Math.S.	Wishart Library, Statistical Laboratory.
Med.	University Medical Library (formerly the Department of Medicine Library).
Met.	Department of Metallurgy and Materials Science.
Min.	Department of Mineralogy and Petrology.
Mol.	Nuttall Library (Parasitology), Molteno Institute.
Nap.	Napier Shaw Library (Meteorology), Department of Physics.
Obs.	Observatories Library, Institute of Astronomy.
P.G.M.S.	University Medical Library (formerly Post-Graduate Medical Library).
P.Math.	Department of Pure Mathematics and Mathematical Statistics.
Path.	Kanthack Library, Department of Pathology.
Philos.	Cambridge Philosophical Society book collection, Scientific Periodicals Library.
Phys.	Department of Physiology.
Psy.	Department of Experimental Psychology.
Radioth.	University Medical Library (formerly Radiotherapeutics Library).
Sco.	Scott Polar Research Institute.
Surg.	Department of Surgery.
T.A.	Institute of Astronomy.
V.A.	Sub-Department of Veterinary Anatomy.
W.S.M.	Whipple Science Museum.

Union Catalogue of Scientific Libraries in the University of Cambridge

Scientific Conference Proceedings
1644-1972

I

Compiled at the Scientific Periodicals Library, University of Cambridge

MANSELL
1975

Mansell Information/Publishing Limited
3 Bloomsbury Place, London WC1A 2QA

International Standard Book Number: 0 7201 0531 5

Text reproduced from computer printout supplied by
the compilers, printed by photolithography and bound at
The Scolar Press Limited, Ilkley, Yorkshire

FOREWORD

When the University of Cambridge set up a Union Catalogue Unit in 1966 as a part of the Scientific Periodicals Library, the great promise of applying computers to the preparation of union catalogues was generally recognized, but the practical problems were largely unknown. The publication of this selective catalogue (a small part of the whole) is a demonstration that the problems are tractable, and that the promise has been fulfilled.

The first tribute must go to the ability and enthusiasm of Miss E. A. Stow, Union Catalogue Librarian, who turned an idea into a catalogue. We are grateful also to the former Office for Scientific and Technical Information for their encouragement and generous financial support and to the Director of the Computer Laboratory for computing facilities. Of the several members of the staff of the Unit we would particularly note the services of Mrs G. Bliss, Mrs J. Downs, Mrs M. Halloran, Mrs M. Lekner, and Mrs P. Robinson. Many others in Cambridge, too numerous to name individually, gave practical help and advice; we mention particularly the staffs of the Computer Laboratory and the Departmental Libraries whose holdings are listed here, but we thank them all.

J.W.S.CASSELS, *Chairman*
D.W.DEWHIRST, *Secretary*
Committee for the Scientific Periodicals Library

PREFACE

The Union Catalogue Unit was set up to compile a central record of the books in the libraries of the scientific research departments of the University. These libraries, some of which date back to the early part of the nineteenth century and one even to 1765, contain important historical collections and at the same time the working material necessary to current research. Together they represent a quarter of a million volumes of periodicals, monographs, and proceedings of conferences. The unit visited the libraries in turn to record their holdings, and the catalogue information was filed and sorted on the Titan computer at the University Computer Laboratory. The catalogues of the various libraries were merged into one alphabetical name sequence and the result printed out in 1973 for use within the University. The Libraries are rich in the published proceedings of conferences, symposia, and lecture meetings. These were specially noted for computer selection and the present catalogue is the result.

Scope

This catalogue includes about 25,000 entries for approximately 6,000 conferences and symposia, ranging from an account of a conference of falconers, published in Rouen in 1644 to a Cold Spring Harbor symposium on the structure and function of proteins at the three-dimensional level held in 1972.

The term conference is used in the widest sense to cover any meeting of scientists for the exchange of ideas so that it includes symposia, lecture meetings, summer institutes, etc. Forty of the scientific departmental libraries and collections of the University are included. These together with the abbreviations adopted for them are listed on the end-papers of this catalogue.

Arrangement

The catalogue is a name and keyword catalogue; each conference is entered under as many headings as are likely to represent its published identity, i.e. official name of conference, title of published proceedings, corporate bodies involved and venues.

Any subject approach is limited to keywords from the titles. Each entry is followed by the symbol denoting the holding library. A number given by the computer to each book catalogued appears at the right hand side of the entry.

Cataloguing rules

The cataloguing follows the general principles laid down in the *Anglo-American Cataloguing Rules, 1967*. The catalogue includes a high proportion of foreign language material, and particular care must be exercised in looking for the names of corporate bodies, since these are not cross-referenced from one form to another. Where the official name of the national or local body appears in a number of different languages the vernacular form is generally preferred. International bodies, on the other hand, are in English if an official English form is available. Publications of a corporate body appear under the official title of the body, e.g. University of Cambridge and not Cambridge University.

Filing order

The filing order is word by word and alphabetical within each word. To achieve this order the computer recognizes all spaces, punctuation and characters and files them in the following order:

. : ; , (space) – A,B, to Z,a,b, to z,0,1,2, to 9

St. and the variant contractions of Mac are filed as though spelt out in full. Special filing rules for certain foreign language modified vowels are ignored. Entries with identical headings are filed alphabetically by the whole title including the definite or indefinite article.

Transliteration of Russian

British Standard BS2978:1958 (with minor modifications)

Аа	a	Дд	d	Ии	i	Нн	n	Тт	t	Чч	ch	Ьь	—
Бб	b	Ее	e	Йй	ï	Оо	o	Уу	u	Шш	sh	Ээ	é
Вв	v	Ёё	ë	Кк	k	Пп	p	Фф	f	Щщ	shch	Юю	yu
Гг	g	Жж	zh	Лл	l	Рр	r	Хх	kh	Ъъ	—	Яя	ya
		Зз	z	Мм	m	Сс	s	Цц	ts	Ыы	y		

Use of asterisk

Characters not available on the typeface of the computer printer are replaced by an asterisk, e.g. in mathematical formulae.

E. A. STOW
Union Catalogue Librarian

Scientific Periodicals Library
University of Cambridge
Bene't Street
Cambridge

A.A.S.-N.A.S.A. SYMPOSIUM ON THE MAGNETIC AND OTHER PECULIAR AND METALLIC-LINE A STARS Proceedings Magnetic and related stars Greenbelt,Md 1965 Nov 8-10 Berkeley,Calif 1965 Dec 29 American astronomical society and National aeronautics and space administration Edited by Robert C. Cameron xi,596p Mono book corporation,Baltimore,1967
A Math 4.1575

A.C.M.-SIGOPS WORKSHOP ON SYSTEM PERFORMANCE EVALUATION Cambridge,Mass. 1971 Apr 5-7 Association for computing machinery.Special interest group in operating systems illus. 378p Association for computing machinery,New York,1971
Math L 5.3564

A.C.T.P.SUMMER SCHOOL ON SYSTEMS PROGRAMMING Lectures London 1966 1969 Collection of typescript copies of lectures
Math L 5.3420

A.G.I. SHORT COURSE ON CHAIN SILICATES Palo Alto,Calif. 1966 Nov 11-13 American geological institute American geological institute,Washington,D.C.,1966 Typescript
Min 10.1286

A.G.I. SHORT COURSE ON LAYER SILICATES New Orleans,La. 1967 Nov 17-19 American geological institute American geological institute,Washington,D.C.,1967 Typescript
Min 10.1287

AACHEN 1964 Statistical mechanics of equilibrum and non-equilibrium The Symposium on statistical mechanics and thermodynamics Proceedings International union of pure and applied physics Edited by J. Meixner 274p North-Holland,Amsterdam,1965
Cav 7.0001

AARHUS 1956 Metabolism of the nervous system International neurochemical symposium 2nd Proceedings Aarhus universitet Edited by Derek Richter Pergamon press, London,1957
Bioch 33.0592

AARHUS 1956 Metabolism of the nervous system International neurochemical symposium 2nd Proceedings Edited by Derek Richter Pergamon,London,1957
Pha 16.0037

AARHUS 1962 Colloquium on algebraic topology Aarhus universitet.Matematisk institut Supported by North Atlantic treaty organization.Scientific affairs division 113p 29cm Aarhus universitet,1962
P. Math 2.0293

AARHUS 1962 Combinatorial methods in probability theory colloquium lectures, mimeograph Aarhus universitet.Matematisk institut Supported by the North Atlantic treaty organization.Scientific affairs division 126p 29cm Aarhus universitet, Aarhus,1962
P. Math 2.2189

AARHUS 1962 Combinatorial methods in probability theory colloquium lectures, mimeograph Aarhus universitet.Matematisk institut Supported by the North Atlantic treaty organization.Scientific affairs division 126p 29cm Aarhus universitet, Aarhus,1962
Math 3.0570

AARHUS 1963 Phonon and phonon interactions Aarhus universitet.Summer school Edited by Thor.A. Bak and others Aarhus summer school lectures,1963 xi,640p W.A.Benjamin,New York;Amsterdam,1964
Cav 7.0002

AARHUS 1969 Summer gathering on function algebras Papers Aarhus universitet. Matematisk institut Aarhus universitet. Matematisk institut.Various publications series, 9 Bibliog. 78p 29cm University of AArhus,AArhus,August,1969
P Math 2.3451

AARHUS 1970 Algebraic topology Proceedings Aarhus universitet.Matematisk institut Aarhus universitet.Matematisk institut.Various publications series, 13 bibliog. 29cm 3 vols Matematisk institut, Aarhus universitet,Aarhus,1970
P Math 2.3822

AARHUS UNIVERSITET Metabolism of the nervous system International neurochemical symposium 2nd Proceedings Aarhus 1956 Jul Edited by Derek Richter Pergamon press, London,1957
Bioch 33.0592

AARHUS UNIVERSITET.MATEMATISK INSTITUT Algebraic topology Proceedings Aarhus 1970 Aug 10-23 Aarhus universitet.Matematisk institut.Various publications series, 13 bibliog. 29cm 3 vols Matematisk institut, Aarhus universitet,Aarhus,1970
P Math 2.3822

AARHUS UNIVERSITET.MATEMATISK INSTITUT Colloquium on algebraic topology Aarhus 1962 Aug 1-10 Supported by North Atlantic treaty organization.Scientific affairs division 113p 29cm Aarhus universitet, 1962
P. Math 2.0293

AARHUS UNIVERSITET.MATEMATISK INSTITUT Combinatorial methods in probability theory colloquium lectures, mimeograph Aarhus 1962 Aug 1-10 Supported by the North Atlantic treaty organization.Scientific affairs division 126p 29cm Aarhus universitet,Aarhus,1962
P. Math 2.2189

AARHUS UNIVERSITET.MATEMATISK INSTITUT Combinatorial methods in probability theory colloquium lectures,mimeograph Aarhus 1962 Aug.1-10 Supported by the North Atlantic treaty organization.Scientific affairs division 126p 29cm Aarhus universitet,Aarhus,1962
Math 3.0570

AARHUS UNIVERSITET.MATEMATISK INSTITUT Papers from the "Open house for algebraists" June-July,1970 at Aarhus universitet... Aarhus universitet.Matematisk institut.Various publications series, 17 bibliog. iv,161p 29cm Aarhus universitet,matematisk institut, Aarhus,1970
P Math 2.3928

AARHUS UNIVERSITET.MATEMATISK INSTITUT Summer gathering on function algebras Papers Aarhus 1969 Jul Aarhus universitet. Matematisk institut.Various publications series, 9 Bibliog. 78p 29cm University of AArhus,AArhus,August,1969
P Math 2.3451

AARHUS UNIVERSITET.SUMMER SCHOOL Phonon and phonon interactions Aarhus 1963 Aug 12-14 Edited by Thor.A. Bak and others Aarhus summer school lectures,1963 xi,640p W.A. Benjamin,New York;Amsterdam,1964
Cav 7.0002

AAS-NASA MAGNETIC STAR SYMPOSIUM The Symposium on the magnetic and other peculiar and metallic-line A stars :magnetic and related stars 1st Proceedings Greenbelt, Md. 1965 Nov 8-10 and Berkeley,Calif. 1965 Dec 29 American astronomical society National aeronautics and space administration Edited by Robert C. Cameron 596p Mono book corporation,Baltimore,1967 Includes the Helen B.Warner prize lecture of the American astronomical society,by George W.Preston
Obs 6.3432

ABELIAN GROUPS Studies on Abelian groups Symposium on the theory of Abelian groups Montpellier 1967 Jun 5-10 Edited by B. Charles Bibliog. x,356p 25cm Dunod, Paris,1968
P Math 2.3507

ABELIAN GROUPS colloquium proceedings Tihany,Hungary 1963 Sep 2-7 Magyar tudomanyos akademia Edited by L. Fuchs and E. T. Schmidt 162p 24cm Akademiai kiado, Budapest,1964
P. Math 2.2028

ABERDEEN 1965 Phonons in perfect lattices and in lattices with point imperfections N.A. T.O.advanced study institute Papers Scottish universities' summer school in physics Edited by R.W.H. Stevenson With financial support of North Atlantic treaty organization Scottish universities' summer school in physics, 6 448p Oliver and Boyd,Edinburgh,1966
TA 15.0212

ABERDEEN 1965 Phonons in perfect lattices and in lattices with point imperfections Scottish universities' summer school 6th Lectures Scottish universities' summer school in physics Edited by R.W.H. Stevenson With financial help from the North Atlantic treaty organization Oliver and Boyd, Edinburgh;London,1966
Chem 18.2643

ABERYSTWYTH 1956 The Biological action of growth substances :a symposium Proceedings Society for experimental biology Society for experimental biology.Symposia, 11 Cambridge university press,Cambridge,1957
Radioth 35.0217

ABERYSTWYTH 1956 The Biological action of growth substances :symposium Society for experimental biology Society for experimental biology.Symposia, 11 Cambridge university press,Cambridge,1957
An 32.0089

ABERYSTWYTH 1956 The Biological replication of macromolecules :a symposium Papers Society for experimental biology Edited by F.K. Sanders Society for experimental biology.Symposia, 12 Cambridge university press,Cambridge,1958
Radioth 35.0237

ABERYSTWYTH 1956 The Biological replication of macromolecules :a symposium Society for experimental biology Edited by K. F. Sanders Society for experimental biology, 12 Cambridge university press,Cambridge, 1957
An 32.5446

ABERYSTWYTH 1965 Biochemistry of chloroplasts :a Nato advanced study institute Proceedings Vol 1-2 North Atlantic treaty organization Edited by T.W. Goodwin 2 vols Academic press,London;New York,1966-67
Bot 42.1429

ABERYSTWYTH 1965 Molecular relaxation processes Symposium on relaxation methods in relation to molecular structure Lectures and papers Chemical society Academic press, London,1966
Pha 16.0369

ABERYSTWYTH 1965 Molecular relaxation processes Symposium on relaxation methods in relation to molecular structure Papers Chemical society Chemical society special publication, 20 Academic press,London,1966
Cav 7.0005

ABERYSTWYTH 1965 Molecular relaxation processes :a symposium Lectures and synopses of papers Chemical society Chemical society.Special publication,20 Chemical society;Academic press,London;New York,1966
Chem 18.1224

ABERYSTWYTH 1966 Terpenoids in plants :a symposium Proceedings Phytochemical group Edited by J.B. Pridham Academic press,London; New York,1967
Chem 18.1510

ABIDJAN 1959 Tropical soils and vegetation Abidjan symposium Proceedings Unesco Jointly organized by the Commission for technical co-operation in Africa south of the Sahara Humid tropics research Unesco,Paris, 1961
Geog 13.1128

ABSORPTION AND DISTRIBUTION OF DRUGS London 1963 Association of medical advisers in the pharmaceutical industry Edited by T.B. Binns Livingstone,Edinburgh; London,1964
Pha 16.0239

ABSORPTION AND DISTRIBUTION OF DRUGS :based on a symposium London 1963 Association of medical advisers in the pharmaceutical industry Edited by T.B. Binns E.and S. Livingstone,Edinburgh;London,1964
PGMS 29.0370

ABSTRACT SPACES AND APPROXIMATION :a conference Proceedings Oberwolfach 1968 Jul 18-27 Edited by P.L. Butzer and Bela Sz-Nagy International series on numerical mathematics, 10 bibliog.,port. 423p 24cm Birkhauser,Basel,1969
P Math 2.3749

ABUNDANCE DETERMINATIONS IN STELLAR SPECTRA : a symposium Utrecht 1964 Aug.10-14 International astronomical union Edited by Hans Hubenet International astronomical union.Symposium, 26 374p Academic press, London;New York,1966
A Math 4.1168

ABUNDANCE DETERMINATIONS IN STELLAR SPECTRA : a symposium Utrecht 1964 Aug.10-14 International astronomical union Edited by Hans Hubenet International astronomical union.Symposium, 26 374p London;New York, 1966
Obs 6.0843

ABUNDANCE DETERMINATIONS IN STELLAR SPECTRA Proceeding :a symposium Utrecht 1964 Aug 10-14 International astronomical union International astronomical union. Symposium, 26 Academic press,London,1966
TA 15.0034

ACADEMIA BRASILEIRA DE CIENCIAS New research techniques in physics :a symposium Proceedings Rio de Janeiro 1952 Jul 15-29 Sao Paulo Sponsored by International union of pure and applied physics illus 447p Unesco,Rio de Janeiro,1954 Under the auspices of Conselho nacional de pesquisas de Brasil and co-sponsored by the university of Sao Paulo
Cav 7.1977

ACADEMIA BRASILEIRA DE CIENCIAS Symposium sobre raios cosmicos Rio de Janeiro 1941 August 4-8 26cm Rio de Janeiro 1943
Philos 1.0379

ACADEMIA BRASILEIRA DE CIENCIAS.CONSELHO NACIONAL DE PESQUISAS International symposium on curare and curare-like agents Proceedings Rio de Janeiro 1957 Aug 5-12 Edited by D. Bovet and others Under the auspices of Unesco illus Elsevier,Amsterdam,1959 Also sponsored by the 'Istituto di sanita',of Rome.Papers in English and French
Chem 26.0103

ACADEMIA REPUBLICII SOCIALISTE ROMANIA Anglo-Romanian conference on mathematics in the archaeological and historical sciences Proceedings Edited by F.R. Hodson and others viii,565p 24cm Edinburgh university press, Edinburgh,1971
Math S 3.1826

ACADEMIA SINICA Recent developments in biochemistry colloquium Proceedings Taipei 1967 Aug 17-19 Sino-American cooperation committee Edited by Hsien-Wen Li and others Academia Sinica,n.p.,1968
Gen 34.0619

ACADEMIE INTERNATIONALE DE PHILOSOPHIE DES SCIENCES Information and prediction in science Symposium de l'Academie internationale de philosophie des sciences Proceedings Brussels 1962 Sep 3-8 Edited by S. Dockx and P. Bernays xi,272p Academic press,London;New York,1965
WSM 43.1136

ACADEMIE INTERNATIONALE D'HISTOIRE DES SCIENCES International congress of the history of science 10th Proceedings Ithaca,N.Y. 1962 Aug 8-Sep 2 Vol 1-2 Co-sponsored by the United States national academy 2 vols Hermann,Paris Congress chairman:H.Guerlac
WSM 43.0030

ACADEMIE ROYALE DE MEDECINE DE BELGIQUE Commemoration solonelle du quatrieme centenaire de la mort d'Andre Vesalius Brussels 1964 Oct 19-24 Brussels,1964
Phys 20.0564

ACCADEMIA DEL CIMENTO Celebrazione della Accademia del cimento nel tricentenario della fondazione Domus galilaeana Pisa 1957 Jun 19 49 pls 79p Domus galilaeana,Pisa,1958 Limited edition.With appendices
WSM 43.4472

ACCADEMIA NAZIONALE DEI LINCEI Evoluzione e genetica :colloquio internazionale Relazioni e discussione Rome 1959 Apr 8-11 Problemi attuali di scienza et di cultura. Quaderno N., 47 Rome,1960 Text in English and Italian.Summaries in English
Bal 39.0927

ACCADEMIA NAZIONALE DEI LINCEI.DONEGANI FOUNDATION International course on materials science 1st Essays Tremezzo 1968 Sep 8-20 Edited by Alan W. Searcy and others illus xxv,715p 24cm Interscience,New York;Chichester,1970 Being the 11th course organized by the Donegani foundation
Met 25.2593

ACCADEMIA NAZIONALE DEL LINCEI Gas chromatographic determination of hormonal steroids :a meeting Proceedings Edited by Filippo Polvani and others Academic press, New York,1967
Inv Med 37.0183

ACCADEMIA PONTIFICA DEI NUOVI LINCEI Study week on nuclei of galaxies Vatican City 1971 Apr 13-18 Edited by D.J.K. O'Connell Accademia pontifica dei nuovi lincei.Scripta varia, 35 795p North-Holland,Amsterdam, 1971
TA 15.0661

The ACCELERATION OF PARTICLES TO HIGH ENERGIES based on a session arranged by the Electronics group Papers London 1949 May Institute of physics Edited by H.R. Lang Physics in industry Institute of physics,London,1950
Cav 7.1166

ACCEPTANCE SAMPLING 105th annual meeting Cleveland,Ohio 1946 Jan.27 American statistical association 155p 23cm American statistical association,Washington,D. C.,1950
Math 3.0816

ACCIDENT PREVENTION The Prevention of accidents in childhood :report on a seminar Spa,Belgium 1958 Jul 16-25 World health organization.Regional office for Europe World health organization,Copenhagen,1960
PGMS 29.0280

ACCRA 1957 C.C.T.A.West-central regional committee for geology :meeting 2nd proceedings Commission for technical co-operation in Africa south of the Sahara London,1958 Bound with proceedings of three other regional meetings
Geol 8.3116

The ACCURACY OF INDUSTRIAL MEASUREMENT OF LENGTH AND DIAMETER :the conference London 1962 Apr 17-18 National physical laboratory National physical laboratory.Conference, 14 H.M.S.O.,London,1963
Met 25.2166

The ACCURACY OF INDUSTRIAL MEASUREMENT OF LENGTH AND DIAMETER :conference Proceedings Teddington 1962 Apr 17-18 National physical laboratory National physical laboratory.Conference, 14 H.M.S.O.,London, 1963
Eng 41.3933

ACETABULARIA International symposium on acetabularia 1st Proceedings Brussels 1969 Jun 18-20 Universite libre de Bruxelles. Centre d'etude de l'energie nucleaire Edited by Jean Brachet and Silvano Bonotto illus. xv,300p Academic press,New York;London,1970
Bot 42.3941

ACETABULARIA International symposium on the biology of acetabularia 1st Proceedings Brussels 1969 Jun 18-20 and Mol 1969 Jun 18-20 Universite libre de Bruxelles Centre d'etude de l'energie nucleaire Edited by Jean Brachet and Silvano Bonotto Academic press,New York,1970
Bioch 33.2261

ACHIEVEMENT OF HIGH FATIGUE RESISTANCE IN METALS AND ALLOYS :a symposium presented at the 72nd annual meeting Atlantic City,N.J. 1969 Jun 22-27 American society for testing materials American society for testing materials.Special technical publication, 467 ASTM,Philadelphia,Pa.,1970
Met 25.2584

ACM CONFERENCE ON PROVING ASSERTIONS ABOUT PROGRAMS Proceedings Las Cruces,N.M. 1972 Jan 6-7 Association for computing machinery.SIGPLAN Association for computing machinery.SIGACT illus 211p ACM,New York, 1972 Published as SIGPLAN notices,vol 7, no 1,Jan 1972 and SIGACT news no 14,Jan 1972
Math L 5.3859

ACM-IEEE SYMPOSIUM ON PROBLEMS IN THE OPTIMIZATION OF DATA COMMUNICATIONS SYSTEMS 2nd Proceedings Palo Alto,Calif. 1971 Oct 20-22 Association for computing machinery Institute of electrical and electronics engineers illus 198p ACM; IEEE,New York,1971
Math L 5.3629

ACM SICFIDET WORKSHOP ON DATA DESCRIPTION AND ACCESS 1970 The Macaims data management system By Robert C. Goldstein and Alois J. Strnad Massachusetts institute of technology. Project MAC Massachusetts institute of technology.MAC technical memorandum, 24 illus 33p Cambridge,Mass.,1971 Paper presented at the 1970 ACM SICFIDET workshop on data description and access,held at Rice university,Houston,Texas,Nov 15-16,1970
Math L 5.3740

ACOUSTICAL FATIGUE Symposium on acoustical fatigue Atlantic City,N.J. 1960 Jun 28 American society for testing materials A.S.T. M.Special technical publication, 284 A.S.T.M. Philadelphia,Pa.,1961
Met 25.0929

ACOUSTICAL SOCIETY OF AMERICA 8TH MEETING 1932 Anecdotal history of the science of sound By Dayton Clarence Miller xii,114p Macmillan, New York,1935 Expanded version of an address delivered at the 8th meeting of the Acoustical society of America 1932
WSM 43.1881

ACQUISITION OF SKILLS :a conference Papers New Orleans,La. 1965 Mar 8-12 United States.Army.Medical research and development command Edited by Edward A. Bilodeau Academic press,New York;London,1966
Psy 31.1005

ACTA PHYSICA AUSTRIACA.SUPPLEMENTUM, 3 Elementary particle theories Internationale universitatswochen fur Kernphysik 5th proceedings Schladming 1966 Feb.24-Mar.9 Edited by Paul Urban Sponsored by the International Atomic Energy Agency Springer-verlag,Vienna;New York,1966
A Math 4.0889

ACTAS Y TRABAJOS DE LAS REUNIONES CELEBRADAS EN MEXICO Association of African geological surveys and commission on the geological map of Africa :joint meeting papers Mexico City 1956 Sep 4-10 International geological congress.Commission on the geological map of Africa,and,Association of African geological surveys Edited by A. Garcia Rojas Mexico City,1959 Text in English,French and Spanish
Geol 8.3038

ACTH Clinical ACTH conference 2nd Proceedings Chicago 1950 Dec 8-9 Vol 1-2 Edited by John R. Mote Sponsored by Armour and company 2 vols Churchill,London, 1951
Bioch 33.0443

ACTINOMYEETALES:MORFOLOGIA,BIOLOGIA E SISTEMATICA Aetinomyeetales :morphology,biology and systematics Edited by E. Baldacci and P. Redaelli International congress of microbiology, 6,1 Rome,1953
Bot 42.0664

ACTIONS EOLIENNES.PHENOMENES DEVAPORATION ET DHYDROLOGIE SUPERFICIELLE DANS LES REGIONS ARIDES Algiers 1951 Mar 27-31 Centre national de la recherche scientifique. Colloques internationaux,35 C.N.R.S.,Paris, 1953
Geog 13.0671

ACTIVATION ANALYSIS PRINCIPLES AND APPLICATIONS NATO advanced study institute Proceedings Glasgow 1964 Aug North Atlantic treaty organization Edited by J.M.A. Lenihan and S. J. Thomson Academic press,London;New York, 1965
Met 25.1689

ACTIVE TRANSPORT AND SECRETION Bangor 1953 Jul Society for experimental biology Society for experimental biology.Symposia, 8 Cambridge university press,Cambridge,1954
Phys 20.0958

ACTIVE TRANSPORT AND SECRETION :a symposium Bangor 1953 Jul Society for experimental biology Society for experimental biology.Symposia, 8 Cambridge university press,Cambridge,1954
Bioch 33.1353

ACTIVE TRANSPORT AND SECRETION :a symposium Papers Bangor 1953 Jul Society for experimental biology Edited by R. Brown and J.F. Danielli Society for experimental biology.Symposia, 8 Cambridge university press,Cambridge,1954
Inv Med 37.0036

ACUTE RENAL FAILURE Symposium on acute renal failure Papers London 1963 Sep Edited by Stanley Shaldon and G.C. Cook Blackwell, Oxford,1964
Inv Med 37.0195

ACUTE RENAL FAILURE :a symposium London 1963 Sep Edited by S. Shaldon and G.C. Cook Held at the Royal free hospital Blackwell, Oxford,1964
PGMS 29.0225

ADANSON BICENTENNIAL SYMPOSIUM Pittsburgh,Pa. 1963 Aug 18-19 Hunt botanical library Hunt botanical library,Pittsburgh,Pa.,1963
BG 38.0952

ADAPTATION IN MICRO-ORGANISMS :symposium London 1953 Apr Society for general microbiology Society for general microbiology.Symposia, 3 Cambridge university press,Cambridge,1953
Gen 34.0694

ADAPTION IN MICPO-ORGANISMS :a symposium Papers London 1953 Apr Society for general microbiology Edited by R. Davies and E.F. Gale Held at the Royal institution Society for general microbiology.Symposia, 3 Cambridge university press,Cambridge,1953
Bioch 33.1166

ADAPTIVE CONTROL International symposium on optimizing and adaptive control 1st Proceedings Rome 1962 Apr 26-28 Edited by Loren E. Bollinger and Emil J. Minnar Sponsored by the International federation of automatic control.Theory committee Instrument society of America,Pittsburgh,Pa., 1962
Eng 41.6017

ADAPTIVE CONTROL SYSTEMS :a symposium Proceedings Garden City,N.Y. 1960 Oct Edited by Felix P. Caruthers and Harold Levenstein illus viii,290p 23cm Pergamon press,Oxford,1963
Math 3.1234

ADDIS ABABA 1961 Economic development for Africa south of the Sahara International economic association :regional conference 3rd Proceedings Edited by E.A.G. Robinson Macmillan,London,1965
Geog 13.4646

ADDISON J.W. and others ed. Theory of models International symposium on the theory of models proceedings Berkeley, Calif. 1963 Jun 25-Jul 11 xv,494p 23cm North-Holland, Amsterdam,1965
P. Math 2.0205

ADDRESSES AND PAPERS,DEDICATION CEREMONIES AND MEDICAL CONFERENCE Peking 1921 Sep 15-22 Peking union medical college illus Peking union medical college,Peking,1922
Path 30.1496

ADELAIDE 1905 Inter-state astronomical and meteorological conference Report By Charles Todd 11p South Australian government,Adelaide,1905
Obs 6.3577

ADELAIDE 1949 and CANBERRA 1949 Plant and animal nutrition in relation to soil and climatic factors British commonwealth scientific official conference.Specialist conference in agriculture 1st Proceedings H.M.S.O.,London,1951
Geog 13.2510

ADELAIDE 1963 Australia-New Zealand conference on soil mechanics and foundation engineering 4th Proceedings University of Adelaide 1963
Eng 41.3169

ADHESION ...Recent developments in adhesion science...Recent advances in adhesives technology...Presented at the 66th annual meeting A.S.T.M. Atlantic city,N.J. 1963 Jun 26 American society for testing materials American society for testing materials.Special technical publication, 360 American society for testing materials, Philadeplphia,Pa.,1964
Eng 41.4027

ADHESION :papers presented at the 66th annual meeting A.S.T.M. Atlantic city,N.J. 1963 Jun 26 American society for testing and materials American society for testing and materials.Special technical publication, 360 American society for testing and materials,Philadelphia,Pa.,1964
Eng 41.7660

ADHESION:FUNDAMENTALS AND PRACTICE ; international conference Report Nottingham 1966 Sep 20-22 Great Britain. Ministry of technology Elsevier,Amsterdam, 1969
Met 25.2629

ADIPOSE TISSUE AS AN ORGAN Deuel conference on lipids Proceedings Edited by Laurance W. Kinsell Thomas,Springfield,Ill.,1962
An 32.3279

ADIPOSE TISSUE METABOLISM AND OBESITY :a conference Papers New York 1964 Dec 14-17 New York academy of sciences Edited by Harold E. Whipple New York academy of sciences.Annals, 131,p 1-683 New York,1965
Bioch 33.0562

ADOPTION D'UN PREMIER MERIDIEN UNIQUE ET D'UNE HEURE UNIVERSELLE :conference Proces-verbaux Washington,D.C. 1884 Oct 1-Nov 1 United States.Department of state 216p Washington,D.C.,1884
Obs 6.3168

ADRENAL CORTEX Conference on adrenal cortex 1st-5th Transactions Edited by Elaine P. Ralli Sponsored by the Josiah Macy jr. foundation 5 vols Josiah Macy foundation, New York,1950-54
An 32.3730

ADRENAL CORTEX :a conference Papers New York 1948 Mar 21-22 New York academy of sciences Edited by Robert Gaunt New York academy of sciences.Annuals, 50,509-678 New York,1949
Inv Med 37.0274

The ADRENAL CORTEX :a symposium Proceedings Chemical pathology in relation to clinical medicine London 1960 Oct 14-15 Association of clinical pathologists Edited by G.K. McGowan and M. Sandler Pitman,London, 1961
An 32.3751

ADRENAL CORTEX 1st :a conference Transactions New York 1949 Nov 21-22 Edited by Elaine P. Ralli and others Sponsored by the Josiah Macy Jr.foundation Josiah Macy Jr.foundation,New York,1950
Bioch 33.0464

ADRENAL CORTEX 2nd :a conference Transactions New York 1950 Nov 16-17 Edited by Elaine P. Ralli Sponsored by the Josiah Macy Jr.foundation Josiah Macy Jr. foundation,New York,1951
Bioch 33.0465

ADRENAL CORTEX 3rd :a conference Transactions New York 1951 Nov 15-16 Edited by Elaine P. Ralli Sponsored by the Josiah Macy Jr.foundation Josiah Macy Jr. foundation,New York,1952
Bioch 33.0466

ADRENAL CORTEX 4th :a conference Transactions New York 1952 Nov 12-14 Edited by Elaine P. Ralli Sponsored by the Josiah Macy Jr.foundation Josiah Macy Jr. foundation,New York,1953
Bioch 33.0467

ADRENAL CORTEX 5th :a conference Transactions Princeton,N.J. 1953 Nov 4-5 Edited by Elaine P. Ralli Sponsored by the Josiah Macy Jr.foundation Josiah Macy Jr. foundation,New York,1954
Bioch 33.0468

ADRENAL GLAND Human adrenal gland and its relation to breast cancer Tenovus workshop 1st Proceedings Cardiff 1969 Jun 26-27 Tenovus institute for cancer research Edited by K. Griffiths and E.H.D. Cameron Alpha omega alpha,Cardiff,1969
Bioch 33.2198

ADRENAL HORMONES Metabolic effects of adrenal hormones in honour of G.W.Thorn London 1960 Jul 15 Ciba foundation Edited by G.E.W. Wolstenholme and Maeve O'Connor Ciba foundation study group, 6 Churchill,London,1960
Bioch 33.0442

ADRENERGIC BLOCKING DRUGS New adrenergic blocking drugs.Their pharmacological, biochemical and clinical actions :a conference Papers New York 1966 Feb 24-26 By N.C. Moran New York academy of sciences New York academy of sciences.Annals, 139,p.541-1009 New York,1967
Bioch 33.0422

ADRENERGIC MECHANISMS London 1960 Mar 28-29 Ciba foundation Committee for symposia on drug action Edited by J.R. Vane and others Ciba foundation.Symposia Churchill, London,1960
PGMS 29.0379

ADRENERGIC MECHANISMS London 1960 Mar 28-31 Ciba foundation Edited by J.R. Vane and others Churchill,London,1960
Pha 16.0070

ADRENERGIC MECHANISMS :a Ciba foundation symposium London 1960 Mar 28-31 Ciba foundation Edited by J.R. Vane and others Churchill,London,1960
Phys 20.0972

ADRENERGIC NEUROTRANSMISSION :Ciba foundation study group London Ciba foundation Edited by G.E.W. Wolstenholme and Maeve O'Connor Ciba foundation study group, 33 Churchill,London,1968
Phys 20.2050

ADRENERGIC NEUROTRANSMISSION 33rd :Ciba foundation study group London Ciba foundation Edited by G.E.W. Wolstenholme and Maeve OConnor Churchill,London,1968
Pha 16.0075

ADRENOCORTICAL HORMONES Bioassay of anterior pituitary and adrenocortical hormones :a colloquium proceedings London 1952 Mar 25-27 Ciba foundation Edited by G.E.W. Wolstenholme Ciba foundation colloquia on endocrinology, 5 illus Churchill,London, 1953
Bioch 33.0435

ADRENOCORTICAL STEROIDS Synthesis and metabolism of adrenocortical steroids :a colloquium Proceedings London 1950 Ciba foundation Edited by W. Klyne Ciba foundation colloquia on endocrinology, 7 Churchill,London,1953
Inv Med 37.0137

ADRENOCORTICAL STEROIDS Synthesis and metabolism of adrenocortical steroids :a colloquium Proceedings London 1952 Jul 7-10 Ciba foundation Edited by W. Klyne and G.E.W. Wolstenholme Ciba foundation colloquia on endocrinology, 7 illus Churchill,London,1953
Bioch 33.0437

ADSORPTION ET CROISSANCE CRISTALLINE : colloque international Nancy 1965 Jun 6-12 Centre national de la recherche scientifique Centre national de la recherche scientifique.Colloques internationaux, 152 Centre national de la recherche scientifique, Paris,1965
Met 25.1621

ADVANCE AND RETREAT OF RURAL SETTLEMENT International geographical congress 19th Papers of the Siljan symposium Norden 1960 Edited by Gerd Enequist and Gunnar Norling Uppsala universitet.Geografiska institution. Meddelanden.Ser A,160 Geografiska annaler,42, no 4 Uppsala,1960
Geog 13.1898

ADVANCE IN TRANSPLANTATION International congress of the Transplantation society 1st Proceedings Paris 1967 Jun 27-30 Transplantation society Edited by J. Dausset and others Munksgaard,Copenhagen,1968
Surg 23.0021

ADVANCED AERO ENGINE TESTING :joint meeting of the Agard combustion and propulsion and wind tunnel and model testing panels Papers Copenhagen 1958 Oct Agard Edited by A.W. Morley and Jean Fabri AGARDograph, 37 Pergamon press,London,1959
Eng 41.7369

ADVANCED AERO ENGINE TESTING :joint meeting of the Agard combustion and propulsion and wind tunnel and model testing panels Papers Copenhagen 1958 Oct Agard Edited by A.W. Morley and Jean Fabri AGARDograph, 37 Pergamon press,London,1959
Eng 41.7377

ADVANCED COURSE ON PROGRAMMING LANGUAGES AND DATA STRUCTURES A Comparative study of some higher programming languages Series of lectures Amsterdam 1972 Jun 12-23 By W.L. van der Poel varp Van der Poel,Delft,1972
Math L 5.3872

ADVANCED MEDICINE A Symposium on advanced medicine 2nd Proceedings London 1965 Nov 29-Dec 3 Edited by J.R. Trounce Held at the Royal college of physicians Pitman medical,London,1966
Inv Med 37.0213

ADVANCED MEDICINE Symposium on advanced medicine :a conference Proceedings London 1964 Nov 16-20 Royal college of physicians of London Edited by Nigel Compston Pitman, London,1965
PGMS 29.0057

ADVANCED NAVIGATIONAL TECHNIQUES :symposium Proceedings Milan 1967 Sep 12-15 AGARD Edited by W.T. Blackband AGARD. Conference proceedings, 28 Technivision, Maidenhead,1970 "...the record of the proceedings of the 14th symposium of the Avionics panel..."-foreward
Eng 41.5726

ADVANCED PRESTRESSED CONCRETE DESIGN National prestressed concrete short course 2nd Papers Dayton beach,Fla. 1958 Jan University of Florida.College of engineering. Bulletin series, 76 University of Florida, Gainesville,Fla.,1958
Eng 41.2847

ADVANCED PROPELLANT CHEMISTRY :a symposium Papers Detroit,Mich. 1965 Apr 6-7 American chemical society.Division of fuel chemistry,and,American institute of aeronautics and astronautics.Propellants and combustion technical committee American chemical society.Advances in chemistry series, 54 American chemical society,Washington,D.C., 1966 Presented at the 149th meeting of the American chemical society.Symposium chairman:Richard T.Holtzmann
Chem 18.0977

ADVANCED PROPULSION SYSTEMS SYMPOSIUM Proceedings Los Angeles,Calif. 1957 Dec 11-13 Edited by Morton Alperin and George P. Sutton International series on aeronautical sciences and space flight.Division 9:symposia, 2 Pergamon press,London,1959
Eng 41.7370

ADVANCED PROPULSION TECHNIQUES :Agard combustion and propulsion panel technical meeting Proceedings Pasadena,Calif. 1960 Aug 24-26 Agard Edited by S.S. Penner Pergamon press,London,1961
Eng 41.7373

ADVANCED RADAR SYSTEMS 19th Avionics panel meeting Istanbul 1970 May 25-29 Agard.Avionics panel AGARD conference proceedings, 66 Agard,Neuilly-sur-Seine, 1970
Eng 41.8322

ADVANCED RESEARCH PROJECTS AGENCY Space trajectories :a symposium Papers Orlando, Fla. 1959 Dec 14-15 Co-sponsored by the American astronautical society Academic press,New York,1960
Geod 9.0014

ADVANCED SCHOOL ENGINEERS.COMBUSTION ENGINES GROUP Air pollution control in transport engines :a symposium Proceedings London 1971 Nov 9-11 IME,London,1972
Eng 41.8645

ADVANCED SCHOOL OF AUTOMOTIVE STUDIES Vibration and noise in motor vehicles :a symposium Proceedings London 1971 Jul 6-7 IME,London,1972
Eng 41.8646

ADVANCES IN AERONAUTICAL SCIENCES International congress for the aeronautical sciences 1st Proceedings Madrid 1958 Sep 8-13 Vol 1-2 Edited by Theodore von Karman International series on aeronautical sciences and space flight. Division 9:symposia 2 vols Pergamon press, London,1959
Eng 41.6963

ADVANCES IN AERONAUTICAL SCIENCES International congress for the aeronautical sciences 2nd Proceedings Zurich 1960 Sep 12-16 Vol 3-4 Edited by Theodore von Karman International series in aeronautics and astronautics.Division 9: symposia, 7-8 2 vols Pergamon press, London,1962
Eng 41.6964

ADVANCES IN AUTOMOBILE ENGINEERING Symposium on automobile engineering Proceedings Cranfield 1963 Jul Pt 2 Edited by N. A. Carter College of aeronautics,Cranfield. Advanced school of automobile engineering Cranfield international symposium series, 5 Pergamon press,London,1964
Eng 41.6190

ADVANCES IN COMPUTER CONTROL United Kingdom automation council.Control convention 2nd Bristol 1967 Apr 11-14 Organized by the Institution of electrical engineers.Control and automation division Institution of electrical engineers.Conference publication, 29 I.E.E.,London,1967
Eng 41.5862

ADVANCES IN CRYOGENIC ENGINEERING Cryogenic engineering conference 6th Proceedings Boulder,Colo. 1960 Aug 23-26 National bureau of standards University of Colorado Edited by K.D. Timmerhaus Advances in cryogenic engineering, 6 Plenum press,New York,1961
Eng 41.7398

ADVANCES IN EARTH SCIENCE International conference on the earth sciences Papers Cambridge 1964 Sep 30-Oct 2 Edited by P.M. Hurley Massachusetts institute of technology press,Cambridge,Mass.,1966
Min 10.0576

ADVANCES IN ELECTRON METALLOGRAPHY Symposium on advances in electron metallography Papers American society for testing materials A.S.T.M.Special technical publication, 245 American society for testing materials, Philadelphia,Pa.,1958
Met 25.1362

ADVANCES IN ELECTRON METALLOGRAPHY a symposium presented at the sixty-eighth annual meeting Lafayette,Ind. 1965 Jun 13-18 Vol 6 American society for testing materials A.S.T.M.Special technical publication, 396 A.S.T.M.,Philadelphia,1966
Met 25.1399

ADVANCES IN ELECTRON TUBE TECHNIQUES National conference on tube techniques 5th Proceedings New York 1960 Sep Edited by David Slater Sponsored by the Advisory group on electron tubes.Working group on tube techniques Pergamon press,Oxford,1961
Eng 41.5621

ADVANCES IN ENZYMIC HYDROLYSIS OF CELLULAR AND RELATED MATERIALS :a symposium Proceedings Washington,D.C. 1962 Mar Edited by Elwyn T. Reese Sponsored by the American chemical society Pergamon,Oxford, 1963 Including a bibliography for the years 1950-61
Bot 42.1728

ADVANCES IN EXTRACTIVE METALLURGY :a symposium Proceedings London 1967 Apr 17-20 Institution of mining and metallurgy Institution of mining and metallurgy,London, 1968
Met 25.0494

ADVANCES IN FLUIDICS Fluidics symposium Chicago,Ill. 1967 May 9-11 American society of mechanical engineers Edited by Forbes T. Brown American society of mechanical engineers,New York,1967
Eng 41.5898

ADVANCES IN GEOPHYSICS, 6 A Symposium on atmospheric diffusion and air pollution proceedings Oxford 1958 Aug 24-29 International union of theoretical and applied mechanics and International union of geodesy and geophysics Edited by F.N. Frenkel and P.A. Sheppard Academic press,New York;London,1959
A Math 4.0576

ADVANCES IN GLASS TECHNOLOGY International congress on glass 4th Technical papers Washington,D.C. 1962 Jul 8-14 American ceramic society Plenum press,New York,1962
Met 25.0116

ADVANCES IN MACHINE TOOL DESIGN AND RESEARCH International M.T.D.R. conference 3rd-9th Proceedings 1963-68 Edited by S.A. Tobias and F. Koenigsberger 9 vols Pergamon press,London,1963-69
Eng 41.3924

ADVANCES IN MAGNETOHYDRODYNAMICS :a colloquium Proceedings Sheffield 1961 Oct Edited by I.A. McGrath and others Organized by the University of Sheffield.Department of fuel technology and chemical engineering Pergamon press,London,1963
Eng 41.4504

ADVANCES IN MAGNETOHYDRODYNAMICS :a colloquium proceedings Sheffield 1961 Oct. Edited by I.A. McGrath and others Organized by University of Sheffield.Department of fuel technology and chemical engineering Pergamon press,London,1963
A Math 4.0622

ADVANCES IN MASS SPECTROMETRY :a conference Proceedings London 1958 Sep 24-26 Institute of petroleum.Hydrocarbon research group,and,American society for testing materials.Committee on mass spectrometry Edited by J.D. Waldron Pergamon press. Symposia publications division,London,1959
Chem 18.0197

ADVANCES IN MATERIALS RESEARCH IN THE N.A.T.O. NATIONS :a symposium Proceedings Washington,D.C. 1961 May North Atlantic treaty organization.Advisory group for aeronautical research and development Edited by H. Brooks and others AGARD 62 Pergamon press,Oxford,1963
Met 25.0096

ADVANCES IN NEUROENDOCRINOLOGY Symposium on neuroendocrinology Miami,Fla. 1961 Dec 6-8 Edited by Andrew V. Nalbandov Supported by a grant from the National institute of neurological diseases and blindness University of Illinois press,Urbana,Ill.,1963
An 32.3766

ADVANCES IN NEUROENDOCRINOLOGY Symposium on neuroendocrinology Miami,Flo. 1961 Dec 6-8 Edited by A.V. Nalbandov Sponsored by the National institute of neurological diseases and blindness University of Illinois press, Urbana,Ill.,1963
Phys 20.1404

ADVANCES IN PROGRAMMING AND NON-NUMERICAL COMPUTATION summer school proceedings Oxford 1963 University of Oxford.Computing laboratory Edited by L. Fox viii,218p 24cm Pergamon press,Oxford,1966
Math 3.0239

ADVANCES IN RADIOBIOLOGY International conference on radiobiology 5th Proceedings Stockholm 1956 Aug 15-19 Edited by George Carl de Hevesy and others port. Oliver and Boyd,Edinburgh;London,1957
Radioth 35.1708

ADVANCES IN TACTICAL ROCKET PROPULSION :a colloquium La Jolla,Calif. 1965 Apr 22-23 Agard Edited by S.S. Penner AGARD conference proceedings, 1 Technivision, Slough,1968
Eng 41.6896

ADVANCES IN THERMOPHYSICAL PROPERTIES AT EXTREME TEMPERATURES AND PRESSURES Symposium on thermophysical properties 3rd Papers Lafayette,Ind. 1965 Mar 22-25 Edited by Serge Gratch Sponsored by the American society of mechanical engineers American society of mechanical engineers,New York,1965
Eng 41.7121

ADVANCES IN TRACER METHODOLOGY, 1 Annual symposium on tracer methodology 5th Proceedings Washington,D.C. 1961 Oct 20 New England nuclear corporation Packard instrument company Edited by Seymour Rothschild Plenum press,New York,1963 Includes selected papers from the first four annual symposia
Gen 34.0633

ADVANCES IN X-RAY ANALYSIS, 12 Annual conference on applications of X-ray analysis 17th Proceedings Estes Park,Colo. 1968 Aug 21-23 Denver research institute. Metallurgy division Edited by Charles S. Barrett and others Plenum press,New York, 1969
Met 25.2824

ADVANCES IN X-RAY ANALYSIS 11 proceedings of the 16th annual conference on applications of X-ray analysis Denver,Colo. 1967 Aug 9-11 Denver research institute Edited by John B. Newkirk Plenum press,New York,1968
Met 25.2437

ADVANCES IN X-RAY ANALYSIS, 7 Annual conference on applications of x-ray analysis 12th 1963 Aug 7-9 Edited by M. Mueller and others Sponsored by the Denver research institute Plenum press,New York,1964
Met 25.1445

ADVISORY GROUP ON ELECTRON TUBES.WORKING GROUP ON SEMICONDUCTOR DEVICES National conference on tube techniques 3rd Proceedings New York 1956 Sep 12-14 New York university press,New York,1958
Eng 41.5619

ADVISORY GROUP ON ELECTRON TUBES.WORKING GROUP ON TUBE TECHNIQUES Advances in electron tube techniques National conference on tube techniques 5th Proceedings New York 1960 Sep Edited by David Slater Pergamon press,Oxford,1961
Eng 41.5621

ADVISORY GROUP ON ELECTRON TUBES.WORKING GROUP ON TUBE TECHNIQUES National conference on tube techniques 4th Proceedings New York 1958 Sep 10-12 New York university press,New York,1959
Eng 41.5620

AERO ENGINE TESTING Advanced aero engine testing :joint meeting of the Agard combustion and propulsion and wind tunnel and model testing panels Papers Copenhagen 1958 Oct Agard Edited by A.W. Morley and Jean Fabri AGARDograph, 37 Pergamon press, London,1959
Eng 41.7369

AERO ENGINE TESTING Advanced aero engine testing :joint meeting of the Agard combustion and propulsion and wind tunnel and model testing panels Papers Copenhagen 1958 Oct Agard Edited by A.W. Morley and Jean Fabri AGARDograph, 37 Pergamon press, London,1959
Eng 41.7377

AERO RESEARCH Structural adhesives:the theory and practice of gluing with synthetic resins lectures Cambridge 1951 Lange, Maxwell and Springer,London,1951 Lectures given at a summer school on the technology of synthetic resin adhesives, Cambridge,1951
Eng 41.4022

AEROARCTIC see INTERNATIONALE STUDIENGESELLSCHAFT ZUR ERFORSCHUNG DER ARKTIS MIT DEM LUFTSCHIFF

AEROBIOLOGY :a symposium Papers American association for the advancement of science.Section on medical sciences National research council.Aerobiology committee Edited by Forest Ray Moulton American association for the advancement of science. Publications, 17 x,289p Washington,D.C., 1942
Bot 42.1228

AERODYNAMIC CAPTURE OF PARTICLES :a conference Proceedings Leatherhead 1960 Edited by E.G. Richardson Pergamon press,London, 1960
Eng 41.6967

AERODYNAMIC CAPTURE OF PARTICLES :a conference Proceedings Leatherhead 1960 Edited by E.G. Richardson viii,200p Pergamon press,Oxford,1960
Chem E 24.0223

AERODYNAMIC PHENOMENA IN STELLAR ATMOSPHERES Symposium on cosmical gas dynamics held at the International school of physics "Enrico Fermi" 4th proceedings Varenna 1960 Aug.18-30 International astronomical union and International union of theoretical and applied mechanics Edited by R.N. Thomas and others Under the auspices of Societa Italiana di fisica International astronomical union. Symposium, 12 515p Bologna,c1961 Reprinted from Supplemento Nuovo Cimento,22,no. 1,1961
A Math 4.1152

AERODYNAMIC PHENOMENA IN STELLAR ATMOSPHERES Symposium on cosmical gas dynamics held at the International school of physics "Enrico Fermi" 4th proceedings Varenna 1960 Aug 18-30 International astronomical union and International union of theoretical and applied mechanics Edited by R.N. Thomas and others under the auspices of the Societa italiana di fisica International astronomical union symposium, 12 515p Zanichelli,Bologna, c1961 Reprinted from the 'Supplemento del nuovo cimento',vol.22,no.1,1961
Obs 6.1736

AERODYNAMICAL PHENOMENA IN STELLAR ATMOSPHERE Nice 1965 Sep International astronomical union Edited by Richard N. Thomas International astronomical union.Symposium, 28 Academic press,London,1967
TA 15.0411

AERODYNAMICS Boundary layer effects in aerodynamics: a symposium proceedings Teddington 1955 March 31 National physical laboratory 28cm London,1955
Philos 1.0437

AERONAUTICAL RESEARCH COUNCIL The Crack propagation symposium Proceedings Cranfield 1961 Sep Vol 1-2 Supported by the Royal aeronautical society 2 vols College of aeronautics,Cranfield,1962
Met 25.0865

AERONAUTICAL SCIENCES Seminar on aeronautical sciences Proceedings 1961 Nov 27-Dec 2 Vol 1 National aeronautical laboratory,Bangalore National aeronautical laboratory,Bangalore,1961
Eng 41.6969

AERONAUTICS Anglo-American aeronautical conference 6th Papers Folkestone 1957 Sep 9-12 Royal aeronautical society Institute of the aeronautical sciences Edited by Joan Bradbrooke and U.A. Libby Royal aeronautical society,London,1959
Eng 41.6962

AERONAUTICS Conference on high-speed aeronautics Proceedings Brooklyn,N.Y. 1955 Jan 20-22 Polytechnic institute of Brooklyn.Department of aeronautical engineering and applied mechanics Edited by Antonio Ferri and others Polytechnic institute of Brooklyn,Brooklyn,N.Y.,1955
Eng 41.6955

AERONAUTICS International council of the aeronautical sciences 4th congress Proceedings Paris 1964 Aug 24-28 Edited by Robert R. Dexter Spartan;Macmillan, Washington,D.C.;London,1965
Eng 41.6966

AERONAUTICS Seminar on aeronautical sciences proceedings Bangalore 1961 Nov 27-Dec.2 Vol 1 India.National aeronautical laboratory National aeronautical laboratory, Bangalore,1961
A Math 4.0526

AEROPAUSE Physics and medicine of the upper atmosphere Symposium on the physics and medicine of the upper atmosphere Proceedings San Antonio,Tex. 1951 Nov 6-9 Lovelace foundation for medical education and research Edited by Clayton S. White and Otis O. Benson Sponsored by School of aviation medicine. Randolph Field illus xxix,611p University of New Mexico press,Albuquerque,N.M. 1952
Nap 11.0635

AEROSPACE MATERIALS Southern metals-materials conference on advances in aerospace materials Proceedings Orlando,Fla. 1964 Apr 16-17 American society of metals.Orlando chapter Edited by Henry M. Otte and Saul R. Locke Plenum press,New York,1965
Met 25.2282

AEROSPACE RESEARCH LABORATORIES Inequalities a symposium Proceedings Ohio 1965 Aug 19-27 Edited by Oved Shisha Bibliog xiv, 360p 24cm Academic press,London;New York, 1967
P Math 2.2925

AEROSPACE RESEARCH LABORATORIES Multivariate analysis :international symposium Proceedings Dayton,Ohio 1965 Jun 14-19 Edited by Paruchuri R. Krishnaiah Bibliog, port xix,592p 23cm Academic press,New York,1966
Math 3.1159

AEROSPACE RESEARCH LABORATORIES Relativity Relativity conference of the Midwest Proceedings Cincinnati,Ohio 1969 Jun 2-6 Edited by M. Carmeli and others 381p Plenum press,New York,1970
TA 15.0506

AEROSPACE RESEARCH LABORATORIES Symposium on the structure of low-medium mass nuclei 3rd Lawrence,Kansas 1968 Apr 18-20 Edited by J.P. Davidson vi,294p 24cm University press of Kansas,Lawrence,Kansas,1968
Cav 7.3099

AEROTHERMOCHEMISTRY Gas dynamics symposium on aerothermochemistry Proceedings Evanston,Ill 1955 Aug 22-24 Northwestern university and American rocket society Edited by Donald K. Fleming Supported by the United States.Air force illus 284p Northwestern university,Evanston,Ill.,1956
Chem E 24.1390

AETINOMYEETALES :morphology,biology and systematics Edited by E. Baldacci and P. Redaelli International congress of microbiology, 6,1 Rome,1953
Bot 42.0664

AETIOLOGY OF DIABETES MELLITUS AND ITS COMPLICATIONS :colloquium London 1963 Oct 8-10 Ciba foundation Edited by Margaret P. Cameron and Maeve O'Connor Ciba foundation colloquia on endocrinology, 15 Churchill,London,1964
Phys 20.1406

AFIPS CONFERENCE PROCEEDINGS INDEX VOLUMES 1-37, 1951-1970 American federation of information processing societies 93p AFIPS press,Montvale,N.J.,1971
Math L 5.3800

AFRICA Actas y trabajos de las reuniones celebradas en Mexico Association of African geological surveys and commission on the geological map of Africa :joint meeting papers Mexico City 1956 Sep 4-10 International geological congress.Commission on the geological map of Africa,and, Association of African geological surveys Edited by A. Garcia Rojas Mexico City,1959 Text in English,French and Spanish
Geol 8.3038

AFRICA Salt basins around Africa Institute of petroleum and Geological society of London : joint meeting proceedings London 1965 Mar 3 Institute of petroleum,and,Geological society of London Institute of petroleum, London,1965
Geol 8.3097

AFRICAN AGRARIAN SYSTEMS International African seminar 2nd Studies Leopoldville 1960 Jan International African institute Edited by Daniel Biebuyck Oxford university press,London,1963
Geog 13.4631

AFRICAN ECOLOGY AND HUMAN EVOLUTION :a symposium Papers and discussions Burg-Wartenstein 1961 Jul 10-22 Wenner-Gren foundation for anthropological research Edited by F.Clark Howell and Francois Bourliere illus,tables 25cm Methuen, London,1964
Bal 44.4056

AFRICAN HYDROBIOLOGY Symposium on African hydrobiology and inland fisheries 2nd Brazzaville 1956 Jul 3-11 Scientific council for Africa south of the Sahara Published under the auspices of the Commission for technical co-operation in Africa south of the Sahara Scientific council for Africa south of the Sahara.Publication, 25 Brazzaville,1956 Title also in French: Symposium sur l'hydriobiologie et la peche en eaux douces en Afrique;text in English and French
Bal 39.1894

AFRICAN PRIMARY PRODUCTS AND INTERNATIONAL TRADE Edinburgh 1964 Sep 20-23 University of Edinburgh.Centre of African studies Edited by I.G. Stewart and H.W. Ord Jointly organized by the University of Edinburgh. Department of political economy Edinburgh university press,Edinburgh,1965
Geog 13.2449

AFRICAN STUDIES ASSOCIATION OF THE UNITED KINGDOM The Soil resources of tropical Africa :a symposium London 1965 Sep 29 Edited by R. P. Moss Cambridge university press,Cambridge, 1968
Geog 13.4644

AGARD AGARD-NATO specialists' meeting The Fluid dynamic aspects of space flight Marseille 1964 Apr 20-24 Vol 1-2 AGARDograph, 87 2 vols Gordon and Breach, New York,1966
Eng 41.6887

AGARD Advanced aero engine testing :joint meeting of the Agard combustion and propulsion and wind tunnel and model testing panels Papers Copenhagen 1958 Oct Edited by A.W. Morley and Jean Fabri AGARDograph, 37 Pergamon press,London,1959
Eng 41.7369

AGARD Advanced aero engine testing :joint meeting of the Agard combustion and propulsion and wind tunnel and model testing panels Papers Copenhagen 1958 Oct Edited by A.W. Morley and Jean Fabri AGARDograph, 37 Pergamon press,London,1959
Eng 41.7377

AGARD Advanced navigational techniques : symposium Proceedings Milan 1967 Sep 12-15 Edited by W.T. Blackband AGARD. Conference proceedings, 28 Technivision, Maidenhead,1970 "...the record of the proceedings of the 14th symposium of the Avionics panel..."-foreward
Eng 41.5726

AGARD Advanced propulsion techniques :Agard combustion and propulsion panel technical meeting Proceedings Pasadena,Calif. 1960 Aug 24-26 Edited by S.S. Penner Pergamon press,London,1961
Eng 41.7373

AGARD Advances in tactical rocket propulsion a colloquium La Jolla,Calif. 1965 Apr 22-23 Edited by S.S. Penner AGARD conference proceedings, 1 Technivision,Slough,1968
Eng 41.6896

AGARD Air intake problems in supersonic propulsion AGARD.Combustion and propulsion panel meeting 11th Paris 1956 Dec Edited by J. Fabri AGARDograph, 27 Pergamon press,London,1958
Eng 41.6753

AGARD Combustion and propulsion:energy sources and energy conversion Agard colloquium 5th Papers Cannes 1964 Mar 16-20 Edited by H.M. DeGroff and others AGARDograph, 81 Gordon and Black,London, 1967
Eng 41.7253

AGARD Combustion researches and reviews 1955 AGARD combustion panel meetings 6th and 7th Papers Scheveningen 1954 May and Paris 1954 Nov Edited by B.P. Mullins AGARDograph, 9 Butterworths,London,1955
Eng 41.7222

AGARD Displays for command and control centers :a conference Proceedings Munich 1966 Nov 11-14 Edited by I.J. Gabelman AGARD.Conference proceedings, 23 Technivision services,Slough,1969 Proceedings of the 11th technical meeting of the AGARD avionics panel
Eng 41.5609

AGARD High frequency radio communications : with emphasis on polar problems Edited by Finn Lied AGARDograph, 104 Technivision, Maidenhead,1967
Eng 41.5677

AGARD History of German guided missiles development :AGARD seminar Munich 1956 Apr Edited by T. Benecke and A.W. Quick AGARDograph, 20 1957
Eng 41.6877

AGARD Navigation systems for aircraft and space vehicles :papers presented at the AGARD avionics panel meeting Istanbul 1960 Oct 3-8 Edited by T.G. Thorne AGARDograph, 55 Pergamon press,London,1962
Eng 41.5706

AGARD New experimental techniques in propulsion and energetics research :a technical meeting Proceedings Munich 1967 Sep 11-15 Edited by David Andrews and Jean Surugue Agard conference proceedings, 38 Technivision,Slough,1970
Eng 41.8359

AGARD Nonsteady flame propagation Edited by George H. Markstein AGARDograph, 75 Pergamon press,London,1964
Eng 41.7246

AGARD Oblique ionospheric radiowave propagation :at frequencies near the lowest usable high frequency:a symposium Rome 1965 AGARD.Conference proceedings, 13 Technivision,Slough,1969 The 11th annual symposium of the AGARD Electromagnetic wave propagation committee
Eng 41.5683

AGARD Phase and frequency instabilities in electromagnetic wave propagation Agard conference proceedings, 33 Technivision, Slough,1970 Contains the proceedings of the Electromagnetic wave propagation committee's thirteenth symposium
Eng 41.8319

AGARD Radar techniques for detection tracking and navigation :8th symposium of the AGARD avionics panel Proceedings London 1964 Sep 21-25 Edited by W.T. Blackband AGARDograph, 100 Gordon and Breach,New York, 1966
Eng 41.5717

AGARD Radio wave propagation :lecture series. held in Leicester...1960 AGARD.Lecture series, 29 1968
Eng 41.5679

AGARD Selected combustion problems Combustion colloquium Cambridge 1953 Dec 7-11 Edited by W.R. Hawthorne and J. Fabri Butterworths,London,1954
A Math 4.0969

AGARD Selected combustion problems: fundamentals and aeronautical applications Combustion colloquium Cambridge 1953 Dec 7-11 Butterworths,London,1954
Eng 41.7219

AGARD Selected combustion problems II: transport phenomena;ignition;altitude behaviour and scaling of aeroengines Combustion colloquium Liege 1959 Dec 5-9 Butterworths,London,1956
Eng 41.7228

AGARD Selected topics on ballistics Cranz centenary colloquium.University of Freiburg Freiburg 1958 Edited by Wilbur C. Nelson NATO-AGARD wind tunnel and model testing panel. Publication AGARDograph, 32 illus,port Pergamon press,London,1959
Eng 41.6772

AGARD Separated flows Pt 1-2 AGARD conference proceedings, 4 2 vols 1966
Eng 41.6531

AGARD.AVIONICS PANEL Advanced radar systems 19th Avionics panel meeting Istanbul 1970 May 25-29 AGARD conference proceedings, 66 Agard,Neuilly-sur-Seine,1970
Eng 41.8322

AGARD.AVIONICS PANEL ANNUAL SYMPOSIUM 8TH Radar techniques for detection tracking and navigation :8th symposium of the AGARD avionics panel Proceedings London 1964 Sep 21-25 Agard Edited by W.T. Blackband AGARDograph, 100 Gordon and Breach,New York, 1966
Eng 41.5717

AGARD.AVIONICS PANEL MEETING Navigation systems for aircraft and space vehicles : papers presented at the AGARD avionics panel meeting Istanbul 1960 Oct 3-8 Agard Edited by T.G. Thorne AGARDograph, 55 Pergamon press,London,1962
Eng 41.5706

AGARD.AVIONICS PANEL SYMPOSIUM 14TH Advanced navigational techniques :symposium Proceedings Milan 1967 Sep 12-15 AGARD Edited by W.T. Blackband AGARD.Conference proceedings, 28 Technivision,Maidenhead, 1970 "...the record of the proceedings of the 14th symposium of the Avionics panel..."-foreward
Eng 41.5726

AGARD.AVIONICS PANEL TECHNICAL MEETING 11TH Displays for command and control centers :a conference Proceedings Munich 1966 Nov 11-14 AGARD Edited by I.J. Gabelman AGARD.Conference proceedings, 23 Technivision services,Slough,1969 Proceedings of the 11th technical meeting of the AGARD avionics panel
Eng 41.5609

AGARD.COMBUSTION AND PROPULSION PANEL Noise, shock tubes,magnetic effects,instability and mixing Agard colloquium 3rd Papers Palermo 1958 Mar 17-21 Edited by M.W. Thring and others Pergamon press,London,1958
Eng 41.7232

AGARD.COMBUSTION AND PROPULSION PANEL MEETING 11th Air intake problems in supersonic propulsion Paris 1956 Dec Agard Edited by J. Fabri AGARDograph, 27 Pergamon press,London,1958
Eng 41.6753

AGARD.CONFERENCE PROCEEDINGS, 13 Oblique ionospheric radiowave propagation :at frequencies near the lowest usable high frequency:a symposium Rome 1965 Agard Technivision,Slough,1969 The 11th annual symposium of the AGARD Electromagnetic wave propagation committee
Eng 41.5683

AGARD.CONFERENCE PROCEEDINGS, 23 Displays for command and control centers :a conference Proceedings Munich 1966 Nov 11-14 AGARD Edited by I.J. Gabelman Technivision services,Slough,1969 Proceedings of the 11th technical meeting of the AGARD avionics panel
Eng 41.5609

AGARD.CONFERENCE PROCEEDINGS, 28 Advanced navigational techniques :symposium Proceedings Milan 1967 Sep 12-15 AGARD Edited by W.T. Blackband Technivision, Maidenhead,1970 "...the record of the proceedings of the 14th symposium of the Avionics panel..."-foreward
Eng 41.5726

AGARD.ELECTROMAGNETIC WAVE PROPAGATION COMMITTEE ANNUAL SYMPOSIUM 11TH Oblique ionospheric radiowave propagation :at frequencies near the lowest usable high frequency:a symposium Rome 1965 Agard AGARD.Conference proceedings, 13 Technivision,Slough,1969 The 11th annual symposium of the AGARD Electromagnetic wave propagation committee
Eng 41.5683

AGARD.PROPULSION AND ENERGETICS AND PANEL MEETING 29TH Performance forecast of selected static energy conversion devices :29th meeting of the AGARD propulsion and energetics panel Liege 1967 Jun 12-16 Edited by G.W. Sherman and L. Devol AGARD,Brussels,1967
Eng 41.4869

AGARD COLLOQUIUM 3rd Papers Noise, shock tubes,magnetic effects,instability and mixing Palermo 1958 Mar 17-21 Agard. Combustion and propulsion panel Edited by M. W. Thring and others Pergamon press,London, 1958
Eng 41.7232

AGARD COLLOQUIUM 5th Papers Combustion and propulsion:energy sources and energy conversion Cannes 1964 Mar 16-20 Agard Edited by H.M. DeGroff and others AGARDograph, 81 Gordon and Black,London, 1967
Eng 41.7253

AGARD COMBUSTION PANEL MEETINGS 6th and 7th Papers Combustion researches and reviews 1955 Scheveningen 1954 May and Paris 1954 Nov Agard Edited by B.P. Mullins AGARDograph, 9 Butterworths,London,1955
Eng 41.7222

AGARD CONFERENCE ON REFRACTORY METALS Proceedings Science and technology of tungsten,tantalum,molybdenum,niobium and their alloys Oslo 1963 Jun 23-26 North Atlantic treaty organization.Advisory group for aeronautical research and development AGARDograph 82 Pergamon press,Oxford,1964
Met 25.1011

AGARD CONFERENCE PROCEEDINGS, 1 Advances in tactical rocket propulsion :a colloquium La Jolla,Calif. 1965 Apr 22-23 Agard Edited by S.S. Penner Technivision,Slough, 1968
Eng 41.6896

AGARD CONFERENCE PROCEEDINGS, 4 Separated flows Pt 1-2 Agard 2 vols 1966
Eng 41.6531

AGARD CONFERENCE PROCEEDINGS, 6 Aircraft flight instrumentation integrated data systems Agard Edited by D. Bosman Technivision, Maidenhead,1967
Eng 41.6867

AGARD CONFERENCE PROCEEDINGS, 33 Phase and frequency instabilities in electromagnetic wave propagation Agard Technivision,Slough, 1970 Contains the proceedings of the Electromagnetic wave propagation committee's thirteenth symposium
Eng 41.8319

AGARD CONFERENCE PROCEEDINGS, 38 New experimental techniques in propulsion and energetics research :a technical meeting Proceedings Munich 1967 Sep 11-15 Agard Edited by David Andrews and Jean Surugue Technivision,Slough,1970
Eng 41.8359

AGARD CONFERENCE PROCEEDINGS, 66 Advanced radar systems 19th Avionics panel meeting Istanbul 1970 May 25-29 Agard.Avionics panel Agard,Neuilly-sur-Seine,1970
Eng 41.8322

AGARD-NATO SPECIALISTS' MEETING The Fluid dynamic aspects of space flight Marseille 1964 Apr 20-24 Vol 1-2 Agard AGARDograph, 87 2 vols Gordon and Breach, New York,1966
Eng 41.6887

AGE HARDENING OF METALS the symposium on precipitation hardening (age hardening) presented before the twenty-first annual convention Papers and discussions Chicago 1939 Oct 23-27 American society for metals American society for metals, Cleveland,1940
Met 25.1249

AGEING Behavior,aging and the nervous system biological determinants of speed of behavior and its changes with age Cambridge 1963 Aug 6-9 Edited by A.T. Welford and James E. Birren American lecture series.Publication, 600,American lectures in geriatrics and gerontology.Bannerstone division monograph Thomas,Springfield,Ill.,1965
Psy 31.2329

AGEING Ciba foundation colloquia on ageing Vol 1-2 Edited by G.E.W. Wolstenholme and others Illus. 2 vols Churchill, London,1955-56 Vol.1:general aspects;vol. 2:ageing in transient tissues
An 32.0344

AGGLOMERATION :based on an international symposium... Philadelphia,Penn. 1961 Apr 12-14 American institute of mining, metallurgical and petroleum engineers Edited by William A. Knepper Interscience,New York; London,1962
Met 25.2288

AGGRESSION The Natural history of aggression a symposium Proceedings London 1963 Oct 21-22 Institute of biology Edited by J.D. Carthy and F.J. Ebling Institute of biology. Symposia, 13 Academic press,London;New York, 1964
Psy 31.0596

AGGRESSION The Natural history of aggression a symposium held at the British museum(natural history) Proceedings London 1963 Oct 21-22 Institute of biology Edited by J.D. Carthy and F.J. Ebling Institute of biology.Symposia, 13 Academic press,London, 1960
Bal 39.1619

AGGRESSION AND DEFENCE:NEURAL MECHANISMS AND SOCIAL PATTERNS (BRAIN FUNCTION,VOL 5) Conference on brain function 5th Proceedings Pacific palisades 1965 Nov 14-17 Edited by Carmine D. Clemente and Donald B. Lindsley UCLA forum in medical sciences, 7 University of California press,Berkeley; Los Angeles,Calif.,1967
An 32.5475

AGGRESSIVE BEHAVIOUR International symposium on the biology of aggressive behaviour Milan 1968 May 2-4 Edited by S. Garattini and E.B. Sigg Excerpta medica foundation,Amsterdam, 1969
Inv Med 37.0086

AGING Symposium on the biology of skin 14th Proceedings Portland,Ore. 1964 May 9-10 Edited by William Montagna Held at the University of Oregon medical school Advances in biology of skin, 6 Oregon regional primate research center, 55 Pergamon press,Oxford,1965
An 32.4041

AGING The Process of aging in the nervous system :a conference Proceedings Bethesda, Md. 1957 Jan 30-Feb 1 Edited by James E. Birren and others Sponsored by the National advisory neurological diseases and blindness council Symposia in neuroanatomical sciences, 5 Blackwell,Oxford,1959
An 32.4171

AGING AROUND THE WORLD International congress of gerontology 5th Proceedings San Francisco 1960 Aug International association of gerontology Edited by Herman T. Blumenthal Columbia university press,New York;London,1962
An 32.0088

AGRICULTURAL CHEMISTRY Liebig and after Liebig:a century of progress in agricultural chemistry A Symposium of papers presented before the sections of chemistry and agriculture of the American association for the advancement of science...in commemoration of the 100th anniversary of the publication of Liebig's 'Organic chemistry in its application to agriculture and physiology' Philadelphia, Pa. 1940 Dec 30 American association for the advancement of science Edited by Forest Ray Moulton American association for the advancement of science.Publication, 16 111p American association for the advancement of science,Washington,D.C.,1942
WSM 43.2238

AGRICULTURAL DEVELOPMENT IN THE SUDAN Philosophical society of the Sudan :annual conference 13th Proceedings and papers Khartoum 1965 Dec 3-6 Vol 1-2 Edited by D.J. Shaw Co-sponsored by the Sudan agricultural society maps 2 vols Philosophical society of the Sudan,Khartoum, 1966 Typescript
Geog 13.4700

AGRICULTURAL RESEARCH COUNCIL Quantitative inheritance colloquium Papers Edinburgh 1950 Apr 4-6 University of Edinburgh. Institute of animal genetics Edited by E.C.R. Reeve and C.H. Waddington H.M.S.O.,London, 1952
Gen 34.1083

AGRICULTURE Plant and animal nutrition in relation to soil and climatic factors British commonwealth scientific official conference.Specialist conference in agriculture 1st Proceedings Adelaide 1949 Aug 22-Sep 15 and Canberra 1949 Aug 22-Sep 15 H.M.S.O.,London,1951
Geog 13.2510

AIR AND WATER POLLUTION IN THE IRON AND STEEL INDUSTRY proceedings of the air pollution meeting 25th and 26th Sept. and the water pollution meeting 11th and 12th Dec Proceedings London 1957 Sep 25-26 London 1957 Dec 11-12 Iron and steel institute Iron and steel institute.Special report, 61 Iron and steel institute,London, 1958
Met 25.0359

AIR FORCE CAMBRIDGE RESEARCH CENTER Conference on extremely high temperatures Boston,Mass. 1958 Mar 18-19 Edited by Heinz Fischer and Lawrence C. Mansur Wiley; Chapman and Hall,New York;London,1958
Met 25.0249

AIR FORCE CAMBRIDGE RESEARCH CENTER.GEOPHYSICAL RESEARCH PAPERS,30 Auroral physics Conference on auroral physics 1st Proceedings London,Ontario 1951 Jul 23-26 Edited by N.C. Gersen and others Sponsored jointly by the University of Western Ontario. Department of physics. illus xxvi,462p Cambridge,Mass.,1954
Sco 14.0613

AIR FORCE CAMBRIDGE RESEARCH CENTER.GEOPHYSICS RESEARCH DIRECTORATE Annual Arctic planning session 2nd Proceedings Bedford, Mass. 1959 Oct Edited by Vivian C. Bushnell Air force Cambridge research center. G.R.D.research notes,29 AFCRC-TN-59-661 illus viii,172p A.F.C.R.C.,Bedford,Mass., 1959
Sco 14.7405

AIR FORCE CAMBRIDGE RESEARCH CENTER.GEOPHYSICS RESEARCH DIRECTORATE Annual arctic planning session 1st Proceedings Boston, Mass. 1958 Nov Edited by J.H. Hartshorn Air force Cambridge research center.G.R.D. research notes,15 AFCRC-TN-59-256,AD 212265 illus viii,65p A.F.C.R.C.,Bedford,Mass., 1959
Sco 14.7404

AIR FORCE CAMBRIDGE RESEARCH CENTER.GEOPHYSICS RESEARCH DIRECTORATE Artificial stimulation of rain Conference on the physics of cloud and precipitation particles 1st Proceedings Woods Hole,Mass. 1955 Sep 7-10 Edited by Helmut Weickmann and Waldo Smith Co-sponsored by United States. Office of naval research.Geophysics branch illus 442p Pergamon,London,1957
Nap 11.0931

AIR FORCE CAMBRIDGE RESEARCH CENTER.GEOPHYSICS RESEARCH DIRECTORATE Cumulus dynamics Conference on cumulus convection 1st Proceedings Portsmouth,N.H. 1959 May 19-22 Edited by Charles E. Anderson ix,211p Pergamon press,Oxford,1960
Nap 11.0012

AIR FORCE CAMBRIDGE RESEARCH CENTER.GEOPHYSICS RESEARCH DIRECTORATE Exploding wires :a conference Papers Boston,Mass. 1959 Apr 2-3 Edited by William G. Chace and Howard K. Moore With the cooperation of Lowell technological institute research foundation Plenum press,New York,1959
Chem 18.0191

AIR FORCE CAMBRIDGE RESEARCH CENTER.GEOPHYSICS RESEARCH DIRECTORATE The Threshold of space :conference on chemical aeronomy Proceedings Cambridge,Mass. 1956 Jun 25-28 Edited by M. Zelikoff Coordinated by Wentworth institute Pergamon press,London, 1957
Chem 18.0478

AIR FORCE CAMBRIDGE RESEARCH CENTER.GEOPHYSICS RESEARCH DIRECTORATE The threshold of space Conference on chemical aeronomy Proceedings Cambridge,Mass. 1956 Jun 25-28 Edited by M. Zelikoff illus xi,342p Pergamon press,London,1957
Nap 11.0654

AIR FORCE CAMBRIDGE RESEARCH CENTER.GEOPHYSICS RESEARCH DIRECTORATE.AEROPHYSICS LABORATORY Recent advances in atmospheric electricity Conference on atmospheric electricity 2nd Proceedings Portsmouth,N.H. 1958 May 20-23 Edited by L.G. Smith illus 646p Pergamon, London,1958
Nap 11.0928

AIR FORCE CAMBRIDGE RESEARCH LABORATORIES
Annual Arctic planning session 3rd Proceedings Bedford,Mass. 1960 Nov Edited by George P. Rigsby and Vivian C. Bushnell Air force Cambridge research laboratories.G.R.D.research notes,55 AFCRL 436 illus viii,148p A.F.C.R.L.,Bedford, Mass.,1961
Sco 14.7406

AIR FORCE CAMBRIDGE RESEARCH LABORATORIES
Crystal growth International conference on crystal growth Proceedings Boston,Mass. 1966 Jun 20-24 Edited by H.Steffen Peiser Physics and chemistry of solids.Supplement Pergamon press,Oxford,1967
Met 25.1520

AIR FORCE CAMBRIDGE RESEARCH LABORATORIES
The Conference on electromagnetic scattering : interdisciplinary conference Papers Potsdam,N.Y. 1962 Aug 13-15 Edited by Milton Kerker Co-sponsored by the American chemical society.Division of colloid and surface chemistry International series of monographs on electromagnetic waves, 5 591p Pergamon press,Oxford,1963
TA 15.0256

AIR FORCE CAMBRIDGE RESEARCH LABORATORIES
The Lunar surface layer materials and characteristics :a conference proceedings Boston 1963 May Edited by John W. Salisbury and Peter E. Glaser 532p Academic Press,New York,1964
Obs 6.1507

AIR FORCE CAMBRIDGE RESEARCH LABORATORIES
The Solar corona :a symposium proceedings Cloudcroft,N.M. 1961 Aug.28-30 International astronomical union Edited by John W. Evans International astronomical union.Symposium, 16 344p Academic press, New York;London,1963
Obs 6.0564

AIR FORCE CAMBRIDGE RESEARCH LABORATORIES.DATA SCIENCE LABORATORY Models for the perception of speech and visual form :a symposium Proceedings Boston,Mass. 1964 Nov 11-14 Edited by Weiant Wathen-Dunn MIT press,Cambridge,Mass.,1967
Psy 31.2951

AIR FORCE CAMBRIDGE RESEARCH LABORATORIES. ELECTROMAGNETIC RADIATION LABORATORY
Electromagnetic aspects of hypersonic flight Symposium on the plasma sheath :its effect upon reentry communication and detection 2nd Proceedings Boston,Mass. 1962 Apr 10-12 Edited by Walter Rotman and others Spartan books;Cleaver-Hume,Baltimore,Md.; London,1964
Eng 41.4544

AIR FORCE CAMBRIDGE RESEARCH LABORATORIES. ELECTRONICS RESEARCH DIRECTORATE
Ultrapurification of semiconductor materials : a conference Proceedings Boston,Mass. 1961 Apr 11-13 Edited by M.S. Brooks and J.K. Kennedy Macmillan,New York,1962
Met 25.1149

AIR FORCE CAMBRIDGE RESEARCH LABORATORIES. GEOPHYSICS RESEARCH DIRECTORATE The Theory of electro-magnetic waves :a symposium Proceedings New York 1950 Jun 6-8 Washington square college of arts and science New York university.Institute for mathematics and mechanics Interscience,New York;London, 1951
Radioth 35.1507

AIR INTAKE PROBLEMS IN SUPERSONIC PROPULSION
AGARD.Combustion and propulsion panel meeting 11th Paris 1956 Dec Agard Edited by J. Fabri AGARDograph, 27 Pergamon press, London,1958
Eng 41.6753

AIR POLLUTION A Symposium on atmospheric diffusion and air pollution proceedings Oxford 1958 Aug 24-29 International union of theoretical and applied mechanics and International union of geodesy and geophysics Edited by F.N. Frenkel and P.A. Sheppard Advances in geophysics, 6 Academic press, New York;London,1959
A Math 4.0576

AIR POLLUTION Air and water pollution in the iron and steel industry proceedings of the air pollution meeting 25th and 26th Sept. and the water pollution meeting 11th and 12th Dec Proceedings London 1957 Sep 25-26 London 1957 Dec 11-12 Iron and steel institute Iron and steel institute.Special report, 61 Iron and steel institute,London,1958
Met 25.0359

AIR POLLUTION Atmospheric diffusion and air pollution :a symposium Proceedings Oxford 1958 Aug 24-29 International union of theoretical and applied mechanics International union of geodesy and geophysics Edited by F.N. Frenkel and P.A. Sheppard Advances in physics, 6 Academic press,New York,1959
Eng 41.6946

AIR POLLUTION Conference on air pollution Proceedings Washington,D.C. 1950 May 3-5 United States.Bureau of mines Edited by Louis C. McCabe Sponsored by United States. Interdepartmental committee on air pollution xiv,847p McGraw-Hill,New York,1952
Nap 11.0388

AIR POLLUTION European congress on the influence of air pollution on plants and animals 1st Proceedings Wageningen 1968 Apr 22-27 Centre for agricultural publishing and documentation illus. 415p Centre for agricultural publishing and documentation,Wageningen,1969
Bot 42.2106

AIR POLLUTION United States technical conference on air pollution Proceedings Washington 1950 May 3-5 Sponsored by the United States.Interdepartmental committee on air pollution xiv,847p McGraw-Hill,New York,1952 Chairman of the committee:Louis C.McCabe
Chem E 24.1542

AIR POLLUTION :based on papers given at a conference held at the University of Sheffield Sheffield 1956 Sep Edited by M.W. Thring 245p Butterworths,London,1957
Nap 11.0607

AIR POLLUTION :based on papers given at a conference held at the University of Sheffield September,1956 Edited by M.W. Thring x, 248p Butterworths,London,1957
Chem M 24.1240

AIR POLLUTION CONTROL IN TRANSPORT ENGINES :a symposium Proceedings London 1971 Nov 9-11 Institution of mechanical engineers. Automobile division Advanced school engineers.Combustion engines group IME, London,1972
Eng 41.8645

AIR RESEARCH AND DEVELOPMENT COMMAND Research on ferroresonant computer and control devices mimeograph Consejo superior de investigaciones cientificas.Instituto de electricidad y automatica Technical note no. 1 bibliog,pl. ix,104p 28cm Instituto de electricidad y automatica,Madrid,1958
Math L 5.0803

AIRBORNE MICROBES :a symposium Papers London 1967 Apr Society for general microbiology Edited by P.H. Gregory and J.L. Monteith Held at the Imperial college of science and technology Society for general microbiology.Symposia, 17 Cambridge university press,Cambridge,1967
Bioch 33.1167

AIRCRAFT STRUCTURES Fatigue in aircraft structures :an international conference New York 1956 Jan 30-Feb 1 United States.Air research and development command Guggenheim institute of flight structures Edited by Alfred M. Freudenthal Academic press,New York,1956
Eng 41.2772

AIRCRAFT STRUCTURES Fatigue of aircraft structures Los Angeles,Calif. 1956 Sep 16-21 American society for testing materials American society for testing materials.Special technical publication, 203 American society for testing materials,Philadelphia,Pa.,1957
Eng 41.2776

AIRCRAFT WAKE TURBULENCE AND ITS DETECTION :a symposium Proceedings Seattle,Wash. 1970 Sep 1-3 Boeing company.Scientific research laboratories Edited by J.H. Olsen and others Plenum press,New York,1971
Eng 41.8333

The AIRGLOW AND THE AURORAE :a symposium Belfast 1955 Sep Edited by E.B. Armstrong and A. Dalgarno Journal of atmospheric and terrestial physics.Special supplement,5 illus x,420p 25cm Pergamon press,London; New York,1956
Sco 14.0598

The AIRGLOW AND THE AURORAE :a symposium Belfast 1955 Sep.6-7 Edited by E.B. Armstrong and A. Dalgarno Journal of atmospheric and terrestrial physics. Supplements., 5 420p Pergamon press, London,1956
Obs 6.0083

The AIRGLOW AND THE AURORAE :a symposium Belfast 1955 Sep.6-7 Edited by E.B. Armstrong and A. Dalgarno Journal of atmospheric and terrestrial physics. Supplements, 5 420p London,1956
Nap 11.0014

AIX-EN-PROVENCE 1961 The International conference on elementary particles Proceedings Vol 1-2 Edited by E. Cremieu-Alcan and others Sponsored by Commissariat a l'energie atomique 2 vols C. E.N.Saclay,Gif-sur-Yvette,1962
Cav 7.0029

AKADEMIE DER WISSENSCHAFTEN Einstein-Symposium Entstehung,Entwicklung und perspektiven der Einsteinschen gravitationstheorie vortrage und diskussionen Berlin 1965 Nov 2-5 Edited by H.J. Treder Akademie Verlag,Berlin,1965 typescript
A Math 4.1031

AKADEMIYA MEDITSINSKIKH NAUK S.S.S.R. Fosforilirovanie i funktsiya :sympozium Leningrad 1958 Jun 13-18 Akademiya meditsinskikh nauk S.S.S.R.Trudy instituta eksperimentalnoi meditsiny Leningrad,1960 Short summaries in English at the end of each chapter
Bioch 33.0561

AKADEMIYA NAUK BELORUSSKOI S.S.R.INSTITUT TORFA Sapropeli ikh ispolzovanie :po matrialam konferentsii 1956 Izdatelstvo akademii nauk B.S.S.R.,Minsk,1958
Bot 42.6733

AKADEMIYA NAUK S.S.S.R Obshchee sobranie... posvyashchennie trudtsatiletiyu velikoi oktyabr'skoi sotsialistichesk revolyutsii Moscow 1947 Akademiya nauk S.S.S.R,Moscow, Leningrad,1948
Philos 1.0032

AKADEMIYA NAUK S.S.S.R. Brain reflexes :the international conference dedicated to the centenary celebration of the publication of I. M.Sechenov's book "Brain reflexes" Moscow Edited by E.A. Asratyan Progress in brain research, 22 Elsevier,Amsterdam,1968 Sponsored also by the International brain research organization
An 32.4243

AKADEMIYA NAUK S.S.S.R. Initial effects of ionizing radiations on cells International symposium on primary and initial effects of ionizing radiations on living cells Papers: discussions Moscow 1960 Oct Edited by R. J.C. Harris Academic press,London;New York, 1961
An 32.3277

AKADEMIYA NAUK S.S.S.R. Initial effects of ionizing radiations on cells The International symposium on primary and initial effects of ionizing radiations on living cells Papers and discussions Moscow 1960 Oct Edited by R.J.C. Harris Academic press, London;New York,1961 A symposium supported by Unesco and the IAEA and sponsored by the Academy of sciences of the U.S.S.R.
Bioch 33.0961

AKADEMIYA NAUK S.S.S.R. International conference on low temperature physics 10th Proceedings Moscow 1966 Aug 31-Sep 6 Vol 1: properties of helium Edited by M.P. Malkov 552p 22cm Moscow,1967
Cav 7.3030

AKADEMIYA NAUK S.S.S.R. International conference on low temperature physics 10th Proceedings Moscow 1966 Aug 31-Sep 6 Vol 3: the electronic properties of metals 404p 22cm Moscow,1967
Cav 7.3032

AKADEMIYA NAUK S.S.S.R. International conference on low temperature physics 10th Proceedings Moscow 1966 Aug 31-Sep 6 Vol 4: antiferromagnetism Edited by M.P. Malkov 361p 22cm Moscow,1967
Cav 7.3033

AKADEMIYA NAUK S.S.S.R. International conference on low temperature physics 10th Proceedings Vol 2 a-b: superconductivity Edited by M.P. Malkov 22cm 2 vols
Cav 7.3031

AKADEMIYA NAUK S.S.S.R. International conference on the physics of semiconductors 9th Proceedings Moscow 1968 Jul 23-29 Vol 1-2 English edition 634p 27cm 2 vols Publishing house 'Nauka',Leningrad, 1968
Cav 7.2955

AKADEMIYA NAUK S.S.S.R. Obshchee sobranie... Moscow 1943 Akademiya nauk S.S.S.R,Moscow, 1944
Philos 1.0031

AKADEMIYA NAUK S.S.S.R. Origin of life on the earth The International symposium on the origin of life on the earth 1st Proceedings Moscow 1957 Aug 19-24 Edited by A.I. Oparin and others Organized under the auspices of the International union of biochemistry I.U.B.symposium series, 1 Pergamon,London,1959 "English-French-German edition edited for the International union of biochemistry by F.Clark and R.L.M. Synge".
Bioch 33.0727

AKADEMIYA NAUK S.S.S.R. Science,technology and application of titanium International conference on titanium Proceedings London 1968 May 21-24 Edited by R.I. Jaffee and N.E. Promisel Pergamon,Oxford,1970
Met 25.2852

AKADEMIYA NAUK S.S.S.R. Sessiya Akademii nauk SSSR po nauchnym problemam avtomatizatsii proizvodstva :plenarnye zasedaniya Moscow 1956 Oct 15-20 Edited by D.Ya. Libenson and others 271p 22cm Akademiya nauk SSSR, Moscow,1957
Math L 5.0775

AKADEMIYA NAUK S.S.S.R. The Initial effects of ionizing radiations on cells :a symposium Moscow 1960 Oct Unesco International atomic energy agency Edited by R.J.C. Harris Academic press,London;New York,1961
Gen 34.2222

AKADEMIYA NAUK S.S.S.R. The Initial effects of ionizing radiations on cells :a symposium Moscow 1960 Oct Unesco International atomic energy agency Edited by R.J.C. Harris Academic press,London;New York,1961
Radioth 35.1127

AKADEMIYA NAUK S.S.S.R.ASTRONOMICAL COUNCIL Optical instability of the earth's atmosphere All-union conference 3rd Proceedings Kiev 1962 July Translated by Z. Lerman from the Russian viii,174p 25cm Israel program for scientific translations,Jerusalem, 1966
Obs 6.3275

AKADEMIYA NAUK S.S.S.R.INSTITUT MERZLOTOVEDENIYA IM V.A.OBRUCHEVA Materially po obshchemu merzlotovedeniyu :mezhduvedomstvennoe coveshchanie po merzlotovedeniyu 7 Papers Moscow 1956 Edited by V.F. Zhukov and I.Ya. Baranov 271p 26cm Moscow,1959 Papers read at the 7th interdepartmental conference on permafrost studies, Moscow,1956
Sco 14.0399

AKADEMIYA NAUK S.S.S.R.INSTITUT MERZLOTOVEDENIYA IM V.A.OBRUCHEVA Materialy po fizike i mekhanike merzlykh gruntov : mezhduvedomstvennoe soveshchanie po merzlotovedeniyu 7 Papers Moscow 1956 Edited by N.A. Tsytovich 107p 26cm Moscow,1959
Sco 14.0398

AKADEMIYA NAUK S.S.S.R.INSTITUT NAUCHNOI INFORMATSII.OTDEL MATEMATIKI Geometriceskogo seminara trudy Moscow 1963 Tom 1 456p 22cm VINITI, Moscow,1966
P. Math 2.0699

AKADEMIYA NAUK S.S.S.R.MINISTERSTVO PO RYBNOGO KHOZYAISTVA S.S.S.R.IKHTIOLOGICHESKAYA KOMISSIYA Morskie mlekopitayushchie Sbornik uklyuchaet materialy tretego vsesoyuznogo soveshchaniya po morskim mlekopitayushchim Vladivostok 1966 Edited by A.V.A. Arsenev and others 342p 27cm Izdatelstvo 'Nauka',Moscow,1969
Sco 14.8244

AKADEMIYA NAUK S.S.S.R.OTDELENIE BIOLOGICHESKIKH NAUK Sessiya Akademii nauk S.S.S.R. po mirnomu ispolzovaniyu atomnoi energii... : zasedaniya otdeleniya biologicheskikh nauk Geneva 1955 Jul 1-5 Izdatelstvo Akademii nauk SSSR,Moscow,1955 Five volumes in box
Chem 18.0878

AKADEMIYA NAUK S.S.S.R.OTDELENIE BIOLOGICHESKIKH NAUK Uspekhi biologicheskikh nauk v S.S.S.R. za 25 let, 1917-42 :symposium Edited by L.A. Orbeli 356p 26cm Moscow,1945
Philos 1.1394

AKADEMIYA NAUK S.S.S.R.OTDELENIE KHIMICHESKIKH NAUK Voprosy khimicheskoi kinetiki kataliza i reaktsionnoi sposobnosti :doklady k vsesoyuznomu soveshchaniyu po khimicheskoi kinetike i reaktsionnoi sposobnosti Doklady Edited by V.N. Kondratev and N.M. Emanuel Izdatelstvo Akademii nauk SSSR,Moscow,1955
Chem 18.0863

AKADEMIYA NAUK S.S.S.R.SOVET PO IZUCHENIYU PROIZVODITELNYKH SIL Razvitie proizvoditelnykh sil Vostochnoi Sibiri : chernaya metallurgiya Trudy 1958 Aug 18-26 Edited by G.I. Lyudogovskii maps 275p 27cm Izdatelstvo Akademii nauk SSSR,Moscow, 1966
Sco 14.8216

AKADEMIYA PEDAGOGICHESKIKH NAUK R.S.F.S.R. INSTITUT PSIKHOLOGII Materialy soveshchaniya po psikhologii 1955 Jul 1-6 Edited by G.N. Voskresenskii and others Izdatekstvo Akademii pedagogicheskikh nauk RSFSR,Moscow,1957
Psy 31.2576

The ALASKAN SCIENCE CONFERENCE 15th Papers Review of research on military problems in cold regions Fairbanks 1964 Aug 31-Sep 4 By Charles R. Kolb and Fritz M.G. Holmstrom University of Alaska Under the auspices of the American association for the advancement of science 1964 Privately printed
Sco 14.0155

ALBANY,N.Y. 1968 National conference on weather modification 1st Proceedings American meteorological society State university of New York 532p American meteorological society,Boston,Mass.,1968
Nap 11.0975

ALBERT EINSTEIN MEDICAL CENTER Conference on murine leukemia Papers Philadelphia,Pa. 1965 Oct 13-15 Edited by Marvin A. Rich and John B. Moloney National cancer institute. Monograph, 22 US government printing office, Washington,D.C.,1966
Path 30.2460

ALBERTA AND NORTHWEST CHAMBER OF MINES AND RESOURCES and EDMONTON CHAMBER OF COMMERCE Last frontier in North America The National northern development conference and post-conference air tour,September 17-21,1958 Proceedings Edmonton,Alta. 1958 Sep 17-21 illus,map 184p 28cm Edmonton,Alta.,c1958
Sco 14.3709

ALBERTA AND NORTHWEST CHAMBER OF MINES AND RESOURCES and EDMONTON CHAMBER OF COMMERCE National northern development conference 3rd Proceedings Edmonton,Alberta 1964 Oct 21-23 illus 170p 28cm Alberta, c1965
Sco 14.7424

ALBERTA SOCIETY OF PETROLEUM GEOLOGISTS Geology of the Arctic First international symposium on Arctic geology Calgary 1960 Jan 11-30 Vol 1-2 Edited by Gilbert O. Raasch illus,maps 26cm 2 vols University of Toronto press,Toronto,1961
Sco 14.2403

ALBERTA SOCIETY OF PETROLEUM GEOLOGISTS Geology of the Arctic International symposium on arctic geology 1st Proceedings Calgary,Alta. 1960 Jan 11-13 Vol 1-2 Edited by Gilbert O. Raasch 2 vols Toronto,1961 Additional maps in a separate container
Geod 9.0436

ALBERTA SOCIETY OF PETROLEUM GEOLOGISTS Geology of the Arctic International symposium on arctic geology 1st proceedings Calgary,Alberta 1960 Jan 11-13 Edited by Gilbert O. Raasch maps 3 vols University of Toronto press,Toronto,1961
Geol 8.2938

ALBERTA SOCIETY OF PETROLEUM GEOLOGISTS Geology of the arctic International symposium on Arctic geology 1st Proceedings Calgary 1960 Jan 11-13 Vol 1-3 Edited by Gilbert O. Raasch 3 vols University of Toronto press,Toronto,1961 Vol.3 consists of maps
Geog 13.5390

ALBUQUERQUE,N.M. 1962 Symposium on dynamic behavior of materials Sponsored by the American society for testing and materials American society for testing and materials. Special technical publication, 336 American society for testing and materials.Materials science series, 5 American society for testing and materials,Philadelphia,Pa.,1963
Eng 41.3788

ALBURQUERQUE,NEW MEXICO 1957 Metals for supersonic aircraft and missiles Heat tolerant metals for aerodynamic applications : a conference Proceedings American society for metals and University of New Mexico Edited by D.W. Grobecker American society for metals,Cleveland,Ohio,1958
Met 25.1003

ALDOSTERONE International symposium on aldosterone Proceedings Prague 1963 Council for international organization of medical sciences Edited by Etienne Emile Baulier and P. Robel Blackwell,Oxford,1964
Inv Med 37.0013

ALDOSTERONE The International symposium on aldosterone Papers and discussions Geneva 1957 Jun 7-8 Edited by Alex F. Muller and Cecilia M. O'Connor Organized by the Universite de Geneve.Clinique therapeutique J.and A.Churchill,London,1958
Bioch 33.0416

ALDOSTERONE :a symposium Prague 1963 Aug 23-25 Edited by E.E. Baulieu and P. Robel Organized by the Council for international organizations of medical sciences Blackwells,Oxford,1964
Phys 20.1398

ALEXANDRA P. and BACQ Z.M. ed. Radiobiology symposium proceedings Liege 1954 Aug-Sep London,1955
Philos 1.1413

ALFRED,N.Y. 1958 Conference on non-crystalline solids National research council and United States.Air research and development command Edited by V.D. Frechette John Wiley,New York;London,1960
Met 25.0126

ALFRED,N.Y. 1967 Kinetics of reactions in ionic systems International symposium on special topics in ceramics Proceedings Edited by T.J. Gray and V.D. Frechette Held at Alfred university Materials science research, 4 Plenum press,New York,1969
Met 25.2750

ALFRED BENZON FOUNDATION Role of nucleotides for the function and conformation of enzymes The Alfred Benzon symposium 1 Proceedings Copenhagen 1968 Sep 9-11 Edited by Herman M. Kalckar Held at the premises of the Royal Danish academy of sciences and letters Munksgaard,Copenhagen,1969
Bioch 33.1913

The ALFRED BENZON SYMPOSIUM 1 Proceedings Role of nucleotides for the function and conformation of enzymes Copenhagen 1968 Sep 9-11 Alfred Benzon foundation Edited by Herman M. Kalckar Held at the premises of the Royal Danish academy of sciences and letters Munksgaard,Copenhagen,1969
Bioch 33.1913

The ALFRED BENZON SYMPOSIUM 2nd Proceedings Capillary permeability:the transfer of molecules and ions between capillary blood and tissue Copenhagen 1969 Jun 23-26 Edited by Christian Crone and Niels A. Lassen Munksgaard,Copenhagen,1970
Inv Med 37.0054

ALFRED P.SLOAN FOUNDATION New directions in mathematics conference Hanover,N.H. 1961 Nov.3-4 Dartmouth college.Department of mathematics Edited by Robert W. Ritchie iv, 124p 22cm Prentice-Hall,Englewood Cliffs,N. J.,1964
P. Math 2.2558

ALGAE The Culture of algae :a symposium Proceedings New York 1969 Sep 7-12 Edited by J.E. Zajic Sponsored by the American chemical society.Division of microbial chemistry and technology illus, tables ix,154p Plenum press,New York; London,1970
Bot 42.3942

ALGAE The Ecology of algae :a symposium Pymatuning,Pa. 1959 Jun 18-19 Edited by C. A. Tryon and R.T. Hartman Pymatuning laboratory of field biology.Special publication, 2 Pittsburgh,Pa.,1960
Bot 42.3900

ALGAE AND MAN Louisville,Ky. 1962 Jul 22-Aug 11 Edited by Daniel F. Jackson illus. x,434p Plenum press,New York,1964 Based on lectures presented at the NATO advanced study institute
Bot 42.3911

ALGEBRA Papers from the "Open house for algebraists" June-July,1970 at Aarhus universitet... Aarhus universitet.Matematisk institut Aarhus universitet.Matematisk institut.Various publications series, 17 bibliog. iv,161p 29cm Aarhus universitet, matematisk institut,Aarhus,1970
P Math 2.3928

ALGEBRAIC GEOMETRY Fondements de la geometrie algebrique moderne Montreal 1964 By Jean Dieudonne North Atlantic treaty organization Societe mathematique du Canada University of Montreal.Seminaire de mathematiques superieures, 8 151p 28cm Universite de Montreal,Montreal,1964
P. Math 2.0507

ALGEBRAIC GEOMETRY Lecture notes prepared in connection with the summer institute on algebraic geometry typescript Woods Hole, Mass. 1964 Jul 6-31 By Shreeram S. Abhyankar and others American mathematical society 29cm American mathematical society, 1964
P. Math 2.0575

ALGEBRAIC GEOMETRY :a colloquium Papers Bombay 1968 Jan 16-23 By S.S. Abhyankar and others viii,426p 26cm Oxford university press,London,1969
P Math 2.3376

ALGEBRAIC GEOMETRY AND TOPOLOGY :a symposium in honor of S.Lefschetz Princeton,N.J. 1954 Apr 8-10 Edited by R.H. Fox and others Princeton mathematical series, 12 viii,399p 24cm Princeton university press,Princeton,N. J.,1957
P. Math 2.0574

ALGEBRAIC GEOMETRY OSLO 1970 Nordic-summer school in mathematics 5th Proceedings Oslo 1970 Aug 5-25 Edited by F. Oort vii, 332p 25cm Wolters-Noordoff,Groningen,1972
P Math 2.4173

ALGEBRAIC GROUP THEORY Colloque sur la theorie des groupes algebriques 20th colloque Brussells 1962 Jun 5-7 Centre Belge de recherches mathematiques 150p 25cm Librairie universitaire,Louvain, 1962
P. Math 2.0573

ALGEBRAIC GROUPS AND DISCONTINUOUS SUBGROUPS Summer mathematical institute 12th Boulder,Colo. 1965 Jul 5-Aug 6 American mathematical society Edited by Armand Borel and George D. Mostow Financed by the National science foundation American mathematical society.Proceedings of symposia in pure mathematics, 9 vii,426p 26cm American mathematical society,Providence,R.I., 1966
P Math 2.2719

ALGEBRAIC K-THEORY AND ITS GEOMETRIC APPLICATIONS Edited by R.M.F. Moss and C.B. Thomas Lecture notes in mathematics, 108 Bibliog. iv,86p 25cm Springer-Verlag,Berlin,1969 Texts of lectures given at a conference held at the University of Hull, Mar 17-21,1969
P Math 2.3355

ALGEBRAIC NUMBER THEORY International symposium on algebraic number theory proceedings Tokyo 1955 Sep 8-13 Nikko Science council of Japan Edited by Shokichi Iyanaga and Yukiyosi Kawada Jointly sponsored by the International mathematical union xx,265p 26cm Science council of Japan,Tokyo,1956
P. Math 2.1771

ALGEBRAIC NUMBER THEORY conference proceedings Brighton 1965 Sep 1-17 London mathematical society Edited by John W. S. Cassels and A. Frohlich With the support of the International mathematical union xviii,366p 24cm Academic press,New York; London,1967
P Math 2.2773

ALGEBRAIC STRUCTURES Les Probabilites sur les structures algebriques Clermont-Ferrand 1969 Jun 30-Jul 5 Centre national de la recherche scientifique Centre national de la recherche scientifique.Colloques internationaux, 186 361p 25cm CNRS,Paris, 1970
P Math 2.4558

ALGEBRAIC STRUCTURES Les Probabilites sur les structures algebriques :colloque international Clermont-Ferrand 1969 Jun 30-Jul 5 Centre national de la recherche scientifique Centre national de la recherche scientifique.Colloques internationaux, 186 361p 25cm Editions du CNRS,Paris,1970 Oeganized by A.Badrikian and P.L.Hennequin
Math S 3.1780

ALGEBRAIC TOPOLOGY Colloquium on algebraic topology Aarhus 1962 Aug 1-10 Aarhus universitet.Matematisk institut Supported by North Atlantic treaty organization.Scientific affairs division 113p 29cm Aarhus universitet,1962
P. Math 2.0293

ALGEBRAIC TOPOLOGY Seminar on algebraic topology Proceedings Seattle,Wash. 1971 Feb 22-26 Battelle memorial institute. Seattle research center Edited by Peter J. Hilton Lecture notes in mathematics, 249 vi,110p 25cm Springer,Berlin,1971
P Math 2.4115

ALGEBRAIC TOPOLOGY Symposium internacional de topologia algebraica Mexico City 1956 Aug Universidad nacional de Mexico.Instituto de matematicas Instituto nacional de la investigacion cientifica,Mexico Sociedad matematica mexicana xii,334p 26cm UNESCO, Mexico City,1958 In memory of Witold Hurewicz
P. Math 2.0290

ALGEBRAIC TOPOLOGY Proceedings Aarhus 1970 Aug 10-23 Aarhus universitet.Matematisk institut Aarhus universitet.Matematisk institut.Various publications series, 13 bibliog. 29cm 3 vols Matematisk institut, Aarhus universitet,Aarhus,1970
P Math 2.3822

ALGEBRAIC VARIETIES Questions on algebraic varieties Centro matematico estivo 3 ciclo Varenna 1969 Sep 7-17 Centro matematico estivo 343p 28cm Edizioni cremonese,Rome,1970 Conference coordinator:E.Marchionna
P Math 2.4172

ALGEBRAISCHE ZAHLENTHEORIE conference an account Oberwolfach 1964 Sep 6-12 Mathematisches forschungsinstitut,Oberwolfach Edited by Helmut Hasse and Peter Roquette Mathematisches forschungsinstitut,Oberwolfach, 2 264p 19cm Bibliographisches institut, Mannheim,1966
P. Math 2.1769

ALGIERS 1951 Actions eoliennes.Phenomenes d'evaporation et d'hydrologie superficielle dans les regions arides :colloque international Centre national de la recherche scientifique.Colloques internationaux,35 C.N.R.S.,Paris,1953
Geog 13.0671

ALGIERS 1952 Genese des roches filoniennes, fasc.6 International geological congress 19th Proceedings,section 6 Algiers,1953
Min 10.0351

ALGIERS 1952 Paleovolcanologie et ses rapports avec la tectonique,fasc.17 International geological congress 19th Proceedings,section 15 Algiers,1954
Min 10.0352

ALGIERS 1952 Symposium sur les gisements de fer du monde International geological congress 19th Tomes 1-2 Edited by F. Blondel and L. Marvier Algiers,1952
Geol 8.3011

ALGIERS 1952 Symposium sur les series de Gondwana International geological congress 19th papers Edited by Curt Teichert Algiers,1952 Text in English,French and German
Geol 8.3012

ALGIERS 1953 Transfusion sanguine et actualites hematologiques :congres national des transfusions sanguines de France et des pays de langue francaise 1st Edited by E. Benhamou and A. Albou Masson et compagnie, Paris,1954
Med 36.0023

ALGOL Informal conference on ALGOL 68 implementations Proceedings Vancouver,B.C. 1969 Aug 29-30 University of British Columbia.Department of computer science Edited by J.E.L. Peck illus 119p University of British Columbia.Department of computer science,Vancouver,B.C.,1969
Math L 5.3845

ALGOL 68 IMPLEMENTATION IFIP working conference on ALGOL and its implementation Proceedings Munich 1970 Jul 20-24 International federation for information processing Edited by J.E.L. Peck illus 375p North-Holland,Amsterdam,1971
Math L 5.3876

ALGUNOS PROBLEMAS MATEMATICOS QUE SE ESTAN ESTUDIANDO EN LATINO AMERICA 2 symposium Villavicencio-Mendoza 1954 Jul.21-25 Universidad nacional,Mendoza Supported by United nations educational,scientific and cultural organisation. Centre for scientific co-operation for Latin America 328p 24cm UNESCO,centro de cooperacion cientifica, Montevideo,1954
P. Math 2.2610

ALIGARH 1967 Symposium on cosmic rays, elementary particle physics and astrophysics 10th Proceedings India.Department of atomic energy.Cosmic ray committee 701p Bombay,1968
TA 15.0553

ALIMENTARY TRACT Tumors of the alimentary tract in Africans :symposium Proceedings Geneva 1965 Nov 29-Dec 5 National cancer institute Edited by J.F. Murray Sponsored by the International union against cancer and the Calouste Gulbenkian foundation National cancer institute.Monograph, 25 National cancer institute,Bethesda,Md.,1967
Bioch 33.0990

The ALKALI METALS :an international symposium Nottingham 1966 Jul 19-22 Chemical society,and,University of Nottingham Chemical society.Special publication,22 Chemical society,London,1966
Chem 18.1226

ALL-UNION CONFERENCE 3rd Proceedings Optical instability of the earth's atmosphere Kiev 1962 July Akademiya nauk S.S.S.R. Astronomical council Glavnaya astronomicheskaya observatoriya,Pulkovo Translated by Z. Lerman from the Russian viii,174p 25cm Israel program for scientific translations,Jerusalem,1966
Obs 6.3275

ALL-UNION SCIENTIFIC AND TECHNICAL CONFERENCE ON THE APPLICATION OF RADIOACTIVE ISOTOPES : a portion of the proceedings...in English translation Radiobiology Moscow 1957 Consultants bureau,New York,1959
Radioth 35.1731

The ALL-BASIC OPEN-HEARTH FURNACE being also the report of the 36th steelmaking conference Leamington-Spa 1951 May 2-3 British ceramic research association and British iron and steel research association Iron and steel institute.Special report, 46 Iron and steel institute,London,1952
Met 25.0330

ALL-UNION SEMINAR ON APPLIED ELECTROCHEMISTRY 5th Reports Electrometallurgy of chloride solutions Dnepropetrovsk 1962 Oct 17-19 Dnepropetrovskii khimiko-tekhnologicheskii institut and Moskovskii khimiko-tekhnologicheskii institut imeni D.I. Mendeleyeva Edited by V.V. Stender Translated by Consultants bureau from the Russian Consultants bureau,New York,1965
Met 25.1843

ALLAHABAD 1958 Recent advances in the study of plant metabolism :a symposium Proceedings Sponsored by the University grants commission Allahabad university studies.Botany section.Symposium supplement port. vi,140p Allahabad,1958
Bot 42.4704

ALLERTON CONFERENCE ON CIRCUIT AND SYSTEM THEORY 1st-6th Proceedings Monticello,Ill. 1963-68 Sponsored by the University of Illinois 6 vols University of Illinois, Evanston,Ill.,1963-68
Eng 41.5240

ALLOYD ELECTRONICS CORPORATION Symposium on electron beam processes 3rd Proceedings Boston,Mass. 1961 Mar 23-24 Edited by Robert Bakish Alloyd,Cambridge,Mass.,1961
Eng 41.5615

ALLOYD ELECTRONICS CORPORATION Symposium on electron beam technology 4th Proceedings Boston,Mass. 1962 Mar 29-30 Edited by R. Bakish Alloyd,Cambridge,Mass.,1962
Eng 41.5616

ALLOYD ELECTRONICS CORPORATION,CAMBRIDGE,MASS The Symposium on electron beam processes 3rd Proceedings Boston,Mass. 1961 Mar 23-24 Edited by R. Bakish 379p Alloyd electronics,Cambridge,Mass.,1961
Cav 7.0113

ALLOYD GENERAL CORPORATION Electron and laser beam symposium 7th Proceedings University Park,Pa. 1965 Mar 31-Apr 2 Pennsylvania state university Pennsylvania state university,University Park,Pa.,1965
Eng 41.5617

ALLOYD RESEARCH CORPORATION Symposium on electron beam melting 1st Proceedings Boston,Mass. 1959 Mar 20 Edited by James S. Hetherington Alloyd,Watertown,Mass.,1959
Eng 41.5613

ALLOYD RESEARCH CORPORATION Symposium on electron beam processes 2nd Proceedings Boston,Mass. 1960 Mar 24-25 Edited by Robert Bakish Alloyd,Cambridge,Mass.,1960
Eng 41.5614

ALLOYING BEHAVIOUR AND EFFECTS IN CONCENTRATED SOLID SOLUTIONS :based on a symposium Cleveland 1963 Oct 21 American institute of mining,metallurgical and petroleum engineers Edited by T.B. Massalski Metallurgical society conferences, 29 Gordon and Breach,Ohio,1963
Met 25.1242

ALLOYS Symposium on internal stresses in metals and alloys Advance copies of papers London 1947 Oct 15-16 Institute of metals Institute of metals,London,1947
Eng 41.3768

ALPBACH 1963 Introduction to solar terrestrial relations Summer school in space physics proceedings European space research organisation.Preparatory commission Edited by J. Ortner and H. Maseland Astrophysics and space science library ix, 506p D.Reidel,Dordrecht-Holland,1965
A Math 4.1162

ALPBACH 1963 Introduction to solar terrestrial relations The Summer school in space physics Proceedings Edited by J. Ortner and H. Maseland Organized by the European preparatory commission for space research Astrophysics and space science library 506p D.Reidel,Dordrecht,1965
Obs 6.1321

ALPBACH 1963 Introduction to solar terrestrial relations The Summer school in space physics Proceedings European preparatory commission for space research Edited by J. Ortner and H. Maseland Astrophysics and space science library Reidel,Dordrecht,1965
TA 15.0025

ALPBACH 1964 Proceedings of a symposium on high latitude particles Proceedings Edited by B. Maehlum illus vii,320p 24cm Logos press,London,1965
Sco 14.7472

ALPBACH 1965 Electromagnetic radiation in space E.S.R.O. summer school in space physics 3rd Proceedings European space research organisation Edited by J.G. Emming Astrophysics and space science library, 9 viii,307p D.Reidel,Dordrecht,1967
A Math 4.1576

ALPINE GARDEN SOCIETY International rock garden plant conference 3rd Report London 1961 Apr 18-28 Edinburgh 1961 Edited by R.C. Elliott and J.L. Mowat Organized by the Scottish rock garden club Alpine garden society;Scottish rock garden club,London;Penicuik,1961
BG 38.0011

ALPINE GARDEN SOCIETY Rock gardens and rock plants conference Reports London 1936 May 5-7 Edited by F.J. Chittenden Royal horticultural society,London,1936
BG 38.3310

ALPINE METEOROLOGIE Internationale tagung fur Alpine meteorologie 9. Wissenschaftliche abhandlungen Brig 1966 Sep14-17 Zermatt 1966 Sep 14-17 Edited by Karin Schram and J.C. Thams Schweizerische meteorologische zentralanstalt. Veroffentlichungen,4 illus 366p City-druck a.g.,Zurich,1967 In French,German, and Italian
Sco 14.7410

ALUMINIUM The Finishing of aluminum : symposium Proceedings Boston,Mass. 1961 Jun 18-23 American electroplaters' society Edited by G.H. Kissin Reinhold,New York,1963
Met 25.2796

ALUMINIUM DEVELOPMENT ASSOCIATION Conference on anodising aluminium Proceedings Nottingham 1961 Sep 12-14 Aluminium development association,London,1961
Met 25.2047

ALUMINIUM DEVELOPMENT ASSOCIATION Conference on anodising aluminium Proceedings Nottingham 1961 Sep 12-14 University of Nottingham Aluminium development association, London,1962
Eng 41.3602

ALUMINIUM DEVELOPMENT ASSOCIATION Welding and riveting larger aluminium structures :a symposium Proceedings London 1951 Nov Aluminium development association,London,1952
Eng 41.3941

ALUMINIUM FEDERATION Aluminium in structural engineering :a symposium Proceedings London 1963 Jun 11-12 Aluminium federation, London,1964
Eng 41.2694

ALUMINIUM FEDERATION Symposium on anodizing aluminium Proceedings Birmingham 1967 Apr 12-13 Aluminium federation,London,1967
Met 25.2048

ALUMINIUM IN STRUCTURAL ENGINEERING :a symposium Proceedings London 1963 Jun 11-12 Institution of structural engineers Aluminium federation Aluminium federation, London,1964
Eng 41.2694

ALUMINIUM-ZINC-MAGNESIUM Conference on weldable Al-Zn-Mg alloys Proceedings 1969 Sep 23-25 Welding institute Welding institute,Abington,1970
Eng 41.8286

AMERICAN ACADEMY FOR CEREBRAL PALSY:ANNUAL MEETING 11th Papers Kernicterus and its importance in cerebral palsy New Orleans Thomas,Springfield,Ill.,1961
An 32.4184

AMERICAN ACADEMY OF ARTS AND SCIENCES Atmospheric exploration The Benjamin Franklin memorial symposium Papers 125p M.I.T.;Wiley,New York,1958
Nap 11.0960

AMERICAN ACADEMY OF NEUROLOGY.SECTION OF NEUROCHEMISTRY Neurochemistry of nucleotides and amino acids :a symposium Papers and discussions presented Philadelphia 1958 Apr 24-25 Edited by Roscoe O. Brady and Donald B. Tower Wiley, New York,1960
Bioch 33.2222

AMERICAN ACADEMY OF PHYSICAL MEDICINE AND REHABILITATION International congress of physical medicine 3rd Proceedings Washington,D.C. 1960 Chicago,Ill.,1960
Phys 20.0990

AMERICAN ANTHROPOLOGICAL ASSOCIATION The Processes of ongoing human evolution Edited by Gabriel W. Lasker Wayne state university press,Detroit,Mich.,1960 Six lectures given at the 58th annual meeting of the American anthropological association
Gen 34.1864

AMERICAN ASSOCIATION FOR THE ADVANCEMENT OF SCIENCE Antimetabolites and cancer :a symposium Boston,Mass 1953 Dec 28-29 Edited by Cornelius P. Rhoades American association for the advancement of science, Washington,D.C.,1955
Radioth 35.0735

AMERICAN ASSOCIATION FOR THE ADVANCEMENT OF SCIENCE Biophysics of physiological and pharmacological actions The American association for the advancement of science meeting 127th Papers presented New York 1960 Dec 26-28 Edited by Abraham M. Shanes and others American association for the advancement of science.Publication, 69 Washington,D.C.,1961
Bal 39.1257

AMERICAN ASSOCIATION FOR THE ADVANCEMENT OF SCIENCE Calcification in biological systems :a symposium presented at the Washington meeting of the American association for the advancement of science Washington,D. C. 1958 Dec 29 Edited by Reidar F. Soggnaes American association for the advancement of science.Publication, 64 American association for the advancement of science,Washington,D.C.,1960
An 32.3237

AMERICAN ASSOCIATION FOR THE ADVANCEMENT OF SCIENCE Congenital heart disease :a symposium Washington,D.C. 1958 Dec 29-30 Edited by Allan D. Bass and Gordon K. Moe American association for the advancement of science.Publications, 63 American association for the advancement of science, Washington,D.C.,1960
An 32.2063

AMERICAN ASSOCIATION FOR THE ADVANCEMENT OF SCIENCE Current issues in the philosophy of science :symposia of scientists and philosophers Proceedings Chicago,Ill. 1959 Dec 27-30 Edited by Herbert Feigl and Grover Maxwell xi,484p Holt,Rinehart and Winston,New York,1961 Proceedings of section L of the Association
WSM 43.0784

AMERICAN ASSOCIATION FOR THE ADVANCEMENT OF SCIENCE Fifty years of Darwinism :modern aspects of evolution Baltimore,Md. 1909 Jan 1 Holt,New York,1909 Centennial addresses in honor of Charles Darwin
Bot 42.6666

AMERICAN ASSOCIATION FOR THE ADVANCEMENT OF SCIENCE Fundamentals of keratinization :a symposium presented at the New York meeting of the New York meeting of the American association for the advancement of science New York 1960 Dec 30 Edited by Earl O. Butcher and Reidar F. Sognnaes American association for the advancement of science. Publications, 70 American association for the advancement of science, Washington, D.C., 1962
An 32.3280

AMERICAN ASSOCIATION FOR THE ADVANCEMENT OF SCIENCE Global effects of environmental pollution :a symposium Dallas, Texas 1968 Dec 218p Reidel, Dordrecht, 1970
Nap 11.0970

AMERICAN ASSOCIATION FOR THE ADVANCEMENT OF SCIENCE Great Lakes basin Syposium Chicago, Ill. 1959 Dec 29-30 Edited by Howard J. Pincus American association for the advancement of science. Publications, 71 American association for the advancement of science, Washington, D.C., 1962
Geog 13.5045

AMERICAN ASSOCIATION FOR THE ADVANCEMENT OF SCIENCE International conference on venoms 1st Papers Berkeley, Calif. 1954 Dec 27-30 Edited by E.E. Buckley and N. Porges American association for the advancement of science. Publication, 44 des plantes et animaux indesirables
Bal 44.6436

AMERICAN ASSOCIATION FOR THE ADVANCEMENT OF SCIENCE International oceanographic congress 1st Preprints of abstracts of papers Washington, D.C. 1959 Aug 31-Sep 12 Edited by Mary Sears illus, maps 1022p 23cm Washington, D.C., 1959
Sco 14.0129

AMERICAN ASSOCIATION FOR THE ADVANCEMENT OF SCIENCE Liebig and after Liebig:a century of progress in agricultural chemistry A Symposium of papers presented before the sections of chemistry and agriculture of the American association for the advancement of science...in commemoration of the 100th anniversary of the publication of Liebig's 'Organic chemistry in its application to agriculture and physiology' Philadelphia, Pa. 1940 Dec 30 Edited by Forest Ray Moulton American association for the advancement of science. Publication, 16 111p American association for the advancement of science, Washington, D.C., 1942
WSM 43.2238

AMERICAN ASSOCIATION FOR THE ADVANCEMENT OF SCIENCE Low level irradiation :a symposium Indianapolis, Ind. 1957 Dec 30 Edited by Austin M. Brues Co-sponsored by the Argonne national laboratory. Division of biological and medical research American association for the advancement of science. Publication, 59 American association for the advancement of science, Washington, D.C., 1959
Radioth 35.1109

AMERICAN ASSOCIATION FOR THE ADVANCEMENT OF SCIENCE Mechanisms of hard tissue destruction :a symposium presented at the Philadelphia meeting of the AAAS... Philadelphia, Pa. 1962 Dec 29-30 Edited by Reidar F. Sognnaes American association for the advancement of science. Publications, 75 American association for the advancement of science, Washington, D.C., 1963
An 32.3661

AMERICAN ASSOCIATION FOR THE ADVANCEMENT OF SCIENCE Molecular mechanisms of temperature adaptation :a symposium Berkeley, Calif. 1965 Dec 27-29 Edited by C.Ladd Prosser American association for the advancement of science. Publication, 84 American association for the advancement of science, Washington, D.C., 1967
Gen 34.0538

AMERICAN ASSOCIATION FOR THE ADVANCEMENT OF SCIENCE Monomolecular layers :a symposium Philadelphia 1951 Dec 27 Edited by Harry Sobotka vii,207p American association for the advancement of science, Washington, D.C., 1954
Chem E 24.0932

AMERICAN ASSOCIATION FOR THE ADVANCEMENT OF SCIENCE Oceanography International oceanographic congress Invited lectures New York 1959 Aug 31-Sep 12 Edited by Mary Sears American association for the advancement of science. Publication, 67 Washington, D.C., 1961
Geod . 9.0576

AMERICAN ASSOCIATION FOR THE ADVANCEMENT OF SCIENCE Oceanography The International oceanographic congress Lectures New York 1959 Aug 31-Sep 12 Edited by Mary Sears American association for the advancement of science. Publication, 67 American association for the advancement of science, Washington, 1961
Bal 39.1951

AMERICAN ASSOCIATION FOR THE ADVANCEMENT OF SCIENCE Outlines of geologic history with especial reference to North America :series of essays ... presented before... the American association for the advancement of science... Baltimore, Md. 1908 Dec Edited by Rollin D. Salisbury University of Chicago press, Chicago, Ill., 1910
Geol 8.1790

AMERICAN ASSOCIATION FOR THE ADVANCEMENT OF SCIENCE Plant biology today:advances and challenges :a symposium Denver, Colo. 1961 Dec Edited by William A. Jensen and Leroy G. Kavaljian illus Macmillan, London, 1963
Bot 42.0298

AMERICAN ASSOCIATION FOR THE ADVANCEMENT OF SCIENCE Review of research on military problems in cold regions The Alaskan science conference 15th Papers Fairbanks 1964 Aug 31-Sep 4 By Charles R. Kolb and Fritz M. G. Holmstrom University of Alaska 1964 Privately printed
Sco 14.0155

AMERICAN ASSOCIATION FOR THE ADVANCEMENT OF SCIENCE Sex in microorganisms :a symposium Philadelphia 1951 Dec 30 Edited by D.H. Wenrich illus v,362p American association for the advancement of science,Washington,D.C.,1954
Bot 42.0758

AMERICAN ASSOCIATION FOR THE ADVANCEMENT OF SCIENCE Spermatozoan motility :a symposium New York 1960 Dec 29-30 Edited by D.W. Bishop Sponsored by the American society of zoologists and the Society of general physiologists American association for the advancement of science.Publication, 72 Washington,D.C.,1962
Bal 39.0391

AMERICAN ASSOCIATION FOR THE ADVANCEMENT OF SCIENCE Symposium on basic research New York 1959 Edited by Dael Wolfle American association for the advancement of science. Publication, 56 Washington,D.C.,1959
Eng 41.1676

AMERICAN ASSOCIATION FOR THE ADVANCEMENT OF SCIENCE The Ontogenesis of grammar :a theoretical symposium Edited by D.I. Slobin Academic press,New York;London,1971 Based on a symposium held in December 1965 in Berkeley at the meeting of the American association for the advancement of science
Psy 31.3299

AMERICAN ASSOCIATION FOR THE ADVANCEMENT OF SCIENCE The Species problem :a symposium Atlanta,Ga 1955 Dec 28-29 Edited by Ernst Mayr American association for the advancement of science.Publication, 50 American association for the advancement of science,Washington,D.C.,1957
Bot 42.3422

AMERICAN ASSOCIATION FOR THE ADVANCEMENT OF SCIENCE The Species problem :a symposium Atlanta,Ga. 1955 Dec 28-29 Edited by Ernst Mayr American association for the advancement of science.Publication, 50 American association for the advancement of science,Washington,D.C.,1957
Gen 34.1294

AMERICAN ASSOCIATION FOR THE ADVANCEMENT OF SCIENCE The Species problem :a symposium presented at the Atlanta meeting of the American association for the advancement of science Atlanta,Ga. 1955 Dec 28-29 Edited by Ernst Mayr American association for the advancement of science.Publication, 50 Washington,D.C.,1957
Bal 39.0904

AMERICAN ASSOCIATION FOR THE ADVANCEMENT OF SCIENCE 128th :annual meeting Symposium Man,culture and animals Denver,Colo. 1961 Dec 30 Edited by Anthony Leeds and Andrew P. Vayda American association for the advancement of science.Publications,78 Washington,D.C.,1963
Geog 13.1313

AMERICAN ASSOCIATION FOR THE ADVANCEMENT OF SCIENCE 132nd :annual meeting Symposium on ground level climatology Ground level climatology Berkeley,Calif. 1965 Dec Edited by Robert H. Shaw American association for the advancement of science. Publications,86 Washington,D.C.,1967
Geog 13.1034

AMERICAN ASSOCIATION FOR THE ADVANCEMENT OF SCIENCE.MISCELLANEOUS PUBLICATION, 70-1 Apollo 11 lunar science conference Papers Houston,Texas 1970 Jan 5-8 National aeronautics and space administration American association for the advancement of science,Washington,D.C.,1970
Min 10.1474

AMERICAN ASSOCIATION FOR THE ADVANCEMENT OF SCIENCE.SECTION E :meetings Papers Outlines of geologic history with especial reference to North America Baltimore,Md. 1908 Dec Edited by Bailey Willis and Rollin D. Salisbury University of Chicago press, Chicago,Ill.,1910
Geog 13.0599

AMERICAN ASSOCIATION FOR THE ADVANCEMENT OF SCIENCE.SECTION ON MEDICAL SCIENCES Evolution of nervous control from primitive organisms to man:a symposium New York 1956 Dec 29-30 Edited by Allan D. Bass American association for the advancement of science. Publications, 52 Washington,D.C.,1954
Phys 20.2015

AMERICAN ASSOCIATION FOR THE ADVANCEMENT OF SCIENCE.SECTION ON MEDICAL SCIENCES Evolution of nervous control from primitive organisms to man :a symposium New York 1956 Dec 29-30 Edited by Allan O. Bass American association for the advancement of science.Publications, 52 American association for the advancement of science, Washington,D.C.,1959
An 32.4357

AMERICAN ASSOCIATION FOR THE ADVANCEMENT OF SCIENCE.SECTION ON MEDICAL SCIENCES Evolution of nervous control from primitive organisms to man :a symposium Proceedings New York 1956 Dec 29-30 Edited by Allan D. Bass American association for the advancement of science.Publication, 52 Washington,D.C.,1959 Arranged by Bernard B.Brodie and Allan D.Bass
Bal 39.1470

AMERICAN ASSOCIATION FOR THE ADVANCEMENT OF SCIENCE.SECTION ON ZOOLOGICAL SCIENCES Beginnings of embryonic development A Symposium on formation and early development of the embryo Atlanta,Ga. 1955 Dec 27 Edited by Albert Tyler and others Co-sponsored by the American society of zoologists and the Association of southeastern biologists American association for the advancement of science.Publication, 48 Washington,D.C.,1957
Bal 39.0401

The AMERICAN ASSOCIATION FOR THE ADVANCEMENT OF SCIENCE MEETING 127th Papers presented Biophysics of physiological and pharmacological actions New York 1960 Dec 26-28 American association for the advancement of science Edited by Abraham M. Shanes and others American association for the advancement of science.Publication, 69 Washington,D.C.,1961
Bal 39.1257

AMERICAN ASSOCIATION OF ANATOMISTS
Microcirculation:Symposium on factors influencing exchange of substances across capillary wall A Sterotaxic atlas of the dog's brain Buffalo,N.Y. 1958 Apr 1 By Robert K.S. Lim and others Edited by S.R.M. Reynolds and Benjamin W. Zweifach Thomas, Springfield,Ill.,1960
VA 19.0362

AMERICAN ASSOCIATION OF AVIAN PATHOLOGISTS
Newcastle disease virus:an evolving pathogen : international symposium Proceedings 1963 Edited by Robert Paul Hanson University of Wisconsin press,Madison,Wis.,1964
Gen 34.0728

AMERICAN ASSOCIATION OF NEUROPATHOLOGISTS
Properties of membranes and diseases of the nervous system By Donald B. Tower and others Springer,New York,1962 Based on a symposium - June 1961 - sponsored jointly by the American neurological association and the American association of neuropathologists
An 32.4197

AMERICAN ASSOCIATION OF PETROLEUM GEOLOGISTS
Classification of carbonate rocks :symposium papers Denver 1961 Apr 27 Edited by William E. Ham American ssociation of petroleum geologists.Memoirs,1 Tulsa,Okla., 1962
Geol 8.3157

AMERICAN ASSOCIATION OF PETROLEUM GEOLOGISTS
Geometry of sandstone bodies :a symposium papers Atlantic city,N.J. 1960 Apr.25-28 Edited by James A. Peterson and John C. Osmond American association of petroleum geologists, Tulsa,Oka.,1961 The symposium was held during the 45th annual meeting of the American association of petroleum geologists
Geol 8.1481

AMERICAN ASSOCIATION OF PETROLEUM GEOLOGISTS
Habitat of oil :symposium papers New York 1955 Mar.28-31 Edited by Lewis G. Weeks illus. viii,1384p 26cm American association of petroleum geologists,Tulsa,Okl., 1958 The symposium was held during the 40th annual meeting of the American association of petroleum geologists
Geol 8.1346

AMERICAN ASSOCIATION OF PETROLEUM GEOLOGISTS
Jurassic and carboniferous of western Canada, with related papers :a symposium papers Jasper,Alta. 1955 Sep.15-16 Edited by A.J. Goodman American association of petroleum geologists,Tulsa,Oka.,1958 John Andrew Allan memorial volume
Geol 8.2712

AMERICAN ASSOCIATION OF PETROLEUM GEOLOGISTS
Theory of continental drift :a symposium on the origin and movement of land masses both intercontinental and intra-continental,as proposed by Alfred Wegener New York 1926 Nov.15 By W.A.J.M. Waterschoot van der Gracht American association of petroleum geologists,Tulsa,Oka.,1928
Geol 8.2130

AMERICAN ASSOCIATION OF PETROLEUM GEOLOGISTS and SOCIETY OF ECONOMIC PALEONTOLOGISTS AND MINERALOGISTS Classification of carbonate rocks :a symposium Papers Denver,Colo. 1961 Apr 27 Edited by William E. Ham American association of petroleum geologists. Memoir,1 Tulsa,Okla.,1962 Held under the joint auspices of the American association of petroleum geologists and the Society of economic paleontologists and mineralogists
Geog 13.0301

AMERICAN ASTRONAUTICAL SOCIETY Space trajectories :a symposium Papers Orlando, Fla. 1959 Dec 14-15 Radiation incorporated, Orlando,Fla. and Advanced research projects agency Academic press,New York,1960
Geod 9.0014

AMERICAN ASTRONOMICAL SOCIETY AAS-NASA magnetic star symposium The Symposium on the magnetic and other peculiar and metallic-line A stars :magnetic and related stars 1st Proceedings Greenbelt,Md. 1965 Nov 8-10 and Berkeley,Calif. 1965 Dec 29 Edited by Robert C. Cameron 596p Mono book corporation,Baltimore,1967 Includes the Helen B.Warner prize lecture of the American astronomical society,by George W.Preston
Obs 6.3432

AMERICAN ASTRONOMICAL SOCIETY Magnetic and related stars A.A.S.-N.A.S.A. symposium on the magnetic and other peculiar and metallic-line A stars Proceedings Greenbelt,Md 1965 Nov 8-10 Berkeley,Calif 1965 Dec 29 Edited by Robert C. Cameron xi,596p Mono book corporation,Baltimore,1967
A Math 4.1575

AMERICAN ASTRONOMICAL SOCIETY Magnetic and related stars The Symposium on the magnetic and other peculiar and metallic-line A stars Greenbelt,Md. 1965 Nov 8-10 Edited by Robert C. Cameron Mono book corporation, Baltimore,Md.,1967
TA 15.0310

AMERICAN ASTRONOMICAL UNION High-energy nuclear reactions in astrophysics Symposium on high-energy nuclear reactions in astrophysics Invted papers Philadelphia,Pa. 1967 Jan Edited by D.S.P. Shen 281p Benjamin,New York,1967 Four of the ten papers were invited after the symposium to complete the coverage
TA 15.0578

AMERICAN CANCER SOCIETY National cancer conference 1st Proceedings Bethesda,Md 1949 Feb 25-27 American cancer society,1949
Med 36.0003

AMERICAN CARBON COMMITTEE Conference on carbon 4th Proceedings Buffalo,N.Y. 1959 Jun 15-19 By S. Mrozowski and others Pergamon press,Oxford,1960
Met 25.0157

AMERICAN CARBON COMMITTEE The Conference on carbon 4th Proceedings Buffalo,N.Y. 1959 Jun 15-19 Edited by S. Mrozowski and others Conference on carbon.Proceedings, 3 778p Pergamon press,Oxford,1960
TA 15.0215

AMERICAN CARBON COMMITTEE The Conference on carbon 5th Proceedings University park, Penn. 1961 Vol 1 Edited by S. Mrozowski Conference on carbon. Proceedings, 4 Pergamon press, Oxford, 1962 Proceedings appeared in two volumes
TA 15.0216

AMERICAN CERAMIC SOCIETY Advances in glass technology International congress on glass 4th Technical papers Washington, D.C. 1962 Jul 8-14 Plenum press, New York, 1962
Met 25.0116

AMERICAN CERAMIC SOCIETY Nucleation and crystallization in glasses and melts :a symposium Toronto 1961 Apr Edited by Margie K. Reser and others American ceramic society, Columbus, Ohio, 1962
Met 25.0127

AMERICAN CHEMICAL SOCIETY Advances in enzymic hydrolysis of cellular and related materials :a symposium Proceedings Washington, D.C. 1962 Mar Edited by Elwyn T. Reese Pergamon, Oxford, 1963 Including a bibliography for the years 1950-61
Bot 42.1728

AMERICAN CHEMICAL SOCIETY American chemical society national meeting 162nd Abstracts of papers Washington, D.C. 1971 Sep 12-17 American chemical society, Washington, D.C., 1971
Chem 18.2812

AMERICAN CHEMICAL SOCIETY American chemical society national meeting 129th Literature of the combustion of petroleum :a symposium Papers Dallas, Texas 1956 Apr Advances in chemistry series, 20 Washington, D.C., 1958
Eng 41.7239

AMERICAN CHEMICAL SOCIETY Fiber spinning and drawing :symposium New York 1966 Sep 13-15 Edited by Myron J. Coplan Applied polymer symposia, 6 181p Interscience publishers, New York, 1967
Chem E 24.0423

AMERICAN CHEMICAL SOCIETY Literature of the combustion of petroleum :a symposium held...at the 129th national meeting of the American chemical society Papers Dallas, Texas 1956 Apr American chemical society. Advances in chemistry series, 20 American chemistry society, Washington, D.C., 1958
Chem 18.0449

AMERICAN CHEMICAL SOCIETY Molecular modification in drug design :a symposium at the 145th meeting of the American chemical society New York 1963 Sep 9-10 Edited by Robert F. Gould American chemical society. Advances in chemistry series, 45 American chemical society, Washington, D.C., 1964
Radioth 35.0429

AMERICAN CHEMICAL SOCIETY Plastic deformation of polymers Symposium on plastic deformation of polymers Selected papers New York 1969 Sep 10-11 Dekker, New York, 1971
Met 25.2576

AMERICAN CHEMICAL SOCIETY Retardation of evaporation by monolayers :transport processes Edited by Victor K. La Mer xx,277p Academic press, New York; London, 1962 Papers presented at a symposium sponsored by the Division of colloid and surface chemistry of the American chemical society, Sep. 1960
Chem E 24.0945

AMERICAN CHEMICAL SOCIETY Solvated electron : a symposium at the 150th meeting of the American chemical society Atlantic City 1965 Sep 15-16 Edited by Robert F. Gould American chemical society. Advances in chemistry series, 50 American chemical society, Washington, D.C., 1965
Radioth 35.0430

AMERICAN CHEMICAL SOCIETY The Mossbauer effect and its application in chemistry New York 1966 Sep 12 Edited by E. Gould American chemical society. Advances in chemistry series, 68 American chemical society, Washington, D.C., 1967 Symposium organized by C.W. Seidel. Chairman: R.H. Herber
Chem 18.1190

AMERICAN CHEMICAL SOCIETY Thermoanalysis of fibers and fiber-forming polymers :symposium Papers Atlantic City, N.J. 1965 Sep 17 Edited by Robert F. Schwenker Applied polymer symposia, 2 Interscience, New York, 1966
Met 25.2705

AMERICAN CHEMICAL SOCIETY Transitions and relaxations in polymers :symposium Papers Atlantic City, N.J. 1965 Sep 13-14 Edited by Raymond F. Boyer Polymer symposia, 14 Interscience, New York, 1966
Met 25.2704

AMERICAN CHEMICAL SOCIETY Werner centennial : a symposium Papers New York 1966 Sep 12-16 American chemical society. Advances in chemistry series, 62 American chemical society, Washington, D.C., 1967 Symposium chairman: G.B. Kauffman
Chem 18.0990

AMERICAN CHEMICAL SOCIETY and WAYNE STATE UNIVERSITY International conference on coordination chemistry Advances in the chemistry of coordination compounds 6th Proceedings Detroit, Mich. 1961 Aug 27-Sep 1 Edited by Stanley Kirschner Macmillan, New York, 1961 Co-sponsored by the United States air force, and the International union of pure and applied physics
Chem 18.0991

AMERICAN CHEMICAL SOCIETY. DIVISION OF AGRICULTURAL AND FOOD CHEMISTRY Gibberellins ...The symposium...presented before the Division of agricultural and food chemistry of the 138th national meeting of the American chemical society Papers New York 1960 Sep Advances in chemistry series, 28 American chemical society, Washington, 1961
Bioch 33.0738

AMERICAN CHEMICAL SOCIETY. DIVISION OF ANALYTICAL CHEMISTRY Fluorescence, theory and practice :a symposium Papers Miami Beach, Fla. 1967 Apr Edited by George G. Guilbault Arnold; Dekker, London; New York, 1967
Col S 12.0091

AMERICAN CHEMICAL SOCIETY.DIVISION OF BIOLOGICAL CHEMISTRY Amino acids,proteins and cancer biochemistry The Jesse P.Greenstein memorial symposium Papers Washington,D.C. 1959 Sep 16 Edited by John T. Edsall Academic press,New York;London,1960 With a biographical article on Dr.Greenstein and a bibliography of his writings
Bioch 33.2345

AMERICAN CHEMICAL SOCIETY.DIVISION OF COLLOID AND SURFACE CHEMISTRY Ice symposium Papers Pittsburgh 1966 Mar Journal of colloid and interface science,25 no 2 illus 131-294p Academic,New York;London,1967 Symposium chairman H.H.G.Jellinek
Sco 14.7934

AMERICAN CHEMICAL SOCIETY.DIVISION OF COLLOID AND SURFACE CHEMISTRY Kendall award symposium Solid surfaces and the gas-solid interface : papers presented at the Kendall award symposium honoring Stephen Brunauer Papers St.Louis,Mo. 1961 Mar 27-29 Edited by Lewellyn E. Copeland and others American chemical society.Advances in chemistry series, 33 American chemical society,Washington,D.C., 1961
Chem 18.0616

AMERICAN CHEMICAL SOCIETY.DIVISION OF COLLOID AND SURFACE CHEMISTRY The Conference on electromagnetic scattering :interdisciplinary conference Papers Potsdam,N.Y. 1962 Aug 13-15 Clarkson college of technology and Air force Cambridge research laboratories Edited by Milton Kerker International series of monographs on electromagnetic waves, 5 591p Pergamon press,Oxford,1963
TA 15.0256

AMERICAN CHEMICAL SOCIETY.DIVISION OF FUEL CHEMISTRY and AMERICAN INSTITUTE OF AERONAUTICS AND ASTRONAUTICS.PROPELLANTS AND COMBUSTION TECHNICAL COMMITTEE Advanced propellant chemistry :a symposium Papers Detroit,Mich. 1965 Apr 6-7 American chemical society.Advances in chemistry series, 54 American chemical society,Washington,D.C., 1966 Presented at the 149th meeting of the American chemical society.Symposium chairman:Richard T.Holtzmann
Chem 18.0977

AMERICAN CHEMICAL SOCIETY.DIVISION OF INDUSTRIAL AND ENGINEERING CHEMISTRY Plasticization and plasticizer processes :a symposium Philadelphia,Pa. 1964 Apr 6-7 Advances in chemistry series, 48 ix,200p American chemical society,Washington,D.C.,1965 Symposium chairman:Norbert A.J.Platzer
Chem E 24.1045

AMERICAN CHEMICAL SOCIETY.DIVISION OF INORGANIC CHEMISTRY Mechanisms of inorganic reactions Lawrence,Kan. 1964 Jun 21-24 Edited by R.Kent Murmann and others American chemical society.Advances in chemistry series, 49 American chemical society,Washington,D.C., 1965 Symposium chairman:J.Kleinberg
Chem 18.1002

AMERICAN CHEMICAL SOCIETY.DIVISION OF INORGANIC CHEMISTRY Reactions of coordinated ligands and homogeneous catalysis :a symposium Papers Washington,D.C. 1962 Mar 22-24 American chemical society.Advances in chemistry series,37 American chemical society,Washington,D.C.,1963 Symposium chairman:D.H.Busch
Chem 18.1010

AMERICAN CHEMICAL SOCIETY.DIVISION OF MICROBIAL CHEMISTRY AND TECHNOLOGY The Culture of algae :a symposium Proceedings New York 1969 Sep 7-12 Edited by J.E. Zajic illus, tables ix,154p Plenum press,New York; London,1970
Bot 42.3942

AMERICAN CHEMICAL SOCIETY.DIVISION OF PHYSICAL CHEMISTRY and AMERICAN INSTITUTE OF PHYSICS.DIVISIONS OF CHEMICAL PHYSICS AND SOLID STATE PHYSICS Chemical physics of nonmetallic crystals :Northwestern university 1961 international conference Papers Urbana,Ill. 1961 Benjamin,New York,1962 First published as a supplement to the Journal of applied physics,January 1962
Chem 18.2385

AMERICAN CHEMICAL SOCIETY.DIVISION OF WATER AND WASTE CHEMISTRY American chemical society national meeting 137th Saline water conversion :a symposium Papers Cleveland, Ohio 1960 Apr Advances in chemistry series American chemical society,Washington,D.C.,1960
Eng 41.7535

AMERICAN CHEMICAL SOCIETY.DIVISIONS OF FUEL CHEMISTRY AND PETROLEUM CHEMISTRY Fuel cells :a symposium 2nd Papers Chicago 1961 Sep Edited by G.J. Young v,225p Reinhold;Chapman and Hall,New York;London,1963 Held during the 140th national meeting of the American chemical society
Chem E 24.1401

AMERICAN CHEMICAL SOCIETY.GAS AND FUEL DIVISION Fuel cells :a symposium Atlantic City 1959 Sep Edited by G.J. Young Reinhold; Chapman and Hall,New York;London,1960
Met 25.1838

AMERICAN CHEMICAL SOCIETY.GAS AND FUEL DIVISION Fuel cells a symposium Proceedings Atlantic City 1959 Sep Edited by G.J. Young v,154p Reinhold;Chapman and Hall,New York;London,1960 Held during the 136th national meeting of the American chemical society
Chem E 24.1400

AMERICAN CHEMICAL SOCIETY.INORGANIC CHEMISTRY DIVISION Bioinorganic chemistry :a symposium Blacksburg,Va. 1970 Jun 22-25 Edited by Robert F. Gould x,436p 23cm American chemical society,Washington,D.C.,1971
Chem 18.2719

AMERICAN CHEMICAL SOCIETY.INORGANIC CHEMISTRY DIVISION Mechanisms of inorganic reactions :symposium Papers Lawrence,Kan 1964 Jun 21-24 Edited by R.Kent Murmann and others American chemical society.Advances in chemistry series, 49 vii,266p American chemical society,Washington,D.C.,1965
Chem E 24.0629

AMERICAN CHEMICAL SOCIETY.NATIONAL MEETING 136TH
Fuel cells :a symposium Atlantic City 1959 Sep American chemical society.Gas and fuel division Edited by G.J. Young Reinhold;Chapman and Hall,New York;London,1960
Met 25.1838

AMERICAN CHEMICAL SOCIETY.NATIONAL MEETING 138TH
Gibberellins ...The symposium...presented before the Division of agricultural and food chemistry of the 138th national meeting of the American chemical society Papers New York 1960 Sep American chemical society.Division of agricultural and food chemistry Advances in chemistry series, 28 American chemical society,Washington,1961
Bioch 33.0738

AMERICAN CHEMICAL SOCIETY MEETING 139TH
Kendall award symposium Solid surfaces and the gas-solid interface :papers presented at the Kendall award symposium honoring Stephen Brunauer Papers St.Louis,Mo. 1961 Mar 27-29 American chemical society.Division of colloid and surface chemistry Edited by Lewellyn E. Copeland and others American chemical society.Advances in chemistry series, 33 American chemical society,Washington,D.C., 1961
Chem 18.0616

AMERICAN CHEMICAL SOCIETY MEETING 141ST
Reactions of coordinated ligands and homogeneous catalysis :a symposium Papers Washington,D.C. 1962 Mar 22-24 American chemical society.Division of inorganic chemistry American chemical society.Advances in chemistry series,37 American chemical society,Washington,D.C.,1963 Symposium chairman:D.H.Busch
Chem 18.1010

AMERICAN CHEMICAL SOCIETY MEETING 149TH
Advanced propellant chemistry :a symposium Papers Detroit,Mich. 1965 Apr 6-7 American chemical society.Division of fuel chemistry,and,American institute of aeronautics and astronautics.Propellants and combustion technical committee American chemical society.Advances in chemistry series, 54 American chemical society,Washington,D.C., 1966 Presented at the 149th meeting of the American chemical society.Symposium chairman:Richard T.Holtzmann
Chem 18.0977

AMERICAN CHEMICAL SOCIETY MEETING 152ND
Werner centennial :a symposium Papers New York 1966 Sep 12-16 American chemical society American chemical society.Advances in chemistry series,62 American chemical society,Washington,D.C.,1967 Symposium chairman:G.B.Kauffman
Chem 18.0990

AMERICAN CHEMICAL SOCIETY NATIONAL MEETING 162nd
Abstracts of papers Washington,D.C. 1971 Sep 12-17 American chemical society American chemical society,Washington,D.C.,1971
Chem 18.2812

AMERICAN CHEMICAL SOCIETY NATIONAL MEETING 129TH
Literature of the combustion of petroleum : a symposium Papers Dallas,Texas 1956 Apr American chemical society Advances in chemistry series, 20 Washington,D.C.,1958
Eng 41.7239

AMERICAN CHEMICAL SOCIETY NATIONAL MEETING 129TH
Literature of the combustion of petroleum : a symposium held...at the 129th national meeting of the American chemical society Papers Dallas,Texas 1956 Apr American chemical society American chemical society. Advances in chemistry series,20 American chemistry society,Washington,D.C.,1958
Chem 18.0449

AMERICAN CHEMICAL SOCIETY NATIONAL MEETING 137TH
Saline water conversion :a symposium Papers Cleveland,Ohio 1960 Apr American chemical society.Division of water and waste chemistry Advances in chemistry series American chemical society,Washington,D.C.,1960
Eng 41.7535

AMERICAN COLLEGE OF NEUROPSYCHOPHARMACOLOGY
Role of cyclic AMP in cell function :symposium San Diego,Calif. 1970 Feb Edited by Paul Greengard and Erminio Costa Advances in biochemical psychopharmacology, 3 Raven press,New York,1970
Bioch 33.2211

AMERICAN CONCRETE INSTITUTE Concrete bridge design 1st international symposium Proceedings 1967 American concrete institute.Publication,SP- 23 ACI,Detroit, Mich.,1969
Eng 41.8488

AMERICAN CONCRETE INSTITUTE Concrete bridge design 2nd international symposium Proceedings American concrete institute. Publication,SP- 26 ACI,Detroit,Mich.,1971
Eng 41.8489

AMERICAN CONCRETE INSTITUTE Symposium on creep of concrete Houston,Tex. 1964 Mar American concrete institute.Publication,SP-9 American concrete institute,Detroit,Mich., 1964
Eng 41.2932

AMERICAN CONGRESS OF PHYSICAL MEDICINE AND REHABILITATION International congress of physical medicine 3rd Proceedings Washington,D.C. 1960 Chicago,Ill.,1960
Phys 20.0990

AMERICAN COUNCIL OF LEARNED SOCIETIES.JOINT COMMITTEE ON LATIN AMERICAN STUDIES see JOINT COMMITTEE ON LATIN AMERICAN STUDIES

AMERICAN CRYSTALLOGRAPHIC ASSOCIATION Small-angle x-ray scattering :the conference Proceedings Syracuse 1965 Jun 24-26 Edited by H. Brumberger Co-sponsored by the United States Army.Research office and the National science foundation Gordon and Breach,New York,1967
Met 25.1466

AMERICAN CRYSTALLOGRAPHIC ASSOCIATION
Symposium on low energy electron diffraction Tucson,Ariz. 1968 Feb 4-7 Edited by David H. Templeton and Gabor A. Somorjai American crystallographic association.Transactions, 4 American crystallographic association, Pittsburgh,1968
Met 25.1476

AMERICAN DENTAL ASSOCIATION.COUNCIL ON DENTAL RESEARCH Genetics and dental health :an international symposium Proceedings Bethesda,Md. 1961 Apr 4-6 Edited by Carl J. Witkop Symposium on genetics related to dental health, 1 McGraw-Hill,New York,1962
Gen 34.2024

AMERICAN ELECTROPLATERS' SOCIETY The Finishing of aluminum :symposium Proceedings Boston,Mass. 1961 Jun 18-23 Edited by G.H. Kissin Reinhold,New York,1963
Met 25.2796

AMERICAN EUGENICS SOCIETY Heredity counseling :a symposium New York Edited by Helen G. Hammons Hoeber-Harper,New York,1959
PGMS 29.0257

AMERICAN FEDERATION OF INFORMATION PROCESSING SOCIETIES AFIPS conference proceedings index volumes 1-37,1951-1970 93p AFIPS press,Montvale,N.J.,1971
Math L 5.3800

AMERICAN FOUNDRYMEN'S SOCIETY Symposium on principles of gating presented at the 55th annual meeting Buffalo 1951 Apr 24 American foundrymen's society,Chicago,Ill., 1951
Met 25.0659

AMERICAN FOUNDRYMEN'S SOCIETY The Symposium on solidification presented at the 64th AFS castings congress and exposition Philadelphia,Pa. 1960 May 9-13 American foundrymen's society,Des Plaines,Ill.,1961
Met 25.0644

AMERICAN FOUNDRYMEN'S SOCIETY.ANNUAL MEETING 55TH Symposium on principles of gating presented at the 55th annual meeting Buffalo 1951 Apr 24 American foundrymen's society American foundrymen's society,Chicago,Ill., 1951
Met 25.0659

AMERICAN FOUNDRYMEN'S SOCIETY.CASTINGS CONGRESS AND EXPOSITION 64TH The Symposium on solidification presented at the 64th AFS castings congress and exposition Philadelphia,Pa. 1960 May 9-13 American foundrymen's society American foundrymen's society,Des Plaines,Ill.,1961
Met 25.0644

AMERICAN FOUNDRYMEN'S SOCIETY.PLANT AND PLANT EQUIPMENT COMMITTEE Symposium on molding machines American foundrymen's society,n.p., n.d.
Met 25.0648

AMERICAN GEOLOGICAL INSTITUTE A.G.I. short course on chain silicates :lecture notes Palo Alto,Calif. 1966 Nov 11-13 American geological institute,Washington,D.C.,1966 Typescript
Min 10.1286

AMERICAN GEOLOGICAL INSTITUTE A.G.I. short course on layer silicates :lecture notes New Orleans,La. 1967 Nov 17-19 American geological institute,Washington,D.C.,1967 Typescript
Min 10.1287

AMERICAN GEOPHYSICAL UNION Antarctica in the International geophysical year Antarctic symposium Proceedings Northfield,Minn. 1956 Apr.26-27 International geophysical year American geophysical union.Publication, 462 Geophysical monograph map 133p Washington,D.C.,1956
Nap 11.0020

AMERICAN GEOPHYSICAL UNION Antarctica in the International geophysical year based on a symposium on the Antarctic Co-sponsored by the National science foundation Geophysical monograph, 1 American geophysical union. Publication, 462 illus,maps v,133p 26cm American geophysical union,Washington,D.C., 1956
Sco 14.8201

AMERICAN GEOPHYSICAL UNION Contemporary geodesy :a conference Proceedings Cambridge,Mass. 1958 Dec 1-2 Harvard college observatory and Smithsonian astrophysical observatory Edited by Charles A. Whitten and Kenneth H. Drummond Geophysical monograph, 4 National research council.Publication, 708 American geophysical union,Washington,D.C.,1959
Geod 9.0028

AMERICAN GEOPHYSICAL UNION Gravity anomalies: unsurveyed areas Extension of gravity anomalies to unsurveyed areas :a symposium Papers Columbus,Ohio 1964 Nov 18-20 Edited by Hyman Orlin Sponsored by the International union of geodesy and geophysics Geophysical monograph, 9 National research council.Publication,1357 American geophysical union,Washington,D.C.,1966
Geod 9.0031

AMERICAN GEOPHYSICAL UNION.CLOUD PHYSICS COMMITTEE and NATIONAL SCIENCE FOUNDATION Physics of precipitation Cloud physics conference 2nd Proceedings Woods Hole, Mass. 1959 Jun 3-5 Edited by Helmut Weickmann Geophysical monograph,5 National research council.Publication,746 illus,map xii,435p 25cm American geophysical union, Washington,1960
Sco 14.0536

AMERICAN GERIATRICS SOCIETY Distinguishing the health care of the aging :...seminar Proceedings Southern Pines,N.C. 1967 Mar 12-14 Edited by E.J. Lorenze American geriatrics society.Journal, 16, no 2 American geriatrics society,New York,1967
PGMS 29.0105

AMERICAN HEART ASSOCIATION Physiology of muscular exercise :a symposium Proceedings Dallas 1966 Feb 7-9 Edited by Carleta B. Chapman Circulation research.Supplement, 20-21 New York,1967
Inv Med 37.0053

AMERICAN HEART ASSOCIATION.COUNCIL ON ARTERIOSCLEROSIS Evolution of the atherosclerotic plaque :an international symposium Presentations Chicago,Ill. 1963 Mar 28-29 Edited by Richard J. Jones University of Chicago press,Chicago,Ill.; London,1963
Path 30.2350

AMERICAN INSTITUTE OF AERONAUTICS AND ASTRONAUTICS Heterogeneous combustion :a conference Palm Beach,Fla. 1963 Dec 11-13 Edited by Hans G. Wolfhard and others Progress in astronautics and aeronautics, 15 Academic press,New York,1964
Eng 41.7241

AMERICAN INSTITUTE OF AERONAUTICS AND ASTRONAUTICS International council of the aeronautical sciences 3rd congress Proceedings Stockholm 1962 Aug 27-31 Spartan;Macmillan,Washington,D.C.;London,1964
Eng 41.6965

AMERICAN INSTITUTE OF AERONAUTICS AND ASTRONAUTICS Joint automatic control conference 7th Preprints of conference papers Washington,D.C. 1966 Aug 17-19 Washington,D.C.,1966
Eng 41.6016

AMERICAN INSTITUTE OF AERONAUTICS AND ASTRONAUTICS.PROPELLANTS AND COMBUSTION TECHNICAL COMMITTEE and AMERICAN CHEMICAL SOCIETY.DIVISION OF FUEL CHEMISTRY Advanced propellant chemistry :a symposium Papers Detroit,Mich. 1965 Apr 6-7 American chemical society.Advances in chemistry series,54 American chemical society,Washington,D.C.,1966 Presented at the 149th meeting of the American chemical society.Symposium chairman:Richard T.Holtzmann
Chem 18.0977

AMERICAN INSTITUTE OF BIOLOGICAL SCIENCES Biochemistry and pharmacology of compounds derived from marine organisms :a conference Papers New York 1960 Apr 6-8 Edited by Ross F. Nigrelli New York academy of sciences.Annals, 90,article 3 New York,1960 Co-sponsored by the New York zoological society and the United States navy,Office of naval research
Chem 26.0302

AMERICAN INSTITUTE OF BIOLOGICAL SCIENCES Brain and behavior Brain function conference 3rd Proceedings Los Angeles 1963 Vol 3: brain and gonadal function Edited by Roger A. Gorski and Richard E. Whalen With the support of the Human ecology fund UCLA forum in medical sciences, 3 University of California press,Berkeley;Los Angeles,1966
Psy 31.0621

AMERICAN INSTITUTE OF BIOLOGICAL SCIENCES Brain function :conference 1st Proceedings Los Angeles,Calif. 1961 Vol 1: cortical excitability and steady potentials;relations of basic research to space biology Edited by Mary A.B. Brazier U.C.L.A. forum in medical sciences, 1 University of California press,Berkeley;Los Angeles,1963
Psy 31.0255

AMERICAN INSTITUTE OF BIOLOGICAL SCIENCES Brain function conference 2nd Proceedings Los Angeles 1962 Vol 2: RNA and brain function memory and learning Edited by Mary A.B. Brazier With the support of the United States.Air force.Office of scientific research U.C.L.A. forum in medical sciences, 2 University of California press,Berkeley;Los Angeles,1964
Psy 31.0246

AMERICAN INSTITUTE OF BIOLOGICAL SCIENCES Conference on learning,remembering and forgetting 1st Proceedings 1963 Vol 1: anatomy of memory Edited by D.P. Kimble Science and behaviour books,Palo Alto,Calif., 1965
An 32.4409

AMERICAN INSTITUTE OF BIOLOGICAL SCIENCES Effects of ionizing radiation on the reproductive system :international symposium Proceedings Fort Collins,Colo. 1962 Edited by William D. Carlson and F.X. Gassner Pergamon press,Oxford,1964
An 32.0153

AMERICAN INSTITUTE OF BIOLOGICAL SCIENCES Genetics in the 20th century :essays on the progress of genetics during its first 50 years Columbus,Ohio 1950 Sep 11-14 Genetics society of America Edited by L.C. Dunn Macmillan,New York,1951
Gen 34.0991

AMERICAN INSTITUTE OF BIOLOGICAL SCIENCES Golden jubilee of genetics Genetics in the 20th century :essays on the progress of genetics during its first 50 years Columbus, Ohio 1950 Sep 11-14 Genetics society of America Edited by L.C. Dunn port Macmillan,New York,1951
Bot 42.0751

AMERICAN INSTITUTE OF BIOLOGICAL SCIENCES Molecular structure and biological specificity a symposium Papers Washington,D.C. 1955 Oct 28-29 United States.Office of naval research Edited by Linus Pauling and Harvey A. Itano American institute of biological sciences.Publication,2 American institute of biological sciences,Washington,D.C.,1957
Chem 18.2137

AMERICAN INSTITUTE OF BIOLOGICAL SCIENCES Molecular structure and biological specificity a symposium Washington,D.C. 1955 Oct 28-29 Edited by Linus Pauling and Harvey A. Hana Sponsored by the United States.Office of naval research American institute of biological sciences.Publications, 2 American institute of biological sciences, Washington,D.C.,1957
An 32.3156

AMERICAN INSTITUTE OF BIOLOGICAL SCIENCES Molecular structure and functional activity of nerve cells :a symposium Washington,D.C. 1955 Jun 3-4 Edited by Robert G. Grenell and L.J. Mullins Sponsored by the United States. Office of naval research American institute of biological sciences.Publications, 1 American institute of biological sciences, Washington,D.C.,1956
An 32.4348

AMERICAN INSTITUTE OF BIOLOGICAL SCIENCES Small blood vessel involvement in diabetes mellitus :conference Proceedings Warrenton, Va. 1963 Mar 25-27 Edited by Marvin D. Siperstein and others American institute of biological sciences,Washington,D.C.,1964
Bioch 33.0482

AMERICAN INSTITUTE OF BIOLOGICAL SCIENCES Surtsey research conference Proceedings Reykjavik 1967 Jun 25-28 Surtsey research society Surtsey research society,Reykjavik, c1967
Sco 14.7944

AMERICAN INSTITUTE OF CHEMICAL ENGINEERS A.I. Ch.E.-Chem.E. joint meeting Proceedings London 1965 Jun 13-17 4: application of mathematical models in chemical engineering research design and production 1965 A.I.Ch.E. I.Chem.E.Symposium series, 4 I.Chem.E., London,1965
Eng 41.5861

AMERICAN INSTITUTE OF CHEMICAL ENGINEERS A.I. Ch.E.-I.Chem.E. joint meeting Proceedings London 1965 Jun 14 10: mixing - theory related to practice 1965 A.I.Ch.E.-I;Chem.E. symposium series, 10 Institution of chemical engineers,London,1965
Chem E 24.1920

AMERICAN INSTITUTE OF CHEMICAL ENGINEERS A.I. Ch.E.-I.Chem.E.joint meeting Proceeding London 1965 Jun 13-17 1: advances in separation techniques 1965 A.I.Ch.E.-I.Chem. E.Symposium series, 1 Institution of chemical engineers,London,1965
Chem E 24.1506

AMERICAN INSTITUTE OF CHEMICAL ENGINEERS A.I. Ch.E.-I.Chem.E.joint meeting Proceedings London 1965 Jun 13-17 2: chemical engineering under extreme conditions 1965 A. I.Ch.E.-I.Chem.E.Symposium series, 2 Institution of chemical engineers,London,1965
Chem E 24.1507

AMERICAN INSTITUTE OF CHEMICAL ENGINEERS A.I. Ch.E.-I.Chem.E.joint meeting Proceedings London 1965 Jun 13-17 4: application of mathematical models in chemical engineering research, design and production 1965 A.I.Ch. E.-I.Chem.E.Symposium, 4 Institution of chemical engineers,London,1965
Chem E 24.1509

AMERICAN INSTITUTE OF CHEMICAL ENGINEERS A.I. Ch.E.-I.Chem.E.joint meeting Proceedings London 1965 Jun 14 3: reaction kinetics in product and process design 1965 A.I.Ch.E.- I.Chem.E.Symposium series, 3 Institution of chemical engineers,London,1965
Chem E 24.1508

AMERICAN INSTITUTE OF CHEMICAL ENGINEERS A.I. Ch.E.-I.Chem.E.joint meeting Proceedings London 1965 Jun 14 6: transport phenomena 1965 A.I.Ch.E.-I.Chem.E.Symposium series, 6 Institution of chemical engineers,London,1965
Chem E 24.1511

AMERICAN INSTITUTE OF CHEMICAL ENGINEERS A.I. Ch.E.-I.Chem.E.joint meeting Proceedings London 1965 Jun 14 7: management oriented topics 1965 A.I.Ch.E.-I.Chem;E. Symposium series, 7 Institution of chemical engineers,London,1965
Chem E 24.1512

AMERICAN INSTITUTE OF CHEMICAL ENGINEERS A.I. Ch.E.-I.Chem.E.joint meeting Proceedings London 1965 Jun 14 9: new chemical engineering problems in the utilisation of water 1965 A.I.Ch.E.-I.Chem.E.Symposium series, 9 Institution of chemical engineers,London,1965
Chem E 24.1513

AMERICAN INSTITUTE OF CHEMICAL ENGINEERS A.I. Ch.E.-I.ChemE.joint meeting London 1965 Jun 14 5: materials associated with direct energy conversion 1965 A.I.Ch.E.-I.Chem.E. Symposium series, 5 Institution of chemical engineers,London,1965
Chem E 24.1510

AMERICAN INSTITUTE OF CHEMICAL ENGINEERS International developments in heat transfer International heat transfer conference Papers Boulder,Col. 1961 Aug 28-Sep 1 London 1962 Jan 8-12 Pt 1-5 Arranged by the American society of mechanical engineers 5 vols A.S.M.E.,New York,1961 Co-sponsored by the American institute of chemical engineers,the Institution of mechanical engineers,and the Institution of chemical engineers
Chem E 24.0595

AMERICAN INSTITUTE OF CHEMICAL ENGINEERS Joint automatic control conference 4th Preprints of technical papers Minneapolis 1963 Jul 19-21 xiv,680p American institute of chemical engineers,New York,1963
Chem E 24.1172

AMERICAN INSTITUTE OF CHEMICAL ENGINEERS.CHEMICAL ENGINEERING PROGRESS SYMPOSIUM SERIES, 50,no 11-13 Nuclear engineering Pt 1-3 American institute of chemical engineers 3 vols New York,1954
Eng 41.7530

AMERICAN INSTITUTE OF CHEMICAL ENGINEERS.WATER COMMITTEE Industrial process design for pollution control :a workshop Proceedings New York 1967 Feb 9-10 Vol 1 iv,59p A.I.Ch.E.,New York,1967
Chem E 24.1575

AMERICAN INSTITUTE OF ELECTRICAL ENGINEERS Solid-state circuit conference digest of technical papers Philadelphia,Pa. 1959 Feb. 12-13 Institute of radio engineers 102p Lewis Winner,1959
Math L 5.2082

AMERICAN INSTITUTE OF ELECTRICAL ENGINEERS Symposium on the engineering aspects of magnetohydrodynamics 2nd Proceedings Philadelphia,Pa. 1961 Mar 9-10 Edited by Clifford Mannal and Norman W. Mather Columbia university press,New York;London,1962
Eng 41.4497

AMERICAN INSTITUTE OF MINING,METALLURGICAL AND MINING ENGINEERS.EXTRACTIVE METALLURGY DIVISION Pyrometallurgical processes in nonferrous metallurgy :a symposium Pittsburgh 1965 Nov 29-Dec 1 Edited by J.N. Anderson and P.E. Queneau Metallurgical society conferences, 39 Gordon and Breach, New York,1967
Met 25.2530

AMERICAN INSTITUTE OF MINING,METALLURGICAL AND PETROLEUM ENGINEERS Agglomeration :based on an international symposium... Philadelphia,Penn. 1961 Apr 12-14 Edited by William A. Knepper Interscience,New York; London,1962
Met 25.2288

AMERICAN INSTITUTE OF MINING,METALLURGICAL AND PETROLEUM ENGINEERS Alloying behaviour and effects in concentrated solid solutions : based on a symposium Cleveland 1963 Oct 21 Edited by T.B. Massalski Metallurgical society conferences, 29 Gordon and Breach, Ohio,1963
Met 25.1242

AMERICAN INSTITUTE OF MINING,METALLURGICAL AND PETROLEUM ENGINEERS Bolton Landing conference oxide dispersion strengthening:a symposium 2nd Bolton Landing,New York 1966 Jun 27-29 Edited by G.S. Ansell Metallurgical society conferences, 47 Gordon and Breach,New York,1968
Met 25.2339

AMERICAN INSTITUTE OF MINING,METALLURGICAL AND PETROLEUM ENGINEERS Columbium metallurgy : a symposium Bolting Landing,N.Y. 1960 Jun 9-10 Edited by D.L. Douglass and F.W. Kunz Metallurgical society conferences, 10 Interscience,New York;London,1958
Met 25.2533

AMERICAN INSTITUTE OF MINING,METALLURGICAL AND PETROLEUM ENGINEERS Continuous processing and process control :a symposium Proceedings Philadelphia,Penn. 1966 Dec 5-8 Edited by Thomas R. Ingraham Metallurgical society conferences, 49 Gordon and Breach,New York, 1968 Symposium sponsored by the Electric furnace conference,the Mechanical working and steel processing conference and the Physical chemistry of steelmaking conference of the Metallurgical society
Met 25.2846

AMERICAN INSTITUTE OF MINING,METALLURGICAL AND PETROLEUM ENGINEERS Decomposition of austenite by diffusional processes :a symposium Proceedings Philadelphia 1960 Oct 19 Edited by V.F. Zackay and H.I. Aaronson Interscience,New York;London,1962
Met 25.1247

AMERICAN INSTITUTE OF MINING,METALLURGICAL AND PETROLEUM ENGINEERS Deformation twinning : a conference Proceedings Gainesville 1963 Mar 21-22 In co-operation with Florida institute for continuing university studies Metallurgical society conferences, 25 Gordon and Breach,New York;London,1964
Met 25.1257

AMERICAN INSTITUTE OF MINING,METALLURGICAL AND PETROLEUM ENGINEERS Extractive and physical metallurgy of plutinium ...based on a symposium San Francisco 1959 Feb 16-17 Edited by W.D. Wilkinson Bibliog Interscience,New York;London,1960 Introduction and annotated bibliography by the editor
Met 25.2364

AMERICAN INSTITUTE OF MINING,METALLURGICAL AND PETROLEUM ENGINEERS International mineral processing congress 7th Technical papers New York 1964 Sep 20-24 Vol 1 Edited by Nathaniel Arbiter Gordon and Breach,New York,1965 Held in conjunction with the centennial of the Henry Krumb school of mines
Met 25.0214

AMERICAN INSTITUTE OF MINING,METALLURGICAL AND PETROLEUM ENGINEERS Iron and its dilute solid solutions :a conference Proceedings Detroit 1961 Oct 23 Edited by C.W. Spencer and F.E. Werner Interscience;Wiley,New York; London,1963
Met 25.1305

AMERICAN INSTITUTE OF MINING,METALLURGICAL AND PETROLEUM ENGINEERS Local atomic arrangements studied by x-ray diffraction :a symposium Proceedings Chicago,Ill. 1965 Feb 15 Edited by J.B. Cohen and J.E. Hilliard Metallurgical society conferences, 36 Gordon and Breach,New York,1966
Met 25.1461

AMERICAN INSTITUTE OF MINING,METALLURGICAL AND PETROLEUM ENGINEERS Mechanical working and steel processing committee 2nd Operating metallurgy conference 9th Proceedings Philadelphia,Pa. 1966 Dec 5-9 Metallurgical society of the AIME.Conference series, 50 Gordon and Breach,New York,1968
Met 25.2572

AMERICAN INSTITUTE OF MINING,METALLURGICAL AND PETROLEUM ENGINEERS Metallurgy of advanced electronic materials a technical conference Philadelphia,Pa. 1962 Aug 27-29 Edited by Geoffrey E. Brock Metallurgical society conferences, 19 Interscience,New York;London,1963
Met 25.1155

AMERICAN INSTITUTE OF MINING,METALLURGICAL AND PETROLEUM ENGINEERS Metallurgy of elemental and compound semiconductors a technical conference Proceedings Boston, Mass. 1960,Aug. 29-31 Edited by Ralph O. Grubel Metallurgical society conferences, 12 Interscience,New York;London,1961
Met 25.1146

AMERICAN INSTITUTE OF MINING,METALLURGICAL AND PETROLEUM ENGINEERS Methods of materials selection :a symposium Metallurgical society conferences, 40 Gordon and Breach,New York; London,1968
Met 25.2502

AMERICAN INSTITUTE OF MINING,METALLURGICAL AND PETROLEUM ENGINEERS Physical chemistry of process metallurgy :an international symposium Pittsburgh,Pa. 1959 Apr 27-May 1 Edited by George R. St.Pierre Metallurgical society conferences, 7,8 Interscience,New York; London,1961
Met 25.0491

AMERICAN INSTITUTE OF MINING,METALLURGICAL AND PETROLEUM ENGINEERS Physical metallurgy of stress-corrosion fracture Pittsburgh 1959 Apr 2-3 Edited by Thor N. Rhodin Metallurgical society conferences, 4 Interscience,New York;London,1959
Met 25.2503

AMERICAN INSTITUTE OF MINING,METALLURGICAL AND PETROLEUM ENGINEERS Process simulation and control in iron and steelmaking :a symposium Proceedings New York 1964 Feb 17-18 Edited by J.M. Uys and H.L. Bishop Metallurgical society conferences, 32 Gordon and Breach,New York,1966
Met 25.0376

AMERICAN INSTITUTE OF MINING,METALLURGICAL AND PETROLEUM ENGINEERS Properties of elemental and compound semiconductors a technical conference Boston,Mass. 1959 Aug 31-Sep 2 Edited by Harry C. Gatos Metallurgical society conferences, 5 Interscience,New York;London,1960
Met 25.1160

AMERICAN INSTITUTE OF MINING,METALLURGICAL AND PETROLEUM ENGINEERS Recovery and recrystallization of metals :a symposium Proceedings New York 1962 Feb 20-21 Edited by L. Himmel Interscience,New York; London,1963
Met 25.1267

AMERICAN INSTITUTE OF MINING,METALLURGICAL AND PETROLEUM ENGINEERS Response of metals to high velocity deformation Technical conference Estes Park,Col. 1960 Jul 11-12 Metallurgical society conferences, 9 Interscience,New York,1961
Met 25.0905

AMERICAN INSTITUTE OF MINING,METALLURGICAL AND PETROLEUM ENGINEERS Science,technology and application of titanium International conference on titanium Proceedings London 1968 May 21-24 Edited by R.I. Jaffee and N.E. Promisel Pergamon,Oxford,1970
Met 25.2852

AMERICAN INSTITUTE OF MINING,METALLURGICAL AND PETROLEUM ENGINEERS Structure and properties of ultrahigh-strength steels :a symposium Cleveland,Ohio 1963 Oct 22 A.S. T.M.Special technical publication, 370 American society for testing and materials, Philadelphia,Pa.,1965
Met 25.2251

AMERICAN INSTITUTE OF MINING,METALLURGICAL AND PETROLEUM ENGINEERS Superconductors technical sessions New York 1962 Feb 18 Edited by M. Tanenbaum and W.V. Wright Interscience,New York;London,1962
Met 25.1181

AMERICAN INSTITUTE OF MINING,METALLURGICAL AND PETROLEUM ENGINEERS. International conference on electron and ion beam science and technology 1st Toronto 1964 May 3-7 Edited by Robert Bakish Electrochemical society series Wiley,New York,1965
Met 25.1678

AMERICAN INSTITUTE OF MINING,METALLURGICAL AND PETROLEUM ENGINEERS.COMMITTEE ON ALLOY PHASES Energy bands in metals and alloys based on a symposium Los Angeles 1967 Edited by L. H. Bennett and J.T. Waber Metallurgical society conferences, 45 Gordon and Breach, New York,1968
Met 25.1060

AMERICAN INSTITUTE OF MINING,METALLURGICAL AND PETROLEUM ENGINEERS.COMMITTEE ON PHYSICAL CHEMISTRY OF STEELMAKING Continuous casting technical sessions Proceedings Detroit 1961 Oct 24 Edited by D.L. McBride and T.E. Dancy Interscience,New York;London, 1962
Met 25.0664

AMERICAN INSTITUTE OF MINING,METALLURGICAL AND PETROLEUM ENGINEERS.COMMITTEE ON PHYSICAL CHEMISTRY OF STEELMAKING Quality requirements of super-duty steels a technical conference Proceedings Pittsburgh 1958 May 5-6 Edited by R.W. Lindsay Metallurgical society conferences, 3 Interscience,New York;London,1959
Met 25.0482

AMERICAN INSTITUTE OF MINING,METALLURGICAL AND PETROLEUM ENGINEERS.EXTRACTIVE METALLURGY DIVISION Extractive metallurgy of aluminum :international symposium 1st New York 1962 Feb 18-22 Vol 1-2: alumina; aluminum Edited by Gary Gerard and P.T. Stroup 2 vols John Wiley,New York,1963
Met 25.0578

AMERICAN INSTITUTE OF MINING,METALLURGICAL AND PETROLEUM ENGINEERS.EXTRACTIVE METALLURGY DIVISION Extractive metallurgy of copper, nickel and cobalt based on an international symposium New York 1960 Feb 15-18 Edited by Paul Quenau Bibliog Interscience,New York;London,1961
Met 25.0510

AMERICAN INSTITUTE OF MINING,METALLURGICAL AND PETROLEUM ENGINEERS.EXTRACTIVE METALLURGY DIVISION Unit processes in hydrometallurgy based on an international symposium Dallas 1963 Feb 24-28 Edited by Milton E. Wadsworth and Franklin T. Davis Metallurgical society conferences, 24 Gordon and Breach,New York;London,1964
Met 25.2289

AMERICAN INSTITUTE OF MINING,METALLURGICAL AND PETROLEUM ENGINEERS.FERROUS METALLURGY COMMITTEE High-temperature high-resolution metallography :a symposium Proceedings Chicago,Ill. 1965 Feb 15 Edited by Hubert I. Aaronson and George S. Ansell Metallurgical society conferences, 38 Gordon and Breach,New York,1967
Met 25.2361

AMERICAN INSTITUTE OF MINING,METALLURGICAL AND PETROLEUM ENGINEERS.FERROUS METALLURGY COMMITTEE Precipitation from iron-base alloys :a symposium Cleveland 1963 Oct 21 Edited by Gilbert R. Speich and John B. Clark Metallurgical society conferences Gordon and Breach,New York;London,1965
Met 25.2525

AMERICAN INSTITUTE OF MINING,METALLURGICAL AND PETROLEUM ENGINEERS.HIGH TEMPERATURE ALLOYS COMMITTEE High temperature materials :a conference Cleveland,Ohio 1957 Apr 16-17 Edited by R.F. Hehemann and G.Mervin Ault John Wiley;Chapman and Hall,New York;London, 1959
Met 25.1004

AMERICAN INSTITUTE OF MINING,METALLURGICAL AND PETROLEUM ENGINEERS.HIGH TEMPERATURE ALLOYS COMMITTEE High temperature materials 11 a technical conference Cleveland,Ohio 1961 Apr 26-27 Metallurgical society conferences, 18 Interscience,New York;London,1963
Met 25.1006

AMERICAN INSTITUTE OF MINING,METALLURGICAL AND PETROLEUM ENGINEERS.INSTITUTE OF METALS DIVISION A Symposium on uranium and uranium dioxide :melting,casting,forging, rolling,coldworking,welding of uranium: preparation and fabrication of uranium dioxide for fuel element use Chicago 1957 Nov 6 Nuclear metallurgy, 4 I.M.D.Special report series, 4 The metallurgical society of American institute of mining,metallurgical,and petroleum engineers,New York,1957
Met 25.1542

AMERICAN INSTITUTE OF MINING,METALLURGICAL AND PETROLEUM ENGINEERS.INSTITUTE OF METALS DIVISION Fracture of solids :an international conference Proceedings Washington,D.C. 1962 Aug 21-24 Edited by D. C. Drucker and J.J. Gilman Metallurgical society conferences, 20 Interscience,New York;London,1963
Met 25.2247

AMERICAN INSTITUTE OF MINING,METALLURGICAL AND PETROLEUM ENGINEERS.INSTITUTE OF METALS DIVISION Symposium on the fabrication of fuel elements :ceramic base elements ;metal base fuels and jacket components Cleveland 1958 Oct 29 Nuclear metallurgy, 5 I.M.D. Special report series, 7 Metallurgical society of American institute of mining, metallurgical and petroleum engineers,New York, 1958
Met 25.1543

AMERICAN INSTITUTE OF MINING,METALLURGICAL AND PETROLEUM ENGINEERS.INSTITUTE OF METALS DIVISION.CHEMISTRY AND PHYSICS COMMITTEE Work hardening :based on a symposium Chicago, Ill. 1966 Nov Edited by J.P. Hirth and J. Weertman Metallurgical society conferences, 46 Gordon and Breach,New York,1968
Met 25.2866

AMERICAN INSTITUTE OF MINING,METALLURGICAL AND PETROLEUM ENGINEERS.MECHANICAL WORKING AND STEEL PROCESSING COMMITTEE Mechanical working of steel 2 The Technical conference 6th Proceedings Chicago,Ill. 1964 Jan 30-31 Edited by T.G. Bradbury Sponsored by the Chicago section Metallurgical society conferences, 26 Gordon and Breach,New York, 1965
Met 25.2244

AMERICAN INSTITUTE OF MINING,METALLURGICAL AND PETROLEUM ENGINEERS.MECHANICAL WORKING COMMITTEE Bar and allied products Technical conference 3rd Proceedings Pittsburgh 1961 Jan 18 Metallurgical society conferences, 13 Interscience,New York;London,1961
Met 25.0729

AMERICAN INSTITUTE OF MINING,METALLURGICAL AND PETROLEUM ENGINEERS.MECHANICAL WORKING COMMITTEE Flat rolled products:rolling and treatment Technical conference 1st Proceedings Chicago 1959 Jan 21 Edited by T.E. Dancy and R.L. Robinson Metallurgical society conferences, 1 Interscience,New York;London,1959
Met 25.0731

AMERICAN INSTITUTE OF MINING,METALLURGICAL AND PETROLEUM ENGINEERS.MECHANICAL WORKING COMMITTEE Flat rolled products III Technical conference 4th Proceedings Chicago,Ill. 1962 Jan 17 Metallurgical society conferences, 16 Interscience,New York;London,n.d.
Met 25.0703

AMERICAN INSTITUTE OF MINING,METALLURGICAL AND PETROLEUM ENGINEERS.MECHANICAL WORKING COMMITTEE Mechanical working of steel 1 Technical conference 5th Proceedings Pittsburgh 1963 Jan 15-16 Metallurgical society conferences, 21 Gordon and Breach, New York,1964
Met 25.0728

AMERICAN INSTITUTE OF MINING,METALLURGICAL AND PETROLEUM ENGINEERS.METALLURGICAL SOCIETY Direct observation of imperfections in crystals :technical conference Proceedings St.Louis,Mo. 1961 Mar 1-2 Interscience,New York,1962
Cav 7.2011

AMERICAN INSTITUTE OF MINING,METALLURGICAL AND PETROLEUM ENGINEERS.METALLURGICAL SOCIETY Electronic structure and alloy chemistry of the transition elements :a symposium New York 1962 Feb 22 Edited by Paul Adams Beck 251p Interscience,New York,1963
Cav 7.1711

AMERICAN INSTITUTE OF MINING,METALLURGICAL AND PETROLEUM ENGINEERS.METALLURGICAL SOCIETY The Sorby centennial symposium on the history of metallurgy :a symposium Cleveland 1963 Oct 22-23 Edited by Cyril Stanley Smith Sponsored by the Society for the history of technology Metallurgical society conferences, 27 Gordon and Breach,New York,1965
Met 25.2141

AMERICAN INSTITUTE OF MINING,METALLURGICAL AND PETROLEUM ENGINEERS.NIAGARA FRONTIER SECTION Reactive metals annual conference 3rd Proceedings Buffalo 1958 May 27-29 Vol 2 Edited by W.R. Clough Interscience, New York;London,1959
Met 25.0505

AMERICAN INSTITUTE OF MINING,METALLURGICAL AND PETROLEUM ENGINEERS.NUCLEAR METALLURGY COMMITTEE Plutonium and its alloys Extractive and physical metallurgy of plutonium and its alloys based on a symposium San Francisco 1959 Feb 16-17 Edited by W.D. Wilkinson Interscience,New York;London,1960 With an introduction and an annotated bibliography by the editor
Met 25.1568

AMERICAN INSTITUTE OF MINING,METALLURGICAL AND PETROLEUM ENGINEERS.PHYSICAL CHEMISTRY OF STEELMAKING COMMITTEE Metallurgy at high pressures and high temperatures :a symposium Proceedings Dallas,Tex. 1963 Feb 25-26 Edited by K.A. Gschneidner and others Metallurgical society conferences, 22 Gordon and Breach,New York;London,1964
Met 25.1639

AMERICAN INSTITUTE OF MINING,METALLURGICAL AND PETROLEUM ENGINEERS.PHYSICAL METALLURGY COMMITTEE Environment-sensitive mechanical behaviour :a conference Proceedings Baltimore 1965 Jun 7-8 Pt 1-2 Edited by A.R.C. Westwood and N.S. Stoloff Sponsored by the Martin company research institute for advanced studies Metallurgical society conferences, 35 2 vols Gordon and Breach,New York,1966
Met 25.1963

AMERICAN INSTITUTE OF MINING,METALLURGICAL AND PETROLEUM ENGINEERS.PHYSICAL METALLURGY COMMITTEE Recovery and recrystallization of metals a symposium Proceedings New York 1962 Feb 20-21 Edited by L. Himmel Interscience,New York,1963
Cav 7.1720

AMERICAN INSTITUTE OF MINING,METALLURGICAL AND PETROLEUM ENGINEERS.POWDER METALLURGY COMMITTEE New types of metal powders :a symposium Proceedings Cleveland,Ohio 1963 Oct 24 Edited by Henry H. Hausner Metallurgical society conferences, 23 Gordon and Breach,New York;London,1964
Met 25.2509

AMERICAN INSTITUTE OF MINING,METALLURGICAL AND PETROLEUM ENGINEERS.POWDER METALLURGY COMMITTEE New types of metal powders :a symposium Proceedings Cleveland,Ohio 1963 Oct 24 Edited by Henry H. Hausner Metallurgical society conferences, 23 Gordon and Breach,New York;London,1964
Met 25.0766

AMERICAN INSTITUTE OF MINING,METALLURGICAL AND PETROLEUM ENGINEERS.REFRACTORY METALS COMMITTEE High temperature refractory metals :a symposium New York 1964 Feb 16-20 1965 Feb Pt 1-2 Metallurgical society conferences, 34 2 vols Gordon and Breach,New York,1966-68 Conference pt 1 1965,pt 2 1964
Met 25.1016

AMERICAN INSTITUTE OF MINING,METALLURGICAL AND PETROLEUM ENGINEERS.REFRACTORY METALS COMMITTEE Refractory metals and alloys :a technical conference Detroit 1960 May 25-26 Edited by M. Semchyshen and J.J. Harwood Metallurgical society conferences, 11 Interscience,New York;London,1961
Met 25.2532

AMERICAN INSTITUTE OF MINING,METALLURGICAL AND PETROLEUM ENGINEERS.REFRACTORY METALS COMMITTEE Refractory metals and alloys 4 research and development:based on a technical conference Proceedings French Lick,Ind. 1965 Oct 3-5 Vol 1-2 Edited by R.I. Jaffee and others Metallurgical society conferences, 41 2 vols Gordon and Breach, New York,1967
Met 25.2250

AMERICAN INSTITUTE OF MINING,METALLURGICAL AND PETROLEUM ENGINEERS.REFRACTORY METALS COMMITTEE Technical conference on refractory metals and alloys 3rd Proceedings Los Angeles 1963 Dec 9-10 Metallurgical society conferences, 30 Gordon and Breach,New York,1966
Met 25.1017

AMERICAN INSTITUTE OF MINING,METALLURGICAL AND PETROLEUM ENGINEERS.SEMICONDUCTORS COMMITTEE Metallurgy of semiconductor materials a technical conference Los Angeles,Calif. 1961 Aug 30-Sep 1 Edited by John B. Schroeder Metallurgical society conferences, 15 Interscience,New York;London,1962
Met 25.1158

AMERICAN INSTITUTE OF MINING,METALLURGICAL AND PETROLEUM ENGINEERS.STRUCTURAL MATERIALS TECHNICAL COMMITTEE Application of fracture toughness parameters to structural metals :a symposium 1964 Oct 20 Edited by Hermann D. Greenberg Metallurgical society conferences, 31 Gordon and Breach,New York; London,1966
Met 25.0867

AMERICAN INSTITUTE OF MINING AND METALLURGICAL ENGINEERS Pyrometry The Papers and discussion of a symposium on pyrometry Chicago,Ill. 1919 Sep illus 701p New York,1920
WSM 43.4920

AMERICAN INSTITUTE OF MINING AND METALLURGICAL ENGINEERS Pyrometry :a symposium Chicago 1919 Sep American institute of mining and metallurgical engineers,New York City,1920
Met 25.0294

AMERICAN INSTITUTE OF MINING AND METALLURGICAL ENGINEERS Symposium on stress-corrosion cracking of metals Philadelphia,Pa. 1944 Nov 29-Dec 1 American society for testing materials;American institute of mining and metallurgical engineers,Philadelphia,Pa.;New York,1945
Eng 41.3717

AMERICAN INSTITUTE OF MINING AND METALLURGICAL ENGINEERS Transactions of the American institute of mining and metallurgical engineers Vol.110: geophysical prospecting 1934:papers and discussions presented at meetings of the Institute.1932,33,34 bibliog. American institute of mining and metallurgical engineers,New York,1934
Geol 8.1789

AMERICAN INSTITUTE OF MINING AND METALLURGICAL ENGINEERS Transactions of the American institute of mining and metallurgical engineers Vol.178: mining geology 1948: papers and discussions presented at meetings of the Institute 1942-48 illus. American institute of mining and metallurgical engineers,New York,1949
Geol 8.2301

AMERICAN INSTITUTE OF MINING AND METALLURGICAL ENGINEERS 1951 :annual meeting Symposium:problems of clay and laterite genesis St.Louis,Mo. 1951 Feb 19-22 Sponsored by the Karl Eilers Memorial Fund A. I.M.E.,New York,1952
Min 10.0319

AMERICAN INSTITUTE OF MINING AND METALLURGICAL ENGINEERS.INSTITUTE OF METALS DIVISION Nuclear metallurgy A Symposium on behavior of materials in reactor environment 1956 Feb 20 By R.C. Dalzell and others I.M.D. Special report series, 2 American institute of mining and metallurgical engineers,New York,1956 Lithoprinted
Met 25.1928

AMERICAN INSTITUTE OF MINING AND METALLURGICAL ENGINEERS.INSTITUTE OF METALS DIVISION Nuclear metallurgy a symposium 1955 Oct 17 I.M.D.Special report series A.I.M.E.,New York,1955 Lithoprinted
Met 25.1573

AMERICAN INSTITUTE OF MINING AND METALLURGICAL ENGINEERS.INSTITUTE OF METALS DIVISION Symposium on stress-corrosion cracking of metals Philadelphia,Pa. 1944 Nov 29-Dec 1 A.S.T.M.,New York,1945
Met 25.1952

AMERICAN INSTITUTE OF PHYSICS Symposium on temperature 3rd Papers Washington,D.C. 1954 Oct 28-30 Edited by Hugh C. Wolfe Reinhold,New York,1955
Eng 41.4126

AMERICAN INSTITUTE OF PHYSICS Temperature : its measurement and control in science and industry :a symposium Papers New York 1939 Nov 2-4 Reinhold,New York,1941
Eng 41.4125

AMERICAN INSTITUTE OF PHYSICS Temperature : its measurement and control in science and industry :symposium Papers New York 1939 Nov Reinhold,New York,1941
Min 10.0599

AMERICAN INSTITUTE OF PHYSICS Temperature: its measurement and control in science and industry :a symposium 4th Proceedings Columbus,Ohio 1961 Mar 27-31 Vol 3,pt 1: basic concepts,standards and methods Edited by C.M. Herzfeld and Ferdinand Graft Brickwedde Co-sponsored by National bureau of standards Reinhold;Chapman and Hall,New York;London,1962
Cav 7.0048

AMERICAN INSTITUTE OF PHYSICS Temperature: its measurement and control in science and industry :a symposium 4th Proceedings Columbus,Ohio 1961 Mar 27-31 Vol 3,pt2: applied methods and instruments Edited by A. I. Dahl and C.M. Herzfeld Co-sponsored by National bureau of standards 1094p Reinhold publishing,New York,1962
Cav 7.0047

AMERICAN INSTITUTE OF PHYSICS Temperature: its measurement and control in science and industry :a symposium New York 1939 Nov 2-4 Edited by C.O. Fairchild 1362p Reinhold publishing,New York,1941
Cav 7.0046

AMERICAN INSTITUTE OF PHYSICS Temperature: its measurement and control in science and industry :a symposium Papers New York 1939 Nov 2-4 With the cooperation of the National bureau of standards Reinhold,New York,1941
Chem 18.0636

AMERICAN INSTITUTE OF PHYSICS Temperature, its measurement and control in science and industry :a symposium Records New York City 1939 Nov 2-4 National bureau of standards Edited by C.O. Fairchild xiii, 1362p 23cm Reinhold publishing,New York, 1941
Sco 14.0234

AMERICAN INSTITUTE OF PHYSICS Temperature - its measurement and control in science and industry 4th Papers Columbus 1961 Mar 27-31 Vol 3,pt.2: applied methods and instruments Edited by A.I. Dahl Reinhold; Chapman and Hall,New York;London,1962
Met 25.0303

AMERICAN INSTITUTE OF PHYSICS Temperature - its measurement and control in science and industry : a symposium 3rd Papers Washington,D.C. 1954 Oct 28-30 Vol 2 Edited by Hugh C. Wolfe Reinhold;Chapman and Hall,New York;London,1955
Met 25.0299

AMERICAN INSTITUTE OF PHYSICS Temperature - its measurement and control in science and industry :a symposium 2nd Papers New York 1939 Nov 2-4 Reinhold,New York,1941
Met 25.0298

AMERICAN INSTITUTE OF PHYSICS Temperature - its measurement and control in science and industry :a symposium Papers New York 1939 Nov 2-4 Reinhold,New York,1941
Radioth 35.1431

AMERICAN INSTITUTE OF PHYSICS Temperature - its measurement and control in science and industry :a symposium Papers New York 1939 Nov 2-4 xiii,1362p Reinhold,New York, 1941
Chem E 24.0478

AMERICAN INSTITUTE OF PHYSICS and INSTRUMENT SOCIETY OF AMERICA Temperature,its measurement and control in science and industry :a symposium 4th Proceedings Columbus,Ohio 1961 Mar 27-31 Vol 3,1: basic concepts,standards and methods Edited by Ferdinand Graft Brickwedde With the cooperation of the National bureau of standards xvi,848p Reinhold,New York,1962 General editor:Herzfeld,Charles
Sco 14.0231

AMERICAN INSTITUTE OF PHYSICS and INSTRUMENT SOCIETY OF AMERICA Temperature,its measurement and control in science and industry :a symposium 4th Proceedings Columbus,Ohio 1961 Mar 27-31 Vol 3,2: applied methods and instruments Edited by A. I. Dahl Co-sponsored by the National bureau of standards xiv,1094p Reinhold,New York, 1962 General editor:Herzfeld,Charles
Sco 14.0233

AMERICAN INSTITUTE OF PHYSICS and INSTRUMENT SOCIETY OF AMERICA Temperature,its measurement and control in science and industry :a symposium 4th Proceedings Columbus,Ohio 1961 Mar 27-31 Vol 3,3: Biology and medicine Edited by James D. Hardy With the cooperation of the National bureau of standards xii,683p Reinhold,New York,1962 General editor:Herzfeld,Charles
Sco 14.0232

AMERICAN INSTITUTE OF PHYSICS.DIVISIONS OF CHEMICAL PHYSICS AND SOLID STATE PHYSICS and AMERICAN CHEMICAL SOCIETY.DIVISION OF PHYSICAL CHEMISTRY Chemical physics of nonmetallic crystals :Northwestern university 1961 international conference Papers Urbana,Ill. 1961 Benjamin,New York,1962 First published as a supplement to the 'Journal of applied physics',January 1962
Chem 18.2385

AMERICAN INSTITUTION OF MINING ENGINEERS. METALLURGICAL SOCIETY Corrosion by liquid metals Proceedings of the sessions on corrosion by liquid metals of the 1969 fall meeting of the Metallurgical society of AIME Philadelphia,Pa. 1969 Oct 13-16 Edited by J.E. Draley and J.R. Weeks Metallurgical society of AIME.Publications Plenum press, New York,1970
Met 25.2790

AMERICAN MANAGEMENT ASSOCIATION Electronics reference handbook :a practical approach to electronics illus. var.p 28cm American management association,New York,1955 Published for distribution at the A.M.A. conference,New York,Feb-Mar,1955
Math L 5.0732

AMERICAN MATHEMATICAL SOCIETY Algebraic groups and discontinuous subgroups Summer mathematical institute 12th Boulder, Colo. 1965 Jul 5-Aug 6 Edited by Armand Borel and George D. Mostow Financed by the National science foundation American mathematical society.Proceedings of symposia in pure mathematics, 9 vii,426p 26cm American mathematical society,Providence,R.I., 1966
P Math 2.2719

AMERICAN MATHEMATICAL SOCIETY Applications of categorical algebra :a symposium Proceedings New York 1968 Apr 10-11 Edited by Alex Heller American mathematical society.Proceedings of symposia in pure mathematics, 17 Bibliog v,231p 26cm American mathematical society,Providence,R.I., 1970
P Math 2.3508

AMERICAN MATHEMATICAL SOCIETY Applied probability Mathematical probability and its applications symposium New York 1955 Apr. 14-15 Edited by L.A. MacColl Cosponsored by the United States.Army.Office of ordnance research American mathematical society. Proceedings of symposia in applied mathematics, 7 v,104p 26cm McGraw-Hill,New York, 1957
Math 3.0682

AMERICAN MATHEMATICAL SOCIETY Applied probability symposium proceedings Brooklyn 1955 Apr 14-15 Edited by L.A. MacColl American mathematical society.Proceedings of symposia in applied mathematics, 7 American mathematical society,Providence,R.I., 1957
A Math 4.0244

AMERICAN MATHEMATICAL SOCIETY Combinatorics Los Angeles,Calif. 1968 Mar 21-22 Edited by Theodore S. Motzkin American mathematical society.Proceedings of symposia in pure mathematics, 19 viii,255p 26cm AMS, Providence,R.I.,1971
P Math 2.4513

AMERICAN MATHEMATICAL SOCIETY Computers in algebra and number theory Symposium in applied mathematics Proceedings New York 1970 Mar 25-26 By Garrett Birkhoff and Marshall Hall SIAM-AMS proceedings, 4 vii,200p 26cm AMS,Providence,R.I.,1971
P Math 2.4476

AMERICAN MATHEMATICAL SOCIETY Convexity symposium Seattle,Wash. 1961 Jun 13-15 Edited by Victor L. Klee American mathematical society.Proceedings of symposia in pure mathematics, 7 xv,516p 26cm American mathematical society,Providence, R.I., 1963
P. Math 2.0487

AMERICAN MATHEMATICAL SOCIETY Differential geometry Tucson, Ariz. 1960 Feb 18-19 Edited by Carl B. Allendoerfer With the support of the National science foundation American mathematical society.Proceedings of symposia in pure mathematics, 3 vii,200p 26cm American mathematical society, Providence,R.I.,1961
P. Math 2.0698

AMERICAN MATHEMATICAL SOCIETY Entire functions and related parts of analysis : summer institute Lecture notes La Jolla 1966 Jun 27-Jul 22 28cm 2 vols 1966 Mimeographed
P Math 2.3682

AMERICAN MATHEMATICAL SOCIETY Entire functions and related parts of analysis : symposium Proceedings La Jolla 1966 Jun 27-Jul 22 Edited by Jacob Korevaar and others American mathematical society. Proceedings of Symposia in pure mathematics, 11 illus. vi,554p 26cm American mathematical society,Providence,R.I.,1968
P Math 2.3199

AMERICAN MATHEMATICAL SOCIETY Finite groups institute report Pasadena,Calif. 1960 Aug 1-28 Edited by Marshall Hall Supported by the National science foundation American mathematical society.Proceedings of symposia in pure mathematics, 6 114p 26cm American mathematical society,Providence,R.I., 1962
P. Math 2.2031

AMERICAN MATHEMATICAL SOCIETY Finite groups symposium New York 1959 Apr 23-24 Edited by A.Adrian Albert and Irving Kaplansky Cosponsored by the Institute for defense analysis.Communications division American mathematical society.Proceedings of symposia in pure mathematics, 1 viii,110p 23cm American mathematical society,Providence,R.I., 1959
P. Math 2.2029

AMERICAN MATHEMATICAL SOCIETY Fluid dynamics a symposium Proceedings College Park 1951 Jun 22-23 Edited by M.H. Martin American mathematical society.Proceedings of symposia in applied mathematics, 4 v,186p McGraw-Hill,New York,1958
Chem E 24.0354

AMERICAN MATHEMATICAL SOCIETY Fluid dynamics symposium College park,Md. 1951 Jun 22-23 Edited by M.H. Martin Co-sponsored by the Naval ordnance laboratory American mathematical society.Proceedings of symposia in applied mathematics, 4 McGraw-Hill,New York,1953
A Math 4.0512

AMERICAN MATHEMATICAL SOCIETY Global analysis :a symposium Proceedings Berkeley, Calif. 1968 Jul 1-26 Edited by Shing-Shen Chern and Stephen Smale American mathematical society.Proceedings of symposia in pure mathematics, 14,16 bibliog. 25cm American mathematical society,Providence,R.I., 1970
P Math 2.3875

AMERICAN MATHEMATICAL SOCIETY Institute in the theory of numbers report Boulder,Colo. 1959 Jun 21-Jul 17 Edited by Donald C.B. Marsh and James H. Jordan With the support of the National science foundation 350p 27cm University of Colorado,Boulder,Colo., 1959
P. Math 2.1773

AMERICAN MATHEMATICAL SOCIETY International congress of mathematicians 11th Proceedings Cambridge,Mass. 1950 Aug 13-Sep 6 Vol 1-2 25cm 2 vols American mathematical society,Providence,R.I., 1952
Math L 5.3246

AMERICAN MATHEMATICAL SOCIETY Lattice theory symposium Monterey,Calif. 1959 Apr 16-18 Edited by R.P. Dilworth With the financial support of the National science foundation American mathematical society.Proceedings of symposia in pure mathematics, 2 viii,208p 26cm American mathematical society, Providence,R.I.,1961
P. Math 2.2030

AMERICAN MATHEMATICAL SOCIETY Lecture notes prepared in connection with the summer institute on algebraic geometry typescript Woods Hole,Mass. 1964 Jul 6-31 By Shreeram S. Abhyankar and others 29cm American mathematical society,1964
P. Math 2.0575

AMERICAN MATHEMATICAL SOCIETY Lectures in applied mathematics :summer seminar proceedings Boulder,Colo. 1957 Jun.23-Jul. 19 Vol. 1: probability and related topics in physical sciences By Mark Kac Supported by United States.Armed services xiii,266p 24cm Interscience publishers,London;New York, 1959 With special lectures by G.E. Uhlenbeck,A.R.Hibbs,and Balth.van der Pol
Math 3.0844

AMERICAN MATHEMATICAL SOCIETY Lectures on fluid mechanics By Sydney Goldstein Lectures in applied mathematics:proceedings of the summer seminar,Boulder, Colo. 1957, 2 xvi,307p Interscience,New York,1960 Seminar held under the joint sponsorship of the University of Colorado, and the American mathematical society.
Chem E 24.0355

AMERICAN MATHEMATICAL SOCIETY Mathematical aspects of computer science :a symposium Proceedings New York 1967 Edited by J.T. Schwartz American mathematical society. Proceedings of symposia in applied mathematics, 19 Bibliog,illus 224p 26cm American mathematical society,Providence,R.I.,1967
P Math 2.3001

AMERICAN MATHEMATICAL SOCIETY Mathematical problems in the geophysical sciences Troy,N. Y. 1970 Vol 1-2: geophysical fluid dynamics;inverse problems,dynamo theory,and tides Edited by W.H. Reid Lectures in applied mathematics, 13-14 illus,maps 24cm 2 vols American mathematical society, Providence,R.I.,1971
A Math 4.1742

AMERICAN MATHEMATICAL SOCIETY Nonlinear functional analysis :a symposium Proceedings Chicago,Ill. 1968 Apr 16-19 Edited by Felix E. Browder American mathematical society.Proceedings of symposia in pure mathematics, 18,part 1 bibliog. v,296p 26cm American mathematical society, Providence,R.I.,1970
P Math 2.3864

AMERICAN MATHEMATICAL SOCIETY Number theory : a symposium Proceedings Houston,Tex. 1967 Jan 24-28 Edited by William J. Leveque and Ernst G. Straus American mathematical society.Proceedings of symposia in pure mathematics, 12 Bibliog. v,98p 26cm American mathematical society,Providence,R.I., 1969 Special session on number theory at the 73rd annual meeting of the A.M.S.
P Math 2.3699

AMERICAN MATHEMATICAL SOCIETY Numerical analysis :symposium 6th Proceedings Santa Monica,Calif. 1953 Aug 26-28 Edited by John H. Curtiss Cosponsored by National bureau of standards American mathematical society.Proceedings of symposia in applied mathematics, 6 vi,303p McGraw-Hill,New York,1956
Math L 5.3245

AMERICAN MATHEMATICAL SOCIETY Partial differential equations :a symposium Proceedings Berkeley,Calif. 1960 Apr 21-22 Edited by Charles B. Morrey American mathematical society.Proceedings of symposia in pure mathematics, 4 Bibliog.,Illus. vi,169p 26cm American mathematical society, Providence,R.I.,1961
A Math 4.1468

AMERICAN MATHEMATICAL SOCIETY Representation theory of finite groups and related topics : symposium Madison,Wisc. 1970 Apr 14-16 Edited by Irving Reiner American mathematical society.Proceedings of symposia in pure mathematics, 21 v,178p 26cm AMS, Providence,R.I.,1971
P Math 2.4432

AMERICAN MATHEMATICAL SOCIETY Singular integrals :symposium Proceedings Chicago 1966 American mathematical society. Proceedings of symposia in pure mathematics, 10 Bibliog,port vi,375p 26cm American mathematical society,Providence,R.I.,1967
P Math 2.3036

AMERICAN MATHEMATICAL SOCIETY Some mathematical problems in biology Symposium on mathematical biology Proceedings Washington 1966 Dec 28 Edited by Murray Gerstenhaber American mathematical society. Lectures on mathematics in the life sciences, 1 Bibliog.,Illus. vi and 117p 22cm American mathematical society,Rhode Island, 1968
P Math 2.3482

AMERICAN MATHEMATICAL SOCIETY Stochastic processes in mathematical physics and engineering symposium 16 proceedings New York 1963 Apr.30-May 2 Edited by Richard Bellman Cosponsored by the Society for industrial and applied mathematics American mathematical society.Proceedings of symposia in applied mathematics, 16 viii,318p 25cm American mathematical society,Providence,R.I., 1964
Math 3.0567

AMERICAN MATHEMATICAL SOCIETY Summer seminar in applied mathematics 1st Proceedings Boulder,Col. 1957 Jun 23-Jul 21 Vol 3: partial differential equations By Lipman Bers and others Edited by Alton S. Householder and others Lectures in applied mathematics, 3 xiii,343p Interscience, New York,1964 Dedicated to Arthur Norman Milgram
A Math 4.1266

AMERICAN MATHEMATICAL SOCIETY Symposium on the theory of finite groups Lectures Urbana,Ill. 1967 Nov 24 Bibliog. 94p 25cm American mathematical society, Providence,R.I.,1969
P Math 2.3581

AMERICAN MATHEMATICAL SOCIETY The Boston colloquium 4th Lectures Boston,Mass. 1903 Sep 2-5 By Edward Burr van Vleck and others American mathematical society. Colloquium lectures, 1 xii,187p 21cm University microfilms,Ann Arbor,Mich.,1967 Facsimile produced by microfilm-xerography of the original volume published in 1905 for the American mathematical society by Macmillan,New York
P Math 2.2865

AMERICAN MATHEMATICAL SOCIETY The General theory of quantized fields proceedings Boulder,Col. 1960 Jul.24-Aug.19 Vol. 4 By Res Jost University of Colorado.Summer institute on applied mathematics Lectures in applied mathematics, 4 American mathematical society,Providence,R.I.,1965
A Math 4.0760

AMERICAN MATHEMATICAL SOCIETY The New Haven mathematical colloquium 5th Lectures New Haven,Conn. 1906 Sep 5-8 By Eliakim Hastings Moore and others American mathematical society.Colloquium lectures, 2 x,222p 24cm University microfilms,Ann Arbor,Mich.,1967 Facsimile produced by microfilm-xerography of the original book published in 1910 by Yale university press,New Haven,Conn.
P Math 2.2864

AMERICAN MATHEMATICAL SOCIETY Theory of numbers symposium Pasadena,Calif. 1963 Nov 21-22 Edited by Albert Leon Whiteman Under a grant from the National science foundation American mathematical society.Proceedings of symposia in pure mathematics, 8 vii,214p 26cm American mathematical society, Providence,R.I.,1965
P. Math 2.1774

AMERICAN MATHEMATICAL SOCIETY Wave motion and vibration theory symposium Washington 1952 June 16-17 Edited by Albert E. Heins American mathematical society.Proceedings of symposia in applied mathematics, 5 McGraw-Hill,New York,1954
A Math 4.0513

AMERICAN MATHEMATICAL SOCIETY 1969 number theory institute Summer institute on number theory Stony Brook, N.Y. 1969 Jul 7-Aug 1 Edited by Donald J. Lewis American mathematical society.Proceedings of symposia in pure mathematics, 20 xiii,451p 26cm AMS,Providence,R.I.,1971
P Math 2.4383

AMERICAN MATHEMATICAL SOCIETY.COLLOQUIUM PUBLICATIONS, 38 Theory of graphs By Oystein Ore Bibliog American mathematical society,Providence,R.I.,1962
Eng 41.1815

AMERICAN MATHEMATICAL SOCIETY.LECTURES ON MATHEMATICS IN THE LIFE SCIENCES, 1 Some mathematical problems in biology Symposium on mathematical biology Proceedings Washington 1966 Dec 28 Edited by Murray Gerstenhaber Prepared by the American mathematical society Bibliog., Illus. vi and 117p 22cm American mathematical society,Rhode Island,1968
P Math 2.3482

AMERICAN MATHEMATICAL SOCIETY.PROCEEDINGS OF SYMPOSIA IN APPLIED MATHEMATICS, 4 Fluid dynamics :a symposium Proceedings College Park 1951 Jun 22-23 American mathematical society Edited by M.H. Martin v,186p McGraw-Hill,New York,1958
Chem E 24.0354

AMERICAN MATHEMATICAL SOCIETY.PROCEEDINGS OF SYMPOSIA IN APPLIED MATHEMATICS, 4 Fluid dynamics symposium College park,Md. 1951 Jun 22-23 American mathematical society Edited by M.H. Martin Co-sponsored by the Naval ordnance laboratory McGraw-Hill,New York,1953
A Math 4.0512

AMERICAN MATHEMATICAL SOCIETY.PROCEEDINGS OF SYMPOSIA IN APPLIED MATHEMATICS, 5 Wave motion and vibration theory symposium Washington 1952 June 16-17 Carnegie institute of technology and American mathematical society Edited by Albert E. Heins McGraw-Hill,New York,1954
A Math 4.0513

AMERICAN MATHEMATICAL SOCIETY.PROCEEDINGS OF SYMPOSIA IN APPLIED MATHEMATICS, 7 Applied probability Mathematical probability and its applications symposium New York 1955 Apr.14-15 American mathematical society Edited by L.A. MacColl Cosponsored by the United States.Army.Office of ordnance research v,104p 26cm McGraw-Hill,New York,1957
Math 3.0682

AMERICAN MATHEMATICAL SOCIETY.PROCEEDINGS OF SYMPOSIA IN APPLIED MATHEMATICS, 7 Applied probability symposium proceedings Brooklyn 1955 Apr 14-15 American mathematical society Edited by L.A. MacColl American mathematical society,Providence,R.I., 1957
A Math 4.0244

AMERICAN MATHEMATICAL SOCIETY.PROCEEDINGS OF SYMPOSIA IN APPLIED MATHEMATICS, 16
Stochastic processes in mathematical physics and engineering symposium 16 proceedings New York 1963 Apr.30-May 2 American mathematical society Edited by Richard Bellman Cosponsored by the Society for industrial and applied mathematics viii,318p 25cm American mathematical society, Providence,R.I.,1964
Math 3.0567

AMERICAN MATHEMATICAL SOCIETY.PROCEEDINGS OF SYMPOSIA IN PURE MATHEMATICS, 1 Finite groups symposium New York 1959 Apr 23-24 American mathematical society Edited by A. Adrian Albert and Irving Kaplansky Cosponsored by the Institute for defense analysis.Communications division viii,110p 23cm American mathematical society, Providence,R.I.,1959
P. Math 2.2029

AMERICAN MATHEMATICAL SOCIETY.PROCEEDINGS OF SYMPOSIA IN PURE MATHEMATICS, 2 Lattice theory symposium Monterey,Calif. 1959 Apr 16-18 American mathematical society Edited by R.P. Dilworth With the financial support of the National science foundation viii,208p 26cm American mathematical society, Providence,R.I.,1961
P. Math 2.2030

AMERICAN MATHEMATICAL SOCIETY.PROCEEDINGS OF SYMPOSIA IN PURE MATHEMATICS, 3
Differential geometry Tucson, Ariz. 1960 Feb 18-19 American mathematical society Edited by Carl B. Allendoerfer With the support of the National science foundation vii,200p 26cm American mathematical society,Providence,R.I.,1961
P. Math 2.0698

AMERICAN MATHEMATICAL SOCIETY.PROCEEDINGS OF SYMPOSIA IN PURE MATHEMATICS, 4 Partial differential equations :a symposium Proceedings Berkeley,Calif. 1960 Apr 21-22 American mathematical society Edited by Charles B. Morrey Bibliog.,Illus. vi,169p 26cm American mathematical society, Providence,R.I.,1961
A Math 4.1468

AMERICAN MATHEMATICAL SOCIETY.PROCEEDINGS OF SYMPOSIA IN PURE MATHEMATICS, 6 Finite groups institute report Pasadena,Calif. 1960 Aug 1-28 American mathematical society Edited by Marshall Hall Supported by the National science foundation 114p 26cm American mathematical society,Providence,R.I., 1962
P. Math 2.2031

AMERICAN MATHEMATICAL SOCIETY.PROCEEDINGS OF SYMPOSIA IN PURE MATHEMATICS, 7
Convexity symposium Seattle,Wash. 1961 Jun 13-15 American mathematical society Edited by Victor L. Klee xv,516p 26cm American mathematical society,Providence, R.I.,1963
P. Math 2.0487

AMERICAN MATHEMATICAL SOCIETY.PROCEEDINGS OF SYMPOSIA IN PURE MATHEMATICS, 8 Theory of numbers symposium Pasadena,Calif. 1963 Nov 21-22 American mathematical society Edited by Albert Leon Whiteman Under a grant from the National science foundation vii, 214p 26cm American mathematical society, Providence,R.I.,1965
P. Math 2.1774

AMERICAN MATHEMATICAL SOCIETY.PROCEEDINGS OF SYMPOSIA IN PURE MATHEMATICS, 10
Singular integrals :symposium Proceedings Chicago 1966 American mathematical society Bibliog,port vi,375p 26cm American mathematical society,Providence,R.I.,1967
P Math 2.3036

AMERICAN MATHEMATICAL SOCIETY.PROCEEDINGS OF SYMPOSIA IN PURE MATHEMATICS, 12 Number theory :a symposium Proceedings Houston, Tex. 1967 Jan 24-28 American mathematical society Edited by William J. Leveque and Ernst G. Straus Bibliog. v,98p 26cm American mathematical society,Providence,R.I., 1969 Special session on number theory at the 73rd annual meeting of the A.M.S.
P Math 2.3699

AMERICAN MATHEMATICAL SOCIETY.PROCEEDINGS OF SYMPOSIA IN PURE MATHEMATICS, 13,pt 1
Axiomatic set theory Symposium in pure mathematics 14th Proceedings Los Angeles, Calif. 1967 Jul 10-Aug 5 Edited by Dana S. Scott Held at the University of California Los Angeles v,474p 26cm American mathematical society,Providence,R.I.,1971
P Math 2.4061

AMERICAN MATHEMATICAL SOCIETY.PROCEEDINGS OF SYMPOSIA IN PURE MATHEMATICS, 17
Applications of categorical algebra :a symposium Proceedings New York 1968 Apr 10-11 American mathematical society Edited by Alex Heller Bibliog v,231p 26cm American mathematical society,Providence,R.I., 1970
P Math 2.3508

AMERICAN MATHEMATICAL SOCIETY.PROCEEDINGS OF SYMPOSIA IN PURE MATHEMATICS, 18,part 1
Nonlinear functional analysis :a symposium Proceedings Chicago,Ill. 1968 Apr 16-19 American mathematical society Edited by Felix E. Browder bibliog. v,296p 26cm American mathematical society,Providence,R.I., 1970
P Math 2.3864

AMERICAN MATHEMATICAL SOCIETY.PROCEEDINGS OF SYMPOSIA IN PURE MATHEMATICS, 19
Combinatorics Los Angeles,Calif. 1968 Mar 21-22 American mathematical society Edited by Theodore S. Motzkin viii,255p 26cm AMS,Providence,R.I.,1971
P Math 2.4513

AMERICAN MATHEMATICAL SOCIETY.PROCEEDINGS OF SYMPOSIA IN PURE MATHEMATICS, 20 1969 number theory institute Summer institute on number theory Stony Brook, N.Y. 1969 Jul 7-Aug 1 American mathematical society Edited by Donald J. Lewis xiii,451p 26cm AMS, Providence,R.I.,1971
P Math 2.4383

AMERICAN MATHEMATICAL SOCIETY.PROCEEDINGS OF SYMPOSIA IN PURE MATHEMATICS, 21
Representation theory of finite groups and related topics :symposium Madison,Wisc. 1970 Apr 14-16 American mathematical society Edited by Irving Reiner v,178p 26cm AMS, Providence,R.I.,1971
P Math 2.4432

AMERICAN MATHEMATICAL SOCIETY.SUMMER SEMINAR IN APPLIED MATHEMATICS Relativity theory and astrophysics 1:relativity and cosmology Lectures Ithaca,N.Y. 1965 Jul 25-Aug 20 By W. Bonner and others Edited by Jurgen Ehlers Lectures in applied mathematics, 8 xvi,292p American mathematical society, Providence,R.I.,1967
TA 15.0350

AMERICAN MATHEMATICAL SOCIETY.SUMMER SEMINAR IN APPLIED MATHEMATICS Relativity theory and astrophysics 2:galactic structure Lectures Ithaca,N.Y. 1965 Jul 25-Aug 20 By E. Burbidge and others Edited by Jurgen Ehlers viii,220p American mathematical society, Providence,R.I.,1967
TA 15.0351

AMERICAN MATHEMATICAL SOCIETY.SYMPOSIUM IN APPLIED MATHEMATICS 3rd Proceedings Elasticity Ann Arbor,Mich. 1949 Jun 14-16 Edited by R.V. Churchill and others Cosponsored by the American society of mechanical engineers.Applied mechanics division McGraw-Hill,New York,1950
Col S 12.0335

AMERICAN MATHEMATICAL SOCIETY.13TH SUMMER MEETING The New Haven mathematical colloquium lectures New Haven 1906 September 3-8 By Eliakim Hastings Moore and others Under the auspices of Yale university Americian mathematical society.Colloquium, 5 Yale university press,New Haven,1910
Philos 1.0198

AMERICAN METEOROLOGICAL SOCIETY Causes of climatic change :a collection of papers derived from the INQUA-NCAR symposium on causes of climatic change,August 30-31,Boulder, Colorado Edited by J.Murray Mitchell American meteorological society.Meteorological monographs, 8,no 30 Boston,Mass.,1968
Bot 42.6236

AMERICAN METEOROLOGICAL SOCIETY Global circulation of the atmosphere :a conference Proceedings London 1969 Aug 25-29 Edited by G.A. Corby bibliog.,illus.,port. 257p 26cm Royal meteorological society,London, 1969
A Math 4.1651

AMERICAN METEOROLOGICAL SOCIETY International conference on cloud physics Proceedings Toronto 1968 873p American meteorological society,Boston,Mass.,1969
Nap 11.0973

AMERICAN METEOROLOGICAL SOCIETY National conference on statistical meteorology 1st Proceedings Hartford,Conn. 1968 May 27-29 American meteorlogical society,Boston,Mass., 1968 Co-Sponsor:A.M.S.committee on statistical meteorology
Math S 3.1583

AMERICAN METEOROLOGICAL SOCIETY National conference on weather modification 1st Proceedings Albany,N.Y. 1968 Apr 28-May 1 532p American meteorological society,Boston, Mass.,1968
Nap 11.0975

AMERICAN METEOROLOGICAL SOCIETY National conference on weather modification 2nd Proceedings Santa Barbara,Calif. 1970 Apr 6-9 436p American meteorological society, Boston,Mass.,1970
Nap 11.0977

AMERICAN METEOROLOGICAL SOCIETY Solar variations,climatic change,and related geophysical problems :a conference Papers New York 1961 Jan 24-28 New York academy of sciences Edited by W.Franklin N. Furness and others New York.Academy of sciences. Annals, 95,1 740p New York,1961 Conference editor:R.W.Fairbridge
Nap 11.0210

AMERICAN METEOROLOGICAL SOCIETY and NEW YORK ACADEMY OF SCIENCES Solar variations, climatic change and related geophysical problems :conference Papers New York 1961 Jan 24-28 Edited by Rhodes W. Fairbridge New York academy of sciences. Annals,95,art.1 New York academy of sciences, New York,1961
Geog 13.1055

AMERICAN MUSEUM OF NATURAL HISTORY Marine bio-acoustics :symposium 1st,2nd Proceedings Bimini,Bahamas 1963 Apr 11-13 New York 1966 Apr 13-15 Vol 1-2 By William N. Tavolga 2 vols Pergamon press,Oxford,1964-67 Vol.1 held at the Lerner marine laboratory Bimini;vol.2 held at the American museum of natural history,New York
Bal 39.1534

AMERICAN NEUROLOGICAL ASSOCIATION Properties of membranes and diseases of the nervous system By Donald B. Tower and others Springer,New York,1962 Based on a symposium - June 1961 - sponsored jointly by the American neurological association and the American association of neuropathologists
An 32.4197

AMERICAN NUCLEAR SOCIETY Neutron physics : symposium Proceedings Troy,N.Y. 1961 May 5-6 Rensselaer polytechnic institute Edited by M.L. Yeater Nuclear science and technology:a series of monographs and text books, 2 xiii,303p Academic press,New York,1962
Cav 7.2543

AMERICAN ORCHID SOCIETY World orchid conference 3rd Proceedings London 1960 May 30-Jun 22 Royal horticultural society,London,1960
BG 38.3118

AMERICAN PHILOSOPHICAL SOCIETY Commemoration of the publication of Gregor Mendel's pioneer experiments in genetics :annual general meeting Papers Philadelphia,Pa. 1965 Apr 23 American philosophical society. Proceedings, 109,no.4 American philosophical society,Philadelphia,Pa.,1965
Gen 34.2207

AMERICAN PHILOSOPHICAL SOCIETY Natural selection and adaptation :annual general meeting Papers Philadelphia,Pa. 1949 Apr American philosophical society.Proceedings, 93,no.6 Philadelphia,Pa.,1949
Gen 34.1305

AMERICAN PHILOSOPHICAL SOCIETY Proceedings commemorative of the 150th anniversary of the foundation of the American philosophical society Philadelphia,Pa. 1893 May 22-26 phylogeny of an acquired characteristic By Alpheus Hyatt American philosophical society. Proceedings,32 Philadelphia,Pa.,1894
Geol 8.0856

AMERICAN PHILOSOPHICAL SOCIETY Symposium on progress in astrophysics Papers 1939 Feb 17 By Donald H. Menzel and others 1939 From the Proceedings of the American philosophical society,vol 81,no 2.Lacks title page
Obs 6.3230

AMERICAN PHYSICAL SOCIETY International conference on II-VI semiconducting compounds Proceedings Providence,R.I. 1967 Sep 6-8 Edited by D.G. Thomas 1490p 20cm Benjamin,New York,1967
Cav 7.2681

AMERICAN PHYSICAL SOCIETY International conference on photoconductivity 3rd Proceedings Stanford,Calif. 1969 Aug 12-15 Edited by E.M. Pell Bibliog,diagrms,graphs xi,410p 26cm Pergamon,Oxford,1971
Cav 7.3073

AMERICAN PHYSICAL SOCIETY.DIVISION OF ELECTRON OPTICS Preparation and characteristics of solid luminescent materials :a conference Ithaca,N.Y. 1946 Oct 24-26 Edited by Gorton R. Fonda and Frederick Seitz Published under the auspices of the National research council graphs xv,459p 22cm Wiley,New York,1948
Cav 7.3061

AMERICAN PHYSICAL SOCIETY.DIVISION OF ELECTRON OPTICS Preparation and characteristics of solid luminescent materials :a symposium Ithaca,N.Y. 1946 Oct 24-26 Wiley;Chapman and Hall,New York;London,1948
Eng 41.5491

AMERICAN PHYSIOLOGICAL SOCIETY Control mechanism in respiration and fermentation :a symposium held during the meeting of the Society of general physiologists at the Marine biological laboratory Woods Hole,Mass. 1961 Sep By Harlyn O. Halvorson Society of general physiologists Edited by Barbara Wright Ronald press,New York,1963
Bot 42.1730

AMERICAN PHYSIOLOGICAL SOCIETY Subcellular particles :a symposium Woods Hole,Mass. 1958 Jun 9-11 Edited by Teru Hayashi Sponsored by the Society of general physiologists Ronald press,New York,1959
Gen 34.0828

AMERICAN PHYSIOLOGICAL SOCIETY Subcellular particles :a symposium held during the meeting of the Society of general physiologists at the Marine biological laboratory Woods Hole,Mass. 1958 Jun 9-11 Society of general physiologists Edited by Teru Hayashi Ronald press,New York,1959
Bioch 33.1014

AMERICAN POLAR EXPLORATIONS Centenary celebration of the Wilkes exploring expedition of the United States navy 1838-1842 and symposium on American polar explorations Papers Philadelphia,Pa. 1940 Feb 23-24 By Edwin G. Conklin and others American philosophical society.Proceedings, 82 519-950p American philosophical society, Philadelphia,Pa.,1940
Sco 14.8094

AMERICAN PSYCHIATRIC ASSOCIATION Language and thought in schizophrenia :collected papers presented at a meeting of the American psychiatric association...and brought up to date Papers Chicago,Ill. 1939 May 12 Edited by J.S. Kasanin University of California press,Berkeley;Los Angeles,1954 Preface by N.D.C.Lewis
Psy 28.0082

AMERICAN PSYCHOLOGICAL ASSOCIATION Behaviour and evolution :conference 2nd Proceedings Princeton,N.J. 1956 Apr 30-May 5 Edited by Anne Roe and George Gaylord Simpson Held in collaboration with the Society for the study of evolution 24cm New Haven,1958
Philos 1.1827

AMERICAN PSYCHOLOGICAL ASSOCIATION Personal and social factors in perception Perception and personality :a symposium Papers Denver, Colo. Edited by Jerome S. Bruner and David Krech Duke university publications Duke university press,Durham,N.C.,1950 Papers originally appeared in the 'Journal of personality',vol 18,no 1-2,Sept and Dec. 1949
Psy 31.0945

AMERICAN PSYCHOLOGICAL ASSOCIATION.DIVISION OF CLINICAL PSYCHOLOGY Research in psychotherapy :a conference Proceedings Washington,D.C. 1958 Apr 9-12 Edited by Eli Rubinstein and Morris B. Parloff National publishing co.,Washington,D.C.,1959
Psy 31.1627

AMERICAN PSYCHOLOGICAL ASSOCIATION MEETINGS SEPTEMBER 1962 Behavior and awareness :a symposium and interpretation Edited by C.W. Eriksen Duke university press,Durham,N.C., 1962 Developed from papers read at the September 1962 meetings of the American psychological association
Psy 31.2985

AMERICAN PSYCHOPATHOLOGICAL ASSOCIATION Comparative psychopathology animal and human Proceedings New York 1965 Feb Edited by Joseph Zubin and Howard F. Hunt Grune and Stratton,New York;London,1967 Proceedings of the 55th annual meeting of the Association
Psy 31.1682

AMERICAN PSYCHOPATHOLOGICAL ASSOCIATION Experimental psychopathology :forty-fifth annual meeting Proceedings New York 1955 Jun Edited by Paul H. Hoch and Joseph Zubin Grune and Stratton,New York;London,1957
Psy 28.0197

AMERICAN PSYCHOPATHOLOGICAL ASSOCIATION Psychiatry and the law :the proceedings of the forty-third annual meeting of the American psychopathological association Proceedings New York 1953 Jun Edited by Paul H. Hoch and Joseph Zubin Grune and Stratton,New York; London,1955
Psy 28.0311

AMERICAN PSYCHOPATHOLOGICAL ASSOCIATION
Psychology of perception :the proceedings of the fifty-third annual meeting of the American psychological association Proceedings New York 1963 Feb Edited by Paul H. Hoch and Joseph Zubin Grune and Stratton,London,1965 Contains the 1963 Samuel Hamilton award lecture by Heinrich Kluver
Psy 31.0958

AMERICAN PSYCHOPATHOLOGICAL ASSOCIATION
Psychopathology of communication :forty-sixth annual meeting Proceedings New York 1956 Jun Edited by Paul H. Hoch and Joseph Zubin Grune and Stratton,New York;London,1958
Psy 28.0196

AMERICAN PSYCHOPATHOLOGICAL ASSOCIATION.ANNUAL MEETING 43RD Psychiatry and the law :the proceedings of the forty-third annual meeting of the American psychopathological association Proceedings New York 1953 Jun American psychopathological association Edited by Paul H. Hoch and Joseph Zubin Grune and Stratton,New York;London,1955
Psy 28.0311

AMERICAN PSYCHOPATHOLOGICAL ASSOCIATION.ANNUAL MEETING 45TH Experimental psychopathology forty-fifth annual meeting Proceedings New York 1955 Jun American psychopathological association Edited by Paul H. Hoch and Joseph Zubin Grune and Stratton,New York;London,1957
Psy 28.0197

AMERICAN PSYCHOPATHOLOGICAL ASSOCIATION.ANNUAL MEETING 46TH Psychopathology of communication :forty-sixth annual meeting Proceedings New York 1956 Jun American psychopathological association Edited by Paul H. Hoch and Joseph Zubin Grune and Stratton,New York;London,1958
Psy 28.0196

AMERICAN PSYCHOPATHOLOGICAL ASSOCIATION.ANNUAL MEETING 55TH Comparative psychopathology animal and human Proceedings New York 1965 Feb American psychopathological association Edited by Joseph Zubin and Howard F. Hunt Grune and Stratton,New York; London,1967 Proceedings of the 55th annual meeting of the Association
Psy 31.1682

AMERICAN PSYCHOPHATHOLOGICAL ASSOCIATION.ANNUAL MEETING 53RD Psychology of perception : the proceedings of the fifty-third annual meeting of the American psychological association Proceedings New York 1963 Feb American psychopathological association Edited by Paul H. Hoch and Joseph Zubin Grune and Stratton,London,1965 Contains the 1963 Samuel Hamilton award lecture by Heinrich Kluver
Psy 31.0958

AMERICAN RED CROSS Conference on the preservation of the formed elements and of the proteins of the blood Transactions Boston 1949 Jan 6-8 1949
Med 36.0005

AMERICAN ROCKET SOCIETY Electric propulsion development... :a selection of technical papers based mainly on the American rocket society electric propulsion conference... Berkeley,Calif. 1962 Mar 14-16 Edited by Ernst Stuhlinger Progress in astronautics and aeronautics, 9 Academic press,New York; London,1963
Eng 41.6889

AMERICAN ROCKET SOCIETY Electrostatic propulsion... a selection of technical papers based mainly on a symposium of the American rocket society... Monterey,Calif. 1960 Nov 3-4 Edited by David B. Langmuir and others Progress in astronautics and rocketry, 5 Academic press,New York;London,1961
Eng 41.6888

AMERICAN ROCKET SOCIETY Energy conversion for space power :a symposium Santa Monica, Calif. 1960 Sep 27-30 Edited by Nathan W. Snyder Progress in astronautics and rocketry, 3 Academic press,New York,1961
Eng 41.4971

AMERICAN ROCKET SOCIETY Gas dynamics symposium on aerothermochemistry Proceedings Evanston,Ill 1955 Aug 22-24 Edited by Donald K. Fleming Supported by the United States.Air force illus 284p Northwestern university,Evanston,Ill.,1956
Chem E 24.1390

AMERICAN ROCKET SOCIETY Hypersonic flow research :a selection of technical papers based mainly a symposium of the American rocket society,held at the Massachusetts institute of technology... Cambridge,Mass. 1961 Aug 16-18 Edited by Frederick R. Riddell Progress in astronautics and rocketry, 7 Academic press,NewYork,1962
Eng 41.6771

AMERICAN ROCKET SOCIETY Hypersonic flow research :a symposium technical papers Cambridge,Mass. 1961 Aug 16-18 Edited by Frederick R. Riddell Sponsored jointly with the United States.Air force.Office of scientific research Progress in astronautics and rocketry, 7 Academic press,New York, London,1962
A Math 4.0533

AMERICAN ROCKET SOCIETY Ionization in high-temperature gases :a selection of technical papers based mainly on the American rocket society conference on ions in flames and rocket exhausts Palm Springs,Calif. 1962 Oct 10-12 Edited by Kurt E. Shuler Progress in astronautics and aeronautics, 12 Academic press,New York,1963
Eng 41.4543

AMERICAN ROCKET SOCIETY Magnetohydrodynamics Biennial gas dynamics symposium 4th proceedings Evanston,Ill. 1961 Aug 23-25 Edited by Ali Bulent Cambel and others Through the generosity of Government agencies and industrial sponsors Northwestern university press,Evanston,Ill.,1962
A Math 4.0623

AMERICAN ROCKET SOCIETY.SPACE POWER SYSTEMS COMMITTEE Energy conversion for space power... based on a symposium Selected technical papers Santa Monica,Calif. 1960 Sep 27-30 Edited by Nathan W. Snyder Progress in astronautics and rocketry, 3 Academic press,New York;London,1961
Met 25.2165

AMERICAN SOCIETY FOR CYBERNETICS 1st : annual symposium Proceedings Purposive systems Edited by Heinz von Foerster and others Spartan books,New York;Washington,D.C. 1968
Psy 31.3068

AMERICAN SOCIETY FOR METALS Age hardening of metals the symposium on precipitation hardening (age hardening) presented before the twenty-first annual convention Papers and discussions Chicago 1939 Oct 23-27 American society for metals,Cleveland,1940
Met 25.1249

AMERICAN SOCIETY FOR METALS Atom movements Chicago 1950 Oct 21-27 American society for metals,Cleveland,1951 A seminar held during the 32nd national metal congress and exposition
Met 25.1272

AMERICAN SOCIETY FOR METALS Atom movements : a seminar proceedings Chicago 1950 Oct 21-27 240p American society for metals, Cleveland,Ohio,1951 Held during the 32nd National metal congress and exposition
Chem E 24.0757

AMERICAN SOCIETY FOR METALS Atomic and electronic structure of metals :a seminar Papers Cleveland,Ohio 1966 Oct 29-30 American society for metals,Cleveland,Ohio, 1967
Met 25.1059

AMERICAN SOCIETY FOR METALS Behaviour of metals at low temperatures :a series of three educational lectures on behaviour of metals at low temperatures presented to members of the ASM during the thirtyfourth National metal congress and exposition,Philadelphia,October 20 t 24,1952 Lecures By R.M. Brick and others diagrs iv,112p 24cm Cleveland, Ohio,1953
Sco 14.1245

AMERICAN SOCIETY FOR METALS Carburizing :a symposium Papers and discussions Atlantic City,N.J. 1937 Oct 18-22 American society for metals,Cleveland,Ohio,1937 The, Symposium presented before the nineteenth annual convention of the American society for Metals...
Met 25.0439

AMERICAN SOCIETY FOR METALS Cold working of metals Philadelphia 1948 Oct 25-29 American society for metals,Cleveland,Ohio, 1949
Met 25.1199

AMERICAN SOCIETY FOR METALS Conference on powder metallurgy held at the Massachusetts institute of technology 1st and 2nd Papers Cambridge,Mass. 1940-41 Edited by John Wulff American society for metals, Cleveland,1942
Met 25.0758

AMERICAN SOCIETY FOR METALS Controlled atmospheres :the symposium...presented before the annual convention Philadelphia,Pa. 1941 Oct.20-24 A.S.M.,Cleveland,Ohio,c1941
Met 25.0250

AMERICAN SOCIETY FOR METALS Creep and recovery a seminar on creep and recovery of metals Cleveland 1956 Oct 6-12 American society for metals,Cleveland,1957 Held during the 38th National metal congress and exposition
Met 25.0950

AMERICAN SOCIETY FOR METALS Ductile chromium and its alloys :a conference...held at the 1955 metal congress and exposition of the American society for metals American society for metals,Cleveland,Ohio,1957
Met 25.2307

AMERICAN SOCIETY FOR METALS Fiber composite materials :a seminar Papers Metals Park, Ohio 1964 Oct 17-18 American society for metals,Metals Park,Ohio,1965
Eng 41.3697

AMERICAN SOCIETY FOR METALS Fracture of engineering materials :a conference Papers New York 1959 Aug American society for metals,Metals Park,Ohio,1964
Eng 41.3796

AMERICAN SOCIETY FOR METALS Fracturing of metals :a seminar...held during the 29th National metal congress and exposition...Co-ordinated by...the metallurgical staff of the Case institute of technology under the direction of George Sachs Chicago,Ill. 1947 Oct 18-24 American society for metals, Cleveland,Ohio,1948
Eng 41.3732

AMERICAN SOCIETY FOR METALS Fracturing of metals a seminar Chicago 1947 Oct 18-24 American society for metals,Cleveland,Ohio,n.d. A seminar held during the 29th National metal congress and exposition
Met 25.0861

AMERICAN SOCIETY FOR METALS Grain-size symposium held during the 16th annual convention New York 1934 Oct 1 American society for metals,Cleveland,c 1934 Damaged copy.No title page
Met 25.1264

AMERICAN SOCIETY FOR METALS High strength steels for aircraft Dallas,Texas 1958 May 12-16 American society for metals,Cleveland, Ohio,1958
Met 25.1002

AMERICAN SOCIETY FOR METALS High-strength steels for the missile industry Golden Gate metals conference Proceedings San Francisco,Calif. 1960 Feb 4 Edited by H.T. Sumsion American society for metals,Novelty, Ohio,1961
Met 25.0487

AMERICAN SOCIETY FOR METALS Impurities and imperfections a seminar held during the 36th national metal congress and exposition American society for metals,Cleveland,1955
Met 25.1230

AMERICAN SOCIETY FOR METALS Induction heating Cleveland 1946 Feb 4-8 By H.B. Osborn and others A.S.M.,Cleveland,Ohio,1946
Met 25.0237

AMERICAN SOCIETY FOR METALS Interpretation of tests and correlation with service Cleveland 1950 Oct 23-27 By M.F. Garwood and others A.S.M.,1950 Lectures presented during the 32nd National metal congress and exposition
Met 25.0883

AMERICAN SOCIETY FOR METALS Interpretation of tests and correlation with service :a series of four educational lectures... presented during the thirty-second National metal congress and exposition Chicago,Ill. 1950 Oct 23-27 By M.F. Garwood and others American society for metals,Cleveland,Ohio, 1951
Eng 41.3837

AMERICAN SOCIETY FOR METALS Liquid metals and solidification a seminar held during the 39th national metal congress and exposition Chicago 1957 Nov 2-8 A.S.M.,Cleveland,Ohio, 1958
Met 25.1510

AMERICAN SOCIETY FOR METALS Machining-theory and practice... :a series of thirteen educational lectures...presented...during the thirty first National metal congress and exposition By Hans Ernst and others American society for metals,Cleveland,Ohio, 1950
Eng 41.3909

AMERICAN SOCIETY FOR METALS Magnetic properties of metals and alloys Cleveland 1958 Oct 25-26 By R.M. Bozorth and others American society for metals,Cleveland,Ohio, 1959 A seminar presented to members of A. S.M.during the National metal congress and exposition
Met 25.2353

AMERICAN SOCIETY FOR METALS Materials engineering exposition and congress Symposium on applications of modern metallographic techniques Philadelphia,Pa. 1969 Oct 13-16 American society for testing and materials.Special technical publication, 480 ASTM,Philadelphia,Pa.,1970 Symposium presented at the exposition and congress
Met 25.2607

AMERICAN SOCIETY FOR METALS Metal interfaces a seminar held during the thirty-third national metal congress and exposition Detroit 1951 Oct 13-19 American society for metals,Cleveland,Ohio,1952
Met 25.1261

AMERICAN SOCIETY FOR METALS Metal thorium The Conference on thorium Proceedings Cleveland,Ohio 1956 Oct 11 Edited by Harley A. Wilhelm American society for metals,Cleveland,Ohio,1958
Met 25.1561

AMERICAN SOCIETY FOR METALS Metals for supersonic aircraft and missiles Heat tolerant metals for aerodynamic applications : a conference Proceedings Albuquerque,New Mexico 1957 Jan 28-29 Edited by D.W. Grobecker American society for metals, Cleveland,Ohio,1958
Met 25.1003

AMERICAN SOCIETY FOR METALS Modern research techniques in physical metallurgy a seminar... held during the thirty-fourth national metal congress and exposition Philadelphia 1952 Oct 18-24 American society for metals, Cleveland,Ohio,1953
Met 25.1640

AMERICAN SOCIETY FOR METALS New fabrication techniques Southwestern metal congress Papers American society for metals,Cleveland, Ohio,n.d.
Met 25.0684

AMERICAN SOCIETY FOR METALS Powder metallurgy in nuclear engineering Conference on powder metallurgy in atomic energy with additional papers Philadelphia 1955 Oct 20 By Henry H. Hausner American society for metals,Cleveland,Ohio,1958
Met 25.0767

AMERICAN SOCIETY FOR METALS Quenching of steels :a compilation of papers presented at the National metal congress Cleveland,Ohio 1948 Oct 27-31 American society for metals, Cleveland,Ohio,1951
Met 25.0446

AMERICAN SOCIETY FOR METALS Recrystallization,grain growth and textures a seminar Papers 1965 Oct 16-17 A.S.M., Metals Park,Ohio,1966
Met 25.1269

AMERICAN SOCIETY FOR METALS Relation of properties to microstructure Cleveland 1953 Oct 17-23 American society for metals, Cleveland,Ohio,1954 A symposium presented at the 35th National metal congress and exposition
Met 25.1201

AMERICAN SOCIETY FOR METALS Relation of properties to microstructure :a seminar held during the thirty-fifth National metal congress and exposition Cleveland,Ohio 1953 Oct 17-23 American society for metals, Cleveland,Ohio,1954
Eng 41.3501

AMERICAN SOCIETY FOR METALS Science, technology and application of titanium International conference on titanium Proceedings London 1968 May 21-24 Edited by R.I. Jaffee and N.E. Promisel Pergamon, Oxford,1970
Met 25.2852

AMERICAN SOCIETY FOR METALS Sheet materials for high temperature service Papers Dallas 1957 May 12-16 American society for metals, Cleveland,1958
Met 25.1001

AMERICAN SOCIETY FOR METALS Short-time high-temperature testing Short-time elevated temperature testing of metals :a symposium Proceedings Los Angeles 1957 Mar 25-29 American society for metals,Cleveland,Ohio, 1958
Met 25.0983

AMERICAN SOCIETY FOR METALS Strength of metals under combined stresses :a series of educational lectures...presented...during the twenty-second National metal congress and exhibition Cleveland,Ohio 1940 Oct 21-25 By Maxwell Gensamer American society for metals,Cleveland,Ohio,1941
Eng 41.3742

AMERICAN SOCIETY FOR METALS Strengthening mechanisms in solids :a seminar 1960 Oct 13-14 American society for metals,Metals Park, Ohio,1962
Met 25.1206

AMERICAN SOCIETY FOR METALS Surface protection against wear and corrosion :two series of educational lectures...presented... during the thirty-fifth National metal congress and exposition Cleveland,Ohio 1953 Oct 19-23 By H.S. Avery and others American society for metals,Cleveland,Ohio, 1954
Eng 41.3715

AMERICAN SOCIETY FOR METALS Surface protection against wear and corrosion... : educational lectures... presented during the thirty-fifth national metal congress and exposition Cleveland 1953 Oct 19-23 By H. S. Avery and others American society for metals,Cleveland,Ohio,1954
Met 25.1977

AMERICAN SOCIETY FOR METALS Surface stressing of metals :a series of five educational lectures...presented...during the twenty-seventh National metal congress and exposition Cleveland,Ohio 1946 Feb 4-8 By H.F. Moore and others American society for metals,Cleveland,Ohio,1947
Eng 41.3731

AMERICAN SOCIETY FOR METALS Surface treatment of metals the symposium...presented before the twenty-second annual convention Papers and discussions Cleveland 1940 Oct 21-25 American society for metals,Cleveland, Ohio,1941
Met 25.1972

AMERICAN SOCIETY FOR METALS The Plastic working metals a symposium Papers and discussions Cleveland 1936 Oct 19-23 American society for metals,Cleveland,Ohio, 1936 The symposium presented before the eighteenth annual convention of the American society for metals.
Met 25.0679

AMERICAN SOCIETY FOR METALS The Sorby centennial symposium on the history of metallurgy :a symposium Cleveland 1963 Oct 22-23 Edited by Cyril Stanley Smith Sponsored by the Society for the history of technology Metallurgical society conferences, 27 Gordon and Breach,New York,1965
Met 25.2141

AMERICAN SOCIETY FOR METALS Theory of alloy phases a seminar Philadelphia 1955 Oct 15-21 American society for metals,Cleveland, Ohio,1956 A seminar held during the 37th National metal congress and exposition
Met 25.1051

AMERICAN SOCIETY FOR METALS Thermodynamics in physical metallurgy :a seminar held during the 31st national metal congress and exposition Cleveland,Ohio 1949 Oct 15-21 American society for metals,Cleveland,Ohio, 1952
Met 25.2372

AMERICAN SOCIETY FOR METALS Thin films papers presented at a seminar of the American society for metals A.M.C.A.,Ohio,1964
Met 25.0175

AMERICAN SOCIETY FOR METALS Ultra high purity metals :a seminar Papers Metals Park,Ohio 1961 Oct 21-22 Edited by R.L. Smith 264p Metals Park,Ohio,1962
Cav 7.0051

AMERICAN SOCIETY FOR METALS Ultra-high-purity metals :a seminar Papers 1961 Oct 21-22 A.S.M.,Metals Park,Ohio,1962
Met 25.1495

AMERICAN SOCIETY FOR METALS Utilization of heat resistant alloys a symposium in honor of Albert Easton White Ann Arbor 1954 Mar 11-12 American society for metals,Cleveland, Ohio,1921
Met 25.0998

AMERICAN SOCIETY FOR METALS World metallurgical congress 1st Proceedings Detroit,Mich. 1951 Oct 14-19 Edited by William Marsh Baldwin 833p American society for metals,Cleveland,Ohio,1952
Met A 61 25.0044

AMERICAN SOCIETY FOR METALS World metallurgical congress 2nd Story of 2nd world metallurgical congress held in the United States of America 1957 Oct-Nov Edited by Kingsley W. Given 225p A.S.M., Cleveland,Ohio,1958
Met 25.0045

AMERICAN SOCIETY FOR METALS Zirconium and zirconium alloys Los Angeles 1953 Mar 23-27 American society for metals,Cleveland, Ohio,1953 A symposium on zirconium and zirconium alloys presented to members of the A. S.M. during the eighth Western metal congress and exposition,Los Angeles, march 23 to 27, 1953
Met 25.0556

AMERICAN SOCIETY FOR METALS.ANNUAL CONVENTION 16TH Grain-size symposium held during the 16th annual convention New York 1934 Oct 1 American society for metals American society for metals,Cleveland,c 1934 Damaged copy. No title page
Met 25.1264

AMERICAN SOCIETY FOR METALS.ANNUAL CONVENTION 18TH The Plastic working metals a symposium Papers and discussions Cleveland 1936 Oct 19-23 American society for metals American society for metals, Cleveland,Ohio,1936 The symposium presented before the eighteenth annual convention of the American society for metals.
Met 25.0679

AMERICAN SOCIETY FOR METALS.ANNUAL CONVENTION 19TH Carburizing :a symposium Papers and discussions Atlantic City,N.J. 1937 Oct 18-22 American society for metals American society for metals,Cleveland,Ohio, 1937 The,Symposium presented before the nineteenth annual convention of the American society for Metals...
Met 25.0439

AMERICAN SOCIETY FOR METALS.ANNUAL CONVENTION 21ST Age hardening of metals the symposium on precipitation hardening (age hardening) presented before the twenty-first annual convention Papers and discussions Chicago 1939 Oct 23-27 American society for metals American society for metals, Cleveland,1940
Met 25.1249

AMERICAN SOCIETY FOR METALS.ANNUAL CONVENTION 23RD Controlled atmospheres :the symposium...presented before the annual convention Philadelphia,Pa. 1941 Oct.20-24 American society for metals A.S.M.,Cleveland, Ohio,c1941
Met 25.0250

AMERICAN SOCIETY FOR METALS.EASTERN NEW YORK CHAPTER Fracture of engineering materials a conference Papers New York 1959 Aug A.S.M.,New York,1959
Met 25.0864

AMERICAN SOCIETY FOR METALS.OAK RIDGE CHAPTER Diffusion in body-centered cubic metals International conference on body-centred cubic materials Papers Gatlinburg 1964 Sep 16-18 American society for metals,Metals Park, Ohio,1965
Met 25.1283

AMERICAN SOCIETY FOR MICROBIOLOGY Spores 4 International spore conference 4th Urbana,Ill. 1968 Oct 4-6 Edited by L.Leon Campbell Under the auspices of the University of Illinois American society for microbiology,1967
Gen 34.0711

AMERICAN SOCIETY FOR PHARMACOLOGY AND EXPERIMENTAL THERAPEUTICS Symposium on catecholamines 2nd Milan 1965 Jul 4-9 Edited by George H. Acheson Held at the Istituto di ricerche farmacologiche "Mario Negri" Williams and Wilkins company, Baltimore,1966 Reprinted from the Pharmacological reviews,18,no.1
Bioch 33.0430

AMERICAN SOCIETY FOR STEEL TREATING.ANNUAL CONVENTION 10TH The Application of science to the steel industry By W.H. Hatfield American society for steel treating, Cleveland,Ohio,1928 Edward De Mille Campbell memorial lecture presented in Philadelphia, Oct.,10,1928 at the tenth annual convention of the American society for steel treating
Met 25.0381

AMERICAN SOCIETY FOR TESTING AND MATERIALS Adhesion :papers presented at the 66th annual meeting A.S.T.M. Atlantic city,N.J. 1963 Jun 26 American society for testing and materials.Special technical publication, 360 American society for testing and materials, Philadelphia,Pa.,1964
Eng 41.7660

AMERICAN SOCIETY FOR TESTING AND MATERIALS Atomic absorption spectroscopy :a symposium presented at the seventy-first annual meeting.. Proceedings San Francisco,Calif. 1968 Jun 23-28 American society for testing and materials.Publication, 443 ASTM,Philadelphia, Pa.,1969
Met 25.2647

AMERICAN SOCIETY FOR TESTING AND MATERIALS Composite materials:testing and design :a symposium New Orleans,La. 1969 Feb 11-13 American society for testing and materials. Special technical publication, 460 ASTM, Philadelphia,Pa.,1969
Eng 41.8290

AMERICAN SOCIETY FOR TESTING AND MATERIALS Determination of stress in rock :a symposium Atlantic City,N.J. 1966 Jun 26-Jul 1 American society for testing and materials. Special technical publication, 429 American society for testing materials,Philadelphia,Pa., 1967
Eng 41.3160

AMERICAN SOCIETY FOR TESTING AND MATERIALS Erosion by cavitation or impingement :a symposium presented at the 69th annual meeting Atlantic City,N.J. 1966 Jun 26-Jul 1 American society for testing and materials. Special technical publications, 408 American society for testing and materials,Philadelphia, Pa.,1967
Eng 41.6561

AMERICAN SOCIETY FOR TESTING AND MATERIALS Evaluation of wear testing :a symposium presented at the 71st annual meeting San Francisco,Calif. 1968 Jun 23-28 American society for testing and materials.Special technical publication, 446 ASTM,Philadelphia, Pa.,1969
Eng 41.8327

AMERICAN SOCIETY FOR TESTING AND MATERIALS Fatigue at high temperature :a symposium presented at the 71st annual meeting... San Francisco,Calif. 1968 Jun 23-28 American society for testing and materials.Special technical publication, 459 ASTM,Philadelphia, Pa.,1969
Eng 41.8293

AMERICAN SOCIETY FOR TESTING AND MATERIALS Fatigue crack propagation :a symposium Atlantic City,N.J. 1966 Jun 26-Jul 1 American society for testing and materials. Special technical publication, 415
Eng 41.3814

AMERICAN SOCIETY FOR TESTING AND MATERIALS Fracture toughness testing and its applications :a symposium presented at the 67th annual meeting A.S.T.M. Chicago,Ill. 1964 American society for testing and materials.Special technical publication, 381
Eng 41.3799

AMERICAN SOCIETY FOR TESTING AND MATERIALS Hot corrosion problems associated with gas turbines :symposium Proceedings Atlantic City,N.J. 1966 Jun 26-Jul 1 American society for testing and materials.Special technical publication, 421 Philadelphia,Pa., 1967 Symposium presented at the annual meeting
Met 25.2788

AMERICAN SOCIETY FOR TESTING AND MATERIALS
Interfaces in composites :a symposium Papers San Francisco,Calif. 1968 Jun 23-28 A.S.T.M.Special technical publication, 452 ASTM, Philadelphia,Pa.,1969 Presented at the 71st annual meeting
Met 25.2749

AMERICAN SOCIETY FOR TESTING AND MATERIALS
Internal friction,damping and cyclic plasticity... :a symposium presented at the 67th annual meeting,A.S.T.M. Chicago,Ill. 1964 Jun 22 American society for testing and materials.Special technical publication, 378
Eng 41.3795

AMERICAN SOCIETY FOR TESTING AND MATERIALS
Materials engineering exposition and congress Symposium on applications of modern metallographic techniques Philadelphia,Pa. 1969 Oct 13-16 American society for testing and materials.Special technical publication, 480 ASTM,Philadelphia,Pa.,1970 Symposium presented at the exposition and congress
Met 25.2607

AMERICAN SOCIETY FOR TESTING AND MATERIALS
Metal corrosion in the atmosphere :symposium Proceedings Boston,Mass. 1967 Jun 25-30 American society for testing and materials. Special technical publication, 435 Philadelphia,Pa.,1968 Symposium presented at the annual meeting
Met 25.2792

AMERICAN SOCIETY FOR TESTING AND MATERIALS
Symposium on dynamic behavior of materials Albuquerque,N.M. 1962 Sep 27-28 American society for testing and materials.Special technical publication, 336 American society for testing and materials.Materials science series, 5 American society for testing and materials,Philadelphia,Pa.,1963
Eng 41.3788

AMERICAN SOCIETY FOR TESTING AND MATERIALS
Symposium on plastics Philadelphia,Pa. 1944 Feb 22-23 American society for testing materials,Philadelphia,Pa.,1944
Eng 41.3642

AMERICAN SOCIETY FOR TESTING AND MATERIALS
Symposium on powder metallurgy :spring meeting Buffalo,N.Y. 1943 American society for testing and materials,Philadelphia,Pa.,1943
Eng 41.3474

AMERICAN SOCIETY FOR TESTING AND MATERIALS
Symposium on stress-strain-time-temperature relationships in materials :presented at the 65th annual meeting A.S.T.M. New York 1962 Jun 27 American society for testing and materials.Special technical publication, 325 American society for testing and materials. Materials science series, 3
Eng 41.3789

AMERICAN SOCIETY FOR TESTING AND MATERIALS
Testing techniques for rock mechanics :a symposium Seattle,Wash. 1965 Oct 31-Nov 5 American society for testing and materials. Special technical publication, 402 American society for testing materials,Philadelphia,Pa., 1966
Eng 41.3161

AMERICAN SOCIETY FOR TESTING AND MATERIALS
Vibration effects of earthquakes on soils and foundations :a symposium San Francsico,Calif. 1968 Jun 23-28 American society for testing and materials.Special technical publication, 450 American society for testing materials,Philadelphia,Pa.,1969
Eng 41.3162

AMERICAN SOCIETY FOR TESTING AND MATERIALS.ANNUAL MEETING 64TH Symposium on major effects of minor constituents on the properties of materials Atlantic City,N.J. 1961 Jun 26 American society for testing materials American society for testing and materials. Materials science series, 2 American society for testing and materials.Special technical publications, 304 American society for testing materials,Philadelphia,Pa.,1962
Eng 41.3429

AMERICAN SOCIETY FOR TESTING AND MATERIALS.ANNUAL MEETING 65TH Symposium on stress-strain-time-temperature relationships in materials : presented at the 65th annual meeting A.S.T.M. New York 1962 Jun 27 American society for testing and materials American society for testing and materials.Special technical publication, 325 American society for testing and materials.Materials science series, 3
Eng 41.3789

AMERICAN SOCIETY FOR TESTING AND MATERIALS.ANNUAL MEETING 66TH Adhesion :papers presented at the 66th annual meeting A.S.T.M. Atlantic city,N.J. 1963 Jun 26 American society for testing and materials American society for testing and materials.Special technical publication, 360 American society for testing and materials,Philadelphia,Pa.,1964
Eng 41.7660

AMERICAN SOCIETY FOR TESTING AND MATERIALS.ANNUAL MEETING 67TH Fracture toughness testing and its applications :a symposium presented at the 67th annual meeting A.S.T.M. Chicago,Ill. 1964 American society for testing and materials American society for testing and materials.Special technical publication, 381
Eng 41.3799

AMERICAN SOCIETY FOR TESTING AND MATERIALS.ANNUAL MEETING 67TH Internal friction,damping and cyclic plasticity... :a symposium presented at the 67th annual meeting,A.S.T.M. Chicago,Ill. 1964 Jun 22 American society for testing and materials American society for testing and materials.Special technical publication, 378
Eng 41.3795

AMERICAN SOCIETY FOR TESTING AND MATERIALS.ANNUAL MEETING 69TH Erosion by cavitation or impingement :a symposium presented at the 69th annual meeting Atlantic City,N.J. 1966 Jun 26-Jul 1 American society for testing and materials American society for testing and materials.Special technical publications, 408 American society for testing and materials, Philadelphia,Pa.,1967
Eng 41.6561

AMERICAN SOCIETY FOR TESTING AND MATERIALS.ANNUAL MEETING 69TH Hot corrosion problems associated with gas turbines :symposium Proceedings Atlantic City,N.J. 1966 Jun 26-Jul 1 American society for testing and materials American society for testing and materials.Special technical publication, 421 Philadelphia,Pa.,1967 Symposium presented at the annual meeting
Met 25.2788

AMERICAN SOCIETY FOR TESTING AND MATERIALS.ANNUAL MEETING 70TH Metal corrosion in the atmosphere :symposium Proceedings Boston, Mass. 1967 Jun 25-30 American society for testing and materials American society for testing and materials.Special technical publication, 435 Philadelphia,Pa.,1968 Symposium presented at the annual meeting
Met 25.2792

AMERICAN SOCIETY FOR TESTING AND MATERIALS.ANNUAL MEETING 71ST Atomic absorption spectroscopy :a symposium presented at the seventy-first annual meeting... Proceedings San Francisco,Calif. 1968 Jun 23-28 American society for testing and materials American society for testing and materials. Publication, 443 ASTM,Philadelphia,Pa.,1969
Met 25.2647

AMERICAN SOCIETY FOR TESTING AND MATERIALS.ANNUAL MEETING 71ST Evaluation of wear testing : a symposium presented at the 71st annual meeting San Francisco,Calif. 1968 Jun 23-28 American society for testing and materials American society for testing and materials.Special technical publication, 446 ASTM,Philadelphia,Pa.,1969
Eng 41.8327

AMERICAN SOCIETY FOR TESTING AND MATERIALS.ANNUAL MEETING 71ST Fatigue at high temperature : a symposium presented at the 71st annual meeting... San Francisco,Calif. 1968 Jun 23-28 American society for testing and materials American society for testing and materials.Special technical publication, 459 ASTM,Philadelphia,Pa.,1969
Eng 41.8293

AMERICAN SOCIETY FOR TESTING AND MATERIALS.ANNUAL MEETING 71ST Interfaces in composites :a symposium Papers San Francisco,Calif. 1968 Jun 23-28 American society for testing and materials A.S.T.M.Special technical publication, 452 ASTM,Philadelphia,Pa.,1969 Presented at the 71st annual meeting
Met 25.2749

AMERICAN SOCIETY FOR TESTING AND MATERIALS.ANNUAL MEETING 73RD Symposium on energy dispersion X-ray analysis :X-ray and electron probe analysis Toronto,Ont. 1970 Jun 21-26 American society for testing and materials. Committee E-4 on metallography ASTM, Philadelphia,Pa.,1971 Symposium presented at the annual meeting.Co-ordinator:J.C.Russ
Met 25.2617

AMERICAN SOCIETY FOR TESTING AND MATERIALS. COMMITTEE E-4 ON METALLOGRAPHY Symposium on energy dispersion X-ray analysis :X-ray and electron probe analysis Toronto,Ont. 1970 Jun 21-26 ASTM,Philadelphia,Pa.,1971 Symposium presented at the annual meeting.Co-ordinator:J.C.Russ
Met 25.2617

AMERICAN SOCIETY FOR TESTING AND MATERIALS. PACIFIC AREA NATIONAL MEETING 5TH Stress-corrosion cracking of titanium :a symposium Seattle 1965 Oct 31-Nov 5 American society for testing materials A.S.T.M.Special technical publication, 397 American society for testing and materials,Philadelphia,Pa., 1966
Met 25.1962

AMERICAN SOCIETY FOR TESTING MATERIALS Achievement of high fatigue resistance in metals and alloys :a symposium presented at the 72nd annual meeting Proceedings Atlantic City,N.J. 1969 Jun 22-27 American society for testing materials.Special technical publication, 467 ASTM,Philadelphia, Pa.,1970
Met 25.2584

AMERICAN SOCIETY FOR TESTING MATERIALS Adhesion ...Recent developments in adhesion science...Recent advances in adhesives technology...Presented at the 66th annual meeting A.S.T.M. Atlantic city,N.J. 1963 Jun 26 American society for testing materials.Special technical publication, 360 American society for testing materials, Philadeplphia,Pa.,1964
Eng 41.4027

AMERICAN SOCIETY FOR TESTING MATERIALS Advances in electron metallography a symposium presented at the sixty-eighth annual meeting Lafayette,Ind. 1965 Jun 13-18 Vol 6 A.S.T.M.Special technical publication, 396 A.S.T.M.,Philadelphia,1966
Met 25.1399

AMERICAN SOCIETY FOR TESTING MATERIALS Applications related phenomena in titanium alloys :a symposium Los Angeles 1967 Apr 19 A.S.T.M.Special technical publication, 432 A.S.T.M.,Philadelphia,Pa.,1968
Met 25.0609

AMERICAN SOCIETY FOR TESTING MATERIALS Atmospheric exposure tests on non-ferrous metals :a symposium Pittsburgh 1946 Feb 27 American society for testing materials, Philadelphia,Pa.,1946
Met 25.1924

AMERICAN SOCIETY FOR TESTING MATERIALS Behaviour of materials at cryogenic temperatures a symposium Lafayette,Ind. 1965 Jun 13-18 A.S.T.M.Special technical publication, 387 American society for testing materials,Philadelphia,Pa.,1916
Met 25.0976

AMERICAN SOCIETY FOR TESTING MATERIALS Cement and concrete Technical papers Los Angeles,Calif. 1956 Sep 19-20 American society for testing materials.Special technical publication, 205 American society for testing materials,Philadelphia,Pa.,1958
Eng 41.2845

AMERICAN SOCIETY FOR TESTING MATERIALS Designs and tests of building structures . Seismic and shock loading...Glued laminated and other constructions Los Angeles,Calif. 1956 Sep 17 American society for testing materials.Special technial publication, 209 American society for testing materials, Philadelphia,Pa.,1957
Eng 41.3017

AMERICAN SOCIETY FOR TESTING MATERIALS
Determination of nonmetallic compounds in steel a symposium presented at the sixty-eighth annual meeting Lafayette 1965 Jun 13-18 A.S.T.M.Special technical publication, 393 A.S.T.M.,Philadelphia,1966
Met 25.2303

AMERICAN SOCIETY FOR TESTING MATERIALS
Effect of cyclic heating and shessing metals Symposium on effect of cyclic heating and stressing on metals at elevated temperatures Chicago 1954 Jun 17 A.S.T.M.Special technical publication, 165 A.S.T.M., Philadelphia,Pa.,1954
Met 25.0980

AMERICAN SOCIETY FOR TESTING MATERIALS
Electron fractography Boston,Mass. 1967 Jun 25-30 A.S.T.M.Special technical publication, 436 A.S.T.M.,Philadelphia,Pa., 1916
Met 25.2248

AMERICAN SOCIETY FOR TESTING MATERIALS
Fatigue crack propagation :a symposium Atlantic City,N.J. 1966 Jun 26-Jul 1 A.S.T.M.Special technical publication, 415 A.S.T.M. Philadelphia,Pa.,1967
Met 25.0938

AMERICAN SOCIETY FOR TESTING MATERIALS
Fatigue crack propagation :a symposium Atlantic City,N.J. 1966 Jun 26-Jul 1 American society for testing and materials American society for testing and materials. Special technical publication, 415
Eng 41.3814

AMERICAN SOCIETY FOR TESTING MATERIALS
Fatigue of aircraft structures Los Angeles, Calif. 1956 Sep 16-21 American society for testing materials.Special technical publication, 203 American society for testing materials,Philadelphia,Pa.,1957
Eng 41.2776

AMERICAN SOCIETY FOR TESTING MATERIALS Fiber-strengthened metallic composites a symposium presented at the 1966 metals congress Chicago 1966 Nov 2-3 A.S.T.M.Special technical publication, 427 A.S.T.M., Philadelphia,1967
Met 25.2338

AMERICAN SOCIETY FOR TESTING MATERIALS
Fracture toughness testing and its applications :a symposium Chicago 1964 Jun 21-25 American society for testing materials, Philadelphia,Pa.,1965
Met 25.0868

AMERICAN SOCIETY FOR TESTING MATERIALS
Fracture toughness testing and its applications :a symposium presented at the 67th annual meeting A.S.T.M. Chicago,Ill. 1964 American society for testing and materials American society for testing and materials.Special technical publication, 381
Eng 41.3799

AMERICAN SOCIETY FOR TESTING MATERIALS
Instruments and apparatus for soil and rock mechanics :a symposium Lafayette,Ind. 1965 Jun 13-18 American society for testing materials.Special technical publication, 392 American society for testing materials, Philadelphia,Pa.,1965
Eng 41.3158

AMERICAN SOCIETY FOR TESTING MATERIALS
Internal friction,damping,and cyclic plasticity a symposium presented before the 67th annual meeting A.S.T.M.Special technical publication, 378 A.S.T.M., Philadelphia,1965
Met 25.2340

AMERICAN SOCIETY FOR TESTING MATERIALS
Internal friction,damping and cyclic plasticity... :a symposium presented at the 67th annual meeting,A.S.T.M. Chicago,Ill. 1964 Jun 22 American society for testing and materials American society for testing and materials.Special technical publication, 378
Eng 41.3795

AMERICAN SOCIETY FOR TESTING MATERIALS Joint international conference on creep session 1-7 New York 1963 Aug 25-29 London 1963 Sep 30-Oct 4 book 1-5 Jointly sponsored by Institution of mechanical engineers I.M.E.,London,c1963 5 books bound together
Met 25.0953

AMERICAN SOCIETY FOR TESTING MATERIALS Large fatigue testing machines and their results :a symposium held at the 1957 annual meeting of the society Atlantic City,N.J. 1957 Jun 18 American society for testing materials, Cleveland,Ohio,1957
Met 25.0922

AMERICAN SOCIETY FOR TESTING MATERIALS Metal matrix composites a symposium presented at the 70th annual meeting... Boston 1967 Jun 25-30 A.S.T.M.Special technical publication, 438 A.S.T.M.,Philadelphia,1968
Met 25.2337

AMERICAN SOCIETY FOR TESTING MATERIALS
Metallic materials at low temperatures Symposium on effect of temperature on the brittle behaviour of metals with particular reference to low temperatures Atlantic City 1953 Jun 28-30 A.S.T.M.Special technical publication, 158 American society for testing materials,Philadelphia,Pa.,1954
Met 25.0850

AMERICAN SOCIETY FOR TESTING MATERIALS
Metals :presented at the 2nd Pacific area national meeting Los Angeles 1956 Sep 17-21 A.S.T.M.Special technical publication, 196 American society for testing materials, Philadelphia,Pa.,1957
Met 25.1036

AMERICAN SOCIETY FOR TESTING MATERIALS Newer metals Symposium on newer metals presented at the third Pacific area national meeting... San Francisco,Calif. 1959 Oct 15 A.S.T.M. Special technical publication, 272 A.S.T.M., Philadelphia,Pa.,1960
Met 25.0514

AMERICAN SOCIETY FOR TESTING MATERIALS
Orientation effects in the mechanical behavior of anisotropic structural materials :a symposium Seattle,Wash. 1965 Oct 31-Nov 5 American society for testing materials.Special technical publication, 405 American society for testing materials,Philadelphia,Pa.,1966
Eng 41.3701

AMERICAN SOCIETY FOR TESTING MATERIALS
Orientation effects in the mechanical behaviour of anisotropic structural materials a symposium presented at the 5th Pacific area national meeting A.S.T.M.Special technical publication, 405 A.S.T.M.,Philadelphia,1966
Met 25.2341

AMERICAN SOCIETY FOR TESTING MATERIALS
Performance of deep foundations :a symposium San Francisco,Calif. 1968 Jun 23-28 American society for testing materials.Special technical publication, 444 American society for testing materials,Philadelphia,Pa.,1969
Eng 41.3159

AMERICAN SOCIETY FOR TESTING MATERIALS
Permeability and capillarity of soils :a symposium Atlantic city,N.J. 1966 Jun 26-Jul 1 American society for testing materials. Special technical publication, 417 American society for testing and materials,Philadelphia, Pa.,1967
Eng 41.3102

AMERICAN SOCIETY FOR TESTING MATERIALS
Procedures for testing soils nomenclatures and definitions,standard methods,suggested methods American society for testing materials,Philadelphia,Pa.,1958
Eng 41.3154

AMERICAN SOCIETY FOR TESTING MATERIALS
Properties of crystalline solids ...Symposium on recent progress in materials sciences; Symposium on nature and origin of strength of materials.Presented at the 63rd annual meeting A.S.T.M. Philadelphia,Pa. 1960 Jun 27 American society for testing materials.Special technical publication, 283 American society for testing materials,Philadelphia,Pa.,1961
Eng 41.3790

AMERICAN SOCIETY FOR TESTING MATERIALS
Stress corrosion testing :a symposium Atlantic City,N.J. 1966 Jun 26-Jul 1 A.S.T. M.Special technical publication, 425 American society for testing and materials, Philadelphia,Pa.,1967
Met 25.1964

AMERICAN SOCIETY FOR TESTING MATERIALS
Stress-corrosion cracking of titanium :a symposium Seattle 1965 Oct 31-Nov 5 A.S. T.M.Special technical publication, 397 American society for testing and materials, Philadelphia,Pa.,1966
Met 25.1962

AMERICAN SOCIETY FOR TESTING MATERIALS
Structural fatigue in aircraft :a symposium A.S.T.M.Special technical publication, 404 American society for testing materials, Philadelphia,Pa.,1966
Met 25.0943

AMERICAN SOCIETY FOR TESTING MATERIALS
Structure and properties of ultrahigh-strength steels :a symposium Cleveland,Ohio 1963 Oct 22 A.S.T.M.Special technical publication, 370 American society for testing and materials,Philadelphia,Pa.,1965
Met 25.2251

AMERICAN SOCIETY FOR TESTING MATERIALS
Symposium on acoustical fatigue Atlantic City,N.J. 1960 Jun 28 A.S.T.M.Special technical publication, 284 A.S.T.M., Philadelphia,Pa.,1961
Met 25.0929

AMERICAN SOCIETY FOR TESTING MATERIALS
Symposium on advances in electron metallography Papers A.S.T.M.Special technical publication, 245 American society for testing materials,Philadelphia,Pa.,1958
Met 25.1362

AMERICAN SOCIETY FOR TESTING MATERIALS
Symposium on atmospheric corrosion of non-ferrous metals Atlantic City,N.J. 1955 Jun 29 A.S.T.M.Special technical publication, 175 A.S.T.M.,Philadelphia,Pa.,1955
Met 25.1918

AMERICAN SOCIETY FOR TESTING MATERIALS
Symposium on basic effects of environment on the strength,scaling,and embrittlement of metals at high temperatures Cincinnati 1955 Feb 2 A.S.T.M.Special technical publication, 171 A.S.T.M.,Philadelphia,Pa., 1955
Met 25.1908

AMERICAN SOCIETY FOR TESTING MATERIALS
Symposium on basic mechanisms of fatigue Papers Boston,Mass. 1958 Jun 23 A.S.T.M. Special technical publication, 237 A.S.T.M., Philadelphia,1959
Met 25.0925

AMERICAN SOCIETY FOR TESTING MATERIALS
Symposium on basic mechanisms of fatigue presented before the 61st annual meeting Boston 1958 Jun 23 A.S.T.M.Special technical publication, 273 A.S.T.M., Philadelphia,1959
Met 25.2328

AMERICAN SOCIETY FOR TESTING MATERIALS
Symposium on brittle failure of rotor forgings presented at special meeting for delegates to world metals congress Philadelphia,Pa. 1957 Nov 11 A.S.T.M.Special technical publication, 231 American society for testing materials,Philadelphia,Pa,1957
Met 25.1325

AMERICAN SOCIETY FOR TESTING MATERIALS
Symposium on consolidation testing of soils Atlantic City,N.J. 1951 Jun 18 American society for testing materials.Special technical publication, 126 American society for testing materials,Philadelphia,Pa.,1952
Eng 41.3149

AMERICAN SOCIETY FOR TESTING MATERIALS
Symposium on corrosion of materials at elevated temperatures Atlantic City,N.J. 1950 Jun 26 A.S.T.M.Special technical publication, 108 American society for testing materials,Philadelphia,Pa.,1951
Met 25.1906

AMERICAN SOCIETY FOR TESTING MATERIALS
Symposium on determination of gases in metals Atlantic City,N.J. 1957 Jun 18 A.S.T.M. Special technical publication, 222 A.S.T.M., Philadelphia,Pa,1958
Met 25.1708

AMERICAN SOCIETY FOR TESTING MATERIALS
Symposium on direct shear testing of soils New York 1952 Jun 26 American society for testing materials.Special technical publication, 131 American society for testing materials,Philadelphia,Pa.,1953
Eng 41.3156

AMERICAN SOCIETY FOR TESTING MATERIALS
Symposium on dynamic testing of soils Atlantic City,N.J. 1953 Jul 2 American society for testing materials.Special technical publication, 156 American society for testing materials,Philadelphia,Pa.,1954
Eng 41.3148

AMERICAN SOCIETY FOR TESTING MATERIALS
Symposium on effect of cyclic heating and stressing on metals at elevated temperatures Chicago 1954 Jun 17 A.S.T.M.Special technical publication, 165 A.S.T.M., Philadelphia,1954
Met 25.2534

AMERICAN SOCIETY FOR TESTING MATERIALS
Symposium on effect of temperature on the brittle behavior of metals with particular reference to low temperatures :presented at the 56th annual meeting A.S.T.M. Atlantic city,N.J. 1953 Jun 28-30 American society for testing materials.Special technical publication, 158 American society for testing materials,Philadelphia,Pa.,1954
Eng 41.3750

AMERICAN SOCIETY FOR TESTING MATERIALS
Symposium on effects of low temperatures on the properties of materials :presented at a meeting of the Philadelphia district,A.S.T.M. Philadelphia,Pa. 1946 Mar 19 American society for testing materials.Special technical publication, 78 American society for testing materials,Philadelphia,Pa.,1950
Eng 41.3778

AMERICAN SOCIETY FOR TESTING MATERIALS
Symposium on electroforming - applications, uses,and properties of electroformed metals presented at the 1962 committee week Dallas 1962 Feb 6-7 A.S.T.M.Special technical publication, 318 A.S.T.M.,Philadelphia,Pa., 1962
Met 25.0733

AMERICAN SOCIETY FOR TESTING MATERIALS
Symposium on electron metallography Atlantic City,N.J. 1959 Jun 23 A.S.T.M.Special technical publication, 262 American society for testing materials,Philadelphia,Pa.,1960
Met 25.1366

AMERICAN SOCIETY FOR TESTING MATERIALS
Symposium on elevated temperature strain gages Philadelphia,Pa. 1957 Dec 4-5 Sponsored by United States.Naval air material center A.S. T.M.Special technical publication, 230 American society for testing materials, Philadelphia,Pa,1958
Met 25.0985

AMERICAN SOCIETY FOR TESTING MATERIALS
Symposium on evaluation of metallic materials in design for low-temperature service presented at the 64th annual meeting Atlantic City,N.J. 1961 Jun 27-28 A.S.T.M. Special technical publication, 302 A.S.T.M., Philadelphia,1962
Met 25.2335

AMERICAN SOCIETY FOR TESTING MATERIALS
Symposium on evaluation tests for stainless steels Atlantic City 1949 Jun 30 A.S.T.M. Special technical publication, 93 American society for testing materials,Philadelphia,Pa., 1950
Met 25.1967

AMERICAN SOCIETY FOR TESTING MATERIALS
Symposium on fatigue tests of aircraft structure:low-cycle,full-scale, and helicopters Los Angeles 1962 Oct 1-3 A.S. T.M.Special technical publication, 338 A.S.T. M.,Philadelphia,Pa.,1963
Met 25.0935

AMERICAN SOCIETY FOR TESTING MATERIALS
Symposium on fatigue with emphasis on statistical approach New York 1952 Jun 24 Pt 2 A.S.T.M.Special technical publication American society for testing materials,Philadelphia 3, Pa.,1953
Met 25.0913

AMERICAN SOCIETY FOR TESTING MATERIALS
Symposium on fretting corrosion New York 1952 Jun 25 A.S.T.M.Special technical publication, 144 American society for testing materials,Philadelphia,Pa.,1953 Fiftieth anniversary publication
Met 25.1953

AMERICAN SOCIETY FOR TESTING MATERIALS
Symposium on full-scale tests on house structures Los Angeles,Calif. 1956 Sep 18 American society for testing materials.Special technical publication, 210 American society for testing materials,Philadelphia,Pa.,1956
Eng 41.3018

AMERICAN SOCIETY FOR TESTING MATERIALS
Symposium on high-purity water corrosion Atlantic City,N.J. 1955 Jun 28 A.S.T.M. Special technical publication, 179 American society for testing materials,Philadelphia,Pa., 1955
Met 25.1927

AMERICAN SOCIETY FOR TESTING MATERIALS
Symposium on impact testing Atlantic City 1955 Jun 27 American society for testing materials,Philadelphia,Pa.,1955
Met 25.0902

AMERICAN SOCIETY FOR TESTING MATERIALS
Symposium on impact testing :presented at the forty-first annual meeting A.S.T.M. Atlantic city,N.J. 1938 Jun 28 American society for testing materials,Philadelphia,Pa.,1938
Eng 41.3751

AMERICAN SOCIETY FOR TESTING MATERIALS
Symposium on impact testing :presented at the 58th A.S.T.M. annual meeting Atlantic city,N. J. 1955 Jun 27 American society for testing materials.Special technical publication, 176 American society for testing materials,Philadelphia,Pa.,1956
Eng 41.3757

AMERICAN SOCIETY FOR TESTING MATERIALS
Symposium on ion exchange and chromatography in analytical chemistry presented at the fifty-ninth annual meeting Atlantic City,N.J. 1956 Jun 18 A.S.T.M.Special technical publication, 195 A.S.T.M.,Philadelphia,Pa., 1958
Met 25.1750

AMERICAN SOCIETY FOR TESTING MATERIALS
Symposium on light microscopy American society for testing materials 55th annual meeting Papers New York 1952 Jun 25 American society for testing materials.Special publication, 143,Fiftieth anniversary publication Philadelphia,Pa.,1953
Min 10.1571

AMERICAN SOCIETY FOR TESTING MATERIALS
Symposium on light microscopy New York 1952 Jun 25 A.S.T.M.Special technical publication, 143 American society for testing materials,Philadelphia,Pa.,1953
Met 25.1333

AMERICAN SOCIETY FOR TESTING MATERIALS
Symposium on load tests of bearing capacity of soils Atlantic City,N.J. 1947 Jun 16-20 American society for testing materials.Special technical publication, 79 American society for testing materials,Philadelphia,Pa.,1948
Eng 41.3145

AMERICAN SOCIETY FOR TESTING MATERIALS
Symposium on low temperature properties of high-strength and missile materials Papers 1960 Jun 30 A.S.T.M.Special technical publication, 287 A.S.T.M.,Philadelphia,1961
Met 25.0974

AMERICAN SOCIETY FOR TESTING MATERIALS
Symposium on lubricants Chicago,Ill. 1937 Mar 3 American society for testing materials, Philadelphia,Pa.,1937
Eng 41.6265

AMERICAN SOCIETY FOR TESTING MATERIALS
Symposium on major effects of minor constituents on the properties of materials Atlantic City,N.J. 1961 Jun 26 American society for testing and materials.Materials science series, 2 American society for testing and materials.Special technical publications, 304 American society for testing materials,Philadelphia,Pa.,1962
Eng 41.3429

AMERICAN SOCIETY FOR TESTING MATERIALS
Symposium on materials for gas turbines : presented at the 49th annual meeting A.S.T.M. Buffalo,N.Y. 1946 Jun 24-28 American society for testing materials,Philadelphia,Pa., 1946
Eng 41.3752

AMERICAN SOCIETY FOR TESTING MATERIALS
Symposium on metallic materials for service at temperatures above 1600 F. Atlantic City,N.J. 1955 Jun 30 A.S.T.M.Special technical publication, 174 American society for testing materials,Philadelphia,Pa.,1956
Met 25.0999

AMERICAN SOCIETY FOR TESTING MATERIALS
Symposium on metallography in color (1948) presented to the 51st annual meeting Detroit 1948 Jun 21 A.S.T.M.Special technical publication, 86 American society for testing materials,Philadelphia,Pa.,1949
Met 25.1342

AMERICAN SOCIETY FOR TESTING MATERIALS
Symposium on methods of metallographic specimen preparation Atlantic City 1960 Jun 28 A.S.T.M.Special technical publication, 285 American society for testing materials, Philadelphia,Pa.,1960
Met 25.1381

AMERICAN SOCIETY FOR TESTING MATERIALS
Symposium on motor lubricants New York 1933 Mar 8 American society for testing materials,Philadelphia,Pa.,1933
Eng 41.6264

AMERICAN SOCIETY FOR TESTING MATERIALS
Symposium on newer structural materials for aerospace vehicles Chicago 1964 Jun 21 A. S.T.M.Special technical publication, 379 Philadelphia,Pa.,1965
Met 25.2516

AMERICAN SOCIETY FOR TESTING MATERIALS
Symposium on nondestructive testing presented to the 2nd Pacific area national meeting Los Angeles 1956 Sep 17-18 A.S.T.M.Special technical publication, 213 A.S.T.M., Philadelphia,1957
Met 25.1323

AMERICAN SOCIETY FOR TESTING MATERIALS
Symposium on nondestructive testing presented to the 55th annual meeting... New York 1952 Jun 26 A.S.T.M.Special technical publication, 145 A.S.T.M.,Philadelphia,1953
Met 25.2359

AMERICAN SOCIETY FOR TESTING MATERIALS
Symposium on nuclear methods for measuring soil density and moisture Atlantic City,N.J. 1960 Jun 27 American society for testing materials.Special technical publication, 293 American society for testing materials, Philadelphia,Pa.,1961
Eng 41.3152

AMERICAN SOCIETY FOR TESTING MATERIALS
Symposium on particle size measurement Papers American society for testing materials.Special technical publication, 234 v,310p American society for testing materials,Philadelphia,Pa.,1959 Presented at the 61st annual meeting of the American society for testing materials
Chem E 24.1440

AMERICAN SOCIETY FOR TESTING MATERIALS
Symposium on porcelain enamels and ceramic coatings as engineering materials presented at the fifty-sixth annual meeting... A.S.T.M. Special technical publication, 153 ASTM, Philadelphia,Pa.,1954
Met 25.0263

AMERICAN SOCIETY FOR TESTING MATERIALS
Symposium on properties of surfaces :presented at the fourth Pacific area national meeting Los Angeles,Calif. 1962 Oct 4 American society for testing materials.Special technical publication, 340 Philadelphia,1963
Met 25.2829

AMERICAN SOCIETY FOR TESTING MATERIALS
Symposium on radiation effects on metals and neutron dosimetry Los Angeles,Calif. 1962 Oct 2-3 A.S.T.M.Special technical publication, 341 American society for testing and materials,Philadelphia,Pa.,1963
Met 25.1585

AMERICAN SOCIETY FOR TESTING MATERIALS
Symposium on radiography 1st and 2nd Atlantic City 1936,1942 A.S.T.M., Philadelphia,1943
Met 25.1317

AMERICAN SOCIETY FOR TESTING MATERIALS
Symposium on radioisotopes presented at the second Pacific area national meeting Los Angeles 1956 Sep 21 A.S.T.M.Special technical publication, 215 A.S.T.M., Philadelphia,1958
Met 25.1559

AMERICAN SOCIETY FOR TESTING MATERIALS
Symposium on radiosotopes in metals analysis and testing presented to the sixty-second annual meeting Atlantic City,N.J. 1959 Hun 22 A.S.T.M.Special technical publication, 261 A.S.T.M.,Philadelphia,Pa.,1960
Met 25.1741

AMERICAN SOCIETY FOR TESTING MATERIALS
Symposium on shear and torsion testing 1960 Jun 28 A.S.T.M.Special technical publication, 289 American society for testing materials, Philadelphia,Pa.,1961
Met 25.0887

AMERICAN SOCIETY FOR TESTING MATERIALS
Symposium on shear testing of soils Atlantic City,N.J. 1939 Jun 28 American society for testing materials,Philadelphia,Pa.,1939
Eng 41.2988

AMERICAN SOCIETY FOR TESTING MATERIALS
Symposium on soil dynamics Atlantic City,N.J. 1961 Jun 26 American society for testing materials.Special technical publication, 305 American society for testing materials, Philadelphia,Pa.,1962
Eng 41.3155

AMERICAN SOCIETY FOR TESTING MATERIALS
Symposium on solder :presented at the fifty-ninth annual meeting A.S.T.M. Atlantic city, N.J. 1956 Jun 19-20 American society for testing materials.Special technical publication, 189 American society for testing materials,Philadelphia,Pa.,1957
Eng 41.4025

AMERICAN SOCIETY FOR TESTING MATERIALS
Symposium on statistical aspects of fatigue Atlantic City,N.J. 1951 Jun 19 Pt 1 A.S.T.M.Special technical publication, 121 American society for testing materials, Philadelphia,Pa.,1951
Met 25.0912

AMERICAN SOCIETY FOR TESTING MATERIALS
Symposium on strength and ductility of metals at elevated temperatures with particular reference to effects of notches and metallurgical changes New York 1952 Jun 23 A.S.T.M.Special technical publication, 128 A.S.T.M.,Philadelphia,Pa.,1953
Met 25.0979

AMERICAN SOCIETY FOR TESTING MATERIALS
Symposium on stress-corrosion cracking of metals Philadelphia,Pa. 1944 Nov 29-Dec 1 A.S.T.M.,New York,1945
Met 25.1952

AMERICAN SOCIETY FOR TESTING MATERIALS
Symposium on stress-corrosion cracking of metals Philadelphia,Pa. 1944 Nov 29-Dec 1 American society for testing materials; American institute of mining and metallurgical engineers,Philadelphia,Pa.;New York,1945
Eng 41.3717

AMERICAN SOCIETY FOR TESTING MATERIALS
Symposium on stress-strain-time-temperature relationships in materials :presented at the 65th annual meeting A.S.T.M. New York 1962 Jun 27 American society for testing and materials American society for testing and materials.Special technical publication, 325 American society for testing and materials. Materials science series, 3
Eng 41.3789

AMERICAN SOCIETY FOR TESTING MATERIALS
Symposium on surface and subsurface reconnaissance Atlantic City,N.J. 1951 Jun 19 American society for testing materials. Special technical publication, 122 American society for testing materials,Philadelphia,Pa., 1952
Eng 41.3153

AMERICAN SOCIETY FOR TESTING MATERIALS
Symposium on techniques for electron metalography Atlantic City,N.J. 1953 Jun 30 A.S.T.M.Special technical publication, 153 American society for testing materials, Philadelphia,Pa.,1954
Met 25.1361

AMERICAN SOCIETY FOR TESTING MATERIALS
Symposium on testing metal powders and metal powder products Papers Cleveland 1952 Mar 5 American society for testing materials. Special technical publication, 140 87p American society for testing materials, Philadelphia,Pa.,1953
Chem E 24.1121

AMERICAN SOCIETY FOR TESTING MATERIALS
Symposium on testing metal powders and metal powder products presented at the Cleveland spring meeting Cleveland 1952 Mar 5 A.S.T.M.Special technical publication, 140 American society for testing materials, Philadelphia,Pa.,1953
Met 25.0762

AMERICAN SOCIETY FOR TESTING MATERIALS
Symposium on the identification and classification of soils Atlantic City,N.J. 1950 Jun 29 American society for testing materials.Special technical publication, 113 American society for testing materials, Philadelphia,Pa.,1951
Eng 41.3146

AMERICAN SOCIETY FOR TESTING MATERIALS
Symposium on the nature,occurrence,and effects of sigma phase presented at the fifty-third annual meeting... Atlantic City,N.J. 1950 Jun 26-27 A.S.T.M.Special technical publication, 110 A.S.T.M.,Philadelphia,Pa., 1951
Met 25.0477

AMERICAN SOCIETY FOR TESTING MATERIALS
Symposium on the use of radioisotopes in soil mechanics Cleveland,Ohio 1952 Mar 5 American society for testing materials.Special technical publication, 134 American society for testing materials,Philadelphia,Pa.,1953
Eng 41.3147

AMERICAN SOCIETY FOR TESTING MATERIALS
Symposium on tin New York 1952 Jun 23 A.S.T.M.Special technical publication, 141 American society for testing materials, Philadelphia,Pa.,1953 Fiftieth anniversary publication. A symposium held at the 55th annual meeting of the American society for testing materials
Met 25.0567

AMERICAN SOCIETY FOR TESTING MATERIALS
Symposium on titanium :presented at the second Pacific area national meeting Los Angeles, Calif. 1956 Sep 17-18 American society for testing materials.Special technical publication, 204 American society for testing materials,Philadelphia,Pa.,1957
Eng 41.3598

AMERICAN SOCIETY FOR TESTING MATERIALS
Symposium on titanium held at the second Pacific area national meeting Los Angeles 1956 Sep 17-18 A.S.T.M.Special technical publication, 204 American society for testing materials,Philadelphia,Pa.,1957
Met 25.0571

AMERICAN SOCIETY FOR TESTING MATERIALS
Symposium on vane shear testing of soils Atlantic City,N.J. 1956 Jun 22 American society for testing materials.Special technical publication, 193 American society for testing materials,Philadelphia,Pa.,1957
Eng 41.3150

AMERICAN SOCIETY FOR TESTING MATERIALS
Techniques of electron microscopy,diffraction and microprobe analysis Papers Atlantic City 1963 Jun 26 A.S.T.M.Special technical publication, 372 Philadelphia,1964
Met 25.2495

AMERICAN SOCIETY FOR TESTING MATERIALS 55th annual meeting Papers Symposium on light microscopy New York 1952 Jun 25 American society for testing materials American society for testing materials.Special publication, 143,Fiftieth anniversary publication Philadelphia,Pa.,1953
Min 10.1571

AMERICAN SOCIETY FOR TESTING MATERIALS 62nd : annual meeting papers Symposium on microscopy Atlantic City,N.J. 1959 Jun 25-26 American society for testing materials. Special technical publications,257 American society for testing materials,Philadelphia,Pa., 1959
Min 10.1046

AMERICAN SOCIETY FOR TESTING MATERIALS.ANNUAL MEETING,59TH Symposium on ion exchange and chromatography in analytical chemistry presented at the fifty-ninth annual meeting Atlantic City,N.J. 1956 Jun 18 American society for testing materials A.S.T.M. Special technical publication, 195 A.S.T.M., Philadelphia,Pa.,1958
Met 25.1750

AMERICAN SOCIETY FOR TESTING MATERIALS.ANNUAL MEETING,62ND Symposium on radiosotopes in metals analysis and testing presented to the sixty-second annual meeting Atlantic City,N. J. 1959 Hun 22 American society for testing materials A.S.T.M.Special technical publication, 261 A.S.T.M.,Philadelphia,Pa., 1960
Met 25.1741

AMERICAN SOCIETY FOR TESTING MATERIALS.ANNUAL MEETING 41ST Symposium on impact testing : presented at the forty-first annual meeting A. S.T.M. Atlantic city,N.J. 1938 Jun 28 American society for testing materials American society for testing materials, Philadelphia,Pa.,1938
Eng 41.3751

AMERICAN SOCIETY FOR TESTING MATERIALS.ANNUAL MEETING 42ND Symposium on shear testing of soils Atlantic City,N.J. 1939 Jun 28 American society for testing materials American society for testing materials, Philadelphia,Pa.,1939
Eng 41.2988

AMERICAN SOCIETY FOR TESTING MATERIALS.ANNUAL MEETING 49TH Symposium on materials for gas turbines :presented at the 49th annual meeting A.S.T.M. Buffalo,N.Y. 1946 Jun 24-28 American society for testing materials American society for testing materials, Philadelphia,Pa.,1946
Eng 41.3752

AMERICAN SOCIETY FOR TESTING MATERIALS.ANNUAL MEETING 50TH Nondestructive testing By Robert C. McMaster Edgar Marburg lecture 1952 American society for testing materials, Philadelphia,Pa.,1952
Met 25.1320

AMERICAN SOCIETY FOR TESTING MATERIALS.ANNUAL MEETING 50TH Symposium on light microscopy New York 1952 Jun 25 American society for testing materials A.S.T.M. Special technical publication, 143 American society for testing materials,Philadelphia,Pa., 1953
Met 25.1333

AMERICAN SOCIETY FOR TESTING MATERIALS.ANNUAL MEETING 51ST Symposium on metallography in color (1948) presented to the 51st annual meeting Detroit 1948 Jun 21 American society for testing materials A.S.T.M. Special technical publication, 86 American society for testing materials,Philadelphia,Pa., 1949
Met 25.1342

AMERICAN SOCIETY FOR TESTING MATERIALS.ANNUAL MEETING 52ND Symposium on evaluation tests for stainless steels Atlantic City 1949 Jun 30 American society for testing materials A.S.T.M.Special technical publication, 93 American society for testing materials,Philadelphia,Pa.,1950
Met 25.1967

AMERICAN SOCIETY FOR TESTING MATERIALS.ANNUAL MEETING 53RD Symposium on corrosion of materials at elevated temperatures Atlantic City,N.J. 1950 Jun 26 American society for testing materials A.S.T.M.Special technical publication, 108 American society for testing materials,Philadelphia,Pa.,1951
Met 25.1906

AMERICAN SOCIETY FOR TESTING MATERIALS.ANNUAL MEETING 53RD Symposium on the nature, occurrence,and effects of sigma phase presented at the fifty-third annual meeting... Atlantic City,N.J. 1950 Jun 26-27 American society for testing materials A.S.T.M. Special technical publication, 110 A.S.T.M., Philadelphia,Pa.,1951
Met 25.0477

AMERICAN SOCIETY FOR TESTING MATERIALS.ANNUAL MEETING 54TH Symposium on statistical aspects of fatigue Atlantic City,N.J. 1951 Jun 19 Pt 1 American society for testing materials A.S.T.M.Special technical publication, 121 American society for testing materials,Philadelphia,Pa.,1951
Met 25.0912

AMERICAN SOCIETY FOR TESTING MATERIALS.ANNUAL MEETING 55TH Symposium on fatigue with emphasis on statistical approach New York 1952 Jun 24 Pt 2 American society for testing materials A.S.T.M.Special technical publication American society for testing materials,Philadelphia 3, Pa.,1953
Met 25.0913

AMERICAN SOCIETY FOR TESTING MATERIALS.ANNUAL MEETING 55TH Symposium on fretting corrosion New York 1952 Jun 25 American society for testing materials A.S.T.M. Special technical publication, 144 American society for testing materials,Philadelphia,Pa., 1953 Fiftieth anniversary publication
Met 25.1953

AMERICAN SOCIETY FOR TESTING MATERIALS.ANNUAL MEETING 55TH Symposium on nondestructive testing presented to the 55th annual meeting... New York 1952 Jun 26 American society for testing materials A.S.T.M.Special technical publication, 145 A.S.T.M., Philadelphia,1953
Met 25.2359

AMERICAN SOCIETY FOR TESTING MATERIALS.ANNUAL MEETING 55TH Symposium on strength and ductility of metals at elevated temperatures with particular reference to effects of notches and metallurgical changes New York 1952 Jun 23 American society for testing materials A.S.T.M.Special technical publication, 128 A.S.T.M.,Philadelphia,Pa., 1953
Met 25.0979

AMERICAN SOCIETY FOR TESTING MATERIALS.ANNUAL MEETING 55TH Symposium on tin New York 1952 Jun 23 American society for testing materials A.S.T.M.Special technical publication, 141 American society for testing materials,Philadelphia,Pa.,1953 Fiftieth anniversary publication. A symposium held at the 55th annual meeting of the American society for testing materials
Met 25.0567

AMERICAN SOCIETY FOR TESTING MATERIALS.ANNUAL MEETING 56TH Metallic materials at low temperatures Symposium on effect of temperature on the brittle behaviour of metals with particular reference to low temperatures Atlantic City 1953 Jun 28-30 American society for testing materials A.S.T.M. Special technical publication, 158 American society for testing materials,Philadelphia,Pa., 1954
Met 25.0850

AMERICAN SOCIETY FOR TESTING MATERIALS.ANNUAL MEETING 56TH Micrometallurgy - the metallurgy of minute additions By Jerome Strauss H.W.Gillett memorial lecture 1953 Presented before the fifty-sixth annual meeting of the American society for testing materials
Met 25.0050

AMERICAN SOCIETY FOR TESTING MATERIALS.ANNUAL MEETING 56TH Symposium on effect of temperature on the brittle behavior of metals with particular reference to low temperatures : presented at the 56th annual meeting A.S.T.M. Atlantic city,N.J. 1953 Jun 28-30 American society for testing materials American society for testing materials.Special technical publication, 158 American society for testing materials,Philadelphia,Pa.,1954
Eng 41.3750

AMERICAN SOCIETY FOR TESTING MATERIALS.ANNUAL MEETING 56TH Symposium on porcelain enamels and ceramic coatings as engineering materials presented at the fifty-sixth annual meeting... American society for testing materials A.S.T.M.Special technical publication, 153 ASTM,Philadelphia,Pa.,1954
Met 25.0263

AMERICAN SOCIETY FOR TESTING MATERIALS.ANNUAL MEETING 56TH Symposium on techniques for electron metalography Atlantic City,N.J. 1953 Jun 30 American society for testing materials A.S.T.M.Special technical publication, 153 American society for testing materials,Philadelphia,Pa.,1954
Met 25.1361

AMERICAN SOCIETY FOR TESTING MATERIALS.ANNUAL MEETING 57TH Effect of cyclic heating and shessing metals Symposium on effect of cyclic heating and stressing on metals at elevated temperatures Chicago 1954 Jun 17 American society for testing materials A.S.T.M.Special technical publication, 165 A.S.T.M. Philadelphia,Pa.,1954
Met 25.0980

AMERICAN SOCIETY FOR TESTING MATERIALS.ANNUAL MEETING 57TH Fatigue of aluminum By R.L. Templin A.S.T.M.,Philadelphia,1954 H.W. Gillett memorial lecture,1954 presented before the 57th annual meeting of the American society for testing materials
Met 25.0916

AMERICAN SOCIETY FOR TESTING MATERIALS.ANNUAL MEETING 58TH Powder metallurgy - now (new techniques,improved properties,wider use) By F.V. Lenel H.W.Gillett memorial lecture 1955 A.S.T.M.,1955
Met 25.0759

AMERICAN SOCIETY FOR TESTING MATERIALS.ANNUAL MEETING 58TH Symposium on atmospheric corrosion of non-ferrous metals Atlantic City,N.J. 1955 Jun 29 American society for testing materials A.S.T.M.Special technical publication, 175 A.S.T.M.,Philadelphia,Pa., 1955
Met 25.1918

AMERICAN SOCIETY FOR TESTING MATERIALS.ANNUAL MEETING 58TH Symposium on high-purity water corrosion Atlantic City,N.J. 1955 Jun 28 American society for testing materials A.S.T.M.Special technical publication, 179 American society for testing materials,Philadelphia,Pa.,1955
Met 25.1927

AMERICAN SOCIETY FOR TESTING MATERIALS.ANNUAL MEETING 58TH Symposium on impact testing Atlantic City 1955 Jun 27 American society for testing materials American society for testing materials,Philadelphia,Pa.,1955
Met 25.0902

AMERICAN SOCIETY FOR TESTING MATERIALS.ANNUAL MEETING 58TH Symposium on impact testing : presented at the 58th A.S.T.M. annual meeting Atlantic city,N.J. 1955 Jun 27 American society for testing materials American society for testing materials.Special technical publication, 176 American society for testing materials,Philadelphia,Pa.,1956
Eng 41.3757

AMERICAN SOCIETY FOR TESTING MATERIALS.ANNUAL MEETING 58TH Symposium on metallic materials for service at temperatures above 1600 F. Atlantic City,N.J. 1955 Jun 30 American society for testing materials A.S.T.M.Special technical publication, 174 American society for testing materials, Philadelphia,Pa.,1956
Met 25.0999

AMERICAN SOCIETY FOR TESTING MATERIALS.ANNUAL MEETING 59TH Structural chemistry and metallurgy of copper By D.K. Crampton H.W. Gillett memorial lecture 1956 A.S.T.M.,1956
Met 25.0534

AMERICAN SOCIETY FOR TESTING MATERIALS.ANNUAL MEETING 59TH Symposium on solder : presented at the fifty-ninth annual meeting A.S.T.M. Atlantic city,N.J. 1956 Jun 19-20 American society for testing materials American society for testing materials.Special technical publication, 189 American society for testing materials,Philadelphia,Pa.,1957
Eng 41.4025

AMERICAN SOCIETY FOR TESTING MATERIALS.ANNUAL MEETING 61ST Symposium on advances in electron metallography Papers American society for testing materials A.S.T.M. Special technical publication, 245 American society for testing materials,Philadelphia,Pa., 1958
Met 25.1362

AMERICAN SOCIETY FOR TESTING MATERIALS.ANNUAL MEETING 61ST Symposium on basic mechanisms of fatigue Papers Boston,Mass. 1958 Jun 23 American society for testing materials A.S.T.M.Special technical publication, 237 A.S.T.M.,Philadelphia,1959
Met 25.0925

AMERICAN SOCIETY FOR TESTING MATERIALS.ANNUAL MEETING 61ST Symposium on basic mechanisms of fatigue presented before the 61st annual meeting Boston 1958 Jun 23 American society for testing materials A.S.T.M.Special technical publication, 273 A.S.T.M. Philadelphia,1959
Met 25.2328

AMERICAN SOCIETY FOR TESTING MATERIALS.ANNUAL MEETING 62ND Symposium on electron metallography Atlantic City,N.J. 1959 Jun 23 American society for testing materials A.S.T.M.Special technical publication, 262 American society for testing materials, Philadelphia,Pa.,1960
Met 25.1366

AMERICAN SOCIETY FOR TESTING MATERIALS.ANNUAL MEETING 63RD Properties of crystalline solids ...Symposium on recent progress in materials sciences;Symposium on nature and origin of strength of materials.Presented at the 63rd annual meeting A.S.T.M. Philadelphia,Pa. 1960 Jun 27 American society for testing materials American society for testing materials.Special technical publication, 283 American society for testing materials,Philadelphia,Pa.,1961
Eng 41.3790

AMERICAN SOCIETY FOR TESTING MATERIALS.ANNUAL MEETING 63RD Symposium on low temperature properties of high-strength and missile materials Papers 1960 Jun 30 American society for testing materials A.S.T.M. Special technical publication, 287 A.S.T.M., Philadelphia,1961
Met 25.0974

AMERICAN SOCIETY FOR TESTING MATERIALS.ANNUAL MEETING 63RD Symposium on methods of metallographic specimen preparation Atlantic City 1960 Jun 28 American society for testing materials A.S.T.M.Special technical publication, 285 American society for testing materials,Philadelphia,Pa.,1960
Met 25.1381

AMERICAN SOCIETY FOR TESTING MATERIALS.ANNUAL MEETING 63RD Symposium on shear and torsion testing 1960 Jun 28 American society for testing materials A.S.T.M. Special technical publication, 289 American society for testing materials,Philadelphia,Pa., 1961
Met 25.0887

AMERICAN SOCIETY FOR TESTING MATERIALS.ANNUAL MEETING 64TH Symposium on evaluation of metallic materials in design for low-temperature service presented at the 64th annual meeting Atlantic City,N.J. 1961 Jun 27-28 American society for testing materials A.S.T.M.Special technical publication, 302 A.S.T.M.,Philadelphia,1962
Met 25.2335

AMERICAN SOCIETY FOR TESTING MATERIALS.ANNUAL MEETING 65TH Fatigue of metals in engineering and design By R.E. Peterson Edgar Marburg lecture,1962 A.S.T.M., Philadelphia,Pa.,1962 Presented before the 65th annual meeting of the A.S.T.M.
Met 25.0937

AMERICAN SOCIETY FOR TESTING MATERIALS.ANNUAL MEETING 65TH Symposium on stress-strain-time-temperature relationships in materials New York 1962 Jun 27 American society for testing materials A.S.T.M.Special technical publication, 325,Materials science series, 3 American society for testing and materials, Philadelphia,Pa.,1962
Met 25.1034

AMERICAN SOCIETY FOR TESTING MATERIALS.ANNUAL MEETING 66TH Adhesion ...Recent developments in adhesion science...Recent advances in adhesives technology...Presented at the 66th annual meeting A.S.T.M. Atlantic city,N.J. 1963 Jun 26 American society for testing materials American society for testing materials.Special technical publication, 360 American society for testing materials,Philadeplphia,Pa.,1964
Eng 41.4027

AMERICAN SOCIETY FOR TESTING MATERIALS.ANNUAL MEETING 66TH Techniques of electron microscopy,diffraction and microprobe analysis Papers Atlantic City 1963 Jun 26 American society for testing materials A.S.T.M.Special technical publication, 372 Philadelphia,1964
Met 25.2495

AMERICAN SOCIETY FOR TESTING MATERIALS.ANNUAL MEETING 67TH Fracture toughness testing and its applications :a symposium Chicago 1964 Jun 21-25 American society for testing materials American society for testing materials,Philadelphia,Pa.,1965
Met 25.0868

AMERICAN SOCIETY FOR TESTING MATERIALS.ANNUAL MEETING 67TH Internal friction,damping, and cyclic plasticity a symposium presented before the 67th annual meeting American society for testing materials A.S.T.M. Special technical publication, 378 A.S.T.M., Philadelphia,1965
Met 25.2340

AMERICAN SOCIETY FOR TESTING MATERIALS.ANNUAL MEETING 67TH Symposium on newer structural materials for aerospace vehicles Chicago 1964 Jun 21 American society for testing materials A.S.T.M.Special technical publication, 379 Philadelphia,Pa.,1965
Met 25.2516

AMERICAN SOCIETY FOR TESTING MATERIALS.ANNUAL MEETING 68TH Behaviour of materials at cryogenic temperatures a symposium Lafayette, Ind. 1965 Jun 13-18 American society for testing materials A.S.T.M.Special technical publication, 387 American society for testing materials,Philadelphia,Pa.,1916
Met 25.0976

AMERICAN SOCIETY FOR TESTING MATERIALS.ANNUAL MEETING 68TH Determination of nonmetallic compounds in steel a symposium presented at the sixty-eighth annual meeting Lafayette 1965 Jun 13-18 American society for testing materials A.S.T.M.Special technical publication, 393 A.S.T.M.,Philadelphia,1966
Met 25.2303

AMERICAN SOCIETY FOR TESTING MATERIALS.ANNUAL MEETING 69TH Fatigue crack propagation :a symposium Atlantic City,N.J. 1966 Jun 26-Jul 1 American society for testing materials A.S.T.M.Special technical publication, 415 A. S.T.M.,Philadelphia,Pa.,1967
Met 25.0938

AMERICAN SOCIETY FOR TESTING MATERIALS.ANNUAL MEETING 69TH Permeability and capillarity of soils :a symposium Atlantic city,N.J. 1966 Jun 26-Jul 1 American society for testing materials American society for testing materials.Special technical publication, 417 American society for testing and materials,Philadelphia,Pa.,1967
Eng 41.3102

AMERICAN SOCIETY FOR TESTING MATERIALS.ANNUAL MEETING 69TH Stress corrosion testing :a symposium Atlantic City,N.J. 1966 Jun 26-Jul 1 American society for testing materials A.S.T.M.Special technical publication, 425 American society for testing and materials, Philadelphia,Pa.,1967
Met 25.1964

AMERICAN SOCIETY FOR TESTING MATERIALS.ANNUAL MEETING 70TH Metal matrix composites a symposium presented at the 70th annual meeting. Boston 1967 Jun 25-30 American society for testing materials A.S.T.M. Special technical publication, 438 A.S.T.M., Philadelphia,1968
Met 25.2337

AMERICAN SOCIETY FOR TESTING MATERIALS.ANNUAL MEETING 71ST Fatigue at high temperature : a symposium presented at the 71st annual meeting Proceedings San Francisco,Calif. 1968 Jun 23-28 American society for testing materials American society for testing materials.Technical publication, 459 ASTM, Philadelphia,Pa.,1969
Met 25.2583

AMERICAN SOCIETY FOR TESTING MATERIALS.ANNUAL MEETING 72ND Achievement of high fatigue resistance in metals and alloys :a symposium presented at the 72nd annual meeting Proceedings Atlantic City,N.J. 1969 Jun 22-27 American society for testing materials American society for testing materials.Special technical publication, 467 ASTM,Philadelphia, Pa.,1970
Met 25.2584

AMERICAN SOCIETY FOR TESTING MATERIALS.CLEVELAND SPRING MEETING 1952 Symposium on testing metal powders and metal powder products presented at the Cleveland spring meeting Cleveland 1952 Mar 5 American society for testing materials A.S.T.M.Special technical publication, 140 American society for testing materials,Philadelphia,Pa.,1953
Met 25.0762

AMERICAN SOCIETY FOR TESTING MATERIALS.COMMITTEE A-I ON STEEL Temper embrittlement in steel a symposium Philadelphia 1967 Oct 3-4 A.S.T.M.Special technical publication, 407 A.S.T.M.,Philadelphia,1968
Met 25.0455

AMERICAN SOCIETY FOR TESTING MATERIALS.COMMITTEE D-30 HIGH MODULUS FIBERS AND THEIR COMPOSITES Symposium on composite materials:testing and design New Orleans,La. 1969 Feb 11-13 A.S.T.M.Special technical publication, 460 Philadelphia,Pa.,1969 Presented at a meeting of committee D-30
Met 25.2859

AMERICAN SOCIETY FOR TESTING MATERIALS.COMMITTEE E-2 ON EMISSION SPECTROSCOPY Symposium on chemical analysis of inorganic solids by means of the mass spectrometer Atlantic City,N.J. 1951 Jun 20 A.S.T.M.Special technical publication, 149 A.S.T.M.,Philadelphia,Pa., 1951
Met 25.1723

AMERICAN SOCIETY FOR TESTING MATERIALS.COMMITTEE E-7 ON NON-DESTRUCTIVE TESTING Papers on non-destructive testing presented at a meeting of committee E-7... 1953 Jul 2 American society for testing materials,Philadelphia,Pa., 1954
Met 25.1318

AMERICAN SOCIETY FOR TESTING MATERIALS.COMMITTEE ON MASS SPECTROMETRY and INSTITUTE OF PETROLEUM.HYDROCARBON RESEARCH GROUP Advances in mass spectrometry :a conference Proceedings London 1958 Sep 24-26 Edited by J.D. Waldron Pergamon press.Symposia publications division,London,1959
Chem 18.0197

AMERICAN SOCIETY FOR TESTING MATERIALS.PACIFIC AREA NATIONAL MEETING 2ND Cement and concrete Technical papers Los Angeles, Calif. 1956 Sep 19-20 American society for testing materials American society for testing materials.Special technical publication, 205 American society for testing materials,Philadelphia,Pa.,1958
Eng 41.2845

AMERICAN SOCIETY FOR TESTING MATERIALS.PACIFIC AREA NATIONAL MEETING 2ND Designs and tests of building structures .Seismic and shock loading...Glued laminated and other constructions Los Angeles,Calif. 1956 Sep 17 American society for testing materials American society for testing materials.Special technial publication, 209 American society for testing materials,Philadelphia,Pa.,1957
Eng 41.3017

AMERICAN SOCIETY FOR TESTING MATERIALS.PACIFIC AREA NATIONAL MEETING 2ND Metals : presented at the 2nd Pacific area national meeting Los Angeles 1956 Sep 17-21 American society for testing materials A.S.T.M.Special technical publication, 196 American society for testing materials, Philadelphia,Pa.,1957
Met 25.1036

AMERICAN SOCIETY FOR TESTING MATERIALS.PACIFIC AREA NATIONAL MEETING 2ND Symposium on full-scale tests on house structures Los Angeles,Calif. 1956 Sep 18 American society for testing materials American society for testing materials.Special technical publication, 210 American society for testing materials,Philadelphia,Pa.,1956
Eng 41.3018

AMERICAN SOCIETY FOR TESTING MATERIALS.PACIFIC AREA NATIONAL MEETING 2ND Symposium on titanium :presented at the second Pacific area national meeting Los Angeles,Calif. 1956 Sep 17-18 American society for testing materials American society for testing materials.Special technical publication, 204 American society for testing materials, Philadelphia,Pa.,1957
Eng 41.3598

AMERICAN SOCIETY FOR TESTING MATERIALS.PACIFIC AREA NATIONAL MEETING 4TH Symposium on fatigue tests of aircraft structure:low-cycle, full-scale, and helicopters Los Angeles 1962 Oct 1-3 American society for testing materials A.S.T.M.Special technical publication, 338 A.S.T.M.,Philadelphia,Pa., 1963
Met 25.0935

AMERICAN SOCIETY FOR TESTING MATERIALS.PACIFIC AREA NATIONAL MEETING 4TH Symposium on properties of surfaces :presented at the fourth Pacific area national meeting Los Angeles,Calif. 1962 Oct 4 American society for testing materials American society for testing materials.Special technical publication, 340 Philadelphia,1963
Met 25.2829

AMERICAN SOCIETY FOR TESTING MATERIALS.PACIFIC AREA NATIONAL MEETING 5TH Orientation effects in the mechanical behavior of anisotropic structural materials :a symposium Seattle,Wash. 1965 Oct 31-Nov 5 American society for testing materials American society for testing materials.Special technical publication, 405 American society for testing materials,Philadelphia,Pa.,1966
Eng 41.3701

AMERICAN SOCIETY FOR TESTING MATERIALS.PACIFIC AREA NATIONAL MEETING 5TH Orientation effects in the mechanical behaviour of anisotropic structural materials a symposium presented at the 5th Pacific area national meeting American society for testing materials A.S.T.M.Special technical publication, 405 A.S.T.M.,Philadelphia,1966
Met 25.2341

AMERICAN SOCIETY FOR TESTING MATERIALS.PACIFIC AREA NATIONAL MEETING 5TH Structural fatigue in aircraft :a symposium American society for testing materials A.S.T.M. Special technical publication, 404 American society for testing materials,Philadelphia,Pa., 1966
Met 25.0943

AMERICAN SOCIETY FOR TESTING MATERIALS. SUBCOMMITTEE IV ON ELECTROPLATING PRACTICE Anodized aluminium :a symposium presented during committee week Papers Cleveland 1965 Feb 9 A.S.T.M.,Philadelphia,1965
Met 25.2049

AMERICAN SOCIETY FOR TESTING MATERIALS. SUBCOMMITTEE ON ELECTRON MICROSTRUCTURE Electron metallography and probe microanalysis 1962 Symposium on advances in electron metallography and electron probe microanalysis Atlantic City,N.J. 1960;1961 A.S.T.M. Special technical publication, 317 A.S.T.M., Philadelphia,Pa.,1962
Met 25.1377

AMERICAN SOCIETY FOR TESTING MATERIALS. SUBCOMMITTEE 11 ON ELECTRON MICROSTRUCTURE OF METALS Techniques in electron metallography 1963 Symposium on advances in techniques in electron metallography New York 1962 Jun 26 A.S.T.M.Special technical publication, 339 American society for testing and materials,Philadelphia,Pa.,1963
Met 25.1378

AMERICAN SOCIETY FOR TESTING MATERIALS,ANNUAL MEETING 60TH Symposium on determination of gases in metals Atlantic City,N.J. 1957 Jun 18 American society for testing materials A.S.T.M.Special technical publication, 222 A.S.T.M.,Philadelphia,Pa, 1958
Met 25.1708

AMERICAN SOCIETY FOR TESTING METALS.ANNUAL MEETING 63RD Symposium on acoustical fatigue Atlantic City,N.J. 1960 Jun 28 American society for testing materials A.S.T.M.Special technical publication, 284 A.S.T.M. Philadelphia,Pa.,1961
Met 25.0929

AMERICAN SOCIETY OF CHEMICAL ENGINEERS. FLUIDS ENGINEERING DIVISION   Biomedical fluid mechanics symposium :presented at Fluids engineering conference   Denver,Col.   1966 Apr 25-27   American society of mechanical engineers,New York,1966
Chem E   24.1797

AMERICAN SOCIETY OF CIVIL ENGINEERS   Coastal engineering :a specialty conference   Santa Barbara,Calif.   1965 Oct   American society of civil engineers,New York,1966
Eng   41.3346

AMERICAN SOCIETY OF CIVIL ENGINEERS
Conference on coastal engineering   10th Proceedings   Tokyo   1966 Sep   2 vols American society of civil engineers,New York, 1967
Eng   41.3348

AMERICAN SOCIETY OF CIVIL ENGINEERS
Conference on coastal engineering   11th Proceedings   London   1968 Sep   Vol   1-2 2 vols   American society of civil engineers, New York,1969
Eng   41.3347

AMERICAN SOCIETY OF CIVIL ENGINEERS
Conference on electronic computation   2nd Proceedings   Pittsburgh,Pa.   1960   American society of civil engineers,Kansas City,Miss., 1960
Eng   41.2718

AMERICAN SOCIETY OF CIVIL ENGINEERS
Conference on electronic computation   3rd Proceedings   Boulder,Colo.   1963 Jun 19-21 Journal of structural division,   89,no ST4 American society of civil engineers,Kansas city,Miss.,1963
Eng   41.2716

AMERICAN SOCIETY OF CIVIL ENGINEERS   Research conference on shear strength of cohesive soils New York   1960   Boulder,Colo.,1960
Eng   41.3171

AMERICAN SOCIETY OF MECHANICAL ENGINEERS
Advances in fluidics   Fluidics symposium Chicago,Ill.   1967 May 9-11   Edited by Forbes T. Brown   American society of mechanical engineers,New York,1967
Eng   41.5898

AMERICAN SOCIETY OF MECHANICAL ENGINEERS
Advances in thermophysical properties at extreme temperatures and pressures   Symposium on thermophysical properties   3rd   Papers Lafayette,Ind.   1965 Mar 22-25   Edited by Serge Gratch   American society of mechanical engineers,New York,1965
Eng   41.7121

AMERICAN SOCIETY OF MECHANICAL ENGINEERS
Condensing heat transfer within horizontal tubes   National heat transfer conference 2nd   Preprints   Chicago,Ill.   1958 Aug 18-21   American society of mechanical engineers, Philadelphia,Pa.,1958
Eng   41.7155

AMERICAN SOCIETY OF MECHANICAL ENGINEERS
Determination of system characteristics from normal operating records   mimeograph   By T.P. Goodman and J.B. Reswick   illus.   17p American society of mechanical engineers,New York,1955   For presentation at the ASME instruments and regulators division conference, Ann Arbor,Mich.,April 1955
Math L   5.1165

AMERICAN SOCIETY OF MECHANICAL ENGINEERS
Engineering developments in energy conversion International conference on energetics Rochester,N.Y.   1965 Aug 18-20   American society of mechanical engineers,New York,1965
Eng   41.4865

AMERICAN SOCIETY OF MECHANICAL ENGINEERS
Experimental techniques in shock and vibration a colloquium   Papers   New York   1962 Nov 27   Edited by W.J. Worley   American society of mechanical engineers,New York,1962
Eng   41.6354

AMERICAN SOCIETY OF MECHANICAL ENGINEERS
Fatigue of metals   International conference on fatigue of metals   Proceedings   London 1956 Sep 10-14   and   New York   1956 Nov 28-30   Institution of mechanical engineers, London,1956
Eng   41.3759

AMERICAN SOCIETY OF MECHANICAL ENGINEERS
General discussion on heat transfer Proceedings   London   1951 Sep 11-12   xiii, 496p   Institution of mechanical engineers, London,1951
Chem E   24.0538

AMERICAN SOCIETY OF MECHANICAL ENGINEERS
General discussion on heat transfer Proceedings   London   1951 Sep 11-13 Institution of mechanical engineers,London, 1952
Eng   41.7146

AMERICAN SOCIETY OF MECHANICAL ENGINEERS
International conference on fatigue of metals Proceedings   London   1956 Sep 10-14   New York   1956   Nov 28-30   Institution of mechanical engineers,London,1956
Met   25.0919

AMERICAN SOCIETY OF MECHANICAL ENGINEERS
International developments in heat transfer Heat transfer conference 1961-62   Proceedings Boulder,Colo.   1961 Aug 28-Sep 1   and London   1962 Jan 8-12   American society of mechanical engineers,New York,1963
Eng   41.7168

AMERICAN SOCIETY OF MECHANICAL ENGINEERS
International developments in heat transfer International heat transfer conference Papers   Boulder,Col.   1961 Aug 28-Sep 1 London   1962   Jan 8-12   Pt   1-5   5 vols   A.S.M.E.,New York,1961   Co-sponsored by the American institute of chemical engineers,the Institution of mechanical engineers,and the Institution of chemical engineers
Chem E   24.0595

AMERICAN SOCIETY OF MECHANICAL ENGINEERS
Joint conference on combustion   Proceedings Boston,Mass.   1955 Jun 15-17   London   1955 Oct 25-27   viii,457p   Institution of mechanical engineers,London,1955
Chem E   24.1376

AMERICAN SOCIETY OF MECHANICAL ENGINEERS
Joint conference on combustion Proceedings Boston,Mass. 1955 Jun 15-17 and London 1955 Oct 25-27 Institute of mechanical engineers,London,1956
Eng 41.7231

AMERICAN SOCIETY OF MECHANICAL ENGINEERS
Joint international conference on creep session 1-7 New York 1963 Aug 25-29 London 1963 Sep 30-Oct 4 book 1-5 Jointly sponsored by Institution of mechanical engineers I.M.E.,London,c1963 5 books bound together
Met 25.0953

AMERICAN SOCIETY OF MECHANICAL ENGINEERS
Mechanical wear :a summer conference Proceedings Cambridge,Mass. 1948 Jun Edited by John T. Burwell Jointly sponsored by the General motors corp. and the Chrysler corp. American society for metals,1950
Met 25.1597

AMERICAN SOCIETY OF MECHANICAL ENGINEERS
Multi-phase flow symposium Papers Philadelphia,Pa. 1963 Nov 17-22 Edited by Norman J. Lipstein illus,diagrms 99p American society of mechanical engineers,New York,1963
Chem E 24.0241

AMERICAN SOCIETY OF MECHANICAL ENGINEERS
Symposium on cavitation in fluid machinery Chicago,Ill. 1965 Nov 7-11 Edited by Glenn M. Wood and others American society of mechanical engineers,New York,1965
Eng 41.6538

AMERICAN SOCIETY OF MECHANICAL ENGINEERS
Symposium on cavitation research.Facilities and techniques New York,1964
Eng 41.6521

AMERICAN SOCIETY OF MECHANICAL ENGINEERS
Symposium on flow visualization :A.S.M.E. annual meeting Presentation summaries New York City 1960 Nov 30 By S.J. Kline and others illus var.p A.S.M.E.,New York, 1960
Chem E 24.0329

AMERICAN SOCIETY OF MECHANICAL ENGINEERS
Symposium on fully separated flows :a conference Papers Philadelphia,Pa. 1964 May 18-20 Edited by Arthur G. Hansen American society of mechanical engineers,New York,1964
Eng 41.6514

AMERICAN SOCIETY OF MECHANICAL ENGINEERS
Symposium on rheology :presented at the ASME applied mechanics and fluids engineering conference Washington,D.C. 1965 Jun 7-9 Edited by A.W. Marris and J.T.S. Wang American society of mechanical engineers,New York,1965
Eng 41.4418

AMERICAN SOCIETY OF MECHANICAL ENGINEERS The Design of heating coils for storage tanks Edited by David Stuhlberg American society of mechanical engineers.Paper, 58-HT-1 American society of mechanical engineers, Philadelphia,Pa.,1958
Eng 41.7154

AMERICAN SOCIETY OF MECHANICAL ENGINEERS :
annual meeting Proceedings Theory and fundamental research in heat transfer New York 1960 Nov Edited by J.A. Clark ix, 220p Pergamon press,Oxford,1963
Chem E 24.0582

AMERICAN SOCIETY OF MECHANICAL ENGINEERS.APPLIED MECHANICS DIVISION Elasticity American mathematical society.Symposium in applied mathematics 3rd Proceedings Ann Arbor, Mich. 1949 Jun 14-16 Edited by R.V. Churchill and others McGraw-Hill,New York, 1950
Col S 12.0335

AMERICAN SOCIETY OF MECHANICAL ENGINEERS.APPLIED MECHANICS DIVISION Shock and vibration instrumentation;a review of the latest developments Urbana,Ill. 1955 Jun 14-16 New York,1956
Eng 41.6324

AMERICAN SOCIETY OF MECHANICAL ENGINEERS.APPLIED MECHANICS DIVISION Structural damping :a colloquium Papers Atlantic City,N.J. 1959 Dec Edited by Jerome E. Ruzicka American society of mechanical engineers,New York,1959
Eng 41.6335

AMERICAN SOCIETY OF MECHANICAL ENGINEERS.
AUTOMATIC CONTROL DIVISION Joint automatic control conference 6th Preprints of conference papers New York 1965 Jun 22-25 New York,1965
Eng 41.6015

AMERICAN SOCIETY OF MECHANICAL ENGINEERS.DESIGN ENGINEERING DIVISION Product design through the engineering analysis of its end use :an approach and a case history.Presented at the Design engineering conference New York 1967 May 15-18 American society of mechanical engineers,New York,1967
Eng 41.0326

AMERICAN SOCIETY OF MECHANICAL ENGINEERS.FLUID MEASUREMENT RESEARCH COMMITTEE Flow measurement symposium :presented at Fluid meters golden anniversary Papers Pittsburgh,Pa. 1966 Sep 26-28 American society of mechanical engineers,New York,1966
Eng 41.6717

AMERICAN SOCIETY OF MECHANICAL ENGINEERS.FLUID MECHANICS COMMISSION Biomedical fluid mechanics symposium Papers Denver,Colo. 1966 Apr 25-27 Co-sponsored by the Rheology society of America American society of mechanical engineers,New York,1966 Papers presented by the Fluid mechanics commission of the ASME at the Fluids engineering conference, co-sponsored by the Rheology society of America
Eng 41.6540

AMERICAN SOCIETY OF MECHANICAL ENGINEERS.
HYDRAULIC DIVISION Symposium on measurement in unsteady flow :presented at the ASME hydraulic division conference Worcester, Mass. 1962 May 2-23 American society of mechanical engineers,New York,1962
Eng 41.6713

AMERICAN SOCIETY OF MECHANICAL ENGINEERS. HYDRAULICS AND HEAT-TRANSFER DIVISION Multi-flow symposium Reports and proceedings Philadelphia,Pa. 1963 Nov 17-22 American society of mechanical engineers,New York,1963
Eng 41.6513

AMERICAN SOCIETY OF MECHANICAL ENGINEERS.STANDING COMMITTEE ON THERMOPHYSICAL PROPERTIES Symposium on thermophysical properties progress in international research on thermodynamic and transport properties 2nd Papers Princeton,N.J. 1962 Jan 24-26 American society of mechanical engineers; Academic press,New York,1962
Chem 18.0326

AMERICAN SOCIETY OF MECHANICAL ENGINEERS:HEAT TRANSFER DIVISION Thermodynamic and transport properties of gases liquids and solids Symposium on thermal properties 3rd Papers Lafayette,Ind 1959 Feb 23-26 x, 472p American society of mechanical engineers;McGraw-Hill,New York,1959
Chem E 24.0984

AMERICAN SOCIETY OF METALS.ORLANDO CHAPTER Southern metals-materials conference on advances in aerospace materials Proceedings Orlando,Fla. 1964 Apr 16-17 Edited by Henry M. Otte and Saul R. Locke Plenum press, New York,1965
Met 25.2282

AMERICAN SOCIETY OF ZOOLOGISTS Behavioral physiology of coelenterates :symposia Papers By L.M. Passano and others American zoologist, 5,3 American society of zoologists,New York,1965
Bal 39.2184

AMERICAN SOCIETY OF ZOOLOGISTS Nervous and hormonal mechanisms of integration St. Andrews 1965 Sep Society for experimental biology.Symposia, 20 Cambridge university press,Cambridge,1966
Psy 31.0137

AMERICAN SOCIETY OF ZOOLOGISTS Refresher course in recent advances in cytogenetics and developmental genetics Corvallis,Ore. 1962 Aug 27-28 American zoologist, 3,no.1 Utica,N.Y.,1963
Gen 34.0856

AMERICAN STATISTICAL ASSOCIATION Acceptance sampling annual meeting 105th Cleveland, Ohio 1946 Jan.27 155p 23cm American statistical association,Washington,D.C.,1950
Math 3.0816

AMERICAN UNIVERSITY IN CAIRO Interaction of radiation with solids Cairo solid state conference Cairo 1966 Sep 3-8 Edited by Adli Bishay Plenum press,New York,1967
TA 15.0429

AMERICAN UNIVERSITY OF BEIRUT Excitons, magnons and phonons in molecular crystals :an international symposium Proceedings Beirut 1968 Jun 15-18 Edited by A.B. Zahlan Cambridge university press,Cambridge,1968
Radioth 35.1689

AMERICAN UNIVERSITY OF BEIRUT The Triplet state an international symposium Proceedings Beirut 1967 Feb 14-19 Edited by A.B. Zahlan and others Cambridge university press, Cambridge,1967
Chem 18.2642

AMERICAN VACUUM SOCIETY National symposium on vacuum technology 7th Transactions Cleveland,Ohio 1960 Oct 12-14 Edited by C. Robert Meissner Pergamon press,Oxford,1961 Earlier symposia published by the Committee on vacuum techniques
Met 25.1669

AMERICAN VACUUM SOCIETY National vacuum symposium 8th Transactions Washington,D. C. 1961 Oct 16-19 Vol 1-2 Edited by Luther E. Preuss 2 vols Pergamon press, Oxford,1962 Combined with the International congress on vacuum science and technology,2nd
Cav 7.2376

AMERICAN VACUUM SOCIETY The National symposium on vacuum technology 7th Transactions Cleveland,Ohio 1960 Oct 12-14 Edited by C.Robert Meissner 427p Pergamon press,Oxford,1961 8th symposium called 'National vacuum symposium' q.v.
Cav 7.2375

AMERICAN ZINC INSTITUTE Zinc as a galvanic anode Navy-industry zinc symposium on cathodic protection development,application and specification 1955 Apr 21-22 U.S.Navy dept.,Washington,D.C.,1955
Met 25.2500

AMERICAN ZOOLOGICAL SOCIETY.COMPARATIVE PHYSIOLOGY DIVISION The Central nervous system and fish behavior :a conference Papers Chicago,Ill. 1967 Apr Edited by David Ingle Jointly sponsored by Perspectives in biology and medicine illus University of Chicago press,Chicago,Ill.; London,1968
Psy 31.2816

AMERSFOORT 1953 International significal summer conference Conference d'ete internationale de linguistique psychologique 9th Proceedings International society for significs Synthese, 9,issue 3,no 3-4 Kroonder,Bussum,c1954
Psy 31.2514

AMES,IOWA 1952 Statistics and mathematics in biology conference Iowa state college Edited by Oscar Kempthorne and others Co-sponsored by Biometrics society xi,632p 23cm Iowa state college press,Ames,Iowa,1954
Math 3.0764

AMES,IOWA 1964 Physiology of digestion in the ruminant International symposium on the physiology of digestion in the ruminant 2nd Papers Edited by R.W. Dougherty Butterworths,London,1965
Phys 20.1337

AMES,IOWA 1964 Physiology of digestion in the ruminant :International symposium 2nd Papers Edited by R.W. Dougherty and others Butterworths,Washington,1965
VA 19.0304

AMHERST,MASS 1955 Cellular mechanisms in differentiation and growth :A SYMPOSIUM Proceedings By M. Delbruck and others Society for the study of development and growth Edited by Dorothea Rudnick Society for the study of development and growth. Symposia, 14 illus. vii,236p Princeton university press,Princeton,N.J.,1956
Bot 42.1474

AMHERST,MASS. 1955 Cellular mechanisms in differentiation and growth :a symposium Edited by Dorothea Rudnick Society for the study of development and growth.Symposia, 14 Princeton university press,Princeton,N.J.,1956
An 32.2478

AMHERST,MASS. 1964 Oxidases and related redox systems :a symposium Proceedings Vol 1-2 Edited by Tsoo E. King and others held at Amherst college 2 vols Wiley,New York,1965
Bioch 33.1085

AMHERST,MASS. 1964 The Role of chromosomes in development Society for the study of development and growth Edited by Michael Locke Society for the study of development and growth.Symposia, 23 illus. Academic press,New York;London,1964
Gen 34.0859

AMHERST,MASS. 1964 The Role of chromosomes in development :a symposium Edited by Michael Locke Society for the study of development and growth.Symposia, 23 Academic press,New York;London,1964
An 32.2506

AMHERST,MASS. 1964 The Role of chromosomes in development :a symposium Society for the study of development and growth Edited by Michael Locke Society for the study of development and growth.Symposia, 23 Academic press,New York;London,1964 The Society for the study of development and growth known later as the Society for developmental biology
Bioch 33.1011

AMHERST,MASS. 1971 The Scientific results from the orbiting astronomical observatory (OAO-2) National aeronautics and space administration Edited by A.D. Code NASA SP-310 590p nasa,washington,D.C.,1972
Obs 6.3626

AMHERST,MASS. 1972 Conference on compact transformation groups 2nd Proceedings Pt 1-2 Held at the University of Massachusetts Lecture notes in mathematics, 298-299 25cm 2 vols Springer,Berlin,1972
P Math 2.4110

AMHERST COLLEGE Oxidases and related redox systems :a symposium Proceedings Amherst, Mass. 1964 Jul 15-19 Vol 1-2 Edited by Tsoo E. King and others 2 vols Wiley, New York,1965
Bioch 33.1085

AMINO-ACID The Relation of optical form to biological activity in the amino-acid series : a symposium London 1947 Feb 15 Biochemical society Edited by R.T. Williams Held at the University college hospital medical school Biochemical society.Symposia, 1 Cambridge university press,Cambridge, 1948
Bioch 33.1363

AMINO ACID ANALYSIS OF PROTEINS :a conference Papers New York 1945 Oct 12-13 By William H. Stein and others New York academy of sciences New York academy of sciences. Annals, 47,p.57-240 New York,1946
Bioch 33.0552

AMINO ACID METABOLISM A Symposium on amino acid metabolism Papers and discussions presented Baltimore,Md. 1954 Jun 14-17 Edited by William D. McElroy and H.Bentley Glass Sponsored by the McCollum-Pratt institute McCollum-Pratt institute. Contribution, 105 Johns Hopkins press, Baltimore,1955
Bioch 33.0599

AMINO ACID METABOLISM A Symposium on amino-acid metabolism Papers Baltimore,Md. 1954 Jun 14-17 McCollum-Pratt institute Edited by William D. McElroy and H.Bentley Glass McCollum-Pratt institute.Contribution, 105 Johns Hopkins,Baltimore,Md.,1955
Chem 18.1532

AMINO ACID METABOLISM :a symposium Papers and discussion Baltimore,Md. 1954 Jun 14-17 Edited by William D. McElroy and H. Bentley Glass Sponsored by the McCollum-Pratt institute McCollum-Pratt institute. Contribution, 108 Johns Hopkins press, Baltimore,Md.,1955
Mol 45.0292

AMINO ACID POOLS :distribution and function of free amino acids.Symposium Proceedings Duarte,Calif. 1961 May 19-22 City of hope medical center Edited by Joseph T. Holden Under the auspices of the Institute for advanced learning in the medical sciences illus Elsevier,Amsterdam,1962 Preface: "The committee undertook the organization of the symposium under the title Conference on free amino acids
Bioch 33.0519

AMINO ACIDS Ciba foundation symposium on amino acids and peptides with antimetabolic activity London 1958 Mar 18-20 Ciba foundation Edited by G.E.W. Wolstenholme and Cecilia M. O'Connor Ciba foundation symposia Churchill,London,1958
Chem 18.1523

AMINO ACIDS International congress of biochemistry 6th Abstracts New York 1964 Jul 26-Aug 1 Vol 2: proteins,peptides, and amino acids Scheduled under the auspices of the International union of biochemistry Washington,D.C.,1964 Title also in French.
Bioch 33.1339

AMINO ACIDS Neurochemistry of nucleotides and amino acids :a symposium Papers and discussions presented Philadelphia 1958 Apr 24-25 American academy of neurology. Section of neurochemistry Edited by Roscoe O. Brady and Donald B. Tower Wiley,New York, 1960
Bioch 33.2222

AMINO ACIDS,PEPTIDES,AND PROTEINS :a conference Papers New York 1960 Feb 25-26 By Karl Folkers and others New York academy of sciences New York academy of sciences.Annals, 88,p.533-770 New York,1960
Bioch 33.0489

AMINO ACIDS,PROTEINS AND CANCER BIOCHEMISTRY Jesse P.Greenstein memorial symposium Papers New York 1959 Sep 16 Edited by John T. Edsall Port Academic press,New York,1960
Radioth 35.0083

AMINO ACIDS,PROTEINS AND CANCER BIOCHEMISTRY The Jesse P.Greenstein memorial symposium Papers Washington,D.C. 1959 Sep 16 American chemical society.Division of biological chemistry Edited by John T. Edsall Academic press,New York;London,1960 With a biographical article on Dr. Greenstein and a bibliography of his writings
Bioch 33.2345

AMINO ACIDS AND PEPTIDES Ciba foundation symposium on amino acids and peptides with antimetabolic activity Proceedings London 1958 Mar 18-20 Ciba foundation Edited by G. E.W. Wolstenholme and Cecilia M. O'Connor illus Churchill,London,1958
Bioch 33.0570

AMINO ACIDS AND PROTEINS :a symposium Papers Cold Spring Harbor 1950 Cold Spring Harbor biological laboratory Cold Spring Harbor symposia on quantitative biology, 14 Long Island biological association,Cold Spring Harbor,1950
Bioch 33.1270

AMORPHOUS AND LIQUID SEMICONDUCTORS :a conference Proceedings Cambridge 1969 Sep 24-27 International union of pure and applied physics.Semiconductor commission Royal society of London Edited by Nevill Francis Mott Bibliog,diagrs,figs 626p 20cm North-Holland,Amsterdam,1970
Cav 7.2678

AMSTERDAM MC-25 information symposium By J. W.de Bakker and others Mathematisch centrum, Amsterdam Mathematical centre tracts, 37 illus varp Mathematisch centrum,Amsterdam, 1971 Dedicated to Prof.van Wijngaarden on the occasion of the centre's 25th anniversary
Math L 5.3857

AMSTERDAM 1923 International tuinbouw-congres 7th Nederlandsche mantschappij voor tuinbouw en plantkunde c1923
Gen 34.1139

AMSTERDAM 1935 Zesde international botanisch congres Proceedings Vol 1-2 Edited by M.J. Sirks 2 vols Brill,Leiden, 1935-36
Bot 42.0657

AMSTERDAM 1938 International geographical congress 15th Comptes rendus;excursions Tom 1-2 International geographical union 17 vols Brill,Leiden,1938 Partly in English
Geog 13.5595

AMSTERDAM 1938 Travaux topographiques et cartographiques International geographical congress Report International geographical union Amsterdam,1938
Geog 13.5708

AMSTERDAM 1946 Zeeman congress Papers Physica, 12,9-10 73p M.Nijhof,The Hague, 1946
Obs 6.2880

AMSTERDAM 1946 Zeeman effect The Zeeman congress,1946 Proceedings Edited by C.J. Bakker 221p The Hague,1946 Reprinted from 'Physica',vol 12,no 9-10
Obs 6.3192

AMSTERDAM 1949 International colloquium on macromolecules Proceedings Internation union of pure and applied chemistry D.B. Ceuten's Uitgerers,Amsterdam,1949
Radioth 35.0014

AMSTERDAM 1949 International colloquium on macromolecules Proceedings Under the auspices of the International union of pure and applied chemistry Amsterdam,1950
Col S 12.0101

AMSTERDAM 1949 International colloquium on macromolecules Proceedings Union internationale de chimie pure et appliquee D. B.Centen's uitgevers-maatschappij N.V., Amsterdam,1950
Chem 18.0503

AMSTERDAM 1950 Congres international d'histoire des sciences et congres de la Societe internationale d'histoire de la medecine Actes Union internationale d'histoire des sciences Genootschap voor geschiedenis der geneeskunde,wiskunde en natuurweteenschappen te Leiden Academie internationale d'histoire des sciences. Collection de travaux, 6 2 vols Academie d'histoire des sciences;Hermann,Paris,1951-53 The 12th congress of the Societe internationale d'histoire de la medecine is section 4 of the 6th Congres international d'histoire des sciences
WSM 43.0067

AMSTERDAM 1950 International conference on spectroscopy at radio-frequencies :a conference Proceedings Netherlands physical society Physica, 17,3-4 169-484p 1951
Cav 7.0055

AMSTERDAM 1954 International congress of mathematicians 15th Proceedings Vol 1-3 Edited by Johan C.H. Gerretson and Johannes de Groot Bibliog 25cm 3 vols North Holland,Amsterdam,1954-57
P Math 2.2875

AMSTERDAM 1955 International conference on coordination compounds Proceedings Nederlandse chemische vereniging Royal Netherlands chemical society,n.p.,c1955
Chem 18.0998

AMSTERDAM 1957 Chemical reaction engineering European symposium on chemical engineering 1st Papers Instituut van ingenieurs Nederlandse chemische vereniging Held during the 12th meeting of the European federation of chemical engineering International series of monographs on chemical engineering, 1 197p Pergamon,London,1957
Chem E 24.1900

AMSTERDAM 1957 Constructivity in mathematics colloquium proceedings Edited by A. Heyting 297p 23cm Amsterdam,1959
P. Math 2.0116

AMSTERDAM 1957 International symposium on isotope separation 1st Proceedings Netherlands physical society and International union of pure and applied physics Edited by J. Kistemaker and others xx,704p North-Holland,Amsterdam,1958
Chem E 24.1654

AMSTERDAM 1958 Gas chromatography :a symposium 2nd Proceedings Institute of petroleum.Gas chromatography discussion group Edited by D.H. Desty Also under the auspices of the Nederlandse chemische vereniging xiii, 383p Butterworths,London,1958
Chem E 24.0849

AMSTERDAM 1958 Gas chromatography 1958 :a symposium 2nd Proceedings Institute of petroleum.Hydrocarbon research group,and, Nederlandse chemische vereniging Edited by D. H. Desty Organised by the Gas chromatography discussion group Butterworths,London,1958
Chem 18.0566

AMSTERDAM 1959 Structure and function of the cerebral cortex International meeting of neurobiologists 2nd Proceedings Edited by D.B. Tower and J.P. Schade Elsevier, Amsterdam,1960
An 32.4646

AMSTERDAM 1959 Structure and function of the cerebral cortex The International meeting of neurobiologists 2nd Proceedings Edited by D.B. Tower and J.F. Schade Elsevier,Amsterdam,1960
Bal 39.1494

AMSTERDAM 1959 The Mechanism of heterogeneous catalysis Proceedings Edited by J.H. De Boer and others vii,179p Elsevier,Amsterdam,1960
Chem E 24.0797

AMSTERDAM 1959 The Mechanism of heterogeneous catalysis :a symposium Proceedings Royal Netherlands chemical society.Section for inorganic and physical chemistry Edited by J.H. de Boer and others Elsevier monographs.Chemistry series Elsevier,Amsterdam,1960
Chem 18.0455

AMSTERDAM 1960 International symposium on the reactivity of solids 4th Nederlandse chemische vereniging and Nederlandse naturkundige vereniging Edited by J.H.de Boer and others Sponsored by the International union of pure and applied physics Elsevier,Amsterdam,1961 Papers in English,French and German
Met 25.1848

AMSTERDAM 1960 International symposium on the reactivity of solids 4th Proceedings Edited by J.H.de Boer and others Elsevier, Amsterdam,1961
Min 10.0648

AMSTERDAM 1960 Reactivity of solids International symposium on the reactivity of solids 4th Proceedings Nederlandse chemische vereniging and International union of pure and applied physics Edited by J.H. DeBoer and others Sponsored also by the International union of pure and applied chemistry illus x,762p
Chem E 24.0882

AMSTERDAM 1960 Reactivity of solids : international symposium 4th Proceedings International union of pure and applied physics,and,International union of pure and applied chemistry Edited by J.H. de Boer and others Elsevier,Amsterdam,1961
Chem 18.0596

AMSTERDAM 1961 Molecular spectroscopy : general and introductory lectures presented at the fifth European congress on molecular spectroscopy 5th International union of pure and applied chemistry.Physical chemistry section,and,Royal Netherlands chemical society Co-sponsored by the Netherlands physical society Butterworths,London,1962 Reprinted from 'Pure and applied chemistry', vol 4,no 1
Chem 18.2315

AMSTERDAM 1962 Nerve,brain and memory models ..a series of lectures delivered during a symposium on cybernetics of the nervous system...held as part of the second international meeting of medical cybernetics at the Royal academy of sciences of Amsterdam.. Lectures Edited by N. Wiener and J.P. Schade Progress in brain research, 2 Elsevier,Amsterdam,1963
Psy 31.0248

AMSTERDAM 1963 Cellular control mechanisms and cancer :a conference Proceedings Edited by P. Emmelot and O. Muhlbock Under the auspices of the International union against cancer Elsevier,Amsterdam,1964 Conference held on the occasion of the 50th anniversary of the Netherlands cancer institute
An 32.2504

AMSTERDAM 1963 Cellular control mechanisms and cancer :conference Proceedings Edited by P. Emmelot and O. Muhlbock Under the auspices of the International union against cancer.Cancer research commision Illus Elsevier,Amsterdam,1964
Radioth 35.0530

AMSTERDAM 1963 Conference on cellular control mechanisms and cancer Proceedings International union against cancer and Netherlands cancer institute Edited by P. Emmelot and O. Muhlbock Elsevier,Amsterdam, 1964 Organized to commemorate the 50th anniversary of the Netherlands cancer institute
Path 30.2459

AMSTERDAM 1963 Degeneration patterns in the nervous system :a workshop ... held as part of the first International summer school of brain research... Edited by M. Singer and J.P. Schade Progress in brain research, 14 Elsevier,Amsterdam,1965
An 32.4234

AMSTERDAM 1963 Mechanisms of neural regeneration :a workshop...held as part of the first International summer school of brain research... Edited by M. Singer and J.P. Schade Progress in brain research, 13 Elsevier,Amsterdam,1964
An 32.4233

AMSTERDAM 1963 Physiology of spinal neurons :a workshop ...held as part of the first International summer school of brain research... Edited by J.C. Eccles and J.P. Schade Progress in brain research, 12 Elsevier,Amsterdam,1964
An 32.4232

AMSTERDAM 1963 Structure and function of the epiphysis cerebri :an international round-table conference Proceedings Edited by J. Ariens Kappers and J.P. Schade Progress in brain research, 10 Elsevier,Amsterdam,1964
An 32.4230

AMSTERDAM 1963 Water and electrolyte metabolism :symposium 2nd Proceedings Edited by J.de Graefe and B. Leijnse West-european symposia on clinical chemistry, 3 Elsevier,Amsterdam,1964
Inv Med 37.0090

AMSTERDAM 1963 Antwerp 1963 May 31-Jun 1 Deltaic and shallow marine deposits International congress of sedimentology 6th Proceedings Edited by L.M.J.U.van Straaten Developments in sedimentology,1 Elsevier, Amsterdam,1964
Geog 13.0304

AMSTERDAM 1964 Chemical reaction engineering European symposium on chemical reaction engineering 3rd Proceedings Instituut van ingenieurs.Afdeling voor chemische techniek and Nederlandse chemische vereniging.Sectie voor chemische technologie Held under the auspices of the European federation of chemical engineering vi,326p Pergamon press,Oxford,1965 Supplement to "Chemical engineering science".
Chem E 24.1269

AMSTERDAM 1964 Sensory mechanisms :a workshop...held during the second International summer school of brain research.. Edited by Y. Zotterman Progress in brain research, 23 Elsevier,Amsterdam,1967
An 32.4244

AMSTERDAM 1965 Flavins and flavoproteins : a symposium Proceedings International union of biochemistry Edited by E.C. Slater organized under the auspices of the Nederlandsche akademie van wetenschappen B.B. A. library, 8 illus Elsevier,Amsterdam, 1966
Bioch 33.1099

AMSTERDAM 1967 Brain barrier systems :a conference Edited by A. Lajtha and D.H. Ford Progress in brain research, 29 Elsevier, Amsterdam,1968
An 32.5233

AMSTERDAM 1967 Logic,methodology and philosophy of science International congress for logic,methodology and philosophy of science Proceedings Nederlandse vereniging voor logica en wijsbegeerte der exacte wetenschappen International union for logic, methodology and philosophy of science.Division of logic,methodology and philosophy of science Edited by B.van Rootselaar and J.F. Staal xiii,554p North-Holland,Amsterdam,1968
WSM 43.0830

AMSTERDAM 1969 International congress on project planning by network analysis 2nd Proceedings North-Holland,Amsterdam,1969
Eng 41.1240

AMSTERDAM 1970 Manifolds - Amsterdam 1970 NUFFIC summer school on manifolds Proceedings Edited by N.H. Kuiper Lecture notes in mathematics, 197 230p 25cm Springer,Berlin,1971
P Math 2.4116

AMSTERDAM 1971 Elementary particles : conference Proceedings Rijksuniversiteit te Amsterdam Edited by A.G. Tenner and H.J.G. Vettman illus viii,472p 26cm North-Holland,Amsterdam,1972
A Math 4.1767

AMSTERDAM 1971 International conference on the physics of electronic and atomic collisions 7th Abstracts of papers Vol 1-2 Edited by L.M. Branscomb and others 23cm 2 vols North-Holland,Amsterdam,1971
Chem 18.2959

AMSTERDAM 1971 Physics of electronics and atomic collisions International conference on the physics of electronic and atomic collisions 7th Invited papers and progress reports Edited by T.R. Govers and F.J. De Heer xi,496p 23cm North Holland,London, 1972
Chem 18.2958

AMSTERDAM 1972 Advanced course on programming languages and data structures A Comparative study of some higher programming languages Series of lectures By W.L.van der Poel varp Van der Poel,Delft,1972
Math L 5.3872

AMTLICHER BERICHT UBER DIE EINUNDZWANZIGSTE VERSAMMLUNG DEUTSCHER NATURFORSCHER UND AERZTE 21st Proceedings Gratz 1843 Sep Deutsche naturforscher und aerzte Edited by L. Langer and A. Schrotter A.Leykam,Gratz, 1844
Hunt B 22.0463

AN :including papers and discussions presented at Conference on brittle fracture mechanics held.. at Massachusetts institute of technology Evaluation of current knowledge of the mechanics of brittle fracture Cambridge, Mass. 1953 Oct 13-16 By D.C. Drucker Sponsored by the National research council. Committee on ship structure design SSC 69 N.A.S.-N.R.C.Division of engineering and industrial research,Washington,D.C.,1954
Eng 41.3743

ANACAPRI 1966 The Fine structure of the solar atmosphere;extended abstracts :a colloquium Edited by K.O. Kiepenheuer Sponsored by the Notgemeinschaft der Deutschen wissenschaft Notgemeinschaft der deutschen wissenschaft.Bericht, 12 149p Franz Steiner,Wiesbaden,1966 German title:Die Feinstruktur der sonenatmosphare
Obs 6.0945

ANACAPRI 1968 Mass motions in solar flares and related phenomena Nobel symposium 9th Proceedings Nobel foundation Edited by Yngve Ohman 245p 24cm Wiley,New York, 1968
Cav 7.2971

ANAEROBIA Role des anaerobies dans la nature Congres international des microbiologistes de langue francaise Brussels 1949 May 23-27 International union of biological sciences International union of biological sciences. Series B.Colloques, 7 Secreteriat general de l'U.I.S.B.,Paris,1949
Bal 39.3919

ANAESTHESIOLOGY European congress of anaesthesiology :post-graduate courses 1st Abstracts of papers Vienna 1962 Sep 3-9 World federation of societies of anaesthesiologists English edition Wiener medizinischen akademie,Vienna,c1962
PGMS 29.0030

ANAESTHESIOLOGY The European congress of anaesthesiology 2nd Proceedings Copenhagen 1966 Aug 8-13 1-3 World federation of societies of anaesthesiologists Acta anaesthesiologica Scandinavica. Supplementum, 23-25 3 vols Universitetsforlaget,Aarhus,1966
PGMS 29.0031

ANALOGUE COMPUTATION International analogue computation meeting 1st,3rd,4th Proceedings Brussels 1955- Brussels,1956-
Eng 41.2242

ANALYSIS IN FUNCTION SPACE Conference on the theory and applications of analysis in function space Proceedings Dedham,Mass. 1963 Jun 9-13 Edited by William T. Martin and Irving Sega vi,218p 24cm MIT press, Cambridge,Mass.,1964
Math S 3.1659

ANALYSIS IN FUNCTION SPACE Conference on the theory and applications of analysis in function space Proceedings Dedham,Mass. 1963 Jun 9-13 Massachusetts institute of technology Edited by William T. Martin and Irving Segal vi,218p 24cm MIT press, Cambridge,Mass.,1964
Cav 7.3154

ANALYSIS IN FUNCTION SPACE Conference on the theory and applications of analysis in function space proceedings Dedham,Mass. 1963 Jun 9-13 Massachusetts institute of technology Edited by William Ted Martin and Irving Segal Financially supported by United States.Air force.Office of scientific research vi,218p 24cm M.I.T.press,Cambridge,Mass., 1964
P. Math 2.1181

ANALYSIS OF CARCINOGENIC AIR POLLUTANTS : symposium Proceedings Cincinnati,Ohio 1961 Aug 29-31 National cancer institute Edited by Eugene Sawicki and Kenneth Cassel National cancer institute.Monograph, 9 illus U.S.Department of health,education and welfare,Bethesda,Md.,1962 Sponsored by the Laboratory of engineering and physical sciences of the Division of air pollution,U.S. Department of health education,and welfare, Public health service.Robert A.Taft sanitary engineering center
Bioch 33.0980

ANALYSIS SITUS By Oswald Veblen 2nd edition revised American mmathematical society.Colloquium publications, 5,pt 2 x, 194p 23cm American mathematical society, New York,1931 First edition entitled 'The Cambridge colloquium,1916.Pt 2:analysis situs'
P Math 2.2836

ANALYSIS SITUS The Cambridge colloquium 1916 pt. 2: analysis situs By Oswald Veblen American mathematical society. Colloquium lectures, 5,pt2 vii,150p 24cm American mathematical society,New York,1922 Second edition called 'Analysis situs'
P. Math 2.0358

ANALYTIC CELL CULTURE Syverton memorial symposium Proceedings Detroit 1961 Jun 6-7 National cancer institute Edited by Robert E. Stevenson Sponsored by Tissue culture association and Cell culture collection committee of the viruses and cancer board,National cancer institute National cancer institute.Monograph, 7 U.S. Department of health,education and welfare, Bethesda,Md.,1962
Bioch 33.0979

ANALYTIC FUNCTIONS Seminars on analytic functions conference typescript Princeton, N.J. 1957 Sep 2-14 Vols. 1-2 Institute for advanced study Edited by Marston Morse and others Under contract with the United States.Air force.Office of scientific research Bibliog. 25cm 2 vols Princeton,N.J.,1957
P. Math 2.1583

ANALYTIC FUNCTIONS Sovremennye problemy teorii analiticheskikh funktsii Mezhdunarodnaya konferentsiya po teorii analiticheskikh funktsii Erevan 1965 Sep 6-14 Edited by M.A. Lavrentev and others 361p 27cm Moscow,1966
P. Math 2.2408

ANALYTIC FUNCTIONS conference papers Princeton,N.J. 1957 Sep 2-14 By Rolf Nevanlinna and others Institute for advanced study With the United States.Air force. Office of scientific research Princeton mathematical series, 24 vii,197p 24cm Princeton university press,Princeton,N.J.,1960
P. Math 2.1492

ANALYTIC THEORY OF DIFFERENTIEL EQUATIONS :a conference Proceedings Kalamazoo,Mich. 1970 Apr 30-May 2 Edited by P.F. Hsieh and A. W.J. Stoddart Held at Western Michigan university Lecture notes in mathematics, 183 bibliog. vi,225p 25cm Springer,Berlin, 1971
P Math 2.3876

ANALYTICAL CHEMISTRY IN NUCLEAR REACTOR TECHNOLOGY:PARTICLE-SIZE ANALYSIS Conference on nuclear reactor technology 2nd Papers Gatlinburg,Tenn. 1958 Sep 29-Oct 1 United States atomic energy commission v,101p Oakridge national laboratory,Oakridge, Tenn.,1959
Chem E 24.1436

ANALYTICAL CONTROL OF RADIOPHARMACEUTICS :a panel Proceedings Vienna 1969 Organized by the International atomic energy agency International atomic energy agency. Panel proceedings series International atomic energy agency,Vienna,1970
Radioth 35.1385

ANATOMICAL SOCIETY OF GREAT BRITAIN AND IRELAND Electron microscopy in anatomy A Symposium on the ultrastructure of cells Proceedings London 1959 Apr 16-17 Edited by J.D. Boyd and others Edward Arnold,London, 1961
Bioch 33.2297

ANATOMICAL SOCIETY OF GREAT BRITAIN AND IRELAND Electron microscopy in anatomy :a symposium Proceedings London 1959 Apr 16-17 Arnold, London,1961
Phys 20.1491

ANATOMICAL SOCIETY OF GREAT BRITAIN AND IRELAND
Electron microscopy in anatomy :a symposium papers London 1959 Apr 16-17 Illus. Arnold,London,1961
VA 19.0356

ANATOMY Congres federatif international d'anatomie 1e Geneva 1905 Aug 6-10 bulletin synthetique Kundig,Geneva,1907
An 32.5266

ANATOMY OF MEMORY Conference on learning, remembering and forgetting 1st Proceedings 1963 Edited by D.P. Kimble Under the auspices of the American institute of biological sciences Science and behaviour books,Palo Alto,Calif.,1965
An 32.4409

AND Photoreception Papers New York 1958 Jan 31-Feb 1 Edited by Jerome J. Wolken New York academy of sciences.Annals, 74,art.2 New York,1958
Phys 20.2225

AND Structure and properties of ultrahigh-strength steels :a symposium Cleveland,Ohio 1963 Oct 22 A.S.T.M.Special technical publication, 370 American society for testing and materials,Philadelphia,Pa.,1965
Met 25.2251

ANDERSON HOSPITAL AND TUMOR INSTITUTE
Biology of normal and atypical pigment cell growth conference 4th Proceedings Houston,Tex. 1957 Nov 14-16 Edited by Myron Gordon Academic press,New York,1959
Gen 34.0601

ANDERSON HOSPITAL AND TUMOR INSTITUTE Breast cancer:early and late Annual clinical conference on cancer 13th Papers Houston, Tex. 1968 Yearbook,Chicago,Ill.,1970
Radioth 35.1914

ANDERSON HOSPITAL AND TUMOR INSTITUTE
Carcinogensis:a broad critique Annual symposium on fundamental cancer research 1966,20th Papers Williams and Wilkins, Baltimore,Md.,1967
Radioth 35.1964

ANDERSON HOSPITAL AND TUMOR INSTITUTE
Carcinoma of the uterine cervix,endometrium and ovary Annual clinical conference on cancer 5th Papers Houston,Tex. 1960 Year book medical,Chicago,Ill.,1962
Radioth 35.0828

ANDERSON HOSPITAL AND TUMOR INSTITUTE
Genetics and cancer Annual symposium on fundamental cancer research 13th Papers 1959 Edited by Russell W. Cumley and others Owen,London,1961
Gen 34.2232

ANDERSON HOSPITAL AND TUMOR INSTITUTE
Leukemia-Lymphoma Annual clinical conference on cancer 14th Papers Houston 1969 Edited by C.C. Shullenberger and R.W. Cumley Yearbook medical publishers,Chicago,1971
PGMS 29.0678

ANDERSON HOSPITAL AND TUMOR INSTITUTE
Molecular basis of neoplasia Annual symposium on fundamental cancer research 15th Papers Houston,Tex. 1961 University of Texas press,Austin,Tex.,1961
Radioth 35.1942

ANDERSON HOSPITAL AND TUMOR INSTITUTE
Pigment cell biology The Conference on biology of normal and atypical pigment cell growth 4th Proceedings Houston,Tex. 1957 Nov 14-16 Edited by Myron Gordon Academic press,New York,1959 Sponsored jointly by the New York zoological society,the M.D.Anderson Hospital and tumor institute and the Damon Runyon memorial fund for cancer research, inc.
Bioch 33.0937

ANDERSON HOSPITAL AND TUMOR INSTITUTE
Proliferation and spread of neoplastic cells Annual symposium on fundamental cancer research 21st 1967 Williams and Wilkins, Baltimore,Md.,1968
Radioth 35.0550

ANDERSON HOSPITAL AND TUMOR INSTITUTE
Radiation biology and cancer Symposium on fundamental cancer research 12th Papers Houston,Tex. 1958 University of Texas press,Austin,Tex.,1958
Phys 20.0988

ANDERSON HOSPITAL AND TUMOR INSTITUTE
Viruses,nucleic acids and cancer Annual symposium on fundamental cancer research,1963 17th Papers Houston,Tex. 1963 Williams and Wilkins,Baltimore,Md.,1963
Radioth 35.0822

The ANDESITE CONFERENCE Proceedings University of Oregon.Department of geology and mineral industries Edited by A.R. McBirney Oregon.Department of geology and mineral industries.Bulletin, 65 Portland,Ore.,1969 Photocopy
Min 10.1477

ANDROGENS IN NORMAL AND PATHOLOGICAL CONDITIONS
Steroid hormones :a symposium 2nd Proceedings Ghent 1965 Jun 17-19 Rijkauniversiteit te Gent Societe belge d'endrocinologie Edited by A. Vermeulen and D. Exley Excerpta medica.International congress series, 101 Excerpta medica, Amsterdam,1966
Inv Med 37.0276

ANESTHESIOLOGY Conference on neurophysiology in relation to anesthesiology Seattle 1966 May 13-14 Anesthesiology,28,no 1 American society of anesthesiologists,Lancaster,Pa., 1967
Pha 16.0052

ANGIOTENSIN SYSTEMS AND EXPERIMENTAL RENAL DISEASES Annual conference on the kidney 14th Proceedings National kidney disease foundation Edited by Jack Metcoff Churchill,London,1963 By 24 authors and other participants
Inv Med 37.0159

ANGLO-AMERICAN AERONAUTICAL CONFERENCE 6th Papers Folkestone 1957 Sep 9-12 Royal aeronautical society Institute of the aeronautical sciences Edited by Joan Bradbrooke and U.A. Libby Royal aeronautical society,London,1959
Eng 41.6962

ANGLO-AMERICAN CONFERENCE ON THE MECHANIZATION OF LIBRARY SERVICES Proceedings Brasenose conference on the automation of libraries Oxford 1966 Jun 30-Jul 3 Edited by John Harrison and Peter Laslett Sponsored by the Old Dominion foundation of New York Mansell, Chicago,Ill.,1967
Eng 41.7672

ANGLO-GERMAN MEDICAL SOCIETY Symposium on drug action and modern therapy London 1961 Sep 28 Anglo-German medical review, 1,no.4 Schattauer,Stuttgart,1962 Text in English and German
Gen 34.1973

ANGLO-ROMANIAN CONFERENCE ON MATHEMATICS IN THE ARCHAEOLOGICAL AND HISTORICAL SCIENCES Proceedings Royal society of London Academia republicii socialiste romania Edited by F.R. Hodson and others viii,565p 24cm Edinburgh university press,Edinburgh, 1971
Math S 3.1826

ANGLO-POLISH SEMINAR 1st Proceedings Problems of applied geography Nieborow 1959 Sep 15-18 Polska akademia nauk.Instytut geografii Geographical studies,25 Panstwowe wydawnictwo naukowe,Warsaw,1961
Geog 13.5617

ANGLO-POLISH SEMINAR 2nd Proceedings Problems of applied geography,2 Keele 1962 Sep 9-20 Polska akademia nauk.Instytut geografii,and,Institute of British geographers Geographia polonica,3 Panstwowe wydawnictwo naukowe,Warsaw,1964
Geog 13.5616

ANIMAL BEHAVIOUR Physiological mechanisms in animal behaviour Cambridge 1949 Jul Society for experimental biology Society for experimental biology.Symposia, 4 Cambridge university press,Cambridge,1950
Phys 20.0945

ANIMAL BEHAVIOUR Physiological mechanisms in animal behaviour Papers Cambridge 1949 Jul Society for experimental biology Society for experimental biology.Symposia, 4 Cambridge university press,Cambridge,1950
Psy 31.0562

ANIMAL BEHAVIOUR AND DRUG ACTION London 1963 Mar 25-28 Ciba foundation,and, Biological council.Co-ordinating committee for symposia on drug action Edited by Hannah Steinberg and others Churchill,London,1964
Pha 16.0030

ANIMAL BEHAVIOUR AND DRUG ACTION :annual meeting Proceedings London 1963 Mar 25-28 Ciba foundation and Biological council.Co-ordinating committee for symposia on drug action Edited by A.V.S.de Reuck and Julie Knight Biological council.Co-ordinating committee for symposia on drug action.Annual meeting, 10 Ciba foundation symposia Churchill,London,1964
Psy 31.0526

ANIMAL CELLS Metabolic control mechanisms in animal cells :a symposium Proceedings Boston,Mass. 1963 May 28-30 National cancer institute Edited by William J. Rutter Sponsored by the Tissue culture association and National cancer institute National cancer institute.Monograph, 13 illus National cancer institute,Bethesda,Md.,1964
Bioch 33.0984

ANIMAL POPULATIONS The Exploitation of natural animal populations :a symposium Durham 1960 Mar 28-31 British ecological society Edited by E.D. Le Cren and M.W. Holdgate British ecological society.Symposia, 2 Oxford,1962
Bal 39.1665

ANIMAL SYMPOSIUM ON FUNDAMENTAL CANCER RESEARCH 12th Papers Radiation biology and cancer 1958 Owen,London,1960
Radioth 35.1101

ANIMAL VIRUS BIOLOGY Basic mechanisms in animal virus biology Papers Cold Spring Harbor 1962 Jun 7-13 Cold Spring Harbor. Biological laboratory Cold Spring Harbor symposia on quantitative biology, 27 Long Island biological association.Biological laboratory,Cold Spring Harbor,L.I.,N.Y.,1962
Path 30.2559

ANIMAL VIRUS BIOLOGY Basic mechanisms in animal virus biology :a symposium Papers Cold Spring Harbor 1962 Cold Spring Harbor biological laboratory Cold Spring Harbor symposia on quantitative biology, 27 Long Island biological association,Cold Spring Harbor,1962
Bioch 33.1283

ANN ARBOR 1940 Lectures in topology : conference University of Michigan Edited by Raymond L. Wilder and William L. Ayres vii,316p 23cm University of Michigan press, Ann Arbor, Mich.,1941
P. Math 2.0288

ANN ARBOR 1952 Heat transfer :a symposium University of Michigan Edited by B.A. Uhlendorf University of Michigan.Engineering research institute special publications: symposia vii,286p Engineering research institute,Michigan,1953
Chem E 24.0544

ANN ARBOR 1954 Utilization of heat resistant alloys a symposium in honor of Albert Easton White American society for metals American society for metals,Cleveland, Ohio,1921
Met 25.0998

ANN ARBOR 1957 The Midwestern conference on fluid mechanics 5th proceedings Edited by R.C.F. Bartels Under the auspices of the University of Michigan.Engineering research institute University of Michigan. Engineering research institute.Publication University of Michigan press,Ann Arbor,1957
A Math 4.0516

ANN ARBOR 1959 Gas chromatography : international symposium 2nd Proceedings Edited by Henry J. Noebels and others Held under the auspices of the Instrument society of America xvi,463p Academic press,New York;London,1961
Chem E 24.0851

ANN ARBOR 1962 Fundamental topics in relativistic fluid dynamics and magnetohydrodynamics :a symposium proceedings Edited by Robert Wasserman and Charles P. Wells Academic press,New York, London,1963
A Math 4.0528

ANN ARBOR 1967 International conference on leukemia lymphoma Proceedings University of Michigan Edited by Chris J.D. Zarafonetis Lea and Febiger;Kimpton,Philadelphia;London, 1968
Med 36.0376

ANN ARBOR, MICH. 1953 Lectures on functions of a complex variable :conference proceedings University of Michigan Edited by Wilfred Kaplan and others Bibliog 435p 24cm University of Michigan,Ann Arbor,Mich., 1955
P. Math 2.1533

ANN ARBOR,MICH 1961 Developments in mechanics Midwestern mechanics conference 7th proceedings Vol 1 Edited by J.E. Lay and L.E. Malvern North Holland, Amsterdam,1961
A Math 4.0313

ANN ARBOR,MICH 1966 Electron and laser beam symposium 8th Proceedings University of Michigan Institute of electrical and electronics engineers Edited by G.I. Haddad University of Michigan,Ann Arbor,Mich.,1966 8th in the series of annual symposia on Electron beam technology initiated in 1959 by Alloyd and held under their sponsorship for seven years.The titles of the symposia vary,therefore those catalogued have been grouped together under the keyword"Electron beams"
Eng 41.5618

ANN ARBOR,MICH. Jun 14-18 United States national congress of applied mechanics 2nd Proceedings Edited by Paul M. Naghdi xx, 821p American society of mechanical engineers,New York,1955
Chem E 24.0231

ANN ARBOR,MICH. 1927 The National symposium on colloid chemistry 5th Papers presented University of Michigan Edited by Harry Boyer Weiser Colloid symposium monograph, 5 Chemical catalog company,New York,1928
Bioch 33.1473

ANN ARBOR,MICH. 1947 Psycholgical diagnosis and counseling of the adult blind : selected papers from the proceedings of the University of Michigan conference for the blind United States.Federal security agency. Office of vocational rehabilitation and Michigan.Department of social welfare.Division of services for the blind Edited by Wilma Donahue and Donald Dabelstein Co-sponsored by the University of Michigan.Institute for human adjustment American foundation for the blind,New York,1950
Psy 31.2118

ANN ARBOR,MICH. 1949 Elasticity American mathematical society.Symposium in applied mathematics 3rd Proceedings Edited by R. V. Churchill and others Cosponsored by the American society of mechanical engineers. Applied mechanics division McGraw-Hill,New York,1950
Col S 12.0335

ANN ARBOR,MICH. 1953 Michigan symposium on astrophysics Proceedings University of Michigan.Observatory Ann Arbor,1953 Typescript
Obs 6.1978

ANN ARBOR,MICH. 1957 Midwestern conference on fluid mechanics 5th Proceedings Held at University of Michigan University of Michigan press,Ann Arbor,Mich.,1957
Eng 41.6958

ANN ARBOR,MICH. 1957 Midwestern conference on fluid mechanics 5th Proceedings Held at the University of Michigan University of Michigan.Engineering research institute publications viii,388p University of Michigan press,Ann Arbor,Mich.,1957
Chem E 24.0405

ANN ARBOR,MICH. 1959 Modern organization theory :a symposium of the Foundation for research on human behavior Foundation for research on human behavior Edited by Mason Haire Wiley,New York,1959 reprinted 1965
Eng 41.0572

ANN ARBOR,MICH. 1962 Categories of human learning :symposium on the psychology of human learning Papers United States.Office of naval research and University of Michigan Edited by Arthur W. Melton Academic press, New York;London,1964
Psy 31.1003

ANN ARBOR,MICH. 1963 Performance appraisals :effects on employees and their performance.A seminar Foundation for research on human behavior Foundation for research on human behavior,Ann Arbor,Mich., 1964
Eng 41.0784

ANN ARBOR,MICH. 1968 Proof techniques in graph theory Ann Arbor graph theory conference 2nd Proceedings Edited by Frank Harary Bibliog.,Illus. xv,330p 24cm Academic press,New York;London,1969
P Math 2.3349

ANN ARBOR,MICH. 1968 Proof techniques in graph theory Ann Arbor graph theory conference 2nd Proceedings University of Michigan Edited by Frank Harary xv,330p 24cm Academic press,New York;London,1969
Math S 3.1658

ANN ARBOR GRAPH THEORY CONFERENCE 2nd Proceedings Proof techniques in graph theory Ann Arbor,Mich. 1968 Feb Edited by Frank Harary Bibliog.,Illus. xv,330p 24cm Academic press,New York;London,1969
P Math 2.3349

ANN ARBOR GRAPH THEORY CONFERENCE 2nd Proceedings Proof techniques in graph theory Ann Arbor,Mich. 1968 Feb University of Michigan Edited by Frank Harary xv,330p 24cm Academic press,New York;London,1969
Math S 3.1658

ANNEALING PROCESSES Recent developments in annealing London 1963 May 2 Iron and steel institute Iron and steel institute. Special report, 79 Iron and steel institute, London,1963
Met 25.0418

ANNEE POLAIRE INTERNATIONAL,1931-32 Rapport de la commission internationale de l'annee polaire 1932-33 Copenhagen 1933 May Vol 3: Compte-rendu des travaux de la commission Octobre 1931-Mai 1933 Organisation meteorologique internationale Ijdo,Leyden, 1934
Sco 14.0491

ANNEE POLAIRE INTERNATIONALE Rapport de la commission internationale de l'annee polaire 1932-33 Innsbruck 1931 Sep Vol 2: compte-rendu des travaux de la commission pendant sa deuxieme annee Organisation meteorologique internationale illus,diagrs 188p Ijdo, Leyden,1932
Sco 14.0492

ANNEE POLAIRE INTERNATIONALE Rapport de la commission internationale de l'annee polaire 1932-33 Leningrad 1930 August Vol 1: compte-rendu des travaux de la commission pendant sa premiere annee de travail... Organization meteorologique internationale illus,diagrms 152p Ijdo,Leyden,1933
Sco 14.0493

ANNUAL ARCTIC PLANNING SESSION 1st Proceedings Boston,Mass. 1958 Nov Air force Cambridge research center.Geophysics research directorate Edited by J.H. Hartshorn Air force Cambridge research center.G.R.D.research notes,15 AFCRC-TN-59-256,AD 212265 illus viii,65p A.F.C.R.C., Bedford,Mass.,1959
Sco 14.7404

ANNUAL ARCTIC PLANNING SESSION 2nd Proceedings Bedford,Mass. 1959 Oct Air force Cambridge research center.Geophysics research directorate Edited by Vivian C. Bushnell Air force Cambridge research center. G.R.D.research notes,29 AFCRC-TN-59-661 illus viii,172p A.F.C.R.C.,Bedford,Mass., 1959
Sco 14.7405

ANNUAL ARCTIC PLANNING SESSION 3rd Proceedings Bedford,Mass. 1960 Nov Air force Cambridge research laboratories Edited by George P. Rigsby and Vivian C. Bushnell Air force Cambridge research laboratories.G.R.D.research notes,55 AFCRL 436 illus viii,148p A.F.C.R.L.,Bedford, Mass.,1961
Sco 14.7406

ANNUAL ASTRONAUTICS SYMPOSIUM 1st Vistas in astronautics United States.Air force. Office of scientific research Edited by Morton Alperin and others Co-sponsored with General dynamic corporation.Convair division Pergamon press,London,1958
Eng 41.6880

ANNUAL CLINICAL CONFERENCE ON CANCER 5th Papers Carcinoma of the uterine cervix, endometrium and ovary Houston,Tex. 1960 Anderson hospital and tumor institute Year book medical,Chicago,Ill.,1962
Radioth 35.0828

ANNUAL CLINICAL CONFERENCE ON CANCER 13th Papers Breast cancer:early and late Houston,Tex. 1968 Anderson hospital and tumor institute Yearbook,Chicago,Ill.,1970
Radioth 35.1914

ANNUAL CLINICAL CONFERENCE ON CANCER 14th Papers Leukemia-Lymphoma Houston 1969 Anderson hospital and tumor institute Edited by C.C. Shullenberger and R.W. Cumley Yearbook medical publishers,Chicago,1971
PGMS 29.0678

ANNUAL COLLOQUIUM ON INFORMATION RETRIEVAL 3rd Information retrieval:a critical view Philadelphia,Pa. 1966 May 12-13 Edited by George Schecter Sponsored by the Association for computing machinery.Special interest group on information retrieval Academic press,New York,1967
Eng 41.7692

ANNUAL CONFERENCE OF THE ASSOCIATION FOR THE STUDY OF MEDICAL EDUCATION Proceedings 2nd Methods of learning and techniques of teaching London 1959 Oct 15-16 Association for the study of medical education Edited by J.R. Ellis Pitman medical publishing company,London,c1960
PGMS 29.0493

ANNUAL CONFERENCE ON APPLICATION OF X-RAY ANALYSIS 18th Proceedings Denver, Colo. 1969 Aug 6-8 Denver research institute.Metallurgy division Edited by Burton L. Henke Advances in X-ray analysis, 13 Plenum press,New York,1970
Met 25.2609

ANNUAL CONFERENCE ON APPLICATIONS OF X-RAY ANALYSIS 7th- Proceedings Denver, Colo. 1958- Denver research institute Edited by William M. Mueller and others Advances in X-ray analysis, 2- Plenum press,New York,1960-
Met 25.2762

ANNUAL CONFERENCE ON APPLICATIONS OF X-RAY ANALYSIS 17th Proceedings Estes Park, Colo. 1968 Aug 21-23 Denver research institute.Metallurgy division Edited by Charles S. Barrett and others Advances in X-ray analysis, 12 Plenum press,New York,1969
Met 25.2824

ANNUAL CONFERENCE ON APPLICATIONS OF X-RAY ANALYSIS 12th 1963 Aug 7-9 Edited by M. Mueller and others Sponsored by the Denver research institute Advances in x-ray analysis, 7 Plenum press,New York,1964
Met 25.1445

ANNUAL CONFERENCE ON APPLICATIONS OF X-RAY ANALYSIS 16TH Advances in X-ray analysis 11 proceedings of the 16th annual conference on applications of X-ray analysis Denver, Colo. 1967 Aug 9-11 Denver research institute Edited by John B. Newkirk Plenum press,New York,1968
Met 25.2437

The ANNUAL CONFERENCE ON PROTEIN METABOLISM 9th six lectures Some conjugated proteins; a symposium New Brunswick,N.J. 1953 Jan 30-31 Rutgers university.Bureau of biological research Edited by William H. Cole Rutgers university press,New Brunswick,N.J.,1953
Bioch 33.0543

ANNUAL CONFERENCE ON THE KIDNEY Proceedings Renal metabolism and epidemiology of some renal diseases Princeton 1961 Oct 12-13 National kidney disease foundation Edited by Jack Metcoff National kidney disease foundation,New York,1964
Inv Med 37.0158

ANNUAL CONFERENCE ON THE KIDNEY 13th Proceedings Hereditary,developmental and immunologic aspects of kidney disease Princeton 1961 Oct 12-13 National kidney disease foundation Edited by Jack Metcoff Northwestern university press,Evanston,1962
Inv Med 37.0157

ANNUAL CONFERENCE ON THE KIDNEY 13th Proceedings Hereditary,developmental and immunologic aspects of kidney disease Princeton,N.J. 1961 Oct 12-13 National kidney disease foundation Edited by Jack Metcoff Northwestern university press, Evanston,Ill.,1962
An 32.2075

ANNUAL CONFERENCE ON THE KIDNEY 14th Proceedings Angiotensin systems and experimental renal diseases National kidney disease foundation Edited by Jack Metcoff Churchill,London,1963 By 24 authors and other participants
Inv Med 37.0159

ANNUAL CONFERENCE ON THE KIDNEY 16th Proceedings Homotransplantation:kidney and other tissues 1964 National kidney disease foundation Edited by Jack Metcoff National kidney disease foundation,New York, 1966
Inv Med 37.0160

ANNUAL GEORGE H.HUDSON SYMPOSIUM 2nd Papers Origin of anorthosite and related topics Plattsburgh,N.Y. 1966 New York state museum and science service.Geological survey State university of New York.College at Plattsburgh Edited by Yngvar W. Isachsen New York state museum and science service. Memoir, 18 New York state museum and science service,Albany,N.Y.,1968
Min 10.1428

ANNUAL MEETING OF JAPAN ENDOCRINOLOGICAL SOCIETY 43rd Proceedings Osaka 1970 Mar 10-12 Japan endocrinological society Kyoto, 1970
PGMS 29.0588

ANNUAL MEETING OF THE ASSOCIATION FOR ACADEMIC SURGERY 2nd Proceedings St.Louis, Miss. 1968 Nov Association for academic surgery Edited by G.D. Zuidema and D.B. Skinner Current topics in surgical research, 1 Academic press,New York;London,1969
Surg 23.0069

ANNUAL RESEARCH CONFERENCE OF THE BUREAU OF BIOLOGICAL RESEARCH, 17 New developments in tissue culture :annual research conference New Brunswick,N.J. 1961 Rutgers university. Bureau of biological research Edited by James W. Green Rutgers university press,New Brunswick,N.J.,1961
An 32.3177

ANNUAL REUNION OF THE SOCIETY OF CHEMICAL PHYSICISTS 11th Proceedings Deoxyribonucleic acid,structure,synthesis and function Col de Voza 1961 Jun 26 - Jul 1 Society of chemical physicists Pergamon press,Oxford,1961
Radioth 35.0111

ANNUAL ROCHESTER CONFERENCE ON HIGH ENERGY NUCLEAR PHYSICS 3rd Proceedings High energy nuclear physics Rochester,N.Y. 1952 Dec 18-20 Edited by H.Pierre Noyes and others Sponsored by the University of Rochester 110p Interscience,New York,c. 1953 Co-sponsored by the National science foundation
Cav 7.1986

ANNUAL ROCHESTER CONFERENCE ON HIGH ENERGY NUCLEAR PHYSICS 4th Proceedings High energy nuclear physics Rochester,N.Y. 1954 Jan 25-27 Edited by H.Pierre Noyes and others Sponsored by the University of Rochester 159p Rochester,N.Y.,1954 Co-sponsored by the National science foundation. Typescript
Cav 7.1987

ANNUAL ROCHESTER CONFERENCE ON HIGH ENERGY NUCLEAR PHYSICS 5th Proceedings High energy nuclear physics Rochester,N.Y. 1955 Jan 31-Feb 2 University of Rochester Edited by H.Pierre Noyes and others Sponsored by the International union of pure and applied physics 196p Interscience,New York,1955 Co-sponsored by the National science foundation in cooperation with the Atomic energy commission and the United States office of naval research
Cav 7.1988

ANNUAL ROCHESTER CONFERENCE ON HIGH ENERGY NUCLEAR PHYSICS 6th Proceedings High energy nuclear physics Rochester,N.Y. 1956 Apr 3-7 Edited by J. Ballam and S.B. Trieman 362p Interscience,New York,1956
Cav 7.1989

ANNUAL SAN FRANCISCO CANCER SYMPOSIUM 1st Proceedings Hyperbolic oxygen and radiation therapy of cancer San Francisco, Calif. 1965 Nov 12-13 Frontiers of radiation therapy and oncology, 1 Karger, Basel;New York,1968
Radioth 35.1970

ANNUAL SAN FRANCISCO CANCER SYMPOSIUM 2nd Proceedings Electron beam therapy San Francisco,Calif. 1966 Frontiers of radiation therapy and oncology, 2 Karger, Basel;New York,1968
Radioth 35.1208

ANNUAL SYMPOSIUM ON CINEFLUOROGRAPHY 1st Proceedings Cinefluorography Rochester, N.Y. 1958 Nov 14-15 Edited by George H.S. Ramsey and others Sponsored by University of Rochester.Department of radiology Thomas, Springfield,Ill.,1960
VA 19.0381

ANNUAL SYMPOSIUM ON COGNITION 1st Papers Problem solving:research,method and theory Pittsburgh 1965 Apr 15-16 Carnegie institute of technology Edited by Benjamin Kleinmutz Wiley,New York,1966
Psy 31.1911

ANNUAL SYMPOSIUM ON FUNDAMENTAL CANCER RESEARCH 21st Proliferation and spread of neoplastic cells 1967 Anderson hospital and tumor institute Williams and Wilkins, Baltimore,Md.,1968
Radioth 35.0550

ANNUAL SYMPOSIUM ON FUNDAMENTAL CANCER RESEARCH 13th Papers Genetics and cancer 1959 Anderson hospital and tumor institute Edited by Russell W. Cumley and others Owen, London,1961
Gen 34.2232

ANNUAL SYMPOSIUM ON FUNDAMENTAL CANCER RESEARCH 15th Papers Molecular basis of neoplasia Houston,Tex. 1961 Anderson hospital and tumor institute University of Texas press,Austin,Tex.,1961
Radioth 35.1942

ANNUAL SYMPOSIUM ON FUNDAMENTAL CANCER RESEARCH 1966,20th Papers Carcinogensis:a broad critique Anderson hospital and tumor institute Williams and Wilkins,Baltimore,Md., 1967
Radioth 35.1964

ANNUAL SYMPOSIUM ON FUNDAMENTAL CANCER RESEARCH, 1963 17th Papers Viruses,nucleic acids and cancer Houston,Tex. 1963 Anderson hospital and tumor institute Williams and Wilkins,Baltimore,Md.,1963
Radioth 35.0822

ANNUAL SYMPOSIUM ON FUNDAMENTAL CANCER RESEARCH, 1964 18th Cellular radiation biology Anderson hospital and tumor institute Williams and Wilkins,Baltimore,Md.,1965
Radioth 35.1742

ANNUAL SYMPOSIUM ON TRACER METHODOLOGY 5th Proceedings Washington,D.C. 1961 Oct 20 New England nuclear corporation Packard instrument company Edited by Seymour Rothschild Advances in tracer methodology, 1 Plenum press,New York,1963 Includes selected papers from the first four annual symposia
Gen 34.0633

ANNUAL TECHNICAL MEETING 16th Proceedings Progress in powder metallurgy Chicago 1960 Apr 25-27 Metal powder industries federation Metal powder industries federation,Chicago, 1960
Met 25.0769

ANNUAL TECHNICAL MEETING 17th Proceedings Progress in powder metallurgy Cleveland 1961 Apr 24-26 Metal powder industries federation Metal powder industries federation,Cleveland,Ohio,1961
Met 25.0770

ANNUAL WORKSHOP ON MICROPROGRAMMING 4th Proceedings Santa Cruz,Calif. 1971 Sep 13-14 Association for computing machinery Association for computing machinery,New York, 1971
Math L 5.3631

ANNUAL WORKSHOP ON MICROPROGRAMMING 5th Preprints Urbana,Ill. 1972 Sep 25-26 Association for computing machinery Institute of electrical and electronics engineers.Computer society Bibliog,diagrs 98p n.p.,1972 Includes 'An annotated bibliography on microprogramming,late 1969-early 1972'
Math L 5.3636

ANODISING ALUMINIUM Conference on anodising aluminium Proceedings Nottingham 1961 Sep 12-14 Aluminium development association and University of Nottingham Aluminium development association,London,1961
Met 25.2047

ANODISING ALUMINIUM Conference on anodising aluminium Proceedings Nottingham 1961 Sep 12-14 University of Nottingham Convened by the Aluminium development association Aluminium development association,London,1962
Eng 41.3602

ANODISING ALUMINIUM Symposium on anodizing aluminium Proceedings Birmingham 1967 Apr 12-13 Aluminium federation and University of Aston in Birmingham Aluminium federation,London,1967
Met 25.2048

ANODIZED ALUMINIUM :a symposium presented during committee week Papers Cleveland 1965 Feb 9 American society for testing materials.Subcommittee IV on electroplating practice A.S.T.M.,Philadelphia,1965
Met 25.2049

ANORTHOSITE Origin of anorthosite and related topics Annual George H.Hudson symposium 2nd Papers Plattsburgh,N.Y. 1966 New York state museum and science service.Geological survey State university of New York.College at Plattsburgh Edited by Yngvar W. Isachsen New York state museum and science service.Memoir, 18 New York state museum and science service,Albany,N.Y.,1968
Min 10.1428

ANTARCTIC BIOLOGY Biologie antarctique : symposium 1er Comptes-rendus Paris 1962 Sep 2-8 Edited by Robert Carrick and others Organized by the Scientific committee on Antarctic research Actualites scientifiques et industrielles,1312 illus, maps 651p 25cm Herman,Paris,1964 Preface by professor P.P.Grasse
Sco 14.6172

ANTARCTIC GEOLOGY 1st :international symposium Proceedings Cape Town 1963 Sep 16-21 Edited by Raymond J. Adie Sponsored by the Scientific committee on Antarctic research illus xx,758p 27cm North-Holland,Amsterdam,1964
Sco 14.6125

ANTARCTIC GEOLOGY 1st :international symposium proceedings Capetown 1963 Sep.16-21 Scientific committee on Antarctic research.,International union of geological sciences Edited by Raymond J. Adie illus. North-Holland,Amsterdam,1964
Geol 8.2509

ANTARCTIC GLACIOLOGY Symposium on Antarctic glaciology International association of scientific hydrology general assembly Papers Helsinki 1960 Jul 25-Aug 6 International association of scientific hydrology. Publication,55 illus 162p 24cm Gentbrugge,1961 Papers in English and French
Sco 14.0130

ANTARCTIC ICE AND WATER MASSES Proceedings Tokyo 1970 Sep 15 Scientific committee on antarctic research Edited by George Deacon illus x,114p 23cm Scientific committee on antarctic research,1971
A Math 4.1746

ANTARCTIC METEOROLOGY :a symposium Proceedings Melbourne 1959 Feb Under the auspices of Australian academy of science Pergamon press,Oxford,1960
Geog 13.0989

ANTARCTIC METEOROLOGY :a symposium Proceedings Melbourne 1959 Feb.18-25 Commonwealth bureau of meteorology and International geophysical year Under the auspices of the Australian academy of science 483p Pergamon press,Oxford;London,1960
Nap 11.0026

ANTARCTIC OCEANOGRAPHY SCAR,SCOR,IAPO,IUBS symposium on Antarctic oceanography Papers Santiago de Chile 1966 Sep 13-16 International council of scientific unions. Scientific committee on Antarctic research International council of scientific unions. Scientific committee on oceanic research International association of physical oceanography International union of biological sciences Held by invitation of the Chilean national committee for antarctic research illus,maps xi,268p 23cm Scott Polar research institute for SCAR,Cambridge, 1968
Sco 14.8300

ANTARCTIC OCEANOGRAPHY Symposium on antarctic oceanography main review papers Santiago,Chile 1966 Sep 13-16 Scientific committee on antarctic research Bibliog., Illus. xi,268p 23cm Scott polar research institute,Cambridge,1966
A Math 4.1501

ANTARCTIC RESEARCH Osnovnye itogi izucheniya Antarktiki za 10 let :doklady vsesoyuznogo soveshchaniya po izucheniyu Antarktiki 1966 god Moscow 1966 Apr 19-20 Mezhduvedomstvennaya kommissiya po izucheniyu Antarktiki map Izdatelstvo 'Nauka',Moscow, 1967
Sco 14.6068

ANTARCTIC RESEARCH Pacific science congress 10th The Matthew Fontaine Maury memorial symposium :papers presented to the tenth Pacific science congress of the Pacific science association 10th Papers Honolulu 1961 Aug 21-Sep 6 Pacific science association,and others Edited by Harry Wexler and others Geophysical monograph,7 American geophysical union.Publication,1036 illus,maps x,228p 25cm American geophysical union,Washington,D.C.,1962
Sco 14.6079

ANTARCTIC RESEARCH Symposium on Antarctic research Wellington,N.Z. 1958 Feb 18-22 New Zealand.National committee for the International geophysical year figures, tables varp Department of scientific and industrial research,Wellington,N.Z.,1958
Sco 14.6067

ANTARCTIC SYMPOSIUM Papers Buenos Aires 1959 Nov 17-25 International union of geodesy and geophysics International union of geodesy and geophysics.Monograph,5 95p Paris,1960
Sco 14.0128

ANTARCTIC SYMPOSIUM Proceedings Antarctica in the International geophysical year Northfield,Minn. 1956 Apr.26-27 International geophysical year Co-sponsored by the American geophysical union American geophysical union.Publication, 462 Geophysical monograph map 133p Washington,D.C.,1956
Nap 11.0020

ANTARCTIC TREATY Antarctic treaty meetings on telecommunications Washington,D.C. 1963 Jun 24-28 varp 37cm Washington,D.C.,1963 Unpublished conference documents
Sco 14.6118

ANTARCTIC TREATY Informe de la segunda reunion consultiva 2nd Proceedings Buenos Aires 1962 Jul 18-28 32p 24cm Buenos Aires,1962
Sco 14.6119

ANTARCTIC TREATY Rapport de la troisieme reunion consultative 3rd Proceedings Brussels 1964 Jun 2-13 39p 21cm Brussels,1964
Sco 14.6117

ANTARCTIC TREATY Report of the first consultative meeting 1st Proceedings Canberra 1961 Jul 10-24 40p 24cm Commonwealth government printer,Canberra,1962
Sco 14.6120

ANTARCTIC TREATY The Conference on Antarctica :conference documents.The Antarctic treaty and related papers Conference documents Washington,D.C. 1959 Oct 15-Dec 1 United States.Department of state. Publication,7060,International organization and conference series,13 78p Washington,D. C.,1960
Sco 14.6121

ANTARCTIC TREATY Tratado antartico informe de la quarte reunion consultiva 4th Proceedings Santiago de Chile 1966 Nov 3-10 plate 86p 20cm Santiago de Chile,n. d.
Sco 14.6116

ANTARCTIC TREATY MEETINGS ON TELECOMMUNICATIONS Washington,D.C. 1963 Jun 24-28 varp 37cm Washington,D.C.,1963 Unpublished conference documents
Sco 14.6118

ANTARCTICA IN THE INTERNATIONAL GEOPHYSICAL YEAR
Antarctic symposium Proceedings Northfield,Minn. 1956 Apr.26-27 International geophysical year Co-sponsored by the American geophysical union American geophysical union.Publication, 462 Geophysical monograph map 133p Washington,D.C.,1956
Nap 11.0020

ANTARCTICA IN THE INTERNATIONAL GEOPHYSICAL YEAR
based on a symposium on the Antarctic American geophysical union and United States national committee for the International geophysical year Co-sponsored by the National science foundation Geophysical monograph, 1 American geophysical union.Publication, 462 illus, maps v,133p 26cm American geophysical union,Washington,D.C.,1956
Sco 14.8201

ANTARTIC BIOLOGY Biologie antarctique : symposium 1er Comptes-rendus Paris 1962 Sep 2-8 Scientific committee for Antarctic research Edited by Robert Carrick and others Hermann,Paris,1964
Bal 39.1669

ANTERIOR PITUITARY HORMONES Bioassay of anterior pituitary and adrenocortical hormones a colloquium proceedings London 1952 Mar 25-27 Ciba foundation Edited by G.E.W. Wolstenholme Ciba foundation colloquia on endocrinology, 5 illus Churchill,London, 1953
Bioch 33.0435

ANTERIOR PITUITARY SECRETION AND HORMONAL INFLUENCES IN WATER METABOLISM :two colloquia London 1951 Jul 9-13 London 1951 Jan 8-10 Ciba foundation Edited by G.E.W. Wolstenholme and Margaret P. Cameron Ciba foundation colloquia on endocrinology, 4 Churhcill,London,1952
Phys 20.1365

ANTERIOR PITUITARY SECRETION AND HORMONAL INFLUENCES IN WATER METABOLISM :two colloquia Proceedings London 1951 Jul 9-13 and London 1951 Jan 8-10 Ciba foundation Edited by G.E.W. Wolstenholme and Margaret P. Cameron Ciba foundation colloquia on endocrinology, 4 illus Churchill,London,1952
Bioch 33.0434

ANTHROPOGENE VEGETATION Internationale vereinigung fur vegetationskunde Bericht Stolzenau 1961 Edited by Reinhold Tuxen illus. Junk,The Hague,1966
Bot 42.4761

ANTHROPOLOGY Congres international d'anthropologie et d'archeologie prehistoriques 9th Compte rendu Lisbon 1880 Academie royale des sciences,Lisbon, 1884
An 32.2637

ANTHROPOLOGY Congres international d'anthropologie et d'archaeologie prehistoriques 13th Compte rendu Monaco 1906 Tome 1-2 2 vols Imprimerie de Monaco,Monaco,1907-08
An 32.2638

ANTHROPOLOGY Congres international d'anthropologie et d'archeologie prehistoriques 8th Compte rendu Budapest 1876 Vol 1 Franklin-Tarsulat,Budapest,1877
An 32.2636

ANTHROPOLOGY Institut international d'anthropologie:session 2e Prague 1924 Sep 14-21 Nourry,Paris,1926
An 32.2639

ANTHROPOLOGY TODAY:AN ENCYCLOPEDIC INVENTORY
The Inventory papers :international symposium on anthropology Edited by A.L. Kroeber xv, 958p University of Chicago press,Chicago,Ill; London,1965 Pre-symposium papers originally prepared as a basis for the symposium held in 1952
WSM 43.3613

ANTIBIOTICS International congress of biochemistry 2nd Paris 1952 Jul 21-27 Vol 6: symposium sur le mode d'action des antibiotiques Council for international organizations of medical sciences Societe d'edition d'enseignement superieur,Paris,1952 Text in English and French
Bioch 33.1309

ANTIBIOTICS International congress of biochemistry 4th Proceedings Vienna 1958 Sep 1-6 Vol 5: symposium 5 - biochemistry of antibiotics International union of biochemistry Edited by K.H. Spitzy and R. Brunner I.U.B.symposium series, 7 Pergamon,London,1959 Added title page in French and German.Text in English,French and German.
Bioch 33.1319

ANTIBIOTICS Selective toxicity and antibiotics Edinburgh 1948 Jul Society for experimental biology Society for experimental biology.Symposia, 3 Cambridge university press,Cambridge,1949
Phys 20.2224

ANTIBIOTICS Selective toxicity and antibiotics :a symposium Edinburgh 1948 Jul Society for experimental biology Society for experimental biology.Symposia, 3 Cambridge university press,Cambridge,1949
Bioch 33.1350

ANTIBIOTICS AND MOULD METABOLITES :a symposium Report Nottingham 1956 Chemical society Chemical society.Special publication, 5 Chemical society,London,1956 Organized by professor Johnson
Chem 18.1215

ANTIBODIES Cell-bound antibodies :a conference Proceedings Washington,D.C. 1963 May 10 National research council. Committee on tissue transplantation Edited by Bernard Amos and Hilary Koprowski Wistar institute press,Philadelphia,1963
Bioch 33.0534

ANTIBODIES :a symposium Papers Cold Spring Harbor 1967 Cold Spring Harbor laboratory of quantitative biology Cold Spring Harbor symposia on quantitative biology, 32 Cold Spring Harbor,1967
Bioch 33.1288

ANTIBODIES :a symposium Proceedings Cold Spring Harbor 1967 Jun 1-7 Cold Spring Harbor laboratory of quantitative biology Edited by Leonora Frisch Cold Spring Harbor symposia on quantitative biology, 32 Cold Spring Harbor laboratory of quantitative biology,Cold Spring Harbor,L.I.,N.Y.,1967
Path 30.2523

ANTIBODIES TO BIOLOGICALLY ACTIVE MOLECULES Federation of European biochemical societies : meeting 2nd Proceedings Vienna 1965 Apr 21-24 Vol 1 Edited by Bernhard Cinader Pergamon,Oxford,1967
Pha 16.0190

ANTIBODY FORMATION Molecular and cellular basis of antibody formation :a symposium Proceedings Prague 1964 Jun 1-5 Ceskoslovenska akademie ved.Institute of microbiology Edited by J. Sterzl Czechoslovak academy of sciences;Academic press,Prague;New York,1965
An 32.3701

ANTIBODY FORMATION Molecular and cellular basis of antibody formation a symposium Proceedings Prague 1964 Jun 1-5 Edited by J. Sterzl Organized by the Ceskoslovenska akademie ved.Institute of microbiology. Immunological department Czechoslovak academy of sciences;Academic press,Prague;New York,1965
Bioch 33.0554

ANTIBODY RESPONSE The Nature and significance of the antibody response Papers and discussions New York 1951 Mar 21-22 New York academy of medicine.Section of microbiology Edited by A.M. Pappenheimer New York academy of sciences.Section of microbiology.Symposia, 5 Columbia university press,New York,1953
Path 30.2532

ANTIBODY RESPONSE The Nature and significance of the antibody response :a symposium Papers and discussions New York 1951 Mar 21-22 New York academy of medicine. Section on microbiology Edited by A.M. Pappenheimer New York academy of medicine. Section on microbiology.Symposia, 5 Columbia university press,New York,1953
Bioch 33.1156

ANTIBODY RESPONSE The Nature and significance of the antibody response; symposium... New York 1951 Mar 21-22 New York academy of medicine.Section on microbiology Edited by Alwin M. Pappenheimer Held at the New York academy of medicine New York academy of medicine.Section on microbiology.Symposia, 5 Columbia university press,New York,1953
Med 36.0272

ANTICHOLINERGIC DRUGS AND BRAIN FUNCTIONS IN ANIMALS AND MAN :a symposium Proceedings Washington,D.C. 1966 Mar Edited by P.B. Bradley and M. Fink Progress in brain research, 28 Elsevier,Amsterdam,1968 Held during the 5th meeting of the Collegium Internationale Neuro-psychopharmacologicum
An 32.5232

ANTIDEPRESSANT DRUGS 1st :international symposium Proceedings Milan 1966 Apr 25-27 Istituto de ricerche farmacologiche "Mario Negri" Edited by S. Garattini and M.N. G. Dukes Excerpta medica international congress series,122 Excerpta medica, Amsterdam,1967
Pha 16.0046

ANTIGENS AND NEOANTIGENS Cross-reacting antigens and neoantigens (with implications for autoimmunity and cancer immunity):a conference Proceedings National research council.Committee on tissue transplantation Edited by John J. Trentin illus,tables Williams and Wilkins,Baltimore,Md.,1967
Path 30.2544

ANTIGUA 1955 Caribbean geological conference 1st report Edited by G.M. Stockley Demerara,1958 Conference organized by G.M.Stockley
Geol 8.2423

ANTIMETABOLITES AND CANCER :a symposium Boston,Mass 1953 Dec 28-29 American association for the advancement of science Edited by Cornelius P. Rhoades American association for the advancement of science, Washington,D.C.,1955
Radioth 35.0735

ANTIMICROBIAL DRUGS Biochemical studies of antimicrobial drugs :a symposium Papers London 1966 Apr Society for general microbiology Edited by B.A. Newton and P.E. Reynolds Held at the Royal institution Society for general microbiology.Symposia, 16 Cambridge university press,Cambridge,1966
Bioch 33.1170

ANTITUMORAL EFFECTS OF VINCA ROSEA ALKALOIDS : first symposium of the G.E.C.A. Proceedings Paris 1965 Jun 21 European cancer chemotherapy group Edited by S. Garattini and E.M. Sproston International congress series, 106 Excerpta medica foundation, Amsterdam,1966
Radioth 35.0875

ANTWERP 1955 Cerebral lipidoses :a symposium Papers Edited by J.N. Cumings and others Blackwell,Oxford,1957
An 32.4165

ANTWERP 1969 Modern diffraction and imaging techniques in material science International summer course on material science Proceedings North Atlantic treaty organization.Scientific affairs division Edited by S. Amelinckx and others North-Holland,Amsterdam,1970
Met 25.2608

APHIDOPHAGOUS INSECTS Ecology of aphidophagous insects :symposium Proceedings Liblice 1965 Sep 27-Oct 1 Ceskoslovenska akademie ved Junk;Academia,The Hague;Prague, 1966
Bal 39.2844

APOLLO 11 LUNAR SCIENCE CONFERENCE Papers Houston,Texas 1970 Jan 5-8 National aeronautics and space administration American association for the advancement of science.Miscellaneous publication, 70-1 American association for the advancement of science,Washington,D.C.,1970
Min 10.1474

APOLLO 11 LUNAR SCIENCE CONFERENCE Proceedings Houston,Texas 1970 Jun 5-8 1-3 National aeronautics and space administration Edited by A.A. Levinson Geochimica et cosmochimica acta, 34,supplement 1 3 vols Pergamon,New York,1970
Min 10.1548

APOLLO 11 LUNAR SCIENCE CONFERENCE 1st Proceedings Houston,Texas 1970 Jan 5-8 Edited by A.A. Levinson Geochimica et cosmochimica acta.Supplement, 1 3 vols Pergamon,New York,1970
Obs 6.3615

APPLICATION OF FRACTURE TOUGHNESS PARAMETERS TO STRUCTURAL METALS :a symposium 1964 Oct 20 American institute of mining, metallurgical and petroleum engineers. Structural materials technical committee Edited by Hermann D. Greenberg Metallurgical society conferences, 31 Gordon and Breach, New York;London,1966
Met 25.0867

APPLICATION OF ISOTOPE TECHNIQUES IN HYDROLOGY a comprehensive report Vienna 1961 Nov 6-9 International atomic energy agency International atomic energy agency.Technical reports series, 11 Vienna,1962
Bot 42.3317

APPLICATION OF MATHEMATICAL MODELS IN CHEMICAL ENGINEERING RESEARCH DESIGN AND PRODUCTION A.I.Ch.E.-Chem.E. joint meeting Proceedings London 1965 Jun 13-17 American institute of chemical engineers Institution of chemical engineers 1965 A.I.Ch.E.-I.Chem.E. Symposium series, 4 I.Chem.E.,London,1965
Eng 41.5861

The APPLICATION OF SCIENTIFIC METHODS TO INDUSTRIAL AND SERVICE MEDICINE :a conference Proceedings London 1950 Mar 29-31 Medical research council H.M.S.O. London,1951
Bioch 33.1542

APPLICATIONS OF CATEGORICAL ALGEBRA :a symposium Proceedings New York 1968 Apr 10-11 American mathematical society Edited by Alex Heller American mathematical society.Proceedings of symposia in pure mathematics, 17 Bibliog v,231p 26cm American mathematical society,Providence,R.I., 1970
P Math 2.3508

APPLICATIONS OF COMMUNICATION THEORY : symposium Papers Communication theory London 1952 Sep 22-26 Edited by Willis Jackson Supported by the British broadcasting corporation xii,532p 25cm Butterworths,London,1953 Also supported by the Ministry of supply
Math L 5.3249

APPLICATIONS OF COMMUNICATION THEORY symposium papers Communication theory London 1952 Sep 22-26 British broadcasting corporation Great Britain.Ministry of supply Edited by Willis Jackson xii,532p Butterworths scientific publications,London, 1953
Math 3.0735

APPLICATIONS OF ELECTRON AND NUCLEAR RESONANCE IN CHEMISTRY Developments in aromatic chemistry.Applications of electron and nuclear resonance in chemistry.Recent work on the inorganic chemistry of sulphur :symposia Bristol 1958 Chemical society Chemical society.Special publication, 12 Chemical society,London,1958
Bioch 33.0630

APPLICATIONS OF FIELD-ION MICROSCOPY IN PHYSICAL METALLURGY AND CORROSION Conference in physical metallurgy and corrosion Proceedings Atlanta,Ga. 1965 May 15-17 Georgia institute of technology Edited by R. F. Hochman and others Georgia institute of technology.Advanced research projects agency, Atlanta,Ga.,1968
Met 25.2631

APPLICATIONS OF FUNDAMENTAL THERMODYNAMICS TO METALLURGICAL PROCESSES Conference on the thermodynamic properties of materials 1st Proceedings Pittsburgh 1964 Nov 29-Dec 1 University of Pittsburgh.Center for the study of thermodynamic properties of materials Edited by G.R. Fitterer Gordon and Breach, New York,1967
Met 25.2443

APPLICATIONS OF MATHEMATICAL PROGRAMMING TECHNIQUES :a conference Cambridge 1968 Jun 24-28 North Atlantic treaty organization.Scientific affairs committee Edited by E.M.L. Beale EUP,London,1970
Eng 41.8254

APPLICATIONS OF MATHEMATICAL PROGRAMMING TECHNIQUES :a conference Proceedings Cambridge 1968 Jun 24-28 North Atlantic treaty organization.Science committee Edited by E.M.L. Beale ix,451p 24cm English universities press,London,1970
Math S 3.1686

APPLICATIONS OF MICRO-ELECTRONICS Symposium on applications of micro-electronics Birmingham 1968 Mar 27 Organized by the Institution of electrical engineers Institution of electrical engineers.Conference publication, 49 Institution of electrical engineers,London,1968
Eng 41.5266

APPLICATIONS OF MICROELECTRONICS :a symposium Southampton 1965 Sep 21-23 Institution of electronic and radio engineers University of Southampton.Department of electronics Sponsored also by the Institution of electrical engineers Institution of electrical engineers.Conference publication, 14 I.E.E.,London,1965 Reproduced from typescript
Eng 41.5576

APPLICATIONS OF MODEL THEORY TO ALGEBRA,ANALYSIS, AND PROBABILITY International conference on the applications of model theory to algebra analysis and probability Proceedings Pasadena 1967 May 23-26 Edited by W.A.J. Luxemburg Held at the California institute of technology Bibliog. v,307p 23cm Holt,Rinehart and Winston,New York,1968
P Math 2.3241

APPLICATIONS OF THE MOSSBAUER EFFECT IN CHEMISTRY AND SOLID-STATE PHYSICS Vienna 1965 Apr 26-30 International atomic energy agency International atomic energy agency.Technical reports series,50 International atomic energy agency,Vienna,1966
Min 10.1131

APPLICATIONS RELATED PHENOMENA IN TITANIUM ALLOYS a symposium Los Angeles 1967 Apr 19 American society for testing materials A.S.T.M.Special technical publication, 432 A.S.T.M. Philadelphia,Pa.,1968
Met 25.0609

APPLIED GEOLOGY International geological congress 20th papers Mexico City 1956 Seccion 13: geologia aplicada a la ingenieria y a la mineria Edited by A. Garcia Rojas and others Mexico City,1959 Text in English, French,German and Spanish
Geol 8.3039

APPLIED LANGUAGE ANALYSIS International conference on machine translation of languages and applied language analysis proceedings Teddington 1961 Sep 5-8 Vol. 1-2 National physical laboratory National physical laboratory.Proceedings of symposia, 13 23cm 2 vols H.M.S.O.,London,1962
Math L 5.0960

APPLIED MATHEMATICS Lectures in applied mathematics :summer seminar proceedings Boulder,Colo. 1957 Jun.23-Jul.19 Vol. 1: probability and related topics in physical sciences By Mark Kac American mathematical society Supported by United States.Armed services xiii,266p 24cm Interscience publishers,London;New York,1959 With special lectures by G.E.Uhlenbeck,A.R.Hibbs, and Balth.van der Pol
Math 3.0844

APPLIED MATHEMATICS Summer seminar in applied mathematics 1st Proceedings Boulder,Col. 1957 Jun 23-Jul 21 Vol 3: partial differential equations By Lipman Bers and others Edited by Alton S. Householder and others Sponsored by the American Mathematical Society Lectures in applied mathematics, 3 xiii,343p Interscience,New York,1964 Dedicated to Arthur Norman Milgram
A Math 4.1266

APPLIED MECHANICS Congres international de mecanique appliquee 9th Actes Tome 1-4 4 vols Universite de Bruxelles, Brussels,1957
Chem E 24.0406

APPLIED MECHANICS International congress for applied mechanics 1st- Proceedings illus Waltman,1925- Lacks 2nd and 6th
Eng 41.1757

APPLIED MECHANICS International congress for applied mechanics 5th proceedings Cambridge,Mass 1938 Sep.12-16 Harvard university and Massachusetts institute of technology Edited by J.P. Den Hartog and H. Peters John Wiley,New York,1939
A Math 4.0325

APPLIED MECHANICS International congress of applied mechanics 10th Proceedings Stresa 1960 Aug 31-Sep 7 Edited by F. Rolla and W.T. Koiter Elsevier,Amsterdam;New York,1962
A Math 4.1280

APPLIED MECHANICS International congress of applied mechanics 12th Proceedings Stanford,Calif. 1968 Aug 26-31 International union of theoretical and applied mechanics Edited by H. Hetenyi and W.G. Vincenti Bibliog,illus xxiv,420p 291cm Springer,Berlin,1969
A Math 4.1722

APPLIED MECHANICS Japan national congress for applied mechanics 18th Proceedings Tokyo 1968 Nov 8-9 Science council of Japan Japan national committee for theoretical and applied mechanics Bibliog, illus 188p 26cm Central scientific publishers,Tokyo,1970
A Math 4.1725

APPLIED MECHANICS National congress of applied mechanics 1st proceedings Chicago 1951 June 11-16 Edited by Eli Sternberg United States national committee on theoretical and applied mechanics American society of mechanical engineers,New York,1952
A Math 4.0327

APPLIED MECHANICS The International congress of applied mechanics 11th Proceedings Munich 1964 International committee for the congresses of applied mechanics Edited by Henry Gortler and Peter Sorger 1184p Springer-verlag,Berlin,1966 Dedicated to Richard Grammel
A Math 4.1279

APPLIED MECHANICS United States national congress of applied mechanics 1st-5th Proceedings Edited by Eli Sternberg American society of mechanical engineers,New York,1952-66
Eng 41.1759

APPLIED MECHANICS United States national congress of applied mechanics 2nd Proceedings Ann Arbor,Mich. Jun 14-18 Edited by Paul M. Naghdi xx,821p American society of mechanical engineers,New York,1955
Chem E 24.0231

APPLIED POLYMER SYMPOSIA, 1 High speed testing Vol 5: 5th international symposium held at Boston,Massachusetts,March 8 and 9, 1965 Sponsored by Plas-tech equipment corporation Interscience,New York,1965
Eng 41.3849

APPLIED POLYMER SYMPOSIA, 2 Thermoanalysis of fibers and fiber-forming polymers : symposium Papers Atlantic City,N.J. 1965 Sep 17 American chemical society Edited by Robert F. Schwenker Interscience,New York, 1966
Met 25.2705

APPLIED POLYMER SYMPOSIA, 12,High speed testing, 7 Rheology of solids International symposium on high speed testing 7th Proceedings Boston,Mass. 1969 Mar 17-18 Edited by R.D. Andrews and F.R. Eirich Interscience,New York,1969
Met 25.2581

APPLIED PSYCHOTHERAPY Conferencia internacional de psicotecnica applicada a l'orientacio professional i a l'organitzacio cientifica del treball 2 Actes Barcelona 1921 Sep 28-30 Conference internationale de psychotechnique appliquee a l'orientation professionelle et a l'organisation scientifique du travail. Secretariat Institut d'orientacio professional,Barcelona,1912 Under the presidency of E.Claparede
Psy 31.2608

APPLIED SCIENCES RESEARCH AND UTILIZATION OF LUNAR RESOURCES Lunar international laboratory (LIL symposium) 4th Proceedings New York 1968 Oct 17 International academy of astronautics Bibliog,illus 200p 20cm Pergamon press, Oxford,1970 LIL symposium organized by the International academy of astronautics at the 19th International astronautical congress
Cav 7.2711

APPROACHES TO THE GENETIC ANALYSIS OF MAMMALIAN CELLS :Michigan conference on genetics Lectures University of Michigan Edited by Donald J. Merchant and James V. Neel University of Michigan press,Ann Arbor,Mich., 1962
Gen 34.0911

APPROXIMATION OF FUNCTIONS Symposium on approximation of functions 8th Proceedings Warren,Mich. 1964 General motors research laboratories Edited by Henry L. Garabedian 220p Elsevier,Amsterdam,1965
TA 15.0097

APPROXIMATION OF FUNCTIONS Symposium on approximation of functions 8th proceedings Warren,Mich. 1964 Aug 31-Sep 2 General motors corporation.Research laboratories Edited by Henry L. Garabedian viii,220p 25cm Elsevier publishing co., Amsterdam,1965
P. Math 2.0978

APPROXIMATION OF FUNCTIONS Symposium on the approximation of functions 8th proceedings Warren,Mich. 1964 Aug.31-Sep.2 General motors corporation.Research laboratories Edited by Henry L. Garabedian bibliog. viii,220p 24cm Elsevier, Amsterdam,1965
Math L 5.0348

APPROXIMATION THEORY :a symposium Proceedings Lancaster 1969 Jul 21-25 Edited by A. Talbot Academic press,London, 1970
Eng 41.2344

APPROXIMATION THEORY :a symposium Proceedings Lancaster 1969 Jul 21-25 Edited by A. Talbot viii,356p 25cm Academic press,New York;London,1970
P Math 2.3850

APPROXIMATIONS WITH SPECIAL EMPHASIS ON SPLINE FUNCTIONS :a symposium Proceedings Madison,Wis. 1969 May 5-7 United States. Army.Mathematics research center Edited by I. J. Schoenberg bibliog. Academic press,New York;London,1969
Math S 3.1504

APPROXIMATIONS WITH SPECIAL EMPHASIS ON SPLINE FUNCTIONS :a symposium Proceedings Madison,Wis. 1970 May 5-7 United States. Army.Mathematics research center Edited by I. J. Schoenberg Held at the University of Wisconsin United States.Army.Mathematics research center.Publications, 23 xi,488p 24cm Academic press,London;New York,1971
Math S 3.1675

AQUEOUS CORROSION A.E.C.-Euratom conference on aqueous corrosion of reactor materials Brussels 1959 Oct 14-17 United States atomic energy commission and European atomic energy society Office of technical services,Washington,D.C.,1959
Met 25.1930

ARABIDOPSIS RESEARCH :an international symposium Report Gottingen 1965 Apr 21-24 Universitat Gottingen.Arabidopsis information service Edited by Gerhard Robbelen Gottingen,1965 Supplement to 'Arabidopsis information service' 1965
Gen 34.1498

ARBEITSGEMEINSCHAFT METALLPHYSIK International conference on vacancies and interstitials in metals Julich 1968 Sep 23-28 Edited by A. Seeger and others North-Holland,Amsterdam,1970
Met 25.2414

ARCHAEOLOGICAL SCIENCE Anglo-Romanian conference on mathematics in the archaeological and historical sciences Proceedings Royal society of London Academia republicii socialiste romania Edited by F.R. Hodson and others viii,565p 24cm Edinburgh university press,Edinburgh, 1971
Math S 3.1826

ARCHAEOLOGY Congres international d'anthropologie et d'archeologie prehistoriques 9th Compte rendu Lisbon 1880 Academie royale des sciences,Lisbon, 1884
An 32.2637

ARCHAEOLOGY Congres international d'anthropologie et d'archaeologie prehistoriques 13th Compte rendu Monaco 1906 Tome 1-2 2 vols Imprimerie de Monaco,Monaco,1907-08
An 32.2638

ARCHAEOLOGY Congres international d'anthropologie et d'archeologie prehistoriques 8th Compte rendu Budapest 1876 Vol 1 Franklin-Tarsulat,Budapest,1877
An 32.2636

ARCHAEOLOGY International congress of prehistoric archaeology 3rd Transactions Norwich 1868 Aug 20-28 and London 1868 Longmans,London,1869
An 32.2635

ARCHAEOLOGY Research seminar on archaeology and related subjects day meeting on statistics and archaeology London 1964 May 30 Institute of archaeology London,1964 Unbound typescript papers
Math 3.1115

ARCHEOLOGIE ET CALCULATEURS:PROBLEMES SEMIOLOGIQUES ET MATHEMATIQUES :colloque international Marseille 1969 Apr 7-12 Centre national de la recherche scientifique Centre national de la recherche scientifique. Colloques internationaux bibliog.,illus. 371p 28cm Centre nationale de la recherche scientifique,Paris,1970
Math S 3.1631

ARCHITECTURAL INSTITUTE OF JAPAN Failure and defects of bridges and structures :a symposium Proceedings Tokyo 1957 Sep 15 Japan society for the promotion of science,Tokyo, 1959
Eng 41.3015

ARCHITECTURAL INSTITUTE OF JAPAN Prestressed structures :a symposium Proceedings Tokyo 1959 Sep 14 Japan society for the promotion of science,Tokyo,1960
Eng 41.3016

ARCS IN INERT ATMOSPHERES AND VACUUM a symposium San Francisco,Calif. 1956 Apr 30-May 1 Electrochemical society. Electrothermics and metallurgy division Edited by W.E. Kuhn Wiley;Chapman and Hall, New York;London,1956
Met 25.0248

ARCTIC Annual Arctic planning session 2nd Proceedings Bedford,Mass. 1959 Oct Air force Cambridge research center.Geophysics research directorate Edited by Vivian C. Bushnell Air force Cambridge research center. G.R.D.research notes,29 AFCRC-TN-59-661 illus viii,172p A.F.C.R.C.,Bedford,Mass., 1959
Sco 14.7405

ARCTIC Annual Arctic planning session 3rd Proceedings Bedford,Mass. 1960 Nov Air force Cambridge research laboratories Edited by George P. Rigsby and Vivian C. Bushnell Air force Cambridge research laboratories.G.R. D.research notes,55 AFCRL 436 illus viii, 148p A.F.C.R.L.,Bedford,Mass.,1961
Sco 14.7406

ARCTIC Annual arctic planning session 1st Proceedings Boston,Mass. 1958 Nov Air force Cambridge research center.Geophysics research directorate Edited by J.H. Hartshorn Air force Cambridge research center.G.R.D.research notes,15 AFCRC-TN-59-256,AD 212265 illus viii,65p A.F.C.R.C., Bedford,Mass.,1959
Sco 14.7404

ARCTIC AND ALPINE ENVIRONMENTS International association for quaternary research congress 7th Proceedings Boulder,Colo. 1965 Aug 14-Sep 19 International association for quaternary research Edited by W.H. Osburn and H.E. Wright Sponsored by the National research council illus,maps xi,308p Indiana university press,Bloomington,Ind.; London,1968 For other volumes of these proceedings,see also International union for quaternary research
Bot 42.2006

ARCTIC AND ALPINE ENVIRONMENTS International association for quaternary research congress 7th Proceedings Boulder,Colo. 1965 Aug 14-Sep 19 10: arctic and alpine environments International association for quaternary research Edited by W.H. Osburn and H.E. Wright Sponsored by the National research council illus,maps xi,308p 25cm Indiana university press,Bloomington,Ind.; London,1968 For other volumes of these proceedings,see also International union for quaternary research
Sco 14.7841

ARCTIC BASIN SYMPOSIUM Proceedings Hershey,Pa. 1962 Oct 8-11 Arctic institute of North America,and,United States.Office of naval research illus,maps 313p 23cm Washington,D.C.,1963
Sco 14.5904

ARCTIC BIOLOGY Oregon state university annual biology colloquium arctic biology:ten papers presented at the 1957 and one at the 1965 biology colloquium 18th Proceedings Cornvallis,Ore. 1957 Oregon state university Edited by Henry P. Hansen 2nd edition Illus,maps 318p 24cm Oregon state university press,Cornvallis,Ore.,1967
Sco 14.7840

ARCTIC BIOLOGY 18 colloquium Biology Colloquium Corvallis 1957 Apr 19-20 Edited by Henry P. Hansen illus,maps 134p 28cm Oregon state college,Corvallis,1957
Sco 14.2434

ARCTIC BIOLOGY AND MEDICINE Symposia on arctic biology and medicine Proceedings Fort Wainwright,Alaska 1962 Aug 28-30 influence of cold on host-parasite interactions By Robert I. McClaughry and others Edited by Eleanor G. Viereck Held under the auspices of the University of Alaska. Geophysical institute illus 455p 23cm Fort Wainwright,Alaska,1963
Sco 14.0882

ARCTIC BIOLOGY AND MEDICINE Symposia on arctic biology and medicine Proceedings Fort Wainwright,Alaska 1963 Feb 17-18 4: frostbite Edited by Eleanor G. Viereck Held under the auspices of the University of Alaska.Geophysical institute illus vi,457p 23cm Arctic aeromedical laboratory,Fort Wainwright,Alaska,1964
Sco 14.0883

ARCTIC CO-OPERATION Conference of the Arctic co-operatives 1st Minutes Frobisher Bay, N.W.T. 1963 Mar 12-18 Canada.Department of northern affairs and national resources 36cm Ottawa,1963 Typescript
Sco 14.3712

ARCTIC CO-OPERATION Conference of the Arctic cooperatives 2nd Minutes Povungnituk,Que. 1966 Apr 19-28 93p 35cm n.p.,1966 Typescript
Sco 14.3711

ARCTIC COMMUNICATIONS 8th proceedings of the eighth meeting of the Agard ionaspheric research committee Proceedings Athens 1963 Jul North Atlantic treaty organisation. Agard ionospheric research committee Edited by B. Landmark tables,graphs 297p Pergamon,Oxford;New York,1964
Sco 14.2468

ARCTIC GEOLOGY Geology of the Arctic First international symposium on Arctic geology Calgary 1960 Jan 11-30 Vol 1-2 Alberta society of petroleum geologists Edited by Gilbert O. Raasch illus,maps 26cm 2 vols University of Toronto press,Toronto,1961
Sco 14.2403

ARCTIC GEOLOGY Geology of the Arctic International symposium on arctic geology 1st Proceedings Calgary,Alta. 1960 Jan 11-13 Vol 1-2 Edited by Gilbert O. Raasch Under the auspices of the Alberta society of petroleum geologists 2 vols Toronto,1961 Additional maps in a separate container
Geod 9.0436

ARCTIC HEAT BUDGET Proceedings of the symposium on the arctic heat budget and atmospheric circulation Lake Arrowhead 1966 Jan-Feb University of California,and, Rand corporation Edited by J.O. Fletcher Prepared for the National science foundation Memorandum RM-5233-NSF illus,diagrs 566p The Rand corporation,Santa Monica,1966
Sco 14.2407

ARCTIC INSTITUTE OF NORTH AMERICA and UNITED STATES.OFFICE OF NAVAL RESEARCH Arctic Basin symposium Proceedings Hershey,Pa. 1962 Oct 8-11 illus,maps 313p 23cm Washington,D.C.,1963
Sco 14.5904

ARCTIC RESEARCH Internationale studiengesellschaft zur erforschung der arktis mit dem luftschiff (aeroarctic) :versammlung 1 Verhandlungen Berlin 1926 Nov 9-13 Edited by Leonid Breitfuss map 115p J. Perthus,Gotha,1927 Contains a paper by Shaw,W.N. 'The influence of the North Polar region upon the meteorology of the northern hemisphere'
Nap 11.0895

The ARCTIC SEA ICE :conference Proceedings Easton,Md. 1958 Feb 24-27 National research council.Division of earth science Supported by United States.Office of naval research National research council. Publication,598 271p 28cm National academy of sciences-National research council, Washington,1958
Sco 14.0359

ARDSLEY-ON-HUDSON,N.Y. 1959 Conference on molecular and radiation biology Edited by R. A. Deering Organised by the National research council.Subcommittee on radiobiology National research council.Nuclear science series.Report, 31 NAS-NRC 823 National research council,Washington,D.C.,1961
Radioth 35.1735

ARECIBO IONOSPHERIC OBSERVATORY Interstellar ionized hydrogen :a symposium Charlottesville,Va. 1967 Dec National radio astronomy observatory Edited by Yervant Jerzian 774p Benjamin,New York, 1968
Obs 6.3447

ARECIBO IONOSPHERIC OBSERVATORY Interstellar ionized hydrogen :a symposium Charlottesville,Va. 1967 Dec National radio astronomy observatory Edited by Yervant Terzian 774p Benjamin,New York, 1968
TA 15.0437

ARGOMENTI SCELTI DI FISICA DELLE PARTICELLE Scuola internazionale di fisica "Enrico Fermi" 41 corso Rendiconti Varenna 1957 Jul 17-29 Societa Italiana di fisica Edited by J. Steinberger Bibliog.,Illus. vii,194p 24cm Academic press,New York,1967
A Math 4.1545

ARGONNE 1964 Symposium on transmethylation and methionine biosynthesis Edited by Stanley K. Shapiro and Fritz Schlenk Sponsored by the Argonne national laboratory. Division of biological and medical research Chicago university press,Chicago,1965
Bioch 33.0607

ARGONNE CANCER RESEARCH HOSPITAL Conference on radioiodine Proceedings Chicago,Ill. 1956 Nov 5-6 Edited by D.E. Clark ACRH-100 University of Chicago,Chicago,Ill.,1956
Radioth 35.1076

ARGONNE NATIONAL LABORATORY Symposium on large scale digital computing machinery proceedings Lemont,Ill. 1953 Aug 3-5 Edited by J.C. Chu illus. 295p 25cm Argonne national laboratory,Lemont,Ill.,1953 Photocopy
Math L 5.0737

ARGONNE NATIONAL LABORATORY.DIVISION OF BIOLOGICAL AND MEDICAL RESEARCH Low level irradiation :a symposium Indianapolis,Ind. 1957 Dec 30 American association for the advancement of science United States atomic energy commission Edited by Austin M. Brues American association for the advancement of science.Publication, 59 American association for the advancement of science, Washington,D.C.,1959
Radioth 35.1109

ARGONNE NATIONAL LABORATORY.DIVISION OF BIOLOGICAL AND MEDICAL RESEARCH Symposium on transmethylation and methionine biosynthesis Argonne 1964 Edited by Stanley K. Shapiro and Fritz Schlenk Chicago university press,Chicago,1965
Bioch 33.0607

ARID LANDS Future of arid lands International arid land meetings Papers and recommendations University of New Mexico 1955 Apr 26-May 4 Edited by Gilbert F. White American association for the advancement of science.Publications,43 American association for the advancement of science,Washington,D.C., 1956
Geog 13.1467

ARITHMETIC FUNCTIONS The Theory of arithmetic functions :conference Proceedings Kalamazoo,Mich. 1971 Apr 29-May 1 Edited by Anthony A. Gioia and Donald L. Goldsmith Held at Western Michigan university Lecture notes in mathematics, 251 287p 25cm Springer,Berlin,1972
P Math 2.4382

ARITHMETICAL ALGEBRAIC GEOMETRY conference proceedings Lafayette,Ind. 1963 Dec 5-7 Purdue university.Division of mathematical sciences Edited by Otto F.G. Schilling Harper's series in modern mathematics Bibliog. vii,200p 22cm Harper and Row, New York,1965
P. Math 2.1770

ARIZONA STATE UNIVERSITY 1961 Researches on meteorites :a symposium Papers Edited by Carleton B. Moore Wiley,New York;London, 1962
Min 10.0499

ARMED FORCES INSTITUTE OF PATHOLOGY Quantitative electron microscopy :a symposium Proceedings Washington,D.C. 1964 Mar30-Apr 3 Edited by Gunter F. Bahr and Elmar H. Zeitler Laboratory investigation, 14,no 6 Williams and Wilkins,Baltimore,Md.,1965
An 32.0162

ARMED FORCES INSTITUTE OF PATHOLOGY Quantitive electron microscopy :a symposium Proceedings Washington,D.C. 1964 Mar 30-Apr 3 Edited by Gunter F. Bahr and Elmar H. Zeitler Laboratory investigation, 14, no 6 Washington,D.C.,1966 Sponsored jointly by the Intersociety committee for research potential in pathology and the Armed forces institute of pathology
Bal 39.0239

ARMED FORCES-NATIONAL RESEARCH COUNCIL COMMITTEE ON VISION Form discrimination as related to military problems :a symposium Proceedings Medford,Mass. 1957 Apr 4-5 Edited by Joseph W. Wulfeck and John H. Taylor National research council.Publication, 561 National research council,Washington,D.C.,1957
Psy 31.0942

ARMED FORCES-NATIONAL RESEARCH COUNCIL COMMITTEE ON VISION Symposium on the measurement of visual function :proceedings of Spring meeting, 1965 Proceedings 1965 Edited by Milton A. Whitcomb and William Benson National research council,Washington,D.C.,1968
Psy 31.0427

ARMED FORCES-NATIONAL RESEARCH COUNCIL COMMITTEE ON VISION Visual search techniques :a symposium Proceedings Washington,D.C. 1959 Apr 7-8 Edited by Ailene Morris and E. Porter Horne National research council. Publication, 712 National research council, Washington,D.C.,1960
Psy 31.0943

ARMOUR AND COMPANY Clinical ACTH conference 2nd Proceedings Chicago 1950 Dec 8-9 Vol 1-2 Edited by John R. Mote 2 vols Churchill,London,1951
Bioch 33.0443

ARMOUR RESEARCH FOUNDATION Computer applications - 1961 Computer applications symposium 8th proceedings Chicago,Ill. 1961 Oct 25-26 Edited by Robert S. Hollitch and Benjamin Mittman illus. vii,198p 23cm McMillan,New York
Math L 5.0854

ARMOUR RESEARCH FOUNDATION Computer applications symposium 4th proceedings Chicago,Ill. 1957 Oct 24-25 illus. x, 126p 23cm Armour research foundation of Illinois institute of technology,Chicago,1958 The Armour research foundation was in 1963 renamed IIT research institute
Math L 5.0853

ARMOUR RESEARCH FOUNDATION Ozone chemistry and technology :international ozone conference Proceedings Chicago,Ill. 1956 Nov 28-30 American chemical society.Advances in chemistry series,21 American chemical society,Washington,D.C.,1959 Edited by the staff of the American chemical society
Chem 18.0031

ARMOUR RESEARCH FOUNDATION Self-organizing systems 1962 conference proceedings Chicago,Ill. 1962 May 22-23 United States. Office of naval research Edited by Marshall C. Yovits and others ix,563p 24cm Spartan books,Washington,D.C.,1962
Math L 5.0992

ARMSTRONG CORK COMPANY Viscoelasticity - phenomenological aspects :a symposium Papers Lancaster,Pa. 1958 Apr 28-29 Edited by J.T. Bergen x,150p Academic press,New York; London,1960
Chem E 24.0413

ARMY INSTITUTE OF RESEARCH Preventive and social psychiatry :a symposium Proceedings Washington 1957 Apr 15-17 1957
Med 36.0011

ARMY MATERIALS RESEARCH AGENCY Strengthening mechanisms,metals and ceramics Sagamore army materials research conference 12th Proceedings Raquette Lake,N.Y. 1965 Aug 24-27 Syracuse university press,Syracuse,N.Y., 1966
Met 25.2746

ARMY MATERIALS RESEARCH AGENCY Ultrafine-grain metals Sagamore army materials research conference 16th Proceedings Raquette Lake,N.Y. 1969 Aug 19-22 Syracuse university press,Syracuse,N.Y.,1970
Met 25.2747

ARNHEM 1964 Soil micromorphology International working meeting on soil micromorphology 2nd Proceedings Edited by A. Jongerius Elsevier,Amsterdam,1964
Eng 41.3115

ARNHEM 1964 Soil micromorphology International working-meeting on soil micromorphology 2nd proceedings Edited by A. Jongerius Elsevier,Amsterdam,1964
Geol 8.1511

AROMATIC CHEMISTRY Developments in aromatic chemistry.Applications of electron and nuclear resonance in chemistry.Recent work on the inorganic chemistry of sulphur :a symposium Bristol 1958 Chemical society Chemical society.Special publication,12 Chemical society,London,1958
Chem 18.1229

AROMATIC RINGS Biological oxidation of aromatic rings :a symposium London 1949 Nov 12 Biochemical society Edited by R.T. Williams Held at the London school of hygiene and tropical medicine Biochemical society.Symposia, 5 Cambridge university press,Cambridge,1950
Bioch 33.1367

AROMATICITY :an international symposium Sheffield 1966 Jul 6-8 Chemical society, and,University of Sheffield Chemical society. Special publication,21 Chemical society, London,1967
Chem 18.1225

ARTERIAL CHEMORECEPTORS The Wates foundation symposium on arterial chemoreceptors Proceedings Oxford 1966 Jul 18-21 Edited by R.W. Torrance Blackwell,Oxford,1968
An 32.3810

ARTERIAL CHEMORECEPTORS The Wates foundation symposium on arterial chemoreceptors Proceedings Oxford 1966 Jul 18-21 Edited by R.W. Torrance pl. xiv,402p Blackwell scientific publications,Oxford,Edinburgh,1968
Phys 20.1968

ARTHROPODS Endocrinologie des arthropodes Colloque international sur l'endocrinologie des arthropodes Paris 1947 Jun Centre national de la recherche scientifique Centre national de la recherche scientifique. Colloques internationaux, 4 Centre national de la recherch scientifique,Paris, 1948
Bal 39.2291

ARTIFICIAL INTELLIGENCE Aspects of the theory of artificial intelligence International symposium on biosimulation 1st Proceedings Locarno 1960 Jun 29-Jul 5 Edited by C.A. Muses Plenum,New York, 1962
Eng 41.5988

ARTIFICIAL INTELLIGENCE International joint conference on artificial intelligence 2nd Advance papers London 1971 Sep 1-3 British computer society illus 658p British computer society,London,1971
Math L 5.3848

ARTIFICIAL INTELLIGENCE AND HEURISTIC PROGRAMMING N.A.T.O. advanced study institute on artificial intelligence and heuristic programming 1st Proceedings Menaggio 1970 Aug North Atlantic treaty organization Edited by N.V. Findler and Bernard Meltzer illus 327p Edinburgh university press, Edinburgh,1971
Math L 5.3853

ARTIFICIAL SATELLITES FOR GEODESY International symposium on the use of artificial satellites for geodesy 1st Proceedings Washington,D.C. 1962 Apr 26-28 Committee on space research and International association of geodesy Edited by George Veis Co-sponsored by the International union of geodesy and geophysics North-Holland,Amsterdam,1963
Geod 9.0055

ARTIFICIAL STIMULATION OF RAIN Conference on the physics of cloud and precipitation particles 1st Proceedings Woods Hole, Mass. 1955 Sep 7-10 Air force Cambridge research center.Geophysics research directorate Edited by Helmut Weickmann and Waldo Smith Co-sponsored by United States. Office of naval research.Geophysics branch illus 442p Pergamon,London,1957
Nap 11.0931

ASIAN REGIONAL CONFERENCE ON SOIL MECHANICS AND FOUNDATION ENGINEERING 1st Proceedings New Delhi 1960 Feb International society of soil mechanics and foundation engineering Indian national society of soil mechanics and foundation engineering New Delhi,1960
Eng 41.3172

ASILOMAR,CALIF 1959 International symposium on high temperature technology 2nd Proceedings Stanford research institute McGraw-Hill,New York,1960
Met 25.2535

ASILOMAR,CALIF. 1954 The Luminescence of biological systems :a conference Proceedings National research council.Committee on photobiology Edited by Frank H. Johnson American assoc.for the advancement of science, Washington,D.C.,1955
Bal 39.0080

ASILOMAR,CALIF. 1962 Cytodifferentiation and macromolecular synthesis Society for the study of development and growth Edited by Michael Locke Society for the study of development and growth.Symposia, 21 Academic press,New York;London,1963
Gen 34.0854

ASILOMAR,CALIF. 1962 Cytodifferentiation and macromolecular synthesis :a symposium Edited by Michael Locke Society for the study of development and growth.Symposia, 21 Academic press,London,1963
An 32.2501

ASILOMAR,CALIF. 1962 Cytodifferentiation and macromolecular synthesis :a symposium Papers Society for the study of development and growth Edited by Michael Locke Society for the study of development and growth. Symposia, 21 Academic press,New York;London, 1963
Bot 42.1496

ASILOMAR,CALIF. 1962 Cytodifferentiation and macromolecular synthesis :a symposium Society for the study of development and growth Edited by Michael Locke Society for the study of development and growth.Symposia, 21 Academic press,New York;London,1963
Bioch 33.1009

ASILOMAR,CALIF. 1964 The Genetics of colonizing species :symposium Proceedings International union of biological sciences Edited by H.G. Baker and G.Ledyard Stebbins International union of biological sciences. Symposia, 1 Academic press,New York;London, 1965
Bot 42.0869

ASILOMAR CONFERENCE The Conference on reactions between complex nuclei Pacific Grove,Calif. 1963 Apr 14-18 Edited by A. Ghiorso and others 456p University of California press,Berkeley,Calif.,1963
Cav 7.0090

ASLIB Conference on classification proceedings London 1962 Apr.6 Aslib proceedings, 14,no.8 215-266p 25cm Aslib,London,1962
Math 3.0776

ASOCIACION GEOLOGICA ARGENTINA Anales de las primeras jornadas geologicas Argentinas San Juan 1960 Nov 4-12 Tom 1-3 Buenos Aires, 1960-62
Geol 8.3165

ASPECTS OF CELL MOTILITY :a symposium Oxford 1967 Sep Society for experimental biology Society for experimental biology. Symposia, 22 Cambridge university press, Cambridge,1968
Bioch 33.1361

ASPECTS OF CELL MOTILITY :a symposium Oxford 1967 Sep Society for experimental biology Society for experimental biology. Symposia, 22 Cambridge university press, Cambridge,1968
An 32.5206

ASPECTS OF CELL MOTILITY :a symposium Papers Oxford 1967 Sep Society for experimental biology Society for experimental biology.Symposia, 22 Cambridge university press,Cambridge,1968
Radioth 35.0473

ASPECTS OF EDUCATIONAL TECHNOLOGY :the proceedings of the programmed learning conference Proceedings Loughborough 1966 Apr 15-18 Association for programmed learning and Leicestershire programmed learning group Edited by Derick Unwin and John Leedham Methuen,London,1967
Psy 28.0233

ASPECTS OF GROWTH Cold Spring Harbor symposia on quantitative biology Papers Cold Spring Harbor 1934 Vol 2 Cold Spring Harbor biological laboratory Long Island biological association,Cold Spring Harbor,1934 Later referred to in vol.9 as "Aspects of growth"
Bioch 33.1258

ASPECTS OF INSECT BIOCHEMISTRY :a symposium London 1965 Apr 1 Biochemical society Edited by T.W. Goodwin Biochemical society. Symposia, 25 Academic press,London;New York, 1965
Bioch 33.1387

ASPECTS OF INSECT BIOCHEMISTRY :symposium London 1965 Apr 1 Biochemical society Edited by T.W. Goodwin Biochemical society. Symposia, 25 Academic press,London;New York, 1965
Gen 34.1646

ASPECTS OF MEDICAL PHYSICS International conference on medical physics Review papers Harrogate 1965 Sep 8-10 Edited by J. Rotblat Convened by the International organization of medical physics Taylor and Francis,London,1966
Radioth 35.1671

ASPECTS OF PROTEIN STRUCTURE The International symposium on protein structure and crystallography Proceedings Madras 1963 Jan 14-18 University of Madras. Department of physics Edited by G.N. Ramachandran Academic press,London,1963
Cav 7.1495

ASPECTS OF PROTEIN STRUCTURE The Symposium on protein structure Proceedings Madras 1963 Jan 14-18 University of Madras. Department of physics Edited by G.N. Ramachandran Academic press,London;New York, 1963 The Symposium on protein structure formed part of an International symposium on protein structure and crystallography organized by the University of Madras
Bioch 33.0542

ASPECTS OF PROTEIN STRUCTURE :a symposium Proceedings Madras 1963 Jan 14-18 Edited by G.N. Ramachandran Organized by the University of Madras Academic press,London; New York,1963
Gen 34.0658

ASPECTS OF SYNTHESIS AND ORDER IN GROWTH :a symposium Hanover,N.H. 1954 Jun 23-26 Edited by Dorothea Rudnick Society for the study of development and growth.Symposia, 13 Princeton university press,Princeton,N.J.,1954
An 32.2482

ASPECTS OF TETHYAN BIOGEOGRAPHY Systematics association symposium 7th papers Leicester 1966 Sep 21-23 Systematics association,and,Palaeontological association Edited by C.G. Adams and D.V. Ager Systematics association.Publication,7 Systematics association,London,1967
Geol 8.4466

ASPECTS OF TETHYAN BIOGEOGRAPHY :a symposium Leicester 1966 Sep 21-23 Systematics association Edited by C.G. Adams and D.V. Ager Systematics association.Publication, 7 illus. vi,336p British museum,London, 1967
Bot 42.3273

ASPECTS OF THE BIOLOGY OF AGEING :a symposium Sheffield 1966 Sep Society for experimental biology Edited by H.W. Woolhouse Society for experimental biology. Symposia, 21 Cambridge university press, Cambridge,1967
An 32.5482

ASPECTS OF THE BIOLOGY OF AGEING :a symposium Sheffield 1966 Sep 5-9 Society for experimental biology Society for experimental biology.Symposia, 21 Cambridge university press,Cambridge,1967
Bioch 33.1360

ASPECTS OF THE THEORY OF ARTIFICIAL INTELLIGENCE International symposium on biosimulation 1st Proceedings Locarno 1960 Jun 29-Jul 5 Edited by C.A. Muses Plenum,New York, 1962
Eng 41.5988

ASPECTS OF YEAST METABOLISM A Guinness symposium held at the Research laboratory St. James's Gate, Dublin Discussions Dublin 1967 Edited by A.K. Mills and Hans Krebs Guinness symposia Blackwell,Oxford;Edinburgh, 1968
Bioch 33.1140

ASPECTS OF YEAST METABOLISM A Guinness symposium held at the research laboratory St. James's Gate Dublin Discussions Dublin 1967 Edited by A.K. Mills and Hans Krebs Guinness symposia Blackwell,Oxford,1968
Bot 42.1751

ASPENASGARDEN,LERUM 1968 Elementary particle theory:relativistic groups and analyticity Nobel symposium 8th Proceedings Nobel foundation Edited by Nils Svartholm illus 399p 24cm Wiley, New York,1968
A Math 4.1765

ASSESSMENT OF RADIOACTIVITY IN MAN Symposium on the assessment of radioactive body burdens in man Proceedings Heidelberg 1964 May 11-16 1-2 International atomic energy agency International labour organization World health organization International atomic energy agency.Proceedings series International atomic energy agency,Vienna,1964
Radioth 35.1373

ASSOCIACION MEXICANA DE GEOLOGOS PETROLEROS Zonificacion microfaunistica de las calizas cretacicas del este de Mexico International geological congress 20th papers By Federico Bonet Mexico City,1956
Geol 8.3050

ASSOCIATION DE PSYCHOLOGIE SCIENTIFIQUE DE LANGUE FRANCAISE Symposium de l'Association de psychologie scientifique de langue francaise : la perception 2eme Textes et communications Louvain 1953 Sep 26-28 By Henri Pieron and others Presses universitaires de France,Paris,1955
Psy 31.0887

ASSOCIATION DES SOCIETES NATIONALES EUROPEENNES ET MEDITERRANEENES DE GASTRO-ENTEROLOGIE Congres international de gastro-enterologie 7e Brussels 1964 Jun 1-6 1-3 3 vols A.S.N.E.M.G.E.,Brussels,c1964
PGMS 29.0357

ASSOCIATION FOR ACADEMIC SURGERY Annual meeting of the Association for academic surgery 2nd Proceedings St.Louis,Miss. 1968 Nov Edited by G.D. Zuidema and D.B. Skinner Current topics in surgical research, 1 Academic press,New York;London,1969
Surg 23.0069

ASSOCIATION FOR APPLIED SOLAR ENERGY The Conference on the use of solar energy :the scientific basis Transactions Tucson,Ariz. 1955 Oct 31-Nov 1 Vol 1-5 University of Arizona and Stanford research institute Edited by Edwin F. Carpenter and others illus 6 vols University of Arizona,Tucson, Ariz.,1958
Nap 11.0111

ASSOCIATION FOR COMPUTING MACHINERY ACM-IEEE symposium on problems in the optimization of data communications systems 2nd Proceedings Palo Alto,Calif. 1971 Oct 20-22 illus 198p ACM;IEEE,New York,1971
Math L 5.3629

ASSOCIATION FOR COMPUTING MACHINERY Annual workshop on microprogramming 4th Proceedings Santa Cruz,Calif. 1971 Sep 13-14 Association for computing machinery,New York,1971
Math L 5.3631

ASSOCIATION FOR COMPUTING MACHINERY Annual workshop on microprogramming 5th Preprints Urbana,Ill. 1972 Sep 25-26 Bibliog,diagrs 98p n.p.,1972 Includes 'An annotated bibliography on microprogramming, late 1969-early 1972'
Math L 5.3636

ASSOCIATION FOR COMPUTING MACHINERY Interactive systems for experimental applied mathematics symposium proceedings Washington,D.C. 1967 Aug Edited by M. Klerer and J. Reinfelds Academic press,New York,1968
Math L 5.3495

ASSOCIATION FOR COMPUTING MACHINERY New computers - a report from the manufacturers symposium Los Angeles 1957 Mar.1 illus. 132p 28cm Association for computing machinery,Los Angeles,1957
Math L 5.0792

ASSOCIATION FOR COMPUTING MACHINERY Program: Association for computing machinery national meeting,September ...1955 64p Association for computing machinery,1955
Math L 5.1079

ASSOCIATION FOR COMPUTING MACHINERY Symposium on operating system principles Gatlinburg 1967 Oct 1-4 ACM,New York,1967
Math L 5.3488

ASSOCIATION FOR COMPUTING MACHINERY Symposium on operating systems principles 3rd Stanford,Calif. 1971 Oct 18-20 illus 163p Association for computing machinery,New York,1971
Math L 5.3824

ASSOCIATION FOR COMPUTING MACHINERY Working conference on mechanical language structures New York 1963 Aug 14-16 Collection of 8 mimeograph abstracts of papers
Math L 5.0928

ASSOCIATION FOR COMPUTING MACHINERY Workshop on microprogramming 3rd Preprints Buffalo,N.Y. 1970 OCT bibliog.,illus. Association for computing machinery,New York, 1970
Math L 5.3574

ASSOCIATION FOR COMPUTING MACHINERY :meeting Proceedings Pittsburgh 1952 May 2-3 Jointly sponsored by the Mellon institute illus. 305p 28cm Rimbach,Pittsburgh,1952
Math L 5.3239

ASSOCIATION FOR COMPUTING MACHINERY :meeting Proceedings Toronto 1952 Sep 8-10 illus. 160p 28cm Association for computing machinery,Washington,D.C.,1952
Math L 5.3238

ASSOCIATION FOR COMPUTING MACHINERY 16th : national meeting Proceedings Los Angeles 1961 Sep 5-8 22cm ACM,New York, 1961
Math L 5.3452

ASSOCIATION FOR COMPUTING MACHINERY 19th- : national conference proceedings 1964- Association for computing machinery,New York,1964-
Math L 5.0972

ASSOCIATION FOR COMPUTING MACHINERY.INFORMATION SYSTEMS BRANCH Large-capacity memory techniques for computing systems :a symposium Washington,D.C. 1961 May 23-25 Edited by Marshall C. Yovits Association for computing machinery.Monograph series Macmillan,New York;London,1962
Eng 41.2196

ASSOCIATION FOR COMPUTING MACHINERY.LOS ANGELES CHAPTER Let the computer decide :a symposium papers Los Angeles 1958 Aug.21 Association for computing machinery,Los Angeles,1958
Math L 5.2373

ASSOCIATION FOR COMPUTING MACHINERY.SIGACT ACM conference on proving assertions about programs Proceedings Las Cruces,N.M. 1972 Jan 6-7 illus 211p ACM,New York, 1972 Published as SIGPLAN notices,vol 7, no 1,Jan 1972 and SIGACT news no 14,Jan 1972
Math L 5.3859

ASSOCIATION FOR COMPUTING MACHINERY.SIGPLAN
ACM conference on proving assertions about programs Proceedings Las Cruces,N.M. 1972 Jan 6-7 illus 211p ACM,New York, 1972 Published as SIGPLAN notices,vol 7, no 1,Jan 1972 and SIGACT news no 14,Jan 1972
Math L 5.3859

ASSOCIATION FOR COMPUTING MACHINERY.SPECIAL INTEREST GROUP IN OPERATING SYSTEMS A.C.M. SIGOPS workshop on system performance evaluation Cambridge,Mass. 1971 Apr 5-7 illus. 378p Association for computing machinery,New York,1971
Math L 5.3564

ASSOCIATION FOR COMPUTING MACHINERY.SPECIAL INTEREST GROUP ON INFORMATION RETRIEVAL Information retrieval:a critical view Annual colloquium on information retrieval 3rd Philadelphia,Pa. 1966 May 12-13 Edited by George Schecter Academic press,New York,1967
Eng 41.7692

ASSOCIATION FOR PROGRAMMED LEARNING Aspects of educational technology :the proceedings of the programmed learning conference Proceedings Loughborough 1966 Apr 15-18 Edited by Derick Unwin and John Leedham Methuen,London,1967
Psy 28.0233

ASSOCIATION FOR RADIATION RESEARCH Energy of transfer in radiation processes:chemical, physical and biological aspects :international symposium Proceedings Cardiff 1965 Jan Edited by Glyn O. Phillips Elsevier, Amsterdam,1966
Radioth 35.1744

ASSOCIATION FOR RADIATION RESEARCH Program of the joint meeting 1969
Radioth 35.1763

ASSOCIATION FOR RESEARCH IN NERVOUS AND MENTAL DISEASE Cerebrovascular disease Proceedings of the Association New York 1961 Dec 8-9 Edited by Clark H. Millikan Association for research in nervous and mental disease.Research publications, 41 Williams and Wilkins,Baltimore,Md.,1966
An 32.4214

ASSOCIATION FOR RESEARCH IN NERVOUS AND MENTAL DISEASE Disorders of communication Proceedings of the Association New York 1962 Dec 7-8 Edited by David McK. Rioch and Edwin A. Weinstein Association for research in nervous and mental disease.Research publications, 42 Williams and Wilkins, Baltimore,Md.,1964
An 32.4561

ASSOCIATION FOR RESEARCH IN NERVOUS AND MENTAL DISEASE Endocrines and the central nervous system Proceedings of the Association New York 1963 Dec 6-7 Edited by Rachmiel Levine Association for research in nervous and mental disease.Research publications, 43 Williams and Wilkins, Baltimore,Md.,1966
An 32.4403

ASSOCIATION FOR RESEARCH IN NERVOUS AND MENTAL DISEASE Genetics,and the inheritance of integrated neurological and psychiatric patterns Proceedings of the Association New York 1953 Dec 11-12 Edited by Davenport Hooker and Clarence C. Hare Association for research in nervous and mental disease.Research publications, 33 Williams and Wilkins,Baltimore,Md.,1954
An 32.4157

ASSOCIATION FOR RESEARCH IN NERVOUS AND MENTAL DISEASE Life stress and bodily disease Proceedings of the association New York 1949 Dec 2-3 Edited by Harold G. Wolff Association for research in nervous and mental disease.Research publications, 29 Hafner, New York,1950 reprinted 1968
An 32.4852

ASSOCIATION FOR RESEARCH IN NERVOUS AND MENTAL DISEASE Localization of function in the cerebral cortex :an investigation of the most recent advances Proceedings of the Association New York 1932 Dec 28-29 Edited by Samuel T. Orton and others Association for research in nervous and mental disease.Research publications, 13 Williams and Wilkins,Baltimore,Md.,1934
An 32.4632

ASSOCIATION FOR RESEARCH IN NERVOUS AND MENTAL DISEASE Mental retardation Proceedings of the Association New York 1959 Dec 11-12 Edited by Lawrence C. Kolb and others Association for research in nervous and mental disease.Research publications, 39 Williams and Wilkins,Baltimore,Md.,1962
An 32.4201

ASSOCIATION FOR RESEARCH IN NERVOUS AND MENTAL DISEASE Metabolic and toxic diseases of the nervous system Proceedings of the Association New York 1952 Dec 12-13 Edited by H.Houston Merritt and Clarence C. Hare Association for research in nervous and mental disease.Research publications, 32 Williams and Wilkins,Baltimore,Md.,1953
An 32.4156

ASSOCIATION FOR RESEARCH IN NERVOUS AND MENTAL DISEASE Neurology and psychiatry in childhood Proceedings of the Association New York 1954 Dec 10-11 Edited by Rustin McIntosh and Clarence C. Hare Association for research in nervous and mental disease. Research publications, 34 Williams and Wilkins,Baltimore,Md.,1954
An 32.4161

ASSOCIATION FOR RESEARCH IN NERVOUS AND MENTAL DISEASE Neuromuscular disorders (the motor unit and its disorders) Proceedings of the Association New York 1958 Dec 12-13 Edited by Raymond D. Adams and others Association for research in nervous and mental disease.Research publications, 38 Williams and Wilkins,Baltimore,Md.,1960
An 32.4185

ASSOCIATION FOR RESEARCH IN NERVOUS AND MENTAL DISEASE Pain Proceedings New York 1942 Dec 18-19 Association for research in nervous and mental disease.Research publications, 23 Baltimore,Md.,1943
Phys 20.2003

ASSOCIATION FOR RESEARCH IN NERVOUS AND MENTAL DISEASE Pain Proceedings of the association New York 1942 Dec 18-19 Edited by Harold G. Wolff and others Association for research in nervous and mental disease.Research publications, 23 Williams and Wilkins,Baltimore,Md.,1945
An 32.4520

ASSOCIATION FOR RESEARCH IN NERVOUS AND MENTAL DISEASE Patterns of organization in the central nervous system Proceedings of the association New York 1950 Dec 15-16 Edited by Philip Bard Association for research in nervous and mental disease. Research publications, 30 Hafner,New York, 1952 reprinted 1968
An 32.4853

ASSOCIATION FOR RESEARCH IN NERVOUS AND MENTAL DISEASE Patterns of recognition in the central nervous system Proceedings New York 1950 Dec 15-16 Association for research in nervous and mental disease. Research publications, 30 Baltimore,Md., 1952
Phys 20.2238

ASSOCIATION FOR RESEARCH IN NERVOUS AND MENTAL DISEASE Sensation:its mechanisms and disturbances :an investigation of the most recent advances Proceedings of the association New York 1934 Dec 27-28 Edited by Clarence A. Patten and others Association for research in nervous and mental disease.Research publications, 15 Williams and Wilkins,Baltimore,Md.,1935
An 32.4523

ASSOCIATION FOR RESEARCH IN NERVOUS AND MENTAL DISEASE The Brain and human behavior : 1956 meeting Proceedings New York 1956 Dec 7-8 Association for research in nervous and mental disease.Research publications, 36 Williams and Wilkins,Baltimore,Md.,1958
Psy 31.0494

ASSOCIATION FOR RESEARCH IN NERVOUS AND MENTAL DISEASE The Brain and human behaviour Proceedings of the Association New York 1956 Dec 7-8 Edited by Harry C. Solomon and others Association for research in nervous and mental disease.Research publications, 36 Williams and Wilkins,Baltimore,Md.,1958
An 32.4353

ASSOCIATION FOR RESEARCH IN NERVOUS AND MENTAL DISEASE The Cerebellum :an investigation of recent advances Proceedings of the Association New York 1926 Dec 28-29 Edited by Frederick Tilney and others Association for research in nervous and mental disease.Research publications, 6 Williams and Wilkins,Baltimore,Md.,1929
An 32.4680

ASSOCIATION FOR RESEARCH IN NERVOUS AND MENTAL DISEASE The Circulation of the brain and spinal cord :a symposium on blood supply Proceedings of the Association New York 1937 Dec 27-28 Edited by Stanley Cobb and others Association for research in nervous and mental disease.Research publications, 18 Hafner,New York,1966 Facsimile of 1938 edition
An 32.4133

ASSOCIATION FOR RESEARCH IN NERVOUS AND MENTAL DISEASE The Diseases of the basal ganglia Proceedings of the association New York 1940 Dec 20-21 Edited by Tracy J. Putnam and others Association for research in nervous and mental disease.Research publications, 21 Williams and Wilkins,Baltimore,Md.,1942
An 32.4511

ASSOCIATION FOR RESEARCH IN NERVOUS AND MENTAL DISEASE The Effect of pharmacologic agents on the nervous system Proceedings of the Association New York 1957 Dec 13-14 Edited by Francis J. Braceland Association for research in nervous and mental disease. Research publications, 37 Williams and Wilkins,Baltimore,Md.,1959
An 32.4169

ASSOCIATION FOR RESEARCH IN NERVOUS AND MENTAL DISEASE The Effect of pharmacologic agents on the nervous system :proceedings of the Association... New York 1957 Dec 13-14 Edited by Francis J. Braceland Association for research in nervous and mental disease. Research publications,37 Williams and Wilkins,Baltimore,Md.,1959
Pha 16.0036

ASSOCIATION FOR RESEARCH IN NERVOUS AND MENTAL DISEASE The Frontal lobes Proceedings New York 1947 Dec 12-13 Association for research in nervous and mental disease. Research publications, 27 Baltimore,Md., 1948
Phys 20.1992

ASSOCIATION FOR RESEARCH IN NERVOUS AND MENTAL DISEASE The Frontal lobes Proceedings of the Association New York 1947 Dec 12-13 Edited by Mary P. Wheeler Association for research in nervous and mental disease. Research publications, 27 Hafner,New York, 1948 reprinted 1966
An 32.5330

ASSOCIATION FOR RESEARCH IN NERVOUS AND MENTAL DISEASE The Hypothalamus and central levels of autonomic function Proceedings of the Association New York 1939 Dec 20-21 Edited by John F. Fulton and others Association for research in nervous and mental disease.Research publications, 22 Baltimore, Md.,1940
Phys 20.1083

ASSOCIATION FOR RESEARCH IN NERVOUS AND MENTAL DISEASE The Hypothalamus and central levels of autonomic function Proceedings of the Association New York 1939 Dec 20-21 Edited by John F. Fulton and others Association for research in nervous and mental disease.Research publications,20 Williams and Wilkins,Baltimore,Md.,1940
Pha 16.0282

ASSOCIATION FOR RESEARCH IN NERVOUS AND MENTAL DISEASE The Hypothalamus and central levels of autonomic function Proceedings of the association New York 1939 Dec 20-21 Edited by John F. Fulton and others Association for research in nervous and mental disease.Research publications, 20 Williams and Wilkins,Baltimore,Md.,1940
An 32.4510

ASSOCIATION FOR RESEARCH IN NERVOUS AND MENTAL DISEASE The Intracranial pressure in health and disease :an investigation of the most recent advances Proceedings of the Association New York 1927 Dec 28-29 Edited by Charles A. Elsberg and others Association for research in nervous and mental disease.Research publications, 8 Williams and Wilkins,Baltimore,Md.,1929
An 32.4155

ASSOCIATION FOR RESEARCH IN NERVOUS AND MENTAL DISEASE The Neurologic and psychiatric aspects of the disorders of aging Proceedings of the Association New York 1955 Dec 9-10 Edited by Joseph Earle Moore and others Association for research in nervous and mental disease.Research publications, 35 Williams and Wilkins, Baltimore,Md.,1956
An 32.4166

ASSOCIATION FOR RESEARCH IN NERVOUS AND MENTAL DISEASE The Pituitary gland :an investigation of the most recent advances Proceedings of the Association New York 1936 Dec 28-29 Edited by Walter Timme and others Association for research in nervous and mental disease.Research publications,17 Williams and Wilkins,Baltimore,Md.,1938
Pha 16.0275

ASSOCIATION FOR RESEARCH IN NERVOUS AND MENTAL DISEASE The Role of nutritional deficiency on nervous and mental disease Proceedings of the Association New York 1941 Dec 19-20 Edited by Stanley Cobb and others Association for research in nervous and mental disease.Research publications, 22 Williams and Wilkins,Baltimore,Md.,1943
An 32.4377

ASSOCIATION FOR RESEARCH IN NERVOUS AND MENTAL DISEASE The Vegetative nervous system :an investigation of the most recent advances Proceedings of the Association New York 1928 Dec 27-28 Edited by Walter Timme and others Association for research in nervous and mental disease.Research publications, 9 Williams and Wilkins,Baltimore,Md.,1930
An 32.4884

ASSOCIATION FOR RESEARCH IN NERVOUS AND MENTAL DISEASE Trauma of the central nervous system Proceedings of the Association New York 1943 Dec 17-18 Association for research in nervous and mental disease. Research publications, 24 Williams and Wilkins,Baltimore,Md.,1945
An 32.4090

ASSOCIATION FOR RESEARCH IN NERVOUS AND MENTAL DISEASE Trauma of the central nervous system Proceedings of the Association New York 1943 Dec 17-18 Association for research in nervous and mental disease. Research publications, 24 Williams and Wilkins,Baltimore,Md.,1945
An 32.5227

ASSOCIATION FOR RESEARCH IN NERVOUS AND MENTAL DISEASE Ultrastructure and metabolism of the nervous system Proceedings of the Association New York 1960 Dec 9-10 Edited by Saul R. Korey and others Association for research in nervous and mental disease.Research publications, 40 Williams and Wilkins,Baltimore,Md.,1962
An 32.4199

ASSOCIATION FOR SYMBOLIC LOGIC Summer school in logic :N.A.T.O. advanced study institute: meeting of the association for symbolic logic Proceedings Leeds 1967 Aug 7-23 Edited by M.H. Lob Lecture notes in mathematics, 70 Bibliog. 331p 28cm Springer,Berlin, 1968
P Math 2.3120

ASSOCIATION FOR THE AID OF CRIPPLED CHILDREN Congenital malformations :a conference on teratology papers Bethesda,Md. 1957 Apr 15-16 Pediatrics, 23,no 1,pt 2,suppt. Thomas,Springfield,Ill.,1959
An 32.2046

ASSOCIATION FOR THE AID OF CRIPPLED CHILDREN Control of ovulation :a conference Proceedings Dedham,Mass. 1960 Feb 26-28 Edited by Claude A. Villee Pergamon press, Oxford,1961
An 32.3823

ASSOCIATION FOR THE AID OF CRIPPLED CHILDREN Genetics of migrant and isolate population Conference on human population genetics in Israel Proceedings Jerusalem 1961 Edited by Elisabeth Goldschmidt Williams and Wilkins,New York,1963
Gen 34.1881

ASSOCIATION FOR THE AID OF CRIPPLED CHILDREN Mechanisms of congenital malformation :second scientific conference Proceedings New York 1954 Jun 15-16 Edited by Harold Wolff Association for the aid of crippled children, New York,1954
An 32.2034

ASSOCIATION FOR THE AID OF CRIPPLED CHILDREN Prematurity,congenital malformation,and birth injury :a conference Proceedings New York 1952 Jun 5-6 Association for the aid of crippled children,New York,1953
An 32.2030

ASSOCIATION FOR THE STUDY OF ANIMAL BEHAVIOUR Evolutionary aspects of animal communication; imprinting and early learning :a symposium Proceedings London 1961 Nov 8-9 Zoological society of London.Symposia, 8 Zoological society of London,London,1962
An 32.4384

ASSOCIATION FOR THE STUDY OF MEDICAL EDUCATION Methods of learning and techniques of teaching Annual conference of the association for the study of medical education 2nd Proceedings London 1959 Oct 15-16 Edited by J.R. Ellis Pitman medical publishing company, London,c1960
PGMS 29.0493

ASSOCIATION FOR THE STUDY OF THE EUROPEAN QUATERNARY,and S.S.S.R.- N.K.T.P.VSESOYUZNOE GEOLOGO-RAZVEDOCHNOE OBEDINENIE Mezhdunarodnaya konferentsiya assotsiatsii po izucheniyu chetvertichnogo perioda Evropy International conference of the Association for the study of the European quaternary 2nd trudy Leningrad 1932 Sep 1-7 Vypusk 1-5 Edited by D.A. Petrovskii and others Gosudarstvennoe nauchno-tekhnicheskoe geologo-razvedochnoe izdatelstvo,Leningrad;Moscow,1932-34
Geol 8.2982

ASSOCIATION FRANCAISE DE CALCUL Congres national de l'Association francaise de calcul 1st mimeograph Grenoble 1960 Sep 14-16 Association francaise de calcul,1960
Math L 5.0869

ASSOCIATION FRANCAISE DE REGULATION D'AUTOMATISME Digital computer applications to process control,2 IFAC-IFIP international conference 2nd Proceedings Menton 1967 Jun 5-9 International federation of automatic control International federation for information processing Edited by W.E. Miller Instrument society of America,Pittsburgh,Pa., 1969
Eng 41.8206

ASSOCIATION FRANCAISE POUR LAVANCEMENT DES SCIENCES.CONGRES 68e Clermont-Ferrand et sa region Clermont-Ferrand 1949 Clermont-Ferrand,c1949
Geog 13.4172

ASSOCIATION INTERNATIONALE D'HYDROLOGIE SCIENTIFIQUE see INTERNATIONAL ASSOCIATION OF SCIENTIFIC HYDROLOGY

ASSOCIATION OF AFRICAN GEOLOGICAL SURVEYS and INTERNATIONAL GEOLOGICAL CONGRESS. COMMISSION ON THE GEOLOGICAL MAP OF AFRICA, Actas y trabajos de las reuniones celebradas en Mexico Association of African geological surveys and commission on the geological map of Africa :joint meeting papers Mexico City 1956 Sep 4-10 Edited by A. Garcia Rojas Mexico City,1959 Text in English, French and Spanish
Geol 8.3038

ASSOCIATION OF AFRICAN GEOLOGICAL SURVEYS AND COMMISSION ON THE GEOLOGICAL MAP OF AFRICA : joint meeting papers Actas y trabajos de las reuniones celebradas en Mexico Mexico City 1956 Sep 4-10 International geological congress.Commission on the geological map of Africa,and,Association of African geological surveys Edited by A. Garcia Rojas Mexico City,1959 Text in English,French and Spanish
Geol 8.3038

ASSOCIATION OF AMERICAN MEDICAL COLLEGES TEACHING INSTITUTE 2nd Report Teaching of pathology,microbiology,immunology,genetics French Lick,Ind. 1954 Oct 10-15 Edited by Lura Street Jackson Association of American medical colleges,Chicago,1955
An 32.0993

ASSOCIATION OF AMERICAN MEDICAL COLLEGES TEACHING INSTITUTE 3rd Report Teaching of anatomy and anthropology in medical education Swampscott,Mass. 1955 Oct 18-22 Edited by Lura Street Jackson Association of American medical colleges,Chicago,1955
An 32.0994

ASSOCIATION OF CLINICAL BIOCHEMISTS International congress on clinical chemistry 4th Proceedings Edinburgh 1960 Aug 14-19 E.and S.Livingstone,Edinburgh;London,1961
Bioch 33.2264

ASSOCIATION OF CLINICAL PATHOLOGISTS Chemical pathology in relation to clinical medicine The Adrenal cortex :a symposium Proceedings London 1960 Oct 14-15 Edited by G.K. McGowan and M. Sandler Pitman,London, 1961
An 32.3751

ASSOCIATION OF CLINICAL PATHOLOGISTS.CHEMICAL PATHOLOGY SUB-COMMITTEE Disorders of carbohydrate metabolism :a symposium Proceedings London 1968 Nov 29-30 Edited by G.K. McGowan and G. Walters held at the Royal society of medicine Chemical pathology in relation to clinical medicine, 5 Journal of clinical pathology.Supplement, 2 British medical association,London,1969
Bioch 33.1205

ASSOCIATION OF CLINICAL SCIENTISTS Measurements of exocrine and endocrine functions of the pancreas... :applied seminar 2nd Proceedings Edited by F.W. Sunderman and F.W. Sunderman,Jr. Lippincott, Philadelphia,Pa.,1961
Pha 16.0094

ASSOCIATION OF MEDICAL ADVISERS IN THE PHARMACEUTICAL INDUSTRY Absorption and distribution of drugs :a symposium London 1963 Edited by T.B. Binns Livingstone, Edinburgh;London,1964
Pha 16.0239

ASSOCIATION OF MEDICAL ADVISERS IN THE PHARMACEUTICAL INDUSTRY Absorption and distribution of drugs :based on a symposium London 1963 Edited by T.B. Binns E.and S. Livingstone,Edinburgh;London,1964
PGMS 29.0370

ASSOCIATION OF SOCIAL ANTHROPOLOGISTS The Social anthropology of complex societies Edited by Michael Banton A.S.A.monographs, 4 Tavistock publications,London,1966 Derived from material presented at a conference on"New approaches in social anthropology"sponsored by the Association of social anthropologists,Cambridge,1963
Eng 41.0885

ASSOCIATIVE LEARNING Fundamental issues in associative learning :a symposium Proceedings Halifax 1968 Jun Dalhousie university Edited by N.J. Mackintosh and W.K. Honig Dalhousie university press,Halifax, 1969
Psy 31.3017

ASSOCIAZZIONE MINERALIA SARDA Remobilization of ores and minerals Convegno sulle rimobilizzazione del minerali metallici e non metallici Rendiconti minerali metallici e non metallici Papers Cagliari 1969 Aug Cagliari,c1969
Min 10.1489

ASSOTSIATSIYA PO IZUCHENIYU CHETVERTICHNOGO PERIODA EVROPY see ASSOCIATION FOR THE STUDY OF THE EUROPEAN QUATERNARY

ASTROFISICA DEL PLASMA Scuola internazionale di fisica 'Enrico Fermi' 39 corso Rendiconti Varenna 1966 Jul 11-30 Societa italiana di fisica Edited by P.A. Sturrock Academic press,New York;London,1967
TA 15.0328

ASTROFISICA DEL PLASMA Scuola internazionale di fisica "Enrico Fermi" 39 corso Rendiconti Varenna 1966 Jul 11-30 Societa italiana di Fisica Edited by P.A. Sturrock xvi,364p Academic press,New York; London,1967
A Math 4.1577

ASTROFISICA DELLE ALTE ENERGIE Scuola internazionale di fisica 'Enrico Fermi' 35 corso Rendiconti Varenna 1965 Jul 12-24 Societa italiana di fisica Edited by L. Gratton Academic press,New York;London,1966
TA 15.0314

ASTROFISICA DELLE ALTE ENERGIE Scuola internazionale di fisica "Enrico Fermi" 35 corso rendiconti Varenna 1965 Jul.12-24 Societa Italiana di Fisica Edited by L. Gratton 463p Academic press,New York,1966
A Math 4.1169

The ASTROMETRIC CONFERENCE 2nd Cincinnati 1959 May 17-21 University of Cincinnati Edited by S.Aa. Strand and O.G. Franz Sponsored by the National science foundation 58p Astronomical journal,1960 Reprinted from the Astronomical journal,vol 65,no 4,May 1960
Obs 6.3243

ASTRONAUTICS International astronautical congress 8th proceedings Barcelona 1957 Oct.6-12 International astronautical federation Edited by F. Hecht International astronautical congress. Proceedings, 8 607p Springer,Vienna,1958
Obs 6.0761

ASTRONAUTICS 1st symposium proceedings Vistas in astronautics San Diego 1957 Feb United States.Air force.Office of scientific research Edited by Morton Alperin and Marvin Stern Co-sponsored by General dynamics corporation.Convair division Vistas in astronautics, 1 International series of monographs on aeronautical sciences and space flight.Astronautics division, 1 329p Pergamon press,London,1958
Nap 11.0022

ASTRONAUTICS 2nd symposium Proceedings Vistas in astronautics Denver,Col. 1958 Apr.28 United States.Air force.Office of scientific research Edited by Morton Alperin and Hollingsworth F. Gregory Co-sponsored by Institute of aeronautical sciences Vistas in astronautics, 2 International series of monographs on aeronautical sciences and space flight.Astronautics division, 2 Pergamon press,London,1959
Nap 11.0023

ASTRONAUTICS 3rd symposium Proceedings Vistas in astronautics Los Angeles,Calif. 1960 Oct.12-14 United States.Air force. Office of scientific research jointly sponsored Society of automative engineers Vistas in astronautics, 3 illus 266p Society of automotive engineers,New York,1960
Nap 11.0024

ASTRONOMICAL CONSTANTS The System of astronomical constants :a symposium proceedings Paris 1963 May 27-31 International astronomical union Edited by J. Kovalevsky International astronomical union. Symposium, 21 330p Gauthier-Villars,Paris, 1965 From the Bulletin astronomique de l'observatoire de Paris,vol.25,fasc.1-3
Obs 6.0986

ASTRONOMICAL OPTICS AND RELATED SUBJECTS A Symposium on astronomical optics and related subjects proceedings Manchester 1955 Apr 19-22 Edited by Zdenek Kopal Under the auspices of Victoria university of Manchester. Astronomy department 428p North-Holland, Amsterdam,1956 Dedicated to the memory of Peter Yorke Millns
Obs 6.0975

ASTRONOMICAL PHOTOELECTRIC CONFERENCE Proceedings Flagstaff,Ariz. 1953 Aug.31 Sep. 1 Lowell observatory Edited by John B. Irwin Sponsored by the National science foundation Indiana university,Bloomington, Ind.,1953
Obs 6.2883

ASTRONOMY Galactic astronomy Stony Brook,N. Y. 1968 Jun 19-Jul 17 Edited by Hong-Yee Chiu and A. Muriel 2 vols Gordon and Breach,New York,1970
TA 15.0656

ASTRONOMY Highlights of astronomy International astronomical union 13th general assembly Prague 1967 Aug Edited by L. Perek illus x,548p 25cm Reidel,Doldrecht,1968
A Math 4.1899

ASTRONOMY Highlights of astronomy International astronomical union 14th general assembly Brighton 1970 Aug Vol 2 International astronomical union Edited by C.de Jager 793p Reidel,Dordrecht, 1971
Obs 6.3688

ASTRONOMY Highlights of astronomy International astronomical union.General assembly 14th Invited discourses Brighton 1970 Aug Vol 2 International astronomical union Edited by C. de Jager 793p Reidel,Dordrecht,1971
TA 15.0655

ASTRONOMY New aspects in the history and philosophy of astronomy Joint symposium of the International astronomical union and the Union internationale d'histoire et de philosophie des sciences 1st Hamburg 1964 Aug 22-24 International astronomical union International union of history and philosophy of science Edited by Arthur Beer Vistas in astronomy, 9 illus,ports,maps 318p 25cm Pergamon press,Oxford,1967
Cav 7.2715

ASTRONOMY New aspects in the history and philosophy of astronomy Joint symposium of the International astronomical union and the Union internationale d'histoire et de philosophie des sciences 1st Hamburg 1964 Aug 22-24 International astronomical union International union of history and philosophy of science Edited by Arthur Beer Vistas in astronomy, 9 xvi,317p Pergamon, Oxford,1967
WSM 43.2439

ASTRONOMY New techniques in astronomy : conferences Proceedings Moscow 1961 Apr 18-20 and Kazan 1964 May 12-14 Edited by H.C. Ingrao 446p Gordon and Breach,New York,1971
Obs 6.3598

ASTRONOMY Reports on astronomy International astronomical union 12th assembly Hamburg 1964 Aug.25-Sep.3 Edited by Jean-Claude Pecker International astronomical union.Transactions, 12a Academic press,London;New York,1965
A Math 4.1114

ASTRONOMY Soleil a la renaissance:sciences et mythes :colloque international Brussels 1963 Apr Federation internationale des instituts et societes pour l'etude de la Renaissance and Belgique.Ministere de l'education et de la culture Universite libre de Bruxelles.Institut pour l'etude de la renaissance et de l'humanisme.Publication, 2 584p Presses universitaires de Bruxelles, Brussels;Paris,1965 Contains Hoskin,M.and Jones,C.'Problems in late renaissance astronomy'
WSM 43.0046

ASTRONOMY,METEOROLOGY AND SEISMOLOGY The Pan American scientific congress 2nd Proceedings Washington,D.C. 1915-16 Dec 27-Jan 8 Pan American union Edited by Glen Levin Swiggett illus xiv,847p Government printing office,Washington,D.C.,1917
Nap 11.0838

ASTROPHYSICS Atomic physics and astrophysics Brandeis university.Summer institute in theoretical physics 12th Proceedings Waltham,Mass. 1969 Jun 16-Jul 25 Edited by Max Chretien and E. Lipworth 216p Gordon and Breach,New York,1971
TA 15.0683

ASTROPHYSICS Etoiles a raies d'emissions Colloque international d'astrophysique 8e Liege Universite de Liege.Institut d'astrophysique 14p Liege,c1957 Summary of contributions
Obs 6.3170

ASTROPHYSICS High energy astrophysics : lectures Les Houches 1966 Vol 3 Universite de Grenoble.Ecole d'ete de physique theorique,Les Houches Edited by C. Dewitt and others Supported by NATO Universite de Grenoble.Ecole d'ete de physique theorique,Les Houches. Lectures,1966 xii,449p Gordon and Breach,New York,1967
A Math 4.1562

ASTROPHYSICS High-energy nuclear reactions in astrophysics Symposium on high-energy nuclear reactions in astrophysics Invted papers Philadelphia,Pa. 1967 Jan Edited by D.S.P. Shen Sponsored by American astronomical union 281p Benjamin,New York, 1967 Four of the ten papers were invited after the symposium to complete the coverage
TA 15.0578

ASTROPHYSICS Instabilite gravitationnelle et formation des etoiles, des galaxies et de leurs structures caracteristiques Colloque international d'astrophysique 4e Communications Liege 1966 Jun 20-22 Universite de Liege.Institut d'astrophysique Institut d'astrophysique,Cointe-Sclessin,1967 Papers in English
A Math 4.1170

ASTROPHYSICS International conference on laboratory astrophysics Selection of papers Lunteren 1968 Sep 2-7 Edited by J. Rosenberg Organised under the auspices of the Nederlandse natuurkundig vereniging Physica, 41,no 1 223p North-Holland, Amsterdam,1969
TA 15.0580

ASTROPHYSICS International conference on laboratory astrophysics Selection of papers Lunteren 1968 Sep 2-7 Nederlandse natuurkundige vereniging Edited by J. Rosenberg Physica, 41,no 1 223p North-Holland,Amsterdam,1969
Obs 6.3347

ASTROPHYSICS Mass loss from stars Trieste colloquium on astrophysics 2nd Proceedings Trieste 1968 Sep 12-17 Edited by Margherita Hack Sponsored by the International astronomical union Astrophysics and space science library, 13 345p Reidel,Dordrecht,1969
Obs 6.3628

ASTROPHYSICS Mass loss from stars Trieste colloquium on astrophysics 2nd Proceedings Trieste 1968 Sep 12-17 Edited by Margherita Hack Sponsored by the International astronomical union Astrophysics and space science library, 13 345p Reidel,Dordrecht,1969
TA 15.0620

ASTROPHYSICS Michigan symposium on astrophysics Proceedings Ann Arbor,Mich. 1953 Jun.29-Jul.24 University of Michigan. Observatory Ann Arbor,1953 Typescript
Obs 6.1978

ASTROPHYSICS Physique fondamentale et astrophysique :colloque Compte-rendu Nice 1969 Centre national de la recherche scientifique Journal de physique.Supplement, 30,no 11-12 Centre national de la recherche scientifique.Colloques nationaux, 925 162p Centre national de la recherche scientifique, Paris,1970
TA 15.0499

ASTROPHYSICS Quasars and high energy astronomy ;including proceedings of 2nd Texas conference on relativistic astrophysics Edited by K.N. Douglas and others 485p Gordon and Breach,New York,1969 Texas conference held in 1964
TA 15.0490

ASTROPHYSICS Relativity theory and astrophysics 1:relativity and cosmology Lectures Ithaca,N.Y. 1965 Jul 25-Aug 20 By W. Bonner and others American mathematical society.Summer seminar in applied mathematics Edited by Jurgen Ehlers Lectures in applied mathematics, 8 xvi, 292p American mathematical society, Providence,R.I.,1967
TA 15.0350

ASTROPHYSICS Relativity theory and astrophysics 2:galactic structure Lectures Ithaca,N.Y. 1965 Jul 25-Aug 20 By E. Burbidge and others American mathematical society.Summer seminar in applied mathematics Edited by Jurgen Ehlers viii,220p American mathematical society,Providence,R.I.,1967
TA 15.0351

ASTROPHYSICS Resonance lines in astrophysics manuscripts presented at a conference... Boulder,Colo. 1968 Sep 9-11 National center for atmospheric research.High altitude observatory Washburn observatory 480p National center for atmospheric research, Boulder,Colo.,1968
Obs 6.3349

ASTROPHYSICS Symposium on cosmic rays, elementary particle physics and astrophysics 10th Proceedings Aligarh 1967 Dec 12-16 India.Department of atomic energy.Cosmic ray committee 701p Bombay,1968
TA 15.0553

ASTROPHYSICS Symposium on progress in astrophysics Papers 1939 Feb 17 By Donald H. Menzel and others American philosophical society Franklin institute 1939 From the Proceedings of the American philosophical society,vol 81,no 2.Lacks title page
Obs 6.3230

ASTROPHYSICS :a topical symposium commemorating the 15th anniversary of the Yerkes observatory Edited by J.A. Hynck 703p McGraw-Hill,New York,1951
TA 15.0021

ASTROPHYSICS AND GENERAL RELATIVITY Brandeis university.Summer institute in theoretical physics 11th Proceedings Waltham,Mass. 1968 Jun 17-Jul 26 Edited by Max Chretien and others 2 vols Gordon and Breach,New York,1971
TA 15.0657

ASTROPHYSICS AND GENERAL RELATIVITY Brandeis university.Summer institute of theoretical physics 11th Waltham,Mass. 1968 Vol 1 Edited by M. Chretien and others illus 300p 23cm 2 vols Gordon and Breach,New York,1969
A Math 4.1791

ASYMPTOTIC DISTRIBUTION Asymptotic distribution modulo 1 Nuffic international summer session in science papers Breukelen 1962 Aug 1-11 Netherlands universities foundation for international cooperation Edited by J.F. Koksma and L. Kuipers With the financial support of North Atlantic treaty organization Bibliog. 203p 24cm Noordhoff,Groningen,1964
P. Math 2.1826

ASYMPTOTIC SOLUTIONS OF DIFFERENTIAL EQUATIONS AND THEIR APPLICATIONS a symposium proceedings Madison,Wis 1964 Edited by C.H. Wilcox United States Army. Mathematics Research Center.Publicatons, 13 John Wiley and sons,New York,1964
A Math 4.0246

ASYMPTOTIC SOLUTIONS OF DIFFERENTIAL EQUATIONS AND THEIR APPLICATIONS symposium proceedings Madison,Wis. 1964 May 4-6 United States.Army.Mathematics research center Edited by Calvin H. Wilcox United States. Army.Mathematics research center.Publications, 13 x,249p 22cm John Wiley and sons,New York,1964 Dedicated to Professor Rudolph E.Langer
Math 3.0291

ATHENS 1936 Congres international de pathologie comparee 3rd Rapports Tome 1 pt 2: rapports:section de pathologie vegetale By A.I. Abricossof and others Flammon,Athens,1936
Bot 42.0658

ATHENS 1936 International congress of comparative pathology 3rd Vol 1: reports.2.Section of plant pathology:immunity of plants Editions 'Flamma',Athens,1936
Bot 42.6540

ATHENS 1959 Congres de la societe europeenne d'opthalmologie 1e Resumees des rapports et des communications European opthalmological society Excerpta medica. International congress series, 25 Excerpta medica,Amsterdam;New York,1960 Papers in English,French and German
PGMS 29.0682

ATHENS 1962 Radioactive dating :a symposium International atomic energy agency Joint commission on applied radioactivity International atomic energy agency.Proceedings series International atomic energy agency, Vienna,1963
Bot 42.3319

ATHENS 1962 Radioactive dating :symposium proceedings sponsored by the International atomic energy agency International atomic energy agency,Vienna,1963
Geol 8.4477

ATHENS 1963 Arctic communications proceedings of the eighth meeting of the Agard ionaspheric research committee 8th Proceedings North Atlantic treaty organisation.Agard ionospheric research committee Edited by B. Landmark tables, graphs 297p Pergamon,Oxford;New York,1964
Sco 14.2468

ATHENS 1964 Medical radioisotope scanning Symposium on medical radioisotope scanning Proceedings Held by the International atomic energy agency International atomic energy agency.Proceedings series International atomic energy agency,Vienna,1964
Radioth 35.1376

ATHENS 1964 Regeneration in animals and related problems :an international symposium Edited by V. Kiortsis and H.A.L. Trampusch Sponsored by the North Atlantic treaty organization.Division of scientific affairs North-Holland,Amsterdam,1965
An 32.5438

ATHENS 1964 Regeneration in animals and related problems :an international symposium Edited by V. Kiortsis and H.A.L. Trampusch Sponsored by the North Atlantic treaty organization.Division of scientific affairs North-Holland,Amsterdam,1965 Text in English and French
Bal 39.1320

ATHENS 1965 Guide to the geology and culture of Greece Petroleum exploration society of Libya :annual field conference 7th guidebook Edited by E.E. Hotz and Peter Norton Amsterdam,1965
Geol 8.4481

ATHENS 1969 Radiation-induced cancer :a symposium Proceedings International atomic energy agency World health organization International atomic energy agency,Vienna,1969
Radioth 35.1917

ATHENS 1969 Structure and evolution of the galaxy North Atlantic treaty organization. Advanced study institute Proceedings Edited by L.N. Mavridis Astrophysics and space science library, 22 312p S.Reidel, Dordrecht,1971
Obs 6.3456

ATHENS 1969 Topology of manifolds University of Georgia topology of manifolds institute Proceedings Edited by J.C. Cantrell and C.H. Edwards xiv,514p 24cm Markham,Chicago,Ill.,1970
P Math 2.3819

ATHENS 1970 Physics of the solar corona NATO advanced study institute on physics of the solar corona Proceedings North Atlantic treaty organization Edited by C.J. Macris Astrophysics and space science library, 27 illus xii,345p 24cm Reidel,Dordrecht,1971
A Math 4.1803

ATHENS 1970 Physics of the solar corona NATO advanced study institute on the physics of the solar corona Proceedings North Atlantic treaty organization Edited by C.J. Macris Astrophysics and space science library, 27 345p Reidel,Dordrecht,1971
Obs 6.3607

ATHENS, GA. 1961 Topology of 3-manifolds and related topics :a symposium proceedings University of Georgia.Topology institute Edited by M.K. Fort supported by United States.Office of naval research and National science foundation viii,256p 24cm Englewood Cliffs, N.J.,1962
P. Math 2.0296

ATHEROSCLEROSIS Evolution of the atherosclerotic plaque :an international symposium Presentations Chicago,Ill. 1963 Mar 28-29 Chicago heart association and American heart association.Council on arteriosclerosis Edited by Richard J. Jones University of Chicago press,Chicago,Ill.; London,1963
Path 30.2350

ATHEROSCLEROSIS Hormones and atherosclerosis a conference Proceedings Brighton,Utah 1958 Mar 11-14 National institutes of health. Endocrinology study section Edited by Gregory Pincus Academic press,New York,1959
Bioch 33.0477

ATHEROSCLEROSIS Symposium on atherosclerosis Papers 1954 Mar 22-23 National research council.Division of medical sciences Held at the request of the United States.Air force. Directorate of research and development.Human factors division National research council. Publication, 338 National research council, Washington,D.C.,1955
Path 30.2349

ATHEROSCLEROSIS Symposium on atherosclerosis Washington,D.C. 1954 Mar 22-23 National research council.Division of medical sciences Sponsored by Human factors division.Air force directorate of research and development National research council.Publication, 338 National research council,Washington,D.C.,1955
Med 36.0266

ATLANTA,GA 1955 The Species problem :a symposium American association for the advancement of science Edited by Ernst Mayr American association for the advancement of science.Publication, 50 American association for the advancement of science, Washington,D.C.,1957
Bot 42.3422

ATLANTA,GA. 1955 Beginnings of embryonic development A Symposium on formation and early development of the embryo American association for the advancement of science. Section on zoological sciences Edited by Albert Tyler and others Co-sponsored by the American society of zoologists and the Association of southeastern biologists American association for the advancement of science.Publication, 48 Washington,D.C., 1957
Bal 39.0401

ATLANTA,GA. 1955 The Beginnings of embryonic development :a symposium American association for the advancement of science. Section of zoological sciences Edited by Albert Tyler and others American association for the advancement of science,Washington,D.C., 1957
An 32.2483

ATLANTA,GA. 1955 The Species problem :a symposium American association for the advancement of science Edited by Ernst Mayr American association for the advancement of science.Publication, 50 American association for the advancement of science, Washington,D.C.,1957
Gen 34.1294

ATLANTA,GA. 1955 The Species problem :a symposium presented at the Atlanta meeting of the American association for the advancement of science Edited by Ernst Mayr American association for the advancement of science. Publication, 50 Washington,D.C.,1957
Bal 39.0904

ATLANTA,GA. 1965 Applications of field-ion microscopy in physical metallurgy and corrosion Conference in physical metallurgy and corrosion Proceedings Georgia institute of technology Edited by R.F. Hochman and others Georgia institute of technology.Advanced research projects agency, Atlanta,Ga.,1968
Met 25.2631

ATLANTIC CITY 1936,1942 Symposium on radiography 1st and 2nd American society for testing materials A.S.T.M., Philadelphia,1943
Met 25.1317

ATLANTIC CITY 1937 Open-hearth steel making By Earnshaw Cook A series of five educational lectures presented to members of the A.S.M. during the nineteenth National metal congress and exposition,Atlantic City, N. J. October 18 to 22,1937
Met 25.0329

ATLANTIC CITY 1946 The Structure of cast iron By Alfred Boyles American society for metals,Cleveland,Ohio,1947, reprinted 1949 A series of three educational lectures... presented to members of the ASM during the twenty-eighth National metal congress and exposition, Atlantic City, November 18 to 22, 1946
Met 25.0401

ATLANTIC CITY 1949 Symposium on evaluation tests for stainless steels American society for testing materials A.S.T.M.Special technical publication, 93 American society for testing materials,Philadelphia,Pa.,1950
Met 25.1967

ATLANTIC CITY 1953 Metallic materials at low temperatures Symposium on effect of temperature on the brittle behaviour of metals with particular reference to low temperatures American society for testing materials A.S.T. M.Special technical publication, 158 American society for testing materials, Philadelphia,Pa.,1954
Met 25.0850

ATLANTIC CITY 1955 Symposium on impact testing American society for testing materials American society for testing materials,Philadelphia,Pa.,1955
Met 25.0902

ATLANTIC CITY 1959 Fuel cells :a symposium American chemical society.Gas and fuel division Edited by G.J. Young Reinhold; Chapman and Hall,New York;London,1960
Met 25.1838

ATLANTIC CITY 1959 Fuel cells a symposium Proceedings American chemical society.Gas and fuel division Edited by G.J. Young v, 154p Reinhold;Chapman and Hall,New York; London,1960 Held during the 136th national meeting of the American chemical society
Chem E 24.1400

ATLANTIC CITY 1960 Polar wandering and continental drift :a symposium papers Edited by Arthur C. Munyan Society of economic paleontologists and mineralogists. Special publication,10 map Society of economic paleontologists and mineralogists, Tulsa,Oka.,1963
Geol 8.1503

ATLANTIC CITY 1960 Symposium on methods of metallographic specimen preparation American society for testing materials A.S.T.M. Special technical publication, 285 American society for testing materials,Philadelphia,Pa., 1960
Met 25.1381

ATLANTIC CITY 1963 Techniques of electron microscopy,diffraction and microprobe analysis Papers American society for testing materials A.S.T.M.Special technical publication, 372 Philadelphia,1964
Met 25.2495

ATLANTIC CITY 1965 Solvated electron :a symposium at the 150th meeting of the American chemical society American chemical society Edited by Robert F. Gould American chemical society.Advances in chemistry series, 50 American chemical society,Washington,D.C.,1965
Radioth 35.0430

ATLANTIC CITY,N.J. 1937 Carburizing :a symposium Papers and discussions American society for metals American society for metals,Cleveland,Ohio,1937 The,Symposium presented before the nineteenth annual convention of the American society for Metals..
Met 25.0439

ATLANTIC CITY,N.J. 1938 Symposium on impact testing :presented at the forty-first annual meeting A.S.T.M. American society for testing materials American society for testing materials,Philadelphia,Pa.,1938
Eng 41.3751

ATLANTIC CITY,N.J. 1939 Symposium on shear testing of soils American society for testing materials American society for testing materials,Philadelphia,Pa.,1939
Eng 41.2988

ATLANTIC CITY,N.J. 1947 Symposium on load tests of bearing capacity of soils American society for testing materials American society for testing materials.Special technical publication, 79 American society for testing materials,Philadelphia,Pa.,1948
Eng 41.3145

ATLANTIC CITY,N.J. 1950 Symposium on corrosion of materials at elevated temperatures American society for testing materials A.S.T.M.Special technical publication, 108 American society for testing materials,Philadelphia,Pa.,1951
Met 25.1906

ATLANTIC CITY,N.J. 1950 Symposium on the identification and classification of soils American society for testing materials American society for testing materials.Special technical publication, 113 American society for testing materials,Philadelphia,Pa.,1951
Eng 41.3146

ATLANTIC CITY,N.J. 1950 Symposium on the nature,occurrence,and effects of sigma phase presented at the fifty-third annual meeting... American society for testing materials A.S.T. M.Special technical publication, 110 A.S.T.M. Philadelphia,Pa.,1951
Met 25.0477

ATLANTIC CITY,N.J. 1951 Symposium on chemical analysis of inorganic solids by means of the mass spectrometer American society for testing materials.Committee E-2 on emission spectroscopy A.S.T.M.Special technical publication, 149 A.S.T.M., Philadelphia,Pa.,1951
Met 25.1723

ATLANTIC CITY,N.J. 1951 Symposium on consolidation testing of soils American society for testing materials American society for testing materials.Special technical publication, 126 American society for testing materials,Philadelphia,Pa.,1952
Eng 41.3149

ATLANTIC CITY,N.J. 1951 Symposium on statistical aspects of fatigue Pt 1 American society for testing materials A.S.T. M.Special technical publication, 121 American society for testing materials, Philadelphia,Pa.,1951
Met 25.0912

ATLANTIC CITY,N.J. 1951 Symposium on surface and subsurface reconnaissance American society for testing materials American society for testing materials.Special technical publication, 122 American society for testing materials,Philadelphia,Pa.,1952
Eng 41.3153

ATLANTIC CITY,N.J. 1953 Symposium on dynamic testing of soils American society for testing materials American society for testing materials.Special technical publication, 156 American society for testing materials,Philadelphia,Pa.,1954
Eng 41.3148

ATLANTIC CITY,N.J. 1953 Symposium on effect of temperature on the brittle behavior of metals with particular reference to low temperatures :presented at the 56th annual meeting A.S.T.M. American society for testing materials American society for testing materials.Special technical publication, 158 American society for testing materials,Philadelphia,Pa.,1954
Eng 41.3750

ATLANTIC CITY,N.J. 1953 Symposium on techniques for electron metalography American society for testing materials A.S.T. M.Special technical publication, 153 American society for testing materials, Philadelphia,Pa.,1954
Met 25.1361

ATLANTIC CITY,N.J. 1954 Photoconductivity conference Edited by R.G. Breckenridge Sponsored by the University of Pennsylvania Wiley;Chapman and Hall,New York;London,1956
Eng 41.5395

ATLANTIC CITY,N.J. 1955 Symposium on atmospheric corrosion of non-ferrous metals American society for testing materials A.S.T. M.Special technical publication, 175 A.S.T.M. Philadelphia,Pa.,1955
Met 25.1918

ATLANTIC CITY,N.J. 1955 Symposium on high-purity water corrosion American society for testing materials A.S.T.M.Special technical publication, 179 American society for testing materials,Philadelphia,Pa.,1955
Met 25.1927

ATLANTIC CITY,N.J. 1955 Symposium on impact testing :presented at the 58th A.S.T.M. annual meeting American society for testing materials American society for testing materials.Special technical publication, 176 American society for testing materials, Philadelphia,Pa.,1956
Eng 41.3757

ATLANTIC CITY,N.J. 1955 Symposium on metallic materials for service at temperatures above 1600 F. American society for testing materials A.S.T.M.Special technical publication, 174 American society for testing materials,Philadelphia,Pa.,1956
Met 25.0999

ATLANTIC CITY,N.J. 1956 Symposium on ion exchange and chromatography in analytical chemistry presented at the fifty-ninth annual meeting American society for testing materials A.S.T.M.Special technical publication, 195 A.S.T.M.,Philadelphia,Pa., 1958
Met 25.1750

ATLANTIC CITY,N.J. 1956 Symposium on vane shear testing of soils American society for testing materials American society for testing materials.Special technical publication, 193 American society for testing materials,Philadelphia,Pa.,1957
Eng 41.3150

ATLANTIC CITY,N.J. 1957 Large fatigue testing machines and their results :a symposium held at the 1957 annual meeting of the society American society for testing materials American society for testing materials,Cleveland,Ohio,1957
Met 25.0922

ATLANTIC CITY,N.J. 1957 Symposium on determination of gases in metals American society for testing materials A.S.T.M. Special technical publication, 222 A.S.T.M., Philadelphia,Pa,1958
Met 25.1708

ATLANTIC CITY,N.J. 1959 Structural damping a colloquium Papers Edited by Jerome E. Ruzicka Sponsored by American society of mechanical engineers.Applied mechanics division American society of mechanical engineers,New York,1959
Eng 41.6335

ATLANTIC CITY,N.J. 1959 Symposium on electron metallography American society for testing materials A.S.T.M.Special technical publication, 262 American society for testing materials,Philadelphia,Pa.,1960
Met 25.1366

ATLANTIC CITY,N.J. 1959 Symposium on microscopy American society for testing materials :annual meeting 62nd papers American society for testing materials.Special technical publications,257 American society for testing materials,Philadelphia,Pa.,1959
Min 10.1046

ATLANTIC CITY,N.J. 1959 Symposium on radiosotopes in metals analysis and testing presented to the sixty-second annual meeting American society for testing materials A.S.T. M.Special technical publication, 261 A.S.T.M. Philadelphia,Pa.,1960
Met 25.1741

ATLANTIC CITY,N.J. 1959 San Francisco,Calif. 1959 Oct 16 Papers on soils Symposium on time rates of loading in soil tension Symposium on Atterberg limits,Session on soils and Symposium on soils for engineering purposes American society for testing materials.Special technical publication, 254 American society for testing materials, Philadelphia,Pa.,1960 Papers presented at various A.S.T.M. meetings during 1959
Eng 41.3151

ATLANTIC CITY,N.J. 1960 Geometry of sandstone bodies :a symposium papers American association of petroleum geologists Edited by James A. Peterson and John C. Osmond American association of petroleum geologists, Tulsa,Oka.,1961 The symposium was held during the 45th annual meeting of the American association of petroleum geologists
Geol 8.1481

ATLANTIC CITY,N.J. 1960 Symposium on nuclear methods for measuring soil density and moisture American society for testing materials American society for testing materials.Special technical publication, 293 American society for testing materials, Philadelphia,Pa.,1961
Eng 41.3152

ATLANTIC CITY,N.J. 1960;1961 Electron metallography and probe microanalysis 1962 Symposium on advances in electron metallography and electron probe microanalysis American society for testing materials. Subcommittee on electron microstructure A.S. T.M.Special technical publication, 317 A.S.T. M.,Philadelphia,Pa.,1962
Met 25.1377

ATLANTIC CITY,N.J. 1960 Jun 28 Symposium on acoustical fatigue American society for testing materials A.S.T.M.Special technical publication, 284 A.S.T.M.,Philadelphia,Pa., 1961
Met 25.0929

ATLANTIC CITY,N.J. 1961 Symposium on evaluation of metallic materials in design for low-temperature service presented at the 64th annual meeting American society for testing materials A.S.T.M.Special technical publication, 302 A.S.T.M.,Philadelphia,1962
Met 25.2335

ATLANTIC CITY,N.J. 1961 Symposium on major effects of minor constituents on the properties of materials American society for testing materials American society for testing and materials.Materials science series, 2 American society for testing and materials.Special technical publications, 304 American society for testing materials, Philadelphia,Pa.,1962
Eng 41.3429

ATLANTIC CITY,N.J. 1961 Symposium on soil dynamics American society for testing materials American society for testing materials.Special technical publication, 305 American society for testing materials, Philadelphia,Pa.,1962
Eng 41.3155

ATLANTIC CITY,N.J. 1963 Adhesion ...Recent developments in adhesion science...Recent advances in adhesives technology...Presented at the 66th annual meeting A.S.T.M. American society for testing materials American society for testing materials.Special technical publication, 360 American society for testing materials,Philadeplphia,Pa.,1964
Eng 41.4027

ATLANTIC CITY,N.J. 1963 Adhesion :papers presented at the 66th annual meeting A.S.T.M. American society for testing and materials American society for testing and materials. Special technical publication, 360 American society for testing and materials,Philadelphia, Pa.,1964
Eng 41.7660

ATLANTIC CITY,N.J. 1963 Nervous control of the heart :symposium Edited by Walter C. Randall Williams and Wilkins,Baltimore,Md., 1965 Held during the annual spring meetings of the American physiological society and the Federation of American societies for experimental biology
An 32.3811

ATLANTIC CITY,N.J. 1965 Thermoanalysis of fibers and fiber-forming polymers :symposium Papers American chemical society Edited by Robert F. Schwenker Applied polymer symposia, 2 Interscience,New York,1966
Met 25.2705

ATLANTIC CITY,N.J. 1965 Transitions and relaxations in polymers :symposium Papers American chemical society Edited by Raymond F. Boyer Polymer symposia, 14 Interscience,New York,1966
Met 25.2704

ATLANTIC CITY,N.J. 1966 Determination of stress in rock :a symposium American society for testing and materials American society for testing and materials.Special technical publication, 429 American society for testing materials,Philadelphia,Pa.,1967
Eng 41.3160

ATLANTIC CITY,N.J. 1966 Erosion by cavitation or impingement :a symposium presented at the 69th annual meeting American society for testing and materials American society for testing and materials. Special technical publications, 408 American society for testing and materials,Philadelphia, Pa.,1967
Eng 41.6561

ATLANTIC CITY,N.J. 1966 Fatigue crack propagation :a symposium American society for testing and materials American society for testing and materials.Special technical publication, 415
Eng 41.3814

ATLANTIC CITY,N.J. 1966 Fatigue crack propagation :a symposium American society for testing materials A.S.T.M.Special technical publication, 415 A.S.T.M., Philadelphia,Pa.,1967
Met 25.0938

ATLANTIC CITY,N.J. 1966 Hot corrosion problems associated with gas turbines : symposium Proceedings American society for testing and materials American society for testing and materials.Special technical publication, 421 Philadelphia,Pa.,1967 Symposium presented at the annual meeting
Met 25.2788

ATLANTIC CITY,N.J. 1966 Permeability and capillarity of soils :a symposium American society for testing materials American society for testing materials.Special technical publication, 417 American society for testing and materials,Philadelphia,Pa., 1967
Eng 41.3102

ATLANTIC CITY,N.J. 1966 Stress corrosion testing :a symposium American society for testing materials A.S.T.M.Special technical publication, 425 American society for testing and materials,Philadelphia,Pa.,1967
Met 25.1964

ATLANTIC CITY,N.J. 1969 Achievement of high fatigue resistance in metals and alloys : a symposium presented at the 72nd annual meeting Proceedings American society for testing materials American society for testing materials.Special technical publication, 467 ASTM,Philadelphia,Pa.,1970
Met 25.2584

ATLAS COMPUTER LABORATORY Computational problems in abstract algebra :a conference Proceedings Oxford 1967 Aug 29-Sep 2 Science research council bibliog.,illus. x, 402p 23cm Pergamon press,Oxford,1970
P Math 2.3758

ATLAS SYMPOSIUM Computers in number theory Science research council Atlas symposium 2nd Proceedings Oxford 1969 Aug 18-23 Science research council Edited by A.O.L. Atkin and B.J. Birch xvii,433p 24cm Academic press,London;New York,1971
P Math 2.4381

ATMOSPHERE Chemical reactions in the lower and upper atmosphere :a symposium Proceedings San Francisco,Calif. 1961 Apr 18-20 Stanford research institute 390p Wiley,New York,1961
Nap 11.0929

ATMOSPHERE Optical instability of the earth's atmosphere All-union conference 3rd Proceedings Kiev 1962 July Akademiya nauk S.S.S.R.Astronomical council Glavnaya astronomicheskaya observatoriya, Pulkovo Translated by Z. Lerman from the Russian viii,174p 25cm Israel program for scientific translations,Jerusalem,1966
Obs 6.3275

ATMOSPHERE OF THE EARTH AND PLANETS A Symposium on planetary atmospheres Papers Williams Bay,Wisc. 1947 Sep 8-10 University of Chicago Edited by Gerard P. Kuiper 366p Chicago,Ill.,1949 Symposium held in connection with the 50th anniversary of the Yerkes observatory
Geod 9.0523

ATMOSPHERES The Origin and evolution of atmospheres and oceans :a conference New York 1963 Apr 8-9 Goddard institute for space studies Edited by Peter J. Brancazio and A.G.W. Cameron Wiley,New York,1964
TA 15.0427

The ATMOSPHERES OF VENUS AND MARS :a symposium Tucson,Ariz. 1967 Feb 28-Mar 2 Goddard institute for space studies and Kitt Peak national observatory.Space division Edited by John C. Brandt and Michael B. McElroy Gordon and Breach,New York,1968
TA 15.0415

ATMOSPHERIC CIRCULATION Problems of atmospheric circulation The Space science symposium 6th Papers of a session Mar del Plata,Argentina 1965 May 11-19 Committee on space research and International union of biochemistry Edited by R.V. Garcia and T.F. Malone Co-sponsored by the International union of physiological sciences 186p Spartan books,Washington,D.C. 1966
Nap 11.0212

ATMOSPHERIC DIFFUSION A Symposium on atmospheric diffusion and air pollution proceedings Oxford 1958 Aug 24-29 International union of theoretical and applied mechanics and International union of geodesy and geophysics Edited by F.N. Frenkel and P.A. Sheppard Advances in geophysics, 6 Academic press,New York; London,1959
A Math 4.0576

ATMOSPHERIC DIFFUSION AND AIR POLLUTION :a symposium Proceedings Oxford 1958 Aug 24-29 International union of theoretical and applied mechanics International union of geodesy and geophysics Edited by F.N. Frenkel and P.A. Sheppard Advances in physics, 6 Academic press,New York,1959
Eng 41.6946

ATMOSPHERIC DIFFUSION AND AIR POLLUTION :a symposium Proceedings Oxford 1958 Aug 24-29 International union of theoretical and applied mechanics and International union of geodesy and geophysics Edited by F.N. Frenkiel and P.A. Sheppard New York,1959
Geod 9.0521

ATMOSPHERIC ELECTRICITY Recent advances in atmospheric electricity Conference on atmospheric electricity 2nd Proceedings Portsmouth,N.H. 1958 May 20-23 Edited by L. G. Smith Sponsored by Air force Cambridge research center.Geophysics research directorate.Aerophysics laboratory illus 646p Pergamon,London,1958
Nap 11.0928

ATMOSPHERIC ELECTRICITY Recent advances in atmospheric electricity Conference on atmospheric electricity 2nd Proceedings Portsmouth,N.H. 1958 May 20-23 Edited by L. G. Smith Sponsored by Air force Cambridge research center,Geophysics research directorate, Aerophysics laboratory illus 646p Pergamon press,London,1958
Nap 11.0562

ATMOSPHERIC EXPLORATION The Benjamin Franklin memorial symposium Papers American academy of arts and sciences 125p M.I.T.;Wiley,New York,1958
Nap 11.0960

ATMOSPHERIC EXPOSURE TESTS ON NON-FERROUS METALS a symposium Pittsburgh 1946 Feb 27 American society for testing materials American society for testing materials, Philadelphia,Pa.,1946
Met 25.1924

ATMOSPHERIC PHYSICS The Emission spectra of the night sky and aurorae :a conference Papers London 1947 Jul.7-10 Royal society of London.Gassiot committee 139p London,1948
Obs 6.1964

ATMOSPHERIC TURBULENCE AND RADIO WAVE PROPAGATION Atmosfernaya turbulentnostui i rasprostranenie radioboli :an international colloquium Proceedings Moscow 1965 Jun 15-22 Edited by A.M. Yaglom and V.I. Tatarsky 370p Moscow,1967
A Math 4.1307

ATOM MOVEMENTS Chicago 1950 Oct 21-27 American society for metals American society for metals,Cleveland,1951 A seminar held during the 32nd national metal congress and exposition
Met 25.1272

ATOM MOVEMENTS :a seminar proceedings Chicago 1950 Oct 21-27 Sponsored by the American society for metals 240p American society for metals,Cleveland,Ohio,1951 Held during the 32nd National metal congress and exposition
Chem E 24.0757

ATOMIC ABSORPTION SPECTROSCOPY :a symposium presented at the seventy-first annual meeting.. Proceedings San Francisco,Calif. 1968 Jun 23-28 American society for testing and materials American society for testing and materials.Publication, 443 ASTM, Philadelphia,Pa.,1969
Met 25.2647

ATOMIC AND ELECTRONIC STRUCTURE OF METALS :a seminar Papers Cleveland,Ohio 1966 Oct 29-30 American society for metals American society for metals,Cleveland,Ohio, 1967
Met 25.1059

ATOMIC COLLISION PHENOMENA Atomic collision processes The International conference on the physics of electronic and atomic collisions 3rd Proceedings London 1963 Jul 22-23 University college,London Edited by M.R.C. McDowell North-Holland, Amsterdam,1964
Cav 7.1461

ATOMIC COLLISION PHENOMENA IN SOLIDS :an international conference Proceedings Brighton 1969 Sep 7-12 Edited by D.W. Palmer and others Bibliog 678p 17cm North-Holland,Amsterdam,1970
Cav 7.2680

ATOMIC COLLISION PROCESSES Symposium on atomic collision processes Boulder summer institute for theoretical physics 11th Proceedings Boulder,Colo. 1968 Jun 17-Aug 23 University of Colorado.Department of physics and astrophysics Edited by Kalyana T. Mahanthappa and others Lectures in theoretical physics, 11c diagrms xiii, 337p 23cm Gordon and Breach,London,1969 Dedicated to George Gamow
Cav 7.3010

ATOMIC ENERGY International conference on the peaceful uses of atomic energy 2nd Proceedings Geneva 1958 Sep 1-13 18,21-26,29 United Nations 8 vols United Nations,Geneva,1958
Radioth 35.1079

ATOMIC ENERGY International conference on the peaceful uses of atomic energy 2nd Proceedings Geneva 1958 Sep 1-13 Vol 31-32 United Nations 2 vols United Nations,Geneva,1958
Eng 41.4466

ATOMIC ENERGY International conference on the peaceful uses of atomic energy Proceedings Geneva 1955 Aug 8-20 10-14 United nations 5 vols United Nations, New York,1955-56
Radioth 35.1009

ATOMIC ENERGY Technology of the gas-cooled power reactor and related subjects United Nations conference on the peaceful uses of atomic energy 2nd Papers United Kingdom atomic energy authority United Kingdom atomic energy authority,Harwell,1958
Eng 41.7494

ATOMIC ENERGY United Nations conference on the peaceful uses of atomic energy Proceedings Geneva 1959 Aug 8-20 Vol 1-15 8 vols United Nations,New York,1956 Lacks vols 6-7.10-14
Eng 41.7540

ATOMIC ENERGY COMMISSION The Chemical basis of heredity :a symposium Baltimore,Md. 1956 Jun 19-22 McCollum-Pratt institute Edited by William D.Bentley McElroy and Bentley Glass McCollum-Pratt institute. Contribution,153 Johns Hopkins press, Baltimore,Md.,1957
Chem 18.1555

ATOMIC ENERGY RESEARCH ESTABLISHMENT Isotope techniques conference Proceedings Oxford 1951 Jul Vol 2: industrial and allied research applications Great Britain.Ministry of supply H.M.S.O.,London,1952
Chem 18.2467

ATOMIC ENERGY RESEARCH ESTABLISHMENT Radioisotope conference 2nd Proceedings Oxford 1954 Jul 19-23 1: medical and physiological applications Edited by J.E. Johnston and others Butterworths,London,1954 First conference called Isotope techniques conference,q.v.
Radioth 35.1349

ATOMIC ENERGY RESEARCH ESTABLISHMENT Radioisotope conference 2nd Proceedings Oxford 1954 Jul 19-23 Vol 1: medical and physiological applications Edited by J.E. Johnston and others Butterworths,London,1954 Previous conference has title "Isotope techniques conference",Oxford,1951
Bioch 33.0300

ATOMIC ENERGY RESEARCH ESTABLISHMENT Radioisotope techniques Isotope techniques conference Proceedings Oxford 1951 Jul 16-20 Vol 1: medical and physiological applications H.M.S.O.,London,1953 Subsequent conference called Radioisotope conference,q.v.
Radioth 35.1348

ATOMIC ENERGY RESEARCH ESTABLISHMENT.CONSULTANT'S SYMPOSIUM 3RD The Interactions between dislocations and point defects :a symposium Proceedings Harwell 1968 Jul 4-12 1-3 Atomic energy research establishment. Metallurgy division United Kingdom atomic energy authority.Research group.Report AERE-R 5944 3 vols harwell,1968
Met 25.2415

ATOMIC ENERGY RESEARCH ESTABLISHMENT.METALLURGY DIVISION The Interactions between dislocations and point defects :a symposium Proceedings Harwell 1968 Jul 4-12 1-3 United Kingdom atomic energy authority. Research group.Report AERE-R 5944 3 vols harwell,1968
Met 25.2415

ATOMIC MECHANISMS OF FRACTURE :an international conference Proceedings Swampscott,Mass. 1959 Apr 12-16 National research council.Committee for the conference on fracture Edited by B.L. Averback and others Technology press books in science and engineering Technology press,John Wiley; Chapman and Hall,New York;London,1959
Met 25.2323

ATOMIC PHYSICS International conference on atomic physics 1st Proceedings New York City 1968 Jun 3-7 Edited by B. Bederson and others Bibliog.,Illus. xiii,620p 26cm Plenum press,New York,1969
A Math 4.1601

ATOMIC PHYSICS AND ASTROPHYSICS Brandeis university.Summer institute in theoretical physics 12th Proceedings Waltham,Mass. 1969 Jun 16-Jul 25 Edited by Max Chretien and E. Lipworth 216p Gordon and Breach,New York,1971
TA 15.0683

ATOMIC SCIENTISTS ASSOCIATION Biological hazards of atomic energy :a conference Papers 1950 Oct Edited by A. Haddow pl. Clarendon press,Oxford,1952
Radioth 35.0180

ATOMIC SCIENTISTS' ASSOCIATION Biological hazards of atomic energy :a conference papers London 1950 Oct Clarendon press, Oxford,1952
HE 27.0097

ATOMIC SPECTRA The Menzel symposium on solar physics,atomic spectra and gaseous nebulae Cambridge,Mass. 1971 Apr 29 Edited by K.B. Gebbie National bureau of standards.Special publication, 353 203p National bureau of standards,Washington,D.C.,1971
Obs 6.3623

ATOMIC SPECTROSCOPY Rydberg centennial conference on atomic spectroscopy Proceedings Lund 1954 Jul 1-5 Edited by Bengt Edlen Arranged by the Swedish national committee for physics Physiographisk sallskaps i Lund.Forhandlingar.N.F., 65,21 108p c.k.w.gleerup,Lund,1954
Obs 6.3285

ATOM'S NUCLEUS AND ITS PROPERTIES Rutherford jubilee international conference Proceedings Manchester 1961 Sep 4-8 Victoria university of Manchester Edited by J.B. Birks Under the sponsorship of International union of pure and applied physics 856p Haywood,London,1961 Includes commemorative session 'Rutherford at Manchester'
Cav 7.1952

ATTENTION AND PERFORMANCE :a symposium Proceedings Driebergen 1966 Aug 17-20 Institute for perception RVO-TNO,Soesterberg Edited by A.F. Sanders North-Holland, Amsterdam,1967
Psy 31.0959

ATTENTION AND PERFORMANCE 2 Donders centenary symposium on reaction time Proceedings Eindhoven 1968 Jul 29-Aug 2 Institute for perception research IPO Edited by W.G. Koster North-Holland,Amsterdam; London,1969
Psy 31.2948

ATTENTION IN NEUROPHYSIOLOGY :an international conference Proceedings Teddington 1967 Oct National physical laboratory Edited by C.R. Evans and T.B. Mulholland port Butterworths,London,1970
Psy 31.2849

AUCKLAND 1949 and CHRISTCHURCH,N.Z. 1949 Pacific science congress 7th proceedings Vol 2: geology Pacific science association Wellington,1953
Geol 8.3069

AUCKLAND N.Z. 1965 Australasian conference on hydraulics and fluid mechanics 2nd proceedings University of Auckland Edited by C.M. Segedin Sponsored jointly with the New Zealand institution of engineers Conference organising committee,Auckland,N.Z., 1966
A Math 4.0531

AUDITION A Discussion held...at the Imperial college of science Report London 1931 Jun 19 Physical society Physical society, London,1931 Bound with 'Report of a joint discussion on vision...'by the Physical and optical society
Psy 31.3361

AUGUSTA,MICH. 1967 Tolbutamide...after ten years The Brook Lodge symposium Proceedings Edited by W.J.H. Butterfield and W.van Westering International congress series, 149 Excerpta,medica,Amsterdam,1967
Bioch 33.0425

AUGUSTA,MICH. 1967 Tolbutamide...after ten years :the Brook Lodge symposium Proceedings Edited by W.J.H. Butterfield and W.van Westering Excerpta medica international congress series,149 Bibliogs 346p Excerpta medica foundation,Amsterdam,1967
Pha 16.0264

AURORA AND AIRGLOW proceedings of the Nato advanced study institute Proceedings Keele,Staffs 1966 Aug 15-26 Nato advanced study institute Edited by Billy M. McCormac illus vii,689p 24cm Reinhold,New York, 1967
Sco 14.7473

AURORAE The Airglow and the aurorae :a symposium Belfast 1955 Sep.6-7 Edited by E.B. Armstrong and A. Dalgarno Journal of atmospheric and terrestrial physics. Supplements., 5 420p Pergamon press, London,1956
Obs 6.0083

AURORAE The Airglow and the aurorae :a symposium Belfast 1955 Sep.6-7 Edited by E.B. Armstrong and A. Dalgarno Journal of atmospheric and terrestrial physics. Supplements, 5 420p London,1956
Nap 11.0014

AURORAL PHYSICS Conference on auroral physics 1st Proceedings London,Ontario 1951 Jul 23-26 Edited by N.C. Gersen and others Sponsored jointly by the University of Western Ontario.Department of physics. Air force Cambridge research center. Geophysical research papers,30 illus xxvi, 462p Cambridge,Mass.,1954
Sco 14.0613

AUSTIN,TEX. 1954 Latin-American geology conference proceedings University of Texas. Department of geology,and,University of Texas Institute of Latin-American studies Edited by Fred M. Bullard map University of Texas, Austin,Tex.,1955
Geol 8.2706

AUSTIN,TEX. 1959 Midwestern conference on solid mechanics 4th proceedings University of Texas,Austin,Tex,1959
A Math 4.0368

AUSTIN,TEXAS 1956 Texas conference on soil mechanics and foundation engineering 8th Proceedings University of Texas.Bureau of engineering research University of Texas, Austin,Texas,1956
Eng 41.3164

AUSTIN,TEXAS 1959 Report on the I.R.E. subcommittee 4.10 round table conference in Austin,Texas mimeograph By Wolfgang J. Poppelbaum University of Illinois.Digital computer laboratory 20p 1959
Math L 5.1676

AUSTIN,TEXAS 1966 Austin symposium on gas phase molecular structure Symposium on machine interpretations of Patterson functions and alternative direct approaches and the Austin symposium on gas phase molecular structure Proceedings Edited by W.F. Bradley and Harold P. Hansen American crystallographic association.Transactions, 2, 1966 Polycrystal book services,Pittsburgh,Pa. 1966
Chem 18.2633

AUSTIN SYMPOSIUM ON GAS PHASE MOLECULAR STRUCTURE Symposium on machine interpretations of Patterson functions and alternative direct approaches and the Austin symposium on gas phase molecular structure Proceedings Austin,Texas 1966 Feb 28-Mar 2 Edited by W. F. Bradley and Harold P. Hansen American crystallographic association.Transactions, 2, 1966 Polycrystal book services,Pittsburgh,Pa. 1966
Chem 18.2633

AUSTRALASIAN CONFERENCE ON HYDRAULICS AND FLUID MECHANICS 1st Proceedings Perth 1962 Dec 6-13 Edited by Richard Silvester Pergamon press,London,1964
Eng 41.6968

AUSTRALASIAN CONFERENCE ON HYDRAULICS AND FLUID MECHANICS 2nd proceedings Auckland N.Z. 1965 Dec.6-11 University of Auckland Edited by C.M. Segedin Sponsored jointly with the New Zealand institution of engineers Conference organising committee,Auckland,N.Z., 1966
A Math 4.0531

AUSTRALASIAN CONFERENCE ON RADIATION BIOLOGY 2nd Proceedings Radiation biology Melbourne 1958 Dec 15-18 Australian radiation society Edited by J.H. Martin Butterworths,London,1959
Gen 34.1110

AUSTRALASIAN CONFERENCE ON RADIOBIOLOGY 3rd Proceedings Radiobiology Sydney 1960 Aug 15-18 Australian radiation society Edited by P.L.T. Ilbery Butterworths,London, 1961
Radioth 35.1727

AUSTRALASIAN INSTITUTE OF MINING AND METALLURGY Empire mining and metallurgical congress 5th Proceedings Melbourne 1953 Apr 14-Jun 12 Vol 4: extractive metallurgy in Australia Edited by J.C. Richards Melbourne,1953
Met 25.2524

AUSTRALASIAN INSTITUTE OF MINING AND METALLURGY Empire mining and metallurgical congress 5th Publications 5: Australian mining and metallurgy,miscellaneous features and practices Edited by A.W. Norrie Australasian institute of mining and metallurgy,Melbourne,1953
Met 25.0069

AUSTRALASIAN INSTITUTE OF MINING AND METALLURGY Extractive metallurgy in Australia:non-ferrous metallurgy Empire mining and metallurgical congress 5th Publications Melbourne 1953 Apr 14-Jun 12 4B Edited by Frank A. Green and others Office of the Congress,Melbourne,1953
Met 25.0498

AUSTRALASIAN INSTITUTE OF MINING AND METALLURGY Geology of Australian ore deposits Empire mining and metallurgical congress 5th Publications Melbourne 1953 Apr 14 Jun 12 1 Edited by A.B. Edwards Office of the congress,and of the Australasian institute of mining and metallurgy,Melbourne, 1953
Met 25.0187

AUSTRALASIAN INSTITUTE OF MINING AND METALLURGY Ore dressing methods in Australia and adjacent territories Empire mining and metallurgical congress 5th Publications Melbourne 1953 Apr 14-Jun 12 3 Edited by H.H. Dunkin Office of the congress, and of the Australasian institute of mining and metallurgy,Melbourne,1953
Met 25.0205

AUSTRALASIAN INSTITUTE OF MINING AND METALLURGY Research in chemical and extraction metallurgy Symposium on 'recent progress in research in chemical and extraction metallurgy' Proceedings Sydney 1965 Feb 24-26 Edited by J.T. Woodcock and others Australasian institute of mining and metallurgy.Monograph series, 2 Melbourne, 1965 reprinted 1967
Met 25.2780

AUSTRALASIAN MEDICAL CONGRESS 8th session Transactions Melbourne 1908 Oct Vol 1-3 Edited by Alan Lewers 3 vols Kemp, Melbourne,1909
An 32.1144

AUSTRALASIAN MEDICAL CONGRESS 9th session Transactions Sydney 1911 Sep Vol 1-2 2 vols Gullick,Sydney,1913
An 32.1145

AUSTRALIA 1965 New Zealand 1965 Commonwealth mining and metallurgical congress 8th publications Vol 1-2: geology of Australian ore deposits,2nd edition; Exploration and mining geology Edited by John McAndrew 2 vols Australasian institute of mining and metallurgy,Melbourne, 1965
Min 10.0222

AUSTRALIA.DEPARTMENT OF SUPPLY.WEAPONS RESEARCH ESTABLISHMENT Data processing and automatic computing machines conference Salisbury,S.Australia 1957 Jun 3-8 Six typescript papers
Math L 5.0776

AUSTRALIA-NEW ZEALAND CONFERENCE ON SOIL MECHANICS AND FOUNDATION ENGINEERING 1st Proceedings Melbourne 1952 Jul University of Melbourne Institution of engineers,Australia 1952
Eng 41.3166

AUSTRALIA-NEW ZEALAND CONFERENCE ON SOIL MECHANICS AND FOUNDATION ENGINEERING 4th Proceedings Adelaide 1963 Aug 19-23 University of Adelaide 1963
Eng 41.3169

AUSTRALIAN ACADEMY OF SCIENCE Antarctic meteorology Proceedings Melbourne Australia.Special committee for the International geophysical year,and,Special committee for Antarctic research illus,maps, tables,diagrs xviii,483p 25cm Pergamon press,London,1960
Sco 14.6145

AUSTRALIAN ACADEMY OF SCIENCE Antarctic meteorology :a symposium Proceedings Melbourne 1959 Feb Pergamon press,Oxford, 1960
Geog 13.0989

AUSTRALIAN ACADEMY OF SCIENCE Antarctic meteorology :a symposium Proceedings Melbourne 1959 Feb.18-25 Commonwealth bureau of meteorology and International geophysical year 483p Pergamon press, Oxford;London,1960
Nap 11.0026

AUSTRALIAN ACADEMY OF SCIENCE Captain Cook, navigator and scientist Cook bicentenary symposium Papers Canberra 1969 May 1 Edited by G.M. Badger ix,143p Hurst; Humanities press,London;New York,1970
WSM 43.1245

AUSTRALIAN ACADEMY OF SCIENCE Haematin enzymes :a symposium Papers and discussions Canberra 1959 Aug 31-Sep 4 Pt 1 International union of biochemistry Edited by J.E. Falk and others I.U.B.Symposium series, 19 Pergamon press,Oxford,1961
Bioch 33.1079

AUSTRALIAN ACADEMY OF SCIENCE International conference on 'Electron diffraction' and 'The nature of defects in crystals' Abstracts of papers Melbourne 1965 Aug 16-21 With the support of the International union of pure and applied physics. Commission on the solid state Pergamon press,Oxford,1966 Extended abstracts of papers to be presented at the conference
Cav 7.1584

AUSTRALIAN ACADEMY OF SCIENCE International conference on"electron diffraction" and "the nature of defects in crystals" Abstracts of papers Melbourne 1965 Aug 16-21 Co-sponsored by the International union of pure and applied physics Pergamon press,Oxford, 1966
Met 25.1463

AUSTRALIAN ACADEMY OF SCIENCE Replication and recombination of genetic material :an international conference Papers Edited by W.J. Peacock and R.D. Brock Australian academy of science,Canberra,1968
Bot 42.6703

AUSTRALIAN ACADEMY OF SCIENCE Symposium on fibrous proteins Papers Canberra 1967 Aug 14-18 Edited by W.G. Crewther Butterworths,Sydney,1968
Bioch 33.0512

AUSTRALIAN ACADEMY OF SCIENCE Water resources use and management :a symposium Proceedings Canberra 1963 Sep 9-13 Melbuorne university press,London;New York, 1964
Geog 13.0787

AUSTRALIAN ACADEMY OF SCIENCES Photosynthesis and photorespiration :a conference Canberra 1970 Nov 23-Dec 5 Edited by M.D. Hatch and others Interscience, New York,1971
Bioch 33.2221

The AUSTRALIAN ENVIRONMENT handbook prepared for the conference on plant and animal nutrition... 1949 Aug Commonwealth scientific and industrial research organization 2nd edition British commonwealth official scientific specialist conferences,Special agricultural conference Commonwealth scientific and industrial research organization,Melbourne,1950
Gen 34.1372

AUSTRALIAN INSTITUTE OF METALS Interfaces conference Papers and abstracts Melbourne 1969 Aug Edited by R.C. Gifkins Butterworth,Sydney,1969
Met 25.2596

AUSTRALIAN MATHEMATICAL SOCIETY Summer research institute of the Australian mathematical society research reports Canberra 1961 Jan. Pt. 1: probability and statistics Australian mathematical society, Canberra,1961 Typescript
Math 3.0676

AUSTRALIAN MINING AND METALLURGY Empire mining and metallurgical congress 5th Publications 5: Australian mining and metallurgy,miscellaneous features and practices Australasian institute of mining and metallurgy Edited by A.W. Norrie Australasian institute of mining and metallurgy,Melbourne,1953
Met 25.0069

AUSTRALIAN NATIONAL UNIVERSITY Data representation :a conference Proceedings Canberra 1969 Dec 10 Edited by R.S. Andersen and M.R. Osborne 117p 24cm University of Queensland press,Brisbane,1970
Math S 3.1800

AUSTRALIAN NATIONAL UNIVERSITY Picture language machines :a conference Proceedings Canberra 1969 Feb 24-28 Edited by S. Kaneff Academic press,London;New York,1970
An 32.5428

AUSTRALIAN NATIONAL UNIVERSITY Picture language machines :a conference Proceedings Canberra 1969 Feb 24-28 Edited by S. Kaneff illus. 425p Academic press,London, 1970
Math L 5.3579

AUSTRALIAN NATIONAL UNIVERSITY The Theory of groups international conference proceedings Canberra 1965 Aug 10-20 Edited by L.G. Kovacs and B.H. Neumann Bibliog xviii,397p 24cm Gordon and Breach,New York,1967
P Math 2.2793

AUSTRALIAN NATIONAL UNIVERSITY.RESEARCH SCHOOL OF PHYSICAL SCIENCES Picture language machines :a conference Proceedings Canberra 1969 Feb 24-28 Edited by S. Kaneff Academic press,London,1970
Psy 31.3083

AUSTRALIAN PHYSICAL EDUCATION ASSOCIATION World congress on physical education Report Melbourne 1956 Nov 16-21 Sponsored by the Commonwealth national fitness council Australian physical education association, Melbourne,1956
HM 27.0219

AUSTRALIAN RADIATION SOCIETY Radiation biology Australasian conference on radiation biology 2nd Proceedings Melbourne 1958 Dec 15-18 Edited by J.H. Martin Butterworths,London,1959
Gen 34.1110

AUSTRALIANN ACADEMY OF SCIENCE Phase transformations and the earth's interior :a conference Canberra 1969 Jan 6-10 Edited by A.E. Ringwood and D.H. Green North-Holland,Amsterdam,1970
Min 10.1459

AUTOIONIZATION Symposium on atomic interactions and space physics Autoionization: astrophysical,theoretical and laboratory experimental aspects Papers Greenbelt,Md. 1965 Aug 10-11 Edited by Aaron Temkin Mono books,Baltimore,Md.,1966
TA 15.0479

AUTOIONIZATION,ASTROPHYSICAL,THEORETICAL AND LABORATORY EXPERIMENTAL ASPECTS Symposium on atomic interactions and space physics Greenbelt,Ma. 1965 Aug.10-11 Goddard space flight center Edited by Aaron Temkin 155p Mono book corp.,Baltimore,Ma.,1966
A Math 4.0911

AUTOMATA Symposium on computers and automata Proceedings New York 1971 Apr 13-15 Microwave research institute Edited by Jerome Fox Microwave research institute. Symposia series, 21 illus 653p Polytechnic press,Brooklyn,N.Y.,1971
Math L 5.3635

AUTOMATA STUDIES Edited by C.E. Shannon and J.M. McCarthy Annals of mathematics studies, 34 viii,285p 26cm Princeton university press,Princeton,N.J.,1956
Math L 5.0749

AUTOMATIC AND MANUAL CONTROL Conference on automatic control papers Cranfield 1951 Jul 18-21 Great Britain.Department of scientific and industrial research xi,584p 25cm Butterworth,London,1951
Math L 5.0021

AUTOMATIC AND MANUAL CONTROL :a conference Papers Cranfield 1951 Jul 16-21 Great Britain.Department of scientific and industrial research Edited by A. Tustin xi, 584p Butterworths,London,1952
Chem E 24.1142

AUTOMATIC AND MANUAL CONTROL Papers Cranfield 1951 Jul 16-21 Institution of electrical engineers Institution of mechanical engineers Edited by A. Tustin Under the auspices of Great Britain.Department of scientific and industrial research Butterworths,London,1952
Psy 31.3398

AUTOMATIC AND MANUAL CONTROL... Conference on automatic control Papers Cranfield 1951 Jul 16-21 Edited by A. Tustin Organized by Great Britain.Department of scientific and industrial research Butterworths,London,1952
Eng 41.6006

AUTOMATIC AND REMOTE CONTROL Congress of the International federation of automatic control 1st-3rd Proceedings 1960,1963,1966 International federation of automatic control Butterworths,London,1961-66
Eng 41.8207

AUTOMATIC CALCULATING MACHINES Conference on high-speed automatic calculating machines Report Cambridge 1949 Jun 22-25 University of Cambridge.Mathematical laboratory Great Britain.Ministry of supply University mathematical laboratory,Cambridge, 1950
Eng 41.2175

AUTOMATIC CODING symposium proceedings Philadelphia,Pa. 1957 Jan 24-25 Edited by Nancy S. Glenn Franklin institute.Journal. Monograph, 3 vii,116p 24cm Journal of the Franklin institute,Philadelphia,Pa.,1957
Math L 5.0781

AUTOMATIC COMPUTING MACHINES :conference proceedings Sydney 1951 Aug 7-9 Commonwealth scientific and industrial research organization Jointly sponsored by University of Sydney.Department of electrical engineering illus. 220p 25cm Commonwealth scientific and industrial research organization,Melbourne,1952
Math L 5.0703

AUTOMATIC CONTROL Joint automatic control conference 4th Preprints of technical papers Minneapolis 1963 Jul 19-21 American institute of chemical engineers xiv, 680p American institute of chemical engineers,New York,1963
Chem E 24.1172

AUTOMATIC CONTROL Joint automatic control conference 5th Preprints of conference papers Stanford,Calif. 1964 Jun 24-26 Under the sponsorship of the Institute of electrical and electronics engineers. Professional technical group in automatic control Stanford,Calif.,1964
Eng 41.6014

AUTOMATIC CONTROL Joint automatic control conference 6th Preprints of conference papers New York 1965 Jun 22-25 Sponsored by American society of mechanical engineers. Automatic control division New York,1965
Eng 41.6015

AUTOMATIC CONTROL Joint automatic control conference 7th Preprints of conference papers Washington,D.C. 1966 Aug 17-19 Sponsored by the American institute of aeronautics and astronautics Washington,D.C., 1966
Eng 41.6016

AUTOMATIC CONTROL Symposium on recent mechanical engineering developments in automatic control Proceedings London 1960 Jan 5-7 Institution of mechanical engineers Institution of mechanical engineers,London,1960
Eng 41.6012

AUTOMATIC CONTROL The Conference on automatic control Papers Cranfield 1951 Jul 16-21 Great Britain.Department of scientific and industrial research Edited by A. Tustin 584p Butterworths,London,1952
Cav 7.2360

AUTOMATIC CONTROL The Teaching of automatic control :a discussion Proceedings London 1962 Jan 29 Institution of mechanical engineers Institution of mechanical engineers,London,1962
Eng 41.5814

AUTOMATIC DATA PROCESSING Economics of automatic data processing :international symposium papers Rome 1965 Oct 19-22 International computation centre Edited by A. B. Frielink pl. xiii,384p 22cm North-Holland,Amsterdam,1965
Math L 5.1022

AUTOMATIC DATA PROCESSING conference proceedings Boston,Mass. 1955 Sep 8-9 Harvard university.Graduate school of business administration Edited by Robert N. Anthony illus. viii,194p 21cm Harvard university, Boston,Mass.,1956
Math L 5.0752

AUTOMATIC DEMONSTRATION Symposium on automatic demonstration Versailles 1968 Dec Edited by M. Laudet and others Organised by the Institut de recherche d'informatique et d'automatique Lecture notes in mathematics, 125 Bibliog 310p 25cm Springer-Verlag,Berlin,1970
P Math 2.3614

AUTOMATIC DIGITAL COMPUTATION International symposium on automatic digital computation 3rd proceedings Teddington 1953 Ot 25-28 National physical laboratory illus vi,296p 26cm National physical laboratory,Teddington, 1954
ATH L 5.0724

The AUTOMATIC FACTORY - WHAT DOES IT MEAN? : conference Margate 1955 Jun 16-19 Institution of production engineers Institution of production engineers,London, 1955
Eng 41.5974

AUTOMATIC MEASUREMENT OF QUALITY IN PROCESS PLANTS :a conference Proceedings Swansea 1957 Sep 23-26 Society of instrument technology xi,320p Butterworths, London,1958
Chem E 24.1160

AUTOMATIC PROGRAMMING Annual review in automatic programming Vol. 1-4 Edited by Richard Goodman International tracts in computer science and technology and their application 25cm 4 vols Pergamon press, Oxford,1960-1964 Vol.1:papers read at a working conference on automatic programming of digital computers,Brighton,April 1959
Math L 5.0893

AUTOMATIC PROGRAMMING FOR DIGITAL COMPUTERS : symposium Washington,D.C. 1954 May 13-14 Organized by United States.Navy mathematical computing advisory panel 152p Office of naval research,Washington,D.C.,1954
Math L 5.1795

AUTOMATION International automation congress Proceedings Paris 1956 Jul 18-24 Presses academiques europeenes,Brussels,1959
Eng 41.6008

AUTOMATION Sessiya Akademii nauk SSSR po nauchnym problemam avtomatizatsii proizvodstva plenarnye zasedaniya Moscow 1956 Oct 15-20 Akademiya nauk S.S.S.R. Edited by D.Ya. Libenson and others 271p 22cm Akademiya nauk SSSR,Moscow,1957
Math L 5.0775

AUTOMATION AND DATA PROCESSING IN PATHOLOGY : a symposium London 1969 Feb. College of pathologists Edited by T.P. Whitehead Journal of clinical pathology,London,1969
HE 27.0161

AUTOMATION '65 British automation conference 1965 Eastbourne 1965 Nov 7-10 Confederation of British industry Trades union congress United Kingdom automation council Organised by the Institution of production engineers Institution of production engineers,London,1965
Eng 41.6018

AUTOREGULATION OF BLOOD FLOW :international symposium Proceedings Indianapolis,Ind. 1963 Jan 10-12 Edited by Paul C. Johnson American heart association monographs, 8 American heart association,New York,1964
Phys 20.0898

AUTOTROPHIC MICRO-ORGANISM :a symposium Papers London 1954 Apr Society for general microbiology Edited by B.A. Fry and J.L. Peel Held at the Institution of electrical engineers Society for general microbiology.Symposia, 4 Cambridge university press,Cambridge,1954
Bioch 33.1168

AUXILIARY INSTRUMENTATION FOR LARGE TELESCOPES a conference Geneva 1972 May 2-5 European southern observatory CERN Edited by S. Lanstsen and A. Reiz 525p European southern observatory,Geneva,1972
Obs 6.3599

AVALANCHES The International symposium on scientific aspects of snow and ice avalanches Rapports et discussions Davos 1965 Apr 5-10 International association of scientific hydrology International association of scientific hydrology.Publication,69 illus 417p Gentbrugge,1966 In English and French
Sco 14.0142

AVIAN TUMOR VIRUSES International conference on Avian tumor viruses Proceedings Durham, N.C. 1964 Mar 31-Apr 3 National cancer institute Duke university.School of medicine Edited by Joseph W. Beard National cancer institute.Monograph, 17 National cancer institute,Bethesda,Md.,1964
Bioch 33.1154

AVIAN TUMOR VIRUSES International conference on avian tumour viruses Proceedings Durham, N.C. 1964 Mar 31-Apr 3 National cancer institute and Duke university.School of medicine Edited by Joseph W. Beard National cancer institute.Monograph, 17 US government printing office,Washington,D.C., 1964
Path 30.2570

The AXIOMATIC METHOD international symposium proceedings Berkeley, Calif. 1957-58 Dec 26 to Jan 4 Edited by Leon Henkin and others Studies in logic and the foundations of mathematics xi,488p 23cm North-Holland,Amsterdam,1959 With special reference to geometry and physics
P. Math 2.0110

AXIOMATIC SET THEORY Symposium in pure mathematics 14th Proceedings Los Angeles, Calif. 1967 Jul 10-Aug 5 Edited by Dana S. Scott Held at the University of California Los Angeles American mathematical society. Proceedings of symposia in pure mathematics, 13,pt 1 v,474p 26cm American mathematical society,Providence,R.I.,1971
P Math 2.4061

AXIOMATICS The Axiomatic method international symposium proceedings Berkeley, Calif. 1957-58 Dec 26 to Jan 4 Edited by Leon Henkin and others Studies in logic and the foundations of mathematics xi, 488p 23cm North-Holland,Amsterdam,1959 With special reference to geometry and physics
P. Math 2.0110

AYRES William L. and WILDER Raymond L. ed. Lectures in topology :conference Ann Arbor 1940 Jun.24-Jul.6 University of Michigan vii,316p 23cm University of Michigan press, Ann Arbor, Mich.,1941
P. Math 2.0288

BAARN 1947 and SCHEVENINGEN 1948 Germany's contribution to European economic life Conference on some aspects of the German problem By Percy W. Bidwell and others Riviere,Paris,1949 Papers in English and French
Geog 13.2112

The BABBAGE MEMORIAL MEETING London 1971 Oct 18 British computer society Royal statistical society illus 40p British computer society,London,1971
Math L 5.3612

The BABBAGE MEMORIAL MEETING Report of the proceedings London 1971 Oct 18 Royal statistical society British computer society iii,39p 30cm British computer society, London,1971
Math S 3.1858

BACKGROUND TO EVOLUTION IN AFRICA Systematic investigation of the African later tertiary and quaternary :symposium Burg Wartenstein 1965 Jul 14-Aug 9 Edited by Walter W. Bishop and J.Desmond Clark Under the auspices of the Wenner-Gren foundation for anthropology and research University of Chicago press,Chicago,Ill.,1967
Geog 13.4660

BACQ Z.M. and ALEXANDRA P. ed. Radiobiology symposium proceedings Liege 1954 Aug-Sep London,1955
Philos 1.1413

BACTERIA Nature of the bacterial surface Society for general microbiology :symposium 1st Proceedings London 1949 Apr 20 Edited by A.A. Miles and N.W. Pirie Blackwell,Oxford,1949
Col S 12.0036

BACTERIAL ANATOMY Society for general microbiology :symposium 6th Papers London 1956 Apr Edited by E.T.C. Spooner and B.A.D. Stocker Cambridge university press,Cambridge,1956
Col S 12.0037

BACTERIAL ANATOMY :a symposium Papers London 1956 Apr Society for general microbiology Edited by E.T.C. Spooner and B. A.D. Stocker Held at the Royal institution Society for general microbiology.Symposia, 6 Cambridge university press,Cambridge,1956
Bioch 33.1169

BACTERIAL CYTOLOGY International congress of microbiology 6th Rome 1953 Vol 1: symposium - bacterial cytology Edited by S. Mudd and G. Penso illus Fondazione Emanuele Paterno,Rome,1953 Title also in Italian;text in English and French
Bioch 33.1800

BACTERIAL ENDOCARDITIS :national symposium Proceedings London 1969 Mar Beecham research laboratories Edited by P.B. Beeson and M. Ridley London,c1970
PGMS 29.0600

BACTERIAL ENDOTOXINS :a symposium Proceedings New Brunswick,N.J. 1963 Sep 4-6 Rutgers university.Institute of microbiology Edited by Maurice Landy and Werner Braun With the support of the National science foundation Rutgers university.Institute of microbiology,New Brunswick,N.J.,c1963
Path 30.2611

BACTERIAL EPISOMES AND PLASMIDS :a Ciba foundation symposium Proceedings London 1968 Sep 9-11 Ciba foundation Edited by G. E.W. Wolstenholme and M. O'Connor Churchill, London,1969
Bioch 33.1118

BACTERIAL PHOTOSYNTHESIS :a symposium Yellow Springs,Ohio 1963 Edited by Howard Gest and others Sponsored by the Charles F. Kettering research laboratory Charles F. Kettering research laboratory.Contribution, 112 illus. xvi,523p Antioch press,Yellow Springs,Ohio,1963
Bot 42.3800

BACTERIAL PHOTOSYNTHESIS :a symposium Papers presented Yellow Springs,Ohio 1963 Mar 18-20 Edited by Howard Gest and others Sponsored by the Charles F.Kettering research laboratory Charles F.Kettering research laboratory.Contribution, 112 Antioch press, Yellow Springs,Ohio,1963
Bioch 33.1116

The BACTERIPHAGE LAMBDA :conference Proceedings Cold Spring Harbor Sep 1970 Cold Spring Harbor laboratory of quantitative biology Edited by A.D. Hershey Cold Spring Harbor symposia on quantitative biology Cold Spring Harbor, N.Y.,1971
Bioch 33.2295

BAD HOMBURG 1963 Natural electromagnetic phenomena below 30 kilocycles NATO advanced study institute Proceedings United States. Office of naval research and Naval ordnance laboratory Edited by D.F. Bleil vii,470p Plenum press,New York,1964
Nap 11.0067

BADEN 1971 Real-time control of electric power systems Symposium of real-time control of electric power systems Brown,Boveri and company Edited by Edmund Handschin Elsevier,Amsterdam,1972 The second Brown, Boveri symposium
Eng 41.8558

BADEN,SWITZERLAND 1969 Brown Boveri symposium 1st Flow research on blading :a symposium Proceedings Brown,Boveri and company Edited by Lang S. Dzung Elsevier, Amsterdam,1970
Eng 41.6828

BAGUIO 1965 Choriocarcinoma :a conference Transactions International union against cancer Edited by James F. Holland and Myroslaw M. Hreshchyshya International union against cancer.Monograph series, 3 Springer,Berlin,1967
Radioth 35.0886

BAHAMA INTERNATIONAL CONFERENCE ON BURNS Report West End,Grand Bahama 1963 Mar 23-25 Sponsored by Colonial research institute Dorrance,Philadelphia,1964
Med 36.0079

BAIKOV INSTITUTE OF METALLURGY.COMMISSION ON THE PHYSICO-CHEMICAL PRINCIPLES OF STEELMAKING The Uses of vacuum in metallurgy Conference proceedings Translated by E. Bishop from the Russian Oliver and Boyd,Edinburgh;London, 1964
Met 25.1674

A BALANCED TEACHING HOSPITAL :a symposium Birmingham 1963 Sep Edited by Thomas McKeown Sponsored by Nuffield provincial hospitals trust viii,131p 28cm Oxford university press,London,1965
PGMS 29.0500

BALATONFURED 1969 Colloquium on combinatorial theory Proceedings Bolyai Janos matematikai tarsulat Edited by P. Erdos and others Societatis Janos Bolyai. Colloquia mathematica, 4 1201p 24cm 3 vols North-Holland,Amsterdam;London,1970
Math S 3.1656

BALATONFURED 1969 Combinatorial theory and its applications :colloquium Proceedings Bolyai Janos matematikai tarsulat Edited by P. Erdos and others Bolyai Janos matematikai tarsulat.Colloquia mathematica, 4 1201p 24cm 3 vols North-Holland,Amsterdam,1970
P Math 2.4509

BALATONVILAGOS,HUNGARY 1961 International symposium on luminescence Papers Roland Eotvos physical society Budapest,1962 Papers in English,German,Russian,reprinted from 'Acta physica Academiae scientiarum hungaricae',tom 14,fasc 2-3
Chem 18.0474

BALLISTICS Selected topics on ballistics Cranz centenary colloquium.University of Freiburg Freiburg 1958 Agard Edited by Wilbur C. Nelson NATO-AGARD wind tunnel and model testing panel.Publication AGARDograph, 32 illus,port Pergamon press,London,1959
Eng 41.6772

BALTIMORE 1929 The Symposium on colloid chemistry 7th Papers presented Johns Hopkins university Edited by Harry Boyer Weiser Colloid symposium annual, 7 Wiley; Chapman and Hall,New York;London,1939 Colloid symposium annual known also as Colloid symposium monograph
Bioch 33.1475

BALTIMORE 1953 A Symposium on the mechanism of enzyme action Papers and discussions Edited by William D. McElroy and Bentley Glass Sponsored by the McCollum-Pratt institute McCollum-Pratt institute. Contribution, 70 Johns Hopkins press, Baltimore,Md.,1954
Bioch 33.1059

BALTIMORE 1956 Chemical basis of heredity A Symposium on the chemical basis of heredity Papers and informal discussions Edited by William D. McElroy and Bentley Glass Sponsored by the McCollum-Pratt institute McCollum-Pratt institute.Contribution, 153 Johns Hopkins press,Baltimore,1957
Bioch 33.0830

BALTIMORE 1958 Chemical basis of development A Symposium on the chemical basis of development Papers and informal discussions Edited by William D. McElroy and Bentley Glass Sponsored by the McCollum-Pratt institute McCollum-Pratt institute. Contribution, 234 Johns Hopkins press, Baltimore,1958
Bioch 33.0829

BALTIMORE 1965 Environment-sensitive mechanical behaviour :a conference Proceedings Pt 1-2 American institute of mining,metallurgical and petroleum engineers.Physical metallurgy committee and United States.Army research office,Durham Edited by A.R.C. Westwood and N.S. Stoloff Sponsored by the Martin company research institute for advanced studies Metallurgical society conferences, 35 2 vols Gordon and Breach,New York,1966
Met 25.1963

BALTIMORE,MD 1958 A Symposium on the chemical basis of development Edited by William D. McElroy and Bentley Glass Sponsored by the McCollum-Pratt institute Johns Hopkins press,Baltimore,Md.,1958
Bal 39.0454

BALTIMORE,MD. 1908 Outlines of geologic history with especial reference to North America American association for the advancement of science.Section E :meetings Papers Edited by Bailey Willis and Rollin D. Salisbury University of Chicago press, Chicago,Ill.,1910
Geog 13.0599

BALTIMORE,MD. 1908 Outlines of geologic history with especial reference to North America :series of essays ... presented before. the American association for the advancement of science... American association for the advancement of science Edited by Rollin D. Salisbury University of Chicago press,Chicago,Ill.,1910
Geol 8.1790

BALTIMORE,MD. 1909 Fifty years of Darwinism :modern aspects of evolution American association for the advancement of science Holt,New York,1909 Centennial addresses in honor of Charles Darwin
Bot 42.6666

BALTIMORE,MD. 1951 Phosphorus metabolism A Symposium on the role of phosphorus in the metabolism of plants and animals Vol 1 Edited by W.D. McElroy and B. Glass Sponsored by McCollum-Pratt institute McCollum-Pratt institute.Contributions, 23 John Hopkins press,Baltimore,Md.,1951
Radioth 35.0028

BALTIMORE,MD. 1951 Phosphorus metabolism A Symposium on the role of phosphorus in the metabolism of plants and animals Vol 2 Edited by W.D. McElroy and B. Glass Sponsored by the McCollum-Pratt institute McCollum-Pratt institute.Contributions, 36 John Hopkins press,Baltimore,Md.,1952
Radioth 35.0029

BALTIMORE,MD. 1951 and BALTIMORE,MD. 1952 Phosphorus metabolism A Symposium on the role of phosphorous in the metabolism of plants and animals Papers and discussions Vol 1-2 Edited by William D. McElroy and H.Bentley Glass Sponsored by the McCollum-Pratt institute McCollum-Pratt institute.Contribution, 23,36 2 vols Johns Hopkins press,Baltimore,Md.,1951-52
Bot 42.1670

BALTIMORE,MD. 1951 and BALTIMORE,MD. 1952 Phosphorus metabolism A Symposium on the role of phosphorus in the metabolism of plants and animals Papers and discussions presented Vol 1-2 Edited by William D. McElroy and H.Bentley Glass Sponsored by the McCollum-Pratt institute McCollum-Pratt institute.Contribution, 23,36 2 vols Johns Hopkins press,Baltimore,Md.,1951-52
Bioch 33.0597

BALTIMORE,MD. 1953 Fluid models in geophysics Symposium on the use of models in geophysical fluid dynamics 1st Proceedings Edited by Robert R. Long Sponsored by United States.Office of naval research illus,diagrms v,162p U.S. Government printing office,Washington,D.C., 1956
Chem E 24.0202

BALTIMORE,MD. 1953 The Mechanism of enzyme action :a symposium Edited by William D. McElroy and Bentley Glass Sponsored by McCollum-Pratt institute McCollum-Pratt institute.Contributions,70 Johns Hopkins press,Baltimore,Md.,1954
Col S 12.0113

BALTIMORE,MD. 1954 A Symposium on amino acid metabolism Papers and discussions presented Edited by William D. McElroy and H. Bentley Glass Sponsored by the McCollum-Pratt institute McCollum-Pratt institute. Contribution, 105 Johns Hopkins press, Baltimore,1955
Bioch 33.0599

BALTIMORE,MD. 1954 A Symposium on amino-acid metabolism Papers McCollum-Pratt institute Edited by William D. McElroy and H. Bentley Glass McCollum-Pratt institute. Contribution,105 Johns Hopkins,Baltimore,Md., 1955
Chem 18.1532

BALTIMORE,MD. 1954 Amino acid metabolism : a symposium Papers and discussion Edited by William D. McElroy and H.Bentley Glass Sponsored by the McCollum-Pratt institute McCollum-Pratt institute.Contribution, 108 Johns Hopkins press,Baltimore,Md.,1955
Mol 45.0292

BALTIMORE,MD. 1955 A Symposium on inorganic nitrogen metabolism:function of metalloflavoproteins Edited by William D. McElroy and Bentley Glass Sponsored by the McCollum-Pratt institute McCollum-Pratt institute.Contributions, 125 Johns Hopkins press,Baltimore,Md.,1956
Gen 34.0659

BALTIMORE,MD. 1955 Inorganic nitrogen metabolism:function of metallo-flavoproteins : a symposium Edited by William D. McElroy and Bentley Glass Sponsored by the McCollum-Pratt institute McCollum-Pratt institute. Contributions, 125 Johns Hopkins press, Baltimore,Md.,1956
Mol 45.0279

BALTIMORE,MD. 1955 21-23 A Symposium on inorganic nitrogen metabolism :function of metallo-flavoproteins Papers and informal discussions presented Edited by William D. McElroy and H.Bentley Glass Sponsored by the McCollum-Pratt institute McCollum-Pratt institute.Contribution, 125 Johns Hopkins press,Baltimore,Md.,1956
Bioch 33.0598

BALTIMORE,MD. 1956 A Symposium on the chemical basis of heredity Edited by William D. McElroy and Bentley Glass Sponsored by the McCollum-Pratt institute McCollum-Pratt institute.Contribution, 153 Johns Hopkins press,Baltimore,Md.,1957
Gen 34.0853

BALTIMORE,MD. 1956 A Symposium on the chemical basis of heredity McCollum-Pratt institute Edited by William D. McElroy and Bentley Glass Johns Hopkins press,Baltimore, Md.,1957
Bot 42.4714

BALTIMORE,MD. 1956 A Symposium on the chemical basis of heredity Papers and informal discussions Edited by William D. McElroy and Bentley Glass Sponsored by the McCollum-Pratt institute McCollum-Pratt institute.Contribution, 153 Johns Hopkins press,Baltimore,Md.,1957
Bal 44.6451

BALTIMORE,MD. 1956 The Chemical basis of heredity :a symposium Edited by W.D. McElroy and Bentley Glass Sponsored by the McCollum-Pratt institute McCollum-Pratt institute. Contributions, 153 John Hopkins press, Baltimore,Md.,1957
Radioth 35.0219

BALTIMORE,MD. 1956 The Chemical basis of heredity :a symposium Edited by William D. McElroy and Bentley Glass Sponsored by the McCollum-Pratt institute McCollum-Pratt institute.Contributions, 153 illus. Johns Hopkins press,Baltimore,Md.,1957
Mol 45.0193

BALTIMORE,MD. 1956 The Chemical basis of heredity :a symposium McCollum-Pratt institute Edited by William D.Bentley McElroy and Bentley Glass With support from the Atomic energy commission McCollum-Pratt institute.Contribution,153 Johns Hopkins press,Baltimore,Md.,1957
Chem 18.1555

BALTIMORE,MD. 1957-58 Researches in geochemistry :seminar series essays Edited by Philip H. Abelson organized by the Carnegie institute of Washington.Geophysical laboratory John Wiley and sons,New York,1959
Geol 8.1495

BALTIMORE,MD. 1958 A Symposium on the chemical basis of development Edited by William D. McElroy and Bentley Glass Sponsored by the McCollum-Pratt institute McCollum-Pratt institute.Symposia illus. Johns Hopkins press,Baltimore,Md.,1958
Gen 34.0852

BALTIMORE,MD. 1958 The Chemical basis of development :a symposium Edited by William D. McElroy and Bentley Glass Sponsored by the McCollum-Pratt institute Johns Hopkins press, Baltimore,Md.,1958
An 32.2488

BALTIMORE,MD. 1958 The Chemical basis of development :a symposium Edited by William D. McElroy and Bentley Glass Sponsored by the McCollum-Pratt institute port Johns Hopkins,Baltimore,Md.,1958
Mol 45.0257

BALTIMORE,MD. 1960 A Symposium on light and life Papers and informal discussions Edited by William D. McElroy and Bentley Glass Sponsored by the McCollum-Pratt institute McCollum-Pratt institute.Contribution, 302 Johns Hopkins university press,Baltimore,Md., 1961
Bal 44.6447

BALTIMORE,MD. 1961 Interhemispheric relations and cerebral dominance :a conference Papers Edited by Vernon B. Mountcastle Johns Hopkins press,Baltimore,Md.,1962
An 32.4652

BALTIMORE,MD. 1961 Reading disability: progress and research needs in dyslexia Johns Hopkins conference on research needs in dyslexia and related aphasic disorders Papers Edited by John Money Johns Hopkins press,Baltimore,Md.,1962
An 32.4203

BALTIMORE,MD. 1963 Drugs in our society :a conference Edited by Paul Talalay Under the auspices of the Johns Hopkins university Johns Hopkins press,Baltimore,Md.,1964
Pha 16.0237

BALTIMORE,MD. 1964 Organogenesis International conference on organogenesis International institute of embryology Edited by Robert L. DeHaan and Heinrich Ursprung Holt,Rinehart and Winston,New York,1965
Bal 39.0490

BALTMORE,MD. 1960 A Symposium on light and life Papers and discussions presented Edited by William D. McElroy and H.Bentley Glass Sponsored by the McCollum-Pratt institute McCollum-Pratt institute. Contribution, 302 John Hopkins press, Baltimore,Md.,1961
Bioch 33.0600

BAMBERG 1965 Position of variable stars in the Hertzsprung-Russell diagram Colloquium on variable stars 3rd International astronomical union.Commissions 27 and 42 Remeis-sternwarte,Bamberg.Kleine veroffentlichungen, 4,no 40 port 304p Astronomisches institut der universitat Erlangen-Nurnberg,Bamberg,1965 I.A.U. combined colloquium of commissions 27 and 42 with the theme 'The position of stars'
TA 15.0540

BANARAS 1967 Recent advances in tropical ecology :a symposium Proceedings International society for tropical ecology Edited by R. Misra and B. Gopal 2 vols Vararasi,1968
Bot 42.2001

BANARAS HINDU UNIVERSITY.DEPARTMENT OF MATHEMATICS Seminar volume in statistics 125p 24cm Banaras Hindu university,1968
Math S 3.1454

BANFF 1933 Canadian papers 1933 Institute of Pacific relations conference 5th Papers Canadian institute of international affairs Canadian institute of international affairs,Toronto,1933
Geog 13.4898

BANFF 1971 IFAC symposium on the control of distributed parameter systems Proceedings International federation for automatic control 2 vols IFAC technical committee on theory, Banff,1971
Eng 41.8593

BANGALORE 1961 Seminar on aeronautical sciences proceedings Vol 1 India. National aeronautical laboratory National aeronautical laboratory,Bangalore,1961
A Math 4.0526

BANGKOK 1960 Radioisotopes in tropical medicine Symposium on the use of radioisotopes in the study of endemic and tropical diseases Proceedings International atomic energy agency World health organization International atomic energy agency.Proceedings series International atomic energy agency,Vienna,1962
Radioth 35.1365

BANGKOK 1968 Nuclear science teaching Panel on nuclear science teaching Report International atomic energy agency Unesco International atomic energy agency.Technical reports series, 94 International atomic energy agency,Vienna,1968
Radioth 35.1777

BANGKOK 1969 Radiation sensitivity of toxins and animal poisons Panel on the radiation sensitivity of toxins and animal poisons Proceedings International atomic energy agency International atomic energy agency.Panel proceedings series International atomic energy agency,Vienna,1970
Radioth 35.1912

BANGKOK 1969 Radiation sensitivity of toxins and animal poisons Panel on the radiation sensitivity of toxins and animal poisons Proceedings Organized by the International atomic energy agency International atomic energy agency.Panel proceedings series International atomic energy agency,Vienna,1970
Radioth 35.1765

BANGOR 1953 Active transport and secretion Society for experimental biology Society for experimental biology.Symposia, 8 Cambridge university press,Cambridge,1954
Phys 20.0958

BANGOR 1953 Active transport and secretion a symposium Papers Society for experimental biology Edited by R. Brown and J.F. Danielli Society for experimental biology.Symposia, 8 Cambridge university press,Cambridge,1954
Inv Med 37.0036

BANGOR 1953 Active transport and secretion a symposium Society for experimental biology Society for experimental biology. Symposia, 8 Cambridge university press, Cambridge,1954
Bioch 33.1353

BANGOR 1962 Grazing in terrestrial and marine environments :a symposium British ecological society Edited by D.J. Crisp British ecological society.Symposia xvi,322p Blackwell scientific,Oxford,1964
Bot 42.1984

BANGOR 1962 Grazing in terrestrial and marine environments :a symposium British ecological society Edited by D.J. Crisp British ecological society.Symposia, 4 Blackwell,Oxford,1964
Bal 39.1671

BANYULS-SUR-MER 1963 Colloque international sur l'histoire de la biologie marine Les Grandes expeditions scientifiques et la creation des laboratoires maritimes International union of history and philosophy of science Vie et milieu.Supplement, 19 370p Laboratoire Arago;Masson,Banyuls-sur-Mer;Paris,1965
WSM 43.2877

BAR AND ALLIED PRODUCTS Technical conference 3rd Proceedings Pittsburgh 1961 Jan 18 American institute of mining,metallurgical and petroleum engineers.Mechanical working committee Metallurgical society conferences, 13 Interscience,New York;London,1961
Met 25.0729

BAR HARBOR,MAINE 1959 Symposium on some problems of normal and abnormal differentiation and development Proceedings National cancer institute Edited by Nathan Kaliss National cancer institute.Monograph, 2 U.S.Department of health,education and welfare,Washington,D.C.,1960
Radioth 35.0808

BARCELONA 1921 Conferencia internacional de psicotecnica applicada a l'orientacio professional i a l'organitzacio cientifica del treball 2 Actes Conference internationale de psychotechnique appliquee a l'orientation professionelle et a l'organisation scientifique du travail. Secretariat Institut d'orientacio professional,Barcelona,1912 Under the presidency of E.Claparede
Psy 31.2608

BARCELONA 1924 World poultry congress 2nd Part 2: informations and communications Arts graphiques,Barcelona, 1924
Gen 34.1767

BARCELONA 1957 International astronautical congress 8th proceedings International astronautical federation Edited by F. Hecht International astronautical congress. Proceedings, 8 607p Springer,Vienna,1958
Obs 6.0761

BARCELONA 1959 and MADRID Congres international d'histoire des sciences 9e Actes Vol 1 Union internationale d'histoire des sciences Academie internationale d'histoire des sciences. Collection de travaux, 12 733p Asociacion para la historia de ciencia espanola;Hermann, Barcelona;Paris,1960
WSM 43.0061

BARCROFT Joseph Haemoglobin :a symposium based on a conference...in memory of Sir Joseph Barcroft Proceedings Cambridge 1948 June Edited by F.J.W. Roughton and J.C. Kendrew Butterworths,London,1949
Med 36.0309

BARI 1965 Regulation of metabolic processes in mitochondria :a symposium Proceedings Edited by J.M. Tager and others organized by the University of Bari,Dept.of Biochemistry and the University of Amsterdam, Laboratory of Biochemistry B.B.A. library, 7 Elsevier,Amsterdam,1966
Bioch 33.1101

BARI 1965 Regulation of metabolic processes in mitochondria :symposium Proceedings Edited by J.M. Tager and others Organized by the Universiteit van Amsterdam and Universita degli studi,Bari B.B.A. library, 7 Elsevier,Amsterdam,1966
Bal 39.0308

BARLEY GENETICS 1 International barley genetics symposium 1st Proceedings Wageningen 1963 Aug 26-31 Centre for agricultural publications and documentation, Wageningen,1964
Gen 34.1560

BASEL 1948 Internationaler kongress uber kernphysik und quantenelektrodynamik Helvetica physica acta, 23,supp.3 Basel, 1950
Radioth 35.1271

BASEL 1960 The International symposium on polarization phenomena of nucleons Universitat Basel Edited by P. Huber and K.P. Meyer Under the sponsorship of the International union of pure and applied physics Helvetica physica acta.Supplementum, 6 Birkhauser,Basel,1961
Cav 7.0135

BASEL 1962 Szintigraphie und radiokardiographie :symposion Schweizerische akademie der medizinischen wissenschaften Schwabe,Basel;Stuttgart,1963
Radioth 35.1371

Les BASES SCIENTIFIQUES D'UNE ORGANISATION INTERNATIONALE POUR LA LUTTE BIOLOGIQUE Biological control Scientific basis of an international biological control organization Stockholm 1948 Aug 5-7 International union of biological sciences International union of biological sciences.Serie B.Colloques, 5 Secretairiat general de I.U.I.S.B.,Paris,1949 Title also in English: The scientific bases of an international biological control organization.Text in English and French
Bal 39.3920

BASIC AERODYNAMIC NOISE RESEARCH :a conference Washington,D.C. 1969 Jul 14-15 National aeronautics and space administration Edited by Ira R. Schwartz NASA SP-207 N.A.S.A.,Washington,D.C.,1969
Eng 41.4224

BASIC CONCEPTS OF INBORN ERRORS AND DEFECTS OF STEROID BIOSYNTHESIS Proceedings Birmingham 1965 Oct 1-2 Society for the study of inborn errors of metabolism Edited by K.S. Holt and D.N. Raine Society for the study of inborn errors of metabolism.Symposia, 3 Livingstone,Edinburgh;London,1968
Gen 34.2000

BASIC CONCEPTS OF INBORN ERRORS AND DEFECTS OF STEROID BIOSYNTHESIS 3rd :a symposium Proceedings Birmingham 1965 Society for the study of inborn errors of metabolism Edited by K.S. Holt and D.N. Raine Livingstone,Edinburgh,1966
Inv Med 37.0271

BASIC MECHANISMS IN ANIMAL VIRUS BIOLOGY :a symposium Papers Cold Spring Harbor 1962 Cold Spring Harbor biological laboratory Cold Spring Harbor symposia on quantitative biology, 27 Long Island biological association,Cold Spring Harbor,1962
Bioch 33.1283

BASIC MECHANISMS IN ANIMAL VIRUS BIOLOGY Papers Cold Spring Harbor 1962 Jun 7-13 Cold Spring Harbor.Biological laboratory Cold Spring Harbor symposia on quantitative biology, 27 Long Island biological association.Biological laboratory,Cold Spring Harbor,L.I.,N.Y.,1962
Path 30.2559

BASIC MECHANISMS IN PHOTOCHEMISTRY AND PHOTOBIOLOGY :an international symposium Caracas 1967 Dec 4-8 Edited by J.W. Longworth Photochemistry and photobiology, 7,no 6 Pergamon press,London,1968
Radioth 35.0279

BASIC MECHANISMS IN RADIOBIOLOGY Cellular aspects of basic mechanisms in radiobiology : an informal conference Proceedings Bear Mountain,N.Y. 1955 May 12-14 Edited by Harvey M. Patt and E.L. Powers Organised by the National research council.Subcommittee on radiobiology National research council. Nuclear science series.Report, 18 NAS-NRC 450 National research council,Washington,D.C. 1956
Radioth 35.1716

BASIC MECHANISMS IN RADIOBIOLOGY.II.PHYSICAL AND CHEMICAL ASPECTS :an informal conference Proceedings Highland Park,Ill. 1953 May 7-9 Edited by John L. Magee and others Organized by the National research council. Subcommittee on radiobiology National research council.Publications, 305 National academy of science,Washington,D.C.,1953
Radioth 35.1546

BASIC MECHANISMS OF FATIGUE Symposium on basic mechanisms of fatigue Papers Boston, Mass. 1958 Jun 23 American society for testing materials A.S.T.M.Special technical publication, 237 A.S.T.M.,Philadelphia,1959
Met 25.0925

BASIC MECHANISMS OF FATIGUE Symposium on basic mechanisms of fatigue presented before the 61st annual meeting Boston 1958 Jun 23 American society for testing materials A.S.T.M.Special technical publication, 273 A.S.T.M. Philadelphia,1959
Met 25.2328

BASIC PROBLEMS IN THIN FILM PHYSICS : international symposium Proceedings Clausthal-Gottingen 1965 Sep 6-11 Edited by R. Niedermayer and H. Mayer VandenHoeck and Ruprecht,Gottingen,1966
Eng 41.5450

BASIC PROBLEMS IN THIN FILM PHYSICS : international symposium Proceedings Clausthal-Gottingen 1965 Sep 6-11 Edited by R. Niedermayer and H. Mayer illus 756p Vandenheck and Ruprecht,Gottingen,1966
Cav 7.0512

BASIC SCIENCE SYMPOSIA, 9 Biological interfaces:flows and exchanges :a symposium Proceedings New York 1967 Dec 8-9 Sponsored by the New York heart association Little,Brown,Boston,Mass.,1968 Simultaneously published in the Journal of general physiology,vol.52,1968
Phys 20.2247

BASLE 1954 International conference on thrombosis and embolism 1 Proceedings Edited by Theo Koller and Willy R. Merz Benno Schwabe,Basle,1955
Med 36.0220

BASLE 1964 Les Fonctions du col uterin : colloque Papers Societe nationale pour l'etude de la sterilite et de la fecondite Edited by R. Moricard and R. Wenner Masson, Paris,1964
An 32.3833

BASLE 1969 Iron deficiency;pathogenesis, clinical aspects and therapy Clinical symposium on iron deficiency Proceedings Edited by L. Hallberg and others Colloquia Geigy Academic press,London,1970
Med 36.0383

BASLE 1969 The Spiral structure of our galaxy International astronomical union Edited by W. Becker and G. Contopoulos International astronomical union.Symposium, 38 478p Reidel,Dordrecht,1970
Obs 6.3320

BASLE 1969 The Spiral structure of our galaxy :symposium Proceedings International astronomical union Edited by W. Becker and G. Contopoulos International astronomical union.Symposium, 38 478p Reidel,Dordrecht,1970
TA 15.0489

BASLE 1971 Challenge of life:biomedical progress and human values Roche anniversary symposium Proceedings Hoffmann-La Roche and company Edited by Robert M. Kunz and Hans Fehr Experientia.Supplementum, 17 456p 24cm Birkhauser,Basle,1972 Symposium held on the occasion of the firm's 75th anniversary.Chairman:Lord Todd
Chem 18.2902

BATON ROUGE,LA. 1967 Symposium on infinite dimensional topology Proceedings Edited by R.D. Anderson Annals of mathematics studies, 69 viii,300p 24cm Princeton university press;University of Tokyo press,Princeton,N.J.; Tokyo,1972
P Math 2.4112

BATON ROUGE,LA. 1970 Louisiana conference on combinatorics,graph theory and computing Proceedings Edited by R.C. Mullin and others Held at Louisiana state university vi,464p 25cm Louisiana state university,Baton Rouge, La.,1970
P Math 2.4503

BATTELLE MATERIALS SCIENCE COLLOQUIUM 1st Proceedings Phase stability in metals and alloys Geneva;Villars 1966 Mar 7-12 Battelle memorial institute Edited by Peter S. Rudman and others McGraw-Hill series in materials science and engineering McGraw-Hill,New York,1967
Met 25.1194

BATTELLE MATERIALS SCIENCE COLLOQUIUM 2nd Dislocation dynamics Seattle,Wash. 1967 May 1-16 Harrison Hot Springs,B.C. Battelle memorial institute Edited by Alan R. Rosenfield and others McGraw-Hill series in materials science McGraw-Hill,New York,1968
Met 25.2253

BATTELLE MEMORIAL INSTITUTE Category theory, homology theory and their application :a conference Proceedings Seattle 1968 Jun 24-Jul 19 Vol 1 Lecture notes in mathematics, 86 Bibliog,Illus vi,216p 28cm Springer-Verlag,Berlin,1969
P Math 2.3343

BATTELLE MEMORIAL INSTITUTE Category theory, homology theory and their applications Seattle 1968 Jun 24-Jul 19 Vol 3 Lecture notes in mathematics, 99 Bibliog., Illus. iv,489p 28cm Springer-Verlag, Berlin,1969
P Math 2.3345

BATTELLE MEMORIAL INSTITUTE Category theory, homology theory and their applications :a conference Seattle 1968 Jun 24-Jul 19 Vol 2 Lecture notes in mathematics, 92 Bibliog.,Illus. vi,216p 28cm Springer-Verlag,Berlin,1969
P Math 2.3344

BATTELLE MEMORIAL INSTITUTE Dislocation dynamics Battelle materials science colloquium 2nd Seattle,Wash. 1967 May 1-16 Harrison Hot Springs,B.C. Edited by Alan R. Rosenfield and others McGraw-Hill series in materials science McGraw-Hill,New York,1968
Met 25.2253

BATTELLE MEMORIAL INSTITUTE Phase stability in metals and alloys Battelle materials science colloquium 1st Proceedings Geneva;Villars 1966 Mar 7-12 Edited by Peter S. Rudman and others McGraw-Hill series in materials science and engineering McGraw-Hill,New York,1967
Met 25.1194

BATTELLE MEMORIAL INSTITUTE The Steenrod algebra and its applications :a conference to celebrate N.E.Steenrod's sixtieth birthday Proceedings Columbus,Ohio 1970 Mar 30-Apr 4 Edited by F.P. Peterson Lecture notes in mathematics, 168 viii,317p 25cm Springer, Berlin,1970
P Math 2.4120

BATTELLE MEMORIAL INSTITUTE.SEATTLE RESEARCH CENTER Seminar on algebraic topology Proceedings Seattle,Wash. 1971 Feb 22-26 Edited by Peter J. Hilton Lecture notes in mathematics, 249 vi,110p 25cm Springer, Berlin,1971
P Math 2.4115

BATTELLE RENCONTRES Edited by Cecile M. DeWitt and John A. Wheeler 1967 lectures in mathematics and physics Bibliog.,illus. xvii,557p 24cm Benjamin,New York,1968
A Math 4.1413

BATTELLE SEATTLE 1968 RENCONTRES Hyperbolic equations and waves Seattle 1968 Edited by M. Froissart Bibliog.,Illus. viii,393p 23cm Springer-Verlag,Berlin,1970
A Math 4.1469

BATTERIES 2 International symposium on batteries :research and development in non-mechanical electrical power sources 4th Brighton 1964 Sep Edited by D.H. Collins Sponsored by the Inter-departmental committee on batteries Pergamon Press,Oxford,1965
Met 25.1841

BCS OCTOBER 71 CONFERENCE ON APRIL 71 REPORT
Collected papers London 1971 Oct British computer society.CODASYL data base task group illus 221p British computer society,London,1972
Math L 5.3865

BEAM TECHNOLOGY Electron,ion and laser beam technology Symposium on electron,ion and laser beam technology 11th Record 1971 Edited by R.F.M. Thornley San Francisco press,San Francisco,Calif.,1971
Eng 41.8582

BEAR MOUNTAIN,N.Y. 1955 Cellular aspects of basic mechanisms in radiobiology :an informal conference Proceedings Edited by Harvey M. Patt and E.L. Powers Organised by the National research council.Subcommittee on radiobiology National research council. Nuclear science series.Report, 18 NAS-NRC 450 National research council,Washington,D.C. 1956
Radioth 35.1716

BEARING CAPACITY OF SOILS Symposium on load tests of bearing capacity of soils Atlantic City,N.J. 1947 Jun 16-20 American society for testing materials American society for testing materials.Special technical publication, 79 American society for testing materials,Philadelphia,Pa.,1948
Eng 41.3145

BECKMAN INSTRUMENTS Simple preparation techniques for liquid scintillation counting The Beckman summer school 1st Proceedings London 1966 Aug 15-18 London,c1966
Bioch 33.2300

BECKMAN INSTRUMENTS Techniques for liquid scintillation counting The Beckman summer school 2nd Proceedings 1967 Glenrothes,c1967
Bioch 33.2301

BECKMAN INSTRUMENTS LIMITED Sample preparation techniques for liquid scintillation counting :1966 summer school Proceedings London 1966 Aug 15-18 Beckman instruments limited,Glenrothes,n.d.
Radioth 35.1778

BECKMAN INSTRUMENTS LIMITED Techniques for liquid scintillation counting :1967 summer school Proceedings Beckman instruments limited,Glenrothes,n.d.
Radioth 35.1779

The BECKMAN SUMMER SCHOOL 1st Proceedings Simple preparation techniques for liquid scintillation counting London 1966 Aug 15-18 Beckman instruments London,c1966
Bioch 33.2300

The BECKMAN SUMMER SCHOOL 2nd Proceedings Techniques for liquid scintillation counting 1967 Beckman instruments Glenrothes,c1967
Bioch 33.2301

BEDFORD,MASS. 1959 Annual Arctic planning session 2nd Proceedings Air force Cambridge research center.Geophysics research directorate Edited by Vivian C. Bushnell Air force Cambridge research center.G.R.D. research notes,29 AFCRC-TN-59-661 illus viii,172p A.F.C.R.C.,Bedford,Mass.,1959
Sco 14.7405

BEDFORD,MASS. 1960 Annual Arctic planning session 3rd Proceedings Air force Cambridge research laboratories Edited by George P. Rigsby and Vivian C. Bushnell Air force Cambridge research laboratories.G.R.D. research notes,55 AFCRL 436 illus viii, 148p A.F.C.R.L.,Bedford,Mass.,1961
Sco 14.7406

BEDFORD,PA. 1966 Cell tissue and organ culture Decennial review conference on cell tissue and organ culture 2nd Proceedings National cancer institute and Tissue culture association Edited by Benton B. Westfall National cancer institute.Monograph, 26 National cancer institute,Bethesda,Md., 1967
Bioch 33.0991

BEDFORD COLLEGE Conference in mathematical logic Proceedings London 1970 Aug 24-28 Edited by Wilfrid Hodges Lecture notes in mathematics, 255 vii,351p 25cm Springer, Berlin,1972
P Math 2.4059

BEDFORD COLLEGE Criticism and the growth of knowledge International colloquium in the philosophy of science Proceedings London 1965 Jul 11-17 Edited by Imre Lakatos and Alan Musgrave vii,282p 23cm Cambridge university press,Cambridge,1970
Math S 3.1850

BEDFORD COLLEGE Oxygen in the animal organism :a symposium Proceedings London 1963 Sep 1-5 International union of biochemistry and International union of physiological sciences Edited by Frank Dickens and Eric Neil I.U.B.Symposium series, 31 Pergamon press,Oxford,1964
Bioch 33.1080

BEECHAM RESEARCH LABORATORIES Bacterial endocarditis :national symposium Proceedings London 1969 Mar Edited by P.B. Beeson and M. Ridley London,c1970
PGMS 29.0600

BEECHAM RESEARCH LABORATORIES National symposium on urinary tract infection 1st Proceedings London 1968 April Edited by Francis O'Grady and William Brumfitt Oxford university press,London,1968
PGMS 29.0238

BEECHAM RESEARCH LABORATORIES Symposium on the adverse effects of drugs 1st Proceedings London 1970 Feb 4-5 Edited by Gillian C. Hanson Beecham research laboratories,London,1971
PGMS 29.0673

BEGINNINGS OF EMBRYONIC DEVELOPMENT A Symposium on formation and early development of the embryo Atlanta,Ga. 1955 Dec 27 American association for the advancement of science.Section on zoological sciences Edited by Albert Tyler and others Co-sponsored by the American society of zoologists and the Association of southeastern biologists American association for the advancement of science.Publication, 48 Washington,D.C.,1957
Bal 39.0401

BEHAVIOR,AGING AND THE NERVOUS SYSTEM : biological determinants of speed of behavior and its changes with age Cambridge 1963 Aug 6-9 Edited by A.T. Welford and James E. Birren American lecture series.Publication, 600,American lectures in geriatrics and gerontology.Bannerstone division monograph Thomas,Springfield,Ill.,1965
Psy 31.2329

BEHAVIOR AND AWARENESS :a symposium and interpretation Edited by C.W. Eriksen Duke university press,Durham,N.C.,1962 Developed from papers read at the September 1962 meetings of the American psychological association
Psy 31.2985

BEHAVIORAL CONSEQUENCES OF GENETIC DIFFERENCES IN MAN Genetic diversity and human behaviour The Burg Wartenstein symposium 27th Papers Burg 1964 Sep 16-28 Social science research council.Committee on genetics and behavior Wenner-Gren foundation for anthropological research Edited by G.N. Spuhler Viking fund publications in anthropology, 45 Aldine,Chicago,Ill.,1967
Gen 34.2032

BEHAVIOUR AND EVOLUTION 2nd :conference Proceedings Princeton,N.J. 1956 Apr 30-May 5 American psychological association Edited by Anne Roe and George Gaylord Simpson Held in collaboration with the Society for the study of evolution 24cm New Haven,1958
Philos 1.1827

BEHAVIOUR OF MATERIALS AT CRYOGENIC TEMPERATURES a symposium Lafayette,Ind. 1965 Jun 13-18 American society for testing materials A.S.T.M.Special technical publication, 387 American society for testing materials, Philadelphia,Pa.,1916
Met 25.0976

BEIRUT 1954 Colloque sur la protection et la conservation de la nature dans le Proche-orient :symposium on the protection and conservation of nature in the Near East Lebanese society of the friends of the trees Unesco-Middle East science cooperation office Beirut,1954 Text in English and French
Bal 39.1787

BEIRUT 1967 The Triplet state an international symposium Proceedings American university of Beirut Edited by A.B. Zahlan and others Cambridge university press, Cambridge,1967
Chem 18.2642

BEIRUT 1968 Excitons,magnons and phonons in molecular crystals :an international symposium Proceedings Edited by A.B. Zahlan Held at the American university of Beirut Cambridge university press,Cambridge, 1968
Radioth 35.1689

BEITOSTOLEN 1966 Physical activity in health and disease :international symposium Proceedings Edited by K. Evang and K.Lange Anderson Williams and Wilkins,Baltimore,1967
HE 27.0178

BEITRAGE ZUR ENTWICKLUNG DER WISSENSCHAFTSTHEORIE IN 19 JAHRHUNDERT :colloquium Vortrage und diskussionen Dusseldorf 1965 Dec 10-11 and Dusseldorf 1966 Dec 9-10 Edited by A. Diemer Studien zur wissenschaftstheorie, 1 234p Hain, Meisenheim am Glan,1968
WSM 43.0983

BEITRAGE ZUR GRAPHENTHEORIE :internationale kolloquium Vortragen Manebach 1967 May 9-12 Technische hochschule Ilmenau. Mathematische institut Mathematische gesellschaft der D.D.R. Edited by Horst Sachs and others Bibliog.,Illus. 394p 23cm Teubner,Leipzig,1968
P Math 2.3352

BELFAST 1955 The Airglow and the aurorae : a symposium Edited by E.B. Armstrong and A. Dalgarno Journal of atmospheric and terrestial physics.Special supplement,5 illus x,420p 25cm Pergamon press,London; New York,1956
Sco 14.0598

BELFAST 1955 The Airglow and the aurorae : a symposium Edited by E.B. Armstrong and A. Dalgarno Journal of atmospheric and terrestrial physics.Supplements., 5 420p Pergamon press,London,1956
Obs 6.0083

BELFAST 1955 The Airglow and the aurorae : a symposium Edited by E.B. Armstrong and A. Dalgarno Journal of atmospheric and terrestrial physics.Supplements, 5 420p London,1956
Nap 11.0014

BELFAST 1971 International seminar on operating systems principles Submitted papers Queen's university of Belfast illus 2 vols Queen's university of Belfast,Belfast, 1971
Math L 5.3846

BELGIQUE.MINISTERE DE L'EDUCATION ET DE LA CULTURE Soleil a la renaissance:sciences et mythes :colloque international Brussels 1963 Apr Universite libre de Bruxelles. Institut pour l'etude de la renaissance et de l'humanisme.Publication, 2 584p Presses universitaires de Bruxelles,Brussels;Paris, 1965 Contains Hoskin,M.and Jones,C. 'Problems in late renaissance astronomy'
WSM 43.0046

BELGRADE 1965 Statistics in the physical sciences :a conference Proceedings Under the auspices of the International association for statistics in physical sciences International association for statistics in physical sciences.Proceedings, 2 Bibliog 24cm International statistitcal institute, Belgrade,c1966 Reprinted from the Bulletin of the International statistical institute. Proceedings of the 35th session, 1965
Math 3.1157

BELGRADE 1966 Solar-terrestrial physics :a symposium Review papers International scientific radio union Edited by J.W. King and W.S. Newman xii,390p Academic press, London;New York,1967
A Math 4.1171

BELLAGIO 1953 International congress of genetics 9th Proceedings Pt 1-2 Edited by G. Montalenti and A. Chiarugi Caryologia, 6,supp. 2 vols Florence,1954
Gen 34.1005

BELLAGIO 1966 Towards a theoretical biology :a collection of essays written as a result of a symposium... International union of biological sciences Edited by C.H. Waddington 253p 26cm Edinburgh university press,Edinburgh,1970
Math S 3.1819

BELLAGIO 1966 Towards a theoretical biology 1.Prolegomena:an I.U.B.S. symposium Essays International union of biological sciences Edited by C.H. Waddington 234p Edinburgh university press,Edinburgh,1968 Essays written after and in the light of the symposium
WSM 43.2897

BELLAGIO,ITALY 1953 International biometric conference 3rd Proceedings Biometric society 1953 Reprinted from 'Biometrica' vol 9,no.4,1953
Gen 34.2147

BELLEVUE 1957 Les Progres recents en spectroscopie interferentielle Proceedings Centre national de la recherche scientifique Edited by Pierre Jacquinot Centre national de la recherche scientifique.Colloques internationaux, 80 252p Editions du CNRS, Paris,1958
Obs 6.2007

BELLEVUE 1957 Proprietes optiques et acoustiques des fluides comprimees et actions intermoleculaires Centre national de la recherche scientifique Edited by B. Vodar Centre national de la recherche scientifique. Colloques internationaux,77 Centre national de la recherche scientifique,Paris,1959
Chem 18.2309

BELLEVUE 1962 Le Bombardement ionique, theories et applications colloque international 113 Actes Centre national de la recherche scientifique 243p Paris, 1962
Cav 7.0169

BELLEVUE,SEINE ET OISE 1962 Le Bombardement ionique theories et applications: colloques internationaux Centre national de la recherche scientifique Centre national de la recherche scientifique.Colloques internationaux, 113 Editions du centre National de la recherche scientifique,Paris, 1962
Met 25.1676

The BENJAMIN FRANKLIN MEMORIAL SYMPOSIUM Papers Atmospheric exploration American academy of arts and sciences 125p M.I.T.: Wiley,New York,1958
Nap 11.0960

BERGEN 1964 Symposium on naval hydrodynamics :ship motions and drag reduction 5th Edited by J.K. Lunde and S.W. Doroff Sponsored by the United States.Office of naval research,and Office of naval research-Department of the navy,Washington,D.C.,1965
A Math 4.1303

BERGEN 1965 Radiation trapped in the earth's magnetic field :the advanced study institute Proceedings Michelson (Chr.) institutt for videnskap og aandsfrihet Edited by Billy M. McCormac Astrophysics and space science library, 5 901p Reidel, Dordrecht,1966
Obs 6.3282

BERING LAND BRIDGE International association for quaternary research congress 7th Proceedings Boulder,Colo. 1965 Aug 30-Sep 5 Edited by D.M. Hopkins Stanford university press,Stanford,Calif.,1967
Bot 42.3247

BERK PHARMACEUTICALS LIMITED A Symposium on carbenoxolone sodium London 1967 Nov 20 Edited by J.M. Robson and F.M. Sullivan Butterworths,London,1968
Surg 23.0013

BERKELEY 1961 Electron microscopy and strength of crystals Berkeley international materials conference :the impact of transmission electron microscopy on theories of the strength of crystals 1st Proceedings Lawrence radiation laboratory. Inorganic materials division and University of California Edited by Gareth Thomas and Jack Washburn Interscience,New York;London,1963
Met 25.1211

BERKELEY 1964 Berkeley international materials conference :high strength materials-present status and anticipated developments 2nd Proceedings Edited by Victor F. Zackay Inorganic materials research division series John Wiley,New York,1965
Met 25.1019

BERKELEY 1966 Ceramic microstructures their analysis,significance and production International materials symposium 3rd Proceedings Edited by Richard M. Fulrath and Joseph A. Pask Wiley,New York,1968
Met 25.0292

BERKELEY, CALIF. 1957-58 The Axiomatic method international symposium proceedings Edited by Leon Henkin and others Studies in logic and the foundations of mathematics xi, 488p 23cm North-Holland,Amsterdam,1959 With special reference to geometry and physics
P. Math 2.0110

BERKELEY, CALIF. 1963 Theory of models International symposium on the theory of models proceedings Edited by J.W. Addison and others xv,494p 23cm North-Holland, Amsterdam,1965
P. Math 2.0205

BERKELEY,CALIF 1961 Impact of transmission electron microscopy on theories of the strength of crystals The International materials conference 1st Proceedings Edited by Gareth Thomas and Jack Washburn xi, 1022p Interscience,John Wiley,New York,1963
Cav 7.0191

BERKELEY,CALIF 1963 The General assembly of the International association of meteorology Proceedings International association of meteorology and atmospheric physics Edited by W.L. Godson 199p Toronto,1963 Part of the 13th general assembly of International union of geodesy and geophysics
Nap 11.0223

BERKELEY,CALIF 1964 The Heat transfer and fluid mechanics institute 1964 proceedings Edited by Warren H. Gledt and Salomon Levy Stanford university press,Stanford,Calif,1964
A Math 4.0522

BERKELEY,CALIF 1968 International materials symposium :the structure and chemistry of solid surfaces 4th Proceedings Lawrence radiation laboratory. Inorganic materials research division and University of California.College of chemistry Edited by Gabor A. Somorjai Inorganic materials research division series, 4 Wiley,New York,1969
Met 25.2476

BERKELEY,CALIF. 1939 San Francisco The Pacific science congress 6th Proceedings Vol 2 Pacific science association illus 451-914p 24cm University of California press,Berkeley,Calif,1940
Sco 14.0115

BERKELEY,CALIF. 1945 Berkeley,Calif. 1946 Jan.27-29 Mathematical statistics and probability Berkeley symposium on mathematical statistics and probability 1st proceedings University of California. Statistical laboratory Edited by Jerzy Neyman viii,501p 26cm University of California press,Berkeley;Los Angeles,Calif., 1949
Math 3.0685

BERKELEY,CALIF. 1945 and 1946 Jan.27-29 Symposium on mathematical statistics and probability Proceedings University of California.Department of mathematics Edited by Jerzy Neyman 501p University of California press,Berkeley;Los Angeles,1949
Nap 11.0048

BERKELEY,CALIF. 1950 Mathematical statistics and probability Berkeley symposium on mathematical statistics and probability 2nd Proceedings University of California.Statistical laboratory Edited by Jerzy Neyman University of California press,Berkeley,Calif.;Los Angeles,Calif.,1951
Radioth 35.0599

BERKELEY,CALIF. 1950 Mathematical statistics and probability Berkeley symposium on mathematical statistics and probability 2nd proceedings University of California.Statistical laboratory Edited by Jerzy Neyman Also supported by the Office of naval research xi,666p 26cm University of California press,Berkeley;Los Angeles,Calif.,1951
Math 3.0686

BERKELEY,CALIF. 1954 International conference on venoms 1st Papers American association for the advancement of science Edited by E.E. Buckley and N. Porges American association for the advancement of science.Publication, 44 des plantes et animaux indesirables
Bal 44.6436

BERKELEY,CALIF. 1954 The Conference for instructors of astronomy :lecture notes Leuschner observatory Edited by Robert Fleischer and others Co-sponsored by National science foundation 265p Berkeley, Calif.,1954 Typescript
Obs 6.2054

BERKELEY,CALIF. 1954 Berkeley,Calif. 1955 Jul,Aug Mathematical statistics and probability Berkeley symposium on mathematical statistics and probability 3rd Proceedings Vol 4: contributions to biology and problems of health University of California.Statistical laboratory Edited by Jerzy Neyman University of California press, Berkeley,Calif.;Los Angeles,Calif.,1956
Radioth 35.0606

BERKELEY,CALIF. 1954 Berkeley,Calif. 1955 Jul-Aug Mathematical statistics and probability Berkeley symposium on mathematical statistics and probability 3r proceedings Vols. 1-5 University of California.Statistical laboratory Edited by Jerzy Neyman 26cm 5 vols University of California press,Berkeley;Los Angeles,Calif., 1956
Math 3.0687

BERKELEY,CALIF. 1956 World conference on earthquake engineering 1st Proceedings Earthquake engineering research institute Earthquake engineering research institute,San Francisco,Calif.,1967
Eng 41.6394

BERKELEY,CALIF. 1958 Heat transfer and fluid mechanics institute 1958 preprints University of California Stanford university press,Stanford,Calif.,1958
A Math 4.0517

BERKELEY,CALIF. 1960 India's urban future Seminar on urbanization in India Edited by Roy Turner University of California press, Berkeley,Calif.;Los Angeles,Calif.,1962 Conference sponsored by Kingsley Davis and others
Geog 13.1947

BERKELEY,CALIF. 1960 Mathematical statistics and probability Berkeley symposium on mathematical statistics and probability 4th proceedings Vols. 1-4 University of California.Statistical laboratory Edited by Jerzy Neyman 26cm 4 vols University of California press,Berkeley; Los Angeles,Calif.,1961
Math 3.0688

BERKELEY,CALIF. 1960 Partial differential equations :a symposium Proceedings American mathematical society Edited by Charles B. Morrey American mathematical society.Proceedings of symposia in pure mathematics, 4 Bibliog.,Illus. vi,169p 26cm American mathematical society, Providence,R.I.,1961
A Math 4.1468

BERKELEY,CALIF. 1960 Rarefied gas dynamics The International symposium on rarefied gas dynamics 2nd Proceedings University of California Edited by L. Talbot Under the sponsorship of International union of theoretical and applied mechanics Advances in applied mechanics.Supplement, 1 748p Academic press,New York,1961 Co-sponsored by National aeronautics and space administration
Cav 7.2254

BERKELEY,CALIF. 1960 Rarefied gas dynamics international symposium 2nd Proceedings Edited by L. Talbot Held at the University of California Advances in applied mechanics. Supplement, 1 x,748p Academic press,New York;London,1961
Chem E 24.0408

BERKELEY,CALIF. 1960 The International symposium on rarefied gas dynamics 2nd proceedings University of California Edited by L. Talbot Advances in applied mechanics.Supplement, 1 Academic press,New York,London,1961
A Math 4.0529

BERKELEY,CALIF. 1961 IAU general assembly 11th News bulletin 1-8 International astronomical union Berkeley, 1961
Obs 6.1945

BERKELEY,CALIF. 1961 Sex and behavior : conferences Papers Committee for research in problems of sex Edited by Frank A. Beach Wiley,New York,1965 Based on conferences held in 1961 and 1962
Psy 31.2709

BERKELEY,CALIF. 1962 Electric propulsion development... :a selection of technical papers based mainly on the American rocket society electric propulsion conference... American rocket society Edited by Ernst Stuhlinger Progress in astronautics and aeronautics, 9 Academic press,New York; London,1963
Eng 41.6889

BERKELEY,CALIF. 1963 Bernoulli 1713:Bayes 1763:Laplace 1818:anniversary volume International research seminar proceedings University of California.Statistical laboratory Edited by Jerzy Neyman and Lucien M. LeCam Supported by the National science foundation ix,262p 23cm Springer-verlag, Berlwn,1965
Math 3.0708

BERKELEY,CALIF. 1963 Ecology of soil-borne plant pathogens.Prelude to biological control Factors determining the behavior of plant pathogens in soil :an international symposium Edited by Kenneth F. Baker and William C. Snyder Sponsored by the National research council illus. Murray,London,1965
Mol 45.0575

BERKELEY,CALIF. 1963 International association of scientific hydrology :general assembly 13th Proceedings Commission of snow and ice International association of scientific hydrology.Publication,61 286p Gentbrugge,1963
Sco 14.0139

BERKELEY,CALIF. 1963 International union of geodesy and geophysics :general assembly 13th comtes-rendus Edited by G.R. Laclavere and G.P. Garland 301p 24cm Paris,1965
Sco 14.0140

BERKELEY,CALIF. 1963 Surface waters : symposium Proceedings World meteorological association,and,International association of scientific hydrology International association of scientific hydrology. Publication,63 illus 615p Gentbrugge, 1964 Papers in English,French,German
Sco 14.0138

BERKELEY,CALIF. 1964 Galapagos The Symposia of the Galapagos international scientific project Edited by Robert I. Bowman illus,tables xvii,318p California university press,Berkeley;Los Angeles,1966 Conference held aboard ship en route to the Islands from Jan 11-18
Bot 42.2246

BERKELEY,CALIF. 1964 and 1964 Jan 11-18 The Galapagos :proceedings of the symposia of the Galapagos international scientific project Proceedings University of California Edited by Robert I. Bowman illus,pls 27cm University of California press,Berkeley;Los Angeles,1966
Bal 44.4211

BERKELEY,CALIF. 1965 Ground level climatology American association for the advancement of science :annual meeting 132nd Symposium on ground level climatology Edited by Robert H. Shaw American association for the advancement of science.Publications,86 Washington,D.C.,1967
Geog 13.1034

BERKELEY,CALIF. 1965 Molecular mechanisms of temperature adaptation :a symposium American association for the advancement of science Edited by C.Ladd Prosser American association for the advancement of science. Publication, 84 American association for the advancement of science,Washington,D.C., 1967
Gen 34.0538

BERKELEY,CALIF. 1965 Berkeley,Calif. 1966 Dec-Jan Mathematical statistics and probability Berkeley symposium on mathematical statistics and probability 5th Proceedings Edited by Lucien M.le Cam and Jerzy Neyman bibliog.,illus. 26cm 6 vols University of California press,Berkeley,Los Angeles,Calif.,1967
Math S 3.1455

BERKELEY,CALIF. 1966 International conference on high-energy physics 13th proceedings Sponsored by International union of pure and applied physics University of California press,Berkeley;Los Angeles,1967
A Math 4.0891

BERKELEY,CALIF. 1966 International symposium on rarefied gas dynamics 2nd Proceedings University of California Edited by L. Talbot Advances in applied mechanics.Supplement, 1 Academic press,New York;London,1961
Eng 41.6764

BERKELEY,CALIF. 1966 Symposium (international) on combustion 11th Papers Combustion institute Combustion institute, Pittsburgh,Pa.,1967
Chem 18.0441

BERKELEY,CALIF. 1966 The International conference on high energy physics 13th Proceedings University of California and United States atomic energy commission Edited by Margaret Alston-Garnjost Sponsored by the International union of pure and applied physics Berkeley,Calif.,1967
Cav 7.0189

BERKELEY,CALIF. 1968 Global analysis :a symposium Proceedings American mathematical society Edited by Shing-Shen Chern and Stephen Smale American mathematical society.Proceedings of symposia in pure mathematics, 14,16 bibliog. 25cm American mathematical society,Providence,R.I., 1970
P Math 2.3875

BERKELEY,CALIF.- 1957- Clays and clay minerals :national conference 6- proceedings International series of monographs on earth sciences,2- Pergamon press,London,1959-
Min 10.0911

BERKELEY CASTLE 1961 Properties of reactor materials and the effects of radiation damage : international conference Proceedings Berkeley nuclear laboratories,Gloucestershire Edited by D.J. Littler Butterworths,London, 1962
Met 25.1574

BERKELEY CASTLE,GLOS. 1961 Properties of reactor materials and the effects of radiation damage :international conference Proceedings Central electricity generating board.Berkeley nuclear laboratories Edited by D.J. Littler xv,562p Butterworths,London,1962
Cav 7.2864

BERKELEY INTERNATIONAL MATERIALS CONFERENCE 1st the impact of transmission electron microscopy on theories of the strength of crystals Proceedings Electron microscopy and strength of crystals Berkeley 1961 Jul 5-8 Lawrence radiation laboratory. Inorganic materials division and University of California Edited by Gareth Thomas and Jack Washburn Interscience,New York;London,1963
Met 25.1211

BERKELEY INTERNATIONAL MATERIALS CONFERENCE 2nd high strength materials- present status and anticipated developments Proceedings Berkeley 1964 Jun 15-18 Edited by Victor F. Zackay Inorganic materials research division series John Wiley,New York,1965
Met 25.1019

BERKELEY NUCLEAR LABORATORIES,GLOUCESTERSHIRE Properties of reactor materials and the effects of radiation damage :international conference Proceedings Berkeley Castle 1961 May 30-Jun 2 Edited by D.J. Littler Butterworths,London,1962
Met 25.1574

BERKELEY SYMPOSIUM ON MATHEMATICAL STATISTICS AND PROBABILITY 2nd Proceedings Mathematical statistics and probability Berkeley,Calif. 1950 Jul 31-Aug 12 University of California.Statistical laboratory Edited by Jerzy Neyman University of California press,Berkeley,Calif.; Los Angeles,Calif.,1951
Radioth 35.0599

BERKELEY SYMPOSIUM ON MATHEMATICAL STATISTICS AND PROBABILITY 2nd proceedings Mathematical statistics and probability Berkeley,Calif. 1950 Jul.31-Aug.12 University of California.Statistical laboratory Edited by Jerzy Neyman Also supported by the Office of naval research xi, 666p 26cm University of California press, Berkeley;Los Angeles,Calif.,1951
Math 3.0686

BERKELEY SYMPOSIUM ON MATHEMATICAL STATISTICS AND PROBABILITY 3rd Proceedings Mathematical statistics and probability Berkeley,Calif. 1954 Dec 26-31 Berkeley, Calif. 1955 Jul,Aug Vol 4: contributions to biology and problems of health University of California.Statistical laboratory Edited by Jerzy Neyman University of California press,Berkeley,Calif.; Los Angeles,Calif.,1956
Radioth 35.0606

BERKELEY SYMPOSIUM ON MATHEMATICAL STATISTICS AND PROBABILITY 4th proceedings Mathematical statistics and probability Berkeley,Calif. 1960 Jun.20-Jul.30 Vols. 1-4 University of California.Statistical laboratory Edited by Jerzy Neyman 26cm 4 vols University of California press,Berkeley; Los Angeles,Calif.,1961
Math 3.0688

BERKELEY SYMPOSIUM ON MATHEMATICAL STATISTICS AND PROBABILITY 5th Proceedings Mathematical statistics and probability Berkeley,Calif. 1965 Jun-Jul Berkeley, Calif. 1966 Dec-Jan Edited by Lucien M. le Cam and Jerzy Neyman bibliog.,illus. 26cm 6 vols University of California press, Berkeley,Los Angeles,Calif.,1967
Math S 3.1455

BERKELEY SYMPOSIUM ON MATHEMATICAL STATISTICS AND PROBABILITY proceedings 3rd Mathematical statistics and probability Berkeley,Calif. 1954 Dec.26-31 Berkeley, Calif. 1955 Jul-Aug Vols. 1-5 University of California.Statistical laboratory Edited by Jerzy Neyman 26cm 5 vols University of California press,Berkeley; Los Angeles,Calif.,1956
Math 3.0687

BERKELEY SYMPOSIUM ON MATHEMATICAL STATISTICS AND PROBABILITY proceedings 1st Mathematical statistics and probability Berkeley,Calif. 1945 Aug.13-18 Berkeley, Calif. 1946 Jan.27-29 University of California.Statistical laboratory Edited by Jerzy Neyman viii,501p 26cm University of California press,Berkeley;Los Angeles,Calif. 1949
Math 3.0685

BERLIN 1874 Commission fur die vorberaschung der beobachtung des Venusdurchgangs von 1874 2 manuscript report 21p Berlin,1871
Obs 6.2161

BERLIN 1897 International scientific conference on leprosy Mittheilungen und verhandlungen Bd 1-3 2 vols Hirschwald,Berlin,1897-98 Papers in French,English,German,Spanish.Bd 2-3 bound together
Path 30.0701

BERLIN 1899 Die Deutsche tiefsee-expedition auf dem schiff 'Valdivia',1898-99 Internationaler geographen-kongress,Berlin 1899 7 By Carl Chun and Gerhard Scott Deutsche tiefsee-expedition,1898-99 Maps, tables 120p 26cm Gesellschaft fur erdkunde zu Berlin,Berlin,1899
Sco 14.6301

BERLIN 1899 Kongress zur bekampfung der tuberkulose als volkskrankheit Bericht Deutsches central-komite zur errichtung von heilstatten fur lungenkranke Edited by Gotthold Pannwitz Berlin,1899
Path 30.0770

BERLIN 1902 International conference on tuberculosis 1st Report Central international bureau for the prevention of consumption Edited by Gotthold Pannwitz Berlin,1903 Text in English,French,German; title also in French and German
Path 30.1025

BERLIN 1904 Versammlung der tuberkulose-arzte 2te Bericht Deutsches central-komite zur errichtung von heilstatten fur lungenkranke Edited by Johannes Nietner Deutsches central-komite zur errichtung von heilstatten fur lungenkranke,Berlin,1905
Path 30.0761

BERLIN 1926 Internationale studiengesellschaft zur erforschung der Arktis mit dem luftschiff :ordentl.versammlung 1ste Verhandlungen Edited by Leonid Breitfuss Petermann's mitteilungen.Erganzungsheft,191 illus,maps 115p 28cm Justus Perthes, Gotha,1927
Sco 14.0106

BERLIN 1926 Internationale studiengesellschaft zur erforschung der arktis mit dem luftschiff (aeroarctic) :versammlung 1 Verhandlungen Edited by Leonid Breitfuss map 115p J.Perthus,Gotha,1927 Contains a paper by Shaw,W.N. 'The influence of the North Polar region upon the meteorology of the northern hemisphere'
Nap 11.0895

BERLIN 1927 Internationaler kongress fur vererbungswissenschaft 5 Verhandlungen Band 1-2 Edited by Hans Nachtsheim Zeitschrift fur induktive abstammungs- und vererbungslehre.Supplementband, 1 2 vols Borntraeger,Leipzig,1928
Gen 34.1001

BERLIN 1930 World power conference 2nd Transactions Bd 5,8 2 vols VDI verlag,Berlin,1930
Eng 41.7438

BERLIN 1935 Bevolkerungsfragen International congress for studies on population 3rd Edited by Hans Harmsen and Franz Lohse illus.,maps Lehmann,Munich, 1936
Geog 13.1489

BERLIN 1935 Bevolkerungsfragen International kongress fur bevolkerungswissenschaft bericht 2nd Internationale vereinigung fur bevolkerungswissenschaft.Deutsche ausschluss Edited by Hans Harmson and Franz Lohse Lehmanns,Munich,1936
Gen 34.2069

BERLIN 1955 Fachkonference radioastronomie jahrestagung der deutschen akademie der wissenschaften Deutsche akademie der wissenschaften zu Berlin Deutsche akademie der wissenschaften zu Berlin.Abhandlungen. Klasse fur mathematik,physik und technik, 3 135p Akademie-verlag,Berlin,1958
Obs 6.2016

BERLIN 1958 International conference on electron microscopy 4th Bd 2: biologisch-medizinischer teil Edited by W. Bargmann and others Springer,Berlin,1960
An 32.3165

BERLIN 1958 Internationaler kongress fur elektronenmikroskopie 4th Verhandlungen Band 1: physikalisch-technischer teil Edited by G. Mollenstedt and others Springer, Berlin,1960
Eng 41.5532

BERLIN 1958 Intionationaler kongress fur elektronmikroskopie 4th Proceedings Bd 1: physikalisch-technischer teil International federation of electron microscope societies Edited by G. Mollenstedt and others Springer,Berlin,1960 General editors W.Bargmann and others. Papers in English,French and German
Met 25.2493

BERLIN 1959 Chemie und biochemie der solanum-alkaloide :international symposium Vortrage und diskussionsbeitrage Deutsche akademie der landwirtschaftswissenschaften zu Berlin Deutsche akademie der landwirtschaftswissenschaften zu Berlin. Tagungsberichte, 27 Berlin,1961 Papers in English and German,with summaries in English,German and Russian
Chem 26.0814

BERLIN 1960 Der Einsatz elektronischer rechengerate fur forschungs- und entwicklungs-aufgaben conference report Hahn-Meitner-Institut fur kernforschung,Berlin Hahn-Meitner-Institut fur kernforschung Berlin. Berichte, 13 var.p 30cm Hahn-Meitner-Institut,Berlin,1960
Math L 5.0870

BERLIN 1962 Internationale tagung uber mathematische statistik und ihre anwendungen Edited by Erna Weber Deutsche akademie der wissenschaften zu Berlin.Abhandlungen,Klasse fur mathematik,physik und technik,4 126p 30cm Akademie-verlag,Berlin,1964
Math 3.0701

BERLIN 1962 Stratospheric and mesospheric circulation The International symposium on stratospheric and mesospheric circulation Proceedings Institut fur meteorologie und geophysik Edited by Richard Scherhag and Gunter Warnecke Institut fur meteorologie und geophysik.Meteorologische abhandlungen,36 illus vi,644p 31cm Dietrich Reimer, Berlin,1963
Sco 14.0527

BERLIN 1965 Einstein symposium
Entstehung, entwicklung und perspektiven der Einsteinschen gravitationstheorie vortrage und diskussionen Deutsche akademie der wissenschaften zu Berlin Edited by H-J. Treder 314p Akademie-verlag,Berlin,1966
Obs 6.1765

BERLIN 1965 Einstein-Symposium
Entstehung,Entwicklung und perspektiven der Einsteinschen gravitationstheorie vortrage und diskussionen Akademie der Wissenschaften Edited by H.J. Treder Akademie Verlag,Berlin, 1965 typescript
A Math 4.1031

BERLIN 1965 Krebs - dokumentation und statistik maligner tumoren Internationale jahrestagung des Arbeitsausschuss medizin 10th Verhandlungsbericht Edited by Gustav Wagner Schattauer,Stuttgart,1966
Radioth 35.0877

BERLIN 1967 Contributions to extension theory :a symposium Proceedings Edited by Jungen Flaschmeyer 279p 24cm VEB Deutscher verlag der wissenschaften,Berlin, 1969
P Math 2.3743

BERLIN 1968 International congress on electron microscopy 4th Verhandlungen Band 2: biologisch-medizinischer teil Edited by D. Bargmann and others Springer, Berlin,1969 In English,French and German
Phys 20.1509

BERLIN 1968 Kristallisation
Gemeinschaftskonferenz der reihe 'Metall' :die kristallisation von metallen aus dem schmelzfluss,der gasphase und durch elektrolytische abscheidung 6 Vortrage Deutsche akademie der wissenschaften zu Berlin. Sektion der physik.Unterkommission metallphysik VEB verlag fur grundstoffindustrie,Leipzig,1969
Met 25.2622

BERLIN 1968 Perinatal medicine European congress of perinatal medicine 1st Edited by Peter John Hungerford and others Thieme;Academic press,Stuttgart;New York,1969
Phys 20.2246

BERN 1955 Funfzig jahre relativitatstheorie...Jubilee of relativity theory Verhandlungen Edited by A. Mercier and M. Kervaire Helvetica physica acta. Supplementum, 5 Birkhauser,Basel,1956 In French,German,English
Radioth 35.1599

BERN 1961 Interpretation of ultrastructure a symposium Edited by R.J.C. Harris International society for cell biology. Symposia, 1 Academic press,New York;London, 1962
Radioth 35.0517

BERN 1961 The Interpretation of ultrastructure :symposium International society for cell biology Edited by R.J.C. Harris International society for cell biology.Symposia, 1 illus Academic press, New York;London,1962
Bioch 33.0964

BERN 1961 The Interpretation of ultrastructure :symposium International society for cell biology Edited by R.J.C. Harris International society for cell biology.Symposia, 1 illus. x,438p Academic press,New York;London,1962
Bot 42.1080

BERN 1964 Symposium uber die markierang der proteine mittels radioaktiver isotopen und deren anwendung in biologie und medizin Schweizerische akademie der medizinischen wissenschaften.Bulletin, 21,fasc.3-4 Schwabe,Basle,Stuttgart,1965 Parallel text in German,French and Italian
Radioth 35.1187

BERNE 1964 Committee on Mediterranean neogene stratigraphy session 3rd proceedings Edited by C.W. Drooger and others illus. E.J.Brill,Leiden,1966
Geol 8.2524

BERNE 1964 Joint European conference of the Institute of mathematical statistics,the International association for statistics in physical sciences,the Biometric society Berne,1964 13 unbound typescript papers
Math 3.1124

BERYLLIUM The Metallurgy of beryllium :an international conference Proceedings London 1961 Oct 16-18 Institute of metals Institute of metals.Monograph and report series, 28 Chapman and Hall,London,1963
Met 25.0605

BETHESDA,MD 1949 National cancer conference 1st Proceedings American cancer society American cancer society,1949
Med 36.0003

BETHESDA,MD 1962 Conference on newer respiratory disease viruses National institutes of health National institute of allergy and infectious diseases Edited by Clayton G. Loosli Sponsored by University of Southern California.School of medicine American review of respiratory diseases, 88,3 pt 2 American thoracic society,1963
PGMS 29.0321

BETHESDA,MD. 1957 Congenital malformations a conference on teratology papers Published under the auspices of the Association for the aid of crippled children Pediatrics, 23,no 1,pt 2,suppt. Thomas, Springfield,Ill.,1959
An 32.2046

BETHESDA,MD. 1957 The Process of aging in the nervous system :a conference Proceedings Edited by James E. Birren and others Sponsored by the National advisory neurological diseases and blindness council Symposia in neuroanatomical sciences, 5 Blackwell,Oxford,1959
An 32.4171

BETHESDA,MD. 1959 Neural mechanisms of the auditory and vertibular systems :a conference Proceedings Edited by G.L. Rasmussen and W.F. Windle Sponsored by the National institute of neurological diseases and blindness Symposia in neuroanatomical sciences, 6 illus. xiv,422p Thomas,Springfield,Ill., 1960
Phys 20.1975

BETHESDA,MD. 1959 Neural mechanisms of the auditory and vestibular systems :a conference Edited by Grant L. Rasmussen and William F. Windle Symposia in neuroanatomical sciences, 6 Thomas,Springfield,Ill.,1960
An 32.4821

BETHESDA,MD. 1959 The Delayed effects of whole body radiation :a symposium Johns Hopkins university.Operation research office Edited by Bernard B. Watson Jointly sponsored by the Walter Reid army institute of research Johns Hopkins press,Baltimore,Md., 1960
Radioth 35.1968

BETHESDA,MD. 1960 Genetic polymorphisms and geographic variations in disease :a conference Proceedings National institute of arthritis and metabolic diseases National heart institute National institutes of health Edited by Barach S. Blumberg Grune and Stratton,New York;London,1961
Gen 34.1984

BETHESDA,MD. 1961 Genetics and dental health :an international symposium Proceedings American dental association. Council on dental research Edited by Carl J. Witkop Symposium on genetics related to dental health, 1 McGraw-Hill,New York,1962
Gen 34.2024

BETHESDA,MD. 1968 Microsomes and drug oxidations :a symposium Edited by J.R. Gillette and others Academic press,New York; London,1969
Bioch 33.1089

BETHSEDA,MD. 1971 Current topics in biochemistry National institutes of health Edited by C.B. Anfinsen National institutes of health.Lectures in biomedical sciences Academic press,New York,1972
Bioch 33.2168

BETTER USE OF THE WORLD'S FAUNA FOR FOOD The Institute of biology symposium 11th London 1961 Edited by J.D. Ovington Institute of biology,London,1963
Geog 13.2660

The BETTER USE OF THE WORLD'S FAUNA FOR FOOD London 1962 Institute of biology Edited by J.D. Ovington Institute of biology. Symposia, 11 21cm London,1963
Bal 44.4059

BEVOLKERUNGSFRAGEN International congress for studies on population 3rd Berlin 1935 Aug 26-Sep 1 Edited by Hans Harmsen and Franz Lohse illus.,maps Lehmann,Munich, 1936
Geog 13.1489

BEVOLKERUNGSFRAGEN International kongress fur bevolkerungswissenschaft bericht 2nd Berlin 1935 Aug 26-Sep 1 Internationale vereinigung fur bevolkerungswissenschaft. Deutsche ausschluss Edited by Hans Harmson and Franz Lohse Lehmanns,Munich,1936
Gen 34.2069

BHAVNAGAR 1965 Sea,salt and plants seminar Proceedings Central salt and marine chemicals research institute Edited by V. Krishnamurthy illus. xv,372p CSMCRI, Bhavnagar,1967
Bot 42.1991

BIBLIOGRAPHY AND NATURAL HISTORY :a conference Essays Lawrence 1964 Jun 25-27 University of Kansas.Linda Hall library of science and technology Edited by Thomas R. Buckman University of Kansas.Publications. Library series, 27 ix,148p University of Kansas libraries,Lawrence,Kansas,1966
WSM 43.3212

BIENNIAL GAS DYNAMICS SYMPOSIUM 4th proceedings Magnetohydrodynamics Evanston,Ill. 1961 Aug 23-25 American rocket society and Northwestern university Edited by Ali Bulent Cambel and others Through the generosity of Government agencies and industrial sponsors Northwestern university press,Evanston,Ill.,1962
A Math 4.0623

BIFURCATION THEORY AND NONLINEAR EIGENVALUE PROBLEMS :a seminar Lectures New York 1966-67 Edited by Joseph B. Keller and Stuart Antman Held at the Courant institute of mathematical sciences Bibliog. xiv,434p 23cm Benjamin,New York;Amsterdam, 1969
P Math 2.3648

BILHARZIASIS :a Ciba foundation symposium Proceedings Cairo 1962 Mar 18-22 Ciba foundation Edited by G.E.W. Wolstenholme and Maeve O'Connor illus. Churchill,London, 1962 Held in commemoration of Theodor Maximilian Bilharz
Mol 45.0266

BIMINI,BAHAMAS 1963 New York 1966 Apr 13-15 Marine bio-acoustics :symposium 1st, 2nd Proceedings Vol 1-2 By William N. Tavolga 2 vols Pergamon press,Oxford, 1964-67 Vol.1 held at the Lerner marine laboratory Bimini;vol.2 held at the American museum of natural history,New York
Bal 39.1534

BINGHAMTON,N.Y. 1970 Conference on monotone mappings and open mappings 1st Proceedings State university of New York, Binghampton Edited by Louis F. McAuley xxii,422p 25cm State university of New York,Binghamton,N.Y.,1971 Proceedings dedicated to the memory of G.T.Whyburn
P Math 2.4118

BIOASSAY OF ANTERIOR PITUITARY AND ADRENOCORTICAL HORMONES :a colloquium proceedings London 1952 Mar 25-27 Ciba foundation Edited by G.E.W. Wolstenholme Ciba foundation colloquia on endocrinology, 5 illus Churchill,London,1953
Bioch 33.0435

BIOCHEMICAL AND HUMAN FACTORS SYMPOSIUM Washington,D.C. 1967 Apr 3-5 Edited by Howard Gage 174p American society of chemical engineers,New York,1967 Presented at the Biomechanical and human factors conference
Chem E 24.1803

BIOCHEMICAL ASPECTS OF GENETICS :a symposium London 1949 Feb 12 Biochemical society Edited by R.T. Williams Biochemical society. Symposia, 4 Cambridge university press, Cambridge,1950
Gen 34.0733

BIOCHEMICAL ASPECTS OF GENETICS :a symposium London 1949 Feb 12 Biochemical society Edited by R.T. Williams Held at the Westminster hospital medical school Biochemical society.Symposia, 4 Cambridge university press,Cambridge,1950
Bioch 33.1366

BIOCHEMICAL ASPECTS OF THE BIOGENESIS OF MITOCHONDRIA The Round table discussion on the biochemical aspects of the biogenesis of mitochondria Proceedings Polignane a Mare 1967 May 15-18 Edited by E.C. Slater and others Adriatica editrice,Bari,1968
Bioch 33.1073

BIOCHEMICAL GENETICS International congress of biochemistry 6th Abstracts New York 1964 Jul 26-Aug 1 Scheduled under the auspices of the International union of biochemistry Washington,D.C.,1964
Bioch 33.1340

BIOCHEMICAL LESIONS Thiamine deficiency : biochemical lesions and their clinical sihnificance London 1966 Nov 10-11 Ciba foundation Edited by G.E.W. Wolstenholme and Maeve O'Connor Ciba foundation study group, 28 Churchill,London,1967 In honour of Sir Rudolph Peters
Bioch 33.0574

BIOCHEMICAL PRINCIPLES OF THE FOOD INDUSTRY International congress of biochemistry 5th Proceedings Moscow 1961 Aug 10-16 International union of biochemistry Edited by V.L. Kretovich and E. Pijanowski I.U.B. symposium series, 28 Pergamon;PWN-Polish scientific publishers,Oxford;Warsaw,1963
Bioch 33.1337

BIOCHEMICAL PROBLEMS OF LIPIDS 2nd :an international conference Proceedings Ghent 1955 Jul 27-30 Vlaamse chemische vereniging Edited by G. Popjak and E. Le Breton Butterworths,London,1956 Conference president:Professor R.Ruyssen
Chem 18.1500

The BIOCHEMICAL REACTIONS OF CHEMICAL WARFARE AGENTS :a symposium London 1947 Dec 13 Biochemical society Edited by R.T. Williams Held at the University college hospital medical school Biochemical society. Symposia, 2 Cambridge university press, Cambridge,1948
Bioch 33.1364

The BIOCHEMICAL REACTIONS OF CHEMICAL WARFARE AGENTS :a symposium London 1947 Dec 13 Edited by R.T. Williams Biochemical society.Symposia, 2 Cambridge university press,Cambridge,1948
Phys 20.1524

BIOCHEMICAL SOCIETY Aspects of insect biochemistry :a symposium London 1965 Apr 1 Edited by T.W. Goodwin Biochemical society.Symposia, 25 Academic press,London; New York,1965
Bioch 33.1387

BIOCHEMICAL SOCIETY Biochemical aspects of genetics :a symposium London 1949 Feb 12 Edited by R.T. Williams Biochemical society. Symposia, 4 Cambridge university press, Cambridge,1950
Gen 34.0733

BIOCHEMICAL SOCIETY Biochemical aspects of genetics :a symposium London 1949 Feb 12 Edited by R.T. Williams Held at the Westminster hospital medical school Biochemical society.Symposia, 4 Cambridge university press,Cambridge,1950
Bioch 33.1366

BIOCHEMICAL SOCIETY Biochemistry of mucopolysaccharides of connective tissue :a symposium London 1960 Feb 3 By J.N. Davidson Edited by F. Clark and J.K. Grant 4th edition Biochemical society.Symposia, 20 Bibliog Cambridge university press, Cambridge,1961
Radioth 35.0098

BIOCHEMICAL SOCIETY Biological oxidation of aromatic rings :a symposium London 1949 Nov 12 Edited by R.T. Williams Held at the London school of hygiene and tropical medicine Biochemical society.Symposia, 5 Cambridge university press,Cambridge,1950
Bioch 33.1367

BIOCHEMICAL SOCIETY Biological transformation of starch and cellulose London 1953 Feb 21 Edited by R.T. Williams Biochemical society.Symposia, 11 Cambridge university press,Cambridge,1953
Bot 42.1690

BIOCHEMICAL SOCIETY Biological transformations of starch and cellulose :a symposium London 1953 Feb 21 Held at the London school of hygiene and tropical medicine Biochemical society.Symposia, 11 Cambridge university press,Cambridge,1953
Bioch 33.1373

BIOCHEMICAL SOCIETY British biochemistry past and present Essays London 1969 Dec Edited by T.W. Goodwin Biochemical society. Symposium, 30 Academic press,London,1970
Bioch 33.2314

BIOCHEMICAL SOCIETY Chemical reactivity and biological role of functional groups in enzymes :a symposium... Proceedings Oxford 1970 Jan Molecular enzymology and protein group Edited by R.M.S. Smellie Biochemical society.Symposia, 31 Academic press,London, 1970 Molecular enzymology and protein group of the Biochemical society and the Chemical society
Chem 18.2745

BIOCHEMICAL SOCIETY Glutathione :a symposium London 1958 Feb 15 Edited by E.M. Crook Held at the University of London.Senate House Biochemical society.Symposia, 17 Cambridge university press,Cambridge,1959
Bioch 33.1379

BIOCHEMICAL SOCIETY Glutathione :a symposium London 1958 Feb 15 Edited by E.M. Crook Held at the University of London.Senate house Biochemical society.Symposia, 17 Cambridge university press,Cambridge,1959
Radioth 35.0078

BIOCHEMICAL SOCIETY Immunochemistry :a symposium London 1952 Nov 15 Edited by R. T. Williams Held at the London school of hygiene and tropical medicine Biochemical society.Symposia, 10 Cambridge university press,Cambridge,1953
Bioch 33.1372

BIOCHEMICAL SOCIETY Instrumentation in biochemistry London 1966 Apr Edited by T. W. Goodwin Biochemical society.Symposia, 26 Academic press,London;New York,1966
Bioch 33.1388

BIOCHEMICAL SOCIETY Instrumentation in biochemistry :a symposium London 1966 Apr Edited by T.W. Goodwin Biochemical society. Symposia, 26 Academic press,London,1966
Radioth 35.0135

BIOCHEMICAL SOCIETY International congress of biochemistry 1st Report of opening and concluding sessions and three lectures Cambridge 1949 Aug 19-25 Glasgow university,Biochemistry department,Glasgow, 1950 Unbound
Bioch 33.1301

BIOCHEMICAL SOCIETY Lipid metabolism :a symposium London 1952 Feb 16 Edited by R. T. Williams Held at the London school of hygiene and tropical medicine Biochemical society.Symposia, 9 Cambridge university press,Cambridge,1952
Bioch 33.1371

BIOCHEMICAL SOCIETY Metabolism and functions in nervous tissue :a symposium London 1951 Nov 10 Edited by R.T. Williams Held at the London school of hygiene and tropical medicine Biochemical society.Symposia, 8 Cambridge university press,Cambridge,1952
Bioch 33.1370

BIOCHEMICAL SOCIETY Metals and enzyme activity :a symposium Leeds 1956 Jul 13 Edited by E.M. Crook Held at the University of Leeds Biochemical society.Symposia, 15 Cambridge university press,Cambridge,1958
Bioch 33.1377

BIOCHEMICAL SOCIETY Metals and enzyme activity :a symposium 15 Leeds 1956 Jul 13 Edited by E.M. Crook Biochemical society symposia,15 Cambridge university press,Cambridge,1958
Chem 18.1269

BIOCHEMICAL SOCIETY Methods of separation of subcellular structural components :a symposium Louvain 1962 May 11 Edited by J.K. Grant Held in the University of Louvain. Physiological institute Biochemical society. Symposia, 23 Cambridge university press, Cambridge,1963 Organized by C.de Duve
Bioch 33.1385

BIOCHEMICAL SOCIETY Methods of separation of subcellular structural components :a symposium Louvain 1962 May 11 Edited by J.K. Grant Held jointly with the Societe belge de biochimie Biochemical society.Symposia, 23 Cambridge university press,Cambridge,1963
Radioth 35.0136

BIOCHEMICAL SOCIETY Natural substances formed biologically from mevalonic acid :a symposium Liverpool 1969 Apr Biochemical society.Symposia, 29 Academic press,London, 1970
Chem 18.2550

BIOCHEMICAL SOCIETY Natural substances formed biologically from mevalonic acid :a symposium Liverpool 1969 Apr Edited by T. W. Goodwin Biochemical society.Symposia, 29 Academic press,London;New York,1970
Bioch 33.1391

BIOCHEMICAL SOCIETY Partition chromatography a symposium London 1948 Oct 30 Edited by R.T. Williams and R.L.M. Synge Biochemical society.Symposia, 3 Cambridge university press,Cambridge,1949
Gen 34.0622

BIOCHEMICAL SOCIETY Partition chromatography a symposium London 1948 Oct 30 Edited by R.T. Williams and R.L.M. Synge Held at the London school of hygiene and tropical medicine Biochemical society.Symposia, 3 Cambridge university press,Cambridge,1949
Bioch 33.1365

BIOCHEMICAL SOCIETY Partition chromatography a symposium 3 London 1948 Oct 30 Edited by R.T. Williams and R.L.M. Synge Biochemical society symposia,3 Cambridge university press,Cambridge,1950
Chem 18.1276

BIOCHEMICAL SOCIETY Porphyrins and related compounds London 1968 Apr Edited by T.W. Goodwin Biochemical society.Symposia, 28 Academic press,London;New York,1968
Bioch 33.1390

BIOCHEMICAL SOCIETY Porphyrins and related compounds London 1968 Apr Edited by T.W. Goodwin Biochemical society.Symposia, 28 Academic press,London;New York,1968
Bot 42.1755

BIOCHEMICAL SOCIETY Porphyrins and related compounds :a symposium London 1968 Apr Edited by T.W. Goodwin Biochemical society. Symposia, 28 Academic press,London,1968
Chem 18.2761

BIOCHEMICAL SOCIETY Steric aspects of the chemistry and biochemistry of natural products London 1959 Jun 30 Edited by J.K. Grant and W. Klyne Biochemical society.Symposia, 19 Cambridge university press,Cambridge, 1960
Chem 18.2613

BIOCHEMICAL SOCIETY Steric aspects of the chemistry and biochemistry of natural products a symposium London 1959 Jun 30 Edited by J.K. Grant and W. Klyne Held at the University of London.Senate House Biochemical society.Symposia, 19 Cambridge university press,Cambridge,1960
Bioch 33.1381

BIOCHEMICAL SOCIETY Structure and function of subcellular components :a symposium London 1957 Feb 23 Edited by E.M. Crook Biochemical society.Symposia, 16 Pl. Cambridge university press,Cambridge,1959
Radioth 35.0069

BIOCHEMICAL SOCIETY Structure and function of subcellular components :a symposium London 1957 Feb 23 Edited by E.M. Crook Held at the University of London.Senate House Biochemical society.Symposia, 16 Cambridge university press,Cambridge,1959
Bot 42.1493

BIOCHEMICAL SOCIETY The Biochemical reactions of chemical warfare agents :a symposium London 1947 Dec 13 Edited by R. T. Williams Held at the University college hospital medical school Biochemical society. Symposia, 2 Cambridge university press, Cambridge,1948
Bioch 33.1364

BIOCHEMICAL SOCIETY The Biochemistry of fertilization and the gametes :a symposium London 1951 Feb 17 Edited by R.T. Williams Held at the London school of hygiene and tropical medicine Biochemical society. Symposia, 7 Cambridge university press, Cambridge,1951
Bal 39.0388

BIOCHEMICAL SOCIETY The Biochemistry of fertilization and the gametes :a symposium London 1951 Feb 17 Edited by R.T. Williams Held at the London school of hygiene and tropical medicine Biochemical society. Symposia, 7 Cambridge university press, Cambridge,1951
Bioch 33.1369

BIOCHEMICAL SOCIETY The Biochemistry of fish a symposium Liverpool 1950 Sep 22 Edited by R.T. Williams Held at the University of Liverpool.Derby Hall Biochemical society.Symposia, 6 Cambridge university press,Cambridge,1961
Bioch 33.1368

BIOCHEMICAL SOCIETY The Biochemistry of mucopolysaccharides of connective tissue :a symposium London 1960 Feb 13 Edited by F. Clark and J.K. Grant Held at the Royal college of surgeons of England Biochemical society.Symposia, 20 Cambridge university press,Cambridge,1961
Bioch 33.1382

BIOCHEMICAL SOCIETY The Biochemistry of vitamin B12 :a symposium London 1955 Feb 19 Edited by R.T. Williams Held at the London school of hygiene and tropical medicine Biochemical society.Symposia, 13 Cambridge university press,Cambridge,1955
Bioch 33.1375

BIOCHEMICAL SOCIETY The Biosynthesis and secretion of adrenocortical steroids :a symposium London 1959 Feb 14 Edited by F. Clark and J.K. Grant Held at the University of London.Senate House Biochemical society. Symposia, 18 Cambridge university press, Cambridge,1960
Bioch 33.1380

BIOCHEMICAL SOCIETY The Chemical pathology of animal pigments :a symposium London 1954 Feb 20 Edited by R.T. Williams Held at the London school of hygiene and tropical medicine Biochemical society.Symposia, 12 Cambridge university press,Cambridge,1954
Bioch 33.1374

BIOCHEMICAL SOCIETY The Control of lipid metabolism :a symposium Oxford 1963 Jul 19 Edited by J.K. Grant Biochemical society. Symposia, 24 Academic press,London;New York, 1963 Organized by G.Popjak
Bioch 33.1386

BIOCHEMICAL SOCIETY The Metabolic roles of citrate :a symposium in honour of...Sir Hans Krebs Oxford 1967 Jul Edited by T.W. Goodwin Biochemical society.Symposia, 27 Academic press,London;New York,1968
Bioch 33.1389

BIOCHEMICAL SOCIETY The Relation of optical form to biological activity in the amino-acid series :a symposium London 1947 Feb 15 Edited by R.T. Williams Held at the University college hospital medical school Biochemical society.Symposia, 1 Cambridge university press,Cambridge,1948
Bioch 33.1363

BIOCHEMICAL SOCIETY The Relation of optical form to biological activity in the amino-acid series :a symposium 1 London 1947 Feb 15 Edited by R.T. Williams Biochemical society symposia,1 Cambridge university press, Cambridge,1948
Chem 18.1275

BIOCHEMICAL SOCIETY The Structure and biosynthesis of macro-molecules London 1961 Mar 27-28 Edited by D.J. Bell and J.K. Grant Held at University of London.Senate house Biochemical society.Symposia, 21 Cambridge,1962 In commemoration of the 50th anniversary of the Biochemical society
Bal 39.0515

BIOCHEMICAL SOCIETY The Structure and biosynthesis of macromolecules :a symposium London 1961 Mar 27-28 Edited by D.J. Bell and J.K. Grant Held at the University of London.Senate House Biochemical society. Symposia, 21 Cambridge university press, Cambridge,1962 In commemoration of the 50th anniversary of the Biochemical society
Bioch 33.1383

BIOCHEMICAL SOCIETY The Structure and biosynthesis of macromolecules :a symposium London 1961 Mar 27-28 Edited by J.K. Grant and D.J. Bell Biochemical society symposia, 21 Cambridge university press,Cambridge,1962
Chem 18.1277

BIOCHEMICAL SOCIETY The Structure and function of subcellular components :a symposium London 1957 Feb 23 Edited by E. M. Crook Held at the University of London. Senate House Biochemical society.Symposia, 16 Cambridge university press,Cambridge,1959
Bioch 33.1378

BIOCHEMICAL SOCIETY The Structure and function of the membranes and surfaces of cells :a symposium London 1962 Mar 9 Edited by D.J. Bell and J.K. Grant Biochemical society.Symposia, 22 Cambridge university press,Cambridge,1963
Bioch 33.1384

BIOCHEMICAL SOCIETY The Structure and function of the membranes and surfaces of cells :a symposium London 1962 Mar 9 Edited by D.J. Bell and J.K. Grant Held at the Middlesex hospital medical school Biochemical society.Symposia, 22 Cambridge university press,Cambridge,1963 Organized by J.K.Grant
Bal 39.0294

BIOCHEMICAL SOCIETY The Structure of nucleic acids and their role in protein synthesis :a symposium London 1956 Feb 18 Edited by E. M. Crook Held at the London school of hygiene and tropical medicine Biochemical society.Symposia, 14 Cambridge university press,Cambridge,1957
Bioch 33.1376

BIOCHEMICAL SOCIETY.SYMPOSIA, 1 The Relation of optical form to biological activity in amino-acid series :a symposium held at University college hospital medical school London 1947 Feb 15 Edited by R.T. Williams Cambridge university press, Cambridge,1948
Radioth 35.0291

BIOCHEMICAL SOCIETY.SYMPOSIA, 1 The Relation of optical form to biological activity in the amino-acid series :a symposium London 1947 Feb 15 Biochemical society Edited by R.T. Williams Held at the University college hospital medical school Cambridge university press,Cambridge,1948
Bioch 33.1363

BIOCHEMICAL SOCIETY.SYMPOSIA, 1 The Relation of optical form to biological activity in the amino-acid series :a symposium London 1947 Feb 15 Edited by R.T. Williams Cambridge university press,Cambridge,1948
Phys 20.1523

BIOCHEMICAL SOCIETY.SYMPOSIA, 2 The Biochemical reactions of chemical warfare agents :a symposium London 1947 Dec 13 Biochemical society Edited by R.T. Williams Held at the University college hospital medical school Cambridge university press, Cambridge,1948
Bioch 33.1364

BIOCHEMICAL SOCIETY.SYMPOSIA, 2 The Biochemical reactions of chemical warfare agents :a symposium London 1947 Dec 13 Edited by R.T. Williams Cambridge university press,Cambridge,1948
Phys 20.1524

BIOCHEMICAL SOCIETY.SYMPOSIA, 3 Partition chromatography :a symposium London 1948 Oct 30 Biochemical society Edited by R.T. Williams and R.L.M. Synge Cambridge university press,Cambridge,1949
Gen 34.0622

BIOCHEMICAL SOCIETY.SYMPOSIA, 3 Partition chromatography :a symposium London 1948 Oct 30 Biochemical society Edited by R.T. Williams and R.L.M. Synge Held at the London school of hygiene and tropical medicine Cambridge university press,Cambridge,1949
Bioch 33.1365

BIOCHEMICAL SOCIETY.SYMPOSIA, 3 Partition chromatography :a symposium London 1948 Oct 30 Edited by R.T. Williams and R.L.M. Synge Cambridge university press,Cambridge, 1951
Phys 20.1525

BIOCHEMICAL SOCIETY.SYMPOSIA, 4 Biochemical aspects of genetics :a symposium London 1949 Feb 12 Biochemical society Edited by R.T. Williams Cambridge university press,Cambridge,1950
Gen 34.0733

BIOCHEMICAL SOCIETY.SYMPOSIA, 4 Biochemical aspects of genetics :a symposium London 1949 Feb 12 Biochemical society Edited by R.T. Williams Held at the Westminster hospital medical school Cambridge university press,Cambridge,1950
Bioch 33.1366

BIOCHEMICAL SOCIETY.SYMPOSIA, 4 Biochemical aspects of genetics :a symposium London 1949 Feb 12 Edited by R.T. Williams Cambridge university press,Cambridge,1950
Phys 20.1526

BIOCHEMICAL SOCIETY.SYMPOSIA, 5 Biological oxidation of aromatic rings :a symposium London 1949 Nov 12 Biochemical society Edited by R.T. Williams Held at the London school of hygiene and tropical medicine Cambridge university press,Cambridge,1950
Bioch 33.1367

BIOCHEMICAL SOCIETY.SYMPOSIA, 5 Biological oxidation of aromatic rings :a symposium London 1949 Nov 12 Edited by R.T. Williams Cambridge university press,Cambridge,1950
Phys 20.1527

BIOCHEMICAL SOCIETY.SYMPOSIA, 6 The Biochemistry of fish ;a symposium Liverpool 1950 Sep 22 Biochemical society Edited by R.T. Williams Held at the University of Liverpool.Derby Hall Cambridge university press,Cambridge,1961
Bioch 33.1368

BIOCHEMICAL SOCIETY.SYMPOSIA, 7 The Biochemistry of fertilization and the gametes : a symposium London 1951 Feb 17 Biochemical society Edited by R.T. Williams Held at the London school of hygiene and tropical medicine Cambridge university press, Cambridge,1951
Bal 39.0388

BIOCHEMICAL SOCIETY.SYMPOSIA, 7 The Biochemistry of fertilization and the gametes : a symposium London 1951 Feb 17 Biochemical society Edited by R.T. Williams Held at the London school of hygiene and tropical medicine Cambridge university press, Cambridge,1951
Bioch 33.1369

BIOCHEMICAL SOCIETY.SYMPOSIA, 8 Metabolism and functions in nervous tissue :a symposium London 1951 Nov 10 Biochemical society Edited by R.T. Williams Held at the London school of hygiene and tropical medicine Cambridge university press,Cambridge,1952
Bioch 33.1370

BIOCHEMICAL SOCIETY.SYMPOSIA, 9 Lipid metabolism :a symposium London 1952 Feb 16 Biochemical society Edited by R.T. Williams Held at the London school of hygiene and tropical medicine Cambridge university press, Cambridge,1952
Bioch 33.1371

BIOCHEMICAL SOCIETY.SYMPOSIA, 10 Immunochemistry :a symposium London 1952 Nov 15 Biochemical society Edited by R.T. Williams Held at the London school of hygiene and tropical medicine Cambridge university press,Cambridge,1953
Bioch 33.1372

BIOCHEMICAL SOCIETY.SYMPOSIA, 11 Biological transformation of starch and cellulose London 1953 Feb 21 Biochemical society Edited by R.T. Williams Cambridge university press,Cambridge,1953
Bot 42.1690

BIOCHEMICAL SOCIETY.SYMPOSIA, 11 Biological transformations of starch and cellulose :a symposium London 1953 Feb 21 Biochemical society Held at the London school of hygiene and tropical medicine Cambridge university press,Cambridge,1953
Bioch 33.1373

BIOCHEMICAL SOCIETY.SYMPOSIA, 12 The Chemical pathology of animal pigments :a symposium London 1954 Feb 20 Biochemical society Edited by R.T. Williams Held at the London school of hygiene and tropical medicine Cambridge university press, Cambridge,1954
Bioch 33.1374

BIOCHEMICAL SOCIETY.SYMPOSIA, 13 The Biochemistry of vitamin B12 :a symposium London 1955 Feb 19 Biochemical society Edited by R.T. Williams Held at the London school of hygiene and tropical medicine Cambridge university press,Cambridge,1955
Bioch 33.1375

BIOCHEMICAL SOCIETY.SYMPOSIA, 14 The Structure of nucleic acids and their role in protein synthesis :a symposium London 1956 Feb 18 Biochemical society Edited by E.M. Crook Held at the London school of hygiene and tropical medicine Cambridge university press,Cambridge,1957
Bioch 33.1376

BIOCHEMICAL SOCIETY.SYMPOSIA, 15 Metals and enzyme activity :a symposium Leeds 1956 Jul 13 Biochemical society Edited by E.M. Crook Held at the University of Leeds Cambridge university press,Cambridge,1958
Bioch 33.1377

BIOCHEMICAL SOCIETY.SYMPOSIA, 16 Structure and function of subcellular components :a symposium London 1957 Feb 23 Biochemical society Edited by E.M. Crook Held at the University of London.Senate House Cambridge university press,Cambridge,1959
Bot 42.1493

BIOCHEMICAL SOCIETY.SYMPOSIA, 16 Structure and function of subcellular components :a symposium London 1957 Feb 23 Biochemical society Edited by E.M. Crook Pl. Cambridge university press,Cambridge,1959
Radioth 35.0069

BIOCHEMICAL SOCIETY.SYMPOSIA, 16 The Structure and function of subcellular components :a symposium London 1957 Feb 23 Biochemical society Edited by E.M. Crook Held at the University of London.Senate House Cambridge university press,Cambridge,1959
Bioch 33.1378

BIOCHEMICAL SOCIETY.SYMPOSIA, 17 Glutathione :a symposium London 1958 Feb 15 Biochemical society Edited by E.M. Crook Held at the University of London. Senate House Cambridge university press, Cambridge,1959
Bioch 33.1379

BIOCHEMICAL SOCIETY.SYMPOSIA, 17 Glutathione :a symposium London 1958 Feb 15 Biochemical society Edited by E.M. Crook Held at the University of London. Senate house Cambridge university press, Cambridge,1959
Radioth 35.0078

BIOCHEMICAL SOCIETY.SYMPOSIA, 18 The Biosynthesis and secretion of adrenocortical steroids :a symposium London 1959 Feb 14 Biochemical society Edited by F. Clark and J. K. Grant Held at the University of London. Senate House Cambridge university press, Cambridge,1960
Bioch 33.1380

BIOCHEMICAL SOCIETY.SYMPOSIA, 19 Steric aspects of the chemistry and biochemistry of natural products London 1959 Jun 30 Biochemical society Edited by J.K. Grant and W. Klyne Cambridge university press Cambridge 1960
Chem 18.2613

BIOCHEMICAL SOCIETY.SYMPOSIA, 19 Steric aspects of the chemistry and biochemistry of natural products :a symposium London 1959 Jun 30 Biochemical society Edited by J.K. Grant and W. Klyne Held at the University of London.Senate House Cambridge university press,Cambridge,1960
Bioch 33.1381

BIOCHEMICAL SOCIETY.SYMPOSIA, 20 Biochemistry of mucopolysaccharides of connective tissue :a symposium London 1960 Feb 3 By J.N. Davidson Biochemical society Edited by F. Clark and J.K. Grant 4th edition Bibliog Cambridge university press, Cambridge,1961
Radioth 35.0098

BIOCHEMICAL SOCIETY.SYMPOSIA, 20 The Biochemistry of mucopolysaccharides of connective tissue :a symposium London 1960 Feb 13 Biochemical society Edited by F. Clark and J.K. Grant Held at the Royal college of surgeons of England Cambridge university press,Cambridge,1961
Bioch 33.1382

BIOCHEMICAL SOCIETY.SYMPOSIA, 21 The Structure and biosynthesis of macro-molecules London 1961 Mar 27-28 Biochemical society Edited by D.J. Bell and J.K. Grant Held at University of London.Senate house Cambridge, 1962 In commemoration of the 50th anniversary of the Biochemical society
Bal 39.0515

BIOCHEMICAL SOCIETY.SYMPOSIA, 21 The Structure and biosynthesis of macromolecules : a symposium London 1961 Mar 27-28 Biochemical society Edited by D.J. Bell and J.K. Grant Held at the University of London. Senate House Cambridge university press, Cambridge,1962 In commemoration of the 50th anniversary of the Biochemical society
Bioch 33.1383

BIOCHEMICAL SOCIETY.SYMPOSIA, 22 The Structure and function of the membranes and surfaces of cells :a symposium London 1962 Mar 9 Biochemical society Edited by D.J. Bell and J.K. Grant Cambridge university press,Cambridge,1963
Bioch 33.1384

BIOCHEMICAL SOCIETY.SYMPOSIA, 22 The Structure and function of the membranes and surfaces of cells :a symposium London 1962 Mar 9 Biochemical society Edited by D.J. Bell and J.K. Grant Held at the Middlesex hospital medical school Cambridge university press,Cambridge,1963 Organized by J.K. Grant
Bal 39.0294

BIOCHEMICAL SOCIETY.SYMPOSIA, 23 Methods of separation of subcellular structural components :a symposium Louvain 1962 May 11 Biochemical society Edited by J.K. Grant Held jointly with the Societe belge de biochimie Cambridge university press, Cambridge,1963
Radioth 35.0136

BIOCHEMICAL SOCIETY.SYMPOSIA, 23 Methods of separation of subcellular structural components :a symposium Louvain 1962 May 11 Biochemical society and Societe belge de biochimie Edited by J.K. Grant Held in the University of Louvain.Physiological institute Cambridge university press, Cambridge,1963 Organized by C.de Duve
Bioch 33.1385

BIOCHEMICAL SOCIETY.SYMPOSIA, 24 The Control of lipid metabolism :a symposium Oxford 1963 Jul 19 Biochemical society Edited by J.K. Grant Academic press,London; New York,1963 Organized by G.Popjak
Bioch 33.1386

BIOCHEMICAL SOCIETY.SYMPOSIA, 25 Aspects of insect biochemistry :a symposium London 1965 Apr 1 Biochemical society Edited by T. W. Goodwin Academic press,London;New York, 1965
Bioch 33.1387

BIOCHEMICAL SOCIETY.SYMPOSIA, 25 Aspects of insect biochemistry :symposium London 1965 Apr 1 Biochemical society Edited by T.W. Goodwin Academic press,London;New York,1965
Gen 34.1646

BIOCHEMICAL SOCIETY.SYMPOSIA, 26 Instrumentation in biochemistry London 1966 Apr Biochemical society Edited by T.W. Goodwin Academic press,London;New York,1966
Bioch 33.1388

BIOCHEMICAL SOCIETY.SYMPOSIA, 26 Instrumentation in biochemistry :a symposium London 1966 Apr Biochemical society Edited by T.W. Goodwin Academic press,London, 1966
Radioth 35.0135

BIOCHEMICAL SOCIETY.SYMPOSIA, 27 The Metabolic roles of citrate :a symposium in honour of...Sir Hans Krebs Oxford 1967 Jul Biochemical society Edited by T.W. Goodwin Academic press,London;New York,1968
Bioch 33.1389

BIOCHEMICAL SOCIETY.SYMPOSIA, 28 Porphyrins and related compounds London 1968 Apr Biochemical society Edited by T.W. Goodwin Academic press,London;New York,1968
Bot 42.1755

BIOCHEMICAL SOCIETY.SYMPOSIA, 28 Porphyrins and related compounds London 1968 Apr Biochemical society Edited by T.W. Goodwin Academic press,London;New York,1968
Bioch 33.1390

BIOCHEMICAL SOCIETY.SYMPOSIA, 28 Porphyrins and related compounds :a symposium London 1968 Apr Biochemical society Edited by T.W. Goodwin Academic press,London,1968
Chem 18.2761

BIOCHEMICAL SOCIETY.SYMPOSIA, 29 Natural substances formed biologically from mevalonic acid :a symposium Liverpool 1969 Apr Biochemical society Academic press,London, 1970
Chem 18.2550

BIOCHEMICAL SOCIETY.SYMPOSIA, 29 Natural substances formed biologically from mevalonic acid :a symposium Liverpool 1969 Apr Biochemical society Edited by T.W. Goodwin Academic press,London;New York,1970
Bioch 33.1391

BIOCHEMICAL SOCIETY.SYMPOSIA, 31 Chemical reactivity and biological role of functional groups in enzymes :a symposium... Proceedings Oxford 1970 Jan Molecular enzymology and protein group Edited by R.M.S. Smellie Academic press,London,1970 Molecular enzymology and protein group of the Biochemical society and the Chemical society
Chem 18.2745

BIOCHEMICAL SOCIETY.SYMPOSIUM, 30 British biochemistry past and present Essays London 1969 Dec Biochemical society Edited by T.W. Goodwin Academic press,London, 1970
Bioch 33.2314

BIOCHEMICAL SOCIETY.SYMPOSIUM, 31 Chemical reactivity and biological role of functional groups in enzymes Oxford 1970 Jan Biochemical society Edited by R.M.S. Smellie Academic press,London,1970
Bioch 33.2315

BIOCHEMICAL SOCIETY.SYMPOSIUM, 32 The Biochemistry of steroid hormone action London 1970 Mar Biochemical society Edited by R.M.S. Smellie Academic press, London,1971
Bioch 33.2316

BIOCHEMICAL SOCIETY.SYMPOSIUM, 33 Plasma lipoproteins London 1971 Apr Biochemical society Edited by R.M.S. Smellie Academic press,London,1971
Bioch 33.2317

BIOCHEMICAL STUDIES OF ANTIMICROBIAL DRUGS : symposium London 1966 Apr Society for general microbiology Society for general microbiology.Symposia, 16 Cambridge university press,Cambridge,1966
Gen 34.0705

BIOCHEMICAL STUDIES OF ANTIMICROBIAL DRUGS :a symposium Papers London 1966 Apr Society for general microbiology Edited by B. A. Newton and P.E. Reynolds Held at the Royal institution Society for general microbiology.Symposia, 16 Cambridge university press,Cambridge,1966
Bioch 33.1170

BIOCHEMIE DES SAUERSTOFFS :colloquium Mosbach 1968 Apr 24-27 Gesellschaft fur biologische chemie Edited by B. Hess and Hj. Staudinger Gesellschaft fur biologische chemie.Colloquia, 19 illus Springer, Berlin,1968 In English and German
Bioch 33.1074

BIOCHEMISCH NACHWEISBARE STRAHLENWIRKUNGEN UND DEREN BEZIEHUNGEN ZUR STRAHLENTHERAPIE : symposium der Arbeitsgemeinschaft fur strahlenbiologie Stuttgart 1969 May 11 Arbeitsgemeinschaft fur strahlenbiologie Edited by G.B. Gerber Thieme,Stuttgart,1970
Radioth 35.1222

BIOCHEMISTRY Current topics in biochemistry Bethseda,Md. 1971 Sep National institutes of health Edited by C.B. Anfinsen National institutes of health.Lectures in biomedical sciences Academic press,New York,1972
Bioch 33.2168

BIOCHEMISTRY Instrumentation in biochemistry London 1966 Apr Biochemical society Edited by T.W. Goodwin Biochemical society. Symposia, 26 Academic press,London;New York, 1966
Bioch 33.1388

BIOCHEMISTRY International congress of biochemistry 1st Cambridge 1949 Aug 19-25 abstracts of communications 1949
Phys 20.1522

BIOCHEMISTRY International congress of biochemistry 4th Proceedings Vienna 1958 Sep 1-6 Vol 7: symposium 7 - biochemistry of viruses International union of biochemistry Edited by E. Broda and W. Frisch-Niggemeyer Pergamon press,London,1960
Radioth 35.0071

BIOCHEMISTRY International congress of biochemistry 4th Proceedings Vienna 1958 Sep 1-6 Vol 8: symposium 8 - proteins International union of biochemistry Edited by H. Neurath and H. Tuppy Pergamon,London, 1960
Radioth 35.0072

BIOCHEMISTRY International congress of biochemistry 4th Proceedings Vienna 1958 Sep 1-6 Vol 9: symposium 9 - physical chemistry of high polymers of biological interest International union of biochemistry Edited by O. Kratky Pergamon press,London, 1960
Radioth 35.1941

BIOCHEMISTRY International congress of biochemistry 4th Proceedings Vienna 1958 Sep 1-6 Vol 10: blood clotting factors By E. Deutsch Pergamon press,Oxford,1959
Med 36.0115

BIOCHEMISTRY International congress of biochemistry 5th Proceedings Moscow 1961 Aug 10-16 Vol 3: evolutionary biochemistry International union of biochemistry I.U.B.symposium series, 23 Pergamon;PWN-Polish scientific publishers,New York;Warsaw,1963
Bioch 33.1332

BIOCHEMISTRY International congress of biochemistry 6th Abstracts New York 1964 Jul 26-Aug 1 Vol 5: special topics in biochemistry Scheduled under the auspices of the International union of biochemistry Washington,1964
Bioch 33.1341

BIOCHEMISTRY Recent developments in biochemistry colloquium Proceedings Taipei 1967 Aug 17-19 Sino-American cooperation committee Edited by Hsien-Wen Li and others Under the auspices of the Academia Sinica Academia Sinica,n.p.,1968
Gen 34.0619

BIOCHEMISTRY AND PHARMACOLOGY OF COMPOUNDS DERIVED FROM MARINE ORGANISMS :a conference Papers New York 1960 Apr 6-8 New York academy of sciences and American institute of biological sciences Edited by Ross F. Nigrelli New York academy of sciences.Annals, 90,article 3 New York,1960 Co-sponsored by the New York zoological society and the United States navy,Office of naval research
Chem 26.0302

BIOCHEMISTRY AND PHARMACOLOGY OF THE BASAL GANGLIA 2nd :symposium Proceedings New York 1965 Nov 29-30 Parkinsons disease information and research center Edited by Erminio Costa and others Raven press,Hewlett, N.Y.,1966
Pha 16.0059

BIOCHEMISTRY AND PHYSIOLOGY OF PLANT GROWTH SUBSTANCES International conference on plant growth substances 6th Ottawa 1967 Jul 24-29 Edited by F. Wightman and G. Setterfield Held at Carleton University Runge press,Ottawa,1968
Bioch 33.0779

BIOCHEMISTRY AND PHYSIOLOGY OF PLANT GROWTH SUBSTANCES International conference on plant growth substances 6th Ottawa 1967 Jul 24-29 Edited by F. Wightman and G. Setterfield Held at the Carleton university xii,1642p Runge press,Ottawa,1968 On the occasion of the centennial of Canadian confederation
Bot 42.1833

BIOCHEMISTRY AND PHYSIOLOGY OF PLANT GROWTH SUBSTANCES International conference on plant growth substances 6th Proceedings Ottawa 1967 Jul 24-29 National research council of Canada Carleton university Edited by F. Wightman and G. Setterfield Runge press,Ottawa,1968
Gen 34.1408

BIOCHEMISTRY OF ANTIBIOTICS International congress of biochemistry 4th Proceedings Vienna 1958 Sep 1-6 Vol 5: symposium 5 - biochemistry of antibiotics International union of biochemistry Edited by K.H. Spitzy and R. Brunner I.U.B.symposium series, 7 Pergamon,London,1959 Added title page in French and German.Text in English,French and German.
Bioch 33.1319

BIOCHEMISTRY OF BLOOD PLATELETS :colloquium held on the occasion of the Third meeting of the Federation of European biochemical societies Warsaw 1966 Apr 4-7 Edited by E. Kowalski and S. Niewiarowski Organized by the Polish biochemical society Academic press,New York,1967
Radioth 35.0148

BIOCHEMISTRY OF CHLOROPLASTS :a Nato advanced study institute Proceedings Aberystwyth 1965 Aug Vol 1-2 North Atlantic treaty organization Edited by T.W. Goodwin 2 vols Academic press, London; New York, 1966-67
Bot 42.1429

BIOCHEMISTRY OF COPPER Symposium on copper in biological systems Proceedings New York 1965 Sep 8-10 Edited by Jack Peisach and others Academic press, New York; London, 1966
Bioch 33.1068

The BIOCHEMISTRY OF FERTILIZATION AND THE GAMETES a symposium London 1951 Feb 17 Biochemical society Edited by R.T. Williams Held at the London school of hygiene and tropical medicine Biochemical society. Symposia, 7 Cambridge university press, Cambridge, 1951
Bioch 33.1369

The BIOCHEMISTRY OF FISH ;a symposium Liverpool 1950 Sep 22 Biochemical society Edited by R.T. Williams Held at the University of Liverpool. Derby Hall Biochemical society. Symposia, 6 Cambridge university press, Cambridge, 1961
Bioch 33.1368

BIOCHEMISTRY OF HUMAN GENETICS Ciba foundation symposium on biochemistry of human genetics Proceedings Naples 1959 May 13-16 Ciba foundation International union of biological sciences Edited by G.E.W. Wolstenholme and Cecilia M. O'Connor illus Churchill, London, 1959
Bioch 33.0890

BIOCHEMISTRY OF HUMAN GENETICS :Ciba foundation symposium London 1959 May 13-16 Ciba foundation Edited by G.E.W. Wolstenholme and Cecilia M. O'Connor Churchill, London, 1959
An 32.0371

BIOCHEMISTRY OF HUMAN GENETICS :a symposium Naples 1959 May 13-16 Ciba foundation and International union of biological sciences Edited by G.E.W. Wolstenholme and C. M. O'Connor Ciba Foundation. Symposia Churchill, London, 1959
PGMS 29.0251

BIOCHEMISTRY OF HUMAN GENETICS symposium Ciba foundation Edited by G.E.W. Wolstenholme and Cecilia M. O'Connor Jointly with the International union of biological sciences Ciba foundation. Symposia Churchill, London, 1959
Gen 34.0602

BIOCHEMISTRY OF INSECTS International congress of biochemistry 4th Proceedings Vienna 1958 Sep 1-6 Vol 12: symposium 12 - biochemistry of insects International union of biochemistry Edited by L. Levenbook I.U.B. symposium series, 14 Pergamon, London, 1959 Added title page in French and German. Text in English and German
Bioch 33.1326

BIOCHEMISTRY OF LIPIDS International conference on the biochemical problems of lipids 5th Proceedings Vienna 1958 Sep Edited by G. Popjak Pergamon press, Oxford, 1960 Section 18 of the 4th international congress of biochemistry
Col S 12.0102

BIOCHEMISTRY OF MITOCHONDRIA; COLLOQUIUM The Federation of European biochemical societies meeting 3rd Warsaw 1966 Apr 4-7 Federation of European biochemical societies Edited by E.C. Slater and others Organized by the Polish biochemical society Academic press, London, 1967
Bioch 33.1394

BIOCHEMISTRY OF MORPHOGENESIS International congress of biochemistry 4th Proceedings Vienna 1958 Sep 1-6 Vol 6: symposium 6 - biochemistry of morphogenesis International union of biochemistry Edited by W.J. Nickerson I.U.B.S. symposium series, 8 Pergamon press, London, 1959
Gen 34.0482

BIOCHEMISTRY OF MORPHOGENESIS International congress of biochemistry 4th Proceedings Vienna 1958 Sep 1-6 Vol 6: symposium 6 - biochemistry of morphogenesis International union of biochemistry Edited by W.J. Nickerson I.U.B. symposium series, 8 Pergamon, London, 1959 Added title page in French and German
Bioch 33.1320

BIOCHEMISTRY OF MUCOPOLYSACCHARIDES OF CONNECTIVE TISSUE :a symposium London 1960 Feb 3 By J.N. Davidson Biochemical society Edited by F. Clark and J.K. Grant 4th edition Biochemical society. Symposia, 20 Bibliog Cambridge university press, Cambridge, 1961
Radioth 35.0098

The BIOCHEMISTRY OF MUCOPOLYSACCHARIDES OF CONNECTIVE TISSUE :a symposium London 1960 Feb 13 Biochemical society Edited by F. Clark and J.K. Grant Held at the Royal college of surgeons of England Biochemical society. Symposia, 20 Cambridge university press, Cambridge, 1961
Bioch 33.1382

BIOCHEMISTRY OF MUSCLE CONTRACTION :a conference Proceedings Dedham, Mass. 1962 May Edited by John Gergely Retina foundation. Institute of biological and medical sciences. Monographs and conferences, 2 Churchill, London, 1964
Bioch 33.2231

The BIOCHEMISTRY OF STEROID HORMONE ACTION London 1970 Mar Biochemical society Edited by R.M.S. Smellie Biochemical society. Symposium, 32 Academic press, London, 1971
Bioch 33.2316

BIOCHEMISTRY OF STEROIDS International congress of biochemistry 4th Proceedings Vienna 1958 Sep 1-6 Vol 4: symposium 4 biochemistry of steroids International union of biochemistry Edited by E. Mosettig I.U.B. symposium series, 6 Pergamon, London, 1959 Added title page in French and German. Text in English, French and German.
Bioch 33.1318

BIOCHEMISTRY OF THE CENTRAL NERVOUS SYSTEM International congress of biochemistry 4th Proceedings Vienna 1958 Sep 1-6 Vol 3: symposium 3: biochemistry of the central nervous system International union of biochemistry Edited by F. Brucke I.U.B. symposium series, 5 Pergamon, London, 1959 Added t.p. in French and German
Bot 42.1715

BIOCHEMISTRY OF THE CENTRAL NERVOUS SYSTEM International congress of biochemistry 4th Proceedings Vienna 1958 Sep 1-6 Vol 3: symposium 3 - biochemistry of the central nervous system International union of biochemistry Edited by Brucke I.U.B. symposium series, 5 Pergamon,London,1959 Added title page in French and German.Text in English,French and German
Bioch 33.1317

BIOCHEMISTRY OF THE DEVELOPING NERVOUS SYSTEM International neurochemical symposium 1st Proceedings Oxford 1954 Jul 13-17 Edited by Heinrich Waelsch Academic press,New York, 1955
An 32.4333

BIOCHEMISTRY OF THE DEVELOPING NERVOUS SYSTEM International neurochemical symposium 1st Proceedings Oxford 1954 Jul 13-17 Edited by Heinrich Waelsch held at Magdalen college, Oxford Academic press,New York,1955
Bal 39.1468

BIOCHEMISTRY OF THE DEVELOPING NERVOUS SYSTEM The International neurochemical symposium 1st Proceedings Oxford 1954 Jul 13-17 Edited by Heinrich Waelsch Academic press, New York,1955
Bioch 33.2233

BIOCHEMISTRY OF VIRUS REPLICATION Federation of European biochemical societies meeting 4th Proceedings Oslo 1967 Jul 4 Federation of European biochemical societies Edited by S.G. Laland and L.O. Froholm Organized by the Norwegian biochemical society Universitetsforlaget;Academic press,Oslo; London,1968 Symposium organizer A.P. Nygaard
Bioch 33.1397

BIOCHEMISTRY OF VIRUSES International congress of biochemistry 4th Proceedings Vienna 1958 Sep 1-6 Vol 7: symposium 7 - biochemistry of viruses International union of biochemistry Edited by E. Broda and W. Frisch-Niggemeyer I.U.B.symposium series, 9 Pergamon,London,1959 Added title page in French and German.Text in German and English
Bioch 33.1321

BIOCHEMISTRY OF VIRUSES International congress of biochemistry 4th Proceedings Vienna 1958 Sep 1-6 Vol 7: symposium 7 - biochemistry of viruses International union of biochemistry Edited by E. Broda and W. Frisch-Niggemeyer Pergamon press,London,1960
Radioth 35.0071

The BIOCHEMISTRY OF VITAMIN B12 :a symposium London 1955 Feb 19 Biochemical society Edited by R.T. Williams Held at the London school of hygiene and tropical medicine Biochemical society.Symposia, 13 Cambridge university press,Cambridge,1955
Bioch 33.1375

BIOCHEMISTRY OF WOOD International congress of biochemistry 4th Proceedings Vienna 1958 Sep 1-6 Vol 2: symposium 2 - biochemistry of wood International union of biochemistry Edited by K. Kratzi and G. Billek I.U.B.symposium series, 4 Pergamon,London,1959 Added title page in French and German.Text in English,French and German
Bioch 33.1316

BIOCHEMISTRY STUDIES OF ANTIMICROBIAL DRUGS 16th symposium London 1966 Apr Society for general microbiology Edited by B.A. Newton and P.E. Reynolds Society for general microbiology.Symposium,15 Cambridge university press,Cambridge,1966
Pha 16.0336

BIOCLIMATOLOGY Biometeorology The International bioclimatological congress 2nd Proceedings London 1960 Sep 4-10 Edited by S.W. Tromp illus xxii,687p 25cm Pergamon,Oxford,1962
Sco 14.0585

BIODETERIORATION OF MATERIALS:MICROBIOLOGICAL AND ALLIED ASPECTS International biodeterioration symposium 1st Proceedings Southampton 1968 Sep 9-14 Edited by A.Harry Walters and John J. Elphick Elsevier,Amsterdam,1968
Met 25.2483

BIODETERIORATION OF MATERIALS:MICROBIOLOGICAL AND ALLIED ASPECTS International biodeterioration symposium 1st Proceedings Southampton 1968 Sep 9-14 Edited by Arthur H. Walters and John J. Elphick Bibliog.,illus. x,740p Elsevier, Amsterdam,1968
Bot 42.3773

BIOENERGETICS International congress of biochemistry 6th Abstracts New York 1964 Jul 26-Aug 1 Scheduled under the auspices of the International union of biochemistry Washington,1964
Bioch 33.1346

BIOENERGETICS :considerations of processes of absorption stabilization, transfer and utilization.A symposium Proceedings Upton,N.Y. 1959 Oct 12-16 Brookhaven national laboratory Edited by Leroy George Augenstine Sponsored by the United States atomic energy commission Radiation research. Supplement, 2 Academic press,New York; London,1960
Bioch 33.0716

BIOGENIC AMINES Edited by Harold E. Himwich and Williamina A. Himwich Progress in brain research, 8 Elsevier,Amsterdam,1964 Derived from a symposium on Binding sites of brain biogenic amines,Galesburg,Ill.,1963
An 32.4228

BIOGENIC AMINES AS PHYSIOLOGICAL REGULATORS : a symposium Woods Hole 1969 Aug 29-Sep 1 Edited by J.J. Blum Held under the auspices of the Society of general physiologists Prentice-Hall,Englewood Cliffs, N.J.,1970
An 32.5479

BIOINORGANIC CHEMISTRY :a symposium Blacksburg,Va. 1970 Jun 22-25 American chemical society.Inorganic chemistry division Chemical institute of Canada Edited by Robert F. Gould x,436p 23cm American chemical society,Washington,D.C.,1971
Chem 18.2719

BIOLOGICAL ACOUSTICS Proceedings London 1961 May 3 Zoological society of London Edited by P.T. Haskell and F.C. Fraser Zoological society of London.Symposia, 7 London,1962
An 32.4826

The BIOLOGICAL ACTION OF GROWTH SUBSTANCES
Cambridge 1956 Sep Society for experimental biology Society for experimental biology.Symposia, 11 Cambridge university press,Cambridge,1956
Bioch 33.1356

The BIOLOGICAL ACTION OF GROWTH SUBSTANCES : symposium Aberystwyth 1956 Sep Society for experimental biology Society for experimental biology.Symposia, 11 Cambridge university press,Cambridge,1957
An 32.0089

The BIOLOGICAL ACTION OF GROWTH SUBSTANCES :a symposium Proceedings Aberystwyth 1956 Sep Society for experimental biology Society for experimental biology.Symposia, 11 Cambridge university press,Cambridge,1957
Radioth 35.0217

The BIOLOGICAL ACTION OF THE VITAMINS :a symposium Chicago 1941 Sep 15-19 University of Chicago Edited by E.A. Evans University of Chicago press,Chicago,1942 Part of the joint Symposium on the respiratory enzymes and the biological action of vitamins arranged with the University of Wisconsin to celebrate the fiftieth anniversary of the University of Chicago.The papers on Respiratory enzymes were presented at Madison, 1941,Sep 11-13
Bioch 33.0343

BIOLOGICAL ACTIVITIES OF STEROIDS IN RELATION TO CANCER :a conference Proceedings Vergennes,Vt. 1959 Sep 27-Oct 2 Edited by Gregory Pincus and Erwin P. Vollmer Sponsored by the Cancer chemotherapy national service center Academic press,New York; London,1960
Radioth 35.0873

BIOLOGICAL AND BIOCHEMICAL EVALUATION OF MALIGNANCY IN EXPERIMENTAL HEPATOMAS The U.S.-Japan joint conference on biological and biochemical evaluation of malignancy in experimental hepatomas Proceedings Kyoto 1965 Nov 4-5 Held under the auspices of the United States-Japan scientific cooperation program G.A.N.N.Monograph, 1 Japanese cancer association;Japanese foundation for cancer research,Tokyo,1966
Bioch 33.0952

BIOLOGICAL AND CHEMICAL ASPECTS OF OXYGENASES The United States-Japan symposium on oxygenases Proceedings Kyoto 1966 May 16-19 Edited by Konrad Bloch and Osamu Hayaishi held under the auspices of the United States-Japan committee on scientific co-operation Maruzen company,Tokyo,1966
Bioch 33.1102

BIOLOGICAL AND CHEMICAL CONTROL OF UNDESIRABLE PLANTS AND ANIMALS Ecological effects of biological and chemical control of undesirable plants and animal symposium International union for the conservation of nature and natural resources. Technical meeting 8th Warsaw 1960 Jul 15-24 Edited by D.J. Kuenen Leiden,1961 Title also in French: Effets ecologiques du controle biologique et chimique des plantes et animaux undesirables
Bal 44.6435

BIOLOGICAL ANTIOXIDANTS 1st :a conference Transactions New York 1946 Oct 10-11 Josiah Macy jr. foundation Edited by Cosmo G. Mackenzie New York,1946
Chem 18.2646

BIOLOGICAL APPLICATION OF TRACER ELEMENTS :a symposium papers Cold Spring Harbor 1948 Jun 8-16 Cold Spring Harbor laboratory of quantitative biology Cold Spring Harbor symposia on quantitative biology, 13 Biological laboratory,New York,1948
Radioth 35.0158

BIOLOGICAL APPLICATIONS OF TRACER ELEMENTS :a symposium Papers Cold Spring Harbor 1948 Cold Spring Harbor biological laboratory Cold Spring Harbor symposia on quantitative biology, 13 Long Island biological association,Cold Spring Harbor,1948
Bioch 33.1269

BIOLOGICAL APPROACHES TO CANCER CHEMOTHERAPY Louvain 1960 Jun Edited by R.J.C. Hams Under the auspices of Unesco Academic press, London;New York,1961
Pha 16.0337

BIOLOGICAL APPROACHES TO CANCER CHEMOTHERAPY a symposium Louvain 1960 Jun Edited by R.J.C. Harris Under the auspices of Unesco Academic press,London;New York,1960
Radioth 35.0238

BIOLOGICAL APPROACHES TO CANCER CHEMOTHERAPY a symposium Louvain 1960 Jun Edited by R.J.C. Harris Under the auspices of Unesco and the World health organization Academic press,London;New York,1961
Bioch 33.0960

BIOLOGICAL ASPECTS OF OCCLUSIVE VASCULAR DISEASE Edited by D.G. Chalmers and G.A. Gresham Cambridge university press,Cambridge,1964 Note on book-jacket:A Cambridge postgraduate medical course at which research workers and clinicians surveyed modern ideas on arterial disease
An 32.3524

BIOLOGICAL ASPECTS OF RACE 5th :a symposium Proceedings Eugenics society Edited by G.Ainsworth Harrison and John Peel Eugenics society.Symposia, 5 Journal of biosocial science.Supplement, 1 Blackwell,Oxford, 1969
Gen 34.2198

BIOLOGICAL ASPECTS OF SOCIAL PROBLEMS London 1964 Oct 1-2 Eugenics society Edited by J.E. Meade and A.S. Parkes Eugenics society.Symposia,1 Oliver and Boyd, Edinburgh;London,1965
Geog 13.1550

BIOLOGICAL CLOCKS : a symposium Papers Cold Spring Harbor 1960 Cold Spring Harbor biological laboratory Cold Spring Harbor symposia on quantitative biology, 25 Long Island biological association,Cold Spring Harbor,1960
Bioch 33.1281

BIOLOGICAL CONTROL International conference of insect pathology and biological control 1st Transactions Prague 1958 Aug 13-18 24cm Prague,1958
Philos 1.1598

BIOLOGICAL CONTROL Les Bases scientifiques d'une organisation internationale pour la lutte biologique Stockholm 1948 Aug 5-7 International union of biological sciences International union of biological sciences. Serie B.Colloques, 5 Secretairiat general de I.U.I.S.B., Paris, 1949 Title also in English: The scientific bases of an international biological control organization. Text in English and French
Bal 39.3920

BIOLOGICAL COUNCIL.CO-ORDINATING COMMITTEE FOR SYMPOSIA ON DRUG ACTION A Symposium on agents affecting fertility London 1964 Edited by C.R. Austin and J.S. Perry Churchill, London, 1965
An 32.3896

BIOLOGICAL COUNCIL.CO-ORDINATING COMMITTEE FOR SYMPOSIA ON DRUG ACTION Animal behaviour and drug action :annual meeting Proceedings London 1963 Mar 25-28 Edited by A.V.S.de Reuck and Julie Knight Biological council.Co-ordinating committee for symposia on drug action.Annual meeting, 10 Ciba foundation symposia Churchill, London, 1964
Psy 31.0526

BIOLOGICAL COUNCIL.CO-ORDINATING COMMITTEE FOR SYMPOSIA ON DRUG ACTION Embryopathic activity of drugs :a symposium Edited by J.M. Robson and others Churchill, London, 1965
An 32.2095

BIOLOGICAL COUNCIL.CO-ORDINATING COMMITTEE FOR SYMPOSIA ON DRUG ACTION Hypotensive drugs and control of vascular tone in hypertension : a symposium Proceedings London 1956 Apr 5-6 Edited by M. Harington Pergamon press, London, 1956
Med 36.0172

BIOLOGICAL COUNCIL.CO-ORDINATING COMMITTEE FOR SYMPOSIA ON DRUG ACTION Scientific basis of drug dependence a symposium London 1968 Apr Edited by Hannah Steinberg J.and A. Churchill, London, 1969
PGMS 29.0398

BIOLOGICAL COUNCIL.CO-ORDINATING COMMITTEE FOR SYMPOSIA ON DRUG ACTION Symposium on calcium and cellular function Essays London 1969 Mar 24-25 Edited by A.W. Cuthbert Macmillan, London, 1969
Bioch 33.2264

BIOLOGICAL COUNCIL.CO-ORDINATING COMMITTEE FOR SYMPOSIA ON DRUG ACTION The Interaction of drugs and subcellular components in animal cells :a symposium London 1967 Apr 10-11 Edited by P.N. Campbell Held in the Middlesex hospital medical school Churchill, London, 1968
Bioch 33.0923

BIOLOGICAL COUNCIL.CO-ORDINATING COMMITTEE FOR SYMPOSIA ON DRUG ACTION.ANNUAL MEETING, 10 Animal behaviour and drug action :annual meeting Proceedings London 1963 Mar 25-28 Ciba foundation and Biological council.Co-ordinating committee for symposia on drug action Edited by A.V.S.de Reuck and Julie Knight Ciba foundation symposia Churchill, London, 1964
Psy 31.0526

BIOLOGICAL COUNCIL.CO-ORDINATING COMMITTEE FOR SYMPOSIA ON DRUG ACTION A Symposium on the interaction of drugs and subcellular components in animal cells London 1967 Edited by P.N. Campbell Pl. viii,355p 21cm Churchill, London, 1968
Pha 16.0252

BIOLOGICAL COUNCIL.CO-ORDINATING COMMITTEE FOR SYMPOSIA ON DRUG ACTION Embryopathic activity of drugs :a symposium London 1965 Edited by J.M. Robson and others Churchill, London, 1965
Pha 16.0320

BIOLOGICAL COUNCIL.CO-ORDINATING COMMITTEE FOR SYMPOSIA ON DRUG ACTION Quantitative methods in human pharmacology and therapeutics a symposium Proceedings London 1958 May 24-25 Edited by D.R. Laurence Biological council.Co-ordinating committee for symposia on drug action.Symposium on drug action series, 3 Pergamon, London, 1959
Pha 16.0419

BIOLOGICAL COUNCIL.CO-ORDINATING COMMITTEE FOR SYMPOSIA ON DRUG ACTION and CIBA FOUNDATION, Animal behaviour and drug action symposium London 1963 Mar 25-28 Edited by Hannah Steinberg and others Churchill, London, 1964
Pha 16.0030

BIOLOGICAL COUNCIL.CO-ORDINATING COMMITTEE FOR SYMPOSIA ON DRUG ACTION and CIBA FOUNDATION, Enzymes and drug action :Ciba foundation symposium 8th London Edited by J.L. Mongar and A.V.S. De Reuck Churchill, London, 1962
Pha 16.0105

BIOLOGICAL COUNCIL.COORDINATING COMMITTEE FOR SYMPOSIA ON DRUG ACTION. SERIES, 4 Polypeptides which affect smooth muscles and blood vessels :a symposium Proceedings London 1959 Mar 23-24 Biological council. Coordinating committee for symposia on drug action Edited by M. Schachter Pergamon press, London, 1960
PGMS 29.0380

BIOLOGICAL EFFECTS OF IONIZING RADIATION AT THE MOLECULAR LEVEL Symposium on the biological effects of ionizing radiation at the molecular level Proceedings Brno 1962 Jul 2-6 Held by the International atomic energy agency International atomic energy agency.Proceedings series I.A.E.A., Vienna, 1962
Radioth 35.1976

BIOLOGICAL EFFECTS OF NEUTRON AND PROTEIN IRRADIATIONS Symposium on biological effects of neutron and proton irradiations Proceedings Upton, N.Y. 1963 Oct 7-11 1-2 International atomic energy agency Held at Brookhaven national laboratory International atomic energy agency.Proceedings series 2 vols International atomic energy agency, Vienna, 1964
Radioth 35.1168

BIOLOGICAL FLUIDS Protides of the biological fluids :colloquia 11th-13th,15th Proceedings Bruges 1963-65,67 Edited by H. Peeters Elsevier, Amsterdam, 1964-68 In English, French and German
Bioch 33.0538

BIOLOGICAL HAZARDS OF ATOMIC ENERGY :a conference Papers 1950 Oct Institute of biology Atomic scientists association Edited by A. Haddow pl. Clarendon press, Oxford,1952
Radioth 35.0180

BIOLOGICAL HAZARDS OF ATOMIC ENERGY :a conference papers London 1950 Oct Institute of biology and Atomic scientists' association Clarendon press, Oxford,1952
HE 27.0097

BIOLOGICAL INTERACTION IN NORMAL AND NEOPLASTIC GROWTH :a contribution to the Host-tumour problem Papers Henry Ford hospital international symposium Detroit,Mich. 1961 May 16-20 Edited by Michael J. Brennan and William L. Simpson Churchill,London,1962
An 32.2497

BIOLOGICAL INTERFACES:FLOWS AND EXCHANGES :a symposium Proceedings New York 1967 Dec 8-9 Sponsored by the New York heart association Basic science symposia, 9 Little,Brown,Boston,Mass.,1968 Simultaneously published in the Journal of general physiology,vol.52,1968
Phys 20.2247

BIOLOGICAL MEMBRANES Regulatory functions of biological membranes The Sigrid Juselius foundation symposium 2nd Proceedings Helsinki 1967 Nov 6-9 Sigrid Juselius foundation Edited by Johan Jarnefelt B.B.A. library, 11 illus. viii,311p Elsevier, Amsterdam,1968
Bot 42.1517

BIOLOGICAL MEMBRANES Regulatory functions of biological membranes The Sigrid Juselius foundation symposium 2nd Proceedings Helsinki 1967 Nov 6-9 Sigrid Juselius foundation Edited by Johan Jarnefelt BBA library, 11 Elsevier,Amsterdam,1968
Bioch 33.0968

BIOLOGICAL MEMBRANES :recent progress New York 1965 Oct 4-7 New York academy of sciences New York academy of sciences.Annals, 137,art.2 New York,1966
Phys 20.0964

BIOLOGICAL MEMBRANES :recent progress: aconference New York 1966 Jul 14 By Kenneth S. Cole and others New York academy of sciences Edited by Werner R. Loewenstein New York academy of sciences.Annals, 137,2 p 403-1048 New York,1966
Bot 42.1508

BIOLOGICAL MEMBRANES:RECENT PROGRESS :a conference New York 1965 Oct 4-6 New York academy of sciences Edited by Edward M. Weyer and others New York academy of science. Annals, 137,p.403-1048 New York,1966
Bioch 33.0921

BIOLOGICAL MEMBRANES:RECENT PROGRESS :a conference New York 1966 Jul 14 By Kenneth S. Cole New York academy of sciences Edited by Werner R. Loewenstein New York academy of sciences.Annals, 137,p 403-1048 New York,1966
Bal 39.0325

BIOLOGICAL MEMBRANES:RECENT PROGRESS :a conference Papers New York 1965 Oct 4-7 Edited by Edward M. Weyer New York academy of sciences.Annals,137,art 2 NNew York,1966
Pha 16.0014

BIOLOGICAL ORGANISATION:CELLULAR AND SUB-CELLULAR a symposium Proceedings Edinburgh 1957 Sep 6-10 Organised by C.H.Waddington on behalf of Unesco Illus Pergamon press, London,1959
Radioth 35.0497

BIOLOGICAL ORGANISATION,CELLULAR AND SUB-CELLULAR a symposium organised on behalf of Unesco by C.H.Waddington Proceedings Edinburgh 1957 Sep 6-10 Unesco held at the University of Edinburgh Pergamon press, London,1959
Bioch 33.1019

BIOLOGICAL ORGANISATION CELLULAR AND SUBCELLULAR a symposium Proceedings Edinburgh 1957 Sep 6-10 Unesco Edited by C.H. Waddington held at the University of Edinburgh Pergamon,London,1959
Bal 39.0455

BIOLOGICAL ORGANIZATION:CELLULAR AND SUB-CELLULAR a symposium Proceedings Edinburgh 1957 Sep 6-10 Organized by C.H.Waddington on behalf of Unesco Pergamon press,London,1959
Gen 34.0829

BIOLOGICAL ORGANIZATION:CELLULAR AND SUBCELLULAR a symposium Proceedings Edinburgh 1957 Sep 6-10 Edited by C.H. Waddington Organised on behalf of Unesco Pergamon press, London,1959
An 32.2485

BIOLOGICAL ORGANIZATION AT THE CELLULAR AND SUPERCELLULAR LEVEL Varenna 1962 Sep 24-27 Edited by R.J.C. Harris Under the auspices of Unesco Academic press,London;New York,1963
Bal 39.0460

BIOLOGICAL ORGANIZATION AT THE CELLULAR AND SUPERCELLULAR LEVEL Varenna 1962 Sep 24-27 Edited by R.J.C. Harris Held under the auspices of Unesco Academic press,London, 1963
Pha 16.0010

BIOLOGICAL ORGANIZATION AT THE CELLULAR AND SUPERCELLULAR LEVEL :a symposium Varenna 1962 Sep 24-27 Edited by R.J.C. Harris held under the auspices of Unesco Academic press,London;New York,1963
Bioch 33.1018

BIOLOGICAL ORGANIZATION AT THE CELLULAR AND SUPERCELLULAR LEVEL :a symposium Papers Varenna 1962 Sep 24-27 Under the auspices of Unesco Academic press,London;New York,1963
Gen 34.0681

BIOLOGICAL ORGANIZATION AT THE CELLULAR AND SUPERCELLULAR LEVEL :a symposium Proceedings Varenna 1962 Sep 24-27 Edited by R.J.C. Harris Under the auspices of Unesco Academic press,London;New York, 1963
An 32.3291

BIOLOGICAL OXIDATION OF AROMATIC RINGS :a symposium London 1949 Nov 12 Biochemical society Edited by R.T. Williams Held at the London school of hygiene and tropical medicine Biochemical society. Symposia, 5 Cambridge university press, Cambridge,1950
Bioch 33.1367

BIOLOGICAL OXIDATION OF AROMATIC RINGS :a symposium London 1949 Nov 12 Edited by R.T. Williams Biochemical society. Symposia, 5 Cambridge university press, Cambridge,1950
Phys 20.1527

BIOLOGICAL OXIDATIONS Cold Spring Harbor symposia on quantitative biology Cold Spring Harbor 1939 Vol 7 Cold Spring Harbor biological laboratory Long Island biological association,Cold Spring Harbor,1939 Later referred to in vol.9 as "Biological oxidations"
Bioch 33.1263

BIOLOGICAL OXYDATION OF NITROGEN IN ORGANIC MOLECULES :symposium Proceedings London 1971 Dec 19-22 Chelsea college Edited by J.W. Bridges and others Taylor and Francis,London,1972
Bioch 33.2286

BIOLOGICAL PROBLEMS ARISING FROM THE CONTROL OF PESTS AND DISEASES :a symposium Proceedings London 1959 Oct Institute of biology Edited by R.K.S. Wood Institute of biology.Symposia, 9 22cm London,1960
Bal 39.2962

BIOLOGICAL PROBLEMS OF GRAFTING :a symposium Proceedings Liege 1959 Mar 18-21 Commission administrative du patrimoine universitaire de Liege and Council for international organizations of medical sciences Edited by F. Albert and G. Lejeune-Ledant Blackwell,Oxford,1959 Symposium chairman P.B.Medawar
PGMS 29.0245

BIOLOGICAL PROBLEMS OF GRAFTING :a symposium Proceedings Liege 1959 Mar 18-21 Commission administrative du patrimoine universitaire de Liege Council for international organizations of medical sciences Edited by F. Albert and G. Lejeune-Ledant Blackwell,Oxford,1959
Bal 39.0458

BIOLOGICAL PSYCHIATRY :the proceedings of the scientific sessions of the Society... San Francisco,Calif. 1958 May Vol 1 Society of biological psychiatry Edited by J. Masserman Grune and Stratton,New York; London,1959
Psy 31.1624

BIOLOGICAL RECEPTOR MECHANISMS Birmingham 1962 Sep 10-16 Society for experimental biology Society for experimental biology. Symposia, 16 Cambridge university press, Cambridge,1962
An 32.4562

BIOLOGICAL RECEPTOR MECHANISMS Contributions Birmingham 1962 Sep 10-16 Society for experimental biology Edited by J.W.L. Beament Society for experimental biology. Symposia, 16 Cambridge university press, Cambridge,1962
Psy 31.0283

The BIOLOGICAL REPLICATION OF MACROMOLECULES a symposium Aberystwyth 1956 Sep Society for experimental biology Edited by K. F. Sanders Society for experimental biology, 12 Cambridge university press,Cambridge, 1957
An 32.5446

The BIOLOGICAL REPLICATION OF MACROMOLECULES a symposium Society for experimental biology Society for experimental biology. Symposia, 12 Cambridge university press, Cambridge,1958
Bioch 33.1357

The BIOLOGICAL REPLICATION OF MACROMOLECULES symposium Society for experimental biology Society for experimental biology. Symposia, 12 Cambridge university press, Cambridge,1958
Gen 34.0736

The BIOLOGICAL REPLICATION OF MACROMOLECULES : a symposium Papers Aberystwyth 1956 Society for experimental biology Edited by F. K. Sanders Society for experimental biology. Symposia, 12 Cambridge university press, Cambridge,1958
Radioth 35.0237

BIOLOGICAL SIGNIFICANCE OF CLIMATIC CHANGES IN BRITAIN Institute of biology symposium 14th Proceedings London 1964 Oct 29-30 Edited by C.G. Johnson and L.P. Smith Academic press,London,1965
Geog 13.1208

The BIOLOGICAL SIGNIFICANCE OF CLIMATIC CHANGES IN BRITAIN :a symposium held at the Royal geographical society Proceedings London 1964 Oct 29-30 Institute of biology Edited by C.G. Johnson and L.P. Smith Institute of biology.Symposia, 14 Academic press,London; New York,1965
Bal 39.1687

The BIOLOGICAL SIGNIFICANCE OF CLIMATIC CHANGES IN BRITAIN Proceedings London 1964 Oct 29-30 Institute of biology Edited by C. G. Johnson and L.P. Smith Held at the Royal geographical society Institute of biology. Symposium, 14 illus x,222p Academic press,London;New York,1965
Bot 42.6678

BIOLOGICAL STRUCTURE AND FUNCTION I.U.B.-I.U.B.S. international symposium 1st Proceedings Stockholm 1960 Sep 12-17 Vol 1 International union of biological sciences and International union of biochemistry Edited by T.W. Goodwin and O. Lindberg Academic press,London,1961
An 32.3276

BIOLOGICAL STRUCTURE AND FUNCTION I.U.B.-I.U.B.S. international symposium 1st Proceedings Stockholm 1960 Sep 12-17 Vol 1-2 International union of biological sciences International union of biochemistry Edited by T.W. Goodwin and O. Lindberg illus 2 vols Academic press,London;New York,1961
Bal 44.6450

BIOLOGICAL STRUCTURE AND FUNCTION IUB-IUBS international symposium 1st Proceedings Stockholm 1960 Sep 12-17 Vol 1-2 International union of biochemistry International union of biological sciences Edited by T.W. Goodwin and O. Lindberg 2 vols Academic press,London,1961
Gen 34.0826

BIOLOGICAL STRUCTURE AND FUNCTION The I.U.B.-I.U.B.S. international symposium 1st Proceedings Stockholm 1960 Sep 12-17 Vol 1-2 International union of biochemistry and International union of biological sciences Edited by T.W. Goodwin and O. Lindberg illus 2 vols Academic press, London;New York,1961
Bioch 33.0584

BIOLOGICAL STRUCTURE AND FUNCTION 1st : international symposium Proceedings Stockholm 1960 Sep 12-17 Vol 1-2 International union of biochemistry and International union of biological sciences Edited by T.W. Goodwin and D. Lindberg Academic press,London;New York,1961
Phys 20.1539

BIOLOGICAL STRUCTURE AND FUNCTION AT THE MOLECULAR LEVEL International congress of biochemistry 5th Proceedings Moscow 1961 Aug 10-16 International union of biochemistry Edited by V.A. Engelhardt I.U.B.symposium series, 21 Pergamon;PWN-Polish scientific publishers,London;Warsaw,1963
Bioch 33.1330

BIOLOGICAL STRUGGLE AND FUNCTION The I.U.B.-I.U.B.S.international symposium 1st Proceedings Stockholm 1960 Sep 12-17 Vol 1-2 International union of biochemistry International union of biological sciences Edited by T.W. Goodwin and O. Lindberg illus. 2 vols Academic press,London;New York, 1961
Bot 42.1344

BIOLOGICAL SYMPOSIA :a series of volumes devoted to symposia in the field of biology Vol 10: frontiers of cytochemistry...Edited by N.L.Hoerr Edited by Jaques Cattell port Cattell press,Lancaster,Pa.,1943
Bot 42.6451

BIOLOGICAL SYMPOSIA, 10 Frontiers in cytochemistry:the physical and chemical organization of the cytoplasm :a symposium Chicago,Ill. 1942 Nov 13 Edited by Normand L. Hoerr Cattell press,Lancaster,Pa.,1943 Held in honour of the 75th birthday of Dr. R.R. Bensley
An 32.3370

BIOLOGICAL SYSTEMS Energy transfer with special reference to biological systems :a general discussion Papers Nottingham 1959 Apr 14-16 Faraday society Faraday society.Discussions, 27 Aberdeen university press,Aberdeen,1960
Bioch 33.0314

BIOLOGICAL SYSTEMS Free radicals in biological systems Symposium on free radicals in biological systems Proceedings Stanford,Calif. 1960 Mar 21-23 Stanford university Edited by M.S. Blois Academic press,New York;London,1961
Bioch 33.0309

BIOLOGICAL SYSTEMS Informative molecules in biological systems International symposium on uptake of informative molecules by living cells Proceedings Mol,Belgium 1970 Aug 31-Sep 2 North Atlantic treaty organization. Scientific affairs division Edited by L. Ledoux North-Holland,Amsterdam,1971
Bioch 33.2240

BIOLOGICAL SYSTEMS Symmetry and function of biological systems at the macromolecular level Nobel symposium 11th Procedings Sodergarn 1968 Aug 26-29 Nobel foundation Edited by Arne Engstrom and Bror Strandberg Wiley,New York,1969
Bioch 33.0327

BIOLOGICAL SYSTEMS The Luminescence of biological systems :a conference Proceedings Asilomar,Calif. 1954 Mar 28-Apr 2 National research council.Committee on photobiology Edited by Frank H. Johnson American assoc. for the advancement of science,Washington,D.C., 1955
Bal 39.0080

BIOLOGICAL TRANSFORMATION OF STARCH AND CELLULOSE London 1953 Feb 21 Biochemical society Edited by R.T. Williams Biochemical society. Symposia, 11 Cambridge university press, Cambridge,1953
Bot 42.1690

BIOLOGICAL TRANSFORMATIONS OF STARCH AND CELLULOSE :a symposium London 1953 Feb 21 Biochemical society Held at the London school of hygiene and tropical medicine Biochemical society.Symposia, 11 Cambridge university press,Cambridge,1953
Bioch 33.1373

BIOLOGIE ANTARCTIQUE 1er :symposium Comptes-rendus Paris 1962 Sep 2-8 Scientific committee for Antarctic research Edited by Robert Carrick and others Hermann, Paris,1964
Bal 39.1669

BIOLOGIE ANTARCTIQUE 1er :symposium Comptes-rendus Paris 1962 Sep 2-8 Edited by Robert Carrick and others Organized by the Scientific committee on Antarctic research Actualites scientifiques et industrielles,1312 illus,maps 651p 25cm Herman,Paris,1964 Preface by professor P.P.Grasse
Sco 14.6172

BIOLOGY Centenary and bicentenary congress of biology :papers delivered in the university of Malaya in commemoration of the works of Darwin,Wallace and Linnaeus Proceedings Singapore 1958 Dec 2-9 Edited by R.D. Purchon Bibliog,illus,port,maps,tables 333p University of Malaya press;Oxford university press,Singapore;London,1960 Includes papers on anthropology
Bal 39.3924

BIOLOGY Comparative biology of calcified tissue papers New York 1962 Oct 12-13 New York academy of sciences New York academy of sciences.Annals,109,art.1 New York,1963
Geol 8.0971

BIOLOGY Concepts of biology Conference report Lee,Mass. 1955 Oct 12-14 National research council.Division of biology and agriculture.Biology council Edited by R.W. Gerard and R.B. Stevens National research council.Publication, 560 Washington,D.C., 1958
Bal 39.0989

BIOLOGY Conference on biology and civil engineering Proceedings London 1948 Sep Institution of civil engineers Institution of civil engineers,London,1949
Geog 13.2688

BIOLOGY Ideas in modern biology International congress of zoology 16th Proceedings Washington,D.C. 1963 Aug 20-27 Edited by J.A. Moore Doubleday,NEW York,1965
Radioth 35.0259

BIOLOGY Models and analogues in biology Bristol 1960 Sep 6-12 Society for experimental biology Edited by J.W.L. Beament Society for experimental biology. Symposia, 14 Cambridge university press, Cambridge,1960
Phys 20.0939

BIOLOGY Models and analogues in biology Bristol 1960 Sep 6-12 Society for experimental biology Society for experimental biology symposia, 14 vii,255p Cambridge university press,Cambridge,1960
WSM 43.2826

BIOLOGY The Social impact of modern biology : an international conference Papers and discussions London 1970 Nov 26-28 British society for social responsibility in science Edited by Watson Fuller 256p Routledge and Kegan Paul,London,1971
WSM 43.0589

BIOLOGY Towards a theoretical biology 4: essays:an IUBS symposium International union of biological sciences Edited by C.H. Waddington 299p 26cm Edinburgh university press,Edinburgh,1972
Math S 3.1866

BIOLOGY Towards a theoretical biology :a collection of essays written as a result of a symposium... Bellagio 1966 Aug 28-Sep 3 International union of biological sciences Edited by C.H. Waddington 253p 26cm Edinburgh university press,Edinburgh,1970
Math S 3.1819

BIOLOGY Towards a theoretical biology :a symposium Papers Lake Como 1967 Aug 3-12 Vol 2: sketches International union of biological sciences Edited by C.H. Waddington I.U.B.S.symposium, 2 Edinburgh university press,Edinburgh,1969
Bal 39.0053

BIOLOGY Towards a theoretical biology :an I. U.B.S. symposium Lake Como 1966 Aug 28-Sep 3 and Lake Como 1967 Aug 3-12 Vol 1-3 International union of biological sciences Edited by C.H. Waddington 3 vols Edinburgh university press,Edinburgh,1970
Bot 42.0320

BIOLOGY Towards a theoretical biology 1. Prolegomena:an I.U.B.S. symposium Essays Bellagio 1966 Aug 28-Sep 3 International union of biological sciences Edited by C.H. Waddington 234p Edinburgh university press, Edinburgh,1968 Essays written after and in the light of the symposium
WSM 43.2897

BIOLOGY Uspekhi biologicheskikh nauk v S.S.S. R. za 25 let, 1917-42 :symposium Akademiya nauk S.S.S.R.Otdelenie biologicheskikh nauk Edited by L.A. Orbeli 356p 26cm Moscow, 1945
Philos 1.1394

BIOLOGY AND BEHAVIOR 3 Environmental influences :a conference Proceedings New York 1967 Apr 21-22 Russell Sage foundation Rockefeller institute Edited by David C. Glass Rockefeller university press; Russell Sage foundation,New York,1968 Third and last in a series of conferences on biology and behavior
Psy 31.3096

BIOLOGY AND ETHICS :a symposium Proceedings London 1968 Sep 26-27 Institute of biology Edited by F.J. Ebling Institute of biology.Symposia, 18 xxix,145p Academic press,London;New York,1969
Bal 39.0055

BIOLOGY AND ETHICS :a symposium Proceedings London 1968 Sep 26-27 Institute of biology Edited by F.J. Ebling held at the Royal geographical society Institute of biology.Symposia, 18 Academic press,London; New York,1969
Bal 39.0018

BIOLOGY COLLOQUIUM Arctic biology 18 colloquium Corvallis 1957 Apr 19-20 Edited by Henry P. Hansen illus,maps 134p 28cm Oregon state college,Corvallis,1957
Sco 14.2434

BIOLOGY OF AGEING Aspects of the biology of ageing :a symposium Sheffield 1966 Sep 5-9 Society for experimental biology Society for experimental biology.Symposia, 21 Cambridge university press,Cambridge,1967
Bioch 33.1360

BIOLOGY OF CUTANEOUS CANCER The International conference on the biology of cutaneous cancer 1st Philadelphia,Pa. 1962 Apr 6-11 Temple university.School of medicine and International union against cancer Edited by Frederick Urbach With financial support from Merck,Sharp and Dohme. Postgraduate program National cancer institute.Monograph, 10 US government printing office,Washington,D.C.,1963
Path 30.2423

BIOLOGY OF DESERTS Biology of hot and cold deserts :a symposium Proceedings London 1952 Sep 25-27 Edited by J.L. Cloudsley-Thompson Organized by the Institute of biology Institute of biology,London,1954
Geog 13.1205

The BIOLOGY OF HAIR GROWTH :a conference Papers London 1957 Aug 7-9 Edited by William Montagna and Richard A. Ellis Sponsored by the British society for research on ageing Academic press,New York,1958
An 32.4034

BIOLOGY OF HOT AND COLD DESERTS :a symposium Proceedings Biology of deserts London 1952 Sep 25-27 Edited by J.L. Cloudsley-Thompson Organized by the Institute of biology Institute of biology,London,1954
Geog 13.1205

The BIOLOGY OF HOT AND COLD DESERTS... :a symposium Proceedings London 1952 Sep 25-27 Institute of biology Edited by J.L. Cloudsley-Thompson Assisted by a grant from Unesco Institute of biology,London,1954
Bal 39.1858

BIOLOGY OF MELANOMAS A Conference on the biology of normal and atypical pigment cell growth New York 1946 Nov 15-16 New York academy of sciences.Section of biology New York academy of sciences.Special publications, 4 pls New York academy of sciences,New York,1948
Path 30.2492

BIOLOGY OF MELANOMAS Conference on the biology of normal and atypical pigment cell growth New York 1946 Nov 15-16 New York academy of sciences.Section of biology New York academy of sciences.Special publications, 4 New York,1948
Phys 20.0952

BIOLOGY OF MELANOMAS Conference on the biology of normal and atypical pigment cell growth Results New York 1946 Nov 15-16 By Myron Gordon and others New York academy of sciences.Special publications, 6 New York academy of sciences,New York,1948
An 32.3374

BIOLOGY OF NEUROGLIA Lectures and discussions Buenos Aires 1963 Oct 17-18 Instituto Torcuato di Tella and International brain research organization Edited by E.D.P.de Robertis and R. Carea Progress in brain research, 15 illus Elsevier,Amsterdam, 1965 Symposium held as part of the 10th Latin-American congress of neurosurgery
Path 30.2466

BIOLOGY OF NORMAL AND ATYPICAL PIGMENT CELL GROWTH 4th conference Proceedings Houston,Tex. 1957 Nov 14-16 New York zoological society Anderson hospital and tumor institute Damon Runyon memorial fund for cancer research Edited by Myron Gordon Academic press,New York,1959
Gen 34.0601

BIOLOGY OF PARASITES;EMPHASIS ON VETERINARY PARASITES The International conference of the World association for the advancement of veterinary parasitology 2nd Proceedings Philadelphia 1965 Sep 7-9 World association for the advancement of veterinary parasitology Edited by E.J.L. Soulsby Held at the University of Pennsylvania Academic press,New York;London,1966 Held...in conjunction with Bicentennial celebrations of medical education in the United States
Bal 39.1737

BIOLOGY OF PHOSPHORUS A Symposium in general biology Papers presented East Lansing,Mich. 1951 Apr 25-27 By G.Evelyn Hutchinson and others Edited by Lester F. Wolterink Held at Michigan state college Michigan state college press,1952
Bioch 33.0590

BIOLOGY OF REPRODUCTION IN MAMMALS :a symposium Proceedings Nairobi,Kenya 1968 Apr Society for the study of fertility Edited by J.S. Perry and I.W. Rowlands Journal of reproduction and fertility. Supplement, 6 Blackwell,Oxford;Edinburgh, 1969
Bal 39.1404

The BIOLOGY OF SURVIVAL 13 :a symposium Proceedings Zoological society of London Edited by O.G. Edholm illus viii,138p Academic press,London;New York,1964
Sco 14.0837

The BIOLOGY OF SURVIVAL 13th :a symposium Proceedings London 1964 May 7-8 Edited by O.G. Edholm Organized by the Physiological society,the society for the study of human biology,and the Zoological society of London Zoological society of London.Symposia, 13 Academic press,London, 1964
Phys 20.2132

BIOLOGY OF THE HYDRA AND OF SOME OTHER COELENTERATES The Physiology and ultrastructure of hydra and of some other coelenterates :a symposium Coral Gables,Fla. 1961 Mar 29-31 Edited by Howard M. Lenhoff and F.Farnsworth Loomis held at the Fairchild tropical gardens University of Miami press,Coral Gables,Fla.,1961
Bal 39.2183

The BIOLOGY OF WEEDS :a symposium Oxford 1959 Apr 2-4 British ecological society Edited by John L. Harper British ecological society.Symposia, 1 Bibliog, maps,2 plates,tables,diagrs xv,256p Blackwell scientific,Oxford,1960
Bot 42.2032

BIOLOGY TODAY:ITS ROLE IN EDUCATION International working session on the teaching of school biology :report of an OECD working session Organization for economic cooperation and development Edited by Paul Duvigneaud English edition prepared by L.S. Comber New thinking in school science OECD,Paris,1966
Gen 34.2234

BIOLUMINESCENCE :a conference New York 1946 Nov 8 By E.Newton Harvey and others New York academy of sciences.Section of biology New York academy of sciences.Annals, 49,p 327-482 New York,1948
Bal 39.0084

BIOMECHANICS Symposium on biomechanics Papers London 1959 Apr 17 Institution of mechanical engineers London,1959
Bal 39.0061

BIOMECHANICS:TECHNIQUE OF DRAWINGS OF MOVEMENT AND MOVEMENT ANALYSIS Zurich 1967 Aug 21-23 Edited by J. Wartenweiler and others Medicine and sport,2 Karger,Basle;NewYork, 1968
VA 19.0128

BIOMEDICAL FLUID MECHANICS SYMPOSIUM : presented at Fluids engineering conference Denver,Col. 1966 Apr 25-27 Sponsored by the American society of chemical engineers. Fluids engineering division American society of mechanical engineers,New York,1966
Chem E 24.1797

BIOMEDICAL FLUID MECHANICS SYMPOSIUM Papers Denver,Colo. 1966 Apr 25-27 American society of mechanical engineers.Fluid mechanics commission Co-sponsored by the Rheology society of America American society of mechanical engineers,New York,1966 Papers presented by the Fluid mechanics commission of the ASME at the Fluids engineering conference,co-sponsored by the Rheology society of America
Eng 41.6540

BIOMEDICAL PROGRESS Challenge of life: biomedical progress and human values Roche anniversary symposium Proceedings Basle 1971 Aug 31-Sep 3 Hoffmann-La Roche and company Edited by Robert M. Kunz and Hans Fehr Experientia.Supplementum, 17 456p 24cm Birkhauser,Basle,1972 Symposium held on the occasion of the firm's 75th anniversary.Chairman:Lord Todd
Chem 18.2902

BIOMETEOROLOGY Biometeorology The International bioclimatological congress 2nd Proceedings London 1960 Sep 4-10 Edited by S.W. Tromp illus xxii,687p 25cm Pergamon,Oxford,1962
Sco 14.0585

BIOMETEOROLOGY International biometeorological congress 3rd Proceedings Pau 1963 Sep 1-7 Vol 2,pt 1-2 International society of biometeorology Edited by S.W. Tromp and W.H. Weihe 2 vols Pergamon press,Oxford,1967
Geog 13.1212

BIOMETEOROLOGY The International bioclimatological congress 2nd Proceedings London 1960 Sep 4-10 Edited by S.W. Tromp illus xxii,687p 25cm Pergamon,Oxford, 1962
Sco 14.0585

BIOMETRIC SOCIETY International biometric conference 1st Woods Hole,Mass. 1947 Sep 5-6 1948 Reprinted from 'Biometrics' December 1947 and March 1948
Gen 34.2145

BIOMETRIC SOCIETY International biometric conference 2nd Proceedings Geneva 1949 Aug 30-Sep 2 1950 Reproduced from 'Biometrics' vol 6,no.1
Gen 34.2146

BIOMETRIC SOCIETY International biometric conference 3rd Proceedings Bellagio, Italy 1953 Sep 1-5 1953 Reprinted from 'Biometrica' vol 9,no.4,1953
Gen 34.2147

BIOMETRIC SOCIETY Joint European conference of the Institute of mathematical statistics, the International association for statistics in physical sciences,the Biometric society Berne 1964 Sep 14-18 Berne,1964 13 unbound typescript papers
Math 3.1124

BIOMETRICAL GENETICS :international symposium Proceedings Ottawa 1958 Aug International union of biological sciences Edited by Oscar Kempthorne Sponsored by the Biometrics society International series of monographs on biometry International union of biological sciences.Series B, 38 Pergamon press,London,1960
Gen 34.0299

BIOMETRICS International biometric conference 1st Woods Hole,Mass. 1947 Sep 5-6 Biometric society 1948 Reprinted from 'Biometrics' December 1947 and March 1948
Gen 34.2145

BIOMETRICS SOCIETY Biometrical genetics : international symposium Proceedings Ottawa 1958 Aug International union of biological sciences Edited by Oscar Kempthorne International series of monographs on biometry International union of biological sciences. Series B, 38 Pergamon press,London,1960
Gen 34.0299

BIOMETRICS SOCIETY Statistics and mathematics in biology conference Ames,Iowa 1952 Jun-Jul Iowa state college Edited by Oscar Kempthorne and others xi,632p 23cm Iowa state college press,Ames,Iowa,1954
Math 3.0764

BIOMETRY International biometric conference 5th Cambridge 1963 Sep 10-13 Cambridge,1963 8 unbound typescript papers
Math 3.1127

BIOMOLECULAR ORGANIZATION Principles of biomolecular organization :a Ciba foundation symposium Proceedings London 1965 Jun 9-11 Ciba foundation Edited by G.E.W. Wolstenholme and Maeve O'Connor Churchill, London,1966
Bot 42.1503

BIOMOLECULAR ORGANIZATION Principles of biomolecular organization :a Ciba foundation symposium Proceedings London 1965 Jun 9-11 Ciba foundation Edited by G.E.W. Wolstenholme and Maeve O'Connor Churchill, London,1966
Bioch 33.0311

BIOPHYSICAL ASPECTS OF RADIATION QUALITY Panel on biophysical aspects of radiation quality 2nd Report Vienna 1967 Apr 10-14 International atomic energy agency International atomic energy agency.Panel proceedings series International atomic energy agency,Vienna,1968
Radioth 35.1422

BIOPHYSICAL ASPECTS OF RADIATION QUALITY Panel on biophysical aspects of radiation quality Report Vienna 1965 Mar 29-Apr 2 International atomic energy agency International atomic energy agency.Technical report series, 58 International atomic energy agency,Vienna,1966
Radioth 35.1421

BIOPHYSICAL SCIENCE :a study program Boulder,Colo. 1958 Jul 20 Aug 16 Edited by J.L. Oncley and others Organized and conducted by the National institutes of health. Biophysics and biophysical chemistry study section Wiley,New York,1959
Bioch 33.0329

BIOPHYSICAL SCIENCE :a study program Boulder,Colo. 1959 Jul 20-Aug 16 Edited by J.L. Oncley and others Organized and conducted by the National institutes of health. Biophysics and biophysical chemistry study section Wiley,New York,1959 Published also in the Reviews of modern physics,Jan.and Apr.1959
Bal 39.0026

BIOPHYSICAL SOCIETY Microsomal particles and protein biosynthesis :symposium Papers Cambridge,Mass. 1958 Feb 5-8 Edited by Richard B. Roberts Biophysical society. Symposia, 1 Washington academy of sciences, Washington,D.C.;London,1958
Gen 34.0657

BIOPHYSICAL SOCIETY Microsomal particles and protein synthesis :a symposium Papers Cambridge,Mass 1958 Feb 5-8 Edited by Richard R. Roberts Biophysical society. Symposia, 1 Pergamon,London;New York,1958
Bot 42.1686

BIOPHYSICAL SOCIETY Microsomal particles and protein synthesis :a symposium Papers presented Cambridge,Mass. 1958 Feb 5-8 Edited by Richard B. Roberts At the Massachusetts institute of technology Biophysical society.Symposia, 1 Pergamon, London,1958
Bioch 33.0924

BIOPHYSICAL SOCIETY 1st :symposium Papers Microsomal particles and protein synthesis Cambridge,Mass. 1958 Feb 5-8 Edited by Richard B. Roberts Pergamon press,London, 1958
Col S 12.0053

BIOPHYSICAL SOCIETY.SYMPOSIA, 1 Microsomal particles and protein biosynthesis :symposium Papers Cambridge,Mass. 1958 Feb 5-8 Biophysical society Edited by Richard B. Roberts Washington academy of sciences, Washington,D.C.;London,1958
Gen 34.0657

BIOPHYSICAL SOCIETY.SYMPOSIA, 1 Microsomal particles and protein synthesis :a symposium Cambridge,Mass. 1958 Feb 5-8 Biophysical society Edited by Richard B. Roberts Pergamon press,London,1958
Radioth 35.0077

BIOPHYSICAL SOCIETY.SYMPOSIA, 1 Microsomal particles and protein synthesis :a symposium Papers Cambridge,Mass 1958 Feb 5-8 Biophysical society Edited by Richard R. Roberts Pergamon,London;New York,1958
Bot 42.1686

BIOPHYSICAL SOCIETY.SYMPOSIA, 1 Microsomal particles and protein synthesis :a symposium Papers presented Cambridge,Mass. 1958 Feb 5-8 Biophysical society Edited by Richard B. Roberts At the Massachusetts institute of technology Pergamon,London,1958
Bioch 33.0924

BIOPHYSICS International biophysics congress Stockholm 1961 Jul 31-Aug 4 abstracts of contributed papers Tryckeri aktiebolaget thule,Stockholm,1961
Radioth 35.1647

BIOPHYSICS International biophysics congress abstracts of contributed papers Stockholm 1961 Jul 31-Aug 4 Tryckeri Aktiebolaget, Stockholm,1961
Radioth 35.1426

BIOPHYSICS National biophysics conference 1st Proceedings Columbus,Ohio 1957 Mar 4-6 Edited by H. Quastler and H.J. Morowitz Yale university press,NEW Haven,Conn.,1959
Radioth 35.0231

BIOPHYSICS CONFERENCE The National biophysics conference 1st Proceedings Columbus,Ohio 1957 Mar 4-6 Edited by Henry Quastler and Harold J. Morowitz With the support of the United States.Air force.Office of scientific research Yale university press, New Haven,1959
Bioch 33.0308

BIOPHYSICS OF PHYSIOLOGICAL AND PHARMACOLOGICAL ACTIONS The American association for the advancement of science meeting 127th Papers presented New York 1960 Dec 26-28 American association for the advancement of science Edited by Abraham M. Shanes and others American association for the advancement of science.Publication, 69 Washington,D.C.,1961
Bal 39.1257

BIOPHYSIK,STRAHLENBIOLOGIE UND STRAHLENTHERAPIE vortrage auf der 39 tagung der Deutschen rontgengesellschaft Proceedings Frankfurt 1957 Oct 20-24 Deutsche rontgengesellschaft Edited by B. Rajewsky Urban and Schwarzenberg,Munich,1958
Radioth 35.1617

BIOPHYSIKALISCHE PROBLEME DER STRAHLEN WIRKUNG jahrestagung der Deutschen gesellschaft fur biophysik Tagungsbericht Hamburg 1965 Apr 23-24 Deutsche gesellschaft fur biophysik Edited by H. Muth Thieme, Stuttgart,1966
Radioth 35.1420

BIOPOLYMERS An International symposium on conformation of biopolymers Madras 1967 Jan 18-21 Vol 1 Edited by G.N. Ramachandran Academic press,London,1967
Col S 12.0132

BIOPOLYMERS Conformation of biopolymers International symposium on conformation of biopolymers Papers read Madras 1967 Jan 18-21 Vol 1-2 University of Madras. Centre of advanced study in biophysics Edited by G.N. Ramachandran Sponsored by the International union of pure and applied physics 2 vols Academic press,London;New York,1967
Bioch 33.0331

BIOSIMULATION Aspects of the theory of artificial intelligence International symposium on biosimulation 1st Proceedings Locarno 1960 Jun 29-Jul 5 Edited by C.A. Muses Plenum,New York,1962
Eng 41.5988

BIOSOZIOLOGIE Internationale vereinigung fur vegetationskunde Bericht Stolzenau 1960 Edited by Reinhold Tuxen illus. Junk,The Hague,1965
Bot 42.4762

BIOSTATISTICS Statistics and mathematics in biology conference Ames,Iowa 1952 Jun-Jul Iowa state college Edited by Oscar Kempthorne and others Co-sponsored by Biometrics society xi,632p 23cm Iowa state college press,Ames,Iowa,1954
Math 3.0764

The BIOSYNTHESIS AND SECRETION OF ADRENOCORTICAL STEROIDS :a symposium London 1959 Feb 14 Biochemical society Edited by F. Clark and J.K. Grant Held at the University of London.Senate House Biochemical society. Symposia, 18 Cambridge university press, Cambridge,1960
Bioch 33.1380

BIOSYNTHESIS OF LIPIDS International congress of biochemistry 5th Proceedings Moscow 1961 Aug 10-16 International union of biochemistry Edited by G. Popjak I.U.B. symposium series, 27 Pergamon;PWN-Polish scientific publishers,Oxford;Warsaw,1963
Bioch 33.1336

BIOSYNTHESIS OF TERPENES AND STEROLS Ciba foundation symposium on the biosynthesis of terpenes and sterols Proceedings London 1958 May 20-22 Ciba foundation Edited by G. E.W. Wolstenholme and Maeve O'Connor illus Churchill,London,1959
Bioch 33.0571

BIOSYNTHETIC PATHWAYS IN HIGHER PLANTS Plant phenolics group symposium Proceedings Leeds 1964 Apr 13-15 Plant phenolics group Edited by J.B. Pridham and T. Swain Academic press,London;New York,1965
Bot 42.1883

BIOSYNTHETIC PATHWAYS IN HIGHER PLANTS Plant phenolics group symposium Proceedings Leeds 1964 Apr 13-15 Plant phenolics group Edited by J.B. Pridham and T. Swain Academic press,London;New York,1965 For later publications of this body see Phytochemical society
Gen 34.1406

BIOSYNTHETIC PATHWAYS IN HIGHER PLANTS The Plant phenolics group symposium Proceedings Leeds 1964 Apr 13-15 Plant phenolics group Edited by J.B. Pridham and T. Swain Academic press,London;New York,1965
Bioch 33.0783

BIRBAL SAHNI INSTITUTE OF PALAEOBOTANY 1964 A Symposium on floristics and stratigraphy of Gondwanaland Proceedings Palaeobotanical society illus. 154p 24cm Birbal Sahni institute of palaeobotany,Lucknow,1966
Bot 42.3271

BIRKBECK COLLEGE Foundations of statistical inference Joint statistics seminar :a discussion London 1959 Jul 27-28 By Leonard J. Savage and others 112p Methuen; Wiley,London;New York,1964
WSM 43.1769

BIRKBECK COLLEGE The Foundations of statistical inference :a discussion opened by Professor L.J. Savage at a meeting of the Joint statistics seminar, Birkbeck and Imperial colleges,in the University of London Methuen's monographs on applied probability and statistics Methuen,London,1962
HE 27.0051

BIRKBECK COLLEGE The Foundations of statistical inference :a discussion opened by Professor L.J.Savage at a meeting of the Joint statistics seminar,Birkbeck and Imperial colleges,in the University of London Methuen's monographs on applied probability and statistics 112p 20cm Methuen,London, 1962
Math 3.1311

BIRMINGHAM 1949 Lipo-proteins :a general discussion Papers Faraday society Faraday society.Discussions, 6 Aberdeen, 1949
Bioch 33.0508

BIRMINGHAM 1950 Birmingham and its regional setting :a scientific survey British association for the advancement of science British association for the advancement of science,Birmingham,1950
Gen 34.1371

BIRMINGHAM 1950 Vacuum physics :a symposium Institute of physics.Midland branch Journal of scientific instruments. Supplement, 1 80,viiip Institute of physics,London,1951
Chem E 24.1353

BIRMINGHAM 1957 Symposium on chemical engineering education Discussion on papers presented Institution of chemical engineers 1957
Chem E 24.1758

BIRMINGHAM 1957 Symposium on chemical engineering education Papers Institution of chemical engineers 61p 1957
Chem E 24.1757

BIRMINGHAM 1962 Biological receptor mechanisms Contributions Society for experimental biology Edited by J.W.L. Beament Society for experimental biology. Symposia, 16 Cambridge university press, Cambridge,1962
Psy 31.0283

BIRMINGHAM 1962 Biological receptor mechanisms Society for experimental biology Society for experimental biology.Symposia, 16 Cambridge university press,Cambridge,1962
An 32.4562

BIRMINGHAM 1963 A Balanced teaching hospital :a symposium Edited by Thomas McKeown Sponsored by Nuffield provincial hospitals trust viii,131p 28cm Oxford university press,London,1965
PGMS 29.0500

BIRMINGHAM 1963 The Symposium on the less common means of separation Proceedings Institution of chemical engineers.Midlands branch Edited by J.M. Pirie 127p I.Chem. E.,London,1964
Chem E 24.1915

BIRMINGHAM 1963-64 Metallurgical achievements :selection of papers presented at the Birmingham metallurgical society's diamond jubilee session 1963-64 Birmingham metallurgical society Edited by W.O. Alexander Pergamon press,London,1965
Eng 41.3485

BIRMINGHAM 1965 Basic concepts of inborn errors and defects of steroid biosynthesis Proceedings Society for the study of inborn errors of metabolism Edited by K.S. Holt and D.N. Raine Society for the study of inborn errors of metabolism.Symposia, 3 Livingstone,Edinburgh;London,1968
Gen 34.2000

BIRMINGHAM 1965 Basic concepts of inborn errors and defects of steroid biosynthesis :a symposium 3rd Proceedings Society for the study of inborn errors of metabolism Edited by K.S. Holt and D.N. Raine Livingstone,Edinburgh,1966
Inv Med 37.0271

BIRMINGHAM 1965 New engineering materials : a conference Institution of mechanical engineers.Process engineering group Institution of mechanical engineers. Proceedings, 180 pt.3D I.Mech.E.,London,1965
Met 25.0105

BIRMINGHAM 1965 Numerical analysis :an introduction Institute of mathematics and its applications Edited by Joan Walsh 212p Academic press,London;New York,1966 Based on a symposium held in Birmingham in 1965
A Math 4.1219

BIRMINGHAM 1965 Numerical analysis:an introduction Symposium on numerical analysis Institute of mathematics and its applications Academic press,London,1966
Radioth 35.0631

BIRMINGHAM 1965 Numerical analysis:an introduction Symposium on numerical analysis Institute of mathematics and its applications Edited by J. Walsh bibliog.,diags. xiv, 212p 24cm Academic press,London,1966
Math L 5.0357

BIRMINGHAM 1965 Reproductive biology and taxonomy of vascular plants :conference Report Botanical society of the British Isles Edited by J.G. Hawkes Botanical society of the British Isles.B.S.B.I. conference reports, 9 183p Pergamon press,Oxford,1966
Bot 42.0307

BIRMINGHAM 1966 The Conference on the human operator in complex systems Proceedings Edited by W.T. Singleton and others Bibliog Taylor and Francis,London, 1967
Eng 41.0678

BIRMINGHAM 1967 Chemotaxonomy and serotaxonomy :a symposium Proceedings Systematics association Edited by J.D. Hawkes Systematics association.Special volume, 2 illus xvi,299p Academic press,London;New York,1968
Bot 42.3449

BIRMINGHAM 1967 Form and function in the human lung :a symposium Proceedings Edited by Gordon Cumming and L.B. Hunt Sponsored by the Lederle laboratories Livingstone, Edinburgh,1968
An 32.3805

BIRMINGHAM 1967 Symposium on anodizing aluminium Proceedings Aluminium federation and University of Aston in Birmingham Aluminium federation,London,1967
Met 25.2048

BIRMINGHAM 1968 Crystal growth 1968 International conference on crystal growth 2nd Proceedings Comite international de croissance crystalline Edited by F.C. Frank and others Jounal of crystal growth,1968,no 3-4 North-Holland,Amsterdam,1968
Min 10.1507

BIRMINGHAM 1968 Stainless steels Conference on stainless steels for the fabricator and user Proceedings Iron and steel institute and Birmingham metallurgical association Iron and steel institute,London,1969
Met 25.2242

BIRMINGHAM 1968 Symposium on applications of micro-electronics Organized by the Institution of electrical engineers Institution of electrical engineers.Conference publication, 49 Institution of electrical engineers,London,1968
Eng 41.5266

BIRMINGHAM 1969 International conference on crystal growth 2nd Proceedings Edited by F.C. Frank and others Sponsored by International union of pure and applied physics xix,842p 28cm North-Holland, Amsterdam,1968 Reprinted from Journal of crystal growth,3-4,1968
Cav 7.2950

BIRMINGHAM,ALA. 1966 Secretory mechanisms of salivary glands International conference on mechanisms of salivary secretion and their regulation Proceedings Edited by Leon H. Schneyer and Charlotte A. Schneyer Held at the University of Alabama medical center Academic press,New York;London,1967
Bal 39.1265

BIRMINGHAM AND ITS REGIONAL SETTING :a scientific survey Birmingham 1950 Aug 30-Sep 6 British association for the advancement of science British association for the advancement of science,Birmingham,1950
Gen 34.1371

BIRMINGHAM METALLURGICAL ASSOCIATION Stainless steels Conference on stainless steels for the fabricator and user Proceedings Birmingham 1968 Sep 10-12 Iron and steel institute,London,1969
Met 25.2242

BIRMINGHAM METALLURGICAL SOCIETY Metallurgical achievements Edited by W.O. Alexander Pergamon press,Oxford,1965 Selection of papers presented at the Birmingham metallurgical society's Diamond Jubilee session, 1963-64
Met 25.0080

BIRMINGHAM METALLURGICAL SOCIETY Metallurgical achievements :selection of papers presented at the Birmingham metallurgical society's diamond jubilee session 1963-64 Birmingham 1963-64 Edited by W.O. Alexander Pergamon press, London,1965
Eng 41.3485

BISHOP (BERNICE P) MUSEUM and UNIVERSITY OF HAWAII Plants and the migrations of Pacific peoples :a symposium Honolulu 1961 Aug 21-Sep 6 Edited by J. Barrau Sponsored by the National science foundation Bishop museum press,Honolulu,1963 Symposium convened at the 10th Pacific science congress
Geog 13.7202

BLACKSBURG,VA. 1961 Physics of the solar system Conference on physics of the solar system and re-entry dynamics proceedings Part 1: physics of the solar system Virginia polytechnic institute.Engineering experiment station. Virginia polytechnic institute.Engineering experiment station series, 149 Virginia polytechnic institute. Bulletin, 55,9 Virginia polytechnic institute,Blacksburg,Va.,1962
A Math 4.1155

BLACKSBURG,VA. 1970 Bioinorganic chemistry a symposium American chemical society. Inorganic chemistry division Chemical institute of Canada Edited by Robert F. Gould x,436p 23cm American chemical society,Washington,D.C.,1971
Chem 18.2719

BLANKETING EFFECT Colloquium on the blanketing effect proceedings Heidelberg 1966 Mar.17-19 Edited by Karl-Heinz Bohm Journal of quantitative spectroscopy and radiative transfer, 6,no.5 Pergamon press, Oxford;London,1966
Obs 6.0207

BLANTYRE 1962 Economic development in Africa Nyasaland economic symposium Papers Edited by E.F. Jackson Blackwell,Oxford,1965
Geog 13.4650

BLARICUM,1966 Computers in mathematical research Edited by R.F. Churchhouse and J.-C. Herz xi,185p 23cm North-Holland, Amsterdam,1968 Mainly on a symposium on Utilization of computers in mathematical research held at the IBM world trade european education centre,Blaricum, Netherlands,August 29-31,1966
Math L 5.3116

BLINDNESS Psycholgical diagnosis and counseling of the adult blind :selected papers from the proceedings of the University of Michigan conference for the blind Ann Arbor, Mich. 1947 United States.Federal security agency.Office of vocational rehabilitation and Michigan.Department of social welfare. Division of services for the blind Edited by Wilma Donahue and Donald Dabelstein Co-sponsored by the University of Michigan. Institute for human adjustment American foundation for the blind,New York,1950
Psy 31.2118

BLOOD CLOTTING AND ALLIED PROBLEMS Conference on blood clotting and allied problems 1-5 Transactions Edited by Joseph Eugene Flynn 5 vols Josiah Macy Jr. foundation,New York,1948-52
Med 36.0141

BLOOD CLOTTING FACTORS International congress of biochemistry 4th Proceedings Vienna 1958 Sep 1-6 Vol 10: symposium 10 - blood clotting factors Edited by E. Deutsch I.U.B.symposium series, 12 Pergamon,London, 1959 Added title page in French and German.Text in English and German
Bioch 33.1324

BLOOD FLOW Autoregulation of blood flow : international symposium Proceedings Indianapolis,Ind. 1963 Jan 10-12 Edited by Paul C. Johnson American heart association monographs, 8 American heart association, New York,1964
Phys 20.0898

BLOOD FLOW Pulsatile blood flow : international symposium 1st Proceedings Philadelphia,Pa. 1963 Apr 11-13 Edited by E.O. Attinger McGraw-Hill,New York,1964
Phys 20.0899

BLOOD FLOW THROUGH ORGANS AND TISSUES :an international conference Proceedings Glasgow 1967 Mar Edited by William H. Bain and A.Murray Harper Livingstone,Edinburgh; London,1968
Inv Med 37.0009

BLOOD GROUPS IN INFRAHUMAN SPECIES :a conference New York 1962 May 3 By Ray D. Owen and others New York academy of sciences Edited by C. Cohen New York academy of sciences.Annals, 97,p1-328 New York,1962
Bal 39.1341

BLOOD GROUPS OF ANIMALS European animal blood group conference 9th Proceedings Prague 1964 Aug 18-22 Ceskoslovenska akademie ved and European society for animal blood group research Edited by Josef Matousek Junk,The Hague,1964
Surg 23.0044

BLOOD OXYGENATION International symposium on blood oxygenation Proceedings Cincinnati, Ohio 1969 Dec 1-3 Edited by Daniel Hershey Plenum press,New York;London,1970
An 32.5457

BLOOD PLATELETS :Henry Ford hospital international symposium Detroit,Mich. 1960 Mar 17-19 Edited by Shirley A. Johnson and others Henry Ford hospital.Symposia, 10 Churchill,London,1961
An 32.3175

BLOOD TRANSFUSION Congres internationale de la societe internationale de transfusion sanguine 4th Rapports et communications Paris 1951 Jul 21-29 International society of blood transfusion Edited by J. Julliard and others L'expansion scientifique francaise,Paris,1952
Med 36.0389

BLOOD TRANSFUSION Europaischen gesellschaft fur hamatologie congress :colloquium uber aktuelle probleme des transfusionswesens und der immunhamatologie 5th Freiburg 1955 Sep 20-24 European society of haematology Edited by Herbert Begemann Springer-verlag,Berlin;Gottingen,1956
Med 36.0022

BLOOD TRANSFUSION International congress of blood transfusion 7th Proceedings Rome 1958 Sep 2-6 International society of blood transfusion Edited by Ludwig Hollander Karger,Basle,1959
Med 36.0200

BLOOD TRANSFUSION International congress of blood transfusion 8th Programme and abstracts Tokyo 1960 Sep International society of blood transfusion Edited by Ludwig Hollander 1960
Med 36.0201

BLOOD TRANSFUSION Transfusion sanguine et actualites hematologiques :congres national des transfusions sanguines de France et des pays de langue francaise 1st Algiers 1953 Apr 1-4 Edited by E. Benhamou and A. Albou Masson et compagnie,Paris,1954
Med 36.0023

BLOOD TRANSFUSIONS International congress of blood transfusion 5th Proceedings Paris 1954 Sep International society of blood transfusion 1955
Med 36.0388

BLOOD VESSELS AND CIRCULATION Brown university symposium on the biology of skin, 1960 Proceedings Providence,R.I. 1960 Jan 30-31 Edited by William Montagna and Richard A. Ellis Advances in biology of skin, 2 Pergamon press,Oxford,1961
An 32.4038

BLOODSTREAM Krebsmetastasierung auf dem blutwege :symposium Geneva 1963 Jun 27-29 Schweizerische nationaliga fur krebsekampfung und krebsforschung Schweizerische akademie der medizinischen wissenschaften Schweizerische hamatologe gesellschaft Schweizerische akademie der medizinischen wissenschaften.Bulletin, 20,fasc 1-3 Schwabe,Basel,1964
Radioth 35.0835

BLOOMINGTON,IND. 1954 Stellar atmospheres : a conference proceedings Edited by Marshal H. Wrubel financed by the National science foundation 183p Indiana university, Bloomington,Ind.,1955
Obs 6.1900

BLOOMINGTON,IND. 1969 Nonparametric techniques in statistical inference International symposium on nonparametric techniques in statistical inference 1st Proceedings Edited by Madan Lal Puri Held at Indiana university bibliog. xiv,623p 24cm Cambridge university press,Cambridge, 1970
Math S 3.1616

BLUE BELL,PA. 1963 International conference on single-crystal films Proceedings Philco scientific laboratories Pergamon press,Oxford,1964
Cav 7.0244

BLUE BELL,PA. 1963 Single crystal films International conference on single crystal films Proceedings Edited by M.H. Francombe and Hiroshi Sato Pergamon press,Oxford,1964
Eng 41.4458

BLUE BELL,PENN. 1963 Single-crystal films : a conference Proceedings United States. Office of naval research and University of Pennsylvania Edited by Maurice H. Francombe and Hiroshi Sato Sponsored jointly by Philco corporation Pergamon press,Oxford,1964
Met 25.0177

BOEING COMPANY.SCIENTIFIC RESEARCH LABORATORIES Aircraft wake turbulence and its detection : a symposium Proceedings Seattle,Wash. 1970 Sep 1-3 Edited by J.H. Olsen and others Plenum press,New York,1971
Eng 41.8333

BOEING SCIENTIFIC RESEARCH LABORATORIES.FLIGHT SCIENCES LABORATORY Clear air turbulence and its detection A Symposium on clear air turbulence and its detection Proceedings Seattle 1968 Aug 14-16 Edited by Yih-Ho Pao and Arnold Goldburg 542p 22cm Plenum press,New York,1969
Cav 7.2921

BOGATSTVA ZEMLI NA SLUZHBU RODINE :materialy soveshchaniya rabotnikov selskogo khozyaistva oblastei i avlovonykh kraynego Severa Materialy Magadan 1962 Feb 22-24 Ministerstvo selskogo khozyaistva,R.S.F.R. 159p 20cm izdatelstvo ministerstva selskogo khozyaistva RSFSR,Moscow,1962
Sco 14.4924

BOLOGNA 1971 La Rotazione come fenomeno e fattore evolutivo nell'universo Edited by G. Righini 212p Societa astronomica italiana, Florence,1972
Obs 6.3715

BOLTING LANDING,N.Y. 1960 Columbium metallurgy :a symposium American institute of mining,metallurgical and petroleum engineers Edited by D.L. Douglass and F.W. Kunz Metallurgical society conferences, 10 Interscience,New York;London,1958
Met 25.2533

BOLTON LANDING,N.Y. 1959 Structure and properties of thin films An International conference on structure and properties of thin films Proceedings General electric company Edited by C.A. Neugebauer and others John Wiley,New York,1959
Cav 7.0265

BOLTON LANDING,N.Y. 1960 Progress in very high pressure research :an international conference Proceedings Wright air development division.Materials central and General electric company research laboratory Edited by F.P. Bundy and others 314p New York,1961
Geod 9.0254

BOLTON LANDING,NEW YORK 1960 Progress in very high pressure research :international conference Proceedings United States.Air research and development command.Materials central and General electric research laboratory Edited by F.P. Bundy and others Wiley,New York;London,1961
Met 25.1638

BOLTON LANDING CONFERENCE 2nd oxide dispersion strengthening:a symposium Bolton Landing,New York 1966 Jun 27-29 American institute of mining,metallurgical and petroleum engineers Edited by G.S. Ansell Metallurgical society conferences, 47 Gordon and Breach,New York,1968
Met 25.2339

BOLYAI JANOS MATEMATIKAI TARSULAT Colloquium on applications of mathematics to economics Budapest 1963 Jun.18-22 Edited by Andras Prekopa 367p 25cm Akademiai kiado, Budapest,1965
Math 3.0917

BOLYAI JANOS MATEMATIKAI TARSULAT Colloquium on combinatorial theory Proceedings Balatonfured 1969 Aug 24 Edited by R. Erdos and others Societatis Janos Bolyai. Colloquia mathematica, 4 1201p 24cm 3 vols North-Holland,Amsterdam;London,1970
Math S 3.1656

BOLYAI JANOS MATEMATIKAI TARSULAT Colloquium on information theory Proceedings Debrecen 1967 Sep 19-24 bibliog.,illus. 24cm 2 vols Janos Bolyai mathematical society, Budapest,1968
Math S 3.1457

BOLYAI JANOS MATEMATIKAI TARSULAT Combinatorial theory and its applications : colloquium Proceedings Balatonfured 1969 Aug 24-29 Edited by P. Erdos and others Bolyai Janos matematikai tarsulat.Colloquia mathematica, 4 1201p 24cm 3 vols North-Holland,Amsterdam,1970
P Math 2.4509

BOLYAI JANOS MATEMATIKAI TARSULAT The Foundations of mathematics,mathematical machines and their applications colloquium Tihany 1962 Sep 11-15 Magyar tudomanyos akademia.Section of mathematics and physics Edited by Laszlo Kalmar 317p 25cm Akademiai kiado,Budapest,1965
P. Math 2.2359

BOLYAI JANOS MATEMATIKAI TARSULAT.COLLOQUIA MATHEMATICA, 4 Combinatorial theory and its applications :colloquium Proceedings Balatonfured 1969 Aug 24-29 Bolyai Janos matematikai tarsulat Edited by P. Erdos and others 1201p 24cm 3 vols North-Holland, Amsterdam,1970
P Math 2.4509

Le BOMBARDEMENT IONIQUE theories et applications:colloques internationaux Bellevue,Seine et Oise 1962 Dec 4-8 Centre national de la recherche scientifique Centre national de la recherche scientifique. Colloques internationaux, 113 Editions du centre National de la recherche scientifique, Paris,1962
Met 25.1676

Le BOMBARDEMENT IONIQUE,THEORIES ET APPLICATIONS 113 colloque international Actes Bellevue 1962 Dec Centre national de la recherche scientifique 243p Paris,1962
Cav 7.0169

BOMBAY 1960 Contributions to function theory international colloquium papers presented Tata institute of fundamental research In joint sponsorship with the Sir Dorabji Tata trust xii,231p 24cm Tata institute of fundamental research,Bombay,1960
P. Math 2.1584

BOMBAY 1960 Radioisotopes and radiation in entomology :a symposium Proceedings By H.J. Bhabha and others Sponsored by the International atomic energy agency International atomic energy agency,Vienna,1962
Bal 39.2390

BOMBAY 1962 Radioisotopes in soil plant nutrition studies The Symposium on the use of radioisotopes in soil-plant nutrition studies Proceedings International atomic energy agency United Nations.Food and agriculture organization International atomic energy agency,Vienna,1962
Bot 42.1368

BOMBAY 1964 Differential analysis International colloquium on differential analysis :the Bombay colloquium papers presented By M.F. Atiyah and others Tata institute of fundamental research Jointly sponsored by the International mathematical union Tata institute of fundamental research. Studies in analysis, 2 viii,253p 25cm Oxford university press,Oxford,1964
P. Math 2.0696

BOMBAY 1964 Inelastic scattering of neutrons Symposium on inelastic scattering of neutrons Vol 1-2 International atomic energy agency Edited by M.M. Brown International atomic energy agency,Vienna,1965
Cav 7.0267

BOMBAY 1966 International council of scientific unions general assembly 11th Record 30cm Rome,1966 Mimeograph
Sco 14.0095

BOMBAY 1967 International conference on spectroscopy India.Department of atomic energy.Spectroscopy division 290p International council of scientific unions, Bombay,1968
TA 15.0636

BOMBAY 1968 Algebraic geometry :a colloquium Papers By S.S. Abhyankar and others viii,426p 26cm Oxford university press,London,1969
P Math 2.3376

BOMBAY 1968 Cosmic ray structures in relation to recent developments in astronomy and astrophysics :a colloquium Proceedings Edited by R.R. Daniel and others 374p Tata institute of fundamental research,Bombay,1969
TA 15.0567

BONDED AIRCRAFT STRUCTURES :a conference Papers Cambridge 1957 Ciba foundation Ciba foundation,Duxford,Cambridge,1958
Eng 41.2774

BONDING IN METALLO-ORGANIC COMPOUNDS Cambridge 1969 Mar 25-27 Faraday society Faraday society.Discussions, 47 208p 20cm Faraday society,London,1969
Cav 7.2717

BONE STRUCTURE The Living bone :symposium on bone structure and metabolism Edited by R. Steendikk and others Folia medica neerlandica, 11,5-6 Amsterdam,1968
PGMS 29.0177

BONN 1955 Probleme der fetalen endokrinologie symposium Edited by H. Nowakowski Deutschen gesellschaft fur endokrinologie.Symposia, 3 Springer,Berlin, 1956
An 32.3750

BONN 1960 International congress of psychology 16th Berichte.Proceedings International union of scientific psychology Deutsche gesellschaft fur psychologie Acta psychologica, 19 North-Holland,Amsterdam, 1961 President of congress:professor Metzger.Conference languages:English,French and German
Psy 31.3396

BONN 1960 Brussels 1960 Internationale arbeitstagung uber die Silur Devon-Grenze und die stratigraphie von Silur und Devon 2nd symposiums-band Edited by H.K. Erben Stuttgart,1962
Geol 8.2417

BONN 1965 Critical rotatory dispersion and circular dichroism in organic chemistry NATO summer school Proceedings North Atlantic treaty organization Edited by G. Snatzke Heyden,London,1967
Chem 18.2791

BORDEAUX 1947 Surface chemistry :a discussion at a joint meeting of the Societe de chimie physique and the Faraday society Papers Societe de chimie physique Faraday society Illus Butterworths scientific publications,London,1949 Papers in English and FRENCH
Radioth 35.0342

BORDEAUX 1947 Surface chemistry discussion at a joint meeting...in honour of professor Henri Devaux Papers Faraday society,and, Societe de chimie physique Butterworths, London,1949 Published as a special supplement to 'Research'
Chem 18.0546

BORON The International symposium on Boron Paris 1964 Jul 17-18 Vol 2: preparation, properties,and applications Edited by Gerhart K. Gaule Plenum press,New York,1965
Met 25.0159

BORON-NITROGEN CHEMISTRY :an international symposium Durham,N.C. 1963 Apr 23-25 United States.Army.Research office American chemical society.Advances in chemistry series, 42 American chemical society,Washington,D.C., 1964 Symposium chairman:Kurt Niedenzu
Chem 18.0986

BORON-NITROGEN CHEMISTRY :an international symposium Durham,N.C. 1963 Apr 23-25 United States.Army research office,Durham American chemical society.Advances in chemistry series, 42 American chemical society,Washington,D.C.,1964 Symposium chairman:K.Niedenzu
Chem 18.2658

BOROUGH POLYTECHNIC,LONDON Quality control in metal finishing based on a symposium at the Borough Polytechnic London Edited by G. Isserlis Columbine press,Manchester;London, 1967
Met 25.1983

BOSTON 1949 Conference on the preservation of the formed elements and of the proteins of the blood Transactions American red cross 1949
Med 36.0005

BOSTON 1956 International congress of hematology 6th Proceedings International society of hematology Edited by A. Richardson Jones Grune and Stratton, New York,1958
Med 36.0202

BOSTON 1958 Symposium on basic mechanisms of fatigue presented before the 61st annual meeting American society for testing materials A.S.T.M.Special technical publication, 273 A.S.T.M.,Philadelphia,1959
Met 25.2328

BOSTON 1963 The Lunar surface layer materials and characteristics :a conference proceedings Edited by John W. Salisbury and Peter E. Glaser Financially supported by Air Force Cambridge Research Laboratories 532p Academic Press,New York,1964
Obs 6.1507

BOSTON 1967 Metal matrix composites a symposium presented at the 70th annual meeting. American society for testing materials A.S.T.M.Special technical publication, 438 A. S.T.M.,Philadelphia,1968
Met 25.2337

BOSTON,MASS 1953 Antimetabolites and cancer :a symposium American association for the advancement of science Edited by Cornelius P. Rhoades American association for the advancement of science,Washington,D.C., 1955
Radioth 35.0735

BOSTON,MASS. High speed testing Vol 1: a symposium held at Boston,Massachusetts, December 8,1958 Sponsored by Plas-tech equipment corporation Interscience,New York, 1968
Eng 41.3845

BOSTON,MASS. High speed testing Vol 3: third annual symposium held at Boston, Massachusetts,October 26 and 27,1961 Sponsored by Plas-tech equipment corporation Interscience,New York,1962
Eng 41.3848

BOSTON,MASS. High speed testing Vol 5: 5th international symposium held at Boston, Massachusetts,March 8 and 9,1965 Sponsored by Plas-tech equipment corporation Applied polymer symposia, 1 Interscience,New York, 1965
Eng 41.3849

BOSTON,MASS. 1954 Vacuum metallurgy symposium Papers Electrochemical society. Electrothermics and metallurgy division Edited by John M. Blocher Electrochemical society inc.,Boston,Mass.,1955
Met 25.2258

BOSTON,MASS. 1955 Automatic data processing conference proceedings Harvard university.Graduate school of business administration Edited by Robert N. Anthony illus. viii,194p 21cm Harvard university, Boston,Mass.,1956
Math L 5.0752

BOSTON,MASS. 1955 London 1955 Oct 25-27 Joint conference on combustion Proceedings Institution of mechanical engineers and American society of mechanical engineers viii,457p Institution of mechanical engineers,London,1955
Chem E 24.1376

BOSTON,MASS. 1955 and LONDON 1955 Joint conference on combustion Proceedings Institution of mechanical engineers American society of mechanical engineers Institute of mechanical engineers,London,1956
Eng 41.7231

BOSTON,MASS. 1958 Annual arctic planning session 1st Proceedings Air force Cambridge research center.Geophysics research directorate Edited by J.H. Hartshorn Air force Cambridge research center.G.R.D.research notes,15 AFCRC-TN-59-256,AD 212265 illus viii,65p A.F.C.R.C.,Bedford,Mass.,1959
Sco 14.7404

BOSTON,MASS. 1958 Conference on extremely high temperatures Air force Cambridge research center Edited by Heinz Fischer and Lawrence C. Mansur Wiley;Chapman and Hall, New York;London,1958
Met 25.0249

BOSTON,MASS. 1958 Symposium on basic mechanisms of fatigue Papers American society for testing materials A.S.T.M. Special technical publication, 237 A.S.T.M., Philadelphia,1959
Met 25.0925

BOSTON,MASS. 1959 Exploding wires :a conference Papers Air force Cambridge research center.Geophysics research directorate Edited by William G. Chace and Howard K. Moore With the cooperation of Lowell technological institute research foundation Plenum press,New York,1959
Chem 18.0191

BOSTON,MASS. 1959 Properties of elemental and compound semiconductors a technical conference American institute of mining, metallurgical and petroleum engineers Edited by Harry C. Gatos Metallurgical society conferences, 5 Interscience,New York; London,1960
Met 25.1160

BOSTON,MASS. 1959 Symposium on electron beam melting 1st Proceedings Edited by James S. Hetherington Sponsored by Alloyd research corporation Alloyd,Watertown,Mass., 1959
Eng 41.5613

BOSTON,MASS. 1960 Symposium on electron beam processes 2nd Proceedings Edited by Robert Bakish Sponsored by the Alloyd research corporation Alloyd,Cambridge,Mass., 1960
Eng 41.5614

BOSTON,MASS. 1960,Aug. 29-31 Metallurgy of elemental and compound semiconductors a technical conference Proceedings American institute of mining,metallurgical and petroleum engineers Edited by Ralph O. Grubel Metallurgical society conferences, 12 Interscience,New York;London,1961
Met 25.1146

BOSTON,MASS. 1961 Sensory communication Symposium on principles of sensory communication Contributions Edited by W.A. Rosenblith Held at the Massachusetts institute of technology M.I.T.press;Wiley, Boston,Mass.;New York,1961
Phys 20.2108

BOSTON,MASS. 1961 Symposium on electron beam processes 3rd Proceedings Edited by Robert Bakish Sponsored by Alloyd electronics corporation Alloyd,Cambridge, Mass.,1961
Eng 41.5615

BOSTON,MASS. 1961 Technological planning on the corporate level a conference Proceedings Harvard business school Edited by James R. Bright Harvard university,Boston, 1962
Met 25.2169

BOSTON,MASS. 1961 The Finishing of aluminum :symposium Proceedings American electroplaters' society Edited by G.H. Kissin Reinhold,New York,1963
Met 25.2796

BOSTON,MASS. 1961 The Symposium on electron beam processes 3rd Proceedings Alloyd electronics corporation,Cambridge,Mass Edited by R. Bakish 379p Alloyd electronics,Cambridge,Mass.,1961
Cav 7.0113

BOSTON,MASS. 1961 Ultrapurification of semiconductor materials :a conference Proceedings Air force Cambridge research laboratories.Electronics research directorate Edited by M.S. Brooks and J.K. Kennedy Macmillan,New York,1962
Met 25.1149

BOSTON,MASS. 1961,1962 Boston studies in the philosophy of science Boston colloquium 1961-62 Proceedings Vol 1 Edited by Marx W. Wartofsky viii,212p Reidel, Dordrecht,1963
WSM 43.0747

BOSTON,MASS. 1962 Electromagnetic aspects of hypersonic flight Symposium on the plasma sheath :its effect upon reentry communication and detection 2nd Proceedings Edited by Walter Rotman and others Sponsored by the Air force Cambridge research laboratories. Electromagnetic radiation laboratory Spartan books;Cleaver-Hume,Baltimore,Md.;London,1964
Eng 41.4544

BOSTON,MASS. 1962 Symposium on electron beam technology 4th Proceedings Edited by R. Bakish Sponsored by the Alloyd electronics corporation Alloyd,Cambridge, Mass.,1962
Eng 41.5616

BOSTON,MASS. 1963 Metabolic control mechanisms in animal cells :a symposium Proceedings National cancer institute Edited by William J. Rutter Sponsored by the Tissue culture association and National cancer institute National cancer institute. Monograph, 13 illus National cancer institute,Bethesda,Md.,1964
Bioch 33.0984

BOSTON,MASS. 1963 Symposium on fundamental phenomena in the material sciences 1st Proceedings Vol 1: sintering and plastic deformation Ilikon corporation,Natick Edited by L.J. Bonis and H.H. Hausner Plenum press,New York,1964
Met 25.2280

BOSTON,MASS. 1964 Models for the perception of speech and visual form :a symposium Proceedings Air force Cambridge research laboratories.Data science laboratory Edited by Weiant Wathen-Dunn MIT press, Cambridge,Mass.,1967
Psy 31.2951

BOSTON,MASS. 1964,1965 Fundamental phenomena in the materials sciences :a symposium 2nd,3rd Vol 2-3: surface phenomena Ilikon corporation,Natick Edited by L.J. Bonis and others 2 vols Plenum press,New York,1966
Met 25.1616

BOSTON,MASS. 1964-66 Boston studies in the philosophy of science Boston colloquium 1964-1966 Proceedings Vol 3: in memory of N.R. Hanson Edited by R.S. Cohen and Marx W. Wartofsky xlix,489p Humanities press,New York,1968
WSM 43.0749

BOSTON,MASS. 1966 Crystal growth International conference on crystal growth Proceedings International union of pure and applied physics and Air force Cambridge research laboratories Edited by H.Steffen Peiser Physics and chemistry of solids. Supplement Pergamon press,Oxford,1967
Met 25.1520

BOSTON,MASS. 1966 Fundamental phenomena in the materials sciences :a symposium 4th Proceedings Vol 4: fracture of metals, polymers,and glasses Ilikon corporation, Natick Edited by L.J. Bonis and others Plenum press,New York,1967
Met 25.0873

BOSTON,MASS. 1966 International conference on operational research 4th Proceedings Edited by David B. Hertz and Jacques Melese Organised by the International foundation of operational research societies Operations research society of America.Publications in operations research, 15 Wiley-Interscience, New York,1966
Eng 41.1219

BOSTON,MASS. 1966-68 Boston studies in the philosophy of science Boston colloquium 1966-1968 Proceedings Vol 4-5 Edited by R.S. Cohen and Marx W. Wartofsky 2 vols Reidel,Dordrecht,1969
WSM 43.0750

BOSTON,MASS. 1967 Electron fractography American society for testing materials A.S.T. M.Special technical publication, 436 A.S.T.M. Philadelphia,Pa.,1916
Met 25.2248

BOSTON,MASS. 1967 Metal corrosion in the atmosphere :symposium Proceedings American society for testing and materials American society for testing and materials.Special technical publication, 435 Philadelphia,Pa., 1968 Symposium presented at the annual meeting
Met 25.2792

BOSTON,MASS. 1968 Summer institute on symbolic mathematical computation Proceedings International business machines corporation Edited by Robert G. Tobey illus 324p IBM Boston programming center, Cambridge,Mass.,1969
Math L 5.3809

BOSTON,MASS. 1969 Rheology of solids International symposium on high speed testing 7th Proceedings Edited by R.D. Andrews and F.R. Eirich Applied polymer symposia, 12,High speed testing, 7 Interscience,New York,1969
Met 25.2581

BOSTON,MASS. 1971 Hardware,software, firmware trade-offs IEEE-International computer society conference 5th Proceedings Institute of electrical and electronics engineers International computer society illus 204p IEEE,New York,1971
Math L 5.3628

BOSTON CITY HOSPITAL Sensory deprivation :a symposium held at Harvard medical school Cambridge,Mass. 1958 Jun 20-21 Edited by Philip Solomon and others Co-sponsored by the United States.Office of naval research. Physiological psychology branch Harvard university press,Cambridge,Mass.,1965
Psy 31.0128

BOSTON CITY HOSPITAL.HARVARD MEDICAL UNIT Training of the physician :celebration of the Centennial of the Boston City hospital and the 40th anniversary of the Thorndike memorial laboratory Papers Boston,Mass.,1964
PGMS 29.0489

The BOSTON COLLOQUIUM 4th Lectures Boston,Mass. 1903 Sep 2-5 By Edward Burr van Vleck and others American mathematical society American mathematical society. Colloquium lectures, 1 xii,187p 21cm University microfilms,Ann Arbor,Mich.,1967 Facsimile produced by microfilm-xerography of the original volume published in 1905 for the American mathematical society by Macmillan,New York
P Math 2.2865

BOSTON COLLOQUIUM 1961-62 Proceedings Boston studies in the philosophy of science Boston,Mass. 1961,1962 Vol 1 Edited by Marx W. Wartofsky viii,212p Reidel, Dordrecht,1963
WSM 43.0747

BOSTON COLLOQUIUM 1964-1966 Proceedings Boston studies in the philosophy of science Boston,Mass. 1964-66 Vol 3: in memory of N.R.Hanson Edited by R.S. Cohen and Marx W. Wartofsky xlix,489p Humanities press,New York,1968
WSM 43.0749

BOSTON COLLOQUIUM 1966-1968 Proceedings Boston studies in the philosophy of science Boston,Mass. 1966-68 Vol 4-5 Edited by R.S. Cohen and Marx W. Wartofsky 2 vols Reidel,Dordrecht,1969
WSM 43.0750

BOSTON STUDIES IN THE PHILOSOPHY OF SCIENCE Boston colloquium 1961-62 Proceedings Boston,Mass. 1961,1962 Vol 1 Edited by Marx W. Wartofsky viii,212p Reidel, Dordrecht,1963
WSM 43.0747

BOSTON STUDIES IN THE PHILOSOPHY OF SCIENCE
Boston colloquium 1964-1966 Proceedings Boston,Mass. 1964-66 Vol 3: in memory of N.R.Hanson Edited by R.S. Cohen and Marx W. Wartofsky xlix,489p Humanities press,New York,1968
WSM 43.0749

BOSTON STUDIES IN THE PHILOSOPHY OF SCIENCE
Boston colloquium 1966-1968 Proceedings Boston,Mass. 1966-68 Vol 4-5 Edited by R.S. Cohen and Marx W. Wartofsky 2 vols Reidel,Dordrecht,1969
WSM 43.0750

BOTANIC GARDENS Multiples fonctions d'un jardin botanique :symposium international de Geneve Actes Geneva 1968 Jul 29-Aug 3 Jardin botanique de Geneve Edited by Jacques Miege Boissiera, 14 Geneva,1969 Symposium held to celebrate the 150th anniversary of the foundation of the botanic garden at Geneva
BG 38.3115

BOTANIC GARDENS Organization scientifique des jardins botaniques Colloque international de l'union international des sciences biologiques Paris 1953 Jun 4-7 Under the auspices of Unesco International union of biological sciences.Series B: Colloquia, 13 International union of biological sciences,Paris,1953
BG 38.3110

BOTANICAL SOCIETY OF AMERICA Plant biology today:advances and challenges :a symposium Denver,Colo. 1961 Dec Edited by William A. Jensen and Leroy G. Kavaljian illus Macmillan,London,1963
Bot 42.0298

BOTANICAL SOCIETY OF THE BRITISH ISLES A Darwin centenary :a conference Report London 1959 Nov 27-28 Edited by P.J. Wanstall Botanical society of the British Isles.B.S.B.I.conference reports, 6 port. London,1961 Centenary of the publication of the Origin of species
Bot 42.4711

BOTANICAL SOCIETY OF THE BRITISH ISLES
British flowering plants and modern systematic methods The Conference on the study of critical British groups Report London 1948 Apr 9-10 Edited by A.J. Wilmott Botanical society of the British Isles.B.S.B.I. conference reports, 1 London,1949
Bot 42.4706

BOTANICAL SOCIETY OF THE BRITISH ISLES Local floras :a conference Proceedings London 1961 Nov 24-25 Edited by P.J. Wanstall Botanical society of the British Isles.B.S.B.I. conference reports, 7 London,1963
Bot 42.4712

BOTANICAL SOCIETY OF THE BRITISH ISLES
Progress in the study of the British flora :a conference Report London 1956 Apr 13-15 Edited by J.E. Lousley Botanical society of the British Isles.B.S.B.I.conference reports, 5 London,1957
Bot 42.4710

BOTANICAL SOCIETY OF THE BRITISH ISLES
Species study in the British flora The Conference on the species concept in its relation to the British flora Report London 1964 Apr 9-11 Edited by J.E. Lousley Botanical society of the British Isles.B.S.B.I.conference reports, 4 London, 1955
Bot 42.4709

BOTANICAL SOCIETY OF THE BRITISH ISLES The Changing flora of Britain :a conference Report London 1962 Apr 4-6 Edited by J.E. Lousley Botanical society of the British Isles.B.S.B.I.conference reports, 3 London, 1953
Bot 42.4708

BOTANICAL SOCIETY OF THE BRITISH ISLES The Study of the distribution of British plants :a conference Report Oxford 1950 Mar 31-Apr 2 Botanical society of the British Isles.B.S.B.I.conference reports, 2 London,1951
Bot 42.4707

BOTANICAL SOCIETY OF THE BRITISH ISLES.B.S.B.I. CONFERENCE REPORTS, 1 British flowering plants and modern systematic methods The Conference on the study of critical British groups Report London 1948 Apr 9-10 Botanical society of the British Isles Edited by A.J. Wilmott London,1949
Bot 42.4706

BOTANICAL SOCIETY OF THE BRITISH ISLES.B.S.B.I. CONFERENCE REPORTS, 2 The Study of the distribution of British plants :a conference Report Oxford 1950 Mar 31-Apr 2 Botanical society of the British Isles London,1951
Bot 42.4707

BOTANICAL SOCIETY OF THE BRITISH ISLES.B.S.B.I. CONFERENCE REPORTS, 3 The Changing flora of Britain :a conference Report London 1962 Apr 4-6 Botanical society of the British Isles Edited by J.E. Lousley London,1953
Bot 42.4708

BOTANICAL SOCIETY OF THE BRITISH ISLES.B.S.B.I. CONFERENCE REPORTS, 4 Species study in the British flora The Conference on the species concept in its relation to the British flora Report London 1964 Apr 9-11 Botanical society of the British Isles Edited by J.E. Lousley London,1955
Bot 42.4709

BOTANICAL SOCIETY OF THE BRITISH ISLES.B.S.B.I. CONFERENCE REPORTS, 5 Progress in the study of the British flora :a conference Report London 1956 Apr 13-15 Botanical society of the British Isles Edited by J.E. Lousley London,1957
Bot 42.4710

BOTANICAL SOCIETY OF THE BRITISH ISLES.B.S.B.I. CONFERENCE REPORTS, 6 A Darwin centenary :a conference Report London 1959 Nov 27-28 Botanical society of the British Isles Edited by P.J. Wanstall port. London,1961 Centenary of the publication of the Origin of species
Bot 42.4711

BOTANICAL SOCIETY OF THE BRITISH ISLES.B.S.B.I. CONFERENCE REPORTS, 7 Local floras :a conference Proceedings London 1961 Nov 24-25 Botanical society of the British Isles Edited by P.J. Wanstall London,1963
Bot 42.4712

BOTANICAL SOCIETY OF THE BRITISH ISLES.B.S.B.I. CONFERENCE REPORTS, 9 Reproductive biology and taxonomy of vascular plants : conference Report Birmingham 1965 Apr 23-24 Botanical society of the British Isles Edited by J.G. Hawkes 183p Pergamon press, Oxford,1966
Bot 42.0307

BOTANY British flowering plants and modern systematic methods The Conference on the study of critical British groups Report London 1948 Apr 9-10 Botanical society of the British Isles Edited by A.J. Wilmott Botanical society of the British Isles.B.S.B.I. conference reports, 1 London,1949
Bot 42.4706

BOTANY International botanical congress 5th Abstracts,programs Cambridge 1930 Aug 16-22 Cambridge university press, Cambridge,1930
BG 38.3127

BOTANY International botanical congress 7th Abstracts Stockholm 1950 section for palaeobotany c1950 Duplicated. Papers in English,French and German.In folder
Bot 42.6225

BOTANY International botanical congress 7th Proceedings Stockholm 1950 Jul 12-20 Edited by Hugo Osvald and Ewert Aberg Almqvist and Wiksell,Uppsala,1953
Bot 42.4696

BOTANY International botanical congress 7th Proposals,Abstracts Stockholm 1950 International union of biological sciences International union of biological sciences, Utrecht,1950
BG 38.3128

BOTANY International botanical congress 8th Comptes rendus des seances et rapports et communications Paris 1954 Sections 3-27 9 vols Paris,1955-57
Bot 42.4699

BOTANY International botanical congress 8th Rapports et communications Paris 1954 Section 2: 2.serie Utrecht,1955 Reprinted from Taxon,vol.4 no.6,pp.121-177
Bot 42.4698

BOTANY International botanical congress 8th Reports of communications Paris 1954 1954
BG 38.3129

BOTANY International botanical congress 9th Proceedings Montreal 1959 Aug 19-29 Vol 1-3 University of Toronto press, Toronto,1959
Bot 42.4703

BOTANY International botanical congress 9th Proceedings Montreal 1959 Aug 19-29 1959
BG 38.3130

BOTANY International botanical congress 10th Proceedings Edinburgh 1964 Aug 1964
BG 38.3131

BOTANY International botanical congress 11th Proceedings Seattle,Wash. 1969 Aug 24-Sep 2 International botanical congress, Washington,D.C.,1970
BG 38.3132

BOTANY International horticultural exhibition and botanical congress :report of proceedings London 1866 May 22-31 London, 1867
Bot 42.4705

BOTANY Local floras :a conference Proceedings London 1961 Nov 24-25 Botanical society of the British Isles Edited by P.J. Wanstall Botanical society of the British Isles.B.S.B.I.conference reports, 7 London,1963
Bot 42.4712

BOTANY Progress in the study of the British flora :a conference Report London 1956 Apr 13-15 Botanical society of the British Isles Edited by J.E. Lousley Botanical society of the British Isles.B.S.B.I. conference reports, 5 London,1957
Bot 42.4710

BOTANY Species study in the British flora The Conference on the species concept in its relation to the British flora Report London 1964 Apr 9-11 Botanical society of the British Isles Edited by J.E. Lousley Botanical society of the British Isles.B.S.B.I. conference reports, 4 London,1955
Bot 42.4709

BOTANY Summer school of botany Proceedings Darjeeling 1960 Jun 2-15 Ministry of scientific research and cultural affairs,New Dehli,1962
BG 38.3112

BOTANY The Changing flora of Britain :a conference Report London 1962 Apr 4-6 Botanical society of the British Isles Edited by J.E. Lousley Botanical society of the British Isles.B.S.B.I.conference reports, 3 London,1953
Bot 42.4708

BOTANY The Study of the distribution of British plants :a conference Report Oxford 1950 Mar 31-Apr 2 Botanical society of the British Isles Botanical society of the British Isles.B.S.B.I.conference reports, 2 London,1951
Bot 42.4707

BOULDER 1963 Lectures in theoretical physics 6th Vol. 6 University of Colorado.Summer institute for theoretical physics Edited by W.E. Brittin and W.R. Chappell Sponsored by the University of Colorado.Department of physics and astrophysics Lectures in theoretical physics, Boulder.,1963,6 Colorado university press, Boulder,Colo.,1964
A Math 4.0834

BOULDER,COL. 1957 Summer seminar in applied mathematics 1st Proceedings Vol 3: partial differential equations By Lipman Bers and others Edited by Alton S. Householder and others Sponsored by the American Mathematical Society Lectures in applied mathematics, 3 xiii,343p Interscience,New York,1964 Dedicated to Arthur Norman Milgram
A Math 4.1266

BOULDER,COL. 1958-62 Lectures in theoretical physics Vol 1-5 University of Colorado.Summer institute for theoretical physics Edited by Wesley E. Brittin and others Lectures in theoretical physics, 1-5 Interscience,New York,1959-63
Cav 7.0292

BOULDER,COL. 1960 The General theory of quantized fields proceedings Vol. 4 By Res Jost University of Colorado.Summer institute on applied mathematics Conducted by the American Mathematical Society Lectures in applied mathematics, 4 American mathematical society,Providence,R.I., 1965
A Math 4.0760

BOULDER,COL. 1961 London 1962 Jan 8-12 International developments in heat transfer International heat transfer conference Papers Pt 1-5 Arranged by the American society of mechanical engineers 5 vols A.S.M.E.,New York,1961 Co-sponsored by the American institute of chemical engineers,the Institution of mechanical engineers,and the Institution of chemical engineers
Chem E 24.0595

BOULDER,COL. 1964 Lectures in theoretical physics 7th 7a: Lorentz group University of Colorado.Summer institute for theoretical physics Edited by Wesley E. Britten and Asim O. Barut Sponsored by the University of Colorado.Department of physics and astrophysics Lectures in theoretical physics,Boulder,1964,7a University of Colorado press,Boulder,Col.,1965 The Lorentz group symposium volume is dedicated to Professor E.Wigner
A Math 4.1335

BOULDER,COL. 1964 Lectures in theoretical physics 7th Vol. 7c: statistical physics,weak interactions,field theory University of Colorado.Summer institute for theoretical physics Edited by Wesley E. Brittin Sponsored by the University of Colorado.Department of physics and astrophysics Lectures in theoretical physics, Boulder,1964,7c University of Colorado press, Boulder,Col.,1965
A Math 4.0835

BOULDER,COLO. 1957 Lectures in applied mathematics :summer seminar proceedings Vol. 1: probability and related topics in physical sciences By Mark Kac American mathematical society Supported by United States.Armed services xiii,266p 24cm Interscience publishers,London;New York,1959 With special lectures by G.E.Uhlenbeck,A.R. Hibbs,and Balth.van der Pol
Math 3.0844

BOULDER,COLO. 1958 Biophysical science :a study program Edited by J.L. Oncley and others Organized and conducted by the National institutes of health.Biophysics and biophysical chemistry study section Wiley, New York,1959
Bioch 33.0329

BOULDER,COLO. 1958 Contemporary approaches to creative thinking :a symposium Edited by Howard E. Gruber and others Held at the University of Colorado Atherton press; Prentice-Hall International,New York;London, 1964
Eng 41.1268

BOULDER,COLO. 1958 Wolf-Rayet stars :a symposium Proceedings Joint institute for laboratory astrophysics Edited by K.B. Gebbie and R.N. Thomas National bureau of standards.Special publication, 307 277p National bureau of standards,Washington,D.C., 1968
Obs 6.3438

BOULDER,COLO. 1959 Biophysical science :a study program Edited by J.L. Oncley and others Organized and conducted by the National institutes of health.Biophysics and biophysical chemistry study section Wiley, New York,1959 Published also in the Reviews of modern physics,Jan.and Apr.1959
Bal 39.0026

BOULDER,COLO. 1959 Institute in the theory of numbers report American mathematical society Edited by Donald C.B. Marsh and James.H. Jordan With the support of the National science foundation 350p 27cm University of Colorado,Boulder,Colo.,1959
P. Math 2.1773

BOULDER,COLO. 1960 Advances in cryogenic engineering Cryogenic engineering conference 6th Proceedings National bureau of standards University of Colorado Edited by K.D. Timmerhaus Advances in cryogenic engineering, 6 Plenum press,New York,1961
Eng 41.7398

BOULDER,COLO. 1961 Curricula in solid mechanics :first study session in theoretical and applied mechanics curricula Edited by H. Liebowitz and J.M. Allen Prentice-Hall, Englewood Cliffs,N.J.,1962
Eng 41.2567

BOULDER,COLO. 1961 and LONDON 1962 International developments in heat transfer Heat transfer conference 1961-62 Proceedings American society of mechanical engineers American society of mechanical engineers,New York,1963
Eng 41.7168

BOULDER,COLO. 1962 Symposium on Antarctic logistics National academy of sciences. Committee on polar research Under the auspices of the Scientific committee on Antarctic research.WOrking party on logistics illus xvl,778p 28cm Washington,D.C.,1963
Sco 14.6199

BOULDER,COLO. 1963 Conference on electronic computation 3rd Proceedings American society of civil engineers Journal of structural division, 89,no ST4 American society of civil engineers,Kansas city,Miss., 1963
Eng 41.2716

BOULDER,COLO. 1963 Number theory conference proceedings University of Colorado With the support of the National science foundation 121p 27cm University of Colorado,Boulder,Colo.,1963
P. Math 2.1772

BOULDER,COLO. 1965 Algebraic groups and discontinuous subgroups Summer mathematical institute 12th American mathematical society Edited by Armand Borel and George D. Mostow Financed by the National science foundation American mathematical society. Proceedings of symposia in pure mathematics, 9 vii,426p 26cm American mathematical society,Providence,R.I.,1966
P Math 2.2719

BOULDER,COLO. 1965 Bering land bridge International association for quaternary research congress 7th Proceedings Edited by D.M. Hopkins Stanford university press,Stanford,Calif.,1967
Bot 42.3247

BOULDER,COLO. 1965 International association for quaternary research congress 7th guidebook for field conference F.Central and south central Alaska Edited by C.B. Schultz and H.T.U. Smith illus 141p 20cm Lincoln,Neb.,1965
Sco 14.0143

BOULDER,COLO. 1965 International association for quaternary research congress 7th Proceedings Vol 10: arctic and alpine environments International association for quaternary research Edited by W.H. Osburn and H.E. Wright Sponsored by the National research council illus,maps xi,308p Indiana university press,Bloomington, Ind.;London,1968 For other volumes of these proceedings,see also International union for quaternary research
Bot 42.2006

BOULDER,COLO. 1965 International association for quaternary research congress 7th Proceedings 10: arctic and alpine environments International association for quaternary research Edited by W.H. Osburn and H.E. Wright Sponsored by the National research council illus,maps xi,308p 25cm Indiana university press,Bloomington,Ind.; London,1968 For other volumes of these proceedings,see also International union for quaternary research
Sco 14.7841

BOULDER,COLO. 1965 International studies on the quaternary Edited by H.E. Wright and David G. Frey Sponsored by the National research council Geological society of America.Special papers, 84 Geological society of America,New York,1965
Bot 42.3133

BOULDER,COLO. 1965 International studies on the quaternary International association for quaternary research :congress 7th Papers Edited by H.E. Wright and David G. Frey Sponsored by the National research council Geological society of America. Special papers,84 Geological society of America,New York,1965
Geog 13.0309

BOULDER,COLO. 1965 International union for quaternary research congress :a collection of papers derived from the INQUA-NCAR symposium 7th Proceedings 5: causes of climatic change International union for quaternary research,and,National center for atmospheric research Edited by J.Murray Mitchell Sponsored by the National research council Meteorological monographs,8,no 30 illus iv, 159p 29cm American meteorological association,Boston,Mass.,1968 For other volumes of these proceedings,see also 'International association for quaternary research'
Sco 14.7750

BOULDER,COLO. 1965 Means of correlation of quaternary successions International association for quaternary research congress 7th Proceedings Edited by R.B. Morrison and H.E. Wright Utah university press,Salt Lake City,Utah,1968
Bot 42.3246

BOULDER,COLO. 1965 Pleistocene extinctions: the search for a cause International association for quaternary research.Congress 7th Proceedings I.N.Q.U.A. Edited by P.S. Martin and H.E. Wright Sponsored by the National research council Yale university press,New Haven,Conn.;London,1967
Geog 13.1320

BOULDER,COLO. 1965 Pleistocene extinctions: the search for a cause International association for quaternary research congress 7th Proceedings Vol 6 Edited by P. S. Martin and H.E. Wright Sponsored by the National research council Yale university press,New Haven,Conn.;London,1967
Bot 42.3244

BOULDER,COLO. 1965 Progress in oceanography International association for quaternary research congress 7th Proceedings Vol 4 Edited by M. Sears Pergamon,Oxford,1967
Bot 42.3248

BOULDER,COLO. 1965 Quaternary paleoecology International association for quaternary research.Congress 7th Proceedings Vol 7 Edited by E.J. Cushing and H.E. Wright Sponsored by the National research council Yale university press,New Haven,Conn.;London, 1967
Geog 13.1318

BOULDER,COLO. 1965 Quaternary paleoecology International association for quaternary research congress 7th Proceedings Vol 7 Edited by E.J. Cushing and H.E. Wright Sponsored by the National research council Yale university press,New Haven,Conn.;London, 1967
Bot 42.3245

BOULDER,COLO. 1965 The Quaternary of the United States :review volume for the 7th Congress of the International association for quaternary research Edited by H.E. Wright and David G. Frey Bibliog. Princeton university press,Princeton,N.J.,1965
Bot 42.6691

BOULDER,COLO. 1965 The Quaternary of the United States :review volume for the 7th congress of the International association for quaternary research Edited by H.E. Wright and David C. Frey bibliog. Princeton university press,Princeton,N.J.,1965
Geol 8.2527

BOULDER,COLO. 1966 Neurosciences :a study program Neurosciences research program. Intensive study program Edited by Gardner C. Quarton and others Rockefeller university press,New York,1967
Bioch 33.2220

BOULDER,COLO. 1968 Elementary particle physics Boulder conference on particle physics held during the first part of the 11th Boulder summer institute for theoretical physics 5th Proceedings Vol 1-2 University of Colorado.Department of physics and astrophysics Edited by Kalyana T. Mahanthappa and others Lectures in theoretical physics, 11a,pt 1-2 Bibliog, diagrms xxi,629p 23cm 2 vols Gordon and Breach,New York,1969 Dedicated to George Gamow
Cav 7.3008

BOULDER,COLO. 1968 Mathematical methods in theoretical physics Boulder summer institute for theoretical physics 11th Proceedings University of Colorado.Department of physics and astrophysics Edited by Kalyana T. Mahanthappa and Wesley E. Brittin Lectures in theoretical physics, 11d diagrms xiv, 648p 23cm Gordon and Breach,New York,1969
Cav 7.3011

BOULDER,COLO. 1968 Quantum fluids and nuclear matter Boulder summer institute for theoretical physics 11th Proceedings University of Colorado.Department of physics and astrophysics Edited by Kalyana T. Mahanthappa and Wesley E. Brittin Lectures in theoretical physics, 11b diagrms xiv, 428p 23cm Gordon and Breach,New York,1969 Dedicated to George Gamow
Cav 7.3009

BOULDER,COLO. 1968 Resonance lines in astrophysics :manuscripts presented at a conference... National center for atmospheric research.High altitude observatory Washburn observatory 480p National center for atmospheric research,Boulder,Colo.,1968
Obs 6.3349

BOULDER,COLO. 1968 Symposium on atomic collision processes Boulder summer institute for theoretical physics 11th Proceedings University of Colorado.Department of physics and astrophysics Edited by Kalyana T. Mahanthappa and others Lectures in theoretical physics, 11c diagrms xiii, 337p 23cm Gordon and Breach,London,1969 Dedicated to George Gamow
Cav 7.3010

BOULDER,COLO. 1968 Wolf Rayet stars :a symposium Proceedings Joint institute for laboratory astrophysics Edited by K.B. Gebbie and R.N. Thomas National bureau of standards.Special publication, 307 277p National bureau of standards,Washington,D.C., 1968
TA 15.0568

BOULDER,COLO. 1969 Communication in development :a symposium Society for developmental biology Edited by Anton Lang Society for developmental biology.Symposia, 28 Developmental biology.Supplement, 3 Academic press,New York;London,1969
An 32.5441

BOULDER,COLO. 1969 Communication in developpment :symposium Society for developmental biology Society for developmental biology.Symposium, 28 Academic press,New York,1969
Bioch 33.2255

BOULDER,COLO.,1957 Lectures on fluid mechanics By Sydney Goldstein Lectures in applied mathematics:proceedings of the summer seminar,Boulder, Colo. 1957, 2 xvi,307p Interscience,New York,1960 Seminar held under the joint sponsorship of the University of Colorado, and the American mathematical society.
Chem E 24.0355

BOULDER CONFERENCE ON PARTICLE PHYSICS 5th held during the first part of the 11th Boulder summer institute for theoretical physics Proceedings Elementary particle physics Boulder,Colo. 1968 Jun 17-Aug 23 Vol 1-2 University of Colorado.Department of physics and astrophysics Edited by Kalyana T. Mahanthappa and others Lectures in theoretical physics, 11a,pt 1-2 Bibliog, diagrms xxi,629p 23cm 2 vols Gordon and Breach,New York,1969 Dedicated to George Gamow
Cav 7.3008

BOULDER SUMMER INSTITUTE FOR THEORETICAL PHYSICS 11th Proceedings Mathematical methods in theoretical physics Boulder,Colo. 1968 Jun 17-Aug 23 University of Colorado. Department of physics and astrophysics Edited by Kalyana T. Mahanthappa and Wesley E. Brittin Lectures in theoretical physics, 11d diagrms xiv,648p 23cm Gordon and Breach,New York,1969
Cav 7.3011

BOULDER SUMMER INSTITUTE FOR THEORETICAL PHYSICS 11th Proceedings Quantum fluids and nuclear matter Boulder,Colo. 1968 Jun 17-Aug 23 University of Colorado.Department of physics and astrophysics Edited by Kalyana T. Mahanthappa and Wesley E. Brittin Lectures in theoretical physics, 11b diagrms xiv, 428p 23cm Gordon and Breach,New York,1969 Dedicated to George Gamow
Cav 7.3009

BOULDER SUMMER INSTITUTE FOR THEORETICAL PHYSICS 11th Proceedings Symposium on atomic collision processes Boulder,Colo. 1968 Jun 17-Aug 23 University of Colorado.Department of physics and astrophysics Edited by Kalyana T. Mahanthappa and others Lectures in theoretical physics, 11c diagrms xiii, 337p 23cm Gordon and Breach,London,1969 Dedicated to George Gamow
Cav 7.3010

BOULDEP SUMMER INSTITUTE FOR THEORETICAL PHYSICS 11TH Elementary particle physics Boulder conference on particle physics held during the first part of the 11th Boulder summer institute for theoretical physics 5th Proceedings Boulder,Colo. 1968 Jun 17-Aug 23 Vol 1-2 University of Colorado.Department of physics and astrophysics Edited by Kalyana T. Mahanthappa and others Lectures in theoretical physics, 11a,pt 1-2 Bibliog, diagrms xxi,629p 23cm 2 vols Gordon and Breach,New York,1969 Dedicated to George Gamow
Cav 7.3008

BOUNDARY LAYER EFFECTS IN AERODYNAMICS :a symposium 1955 National physical laboratory London,1955
Eng 41.6960

BOUNDARY LAYER EFFECTS IN AERODYNAMICS 5th : a symposium Proceedings Teddington 1955 Mar 31-Apr 1 National physical laboratory illus,diagrms H.M.S.O.,London, 1955 Mimeograph
Chem E 24.0229

BOUNDARY LAYER EFFECTS IN AERODYNAMICS: A SYMPOSIUM proceedings Teddington 1955 March 31 National physical laboratory 28cm London,1955
Philos 1.0437

BOUNDARY LAYER RESEARCH (GRENZSCHICHTFORSCHUNG) a symposium Proceedings Freiburg 1957 Aug 26-29 International union of theoretical and applied mechanics Edited by H. Gortler Springer,Berlin,1958
Eng 41.6961

BOUNDARY PROBLEMS IN DIFFERENTIAL EQUATIONS symposium proceedings Madison,Wis. 1959 Apr.20-22 United States.Army. Mathematics research center Edited by Rudolph E. Langer United States army. Mathematics research center.Publications, 2 x,320p 23cm University of Wisconsin press, Madison,Wis.,1960
Math L 5.0281

BOYCE THOMPSON INSTITUTE FOR PLANT RESEARCH Plant growth regulation International conference on plant growth regulation 4th Yonkers 1959 Aug 10-14 Iowa state university press,Ames,Iowa,1961
Bot 42.1830

BOZEMAN,MONTANA 1962 Symposium on listeric infection 2nd Papers and discussions Montana state college Edited by M.L. Gray With the support of the National institute of allergy and infectious diseases Montana state college,Bozeman,Montana,1963
Path 30.2590

BRADYKININ AND RELATED KININS;CARDIOVASCULAR, BIOCHEMICAL,AND NEURAL ACTIONS An International symposium on the cardiovascular and neural actions of bradykinin and related kinins Castel di Poggio 1969 Jul 21-25 Edited by F. Sicuteri and others Advances in experimental medicine and biology, 8 Plenum,New York;London,1970
Inv Med 37.0196

BRAIN AND BEHAVIOR Brain and behavior,vol.1 Conference on brain and behavior 1st Proceedings Los Angeles,Calif. 1961 Vol 1 Edited by Mary A.B. Brazier American institute of biological sciences,Washington,D. C.,1961
An 32.4382

BRAIN AND BEHAVIOR Brain and behavior,vol.2: the internal environment and alimentary behavior Conference on brain and behavior 2nd Proceedings Los Angeles,Calif. 1962 Feb 18-21 Edited by Mary A.B. Brazier American institute of biological sciences, Washington,D.C.,1963
An 32.4383

BRAIN AND BEHAVIOR Brain function conference 3rd Proceedings Los Angeles 1963 Vol 3: brain and gonadal function Brain research institute American institute of biological sciences Edited by Roger A. Gorski and Richard E. Whalen With the support of the Human ecology fund UCLA forum in medical sciences, 3 University of California press,Berkeley;Los Angeles,1966
Psy 31.0621

BRAIN AND BEHAVIOR,VOL.1 Conference on brain and behavior 1st Proceedings Los Angeles,Calif. 1961 Vol 1 Edited by Mary A.B. Brazier American institute of biological sciences,Washington,D.C.,1961
An 32.4382

BRAIN AND CONSCIOUS EXPERIENCE :study week Vatican 1964 Sep 28-Oct 4 Pontificia academia scientiarum Edited by John Carew Eccles Springer,Berlin,1966
Phys 20.2041

BRAIN AND CONSCIOUS EXPERIENCE Vatican 1964 Sep 28-Oct 4 Pontificia academia scientiarum Edited by John C. Eccles Springer,Berlin,1966
Pha 16.0309

BRAIN AND GONADAL FUNCTION Brain and behavior Brain function conference 3rd Proceedings Los Angeles 1963 Brain research institute American institute of biological sciences Edited by Roger A. Gorski and Richard E. Whalen With the support of the Human ecology fund UCLA forum in medical sciences, 3 University of California press,Berkeley;Los Angeles,1966
Psy 31.0621

The BRAIN AND HUMAN BEHAVIOR :1956 meeting Proceedings New York 1956 Dec 7-8 Association for research in nervous and mental disease Association for research in nervous and mental disease.Research publications, 36 Williams and Wilkins,Baltimore,Md.,1958
Psy 31.0494

The BRAIN AND HUMAN BEHAVIOUR Proceedings of the Association New York 1956 Dec 7-8 Association for research in nervous and mental disease Edited by Harry C. Solomon and others Association for research in nervous and mental disease.Research publications, 36 Williams and Wilkins,Baltimore,Md.,1958
An 32.4353

BRAIN BARRIER SYSTEMS :a conference Amsterdam 1967 Sep 26-30 Edited by A. Lajtha and D.H. Ford Progress in brain research, 29 Elsevier,Amsterdam,1968
An 32.5233

BRAIN FUNCTION Aggression and defence:neural mechanisms and social patterns (brain function, vol 5) Conference on brain function 5th Proceedings Pacific palisades 1965 Nov 14-17 Edited by Carmine D. Clemente and Donald B. Lindsley UCLA forum in medical sciences, 7 University of California press,Berkeley; Los Angeles,Calif.,1967
An 32.5475

BRAIN FUNCTION 1st :conference Proceedings Los Angeles,Calif. 1961 Vol 1: cortical excitability and steady potentials;relations of basic research to space biology Brain research institute and American institute of biological sciences Edited by Mary A.B. Brazier U.C.L.A. forum in medical sciences, 1 University of California press,Berkeley;Los Angeles,1963
Psy 31.0255

BRAIN FUNCTION 2nd conference Proceedings Los Angeles 1962 Vol 2: RNA and brain function memory and learning Brain research institute and American institute of biological sciences Edited by Mary A.B. Brazier With the support of the United States.Air force.Office of scientific research U.C.L.A. forum in medical sciences, 2 University of California press,Berkeley;Los Angeles,1964
Psy 31.0246

BRAIN FUNCTION 3rd :a conference Proceedings Los Angeles,Calif. 1963 Nov Vol 3: speech,language and communication Brain research institute Edited by Edward C. Carterette With the support of the United States.Air force.Office of scientific research UCLA forum in medical sciences, 4 University of California press,Berkeley;Los Angeles,1966
Psy 31.0458

BRAIN FUNCTION 3rd conference Proceedings Brain and behavior Los Angeles 1963 Vol 3: brain and gonadal function Brain research institute American institute of biological sciences Edited by Roger A. Gorski and Richard E. Whalen With the support of the Human ecology fund UCLA forum in medical sciences, 3 University of California press,Berkeley;Los Angeles,1966
Psy 31.0621

BRAIN MECHANISM AND LEARNING :a symposium Montevideo 1959 Aug 2-8 Council for international organizations of medical sciences Edited by J.F. Delafresnaye Under the auspices of Unesco Blackwell,Oxford,1960 Under the joint auspices of Unesco and WHO
Psy 31.0608

BRAIN MECHANISM AND LEARNING :a symposium Montevideo 1959 Aug 2-8 Council for international organizations of medical sciences Under the joint auspices of Unesco World health organization Blackwell,Oxford, 1961
Bal 39.1604

BRAIN MECHANISMS :international colloquium... on specific and unspecific mechanisms of sensory motor integration Papers Pisa 1961 Edited by Giuseppe Moruzzi and others Sponsored by the International brain research organisation Progress in brain research, 1 Elsevier,Amsterdam,1963
An 32.5230

BRAIN MECHANISMS AND CONSCIOUSNESS :a symposium Laurentian symposium Council for international organizations of medical sciences Edited by J.F. Delafresnaye Blackwell,Oxford,1954
Psy 31.0519

BRAIN MECHANISMS AND CONSCIOUSNESS :a symposium Ste.Marguerite,Que. 1953 Aug 23-28 Council for international organizations of medical sciences Blackwell, Oxford,1954
Phys 20.2009

BRAIN MECHANISMS AND CONSCIOUSNESS :a symposium Ste.Marguerite,Que. 1953 Aug 23-28 Council for international organizations of medical sciences Edited by H.H. Jasper and others Blackwell,Oxford,1954
An 32.4625

BRAIN MECHANISMS AND DRUG ACTION :a symposium: fourth annual general meeting of the Houston neurological society Papers Houston, Texas 1956 Mar 16 Houston neurological society Edited by William S. Fields C.C. Thomas,Springfield,Ill.,1957
Psy 31.3371

BRAIN MECHANISMS AND LEARNING :a symposium Montevideo 1959 Aug 2-8 By J.F. Delafresnaye Council for international organizations of medical sciences Blackwell, Oxford,1961
An 32.4647

BRAIN MECHANISMS AND LEARNINGS Montevideo 1959 Aug 2-8 Edited by J.F. Pelafresnaye Organised by Council for international organisations of medical sciences Blackwell, Oxford,1961
VA 19.0236

BRAIN MECHANISMS UNDERLYING SPEECH AND LANGUAGE Princeton,N.J. 1965 Nov Edited by F.L. Darley and Millikan Grune and Stratton,New York,1967
Psy 31.0460

BRAIN REFLEXES :the international conference dedicated to the centenary celebration of the publication of I.M.Sechenov's book "Brain reflexes" Moscow Edited by E.A. Asratyan Sponsored by Akademiya nauk S.S.S.R. Progress in brain research, 22 Elsevier, Amsterdam,1968 Sponsored also by the International brain research organization
An 32.4243

BRAIN RESEARCH INSTITUTE Brain and behavior Brain function conference 3rd Proceedings Los Angeles 1963 Vol 3: brain and gonadal function Edited by Roger A. Gorski and Richard E. Whalen With the support of the Human ecology fund UCLA forum in medical sciences, 3 University of California press, Berkeley;Los Angeles,1966
Psy 31.0621

BRAIN RESEARCH INSTITUTE Brain function :a conference 3rd Proceedings Los Angeles, Calif. 1963 Nov Vol 3: speech,language and communication Edited by Edward C. Carterette With the support of the United States.Air force.Office of scientific research UCLA forum in medical sciences, 4 University of California press,Berkeley;Los Angeles,1966
Psy 31.0458

BRAIN RESEARCH INSTITUTE Brain function : conference 1st Proceedings Los Angeles, Calif. 1961 Vol 1: cortical excitability and steady potentials;relations of basic research to space biology Edited by Mary A.B. Brazier U.C.L.A. forum in medical sciences, 1 University of California press,Berkeley; Los Angeles,1963
Psy 31.0255

BRAIN RESEARCH INSTITUTE Brain function conference 2nd Proceedings Los Angeles 1962 Vol 2: RNA and brain function memory and learning Edited by Mary A.B. Brazier With the support of the United States.Air force.Office of scientific research U.C.L.A. forum in medical sciences, 2 University of California press,Berkeley;Los Angeles,1964
Psy 31.0246

BRAIN RESEARCH INSTITUTE Frontiers in brain research Edited by John D. French Columbia university press,New York,1962 Speeches given at the inauguration of the Brain Research institute,Los Angeles,Oct.1961
An 32.4379

BRAIN RESEARCH INSTITUTE The Interneuron :a conference Proceedings Los Angeles,Calif. 1967 Sep Edited by Mary A.B. Brazier UCLA forum in medical sciences, 11 University of California press,Berkeley;Los Angeles,Calif., 1969
An 32.5473

BRANDEIS UNIVERSITY Synthesis of molecular and cellular structure :a symposium Papers Waltham,Mass. 1960 Society for the study of development and growth Edited by Dorothea Rudnick Society for the study of development and growth.Symposia, 19 Ronald press,New York,1961
Bioch 33.1012

BRANDEIS UNIVERSITY.SUMMER INSTITUTE IN THEORETICAL PHYSICS Lectures in theoretical physics Waltham,Mass 1961 Vol. 2 By M.E. Rose and E.C.G. Sudarshan Lectures in theoretical physics,1961,2 Benjamin,New York,1962
A Math 4.0830

BRANDEIS UNIVERSITY.SUMMER INSTITUTE IN THEORETICAL PHYSICS Lectures in theoretical physics Waltham,Mass. 1961 Pt 1 By R.J. Eden Lectures in theoretical physics,1961,1 Benjamin,New York, 1962
Cav 7.0315

BRANDEIS UNIVERSITY.SUMMER INSTITUTE IN THEORETICAL PHYSICS Lectures in theoretical physics Waltham,Mass. 1961 Pt 2 By M.E. Rose and E.C.G. Sudarshan Lectures in theoretical physics,1961,2 Benjamin,New York,1962
Cav 7.0316

BRANDEIS UNIVERSITY.SUMMER INSTITUTE IN THEORETICAL PHYSICS Lectures in theoretical physics Waltham,Mass. 1962 Vol 2: astrophysics and the many-body problem By E.N. Parker and J.S. Goldstein Lectures in theoretical physics,1962,2 Benjamin,New York,1963
Cav 7.0318

BRANDEIS UNIVERSITY.SUMMER INSTITUTE IN THEORETICAL PHYSICS Lectures on astrophysics and weak interactions Waltham, Mass. 1963 By S. Hayakawa and others Lectures in theoretical physics,1963,2 Brandeis university,Waltham,Mass.,1963
A Math 4.1087

BRANDEIS UNIVERSITY.SUMMER INSTITUTE IN THEORETICAL PHYSICS Lectures on astrophysics and weak interactions Waltham, Mass. 1963 Vol. 2 By S. Hayakawa and others Lectures in theoretical physics,1963, 2 Brandeis university,Waltham,Mass.,1964
A Math 4.0832

BRANDEIS UNIVERSITY.SUMMER INSTITUTE IN THEORETICAL PHYSICS Lectures on general relativity Waltham,Mass. 1964 By A. Trautman Lectures in theoretical physics, 1964,1 Prentice-Hall,Englewood Cliffs,N.J., 1965
A Math 4.1025

BRANDEIS UNIVERSITY.SUMMER INSTITUTE IN THEORETICAL PHYSICS Lectures on general relativity Waltham,Mass. 1964 By A. Trautman and others Lectures in theoretical physics,1964,1 Prentice-Hall,Englewood Cliffs,N.J.,1965
Cav 7.0320

BRANDEIS UNIVERSITY.SUMMER INSTITUTE IN THEORETICAL PHYSICS Lectures on particle and field theory Waltham,Mass 1962 Vol 1: elementary particle physics and field theory By I. Fulton and others Lectures in theoretical physics,1962,1 Benjamin,New York, 1963
Cav 7.0317

BRANDEIS UNIVERSITY.SUMMER INSTITUTE IN THEORETICAL PHYSICS Lectures on particle symmetries and axiomatic field theory Papers Waltham,Mass. 1965 1: axiomatic field theory Edited by M. Chretien and S. Deser Lectures in theoretical physics,1965,1 Gordon and Breach,New York,1966
A Math 4.1334

BRANDEIS UNIVERSITY.SUMMER INSTITUTE IN THEORETICAL PHYSICS Lectures on particles and field theory Waltham,Mass 1964 Vol. 2 By K. Johnson and others Lectures on theoretical physics,1964,2 Prentice-Hall,New York,1965
A Math 4.0833

BRANDEIS UNIVERSITY.SUMMER INSTITUTE IN THEORETICAL PHYSICS Lectures on particles and field theory Waltham,Mass. 1962 Vol. 1: elementary particle physics and field theory By I. Fulton and others Lectures in theoretical physics,1962,1 Benjamin,New York, 1963
A Math 4.0831

BRANDEIS UNIVERSITY.SUMMER INSTITUTE IN THEORETICAL PHYSICS Lectures on particles and field theory Waltham,Mass. 1964 By K. Johnson and others Lectures in theoretical physics,1964,2 Prentice-Hall,Englewood Cliffs,N.J.,1965
Cav 7.0321

BRANDEIS UNIVERSITY.SUMMER INSTITUTE IN THEORETICAL PHYSICS Lectures on strong and electromagnetic interactions Waltham, Mass. 1963 By P.T. Matthews and others Lectures in theoretical physics,1963,1 Brandeis University,Waltham,Mass.,1964
A Math 4.0036

BRANDEIS UNIVERSITY.SUMMER INSTITUTE IN THEORETICAL PHYSICS Statistical physics Waltham,Mass. 1962 By G.E. Uhlenbeck and others Lectures in theoretical physics,1962, 3 252p W.A.Benjamin,New York,1963
Cav 7.2414

BRANDEIS UNIVERSITY.SUMMER INSTITUTE IN THEORETICAL PHYSICS Theoretical physics : lecture notes Typescript Waltham,Mass. 1960 By C. Moller and others Edited by J. Stachel and others Lectures in theoretical physics,1960 Brandeis university,Waltham, Mass.,1960
Cav 7.0314

BRANDEIS UNIVERSITY.SUMMER INSTITUTE IN THEORETICAL PHYSICS Theoretical physics : lecture notes Waltham,Mass. 1960 By C. Moller and others Edited by J. Stachel and others Lectures in theoretical physics,1960 Brandeis university,Waltham,Mass,c1961
A Math 4.0829

BRANDEIS UNIVERSITY.SUMMER INSTITUTE IN THEORETICAL PHYSICS Theoretical physics: lecture notes Waltham,Mass. 1959 By F.E. Low Edited by Wayne A. Mills Lectures in theoretical physics,1959 Brandeis university, Waltham,Mass.,1959
A Math 4.0828

BRANDEIS UNIVERSITY.SUMMER INSTITUTE IN THEORETICAL PHYSICS 11th Proceedings Astrophysics and general relativity Waltham, Mass. 1968 Jun 17-Jul 26 Edited by Max Chretien and others 2 vols Gordon and Breach,New York,1971
TA 15.0657

BRANDEIS UNIVERSITY.SUMMER INSTITUTE IN THEORETICAL PHYSICS 12th Proceedings Atomic physics and astrophysics Waltham,Mass. 1969 Jun 16-Jul 25 Edited by Max Chretien and E. Lipworth 216p Gordon and Breach,New York,1971
TA 15.0683

BRANDEIS UNIVERSITY.SUMMER INSTITUTE OF THEORETICAL PHYSICS 11th Astrophysics and general relativity Waltham,Mass. 1968 Vol 1 Edited by M. Chretien and others illus 300p 23cm 2 vols Gordon and Breach,New York,1969
A Math 4.1791

BRASENOSE CONFERENCE ON THE AUTOMATION OF LIBRARIES Anglo-American conference on the mechanization of library services Proceedings Oxford 1966 Jun 30-Jul 3 Edited by John Harrison and Peter Laslett Sponsored by the Old Dominion foundation of New York Mansell,Chicago,Ill.,1967
Eng 41.7672

BRASOV 1969 Romanian-Finnish seminar on Teichmuller spaces and quasiformal mappings Proceedings Edited by Cabiria Andreian Cazacu 307p 21cm Publishing house of the Academy of the socialist republic of Romania, Bucharest,1971
P Math 2.4355

BRAZIL 1956 International geographical congress 18th Excursion guides 1-9 International geographical union.Brazilian national committee 3 vols International geographical union,Rio de Janeiro,1956
Geog 13.5349

BRAZZAVILLE 1956 Symposium on African hydrobiology and inland fisheries 2nd Scientific council for Africa south of the Sahara Published under the auspices of the Commission for technical co-operation in Africa south of the Sahara Scientific council for Africa south of the Sahara. Publication, 25 Brazzaville,1956 Title also in French:Symposium sur l'hydriobiologie et la peche en eaux douces en Afrique;text in English and French
Bal 39.1894

BRAZZAVILLE 1956 Symposium on African hydrobiology and inland fisheries 2nd Scientific council for Africa south of the Sahara Scientific council for Africa south of the Sahara.Publications,25 Commission for technical co-operation in Africa,London,1957
Geog 13.2726

BREAST CANCER Endocrine aspects of breast cancer :a conference Proceedings Glasgow 1957 Jul 8-10 University of Glasgow Edited by Alastair R. Currie Livingstone,Edinburgh: London,1958
Path 30.2082

BREAST CANCER Human adrenal gland and its relation to breast cancer Tenovus workshop 1st Proceedings Cardiff 1969 Jun 26-27 Tenovus institute for cancer research Edited by K. Griffiths and E.H.D. Cameron Alpha omega alpha,Cardiff,1969
Bioch 33.2198

BREAST CANCER:EARLY AND LATE Annual clinical conference on cancer 13th Papers Houston, Tex. 1968 Anderson hospital and tumor institute Yearbook,Chicago,Ill.,1970
Radioth 35.1914

BREATHING Hering-Breur centenary symposium Proceedings London 1969 Jul 8-10 Ciba foundation Edited by Ruth Porter Ciba foundation.Symposia Churchill,London,1970
Inv Med 37.0184

BRESSANONE 1960 Calculus of variations, classical and modern Centro internazionale matematico estivo Edited by R. Conti bibliog.,illus. 369p 27cm Edizione Cremonese,Rome,1967 In English and French
P Math 2.3729

BREST 1970 Theorie des matroides Rencontre Franco-britannique Proceedings Edited by C.P. Bruter Lecture notes in mathematics, 211 108p 25cm Springer, Berlin,1970
P Math 2.4508

BREUKELEN 1960 Present problems concerning the structure and evolution of the galactic system NUFFIC international summer course in science Lecture notes Netherlands universities foundation for international cooperation Edited by J.H. Oort and H.G. Quik Supported by North atlantic treaty organization The Hague,1960
Obs 6.2067

BREUKELEN 1962 Asymptotic distribution modulo 1 Nuffic international summer session in science papers Netherlands universities foundation for international cooperation Edited by J.F. Koksma and L. Kuipers With the financial support of North Atlantic treaty organization Bibliog. 203p 24cm Noordhoff,Groningen,1964
P. Math 2.1826

BREUKELEN,NETHERLANDS 1963 Selected topics in nuclear spectroscopy :N.U.F.F.I.C. international summer course in science, Nijenrode Castle... Proceedings Netherlands universities foundation for international cooperation,and,Dutch physical society Edited by B.J. Verhaar With financial support from the North Atlantic treaty organization North-Holland,Amsterdam, 1964
Chem 18.2279

BRICKWORK Load-bearing brickwork :a conference Proceedings London 1964 Nov 18-19 British ceramic society British ceramic society.Proceedings, 4 British ceramic society,Stoke-on-Trent,1965
Eng 41.2941

BRIDGE AND STRUCTURAL ENGINEERING International association for bridge and structural engineering :congress 1st-8th Preliminary and final reports 1932-1967 International association for bridge and structural engineering,Zurich,1932-67
Eng 41.2927

BRIDGE DESIGN Concrete bridge design 1st international symposium Proceedings 1967 American concrete institute American concrete institute.Publication,SP- 23 ACI, Detroit,Mich.,1969
Eng 41.8488

BRIDGE DESIGN Concrete bridge design 2nd international symposium Proceedings American concrete institute American concrete institute.Publication,SP- 26 ACI, Detroit,Mich.,1971
Eng 41.8489

BRIDGES AND STRUCTURES Failure and defects of bridges and structures :a symposium Proceedings Tokyo 1957 Sep 15 Japan society of civil engineers Architectural institute of Japan Japan society for the promotion of science,Tokyo,1959
Eng 41.3015

BRIG 1966 Zermatt 1966 Sep 14-17 Internationale tagung fur Alpine meteorologie 9. Wissenschaftliche abhandlungen Edited by Karin Schram and J.C. Thams Schweizerische meteorologische zentralanstalt. Veroffentlichungen,4 illus 366p City-druck a.g.,Zurich,1967 In French,German, and Italian
Sco 14.7410

BRIGHTON 1955 The Scope for electronic computers in the office National conference of the Office management association papers Office management association illus. 102p 26cm Office management association,London, 1955
Math L 5.0727

BRIGHTON 1960 International symposium on distillation European federation of chemical engineering 281p Institution of chemical engineers,London,1960 24th meeting of the European Federation of Chemical Engineering
Chem E 24.1303

BRIGHTON 1962 One-day symposium on structural lightweight concrete Vol 1-2 Reinforced concrete association 2 vols RCA, London,1963
Eng 41.2968

BRIGHTON 1964 Batteries 2 International symposium on batteries :research and development in non-mechanical electrical power sources 4th Edited by D.H. Collins Sponsored by the Inter-departmental committee on batteries Pergamon Press,Oxford,1965
Met 25.1841

BRIGHTON 1964 Gas chromatography 1964 :a symposium 5th Proceedings Institute of petroleum.Gas chromatography discussion group Edited by A. Goldop Institute of petroleum, London,1965
Chem 18.0568

BRIGHTON 1965 Algebraic number theory conference proceedings London mathematical society Edited by John W.S. Cassels and A. Frohlich With the support of the International mathematical union xviii,366p 24cm Academic press,New York;London,1967
P Math 2.2773

BRIGHTON 1965 Quantum fluids The International symposium on quantum fluids University of Sussex Edited by D.F. Brewer viii,360p 23cm North-Holland,Amsterdam, 1966
Cav 7.2935

BRIGHTON 1966 Computer typesetting conference preprints and summaries Institute of printing 29cm 2 vols Institute of printing,1966
Math L 5.1023

BRIGHTON 1967 The Solidification of metals the conference Proceedings Iron and steel institute and Institute of metals Jointly organized with the Institute of physics and the physical society and the Institution of metallurgy Iron and steel institute.Publication, 110 Iron and steel institute,London,1968
Met 25.1529

BRIGHTON 1968 I.Chem.E.,VTG-VDI joint meeting on the engineering of gas-solid reactions Preprints characteristics of solid particles which are relevant to gas-solid reactions Institution of chemical engineers and Verein deutscher ingenieure 43p Institution of chemical engineers,London, 1968 In English and German
Chem E 24.1516

BRIGHTON 1968 I.Chem.E.,VTG-VDI joint meeting on the engineering of gas-solid reactions Preprints experiences with the operation of industrial process equipment Institution of chemical engineers and Verein deutscher ingenieure 51p Institution of chemical engineers,London,1968 In English and German
Chem E 24.1519

BRIGHTON 1968 I.Chem.E.,VTG-VDI joint meeting on the engineering of gas-solid reactions Preprints physical behaviour of gas-solid systems Institution of chemical engineers and Verein deutscher ingenieure 69p Institution of chemical engineers,London, 1968 In English and German
Chem E 24.1517

BRIGHTON 1968 I.Chem.E.,VTG-VDI joint meeting on the engineering of gas-solid reactions Preprints the technical design and performance of industrial process equipment Institution of chemical engineers and Verein deutscher ingenieure 52p Institution of chemical engineers,London,1968 In English and German
Chem E 24.1518

BRIGHTON 1968 The Engineering of gas-solid reactions :I.Chem.E.-VTG-VDI joint meeting Proceedings Institution of chemical engineers Verein deutscher ingenieure Edited by J.M. Pirie Institution of chemical engineers.Symposium series, 27 illus 255p Institution of chemical engineers,London,1968
Chem E 24.1873

BRIGHTON 1969 Atomic collision phenomena in solids :an international conference Proceedings Edited by D.W. Palmer and others Bibliog 678p 17cm North-Holland, Amsterdam,1970
Cav 7.2680

BRIGHTON 1969 Fracture International conference on fracture 2nd Proceedings Edited by P.L. Pratt and others Chapman and Hall,London,1969
Eng 41.3808

BRIGHTON 1969 International conference on fracture 2nd Proceedings Edited by P.L. Pratt and others Chapman and Hall,London, 1969
Met 25.2327

BRIGHTON 1970 Highlights of astronomy International astronomical union 14th general assembly Vol 2 International astronomical union Edited by C. de Jager 793p Reidel,Dordrecht,1971
Obs 6.3688

BRIGHTON 1970 Highlights of astronomy International astronomical union.General assembly 14th Invited discourses Vol 2 International astronomical union Edited by C.de Jager 793p Reidel,Dordrecht,1971
TA 15.0655

BRIGHTON 1970 Lunar nomenclature Report of the working group of I.A.U. commission 17: the moon:proposed names for craters on the moon's far side...presented at the 14th assembly... International astronomical union. Commission 17 43p Harvard college observatory,Cambridge,Mass.,1970 Duplicated
Obs 6.3342

BRIGHTON,UTAH 1958 Hormones and atherosclerosis :a conference Proceedings National institutes of health.Endocrinology study section Edited by Gregory Pincus Academic press,New York,1959
Bioch 33.0477

BRISBANE 1899 Intercolonial medical congress of Australasia 5th session Transactions Edited by Wilton Love E. Gregory,Brisbane,1901
An 32.1143

BRISTOL 1947 A Conference on strength of solids :held at the H.H.Wills physical laboratory Report University of Bristol Edited by N.F. Mott 162p Physical sciety, London,1948
Cav 7.0346

BRISTOL 1947 Strength of solids :a conference Report Wills (H.H.) physical laboratory Physical society,London,1948
Met 25.1197

BRISTOL 1948 Cosmic radiation based on a symposium promoted by the Colston research society 1st Papers Colston research society and University of Bristol Edited by F.C. Frank and D.R. Rexworthy Colston research society.Colston papers, 1 illus viii,189p Butterworths,London,1949
Cav 7.0348

BRISTOL 1949 A Discussion of the Faraday society 5th Papers University of Bristol.Department of physics and Faraday society Faraday society.Discussions, 5, 1949 Butterworth,London,1959
Cav 7.0760

BRISTOL 1949 Crystal growth :a general discussion Faraday society Faraday society. Discussions, 5 Gurney and Jackson,London, 1949
Eng 41.4424

BRISTOL 1949 Crystal growth :a general discussion Faraday society Faraday society. Discussions,5 Gurney and Jackson,London,1949
Min 10.1087

BRISTOL 1949 Engineering structures :a symposium Papers Colston research society University of Bristol Colston research society.Research special supplement, 49 Butterworths,London,1949
Eng 41.3022

BRISTOL 1950 Fundamental mechanisms of photographic sensitivity A Conference on fundamental mechanisms of photographic sensitivity :held in the H.H.Wills physical laboratory Proceedings University of Bristol Edited by J.W. Mitchell Butterworths,London,1951
Cav 7.0347

BRISTOL 1950 Fundamental mechanisms of photographic sensitivity :a symposium Proceedings Butterworths,London,1951
Radioth 35.1506

BRISTOL 1950 Principles and methods of colonial administration :a symposium Colston research society Colston papers Butterworths,London,1950
Geog 13.3107

BRISTOL 1951 Structural aspects of cell physiology Society for experimental biology Society for experimental biology.Symposia, 6 Cambridge university press,Cambridge,1952
Phys 20.0954

BRISTOL 1951 Structural aspects of cell physiology :a symposium Society for experimental biology Edited by J.F. Danielli and R. Brown Society for experimental biology, 6 Cambridge university press, Cambridge,1952
An 32.5445

BRISTOL 1951 Structural aspects of cell physiology :a symposium Society for experimental biology Society for experimental biology.Symposia, 6 Cambridge university press,Cambridge,1952
Bioch 33.1352

BRISTOL 1951 Structural aspects of cell physiology :a symposium papers Society for experimental biology Society for experimental biology.Symposia,6 Cambridge university press,Cambridge,1952
VA 19.0361

BRISTOL 1951 Structural aspects of cell physiology :symposium papers Society for experimental biology Society for experimental biology.Symposia, 6 Cambridge university press,Cambridge,1952
Radioth 35.0456

BRISTOL 1952 The Suprarenal cortex :a symposium Proceedings Colston research society Edited by J.M. Yoffey Colston research society.Symposia, 5 Colston papers, 5 Butterworth,London,1953
Bioch 33.0445

BRISTOL 1954 Conference on defects in crystalline solids Report Wills (H.H.) physical laboratory The Physcial society, London,1955
Met 25.2358

BRISTOL 1954 Conference on the defects in crystalline solids :held at the H.H.Wills physical laboratory Report By N. Bloembergen and others University of Bristol and Institute of physics With co-operation of the International union of pure and applied physics.Commission on the solid state 429p Physical society,London,1955
Cav 7.0345

BRISTOL 1954 Defects in crystalline solids Conference on defects in crystalline solids Report Physical society,London,1955
Eng 41.4311

BRISTOL 1954 Recent developments in cell physiology :a symposium Colston research society Edited by J.A. Kitching Held at the University of Bristol Colston papers, 7 London,1954
Bal 39.0332

BRISTOL 1954 Recent developments in cell physiology :symposium 7th Proceedings Colston research society Edited by J.A. Kitching Colston papers, 7 Butterworths, London,1954
Phys 20.1481

BRISTOL 1954 Recent developments in cell physiology :symposium Proceedings Colston research society Edited by J.A. Kitching Colston research society.Symposia, 7 Colston papers, 7 Butterworth,London,1954
Radioth 35.0463

BRISTOL 1954 Recent developments in cell physiology :symposium Proceedings Colston research society Edited by J.A. Kitching Colston research society.Symposia, 7 Colston papers, 7 Butterworths,London,1954
Bioch 33.0936

BRISTOL 1956 Neurohypophysis :symposium Proceedings Colston research society Edited by H. Heller Colston research society. Symposia, 8 Colston papers, 8 Butterworths,London,1957
Bioch 33.0444

BRISTOL 1956 The Neurohypophysis : symposium Proceedings Colston research society Edited by H. Heller Colston papers, 8 Butterworth,London,1957
An 32.4689

BRISTOL 1956 The Neurohypophysis : symposium Proceedings Colston research society Edited by H. Heller Colston papers, 8 Butterworths,London,1957
VA 19.0224

BRISTOL 1957 Observation and interpretation... symposium 9th proceedings Colston research society Edited by S. Korner Colston papers, 9 xiii,218p 25cm Butterworths,London,1957
Math 3.0972

BRISTOL 1957 Observation and interpretation:a symposium of philosophers and physicists Colston research society symposium 9th Proceedings Edited by S. Korner xiv,218p Butterworths,London,1957
WSM 43.1022

BRISTOL 1957 Observation and interpretation:a symposium of philosophers and physicists Proceedings Colston research society Edited by S. Korner Colston research society.Symposia, 9 Butterworths, London,1957
Psy 31.2431

BRISTOL 1958 Developments in aromatic chemistry.Applications of electron and nuclear resonance in chemistry.Recent work on the inorganic chemistry of sulphur :a symposium Chemical society Chemical society.Special publication,12 Chemical society,London,1958
Chem 18.1229

BRISTOL 1958 Developments in aromatic chemistry.Applications of electron and nuclear resonance in chemistry.Recent work on the inorganic chemistry of sulphur :symposia Chemical society Chemical society.Special publication, 12 Chemical society,London, 1958
Bioch 33.0630

BRISTOL 1958 Structure and properties of porous materials Colston research society : symposium 10th Proceedings Edited by D.H. Everett and F.S. Stone Butterworths,London, 1958
Min 40 77 10.0655

BRISTOL 1958 The Structure and properties of porous materials :a symposium Proceedings a symposium 10th Proceedings Colston research society Edited by D.H. Everett and F.S. Stone Colston papers, 10 Butterworths,London,1958
Eng 41.3084

BRISTOL 1958 The Structure and properties of porous materials :a symposium Proceedings Colston research society Edited by D.H. Everett and F.S. Stone Colston papers, 10 xiv,389p Butterworths,London,1958
Chem E 24.0854

BRISTOL 1958 The Structure and properties of porous materials :a symposium 10th Proceedings Colston research society Edited by D.H. Everett and F.S. Stone Butterworths,London,1958
Chem 18.0558

BRISTOL 1959 Hypersonic flow 11th symposium Colston research society Edited by A.R. Collar and J. Tinkler Colston papers, 11 Butterworths,London,1960
Eng 41.6759

BRISTOL 1960 Models and analogues in biology Society for experimental biology Edited by J.W.L. Beament Society for experimental biology.Symposia, 14 Cambridge university press,Cambridge,1960
Phys 20.0939

BRISTOL 1960 Models and analogues in biology Society for experimental biology Society for experimental biology symposia, 14 vii,255p Cambridge university press, Cambridge,1960
WSM 43.2826

BRISTOL 1960 Models and analogues in biology :a symposium Society for experimental biology.Symposia, 14 Cambridge university press,Cambridge,1960
An 32.0058

BRISTOL 1965 Endogenous substances affecting the myometrium :a symposium Proceedings Society for endocrinology Edited by V.R. Pickles and R.J. Fitzpatrick Society for endocrinology.Memoirs, 14 Cambridge university press,Cambridge,1966
Inv Med 37.0180

BRISTOL 1965 Endogenous substances affecting the myometrium :a symposium Proceedings Society for endocrinology Edited by V.R. Pickles and R.J. Fitzpatrick Society for endocrinology.Memoirs,14 Cambridge university press,Cambridge,1966
Pha 16.0083

BRISTOL 1965 Science of ceramics :a conference 3rd Proceedings British ceramic society and Nederlandse keramische vereniging Edited by G.H. Stewart Under the auspices of the European ceramic association Academic press,London;New York, 1967
Met 25.0286

BRISTOL 1965 Submarine geology and geophysics Colston research society : symposium 17th proceedings Edited by W.F. Whittard and R. Bradshaw Butterworths, London,1965
Min 10.0573

BRISTOL 1965 Submarine geology and geophysics Colston research society symposium 17th Proceedings Edited by W.F. Whittard and R. Bradshaw Colston papers,17 Butterworths,London,1965
Geog 13.0740

BRISTOL 1965 Submarine geology and geophysics :a symposium 17th Proceedings Colston research society Edited by W.F. Whittard and R. Bradshaw Colston research society.Colston papers, 17 London,1965
Geod 9.0580

BRISTOL 1965 Submarine geology and geophysics :symposium 17th proceedings Colston research society Edited by W.F. Whittard and R. Bradshaw Colston papers,17 illus. map Butterworths,London,1965
Geol 8.1358

BRISTOL 1966 Fungus spore :a symposium Proceedings Colston research society Edited by M.F. Madelin Colston research society.Symposia, 18 Colston papers, 18 Butterworths,London,1966
Bot 42.4006

BRISTOL 1966 Wetting :joint symposium Papers and discussion Society of chemical industry.Bristol section,and,Society of chemical industry.Colloid and surface chemistry group Society of chemical industry. Monographs,25 Society of chemical industry, London,1967
Col S 12.0147

BRISTOL 1967 Advances in computer control United Kingdom automation council.Control convention 2nd Organized by the Institution of electrical engineers.Control and automation division Institution of electrical engineers.Conference publication, 29 I.E.E.,London,1967
Eng 41.5862

BRISTOL 1967 Nitrogen metabolism in plants Recent aspects of nitrogen metabolism in plants :a symposium Proceedings Long Aston research station Edited by E.J. Hewitt and C. V. Cutting Long Ashton research station. Symposia, 1 Academic press,London;New York, 1968
Bot 42.1893

BRISTOL 1967 The Liver :...symposium Proceedings Colston research society Edited by A.E. Read Colston research society. Symposium, 19 Colston papers, 19 Butterworths,London,1967
PGMS 29.0367

BRISTOL 1969 Phytochemical phylogeny :a symposium Proceedings Phytochemical society Edited by Jeffrey B. Harborne Bibliog.,illus,maps xiii,334p Academic press,London,1970
Bot 42.3453

BRISTOL 1972 Conference on teaching computing Proceedings University grants committee.Mathematical sciences sub-committee illus 165p University grants committee, London,1972
Math L 5.3878

BRITISH ASSOCIATION FOR THE ADVANCEMENT OF SCIENCE A Scientific survey of Dundee and district :a scientific survey prepared for the Dundee meeting,1939 Edited by R.L. Mackie London,1939
Geog 13.4063

BRITISH ASSOCIATION FOR THE ADVANCEMENT OF SCIENCE Address delivered in the Senate house Cambridge 1833 Jun 25 By William Whewell Cambridge,1833 One of nineteen tracts bound together
Philos 1.0880

BRITISH ASSOCIATION FOR THE ADVANCEMENT OF SCIENCE Birmingham and its regional setting :a scientific survey Birmingham 1950 Aug 30-Sep 6 British association for the advancement of science,Birmingham,1950
Gen 34.1371

BRITISH ASSOCIATION FOR THE ADVANCEMENT OF SCIENCE Discussion on the teaching of mathematics... to which is added a report of the British association committee drawn up by the chairman,professor Forsyth Glasgow 1901 Sep 14 Edited by John Perry 2nd edition Macmillan,London;New York,1902
Psy 31.3384

BRITISH ASSOCIATION FOR THE ADVANCEMENT OF SCIENCE Lithographed signatures Cambridge 1833 Cambridge,1833 Two copies
Philos 1.0098

BRITISH ASSOCIATION FOR THE ADVANCEMENT OF SCIENCE Mathematical tables Vol 1: circular and hyperbolic functions,exponential and sine cosine integrals,factorial function and allied functions,hermition probability functions 2nd edition University press, Cambridge,1946
Cav 7.0350

BRITISH ASSOCIATION FOR THE ADVANCEMENT OF SCIENCE The Botanical features of the southwestern Cape Province :essays By Robert S. Adamson and others 127p Speciality press,Cape Town,1929 Published on the occasion of the visit to South Africa of the British association for the advancement of science,July 1929
Bot 42.2342

BRITISH ASSOCIATION FOR THE ADVANCEMENT OF SCIENCE The Cambridge region 1965 : meeting of the British Association Cambridge 1965 Sep 1-8 Edited by J.A. Steers illus., maps. British association for the advancement of science,London,1965
Gen 34.0131

BRITISH ASSOCIATION FOR THE ADVANCEMENT OF SCIENCE 105th :annual meeting Report Norwich 1935 Sep 4-11 London,1935
Geog 13.3987

BRITISH ASSOCIATION FOR THE ADVANCEMENT OF SCIENCE 108th :annual meeting Report Cambridge 1938 Aug 17-24 London,1938 With appendix:a scientific survey of Cambridge and district,edited by H.C.Darby
Geog 13.3988

BRITISH ASSOCIATION FOR THE ADVANCEMENT OF SCIENCE.CAMBRIDGE MEETING 1833 Lithographed signatures of the members of the British association for the advancement of science who met at Cambridge June 1833.With a report of the proceedings at the public meetings... British association for the advancement of science iv,125 Pitt press, Cambridge,1833
WSM 43.0435

BRITISH ASSOCIATION FOR THE ADVANCEMENT OF SCIENCE.CAMBRIDGE MEETING 1965 The Cambridge region 1965 :prepared for the meeting of the British association...1965 British association for the advancement of science Edited by J.A. Steers British association for the advancement of science, London,1965
Eng 41.3266

BRITISH AUTOMATION CONFERENCE 1965 Automation '65 Eastbourne 1965 Nov 7-10 Confederation of British industry Trades union congress United Kingdom automation council Organised by the Institution of production engineers Institution of production engineers,London,1965
Eng 41.6018

BRITISH BIOCHEMISTRY PAST AND PRESENT Essays London 1969 Dec Biochemical society Edited by T.W. Goodwin Biochemical society. Symposium, 30 Academic press,London,1970
Bioch 33.2314

BRITISH BIOPHYSICAL SOCIETY Cell electrophoresis :a symposium London 1963 May Edited by E.J. Ambrose Churchill, London,1965
Radioth 35.1414

BRITISH BIOPHYSICAL SOCIETY Cell electrophoresis :a symposium London 1963 May Edited by E.J. Ambrose held at the Chester Beatty research institute illus Churchill,London,1965
Bal 39.0034

BRITISH BROADCASTING CORPORATION Communication theory Applications of communication theory :symposium Papers London 1952 Sep 22-26 Edited by Willis Jackson xii,532p 25cm Butterworths, London,1953 Also supported by the Ministry of supply
Math L 5.3249

BRITISH BROADCASTING CORPORATION Communication theory Applications of communication theory symposium papers London 1952 Sep 22-26 Edited by Willis Jackson xii,532p Butterworths scientific publications,London,1953
Math 3.0735

The BRITISH CALEDONIDES symposium papers Edinburgh 1961 Dec. Edited by M.R.W. Johnson and F.H. Stewart Sponsored by the Edinburgh geological society illus. map Oliver and Boyd,Edinburgh;London,1963
Geol 8.1260

BRITISH CERAMIC RESEARCH ASSOCIATION Special ceramics :a symposium 4th Proceedings Stoke-on-Trent 1967 Jul 11-13 Edited by P. Popper British ceramic research association, Stoke-on-Trent,1968
Met 25.2554

BRITISH CERAMIC RESEARCH ASSOCIATION Special ceramics :a symposium Proceedings Stoke-on-Trent 1959 Jul 13-15 By P. Popper Heywood,London,1960
Met 25.0267

BRITISH CERAMIC RESEARCH ASSOCIATION Special ceramics 1962 :a symposium 2nd Proceedings Stoke-on-Trent 1962 Jul 11-13 Edited by P. Popper Academic press,London; New York,1963
Met 25.0280

BRITISH CERAMIC RESEARCH ASSOCIATION Special ceramics 1964 :a symposium 3rd Proceedings Stoke-on-Trent 1964 Jul 8-10 By P. Popper Edited by P. Popper Academic press,London;New York,1965
Met 25.0281

BRITISH CERAMIC RESEARCH ASSOCIATION The All-basic open-hearth furnace being also the report of the 36th steelmaking conference Leamington-Spa 1951 May 2-3 Iron and steel institute.Special report, 46 Iron and steel institute,London,1952
Met 25.0330

BRITISH CERAMIC RESEARCH ASSOCIATION 3rd : symposium Proceedings Special ceramics 1964 Stoke-on-Trent 1964 Jul 8-10 Edited by P. Popper Academic press,London;New York, 1965
Min 10.0656

BRITISH CERAMIC SOCIETY Load-bearing brickwork :a conference Proceedings London 1964 Nov 18-19 British ceramic society. Proceedings, 4 British ceramic society, Stoke-on-Trent,1965
Eng 41.2941

BRITISH CERAMIC SOCIETY Science of ceramics : a conference 1st Proceedings Oxford 1961 Jun 26-30 Edited by G.H. Stewart Academic press,London;New York,1962
Met 25.0274

BRITISH CERAMIC SOCIETY Science of ceramics : a conference 2nd Proceedings Noordwijk aan Zee 1963 May 13-17 Edited by G.H. Stewart Academic press,London;New York,1965
Met 25.0278

BRITISH CERAMIC SOCIETY Science of ceramics : a conference 3rd Proceedings Bristol 1965 Jul 5-8 Edited by G.H. Stewart Under the auspices of the European ceramic association Academic press,London;New York, 1967
Met 25.0286

BRITISH COLUMBIA NATURAL RESOURCES CONFERENCE 8th Transactions Victoriia,B.C. 1955 Feb 23-25 Edited by D,B, Turner and J.D. Chapman maps 338p 25cm British Columbia natural resources conference,Victoria,B.C.,1955
Sco 14.4292

BRITISH COLUMBIA NATURAL RESOURCES CONFERENCE 9TH British Columbia atlas of resources Atlas Edited by J.D. Chapman and others British Columbia natural resources conference, Vancouver,1956
Geog 13.6297

BRITISH COMMONWEALTH OFFICIAL SCIENTIFIC SPECIALIST CONFERENCES,Special agricultural conference The Australian environment handbook prepared for the conference on plant and animal nutrition... 1949 Aug Commonwealth scientific and industrial research organization 2nd edition Commonwealth scientific and industrial research organization,Melbourne,1950
Gen 34.1372

BRITISH COMMONWEALTH SCIENTIFIC OFFICIAL CONFERENCE.SPECIALIST CONFERENCE IN AGRICULTURE 1st Proceedings Plant and animal nutrition in relation to soil and climatic factors Adelaide 1949 Aug 22-Sep 15 and Canberra 1949 Aug 22-Sep 15 H.M. S.O.,London,1951
Geog 13.2510

BRITISH COMPUTER SOCIETY International joint conference on artificial intelligence 2nd Advance papers London 1971 Sep 1-3 illus 658p British computer society,London,1971
Math L 5.3848

BRITISH COMPUTER SOCIETY Joint symposium on instrumentation and computation in process development and plant design Papers London 1959 May 11-12 British conference on automation and computation Sponsored by the Institution of chemical engineers Institution of chemical engineers,London,1959 Sponsored also by the Society of instrument technology,and the British computer society
Math L 5.3290

BRITISH COMPUTER SOCIETY Joint symposium on instrumentation and computation in process development and plant design Proceedings London 1959 May 11-13 Institution of chemical engineers,London,1959
Eng 41.6011

BRITISH COMPUTER SOCIETY Models for decision a conference London 1960 Oct 13-144 Under the auspices of the United Kingdom automation council English universities press,London,1965
Eng 41.1270

BRITISH COMPUTER SOCIETY The Babbage memorial meeting London 1971 Oct 18 illus 40p British computer society,London, 1971
Math L 5.3612

BRITISH COMPUTER SOCIETY The Babbage memorial meeting Report of the proceedings London 1971 Oct 18 iii,39p 30cm British computer society,London,1971
Math S 3.1858

BRITISH COMPUTER SOCIETY The Impact of users' needs on the design of data processing systems : conference Edinburgh 1964 Mar 31-Apr 3 Held under the aegis of the United Kingdom automation council 1964 Organized jointly by the British computer society,the British institution of radio engineers and the Institution of electrical engineers
Math L 5.3265

BRITISH COMPUTER SOCIETY The Reliability and maintenance of digital computer systems: managerial and engineering aspects :discussion meetings London 1960 Jan 20-21 70p 28cm British computer society;Institution of electrical engineers,London,1960
Math L 5.3255

BRITISH COMPUTER SOCIETY.CODASYL DATA BASE TASK GROUP BCS October 71 conference on April 71 report Collected papers London 1971 Oct illus 221p British computer society, London,1972
Math L 5.3865

BRITISH CONFERENCE ON AUTOMATION AND COMPUTATION Joint symposium on instrumentation and computation in process development and plant design Papers London 1959 May 11-12 Sponsored by the Institution of chemical engineers Institution of chemical engineers, London,1959 Sponsored also by the Society of instrument technology,and the British computer society
Math L 5.3290

BRITISH CONGRESS ON TUBERCULOSIS Descriptive catalogue of the museum of the British congress on tuberculosis held in London,July 22nd to 26th,1901 Edited by W.Jobson Horne illus Adlard,London,1901
Path 30.0762

BRITISH CONSTRUCTIONAL STEELWORK ASSOCIATION Conference on steel in architecture Proceedings London 1969 Nov 24-26 British constructional steelwork association, London,1970
Eng 41.0269

BRITISH CONSTRUCTIONAL STEELWORK ASSOCIATION Structural steelwork :a conference Proceedings London 1966 Sep 26-28 British constructional steelwork association, London,1966
Eng 41.3024

BRITISH COUNCIL Evolution symposium 7th Oxford 1952 Jul. Society for experimental biology Edited by R. Brown and J.F. Danielli Society for experimental biology.Symposia, 7 xix,448p 25cm Cambridge university press,Cambridge,1953
Math 3.0773

BRITISH ECOLOGICAL SOCIETY Ecological aspects of the mineral nutrition of plants :a symposium Sheffield 1968 Apr 1-5 Edited by Ian H. Rorison and others British ecological society.Symposia, 9 illus, plates xxi,484p Blackwell scientific, Oxford,1969
Bot 42.2107

BRITISH ECOLOGICAL SOCIETY Grazing in terrestrial and marine environments :a symposium Bangor 1962 Apr 11-14 Edited by D.J. Crisp British ecological society. Symposia, 4 Blackwell,Oxford,1964
Bal 39.1671

BRITISH ECOLOGICAL SOCIETY Light as a ecological factor :a symposium Cambridge 1965 Mar 30-Apr 1 Edited by Richard Bainbridge and G. Evans British ecological society.Symposia, 6 Blackwell,Oxford,1966
Bal 39.1683

BRITISH ECOLOGICAL SOCIETY Light as an ecological factor British ecological society a symposium 6th Proceedings Cambridge 1965 Mar 30-Apr 1 Edited by Richard Bainbridge and others British ecological society.Symposia, 6 Bibliog.,tables,diagrs 452p Blackwell,Oxford,1966
Bot 42.2093

BRITISH ECOLOGICAL SOCIETY The Biology of weeds :a symposium Oxford 1959 Apr 2-4 Edited by John L. Harper British ecological society.Symposia, 1 Bibliog,maps,2 plates, tables,diagrs xv,256p Blackwell scientific, Oxford,1960
Bot 42.2032

BRITISH ECOLOGICAL SOCIETY The Exploitation of natural animal populations :a symposium Durham 1960 Mar 28-31 Edited by E.D. Le Cren and M.W. Holdgate British ecological society.Symposia, 2 Oxford,1962
Bal 39.1665

BRITISH ECOLOGICAL SOCIETY The Exploitation of natural animal populations :a symposium 2 Durham 1960 Mar 28-31 Edited by E.David Le Cren and Martin Wyatt Holdgate tables,diagrs, maps x,399p Blackwell scientific publications,Oxford,1962
Sco 14.0721

BRITISH ECOLOGICAL SOCIETY The Water relations of plants :a symposium London 1961 Apr 5-8 Edited by A.J. Rutter and F.H. Whitehead British ecological society, 3 illus. x,349p Blackwell,London,1963
Bot 42.1879

BRITISH ECOLOGICAL SOCIETY 6th :a symposium Proceedings Light as an ecological factor Cambridge 1965 Mar 30-Apr 1 British ecological society Edited by Richard Bainbridge and others British ecological society.Symposia, 6 Bibliog.,tables,diagrs 452p Blackwell,Oxford,1966
Bot 42.2093

BRITISH ECOLOGICAL SOCIETY.SYMPOSIA Grazing in terrestrial and marine environments :a symposium Bangor 1962 Apr 11-14 British ecological society Edited by D.J. Crisp xvi,322p Blackwell scientific,Oxford,1964
Bot 42.1984

BRITISH ECOLOGICAL SOCIETY.SYMPOSIA, 2 The Exploitation of natural animal populations :a symposium Durham 1960 Mar 28-31 British ecological society Edited by E.D. Le Cren and M.W. Holdgate Oxford,1962
Bal 39.1665

BRITISH ECOLOGICAL SOCIETY.SYMPOSIA, 4 Grazing in terrestrial and marine environments a symposium Bangor 1962 Apr 11-14 British ecological society Edited by D.J. Crisp Blackwell,Oxford,1964
Bal 39.1671

BRITISH ECOLOGICAL SOCIETY.SYMPOSIA, 5 Ecology and the industrial society :a symposium Swansea 1964 Apr 13-16 British ecological society Edited by Gordon T. Goodman and others illus. viii,395p Blackwell scientific,Oxford,1965
Bot 42.1974

BRITISH ECOLOGICAL SOCIETY.SYMPOSIA, 6 Light as a ecological factor :a symposium Cambridge 1965 Mar 30-Apr 1 British ecological society Edited by Richard Bainbridge and G. Evans Blackwell,Oxford, 1966
Bal 39.1683

BRITISH ECOLOGICAL SOCIETY.SYMPOSIA, 6 Light as an ecological factor British ecological society :a symposium 6th Proceedings Cambridge 1965 Mar 30-Apr 1 British ecological society Edited by Richard Bainbridge and others Bibliog.,tables,diagrs 452p Blackwell,Oxford,1966
Bot 42.2093

BRITISH ECOLOGICAL SOCIETY.SYMPOSIA, 7 The Teaching of ecology :a symposium London 1966 Apr 13-16 British ecological society Edited by J.M. Lambert illus. xi,294p Blackwell scientific,Oxford,1967
Bot 42.1988

BRITISH ECOLOGICAL SOCIETY.TROPICAL GROUP Speciation in tropical environments symposium Papers London 1968 Oct 31-Nov 1 Linnean society of London.Biological journal, 1,no.1-2 Academic press,London,1969
Gen 34.1298

BRITISH ECOLOGICAL SOCIETY JUBILEE SYMPOSIUM London 1963 Mar 28-30 Edited by A. Macfadyen and P.J. Newbound Blackwell,Oxford, 1964 Supplement to "Journal of ecology", vol 52 and "Journal of animal ecology",vol 33
Geog 13.1309

BRITISH ECOLOGICAL SOCIETY SYMPOSIUM 5th Ecology and the industrial society Swansea 1964 Apr 13-16 Edited by Gordon T. Goodman and others Blackwell,Oxford,1965
Geog 13.1216

BRITISH ECOLOGICAL SOCIETY SYMPOSIUM 6th Light as an ecological factor Cambridge 1965 Mar 30-Apr 1 Edited by Richard Bainbridge and others Blackwell,Oxford,1966
Geog 13.1217

BRITISH EGG MARKETING BOARD Physiology of the domestic fowl :a symposium Proceedings Sutton Bonington 1964 Dec 16-18 Edited by C. Horton-Smith and E.C. Amoroso British egg marketing board symposium,1 Oliver and Boyd, Edinburgh;London,1966
VA 19.0308

BRITISH EGG MARKETING BOARD Protein utilization by poultry :a symposium Proceedings Sutton Bonington 1965 Sep 22-23 Edited by R.A. Morton and E.C. Amoroso British egg marketing board symposium,2 Oliver and Boyd,Edinburgh;London,1967
VA 19.0309

BRITISH EGG MARKETING BOARD SYMPOSIUM,1 Physiology of the domestic fowl :a symposium Proceedings Sutton Bonington 1964 Dec 16-18 British Egg marketing board Edited by C. Horton-Smith and E.C. Amoroso Oliver and Boyd,Edinburgh;London,1966
VA 19.0308

BRITISH EGG MARKETING BOARD SYMPOSIUM,2 Protein utilization by poultry :a symposium Proceedings Sutton Bonington 1965 Sep 22-23 British egg marketing board Edited by R. A. Morton and E.C. Amoroso Oliver and Boyd, Edinburgh;London,1967
VA 19.0309

BRITISH EMPIRE EXHIBITION Phases of modern science :collection of essays and the 'Guide to the exhibits in the science galleries' Wembley 1924-25 By Oliver Lodge and others Royal society of London 2nd edition 231p A.and F.Denny,London,1925 1st edition entitled 'Handbook to the exhibition of pure science arranged by the Royal society'
Nap 11.0962

BRITISH FLOWERING PLANTS AND MODERN SYSTEMATIC METHODS The Conference on the study of critical British groups Report London 1948 Apr 9-10 Botanical society of the British Isles Edited by A.J. Wilmott Botanical society of the British Isles.B.S.B.I. conference reports, 1 London,1949
Bot 42.4706

BRITISH GELATINE AND GLUE RESEARCH ASSOCIATION Recent advances in gelatin and glue research : a conference Proceedings Cambridge 1957 Jul 1-5 Edited by G. Stainsby Pergamon, London,1958
Bioch 33.2247

BRITISH GRASSLAND SOCIETY International grassland congress 8th Proceedings Reading 1960 Jul 11-21 Grassland research institute,Hurley,1961
BG 38.3114

BRITISH HYDROMECHANICS RESEARCH ASSOCIATION Cranfield fluidics conference 2nd Proceedings Cambridge 1967 Turin 1968 British hydromechanics research association, Cranfield,1967-68
Eng 41.5906

BRITISH HYDROMECHANICS RESEARCH ASSOCIATION Cranfield fluidics conference 3rd Proceedings Turin 1968 May 8-10 Vol 1-3 3 vols British hydromechanics research association,Cranfield,1968
Eng 41.5907

BRITISH HYDROMECHANICS RESEARCH ASSOCIATION International conference on fluid logic and amplification 1st Proceedings Cranfield 1965 Sep Edited by C.J. Charnley and H.S. Stephens British hydromechanics research association,Cranfield,1965
Eng 41.5895

BRITISH INSTITUTE OF MANAGEMENT Conference on work study Papers Harrogate 1954 May 27-29 British institute of management,London, 1954
Eng 41.1544

BRITISH INSTITUTE OF MANAGEMENT.WINTER PROCEEDINGS 1948-49, 2 Problems of growth in industrial undertakings By L. Urwick British institute of management, London,1949
Eng 41.0531

BRITISH INSTITUTION OF RADIO ENGINEERS The Impact of users' needs on the design of data processing systems : conference Edinburgh 1964 Mar 31-Apr 3 British computer society Held under the aegis of the United Kingdom automation council 1964 Organized jointly by the British computer society,the British institution of radio engineers and the Institution of electrical engineers
Math L 5.3265

BRITISH INSULIN MANUFACTURERS The Mechanism of action of insulin :a symposium London 1958 Sep 9-10 Edited by F.G. Young and others Blackwell scientific publication, Oxford,1960
PGMS 29.0378

BRITISH INSULIN MANUFACTURERS The Mechanism of action of insulin a symposium Discussions London 1958 Sep 9-10 Edited by F.G. Young Blackwell,Oxford,1960
Bioch 33.0474

BRITISH IRON AND STEEL RESEARCH ASSOCIATION
Design in high-strength structural steels :a conference Proceedings Eastbourne 1969 Iron and steel institute.Publication, 122 Iron and steel institute,London,1969
Eng 41.2761

BRITISH IRON AND STEEL RESEARCH ASSOCIATION
Energy requirements in the iron and steel industry a conference Harrogate 1959 Jun 24-25 BISRA,London,1959
Met 25.0409

BRITISH IRON AND STEEL RESEARCH ASSOCIATION
High-temperature properties of steels :joint conference Proceedings Eastbourne 1966 Apr 4-7 Iron and steel institute. Publications, 97 Iron and steel institute, London,1967
Eng 41.3827

BRITISH IRON AND STEEL RESEARCH ASSOCIATION
High-temperature properties of steels the joint conference Proceedings Eastbourne 1966 Apr 4-7 Iron and steel institute. Publication, 97 Eyre and Spottiswoode, Margate,1967
Met 25.0429

BRITISH IRON AND STEEL RESEARCH ASSOCIATION
Joint conference on low-alloy steels organized by BISRA and the Iron and steel institute Proceedings Scarborough 1965 Apr 2-4 Iron and steel institute.Publication, 114 illus 269p London,1969
Met 25.2566

BRITISH IRON AND STEEL RESEARCH ASSOCIATION
Metallurgical developments in carbon steels :a conference Harrogate 1963 May 15-16 Iron and steel institute Iron and steel institute. Special report, 81 Iron and steel institute, London,1963
Met 25.0417

BRITISH IRON AND STEEL RESEARCH ASSOCIATION
Metallurgical developments in high-alloy steels A Joint conference on high-alloy steels Proceedings Scarborough 1964 Jun 2-4 Iron and steel institute.Special report, 86 Iron and steel institute,London,1964
Met 25.0488

BRITISH IRON AND STEEL RESEARCH ASSOCIATION
Strong tough structural steels joint conference... Proceedings Scarborough 1967 Apr 4-6 Iron and steel institute. Publication, 104 Eyre and Spottiswoode, Margate,1967
Met 25.0428

BRITISH IRON AND STEEL RESEARCH ASSOCIATION
The All-basic open-hearth furnace being also the report of the 36th steelmaking conference Leamington-Spa 1951 May 2-3 Iron and steel institute.Special report, 46 Iron and steel institute,London,1952
Met 25.0330

BRITISH JOINT CORROSION GROUP Corrosion and its prevention in motor vehicles a symposium Proceedings London 1968 Mar 28-29 Institution of mechanical engineers. Proceedings, 182 pt 3J Inst.of mechanical engineers,London,1968
Met 25.1943

BRITISH MATHEMATICAL COLLOQUIUM 10 lectures, typescript Reading 1958 Mar 84p 33cm University of Reading,Reading,1958
P. Math 2.2600

BRITISH MEDICAL ASSOCIATION ANNUAL MEETING
Proceedings London 1948 Butterworth's, London,1949
Radioth 35.0666

BRITISH MEDICAL ASSOCIATION ANNUAL MEETING 104TH
The Book of Oxford British medical association xvii,243p London,1936 Printed for the 104th meeting of the B.M.A.
WSM 43.0296

BRITISH MICROCIRCULATION SOCIETY Symposium on techniques used in the study of microcirculation Papers Oxford 1968 Sep Edited by D.R. Chambers and P.A.G. Monro British microcirculation society,Cambridge, 1969
An 32.5223

BRITISH NATIONAL ASSOCIATION FOR THE PREVENTION OF TUBERCULOSIS International conference of the International union against tuberculosis 2nd Transactions London 1921 Jul 26-28 International union against tuberculosis Adlard and Newman,London,1921
An 32.2990

BRITISH NATIONAL COMMITTEE FOR NON-DESTRUCTIVE TESTING International conference on non-destructive testing 4th Proceedings London 1963 Sep 9-13 Butterworths,London, 1964
Eng 41.3847

BRITISH NATIONAL COMMITTEE ON MATERIALS
Machines for materials and environmental testing Symposium on techniques and equipment for environmental testing; symposium on developments in materials testing machine design :joint conference Manchester 1965 Sep 6-10 Institution of mechanical engineers and Society of environmental engineers Institution of mechanical engineers. Proceedings, 180,pt 3a I.M.E.,London,1966
Met 25.0891

BRITISH NATIONAL COMMITTEE ON SPACE RESEARCH
A Discussion on solar studies with special reference to space observations London 1970 Apr 21-22 Philosophical transactions, 270A,1202 195p royal society of London, London,1971
Obs 6.3606

BRITISH ORNITHOLOGISTS' UNION Progress and prospects in ornithology being the centenary symposium Proceedings Cambridge 1959 Mar 20-23 Ibis, 101,no 3-4,Centenary celebration number 25cm London,1959
Bal 44.1519

BRITISH ORTHOPAEDIC ASSOCIATION Lubrication and wear in living and artificial human joints a symposium London 1967 Apr 7 Institution of mechanical engineers. Proceedings, 181, pt.3 Institution of mechanical engineers,London,1967
Met 25.1626

BRITISH PHARMACOLOGICAL SOCIETY Metabolism of amines in the brain :a symposium Proceedings Edinburgh 1968 Jul 11 Edited by G. Hooper Macmillan,London,1969
An 32.5040

BRITISH RAILWAYS ELECTRIFICATION CONFERENCE
Proceeding Railway electrification at industrial frequency London 1960 Oct 3 British transport commission Illus British transport commission,London,1960
Eng 41.1605

BRITISH ROAD FEDERATION Urban motorways :a conference British road federation,London, 1957
Eng 41.3298

BRITISH SMALL ANIMALS VETERINARY ASSOCIATION
Small animal anaesthesia :a symposium Proceedings London 1963 Jul Edited by Oliver Graham-Jones Pergamon press,London, 1964
Radioth 35.0270

BRITISH SMALL ANIMALS VETERINARY ASSOCIATION and UNIVERSITIES FEDERATION FOR ANIMAL WELFARE Small animal anaesthesia :a symposium Proceedings London 1963 Jul Edited by Oliver Graham-Jones Illus. Pergamon press,London,1964
VA 19.0264

BRITISH SOCIETY FOR IMMUNOLOGY Immunity to protozoa :a symposium Papers Edited by P.C. C. Garnham and others Blackwell,Oxford,1963
Path 30.2335

BRITISH SOCIETY FOR RESEARCH ON AGEING The Biology of hair growth :a conference Papers London 1957 Aug 7-9 Edited by William Montagna and Richard A. Ellis Academic press, New York,1958
An 32.4034

BRITISH SOCIETY FOR SOCIAL RESPONSIBILITY IN SCIENCE The Social impact of modern biology :an international conference Papers and discussions London 1970 Nov 26-28 Edited by Watson Fuller 256p Routledge and Kegan Paul,London,1971
WSM 43.0589

BRITISH SOCIETY FOR THE PHILOSOPHY OF SCIENCE
Problems in the philosophy of mathematics International colloquium in the philosophy of science Proceedings London 1965 Jul 11-17 Vol 1 Edited by Imre Lakatos Under the auspices of the International union of history and philosophy of science.Division of logic,methodology and philosophy of science Studies in logic and the foundations of mathematics xv,241p North-Holland, Amsterdam,1967
WSM 43.1813

BRITISH SOCIETY FOR THE PHILOSOPHY OF SCIENCE
Problems in the philosophy of science International colloquium in the philosophy of science Proceedings London 1965 Jul 11-17 Vol 3 Edited by Imre Lakatos and Alan Musgrave Under the auspices of the International union of history and philosophy of science.Division of logic,methodology and philosophy of science Studies in logic and the foundations of mathematics ix,448p North-Holland,Amsterdam,1968
WSM 43.0931

BRITISH SOCIETY OF RHEOLOGY Flow properties of blood and other biological systems :an informal discussion Proceedings Oxford 1959 Sep 23-24 Edited by A.L. Copley and G. Stainsby Pergamon,London,1960
PGMS 29.0269

BRITISH SOCIETY OF RHEOLOGY Flow properties of blood and other biological systems :an informal discussion... Proceedings Oxford 1959 Sep 23-24 Edited by A.L. Copley and G. Stainsby Pergamon press,Oxford,1960
An 32.3471

BRITISH SOCIETY OF SOIL SCIENCE Sorption and transport processes in soils... :a symposium 1970 Society of chemical industry.Monograph, 37 Society of chemical industry,London,1970
Eng 41.8281

BRITISH SOCIOLOGICAL ASSOCIATION The Development of industrial societies :papers read at the Nottingham conference of the British sociological association Nottingham 1964 Apr Edited by Paul Halmos Sociological review monograph, 8 University of Keele,Keele,1964
Eng 41.0914

BRITISH TRANSPORT COMMISSION Railway electrification at industrial frequency British railways electrification conference Proceedings London 1960 Oct 3 Illus British transport commission,London,1960
Eng 41.1605

BRITISH WELDING RESEARCH ASSOCIATION
Pressure vessel research towards better design a symposium Proceedings London 1961 Jan Institution of mechanical engineers,London, 1962
Eng 41.2875

BRITTLE FAILURE OF ROTOR FORGINGS Symposium on brittle failure of rotor forgings presented at special meeting for delegates to world metals congress Philadelphia,Pa. 1957 Nov 11 American society for testing materials A.S.T.M.Special technical publication, 231 American society for testing materials, Philadelphia,Pa,1957
Met 25.1325

BRITTLE FRACTURE IN STEEL :conference Proceedings Cambridge 1959 Sep 28-30 Great Britain.Admiralty advisory committee on structural steel P.3 H.M.S.O.,London,1962
Eng 41.3775

BRITTLENESS IN METALS :a conference Culcheth,Lancs 1957 Nov 1 United Kingdom atomic energy authority.Industrial group I.G. Report 145 (RDC) Industrial group headquarters,Risley;Warrington,1959
Met 25.0859

BRNO 1962 Biological effects of ionizing radiation at the molecular level Symposium on the biological effects of ionizing radiation at the molecular level Proceedings Held by the International atomic energy agency International atomic energy agency.Proceedings series I.A.E.A.,Vienna,1962
Radioth 35.1976

BRNO 1965 Prague 1965 Aug 9-11 Symposia Csav The Physiology of gene and mutation expression :symposium on the mutational process Proceedings Edited by Margita Kohoutova and J. Hubacek Organized by the Ceskoslovenska akademie ved.Institute of microbiology Academia,Prague,1967
Gen 34.0770

BROOK LODGE,MICH. 1961 Mechanism of cell and tissue damage produced by immune reactions International symposium on immunopathology 2nd Papers International committee on immunology and Upjohn company Edited by Pierre Grabar and Peter Miescher Schwabe, Basle;Stuttgart,1962
Path 30.2336

The BROOK LODGE SYMPOSIUM Proceedings Tolbutamide...after ten years Augusta,Mich. 1967 Mar 6-7 Edited by W.J.H. Butterfield and W.van Westering International congress series, 149 Excerpta,medica,Amsterdam,1967
Bioch 33.0425

BROOKHAVEN 1951 Chemistry and physiology of the nucleus :symposium Proceedings Held by the Brookhaven national laboratory.Biology department Experimental cell research. Supplement, 2 Academic press,New York,1952
Radioth 35.0459

BROOKHAVEN NATIONAL LABORATORY Bioenergetics considerations of processes of absorption stabilization, transfer and utilization.A symposium Proceedings Upton,N.Y. 1959 Oct 12-16 Edited by Leroy George Augenstine Sponsored by the United States atomic energy commission Radiation research.Supplement, 2 Academic press,New York;London,1960
Bioch 33.0716

BROOKHAVEN NATIONAL LABORATORY Biological effects of neutron and protein irradiations Symposium on biological effects of neutron and proton irradiations Proceedings Upton,N.Y. 1963 Oct 7-11 1-2 International atomic energy agency International atomic energy agency.Proceedings series 2 vols International atomic energy agency,Vienna,1964
Radioth 35.1168

BROOKHAVEN NATIONAL LABORATORY Meristems and differentiation report of symposium Upton,N. Y. 1963 Jan 3-5 Brookhaven symposia in biology, 16 BNL 805 Brookhaven national laboratory,Upton,N.Y.,1964
Gen 34.1414

BROOKHAVEN NATIONAL LABORATORY Subunit structure of proteins,biochemical and genetic aspects report of symposium Upton,N.Y. 1964 Jun 1-3 Brookhaven symposia in biology, 17 Brookhaven national laboratory,Upton,N.Y. 1964
Gen 34.0769

BROOKHAVEN NATIONAL LABORATORY.BIOLOGICAL DEPARTMENT Structure and function of genetic elements report of a symposium Upton, N.Y. 1959 Jun 1-3 Brookhaven symposia in biology, 12 BNL 558 Brookhaven national laboratory,Upton,N.Y.,1959
Gen 34.0674

BROOKHAVEN NATIONAL LABORATORY.BIOLOGY DEPARTMENT Chemistry and physiology of the nucleus : symposium Proceedings Brookhaven 1951 Aug Experimental cell research.Supplement, 2 Academic press,New York,1952
Radioth 35.0459

BROOKHAVEN NATIONAL LABORATORY.BIOLOGY DEPARTMENT Energy conversion by the photosynthetic apparatus :symposium Report Upton,N.Y. 1966 Jun 6-9 Brookhaven symposia in biology, 19 BNL989(C-48) Brookhaven national laboratory.Biology department,Upton,N.Y.,1967
Bioch 33.1297

BROOKHAVEN NATIONAL LABORATORY.BIOLOGY DEPARTMENT Enzyme models and enzyme structure : symposium Report Upton,N.Y. 1962 Jun 4-6 Brookhaven symposia in biology, 15 BNL738(C-34) Brookhaven national laboratory.Biology department,Upton,N.Y.,1962
Bioch 33.1293

BROOKHAVEN NATIONAL LABORATORY.BIOLOGY DEPARTMENT Fundamental aspects of radiosensitivity report of a symposium Upton,N.Y. 1961 Jun 5-7 Brookhaven symposia in biology, 14 BNL 675 Brookhaven national laboratory,Upton, N.Y.,1961
Gen 34.1115

BROOKHAVEN NATIONAL LABORATORY.BIOLOGY DEPARTMENT Genetic control of differentiation :report of a symposium Upton,N.Y. 1965 Jun 7-9 Brookhaven symposia in biology, 18 Brookhaven national laboratory,Upton,N.Y.,1965
Gen 34.0499

BROOKHAVEN NATIONAL LABORATORY.BIOLOGY DEPARTMENT Genetic control of differentiation : symposium Report Upton,N.Y. 1965 Jun 7-9 Brookhaven symposia in biology, 18 BNL931(C-44) Brookhaven national laboratory.Biology department,Upton,N.Y.,1965
Bioch 33.1296

BROOKHAVEN NATIONAL LABORATORY.BIOLOGY DEPARTMENT Genetics in plant breeding report of symposium Upton,N.Y. 1956 May 21-23 Brookhaven symposia in biology, 9 Brookhaven national laboratory,Upton,N.Y.,1956
Gen 34.1152

BROOKHAVEN NATIONAL LABORATORY.BIOLOGY DEPARTMENT Meristems and differentiation :symposium Report Upton,N.Y. 1963 Jun 3-5 Brookhaven symposia in biology, 16 BNL805(C-38) Brookhaven national laboratory.Biology department,Upton,N.Y.,1964
Bioch 33.1294

BROOKHAVEN NATIONAL LABORATORY.BIOLOGY DEPARTMENT Protein structure and function :a symposium Report Upton,N.Y. 1960 Jun 6-8 Brookhaven symposia in biology, 13 BNL608(C-30) Brookhaven national laboratory.Biology department,Upton,N.Y.,1960
Bioch 33.1292

BROOKHAVEN NATIONAL LABORATORY.BIOLOGY DEPARTMENT Protein structure and function :symposium Upton,N.Y. 1960 Jun 6-8 Brookhaven symposia in biology,13 BNL-608 Brookhaven national laboratory,biology dep.,Upton,N.Y., 1960
Pha 16.0120

BROOKHAVEN NATIONAL LABORATORY.BIOLOGY DEPARTMENT Recovery and repair mechanics in radiography :symposium Report Upton,N.Y. 1967 Jun 5-7 Brookhaven symposia in biology, 20 BNL50058)C-51) Brookhaven national laboratory.Biology department,Upton,N.Y.,1968
Bioch 33.1298

BROOKHAVEN NATIONAL LABORATORY.BIOLOGY DEPARTMENT Recovery and repair mechanisms in radiobiology report of symposium Upton,N.Y. 1967 Jun 5-7 Brookhaven symposia in biology, 20 BNL 50058 Brookhaven national laboratory,Upton,N.Y.,1968
Gen 34.1128

BROOKHAVEN NATIONAL LABORATORY.BIOLOGY DEPARTMENT
Structure,function and evolution in proteins :symposium Report Upton,N.Y. 1968 Jun 3-5 Brookhaven symposia in biology, 21 BNL50116(C-53) Brookhaven national laboratory.Biology department,Upton,N.Y.,1969
Bioch 33.1299

BROOKHAVEN NATIONAL LABORATORY.BIOLOGY DEPARTMENT
Structure and function of genetic elements : symposium Report Upton,N.Y. 1959 Jun 1-3 Brookhaven symposia in biology, 12 BNL558(C-29) Brookhaven national laboratory.Biology department,Upton,N.Y.,1959
Bioch 33.1291

BROOKHAVEN NATIONAL LABORATORY.BIOLOGY DEPARTMENT
Subunit structure of proteins :biochemical and genetic aspects.Symposium Report Upton, N.Y. 1964 Jun 1-3 Brookhaven symposia in biology, 17 BNL869(C-40) Brookhaven national laboratory.Biology department,Upton,N.Y.,1964
Bioch 33.1295

BROOKHAVEN NATIONAL LABORATORY.BIOLOGY DEPARTMENT
The Photochemical apparatus;its structure and function :a symposium Report Upton,N.Y. 1958 Jun 16-18 Brookhaven symposia in biology, 11 BWL512(C-28) Brookhaven national laboratory.Biology department,Upton,N.Y.,1959
Bioch 33.1290

BROOKHAVEN NATIONAL LABORATORY.MEDICAL DEPARTMENT
Structure and function of polypeptide hormones;insulin Symposium on insulin Proceedings Upton,N.Y. 1965 Nov 8-10 Edited by P.G. Katsoyannis and I.L. Schwartz American journal of medicine, 40, no. 5 Yorke medical group.Publication New York, 1966
Bioch 33.0557

BROOKHAVEN SYMPOSIA IN BIOLOGY, 9 Genetics in plant breeding report of symposium Upton, N.Y. 1956 May 21-23 Brookhaven national laboratory.Biology department Brookhaven national laboratory,Upton,N.Y.,1956
Gen 34.1152

BROOKHAVEN SYMPOSIA IN BIOLOGY, 11 The Photochemical apparatus;its structure and function :a symposium Report Upton,N.Y. 1958 Jun 16-18 Brookhaven national laboratory.Biology department BWL512(C-28) Brookhaven national laboratory.Biology department,Upton,N.Y.,1959
Bioch 33.1290

BROOKHAVEN SYMPOSIA IN BIOLOGY, 12
Structure and function of genetic elements : symposium Report Upton,N.Y. 1959 Jun 1-3 Brookhaven national laboratory.Biology department BNL558(C-29) Brookhaven national laboratory.Biology department,Upton,N.Y.,1959
Bioch 33.1291

BROOKHAVEN SYMPOSIA IN BIOLOGY, 12
Structure and function of genetic elements report of a symposium Upton,N.Y. 1959 Jun 1-3 Brookhaven national laboratory. Biological department BNL 558 Brookhaven national laboratory,Upton,N.Y.,1959
Gen 34.0674

BROOKHAVEN SYMPOSIA IN BIOLOGY, 13 Protein structure and function :a symposium Report Upton,N.Y. 1960 Jun 6-8 Brookhaven national laboratory.Biology department BNL608(C-30) Brookhaven national laboratory. Biology department,Upton,N.Y.,1960
Bioch 33.1292

BROOKHAVEN SYMPOSIA IN BIOLOGY, 14
Fundamental aspects of radiosensitivity report of a symposium Upton,N.Y. 1961 Jun 5-7 Brookhaven national laboratory.Biology department BNL 675 Brookhaven national laboratory,Upton,N.Y.,1961
Gen 34.1115

BROOKHAVEN SYMPOSIA IN BIOLOGY, 15 Enzyme models and enzyme structure :symposium Report Upton,N.Y. 1962 Jun 4-6 Brookhaven national laboratory.Biology department BNL738(C-34) Brookhaven national laboratory.Biology department,Upton,N.Y.,1962
Bioch 33.1293

BROOKHAVEN SYMPOSIA IN BIOLOGY, 16
Meristems and differentiation :symposium Report Upton,N.Y. 1963 Jun 3-5 Brookhaven national laboratory.Biology department BNL805(C-38) Brookhaven national laboratory.Biology department,Upton,N.Y.,1964
Bioch 33.1294

BROOKHAVEN SYMPOSIA IN BIOLOGY, 16
Meristems and differentiation report of symposium Upton,N.Y. 1963 Jan 3-5 Brookhaven national laboratory BNL 805 Brookhaven national laboratory,Upton,N.Y.,1964
Gen 34.1414

BROOKHAVEN SYMPOSIA IN BIOLOGY, 17 Subunit structure of proteins :biochemical and genetic aspects.Symposium Report Upton,N.Y. 1964 Jun 1-3 Brookhaven national laboratory. Biology department BNL869(C-40) Brookhaven national laboratory.Biology department,Upton,N.Y.,1964
Bioch 33.1295

BROOKHAVEN SYMPOSIA IN BIOLOGY, 17 Subunit structure of proteins,biochemical and genetic aspects report of symposium Upton,N.Y. 1964 Jun 1-3 Brookhaven national laboratory Brookhaven national laboratory,Upton,N.Y.,1964
Gen 34.0769

BROOKHAVEN SYMPOSIA IN BIOLOGY, 18 Genetic control of differentiation :report of a symposium Upton,N.Y. 1965 Jun 7-9 Brookhaven national laboratory.Biology department Brookhaven national laboratory, Upton,N.Y.,1965
Gen 34.0499

BROOKHAVEN SYMPOSIA IN BIOLOGY, 18 Genetic control of differentiation :symposium Report Upton,N.Y. 1965 Jun 7-9 Brookhaven national laboratory.Biology department BNL931(C-44) Brookhaven national laboratory. Biology department,Upton,N.Y.,1965
Bioch 33.1296

BROOKHAVEN SYMPOSIA IN BIOLOGY, 19 Energy conversion by the photosynthetic apparatus : symposium Report Upton,N.Y. 1966 Jun 6-9 Brookhaven national laboratory.Biology department BNL989(C-48) Brookhaven national laboratory.Biology department,Upton,N. Y.,1967
Bioch 33.1297

BROOKHAVEN SYMPOSIA IN BIOLOGY, 20 Recovery and repair mechanics in radiography :symposium Report Upton,N.Y. 1967 Jun 5-7 Brookhaven national laboratory.Biology department BNL50058)C-51) Brookhaven national laboratory.Biology department,Upton,N. Y.,1968
Bioch 33.1298

BROOKHAVEN SYMPOSIA IN BIOLOGY, 20 Recovery and repair mechanisms in radiobiology report of symposium Upton,N.Y. 1967 Jun 5-7 Brookhaven national laboratory.Biology department BNL 50058 Brookhaven national laboratory,Upton,N.Y.,1968
Gen 34.1128

BROOKHAVEN SYMPOSIA IN BIOLOGY, 21 Structure,function and evolution in proteins : symposium Report Upton,N.Y. 1968 Jun 3-5 Brookhaven national laboratory.Biology department BNL50116(C-53) Brookhaven national laboratory.Biology department,Upton,N. Y.,1969
Bioch 33.1299

BROOKHAVEN SYMPOSIA IN BIOLOGY, 21 Structure,function and evolution in proteins report of symposium Upton.N?Y. 1968 Jun 3-5 Vol 1-2 Brookhaven national laboratory.Biology department BNL 50116 2 vols brookhaven national laboratory,Upton,N. Y.,1969
Gen 34.0671

BROOKHAVEN SYMPOSIA IN BIOLOGY,13 Protein structure and function :symposium Upton,N.Y. 1960 Jun 6-8 Brookhaven national laboratory. Biology department BNL-608 Brookhaven national laboratory,biology dep.,Upton,N.Y., 1960
Pha 16.0120

BROOKLINE,MASS. 1962 Human ovulation :a symposium Edited by Chester S. Keefer Sponsored by the Lowell M.Palmer foundation Churchill,London,1965
An 32.3886

BROOKLYN 1955 Applied probability symposium proceedings American mathematical society Edited by L.A. MacColl American mathematical society.Proceedings of symposia in applied mathematics, 7 American mathematical society,Providence,R.I., 1957
A Math 4.0244

BROOKLYN,N.Y. 1935 Brooklyn botanic garden memoirs twenty-fifth anniversary papers Vol 4 Brooklyn botanic garden Brooklyn botanic garden,Brooklyn,N.Y.,1936
Bot 42.0659

BROOKLYN,N.Y. 1955 Conference on high-speed aeronautics Proceedings Polytechnic institute of Brooklyn.Department of aeronautical engineering and applied mechanics Edited by Antonio Ferri and others Polytechnic institute of Brooklyn,Brooklyn,N.Y. 1955
Eng 41.6955

BROOKLYN,N.Y. 1964 Cerebrospinal fluid and the regulation of ventilation :a symposium Proceedings Edited by Chandler McC. Brooks and others Held at the Downstate medical centre Blackwell,Oxford,1965
An 32.4209

BROWN,BOVERI AND COMPANY Brown Boveri symposium 1st Flow research on blading :a symposium Proceedings Baden,Switzerland 1969 Mar 10-11 Edited by Lang S. Dzung Elsevier,Amsterdam,1970
Eng 41.6828

BROWN,BOVERI AND COMPANY Real-time control of electric power systems Symposium of real-time control of electric power systems Baden 1971 Sep 27-28 Edited by Edmund Handschin Elsevier,Amsterdam,1972 The second Brown, Boveri symposium
Eng 41.8558

BROWN BOVERI SYMPOSIUM 1ST Flow research on blading :a symposium Proceedings Baden, Switzerland 1969 Mar 10-11 Brown,Boveri and company Edited by Lang S. Dzung Elsevier,Amsterdam,1970
Eng 41.6828

BROWN UNIVERSITY Cutaneous innervation : Brown university symposium Proceedings Providence,R.I. 1959 Jan 24-25 Edited by William Montagna Advances in the biology of skin,1 Pergamon press,London,1966
VA 19.0232

BROWN UNIVERSITY Plasticity Naval structural mechanics :a symposium 2nd Proceedings Rhode Island 1960 Apr 5-7 Edited by E.H. Lee and P.S. Symonds Pergamon press,New York,1960
Met 25.0845

BROWN UNIVERSITY Plasticity Naval structural mechanics symposium 2nd Proceedings Providence,R.I. 1960 Apr 5-7 Edited by E.H. Lee and P.S. Symonds United States.Office of naval research.Structural mechanics series Pergamon press,London,1960
Eng 41.2479

BROWN UNIVERSITY Symposium on time series analysis proceedings Providence,R.I. 1962 Jun.11-14 Edited by Murray Rosenblatt Also sponsored by United States.Office of naval research xiv,497p 25cm John Wiley and sons,New York,1963
Math 3.0695

BROWN UNIVERSITY The Symposium on time series analysis Proceedings Providence,R.I. 1962 Jun 11-14 Edited by Murray Rosenblatt xiv,497p 26cm Wiley,New York; London,1963
Cav 7.2744

BROWN UNIVERSITY.ADVANCED INSTRUCTION AND RESEARCH IN MECHANICS,1941 Fluid dynamics manuscript By Richard von Mises and Kurt O. Friedrichs Providence,R.I.,1942 Containing'The Hodograph method in the theory of compressible fluid'by Stefan Bergman
A Math 4.1292

BROWN UNIVERSITY.ADVANCED INSTRUCTION AND RESEARCH IN MECHANICS,1942 The Hodograph method in the theory of compressible fluid By Stefan Bergman Providence,R.I.,1942 Supplementary notes contained in Mises,Richard von and Friedrich,Kurt O., 'Fluid dynamics'
A Math 4.1293

BROWN UNIVERSITY SYMPOSIUM ON THE BIOLOGY OF SKIN, 1959 Proceedings Cutaneous innervation Providence,R.I. 1959 Jan 24-25 Edited by William Montagna Advances in biology of skin, 1 Pergamon press,Oxford,1960
An 32.4037

BROWN UNIVERSITY SYMPOSIUM ON THE BIOLOGY OF SKIN, 1960 Proceedings Blood vessels and circulation Providence,R.I. 1960 Jan 30-31 Edited by William Montagna and Richard A. Ellis Advances in biology of skin, 2 Pergamon press,Oxford,1961
An 32.4038

BROWN UNIVERSITY SYMPOSIUM ON THE BIOLOGY OF SKIN, 1961 Proceedings Eccrine sweat glands and eccrine sweating Providence,R.I. 1961 Jan 28-29 Edited by William Montagna and others Advances in biology of skin, 3 Pergamon press,Oxford,1962
An 32.4039

BROWN UNIVERSITY SYMPOSIUM ON THE BIOLOGY OF SKIN, 1962 Proceedings Sebaceous glands Providence,R.I. 1962 Jan 27-28 Edited by William Montagna and others Advances in biology of skin, 4 Pergamon press,Oxford, 1963
An 32.4040

BROWN UNIVERSITY SYMPOSIUM ON THE BIOLOGY OF SKIN, 1963 Proceedings Wound healing Providence,R.I. 1963 Jan 26-27 Edited by William Montagna and Rupert E. Billingham Advances in biology of skin, 5 Pergamon press,Oxford,1964
An 32.4044

BROWNIAN MOTION Promenades aleatoires et mouvement brownien :introduction a la theorie des probabilites mimeograph Montreal 1963 By Anatole Joffe North Atlantic treaty organization Societe mathematique du Canada Montreal.University.Seminaire de mathematiques superieures, 7 vii,143p 28cm Universite de Montreal,Montreal,1964
P. Math 2.2145

BRUGES 1957 Protides of the biological fluids :colloquium 5th Proceedings Edited by H. Peeters Elsevier,Amsterdam,1958
Col S 12.0126

BRUGES 1963-65,67 Protides of the biological fluids :colloquia 11th-13th,15th Proceedings Edited by H. Peeters Elsevier, Amsterdam,1964-68 In English,French and German
Bioch 33.0538

BRUGES 1966 Summer school on topological algebra theory Bibliog 285p 27cm University of Brussels,Brussels,1966
P Math 2.2745

BRUNSWICK 1897 Versammlung Deutscher Naturforscher und Aerzte 69 Festgruss Verein fur naturwissenschaft zu Braunschweig Schulbuchhandlung,Brunswick,1897
Philos 1.0012

BRUNSWICK 1897 Versammlung Deutscher naturforscher und aertze 69 Festschrift Gesellschaft deutscher naturforscher und arzte Harald Bruhn,Brunswick,1897
Philos 1.0008

BRUSSELLS 1962 Colloque sur la theorie des groupes algebriques 20th colloque Centre Belge de recherches mathematiques 150p 25cm Librairie universitaire,Louvain, 1962
P. Math 2.0573

BRUSSELS 1906 Congres international pour l'etude des regions polaires Rapport d'ensemble documents preliminaires et comte rendu des seances Hayez,Brussels,1906
Sco 14.0109

BRUSSELS 1906 Congres international pour l'etude des regions polaires programme 40p Hayez,Brussels,1906 In English,French and German
Sco 14.0108

BRUSSELS 1908 Commission polaire internationale :session Proces-verbaux des seances International polar commission Edited by Georges Lecointe 110,clxiip 25cm Brussels,1908 In English,French,German, Russian
Sco 14.0110

BRUSSELS 1910 Congress international de botanique 3rd Actes Vol 2: conferences et memoires Edited by E.de Wildeman Boeck,Brussels,n.d.
Bot 42.0646

BRUSSELS 1910 International tuberculosis conference 9th Report International anti-tuberculosis association Edited by Gotthold Pannwitz port Berlin-Charlottenburg,1911 Text in English, French,German;title also in French and German
Path 30.1026

BRUSSELS 1921 Atomes et electrons Conseil de physique 3eme Rapports Institut international de physique Solvay Solvay conference,1921,3 Paris,1923
Cav 7.2080

BRUSSELS 1922 Congres geologique international International geological congress 13th comptes rendus Fasc 1-3 3 vols Liege,1924-26
Geog 13.0200

BRUSSELS 1927 Electrons et photons Conseil de physique 5eme Rapports Institut international de physique Solvay Solvay conference,1927,5 Paris,1928
Cav 7.2081

BRUSSELS 1930 Congres national des sciences :section d'astronomie Comptes rendus Federation belge des societes scientifiques 103p Societe d'astronomie d'Anvers,Brussels,1930
Obs 6.2021

BRUSSELS 1933 Structure et proprietes des noyaux atomiques Conseil de physique 7eme Rapports Institut international de physique Solvay and Universite libre de Bruxelles Solvay conference,1933,7 Paris,1934
Cav 7.2083

BRUSSELS 1949 Role des anaerobies dans la nature Congres international des microbiologistes de langue francaise International union of biological sciences International union of biological sciences. Series B.Colloques, 7 Secreteriat general de l'U.I.S.B.,Paris,1949
Bal 39.3919

BRUSSELS 1950 Colloque de topologie (espaces fibres) Centre Belge de recherches mathematiques 136p 24cm Centre Belge de recherches mathematiques,Brussels,1950
P Math 2.3126

BRUSSELS 1951 Association internationale d'hydrologie scientifique :assemblee generale 9th Rapports et comtes-rendus Tome 1: 1... 2...3:rapports et comtes-rendus des seances de la commission de neige et des glaciers International association of scientific hydrology.Publication,32 Louvain,1951 Papers in English,French and German
Sco 14.0120

BRUSSELS 1951 Etat solide Conseil de physique 9 eme Rapports et discussions Institut international de physique Solvay R. Stoops,Brussels,1952
Met 25.1135

BRUSSELS 1951 Symposium on the general circulation of the oceans and the atmosphere International geodetic and geophysical union : general assembly 9th Symposium proceedings International association of meteorology and atmospheric physics Unesco,Brussels,1951
Geog 13.0960

BRUSSELS 1953 Proteines Conseil de chimie 9e Rapports et discussions Institut international de chimie Solvay and Universite libre de Bruxelles Edited by R. Stoops Brussels,1953
Bioch 33.0522

BRUSSELS 1955 International congress of biochemistry 3rd Proceedings Edited by Claude Liebecq Academic press,New York,1956 Text in English,French and German
Bioch 33.1311

BRUSSELS 1955 International congress of biochemistry 3rd Programme et liste des membres Council for international organizations of medical sciences Secretariat general;Agence de voyage,Liege; Brussels,c1956 Bound with 3 eme Congres international de biochimie.Resumes des communications
Bioch 33.1312

BRUSSELS 1955 International congress of biochemistry 3rd Rapports (volume provisoire) Council for international organizations of medical sciences Imprimerie Vaillant-Carmanne,Liege,1955
Bioch 33.1314

BRUSSELS 1955 International congress of biochemistry 3rd Resumes des communications Council for international organizations of medical sciences c1956 Text in English,French and German.Bound with 3 eme Congres international de biochimie. Programme et liste des membres
Bioch 33.1313

BRUSSELS 1955- International analogue computation meeting 1st,3rd,4th Proceedings Brussels,1956-
Eng 41.2242

BRUSSELS 1956 International physiological congress 20th abstracts of communications 1956
Phys 20.2101

BRUSSELS 1956 International physiological congress 20th abstracts of reviews 1956
Phys 20.2102

BRUSSELS 1956 Quelques problemes de chimie minerale :conseil de chimie 10e Rapports et discussions Institut international de chimie Solvay.Comite scientifique Solvay conference on chemistry,1956 R.Stoops, Brussels,1956
Chem 18.1004

BRUSSELS 1958 Brussels conference on earth pressure problems Proceedings Vol 1-3 International society of soil mechanics and foundation engineering Brussels,1958
Eng 41.3165

BRUSSELS 1958 Optics in metrology Colloquia of the International commission for optics Papers International commission for optics Edited by Pol Mollet Pergamon press, Oxford,1960 Papers presented in English, French or German
Cav 7.1631

BRUSSELS 1958 Semaine internationale de geographie :international geographical week Report Federation Belge des geographes professeurs de l'enseignement moyen,normal et technique Ghent,c1958
Geog 13.5602

BRUSSELS 1958 Solid state physics in electronics and telecommunications : international conference Proceedings Vol 3,pt 1: magnetic and optical properties International union of pure and applied physics Edited by M. Desirant and J.L. Michiels Academic press,London,1960
Cav 7.0387

BRUSSELS 1958 Solid state physics in electronics and telecommunications : international conference Proceedings Vol 1-2,pt 1-2: semiconductors International union of pure and applied physics Edited by M. Desirant and J.L. Michiels 2 vols Academic press,London,1960
Cav 7.0388

BRUSSELS 1958 Structure et evolution de l'univers Conseil de physique 11e Rapports Universite libre de Bruxelles Institut de physique Solvay Edited by R. Stoops 309p Brussels,1958
Obs 6.3255

BRUSSELS 1958 Structure et evolution de l'univers Conseil de physique 11eme Rapports Institut international de physique Solvay and Universite libre de Bruxelles Edited by R. Stoops Solvay conference,1958, 11 Brussels,1958
Cav 7.2084

BRUSSELS 1958 The Relationship between nucleus and cytoplasm :a symposium Proceedings Organised by the International society for cell biology Experimental cell research.Supplement, 6 Academic press,New York,1959
An 32.2486

BRUSSELS 1958 The Relationship between nucleus and cytoplasm :symposium Proceedings Organized by the International society for cell biology Experimental cell research,Supp. 6,1959 illus. Academic press,New York, 1959
Gen 34.0837

BRUSSELS 1959 A.E.C.-Euratom conference on aqueous corrosion of reactor materials United States atomic energy commission and European atomic energy society Office of technical services,Washington,D.C.,1959
Met 25.1930

BRUSSELS 1959 Nucleoproteines Conseil de chimie 11eme Institut international de chimie Solvay Edited by R. Stoops Interscience;Stoops,New York;Brussels,1960
Radioth 35.0100

BRUSSELS 1959 Some ionospheric results obtained during the international geophysical year :a symposium Proceedings International scientific radio union.Special commission for the International Geophysical Year Edited by W.J.G. Beynon xi,399p Elsevier,Amsterdam,1960
Nap 11.0053

BRUSSELS 1960 Les Fonctions de nidation uterine et leurs troubles :colloque Papers Societe nationale pour l'etude de la sterilite et de la fecondite Edited by J. Ferin and M. Gaudefroy Masson,Paris,1960
An 32.3834

BRUSSELS 1961 Simplified calculation methods of shell structures :colloquium Proceedings International association for shell structures Edited by A. Paduart and R. Dutron North-Holland,Amsterdam,1962
Eng 41.2864

BRUSSELS 1961 Symposium international sur les marees terrestres :symposium on earth tides 4th Proceedings Edited by P. Melchior Observatoire royal de Belgique. Communication., 188,Serie geophysique, 58 R. Louis,Ixelles,i961
Geod 9.0025

BRUSSELS 1961 Theorie quantique des champs Conseil de physique 12eme Rapports Institut international de physique Solvay and Universite libre de Bruxelles Edited by R. Stoops Solvay conference,1961,12 Interscience,New York,c 1961 13th Solvay conference issued as Conference on physics,q.v.
Cav 7.2077

BRUSSELS 1962 Hortus belgicus :catalogue de l'exposition Bibliotheque Albert I, Belgium Edited by Jan Balis Bibliotheque royale de Belgique.Catalogues des expositions, 10 Brussels,1962 Exhibition organised to coincide with the 16th International horticultural congress Book wrongly calls it Botanical congress
BG 38.1857

BRUSSELS 1962 Information and prediction in science Symposium de l'Academie internationale de philosophie des sciences Proceedings Academie internationale de philosophie des sciences Edited by S. Dockx and P. Bernays xi,272p Academic press, London;New York,1965
WSM 43.1136

BRUSSELS 1962 International horticultural congress 16th Vol 1-5 International society for horticultural sciences 5 vols Duculot,Gembloux,1962
BG 38.3126

BRUSSELS 1962 Transfert d'energie dans les gaz 12th Institut international de chimie Solvay Solvay conference on chemistry,1962 Interscience;R.Stoops,New York;Brussels,c1962
Chem 18.0514

BRUSSELS 1963 European anatomical meeting 2nd lectures on morphology Archives de biologie,Liege,1965 Most papers in French
An 32.1195

BRUSSELS 1963 Soleil a la renaissance: sciences et mythes :colloque international Federation internationale des instituts et societes pour l'etude de la Renaissance and Belgique.Ministere de l'education et de la culture Universite libre de Bruxelles. Institut pour l'etude de la renaissance et de l'humanisme.Publication, 2 584p Presses universitaires de Bruxelles,Brussels;Paris, 1965 Contains Hoskin,M.and Jones,C. 'Problems in late renaissance astronomy'
WSM 43.0046

BRUSSELS 1964 Colloque de topologie Centre belge de recherches mathematiques 235p 25cm Louvain, Paris,1966
P. Math 2.0294

BRUSSELS 1964 Commemoration solonelle du quatrieme centenaire de la mort d'Andre Vesalius Academie royale de medecine de Belgique Brussels,1964
Phys 20.0564

BRUSSELS 1964 Congres international de gastro-enterologie 7e 1-3 Association des societes nationales europeennes et mediterraneenes de gastro-enterologie 3 vols A.S.N.E.M.G.E.,Brussels, c1964
PGMS 29.0357

BRUSSELS 1964 Rapport de la troisieme reunion consultative 3rd Proceedings Antarctic treaty 39p 21cm Brussels,1964
Sco 14.6117

BRUSSELS 1964 Structure and evolution of galaxies Conference on physics 13th Proceedings Universite libre de Bruxelles and Institut international de physique Solvay 174p Interscience,London,1965
Obs 6.2134

BRUSSELS 1964 Structure and evolution of galaxies The Conference on physics 13th Proceedings Institut international de physique Solvay and Universite libre de Bruxelles Solvay conference,1964,13 Interscience,London,1965 Other Solvay conferences issued as Conseil de physique,q.v.
Cav 7.2078

BRUSSELS 1964 Symposium "Vesale" cellule, forme et fonction Archives de biologie,Liege, 1965 In English and French
An 32.3303

BRUSSELS 1965 Reactivity of photo-excited organic molecula Solvay conference on chemistry 13th Proceedings International institute of chemistry,and,Universite libre de Bruxelles Interscience,London,1967
Col S 12.0568

BRUSSELS 1968 International congress of cell biology 12th Summaries of reports and communications Edited by P. Dustin and others International congress series, 166 Excerpta medica,Amsterdam,1968
Bioch 33.0963

BRUSSELS 1968 International congress of cell biology 12th Summaries of reports and communications Edited by P. Dustin and others Under the auspices of the International society for cell biology International congress series, 166 Excerpta medica,Amsterdam,1968
Bot 42.1515

BRUSSELS 1968 International congress on cell biology 12th Summaries of reports and communications Edited by P. Dustin Excerpta medica.International congress series, 166 Excerpta medica,Amsterdam,1968
Inv Med 37.0250

BRUSSELS 1969 Conformational analysis: scope and present limitations International symposium on conformational analysis Papers Edited by Gregoire Chiurdoglu Organic chemistry, 21
Chem 18.2788

BRUSSELS 1969 International symposium on acetabularia 1st Proceedings Universite libre de Bruxelles.Centre d'etude de l'energie nucleaire Edited by Jean Brachet and Silvano Bonotto illus. xv,300p Academic press, New York;London,1970
Bot 42.3941

BRUSSELS 1969 and MOL 1969 International symposium on the biology of acetabularia 1st Proceedings Universite libre de Bruxelles Centre d'etude de l'energie nucleaire Edited by Jean Brachet and Silvano Bonotto Academic press,New York, 1970
Bioch 33.2261

BRUSSELS CONFERENCE ON EARTH PRESSURE PROBLEMS Proceedings Brussels 1958 Vol 1-3 International society of soil mechanics and foundation engineering Brussels,1958
Eng 41.3165

BUCHAREST 1967 Les Travaux colloque sur les techniques de calcul et les calculations Travaux Societate de Stiinte matematice din R.S.Romania.Bulletin mathematique, 12,1 Societate de Stiinte matematice din R.S. Romania,Bucharest,1968
Math L 5.3513

BUCHAREST 1968 International colloquium on the modernization of mathematics teaching in European countries Proceedings Unesco 572p 25cm Editions didactiques et pedagogiques,Bucharest,1968
P Math 2.4020

BUCHAREST 1968 Symposium on modernization of mathematics teaching in European countries Proceedings Unesco Bibliog 573p 24cm Editions didactiques et pedagogiques,Bucharest, 1969
A Math 4.1688

BUCHAREST 1969 International conference on phenomena in ionized gases 9th Proceedings Institutul de fizica,Bucharest Institute of physics,Bucharest,1969
Eng 41.8553

BUDAPEST 1876 Congres international d'anthropologie et d'archeologie prehistoriques 8th Compte rendu Vol 1 Franklin-Tarsulat,Budapest,1877
An 32.2636

BUDAPEST 1891 Ausstellungen der ungarischen vogelfauna... Edited by Gyula Madarasz Budapest,1891 One of three papers bound together under spine title 'Vom ornithologi congresse in Budapest 1890'
Bal 44.2249

BUDAPEST 1891 Aves Hungariae... : enumeratio systematica avium Hungariae cum notis... By Janos Frivaldszky 22cm Franklin,Budapest,1891 One of three papers bound together under spine title 'Vom ornithologi congresse in Budapest 1890'
Bal 44.2247

BUDAPEST 1891 Die Vogelsammlung des Bosnisch-Herzegovinischen landesmuseum in Sarajevo By C.O. Reiser Budapest,1891 One of three papers bound together under spine title 'Vom ornithologi congresse in Budapest 1890'
Bal 44.2248

BUDAPEST 1891 Die Vogelsammlung des bosnisch-hercegovinischen landesmuseums in Sarajevo Edited by Custos O. Reiser 2 pls 22cm Budapest,1891 Paper read at the conference
Bal 44.2369

BUDAPEST 1918 Zur psychoanalyse der kriegsneurosen :diskussion gehalten auf dem V internationalen psychoanalytischen kongress Beitrage By Sigmund Freud and others Internationale psychoanalytische bbibliothek, 1 Internationaler psychoanelytischer verlag, Leipzig;Vienna,1919
Psy 28.0171

BUDAPEST 1960 Magyar matematikai kongresszus 2 abstracts of lectures, mimeograph Vols. 1-2 and suppt. 29cm 3 vols Budapest,1960
P. Math 2.2607

BUDAPEST 1960 Physiologie (bewegung) der spermien :symposium Edited by I. Toro Symposia biologica hungarica, 4 Akademia Kiado,Budapest,1964
An 32.3880

BUDAPEST 1960 Regeneration and wound healing Proceedings Magyar tudonanyos akademia Symposia biologica hungarica, 3 Akademiai kiado,Budapest,1960
Phys 20.1001

BUDAPEST 1960 Regeneration and wound healing :symposium Edited by G. Szanto Symposia biologica Hungarica, 3 Akademiai Kiado,Budapest,1964
An 32.0093

BUDAPEST 1963 Colloquium on applications of mathematics to economics Bolyai Janos matematikai tarsulat Edited by Andras Prekopa 367p 25cm Akademiai kiado, Budapest,1965
Math 3.0917

BUDAPEST 1964 Host-parasite relations in plant pathology :a symposium Hungarian academy of sciences.Research institute for plant protection Edited by Z. Kiraly and G. Ubrizsy Sponsored by the Ministry of agriculture 257p Budapest,c1965
Bot 42.4628

BUDAPEST 1966 International conference on luminescence Proceedings International union of pure and applied physics Magyar tudomanyos akademia Edited by G. Szigeti diagrms 24cm 2 vols Akademiai kiado, Budapest,1968
Cav 7.3097

BUDAPEST 1966 Symposium on muscle Edited by E. Ernst and F.B. Straub Symposia biologica hungarica, 8 Akademia Kaido, Budapest,1968
An 32.4011

BUDAPEST 1967 Structure and development of solar active regions International astronomical union Edited by K.O. Kiepenheuer International astronomical union. Symposium, 35 Reidel,Dordrecht,1968
TA 15.0420

BUDAPEST 1967 Structure and development of solar active regions :a symposium International astronomical union Edited by K. O. Kiepenheuer International astronomical union.Symposia, 35 Bibliog.,Illus. xvii, 608p 25cm D.Reidel,Dordrecht,1968
A Math 4.1602

BUDAPEST 1967 Structure and development of solar active regions :a symposium Proceedings International astronomical union Edited by K.O. Kiepenheuer International astronomical union.Symposium, 35 608p Reidel,Dordrecht,1968
Obs 6.3266

BUDAPEST 1968 Budapest conference on soil mechanics and foundation engineering 3rd Proceedings Magyar tudomanyos akademia Akademiai kiado,Budapest,1968
Eng 41.3137

BUDAPEST 1968 Colloquium on variable stars Non-periodic phenomena in variable stars International astronomical union.Commissions 27 and 42 International astronomical union. Colloquium, 4 490p Academic press, Budapest,1969
Obs 6.3360

BUDAPEST 1968 Conference on dimensioning and strength calculations 3rd Proceedings Hungarian academy of sciences Edited by E. Czoboly Akademiai kiado,Budapest,1968
Met 25.2407

BUDAPEST 1968 I.F.A.C. symposium on pulse-rate and pulse-number signals in automatic control Proceedings Sponsored by the International federation of automatic control International federation of automatic control, Debrecen,1968
Eng 41.6020

BUDAPEST 1969 Conference on constructive theory of functions Proceedings Edited by G. Alexits and S.B. Stechkin 538p 25cm Akademiai kiado,Budapest,1972
P Math 2.4225

BUDAPEST 1969 International conference on cosmic rays 11th Invited papers and rapporteur talks International union of pure and applied physics Magyar tudomanyos akademia 612p Central research institute of physics,Budapest,1969
TA 15.0551

BUDAPEST 1969 International conference on cosmic rays 11th Proceedings 4: muons and neutrons techniques Edited by A. Somogyi Hungarica acta physica.Supplement, 29 Akademiai kiado,Budapest,1970
TA 15.0635

BUDAPEST 1969 Sterilisation and preservation of biological tissues by ionizing radiations Panel on radiation sterilisation of biological tissues for transplantation International atomic energy agency International atomic energy agency.Panel proceedings series International atomic energy agency,Vienna,1970
Radioth 35.1915

BUDAPEST CONFERENCE ON SOIL MECHANICS AND FOUNDATION ENGINEERING 3rd Proceedings Budapest 1968 Oct 15-18 Magyar tudomanyos akademia Akademiai kiado,Budapest, 1968
Eng 41.3137

BUENOS AIRES 1910 Congreso internacional de americanistas 17o Actas Edited by Robert Lehmann-Nitsche Hermanos,Buenos Aires, 1912
An 32.2621

BUENOS AIRES 1959 Antarctic symposium Papers International union of geodesy and geophysics International union of geodesy and geophysics.Monograph,5 95p Paris,1960
Sco 14.0128

BUENOS AIRES 1959 Human pituitary hormones a colloquium in honour of B.A.Houssay Proceedings Ciba foundation Edited by G.E. W. Wolstenholme and Cecilia M. O'Connor Ciba foundation colloquia on endocrinology, 13 illus Churchill,London,1960
Bioch 33.0440

BUENOS AIRES 1959 International congress of physiological sciences 21st abstracts of communications 1959
Phys 20.2105

BUENOS AIRES 1959 International congress of physiological sciences 21st symposia and special lectures 1959
Phys 20.2106

BUENOS AIRES 1959 International symposium on cold acclimation Proceedings Edited by Robert E. Smith and others Federation of American societies for experimental biology. Federation proceedings,19,4,pt 2,Supplement,5 illus viii,165p Buenos Aires,1966
Sco 14.0862

BUENOS AIRES 1959 Simposio internacional de acclimation al frio Actas By Robert E. Smith and others Institute Antartico Argentino.Publicacion,9 illus 414p 1962
Sco 14.0863

BUENOS AIRES 1962 Informe de la segunda reunion consultiva 2nd Proceedings Antarctic treaty 32p 24cm Buenos Aires, 1962
Sco 14.6119

BUENOS AIRES 1963 Biology of neuroglia Lectures and discussions Instituto Torcuato di Tella and International brain research organization Edited by E.D.P.de Robertis and R. Carea Progress in brain research, 15 illus Elsevier,Amsterdam,1965 Symposium held as part of the 10th Latin-American congress of neurosurgery
Path 30.2466

BUENOS AIRES 1963 Biology of neuroglia :a symposium... held as part of the 10th Latin-American congress of neurosurgery Edited by E.D.P.de Robertis and R. Carrea Progress in brain research, 15 Elsevier,Amsterdam,1965
An 32.4235

BUENOS AIRES 1964 International symposium on genes and chromosomes:structure and function Proceedings National cancer institute Edited by Juan I. Valencia and Rhoda F. Grell National cancer institute. Monograph, 18 National cancer institute, Bethesda,Md.,1965
Bioch 33.0987

BUENOS AIRES 1964 International symposium on genes and chromosomes structure and function Proceedings Edited by Juan I. Valencia and Rhoda F. Grell National cancer institute.Monograph, 18 illus 354p U.S. Dept. of health,education and welfare,Bethesda, Md.,1965
Bot 42.0792

BUENOS AIRES 1964 International symposium on genes and chromosomes structure and function Proceedings Edited by Juan I. Valencia and others National cancer institute.Monographs, 18 U.S.Department of health,education and welfare,Washington,D.C., 1965
Gen 34.0833

BUFFALO 1951 Symposium on principles of gating presented at the 55th annual meeting American foundrymen's society American foundrymen's society,Chicago,Ill.,1951
Met 25.0659

BUFFALO 1958 Reactive metals annual conference 3rd Proceedings Vol 2 American institute of mining,metallurgical and petroleum engineers.Niagara frontier section Edited by W.R. Clough Interscience,New York; London,1959
Met 25.0505

BUFFALO,N.Y. Immunity,cancer,and chemotherapy :basic relationships on the cellular level 1st Report Universita di Milano.Istituto di farmacologia and State university of New York.School of pharmacy Edited by Enrico Mihich Academic press,New York;London,1967
Path 30.2744

BUFFALO,N.Y. 1943 Symposium on powder metallurgy :spring meeting American society for testing and materials American society for testing and materials,Philadelphia,Pa., 1943
Eng 41.3474

BUFFALO,N.Y. 1946 Symposium on materials for gas turbines :presented at the 49th annual meeting A.S.T.M. American society for testing materials American society for testing materials,Philadelphia,Pa.,1946
Eng 41.3752

BUFFALO,N.Y. 1953,1955 Conference on carbon 1st and 2nd Proceedings By S. Mrozowski and others University of Buffalo University of Buffalo,New York,1956
Met 25.0154

BUFFALO,N.Y. 1957 Conference on carbon 3rd Proceedings By S. Morzowski and others University of Buffalo and United States. Office of naval research Co-sponsored by the National science foundation Pergamon press, London,1959
Met 25.0156

BUFFALO,N.Y. 1957 Conference on carbon 3rd Proceedings State university of New York,Buffalo Edited by S. Mrozowski and others Co-sponsored by the United States. Office of naval research Conference on carbon.Proceedings, 2 718p Pergamon press,Oxford,1959
TA 15.0214

BUFFALO,N.Y. 1958 Conference on microcirculatory physiology and pathology,5th The Microcirculation :symposium on factors influencing exchange of substances across capillary wall Edited by S.R.M. Reynolds and Benjamin W. Zweifach University of Illinois press,Urbana,Ill.,1959
Phys 20.0888

BUFFALO,N.Y. 1958 Microcirculation: Symposium on factors influencing exchange of substances across capillary wall A Sterotaxic atlas of the dog's brain By Robert K.S. Lim and others American association of anatomists Edited by S.R.M. Reynolds and Benjamin W. Zweifach Thomas, Springfield,Ill.,1960
VA 19.0362

BUFFALO,N.Y. 1959 Conference on carbon 4th Proceedings By S. Mrozowski and others University of Buffalo and American carbon committee Pergamon press,Oxford,1960
Met 25.0157

BUFFALO,N.Y. 1959 The Conference on carbon 4th Proceedings State university of New York,Buffalo and American carbon committee Edited by S. Mrozowski and others Conference on carbon.Proceedings, 3 778p Pergamon press,Oxford,1960
TA 15.0215

BUFFALO,N.Y. 1968 Information processing in the nervous system :a symposium Proceedings Edited by K.N. Leibovic held at the State university of New York Springer, Berlin,1969
Bal 39.1019

BUFFALO,N.Y. 1968 Intuitionism and proof theory Conference on intuitionism and proof theory Proceedings Edited by A. Kino Studies in logic and the foundations of mathematics Bibliog. viii,516p 23cm North Holland,Amsterdam;London,1970
P Math 2.3613

BUFFALO,N.Y. 1968 Intuitionism and proof theory :summer conference Proceedings State university of New York Edited by A. Kino and others Studies in logic and the foundations of mathematics viii,516p North-Holland,Amsterdam;London,1970
WSM 43.1239

BUFFALO,N.Y. 1969 Set-valued mappings, selections and topological properties Proceedings Edited by W.M. Fleischman Held at the State university of New York Lecture notes in mathematics, 171 Springer,Berlin, 1970
Math S 3.1657

BUFFALO,N.Y. 1969 Set-valued mappings, selections and topological properties : conference Proceedings State university of New York Lecture notes in mathematics, 171 x,109p 25cm Springer,Berlin,1970
P Math 2.4114

BUFFALO,N.Y. 1970 Workshop on microprogramming 3rd Preprints Association for computing machinery bibliog., illus. Association for computing machinery, New York,1970
Math L 5.3574

BUHL FOUNDATION Buhl international conference on materials :transition metal compounds,transport and magnetic properties 1st Informal proceedings Pittsburgh 1963 Oct 31-Nov 1 Edited by E.R. Schatz Gordon and Breach,New York,1964
Chem 18.2630

BUHL FOUNDATION Buhl international conference on materials :transition metal compounds ;transport and magnetic properties 1st Informal proceedings Pittsburgh,Penn. 1963 Oct 31-Nov 1 Edited by E.R. Schatz Gordon and Breach,New York;London,1964
Met 25.2351

BUHL FOUNDATION Metal transformations Buhl international conference on materials 2nd Informal proceedings Pittsburgh,Pa. 1966 Mar 16-17 Edited by William W. Mullins and Milton C. Shaw Gordon and Breach,London;New York,1968
Met 25.2514

BUHL INTERNATIONAL CONFERENCE ON MATERIALS 1st transition metal compounds,transport and magnetic properties Informal proceedings Pittsburgh 1963 Oct 31-Nov 1 Buhl foundation Edited by E.R. Schatz Gordon and Breach,New York,1964
Chem 18.2630

BUHL INTERNATIONAL CONFERENCE ON MATERIALS 1st transition metal compounds ;transport and magnetic properties Informal proceedings Pittsburgh,Penn. 1963 Oct 31-Nov 1 Buhl foundation and Carnegie institute of technology,Pittsburgh Edited by E.R. Schatz Gordon and Breach,New York;London,1964
Met 25.2351

BUHL INTERNATIONAL CONFERENCE ON MATERIALS 2nd Informal proceedings Metal transformations Pittsburgh,Pa. 1966 Mar 16-17 Buhl foundation Edited by William W. Mullins and Milton C. Shaw Gordon and Breach, London;New York,1968
Met 25.2514

BUILDING RESEARCH ADVISORY BOARD Condensation control in buildings as related to paints,papers and insulating materials : conference Report 1952 Feb 26-27 Building research advisory board.Research conference report,4 illus ii,118p 28cm Washington,D.C.,1952
Sco 14.1273

BUILDING RESEARCH ADVISORY BOARD Permafrost international conference 1st Proceedings Lafayette,Indiana 1963 Nov 11-15 National research council National research council. Publication,1287 DLC,GB641.16 1963 illus, maps,tables,graphs 563p National academy of sciences,Washington,1966
Sco 14.0390

BUILDING RESEARCH ADVISORY BOARD World conference on shell structures Proceedings San Francisco,Calif. 1962 Oct 1-4 National research council.Publication,1187 National academy of sciences,Washington,D.C.,1064
Eng 41.3025

BUILDING RESEARCH CONGRESS 1951 Papers presented in division 1-3 London 1951 London,1951
Eng 41.3372

BUILDINGS AND STRUCTURES Wind effects on buildings and structures :a conference Proceedings Teddington 1963 Jun 26-28 Vol 1-2 Great Britain.Department of scientific and industrial research National physical laboratory National physical laboratory.Symposium, 16 H.M.S.O.,London, 1963
Eng 41.6904

BUILDINGS AND STRUCTURES Wind effects on buildings and structures :a symposium Proceedings Loughborough 1968 Apr 2-4 Vol 1-2 Loughborough university of technology Edited by D.J. Johns 2 vols Loughborough,1968
Eng 41.2762

BUILDINGS AND STRUCTURES Wind effects on buildings and structures :international research seminar Proceedings Ottawa 1967 Sep 11-15 Vol 1-2 National research council of Canada 2 vols University of Toronto press,Toronto,1968
Eng 41.2746

BULGARSKO KNIZHOVNO DRUZHESTVO.NATIONAL COMMITTEE ON THEORETICAL AND APPLIED MECHANICS National congress of theoretical and applied mechanics 1st Proceedings Varna 1969 Nov 3-6 Bibliog,illus 632p 24cm Bulgarsko knizhovno druzhestvo,Sofia,1971 Papers in English,French and Russian
A Math 4.1723

BULLETIN DE L'INSTITUT INTERNATIONAL DE STATISTIQUE:COMPTE RENDUS DES SESSIONS DE L'INSTITUT 19e 1930- Vol 25- Institut international de statistique v.p., 1931- Wants Vol 31:25th session
Math 3.1329

BUNKER-RAMO CORPORATION Symposium on computer augmentation of human reasoning proceedings Washington,D.C. 1964 Jun.16-17 United States.Office of naval research Edited by Margo A. Sass and William D. Wilkinson vi,235p 24cm Spartan books; Macmillan,Washington,D.C.;London,1965
Math 3.0244

BUNKER-RAMO CORPORATION Symposium on computer augmentation of human reasoning Proceedings Washington,D.C. 1964 Jun 16-17 United States.Office of naval research Edited by Margo A. Sass and William D. Wilkinson Spartan books,Washington,D.C.,1965
Eng 41.2202

BUREAU DES LONGITUDES,PARIS Conference internationale des etoiles fondamentales Proceedings Paris 1896 May 18-21 90p Gauthier-Villars,Paris,1896
Obs 6.2315

BUREAU DES LONGITUDES,PARIS Congres international des ephemerides astronomiques Rapports Paris 1911 Oct. 23-26 53p Gauthier-Villars,Paris,1912
Obs 6.2483

BURG Wartenstein Jul 14-Aug 9 Background to evolution in Africa Systematic investigation of the African later tertiary and quaternary :symposium Edited by Walter W. Bishop and J.Desmond Clark Under the auspices of the Wenner-Gren foundation for anthropology and research University of Chicago press,Chicago,Ill.,1967
Geog 13.4660

BURG 1964 Genetic diversity and human behaviour The Burg Wartenstein symposium 27th Papers behavioral consequences of genetic differences in man Social science research council.Committee on genetics and behavior Wenner-Gren foundation for anthropological research Edited by G.N. Spuhler Viking fund publications in anthropology, 45 Aldine,Chicago,Ill.,1967
Gen 34.2032

BURG-WARTENSTEIN 1961 African ecology and human evolution :a symposium Papers and discussions Wenner-Gren foundation for anthropological research Edited by F.Clark Howell and Francois Bourliere illus,tables 25cm Methuen,London,1964
Bal 44.4056

The BURG WARTENSTEIN SYMPOSIUM 27th Papers Genetic diversity and human behaviour Burg 1964 Sep 16-28 behavioral consequences of genetic differences in man Social science research council.Committee on genetics and behavior Wenner-Gren foundation for anthropological research Edited by G.N. Spuhler Viking fund publications in anthropology, 45 Aldine,Chicago,Ill.,1967
Gen 34.2032

BURNS Bahama international conference on burns Report West End,Grand Bahama 1963 Mar 23-25 Sponsored by Colonial research institute Dorrance,Philadelphia,1964
Med 36.0079

BUSHVELD IGNEOUS COMPLEX Symposium on the Bushveld igneous complex and other layered intrusions Proceedings Pretoria 1969 Jul 7-14 Geological society of South Africa Edited by D.J.L. Visser and G.von Gruenewaldt Geological society of South Africa.Special paper, 1 Geological society of South Africa,Johannesburg,1970
Min 10.1476

The BUSINESS COMPUTER SYMPOSIUM Papers London 1958 Dec 1-3 No 1-22 Electronic engineering association and Office appliance and business equipment Collection of 22 papers and other mimeographed material
Math L 5.3252

BUXTON 1953 The Post-war expansion of the U.K.petroleum industry :summer meeting Report Institute of petroleum Edited by George Sell vii,220,xxxvp Institute of petroleum,London,1954
Chem E 24.1705

BUXTON 1962 Spectroscopy in the metallurgical industry the symposium 2nd Papers Institute of physics and the physical society.Spectroscopy group Edited by L. Bovey Hilger and Watts,London,1963
Met 25.1752

BYURAKAN 1964 Extraterrestrial civilizations All-Union conference on extraterrestrial civilizations and interstellar communications 1st Proceedings Byurakanskaya astrofizicheskaya observatoriya Edited by S.M. Tovmasyan Translated by Z. Lerman from the Russian Israel program for scientific translations, Jerusalem,1967
TA 15.0322

BYURAKAN 1966 Non-stable phenomena in galaxies Nestasionarnye yavleniya v galaktikakh International astronomical union International astronomical union.Symposium, 29 480p Akademiya nauk Armyanskey S.S.R., Erevan,1968
Obs 6.3269

BYURAKANSKAYA ASTROFIZICHESKAYA OBSERVATORIYA Extraterrestrial civilizations All-Union conference on extraterrestrial civilizations and interstellar communications 1st Proceedings Byurakan 1964 May 20-23 Edited by S.M. Tovmasyan Translated by Z. Lerman from the Russian Israel program for scientific translations,Jerusalem,1967
TA 15.0322

C.C.T.A.EAST-CENTRAL AND SOUTHERN REGIONAL COMMITTEE FOR GEOLOGY 2nd :meeting proceedings Tananarive 1957 Apr 11-13 Commission for technical co-operation in Africa south of the Sahara London,1958 Bound with proceedings of three other regional meetings
Geol 8.3117

C.C.T.A.EAST-CENTRAL REGIONAL COMMITTEE FOR GEOLOGY 1st :meeting proceedings Dar-es-Salaam 1956 Commission for technical co-operation in Africa south of the Sahara London,1957 Bound with proceedings of three other regional meetings
Geol 8.3115

C.C.T.A.SOUTHERN REGIONAL COMMITTEE FOR GEOLOGY 1st :meeting proceedings Salisbury 1955 Sep 19-24 Commission for technical co-operation in Africa south of the Sahara London,1956 Bound with proceedings of three other regional meetings
Geol 8.3114

C.C.T.A.WEST-CENTRAL REGIONAL COMMITTEE FOR GEOLOGY 2nd :meeting proceedings Accra 1957 Jan 29-Feb 2 Commission for technical co-operation in Africa south of the Sahara London,1958 Bound with proceedings of three other regional meetings
Geol 8.3116

C.I.T.C.E. MEETING 6th-9th Proceedings 1954-57 International committee of electrochemical thermodynamics and kinetics 4 vols Butterworths,London,1955-59 Papers in English,French and German
Met 25.1833

CAGLIARI 1969 Remobilization of ores and minerals Convegno sulle rimobilizzazione del minerali metallici e non metallici Rendiconti minerali metallici e non metallici Papers Associazzione mineralia sarda Istituto di giacimenti minerali universitaria Cagliari,c1969
Min 10.1489

CAIRNS Stewart S. ed. Differential and combinatorial topology :a symposium in honor of Marston Morse Princeton, N.J. 1963 Supported by United States.Air force.Office of scientific research.Mathematics division vi, 265p 24cm Princeton university press, Princeton, N.J.,1965
P. Math 2.0295

CAIRO 1925 International geographical congress 11th Compte rendu International geographical union 2 vols Societe royale de geographie d'Egypte,Cairo,1925
Geog 13.5593

CAIRO 1962 Bilharziasis :a Ciba foundation symposium Proceedings Ciba foundation Edited by G.E.W. Wolstenholme and Maeve O'Connor illus. Churchill,London,1962 Held in commemoration of Theodor Maximilian Bilharz
Mol 45.0266

CAIRO 1962 Markets and marketing as factors of development in the Mediterranean basin Mediterranean social sciences research council.Assembly 2nd Papers Edited by C. A.O.van Nieuwenhuijze Institute of social studies,The Hague.Publications.Series maior,11 Mouton,The Hague,1963
Geog 13.4117

CAIRO 1966 Interaction of radiation with solids Cairo solid state conference American university in Cairo Edited by Adli Bishay Plenum press,New York,1967
TA 15.0429

CAIRO SOLID STATE CONFERENCE Interaction of radiation with solids Cairo 1966 Sep 3-8 American university in Cairo Edited by Adli Bishay Plenum press,New York,1967
TA 15.0429

CALCIFICATION IN BIOLOGICAL SYSTEMS :a symposium presented at the Washington meeting of the American association for the advancement of science Washington,D.C. 1958 Dec 29 American association for the advancement of science Edited by Reidar F. Soggnaes American association for the advancement of science.Publication, 64 American association for the advancement of science,Washington,D.C.,1960
An 32.3237

CALCIFIED TISSUE Comparative biology of calcified tissue papers New York 1962 Oct 12-13 New York academy of sciences New York academy of sciences.Annals,109,art.1 New York,1963
Geol 8.0971

CALCITONIN Symposium on thyrocalcitonin and the C cells Proceedings London 1967 Jul 17-20 Edited by Selwyn Taylor Heinemann, London,1968
An 32.3778

CALCIUM Symposium on calcium and cellular function Essays London 1969 Mar 24-25 Biological council.Co-ordinating committee for symposia on drug action Edited by A.W. Cuthbert Macmillan,London,1969
Bioch 33.2264

CALCIUM The Transfer of calcium and strontium across biological membranes :a conference Proceedings Ithaca,N.Y. 1962 May 13-16 Edited by R.H. Wasserman held at Cornell university Academic press,New York; London,1963
Bioch 33.1021

Le CALCUL DE PROBABILITES ET SES APPLICATIONS colloque international proceedings Lyon 1948 Jun.28-Jul.3 Centre national de la recherche scientifique Edited by Maurice Frechet Centre national de la recherche scientifique.Colloques internationaux 130p 27cm Centre national de la recherche scientifique,Paris,1949
Math 3.0678

Le CALCUL DES PROBABILITES ET SES APPLICATIONS colloque international proceedings Paris 1958 Jul.15-20 Centre national de la recherche scientifique Edited by Georges Darmois Centre national de la recherche scientifique.Colloques internationaux, 87 196p 24cm Centre national de la recherche scientifique,Paris,1959
Math 3.0679

Le CALCUL DS VARIATIONS :son evolution et ses progres son role dans la physique mathematique: conferences Prague 1931 and Brno By Vito Volterra Prague,1932
Philos 1.1949

CALCULATING MACHINES Conference Konferentsiya puti razvitiya sovetskogo matematicheskogo mashchinostroeniya i priborostroeniya :plenarnye zasedaniya Moscow 1956 Mar 12-17 illus. 131p 22cm Moscow,1956
Math L 5.0774

CALCULUS OF VARIATIONS Control theory and the calculus of variations based on the lectures presented at the workshop on calculus of variations and control theory Edited by A. V. Balakrishnan Academic press,New York; London,1969
Math S 3.1515

CALCUTTA 1938 Indian science congress association,25th session The Second city of the empire Edited by P.C. Bagchi illus. Calcutta,1938
Geog 13.4484

CALCUTTA 1968 International seminar on chromosome :its structure and function Proceedings University of Calcutta Ramakrisha mission institute of culture University grants commission Edited by Arunkumar Sharma and Archana Sharma Nucleus, 1968,Supplementary volume Nucleus,Calcutta, 1968
Gen 34.2173

CALEDONIDES The British Caledonides :a symposium Papers Edinburgh 1961 Dec Edinburgh university and Edinburgh geological society Edited by M.R.W. Johnson and F.H. Stewart Edinburgh,1963
Geod 9.0414

CALEDONIDES The British Caledonides symposium papers Edinburgh 1961 Dec. Edited by M.R.W. Johnson and F.H. Stewart Sponsored by the Edinburgh geological society illus. map Oliver and Boyd,Edinburgh;London, 1963
Geol 8.1260

CALGARY 1960 Geology of the Arctic First international symposium on Arctic geology Vol 1-2 Alberta society of petroleum geologists Edited by Gilbert O. Raasch illus,maps 26cm 2 vols University of Toronto press,Toronto,1961
Sco 14.2403

CALGARY 1960 Geology of the arctic International symposium on Arctic geology 1st Proceedings Vol 1-3 Edited by Gilbert O. Raasch Under the auspices of the Alberta society of petroleum geologists 3 vols University of Toronto press,Toronto, 1961 Vol.3 consists of maps
Geog 13.5390

CALGARY,ALBERTA 1960 Geology of the Arctic International symposium on arctic geology 1st proceedings Edited by Gilbert O. Raasch Under the auspices of Alberta society of petroleum geologists maps 3 vols University of Toronto press,Toronto,1961
Geol 8.2938

CALGARY,ALTA. 1960 Geology of the Arctic International symposium on arctic geology 1st Proceedings Vol 1-2 Edited by Gilbert O. Raasch Under the auspices of the Alberta society of petroleum geologists 2 vols Toronto,1961 Additional maps in a separate container
Geod 9.0436

CALGARY,ALTA. 1969 International conference on combinatorial structures Proceedings Edited by Richard Guy and others xvi,508p 24cm Gordon and Breach,New York, 1970
P Math 2.4511

CALIFORNIA INSTITUTE OF TECHNOLOGY Applications of model theory to algebra, analysis,and probability International conference on the applications of model theory to algebra analysis and probability Proceedings Pasadena 1967 May 23-26 Edited by W.A.J. Luxemburg Bibliog. v,307p 23cm Holt,Rinehart and Winston,New York,1968
P Math 2.3241

CALIFORNIA INSTITUTE OF TECHNOLOGY Proceedings of the symposium on information processing in sight sensory systems Proceedings Pasadena,Calif. 1965 Nov 1-3 Edited by P.W. Nye California institute of technology,Pasadena,Calif.,1965
Psy 31.0419

CALIFORNIA INSTITUTE OF TECHNOLOGY.DIVISION OF PHYSICS,MATHEMATICS AND ASTRONOMY Phenomenology in particle physics :conference Proceedings Pasadena,Calif. 1971 Mar 25-26 Edited by C.B. Chiu and others illus xvi, 814p 28cm California institute of technology,Pasadena,Calif.,1971
A Math 4.1766

CALIFORNIA INSTITUTE OF TECHNOLOGY.HIXON COMMITTEE Hixon symposium Cerebral mechanisms in behaviour :the Hixon symposium Report Pasadena,Calif. 1948 Sep 20-25 Edited by Lloyd A. Jeffress Wiley;Chapman and Hall,New York;London,1951
Psy 31.0522

CALIFORNIA INSTITUTE OF TECHNOLOGY.JET PROPULSION LABORATORY Solar wind The Conference on the solar wind Proceedings Pasadena,Calif. 1964 Apr 1-4 Edited by Robert J. Mackin and Marcia Neugebauer JPL TR 32-630 400p pergamon press,Oxford,1966
TA 15.0201

CALIFORNIA INSTITUTE OF TECHNOLOGY.JET PROPULSION LABORATORY The Solar wind :a conference proceedings Pasadena,Calif. 1964 Apr.1-4 Edited by Robert J. Mackin and Marcia Neugebauer Pergamon press,Oxford,1966
A Math 4.1165

CALIFORNIA INSTITUTE OF TECHNOLOGY.OFFICE FOR INDUSTRIAL ASSOCIATION New methods of thought and procedure Symposium on methodologies Contributions Pasadena,Calif. 1967 May 22-24 Edited by F. Zwicky and A. G. Wilson viii,338p Springer,Berlin,1967
WSM 43.2639

CALLUS FORMATION :symposium on the biology of fracture healing Debrecen 1965 Jul 5-8 Edited by St. Krompecher and E. Kerner Translated by E. Kerner Symposia biologica hungarica, 7 Akademiai Kaido,Budapest,1967
An 32.3921

CAMBRIAN Sistema cambrico,su paleografia y el problema de su base International geological congress 20th papers of a symposium Mexico city 1956 Pt 1-2 Mexico city,1956 Text in English,French and Spanish
Geol 8.3024

CAMBRIDGE 1833 Address delivered in the Senate house By William Whewell British association for the advancement of science Cambridge,1833 One of nineteen tracts bound together
Philos 1.0880

CAMBRIDGE 1833 Lithographed signatures British association for the advancement of science Cambridge,1833 Two copies
Philos 1.0098

CAMBRIDGE 1903 Air currents and the law of ventilation :lectures on the physics of the ventilation of buildings By W.H. Shaw Cambridge physical series Cambridge university press,Cambridge,1907
BG 38.0440

CAMBRIDGE 1909 Order of the proceedings at the Darwin celebration... :with a sketch of Darwin's life University of Cambridge 23p Cambridge university press,Cambridge,1909
WSM 43.2842

CAMBRIDGE 1912 International congress of mathematicians 5th proceedings Vol. 2. Communications to sections 2-4 International mathematical union Edited by E. W. Hobson and A.E.H. Love 657p 26cm Cambridge university press,Cambridge,1913
Math L 5.0751

CAMBRIDGE 1928 International geographical congress Esplorazione scientifica del Mar Rosso :comunicazione al congresso geografice internazionale Italy.Navy.Hydrographic institute Genoa,1928 Bound with English translation
Geog 13.0854

CAMBRIDGE 1928 International geographical congress 12th Report of the proceedings Cambridge university press,Cambridge,1930
Geog 13.5594

CAMBRIDGE 1928 Triennial congress...on the subject of fundamental relationships between all sections of the industrial community 1st Report International association for the study and improvement of human relations and conditions in industry Secretariat I.R.I. The Hague,1928
Psy 31.2516

CAMBRIDGE 1930 International botanical congress 5th Abstracts International botanical congress Cambridge university press,Cambridge,1930 President of congress:prof.Seward.Bound with complete set of documents relating to this congress
Bot 42.5914

CAMBRIDGE 1930 International botanical congress 5th Abstracts,programs Cambridge university press,Cambridge,1930
BG 38.3127

CAMBRIDGE 1930 International botanical congress :a report 5th Proceedings Edited by F.T. Brooks and T.F. Chipp port Cambridge university press,Cambridge,1931
Bot 42.0655

CAMBRIDGE 1930 International botanical congress... :minutes of the executive countil 5th International botanical congress Cambridge,1927-35
Bot 42.5961

CAMBRIDGE 1930 International rules of botanical nomenclature revised by the International botanical congress,Cambridge 1930,compiled from the report of J.Briquet Fischer,Jena,1935
Bot 42.5221

CAMBRIDGE 1931 Commission internationale de l'eclairage en succession a la Commission internationale de photometrie 8e session Recueil des travaux et compte rendu Published under the direction of the National physical laboratory Cambridge university press,Cambridge,1932
Psy 31.2562

CAMBRIDGE 1938 British association for the advancement of science :annual meeting 108th Report London,1938 With appendix:a scientific survey of Cambridge and district, edited by H.C.Darby
Geog 13.3988

CAMBRIDGE 1946 Nucleic acid Society for experimental biology Society for experimental biology.Symposia, 1 Cambridge university press,Cambridge,1947
Phys 20.0942

CAMBRIDGE 1946 Nucleic acid :a symposium Society for experimental biology Society for experimental biology.Symposia, 1 Cambridge university press,Cambridge,1947
Bioch 33.1348

CAMBRIDGE 1946 Nucleic acid :a symposium Society for experimental biology Society for experimental biology.Symposia, 1 Cambridge university press,Cambridge,1947
Bot 42.1722

CAMBRIDGE 1946 Nucleic acid :a symposium Society for experimental biology Society for experimental biology.Symposia, 1 Cambridge university press,Cambridge,1947
Radioth 35.0009

CAMBRIDGE 1946 Nucleic acid :a symposium 1 Papers Society for experimental biology Society for experimental biology.Symposia,1 Cambridge university press,Cambridge,1947
Chem 18.1554

CAMBRIDGE 1946 Nucleic acid :symposium Society for experimental biology Society for experimental biology.Symposia, 1 Cambridge university press,Cambridge,1947
Gen 34.0683

CAMBRIDGE 1946 Nucleic acid symposium 1st Papers Society for experimental biology Society for experimental biology.Symposia, 1 plates 290p Cambridge university press, Cambridge,1947
Cav 7.0459

CAMBRIDGE 1946 The International conference on fundamental particles and low temperature Report Vol 1: fundamental particles Physical society and Cavendish laboratory Physical society,London,1947
Cav 7.0460

CAMBRIDGE 1946 The International conference on fundamental particles and low temperatures Report Vol 2: low temperatures Physical society and Cavendish laboratory Physical society,London, 1947
Cav 7.0458

CAMBRIDGE 1947 Conference on colour vision Papers Documenta ophthalmologica, 3 Junk, The Hague,1949
Phys 20.1021

CAMBRIDGE 1947 Documenta ophthalmologica : advances in ophthalmology Vol 3 Edited by F.P. Fisher and others W.Junk,'S. Gravenhage,1949 Contains most of the papers read at the conference on colour vision held in Cambridge,July 28-August 2,1947
Psy 31.0409

CAMBRIDGE 1947 Gorwth in relation to differentiation and morphogenesis Society for experimental biology Edited by J.F. Danielli and R. Brown Society for experimental biology, 2 Cambridge university press,Cambridge,1948
An 32.5434

CAMBRIDGE 1947 Growth :in relation to differentiation and morphogenesis Society for experimental biology Society for experimental biology.Symposia, 2 Cambridge university press,Cambridge,1958
Phys 20.0943

CAMBRIDGE 1947 Growth in relation to differentiation and morphogenesis.A symposium Society for experimental biology Society for experimental biology.Symposia, 2 Cambridge university press,Cambridge,1948
Bioch 33.1349

CAMBRIDGE 1947 Growth,in relation to differentiation and morphogenesis Society for experimental biology Society for experimental biology.Symposia, 2 Cambridge university press,Cambridge,1948
Chem 18.2652

CAMBRIDGE 1948 Haemoglobin :a symposium Edited by F.J.W. Roughton and J.C. Kendrew port. Butterworths,London,1949 In memory of Sir Joseph Barcroft
Phys 20.0881

CAMBRIDGE 1948 Haemoglobin :a symposium based on a conference...held in memory of Sir Joseph Barcroft Edited by F.J.W. Roughton and J.C. Kendrew Butterworths,London,1949
Col S 12.0049

CAMBRIDGE 1948 Haemoglobin :a symposium based on a conference...in memory of Sir Joseph Barcroft Proceedings Edited by F.J. W. Roughton and J.C. Kendrew Butterworths, London,1949
Med 36.0309

CAMBRIDGE 1948 Haemoglobin :a symposium based on a conference held in memory of Sir Joseph Barcroft Edited by Francis John Worsley Roughton and John Cowdery Kendrew Butterworths,London,1949
Bioch 33.0490

CAMBRIDGE 1949 Conference on high speed automatic calculating-machines report University of Cambridge.Mathematical laboratory Bibliog 31cm Cambridge,1950
Philos 1.0336

CAMBRIDGE 1949 Conference on high-speed automatic calculating machines Report University of Cambridge.Mathematical laboratory Great Britain.Ministry of supply University mathematical laboratory,Cambridge, 1950
Eng 41.2175

CAMBRIDGE 1949 Conference on high-speed automatic calculating machines Report University of Cambridge.Mathematical laboratory Great Britain.Ministry of supply University mathematical laboratory,Cambridge, 1950
Math L 5.3236

CAMBRIDGE 1949 International congress of bio-chemistry 1st abstracts of communications 1949
Phys 20.1522

CAMBRIDGE 1949 International congress of biochemistry 1st Abstracts of communication 1949 Text in English and French
Bioch 33.1300

CAMBRIDGE 1949 International congress of biochemistry 1st Report of opening and concluding sessions and three lectures Printed for the Biochemical society Glasgow university,Biochemistry department,Glasgow, 1950 Unbound
Bioch 33.1301

CAMBRIDGE 1949 Physiological mechanisms in animal behaviour Papers Society for experimental biology Society for experimental biology.Symposia, 4 Cambridge university press,Cambridge,1950
Psy 31.0562

CAMBRIDGE 1949 Physiological mechanisms in animal behaviour Society for experimental biology Edited by J.F. Danielli and R. Brown Society for experimental biology.Symposia, 4 Cambridge university press,Cambridge,1950
An 32.4327

CAMBRIDGE 1949 Physiological mechanisms in animal behaviour Society for experimental biology Society for experimental biology. Symposia, 4 Cambridge university press, Cambridge,1950
Phys 20.0945

CAMBRIDGE 1950 International congress of hematology 3rd Proceedings International society of hematology Edited by Carl V. Moore Grune and Stratton,New York, 1951
Med 36.0256

CAMBRIDGE 1950 Spectroscopy and molecular structure,and optical methods of investigating cell structure :a general discussion Faraday society Faraday society.Discussions, 9 Aberdeen,1951
Bal 39.0230

CAMBRIDGE 1950 Spectroscopy and molecular structure and optical methods of investigating cell structure :general discussion Faraday society Faraday society.Discussions, 9 Faraday society,Aberdeen,1951
Radioth 35.0453

CAMBRIDGE 1951 Structural adhesives:the theory and practice of gluing with synthetic resins lectures Aero research Lange, Maxwell and Springer,London,1951 Lectures given at a summer school on the technology of synthetic resin adhesives, Cambridge,1951
Eng 41.4022

CAMBRIDGE 1952 The Physical chemistry of proteins :a general discussion Faraday society Faraday society.Discussions, 13 Faraday society,Aberdeen,1953
Radioth 35.0336

CAMBRIDGE 1952 The Physical chemistry of proteins :a general discussion Papers Faraday society Faraday society.Discussions, 13 Aberdeen university press,Aberdeen,1953
Bioch 33.0510

CAMBRIDGE 1953 Gas dynamics of cosmic clouds Symposium on cosmical gas dynamics 2nd International union of theoretical and applied mechanics International astronomical union International astronomical union.Symposium series, 2 North-Holland,Amsterdam,1955
Eng 41.4468

CAMBRIDGE 1953 Gas dynamics of cosmic clouds The Symposium on cosmical gas dynamics Proceedings International astronomical union and International union of theoretical and applied mechanics Edited by J.M. Burgess and H.C.van de Hulst International astronomical union.Symposium, 2 North Holland,Amsterdam,1955
TA 15.0217

CAMBRIDGE 1953 Gas dynamics of cosmic clouds :a symposium International union of theoretical and applied mechanics and International astronomical union Edited by J. M. Burgers and H.C.van de Hulst International astronomical union.Symposium, 2 247p North-Holland,Amsterdam,1955
Obs 6.0308

CAMBRIDGE 1953 Selected combustion problems Combustion colloquium Agard Edited by W.R. Hawthorne and J. Fabri Butterworths,London,1954
A Math 4.0969

CAMBRIDGE 1953 Selected combustion problems:fundamentals and aeronautical applications Combustion colloquium Agard Butterworths,London,1954
Eng 41.7219

CAMBRIDGE 1953 Selected combustion problems:fundamentals and aeronautical applications Combustion colloquium Proceedings North Atlantic treaty organization.Advisory group for aeronautical research and development viii,534p Butterworths,London,1954
Chem E 24.1912

CAMBRIDGE 1954 Conference on the physics of the ionosphere Report Physical society Cavendish laboratory 406p Physical society, London,1955
Nap 11.0926

CAMBRIDGE 1955 Progress in radiobiology International conference on radiobiology 4th Proceedings Edited by Joseph S. Mitchell and others port. Oliver and Boyd, Edinburgh;London,1956
Radioth 35.1707

CAMBRIDGE 1956 Plant and process dynamic characteristics :a conference Proceedings Society of instrument technicians Academic press;Butterworths,New York;London,1957
Eng 41.5908

CAMBRIDGE 1956 Plant and process dynamic characteristics :a conference Proceedings Society of instrument technology xii,246p Butterworths,London,1957
Chem E 24.1157

CAMBRIDGE 1956 The Biological action of growth substances Society for experimental biology Society for experimental biology. Symposia, 11 Cambridge university press, Cambridge,1956
Bioch 33.1356

CAMBRIDGE 1956 X-ray microscopy and microradiography Symposium on x-ray microscopy and microradiography Proceedings Edited by V.E. Coslett and others Sponsored by the International union of pure and applied physics Academic press,New York,1957
Radioth 35.1600

CAMBRIDGE 1956 X-ray microscopy and microradiography The Symposium on X-ray microscopy and microradiography Proceedings Cavendish laboratory Edited by V.E. Cosslett and others Sponsored by the International union of pure and applied physics Academic press,New York,1957
Cav 7.0461

CAMBRIDGE 1957 Bonded aircraft structures : a conference Papers Ciba foundation Ciba foundation,Duxford,Cambridge,1958
Eng 41.2774

CAMBRIDGE 1957 Entrepreneur Economic history society.Annual conference Papers Edited by B.E. Supple 63p Harvard university.Research center in entrepreneurial history,Cambridge,Mass.,1957
Geog 13.3190

CAMBRIDGE 1957 Phosphoric esters and related compounds :a symposium Report Chemical society Chemical society.Special publication,8 Chemical society,London,1957 Organized by Dr.Kenner and Dr.Brown
Chem 18.1217

CAMBRIDGE 1957 Phosphoric esters and related compounds :a symposium held at the Chemical society anniversary meeting Report Chemical society Chemical society.Special publication, 8 Chemical society,London, 1957
Bioch 33.0842

CAMBRIDGE 1957 Reactions of free radicals in the gas phase :a symposium Report Chemical society Chemical society.Special publication,9 Chemical society,London,1957 Organized by Dr.Sugden
Chem 18.1218

CAMBRIDGE 1957 Recent advances in gelatin and glue research :a conference Proceedings Edited by G. Stainsby Sponsored by the British gelatine and glue research association Pergamon,London,1958
Bioch 33.2247

CAMBRIDGE 1957 Recent aspects of the inorganic chemistry of nitrogen :a symposium Report Chemical society Chemical society. Special publication,10 Chemical society, London,1957 Organized by Dr.Maddock and Dr.Sharpe
Chem 18.1219

CAMBRIDGE 1959 Brittle fracture in steel : conference Proceedings Great Britain. Admiralty advisory committee on structural steel P.3 H.M.S.O.,London,1962
Eng 41.3775

CAMBRIDGE 1959 Depression :a symposium Proceedings University of Cambridge.School of clinical research and postgraduate medical training Edited by E.Beresford Davies Cambridge university press,Cambridge,1964
Psy 31.1679

CAMBRIDGE 1959 Depression :the symposium Proceedings University of Cambridge.School of clinical research and postgraduate medical teaching Edited by E.B. Davies University of Cambridge.School of clinical research and postgraduate medical teaching.Course,1959 Cambridge university press,Cambridge,1964
PGMS 29.0330

CAMBRIDGE 1959 Progress and prospects in ornithology being the centenary symposium Proceedings British ornithologists' union Ibis, 101,no 3-4,Centenary celebration number 25cm London,1959
Bal 44.1519

CAMBRIDGE 1959 The Cell nucleus :an informal meeting Proceedings Faraday society Edited by J.S. Mitchell Butterworth,London,1960
An 32.3234

CAMBRIDGE 1959 The Cell nucleus :an informal meeting Proceedings Faraday society Held at the University of Cambridge. Department of radiotherapeutics Butterworths, London,1960 Chairman:J.S.Mitchell
Bal 39.0290

CAMBRIDGE 1959 The Cell nucleus :an informal meeting Proceedings Faraday society held at the University of Cambridge. Department of radio-therapeutics Butterworths,London,1960
Bioch 33.0946

CAMBRIDGE 1959 The Cell nucleus :an informal meeting held at the Department of radiotherapeutics,University of Cambridge...by the Faraday society Proceedings Faraday society Illus Butterworths,London,1960 Chairman:Professor J.S.Mitchell
Radioth 35.0501

CAMBRIDGE 1959 The Cell nucleus informal meeting Proceedings Faraday society Edited by J.S. Mitchell Butterworths,London, 1960
Gen 34.0839

CAMBRIDGE 1962 Hormones and the kidney Society for endocrinology :a meeting 89th Proceedings Society for endocrinology Edited by Peter C. Williams Society for endocrinology.Memoirs, 13 Academic press, London;New York,1963
Inv Med 37.0233

CAMBRIDGE 1962 Hormones and the kidney : 89th meeting Proceedings Society for endocrinology Edited by Peter C. Williams Society for endocrinology.Memoirs, 13 Academic press,London;New York,1963
Gen 34.0557

CAMBRIDGE 1963 Behavior,aging and the nervous system :biological determinants of speed of behavior and its changes with age Edited by A.T. Welford and James E. Birren American lecture series.Publication, 600, American lectures in geriatrics and gerontology.Bannerstone division monograph Thomas,Springfield,Ill.,1965
Psy 31.2329

CAMBRIDGE 1963 Food supply and nature conservation :a symposium Cambridgeshire college of arts and technology Edited by R.I. B. Cooper Cambridge,1964
Bal 39.2963

CAMBRIDGE 1963 Homeostasis and feedback mechanisms Society for experimental biology Society for experimental biology.Symposia, 18 Cambridge university press,Cambridge,1964
Phys 20.0962

CAMBRIDGE 1963 International biometric conference 5th Cambridge,1963 8 unbound typescript papers
Math 3.1127

CAMBRIDGE 1963 International symposium on free radicals 6th Papers University of Cambridge Sponsored in part by Great Britain. Air force.Office of scientific research,OAR Cambridge,1963 Mimeographed
Chem 18.0414

CAMBRIDGE 1963 Metabolism and physiological significance of lipids :advanced study course Proceedings Edited by R.M.C. Dawson and D.N. Rhodes Wiley,London,1964
Pha 16.0127

CAMBRIDGE 1963 Metabolism and physiological significance of lipids :advanced study course Proceedings Edited by R.M.C. Dawson and Douglas N. Rhodes Wiley,London, 1964
Bioch 33.1235

CAMBRIDGE 1964 A Seals symposium Proceedings Nature conservancy.Consultative committee on grey seals Edited by E.A. Smith illus 101p 28cm Edinburgh,1965
Sco 14.1172

CAMBRIDGE 1964 A Seals symposium Proceedings Nature conservancy,Edinburgh. Consultative committee on grey seals and fisheries Edited by E.A. Smith maps,tables 28cm Edinburgh,1965 Number 84 of 100 mimeographed copies.In folder
Bal 44.5093

CAMBRIDGE 1964 Advances in earth science International conference on the earth sciences Papers Edited by P.M. Hurley Massachusetts institute of technology press,Cambridge,Mass., 1966
Min 10.0576

CAMBRIDGE 1964 Cereal rust conferences Plant breeding institute Cambridge,1966
Bot 42.4125

CAMBRIDGE 1964 Comparative physiology and pathology of the skin :a symposium Proceedings Edited by Arthur J. Rook and G.S. Walton Blackwell,Oxford,1965
An 32.4050

CAMBRIDGE 1964 Natural rubber producers' research association jubilee conference Proceedings Natural rubber producers' research association Edited by L. Mullins Maclaren,London,1965
Bioch 33.1740

CAMBRIDGE 1964 Some factors affecting drug toxicity European society for the study of drug toxicity :meeting 4th Proceedings Edited by D.G. Davey European society for the study of drug toxicity.Proceedings,4 International congress series,81 Excerpta medica,Amsterdam,1964
Pha 16.0326

CAMBRIDGE 1964 Symposium (international) on combustion 10th Combustion institute Combustion institute,Pittsburgh,Pa.,1965
Chem 18.0440

CAMBRIDGE 1964 The Natural rubber producers' research association Jubillee conference Proceedings Natural rubber producers' research association Edited by L. Mullins Bibliog.,illus,plate viii,250p McLaren,London,1965
Bot 42.1427

CAMBRIDGE 1965 Light as a ecological factor :a symposium British ecological society Edited by Richard Bainbridge and G. Evans British ecological society.Symposia, 6 Blackwell,Oxford,1966
Bal 39.1683

CAMBRIDGE 1965 Light as an ecological factor British ecological society :a symposium 6th Proceedings British ecological society Edited by Richard Bainbridge and others British ecological society.Symposia, 6 Bibliog.,tables,diagrs 452p Blackwell,Oxford,1966
Bot 42.2093

CAMBRIDGE 1965 Light as an ecological factor British ecological society symposium 6th Edited by Richard Bainbridge and others Blackwell,Oxford,1966
Geog 13.1217

CAMBRIDGE 1965 The Cambridge region 1965 : meeting of the British Association British association for the advancement of science Edited by J.A. Steers illus.,maps. British association for the advancement of science, London,1965
Gen 34.0131

CAMBRIDGE 1966 Endocrine genetics :a symposium Proceedings Soceity for endocrinology Edited by S.G. Spicket and J.G. M. Shire Society for endocrinology.Memoirs, 15 Cambridge university press,Cambridge,1967
Pha 16.0093

CAMBRIDGE 1966 Endocrine genetics :a symposium Proceedings Society for endocrinology Edited by S.G. Spickett and J. G.M. Shire Society for endocrinology.Memoirs, 15 Cambridge university press,Cambridge, 1967
Gen 34.0555

CAMBRIDGE 1966 Endocrine genetics a symposium Proceedings Society for endocrinology Edited by S.G. Spickett and J. G.M. Shire Society for endocrinology.Memoir, 15 Cambridge university press,Cambridge, 1967
Inv Med 37.0203

CAMBRIDGE 1966 International conference on the use of computers in therapeutic radiology Summary report British institute of radiology,London,1967
Radioth 35.1198

CAMBRIDGE 1966 International symposium on geophysical theory and computers 3rd Proceedings International upper mantle committee Sponsored by the International union of geodesy and geophysics Geophysical journal, 13;1-3 375p Blackwell,Oxford, 1967
Geod 9.0694

CAMBRIDGE 1966 Microwave and optical generation and amplification International conference on microwave tubes 6th Institution of electrical engineers Institution of electrical engineers.Conference publication, 27 I.E.E.,London,1967
Eng 41.5588

CAMBRIDGE 1967 Internal aerodynamics (turbomachinery) :a conference Royal society of London Institution of mechanical engineers,London,1970
Eng 41.6829

CAMBRIDGE 1967 The Treatment of carcinoma of the breast :a symposium Proceedings Institute of hormone biology Syntex pharmaceuticals Edited by Antony S. Jarrett Excerpta medica foundation,1968
PGMS 29.0460

CAMBRIDGE 1967 The Treatment of carcinoma of the breast :a symposium Proceedings Syntex pharmaceuticals Edited by Anthony S. Jarrett Under the auspices of the Institute of hormone biology Lewis,Cambridge,1968
Inv Med 37.0122

CAMBRIDGE 1967 The Treatment of carcinoma of the breat :a symposium Proceedings Syntex pharmaceuticals Edited by Antony S. Jarrett Under the auspices of the Institute of hormone biology Excerpta medica,Amsterdam, 1967
Radioth 35.0926

CAMBRIDGE 1967 1968 Cranfield fluidics conference 2nd Proceedings Organized by the British hydromechanics research association British hydromechanics research association,Cranfield,1967-68
Eng 41.5906

CAMBRIDGE 1968 Applications of mathematical programming techniques :a conference North Atlantic treaty organization.Scientific affairs committee Edited by E.M.L. Beale EUP,London,1970
Eng 41.8254

CAMBRIDGE 1968 Applications of mathematical programming techniques :a conference Proceedings North Atlantic treaty organization.Science committee Edited by E.M.L. Beale ix,451p 24cm English universities press,London,1970
Math S 3.1686

CAMBRIDGE 1968 Conference on electronics design Sponsored by the Institution of electrical engineers Institution of electrical engineers.Conference publication, 45 IEE,London,1968
Math L 5.3468

CAMBRIDGE 1968 Engineering plasticity :a conference Papers Edited by J. Heyman and F.A. Leckie Cambridge university press, Cambridge,1968
Eng 41.2501

CAMBRIDGE 1968 Perspectives in phytochemistry Phytochemical society symposium Proceedings Phytochemical society Edited by J.B. Harborne and T. Swain Academic press,London;New York,1969 For earlier publications of this body see Plant phenolics group
Gen 34.1407

CAMBRIDGE 1968 Perspectives in phytochemistry :a symposium Proceedings Phytochemical society Edited by J.B. Harborne and T. Swain Bibliog.,illus. Academic press,London;New York,1969
Bot 42.1436

CAMBRIDGE 1968 Quantum theory and beyond : essays and discussions arising from a colloquium Edited by Ted Bastin ix,345p Cambridge university press,Cambridge,1971 Based on papers read at an informal colloquium held in Cambridge 1968
WSM 43.1119

CAMBRIDGE 1968 Quantum theory and beyond : essays and discussions arising from a colloquium Royal society of London Theoria incorporated Carnegie institution of Washington illus viii,345p 24cm Cambridge university press,Cambridge,1971 Based on papers read at an informal colloquium
A Math 4.1759

CAMBRIDGE 1969 Amorphous and liquid semiconductors :a conference Proceedings International union of pure and applied physics.Semiconductor commission Royal society of London Edited by Nevill Francis Mott Bibliog,diagrs,figs 626p 20cm North-Holland,Amsterdam,1970
Cav 7.2678

CAMBRIDGE 1969 Bonding in metallo-organic compounds Faraday society Faraday society. Discussions, 47 208p 20cm Faraday society,London,1969
Cav 7.2717

CAMBRIDGE 1969 Charged particle tracks in solids and liquids Gray (L.H.) conference 2nd Proceedings Institute of physics and the physical society.Conference series Institute of physics,London,1970
Radioth 35.1702

CAMBRIDGE 1969 Short-term changes in neural activity and behaviour :a conference Edited by Gabriel Horn and Robert A. Hinde Cambridge university press,Cambridge,1970
An 32.5239

CAMBRIDGE 1970 Gravitational N-body problem Proceedings International astronomical union Edited by M. Lecar International astronomical union.Colloquium, 10 441p Reidel,Dordrecht,1972
TA 15.0667

CAMBRIDGE 1970 Teaching of programming at university level Joint IBM-University of Newcastle upon Tyne seminar 3rd Proceedings University of Newcastle upon Tyne.Computing laboratory International business machines corporation Edited by B. Shaw illus. 153p University of Newcastle, Newcastle,1971
Math L 5.3590

CAMBRIDGE 1970 Tissue proteinases Wates symposium Proceedings Royal society of London Edited by A.J. Barrett and J.T. Dingle Held at the Strangeways research laboratory North-Holland,Amsterdam,1971
Bioch 33.2284

CAMBRIDGE 1971 Electron microscopy and analysis Institute of physics.Electron microscopy and analysis group 25th anniversary meeting Proceedings Institute of physics Edited by W.C. Nixon Institute of physics.Conference series, 10 London, 1971
Met 25.2753

CAMBRIDGE,MASS 1938 International congress for applied mechanics 5th proceedings Harvard university and Massachusetts institute of technology Edited by J.P. Den Hartog and H. Peters John Wiley,New York, 1939
A Math 4.0325

CAMBRIDGE,MASS 1958 Microsomal particles and protein synthesis :a symposium Papers Biophysical society Edited by Richard B. Roberts Biophysical society.Symposia, 1 Pergamon,London;New York,1958
Bot 42.1686

CAMBRIDGE,MASS 1968 Problems in multiple access computer system design :special summer session Massachusetts institute of technology.Department of electrical engineering 1968 Administrative material,class notes and other course material in loose-leaf binder
Math L 5.3474

CAMBRIDGE,MASS. 1926 The National symposium on colloid chemistry 4th Papers presented Massachusetts institute of technology Edited by Harry Boyer Weiser Colloid symposium monograph, 4 Chemical catalog company,New York,1926
Bioch 33.1472

CAMBRIDGE,MASS. 1936 Factors determining human behavior Harvard tercentenary conference of arts and sciences Papers Harvard university Harvard tercentenary publications Harvard university press, Cambridge,Mass.,1937
Psy 31.0770

CAMBRIDGE,MASS. 1936 International conference on soils mechanics and foundation engineering Proceedings Vol 1-3 3 vols Cambridge,Mass.,1936
Eng 41.3138

CAMBRIDGE,MASS. 1940-41 Conference on powder metallurgy held at the Massachusetts institute of technology 1st and 2nd Papers American society for metals Edited by John Wulff American society for metals, Cleveland,1942
Met 25.0758

CAMBRIDGE,MASS. 1947 Symposium on large-scale digital calculating machinery proceedings Harvard university.Computation laboratory Jointly sponsored by United States.Bureau of ordnance Harvard university. Computation laboratory.Annals, 16 illus. xxix,302p 27cm Harvard university press, Cambridge,Mass.,1948
Math L 5.0681

CAMBRIDGE,MASS. 1948 Mechanical wear :a summer conference Proceedings Massachusetts institute of technology and American society of mechanical engineers Edited by John T. Burwell Jointly sponsored by the General motors corp. and the Chrysler corp. American society for metals,1950
Met 25.1597

CAMBRIDGE,MASS. 1949 Conference on physical electronics Report By E.B. Callick 86p 27cm British joint services mission,Washington,D.C.,1949
Math L 5.3263

CAMBRIDGE,MASS. 1949 Mid-century Convocation on the social implications of scientific progress Verbatim account Massachusetts institute of technology Edited by John Ely Burchard xx,549p Technology press of the Massachusetts institute of technology,Cambridge,Mass.,1950
Cav 7.0406

CAMBRIDGE,MASS. 1949 Symposium on large-scale digital calculating machinery 2nd proceedings Harvard university.Computation laboratory Jointly sponsored by United States.Bureau of ordnance Harvard university. Computation laboratory Annals, 26 illus. xxxviii,393p 27cm Harvard university press, Cambridge,Mass.,1951
Math L 5.0684

CAMBRIDGE,MASS. 1950 Fatigue and fracture of metals :a symposium Edited by William M. Murray Held at the Massachusetts institute of technology Technology press books M.I.T. press;Wiley,Cambridge,Mass.;New York,1952
Eng 41.3740

CAMBRIDGE,MASS. 1950 Fatigue and fracture of metals :a symposium Massachusetts institute of technology Edited by W.M. Murray Technology press;Wiley,New York,1952
Met 25.0911

CAMBRIDGE,MASS. 1950 International congress of mathematicians 11th Proceedings Vol 1-2 International mathematical union and American mathematical society 25cm 2 vols American mathematical society,Providence,R.I., 1952
Math L 5.3246

CAMBRIDGE,MASS. 1951 United States conference on prestressed concrete 1st Proceedings Massachusetts institute of technology Cambridge,Mass.,1951
Eng 41.3012

CAMBRIDGE,MASS. 1952 Conference on soil stabilization Proceedings Massachusetts institute of technology M.I.T.press, Cambridge,Mass.,1952
Eng 41.3134

CAMBRIDGE,MASS. 1952 Symposium (International) on combustion (combustion and detonation waves) 4th Standing committee on combustion symposia xx,926p Williams and Wilkins company,Baltimore,Md., 1953
Chem E 24.1405

CAMBRIDGE,MASS. 1952 Symposium (international) on combustion,flame and explosion phenomena (combustion and detonation waves) 4th Papers Standing committee on combustion symposia Held at the Massachusetts institue of technology Williams and Wilkins,Baltimore,Md.,1953 Committee chairman:B.Lewis.Earlier volumes entitled 'Symposium on combustion,flame and explosion phenomena'
Chem 18.0434

CAMBRIDGE,MASS. 1953 An Evaluation of current knowledge of the mechanics of brittle fracture :including papers and discussions presented at Conference on brittle fracture mechanics held...at Massachusetts institute of technology By D.C. Drucker Sponsored by the National research council.Committee on ship structure design SSC 69 N.A.S.-N.R.C. Division of engineering and industrial research,Washington,D.C.,1954
Eng 41.3743

CAMBRIDGE,MASS. 1956 The Threshold of space :conference on chemical aeronomy Proceedings Air force Cambridge research center.Geophysics research directorate Edited by M. Zelikoff Coordinated by Wentworth institute Pergamon press,London, 1957
Chem 18.0478

CAMBRIDGE,MASS. 1956 The threshold of space Conference on chemical aeronomy Proceedings Air force Cambridge research center.Geophysics research directorate Edited by M. Zelikoff illus xi,342p Pergamon press,London,1957
Nap 11.0654

CAMBRIDGE,MASS. 1957 Symposium on cosmical gas dynamics 3rd Proceedings International union of theoretical and applied mechanics International astronomical union Edited by J.M. Burgers and R.N. Thomas Reviews of modern physics, 30,no 3 American physical society,New York,1958
Eng 41.4467

CAMBRIDGE,MASS. 1957 Symposium on cosmical gas dynamics 3rd proceedings International union of theoretical and applied mechanics and International astronomical union Edited by J.M. Burgers and others Financial grant from the Smithsonian institution International astronomical union. Symposium, 8 Smithsonian Institution. Smithsonian symposium publications, 1 American physical society,New York,1958 Reviews of modern physics.Vol.30,no.3,July 1958
Obs 6.0306

CAMBRIDGE,MASS. 1957 The International symposium on the theory of switching proceedings Part 1-2 Harvard university.Computation laboratory Harvard university.Computation laboratory.Annals, 29 27cm 2 vols Harvard university press, Cambridge,Mass.,1959
Math L 5.0686

CAMBRIDGE,MASS. 1958 Contemporary geodesy : a conference Proceedings Harvard college observatory and Smithsonian astrophysical observatory Edited by Charles A. Whitten and Kenneth H. Drummond Co-sponsored by the American geophysical union Geophysical monograph, 4 National research council. Publication, 708 American geophysical union, Washington,D.C.,1959
Geod 9.0028

CAMBRIDGE,MASS. 1958 Kinetics of high-temperature processes :conference Edited by W.D. Kingery Supported in part by United States air force.Aeronautical research laboratory Technology press books in science and engineering Massachusetts institute of technology press,Cambridge,Mass.,1959
Min 10.1127

CAMBRIDGE,MASS. 1958 Microsomal particles and protein biosynthesis :symposium Papers Biophysical society Edited by Richard B. Roberts Biophysical society.Symposia, 1 Washington academy of sciences,Washington,D.C.; London,1958
Gen 34.0657

CAMBRIDGE,MASS. 1958 Microsomal particles and protein synthesis Biophysical society : symposium 1st Papers Edited by Richard B. Roberts Pergamon press,London,1958
Col S 12.0053

CAMBRIDGE,MASS. 1958 Microsomal particles and protein synthesis :a symposium Biophysical society Edited by Richard B. Roberts Biophysical society.Symposia, 1 Pergamon press,London,1958
Radioth 35.0077

CAMBRIDGE,MASS. 1958 Microsomal particles and protein synthesis :a symposium Papers presented Biophysical society Edited by Richard B. Roberts At the Massachusetts institute of technology Biophysical society. Symposia, 1 Pergamon,London,1958
Bioch 33.0924

CAMBRIDGE,MASS. 1958 Sensory deprivation : a symposium held at Harvard medical school Harvard medical school and Boston city hospital Edited by Philip Solomon and others Co-sponsored by the United States.Office of naval research.Physiological psychology branch Harvard university press,Cambridge,Mass.,1965
Psy 31.0128

CAMBRIDGE,MASS. 1959 Sensory communication Symposium on principles of sensory communication Contributions Edited by Walter Rosenblith M.I.T.press;Wiley,New York, 1961
An 32.4395

CAMBRIDGE,MASS. 1961 Harvard symposium on digital computers and their applications proceedings Harvard university.Computation laboratory Harvard university.Computation laboratory.Annals, 31 xiv,332p 27cm Harvard university press,Cambridge,Mass.,1962
Math L 5.0687

CAMBRIDGE,MASS. 1961 High magnetic fields International conference on high magnetic fields Proceedings Edited by Henry Kolm Sponsored by the United States.Air force. Office of scientific research.Solid state sciences division M.I.T.press;Wiley, Cambridge,Mass.;New York,1962
Eng 41.5331

CAMBRIDGE,MASS. 1961 Hypersonic flow research :a selection of technical papers based mainly a symposium of the American rocket society,held at the Massachusetts institute of technology... American rocket society Edited by Frederick R. Riddell Progress in astronautics and rocketry, 7 Academic press,NewYork,1962
Eng 41.6771

CAMBRIDGE,MASS. 1961 Hypersonic flow research :a symposium technical papers American rocket society Edited by Frederick R. Riddell Sponsored jointly with the United States.Air force.Office of scientific research Progress in astronautics and rocketry, 7 Academic press,New York,London,1962
A Math 4.0533

CAMBRIDGE,MASS. 1962 Ice and snow: properties,processes,and application :a conference held at the Massachusetts Institute of technology Proceedings Air force Cambridge research laboratories.Terrestrial sciences laboratory Edited by W.D. Kingery Sponsored also by the United States naval civil engineering laboratory illus xv,684p 24cm M.I.T.press,Cambridge,Mass.,1963
Sco 14.0269

CAMBRIDGE,MASS. 1962 Ice and snow: properties,processes and applications : conference Proceedings Edited by W.D. Kingery Massachusetts institute of technology press,Cambridge,Mass.,1963
Geog 13.0652

CAMBRIDGE,MASS. 1962 Phylogeny and evolution of crustacea :a conference Proceedings Harvard university.Museum of comparative zoology Edited by H.B. Whittington and W.D.I. Rolfe Supported by the National science foundation Museum of comparative zoology,Cambridge,Mass.,1963
Bal 39.2305

CAMBRIDGE,MASS. 1963 Symposium on group theory abstracts of lectures Harvard university 59p 23cm Harvard university, Cambridge,Mass.,1963
P Math 2.2792

CAMBRIDGE,MASS. 1964 Advances in earth sciences International conference on the earth sciences contributions Massachusetts institute of technology Edited by P.M. Hurley M.I.T. press,Cambridge,Mass.,1966
Geol 8.4429

CAMBRIDGE,MASS. 1964 Mycotoxins in foodstuffs :a symposium Proceedings Edited by Gerald N. Wogan Held at the Massachusetts institute of technology M.I.T.press, Cambridge,Mass.,1965 "This volume comprises the proceedings of an International symposium on mycotoxins in foodstuffs".
Bioch 33.0708

CAMBRIDGE,MASS. 1964 Recent research on carnitine:its relation to lipid metabolism :a symposium Papers presented Edited by George Wolf at the Massachusetts institute of technology M.I.T. press,Cambridge,Mass., 1965
Bioch 33.1249

CAMBRIDGE,MASS. 1965 Formation of spectrum lines Harvard-Smithsonian conference on stellar atmospheres 2nd Proceedings Smithsonian institution.Astrophysical laboratory and Harvard university. Astronomical observatory Smithsonian institution.Astrophysical observatory,Research in space. Special report, 174 illus 461p Cambridge,Mass.,1965
Obs 6.2121

CAMBRIDGE,MASS. 1966 Factors in the transfer of technology Massachusetts institute of technology conference on the human factor in the transfer of technology Edited by W.H. Gruber and D.G. Marquis M.I.T. press,Cambridge,Mass.,1969
Eng 41.1687

CAMBRIDGE,MASS. 1968 International symposium on rarefied gas dynamics 6th Proceedings Vol 1-2 United States.Air force.Office of scientific research National aeronautics and space administration United States.Office of naval research Edited by Leon Trilling and Harold Y. Wachman Advances in applied mechanics.Supplement, 5 2 vols Academic press,New York,1969
Eng 41.6797

CAMBRIDGE,MASS. 1968 Rarefied gas dynamics international symposium 6th Proceedings Vol 1-2 Edited by Leon Trilling and Harold Y. Wachman Advances in applied mechanics.Supplement, 5 2 vols Academic press,New York;London,1969
Chem E 24.1814

CAMBRIDGE,MASS. 1968 Theory and observation of normal stellar atmospheres Harvard-Smithsonian conference on stellar atmospheres 3rd Proceedings Harvard university.Astronomical observatory Smithsonian institution.Astrophysical observatory Edited by O. Gingerich 472p Massachusetts institute of technology press, Cambridge,Mass.,1969
Obs 6.3354

CAMBRIDGE,MASS. 1969 Neurosciences research symposium summaries :an anthology from the 'Neurosciences research program bulletin' Neurosciences research program. Intensive study program Edited by Francis O. Schmitt and others Sponsored by the Massachusetts institute of technology Rockefeller university press;MIT press,New York;Cambridge,Mass.,1970
Bioch 33.2164

CAMBRIDGE,MASS. 1971 A.C.M.-SIGOPS workshop on system performance evaluation Association for computing machinery.Special interest group in operating systems illus. 378p Association for computing machinery,New York,1971
Math L 5.3564

CAMBRIDGE,MASS. 1971 The Menzel symposium on solar physics,atomic spectra and gaseous nebulae Edited by K.B. Gebbie National bureau of standards.Special publication, 353 203p National bureau of standards,Washington, D.C.,1971
Obs 6.3623

CAMBRIDGE CONFERENCE ON SCHOOL MATHEMATICS The Ditchley mathematical conference :a report by the British co-chairman (Prof.B.Thwaites) of the Anglo-American conference on mathematical education Ditchley Park 1966 Sep 9-12 S.M.P.,London,1966
Eng 41.1829

The CAMBRIDGE REGION 1965 :meeting of the British Association Cambridge 1965 Sep 1-8 British association for the advancement of science Edited by J.A. Steers illus., maps. British association for the advancement of science,London,1965
Gen 34.0131

CAMBRIDGESHIRE COLLEGE OF ARTS AND TECHNOLOGY Food supply and nature conservation :a symposium Cambridge 1963 Nov 2 Edited by R.I.B. Cooper Cambridge,1964
Bal 39.2963

CAMELLIAS AND MAGNOLIAS conference Report London 1950 Apr 4-5 Royal horticultural society illus. Royal horticultural society, London,1950
BG 38.2668

CANADA 1962 Conference across a continent The Duke of Edinburgh's commonwealth study conference 2nd Macmillan,Toronto,1963
Geog 13.5039

CANADA.DEFENCE RESEARCH BOARD The Symposium on physical processes in the sun-earth environment Proceedings Ottawa 1959 Jul 20-21 Edited by C. Collins DRTE no 1025 xii,392p Ottawa,1960
Nap 11.0140

CANADA.DEPARTMENT OF MINES AND TECHNICAL SURVEYS The World rift system symposium report Ottawa 1965 Sep 4-5 International upper mantle committee Geological survey of Canada. Paper,66-14 Ottawa,1967
Geol 8.4451

CANADA.DEPARTMENT OF NORTHERN AFFAIRS AND NATIONAL RESOURCES Conference of the Arctic co-operatives 1st Minutes Frobisher Bay,N.W.T. 1963 Mar 12-18 36cm Ottawa,1963 Typescript
Sco 14.3712

CANADIAN ASSOCIATION FOR APPLIED SPECTROSCOPY Colloquium spectroscopicum internationale 13th Proceedings Ottawa 1967 Jun 19-23 Co-sponsored by the Society for applied spectroscopy xix,460p 23cm Hilger,London, 1968
Chem 18.2845

CANADIAN ASSOCIATION OF PHYSICISTS The International conference on low temperature physics 7th Proceedings Toronto 1960 Aug 29-Sep 3 International union of pure and applied physics Edited by J.M. Graham and A. C. Hollis-Hallett University of Toronto press;North-Holland,Toronto;Amsterdam,1961
Cav 7.2330

CANADIAN ASSOCIATION OF PHYSICISTS.THEORETICAL PHYSICS DIVISION Summer institute in nuclear and particle physics Proceedings Toronto 1967 Edited by B. Margolis and C.S. Lam 547p 24cm Benjamin,New York,1968
Cav 7.3100

CANADIAN CANCER CONFERENCE Canadian cancer research conference 1st,3rd Proceedings Honey Habor,Ont. 1954,1958 Jun Edited by R. W. Begg illus. 2 vols Academic press,New York,1955,1959
Radioth 35.1958

CANADIAN CANCER CONFERENCE 5th :Canadian cancer research conference Proceedings Honey harbor,Ont. 1962 Jun 10-14 National cancer institute of Canada Academic press, New York;London,1963
Gen 34.1993

CANADIAN CANCER RESEARCH CONFERENCE 1st,3rd Proceedings Canadian cancer conference Honey Habor,Ont. 1954,1958 Jun Edited by R. W. Begg illus. 2 vols Academic press,New York,1955,1959
Radioth 35.1958

CANADIAN CONFERENCE FOR COMPUTING AND DATA PROCESSING proceedings Toronto 1958 June 9-10 illus. xii,383p University of Toronto press,Toronto,1958
Math L 5.0845

CANADIAN INSTITUTE OF INTERNATIONAL AFFAIRS Canadian papers 1933 Institute of Pacific relations conference 5th Papers Banff 1933 Aug 14-28 Canadian institute of international affairs,Toronto,1933
Geog 13.4898

CANADIAN MATHEMATICAL CONGRESS Counting labelled trees Canadian mathematical congress 12th biennial seminar Lectures Vancouver,B.C. 1969 Canadian mathematical monographs, 1 x,113p 23cm Canadian mathematical congress,Montreal,Que.,1970
P Math 2.4447

CANADIAN MATHEMATICAL CONGRESS International symposium on classical and contagious discrete distributions proceedings Montreal 1963 Aug 15-20 Edited by Ganapati P. Patil Supported by the National research council of Canada Bibliog. xiv,552p 27cm Statistical publishing society,Calctta,1965
Math 3.0709

CANADIAN MATHEMATICAL CONGRESS :summer seminar Notes of lectures Nilpotent groups Edmonton,Alta. 1957 Aug 12-30 By Philip Hall 2nd edition iii,76p 25cm Queen Mary college,London,1957
P Math 2.4544

CANADIAN MATHEMATICAL CONGRESS 12th biennial seminar Lectures Counting labelled trees Vancouver,B.C. 1969 Canadian mathematical congress Canadian mathematical monographs, 1 x,113p 23cm Canadian mathematical congress,Montreal,Que.,1970
P Math 2.4447

CANADIAN MATHEMATICAL CONGRESS.LECTURE SERIES, 1 Introduction to the theory of distributions By Israel Halperin 33p 23cm University of Toronto press,Toronto, 1952 Based on lectures given by Laurent Schwartz
P. Math 2.1089

CANADIAN NATIONAL COMMITTEE FOR THE INTERNATIONAL BIOLOGICAL PROGRAM and UNITED STATES NATIONAL COMMITTEE FOR THE INTERNATIONAL BIOLOGICAL PROGRAM Working party conference for the international biological program human adaptability study of circumpolar populations Report Point Barrow 1967 Nov 17-22 variousp 29cm Naval arctic research laboratory,Point Barrow, 1967
Sco 14.7529

The CANADIAN NORTHWEST:ITS POTENTIALITIES Edmonton 1958 Jun Royal society of Canada Edited by Frank H. Underhill Royal society of Canada."Studia varia" series,3 University of Toronto press,Toronto,1959
Geog 13.5038

The CANADIAN NORTHWEST:ITS POTENTIALITIES : symposium presented to the Royal society of Canada in 1958 Papers Edmonton 1958 Jun 3 Royal society of Canada Edited by Frank H. Underhill Royal society of Canada. Studia varia.Series 3 map 104p 24cm University of Toronto press,Toronto,1959
Sco 14.4228

CANADIAN PAPERS 1933 Institute of Pacific relations conference 5th Papers Banff 1933 Aug 14-28 Canadian institute of international affairs Canadian institute of international affairs,Toronto,1933
Geog 13.4898

CANADIAN PSYCHIATRIC ASSOCIATION The World congress of psychiatry 3rd Proceedings Montreal 1961 Jun 4-10 2 vols University of Toronto press;McGill university press, Toronto;Montreal,1961 Conference languages:English,French,German and Spanish
Psy 31.2540

CANADIAN PSYCHIATRIC ASSOCIATION World congress of psychiatry Proceedings Montreal 1961 Jun 4-10 Vol 1-2 Edited by R.A. Cleghorn and others 2 vols University of Toronto press;McGill university press,Toronto;Montreal,1961 Papers in English,French,German and Spanish
PGMS 29.0322

CANADIAN SOCIETY FOR CHEMICAL ENGINEERING. SYMPOSIUM SERIES, 1 Cocurrent gas-liquid flow International symposium on research in cocurrent gas-liquid flow Proceedings Waterloo,Ont. 1968 Sep 18-19 Edited by Edward Rhodes and Donald S. Scott illus ix,698p Plenum press,New York,1969
Chem E 24.1842

CANADIAN SOCIETY OF MICROBIOLOGISTS Recent progress in microbiology :symposia presented at the 8th international congress for microbiology Montreal 1962 Aug 20-24 Edited by N.E. Gibbons University of Toronto press,Toronto,1963
Gen 34.0701

CANBERRA 1952 Mixed commission on the ionosphere meeting 3rd Proceedings International scientific radio union Edited by W.J.G. Beynon Organized by the International council of scientific unions 194p General secretariat,Brussels,1953
Obs 6.2579

CANBERRA 1956 Climatology and microclimatology :symposium Proceedings Unesco,and,Commonwealth scientific and industrial research organization Arid zone research,11 Unesco,Paris,1958
Geog 13.1018

CANBERRA 1959 Haematin enzymes :a symposium Papers and discussions Pt 1 International union of biochemistry Edited by J.E. Falk and others Organized by the Australian academy of science I.U.B. Symposium series, 19 Pergamon press,Oxford, 1961
Bioch 33.1079

CANBERRA 1961 Report of the first consultative meeting 1st Proceedings Antarctic treaty 40p 24cm Commonwealth government printer,Canberra,1962
Sco 14.6120

CANBERRA 1961 Summer research institute of the Australian mathematical society research reports Pt. 1: probability and statistics Australian mathematical society Australian mathematical society,Canberra,1961 Typescript
Math 3.0676

CANBERRA 1962 Environmental control of plant growth :a symposium Proceedings Edited by L.T. Evans Academic press,New York; London,1963
Bot 42.1351

CANBERRA 1963 The Galaxy and the magellanic clouds :a symposium International astronomical union and International scientific radio union Edited by F.J. Kerr and A.W. Rodgers International astronomical union.Symposium, 20 393p Australian academy of science,Canberra,1964 Dedicated to Joseph Lade Pawsey
A Math 4.1158

CANBERRA 1963 The Galaxy and the magellanic clouds :a symposium Proceedings International astronomical union and International scientific radio union Edited by F.J. Kerr and A.W. Rodgers International astronomical union.Symposium, 20 393p Australian academy of science,Canberra,1964 Publication dedicated to memory of Joseph Lade Pawsey
Obs 6.0941

CANBERRA 1963 Water resources use and management :a symposium Proceedings Australian academy of science Melbuorne university press,London;New York,1964
Geog 13.0787

CANBERRA 1965 The Theory of groups international conference proceedings Australian national university Edited by L.G. Kovacs and B.H. Neumann Bibliog xviii, 397p 24cm Gordon and Breach,New York,1967
P Math 2.2793

CANBERRA 1967 Symposium on fibrous proteins Papers Edited by W.G. Crewther Sponsored by the Australian academy of science Butterworths,Sydney,1968
Bioch 33.0512

CANBERRA 1969 Captain Cook,navigator and scientist Cook bicentenary symposium Papers Australian academy of science Edited by G.M. Badger ix,143p Hurst; Humanities press,London;New York,1970
WSM 43.1245

CANBERRA 1969 Data representation :a conference Proceedings Edited by R.S. Andersen and M.R. Osborne Held at the Australian national university 117p 24cm University of Queensland press,Brisbane,1970
Math S 3.1800

CANBERRA 1969 Phase transformations and the earth's interior :a conference International upper mantle committee Australiann academy of science Edited by A.E. Ringwood and D.H. Green North-Holland, Amsterdam,1970
Min 10.1459

CANBERRA 1969 Picture language machines :a conference Proceedings Australian national university Edited by S. Kaneff illus. 425p Academic press,London,1970
Math L 5.3579

CANBERRA 1969 Picture language machines :a conference Proceedings Commonwealth scientific and industrial research organisation Australian national university. Research school of physical sciences Edited by S. Kaneff Academic press,London,1970
Psy 31.3083

CANBERRA 1969 Picture language machines :a conference Proceedings Edited by S. Kaneff Held at the Australian national university Academic press,London;New York,1970
An 32.5428

CANBERRA 1970 International conference on plant growth substances 7th Proceedings Edited by Denis J. Carr Springer,Berlin,1972
Bioch 33.2231

CANBERRA 1970 Photosynthesis and photorespiration :a conference Australian academy of sciences Edited by M.D. Hatch and others Interscience,New York,1971
Bioch 33.2221

CANCER Biological activities of steroids in relation to cancer :a conference Proceedings Vergennes,Vt. 1959 Sep 27-Oct 2 Edited by Gregory Pincus and Erwin P. Vollmer Sponsored by the Cancer chemotherapy national service center Academic press,New York; London,1960
Radioth 35.0873

CANCER Carcinoma of the uterine cervix, endometrium and ovary Annual clinical conference on cancer 5th Papers Houston, Tex. 1960 Anderson hospital and tumor institute Year book medical,Chicago,Ill., 1962
Radioth 35.0828

CANCER Conference on cellular control mechanisms and cancer Proceedings Amsterdam 1963 Sep 9-13 International union against cancer and Netherlands cancer institute Edited by P. Emmelot and O. Muhlbock Elsevier,Amsterdam,1964 Organized to commemorate the 50th anniversary of the Netherlands cancer institute
Path 30.2459

CANCER Electron beam therapy Annual San Francisco cancer symposium 2nd Proceedings San Francisco,Calif. 1966 Frontiers of radiation therapy and oncology, 2 Karger,Basel;New York,1968
Radioth 35.1208

CANCER Hyperbolic oxygen and radiation therapy of cancer Annual San Francisco cancer symposium 1st Proceedings San Francisco,Calif. 1965 Nov 12-13 Frontiers of radiation therapy and oncology, 1 Karger,Basel;New York,1968
Radioth 35.1970

CANCER International cancer congress 7th London 1958 Jul 6-12 Abstract of papers Under the auspices of the International union against cancer 1958
Radioth 35.0777

CANCER International cancer congress 8th abstracts of papers Moscow 1962 Jul 22-28 Moscow,1962
Radioth 35.0785

CANCER International conference on radiation biology and cancer Proceedings Kyoto 1966 Nov 1-2 Edited by Tsutomu Sugahara Under the auspices of Nippon societas radiologica.Biological subgroup Radiation society of Japan,1967
Radioth 35.1757

CANCER International symposium on chemotherapy of cancer Proceedings Lugano 1964 Apr 28-May 1 Swiss academy of medical sciences Edited by Placidus A. Plattner Sponsored by Hoffmann-La Roche and co ltd. Elsevier,Amsterdam,1964
Pha 16.0335

CANCER Krebs - dokumentation und statistik maligner tumoren Internationale jahrestagung des Arbeitsausschuss medizin 10th Verhandlungsbericht Berlin 1965 Oct 25-28 Edited by Gustav Wagner Schattauer,Stuttgart, 1966
Radioth 35.0877

CANCER Krebsmetastasierung auf dem blutwege : symposium Geneva 1963 Jun 27-29 Schweizerische nationaliga fur krebsekampfung und krebsforschung Schweizerische akademie der medizinischen wissenschaften Schweizerische hamatologe gesellschaft Schweizerische akademie der medizinischen wissenschaften.Bulletin, 20,fasc 1-3 Schwabe,Basel,1964
Radioth 35.0835

CANCER Leukemia-Lymphoma Annual clinical conference on cancer 14th Papers Houston 1969 Anderson hospital and tumor institute Edited by C.C. Shullenberger and R.W. Cumley Yearbook medical publishers,Chicago,1971
PGMS 29.0678

CANCER Radiation biology and cancer Symposium on fundamental cancer research 12th Papers Houston,Tex. 1958 Anderson hospital and tumor institute University of Texas press,Austin,Tex.,1958
Phys 20.0988

CANCER Some aspects of the aetiology and biochemistry of prostatic cancer Tenovus workshop 3rd Proceedings Cardiff 1970 Tenovus institute for cancer research Edited by K. Griffiths and C.G. Pierrepoint Alpha omega alpha,Cardiff,1971
Radioth 35.1928

CANCER Symposion uber krebsprobleme : arbeitstagung des beratungsaussschusses fur krebsforschung beim Kultusministerium des landes Nordrhein-Westfalen Dusseldorf 1960 Jun 27-28 Edited by K.G. Ober and others Illus Springer,Berlin,1961
Radioth 35.0515

CANCER Symposium on chemotherapy of cancer Oslo 1956 May 21-25 International union against cancer International union against cancer.Acta, 13,no 3 U.I.C.C.,Louvain,1957
Radioth 35.0814

CANCER Symposium on virus and cancer Papers Saltsjobaden 1964 Jun 8-9 Swedish cancer society and Unio nordica contra cancrum Edited by H. Bergstrand and K.E. Hellstrom illus Balder,Stockholm,1965 Symposium arranged by the Swedish cancer society in conjunction with the annual general meeting of the Unio nordica contra cancrum
Path 30.2363

CANCER The International conference on the biology of cutaneous cancer 1st Philadelphia,Pa. 1962 Apr 6-11 Temple university.School of medicine and International union against cancer Edited by Frederick Urbach With financial support from Merck,Sharp and Dohme.Postgraduate program National cancer institute.Monograph, 10 US government printing office,Washington,D.C., 1963
Path 30.2423

CANCER The Surgical clinics of North America New York 1953 April New York number: symposium on the surgery of cancer Saunders, Philadelphia,Pa.;London,1953
Radioth 35.0754

CANCER BIOCHEMISTRY Amino acids,proteins and cancer biochemistry The Jesse P.Greenstein memorial symposium Papers Washington,D.C. 1959 Sep 16 American chemical society. Division of biological chemistry Edited by John T. Edsall Academic press,New York; London,1960 With a biographical article on Dr.Greenstein and a bibliography of his writings
Bioch 33.2345

CANCER CHEMOTHERAPY Biological approaches to cancer chemotherapy :a symposium Louvain 1960 Jun Edited by R.J.C. Harris Under the auspices of Unesco and the World health organization Academic press,London;New York, 1961
Bioch 33.0960

CANCER CHEMOTHERAPY :basic and clinical application Edited by I. Brodsky and S.B. Kahn Hahnemann symposia, 15 Grune and Stratton,New York,1967
Radioth 35.0899

CANCER CHEMOTHERAPY NATIONAL SERVICE CENTER Biological activities of steroids in relation to cancer :a conference Proceedings Vergennes,Vt. 1959 Sep 27-Oct 2 Edited by Gregory Pincus and Erwin P. Vollmer Academic press,New York;London,1960
Radioth 35.0873

CANCER CONTROL :Latin American regional conference Papers Santiago 1967 Nov 25-28 International union against cancer Sponsored by the Committee on national cancer control programmes of the commission on cqncer control International union against cancer. Technical report series, 1 International union against cancer,Geneva,1968
Radioth 35.0888

CANCER OF THE CERVIX :diagnosis of early forms London 1959 May 8 Ciba foundation Edited by G.E.W. Wolstenholme and Maeve O'Connor Ciba foundation study group, 3 Churchill,London,1959
Radioth 35.1960

CANCER RESEARCH Carcinogensis:a broad critique Annual symposium on fundamental cancer research 1966,20th Papers Anderson hospital and tumor institute Williams and Wilkins,Baltimore,Md.,1967
Radioth 35.1964

CANCER RESEARCH Molecular basis of neoplasia Annual symposium on fundamental cancer research 15th Papers Houston,Tex. 1961 Anderson hospital and tumor institute University of Texas press,Austin,Tex.,1961
Radioth 35.1942

CANCER RESEARCH Proliferation and spread of neoplastic cells Annual symposium on fundamental cancer research 21st 1967 Anderson hospital and tumor institute Williams and Wilkins,Baltimore,Md.,1968
Radioth 35.0550

CANCER RESEARCH Radiation biology and cancer Animal symposium on fundamental cancer research 12th Papers 1958 Owen,London, 1960
Radioth 35.1101

CANCER RESEARCH Viruses,nucleic acids and cancer Annual symposium on fundamental cancer research,1963 17th Papers Houston, Tex. 1963 Anderson hospital and tumor institute Williams and Wilkins,Baltimore,Md., 1963
Radioth 35.0822

CANCER THERAPY End results of cancer therapy International symposium on end results of cancer therapy Proceedings Sandefjord 1963 Sep 16-20 National cancer institute Edited by Sidney J. Cutler National cancer institute.Monograph, 15 National cancer institute,Bethesda,Md.,1964
Bioch 33.0986

CANCER THERAPY International symposium on end results of cancer therapy Sandefjord 1963 Sep 16-20 National cancer institute Edited by Sidney J. Cutler National cancer institute.Monograph, 15 U.S.Department of health,education and welfare,Washington,D.C., 1964
Radioth 35.0844

CANNES 1964 Combustion and propulsion: energy sources and energy conversion Agard colloquium 5th Papers Agard Edited by H.M. DeGroff and others AGARDograph, 81 Gordon and Black,London,1967
Eng 41.7253

CANTERBURY 1969 Joint conference on digital methods of measurement Proceedings Institution of electronic and radio engineers Institution of electrical engineers Institute of electrical and electronics engineers Institution of electronic and radio engineers.Conference proceedings, 15 IERE,London,1969
Eng 41.8305

CANTON 1935 Pathology and microbiology Chinese society of pathology and microbiolgy Proceedings Chinese medical journal. Supplement, 1 Chinese medical journal, Peiping,1935
An 32.1220

CAPE TOWN 1963 Antarctic geology : international symposium 1st Proceedings Edited by Raymond J. Adie Sponsored by the Scientific committee on Antarctic research illus xx,758p 27cm North-Holland, Amsterdam,1964
Sco 14.6125

CAPETOWN 1963 Antarctic geology : international symposium 1st proceedings Scientific committee on Antarctic research., International union of geological sciences Edited by Raymond J. Adie illus. North-Holland,Amsterdam,1964
Geol 8.2509

CAPILLARY PERMEABILITY:THE TRANSFER OF MOLECULES AND IONS BETWEEN CAPILLARY BLOOD AND TISSUE The Alfred Benzon symposium 2nd Proceedings Copenhagen 1969 Jun 23-26 Edited by Christian Crone and Niels A. Lassen Munksgaard,Copenhagen,1970
Inv Med 37.0054

CAPRI 1969 Pharmacokinetics and mode of action of oral hypoglycaemic agents conference 3rd Edited by A. Loubatieres and A.E. Renold Acta diabetologica latina, 6 1969
PGMS 29.0400

CAPTAIN COOK,NAVIGATOR AND SCIENTIST Cook bicentenary symposium Papers Canberra 1969 May 1 Australian academy of science Edited by G.M. Badger ix,143p Hurst; Humanities press,London;New York,1970
WSM 43.1245

CARACAS 1967 Basic mechanisms in photochemistry and photobiology :an international symposium Edited by J.W. Longworth Photochemistry and photobiology, 7,no 6 Pergamon press,London,1968
Radioth 35.0279

CARBOHYDRATE CHEMISTRY OF SUBSTANCES OF BIOLOGICAL INTEREST International congress of biochemistry 4th Vol 1: symposium 1- carbohydrate chemistry of substances of biological interest International union of biochemistry Edited by M.L. Wolfrom I.U.B.symposium series, 3 Pergamon,London,1959 Added title page in French and German.Text in English,French and German
Bioch 33.1315

CARBOHYDRATE METABOLISM Comparative physiology of carbohydrate metabolism in heterothermic animals :symposium 1959 Western society of naturalists Edited by Arthur W. Martin University of Washington press,Seattle,1961
Bal 39.1258

CARBOHYDRATE METABOLISM Disorders of carbohydrate metabolism :a symposium Proceedings London 1968 Nov 29-30 Association of clinical pathologists.Chemical pathology sub-committee Edited by G.K. McGowan and G. Walters held at the Royal society of medicine Chemical pathology in relation to clinical medicine, 5 Journal of clinical pathology.Supplement, 2 British medical association,London,1969
Bioch 33.1205

CARBOHYDRATE METABOLISM Hormonal factors in carbohydrate metabolism :a colloquium Proceedings London 1952 Jun 30-Jul 3 Ciba foundation Edited by G.E.W. Wolstenholme Ciba foundation colloquia on endocrinology, 6 illus Churchill,London, 1953
Bioch 33.0436

CARBOHYDRATES Neuere ergebnisse aus chemie und stoffwechsel der kohlenhydrate colloquium Mosbach,Baden 1957 May 2-4 Gesellschaft fur biologische chemie Gesellschaft fur biologische chemie.Colloquia, 8 illus Springer,Berlin,1958
Bioch 33.2347

CARBON Conference on carbon 1st and 2nd Proceedings Buffalo,N.Y. 1953,1955 By S. Mrozowski and others University of Buffalo University of Buffalo,New York,1956
Met 25.0154

CARBON Conference on carbon 3rd Proceedings Buffalo,N.Y. 1957 Jun 17-21 By S. Morzowski and others University of Buffalo and United States.Office of naval research Co-sponsored by the National science foundation Pergamon press,London, 1959
Met 25.0156

CARBON Conference on carbon 3rd Proceedings Buffalo,N.Y. 1957 Jun 17-21 State university of New York,Buffalo Edited by S. Mrozowski and others Co-sponsored by the United States.Office of naval research Conference on carbon.Proceedings, 2 718p Pergamon press,Oxford,1959
TA 15.0214

CARBON Conference on carbon 4th Proceedings Buffalo,N.Y. 1959 Jun 15-19 By S. Mrozowski and others University of Buffalo and American carbon committee Pergamon press,Oxford,1960
Met 25.0157

CARBON Conference on carbon 5th Proceedings University Park,Penn. 1961 Jun 19-23 Clarendon press,Oxford,1965
Met 25.0158

CARBON The Conference on carbon 4th Proceedings Buffalo,N.Y. 1959 Jun 15-19 State university of New York,Buffalo and American carbon committee Edited by S. Mrozowski and others Conference on carbon. Proceedings, 3 778p Pergamon press, Oxford,1960
TA 15.0215

CARBON The Conference on carbon 5th Proceedings University park,Penn. 1961 Vol 1 Pennsylvania state university and American carbon committee Edited by S. Mrozowski Conference on carbon.Proceedings, 4 Pergamon press,Oxford,1962 Proceedings appeared in two volumes
TA 15.0216

CARBON DIOXIDE FIXATION AND PHOTOSYNTHESIS Sheffield 1950 Jul Society for experimental biology Society for experimental biology.Symposia, 5 Cambridge university press,Cambridge,1951
Phys 20.0953

CARBON DIOXIDE FIXATION AND PHOTOSYNTHESIS : a symposium Sheffield 1950 Jul Society for experimental biology Society for experimental biology.Symposia, 5 Cambridge university press,Cambridge,1951
Bioch 33.1351

CARBON FLUORINE COMPOUNDS Ciba foundation symposium on carbon fluorine compounds : chemistry biochemistry and biological London 1971 Sep 13-15 Ciba foundation Edited by Katherine Elliott and John Birch Elsevier, Amsterdam,1972
Bioch 33.2244

CARBON STEELS Metallurgical developments in carbon steels :a conference Harrogate 1963 May 15-16 Iron and steel institute Sponsored by the British iron and steel research association Iron and steel institute.Special report, 81 Iron and steel institute,London,1963
Met 25.0417

CARBONATE ROCKS Classification of carbonate rocks :a symposium Papers Denver,Colo. 1961 Apr 27 Edited by William E. Ham American association of petroleum geologists. Memoir,1 Tulsa,Okla.,1962 Held under the joint auspices of the American association of petroleum geologists and the Society of economic paleontologists and mineralogists
Geog 13.0301

CARBONDALE,ILL. 1969 Teaching of probability and statistics Comprehensive school mathematics program international conference 1st Proceedings Southern Illinois university Central Midwestern regional educational laboratory Edited by Lennart Rade Wiley;Almqvist and Wiksell,New York;Stockholm,1970
An 32.5410

CARBURIZING :a symposium Papers and discussions Atlantic City,N.J. 1937 Oct 18-22 American society for metals American society for metals,Cleveland,Ohio,1937 The,Symposium presented before the nineteenth annual convention of the American society for Metals...
Met 25.0439

CARCINOGENESIS Ciba foundation symposium on carcinogenesis:mechanisms of action London 1958 Jun 24-26 Ciba foundation Edited by G. E.W. Wolstenholme and Maeve O'Connor Illus Churchill,London,1959
Radioth 35.0107

CARCINOGENESIS Subviral carcinogenesis International symposium on tumor viruses 1st 1966 Edited by Yohei Ito Sponsored by the Aichi cancer center.Research institute and the Japanese cancer association Aichi cancer center,Nagoya,1967
Bioch 33.1191

CARCINOGENESIS Symposium on the biology of skin 15th Proceedings Portland,Ore. 1965 Edited by William Montagna and Richard L. Dobson Held at the University of Oregon medical school Advances in biology of skin, 7 Oregon regional primate research center, 87 Pergamon press,Oxford,1966
An 32.4042

CARCINOGENIC AIR POLLUTANTS Analysis of carcinogenic air pollutants :symposium Proceedings Cincinnati,Ohio 1961 Aug 29-31 National cancer institute Edited by Eugene Sawicki and Kenneth Cassel National cancer institute.Monograph, 9 illus U.S. Department of health,education and welfare, Bethesda,Md.,1962 Sponsored by the Laboratory of engineering and physical sciences of the Division of air pollution,U.S. Department of health education,and welfare, Public health service.Robert A.Taft sanitary engineering center
Bioch 33.0980

CARCINOGENSIS:A BROAD CRITIQUE Annual symposium on fundamental cancer research 1966,20th Papers Anderson hospital and tumor institute Williams and Wilkins, Baltimore,Md.,1967
Radioth 35.1964

CARCINOMA OF THE BREAST The Treatment of carcinoma of the breast :a symposium Proceedings Cambridge 1967 Sep 9 Syntex pharmaceuticals Edited by Anthony S. Jarrett Under the auspices of the Institute of hormone biology Lewis,Cambridge,1968
Inv Med 37.0122

CARCINOMA OF THE UTERINE CERVIX,ENDOMETRIUM AND OVARY Annual clinical conference on cancer 5th Papers Houston,Tex. 1960 Anderson hospital and tumor institute Year book medical,Chicago,Ill.,1962
Radioth 35.0828

CARDIFF 1964 Royal statistical society. Research section and industrial applications joint conference Royal statistical society Cardiff,1964 Unbound mimeograph papers
Math 3.1114

CARDIFF 1964 and LONDON 1964 Rheomacrodex :a symposium Proceedings 1 Edited by H. Scarborough and H.G. Melrose Pharmacia limited,London,1965
PGMS 29.0610

CARDIFF 1965 Energy of transfer in radiation processes:chemical,physical and biological aspects :international symposium Proceedings Association for radiation research Edited by Glyn O. Phillips Elsevier,Amsterdam,1966
Radioth 35.1744

CARDIFF 1967 Prognostic factors in breast cancer :symposium Proceedings Tenovus institute for cancer research Tenovus symposia, 1 Livingstone,Edinburgh;London, 1968
Radioth 35.0935

CARDIFF 1969 Human adrenal gland and its relation to breast cancer Tenovus workshop 1st Proceedings Tenovus institute for cancer research Edited by K. Griffiths and E. H.D. Cameron Alpha omega alpa,Cardiff,1969
Inv Med 37.0270

CARDIFF 1969 Human adrenal gland and its relation to breast cancer Tenovus workshop 1st Proceedings Tenovus institute for cancer research Edited by K. Griffiths and E. H.D. Cameron Alpha omega alpha,Cardiff,1969
Bioch 33.2198

CARDIFF 1969 The Human adrenal gland and its relation to breast cancer Tenovus institute for cancer research Edited by K. Griffiths and E.H.D. Cameron Tenovus workshops, 1 Alpha omega alpha publishing, Cardiff,1969
Radioth 35.0940

CARDIFF 1970 Clinical management of advanced breast cancer Tenovus workshop 2nd Proceedings Tenovus institute for cancer research Edited by C.A.F. Joslin and E.N. Gleave Alpha omega alpha,Cardiff,1971
Radioth 35.1927

CARDIFF 1970 Some aspects of the aetiology and biochemistry of prostatic cancer Tenovus workshop 3rd Proceedings Tenovus institute for cancer research Edited by K. Griffiths and C.G. Pierrepoint Alpha omega alpha,Cardiff,1971
Radioth 35.1928

CARGESE,CORSICA 1962 Lectures in theoretical physics :notes from the French summer school Edited by Maurice Levy Cargese lectures in theoretical physics,1962 W.A.Benjamin,New York,1963
Cav 7.0437

CARIBBEAN:MEXICO TODAY Conference on the Caribbean 14th Papers Gainesville,Fla. 1963 Dec 4-7 University of Florida.Center for Latin American studies Edited by A. Curtis Wilgus University of Florida.Center for Latin American studies.Publications.Series 1,14 University of Florida press,Gainesville, Fla.,1964
Geog 13.5202

CARIBBEAN:VENEZUELAN DEVELOPMENT,A CASE HISTORY Conference on the Caribbean 13th Papers Gainesville,Fla. 1962 Dec 6-8 Edited by A. Curtis Wilgus University of Florida.School of Inter-American studies.Publications.Series 1,13 University of Florida press,Gainesville, Fla.,1963
Geog 13.5242

CARIBBEAN GEOLOGICAL CONFERENCE 1st report Antigua 1955 Dec.5-8 Edited by G.M. Stockley Demerara,1958 Conference organized by G.M.Stockley
Geol 8.2423

CARIBBEAN GEOLOGICAL CONFERENCE 2nd transactions Mayaguez 1959 Jan.4-9 Edited by John D. Weaver University of Puerto Rico,Mayaguez,1960
Geol 8.2424

CARIBBEAN GEOLOGICAL CONFERENCE 3rd transactions Kingston,Jamaica 1962 Apr. 2-4 Edited by Edward Robinson Jamaica geological survey department,Kingston,Jamaica, 1966
Geol 8.2425

CARIES-RESISTANT TEETH :Ciba foundation symposium Ciba foundation Edited by G.E. W. Wolstenholme and M. O'Connor Churchill, London,1965
Phys 20.2131

CARLETON,MINN. 1965 Reproduction;molecular, subcellular and cellular :a symposium Society for developmental biology Edited by Michael Locke Society for developmental biology.Symposia, 24 Academic press,London, 1965 The Society for developmental biology formerly known as the Society for the study of development
Bioch 33.0900

CARLETON UNIVERSITY Biochemistry and physiology of plant growth substances International conference on plant growth substances 6th Ottawa 1967 Jul 24-29 Edited by F. Wightman and G. Setterfield Runge press,Ottawa,1968
Bioch 33.0779

CARLETON UNIVERSITY Biochemistry and physiology of plant growth substances International conference on plant growth substances 6th Ottawa 1967 Jul 24-29 Edited by F. Wightman and G. Setterfield xii, 1642p Runge press,Ottawa,1968 On the occasion of the centennial of Canadian confederation
Bot 42.1833

CARLETON UNIVERSITY Biochemistry and physiology of plant growth substances International conference on plant growth substances 6th Proceedings Ottawa 1967 Jul 24-29 Edited by F. Wightman and G. Setterfield Runge press,Ottawa,1968
Gen 34.1408

CARLETON UNIVERSITY Colloquium spectroscopicum internationale 13th Proceedings Ottawa 1967 Jun 19-23 Co-sponsored by the Society for applied spectroscopy xix,460p 23cm Hilger,London, 1968
Chem 18.2845

CARNEGIE INSTITUTE OF TECHNOLOGY Concepts and the structure of memory Symposium on cognition 2nd Papers Pittsburgh,Pa. 1966 Apr 7-8 Edited by Benjamin Kleinmuntz Wiley,New York;London,1967
Eng 41.0652

CARNEGIE INSTITUTE OF TECHNOLOGY Problem solving:research,method and theory Annual symposium on cognition 1st Papers Pittsburgh 1965 Apr 15-16 Edited by Benjamin Kleinmutz Wiley,New York,1966
Psy 31.1911

CARNEGIE INSTITUTE OF TECHNOLOGY Problem solving:research,method and theory Symposium on cognition 1st Pittsburgh,Pa. 1966 Edited by Benjamin Kleinmuntz Wiley,New York; London,1966
Eng 41.1291

CARNEGIE INSTITUTE OF TECHNOLOGY Wave motion and vibration theory symposium Washington 1952 June 16-17 Edited by Albert E. Heins American mathematical society.Proceedings of symposia in applied mathematics, 5 McGraw-Hill,New York,1954
A Math 4.0513

CARNEGIE INSTITUTE OF TECHNOLOGY,PITTSBURGH
Buhl international conference on materials : transition metal compounds ;transport and magnetic properties 1st Informal proceedings Pittsburgh,Penn. 1963 Oct 31-Nov 1 Edited by E.R. Schatz Gordon and Breach,New York;London,1964
Met 25.2351

CARNEGIE INSTITUTE OF WASHINGTON.GEOPHYSICAL LABORATORY Researches in geochemistry : seminar series essays Baltimore,Md. 1957-58 Edited by Philip H. Abelson John Wiley and sons,New York,1959
Geol 8.1495

CARNEGIE INSTITUTION OF WASHINGTON
Conference on cycles 1st-2nd Report Washington 1922 and 1928 Washington, 1929
Philos 1.0096

CARNEGIE INSTITUTION OF WASHINGTON Quantum theory and beyond :essays and discussions arising from a colloquium Cambridge 1968 Jul illus viii,345p 24cm Cambridge university press,Cambridge,1971 Based on papers read at an informal colloquium
A Math 4.1759

CARNEGIE INSTITUTION OF WASHINGTON.PUBLICATIONS, 490 Leaf xanthophylls By Harold H. Strain Carnegie institution of Washington, Washington,D.C.,1938
Bot 42.1694

CARNEGIE-MELLON UNIVERSITY Cognition and the development of language :annual symposium 4th Papers Pittsburgh,Pa. 1968 Apr 11-12 Edited by John R. Hayes Wiley,New York,1970
Psy 31.3113

CARNEGIE-MELLON UNIVERSITY Formal representation of human judgement Symposium on cognition 3rd Papers Pittsburgh,Pa. 1967 Apr 13-14 Edited by Benjamin Kleinmuntz Wiley,New York;London,1968
Eng 41.0653

CARNITINE Recent research on carnitine:its relation to lipid metabolism :a symposium Papers presented Cambridge,Mass. 1964 Jul 24-25 Edited by George Wolf at the Massachusetts institute of technology M.I.T. press,Cambridge,Mass.,1965
Bioch 33.1249

CAROLINA CONFERENCE Proceedings Chapel Hill,N.C. 1970 Mar 30-Apr 3 University of North Carolina.Department of mathematics Edited by Donald A. Eisenman and Laird E. Taylor 107p 28cm University of North Carolina,Chapel Hill,N.C.,1970
P Math 2.4356

CARTAN H. and others Structures algebriques et structures topologiques :conferences a l'Institut Henri Poincare de la Sorbonne Paris 1956 Feb.9 and 1957 Jun.6 Societe mathematique de France en accord avec L' Association des Professeurs de mathematique de l'enseignement public Enseignement mathematique.Monographies, 7 198p Geneva, Paris,1958
P. Math 2.0241

CARTE DU MONDE AU MILLIONIEME 2nd :
Conference internationale Comptes rendus Paris 1913 Dec France.Service geographique de larmee Service geographique de larmee, Paris,1914
Geog 13.5472

CARTOGRAPHY United nations regional cartographic conference for Asia and the Far East 4th Manila 1964 Nov 21-Dec 5 2: proceedings of the conference and technical papers United nations.Department of economic and social affairs United nations,New York, 1966
Geog 13.7226

CASE INSTITUTE OF TECHNOLOGY Fracturing of metals a seminar Chicago 1947 Oct 18-24 American society for metals,Cleveland,Ohio,n.d. A seminar held during the 29th National metal congress and exposition
Met 25.0861

CASE INSTITUTE OF TECHNOLOGY Operations research in research development :a conference Proceedings Cleveland,Ohio 1962 Edited by Barton V. Dean Wiley,New York;London,1963
Eng 41.1173

CASE INSTITUTE OF TECHNOLOGY The Annealing of low carbon steel :the international symposium Proceedings Cleveland 1957 Oct 29-30 Lee Wilson engineering company, Cleveland,Ohio,1958
Met 25.0447

CASE INSTITUTE OF TECHNOLOGY Views on general systems theory Systems symposium 2nd Proceedings Cleveland,Ohio 1963 Apr Edited by Mihajlo D. Mesarovic Case institute of technology.Systems research center.Publications Wiley,New York,1964
Eng 41.5967

CASE INSTITUTE OF TECHNOLOGY Views on general systems theory The Second systems symposium proceedings Cleveland,Ohio 1963 Apr Edited by Mihajlo D. Mesarovic xvii,178p 24cm John Wiley and sons,New York;London,1964
Math 3.0880

CASE INSTITUTE OF TECHNOLOGY.METALLURGICAL STAFF Fracturing of metals :a seminar...held during the 29th National metal congress and exposition...Co-ordinated by...the metallurgical staff of the Case institute of technology under the direction of George Sachs Chicago,Ill. 1947 Oct 18-24 Sponsored by the American society for metals American society for metals,Cleveland,Ohio,1948
Eng 41.3732

CASE INSTITUTE OF TECHNOLOGY.SYSTEMS RESEARCH CENTER Systems theory and biology Systems symposium 3rd Proceedings Cleveland,Ohio 1966 Oct Edited by M.D. Mesarovic Springer,Berlin,1968
Gen 34.0226

CASE WESTERN RESERVE UNIVERSITY Conference on global differentiable dynamics Proceedings Cleveland,Ohio 1969 Jun 2-6 Edited by O. Hajek and others Lecture notes in mathematics, 235 x,140p 25cm Springer, Berlin,1971
P Math 2.4317

CASE WESTERN RESERVE UNIVERSITY.SYSTEMS RESEARCH CENTER Theoretical approaches to non-numerical problem solving Systems symposium 4th Proceedings Cleveland,Ohio 1968 Nov 19-20 Edited by R.B. Banerji and M.D. Mesarovic Lecture notes in operations research and mathematical systems, 28 bibliog.,illus. vi,466p 25cm Springer, Berlin,1970
Math S 3.1617

CASTEL DI POGGIO 1969 Bradykinin and related kinins;cardiovascular,biochemical,and neural actions An International symposium on the cardiovascular and neural actions of bradykinin and related kinins Edited by F. Sicuteri and others Advances in experimental medicine and biology, 8 Plenum,New York; London,1970
Inv Med 37.0196

CATALINA ISLAND 1957 Radiobiology at the intra-cellular level U.C.L.A.conference on radiobiology 1st Proceedings University of California Edited by T.G. Hennessy and others ports. Pergamon press,London,1959
Radioth 35.1722

CATALYSIS International congress on catalysis Philadelphia 1956 Sep Edited by Adalbert Farkas Advances in chemistry and related subjects, 9 xviii,847p Academic press,New York,1957
Chem E 24.0800

CATALYSIS Symposium on catalysis in practice Proceedings Harrogate 1963 Jun 20-21 Edited by J.M. Pirie The Yorkshire meeting of the Institution of chemical engineers 83p Institution of chemical engineers,London,1963
Chem E 24.0803

CATALYSIS The Mechanism of heterogeneous catalysis :a symposium Proceedings Amsterdam 1959 Nov 12-13 Royal Netherlands chemical society.Section for inorganic and physical chemistry Edited by J.H. de Boer and others Elsevier monographs.Chemistry series Elsevier,Amsterdam,1960
Chem 18.0455

CATALYSIS ,WITH SPECIAL REFERENCE TO NEWER THEORIES OF CHEMICAL ACTION :a general discussion London 1921 Sep 28 Pt 1-2 Faraday society London,1922 Reprinted from the 'Transactions' of the society,vol 17,pt 3.Chairman of the discussion A.W.Porter
Bot 42.6477

CATAPRES IN HYPERTENSION :a symposium Proceedings London 1969 Mar Royal college of surgeons Edited by Matthew E. Conolly Butterworths,London,1970
PGMS 29.0608

CATECHOLAMINES Symposium on catecholamines 2nd Milan 1965 Jul 4-9 American society for pharmacology and experimental therapeutics Edited by George H. Acheson Held at the Istituto di ricerche farmacologiche "Mario Negri" Williams and Wilkins company,Baltimore,1966 Reprinted from the Pharmacological reviews,18,no.1
Bioch 33.0430

CATECHOLAMINES Symposium on catecholamines 2nd Milan 1965 Jul 4-9 Edited by George H. Acheson Pharmacological reviews, 18,no.1 Williams and Wilkins,Baltimore,Md., 1966 Reprint
An 32.3777

CATEGORICAL ALGEBRA conference proceedings La Jolla,Calif. 1965 Jun 7-12 Edited by S. Eilenberg and others Supported by the United States.Air force.Office of scientific research 562p 24cm Springer-Verlag, Berlin,1966
P. Math 2.2027

CATEGORIES OF HUMAN LEARNING :symposium on the psychology of human learning Papers Ann Arbor,Mich. 1962 Jan 31-Feb 1 United States.Office of naval research and University of Michigan Edited by Arthur W. Melton Academic press,New York;London,1964
Psy 31.1003

CATEGORY THEORY,HOMOLOGY THEORY AND THEIR APPLICATION :a conference Proceedings Seattle 1968 Jun 24-Jul 19 Vol 1 Supported by the Battelle memorial institute Lecture notes in mathematics, 86 Bibliog, Illus vi,216p 28cm Springer-Verlag, Berlin,1969
P Math 2.3343

CATEGORY THEORY,HOMOLOGY THEORY AND THEIR APPLICATIONS Seattle 1968 Jun 24-Jul 19 Vol 3 Supported by the Battelle memorial institute Lecture notes in mathematics, 99 Bibliog.,Illus. iv,489p 28cm Springer-Verlag,Berlin,1969
P Math 2.3345

CATEGORY THEORY,HOMOLOGY THEORY AND THEIR APPLICATIONS :a conference Seattle 1968 Jun 24-Jul 19 Vol 2 Supported by the Battelle memorial institute Lecture notes in mathematics, 92 Bibliog.,Illus. vi,216p 28cm Springer-Verlag,Berlin,1969
P Math 2.3344

CATHOLIC UNIVERSITY OF AMERICA The Plasma space science symposium proceedings Washington D.C. 1963 Jun 11-14 Edited by C. C. Chang and S.S. Huang In co-operation with the National aeronautics and space administration Astrophysics and space science library, 3 D.Reidel,Dordrecht,1965
A Math 4.0625

CAUSES OF CLIMATIC CHANGE International union for quaternary research congress :a collection of papers derived from the INQUA-NCAR symposium 7th Proceedings Boulder, Colo. 1965 Aug 14-Sep 19 5: causes of climatic change International union for quaternary research,and,National center for atmospheric research Edited by J.Murray Mitchell Sponsored by the National research council Meteorological monographs,8,no 30 illus iv,159p 29cm American meteorological association,Boston,Mass.,1968 For other volumes of these proceedings,see also 'International association for quaternary research'
Sco 14.7750

CAVENDISH LABORATORY Conference on the physics of the ionosphere Report Cambridge 1954 Sep 6-9 406p Physical society,London, 1955
Nap 11.0926

CAVENDISH LABORATORY The International conference on fundamental particles and low temperature Report Cambridge 1946 Jul 22-27 Vol 1: fundamental particles Physical society,London,1947
Cav 7.0460

CAVENDISH LABORATORY The International conference on fundamental particles and low temperatures Report Cambridge 1946 Jul 22-27 Vol 2: low temperatures Physical society,London,1947
Cav 7.0458

CAVENDISH LABORATORY X-ray microscopy and microradiography The Symposium on X-ray microscopy and microradiography Proceedings Cambridge 1956 Aug 16-25 Edited by V.E. Cosslett and others Sponsored by the International union of pure and applied physics Academic press,New York,1957
Cav 7.0461

CAVITATION Cavitation in real liquids :a symposium Proceedings Warren,Mich. 1962 General motors research laboratories Edited by Robert Davies Elsevier,Amsterdam,1964
Eng 41.6512

CAVITATION Erosion by cavitation or impingement :a symposium presented at the 69th annual meeting Atlantic City,N.J. 1966 Jun 26-Jul 1 American society for testing and materials American society for testing and materials.Special technical publications, 408 American society for testing and materials, Philadelphia,Pa.,1967
Eng 41.6561

CAVITATION International association for hydraulic research symposium Cavitation and hydraulic machinery Proceedings Sendai 1962 Sep 3-5 Edited by F. Numachi Tohoku university,Sendai,163
Eng 41.6518

CAVITATION Symposium on cavitation research. Facilities and techniques American society of mechanical engineers New York,1964
Eng 41.6521

CAVITATION AND HYDRAULIC MACHINERY Proceedings International association for hydraulic research symposium Sendai 1962 Sep 3-5 Edited by F. Numachi Tohoku university, Sendai,163
Eng 41.6518

CAVITATION IN FLUID MACHINERY Symposium on cavitation in fluid machinery Chicago,Ill. 1965 Nov 7-11 American society of mechanical engineers Edited by Glenn M. Wood and others American society of mechanical engineers,New York,1965
Eng 41.6538

CAVITATION IN REAL LIQUIDS :a symposium Proceedings Warren,Mich. 1962 General motors research laboratories Edited by Robert Davies Elsevier,Amsterdam,1964
Eng 41.6512

CAVITATION IN REAL LIQUIDS... 6th :a symposium Proceedings Warren,Mich. 1962 Sep 24-25 Edited by Robert Davies Sponsored by General motors research laboratories port vii,189p Elsevier, Amsterdam,1964
Chem E 24.0240

CAYENNE 1951 Conference geologique des trois Guyanes 2nd report 1951 Typescript
Geol 8.2429

CAYENNE 1957 Conference geologique des Guyanes 4me communications presentees Memoires pour servir a l'explication de la carte geologique detaillee de la France: departement de la Guyane Francaise Imprimerie nationale,Paris,1959
Geol 8.2427

CELEBRAZIONE DELLA ACCADEMIA DEL CIMENTO NEL TRICENTENARIO DELLA FONDAZIONE DOMUS GALILAEANA Pisa 1957 Jun 19 Accademia del cimento 49 pls 79p Domus galilaeana, Pisa,1958 Limited edition.With appendices
WSM 43.4472

CELEBRAZIONI ARCHIMEDEE DEL SECOLO XX Syracuse 1961 Apr 11-16 Vol. 1: conferenze generali e simposio di geometria differenziale Universita di Messina 68, 190p 25cm Oderisi,Gubbio,1962
P. Math 2.2608

CELEBRAZIONI ARCHIMEDEE DEL SECOLO XX Syracuse 1961 Apr 11-16 Vol. 2: simposio di analisi Universita di Messina 108p 25cm Oderisi,Gubbio,1962
P. Math 2.2609

CELERINA 1962 Gyrodynamics symposium International union of theoretical and applied mechanics Edited by Hans Ziegler Springer-Verlag,Berlin,1963
A Math 4.0314

CELERINA,SWITZERLAND 1962 Kreisel probleme internationale symposium Vortrage International union of theoretical and applied mechanics and Swiss federal institute of technology Edited by Hans Ziegler Berlin, 1963 Papers in English,French and German
Geod 9.0236

CELL,ORGANISM AND MILIEU :a symposium South Hadley,Mass. 1958 Jun 9-11 Edited by Dorothea Rudnick Society for the study of development and growth.Symposia, 17 Ronald press,New York,1959
An 32.3226

CELL,ORGANISM AND MILIEU :a symposium South Hadley,Mass. 1958 Jun 9-11 Society for the study of development and growth Edited by Dorothea Rudnick held at Mount Holyoke college Society for the study of development and growth.Symposia, 17 Ronald press,New York,1959
Bioch 33.1008

CELL AND TISSUE DAMAGE Mechanism of cell and tissue damage produced by immune reactions International symposium on immunopathology 2nd Papers Brook Lodge,Mich. 1961 International committee on immunology and Upjohn company Edited by Pierre Grabar and Peter Miescher Schwabe,Basle;Stuttgart,1962
Path 30.2336

CELL BIOLOGY Fine structure of cells :a symposium held at the VIIIth congress of cell biology Papers Leiden 1954 International union of biological sciences With financial support of Unesco International union of biological sciences. Publications.Series B, 21 Noordhoff, Groningen,1955
Path 30.2410

CELL BIOLOGY International congress of cell biology 12th Summaries of reports and communications Brussels 1968 Aug 25-31 Edited by P. Dustin and others International congress series, 166 Excerpta medica, Amsterdam,1968
Bioch 33.0963

CELL BIOLOGY International congress of cell biology 12th Summaries of reports and communications Brussels 1968 Aug 25-31 Edited by P. Dustin and others Under the auspices of the International society for cell biology International congress series, 166 Excerpta medica,Amsterdam,1968
Bot 42.1515

CELL BIOLOGY International congress on cell biology 12th Summaries of reports and communications Brussels 1968 Aug 25-31 Edited by P. Dustin Excerpta medica. International congress series, 166 Excerpta medica,Amsterdam,1968
Inv Med 37.0250

CELL BIOLOGY New approaches in cell biology : a symposium Proceedings London 1958 Jul Edited by P.M.B. Walker Academic press, London,1960 Held during the fifteenth international congress of zoology
An 32.3236

CELL BIOLOGY New approaches in cell biology; a symposium The International congress of zoology 15th Proceedings London 1958 Jul Edited by P.M.B. Walker Held at Imperial college of science and technology Academic press,London,1960 A section of the Proceedings of the 15th International congress of zoology
Bal 39.0289

CELL BIOLOGY Primitive motile systems in cell biology A Symposium on the mechanism of cytoplasmic streaming,cell movement,and the saltatory motion of subcellular particles Princeton,N.J. 1963 Apr 2-5 Edited by Robert D. Allen and Noburo Kamiya Held at Princeton university Academic press,New York; London,1964
Bal 39.0297

CELL-BOUND ANTIBODIES :a conference Proceedings Washington,D.C. 1963 May 10 National research council.Committee on tissue transplantation Edited by Bernard Amos and Hilary Koprowski Wistar institute press, Philadelphia,1963
Bioch 33.0534

CELL-BOUND ANTIBODIES :conference Proceedings Washington,D.C. 1963 May 10 Edited by Bernard Amos and Hilary Koprowski Sponsored by the National research council. Committee on tissue transplantation Wistar institute press,Philadelphia,Pa.,1963
An 32.3296

CELL CULTURE Analytic cell culture Syverton memorial symposium Proceedings Detroit 1961 Jun 6-7 National cancer institute Edited by Robert E. Stevenson Sponsored by Tissue culture association and Cell culture collection committee of the viruses and cancer board,National cancer institute National cancer institute. Monograph, 7 U.S.Department of health, education and welfare,Bethesda,Md.,1962
Bioch 33.0979

CELL CULTURE Ciba foundation symposium on growth control in cell culture Papers London 1971 Oct 14-16 Ciba foundation Edited by G.E.W. Wolstenholme and Julie Knight Churchill;Livingstone,Edinburgh,1971
Bioch 33.2263

CELL DIFFERENTATION AND MORPHOGENESIS : international lecture course Wageningen 1965 Apr 26-29 By W. Beermann and others Organized by the Landbouwhoogeschool, Wageningen North-Holland,Amsterdam,1966 "The fourth international symposium organized by the Landbouwhoogeschool"
Bal 39.0309

CELL DIFFERENTIATION Edinburgh 1962 Sep 3-8 Society for experimental biology Society for experimental biology.Symposia, 17 Cambridge university press,Cambridge,1963
Bioch 33.1359

CELL DIFFERENTIATION :a Ciba foundation symposium London 1967 Jan 31-Feb 2 Ciba foundation Edited by A.V.S. De Reuck and Julie Knight Churchill,London,1967
An 32.2531

CELL DIFFERENTIATION :a Ciba foundation symposium London 1967 Jan 31-Feb 2 Ciba foundation Edited by A.V.S. De Reuck and Julie Knight Churchill,London,1967
Phys 20.1511

CELL DIFFERENTIATION :a Ciba foundation symposium London 1967 Jan 31-Feb 2 Ciba foundation Edited by A.V.S. De Reuck and Julie Knight Ciba foundation.Symposia Churchill,London,1967
Bal 39.0317

CELL DIFFERENTIATION :a symposium Edinburgh 1962 Sep 3-8 Edited by G.E. Fogg Society for experimental biology.Symposia, 17 Cambridge university press,Cambridge,1963
An 32.2500

CELL DIFFERENTIATION :symposium Edinburgh 1962 Sep 3-8 Society for experimental biology Society for experimental biology.Symposia, 17 Cambridge university press,Cambridge,1963
Phys 20.1494

CELL DIFFERENTIATION :a Ciba foundation symposium Proceedings London 1967 Jan 31-Feb 2 Ciba foundation Edited by A.V.S. De Reuck and Julie Knight Churchill,London, 1967
Bioch 33.0889

CELL DIFFERENTIATION AND MORPHOGENESIS : international lecture course Wageningen 1965 Apr 26-29 By W. Beermann and others Organized by the Agricultural university of Wageningen North Holland,Amsterdam,1966 The fourth international symposium organized by the Agricultural university of Wageningen
Bot 42.1098

CELL DIFFERENTIATION AND MORPHOGENESIS : international lecture course Wageningen 1965 Apr 26-29 Edited by W. Beermann and others Organized by the Agricultural university of Wageningen North-Holland, Amsterdam,1966 "The fourth international symposium organized by the Agricultural university of Wageningen"
Bioch 33.1020

CELL DIFFERENTIATION AND MORPHOGENESIS... : international lecture course Wageningen 1965 Apr 26-29 By W. Beerman Landbouwhogeschool,Wageningen North-Holland, Amsterdam,1966
Gen 34.2211

CELL DIVISION Conference on the mechanisms of cell divisions 2nd Proceedings New York 1960 Oct 7 By M.J. Kopac and others New York academy of sciences Edited by Paul R. Gross New York academy of sciences.Annals, 90,p.345-613 New York,1960
Bal 39.0291

CELL DIVISION Control of cell division and the induction of cancer International symposium on the control of cell division and the induction of cancer Proceedings Lima 1963 Jul 1-6 and Cali,Columbia 1963 Jul 1-6 National cancer institute National cancer institute.Monograph, 14 illus National cancer institute,Bethesda,Md.,1964 Introduction,abstracts of each paper and summary in English and Spanish
Bioch 33.0985

CELL ELECTROPHORESIS :a symposium London 1963 May British biophysical society Edited by E.J. Ambrose held at the Chester Beatty research institute illus Churchill,London,1965
Bal 39.0034

CELL ELECTROPHORESIS :a symposium London 1963 May Edited by E.J. Ambrose Convened by the British biophysical society Churchill,London,1965
Radioth 35.1414

CELL GROWTH AND CELL DIVISION :a symposium Liege 1962 May 19-24 Edited by R.J.C. Harris International society for cell biology.Symposia, 2 Academic press,New York,1963
An 32.3288

CELL GROWTH AND CELL DIVISION :a symposium Liege 1962 May 19-24 International society for cell biology Edited by R.J.C. Harris International society for cell biology.Symposia, 2 illus. Academic press,New York;London,1963
Gen 34.0863

CELL GROWTH AND CELL DIVISION :symposium Liege 1962 May 19-24 Edited by R.J.C. Harris International society for cell biology.Symposia, 2 Illus Academic press, New York;London,1963
Radioth 35.0518

CELL GROWTH AND CELL DIVISION Report Liege 1962 May 19-24 International society for cell biology Edited by R.J.C. Harris held at the Institute of histology,Liege International society for cell biology. Symposia, 2 illus Academic press,New York;London,1963
Bioch 33.0965

CELL GROWTH AND CELL DIVISION;A SYMPOSIUM Report Liege 1962 May 19-24 International society for cell biology Edited by R.J.C. Harris International society for cell biology.Symposia, 2 Academic press,New York,1963
Med 36.0178

The CELL IN MITOSIS :a symposium Proceedings Detroit 1961 Nov 6-8 Wayne state university Edited by Laurence Levine Wayne state fund research recognition award.Annual symposia, 1 Academic press,New York,1963
An 32.3412

The CELL IN MITOSIS annual symposium Proceedings Detroit,Mich. 1961 Nov 6-8 Wayne state university Edited by Laurence Levine Wayne state fund research recognition award.Annual symposia, 1 illus. Academic press,New York;London,1963
Gen 34.0862

The CELL IN MITOSIS symposium Proceedings Detroit 1961 Nov 6-8 Edited by Laurence Levine Held at Wayne state university Wayne State fund research recognition award. Annual symposia, 1 Academic press,New York; London,1963
Bioch 33.0973

CELL INTERACTIONS Lepetit colloquium on biology and medicine 3rd Proceedings London 1971 Nov Edited by Luigi G. Silvestri North-Holland,Amsterdam,1972
Bioch 33.2267

CELL MECHANISM IN HORMONE PRODUCTION AND ACTION a symposium Proceedings London 1960 May 3-4 Society for endocrinology Edited by P.C. Williams and C.R. Austin Society for endocrinology.Memoirs, 11 Cambridge university press,Cambridge,1961
Phys 20.1363

CELL MECHANISMS IN HORMONE PRODUCTION AND ACTION a symposium Proceedings London 1960 May 3-4 Society for endocrinology Edited by P.C. Williams and C.R. Austin Society for endocrinology.Memoirs,11 Cambridge university press,Cambridge,1961
Pha 16.0091

CELL METABOLISM Ciba foundation symposium on ionizing radiations and cell metabolism Proceedings London 1956 Mar 6-9 Ciba foundation Edited by G.E.W. Wolstenholme and C.M. O'Connor illus. Churchill,London,1956
Radioth 35.1355

CELL METABOLISM Ciba foundation symposium on the regulation of cell metabolism London 1958 Jul 28-30 Ciba foundation Edited by G. E.W. Wolstenholme and Cecilia M. O'Connor Illus J.and A.Churchill,London,1959
Radioth 35.0493

CELL METABOLISM Ciba foundation symposium on the regulation of cell metabolism Proceedings London 1958 Jul 28-30 Ciba foundation Edited by G.E.W. Wolstenholme and Cecilia M. O'Connor illus Churchill,London, 1959
Bioch 33.0575

CELL NUCLEUS International symposium on the cell nucleus:metabolism and radiosensitivity Proceedings Rijswijk 1966 May 9-12 Taylor and Francis,London,1966
Radioth 35.1751

The CELL NUCLEUS :an informal meeting Proceedings Cambridge 1959 Aug 31-Sep 1 Faraday society Edited by J.S. Mitchell Butterworth,London,1960
An 32.3234

The CELL NUCLEUS :an informal meeting Proceedings Cambridge 1959 Aug 31-Sep 1 Faraday society Held at the University of Cambridge.Department of radiotherapeutics Butterworths,London,1960 Chairman:J.S. Mitchell
Bal 39.0290

The CELL NUCLEUS :an informal meeting Proceedings Cambridge 1959 Aug 31-Sep 1 Faraday society held at the University of Cambridge.Department of radio-therapeutics Butterworths,London,1960
Bioch 33.0946

The CELL NUCLEUS informal meeting Proceedings Cambridge 1959 Aug 31-Sep 1 Faraday society Edited by J.S. Mitchell Butterworths,London,1960
Gen 34.0839

The CELL NUCLEUS Proceedings :an informal meeting held at the Department of radiotherapeutics,University of Cambridge...by the Faraday society Cambridge 1959 Aug 31 - Sep 1 Faraday society Illus Butterworths,London,1960 Chairman: Professor J.S.Mitchell
Radioth 35.0501

CELL PHYSIOLOGY Recent developments in cell physiology :a symposium Bristol 1954 Mar 29-Apr 1 Colston research society Edited by J.A. Kitching Held at the University of Bristol Colston papers, 7 London,1954
Bal 39.0332

CELL PHYSIOLOGY Recent developments in cell physiology :symposium Proceedings Bristol 1954 Mar 29-Apr 1 Colston research society Edited by J.A. Kitching Colston research society.Symposia, 7 Colston papers, 7 Butterworths,London,1954
Bioch 33.0936

CELL PHYSIOLOGY Structural aspects of cell physiology Bristol 1951 Jul Society for experimental biology Society for experimental biology.Symposia, 6 Cambridge university press,Cambridge,1952
Phys 20.0954

CELL PHYSIOLOGY Structural aspects of cell physiology :a symposium Bristol 1951 Jul Society for experimental biology Society for experimental biology.Symposia, 6 Cambridge university press,Cambridge,1952
Bioch 33.1352

CELL RESEARCH Inhibitions in cell research : colloquium Mosbach 1969 Apr 14-16 Gesellschaft fur biologische chemie Edited by T. Bucher and H. Sies Gesellschaft fur biologische chemie.Colloquia, 20 Springer, Berlin,1969
Bioch 33.1908

CELL RESEARCH Inhibitions in cell research : colloquium Mosbach 1969 Apr 14-16 Gesellschaft fur biologische chemie Edited by T. Bucher and H. Sies Gesellschaft fur biologische chemie.Colloquia, 20 Springer, Berlin,1969
Bot 42.1758

CELL STRUCTURE Spectroscopy and molecular structure,and optical methods of investigating cell structure :a general discussion Cambridge 1950 Sep 25-28 Faraday society Faraday society.Discussions, 9 Aberdeen, 1951
Bal 39.0230

CELL SYSTEMS Developing cell systems and their control :a symposium Madison,Wis. 1959 Society for the study of development and growth Edited by Dorothea Rudnick at the University of Wisconsin Society for the study of development and growth.Symposia, 18 Ronald press,New York,1960
Bioch 33.1010

CELL TISSUE AND ORGAN CULTURE Decennial review conference on cell tissue and organ culture 2nd Proceedings Bedford,Pa. 1966 Sep 11-15 National cancer institute and Tissue culture association Edited by Benton B. Westfall National cancer institute. Monograph, 26 National cancer institute, Bethesda,Md.,1967
Bioch 33.0991

CELLS Fine structure of cells :symposium held at the 8th congress of cell biology Leiden 1954 With the financial support of Unesco International union of biological sciences.Publications.Series B, 21 Noordhoff,Groningen,1955
An 32.3219

CELLS Fine structure of cells;symposium The Congress of cell biology 8th Papers Leiden 1954 International union of biological sciences With financial support of Unesco International union of biological sciences.Ser.B, 21 Noordhoff,Groningen,1955
Bal 39.0283

CELLS International congress of microbiology 6th Rome 1953 Vol 6: symposium - interaction of viruses and cells Edited by F. C. Bawden and G. Penso illus Fondazione Emanuele Paterno,Rome,1953 Title also in Italian;text in English and French
Bioch 33.1804

CELLULAR AND HUMORAL ASPECTS OF THE HYPERSENSITIVE STATES New York Edited by H.Sherwood Lawrence New York academy of medicine.Section microbiology. Symposia,9 Cassell,London,1959
Pha 16.0181

CELLULAR AND HUMORAL ASPECTS OF THE HYPERSENSITIVE STATES :a symposium New York 1959 New York academy of medicine. Section on microbiology Edited by H.Sherwood Lawrence New York academy of medicine. Section of microbiology.Symposia, 9 illus, diagrs Cassell,London,1959
Path 30.2526

CELLULAR AND MOLECULAR ASPECTS OF FLORAL INDUCTION :a symposium Proceedings Liege 1967 Sep 4-8 Edited by Georges Bernier Bibliog.,illus,plates xv,492p Longman,London,1970
Bot 42.1440

CELLULAR ASPECTS OF BASIC MECHANISMS IN RADIOBIOLOGY :an informal conference Proceedings Bear Mountain,N.Y. 1955 May 12-14 Edited by Harvey M. Patt and E.L. Powers Organised by the National research council.Subcommittee on radiobiology National research council.Nuclear science series.Report, 18 NAS-NRC 450 National research council,Washington,D.C.,1956
Radioth 35.1716

CELLULAR ASPECTS OF IMMUNITY Ciba foundation symposium on cellular aspects of immunity Proceedings Royaumont 1959 Jun 3-5 Ciba foundation Edited by G.E.W. Wolstenholme and Maeve O'Connor Churchill,London,1960
Bioch 33.1117

CELLULAR BASIS AND AETIOLOGY OF LATE SOMATIC EFFECTS OF IONIZING RADIATION :a symposium London 1962 Mar 27-30 Edited by R.J.C. Harris Under the auspices of Unesco Academic press,London;New York, 1963
Radioth 35.1750

CELLULAR BIOLOGY NUCLEIC ACIDS AND VIRUSES :a conference Papers New York 1957 Jun 7-9 Edited by Thomas M. Rivers Held at the New York academy of sciences New York academy of sciences.Special publications, 5 Port New York academy of sciences,New York, 1957
Radioth 35.0496

CELLULAR BIOLOGY NUCLEIC ACIDS AND VIRUSES : conference Papers New York 1957 Jan 7-9 New York academy of sciences Edited by Thomas M. Rivers New York academy of sciences.Special publications, 5 New York academy of sciences,New York,1957
Gen 34.0724

CELLULAR BIOLOGY OF MYXOVIRUS INFECTIONS London 1964 Feb 4-6 Ciba foundation Edited by G.E.W. Wolstenholme and Julie Knight Ciba foundation.Symposia Churchill,London, 1964
Gen 34.0727

CELLULAR CHEMISTRY Nucleic acid metabolism, cell differentiation and cancer growth International symposium for cellular chemistry 2nd Proceedings Ohtsu 1966 Oct 17-21 Edited by E.V. Cowdry and S. Seno Organised by the Japan society for cell biology Pergamon,London,1969
Radioth 35.0561

CELLULAR COMPARTMENTALIZATION AND CONTROL OF FATTY ACID METABOLISM Federation of European biochemical societies meeting 4th Proceedings Oslo 1967 Jul 5 Federation of European biochemical societies Edited by F.C. Gran Organized by the Norwegian biochemical society Universitetsforlaget; Academic press,Oslo;London,1968 Symposium organizer J. Bremer
Bioch 33.1399

CELLULAR CONTROL MECHANISMS AND CANCER Conference on cellular control mechanisms and cancer Proceedings Amsterdam 1963 Sep 9-13 International union against cancer and Netherlands cancer institute Edited by P. Emmelot and O. Muhlbock Elsevier,Amsterdam, 1964 Organized to commemorate the 50th anniversary of the Netherlands cancer institute
Path 30.2459

CELLULAR CONTROL MECHANISMS AND CANCER :a conference Proceedings Amsterdam 1963 Sep 9-13 Edited by P. Emmelot and O. Muhlbock Under the auspices of the International union against cancer Elsevier, Amsterdam,1964 Conference held on the occasion of the 50th anniversary of the Netherlands cancer institute
An 32.2504

CELLULAR CONTROL MECHANISMS AND CANCER : conference Proceedings Amsterdam 1963 Sep 9-13 Edited by P. Emmelot and O. Muhlbock Under the auspices of the International union against cancer.Cancer research commision Illus Elsevier, Amsterdam,1964
Radioth 35.0530

CELLULAR FUNCTION Symposium on calcium and cellular function Essays London 1969 Mar 24-25 Biological council.Co-ordinating committee for symposia on drug action Edited by A.W. Cuthbert Macmillan,London,1969
Bioch 33.2264

The CELLULAR FUNCTIONS OF MEMBRANE TRANSPORT Woods Hole,Mass. 1963 Sep 4-7 Edited by J.F. Hoffmann Held under the auspices of the Society of general physiologists Prentice-Hall,Englewood Cliffs,N.J.,1964
Pha 16.0280

The CELLULAR FUNCTIONS OF MEMBRANE TRANSPORT : a symposium Papers Woods Hole,Mass. 1963 Sep 4-7 Edited by J.F. Hoffman Held under the auspices of the Society of general physiologists illus. viii,291p Prentice-Hall,New Jersey,1964 Held at the annual meeting of the society
Bot 42.1500

CELLULAR GROWTH Control of cellular growth in adult organisms A Sigrid Juselius foundation symposium Helsinki 1965 Oct 4-6 Sigrid Juselius foundation Edited by Harald Teir and Tapio Rytomaa Academic press,London; New York,1967
Bioch 33.1007

CELLULAR INJURY :a Ciba foundation symposium Papers London 1963 Jul 2-4 Ciba foundation Edited by A.V.S. De Reuck and Julie Knight Churchill,London,1964
An 32.5317

CELLULAR INJURY AND RESISTANCE IN FREEZING ORGANISMS International conference on low temperature science 2nd Proceedings Sapporo 1966 Aug 14-19 Vol 2: conference on cryobiology Hokkaido university.Institute of low temperature science Edited by Eizo Asahina illus Hokkaido university. Institute of low temperature science,Sapporo, 1967 Title page in English and Japanese Conference held in commemoration of the 25th anniversary of the establishment of the Institute
Path 30.2625

CELLULAR MECHANISMS IN DIFFERENTIATION AND GROWTH a symposium Amherst,Mass. 1955 Jun 15-18 Edited by Dorothea Rudnick Society for the study of development and growth. Symposia, 14 Princeton university press, Princeton,N.J.,1956
An 32.2478

CELLULAR MECHANISMS IN DIFFERENTIATION AND GROWTH A SYMPOSIUM Proceedings Amherst, Mass 1955 Jun 15-18 By M. Delbruck and others Society for the study of development and growth Edited by Dorothea Rudnick Society for the study of development and growth.Symposia, 14 illus. vii,236p Princeton university press,Princeton,N.J.,1956
Bot 42.1474

CELLULAR MEMBRANES IN DEVELOPMENT :a symposium Storrs,Conn. 1963 Jun 17-19 Society for the study of development and growth Edited by Michael Locke Society for the study of development and growth.Symposia, 22 Academic press,New York,1964
Bioch 33.0910

CELLULAR MEMBRANES IN DEVELOPMENT :symposium Storrs,Conn. 1963 Jun Society for the study of development and growth Edited by Michael Locke Society for the study of development and growth.Symposia, 22 Academic press,New York,1964
Gen 34.0480

CELLULAR MEMBRANES IN DEVELOPMENT :symposium Storrs,Conn. 1963 Jun 17-19 Society for the study of development and growth Edited by Michael Locke Society for the study of development and growth.Symposia, 22 illus. xvi,382p Academic press,New York;London,1964
Bot 42.4733

CELLULAR MEMBRANES IN DEVELOPMENT a symposium Proceedings Storrs,Conn. 1963 Jun Edited by Michael Locke Society for the study of development and growth.Symposia, 22 Academic press,New York;London,1964
An 32.3295

CELLULAR ORGANIZATION International congress of biochemistry 6th Abstracts New York 1964 Jul 26-Aug 1 Scheduled under the auspices of the International union of biochemistry Washington,1964
Bioch 33.1344

CELLULAR RADIATION BIOLOGY Annual symposium on fundamental cancer research,1964 18th Anderson hospital and tumor institute Williams and Wilkins,Baltimore,Md.,1965
Radioth 35.1742

CELLULAR REGULATORY MECHANISMS :a symposium Papers Cold Spring Harbor 1961 Cold Spring Harbor biological laboratory Cold Spring Harbor symposia on quantitative biology, 26 Long Island biological association,Cold Spring Harbor,1961
Bioch 33.1282

CELLULAR ULTRASTRUCTURE OF WOODY PLANTS : advanced science seminar Proceedings Syracuse university.Pinebrook conference center 1964 Sep 20-26 Edited by Wilfred A. Cote xii,603p Syracuse university press, Syracuse,N.Y.,1965
Bot 42.4732

CELLULOSE Advances in enzymic hydrolysis of cellular and related materials :a symposium Proceedings Washington,D.C. 1962 Mar Edited by Elwyn T. Reese Sponsored by the American chemical society Pergamon,Oxford, 1963 Including a bibliography for the years 1950-61
Bot 42.1728

CEMENT Chemistry of cement International symposium on the chemistry of cement 4th Proceedings Washington,D.C. 1960 Oct 2-7 Vol 1-2 National bureau of standards National bureau of standards.Monographs,43 2 vols Washington,D.C.,1962
Min 10.0647

CEMENT International symposium on the chemistry of cement 2nd Proceedings Stockholm 1938 July 6-8 Under the auspices of the ingeniorsvetenskapsakademien and Svenska cementforeningen Ingeniorsvetenskapsakademien,Stockholm,1939
Min 10.0598

CEMENT International symposium on the chemistry of cement 3rd Proceedings London 1952 Under the auspices of the Great Britain.Building research station...and Cement and concrete association Cement and concrete association,London,1954
Min 10.0615

CEMENT AND CONCRETE Technical papers Los Angeles,Calif. 1956 Sep 19-20 American society for testing materials American society for testing materials.Special technical publication, 205 American society for testing materials,Philadelphia,Pa.,1958
Eng 41.2845

CEMENT AND CONCRETE ASSOCIATION Concrete shell roof construction :a symposium Proceedings London 1952 Jul 2-4 Cement and concrete association,London,1954
Eng 41.3008

CEMENT AND CONCRETE ASSOCIATION Mix design and quality control of concrete :a symposium Proceedings London 1954 May Cement and concrete association,London,1955
Eng 41.2929

CEMENT AND CONCRETE ASSOCIATION Model testing :one-day meeting Proceedings London 1964 Mar 17 Cement and concrete association,London,1964
Eng 41.2934

CEMENT AND CONCRETE ASSOCIATION Prestressed concrete statically indeterminate structures : a symposium London 1951 Sep 24-25 Cement and concrete association,London,1953
Eng 41.3006

CEMENT AND CONCRETE ASSOCIATION The Strength of concrete structures :a symposium Proceedings London 1956 May Cement and concrete association,London,1958
Eng 41.2930

CEMENT AND CONCRETE ASSOCIATION The Structure of concrete and its behaviour under load :an international conference Proceedings London 1965 Sep Cement and concrete association,London,1968
Eng 41.3007

CEMENT INDUSTRY RESEARCH AND INFORMATION ASSOCIATION The Modern design of wind-sensitive structures :a seminar Proceedings London 1970 Jun 18 CIRIA,London,1971
Eng 41.8276

CENOZOIC International geological congress 20th papers Mexico City 1956 Seccion 1: vulcanologia del cenozoico,tom 1-2 Edited by A. Garcia Rojas and others Mexico City, 1957 Text in English,French,Russian and Spanish
Geol 8.3036

CENTENARY AND BICENTENARY CONGRESS OF BIOLOGY : papers delivered in the university of Malaya in commemoration of the works of Darwin, Wallace and Linnaeus Proceedings Singapore 1958 Dec 2-9 Edited by R.D. Purchon Bibliog,illus,port,maps,tables 333p University of Malaya press;Oxford university press,Singapore;London,1960 Includes papers on anthropology
Bal 39.3924

CENTENNIAL OF ENTOMOLOGY IN CANADA,1863-1963 : a commemorative joint meeting Proceedings Ottawa 1963 Sep 3-6 Entomological society of Canada,and,Entomological society of Ontario Canadian entomologist,96,pt.1-2 475p Carelton university,Ottawa,1963
Sco 14.2471

CENTRA NATIONAL DE LA RECHERCHE SCIENTIFIQUE. COLLOQUES INTERNATIONAUX.ACTES, 104 Problemes actuels de paleontologie (evolution des vertebres) Paris 1961 May 29-Jun 3 Centre national de la recherche scientifique 27cm C.N.R.S.,Paris,1962 Organized by J. P.Lehmann
Bal 44.4639

CENTRAL CHEMICAL ASSOCIATION Current corrosion research in Scandinavia Scandinavian corrosion congress (NKM) : lectures 4th Helsinki 1964 Nov 24-27 Edited by Jori Larinkari and others Organized by the Scandinavian council for applied research Kemian keskusliton julkaisuja, 24 Sanoma Osakeyhtio,Helsinki, 1965
Met 25.1901

CENTRAL ELECTRICITY GENERATING BOARD.BERKELEY NUCLEAR LABORATORIES Properties of reactor materials and the effects of radiation damage :international conference Proceedings Berkeley Castle,Glos. 1961 May 30-Jun 2 Edited by D.J. Littler xv,562p Butterworths,London,1962
Cav 7.2864

CENTRAL ELECTRICITY RESEARCH LABORATORIES Gas discharges and the electricity supply industry International conference on gas discharges and the electricity supply industry Leatherhead 1962 May 7-11 Edited by J.S. Forrest and others Butterworths,London,1962
Eng 41.5415

CENTRAL ELECTRICITY RESEARCH LABORATORIES Gas discharges and the electricity supply industry :a conference Proceedings Leatherhead,Surrey 1962 May 7-11 Edited by J.S. Forrest and others 677p Butterworths, London,1962
Nap 11.0123

CENTRAL INTERNATIONAL BUREAU FOR THE PREVENTION OF CONSUMPTION International conference on tuberculosis 1st Report Berlin 1902 Oct 22-26 Edited by Gotthold Pannwitz Berlin,1903 Text in English,French,German; title also in French and German
Path 30.1025

CENTRAL LONDON PRODUCTIVITY ASSOCIATION The Use of computers in production control conference proceedings London 1959 Apr 2 69p 32cm Central London productivity association,London,1959
Math L 5.0843

CENTRAL METEOROLOGICAL OBSERVATORY,TOKYO Symposium on typhoons Proceedings Tokyo 1954 Nov 9-12 illus 257p Japanese national commission for Unesco,Tokyo,1955
Nap 11.0612

CENTRAL MIDWESTERN REGIONAL EDUCATIONAL LABORATORY Teaching of probability and statistics Comprehensive school mathematics program international conference 1st Proceedings Carbondale,Ill. 1969 Mar 18-27 Edited by Lennart Rade Wiley;Almqvist and Wiksell,New York;Stockholm,1970
An 32.5410

CENTRAL NERVOUS SYSTEM International congress of biochemistry 4th Proceedings Vienna 1958 Sep 1-6 Vol 3: symposium 3: biochemistry of the central nervous system International union of biochemistry Edited by F. Brucke I.U.B.symposium series, 5 Pergamon,London,1959 Added t.p.in French and German
Bot 42.1715

CENTRAL NERVOUS SYSTEM International congress of biochemistry 4th Proceedings Vienna 1958 Sep 1-6 Vol 3: symposium 3 - biochemistry of the central nervous system International union of biochemistry Edited by Brucke I.U.B.symposium series, 5 Pergamon,London,1959 Added title page in French and German.Text in English,French and German
Bioch 33.1317

CENTRAL NERVOUS SYSTEM Patterns of recognition in the central nervous system Proceedings New York 1950 Dec 15-16 Association for research in nervous and mental disease Association for research in nervous and mental disease.Research publications, 30 Baltimore,Md.,1952
Phys 20.2238

The CENTRAL NERVOUS SYSTEM AND BEHAVIOR 3rd a conference Transactions Princeton,N. J. 1960 Feb 21-24 Josiah Macy jr. foundation Edited by Mary A.B. Brazier With the cooperation of the National science foundation Josiah Macy,jr.foundation,New York,1960
Psy 31.0063

CENTRAL NERVOUS SYSTEM AND BEHAVIOUR Conference on the central nervous system and behaviour 1st Transactions 1958 Feb 23-26 Edited by Mary A.B. Brazier Sponsored by the Josiah Macy jr.foundation New York, 1959
Phys 20.2239

CENTRAL NERVOUS SYSTEM AND BEHAVIOUR The Conference on the central nervous system and behaviour 1st-3rd Transactions 1958-60 Edited by Mary A. Brazier Josiah Macy jr. foundation 3 vols Josiah Macy jr. foundation,New York,1959-61
An 32.5329

The CENTRAL NERVOUS SYSTEM AND FISH BEHAVIOR : a conference Papers Chicago,Ill. 1967 Apr United States.Air force.Office of scientific research American zoological society.Comparative physiology division Edited by David Ingle Jointly sponsored by Perspectives in biology and medicine illus University of Chicago press,Chicago,Ill.; London,1968
Psy 31.2816

The CENTRAL NERVOUS SYSTEM AND FISH BEHAVIOR : symposium Papers presented Chicago 1967 Apr Edited by D. Ingle at the University of Chicago University of Chicago press,Chicago;London,1968 Sponsored jointly by the United States Air force office of scientific research, comparative physiology division of the American zoological society, and the academic journal Perspectives in biology and medicine
Bal 39.3266

CENTRAL SALT AND MARINE CHEMICALS RESEARCH INSTITUTE Sea,salt and plants seminar Proceedings Bhavnagar 1965 Dec 20-23 Edited by V. Krishnamurthy illus. xv,372p CSMCRI,Bhavnagar,1967
Bot 42.1991

CENTRE BELGE DE RECHERCHES MATHEMATIQUES Colloque de topologie Brussels 1964 Sep 7-10 235p 25cm Louvain, Paris,1966
P. Math 2.0294

CENTRE BELGE DE RECHERCHES MATHEMATIQUES Colloque de topologie (espaces fibres) Brussels 1950 Jun 5-8 136p 24cm Centre Belge de recherches mathematiques,Brussels, 1950
P Math 2.3126

CENTRE BELGE DE RECHERCHES MATHEMATIQUES Colloque sur l'analyse fonctionelle 2nd Liege 1964 May 4-6 166p 25cm Librairie universitaire,Louvain,1964
P. Math 2.1182

CENTRE BELGE DE RECHERCHES MATHEMATIQUES Colloque sur la theorie des groupes algebriques 20th colloque Brussells 1962 Jun 5-7 150p 25cm Librairie universitaire,Louvain,1962
P. Math 2.0573

CENTRE BELGE DE RECHERCHES MATHEMATIQUES Groupement des mathematiciens d'expression latine 3rd reunion comptes rendus Namur 1965 Sep.20-23 153p 25cm Librairie universitaire,Louvain,1966
P. Math 2.2611

CENTRE CULTUREL INTERNATIONAL DE CERISY-LA-SALLE Entretiens sur le temps Cerisy-la-Salle 1964 Jul 14-23 Edited by Jeanne Hersch and Rene Poirier Centre culturel international de Cerisy-la-Salle.Decade nouvelle serie, 5 351p Mouton,Paris;The Hague,1967
WSM 43.1000

CENTRE D'ETUDE DE L'ENERGIE NUCLEAIRE International symposium on the biology of acetabularia 1st Proceedings Brussels 1969 Jun 18-20 and Mol 1969 Jun 18-20 Edited by Jean Brachet and Silvano Bonotto Academic press,New York,1970
Bioch 33.2261

CENTRE D'ETUDES DE BRUYERES-LE-CHATEL L'Hydrogene dans les metaux :colloque Valduc 1967 Sep 27-28 Paris,1969
Met 25.2624

CENTRE D'ETUDES NUCLEAIRES,SACLAY Colloque de metallurgie proprietes des joints de grains 4th Saclay 1960 Jun 27-28 Presses universitaires de France,Paris,1961 Papers in English and French
Met 25.1266

CENTRE D'ETUDES NUCLEAIRES,SACLAY Symposium de metallurgie speciale Saclay 1957 Jun 27-28 Presses universitaires de France,Paris, 1958
Met 25.2517

CENTRE D'ETUDES NUCLEAIRES,SACLAY Colloque de metallurgie corrosion,seche et aqueuse 3e Saclay 1959 Jun 29-Jul 1 Sponsored by the Institut national des sciences et techniques nucleaires Centre d'etudes nucleaires de Saclay;North Holland,Paris; Amsterdam,1960
Met 25.1891

CENTRE D'ETUDES NULEAIRES,SACLAY Diffusion a l'etat solide Colloque sur la diffusion a l'etat solide Saclay 1958 Jul 3-5 Centre d'etudes nucleaires de Saclay;North Holland, Gif-sur-Yvette;Amsterdam,1959
Met 25.1277

CENTRE D'ETUDES SOCIOLOGIQUES Villes et campagnes Semaine sociologique 2nd Compte rendu Paris 1951 Mar Edited by Georges Friedmann Ecole pratique des hautes etudes,Paris.Bibliotheque generale.Section 6 Colin,Paris,1953
Geog 13.4272

CENTRE FOR ADVANCED STUDY IN THE DEVELOPMENTAL SCIENCES STUDY GROUP ON 'MECHANISMS OF MOTOR SKILL DEVELOPMENT' Proceedings Mechanisms of motor skill development London 1968 Nov Ciba foundation Edited by Kevin Connolly Academic press,London;New York,1970 Being the 4th study group in a C.A.S.D.S. programme on 'The origins of human behaviour' held jointly with the Ciba foundation
Psy 31.3101

CENTRE INTERNATIONAL DE SYNTHESE Science et loi... semaine internationale de synthese 5e Discussions et conclusions Paris 1963 May 29-Jun 3 By Abel Rey and others vi, 228p Alcan,Paris,1934
WSM 43.0874

CENTRE INTERNATIONAL D'ETUDE DES PROBLEMES HUMAINS Human displacements;les deplacements humains 1962 May 24-29 Edited by Jean Sutter Entretiens de Monaco en sciences humaines, 1 Hachette,1963 Papers and title page in English and French
Gen 34.1883

CENTRE NATIONAL DE LA RECHERCHE SCIENTIFIQUE Adsorption et croissance cristalline :colloque international Nancy 1965 Jun 6-12 Centre national de la recherche scientifique. Colloques internationaux, 152 Centre national de la recherche scientifique,Paris, 1965
Met 25.1621

CENTRE NATIONAL DE LA RECHERCHE SCIENTIFIQUE Archeologie et calculateurs:problemes semiologiques et mathematiques :colloque international Marseille 1969 Apr 7-12 Centre national de la recherche scientifique. Colloques internationaux bibliog.,illus. 371p 28cm Centre nationale de la recherche scientifique,Paris,1970
Math S 3.1631

CENTRE NATIONAL DE LA RECHERCHE SCIENTIFIQUE Cinetique et mecanisme des reactions d'inflammation et de combustion en phase gazeuse Paris 1948 Apr 26-May 1 With the financial support of the Rockefeller foundation Centre national de la recherche scientifique.Colloques internationaux,16 Centre national de la recherche scientifique, Paris,1949 With summaries in English
Chem 18.0431

CENTRE NATIONAL DE LA RECHERCHE SCIENTIFIQUE Cytologie de l'adenohypophyse :colloque international Paris 1963 Sep 9-14 Edited by Jacques Benoit and Christian Da Lage Centre national de la recherche scientifique. Colloques internationaux, 128 C.N.R.S.,Paris, 1963
An 32.3770

CENTRE NATIONAL DE LA RECHERCHE SCIENTIFIQUE Dosage des elements a l'etat de traces dans les roches et autres substances :colloque Comptes-rendus Nancy 1968 Dec 4-6 Centre national de la recherche scientifique. Colloques nationaux, 923 CNRS,Paris,1970
Min 10.1475

CENTRE NATIONAL DE LA RECHERCHE SCIENTIFIQUE Ecologie des algues marines Dinard 1957 Sep 20-28 By R. Biebl and others Centre national de la recherche scientifique. Colloques internationaux, 81 illus. 276p Paris,1959
Bot 42.3814

CENTRE NATIONAL DE LA RECHERCHE SCIENTIFIQUE Endocrinologie des arthropodes Colloque international sur l'endocrinologie des arthropodes Paris 1947 Jun Centre national de la recherche scientifique. Colloques internationaux, 4 Centre national de la recherch scientifique,Paris, 1948
Bal 39.2291

CENTRE NATIONAL DE LA RECHERCHE SCIENTIFIQUE La Structure des solutions solides metalliques Orsay 1962 Jul 9-11 Centre national de la recherche scientifique.Colloques internationaux, 118 Centre national de la recherche scientifique,Paris,1962 Papers in English and French
Met 25.1210

CENTRE NATIONAL DE LA RECHERCHE SCIENTIFIQUE La Theorie des images optiques Memoires Paris 1946 Oct 21-26 Edited by Pierre Fleury Financed by the Rockefeller foundation Colloques internationaux du centre de la recherche scientifique Editions de la revue d'optique,Paris,1949
Cav 7.1758

CENTRE NATIONAL DE LA RECHERCHE SCIENTIFIQUE La Theorie du potentiel :colloque international Orsay 1964 Jun.22-26 Edited by M. Brelot and others Centre national de la recherche scientifique. Colloques internationaux, 146 x,312p 25cm Centre national de la recherche scientifique, Paris,1965
Math 3.0050

CENTRE NATIONAL DE LA RECHERCHE SCIENTIFIQUE Le Bombardement ionique theories et applications:colloques internationaux Bellevue,Seine et Oise 1962 Dec 4-8 Centre national de la recherche scientifique. Colloques internationaux, 113 Editions du centre National de la recherche scientifique, Paris,1962
Met 25.1676

CENTRE NATIONAL DE LA RECHERCHE SCIENTIFIQUE Le Calcul de probabilites et ses applications colloque international proceedings Lyon 1948 Jun.28-Jul.3 Edited by Maurice Frechet Centre national de la recherche scientifique. Colloques internationaux 130p 27cm Centre national de la recherche scientifique, Paris,1949
Math 3.0678

CENTRE NATIONAL DE LA RECHERCHE SCIENTIFIQUE Le Calcul des probabilites et ses applications colloque international proceedings Paris 1958 Jul.15-20 Edited by Georges Darmois Centre national de la recherche scientifique. Colloques internationaux, 87 196p 24cm Centre national de la recherche scientifique, Paris,1959
Math 3.0679

CENTRE NATIONAL DE LA RECHERCHE SCIENTIFIQUE Le Magnetisme Travaux Strasbourg 1939 May 21-25 Universite de Strasbourg.Institut de physique and Institut international de cooperation intellectuelle Collection scientifique,Paris,1940
Cav 7.2241

CENTRE NATIONAL DE LA RECHERCHE SCIENTIFIQUE Les Cultures de tissus de plantes Proceedings Strasburg 1970 Jul 6-10 Centre national de la recherche scientifique. Colloques internationaux, 193 CNRS,Paris, 1971
Bioch 33.2223

CENTRE NATIONAL DE LA RECHERCHE SCIENTIFIQUE
Les Machines a calculer et la pensee humaine colloque international Paris 1951 Jan 8-13 Centre national de la recherche scientifique., Colloques internationaux,37 illus. xix, 570p 24cm Centre national de la recherche scientifique,Paris,1953
Math L 5.0716

CENTRE NATIONAL DE LA RECHERCHE SCIENTIFIQUE
Les Probabilites sur les structures algebriques Clermont-Ferrand 1969 Jun 30-Jul 5 Centre national de la recherche scientifique.Colloques internationaux, 186 361p 25cm CNRS,Paris,1970
P Math 2.4558

CENTRE NATIONAL DE LA RECHERCHE SCIENTIFIQUE
Les Probabilites sur les structures algebriques :colloque international Clermont-Ferrand 1969 Jun 30-Jul 5 Centre national de la recherche scientifique.Colloques internationaux, 186 361p 25cm Editions du CNRS,Paris,1970 Oeganized by A. Badrikian and P.L.Hennequin
Math S 3.1780

CENTRE NATIONAL DE LA RECHERCHE SCIENTIFIQUE
Les Progres recents en spectroscopie interferentielle Proceedings Bellevue 1957 Sep 9 - 13 Edited by Pierre Jacquinot Centre national de la recherche scientifique. Colloques internationaux, 80 252p Editions du CNRS,Paris,1958
Obs 6.2007

CENTRE NATIONAL DE LA RECHERCHE SCIENTIFIQUE
Les Proprietes optiques des lames minces solides Proceedings Marseille 1949 Apr 19 -23 Supported by the Rockefeller foundation Centre national de la recherche scientifique.Colloques internationaux, 23 176p Centre national de la recherche scientifique,Paris,1950
Obs 6.2009

CENTRE NATIONAL DE LA RECHERCHE SCIENTIFIQUE
Les Relations entre precambrien et cambrien problemes des series intermediaires colloque international Paris 1957 Jun 27- Jul 4 Centre national de la recherche scientifique. Colloques internationaux,66 bibliog. Centre national de la recherche scientifique, Paris,1958
Geol 8.2716

CENTRE NATIONAL DE LA RECHERCHE SCIENTIFIQUE
Les Tendances geometriques en algebre et theorie des nombres colloque international Clermont-Ferrand 1964 Apr 2-9 Edited by Marc Krasner Centre nationale de la recherche scientifique.Colloques internationaux, 143 Bibliog. 256p 25cm Centre national de la recherche scientifique, Paris,1966
P. Math 2.1775

CENTRE NATIONAL DE LA RECHERCHE SCIENTIFIQUE
Les techniques recentes en microscopie electronique et corpusculaire :colloque international Toulouse 1955 Apr 4-8 Edited by Ch. Fert Centre national de la recherche scientifique.Colloques internationaux, 58 Paris,1956
Cav 7.0525

CENTRE NATIONAL DE LA RECHERCHE SCIENTIFIQUE
Low energy nuclear interactions and nuclear structure International conference on nuclear physics Proceedings Paris 1958 Jul 7-12 Under the auspices of International union of pure and applied physics Crosby Lockwood,London,1959
Cav 7.1759

CENTRE NATIONAL DE LA RECHERCHE SCIENTIFIQUE
Mecanique de la turbulence :colloque international Marseille 1961 Aug 28-Sep 2 Centre national de la recherche scientifique. Colloques internationaux, 108 470p Editions du centre national de la recherche scientifique,Paris,1962
Chem E 24.0370

CENTRE NATIONAL DE LA RECHERCHE SCIENTIFIQUE
Mecanique de la turbulence colloque international Marseille 1961 Aug 28-Sep 2 Mecanique de la turbulence Centre national de la recherche scientifique.Colloques internationaux, 108 Centre national de la recherche scientifique,Paris,1962
A Math 4.0525

CENTRE NATIONAL DE LA RECHERCHE SCIENTIFIQUE
Mecanisme physiologique de la secretion lactee colloque international Strasbourg 1950 Aug 22-29 Centre national de la recherche scientifique.Colloques internationaux, 32 C. N.R.S.,Paris,1950
Phys 20.2212

CENTRE NATIONAL DE LA RECHERCHE SCIENTIFIQUE
Methodes nouvelles de spectroscopie instrumentale Exposes Orsay 1966 Apr 25-29 2nd edition Centre national de la recherche scientifique.Colloques internationaux, 161 344p CNRS,Paris,1969
Obs 6.3376

CENTRE NATIONAL DE LA RECHERCHE SCIENTIFIQUE
Microphysiologie comparee des elements excitables :colloque international Gif-sur-Yvette 1955 Jul 19-23 Centre national de la recherche scientifique.Colloques internationaux, 67
Phys 20.2022

CENTRE NATIONAL DE LA RECHERCHE SCIENTIFIQUE
Morphogenese :colloque international Strasbourg 1949 Jul 4-11 Centre national de la recherche scientifique.Colloques internationaux, 28 C.N.R.S.,Paris,1951
An 32.2463

CENTRE NATIONAL DE LA RECHERCHE SCIENTIFIQUE
New physical and chemical properties of metals of very high purity :international symposia Paris 1959 Oct 12-14 Centre national de la recherche scientifique.International symposia, 40 Gordon and Breach,New York,1965
Met 25.1500

CENTRE NATIONAL DE LA RECHERCHE SCIENTIFIQUE
Nouvelles proprietes physiques, mecaniques et chimiques Colloque international Paris 1966 Sep 26-Oct 1 Centre national de la recherche scientifique.Colloques internationaux, 167 Editions du centre national de la recherche scientifique,Paris, 1968
Met 25.1503

CENTRE NATIONAL DE LA RECHERCHE SCIENTIFIQUE
Nouvelles proprietes physiques et chimiques des metaux de tres haut purete colloque international Paris 1959 Oct 12-14 Centre national de la recherche scientifique. Colloques internationaux, 40 Centre national de la recherche scientifique,Paris, 1960 English translation also available
Met 25.1499

CENTRE NATIONAL DE LA RECHERCHE SCIENTIFIQUE
Oceanographie geologique et geophysique de la Mediterrane occidentale :colloque international Travaux Villefranche sur Mer 1961 Apr 4-8 Paris,1962
Geod 9.0581

CENTRE NATIONAL DE LA RECHERCHE SCIENTIFIQUE
Paleontologie :colloque internationale Paris 1947 Apr Centre national de la recherche scientifique.Colloques internationaux, 21 C. N.R.S.,Paris,1950
An 32.2180

CENTRE NATIONAL DE LA RECHERCHE SCIENTIFIQUE
Physiologie comportement et ecologie des acridiens en rapport avec la phase :colloque international Actes Paris 1962 Apr 9-13 Centre national de la recherche scientifique. Colloques internationaux, 114 Paris,1962 Organised by F.O.Albrecht
Bal 39.2555

CENTRE NATIONAL DE LA RECHERCHE SCIENTIFIQUE
Physique fondamentale et astrophysique : colloque Compte-rendu Nice 1969 Journal de physique.Supplement, 30,no 11-12 Centre national de la recherche scientifique. Colloques nationaux, 925 162p Centre national de la recherche scientifique,Paris, 1970
TA 15.0499

CENTRE NATIONAL DE LA RECHERCHE SCIENTIFIQUE
Phytochimie et plantes medicinales des terres du Pacifique :colloque international Noumea 1964 Apr 28-May 5 Centre national de la recherche scientifique.Colloques internationaux,144 Editions du Centre national de la recherche scientifique,Paris, 1966
Chem 18.1296

CENTRE NATIONAL DE LA RECHERCHE SCIENTIFIQUE
Principes fondamentaux de classification stellaire Proceedings Paris 1953 Jun 29 - Jul 4 With the support of the Rockefeller foundation Centre national de la recherche scientifique.Colloques internationaux, 55 190p Centre national de la recherche scientifique,Paris,1955
Obs 6.2008

CENTRE NATIONAL DE LA RECHERCHE SCIENTIFIQUE
Problemes actuels de paleontologie (evolution des vertebres) Paris 1961 May 29-Jun 3 Centra national de la recherche scientifique. Colloques internationaux.Actes, 104 27cm C. N.R.S.,Paris,1962 Organized by J.P. Lehmann
Bal 44.4639

CENTRE NATIONAL DE LA RECHERCHE SCIENTIFIQUE
Problemes actuels de paleontologie (evolution des vertebres) Paris 1966 Jun 6-11 Centre national de la recherche scientifique. Colloques internationaux.Actes, 163 pls 27cm C.N.R.S.,Paris,1967 Organized by J. P.Lehman.Papers in English,French and German
Bal 44.4640

CENTRE NATIONAL DE LA RECHERCHE SCIENTIFIQUE
Proprietes optiques et acoustiques des fluides comprimees et actions intermoleculaires Bellevue 1957 Jul 1-6 Edited by B. Vodar Centre national de la recherche scientifique. Colloques internationaux,77 Centre national de la recherche scientifique,Paris,1959
Chem 18.2309

CENTRE NATIONAL DE LA RECHERCHE SCIENTIFIQUE
Relations entre les phenomenes solaires et geophysiques Proceedings Lyon 1947 Sep 1 7 Edited by Jean Dufay Supported by the Rockefeller foundation Centre national de la recherche scientifique.Colloques internationaux, 9 312p Centre national de la recherche scientifique,Lyon,1947
Obs 6.2010

CENTRE NATIONAL DE LA RECHERCHE SCIENTIFIQUE
Regulateurs naturels de la croissance vegetales Colloque international sur les substances de croissance vegetales 5e Actes Gif sur Yvette 1963 Jul 15-20 Centre national de la recherche scientifique. Colloques internationaux, 123 Paris,1964 Text in English and French
Bioch 33.0747

CENTRE NATIONAL DE LA RECHERCHE SCIENTIFIQUE
Structure et physiologie des societes animales colloque international Paris 1950 Mar Centre national de la recherche scientifique. Colloques internationaux, 34 C.N.R.S.,Paris, 1952
Phys 20.0540

CENTRE NATIONAL DE LA RECHERCHE SCIENTIFIQUE
Structure et proprietes des surfaces des solides :colloque international Paris 1969 Jul 7-11 Centre national de la recherche scientifique.Colloques internationaux, 187 Paris,1970
Met 25.2626

CENTRE NATIONAL DE LA RECHERCHE SCIENTIFIQUE
Theorie du potentiel colloque international 1964 Centre national de la recherche scientifique.Colloques internationaux 15 unbound typescript papers
Math 3.1110

CENTRE NATIONAL DE LA RECHERCHE SCIENTIFIQUE
Unites biologiques douees de continuite genetique colloque international Paris 1948 Jul 25-Jul 3 Centre national de la recherche scientifique.Colloques internationaux, 8 Centre national de la recherche scientifique,Paris,1949
Gen 34.0988

CENTRE NATIONAL DE LA RECHERCHE SCIENTIFIQUE
Regionalisation et developpement Strasbourg 1967 Jun 26-30 Centre national de la recherche scientifique Centre national de la recherche scientifique,Paris,1968 Papers in English and French
Geog 13.7207

CENTRE NATIONAL DE LA RECHERCHE SCIENTIFIQUE 68th colloque international Proceedings Echanges de matieres au cours de la genese des roches grenues acides et basiques Nancy 1955 Sep 4-11 Edited by M. Roubault Sciences de la terre,no.hors serie Ecole superieure de geologie,Nancy,1955
Min 10.0713

CENTRE NATIONAL DE LA RECHERCHE SCIENTIFIQUE. COLLOQUES INTERNATIONAUX Le Calcul de probabilites et ses applications colloque international proceedings Lyon 1948 Jun. 28-Jul.3 Centre national de la recherche scientifique Edited by Maurice Frechet 130p 27cm Centre national de la recherche scientifique,Paris,1949
Math 3.0678

CENTRE NATIONAL DE LA RECHERCHE SCIENTIFIQUE. COLLOQUES INTERNATIONAUX.ACTES, 163 Problemes actuels de paleontologie (evolution des vertebres) Paris 1966 Jun 6-11 Centre national de la recherche scientifique pls 27cm C.N.R.S.,Paris,1967 Organized by J.P.Lehman.Papers in English, French and German
Bal 44.4640

CENTRE NATIONAL DE LA RECHERCHE SCIENTIFIQUE. COLLOQUES INTERNATIONAUX, 4 Endocrinologie des arthropodes Colloque international sur l'endocrinologie des arthropodes Paris 1947 Jun Centre national de la recherche scientifique Centre national de la recherch scientifique,Paris, 1948
Bal 39.2291

CENTRE NATIONAL DE LA RECHERCHE SCIENTIFIQUE. COLLOQUES INTERNATIONAUX, 21 Paleontologie :colloque internationale Paris 1947 Apr Centre national de la recherche scientifique C.N.R.S.,Paris,1950
An 32.2180

CENTRE NATIONAL DE LA RECHERCHE SCIENTIFIQUE. COLLOQUES INTERNATIONAUX, 28 Morphogenese :colloque international Strasbourg 1949 Jul 4-11 Centre national de la recherche scientifique C.N.R.S.,Paris, 1951
An 32.2463

CENTRE NATIONAL DE LA RECHERCHE SCIENTIFIQUE. COLLOQUES INTERNATIONAUX, 32 Mecanisme physiologique de la secretion lactee :colloque international Strasbourg 1950 Aug 22-29 Centre national de la recherche scientifique C.N.R.S.,Paris,1950
Phys 20.2212

CENTRE NATIONAL DE LA RECHERCHE SCIENTIFIQUE. COLLOQUES INTERNATIONAUX, 34 Structure et physiologie des societes animales :colloque international Paris 1950 Mar Centre national de la recherche scientifique C.N.R. S.,Paris,1952
Phys 20.0540

CENTRE NATIONAL DE LA RECHERCHE SCIENTIFIQUE. COLLOQUES INTERNATIONAUX, 40 Nouvelles proprietes physiques et chimiques des metaux de tres haut purete colloque international Paris 1959 Oct 12-14 Centre national de la recherche scientifique Centre national de la recherche scientifique,Paris,1960 English translation also available
Met 25.1499

CENTRE NATIONAL DE LA RECHERCHE SCIENTIFIQUE. COLLOQUES INTERNATIONAUX, 58 Les techniques recentes en microscopie electronique et corpusculaire :colloque international Toulouse 1955 Apr 4-8 Centre national de la recherche scientifique Edited by Ch. Fert Paris,1956
Cav 7.0525

CENTRE NATIONAL DE LA RECHERCHE SCIENTIFIQUE. COLLOQUES INTERNATIONAUX, 81 Ecologie des algues marines Dinard 1957 Sep 20-28 By R. Biebl and others Centre national de la recherche scientifique illus. 276p Paris, 1959
Bot 42.3814

CENTRE NATIONAL DE LA RECHERCHE SCIENTIFIQUE. COLLOQUES INTERNATIONAUX, 87 Le Calcul des probabilites et ses applications colloque international proceedings Paris 1958 Jul. 15-20 Centre national de la recherche scientifique Edited by Georges Darmois 196p 24cm Centre national de la recherche scientifique,Paris,1959
Math 3.0679

CENTRE NATIONAL DE LA RECHERCHE SCIENTIFIQUE. COLLOQUES INTERNATIONAUX, 108 Mecanique de la turbulence colloque international Marseille 1961 Aug 28-Sep 2 Mecanique de la turbulence Centre national de la recherche scientifique Centre national de la recherche scientifique,Paris,1962
A Math 4.0525

CENTRE NATIONAL DE LA RECHERCHE SCIENTIFIQUE. COLLOQUES INTERNATIONAUX, 113 Le Bombardement ionique theories et applications: colloques internationaux Bellevue,Seine et Oise 1962 Dec 4-8 Centre national de la recherche scientifique Editions du centre National de la recherche scientifique,Paris, 1962
Met 25.1676

CENTRE NATIONAL DE LA RECHERCHE SCIENTIFIQUE. COLLOQUES INTERNATIONAUX, 114 Physiologie comportement et ecologie des acridiens en rapport avec la phase :colloque international Actes Paris 1962 Apr 9-13 Centre national de la recherche scientifique Paris, 1962 Organised by F.O.Albrecht
Bal 39.2555

CENTRE NATIONAL DE LA RECHERCHE SCIENTIFIQUE. COLLOQUES INTERNATIONAUX, 118 La Structure des solutions solides metalliques Orsay 1962 Jul 9-11 Centre national de la recherche scientifique Centre national de la recherche scientifique,Paris,1962 Papers in English and French
Met 25.1210

CENTRE NATIONAL DE LA RECHERCHE SCIENTIFIQUE. COLLOQUES INTERNATIONAUX, 123 Regulateurs naturels de la croissance vegetales Colloque international sur les substances de croissance vegetales 5e Actes Gif sur Yvette 1963 Jul 15-20 Centre national de la recherche scientifique Paris,1964 Text in English and French
Bioch 33.0747

CENTRE NATIONAL DE LA RECHERCHE SCIENTIFIQUE. COLLOQUES INTERNATIONAUX, 124 Mecanismes de regulation des activites cellulaires chez les microorganismes Colloque international Marseilles 1963 Jul 23-27 Centre national de la recherche scientifique Paris,1965
Gen 34.0707

CENTRE NATIONAL DE LA RECHERCHE SCIENTIFIQUE. COLLOQUES INTERNATIONAUX, 128 Cytologie de l'adenohypophyse :colloque international Paris 1963 Sep 9-14 Centre national de la recherche scientifique Edited by Jacques Benoit and Christian Da Lage C.N.R.S.,Paris, 1963
An 32.3770

CENTRE NATIONAL DE LA RECHERCHE SCIENTIFIQUE. COLLOQUES INTERNATIONAUX, 144 Phytochimie et plantes medicinales des terres du Pacifique colloque Proceedings Noumea,New Caledonia 1964 Apr 28-May 5 Centre nationale de la recherche scientifique Editions du centre national de la recherche scientifique,Paris, 1966
Chem 26.0137

CENTRE NATIONAL DE LA RECHERCHE SCIENTIFIQUE. COLLOQUES INTERNATIONAUX, 146 La Theorie du potentiel :colloque international Orsay 1964 Jun.22-26 Centre national de la recherche scientifique Edited by M. Brelot and others x,312p 25cm Centre national de la recherche scientifique,Paris,1965
Math 3.0050

CENTRE NATIONAL DE LA RECHERCHE SCIENTIFIQUE. COLLOQUES INTERNATIONAUX, 161 Methodes nouvelles de spectroscopie instrumentale Exposes Orsay 1966 Apr 25-29 Centre national de la recherche scientifique 2nd edition 344p CNRS,Paris,1969
Obs 6.3376

CENTRE NATIONAL DE LA RECHERCHE SCIENTIFIQUE. COLLOQUES INTERNATIONAUX, 167 Nouvelles proprietes physiques, mecaniques et chimiques Colloque international Paris 1966 Sep 26-Oct 1 Centre national de la recherche scientifique Editions du centre national de la recherche scientifique,Paris,1968
Met 25.1503

CENTRE NATIONAL DE LA RECHERCHE SCIENTIFIQUE. COLLOQUES INTERNATIONAUX, 186 Les Probabilites sur les structures algebriques Clermont-Ferrand 1969 Jun 30-Jul 5 Centre national de la recherche scientifique 361p 25cm CNRS,Paris,1970
P Math 2.4558

CENTRE NATIONAL DE LA RECHERCHE SCIENTIFIQUE. COLLOQUES INTERNATIONAUX, 186 Les Probabilites sur les structures algebriques : colloque international Clermont-Ferrand 1969 Jun 30-Jul 5 Centre national de la recherche scientifique 361p 25cm Editions du CNRS,Paris,1970 Oeganized by A.Badrikian and P.L.Hennequin
Math S 3.1780

CENTRE NATIONAL DE LA RECHERCHE SCIENTIFIQUE. COLLOQUES INTERNATIONAUX, 187 Structure et proprietes des surfaces des solides : colloque international Paris 1969 Jul 7-11 Centre national de la recherche scientifique Paris,1970
Met 25.2626

CENTRE NATIONAL DE LA RECHERCHE SCIENTIFIQUE. COLLOQUES INTERNATIONAUX, 193 Les Cultures de tissus de plantes Proceedings Strasburg 1970 Jul 6-10 Centre national de la recherche scientifique CNRS,Paris,1971
Bioch 33.2223

CENTRE NATIONAL DE LA RECHERCHE SCIENTIFIQUE. COLLOQUES INTERNATIONAUX,35 Actions eoliennes.Phenomenes d'evaporation et d'hydrologie superficielle dans les regions arides :colloque international Algiers 1951 Mar 27-31 C.N.R.S.,Paris,1953
Geog 13.0671

CENTRE NATIONAL DE LA RECHERCHE SCIENTIFIQUE. COLLOQUES INTERNATIONAUX,66 Les Relations entre precambrien et cambrien problemes des series intermediaires colloque international Paris 1957 Jun 27- Jul 4 Centre national de la recherche scientifique bibliog. Centre national de la recherche scientifique, Paris,1958
Geol 8.2716

CENTRE NATIONAL DE LA RECHERCHE SCIENTIFIQUE. COLLOQUES INTERNATIONAUX,77 Proprietes optiques et acoustiques des fluides comprimees et actions intermoleculaires Bellevue 1957 Jul 1-6 Centre national de la recherche scientifique Edited by B. Vodar Centre national de la recherche scientifique,Paris, 1959
Chem 18.2309

CENTRE NATIONAL DE LA RECHERCHE SCIENTIFIQUE. COLLOQUES INTERNATIONAUX,83 Topographie et la geologie des profondeurs oceaniques : colloque international Nice 1958 May 5-12 C.N.R.S.,Paris,1959
Geog 13.0734

CENTRE NATIONAL DE LA RECHERCHE SCIENTIFIQUE. COLLOQUES NATIONAUX, 923 Dosage des elements a l'etat de traces dans les roches et autres substances :colloque Comptes-rendus Nancy 1968 Dec 4-6 Centre national de la recherche scientifique CNRS,Paris,1970
Min 10.1475

CENTRE NATIONAL DE LA RECHERCHE SCIENTIFIQUE. INTERNATIONAL SYMPOSIA, 40 New physical and chemical properties of metals of very high purity :international symposia Paris 1959 Oct 12-14 Centre national de la recherche scientifique Gordon and Breach,New York,1965
Met 25.1500

CENTRE NATIONAL DE LA RECHERCHE SCIENTIFIQUE. SCIENCES HUMAINES Leonard de Vinci et l'experience scientifique au XVIe siecle : colloque international Paris 1952 Jul 4-7 viii,273p Presses universitaires de France, Paris,1953
WSM 43.1938

CENTRE NATIONAL DE RECHERCHES METALLURGIQUES Le Frottement interieur des metaux Comptes rendus Saint-Germain-en-Laye 1960 Oct 13-14 Edited by C. Crussard and others Editions metaux,n.p.,c1961
Met 25.1026

CENTRE NATIONALE DE LA RECHERCHE SCIENTIFIQUE Phytochimie et plantes medicinales des terres du Pacifique colloque Proceedings Noumea, New Caledonia 1964 Apr 28-May 5 Centre national de la recherche scientifique. Colloques internationaux, 144 Editions du centre national de la recherche scientifique, Paris,1966
Chem 26.0137

CENTRE NATIONALE DE LA RECHERCHE SCIENTIFIQUE. COLLOQUES INTERNATIONAUX, 143 Les Tendances geometriques en algebre et theorie des nombres colloque international Clermont-Ferrand 1964 Apr 2-9 Centre national de la recherche scientifique Edited by Marc Krasner Bibliog. 256p 25cm Centre national de la recherche scientifique,Paris, 1966
P. Math 2.1775

CENTRO DE CALCULO CIENTIFICO Theory of distributions :international summer institute Proceedings Lisbon 1964 Sep Sponsored by the North Atlantic treaty organization.Science committee Illus. xviii,390p 25cm Centro de calculo cientifico,Lisbon,1964 In English and French
P Math 2.3180

CENTRO INTERNAZIONALE MATEMATICO ESTIVO Calculus of variations,classical and modern Bressanone 1960 Jun 10-18 Edited by R. Conti bibliog.,illus. 369p 27cm Edizione Cremonese,Rome,1967 In English and French
P Math 2.3729

CENTRO INTERNAZIONALE MATEMATICO ESTIVO Functional equations and inequalities Centro internazionale matematico estivo 3 ciclo La Mendola,Trento 1970 Aug 20-28 426p 29cm Edizioni cremonese,Rome,1971 Conference coordinator:B.Forte
P Math 2.4213

CENTRO INTERNAZIONALE MATEMATICO ESTIVO Topologia differenziale 1 ciclo Urbino 1962 Jul 2-12 By J. Cerf and others 27cm Institute matematica dell'universita,Rome,1962
P. Math 2.0243

CENTRO MATEMATICO ESTIVO Questions on algebraic varieties Centro matematico estivo 3 ciclo Varenna 1969 Sep 7-17 343p 28cm Edizioni cremonese,Rome,1970 Conference coordinator:E.Marchionna
P Math 2.4172

CERAMIC EDUCATION COUNCIL Mechanical behavior of crystalline solids a symposium Proceedings New York 1962 Apr 28-29 Sponsored by the Edward Orton junior ceramic foundation National bureau of standards monograph, 59 U.S.Government printing office,Washington,1963 Lithographed
Met 25.1213

CERAMIC EDUCATION COUNCIL Microstructure of ceramic materials :a symposium Proceedings Pittsburgh 1963 Apr 27-28 Sponsored by Edward Orton junior ceramic foundation National bureau of standards.Miscellaneous publication, 257 U.S.dept of commerce, Washington D.C.,1964
Met 25.0285

CERAMIC MICROSTRUCTURES THEIR ANALYSIS, SIGNIFICANCE AND PRODUCTION International materials symposium 3rd Proceedings Berkeley 1966 Jun 13-16 Edited by Richard M. Fulrath and Joseph A. Pask Wiley,New York, 1968
Met 25.0292

CERAMICS Kinetics of reactions in ionic systems International symposium on special topics in ceramics Proceedings Alfred,N.Y. 1967 Jun 18-23 Edited by T.J. Gray and V.D. Frechette Held at Alfred university Materials science research, 4 Plenum press, New York,1969
Met 25.2750

CERAMICS Materials science research :a conference Proceedings Raleigh 1964 Nov 16-18 Vol 3: role of grain boundaries and surfaces in ceramics North Carolina state university Edited by W.Wurth Kriegel and Hayne Palmour Plenum press,New York,1966
Met 25.0288

CERAMICS Microstructure of ceramic materials a symposium Proceedings Pittsburgh 1963 Apr 27-28 Ceramic education council and National bureau of standards Sponsored by Edward Orton junior ceramic foundation National bureau of standards.Miscellaneous publication, 257 U.S.dept of commerce, Washington D.C.,1964
Met 25.0285

CERAMICS Science of ceramics :a conference 1st Proceedings Oxford 1961 Jun 26-30 British ceramic society Edited by G.H. Stewart Academic press,London;New York,1962
Met 25.0274

CERAMICS Science of ceramics :a conference 2nd Proceedings Noordwijk aan Zee 1963 May 13-17 British ceramic society Edited by G.H. Stewart Academic press,London;New York,1965
Met 25.0278

CERAMICS Science of ceramics :a conference 3rd Proceedings Bristol 1965 Jul 5-8 British ceramic society and Nederlandse keramische vereniging Edited by G.H. Stewart Under the auspices of the European ceramic association Academic press,London;New York, 1967
Met 25.0286

CERAMICS Special ceramics :a symposium 4th Proceedings Stoke-on-Trent 1967 Jul 11-13 British ceramic research association Edited by P. Popper British ceramic research association,Stoke-on-Trent,1968
Met 25.2554

CERAMICS Special ceramics :a symposium Proceedings Stoke-on-Trent 1959 Jul 13-15 By P. Popper British ceramic research association Heywood,London,1960
Met 25.0267

CERAMICS Special ceramics 1962 :a symposium 2nd Proceedings Stoke-on-Trent 1962 Jul 11-13 British ceramic research association Edited by P. Popper Academic press,London; New York,1963
Met 25.0280

CERAMICS Special ceramics 1964 British ceramic research association :symposium 3rd Proceedings Stoke-on-Trent 1964 Jul 8-10 Edited by P. Popper Academic press,London; New York,1965
Min 10.0656

CERAMICS Special ceramics 1964 :a symposium 3rd Proceedings Stoke-on-Trent 1964 Jul 8-10 By P. Popper British ceramic research association Edited by P. Popper Academic press,London;New York,1965
Met 25.0281

CERAMICS The Physics and chemistry of ceramics :a symposium Proceedings Philadelphia,Pa. 1962 May 28-30 United States.Office of naval research Edited by Cyrus Klingsberg Gordon and Breach,New York; London,1963
Met 25.0277

CEREAL RUST CONFERENCES Cambridge 1964 Jun-Jul Plant breeding institute Cambridge, 1966
Bot 42.4125

CEREALS Growth of cereals and grasses Easter school in agricultural science 12th Proceedings Nottingham University of Nottingham Edited by F.L. Milthorpe and J.D. Ivins Bibliog.,illus,diagrs,tables xii, 359p Butterworths,London,1966
Bot 42.2035

The CEREBELLUM Edited by C.A. Fox and R.S. Snider Progress in brain research, 25 Elsevier,Amsterdam,1967 Derived from a meeting held in the Netherlands,1965
An 32.4246

The CEREBELLUM :an investigation of recent advances Proceedings of the Association New York 1926 Dec 28-29 Association for research in nervous and mental disease Edited by Frederick Tilney and others Association for research in nervous and mental disease.Research publications, 6 Williams and Wilkins,Baltimore,Md.,1929
An 32.4680

CEREBRAL CIRCULATION European congress of neurosurgery 3rd Madrid 1967 Apr Edited by W. Luyendijk Progress in brain research, 30 Elsevier,Amsterdam,1968
An 32.4248

CEREBRAL LIPIDOSES :a symposium Papers Antwerp 1955 Jul 26-27 Edited by J.N. Cumings and others Blackwell,Oxford,1957
An 32.4165

CEREBRAL LOCALIZATION AND ORGANIZATION :a symposium Proceedings Lisbon 1960 Oct 21-23 World federation of neurology Edited by Georges Schattenbrand and Clinton N. Woolsey University of Wisconsin press, Madison,Wis.;Milwaukee,Wis.,1964
Psy 31.0254

CEREBRAL MECHANISMS IN BEHAVIOUR :the Hixon symposium Pasadena,Calif. 1948 Sep Edited by Lloyd A. Jeffress Under the auspices of the Hixon fund committee Wiley; Chapman and Hall,New York;London,1951
An 32.4627

CEREBRAL MECHANISMS IN BEHAVIOUR :the Hixon symposium Report Hixon symposium Pasadena,Calif. 1948 Sep 20-25 California institute of technology.Hixon committee Edited by Lloyd A. Jeffress Wiley;Chapman and Hall,New York;London,1951
Psy 31.0522

CEREBROSPINAL FLUID AND THE REGULATION OF VENTILATION :a symposium Proceedings Brooklyn,N.Y. 1964 Apr 9-10 Edited by Chandler McC. Brooks and others Held at the Downstate medical centre Blackwell,Oxford, 1965
An 32.4209

CEREBROSPINAL FLUID AND THE REGULATION OF VENTILATION 6 :a symposium Proceedings New York 1964 Apr 9-10 Downstate medical center Edited by Chandler McC. Brooks and others Blackwell,Oxford,1965
Pha 16.0278

CEREBROVASCULAR DISEASE Proceedings of the Association New York 1961 Dec 8-9 Association for research in nervous and mental disease Edited by Clark H. Millikan Association for research in nervous and mental disease.Research publications, 41 Williams and Wilkins,Baltimore,Md.,1966
An 32.4214

CERISY-LA-SALLE 1964 Entretiens sur le temps Centre culturel international de Cerisy-la-Salle Edited by Jeanne Hersch and Rene Poirier Centre culturel international de Cerisy-la-Salle.Decade nouvelle serie, 5 351p Mouton,Paris;The Hague,1967
WSM 43.1000

CERN Auxiliary instrumentation for large telescopes :a conference Geneva 1972 May 2-5 Edited by S. Lanstsen and A. Reiz 525p European southern observatory,Geneva,1972
Obs 6.3599

CERN Large telescope design :a conference Proceedings Geneva 1971 Mar 1-5 Edited by R.M. West 499p European southern observatory,Geneva,1971
Obs 6.3601

CESKOSLOVENSKA AKADEMIE VED Blood groups of animals European animal blood group conference 9th Proceedings Prague 1964 Aug 18-22 Edited by Josef Matousek Junk,The Hague,1964
Surg 23.0044

CESKOSLOVENSKA AKADEMIE VED Ecology of aphidophagous insects :symposium Proceedings Liblice 1965 Sep 27-Oct 1 Junk;Academia, The Hague;Prague,1966
Bal 39.2844

CESKOSLOVENSKA AKADEMIE VED Electronics and vacuum physics Czechoslovak conference 3rd Transactions Prague 1965 Sep 23-28 Edited by Libor Paty Sponsored by the Universita Karlova Academia,Prague,1967
Eng 41.5453

CESKOSLOVENSKA AKADEMIE VED General topology and its relations to modern analysis and algebra Prague topological symposium 2nd Proceedings Prague 1966 Aug 30-Sep 4 365p 24cm Academic press;Academia,New York; Prague,1967
P Math 2.3346

CESKOSLOVENSKA AKADEMIE VED General topology and its relations to modern analysis and algebra :symposium proceedings Prague 1961 Sep 1-8 Edited by J. Novak jointly with International mathematical union 363p 25cm Academic press,New York,1962
P. Math 2.0291

CESKOSLOVENSKA AKADEMIE VED Information processing machines symposium proceedings Prague 1964 Sep 7-9 Edited by Vladimir Bubenik Czechoslovak academy of sciences, Prague,1965
Math L 5.1011

CESKOSLOVENSKA AKADEMIE VED Prague conference on information theory,statistical decision functions, random processes 3rd transactions Liblice,Czech. 1962 Jun 5-13 Edited by Jaroslav Kozesnik Organized by the Institute of information theory and automation 84625 Czechoslovak academy of sciences, Prague,1964 Dedicated to the memory of Antonin Spacek
Math 3.0697

CESKOSLOVENSKA AKADEMIE VED Purkyne symposium Vortrage und diskussionsbeitrage Halle 1959 Oct 31-Nov 1 Edited by Rudolph Zaunick Nova acta leopoldina.Neue folge, 24 230p Barth,Leipzig,1961
WSM 43.2940

CESKOSLOVENSKA AKADEMIE VED Regulation of body fluid volumes by the kidney A Symposium on natriuretic hormone Smolenice castle 1969 Jun Edited by J.H. Cort and B. Lichardus Karger,Basle,1970
Inv Med 37.0268

CESKOSLOVENSKA AKADEMIE VED Symposium on mechanisms of immunological tolerance Proceedings Liblice 1961 Nov 8-10 Edited by M. Hasek and others illus. Academic press,New York,1963
EG 23.0056

CESKOSLOVENSKA AKADEMIE VED Water stress in plants :a symposium Proceedings Prague 1963 Sep 30-Oct 4 Edited by Bohdan Slavik 318p Junk,The Hague,1965
Bot 42.1881

CESKOSLOVENSKA AKADEMIE VED Structure of organic solids Microsymposium 'Structure of organic solids' 2nd Lectures Prague 1968 Sep 16-19 International union of pure and applied chemistry.Macromolecular division Czechoslovak chemical society illus 550p 25cm Butterworths,London,1969 Simultaneously published in 'Pure and applied chemistry',vol 18,no 4,1969
Cav 7.2659

CESKOSLOVENSKA AKADEMIE VED.INSTITUTE OF BIOLOGY Membrane transport and metabolism :a symposium Proceedings Prague 1960 Aug 22-27 Edited by A. Kleinzeller and A. Kotyk Academic press,London;New York,1961
Bot 42.1483

CESKOSLOVENSKA AKADEMIE VED.INSTITUTE OF BIOLOGY Membrane transport and metabolism :a symposium Proceedings Prague 1960 Aug 22-27 Edited by A. Kleinzeller and A. Kotyk Academic press,London;New York,1961
Bioch 33.0972

CESKOSLOVENSKA AKADEMIE VED.INSTITUTE OF INFORMATION THEORY AND AUTOMATION Identification in automatic control systems I.F.A.C. symposium on identification in automatic control systems Preprints Prague 1967 Jun 12-17 Pt 1-2 Sponsored by International federation of automatic control 2 vols Academia,Prague,1967
Eng 41.6019

CESKOSLOVENSKA AKADEMIE VED.INSTITUTE OF MICROBIOLOGY Molecular and cellular basis of antibody formation :a symposium Proceedings Prague 1964 Jun 1-5 Edited by J. Sterzl Czechoslovak academy of sciences;Academic press,Prague;New York,1965
An 32.3701

CESKOSLOVENSKA AKADEMIE VED.INSTITUTE OF MICROBIOLOGY Symposia Csav The Physiology of gene and mutation expression : symposium on the mutational process Proceedings Brno 1965 Aug 4-7 Prague 1965 Aug 9-11 Edited by Margita Kohoutova and J. Hubacek Academia,Prague,1967
Gen 34.0770

CESKOSLOVENSKA AKADEMIE VED.INSTITUTE OF MICROBIOLOGY. IMMUNOLOGICAL DEPARTMENT Molecular and cellular basis of antibody formation a symposium Proceedings Prague 1964 Jun 1-5 Edited by J. Sterzl Czechoslovak academy of sciences;Academic press,Prague;New York,1965
Bioch 33.0554

CESKOSLOVENSKA AKADEMIE VED.SEKCE BIOLOGICKO - LEKARSKA Ontogeny of insects :a symposium Acta Prague 1959 Edited by Ivan Hrdy Czechoslovak academy of sciences,Prague,1960 Text in English,French,German and Russian
Bal 39.2433

CESKOSLOVENSKA AKADEMIE VED.USTAV JADERNEHO VYZKUMU Nuclear theory International summer school on selected topics in nuclear theory lectures Low Tatra mountains, Czechoslovakia 1962 Aug 20-Sep 8 By N. Austern and others Edited by F. Janouch With the co-operation of the International atomic energy agency 452p International atomic energy agency,Vienna,1963
A Math 4.0885

CEYLON ASSOCIATION FOR THE ADVANCEMENT OF SCIENCE The Development of agriculture in the dry zone :a symposium Proceedings Colombo 1967 Jul 30-31 Edited by O.S. Peries bibliog. Colombo,1967
Geog 13.4482

CHAD South-central Libya and northern Chad Petroleum exploration society of Libya :annual field conference 8th Guidebook Libya 1966 Edited by James J. Williams Amsterdam, 1966
Geog 13.0598

CHAIN SILICATES A.G.I. short course on chain silicates :lecture notes Palo Alto,Calif. 1966 Nov 11-13 American geological institute American geological institute,Washington,D.C., 1966 Typescript
Min 10.1286

CHALLENGE FOR STEEL The Conference on 'The challenge for steel' :I.S.I. autumn general meeting Proceedings 1968 Nov Iron and steel institute Iron and steel institute. Publication, 119 Iron and steel institute, London,1969
Met 25.2241

CHALLENGE OF LIFE:BIOMEDICAL PROGRESS AND HUMAN VALUES   Roche anniversary symposium   Proceedings   Basle   1971 Aug 31-Sep 3   Hoffmann-La Roche and company   Edited by Robert M. Kunz and Hans Fehr   Experientia. Supplementum, 17   456p   24cm   Birkhauser, Basle,1972   Symposium held on the occasion of the firm's 75th anniversary.Chairman:Lord Todd
Chem   18.2902

CHALLENGE TO PHARMACY IN THE SEVENTIES   Invitational conference on pharmacy manpower   Proceedings   San Francisco   1970 Sep 10-12   Edited by Joe B. Graber and Donald C. Brodie   Sponsored by the University of California. School of pharmacy   U.S.Department of health, education and welfare,Washington,D.C.,1972
PGMS   29.0674

CHALMERS UNIVERSITY OF TECHNOLOGY   International conference on solvent extraction chemistry   Proceedings   Gothenburg   1966 Aug 27-Sep 1   By D. Dyrssen and others   North-Holland,Amsterdam,1967
Met   25.2292

CHAMBRE SYNDICATE DE LA GROSSE FORGE FRANCAISE   Forgemasters' meeting 1954   Proceedings   London   1954 Oct 11-15   Iron and steel institute   and   National forgemasters' association   Iron and steel institute.Special report, 60   Iron and steel institute,London, 1957
Met   25.0715

CHAMONIX   1958   Physics of the movement of the ice   The International association of scientific hydrology :a symposium   Papers   International association of scientific hydrology.Publication,47   illus.   393p   22cm   Gentbrugge,1958   Papers in English and French
Sco   14.0127

The CHANGING FLORA OF BRITAIN   :a conference   Report   London   1962 Apr 4-6   Botanical society of the British Isles   Edited by J.E. Lousley   Botanical society of the British Isles.B.S.B.I.conference reports, 3   London, 1953
Bot   42.4708

CHAPEL HILL,N.C.   1963   Research methodology and needs in perinatal studies :conference   Proceedings   Edited by Sidney S. Chipman and others   Thomas,Springfield,Ill.,1966
An   32.2096

CHAPEL HILL,N.C.   1964   Symposium on congestion theory   Proceedings   Edited by Walter L. Smith and William E. Wilkinson   Held at the University of North Carolina   University of North Carolina.Monograph series in probability and statistics, 2   University of North Carolina press,Chapel Hill, N.C.,1965
Eng   41.2150

CHAPEL HILL,N.C.   1964   Symposium on congestion theory   proceedings   University of North Carolina.Department of statistics   Edited by Walter L. Smith and William E. Wilkinson   Financially supported by the United States.Office of naval research   North Carolina.University.Monograph series in probability and statistics, 2   xv,457p   24cm   University of North Carolina press, Chapel Hill,N.C.,1965
Math   3.0568

CHAPEL HILL,N.C.   1970   Carolina conference   Proceedings   University of North Carolina. Department of mathematics   Edited by Donald A. Eisenman and Laird E. Taylor   107p   28cm   University of North Carolina,Chapel Hill,N.C., 1970
P Math   2.4356

CHAPEL HILL,N.C.   1970   Combinatorial mathematics and its applications   2nd conference   Proceedings   University of North Carolina   United States.Air force.Office of scientific research   Edited by R.C. Bose and others   548p   25cm   University of North Carolina,Chapel Hill,N.C.,1970
Math S   3.1655

CHAPEL HILL,N.C.   1970   Combinatorial mathematics and its applications   Chapel Hill conference   2nd   Proceedings   University of North Carolina   United States.Air force. Office of scientific research   Edited by R.C. Bose and others   548p   25cm   University of North Carolina,Chapel Hill,N.C.,1970
P Math   2.4507

CHAPEL HILL CONFERENCE   2nd   Proceedings   Combinatorial mathematics and its applications   Chapel Hill,N.C.   1970 May 8-13   University of North Carolina   United States.Air force. Office of scientific research   Edited by R.C. Bose and others   548p   25cm   University of North Carolina,Chapel Hill,N.C.,1970
P Math   2.4507

CHAPELCROSS   1966   Strontium metabolism   International symposium on some aspects of strontium metabolism   Proceedings   Edited by J.M.A. Lenihan and others   Academic press, London;New York,1967
Radioth   35.1719

CHARACTERISTICS AND APPLICATIONS OF RESISTANCE STRAIN GAGES   :a symposium   Proceedings   Washington,D.C.   1951 Nov 8-9   National bureau of standards   National bureau of standards.Circular, 528   Washington,D.C.,1954
Eng   41.2902

CHARGED PARTICLE TRACKS IN SOLIDS AND LIQUIDS   Gray (L.H.) conference   2nd   Proceedings   Cambridge   1969 Apr   Institute of physics and the physical society.Conference series   Institute of physics,London,1970
Radioth   35.1702

CHARLES DARWIN AS GEOLOGIST   By Archibald Geikie   Cambridge university press,Cambridge, 1909
Geol   8.1950

CHARLES F.KETTERING RESEARCH LABORATORY   Bacterial photosynthesis :a symposium   Papers presented   Yellow Springs,Ohio   1963 Mar 18-20   Edited by Howard Gest and others   Charles F.Kettering research laboratory. Contribution, 112   Antioch press,Yellow Springs,Ohio,1963
Bioch   33.1116

CHARLES F.KETTERING RESEARCH LABORATORY   Bacterial photosynthesis :a symposium   Yellow Springs,Ohio   1963   Edited by Howard Gest and others   Charles F.Kettering research laboratory.Contribution, 112   illus.   xvi, 523p   Antioch press,Yellow Springs,Ohio,1963
Bot   42.3800

CHARLES F.KETTERING RESEARCH LABORATORY Non-heme iron proteins;role in energy conversion : a symposium Proceedings Yellow Springs, Ohio 1965 Mar 22-24 Edited by Anthony San Pietro Charles F.Kettering research laboratory.Contribution, 201 Antioch press, Yellow Springs,Ohio,1965
Bioch 33.0546

CHARLES SALT RESEARCH CENTRE 1968 Transplantation antigens and tissue typing symposium Proceedings Robert Jones and Agnes Hunt orthopaedic hospital management committee Edited by M.W. Elves and N.W. Nisbet Oswestry,1969
Surg 23.0031

CHARLOTTENLUND 1966 Symposium on the ecology of pelagic fish species in Arctic waters and adjacent seas Papers International council for the exploration of the sea.Distant northern seas committee and International council for the exploration of the sea.Gadoid fish committee Edited by R.W. Blacker Conseil permanent international pour l'exploration de la mer.Rapports et proces-verbaux des reunions, 158 illus,maps Host, Copenhagen,1968
Sco 14.8291

CHARLOTTESVILLE,VA. 1967 Interstellar ionized hydrogen :a symposium National radio astronomy observatory Edited by Yervant Jerzian Co-sponsored by the Arecibo ionospheric observatory 774p Benjamin,New York,1968
Obs 6.3447

CHARLOTTESVILLE,VA. 1967 Interstellar ionized hydrogen :a symposium National radio astronomy observatory Edited by Yervant Terzian Jointly sponsored by Arecibo ionospheric observatory 774p Benjamin,New York,1968
TA 15.0437

CHARLOTTESVILLE,VA. 1968 Low-luminosity stars Symposium on low-luminosity stars Proceedings University of Virginia Edited by Shiv S. Kumar 542p Gordon and Breach, New York,1969
Obs 6.3627

CHELSEA COLLEGE Biological oxydation of nitrogen in organic molecules :symposium Proceedings London 1971 Dec 19-22 Edited by J.W. Bridges and others Taylor and Francis,London,1972
Bioch 33.2286

CHELTENHAM 1956 Chemical engineering in the coal industry :an international conference Edited by Forbes W. Sharpley Organised by the National coal board 141p Pergamon press,London,n.d.
Chem E 24.1603

CHEMICAL AERONOMY The Threshold of space : conference on chemical aeronomy Proceedings Cambridge,Mass. 1956 Jun 25-28 Air force Cambridge research center.Geophysics research directorate Edited by M. Zelikoff Coordinated by Wentworth institute Pergamon press,London,1957
Chem 18.0478

CHEMICAL AERONOMY The threshold of space Conference on chemical aeronomy Proceedings Cambridge,Mass. 1956 Jun 25-28 Air force Cambridge research center.Geophysics research directorate Edited by M. Zelikoff illus xi,342p Pergamon press,London,1957
Nap 11.0654

CHEMICAL AND BIOLOGICAL ASPECTS OF PYRIDOXAL CATALYSIS Pyridoxal catalysis:enzymes and model systems The International symposium on chemical and biological aspects of pyridoxical catalysis 2nd Proceedings Moscow 1966 Sep 15-21 Edited by E.E. Snell and others sponsored by the International union of biochemistry I.U.B.Symposium series, 35 Interscience,New York,1968 First International symposium on chemical and biological aspects of pyridoxal catalysis published under title chemical and biological aspects of pyridoxical catalysis
Bioch 33.1038

CHEMICAL AND BIOLOGICAL ASPECTS OF PYRIDOXAL CATALYSIS :a symposium of the International union of biochemistry Proceedings Rome 1960 Oct Edited by E.E. Snell International union of biochemistry.Symposium series, 30 Pergamon,Oxford,1963
Radioth 35.0131

CHEMICAL BASIS OF DEVELOPMENT A Symposium on the chemical basis of development Baltimore, Md 1958 Mar 24-27 Edited by William D. McElroy and Bentley Glass Sponsored by the McCollum-Pratt institute Johns Hopkins press, Baltimore,Md.,1958
Bal 39.0454

CHEMICAL BASIS OF DEVELOPMENT A Symposium on the chemical basis of development Papers and informal discussions Baltimore 1958 Mar 24-27 Edited by William D. McElroy and Bentley Glass Sponsored by the McCollum-Pratt institute McCollum-Pratt institute. Contribution, 234 Johns Hopkins press, Baltimore,1958
Bioch 33.0829

The CHEMICAL BASIS OF DEVELOPMENT :a symposium Baltimore,Md. 1958 Mar 24-27 Edited by William D. McElroy and Bentley Glass Sponsored by the McCollum-Pratt institute Johns Hopkins press,Baltimore,Md.,1958
An 32.2488

The CHEMICAL BASIS OF DEVELOPMENT :a symposium Baltimore,Md. 1958 Mar 24-27 Edited by William D. McElroy and Bentley Glass Sponsored by the McCollum-Pratt institute port Johns Hopkins,Baltimore,Md.,1958
Mol 45.0257

CHEMICAL BASIS OF HEREDITY A Symposium on the chemical basis of heredity Papers and informal discussions Baltimore 1956 Jun 19-22 Edited by William D. McElroy and Bentley Glass Sponsored by the McCollum-Pratt institute McCollum-Pratt institute. Contribution, 153 Johns Hopkins press, Baltimore,1957
Bioch 33.0830

CHEMICAL BASIS OF HEREDITY A Symposium on the chemical basis of heredity Papers and informal discussions Baltimore,Md. 1956 Jun 19-22 Edited by William D. McElroy and Bentley Glass Sponsored by the McCollum-Pratt institute McCollum-Pratt institute. Contribution, 153 Johns Hopkins press, Baltimore,Md.,1957
Bal 44.6451

The CHEMICAL BASIS OF HEREDITY :a symposium Baltimore,Md. 1956 Jun 19-22 Edited by W.D. McElroy and Bentley Glass Sponsored by the McCollum-Pratt institute McCollum-Pratt institute.Contributions, 153 John Hopkins press,Baltimore,Md.,1957
Radioth 35.0219

The CHEMICAL BASIS OF HEREDITY :a symposium Baltimore,Md. 1956 Jun 19-22 McCollum-Pratt institute Edited by William D.Bentley McElroy and Bentley Glass With support from the Atomic energy commission McCollum-Pratt institute.Contribution,153 Johns Hopkins press,Baltimore,Md.,1957
Chem 18.1555

CHEMICAL-BIOLOGICAL CO-ORDINATION CENTRE Symposium on chemical-biological correlation 1st Washington,D.C. 1950 May 26-27 National academy of sciences.National research council,Washington,D.C.,1951
Radioth 35.0329

CHEMICAL EFFECTS OF NUCLEAR TRANSFORMATIONS :a symposium Proceedings Prague 1960 Oct 24-27 Vol 1-2 International atomic energy agency International atomic energy agency. Proceedings series STI-PUB 34 2 vols international atomic energy agency,Vienna,1961
Chem 18.1150

CHEMICAL EFFECTS OF NUCLEAR TRANSFORMATIONS : symposium on chemical effects associated with nuclear reactions and radioactive transformations Proceedings Vienna 1964 Dec 7-11 Vol 1-2 International atomic energy agency International atomic energy agency.Proceedings series STI-PUB 91 International atomic energy agency,Vienna,1965
Chem 18.1151

CHEMICAL EFFECTS OF RADIATION International conference on the peaceful uses of atomic energy 2nd Geneva 1958 Sep 1-13 Vol 29: chemical effects of radiation United Nations United Nations,Geneva,1958
Met 25.1567

CHEMICAL ENGINEER Function and training of the chemical engineer :international conference London 1955 Mar 21-23 Organisation for European economic co-operation 85p O.E.E.C.,Paris,1955
Chem E 24.1760

CHEMICAL ENGINEERING International congress of chemical engineering,chemical equipment construction and automation 3rd Lecture summaries Marienbad 1969 Sep Pt A-H Scientific committee Chisa '69,1969 Cover title:Chisa '69
Chem E 24.1872

CHEMICAL ENGINEERING EDUCATION Symposium on chemical engineering education Discussion on papers presented Birmingham 1957 Apr 9-11 Institution of chemical engineers 1957
Chem E 24.1758

CHEMICAL ENGINEERING EDUCATION Symposium on chemical engineering education Papers Birmingham 1957 Apr 9-11 Institution of chemical engineers 61p 1957
Chem E 24.1757

CHEMICAL ENGINEERING GROUP Corrosion problems of the petroleum industry a joint symosium Papers London 1959 Nov 26-27 Joint symposium with the Corrosion group S.C.I.Monograph, 10 Society of chemical industry,London,1960
Met 25.1934

CHEMICAL ENGINEERING IN THE COAL INDUSTRY : an international conference Cheltenham 1956 June Edited by Forbes W. Sharpley Organised by the National coal board 141p Pergamon press,London,n.d.
Chem E 24.1603

CHEMICAL ENGINEERING IN THE METALLURGICAL INDUSTRIES a symposium Proceedings Edinburgh 1963 Sep 25-26 Institution of chemical engineers Institution of chemical engineers,London,1963
Met 25.0316

CHEMICAL ENGINEERING PROGRESS Safety in air and ammonia plants :a symposium New York 1966 9 96p A.I.Ch.E.,New York,1967 A C.E.P.technical manual
Chem E 24.1576

CHEMICAL EVOLUTION AND THE ORIGIN OF LIFE International conference on the origin of life Proceedings Vol 1: molecular evolution Edited by Rene Buvet and C. Ponnamperuma 560p. North-Holland,Amsterdam,1971
TA 15.0690

CHEMICAL INSTITUTE OF CANADA Bioinorganic chemistry :a symposium Blacksburg,Va. 1970 Jun 22-25 Edited by Robert F. Gould x,436p 23cm American chemical society,Washington,D.C.,1971
Chem 18.2719

CHEMICAL INSTITUTE OF CANADA.PHYSICAL CHEMISTRY DIVISION The Structure and reactivity of electronically-excited species :a symposium Preprint of papers Ottawa 1957 Sep 5-6 Part-financed by the National research council of Canada Chemical institute of Canada. Physical chemistry division,1957 Chairman: Professor K.J.Laidler
Chem 18.0227

CHEMICAL KINETICS Fast reactions and primary processes in chemical kinetics Nobel symposium 5th Proceedings Sodergarn 1967 Aug 28 Sep 2 Nobel foundation Edited by Stig Claesson Wiley,New York;London,1967
Bioch 33.0328

CHEMICAL KINETICS Voprosy khimicheskoi kinetiki kataliza i reaktsionnoi sposobnosti : doklady k vsesoyuznomu soveshchaniyu po khimicheskoi kinetike i reaktsionnoi sposobnosti Doklady Akademiya nauk S.S.S.R. Otdelenie khimicheskikh nauk Edited by V.N. Kondratev and N.M. Emanuel Izdatelstvo Akademii nauk SSSR,Moscow,1955
Chem 18.0863

CHEMICAL PATHOLOGY IN RELATION TO CLINICAL MEDICINE, 5 Disorders of carbohydrate metabolism :a symposium Proceedings London 1968 Nov 29-30 Association of clinical pathologists.Chemical pathology sub-committee Edited by G.K. McGowan and G. Walters held at the Royal society of medicine Journal of clinical pathology.Supplement, 2 British medical association,London,1969
Bioch 33.1205

The CHEMICAL PATHOLOGY OF ANIMAL PIGMENTS :a symposium London 1954 Feb 20 Biochemical society Edited by R.T. Williams Held at the London school of hygiene and tropical medicine Biochemical society. Symposia, 12 Cambridge university press, Cambridge,1954
Bioch 33.1374

CHEMICAL PATHOLOGY OF THE NERVOUS SYSTEM International neurochemical symposium 3rd Proceedings Strasbourg 1958 Aug Edited by Jordi Folch-Pi Pergamon press,London,1961
An 32.4186

CHEMICAL PHYSICS OF IONIC SOLUTIONS :a selction of invited papers and discussions presented at the international symposium of the Electrochemical society...under the chairmanship of B.E.Conway Toronto 1964 May 4-6 Electrochemical society Edited by B.E. Conway and R.G. Barradas J.Wiley,New York,1966
Chem 18.0342

CHEMICAL PHYSICS OF NONMETALLIC CRYSTALS : Northwestern university 1961 international conference Papers Urbana,Ill. 1961 American institute of physics.Divisions of chemical physics and solid state physics,and, American chemical society.Division of physical chemistry Benjamin,New York,1962 First published as a supplement to the 'Journal of applied physics',January 1962
Chem 18.2385

CHEMICAL PROCESS HAZARDS Symposium on chemical process hazards with special reference to plant design Proceedings Manchester 1960 Mar 29-31 Edited by D.M. Pirie 117p Institution of chemical engineers,London,1960
Chem E 24.1395

CHEMICAL REACTION ENGINEERING European symposium on chemical engineering 1st Papers Amsterdam 1957 May 7-9 Instituut van ingenieurs Nederlandse chemische vereniging Held during the 12th meeting of the European federation of chemical engineering International series of monographs on chemical engineering, 1 197p Pergamon,London,1957
Chem E 24.1900

CHEMICAL REACTION ENGINEERING European symposium on chemical reaction engineering 3rd Proceedings Amsterdam 1964 Sep 15-17 Instituut van ingenieurs.Afdeling voor chemische techniek and Nederlandse chemische vereniging.Sectie voor chemische technologie Held under the auspices of the European federation of chemical engineering vi,326p Pergamon press,Oxford,1965 Supplement to "Chemical engineering science".
Chem E 24.1269

CHEMICAL REACTIONS IN THE LOWER AND UPPER ATMOSPHERE :a symposium Proceedings San Francisco,Calif. 1961 Apr 18-20 Stanford research institute 390p Wiley,New York,1961
Nap 11.0929

CHEMICAL REACTIVITY AND BIOLOGICAL ROLE OF FUNCTIONAL GROUPS IN ENZYMES Oxford 1970 Jan Biochemical society Edited by R.M. S. Smellie Biochemical society.Symposium, 31 Academic press,London,1970
Bioch 33.2315

CHEMICAL REACTIVITY AND BIOLOGICAL ROLE OF FUNCTIONAL GROUPS IN ENZYMES :a symposium... Proceedings Oxford 1970 Jan Molecular enzymology and protein group Edited by R.M.S. Smellie Biochemical society. Symposia, 31 Academic press,London,1970 Molecular enzymology and protein group of the Biochemical society and the Chemical society
Chem 18.2745

CHEMICAL RESEARCH Conference on chemical research Proceedings Houston,Texas 1969 Nov 17-19 13: the transuranium elements - the Mendeleev centennial Robert A.Welch foundation Edited by W.O. Milligan 494p R.A.Welch foundation,Houston,Texas,1970
TA 15.0699

CHEMICAL RESEARCH Conference on chemical research Proceedings Houston,Texas 1969 Nov 17-19 Robert A.Welsh foundation Robert A.Welch foundation.Conference, 13 photos, graphs,diagrms xii,494p 23cm Robert A. Welch foundation,Houston,Texas,1970
Cav 7.3126

CHEMICAL SOCIETY Antibiotics and mould metabolites :a symposium Report Nottingham 1956 Chemical society.Special publication,5 Chemical society,London,1956 Organized by professor Johnson
Chem 18.1215

CHEMICAL SOCIETY Chemical reactivity and biological role of functional groups in enzymes :a symposium... Proceedings Oxford 1970 Jan Molecular enzymology and protein group Edited by R.M.S. Smellie Biochemical society.Symposia, 31 Academic press,London, 1970 Molecular enzymology and protein group of the Biochemical society and the Chemical society
Chem 18.2745

CHEMICAL SOCIETY Chemisorption :a symposium Proceedings Keele 1956 Jul 16-19 Edited by W.E. Garner Butterworths,London,1957
Chem 18.0618

CHEMICAL SOCIETY Chemisorption :symposium Proceedings Keele 1956 Jul 16-19 Edited by W.E. Garner Butterworths,London,1957
Col S 12.0266

CHEMICAL SOCIETY Chemistry of natural products The International symposium on the chemistry of natural products 5th Plenary lectures London 1968 Jul 8-13 Butterworths,London,1968 "The contents of this book appear in 'Pure and applied chemistry,vol.17, nos.3-4 (196819".Added title page in French
Bioch 33.1392

CHEMICAL SOCIETY Developments in aromatic chemistry.Applications of electron and nuclear resonance in chemistry.Recent work on the inorganic chemistry of sulphur :a symposium Bristol 1958 Chemical society.Special publication,12 Chemical society,London,1958
Chem 18.1229

CHEMICAL SOCIETY Developments in aromatic chemistry.Applications of electron and nuclear resonance in chemistry.Recent work on the inorganic chemistry of sulphur :symposia Bristol 1958 Chemical society.Special publication, 12 Chemical society,London, 1958
Bioch 33.0630

CHEMICAL SOCIETY Molecular relaxation processes Symposium on relaxation methods in relation to molecular structure Lectures and papers Aberystwyth 1965 Jul 7-9 Academic press,London,1966
Pha 16.0369

CHEMICAL SOCIETY Molecular relaxation processes Symposium on relaxation methods in relation to molecular structure Papers Aberystwyth 1965 Jul 7-9 Chemical society special publication, 20 Academic press, London,1966
Cav 7.0005

CHEMICAL SOCIETY Molecular relaxation processes :a symposium Lectures and synopses of papers Aberystwyth 1965 Jul 7-9 Chemical society.Special publication,20 Chemical society;Academic press,London;New York,1966
Chem 18.1224

CHEMICAL SOCIETY Peptide chemistry :a symposium Report London 1955 Mar 30 Edited by A.D. Jenkins and D.F. Elliott Chemical society.Special publication, 2 London,1955 Organised by D.H.Hey
Bioch 33.0498

CHEMICAL SOCIETY Peptide chemistry :a symposium Report London 1955 Mar 30 Edited by A.D. Jenkins and D.F. Elliott Chemical society.Special publication,2 Chemical society,London,1955 Symposium organized by professor Hey
Chem 18.1524

CHEMICAL SOCIETY Phosphoric esters and related compounds :a symposium Report Cambridge 1957 Apr 9-12 Chemical society. Special publication,8 Chemical society, London,1957 Organized by Dr.Kenner and Dr. Brown
Chem 18.1217

CHEMICAL SOCIETY Phosphoric esters and related compounds :a symposium held at the Chemical society anniversary meeting Report Cambridge 1957 Apr 9-12 Chemical society. Special publication, 8 Chemical society, London,1957
Bioch 33.0842

CHEMICAL SOCIETY Reactions of free radicals in the gas phase :a symposium Report Cambridge 1957 Apr 9-12 Chemical society. Special publication,9 Chemical society, London,1957 Organized by Dr.Sugden
Chem 18.1218

CHEMICAL SOCIETY Recent advances in the chemistry of colouring matter :a symposium Report London 1956 Jan 19 Chemical society.Special publication,4 Chemical society,London,1956 Organized by professor Braude
Chem 18.1214

CHEMICAL SOCIETY Recent aspects of the inorganic chemistry of nitrogen :a symposium Report Cambridge 1957 Apr 10-12 Chemical society.Special publication,10 Chemical society,London,1957 Organized by Dr. Maddock and Dr.Sharpe
Chem 18.1219

CHEMICAL SOCIETY Recent work on naturally occuring nitrogen heterocyclic compounds :a symposium Report Exeter 1955 Jul 13-15 Edited by K. Schofield Chemical society. Special publication,3 Chemical society, London,1955
Chem 18.1329

CHEMICAL SOCIETY Recent work on naturally occurring nitrogen heterocyclic compounds :a symposium Report Exeter 1955 Jul 13-15 Edited by K. Schofield Chemical society. Special publication, 3 London,1955
Bioch 33.2228

CHEMICAL SOCIETY Solution properties of natural polymers :an international symposium Proceedings Edinburgh 1967 Jul 25-28 Chemical society.Special publication, 23 Chemical society,London,1968
Bioch 33.0632

CHEMICAL SOCIETY Steric effects in conjugated systems :a symposium Hull 1958 Jul 15-17 Edited by G.W. Gray Butterworths, London,1958
Chem 18.1687

CHEMICAL SOCIETY The Kinetics and mechanism of inorganic reaction in solution;a survey of recent work :a symposium Report London 1954 Feb 4 Edited by K.W. Sykes Chemical society.Special publication, 1 London,1954
Bioch 33.0631

CHEMICAL SOCIETY Theoretical organic chemistry :Kekule symposium Papers London 1958 Sep 15-17 International union of pure and applied chemistry.Section of organic chemistry Butterworths,London,1959
Chem 18.2165

CHEMICAL SOCIETY and INTERNATIONAL UNION OF PURE AND APPLIED CHEMISTRY International conference on co-ordination chemistry Lectures delivered and abstracts of papers London 1959 Apr 6-11 Chemical society. Special publication,13 Chemical society, London,1959
Chem 18.1220

CHEMICAL SOCIETY and INTERNATIONAL UNION OF PURE AND APPLIED CHEMISTRY.SECTION OF ORGANIC CHEMISTRY Theoretical organic chemistry : Kekule symposium Papers London 1958 Sep 15-17 Butterworths,London,1959
Chem 18.1604

CHEMICAL SOCIETY and UNIVERSITY COLLEGE,CORK
Organic reaction mechanisms :an international symposium Lectures Cork, Ireland 1964 Jul 20-25 Chemical society. Special publication,19 Chemical society, London,1965
Chem 18.1223

CHEMICAL SOCIETY and UNIVERSITY OF NOTTINGHAM
Inorganic polymers :an international conference Lectures Nottingham 1961 Jul 18-21 Chemical society.Special publication, 15 Chemical society,London,1961
Chem 18.1221

CHEMICAL SOCIETY and UNIVERSITY OF NOTTINGHAM
The Alkali metals :an international symposium Nottingham 1966 Jul 19-22 Chemical society.Special publication,22 Chemical society,London,1966
Chem 18.1226

CHEMICAL SOCIETY and UNIVERSITY OF SHEFFIELD
Aromaticity :an international symposium Sheffield 1966 Jul 6-8 Chemical society. Special publication,21 Chemical society, London,1967
Chem 18.1225

CHEMICAL SOCIETY and UNIVERSITY OF SHEFFIELD
The Transition state :a symposium Sheffield 1962 Apr 3-4 Chemical society. Special publication,16 Chemical society, London,1962
Chem 18.1222

CHEMICAL STABILITY OF TRITIATED PRYIMIDINE NUCLEOSIDES AND SOME PROBLEMS RAISED BY THEIR USE IN BIOLOGY :a colloquium Proceedings Mol 1965 Mar 16-17 Edited by R. Goutier and others EUR 3290 f-e Euratom,Brussels, 1967
Radioth 35.1754

The CHEMICAL STRUCTURE OF PROTEINS :a Ciba foundation symposium Proceedings London 1952 Dec 1-3 Ciba foundation Edited by G.E. W. Wolstenholme and Margaret P. Cameron illus Churchill,London,1953
Bioch 33.0500

CHEMICAL WARFARE AGENTS The Biochemical reactions of chemical warfare agents :a symposium London 1947 Dec 13 Biochemical society Edited by R.T. Williams Held at the University college hospital medical school Biochemical society.Symposia, 2 Cambridge university press,Cambridge,1948
Bioch 33.1364

CHEMIE UND BIOCHEMIE DER SOLANUM-ALKALOIDE : international symposium Vortrage und diskussionsbeitrage Berlin 1959 Jun 2-4 Deutsche akademie der landwirtschaftswissenschaften zu Berlin Deutsche akademie der landwirtschaftswissenschaften zu Berlin. Tagungsberichte, 27 Berlin,1961 Papers in English and German,with summaries in English,German and Russian
Chem 26.0814

CHEMIE UND STOFFWECHSEL VON BINDE-UND KNOCHENGEWEBE :colloquium Mosbach Baden 1956 Apr 12-14 Gesellschaft fur physiologische chemie Gesellschaft fur physiologische chemie.Colloquium, 7 Springer,Berlin,1956
An 32.3221

CHEMISORPTION :a symposium Proceedings Keele 1956 Jul 16-19 Chemical society Edited by W.E. Garner Butterworths,London, 1957
Chem 18.0618

CHEMISORPTION :symposium Proceedings Keele 1956 Jul 16-19 Chemical society Edited by W.E. Garner Butterworths,London, 1957
Col S 12.0266

CHEMISTRY Reactivity of photo-excited organic molecula Solvay conference on chemistry 13th Proceedings Brussels 1965 Oct 25-30 International institute of chemistry,and,Universite libre de Bruxelles Interscience,London,1967
Col S 12.0568

The CHEMISTRY AND BIOCHEMISTRY OF FUNGI AND YEASTS :a symposium Proceedings Dublin 1963 Jul 18-20 International union of pure and applied chemistry sponsored by the International union of pure and applied chemistry and Royal Irish academy.Irish national committee for chemistry Butterworths,London,1963 Reprinted from 'Pure and applied chemistry',vol.7,no.,4.Added title-page in French:Chimie et biochimie des champignons inferieurs et des levures
Bioch 33.1146

The CHEMISTRY AND BIOCHEMISTRY OF FUNGI AND YEASTS :symposium Proceedings Dublin 1963 Jul 18-20 International union of pure and applied chemistry Royal Irish academy Irish national committee for chemistry Butterworths,London,1963 Reprinted from 'Pure and applied chemistry' vol.7,no.4
Gen 34.0729

CHEMISTRY AND BIOLOGY OF MUCOPOLYSACCHARIDES
Ciba foundation ssymposium on the chemistry and biology of mucopolysaccharides Proceedings London 1957 Apr 23-25 Ciba foundation Edited by G.E.W. Wolstenholme and Maeve O'Connor Churchill,London,1958
Bioch 33.1202

CHEMISTRY AND BIOLOGY OF MUCOPOLYSACCHARIDES
London 1967 Apr 23-25 Ciba foundation Edited by G.E.W. Wolstenholme and Maeve O'Connor Ciba foundation.Symposia Churchill,London,1958
PGMS 29.0401

CHEMISTRY AND BIOLOGY OF PURINES Ciba foundation symposium on the chemistry and biology of purines Papers London 1956 May 8-10 Ciba foundation Edited by G.E.W. Wolstenholme and Cecilia M. O'Connor illus Churchill,London,1957
Bioch 33.0814

CHEMISTRY AND METABOLISM OF L- AND D- LACTIC ACIDS :a conference Papers New York 1964 Nov 12-14 New York academy of sciences Edited by Harold E. Whipple New York academy of sciences.Annals, 119,p.851-1165 New York, 1965
Bioch 33.0563

The CHEMISTRY AND MODE OF ACTION OF PLANT GROWTH SUBSTANCES :a symposium London 1955 Jul Edited by R.L. Wain and F. Wightman Butterworth scientific,London,1956
Radioth 35.0053

The CHEMISTRY AND MODE OF ACTION OF PLANT GROWTH SUBSTANCES :a symposium Proceedings Plant growth substances London 1955 Jul Edited by R.L. Wain and F. Wightman Butterworths,London,1956
BG 38.0165

CHEMISTRY AND PHYSIOLOGY OF THE NUCLEUS : symposium Proceedings Brookhaven 1951 Aug Held by the Brookhaven national laboratory.Biology department Experimental cell research.Supplement, 2 Academic press, New York,1952
Radioth 35.0459

CHEMISTRY OF CEMENT International symposium in the chemistry of cement 4th Proceedings Washington,D.C. 1960 Oct 2-7 Vol 1-2 National bureau of standards National bureau of standards.Monograph, 43, vol 1-2 2 vols U.S.government printing office,Washington,D.C.,1962
Eng 41.3026

CHEMISTRY OF CEMENT International symposium on the chemistry of cement 4th Proceedings Washington,D.C. 1960 Oct 2-7 Vol 1-2 National bureau of standards National bureau of standards.Monographs,43 2 vols Washington,D.C.,1962
Min 10.0647

CHEMISTRY OF COLOURING MATTER Recent advances in the chemistry of colouring matter : a symposium Report London 1956 Jan 19 Chemical society Chemical society.Special publication,4 Chemical society,London,1956 Organized by professor Braude
Chem 18.1214

The CHEMISTRY OF EXTENDED DEFECTS IN NON-METALLIC SOLIDS Proceedings Scotsdale,Ariz. 1969 Apr 16-26 Institute of advanced study on the chemistry of extended defects in non-metallic solids Edited by LeRoy Eyring and Michael O'Keefe North-Holland,Amsterdam,1970
Min 10.1545

CHEMISTRY OF NATURAL PRODUCTS The International symposium on the chemistry of natural products 5th Plenary lectures London 1968 Jul 8-13 International union of pure and applied chemistry.Division of organic chemistry and Chemical society Butterworths,London,1968 "The contents of this book appear in 'Pure and applied chemistry,vol.17, nos.3-4 (196819".Added title page in French
Bioch 33.1392

CHEMOTAXONOMY AND SEROTAXONOMY :a symposium Proceedings Birmingham 1967 Sep 15-16 Systematics association Edited by J.D. Hawkes Systematics association.Special volume, 2 illus xvi,299p Academic press,London;New York,1968
Bot 42.3449

CHEMOTHERAPEUTISCHE PROBLEME MALIGNER TUMOREN Kolloquium uber cytostatica Hemer 1958 Jul 28 Edited by F. Meythaler Symposien aktueller therapeutischer probleme, 1 Enke, Stuttgart,1958
Radioth 35.1082

CHEMOTHERAPY International congress of microbiology 6th Rome 1953 Vol 4: growth inhibition and chemotherapy Edited by H. Eagle and E.B. Chain Fondazione Emanuele Paterno,Rome,1953 Title also in Italian;text in English and French
Bioch 33.1803

CHEMOTHERAPY Strategy of chemotherapy :a symposium Papers London 1958 Apr Society for general microbiology Edited by S. T. Cowan and Elizabeth Rowatt Held at the Royal institution Society for general microbiology.Symposia, 8 Cambridge university press,Cambridge,1958
Bioch 33.1180

CHEMOTHERAPY OF CANCER International symposium on chemotherapy of cancer Proceedings Lugano 1964 Apr 28-May 1 Swiss academy of medical sciences Edited by Placidus A. Plattner Sponsored by Hoffmann-La Roche and co ltd. Elsevier,Amsterdam,1964
Pha 16.0335

CHEMOTHERAPY OF CANCER International symposium on the chemotherapy of cancer Proceedings Lugano 1964 Apr 28-May 1 Edited by Placidus Plattner Organised by the Swiss academy of medical sciences Elsevier, Amsterdam,1964 Sponsored by F.Hoffman-Laroche and co.ltd.
Radioth 35.0837

CHEMOTHERAPY OF FILARIASIS :a conference New York 1948 May 25 By L.L. Ashburn and others New York academy of sciences Edited by Roy Waldo Miner New York academy of sciences.Annals, 50,art.2 New York,1948
Mol 45.0271

CHESTER BEATTY RESEARCH INSTITUTE Cell electrophoresis :a symposium London 1963 May British biophysical society Edited by E.J. Ambrose illus Churchill,London,1965
Bal 39.0034

CHICAGO 1919 Pyrometry :a symposium American institute of mining and metallurgical engineers and National bureau of standards American institute of mining and metallurgical engineers,New York City,1920
Met 25.0294

CHICAGO 1935 Principles of heat treatment By M.A. Grossman American society for metals, Cleveland,Ohio,1953,reprinted 1955 A series of educational lectures...first presented to members of the ASM during the 17th National Metal Congress and Exposition, Chicago, 1935, and later extended to include the more recent developments
Met 25.0434

CHICAGO 1935 Principles of heat treatment By M.A. Grossman 3rd edition American society for metals,Cleveland,Ohio,1940 Lectures first presented to members of the A.S. M. during the 17th National Metal Congress and exposition,Chicago, 1935 and later extended to include the more recent developments
Met 25.0433

CHICAGO 1939 Age hardening of metals the symposium on precipitation hardening (age hardening) presented before the twenty-first annual convention Papers and discussions American society for metals American society for metals,Cleveland,1940
Met 25.1249

CHICAGO 1941 The Biological action of the vitamins :a symposium University of Chicago Edited by E.A. Evans University of Chicago press,Chicago,1942 Part of the joint Symposium on the respiratory enzymes and the biological action of vitamins arranged with the University of Wisconsin to celebrate the fiftieth anniversary of the University of Chicago.The papers on Respiratory enzymes were presented at Madison,1941,Sep 11-13
Bioch 33.0343

CHICAGO 1947 Copper and copper alloys By Owen W. Ellis American society for metals, Cleveland,Ohio,1948 A series of lectures on copper and copper alloys presented to members of the A.S.M. during the twenty-ninth National Metal Congress and Exposition, Chicago,October 18 to 24,1947
Met 25.0532

CHICAGO 1947 Fracturing of metals a seminar American society for metals and Case institute of technology American society for metals,Cleveland,Ohio,n.d. A seminar held during the 29th National metal congress and exposition
Met 25.0861

CHICAGO 1947 Introductory physical metallurgy By Clyde W. Mason American society for metals A.S.M.,Ohio,1947 A series of lectures...presented to members of the A.S.M.,during the twenty-ninth National metal congress and exposition,Chicago October 18 to 24,1
Met 25.0030

CHICAGO 1950 Atom movements American society for metals American society for metals,Cleveland,1951 A seminar held during the 32nd national metal congress and exposition
Met 25.1272

CHICAGO 1950 Atom movements :a seminar proceedings Sponsored by the American society for metals 240p American society for metals,Cleveland,Ohio,1951 Held during the 32nd National metal congress and exposition
Chem E 24.0757

CHICAGO 1950 Clinical ACTH conference 2nd Proceedings Vol 1-2 Edited by John R. Mote Sponsored by Armour and company 2 vols Churchill,London,1951
Bioch 33.0443

CHICAGO 1950 High temperature properties of metals By E.R. Parker and others American society for metals,Cleveland,Ohio, 1951 A series of five educational lectures presented to members of the A.S.M. during the 32nd National metal congress and exposition
Met 25.0978

CHICAGO 1950 Interpretation of tests and correlation with service By M.F. Garwood and others American society for metals,Cleveland, Ohio,1951 Lectures presented during the thirty-second National metal congress and exposition.
Met 25.0811

CHICAGO 1951 National congress of applied mechanics 1st proceedings Edited by Eli Sternberg United States national committee on theoretical and applied mechanics American society of mechanical engineers,New York,1952
A Math 4.0327

CHICAGO 1953 Low temperature test methods and standards for containers :a symposium National research council Edited by Earl C. Myers and Norbert J. Leiner Sponsored by the Quartermaster food and container institute illus vi,126p 23cm Washington,1954
Sco 14.0889

CHICAGO 1954 Effect of cyclic heating and shessing metals Symposium on effect of cyclic heating and stressing on metals at elevated temperatures American society for testing materials A.S.T.M.Special technical publication, 165 A.S.T.M.,Philadelphia,Pa., 1954
Met 25.0980

CHICAGO 1954 Symposium on effect of cyclic heating and stressing on metals at elevated temperatures American society for testing materials A.S.T.M.Special technical publication, 165 A.S.T.M.,Philadelphia,1954
Met 25.2534

CHICAGO 1956-57 A Symposium on molecular biology :a seminar series and a symposium Lectures University of Chicago Edited by Raymond E. Zirkle Under the auspices of the Chicago-Frankfurt inter-university program University of Chicago press,Chicago,1959
Bioch 33.0839

CHICAGO 1956-57 A Symposium on molecular biology :a seminar series and a symposium lectures University of Chicago Edited by Raymond E. Zirkle Under the auspices of the Chicago-Frankfurt inter-university program University of Chicago press,Chicago,1959
Bal 39.0617

CHICAGO 1957 A Symposium on uranium and uranium dioxide :melting,casting,forging, rolling,coldworking,welding of uranium: preparation and fabrication of uranium dioxide for fuel element use American institute of mining,metallurgical and petroleum engineers. Institute of metals division Nuclear metallurgy, 4 I.M.D.Special report series, 4 The metallurgical society of American institute of mining,metallurgical,and petroleum engineers,New York,1957
Met 25.1542

CHICAGO 1957 Liquid metals and solidification a seminar held during the 39th national metal congress and exposition American society for metals A.S.M.,Cleveland, Ohio,1958
Met 25.1510

CHICAGO 1959 Flat rolled products:rolling and treatment Technical conference 1st Proceedings American institute of mining, metallurgical and petroleum engineers. Mechanical working committee Edited by T.E. Dancy and E.L. Robinson Metallurgical society conferences, 1 Interscience,New York;London,1959
Met 25.0731

CHICAGO 1959 Medicine,a life-long study World conference on medical education 2 Proceedings World medical association British medical journal,London,1961
Med 36.0371

CHICAGO 1960 Progress in powder metallurgy Annual technical meeting 16th Proceedings Metal powder industries federation Metal powder industries federation,Chicago,1960
Met 25.0769

CHICAGO 1960 Rhenium :a symposium Electrochemical society.Electrothermics and metallurgy division Edited by B.W. Gonser Elsevier,Amsterdam;New York,1962
Met 25.0585

CHICAGO 1961 Fuel cells :a symposium 2nd Papers American chemical society. Divisions of fuel chemistry and petroleum chemistry Edited by G.J. Young v,225p Reinhold;Chapman and Hall,New York;London,1963 Held during the 140th national meeting of the American chemical society
Chem E 24.1401

CHICAGO 1961 The Progress of ALGOL in Europe By Peter Naur ALGOL bulletin. Supplement, 18 1961 A paper presented at the Computer applications symposium,Chicago, Oct 1961
Math L 5.0835

CHICAGO 1962 NASA-University conference on the science and technology of space exploration proceedings Vol 1-2 United States.National aeronautics and space administration.NASA SP, 11 NASA SP 11 2 vols office of scientific and technical information,Washington,D.C.,1962
A Math 4.1157

CHICAGO 1964 Fracture toughness testing and its applications :a symposium American society for testing materials American society for testing materials,Philadelphia,Pa., 1965
Met 25.0868

CHICAGO 1964 Lattice defects in quenched metals :an international conference Edited by R.M.J. Cotterill and others Sponsored by the United States atomic energy commission Academic press,New York;London,1965
Met 25.1240

CHICAGO 1964 Symposium on newer structural materials for aerospace vehicles American society for testing materials A.S.T.M. Special technical publication, 379 Philadelphia,Pa.,1965
Met 25.2516

CHICAGO 1965 A Symposium on non-equilibrium thermodynamics, variational techniques and stability University of Chicago General motors corporation.Research laboratories Universite libre de Bruxelles Edited by Russell J. Donnelly and I. Prigogine University of Chicago,Chicago,1966
A Math 4.0712

CHICAGO 1966 Fiber-strengthened metallic composites a symposium presented at the 1966 metals congress American society for testing materials A.S.T.M.Special technical publication, 427 A.S.T.M.,Philadelphia,1967
Met 25.2338

CHICAGO 1966 Harvesting the sun: photosynthesis in plant life :a symposium Edited by A. San Pietro and others Sponsored by the International minerals and chemical corporation Academic press,New York;London, 1967
Bioch 33.0789

CHICAGO 1966 Singular integrals :symposium Proceedings American mathematical society American mathematical society.Proceedings of symposia in pure mathematics, 10 Bibliog, port vi,375p 26cm American mathematical society,Providence,R.I.,1967
P Math 2.3036

CHICAGO 1967 Projective geometry conference Proceedings Bibliog v,162p 28cm University of Illinois at Chicago Circle,Chicago,1967 Mimeographed
P Math 2.3367

CHICAGO 1967 The Central nervous system and fish behavior :symposium Papers presented Edited by D. Ingle at the University of Chicago University of Chicago press,Chicago;London,1968 Sponsored jointly by the United States Air force office of scientific research, comparative physiology division of the American zoological society, and the academic journal Perspectives in biology and medicine
Bal 39.3266

CHICAGO 1968 Scanning electron microscopy 1968 Scanning electron microscope :the instrument and its applications Illinois institute of technology.Research institute Chicago,1968
Met 25.2869

CHICAGO,ILL 1956 National symposium on vacuum technology Transactions Committee on vacuum techniques Edited by Edmond S. Perry and John H. Durant Pergamon press, London,1957 Later symposia published by the American vacuum society
Met 25.1629

CHICAGO,ILL. 1965 Local atomic arrangements studied by x-ray diffraction :a symposium Proceedings American institute of mining,metallurgical and petroleum engineers Edited by J.B. Cohen and J.E. Hilliard Metallurgical society conferences, 36 Gordon and Breach,New York,1966
Met 25.1461

CHICAGO,ILL. 1919 Pyrometry The Papers and discussion of a symposium on pyrometry American institute of mining and metallurgical engineers National research council National bureau of standards illus 701p New York,1920
WSM 43.4920

CHICAGO,ILL. 1937 Symposium on lubricants American society for testing materials American society for testing materials, Philadelphia,Pa.,1937
Eng 41.6265

CHICAGO,ILL. 1939 Language and thought in schizophrenia :collected papers presented at a meeting of the American psychiatric association...and brought up to date Papers American psychiatric association Edited by J. S. Kasanin University of California press, Berkeley;Los Angeles,1954 Preface by N.D. C.Lewis
Psy 28.0082

CHICAGO,ILL. 1941 Visual mechanisms :a symposium Edited by Heinrich Kluver Biological symposia, 7 Cattell press, Lancaster,Pa.,1942
An 32.4752

CHICAGO,ILL. 1942 Frontiers in cytochemistry:the physical and chemical organization of the cytoplasm :a symposium Edited by Normand L. Hoerr Biological symposia, 10 Cattell press,Lancaster,Pa., 1943 Held in honour of the 75th birthday of Dr. R.R.Bensley
An 32.3370

CHICAGO,ILL. 1947 Fracturing of metals :a seminar...held during the 29th National metal congress and exposition...Co-ordinated by... the metallurgical staff of the Case institute of technology under the direction of George Sachs Sponsored by the American society for metals American society for metals,Cleveland, Ohio,1948
Eng 41.3732

CHICAGO,ILL. 1949 Genetic neurology International conference on the development, growth and regeneration of the nervous system Edited by Paul Weiss Sponsored by the International union of biological sciences University of Chicago,Ill.,Chicago,Ill.,1950
An 32.4329

CHICAGO,ILL. 1949 Genetic neurology : problems of the development,growth,and regeneration of the nervous system and of its functions:an international conference Essays International union of biological sciences Edited by Paul Weiss Subsidized by Unesco University of Chicago press,Chicago,Ill.,1950
Psy 31.0170

CHICAGO,ILL. 1950 Interpretation of tests and correlation with service :a series of four educational lectures...presented during the thirty-second National metal congress and exposition By M.F. Garwood and others American society for metals American society for metals,Cleveland,Ohio,1951
Eng 41.3837

CHICAGO,ILL. 1956 Conference on radioiodine Proceedings Argonne cancer research hospital United States atomic energy commission Edited by D.E. Clark ACRH-100 University of Chicago,Chicago,Ill., 1956
Radioth 35.1076

CHICAGO,ILL. 1956 Ozone chemistry and technology :international ozone conference Proceedings Armour research foundation American chemical society.Advances in chemistry series,21 American chemical society,Washington,D.C.,1959 Edited by the staff of the American chemical society
Chem 18.0031

CHICAGO,ILL. 1956-57 A Symposium on molecular biology Edited by Raymond E. Zirkle Held under the auspices of the Chicago-Frankfurt inter-university program illus. University of Chicago press,Chicago, Ill.,1959
Gen 34.0647

CHICAGO,ILL. 1957 Computer applications symposium 4th proceedings Sponsored by Armour research foundation illus. x,126p 23cm Armour research foundation of Illinois institute of technology,Chicago,1958 The Armour research foundation was in 1963 renamed IIT research institute
Math L 5.0853

CHICAGO,ILL. 1958 Condensing heat transfer within horizontal tubes National heat transfer conference 2nd Preprints American society of mechanical engineers American society of mechanical engineers, Philadelphia,Pa.,1958
Eng 41.7155

CHICAGO,ILL. 1959 Current issues in the philosophy of science :symposia of scientists and philosophers Proceedings American association for the advancement of science Edited by Herbert Feigl and Grover Maxwell xi,484p Holt,Rinehart and Winston,New York, 1961 Proceedings of section L of the Association
WSM 43.0784

CHICAGO,ILL. 1959 Great Lakes basin Syposium American association for the advancement of science Edited by Howard J. Pincus American association for the advancement of science.Publications,71 American association for the advancement of science,Washington,D.C.,1962
Geog 13.5045

CHICAGO,ILL. 1959 Self-organizing systems : an interdisciplinary conference Proceedings Edited by Marshall C. Yovits and Scott Cameron Sponsored by the United States.Office of naval research.Information systems branch International tracts in computer science and technology and their application, 2 Pergamon press,New York,1960
Eng 41.1162

CHICAGO,ILL. 1960 Industrialization and society Conference on social implications of industrialization and technical change Papers Edited by Bert F. Hoselitz and Wilbert E. Moore Sponsored by Unesco Unesco,Mouton,1963
Eng 41.1052

CHICAGO,ILL. 1960 Industrialization and society Conference on the social implications of industrialization and technological change Papers University of Chicago Edited by Bert F. Hoselitz and Wilbert E. Moore Sponsored jointly by Unesco Mouton,The Hague,1963
Geog 13.1608

CHICAGO,ILL. 1960 Response of the nervous system to ionizing radiation :an international symposium Proceedings Edited by Thomas J. Haley and Ray S. Snider Academic press,New York;London,1962
Radioth 35.1730

CHICAGO,ILL. 1960 Response of the nervous system to ionizing radiation :an international symposium Proceedings Edited by Thomas J. Hayley and Ray S. Snider Academic press,New York,1902
An 32.4191

CHICAGO,ILL. 1961 Computer applications - 1961 Computer applications symposium 8th proceedings Edited by Robert S. Hollitch and Benjamin Mittman Sponsored by Armour research foundation illus. vii,198p 23cm McMillan,New York
Math L 5.0854

CHICAGO,ILL. 1961 Photoelasticity :an international symposium Proceedings United States.Army.Research office Edited by M.M. Frocht Pergamon press,London,1963
Eng 41.2913

CHICAGO,ILL. 1962 Flat rolled products III Technical conference 4th Proceedings American institute of mining,metallurgical and petroleum engineers.Mechanical working committee Metallurgical society conferences, 16 Interscience,New York;London,n.d.
Met 25.0703

CHICAGO,ILL. 1962 Self-organizing systems 1962 conference proceedings United States. Office of naval research Edited by Marshall C. Yovits and others Jointly sponsored by Armour research foundation ix,563p 24cm Spartan books,Washington,D.C.,1962
Math L 5.0992

CHICAGO,ILL. 1963 Evolution of the atherosclerotic plaque :an international symposium Presentations Chicago heart association and American heart association. Council on arteriosclerosis Edited by Richard J. Jones University of Chicago press, Chicago,Ill.;London,1963
Path 30.2350

CHICAGO,ILL. 1964 Fracture toughness testing and its applications :a symposium presented at the 67th annual meeting A.S.T.M. American society for testing and materials American society for testing and materials. Special technical publication, 381
Eng 41.3799

CHICAGO,ILL. 1964 Internal friction, damping and cyclic plasticity... :a symposium presented at the 67th annual meeting,A.S.T.M. American society for testing and materials American society for testing and materials. Special technical publication, 378
Eng 41.3795

CHICAGO,ILL. 1964 Isotopes in experimental pharmacology International conference on the uses of isotopically labeled drugs in experimental pharmacology Lectures Edited by Lloyd J. Roth University of Chicago press, Chicago,Ill.,1965
Pha 16.0155

CHICAGO,ILL. 1964 Mechanical working of steel 2 The Technical conference 6th Proceedings American institute of mining, metallurgical and petroleum engineers. Mechanical working and steel processing committee Edited by T.G. Bradbury Sponsored by the Chicago section Metallurgical society conferences, 26 Gordon and Breach,New York,1965
Met 25.2244

CHICAGO,ILL. 1965 High-temperature high-resolution metallography :a symposium Proceedings American institute of mining, metallurgical and petroleum engineers.Ferrous metallurgy committee Edited by Hubert I. Aaronson and George S. Ansell Metallurgical society conferences, 38 Gordon and Breach, New York,1967
Met 25.2361

CHICAGO,ILL. 1965 Non-equilibrium thermodynamics variational techniques and stability :a symposium Proceedings Edited by Russell J. Donnelly and others University of Chicago press,Chicago,Ill.;London,1966
Eng 41.7117

CHICAGO,ILL. 1965 Symposium on cavitation in fluid machinery American society of mechanical engineers Edited by Glenn M. Wood and others American society of mechanical engineers,New York,1965
Eng 41.6538

CHICAGO,ILL. 1966 Chicago conference Standardization in human cytogenetics Sponsored by the National foundation-March of dimes Birth defects:original article series, 2,no.2 National foundation -March of dimes, New York,1966
Gen 34.0895

CHICAGO,ILL. 1966 Harvesting the sun: photosynthesis in plant life :a symposium Papers Edited by Anthony San Pietro and others Sponsored by the International minerals and chemical corporation illus. ix,342p Academic press,New York;London,1967
Bot 42.1899

CHICAGO,ILL. 1966 International congress of human genetics :plenary sessions and symposia 3rd Proceedings Edited by James F. Crow and James V. Neel John Hopkins press,Baltimore,Md.,1967
PGMS 29.0264

CHICAGO,ILL. 1966 International congress on human genetics 3rd Proceedings plenary sessions and symposia Edited by James F. Crow and James V. Neel Johns Hopkins press,Baltimore,Md.,1967
An 32.0057

CHICAGO,ILL. 1966 Malignant transformation by viruses :a symposium Edited by W.H. Kirsten Recent advances in cancer research, 6 Springer,Berlin,1966
Radioth 35.0856

CHICAGO,ILL. 1966 The International congress of human genetics 3rd Plenary sessions and symposia Edited by James F. Crow and James V. Neel Hopkins,Baltimore,Md., 1967
Gen 34.1858

CHICAGO,ILL. 1966 Work hardening :based on a symposium American institute of mining, metallurgical and petroleum engineers. Institute of metals division.Chemistry and physics committee Edited by J.P. Hirth and J. Weertman Metallurgical society conferences, 46 Gordon and Breach,New York,1968
Met 25.2866

CHICAGO,ILL. 1967 Advances in fluidics Fluidics symposium American society of mechanical engineers Edited by Forbes T. Brown American society of mechanical engineers,New York,1967
Eng 41.5898

CHICAGO,ILL. 1967 Stereology International congress for stereology 2nd Proceedings International society for stereology Edited by Hans Elias Sponsored by the National science foundation Springer, Berlin,1967
An 32.0107

CHICAGO,ILL. 1967 The Central nervous system and fish behavior :a conference Papers United States.Air force.Office of scientific research American zoological society.Comparative physiology division Edited by David Ingle Jointly sponsored by Perspectives in biology and medicine illus University of Chicago press,Chicago,Ill.; London,1968
Psy 31.2816

CHICAGO,ILL. 1968 Functional analysis and related fields :a conference held in honour of Professor Marshall Harvey Stone Proceedings Edited by Felix E. Browder Supported by United States.Air Force.Office of scientific research port. 241p 24cm Springer, Berlin,1970
P Math 2.3863

CHICAGO,ILL. 1968 Nonlinear functional analysis :a symposium Proceedings American mathematical society Edited by Felix E. Browder American mathematical society. Proceedings of symposia in pure mathematics, 18,part 1 bibliog. v,296p 26cm American mathematical society,Providence,R.I., 1970
P Math 2.3864

CHICAGO,ILL. 1968 Normal and malignant cell growth :symposium Proceedings Edited by R.J.M. Fry and others Sponsored by the University of Chicago.Cancer training grant Recent results in cancer research, 17 Springer,New York,1969
Radioth 35.0865

CHICAGO,ILL. 1968-69 Scanning electron microscopy symposia 1st-2nd Proceedings Sponsored by the Illinois institute of technology.Research institute 2cm I.I.T., Chicago,Ill.,1968-69
Eng 41.5623

CHICAGO,ILL. 1969 Late effects of radiation :a colloquium Proceedings Edited by R.J.M. Fry and others Taylor and Francis, London,1970
Radioth 35.1221

CHICAGO,ILL. 1969 Scanning electron microscope 1969 Annual scanning electron microscope symposium 2nd Illinois institute of technology.Research institute Chicago,c1969
Met 25.2870

CHICAGO,ILL. 1970 Scanning electron microscopy 1970 Scanning electron microscope symposium 3rd Proceedings Sponsored by Illinois institute of technology.Research institute I.I.T.research institute,Chicago, Ill.,1970
Eng 41.8204

CHICAGO CONFERENCE Standardization in human cytogenetics Chicago,Ill. 1966 Sep 3-4,10 Sponsored by the National foundation-March of dimes Birth defects:original article series, 2,no.2 National foundation -March of dimes, New York,1966
Gen 34.0895

CHICAGO-FRANKFURT INTER-UNIVERSITY PROGRAM A Symposium on molecular biology Chicago,Ill. 1956-57 Nov-Mar Edited by Raymond E. Zirkle illus. University of Chicago press,Chicago, Ill.,1959
Gen 34.0647

CHICAGO-FRANKFURT INTER-UNIVERSITY PROGRAM A Symposium on molecular biology :a seminar series and a symposium Lectures Chicago 1956-57 Nov-Mar University of Chicago Edited by Raymond E. Zirkle University of Chicago press,Chicago,1959
Bioch 33.0839

CHICAGO-FRANKFURT INTER-UNIVERSITY PROGRAM A Symposium on molecular biology :a seminar series and a symposium lectures Chicago 1956-57 Nov-Mar University of Chicago Edited by Raymond E. Zirkle University of Chicago press,Chicago,1959
Bal 39.0617

CHICAGO HEART ASSOCIATION Evolution of the atherosclerotic plaque :an international symposium Presentations Chicago,Ill. 1963 Mar 28-29 Edited by Richard J. Jones University of Chicago press,Chicago,Ill.; London,1963
Path 30.2350

The CHICK EMBRYO IN BIOLOGICAL RESEARCH :a conference Papers New York 1951 Dec 7-8 Edited by Roy Waldo Miner and D.A. Karnofsky Held by the New York academy of sciences.Section of biology New York academy of sciences.Annals, 55,art.2 New York academy of sciences,New York,1952
An 32.2265

CHIHUAHUA Memoria geologico-minera del estado de Chihuahua International geological congress 20th papers Mexico City 1956 By Jenaro Gonzalez Reyna Mexico City,1956 Bound with two other publications of the congress
Geol 8.3029

CHILD AND EDUCATION International congress of applied psychology 14th Proceedings Copenhagen 1961 Aug 13-19 Vol 3 International association of applied psychology and Danish psychological association Edited by Gerhard S. Nielsen Munksgaard,Copenhagen,1962
Psy 31.1198

CHILD DEVELOPMENT Discussions on child development :consideration of the biological, psychological and cultural approaches to the understanding of human development and behaviour :meeting Geneva 1956 Vol 4 By Jean Piaget and others World health organization.Study group on the psychobiological development of the child Edited by J.M. Tanner and Barbel Inhelder Tavistock publications,London,1960
Psy 28.0224

CHILEAN NATIONAL COMMITTEE FOR ANTARCTIC RESEARCH SCAR,SCOR,IAPO,IUBS symposium on Antarctic oceanography Papers Santiago de Chile 1966 Sep 13-16 International council of scientific unions.Scientific committee on Antarctic research International council of scientific unions.Scientific committee on oceanic research International association of physical oceanography International union of biological sciences illus,maps xi,268p 23cm Scott Polar research institute for SCAR, Cambridge,1968
Sco 14.8300

CHILTON,BERKS. 1969 Conference on radiation protection in accelerator environments Proceedings Edited by G.R. Stevenson Sponsored by the Science research council Rutherford laboratory,Chilton,1969 Organized by and held at the Rutherford laboratory
Radioth 35.1393

CHIMIE ET BIOCHIMIE DES CHAMPIGNONS INFERIEURS ET DES LEVURES The Chemistry and biochemistry of fungi and yeasts :a symposium Proceedings Dublin 1963 Jul 18-20 International union of pure and applied chemistry sponsored by the International union of pure and applied chemistry and Royal Irish academy.Irish national committee for chemistry Butterworths,London,1963 Reprinted from 'Pure and applied chemistry', vol.7,no.,4.Added title-page in French:Chimie et biochimie des champignons inferieurs et des levures
Bioch 33.1146

CHINESE SOCIETY OF PATHOLOGY AND MICROBIOLGY Proceedings Pathology and microbiology Canton 1935 Nov 5-8 Chinese medical journal.Supplement, 1 Chinese medical journal,Peiping,1935
An 32.1220

CHISA 69 International congress of chemical engineering,chemical equipment construction and automation 3rd Lecture summaries Marienbad 1969 Sep Pt A-H Scientific committee Chisa '69,1969 Cover title: Chisa '69
Chem E 24.1872

CHLOROPLASTS Biochemistry of chloroplasts :a Nato advanced study institute Proceedings Aberystwyth 1965 Aug Vol 1-2 North Atlantic treaty organization Edited by T.W. Goodwin 2 vols Academic press,London;New York,1966-67
Bot 42.1429

CHLORPROPAMIDE AND DIABETES MELLITUS :a conference Papers New York 1958 Sep 25-27 By Martin G. Goldner and others New York academy of sciences New York academy of sciences.Annals, 74,p.407-1028 New York, 1959
Bioch 33.0419

Le CHOIX DES SITES D'OBSERVATOIRES ASTRONOMIQUES (SITE TESTING) :a symposium proceedings Rome 1962 Oct 1-2 Edited by J. Rosch and others International astronomical union. Symposium, 19 382p Gauthier-Villars,Paris, 1964 Reprinted from 'Bulletin astronomique de l'observatoire de Paris, vol 24, fasc.2 et 3, 1963
Obs 6.1467

CHOLINERGIC MECHANISMS :a conference New York 1966 May 19-21 Edited by S. Ehrenpreis New York academy of sciences. Annals, 144,art.2 New York academy of sciences,New York,1967
An 32.5248

CHOLINERGIC MECHANISMS :a conference Papers New York 1966 May 19-21 Edited by S. Ehrenpreis New York academy of sciences. Annals,144,art 2 New York academy of sciences,New York,1967
Pha 16.0311

CHOLINERGIC MECHANISMS :a conference Papers New York 1966 May 19-21 New York academy of sciences Edited by Edward M. Weyer New York academy of sciences.Annals, 144,p. 383-936 New York,1967
Bioch 33.0564

CHORIOCARCINOMA :a conference Transactions Baguio 1965 Jan International union against cancer Edited by James F. Holland and Myroslaw M. Hreshchyshya International union against cancer.Monograph series, 3 Springer,Berlin,1967
Radioth 35.0886

CHRISTIANA 1914 Tableau synoptique des dispositions adoptees et reservees Conference internationale du Spitsbergen 35p 33cm Grondahl,Christiana,1914
Sco 14.2859

CHRISTIANIA 1856 Skandinaviske naturforskeres syvende mode Forhandlingar Skandinaviske naturforskere Werner, Christiania,1857 Includes two papers by J. Wolley
Bal 44.1980

CHROMATIC DISCRIMINATION IN ANIMALS AND MAN Mechanisms of colour discrimination :an international symposium on the fundamental mechanisms of the chromatic discrimination in animals and man Proceedings Paris 1958 Jul 25-29 International council of scientific unions Pergamon press:symposium publications division,London,1960
Psy 31.0384

CHROMATOGRAPHY Gas chromatography :a symposium 2nd Proceedings Amsterdam 1958 May 19-23 Institute of petroleum.Gas chromatography discussion group Edited by D. H. Desty Also under the auspices of the Nederlandse chemische vereniging xiii,383p Butterworths,London,1958
Chem E 24.0849

CHROMATOGRAPHY Gas chromatography :a symposium 3rd Proceedings Edinburgh 1960 Jun 8-10 Institute of petroleum.Gas chromatography discussion group Edited by R. P.W. Scott Organized also by the Society for physical chemistry xvii,466p Butterworths, London,1960
Chem E 24.0850

CHROMATOGRAPHY Gas chromatography : international symposium 2nd Proceedings Ann Arbor 1959 Jun 10-13 Edited by Henry J. Noebels and others Held under the auspices of the Instrument society of America xvi, 463p Academic press,New York;London,1961
Chem E 24.0851

CHROMATOGRAPHY Partition chromatography :a symposium London 1948 Oct 30 Biochemical society Edited by R.T. Williams and R.L.M. Synge Held at the London school of hygiene and tropical medicine Biochemical society. Symposia, 3 Cambridge university press, Cambridge,1949
Bioch 33.1365

CHROMATOGRAPHY Vapour phase chromatography : a symposium Proceedings London 1956 May 30-Jun 1 Edited by D.H. Desty and C.L.A. Harbourn Sponsored by the Institute of petroleum.Hydrocarbon research group xv,435p Butterworths,London,1957
Chem E 24.0845

CHROMATOGRAPHY :a conference Papers New York 1946 Nov 29-30 By Harold G, Cassidy and others New York academy of sciences. Section of physics and chemistry New York academy of sciences.Annals, 49,p.141-326 New York,1948
Bioch 33.2362

CHROMATOGRAPHY Conference papers 1946 Nov 29-30 By Harold G. Cassidy and others New York academy of sciences.Section of physics and chemistry New York academy of sciences. Annals, 49;2 New York academy of sciences, New York,1948
Chem 18.2636

CHROMOSOME International seminar on chromosome :its structure and function Proceedings Calcutta 1968 Aug 11-13 University of Calcutta Ramakrisha mission institute of culture University grants commission Edited by Arunkumar Sharma and Archana Sharma Nucleus,1968,Supplementary volume Nucleus,Calcutta,1968
Gen 34.2173

CHROMOSOME:STRUCTURAL AND FUNCTIONAL ASPECTS Tissue culture association symposia 5th Miami,Fla. 1965 Jun 1-4 Edited by George Yerganian illus. xiii,107p 26cm Tissue cultural association,n.p.,1965
Bot 42.4734

The CHROMOSOME:STRUCTURAL AND FUNCTIONAL ASPECTS a symposium Miami,Fla. 1965 Jun 1-4 Edited by Clyde J. Dawe Tissue culture association.Publications, 1 Tissue culture association,1965
An 32.3302

The CHROMOSOME:STRUCTURAL AND FUNCTIONAL ASPECTS a symposium Miami,Fla. 1965 Jun 1-4 Tissue culture association Edited by Clyde Johnson-Dawe Williams and Wilkins,Baltimore, Md.,1965 Organized by George Yerganian
Gen 34.0783

The CHROMOSOME;STRUCTURAL AND FUNCTIONAL ASPECTS a symposium Miami 1965 Jun 1-4 Vol 1 Tissue culture association Edited by Clyde Johnson Dawe Williams and Wilkins,Baltimore,Md.,1966 Organized by George Yerganian
Bal 39.0339

CHROMOSOME BREAKAGE Symposium on chromosome breakage Contributions London 1952 Jun 9-11 By C.D. Darlington and others John Innes horticultural institution Heredity, 5,supplement Oliver and Boyd,London; Edinburgh,1953
Path 30.0219

CHROMOSOME MANIPULATIONS AND PLANT GENETICS : a symposium held during the 10th international botanical congress Edinburgh 1964 Aug Edited by Ralph Riley and K.R. Lewis Oliver and Boyd,Edinburgh,1966
Bot 42.0789

CHROMOSOME MANIPULATIONS AND PLANT GENETICS : a symposium held during the 10th international botanical congress Edinburgh 1964 Aug Edited by Ralph Riley and K.R. Lewis Oliver and Boyd,Edinburgh;London,1966
Gen 34.0894

CHROMOSOME MECHANICS AT THE MOLECULAR LEVEL Symposium on chromosome mechanics at the molecular level :given at research conference for biology and medicine of the atomic energy commission Gatlinburg,Tenn. 1967 Apr 10-13 Sponsored by the Oak Ridge national laboratory. Biology division Oak Ridge national laboratory.Symposia, 20 Journal of cellular physiology, 70,suppl Wistar institute of anatomy and biology,Philadelphia,Pa.,1967
Gen 34.0846

CHROMOSOMES Chromosome:structural and functional aspects Tissue culture association symposia 5th Miami,Fla. 1965 Jun 1-4 Edited by George Yerganian illus. xiii,107p 26cm Tissue cultural association,n.p.,1965
Bot 42.4734

CHROMOSOMES Chromosomes today The Oxford chromosome conference 1st Proceedings Oxford 1964 Jul 28-31 Edited by C.D. Darlington and K.R. Lewis Oliver and Boyd, Edinburgh;London,1966
Mol 45.0190

CHROMOSOMES Conference on chromosomes Lectures Wageningen 1956 Apr 16-19 Willink,Zwolle,1956 Papers in English and German
An 32.3379

CHROMOSOMES International symposium on genes and chromosomes:structure and function Proceedings Buenos Aires 1964 Nov 30-Dec 4 National cancer institute Edited by Juan I. Valencia and Rhoda F. Grell National cancer institute.Monograph, 18 National cancer institute,Bethesda,Md.,1965
Bioch 33.0987

CHROMOSOMES Radiation-induced chromosome aberrations :report of a conference on biochemical and biophysical mechanisms in the production of radiation-induced chromosome aberrations San Juan,P.R. 1961 Nov 16-18 Edited by Sheldon Wolff Sponsored by the National research council illus. xii,304p Columbia university press,New York;London,1963
Bot 42.4730

CHROMOSOMES The Role of chromosomes in development :a symposium Amherst,Mass. 1964 Jun Society for the study of development and growth Edited by Michael Locke Society for the study of development and growth.Symposia, 23 Academic press,New York;London,1964 The Society for the study of development and growth known later as the Society for developmental biology
Bioch 33.1011

CHROMOSOMES TODAY The Oxford chromosome conference 1st Proceedings Oxford 1964 Jul 28-31 Edited by C.D. Darlington and K.R. Lewis Oliver and Boyd,Edinburgh;London, 1966
Mol 45.0190

CHROMOSOMES TODAY :Oxford chromosome conference Proceedings Oxford 1964 Jul 28-31 and Oxford 1967 Sep 5-8 Vol 1-2 Edited by C.D. Darlington and K.R. Lewis Heredity.Supplement series, 19,24 illus Oliver and Boyd,Edinburgh,1966-68
Bot 42.0788

CHROMOSOMES TODAY :Oxford chromosome conference Proceedings Oxford 1964 Jun 28-31 Oxford 1967 Sep 5-8 Vol 1-2 Edited by C.D. Darlington and K.R. Lewis illus. 2 vols Oliver and Boyd, Edinburgh.London,1966-68
Gen 34.0893

CHROMOSOMES TODAY 2 Oxford chromosome conference 2nd Proceedings Oxford 1967 Sep 5-8 Edited by C.D. Darlington and K. R. Lewis Oliver and Boyd,Edinburgh,1969
Gen 34.2179

CIBA Protein metabolism :influence of growth hormone anabolic steroids, and nutrition in health and disease :an international symposium Leyden 1962 Jun 25-29 Edited by F. Gross Springer,Berlin;Gottingen,1967 Chairman A. Querido
PGMS 29.0406

CIBA Shock pathogenesis and therapy :an international symposium Stockholm Jun 27-30 Edited by K.D. Bock Springer-verlag, Berlin,1962
PGMS 29.0080

CIBA,HORSHAM The Investigation of hypothalamic-pituitary-adrenal function :a symposium Proceedings London 1967 Feb 22-23 Society for endocrinology Royal society of medicine.Endocrine section Edited by V.H. T. James and J. Landon Society for endocrinology.Memoirs, 17 Cambridge university press,Cambridge,1968
Gen 34.0558

CIBA FOUNDATION Adrenergic mechanisms London 1960 Mar 28-29 Edited by J.R. Vane and others Ciba foundation.Symposia Churchill,London,1960
PGMS 29.0379

CIBA FOUNDATION Adrenergic mechanisms :Ciba foundation symposium London 1960 Mar 28-31 Edited by J.R. Vane and others Churchill, London,1960
Pha 16.0070

CIBA FOUNDATION Adrenergic mechanisms :a Ciba foundation symposium London 1960 Mar 28-31 Edited by J.R. Vane and others Churchill,London,1960
Phys 20.0972

CIBA FOUNDATION Adrenergic neurotransmission Edited by G.E.W. Wolstenholme and Maeve O'Connor Ciba foundation study group, 33 Churchill,London,1968
An 32.5042

CIBA FOUNDATION Adrenergic neurotransmission Ciba foundation study group London Edited by G.E.W. Wolstenholme and Maeve O'Connor Ciba foundation study group, 33 Churchill, London,1968
Phys 20.2050

CIBA FOUNDATION Adrenergic neurotransmission Ciba foundation study group 33rd London Edited by G.E.W. Wolstenholme and Maeve O'Connor Churchill,London,1968
Pha 16.0075

CIBA FOUNDATION Aetiology of diabetes mellitus and its complications :colloquium London 1963 Oct 8-10 Edited by Margaret P. Cameron and Maeve O'Connor Ciba foundation colloquia on endocrinology, 15 Churchill, London,1964
Phys 20.1406

CIBA FOUNDATION Animal behaviour and drug action :annual meeting Proceedings London 1963 Mar 25-28 Edited by A.V.S.de Reuck and Julie Knight Biological council.Co-ordinating committee for symposia on drug action.Annual meeting, 10 Ciba foundation symposia Churchill,London,1964
Psy 31.0526

CIBA FOUNDATION Anterior pituitary secretion and Hormonal influences in water metabolism : two colloquia Proceedings London 1951 Jul 9-13 and London 1951 Jan 8-10 Edited by G.E.W. Wolstenholme and Margaret P. Cameron Ciba foundation colloquia on endocrinology, 4 illus Churchill,London, 1952
Bioch 33.0434

CIBA FOUNDATION Anterior pituitary secretion and hormonal influences in water metabolism : two colloquia London 1951 Jul 9-13 London 1951 Jan 8-10 Edited by G.E.W. Wolstenholme and Margaret P. Cameron Ciba foundation colloquia on endocrinology, 4 Churhcill,London,1952
Phys 20.1365

CIBA FOUNDATION Bacterial episomes and plasmids :a Ciba foundation symposium Proceedings London 1968 Sep 9-11 Edited by G.E.W. Wolstenholme and M. O'Connor Churchill,London,1969
Bioch 33.1118

CIBA FOUNDATION Bilharziasis :a Ciba foundation symposium Proceedings Cairo 1962 Mar 18-22 Edited by G.E.W. Wolstenholme and Maeve O'Connor illus. Churchill,London, 1962 Held in commemoration of Theodor Maximilian Bilharz
Mol 45.0266

CIBA FOUNDATION Bioassay of anterior pituitary and adrenocortical hormones :a colloquium proceedings London 1952 Mar 25-27 Edited by G.E.W. Wolstenholme Ciba foundation colloquia on endocrinology, 5 illus Churchill,London,1953
Bioch 33.0435

CIBA FOUNDATION Biochemistry of human genetics :Ciba foundation symposium London 1959 May 13-16 Edited by G.E.W. Wolstenholme and Cecilia M. O'Connor Churchill,London, 1959
An 32.0371

CIBA FOUNDATION Biochemistry of human genetics symposium Edited by G.E.W. Wolstenholme and Cecilia M. O'Connor Jointly with the International union of biological sciences Ciba foundation.Symposia Churchill,London,1959
Gen 34.0602

CIBA FOUNDATION Cancer of the cervix : diagnosis of early forms London 1959 May 8 Edited by G.E.W. Wolstenholme and Maeve O'Connor Ciba foundation study group, 3 Churchill,London,1959
Radioth 35.1960

CIBA FOUNDATION Caries-resistant teeth :Ciba foundation symposium Edited by G.E.W. Wolstenholme and M. O'Connor Churchill, London,1965
Phys 20.2131

CIBA FOUNDATION Cell differentiation :a Ciba foundation symposium London 1967 Jan 31-Feb 2 Edited by A.V.S. De Reuck and Julie Knight Churchill,London,1967
An 32.2531

CIBA FOUNDATION Cell differentiation :a Ciba foundation symposium London 1967 Jan 31-Feb 2 Edited by A.V.S. De Reuck and Julie Knight Churchill,London,1967
Phys 20.1511

CIBA FOUNDATION Cell differentiation :a Ciba foundation symposium London 1967 Jan 31-Feb 2 Edited by A.V.S. De Reuck and Julie Knight Ciba foundation.Symposia Churchill, London,1967
Bal 39.0317

CIBA FOUNDATION Cell differentiation :a Ciba foundation symposium Proceedings London 1967 Jan 31-Feb 2 Edited by A.V.S. De Reuck and Julie Knight Churchill,London,1967
Bioch 33.0889

CIBA FOUNDATION Cellular biology of myxovirus infections London 1964 Feb 4-6 Edited by G.E.W. Wolstenholme and Julie Knight Ciba foundation.Symposia Churchill,London, 1964
Gen 34.0727

CIBA FOUNDATION Cellular injury :a Ciba foundation symposium Papers London 1963 Jul 2-4 Edited by A.V.S. De Reuck and Julie Knight Churchill,London,1964
An 32.5317

CIBA FOUNDATION Chemistry and biology of mucopolysaccharides London 1967 Apr 23-25 Edited by G.E.W. Wolstenholme and Maeve O'Connor Ciba foundation.Symposia Churchill,London,1958
PGMS 29.0401

CIBA FOUNDATION Ciba foundation conference on isotopes in biochemistry London 1951 Mar 12-15 Edited by G.E.W. Wolstenholme Illus J.and A.Churchill,London,1951
Radioth 35.0023

CIBA FOUNDATION Ciba foundation conference on isotopes in biochemistry Papers London 1951 Mar 12-15 Edited by J.N. Davidson Churchill,London,1951
Bioch 33.0274

CIBA FOUNDATION Ciba foundation ssymposium on the chemistry and biology of mucopolysaccharides Proceedings London 1957 Apr 23-25 Edited by G.E.W. Wolstenholme and Maeve O'Connor Churchill,London,1958
Bioch 33.1202

CIBA FOUNDATION Ciba foundation symposium... on sensory function :touch,heat and pain 2nd Proceedings London 1965 Sep 21-23 Edited by A.V.S. de Reuck and Julie Knight J. and A.Churchill,London,1966
Psy 31.0286

CIBA FOUNDATION Ciba foundation symposium on amino acids and peptides with antimetabolic activity London 1958 Mar 18-20 Edited by G.E.W. Wolstenholme and Cecilia M. O'Connor Ciba foundation symposia Churchill,London, 1958
Chem 18.1523

CIBA FOUNDATION Ciba foundation symposium on amino acids and peptides with antimetabolic activity Proceedings London 1958 Mar 18-20 Edited by G.E.W. Wolstenholme and Cecilia M. O'Connor illus Churchill,London,1958
Bioch 33.0570

CIBA FOUNDATION Ciba foundation symposium on biochemistry of human genetics Proceedings Naples 1959 May 13-16 Edited by G.E.W. Wolstenholme and Cecilia M. O'Connor illus Churchill,London,1959
Bioch 33.0890

CIBA FOUNDATION Ciba foundation symposium on carbon fluorine compounds :chemistry biochemistry and biological London 1971 Sep 13-15 Edited by Katherine Elliott and John Birch Elsevier,Amsterdam,1972
Bioch 33.2244

CIBA FOUNDATION Ciba foundation symposium on carcinogenesis:mechanisms of action London 1958 Jun 24-26 Edited by G.E.W. Wolstenholme and Maeve O'Connor Illus Churchill,London, 1959
Radioth 35.0107

CIBA FOUNDATION Ciba foundation symposium on cellular aspects of immunity Proceedings Royaumont 1959 Jun 3-5 Edited by G.E.W. Wolstenholme and Maeve O'Connor Churchill, London,1960
Bioch 33.1117

CIBA FOUNDATION Ciba foundation symposium on chemistry and biology of pteridines Papers London 1954 Mar 22-26 Edited by G.E.W. Wolstenholme and Margaret P. Cameron Ciba foundation symposia J. and A. Churchill, London,1954
Chem 18.1327

CIBA FOUNDATION Ciba foundation symposium on haemopoiesis :cell production and its regulation London 1960 Feb 2-4 Edited by G.E.W. Wolstenholme and Maeve O'Connor Churchill,London,1960
An 32.3166

CIBA FOUNDATION Ciba foundation symposium on ionizing radiations and cell metabolism Proceedings London 1956 Mar 6-9 Edited by G.E.W. Wolstenholme and C.M. O'Connor illus. Churchill,London,1956
Radioth 35.1355

CIBA FOUNDATION Ciba foundation symposium on leukaemia research London 1953 Nov 16-19 Edited by G.E.W. Wolstenholme and Margaret P. Cameron Ciba foundation.Symposia Churchill, London,1954
Radioth 35.0800

CIBA FOUNDATION Ciba foundation symposium on paper electrophoresis Proceedings London 1955 Jul 27-29 Edited by G.E.W. Wolstenholme and Elaine C.P. Millar Ciba foundation symposia Churchill,London,1956
Chem 18.1445

CIBA FOUNDATION Ciba foundation symposium on paper electrophoresis proceedings London 1955 Jul 27-29 Edited by G.E.W. Wolstenholme and Elaine C.P. Miller Illus Churchill, London,1956
Radioth 35.0361

CIBA FOUNDATION Ciba foundation symposium on pathogenic mycoplasma Proceeings London 1972 Jan 25-27 Edited by Katherine Elliott and Joan Birch Associated scientific publishers,Amsterdam,1972
Bioch 33.2296

CIBA FOUNDATION Ciba foundation symposium on porphyrin biosynthesis and metabolism London 1955 Feb 8-10 Edited by G.E.W. Wolstenholme and E. Miller Churchill,London,1955
Radioth 35.0088

CIBA FOUNDATION Ciba foundation symposium on porphyrin biosynthesis and metabolism Papers London 1955 Feb 8-10 Churchill,London, 1955
Chem 18.2637

CIBA FOUNDATION Ciba foundation symposium on porphyrin biosynthesis and metabolism Proceedings London 1955 Feb 8-10 Edited by G.E.W. Wolstenholme and Elaine C.P. Millar illus Churchill,London,1955
Bioch 33.0572

CIBA FOUNDATION Ciba foundation symposium on quinones in electron transport London 1960 May 11-13 Edited by G.E.W. Wolstenholme and Cecilia M. O'Connor Churchill,London,1961
Radioth 35.0099

CIBA FOUNDATION Ciba foundation symposium on quinones in electron transport Papers London 1960 May 11-13 Edited by G.E.W. Wolstenholme and Cecilia M. O'Connor Churchill,London,1961
Chem 18.1489

CIBA FOUNDATION Ciba foundation symposium on quinones in electron transport Proceedings London 1960 May 11-13 Edited by G.E.W. Wolstenholme and Cecilia M. O'Connor illus Churchill,London,1961
Bioch 33.0573

CIBA FOUNDATION Ciba foundation symposium on the biosynthesis of terpenes and sterols Papers London 1958 May 20-22 Edited by G. E.W. Wolstenholme and Maeve O'Connor Ciba foundation symposia Churchill,London,1959
Chem 18.1511

CIBA FOUNDATION Ciba foundation symposium on the biosynthesis of terpenes and sterols Proceedings London 1958 May 20-22 Edited by G.E.W. Wolstenholme and Maeve O'Connor illus Churchill,London,1959
Bioch 33.0571

CIBA FOUNDATION Ciba foundation symposium on the chemistry and biology of mucopolysaccharides London 1957 Apr 23-25 Edited by G.E.W. Wolstenholme and Maeve O'Connor Churchill,London,1958
An 32.1490

CIBA FOUNDATION Ciba foundation symposium on the chemistry and biology of purines Papers London 1956 May 8-10 Edited by G.E.W. Wolstenholme and Cecilia M. O'Connor Ciba foundation symposia J. and A. Churchill, London,1957
Chem 18.1326

CIBA FOUNDATION Ciba foundation symposium on the chemistry and biology of purines Papers London 1956 May 8-10 Edited by G.E.W. Wolstenholme and Cecilia M. O'Connor illus Churchill,London,1957
Bioch 33.0814

CIBA FOUNDATION Ciba foundation symposium on the nature of sleep London 1960 Jun 27-29 Edited by G.E.W. Wolstenholme and Maeve O'Connor Churchill,London,1961 Under the chairmanship of sir John Eccles
Psy 31.3387

CIBA FOUNDATION Ciba foundation symposium on the neurological basis of behaviour :in commemoration of Sir Charles Sherrington,O.M., G.B.E.,F.R.S.,1857-1952 Papers London 1957 Jul 2-4 Edited by G.E.W. Wolstenholme and Cecilia M. O'Connor Churchill,London, 1958
Psy 31.0512

CIBA FOUNDATION Ciba foundation symposium on the regulation of cell metabolism London 1958 Jul 28-30 Edited by G.E.W. Wolstenholme and Cecilia M. O'Connor Illus J.and A. Churchill,London,1959
Radioth 35.0493

CIBA FOUNDATION Ciba foundation symposium on the regulation of cell metabolism Proceedings London 1958 Jul 28-30 Edited by G.E.W. Wolstenholme and Cecilia M. O'Connor illus Churchill,London,1959
Bioch 33.0575

CIBA FOUNDATION Circulatory and respiratory mass transport :a Ciba foundation symposium London 1968 Jul Edited by G.E.W. Wolstenholme and Julie Knight Churchill, London,1969
Inv Med 37.0236

CIBA FOUNDATION Circulatory and respiratory mass transport :a Ciba foundation symposium London 1968 Jul 16-18 Edited by G.E.W. Wolstenholme and Julie Knight Churchill, London,1969
An 32.5222

CIBA FOUNDATION Circulatory and respiratory mass transport :a Ciba foundation symposium London 1968 Jul 16-18 Edited by G.E.W. Wolstenholme and Julie Knight illus.,pl. x, 310p Churchill,London,1969
Phys 20.1969

CIBA FOUNDATION Colour vision :physiology and experimental psychology London Edited by A.V.S.de Reuck and Julie Knight Churchill, London,1965
Psy 31.3373

CIBA FOUNDATION Complement :a Ciba foundation symposium London 1964 May 26-28 Edited by G.E.W. Wolstenholme and Julie Knight Churchill,London,1965
Pha 16.0189

CIBA FOUNDATION Conflict in society :a Ciba foundation symposium Edited by A.V.S. De Reuck and Julie Knight Churchill,London,1966
Eng 41.0909

CIBA FOUNDATION Congenital malformation :a symposium 1960 Jan 19-21 Edited by G.E.W. Wolstenholme and C.M. O'Connor Ciba foundation.Symposia Churchill,London,1960
PGMS 29.0216

CIBA FOUNDATION Congenital malformations : Ciba foundation symposium London 1960 Jan 19-21 Edited by G.E.W. Wolstenholme and Cecilia M. O'Connor Churchill,London,1960
An 32.2079

CIBA FOUNDATION Congential malformations : symposium London 1960 Jun 19-21 Edited by G.E.W. Wolstenholme and Cecilia M. O'Connor Ciba foundation.Symposia Churchill,London, 1960
Gen 34.0517

CIBA FOUNDATION Control of glycogen metabolism :Ciba foundation symposium Edited by W.J. Whelan and M.P. Cameron Churchill, London,1966
Phys 20.1339

CIBA FOUNDATION Control of glycogen metabolism :a Ciba foundation symposium Proceedings London 1963 Jul 23-25 Edited by W.J. Whelan and Margaret P. Cameron illus Churchill,London,1964
Bioch 33.0576

CIBA FOUNDATION Control processes in multicellular organisms :a Ciba foundation symposium New Delhi 1969 Mar 17-21 Edited by G.E.W. Wolstenholme and Julie Knight Churchill,London,1970
An 32.5436

CIBA FOUNDATION Control processes in multicellular organisms :a Ciba foundation symposium New Delhi 1969 Mar 17-21 Edited by G.E.W. Wolstenholme and Julie Knight Churchill,London,1970
Phys 20.2271

CIBA FOUNDATION Curare and curare-like agents :Ciba foundation study group 12th London 1961 Jul 4 Edited by A.V.S. De Reuck Churchill,London,1962
Pha 16.0044

CIBA FOUNDATION Development of the lung :a Ciba foundation symposium London 1965 Nov 1-3 Edited by A.V.S. De Reuck and Ruth Porter Churchill,London,1967
Phys 20.2220

CIBA FOUNDATION Development of the lung :a Ciba foundation symposium London 1965 Nov 1-3 Edited by A.V.S. Reuck and Ruth Porter Churchill,London,1967
An 32.3803

CIBA FOUNDATION Disorders of language Edited by A.V.S. de Reuck and Maeve O'Connor Ciba foundation symposia Churchill,London, 1964
Psy 31.0456

CIBA FOUNDATION Diurese und diuretica :ein internationales symposium Herrenchiemsee 1959 Jun 17-20 Edited by H. Schwiegk and others Springer,Berlin,1959 t.p. also in English "Diuresis and diuretics"
Inv Med 37.0193

CIBA FOUNDATION Drug responses in man :a symposium Papers London 1966 Jun 14-16 Edited by Gordon Wolstenholme and Ruth Porter Churchill,London,1967
Pha 16.0253

CIBA FOUNDATION Egg implantation London 1965 Oct 21 Edited by G.E.W. Wolstenholme and Maeve O'Connor Ciba foundation study group, 23 Churchill,London,1966
An 32.3889

CIBA FOUNDATION Egg implantation London 1965 Oct 21 Edited by G.E.W. Wolstenholme and Maeve O'Connor Ciba foundation study group, 23 Churchill,London,1966
Phys 20.1408

CIBA FOUNDATION Endocrinology of the testis : colloquium London 1966 May 18-20 Edited by G.E.W. Wolstenholme and Maeve O'Connor Ciba foundation colloquia on endocrinology, 16 Churchill,London,1967
An 32.3893

CIBA FOUNDATION Endocrinology of the testis : colloquium London 1966 May 18-22 Ciba foundation colloquia on endocrinology, 16 Churchill,London,1967
Phys 20.1412

CIBA FOUNDATION Enzymes and drug action London 1961 Mar 20-23 Edited by J.L. Mongar and A.V.S. De Reuck Churchill,London, 1962
Bioch 33.1872

CIBA FOUNDATION Foetal autonomy :a Ciba foundation symposium London 1968 Dec 3-5 Edited by G.E.W. Wolstenholme and Maeve O'Connor Churchill,London,1969
An 32.2543

CIBA FOUNDATION Foetal autonomy :a Ciba foundation symposium London 1969 Edited by G.E.W. Wolstenholme and Maeve O'Connor Churchill,London,1969
Phys 20.2262

CIBA FOUNDATION Foetal autonomy :a Ciba foundation symposium Proceedings London 1968 Dec 3-5 Edited by G.E.W. Wolstenholme and Maeve O'Connor Churchill,London,1969
Inv Med 37.0237

CIBA FOUNDATION Functions of the corpus callosum London 1964 Sep 4 Edited by E.G. Ettlinger Ciba foundation study group, 20 Churchill,London,1965
An 32.4657

CIBA FOUNDATION Functions of the corpus callosum London 1964 Sep 4 Edited by E.G. Ettlinger Ciba foundation study groups, 20 Churchill,London,1965
Phys 20.1996

CIBA FOUNDATION Functions of the corpus callosum :a symposium Proceedings London 1964 Sep 4 Edited by E.George Ettlinger Ciba foundation study group, 20 J.and A. Churchill,London,1965 Proceedings dedicated to the Rt.hon.lord Adrian,O.M.,on his retirement
Psy 31.0193

CIBA FOUNDATION Gas chromotography in biology and medicine :a Ciba foundation symposium London 1969 Feb 5-6 Edited by Ruth Porter Churchill,London,1969
Phys 20.2256

CIBA FOUNDATION Gonadotropins : physiochemical and immunological properties Edited by G.E.W. Wolstenholme and Julie Knight Ciba foundation study group, 22 Churchill, London,1965 In honour of Dr.C.Hamburger
Inv Med 37.0243

CIBA FOUNDATION Gonadotropins: physiocochemical and immunological properties London 1965 Feb 5 Edited by G.E.W. Wolstenholme and Julie Knight Ciba foundation study group, 22 Churchill,London, 1965
Phys 20.1409

CIBA FOUNDATION Growth of the nervous system Ciba foundation symposium Edited by G.E.W. Wolstenholme and M. O'Connor Churchill, London,1968
Phys 20.2046

CIBA FOUNDATION Growth of the nervous system a Ciba foundation symposia London 1967 Edited by G.E.W. Wolstenholme and Maeve O'Connor Churchill,London,1968
An 32.2536

CIBA FOUNDATION Hashish:its chemistry and pharmacology :a Ciba foundation study group 21st London 1965 Churchill,London,1965
Pha 16.0431

CIBA FOUNDATION Health of mankind :a Ciba foundation symposium London 1967 Mar 8-10 Edited by G.E.W. Wolstenholme and Maeve O'Connor Churchill,London,1967 Ciba foundation 100th symposium
Inv Med 37.0267

CIBA FOUNDATION Hearing mechanisms in vertebrates Ciba foundation symposium...on sensory function 4th Proceedings London 1967 Sep 26-28 Edited by A.V.S.de Reuck and Julie Knight J.and A.Churchill,London,1968
Psy 31.3372

CIBA FOUNDATION Hearing mechanisms in vertebrates :a Ciba foundation symposium London Edited by A.V.S.de Reuck and Julie Knight London,1968
Bal 44.6455

CIBA FOUNDATION Hearing mechanisms in vertebrates :a Ciba foundation symposium London 1967 Edited by A.V.S. De Reuck and Julie Knight Churchill,London,1968
Phys 20.1050

CIBA FOUNDATION Hering-Breur centenary symposium Proceedings London 1969 Jul 8-10 Edited by Ruth Porter Ciba foundation. Symposia Churchill,London,1970
Inv Med 37.0184

CIBA FOUNDATION Histamine :Ciba foundation symposium Edited by G.E.W. Wolstenholme and Cecilia M. O'Connor Jointly with the Physiological society and the British pharmacological society Churchill,London, 1956
Radioth 35.0087

CIBA FOUNDATION Histamine :Ciba foundation symposium London 1955 Apr 6-7 Edited by G.E.W. Wolstenholme and Cecilia M. O'Connor Churchill,London,1956
Pha 16.0076

CIBA FOUNDATION Histones :their role in the transfer of genetic information London 1965 Dec 3 Edited by A.V.S. De Reuck and Julie Knight Ciba foundation study group, 24 Little,Brown and company,Boston,1966 In honour of W.A.Engelhardt
Bioch 33.0888

CIBA FOUNDATION Homeostatic regulators :a Ciba foundation symposium Proceedings London 1969 Jan 28-30 Edited by G.E.W. Wolstenholme and Julie Knight Churchill, London,1969
Bioch 33.0934

CIBA FOUNDATION Homeostatic regulators :a Ciba foundation symposium held jointly with the Wellcome trust London 1969 Jan 28-30 Edited by G.E.W. Wolstenholme and Julie Knight Churchill,London,1969
Phys 20.2257

CIBA FOUNDATION Hormonal factors in carbohydrate metabolism :a colloquium Proceedings London 1952 Jun 30-Jul 3 Edited by G.E.W. Wolstenholme Ciba foundation colloquia on endocrinology, 6 illus Churchill,London,1953
Bioch 33.0436

CIBA FOUNDATION Hormone production in endocrine tumours :a colloquium Proceedings Edited by G.E.W. Wolstenholme and Maeve O'Connor Ciba foundation colloquia on endocrinology, 12 Little,Brown,Boston,1958
Inv Med 37.0242

CIBA FOUNDATION Hormones,psychology and behaviour,and Steroid hormone administration : two colloquia Proceedings Edited by G.E.W. Wolstenholme Ciba foundation colloquia on endocrinology,3 Illus. Churchill,London, 1952
VA 19.0225

CIBA FOUNDATION Hormones,psychology and behaviour,and steroid hormone administration : two colloquia Proceedings London 1951 Apr 9-12 and London 1950 Feb 23-24 Edited by G.E.W. Wolstenholme and Margaret P. Cameron Ciba foundation colloquia on endocrinology, 3 illus Churchill,London, 1952
Bioch 33.0433

CIBA FOUNDATION Hormones and the immune response :Ciba foundation study group London 1970 May 1 Edited by G.E.W. Wolstenholme and Julie Knight Ciba foundation study group, 36 Churchill,London,1970
An 32.5453

CIBA FOUNDATION Hormones in blood :a colloquium Proceedings London 1950 Edited by G.E.W. Wolstenholme and Elaine C.P. Millar Ciba foundation colloquia on endocrinology, 11 Churchill,London,1957
Inv Med 37.0241

CIBA FOUNDATION Human pituitary hormones :a colloquium in honour of B.A.Houssay Proceedings Buenos Aires 1959 Aug 6-8 Edited by G.E.W. Wolstenholme and Cecilia M. O'Connor Ciba foundation colloquia on endocrinology, 13 illus Churchill,London, 1960
Bioch 33.0440

CIBA FOUNDATION Immunoassay of hormones London 1961 Jul 25-27 Edited by G.E.W. Wolstenholme and Margaret P. Cameron Ciba foundation colloquia on endocrinology, 14 Churchill,London,1962
Gen 34.0654

CIBA FOUNDATION Immunoassay of hormones :a colloquium Proceedings London 1961 Jul 25-27 Edited by G.E.W. Wolstenholme and Margaret P. Cameron Ciba foundation colloquia on endocrinology, 14 Churchill, London,1962
Bioch 33.0441

CIBA FOUNDATION Immunoassay of hormones : colloquium London 1961 Jul 25-27 Ciba foundation colloquia on endocrinology, 14 Churchill,London,1962
Phys 20.1400

CIBA FOUNDATION Interferon :a Ciba foundation symposium London 1967 Apr 19 Edited by G.E.W. Wolstenholme and Maeve O'Connor Churchill,London,1968
Phys 20.0999

CIBA FOUNDATION Interferon a Ciba foundation symposium dedicated to Alick Isaacs Proceedings London 1967 Apr 19-21 Edited by G.E.W. Wolstenholme and Maeve O'Connor illus Churchill,London,1968
Bioch 33.1119

CIBA FOUNDATION Internal secretion of the pancreas :colloquium London 1955 Jun 21-23 Edited by G.E.W. Wolstenholme and Cecilia M. O'Connor Ciba foundation colloquia on endocrinology, 9 Churchill,London,1956
An 32.3739

CIBA FOUNDATION Internal secretions of the pancreas The Nature and actions of the internal secretions of the pancreas :a colloquium Proceedings London 1955 Jun 21-23 Edited by G.E.W. Wolstenholme and Cecilia M. O'Connor Ciba foundation colloquia on endocrinology, 9 illus Churchill,London,1956
Bioch 33.0438

CIBA FOUNDATION Leukaemia research :a Ciba foundation symposium Edited by G.E.W. Wolstenholme Churchill,London,1954
Med 36.0366

CIBA FOUNDATION Liver disease :a Ciba foundation symposium Proceedings Edited by Sheila Sherlock Churchill,London,1955
Med 36.0318

CIBA FOUNDATION Lysosmes :a Ciba foundation symposium London 1963 Feb 12-14 Edited by A.V.S. De Reuck and Margaret P. Cameron Churchill,London,1963
An 32.5316

CIBA FOUNDATION Lysosomes :a Ciba foundation symposium Papers London 1963 Feb 12-14 Edited by A.V.S. De Reuck and Margaret P. Cameron Ciba foundation.Symposia Churchill, London,1963
Bal 39.0298

CIBA FOUNDATION Lysosomes :a Ciba foundation symposium Papers London 1963 Feb 12-14 Edited by A.V.S. De Reuck and Margaret P. Cameron illus Churchill,London,1963
Bioch 33.0935

CIBA FOUNDATION Mammalian germ cells :Ciba foundation symposium London 1952 Jun 17-20 Edited by G.E.W. Wolstenholme and others Churchill,London,1953
Phys 20.1478

CIBA FOUNDATION Mechanisms of motor skill development Centre for advanced study in the developmental sciences study group on 'Mechanisms of motor skill development' Proceedings London 1968 Nov Edited by Kevin Connolly Academic press,London;New York,1970 Being the 4th study group in a C.A.S.D.S. programme on 'The origins of human behaviour' held jointly with the Ciba foundation
Psy 31.3101

CIBA FOUNDATION Metabolic effects of adrenal hormones London 1960 Jul 18 Edited by G. E.W. Wolstenholme and Maeve O'Connor Ciba foundation study group, 6 Churchill,London, 1960
Gen 34.0540

CIBA FOUNDATION Metabolic effects of adrenal hormones in honour of G.W.Thorn London 1960 Jul 15 Edited by G.E.W. Wolstenholme and Maeve O'Connor Ciba foundation study group, 6 Churchill,London,1960
Bioch 33.0442

CIBA FOUNDATION Molecular properties of drug receptors :a Ciba foundation symposium London 1970 Jan 27-29 Edited by Ruth Pater and Maeve O'Connor Churchill,London,1970
An 32.5477

CIBA FOUNDATION Mutation as cellular process a Ciba foundation symposium London 1969 Feb 11-13 Edited by G.E.W. Wolstenholme and M. O'Connor illus Chuchill,London,1969
Bioch 33.0891

CIBA FOUNDATION Myotatic,kinesthetic and vestibular mechanisms :Ciba foundation symposium Edited by A.V.S. De Reuck and Julie Knight Churchill,London,1967
Phys 20.2039

CIBA FOUNDATION Pain and itch :nervous mechanisms London 1959 Mar 10 Edited by G.E.W. Wolstenholme and Maeve O'Connor Ciba foundation study groups, 1 Churchill, London,1959
Phys 20.2017

CIBA FOUNDATION Pain and itch:nervous mechanisms London 1959 Mar 10 Edited by G.E.W. Wolstenholme and Maeve O'Connor Ciba foundation study group, 1 Churchill,London, 1959
An 32.4535

CIBA FOUNDATION Peripheral circulation in man :Ciba foundation symposium London 1953 May 11-13 Edited by G.E.W. Wolstenholme and Jessie S. Freeman Churchill,London,1954
An 32.3459

CIBA FOUNDATION Physiology and experimental psychology :a Ciba foundation symposium London Edited by A.V.S. De Reuck and Julie Knight Churchill,London,1965
Phys 20.2226

CIBA FOUNDATION Preimplantation stages of pregnancy :Ciba foundation symposium London 1965 Apr 13-15 Edited by G.E.W. Wolstenholme and M. O'Connor Churchill,London,1965
Phys 20.1407

CIBA FOUNDATION Preimplantation stages of pregnancy :Ciba foundation symposium London 1965 Apr 13-15 Edited by G.E.W. Wolstenholme and M. O'Connor London,1965
Bal 39.3558

CIBA FOUNDATION Preimplantation stages of pregnancy :a Ciba foundation symposium London 1965 Edited by G.E.W. Wolstenholme and Maeve O'Connor Churchill,London,1965
An 32.2517

CIBA FOUNDATION Preservation and transplantation of normal tissues :Ciba foundation symposium London 1953 Mar 16-18 Edited by G.E.W. Wolstenholme and Margaret P. Cameron Churchill,London,1954
An 32.3211

CIBA FOUNDATION Principles of biomolecular organisation :a CIBA foundation symposium London 1965 Jan 9-11 Edited by G.E.W. Wolstenholme and Maeve O'Connor Churchill, London,1966
An 32.1523

CIBA FOUNDATION Principles of biomolecular organization :Ciba foundation symposium London 1965 Jul 9-11 Edited by G.E.W. Wolstenholme and Maeve O'Connor Churchill, London,1966
Phys 20.1505

CIBA FOUNDATION Principles of biomolecular organization :a Ciba foundation symposium Proceedings London 1965 Jun 9-11 Edited by G.E.W. Wolstenholme and Maeve O'Connor Churchill,London,1966
Bioch 33.0311

CIBA FOUNDATION Principles of biomolecular organization :a Ciba foundation symposium Proceedings London 1965 Jun 9-11 Edited by G.E.W. Wolstenholme and Maeve O'Connor Churchill,London,1966
Bot 42.1503

CIBA FOUNDATION Progesterone :its regulatory effect on the myometrium Edited by G.E.W. Wolstenholme and Julie Knight Ciba foundation study group, 34 Churchill,London, 1969 In memory of Brenda M.Schofield
Inv Med 37.0249

CIBA FOUNDATION Progesterone and the defence mechanism of pregnancy London 1961 Feb 10 Edited by G.E.W. Wolstenholme and Margaret P. Cameron Ciba foundation study group, 9 Churchill,London,1961
An 32.3753

CIBA FOUNDATION Progesterone and the defence mechanism of pregnancy 9th London 1961 Feb 10 Edited by G.E.W. Wolstenholme and Margaret Cameron Ciba foundation study group, 9 Churchill,London,1961
Phys 20.1388

CIBA FOUNDATION Protein metabolism : influence of growth hormone,anabolic steroids, and nutrition in health and disease:an international symposium Leyden 1962 Jun 25-29 Edited by F. Gross Springer,Berlin,1967
Inv Med 37.0255

CIBA FOUNDATION Regulation and mode of action of thyroid hormones :a colloquium Proceedings London 1956 Jun 20-22 Edited by G.E.W. Wolstenholme and Elaine C.P. Millar Ciba foundation colloquia on endocrinology, 10 Churchill,London,1957
Inv Med 37.0240

CIBA FOUNDATION Regulation and mode of action of thyroid hormones :a colloquium Proceedings London 1956 Jun 20-22 Edited by G.E.W. Wolstenholme and Elaine C.P. Millar Ciba foundation colloquia on endocrinology, 10 illus Churchill,London,1957
Bioch 33.0439

CIBA FOUNDATION Somatic stability in the newly born :Ciba foundation symposium Edited by G.E.W. Wolstenholme and M. O'Connor Churchill,London,1961
Phys 20.2127

CIBA FOUNDATION Steric course of microbiological reactions Edited by G.E.W. Wolstenholme and Cecilia M. O'Connor Ciba foundation study group, 2 Churchill,London, 1959
Gen 34.0695

CIBA FOUNDATION Steric course of microbiological reactions :in honour of Dr.V. Prelog London 1959 Mar 17 Edited by G.E. W. Wolstenholme and Cecilia M. O'Connor Ciba foundation study group, 2 illus Churchill,London,1959
Bioch 33.1120

CIBA FOUNDATION Steroid hormones and tumour growth,and Steroid hormones and enzymes :two colloquia Proceedings London 1950 Edited by G.E.W. Wolstenholme and Margaret P. Cameron Ciba foundation colloquia on endocrinology, 1 Churchill,London,1952
Phys 20.1366

CIBA FOUNDATION Steroid metabolism and estimation :a colloquium Proceedings London 1950 Jul 31-Aug 2 Edited by G.E.W. Wolstenholme Ciba foundation colloquia on endocrinology, 2 illus Churchill,London, 1952
Bioch 33.0432

CIBA FOUNDATION Steroid metabolism and estimation :two colloquia London 1950 Jul 31-Aug 2 London 1950 Aug 9-10 Edited by G.E.W. Wolstenholme and Margaret Cameron Ciba foundation colloquia on endocrinology, 2 Churchill,London,1952
Phys 20.1367

CIBA FOUNDATION Symposium on haemopoiesis cell production and its regulation London 1960 Feb 2-4 Edited by G.E.W. Wolstenholme and Maeve O'Connor Ciba foundation.Symposia Churchill,London,1960
PGMS 29.0273

CIBA FOUNDATION Synthesis and metabolism of adrenocortical steroids :a colloquium Proceedings London 1950 Edited by W. Klyne Ciba foundation colloquia on endocrinology, 7 Churchill,London,1953
Inv Med 37.0137

CIBA FOUNDATION Synthesis and metabolism of adrenocortical steroids :a colloquium Proceedings London 1952 Jul 7-10 Edited by W. Klyne and G.E.W. Wolstenholme Ciba foundation colloquia on endocrinology, 7 illus Churchill,London,1953
Bioch 33.0437

CIBA FOUNDATION Systemic mycoses :a Ciba foundation symposium Ibadan 1967 Mar 29-31 Edited by G.E.W. Wolstenholme and Ruth Porter Churchill,London,1968
Phys 20.0998

CIBA FOUNDATION Systemic mycoses :a Ciba foundation symposium in commemoration of William Balfour Baikie Ibadan 1967 Mar 29-31 Edited by G.E.W. Wolstenholme and Ruth Porter illus,port Churchill,London,1968
Path 30.2632

CIBA FOUNDATION The Chemical structure of proteins :a Ciba foundation symposium Proceedings London 1952 Dec 1-3 Edited by G.E.W. Wolstenholme and Margaret P. Cameron illus Churchill,London,1953
Bioch 33.0500

CIBA FOUNDATION The Ciba foundation symposium on the nature of viruses London 1956 Edited by G.E.W. Wolstenholme and E.C.P. Millar Ciba foundation.Symposia Churchill, London,1957
Radioth 35.0752

CIBA FOUNDATION The Exocrine pancreas :Ciba foundation symposium Edited by A.V.S. De Reuck and M.P. Cameron Churchill,London,1962
Phys 20.1340

CIBA FOUNDATION The Frozen cell :a Ciba foundation symposium London 1969 May 20-22 Edited by G.E.W. Wolstenholme and Maeve O'Connor Churchill,London,1970
An 32.5444

CIBA FOUNDATION The Frozen cell :a Ciba foundation symposium Proceedings London 1969 May 20-22 Edited by G.E.W. Wolstenholme and Maeve O'Connor illus. ix,316p Churchill,London,1970
Bot 42.1383

CIBA FOUNDATION The Human adrenal cortex :a colloquium Proceedings Edited by G.E.W. Wolstenholme and others Ciba foundation colloquia on endocrinology, 8 Churchill, London,1955
Inv Med 37.0239

CIBA FOUNDATION The Human adrenal cortex : colloqium London 1954 Apr 21-25 Edited by G.E.W. Wolstenholme and Margaret Cameron Ciba foundation colloquia on endocrinology, 14 Churchill,London,1962
Phys 20.1380

CIBA FOUNDATION The Mechanism of action of water soluble vitamins :a meeting 11 London 1961 Apr 28 Edited by A.V.S. de Reuck and Maeve O'Connor Ciba foundation study group,11 J.and A.Churchill,London,1961
Chem 18.1283

CIBA FOUNDATION The Nature of sleep :a Ciba foundation symposium Proceedings London 1960 Jun 27-29 Edited by G.E.W. Wolstenholme and Maeve O'Connor Churchill,London,1961
An 32.4378

CIBA FOUNDATION The Pineal gland :a Ciba foundation symposium London 1970 Jun 30-Jul 2 Edited by G.E.W. Wolstenholme and Julie Knight Churchill;Livingstone,Edinburgh; London,1971
An 32.5456

CIBA FOUNDATION The Spinal cord :a Ciba foundation symposium Churchill,London,1953
Phys 20.2005

CIBA FOUNDATION The Spinal cord :a Ciba foundation symposium London 1952 Feb 26-28 Edited by G.E.W. Wolstenholme Churchill, London,1953
An 32.4129

CIBA FOUNDATION The Thymus :experimental and clinical studies:a Ciba foundation symposium Melbourne 1965 Aug 25-27 Edited by G.E.W. Wolstenholme and Ruth Porter Churchill, London,1966
Phys 20.0905

CIBA FOUNDATION The Thymus:experimental and clinical studies :a Ciba foundation symposium Melbourne 1965 Aug 25-27 Edited by G.E.W. Wolstenholme and Ruth Porter Churchill, London,1966
An 32.3703

CIBA FOUNDATION The Thymus:experimental and clinical studies Ciba foundation symposium Melbourne 1965 Aug 25-27 Edited by G.E.W. Wolstenholme and Ruth Porter illus Churchill,London,1966 In honour of sir Macfarlane Burnet,chairman of the conference
Path 30.2461

CIBA FOUNDATION Thiamine deficiency : biochemical lesions and their clinical sihnificance London 1966 Nov 10-11 Edited by G.E.W. Wolstenholme and Maeve O'Connor Ciba foundation study group, 28 Churchill,London,1967 In honour of Sir Rudolph Peters
Bioch 33.0574

CIBA FOUNDATION Touch,heat and pain :Ciba foundation symposium Edited by A.V.S. De Reuck and J. Knight Churchill,London,1966
Phys 20.2036

CIBA FOUNDATION Touch,heat and pain :Ciba foundation symposium Proceedings London 1965 Sep 21-23 Edited by A.V.S. De Reuck and Julie Knight Churchill,London,1966
An 32.5335

CIBA FOUNDATION.SYMPOSIA Biochemistry of human genetics :a symposium Naples 1959 May 13-16 Ciba foundation and International union of biological sciences Edited by G.E.W. Wolstenholme and C.M. O'Connor Churchill,London,1959
PGMS 29.0251

CIBA FOUNDATION.SYMPOSIA, 50 Medical biology and etruscan origins :a Ciba foundation symposium London 1958 Apr Ciba foundation Edited by G.E.W. Wolstenholme and Cecilia M. O'Connor illus. Churchill,London,1959
Gen 34.1877

CIBA FOUNDATION,and BIOLOGICAL COUNCIL.CO-ORDINATING COMMITTEE FOR SYMPOSIA ON DRUG ACTION Animal behaviour and drug action : symposium London 1963 Mar 25-28 Edited by Hannah Steinberg and others Churchill, London,1964
Pha 16.0030

CIBA FOUNDATION,and BIOLOGICAL COUNCIL.CO-ORDINATING COMMITTEE FOR SYMPOSIA ON DRUG ACTION Enzymes and drug action :Ciba foundation symposium 8th London Edited by J.L. Mongar and A.V.S. De Reuck Churchill, London,1962
Pha 16.0105

CIBA FOUNDATION COLLOQUIA ON AGEING Vol 1-2 Edited by G.E.W. Wolstenholme and others Illus. 2 vols Churchill,London, 1955-56 Vol.1:general aspects;vol.2: ageing in transient tissues
An 32.0344

CIBA FOUNDATION COLLOQUIA ON ENDOCRINOLOGY, 1 Steroid hormones and tumour growth and steroid hormones and enzymes :two colloquia Proceedings London 1950 Jul 10-12 and London 1950 Mar 8-10 Ciba foundation Edited by G.E.W. Wolstenholme illus Churchill,London,1952
Bioch 33.0431

CIBA FOUNDATION COLLOQUIA ON ENDOCRINOLOGY, 2 Steroid metabolism and estimation :a colloquium Proceedings London 1950 Jul 31-Aug 2 Ciba foundation Edited by G.E.W. Wolstenholme illus Churchill,London,1952
Bioch 33.0432

CIBA FOUNDATION COLLOQUIA ON ENDOCRINOLOGY, 3 Hormones,psychology and behaviour,and steroid hormone administration :two colloquia Proceedings London 1951 Apr 9-12 and London 1950 Feb 23-24 Ciba foundation Edited by G.E.W. Wolstenholme and Margaret P. Cameron illus Churchill,London,1952
Bioch 33.0433

CIBA FOUNDATION COLLOQUIA ON ENDOCRINOLOGY, 4 Anterior pituitary secretion and Hormonal influences in water metabolism :two colloquia Proceedings London 1951 Jul 9-13 and London 1951 Jan 8-10 Ciba foundation Edited by G.E.W. Wolstenholme and Margaret P. Cameron illus Churchill,London,1952
Bioch 33.0434

CIBA FOUNDATION COLLOQUIA ON ENDOCRINOLOGY, 4 Anterior pituitary secretion and hormonal influences in water metabolism :two colloquia London 1951 Jul 9-13 London 1951 Jan 8-10 Ciba foundation Edited by G.E.W. Wolstenholme and Margaret P. Cameron Churhcill,London,1952
Phys 20.1365

CIBA FOUNDATION COLLOQUIA ON ENDOCRINOLOGY, 5 Bioassay of anterior pituitary and adrenocortical hormones :a colloquium proceedings London 1952 Mar 25-27 Ciba foundation Edited by G.E.W. Wolstenholme illus Churchill,London,1953
Bioch 33.0435

CIBA FOUNDATION COLLOQUIA ON ENDOCRINOLOGY, 6 Hormonal factors in carbohydrate metabolism a colloquium Proceedings London 1952 Jun 30-Jul 3 Ciba foundation Edited by G.E. W. Wolstenholme illus Churchill,London, 1953
Bioch 33.0436

CIBA FOUNDATION COLLOQUIA ON ENDOCRINOLOGY, 7 Synthesis and metabolism of adrenocortical steroids :a colloquium Proceedings London 1950 Ciba foundation Edited by W. Klyne Churchill,London,1953
Inv Med 37.0137

CIBA FOUNDATION COLLOQUIA ON ENDOCRINOLOGY, 7 Synthesis and metabolism of adrenocortical steroids :a colloquium Proceedings London 1952 Jul 7-10 Ciba foundation Edited by W. Klyne and G.E.W. Wolstenholme illus Churchill,London,1953
Bioch 33.0437

CIBA FOUNDATION COLLOQUIA ON ENDOCRINOLOGY, 8 The Human adrenal cortex :a colloquium Proceedings Ciba foundation Edited by G.E. W. Wolstenholme and others Churchill,London, 1955
Inv Med 37.0239

CIBA FOUNDATION COLLOQUIA ON ENDOCRINOLOGY, 9 Internal secretion of the pancreas : colloquium London 1955 Jun 21-23 Ciba foundation Edited by G.E.W. Wolstenholme and Cecilia M. O'Connor Churchill,London,1956
An 32.3739

CIBA FOUNDATION COLLOQUIA ON ENDOCRINOLOGY, 9 Internal secretions of the pancreas The Nature and actions of the internal secretions of the pancreas :a colloquium Proceedings London 1955 Jun 21-23 Ciba foundation Edited by G.E.W. Wolstenholme and Cecilia M. O'Connor illus Churchill,London,1956
Bioch 33.0438

CIBA FOUNDATION COLLOQUIA ON ENDOCRINOLOGY, 10 Regulation and mode of action of thyroid hormones :a colloquium Proceedings London 1956 Jun 20-22 Ciba foundation Edited by G. E.W. Wolstenholme and Elaine C.P. Millar Churchill,London,1957
Inv Med 37.0240

CIBA FOUNDATION COLLOQUIA ON ENDOCRINOLOGY, 10 Regulation and mode of action of thyroid hormones :a colloquium Proceedings London 1956 Jun 20-22 Ciba foundation Edited by G. E.W. Wolstenholme and Elaine C.P. Millar illus Churchill,London,1957
Bioch 33.0439

CIBA FOUNDATION COLLOQUIA ON ENDOCRINOLOGY, 11
Hormones in blood :a colloquium Proceedings London 1950 Ciba foundation Edited by G.E.W. Wolstenholme and Elaine C.P. Millar Churchill,London,1957
Inv Med 37.0241

CIBA FOUNDATION COLLOQUIA ON ENDOCRINOLOGY, 12
Hormone production in endocrine tumours :a colloquium Proceedings Ciba foundation Edited by G.E.W. Wolstenholme and Maeve O'Connor Little,Brown,Boston,1958
Inv Med 37.0242

CIBA FOUNDATION COLLOQUIA ON ENDOCRINOLOGY, 13
Human pituitary hormones :a colloquium in honour of B.A.Houssay Proceedings Buenos Aires 1959 Aug 6-8 Ciba foundation Edited by G.E.W. Wolstenholme and Cecilia M. O'Connor illus Churchill,London,1960
Bioch 33.0440

CIBA FOUNDATION COLLOQUIA ON ENDOCRINOLOGY, 14
Immunoassay of hormones London 1961 Jul 25-27 Ciba foundation Edited by G.E.W. Wolstenholme and Margaret P. Cameron Churchill,London,1962
Gen 34.0654

CIBA FOUNDATION COLLOQUIA ON ENDOCRINOLOGY, 14
Immunoassay of hormones :a colloquium Proceedings London 1961 Jul 25-27 Ciba foundation Edited by G.E.W. Wolstenholme and Margaret P. Cameron Churchill,London,1962
Bioch 33.0441

CIBA FOUNDATION COLLOQUIA ON ENDOCRINOLOGY, 16
Endocrinology of the testis :colloquium London 1966 May 18-20 Ciba foundation Edited by G.E.W. Wolstenholme and Maeve O'Connor Churchill,London,1967
An 32.3893

CIBA FOUNDATION COLLOQUIA ON ENDOCRINOLOGY,3
Hormones,psychology and behaviour,and Steroid hormone administration :two colloquia Proceedings Ciba foundation Edited by G.E. W. Wolstenholme Illus. Churchill,London, 1952
VA 19.0225

CIBA FOUNDATION CONFERENCE ON ISOTOPES IN BIOCHEMISTRY London 1951 Mar 12-15 Ciba foundation Edited by G.E.W. Wolstenholme Illus J.and A.Churchill, London,1951
Radioth 35.0023

CIBA FOUNDATION CONFERENCE ON ISOTOPES IN BIOCHEMISTRY Papers London 1951 Mar 12-15 Ciba foundation Edited by J.N. Davidson Churchill,London,1951
Bioch 33.0274

CIBA FOUNDATION SSYMPOSIUM ON THE CHEMISTRY AND BIOLOGY OF MUCOPOLYSACCHARIDES Proceedings London 1957 Apr 23-25 Ciba foundation Edited by G.E.W. Wolstenholme and Maeve O'Connor Churchill,London,1958
Bioch 33.1202

CIBA FOUNDATION STUDY GROUP, 1 Pain and itch nervous mechanisms London 1959 Mar 10 Ciba foundation Edited by G.E.W. Wolstenholme and M. O'Connor Churchill, London,1959
PGMS 29.0539

CIBA FOUNDATION STUDY GROUP, 2 Steric course of microbiological reactions Ciba foundation Edited by G.E.W. Wolstenholme and Cecilia M. O'Connor Churchill,London,1959
Gen 34.0695

CIBA FOUNDATION STUDY GROUP, 2 Steric course of microbiological reactions London 1959 Mar 17 Ciba foundation Edited by G.E. W. Wolstenholme and C.M. O'Connor Churchill, London,1959
PGMS 29.0540

CIBA FOUNDATION STUDY GROUP, 2 Steric course of microbiological reactions :in honour of Dr.V.Prelog London 1959 Mar 17 Ciba foundation Edited by G.E.W. Wolstenholme and Cecilia M. O'Connor illus Churchill,London, 1959
Bioch 33.1120

CIBA FOUNDATION STUDY GROUP, 3 Cancer of the cervix :diagnosis of early forms London 1959 May 8 Ciba foundation Edited by G.E.W. Wolstenholme and Maeve O'Connor Churchill, London,1959
Radioth 35.1960

CIBA FOUNDATION STUDY GROUP, 4 Virus virulence and pathogenicity London 1959 Jun 15 Ciba foundation Edited by G.E.W. Wolstenholme and C.M. O'Connor Churchill, London,1960
PGMS 29.0541

CIBA FOUNDATION STUDY GROUP, 6 Metabolic effects of adrenal hormones London 1960 Jul 18 Ciba foundation Edited by G.E.W. Wolstenholme and Maeve O'Connor Churchill, London,1960
Gen 34.0540

CIBA FOUNDATION STUDY GROUP, 6 Metabolic effects of adrenal hormones in honour of G.W. Thorn London 1960 Jul 15 Ciba foundation Edited by G.E.W. Wolstenholme and Maeve O'Connor Churchill,London,1960
Bioch 33.0442

CIBA FOUNDATION STUDY GROUP, 9
Progesterone and the defence mechanism of pregnancy London 1961 Feb 10 Ciba foundation Edited by G.E.W. Wolstenholme and Margaret P. Cameron Churchill,London,1961
An 32.3753

CIBA FOUNDATION STUDY GROUP, 20 Functions of the corpus callosum London 1964 Sep 4 Ciba foundation Edited by E.G. Ettlinger Churchill,London,1965
An 32.4657

CIBA FOUNDATION STUDY GROUP, 20 Functions of the corpus callosum :a symposium Proceedings London 1964 Sep 4 Ciba foundation Edited by E.George Ettlinger J. and A.Churchill,London,1965 Proceedings dedicated to the Rt.hon.lord Adrian,O.M.,on his retirement
Psy 31.0193

CIBA FOUNDATION STUDY GROUP, 22
Gonadotropins :physiochemical and immunological properties Ciba foundation Edited by G.E.W. Wolstenholme and Julie Knight Churchill,London,1965 In honour of Dr.C. Hamburger
Inv Med 37.0243

CIBA FOUNDATION STUDY GROUP, 23 Egg implantation London 1965 Oct 21 Ciba foundation Edited by G.E.W. Wolstenholme and Maeve O'Connor Churchill,London,1966
An 32.3889

CIBA FOUNDATION STUDY GROUP, 24 Histones : their role in the transfer of genetic information London 1965 Dec 3 Ciba foundation Edited by A.V.S. De Reuck and Julie Knight Little,Brown and company,Boston, 1966 In honour of W.A.Engelhardt
Bioch 33.0888

CIBA FOUNDATION STUDY GROUP, 26 Effects of external stimuli on reproduction :a Ciba foundation study group London 1966 Jul 27 Edited by G.E.W. Wolstenholme and Maeve O'Connor Churchill,London,1967
An 32.3895

CIBA FOUNDATION STUDY GROUP, 28 Thiamine deficiency :biochemical lesions and their clinical sihnificance London 1966 Nov 10-11 Ciba foundation Edited by G.E.W. Wolstenholme and Maeve O'Connor Churchill, London,1967 In honour of Sir Rudolph Peters
Bioch 33.0574

CIBA FOUNDATION STUDY GROUP, 33 Adrenergic neurotransmission Ciba foundation Edited by G.E.W. Wolstenholme and Maeve O'Connor Churchill,London,1968
An 32.5042

CIBA FOUNDATION STUDY GROUP, 34 Progesterone :its regulatory effect on the myometrium Ciba foundation Edited by G.E.W. Wolstenholme and Julie Knight Churchill, London,1969 In memory of Brenda M. Schofield
Inv Med 37.0249

CIBA FOUNDATION STUDY GROUP, 36 Hormones and the immune response :Ciba foundation study group London 1970 May 1 Ciba foundation Edited by G.E.W. Wolstenholme and Julie Knight Churchill,London,1970
An 32.5453

CIBA FOUNDATION SYMPOSIUM Paper electrophoresis London 1955 Jul 27-29 Edited by G.E.W. Wolstenholme and Elaine C.S. Millar Churchill,London,1958
Col S 12.0070

CIBA FOUNDATION SYMPOSIUM...ON SENSORY FUNCTION 2nd :touch,heat and pain Proceedings London 1965 Sep 21-23 Ciba foundation Edited by A.V.S. de Reuck and Julie Knight J. and A.Churchill,London,1966
Psy 31.0286

CIBA FOUNDATION SYMPOSIUM...ON SENSORY FUNCTION 4th Proceedings Hearing mechanisms in vertebrates London 1967 Sep 26-28 Ciba foundation Edited by A.V.S.de Reuck and Julie Knight J.and A.Churchill,London,1968
Psy 31.3372

CIBA FOUNDATION SYMPOSIUM ON AMINO ACIDS AND PEPTIDES WITH ANTIMETABOLIC ACTIVITY Proceedings London 1958 Mar 18-20 Ciba foundation Edited by G.E.W. Wolstenholme and Cecilia M. O'Connor illus Churchill,London,1958
Bioch 33.0570

CIBA FOUNDATION SYMPOSIUM ON BIOCHEMISTRY OF HUMAN GENETICS Proceedings Naples 1959 May 13-16 Ciba foundation International union of biological sciences Edited by G.E.W. Wolstenholme and Cecilia M. O'Connor illus Churchill,London,1959
Bioch 33.0890

CIBA FOUNDATION SYMPOSIUM ON CARBON FLUORINE COMPOUNDS :chemistry biochemistry and biological London 1971 Sep 13-15 Ciba foundation Edited by Katherine Elliott and John Birch Elsevier,Amsterdam,1972
Bioch 33.2244

CIBA FOUNDATION SYMPOSIUM ON CARCINOGENESIS: MECHANISMS OF ACTION London 1958 Jun 24-26 Ciba foundation Edited by G.E.W. Wolstenholme and Maeve O'Connor Illus Churchill,London,1959
Radioth 35.0107

CIBA FOUNDATION SYMPOSIUM ON CELLULAR ASPECTS OF IMMUNITY Proceedings Royaumont 1959 Jun 3-5 Ciba foundation Edited by G.E.W. Wolstenholme and Maeve O'Connor Churchill, London,1960
Bioch 33.1117

CIBA FOUNDATION SYMPOSIUM ON GROWTH CONTROL IN CELL CULTURE Papers London 1971 Oct 14-16 Ciba foundation Edited by G.E.W. Wolstenholme and Julie Knight Churchill; Livingstone,Edinburgh,1971
Bioch 33.2263

CIBA FOUNDATION SYMPOSIUM ON HAEMOPOIESIS : cell production and its regulation London 1960 Feb 2-4 Ciba foundation Edited by G.E. W. Wolstenholme and Maeve O'Connor Churchill, London,1960
An 32.3166

CIBA FOUNDATION SYMPOSIUM ON IONIZING RADIATIONS AND CELL METABOLISM Proceedings London 1956 Mar 6-9 Ciba foundation Edited by G.E. W. Wolstenholme and C.M. O'Connor illus. Churchill,London,1956
Radioth 35.1355

CIBA FOUNDATION SYMPOSIUM ON LEUKAEMIA RESEARCH London 1953 Nov 16-19 Ciba foundation Edited by G.E.W. Wolstenholme and Margaret P. Cameron Ciba foundation.Symposia Churchill, London,1954
Radioth 35.0800

CIBA FOUNDATION SYMPOSIUM ON PAPER ELECTROPHORESIS London 1955 Jul 27-29 Ciba foundation Edited by G.E. Wolstenholme and E.C.P. Millar Ciba foundation.Symposia Churchill,London,1958
PGMS 29.0516

CIBA FOUNDATION SYMPOSIUM ON PAPER ELECTROPHORESIS proceedings London 1955 Jul 27-29 Ciba foundation Edited by G. E.W. Wolstenholme and Elaine C.P. Miller Illus Churchill,London,1956
Radioth 35.0361

CIBA FOUNDATION SYMPOSIUM ON PATHOGENIC MYCOPLASMA Proceeings London 1972 Jan 25-27 Ciba foundation Edited by Katherine Elliott and Joan Birch Associated scientific publishers,Amsterdam,1972
Bioch 33.2296

CIBA FOUNDATION SYMPOSIUM ON PORPHYRIN BIOSYNTHESIS AND METABOLISM London 1955 Feb 8-10 Ciba foundation Edited by G.E.W. Wolstenholme and E. Miller Churchill,London, 1955
Radioth 35.0088

CIBA FOUNDATION SYMPOSIUM ON PORPHYRIN BIOSYNTHESIS AND METABOLISM Proceedings London 1955 Feb 8-10 Ciba foundation Edited by G.E.W. Wolstenholme and Elaine C.P. Millar illus Churchill,London,1955
Bioch 33.0572

CIBA FOUNDATION SYMPOSIUM ON PORPHYRIN BIOSYNTHESIS AND METABOLISM Papers London 1955 Feb 8-10 Ciba foundation Churchill,London,1955
Chem 18.2637

CIBA FOUNDATION SYMPOSIUM ON QUINONES IN ELECTRON TRANSPORT London 1960 May 11-13 Ciba foundation Edited by G.E.W. Wolstenholme and Cecilia M. O'Connor Churchill,London,1961
Radioth 35.0099

CIBA FOUNDATION SYMPOSIUM ON QUINONES IN ELECTRON TRANSPORT Proceedings London 1960 May 11-13 Ciba foundation Edited by G. E.W. Wolstenholme and Cecilia M. O'Connor illus Churchill,London,1961
Bioch 33.0573

CIBA FOUNDATION SYMPOSIUM ON THE BIOSYNTHESIS OF TERPENES AND STEROLS Proceedings London 1958 May 20-22 Ciba foundation Edited by G.E.W. Wolstenholme and Maeve O'Connor illus Churchill,London,1959
Bioch 33.0571

The CIBA FOUNDATION SYMPOSIUM ON THE CEREBROSPINAL FLUID :production, circulation and absorption London 1957 May 27-29 Edited by G.E.W. Wolstenholme and Cecilia M. O'Connor Churchill,London,1958
An 32.4167

CIBA FOUNDATION SYMPOSIUM ON THE CHEMISTRY AND BIOLOGY OF MUCOPOLYSACCARIDES London 1957 Apr 23-25 Edited by G.E.W. Wolstenholme and M. O'Connor Illus J.and A.Churchill, London,1958
Radioth 35.0057

CIBA FOUNDATION SYMPOSIUM ON THE CHEMISTRY AND BIOLOGY OF PURINES London 1954 Mar 22-26 Edited by G.E.W. Wolstenholme and M. Cameron Illus J.and A.Churchill,London, 1954
Radioth 35.0058

CIBA FOUNDATION SYMPOSIUM ON THE CHEMISTRY AND BIOLOGY OF PURINES London 1956 May 8-10 Edited by G.E.W. Wolstenholme and C.M. O'Connor Illus Churchill,London,1957
Radioth 35.0059

CIBA FOUNDATION SYMPOSIUM ON THE CHEMISTRY AND BIOLOGY OF PURINES Papers London 1956 May 8-10 Ciba foundation Edited by G. E.W. Wolstenholme and Cecilia M. O'Connor illus Churchill,London,1957
Bioch 33.0814

CIBA FOUNDATION SYMPOSIUM ON THE EXOCRINE PANCREAS :normal and abnormal functions London 1961 May 30-Jun 1 Edited by A.V. S. De Reuck and Margaret P. Cameron Churchill,London,1962
An 32.3659

CIBA FOUNDATION SYMPOSIUM ON THE NATURE OF SLEEP London 1960 Jun 27-29 Ciba foundation Edited by G.E.W. Wolstenholme and Maeve O'Connor Churchill,London,1961 Under the chairmanship of sir John Eccles
Psy 31.3387

The CIBA FOUNDATION SYMPOSIUM ON THE NATURE OF VIRUSES London 1956 Ciba foundation Edited by G.E.W. Wolstenholme and E.C.P. Millar Ciba foundation.Symposia Churchill, London,1957
Radioth 35.0752

CIBA FOUNDATION SYMPOSIUM ON THE NEUROLOGICAL BASIS OF BEHAVIOUR :in commemoration of Sir Charles Sherrington,O.M.,G.B.E.,F.R.S.,1857-1952 Papers London 1957 Jul 2-4 Ciba foundation Edited by G.E.W. Wolstenholme and Cecilia M. O'Connor Churchill,London,1958
Psy 31.0512

CIBA FOUNDATION SYMPOSIUM ON THE REGULATION OF CELL METABOLISM London 1958 Jul 28-30 Ciba foundation Edited by G.E.W. Wolstenholme and Cecilia M. O'Connor Illus J.and A.Churchill,London,1959
Radioth 35.0493

CIBA FOUNDATION SYMPOSIUM ON THE REGULATION OF CELL METABOLISM Proceedings London 1958 Jul 28-30 Ciba foundation Edited by G. E.W. Wolstenholme and Cecilia M. O'Connor illus Churchill,London,1959
Bioch 33.0575

CIBA FOUNDATION SYMPOSIUM ON TRANSPLANTATION London 1961 Edited by G.E.W. Wolstenholme and Margaret P. Cameron Churchill,London, 1962
An 32.3281

CIBA LABORATORIES LTD The Investigation of hypothalamic-pituitary-adrenal function :a symposium Proceedings London 1967 Feb 22-23 Edited by V.H.T. James and J. Landon Sponsored also by the Royal society of medicine.Endocrine section Society for endocrinology.Memoirs, 17 Cambridge university press,Cambridge,1968
Phys 20.1421

CIGARETTE Toward a less harmful cigarette The World conference on smoking and health :a workshop Proceedings 1967 Sep 11-13 National cancer institute Edited by Ernest L. Wynder and Dietrich Hoffmann National cancer institute.Monograph, 28 National cancer institute,Bethesda,Md.,1968
Bioch 33.0992

CINCINNATI 1955 Symposium on basic effects of environment on the strength,scaling,and embrittlement of metals at high temperatures American society for testing materials A.S.T. M.Special technical publication, 171 A.S.T.M. Philadelphia,Pa.,1955
Met 25.1908

CINCINNATI 1959 The Astrometric conference 2nd University of Cincinnati Edited by S.Aa. Strand and O.G. Franz Sponsored by the National science foundation 58p Astronomical journal,1960 Reprinted from the Astronomical journal,vol 65,no 4,May 1960
Obs 6.3243

CINCINNATI,OHIO 1961 Analysis of carcinogenic air pollutants :symposium Proceedings National cancer institute Edited by Eugene Sawicki and Kenneth Cassel National cancer institute.Monograph, 9 illus U.S.Department of health,education and welfare,Bethesda,Md.,1962 Sponsored by the Laboratory of engineering and physical sciences of the Division of air pollution,U.S. Department of health education,and welfare, Public health service.Robert A.Taft sanitary engineering center
Bioch 33.0980

CINCINNATI,OHIO 1969 Blood oxygenation International symposium on blood oxygenation Proceedings Edited by Daniel Hershey Plenum press,New York;London,1970
An 32.5457

CINCINNATI,OHIO 1969 Relativity Relativity conference in the midwest Proceedings Edited by Moshe Carmel and others Sponsored jointly by the Aerospace research laboratories and the University of Cincinnati bibliog.,illus. xii,381p 26cm Plenum,New York,1970
A Math 4.1622

CINCINNATI,OHIO 1969 Relativity Relativity conference of the Midwest Proceedings Aerospace research laboratories University of Cincinnati Edited by M. Carmeli and others 381p Plenum press,New York,1970
TA 15.0506

CINEFLUOROGRAPHY Annual symposium on cinefluorography 1st Proceedings Rochester,N.Y. 1958 Nov 14-15 Edited by George H.S. Ramsey and others Sponsored by University of Rochester.Department of radiology Thomas,Springfield,Ill.,1960
VA 19.0381

CINEMATOGRAPHY Internationaler kongress fur medizinische photographie und Kinematographie 1st Abhandlungen Dusseldorf 1960 Sep 27-30 Edited by Heinz Orbach Thieme,Stuttgart, 1962
An 32.0311

CINETIQUE ET MECANISME DES REACTIONS D'INFLAMMATION ET DE COMBUSTION EN PHASE GAZEUSE Paris 1948 Apr 26-May 1 Centre national de la recherche scientifique With the financial support of the Rockefeller foundation Centre national de la recherche scientifique.Colloques internationaux,16 Centre national de la recherche scientifique, Paris,1949 With summaries in English
Chem 18.0431

CIRCADIAN CLOCKS The Feldafing summer school Proceedings 1964 Sep 7-18 Edited by Jurgen Aschoff Sponsored by the North Atlantic treaty organization.Scientific affairs division North Holland,Amsterdam,1965
Bal 39.1766

CIRCADIAN CLOCKS The Feldafing summer school Proceedings 1964 Sep 7-18 Edited by Jurgen Aschoff Sponsored by the North Atlantic treaty organization North-Holland,Amsterdam, 1965
Bot 42.1365

CIRCADIAN CLOCKS :summer school Proceedings Feldafing 1964 Sep 7-18 Edited by Jurgen Aschoff Sponsored by the North Atlantic treaty organization.Scientific affairs division North-Holland,Amsterdam, 1965
An 32.5375

CIRCUIT ANALYSIS Symposium on circuit analysis Proceedings Monticello,Ill. 1955 University of Illinois 1955
Eng 41.5160

CIRCUIT AND SYSTEM THEORY Allerton conference on circuit and system theory 1st-6th Proceedings Monticello,Ill. 1963-68 Sponsored by the University of Illinois 6 vols University of Illinois,Evanston,Ill., 1963-68
Eng 41.5240

CIRCUIT THEORY International symposium on circuit and information theory transactions Los Angeles 1959 Jun.16-18 Institute of radio engineers Co-sponsored by the International scientific radio union Institute of radio engineers.Transactions on information theory,IT, 5,special supplement xiv,298p Institute of radio engineers,New York,1959 Papers published in advance of the Symposium
Math L 5.2071

The CIRCULATION OF THE BRAIN AND SPINAL CORD : a symposium on blood supply Proceedings of the Association New York 1937 Dec 27-28 Association for research in nervous and mental disease Edited by Stanley Cobb and others Association for research in nervous and mental disease.Research publications, 18 Hafner, New York,1966 Facsimile of 1938 edition
An 32.4133

CIRCULATORY AND RESPIRATORY MASS TRANSPORT : a Ciba foundation symposium London 1968 Jul Ciba foundation Edited by G.E.W. Wolstenholme and Julie Knight Churchill, London,1969
Inv Med 37.0236

CIRCULATORY AND RESPIRATORY MASS TRANSPORT : a Ciba foundation symposium London 1968 Jul 16-18 Ciba foundation Edited by G.E.W. Wolstenholme and Julie Knight Churchill, London,1969
An 32.5222

CIRCULATORY AND RESPIRATORY MASS TRANSPORT : a Ciba foundation symposium London 1968 Jul 16-18 Ciba foundation Edited by G.E.W. Wolstenholme and Julie Knight illus.,pl. x, 310p Churchill,London,1969
Phys 20.1969

CITLOGIA BATTERICA:BACTERIAL CYTOLOGY Congresso internazionale de microbiologia symposium 6e Edited by S. Mudd and G. Penso International congress of microbiology Paterno,Rome,1953
Bot 42.0666

CITRATE The Metabolic roles of citrate :a symposium in honour of...Sir Hans Krebs Oxford 1967 Jul Biochemical society Edited by T.W. Goodwin Biochemical society. Symposia, 27 Academic press,London;New York, 1968
Bioch 33.1389

CITY OF HOPE MEDICAL CENTER Amino acid pools distribution and function of free amino acids. Symposium Proceedings Duarte,Calif. 1961 May 19-22 Edited by Joseph T. Holden Under the auspices of the Institute for advanced learning in the medical sciences illus Elsevier,Amsterdam,1962 Preface:"The committee undertook the organization of the symposium under the title Conference on free amino acids
Bioch 33.0519

CITY OF HOPE MEDICAL CENTER Inhibition in the nervous system and gamma-aminobutyric acid an international symposium held at the City of Hope medical center Proceedings Duarte, Calif. 1959 May 22-24 Edited by Eugene Roberts and others Sponsored by the United States.Air force.Office of scientific research Pergamon press,Oxford,1960
Bal 39.1471

CITY UNIVERSITY.DEPARTMENT OF AERONAUTICS Road vehicle aerodynamics Symposium on road vehicle aerodynamics 1st Proceedings London 1969 Nov 6-7 Edited by A.J. Scibor-Rylski City university,London,1970 Organised in cooperation with the National physical laboratory,the Motor industry research association and the Institution of mechanical engineers
Eng 41.8336

CIVIL ENGINEERING Conference on biology and civil engineering Proceedings London 1948 Sep Institution of civil engineers Institution of civil engineers,London,1949
Geog 13.2688

CIVIL ENGINEERING Vibration in civil engineering :a symposium Proceedings London 1965 Apr Edited by B.O. Skipp Organised by the International association for earthquake engineering Butterworths,London, 1966
Eng 41.6363

The CIVITAS CAPITALS OF ROMAN BRITAIN :a conference Papers Leicester 1963 Dec 13-15 Edited by J.S. Wacher Bibliog,illus 124p Leicester university press,Leicester, 1966
Geog 13.3213

CLARKSON COLLEGE OF TECHNOLOGY The Conference on electromagnetic scattering : interdisciplinary conference Papers Potsdam,N.Y. 1962 Aug 13-15 Edited by Milton Kerker Co-sponsored by the American chemical society.Division of colloid and surface chemistry International series of monographs on electromagnetic waves, 5 591p Pergamon press,Oxford,1963
TA 15.0256

CLASSICAL CONDITIONING :a symposium Papers University Park,Pa. 1963 Aug Edited by William F. Prokasy Century psychology series Appleton-Century-Crofts,New York,1965 Organised by the editor
Psy 31.3015

CLASSIFICATION Conference on classification proceedings London 1962 Apr.6 Aslib Aslib proceedings, 14,no.8 215-266p 25cm Aslib,London,1962
Math 3.0776

CLASSIFICATION OF BRACKISH WATERS Symposium on the classification of brackish waters Venice 1958 Apr 8-14 By G.Sven Segerstrale International association of theoretical and applied limnology Archivio di oceanografia e limnologia, 11,suppl. Centro nazionale di studi talassografafici del consiglio nazionale delle ricerche,Venice,1959 Title also in Italian:"Simposio sulla classificazione della acque salmastre" Text in English and Italian.
Bal 39.1962

CLASSIFICATION OF CARBONATE ROCKS :a symposium Papers Denver,Colo. 1961 Apr 27 Edited by William E. Ham American association of petroleum geologists.Memoir,1 Tulsa,Okla.,1962 Held under the joint auspices of the American association of petroleum geologists and the Society of economic paleontologists and mineralogists
Geog 13.0301

CLASSIFICATION OF CARBONATE ROCKS :symposium papers Denver 1961 Apr 27 American association of petroleum geologists Edited by William E. Ham American ssociation of petroleum geologists.Memoirs,1 Tulsa,Okla., 1962
Geol 8.3157

CLAUSTHAL-GOTTINGEN 1965 Basic problems in thin film physics :international symposium Proceedings Edited by R. Niedermayer and H. Mayer VandenHoeck and Ruprecht,Gottingen, 1966
Eng 41.5450

CLAUSTHAL-GOTTINGEN 1965 Basic problems in thin film physics :international symposium Proceedings Edited by R. Niedermayer and H. Mayer illus 756p Vandenheck and Ruprecht, Gottingen,1966
Cav 7.0512

CLAUSTHAL-ZELLERFELD 1968 Texturen in forschung und praxis international symposium Proceedings Institut fur metallkunde und metallphysik der technischen universitat Clausthal Edited by J. Grewen and G. Wassermann Springer-Verlag,Berlin;Heidelberg, 1969
Met 25.2363

CLAY American institute of mining and metallurgical engineers :annual meeting 1951 Symposium:problems of clay and laterite genesis St.Louis,Mo. 1951 Feb 19-22 Sponsored by the Karl Eilers Memorial Fund A. I.M.E.,New York,1952
Min 10.0319

CLAYS AND CLAY MINERALS 6- :national conference proceedings Berkeley,Calif.- 1957- International series of monographs on earth sciences,2- Pergamon press,London,1959-
Min 10.0911

CLEAN STEEL Iron and steel institute annual general meeting,1962 Papers and discussions London 1962 Nov 28-29 Iron and steel institute Iron and steel institute.Special report, 77 Iron and steel institute,London, 1963
Met 25.0412

CLEAN SURFACES THEIR PREPARATION AND CHARACTERIZATION FOR INTERFACIAL STUDIES : based on a symposium Raleigh,N.C. 1968 Apr 8-10 North Carolina State university Edited by George Goldfinger Dekker,New York, 1970
Met 25.2828

CLEAR AIR TURBULENCE AND ITS DETECTION A Symposium on clear air turbulence and its detection Proceedings Seattle 1968 Aug 14-16 Boeing scientific research laboratories.Flight sciences laboratory Edited by Yih-Ho Pao and Arnold Goldburg 542p 22cm Plenum press,New York,1969
Cav 7.2921

CLERMONT-FERRAND 1962 Colloque de mathematiques proceedings Tome 2: calcul des probabilites;analyse numerique et calcul automatique; geometrie et physique mathematique Clermont-Ferrand.University. Faculte des sciences.Annales, 8, Mathematiques, 2 189p 1962 A l'occasion du tricentenaire de la mort de Blaise Pascal
Math 3.0680

CLERMONT-FERRAND 1964 Les Tendances geometriques en algebre et theorie des nombres colloque international Centre national de la recherche scientifique Edited by Marc Krasner Centre nationale de la recherche scientifique.Colloques internationaux, 143 Bibliog. 256p 25cm Centre national de la recherche scientifique,Paris,1966
P. Math 2.1775

CLERMONT-FERRAND 1969 Les Probabilites sur les structures algebriques Centre national de la recherche scientifique Centre national de la recherche scientifique.Colloques internationaux, 186 361p 25cm CNRS,Paris, 1970
P Math 2.4558

CLERMONT-FERRAND 1969 Les Probabilites sur les structures algebriques :colloque international Centre national de la recherche scientifique Centre national de la recherche scientifique.Colloques internationaux, 186 361p 25cm Editions du CNRS,Paris,1970 Oeganized by A. Badrikian and P.L.Hennequin
Math S 3.1780

CLERMONT-FERRAND 1949 Clermont-Ferrand et sa region Association francaise pour l'avancement des sciences.Congres 68e Clermont-Ferrand,c1949
Geog 13.4172

CLERMONT-FERRAND ET SA REGION Association francaise pour l'avancement des sciences. Congres 68e Clermont-Ferrand 1949 Clermont-Ferrand,c1949
Geog 13.4172

CLEVELAND 1936 The Plastic working metals a symposium Papers and discussions American society for metals American society for metals,Cleveland,Ohio,1936 The symposium presented before the eighteenth annual convention of the American society for metals.
Met 25.0679

CLEVELAND 1940 Strength of metals under combined stresses By Maxwell Gensamer American society for metals,Cleveland,Ohio, 1940 A series of educational lectures presented to members of the A.S.M. during the twenty-second National metal congress and exposition Cleveland, Ohio, October 2-25,1940
Met 25.0815

CLEVELAND 1940 Surface treatment of metals the symposium...presented before the twenty-second annual convention Papers and discussions American society for metals American society for metals,Cleveland,Ohio, 1941
Met 25.1972

CLEVELAND 1946 Induction heating By H.B. Osborn and others American society for metals A.S.M.,Cleveland,Ohio,1946
Met 25.0237

CLEVELAND 1948 Magnesium By L.M. Pidgeon and others A series of five educational lectures on magnesium presented to members of the A.S.M. during the twenty-seventh National metal congress and exposition,Cleveland, February 4 to 8.1946
Met 25.0542

CLEVELAND 1950 Interpretation of tests and correlation with service By M.F. Garwood and others American society for metals A.S.M., 1950 Lectures presented during the 32nd National metal congress and exposition
Met 25.0883

CLEVELAND 1952 Symposium on testing metal powders and metal powder products Papers American society for testing materials American society for testing materials.Special technical publication, 140 87p American society for testing materials,Philadelphia,Pa., 1953
Chem E 24.1121

CLEVELAND 1952 Symposium on testing metal powders and metal powder products presented at the Cleveland spring meeting American society for testing materials A.S.T.M. Special technical publication, 140 American society for testing materials,Philadelphia,Pa., 1953
Met 25.0762

CLEVELAND 1953 Relation of properties to microstructure American society for metals American society for metals,Cleveland,Ohio, 1954 A symposium presented at the 35th National metal congress and exposition
Met 25.1201

CLEVELAND 1953 Surface protection against wear and corrosion... :educational lectures... presented during the thirty-fifth national metal congress and exposition By H.S. Avery and others American society for metals American society for metals,Cleveland,Ohio, 1954
Met 25.1977

CLEVELAND 1956 Creep and recovery a seminar on creep and recovery of metals American society for metals American society for metals,Cleveland,1957 Held during the 38th National metal congress and exposition
Met 25.0950

CLEVELAND 1957 The Annealing of low carbon steel :the international symposium Proceedings Case institute of technology and Lee Wilson engineering company,inc. Lee Wilson engineering company,Cleveland,Ohio, 1958
Met 25.0447

CLEVELAND 1958 Magnetic properties of metals and alloys By R.M. Bozorth and others American society for metals American society for metals,Cleveland,Ohio,1959 A seminar presented to members of A.S.M.during the National metal congress and exposition
Met 25.2353

CLEVELAND 1958 Symposium on the fabrication of fuel elements :ceramic base elements ;metal base fuels and jacket components American institute of mining, metallurgical and petroleum engineers. Institute of metals division Nuclear metallurgy, 5 I.M.D.Special report series, 7 Metallurgical society of American institute of mining,metallurgical and petroleum engineers,New York,1958
Met 25.1543

CLEVELAND 1961 Progress in powder metallurgy Annual technical meeting 17th Proceedings Metal powder industries federation Metal powder industries federation,Cleveland,Ohio,1961
Met 25.0770

CLEVELAND 1963 Alloying behaviour and effects in concentrated solid solutions :based on a symposium American institute of mining, metallurgical and petroleum engineers Edited by T.B. Massalski Metallurgical society conferences, 29 Gordon and Breach,Ohio,1963
Met 25.1242

CLEVELAND 1963 Precipitation from iron-base alloys :a symposium American institute of mining,metallurgical and petroleum engineers.Ferrous metallurgy committee Edited by Gilbert R. Speich and John B. Clark Metallurgical society conferences Gordon and Breach,New York;London,1965
Met 25.2525

CLEVELAND 1963 The Sorby centennial symposium on the history of metallurgy :a symposium American society for metals and American institute of mining,metallurgical and petroleum engineers.Metallurgical society Edited by Cyril Stanley Smith Sponsored by the Society for the history of technology Metallurgical society conferences, 27 Gordon and Breach,New York,1965
Met 25.2141

CLEVELAND 1965 Anodized aluminium :a symposium presented during committee week Papers American society for testing materials.Subcommittee IV on electroplating practice A.S.T.M.,Philadelphia,1965
Met 25.2049

CLEVELAND 1965 International conference on sintering and related phenomena University of Notre Dame Edited by G.C. Kuczynski and others Gordon and Breach,New York,1967
Met 25.1285

CLEVELAND,OHIO 1940 Strength of metals under combined stresses :a series of educational lectures...presented...during the twenty-second National metal congress and exhibition By Maxwell Gensamer American society for metals American society for metals,Cleveland,Ohio,1941
Eng 41.3742

CLEVELAND,OHIO 1946 Acceptance sampling annual meeting 105th American statistical association 155p 23cm American statistical association,Washington,D. C.,1950
Math 3.0816

CLEVELAND,OHIO 1946 Surface stressing of metals :a series of five educational lectures.. presented...during the twenty-seventh National metal congress and exposition By H. F. Moore and others American society for metals American society for metals,Cleveland, Ohio,1947
Eng 41.3731

CLEVELAND,OHIO 1948 Quenching of steels :a compilation of papers presented at the National metal congress American society for metals American society for metals,Cleveland, Ohio,1951
Met 25.0446

CLEVELAND,OHIO 1949 Thermodynamics in physical metallurgy :a seminar held during the 31st national metal congress and exposition American society for metals American society for metals,Cleveland,Ohio,1952
Met 25.2372

CLEVELAND,OHIO 1952 Symposium on the use of radioisotopes in soil mechanics American society for testing materials American society for testing materials.Special technical publication, 134 American society for testing materials,Philadelphia,Pa.,1953
Eng 41.3147

CLEVELAND,OHIO 1953 Relation of properties to microstructure :a seminar held during the thirty-fifth National metal congress and exposition Sponsored by the American society for metals American society for metals, Cleveland,Ohio,1954
Eng 41.3501

CLEVELAND,OHIO 1953 Surface protection against wear and corrosion :two series of educational lectures...presented...during the thirty-fifth National metal congress and exposition By H.S. Avery and others American society for metals American society for metals,Cleveland,Ohio,1954
Eng 41.3715

CLEVELAND,OHIO 1956 A Symposium on the effects of radiation on metals American institute of mining,metallurgical and petroleum engineers.Nuclear metallurgy committee I.M.D.Special report series, 3 Nuclear metallurgy, 3 American institute of mining,metallurgical,and petroleum engineers,New York,1956
Met 25.1576

CLEVELAND,OHIO 1956 Metal thorium The Conference on thorium Proceedings American society for metals and United States atomic energy commission Edited by Harley A. Wilhelm American society for metals, Cleveland,Ohio,1958
Met 25.1561

CLEVELAND,OHIO 1957 High temperature materials :a conference American institute of mining,metallurgical and petroleum engineers.High temperature alloys committee Edited by R.F. Hehemann and G.Mervin Ault John Wiley;Chapman and Hall,New York;London, 1959
Met 25.1004

CLEVELAND,OHIO 1960 American chemical society national meeting 137th Saline water conversion :a symposium Papers American chemical society.Division of water and waste chemistry Advances in chemistry series American chemical society,Washington,D.C.,1960
Eng 41.7535

CLEVELAND,OHIO 1960 National symposium on vacuum technology 7th Transactions American vacuum society Edited by C.Robert Meissner Pergamon press,Oxford,1961 Earlier symposia published by the Committee on vacuum techniques
Met 25.1669

CLEVELAND,OHIO 1960 The National symposium on vacuum technology 7th Transactions American vacuum society Edited by C.Robert Meissner 427p Pergamon press,Oxford,1961 8th symposium called 'National vacuum symposium' q.v.
Cav 7.2375

CLEVELAND,OHIO 1961 High temperature materials 11 a technical conference American institute of mining,metallurgical and petroleum engineers.High temperature alloys committee Metallurgical society conferences, 18 Interscience,New York;London,1963
Met 25.1006

CLEVELAND,OHIO 1962 Operations research in research development :a conference Proceedings Edited by Barton V. Dean Held at the Case institute of technology Wiley, New York;London,1963
Eng 41.1173

CLEVELAND,OHIO 1963 Developments in mechanics Midwestern mechanics conference 6th Proceedings Vol 2,pt 1: fluid mechanics Edited by Simon Ostrach and Robert H. Scanlan Pergamon,London,1965 Combining the 6th Midwestern conference on soil mechanics and the 8th Midwestern conference on fluid mechanics
Eng 41.6959

CLEVELAND,OHIO 1963 New types of metal powders :a symposium Proceedings American institute of mining,metallurgical and petroleum engineers.Powder metallurgy committee Edited by Henry H. Hausner Metallurgical society conferences, 23 Gordon and Breach,New York;London,1964
Met 25.2509

CLEVELAND,OHIO 1963 New types of metal powders :a symposium Proceedings American institute of mining,metallurgical and petroleum engineers.Powder metallurgy committee Edited by Henry H. Hausner Metallurgical society conferences, 23 Gordon and Breach,New York;London,1964
Met 25.0766

CLEVELAND,OHIO 1963 Structure and properties of ultrahigh-strength steels :a symposium American society for testing materials and American institute of mining, metallurgical and petroleum engineers A.S.T.M.Special technical publication, 370 American society for testing and materials, Philadelphia,Pa.,1965
Met 25.2251

CLEVELAND,OHIO 1963 Views on general systems theory Systems symposium 2nd Proceedings Edited by Mihajlo D. Mesarovic Held at the Case institute of technology Case institute of technology.Systems research center.Publications Wiley,New York,1964
Eng 41.5967

CLEVELAND,OHIO 1963 Views on general systems theory The Second systems symposium proceedings Case institute of technology Edited by Mihajlo D. Mesarovic xvii,178p 24cm John Wiley and sons,New York;London, 1964
Math 3.0880

CLEVELAND,OHIO 1966 Atomic and electronic structure of metals :a seminar Papers American society for metals American society for metals,Cleveland,Ohio,1967
Met 25.1059

CLEVELAND,OHIO 1966 Systems theory and biology Systems symposium 3rd Proceedings Edited by M.D. Mesarovic Sponsored by the Case institute of technology. Systems research center Springer,Berlin,1968
Gen 34.0226

CLEVELAND,OHIO 1968 Theoretical approaches to non-numerical problem solving Systems symposium 4th Proceedings Edited by R.B. Banerji and M.D. Mesarovic Sponsored by Case western reserve university.Systems research center Lecture notes in operations research and mathematical systems, 28 bibliog.,illus. vi,466p 25cm Springer, Berlin,1970
Math S 3.1617

CLEVELAND,OHIO 1969 Conference on global differentiable dynamics Proceedings Case western reserve university Edited by O. Hajek and others Lecture notes in mathematics, 235 x,140p 25cm Springer, Berlin,1971
P Math 2.4317

CLEVELAND ELECTRONICS CONFERENCE A High speed magnetic core storage buffer By Ben T. Goda and Waclaw Pryciak Telemeter magnetics illus. 12p Telemeter magnetics,Los Angeles, Calif.,1959 Given by invitation at the Cleveland electronics conference,1959
Math L 5.1172

CLIMATE Biological significance of climatic changes in Britain Institute of biology symposium 14th Proceedings London 1964 Oct 29-30 Edited by C.G. Johnson and L.P. Smith Academic press,London,1965
Geog 13.1208

CLIMATE Solar variations,climatic change and related geophysical problems :conference Papers New York 1961 Jan 24-28 New York academy of sciences,and,American meteorological society Edited by Rhodes W. Fairbridge New York academy of sciences. Annals,95,art.1 New York academy of sciences, New York,1961
Geog 13.1055

CLIMATE :an inquiry into the causes of its differences,and its influence on vegetable life Torquay 1863 Feb By C. Daubeny Natural history society Bohn,Oxford,1863 Being the substance of four lectures delivered before the Natural history society
Bot 42.2042

CLIMATIC CHANGE Causes of climatic change :a collection of papers derived from the INQUA-NCAR symposium on causes of climatic change, August 30-31,Boulder,Colorado American meteorological society Edited by J.Murray Mitchell American meteorological society. Meteorological monographs, 8,no 30 Boston, Mass.,1968
Bot 42.6236

CLIMATIC CHANGES IN BRITAIN The Biological significance of climatic changes in Britain :a symposium held at the Royal geographical society Proceedings London 1964 Oct 29-30 Institute of biology Edited by C.G. Johnson and L.P. Smith Institute of biology. Symposia, 14 Academic press,London;New York, 1965
Bal 39.1687

CLIMATIC CHANGES IN GREAT BRITAIN The Biological significance of climatic changes in Britain Proceedings London 1964 Oct 29-30 Institute of biology Edited by C.G. Johnson and L.P. Smith Held at the Royal geographical society Institute of biology. Symposium, 14 illus x,222p Academic press,London;New York,1965
Bot 42.6678

CLIMATOLOGY Ground level climatology American association for the advancement of science :annual meeting 132nd Symposium on ground level climatology Berkeley,Calif. 1965 Dec Edited by Robert H. Shaw American association for the advancement of science. Publications,86 Washington,D.C.,1967
Geog 13.1034

CLIMATOLOGY AND MICROCLIMATOLOGY :symposium Proceedings Canberra 1956 Oct Unesco, and,Commonwealth scientific and industrial research organization Arid zone research,11 Unesco,Paris,1958
Geog 13.1018

CLINICAL ACTH CONFERENCE 2nd Proceedings Chicago 1950 Dec 8-9 Vol 1-2 Edited by John R. Mote Sponsored by Armour and company 2 vols Churchill,London,1951
Bioch 33.0443

CLINICAL ASPECTS OF GENETICS Royal college of physicians scientific conference 5th Proceedings London 1961 Mar 17-18 Royal college of physicians Edited by F.Avery Jones Pitman medical,London,1961
Gen 34.1981

CLINICAL ASPECTS OF GENETICS :a conference Proceedings London 1961 Mar 17-18 Royal college of physicians of London Edited by F.Avery Jones Pitman,London,1961
PGMS 29.0255

CLINICAL CHEMISTRY International congress on clinical chemistry 4th Proceedings Edinburgh 1960 Aug 14-19 Organised by the Association of clinical biochemists E.and S. Livingstone,Edinburgh;London,1961
Bioch 33.2264

CLINICAL ELECTRORETINOGRAPHY 3rd : international symposium Proceedings Highland Park,Ill. 1964 Oct 14-16 International society for clinical electroretinography Edited by Hermann M. Burian and Jerry Hart Jacobsen Pergamon press.Symposium publication division,Oxford, 1966 Supplement to 'Visual research'
Psy 31.0383

CLINICAL ELECTRORETINOGRAPHY INTERNATIONAL SYMPOSIUM 3rd Proceedings Highland Park,Ill. 1964 Oct 14-16 International society for clinical electroretinography Edited by Hermann M. Burian and Jerry Hart Jacobson Vision research.Supplement Pergamon,Oxford,1966
PGMS 29.0340

CLINICAL EVALUATION IN BREAST CANCER Imperial cancer research fund symposium 1st Proceedings London 1965 Oct 21-23 Edited by J.L. Haywood and R.D. Bulbrook illus. Academic press,London;New York,1966
Radioth 35.0870

CLINICAL MANAGEMENT OF ADVANCED BREAST CANCER Tenovus workshop 2nd Proceedings Cardiff 1970 Tenovus institute for cancer research Edited by C.A.F. Joslin and E.N. Gleave Alpha omega alpha,Cardiff,1971
Radioth 35.1927

CLINICAL SYMPOSIUM ON IRON DEFICIENCY Proceedings Iron deficiency;pathogenesis, clinical aspects and therapy Basle 1969 Edited by L. Hallberg and others Colloquia Geigy Academic press,London,1970
Med 36.0383

CLINICAL TRIALS :a symposium Report London 1962 Apr 5 Pharmaceutical society of Great Britain Pharmaceutical press,London, 1962
Pha 16.0130

CLINICAL USES OF WHOLE-BODY COUNTING Panel on clinical uses of whole-body counting Proceedings Vienna 1965 Jun 28-Jul 2 International atomic energy agency International atomic energy agency.Panel proceedings series International atomic energy agency,Vienna,1966
Radioth 35.1383

CLOSE BINARIES Mass loss and evolution in close binaries Elsinore 1969 International astronomical union Edited by K. Gyldenkerne and R.M. West International astronomical union.Colloquium, 6 238p Copenhagen university observatory,Copenhagen, 1970
Obs 6.3358

CLOUD PHYSICS International conference on cloud physics Proceedings Toronto 1968 American meteorological society 873p American meteorological society,Boston,Mass., 1969
Nap 11.0973

CLOUD PHYSICS CONFERENCE 2nd Proceedings Physics of precipitation Woods Hole,Mass. 1959 Jun 3-5 American geophysical union. Cloud physics committee,and,National science foundation Edited by Helmut Weickmann Geophysical monograph,5 National research council.Publication,746 illus,map xii,435p 25cm American geophysical union,Washington, 1960
Sco 14.0536

CLOUDCROFT,N.M. 1961 The Solar corona :a symposium proceedings International astronomical union Edited by John W. Evans Sponsored by the Air Force Cambridge research laboratories International astronomical union.Symposium, 16 344p Academic press, New York;London,1963
Obs 6.0564

CO-ORDINATING COMMITTEE FOR SYMPOSIA ON DRUG ACTION A Symposium on the interaction of drugs and subcellular components in animal cells London 1967 Apr 10-11 Edited by P. N. Campbell Illus Churchill,London,1968
Radioth 35.0273

CO-ORDINATING COMMITTEE FOR SYMPOSIA ON DRUG ACTION Enzymes and drug action London 1961 Mar 20-23 Edited by J.L. Mongar and A.V. S. De Reuck Churchill,London,1962
Bioch 33.1872

CO-ORDINATION OF GALACTIC RESEARCH a symposium International astronomical union Edited by A. Blaauw International astronomical union.Symposium, 1 59p Cambridge university press,Cambridge,1955
Obs 6.0191

CO-ORDINATION OF GALACTIC RESEARCH 2nd : conference Saltsjobaden 1957 Jun.17-22 International astronomical union Edited by A. Blaauw and others International astronomical union.Symposium, 7 93p Cambridge university press,Cambridge,1959
Obs 6.0192

CO-ORDINATION CHEMISTRY International conference on co-ordination chemistry Lectures delivered and abstracts of papers London 1959 Apr 6-11 Chemical society,and, International union of pure and applied chemistry Chemical society.Special publication,13 Chemical society,London,1959
Chem 18.1220

COAL PETROLOGY International congress on coal petrology 1st proceedings Heerlen 1958 Sep 10-13 International committee for coal petrology.Proceedings,3 Ernest van Aelst,Maastricht,1960
Geol 8.3137

COAL PREPARATION Symposium on coal preparation 2nd Papers Leeds 1957 Oct 21-25 University of Leeds.Department of mining xiii,513p University of Leeds,Leeds, 1957
Chem E 24.1392

COASTAL ENGINEERING Conference on coastal engineering 10th Proceedings Tokyo 1966 Sep American society of civil engineers 2 vols American society of civil engineers, New York,1967
Eng 41.3348

COASTAL ENGINEERING Conference on coastal engineering 11th Proceedings London 1968 Sep Vol 1-2 American society of civil engineers Institution of civil engineers 2 vols American society of civil engineers,New York,1969
Eng 41.3347

COASTAL ENGINEERING :a specialty conference Santa Barbara,Calif. 1965 Oct American society of civil engineers American society of civil engineers,New York,1966
Eng 41.3346

COASTAL GEOGRAPHY CONFERENCE 2nd Louisiana state university 1959 Apr 6-9 By Richard J. Russell Sponsored by the Office of naval research.Geography branch,Washington, D.C.,1959
Geog 13.0712

COBALT-60 BEAM THERAPY Symposium on cobalt-60 beam therapy in malignant disease Edited by S. Krishnamurthi and V. Shanta Associated printers,Madras,c1962
Radioth 35.1134

COCURRENT GAS-LIQUID FLOW International symposium on research in cocurrent gas-liquid flow Proceedings Waterloo,Ont. 1968 Sep 18-19 Edited by Edward Rhodes and Donald S. Scott Canadian society for chemical engineering.Symposium series, 1 illus ix, 698p Plenum press,New York,1969
Chem E 24.1842

COENZYMES Vitamin B 12 coenzymes :a conference Papers New York 1963 Apr 10-11 New York academy of sciences Edited by Harold E. Whipple New York academy of sciences.Annals, 112,p.547-921 New York,1964
Bioch 33.0565

COGNITION Concepts and the structure of memory Symposium on cognition 2nd Papers Pittsburgh,Pa. 1966 Apr 7-8 Edited by Benjamin Kleinmuntz Sponsored by the Carnegie institute of technology Wiley, New York;London,1967
Eng 41.0652

COGNITION Formal representation of human judgement Symposium on cognition 3rd Papers Pittsburgh,Pa. 1967 Apr 13-14 Edited by Benjamin Kleinmuntz Sponsored by the Carnegie-Mellon university Wiley,New York;London,1968
Eng 41.0653

COGNITION Problem solving:research,method and theory Annual symposium on cognition 1st Papers Pittsburgh 1965 Apr 15-16 Carnegie institute of technology Edited by Benjamin Kleinmutz Wiley,New York,1966
Psy 31.1911

COGNITION Problem solving:research,method and theory Symposium on cognition 1st Pittsburgh,Pa. 1966 Edited by Benjamin Kleinmuntz Sponsored by the Carnegie institute of technology Wiley,New York; London,1966
Eng 41.1291

COGNITION,THEORY,PROMISE :papers read at the Martin Scheerer memorial meetings on cognitive psychology Papers University of Kansas 1962 May 7-9 Edited by Constance Scheerer Harper and Row,New York,1964
Psy 31.1919

COGNITION AND THE DEVELOPMENT OF LANGUAGE 4th annual symposium Papers Pittsburgh,Pa. 1968 Apr 11-12 Carnegie-Mellon university Edited by John R. Hayes Wiley,New York,1970
Psy 31.3113

COHESIVE SOILS Research conference on shear strength of cohesive soils New York 1960 University of Colorado American society of civil engineers Boulder,Colo.,1960
Eng 41.3171

COHOMOLOGY Homotopie et cohomologie manuscript Montreal 1964 By Beno Eckmann North Atlantic treaty organization Societe mathematique du Canada University of Montreal.Seminaire de mathematiques superieures, 11 129p 28cm Universite de Montreal,Montreal,1965
P. Math 2.0297

COINS, 3 Software engineering Symposium on computer and information sciences 3rd Proceedings Miami,Fla. 1969 Dec Vol 1-2 Edited by Julius T. Tou illus 2 vols Academic press,New York,1970
Math L 5.3830

COL DE VOZA 1961 Deoxyribonucleic acid, structure,synthesis and function Annual reunion of the society of chemical physicists 11th Proceedings Society of chemical physicists Pergamon press,Oxford,1961
Radioth 35.0111

COLD INJURY Conference on cold injury 1st Transactions 1-6 Sponsored by the Josiah Macy foundation 248p 23cm 6 vols New York,1952-60
Sco 14.0375

COLD SPRING HARBOR Sep 1970 The Bacteriphage lambda :conference Proceedings Cold Spring Harbor laboratory of quantitative biology Edited by A.D. Hershey Cold Spring Harbor symposia on quantitative biology Cold Spring Harbor, N.Y.,1971
Bioch 33.2295

COLD SPRING HARBOR 1933 Cold Spring Harbor symposia on quantitative biology Papers Vol 1 Cold Spring Harbor biological laboratory Long Island biological association,Cold Spring Harbor,1933 Later referred to in vol.9 as "Surface phenomena"
Bioch 33.1257

COLD SPRING HARBOR 1934 Cold Spring Harbor symposia on quantitative biology Papers Vol 2 Cold Spring Harbor biological laboratory Long Island biological association,Cold Spring Harbor,1934 Later referred to in vol.9 as "Aspects of growth"
Bioch 33.1258

COLD SPRING HARBOR 1935 Cold Spring Harbor symposia on quantitative biology Papers Vol 3 Cold Spring Harbor biological laboratory Long Island biological association,Cold Spring Harbor,1935 Later referred to in vol.9 as "Photochemical reactions"
Bioch 33.1259

COLD SPRING HARBOR 1936 Cold Spring Harbor symposia on quantitative biology Papers Vol 4 Cold Spring Harbor biological laboratory Long Island biological association,Cold Spring Harbor,1936 Later referred to in vol.9 as "Excitation phenomena"
Bioch 33.1260

COLD SPRING HARBOR 1937 Cold Spring Harbor symposia on quantitative biology Papers Vol 5 Cold Spring Harbor biological laboratory Long Island biological association,Cold Spring Harbor,1937 Later referred to in vol. 9 as "Internal secretions"
Bioch 33.1261

COLD SPRING HARBOR 1938 Cold Spring Harbor symposia on quantitative biology Papers Cold Spring Harbor biological laboratory Long Island biological association,Cold Spring Harbor,1938 Later referred to in vol.9 as "Protein chemistry"
Bioch 33.1262

COLD SPRING HARBOR 1939 Cold Spring Harbor symposia on quantitative biology Vol 7 Cold Spring Harbor biological laboratory Long Island biological association,Cold Spring Harbor,1939 Later referred to in vol.9 as "Biological oxidations"
Bioch 33.1263

COLD SPRING HARBOR 1941 Genes and chromosones :structure and organization.A symposium Papers Cold Spring Harbor biological laboratory Cold Spring Harbor symposia on quantitative biology, 9 Long Island biological association,Cold Spring Harbor,1941
Bioch 33.1265

COLD SPRING HARBOR 1942 The Relation of hormones to development :a symposium Papers Cold Spring Harbor biological laboratory Cold Spring Harbor symposia on quantitative biology, 10 Long Island biological association,Cold Spring Harbor,1942
Bioch 33.1266

COLD SPRING HARBOR 1946 Heredity and variation in microorganisms :a symposium Papers Cold Spring Harbor biological laboratory Cold Spring Harbor symposia on quantitative biology, 11 Long Island biological association,Cold Spring Harbor,1946
Bioch 33.1267

COLD SPRING HARBOR 1947 Nucleic acids and nucleoproteins Cold Spring Harbor laboratory of quantitative biology Cold Spring Harbor symposia on quantitative biology, 12 Port Biological laboratory,New York,1947
Radioth 35.0157

COLD SPRING HARBOR 1947 Nucleic acids and nucleoproteins :a symposium Papers Cold Spring Harbor biological laboratory Cold Spring Harbor symposia on quantitative biology, 12 Long Island biological association,Cold Spring Harbor,1947
Bioch 33.1268

COLD SPRING HARBOR 1948 Biological application of tracer elements :a symposium papers Cold Spring Harbor laboratory of quantitative biology Cold Spring Harbor symposia on quantitative biology, 13 Biological laboratory,New York,1948
Radioth 35.0158

COLD SPRING HARBOR 1948 Biological applications of tracer elements :a symposium Papers Cold Spring Harbor biological laboratory Cold Spring Harbor symposia on quantitative biology, 13 Long Island biological association,Cold Spring Harbor,1948
Bioch 33.1269

COLD SPRING HARBOR 1950 Amino acids and proteins :a symposium Papers Cold Spring Harbor biological laboratory Cold Spring Harbor symposia on quantitative biology, 14 Long Island biological association,Cold Spring Harbor,1950
Bioch 33.1270

COLD SPRING HARBOR 1950 Origin and evolution of man :a symposium Edited by Katherine Brehme Warren Cold Spring Harbor symposia on quantitative biology, 15 Long Island biological association,Cold Spring Harbor,N.Y.,1950
An 32.2807

COLD SPRING HARBOR 1950 Origin and evolution of man :a symposium Papers Cold Spring Harbor biological laboratory Cold Spring Harbor symposia on quantitative biology, 15 Long Island biological association,Cold Spring Harbor,1950
Bioch 33.1271

COLD SPRING HARBOR 1951 Genes and mutation a symposium Papers Cold Spring Harbor biological laboratory Cold Spring Harbor symposia on quantitative biology, 16 Long Island biological association,Cold Spring Harbor,1951
Bioch 33.1272

COLD SPRING HARBOR 1952 The Neuron :a symposium Cold Spring Harbor symposia on quantitative biology, 17 Cold Spring Harbor, N.Y.,1952
Phys 20.0956

COLD SPRING HARBOR 1952 The Neuron :a symposium Papers Cold Spring Harbor biological laboratory Cold Spring Harbor symposia on quantitative biology, 17 Long Island biological association,Cold Spring Harbor,1952
Bioch 33.1273

COLD SPRING HARBOR 1952 The Neuron :a ymposium Cold Spring Harbor biological laboratory Edited by Katherine Brehme Warren Cold Spring Harbor symposia on quantitative biology, 17 Biological laboratory,Cold Spring Harbor,1952
An 32.4330

COLD SPRING HARBOR 1953 Viruses :a symposium Papers Cold Spring Harbor biological laboratory Cold Spring Harbor symposia on quantitative biology, 18 Long Island biological association,Cold Spring Harbor,1953
Bioch 33.1274

COLD SPRING HARBOR 1954 The Mammalian fetus :physiological aspects of development.A symposium Papers Cold Spring Harbor biological laboratory Cold Spring Harbor symposia on quantitative biology, 19 Long Island biological association,Cold Spring Harbor,1954
Bioch 33.1275

COLD SPRING HARBOR 1954 The Mammalian fetus:physiological aspects of development : symposium Cold Spring Harbor symposia on quantitative biology, 19 Long Island biological association,New York,1954
Phys 20.1376

COLD SPRING HARBOR 1955 Population genetics :the nature and causes of genetic variability in populations.A symposium Papers Cold Spring Harbor biological laboratory Cold Spring Harbor symposia on quantitative biology, 20 Long Island biological association,Cold Spring Harbor,1955 Dedicated to Vannevar Bush,President of the Carnegie Institute of Washington, 1939-1955
Bioch 33.1276

COLD SPRING HARBOR 1956 Genetic mechanisms structure and function.A symposium Papers Cold Spring Harbor biological laboratory Cold Spring Harbor symposia on quantitative biology, 21 Long Island biological association,Cold Spring Harbor,1956
Bioch 33.1277

COLD SPRING HARBOR 1957 Population studies animal ecology and demography.A symposium Papers Cold Spring Harbor biological laboratory Cold Spring Harbor symposia on quantitative biology, 22 Long Island biological association,Cold Spring Harbor,1957
Bioch 33.1278

COLD SPRING HARBOR 1958 Comparative endocrinology :symposium Proceedings Edited by A. Gorbman Wiley;Chapman and Hall, New York;London,1959
VA 19.0286

COLD SPRING HARBOR 1958 Comparative endocrinology :symposium Proceedings Edited by A. Gorbman Wiley;Chapman and Hall, New York;London,1959
Phys 20.1386

COLD SPRING HARBOR 1958 Exchange of genetic material :mechanisms and consequences. A symposium Papers Cold Spring Harbor biological laboratory Cold Spring Harbor symposia on quantitative biology, 23 Long Island biological association,Cold Spring Harbor,1958
Bioch 33.1279

COLD SPRING HARBOR 1959 Genetics and twentieth century Darwinism :a symposium Edited by Clara Wooldridge Cold Spring Harbor symposia on quantitative biology, 24 Biological laboratory,Cold Spring Harbor,L.I., 1959
An 32.0043

COLD SPRING HARBOR 1959 Genetics and twentieth century Darwinism :a symposium Papers Cold Spring Harbor biological laboratory Cold Spring Harbor symposia on quantitative biology, 24 Long Island biological association,Cold Spring Harbor,1959
Bioch 33.1280

COLD SPRING HARBOR 1960 Biological clocks : a symposium Papers Cold Spring Harbor biological laboratory Cold Spring Harbor symposia on quantitative biology, 25 Long Island biological association,Cold Spring Harbor,1960
Bioch 33.1281

COLD SPRING HARBOR 1961 Cellular regulatory mechanisms :a symposium Papers Cold Spring Harbor biological laboratory Cold Spring Harbor symposia on quantitative biology, 26 Long Island biological association,Cold Spring Harbor,1961
Bioch 33.1282

COLD SPRING HARBOR 1962 Basic mechanisms in animal virus biology Papers Cold Spring Harbor.Biological laboratory Cold Spring Harbor symposia on quantitative biology, 27 Long Island biological association.Biological laboratory,Cold Spring Harbor,L.I.,N.Y.,1962
Path 30.2559

COLD SPRING HARBOR 1962 Basic mechanisms in animal virus biology :a symposium Papers Cold Spring Harbor biological laboratory Cold Spring Harbor symposia on quantitative biology, 27 Long Island biological association,Cold Spring Harbor,1962
Bioch 33.1283

COLD SPRING HARBOR 1963 Synthesis and structure of macromolecules :a symposium Papers Cold Spring Harbor laboratory of quantitative biology Cold Spring Harbor symposia on quantitative biology, 28 Cold Spring Harbor,1963
Bioch 33.1284

COLD SPRING HARBOR 1963 Synthesis and structure of macromolecules :a symposium Papers Cold Spring Harbor laboratory of quantitative biology Cold Spring Harbor symposia on quantitative biology, 28 Cold Spring Harbor laboratory of quantitative biology,Cold Spring Harbor,1963
Radioth 35.0245

COLD SPRING HARBOR 1964 Human genetics :a symposium Edited by Leonora Frisch Cold Spring Harbor symposia on quantitative biology, 29 Cold Spring Harbor laboratory on quantitative biology,Cold Spring Harbor, L.I., 1964
An 32.0051

COLD SPRING HARBOR 1964 Human genetics :a symposium Papers Cold Spring Harbor laboratory of quantitative biology Cold Spring Harbor symposia on quantitative biology, 29 Cold Spring Harbor,1964
Bioch 33.1285

COLD SPRING HARBOR 1965 Sensory receptors : a symposium Cold Spring Harbor symposia on quantitative biology, 30 Cold Spring Harbor, N.Y.,1965
Phys 20.1058

COLD SPRING HARBOR 1965 Sensory receptors : a symposium Papers Cold Spring Harbor laboratory of quantitative biology Cold Spring Harbor symposia on quantitative biology, 30 Cold Spring Harbor,1965
Bioch 33.1286

COLD SPRING HARBOR 1966 The Genetic code : a symposium Papers Cold Spring Harbor laboratory of quantitative biology Cold Spring Harbor symposia on quantitative biology, 31 Cold Spring Harbor,1966
Bioch 33.1287

COLD SPRING HARBOR 1967 Antibodies :a symposium Papers Cold Spring Harbor laboratory of quantitative biology Cold Spring Harbor symposia on quantitative biology, 32 Cold Spring Harbor,1967
Bioch 33.1288

COLD SPRING HARBOR 1967 Antibodies :a symposium Proceedings Cold Spring Harbor laboratory of quantitative biology Edited by Leonora Frisch Cold Spring Harbor symposia on quantitative biology, 32 Cold Spring Harbor laboratory of quantitative biology,Cold Spring Harbor,L.I.,N.Y.,1967
Path 30.2523

COLD SPRING HARBOR 1968 Replication of DNA in micro-organism :a symposium Papers Cold Spring Harbor laboratory of quantitative biology Cold Spring Harbor symposia on quantitative biology, 33 Cold Spring Harbor, 1968
Bioch 33.1289

COLD SPRING HARBOR 1968 Replication of DNA in micro-organisms :a symposium Papers Cold Spring Harbor laboratory of quantitative biology Cold Spring Harbor symposia on quantitative biology, 33 2 vols Cold Spring Harbor,1968
Radioth 35.0246

COLD SPRING HARBOR 1969 Lactose operon : conference Proceedings Cold Spring Harbor laboratory of quantitative biology Cold Spring harbor symposia on quantitative biology Cold Spring Harbor,L.I.,N.Y.,1970
Bioch 33.2251

COLD SPRING HARBOR 1969 The Mechanism of protein synthesis :a symposium Proceedings Cold Spring Harbor laboratory of quantitative biology Cold Spring Harbor symposia on quantitative biology, 34 Cold Spring Harbor, 1969
Bioch 33.1887

COLD SPRING HARBOR 1970 Transcription of genetic material :a symposium Papers Cold Spring Harbor laboratory of quantitative biology Cold Spring Harbor symposia on quantitative biology, 35 Cold Spring Harbor, 1971
Bioch 33.2335

COLD SPRING HARBOR 1971 Replication of DNA a symposium Abstracts Cold Spring Harbor laboratory of quantitative biology Cold Spring Harbor symposia on quantitative biology Cold Spring Harbor,1971
Bioch 33.2333

COLD SPRING HARBOR 1972 Structure and function of proteins at the three-dimensional level :a symposium Papers Cold Spring Harbor laboratory of quantitative biology Cold Spring Harbor symposia on quantitative biology, 36 Cold Spring Harbor,1972
Bioch 33.2334

COLD SPRING HARBOR.BIOLOGICAL LABORATORY
Basic mechanisms in animal virus biology Papers Cold Spring Harbor 1962 Jun 7-13 Cold Spring Harbor symposia on quantitative biology, 27 Long Island biological association.Biological laboratory,Cold Spring Harbor,L.I.,N.Y.,1962
Path 30.2559

COLD SPRING HARBOR,L.I.,N.Y. 1935 Cold Spring Harbor symposia on quantitative biology Papers Vol 3 Long Island biological association.Biological laboratory Long Island biological association,Cold Spring Harbor,L.I.,N.Y.,1935
Chem 18.2002

COLD SPRING HARBOR,L.I.,N.Y. 1965 Sensory receptors :a symposium Cold Spring Harbor laboratory of quantitative biology Edited by John Cairns Cold Spring Harbor symposia on quantitative biology, 30 Cold Spring Harbor laboratory of quantitative biology,Cold Spring Harbor,L.I.,N.Y.,1965
Psy 31.0298

COLD SPRING HARBOR BIOLOGICAL ASSOCIATION
Cold Spring Harbor symposia on quantitative biology Papers Cold Spring Harbor,L.I.,N.Y. 1938 Vol 6 Long Island biological laboratory,Cold Spring Harbor,L.I.,N.Y.,1938
Bot 42.6450

COLD SPRING HARBOR BIOLOGICAL LABORATORY
Amino acids and proteins :a symposium Papers Cold Spring Harbor 1950 Cold Spring Harbor symposia on quantitative biology, 14 Long Island biological association,Cold Spring Harbor,1950
Bioch 33.1270

COLD SPRING HARBOR BIOLOGICAL LABORATORY
Basic mechanisms in animal virus biology :a symposium Papers Cold Spring Harbor 1962 Cold Spring Harbor symposia on quantitative biology, 27 Long Island biological association,Cold Spring Harbor,1962
Bioch 33.1283

COLD SPRING HARBOR BIOLOGICAL LABORATORY
Biological applications of tracer elements :a symposium Papers Cold Spring Harbor 1948 Cold Spring Harbor symposia on quantitative biology, 13 Long Island biological association,Cold Spring Harbor,1948
Bioch 33.1269

COLD SPRING HARBOR BIOLOGICAL LABORATORY
Biological clocks : a symposium Papers Cold Spring Harbor 1960 Cold Spring Harbor symposia on quantitative biology, 25 Long Island biological association,Cold Spring Harbor,1960
Bioch 33.1281

COLD SPRING HARBOR BIOLOGICAL LABORATORY
Cellular regulatory mechanisms :a symposium Papers Cold Spring Harbor 1961 Cold Spring Harbor symposia on quantitative biology, 26 Long Island biological association,Cold Spring Harbor,1961
Bioch 33.1282

COLD SPRING HARBOR BIOLOGICAL LABORATORY
Cold Spring Harbor symposia on quantitative biology Cold Spring Harbor 1939 Vol 7 Long Island biological association,Cold Spring Harbor,1939 Later referred to in vol.9 as "Biological oxidations"
Bioch 33.1263

COLD SPRING HARBOR BIOLOGICAL LABORATORY
Cold Spring Harbor symposia on quantitative biology Papers Cold Spring Harbor 1933 Vol 1 Long Island biological association,Cold Spring Harbor,1933 Later referred to in vol.9 as "Surface phenomena"
Bioch 33.1257

COLD SPRING HARBOR BIOLOGICAL LABORATORY
Cold Spring Harbor symposia on quantitative biology Papers Cold Spring Harbor 1934 Vol 2 Long Island biological association,Cold Spring Harbor,1934 Later referred to in vol.9 as "Aspects of growth"
Bioch 33.1258

COLD SPRING HARBOR BIOLOGICAL LABORATORY
Cold Spring Harbor symposia on quantitative biology Papers Cold Spring Harbor 1935 Vol 3 Long Island biological association,Cold Spring Harbor,1935 Later referred to in vol.9 as "Photochemical reactions"
Bioch 33.1259

COLD SPRING HARBOR BIOLOGICAL LABORATORY
Cold Spring Harbor symposia on quantitative biology Papers Cold Spring Harbor 1936 Vol 4 Long Island biological association,Cold Spring Harbor,1936 Later referred to in vol.9 as "Excitation phenomena"
Bioch 33.1260

COLD SPRING HARBOR BIOLOGICAL LABORATORY
Cold Spring Harbor symposia on quantitative biology Papers Cold Spring Harbor 1937 Vol 5 Long Island biological association,Cold Spring Harbor,1937 Later referred to in vol. 9 as "Internal secretions"
Bioch 33.1261

COLD SPRING HARBOR BIOLOGICAL LABORATORY
Cold Spring Harbor symposia on quantitative biology Papers Cold Spring Harbor 1938 Long Island biological association,Cold Spring Harbor,1938 Later referred to in vol.9 as "Protein chemistry"
Bioch 33.1262

COLD SPRING HARBOR BIOLOGICAL LABORATORY
Cold Spring Harbor symposia on quantitative biology Papers Vol 8 Long Island biological association,Cold Spring Harbor,1940 Later referred to in vol.9 as "Permeability and the nature of cell membranes"
Bioch 33.1264

COLD SPRING HARBOR BIOLOGICAL LABORATORY
Exchange of genetic material :mechanisms and consequences.A symposium Papers Cold Spring Harbor 1958 Cold Spring Harbor symposia on quantitative biology, 23 Long Island biological association,Cold Spring Harbor,1958
Bioch 33.1279

COLD SPRING HARBOR BIOLOGICAL LABORATORY
Genes and chromosones :structure and organization.A symposium Papers Cold Spring Harbor 1941 Cold Spring Harbor symposia on quantitative biology, 9 Long Island biological association,Cold Spring Harbor,1941
Bioch 33.1265

COLD SPRING HARBOR BIOLOGICAL LABORATORY Genes and mutation :a symposium Papers Cold Spring Harbor 1951 Cold Spring Harbor symposia on quantitative biology, 16 Long Island biological association,Cold Spring Harbor,1951
Bioch 33.1272

COLD SPRING HARBOR BIOLOGICAL LABORATORY Genetic mechanisms :structure and function.A symposium Papers Cold Spring Harbor 1956 Cold Spring Harbor symposia on quantitative biology, 21 Long Island biological association,Cold Spring Harbor,1956
Bioch 33.1277

COLD SPRING HARBOR BIOLOGICAL LABORATORY Genetics and twentieth century Darwinism :a symposium Papers Cold Spring Harbor 1959 Cold Spring Harbor symposia on quantitative biology, 24 Long Island biological association,Cold Spring Harbor,1959
Bioch 33.1280

COLD SPRING HARBOR BIOLOGICAL LABORATORY Heredity and variation in microorganisms :a symposium Papers Cold Spring Harbor 1946 Cold Spring Harbor symposia on quantitative biology, 11 Long Island biological association,Cold Spring Harbor,1946
Bioch 33.1267

COLD SPRING HARBOR BIOLOGICAL LABORATORY Nucleic acids and nucleoproteins :a symposium Papers Cold Spring Harbor 1947 Cold Spring Harbor symposia on quantitative biology, 12 Long Island biological association,Cold Spring Harbor,1947
Bioch 33.1268

COLD SPRING HARBOR BIOLOGICAL LABORATORY Origin and evolution of man :a symposium Papers Cold Spring Harbor 1950 Cold Spring Harbor symposia on quantitative biology, 15 Long Island biological association,Cold Spring Harbor,1950
Bioch 33.1271

COLD SPRING HARBOR BIOLOGICAL LABORATORY Population genetics :the nature and causes of genetic variability in populations.A symposium Papers Cold Spring Harbor 1955 Cold Spring Harbor symposia on quantitative biology, 20 Long Island biological association,Cold Spring Harbor,1955 Dedicated to Vannevar Bush,President of the Carnegie Institute of Washington, 1939-1955
Bioch 33.1276

COLD SPRING HARBOR BIOLOGICAL LABORATORY Population studies :animal ecology and demography.A symposium Papers Cold Spring Harbor 1957 Cold Spring Harbor symposia on quantitative biology, 22 Long Island biological association,Cold Spring Harbor,1957
Bioch 33.1278

COLD SPRING HARBOR BIOLOGICAL LABORATORY The Mammalian fetus :physiological aspects of development.A symposium Papers Cold Spring Harbor 1954 Cold Spring Harbor symposia on quantitative biology, 19 Long Island biological association,Cold Spring Harbor,1954
Bioch 33.1275

COLD SPRING HARBOR BIOLOGICAL LABORATORY The Neuron :a symposium Papers Cold Spring Harbor 1952 Cold Spring Harbor symposia on quantitative biology, 17 Long Island biological association,Cold Spring Harbor,1952
Bioch 33.1273

COLD SPRING HARBOR BIOLOGICAL LABORATORY The Relation of hormones to development :a symposium Papers Cold Spring Harbor 1942 Cold Spring Harbor symposia on quantitative biology, 10 Long Island biological association,Cold Spring Harbor,1942
Bioch 33.1266

COLD SPRING HARBOR BIOLOGICAL LABORATORY Viruses :a symposium Papers Cold Spring Harbor 1953 Cold Spring Harbor symposia on quantitative biology, 18 Long Island biological association,Cold Spring Harbor,1953
Bioch 33.1274

COLD SPRING HARBOR LABORATORY OF QUANTITATIVE BIOLOGY Antibodies :a symposium Papers Cold Spring Harbor 1967 Cold Spring Harbor symposia on quantitative biology, 32 Cold Spring Harbor,1967
Bioch 33.1288

COLD SPRING HARBOR LABORATORY OF QUANTITATIVE BIOLOGY Antibodies :a symposium Proceedings Cold Spring Harbor 1967 Jun 1-7 Edited by Leonora Frisch Cold Spring Harbor symposia on quantitative biology, 32 Cold Spring Harbor laboratory of quantitative biology,Cold Spring Harbor,L.I.,N.Y.,1967
Path 30.2523

COLD SPRING HARBOR LABORATORY OF QUANTITATIVE BIOLOGY Human genetics :a symposium Papers Cold Spring Harbor 1964 Cold Spring Harbor symposia on quantitative biology, 29 Cold Spring Harbor,1964
Bioch 33.1285

COLD SPRING HARBOR LABORATORY OF QUANTITATIVE BIOLOGY Phage and origins of molecular biology symposium Edited by John Cairns and others Cold Spring harbor symposia, 32 Cold Spring harbor laboratory of quantitative biology,Long Island,N.Y.,1966
Gen 34.0908

COLD SPRING HARBOR LABORATORY OF QUANTITATIVE BIOLOGY Replication of DNA :a symposium Abstracts Cold Spring Harbor 1971 Sep Cold Spring Harbor symposia on quantitative biology Cold Spring Harbor,1971
Bioch 33.2333

COLD SPRING HARBOR LABORATORY OF QUANTITATIVE BIOLOGY Replication of DNA in micro-organism :a symposium Papers Cold Spring Harbor 1968 Cold Spring Harbor symposia on quantitative biology, 33 Cold Spring Harbor, 1968
Bioch 33.1289

COLD SPRING HARBOR LABORATORY OF QUANTITATIVE BIOLOGY Sensory receptors :a symposium Papers Cold Spring Harbor 1965 Cold Spring Harbor symposia on quantitative biology, 30 Cold Spring Harbor,1965
Bioch 33.1286

COLD SPRING HARBOR LABORATORY OF QUANTITATIVE BIOLOGY Structure and function of proteins at the three-dimensional level :a symposium Papers Cold Spring Harbor 1972 Cold Spring Harbor symposia on quantitative biology, 36 Cold Spring Harbor,1972
Bioch 33.2334

COLD SPRING HARBOR LABORATORY OF QUANTITATIVE BIOLOGY Synthesis and structure of macromolecules :a symposium Papers Cold Spring Harbor 1963 Cold Spring Harbor symposia on quantitative biology, 28 Cold Spring Harbor,1963
Bioch 33.1284

COLD SPRING HARBOR LABORATORY OF QUANTITATIVE BIOLOGY The Bacteriphage lambda : conference Proceedings Cold Spring Harbor Sep 1970 Edited by A.D. Hershey Cold Spring Harbor symposia on quantitative biology Cold Spring Harbor, N.Y.,1971
Bioch 33.2295

COLD SPRING HARBOR LABORATORY OF QUANTITATIVE BIOLOGY The Genetic code :a symposium Papers Cold Spring Harbor 1966 Cold Spring Harbor symposia on quantitative biology, 31 Cold Spring Harbor,1966
Bioch 33.1287

COLD SPRING HARBOR LABORATORY OF QUANTITATIVE BIOLOGY The Mechanism of protein synthesis :a symposium Proceedings Cold Spring Harbor 1969 Cold Spring Harbor symposia on quantitative biology, 34 Cold Spring Harbor,1969
Bioch 33.1887

COLD SPRING HARBOR LABORATORY OF QUANTITATIVE BIOLOGY Transcription of genetic material a symposium Papers Cold Spring Harbor 1970 Jun 4-11 Cold Spring Harbor symposia on quantitative biology, 35 Cold Spring Harbor, 1971
Bioch 33.2335

COLD SPRING HARBOR SYMPOSIA, 32 Phage and origins of molecular biology symposium Cold Spring harbor laboratory of quantitative biology Edited by John Cairns and others Cold Spring harbor laboratory of quantitative biology,Long Island,N.Y.,1966
Gen 34.0908

COLD SPRING HARBOR SYMPOSIA ON QUANTITATIVE BIOLOGY Cold Spring Harbor 1939 Vol 7 Cold Spring Harbor biological laboratory Long Island biological association,Cold Spring Harbor,1939 Later referred to in vol.9 as "Biological oxidations"
Bioch 33.1263

COLD SPRING HARBOR SYMPOSIA ON QUANTITATIVE BIOLOGY Lactose operon :conference Proceedings Cold Spring Harbor 1969 Sep Cold Spring Harbor laboratory of quantitative biology Cold Spring Harbor,L.I.,N.Y.,1970
Bioch 33.2251

COLD SPRING HARBOR SYMPOSIA ON QUANTITATIVE BIOLOGY Replication of DNA :a symposium Abstracts Cold Spring Harbor 1971 Sep Cold Spring Harbor laboratory of quantitative biology Cold Spring Harbor,1971
Bioch 33.2333

COLD SPRING HARBOR SYMPOSIA ON QUANTITATIVE BIOLOGY The Bacteriphage lambda : conference Proceedings Cold Spring Harbor Sep 1970 Cold Spring Harbor laboratory of quantitative biology Edited by A.D. Hershey Cold Spring Harbor, N.Y.,1971
Bioch 33.2295

COLD SPRING HARBOR SYMPOSIA ON QUANTITATIVE BIOLOGY Papers Cold Spring Harbor 1933 Vol 1 Cold Spring Harbor biological laboratory Long Island biological association,Cold Spring Harbor,1933 Later referred to in vol.9 as "Surface phenomena"
Bioch 33.1257

COLD SPRING HARBOR SYMPOSIA ON QUANTITATIVE BIOLOGY Papers Cold Spring Harbor 1934 Vol 2 Cold Spring Harbor biological laboratory Long Island biological association,Cold Spring Harbor,1934 Later referred to in vol.9 as "Aspects of growth"
Bioch 33.1258

COLD SPRING HARBOR SYMPOSIA ON QUANTITATIVE BIOLOGY Papers Cold Spring Harbor 1935 Vol 3 Cold Spring Harbor biological laboratory Long Island biological association,Cold Spring Harbor,1935 Later referred to in vol.9 as "Photochemical reactions"
Bioch 33.1259

COLD SPRING HARBOR SYMPOSIA ON QUANTITATIVE BIOLOGY Papers Cold Spring Harbor 1936 Vol 4 Cold Spring Harbor biological laboratory Long Island biological association,Cold Spring Harbor,1936 Later referred to in vol.9 as "Excitation phenomena"
Bioch 33.1260

COLD SPRING HARBOR SYMPOSIA ON QUANTITATIVE BIOLOGY Papers Cold Spring Harbor 1937 Vol 5 Cold Spring Harbor biological laboratory Long Island biological association,Cold Spring Harbor,1937 Later referred to in vol. 9 as "Internal secretions"
Bioch 33.1261

COLD SPRING HARBOR SYMPOSIA ON QUANTITATIVE BIOLOGY Papers Cold Spring Harbor 1938 Cold Spring Harbor biological laboratory Long Island biological association,Cold Spring Harbor,1938 Later referred to in vol.9 as "Protein chemistry"
Bioch 33.1262

COLD SPRING HARBOR SYMPOSIA ON QUANTITATIVE BIOLOGY Papers Cold Spring Harbor,L.I., N.Y. 1935 Summer Vol 3 Long Island biological association.Biological laboratory Long Island biological association,Cold Spring Harbor,L.I.,N.Y.,1935
Chem 18.2002

COLD SPRING HARBOR SYMPOSIA ON QUANTITATIVE BIOLOGY Papers Cold Spring Harbor,L.I., N.Y. 1938 Vol 6 Cold Spring Harbor biological association Long Island biological laboratory,Cold Spring Harbor,L.I., N.Y.,1938
Bot 42.6450

COLD SPRING HARBOR SYMPOSIA ON QUANTITATIVE BIOLOGY Papers Vol 8 Cold Spring Harbor biological laboratory Long Island biological association,Cold Spring Harbor,1940 Later referred to in vol.9 as "Permeability and the nature of cell membranes"
Bioch 33.1264

COLD SPRING HARBOR SYMPOSIA ON QUANTITATIVE BIOLOGY, 9 Genes and chromosones : structure and organization.A symposium Papers Cold Spring Harbor 1941 Cold Spring Harbor biological laboratory Long Island biological association,Cold Spring Harbor,1941
Bioch 33.1265

COLD SPRING HARBOR SYMPOSIA ON QUANTITATIVE BIOLOGY, 10 The Relation of hormones to development :a symposium Papers Cold Spring Harbor 1942 Cold Spring Harbor biological laboratory Long Island biological association,Cold Spring Harbor,1942
Bioch 33.1266

COLD SPRING HARBOR SYMPOSIA ON QUANTITATIVE BIOLOGY, 11 Heredity and variation in microorganisms :a symposium Papers Cold Spring Harbor 1946 Cold Spring Harbor biological laboratory Long Island biological association,Cold Spring Harbor,1946
Bioch 33.1267

COLD SPRING HARBOR SYMPOSIA ON QUANTITATIVE BIOLOGY, 12 Nucleic acids and nucleoproteins Cold Spring Harbor 1947 Jun 11-20 Cold Spring Harbor laboratory of quantitative biology Port Biological laboratory,New York,1947
Radioth 35.0157

COLD SPRING HARBOR SYMPOSIA ON QUANTITATIVE BIOLOGY, 12 Nucleic acids and nucleoproteins :a symposium Papers Cold Spring Harbor 1947 Cold Spring Harbor biological laboratory Long Island biological association,Cold Spring Harbor,1947
Bioch 33.1268

COLD SPRING HARBOR SYMPOSIA ON QUANTITATIVE BIOLOGY, 13 Biological application of tracer elements :a symposium papers Cold Spring Harbor 1948 Jun 8-16 Cold Spring Harbor laboratory of quantitative biology Biological laboratory,New York,1948
Radioth 35.0158

COLD SPRING HARBOR SYMPOSIA ON QUANTITATIVE BIOLOGY, 13 Biological applications of tracer elements :a symposium Papers Cold Spring Harbor 1948 Cold Spring Harbor biological laboratory Long Island biological association,Cold Spring Harbor,1948
Bioch 33.1269

COLD SPRING HARBOR SYMPOSIA ON QUANTITATIVE BIOLOGY, 14 Amino acids and proteins :a symposium Papers Cold Spring Harbor 1950 Cold Spring Harbor biological laboratory Long Island biological association,Cold Spring Harbor,1950
Bioch 33.1270

COLD SPRING HARBOR SYMPOSIA ON QUANTITATIVE BIOLOGY, 15 Origin and evolution of man : a symposium Cold Spring Harbor 1950 Jun 9-17 Edited by Katherine Brehme Warren Long Island biological association,Cold Spring Harbor,N.Y.,1950
An 32.2807

COLD SPRING HARBOR SYMPOSIA ON QUANTITATIVE BIOLOGY, 15 Origin and evolution of man : a symposium Papers Cold Spring Harbor 1950 Cold Spring Harbor biological laboratory Long Island biological association,Cold Spring Harbor,1950
Bioch 33.1271

COLD SPRING HARBOR SYMPOSIA ON QUANTITATIVE BIOLOGY, 16 Genes and mutation :a symposium Papers Cold Spring Harbor 1951 Cold Spring Harbor biological laboratory Long Island biological association,Cold Spring Harbor,1951
Bioch 33.1272

COLD SPRING HARBOR SYMPOSIA ON QUANTITATIVE BIOLOGY, 17 The Neuron :a symposium Cold Spring Harbor 1952 Jun 6-13 Cold Spring Harbor,N.Y.,1952
Phys 20.0956

COLD SPRING HARBOR SYMPOSIA ON QUANTITATIVE BIOLOGY, 17 The Neuron :a symposium Papers Cold Spring Harbor 1952 Cold Spring Harbor biological laboratory Long Island biological association,Cold Spring Harbor,1952
Bioch 33.1273

COLD SPRING HARBOR SYMPOSIA ON QUANTITATIVE BIOLOGY, 17 The Neuron :a ymposium Cold Spring Harbor 1952 Jun 6-13 Cold Spring Harbor biological laboratory Edited by Katherine Brehme Warren Biological laboratory,Cold Spring Harbor,1952
An 32.4330

COLD SPRING HARBOR SYMPOSIA ON QUANTITATIVE BIOLOGY, 18 Viruses :a symposium Papers Cold Spring Harbor 1953 Cold Spring Harbor biological laboratory Long Island biological association,Cold Spring Harbor,1953
Bioch 33.1274

COLD SPRING HARBOR SYMPOSIA ON QUANTITATIVE BIOLOGY, 19 The Mammalian fetus : physiological aspects of development.A symposium Papers Cold Spring Harbor 1954 Cold Spring Harbor biological laboratory Long Island biological association,Cold Spring Harbor,1954
Bioch 33.1275

COLD SPRING HARBOR SYMPOSIA ON QUANTITATIVE BIOLOGY, 19 The Mammalian fetus: physiological aspects of development : symposium Cold Spring Harbor 1954 Jun 7-14 Long Island biological association,New York, 1954
Phys 20.1376

COLD SPRING HARBOR SYMPOSIA ON QUANTITATIVE BIOLOGY, 20 Population genetics :the nature and causes of genetic variability in populations.A symposium Papers Cold Spring Harbor 1955 Cold Spring Harbor biological laboratory Long Island biological association,Cold Spring Harbor,1955 Dedicated to Vannevar Bush,President of the Carnegie Institute of Washington, 1939-1955
Bioch 33.1276

COLD SPRING HARBOR SYMPOSIA ON QUANTITATIVE BIOLOGY, 21 Genetic mechanisms : structure and function.A symposium Papers Cold Spring Harbor 1956 Cold Spring Harbor biological laboratory Long Island biological association,Cold Spring Harbor,1956
Bioch 33.1277

COLD SPRING HARBOR SYMPOSIA ON QUANTITATIVE BIOLOGY, 22 Population studies :animal ecology and demography.A symposium Papers Cold Spring Harbor 1957 Cold Spring Harbor biological laboratory Long Island biological association,Cold Spring Harbor,1957
Bioch 33.1278

COLD SPRING HARBOR SYMPOSIA ON QUANTITATIVE BIOLOGY, 23 Exchange of genetic material mechanisms and consequences.A symposium Papers Cold Spring Harbor 1958 Cold Spring Harbor biological laboratory Long Island biological association,Cold Spring Harbor,1958
Bioch 33.1279

COLD SPRING HARBOR SYMPOSIA ON QUANTITATIVE BIOLOGY, 24 Genetics and twentieth century Darwinism :a symposium Cold Spring Harbor 1959 Jun 3-10 Edited by Clara Wooldridge Biological laboratory,Cold Spring Harbor,L.I.,1959
An 32.0043

COLD SPRING HARBOR SYMPOSIA ON QUANTITATIVE BIOLOGY, 24 Genetics and twentieth century Darwinism :a symposium Papers Cold Spring Harbor 1959 Cold Spring Harbor biological laboratory Long Island biological association,Cold Spring Harbor,1959
Bioch 33.1280

COLD SPRING HARBOR SYMPOSIA ON QUANTITATIVE BIOLOGY, 25 Biological clocks : a symposium Papers Cold Spring Harbor 1960 Cold Spring Harbor biological laboratory Long Island biological association,Cold Spring Harbor,1960
Bioch 33.1281

COLD SPRING HARBOR SYMPOSIA ON QUANTITATIVE BIOLOGY, 26 Cellular regulatory mechanisms :a symposium Papers Cold Spring Harbor 1961 Cold Spring Harbor biological laboratory Long Island biological association,Cold Spring Harbor,1961
Bioch 33.1282

COLD SPRING HARBOR SYMPOSIA ON QUANTITATIVE BIOLOGY, 27 Basic mechanisms in animal virus biology Papers Cold Spring Harbor 1962 Jun 7-13 Cold Spring Harbor.Biological laboratory Long Island biological association.Biological laboratory,Cold Spring Harbor,L.I.,N.Y.,1962
Path 30.2559

COLD SPRING HARBOR SYMPOSIA ON QUANTITATIVE BIOLOGY, 27 Basic mechanisms in animal virus biology :a symposium Papers Cold Spring Harbor 1962 Cold Spring Harbor biological laboratory Long Island biological association,Cold Spring Harbor,1962
Bioch 33.1283

COLD SPRING HARBOR SYMPOSIA ON QUANTITATIVE BIOLOGY, 28 Synthesis and structure of macromolecules :a symposium Papers Cold Spring Harbor 1963 Cold Spring Harbor laboratory of quantitative biology Cold Spring Harbor,1963
Bioch 33.1284

COLD SPRING HARBOR SYMPOSIA ON QUANTITATIVE BIOLOGY, 28 Synthesis and structure of macromolecules :a symposium Papers Cold Spring Harbor 1963 Jun 7-13 Cold Spring Harbor laboratory of quantitative biology Cold Spring Harbor laboratory of quantitative biology,Cold Spring Harbor,1963
Radioth 35.0245

COLD SPRING HARBOR SYMPOSIA ON QUANTITATIVE BIOLOGY, 29 Human genetics :a symposium Cold Spring Harbor 1964 Jun 5-11 Edited by Leonora Frisch Cold Spring Harbor laboratory on quantitative biology,Cold Spring Harbor, L. I.,1964
An 32.0051

COLD SPRING HARBOR SYMPOSIA ON QUANTITATIVE BIOLOGY, 29 Human genetics :a symposium Papers Cold Spring Harbor 1964 Cold Spring Harbor laboratory of quantitative biology Cold Spring Harbor,1964
Bioch 33.1285

COLD SPRING HARBOR SYMPOSIA ON QUANTITATIVE BIOLOGY, 30 Sensory receptors :a symposium Cold Spring Harbor,L.I.,N.Y. 1965 Jun 4-11 Cold Spring Harbor laboratory of quantitative biology Edited by John Cairns Cold Spring Harbor laboratory of quantitative biology,Cold Spring Harbor,L.I.,N. Y.,1965
Psy 31.0298

COLD SPRING HARBOR SYMPOSIA ON QUANTITATIVE BIOLOGY, 30 Sensory receptors :a symposium Papers Cold Spring Harbor 1965 Cold Spring Harbor laboratory of quantitative biology Cold Spring Harbor,1965
Bioch 33.1286

COLD SPRING HARBOR SYMPOSIA ON QUANTITATIVE BIOLOGY, 31 The Genetic code :a symposium Papers Cold Spring Harbor 1966 Cold Spring Harbor laboratory of quantitative biology Cold Spring Harbor,1966
Bioch 33.1287

COLD SPRING HARBOR SYMPOSIA ON QUANTITATIVE BIOLOGY, 32 Antibodies :a symposium Papers Cold Spring Harbor 1967 Cold Spring Harbor laboratory of quantitative biology Cold Spring Harbor,1967
Bioch 33.1288

COLD SPRING HARBOR SYMPOSIA ON QUANTITATIVE BIOLOGY, 32 Antibodies :a symposium Proceedings Cold Spring Harbor 1967 Jun 1-7 Cold Spring Harbor laboratory of quantitative biology Edited by Leonora Frisch Cold Spring Harbor laboratory of quantitative biology,Cold Spring Harbor,L.I.,N. Y.,1967
Path 30.2523

COLD SPRING HARBOR SYMPOSIA ON QUANTITATIVE BIOLOGY, 33 Replication of DNA in micro-organism :a symposium Papers Cold Spring Harbor 1968 Cold Spring Harbor laboratory of quantitative biology Cold Spring Harbor, 1968
Bioch 33.1289

COLD SPRING HARBOR SYMPOSIA ON QUANTITATIVE BIOLOGY, 33 Replication of DNA in micro-organisms :a symposium Papers Cold Spring Harbor 1968 Jun 5-12 Cold Spring Harbor laboratory of quantitative biology 2 vols Cold Spring Harbor,1968
Radioth 35.0246

COLD SPRING HARBOR SYMPOSIA ON QUANTITATIVE BIOLOGY, 34 The Mechanism of protein synthesis :a symposium Proceedings Cold Spring Harbor 1969 Cold Spring Harbor laboratory of quantitative biology Cold Spring Harbor,1969
Bioch 33.1887

COLD SPRING HARBOR SYMPOSIA ON QUANTITATIVE BIOLOGY, 35 Transcription of genetic material :a symposium Papers Cold Spring Harbor 1970 Jun 4-11 Cold Spring Harbor laboratory of quantitative biology Cold Spring Harbor,1971
Bioch 33.2335

COLD SPRING HARBOR SYMPOSIA ON QUANTITATIVE BIOLOGY, 36 Structure and function of proteins at the three-dimensional level :a symposium Papers Cold Spring Harbor 1972 Cold Spring Harbor laboratory of quantitative biology Cold Spring Harbor,1972
Bioch 33.2334

COLD WORKING OF METALS Philadelphia 1948 Oct 25-29 American society for metals American society for metals,Cleveland,Ohio, 1949
Met 25.1199

COLLAGEN Nature and structure of collagen :a discussion Papers London 1953 Mar 26-27 Faraday society.Colloid and biophysics committee Edited by J.T. Randall and Sylvia Fitton Jackson Butterworths,London,1953
Bioch 33.2346

COLLAGEN Nature and structure of collagen :a discussion Papers presented London 1953 Mar 26-27 Faraday society.Colloid and biophysics committee Edited by J.T. Randall and Sylvia Fitton Jackson Butterworths, London,1953
Bal 39.1202

COLLECTIVIZATION OF AGRICULTURE IN EASTERN EUROPE Conference on collectivization in eastern Europe Papers Lexington,Ky. 1955 Apr 14-16 Edited by Irwin T. Sanders maps University of Kentucky press,Lexington,Ky., 1958
Geog 13.2582

COLLEGE OF AERONAUTICS,CRANFIELD.ADVANCED SCHOOL OF AUTOMOBILE ENGINEERING Advances in automobile engineering Symposium on automobile engineering Proceedings Cranfield 1963 Jul Pt 2 Edited by N. A. Carter Cranfield international symposium series, 5 Pergamon press,London,1964
Eng 41.6190

COLLEGE OF AERONAUTICS,CRANFIELD.DEPARTMENT OF ELECTRICAL AND CONTROL ENGINEERING Recent developments in network theory a symposium Proceedings Cranfield 1961 Sep Edited by S.R. Deards Pergamon press,Oxford,1963
Cav 7.0582

COLLEGE OF GENERAL PRACTITIONERS Design guide for medical group practice centres :a study...incorporating papers given at a symposium London 1966 Jun 3 College of general practitioners,London,c1966
PGMS 29.0555

COLLEGE OF PATHOLOGISTS Automation and data processing in pathology :a symposium London 1969 Feb. Edited by T.P. Whitehead Journal of clinical pathology,London,1969
HE 27.0161

COLLEGE PARK 1951 Fluid dynamics :a symposium Proceedings American mathematical society Edited by M.H. Martin American mathematical society.Proceedings of symposia in applied mathematics, 4 v,186p McGraw-Hill,New York,1958
Chem E 24.0354

COLLEGE PARK 1961 Fluid dynamics and applied mathematics :a symposium Proceedings Edited by J.B. Diaz and S.I. Pai Sponsored by the Institute for fluid dynamics and applied mathematics 207p Gordon and Breach, New York;London,1962
Chem E 24.0371

COLLEGE PARK, MD 1970 Seminar on differential equations and dynamical systems 2n Lectures Edited by J.A. Yorke Held at the University of Maryland Lecture notes in mathematics, 144 viii,268p 17cm Springer-Verlag,Berlin,1970
P Math 2.3534

COLLEGE PARK,MD. 1951 Fluid dynamics symposium American mathematical society Edited by M.H. Martin Co-sponsored by the Naval ordnance laboratory American mathematical society.Proceedings of symposia in applied mathematics, 4 McGraw-Hill,New York,1953
A Math 4.0512

COLLEGE PARK,MD. 1965 Numerical solution of partial differential equations :a symposium Proceedings Edited by James H. Bramble Academic press,New York;London,1966
Eng 41.2331

COLLEGE PARK,MD. 1965 Numerical solution of partial differential equations :symposium Proceedings Edited by James H. Bramble 373p 23cm Academic press,New York,1966
Math L 5.3110

COLLEGE PARK,MD. 1966 International horticultural congress 17th Proceedings Vol 1-4 International society for horticultural sciences Sponsored by the American horticultural society 4 vols 1966-67
BG 38.3134

COLLEGE PARK,MD. 1967 Seminar on differential equations and dynamical systems Lectures Edited by G.Stephen Jones Lecture notes in mathematics, 60 Bibliog v,106p 28cm Springer-verlag,Berlin,1968
P Math 2.3046

COLLEGE PARK,MD. 1970 Several complex variables International mathematical conference Proceedings 1 University of Maryland Edited by John Horvath Lecture notes in mathematics, 155 bibliog. 214p 25cm Springer,Berlin,1970
P Math 2.3890

COLLEGE PARK,MD. 1970 Several complex variables International mathematical conference Proceedings 2 University of Maryland Edited by John Horvath Lecture notes in mathematics, 185 214p 25cm Springer,Berlin,1971
P Math 2.3891

COLLEGE PARK,MD. 1971 International conference on harmonic analysis Proceedings Edited by Denny Gulick and Ronald L. Lipsman Held at University of Maryland Lecture notes in mathematics, 266 vi,323p 25cm Springer,Berlin,1972
P Math 2.4283

COLLEGIUM FOR NEURO-PSYCHOPHARMACOLOGY Neuro-psychopharmacology International congress of neuro-pharmacology 1st Proceedings Rome 1958 Sep Edited by P.B. Bradley and others Elsevier,London,1959
PGMS 29.0192

COLLEGIUM INTERNATIONALE NEURO-PSYCHOPHARMACOLOGICUM Present status of psychotropic drugs,pharmacological and clinical aspects International congress of the Collegium internationale neuro-psychopharmacologicum 6th Proceedings Tarragona 1968 Apr 24-27 Edited by A. Cerletti and F.J. Bove Excerpta medica foundation.International congress series, 180 Excerpta medica foundation,Amsterdam,1969
Psy 31.2902

COLLISIONS Physics of electronic and atomic collisions International conference on the physics of electronic and atomic collisions 5th Invited papers Leningrad 1967 Jul 17-23 Edited by Lewis M. Branscomb Sponsored by the International union of pure and applied physics Bibliog,illus xvi,200p 23cm University of Colorado press,Boulder,Colo., 1968
A Math 4.1906

COLLOID CHEMISTRY The National symposium on colloid chemistry 1st Papers and discussions Madison,Wis. 1923 Jun 12-15 University of Wisconsin Edited by J.Howard Mathews Colloid symposium monograph, 1 Department of chemistry,university of Wisconsin,Madison,Wis.,1923
Bioch 33.1469

COLLOID CHEMISTRY The National symposium on colloid chemistry 2nd Papers presented Evanston,Ill. 1924 Jun 18-21 Northwestern university Edited by Harry N. Holmes Colloid symposium monograph, 2 Chemical catalog company,New York,1925
Bioch 33.1470

COLLOID CHEMISTRY The National symposium on colloid chemistry 3rd Papers presented Minneapolis 1925 Jun 17-19 University of Minnesota Edited by Harry N. Holmes and Harry B. Weiser Colloid symposium monograph, 3 Chemical catalog company,New York,1925
Bioch 33.1471

COLLOID CHEMISTRY The National symposium on colloid chemistry 4th Papers presented Cambridge,Mass. 1926 Jun 23-25 Massachusetts institute of technology Edited by Harry Boyer Weiser Colloid symposium monograph, 4 Chemical catalog company,New York,1926
Bioch 33.1472

COLLOID CHEMISTRY The National symposium on colloid chemistry 5th Papers presented Ann Arbor,Mich. 1927 Jun 22-24 University of Michigan Edited by Harry Boyer Weiser Colloid symposium monograph, 5 Chemical catalog company,New York,1928
Bioch 33.1473

COLLOID CHEMISTRY The Symposium on colloid chemistry 6th Papers presented Toronto 1928 Jun 14-16 University of Toronto Edited by Harry Boyer Weiser Colloid symposium monograph, 6 Chemical catalog company,New York,1928 Colloid symposium monograph known also as Colloid symposium annual
Bioch 33.1474

COLLOID CHEMISTRY The Symposium on colloid chemistry 7th Papers presented Baltimore 1929 Jun 20-22 Johns Hopkins university Edited by Harry Boyer Weiser Colloid symposium annual, 7 Wiley;Chapman and Hall,New York;London,1939 Colloid symposium annual known also as Colloid symposium monograph
Bioch 33.1475

COLLOID CHEMISTRY The Symposium on colloid chemistry 9th Papers presented Columbus, Ohio 1931 Jun Ohio State university Edited by Harry Boyer Weiser Colloid symposium monograph, 9 Journal of physical chemistry,New York,1931
Bioch 33.1476

COLLOID CHEMISTRY The Symposium on colloid chemistry 10th Papers presented Ottawa 1932 Jun Edited by Harry Boyer Weiser Colloid symposium monograph, 10 Journal of physical chemistry,New York,1932
Bioch 33.1477

COLLOID CHEMISTRY The Symposium on colloid chemistry 11th Papers presented Madison, Wis. 1934 Jun Edited by Harry Boyer Weiser Colloid symposium monograph, 11 Williams and Wilkins,Baltimore,1935
Bioch 33.1478

COLLOID CHEMISTRY The Symposium on colloid chemistry 12th Papers presented Ithaca,N. Y. 1935 Jun Edited by Harry Boyer Weiser Colloid symposium monograph, 12 Williams and Wilkins,Baltimore,1936
Bioch 33.1479

COLLOID CHEMISTRY The Symposium on colloid chemistry 13th Papers presented St.Louis, Mo. 1936 Jun Edited by Harry Boyer Weiser Colloid symposium monograph, 13 William and Wilkins,Baltimore,1937
Bioch 33.1480

COLLOIDAL SYSTEMS The Size and shape factor in colloidal systems :a discussion Leamington Spa 1951 Jul 18-20 Faraday society Faraday society.Discussions, 11 Faraday society,Aberdeen,1952
Bioch 33.1529

COLLOIDS The Physics and chemistry of colloids and their bearing on industrial questions :a general discussion Report London 1920 Oct 25 Faraday society Physical society of London H.M.S.O.,London, 1921
Bioch 33.1522

COLLOQUE AMPERE 11th :international conference Proceedings Magnetic and electric resonance and relaxation Eindhoven 1962 Jul 2-7 Netherlands physical society Edited by J. Smidt Under the auspices of the International union of pure and applied physics 789p North-Holland,Amsterdam,1963
Cav 7.0712

COLLOQUE DE L'U.R.S.S. 3rd selection of papers presented Fonctions d'une variable complexe:problemes contemporaines Moscow 1957 Edited by A.I. Markushevitch Translated by L. Nicolas from the Russian Monographies internationales de mathematiques modernes, 1 271p 24cm Gauthier-Villars, Paris,1962
P. Math 2.1557

COLLOQUE DE MATHEMATIQUES proceedings Clermont-Ferrand 1962 Jun.4-8 Tome 2: calcul des probabilites;analyse numerique et calcul automatique; geometrie et physique mathematique Clermont-Ferrand.University. Faculte des sciences.Annales, 8, Mathematiques, 2 189p 1962 A l'occasion du tricentenaire de la mort de Blaise Pascal
Math 3.0680

COLLOQUE DE METALLURGIE 3e corrosion,seche et aqueuse Saclay 1959 Jun 29-Jul 1 Centre d'etudes nucleaires,Saclay Sponsored by the Institut national des sciences et techniques nucleaires Centre d'etudes nucleaires de Saclay;North Holland,Paris; Amsterdam,1960
Met 25.1891

COLLOQUE DE METALLURGIE 4th proprietes des joints de grains Saclay 1960 Jun 27-28 Centre d'etudes nucleaires,Saclay Presses universitaires de France,Paris,1961 Papers in English and French
Met 25.1266

COLLOQUE DE TOPOLOGIE Brussels 1964 Sep 7-10 Centre belge de recherches mathematiques 235p 25cm Louvain, Paris,1966
P. Math 2.0294

COLLOQUE DE TOPOLOGIE (ESPACES FIBRES) Brussels 1950 Jun 5-8 Centre Belge de recherches mathematiques 136p 24cm Centre Belge de recherches mathematiques, Brussels,1950
P Math 2.3126

COLLOQUE D'ENDOCRINOLOGIE 1959 Thyroide et exploration thyroidienne medullo-surrenale : rapports de la vie reunion des endocrinologistes de langue francaise Paris 1959 Jun 29-Jul 1 Endocrinologistes de langue francaise Annales d'endocrinologie, 19,no.4 Masson;Doin,Paris,1959 At head of title:Colloque d'endocrinologie 1959
An 32.3755

COLLOQUE DU JURASSIQUE Luxembourg 1962 Aug. 1- International geological congress. International commission on stratigraphy,and, International palaeontological union Ministere des arts et des sciences,Luxembourg, 1964
Geol 8.2510

COLLOQUE INTERNATIONAL D'ASTROPHYSIQUE Reports Novae and white dwarfs Paris 1939 Jul. 17-23 1: observations of novae By Knut Lundmark and others Actualites scientifiques et industrielles, 901 College de France. Conferences, 13 140p Hermann,Paris,1941
Obs 6.1146

COLLOQUE INTERNATIONAL D'ASTROPHYSIQUE reports Novae and white dwarfs Paris 1939 Jul. 17-23 3: white dwarfs By G.P. Kuiper and others Actualites scientifiques et industrielles, 903 College de France. Conferences, 15 267p Hermann,Paris,1941
Obs 6.0996

COLLOQUE INTERNATIONAL D'ASTROPHYSIQUE reports Novae and white dwarfs Paris 1939 Jul 17-23 2: nova theory supernovae By P. Swings and others Actualites scientifiques et industrielles, 902 College de France. Conferences, 14 267p Hermann,Paris,1941
Obs 6.1720

COLLOQUE INTERNATIONAL D'ASTROPHYSIQUE 4e Communications Instabilite gravitationnelle et formation des etoiles, des galaxies et de leurs structures caracteristiques Liege 1966 Jun 20-22 Universite de Liege.Institut d'astrophysique Institut d'astrophysique,Cointe-Sclessin,1967 Papers in English
A Math 4.1170

COLLOQUE INTERNATIONAL D'ASTROPHYSIQUE 8e Etoiles a raies d'emissions Liege Universite de Liege.Institut d'astrophysique 14p Liege,c1957 Summary of contributions
Obs 6.3170

COLLOQUE INTERNATIONAL D'ASTROPHYSIQUE 15ieme Communications Transitions interdites dans les spectres des astres Liege 1968 Jun 24-26 Societe royale des sciences de Liege.Memoires,5s, 17 410p Institut d'astrophysique,Liege,1969
TA 15.0576

COLLOQUE INTERNATIONAL DE GEOGRAPHIE APPLIQUEE 3rd comptes rendus Liege 1967 Sep 7-13 Commission de geographie appliquee Christans,Liege,c1967
Geog 13.7167

COLLOQUE INTERNATIONAL DE GEOGRAPHIE ET D'HISTOIRE AGRAIRES Structures agraires et paysages ruraux :un quart de siecle de recherches francaises By Etienne Juillard and others Annales de l'Est.Memoires,17 Faculte des lettres de l'universite de Nancy, Nancy,1957 Published on the occasion of the colloque international de geographie et d'histoire agraires,Nancy,1957
Geog 13.4206

COLLOQUE INTERNATIONAL DE L'UNION INTERNATIONAL DES SCIENCES BIOLOGIQUES Organization scientifique des jardins botaniques Paris 1953 Jun 4-7 Under the auspices of Unesco International union of biological sciences. Series B:Colloquia, 13 International union of biological sciences,Paris,1953
BG 38.3110

COLLOQUE INTERNATIONAL DES VIBRATIONS NON LINEAIRES Actes Vibrations non lineaires Ile de Porquerolles 1951 Sep 18-21 Union internationale de radio-science Organised by the Union internationale de mecanique theorique et appliquee France. Ministere de l'air.Publications scientifiques et techniques, 281 Information technique de l'aeronautique,Paris,1953
Eng 41.6372

COLLOQUE INTERNATIONAL SUR LA FORMATION DES INGENIEURS EN EUROPE Comte-rendu Paris 1961 Jun 21-24 Paris,1961
Chem E 24.1779

COLLOQUE INTERNATIONAL SUR LA RETENTION ET LA MIGRATION DES IONS RADIOACTIFS DANS LES SOLS Saclay 1962 Oct 16-18 Institut national des sciences et techniques nucleaires Centre d'etudes nucleaires de Saclay;Presses universitaires de France,Gif-sur-Yvette;Paris, 1963
Eng 41.3174

COLLOQUE INTERNATIONAL SUR LE PHOTOTHERMOPERIODISME ACTION DES DIVERSES RADIATIONS ON GIBBERELLINES ET DE QUELQUES AUTRES SUBSTANCES Photo-thermoperiodism, the action of different radiations on gibberellins and other active substances Parma 1957 Jun International union of biological sciences International union of biological sciences.Series B, 34 U.I.S.B., Paris,1958
Bot 42.1420

COLLOQUE INTERNATIONAL SUR L'ENDOCRINOLOGIE DES ARTHROPODES Endocrinologie des arthropodes Paris 1947 Jun Centre national de la recherche scientifique Centre national de la recherche scientifique. Colloques internationaux, 4 Centre national de la recherch scientifique,Paris, 1948
Bal 39.2291

COLLOQUE INTERNATIONAL SUR LES SUBSTANCES DE CROISSANCE VEGETALES 5e Actes Regulateurs naturels de la croissance vegetales Gif sur Yvette 1963 Jul 15-20 Centre national de la recherche scientifique Centre national de la recherche scientifique. Colloques internationaux, 123 Paris,1964 Text in English and French
Bioch 33.0747

COLLOQUE INTERNATIONAL SUR L'HISTOIRE DE LA BIOLOGIE MARINE Les Grandes expeditions scientifiques et la creation des laboratoires maritimes Banyuls-sur-Mer 1963 Sep 2-6 International union of history and philosophy of science Vie et milieu.Supplement, 19 370p Laboratoire Arago;Masson,Banyuls-sur-Mer;Paris,1965
WSM 43.2877

COLLOQUE INTERNATIONAL SUR "LE MARCHE DES BOIS DU NORD ET LA REGION ECONOMIQUE DE HAUTE-NORMANDIE Rouen 1964 Nov 17-18 Fondation Francaise d'etudes nordiques Edited by Jean Malaurie Fondation Francaise d'etudes nordiques.Actes et documents,1 256p Rouen,1964 Mimeograph
Sco 14.2469

COLLOQUE SUR LA CHIMIE DU BOIS The Wood chemistry symposium Proceedings Montreal 1961 Aug 9-11 International union of pure and applied chemistry.Applied chemistry section Pure and applied chemistry, 5,nos 1-2 illus Butterworths,London,1962 Added title page in French "Colloque sur la chimie du bois".
Bioch 33.1212

COLLOQUE SUR LA CULTURE DES TISSUS 1963 Oct 26 By R.J. Gautheret and others Organise au nom de la Societe botanique de France Revue de cytologie et de biologie vegetale, 27,fasc.,2,3,4 Paris,1964
Bioch 33.2364

COLLOQUE SUR LA DIFFUSION Actes Montpellier 1955 Jun By J. Salvinien and others France.Ministere de l'air. Publications scientifiques et techniques de l'air.Notes techniques, 59 96p Service de documentation et d'information technique de l'aeronautique,Paris,1956
Chem E 24.0760

COLLOQUE SUR LA DIFFUSION A L'ETAT SOLIDE Diffusion a l'etat solide Saclay 1958 Jul 3-5 Centre d'etudes nuleaires,Saclay and Commissariat a l'energie atomique Centre d'etudes nucleaires de Saclay;North Holland, Gif-sur-Yvette;Amsterdam,1959
Met 25.1277

La COLLOQUE SUR LA PHYSIQUE DES METAUX Comptes rendus Diffusion dans les metaux Eindhoven 1956 Sep 10-11 Edited by J.D. Fast and others Bibliotheque technique Philips,Eindhoven,1957
Met 25.1276

COLLOQUE SUR LA PROTECTION ET LA CONSERVATION DE LA NATURE DANS LE PROCHE-ORIENT : symposium on the protection and conservation of nature in the Near East Beirut 1954 Jun 3-8 Lebanese society of the friends of the trees Unesco-Middle East science cooperation office Beirut,1954 Text in English and French
Bal 39.1787

COLLOQUE SUR LA THEORIE DES GROUPES ALGEBRIQUES 20th colloque Brussells 1962 Jun 5-7 Centre Belge de recherches mathematiques 150p 25cm Librairie universitaire,Louvain, 1962
P. Math 2.0573

COLLOQUE SUR L'ANALYSE FONCTIONELLE 2nd Liege 1964 May 4-6 Centre Belge de recherches mathematiques 166p 25cm Librairie universitaire,Louvain,1964
P. Math 2.1182

COLLOQUIUM IN NUMERICAL TAXONOMY Proceedings Numerical taxonomy St.Andrews 1968 Sep Edited by A.J. Cole Academic press,London, 1969
HE 27.0022

COLLOQUIUM IN NUMERICAL TAXONOMY Proceedings Numerical taxonomy St.Andrews 1968 Sep. Edited by A.J. Cole Academic press,London, 1969
Math L 5.3440

COLLOQUIUM IN NUMERICAL TAXONOMY Proceedings Numerical taxonomy St.Andrews 1968 Sep 2-4 Edited by A.J. Cole Academic press, London;New York,1969
Gen 34.2149

COLLOQUIUM IN THE PHILOSOPHY OF SCIENCE Proceedings and discussions Induction, physics,and ethics Salzburg 1968 Aug 28-31 International union of history and philosophy of science.Division of logic,methodology and philosophy of science Institut fur wissenschaftstheorie,Salzburg Edited by Paul Weingartner and Gerhard Zecha Synthese library x,382p Reidel,Dordrecht,1970
WSM 43.0992

COLLOQUIUM LECTURES IN PURE AND APPLIED SCIENCE 8 Dynamic programming Dallas,Texas 1963 Dec 9-13 By Rutherford Aris Socony mobile oil company.Field research laboratory 310p 1965
Chem E 24.0172

COLLOQUIUM OF THE JOHNSON RESEARCH FOUNDATION 5th Proceedings Probes of structure and function of macromolecules and membranes Philadelphia,Pa. 1969 Apr 19-21 Vol 1-2: probes and membrane function;probes of enzymes and hemoproteins Johnson research foundation Edited by Britton Chance and others Academic press,New York,1971
Bioch 33.2190

COLLOQUIUM ON ALGEBRAIC TOPOLOGY Aarhus 1962 Aug 1-10 Aarhus universitet.Matematisk institut Supported by North Atlantic treaty organization.Scientific affairs division 113p 29cm Aarhus universitet,1962
P. Math 2.0293

COLLOQUIUM ON APPLICATIONS OF MATHEMATICS TO ECONOMICS Budapest 1963 Jun.18-22 Bolyai Janos matematikai tarsulat Edited by Andras Prekopa 367p 25cm Akademiai kiado, Budapest,1965
Math 3.0917

COLLOQUIUM ON COMBINATORIAL THEORY Proceedings Balatonfured 1969 Aug 24 Bolyai Janos matematikai tarsulat Edited by P. Erdos and others Societatis Janos Bolyai.Colloquia mathematica, 4 1201p 24cm 3 vols North-Holland,Amsterdam;London,1970
Math S 3.1656

COLLOQUIUM ON CONVEXITY Proceedings Copenhagen 1965 Aug 12-19 Sponsored under the Nato advanced study institute programme Bibliog,illus xii,325p 25cm Kobenhavens universitets matematiske inst.,Copenhagen,1967 In English,French and German
P Math 2.3016

COLLOQUIUM ON FATIGUE Proceedings Stockholm 1955 May 25-27 International union of theoretical and applied mechanics Edited by Waloddi Weibull and Folke K.G. Odqvist Springer,Berlin,1956
Eng 41.3754

COLLOQUIUM ON FATIGUE Proceedings Stockholm 1955 May 25-27 International union of theoretical and applied mechanics Edited by Waloddi Weibull and Folke K.G. Odqvist Springer-verlag,Gottingen;Heidelberg, 1956 Lectures in English,French and German
Met 25.0917

COLLOQUIUM ON INFORMATION THEORY Proceedings Debrecen 1967 Sep 19-24 Bolyai Janos matematikai tarsulat bibliog.,illus. 24cm 2 vols Janos Bolyai mathematical society, Budapest,1968
Math S 3.1457

COLLOQUIUM ON LATE-TYPE STARS Proceedings Trieste 1966 Jun 13-17 Osservatorio astronomico di Trieste Edited by Margherita Hack 465p Trieste,1966
TA 15.0315

COLLOQUIUM ON LATE-TYPE STARS proceedings Trieste 1966 Jun.13-17 Osservatorio astronomico di Trieste Edited by Margherita Hack 465p Trieste,1967
Obs 6.0712

A COLLOQUIUM ON METABOLIC CONTROL and a Symposium on control of energy metabolism Proceedings Control of energy metabolism Philadelphia 1965 May 20-21 Johnson research foundation Edited by Britton Chance and others Johnson research foundation. Colloquia Academic press,New York;London, 1965 "In celebration of the bicentennial of the University of Pennsylvania school of medicine"
Bioch 33.1077

COLLOQUIUM ON METABOLIC CONTROL :and symposium on control of metabolism Control of energy metabolism Philadelphia,Pa. 1965 May 20-21 Johnson research foundation Edited by Britton Chance and others Academic press,New York;London,1965
Gen 34.0614

COLLOQUIUM ON METHODS OF OPTIMIZATION Novosibirsk 1968 Jun Edited by N.N. Moiseev Lecture notes in mathematics, 112 Illus. 293p 25cm Springer-Verlag,Berlin, 1970
P Math 2.3531

COLLOQUIUM ON METHODS OF OPTIMIZATION Novosibirsk 1968 Jun Edited by N.N. Moiseev Lecture notes in mathematics, 112 293p 26cm Springer,Berlin,1970
Math S 3.1688

COLLOQUIUM ON SOIL FAUNA AND SOIL MICROFLORA AND THEIR RELATIONSHIPS Proceedings Soil organisms Oosterbeck 1962 Sep 10-16 Edited by J. Doeksen and Joseph van der Drift North Holland,Amsterdam,1963
Bot 42.4690

COLLOQUIUM ON SPECTRA OF METEOROLOGICAL VARIABLES Proceedings Stockholm 1969 Jun 9-19 Vol 3: radio science Inter-union commission on radio astronomy International union of radio science.Swedish national committee illus 28cm American geophysical union,Washington,D.C.,1969
A Math 4.1745

COLLOQUIUM ON THE BLANKETING EFFECT proceedings Heidelberg 1966 Mar.17-19 Edited by Karl-Heinz Bohm Journal of quantitative spectroscopy and radiative transfer, 6,no.5 Pergamon press,Oxford; London,1966
Obs 6.0207

COLLOQUIUM ON THE THEORY OF STELLAR ATMOSPHERES 3rd Transfer of radiation in stellar atmospheres Herstmonceux 1962 Aug.15-16 International astronomical union Edited by C. de Jager and A.B. Underhill 201p Pergamon press,London,1963 Vol 3,no.2 of Journal of quantitative spectroscopy and radiative transfer. AprJun.1963
Obs 6.0875

COLLOQUIUM ON VARIABLE STARS Non-periodic phenomena in variable stars Budapest 1968 Sep 5-9 International astronomical union. Commissions 27 and 42 International astronomical union.Colloquium, 4 490p Academic press,Budapest,1969
Obs 6.3360

COLLOQUIUM ON VARIABLE STARS 3rd Position of variable stars in the Hertzsprung-Russell diagram Bamberg 1965 Aug 11-14 International astronomical union.Commissions 27 and 42 Remeis-sternwarte,Bamberg.Kleine veroffentlichungen, 4,no 40 port 304p Astronomisches institut der universitat Erlangen-Nurnberg,Bamberg,1965 I.A.U. combined colloquium of commissions 27 and 42 with the theme 'The position of stars'
TA 15.0540

COLLOQUIUM SPECTROSCOPICUM INTERNATIONALE 13th Proceedings Ottawa 1967 Jun 19-23 Canadian association for applied spectroscopy Carleton university Co-sponsored by the Society for applied spectroscopy xix,460p 23cm Hilger,London,1968
Chem 18.2845

COLLOQUIUM SPECTROSCOPICUM INTERNATIONALE 14th Proceedings Debrecen 1967 Aug 7-12 Vol 1-3 Scientific society of mechanical engineers 1579p 24cm 3 vols Hilger,London,1968
Chem 18.2846

COLLOQUIUM SPECTROSCOPICUM INTERNATIONALE 15th Madrid 1969 May 26-30 Vol 1-2: plenary conferences;summary of papers 22cm 2 vols Hilger,London,1969-70
Chem 18.2847

COLOGNE 1927 Deutsche anthropologische gesellschaft:versammlung 49th Bericht Edited by Walter Venn-Koln Im auftrage der Kolner anthropologische gesellschaft Kabitzsch,Leipzig,1928
An 32.2563

COLOGNE 1960 International congress of surface activity 3rd Proceedings Vol 1-4 Universitaetsdruckerei,Mainz,1961
Col S 12.0279

COLOGNE 1961 Tagungsbericht und wissenschaftliche abhandlungen Papers Deutscher geographentag,Cologne Edited by Wolfgang Hartke and Friedrich Wilhelm Deutscher geographentag.Verhandlungen,33 illus,maps,15 plates 407p 26cm Franz Steiner,Wiesbaden,1962
Sco 14.2860

COLOMBO 1967 The Development of agriculture in the dry zone :a symposium Proceedings Ceylon association for the advancement of science Edited by O.S. Peries bibliog. Colombo,1967
Geog 13.4482

COLONIAL ADMINISTRATION Principles and methods of colonial administration :a symposium Bristol 1950 Apr Colston research society Colston papers Butterworths,London,1950
Geog 13.3107

COLONIAL RESEARCH INSTITUTE Bahama international conference on burns Report West End,Grand Bahama 1963 Mar 23-25 Dorrance,Philadelphia,1964
Med 36.0079

COLOQUIO INTERNATIONAL SOBRE GEOMETRIA ALGEBRAICA Actas Madrid 1965 Sep Consejo superior de investigaciones cientificas. Instituto "Jorge Juan" de matematicas and International mathematical union Edited by Pedro Abellanas Bibliog. 190p 24cm Madrid,1966 Papers mainly in English and Spanish
P Math 2.3137

COLOQUIO SOBRE QUIMICA FISICA DE PROCESSOS EN SUPERFICIES SOLIDAS Trabajos Madrid 1964 Oct 19-25 Consejo superior de investigaciones cientificas Consejo superior de investigaciones cientificas,Madrid,1965 Conference celebrating the 25th anniversary of C.S.I.C.
Chem 18.0609

COLORADO SPRINGS 1965 Conference on radiobiology and radiotherapy Proceedings National cancer institute Edited by J.A. Del Regato National cancer institute.Monograph, 24 National cancer institute,Bethesda,Md., 1967
Bioch 33.0989

COLORADO SPRINGS,COL 1961 International symposium on nonlinear differential equations and nonlinear mechanics proceedings Edited by Joseph P. La Salle and Solomon Lefschetz Sponsored by the United States.Air force. Office of scientific research Academic press, New York,1963
A Math 4.0245

COLORADO SPRINGS,COLO. 1961 International symposium on nonlinear differential equations and nonlinear mechanics Edited by Joseph P. LaSalle and Solomon Lefschetz Sponsored by United States.Air force.Office of scientific research Academic press,New York;London,1963
Eng 41.6385

COLORADO STATE UNIVERSITY Effects of ionizing radiation on the reproductive system : international symposium Proceedings Fort Collins,Colo. 1962 Edited by William D. Carlson and F.X. Gassner Pergamon press, Oxford,1964
An 32.0153

COLORADO STATE UNIVERSITY.AGRICULTURAL EXPERIMENT STATION Reproduction and infertility :a symposium 3rd Papers Fort Collins,Colo. 1957 Jul 1-4 Edited by F.X. Gassner and others Pergamon press,London,1958
Chem 26.0313

COLORADO STATE UNIVERSITY.COLLEGE OF VETERINARY MEDICINE Reproduction and infertility :a symposium 3rd Papers Fort Collins,Colo. 1957 Jul 1-4 Edited by F.X. Gassner and others Pergamon press,London,1958
Chem 26.0313

COLOUR VISION Conference on colour vision Papers Cambridge 1947 Jul 28-Aug 2 Documenta ophthalmologica, 3 Junk,The Hague,1949
Phys 20.1021

COLOUR VISION Documenta ophthalmologica : advances in ophthalmology Cambridge 1947 Jul 28-Aug 2 Vol 3 Edited by F.P. Fisher and others W.Junk,'S.Gravenhage,1949 Contains most of the papers read at the conference on colour vision held in Cambridge, July 28-August 2,1947
Psy 31.0409

COLOUR VISION Visual problems of colour :a symposium Teddington 1957 Sep 23-25 Vol 1-2 National physical laboratory National physical laboratory.Symposia, 8 2 vols H.M.S.O.,London,1958
Phys 20.1033

COLOUR VISION :physiology and experimental psychology London Ciba foundation Edited by A.V.S.de Reuck and Julie Knight Churchill,London,1965
Psy 31.3373

COLSTON PAPERS, 7 Recent developments in cell physiology :a symposium Bristol 1954 Mar 29-Apr 1 Colston research society Edited by J.A. Kitching Held at the University of Bristol London,1954
Bal 39.0332

COLSTON PAPERS, 11 Hypersonic flow 11th symposium Bristol 1959 Apr 6-8 Colston research society Edited by A.R. Collar and J. Tinkler Butterworths,London, 1960
Eng 41.6759

COLSTON RESEARCH SOCIETY Cosmic radiation based on a symposium promoted by the Colston research society 1st Papers Bristol 1948 Sep Edited by F.C. Frank and D.R. Rexworthy Colston research society.Colston papers, 1 illus viii,189p Butterworths, London,1949
Cav 7.0348

COLSTON RESEARCH SOCIETY Engineering structures :a symposium Papers Bristol 1949 Sep Colston research society.Research special supplement, 49 Butterworths,London, 1949
Eng 41.3022

COLSTON RESEARCH SOCIETY Fungus spore :a symposium Proceedings Bristol 1966 Mar 28-Apr 1 Edited by M.F. Madelin Colston research society.Symposia, 18 Colston papers, 18 Butterworths,London,1966
Bot 42.4006

COLSTON RESEARCH SOCIETY Hypersonic flow 11th symposium Bristol 1959 Apr 6-8 Edited by A.R. Collar and J. Tinkler Colston papers, 11 Butterworths,London,1960
Eng 41.6759

COLSTON RESEARCH SOCIETY Neurohypophysis : symposium Proceedings Bristol 1956 Apr 9-12 Edited by H. Heller Colston research society.Symposia, 8 Colston papers, 8 Butterworths,London,1957
Bioch 33.0444

COLSTON RESEARCH SOCIETY Observation and interpretation... symposium 9th proceedings Bristol 1957 Apr 1-4 Edited by S. Korner Colston papers, 9 xiii,218p 25cm Butterworths,London,1957
Math 3.0972

COLSTON RESEARCH SOCIETY Observation and interpretation:a symposium of philosophers and physicists Proceedings Bristol 1957 Apr Edited by S. Korner Colston research society.Symposia, 9 Butterworths,London, 1957
Psy 31.2431

COLSTON RESEARCH SOCIETY Principles and methods of colonial administration :a symposium Bristol 1950 Apr Colston papers Butterworths,London,1950
Geog 13.3107

COLSTON RESEARCH SOCIETY Recent developments in cell physiology :a symposium Bristol 1954 Mar 29-Apr 1 Edited by J.A. Kitching Held at the University of Bristol Colston papers, 7 London,1954
Bal 39.0332

COLSTON RESEARCH SOCIETY Recent developments in cell physiology :symposium 7th Proceedings Bristol 1954 Mar 29-Apr 1 Edited by J.A. Kitching Colston papers, 7 Butterworths,London,1954
Phys 20.1481

COLSTON RESEARCH SOCIETY Recent developments in cell physiology :symposium Proceedings Bristol 1954 Mar 29-Apr 1 Edited by J.A. Kitching Colston research society.Symposia, 7 Colston papers, 7 Butterworths,London, 1954
Bioch 33.0936

COLSTON RESEARCH SOCIETY Submarine geology and geophysics :a symposium 17th Proceedings Bristol 1965 Apr 5-9 Edited by W.F. Whittard and R. Bradshaw Colston research society.Colston papers, 17 London, 1965
Geod 9.0580

COLSTON RESEARCH SOCIETY Submarine geology and geophysics :symposium 17th proceedings Bristol 1965 Apr.5-9 Edited by W.F. Whittard and R. Bradshaw Colston papers,17 illus. map Butterworths,London,1965
Geol 8.1358

COLSTON RESEARCH SOCIETY The Liver :... symposium Proceedings Bristol 1967 Apr 3-7 Edited by A.E. Read Colston research society.Symposium, 19 Colston papers, 19 Butterworths,London,1967
PGMS 29.0367

COLSTON RESEARCH SOCIETY The Neurohypophysis symposium Proceedings Bristol 1956 Apr 9-12 Edited by H. Heller Colston papers, 8 Butterworth,London,1957
An 32.4689

COLSTON RESEARCH SOCIETY The Neurohypophysis symposium Proceedings Bristol 1956 Apr 9-12 Edited by H. Heller Colston papers,8 Butterworths,London,1957
VA 19.0224

COLSTON RESEARCH SOCIETY The Structure and properties of porous materials :a symposium Proceedings :a symposium 10th Proceedings Bristol 1958 Mar 24-27 Edited by D.H. Everett and F.S. Stone Colston papers, 10 Butterworths,London,1958
Eng 41.3084

COLSTON RESEARCH SOCIETY The Structure and properties of porous materials :a symposium Proceedings Bristol 1958 Mar 24-27 Edited by D.H. Everett and F.S. Stone Colston papers, 10 xiv,389p Butterworths, London,1958
Chem E 24.0854

COLSTON RESEARCH SOCIETY The Structure and properties of porous materials :a symposium 10th Proceedings Bristol 1958 Mar 24-27 Edited by D.H. Everett and F.S. Stone Butterworths,London,1958
Chem 18.0558

COLSTON RESEARCH SOCIETY The Suprarenal cortex :a symposium Proceedings Bristol 1952 Apr 1-4 Edited by J.M. Yoffey Colston research society.Symposia, 5 Colston papers, 5 Butterworth,London,1953
Bioch 33.0445

COLSTON RESEARCH SOCIETY 10th :symposium Proceedings Structure and properties of porous materials Bristol 1958 Mar 24-27 Edited by D.H. Everett and F.S. Stone Butterworths,London,1958
Min 40 77 10.0655

COLSTON RESEARCH SOCIETY 17th :symposium proceedings Submarine geology and geophysics Bristol 1965 Apr 5-9 Edited by W.F. Whittard and R. Bradshaw Butterworths,London,1965
Min 10.0573

COLSTON RESEARCH SOCIETY.SYMPOSIA, 5 The Suprarenal cortex :a symposium Proceedings Bristol 1952 Apr 1-4 Colston research society Edited by J.M. Yoffey Colston papers, 5 Butterworth,London,1953
Bioch 33.0445

COLSTON RESEARCH SOCIETY.SYMPOSIA, 7 Recent developments in cell physiology : symposium Proceedings Bristol 1954 Mar 29-Apr 1 Colston research society Edited by J.A. Kitching Colston papers, 7 Butterworth,London,1954
Radioth 35.0463

COLSTON RESEARCH SOCIETY.SYMPOSIA, 7 Recent developments in cell physiology : symposium Proceedings Bristol 1954 Mar 29-Apr 1 Colston research society Edited by J.A. Kitching Colston papers, 7 Butterworths,London,1954
Bioch 33.0936

COLSTON RESEARCH SOCIETY.SYMPOSIA, 8 Neurohypophysis :symposium Proceedings Bristol 1956 Apr 9-12 Colston research society Edited by H. Heller Colston papers, 8 Butterworths,London,1957
Bioch 33.0444

COLSTON RESEARCH SOCIETY.SYMPOSIA, 9 Observation and interpretation:a symposium of philosophers and physicists Proceedings Bristol 1957 Apr Colston research society Edited by S. Korner Butterworths,London,1957
Psy 31.2431

COLSTON RESEARCH SOCIETY.SYMPOSIA, 18 Fungus spore :a symposium Proceedings Bristol 1966 Mar 28-Apr 1 Colston research society Edited by M.F. Madelin Colston papers, 18 Butterworths,London,1966
Bot 42.4006

COLSTON RESEARCH SOCIETY.SYMPOSIUM, 19 The Liver :...symposium Proceedings Bristol 1967 Apr 3-7 Colston research society Edited by A.E. Read Colston papers, 19 Butterworths,London,1967
PGMS 29.0367

COLSTON RESEARCH SOCIETY SYMPOSIUM 9th Proceedings Observation and interpretation:a symposium of philosophers and physicists Bristol 1957 Apr 1-4 Edited by S. Korner xiv,218p Butterworths,London, 1957
WSM 43.1022

COLSTON RESEARCH SOCIETY SYMPOSIUM 17th Proceedings Submarine geology and geophysics Bristol 1965 Apr 5-9 Edited by W.F. Whittard and R. Bradshaw Colston papers,17 Butterworths,London,1965
Geog 13.0740

COLUMBIA UNIVERSITY Crust of the earth :a symposium Proceedings New York 1954 Oct 14-16 Edited by Arie Poldervaart Under the auspices of the Geological society of America Geological society of America.Special paper, 62 New York,1955
Geod 9.0100

COLUMBIA UNIVERSITY Quantum electronics Conference on quantum electronics-resonance phenomena Papers New York 1959 Sep 14-16 Edited by Charles H. Townes 601p Columbia university press,New York,1960
Cav 7.2332

COLUMBIA UNIVERSITY.COLLEGE OF PHYSICIANS AND SURGEONS Ion transport across membranes : a symposium Papers New York 1953 Oct 2-3 Edited by Hans T. Clarke and David Nachmansohn Academic press,New York,1954 Incorporating papers presented at a symposium on the role of proteins in ion transport across membranes together with six invited articles on allied topics
Bal 39.0560

COLUMBIA UNIVERSITY.DEPARTMENT OF GEOLOGY Crust of the earth :a symposium New York 1954 Oct 14-16 Edited by Arie Poldervaart Geological society of America.Special papers, 62 Geological society of America,New York, 1955
Geog 13.0219

COLUMBIUM Technology of columbium (Niobium) : a symposium Washington,D.C. 1958 May 15-16 Electrochemical society.Electrothermics and metallurgy division Wiley;Chapman and Hall, London,1958
Met 25.0587

COLUMBIUM METALLURGY :a symposium
Bolting Landing,N.Y. 1960 Jun 9-10
American institute of mining,metallurgical and petroleum engineers Edited by D.L. Douglass and F.W. Kunz Metallurgical society conferences, 10 Interscience,New York; London,1958
Met 25.2533

COLUMBUS 1961 Temperature - its measurement and control in science and industry 4th Papers Vol 3,pt.2: applied methods and instruments American institute of physics Edited by A.I. Dahl Reinhold;Chapman and Hall,New York;London,1962
Met 25.0303

COLUMBUS,OHIO 1931 The Symposium on colloid chemistry 9th Papers presented Ohio State university Edited by Harry Boyer Weiser Colloid symposium monograph, 9 Journal of physical chemistry,New York,1931
Bioch 33.1476

COLUMBUS,OHIO 1950 Genetics in the 20th century :essays on the progress of genetics during its first 50 years Genetics society of America Edited by L.C. Dunn Sponsored by the American institute of biological sciences Macmillan,New York,1951
Gen 34.0991

COLUMBUS,OHIO 1950 Golden jubilee of genetics Genetics in the 20th century : essays on the progress of genetics during its first 50 years Genetics society of America Edited by L.C. Dunn Sponsored by the American institute of biological sciences port Macmillan,New York,1951
Bot 42.0751

COLUMBUS,OHIO 1952 Midwestern conference on fluid mechanics 2nd Proceedings Sponsored by Ohio state university.Graduate school Ohio state university.Engineering experiment station.Bulletin, 149 Ohio state university studies:engineering series, 21,no 3 v,529p Ohio state university,Columbus, Ohio,1952
Chem E 24.0402

COLUMBUS,OHIO 1957 Biophysics conference The National biophysics conference 1st Proceedings Edited by Henry Quastler and Harold J. Morowitz With the support of the United States.Air force.Office of scientific research Yale university press,New Haven, 1959
Bioch 33.0308

COLUMBUS,OHIO 1957 National biophysics conference 1st Proceedings Edited by H. Quastler and H.J. Morowitz Yale university press,NEW Haven,Conn.,1959
Radioth 35.0231

COLUMBUS,OHIO 1961 Geodesy in the space age :a symposium Papers Ohio state university Edited by Simo Laurila and others Ohio state university.Institute of geodesy, photogrammetry,and cartography.Publication, 15 Columbus,Ohio,1961 Privately printed
Geod 9.0032

COLUMBUS,OHIO 1961 Temperature:its measurement and control in science and industry :a symposium 4th Proceedings Vol 3,pt 1: basic concepts,standards and methods American institute of physics and Instrument society of America Edited by C.M. Herzfeld and Ferdinand Graft Brickwedde Co-sponsored by National bureau of standards Reinhold;Chapman and Hall,New York;London,1962
Cav 7.0048

COLUMBUS,OHIO 1961 Temperature:its measurement and control in science and industry :a symposium 4th Proceedings Vol 3,pt2: applied methods and instruments American institute of physics and Instrument society of America Edited by A.I. Dahl and C.M. Herzfeld Co-sponsored by National bureau of standards 1094p Reinhold publishing,New York,1962
Cav 7.0047

COLUMBUS,OHIO 1961 Temperature,its measurement and control in science and industry :a symposium 4th Proceedings Vol 3,1: basic concepts,standards and methods American institute of physics,and,Instrument society of America Edited by Ferdinand Graft Brickwedde With the cooperation of the National bureau of standards xvi,848p Reinhold,New York,1962 General editor: Herzfeld,Charles
Sco 14.0231

COLUMBUS,OHIO 1961 Temperature,its measurement and control in science and industry :a symposium 4th Proceedings Vol 3,2: applied methods and instruments American institute of physics,and,Instrument society of America Edited by A.I. Dahl Co-sponsored by the National bureau of standards xiv,1094p Reinhold,New York,1962 General editor:Herzfeld,Charles
Sco 14.0233

COLUMBUS,OHIO 1961 Temperature,its measurement and control in science and industry :a symposium 4th Proceedings Vol 3,3: Biology and medicine American institute of physics,and,Instrument society of America Edited by James D. Hardy With the cooperation of the National bureau of standards xii,683p Reinhold,New York,1962 General editor:Herzfeld,Charles
Sco 14.0232

COLUMBUS,OHIO 1964 Gravity anomalies: unsurveyed areas Extension of gravity anomalies to unsurveyed areas :a symposium Papers Ohio state university and American geophysical union Edited by Hyman Orlin Sponsored by the International union of geodesy and geophysics Geophysical monograph, 9 National research council. Publication,1357 American geophysical union, Washington,D.C.,1966
Geod 9.0031

COLUMBUS,OHIO 1966 Foundations of mathematics :symposium papers commemorating the sixtieth birthday of Kurt Godel Edited by Jack J. Bulloff and others Bibliog.,Port. xii,195p 24cm Springer-Verlag,Berlin,1969 Some papers in German
P Math 2.3309

COLUMBUS,OHIO 1970 Contributions to ergodic theory and probability Midwestern conference on ergodic theory 1st Proceedings Held at Ohio state university Lecture notes in mathematics, 160 bibliog. vii,278p 25cm Springer,Berlin,1970
P Math 2.3939

COLUMBUS,OHIO 1970 Contributions to ergodic theory and probability Midwestern conference on ergodic theory and probability 1st Proceedings Held at the Ohio state university Lecture notes in mathematics, 160 vii,278p 26cm Springer,Berlin,1970
Math S 3.1779

COLUMBUS,OHIO 1970 The Steenrod algebra and its applications :a conference to celebrate N.E.Steenrod's sixtieth birthday Proceedings Battelle memorial institute Edited by F.P. Peterson Lecture notes in mathematics, 168 viii,317p 25cm Springer, Berlin,1970
P Math 2.4120

COMBINATORIAL MATHEMATICS International conference on combinatorial mathematics Papers New York 1970 Apr 1-4 New York academy of sciences Edited by Allan Gewirtz and Louis Quintas New York academy of sciences.Annals, 175,art 1 412p 23cm New York academy of sciences,New York,1970
P Math 2.4510

COMBINATORIAL MATHEMATICS AND ITS APPLICATION : conference Proceedings Oxford 1969 Jul 7-10 Edited by J.J.A. Welsh Held at the Mathematical institute,Oxford x,364p 24cm Academic press,New York;London,1971
Math S 3.1661

COMBINATORIAL MATHEMATICS AND ITS APPLICATIONS Chapel Hill conference 2nd Proceedings Chapel Hill,N.C. 1970 May 8-13 University of North Carolina United States.Air force. Office of scientific research Edited by R.C. Bose and others 548p 25cm University of North Carolina,Chapel Hill,N.C.,1970
P Math 2.4507

COMBINATORIAL MATHEMATICS AND ITS APPLICATIONS conference Proceedings Oxford 1969 Jul 7-10 University of Oxford.Mathematics institute Edited by D.J.A. Welsh x,364p 24cm Academic press,London;New York,1971
P Math 2.4514

COMBINATORIAL MATHEMATICS AND ITS APPLICATIONS 2nd conference Proceedings Chapel Hill,N.C. 1970 May University of North Carolina United States.Air force.Office of scientific research Edited by R.C. Bose and others 548p 25cm University of North Carolina,Chapel Hill,N.C.,1970
Math S 3.1655

COMBINATORIAL METHODS IN PROBABILITY THEORY colloquium lectures, mimeograph Aarhus 1962 Aug 1-10 Aarhus universitet.Matematisk institut Supported by the North Atlantic treaty organization.Scientific affairs division 126p 29cm Aarhus universitet, Aarhus,1962
P. Math 2.2189

COMBINATORIAL METHODS IN PROBABILITY THEORY colloquium lectures,mimeograph Aarhus 1962 Aug.1-10 Aarhus universitet.Matematisk institut Supported by the North Atlantic treaty organization.Scientific affairs division 126p 29cm Aarhus universitet, Aarhus,1962
Math 3.0570

COMBINATORIAL STRUCTURES International conference on combinatorial structures Proceedings Calgary,Alta. 1969 Jun Edited by Richard Guy and others xvi,508p 24cm Gordon and Breach,New York,1970
P Math 2.4511

COMBINATORIAL THEORY Colloquium on combinatorial theory Proceedings Balatonfured 1969 Aug 24 Bolyai Janos matematikai tarsulat Edited by R. Erdos and others Societatis Janos Bolyai.Colloquia mathematica, 4 1201p 24cm 3 vols North-Holland,Amsterdam;London,1970
Math S 3.1656

COMBINATORIAL THEORY AND ITS APPLICATIONS : colloquium Proceedings Balatonfured 1969 Aug 24-29 Bolyai Janos matematikai tarsulat Edited by P. Erdos and others Bolyai Janos matematikai tarsulat.Colloquia mathematica, 4 1201p 24cm 3 vols North-Holland,Amsterdam,1970
P Math 2.4509

COMBINATORIAL TOPOLOGY Differential and combinatorial topology :a symposium in honor of Marston Morse Princeton, N.J. 1963 Edited by Stewart S. Cairns Supported by United States.Air force.Office of scientific research.Mathematics division vi,265p 24cm Princeton university press,Princeton, N.J., 1965
P. Math 2.0295

COMBINATORICS Los Angeles,Calif. 1968 Mar 21-22 American mathematical society Edited by Theodore S. Motzkin American mathematical society.Proceedings of symposia in pure mathematics, 19 viii,255p 26cm AMS, Providence,R.I.,1971
P Math 2.4513

COMBINATORICS Louisiana conference on combinatorics,graph theory and computing Proceedings Baton Rouge,La. 1970 Mar 1-5 Edited by R.C. Mullin and others Held at Louisiana state university vi,464p 25cm Louisiana state university,Baton Rouge,La., 1970
P Math 2.4503

COMBINATORICS Recent progress in combinatorics Waterloo conference on combinatorics 3rd Proceedings Waterloo, Ont. 1969 May 20-31 Edited by W.T. Tutte Supported by N.A.T.O. and National research council of Canada xiv,347p 24cm Academic press,New York;London,1969
P Math 2.3354

COMBINATORICS Recent progress in combinatorics Waterloo conference on combinatorics 3rd Proceedings Waterloo, Ont. 1969 May 21-30 University of Waterloo xiv,341p 24cm Academic press,New York; London,1969
Math S 3.1660

COMBUSTION Heterogeneous combustion :a conference Palm Beach,Fla. 1963 Dec 11-13 American institute of aeronautics and astronautics Edited by Hans G. Wolfhard and others Progress in astronautics and aeronautics, 15 Academic press,New York, 1964
Eng 41.7241

COMBUSTION International symposia on combustion 1st-2nd Combustion institute Combustion institute,Pittsburgh,Pa. 1965
Eng 41.7257

COMBUSTION International symposia on combustion 4th-12th Combustion institute 9 vols Combustion institute, Pittsburgh,Pa.,1953-69
Eng 41.7258

COMBUSTION Joint conference on combustion Proceedings Boston,Mass. 1955 Jun 15-17 London 1955 Oct 25-27 Institution of mechanical engineers and American society of mechanical engineers viii,457p Institution of mechanical engineers,London, 1955
Chem E 24.1376

COMBUSTION Joint conference on combustion Proceedings Boston,Mass. 1955 Jun 15-17 and London 1955 Oct 25-27 Institution of mechanical engineers American society of mechanical engineers Institute of mechanical engineers,London,1956
Eng 41.7231

COMBUSTION Symposium on combustion 9th proceedings Ithaca,New York 1962 Aug.27-Sep.1 Organized by the Combustion Institute Academic press,New York,1963
A Math 4.0970

COMBUSTION Symposium on combustion and flame and explosion phenomena 3rd Madison, Wis. 1948 Sep 7-11 Standing committee on combustion symposia xiii,748p Williams and Wilkins,Baltimore,Md.,1949
Chem E 24.1404

COMBUSTION Symposium (International) on combustion 6th New Haven,Conn. 1956 Aug 19-24 Combustion institute xxv,943p Reinhold;Chapman and Hall,New York;London,1957
Chem E 24.1407

COMBUSTION Symposium (International) on combustion 12 Abstracts of papers Poitiers 1968 Jul 14-20 Combustion institute xxx,228p Combustion institute, Pittsburgh,Pa.,1968
Chem E 24.1408

COMBUSTION Symposium (International) on combustion (combustion and detonation waves) 4th Cambridge,Mass. 1952 Sep 1-5 Standing committee on combustion symposia xx, 926p Williams and Wilkins company,Baltimore, Md.,1953
Chem E 24.1405

COMBUSTION Symposium (international) on combustion 5th Pittsburgh,Pa 1954 Aug 30-Sep 3 combustion in engines and combustion kinetics Combustion institute Under the auspices of the Standing committee on combustion symposia v,802p Reinhold; Chapman and Hall,New York;London,1955
Chem E 24.1914

COMBUSTION Symposium (international) on combustion 10th Cambridge 1964 Aug 17-21 Combustion institute Combustion institute, Pittsburgh,Pa.,1965
Chem 18.0440

COMBUSTION Symposium (international) on combustion 11th Papers Berkeley,Calif. 1966 Aug 14-20 Combustion institute Combustion institute,Pittsburgh,Pa.,1967
Chem 18.0441

COMBUSTION Symposium (international) on combustion 7th Papers London 1958 Aug 28-Sep 3 Oxford Combustion institute Butterworths scientific publications,London, 1959 Earlier symposia entitled 'Symposium (international) on combustion,flame and explosion phenomena'
Chem 18.0437

COMBUSTION Symposium (international) on combustion 8th Papers Pasadena,Calif. Aug 28-Sep 3 Combustion institute Williams and Wilkins,Baltimore,Md.,1962
Chem 18.0438

COMBUSTION Symposium (international) on combustion 9th Papers Ithaca,N.Y. 1962 Aug 27-Sep 1 Combustion institute Academic press,New York;London,1963
Chem 18.0439

COMBUSTION,FLAME AND EXPLOSION PHENOMENA Symposium on combustion,flame and explosion phenomena 3rd Papers Madison,Wisc. 1948 Sep 7-11 Standing committee on combustion symposia Williams and Wilkins, Baltimore,Md.,1949 Chairman of committee: B.Lewis.Later volumes entitled 'Symposium (international) on combustion,flame and explosion phenomena'
Chem 18.0433

COMBUSTION,FLAME AND EXPLOSION PHENOMENA Symposium on combustion,flame and explosion phenomena abstracts of papers 3rd Madison, Wisc. 1948 Sep 7-11 Standing committee on combustion symposia University of Wisconsin, Madison,Wisc.,1948
Chem 18.0432

COMBUSTION,FLAME AND EXPLOSION PHENOMENA Symposium (international) on combustion,flame and explosion phenomena 6th Papers New Haven,Conn. 1956 Aug 19-24 Combustion institute Reinhold;Chapman and Hall,New York; London,1957 Earlier symposia organized by the Standing committee on combustion.Later symposia entitled 'Symposium (international) on combustion'
Chem 18.2602

COMBUSTION,FLAME AND EXPLOSION PHENOMENA Symposium (international) on combustion,flame and explosion phenomena :combustion in engines and combustion kinetics 5th Papers Pittsburgh,Pa. 1954 Aug 30-Sep 3 Standing committee on combustion symposia Published for the Combustion institute Reinhold: Chapman and Hall,New York;London,1955 Later symposia organized by the Combustion institute
Chem 18.0435

COMBUSTION,FLAME AND EXPLOSION PHENOMENA
Symposium (international) on combustion,flame and explosion phenomena (combustion and detonation waves) 4th Papers Cambridge, Mass. 1952 Sep 1-5 Standing committee on combustion symposia Held at the Massachusetts institue of technology Williams and Wilkins,Baltimore,Md.,1953 Committee chairman:B.Lewis.Earlier volumes entitled 'Symposium on combustion,flame and explosion phenomena'
Chem 18.0434

COMBUSTION AND PROPULSION Advanced aero engine testing :joint meeting of the Agard combustion and propulsion and wind tunnel and model testing panels Papers Copenhagen 1958 Oct Agard Edited by A.W. Morley and Jean Fabri AGARDograph, 37 Pergamon press, London,1959
Eng 41.7369

COMBUSTION AND PROPULSION Advanced aero engine testing :joint meeting of the Agard combustion and propulsion and wind tunnel and model testing panels Papers Copenhagen 1958 Oct Agard Edited by A.W. Morley and Jean Fabri AGARDograph, 37 Pergamon press, London,1959
Eng 41.7377

COMBUSTION AND PROPULSION Advanced propulsion techniques :Agard combustion and propulsion panel technical meeting Proceedings Pasadena,Calif. 1960 Aug 24-26 Agard Edited by S.S. Penner Pergamon press, London,1961
Eng 41.7373

COMBUSTION AND PROPULSION Combustion and propulsion:energy sources and energy conversion Agard colloquium 5th Papers Cannes 1964 Mar 16-20 Agard Edited by H. M. DeGroff and others AGARDograph, 81 Gordon and Black,London,1967
Eng 41.7253

COMBUSTION AND PROPULSION Combustion and propulsion:high temperature phenomena Agard colloquium 5th Papers Brunswick 1962 Apr 9-13 Agard Edited by R.P. Hagerty and others Pergamon press,London,1963
Eng 41.7245

COMBUSTION AND PROPULSION Noise,shock tubes, magnetic effects,instability and mixing Agard colloquium 3rd Papers Palermo 1958 Mar 17-21 Agard.Combustion and propulsion panel Edited by M.W. Thring and others Pergamon press,London,1958
Eng 41.7232

COMBUSTION AND PROPULSION:ENERGY SOURCES AND ENERGY CONVERSION Agard colloquium 5th Papers Cannes 1964 Mar 16-20 Agard Edited by H.M. DeGroff and others AGARDograph, 81 Gordon and Black,London, 1967
Eng 41.7253

COMBUSTION AND PROPULSION:HIGH TEMPERATURE PHENOMENA Agard colloquium 5th Papers Brunswick 1962 Apr 9-13 Agard Edited by R.P. Hagerty and others Pergamon press, London,1963
Eng 41.7245

COMBUSTION COLLOQUIUM Selected combustion problems Cambridge 1953 Dec 7-11 Agard Edited by W.R. Hawthorne and J. Fabri Butterworths,London,1954
A Math 4.0969

COMBUSTION COLLOQUIUM Selected combustion problems:fundamentals and aeronautical applications Cambridge 1953 Dec 7-11 Agard Butterworths,London,1954
Eng 41.7219

COMBUSTION COLLOQUIUM Selected combustion problems II:transport phenomena;ignition; altitude behaviour and scaling of aeroengines Liege 1959 Dec 5-9 Agard Butterworths, London,1956
Eng 41.7228

COMBUSTION COLLOQUIUM Proceedings Selected combustion problems:fundamentals and aeronautical applications Cambridge 1953 Dec 7-11 North Atlantic treaty organization. Advisory group for aeronautical research and development viii,534p Butterworths,London, 1954
Chem E 24.1912

COMBUSTION INSTITUTE International symposia on combustion 1st-2nd Combustion institute,Pittsburgh,Pa.,1965
Eng 41.7257

COMBUSTION INSTITUTE International symposia on combustion 4th-12th 9 vols Combustion institute,Pittsburgh,Pa.,1953-69
Eng 41.7258

COMBUSTION INSTITUTE Symposium on combustion 9th proceedings Ithaca,New York 1962 Aug.27-Sep.1 Academic press,New York,1963
A Math 4.0970

COMBUSTION INSTITUTE Symposium (International) on combustion 6th New Haven,Conn. 1956 Aug 19-24 xxv,943p Reinhold;Chapman and Hall,New York;London,1957
Chem E 24.1407

COMBUSTION INSTITUTE Symposium (International) on combustion 12 Abstracts of papers Poitiers 1968 Jul 14-20 xxx, 228p Combustion institute,Pittsburgh,Pa., 1968
Chem E 24.1408

COMBUSTION INSTITUTE Symposium (international) on combustion 5th Pittsburgh,Pa 1954 Aug 30-Sep 3 combustion in engines and combustion kinetics Under the auspices of the Standing committee on combustion symposia v,802p Reinhold; Chapman and Hall,New York;London,1955
Chem E 24.1914

COMBUSTION INSTITUTE Symposium (international) on combustion 10th Cambridge 1964 Aug 17-21 Combustion institute,Pittsburgh,Pa.,1965
Chem 18.0440

COMBUSTION INSTITUTE Symposium (international) on combustion 11th Papers Berkeley,Calif. 1966 Aug 14-20 Combustion institute,Pittsburgh,Pa.,1967
Chem 18.0441

COMBUSTION INSTITUTE Symposium (international) on combustion 13th Proceedings Salt Lake City,Utah 1970 Aug 23-29 xix,1190p 26cm Combustion institute,Pittsburgh,Pa.,1971
Chem 18.2863

COMBUSTION INSTITUTE Symposium (international) on combustion 7th Papers London 1958 Aug 28-Sep 3 Oxford Butterworths scientific publications,London, 1959 Earlier symposia entitled 'Symposium (international) on combustion,flame and explosion phenomena'
Chem 18.0437

COMBUSTION INSTITUTE Symposium (international) on combustion 8th Papers Pasadena,Calif. Aug 28-Sep 3 Williams and Wilkins,Baltimore,Md.,1962
Chem 18.0438

COMBUSTION INSTITUTE Symposium (international) on combustion 9th Papers Ithaca,N.Y. 1962 Aug 27-Sep 1 Academic press,New York;London,1963
Chem 18.0439

COMBUSTION INSTITUTE Symposium (international) on combustion,flame and explosion phenomena 6th Papers New Haven,Conn. 1956 Aug 19-24 Reinhold; Chapman and Hall,New York;London,1957 Earlier symposia organized by the Standing committee on combustion.Later symposia entitled 'Symposium (international) on combustion'
Chem 18.2602

COMBUSTION INSTITUTE Symposium (international) on combustion,flame and explosion phenomena :combustion in engines and combustion kinetics 5th Papers Pittsburgh,Pa. 1954 Aug 30-Sep 3 Standing committee on combustion symposia Reinhold; Chapman and Hall,New York;London,1955 Later symposia organized by the Combustion institute
Chem 18.0435

COMBUSTION INSTITUTE.WESTERN STATES SECTION Kinetics,equilibria and performance of high temperature systems :a conference 1st Proceedings Los Angeles 1959 Nov 2-5 Edited by Gilbert S. Bahn and Edward E. Zukoski x,255p Combustion institute,Los Angeles,1960
Chem E 24.1394

COMBUSTION RESEARCHES AND REVIEWS,1955 AGARD combustion panel meetings 6th-7th Invited papers Scheveningen 1954 May Paris 1954 Nov North Atlantic treaty organisation.Advisory group for aeronautical research and development illus xv,187p Butterworths scientific publications,London, 1955
Chem E 24.1386

COMBUSTION RESEARCHES AND REVIEWS 1955 AGARD combustion panel meetings 6th and 7th Papers Scheveningen 1954 May and Paris 1954 Nov Agard Edited by B.P. Mullins AGARDograph, 9 Butterworths,London,1955
Eng 41.7222

COMETS The Motion,evolution of orbits,and origin of comets Leningrad 1970 Aug 4-11 International astronomical union Edited by G. A. Chebotarev and others International astronomical union.Symposium, 45 521p Reidel,Dordrecht,1972
Obs 6.3591

COMETS The Motion,evolution of orbits,and origin of comets Leningrad 1970 Aug 4-11 International astronomical union Edited by G. A. Chebotarev and others International astronomical union.Symposium, 45 521p Reidel,Dordrecht,1972
TA 15.0687

COMETS :a conference :scientific data and missions Tucson,Ariz. 1970 Apr 8-9 University of Arizona.Lunar and planetary laboratory Revised edition 222p University of Arizona.Lunar and planetary laboratory,Tucson,Ariz.,1972
Obs 6.3619

COMITATO NAZIONALE PER LE RICCERCHE NUCLEARI Immediate and low level effects of ionizing radiations :symposium Proceedings Venice 1959 Jun 22-26 Under the auspices of Unesco International journal of radiation biology. Supplement Taylor and Francis,London,1960
Gen 34.2221

COMITE INTERNATIONAL DE CROISSANCE CRYSTALLINE Crystal growth 1968 International conference on crystal growth 2nd Proceedings Birmingham 1968 Jul 15-19 Edited by F.C. Frank and others Jounal of crystal growth, 1968,no 3-4 North-Holland,Amsterdam,1968
Min 10.1507

COMITE INTERNATIONAL DE PHOTOBIOLOGIE Recent progress in photobiology International congress of photobiology 4th Proceedings Oxford 1964 Jul Edited by E.J. Bowen Bibliog.,illus. vii,400p Blackwell scientific publications,Oxford,1965
Bot 42.1362

COMITE INTERNATIONAL DES POIDS ET MESURES La Conference generale des poids et mesures 9eme comptes rendus Paris 1948 Oct. 12-21 127p Gauthier-Villars,Paris,1949
Obs 6.2388

COMITE INTERNATIONALE DE PHOTOBIOLOGIE Recent progress in photobiology International photobiology congress 4th Proceedings Oxford 1964 Jul 26-30 Edited by E.J. Bowen Blackwell,Oxford,1965
Radioth 35.0253

COMITE METEOROLOGIQUE INTERNATIONALE :seances.. et rapport de 3 commissions et d'une sous-commission Proces-verbaux Innsbruck 1931 Oct 5-8 Locarno 1931 Sep Comite meteorologique internationale,10 385p 24cm Leyden,1932 In English,French and German
Sco 14.0113

COMITE METEROLOGIQUE INTERNATIONALE 7th : seances de la conference internationale des directeurs...et de diverses commissions Proces-verbaux Copenhagen 1929 Sep Comite meteorologique international,3 401p 24cm Kemink,Utrecht,1930 In French and English
Sco 14.0112

COMITE NATIONAL DES CELEBRATIONS DU 1ER CENTENAIRE DE L'UNITE D'ITALIE Symposium international d'histoire des sciences Actes Turin 1961 Jul 28-30 With the participation of the Union internationale d'histoire et philosophie des sciences Academie internationale d'histoire des sciences.Collections de travaux, 14 189p Gruppe italiano di storia delle science,Vinci (Florence),1964
WSM 43.0059

COMITE NATIONAL FRANCAIS D'ASTRONOMIE Congres national d'astronomie rapports Paris 1931 Jul. 21-23 203p Paris,1932
Obs 6.2481

COMITE NATIONAL FRANCAIS D'ASTRONOMIE Reunions scientifiques...pour le centenaire de la decouverte de Neptune reports Paris 1946 Oct.22-24 Edited by P. Couderc illus. 164p Imprimerie nouvelle,Orleans,1948
Obs 6.0418

COMMEMORATION OF THE CENTENNIAL OF THE PUBLICATION OF THE ORIGIN OF THE SPECIES : annual general meeting Philadelphia,Pa. 1959 Apr American philosophical society American philosophical society.Proceedings, 103,no.2 Philadelphia,Pa.,1959
Gen 34.1265

COMMEMORATION OF THE PUBLICATION OF GREGOR MENDEL'S PIONEER EXPERIMENTS IN GENETICS : annual general meeting Papers Philadelphia,Pa. 1965 Apr 23 American philosophical society American philosophical society.Proceedings, 109,no.4 American philosophical society,Philadelphia,Pa.,1965
Gen 34.2207

COMMEMORATION SOLENELLE DU QUATRIEME CENTENAIRE DE LA MORT D'ANDRE VESALIUS Brussels 1964 Oct 19-24 Academie royale de medecine de Belgique Brussels,1964
Phys 20.0564

COMMERCIALLY AVAILABLE GENERAL-PURPOSE ELECTRONIC DIGITAL COMPUTERS OF MODERATE PRICE symposium Washington,D.C. 1952 May 14 Sponsored by United States.Navy mathematical computing advisory panel 41p Washington,D. C.,1952
Math L 5.1790

COMMISSARIAT A L'ENERGIE ATOMIQUE Diffusion a l'etat solide Colloque sur la diffusion a l'etat solide Saclay 1958 Jul 3-5 Centre d'etudes nucleaires de Saclay;North Holland, Gif-sur-Yvette;Amsterdam,1959
Met 25.1277

COMMISSARIAT A L'ENERGIE ATOMIQUE Plutonium 1960 International conference on plutonium metallurgy 2nd Proceedings Grenoble 1960 Apr 19-22 Edited by Emmanuel Grison and others Cleaver-Hume,London,1961 Papers in English and French
Met 25.1549

COMMISSARIAT A L'ENERGIE ATOMIQUE The International conference on elementary particles Proceedings Aix-en-Provence 1961 Sep 14-20 Vol 1-2 Edited by E. Cremieu-Alcan and others 2 vols C.E.N. Saclay,Gif-sur-Yvette,1962
Cav 7.0029

COMMISSION ADMINISTRATIVE DU PATRIMOINE UNIVERSITAIRE DE LIEGE Biological problems of grafting :a symposium Proceedings Liege 1959 Mar 18-21 Edited by F. Albert and G. Lejeune-Ledant Blackwell, Oxford,1959 Symposium chairman P.B. Medawar
PGMS 29.0245

COMMISSION ADMINISTRATIVE DU PATRIMOINE UNIVERSITAIRE DE LIEGE Biological problems of grafting :a symposium Proceedings Liege 1959 Mar 18-21 Edited by F. Albert and G. Lejeune-Ledant Blackwell, Oxford,1959
Bal 39.0458

COMMISSION DE GEOGRAPHIE APPLIQUEE Colloque international de geographie appliquee comptes rendus 3rd Liege 1967 Sep 7-13 Christans,Liege,c1967
Geog 13.7167

COMMISSION FOR MARITIME METEOROLOGY abridged final report of the first session Report London 1952 Jul 14-29 Commission for maritime meteorology 108p 28cm World meteorological organization,Geneva,1952
Sco 14.0349

COMMISSION FOR TECHNICAL CO-OPERATION IN AFRICA SOUTH OF THE SAHARA Symposium on African hydrobiology and inland fisheries 2nd Brazzaville 1956 Jul 3-11 Scientific council for Africa south of the Sahara Scientific council for Africa south of the Sahara.Publication, 25 Brazzaville,1956 Title also in French:Symposium sur l'hydriobiologie et la peche en eaux douces en Afrique;text in English and French
Bal 39.1894

COMMISSION FOR TECHNICAL CO-OPERATION IN AFRICA SOUTH OF THE SAHARA Tropical soils and vegetation :Abidjan symposium Proceedings Abidjan 1959 Oct 20-24 Unesco Humid tropics research Unesco,Paris,1961
Geog 13.1128

COMMISSION INTERNATIONALE DE L'ECLAIRAGE 7e session en succession a la Commission internationale de photometrie Recueil des travaux et compte rendu Saranac Inn,N.Y. 1928 Sep National physical laboratory, Teddington,c1929
Psy 31.2561

COMMISSION INTERNATIONALE DE L'ECLAIRAGE 8e session en succession a la Commission internationale de photometrie Recueil des travaux et compte rendu Cambridge 1931 Sep Published under the direction of the National physical laboratory Cambridge university press,Cambridge,1932
Psy 31.2562

COMMISSION ON ATMOSPHERIC CHEMISTRY AND RADIOACTIVITY and WORLD METEOROLOGICAL ORGANIZATION Symposium:atmospheric chemistry,circulation and aerosols Proceedings Visby,Sweden 1965 Aug Tellus, 18,pt 2-3 Svenska geofysiska foreningen, Stockholm,1966
Chem 18.0571

COMMITTEE FOR RESEARCH IN PROBLEMS OF SEX
Sex and behavior :conferences Papers Berkeley,Calif. 1961 Edited by Frank A. Beach Wiley,New York,1965 Based on conferences held in 1961 and 1962
Psy 31.2709

COMMITTEE FOR SYMPOSIA ON DRUG ACTION
Adrenergic mechanisms London 1960 Mar 28-29 Edited by J.R. Vane and others Ciba foundation.Symposia Churchill,London,1960
PGMS 29.0379

COMMITTEE OF MUSCLE CHEMISTRY OF JAPAN
Conference on the chemistry of muscular contraction Paper and discussions presented Tokyo 1957 Oct 12-14 Igaku Shoin ltd., Tokyo;Osaka,1957
Bioch 33.2229

COMMITTEE ON MEDITERRANEAN NEOGENE STRATIGRAPHY
3rd session proceedings Berne 1964 Jun.8-13 Edited by C.W. Drooger and others illus. E.J.Brill,Leiden,1966
Geol 8.2524

COMMITTEE ON SPACE RESEARCH International symposium on rocket and satellite meteorology Washington,D.C. 1962 Apr 23-25 Edited by H. Wexler and J.E. Caskey,jnr ix,440p North-Holland,Amsterdam,1963
Sco 14.0516

COMMITTEE ON SPACE RESEARCH International symposium on the use of artificial satellites for geodesy 1st Proceedings Washington, D.C. 1962 Apr 26-28 Edited by George Veis Co-sponsored by the International union of geodesy and geophysics North-Holland, Amsterdam,1963
Geod 9.0055

COMMITTEE ON SPACE RESEARCH Moon and planets International space science :symposium 7th Papers Vienna 1966 May 10-18 International astronomical union Edited by A. Dollfus 313p North-Holland,Amsterdam,1967 Papers of one session of the symposium
Obs 6.3206

COMMITTEE ON SPACE RESEARCH Moon and planets, 2 Cospar plenary meeting :joint open meeting of working groups 1,2 and 5 10th London 1967 Jul 26-27 Edited by A. Dollfus 196p North-Holland,Amsterdam,1968
Obs 6.3419

COMMITTEE ON SPACE RESEARCH Problems of atmospheric circulation The Space science symposium 6th Papers of a session Mar del Plata,Argentina 1965 May 11-19 Edited by R.V. Garcia and T.F. Malone Co-sponsored by the International union of physiological sciences 186p Spartan books,Washington,D.C. 1966
Nap 11.0212

COMMITTEE ON SPACE RESEARCH Solar flares and space research Proceedings of the symposium held on the occasion of the 11th plenary meeting of the Committee on space research Tokyo 1968 May 9-11 Edited by C.de Jager and Z. Svestka 419p North-Holland, Amsterdam,1969 Co-sponsored by International astronomical union,International union of geodesy and geophysics and International union of radio scientists
Obs 6.3335

COMMITTEE ON SPACE RESEARCH Solar-terrestrial physics:solar aspects Proceedings of joint I.Q.S.Y.-Cospar symposium London 1967 Jul 17-22 Pt 1 Edited by A.C. Stickland International quiet sun year.Annals, 4 414p MIT press,Cambridge, Mass.,1969
Obs 6.3341

COMMITTEE ON SPACE RESEARCH Space research The Space science symposium 1st Proceedings Nice 1960 Jan 11-16 Edited by Hilde Kallmann-Bijl xvi,1195p North-Holland,Amsterdam,1960
Nap 11.0317

COMMITTEE ON SPACE RESEARCH The Theory of orbits in the solar system and in stellar systems a symposium Thessaloniki 1964 Aug. 17-22 International astronomical union Edited by George Contopoulos International Astronomical Union.Symposium, 25 380p Academic press,London;New York,1966
Obs 6.0409

COMMITTEE ON SPACE RESEARCH The Theory of orbits in the solar system and in stellar systems :a symposium Thessaloniki 1964 Aug. 17-22 International astronomical union Edited by George Contopoulos International astronomical union.Symposium, 25 380p Academic press,London;New York,1966
A Math 4.1167

COMMITTEE ON SPACE RESEARCH 11TH PLENARY MEETING
Solar flares and space research Proceedings of the symposium held on the occasion of the 11th plenary meeting of the Committee on space research Tokyo 1968 May 9-11 Committee on space research International astronomical union Edited by C. de Jager and Z. Svestka 419p North-Holland, Amsterdam,1969 Co-sponsored by International astronomical union,International union of geodesy and geophysics and International union of radio scientists
Obs 6.3335

COMMITTEE ON VACUUM TECHNIQUES National symposium on vacuum technology Transactions Chicago,Ill 1956 Oct 10-11 Edited by Edmond S. Perry and John H. Durant Pergamon press,London,1957 Later symposia published by the American vacuum society
Met 25.1629

COMMONWEALTH AND EMPIRE HEALTH AND TUBERCULOSIS CONFERENCE Transactions Tuberculosis in the commonwealth 1947 London 1947 National association for the prevention of tuberculosis N.A.P.T.,London,1947
HE 27.0132

COMMONWEALTH BUREAU OF SOIL SCIENCE
Commonwealth conference on tropical and sub-tropical soils 1st proceedings Harpenden 1948 Jun Commonwealth bureau of soil science. Technical communications,46 Commonwealth bureau of soil science,Harpenden,1949
Geog 13.1093

COMMONWEALTH CONFERENCE ON TROPICAL AND SUB-TROPICAL SOILS 1st proceedings Harpenden 1948 Jun Commonwealth bureau of soil science Commonwealth bureau of soil science.Technical communications,46 Commonwealth bureau of soil science,Harpenden, 1949
Geog 13.1093

COMMONWEALTH COUNCIL OF MINING AND METALLURGICAL INSTITUTIONS The Commonwealth mining and metallurgical congress 9th Proceedings London 1969 May 3-24 Vol 4: physical and fabrication metallurgy Edited by M.J. Jones Institution of mining and metallurgy,London, 1970
Met 25.2847

COMMONWEALTH ENTOMOLOGICAL CONFERENCE... 6th Report London 1954 Jul 7-16 Commonwealth institute of entomology London, 1954
Bal 39.2874

COMMONWEALTH INSTITUTE OF ENTOMOLOGY Commonwealth entomological conference... 6th Report London 1954 Jul 7-16 London, 1954
Bal 39.2874

The COMMONWEALTH MINING AND METALLURGICAL CONGRESS 9th Proceedings London 1969 May 3-24 Vol 4: physical and fabrication metallurgy Commonwealth council of mining and metallurgical institutions Edited by M.J. Jones Institution of mining and metallurgy,London,1970
Met 25.2847

COMMONWEALTH MINING AND METALLURGICAL CONGRESS 6th Papers Geology of Canadian industrial mineral deposits Montreal 1957 Sep 8-Oct 9 Canadian institute of mining and metallurgy Montreal,1957
Min 10.0716

COMMONWEALTH MINING AND METALLURGICAL CONGRESS 8th publications Australia 1965 and New Zealand 1965 Vol 1-2: geology of Australian ore deposits,2nd edition; Exploration and mining geology Edited by John McAndrew 2 vols Australasian institute of mining and metallurgy,Melbourne, 1965
Min 10.0222

COMMONWEALTH MYCOLOGICAL INSTITUTE Plant pathologist's pocketbook International congress of plant pathology 1st Proceedings London 1968 Jul Commonwealth mycological institute,Kew,1968
BG 38.1757

COMMONWEALTH NATIONAL FITNESS COUNCIL World congress on physical education Report Melbourne 1956 Nov 16-21 Australian physical education association Australian physical education association,Melbourne,1956
HE 27.0219

The COMMONWEALTH OCEANOGRAPHIC CONFERENCE Proceedings Wormley,Surrey 1954 Oct 18-22 Arranged by the National oceanographic council vi,93p 25cm Cambridge university press,Cambridge,1955
Sco 14.0123

COMMONWEALTH SCIENTIFIC AND INDUSTRIAL RESEARCH ORGANISATION Picture language machines :a conference Proceedings Canberra 1969 Feb 24-28 Edited by S. Kaneff Academic press, London,1970
Psy 31.3083

COMMONWEALTH SCIENTIFIC AND INDUSTRIAL RESEARCH ORGANIZATION Automatic computing machines conference proceedings Sydney 1951 Aug 7-9 Jointly sponsored by University of Sydney.Department of electrical engineering illus. 220p 25cm Commonwealth scientific and industrial research organization,Melbourne, 1952
Math L 5.0703

COMMONWEALTH SCIENTIFIC AND INDUSTRIAL RESEARCH ORGANIZATION Conference on applications of isotopes in scientific research Proceedings Melbourne 1950 Aug 14-17 University of Melbourne,Melbourne,1951
Radioth 35.1495

COMMONWEALTH SCIENTIFIC AND INDUSTRIAL RESEARCH ORGANIZATION The Australian environment handbook prepared for the conference on plant and animal nutrition... 1949 Aug 2nd edition British commonwealth official scientific specialist conferences,Special agricultural conference Commonwealth scientific and industrial research organization,Melbourne,1950
Gen 34.1372

COMMONWEALTH SCIENTIFIC AND INDUSTRIAL RESEARCH ORGANIZATION and UNESCO Climatology and microclimatology :symposium Proceedings Canberra 1956 Oct Arid zone research,11 Unesco,Paris,1958
Geog 13.1018

COMMONWEALTH STUDY CONFERENCE see DUKE OF EDINBURGH'S COMMONWEALTH STUDY CONFERENCE

COMMUNICABLE DISEASES:METHODS OF SURVEILLANCE report on a seminar The Hague 1969 May 21-30 Convened by the World Health Organisation.Regional office for Europe Regional office for Europe,W.H.O.,Copenhagen, 1969
PGMS 29.0117

COMMUNICATION Psychopathology of communication :forty-sixth annual meeting Proceedings New York 1956 Jun American psychopathological association Edited by Paul H. Hoch and Joseph Zubin Grune and Stratton,New York;London,1958
Psy 28.0196

COMMUNICATION IN DEVELOPMENT :a symposium Boulder,Colo. 1969 Jun 16-18 Society for developmental biology Edited by Anton Lang Society for developmental biology.Symposia, 28 Developmental biology.Supplement, 3 Academic press,New York;London,1969
An 32.5441

COMMUNICATION IN DEVELOPMENT :a symposium Proceedings Society for developmental biology Edited by A. Lang Society for developmental biology.Symposia, 28 Developmental biology.Supplement, 3 Academic press,New York,1969
Gen 34.2155

COMMUNICATION IN DEVELOPPMENT :symposium Boulder,Colo. 1969 Jun 16-18 Society for developmental biology Society for developmental biology.Symposium, 28 Academic press,New York,1969
Bioch 33.2255

COMMUNICATION IN ORGANIZATIONS :some new research findings.A meeting 1958 Oct Foundation for research on human behavior Foundation for research on human behavior,Ann Arbor,Mich.,1959 One of four publications of the Foundation bound together
Eng 41.0729

COMMUNICATION OR CONFLICT :an international conference :conferences.Conferences;their nature,dynamics,and planning Eastbourne 1956 Edited by Mary Capes Tavistock publications,London,1960
Eng 41.0699

COMMUNICATION PROCESSES NATO symposium on communication processes Proceedings Washington,D.C. 1963 North Atlantic treaty organization.Advisory group on human factors Edited by Frank A. Geldard and others NATO conference series, 4 298p Pergamon press, Oxford,1965
Cav 7.2423

COMMUNICATION THEORY Applications of communication theory :symposium Papers London 1952 Sep 22-26 Edited by Willis Jackson Supported by the British broadcasting corporation xii,532p 25cm Butterworths,London,1953 Also supported by the Ministry of supply
Math L 5.3249

COMMUNICATION THEORY Applications of communication theory symposium papers London 1952 Sep 22-26 British broadcasting corporation Great Britain.Ministry of supply Edited by Willis Jackson xii,532p Butterworths scientific publications,London, 1953
Math 3.0735

COMMUNICATION THEORY Applications of communication theory symposium Papers London 1952 Sep 22-26 Institution of electrical engineers Edited by Willis Jackson Butterworths,London,1953
Psy 31.2016

COMMUNICATION THEORY AND RESEARCH:INTERNATIONAL SYMPOSIUM 1st Proceedings Kansas City 1965 Mar 24-27 University of Missouri and National society for the study of communication Edited by Lee Thayer Thomas, Springfield,Ill.,1967
Psy 31.1068

COMO,PAVIA,ROMA 1927 Congresso internazionale dei fisici Atti Vol 1-2 Nicola Zanichelli,Bologna,1928 Polyglot publication
Cav 7.2404

A COMPANION TO MEDICAL STUDIES Vol 1: anatomy,biochemistry,physiology and related subjects Edited by R. Passmore and J.S. Robson Blackwell scientific,Oxford;Edinburgh, 1968
An 32.1536

COMPARATIVE ASPECTS OF NEUROHYPOPHYSICAL MORPHOLOGY AND FUNCTION :a symposium Proceedings London 1962 Feb 8 Zoological society of London Edited by H. Heller Zoological society of London.Symposia, 9 Zoological society of London,1963
An 32.3765

COMPARATIVE ASPECTS OF REPRODUCTIVE FAILURE : an international conference Hanover,N.H. 1966 Jul 25-29 Edited by Kurt Benirschke Springer,Berlin,1967
Inv Med 37.0017

COMPARATIVE BIOCHEMISTRY AND BIOPHYSICS OF PHOTOSYNTHESIS :a conference Papers Hakone 1967 Aug 12-15 Edited by K. Shibata and others Sponsored by the Japan-U.S. cooperative science program illus viii, 445p University of Tokyo press;University Park press,Tokyo;University Park,Pa.,1968
Bot 42.1903

COMPARATIVE BIOCHEMISTRY AND BIOPHYSICS OF PHOTOSYNTHESIS :a conference Papers presented Hakone,Japan 1967 Aug 12-15 Edited by K. Shibata and others Sponsored by the Japan-United States cooperative science program University of Tokyo press;University park press,State college,Tokyo;University Park, Pa.,1968
Bioch 33.2349

COMPARATIVE BIOCHEMISTRY OF PHOTOREACTIVE SYSTEMS Symposium on comparative biology 1st Papers Richmond,Calif. Kaiser foundation research institute Edited by Mary Belle Allen Kaiser foundation research institute. Symposia on comparative biology, 1 Academic press,New York;London,1960
Bioch 33.0334

COMPARATIVE BIOCHEMISTRY OF PHOTOREACTIVE SYSTEMS 1st :annual symposium on comparative biology of the Kaiser foundation research institute Papers Kaiser foundation research institute Edited by Mary Belle Allen Kaiser foundation research institute. Symposia on comparative biology,1 Academic press,New York;London,1960
Chem 18.2139

COMPARATIVE BIOLOGY OF CALCIFIED TISSUE papers New York 1962 Oct 12-13 New York academy of sciences New York academy of sciences.Annals,109,art.1 New York,1963
Geol 8.0971

COMPARATIVE BIOLOGY OF REPRODUCTION IN MAMMALS international symposium Proceedings London 1964 Nov 24-26 Society for the study of fertility,and,Zoological society of London Edited by I.W. Rowlands Zoological society of London.Symposia,15 Academic press, London,1966
VA 19.0088

COMPARATIVE BIOLOGY OF REPRODUCTION IN MAMMALS symposium Proceedings London 1964 Nov 24-26 Society for the study of fertility Zoological society of London Edited by I.W. Rowlands Zoological society of London. Symposia, 15 Academic press,London,1966
An 32.3887

COMPARATIVE CELLULAR AND SPECIES RADIOSENSITIVITY international seminar Proceedings Kyoto 1968 May 20-23 Edited by Victor P. Bond and Tsutomu Suyahara Iyaku Shoin,Tokyo, 1969
Radioth 35.0554

COMPARATIVE EFFECTS OF RADIATION :a conference Report San Juan 1960 Feb 15-19 Edited by Milton Burton and others Wiley,New York, 1960
Radioth 35.0947

COMPARATIVE ENDOCRINOLOGY Progress in comparative endocrinology International symposium on comparative endocrinology 3rd Proceedings Oiso,Japan 1961 Jun 5-11 Edited by Kiyoshi Takewaki Sponsored by the Zoological society of Japan General and comparative endocrinology.Suppl., 1,1962 Academic press,New York,1962
Bal 39.1381

COMPARATIVE ENDOCRINOLOGY :symposium Proceedings Cold Spring Harbor 1958 May 25-29 Edited by A. Gorbman Wiley;Chapman and Hall,New York;London,1959
VA 19.0286

The COMPARATIVE ENDOCRINOLOGY OF VERTEBRATES : a conference Proceedings Liverpool 1954 Jul 12-16 Part 1: the comparative physiology of reproduction and the effects of sex hormones in vertebrates Society for endocrinology Edited by I.Chester Jones and P. Eckstein Society for endocrinology. Memoirs, 4 Cambridge university press, Cambridge,1955
Phys 20.1378

The COMPARATIVE ENDOCRINOLOGY OF VERTEBRATES : a conference Proceedings Liverpool 1954 Jul 12-16 Part 2: the hormonal control of water and salt electrolyte metabolism in vertebrates Society for endocrinology Edited by I.Chester Jones and P. Eckstein Society for endocrinology. Memoirs, 5 Cambridge university press, Cambridge,1955
Phys 20.1379

The COMPARATIVE ENDOCRINOLOGY OF VERTEBRATES : a conference Proceedings Liverpool 1954 Jul 12-16 Pt 1: the comparative physiology of reproduction and the effects of sex hormones in vertebrates Society for endocrinology Edited by I.Chester Jones and P. Eckstein held at the University of Liverpool.Department of zoology Society for endocrinology.Memoirs, 4 Cambridge university press,Cambridge,1955
Bal 39.1400

COMPARATIVE MORPHOLOGY OF HEMATOPOIETIC NEOPLASMS a symposium Proceedings Washington, D.C. 1968 Mar 11-12 National cancer institute Edited by Carolyn H. Lingeman and F.M. Garner held at United States.Armed forces institute of pathology National cancer institute.Monograph, 32 National cancer institute,Bethesda,Md.,1969
Bioch 33.0995

COMPARATIVE PHYSIOLOGY AND PATHOLOGY OF THE SKIN a symposium Proceedings Cambridge 1964 Apr Edited by Arthur J. Rook and G.S. Walton Blackwell,Oxford,1965
An 32.4050

COMPARATIVE PHYSIOLOGY OF CARBOHYDRATE METABOLISM IN HETEROTHERMIC ANIMALS :symposium 1959 Western society of naturalists Edited by Arthur W. Martin University of Washington press,Seattle,1961
Bal 39.1258

COMPARATIVE PSYCHOPATHOLOGY ANIMAL AND HUMAN Proceedings New York 1965 Feb American psychopathological association Edited by Joseph Zubin and Howard F. Hunt Grune and Stratton,New York;London,1967 Proceedings of the 55th annual meeting of the Association
Psy 31.1682

A COMPARATIVE STUDY OF SOME HIGHER PROGRAMMING LANGUAGES Series of lectures Advanced course on programming languages and data structures Amsterdam 1972 Jun 12-23 By W. L.van der Poel varp Van der Poel,Delft, 1972
Math L 5.3872

COMPARISON OF THE LARGE-SCALE STRUCTURE OF THE GALACTIC SYSTEM WITH THAT OF OTHER STELLAR SYSTEMS :a symposium papers Dublin 1955 Sep 2 International astronomical union Edited by N.G. Roman International astronomical union. Symposium, 5 illus 72p Cambridge university press,Cambridge, 1958
Obs 6.1466

COMPARTMENTS POOLS AND SPACES IN MEDICAL PHYSIOLOGY :a symposium Oak Ridge, Tenn. 1966 Oct 24-27 Edited by Per.Erik E. Bergner and others Held at the Oak Ridge institute of nuclear studies United States atomic energy commission.Symposium series, 11 U.S.atomic energy commission,1967
Radioth 35.0927

COMPLEMENT London 1964 May 26-28 Ciba foundation Edited by G.E.W. Wolstenholme and Julie Knight Churchill,London,1965
Pha 16.0189

COMPLEX ANALYSIS Conference on complex analysis Proceedings Houston,Tex. 1967 Apr 13-15 Rice university Edited by H.L. Resnikoff and R.O. Wells Rice university studies, 54,no.4 Bibliog. 84p 23cm William Marsh Rice University,Houston,Tex., 1968
P Math 2.3684

COMPLEX ANALYSIS Conference on complex analysis proceedings Minneapolis,Minn. 1964 Mar 16-21 United States.Air force. Office of scientific research Edited by A. Aeppli and others Bibliog. vii,308p 24cm Springer-Verlag,Berlin,Heidelberg,1965
P. Math 2.1582

COMPLEX ANALYSIS 1969 :conference Proceedings Houston,Texas 1969 Mar 26-29 Rice university Edited by H.L. Resnikoff and R.O. Wells Rice university studies, 56,no 2 vi,222p 23cm Rice university,Houston,Texas,1971
P Math 2.4357

COMPLEX MANIFOLDS Katata conference on the theory of partial differential equations and on the theory of complex manifolds Proceedings Katata 1966 Sep 18-22 Bibliog.,Illus. vi,107p 25cm Research institute for mathematical science,Kyoto university,Kyoto,1966
P Math 2.3188

COMPLEX MULTIPLICATION Seminar on complex multiplication Princeton,N.J. 1957-58 Institute for advanced study Lecture notes in mathematics, 21 28cm Springer-verlag, Berlin,1966
P Math 2.2857

COMPLEX NUCLEI Asilomar conference The Conference on reactions between complex nuclei Pacific Grove,Calif. 1963 Apr 14-18 Edited by A. Ghiorso and others 456p University of California press,Berkeley,Calif.,1963
Cav 7.0090

COMPLEX VARIABLES Fonctions d'une variable complexe:problemes contemporaines Colloque de l'U.R.S.S. 3rd selection of papers presented Moscow 1957 Edited by A.I. Markushevitch Translated by L. Nicolas from the Russian Monographies internationales de mathematiques modernes, 1 271p 24cm Gauthier-Villars,Paris,1962
P. Math 2.1557

COMPLEX VARIABLES Lectures on functions of a complex variable :conference proceedings Ann Arbor, Mich. 1953 Jun 17-Jul1 University of Michigan Edited by Wilfred Kaplan and others Bibliog 435p 24cm University of Michigan,Ann Arbor,Mich.,1955
P. Math 2.1533

COMPLEX VARIABLES Symposium on several complex variables Proceedings Park City, Utah 1970 Mar 30-Apr 3 Edited by R.M. Brooks Lecture notes in mathematics, 184 234p 25cm Springer,Berlin,1971
P Math 2.4354

COMPLEXES OF BIOLOGICALLY ACTIVE SUBSTANCES WITH NUCLEIC ACID AND THEIR MODE OF ACTION :a research symposium Proceedings Washington,D.C. 1970 Mar 16-19 Walter Reed army institute of research Edited by F.E. Hahn Progress in molecular and subcellular biology, 2 Springer,Berlin,1971
Bioch 33.2242

COMPOSITE MATERIALS Symposium on composite materials:testing and design New Orleans,La. 1969 Feb 11-13 American society for testing materials.Committee D-30 high modulus fibers and their composites A.S.T.M.Special technical publication, 460 Philadelphia,Pa., 1969 Presented at a meeting of committee D-30
Met 25.2859

COMPOSITE MATERIALS:TESTING AND DESIGN :a symposium New Orleans,La. 1969 Feb 11-13 American society for testing and materials American society for testing and materials.Special technical publication, 460 ASTM,Philadelphia,Pa.,1969
Eng 41.8290

COMPOSITE MATERIALS AND COMPOSITE STRUCTURES Sagamore ordnance materials research conference 6th Proceedings New York 1959 Aug 8-21 United States.Army.Ordnance materials research office and United States.Army.Office of ordnance research Arrangements by Syracuse university research institute New York,1959 Mimeograph
Met 25.0093

COMPREHENSIVE SCHOOL MATHEMATICS PROGRAM INTERNATIONAL CONFERENCE 1st Proceedings Teaching of probability and statistics Carbondale,Ill. 1969 Mar 18-27 Southern Illinois university Central Midwestern regional educational laboratory Edited by Lennart Rade Wiley;Almqvist and Wiksell,New York;Stockholm,1970
An 32.5410

COMPUTATION Joint symposium on instrumentation and computation in process development and plant design Papers London 1959 May 11-12 British conference on automation and computation Sponsored by the Institution of chemical engineers Institution of chemical engineers,London,1959 Sponsored also by the Society of instrument technology,and the British computer society
Math L 5.3290

COMPUTATION OF TURBULENT BOUNDARY LAYERS :1968 AFOSR-IFP-Stanford conference Proceedings Stanford,Calif 1968 Vol 1-2 United States.Air force.Office of scientific research Edited by S.J. Kline and others Bibliog., Illus,port 26cm 2 vols Stanford university,Stanford,1969
A Math 4.1481

COMPUTATION SEMINAR proceedings Endicott, N.Y. 1949 Dec 5-9 Sponsored by International business machines corporation 173p 28cm International business machines corporation,New York,1951
Math L 5.0700

COMPUTATION SEMINAR proceedings Endicott, N.Y. 1951 Aug 13-17 Sponsored by International business machines corporation 148p 28cm International business machines corporation,New York,1951
Math L 5.0702

COMPUTATIONAL PROBLEMS IN ABSTRACT ALGEBRA :a conference Proceedings Oxford 1967 Aug 29-Sep 2 Science research council Held under the auspices of the Atlas computer laboratory bibliog.,illus. x,402p 23cm Pergamon press,Oxford,1970
P Math 2.3758

COMPUTER AIDED CIRCUIT DESIGN Conference on CACD Proceedings 1968 Mar 26-28 Sheffield university Kingston college of technology Jointly sponsored by Design electronics Design electronics,London,1968
Eng 41.2226

COMPUTER APPLICATIONS SYMPOSIUM 4th proceedings Chicago,Ill. 1957 Oct 24-25 Sponsored by Armour research foundation illus. x,126p 23cm Armour research foundation of Illinois institute of technology, Chicago,1958 The Armour research foundation was in 1963 renamed IIT research institute
Math L 5.0853

COMPUTER APPLICATIONS SYMPOSIUM 8th proceedings Computer applications - 1961 Chicago,Ill. 1961 Oct 25-26 Edited by Robert S. Hollitch and Benjamin Mittman Sponsored by Armour research foundation illus. vii,198p 23cm McMillan,New York
Math L 5.0854

COMPUTER APPLICATIONS SYMPOSIUM 9th
proceedings Computer applications - 1962 Edited by Milton M. Gutterman and Robert S. Hollitch Sponsored by Illinois institute of technology.Research institute illus. 224p 23cm Spartan books,Baltimore,Md.,1964 The IIT research institute was formerly called the Armour research foundation
Math L 5.0855

COMPUTER AUGMENTATION OF HUMAN REASONING
Symposium on computer augmentation of human reasoning proceedings Washington,D.C. 1964 Jun.16-17 United States.Office of naval research Edited by Margo A. Sass and William D. Wilkinson In joint sponsorship with the Bunker-Ramo corporation vi,235p 24cm Spartan books;Macmillan,Washington,D.C.;London, 1965
Math 3.0244

COMPUTER AUGMENTATION OF HUMAN REASONING
Symposium on computer augmentation of human reasoning Proceedings Washington,D.C. 1964 Jun 16-17 United States.Office of naval research Edited by Margo A. Sass and William D. Wilkinson In joint sponsorship with the Bunker-Ramo corporation Spartan books, Washington,D.C.,1965
Eng 41.2202

COMPUTER AUGMENTATION OF HUMAN REASONING
Symposium on computer augmentation of human reasoning proceedings Washington,D.C. 1964 Jun 16-17 United States.Office of naval research Edited by Margo A. Sass and William D. Wilkinson Jointly sponsored by Bunker-Ramo corporation bibliog,illus. vi,235p 24cm Spartan books,Washington,D.C.,1965
Math L 5.1005

COMPUTER CONTROL OF NATURAL RESOURCES
Symposium on computer control of natural resources and public utilities Preprints Haifa 1967 Sep 11-14 Sponsored by International federation of automatic control. Committee on applications Haifa,1967
Eng 41.5866

COMPUTER DESIGN Teaching of computer design
Joint IBM-University of Newcastle-upon-Tyne seminar Proceedings Newcastle upon Tyne 1971 Sep 7-10 International business machines corporation University of Newcastle-upon-Tyne.Computing laboratory illus 121p University of Newcastle upon Tyne.Computing laboratory,Newcastle upon Tyne,1972
Math L 5.3723

COMPUTER GRAPHICS Computer graphics:utility, production,art The Graphic symposium 3rd Papers Los Angeles,Calif 1966 University of California,Los Angeles and Informatics inc. Edited by Fred Gruenberger 225p Thompson book co.;Academic press,Washington,D. C.;London,1967
TA 15.0098

COMPUTER GRAPHICS:UTILITY,PRODUCTION,ART
Graphics symposium Los Angeles n.d. University of California Edited by Fred Gruenberger Jointly sponsored by Informatics incorporated 225p 22cm Academic press, London,1967
Math L 5.3273

COMPUTER GRAPHICS:UTILITY,PRODUCTION,ART
Graphics symposium Los Angeles,Calif. 1967 University of California Edited by Fred Gruenberger Jointly sponsored by Informatics incorporated Academic press,London,1967
Eng 41.2210

COMPUTER NETWORKS Courantter science symposium 3rd New York 1970 Nov 30-Dec 1 Courant institute of mathematical sciences Edited by Randall Rustin diagrs 205p Prentice-Hall,Englewood Cliffs,N.J., 1972
Math L 5.3638

COMPUTER PROCESSING Symposium on computer processing in communications New York 1969 Microwave research institute Microwave research institute.Symposium series, 19 Polytechnic press,Brooklyn,N.Y.,1970
Eng 41.8313

COMPUTER PROCESSING IN COMMUNICATIONS
Symposium on computer processing in communications Proceedings New York 1969 Microwave research institute Edited by Jerome Fox Microwave research institute. Symposia series, 19 illus. 850p Polytechnic institute of Brooklyn,Brooklyn,N.Y. 1970
Math L 5.3558

COMPUTER SCIENCE :a symposium Proceedings Girton 1969 Aug Institute of mathematics and its applications.Bulletin, 6,no.1 I.M. A.,Southend,1970 Held as part of the centenary celebrations of Girton college, Cambridge
Math L 5.3553

COMPUTER SCIENCES Software engineering
Symposium on computer and information sciences 3rd Proceedings Miami,Fla. 1969 Dec Vol 1-2 Edited by Julius T. Tou COINS, 3 illus 2 vols Academic press,New York,1970
Math L 5.3830

COMPUTER TYPESETTING CONFERENCE preprints and summaries Brighton 1966 July 15-16 Institute of printing 29cm 2 vols Institute of printing,1966
Math L 5.1023

COMPUTERS Conference on high speed automatic calculating-machines report Cambridge 1949 June 22-25 University of Cambridge. Mathematical laboratory Bibliog 31cm Cambridge,1950
Philos 1.0336

COMPUTERS Symposium on computers and automata Proceedings New York 1971 Apr 13-15 Microwave research institute Edited by Jerome Fox Microwave research institute. Symposia series, 21 illus 653p Polytechnic press,Brooklyn,N.Y.,1971
Math L 5.3635

COMPUTERS AND COMMUNICATIONS - TOWARD A COMPUTER UTILITY Computer communications symposium Los Angeles 1967 Mar 20-22 University of California Informatics incorporated Edited by Fred Gruenberger 219p 23cm Prentice-Hall,Englewood Cliffs,N.J.,1968
Math L 5.3117

COMPUTERS IN ALGEBRA AND NUMBER THEORY Symposium in applied mathematics Proceedings New York 1970 Mar 25-26 By Garrett Birkhoff and Marshall Hall American mathematical society Society for industrial and applied mathematics SIAM-AMS proceedings, 4 vii,200p 26cm AMS,Providence,R.I., 1971
P Math 2.4476

COMPUTERS IN MATHEMATICAL RESEARCH Edited by R.F. Churchhouse and J.-C. Herz xi,185p 23cm North-Holland,Amsterdam,1968 Mainly on a symposium on Utilization of computers in mathematical research held at the IBM world trade european education centre, Blaricum, Netherlands,August 29-31,1966
Math L 5.3116

COMPUTERS IN MATHEMATICAL RESEARCH Edited by R.F. Churchhouse and J.C. Herz Bibliog,illus xi,185p 23cm North-Holland,Amsterdam,1968 Based mainly on the Symposium on Utilization of computers in mathematical research held at Blaricum,Netherlands,August 1966
P Math 2.3000

COMPUTERS IN NUMBER THEORY Science research council Atlas symposium 2nd Proceedings Oxford 1969 Aug 18-23 Science research council Edited by A.O.L. Atkin and B.J. Birch xvii,433p 24cm Academic press, London;New York,1971
P Math 2.4381

COMPUTING Canadian conference for computing and data processing proceedings Toronto 1958 June 9-10 illus. xii,383p University of Toronto press,Toronto,1958
Math L 5.0845

COMPUTING Conference on teaching computing Proceedings Bristol 1972 Mar 28 University grants committee.Mathematical sciences sub-committee illus 165p University grants committee,London,1972
Math L 5.3878

COMPUTING International computing symposium Proceedings Venice 1972 Apr 12-14 Consiglio nazionale delle ricerche.Cybernetics group illus 634p Venice,1972
Math L 5.3862

COMPUTING Louisiana conference on combinatorics,graph theory and computing Proceedings Baton Rouge,La. 1970 Mar 1-5 Edited by R.C. Mullin and others Held at Louisiana state university vi,464p 25cm Louisiana state university,Baton Rouge,La., 1970
P Math 2.4503

The COMPUTING AND DATA PROCESSING SOCIETY OF CANADA 2nd conference proceedings Toronto 1960 June 6-7 vi,365p 25cm University of Toronto press,Toronto,1960
Math L 5.0896

The COMPUTING LABORATORY IN THE UNIVERSITY conference papers Madison,Wis. 1955 Aug Edited by Preston C. Hammer pl. xv, 236p 23cm University of Wisconsin press, Madison,Wis.,1957
Math L 5.0784

COMPUTING METHODS Computing methods and the phase problem in X-ray crystal analysis : conference report State college,Pa. 1950 Apr 6-8 Edited by Ray Pepinsky sponsored by the Rockefeller foundation Pennsylvania state college,State college,Pa.,1952
Min 10.0756

COMPUTING METHODS AND THE PHASE PROBLEM IN X-RAY CRYSTAL ANALYSIS :a conference Report Pennsylvania state college 1950 Apr 6-8 Rockefeller foundation and United States. Office of naval research Pennsylvania state college,State College,Pa.,1952
Cav 7.1789

COMPUTING METHODS AND THE PHASE PROBLEM IN X-RAY CRYSTAL ANALYSIS conference report Glasgow 1960 Aug.10-12 Edited by Ray Pepinsky and others International tracts in computer science and technology and their application, 4 viii,326p 23cm Pergamon press,Oxford,1961
Math L 5.0303

COMPUTING METHODS AND THE PHASE PROBLEM IN X-RAY CRYSTAL ANALYSIS :conference report State college,Pa. 1950 Apr 6-8 Edited by Ray Pepinsky sponsored by the Rockefeller foundation Pennsylvania state college,State college,Pa.,1952
Min 10.0756

COMPUTING METHODS AND THE PHASE PROBLEM IN X-RAY CRYSTAL ANALYSIS 2nd :conference report Glasgow 1960 Aug 10-12 Edited by Ray Pepinsky and others International tracts in computer science and technology and their application,4 port. Pergamon press,Oxford, 1961 The first conference was held in 1950
Min 10.1207

COMPUTING METHODS IN OPTIMIZATION PROBLEMS International conference on computing methods in optimization problems 2nd Papers San Remo 1968 Sep 9-13 Sponsored by the Society for industrial and applied mathematics Lecture notes in operations research and mathematical economics, 14 bibliog.,illus. Springer,Berlin,1968
Math S 3.1516

COMPUTING METHODS IN OPTIMIZATION PROBLEMS :a conference Proceedings Los Angeles 1964 Jan 30-31 Edited by A.V. Balakrishnan and Lucien W. Neustadt illus x,327p 24cm Academic press,New York;London,1964
Math 3.1205

COMPUTING METHODS IN OPTIMIZATION PROBLEMS-2 International conference on computing methods in optimization problems 2nd Papers San Remo 1968 Sep 9-13 Edited by Lofti A. Zadeh and others Sponsored by the Society for industrial and applied mathematics Academic press,New York,1969
Eng 41.5865

COMPUTING PROBLEMS IN OPTIMIZATION PROBLEMS :a conference Proceedings Los Angeles, Calif. 1964 Edited by A.V. Balakrishnan Academic press,New York,1964
Eng 41.5857

COMPUTING SCIENCE University education in computing science :a conference on graduate academic and related research programs in computing science Proceedings New York 1967 Jun 5-8 Edited by Aaron Finerman Sponsored by New York state science and technology foundation ACM monograph series xvi,237p 23cm Academic press,New York,1968
Math L 5.3107

CONCEPT AND THE ROLE OF THE MODEL IN MATHEMATICS AND NATURAL AND SOCIAL SCIENCES International union of history and philosophy of science colloquium Proceedings Utrecht 1960 International union of history and philosophy of science.Division of philosophy of science Synthese library 194p Reidel, Dordrecht,1961 Colloquium organized by H. Freudenthal
WSM 43.0785

The CONCEPT AND THE ROLE OF THE MODEL IN MATHEMATICS AND NATURAL AND SOCIAL SCIENCES a colloquium Proceedings Utrecht 1960 Jan International union of history and philosophy of science.Division of philosophy of science Edited by B.H. Kazemier and D. Vuysje 164p Reidel,Dordrecht,1961
WSM 43.1797

CONCEPTS AND THE STRUCTURE OF MEMORY Symposium on cognition 2nd Papers Pittsburgh,Pa. 1966 Apr 7-8 Edited by Benjamin Kleinmuntz Sponsored by the Carnegie institute of technology Wiley,New York;London,1967
Eng 41.0652

Les CONCEPTS DE CLAUDE BERNARD SUR LE MILIEU INTERIEUR :colloque international Actes Paris 1965 Jun 29-Jul 2 Organise par la Fondation Singer-Polignac Masson,Paris,1967
Phys 20.1343

CONCEPTS OF BIOLOGY Conference report Lee, Mass. 1955 Oct 12-14 National research council.Division of biology and agriculture. Biology council Edited by R.W. Gerard and R. B. Stevens National research council. Publication, 560 Washington,D.C.,1958
Bal 39.0989

CONCEPTUAL ADVANCES IN IMMUNOLOGY AND ONCOLOGY Symposium on fundamental cancer research 16th Papers Houston,Tex. 1962 Harper, New York,1962
Pha 16.0185

CONCRETE Mix design and quality control of concrete :a symposium Proceedings London 1954 May Cement and concrete association Cement and concrete association,London,1955
Eng 41.2929

CONCRETE One-day symposium on structural lightweight concrete Brighton 1962 Jun 26 Vol 1-2 Reinforced concrete association 2 vols RCA,London,1963
Eng 41.2968

CONCRETE Symposium on creep of concrete Houston,Tex. 1964 Mar American concrete institute American concrete institute. Publication,SP- 9 American concrete institute,Detroit,Mich.,1964
Eng 41.2932

CONCRETE BRIDGE DESIGN 1st international symposium Proceedings 1967 American concrete institute American concrete institute.Publication,SP- 23 ACI, Detroit,Mich.,1969
Eng 41.8488

CONCRETE BRIDGE DESIGN 2nd international symposium Proceedings American concrete institute American concrete institute.Publication,SP- 26 ACI,Detroit, Mich.,1971
Eng 41.8489

CONCRETE SHELL ROOF CONSTRUCTION :a symposium Proceedings London 1952 Jul 2-4 Cement and concrete association Cement and concrete association,London,1954
Eng 41.3008

CONCRETE SHELL ROOF CONSTRUCTION 2nd : symposium Proceedings Oslo 1957 Jul 1-3 Teknisk ukeblad,Norway Teknisk ukeblad, Oslo,1958
Eng 41.3009

CONCRETE STRUCTURES Prestressed concrete statically indeterminate structures :a symposium London 1951 Sep 24-25 Cement and concrete association Cement and concrete association,London,1953
Eng 41.3006

CONCRETE STRUCTURES The Strength of concrete structures :a symposium Proceedings London 1956 May Cement and concrete association Cement and concrete association,London,1958
Eng 41.2930

CONDENSATION Condensation control in buildings as related to paints,papers and insulating materials :conference Report 1952 Feb 26-27 Building research advisory board Building research advisory board. Research conference report,4 illus ii,118p 28cm Washington,D.C.,1952
Sco 14.1273

CONDENSATION AND EVAPORATION OF SOLIDS International symposium on condensation and evaporation of solids Proceedings Dayton, Ohio 1962 Sep 12-14 United States.Air force.Directorate of materials and processes Edited by Emile Rutner and others Gordon and Breach,New York,1964
Met 25.1518

CONDENSATION CONTROL IN BUILDINGS AS RELATED TO PAINTS,PAPERS AND INSULATING MATERIALS : conference Report 1952 Feb 26-27 Building research advisory board Building research advisory board.Research conference report,4 illus ii,118p 28cm Washington, D.C.,1952
Sco 14.1273

CONDENSING HEAT TRANSFER WITHIN HORIZONTAL TUBES National heat transfer conference 2nd Preprints Chicago,Ill. 1958 Aug 18-21 American society of mechanical engineers American society of mechanical engineers, Philadelphia,Pa.,1958
Eng 41.7155

CONDITIONING Classical conditioning :a symposium Papers University Park,Pa. 1963 Aug Edited by William F. Prokasy Century psychology series Appleton-Century-Crofts,New York,1965 Organised by the editor
Psy 31.3015

CONDUCTING FLUIDS Magnetodynamics of conducting fluids Lockheed symposium on magnetohydrodynamics 3rd proceedings Palo Alto 1958 Nov.21-22 Edited by Daniel Bershader Sponsored by Lockheed aircraft corporation.Missiles and space division Stanford university press,Stanford,Calif.,1959
A Math 4.0616

CONFEDERATION OF BRITISH INDUSTRY Automation '65 British automation conference 1965 Eastbourne 1965 Nov 7-10 Organised by the Institution of production engineers Institution of production engineers,London, 1965
Eng 41.6018

CONFERENCE Konferentsiya puti razvitiya sovetskogo matematicheskogo mashchinostroeniya i priborostroeniya :plenarnye zasedaniya Moscow 1956 Mar 12-17 illus. 131p 22cm Moscow,1956
Math L 5.0774

CONFERENCE ACROSS A CONTINENT The Duke of Edinburgh's commonwealth study conference 2nd Canada 1962 May 13-Jun 6 Macmillan, Toronto,1963
Geog 13.5039

CONFERENCE AND WORKSHOP ON HISTOCOMPATIBILITY TESTING Report 3rd Histocompatibility testing 1967 Turin 1967 Jun 14-24 St. Vincent Edited by E.S. Curtoni and others Sponsored by the Consiglio nazionale delle ricerche Munksgaard,Copenhagen,1967
Surg 23.0032

CONFERENCE BOARD OF THE MATHEMATICAL SCIENCES Recent advances in homotopy theory By George W. Whitehead American mathematical society.C. M.B.S. regional conference series in mathematics, 5 ix,82p 25cm Providence, R.I.,1970 Expository papers developed from lectures given by the author at one of a series of regional conferences
P Math 2.4105

CONFERENCE DE L'UNION INTERNATIONALE DE CHIMIE 10th :rapports sur les hydrates de carbone (Glucides) Rapports Liege 1930 Sep 14-20 Union internationale de chimie Union internationale de chimie,Paris, 1930
Chem 18.0066

CONFERENCE DES FAUCONNIERS La Fauconnerie... avec les portraicts au naturel de tous les oyseaux By Charles d' Arcussia,seigneur d'Esparron 21cm Vaultier;Besongne,Rouen, 1644 Parts 1-10,including an account of the conference and discussions of falconers
Bal 44.4090

The CONFERENCE FOR INSTRUCTORS OF ASTRONOMY : lecture notes Berkeley,Calif. 1954 Aug 12-Sep 8 Leuschner observatory Edited by Robert Fleischer and others Co-sponsored by National science foundation 265p Berkeley, Calif.,1954 Typescript
Obs 6.2054

La CONFERENCE GENERALE DES POIDS ET MESURES 9eme comptes rendus Paris 1948 Oct. 12-21 Comite international des poids et mesures 127p Gauthier-Villars,Paris,1949
Obs 6.2388

CONFERENCE GEOLOGIQUE DES GUYANES 4me communications presentees Cayenne 1957 Sep. Memoires pour servir a lexplication de la carte geologique detaillee de la France: departement de la Guyane Francaise Imprimerie nationale,Paris,1959
Geol 8.2427

CONFERENCE GEOLOGIQUE DES GUYANES see also CONFERENCE GEOLOGIQUE DES TROIS GUYANES

CONFERENCE GEOLOGIQUE DES GUYANES see also GEOLOGISCHE CONFERENTIE

CONFERENCE GEOLOGIQUE DES TROIS GUYANES 2nd report Cayenne 1951 Mar.14-28 1951 Typescript
Geol 8.2429

CONFERENCE GEOLOGIQUE DES TROIS GUYANES see also CONFERENCE GEOLOGIQUE DES GUYANES

CONFERENCE GEOLOGIQUE DES TROIS GUYANES see also GEOLOGISCHE CONFERENTIE

CONFERENCE GEOLOGIQUE DES TROIS GUYANES see also INTER-GUIANA GEOLOGICAL CONFERENCE

CONFERENCE GEOLOGIQUES DES GUYANES see also INTER-GUIANA GEOLOGICAL CONFERENCE

CONFERENCE HYDROLOGIQUE DES ETATS BALTIQUES 4th Compte-rendu Leningrad 1933 Sep 6-22 Union of Soviet socialist republics.Service hydro-meteorologique Leningrad,1934
Geog 13.0767

CONFERENCE IN MATHEMATICAL LOGIC Proceedings London 1970 Aug 24-28 Edited by Wilfrid Hodges Held at Bedford college Lecture notes in mathematics, 255 vii,351p 25cm Springer,Berlin,1972
P Math 2.4059

CONFERENCE IN PHYSICAL METALLURGY AND CORROSION Proceedings Applications of field-ion microscopy in physical metallurgy and corrosion Atlanta,Ga. 1965 May 15-17 Georgia institute of technology Edited by R. F. Hochman and others Georgia institute of technology.Advanced research projects agency, Atlanta,Ga.,1968
Met 25.2631

CONFERENCE INTERNATIONAL DE GENETIQUE 4e Comptes rendus Paris 1911 Sep 18-23 Societe nationale d'horticulture de France Edited by Ph.de Vilmorin 2 vols Masson, Paris,1913
Gen 34.1000

CONFERENCE INTERNATIONALE DE PSYCHOTECHNIQUE 4th Comptes rendues Paris 1927 Oct 10-14 Conferences internationales de psychotechnique and Institut international de cooperation intellectuelle Alcan,Paris,1929 Papers in English,French,German and Italian
Psy 31.2517

CONFERENCE INTERNATIONALE DE PSYCHOTECHNIQUE APPLIQUEE A L'ORIENTATION PROFESSIONELLE ET A L'ORGANISATION SCIENTIFIQUE DU TRAVAIL. SECRETARIAT Conferencia internacional de psicotecnica applicada a l'orientacio professional i a l'organitzacio cientifica del treball 2 Actes Barcelona 1921 Sep 28-30 Institut d'orientacio professional, Barcelona,1912 Under the presidency of E. Claparede
Psy 31.2608

CONFERENCE INTERNATIONALE DES ETOILES FONDAMENTALES Proceedings Paris 1896 May 18-21 Sponsored by the Bureau des longitudes,Paris 90p Gauthier-Villars, Paris,1896
Obs 6.2315

CONFERENCE INTERNATIONALE DES IRIS 1ere Actes et comptes-rendus Iris cultivees Paris 1922 Societe nationale d'horticulture de France Societe nationale d'horticulture,Paris,1923
BG 38.2574

CONFERENCE INTERNATIONALE DU SPITSBERGEN Tableau synoptique des dispositions adoptees et reservees Christiana 1914 Jun 16-Jul 30 35p 33cm Grondahl,Christiana,1914
Sco 14.2859

CONFERENCE OF BRITISH COMMONWEALTH SURVEY OFFICERS 1947- British commonwealth survey officers H.M.S.O.,London,1951-
Geog 13.6270

CONFERENCE OF ENGINEERING SOCIETIES OF WESTERN EROPE AND THE UNITED STATES OF AMERICA EUSEC conference on engineering education 1st Proceedings London 1953 Jan 12-17 Institution of civil engineers,London,1953
Eng 41.1543

CONFERENCE OF ENGINEERING SOCIETIES OF WESTERN EUROPE AND THE UNITED STATES OF AMERICA EUSEC conference on engineering education 3rd Proceedings Paris 1957 Sep Institution of mechanical engineers,London, 1958
Eng 41.1546

CONFERENCE OF ENGINEERING SOCIETIES OF WESTERN EUROPE AND THE UNITED STATES OF AMERICA EUSEC meeting on engineering education and training 4th Proceedings London 1962 Jan Institution of electrical engineers, London,1963
Eng 41.1547

CONFERENCE OF THE ARCTIC CO-OPERATIVES 1st Minutes Frobisher Bay,N.W.T. 1963 Mar 12-18 Canada.Department of northern affairs and national resources 36cm Ottawa,1963 Typescript
Sco 14.3712

CONFERENCE OF THE ARCTIC COOPERATIVES 2nd Minutes Povungnituk,Que. 1966 Apr 19-28 93p 35cm n.p.,1966 Typescript
Sco 14.3711

CONFERENCE OF THE GESELLSCHAFT FUR BIOLOGISCHE CHEMIE 9th Proceedings Regulation of gluconeogenesis Gottingen 1971 Jan 15 Gesellschaft fur biologische chemie Edited by Hans-Dieter Soling and Berend Willms Thieme;Academic press,Stuttgart;London,1971
Bioch 33.2210

CONFERENCE OF THE INSTITUTE OF MATHEMATICS AND ITS APPLICATIONS Proceedings Large sparse sets of linear equations Oxford 1970 Apr Institute of mathematics and its applications Edited by J.K. Reid Illus 284p Academic press,London,1971
Math L 5.3804

CONFERENCE ON ADRENAL CORTEX 1st-5th Transactions Adrenal cortex Edited by Elaine P. Ralli Sponsored by the Josiah Macy jr.foundation 5 vols Josiah Macy foundation,New York,1950-54
An 32.3730

CONFERENCE ON AIR POLLUTION Proceedings Air pollution Washington,D.C. 1950 May 3-5 United States.Bureau of mines Edited by Louis C. McCabe Sponsored by United States. Interdepartmental committee on air pollution xiv,847p McGraw-Hill,New York,1952
Nap 11.0388

CONFERENCE ON ANALYTIC FUNCTIONS Proceedings Lodz 1966 Sep 1-7 Polska akademia nauk. Instytut matematyczny Bibliog 36p 24cm Lodz,1966
P Math 2.3052

CONFERENCE ON ANODISING ALUMINIUM Proceedings Nottingham 1961 Sep 12-14 Aluminium development association and University of Nottingham Aluminium development association, London,1961
Met 25.2047

CONFERENCE ON ANODISING ALUMINIUM Proceedings Nottingham 1961 Sep 12-14 University of Nottingham Convened by the Aluminium development association Aluminium development association,London,1962
Eng 41.3602

CONFERENCE ON APPLICATIONS OF ISOTOPES IN SCIENTIFIC RESEARCH Proceedings Melbourne 1950 Aug 14-17 Commonwealth scientific and industrial research organization University of Melbourne. Chemistry department University of Melbourne, Melbourne,1951
Radioth 35.1495

CONFERENCE ON APPLICATIONS OF NUMERICAL ANALYSIS Proceedings Dundee 1971 Mar 23-26 Edited by John L. Morris Lecture notes in mathematics, 228 x,358p 25cm Springer, Berlin,1971
P Math 2.4463

CONFERENCE ON APPLICATIONS OF X-RAY ANALYSIS 15- Proceedings Denver,Colo. 1966- Denver research institute Advances in X-ray analysis, 10- Plenum press,New York,1967-
Min 10.1549

CONFERENCE ON ATMOSPHERIC ELECTRICITY 2nd Proceedings Recent advances in atmospheric electricity Portsmouth,N.H. 1958 May 20-23 Edited by L.G. Smith Sponsored by Air force Cambridge research center.Geophysics research directorate. Aerophysics laboratory illus 646p Pergamon,London,1958
Nap 11.0928

CONFERENCE ON AURORAL PHYSICS 1st
Proceedings Auroral physics London, Ontario 1951 Jul 23-26 Edited by N.C. Gersen and others Sponsored jointly by the University of Western Ontario.Department of physics. Air force Cambridge research center. Geophysical research papers,30 illus xxvi, 462p Cambridge,Mass.,1954
Sco 14.0613

CONFERENCE ON AUTOMATIC CONTROL Papers
Automatic and manual control... Cranfield 1951 Jul 16-21 Edited by A. Tustin Organized by Great Britain.Department of scientific and industrial research Butterworths,London,1952
Eng 41.6006

CONFERENCE ON AUTOMATIC CONTROL papers
Automatic and manual control Cranfield 1951 Jul 18-21 Great Britain.Department of scientific and industrial research xi,584p 25cm Butterworth,London,1951
Math L 5.0021

The CONFERENCE ON AUTOMATIC CONTROL Papers
Cranfield 1951 Jul 16-21 Great Britain. Department of scientific and industrial research Edited by A. Tustin 584p Butterworths,London,1952
Cav 7.2360

CONFERENCE ON BIOCHEMICAL AND BIOPHYSICAL MECHANISMS IN THE PRODUCTION OF RADIATION-INDUCED CHROMOSOME ABERRATIONS Radiation-induced chromosome aberrations San Juan, Puerto Rico 1961 Nov 16-18 Edited by Sheldon Wolff Sponsored by the National research council Columbia university press, New York;London,1963
Gen 34.1122

CONFERENCE ON BIOLOGY AND CIVIL ENGINEERING
Proceedings London 1948 Sep Institution of civil engineers Institution of civil engineers,London,1949
Geog 13.2688

The CONFERENCE ON BIOLOGY OF NORMAL AND ATYPICAL PIGMENT CELL GROWTH 4th Proceedings
Pigment cell biology Houston,Tex. 1957 Nov 14-16 Edited by Myron Gordon Held at the Anderson hospital and tumor institute Academic press,New York,1959 Sponsored jointly by the New York zoological society,the M.D.Anderson Hospital and tumor institute and the Damon Runyon memorial fund for cancer research, inc.
Bioch 33.0937

CONFERENCE ON BLOOD CLOTTING AND ALLIED PROBLEMS 1-5 Transactions Blood clotting and allied problems Edited by Joseph Eugene Flynn 5 vols Josiah Macy Jr.foundation,New York,1948-52
Med 36.0141

CONFERENCE ON BRAIN AND BEHAVIOR 1st
Proceedings Brain and behavior,vol.1 Los Angeles,Calif. 1961 Vol 1 Edited by Mary A.B. Brazier American institute of biological sciences,Washington,D. C.,1961
An 32.4382

CONFERENCE ON BRAIN AND BEHAVIOR 2nd
Proceedings Brain and behavior,vol.2:the internal environment and alimentary behavior Los Angeles,Calif. 1962 Feb 18-21 Edited by Mary A.B. Brazier American institute of biological sciences,Washington,D.C.,1963
An 32.4383

CONFERENCE ON BRAIN FUNCTION 5th
Proceedings Aggression and defence:neural mechanisms and social patterns (brain function, vol 5) Pacific palisades 1965 Nov 14-17 Edited by Carmine D. Clemente and Donald B. Lindsley UCLA forum in medical sciences, 7 University of California press,Berkeley;Los Angeles,Calif.,1967
An 32.5475

CONFERENCE ON BRITTLE FRACTURE MACHINES An
Evaluation of current knowledge of the mechanics of brittle fracture :including papers and discussions presented at Conference on brittle fracture mechanics held...at Massachusetts institute of technology Cambridge,Mass. 1953 Oct 13-16 By D.C. Drucker Sponsored by the National research council.Committee on ship structure design SSC 69 N.A.S.-N.R.C.Division of engineering and industrial research,Washington,D.C.,1954
Eng 41.3743

CONFERENCE ON CACD Proceedings Computer
aided circuit design 1968 Mar 26-28 Sheffield university Kingston college of technology Jointly sponsored by Design electronics Design electronics,London,1968
Eng 41.2226

CONFERENCE ON CARBON 1st and 2nd
Proceedings Buffalo,N.Y. 1953,1955 By S. Mrozowski and others University of Buffalo University of Buffalo,New York,1956
Met 25.0154

CONFERENCE ON CARBON 3rd Proceedings
Buffalo,N.Y. 1957 Jun 17-21 By S. Morzowski and others University of Buffalo and United States.Office of naval research Co-sponsored by the National science foundation Pergamon press,London,1959
Met 25.0156

CONFERENCE ON CARBON 3rd Proceedings
Buffalo,N.Y. 1957 Jun 17-21 State university of New York,Buffalo Edited by S. Mrozowski and others Co-sponsored by the United States.Office of naval research Conference on carbon.Proceedings, 2 718p Pergamon press,Oxford,1959
TA 15.0214

CONFERENCE ON CARBON 4th Proceedings
Buffalo,N.Y. 1959 Jun 15-19 By S. Mrozowski and others University of Buffalo and American carbon committee Pergamon press,Oxford,1960
Met 25.0157

The CONFERENCE ON CARBON 4th Proceedings
Buffalo,N.Y. 1959 Jun 15-19 State university of New York,Buffalo and American carbon committee Edited by S. Mrozowski and others Conference on carbon. Proceedings, 3 778p Pergamon press, Oxford,1960
TA 15.0215

CONFERENCE ON CARBON 5th Proceedings University Park,Penn. 1961 Jun 19-23 Clarendon press,Oxford,1965
Met 25.0158

The CONFERENCE ON CARBON 5th Proceedings University park,Penn. 1961 Vol 1 Pennsylvania state university and American carbon committee Edited by S. Mrozowski Conference on carbon.Proceedings, 4 Pergamon press,Oxford,1962 Proceedings appeared in two volumes
TA 15.0216

CONFERENCE ON CAUSES OF SUDDEN DEATH IN INFANTS Proceedings Sudden death in infants Seattle 1963 Sep 9-10 United States. Department of health,education and welfare Edited by Ralph J. Wedgwood and others United States.Public health service publication,1412 U.S.govt.printing office, Washington,D.C.,1965
HM 27.0158

CONFERENCE ON CELL TISSUE AND ORGAN CULTURE Cell tissue and organ culture Decennial review conference on cell tissue and organ culture 2nd Proceedings Bedford,Pa. 1966 Sep 11-15 National cancer institute and Tissue culture association Edited by Benton B. Westfall National cancer institute. Monograph, 26 National cancer institute, Bethesda,Md.,1967
Bioch 33.0991

CONFERENCE ON CELLULAR CONTROL MECHANISMS AND CANCER Proceedings Amsterdam 1963 Sep 9-13 International union against cancer and Netherlands cancer institute Edited by P. Emmelot and O. Muhlbock Elsevier, Amsterdam,1964 Organized to commemorate the 50th anniversary of the Netherlands cancer institute
Path 30.2459

CONFERENCE ON CHEMICAL AERONOMY Proceedings The threshold of space Cambridge,Mass. 1956 Jun 25-28 Air force Cambridge research center.Geophysics research directorate Edited by M. Zelikoff illus xi,342p Pergamon press,London,1957
Nap 11.0654

CONFERENCE ON CHEMICAL RESEARCH Proceedings Houston,Texas 1969 Nov 17-19 13: the transuranium elements - the Mendeleev centennial Robert A.Welch foundation Edited by W.O. Milligan 494p R.A.Welch foundation,Houston,Texas,1970
TA 15.0699

CONFERENCE ON CHEMICAL RESEARCH Proceedings Houston,Texas 1969 Nov 17-19 Robert A. Welsh foundation Robert A.Welch foundation. Conference, 13 photos,graphs,diagrms xii, 494p 23cm Robert A.Welch foundation, Houston,Texas,1970
Cav 7.3126

CONFERENCE ON CHROMOSOMES Lectures Chromosomes Wageningen 1956 Apr 16-19 Willink,Zwolle,1956 Papers in English and German
An 32.3379

CONFERENCE ON CLASSIFICATION proceedings London 1962 Apr.6 Aslib Aslib proceedings, 14,no.8 215-266p 25cm Aslib,London,1962
Math 3.0776

CONFERENCE ON COASTAL ENGINEERING 10th Proceedings Tokyo 1966 Sep American society of civil engineers 2 vols American society of civil engineers,New York,1967
Eng 41.3348

CONFERENCE ON COASTAL ENGINEERING 11th Proceedings London 1968 Sep Vol 1-2 American society of civil engineers Institution of civil engineers 2 vols American society of civil engineers,New York, 1969
Eng 41.3347

CONFERENCE ON COLD INJURY 1st Transactions 1-6 Sponsored by the Josiah Macy foundation 248p 23cm 6 vols New York, 1952-60
Sco 14.0875

CONFERENCE ON COLLECTIVIZATION IN EASTERN EUROPE Papers Collectivization of agriculture in eastern Europe Lexington,Ky. 1955 Apr 14-16 Edited by Irwin T. Sanders maps University of Kentucky press,Lexington,Ky., 1958
Geog 13.2582

CONFERENCE ON COLOUR VISION Documenta ophthalmologica :advances in ophthalmology Cambridge 1947 Jul 28-Aug 2 Vol 3 Edited by F.P. Fisher and others W.Junk,'S. Gravenhage,1949 Contains most of the papers read at the conference on colour vision held in Cambridge,July 28-August 2,1947
Psy 31.0409

CONFERENCE ON COLOUR VISION Papers Cambridge 1947 Jul 28-Aug 2 Documenta ophthalmologica, 3 Junk,The Hague,1949
Phys 20.1021

CONFERENCE ON COMPACT TRANSFORMATION GROUPS 2nd Proceedings Amherst,Mass. 1972 Pt 1-2 Held at the University of Massachusetts Lecture notes in mathematics, 298-299 25cm 2 vols Springer,Berlin,1972
P Math 2.4110

CONFERENCE ON COMPLEX ANALYSIS Proceedings Complex analysis Houston,Tex. 1967 Apr 13-15 Rice university Edited by H.L. Resnikoff and R.O. Wells Rice university studies, 54,no.4 Bibliog. 84p 23cm William Marsh Rice University,Houston,Tex., 1968
P Math 2.3684

CONFERENCE ON COMPLEX ANALYSIS proceedings Minneapolis,Minn. 1964 Mar 16-21 United States.Air force.Office of scientific research Edited by A. Aeppli and others Bibliog. vii,308p 24cm Springer-Verlag,Berlin, Heidelberg,1965
P. Math 2.1582

CONFERENCE ON COMPUTER SCIENCE AND TECHNOLOGY Manchester 1969 Jun 30-Jul 3 Sponsored by the institution of electrical engineers. Electronics division.Computer design professional group Institution of electrical engineers.Conference publication, 55 The institution of electrical engineers,London, 1969
Math L 5.3503

CONFERENCE ON CONNECTIONS BETWEEN CATEGORY THEORY AND ALGEBRAIC GEOMETRY AND INTUITIONISTIC LOGIC Partial report Toposes,algebraic geometry and logic Halifax,Nova Scotia 1971 Jan 16-19 Edited by F.W. Lawvere Lecture notes in mathematics, 274 189p 25cm Springer,Berlin,1972
P Math 2.4171

CONFERENCE ON CONNECTIVE TISSUES 1st Transactions Connective tissues New York 1950 Apr 24-25 Josiah Macy jr. foundation Edited by Charles Ragan Josiaph Macy jr.foundation,New York,1951
Path 30.2487

CONFERENCE ON CONNECTIVE TISSUES 1st-5th Transactions Connective tissues New York 1950 Princeton,N.J. 1953-54 Edited by Charles Ragan Sponsored by the Josiah Macy jr.foundation Josiah Macy,jr. foundation,New York,1951-54
An 32.3218

CONFERENCE ON CONSTRUCTIVE THEORY OF FUNCTIONS Proceedings Budapest 1969 Aug 24-Sep 3 Edited by G. Alexits and S.B. Stechkin 538p 25cm Akademiai kiado,Budapest,1972
P Math 2.4225

CONFERENCE ON CRYOBIOLOGY Cellular injury and resistance in freezing organisms International conference on low temperature science 2nd Proceedings Sapporo 1966 Aug 14-19 Hokkaido university.Institute of low temperature science Edited by Eizo Asahina illus Hokkaido university. Institute of low temperature science,Sapporo, 1967 Title page in English and Japanese Conference held in commemoration of the 25th anniversary of the establishment of the Institute
Path 30.2625

CONFERENCE ON CRYSTAL GROWTH 1st- Proceedings Growth of crystals Moscow 1956- 1- Institut kristallografii, Moscow Edited by A.V. Shubnikov and N.N. Sheftal Translated by Consultants' bureau from the Russian Rost kristallov, 1- Consultants bureau,inc.,New York,1958- Includes interim reports published between conferences.Held as part of the international crystallographic congresses
Met 25.1508

CONFERENCE ON CUMULUS CONVECTION 1st Proceedings Cumulus dynamics Portsmouth, N.H. 1959 May 19-22 Edited by Charles E. Anderson Sponsored by the Air Force Cambridge research center.Geophysics research directorate ix,211p Pergamon press,Oxford, 1960
Nap 11.0012

CONFERENCE ON CYBERNETICS 7th Transactions Cybernetics:circular causal and feedback mechanisms in biological and social systems New York 1950 Mar 23-24 Josiah Macy jr. foundation Edited by Heinz von Foerster Josiah Macy jr. foundation,New York,1951
Psy 31.1064

CONFERENCE ON CYBERNETICS 9th Transactions Cybernetics:circular causal and feedback mechanisms in biological and social systems New York 1952 Mar 20-21 Josiah Macy jr. foundation Edited by Heinz von Foerster Josiah Macy jr. foundation,New York,1953
Psy 31.1063

CONFERENCE ON CYBERNETICS 8TH Cybernetics : circular causal and feedback mechanisms in biological and social systems Transactions New York 1951 Mar 15-16 Josiah Macy jr. foundation Edited by Heinz von Foerster Josiah Macy,jr. foundation,New York,1952
Psy 31.2046

CONFERENCE ON CYCLES 1st-2nd Report Washington 1922 and 1928 Carnegie Institution of Washington Washington,1929
Philos 1.0096

CONFERENCE ON DEFECTS IN CRYSTALLINE SOLIDS Report Bristol 1954 July Wills (H.H.) physical laboratory The Physcial society, London,1955
Met 25.2358

CONFERENCE ON DEFECTS IN CRYSTALLINE SOLIDS Report Defects in crystalline solids Bristol 1954 Jul Physical society,London, 1955
Eng 41.4311

CONFERENCE ON DESIGN METHODS :conference on systematic and intuitive methods in engineering industrial design,architecture and communications Papers London 1962 Sep 19-21 Edited by J.Christopher Jones and D.G. Thornley Pergamon press,London,1963
Eng 41.0304

CONFERENCE ON DIMENSIONING AND STRENGTH CALCULATIONS 3rd Proceedings Budapest 1968 Nov Hungarian academy of sciences Edited by E. Czoboly Akademiai kiado,Budapest,1968
Met 25.2407

CONFERENCE ON DUST COOLING Proceedings London 1964 Dec 21-22 Vol 1-2 Queen Mary college.Nuclear engineering department 2 vols London,1964 Mimeograph
Chem E 24.1464

CONFERENCE ON ELECTROCHEMISTRY 4th Proceedings Soviet electrochemistry Moscow 1956 Oct 1-6 Vol 1-3 Translated by Consultants bureau from the Russian New York,1961
Met 25.1821

The CONFERENCE ON ELECTROMAGNETIC SCATTERING : interdisciplinary conference Papers Potsdam,N.Y. 1962 Aug 13-15 Clarkson college of technology and Air force Cambridge research laboratories Edited by Milton Kerker Co-sponsored by the American chemical society.Division of colloid and surface chemistry International series of monographs on electromagnetic waves, 5 591p Pergamon press,Oxford,1963
TA 15.0256

CONFERENCE ON ELECTRONIC COMPUTATION 2nd Proceedings Pittsburgh,Pa. 1960 American society of civil engineers American society of civil engineers,Kansas City,Miss., 1960
Eng 41.2718

CONFERENCE ON ELECTRONIC COMPUTATION 3rd Proceedings Boulder,Colo. 1963 Jun 19-21 American society of civil engineers Journal of structural division, 89,no ST4 American society of civil engineers,Kansas city,Miss.,1963
Eng 41.2716

CONFERENCE ON ELECTRONIC TELEPHONE EXCHANGES papers London 1960 Nov 22-24 Institution of electrical engineers Collection of typescript papers and reprints
Math L 5.0873

CONFERENCE ON ELECTRONICS DESIGN Cambridge 1968 Sep 23-27 Sponsored by the Institution of electrical engineers Institution of electrical engineers.Conference publication, 45 IEE,London,1968
Math L 5.3468

The CONFERENCE ON EMISSION SPECTRA OF THE NIGHT SKY AND AURORAE Papers London 1947 Jul 7-10 Royal society of London.Gassiot committee 139p Physical society,London, 1948
Cav 7.1453

The CONFERENCE ON EMISSION SPECTRA OF THE NIGHT SKY AND AURORAE Papers London 1947 Jul 7-10 Royal society of London.Gassiot committee 139p Physical society,London, 1948
TA 15.0039

CONFERENCE ON ESTUARIES Papers Estuaries Jekyll Island,Ga. 1964 University of Georgia.Marine institute Edited by George H. Lauff American association for the advancement of science.Publications,83 New York,1967
Geog 13.0743

CONFERENCE ON EXPERIMENTAL CLINICAL CANCER CHEMOTHERAPY Washington,D.C. 1959 Nov 11-12 National cancer institute Edited by B.H. Morrison National cancer institute. Monograph, 3 U.S.Department of health, education and welfare,Washington,D.C.,1960
Radioth 35.0809

CONFERENCE ON EXTREMELY HIGH TEMPERATURES Boston,Mass. 1958 Mar 18-19 Air force Cambridge research center Edited by Heinz Fischer and Lawrence C. Mansur Wiley;Chapman and Hall,New York;London,1958
Met 25.0249

CONFERENCE ON FATIGUE OF WELDED STRUCTURES Proceedings 1970 Vol 1-2 Welding institute Welding institute,Abington,1971 Conference director:T.R.Gurney
Eng 41.8534

CONFERENCE ON FETAL HOMEOSTASIS 1st Proceedings Fetal homeostasis Edited by Ralph M. Wynn Sponsored by the New York academy of sciences New York academy of sciences,New York,1965
An 32.2519

CONFERENCE ON FETAL HOMEOSTASIS 2nd Proceedings Fetal homeostasis Princeton, N.J. 1966 Apr 3-6 Edited by Ralph M. Wynn Sponsored by the New York academy of sciences New York academy of sciences,New York,1967
An 32.2520

The CONFERENCE ON FLAVINS AND FLAVOPROTEINS 2nd Proceedings Flavins and flavoproteins Nagoya 1967 Aug 14-17 Edited by Kunio Yagi University of Tokyo press,Tokyo,1968
Bioch 33.1076

CONFERENCE ON FREE AMINO ACIDS Amino acid pools :distribution and function of free amino acids.Symposium Proceedings Duarte,Calif. 1961 May 19-22 City of hope medical center Edited by Joseph T. Holden Under the auspices of the Institute for advanced learning in the medical sciences illus Elsevier,Amsterdam,1962 Preface:"The committee undertook the organization of the symposium under the title Conference on free amino acids
Bioch 33.0519

A CONFERENCE ON FUNDAMENTAL MECHANISMS OF PHOTOGRAPHIC SENSITIVITY :held in the H.H. Wills physical laboratory Proceedings Fundamental mechanisms of photographic sensitivity Bristol 1950 March University of Bristol Edited by J.W. Mitchell Butterworths,London,1951
Cav 7.0347

CONFERENCE ON FUNDAMENTAL PROBLEMS AND TECHNICS FOR THE STUDY OF THE KINETICS OF CELLULAR PROLIFERATION Proceedings Kinetics of cellular proliferation Salt Lake City,Utah 1959 Jan 19-21 Edited by Frederick Stohlman Grune and Stratton,New York;London,1959
An 32.3235

CONFERENCE ON GAS DISCHARGES 1972 Institution of electrical engineers Institute of electrical and electronics engineers Institution of electrical engineers.Conference publication, 90 IEE, London,1972
Eng 41.8570

CONFERENCE ON GENERAL TOPOLOGY Proceedings Pullman,Wash. 1970 Mar 25-27 Washington state university.Department of mathematics iv,136p 27cm Washington state university. Department of mathematics,Pullman,Wash.,1970
P Math 2.4108

CONFERENCE ON GENETICS 1st Transactions Genetics:genetic information and the control of protein structure and function Princeton, N.J. 1959 Oct 19-22 Edited by H.Eldon Sutton Sponsored by the Josiah Macy jr. foundation Macy Jr.foundation,New York,1960
An 32.0360

CONFERENCE ON GENETICS 2nd Mutations Princeton,N.J. 1960 Oct 16-19 Edited by William J. Schull Sponsored by the Josiah Macy jr.foundation University of Michigan press,Ann Arbor,Mich.,1962
An 32.0361

CONFERENCE ON GENETICS 3rd Genetic selection in man Princeton,N.J. 1961 Oct 15-18 Edited by William J. Schull Sponsored by the Josiah Macy jr.foundation University of Michigan press,Ann Arbor,Mich., 1963
An 32.0362

CONFERENCE ON GESTATION 1st Transactions Gestation Princeton,N.J. 1954 Mar 9-11 Edited by Louis B. Flexner Sponsored by the Josiah Macy jr.foundation Josiah Macy jr. foundation,New York,1955
An 32.3815

CONFERENCE ON GESTATION 2nd-5th
Transactions Gestation Princeton,N.J. 1955-58 Mar Edited by Claude A. Villee Sponsored by the Josiah Macy jr.foundation 3 vols Josiah Macy jr. foundation,New York, 1956-58
An 32.3816

CONFERENCE ON GLOBAL DIFFERENTIABLE DYNAMICS
Proceedings Cleveland,Ohio 1969 Jun 2-6 Case western reserve university Edited by O. Hajek and others Lecture notes in mathematics, 235 x,140p 25cm Springer, Berlin,1971
P Math 2.4317

CONFERENCE ON GRADUATE ACADEMIC AND RELATED RESEARCH PROGRAMS IN COMPUTING SCIENCE
Proceedings University education in computing science Stony Brook,N.Y. 1967 Jun Edited by Aaron Finerman Association for computing machinery.Monograph series Bibliog.,Illus.Port. xvi,237p 24cm Academic press,New York;London,1968
P Math 2.3505

CONFERENCE ON GRADUATE ACADEMIC AND RELATED RESEARCH PROGRAMS IN COMPUTING SCIENCE
Proceedings University education in computing science Stony Brook,N.Y. 1967 Jun Edited by Aaron Finerman Held at the State university of New York Association for computing machinery.Monograph series Academic press,New York,1968
Eng 41.2223

CONFERENCE ON GRAPH THEORY Proceedings
Many facets of graph theory Kalamazoo 1969 Oct 31-Nov 2 Edited by G. Chartrand and S.F. Kapoor Lecture notes in mathematics, 110 Illus viii,290p 24cm Springer-Verlag, Berlin,1969
P Math 2.3347

A CONFERENCE ON GRAVITATION Proceedings
Relativistic theories of gravitation Jablonna 1962 Jul 25-31 Polska akademia nauk Edited by Leopold Infeld With financial support of the International union of pure and applied physics Pergamon press, Oxford,1964 Binder's title in French
TA 15.0085

CONFERENCE ON HIGH-SPEED AERONAUTICS
Proceedings Brooklyn,N.Y. 1955 Jan 20-22 Polytechnic institute of Brooklyn. Department of aeronautical engineering and applied mechanics Edited by Antonio Ferri and others Polytechnic institute of Brooklyn, Brooklyn,N.Y.,1955
Eng 41.6955

CONFERENCE ON HIGH SPEED AUTOMATIC CALCULATING-MACHINES report Cambridge 1949 June 22-25 University of Cambridge.Mathematical laboratory Bibliog 31cm Cambridge,1950
Philos 1.0336

CONFERENCE ON HIGH-SPEED AUTOMATIC CALCULATING MACHINES Report Cambridge 1949 Jun 22-25 University of Cambridge.Mathematical laboratory Great Britain.Ministry of supply University mathematical laboratory,Cambridge, 1950
Eng 41.2175

CONFERENCE ON HIGH-SPEED AUTOMATIC CALCULATING MACHINES Report Cambridge 1949 Jun 22-25 University of Cambridge.Mathematical laboratory Great Britain.Ministry of supply University mathematical laboratory,Cambridge, 1950
Math L 5.3236

CONFERENCE ON HIGH-STRENGTH STEELS... Papers and discussions Harrogate 1962 May 23-24 Iron and steel institute Iron and steel institute.Special report, 76 Iron and steel institute,London,1962
Met 25.0486

CONFERENCE ON HUMAN POPULATION GENETICS IN ISRAEL
Proceedings Genetics of migrant and isolate population Jerusalem 1961 Edited by Elisabeth Goldschmidt Published for the Association for the aid of crippled children Williams and Wilkins,New York,1963
Gen 34.1881

CONFERENCE ON HYDRAULIC MECHANISMS Proceedings
London 1954 Mar 26 Institution of mechanical engineers Institution of mechanical engineers,London,1954
Eng 41.6686

CONFERENCE ON HYDRAULIC SERVO-MECHANISMS
Proceedings London 1953 Feb 13 Institution of mechanical engineers Institution of mechanical engineers,London, 1963
Eng 41.5891

CONFERENCE ON HYDROGEN IN STEEL papers and discussions Hydrogen in steel Harrogate 1961 Oct Iron and steel institute Iron and steel institute.Special report, 73 Iron and steel institute,London,1962 No title page
Met 25.1538

CONFERENCE ON IMMUNO-REPRODUCTION Proceedings
La Jolla,Calif. 1962 Sep 9-11 Population council and Ford foundation Population council,New York,c1962 Organized by A.Tyler and K.A.Laurence and edited by the latter
Path 30.2340

CONFERENCE ON INDUSTRIAL MEASUREMENT TECHNIQUES FOR ON-LINE COMPUTERS London 1968 Jun 11-13 Institution of electrical engineers Institution of electrical engineers.Conference publications, 43 I.E.E.,London,1968
Eng 41.2220

CONFERENCE ON INFORMATION AND CONTROL PROCESSES IN LIVING SYSTEMS 1ST Information and control processes in living systems :first interdisciplinary conference Proceedings Princeton,N.J. 1965 Feb 28-Mar 3 New York academy of sciences Edited by Diane M. Ramsey New York academy of sciences interdisciplinary communications program,New York,1967
Psy 31.0142

CONFERENCE ON INTEGRATED CIRCUITS Integrated circuits Eastbourne 1967 May 2-4 Sponsored by the Institution of electrical engineers.Electronics division I.E.E. Conference publication, 30 Institution of electrical engineers,London,1967 Joint sponsors;Institution of electronic and radio engineers and the Institute of electrical and electronics engineers
Eng 41.5461

CONFERENCE ON INTUITIONISM AND PROOF THEORY Proceedings Intuitionism and proof theory Buffalo,N.Y. 1968 Aug Edited by A. Kino Studies in logic and the foundations of mathematics Bibliog. viii,516p 23cm North Holland,Amsterdam;London,1970
P Math 2.3613

The CONFERENCE ON INVERTEBRATE NERVOUS SYSTEMS their significance for mammalian neurophysiology Papers Invertebrate nervous systems Pasadena,Calif. 1966 Jan 10-12 Edited by C.A.G. Wiersma Sponsored by the National institute of health University of Chicago press,Chicago,Ill.; London,1967 Dedicated to the memory of George Howard Parker
Bal 44.6453

CONFERENCE ON LATIN AMERICAN STUDIES Papers Social science in Latin America Rio de Janeiro 1965 Mar 29-31 Joint committee on Latin American studies Edited by Manuel Diegues and Bryce Wood Jointly sponsored by the Latin American center for research in the social sciences,Rio de Janeiro Columbia university press,New York,1967
Geog 13.5108

CONFERENCE ON LEARNING,REMEMBERING AND FORGETTING 1st Proceedings 1963 Vol 1: anatomy of memory Edited by D.P. Kimble Under the auspices of the American institute of biological sciences Science and behaviour books,Palo Alto,Calif.,1965
An 32.4409

CONFERENCE ON LEARNING,REMEMBERING AND FORGETTING 3rd Proceedings Readiness to remember Princeton,N.J. 1965 Vol 1-2 New York academy of sciences Edited by Daniel P. Kimble Gordon and Breach,New York, 1969
Psy 31.3281

CONFERENCE ON LIVER INJURY 10th Transactions Liver injury New York 1951 May 21-22 Edited by F.W. Hoffbauer Churchill,London, 1951
Med 36.0189

CONFERENCE ON LOCALIZED EXCITATIONS Elementary excitations in solids Cortina lectures and four lectures from the Conference on localized excitations Proceedings Milan 1966 Jul 25-26 Edited by A.A. Maradudin and G.F. Nardelli Bibliog 526p 20cm Plenum press,New York,1969
Cav 7.2677

CONFERENCE ON LUBRICATION AND WEAR 1957 Dec Scientific lubrication Scientific publications,Broseley,1957
Eng 41.6294

CONFERENCE ON LUBRICATION AND WEAR Proceedings London 1957 Oct 1-3 Institution of mechanical engineers Institution of mechanical engineers,London,1957
Eng 41.6295

CONFERENCE ON MACHINE PERCEPTION OF PATTERNS AND PICTURES Proceedings Teddington 1972 Apr 12-14 Institute of physics National physical laboratory Institute of electrical engineers Institute of physics.Conference series, 13 illus ix,362p Institute of physics,London,1972
Math L 5.3882

CONFERENCE ON MATHEMATICS AND COMPUTER SCIENCE IN BIOLOGY AND MEDICINE Proceedings Mathematics and computer science in biology and medicine Oxford 1964 Jul Medical research council 317p H.M.S.O.,London,1965
Bot 42.0512

CONFERENCE ON MICROCIRCULATION,PERFUSION AND TRANSPLANTATION OF ORGANS Proceedings Microcirculation,perfusion,and transplantation of organs Miami,Fla. 1969 Oct 30 University of Miami school of medicine United States.Office of naval research Edited by Theodore I. Malinin and others Academic press,London;New York,1970
An 32.5458

CONFERENCE ON MICROCIRCULATORY PHYSIOLOGY AND PATHOLOGY 2nd Vascular patterns as related to functions Philadelphia,Pa. 1955 Apr 5 Williams and Wilkins,Baltimore,Md., 1955
An 32.3461

CONFERENCE ON MICROCIRCULATORY PHYSIOLOGY AND PATHOLOGY 5th Proceedings Microcirculation:Symposium on factors influencing exchange of substances across capillary wall A Sterotaxic atlas of the dogs brain Buffalo,N.Y. 1958 Apr 1 By Robert K.S. Lim and others American association of anatomists Edited by S.R.M. Reynolds and Benjamin W. Zweifach Thomas, Springfield,Ill.,1960
VA 19.0362

CONFERENCE ON MICROCIRCULATORY PHYSIOLOGY AND PATHOLOGY,5TH The Microcirculation : symposium on factors influencing exchange of substances across capillary wall Buffalo,N.Y. 1958 Apr 1 Edited by S.R.M. Reynolds and Benjamin W. Zweifach University of Illinois press,Urbana,Ill.,1959
Phys 20.0888

CONFERENCE ON MOLECULAR AND RADIATION BIOLOGY Ardsley-on-Hudson,N.Y. 1959 Dec 2-4 Edited by R.A. Deering Organised by the National research council.Subcommittee on radiobiology National research council.Nuclear science series.Report, 31 NAS-NRC 823 National research council,Washington,D.C.,1961
Radioth 35.1735

CONFERENCE ON MONONUCLEAR PHAGOCYTES Mononuclear phagocytes Leiden 1969 Sep 2-5 Edited by Ralph van Furth Blackwell scientific,Oxford;Edinburgh,1970
An 32.5450

CONFERENCE ON MONOTONE MAPPINGS AND OPEN MAPPINGS 1st Proceedings Binghamton,N.Y. 1970 Oct 8-11 State university of New York, Binghampton Edited by Louis F. McAuley xxii,422p 25cm State university of New York,Binghamton,N.Y.,1971 Proceedings dedicated to the memory of G.T.Whyburn
P Math 2.4118

CONFERENCE ON MURINE LEUKEMIA Papers Philadelphia,Pa. 1965 Oct 13-15 National cancer institute.Virology research branch and Albert Einstein medical center Edited by Marvin A. Rich and John B. Moloney National cancer institute.Monograph, 22 US government printing office,Washington,D.C., 1966
Path 30.2460

CONFERENCE ON MUSCULAR CONTRACTION 2ND
Conference on physicochemical mechanism of nerve activity and Conference on muscular contraction 2nd New York academy of sciences Edited by David A. Nachmansohn New York academy of sciences.Annals, 81,p 215-510 New York,1959
Bal 39.1450

CONFERENCE ON NERVE IMPULSE 3rd
Transactions Nerve impulse New York 1952 Mar 3-4 Edited by H.Houston Merritt Sponsored by the Josiah Macy jr.foundation Josiah Macy jr.foundation,New York,1952
An 32.4315

CONFERENCE ON NEUROPHYSIOLOGY IN RELATION TO ANESTHESIOLOGY Seattle 1966 May 13-14 Anesthesiology,28,no 1 American society of anesthesiologists,Lancaster,Pa.,1967
Pha 16.0052

CONFERENCE ON NEWER RESPIRATORY DISEASE VIRUSES
Bethesda,Md 1962 Oct 3-5 National institutes of health National institute of allergy and infectious diseases Edited by Clayton G. Loosli Sponsored by University of Southern California.School of medicine American review of respiratory diseases, 88,3 pt 2 American thoracic society,1963
PGMS 29.0321

CONFERENCE ON NON-DESTRUCTIVE EXAMINATION APPLIED TO PROCESS CONTROL IN THE STEEL INDUSTRY
Papers and discussions Non-destructive examination in the steel industry Swansea 1967 Jan 3-5 Swansea and district metallurgical society Iron and steel institute Society of non-destructive examination Iron and steel institute. Publication, 103 illus 193p London,1967
Met 25.2601

CONFERENCE ON NONLINEAR OSCILLATIONS 4th
Proceedings Prague 1967 Sep 5-9 Edited by Jan Gonda Academia kiado,Prague, 1968
Eng 41.6391

CONFERENCE ON NUCLEAR AND MESON PHYSICS
Proceedings Glasgow 1954 Jul 13-17 Edited by E.H. Bellamy and R.G. Moorhouse Sponsored by the International union of pure and applied physics Pergamon press,London; New York,1955
Radioth 35.1563

CONFERENCE ON NUCLEAR PROCESSES IN GEOLOGIC SETTINGS Proceedings Williamsbay 1953 Sep,21-23 University of Chicago Co-sponsored by National research council Williams Bay,Wisconsin,1953
Philos 1.0514

CONFERENCE ON NUCLEAR REACTOR TECHNOLOGY 2nd
Papers Analytical chemistry in nuclear reactor technology:particle-size analysis Gatlinburg,Tenn. 1958 Sep 29-Oct 1 United States atomic energy commission v,101p Oakridge national laboratory,Oakridge,Tenn., 1959
Chem E 24.1436

CONFERENCE ON OCEANWAVE SPECTRA 1961
National academy of sciences Englewood Cliffs,N.J.,1963
Eng 41.6696

A CONFERENCE ON OIL HYDRAULIC POWER TRANSMISSION
London 1961 Nov 29-30 Institution of mechanical engineers.Hydraulic plant and machinery group Institution of mechanical engineers,London,1962
Eng 41.6685

CONFERENCE ON OPTICAL INSTRUMENTS AND TECHNIQUES
Proceedings London 1961 Jul 11-14 Edited by K.J. Habell Chapman and Hall, London,1962 Held at Imperial college
Cav 7.1457

CONFERENCE ON OPTICAL INSTRUMENTS AND TECHNIQUES
Proceedings London 1961 Jul 11-14 Edited by K.J. Habell Under the auspices of the International commission for optics Chapman and Hall,London,1962
Obs 6.3215

CONFERENCE ON OPTIMISATION TECHNIQUES IN CIRCUIT AND CONTROL APPLICATIONS London 1970 Jun 29-30 Institution of electrical engineers Institution of electrical engineers.Conference publication, 66 IEE, London,1970 Organised in association with the Institute of electrical and electronics engineers and the Institute of physics and the Physical society
Eng 41.8587

CONFERENCE ON OPTIMIZATION OF STEEL PRODUCT YIELD
Proceedings Optimization of steel product yield London 1967 May 3-4 Iron and steel institute Iron and steel institute. Publication, 107 London,1967 Conference held at the Institute's annual general meeting
Met 25.2561

CONFERENCE ON PARTIAL DIFFERENTIAL EQUATIONS
typescript Contributions to the theory of partial differential equations Harriman,N.Y. 1952 Oct By S. Bergman and others Edited by Lipman Bers and others Sponsored by National research council Annals of mathematics studies, 33 Bibliog 257p 26cm Princeton university press,Princeton,N. J.,1954
P. Math 2.1228

CONFERENCE ON PATTERN RECOGNITION Teddington 1968 Jul 29-31 Organised by Institution of electrical engineers.Control and automation division Institution of electrical engineers. Conference publication, 42 Institution of electrical engineers,London,1968
Eng 41.5311

CONFERENCE ON PHENOMENA IN THE NEIGHBOURHOOD OF CRITICAL POINTS Proceedings Critical phenomena Washington,D.C. 1965 Apr 5-8 National bureau of standards Edited by M.S. Green and J.V. Sengers National bureau of standards.Miscellaneous publication, 273 xiv, 242p 27cm National bureau of standards, Washington,D.C.,1966
Cav 7.2931

The CONFERENCE ON PHOTOGRAPHIC AND SPECTROGRAPHIC OPTICS Proceedings Tokyo 1964 Sep 1-5 and Kyoto 1964 Sep 1-5-,7-8 International commission for optics Under the auspices of the Science council of Japan Japanese journal of applied optics, 4,suppl. 1 669p Tokyo,1965
Obs 6.2050

CONFERENCE ON PHOTOPERIODISM Photoperiodism and related phenomena in plants and animals Gatlinburg 1957 American association for the advancement of science Edited by Robert B. Withrow and others American association for the advancement of science.Publication, 55 Washington,D.C.,1959
Bot 42.1826

CONFERENCE ON PHOTOSYNTHESIS Currents in photosyntheses Western-Europe conference on photosynthesis 2nd Proceedings Woudschoten,Zeist 1965 Sep 6-12 Edited by J.B. Thomas and J.C. Goedheer A.D.Donker, Rotterdam,1966
Bioch 33.0805

CONFERENCE ON PHYSICAL AND CHEMICAL PROPERTIES OF SEA WATER Report Easton,Md. 1958 Sep 4-5 United States.Office of naval research and National research council. Committee on oceanography National research council. Publication, 600 Washington,D.C.,1959
Geod 9.0610

CONFERENCE ON PHYSICAL ASPECTS OF NOISE IN ELECTRONIC DEVICES Papers Nottingham 1968 Sep 11-13 Institute of physics and the Physical society.Electronics group iv,248p 31cm Peregrinus,Stevenage,1968
Cav 7.2840

CONFERENCE ON PHYSICAL ELECTRONICS Report Cambridge,Mass. 1949 Apr 7-9 By E.B. Callick 86p 27cm British joint services mission,Washington,D.C.,1949
Math L 5.3263

CONFERENCE ON PHYSICOCHEMICAL MECHANISM OF NERVE ACTIVITY 2nd and Conference on muscular contraction New York academy of sciences Edited by David A. Nachmansohn New York academy of sciences.Annals, 81,p 215-510 New York,1959
Bal 39.1450

CONFERENCE ON PHYSICS 13th Proceedings Structure and evolution of galaxies Brussels 1964 Sep Universite libre de Bruxelles and Institut international de physique Solvay 174p Interscience,London,1965
Obs 6.2134

The CONFERENCE ON PHYSICS 13th Proceedings Structure and evolution of galaxies Brussels 1964 Sep Institut international de physique Solvay and Universite libre de Bruxelles Solvay conference,1964,13 Interscience,London,1965 Other Solvay conferences issued as Conseil de physique,q.v.
Cav 7.2078

CONFERENCE ON PHYSICS OF THE SOLAR SYSTEM AND RE-ENTRY DYNAMICS proceedings Physics of the solar system Blacksburg,Va. 1961 Jul. 31-Aug.11 Part 1: physics of the solar system Virginia polytechnic institute. Engineering experiment station. Virginia polytechnic institute.Engineering experiment station series, 149 Virginia polytechnic institute.Bulletin, 55,9 Virginia polytechnic institute,Blacksburg,Va.,1962
A Math 4.1155

CONFERENCE ON PHYSIOLOGY OF PREMATURITY 1st-4th Transactions Physiology of pre-maturity Princeton,N.J. 1956-59 Mar Edited by Jonathan T. Lanman Sponsored by the Josiah Macy jr.foundation 4 vols Josiah Macy jr. foundation,New York,1957-60
An 32.3819

CONFERENCE ON POLYSACCHARIDES IN BIOLOGY 1st-5th Transactions Polysaccarides in biology Princeton,N.J. 1955-1959 Edited by Georg F. Springer Sponsored by Josiah Macy Jr. foundation 5 vols Josiah Macy,Jr. foundation,New York,1955-59
An 32.1493

CONFERENCE ON POWDER METALLURGY 1st and 2nd held at the Massachusetts institute of technology Papers Cambridge,Mass. 1940-41 American society for metals Edited by John Wulff American society for metals, Cleveland,1942
Met 25.0758

CONFERENCE ON POWDER METALLURGY IN ATOMIC ENERGY with additional papers Powder metallurgy in nuclear engineering Philadelphia 1955 Oct 20 By Henry H. Hausner United States atomic energy commission and American society for metals American society for metals,Cleveland,Ohio, 1958
Met 25.0767

CONFERENCE ON POWER APPLICATION OF CONTROLLABLE SEMICONDUCTOR DEVICES London 1965 Pt 1 Institution of electrical engineers Institution of electrical engineers.Conference publications, 17 I.E.E.,London,1965
Eng 41.4868

CONFERENCE ON PRESTRESSED CONCRETE PRESSURE VESSELS London 1967 Mar 13-17 Institution of civil engineers Institution of civil engineers,London,1968
Eng 41.2891

CONFERENCE ON PROBLEMS CONNECTED WITH THE PREPARATION AND USE OF LABELLED PROTEINS IN TRACER STUDIES Proceedings Labelled proteins in tracer studies Pisa 1966 Jan 17-19 Edited by L. Donato and others Sponsored by the Universita di Pisa.Nuclear medicine center EUR 2950 d,f,e European atomic energy community,Brussels,1966
Radioth 35.1752

CONFERENCE ON PROBLEMS IN ECONOMIC DEVELOPMENT Proceedings Problems in economic development International economic association Edited by E.A.G. Robinson Macmillan,London,1965
Geog 13.2413

The CONFERENCE ON PROBLEMS OF THE DISTRIBUTION AND MOTION OF INTERSTELLAR MATTER IN GALAXIES Proceedings Distribution and motion of interstellar matter in galaxies Princeton,N. J. 1961 Apr 10-20 Institute for advanced study Edited by L. Woltjer 330p Benjamin, New York,1962
TA 15.0213

CONFERENCE ON PULMONARY CIRCULATION Oslo 1965 Sep 10-11 Edited by Carsten Muller Universitetsforlaget,Oslo,1966
An 32.3575

CONFERENCE ON QUANTUM ELECTRONICS - RESONANCE PHENOMENA Quantum electronics High View, N.Y. 1959 Sep 14-16 Edited by Charles H. Townes Columbia university press, New York, 1960
Eng 41.5411

CONFERENCE ON QUANTUM ELECTRONICS-RESONANCE PHENOMENA Papers Quantum electronics New York 1959 Sep 14-16 Columbia university Edited by Charles H. Townes 601p Columbia university press, New York, 1960
Cav 7.2332

CONFERENCE ON RADIATION PROTECTION IN ACCELERATOR ENVIRONMENTS Proceedings Chilton, Berks. 1969 Mar Edited by G.R. Stevenson Sponsored by the Science research council Rutherford laboratory, Chilton, 1969 Organized by and held at the Rutherford laboratory
Radioth 35.1393

CONFERENCE ON RADIOBIOLOGY AND RADIATION PROTECTION Stockholm 1952 Sep 15-20 Acta radiologica, 41, fasc.1 Stockholm, 1954
Radioth 35.1228

CONFERENCE ON RADIOBIOLOGY AND RADIOTHERAPY Proceedings Colorado Springs 1965 Nov 1-3 National cancer institute Edited by J.A. Del Regato National cancer institute. Monograph, 24 National cancer institute, Bethesda, Md., 1967
Bioch 33.0989

CONFERENCE ON RADIOIODINE Proceedings Chicago, Ill. 1956 Nov 5-6 Argonne cancer research hospital United States atomic energy commission Edited by D.E. Clark ACRH-100 University of Chicago, Chicago, Ill., 1956
Radioth 35.1076

The CONFERENCE ON REACTIONS BETWEEN COMPLEX NUCLEI Asilomar conference Pacific Grove, Calif. 1963 Apr 14-18 Edited by A. Ghiorso and others 456p University of California press, Berkeley, Calif., 1963
Cav 7.0090

CONFERENCE ON REFRACTORY METALS 4TH Refractory metals and alloys 4 research and development: based on a technical conference Proceedings French Lick, Ind. 1965 Oct 3-5 Vol 1-2 American institute of mining, metallurgical and petroleum engineers. Refractory metals committee Edited by R.I. Jaffee and others Metallurgical society conferences, 41 2 vols Gordon and Breach, New York, 1967
Met 25.2250

CONFERENCE ON RELATIVISTIC THEORIES OF GRAVITATION proceedings Relativistic theories of gravitation Jablonna 1962 Jul 25-31 Edited by Leopold Infeld Organised and sponsored by Polska akademia nauk Pergamon press, Oxford, 1964
A Math 4.1029

The CONFERENCE ON RESEARCH ON THE RADIOTHERAPY OF CANCER Proceedings Madison, Wis. 1960 Jan 16-18 American cancer society, New York, 1961
Radioth 35.1133

CONFERENCE ON SCIENCE AND TECHNOLOGY FOR DEANS OF ENGINEERING Proceedings Recent advances in the engineering sciences: their impact on engineering education Lafayette, Ind. 1957 Sep 9-12 Purdue research foundation and Purdue university viii, 257p McGraw-Hill, New York, 1958
Chem E 24.1074

CONFERENCE ON SERVOCOMPONENTS London 1967 Nov 21-23 Sponsored by Institution of electrical engineers. Professional group on measuring and control equipment Institution of electrical engineers. Conference publication, 37 I.E.E., London, 1967
Eng 41.5879

CONFERENCE ON SHOCK METAMORPHOSIS OF NATURAL MATERIALS 1st Proceedings Greenbelt, Md. 1966 Apr 14-16 National aeronautics and space administration. Goddard space flight center Edited by Bevan M. French and Nicholas M. Short Held at the Goddard space flight center Mono book corporation, Baltimore, Md., 1968
Min 10.1443

CONFERENCE ON SMALL BLOOD VESSEL INVOLVEMENT IN DIABETES MELLITUS Small blood vessel involvement in diabetes mellitus :conference Proceedings Warrenton, Va. 1963 Mar 25-27 American institute of biological sciences Edited by Marvin D. Siperstein and others American institute of biological sciences, Washington, D.C., 1964
Bioch 33.0482

CONFERENCE ON SOCIAL IMPLICATIONS OF INDUSTRIALIZATION AND TECHNICAL CHANGE Papers Industrialization and society Chicago, Ill. 1960 Sep 15-22 Edited by Bert F. Hoselitz and Wilbert E. Moore Sponsored by Unesco Unesco, Mouton, 1963
Eng 41.1052

CONFERENCE ON SOIL SCIENCE PROBLEMS 1st proceedings Rothamsted 1930 Sep 16-18 Imperial bureau of soil science Imperial bureau of soil science. Technical communications, 17 H.M.S.O., London, 1931
Geog 13.1071

CONFERENCE ON SOIL STABILIZATION Proceedings Cambridge, Mass. 1952 Jun 18-20 Massachusetts institute of technology M.I.T. press, Cambridge, Mass., 1952
Eng 41.3134

CONFERENCE ON SOLID CIRCUITS AND MICROMINIATURIZATION :a conference Proceedings Solid circuits and microminiaturization West Ham 1963 Jun Edited by G.W.A. Dummer Pergamon press, London, 1964
Eng 41.5567

CONFERENCE ON SOME ASPECTS OF THE GERMAN PROBLEM Germanys contribution to European economic life Baarn 1947 Oct 6-11 and Scheveningen 1948 Apr 11-17 By Percy W. Bidwell and others Riviere, Paris, 1949 Papers in English and French
Geog 13.2112

CONFERENCE ON SOVIET AGRICULTURAL AFFAIRS 2nd Papers Soviet and east European agriculture Santa Barbara,Calif. 1965 Aug 26-28 University of California,Berkeley. Center of Slavic and east European studies Edited by Jerzy F. Karcz Russian and east European studies University of California press,Berkeley,Calif.;Los Angeles,Calif.,1967
Geog 13.2584

CONFERENCE ON STAINLESS STEELS FOR THE FABRICATOR AND USER Proceedings Stainless steels Birmingham 1968 Sep 10-12 Iron and steel institute and Birmingham metallurgical association Iron and steel institute,London, 1969
Met 25.2242

CONFERENCE ON STATISTICAL MECHANICS AND THERMODYNAMICS.FOUNDATIONS AND APPLICATIONS International union of pure and applied physics :meeting Proceedings Copenhagen 1966 Edited by Thor A. Bak table 582p 25cm Benjamin,New York,1967
Cav 7.3022

CONFERENCE ON STEAM TURBINE RESEARCH AND DEVELOPMENT Proceedings London 1953 Mar 6 Institution of mechanical engineers Institution of mechanical engineers,London, 1953
Eng 41.7296

CONFERENCE ON STEEL IN ARCHITECTURE Proceedings London 1969 Nov 24-26 Organized by the British constructional steelwork association British constructional steelwork association,London,1970
Eng 41.0269

A CONFERENCE ON STRENGTH OF SOLIDS :held at the H.H.Wills physical laboratory Report Bristol 1947 Jul 7-9 University of Bristol Edited by N.F. Mott 162p Physical sciety, London,1948
Cav 7.0346

The CONFERENCE ON STRUCTURE AND REACTIONS OF DFP SENSITIVE ENZYMES Proceedings Stockholm 1966 Sep 5-7 Edited by Edith Heilbronn organized by the Forsvarets forskningsanstalt,Stockholm Research institute of national defence,Stockholm,1967
Bioch 33.1065

CONFERENCE ON SUPERNOVAE Proceedings Supernovae and their remnants Greenbelt,Md. 1967 Goddard institute for space studies Edited by P.J. Brancazio and A.G.W. Cameron 240p Gordon and Breach,New York,1969
Obs 6.3361

CONFERENCE ON SWITCHING TECHNIQUES FOR TELECOMMUNICATIONS NETWORKS London 1969 21-25 Apr Sponsored by the Institution of electrical engineers Institution of electrical engineers.Conference publications, 52 Institution of electrical engineers, London,1969
Math L 5.3531

CONFERENCE ON SYMMETRY PRINCIPLES AT HIGH ENERGY Coral Gables,Fla. 1964 Jan 30-31 University of Miami Edited by Behram Kursunoglu and Arnold Perlmutter Freeman,San Francisco,1964
Cav 7.0543

CONFERENCE ON SYSTEMS AND COMPUTER SCIENCE : held at the University of Western Ontario Proceedings Systems and computer science London,Ont. 1965 Sep 10-11 Edited by John F. Hart and Satoru Takasu Bibliog.,Illus. xiv,249p 25cm Toronto university press, Toronto,1967
P Math 2.3494

CONFERENCE ON TEACHING COMPUTING Proceedings Bristol 1972 Mar 28 University grants committee.Mathematical sciences sub-committee illus 165p University grants committee, London,1972
Math L 5.3878

CONFERENCE ON TECHNOLOGY OF ENGINEERING MANUFACTURE Proceedings London 1958 Mar 25-27 Institution of mechanical engineers Institution of mechanical engineers,London,1958
Eng 41.3870

CONFERENCE ON THE APPLICATION OF NUMERICAL INTEGRATION TECHNIQUES TO THE PROBLEM OF THE GENERAL CIRCULATION proceedings Dynamics of climate 1955 Oct 26-28 Institute for advanced study Edited by Richard L. Pfeffer Sponsored jointly by Air force Cambridge research center Pergamon press,Oxford;London,1960
A Math 4.0560

CONFERENCE ON THE BIOLOGY OF CUTANEOUS CANCER The International conference on the biology of cutaneous cancer 1st Proceedings Philadelphia 1962 Apr 6-11 National cancer institute Edited by Frederick Urbach National cancer institute.Monograph, 10 U.S. Department of health,education,and welfare, Bethesda,Md.,1963 Sponsored by the skin and cancer hospital,Dept.of dermatology,Temple university school of medicine and The Committee on geographie pathology,Unio internationalis contra cancrum.A Merck Sharp and Dohme medical research conference
Bioch 33.0981

A CONFERENCE ON THE BIOLOGY OF NORMAL AND ATYPICAL PIGMENT CELL GROWTH Biology of melanomas New York 1946 Nov 15-16 New York academy of sciences.Section of biology New York academy of sciences.Special publications, 4 pls New York academy of sciences,New York,1948
Path 30.2492

CONFERENCE ON THE BIOLOGY OF NORMAL AND ATYPICAL PIGMENT CELL GROWTH Biology of melanomas New York 1946 Nov 15-16 New York academy of sciences.Section of biology New York academy of sciences.Special publications, 4 New York,1948
Phys 20.0952

CONFERENCE ON THE BIOLOGY OF NORMAL AND ATYPICAL PIGMENT CELL GROWTH Results Biology of melanomas New York 1946 Nov 15-16 By Myron Gordon and others New York academy of sciences.Special publications, 6 New York academy of sciences,New York,1948
An 32.3374

CONFERENCE ON THE BIOLOGY OF NORMAL AND ATYPICAL PIGMENT CELL GROWTH 3rd Proceedings Pigment cell growth New York 1951 Nov 15-17 Edited by Myron Gordon Sponsored by the New York zoological society Academic press, New York,1953
An 32.3368

CONFERENCE ON THE BIOLOGY OF NORMAL AND ATYPICAL PIGMENT CELL GROWTH 4th Proceedings Pigment cell biology Houston,Tex. 1957 Nov 14-16 Edited by Myron Gordon Academic press,New York,1959 For 5th and further conferences see:International pigment cell conference
An 32.3388

CONFERENCE ON THE CARIBBEAN 13th Papers Caribbean:Venezuelan development,a case history Gainesville,Fla. 1962 Dec 6-8 Edited by A.Curtis Wilgus University of Florida.School of Inter-American studies. Publications.Series 1,13 University of Florida press,Gainesville,Fla.,1963
Geog 13.5242

CONFERENCE ON THE CARIBBEAN 14th Papers Caribbean:Mexico today Gainesville,Fla. 1963 Dec 4-7 University of Florida.Center for Latin American studies Edited by A. Curtis Wilgus University of Florida.Center for Latin American studies.Publications.Series 1,14 University of Florida press,Gainesville, Fla.,1964
Geog 13.5202

CONFERENCE ON THE CENTRAL NERVOUS SYSTEM AND BEHAVIOUR 1st Transactions Central nervous system and behaviour 1958 Feb 23-26 Edited by Mary A.B. Brazier Sponsored by the Josiah Macy jr.foundation New York,1959
Phys 20.2239

The CONFERENCE ON THE CENTRAL NERVOUS SYSTEM AND BEHAVIOUR 1st-3rd Transactions Central nervous system and behaviour 1958-60 Edited by Mary A. Brazier Josiah Macy jr. foundation 3 vols Josiah Macy jr. foundation,New York,1959-61
An 32.5329

The CONFERENCE ON 'THE CHALLENGE FOR STEEL' :I. S.I. autumn general meeting Proceedings Challenge for steel 1968 Nov Iron and steel institute Iron and steel institute. Publication, 119 Iron and steel institute, London,1969
Met 25.2241

CONFERENCE ON THE CHEMISTRY OF MUSCULAR CONTRACTION Paper and discussions presented Tokyo 1957 Oct 12-14 Committee of muscle chemistry of Japan Igaku Shoin ltd., Tokyo;Osaka,1957
Bioch 33.2229

CONFERENCE ON THE DEFECTS IN CRYSTALLINE SOLIDS held at the H.H.Wills physical laboratory Report Bristol 1954 Jul By N. Bloembergen and others University of Bristol and Institute of physics With co-operation of the International union of pure and applied physics.Commission on the solid state 429p Physical society,London,1955
Cav 7.0345

The CONFERENCE ON THE EDUCATION AND TRAINING OF ENGINEERING TECHNICIANS Proceedings London 1963 Mar 21-22 Sponsored by Institution of mechanical engineers.Education and training group Institution of mechanical engineers,London,1963
Eng 41.1503

CONFERENCE ON THE FATIGUE OF METALS 2nd Papers Institut metallurgii imini A.A. Baykov Translated by Leo Ronson from the Russian N.L.L.S.T.,Boston Spa,1964 Translation edited by P.G.Forrest
Met 25.2505

CONFERENCE ON THE HIGHWAY NEEDS OF GREAT BRITAIN Proceedings London 1957 Nov 13-15 Institution of civil engineers Institution of civil engineers,London,1958
Eng 41.3260

CONFERENCE ON THE HUMAN ENVIRONMENT 1972 Air pollution across national boundaries :the impact on the environment of sulfur in air and precipitation.Sweden's case study for the United Nations conference on human environment Sweden.Utrikesdepartementet Nordstedt, Stockholm,1971 Case prepared with the Swedish ministry of agriculture
Eng 41.8613

The CONFERENCE ON THE HUMAN OPERATOR IN COMPLEX SYSTEMS Proceedings Birmingham 1966 Edited by W.T. Singleton and others Bibliog Taylor and Francis,London,1967
Eng 41.0678

CONFERENCE ON THE MECHANISMS OF CELL DIVISIONS 2nd Proceedings New York 1960 Oct 7 By M.J. Kopac and others New York academy of sciences Edited by Paul P. Gross New York academy of sciences.Annals, 90,p.345-613 New York,1960
Bal 39.0291

The CONFERENCE ON THE NATURE OF THE SURFACE OF THE MOON Proceedings IAU-NASA symposium Greenbelt,Md. 1965 Apr 15-16 Goddard space flight center and International astronomical union Edited by Wilmot N. Hess and others Sponsored bt the National aeronautics and space administration 320p Johns Hoplins press,Baltimore,1966
TA 15.0203

CONFERENCE ON THE NORTH SEA FLOODS OF 31 JANUARY-1 FEBRUARY,1953 Papers London 1953 Dec Institution of civil engineers Institution of civil engineers,London,1954
Geog 13.0706

CONFERENCE ON THE NUMERICAL SOLUTION OF DIFFERENTIAL EQUATIONS Dundee 1969 Jun 23-27 Edited by J.L. Morris Lecture notes in mathematics, 109 Bibliog. vi,275p 25cm Springer-Verlag,Berlin,1969
P Math 2.3474

A CONFERENCE ON THE ORIGINS OF PREBIOLOGICAL SYSTEMS AND OF THEIR MOLECULAR MATRICES Proceedings Origins of prebiological systems and of their molecular matrices Wakulla Springs,Fla 1963 Oct 27-30 Institute for space biosciences Florida state university National aeronautics and space administration Edited by Sidney W. Fox Bibliog.,illus. xx,482p Academic press,New York;London,1965
Bot 42.1364

CONFERENCE ON THE PHYSICS OF CLOUD AND PRECIPITATION PARTICLES 1st Proceedings Artificial stimulation of rain Woods Hole, Mass. 1955 Sep 7-10 Air force Cambridge research center.Geophysics research directorate Edited by Helmut Weickmann and Waldo Smith Co-sponsored by United States. Office of naval research.Geophysics branch illus 442p Pergamon,London,1957
Nap 11.0931

CONFERENCE ON THE PHYSICS OF SEMICONDUCTOR SURFACES Proceedings Semiconductor surface physics Philadelphia,Pa. 1956 Jun 4-6 Edited by R.H. Kingston Sponsored by the United States.Office of naval research,the University of Pennsylvania and the Lincoln laboratory University of Pennsylvania, Philadelphia,Pa.,1957
Eng 41.5399

CONFERENCE ON THE PHYSICS OF THE IONOSPHERE Report Cambridge 1954 Sep 6-9 Physical society Cavendish laboratory 406p Physical society,London,1955
Nap 11.0926

CONFERENCE ON THE PRESERVATION OF THE FORMED ELEMENTS AND OF THE PROTEINS OF THE BLOOD Transactions Boston 1949 Jan 6-8 American red cross 1949
Med 36.0005

CONFERENCE ON THE PROBLEM 'INTERACTIONS BETWEEN THE ATMOSPHERE AND THE HYDROSPHERE IN THE NORTH ATLANTIC 2nd Papers Konferentsiya po probleme 'Vsaimodeistvie atmosfery i gidrosfery v severnoi chasti Atlanticheskogo okeana' Leningrad 1961 May 25-30 Leningradskii gidrometeorologicheskii institut 281p Izdatelstvo Leningradskogo universiteta,Leningrad,1964
Sco 14.5941

CONFERENCE ON THE PROBLEMS OF TECTONICS Stroenie i razvitie zemnoii kory conference proceedings Moscow 1963 Edited by P.N. Kropokkim Akademiya nauk SSSR,Mowcow,1964 Includes English title-page,Structure and evolution of the crust.(Proceedings of a conference on the problems of tectonics in Moscow)
Geol 8.1367

CONFERENCE ON THE PROBLEMS OF TECTONICS Voprosy sravnitelnoi tektoniki drevnikh platform conference proceedings Moscow 1963 Edited by A.A. Bogdanov and others Akademiya nauk SSSR,Moscow,1964 Includes English title-page:Problems of comparative tectonics of old platforms (Proceedings of the conference on the problems of tectonics in Moscow)
Geol 8.1366

CONFERENCE ON THE SOCIAL IMPLICATIONS OF INDUSTRIALIZATION AND TECHNOLOGICAL CHANGE Papers Industrialization and society Chicago,Ill. 1960 Sep 15-22 University of Chicago Edited by Bert F. Hoselitz and Wilbert E. Moore Sponsored jointly by Unesco Mouton,The Hague,1963
Geog 13.1608

The CONFERENCE ON THE SOLAR WIND Proceedings Solar wind Pasadena,Calif. 1964 Apr 1-4 California institute of technology.Jet propulsion laboratory Edited by Robert J. Mackin and Marcia Neugebauer JPL TR 32-630 400p pergamon press,Oxford,1966
TA 15.0201

The CONFERENCE ON THE SPECIES CONCEPT IN ITS RELATION TO THE BRITISH FLORA Report Species study in the British flora London 1964 Apr 9-11 Botanical society of the British Isles Edited by J.E. Lousley Botanical society of the British Isles.B.S.B.I. conference reports, 4 London,1955
Bot 42.4709

The CONFERENCE ON THE STUDY OF CRITICAL BRITISH GROUPS Report British flowering plants and modern systematic methods London 1948 Apr 9-10 Botanical society of the British Isles Edited by A.J. Wilmott Botanical society of the British Isles.B.S.B.I. conference reports, 1 London,1949
Bot 42.4706

CONFERENCE ON THE TEACHING OF ENGINEERING DESIGN Papers and summary Scarborough 1964 Apr 1-4 Organised by the Enfield college of technology Illus Institution of engineering designers,London,1964 Co-organized by the Institution of engineering designers and the Hornsey College of art
Eng 41.0309

CONFERENCE ON THE THEORY AND APPLICATIONS OF ANALYSIS IN FUNCTION SPACE Proceedings Dedham,Mass. 1963 Jun 9-13 Massachusetts institute of technology Edited by William T. Martin and Irving Segal vi,218p 24cm MIT press,Cambridge,Mass.,1964
Cav 7.3154

CONFERENCE ON THE THEORY AND APPLICATIONS OF ANALYSIS IN FUNCTION SPACE Proceedings Analysis in function space Dedham,Mass. 1963 Jun 9-13 Edited by William T. Martin and Irving Sega vi,218p 24cm MIT press, Cambridge,Mass.,1964
Math S 3.1659

CONFERENCE ON THE THEORY AND APPLICATIONS OF ANALYSIS IN FUNCTION SPACE proceedings Analysis in function space Dedham,Mass. 1963 Jun 9-13 Massachusetts institute of technology Edited by William Ted Martin and Irving Segal Financially supported by United States.Air force.Office of scientific research vi,218p 24cm M.I.T.press,Cambridge,Mass., 1964
P. Math 2.1181

CONFERENCE ON THE THEORY OF ORDINARY AND PARTIAL DIFFERENTIAL EQUATIONS Proceedings Dundee 1972 Mar 28-31 Edited by W.N. Everitt and B.D. Sleeman Lecture notes in mathematics, 280 xv,367p 25cm Springer, Berlin,1972
P Math 2.4316

CONFERENCE ON THE THERMODYNAMIC PROPERTIES OF MATERIALS 1st Proceedings Applications of fundamental thermodynamics to metallurgical processes Pittsburgh 1964 Nov 29-Dec 1 University of Pittsburgh.Center for the study of thermodynamic properties of materials Edited by G.R. Fitterer Gordon and Breach,New York,1967
Met 25.2443

CONFERENCE ON THE THYMUS;THE THYMUS IN IMMUNOLOGY structure,function and role in disease Proceedings Minneapolis,Minn. 1962 Nov Edited by Robert A. Good and Ann E. Gabrielson Harper and Row,New York,1964
Med 36.0382

CONFERENCE ON THE ULTRA-FINE STRUCTURE OF COAL AND COKES held at the Royal Institution Proceedings London 1943 June 24-25 British coal utilisation research association 23cm London,1944
Philos 1.0687

The CONFERENCE ON THE USE OF SOLAR ENERGY :the scientific basis Transactions Tucson, Ariz. 1955 Oct 31-Nov 1 Vol 1-5 University of Arizona and Stanford research institute Edited by Edwin F. Carpenter and others Co-sponsored by the Association for applied solar energy illus 6 vols University of Arizona,Tucson,Ariz., 1958
Nap 11.0111

CONFERENCE ON THERMAL CONDUCTIVITY 8th Proceedings West Lafayette,Ind. 1968 Oct 7-10 Edited by C.Y. Ho and R.E. Taylor Sponsored by Purdue university.Thermophysical properties research center xx,1169p 29cm Plenum press,New York,1969
Cav 7.3075

The CONFERENCE ON THORIUM Proceedings Metal thorium Cleveland,Ohio 1956 Oct 11 American society for metals and United States atomic energy commission Edited by Harley A. Wilhelm American society for metals,Cleveland,Ohio,1958
Met 25.1561

CONFERENCE ON TRAINING IN APPLIED MATHEMATICS proceedings New York 1953 Oct 22-24 American mathematical society Sponsored also by the National research council 97p 27cm Columbia university,New York,1953
Math L 5.0709

CONFERENCE ON TRAINING PERSONNEL FOR THE COMPUTING MACHINE FIELD 1st proceedings Detroit,Mich. 1954 June 22-23 Wayne state university Edited by Arvid W. Jacobson Sponsored also by the Association for computing machinery,the Industrial mathematics society and the Institute of radio engineers 104p 23cm Wayne university press,Detroit, Mich.,1955
Math L 5.0730

CONFERENCE ON TRANSFORMATION GROUPS Proceedings New Orleans 1967 May 8-Jun 2 Edited by Paul S. Mostert Sponsored by the National science foundation.Advanced science seminar projects Bibliog.,Illus. xiii,456p 24cm Springer-Verlag,Berlin,1968
P Math 2.3395

CONFERENCE ON UNIVERSAL ALGEBRA Proceedings Kingston,Ont 1969 Oct Held at Queen's university,Kingston,Ont. Queen's papers in pure and applied mathematics, 25 Bibliog. 273p 27cm Queen's university,Kingston,Ont., 1970
P Math 2.3712

CONFERENCE ON VERBAL LEARNING 2nd Proceedings Verbal behavior and learning United States.Office of naval research New York university Edited by C.N. Cofer and Barbara S. Musgrave McGraw-Hill,New York, 1963
Psy 31.2737

CONFERENCE ON WELDING CREEP-RESISTANT STEELS Proceedings Newcastle-upon-Tyne 1970 Feb 17-18 Welding institute Welding institute,Abington,1970 Conference director:B.Phelps
Met 25.2725

CONFERENCE ON WORK STUDY Papers Harrogate 1954 May 27-29 British institute of management British institute of management, London,1954
Eng 41.1544

CONFERENCES Communication or conflict : conferences.Conferences;their nature,dynamics, and planning :an international conference Eastbourne 1956 Edited by Mary Capes Tavistock publications,London,1960
Eng 41.0699

CONFERENCES DU COLLEGE DE FRANCE Les Hormones sexuelles colloque international Comptes-rendus Paris 1937 Jun 10-19 Edited by L. Brouha Under the auspices of Fondation Singer-Polignac Hermann,Paris,1938
Gen 34.1603

CONFERENCES HELD IN CONNECTION WITH SPECIAL LOAN COLLECTION OF SCIENTIFIC APPARATUS : physics and mechanics London 1876 South Kensington museum 420p Chapman and Hall,London,1876
Obs 6.1966

CONFERENCES INTERNATIONALES DE PSYCHOTECHNIQUE Conference internationale de psychotechnique 4th Comptes rendues Paris 1927 Oct 10-14 Alcan,Paris,1929 Papers in English, French,German and Italian
Psy 31.2517

CONFERENCES ON CELLULAR DYNAMICS 3rd and 4th interdisciplinary conferences Proceedings Princeton,N.J. 1965 Feb Princeton,N.J. 1966 Feb Edited by Lee D. Peachey New York academy of sciences,New York,1967
Phys 20.1513

CONFERENCIA AMERICANA DA LEPRA Prophylaxia da lepra e das doencas venereas no estado do Para Vol 2 By Heraclides C.de Souza Araujo Brazil.Departamento nacional de saude publica Livreria classica,Belem-Para,1922 Publication issued for the Conferencia americana da lepra and to commemorate the centenary of independence
Path 30.0063

CONFERENCIA INTERNACIONAL DE PSICOTECNICA APPLICADA A L'ORIENTACIO PROFESSIONAL I A L'ORGANITZACIO CIENTIFICA DEL TREBALL 2 Actes Barcelona 1921 Sep 28-30 Conference internationale de psychotechnique appliquee a l'orientation professionelle et a l'organisation scientifique du travail. Secretariat Institut d'orientacio professional,Barcelona,1912 Under the presidency of E.Claparede
Psy 31.2608

CONFERENZA INTERNAZIONALE DI INFORMAZIONE VISIVA 1mo :modalita dell'informazione visiva: ricerca scientifica e azione politica 1-2 Istituto per lo studio sperimentale di problemi sociali con tecniche filmologiche 2 vols Istituto per lo studio sperimentale di problemi sociali con tecniche filmologiche, Milan,1961 Papers in English,French and Italian
Psy 31.2545

CONFIRMATION Aspects of inductive logic Edited by Jaakko Hintikka and Patrick Suppes Studies in logic and the foundations of mathematics vii,320p 23cm North-Holland, Amsterdam,1966 Includes papers read at the International symposium on confirmation and information,Helsinki,1965
P Math 2.2678

CONFIRMATION AND INFORMATION Aspects of inductive logic Edited by Jaakko Hintikka and Patrick Suppes Studies in logic and the foundations of mathematics vii,320p North-Holland,Amsterdam,1966 Eight of the fourteen papers based on those read at an international symposium on confirmation and information held at Helsinki,Sept.30-Oct.2, 1965
WSM 43.1205

CONFORMATION OF BIOPOLYMERS International symposium on conformation of biopolymers Papers read Madras 1967 Jan 18-21 Vol 1-2 University of Madras.Centre of advanced study in biophysics Edited by G.N. Ramachandran Sponsored by the International union of pure and applied physics 2 vols Academic press,London;New York,1967
Bioch 33.0331

CONFORMATIONAL ANALYSIS:SCOPE AND PRESENT LIMITATIONS International symposium on conformational analysis Papers Brussels 1969 Sep Edited by Gregoire Chiurdoglu Organic chemistry, 21
Chem 18.2788

CONGENITAL ANOMALIES OF THE FACE AND ASSOCIATED STRUCTURES :an international symposium Proceedings Gatlinburg,Tenn. 1959 Dec 6-9 Edited by Samuel Pruzansky Thomas, Springfield,Ill.,1961
An 32.2054

CONGENITAL HEART DISEASE :a symposium Washington,D.C. 1958 Dec 29-30 American association for the advancement of science Edited by Allan D. Bass and Gordon K. Moe American association for the advancement of science.Publications, 63 American association for the advancement of science, Washington,D.C.,1960
An 32.2063

CONGENITAL HEART DISEASE... :an international symposium Philadelphia,Pa. 1960 Apr Edited by Dryden P. Morse Sponsored by Deborah hospital Blackwell, Oxford,1962
An 32.2064

CONGENITAL MALFORMATION :a symposium 1960 Jan 19-21 Ciba foundation Edited by G. E.W. Wolstenholme and C.M. O'Connor Ciba foundation.Symposia Churchill,London,1960
PGMS 29.0216

CONGENITAL MALFORMATIONS International conference on congenital malformations 2nd Papers New York 1963 Jul 14-19 International medical congress Edited by Morris Fishbein International medical congress,New York,1864
An 32.2091

CONGENITAL MALFORMATIONS :Ciba foundation symposium London 1960 Jan 19-21 Ciba foundation Edited by G.E.W. Wolstenholme and Cecilia M. O'Connor Churchill,London,1960
An 32.2079

CONGENITAL MALFORMATIONS :a conference on teratology papers Bethesda,Md. 1957 Apr 15-16 Published under the auspices of the Association for the aid of crippled children Pediatrics, 23,no 1,pt 2,suppt. Thomas,Springfield,Ill.,1959
An 32.2046

CONGESTION THEORY Symposium on congestion theory Proceedings Chapel Hill,N.C. 1964 Aug 24-26 Edited by Walter L. Smith and William E. Wilkinson Held at the University of North Carolina University of North Carolina.Monograph series in probability and statistics, 2 University of North Carolina press,Chapel Hill,N.C.,1965
Eng 41.2150

CONGESTION THEORY Symposium on congestion theory proceedings Chapel Hill,N.C. 1964 Aug.24-26 University of North Carolina. Department of statistics Edited by Walter L. Smith and William E. Wilkinson Financially supported by the United States.Office of naval research North Carolina.University.Monograph series in probability and statistics, 2 xv, 457p 24cm University of North Carolina press,Chapel Hill,N.C.,1965
Math 3.0568

CONGRES DE LA SOCIETE EUROPEENNE D'OPTHALMOLOGIE 1e Resumees des rapports et des communications Athens 1959 Apr 18-22 European opthalmological society Excerpta medica.International congress series, 25 Excerpta medica,Amsterdam;New York,1960 Papers in English,French and German
PGMS 29.0682

CONGRES DE L'ALPINISME Comptes rendus Monaco 1920 May 1-10 Tomes 1-2 22cm 2 vols Paris,1921
Sco 14.1300

CONGRES DE PSYCHIATRIE ET DE NEUROLOGIE DE LANGUE FRANCAISE Les Agenesies du corps calleux : rapport de neurologie presente au Congres de psychiatrie et de neurologie de langue franciase Tours 1959 Jun 8-13 By Francis Rohmer Masson,Paris,1959
An 32.4178

CONGRES DES GEOGRAPHES ALLEMANDS notice Halle 1882 Apr 12-14 By Alexis de Tillo Halle,1882 One of fifteen tracts bound together
Philos 1.0875

CONGRES DES MATHEMATICIENS SCANDINAVES,1953 On the machine calculation of characters of the symmetric group :comptes rendus du douzieme Conges des Mathematiciens Scandinaves Lund 1953 Aug 10-15 By Stig Comet Swedish board for computing machinery, Stockholm,1954
Math L 5.1123

CONGRES FEDERATIF INTERNATIONAL D'ANATOMIE 1e
Geneva 1905 Aug 6-10 bulletin synthetique Kundig,Geneva,1907
An 32.5266

CONGRES GEOLOGIQUE INTERNATIONAL
International geological congress 13th comptes rendus Brussels 1922 Aug Fasc 1-3 3 vols Liege,1924-26
Geog 13.0200

CONGRES INTERNACIONAL DE CIENCIAS FISIOLOGICAS
see INTERNATIONAL CONGRESS OF PHYSIOLOGICAL SCIENCES

CONGRES INTERNATIONAL D'ANTHROPOLOGIE ET D'ARCHAEOLOGIE PREHISTORIQUES 13th Compte rendu Monaco 1906 Tome 1-2 2 vols Imprimerie de Monaco,Monaco,1907-08
An 32.2638

CONGRES INTERNATIONAL D'ANTHROPOLOGIE ET D'ARCHEOLOGIE PREHISTORIQUES 8th Compte rendu Budapest 1876 Vol 1 Franklin-Tarsulat,Budapest,1877
An 32.2636

CONGRES INTERNATIONAL D'ANTHROPOLOGIE ET D'ARCHEOLOGIE PREHISTORIQUES 9th Compte rendu Lisbon 1880 Academie royale des sciences,Lisbon,1884
An 32.2637

CONGRES INTERNATIONAL D'AVICULTURE ET DE PECHE
Memoires et comptes-rendus Paris 1900 France.Ministere de l'industrie,des postes et des telegraphes Edited by Joseph Perard and Maire 24cm Challamel,Paris,1901
Bal 44.1241

CONGRES INTERNATIONAL DE BIOCHEMIE, 2
International congress of biochemistry 2nd Proceedings Paris 1952 Jul 21-27 Vol 3: symposium sur le cycle tricarboxylique Council for international organizations of medical science Edited by Michel Polonovski Societe d'edition d'enseignement superieur, Paris,1952
Bot 42.1697

CONGRES INTERNATIONAL DE BOTANIQUE Actes
Paris 1900 Edited by Emile Perrot Lons-le-Saunier,1900 On the occasion of 'L'exposition universelle de 1900'
Bot 42.0648

CONGRES INTERNATIONAL DE CHIMIE INDUSTRIELLE
und der Dechema-hauptversammlung 1952 Vortrage Frankfurt am Main 1952 Dechema-monographie, 21,p245-268 Verlag chemie GMBH,Weinheim,1952
Chem E 24.1906

CONGRES INTERNATIONAL DE GASTRO-ENTEROLOGIE 7e
Brussels 1964 Jun 1-6 1-3 Association des societes nationales europeennes et mediterraneenes de gastro-enterologie 3 vols A.S.N.E.M.G.E.,Brussels, c1964
PGMS 29.0357

CONGRES INTERNATIONAL DE GEOGRAPHIE
International geographical congress Excursion guide Paris 1931 Excursion A2: le sud-est du Massif central Edited by Henri Baulig Colin,Paris,1931
Geog 13.0525

CONGRES INTERNATIONAL DE GEOGRAPHIE
International geographical congress Guidebooks Paris 1931 Colin,Paris,1931
Geog 13.4158

CONGRES INTERNATIONAL DE LA POPULATION
International congress for studies on population 4th Paris 1937 July 29-Aug 1 Under the auspices of l' Union internationale pour l'etude scientifique des problemes de la population 8 vols Hermann,Paris,1938
Geog 13.1494

CONGRES INTERNATIONAL DE L'INDUSTRIE MORUTIERE DANS L'ATLANTIQUE-NORD 1st :tradition et avenir Rouen 1966 Jan 27-29 Fecamp 1966 Jan 27-29 Tome 1 Fondation francaise d'etudes nordiques Fondation francaise d'etudes nordiques.Actes et documents,2 Rouen;Fecamp,1966
Sco 14.1037

CONGRES INTERNATIONAL DE MECANIQUE APPLIQUEE 9th
Actes Tome 1-4 4 vols Universite de Bruxelles,Brussels,1957
Chem E 24.0406

CONGRES INTERNATIONAL DE METEOROLOGIE ALPINE,4TH
La Meteorologie... Numero special consacre a la meteorologie Alpine 1957,Janvier-Juin Societe meteorologigne de France,Paris,1957 Ce numero contient les communications presentees au 1Ve. Congres international de meteorologie Alpine
Sco 14.0288

CONGRES INTERNATIONAL DE MICROSCOPIE ELECTRONIQUE 7e Resumes des communications Microscopie electronique 1970 Grenoble 1970 Aug 30-Sep 3 International federation of societies for electron microscopy Edited by Pierre Favard Societe francaise de microscopie electronique,Paris,1970
Met 25.2604

CONGRES INTERNATIONAL DE NEPHROLOGIE
International congress of nephrology 1st Proceedings Geneva 1969 Sep 1-4 Evian 1960 Sep 1-4 Edited by E. Richet S. Karger,Basle;New York,1961
Inv Med 37.0187

CONGRES INTERNATIONAL DE NEPHROLOGIE 2
Comptes rendus Prague 1963 Aug 26-30 Societe internationale de nephrologie Societe tchecoslovaque de medecine Edited by J. Vostal and G. Richet Excerpta medica. International congress series, 78 Excerpta medica,Amsterdam,1964
Inv Med 37.0220

CONGRES INTERNATIONAL DE PHILOSOPHIE DES SCIENCES
Actes Paris 1949 Oct 17-22 Tome 1- Institut international de philosophie Actualites scientifiques et industrielles,1126, 1134,1137,1146,1153,1155-56,1166-67- Hermann, Paris,1951-
WSM 43.0981

CONGRES INTERNATIONAL DE PHOTOGRAPHIE SCIENTIFIQUE ET APPLIQUEE 8e Comptes rendus Dresden 1931 Aug 3-7 Edited by L.P. Clerc Revue d'optique theorique et instrumentale,Paris,1933
Chem 18.0248

CONGRES INTERNATIONAL DE PSYCHOLOGIE PHYSIOLOGIQUE 1st :premiere session Compte-rendu Paris 1889 Societe de psychologie physiologique de Paris Bureau des revues,Paris,1890 Congresses continued as 'International congress of experimental psychology' q.v.
Psy 31.3390

CONGRES INTERNATIONAL DE STRATIGRAPHIE ET DE GEOLOGIE DU CARBONIFERE 5e compte rendu Paris 1963 Sep 9-12 Tom 1-3 3 vols Paris,1964 Text in English,French and German
Geol 8.2988

CONGRES INTERNATIONAL DES MATHEMATICIENS : les 265 communications individuelles Nice 1970 Sep vii,290p 24cm Gauthier-Villars, Paris,1970
P Math 2.3775

CONGRES INTERNATIONAL DES MATHEMATICIENS 1970 Actes Nice 1970 Sep 1-10 Vol 1-3 25cm 3 vols Gauthier-Villars,Paris,1971
P Math 2.3995

CONGRES INTERNATIONAL DES MICROBIOLOGISTES DE LANGUE FRANCAISE Role des anaerobies dans la nature Brussels 1949 May 23-27 International union of biological sciences International union of biological sciences. Series B.Colloques, 7 Secreteriat general de l'U.I.S.B.,Paris,1949
Bal 39.3919

CONGRES INTERNATIONAL D'HISTOIRE DES SCIENCES 7e Actes Jerusalem 1953 Aug 4-12 International union of history and philosophy of science Edited by F.S. Bodenheimer Academie internationale d'histoire des sciences.Collection de travaux, 8 xii,662p Academie internationale d'histoire des sciences;Hermann,Paris,c1954
WSM 43.0058

CONGRES INTERNATIONAL D'HISTOIRE DES SCIENCES 8e Actes Florence 1956 Sep 3-9 and Milan Vol 1-3 International union of the history and philosophy of science Academie internationale d'histoire des sciences.Collection de travaux, 9 3 vols Gruppe italiano di storia delle scienze; Hermann,Vinci (Florence);Paris,1958
WSM 43.0057

CONGRES INTERNATIONAL D'HISTOIRE DES SCIENCES 9e Actes Barcelona 1959 Sep 1-7 and Madrid Vol 1 Union internationale d'histoire des sciences Academie internationale d'histoire des sciences. Collection de travaux, 12 733p Asociacion para la historia de ciencia espanola;Hermann, Barcelona;Paris,1960
WSM 43.0061

CONGRES INTERNATIONAL D'HISTOIRE DES SCIENCES Actes 5e Lausanne 1947 Sep 30-Oct 6 Union internationale d'histoire des sciences Academie internationale d'histoire des sciences.Collections de travaux, 2 288p Academie internationale d'histoire des sciences;Hermann,Paris,1948
WSM 43.0060

CONGRES INTERNATIONAL D'HISTOIRE DES SCIENCES 11e Actes Warsaw 1965 Aug 24-31 Cracow 1-6 Polska akademia nauk.Comite et institut d'histoire de la science et de la technique 6 vols Ossolineum.Maison d'edition de l'Academie polonaise des sciences, Wroclaw,1968
WSM 43.0079

CONGRES INTERNATIONAL D'HISTOIRE DES SCIENCES ET CONGRES DE LA SOCIETE INTERNATIONALE D'HISTOIRE DE LA MEDECINE Actes Amsterdam 1950 Aug 14-21 Union internationale d'histoire des sciences Genootschap voor geschiedenis der geneeskunde, wiskunde en natuurweteenschappen te Leiden Academie internationale d'histoire des sciences.Collection de travaux, 6 2 vols Academie d'histoire des sciences;Hermann,Paris, 1951-53 The 12th congress of the Societe internationale d'histoire de la medecine is section 4 of the 6th Congres international d'histoire des sciences
WSM 43.0067

CONGRES INTERNATIONAL POUR L'ETUDE DES REGIONS POLAIRES Brussels 1906 Sep 7-11 programme 40p Hayez,Brussels,1906 In English,French and German
Sco 14.0108

CONGRES INTERNATIONAL POUR L'ETUDE DES REGIONS POLAIRES Rapport d'ensemble Brussels 1906 Sep 7-11 documents preliminaires et comte rendu des seances Hayez,Brussels,1906
Sco 14.0109

CONGRES INTERNATIONAL SUR L'OPTIQUE DES RAYONS X ET LA MICROANALYSE 4e Optique des rayons X et microanalyse Orsay 1965 Sep 7-10 Edited by Raymond Castaing Hermann, Paris,1966
Met 25.1460

CONGRES INTERNATIONALE D'ANTHROPOLOGIE et d'archeologie prehistorique et de zoologie Moscow 1892 Aug 10-18,22-30 1e-2e partie: materiaux...concernant les expositions,les excursions et les rapports sur des questions touchant les congres illus.,pl. Moscow, 1893 2 volumes bound together
Phys 20.0376

CONGRES INTERNATIONALE DE LA SOCIETE INTERNATIONALE DE TRANSFUSION SANGUINE 4th Rapports et communications Paris 1951 Jul 21-29 International society of blood transfusion Edited by J. Julliard and others L'expansion scientifique francaise,Paris,1952
Med 36.0389

CONGRES MONDIAL DES MEDECINS POUR L'ETUDE DES CONDITIONS ACTUELLES DE VIE 1953 Atomic bomb injuries Japanese preparatory committee for le Congres mondial des medicins pour l'etude de conditions actuelles de vie Edited by Nobuo Kusano Illus Tsukiji; Shokan,Tokyo,1953 Based on Dr.Kusano's report to the Congres mondial des medecins pour l'etude des conditions actuelles de vie, Vienna June 1953
Radioth 35.0728

CONGRES MONDIALE D'AVICULTURE,2ND World poultry congress 2nd Barcelona 1924 Part 2: informations and communications Arts graphiques,Barcelona,1924
Gen 34.1767

CONGRES NATIONAL D'ASTRONOMIE rapports Paris 1931 Jul. 21-23 Comite national francais d'astronomie 203p Paris,1932
Obs 6.2481

CONGRES NATIONAL DE L'ASSOCIATION FRANCAISE DE CALCUL 1st mimeograph Grenoble 1960 Sep 14-16 Association francaise de calcul Association francaise de calcul,1960
Math L 5.0869

CONGRES NATIONAL DES SCIENCES :section d'astronomie Comptes rendus Brussels 1930 Jun.29-Jul.2 Federation belge des societes scientifiques 103p Societe d'astronomie d'Anvers,Brussels,1930
Obs 6.2021

CONGRES POUR LAVANCEMENT DES ETUDES DE STRATIGRAPHIE CARBONIFERE 1er compte rendu Heerlen 1927 Jun 7-11 Edited by W.J. Jongmans Published by the geologisch-mijnbouwkundig genootschap voor Nederland en kolonien Liege,1928 Text in English, French and German
Geol 8.2984

CONGRES POUR LAVANCEMENT DES ETUDES DE STRATIGRAPHIE CARBONIFERE 2e compte rendu Heerlen 1935 Sep Tom 1-3 Edited by W. J. Jongmans 3 vols Maestricht,1937-38 Text in English,French and German
Geol 8.2985

CONGRES POUR LAVANCEMENT DES ETUDES DE STRATIGRAPHIE ET DE GEOLOGIE DU CARBONIFERE 3e compte rendu Heerlen 1951 Jun 25-31 Tom 1-2 Edited by Geologisch bureau Heerlen Maestricht,1952 Text in English,French and German
Geol 8.2986

CONGRES POUR LAVANCEMENT DES ETUDES DE STRATIGRAPHIE ET DE GEOLOGIE DU CARBONIFERE 4e compte rendu Heerlen 1958 Sep 15-20 Tom 1-3 3 vols Maestricht,1960-62 Text in English,French and German
Geol 8.2987

CONGRESO GEOLOGICO INTERNACIONAL see INTERNATIONAL GEOLOGICAL CONGRESS

CONGRESO INTERNACIONAL DE AMERICANISTAS 17o Actas Buenos Aires 1910 May 17-32 Edited by Robert Lehmann-Nitsche Hermanos, Buenos Aires,1912
An 32.2621

CONGRESO PANAMERICANO DE MECANICA DE SUELOS Y CIMENTACIONES 1st Proceedings Mexico,D.F. 1959 Sep 7-12 Vol 1-3 Sociedad mexicana de mecanica de suelos 3 vols Mexico,D.F.,1960
Eng 41.3178

CONGRESS FUR PSYCHOLOGIE.INTERNATIONALES ORGANISATIONS-COMITE International congress of psychology 3rd Munich 1896 Aug 4-7 Lehmann,Munich,1897 Papers in English,French,German and Italian
Psy 31.3391

CONGRESS FUR PSYCHOLOGIE.INTERNATIONALES ORGANISATIONS-COMITE Internationaler congress fur psychologie 3rd Munich 1896 Aug 4-7 Lehmann,Munich,1897 Papers in English,French,German and Italian
Psy 31.2518

CONGRESS G.A.M.S. 4th Paris 1951 June 20-22 Groupement pour l'advancement des methodes d'analyse spectrographique des produits metallurgiques 24cm Paris,1951
Philos 1.0678

CONGRESS INTERNATIONAL DE BOTANIQUE 3rd Actes Brussels 1910 Vol 2: conferences et memoires Edited by E.de Wildeman Boeck,Brussels,n.d.
Bot 42.0646

CONGRESS INTERNATIONAL POMOLOGIE Ghent 1862 Federation des societes d'horticulture de Belgique Federation des societes d'horticulture de Belgique.Bulletin, 5 Ghent,1865
Bot 42.0644

CONGRESS OF AMERICAN PHYSICANS AND SURGEONS 4th Transactions Washington,D.C. 1897 May 4-6 New Haven,Conn.,1897
An 32.1146

CONGRESS OF AMERICAN PHYSICIANS AND SURGEONS 6th triennial session Transactions Washington,D.C. 1903 May 12-14 New Haven, Conn.,1903
An 32.1147

The CONGRESS OF CELL BIOLOGY 8th Papers Fine structure of cells;symposium Leiden 1954 International union of biological sciences With financial support of Unesco International union of biological sciences.Ser. B, 21 Noordhoff,Groningen,1955
Bal 39.0283

CONGRESS OF CELL BIOLOGY,8TH Fine structure of cells :symposium held at the VIIIth congress of cell biology Leiden 1954 With financial support of Unesco International union of biological sciences. Series B, 21 Noordhoff,Groningen,1955
Phys 20.1483

CONGRESS OF CELL BIOLOGY,8TH Fine structure of cells :symposium held at the 8th congress of cell biology Leiden 1954 With the financial support of Unesco International union of biological sciences.Publications. Series B, 21 Noordhoff,Groningen,1955
An 32.3219

CONGRESS OF CELL BIOLOGY 8TH Fine structure of cells :a symposium held at the VIIIth congress of cell biology Papers Leiden 1954 International union of biological sciences With financial support of Unesco International union of biological sciences. Publications.Series B, 21 Noordhoff, Groningen,1955
Path 30.2410

CONGRESS OF CELL BIOLOGY 8TH Fine structure of cells :a symposium held at the 8th congress of cell biology Leiden 1954 International union of biological sciences With financial support of Unesco International union of biological sciences. Series B., 21 Noordhoff,Groningen,1955
Radioth 35.0467

CONGRESS OF SCANDINAVIAN MATHEMATICIANS 15th Proceedings Oslo 1968 Aug 12-16 Edited by K.E. Aubert and W. Ljunggren Lecture notes in mathematics, 118 Bibliog. 162p 25cm Springer-verlag,Berlin,1970
P Math 2.3240

CONGRESS OF SCANDINAVIAN NEUROBIOLOGISTS 16TH The So-called extrapyramidal system :symposium from the sixteenth congress of Scandinavian neurobiologists Oslo 1962 Scandinavian neurological society and Norwegian neurological association Edited by Sigvald Refsum and others Universitetsforlaget,Oslo, 1963
An 32.4655

CONGRESS OF THE EUROPEAN SOCIETY OF HAEMATOLOGY 7th Proceedings London 1959 2, pt 1: papers European society of haematology Edited by E. Neumark and others Karger,Basle, 1960
Med 36.0132

CONGRESS OF THE INTERNATIONAL ASSOCIATION FOR THE SCIENTIFIC STUDY OF MENTAL DEFIENCY 1st Proceedings Montpellier 1967 Sep 12-20 International association for the scientific study of mental deficiency Edited by B.W. Richards Jackson,Reigate,1968 Title page and papers in English and French
Psy 28.0269

CONGRESS OF THE INTERNATIONAL FEDERATION OF AUTOMATIC CONTROL 1st-3rd Proceedings Automatic and remote control 1960,1963,1966 International federation of automatic control Butterworths,London,1961-66
Eng 41.8207

CONGRESS OF THE INTERNATIONAL SOCIETY OF ROCK MECHANICS 1st Proceeedings Lisbon 1966 Sep 25-Oct 1 Vol 1-3 International society of rock mechanics 3 vols Laboratorio nacional de engenharia civil,Lisbon,1966-67
Eng 41.3176

CONGRESSO BOTANICO INTERNAZIONALE Atti Genoa 1892 Edited by Ottone Penzig Genoa, 1893
Bot 42.0649

CONGRESSO DE SCIENZIATI ITALIANI 7TH Vocabolario dei sinonimi classici dell'ornithologia Europea presentata in mss al settimo congresso dei scienziati italiani By Antonio Schembri 22cm Bologna,1846 From 'Annali delle scienze naturali'
Bal 44.2051

CONGRESSO DELL'UNIONE MATEMATICA ITALIANA 6to Atti Naples 1959 Sep 11-16 Unione matematica italiana Bibliog,illus 497p 24cm Edizioni Cremonese,Rome,1960
P Math 2.3008

CONGRESSO INTERNAZIONALE DE MICROBIOLOGIA 6e symposium Citlogia batterica:Bacterial cytology Edited by S. Mudd and G. Penso International congress of microbiology Paterno,Rome,1953
Bot 42.0666

CONGRESSO INTERNAZIONALE DE MICROBIOLOGIA 6e symposia Abstracts Rome 1953 Instituto superiore di Sanita,Rome,1953
Bot 42.0667

CONGRESSO INTERNAZIONALE DEI FISICI Atti Como,Pavia,Roma 1927 Sep 11-20 Vol 1-2 Nicola Zanichelli,Bologna,1928 Polyglot publication
Cav 7.2404

CONGRESSO INTERNAZIONALE DI PATOLOGIA 7 Atti Problemi attuali di scienza e di cultura Milan 1968 Sep 5-11 International congress of pathology Edited by Alfonso Giordano Accademia nazionale dei Lincei,Rome,1970
PGMS 29.0618

CONGRESSO NAZIONALE DELLA SOCIETA ITALIANA DI EMATOLOGIA 11th Atti Rome 1953 May 10-11 Societa italiana di ematologia Edited by L. Villa and others Edizioni mediche e scientifiche,Rome,1953
Med 36.0341

CONGRESSO NAZIONALE DELLA SOCIETA ITALIANA DI EMATOLOGIA 13th Atti Rome 1955 Oct 6-7 Societa italiana di ematologia Edited by G. Dominici and others Edizioni mediche e scientifiche,Rome,1956
Med 36.0380

CONGRESSO SCIENTIFICO ITALIANO 5th Atti Lucca 1843 Sep 15-Sep 29 Unione degli scienzati italiani 30cm Lucca,1844 Secretary general of conference:prof.Pacini
Bal 44.6331

CONIFER CONFERENCE Conifers in cultivation London 1931 Nov 10-12 Royal horticultural society Edited by F.J. Chittenden Royal horticultural society,London,1932
BG 38.2424

CONIFER CONFERENCE Report London 1891 Edited by W. Wilks and John Weathers Royal horticultural society.Journal, 14 Spottiswoode,London,1892
BG 38.2423

CONIFERS IN CULTIVATION Conifer conference London 1931 Nov 10-12 Royal horticultural society Edited by F.J. Chittenden Royal horticultural society,London,1932
BG 38.2424

CONJUGATED PROTEINS Some conjugated proteins; a symposium The Annual conference on protein metabolism 9th six lectures New Brunswick,N.J. 1953 Jan 30-31 Rutgers university.Bureau of biological research Edited by William H. Cole Rutgers university press,New Brunswick,N.J.,1953
Bioch 33.0543

CONNECTICUT.PUBLIC DOCUMENT, 47 Guide to the insects of Connecticut Pt 5: Odonata or dragonflies of Connecticut By Philip Garman Connecticut.State geological and natural history survey.Bulletin, 39 Hartford,Conn.,1927 Printed for the State geological and Natural history survey
Bal 39.2533

CONNECTIVE AND SKELETAL TISSUE Structure and functions of connective and skeletal tissue an advanced study institute Proceedings St. Andrews 1964 Jun 15-25 North Atlantic treaty organization.Scientific affairs division Butterworths,London,1964
Path 30.2421

CONNECTIVE TISSUE :a symposium Papers and discussions London 1956 Jul 22-26 Edited by R.E. Tunbridge and others Organized by the Council for international organizations of medical sciences Blackwell, Oxford,1957
Path 30.2490

CONNECTIVE TISSUE,THROMBOSIS AND ATHEROSCLEROSIS a conference Proceedings Princeton,N.J. 1958 May 12-14 Edited by Irvine H. Page Academic press,New York,1959
An 32.3232

CONNECTIVE TISSUES Conference on connective tissues 1st Transactions New York 1950 Apr 24-25 Josiah Macy jr.foundation Edited by Charles Ragan Josiaph Macy jr. foundation,New York,1951
Path 30.2487

CONNECTIVE TISSUES Conference on connective tissues 1st-5th Transactions New York 1950 Princeton,N.J. 1953-54 Edited by Charles Ragan Sponsored by the Josiah Macy jr.foundation Josiah Macy,jr.foundation,New York,1951-54
An 32.3218

CONSCIOUSNESS Brain mechanisms and consciousness :a symposium Ste.Marguerite, Que. 1953 Aug 23-28 Council for international organizations of medical sciences Blackwell,Oxford,1954
Phys 20.2009

CONSEIL DE CHIMIE 9e Rapports et discussions Proteines Brussels 1953 Apr 6-14 Institut international de chimie Solvay and Universite libre de Bruxelles Edited by R. Stoops Brussels,1953
Bioch 33.0522

CONSEIL DE CHIMIE 11eme Nucleoproteines Brussels 1959 Jun 1-6 Institut international de chimie Solvay Edited by R. Stoops Interscience;Stoops,New York;Brussels, 1960
Radioth 35.0100

CONSEIL DE PHYSIQUE 3eme Rapports Atomes et electrons Brussels 1921 Apr 1-6 Institut international de physique Solvay Solvay conference,1921,3 Paris,1923
Cav 7.2080

CONSEIL DE PHYSIQUE 5eme Rapports Electrons et photons Brussels 1927 Oct 24-27 Institut international de physique Solvay Solvay conference,1927,5 Paris,1928
Cav 7.2081

CONSEIL DE PHYSIQUE 7eme Rapports Structure et proprietes des noyaux atomiques Brussels 1933 Oct 22-29 Institut international de physique Solvay and Universite libre de Bruxelles Solvay conference,1933,7 Paris,1934
Cav 7.2083

CONSEIL DE PHYSIQUE 9 eme Rapports et discussions Etat solide Brussels 1951 Sep 25-29 Institut international de physique Solvay R.Stoops,Brussels,1952
Met 25.1135

CONSEIL DE PHYSIQUE 11e Rapports Structure et evolution de l'univers Brussels 1958 Jun 9-13 Universite libre de Bruxelles Institut de physique Solvay Edited by R. Stoops 309p Brussels,1958
Obs 6.3255

CONSEIL DE PHYSIQUE 11eme Rapports Structure et evolution de l'univers Brussels 1958 Jun 9-13 Institut international de physique Solvay and Universite libre de Bruxelles Edited by R. Stoops Solvay conference,1958,11 Brussels,1958
Cav 7.2084

CONSEIL DE PHYSIQUE 12eme Rapports Theorie quantique des champs Brussels 1961 Oct 9-14 Institut international de physique Solvay and Universite libre de Bruxelles Edited by R. Stoops Solvay conference,1961, 12 Interscience,New York,c 1961 13th Solvay conference issued as Conference on physics,q.v.
Cav 7.2077

CONSEIL LAPON INTER-NORDIQUE see NORDIC LAPP COUNCIL

CONSEIL PERMANENT INTERNATIONAL POUR L'EXPLORATION DE LA MER Rapports et proces-verbaux des reunions Vol 149: contributions to special IGY meeting 1959 TR-58 illus,maps 218p And.Fred Host, Copenhagen,1961 Meeting held in Copenhagen
Sco 14.5911

CONSEIL PERMANENT INTERNATIONAL POUR L'EXPLORATION DE LA MER Reports of the British delegates attending the international conferences held at Stockholm,Christiania,and Copenhagen with respect to fishery and hydrographical investigations in the North Sea. Great Britain.North Sea fishery investigations Cd 1313 32cm H.M.S.O., London,1903 Bound with continuations of above,covering conferences held from 1903 to 1909,in four parts:Cd 2966,3033,3165 and 5032
Bal 44.4999

CONSEIL PERMANENT INTERNATIONAL POUR L'EXPLORATION DE LA MER see also INTERNATIONAL COUNCIL FOR THE EXPLORATION OF THE SEA

CONSEIL POUR LA CO-ORDINATION DES CONGRES INTERNATIONAUX DES SCIENCES MEDICALES Le Muscle:etude de biologie et de pathologie : colloque Compte rendu Royaumont 1950 Aug 31-Sep 6 L'Expansion scientifique francaise, 1950 In French and English
Phys 20.1557

CONSEJO SUPERIOR DE INVESTIGACIONES CIENTIFICAS Coloquio sobre quimica fisica de processos en superficies solidas Trabajos Madrid 1964 Oct 19-25 Consejo superior de investigaciones cientificas,Madrid,1965 Conference celebrating the 25th anniversary of C.S.I.C.
Chem 18.0609

CONSEJO SUPERIOR DE INVESTIGACIONES CIENTIFICAS. INSTITUTO DE ELECTRICIDAD Y AUTOMATICA International automation congress proceedings Madrid 1958 Oct 13-18 illus. 411p 30cm Instituto de electricidad y automatica,Madrid,1958
Math L 5.0894

CONSEJO SUPERIOR DE INVESTIGACIONES CIENTIFICAS. INSTITUTO "JORGE JUAN" DE MATEMATICAS Coloquio international sobre geometria algebraica Actas Madrid 1965 Sep Edited by Pedro Abellanas Bibliog. 190p 24cm Madrid,1966 Papers mainly in English and Spanish
P Math 2.3137

CONSERVATION Colloque sur la protection et la conservation de la nature dans le Proche-orient :symposium on the protection and conservation of nature in the Near East Beirut 1954 Jun 3-8 Lebanese society of the friends of the trees Unesco-Middle East science cooperation office Beirut,1954 Text in English and French
Bal 39.1787

CONSERVATION OF WATER RESOURCES IN THE UNITED KINGDOM :a symposium Proceedings London 1962 Oct 30-Nov 1 Institution of civil engineers Institution of civil engineers,London,1963
Geog 13.0786

CONSIGLIO NAZIONALE DELLA RICERCHE Teoria dei gruppi finiti e applicazioni :convegno internazionale Florence 1960 Apr 11-13 Unione matematica Italiana Universita di Firenze vii,156p 24cm Edizioni Cremonese, Rome,1960
P Math 2.2858

CONSIGLIO NAZIONALE DELLE RICERCHE Histocompatibility testing 1967 Conference and workshop on histocompatibility testing 3r Report Turin 1967 Jun 14-24 St. Vincent Edited by E.S. Curtoni and others Munksgaard,Copenhagen,1967
Surg 23.0032

CONSIGLIO NAZIONALE DELLE RICERCHE Symposium on solar seeing 2nd Proceedings Rome 1961 Feb 20-25 International electronic and nuclear exhibition,Rome 158p Rome,1962
Obs 6.2051

CONSIGLIO NAZIONALE DELLE RICERCHE.CYBERNETICS GROUP International computing symposium Proceedings Venice 1972 Apr 12-14 illus 634p Venice,1972
Math L 5.3862

CONSIGLIO NAZIONALE DELLE RICHERCHE Differential games and related topics International summer school on mathematical models of action and reaction Proceedings Varenna 1970 Jun 15-27 Edited by H.W. Kuhn and G.P. Szego x,489p 23cm North-Holland, Amsterdam,1971
P Math 2.4574

CONSIGLIO NAZIONALE PER LA RICERCA Embriologia e genetica congress Naples 1948 Jun 1-318 Per iniziativa del Centre di studio di biologia,di citologia genetica e di enzimologia Stazione zoologica di Napoli. Publicazioni, 21 Naples,1949
Gen 34.0525

The CONSTRUCTION OF LARGE TELESCOPES :a symposium International astronomical union Edited by D.L. Crawford International astronomical union.Symposium, 27 Academic press,New York,1966
TA 15.0311

The CONSTRUCTION OF LARGE TELESCOPES :a symposium Tucson,Ariz. 1965 Apr 5-12 Pasadena,Calif. International astronomical union Edited by D.L. Crawford International astronomical union.Symposium, 27 234p Academic press,New York;London, 1966
Obs 6.3203

CONSTRUCTIVE ASPECTS OF THE FUNDAMENTAL THEOREM OF ALGEBRA :a symposium Proceedings Ruschlikon 1967 Jun 5-7 Edited by Bruno Dejon and Peter Henrici Bibliog.,illus. vii,337p 24cm Wiley,New York,1969
P Math 2.3485

CONSTRUCTIVE THEORY OF FUNCTIONS Conference on constructive theory of functions Proceedings Budapest 1969 Aug 24-Sep 3 Edited by G. Alexits and S.B. Stechkin 538p 25cm Akademiai kiado,Budapest,1972
P Math 2.4225

CONSTRUCTIVITY IN MATHEMATICS colloquium proceedings Amsterdam 1957 Aug 26-31 Edited by A. Heyting 297p 23cm Amsterdam, 1959
P. Math 2.0116

CONSULTANTS SYMPOSIUM 2nd Report Nature of small defect clusters Harwell 1966 Jul 4-6 Vol 1-2 United Kingdom atomic energy authority.Research group Edited by M.J. Makin 2 vols Atomic energy research establishment,Harwell,1966
Met 25.1243

CONTEMPORARY APPROACHES TO CREATIVE THINKING a symposium Boulder,Colo. 1958 Edited by Howard E. Gruber and others Held at the University of Colorado Atherton press; Prentice-Hall International,New York;London, 1964
Eng 41.1268

CONTEMPORARY GEODESY :a conference Proceedings Cambridge,Mass. 1958 Dec 1-2 Harvard college observatory and Smithsónian astrophysical observatory Edited by Charles A. Whitten and Kenneth H. Drummond Co-sponsored by the American geophysical union Geophysical monograph, 4 National research council.Publication, 708 American geophysical union,Washington,D.C.,1959
Geod 9.0028

CONTEMPORARY NEUROLOGY SYMPOSIA, 1 The Remote effects of cancer on the nervous system a symposium Proceedings Rochester N.Y. 1964 Sep 30-Oct 1 Edited by Russell Brain and Forbes H. Norris Sponsored by the University of Rochester school of medicine and dentistry. Division of neurology Grune and Stratton,New York;London,1965
PGMS 29.0451

CONTINENTAL DRIFT A Symposium on Continental drift papers London 1964 Mar 19-20 By P.M.S. Blackett and others Royal society of London Royal society of London.Philosophical transactions.Series A,no.258,no.1088 map Royal society of London,London,1965
Geol 8.1262

CONTINENTAL DRIFT A Symposium on continental drift London 1964 Mar Royal society Edited by P.M.S. Blackett and others Royal society.Philosophical transactions,1088 Royal society,London,1965
Geog 13.5647

CONTINENTAL DRIFT A Symposium on continental drift Papers London 1964 Mar 19-20 By P.M.S. Blackett and others Organised for the Royal society of London Royal society of London.Philosophical transactions.Series A,258, no.1088 vi,323p Royal society of London, London,1965
Sco 14.0260

CONTINENTAL DRIFT A Symposium on continental drift Papers London 1964 Mar 19-20 Royal society of London Royal society of London.Philosophical transactions.Series A, 258 London,1965 Symposium organized for the Royal society by P.M.S.Blackett,Sir Edward Bullard, and S.K.Runcorn
Geod 9.0129

CONTINENTAL DRIFT Continental drift,secular motion of the pole,and rotation of the earth : a symposium Proceedings Stresa 1967 Mar International astronomical union International union of geodesy and geophysics Edited by W. Markowitz and B. Guinot International astronomical union.Symposium, 32 108p Reidel,Dordrecht,1968
Obs 6.3268

CONTINENTAL DRIFT Stresa symposium on continental drift,secular motion of the pole and rotation of the earth Stresa,Italy 1967 Mar 21-25 International astronomical union and International union of geodesy and geophysics Edited by W. Markowitz and B. Guinot International astronomical union. Symposium, 32 Reidel,Dordrecht,1968
A Math 4.1308

CONTINENTAL DRIFT Theory of continental drift :a symposium on the origin and movement of land masses both intercontinental and intra-continental,as proposed by Alfred Wegener New York 1926 Nov.15 By W.A.J.M. Waterschoot van der Gracht American association of petroleum geologists American association of petroleum geologists,Tulsa,Oka., 1928
Geol 8.2130

CONTINENTAL DRIFT University of Tasmania. Geology department :symposium 2nd Papers Hobart 1956 Mar Edited by S.Warren Carey University of Tasmania,Hobart,1958
Min 10.0561

CONTINENTAL DRIFT :a symposium Papers Hobart,Tasmania 1956 Mar University of Tasmania Edited by S.Warren Carey University of Tasmania.Geology department. Symposia, 2 Hobart,1958
Geod 9.0083

CONTINENTAL DRIFT Papers Hobart 1956 Mar Edited by S.Warren Carey University of Tasmania.Geology department.Symposia,2 University of Tasmania,Hobart,1958
Geog 13.0367

CONTINENTAL DRIFT,SECULAR MOTION OF THE POLE,AND ROTATION OF THE EARTH Stresa 1967 Mar 21-25 International astronomical union and International union of geodesy and geophysics Edited by Wm. Markowitz and B. Guinot International astronomical union.Symposium, 32 Reidel,Dordrecht,1968
TA 15.0424

CONTINENTAL DRIFT,SECULAR MOTION OF THE POLE,AND ROTATION OF THE EARTH :a symposium Proceedings Stresa 1967 Mar International astronomical union International union of geodesy and geophysics Edited by W. Markowitz and B. Guinot International astronomical union.Symposium, 32 108p Reidel,Dordrecht,1968
Obs 6.3268

CONTINENTAL DRIFT,SECULAR MOTION OF THE POLE AND ROTATION OF THE EARTH I.A.U. symposium 32nd Papers Stresa 1967 Mar 21-25 International astronomical union and International union of geodesy and geophysics Edited by William Markowitz and B. Guinot illus 107p 25cm D.Riedel,Dordrecht,1968
Sco 14.8118

CONTINUOUS CASTING technical sessions Proceedings Detroit 1961 Oct 24 American institute of mining,metallurgical and petroleum engineers.Committee on physical chemistry of steelmaking Edited by D.L. McBride and T.E. Dancy Interscience,New York; London,1962
Met 25.0664

CONTINUOUS CASTING OF STEEL Iron and steel institute :autumn general meeting Proceedings London 1964 Nov 25-26 Iron and steel institute Iron and steel institute. Special report, 89 London,1965
Met 25.2569

CONTINUOUS PROCESSING AND PROCESS CONTROL :a symposium Proceedings Philadelphia,Penn. 1966 Dec 5-8 American institute of mining, metallurgical and petroleum engineers Edited by Thomas R. Ingraham Metallurgical society conferences, 49 Gordon and Breach,New York, 1968 Symposium sponsored by the Electric furnace conference,the Mechanical working and steel processing conference and the Physical chemistry of steelmaking conference of the Metallurgical society
Met 25.2846

CONTRIBUTIONS TO DYNAMICS AND THEORY OF CURRENTS dissertation By M.O. Taha Cambridge, 1967
A Math 4.0800

CONTRIBUTIONS TO ERGODIC THEORY AND PROBABILITY Midwestern conference on ergodic theory 1st Proceedings Columbus,Ohio 1970 Mar 27-30 Held at Ohio state university Lecture notes in mathematics, 160 bibliog. vii,278p 25cm Springer,Berlin,1970
P Math 2.3939

CONTRIBUTIONS TO ERGODIC THEORY AND PROBABILITY Midwestern conference on ergodic theory and probability 1st Proceedings Columbus, Ohio 1970 Mar 27-30 Held at the Ohio state university Lecture notes in mathematics, 160 vii,278p 26cm Springer,Berlin,1970
Math S 3.1779

CONTRIBUTIONS TO EXTENSION THEORY :a symposium Proceedings Berlin 1967 Aug 14-19 Edited by Jungen Flaschmeyer 279p 24cm VEB Deutscher verlag der wissenschaften,Berlin, 1969
P Math 2.3743

CONTRIBUTIONS TO FUNCTION THEORY international colloquium papers presented Bombay 1960 Jan 12-19 Tata institute of fundamental research In joint sponsorship with the Sir Dorabji Tata trust xii,231p 24cm Tata institute of fundamental research,Bombay,1960
P. Math 2.1584

CONTRIBUTIONS TO MARINE BIOLOGY :lectures and symposia given at the Hopkins marine station...at The midwinter meeting of the Western society of naturalists Pacific Grove,Calif. 1929 Dec 20-21 Western society of naturalists Stanford university press,Stanford,Calif.,1930
Bal 39.1933

CONTRIBUTIONS TO MATHEMATICAL LOGIC Logic colloquium Proceedings Hanover 1966 Aug Edited by H.Arnold Schmidt and others Studies in logic and the foundations of mathematics Bibliog. ix,298p 23cm North Holland,Amsterdam,1968
P Math 2.3312

CONTRIBUTIONS TO NON-STANDARD ANALYSIS :a symposium Papers Oberwolfach 1970 Jul 19-25 Edited by W.A.J. Luxemburg and A. Robinson Studies in logic and the foundations of mathematics, 69 vi,289p 23cm North-Holland,Amsterdam;London,1972
P Math 2.4060

CONTRIBUTIONS TO THE THEORY OF PARTIAL DIFFERENTIAL EQUATIONS Conference on partial differential equations typescript Harriman,N.Y. 1952 Oct By S. Bergman and others Edited by Lipman Bers and others Sponsored by National research council Annals of mathematics studies, 33 Bibliog 257p 26cm Princeton university press, Princeton,N.J.,1954
P. Math 2.1228

CONTRIBUTIONS TO THE THEORY OF RIEMANN SURFACES conference typescript Princeton,N.J. 1951 Dec 14-15 Institute for advanced study Edited by Lars V. Ahlfors and others In joint sponsorship with Princeton university Annals of mathematics studies, 30 264p 26cm Princeton university press,Princeton,N. J.,1953 Centennial celebration of Riemann's dissertation
P. Math 2.1434

CONTROL Automatic and manual control Conference on automatic control papers Cranfield 1951 Jul 18-21 Great Britain. Department of scientific and industrial research xi,584p 25cm Butterworth,London, 1951
Math L 5.0021

CONTROL Mathematical theory of control :a conference Proceedings Los Angeles,Calif. 1967 Jan 30-Feb 1 Edited by A.V. Balakrishnan and Lucien W. Neustadt Academic press,New York;London,1967
Eng 41.5763

CONTROL AND INNERVATION OF SKELETAL MUSCLE : a symposium Dundee 1965 Sep Edited by B.L. Andrew University of St.Andrews,St. Andrews,1966
An 32.4009

CONTROL AND INNERVATION OF SKELETAL MUSCLE : a symposium Dundee 1965 Sep Edited by B.L. Andrew University of St Andrews,St. Andrews,1966
Phys 20.1560

CONTROL IN ELECTROPLATING :a one-day symposium London 1958 Nov 19 Institute of metal finishing.London branch Draper,Teddington,1959
Met 25.2670

CONTROL MECHANISM IN RESPIRATION AND FERMENTATION a symposium held during the meeting of the Society of general physiologists at the Marine biological laboratory Woods Hole,Mass. 1961 Sep By Harlyn O. Halvorson Society of general physiologists Edited by Barbara Wright Published for the American physiological society Ronald press,New York, 1963
Bot 42.1730

CONTROL MECHANISMS IN CELLULAR PROCESSES :a symposium Woods Hole,Mass 1960 By Sigmund R. Suskind Edited by David M. Bonner Under the auspices of the Society of general physiologists Ronald press,New York,1961 Held at the 7th annual meeting of the society
Bot 42.1492

CONTROL MECHANISMS IN DEVELOPMENTAL PROCESSES a symposium La Jolla,Calif 1967 Jun Society for developmental biology Edited by Michael Locke Society for developmental biology.Symposia, 26 Developmental biology. Supplement, 1 Academic press,New York,1967
Bioch 33.0911

CONTROL MECHANISMS IN DEVELOPMENTAL PROCESSES a symposium La Jolla,Calif. 1967 Jun Edited by Michael Locke Society for developmental biology.Symposia, 26,supp.1 Academic press,New York,1967 Formerly the Society for the study of development and growth
An 32.2535

CONTROL MECHANISMS IN DEVELOPMENTAL PROCESSES : a symposium Proceedings Society for developmental biology Edited by Michael Locke Society for developmental biology. Symposia, 26 Developmental biology. Supplement, 1 Academic press,New York; London,1967
Gen 34.2153

CONTROL OF CELL DIVISION AND THE INDUCTION OF CANCER International symposium on the control of cell division and the induction of cancer Proceedings Lima 1963 Jul 1-6 and Cali,Columbia 1963 Jul 1-6 National cancer institute National cancer institute. Monograph, 14 illus National cancer institute,Bethesda,Md.,1964 Introduction, abstracts of each paper and summary in English and Spanish
Bioch 33.0985

The CONTROL OF CELL DIVISION AND THE INDUCTION OF CANCER :international symposium Proceedings Lima 1963 Jul 1-6 and Cali,Columbia 1963 Edited by C.C. Congdon and Pablo Mori-Chavez National cancer institute.Monographs, 14 Illus U.S. department of health,education and welfare; National cancer institute,Bethesda,Md.,1964 In Spanish and English
Radioth 35.0532

CONTROL OF CELLULAR GROWTH IN ADULT ORGANISMS A Sigrid Juselius foundation symposium Helsinki 1965 Oct 4-6 Sigrid Juselius foundation Edited by Harald Teir and Tapio Rytomaa Academic press,London;New York,1967
Bioch 33.1007

CONTROL OF CELLULAR GROWTH IN ADULT ORGANISMS Helsinki 1965 Oct 4-6 Edited by Harald Teir and Tapio Rytomaa Academic press,London; New York,1967
Pha 16.0177

CONTROL OF CELLULAR GROWTH IN ADULT ORGANISMS a Sigrid Juselius foundation symposium Helsinki 1965 Oct 4-6 Edited by Harald Tier and Tapio Rytomaa Academic press,London; New York,1967
Phys 20.2232

CONTROL OF COMPOSITION IN STEELMAKING :annual general meeting Papers and discussions London 1966 May 4-5 Iron and steel institute Iron and steel institute.Special publication, 99 London,1967
Met 25.2557

CONTROL OF ENERGY METABOLISM A Colloquium on metabolic control and a Symposium on control of energy metabolism Proceedings Philadelphia 1965 May 20-21 Johnson research foundation Edited by Britton Chance and others Johnson research foundation. Colloquia Academic press,New York;London, 1965 "In celebration of the bicentennial of the University of Pennsylvania school of medicine"
Bioch 33.1077

CONTROL OF ENERGY METABOLISM Colloquium on metabolic control :and symposium on control of metabolism Philadelphia,Pa. 1965 May 20-21 Johnson research foundation Edited by Britton Chance and others Academic press,New York;London,1965
Gen 34.0614

CONTROL OF GLYCOGEN METABOLISM Federation of European biochemical societies meeting 4th Proceedings Oslo 1967 Jul 7 Federation of European biochemical societies Edited by W.J. Whelan Organized by the Norwegian biochemical society Universitetsforlaget; Academic press,Oslo;London,1968 Symposium organizer W.J. Whelan
Bioch 33.1400

CONTROL OF GLYCOGEN METABOLISM :Ciba foundation symposium Ciba foundation Edited by W.J. Whelan and M.P. Cameron Churchill,London,1966
Phys 20.1339

CONTROL OF GLYCOGEN METABOLISM :a Ciba foundation symposium Proceedings London 1963 Jul 23-25 Ciba foundation Edited by W. J. Whelan and Margaret P. Cameron illus Churchill,London,1964
Bioch 33.0576

The CONTROL OF LIPID METABOLISM :a symposium Oxford 1963 Jul 19 Biochemical society Edited by J.K. Grant Biochemical society. Symposia, 24 Academic press,London;New York, 1963 Organized by G.Popjak
Bioch 33.1386

CONTROL OF LIVESTOCK INSECT PESTS BY THE STERILE-MALE TECHNIQUE Panel on the control of livestock insect pests by the sterile-male technique Proceedings Vienna 1967 Jan 23-27 Organized bythe Joint FAO-IAEA division of atomic energy in food and agriculture International atomic energy agency.Panel proceedings series International atomic energy agency,Vienna,1968
Radioth 35.0263

The CONTROL OF NOISE :a conference Proceedings Teddington 1961 Jun 26-28 National physical laboratory National physical laboratory.Symposia, 12 H.M.S.O., London,1962
Eng 41.4218

The CONTROL OF NUCLEAR ACTIVITY :a symposium held under the auspices of the Society of general physiologists at its annual meeting at the Marine biological laboratory Woods Hole,Mass. 1966 Aug 31-Sep 3 Society of general physiologists Edited by Lester Goldstein Prentice-Hall,Englewood Cliffs,N.J. 1967
Bioch 33.1013

The CONTROL OF NUCLEAR ACTIVITY :a symposium Papers Woods Hole,Mass. 1966 Aug 31-Sep 3 Edited by Lester Goldstein Held under the auspices of the Society of general physiologists Prentice-Hall,Englewood Cliffs, N.J.,1967 Original title of the symposium 'Cytoplasmic and environmental influences on nuclear behavior'
Bot 42.4736

CONTROL OF ORGANELLE DEVELOPMENT :a symposium London 1969 Sep Society for experimental biology Society for experimental biology.Symposia, 24 Cambridge university press,Cambridge,1970
Bioch 33.2340

CONTROL OF ORGANELLE DEVELOPMENT :a symposium Proceedings London 1969 Sep 8-12 Society for experimental biology Edited by P. L. Miller Society for experimental biology. Symposia, 24 Cambridge university press, Cambridge,1970
Gen 34.2205

CONTROL OF OVULATION :a conference Proceedings Dedham,Mass. 1960 Feb 26-28 Edited by Claude A. Villee Sponsored by the Association for the aid of crippled children Pergamon press,Oxford,1961
An 32.3823

The CONTROL OF QUALITY IN THE PRODUCTION OF WROUGHT NON-FERROUS METALS AND ALLOYS :a symposium London 1954 Apr 28 2: the control of quality in working operations Institute of metals Institute of metals. Monograph and report series, 16 Institute of metals,London,1954
Met 25.0711

The CONTROL OF QUALITY IN THE PRODUCTION OF WROUGHT NON-FERROUS METALS AND ALLOYS a symposium held... on the occasion of the annual general meeting of the Institute of metals London 1955 Mar 31 Institute of metals Institute of metals.Monograph and report series, 17 Institute of metals, London,1955
Met 25.0504

CONTROL OF THE PLANT ENVIRONMENT Easter school in agricultural science 4th Proceedings Nottingham 1957 University of Nottingham Edited by J.P. Hudson Butterworths,London,1957
BG 38.0605

CONTROL OF THE PLANT ENVIRONMENT Easter school in agricultural science 4th Proceedings Nottingham 1957 University of Nottingham Edited by J.P. Hudson illus. xvi,240p Butterworths,London,1957
Bot 42.1943

CONTROL PROCESSES IN MULTICELLULAR ORGANISMS a Ciba foundation symposium New Delhi 1969 Mar 17-21 Ciba foundation Edited by G. E.W. Wolstenholme and Julie Knight Churchill, London,1970
An 32.5436

CONTROL PROCESSES IN MULTICELLULAR ORGANISMS a Ciba foundation symposium New Delhi 1969 Mar 17-21 Ciba foundation Edited by G. E.W. Wolstenholme and Julie Knight Churchill, London,1970
Phys 20.2271

CONTROL SYSTEM DESIGN Multivariable control system design and application U.K.A.C. control convention 4th Manchester 1971 Sep 1-3 United Kingdom automation council Organized by the Institution of electrical engineers Institution of electrical engineers.Conference publication, 78 IEE,London,1971
Eng 41.8590

CONTROL SYSTEMS I.F.A.C.symposium on the theory of self-adaptive control systems 2nd Proceedings Teddington 1965 Sep 14-17 International federation of automatic control Edited by P.H. Hammond Plenum press,New York, 1966
Eng 41.5888

CONTROL THEORY Sensitivity methods in control theory International symposium on sensitivity analysis Dubrovnik 1964 Aug 31-Sep 5 Yugoslav committee for electronics and automation Edited by L. Radanovic Sponsored by the International federation of automatic control.Theory committee Pergamon press,Oxford,1965
Eng 41.5835

CONTROL THEORY AND THE CALCULUS OF VARIATIONS based on the lectures presented at the workshop on calculus of variations and control theory Edited by A.V. Balakrishnan Academic press,New York;London,1969
Math S 3.1515

CONTROLLED ATMOSPHERES :the symposium... presented before the annual convention Philadelphia,Pa. 1941 Oct.20-24 American society for metals A.S.M.,Cleveland,Ohio, c1941
Met 25.0250

CONTROLLED CLINICAL TRIALS :a conference Papers Vienna 1959 Mar 23-27 Convened by the Council for international organizations of medical sciences Blackwell,Oxford,1960
Radioth 35.0799

CONTROLLED CLINICAL TRIALS :a conference Papers Vienna 1959 Mar 23-27 Council for international organizations of medical sciences Blackwell,Oxford,1960 Organized under the direction of A. Bradford Hill
HE 27.0076

CONTROLS OF METAMORPHISM Liverpool 1964 Jan Edited by Wallace S. Pitcher and Glenys W. Flinn Under the auspices of the Liverpool geological society Liverpool and Manchester geological journal,special issue,1965 Oliver and Boyd,Edinburgh;London,1965
Geol 8.3168

CONVECTION Heat and mass transfer by combined forced and natural convection :a symposium Proceedings London 1971 Sep 15 Institution of mechanical engineers. Thermodynamics and fluid mechanics group IME, London,1972
Eng 41.8647

CONVEGNO SULLA COSMOLOGIA;MEETING ON COSMOLOGY Atti;Proceedings Padua 1964 Sep 14-16 and Venice Quatro centenario della nascita di Galileo Galilei 1564-1964.Comitato nazionale per le manifestazioni celebrative Edited by L. Rosino Organized by Universita di Padua Quatro centenario della nascita di Galileo Galilei.Comitato nazionale per le manifestazioni celebrative.Pubblicazioni, 2, Atti dei convegni, 3 206p Florence,1966
Obs 6.3451

CONVEGNO SULLA RELATIVITA:PROBLEMI DI ENERGIA E ONDE GRAVITAZIONALI;MEETING ON GENERAL RELATIVITY:PROBLEMS OF ENERGY AND GRAVITATIONAL WAVES Atti;proceedings Florence 1964 Sep 9-12 Quatro centenario della nascita di Galileo Galilei,1564-1964. Comitato nazionale per le manifestazioni celebrative Edited by Giorgio Sestini Quatro centenario della nascita di Galileo Galilei,1564-1964.Comitato nazionale per le manifestazioni celebrative.Pubblicazioni, 2, Atti dei convegni, 1 267p Florence,1966
Obs 6.3276

CONVEGNO SULLE RIMOBILIZZAZIONE DEL MINERALI METALLICI E NON METALLICI Rendiconti minerali metallici e non metallici Remobilization of ores and minerals Cagliari 1969 Aug Associazzione mineralia sarda Istituto di giacimenti minerali universitaria Cagliari,c1969
Min 10.1489

CONVENTION ON DIGITAL-COMPUTER TECHNIQUES proceedings London 1956 Apr 9-14 Institution of electrical engineers Edited by W.K. Brasher Institution of electrical engineers.Proceedings, 103,Part B,Suppts.1-3 illus 342p 27cm Institution of electrical engineers,London,1956
Math L 5.0762

CONVENTION ON ELECTRIC TRACTION BY SINGLE-PHASE CURRENT AT INDUSTRIAL FREQUENCY Documents Lille 1955 May 11-14 Lille,1955
Eng 41.4842

CONVEXITY Colloquium on convexity Proceedings Copenhagen 1965 Aug 12-19 Sponsored under the Nato advanced study institute programme Bibliog,illus xii,325p 25cm Kobenhavens universitets matematiske inst.,Copenhagen,1967 In English,French and German
P Math 2.3016

CONVEXITY symposium Seattle,Wash. 1961 Jun 13-15 American mathematical society Edited by Victor L. Klee American mathematical society.Proceedings of symposia in pure mathematics, 7 xv,516p 26cm American mathematical society,Providence, R.I., 1963
P. Math 2.0487

CONVOCATION ON THE SOCIAL IMPLICATIONS OF SCIENTIFIC PROGRESS Verbatim account Mid-century Cambridge,Mass. 1949 Mar 31-Apr 2 Massachusetts institute of technology Edited by John Ely Burchard xx,549p Technology press of the Massachusetts institute of technology,Cambridge,Mass.,1950
Cav 7.0406

COOK James Captain Cook,navigator and scientist Cook bicentenary symposium Papers Canberra 1969 May 1 Australian academy of science Edited by G.M. Badger ix,143p Hurst;Humanities press,London;New York,1970
WSM 43.1245

COOK BICENTENARY SYMPOSIUM Papers Captain Cook,navigator and scientist Canberra 1969 May 1 Australian academy of science Edited by G.M. Badger ix,143p Hurst;Humanities press,London;New York,1970
WSM 43.1245

COOLING TOWERS Natural draught cooling towers - Ferrybridge and after :a conference Proceedings London 1967 Jun 12 Institution of civil engineers Institution of civil engineers,London,1967
Eng 41.2889

COOPERATIVE SUMMER SEMINAR 1st Metric geometry over affine spaces Ithaca,N.Y. 1964 By Ernst Snapper Mathematical association of America Held at Cornell university bibliog. iv,166p 28cm Mathematical association of America,Buffalo,N.Y.,1964 Duplicated notes
P Math 2.3829

COOPERSTOWN,N.Y. 1958 Growth and perfection of crystals International conference on crystal growth Proceedings General electric company Edited by R.H. Doremus and others Co-sponsored by United States.Air force.Office of scientific research Wiley;Chapman and Hall,New York;London,1958
Eng 41.4438

COOPERSTOWN,N.Y. 1960 The Fermi surface : an international conference Proceedings General electric research laboratory Edited by W.A. Harrison and M.B. Webb Sponsored by the United States.Air force.Office of scientific research John Wiley,New York,1960
Cav 7.0541

COOPERSTOWN,N.Y. 1960 The Fermi surface : an international conference Proceedings United States.Air force.Office of scientific research,and,General electric research laboratory J.Wiley,New York;London,1960
Chem 18.2382

COOPERSTOWN,NEW YORK 1958 Growth and perfection of crystals International conference on crystal growth Proceedings United States.Air research and development command and General electric research laboratory Wiley;Chapman and Hall,New York;London,1958
Met 25.1507

COORDINATED LIGANDS Reactions of coordinated ligands and homogeneous catalysis :a symposium Papers Washington,D.C. 1962 Mar 22-24 American chemical society.Division of inorganic chemistry American chemical society.Advances in chemistry series,37 American chemical society,Washington,D.C.,1963 Symposium chairman:D.H.Busch
Chem 18.1010

COORDINATION CHEMISTRY International conference on coordination chemistry Advances in the chemistry of coordination compounds 6th Proceedings Detroit,Mich. 1961 Aug 27-Sep 1 American chemical society,and,Wayne state university Edited by Stanley Kirschner Macmillan,New York,1961 Co-sponsored by the United States air force,and the International union of pure and applied physics
Chem 18.0991

COORDINATION COMPOUNDS International conference on coordination compounds Proceedings Amsterdam 1955 Apr 28-May 3 Nederlandse chemische vereniging Royal Netherlands chemical society,n.p.,c1955
Chem 18.0998

COPENHAGEN 1928 Reunion geologique internationale compte-rendu Edited by V. Nordmann Copenhagen,1930 Sponsored by Danmarks geologiske undersogelse in celebration of its 40th anniversary
Geol 8.2917

COPENHAGEN 1929 Comite meterologique internationale :seances de la conference internationale des directeurs...et de diverses commissions 7th Proces-verbaux Comite meteorologique international,3 401p 24cm Kemink,Utrecht,1930 In French and English
Sco 14.0112

COPENHAGEN 1932 International congress of psychology 10th Report Levin and Munksgaard,Copenhagen,1935 Under the presidency of professor Rubin
Psy 31.2537

COPENHAGEN 1933 Rapport de la commission internationale de l'annee polaire 1932-33 Vol 3: Compte-rendu des travaux de la commission Octobre 1931-Mai 1933 Organisation meteorologique internationale Ijdo,Leyden,1934
Sco 14.0491

COPENHAGEN 1947 International congress of microbiology 4th Report of proceedings Edited by Mogens Bjorneboe Rosenkilde and Bagger,Copenhagen,1949
Bot 42.0663

COPENHAGEN 1950 International physiological congress 18th abstracts of communications 1950
Phys 20.2086

COPENHAGEN 1953 International congress of radiology 7th Invited papers Acta radiologica.Supplementum, 116 Stockholm,1954
Radioth 35.1311

COPENHAGEN 1953 Zoological nomenclature Copenhagen decisions on zoological nomenclature additions to...les regles internationales de la nomenclature zoologique International trust for zoological nomenclature Edited by Francis Hemming International trust for zoological nomenclature,London,1953
BG 38.3107

COPENHAGEN 1955 International congress on the application of the theory of probability in telephone engineering and administration 1st proceedings Kobnhavns telefon aktieselskab Edited by Arne Jensen 130p 30cm Copenhagen,1957 Published simultaneously as vol.1,1957,no.1 of "Teleteknik"
Math 3.0881

COPENHAGEN 1956 Effect of radiation on human heredity :a study group Report World health organization World health organization,Geneva,1957
Radioth 35.1055

COPENHAGEN 1956 Effect of radiation on human heredity report of a study group World health organization.Study group World health organization,Geneva,1957
Gen 34.1106

COPENHAGEN 1956 International congress of Americanists 32nd Proceedings Published with the support of the International council of philosophy and human sciences illus,maps 743p 26cm Munksgaard,Copenhagen,1958
Sco 14.2361

COPENHAGEN 1956 International congress of human genetics 1st Proceedings Edited by Tage Kemp and others Karger,Basle,1957
An 32.0355

COPENHAGEN 1957 European society of haematology.Congress 6th Transactions Vol 1: principal papers European society of haematology Edited by Aage Videbaek and others S.Karger,Basle;New York,1958
Med 36.0128

COPENHAGEN 1957 European society of haematology congress 6th Transactions Vol 2: short papers - communications - kurzreferate European society of haematology Edited by Aage Videbaek and others 3 vols Karger,Basle;New York,1958
Med 36.0129

COPENHAGEN 1957 European society of haematology congress 6th Transactions Vol 3: panel discussion on "Efficiency and limitations of anticoagulant therapy in arterial thrombosis European society of haematology Edited by T. Astrup Karger, Basle;New York,1958
Med 36.0130

COPENHAGEN 1958 Advanced aero engine testing :joint meeting of the Agard combustion and propulsion and wind tunnel and model testing panels Papers Agard Edited by A. W. Morley and Jean Fabri AGARDograph, 37 Pergamon press,London,1959
Eng 41.7369

COPENHAGEN 1958 Advanced aero engine testing :joint meeting of the Agard combustion and propulsion and wind tunnel and model testing panels Papers Agard Edited by A. W. Morley and Jean Fabri AGARDograph, 37 Pergamon press,London,1959
Eng 41.7377

COPENHAGEN 1960 Finsen memorial congress International congress on photobiology 3rd Proceedings Edited by B.Chr. Christensen and B. Buchmann illus. Elsevier,Amsterdam,1961
Bot 42.1347

COPENHAGEN 1960 International geological congress 21st Norden-Denmark,Finland, Iceland,Norway,Sweden,Volume of abstracts Edited by Theodor Sorgenfrei 252p 24cm Copenhagen,1960
Sco 14.0134

COPENHAGEN 1960 International geological congress 21st reports 1-22 Edited by Theodor Sorgenfrei Copenhagen,1960
Geol 8.3123

COPENHAGEN 1960 Physical geography of Greenland International geographical congress 19th Edited by Niels Kingo Jacobsen Arranged at the University of Copenhagen.Geographical institute Folia geographica Danica, 9 illus,maps 234p C.A.Reitzel,Copenhagen,1961
Sco 14.8357

COPENHAGEN 1961 Child and education International congress of applied psychology 14th Proceedings Vol 3 International association of applied psychology and Danish psychological association Edited by Gerhard S. Nielsen Munksgaard,Copenhagen,1962
Psy 31.1198

COPENHAGEN 1961 International congress of applied psychology 14th Proceedings Vol 2: personality research International association of applied psychology and Danish psychological association Edited by Stanley Coopersmith Munksgaard,Copenhagen, 1952
Psy 31.1968

COPENHAGEN 1961 International congress of applied psychology 14th Proceedings Vol 5: industrial and business psychology International association of applied psychology and Danish psychological association Edited by Gerhard S. Nielsen Munksgaard,Copenhagen,1962
Psy 31.2252

COPENHAGEN 1963 Lattice dynamics The International conference on lattice dynamics Proceedings International union of pure and applied physics and University of Pennsylvania Edited by R.F. Wallis With the support of the International atomic energy agency 730p Pergamon press,Oxford,1965
Cav 7.0542

COPENHAGEN 1963 Nordisk meteorologmode 3 Foredragenes resumeer Danske meteorologiske institut.Meddeleser,17 illus,maps xi,87p Charlottenlund,1964 Mimeographed
Sco 14.0141

COPENHAGEN 1965 Colloquium on convexity Proceedings Sponsored under the Nato advanced study institute programme Bibliog, illus xii,325p 25cm Kobenhavens universitets matematiske inst.,Copenhagen,1967 In English,French and German
P Math 2.3016

COPENHAGEN 1966 Conference on statistical mechanics and thermodynamics.Foundations and applications International union of pure and applied physics :meeting Proceedings Edited by Thor A. Bak table 582p 25cm Benjamin,New York,1967
Cav 7.3022

COPENHAGEN 1966 The European congress of anaesthesiology 2nd Proceedings 1-3 World federation of societies of anaesthesiologists Acta anaesthesiologica Scandinavica.Supplementum, 23-25 3 vols Universitetsforlaget,Aarhus,1966
PGMS 29.0031

COPENHAGEN 1967 Theory of thin shells :a symposium 2nd Proceedings International union of theoretical and applied mechanics Edited by F.I. Niordson Springer,Berlin,1969
Eng 41.2892

COPENHAGEN 1968 Role of nucleotides for the function and conformation of enzymes The Alfred Benzon symposium 1 Proceedings Alfred Benzon foundation Edited by Herman M. Kalckar Held at the premises of the Royal Danish academy of sciences and letters Munksgaard,Copenhagen,1969
Bioch 33.1913

COPENHAGEN 1969 Capillary permeability:the transfer of molecules and ions between capillary blood and tissue The Alfred Benzon symposium 2nd Proceedings Edited by Christian Crone and Niels A. Lassen Munksgaard,Copenhagen,1970
Inv Med 37.0054

COPENHAGEN,1959 Rapports et proces-verbaux des reunions Vol 149: contributions to special IGY meeting 1959 Conseil permanent international pour l'exploration de la mer TR-58 illus,maps 218p And.Fred Host, Copenhagen,1961 Meeting held in Copenhagen
Sco 14.5911

COPENHAGEN DECISIONS ON ZOOLOGICAL NOMENCLATURE ADDITIONS TO...LES REGLES INTERNATIONALES DE LA NOMENCLATURE ZOOLOGIQUE Zoological nomenclature Copenhagen 1953 Aug International trust for zoological nomenclature Edited by Francis Hemming International trust for zoological nomenclature,London,1953
BG 38.3107

COPPER Biochemistry of copper Symposium on copper in biological systems Proceedings New York 1965 Sep 8-10 Edited by Jack Peisach and others Academic press,New York; London,1966
Bioch 33.1068

COPPER RESOURCES OF THE WORLD International geological congress 16th papers Washington,D.C. 1935 Vol 1-2 2 vols Washington,D.C.,1935
Min 10.0202

CORAL GABLES,FLA. 1961 Biology of the hydra and of some other coelenterates The Physiology and ultrastructure of hydra and of some other coelenterates :a symposium Edited by Howard M. Lenhoff and F.Farnsworth Loomis held at the Fairchild tropical gardens University of Miami press,Coral Gables,Fla., 1961
Bal 39.2183

CORAL GABLES,FLA. 1964 Conference on symmetry principles at high energy University of Miami Edited by Behram Kursunoglu and Arnold Perlmutter Freeman,San Francisco,1964
Cav 7.0543

CORAL GABLES,FLA. 1967 Coral Gables conferences on symmetry principles at high energy 4 Proceedings University of Miami.Center for theoretical studies Edited by Arnold Perlmutter and Behram Kursunoglu Sponsored jointly by the United States atomic energy commission 253p Freeman,San Francisco;London,1967
A Math 4.1347

CORAL GABLES CONFERENCES ON SYMMETRY PRINCIPLES AT HIGH ENERGY 4 Proceedings Coral Gables,Fla. 1967 Jan 25-27 University of Miami.Center for theoretical studies Edited by Arnold Perlmutter and Behram Kursunoglu Sponsored jointly by the United States atomic energy commission 253p Freeman,San Francisco;London,1967
A Math 4.1347

CORFU 1961 Meteorological and astronomical influences on radio wave propagation NATO Advanced study institute Papers North Atlantic treaty organization.Division of scientific affairs NATO conference series, 3 318p Pergamon press,Oxford,1963
Nap 11.0915

CORK,IRELAND 1964 Organic reaction mechanisms :an international symposium Lectures Chemical society,and,University college,Cork Chemical society.Special publication,19 Chemical society,London,1965
Chem 18.1223

CORNEA... World congress on the cornea 1st Papers Washington,D.C. 1964 Oct Edited by John Harry Kind and John W. McTigue Butterworths,London,1965
An 32.4777

CORNELL UNIVERSITY Metric geometry over affine spaces Cooperative summer seminar 1st Ithaca,N.Y. 1964 By Ernst Snapper Mathematical association of America bibliog. iv,166p 28cm Mathematical association of America,Buffalo,N.Y.,1964 Duplicated notes
P Math 2.3829

CORNELL UNIVERSITY The Nature of time :a meeting convened by T.Gold and H.Bondi Ithaca,N.Y. 1963 May 30-Jun 1 Edited by T. Gold xiv,248p Cornell university press, Ithaca,N.Y.,1967
WSM 43.1104

CORNELL UNIVERSITY The Nature of time : report of a meeting Ithaca,N.Y. 1963 May 30-Jun 1 Edited by T. Gold and D.L. Schumacher ix,248p 23cm Cornell university,Ithaca,N.Y.,1967
Cav 7.2764

CORNELL UNIVERSITY The Transfer of calcium and strontium across biological membranes :a conference Proceedings Ithaca,N.Y. 1962 May 13-16 Edited by R.H. Wasserman Academic press,New York;London,1963
Bioch 33.1021

CORNELL UNIVERSITY 1948 Phase transformations in solids :a symposium Papers Edited by R. Smoluchowski and others Sponsored by the National research council. Division of physical sciences. Committee on solids J.Wiley;Chapman and Hall,New York; London,1951
Chem 18.2149

CORNVALLIS,ORE. 1957 Oregon state university annual biology colloquium arctic biology:ten papers presented at the 1957 and one at the 1965 biology colloquium 18th Proceedings Oregon state university Edited by Henry P. Hansen 2nd edition Illus,maps 318p 24cm Oregon state university press, Cornvallis,Ore.,1967
Sco 14.7840

CORPORATION FOR ECONOMIC AND INDUSTRIAL RESEARCH Mathematical model building in economics and industry :a conference Papers London 1967 Jul 4-6 Griffin,London,1968
Eng 41.1223

CORPORATION FOR ECONOMIC AND INDUSTRIAL RESEARCH Mathematical model building in economics and industry :a conference Papers London 1967 Jul 4-6 vii,165p 23cm Griffin, London,1968
Math S 3.1407

CORPUS CALLOSUM Functions of the corpus callosum London 1964 Sep 4 Ciba foundation Edited by E.G. Ettlinger Ciba foundation study group, 20 Churchill,London, 1965
An 32.4657

CORPUS CALLOSUM Functions of the corpus callosum London 1964 Sep 4 Ciba foundation Edited by E.G. Ettlinger Ciba foundation study groups, 20 Churchill, London,1965
Phys 20.1996

CORRELATION BETWEEN CALCULATED AND OBSERVED STRESSES AND DISPLACEMENTS IN STRUCTURES : a conference London 1955 Sep 21-22 Institution of civil engineers 2 vols Institution of civl engineers,London,1955
Eng 41.3020

CORROSION Colloque de metallurgie corrosion, seche et aqueuse 3e Saclay 1959 Jun 29-Jul 1 Centre d'etudes nucleaires,Saclay Sponsored by the Institut national des sciences et techniques nucleaires Centre d'etudes nucleaires de Saclay;North Holland, Paris;Amsterdam,1960
Met 25.1891

CORROSION Current corrosion research in Scandinavia Scandinavian corrosion congress (NKM) :lectures 4th Helsinki 1964 Nov 24-27 Teknillisten tieteiden akatemia and Central chemical association Edited by Jori Larinkari and others Organized by the Scandinavian council for applied research Kemian keskusliton julkaisuja, 24 Sanoma Osakeyhtio,Helsinki,1965
Met 25.1901

CORROSION International congress on metallic corrosion 3rd Proceedings Moscow 1966 Vol 1-4 Edited by Ya.M. Kolotyrkin and others 4 vols Mir,Moscow,1969
Met 25.2662

CORROSION Symposium on corrosion fundamentals :a series of lectures... Knoxsville 1955 Mar 1-3 University of Tennessee.Chemical engineering department and National association of corrosion engineers.Southeast region Edited by Anton De S. Brasunas and E.E. Stansbury University of Tennessee press,Knoxville,Tenn.,1956
Met 25.1884

CORROSION AND ITS PREVENTION IN MOTOR VEHICLES a symposium Proceedings London 1968 Mar 28-29 Institution of mechanical engineers.Automobile division and British joint corrosion group Institution of mechanical engineers.Proceedings, 182 pt 3J Inst.of mechanical engineers,London,1968
Met 25.1943

CORROSION BY LIQUID METALS Proceedings of the sessions on corrosion by liquid metals of the 1969 fall meeting of the Metallurgical society of AIME Philadelphia,Pa. 1969 Oct 13-16 American institution of mining engineers.Metallurgical society Edited by J. E. Draley and J.R. Weeks Metallurgical society of AIME.Publications Plenum press, New York,1970
Met 25.2790

CORROSION CONVENTION Proceedings Industry fights corrosion London 1937 Oct 15-16 Sponsored by Corrosion technology Corrosion technology,London,1957
Eng 41.3719

The CORROSION CONVENTION Proceedings Industry fights corrosion London 1957 Oct 15-16 Corrosion technology Corrosion technology,London,c1957
Met 25.1890

CORROSION GROUP Corrosion problems of the petroleum industry a joint symosium Papers London 1959 Nov 26-27 Institute of petroleum and Chemical engineering group S.C.I.Monograph, 10 Society of chemical industry,London,1960
Met 25.1934

CORROSION GROUP The Protection of motor vehicles from corrosion :a symposium London 1958 Mar 11-12 S.C.I.monograph, 4 Society of chemical industry,London,1958
Met 25.1940

CORROSION OF REACTOR MATERIALS the conference Proceedings Salzburg 1962 Jun 4-8 Vol 1-2 International atomic energy agency 2 vols International atomic energy agency, Vienna,1962
Met 25.1937

CORROSION PROBLEMS OF THE PETROLEUM INDUSTRY a joint symosium Papers London 1959 Nov 26-27 Institute of petroleum and Chemical engineering group Joint symposium with the Corrosion group S.C.I.Monograph, 10 Society of chemical industry,London,1960
Met 25.1934

CORROSION TECHNOLOGY Industry fights corrosion Corrosion convention Proceedings London 1937 Oct 15-16 Corrosion technology, London,1957
Eng 41.3719

CORROSION TECHNOLOGY Industry fights corrosion The Corrosion convention Proceedings London 1957 Oct 15-16 Corrosion technology,London,c1957
Met 25.1890

CORTICAL EXCITABILITY AND STEADY POTENTIALS; RELATIONS OF BASIC RESEARCH TO SPACE BIOLOGY Brain function :conference 1st Proceedings Los Angeles,Calif. 1961 Brain research institute and American institute of biological sciences Edited by Mary A.B. Brazier U.C.L.A. forum in medical sciences, 1 University of California press, Berkeley;Los Angeles,1963
Psy 31.0255

CORTINA D'AMPEZZO 1966 International congress of radiation research 3rd book of abstracts Cortina d'Ampezzo,1966
Radioth 35.1969

CORTINA D'AMPEZZO 1966 Radiation research International congress of radiation research 3rd Proceedings Edited by G. Silini Under the auspices of the International association for radiation research North-Holland,Amsterdam,1967
Radioth 35.1173

CORTINA LECTURES and four lectures from the Conference on localized excitations Proceedings Elementary excitations in solids Milan 1966 Jul 25-26 Edited by A. A. Maradudin and G.F. Nardelli Bibliog 526p 20cm Plenum press,New York,1969
Cav 7.2677

CORVALLIS 1957 Biology Colloquium Arctic biology 18 colloquium Edited by Henry P. Hansen illus,maps 134p 28cm Oregon state college,Corvallis,1957
Sco 14.2434

CORVALLIS,ORE. 1962 Refresher course in recent advances in cytogenetics and developmental genetics American society of zoologists American zoologist, 3,no.1 Utica,N.Y.,1963
Gen 34.0856

COSMIC CLOUDS Gas dynamics of cosmic clouds Symposium on cosmical gas dynamics 2nd Cambridge 1953 Jul 6-11 International union of theoretical and applied mechanics International astronomical union International astronomical union.Symposium series, 2 North-Holland,Amsterdam,1955
Eng 41.4468

COSMIC CLOUDS Gas dynamics of cosmic clouds : a symposium Cambridge 1953 Jul.6-11 International union of theoretical and applied mechanics and International astronomical union Edited by J.M. Burgers and H.C.van de Hulst International astronomical union. Symposium, 2 247p North-Holland, Amsterdam,1955
Obs 6.0308

COSMIC PLASMA PHYSICS :a conference Frascati 1971 Sep 20-24 Edited by K. Schindler 369p Plenum press,New York,1972
Obs 6.3625

COSMIC RADIATION 1st based on a symposium promoted by the Colston research society Papers Bristol 1948 Sep Colston research society and University of Bristol Edited by F.C. Frank and D.P. Rexworthy Colston research society.Colston papers, 1 illus viii,189p Butterworths,London,1949
Cav 7.0348

COSMIC RAY STRUCTURES IN RELATION TO RECENT DEVELOPMENTS IN ASTRONOMY AND ASTROPHYSICS : a colloquium Proceedings Bombay 1968 Nov 11-16 Edited by R.R. Daniel and others 374p Tata institute of fundamental research, Bombay,1969
TA 15.0567

COSMIC RAYS International conference on cosmic rays 8th proceedings Jaipur 1963 Dec.2-14 Vols. 1-6 International union of pure and applied physics.Cosmic ray commission Edited by R.R. Daniel and others Sponsored by India.Department of atomic energy 6 vols Bombay,1964
A Math 4.1159

COSMIC RAYS International conference on cosmic rays 11th Invited papers and rapporteur talks Budapest 1969 Aug 25-Sep 4 International union of pure and applied physics Magyar tudomanyos akademia 612p Central research institute of physics,Budapest, 1969
TA 15.0551

COSMIC RAYS International conference on cosmic rays 11th Proceedings 3: high energy interactions - extensive air showers Edited by A. Somogyi Hungarica acta physica. Supplement, 29 766p Akademiai kiado, Budapest,1970
TA 15.0634

COSMIC RAYS International conference on cosmic rays 11th Proceedings Budapest 1969 Aug 25-Sep 4 4: muons and neutrons techniques Edited by A. Somogyi Hungarica acta physica.Supplement, 29 Akademiai kiado, Budapest,1970
TA 15.0635

COSMIC RAYS Soveshchaniya po voprosam kosmogonii, proiskhozhdenie kosmicheskikh luchei 3 trudy Moscow 1953 May 14-15 Edited by Viktor Amazaspovich Ambartsumyan 319p Akademia Nauk SSSR,Moscow,1954
Obs 6.0060

COSMIC RAYS Symposium on cosmic rays, elementary particle physics and astrophysics 10th Proceedings Aligarh 1967 Dec 12-16 India.Department of atomic energy.Cosmic ray committee 701p Bombay,1968
TA 15.0553

COSMIC RAYS Symposium sobre raios cosmicos Rio de Janeiro 1941 August 4-8 Academia Brasileira de Ciencias 26cm Rio de Janeiro 1943
Philos 1.0379

COSMIC RAYS The International conference on cosmic rays and the earth storm :combined meeting of the International symposium on the earth storm and the International conference on cosmic rays Proceedings Kyoto 1961 Sep 4-15 Vol 2-3 International union of geodesy and geophysics and International astronomical union Edited by Ken-ichi Maeda and Osamu Minakawa Co-sponsored by the International scientific radio union Physical society of Japan. Journal, 17,suppl.A-2,3 2 vols physical society of Japan,Tokyo,1962
TA 15.0038

COSMICAL AERODYNAMICS Problems of cosmical aerodynamics Symposium on the motion of gaseous masses of cosmical dimensions Proceedings Paris 1949 Aug 16-19 International union of theoretical and applied mechanics International astronomical union Edited by J.M. Burgess and H.C.van de Hulst 237p Central air documents office,Dayton, Ohio,1951
Obs 6.3550

COSMICAL GAS DYNAMICS Aerodynamic phenomena in stellar atmospheres Symposium on cosmical gas dynamics held at the International school of physics "Enrico Fermi" 4th proceedings Varenna 1960 Aug.18-30 International astronomical union and International union of theoretical and applied mechanics Edited by R.N. Thomas and others Under the auspices of Societa Italiana di fisica International astronomical union.Symposium, 12 515p Bologna,c1961 Reprinted from Supplemento Nuovo Cimento,22,no.1,1961
A Math 4.1152

COSMICAL GAS DYNAMICS Aerodynamic phenomena in stellar atmospheres Symposium on cosmical gas dynamics held at the International school of physics "Enrico Fermi" 4th proceedings Varenna 1960 Aug 18-30 International astronomical union and International union of theoretical and applied mechanics Edited by R.N. Thomas and others under the auspices of the Societa italiana di fisica International astronomical union symposium, 12 515p Zanichelli,Bologna,c1961 Reprinted from the 'Supplemento del nuovo cimento',vol.22,no.1,1961
Obs 6.1736

COSMICAL GAS DYNAMICS Gas dynamics of cosmic clouds The Symposium on cosmical gas dynamics Proceedings Cambridge 1953 Jul 6-11 International astronomical union and International union of theoretical and applied mechanics Edited by J.M. Burgess and H.C.van de Hulst International astronomical union. Symposium, 2 North Holland,Amsterdam,1955
TA 15.0217

COSMICAL GAS DYNAMICS Interstellar gas dynamics Symposium on cosmical gas dynamics 6th Proceedings Yalta 1969 Sep 8-18 International astronomical union International union of theoretical and applied mechanics International astronomical union. Symposium, 39 388p Reidel,Dordrecht,1970
Obs 6.3323

COSMICAL GAS DYNAMICS Interstellar gas dynamics Symposium on cosmical gas dynamics 6th Proceedings Yalta 1969 Sep 8-18 International astronomical union International union of theoretical and applied mechanics International astronomical union. Symposium, 39 388p Reidel,Dordrecht,1970
TA 15.0541

COSMICAL GAS DYNAMICS Interstellar gas dynamics Symposium on cosmical gas dynamics 6th Proceedings Yalta 1969 Sep 8-18 International union of theoretical and applied mechanics and International astronomical union Edited by H.J. Habing International astronomical union.Symposium, 39 diagrs, photos,graphs xii,388p 24cm Reidel, Dordrecht,1970
Cav 7.2710

COSMICAL GAS DYNAMICS Symposium on cosmical gas dynamics 3rd Proceedings Cambridge, Mass. 1957 Jun 24-29 International union of theoretical and applied mechanics International astronomical union Edited by J. M. Burgers and R.N. Thomas Reviews of modern physics, 30,no 3 American physical society, New York,1958
Eng 41.4467

COSMICAL GAS DYNAMICS Symposium on cosmical gas dynamics 3rd proceedings Cambridge, Mass. 1957 Jun.24-29 International union of theoretical and applied mechanics and International astronomical union Edited by J. M. Burgers and others Financial grant from the Smithsonian institution International astronomical union.Symposium, 8 Smithsonian Institution.Smithsonian symposium publications, 1 American physical society, New York,1958 Reviews of modern physics. Vol.30,no.3,July 1958
Obs 6.0306

COSMICAL PHYSICS Electromagnetic phemomena in cosmical physics a symposium Proceedings Stockholm 1956 Aug.27-31 International astronomical union Edited by B. Lehnert In co-operation with the International union of theoretical and applied mechanics International astronomical union.Symposium, 6 544p Cambridge university press, Cambridge,1958
Obs 6.1054

COSMICAL PHYSICS Electromagnetic phenomena in cosmical physics Stockholm 1956 Aug Edited by B. Lehnert International astronomical union.Symposium, 6 Cambridge university press,Cambridge,1958
Eng 41.4475

COSMICAL PHYSICS Electromagnetic phenomena in cosmical physics Stockholm 1956 Aug 27-31 International astronomical union Edited by B. Lehnert In co-operation with the International union of theoretical and applied mechanics International astronomical union. Symposium, 6 13,544p Cambridge university press,Cambridge,1958
Nap 11.0916

COSMICAL PHYSICS Electromagnetic phenomena in cosmical physics :a symposium proceedings Stockholm 1956 Aug.27-31 International astronomical union Edited by B. Lehnert International astronomical union.Symposia, 6 Cambridge university press,Cambridge,1958
A Math 4.0615

COSMOLOGY Convegno sulla cosmologia;meeting on cosmology Atti;Proceedings Padua 1964 Sep 14-16 and Venice Quatro centenario della nascita di Galileo Galilei 1564-1964. Comitato nazionale per le manifestazioni celebrative Edited by L. Rosino Organized by Universita di Padua Quatro centenario della nascita di Galileo Galilei.Comitato nazionale per le manifestazioni celebrative. Pubblicazioni, 2,Atti dei convegni, 3 206p Florence,1966
Obs 6.3451

COSPAR MEETING,FLORENCE,1961 CIRA 1961 COSPAR international reference atmosphere 1961 report of the preparatory group for an International reference atmosphere accepted at the Cospar meeting in Florence,1961 Report Committee on space research Edited by H. Kallmann-Bijl and others 1st edition 177p North Holland,Amsterdam,1961
Obs 6.0925

COSPAR PLENARY MEETING 10th :joint open meeting of working groups 1,2 and 5 Moon and planets,2 London 1967 Jul 26-27 Committee on space research Royal society of London Edited by A. Dollfus 196p North-Holland,Amsterdam,1968
Obs 6.3419

COUNCIL FOR INTERNATIONAL ORGANISATIONS OF MEDICAL SCIENCES Brain mechanisms and learnings :a symposium Montevideo 1959 Aug 2-8 Edited by J.F. Pelafresnaye Blackwell, Oxford,1961
VA 19.0236

COUNCIL FOR INTERNATIONAL ORGANISATIONS OF MEDICAL SCIENCES The Support of medical research :a symposium London 1954 Oct 4-8 Edited by H. Himsworth and J.F. Delafresnaye Blackwell,Oxford,1956
PGMS 29.0496

COUNCIL FOR INTERNATIONAL ORGANIZATION OF MEDICAL SCIENCES International symposium on aldosterone Proceedings Prague 1963 Edited by Etienne Emile Baulier and P. Robel Blackwell,Oxford,1964
Inv Med 37.0013

COUNCIL FOR INTERNATIONAL ORGANIZATIONS OF MEDICAL SCIENCE International congress of biochemistry 2nd Proceedings Paris 1952 Jul 21-27 Vol 3: symposium sur le cycle tricarboxylique Edited by Michel Polonovski Congres international de biochemie, 2 Societe d'edition d'enseignement superieur,Paris,1952
Bot 42.1697

COUNCIL FOR INTERNATIONAL ORGANIZATIONS OF MEDICAL SCIENCES Aldosterone :a symposium Prague 1963 Aug 23-25 Edited by E.E. Baulieu and P. Robel Blackwells,Oxford,1964
Phys 20.1398

COUNCIL FOR INTERNATIONAL ORGANIZATIONS OF MEDICAL SCIENCES Biological problems of grafting :a symposium Proceedings Liege 1959 Mar 18-21 Edited by F. Albert and G. Lejeune-Ledant Blackwell,Oxford,1959
Bal 39.0458

COUNCIL FOR INTERNATIONAL ORGANIZATIONS OF MEDICAL SCIENCES Biological problems of grafting :a symposium Proceedings Liege 1959 Mar 18-21 Edited by F. Albert and G. Lejeune-Ledant Blackwell,Oxford,1959 Symposium chairman P.B.Medawar
PGMS 29.0245

COUNCIL FOR INTERNATIONAL ORGANIZATIONS OF MEDICAL SCIENCES Brain mechanism and learning :a symposium Montevideo 1959 Aug 2-8 Edited by J.F. Delafresnaye Under the auspices of Unesco Blackwell,Oxford,1960 Under the joint auspices of Unesco and WHO
Psy 31.0608

COUNCIL FOR INTERNATIONAL ORGANIZATIONS OF MEDICAL SCIENCES Brain mechanism and learning :a symposium Montevideo 1959 Aug 2-8 Under the joint auspices of Unesco World health organization Blackwell,Oxford,1961
Bal 39.1604

COUNCIL FOR INTERNATIONAL ORGANIZATIONS OF MEDICAL SCIENCES Brain mechanisms and consciousness :a symposium Ste.Marguerite, Que. 1953 Aug 23-28 Blackwell,Oxford,1954
Phys 20.2009

COUNCIL FOR INTERNATIONAL ORGANIZATIONS OF MEDICAL SCIENCES Brain mechanisms and consciousness :a symposium Ste.Marguerite, Que. 1953 Aug 23-28 Edited by H.H. Jasper and others Blackwell,Oxford,1954
An 32.4625

COUNCIL FOR INTERNATIONAL ORGANIZATIONS OF MEDICAL SCIENCES Brain mechanisms and learning :a symposium Montevideo 1959 Aug 2-8 By J.F. Delafresnaye Blackwell,Oxford, 1961
An 32.4647

COUNCIL FOR INTERNATIONAL ORGANIZATIONS OF MEDICAL SCIENCES Connective tissue :a symposium Papers and discussions London 1956 Jul 22-26 Edited by R.E. Tunbridge and others Blackwell,Oxford,1957
Path 30.2490

COUNCIL FOR INTERNATIONAL ORGANIZATIONS OF MEDICAL SCIENCES Controlled clinical trials :a conference Papers Vienna 1959 Mar 23-27 Blackwell,Oxford,1960
Radioth 35.0799

COUNCIL FOR INTERNATIONAL ORGANIZATIONS OF MEDICAL SCIENCES Controlled clinical trials :a conference Papers Vienna 1959 Mar 23-27 Blackwell,Oxford,1960 Organized under the direction of A. Bradford Hill
HE 27.0076

COUNCIL FOR INTERNATIONAL ORGANIZATIONS OF MEDICAL SCIENCES Experimental diabetes and its relation to the clinical disease :a symposium Leiden 1952 Jul 14-18 Edited by J.F. Delafresnaye and G.Howard Smith Blackwell,Oxford,1954
Bioch 33.0448

COUNCIL FOR INTERNATIONAL ORGANIZATIONS OF MEDICAL SCIENCES Immunological methods :a symposium Edited by J.A. Ackroyd Blackwell, Oxford,1964
Pha 16.0188

COUNCIL FOR INTERNATIONAL ORGANIZATIONS OF MEDICAL SCIENCES Immunological methods :a symposium Papers and discussions 1963 Edited by J.F. Ackroyd Blackwell,Oxford,1964
Path 30.2539

COUNCIL FOR INTERNATIONAL ORGANIZATIONS OF MEDICAL SCIENCES International congress biochemistry 2nd Paris 1952 Jul 21-27 Vol 7: symposium sur la biochemie des steroides Societe d'edition d'enseignement superieur,Paris,1952 Text in English and French
Bioch 33.1310

COUNCIL FOR INTERNATIONAL ORGANIZATIONS OF MEDICAL SCIENCES International congress of biochemistry 2nd Paris 1952 Jul 21-27 Vol 1: symposium sur la biochimie de l'hematopoiese Societe d'edition d'enseignement superieur,Paris,1952 Text in English and French
Bioch 33.1304

COUNCIL FOR INTERNATIONAL ORGANIZATIONS OF MEDICAL SCIENCES International congress of biochemistry 2nd Paris 1952 Jul 21-27 Vol 2: symposium sur la biogenese des proteines Societe d'edition d'enseignement superieur,Paris,1952 Text in English and French
Bioch 33.1305

COUNCIL FOR INTERNATIONAL ORGANIZATIONS OF MEDICAL SCIENCES International congress of biochemistry 2nd Paris 1952 Jul 21-27 Vol 3: symposium sur le cycle tricarboxylique Societe d'edition d'enseignement superieur,Paris,1952 Text in English,French and German
Bioch 33.1306

COUNCIL FOR INTERNATIONAL ORGANIZATIONS OF MEDICAL SCIENCES International congress of biochemistry 2nd Paris 1952 Jul 21-27 Vol 4: symposium sur les hormones proteiques et derivees des proteines Societe d'edition d'enseignement superieur,Paris,1952 Text in English and French
Bioch 33.1307

COUNCIL FOR INTERNATIONAL ORGANIZATIONS OF MEDICAL SCIENCES International congress of biochemistry 2nd Paris 1952 Jul 21-27 Vol 5: symposium sur le metabolisme microbien Societe d'edition d'enseignement superieur,Paris,1952 Text in English and French
Bioch 33.1308

COUNCIL FOR INTERNATIONAL ORGANIZATIONS OF MEDICAL SCIENCES International congress of biochemistry 2nd Paris 1952 Jul 21-27 Vol 6: symposium sur le mode d'action des antibiotiques Societe d'edition d'enseignement superieur,Paris,1952 Text in English and French
Bioch 33.1309

COUNCIL FOR INTERNATIONAL ORGANIZATIONS OF MEDICAL SCIENCES International congress of biochemistry 2nd Compte rendu Paris 1952 Jul 21-27 Societe de chimie biologique. Bulletin, 35,no 1-2 Masson,Paris,1953 Supplement.Text in English and French
Bioch 33.1302

COUNCIL FOR INTERNATIONAL ORGANIZATIONS OF MEDICAL SCIENCES International congress of biochemistry 2nd Resumes des communication Paris 1952 Jul 21-27 Masson,Paris,1952 Text in English and French
Bioch 33.1303

COUNCIL FOR INTERNATIONAL ORGANIZATIONS OF MEDICAL SCIENCES International congress of biochemistry 3rd Programme et liste des membres Brussels 1955 Aug 1-6 Secretariat general;Agence de voyage,Liege; Brussels,c1956 Bound with 3 eme Congres international de biochimie.Resumes des communications
Bioch 33.1312

COUNCIL FOR INTERNATIONAL ORGANIZATIONS OF MEDICAL SCIENCES International congress of biochemistry 3rd Rapports (volume provisoire) Brussels 1955 Aug 1-6 Imprimerie Vaillant-Carmanne,Liege,1955
Bioch 33.1314

COUNCIL FOR INTERNATIONAL ORGANIZATIONS OF MEDICAL SCIENCES International congress of biochemistry 3rd Resumes des communications Brussels 1955 Aug 1-6 c1956 Text in English,French and German. Bound with 3 eme Congres international de biochimie.Programme et liste des membres
Bioch 33.1313

COUNCIL FOR INTERNATIONAL ORGANIZATIONS OF MEDICAL SCIENCES Laurentian symposium Brain mechanisms and consciousness :a symposium Edited by J.F. Delafresnaye Blackwell,Oxford,1954
Psy 31.0519

COUNCIL FOR INTERNATIONAL ORGANIZATIONS OF MEDICAL SCIENCES Le Muscle :etude de biologie et de pathologie.Colloque Compte rendu Royaumont 1950 Aug 31-Sep 6 L'expansion scientifique,1952 In English and French
Bioch 33.2245

COUNCIL FOR INTERNATIONAL ORGANIZATIONS OF MEDICAL SCIENCES Oxygen supply to the human foetus :a symposium Princeton,N.J. 1957 Dec Edited by James Walker and Alec C. Turnbull Blackwell,Oxford,1959
An 32.2514

COUNCIL FOR INTERNATIONAL ORGANIZATIONS OF MEDICAL SCIENCES Oxygen supply to the human foetus :a symposium Princeton,N.J. 1957 Dec Edited by James Walker and Alec C. Turnbull Blackwell,Oxford,1959
Phys 20.1003

COUNCIL FOR INTERNATIONAL ORGANIZATIONS OF MEDICAL SCIENCES Physiopathology of the reticular endothelial system :a symposium Paris Edited by B.N. Halpern Blackwell, Oxford,1957 Summary of each paper in French
An 32.3154

COUNCIL FOR INTERNATIONAL ORGANIZATIONS OF MEDICAL SCIENCES Protein biosynthesis :a symposium Wassenaar 1960 Aug 29-Sep 2 Edited by R.J.C. Harris Under the auspice of Unesco Academic press,London;New York,1961 Co-sponsored by the Council for international organizations of medical sciences
Radioth 35.0102

COUNCIL FOR INTERNATIONAL ORGANIZATIONS OF MEDICAL SCIENCES Protein biosynthesis :a symposium Wassenaar 1960 Aug 29-Sep 2 Unesco Edited by R.J.C. Harris Academic press,London;New York,1961
An 32.1509

COUNCIL FOR INTERNATIONAL ORGANIZATIONS OF MEDICAL SCIENCES Protein biosynthesis :a symposium Wassennaar 1960 Aug 29-Sep 2 Edited by R.J.C. Harris Academic press, London;New York,1961
Gen 34.0652

COUNCIL FOR INTERNATIONAL ORGANIZATIONS OF MEDICAL SCIENCES Sensitivity reactions to drugs :a symposium Liege 1957 Jul 9-12 Edited by M.L. Rosenheim and R. Moulton Blackwell,Oxford,1958
Pha 16.0187

COUNCIL FOR INTERNATIONAL ORGANIZATIONS OF MEDICAL SCIENCES The Transparancy of the cornea :a symposium Knokke 1959 Aug 31-Sep 5 Edited by William Stewart Duke-Elder and E. S. Perkins Blackwell,Oxford,1960
An 32.4772

COUNCIL FOR NATURE and ROYAL SOCIETY OF ARTS The Countryside in 1970 :conference 2nd Proceedings London 1965 Nov 10-12 Jointly sponsored by the Nature conservancy Royal society of arts;Nature conservancy, London,1966
Geog 13.1899

COUNCIL FOR SCIENCES OF INDONESIA The Vegetation of the humid tropics :symposium 2nd Proceedings Tjiawi 1953 Dec Unesco. Science cooperation office for south-east Asia Unesco,c1953
Bot 42.4770

COUNCIL OF SCIENTIFIC AND INDUSTRIAL RESEARCH. BIOLOGICAL RESEARCH COMMITTEE Plant embryology :a symposium New Delhi 1960 Nov 11-14 Council of scientific and industrial research,New Delhi,1962
Bot 42.1240

COUNTING LABELLED TREES Canadian mathematical congress 12th biennial seminar Lectures Vancouver,B.C. 1969 Canadian mathematical congress Canadian mathematical monographs, 1 x,113p 23cm Canadian mathematical congress,Montreal,Que.,1970
P Math 2.4447

The COUNTRYSIDE IN 1970 2nd :conference Proceedings London 1965 Nov 10-12 Council for nature,and,Royal society of arts Jointly sponsored by the Nature conservancy Royal society of arts;Nature conservancy, London,1966
Geog 13.1899

COURANT INSTITUTE OF MATHEMATICAL SCIENCES Bifurcation theory and nonlinear eigenvalue problems :a seminar Lectures New York 1966-67 Edited by Joseph B. Keller and Stuart Antman Bibliog. xiv,434p 23cm Benjamin,New York;Amsterdam,1969
P Math 2.3648

COURANT INSTITUTE OF MATHEMATICAL SCIENCES Computer networks Courantter science symposium 3rd New York 1970 Nov 30-Dec 1 Edited by Randall Rustin diagrs 205p Prentice-Hall,Englewood Cliffs,N.J., 1972
Math L 5.3638

COURANT INSTITUTE OF MATHEMATICAL SCIENCES Lectures on advanced ordinary differential equations By K.O. Friedrichs Notes on mathematics and its applications 205p Nelson,London,1967
A Math 4.1235

COURANTTER SCIENCE SYMPOSIUM 3rd Computer networks New York 1970 Nov 30-Dec 1 Courant institute of mathematical sciences Edited by Randall Rustin diagrs 205p Prentice-Hall,Englewood Cliffs,N.J.,1972
Math L 5.3638

COVENTRY 1972 Stability of stochastic dynamical systems Proceedings Edited by Ruth E. Curtain illus ix,334p 25cm Springer,Berlin,1972
A Math 4.1870

COVENTRY 1972 Stability of stochastic dynamical systems :international symposium Proceedings University of Warwick.Control theory centre Edited by Ruth F. Curtain Lecture notes in mathematics, 254 ix,332p 26cm Springer,Berlin,1972
P Math 2.4569

The CRAB NEBULA Jodrell Bank 1970 Aug 5-7 International astronomical union Edited by R. D. Davies and F.G. Smith International astronomical union.Symposium, 46 470p Reidel,Dordrecht,1971
Obs 6.3588

The CRAB NEBULA Jodrell Bank 1970 Aug 5-7 International astronomical union Edited by R. D. Davies and F.G. Smith International astronomical union.Symposium, 46 470p Reidel,Dordrecht,1971
TA 15.0689

CRACK PROPAGATION SYMPOSIUM Proceedings Cranfield 1961 Sep Vol 1-2 Cranfield college of aeronautics Cranfield, 1962
Eng 41.2779

The CRACK PROPAGATION SYMPOSIUM Proceedings Cranfield 1961 Sep Vol 1-2 Aeronautical research council Supported by the Royal aeronautical society 2 vols College of aeronautics,Cranfield,1962
Met 25.0865

CRANFIELD 1951 Automatic and manual control Conference on automatic control papers Great Britain.Department of scientific and industrial research xi,584p 25cm Butterworth,London,1951
Math L 5.0021

CRANFIELD 1951 Automatic and manual control Papers Institution of electrical engineers Institution of mechanical engineers Edited by A. Tustin Under the auspices of Great Britain.Department of scientific and industrial research Butterworths,London,1952
Psy 31.3398

CRELTON,MINN. 1965 Reproduction molecular, subcellular and cellular :symposium Society for developmental biology Edited by Michael Locke Society for developmental biology. Symposia, 24 Academic press,New York;London, 1965
Gen 34.0485

CRETACIOUS SYMPOSIUM Sistema cretacico Un Symposium sobre el cretacico en el hemisferio occidental y su correlacio mundial papers Mexico City 1956 Tom 1-2 Edited by Lewis B. Kellum 2 vols Mexico City,1959 Prepared under the auspices of the International commission on stratigraphy and the International palaeontological union for the 20th International geological congress
Geol 8.3049

CRITICAL PATH ANALYSIS PLANNING TECHNIQUES Discussion on critical path analysis planning techniques for engineering management Proceedings London 1963 Apr 23 Institution of mechanical engineers Institution of mechanical engineers,London, 1964
Eng 41.0499

CRITICAL PHENOMENA Conference on phenomena in the neighbourhood of critical points Proceedings Washington,D.C. 1965 Apr 5-8 National bureau of standards Edited by M.S. Green and J.V. Sengers National bureau of standards.Miscellaneous publication, 273 xiv, 242p 27cm National bureau of standards, Washington,D.C.,1966
Cav 7.2931

CRITICAL ROTATORY DISPERSION AND CIRCULAR DICHROISM IN ORGANIC CHEMISTRY NATO summer school Proceedings Bonn 1965 Sep 24-Oct 10 North Atlantic treaty organization Edited by G. Snatzke Heyden,London,1967
Chem 18.2791

CRITICISM AND THE GROWTH OF KNOWLEDGE International colloquium in the philosophy of science Proceedings London 1965 Jul 11-17 Edited by Imre Lakatos and Alan Musgrave Held at Bedford college vii,282p 23cm Cambridge university press,Cambridge,1970
Math S 3.1850

CRITICISM AND THE GROWTH OF KNOWLEDGE International colloquium in the philosophy of science Proceedings London 1965 Jul 11-17 Vol 4: criticism and the growth of knowledge Edited by Imre Lakatos and Alan Musgrave Cambridge university press, Cambridge,1970
Eng 41.1688

CROSS-REACTING ANTIGENS AND NEOANTIGENS (with implications for autoimmunity and cancer immunity):a conference Proceedings National research council.Committee on tissue transplantation Edited by John J. Trentin illus,tables Williams and Wilkins,Baltimore, Md.,1967
Path 30.2544

CROSSLEY J.N. and DUMMETT M.A.E. ed. Formal systems and recursive functions The Logic colloquium 8th proceedings Oxford 1963 Jul 320p 23cm North-Holland, Amsterdam,1965
P. Math 2.0207

CRUST OF THE EARTH New York 1954 Oct 14-16 Edited by Arie Poldervaart Organized by the Columbia university.Department of geology Geological society of America.Special papers, 62 Geological society of America,New York, 1955
Geog 13.0219

CRUST OF THE EARTH :a symposium Proceedings New York 1954 Oct 14-16 Columbia university Edited by Arie Poldervaart Under the auspices of the Geological society of America Geological society of America. Special paper, 62 New York,1955
Geod 9.0100

The CRUST OF THE PACIFIC BASIN :two symposia, presented at the 10th Pacific Honolulu 1961 Aug 21-Sep 6 Edited by Gordon A. MacDonald and Hisashi Kuno Geophysical monograph, 6 National research council. Publication,1035 Washington,D.C.,1961 Based on the papers presented at the symposia
Geod 9.0423

CRYOBIOLOGY Cellular injury and resistance in freezing organisms International conference on low temperature science 2nd Proceedings Sapporo 1966 Aug 14-19 Vol 2: conference on cryobiology Hokkaido university.Institute of low temperature science Edited by Eizo Asahina illus Hokkaido university.Institute of low temperature science,Sapporo,1967 Title page in English and Japanese Conference held in commemoration of the 25th anniversary of the establishment of the Institute
Path 30.2625

CRYOBIOLOGY International conference on low temperature science Papers Sapporo,Japan 1966 Aug 14-19 conference on cryobiology. Cellular injury and resistance in freezing organisms Edited by Eizo Asahina Sponsored by the Institute of low temperature science illus xxiii,257p 27cm Sapporo,1967
Sco 14.0144

CRYOGENIC ENGINEERING CONFERENCE 6th Proceedings Advances in cryogenic engineering Boulder,Colo. 1960 Aug 23-26 National bureau of standards University of Colorado Edited by K.D. Timmerhaus Advances in cryogenic engineering, 6 Plenum press,New York,1961
Eng 41.7398

CRYSTAL GROWTH A Discussion of the Faraday society 5th Papers Bristol 1949 Apr 12-14 University of Bristol.Department of physics and Faraday society Faraday society.Discussions, 5,1949 Butterworth, London,1959
Cav 7.0760

CRYSTAL GROWTH Growth and perfection of crystals International conference on crystal growth Proceedings Cooperstown,N.Y. 1958 Aug 27-29 General electric company Edited by R.H. Doremus and others Co-sponsored by United States.Air force.Office of scientific research Wiley;Chapman and Hall,New York; London,1958
Eng 41.4438

CRYSTAL GROWTH Growth of crystals Conference on crystal growth 1st- Proceedings Moscow 1956- 1- Institut kristallografii,Moscow Edited by A. V. Shubnikov and N.N. Sheftal Translated by Consultants' bureau from the Russian Rost kristallov, 1- Consultants bureau,inc.,New York,1958- Includes interim reports published between conferences.Held as part of the international crystallographic congresses
Met 25.1508

CRYSTAL GROWTH International conference on crystal growth 2nd Proceedings Birmingham 1969 Jul 15-19 Edited by F.C. Frank and others Sponsored by International union of pure and applied physics xix,842p 28cm North-Holland,Amsterdam,1968 Reprinted from Journal of crystal growth,3-4, 1968
Cav 7.2950

CRYSTAL GROWTH International conference on crystal growth Proceedings Boston,Mass. 1966 Jun 20-24 International union of pure and applied physics and Air force Cambridge research laboratories Edited by H. Steffen Peiser Physics and chemistry of solids.Supplement Pergamon press,Oxford,1967
Met 25.1520

CRYSTAL GROWTH Bristol 1949 Apr 12-14 Faraday society Faraday society.Discussions, 5 Gurney and Jackson,London,1949
Min 10.1087

CRYSTAL GROWTH :a general discussion Bristol 1949 Apr 12-14 Faraday society Faraday society.Discussions, 5 Gurney and Jackson,London,1949
Eng 41.4424

CRYSTAL GROWTH AND EPITAXY FROM THE VAPOUR PHASE International conference on crystal growth and epitaxy from the vapour phase 1st Proceedings Zurich 1970 Sep 23-26 Edited by E. Kaldis and M. Schieber North-Holland, Amsterdam,1971
Met 25.2765

CRYSTAL GROWTH 1968 International conference on crystal growth 2nd Proceedings Birmingham 1968 Jul 15-19 Comite international de croissance crystalline Edited by F.C. Frank and others Journal of crystal growth,1968,no 3-4 North-Holland, Amsterdam,1968
Min 10.1507

CRYSTAL LATTICE DEFECTS International conference on crystal lattice defects Proceedings Tokyo 1962 Sep 3-4 and Kyoto 1962 Sep 7-12 Physical society of Japan Physical society of Japan.Journal, 18, Supplements, 1-3 photos,diagrms 1000p 26cm Physical society of Japan,Tokyo,1963
Cav 7.3058

CRYSTAL LATTICE DEFECTS The International conference on crystal lattice defects Proceedings Tokyo 1962 Sep 3-4 Kyoto 1962 Sep 7-12 Pt 1-3 Physical society of Japan Under the auspices of the International union of pure and applied physics Physical society of Japan.Journal, 18; Suppl.1-3 3 vols Physical society of Japan,Tokyo,1963
Cav 7.1343

CRYSTAL STRUCTURE AND CHEMICAL CONSTITUTION : a general discussion London 1929 May 14 Faraday society London,1929
Bot 42.6478

CRYSTAL STRUCTURES Computing methods and the phase problem in X-ray crystal analysis : conference report State college,Pa. 1950 Apr 6-8 Edited by Ray Pepinsky sponsored by the Rockefeller foundation Pennsylvania state college,State college,Pa.,1952
Min 10.0756

CRYSTALLINE SOLIDS Conference on the defects in crystalline solids :held at the H.H.Wills physical laboratory Report Bristol 1954 Jul By N. Bloembergen and others University of Bristol and Institute of physics With co-operation of the International union of pure and applied physics.Commission on the solid state 429p Physical society,London,1955
Cav 7.0345

CRYSTALLINE SOLIDS Defects in crystalline solids Conference on defects in crystalline solids Report Bristol 1954 Jul Physical society,London,1955
Eng 41.4311

CRYSTALLINE SOLIDS The Mechanism of phase transformations in crystalline solids :an international symposium Proceedings Manchester 1968 Jul 3-5 Institute of metals Institute of metals.Monograph and report series, 33 illus,diagrs 324p 25cm London,1969
Cav 7.2666

CRYSTALLINE SURFACES Interactions des electrons,phonons et magnons avec les surfaces cristallines :colloque Exposes et communications presentes Lille 1969 Sep 17-19 Societe francaise de physique Journal de physique, 31,no 4,supplement,Colloque,C-1 Paris,1970
Met 25.2627

CRYSTALLISATION Kristallisation Gemeinschaftskonferenz der reihe 'Metall' :die kristallisation von metallen aus dem schmelzfluss,der gasphase und durch elektrolytische abscheidung 6 Vortrage Berlin 1968 Mar 28-29 Deutsche akademie der wissenschaften zu Berlin.Sektion der physik.Unterkommission metallphysik VEB verlag fur grundstoffindustrie,Leipzig,1969
Met 25.2622

CRYSTALLIZATION Nucleation and crystallization in glasses and melts :a symposium Toronto 1961 Apr American ceramic society Edited by Margie K. Reser and others American ceramic society,Columbus, Ohio,1962
Met 25.0127

CRYSTALLOGRAPHIC COMPUTATION International summer school on crystallographic computation Proceedings Ottawa 1969 Aug 4-11 International union of crystallography. Commission on crystallographic computation Edited by F.R. Ahmed Munksgaard,Copenhagen, 1970
Min 10.1527

CRYSTALLOGRAPHIC COMPUTING :a conference Proceedings Ottawa 1969 Aug 4-11 International union of crystallography. Commission on crystallographic computing Edited by F.R. Ahmed and others Bibliog 384p 20cm Munksgaard,Copenhagen,1970
Cav 7.2655

CRYSTALLOGRAPHIC COMPUTING :international summer school Proceedings Ottawa 1969 Aug 4-11 International union of crystallographers.Commission on crystallographic computing Edited by F.R. Ahmed and others Munksgaard,Copenhagen,1970
Chem 18.2733

CRYSTALLOGRAPHIC SOCIETY OF JAPAN International conference on magnetism and crystallography Proceedings Kyoto 1961 Sep 25-30 Pt 2: electron and neutron diffraction Edited by Shizuo Miyake and others Sponsored by the International union of pure and applied physics Physical society of Japan.Journal, 17,B 2 396p physical society of Japan,Tokyo,1962
Cav 7.1338

CRYSTALLOGRAPHIC SOCIETY OF JAPAN International conference on magnetism and crystallography Proceedings Kyoto 1961 Sep 25-30 Pt 3: neutron diffraction:study of magnetic materials Edited by Takeo Nagamiya and Shizno Miyake Under the auspices of International union of pure and applied physics Physical society of Japan. Journal, 17,B 3 71p physical society of Japan,Tokyo,1962
Cav 7.1339

CRYSTALLOGRAPHIC SOCIETY OF JAPAN International conference on magnetism and crystallography Proceedings Kyoto 1961 Sep 25-30 Pt 1: magnetism Edited by Kei Josida and others Held under the auspices of International union of pure and applied physics Physical society of Japan.Journal, 17,B 1 xi,718p Physical society of Japan, Tokyo,1962
Cav 7.1337

CRYSTALLOGRAPHY Aspects of protein structure The International symposium on protein structure and crystallography Proceedings Madras 1963 Jan 14-18 University of Madras. Department of physics Edited by G.N. Ramachandran Academic press,London,1963
Cav 7.1495

CRYSTALLOGRAPHY International conference on magnetism and crystallography Proceedings Kyoto 1961 Sep 25-30 Pt 2: electron and neutron diffraction Science council of Japan and Crystallographic society of Japan Edited by Shizuo Miyake and others Sponsored by the International union of pure and applied physics Physical society of Japan.Journal, 17,B 2 396p physical society of Japan, Tokyo,1962
Cav 7.1338

CRYSTALLOGRAPHY International conference on magnetism and crystallography Proceedings Kyoto 1961 Sep 25-30 Pt 3: neutron diffraction:study of magnetic materials Science council of Japan and Crystallographic society of Japan Edited by Takeo Nagamiya and Shizno Miyake Under the auspices of International union of pure and applied physics Physical society of Japan. Journal, 17,B 3 71p physical society of Japan,Tokyo,1962
Cav 7.1339

CRYSTALLOGRAPHY International conference on magnetism and crystallography Proceedings Kyoto 1961 Sep 25-30 Pt 1: magnetism Science council of Japan and Crystallographic society of Japan Edited by Kei Josida and others Held under the auspices of International union of pure and applied physics Physical society of Japan. Journal, 17,B 1 xi,718p Physical society of Japan,Tokyo,1962
Cav 7.1337

CRYSTALLOGRAPHY AND CRYSTAL PERFECTION The International symposium on protein structure and crystallography Proceedings Madras 1963 Jan 14-18 session on crystallography and crystal perfection University of Madras. Department of physics Edited by G.N. Ramachandran Academic press,London,1963
Cav 7.1494

CRYSTALLOGRAPHY AND CRYSTAL PERFECTION Madras 1963 Jan 14-18 Edited by G.N. Ramachandran Organized by University of Madras Academic press,London;New York,1963
Min 10.1193

CRYSTALLOGRAPHY AND CRYSTAL PERFECTION :a symposium Proceedings Madras 1963 Jan 14-18 Edited by G.N. Ramachandran Academic press,London;New York,1963 Papers of a symposium which formed part of the International symposium on protein structure and crystallography
Met 25.1448

CRYSTALS Dislocations and mechanical properties of crystals :international conference Proceedings Lake Placid,N.Y. 1956 Sep 6-8 United States.Air force.Office of scientific research and General electric research laboratory Edited by J.C. Fisher and others John Wiley,New York,1957
Cav 7.0804

CRYSTALS Growth and perfection of crystals International conference on crystal growth Proceedings Cooperstown,N.Y. 1958 Aug 27-29 General electric company Edited by R.H. Doremus and others Co-sponsored by United States.Air force.Office of scientific research Wiley;Chapman and Hall,New York;London,1958
Eng 41.4438

CRYSTALS Imperfections in nearly perfect crystals :a symposium Pocono Manor,Pa. 1950 Oct 12-14 Edited by W. Shockley and others Sponsored by the National research council.Committee on solids Wiley;Chapman and Hall,New York;London,1952
Eng 41.4427

CUERNAVACA,MEXICO 1951 Symposium on steroids in experimental and clinical practice Edited by Abraham White Churchill,London, 1951
Radioth 35.0697

CULCHETH,LANCS 1957 Brittleness in metals : a conference United Kingdom atomic energy authority.Industrial group I.G.Report 145 (RDC) Industrial group headquarters,Risley; Warrington,1959
Met 25.0859

CULTURE AND PERSONALITY :an interdisciplinary conference Proceedings New York 1947 Nov 7-8 Viking fund Edited by S.Stansfeld Sargent and Marian W. Smith Viking fund,New York,1949
Psy 31.1931

The CULTURE OF ALGAE :a symposium Proceedings New York 1969 Sep 7-12 Edited by J.E. Zajic Sponsored by the American chemical society.Division of microbial chemistry and technology illus, tables ix,154p Plenum press,New York; London,1970
Bot 42.3942

Les CULTURES DE TISSUS DE PLANTES Proceedings Strasburg 1970 Jul 6-10 Centre national de la recherche scientifique Centre national de la recherche scientifique.Colloques internationaux, 193 CNRS,Paris,1971
Bioch 33.2223

CUMULUS DYNAMICS Conference on cumulus convection 1st Proceedings Portsmouth,N. H. 1959 May 19-22 Edited by Charles E. Anderson Sponsored by the Air Force Cambridge research center.Geophysics research directorate ix,211p Pergamon press,Oxford, 1960
Nap 11.0012

CURARE AND CURARE-LIKE AGENTS International symposium on curare and curare-like agents Proceedings Rio de Janeiro 1957 Aug 5-12 Academia brasileira de ciencias.Conselho nacional de pesquisas and Universidade do Brasil Edited by D. Bovet and others Under the auspices of Unesco illus Elsevier, Amsterdam,1959 Also sponsored by the 'Istituto di sanita',of Rome.Papers in English and French
Chem 26.0103

CURARE AND CURARE-LIKE AGENTS International symposium on curare and curare-like agents Proceedings Rio de Janeiro 1957 Aug 5-12 Edited by D. Bovet and others Organized under the auspices of Unesco Elsevier, Amsterdam,1959
Pha 16.0243

CURARE AND CURARE-LIKE AGENTS 12th :Ciba foundation study group London 1961 Jul 4 Ciba foundation Edited by A.V.S. De Reuck Churchill,London,1962
Pha 16.0044

CURRENT CORROSION RESEARCH IN SCANDINAVIA Scandinavian corrosion congress (NKM) : lectures 4th Helsinki 1964 Nov 24-27 Teknillisten tieteiden akatemia and Central chemical association Edited by Jori Larinkari and others Organized by the Scandinavian council for applied research Kemian keskusliton julkaisuja, 24 Sanoma Osakeyhtio,Helsinki,1965
Met 25.1901

CURRENT ISSUES IN THE PHILOSOPHY OF SCIENCE : symposia of scientists and philosophers Chicago,Ill. 1959 Dec 27-30 American association for the advancement of science Edited by Herbert Feigl and Grover Maxwell xi,484p Holt,Rinehart and Winston,New York, 1961 Proceedings of section L of the Association
WSM 43.0784

CURRENT PROBLEMS OF LOWER VERTEBRATE PHYLOGENY Nobel symposium 4th Proceedings Stockholm 1967 Jun Edited by Tor Orvig Wiley,New York,1968
Bal 39.2054

CURRENT STATUS OF SOME MAJOR PROBLEMS IN DEVELOPMENTAL BIOLOGY :symposium Haverford,Penn. 1966 Jun Society for developmental biology Edited by Michael Locke Society for developmental biology. Symposia, 25 Academic press,New York;London, 1966 Formerly the Society for the study of development and growth
Gen 34.0487

CURRENT THEORY AND RESEARCH IN MOTIVATION :a symposium Papers Lincoln,Nebraska 1953 Jan 15-16 Mar 26-27 By Judson S. Brown and others University of Nebraska. Department of psychology University of Nebraska press,Lincoln,Nebraska,1953
Psy 31.1857

CURRENT TOPICS IN BIOCHEMISTRY Bethseda,Md. 1971 Sep National institutes of health Edited by C.B. Anfinsen National institutes of health.Lectures in biomedical sciences Academic press,New York,1972
Bioch 33.2168

CURRENT TOPICS IN SURGICAL RESEARCH, 1 Annual meeting of the Association for academic surgery 2nd Proceedings St.Louis,Miss. 1968 Nov Association for academic surgery Edited by G.D. Zuidema and D.B. Skinner Academic press,New York;London,1969
Surg 23.0069

CURRENT TRENDS IN INFORMATION THEORY 7th conference lectures Pittsburgh,Pa. 1953 Feb.20-21 By Brockway McMillan and others University of Pittsburgh.Department of psychology Current trends in psychology series xii,188p 22cm University of Pittsburgh press,Pittsburgh,Pa.,1953 7th annual conference on Current trends in psychology
Math 3.0767

CURRENT TRENDS IN PSYCHOLOGY AND THE BEHAVIORAL SCIENCES... :six lectures Lectures Pittsburgh,Pa. 1954 Mar 11-12 By John T. Wilson and others University of Pittsburgh. Department of psychology University of Pittsburgh press,Pittsburgh,1954
Psy 31.2597

CURRENT TRENDS IN RESEARCH AND CLINICAL MANAGEMENT IN DIABETES :a conference Papers New York 1959 Apr 10-11 By Peter H. Forsham and others New York academy of sciences New York academy of sciences. Annals, 82,p.191-644 New York,1959
Bioch 33.0420

CURRENTS IN PHOTOSYNTHESES Western-Europe conference on photosynthesis 2nd Proceedings Woudschoten,Zeist 1965 Sep 6-12 Edited by J.B. Thomas and J.C. Goedheer A.D.Donker,Rotterdam,1966
Bioch 33.0805

CURRENTS IN PHOTOSYNTHESIS The Western-Europe conference on photosynthesis 2nd Proceedings Woudschoten,Zeist 1965 Sep Edited by J.B. Thomas and J.C. Goedheer 486, 23p Donker,Rotterdam,1966
Bot 42.1890

CURRICULA IN SOLID MECHANICS :first study session in theoretical and applied mechanics curricula Boulder,Colo. 1961 Jun 19-24 Edited by H. Liebowitz and J.M. Allen Prentice-Hall,Englewood Cliffs,N.J.,1962
Eng 41.2567

CUTANEOUS CANCER The International conference on the biology of cutaneous cancer 1st Proceedings Philadelphia 1962 Apr 6-11 National cancer institute Edited by Frederick Urbach National cancer institute. Monograph, 10 U.S.Department of health, education,and welfare,Bethesda,Md.,1963 Sponsored by the skin and cancer hospital,Dept. of dermatology,Temple university school of medicine and The Committee on geographie pathology,Unio internationalis contra cancrum. A Merck Sharp and Dohme medical research conference
Bioch 33.0981

CUTANEOUS INNERVATION Brown university symposium on the biology of skin,1959 Proceedings Providence,R.I. 1959 Jan 24-25 Edited by William Montagna Advances in biology of skin, 1 Pergamon press,Oxford, 1960
An 32.4037

CUTANEOUS INNERVATION :Brown university symposium Proceedings Providence,R.I. 1959 Jan 24-25 Brown university Edited by William Montagna Advances in the biology of skin,1 Pergamon press,London,1966
VA 19.0232

CUTTACK 1961-62 Plant and animal viruses : a symposium Proceedings National institute of sciences of India National institute of sciences of India.Bulletin, 24 illus. vi, 248p National institute of sciences of India, New Delhi,1963
Bot 42.3739

CUTTACK 1961-62 The Symposium on plant and animal viruses Proceedings National institute of sciences of India National institute of sciences of India.Bulletin, 24, 1963 National institute of sciences of India, New Delhi,1963 Held under the joint auspices of the National institute and the Council of scientific and industrial research
Bal 39.1071

CYBERNETICS La Cybernetique:theorie du signal et de l'information :reunions d'etudes et de mises au point Paris 1951 Apr. Institut Henri Poincare.Seminaire de theories physiques Edited by Louis de Broglie vi, 317p 22cm Revue d'optique theorique et instrumentale,Paris,1951
Math L 5.0169

CYBERNETICS :circular causal and feedback mechanisms in biological and social systems Transactions New York 1951 Mar 15-16 Josiah Macy jr.foundation Edited by Heinz von Foerster Josiah Macy,jr. foundation,New York,1952
Psy 31.2046

CYBERNETICS:CIRCULAR CAUSAL AND FEEDBACK MECHANISMS IN BIOLOGICAL AND SOCIAL SYSTEMS Conference on cybernetics 7th Transactions New York 1950 Mar 23-24 Josiah Macy jr.foundation Edited by Heinz von Foerster Josiah Macy jr. foundation,New York,1951
Psy 31.1064

CYBERNETICS:CIRCULAR CAUSAL AND FEEDBACK MECHANISMS IN BIOLOGICAL AND SOCIAL SYSTEMS Conference on cybernetics 9th Transactions New York 1952 Mar 20-21 Josiah Macy jr.foundation Edited by Heinz von Foerster Josiah Macy jr. foundation,New York,1953
Psy 31.1063

CYBERNETICS OF THE NERVOUS SYSTEM Nerve, brain and memory models ..a series of lectures delivered during a symposium on cybernetics of the nervous system...held as part of the second international meeting of medical cybernetics at the Royal academy of sciences of Amsterdam... Lectures Amsterdam 1962 Apr 16-18 Edited by N. Wiener and J.P. Schade Progress in brain research, 2 Elsevier,Amsterdam,1963
Psy 31.0248

CYCLES Conference on cycles 1st-2nd Report Washington 1922 and 1928 Carnegie Institution of Washington Washington,1929
Philos 1.0096

CYCLIC AMP Role of cyclic AMP in cell function :symposium San Diego,Calif. 1970 Feb American college of neuropsychopharmacology Edited by Paul Greengard and Erminio Costa Advances in biochemical psychopharmacology, 3 Raven press,New York,1970
Bioch 33.2211

CYCLICAL ACTIVITY IN ENDOCRINE SYSTEMS :a symposium Proceedings London 1959 Dec 8 Zoological society of London Edited by F. J.W. Barrington Zoological society of London. Symposia, 2 Zoological society of London, London,1960
An 32.3764

CYCLITOLS AND PHOSPHOINOSITIDES The Federation of European biochemical societies meeting 2nd Proceedings Vienna 1965 Apr 21-24 Federation of European biochemical societies Edited by H. Kindl Pergamon press,Oxford,1966
Bioch 33.1393

CYRENAICA 1968 Geology and archaeology of northern Cyrenaica,Libya Petroleum exploration society of Libya :annual field conference 10th guidebook Edited by F.T. Barr Amsterdam,1968
Geol 8.4487

CYTOCHEMICAL PROGRESS IN ELECTRON MICROSCOPY Symposium on cytochemical progress in electron microscopy Oxford 1962 Jul 2-4 Royal microscopical society Journal of the Royal microscopical society.Ser.3, 81,pt 3-4,p.107-117 London,1963
Bal 39.0235

CYTOCHEMICAL PROGRESS IN ELECTRON MICROSCOPY a symposium Oxford 1962 Jul 2-4 Royal microscopical society Royal microscopical society.Journal, 81,pt 3-4 London,1963
Bot 42.4729

CYTOCHEMISTRY International congress of histochemistry and cytochemistry 3rd Summary reports New York 1968 Aug 18-22 Springer,New York,1968
Radioth 35.0547

CYTOCHEMISTRY International congress of histochemistry and cytochemistry 3rd summary reports New York 1968 Aug 18-22 United States.Office of the secretariat 317p Springer-verlag,New York,1968
Bot 42.0703

CYTOCHEMISTRY International congress of histochemistry and cytochemistry 3rd Summary reports New York 1968 Aug 18-22 Springer,New York,1968
An 32.0438

CYTOCHROMES Structure and function of cytochromes The Symposium on structural and chemical aspects of cytochromes Proceedings Osaka 1967 Aug 16-18 Edited by Kazuo Okunuki and others University of Tokyo press, Tokyo,1968
Bioch 33.1100

CYTODIFFERENTIATION AND MACROMOLECULAR SYNTHESIS Asilomar,Calif. 1962 Jun Society for the study of development and growth Edited by Michael Locke Society for the study of development and growth.Symposia, 21 Academic press,New York;London,1963
Gen 34.0854

CYTODIFFERENTIATION AND MACROMOLECULAR SYNTHESIS a symposium Asilomar,Calif. 1962 Jun Edited by Michael Locke Society for the study of development and growth.Symposia, 21 Academic press,London,1963
An 32.2501

CYTODIFFERENTIATION AND MACROMOLECULAR SYNTHESIS a symposium Asilomar,Calif. 1962 Jun Society for the study of development and growth Edited by Michael Locke Society for the study of development and growth.Symposia, 21 Academic press,New York;London,1963
Bioch 33.1009

CYTODIFFERENTIATION AND MACROMOLECULAR SYNTHESIS a symposium Papers Asilomar,Calif. 1962 Jun Society for the study of development and growth Edited by Michael Locke Society for the study of development and growth.Symposia, 21 Academic press,New York;London,1963
Bot 42.1496

CYTOGENETICS OF CELLS IN CULTURE symposium Pasadena,Calif. 1964 Jun 17 International society for cell biology Edited by R.J.C. Harris International society for cell biology.Symposia, 3 Academic press,New York,1964
Gen 34.0892

CYTOGENTICS OF CELLS IN CULTURE :a symposium Papers Pasadena,Calif. 1964 Jul 17 International society for cell biology Edited by R.J.C. Harris International society for cell biology.Symposia, 3 illus, port ix,307p Academic press,New York; London,1964
Bot 42.0784

CYTOLOGIE DE L'ADENOHYPOPHYSE :colloque international Paris 1963 Sep 9-14 Centre national de la recherche scientifique Edited by Jacques Benoit and Christian Da Lage Centre national de la recherche scientifique. Colloques internationaux, 128 C.N.R.S.,Paris, 1963
An 32.3770

CYTOLOGY International congress of microbiology 6th Rome 1953 Vol 1: symposium - bacterial cytology Edited by S. Mudd and G. Penso illus Fondazione Emanuele Paterno,Rome,1953 Title also in Italian;text in English and French
Bioch 33.1800

CYTOLOGY,GENETICS AND EVOLUTION University of Pennsylvania bicentennial conference By M. Demerec and others University of Pennsylvania University of Pennsylvania press,Philadelphia,Pa.,1941
Gen 34.0910

CYTOLOGY,GENETICS AND EVOLUTION University of Pennsylvania bicentennial conference By M. Demerec and others University of Pennsylvania press,Philadelphia,Pa.,1941
An 32.3347

CYTOLOGY,GENETICS AND EVOLUTION :university of Pennsylvania bicentennial conference By M. Demerec and others University of Pennsylvania.Bicentennial conference University of Pennsylvania press,Philadelphia, Pa.,1941
Bot 42.6465

CYTOPLASM The Relationship between nucleus and cytoplasm :a symposium Proceedings Brussels 1958 Jan 9-13 Organised by the International society for cell biology Experimental cell research.Supplement, 6 Academic press,New York,1959
An 32.2486

CYTOPLASMIC AND ENVIRONMENTAL INFLUENCES ON NUCLEAR BEHAVIOR The Control of nuclear activity :a symposium Papers Woods Hole, Mass. 1966 Aug 31-Sep 3 Edited by Lester Goldstein Held under the auspices of the Society of general physiologists Prentice-Hall,Englewood Cliffs,N.J.,1967 Original title of the symposium 'Cytoplasmic and environmental influences on nuclear behavior'
Bot 42.4736

CZECHOSLOVAK CHEMICAL SOCIETY International congress of chemical engineering,chemical equipment and automation 2nd Lecture summaries Marienbad 1965 Sep 12-19 Edited by Tomas Misek and Pavel Mitschka 1965
Chem E 24.1907

CZECHOSLOVAK CHEMICAL SOCIETY Structure of organic solids Microsymposium 'Structure of organic solids' 2nd Lectures Prague 1968 Sep 16-19 In conjunction with Ceskoslovenska akademie ved illus 550p 25cm Butterworths,London,1969 Simultaneously published in 'Pure and applied chemistry',vol 18,no 4,1969
Cav 7.2659

CZECHOSLOVAK CONFERENCE 3rd Transactions Electronics and vacuum physics Prague 1965 Sep 23-28 Ceskoslovenska akademie ved Edited by Libor Paty Sponsored by the Universita Karlova Academia,Prague,1967
Eng 41.5453

CZECHOSLOVAK PLANT VIROLOGISTS The Czechoslovak plant virologists conference 5th Proceedings Prague 1962 Jul 26-30 illus 364p Publishing house of the Czechoslovak academy of sciences,Prague,1964
Bot 42.3745

CZECHOSLOVAK SCIENTIFIC-TECHNICAL SOCIETY International congress of chemical engineering, chemical equipment and automation 2nd Lecture summaries Marienbad 1965 Sep 12-19 Edited by Tomas Misek and Pavel Mitschka 1965
Chem E 24.1907

CZECHOSLOVAK SCIENTIFIC-TECHNICAL SOCIETY Symposium on fatigue Papers and discussions Prague 1960 Sep 8-10 Ceskoslovenska vedecko-technicka spolecnost,1961 Papers in English,French,German and Russian
Met 25.0936

DACCA 1964 Scientific problems of the humid tropical zone deltas and their implication :a symposium Proceedings Unesco Jointly sponsored by the government of Pakistan Humid tropics research Unesco, Paris,1966 Partly in French
Geog 13.5396

DAINTON REPORT Framework for government research and development ;presented to Parliament by the Lord Privy Seal Great Britain.Lord Privy Seal Cmnd 4814 HMSO, London,1971 Contains the Rothschild and Dainton reports
Met 25.2797

DALHOUSIE UNIVERSITY Fundamental issues in associative learning :a symposium Proceedings Halifax 1968 Jun Edited by N. J. Mackintosh and W.K. Honig Dalhousie university press,Halifax,1969
Psy 31.3017

DALLAS 1957 Sheet materials for high temperature service Papers American society for metals American society for metals,Cleveland,1958
Met 25.1001

DALLAS 1962 Symposium on electroforming - applications,uses,and properties of electroformed metals presented at the 1962 committee week American society for testing materials A.S.T.M.Special technical publication, 318 A.S.T.M.,Philadelphia,Pa., 1962
Met 25.0733

DALLAS 1963 Unit processes in hydrometallurgy based on an international symposium American institute of mining, metallurgical and petroleum engineers. Extractive metallurgy division and Society of mining engineers.Minerals benefication division Edited by Milton E. Wadsworth and Franklin T. Davis Metallurgical society conferences, 24 Gordon and Breach,New York; London,1964
Met 25.2289

DALLAS 1966 Physiology of muscular exercise :a symposium Proceedings American heart association Edited by Carleta B. Chapman Circulation research.Supplement, 20-21 New York,1967
Inv Med 37.0053

DALLAS,TEX. 1963 25-26 Metallurgy at high pressures and high temperatures :a symposium Proceedings American institute of mining,metallurgical and petroleum engineers.Physical chemistry of steelmaking committee Edited by K.A. Gschneidner and others Metallurgical society conferences, 22 Gordon and Breach,New York;London,1964
Met 25.1639

DALLAS,TEXAS 1956 American chemical society national meeting 129th Literature of the combustion of petroleum :a symposium Papers American chemical society Advances in chemistry series, 20 Washington,D.C., 1958
Eng 41.7239

DALLAS,TEXAS 1956 Literature of the combustion of petroleum :a symposium held...at the 129th national meeting of the American chemical society Papers American chemical society American chemical society.Advances in chemistry series,20 American chemistry society,Washington,D.C.,1958
Chem 18.0449

DALLAS,TEXAS 1958 High strength steels for aircraft American society for metals American society for metals,Cleveland,Ohio, 1958
Met 25.1002

DALLAS,TEXAS 1963 Dynamic programming Colloquium lectures in pure and applied science 8 By Rutherford Aris Socony mobile oil company.Field research laboratory 310p 1965
Chem E 24.0172

DALLAS,TEXAS 1963 Quasi stellar sources and gravitational collapse Texas symposium on relativistic astrophysics 1st Proceedings By Ivor Robinson and others Southwest center for advanced studies Sponsored by University of Texas 475p 25cm University of Chicago press,Chicago,Ill.,1965
Cav 7.1985

DALLAS,TEXAS 1963 Quasi-stellar sources and gravitational collapse Texas symposium on relativistic astrophysics 1st proceedings Southwest center for advanced studies Edited by Ivor Robinson and others Sponsored by University of Texas xvii,475p 25cm University of Chicago press,Chicago, 1965
A Math 4.1161

DALLAS,TEXAS 1963 Quasi-stellar sources and gravitational collapse Texas symposium on relativistic astrophysics 1st proceedings Southwest center for advanced studies Edited by Ivor Robinson and others sponsored by University of Texas xvii,475p 25cm University of Chicago press,Chicago, 1965
Obs 6.1457

DALLAS,TEXAS 1968 Global effects of environmental pollution :a symposium American association for the advancement of science 218p Reidel,Dordrecht,1970
Nap 11.0970

DALLAS,TEXAS 1968 Symposium on viscous drag reduction Proceedings United States. Office of naval research.Naval ship research and development center National aeronautics and space administration Edited by C.S. Wells illus xi,500p 25cm Plenum press, New York,1969
A Math 4.1736

DALLAS,TEXAS 1968 Viscous drag reduction A Symposium on viscous drag reduction Proceedings Edited by C.Sinclair Wells Sponsored by United States.Office of naval research 500p 25cm Plenum press,New York, 1969
Cav 7.2923

DAMON RUNYON MEMORIAL FUND FOR CANCER RESEARCH Biology of normal and atypical pigment cell growth conference 4th Proceedings Houston,Tex. 1957 Nov 14-16 Edited by Myron Gordon Academic press,New York,1959
Gen 34.0601

DAMS International congress on large dams 5th Transactions Paris 1955 May 31-Jun 4 Vol 1-5 International commission on large dams 5 vols Paris,1956
Eng 41.3334

DANESBURY NUCLEAR PHYSICS LABORATORY International symposium on electron and photon interactions at high energies 4th Proceedings Liverpool 1969 Sep 14-20 illus 30cm Danesbury nuclear physics laboratory,Danesbury,1969
A Math 4.1760

DANISH ATOMIC ENERGY CENTRE Health physics in nuclear installations :a symposium Riso 1959 May 25-28 Organization for European economic cooperation European nuclear energy agency European nuclear energy agency,Paris, 1959
Radioth 35.1110

DANISH PSYCHOLOGICAL ASSOCIATION Child and education International congress of applied psychology 14th Proceedings Copenhagen 1961 Aug 13-19 Vol 3 Edited by Gerhard S. Nielsen Munksgaard,Copenhagen, 1962
Psy 31.1198

DANISH PSYCHOLOGICAL ASSOCIATION International congress of applied psychology 14th Proceedings Copenhagen 1961 Aug 13-19 Vol 2: personality research Edited by Stanley Coopersmith Munksgaard,Copenhagen, 1952
Psy 31.1968

DANISH PSYCHOLOGICAL ASSOCIATION International congress of applied psychology 14th Proceedings Copenhagen 1961 Aug 13-19 Vol 5: industrial and business psychology Edited by Gerhard S. Nielsen Munksgaard,Copenhagen,1962
Psy 31.2252

DANMARKS GEOLOGISKA UNDERSOGELSE Reunion geologique internationale compte-rendu Copenhagen 1928 Jun 25-28 Edited by V. Nordmann Copenhagen,1930 Sponsored by Danmarks geologiske undersogelse in celebration of its 40th anniversary
Geol 8.2917

DAR-ES-SALAAM 1956 C.C.T.A.East-central regional committee for geology :meeting 1st proceedings Commission for technical co-operation in Africa south of the Sahara London,1957 Bound with proceedings of three other regional meetings
Geol 8.3115

DARJEELING 1960 Summer school of botany Proceedings Ministry of scientific research and cultural affairs,New Dehli,1962
BG 38.3112

DARK NEBULAE,GLOBULES AND PROTOSTARS :a conference Tucson,Ariz. 1970 Mar 26-27 Edited by Beverly T. Lynds 150p University of Arizona press,Tucson,Ariz.
Obs 6.3638

DARMSTADT 1956 Electronic digital computers and information processing : conference Proceedings Gesellschaft fur angewandte mathematik und mechanik Nachrichtentechnische gesellschaft Edited by A. Walther and W. Hoffmann Nachrichtentechnische fachberichte, 4 illus. viii,229p 30cm Friedr.Vieweg und sohn,Brunswick,1956
Math L 5.3247

DARTMOUTH COLLEGE Tissue elasticity :a conference Papers Hanover,N.H. 1955 Sep 1-3 Edited by John W. Remington American physiological society,Washington,D.C.,1957
Bal 39.1204

DARTMOUTH COLLEGE.DEPARTMENT OF MATHEMATICS New directions in mathematics conference Hanover,N.H. 1961 Nov.3-4 Edited by Robert W. Ritchie Supported by the Alfred P.Sloan foundation iv,124p 22cm Prentice-Hall, Englewood Cliffs,N.J.,1964
P. Math 2.2558

DARWIN Charles A Darwin centenary :a conference Report London 1959 Nov 27-28 Botanical society of the British Isles Edited by P.J. Wanstall Botanical society of the British Isles.B.S.B.I.conference reports, 6 port. London,1961 Centenary of the publication of the Origin of species
Bot 42.4711

DARWIN Charles The Evolution of living organisms :a symposium to mark the centenary of Darwin's 'Origin of species' and of the Royal society of Victoria Melbourne 1959 Dec 8-11 Edited by G.W. Leiper Melbourne university press,Melbourne,1962
Gen 34.1277

DARWIN CELEBRATION 1909 Order of the proceedings at the Darwin celebration... :with a sketch of Darwin's life Cambridge 1909 Jun 22-24 University of Cambridge 23p Cambridge university press,Cambridge,1909
WSM 43.2842

A DARWIN CENTENARY :a conference Report London 1959 Nov 27-28 Botanical society of the British Isles Edited by P.J. Wanstall Botanical society of the British Isles.B.S.B.I. conference reports, 6 port. London,1961 Centenary of the publication of the Origin of species
Bot 42.4711

The DARWIN-WALLACE CELEBRATION London 1908 Linnean society of London London,1908
Philos 1.0757

DATA ACQUISITION AND PROCESSING IN BIOLOGY AND MEDICINE :a conference Proceedings Rochester 1961 Jul 13-14 Edited by K. Enslein Pergamon,Oxford;London,1962
PGMS 29.0509

DATA COMMUNICATION New York 1970 Mar 27 Institute of electrical and electronics engineers.Communication technology group.Data communication committee illus 24p IEEE, New York,1970
Math L 5.3626

DATA COMMUNICATION SYSTEMS ACM-IEEE symposium on problems in the optimization of data communications systems 2nd Proceedings Palo Alto,Calif. 1971 Oct 20-22 Association for computing machinery Institute of electrical and electronics engineers illus 198p ACM;IEEE,New York, 1971
Math L 5.3629

DATA PROCESSING Canadian conference for computing and data processing proceedings Toronto 1958 June 9-10 illus. xii,383p University of Toronto press,Toronto,1958
Math L 5.0845

DATA PROCESSING MACHINES International symposium on data processing machines Papers Prague 1964 Sep 1964 Collection of typescript papers in folder
Math L 5.3266

DATA PROCESSING SEMINAR ON STATUS OF DIGITAL COMPUTER AND DATA PROCESSING DEVELOPMENTS IN THE SOVIET UNION Washington,D.C. 1958 Nov 12 United States.Office of Naval research 179p 26cm United States Office of Naval research,Washington,D.C.,1958
Math L 5.0821

DATA PROCESSING SYSTEMS The Impact of users' needs on the design of data processing systems conference Edinburgh 1964 Mar 31-Apr 3 British computer society Held under the aegis of the United Kingdom automation council 1964 Organized jointly by the British computer society,the British institution of radio engineers and the Institution of electrical engineers
Math L 5.3265

DATA REPRESENTATION :a conference Proceedings Canberra 1969 Dec 10 Edited by R.S. Andersen and M.R. Osborne Held at the Australian national university 117p 24cm University of Queensland press, Brisbane,1970
Math S 3.1800

DATA TRANSMISSION International symposium on data transmission transactions Delft 1960 Sep 19-21 Institute of radio engineers. Benelux section Institute of radio engineers. Transactions on communications systems, 9,no. 1 99p I.R.E.professional group on communications systems,New York,1961
Math L 5.0868

DAVOS 1925 Verhandlungen der klimatologischen tagung in Davos 1925 Schweizerisches institut fur hochgebirgesphysiologie und tuberkuloseforschung illus vii,576p Benno Schwabe,Basel,c.1925
Nap 11.0894

DAVOS 1965 The International symposium on scientific aspects of snow and ice avalanches Rapports et discussions International association of scientific hydrology International association of scientific hydrology.Publication,69 illus 417p Gentbrugge,1966 In English and French
Sco 14.0142

DAVOS 1968 The Immune response and its suppression :an international symposium Forschungsinstitut,Davos Edited by E. Sorkin Antibiotica et chematherapia, 15 illus, tables Karger,Basle;New York,1969
Path 30.2545

DAYTON,OHIO 1962 Condensation and evaporation of solids International symposium on condensation and evaporation of solids Proceedings United States.Air force. Directorate of materials and processes Edited by Emile Rutner and others Gordon and Breach,New York,1964
Met 25.1518

DAYTON,OHIO 1965 International symposium on multivariate analysis Proceedings Edited by P.R. Krishnaiah Sponsored by United States.Air force.Aerospace research laboratories Academic press,New York,1966
Eng 41.2206

DAYTON,OHIO 1965 Matrix methods in structural mechanics :a conference Proceedings United States.Air force systems command United States.Air university 1966
Eng 41.2732

DAYTON,OHIO 1965 Multivariate analysis : international symposium Proceedings Edited by Paruchuri R. Krishnaiah Sponsored by Aerospace research laboratories Bibliog,port xix,592p 23cm Academic press,New York,1966
Math 3.1159

DAYTON,OHIO 1968 International symposium on multivariate analysis Proceedings Edited by Parachuri R. Krishnaiah Held at the Wright state university xx,696p 24cm Academic press,New York;London,1969
Math S 3.1802

DAYTON BEACH,FLA. 1958 Advanced prestressed concrete design National prestressed concrete short course 2nd Papers University of Florida.College of engineering.Bulletin series, 76 University of Florida,Gainesville,Fla.,1958
Eng 41.2847

DEBORAH HOSPITAL Congenital heart disease... an international symposium Philadelphia,Pa. 1960 Apr Edited by Dryden P. Morse Blackwell,Oxford,1962
An 32.2064

DEBRECEN 1965 Callus formation :symposium on the biology of fracture healing Edited by St. Krompecher and E. Kerner Translated by E. Kerner Symposia biologica hungarica, 7 Akademiai Kaido,Budapest,1967
An 32.3921

DEBRECEN 1967 Colloquium on information theory Proceedings Bolyai Janos matematikai tarsulat bibliog.,illus. 24cm 2 vols Janos Bolyai mathematical society, Budapest,1968
Math S 3.1457

DEBRECEN 1967 Colloquium spectroscopicum internationale 14th Proceedings Vol 1-3 Scientific society of mechanical engineers 1579p 24cm 3 vols Hilger, London,1968
Chem 18.2846

DEBRECEN 1968 Conference on the electron capture and higher processes in nuclear decays Proceedings Vol 1-3 International union of pure and applied physics Magyar tudomanyos akademia.Institute of nuclear research diagrs 20cm 3 vols Eotvos Lorand physical society,Budapest,1968
Cav 7.2700

DEBRECEN 1968 Number theory :a colloquium held at Edited by Paul Turan Colloquia mathematica societatis Janos Bolyai, 2 bibliog. 244p 24cm North-Holland, Amsterdam,1970
P Math 2.3898

DECADE OF PROGRESS IN EUGENICS International congress of eugenics 3rd Papers New York 1932 Aug 21-23 port.pl. Williams and Wilkins,Baltimore,Md.,1934
Gen 34.1927

DECENNIAL REVIEW CONFERENCE ON CELL TISSUE AND ORGAN CULTURE 2nd Proceedings Cell tissue and organ culture Bedford,Pa. 1966 Sep 11-15 National cancer institute and Tissue culture association Edited by Benton B. Westfall National cancer institute. Monograph, 26 National cancer institute, Bethesda,Md.,1967
Bioch 33.0991

DECHEMA see DEUTSCHE GESELLSCHAFT FUR CHEMISCHE APPARATEWESEN

DECISION MAKING AND AGE :N.A.T.O. advanced study institute Papers Thessaloniki 1967 Aug 14-19 North Atlantic treaty organization Edited by A.T. Welford and J.E. Birren Interdisciplinary topics in gerontology, 4 Karger,Basle;New York,c1968
Psy 31.3208

DECISION PROCESSES seminar proceedings Santa Monica,Calif. 1952 summer University of Michigan Edited by R.M. Thrall and others Supported by Ford foundation viii,332p 22cm John Wiley and sons,New York,1954
Math 3.0771

DECOMPOSITION OF AUSTENITE BY DIFFUSIONAL PROCESSES :a symposium Proceedings Philadelphia 1960 Oct 19 American institute of mining,metallurgical and petroleum engineers Edited by V.F. Zackay and H.I. Aaronson Interscience,New York; London,1962
Met 25.1247

DEDHAM,MASS. 1956 The Physical chemistry of steelmaking :a conference Proceedings Edited by John F. Elliott Sponsored by Massachusetts institute of technology. Department of metallurgy Wiley;Chapman and Hall,New York;London,1958
Met 25.1789

DEDHAM,MASS. 1959 Sensory communication The Symposium on principles of sensory communication Contributions Massachusetts institute of technology Edited by W.A. Rosenblith M.I.T.press;Wiley,New York;London, 1961 Held at M.I.T.'s Endicott House,1961
Bal 39.1518

DEDHAM,MASS. 1959 Sensory communications The Symposium on principles of sensory communications Contributions United States. Air force.Office of scientific research Edited by Walter A. Rosenblith MIT press, Cambridge,Mass.,1961
Psy 31.2842

DEDHAM,MASS. 1960 Control of ovulation :a conference Proceedings Edited by Claude A. Villee Sponsored by the Association for the aid of crippled children Pergamon press, Oxford,1961
An 32.3823

DEDHAM,MASS. 1960 Mechanism of action of steroid hormones :a conference Proceedings Edited by Claude A. Villee and Lewis L. Engel International series of monographs on pure and applied biology Pergamon,Oxford,1961
Inv Med 37.0219

DEDHAM,MASS. 1960 Mechanism of action of steroid hormones :conference Proceedings Edited by Claude A. Villee and Lewis L. Engel International series of monographs on pure and applied biology.Symposium division,1 Pergamon,Oxford,1961
Pha 16.0088

DEDHAM,MASS. 1962 Biochemistry of muscle contraction :a conference Proceedings Edited by John Gergely Retina foundation. Institute of biological and medical sciences. Monographs and conferences, 2 Churchill, London,1964
Bioch 33.2231

DEDHAM,MASS. 1963 Analysis in function space Conference on the theory and applications of analysis in function space Proceedings Edited by William T. Martin and Irving Sega vi,218p 24cm MIT press, Cambridge,Mass.,1964
Math S 3.1659

DEDHAM,MASS. 1963 Analysis in function space Conference on the theory and applications of analysis in function space proceedings Massachusetts institute of technology Edited by William Ted Martin and Irving Segal Financially supported by United States.Air force.Office of scientific research vi,218p 24cm M.I.T.press,Cambridge,Mass., 1964
P. Math 2.1181

DEDHAM,MASS. 1963 Conference on the theory and applications of analysis in function space Proceedings Massachusetts institute of technology Edited by William T. Martin and Irving Segal vi,218p 24cm MIT press, Cambridge,Mass.,1964
Cav 7.3154

DEFECTS IN CRYSTALLINE SOLIDS Conference on defects in crystalline solids Report Bristol 1954 Jul Physical society,London, 1955
Eng 41.4311

DEFECTS IN CRYSTALLINE SOLIDS Conference on defects in crystalline solids Report Bristol 1954 July Wills (H.H.)physical laboratory The Physcial society,London,1955
Met 25.2358

DEFORMATION AND FLOW OF SOLIDS :colloquium Verformung und fliessen des festkorpers Madrid 1955 Sep 26-30 International union of theoretical and applied mechanics Edited by R. Grammel Springer,Berlin,1956 Papers in English,French and German
Met 25.1203

DEFORMATION OF CRYSTALLINE SOLIDS International conference on the deformation of crystalline solids Ottawa 1966 Aug 22-26 National research council of Canada Canadian journal of physics, 2,pt 2-3 Ottawa,1967
Met 25.2491

DEFORMATION TWINNING :a conference Proceedings Gainesville 1963 Mar 21-22 American institute of mining,metallurgical and petroleum engineers and University of Florida.College of engineering In co-operation with Florida institute for continuing university studies Metallurgical society conferences, 25 Gordon and Breach, New York;London,1964
Met 25.1257

DEFORMATION UNDER HOT WORKING CONDITIONS :the conference Proceedings Sheffield 1966 Jul 5-6 University of Sheffield.Department of metallurgy Edited by C.M. Sellars and W.J. McG. Tegart Iron and Steel institute,London, 1968
Met 25.2402

DEGENERATION PATTERNS IN THE NERVOUS SYSTEM : a workshop ... held as part of the first International summer school of brain research.. Amsterdam 1963 Jul 15-26 Edited by M. Singer and J.P. Schade Progress in brain research, 14 Elsevier,Amsterdam,1965
An 32.4234

DEHYDROGENASES Pyridine-nucleotic-dependent dehydrogenases :advanced study institute Proceedings Konstanz 1969 Sep 15-20 Universitat Konstanz Edited by Horst Sund Springer,Berlin,1970
Bioch 33.2292

DEISTVIE OBLUCHENIYA NA ORGANIZM :doklady sovetskoi delegatsii na Mezhdunarodnoi konferentsii po mirnomu ispolzovaniyu atomnoi energii Doklady Geneva 1955 Jul 1-5 By A.A. Letavet and others Izdatelstvo Akademii nauk SSSR,Moscow,1955
Chem 18.0879

DELAWARE SEMINAR IN THE PHILOSOPHY OF SCIENCE Philosophy of science :the Delaware seminar Vol 1-: 1961- University of Delaware Edited by Bernard Baumrin Interscience,New York;London,1963-
WSM 43.0757

The DELAYED EFFECTS OF WHOLE BODY RADIATION : a symposium Bethesda,Md. 1959 Oct 29 Johns Hopkins university.Operation research office Edited by Bernard B. Watson Jointly sponsored by the Walter Reid army institute of research Johns Hopkins press,Baltimore,Md., 1960
Radioth 35.1968

DELAYED IMPLANTATION :symposium Houston, Tex. 1963 Jan 23-26 Rice university Edited by Allen C. Enders Rice university semicentennial series University of Chicago press,Chicago,Ill.,1963
An 32.3890

DELFT 1959 Symposium on the theory of thin elastic shells proceedings International union of theoretical and applied mechanics Edited by W.T. Koiter North-Holland, Amsterdam,1960
A Math 4.0366

DELFT 1959 The Theory of thin elastic shells :a symposium Proceedings International union of theoretical and applied mechanics North-Holland,Amsterdam,1960
Eng 41.2854

DELFT 1960 European regional conference on electron microscopy Proceedings Vol 1 Edited by A.L. Houwink and B.J. Spit Nederlandse vereniging voor electronenmicroscopie,Delft,1960
Met 25.1370

DELFT 1960 European regional conference on electron microscopy Proceedings Vols 1-2 Edited by A.L. Houwink and B.J. Spit 2 vols Delft,1961
Eng 41.5543

DELFT 1960 International symposium on data transmission transactions Institute of radio engineers.Benelux section Institute of radio engineers.Transactions on communications systems, 9,no.1 99p I.R.E.professional group on communications systems,New York,1961
Math L 5.0868

DELFT 1960 The European regional conference on electron microscopy Proceedings Vol 2: cytology Edited by A. L. Houwink and others Delft,1960
Phys 20.1508

DELFT 1961 Shell research :a symposium Proceedings International union of testing and research laboratories for materials and structure International association for shell structures Edited by A.M. Haas and A.L. Bouma North-Holland,Amsterdam,1961
Eng 41.2860

DELFT 1965 Performance of the eye at low luminances :colloquium Proceedings Institute for perception RVO-TNO N.V. Optische industrie "De oude Delft'.Scientific fund Scientiae serviens Edited by M.A. Bouman and J.J. Vos International congress series, 125 Excerpta medica,Amsterdam,1966
Psy 31.2862

DELTAIC AND SHALLOW MARINE DEPOSITS International congress of sedimentology 6th Proceedings Amsterdam 1963 May 29-30 Antwerp 1963 May 31-Jun 1 Edited by L.M. J.U.van Straaten Developments in sedimentology,1 Elsevier,Amsterdam,1964
Geog 13.0304

DEMONSTRATION,VERIFICATION,JUSTIFICATION : entretiens de l'Institut international de philosophie Liege 1967 Sep Institut international de philosophie Edited by Philippe Devaux xii,352p Nauwelaerts, Louvain;Paris,1968
WSM 43.0967

DENTAL ANTHROPOLOGY :a symposium London 1962 Edited by D.R. Brothwell Society for the study of human biology.Symposia, 5 Pergamon press,Oxford,1963
An 32.2883

DENVER 1961 Classification of carbonate rocks :symposium papers American association of petroleum geologists Edited by William E. Ham American ssociation of petroleum geologists.Memoirs,1 Tulsa,Okla., 1962
Geol 8.3157

DENVER 1962-65 Energetics in metallurgical phenomena :seminar 1st-4th Proceedings University of Denver Edited by William M. Mueller Energetics in metallurgical phenomena, 1-4 Gordon and Breach,New York, 1965-68
Met 25.1287

DENVER 1971 Atlantic City 1971 May Renal disease and hypertension :tentn annual conference... Proceedings United States. Public health service Edited by Richard H. Thurm Proceedings of the annual conferences of the U.S.Public health service co-operative study (antihypertensive agents), 8 U.S. Dept. of health,education and welfare,Bethesda, 1971
PGMS 29.0659

DENVER,COL. 1958 Vistas in astronautics Astronautics symposium 2nd Proceedings United States.Air force.Office of scientific research Edited by Morton Alperin and Hollingsworth F. Gregory Co-sponsored by Institute of aeronautical sciences Vistas in astronautics, 2 International series of monographs on aeronautical sciences and space flight.Astronautics division, 2 Pergamon press,London,1959
Nap 11.0023

DENVER,COL. 1966 Biomedical fluid mechanics symposium :presented at Fluids engineering conference Sponsored by the American society of chemical engineers.Fluids engineering division American society of mechanical engineers,New York,1966
Chem E 24.1797

DENVER,COLO. Personal and social factors in perception Perception and personality :a symposium Papers American psychological association Edited by Jerome S. Bruner and David Krech Duke university publications Duke university press,Durham,N.C.,1950 Papers originally appeared in the 'Journal of personality',vol 18,no 1-2,Sept and Dec. 1949
Psy 31.0945

DENVER,COLO. 1958- Annual conference on applications of X-ray analysis 7th- Proceedings Denver research institute Edited by William M. Mueller and others Advances in X-ray analysis, 2- Plenum press,New York,1960-
Met 25.2762

DENVER,COLO. 1961 Classification of carbonate rocks :a symposium Papers Edited by William E. Ham American association of petroleum geologists.Memoir,1 Tulsa,Okla., 1962 Held under the joint auspices of the American association of petroleum geologists and the Society of economic paleontologists and mineralogists
Geog 13.0301

DENVER,COLO. 1961 Man,culture and animals American association for the advancement of science :annual meeting 128th Symposium Edited by Anthony Leeds and Andrew P. Vayda American association for the advancement of science.Publications,78 Washington,D.C.,1963
Geog 13.1313

DENVER,COLO. 1961 Plant biology today: advances and challenges :a symposium American association for the advancement of science Botanical society of America Edited by William A. Jensen and Leroy G. Kavaljian illus Macmillan,London,1963
Bot 42.0298

DENVER,COLO. 1966 Biomedical fluid mechanics symposium Papers American society of mechanical engineers.Fluid mechanics commission Co-sponsored by the Rheology society of America American society of mechanical engineers,New York,1966 Papers presented by the Fluid mechanics commission of the ASME at the Fluids engineering conference,co-sponsored by the Rheology society of America
Eng 41.6540

DENVER,COLO. 1966 Physics,logic and history International colloquium on logic, physical reality and history Edited by Wolfgang Yourgrau and Allen D. Breck xiv, 336p Plenum press,New York;London,1970 Based on the 1st colloquium
WSM 43.1116

DENVER,COLO. 1966- Conference on applications of X-ray analysis 15- Proceedings Denver research institute Advances in X-ray analysis, 10- Plenum press,New York,1967-
Min 10.1549

DENVER,COLO. 1967 Advances in X-ray analysis 11 proceedings of the 16th annual conference on applications of X-ray analysis Denver research institute Edited by John B. Newkirk Plenum press,New York,1968
Met 25.2437

DENVER,COLO. 1969 Annual conference on application of X-ray analysis 18th Proceedings Denver research institute. Metallurgy division Edited by Burton L. Henke Advances in X-ray analysis, 13 Plenum press,New York,1970
Met 25.2609

DENVER RESEARCH INSTITUTE Advances in X-ray analysis 11 proceedings of the 16th annual conference on applications of X-ray analysis Denver,Colo. 1967 Aug 9-11 Edited by John B. Newkirk Plenum press,New York,1968
Met 25.2437

DENVER RESEARCH INSTITUTE Annual conference on applications of X-ray analysis 7th- Proceedings Denver,Colo. 1958- Edited by William M. Mueller and others Advances in X-ray analysis, 2- Plenum press,New York, 1960-
Met 25.2762

DENVER RESEARCH INSTITUTE Annual conference on applications of x-ray analysis 12th 1963 Aug 7-9 Edited by M. Mueller and others Advances in x-ray analysis, 7 Plenum press, New York,1964
Met 25.1445

DENVER RESEARCH INSTITUTE Conference on applications of X-ray analysis 15- Proceedings Denver,Colo. 1966- Advances in X-ray analysis, 10- Plenum press,New York,1967-
Min 10.1549

DENVER RESEARCH INSTITUTE.METALLURGY DIVISION Annual conference on application of X-ray analysis 18th Proceedings Denver,Colo. 1969 Aug 6-8 Edited by Burton L. Henke Advances in X-ray analysis, 13 Plenum press, New York,1970
Met 25.2609

DENVER RESEARCH INSTITUTE.METALLURGY DIVISION Annual conference on applications of X-ray analysis 17th Proceedings Estes Park, Colo. 1968 Aug 21-23 Edited by Charles S. Barrett and others Advances in X-ray analysis, 12 Plenum press,New York,1969
Met 25.2824

DEOXYRIBONUCLEIC ACID,STRUCTURE,SYNTHESIS AND FUNCTION Annual reunion of the society of chemical physicists 11th Proceedings Col de Voza 1961 Jun 26 - Jul 1 Society of chemical physicists Pergamon press,Oxford, 1961
Radioth 35.0111

DEPRESSION :a symposium Proceedings Cambridge 1959 Sep 22-26 University of Cambridge.School of clinical research and postgraduate medical training Edited by E. Beresford Davies Cambridge university press, Cambridge,1964
Psy 31.1679

DEPRESSION :the symposium Proceedings Cambridge 1959 Sep 22-26 University of Cambridge.School of clinical research and postgraduate medical teaching Edited by E.B. Davies University of Cambridge.School of clinical research and postgraduate medical teaching.Course,1959 Cambridge university press,Cambridge,1964
PGMS 29.0330

DEPRESSION AND ALLIED STATES McGill university conference on depression and allied states Papers Montreal 1959 Mar 19-21 McGill university.Department of psychiatry Canadian psychiatric association journal, 4, Special supplement, 1 Canadian psychiatric association journal,Ottawa,1959
Psy 28.0086

DESERT RESEARCH :international symposium Proceedings Jerusalem 1952 May 7-14 Jointly sponsored by the Research council of Israel Research council of Israel.Special publications,2 Research council of Israel, Jerusalem,1953
Geog 13.1458

DESERTS Biology of deserts Biology of hot and cold deserts :a symposium Proceedings London 1952 Sep 25-27 Edited by J.L. Cloudsley-Thompson Organized by the Institute of biology Institute of biology, London,1954
Geog 13.1205

DESERTS The Biology of hot and cold deserts.. a symposium Proceedings London 1952 Sep 25-27 Institute of biology Edited by J. L. Cloudsley-Thompson Assisted by a grant from Unesco Institute of biology,London,1954
Bal 39.1858

DESIGN Conference on design methods : conference on systematic and intuitive methods in engineering industrial design,architecture and communications Papers London 1962 Sep 19-21 Edited by J.Christopher Jones and D.G. Thornley Pergamon press,London,1963
Eng 41.0304

DESIGN ELECTRONICS Computer aided circuit design Conference on CACD Proceedings 1968 Mar 26-28 Sheffield university Kingston college of technology Design electronics,London,1968
Eng 41.2226

DESIGN ENGINEERING Product design through the engineering analysis of its end use :an approach and a case history.Presented at the Design engineering conference New York 1967 May 15-18 Sponsored by the American society of mechanical engineers.Design engineering division American society of mechanical engineers,New York,1967
Eng 41.0326

DESIGN GUIDE FOR MEDICAL GROUP PRACTICE CENTRES a study...incorporating papers given at a symposium London 1966 Jun 3 College of general practitioners National building agency College of general practitioners, London,c1966
PGMS 29.0555

DESIGN IN HIGH-STRENGTH STRUCTURAL STEELS :a conference Proceedings Eastbourne 1969 British iron and steel research association Iron and steel institute. Publication, 122 Iron and steel institute, London,1969
Eng 41.2761

The DESIGN OF HEATING COILS FOR STORAGE TANKS American society of mechanical engineers Edited by David Stuhlberg American society of mechanical engineers.Paper, 58-HT-1 American society of mechanical engineers, Philadelphia,Pa.,1958
Eng 41.7154

DESIGN OF HIGH BUILDINGS Symposium on the design of high buildings Proceedings of a meeting...held as part of the Golden jubilee congress of the University of Hong Kong Hong Kong 1961 Sep University of Hong Kong Edited by Sean Mackey Hong Kong university press,Hong Kong,1962
Eng 41.2685

DESIGN OF INDUSTRIAL EXPERIMENTS Experimental designs in industry Symposium on design of industrial experiments proceedings Raleigh,N.C. 1956 Nov.5-9 North Carolina state college.Institute of statistics Edited by Victor Chew Supported by the United States.Air force.Office of scientific research Wiley publications in statistics Bibliog. xi,268p 28cm John Wiley and sons,New York,1958 Being parts A and B of Symposium on design of industrial experiments, proceedings,edited by Victor Chew, 1957
Math 3.0583

DESIGN OF INDUSTRIAL EXPERIMENTS Symposium on design of industrial experiments proceedings Raleigh,N.C. 1956 Nov.5-9 North Carolina state college.Institute of statistics Edited by Victor Chew Supported by the United States.Air force.Office of scientific research Bibliog. viii,374p 27cm Institute of statistics,University of North Carolina,Raleigh,N.C.,1957
Math 3.0584

The DESIGN OF PHYSICS RESEARCH LABORATORIES :a symposium Proceedings London 1957 Nov 27 Held by the Institute of physics.London and home counties branch bibliog. Chapman and Hall;Reinhold,London;New York,1959
Radioth 35.1634

DESIGNS AND TESTS OF BUILDING STRUCTURES Seismic and shock loading...Glued laminated and other constructions Los Angeles,Calif. 1956 Sep 17 American society for testing materials American society for testing materials.Special technial publication, 209 American society for testing materials, Philadelphia,Pa.,1957
Eng 41.3017

DESTITUTION National conference on the prevention of destitution 1st Proceedings London 1911 May 30-Jun 2 Great Britain. Local government board King,London,1911 President of conference:the Rt.Hon.the Lord Mayor of London
Path 30.1285

DESTITUTION National conference on the prevention of destitution 1st Proceedings of the Public health section London 1911 May 30-Jun 2 Great Britain.Local government board King,London,1911 President of the Public health section of the Local government board:T.C.Allbutt
Path 30.1284

The DETECTION AND USE OF TRITIUM IN THE PHYSICAL AND BIOLOGICAL SCIENCES symposium Proceedings Vienna 1961 May 3-10 Vol 1-2 Joint commission on applied radioactivity Sponsored by the International atomic energy agency International atomic energy agency.Proceedings series 2 vols International atomic energy agency,Vienna,1962 Papers in English and Russian.Summaries in English,French,Russian and Spanish
Gen 34.0470

DETERGENTS,WETTING AND EMULSIFYING AGENTS :a symposium Papers London 1948 Apr 5-6 Organized by the Society of chemical industry. London section 48p Society of chemical industry,London,1950
Chem E 24.0923

DETERMINANTS OF INFANT BEHAVIOUR : proceedings of a Tavistock study group on mother-infant interaction London 1959 Sep Tavistock clinic.Child development research unit Edited by B.M. Foss Methuen; Wiley,London;New York,1961
Psy 31.1203

DETERMINANTS OF INFANT BEHAVIOUR 2 Tavistock study group on mother-infant interaction 2nd Proceedings London 1961 Sep Tavistock clinic.Child development research unit Edited by B.M. Foss Methuen;Wiley, London;New York,1963
Psy 31.1208

DETERMINANTS OF INFANT BEHAVIOUR 3 Tavistock study group on mother-infant interaction 3rd Proceedings London 1963 Sep Tavistock clinic.Child development research unit Edited by B.M. Foss Methuen;Wiley, London;New York,1965
Psy 31.1205

The DETERMINATION OF ADRENOCORTICAL STEROIDS AND THEIR METABOLITES :a conference Proceedings London 1953 May 21 Society for endocrinology Royal society of medicine.Endocrinological section Edited by P. Eckstein and S. Zuckerman Society for endocrinology.Memoirs, 2 Cambridge university press,Cambridge,1955
Radioth 35.1940

The DETERMINATION OF ADRENOCORTICAL STEROIDS AND THEIR METABOLITES Proceedings London 1953 May 21 Society for endocrinology Edited by P. Eckstein and S. Zuckermann Society for endocrinology.Memoirs Dobson, London,1953
Med 36.0126

DETERMINATION OF GASES IN METALS Symposium on determination of gases in metals Atlantic City,N.J. 1957 Jun 18 American society for testing materials A.S.T.M.Special technical publication, 222 A.S.T.M.,Philadelphia,Pa, 1958
Met 25.1708

The DETERMINATION OF GASES IN METALS :a symposium Report Iron and steel institute and Society for analytical chemistry Organised in conjunction with the Institute of metals Iron and steel institute. Special report, 68 Iron and steel institute, London,1960
Met 25.1747

DETERMINATION OF NONMETALLIC COMPOUNDS IN STEEL a symposium presented at the sixty-eighth annual meeting Lafayette 1965 Jun 13-18 American society for testing materials A.S.T.M.Special technical publication, 393 A.S.T.M. Philadelphia,1966
Met 25.2303

DETERMINATION OF RADIAL VELOCITIES AND THEIR APPLICATIONS Toronto 1966 Jun 20-24 International astronomical union and International council of scientific unions Edited by A.H. Batten and J.F. Heard International astronomical union.Symposium, 30 Academic press,London;New York,1967
TA 15.0307

DETERMINATION OF RADIAL VELOCITIES AND THEIR APPLICATIONS :a symposium Toronto 1966 Jun.20-24 International astronomical union Edited by A.H. Batten and J.F. Heard International astronomical union.Symposium, 30 262p academic press,London,1967 publication dedicated to the memory of R. Methven Petrie
Obs 6.0147

DETERMINATION OF STRESS IN ROCK :a symposium Atlantic City,N.J. 1966 Jun 26-Jul 1 American society for testing and materials American society for testing and materials. Special technical publication, 429 American society for testing materials,Philadelphia,Pa., 1967
Eng 41.3160

DETERMINISM AND FREEDOM IN THE AGE OF MODERN SCIENCE:A PHILOSOPHICAL SYMPOSIUM New York university institute of philosophy : annual symposium 1st Proceedings New York 1957 Feb 9-10 Edited by Sidney Hook xv,237p New York university press,New York, 1965
WSM 43.1662

DETROIT 1938 The Pyrometry of solids and surfaces By Robert B. Sosman American society for metals,Cleveland,1940 A series of three educational lectures presented to members of the American society for metals during the twentieth National metal congress and exposition Detroit,Michigan,Oct.17 to 21. 1938
Met 25.0297

DETROIT 1948 Symposium on metallography in color (1948) presented to the 51st annual meeting American society for testing materials A.S.T.M.Special technical publication, 86 American society for testing materials,Philadelphia,Pa.,1949
Met 25.1342

DETROIT 1951 Metal interfaces a seminar held during the thirty-third national metal congress and exposition American society for metals American society for metals,Cleveland, Ohio,1952
Met 25.1261

DETROIT 1951 Residual stress measurements By R.G. Treuting and others A series of four educational lectures presented to members of the A.S.M. during the 33rd National metal congress and exposition
Met 25.0962

DETROIT 1954 The Hypophyseal growth hormone;nature and actions :an international symposium Proceedings By Richmond W. Smith and others Sponsored by the Henry Ford Hospital and Edsel B.Ford institute for medical research New York,1955
Bioch 33.0483

DETROIT 1955 Enzymes:units of biological structure and function Henry Ford hospital international symposium Proceedings Edited by Oliver H. Gaebler held at Henry Ford hospital Henry Ford hospital.Symposia Academic press,New York,1956
Bioch 33.1050

DETROIT 1955 Enzymes:units of biological structure and function :international symposium Proceedings Henry Ford hospital Edited by Oliver H. Guebler Henry Ford hospital symposia, 4 Academic press,New York,1956
Radioth 35.0052

DETROIT 1956 International symposium on the leukemias :etiology,pathophysiology and treatment Proceedings Edited by John W. Rebuck and others Sponsored by Henry Ford hospital Academic press,New York,1957
Med 36.0299

DETROIT 1958 Warren,Mich. 1958 Internal stresses and fatigue in metals the symposium Proceedings General motors corporation.Research laboratories Edited by Gerald M. Rasswiler and William L. Grube Elsevier,Amsterdam,1959
Met 25.0965

DETROIT 1959 General session on powder metallurgy Metal powder industries federation Metal powder industries federation,New York,1959
Met 25.0764

DETROIT 1960 National die casting exposition and congress 1st Technical papers Society of die casting engineers Society of die casting engineers,Detroit,1960
Met 25.0630

DETROIT 1960 Refractory metals and alloys : a technical conference American institute of mining,metallurgical and petroleum engineers. Refractory metals committee Edited by M. Semchyshen and J.J. Harwood Metallurgical society conferences, 11 Interscience,New York;London,1961
Met 25.2532

DETROIT 1961 Analytic cell culture Syverton memorial symposium Proceedings National cancer institute Edited by Robert F. Stevenson Sponsored by Tissue culture association and Cell culture collection committee of the viruses and cancer board, National cancer institute National cancer institute.Monograph, 7 U.S.Department of health,education and welfare,Bethesda,Md.,1962
Bioch 33.0979

DETROIT 1961 Continuous casting technical sessions Proceedings American institute of mining,metallurgical and petroleum engineers. Committee on physical chemistry of steelmaking Edited by D.L. McBride and T.E. Dancy Interscience,New York;London,1962
Met 25.0664

DETROIT 1961 Iron and its dilute solid solutions :a conference Proceedings American institute of mining,metallurgical and petroleum engineers Edited by C.W. Spencer and F.E. Werner Interscience;Wiley,New York; London,1963
Met 25.1305

DETROIT 1961 The Cell in mitosis :a symposium Proceedings Wayne state university Edited by Laurence Levine Wayne state fund research recognition award.Annual symposia, 1 Academic press,New York,1963
An 32.3412

DETROIT 1961 The Cell in mitosis symposium Proceedings Edited by Laurence Levine Held at Wayne state university Wayne State fund research recognition award.Annual symposia, 1 Academic press,New York;London,1963
Bioch 33.0973

DETROIT,MICH. 1951 World metallurgical congress 1st Proceedings American society for metals Edited by William Marsh Baldwin 833p American society for metals, Cleveland,Ohio,1952
Met A 61 25.0044

DETROIT,MICH. 1953 The Dynamics of virus and rickettsial infections :an international symposium Proceedings Henry Ford hospital Edited by Frank W. Hartmann and others illus, tables,diagrs Blakiston,New York;Toronto, 1954
Path 30.2548

DETROIT,MICH. 1954 Conference on training personnel for the computing machine field 1st proceedings Wayne state university Edited by Arvid W. Jacobson Sponsored also by the Association for computing machinery,the Industrial mathematics society and the Institute of radio engineers 104p 23cm Wayne university press,Detroit,Mich.,1955
Math L 5.0730

DETROIT,MICH. 1954 The Hypophyseal growth hormone,nature and actions :international symposium Edited by Richmond W. Smith and others Sponsored by the Henry Ford hospital McGraw-Hill,New York,1955
An 32.3736

DETROIT,MICH. 1955 Enzymes:units of biological structure and function : international symposium Proceedings Henry Ford hospital Edited by Oliver H. Gaebler Henry Ford hospital.Symposia, 4 Academic press,New York,1956
Gen 34.0646

DETROIT,MICH. 1957 Reticular formation of the brain Henry Ford hospital international symposium Contributions Henry Ford hospital Edited by Herbert H. Jasper and others J.and A.Churchill,London,1957
Psy 31.0230

DETROIT,MICH. 1957 Symposium on friction and wear Proceedings Edited by Robert Davies port Elsevier,Amsterdam,1959
Eng 41.6297

DETROIT,MICH. 1960 Blood platelets :Henry Ford hospital international symposium Edited by Shirley A. Johnson and others Henry Ford hospital.Symposia, 10 Churchill,London,1961
An 32.3175

DETROIT,MICH. 1961 Henry Ford hospital international symposium Biological interaction in normal and neoplastic growth :a contribution to the Host-tumour problem Papers Edited by Michael J. Brennan and William L. Simpson Churchill,London,1962
An 32.2497

DETROIT,MICH. 1961 International conference on coordination chemistry Advances in the chemistry of coordination compounds 6th Proceedings American chemical society, and,Wayne state university Edited by Stanley Kirschner Macmillan,New York,1961 Co-sponsored by the United States air force,and the International union of pure and applied physics
Chem 18.0991

DETROIT,MICH. 1961 The Cell in mitosis annual symposium Proceedings Wayne state university Edited by Laurence Levine Wayne state fund research recognition award.Annual symposia, 1 illus. Academic press,New York;London,1963
Gen 34.0862

DETROIT,MICH. 1965 Advanced propellant chemistry :a symposium Papers American chemical society.Division of fuel chemistry, and,American institute of aeronautics and astronautics.Propellants and combustion technical committee American chemical society.Advances in chemistry series,54 American chemical society,Washington,D.C.,1966 Presented at the 149th meeting of the American chemical society.Symposium chairman: Richard T.Holtzmann
Chem 18.0977

DETROIT,MICH. 1965 Positron annihilation conference :conference Proceedings Wayne state university Edited by A.T. Stewart and L.O. Roellig xvi,438p 24cm Academic press,New York,1967
Cav 7.2855

DETROIT,MICH. 1968 Semigroups Symposium on semigroups Proceedings Wayne state university Edited by Karl W. Folley Sponsored by the National science foundation bibliog. xi,277p 24cm Academic press,New York;London,1969
P Math 2.3927

DEUCE USERS' COLLOQUIUM ON MONTE CARLO METHODS papers London 1960 May 17 English electric company DEUCE news, 52 21p English electric,Nelson,Staffs.,1960 Typescript
Math L 5.2410

DEUCE USERS' COLLOQUIUM ON PARTIAL DIFFERENTIAL EQUATIONS papers London 1959 Oct 6 English electric company ltd DEUCE news, 45 91p English electric,Nelson,Staffs., 1959
Math L 5.2408

DEUEL CONFERENCE ON LIPIDS Proceedings
Adipose tissue as an organ Edited by Laurance W. Kinsell Thomas,Springfield,Ill., 1962
An 32.3279

DEUTSCHE AKADEMIE DER LANDWIRTSCHAFTSWISSENSCHAFTEN ZU BERLIN
Chemie und biochemie der solanum-alkaloide : international symposium Vortrage und diskussionsbeitrage Berlin 1959 Jun 2-4 Deutsche akademie der landwirtschaftswissenschaften zu Berlin. Tagungsberichte, 27 Berlin,1961 Papers in English and German,with summaries in English,German and Russian
Chem 26.0814

DEUTSCHE AKADEMIE DER NATURFORSCHER LEOPOLDINA
Purkyne symposium Vortrage und diskussionsbeitrage Halle 1959 Oct 31-Nov 1 Edited by Rudolph Zaunick Nova acta leopoldina.Neue folge, 24 230p Barth, Leipzig,1961
WSM 43.2940

DEUTSCHE AKADEMIE DER WISSENSCHAFTEN ZU BERLIN
Einstein symposium Entstehung, entwicklung und perspektiven der Einsteinschen gravitationstheorie vortrage und diskussionen Berlin 1965 Nov 2-5 Edited by H-J. Treder 314p Akademie-verlag,Berlin, 1966
Obs 6.1765

DEUTSCHE AKADEMIE DER WISSENSCHAFTEN ZU BERLIN
Fachkonference radioastronomie :jahrestagung der deutschen akademie der wissenschaften Berlin 1955 Mar 28- Apr 2 Deutsche akademie der wissenschaften zu Berlin. Abhandlungen.Klasse fur mathematik,physik und technik, 3 135p Akademie-verlag,Berlin, 1958
Obs 6.2016

DEUTSCHE AKADEMIE DER WISSENSCHAFTEN ZU BERLIN. SEKTION DER PHYSIK.UNTERKOMMISSION METALLPHYSIK Kristallisation
Gemeinschaftskonferenz der reihe 'Metall' :die kristallisation von metallen aus dem schmelzfluss,der gasphase und durch elektrolytische abscheidung 6 Vortrage Berlin 1968 Mar 28-29 VEB verlag fur grundstoffindustrie,Leipzig,1969
Met 25.2622

DEUTSCHE ANTHROPOLOGISCHE GESELLSCHAFT: VERSAMMLUNG 49th Bericht Cologne
1927 Sep 11-17 Edited by Walter Venn-Koln Im auftrage der Kolner anthropologische gesellschaft Kabitzsch,Leipzig,1928
An 32.2563

DEUTSCHE BUNSEN-GESELLSCHAFT FUR PHYSIKALISCHE CHEMIE International symposium on the reactivity of solids 5th Munich 1964 Aug 2-8 Edited by G.M. Schwab Sponsored by the International union of pure and applied chemistry Elsevier,Amsterdam,1965 Papers in English,French and German
Met 25.1851

DEUTSCHE BUNSENGESELLSCHAFT FUR PHYSIKALISCHE CHEMIE and INTERNATIONAL UNION OF PURE AND APPLIED CHEMISTRY, International symposium on the reactivity of solids 5th Plenary lectures Munich 1964 Aug 2-8 Butterworths,London,1965
Min 10.0662

DEUTSCHE FORSCHUNGSGEMEINSCHAFT Vortrage uber rechenanlagen Gottingen 1953 Mar 19-21 Edited by L. Biermann 145p 21cm Max-Planck institut fur physik,Gottingen,1953
Math L 5.1087

DEUTSCHE GESELLSCHAFT FUR BIOPHYSIK
Biophysikalische probleme der strahlen wirkung jahrestagung der Deutschen gesellschaft fur biophysik Tagungsbericht Hamburg 1965 Apr 23-24 Edited by H. Muth Thieme, Stuttgart,1966
Radioth 35.1420

DEUTSCHE GESELLSCHAFT FUR BIOPHYSIK
Molekulare struktur und strahlenwirkung : jahrestagung der Deutschen gesellschaft fur biophysik Tagungsbericht Hanover 1966 Jan 2-3 Edited by H. Glubrecht illus. Thieme,Stuttgart,1968
Radioth 35.1423

DEUTSCHE GESELLSCHAFT FUR CHEMISCHES APPARATEWESEN Congres international de chimie industrielle und der Dechema-hauptversammlung 1952 Vortrage Frankfurt am Main 1952 Dechema-monographie, 21,p245-268 Verlag chemie GMBH,Weinheim,1952
Chem E 24.1906

DEUTSCHE GESELLSCHAFT FUR METALLKUNDE
Passivierend filme und deckschichten anlaufschichten mechanismus ihrer entstehung und ihre schutzwirkung gegen korrosion Frankfurt 1955 Oct 13-14 Edited by H. Fischer and K. Wiederholt Springer-verlag, Berlin,1956
Met 25.1944

DEUTSCHE GESELLSCHAFT FUR PSYCHOLOGIE
International congress of psychology 16th Berichte.Proceedings Bonn 1960 Jul 31-Aug 5 Acta psychologica, 19 North-Holland, Amsterdam,1961 President of congress: professor Metzger.Conference languages:English, French and German
Psy 31.3396

DEUTSCHE GESELLSCHAFT FUR VAKUUMTECHNIK
Vacuum technique Meeting of the German society for vacuum technique Proceedings Heidelberg 1962 Sep 18-21 Macmillan; Pergamon press,London;Oxford,1965
Cav 7.1062

DEUTSCHE NATURFORSCHER UND AERZTE Amtlicher bericht uber die einundzwanzigste versammlung deutscher naturforscher und aerzte 21st Proceedings Gratz 1843 Sep Edited by L. Langer and A. Schrotter A.Leykam,Gratz,1844
Hunt B 22.0463

DEUTSCHE NATURFORSCHER UND AERZTE VERSAMMLUNG 60
Katalog zur wissenschaftlichen ausstellung der 60 versammlung deutscher naturforscher und aerzte Edited by Ludwig Dreyfus 224p Bergmann,Wiesbaden,1887
WSM 43.5009

DEUTSCHE RONTGENGESELLSCHAFT Biophysik, strahlenbiologie und strahlentherapie : vortrage auf der 39 tagung der Deutschen rontgengesellschaft Proceedings Frankfurt 1957 Oct 20-24 Edited by B. Rajewsky Urban and Schwarzenberg,Munich,1958
Radioth 35.1617

DEUTSCHE RONTGENGESELLSCHAFT 70 jahre radiologie am Katharinen hospital der stadt : kongress der deutschen rontgengesellschaft Stuttgart 1969 May 8-11 Edited by Friedrich Heuck Stuttgart,1970
Radioth 35.1913

DEUTSCHE RONTGENKONGRESS 1970 Vortrage Pra-operative tumorbestrahlung Edited by Otto Hugg Urban and Schwarzenberg,Munich, 1971
Radioth 35.1939

DEUTSCHE TIEFSEE-EXPEDITION,1898-99 Die Deutsche tiefsee-expedition auf dem schiff 'Valdivia',1898-99 Internationaler geographen-kongress,Berlin 1899 7 Berlin 1899 By Carl Chun and Gerhard Scott Maps,tables 120p 26cm Gesellschaft fur erdkunde zu Berlin,Berlin,1899
Sco 14.6301

DEUTSCHEN GESELLSCHAFT FUR ENDOKRINOLOGIE. SYMPOSIA, 3 Probleme der fetalen endokrinologie symposium Bonn 1955 Mar 4-5 Edited by H. Nowakowski Springer,Berlin,1956
An 32.3750

DEUTSCHES CENTRAL-KOMITE ZUR ERRICHTUNG VON HEILSTATTEN FUR LUNGENKRANKE Kongress zur bekampfung der tuberkulose als volkskrankheit Bericht Berlin 1899 May 24-27 Edited by Gotthold Pannwitz Berlin,1899
Path 30.0770

DEUTSCHES CENTRAL-KOMITE ZUR ERRICHTUNG VON HEILSTATTEN FUR LUNGENKRANKE Versammlung der tuberkulose-arzte 2te Bericht Berlin 1904 Nov 24-26 Edited by Johannes Nietner Deutsches central-komite zur errichtung von heilstatten fur lungenkranke, Berlin,1905
Path 30.0761

The DEVELOPING BRAIN Edited by Williamina A. Himwich and Harold E. Himwich Progress in brain research, 9 Elsevier,Amsterdam,1964 Derived from a Symposium held at Galesburg, Ill.,April 1963
An 32.4229

DEVELOPING CELL SYSTEMS AND THEIR CONTROL 1959 Society for the study of development and growth Edited by Dorothea Rudnick Society for the study of development and growth.Symposia, 18 Ronald press,New York, 1960
Gen 34.0850

DEVELOPING CELL SYSTEMS AND THEIR CONTROL :a symposium Madison,Wis. 1959 Society for the study of development and growth Edited by Dorothea Rudnick at the University of Wisconsin Society for the study of development and growth.Symposia, 18 Ronald press,New York,1960
Bioch 33.1010

DEVELOPING CELL SYSTEMS AND THEIR CONTROL :a symposium Papers Edited by Dorothea Rudnick Society for the study of development and growth.Symposia, 18 Ronald press,New York,1960
An 32.2495

DEVELOPMENT Problemi di sviluppo :un simposio Milan 1952 Sep Unione zoologie italiana Casa editrice ambrosiana,Milan,1954
An 32.2477

The DEVELOPMENT OF AGRICULTURE IN THE DRY ZONE a symposium Proceedings Colombo 1967 Jul 30-31 Ceylon association for the advancement of science Edited by O.S. Peries bibliog. Colombo,1967
Geog 13.4482

The DEVELOPMENT OF INDUSTRIAL SOCIETIES : papers read at the Nottingham conference of the British sociological association Nottingham 1964 Apr British sociological association Edited by Paul Halmos Sociological review monograph, 8 University of Keele,Keele,1964
Eng 41.0914

DEVELOPMENT OF THE LUNG :a Ciba foundation symposium London 1965 Nov 1-3 Ciba foundation Edited by A.V.S. De Reuck and Ruth Porter Churchill,London,1967
Phys 20.2220

DEVELOPMENT OF THE LUNG :a Ciba foundation symposium London 1965 Nov 1-3 Ciba foundation Edited by A.V.S. Reuck and Ruth Porter Churchill,London,1967
An 32.3803

DEVELOPMENTAL BIOLOGY Major problems in developmental biology :a symposium Haverford, Pa. 1966 Jun Society for developmental biology Edited by Michael Locke Society for developmental biology.Symposia, 25 Academic press,New York;London,1966
Bioch 33.0912

DEVELOPMENTAL BIOLOGY The Emergence of order in developing systems Ithaca,N.Y. 1968 Jun Society for developmental biology Edited by Michael Locke Society for developmental biology.Symposia, 27 350p Academic press, New York,1968
Bot 42.1097

DEVELOPMENTAL BIOLOGY CONFERENCE SERIES,1956 Dynamics of proliferating tissues :a symposium Upton,N.Y. 1956 Sep 5-8 Edited by Dorothy Price University of Chicago press,Chicago, Ill.,1958
An 32.3285

DEVELOPMENTAL BIOLOGY CONFERENCE SERIES,1956 Endocrines in development Shelter Island,N.Y. 1956 Sep 11-13 National research council Edited by Ray L. Watterson University of Chicago.Committee on publications in biology and medicine University of Chicago press, Chicago,Ill.,1959
Gen 34.0553

DEVELOPMENTAL BIOLOGY CONFERENCES SERIES,1956 Wound healing and tissue repair :a symposium New York 1956 Oct 2-4 Edited by W.Bradford Patterson University of Chicago press, Chicago,Ill.,1959
An 32.3161

DEVELOPMENTAL CYTOLOGY Kingston,R.I. 1957 Jun 19-24 Society for the study of development and growth Edited by Dorothea Rudnick Society for the study of development and growth.Symposia, 16 Ronald press,New York,1957
Gen 34.0849

DEVELOPMENTS IN AROMATIC CHEMISTRY
Developments in aromatic chemistry. Applications of electron and nuclear resonance in chemistry.Recent work on the inorganic chemistry of sulphur :symposia Bristol 1958 Chemical society Chemical society. Special publication, 12 Chemical society, London,1958
Bioch 33.0630

DEVELOPMENTS IN MECHANICS Midwestern mechanics conference 6th Proceedings Cleveland,Ohio 1963 Apr 1-3 Vol 2,pt 1: fluid mechanics Edited by Simon Ostrach and Robert H. Scanlan Pergamon,London,1965 Combining the 6th Midwestern conference on soil mechanics and the 8th Midwestern conference on fluid mechanics
Eng 41.6959

DEVELOPMENTS IN MECHANICS Midwestern mechanics conference 7th proceedings Ann Arbor,Mich 1961 Sep. 6-8 Vol 1 Edited by J.E. Lay and L.E. Malvern North Holland,Amsterdam,1961
A Math 4.0313

DEVELOPMENTS IN THE PHARMACOLOGY AND CLINICAL USES OF HUMAN GONADOTROPHINS :a private scientific meeting Proceedings London 1968 Mar 15-16 Sponsored by Searle (G.D.) and company Searle,High Wycombe,1970
Inv Med 37.0261

DEVELOPMENTS IN THEORETICAL AND APPLIED MECHANICS Southeastern conference on theoretical and applied technics 4th Proceedings New Orleans,La. 1968 Feb 29-Mar 1 Edited by D. Frederick Sponsored by Tulane university illus 637p 26cm Pergamon,Oxford,1970
A Math 4.1903

DFP SENSITIVE ENZYMES The Conference on structure and reactions of DFP sensitive enzymes Proceedings Stockholm 1966 Sep 5-7 Edited by Edith Heilbronn organized by the Forsvarets forskningsanstalt,Stockholm Research institute of national defence, Stockholm,1967
Bioch 33.1065

DIABETES Current trends in research and clinical management in diabetes :a conference Papers New York 1959 Apr 10-11 By Peter H. Forsham and others New York academy of sciences New York academy of sciences.Annals, 82,p.191-644 New York,1959
Bioch 33.0420

DIABETES Experimental diabetes and its relation to the clinical disease :a symposium Leiden 1952 Jul 14-18 Council for international organizations of medical sciences Edited by J.F. Delafresnaye and G. Howard Smith Blackwell,Oxford,1954
Bioch 33.0448

DIABETES International diabetes federation congress 6th Proceedings Stockholm 1967 Jul 30-Aug 4 International diabetes federation Edited by J. Ostman and R.D.G. Milner Excerpta medica,Amsterdam,1969
Bioch 33.0462

DIABETES MELITUS :the opening symposium of the Pfizer foundation of the post-graduate medical school,University of Edinburgh Edinburgh 1965 Pfizer foundation Edited by L.J.P. Duncan University of Edinburgh. Pfizer medical monographs Edinburgh university press,Edinburgh,1966
PGMS 29.0179

DIABETES MELLITUS Chlorpropamide and diabetes mellitus :a conference Papers New York 1958 Sep 25-27 By Martin G. Goldner and others New York academy of sciences New York academy of sciences.Annals, 74,p.407-1028 New York,1959
Bioch 33.0419

DIABETES MELLITUS Small blood vessel involvement in diabetes mellitus :conference Proceedings Warrenton,Va. 1963 Mar 25-27 American institute of biological sciences Edited by Marvin D. Siperstein and others American institute of biological sciences, Washington,D.C.,1964
Bioch 33.0482

DIAGNOSIS AND TREATMENT OF RADIOACTIVE POISONING Scientific meeting on the diagnosis and treatment of radioactive poisoning Proceedings Vienna 1962 Oct 15-18 World health organization International atomic energy agency International atomic energy agency.Proceedings series International atomic energy agency,Vienna,1963
Radioth 35.1368

DIALYSIS AND RENAL TRANSPLANTATION European dialysis and transplant association conference 4th Proceedings Paris 1967 Jun Edited by David S. Kerr and others European dialysis and transplant association. Proceedings, 4 International congress series, 155 Excerpta medica,Amsterdam,1968
Surg 23.0043

DIAMOND ORDNANCE FUZE LABORATORIES Symposium on microminiaturization of electronic assemblies Proceedings Washington,D.C. 1958 Sep 30-Oct 1 Edited by Eleonor F. Horsey and Laurence D. Shergalis Hayden,New York,1960
Eng 41.5541

DIAZEPAM IN ANAESTHESIA :a symposium Proceedings London 1967 Jun 30 Edited by Peter F. Knight and C.G. Burgess Sponsored by Roche Products Wright,Bristol, 1968
PGMS 29.0039

DIE CASTING National die casting exposition and congress 1st Technical papers Detroit 1960 Nov 8-11 Society of die casting engineers Society of die casting engineers,Detroit,1960
Met 25.0630

DIFFERENTIAL ANALYSIS International colloquium on differential analysis :the Bombay colloquium papers presented Bombay 1964 Jan 7-14 By M.F. Atiyah and others Tata institute of fundamental research Jointly sponsored by the International mathematical union Tata institute of fundamental research.Studies in analysis, 2 viii,253p 25cm Oxford university press, Oxford,1964
P. Math 2.0696

DIFFERENTIAL AND COMBINATORIAL TOPOLOGY :a symposium in honor of Marston Morse Princeton, N.J. 1963 Edited by Stewart S. Cairns Supported by United States.Air force. Office of scientific research.Mathematics division vi,265p 24cm Princeton university press,Princeton, N.J.,1965
P. Math 2.0295

DIFFERENTIAL AND FUNCTIONAL EQUATIONS Japan-United States seminar on ordinary differential functional equations Proceedings Kyoto 1971 Sep 6-11 Edited by Minoru Urabe Lecture notes in mathematics, 243 viii,332p 25cm Springer,Berlin,1971
P Math 2.4318

DIFFERENTIAL EQUATIONS Asymptotic solutions of differential equations and their applications a symposium proceedings Madison,Wis 1964 Edited by C.H. Wilcox United States Army.Mathematics Research Center. Publicatons, 13 John Wiley and sons,New York,1964
A Math 4.0246

DIFFERENTIAL EQUATIONS Boundary problems in differential equations symposium proceedings Madison,Wis. 1959 Apr.20-22 United States. Army.Mathematics research center Edited by Rudolph E. Langer United States army. Mathematics research center.Publications, 2 x,320p 23cm University of Wisconsin press, Madison,Wis.,1960
Math L 5.0281

DIFFERENTIAL EQUATIONS Conference on the numerical solution of differential equations Dundee 1969 Jun 23-27 Edited by J.L. Morris Lecture notes in mathematics, 109 Bibliog. vi,275p 25cm Springer-Verlag, Berlin,1969
P Math 2.3474

DIFFERENTIAL EQUATIONS Conference on the theory of ordinary and partial differential equations Proceedings Dundee 1972 Mar 28-31 Edited by W.N. Everitt and B.D. Sleeman Lecture notes in mathematics, 280 xv,367p 25cm Springer,Berlin,1972
P Math 2.4316

DIFFERENTIAL EQUATIONS Instructional conference on differential equations Edinburgh 1967 Sep 18-22 notes for the introductory lectures on Hilbert spaces and ordinary differential equations Illus. 28, 35cm 2 vols Edinburgh,1967 Typescript
P Math 2.3476

DIFFERENTIAL EQUATIONS Instructional conference on differential equations Lectures Edinburgh 1967 Sep 18-22 Edinburgh,1967 Typescript
P Math 2.3477

DIFFERENTIAL EQUATIONS Numerical solutions of ordinary and partial differential equations summer school Lectures Oxford 1961 Aug-Sep University of Oxford.Computing laboratory Edited by L. Fox In collaboration with the University of Oxford. Delegacy for extra-mural studies Pergamon press,Oxford,1962
Cav 7.0830

DIFFERENTIAL EQUATIONS Seminar on differential equations and dynamical systems Lectures College Park,Md. 1967 Aug Edited by G.Stephen Jones Lecture notes in mathematics, 60 Bibliog v,106p 28cm Springer-verlag,Berlin,1968
P Math 2.3046

DIFFERENTIAL EQUATIONS United States-Japan seminar on differential and functional equations Proceedings Minneapolis,Ma. 1967 Jun 26-30 Edited by William A. Harris and Yasutaka Sibuya Illus. xvi,585p 24cm Benjamin,New York;Amsterdam,1967
P Math 2.3164

DIFFERENTIAL EQUATIONS AND DYNAMICAL SYSTEMS Symposium on differential equations and dynamical systems Proceedings Coventry 1968-69 and 1969 Jul 15-25 University of Warwick.Mathematics institute Edited by David Chillingworth Lecture notes in mathematics, 206 x,173p 25cm Springer, Berlin,1971 Notes taken from the seminars and lectures given in the Mathematics institute during the symposium and also from the summer school held in July
P Math 2.4315

DIFFERENTIAL EQUATIONS AND DYNAMICAL SYSTEMS : an international symposium Proceedings Mayaguez,Puerto Rico 1965 Dec 27-30 Edited by Jack K. Hale and Joseph P. LaSalle Bibliog,illus,port xvii,544p 24cm Academic press,New York;London,1967 Dedicated to Solomon Lefschetz
P Math 2.2957

DIFFERENTIAL GAMES AND RELATED TOPICS International summer school on mathematical models of action and reaction Proceedings Varenna 1970 Jun 15-27 Consiglio nazionale delle richerche Ente per gli studi monetari, bancari et finanziari 'Luigi Einaudi' Edited by H.W. Kuhn and G.P. Szego x,489p 23cm North-Holland,Amsterdam,1971
P Math 2.4574

DIFFERENTIAL GAMES AND RELATED TOPICS International summer school on mathematical models of action and reaction Proceedings Varenna 1970 Jun 15-27 Edited by H.W. Kuhn and G.P. Szego x,489p 23cm North-Holland, Amsterdam;London,1971
Math S 3.1721

DIFFERENTIAL GEOMETRY Tucson, Ariz. 1960 Feb 18-19 American mathematical society Edited by Carl B. Allendoerfer With the support of the National science foundation American mathematical society.Proceedings of symposia in pure mathematics, 3 vii,200p 26cm American mathematical society, Providence,R.I.,1961
P. Math 2.0698

DIFFERENTIALGEOMETRIE UND TOPOLOGIE Internationales kolloquium Zurich 1960 Jun 20-25 Schweizerische mathematische gesellschaft Sponsored by the International mathematical union Enseignement mathematique. Monographies, 11 159p 24cm L'enseignement mathematique universite,Geneva, 1962
P. Math 2.0697

DIFFERENTIATION Genetic control of differentiation :report of a symposium Upton, N.Y. 1965 Jun 7-9 Brookhaven national laboratory.Biology department Brookhaven symposia in biology, 18 Brookhaven national laboratory,Upton,N.Y.,1965
Gen 34.0499

DIFFERENTIATION Genetic control of differentiation :symposium Report Upton,N. Y. 1965 Jun 7-9 Brookhaven national laboratory.Biology department Brookhaven symposia in biology, 18 BNL931(C-44) Brookhaven national laboratory.Biology department,Upton,N.Y.,1965
Bioch 33.1296

DIFFERENTIATION Growth,in relation to differentiation and morphogenesis Cambridge 1947 Jul Society for experimental biology Society for experimental biology.Symposia, 2 Cambridge university press,Cambridge,1948
Chem 18.2652

DIFFERENTIATION Meristems and differentiation :symposium Report Upton,N. Y. 1963 Jun 3-5 Brookhaven national laboratory.Biology department Brookhaven symposia in biology, 16 BNL805(C-38) Brookhaven national laboratory.Biology department,Upton,N.Y.,1964
Bioch 33.1294

DIFFERENTIATION AND DEVELOPMENT :a symposium Proceedings New York 1964 Feb 7-8 New York heart association Journal of experimental zoology, 157,no.1 Wistar institute of anatomy and biology,Philadelphia, Pa.,1964
Gen 34.0530

DIFFERENTIATION AND GROWTH OF HAEMOGLOBIN AND IMMUNOGLOBIN-SYNTHESISING CELLS Symposium on differentiation and growth of haemoglobin- and immunoglobin-synthesising cells :given at research conference for biology and medicine of the Atomic energy commission Gatlinburg, Tenn. 1966 Apr 4-7 Sponsored by the Oak Ridge national laboratory.Biology division Oak Ridge national laboratory.Biology division. Symposia, 19 Journal of cellular physiology, 67,supp.1 Wistar institute of anatomy and biology,Philadelphia,Pa.,1966
Gen 34.0504

DIFFERENTIATION AND IMMUNOLOGY International society for cell biology Edited by Katherine Brehme Warren International society for cell biology.Symposia, 7 Academic press,New York;London,1968
Gen 34.0503

DIFFUSION Colloque sur la diffusion Actes Montpellier 1955 Jun By J. Salvinien and others France.Ministere de l'air. Publications scientifiques et techniques de l'air.Notes techniques, 59 96p Service de documentation et d'information technique de l'aeronautique,Paris,1956
Chem E 24.0760

DIFFUSION A L'ETAT SOLIDE Colloque sur la diffusion a l'etat solide Saclay 1958 Jul 3-5 Centre d'etudes nuleaires,Saclay and Commissariat a l'energie atomique Centre d'etudes nucleaires de Saclay;North Holland, Gif-sur-Yvette;Amsterdam,1959
Met 25.1277

DIFFUSION DANS LES METAUX La Colloque sur la physique des metaux Comptes rendus Eindhoven 1956 Sep 10-11 Edited by J.D. Fast and others Bibliotheque technique Philips,Eindhoven,1957
Met 25.1276

DIFFUSION IN BODY-CENTERED CUBIC METALS International conference on body-centred cubic materials Papers Gatlinburg 1964 Sep 16-18 American society for metals.Oak Ridge chapter and Oak Ridge national laboratory American society for metals,Metals Park,Ohio, 1965
Met 25.1283

DIFFUSION PROCESSES Thomas Graham memorial symposium Proceedings Glasgow 1969 Sep 16 University of Strathclyde Edited by John N. Sherwood and others 2 vols Gordon and Breach,London,1971
Met 25.2597

DIGESTIVE PHYSIOLOGY AND NUTRITION OF THE RUMINANT Easter school in agricultural sciences 7th Proceedings Nottingham 1960 University of Nottingham Edited by D. Lewis Butterworth,London,1961
VA 19.0296

DIGITAL COMPUTATION Error in digital computation advanced seminar proceedings Madison,Wis. 1964 Oct.5-7 Vol. 1 United States.Army.Mathematics research center Edited by Louis B. Rall United States army. Mathematics research center.Publications, 14 Bibliog. ix,324p 24cm John Wiley and sons,New York,1965
Math 3.0242

DIGITAL COMPUTATION Error in digital computation seminar proceedings Wisconsin 1964 Oct.5-7 Vols. 1-2 Edited by Louis B. Rall United States Army.Mathematics research center.publication, 14 2 vols John Wiley,New York,1965
A Math 4.1209

DIGITAL COMPUTATION Error in digital computation symposium proceedings Madison, Wis. 1965 Apr.26-28 Vol. 2 United States.Army.Mathematics research center Edited by Louis B. Rall United States army. Mathematics research center.Publications, 15 x,288p 24cm John Wiley and sons,New York, 1965
Math 3.0243

DIGITAL COMPUTER APPLICATIONS TO PROCESS CONTROL, 2 IFAC-IFIP international conference 2nd Proceedings Menton 1967 Jun 5-9 International federation of automatic control International federation for information processing Edited by W.E. Miller Sponsored also by the Association francaise de regulation d'automatisme Instrument society of America,Pittsburgh,Pa.,1969
Eng 41.8206

DIGITAL COMPUTER TECHNIQUES Convention on digital-computer techniques proceedings London 1956 Apr 9-14 Institution of electrical engineers Edited by W.K. Brasher Institution of electrical engineers. Proceedings, 103,Part B,Suppts.1-3 illus 342p 27cm Institution of electrical engineers,London,1956
Math L 5.0762

DIGITAL COMPUTERS Harvard symposium on digital computers and their applications proceedings Cambridge,Mass. 1961 Apr 3-6 Harvard university.Computation laboratory Harvard university.Computation laboratory. Annals, 31 xiv,332p 27cm Harvard university press,Cambridge,Mass.,1962
Math L 5.0687

DIGITAL COMPUTING Symposium on digital computing in the aircraft industry proceedings New York 1957 Jan 31-Feb 1 New York university Jointly sponsored by International business machines corporation illus. 399p 28cm International business machines corporation,New York,1957
Math L 5.0777

DIGITAL EQUIPMENT COMPUTER USERS'SOCIETY DECUS proceedings 1967 :European seminar Ijmuiden 1967 Oct.19-20 iii,85p Digital equipment computer users' society,Ijmuiden, 1967
Math L 5.3103

DIGITAL EQUIPMENT COMPUTER USERS'SOCIETY DECUS proceedings 1963:papers and presentations... illus. xi,273,A 10p 28cm digital equipment computer users' society,Maynard,Mass.,1964
Math L 5.0978

DIGITAL METHODS OF MEASUREMENT Joint conference on digital methods of measurement Proceedings Canterbury 1969 Jul 23-25 Institution of electronic and radio engineers Institution of electrical engineers Institute of electrical and electronics engineers Institution of electronic and radio engineers.Conference proceedings, 15 IERE,London,1969
Eng 41.8305

DIGITAL PROCESSING Joint conference on digital processing of signals in communications Proceedings 1972 Institution of electronic and radio engineers Institution of electrical engineers Institution of electronic and radio engineers. Conference proceedings, 23 IERE,London,1972
Eng 41.8565

DIGITAL SIMULATION IN OPERATIONAL RESEARCH :a conference Proceedings Hamburg 1965 Sep 6-10 Edited by S.H. Hollingdale Under the aegis of the North Atlantic Treaty Organization.Scientific affairs division English universities press,London,1967
Eng 41.1192

DIMENSIONING AND STRENGTH CALCULATIONS Conference on dimensioning and strength calculations 3rd Proceedings Budapest 1968 Nov Hungarian academy of sciences Edited by F. Czoboly Akademiai kiado, Budapest,1968
Met 25.2407

DINARD 1957 Ecologie des algues marines By R. Biebl and others Centre national de la recherche scientifique Centre national de la recherche scientifique.Colloques internationaux, 81 illus. 276p Paris, 1959
Bot 42.3814

DISCONTINUOUS SUBGROUPS Algebraic groups and discontinuous subgroups Summer mathematical institute 12th Boulder,Colo. 1965 Jul 5-Aug 6 American mathematical society Edited by Armand Borel and George D. Mostow Financed by the National science foundation American mathematical society.Proceedings of symposia in pure mathematics, 9 vii,426p 26cm American mathematical society, Providence,R.I.,1966
P Math 2.2719

A DISCUSSION HELD...AT THE IMPERIAL COLLEGE OF SCIENCE Report London 1931 Jun 19 Physical society Physical society,London, 1931 Bound with 'Report of a joint discussion on vision...'by the Physical and optical society
Psy 31.3361

DISCUSSION ON CRITICAL PATH ANALYSIS PLANNING TECHNIQUES FOR ENGINEERING MANAGEMENT Proceedings London 1963 Apr 23 Institution of mechanical engineers Institution of mechanical engineers,London, 1964
Eng 41.0499

A DISCUSSION ON DETERMINATION OF SEX 1969 May 22-23 Royal society of London Edited by G.W. Harris and R.G. Edwards Royal society of London.Philosophical transactions. Biological sciences. Series B, 259,p.1-206 Royal society,London,1970
Inv Med 37.0102

A DISCUSSION ON INFRARED ASTRONOMY London 1966 May 1-2 Edited by H. Massey and J. Ring Philosophical transactions,A., 264,p 107-320 Royal society,London,1969
Obs 6.3411

DISCUSSION ON NEW MATERIALS Proceedings of the Royal society Series A:mathematical and physical sciences Royal society of London Royal society,London,1964 Organized by J. D.Bernal and others
Eng 41.3430

A DISCUSSION ON SOLAR STUDIES with special reference to space observations London 1970 Apr 21-22 British national committee on space research Philosophical transactions, 270A,1202 195p royal society of London, London,1971
Obs 6.3606

A DISCUSSION ON THE PHYSICS OF THE MOON AND ITS ENVIRONMENT :a meeting Papers London 1965 Jun 3-4 Royal society of London Royal society of London.Proceedings.Series A. Mathematical and physical sciences, 296,1446 Royal society,London,1967
Geod 9.0015

DISCUSSION ON THE PRESENT STATUS OF RADIATION GENETICS information meeting... Oak Ridge,Tenn. 1948 Mar 26-27 Sponsored by the Oak Ridge national laboratory Oak Ridge national laboratory.Biology division.Symposia, 1 Journal of cellular and comparative physiology, 35,supp Wistar institute of anatomy and biology,Philadelphia,Pa.,1950
Gen 34.1103

DISCUSSION ON THE TEACHING OF MATHEMATICS... to which is added a report of the British association committee drawn up by the chairman, professor Forsyth Glasgow 1901 Sep 14 British association for the advancement of science Edited by John Perry 2nd edition Macmillan,London;New York,1902
Psy 31.3384

DISCUSSIONS ON CHILD DEVELOPMENT : consideration of the biological,psychological and cultural approaches to the understanding of human development and behaviour :meeting Geneva 1956 Vol 4 By Jean Piaget and others World health organization.Study group on the psychobiological development of the child Edited by J.M. Tanner and Barbel Inhelder Tavistock publications,London,1960
Psy 28.0224

DISCUSSIONS ON CHILD DEVELOPMENT ;a consideration of the biological,psychological, and cultural approaches to the understanding of human development and behaviour Geneva, Switzerland 1953 Vol 1: proceedings World health organization.Study group on the psychobiological development of the child Edited by J.M. Tanner and Barbel Inhelder Tavistock publications,London,1956
Psy 31.1178

The DISEASES OF THE BASAL GANGLIA Proceedings of the association New York 1940 Dec 20-21 Association for research in nervous and mental disease Edited by Tracy J. Putnam and others Association for research in nervous and mental disease.Research publications, 21 Williams and Wilkins,Baltimore,Md.,1942
An 32.4511

DISLOCATION DYNAMICS Battelle materials science colloquium 2nd Seattle,Wash. 1967 May 1-16 Harrison Hot Springs,B.C. Battelle memorial institute Edited by Alan R. Rosenfield and others McGraw-Hill series in materials science McGraw-Hill,New York,1968
Met 25.2253

DISLOCATION THEORY Fundamental aspects of dislocation theory Conference on the fundamental aspects of dislocation theory Proceedings Gaithersburg,Md. 1969 Apr 21-25 Institute of materials research National bureau of standards.Special publication, 317 diagrms 1338p 24cm 2 vols National bureau of standards,Washington, D.C.,1970
Cav 7.3060

DISLOCATIONS AND MECHANICAL PROPERTIES OF CRYSTALS an international conference Lake Placid 1956 Sep 6-8 United States.Air force.Office of scientific research and General electric research laboratory Edited by J.C. Fisher and others Sponsored by United States.Air research and development command Wiley;Chapman and Hall,New York; London,1957
Met 25.1232

DISLOCATIONS AND MECHANICAL PROPERTIES OF CRYSTALS :international conference Proceedings Lake Placid,N.Y. 1956 Sep 6-8 United States.Air force.Office of scientific research and General electric research laboratory Edited by J.C. Fisher and others John Wiley,New York,1957
Cav 7.0804

DISLOCATIONS IN SOLIDS :a general discussion Gottingen 1964 Sep 15-17 Faraday society Faraday society.Discussions, 38 Faraday society,London,1964
Eng 41.3532

DISORDERS OF CARBOHYDRATE METABOLISM :a symposium Proceedings London 1968 Nov 29-30 Association of clinical pathologists. Chemical pathology sub-committee Edited by G. K. McGowan and G. Walters held at the Royal society of medicine Chemical pathology in relation to clinical medicine, 5 Journal of clinical pathology.Supplement, 2 British medical association,London,1969
Bioch 33.1205

DISORDERS OF COMMUNICATION Proceedings of the Association New York 1962 Dec 7-8 Association for research in nervous and mental disease Edited by David McK. Rioch and Edwin A. Weinstein Association for research in nervous and mental disease.Research publications, 42 Williams and Wilkins, Baltimore,Md.,1964
An 32.4561

DISORDERS OF THE DEVELOPING NERVOUS SYSTEM Houston neurological society:annual scientific meeting 8th Houston Tex. 1960 Mar Edited by Williams S. Fields and Murdina M. Desmond Houston neurological society. Symposia, 6 Thomas,Springfield,Ill.,1961
An 32.4187

DISPERSION AND ABSOPTION OF SOUND BY MOLECULAR PROCESSES Dispersione ed assorbimento del suono Scuola internazionale de fisica 'Enrico Fermi' 27 corso Rendiconti Varenna 1962 Aug 6-18 Societa italiana di fisica Edited by D. Sette Academic press, New York;London,1963 Papers in English and French
Chem 18.0517

DISPERSIONE ED ASSORBIMENTO DEL SUONO Scuola internazionale de fisica 'Enrico Fermi' 27 corso Rendiconti Varenna 1962 Aug 6-18 Societa italiana di fisica Edited by D. Sette Academic press,New York;London,1963 Papers in English and French
Chem 18.0517

DISPLAYS FOR COMMAND AND CONTROL CENTERS :a conference Proceedings Munich 1966 Nov 11-14 AGARD Edited by I.J. Gabelman AGARD.Conference proceedings, 23 Technivision services,Slough,1969 Proceedings of the 11th technical meeting of the AGARD avionics panel
Eng 41.5609

DISTANCE MEASUREMENT Electromagnetic distance measurement :a symposium Oxford 1965 Sep 6-11 International association of geodesy.Special duty group no 19 Hilger and Watts,London,1967
Eng 41.3278

DISTILLATION International symposium on distillation Brighton 1960 May 4-6 European federation of chemical engineering 281p Institution of chemical engineers, London,1960 24th meeting of the European Federation of Chemical Engineering
Chem E 24.1303

DISTINGUISHING THE HEALTH CARE OF THE AGING :.. seminar Proceedings Southern Pines,N.C. 1967 Mar 12-14 American geriatrics society Edited by E.J. Lorenze American geriatrics society.Journal, 16, no 2 American geriatrics society,New York,1967
PGMS 29.0105

DISTRIBUTION AND MOTION OF INTERSTELLAR MATTER IN GALAXIES :a conference Proceedings Princeton,N.J. 1961 Apr 10-20 Institute for advanced study Edited by L. Woltjer 330p W.A.Benjamin,New York,1962
Obs 6.3253

The DITCHLEY MATHEMATICAL CONFERENCE :a report by the British co-chairman (Prof.B. Thwaites) of the Anglo-American conference on mathematical education Ditchley Park 1966 Sep 9-12 School mathematics project Cambridge conference on school mathematics S. M.P.,London,1966
Eng 41.1829

DITCHLEY PARK 1966 The Ditchley mathematical conference :a report by the British co-chairman (Prof.B.Thwaites) of the Anglo-American conference on mathematical education School mathematics project Cambridge conference on school mathematics S. M.P.,London,1966
Eng 41.1829

DIURESE UND DIURETICA :ein internationales symposium Herrenchiemsee 1959 Jun 17-20 Edited by H. Schwiegk and others Sponsored by Ciba foundation Springer,Berlin,1959 t.p. also in English "Diuresis and diuretics"
Inv Med 37.0193

DIURESIS AND DIURETICS Diurese und diuretica ein internationales symposium Herrenchiemsee 1959 Jun 17-20 Edited by H. Schwiegk and others Sponsored by Ciba foundation Springer,Berlin,1959 t.p. also in English "Diuresis and diuretics"
Inv Med 37.0193

Les DIVISIONS ECOLOGIQUES DU MONDE MOYENS D'EXPRESSION,NOMENCLATURE CARTOGRAPHIE : colloques internationaux Paris 1954 Jun 28-Jul 3 Centre national de la recherche scientifique illus xii,235p Centre national de la recherche scientifique,Paris, 1955
Bot 42.2071

DNA Replication of DNA :a symposium Abstracts Cold Spring Harbor 1971 Sep Cold Spring Harbor laboratory of quantitative biology Cold Spring Harbor symposia on quantitative biology Cold Spring Harbor,1971
Bioch 33.2333

DNA Replication of DNA in micro-organism :a symposium Papers Cold Spring Harbor 1968 Cold Spring Harbor laboratory of quantitative biology Cold Spring Harbor symposia on quantitative biology, 33 Cold Spring Harbor, 1968
Bioch 33.1289

DNEPROPETROVSK 1962 Electrometallurgy of chloride solutions All-union seminar on applied electrochemistry 5th Reports Dnepropetrovskii khimiko-tekhnologicheskii institut and Moskovskii khimiko-tekhnologicheskii institut imeni D.I. Mendeleyeva Edited by V.V. Stender Translated by Consultants bureau from the Russian Consultants bureau,New York,1965
Met 25.1843

DNEPROPETROVSKII KHIMIKO-TEKHNOLOGICHESKII INSTITUT Electrometallurgy of chloride solutions All-union seminar on applied electrochemistry 5th Reports Dnepropetrovsk 1962 Oct 17-19 Edited by V. V. Stender Translated by Consultants bureau from the Russian Consultants bureau,New York, 1965
Met 25.1843

DOKLADY 6 RASSHIRENNOI SESSII UCHENOGO SOVETA INSTITUTA 1953 Mar 2-9 Vyp 2-3 Nauchno-issledovatelskii institut polyarnogo zemledeliya,zhivotnovodstva i promyslovogo khozyaistva Leningrad,1953-54
Sco 14.4902

DOLERITE University of Tasmania.Geology department :symposium 4th Hobart 1957 July 17-22 Hobart,1958 Typescript
Min 10.0334

DOMICILIARY CARE OF THE PATIENT WITH CANCER : symposium London 1969 Marie Curie foundation fund Edited by R.W. Raven Heinemann,London,1970
PGMS 29.0561

DOMUS GALILAEANA Celebrazione della Accademia del cimento nel tricentenario della fondazione Domus galilaeana Pisa 1957 Jun 19 Accademia del cimento 49 pls 79p Domus galilaeana,Pisa,1958 Limited edition.With appendices
WSM 43.4472

DONDERS CENTENARY SYMPOSIUM ON REACTION TIME Proceedings Attention and performance 2 Eindhoven 1968 Jul 29-Aug 2 Institute for perception research IPO Edited by W.G. Koster North-Holland,Amsterdam;London,1969
Psy 31.2948

DONEGANI FOUNDATION.COURSE 11 International course on materials science 1st Essays Tremezzo 1968 Sep 8-20 Accademia nazionale dei lincei.Donegani foundation Edited by Alan W. Searcy and others illus xxv,715p 24cm Interscience,New York;Chichester,1970 Being the 11th course organized by the Donegani foundation
Met 25.2593

DORADO,PUERTO RICO 1967 The Transmission of schizophrenia:research conference 2nd Proceedings Foundations fund for research in psychiatry Edited by David Rosenthal and, Seymour S. Kety Pergamon press,Oxford,1968 The second in a series of three psychiatric research conferences
Psy 28.0084

DORADO,PUERTO RICO 1967 Transmission of schizophrenia :research conference Proceedings Foundations' fund for research in psychiatry Edited by Dvid Rosenthal and Seymour S. Kety Foundations' fund for research in psychiatry.Research conference, 2 Pergamon press,Oxford,1968 reprinted 1969
PGMS 29.0566

DORMANCY AND SURVIVAL :a symposium Norwich 1968 Sep 2-6 Society for experimental biology Society for experimental biology.Symposia, 23 Cambridge university press,Cambridge,1969
Bioch 33.1362

DORMANCY AND SURVIVAL :a symposium Papers Norwich 1968 Sep 2-6 Society for experimental biology Society for experimental biology.Symposia, 23 Cambridge university press,Cambridge,1969
Gen 34.2224

DORMANCY AND SURVIVAL :symposium Papers Norwich 1968 Sep 2-6 Society for experimental biology Edited by H.W. Woolhouse Society for experimental biology. Symposia, 23 illus vi,598p 23cm Cambridge university press,Cambridge,1969
Sco 14.8232

DOSAGE DES ELEMENTS A L'ETAT DE TRACES DANS LES ROCHES ET AUTRES SUBSTANCES :colloque Comptes-rendus Nancy 1968 Dec 4-6 Centre national de la recherche scientifique Centre national de la recherche scientifique. Colloques nationaux, 923 CNRS,Paris,1970
Min 10.1475

DOUGLAS,I.O.M. 1971 Tribology convention Institution of mechanical engineers.Tribology group Institution of mechanical engineers. Proceedings, 19 IME,London,1971
Eng 41.8642

La DOULEUR ET LES DOULEURS :a symposium Paris 1955 Hopital de la Salpetriere. Clinique des maladies du systeme nerveux Edited by Th. Alajouanine and others Masson, Paris,1957
Psy 31.0214

DOWNSTATE MEDICAL CENTER Cerebrospinal fluid and the regulation of ventilation :a symposium 6 Proceedings New York 1964 Apr 9-10 Edited by Chandler McC. Brooks and others Blackwell,Oxford,1965
Pha 16.0278

DOWNSTATE MEDICAL CENTRE Cerebrospinal fluid and the regulation of ventilation :a symposium Proceedings Brooklyn,N.Y. 1964 Apr 9-10 Edited by Chandler McC. Brooks and others Blackwell,Oxford,1965
An 32.4209

DRAMMEN 1965 Molecular basis of some aspects of mental activity :a NATO advanced study institute Proceedings North Atlantic treaty organization Edited by Otto Walaas 2 vols Academic press,London,1966-67
Pha 16.0053

DRESDEN 1931 Congres international de photographie scientifique et appliquee 8e Comptes rendus Edited by L.P. Clerc Revue d'optique theorique et instrumentale,Paris, 1933
Chem 18.0248

DRIEBERGEN 1966 Attention and performance : a symposium Proceedings Institute for perception RVO-TNO,Soesterberg Edited by A.F. Sanders North-Holland,Amsterdam,1967
Psy 31.0959

DRIEBERGEN 1966 Local fields summer school proceedings Netherlands universities foundation for international co-operation Edited by T.A. Springer viii,214p 24cm Springer,Berlin,1967
P Math 2.2774

DRUG ACTION Enzymes and drug action London 1961 Mar 20-23 Ciba foundation and Co-ordinating committee for symposia on drug action Edited by J.L. Mongar and A.V.S. De Reuck Churchill,London,1962
Bioch 33.1872

DRUG ACTION AND MODERN THERAPY Symposium on drug action and modern therapy London 1961 Sep 28 Anglo-German medical society Anglo-German medical review, 1,no.4 Schattauer, Stuttgart,1962 Text in English and German
Gen 34.1973

DRUG RESPONSES IN MAN :a symposium Papers London 1966 Jun 14-16 Ciba foundation Edited by Gordon Wolstenholme and Ruth Porter Churchill,London,1967
Pha 16.0253

DRUG TOXICITY A Symposium on the evaluation of drug toxicity Macclesfield 1957 Oct 1 Imperial chemical industries.Pharmaceuticals division Edited by A.L. Walpole and A. Spinks Churchill,London,1958
Pha 16.0327

DRUG TOXICITY Some factors affecting drug toxicity European society for the study of drug toxicity :meeting 4th Proceedings Cambridge 1964 Jul 2-3 Edited by D.G. Davey European society for the study of drug toxicity.Proceedings,4 International congress series,81 Excerpta medica,Amsterdam, 1964
Pha 16.0326

DRUGS Symposium on the adverse effects of drugs 1st Proceedings London 1970 Feb 4-5 Edited by Gillian C. Hanson Sponsored by Beecham research laboratories Beecham research laboratories,London,1971
PGMS 29.0673

DRUGS AFFECTING LIPID MATABOLISM The Symposium on drugs affecting lipid metabolism Proceedings Milan 1960 Edited by S. Garattini and R. Paoletti illus Elsevier, Amsterdam,1961
Bioch 33.1237

DRUGS AFFECTING LIPID METABOLISM International symposium on drugs affecting lipid metabolism 2nd Papers Milan 1965 Pt 1-2 Edited by D. Kritchevsky and others organised by the European society for biochemical pharmacology Progress in biochemical pharmacology, 2-3 illus 2 vols Karger,Basle;New York,1967
Bioch 33.1250

DRUGS AFFECTING LIPID METABOLISM :symposium Proceedings Milan 1960 Edited by S. Garattini and R. Paoletti Elsevier,Amsterdam, 1961
Pha 16.0128

DRUGS AND THE MIND :a symposium Proceedings Dunedin 1967 Jul 14-15 University of Otago medical school.Department of psychological medicine Edited by Basil James University of Otago medical school,Dunedin, 1967
PGMS 29.0394

DRUGS AND YOUTH Rutgers symposium on drug abuse Proceedings Rutgers interdisciplinary research center Edited by J.R. Wittenborn and others Held at Rutgers university Charles C.Thomas,Springfield,Ill., 1969
Inv Med 37.0235

DRUGS IN OUR SOCIETY Baltimore,Md. 1963 Nov Edited by Paul Talalay Under the auspices of the Johns Hopkins university Johns Hopkins press,Baltimore,Md.,1964
Pha 16.0237

DRY ZONE AGRICULTURE The Development of agriculture in the dry zone :a symposium Proceedings Colombo 1967 Jul 30-31 Ceylon association for the advancement of science Edited by O.S. Peries bibliog. Colombo,1967
Geog 13.4482

DUARTE,CALIF. 1959 Inhibition in the nervous system and gamma-aminobutyric acid :an international symposium Proceedings Edited by Eugene Roberts Sponsored by the United States.Air force.Office of scientific research Pergamon press,London,1960
An 32.4362

DUARTE,CALIF. 1959 Inhibition in the nervous system and gamma-aminobutyric acid :an international symposium held at the City of Hope medical center Proceedings Edited by Eugene Roberts and others Sponsored by the United States.Air force.Office of scientific research Pergamon press,Oxford,1960
Bal 39.1471

DUARTE,CALIF. 1961 Amino acid pools : distribution and function of free amino acids. Symposium Proceedings City of hope medical center Edited by Joseph T. Holden Under the auspices of the Institute for advanced learning in the medical sciences illus Elsevier,Amsterdam,1962 Preface:"The committee undertook the organization of the symposium under the title Conference on free amino acids
Bioch 33.0519

DUBLIN 1955 Comparison of the large-scale structure of the galactic system with that of other stellar systems :a symposium papers International astronomical union Edited by N. G. Roman International astronomical union. Symposium, 5 illus 72p Cambridge university press,Cambridge,1958
Obs 6.1466

DUBLIN 1955 Non-stable stars :a symposium Proceedings International astronomical union Edited by George H. Herbig International astronomical union.Symposium, 3 Cambridge university press,Cambridge,1957
TA 15.0033

DUBLIN 1955 Symposium of recent advances in the chemistry of naturally occurring pyrones and related compounds Proceedings University college,Dublin,and,Institute of chemistry of Ireland Royal Dublin society. Scientific proceedings,27,no 6 Royal Dublin society,Dublin,1956
Chem 18.1494

DUBLIN 1963 The Chemistry and biochemistry of fungi and yeasts :a symposium Proceedings International union of pure and applied chemistry sponsored by the International union of pure and applied chemistry and Royal Irish academy.Irish national committee for chemistry Butterworths,London,1963 Reprinted from 'Pure and applied chemistry', vol.7,no.,4.Added title-page in French:Chimie et biochimie des champignons inferieurs et des levures
Bioch 33.1146

DUBLIN 1963 The Chemistry and biochemistry of fungi and yeasts :symposium Proceedings International union of pure and applied chemistry Royal Irish academy Irish national committee for chemistry Butterworths,London,1963 Reprinted from 'Pure and applied chemistry' vol.7,no.4
Gen 34.0729

DUBLIN 1967 Aspects of yeast metabolism A Guinness symposium held at the Research laboratory St.James's Gate, Dublin Discussions Edited by A.K. Mills and Hans Krebs Guinness symposia Blackwell,Oxford; Edinburgh,1968
Bioch 33.1140

DUBLIN 1967 Aspects of yeast metabolism A Guinness symposium held at the research laboratory St.James's Gate Dublin Discussions Edited by A.K. Mills and Hans Krebs Guinness symposia Blackwell,Oxford, 1968
Bot 42.1751

DUBLIN 1970 Gas chromatography 1970 International symposium on gas chromatography 8th Proceedings Institute of petroleum. Gas chromatography discussion group Edited by R. Stock viii,445p 25cm Institute of petroleum,London,1971
Chem 18.2697

DUBNA,U.S.S.R. 1966 Research applications of nuclear pulsed systems Panel on research applications of repetitively pulsed reactors and boosters Proceedings International atomic energy agency International atomic energy agency.Panel proceedings series International atomic energy agency,Vienna,1967
Radioth 35.1682

DUBROVNIK 1964 Sensitivity methods in control theory International symposium on sensitivity analysis Yugoslav committee for electronics and automation Edited by L. Radanovic Sponsored by the International federation of automatic control.Theory committee Pergamon press,Oxford,1965
Eng 41.5835

DUBROVNIK 1968 I.F.A.C. symposium on system sensitivity and adaptivity 2nd Preprints International federation of automatic control Organised by the Yugoslav committee for electronics and automation I.F. A.C.,Dubrovnik,1968
Eng 41.5889

DUCTILE CHROMIUM AND ITS ALLOYS :a conference...held at the 1955 metal congress and exposition of the American society for metals United States.Army.Office of ordnance research and American society for metals American society for metals,Cleveland, Ohio,1957
Met 25.2307

The DUKE OF EDINBURGHS COMMONWEALTH STUDY CONFERENCE 2nd Conference across a continent Canada 1962 May 13-Jun 6 Macmillan,Toronto,1963
Geog 13.5039

DUKE UNIVERSITY.SCHOOL OF MEDICINE International conference on Avian tumor viruses Proceedings Durham,N.C. 1964 Mar 31-Apr 3 Edited by Joseph W. Beard National cancer institute.Monograph, 17 National cancer institute,Bethesda,Md.,1964
Bioch 33.1154

DUKE UNIVERSITY.SCHOOL OF MEDICINE International conference on avian tumour viruses Proceedings Durham,N.C. 1964 Mar 31-Apr 3 Edited by Joseph W. Beard National cancer institute.Monograph, 17 US government printing office,Washington,D.C., 1964
Path 30.2570

DUMMETT M.A.E. and CROSSLEY J.N. ed. Formal systems and recursive functions The Logic colloquium 8th proceedings Oxford 1963 Jul 320p 23cm North-Holland, Amsterdam,1965
P. Math 2.0207

DUNDEE 1955 Pulmonary circulation and respiratory function :a symposium held at Queen's college Supported by the University of St.Andrews University of St.Andrews,St. Andrews,1956
An 32.3562

DUNDEE 1961 Summer school in mathematics, geometry and topology proceedings Queen's college,Dundee 33cm Queen's College,Dundee, 1961
P. Math 2.0292

DUNDEE 1964 The Early conceptus,normal and abnormal :a symposium Papers and discussions Edited by W.Wallace Park University of St. Andrews,St.Andrews,1965
An 32.2521

DUNDEE 1965 Control and innervation of skeletal muscle :a symposium Edited by B.L. Andrew University of St.Andrews,St.Andrews, 1966
An 32.4009

DUNDEE 1965 Control and innervation of skeletal muscle :a symposium Edited by B.L. Andrew University of St Andrews,St. Andrews, 1966
Phys 20.1560

DUNDEE 1969 Conference on the numerical solution of differential equations Edited by J.L. Morris Lecture notes in mathematics, 109 Bibliog. vi,275p 25cm Springer-Verlag,Berlin,1969
P Math 2.3474

DUNDEE 1970 Symposium on the theory of numerical numbers Proceedings Edited by John L. Morris Lecture notes in mathematics, 193 vi,152p 25cm Springer,Berlin,1971
P Math 2.4462

DUNDEE 1971 Conference on applications of numerical analysis Proceedings Edited by John L. Morris Lecture notes in mathematics, 228 x,358p 25cm Springer,Berlin,1971
P Math 2.4463

DUNDEE 1972 Conference on the theory of ordinary and partial differential equations Proceedings Edited by W.N. Everitt and B.D. Sleeman Lecture notes in mathematics, 280 xv,367p 25cm Springer,Berlin,1972
P Math 2.4316

DUNEDIN 1865 New Zealand exhibition reports and awards of the jurors New Zealand commissioners,Dunedin,1866
Philos 1.0056

DUNEDIN 1896 Intercolonial medical congress of Australasia 4th session Transactions Otago daily times and witness newspapers,Dunedin,1897
An 32.1142

DUNEDIN 1967 Drugs and the mind :a symposium Proceedings University of Otago medical school.Department of psychological medicine Edited by Basil James University of Otago medical school,Dunedin,1967
PGMS 29.0394

DURHAM 1960 Symposium on electrical conductivity in organic solids Proceedings United States.Army.Office of ordnance research United States.Office of naval research Edited by H. Kallman and M. Silver Sponsored also by United States.Air force.Office of scientific research Interscience,New York, 1961
Radioth 35.1646

DURHAM 1960 The Exploitation of natural animal populations :a symposium British ecological society Edited by E.D. Le Cren and M.W. Holdgate British ecological society. Symposia, 2 Oxford,1962
Bal 39.1665

DURHAM 1960 The Exploitation of natural animal populations :a symposium 2 British ecological society Edited by E.David Le Cren and Martin Wyatt Holdgate tables,diagrs,maps x,399p Blackwell scientific publications, Oxford,1962
Sco 14.0721

DURHAM,N.C. 1963 Boron-nitrogen chemistry : an international symposium United States. Army.Research office American chemical society.Advances in chemistry series,42 American chemical society,Washington,D.C.,1964 Symposium chairman:Kurt Niedenzu
Chem 18.0986

DURHAM,N.C. 1963 Boron-nitrogen chemistry : an international symposium United States. Army research office,Durham American chemical society.Advances in chemistry series, 42 American chemical society,Washington,D.C. 1964 Symposium chairman:K.Niedenzu
Chem 18.2658

DURHAM,N.C. 1964 International conference on Avian tumor viruses Proceedings National cancer institute Duke university. School of medicine Edited by Joseph W. Beard National cancer institute.Monograph, 17 National cancer institute,Bethesda,Md.,1964
Bioch 33.1154

DURHAM,N.C. 1964 International conference on avian tumour viruses Proceedings National cancer institute and Duke university.School of medicine Edited by Joseph W. Beard National cancer institute. Monograph, 17 US government printing office, Washington,D.C.,1964
Path 30.2570

DURHAM N.C. 1970 International symposium on flaveins and flavoproteins 3rd Proceedings International union of biochemistry International union of biochemistry.Symposium, 39 University park press;Butterworths,Baltimore,Md.;London,1971
Bioch 33.2373

DUSSELDORF 1960 Internationaler kongress fur medizinische photographie und Kinematographie 1st Abhandlungen Edited by Heinz Orbach Thieme,Stuttgart,1962
An 32.0311

DUSSELDORF 1960 Symposion uber krebsprobleme :arbeitstagung des beratungsaussschusses fur krebsforschung beim Kultusministerium des landes Nordrhein-Westfalen Edited by K.G. Ober and others Illus Springer,Berlin,1961
Radioth 35.0515

DUSSELDORF 1965 and DUSSELDORF 1966 Beitrage zur entwicklung der wissenschaftstheorie in 19 jahrhundert : colloquium Vortrage und diskussionen Edited by A. Diemer Studien zur wissenschaftstheorie, 1 234p Hain, Meisenheim am Glan,1968
WSM 43.0983

DUSSELDORF 1967 System und klassifikkation in wissenschaft und dokumentation :colloquium Vortrage und diskussionen Edited by A. Diemer Studien zur wissenschaftstheorie, 2 183p Hain,Meisenheim am Glan,1968
WSM 43.0984

DUST COOLING Conference on dust cooling Proceedings London 1964 Dec 21-22 Vol 1-2 Queen Mary college.Nuclear engineering department 2 vols London,1964 Mimeogra?h
Chem E 24.1464

DUTCH PHYSICAL SOCIETY nd NETHERLANDS UNIVERSITIES FOUNDATION FOR INTERNATIONAL COOPERATION Selected topics in nuclear spectroscopy :N.U.F.F.I.C.international summer course in science,Nijenrode Castle... Proceedings Breukelen,Netherlands 1963 Jul 30-Aug 17 Edited by B.J. Verhaar With financial support from the North Atlantic treaty organization North-Holland,Amsterdam, 1964
Chem 18.2279

DYNAMIC BEHAVIOR OF MATERIALS Symposium on dynamic behavior of materials Albuquerque,N. M. 1962 Sep 27-28 Sponsored by the American society for testing and materials American society for testing and materials. Special technical publication, 336 American society for testing and materials.Materials science series, 5 American society for testing and materials,Philadelphia,Pa.,1963
Eng 41.3788

DYNAMIC CLINICAL STUDIES WITH RADIOISOTOPES :a symposium held at the Oak Ridge institute of nuclear studies Proceedings Oak Ridge, Tenn. 1963 Oct 21-25 Edited by R.M. Kniseley and W.N. Tauke Sponsored by the United States atomic energy commission. Division of technical information U.S.A.E.C., Washington,D.C.,1964
Radioth 35.1403

DYNAMIC MASS SPECTROMETRY European symposium on the time-of-flight mass spectrometry 2nd Proceedings Salford 1969 Jul 1 University of Salford Edited by D. Price and J.E. Williams Heyden,London,1970
Chem 18.2669

DYNAMIC PROGRAMMING Colloquium lectures in pure and applied science 8 Dallas, Texas 1963 Dec 9-13 By Rutherford Aris Socony mobile oil company.Field research laboratory 310p 1965
Chem E 24.0172

The DYNAMIC ROLE OF MOLECULAR CONSTITUENTS IN PLANT-PARASITE INTERACTION :a conference Proceedings Gamagori 1966 May 15-21 American phytopathological society Edited by Chester J. Mirocha and Ikuzo Uritani American phytopathological society,St.Paul, Minn.,1967
Bot 42.3761

DYNAMIC STABILITY OF STRUCTURES :an international conference Proceedings Evanston,Ill. 1965 Oct 18-20 Edited by George Herrmann Sponsored by Northwestern university Pergamon press,London,1967
Eng 41.2531

DYNAMICS OF CLIMATE Conference on the application of numerical integration techniques to the problem of the general circulation proceedings 1955 Oct 26-28 Institute for advanced study Edited by Richard L. Pfeffer Sponsored jointly by Air force Cambridge research center Pergamon press,Oxford;London,1960
A Math 4.0560

DYNAMICS OF GROWTH PROCESSES :a symposium Williamstown,Mass. 1952 Jun Edited by Edgar J. Boell Society for the study of development and growth.Symposia, 11 Princeton university press,Princeton,N.J.,1954
An 32.2467

DYNAMICS OF GROWTH PROCESSES :a symposium Papers Massachusetts 1952 Jun Society for the study of development and growth Edited by Edgar J. Boell Society for the study of development and growth.Symposia, 11 Princeton university press,Princeton,N.J.,1954
Radioth 35.0197

DYNAMICS OF PROLIFERATING TISSUES :a symposium Upton,N.Y. 1956 Sep 5-8 Edited by Dorothy Price Developmental biology conference series,1956 University of Chicago press,Chicago,Ill.,1958
An 32.3285

DYNAMICS OF SATELLITES :a symposium Paris 1962 May 28-30 International union of theoretical and applied mechanics Edited by Maurice Roy Springer,Berlin,1963
Eng 41.6890

The DYNAMICS OF THE EARTH-MOON SYSTEM conference Proceedings Earth-moon system New York 1964 Jan.20-21 Goddard institute for space studies Edited by B.G. Marsden and A.G.W. Cameron 288p Plenum press,New York,1966
Obs 6.1169

The DYNAMICS OF VIRUS AND RICKETTSIAL INFECTIONS an international symposium Proceedings Detroit,Mich. 1953 Oct 21-23 Henry Ford hospital Edited by Frank W. Hartmann and others illus,tables,diagrs Blakiston,New York;Toronto,1954
Path 30.2548

DYSLEXIA Reading disability:progress and research needs in dyslexia Johns Hopkins conference on research needs in dyslexia and related aphasic disorders Papers Baltimore, Md. 1961 Nov 15-17 Edited by John Money Johns Hopkins press,Baltimore,Md.,1962
An 32.4203

E.S.R.O. SUMMER SCHOOL IN SPACE PHYSICS 3rd Proceedings Electromagnetic radiation in space Alpbach 1965 Jul 19-Aug 13 European space research organisation Edited by J.G. Emming Astrophysics and space science library, 9 viii,307p D.Reidel, Dordrecht,1967
A Math 4.1576

The EARLY CONCEPTUS,NORMAL AND ABNORMAL :a symposium Papers and discussions Dundee 1964 Sep 17-19 Edited by W.Wallace Park University of St.Andrews,St.Andrews,1965
An 32.2521

EARLY MAN... International symposium on early man Papers Philadelphia,Pa. 1937 Mar 17-20 Edited by George Grant MacCurdy illus. Lippincott,Philadelphia,Pa.,1937
Geog 13.1760

EARTH-MOON SYSTEM The Dynamics of the earth-moon system conference Proceedings New York 1964 Jan.20-21 Goddard institute for space studies Edited by B.G. Marsden and A.G. W. Cameron 288p Plenum press,New York,1966
Obs 6.1169

EARTH PRESSURE PROBLEMS Brussels conference on earth pressure problems Proceedings Brussels 1958 Vol 1-3 International society of soil mechanics and foundation engineering Brussels,1958
Eng 41.3165

EARTH SCIENCES Advances in earth sciences International conference on the earth sciences contributions Cambridge,Mass. 1964 Sep 30-Oct 2 Massachusetts institute of technology Edited by P.M. Hurley M.I.T. press,Cambridge, Mass.,1966
Geol 8.4429

EARTH STORM The International conference on cosmic rays and the earth storm :combined meeting of the International symposium on the earth storm and the International conference on cosmic rays Proceedings Kyoto 1961 Sep 4-15 Vol 2-3 International union of geodesy and geophysics and International astronomical union Edited by Ken-ichi Maeda and Osamu Minakawa Co-sponsored by the International scientific radio union Physical society of Japan. Journal, 17,suppl.A-2,3 2 vols physical society of Japan,Tokyo,1962
TA 15.0038

EARTH TIDES Symposium international sur les marees terrestres :symposium on earth tides 4th Proceedings Brussels 1961 Jun 5-10 Edited by P. Melchior Observatoire royal de Belgique.Communication., 188,Serie geophysique, 58 R.Louis,Ixelles,1961
Geod 9.0025

EARTHQUAKE ENGINEERING World conference on earthquake engineering 1st Proceedings Berkeley,Calif. 1956 Jun Earthquake engineering research institute Earthquake engineering research institute,San Francisco, Calif.,1967
Eng 41.6394

EARTHQUAKE ENGINEERING World conference on earthquake engineering 2nd Proceedings Tokyo 1960 Jul 11-18 and Kyoto 1960 Jul 11-18 Vol 1-3 Japan society of civil engineers Organised by the Science council of Japan 3 vols Science council of Japan,Tokyo,1960
Eng 41.6395

EARTHQUAKE ENGINEERING RESEARCH INSTITUTE World conference on earthquake engineering 1st Proceedings Berkeley,Calif. 1956 Jun Earthquake engineering research institute,San Francisco,Calif.,1967
Eng 41.6394

EARTHQUAKES Vibration effects of earthquakes on soils and foundations :a symposium San Francsico,Calif. 1968 Jun 23-28 American society for testing and materials American society for testing and materials.Special technical publication, 450 American society for testing materials,Philadelphia,Pa.,1969
Eng 41.3162

EAST AFRICA HIGH COMMISSION Interterritorial conference on hydrology and water resources 1st Nairobi 1950 Nov Nairobi,1950
Geog 13.5670

The EAST AFRICAN RIFT SYSTEM :a seminar Papers Nairobi 1965 Apr 12-17 International upper mantle committee Sponsored by the International union of geodesy and geophysics Nairobi,1965
Geod 9.0451

EAST LANSING,MICH. 1951 Biology of phosphorus A Symposium in general biology Papers presented By G.Evelyn Hutchinson and others Edited by Lester F. Wolterink Held at Michigan state college Michigan state college press,1952
Bioch 33.0590

EAST LANSING,MICH. 1962 Fundamental topics in relativistic fluid mechanics and magnetohydrodynamics :a symposium Proceedings Edited by Robert Wasserman and Charles P. Wells Held at Michigan state university Academic press,New York,1963
Eng 41.4547

EAST LANSING,MICH. 1967 Topology of manifolds :a conference Proceedings Edited by John G. Hocking Held at Michigan state university Bibliog. viii,161p 21cm Prindle,Weber and Schmidt,Boston,Mass.,1968
P Math 2.3351

EASTBOURNE 1956 Communication or conflict : conferences.Conferences;their nature,dynamics, and planning :an international conference Edited by Mary Capes Tavistock publications, London,1960
Eng 41.0699

EASTBOURNE 1962 Pattern of research in British industry :a conference Report Federation of British industries Federation of British industries,London,1962
Eng 41.1671

EASTBOURNE 1965 Automation '65 British automation conference 1965 Confederation of British industry Trades union congress United Kingdom automation council Organised by the Institution of production engineers Institution of production engineers,London, 1965
Eng 41.6018

EASTBOURNE 1966 High-temperature properties of steels :joint conference Proceedings British iron and steel research association Iron and steel institute Iron and steel institute.Publications, 97 Iron and steel institute,London,1967
Eng 41.3827

EASTBOURNE 1966 High-temperature properties of steels the joint conference Proceedings British iron and steel research association and Iron and steel institute Iron and steel institute.Publication, 97 Eyre and Spottiswoode,Margate,1967
Met 25.0429

EASTBOURNE 1967 Integrated circuits Conference on integrated circuits Sponsored by the Institution of electrical engineers. Electronics division I.E.E.Conference publication, 30 Institution of electrical engineers,London,1967 Joint sponsors; Institution of electronic and radio engineers and the Institute of electrical and electronics engineers
Eng 41.5461

EASTBOURNE 1968 Modern chemistry in industry An International union of pure and applied chemistry symposium Proceedings Society of chemical industry Edited by J.G. Gregory London,1968
Bioch 33.1874

EASTBOURNE 1968 Modern chemistry in industry :a symposium Proceedings International union of pure and applied chemistry Edited by J.G. Gregory 311p Society of chemical industry,London,1968
Chem E 24.1551

EASTBOURNE 1968 Modern chemistry in industry :an International union of pure and applied chemistry symposium Proceedings International union of pure and applied chemistry Edited by J.G. Gregory illus 311p Society of chemical industry,London, 1968
Chem E 24.1874

EASTBOURNE 1969 Design in high-strength structural steels :a conference Proceedings British iron and steel research association Iron and steel institute.Publication, 122 Iron and steel institute,London,1969
Eng 41.2761

EASTER SCHOOL IN AGRICULTURAL SCIENCE 2nd Proceedings Soil zoology Nottingham 1955 University of Nottingham Edited by D. Keith McE. Kevan Butterworths,London,1955
Bal 39.2006

EASTER SCHOOL IN AGRICULTURAL SCIENCE 2nd Proceedings Soil zoology Nottingham 1955 University of Nottingham Edited by D. Keith McE. Kevan Butterworths scientific publications illus. xiv,512p Butterworths,London,1955
Bot 42.2070

EASTER SCHOOL IN AGRICULTURAL SCIENCE 3rd Proceedings Growth of leaves Nottingham 1956 University of Nottingham Edited by F. L. Milthorpe illus. x,223p Butterworths, London,1956
Bot 42.1825

EASTER SCHOOL IN AGRICULTURAL SCIENCE 4th Proceedings Control of the plant environment Nottingham 1957 University of Nottingham Edited by J.P. Hudson Butterworths,London,1957
BG 38.0605

EASTER SCHOOL IN AGRICULTURAL SCIENCE 4th Proceedings Control of the plant environment Nottingham 1957 University of Nottingham Edited by J.P. Hudson illus. xvi,240p Butterworths,London,1957
Bot 42.1943

EASTER SCHOOL IN AGRICULTURAL SCIENCE 5th Proceedings Nutrition of the legumes Nottingham 1958 University of Nottingham Edited by E.G. Hallsworth Butterworths, London,1958
Bioch 33.0776

EASTER SCHOOL IN AGRICULTURAL SCIENCE 6th Proceedings Measurement of grassland productivity Nottingham 1959 University of Nottingham Edited by J.D. Ivins Butterworths,London,1959
Bot 42.1951

EASTER SCHOOL IN AGRICULTURAL SCIENCE 11th Proceedings Experimental pedology Nottingham 1964 University of Nottingham Edited by E.G. Hallsworth and D.V. Crawford xi,413p Butterworths,London,1965
Bot 42.2090

EASTER SCHOOL IN AGRICULTURAL SCIENCE 12th Proceedings Growth of cereals and grasses Nottingham University of Nottingham Edited by F.L. Milthorpe and J.D. Ivins Bibliog., illus,diagrs,tables xii,359p Butterworths, London,1966
Bot 42.2035

EASTER SCHOOL IN AGRICULTURAL SCIENCE 13th
Proceedings Reproduction in the female mammals Nottingham 1966 University of Nottingham Edited by G.E. Lamming and E.C. Amoroso Butterworths,London,1967
An 32.5323

EASTER SCHOOL IN AGRICULTURAL SCIENCE 14th
Proceedings Growth and development of mammals Nottingham 1967 University of Nottingham.School of agriculture Edited by G. A. Lodge and G.E. Lamming Butterworths, London,1968
Gen 34.1592

EASTER SCHOOL IN AGRICULTURAL SCIENCE 15th
Proceedings Root growth Nottingham 1968 University of Nottingham Edited by William J. Whittington Bibliog.,illus. xi, 450p Butterworths,London,1969
Bot 42.1437

EASTER SCHOOL IN AGRICULTURAL SCIENCES 5th
Proceedings Nutrition of the legumes Nottingham 1958 Edited by E.G. Hallsworth Butterworths,London,1958
Bot 42.1781

EASTER SCHOOL IN AGRICULTURAL SCIENCES 7th
Proceedings Digestive physiology and nutrition of the ruminant Nottingham 1960 University of Nottingham Edited by D. Lewis Butterworth,London,1961
VA 19.0296

EASTON,MD. 1958 Conference on physical and chemical properties of sea water Report United States.Office of naval research and National research council. Committee on oceanography National research council. Publication, 600 Washington,D.C.,1959
Geod 9.0610

EASTON,MD. 1958 The Arctic sea ice : conference Proceedings National research council.Division of earth science Supported by United States.Office of naval research National research council.Publication,598 271p 28cm National academy of sciences-National research council,Washington,1958
Sco 14.0359

EATON LABORATORIES DIVISION Symposium on pyelonephritis Edinburgh 1966 Livingstone,London,1967
PGMS 29.0236

ECCRINE SWEAT GLANDS AND ECCRINE SWEATING
Brown university symposium on the biology of skin,1961 Proceedings Providence,R.I. 1961 Jan 28-29 Edited by William Montagna and others Advances in biology of skin, 3 Pergamon press,Oxford,1962
An 32.4039

ECHANGES DE MATIERES AU COURS DE LA GENESE DES ROCHES GRENUES ACIDES ET BASIQUES Centre national de la recherche scientifique : colloque international 68th Proceedings Nancy 1955 Sep 4-11 Edited by M. Roubault Sciences de la terre,no.hors serie Ecole superieure de geologie,Nancy,1955
Min 10.0713

ECLIPSES Solar eclipses and the ionosphere : a symposium London 1955 Aug.22-24 Edited by W.J.G. Beynon and G.M. Brown Under the auspices of the Mixed commission on the ionosphere Pergamon press,London,1956 Vol.6 of special supplements to Journal of atmospheric and terrestrial physics
Obs 6.0176

ECOLE NORMALE SUPERIEURE DE SAINT-CLOUD
L'histoire sociale.Sources et methodes : colloque Saint-Cloud 1965 May 15-16 Presses universitaires de France,Paris,1967
Geog 13.1610

ECOLE PRATIQUE DES HAUTES ETUDES,PARIS.CENTRE DE SOCIOLOGIE EUROPEENNE Problemes du developpement economique dans les pays mediterraneens colloque international Actes Naples 1962 Oct 28-Nov 2 Edited by Jean Cuisenier Maison des sciences de l'homme. Recherches mediterraneennes,5 Mouton,The Hague,1963
Geog 13.4116

ECOLOGICAL ASPECTS OF THE MINERAL NUTRITION OF PLANTS :a symposium Sheffield 1968 Apr 1-5 British ecological society Edited by Ian H. Rorison and others British ecological society.Symposia, 9 illus, plates xxi,484p Blackwell scientific, Oxford,1969
Bot 42.2107

ECOLOGICAL EFFECTS OF BIOLOGICAL AND CHEMICAL CONTROL OF UNDESIRABLE PLANTS AND ANIMAL SYMPOSIUM International union for the conservation of nature and natural resources. Technical meeting 8th Warsaw 1960 Jul 15-24 Edited by D.J. Kuenen Leiden, 1961 Title also in French:Effets ecologiques du controle biologique et chimique des plantes et animaux undesirables
Bal 44.6435

ECOLOGIE DES ALGUES MARINES Dinard 1957 Sep 20-28 By R. Biebl and others Centre national de la recherche scientifique Centre national de la recherche scientifique. Colloques internationaux, 81 illus. 276p Paris,1959
Bot 42.3814

ECOLOGY Man's place in the island ecosystem Pacific science congress 10th Symposium Honolulu 1961 Aug 21-Sep 6 Edited by F.R. Fosberg Bishop museum press,Honolulu,1963
Geog 13.1311

ECOLOGY Phytochemical ecology Phytochemical society symposium 8th Annual proceedings Englefield Green,Surrey 1971 Apr Phytochemical society Academic press,London,1972
Bioch 33.2229

ECOLOGY Recent advances in tropical ecology : a symposium Proceedings Banaras 1967 Jan 16-21 International society for tropical ecology Edited by R. Misra and B. Gopal 2 vols Vararasi,1968
Bot 42.2001

ECOLOGY Symposium on the ecology of pelagic fish species in Arctic waters and adjacent seas Papers Charlottenlund 1966 Sep 30-Oct 1 International council for the exploration of the sea.Distant northern seas committee and International council for the exploration of the sea.Gadoid fish committee Edited by R.W. Blacker Conseil permanent international pour l'exploration de la mer.Rapports et proces-verbaux des reunions, 158 illus,maps Host,Copenhagen,1968
Sco 14.8291

ECOLOGY The Teaching of ecology :a symposium London 1966 Apr 13-16 British ecological society Edited by J.M. Lambert British ecological society.Symposia, 7 illus. xi, 294p Blackwell scientific,Oxford,1967
Bot 42.1988

ECOLOGY AND THE INDUSTRIAL SOCIETY British ecological society symposium 5th Swansea 1964 Apr 13-16 Edited by Gordon T. Goodman and others Blackwell,Oxford,1965
Geog 13.1216

The ECOLOGY OF ALGAE :a symposium Pymatuning,Pa. 1959 Jun 18-19 Edited by C. A. Tryon and R.T. Hartman Pymatuning laboratory of field biology.Special publication, 2 Pittsburgh,Pa.,1960
Bot 42.3900

ECOLOGY OF APHIDOPHAGOUS INSECTS :symposium Proceedings Liblice 1965 Sep 27-Oct 1 Ceskoslovenska akademie ved Junk;Academia, The Hague;Prague,1966
Bal 39.2844

ECOLOGY OF HEALTH :institute on public health Papers New York 1947 Apr 1-3 New York academy of medicine Edited by E.H.L. Corwin Commonwealth fund,New York,1949 Issued for the centennial celebration of the New York academy of medicine
An 32.2035

The ECOLOGY OF SOIL BACTERIA :an international symposium Edited by T.R.G. Gray and D. Parkinson illus. xv,681p Liverpool university press,Liverpool,1968
Bot 42.3764

ECOLOGY OF SOIL-BORNE PLANT PATHOGENS.PRELUDE TO BIOLOGICAL CONTROL Factors determining the behavior of plant pathogens in soil :an international symposium Berkeley,Calif. 1963 Apr 7-13 Edited by Kenneth F. Baker and William C. Snyder Sponsored by the National research council illus. Murray,London,1965
Mol 45.0575

The ECOLOGY OF SOIL FUNGI :an international symposium Edited by D. Parkinson and J.S. Waid illus. xi,324p Liverpool university press,Liverpool,1960
Bot 42.3717

ECONOMIC CONSEQUENCES OF THE SIZE OF NATIONS International economic association conference Proceedings Lisbon 1957 Edited by E.A.G. Robinson Macmillan,London,1960
Geog 13.2036

ECONOMIC DEVELOPMENT FOR AFRICA SOUTH OF THE SAHARA International economic association regional conference 3rd Proceedings Addis Ababa 1961 Edited by E.A.G. Robinson Macmillan,London,1965
Geog 13.4646

ECONOMIC DEVELOPMENT IN AFRICA Nyasaland economic symposium Papers Blantyre 1962 Jul 18-28 Edited by E.F. Jackson Blackwell, Oxford,1965
Geog 13.4650

ECONOMIC DEVELOPMENT WITH SPECIAL REFERENCE TO EAST ASIA International economic association.Round table conference Proceedings Gamagori,Japan 1963 Edited by Kenneth Berrill Macmillan,London,1964
Geog 13.2407

ECONOMIC HISTORY SOCIETY.ANNUAL CONFERENCE Papers Entrepreneur Cambridge 1957 Apr Edited by B.E. Supple 63p Harvard university.Research center in entrepreneurial history,Cambridge,Mass.,1957
Geog 13.3190

ECONOMICS OF AUTOMATIC DATA PROCESSING : international symposium papers Rome 1965 Oct 19-22 International computation centre Edited by A.B. Frielink pl. xiii, 384p 22cm North-Holland,Amsterdam,1965
Math L 5.1022

ECONOMICS OF TAKE-OFF INTO SUSTAINED GROWTH International economic association conference Proceedings Konstanz 1960 Edited by W.W. Rostow Macmillan,London,1963
Geog 13.2064

EDINBURGH 1923 International physiological congress 11th Abstracts abstracts of communications 1923 Proof copy
Phys 20.2067

EDINBURGH 1936 Association internationale d'hydrologie scientifique :assemblee generale 6me comtes-rendus meetings of the International commissions of snow and of glaciers International association of scientific hydrology International association of scientific hydrology.Bulletin, 23 Riga,1938
Sco 14.0116

EDINBURGH 1936 International union of geodesy and geophysics :assemblee generale 6me 184p 25cm Camelot press,London,1937 In English and French
Sco 14.0114

EDINBURGH 1939 Genetics International genetical congress 7th Proceedings International genetical congress.Permanent international committee Edited by R.C. Punnett Cambridge university press,Cambridge, 1939 Issued as a supplementary volume to the 'Journal of genetics'
Bot 42.6483

EDINBURGH 1939 International genetical congress 7th Proceedings Edited by R.C. Punnett Cambridge university press,Cambridge, 1941
Bot 42.0661

EDINBURGH 1939 International genetical congress 7th Proceedings Genetical society Edited by Reginald Crundall Punnett Journal of genetics.Supplement Cambridge university press,Cambridge,1941
Gen 34.2219

EDINBURGH 1948 International congress of psychology 12th Proceedings and papers Edited by Mary Collins Oliver and Boyd, Edinburgh;London,1950 Under the presidency of J.Drever
Psy 31.3397

EDINBURGH 1948 Selective toxicity and antibiotics Society for experimental biology Edited by J.F. Danielli and R. Brown Society for experimental biology.Symposia, 3 Cambridge university press,Cambridge,1949
Radioth 35.1931

EDINBURGH 1948 Selective toxicity and antibiotics Society for experimental biology Society for experimental biology.Symposia, 3 Cambridge university press,Cambridge,1949
Phys 20.2224

EDINBURGH 1948 Selective toxicity and antibiotics :a symposium Society for experimental biology Society for experimental biology.Symposia, 3 Cambridge university press,Cambridge,1949
Bioch 33.1350

EDINBURGH 1950 Quantitative inheritance colloquium Papers University of Edinburgh. Institute of animal genetics Edited by E.C.R. Reeve and C.H. Waddington Under the auspices of the Agricultural research council H.M.S.O.,London,1952
Gen 34.1083

EDINBURGH 1953 Lind bicentenary symposium : a conference on scurvy and vitamin C in honour of James Lind Proceedings Nutrition society Nutrition society.Proceedings,12,3 Cambridge university press,London,1953
Sco 14.0848

EDINBURGH 1956 International union for the conservation of nature and natural resources : technical meeting 6th Proceedings and papers Society for the promotion of nature reserves,London,1956 Held simultaneously with the Union's 5th general assembly
Geog 13.2698

EDINBURGH 1957 Biological organisation: cellular and sub-cellular :a symposium Proceedings Organised by C.H.Waddington on behalf of Unesco Illus Pergamon press, London,1959
Radioth 35.0497

EDINBURGH 1957 Biological organisation, cellular and sub-cellular :a symposium organised on behalf of Unesco by C.H. Waddington Proceedings Unesco held at the University of Edinburgh Pergamon press, London,1959
Bioch 33.1019

EDINBURGH 1957 Biological organisation cellular and subcellular :a symposium Proceedings Unesco Edited by C.H. Waddington held at the University of Edinburgh Pergamon,London,1959
Bal 39.0455

EDINBURGH 1957 Biological organization: cellular and sub-cellular :a symposium Proceedings Organized by C.H.Waddington on behalf of Unesco Pergamon press,London,1959
Gen 34.0829

EDINBURGH 1957 Biological organization: cellular and subcellular :a symposium Proceedings Edited by C.H. Waddington Organised on behalf of Unesco Pergamon press, London,1959
An 32.2485

EDINBURGH 1959 Progress in endocrinology Proceedings Vol 1-2 Society for endocrinology Edited by K. Fotherby and others Society for endocrinology.Memoirs, 9-10 2 vols Cambridge university press, Cambridge,1964
Inv Med 37.0082

EDINBURGH 1959 Progress in endocrinology The Edinburgh meeting on endocrinology Proceedings Pt 1-2 Society for endocrinology Edited by K. Fotherby and others Society for endocrinology.Memoirs, 9-10 2 vols Cambridge university press, Cambridge,1960-1961
Bioch 33.0484

EDINBURGH 1960 Gas chromatography :a symposium 3rd Proceedings Institute of petroleum.Gas chromatography discussion group Edited by R.P.W. Scott Organized also by the Society for physical chemistry xvii,466p Butterworths,London,1960
Chem E 24.0850

EDINBURGH 1960 Gas chromatography 1960 :a symposium 3rd Proceedings Society for analytical chemistry,and,Institute of petroleum.Gas chromatography discussion group Edited by R.P.W. Scott Butterworths,London, 1960
Chem 18.0567

EDINBURGH 1960 International congress on clinical chemistry 4th Proceedings Organised by the Association of clinical biochemists E.and S.Livingstone,Edinburgh; London,1961
Bioch 33.2264

EDINBURGH 1960 Natural resources in Scotland :symposium Royal society of Edinburgh 25cm Edinburgh,1961
Philos 1.1867

EDINBURGH 1961 Human radiation cytogenetics :an international symposium Proceedings United Kingdom atomic energy authority Medical research council University of Aberdeen.Department of genetics Edited by H.J. Evans and others North-Holland,Amsterdam,1967
Gen 34.1127

EDINBURGH 1961 The British Caledonides :a symposium Papers Edinburgh university and Edinburgh geological society Edited by M.R.W. Johnson and F.H. Stewart Edinburgh, 1963
Geod 9.0414

EDINBURGH 1961 The British Caledonides symposium papers Edited by M.R.W. Johnson and F.H. Stewart Sponsored by the Edinburgh geological society illus. map Oliver and Boyd,Edinburgh;London,1963
Geol 8.1260

EDINBURGH 1962 Cell differentiation Society for experimental biology Society for experimental biology.Symposia, 17 Cambridge university press,Cambridge,1963
Bioch 33.1359

EDINBURGH 1962 Cell differentiation :a symposium Edited by G.E. Fogg Society for experimental biology.Symposia, 17 Cambridge university press,Cambridge,1963
An 32.2500

EDINBURGH 1962 Cell differentiation : symposium Society for experimental biology Society for experimental biology.Symposia, 17 Cambridge university press,Cambridge,1963
Phys 20.1494

EDINBURGH 1963 Chemical engineering in the metallurgical industries a symposium Proceedings Institution of chemical engineers Institution of chemical engineers, London,1963
Met 25.0316

EDINBURGH 1963 Symposium on chemical engineering in the metallurgical industries Proceedings Institution of chemical engineers 147p Institution of chemical engineers,London,1963
Chem E 24.1131

EDINBURGH 1964 African primary products and international trade :international seminar University of Edinburgh.Centre of African studies Edited by I.G. Stewart and H.W. Ord Jointly organized by the University of Edinburgh.Department of political economy Edinburgh university press,Edinburgh,1965
Geog 13.2449

EDINBURGH 1964 Chromosome manipulations and plant genetics :a symposium held during the 10th international botanical congress Edited by Ralph Riley and K.R. Lewis Oliver and Boyd,Edinburgh;London,1966
Gen 34.0894

EDINBURGH 1964 Chromosome manipulations and plant genetics :a symposium held during the 10th international botanical congress Edited by Ralph Riley and K.R. Lewis Oliver and Boyd,Edinburgh,1966
Bot 42.0789

EDINBURGH 1964 Incompatibility in fungi :a symposium Proceedings Edited by Karl Esser and John R. Raper viii,124p 24cm Springer,Berlin,1965 Symposium held at the 10th International botanical congress
Bot 42.4713

EDINBURGH 1964 International botanical congress 10th Proceedings 1964
BG 38.3131

EDINBURGH 1964 The Impact of users' needs on the design of data processing systems : conference British computer society Held under the aegis of the United Kingdom automation council 1964 Organized jointly by the British computer society,the British institution of radio engineers and the Institution of electrical engineers
Math L 5.3265

EDINBURGH 1965 Diabetes melitus :the opening symposium of the Pfizer foundation of the post-graduate medical school,University of Edinburgh Pfizer foundation Edited by L.J. P. Duncan University of Edinburgh.Pfizer medical monographs Edinburgh university press,Edinburgh,1966
PGMS 29.0179

EDINBURGH 1965- Machine intelligence workshop 1st- Essays University of Edinburgh.Experimental programming unit Edited by Donald Michie and others Oliver and Boyd;Edinburgh university press,Edinburgh, 1967- The 5th contains a previously unpublished report by Alan Turing
Math S 3.1818

EDINBURGH 1966 Human radiation cytogenetics :an international symposium Proceedings Edited by H.J. Evans and others Sponsored by the Medical research council North-Holland,Amsterdam,1967
Radioth 35.0540

EDINBURGH 1966 Psycholinguistic papers :a conference Proceedings Edinburgh university.Faculty of social sciences.Research board Edited by J. Lyons and R.J. Wales Edinburgh university press,Edinburgh,1966
Psy 31.2745

EDINBURGH 1966 Recent research on gonadotrophic hormones Gonadotrophin club meeting 5th Proceedings Gonadotrophin club Edited by E.Trevor Bell and John A. Loraine Livingstone,Edinburgh;London,1967
Inv Med 37.0015

EDINBURGH 1966 Symposium on pyelonephritis Norwich pharmacal company and Eaton Laboratories division Livingstone,London, 1967
PGMS 29.0236

EDINBURGH 1967 Instructional conference on differential equations Lectures Edinburgh, 1967 Typescript
P Math 2.3477

EDINBURGH 1967 Instructional conference on differential equations notes for the introductory lectures on Hilbert spaces and ordinary differential equations Illus. 28, 35cm 2 vols Edinburgh,1967 Typescript
P Math 2.3476

EDINBURGH 1967 Solution properties of natural polymers :an international symposium Proceedings Chemical society Chemical society.Special publication, 23 Chemical society,London,1968
Bioch 33.0632

EDINBURGH 1968 IFIP congress 1968 invited papers International federation for information processing 1600p 30cm North-Holland,Amsterdam,1968
Math L 5.3123

EDINBURGH 1968 Metabolism of amines in the brain :a symposium Proceedings British pharmacological society Scandinavian pharmacological society Edited by G. Hooper Macmillan,London,1969
An 32.5040

EDINBURGH GEOLOGICAL SOCIETY The British Caledonides :a symposium Papers Edinburgh 1961 Dec Edited by M.R.W. Johnson and F.H. Stewart Edinburgh,1963
Geod 9.0414

EDINBURGH GEOLOGICAL SOCIETY The British Caledonides symposium papers Edinburgh 1961 Dec. Edited by M.R.W. Johnson and F.H. Stewart illus. map Oliver and Boyd, Edinburgh;London,1963
Geol 8.1260

The EDINBURGH MEETING ON ENDOCRINOLOGY Proceedings Progress in endocrinology Edinburgh 1959 Aug 16-20 Pt 1-2 Society for endocrinology Edited by K. Fotherby and others Society for endocrinology.Memoirs, 9-10 2 vols Cambridge university press,Cambridge,1960-1961
Bioch 33.0484

EDINBURGH UNIVERSITY The British Caledonides a symposium Papers Edinburgh 1961 Dec Edited by M.R.W. Johnson and F.H. Stewart Edinburgh,1963
Geod 9.0414

EDINBURGH UNIVERSITY.FACULTY OF SOCIAL SCIENCES. RESEARCH BOARD Psycholinguistic papers :a conference Proceedings Edinburgh 1966 Mar 18-20 Edited by J. Lyons and R.J. Wales Edinburgh university press,Edinburgh,1966
Psy 31.2745

EDMONTON 1958 The Canadian Northwest:its potentialities :symposium presented to the Royal society of Canada in 1958 Papers Royal society of Canada Edited by Frank H. Underhill Royal society of Canada.Studia varia.Series 3 map 104p 24cm University of Toronto press,Toronto,1959
Sco 14.4228

EDMONTON 1958 The Canadian northwest:its potentialities :symposium Royal society of Canada Edited by Frank H. Underhill Royal society of Canada."Studia varia" series,3 University of Toronto press,Toronto,1959
Geog 13.5038

EDMONTON 1965 Gastric secretion:mechanisms and control :international symposium Proceedings Edited by T.K. Shnitka and others Pergamon press,Oxford,1967
Phys 20.1428

EDMONTON,ALBERTA 1964 Muscle :a symposium Proceedings University of Alberta.Faculty of medicine Edited by W.M. Paul and others Pergamon press,Oxford,1965 First symposium of the University of Alberta medical school held as part of the celebrations of the 50th year of this medical faculty
Bioch 33.2244

EDMONTON,ALBERTA 1964 National northern development conference 3rd Proceedings Alberta and northwest chamber of mines and resources,and,Edmonton chamber of commerce illus 170p 28cm Alberta,c1965
Sco 14.7424

EDMONTON,ALBERTA 1966 Molecular biology of viruses Symposium of the molecular biology of viruses Proceedings University of Alberta.Faculty of medicine Edited by John S. Colter and William Paranchych Academic press,New York;London,1967
Path 30.2745

EDMONTON,ALTA. 1957 Nilpotent groups Canadian mathematical congress :summer seminar Notes of lectures By Philip Hall 2nd edition iii,76p 25cm Queen Mary college, London,1957
P Math 2.4544

EDMONTON,ALTA. 1958 Last frontier in North America The National northern development conference and post-conference air tour, September 17-21,1958 Proceedings Edmonton chamber of commerce,and,Alberta and Northwest chamber of mines and resources illus,map 184p 28cm Edmonton,Alta.,c1958
Sco 14.3709

EDMONTON,ALTA. 1964 Muscle :symposium Proceedings University of Alberta.Faculty of medicine Edited by W.M. Paul and others Pergamon press,Oxford,1965 50th anniversary of the University of Alberta medical school
An 32.4007

EDMONTON CHAMBER OF COMMERCE and ALBERTA AND NORTHWEST CHAMBER OF MINES AND RESOURCES Last frontier in North America The National northern development conference and post-conference air tour,September 17-21,1958 Proceedings Edmonton,Alta. 1958 Sep 17-21 illus,map 184p 28cm Edmonton,Alta.,c1958
Sco 14.3709

EDMONTON CHAMBER OF COMMERCE and ALBERTA AND NORTHWEST CHAMBER OF MINES AND RESOURCES National northern development conference 3rd Proceedings Edmonton,Alberta 1964 Oct 21-23 illus 170p 28cm Alberta,c1965
Sco 14.7424

EDUCATION University education in computing science :a conference on graduate academic and related research programs in computing science Proceedings New York 1967 Jun 5-8 Edited by Aaron Finerman Sponsored by New York state science and technology foundation ACM monograph series xvi,237p 23cm Academic press,New York,1968
Math L 5.3107

EDUCATIONAL RESEARCH Improving experimental design and statistical analysis Phi Delta Kappa annual symposium on educational research 7th Discussions Madison,Wisc. 1966 Phi Delta Kappa and University of Wisconsin.Phi Delta Kappa chapter Rand-McNally education series Rand-McNally, Chicago,Ill.,1967
Psy 31.1123

EDUCATIONAL TESTING SERVICE Measurement in personality and cognition :a conference Princeton,N.J. 1960 Oct Edited by Samuel Messick and John Ross Wiley,New York;London, 1962
Psy 31.2152

EDWARD ORTON JUNIOR CERAMIC FOUNDATION Mechanical behavior of crystalline solids a symposium Proceedings New York 1962 Apr 28-29 Ceramic education council and National bureau of standards National bureau of standards monograph, 59 U.S.Government printing office,Washington,1963 Lithographed
Met 25.1213

EDWARD ORTON JUNIOR CERAMIC FOUNDATION Microstructure of ceramic materials :a symposium Proceedings Pittsburgh 1963 Apr 27-28 Ceramic education council and National bureau of standards National bureau of standards.Miscellaneous publication, 257 U.S.dept of commerce,Washington D.C.,1964
Met 25.0285

EDWARDSVILLE 1967 Orthogonal expansions and their continuous analogues :a conference Proceedings Edited by Deborah Tepper Haimo Bibliog. xx,307p 24cm Feffer and Simons Inc.,London;Amsterdam,1968
P Math 2.3408

EFFECT DU GEL DANS LES MATERIAUX DE CONSTRUCTION colloque Compte rendu Paris 1957 May Pt 1-2 Association francaise de recherches et d'essais sur les materiaux et les constructions Cahiers de la recherche theorique et experimentale sur les materiaux et les structures,8,9 illus,diagrs 81p 27cm Paris,1959
Sco 14.1263

EFFECT OF CYCLIC HEATING AND SHESSING METALS Symposium on effect of cyclic heating and stressing on metals at elevated temperatures Chicago 1954 Jun 17 American society for testing materials A.S.T.M.Special technical publication, 165 A.S.T.M.,Philadelphia,Pa., 1954
Met 25.0980

EFFECT OF CYCLIC HEATING AND STRESSING ON METALS Symposium on effect of cyclic heating and stressing on metals at elevated temperatures Chicago 1954 Jun 17 American society for testing materials A.S.T.M.Special technical publication, 165 A.S.T.M.,Philadelphia,1954
Met 25.2534

The EFFECT OF PHARMACOLOGIC AGENTS ON THE NERVOUS SYSTEM New York 1957 Dec 13-14 Association for research in nervous and mental disease Edited by Francis J. Braceland Association for research in nervous and mental disease.Research publications,37 Williams and Wilkins,Baltimore,Md.,1959
Pha 16.0036

The EFFECT OF PHARMACOLOGIC AGENTS ON THE NERVOUS SYSTEM Proceedings of the Association New York 1957 Dec 13-14 Association for research in nervous and mental disease Edited by Francis J. Braceland Association for research in nervous and mental disease. Research publications, 37 Williams and Wilkins,Baltimore,Md.,1959
An 32.4169

EFFECT OF RADIATION ON HUMAN HEREDITY report of a study group Copenhagen 1956 Aug 7-11 World health organization.Study group World health organization,Geneva,1957
Gen 34.1106

EFFECT OF RADIATION ON HUMAN HEREDITY :a study group Report Copenhagen 1956 Aug 7-11 World health organization World health organization,Geneva,1957
Radioth 35.1055

EFFECT OF TEMPERATURE Symposium on effect of temperature on the brittle behavior of metals with particular reference to low temperatures : presented at the 56th annual meeting A.S.T.M. Atlantic city,N.J. 1953 Jun 28-30 American society for testing materials American society for testing materials.Special technical publication, 158 American society for testing materials,Philadelphia,Pa.,1954
Eng 41.3750

EFFECT OF THE VARIOUS STEELMAKING PROCESSES ON THE ENERGY BALANCES OF INTEGRATED IRON AND STEELWORKS The Iron and steel engineers group meeting 43rd Report London 1961 Dec 1 Iron and steel institute.Engineers group London,1962
Met 25.2522

EFFECTS OF EARTHQUAKES ON SOILS AND FOUNDATIONS Vibration effects of earthquakes on soils and foundations :a symposium San Francsico, Calif. 1968 Jun 23-28 American society for testing and materials American society for testing and materials.Special technical publication, 450 American society for testing materials,Philadelphia,Pa.,1969
Eng 41.3162

EFFECTS OF EXTERNAL STIMULI ON REPRODUCTION : a Ciba foundation study group London 1966 Jul 27 Edited by G.E.W. Wolstenholme and Maeve O'Connor Ciba foundation study group, 26 Churchill,London,1967
An 32.3895

EFFECTS OF IONIZING RADIATION ON THE NERVOUS SYSTEM Symposium on ionizing radiation on the nervous system Proceedings Vienna 1961 Jun 5-9 Sponsored by the International atomic energy agency International atomic energy agency.Proceedings series International atomic energy agency,Vienna,1962
Radioth 35.1130

EFFECTS OF IONIZING RADIATION ON THE NERVOUS SYSTEM Symposium on the effects of ionizing radiation on the nervous system Vienna 1961 Jun 5-9 Sponsored by the International atomic energy agency International atomic energy agency,Vienna,1962
An 32.4190

EFFECTS OF IONIZING RADIATION ON THE REPRODUCTIVE SYSTEM :international symposium Proceedings Fort Collins,Colo. 1962 Colorado state university and American institute of biological sciences Edited by William D. Carlson and F.X. Gassner Pergamon press,Oxford,1964
An 32.0153

EFFECTS OF IONIZING RADIATIONS ON THE HAEMATOPOIETIC TISSUE Panel on the effects of ionizing radiations from different sources on haematopoietic tissue Proceedings Vienna 1966 May 17-20 International atomic energy agency International atomic energy agency.Panel proceedings series International atomic energy agency,Vienna,1967
Radioth 35.1200

EFFECTS OF LOW DOSES OF RADIATION ON CROP PLANTS a panel Report Rome 1964 Jun 1 Food and agriculture organization International atomic enérgy agency.Division of atomic energy in agriculture International atomic energy agency.Technical reports series, 64 International atomic energy agency, Vienna,1966
Radioth 35.1151

EFFECTS OF LOW TEMPERATURES Symposium on effects of low temperatures on the properties of materials :presented at a meeting of the Philadelphia district,A.S.T.M. Philadelphia, Pa. 1946 Mar 19 American society for testing materials American society for testing materials.Special technical publication, 78 American society for testing materials,Philadelphia,Pa.,1950
Eng 41.3778

EFFECTS OF RADIATION ON CELLULAR PROLIFERATION AND DIFFERENTIATION Symposium on the effects of radiation on cellular proliferation and differentiation Proceedings Monaco 1968 Apr 1-5 International atomic energy agency Organized in co-operation with the Joint commission on applied radioactivity International atomic energy agency,Vienna,1968
Radioth 35.1760

EFFECTS OF RADIATION ON MEIOTIC SYSTEMS Study group on the effects of radiation on meiotic systems Vienna 1967 May 8-11 Organized by the International atomic energy agency International atomic energy agency. Panel proceedings series International atomic energy agency,Vienna,1968
Radioth 35.1201

EFFECTS OF RADIATION ON METALS A Symposium on the effects of radiation on metals Cleveland,Ohio 1956 Oct 8 American institute of mining,metallurgical and petroleum engineers.Nuclear metallurgy committee I.M.D.Special report series, 3 Nuclear metallurgy, 3 American institute of mining,metallurgical,and petroleum engineers,New York,1956
Met 25.1576

EFFICIENT COMPUTER METHODS FOR THE PRACTISING CHEMICAL ENGINEER :a symposium Proceedings Nottingham 1967 Apr 19 Edited by J.M. Pine Organised by the Institution of chemical engineers.Midlands branch Institution of chemical engineers. Symposium series, 23 illus 194p I.Chem. E.,London,1967
Chem E 24.1809

EFFICIENT COMPUTER METHODS FOR THE PRACTISING CHEMICAL ENGINEER :a symposium Proceedings Nottingham 1967 Apr 19 Organised by the Institution of chemical engineers.Midlands branch Institution of chemical engineers.Symposium series, 23 Institution of chemical engineers,London,1967
Eng 41.5860

EGG IMPLANTATION London 1965 Oct 21 Ciba foundation Edited by G.E.W. Wolstenholme and Maeve O'Connor Ciba foundation study group, 23 Churchill,London,1966
An 32.3889

EGG IMPLANTATION London 1965 Oct 21 Ciba foundation Edited by G.E.W. Wolstenholme and Maeve O'Connor Ciba foundation study group, 23 Churchill,London,1966
Phys 20.1408

EINDHOVEN 1956 Diffusion dans les metaux La Colloque sur la physique des metaux Comptes rendus Edited by J.D. Fast and others Bibliotheque technique Philips, Eindhoven,1957
Met 25.1276

EINDHOVEN 1962 Magnetic and electric resonance and relaxation Colloque Ampere : international conference 11th Proceedings Netherlands physical society Edited by J. Smidt Under the auspices of the International union of pure and applied physics 789p North-Holland,Amsterdam,1963
Cav 7.0712

EINDHOVEN 1967 International symposium on fluidization Proceedings Edited by A.A.H. Drinkenburg illus 829p Netherlands university press,Amsterdam,1967
Chem E 24.1918

EINDHOVEN 1968 Attention and performance 2 Donders centenary symposium on reaction time Proceedings Institute for perception research IPO Edited by W.G. Koster North-Holland,Amsterdam;London,1969
Psy 31.2948

EINSTEIN SYMPOSIUM Entstehung, entwicklung und perspektiven der Einsteinschen gravitationstheorie vortrage und diskussionen Berlin 1965 Nov 2-5 Deutsche akademie der wissenschaften zu Berlin Edited by H-J. Treder 314p Akademie-verlag, Berlin,1966
Obs 6.1765

EINSTEIN-SYMPOSIUM Entstehung,Entwicklung und perspektiven der Einsteinschen gravitationstheorie vortrage und diskussionen Berlin 1965 Nov 2-5 Akademie der Wissenschaften Edited by H.J. Treder . Akademie Verlag,Berlin,1965 typescript
A Math 4.1031

EL JURASICO INFERIOR DE MEXICO Y SUS AMONITAS International geological congress 20th papers Mexico City 1956 By Heinrich Karl Erben Translated by U. Hungsberg from the German Mexico City,1956
Geol 8.3033

EL JURASICO MEDIO Y EL CALLOVIANO DE MEXICO International geological congress 20th papers Mexico City 1956 By Heinrich Karl Erben maps Mexico City,1956
Geol 8.3032

ELABORAZIONE DI DATI OFFICI DE PARTE DI ORGANISM E DI MACCHINE Scuola internazionale di fisica "Enrico Fermi" 43 corso Rendaconti Varenna 1968 Jul 15-27 Societa italiana di fisica Edited by W. Reichardt Academic press,New York;London,1969
An 32.5470

ELASTIC SHELLS The Theory of thin elastic shells :a symposium Proceedings Delft 1959 Aug 24-28 International union of theoretical and applied mechanics North-Holland,Amsterdam,1960
Eng 41.2854

ELASTICITY American mathematical society. Symposium in applied mathematics 3rd Proceedings Ann Arbor,Mich. 1949 Jun 14-16 Edited by R.V. Churchill and others Cosponsored by the American society of mechanical engineers.Applied mechanics division McGraw-Hill,New York,1950
Col S 12.0335

ELASTOHYDRODYNAMIC LUBRICATION :a symposium Proceedings Leeds 1972 Apr 11-13 Institution of mechanical engineers.Tribology group IME,London,1972
Eng 41.8650

ELECTRIC POWER SYSTEMS Real-time control of electric power systems Symposium of real-time control of electric power systems Baden 1971 Sep 27-28 Brown,Boveri and company Edited by Edmund Handschin Elsevier, Amsterdam,1972 The second Brown,Boveri symposium
Eng 41.8558

ELECTRIC PROPULSION DEVELOPMENT... :a selection of technical papers based mainly on the American rocket society electric propulsion conference... Berkeley,Calif. 1962 Mar 14-16 American rocket society Edited by Ernst Stuhlinger Progress in astronautics and aeronautics, 9 Academic press,New York;London,1963
Eng 41.6889

ELECTRIC TRACTION Convention on electric traction by single-phase current at industrial frequency Documents Lille 1955 May 11-14 Lille,1955
Eng 41.4842

ELECTRICAL STUDIES ON THE UNANESTHETIZED BRAIN Georgetown Edited by Estelle R. Ramey and Desmond S. ODoherty Hoeber,New York,1960
VA 19.0238

ELECTRICAL STUDIES ON THE UNANESTHETIZED BRAIN a symposium Washington,D.C. 1960 Edited by Estelle R. Ramey and Desmond S. O'Doherty Hoeber,New York,1960
An 32.4648

ELECTRICITY SUPPLY INDUSTRY Gas discharges and the electricity supply industry :a conference Proceedings Leatherhead,Surrey 1962 May 7-11 Central electricity research laboratories Edited by J.S. Forrest and others 677p Butterworths,London,1962
Nap 11.0123

ELECTROCHEMICAL CONSTANTS NBS semicentennial symposium 7th Proceedings Washington,D. C. 1951 Sep 19-21 National bureau of standards National bureau of standards. Circular, 524 U.S.Government printing office, Washington,D.C.,1953 One of 12 semicentennial symposia
Met 25.2401

ELECTROCHEMICAL SOCIETY Chemical physics of ionic solutions :a selction of invited papers and discussions presented at the international symposium of the Electrochemical society... under the chairmanship of B.E.Conway Toronto 1964 May 4-6 Edited by B.E. Conway and R.G. Barradas J.Wiley,New York,1966
Chem 18.0342

ELECTROCHEMICAL SOCIETY Mechanical properties of intermetallic compounds a symposium held during the 115th meeting of the Electrochemical society Philadelphia 1959 May 3-7 Edited by J.H. Westbrook John Wiley,London,1960
Met 25.2277

ELECTROCHEMICAL SOCIETY The Electron microprobe :a symposium Proceedings Washington,D.C. 1964 Oct 12-15 Edited by T. D. McKinley and others Electrochemical society series John Wiley,New York,New York, 1966
Cav 7.2425

ELECTROCHEMICAL SOCIETY.CORROSION DIVISION Stress corrosion cracking and embrittlement Boston,Mass. 1954 Oct Edited by William D. Robertson Electrochemical society series Wiley;Chapman and Hall,New York;London,1956
Met 25.2262

ELECTROCHEMICAL SOCIETY.ELECTROTHERMICS AND METALLURGY DIVISION Arcs in inert atmospheres and vacuum a symposium San Francisco,Calif. 1956 Apr 30-May 1 Edited by W.E. Kuhn Wiley;Chapman and Hall,New York; London,1956
Met 25.0248

ELECTROCHEMICAL SOCIETY.ELECTROTHERMICS AND METALLURGY DIVISION Rhenium :a symposium Chicago 1960 May 3-4 Edited by B.W. Gonser Elsevier,Amsterdam;New York,1962
Met 25.0585

ELECTROCHEMICAL SOCIETY.ELECTROTHERMICS AND METALLURGY DIVISION Technology of columbium (Niobium) :a symposium Washington, D.C. 1958 May 15-16 Wiley;Chapman and Hall, London,1958
Met 25.0587

ELECTROCHEMICAL SOCIETY.ELECTROTHERMICS AND METALLURGY DIVISION The Electron microprobe :a symposium Proceedings Washington 1964 Oct Edited by T.D. McKinley and others Wiley,New York,1966
Met 25.2362

ELECTROCHEMICAL SOCIETY.ELECTROTHERMICS AND METALLURGY DIVISION Vacuum metallurgy symposium Papers Boston,Mass. 1954 Oct 6-7 Edited by John M. Blocher Electrochemical society inc.,Boston,Mass.,1955
Met 25.2258

ELECTROCHEMICAL SOCIETY.ELECTROTHERMICS AND METALLURGY DIVISIONS International conference on electron and ion beam science and technology 1st Toronto 1964 May 3-7 Edited by Robert Bakish Electrochemical society series Wiley,New York,1965
Eng 41.5564

ELECTROCHEMICAL SOCIETY.ELECTROTHERMICS AND METALLURGY DIVISIONS International conference on electron and ion beam science and technology 1st Toronto 1964 May 3-7 Edited by Robert Bakish Electrochemical society series Wiley,New York,1965
Met 25.1678

ELECTROCHEMICAL SOCIETY.THEORETICAL ELECTROCHEMISTRY DIVISION Symposium on electrode processes Transactions Philadelphia 1959 May 4-6 Edited by Ernest Yeager Wiley,New York;London,1961
Met 25.1820

ELECTROCHEMICAL SOCIETY.THEORETICAL ELECTROCHEMISTRY DIVISION Symposium on electrode processes Transactions Philadelphia,Pa. 1959 May 4-6 Wiley,New York,1961
Chem 18.2624

ELECTRODE PROCESSES Symposium on electrode processes Transactions Philadelphia 1959 May 4-6 Electrochemical society.Theoretical electrochemistry division Edited by Ernest Yeager Wiley,New York;London,1961
Met 25.1820

ELECTRODE PROCESSES Symposium on electrode processes Transactions Philadelphia,Pa. 1959 May 4-6 Electrochemical society. Theoretical electrochemistry division and United States.Air force.Office of scientific research Wiley,New York,1961
Chem 18.2624

ELECTRODE PROCESSES :a general discussion Manchester 1947 Apr 9 Faraday society Faraday society.Discussions, 1 Butterworths,London,1961
Met 25.2396

ELECTRODE REACTIONS OF ORGANIC COMPOUNDS Newcastle-upon-Tyne 1968 Faraday society Faraday society.Discussions, 45 graphs, photos,diagrms 281p 26cm Faraday society, London,1968
Cav 7.3124

ELECTRODYNAMICS Planetary electrodynamics International conference on the universal aspects of atmospheric electricity 4th Proceedings Tokyo 1968 Vol 1-2 Edited by Samuel C. Coroniti and James Hughes 2 vols Gordon and Breach,New York;London, 1969
Nap 11.0974

ELECTROENCEPHALOGRAPHIC CORRELATES OF BEHAVIOR EEG and behavior Edited by Gilbert H. Glaser Basi books,New York;London,1963 Based upon a research conference, "Electroencephalographic correlates of behavior",held at the Yale university school of medicine,November 27 and 28, 1961"
Bal 39.1498

ELECTROENCEPHALOGRAPHY Recent advances in clinical neurophysiology International congress of eletroencephalography and clinical neurophysiology 6th Proceedings Vienna 1965 Sep 5-10 functions of the spinal cord; neurophysiological investigations of brain diseases;EEG in stress Edited by L. Widen Electroencephalography and clinical neurophysiology.Supplement,25 Elsevier, Amsterdam,1967
Pha 16.0304

ELECTROFORMED METALS Symposium on electroforming - applications,uses,and properties of electroformed metals presented at the 1962 committee week Dallas 1962 Feb 6-7 American society for testing materials A.S.T.M.Special technical publication, 318 A. S.T.M.,Philadelphia,Pa.,1962
Met 25.0733

ELECTROMAGNETIC ASPECTS OF HYPERSONIC FLIGHT Symposium on the plasma sheath :its effect upon reentry communication and detection 2nd Proceedings Boston,Mass. 1962 Apr 10-12 Edited by Walter Rotman and others Sponsored by the Air force Cambridge research laboratories.Electromagnetic radiation laboratory Spartan books;Cleaver-Hume, Baltimore,Md.;London,1964
Eng 41.4544

ELECTROMAGNETIC DISTANCE MEASUREMENT Oxford 1965 Sep 6-11 International association of geodesy Hilger and Watts, London,1967
Geog 13.5530

ELECTROMAGNETIC DISTANCE MEASUREMENT :a symposium Oxford 1965 Sep 6-11 International association of geodesy.Special duty group no 19 Hilger and Watts,London, 1967
Eng 41.3278

ELECTROMAGNETIC PHENOMENA IN COSMICAL PHYSICS Stockholm 1956 Aug Edited by B. Lehnert International astronomical union.Symposium, 6 Cambridge university press,Cambridge,1958
Eng 41.4475

ELECTROMAGNETIC PHENOMENA IN COSMICAL PHYSICS Stockholm 1956 Aug 27-31 International astronomical union Edited by B. Lehnert In co-operation with the International union of theoretical and applied mechanics International astronomical union.Symposium, 6 13,544p Cambridge university press, Cambridge,1958
Nap 11.0916

ELECTROMAGNETIC PHENOMENA IN COSMICAL PHYSICS : a symposium proceedings Stockholm 1956 Aug.27-31 International astronomical union Edited by B. Lehnert International astronomical union.Symposia, 6 Cambridge university press,Cambridge,1958
A Math 4.0615

ELECTROMAGNETIC RADIATION IN SPACE E.S.R.O. summer school in space physics 3rd Proceedings Alpbach 1965 Jul 19-Aug 13 European space research organisation Edited by J.G. Emming Astrophysics and space science library, 9 viii,307p D.Reidel, Dordrecht,1967
A Math 4.1576

ELECTROMAGNETIC SCATTERING The Conference on electromagnetic scattering :interdisciplinary conference Papers Potsdam,N.Y. 1962 Aug 13-15 Clarkson college of technology and Air force Cambridge research laboratories Edited by Milton Kerker Co-sponsored by the American chemical society.Division of colloid and surface chemistry International series of monographs on electromagnetic waves, 5 591p Pergamon press,Oxford,1963
TA 15.0256

ELECTROMAGNETIC WAVES :a symposium proceedings Madison,Wis. 1961 Apr.10-12 United States.Army.Mathematics research center Edited by Rudolph E. Langer 396p University of Wisconsin press,Madison,Wis., 1962
Obs 6.1032

ELECTROMAGNETIC WAVES symposium proceedings Madison,Wis. 1961 Apr. 10-12 United States.Army.Mathematics research center Edited by Rudolph E. Langer University of Wisconsin press,Madison,Wis.,1962 photolith
A Math 4.0684

ELECTROMETALLURGY OF CHLORIDE SOLUTIONS All-union seminar on applied electrochemistry 5th Reports Dnepropetrovsk 1962 Oct 17-19 Dnepropetrovskii khimiko-tekhnologicheskii institut and Moskovskii khimiko-tekhnologicheskii institut imeni D.I. Mendeleyeva Edited by V.V. Stender Translated by Consultants bureau from the Russian Consultants bureau,New York,1965
Met 25.1843

ELECTRON,ION AND LASER BEAM TECHNOLOGY Symposium on electron,ion and laser beam technology 11th Record 1971 Edited by R.F.M. Thornley San Francisco press,San Francisco,Calif.,1971
Eng 41.8582

ELECTRON AND ION BEAM SCIENCE International conference on electron and ion beam science and technology 1st Toronto 1964 May 3-7 American institute of mining, metallurgical and petroleum engineers. and Electrochemical society.Electrothermics and metallurgy divisions Edited by Robert Bakish Electrochemical society series Wiley,New York,1965
Met 25.1678

ELECTRON AND ION BEAM SCIENCE AND TECHNOLOGY International conference on electron and ion beam science and technology 1st Toronto 1964 May 3-7 Edited by Robert Bakish Sponsored by the Electrochemical society.Electrothermics and metallurgy divisions Electrochemical society series Wiley,New York,1965
Eng 41.5564

ELECTRON AND LASER BEAM SYMPOSIUM 7th Proceedings University Park,Pa. 1965 Mar 31-Apr 2 Pennsylvania state university Sponsored also by the Alloyd general corporation Pennsylvania state university, University Park,Pa.,1965
Eng 41.5617

ELECTRON AND LASER BEAM SYMPOSIUM 8th Proceedings Ann Arbor,Mich 1966 Apr 6-8 University of Michigan Institute of electrical and electronics engineers Edited by G.I. Haddad University of Michigan,Ann Arbor,Mich.,1966 8th in the series of annual symposia on Electron beam technology initiated in 1959 by Alloyd and held under their sponsorship for seven years.The titles of the symposia vary,therefore those catalogued have been grouped together under the keyword"Electron beams"
Eng 41.5618

ELECTRON AND NUCLEAR RESONANCE IN CHEMISTRY Developments in aromatic chemistry. Applications of electron and nuclear resonance in chemistry.Recent work on the inorganic chemistry of sulphur :a symposium Bristol 1958 Chemical society Chemical society. Special publication,12 Chemical society, London,1958
Chem 18.1229

ELECTRON AND PHOTON INTERACTIONS International symposium on electron and photon interactions at high energies 4th Proceedings Liverpool 1969 Sep 14-20 Danesbury nuclear physics laboratory International union of pure and applied physics University of Liverpool illus 30cm Danesbury nuclear physics laboratory, Danesbury,1969
A Math 4.1760

ELECTRON BEAM THERAPY Annual San Francisco cancer symposium 2nd Proceedings San Francisco,Calif. 1966 Frontiers of radiation therapy and oncology, 2 Karger, Basel;New York,1968
Radioth 35.1208

ELECTRON BEAMS Electron and laser beam symposium 7th Proceedings University Park,Pa. 1965 Mar 31-Apr 2 Pennsylvania state university Sponsored also by the Alloyd general corporation Pennsylvania state university,University Park,Pa.,1965
Eng 41.5617

ELECTRON BEAMS Electron and laser beam symposium 8th Proceedings Ann Arbor, Mich 1966 Apr 6-8 University of Michigan Institute of electrical and electronics engineers Edited by G.I. Haddad University of Michigan,Ann Arbor,Mich.,1966 8th in the series of annual symposia on Electron beam technology initiated in 1959 by Alloyd and held under their sponsorship for seven years. The titles of the symposia vary,therefore those catalogued have been grouped together under the keyword"Electron beams"
Eng 41.5618

ELECTRON BEAMS Symposium on electron beam melting 1st Proceedings Boston,Mass. 1959 Mar 20 Edited by James S. Hetherington Sponsored by Alloyd research corporation Alloyd,Watertown,Mass.,1959
Eng 41.5613

ELECTRON BEAMS Symposium on electron beam processes 2nd Proceedings Boston,Mass. 1960 Mar 24-25 Edited by Robert Bakish Sponsored by the Alloyd research corporation Alloyd,Cambridge,Mass.,1960
Eng 41.5614

ELECTRON BEAMS Symposium on electron beam processes 3rd Proceedings Boston,Mass. 1961 Mar 23-24 Edited by Robert Bakish Sponsored by Alloyd electronics corporation Alloyd,Cambridge,Mass.,1961
Eng 41.5615

ELECTRON BEAMS Symposium on electron beam technology 4th Proceedings Boston,Mass. 1962 Mar 29-30 Edited by R. Bakish Sponsored by the Alloyd electronics corporation Alloyd,Cambridge,Mass.,1962
Eng 41.5616

ELECTRON DENSITY PROFILES IN THE IONOSPHERE AND EXOSPHERE :a symposium Proceedings N. A.T.O. Advanced study institute Skeikampen, Norway 1961 Apr 17-26 Norwegian defence research establishment Edited by B. Maehlum Sponsored by N.A.T.O. science committee N.A. T.O. conference series, 2 428p Pergamon press,Oxford,1962
Nap 11.0391

ELECTRON DIFFRACTION International conference on 'Electron diffraction' and 'The nature of defects in crystals' Abstracts of papers Melbourne 1965 Aug 16-21 Australian academy of science and International union of crystallography With the support of the International union of pure and applied physics. Commission on the solid state Pergamon press,Oxford,1966
Extended abstracts of papers to be presented at the conference
Cav 7.1584

ELECTRON DIFFRACTION International conference on"electron diffraction" and "the nature of defects in crystals" Abstracts of papers Melbourne 1965 Aug 16-21 Australian academy of science and International union of crystallography Co-sponsored by the International union of pure and applied physics Pergamon press,Oxford, 1966
Met 25.1463

ELECTRON FRACTOGRAPHY Boston,Mass. 1967 Jun 25-30 American society for testing materials A.S.T.M.Special technical publication, 436 A.S.T.M.,Philadelphia,Pa., 1916
Met 25.2248

ELECTRON METALLOGRAPHY Symposium on advances in electron metallography Papers American society for testing materials A.S.T.M. Special technical publication, 245 American society for testing materials,Philadelphia,Pa., 1958
Met 25.1362

ELECTRON METALLOGRAPHY Symposium on electron metallography Atlantic City,N.J. 1959 Jun 23 American society for testing materials A.S.T.M.Special technical publication, 262 American society for testing materials, Philadelphia,Pa.,1960
Met 25.1366

ELECTRON METALLOGRAPHY Symposium on techniques for electron metalography Atlantic City,N.J. 1953 Jun 30 American society for testing materials A.S.T.M. Special technical publication, 153 American society for testing materials,Philadelphia,Pa., 1954
Met 25.1361

ELECTRON METALLOGRAPHY Techniques in electron metallography 1963 Symposium on advances in techniques in electron metallography New York 1962 Jun 26 American society for testing materials. Subcommittee 11 on electron microstructure of metals A.S.T.M.Special technical publication, 339 American society for testing and materials,Philadelphia,Pa.,1963
Met 25.1378

ELECTRON METALLOGRAPHY Techniques of electron microscopy,diffraction and microprobe analysis Papers Atlantic City 1963 Jun 26 American society for testing materials A.S.T.M.Special technical publication, 372 Philadelphia,1964
Met 25.2495

ELECTRON METALLOGRAPHY AND PROBE MICROANALYSIS 1962 Symposium on advances in electron metallography and electron probe microanalysis Atlantic City,N.J. 1960;1961 American society for testing materials.Subcommittee on electron microstructure A.S.T.M.Special technical publication, 317 A.S.T.M., Philadelphia,Pa.,1962
Met 25.1377

The ELECTRON MICROPROBE :a symposium Proceedings Washington 1964 Oct Electrochemical society.Electrothermics and metallurgy division Edited by T.D. McKinley and others Wiley,New York,1966
Met 25.2362

The ELECTRON MICROPROBE :a symposium Proceedings Washington,D.C. 1964 Oct 12-15 Electrochemical society Edited by T.D. McKinley and others Electrochemical society series John Wiley,New York,New York,1966
Cav 7.2425

ELECTRON MICROSCOPE SOCIETY OF AMERICA Electron microscopy International congress for electron microscopy 5th Philadelphia 1962 Aug 29-Sep 5 Vol 1: non-biology Edited by Sydney S. Breese Academic press,New York;London,1962
Met 25.1368

ELECTRON MICROSCOPY Cytochemical progress in electron microscopy :a symposium Oxford 1962 Jul 2-4 Royal microscopical society Royal microscopical society.Journal, 81,pt 3-4 London,1963
Bot 42.4729

ELECTRON MICROSCOPY Electron microscopy 1966 The International congress for electron microscopy 6th Extended abstracts Kyoto Aug 28-Sep 4 Vol 1-2 International federation of societies for electron microscopy Edited by Ryozi Uyeda 2 vols Maruzen company,Tokyo,1966
Cav 7.1340

ELECTRON MICROSCOPY European congress on electron microscopy 5th Proceedings Manchester 1972 Institute of physics Institute of physics.Conference series, 14 illus xxviii,683p 26cm Institute of physics,London,1972
Cav 7.3182

ELECTRON MICROSCOPY European regional conference on electron microscopy Proceedings Delft 1960 Vol 1 Edited by A.L. Houwink and B.J. Spit Nederlandse vereniging voor electronenmicroscopie,Delft,1960
Met 25.1370

ELECTRON MICROSCOPY European regional conference on electron microscopy Proceedings Delft 1960 Aug 29-Sep 3 Vols 1-2 Edited by A.L. Houwink and B.J. Spit 2 vols Delft,1961
Eng 41.5543

ELECTRON MICROSCOPY Europees congres toegepaste elektronmicroscopie Proceedings Ghent 1954 Apr 7-10 Edited by G. Van der Meersche Rijksuniversitet,Ghent,1954
Papers in English,Dutch,German,French and Spanish
Radioth 35.1583

ELECTRON MICROSCOPY International conference on electron microscopy 4th Berlin 1958 Sep 10-17 Bd 2: biologisch-medizinischer teil Edited by W. Bargmann and others Springer,Berlin,1960
An 32.3165

ELECTRON MICROSCOPY International congress for electron microscopy 5th Philadelphia 1962 Aug 29-Sep 5 Vol 1: non-biology International federation of electron microscope societies and Electron microscope society of America Edited by Sydney S. Breese Academic press,New York; London,1962
Met 25.1368

ELECTRON MICROSCOPY International congress for electron microscopy 5th Philadelphia,Pa. 1962 Aug 29-Sep 5 Vol 2: biology Edited by Sidney S. Breese Academic press,New York,1962
An 32.3183

ELECTRON MICROSCOPY Internationaler kongress fur elektronenmikroskopie 4th Verhandlungen Berlin 1958 Sep 10-17 Band 1: physikalisch-technischer teil Edited by G. Mollenstedt and others Springer,Berlin, 1960
Eng 41.5532

ELECTRON MICROSCOPY Intionationaler kongress fur elektronmikroskopie 4th Proceedings Berlin 1958 Sep 10-17 Bd 1: physikalisch-technischer teil International federation of electron microscope societies Edited by G. Mollenstedt and others Springer,Berlin,1960 General editors W.Bargmann and others. Papers in English,French and German
Met 25.2493

ELECTRON MICROSCOPY Les techniques recentes en microscopie electronique et corpusculaire : colloque international Toulouse 1955 Apr 4-8 Centre national de la recherche scientifique Edited by Ch. Fert Centre national de la recherche scientifique. Colloques internationaux, 58 Paris,1956
Cav 7.0525

ELECTRON MICROSCOPY Microscopie electronique 1970 Congres international de microscopie electronique 7e Resumes des communications Grenoble 1970 Aug 30-Sep 3 International federation of societies for electron microscopy Edited by Pierre Favard Societe francaise de microscopie electronique, Paris,1970
Met 25.2604

ELECTRON MICROSCOPY Quantitive electron microscopy :a symposium Proceedings Washington,D.C. 1964 Mar 30 - Apr 3 Edited by Gunter F. Bahr and Elmar H. Zeitler Sponsored by the Intersociety committee for research potential in pathology,inc. Laboratory investigation, 14 no.6 illus. viii,605p Baltimore,Md.,1965
Bot 42.4731

ELECTRON MICROSCOPY Quantitive electron microscopy :a symposium Proceedings Washington,D.C. 1964 Mar 30-Apr 3 Edited by Gunter F. Bahr and Elmar H. Zeitler held at the Armed forces institute of pathology Laboratory investigation, 14, no 6 Washington,D.C.,1966 Sponsored jointly by the Intersociety committee for research potential in pathology and the Armed forces institute of pathology
Bal 39.0239

ELECTRON MICROSCOPY Symposium on cytochemical progress in electron microscopy Oxford 1962 Jul 2-4 Royal microscopical society Journal of the Royal microscopical society.Ser.3, 81,pt 3-4,p.107-117 London, 1963
Bal 39.0235

ELECTRON MICROSCOPY Symposium on cytochemical progress in electron microscopy Oxford 1962 Jul 2-4 Royal microscopical society Journal of the Royal microscopical society,1962,pt.3,4 Royal microscopical society,London,1963
An 32.3182

ELECTRON MICROSCOPY The International conference on electron microscopy 3rd Proceedings London 1954 Jul 15-21 Edited by R. Ross Royal microscopical society, London,1956
Radioth 35.1582

ELECTRON MICROSCOPY The International conference on electron microscopy 3rd Proceedings London 1954 Jul 15-21 International council of scientific unions. Joint commission on electron microscopy Edited by V.E. Cosslett and others Royal microscopical society,London,1956
Cav 7.1455

ELECTRON MICROSCOPY :a conference Proceedings Stockholm 1956 Sep 17-20 Edited by F.S. Sjostrand and J. Rhodin Almqvist and Wiksell,Stockholm,1957 Some papers in German
An 32.3155

ELECTRON MICROSCOPY,1968 European regional conference on electron microscopy 4th Pre-congress abstracts Rome 1968 Sep 1-7 Edited by Daria Steve Bocciarelli Tipografia poliglotta vaticana,Rome,1968
Met 25.1395

ELECTRON MICROSCOPY AND ANALYSIS Institute of physics.Electron microscopy and analysis group 25th anniversary meeting Proceedings Cambridge 1971 Jun 29-Jul 1 Institute of physics Edited by W.C. Nixon Institute of physics.Conference series, 10 London,1971
Met 25.2753

ELECTRON MICROSCOPY AND STRENGTH OF CRYSTALS Berkeley international materials conference : the impact of transmission electron microscopy on theories of the strength of crystals 1st Proceedings Berkeley 1961 Jul 5-8 Lawrence radiation laboratory.Inorganic materials division and University of California Edited by Gareth Thomas and Jack Washburn Interscience,New York;London,1963
Met 25.1211

ELECTRON MICROSCOPY IN ANATOMY A Symposium on the ultrastructure of cells Proceedings London 1959 Apr 16-17 Anatomical society of Great Britain and Ireland Edited by J.D. Boyd and others Edward Arnold,London,1961
Bioch 33.2297

ELECTRON MICROSCOPY IN ANATOMY :a symposium Proceedings London 1959 Apr 16-17 Anatomical society of Great Britain and Ireland Arnold,London,1961
Phys 20.1491

ELECTRON MICROSCOPY IN ANATOMY :a symposium papers London 1959 Apr 16-17 Anatomical society of Great Britain and Ireland Illus. Arnold,London,1961
VA 19.0356

ELECTRON MICROSCOPY SOCIETY OF AMERICA ANNUAL MEETING 26th- Electron microscopy society of America Edited by Claude J. Arcencaux Claitor,Baton Rouge,La.,1968-
Eng 41.5622

ELECTRON MICROSCOPY 1964 European regional conference on electron microscopy 3rd Proceedings Prague 1964 Aug 26-Sep 3 Vol A: non-biology International federation of electron microscope societies Edited by M. Titlbach Czech.academy of sciences,Prague, 1964
Met 25.2494

ELECTRON MICROSCOPY 1966 International congress for electron microscopy 6th Kyoto 1966 Aug 28-Sep 4 Vol 1: non-biology International federation of electron microscope societies Edited by Ryozi Uyeda Maruzen,Tokyo,1966
Met 25.1388

ELECTRON MICROSCOPY 1966 International congress for electron microscopy 6th Papers and extended abstracts Kyoto 1966 Aug 28-Sep 4 International federation of electron microscope societies Edited by Ryozi Uyeda illus Maruzen,Tokyo,1966
Path 30.2615

ELECTRON MICROSCOPY 1968 European regional conference on electron microscopy 4th Pre-congress abstracts Rome 1968 Sep 1-7 Vol 2 Edited by D.S. Bocciarelli Tipografia polyglotta vaticana,Rome,1968 Chairman:J.S.Mitchell
Radioth 35.0559

ELECTRON MICROSCOPY 1972 European congress on electron microscopy 5th Proceedings Manchester 1972 International federation of societies for electron microscopy Institute of physics.Conference series, 14 Institute of physics,London,1972
Eng 41.8576

ELECTRON SPECTROSCOPY International conference on electron spectroscopy Proceedings Pacific Grove,Calif. 1971 Sep 7-10 Edited by D.A. Shirley xii,916p 23cm North-Holland,London,1972
Chem 18.2871

ELECTRON SPIN RESONANCE AND THE EFFECTS OF RADIATION ON BIOLOGICAL SYSTEMS :a conference Gatlinburg 1965 May 10-12 Edited by Wallace Snipes Organised by the National research council.Subcommittee on radiobiology National research council. Nuclear science series.Report, 43 National research council,Washington,D.C.,1966
Radioth 35.1748

ELECTRON TRANSPORT AND ENERGY CONSERVATION Round table discussion Fasano 1969 May 12-15 Edited by J.M. Tager and others Adriatica editrice,Bari,1970
Bioch 33.2288

ELECTRONIC AND ATOMIC COLLISIONS International conference on the physics of electronic and atomic collisions 7th Abstracts of papers Amsterdam 1971 Jul 26-30 Vol 1-2 Edited by L.M. Branscomb and others 23cm 2 vols North-Holland, Amsterdam,1971
Chem 18.2959

ELECTRONIC AND ATOMIC COLLISIONS Physics of electronics and atomic collisions International conference on the physics of electronic and atomic collisions 7th Invited papers and progress reports Amsterdam 1971 Jul 26-30 Edited by T.R. Govers and F.J. De Heer xi,496p 23cm North Holland,London,1972
Chem 18.2958

ELECTRONIC ASPECTS OF BIOCHEMISTRY The International conference on the electronic aspects of biochemistry New York 1967 Dec 11-13 By Bernard Pullman and others New York academy of sciences Edited by Philip Feigelson New York academy of sciences. Annals, 158,p.1-438 New York,1969
Bioch 33.0313

ELECTRONIC CALCULATING MACHINES Der Einsatz elektronischer rechengerate fur forschungs- und entwicklungs- aufgaben conference report Berlin 1960 Mai 9-11 Hahn-Meitner-Institut fur kernforschung,Berlin Hahn-Meitner-Institut fur kernforschung Berlin.Berichte, 13 var.p 30cm Hahn-Meitner-Institut, Berlin,1960
Math L 5.0870

ELECTRONIC COMPONENTS SYMPOSIUM,1953 Rudiments of good circuit design :talk delivered at Electronic components symposium, Pasadena ...1953 By N.H. Taylor Massachusetts institute of technology.Digital computer laboratory P 224 13p massachusetts institute of technology, Cambridge,Mass.,1953
Math L 5.1569

ELECTRONIC COMPUTATION Conference on electronic computation 2nd Proceedings Pittsburgh,Pa. 1960 American society of civil engineers American society of civil engineers,Kansas City,Miss.,1960
Eng 41.2718

ELECTRONIC COMPUTATION Conference on electronic computation 3rd Proceedings Boulder,Colo. 1963 Jun 19-21 American society of civil engineers Journal of structural division, 89,no ST4 American society of civil engineers,Kansas city,Miss., 1963
Eng 41.2716

ELECTRONIC DIGITAL COMPUTERS AND INFORMATION PROCESSING :conference Proceedings Darmstadt 1956 Oct 25-27 Gesellschaft fur angewandte mathematik und mechanik Nachrichtentechnische gesellschaft Edited by A. Walther and W. Hoffmann Nachrichtentechnische fachberichte, 4 illus. viii,229p 30cm Friedr.Vieweg und sohn,Brunswick,1956
Math L 5.3247

ELECTRONIC ENGINEERING ASSOCIATION The Business computer symposium Papers London 1958 Dec 1-3 No 1-22 Collection of 22 papers and other mimeographed material
Math L 5.3252

ELECTRONIC EQUIPMENT RELIABILITY Symposium digest Symposium on electronic equipment reliability 2nd London 1962 Oct 24-26 Institution of electrical engineers illus. 47p 28cm Institution of electrical engineers,London,1962
Math L 5.0926

ELECTRONIC PHENOMENA Symposium on electronic phenomena in chemisorption and catalysis on semiconductors Moscow 1968 Jul 2-4 Edited by K. Hauffe and T. Wolkenstein De Gruyter,Berlin,1969
Chem 18.2698

ELECTRONIC PHENOMENA IN CHEMISORPTION AND CATALYSIS ON SEMICONDUCTORS Symposium on electronic phenomena in chemisorption and catalysis on semiconductors Moscow 1968 Jul 2-4 Edited by K. Hauffe and Th. Wolkenstein English edition De Gruyter, Berlin,1969 Held during the fourth international congress on catalysis
Met 25.2475

ELECTRONIC PROPERTIES OF METALS The International conference on the electronic properties of metals at low temperatures Report Geneva,N.Y. 1958 Aug 25-29 Hobart college and International union of pure and applied physics Co-sponsored by General electric research laboratory 244p c1958 Privately printed report to the sponsors
Cav 7.0895

ELECTRONIC STRUCTURE AND ALLOY CHEMISTRY OF THE TRANSITION ELEMENTS :a symposium New York 1962 Feb 22 American institute of mining,metallurgical and petroleum engineers. Metallurgical society Edited by Paul Adams Beck 251p Interscience,New York,1963
Cav 7.1711

ELECTRONIC WAVEGUIDES :symposium Proceedings New York 1958 Apr 8-10 Polytechnic institute of Brooklyn Microwave research institute Microwave research institute. Symposia series, 8 Polytechnic press, Brooklyn,N.Y.,1958
Eng 41.5649

ELECTRONICS Medical electronics International conference on medical electronics 2nd Proceedings Paris 1959 Jun 24-27 International federation for medical electronics Edited by C.N. Smyth Iliffe,London,1960
Psy 31.3062

ELECTRONICS AND VACUUM PHYSICS Czechoslovak conference 3rd Transactions Prague 1965 Sep 23-28 Ceskoslovenska akademie ved Edited by Libor Paty Sponsored by the Universita Karlova Academia,Prague,1967
Eng 41.5453

ELECTRONICS DESIGN Conference on electronics design Cambridge 1968 Sep 23-27 Sponsored by the Institution of electrical engineers Institution of electrical engineers.Conference publication, 45 IEE, London,1968
Math L 5.3468

ELECTRONS Interactions des electrons,phonons et magnons avec les surfaces cristallines : colloque Exposes et communications presentes Lille 1969 Sep 17-19 Societe francaise de physique Journal de physique, 31,no 4, supplement,Colloque,C- 1 Paris,1970
Met 25.2627

ELECTRONS DENSITY PROFILES N.A.T.O. Advanced study institute Electron density profiles in the ionosphere and exosphere :a symposium Proceedings Skeikampen,Norway 1961 Apr 17-26 Norwegian defence research establishment Edited by B. Maehlum Sponsored by N.A.T.O. science committee N.A.T.O. conference series, 2 428p Pergamon press,Oxford,1962
Nap 11.0391

ELECTROPHYSIOLOGY OF THE HEART International symposium on the electrophysiology of the heart Proceedings Milan 1963 Edited by B. Taccardi and G. Marchetti Macmillan; Pergamon,New York;Oxford,1965
Pha 16.0286

ELECTROPLATING Control in electroplating :a one-day symposium London 1958 Nov 19 Institute of metal finishing.London branch Draper,Teddington,1959
Met 25.2670

ELECTRORETINOGRAPHY Clinical electroretinography international symposium 3rd Proceedings Highland Park,Ill. 1964 Oct 14-16 International society for clinical electroretinography Edited by Hermann M. Burian and Jerry Hart Jacobson Vision research.Supplement Pergamon,Oxford,1966
PGMS 29.0340

ELECTROSTATIC PROPULSION... a selection of technical papers based mainly on a symposium of the American rocket society... Monterey,Calif. 1960 Nov 3-4 American rocket society Edited by David B. Langmuir and others Progress in astronautics and rocketry, 5 Academic press,New York;London, 1961
Eng 41.6888

ELECTROTHERAPEUTIC SLEEP AND ELECTROANAESTHESIA 1st :international symposium Proceedings Graz 1966 Sep 12-17 Edited by F.M. Wageneder and others Excerpta medica.International congress series, 136 Excerpta medica,Amsterdam,1967
Inv Med 37.0221

ELECTROTHERAPEUTIC SLEEP AND ELECTROANESTHESIA 2nd :international symposium Proceedings Graz 1969 Sep 8-13 Edited by F.M. Wageneder and others Excerpta medica. International congress series, 212 Excerpta medica,Amsterdam,1970
Inv Med 37.0222

ELEMENTARY EXCITATIONS IN SOLIDS Cortina lectures and four lectures from the Conference on localized excitations Proceedings Milan 1966 Jul 25-26 Edited by A.A. Maradudin and G.F. Nardelli Bibliog 526p 20cm Plenum press,New York,1969
Cav 7.2677

ELEMENTARY PARTICLE PHYSICS Boulder conference on particle physics held during the first part of the 11th Boulder summer institute for theoretical physics 5th Proceedings Boulder,Colo. 1968 Jun 17-Aug 23 Vol 1-2 University of Colorado. Department of physics and astrophysics Edited by Kalyana T. Mahanthappa and others Lectures in theoretical physics, 11a,pt 1-2 Bibliog,diagrms xxi,629p 23cm 2 vols Gordon and Breach,New York,1969 Dedicated to George Gamow
Cav 7.3008

ELEMENTARY PARTICLE PHYSICS Symposium on cosmic rays,elementary particle physics and astrophysics 10th Proceedings Aligarh 1967 Dec 12-16 India.Department of atomic energy.Cosmic ray committee 701p Bombay, 1968
TA 15.0553

ELEMENTARY PARTICLE THEORIES Internationale universitatswochen fur Kernphysik 5th proceedings Schladming 1966 Feb.24-Mar.9 Edited by Paul Urban Sponsored by the International Atomic Energy Agency Acta physica austriaca.Supplementum, 3 Springer-verlag,Vienna;New York,1966
A Math 4.0889

ELEMENTARY PARTICLE THEORY:RELATIVISTIC GROUPS AND ANALYTICITY Nobel symposium 8th Proceedings Aspenasgarden,Lerum 1968 May 19-25 Nobel foundation Edited by Nils Svartholm illus 399p 24cm Wiley,New York,1968
A Math 4.1765

ELEMENTARY PARTICLES European conference on elementary particles 4th Proceedings Heidelberg 1967 Sep 20-27 Ruprecht-Karl-universitat Heidelberg and European organization for nuclear research Edited by H. Filthuth viii,550p North-Holland, Amsterdam,1968
A Math 4.1348

ELEMENTARY PARTICLES High-energy physics and elementary particles Seminar on elementary particles lectures Trieste 1965 May 3-Jun.30 International atomic energy agency International atomic energy agency.Proceedings series STIPUB117 Vienna,1965
A Math 4.0888

ELEMENTARY PARTICLES International conference on elementary particles proceedings Oxford 1965 Sep.19-25 Rutherford high energy laboratory Sponsored by the International union of pure and applied physics Rutherford high energy laboratory, Oxford,1965
A Math 4.0887

ELEMENTARY PARTICLES Lund international conference on elementary particles Proceedings Lund 1969 Jun 25-Jul 1 Edited by G.von Dardel Bibliog. 452p 30cm Berlingska Boktryckeriet,Lund,1969
A Math 4.1600

ELEMENTARY PARTICLES The International conference on elementary particles Proceedings Aix-en-Provence 1961 Sep 14-20 Vol 1-2 Edited by E. Cremieu-Alcan and others Sponsored by Commissariat a l'energie atomique 2 vols C.E.N.Saclay,Gif-sur-Yvette,1962
Cav 7.0029

ELEMENTARY PARTICLES :conference Proceedings Amsterdam 1971 Jun 30-Jul 6 Rijksuniversiteit te Amsterdam Edited by A.G. Tenner and H.J.G. Vettman illus viii,472p 26cm North-Holland,Amsterdam,1972
A Math 4.1767

ELEMENTARY PARTICLES PHYSICS Tokyo summer lectures in theoretical physics 2nd Papers Tokyo 1966 Aug 29-Sep 3 Pt 2: elementary particle physics University of Tokyo.Summer institute of theoretical physics Edited by Gyo Takeda and Akihiko Fujii Tokyo; New York,1967 Course under chairmanship of Kubo,Ryogo
Cav 7.2317

ELEMENTE DER MINERALOGIE By Carl Friedrich Naumann 4th edition revised,expanded illus. Engelmann,Leipzig,1855
Min 10.1370

ELEMENTS Origin and distribution of the elements Symposium on the origin and distribution of the elements Paris 1967 International union of geological sciences Edited by L.H. Ahrens International series of monographs on earth sciences, 30 xvii, 1178p 24cm Pergamon press,Oxford,1968
TA 15.0565

ELEMENTS Origin and distribution of the elements :a symposium Proceedings Paris 1967 May 8-11 International association of geochemistry and cosmochemistry Edited by L.H. Ahrens International series of monographs on earth sciences, 30 1178p Pergamon,Oxford,1968
Obs 6.3444

ELEVATED TEMPERATURE STRAIN GAGES Symposium on elevated temperature strain gages Philadelphia,Pa. 1957 Dec 4-5 American society for testing materials Sponsored by United States.Naval air material center A.S. T.M.Special technical publication, 230 American society for testing materials, Philadelphia,Pa,1958
Met 25.0985

ELGA GROUP OF COMPANIES Water requirements for kidney dialysis London 1969 May 28 Elga Publications,Lane End,Bucks,1969
PGMS 29.0239

ELIMINATION OF HARMFUL ORGANISMS FROM FOOD AND FEED BY IRRADIATION Panel on elimination of harmful organisms from food and feed by irradiation Report Zeist 1967 Jun 12-16 Organized by the Joint FAO-IAEA division of atomic energy in food and agriculture International atomic energy agency.Panel proceedings series International atomic energy agency,Vienna,1968
Radioth 35.1761

ELLIPSOMETRY IN THE MEASUREMENT OF SURFACES AND THIN FILMS symposium Proceedings Washington,D.C. 1963 Sep 5-6 National bureau of standards Edited by E. Passaglia and others National bureau of standards. Miscellaneous publication, 256 U.S. Government printing office,Washington,D.C., 1964
Met 25.1687

ELSINORE 1969 Mass loss and evolution in close binaries International astronomical union Edited by K. Gyldenkerne and R.M. West International astronomical union.Colloquium, 6 238p Copenhagen university observatory, Copenhagen,1970
Obs 6.3358

EMBRIOLOGIA E GENETICA congress Naples 1948 Jun 1-318 Consiglio nazionale per la ricerca Per iniziativa del Centre di studio di biologia,di citologia genetica e di enzimologia Stazione zoologica di Napoli. Publicazioni, 21 Naples,1949
Gen 34.0525

EMBRYONIC DEVELOPMENT Beginnings of embryonic development A Symposium on formation and early development of the embryo Atlanta,Ga. 1955 Dec 27 American association for the advancement of science. Section on zoological sciences Edited by Albert Tyler and others Co-sponsored by the American society of zoologists and the Association of southeastern biologists American association for the advancement of science.Publication, 48 Washington,D.C., 1957
Bal 39.0401

EMBRYONIC DEVELOPMENT Symposium on effects of radiation and other deleterious agents on embryonic development Oak Ridge,Tenn. 1953 Apr 20-21 Sponsored by the Oak Ridge national laboratory.Biology division Wistar institute of anatomy and biology,Philadelphia, Pa.,1954 Reprinted from the Journal of cellular and comparative physiology Vol 43, Suppt 1,May 1954
An 32.2027

EMBRYOPATHIC ACTIVITY OF DRUGS London 1965 Biological council.Co-ordinating committee for symposia on drug action Edited by J.M. Robson and others Churchill,London, 1965
Pha 16.0320

EMBRYOPATHIC ACTIVITY OF DRUGS :a symposium Biological council.Co-ordinating committee for symposia on drug action Edited by J.M. Robson and others Churchill,London,1965
An 32.2095

The EMERGENCE OF ORDER IN DEVELOPING SYSTEMS Ithaca,N.Y. 1968 Jun Society for developmental biology Edited by Michael Locke Society for developmental biology. Symposia, 27 350p Academic press,New York, 1968
Bot 42.1097

EMERGENCE OF ORDER IN DEVELOPING SYSTEMS :a symposium Proceedings Society for developmental biology Edited by Michael Locke Society for developmental biology. Symposia, 27 Developmental biology. Supplement, 2 Academic press,New York; London,1968
Gen 34.2154

EMISSION SPECTRA The Conference on emission spectra of the night sky and aurorae Papers London 1947 Jul 7-10 Royal society of London.Gassiot committee 139p Physical society,London,1948
Cav 7.1453

EMISSION SPECTRA OF AURORAE The Conference on emission spectra of the night sky and aurorae Papers London 1947 Jul 7-10 Royal society of London.Gassiot committee 139p Physical society,London,1948
TA 15.0039

The EMISSION SPECTRA OF THE NIGHT SKY AND AURORAE a conference Papers London 1947 Jul.7-10 Royal society of London.Gassiot committee 139p London,1948
Obs 6.1964

EMPIRE MINING AND METALLURGICAL CONGRESS 5th Proceedings Melbourne 1953 Apr 14-Jun 12 Vol 4: extractive metallurgy in Australia Australasian institute of mining and metallurgy Edited by J.C. Richards Melbourne,1953
Met 25.2524

EMPIRE MINING AND METALLURGICAL CONGRESS 5th Publications 5: Australian mining and metallurgy,miscellaneous features and practices Australasian institute of mining and metallurgy Edited by A.W. Norrie Australasian institute of mining and metallurgy,Melbourne,1953
Met 25.0069

EMPIRE MINING AND METALLURGICAL CONGRESS 5th Publications Ore dressing methods in Australia and adjacent territories Melbourne 1953 Apr 14-Jun 12 3 Australasian institute of mining and metallurgy Edited by H.H. Dunkin Office of the congress,and of the Australasian institute of mining and metallurgy,Melbourne,1953
Met 25.0205

EMPIRE MINING AND METALLURGICAL CONGRESS Publications 5th Extractive metallurgy in Australia:non-ferrous metallurgy Melbourne 1953 Apr 14-Jun 12 4B Australasian institute of mining and metallurgy Edited by Frank A. Green and others Office of the Congress,Melbourne,1953
Met 25.0498

EMPIRE MINING AND METALLURGICAL CONGRESS 4th Proceedings London 1943 Jul 9-23 Pt 1-2 Edited by F. Higham 2 vols London,1950
Min 10.0726

EMPIRE SCIENTIFIC CONFERENCE Report London 1946 Jun-Jul Vol 1-2 Royal society of London 2 vols London,1948
Bioch 33.1683

The EMPIRE SCIENTIFIC CONFERENCE Reports London;Cambridge;Oxford 1946 Jun-Jul Vol 1-2 Royal society of London 2 vols Royal society,London,1948
Gen 34.0053

EMULSION TECHNOLOGY,THEORETICAL AND APPLIED : including the Symposium on technical aspects of emulsions 2nd edition expanded xiii,360p Chemical publishing company, Brooklyn,N.Y.,1946
Chem E 24.0910

END RESULTS OF CANCER THERAPY International symposium on end results of cancer therapy Proceedings Sandefjord 1963 Sep 16-20 National cancer institute Edited by Sidney J. Cutler National cancer institute.Monograph, 15 National cancer institute,Bethesda,Md., 1964
Bioch 33.0986

ENDICOTT,N.Y. 1949 Computation seminar proceedings Sponsored by International business machines corporation 173p 28cm International business machines corporation, New York,1951
Math L 5.0700

ENDICOTT,N.Y. 1950 Industrial computation seminar proceedings Sponsored by International business machines corporation 103p 28cm International business machines corporation,New York,1951
Math L 5.0701

ENDICOTT,N.Y. 1951 Computation seminar proceedings Sponsored by International business machines corporation 148p 28cm International business machines corporation, New York,1951
Math L 5.0702

ENDICOTT HOUSE,M.I.T. 1959 Sensory communication :symposium on principles of sensory communication Contributions Edited by Walter A. Rosenblith M.I.T.press, Cambridge,Mass.
Psy 31.0282

ENDOCRINE ASPECTS OF BREAST CANCER :a conference Proceedings Glasgow 1957 Jul 8-10 University of Glasgow Edited by Alastair R. Currie Livingstone,Edinburgh; London,1958
Path 30.2082

ENDOCRINE GENETICS :a symposium Proceedings Cambridge 1966 Mar 29-31 Soceity for endocrinology Edited by S.G. Spicket and J.G. M. Shire Society for endocrinology.Memoirs, 15 Cambridge university press,Cambridge,1967
Pha 16.0093

ENDOCRINE GENETICS :a symposium Proceedings Cambridge 1966 Mar 29-31 Society for endocrinology Edited by S.G. Spickett and J. G.M. Shire Society for endocrinology.Memoirs, 15 Cambridge university press,Cambridge, 1967
Gen 34.0555

ENDOCRINE GENETICS a symposium Proceedings Cambridge 1966 Mar 29-30 Society for endocrinology Edited by S.G. Spickett and J. G.M. Shire Society for endocrinology.Memoir, 15 Cambridge university press,Cambridge, 1967
Inv Med 37.0203

ENDOCRINES AND THE CENTRAL NERVOUS SYSTEM Proceedings of the Association New York 1963 Dec 6-7 Association for research in nervous and mental disease Edited by Rachmiel Levine Association for research in nervous and mental disease.Research publications, 43 Williams and Wilkins, Baltimore,Md.,1966
An 32.4403

ENDOCRINES IN DEVELOPMENT Shelter Island,N.Y. 1956 Sep 11-13 National research council Edited by Ray L. Watterson Developmental biology conference series,1956 University of Chicago.Committee on publications in biology and medicine University of Chicago press, Chicago,Ill.,1959
Gen 34.0553

ENDOCRINOLOGIC AND MORPHOLOGIC CORRELATIONS OF THE OVARY :the Florentine conference Florence Edited by W. Ingiulla and Robert B. Greenblatt
An 32.5225

ENDOCRINOLOGIE DES ARTHROPODES Colloque international sur l'endocrinologie des arthropodes Paris 1947 Jun Centre national de la recherche scientifique Centre national de la recherche scientifique. Colloques internationaux, 4 Centre national de la recherch scientifique,Paris, 1948
Bal 39.2291

ENDOCRINOLOGISTES DE LANGUE FRANCAISE Thyroide et exploration thyroidienne medullo-surrenale :rapports de la vie reunion des endocrinologistes de langue francaise Paris 1959 Jun 29-Jul 1 Annales d'endocrinologie, 19,no.4 Masson;Doin,Paris,1959 At head of title:Colloque d'endocrinologie 1959
An 32.3755

ENDOCRINOLOGY International congress of endocrinology 2nd Proceedings London 1964 Aug 17-22 Pt 1-2 Edited by S. Taylor Exerpta medica.International congress series, 83 2 vols Excerpta medica foundation,London,1965
PGMS 29.0157

ENDOCRINOLOGY International congress of endocrinology 2nd Proceedings London 1964 Aug 17-22 Pt 1-2 Excerpta medica. International congress series, 83 2 vols Excerpta medica,Amsterdam,1966
Inv Med 37.0114

ENDOCRINOLOGY Pan-American congress of endocrinology 6th Proceedings Mexico City 1965 Oct 10-13 Edited by C. Gual Excerpta medica.International congress series, 112 Excerpta medica,Amsterdam,1966
Inv Med 37.0096

ENDOCRINOLOGY Pan-American congress of endocrinology 6th Proceedings Mexico City 1965 Oct 10-15 Edited by C. Gual Excerpta medica.International congress series, 112 Excerpta medica foundation,Amsterdam, 1966
Inv Med 37.0254

ENDOCRINOLOGY Pan-American congress of endocrinology 6th Proceedings Mexico City 1965 Oct 10-15 Edited by C. Gual Excerpta medica.International congress series, 112 Excerpta medica foundation,New York, c1965
PGMS 29.0167

ENDOCRINOLOGY Progress in endocrinology International congress of endocrinology Proceedings Mexico City 1968 Jun 30-Jul 5 Edited by Carlos Gual and F.J.C. Ebling Excerpta medica.International congress series, 184 Excerpta medica,Amsterdam,1969
Inv Med 37.0095

ENDOCRINOLOGY Progress in endocrinology Proceedings Edinburgh 1959 Aug 16-20 Vol 1-2 Society for endocrinology Edited by K. Fotherby and others Society for endocrinology.Memoirs, 9-10 2 vols Cambridge university press,Cambridge,1964
Inv Med 37.0082

ENDOCRINOLOGY Progress in endocrinology The Edinburgh meeting on endocrinology Proceedings Edinburgh 1959 Aug 16-20 Pt 1-2 Society for endocrinology Edited by K. Fotherby and others Society for endocrinology.Memoirs, 9-10 2 vols Cambridge university press,Cambridge,1960-1961
Bioch 33.0484

ENDOCRINOLOGY OF THE TESTIS :colloquium London 1966 May 18-20 Ciba foundation Edited by G.E.W. Wolstenholme and Maeve O'Connor Ciba foundation colloquia on endocrinology, 16 Churchill,London,1967
An 32.3893

ENDOCRINOLOGY OF VERTEBRATES The Comparative endocrinology of vertebrates :a conference Proceedings Liverpool 1954 Jul 12-16 Pt 1: the comparative physiology of reproduction and the effects of sex hormones in vertebrates Society for endocrinology Edited by I. Chester Jones and P. Eckstein held at the University of Liverpool.Department of zoology Society for endocrinology.Memoirs, 4 Cambridge university press,Cambridge,1955
Bal 39.1400

ENDOGENOUS METABOLISM WITH SPECIAL REFERENCE TO BACTERIA New York 1963 Jan 21 New York academy of sciences Edited by Carl Lamanna New York academy of sciences.Annals, 102,3 New York,1963
Bot 42.3726

ENDOGENOUS SUBSTANCES AFFECTING THE MYOMETRIUM a symposium Proceedings Bristol 1965 Jul 19-20 Society for endocrinology Edited by V.R. Pickles and R.J. Fitzpatrick Society for endocrinology.Memoirs, 14 Cambridge university press,Cambridge,1966
Inv Med 37.0180

ENDOGENOUS SUBSTANCES AFFECTING THE MYOMETRIUM a symposium Proceedings Bristol 1965 Jul 19-20 Society for endocrinology Edited by V.R. Pickles and R.J. Fitzpatrick Society for endocrinology.Memoirs,14 Cambridge university press,Cambridge,1966
Pha 16.0083

ENDOPLASMIC RETICULUM The Federation of European biochemical societies meeting 4th Proceedings Oslo 1967 Jul 7 Vol 6: structure and function of the endoplasmic reticulum in animal cells Federation of European biochemical societies Edited by F.C. Gran Organized by the Norwegian biochemical society Universitetsforlaget;Academic press, Oslo;London,1968 Symposium organizer P.N. Campbell
Bioch 33.1401

ENERGETICS Engineering developments in energy conversion International conference on energetics Rochester,N.Y. 1965 Aug 18-20 American society of mechanical engineers American society of mechanical engineers,New York,1965
Eng 41.4865

ENERGETICS New experimental techniques in propulsion and energetics research :a technical meeting Proceedings Munich 1967 Sep 11-15 Agard Edited by David Andrews and Jean Surugue Agard conference proceedings, 38 Technivision,Slough,1970
Eng 41.8359

ENERGETICS IN METALLURGICAL PHENOMENA 1st-4th seminar Proceedings Denver 1962-65 University of Denver Edited by William M. Mueller Energetics in metallurgical phenomena, 1-4 Gordon and Breach,New York, 1965-68
Met 25.1287

ENERGETICS IN METALLURGICAL PHENOMENA, 1-4 Energetics in metallurgical phenomena :seminar 1st-4th Proceedings Denver 1962-65 University of Denver Edited by William M. Mueller Gordon and Breach,New York,1965-68
Met 25.1287

ENERGY AND GRAVITATIONAL WAVES Centenario della nascita di Galileo Galilei 1564-1964 4th Atti Florence 1964 Sep 9-12 Vol 2,tomo 1: atti del convegno sulla relativita generale,problemi di energia e onde gravitazionali.Proceedings of the meeting on general electricity problems and gravitational waves Galileo Galilei.Comitato nazionale per le manifestazioni celebrative Galileo Galilei.Comitato nazionale per le manifestazioni celebrative.Publicazioni 267p Barbera,Florence,1965
WSM 43.4484

ENERGY BANDS IN METALS AND ALLOYS based on a symposium Los Angeles 1967 American institute of mining,metallurgical and petroleum engineers.Committee on alloy phases Edited by L.H. Bennett and J.T. Waber Metallurgical society conferences, 45 Gordon and Breach,New York,1968
Met 25.1060

ENERGY CONSERVATION Electron transport and energy conservation Round table discussion Fasano 1969 May 12-15 Edited by J.M. Tager and others Adriatica editrice,Bari,1970
Bioch 33.2288

ENERGY CONVERSION Engineering developments in energy conversion International conference on energetics Rochester,N.Y. 1965 Aug 18-20 American society of mechanical engineers American society of mechanical engineers,New York,1965
Eng 41.4865

ENERGY CONVERSION Non-heme iron proteins; role in energy conversion :a symposium Proceedings Yellow Springs,Ohio 1965 Mar 22-24 Edited by Anthony San Pietro Sponsored by the Charles F.Kettering research laboratory Charles F.Kettering research laboratory.Contribution, 201 Antioch press, Yellow Springs,Ohio,1965
Bioch 33.0546

ENERGY CONVERSION BY THE PHOTOSYNTHETIC APPARATUS symposium Report Upton,N.Y. 1966 Jun 6-9 Brookhaven national laboratory. Biology department Brookhaven symposia in biology, 19 BNL989(C-48) Brookhaven national laboratory.Biology department,Upton,N. Y.,1967
Bioch 33.1297

ENERGY CONVERSION FOR SPACE POWER :a symposium Santa Monica,Calif. 1960 Sep 27-30 American rocket society Edited by Nathan W. Snyder Progress in astronautics and rocketry, 3 Academic press,New York, 1961
Eng 41.4971

ENERGY CONVERSION FOR SPACE POWER... based on a symposium Selected technical papers Santa Monica,Calif. 1960 Sep 27-30 American rocket society.Space power systems committee Edited by Nathan W. Snyder Progress in astronautics and rocketry, 3 Academic press,New York;London,1961
Met 25.2165

ENERGY LEVEL AND METABOLIC CONTROL IN MITOCHONDRIA The Round table discussion on the energy level and metabolic control in mitochondria Proceedings Polignano a Mare 1968 May 13-16 Edited by S. Papa and others Organized by the University of Bari.Department of biochemistry and the University of Amsterdam.Laboratory of biochemistry Adriatica editrice,Bari,1969
Bioch 33.1075

ENERGY-LINKED FUNCTIONS OF MITOCHONDRIA : colloquium Papers presented Philadelphia 1963 Apr 13 Johnson research foundation Edited by Britton Chance Johnson research foundation.Colloquia, 1 Academic press,New York;London,1963
Bioch 33.1081

ENERGY-LINKED FUNCTIONS OF MITOCHONDRIA :first colloquium of the Johnson research foundation of the University of Pennsylvania Papers Philadelphia 1963 Apr 13 Johnson research foundation Edited by Britton Chance Academic press,NEW York,1963
Radioth 35.0412

ENERGY METABOLISM Hormonal regulation of energy metabolism :a conference Proceedings 1956 Feb 5 Edited by Laurance W. Kinsell Charles C.Thomas,Springfield,Ill.,1957
Bioch 33.0461

ENERGY OF TRANSFER IN RADIATION PROCESSES: CHEMICAL,PHYSICAL AND BIOLOGICAL ASPECTS : international symposium Proceedings Cardiff 1965 Jan Association for radiation research Edited by Glyn O. Phillips Elsevier,Amsterdam,1966
Radioth 35.1744

ENERGY RELATIONSHIPS IN ENZYME REACTIONS :a conference New York 1944 Feb 11-12 New York academy of sciences.Section of physics and chemistry New York academy of sciences.Annals, 45,p.357-436 New York,1944
Bioch 33.1027

ENERGY REQUIREMENTS IN THE IRON AND STEEL INDUSTRY a conference Harrogate 1959 Jun 24-25 British iron and steel research association BISRA,London,1959
Met 25.0409

ENERGY TRANSFER IN HOT GASES NBS semicentennial symposium on energy transfer in hot gases Proceedings 1951 Sep 17-18 National bureau of standards National bureau of standards.Circular, 523 U.S.government printing office,Washington,D.C.,1954
Eng 41.7220

ENERGY TRANSFER WITH SPECIAL REFERENCE TO BIOLOGICAL SYSTEMS :a general discussion Nottingham 1959 Apr 14-16 Faraday society Faraday society.Discussions, 27 Faraday society,London,1959
Radioth 35.1406

ENERGY TRANSFER WITH SPECIAL REFERENCE TO BIOLOGICAL SYSTEMS :a general discussion Nottingham 1959 Apr 14-16 Faraday society Faraday society.Discussions, 27 The Faraday society,London,1960
Radioth 35.0249

ENERGY TRANSFER WITH SPECIAL REFERENCE TO BIOLOGICAL SYSTEMS :a general discussion Papers Nottingham 1959 Apr 14-16 Faraday society Faraday society.Discussions, 27 Aberdeen university press,Aberdeen,1960
Bioch 33.0314

ENFIELD COLLEGE OF TECHNOLOGY Conference on the teaching of engineering design Papers and summary Scarborough 1964 Apr 1-4 Illus Institution of engineering designers, London,1964 Co-organized by the Institution of engineering designers and the Hornsey College of art
Eng 41.0309

ENGINEERING Coastal engineering :a specialty conference Santa Barbara,Calif. 1965 Oct American society of civil engineers American society of civil engineers,New York,1966
Eng 41.3346

ENGINEERING Conference on coastal engineering 10th Proceedings Tokyo 1966 Sep American society of civil engineers 2 vols American society of civil engineers, New York,1967
Eng 41.3348

ENGINEERING Conference on coastal engineering 11th Proceedings London 1968 Sep Vol 1-2 American society of civil engineers Institution of civil engineers 2 vols American society of civil engineers,New York,1969
Eng 41.3347

ENGINEERING International engineering congress :report of the proceedings and abstracts of the papers read Glasgow 1901 Sep 2-3 Edited by J.D. Cormack Asher, Glasgow,1902
Eng 41.1563

ENGINEERING Joint engineering conference Proceedings London 1951 Jun 4-15 Institution of civil engineers Institution of mechanical engineers Institution of electrical engineers Institution of civil engineers,London,1951
Eng 41.1761

ENGINEERING ASPECTS OF MAGNETOHYDRODYNAMICS Symposium on the engineering aspects of magnetohydrodynamics 2nd Proceedings Philadelphia,Pa. 1961 Mar 9-10 Edited by Clifford Mannal and Norman W. Mather Sponsored by the American institute of electrical engineers Columbia university press,New York;London,1962
Eng 41.4497

ENGINEERING ASPECTS OF MAGNETOHYDRODYNAMICS Symposium on the engineering aspects of magnetohydrodynamics 3rd Proceedings Rochester,N.Y. 1962 Mar 28-29 Edited by Norman W. Mather and George W. Sutton Gordon and Breach,New York,1964
Eng 41.4546

ENGINEERING CERAMICS Mechanical properties of engineering ceramics :a conference Proceedings Raleigh,N.C. 1960 North Carolina state college and United States. Army.Office of ordnance research Edited by W. Wurth Kriegel and Hayne Palmour Interscience, New York;London,1961
Met 25.0853

ENGINEERING DESIGN Conference on the teaching of engineering design Papers and summary Scarborough 1964 Apr 1-4 Organised by the Enfield college of technology Illus Institution of engineering designers, London,1964 Co-organized by the Institution of engineering designers and the Hornsey College of art
Eng 41.0309

ENGINEERING DEVELOPMENTS IN ENERGY CONVERSION International conference on energetics Rochester,N.Y. 1965 Aug 18-20 American society of mechanical engineers American society of mechanical engineers,New York,1965
Eng 41.4865

ENGINEERING DIMENSIONAL METROLOGY :a symposium Proceedings Teddington 1953 Oct 21-24 Vol 1-2 National physical laboratory 2 vols H.M.S.O.,London,1955
Eng 41.3927

ENGINEERING DIMENSIONAL METROLOGY:A SYMPOSIUM proceedings Teddington 1953 October 21-24 Vols 1-2 National physical laboratory 26cm London,1955
Philos 1.0402

ENGINEERING EDUCATION EUSEC conference on engineering education 1st Proceedings London 1953 Jan 12-17 Conference of engineering societies of western Erope and the United States of America Institution of civil engineers,London,1953
Eng 41.1543

ENGINEERING EDUCATION EUSEC conference on engineering education 3rd Proceedings Paris 1957 Sep Conference of engineering societies of western Europe and the United States of America Institution of mechanical engineers,London,1958
Eng 41.1546

ENGINEERING EDUCATION EUSEC meeting on engineering education and training 4th Proceedings London 1962 Jan Conference of engineering societies of western Europe and the United States of America Institution of electrical engineers,London, 1963
Eng 41.1547

ENGINEERING HYDRAULICS Hydraulics conference 4th proceedings Iowa city 1949 Jun 12-15 Iowa institute of hydraulic research Edited by Hunter Rouse John Wiley and sons, New York;London,1950
A Math 4.0511

ENGINEERING HYDRAULICS 4th conference Proceedings Iowa city,Iowa 1949 Jun 12-15 Iowa institute of hydraulic research Edited by Hunter Rouse Wiley;Chapman and Hall,New York;London,1950
Eng 41.6662

ENGINEERING MANUFACTURE Conference on technology of engineering manufacture Proceedings London 1958 Mar 25-27 Institution of mechanical engineers Institution of mechanical engineers,London, 1958
Eng 41.3870

The ENGINEERING OF GAS-SOLID REACTIONS :I.Chem. E.-VTG-VDI joint meeting Proceedings Brighton 1968 Apr 24-26 Institution of chemical engineers Verein deutscher ingenieure Edited by J.M. Pirie Institution of chemical engineers.Symposium series, 27 illus 255p Institution of chemical engineers,London,1968
Chem E 24.1873

ENGINEERING PLASTICITY :a conference Papers Cambridge 1968 Mar Edited by J. Heyman and F.A. Leckie Cambridge university press, Cambridge,1968
Eng 41.2501

ENGINEERING SCIENCE Recent advances in the engineering sciences:their impact on engineering education Conference on science and technology for deans of engineering Proceedings Lafayette,Ind. 1957 Sep 9-12 Purdue research foundation and Purdue university viii,257p McGraw-Hill,New York, 1958 .
Chem E 24.1074

ENGINEERING STRUCTURES :a symposium Papers Bristol 1949 Sep Colston research society University of Bristol Colston research society.Research special supplement, 49 Butterworths,London,1949
Eng 41.3022

The ENGINEERING USES OF HOLOGRAPHY :a symposium Proceedings Glasgow 1968 Sep 17-20 University of Strathclyde Edited by E.R. Robertson and J.M. Harvey Organized in association with the National physical laboratory Cambridge university press, Cambridge,1970
Eng 41.4290

The ENGINEERING USES OF HOLOGRAPHY :a symposium Proceedings Glasgow 1968 Sep 17-20 University of Strathclyde National physical laboratory Edited by Elliot R. Robertson and James M. Harvey Cambridge university press,Cambridge,1970
An 32.5364

ENGLEFIELD GREEN,SURREY 1971 Phytochemical ecology Phytochemical society symposium 8th Annual proceedings Phytochemical society Academic press,London,1972
Bioch 33.2229

ENTE PER GLI STUDI MONETARI,BANCARI ET FINANZIARI 'LUIGI EINAUDI' Differential games and related topics International summer school on mathematical models of action and reaction Proceedings Varenna 1970 Jun 15-27 Edited by H.W. Kuhn and G.P. Szego x,489p 23cm North-Holland,Amsterdam,1971
P Math 2.4574

ENTIRE FUNCTIONS AND RELATED PARTS OF ANALYSIS summer institute Lecture notes La Jolla 1966 Jun 27-Jul 22 American mathematical society 28cm 2 vols 1966 Mimeographed
P Math 2.3682

ENTIRE FUNCTIONS AND RELATED PARTS OF ANALYSIS symposium Proceedings La Jolla 1966 Jun 27-Jul 22 American mathematical society Edited by Jacob Korevaar and others American mathematical society.Proceedings of Symposia in pure mathematics, 11 illus. vi,554p 26cm American mathematical society, Providence,R.I.,1968
P Math 2.3199

ENTOMOLOGICAL SOCIETY OF CANADA and ENTOMOLOGICAL SOCIETY OF ONTARIO Centennial of entomology in Canada,1863-1963 : a commemorative joint meeting Proceedings Ottawa 1963 Sep 3-6 Canadian entomologist, 96,pt.1-2 475p Carelton university,Ottawa, 1963
Sco 14.2471

ENTOMOLOGICAL SOCIETY OF ONTARIO and ENTOMOLOGICAL SOCIETY OF CANADA Centennial of entomology in Canada,1863-1963 : a commemorative joint meeting Proceedings Ottawa 1963 Sep 3-6 Canadian entomologist, 96,pt.1-2 475p Carelton university,Ottawa, 1963
Sco 14.2471

ENTOMOLOGY Physiology of the insect central nervous system International congress of entomology 12th Papers London 1964 Jul Edited by J.E. Treherne and J.W.L. Beament Academic press,London,1965
Pha 16.0276

ENTOMOLOGY Radioisotopes and radiation in entomology :a symposium Proceedings Bombay 1960 Dec 5-9 By H.J. Bhabha and others Sponsored by the International atomic energy agency International atomic energy agency, Vienna,1962
Bal 39.2390

ENTREPRENEUR Economic history society.Annual conference Papers Cambridge 1957 Apr Edited by B.E. Supple 63p Harvard university.Research center in entrepreneurial history,Cambridge,Mass.,1957
Geog 13.3190

ENTRETIENS DE MONACO EN SCIENCES HUMAINES 1ere session Human displacements: measurement,methodological aspects Monaco 1962 May 24-29 Centre international detude des problemes humains Edited by Jean Sutter Hachette,Paris,1962 Text in English and French
Geog 13.1566

ENTRETIENS SUR LE TEMPS Cerisy-la-Salle 1964 Jul 14-23 Centre culturel international de Cerisy-la-Salle Edited by Jeanne Hersch and Rene Poirier Centre culturel international de Cerisy-la-Salle.Decade nouvelle serie, 5 351p Mouton,Paris;The Hague,1967
WSM 43.1000

ENTSTEHUNG, ENTWICKLUNG UND PERSPEKTIVEN DER EINSTEINSCHEN GRAVITATIONSTHEORIE vortrage und diskussionen Einstein symposium Berlin 1965 Nov 2-5 Deutsche akademie der wissenschaften zu Berlin Edited by H-J. Treder 314p Akademie-verlag,Berlin,1966
Obs 6.1765

ENTSTEHUNG,ENTWICKLUNG UND PERSPEKTIVEN DER EINSTEINSCHEN GRAVITATIONSTHEORIE vortrage und diskussionen Einstein-Symposium Berlin 1965 Nov 2-5 Akademie der Wissenschaften Edited by H.J. Treder Akademie Verlag,Berlin,1965 typescript
A Math 4.1031

ENVIRONMENTAL CONTROL OF PLANT GROWTH :a symposium Proceedings Canberra 1962 Aug Edited by L.T. Evans Academic press, New York;London,1963
Bot 42.1351

ENVIRONMENTAL INFLUENCES :a conference Proceedings Biology and behavior 3 New York 1967 Apr 21-22 Russell Sage foundation Rockefeller institute Edited by David C. Glass Rockefeller university press; Russell Sage foundation,New York,1968 Third and last in a series of conferences on biology and behavior
Psy 31.3096

ENVIRONMENTAL PROTECTION Symposium on multiple-source urban diffusion models Proceedings 1970 United States. Environmental protection agency National air pollution control administration North Carolina consortium on air pollution Air pollution control office.Publication,AP- 86 US government printing office,Washington,D.C., 1970
Eng 41.8616

ENVIRONMENTAL STUDIES OF THE GLACIAL LAKE AGASSIZ REGION :a conference Proceedings Life,land and water 1966 Edited by William J. Mayer-Oakes University of Manitoba. Department of anthropology.Occasional papers, 1 maps xvi,414p University of Manitoba press,Winnipeg,1967
Bot 42.2101

ENZYMATIC ASPECTS OF METABOLIC REGULATION International symposium.Enzymatic aspects of metabolic regulation Proceedings Mexico City 1966 Nov 28-Dec 1 National cancer institute Edited by M.P. Stulberg National cancer institute.Monograph, 27 National cancer institute,Betheseda,Md.,1967
Bioch 33.1062

ENZYME ACTION A Symposium on the mechanism of enzyme action Papers and discussions Baltimore 1953 Jun 16-19 Edited by William D. McElroy and Bentley Glass Sponsored by the McCollum-Pratt institute McCollum-Pratt institute.Contribution, 70 Johns Hopkins press,Baltimore,Md.,1954
Bioch 33.1059

ENZYME ACTION A symposium on the mechanism of enzyme activity McCollum-Pratt institute Edited by William D. McElroy McCollum-Pratt institute.Contribution,70 Johns Hopkins press,Baltimore,Md.,1954
Chem 18.1270

ENZYME ACTION Federation of European biochemical societies meeting 6th Proceedings Madrid 1969 Apr metabolic regulation and enzyme action Federation of European biochemical societies Edited by A. Sols and S. Grisolia Federation of European biochemical societies.Publications, 19 Academic press,London,1970
Bioch 33.1889

ENZYME ACTION International congress of biochemistry 5th Proceedings Moscow 1961 Aug 10-16 Vol 4: molecular basis of enzyme action and inhibition International union of biochemistry Edited by P.A.E. Desnuelle I.U.B.symposium series, 24 Pergamon;PWN-Polish scientific publishers, Oxford;Warsaw,1963
Bioch 33.1333

ENZYME ACTIVITY Symposium on regulation of enzyme activity and synthesis in normal and neoplastic liver 1st Proceedings Indianapolis,Ind. 1962 Oct 1-2 Edited by George Weber Held at Indiana university school of medicine Advances in enzyme regulation, 1 Pergamon,Oxford,1963 Technical editor Catherine E.Forrest Weber
Bot 42.1738

ENZYME ACTIVITY Symposium on regulation of enzyme activity and synthesis in normal and neoplastic liver 2nd Proceedings Indianapolis,Ind. 1963 Sep 30-Oct 1 Edited by George Weber Held at Indiana university school of medicine Advances in enzyme regulation, 2 Pergamon,Oxford,1964 Technical editor Catherine E.Forrest,Weber
Bot 42.1739

ENZYME ACTIVITY The Federation of European biochemical societies meeting 4th Proceedings Oslo 1976 Jul Vol 1: regulation of enzyme activity and allosteric interactions Federation of European biochemical societies Edited by E. Kvamme and A. Phil Organized by the Norwegian biochemical society Universitetsforlaget; Academic press,Oslo;London,1968 Symposium organizer J.P.Changeux
Bioch 33.1396

ENZYME CHEMISTRY The International symposium on enzyme chemistry Proceedings Tokyo 1957 Oct 15-23 and Kyoto 1957 Oct 15-23 International union of biochemistry Organized by Science council of Japan I.U.B. Symposium series, 2 Pergamon,London,1958
Bioch 33.1053

ENZYME CHEMISTRY OF PHENOLIC COMPOUNDS :a symposium Proceedings Liverpool 1962 Apr 11-12 Plant phenolics group Edited by J.B. Pridham Pergamon press.Symposium publications division,Oxford,1963
Chem 18.1247

ENZYME MODELS AND ENZYME STRUCTURE :symposium Report Upton,N.Y. 1962 Jun 4-6 Brookhaven national laboratory.Biology department Brookhaven symposia in biology, 15 BNL738(C-34) Brookhaven national laboratory.Biology department,Upton,N.Y.,1962
Bioch 33.1293

ENZYME REACTION MECHANISMS Symposium on enzyme reaction mechanisms Research conference for biology and medicine 12th Gatlinburg,Tenn. 1959 Apr 1-4 Sponsored by the Oak Ridge national laboratory.Biology division Journal of cellular and comparative physiology, 54,suppl.1 Wistar institute of anatomy and biology,Philadelphia,Pa.,1959
Bioch 33.1046

ENZYME REACTIONS Energy relationships in enzyme reactions :a conference New York 1944 Feb 11-12 New York academy of sciences. Section of physics and chemistry New York academy of sciences.Annals, 45,p.357-436 New York,1944
Bioch 33.1027

ENZYMES Chemical reactivity and biological role of functional groups in enzymes Oxford 1970 Jan Biochemical society Edited by R.M. S. Smellie Biochemical society.Symposium, 31 Academic press,London,1970
Bioch 33.2315

ENZYMES Chemical reactivity and biological role of functional groups in enzymes :a symposium... Proceedings Oxford 1970 Jan Molecular enzymology and protein group Edited by R.M.S. Smellie Biochemical society. Symposia, 31 Academic press,London,1970 Molecular enzymology and protein group of the Biochemical society and the Chemical society
Chem 18.2745

ENZYMES Haematin enzymes :a symposium Papers and discussions Canberra 1959 Aug 31-Sep 4 Pt 1 International union of biochemistry Edited by J.E. Falk and others Organized by the Australian academy of science I.U.B.Symposium series, 19 Pergamon press, Oxford,1961
Bioch 33.1079

ENZYMES Multiple molecular forms of enzymes : a conference 1st Papers New York 1961 Feb 1-3 New York academy of sciences Edited by Felix Wroblewski and others New York academy of sciences.Annals, 94,p 655-1030 New York,1961
Bioch 33.1028

ENZYMES Multiple molecular forms of enzymes : a conference 2nd Papers New York 1966 Dec 1-3 New York academy of sciences Edited by Edward M. Weyer New York academy of sciences.Annals, 151,p 1-689 New York, 1968
Bioch 33.1029

ENZYMES Role of nucleotides for the function and conformation of enzymes The Alfred Benzon symposium 1 Proceedings Copenhagen 1968 Sep 9-11 Alfred Benzon foundation Edited by Herman M. Kalckar Held at the premises of the Royal Danish academy of sciences and letters Munksgaard, Copenhagen,1969
Bioch 33.1913

ENZYMES Structure and activity of enzymes Federation of European biochemical societies symposium 1st London 1914 Mar 24 Edited by T.W. Goodwin and others Academic press,London;New York,1964
Radioth 35.0119

ENZYMES Structure and activity of enzymes :a symposium Proceedings London 1964 Mar 24 Federation of European biochemical societies Edited by T.W. Goodwin and others Federation of European biochemical societies.Symposium, 1 Academic press,London;New York,1964
Bioch 33.1404

ENZYMES The Physical chemistry of enzymes :a discussion Papers Oxford 1955 Aug 10-12 Faraday society Faraday society.Discussions, 20 Aberdeen,1956
Bioch 33.1047

ENZYMES:UNITS OF BIOLOGICAL STRUCTURE AND FUNCTION Henry Ford hospital international symposium Proceedings Detroit 1955 Nov 1-3 Edited by Oliver H. Gaebler held at Henry Ford hospital Henry Ford hospital.Symposia Academic press,New York,1956
Bioch 33.1050

ENZYMES:UNITS OF BIOLOGICAL STRUCTURE AND FUNCTION :international symposium Proceedings Detroit 1955 Nov 1-3 Henry Ford hospital Edited by Oliver H. Guebler Henry Ford hospital symposia, 4 Academic press,New York,1956
Radioth 35.0052

ENZYMES:UNITS OF BIOLOGICAL STRUCTURE AND FUNCTION :international symposium Proceedings Detroit,Mich. 1955 Nov 1-3 Henry Ford hospital Edited by Oliver H. Gaebler Henry Ford hospital.Symposia, 4 Academic press,New York,1956
Gen 34.0646

ENZYMES AND DRUG ACTION London 1961 Mar 20-23 Ciba foundation and Co-ordinating committee for symposia on drug action Edited by J.L. Mongar and A.V.S. De Reuck Churchill, London,1962
Bioch 33.1872

ENZYMES AND ISOENZYMES:STRUCTURE,PROPERTIES AND FUNCTION Federation of European biochemical societies meeting 5th Prague 1968 Jul Federation of European biochemical societies Federation of European biochemical societies.Publications, 18 Academic press,London,1970
Bioch 33.1888

ENZYMOLOGICAL ASPECTS OF FOOD IRRADIATION Panel on enzymological aspects of the application of ionizing radiation to food preservation Proceedings Vienna 1968 Apr 8-12 Organized by the Joint FAO-IAEA division of atomic energy in food and agriculture International atomic energy agency.Panel proceedings series International atomic energy agency,Vienna,1969
Radioth 35.1762

EPIDEMIOLOGY International epidemiological association 5th Proceedings Primosten 1968 Aug 25-31 Edited by Milutin Srdic and others Savremena administracija,Belgrade, 1970
PGMS 29.0635

EPITHELIAL-MESENCHYMAL INTERACTIONS The Hahnemann symposium 18th Philadelphia 1968 Hahnemann medical college and hospital Edited by Raul Fleischmajor and Rupert E. Billingham Williams and Wilkins,Baltimore, 1968 Dedicated to Dr. Johannes Holtfreter
Bal 39.0491

EQUATIONS Symposium on the numerical treatment of ordinary differential equations, integral and integro-differential equations 4th proceedings Rome 1960 Sep.20-24 Provisional international computation centre bibliog. 679p 26cm Birkhauser,Basel,1960
Math L 5.0299

EQUIPMENT FOR THE THERMAL TREATMENT OF NON-FERROUS METALS AND ALLOYS a symposium on the metallurgical aspects of the subject London 1952 Mar 26 Institute of metals Institute of metals.Monograph and report series, 14 Institute of metals,London,1953
Met 25.2306

EREVAN 1965 Sovremennye problemy teorii analiticheskikh funktsii Mezhdunarodnaya konferentsiya po teorii analiticheskikh funktsii Edited by M.A. Lavrentev and others 361p 27cm Moscow,1966
P. Math 2.2408

ERGODIC THEORIES Teorie ergodiche Scuola internationale di fisica 'Enrico Fermi' 14 corso Rendiconti Varenna 1960 May 23-31 Societa Italiana di Fisica Edited by P. Caldirola Academic press,New York;London, 1961
A Math 4.0950

ERGODIC THEORY international symposium papers New Orleans,La. 1961 Oct.23-27 Tulane university Edited by Fred B. Wright Sponsored by the National science foundation xii,316p Academic press,New York;London,1963
Math 3.0044

ERGODIC THEORY international symposium proceedings New Orleans,La. 1961 Oct 23-27 Edited by Fred B. Wright Sponsored by the National science foundation xii,316p 24cm Academic press,New York,London,1963
P. Math 2.2188

ERGODIC THEORY AND PROBABILITY Contributions to ergodic theory and probability Midwestern conference on ergodic theory and probability 1st Proceedings Columbus,Ohio 1970 Mar 27-30 Held at the Ohio state university Lecture notes in mathematics, 160 vii,278p 26cm Springer,Berlin,1970
Math S 3.1779

ERGONOMICS RESEARCH SOCIETY Symposium on fatigue Contributions Cranfield 1952 Mar Edited by W.F. Floyd and A.T. Welford Lewis, London,1953
Psy 31.1027

ERICE 1963 Strong,electromagnetic,and weak interactions International school of physics 'Ettore Majorana' 1st course proceedings Edited by A. Zichichi Sponsored by the European organization for nuclear research W. A.Benjamin,New York;Amsterdam,1964
A Math 4.0846

ERICE 1964 Symmetries in elementary particle physics International school of physics 'Ettore Majorana' 2nd course proceedings Edited by A. Zichichi Sponsored by the European organization for nuclear research 429p Academic press,New York;London,1965
A Math 4.0847

ERICE 1965 Recent developments in particle symmetries International school of physics 'Ettore Majorana 3rd course proceedings Edited by A. Zichichi Sponsored by the European organization for nuclear research 460p Academic press,New York,1966
A Math 4.0848

ERICE 1966 strong and weak interactions, present problems International school of physics 'Ettore Majorana' 4th course proceedings Edited by A. Zichichi Sponsored by the European organization for nuclear research 859p Academic press,New York,1966
A Math 4.0849

ERICE 1968 Theory and phenomonology in particle physica "Ettore Majorana" International school of physics 6th course Part A-B European organization for nuclear research Edited by A. Zichichi Co-sponsored by the Weizmann institute of science Bibliog.,Illus. 23cm 2 vols Academic press,New York,1969
A Math 4.1541

EROSION BY CAVITATION OR IMPINGEMENT :a symposium presented at the 69th annual meeting Atlantic City,N.J. 1966 Jun 26-Jul 1 American society for testing and materials American society for testing and materials. Special technical publications, 408 American society for testing and materials,Philadelphia, Pa.,1967
Eng 41.6561

ERROR CORRECTING CODES Madison,Wis. 1968 May 6-8 United States.Army.Mathematics research center Edited by Henry B. Mann United States.Army.Mathematics research center, 21 Bibliog.,Illus. xii,231p 23cm Wiley,New York,1968
P Math 2.3700

ERROR CORRECTING CODES :a symposium Proceedings Madison 1968 May 6-8 United States.Army.Mathematics research center Edited by Henry B. Mann United States.Army. Mathematics research center.Publications, 22 Wiley,New York,1968
Math L 5.3514

ERROR IN DIGITAL COMPUTATION advanced seminar proceedings Madison,Wis. 1964 Oct.5-7 Vol. 1 United States.Army.Mathematics research center Edited by Louis B. Rall United States army.Mathematics research center. Publications, 14 Bibliog. ix,324p 24cm John Wiley and sons,New York,1965
Math 3.0242

ERROR IN DIGITAL COMPUTATION seminar proceedings Wisconsin 1964 Oct.5-7 Vols. 1-2 Edited by Louis B. Rall United States Army.Mathematics research center. publication, 14 2 vols John Wiley,New York,1965
A Math 4.1209

ERROR IN DIGITAL COMPUTATION symposium proceedings Madison,Wis. 1965 Apr.26-28 Vol. 2 United States.Army.Mathematics research center Edited by Louis B. Rall United States army.Mathematics research center. Publications, 15 x,288p 24cm John Wiley and sons,New York,1965
Math 3.0243

ESPLORAZIONE SCIENTIFICA DEL MAR ROSSO International geographical congress Cambridge 1928 Jul Italy.Navy.Hydrographic institute Genoa,1928 Bound with English translation
Geog 13.0854

ESSO PETROLEUM COMPANY Control and simulation language :introductory manual 39p 28cm Esso;I.B.M.,London,1963
Math L 5.3262

ESSO PETROLEUM COMPANY Control and simulation language reference manual 95p 28cm Esso;I.B.M.,London,1963
Math L 5.3261

ESTES PARK,COL. 1960 Response of metals to high velocity deformation Technical conference American institute of mining, metallurgical and petroleum engineers Metallurgical society conferences, 9 Interscience,New York,1961
Met 25.0905

ESTES PARK,COLO. 1968 Annual conference on applications of X-ray analysis 17th Proceedings Denver research institute. Metallurgy division Edited by Charles S. Barrett and others Advances in X-ray analysis, 12 Plenum press,New York,1969
Met 25.2824

ESTUARIES Conference on estuaries Papers Jekyll Island,Ga. 1964 University of Georgia.Marine institute Edited by George H. Lauff American association for the advancement of science.Publications,83 New York,1967
Geog 13.0743

ESTUARIES :a symposium Papers Jekyll island,Ga 1964 Mar-Apr University of Georgia.Marine institute Edited by George H. Lauff American association for the advance of science.Publication, 83 Washington,D.C., 1967
Bot 42.3345

ESTUDIO SOBRE ALGUNOS DE LOS CONGLOMERADOS ROJOS DEL TERCIARIO INFERIOR DEL CENTRO DE MEXICO International geological congress 20th papers Mexico City 1956 By John D. Edwards Translated by Salvador Ulloa from the English Mexico City,1956 Bound with two other publications of the congress
Geol 8.3042

ETAT SOLIDE Conseil de physique 9 eme Rapports et discussions Brussels 1951 Sep 25-29 Institut international de physique Solvay R.Stoops,Brussels,1952
Met 25.1135

ETHNOPHARMACOLOGIC SEARCH FOR PSYCHOACTIVE DRUGS a symposium Proceedings San Francisco 1967 Jan 28-30 Edited by Daniel H. Efron and others Sponsored by the National institute of mental health National institute of mental health.Pharmacology section.Workshop series,2 United States.Public health service publication,1645 U.S.Dept.of public health, Washington,D.C.,1967
Pha 16.0065

ETOILES A RAIES D'EMISSIONS Colloque international d'astrophysique 8e Liege Universite de Liege.Institut d'astrophysique 14p Liege,c1957 Summary of contributions
Obs 6.3170

ETUDES ET RECHERCHES SUR LES PHYTOHORMONES Paris 1937 Oct 1-2 By Peter Boysen-Jensen and others League of Nations.Committee on intellectual co-operation Organized in collaboration with Union internationale des sciences biologiques League of Nations,Paris, 1937
Bot 42.1815

ETUDES SUR LES SCHISTES CRISTALLINS International geological congress 4th papers London 1888 London,1888
Min 10.0237

ETUDES SUR LES TERMITES AFRICAINS :colloque international Comptes rendus Leopoldville 1964 May 11-16 Universite de Louvain Edited by A. Bouillon Under the auspices of Unesco figs,pls,tables 27cm Masson,Paris,1964
Bal 44.4805

EUGENE 1955 Recent advances in invertebrate physiology :a symposium Edited by B.T. Scheer and others held at the University of Oregon University of Oregon publications,Eugene,Oregon,1957 Sponsored by the National science foundation,Tektronix foundation,the University of Oregon
Bal 39.2107

EUGENICS Eugenics,genetics and the family International congress of eugenics 2nd Papers New York 1921 Sep 22-28 Vol 1 Port 2 vols Williams and Wilkins, Baltimore,Md.,1923
An 32.0031

EUGENICS Eugenics in race and state The International congress of eugenics 2nd Scientific papers New York 1921 Sep 22-28 Vol 2 Port Williams and Wilkins, Baltimore,1923
An 32.0032

EUGENICS Problems in eugenics International eugenics congress 1st Papers London 1912 Jul 24-20 Eugenics education society Eugenics education society, London,1912
Bal 44.6430

EUGENICS Problems in eugenics International eugenics congress 1st Papers London 1912 Jul 24-30 Eugenics education society Held at the University of London Eugenics education society,London, 1912
Bot 42.6705

EUGENICS,GENETICS AND THE FAMILY International congress of eugenics 2nd Papers New York 1921 Sep 22-28 Vol 1 Port 2 vols Williams and Wilkins, Baltimore,Md.,1923
An 32.0031

EUGENICS,GENETICS AND THE FAMILY International congress of eugenics 2nd Papers New York 1921 Sep 22-28 Vol 1-2 port. 2 vols Williams and Wilkins, Baltimore,Md.,1923
Gen 34.1926

EUGENICS EDUCATION SOCIETY Problems in eugenics International eugenics conference 1st Papers London 1912 Jul 24-30 2 vols Eugenics education society,London,1913 Later known as the 1st international congress of eugenics
Gen 34.1925

EUGENICS EDUCATION SOCIETY Problems in eugenics International eugenics congress 1st Papers London 1912 Jul 24-20 Eugenics education society,London,1912
Bal 44.6430

EUGENICS IN RACE AND STATE The International congress of eugenics 2nd Scientific papers New York 1921 Sep 22-28 Vol 2 Port Williams and Wilkins,Baltimore,1923
An 32.0032

EUGENICS SOCIETY Biological aspects of race : a symposium 5th Proceedings Edited by G. Ainsworth Harrison and John Peel Eugenics society.Symposia, 5 Journal of biosocial science.Supplement, 1 Blackwell,Oxford, 1969
Gen 34.2198

EUGENICS SOCIETY Biological aspects of social problems :a symposium London 1964 Oct 1-2 Edited by J.E. Meade and A.S. Parkes Eugenics society.Symposia,1 Oliver and Boyd, Edinburgh;London,1965
Geog 13.1550

EUGENICS SOCIETY Genetic and environmental influences on behaviour :a symposium 1967 Sep Edited by J.M. Thoday and Alan S. Parkes Eugenics society.Symposium, 4 bibliog., illus. x,27p 22cm Oliver and Boyd, Edinburgh,1968
Math S 3.1468

EUGENICS SOCIETY.SYMPOSIA, 2 Genetic and environmental factors in human ability : Eugenics society symposium 1965 Sep-Oct Eugenics society Edited by G.E. Meade and A. S. Parkes Oliver and Boyd,Edinburgh,1966
Gen 34.2042

EUGENICS SOCIETY.SYMPOSIA, 3 Social and genetic influences on life and death :a symposium Papers 1966 Sep Eugenics society Edited by Robert Platt and A.S. Parkes Oliver and Boyd,Edinburgh;London,1967
Gen 34.1911

EUGENICS SOCIETY.SYMPOSIA, 5 Biological aspects of race :a symposium 5th Proceedings Eugenics society Edited by G. Ainsworth Harrison and John Peel Journal of biosocial science.Supplement, 1 Blackwell, Oxford,1969
Gen 34.2198

EUGENICS SOCIETY.SYMPOSIUM, 4 Genetic and environmental influences on behaviour :a symposium 1967 Sep Eugenics society Edited by J.M. Thoday and Alan S. Parkes bibliog.,illus. x,27p 22cm Oliver and Boyd,Edinburgh,1968
Math S 3.1468

EUROPAISCHEN GESELLSCHAFT FUR HAMATOLOGIE CONGRESS 5th :colloquium uber aktuelle probleme des transfusionswesens und der immunhamatologie Freiburg 1955 Sep 20-24 European society of haematology Edited by Herbert Begemann Springer-verlag,Berlin; Gottingen,1956
Med 36.0022

EUROPEAN ANATOMICAL MEETING 2nd Brussels 1963 Sep lectures on morphology Archives de biologie,Liege,1965 Most papers in French
An 32.1195

EUROPEAN ANIMAL BLOOD GROUP CONFERENCE 9th Proceedings Blood groups of animals Prague 1964 Aug 18-22 Ceskoslovenska akademie ved and European society for animal blood group research Edited by Josef Matousek Junk,The Hague,1964
Surg 23.0044

EUROPEAN ASSOCIATION OF RADIOLOGISTS Symposium on high-energy electrons Proceedings Montreux 1964 Sep 7-11 Edited by A. Zuppinger and G. Poretti Springer,Berlin,1965
Chem 18.0123

EUROPEAN ATOMIC ENERGY SOCIETY A.E.C.-Euratom conference on aqueous corrosion of reactor materials Brussels 1959 Oct 14-17 Office of technical services,Washington,D.C., 1959
Met 25.1930

EUROPEAN BROWN RUST CONFERENCE,1ST Cereal rust conferences Cambridge 1964 Jun-Jul Plant breeding institute Cambridge,1966
Bot 42.4125

EUROPEAN CANCER CHEMOTHERAPY GROUP Antitumoral effects of vinca rosea alkaloids : first symposium of the G.E.C.A. Proceedings Paris 1965 Jun 21 Edited by S. Garattini and E.M. Sproston International congress series, 106 Excerpta medica foundation, Amsterdam,1966
Radioth 35.0875

EUROPEAN CERAMIC ASSOCIATION Science of ceramics :a conference 3rd Proceedings Bristol 1965 Jul 5-8 British ceramic society and Nederlandse keramische vereniging Edited by G.H. Stewart Academic press,London;New York,1967
Met 25.0286

EUROPEAN COLLOQUIUM ON BLACK RUST OF CEREALS,3RD Cereal rust conferences Cambridge 1964 Jun-Jul Plant breeding institute Cambridge, 1966
Bot 42.4125

EUROPEAN CONFERENCE ON ELEMENTARY PARTICLES 4th Proceedings Heidelberg 1967 Sep 20-27 Ruprecht-Karl-universitat Heidelberg and European organization for nuclear research Edited by H. Filthuth viii,550p North-Holland,Amsterdam,1968
A Math 4.1348

EUROPEAN CONFERENCE ON SOIL MECHANICS AND FOUNDATION ENGINEERING Proceedings Problems of settlements and compressibility of soils Wiesbaden 1963 Vol 1-2 2 vols 1963
Eng 41.3175

EUROPEAN CONFERENCE ON TUMOUR BIOLOGY Warsaw 1961 May 22-27 International union against cancer International union against cancer. Acta, 18,no 1-2 U.I.C;C.,Louvain,1962
Radioth 35.0817

EUROPEAN CONGRESS OF ACCOUNTANTS 1963 European accounting history :exhibition September 2-14 Institute of chartered accountants of Scotland pls 72p Edinburgh,1963 Exhibition held on the occasion of the congress
WSM 43.5043

EUROPEAN CONGRESS OF ANAESTHESIOLOGY 1st : post-graduate courses Abstracts of papers Vienna 1962 Sep 3-9 World federation of societies of anaesthesiologists English edition Wiener medizinischen akademie,Vienna, c1962
PGMS 29.0030

The EUROPEAN CONGRESS OF ANAESTHESIOLOGY 2nd Proceedings Copenhagen 1966 Aug 8-13 1-3 World federation of societies of anaesthesiologists Acta anaesthesiologica Scandinavica.Supplementum, 23-25 3 vols Universitetsforlaget,Aarhus,1966
PGMS 29.0031

EUROPEAN CONGRESS OF NEUROSURGERY 3rd Cerebral circulation Madrid 1967 Apr Edited by W. Luyendijk Progress in brain research, 30 Elsevier,Amsterdam,1968
An 32.4248

EUROPEAN CONGRESS OF PERINATAL MEDICINE 1st Perinatal medicine Berlin 1968 Mar 28-30 Edited by Peter John Hungerford and others Thieme;Academic press,Stuttgart;New York,1969
Phys 20.2246

EUROPEAN CONGRESS ON ELECTRON MICROSCOPY 5th Proceedings Electron microscopy 1972 Manchester 1972 International federation of societies for electron microscopy Institute of physics.Conference series, 14 Institute of physics,London,1972
Eng 41.8576

EUROPEAN CONGRESS ON ELECTRON MICROSCOPY 5th Proceedings Manchester 1972 Institute of physics Institute of physics.Conference series, 14 illus xxviii,683p 26cm Institute of physics,London,1972
Cav 7.3182

EUROPEAN CONGRESS ON MOLECULAR SPECTROSCOPY 5TH Molecular spectroscopy :general and introductory lectures presented at the fifth European congress on molecular spectroscopy 5th Amsterdam 1961 May 29-Jun 3 International union of pure and applied chemistry.Physical chemistry section,and,Royal Netherlands chemical society Co-sponsored by the Netherlands physical society Butterworths,London,1962 Reprinted from 'Pure and applied chemistry',vol 4,no 1
Chem 18.2315

EUROPEAN CONGRESS ON THE INFLUENCE OF AIR POLLUTION ON PLANTS AND ANIMALS 1st Proceedings Air pollution Wageningen 1968 Apr 22-27 Centre for agricultural publishing and documentation illus. 415p Centre for agricultural publishing and documentation,Wageningen,1969
Bot 42.2106

EUROPEAN DIALYSIS AND TRANSPLANT ASSOCIATION
Renal failure and replacement of renal function European dialysis and transport association :a conference 2nd Proceedings Newcastle-upon-Tyne 1965 Vol 2 Edited by David Kerr Excerpta medica. International congress series, 103 Excerpta medica,Amsterdam,1966
Inv Med 37.0130

EUROPEAN DIALYSIS AND TRANSPLANT ASSOCIATION
CONFERENCE 4th Proceedings Dialysis and renal transplantation Paris 1967 Jun Edited by David S. Kerr and others European dialysis and transplant association. Proceedings, 4 International congress series, 155 Excerpta medica,Amsterdam,1968
Surg 23.0043

EUROPEAN DIALYSIS AND TRANSPORT ASSOCIATION 2nd
a conference Proceedings Renal failure and replacement of renal function Newcastle-upon-Tyne 1965 Vol 2 European dialysis and transplant association Edited by David Kerr Excerpta medica. International congress series, 103 Excerpta medica,Amsterdam,1966
Inv Med 37.0130

EUROPEAN FEDERATION OF CHEMICAL ENGINEERING
Chemical reaction engineering European symposium on chemical engineering 1st Papers Amsterdam 1957 May 7-9 Instituut van ingenieurs Nederlandse chemische vereniging International series of monographs on chemical engineering, 1 197p Pergamon,London,1957
Chem E 24.1900

EUROPEAN FEDERATION OF CHEMICAL ENGINEERING
Chemical reaction engineering European symposium on chemical reaction engineering 3rd Proceedings Amsterdam 1964 Sep 15-17 Instituut van ingenieurs.Afdeling voor chemische techniek and Nederlandse chemische vereniging.Sectie voor chemische technologie vi,326p Pergamon press,Oxford, 1965 Supplement to "Chemical engineering science".
Chem E 24.1269

EUROPEAN FEDERATION OF CHEMICAL ENGINEERING
International symposium on distillation Brighton 1960 May 4-6 281p Institution of chemical engineers,London,1960 24th meeting of the European Federation of Chemical Engineering
Chem E 24.1303

EUROPEAN FEDERATION OF CHEMICAL ENGINEERING
Symposium on process optimisation Proceedings London 1962 Jun 26 Institution of chemical engineers Edited by J.M. Pirie 70p Institution of chemical engineers,London,1962 3rd congress of the European federation of chemical engineering
Chem E 24.1630

EUROPEAN FEDERATION OF CHEMICAL ENGINEERING
Symposium on the handling of solids Proceedings London 1962 Jun 25 Institution of chemical engineers Edited by P.A. Rottenburg Institution of chemical engineers,London,1962 3rd congress of the European federation of chemical engineering
Chem E 24.1446

EUROPEAN FEDERATION OF CHEMICAL ENGINEERING
Symposium on the interaction between fluids and particles Proceedings London 1962 Jun 20-22 Institution of chemical engineers illus 351p Institution of chemical engineers,London,n.d. Third congress on the European federation of chemical engineering
Chem E 24.1447

EUROPEAN FEDERATION OF CHEMICAL ENGINEERING
The Organization of chemical engineering projects :joint symposium Proceedings London 1958 Jun 24-26 Institute of petroleum and Institution of chemical engineers Institution of chemical engineers, London,1958 17th meeting of the European federation of chemical engineering
Chem E 24.1543

EUROPEAN NUCLEAR ENERGY AGENCY Health physics in nuclear installations :a symposium Riso 1959 May 25-28 Held at the Danish atomic energy centre European nuclear energy agency,Paris,1959
Radioth 35.1110

EUROPEAN OPTHALMOLOGICAL SOCIETY Congres de la societe europeenne d'opthalmologie 1e Resumees des rapports et des communications Athens 1959 Apr 18-22 Excerpta medica. International congress series, 25 Excerpta medica,Amsterdam;New York,1960 Papers in English,French and German
PGMS 29.0682

EUROPEAN ORCHID CONGRESS 2nd Proceedings Paris 1969 Apr 24-26 Comite national interprofessionnel de l'horticulture et des pepinieres.Monographie, 2 C.N.I.H.,Rungis, 1969
BG 38.3116

EUROPEAN ORGANIZATION FOR NUCLEAR RESEARCH
European conference on elementary particles 4th Proceedings Heidelberg 1967 Sep 20-27 Edited by H. Filthuth viii,550p North-Holland,Amsterdam,1968
A Math 4.1348

EUROPEAN ORGANIZATION FOR NUCLEAR RESEARCH
International conference on high-energy physics at CERN 11th Proceedings Geneva 1962 Jul 4-11 Edited by J. Prentki Sponsored by the International union of pure and applied physics Bibliog.,Illus. xxiv, 949p 28cm CERN,Geneva,1962
A Math 4.1544

EUROPEAN ORGANIZATION FOR NUCLEAR RESEARCH
Recent developments in particle symmetries International school of physics 'Ettore Majorana 3rd course proceedings Erice 1965 Sep.-Oct. Edited by A. Zichichi 460p Academic press,New York,1966
A Math 4.0848

EUROPEAN ORGANIZATION FOR NUCLEAR RESEARCH
Strong,electromagnetic,and weak interactions International school of physics 'Ettore Majorana' 1st course proceedings Erice 1963 May-June Edited by A. Zichichi W.A. Benjamin,New York;Amsterdam,1964
A Math 4.0846

EUROPEAN ORGANIZATION FOR NUCLEAR RESEARCH Symmetries in elementary particle physics International school of physics 'Ettore Majorana' 2nd course proceedings Erice 1964 Aug.-Sep. Edited by A. Zichichi 429p Academic press, New York; London, 1965
A Math 4.0847

EUROPEAN ORGANIZATION FOR NUCLEAR RESEARCH Theory and phenomonology in particle physica "Ettore Majorana" International school of physics 6th course Erice 1968 Jul 13-28 Part A-B Edited by A. Zichichi Co-sponsored by the Weizmann institute of science Bibliog., Illus. 23cm 2 vols Academic press, New York, 1969
A Math 4.1541

EUROPEAN ORGANIZATION FOR NUCLEAR RESEARCH strong and weak interactions, present problems International school of physics 'Ettore Majorana' 4th course proceedings Erice 1966 Jun.19-Jul.4 Edited by A. Zichichi 859p Academic press, New York, 1966
A Math 4.0849

EUROPEAN ORGANIZATION FOR RESEARCH INTO CANCER TREATMENT Scientific basis of cancer chemotherapy :seminar Paris 1968 Mar 22-23 Edited by Georges Mathe Recent results in cancer research, 21 Heinemann medical, London, 1969
Radioth 35.0866

EUROPEAN PEPTIDE SYMPOSIUM 8th Proceedings Peptides Noordwijk 1966 Sep Edited by H.C. Beyerman and others North-Holland, Amsterdam, 1967
Bioch 33.0539

EUROPEAN PREPARATORY COMMISSION FOR SPACE RESEARCH Introduction to solar terrestrial relations The Summer school in space physics Proceedings Alpbach 1963 Jul.15-Aug.10 Edited by J. Ortner and H. Maseland Astrophysics and space science library 506p D.Reidel, Dordrecht, 1965
Obs 6.1321

EUROPEAN PREPARATORY COMMISSION FOR SPACE RESEARCH Introduction to solar terrestrial relations The Summer school in space physics Proceedings Alpbach 1963 Jul 15-Aug 10 Edited by J. Ortner and H. Maseland Astrophysics and space science library Reidel, Dordrecht, 1965
TA 15.0425

EUROPEAN QUATERNARY Mezhdunarodnaya konferentsiya assotsiatsii po izucheniyu chetvertichnogo perioda Evropy International conference of the Association for the study of the European quaternary 2nd trudy Leningrad 1932 Sep 1-7 Vypusk 1-5 Association for the study of the European quaternary, and, S.S.S.R.- N.K.T.P. Vsesoyuznoe geologo-razvedochnoe obedinenie Edited by D. A. Petrovskii and others Gosudarstvennoe nauchno-tekhnicheskoe geologo-razvedochnoe izdatelstvo, Leningrad; Moscow, 1932-34
Geol 8.2982

EUROPEAN REGIONAL CONFERENCE ON ELECTRON MICROSCOPY Proceedings Delft 1960 Vol 1 Edited by A.L. Houwink and B.J. Spit Nederlandse vereniging voor electronenmicroscopie, Delft, 1960
Met 25.1370

EUROPEAN REGIONAL CONFERENCE ON ELECTRON MICROSCOPY Proceedings Delft 1960 Aug 29-Sep 3 Vols 1-2 Edited by A.L. Houwink and B.J. Spit 2 vols Delft, 1961
Eng 41.5543

The EUROPEAN REGIONAL CONFERENCE ON ELECTRON MICROSCOPY Proceedings Delft 1960 Vol 2: cytology Edited by A.L. Houwink and others Delft, 1960
Phys 20.1508

EUROPEAN REGIONAL CONFERENCE ON ELECTRON MICROSCOPY 3rd Proceedings Electron microscopy 1964 Prague 1964 Aug 26-Sep 3 Vol A: non-biology International federation of electron microscope societies Edited by M. Titlbach Czech.academy of sciences, Prague, 1964
Met 25.2494

EUROPEAN REGIONAL CONFERENCE ON ELECTRON MICROSCOPY 4th Pre-congress abstracts Electron microscopy, 1968 Rome 1968 Sep 1-7 Edited by Daria Steve Bocciarelli Tipografia poliglotta vaticana, Rome, 1968
Met 25.1395

EUROPEAN REGIONAL CONFERENCE ON ELECTRON MICROSCOPY 4th Pre-congress abstracts Electron microscopy 1968 Rome 1968 Sep 1-7 Vol 2 Edited by D.S. Bocciarelli Tipografia polyglotta vaticana, Rome, 1968 Chairman: J.S. Mitchell
Radioth 35.0559

EUROPEAN SELENIUM-TELLURIUM COMMITTEE Recent advances in selenium physics A Symposium on solid and liquid state selenium physics... held at the Chemical society Proceedings London 1964 June Edited by Sven Walden and others Pergamon press, Oxford, 1965
Cav 7.1462

EUROPEAN SOCIETY FOR ANIMAL BLOOD GROUP RESEARCH Blood groups of animals European animal blood group conference 9th Proceedings Prague 1964 Aug 18-22 Edited by Josef Matousek Junk, The Hague, 1964
Surg 23.0044

EUROPEAN SOCIETY FOR BIOCHEMICAL PHARMACOLOGY Drugs affecting lipid metabolism International symposium on drugs affecting lipid metabolism 2nd Papers Milan 1965 Pt 1-2 Edited by D. Kritchevsky and others Progress in biochemical pharmacology, 2-3 illus 2 vols Karger, Basle; New York, 1967
Bioch 33.1250

EUROPEAN SOCIETY FOR BIOCHEMICAL PHARMACOLOGY Symposium on radiosensitizers and radioprotective drugs 1st Milan 1964 Edited by R. Paoletti and R. Vertus Progress in biochemical pharmacology, 1 Karger, Basel, 1965
Radioth 35.1943

EUROPEAN SOCIETY FOR THE STUDY OF DRUG TOXICITY 4th :meeting Proceedings Some factors affecting drug toxicity Cambridge 1964 Jul 2-3 Edited by D.G. Davey European society for the study of drug toxicity. Proceedings, 4 International congress series, 81 Excerpta medica, Amsterdam, 1964
Pha 16.0326

EUROPEAN SOCIETY FOR THE STUDY OF DRUG TOXICITY 6th :meeting Proceedings Experimental studies and clinical experience - the assessment of risk Stockholm 1965 Jun Edited by D.G. Davey European society for the study of drug toxicity.Proceedings,6 International congress series,97 Excerpta medica,Amsterdam,1965
Pha 16.0331

EUROPEAN SOCIETY FOR THE STUDY OF DRUG TOXICITY 8th :meeting Proceedings Neurotoxicity of drugs Prague 1966 Jun-Jul European society for the study of drug toxicity.Proceedings,8 International congress series,118 Excerpta medica, Amsterdam,1967
Pha 16.0332

EUROPEAN SOCIETY OF HAEMATOLOGY Congress of the European society of haematology 7th Proceedings London 1959 2,pt 1: papers Edited by E. Neumark and others Karger,Basle, 1960
Med 36.0132

EUROPEAN SOCIETY OF HAEMATOLOGY Europaischen gesellschaft fur hamatologie congress : colloquium uber aktuelle probleme des transfusionswesens und der immunhamatologie 5th Freiburg 1955 Sep 20-24 Edited by Herbert Begemann Springer-verlag,Berlin; Gottingen,1956
Med 36.0022

EUROPEAN SOCIETY OF HAEMATOLOGY European society of haematology.Congress 6th Transactions Copenhagen 1957 Vol 1: principal papers Edited by Aage Videbaek and others S.Karger,Basle;New York,1958
Med 36.0128

EUROPEAN SOCIETY OF HAEMATOLOGY European society of haematology congress 6th Transactions Copenhagen 1957 Vol 2: short papers - communications - kurzreferate Edited by Aage Videbaek and others 3 vols Karger,Basle;New York,1958
Med 36.0129

EUROPEAN SOCIETY OF HAEMATOLOGY European society of haematology congress 6th Transactions Copenhagen 1957 Vol 3: panel discussion on "Efficiency and limitations of anticoagulant therapy in arterial thrombosis Edited by T. Astrup Karger,Basle;New York,1958
Med 36.0130

EUROPEAN SOCIETY OF HAEMATOLOGY European society of haematology congress 7th Papers and proceedings London 1959 1-2 Edited by E. Neumark and others 2 vols Karger,Basle;New York,1960
Med 36.0131

EUROPEAN SOCIETY OF HAEMATOLOGY European society of haematology congress 8th Proceedings Vienna 1961 1-2 Edited by Adolf Scharf 2 vols Karger,Basle; New York,1962
Med 36.0134

EUROPEAN SOCIETY OF HAEMATOLOGY.CONGRESS 6th Transactions Copenhagen 1957 Vol 1: principal papers European society of haematology Edited by Aage Videbaek and others S.Karger,Basle;New York,1958
Med 36.0128

EUROPEAN SOCIETY OF HAEMATOLOGY CONGRESS 6th Transactions Copenhagen 1957 Vol 2: short papers - communications - kurzreferate European society of haematology Edited by Aage Videbaek and others 3 vols Karger, Basle;New York,1958
Med 36.0129

EUROPEAN SOCIETY OF HAEMATOLOGY CONGRESS 6th Transactions Copenhagen 1957 Vol 3: panel discussion on "Efficiency and limitations of anticoagulant therapy in arterial thrombosis European society of haematology Edited by T. Astrup Karger, Basle;New York,1958
Med 36.0130

EUROPEAN SOCIETY OF HAEMATOLOGY CONGRESS 7th Papers and proceedings London 1959 1-2 European society of haematology Edited by E. Neumark and others 2 vols Karger,Basle;New York,1960
Med 36.0131

EUROPEAN SOCIETY OF HAEMATOLOGY CONGRESS 8th Proceedings Vienna 1961 1-2 European society of haematology Edited by Adolf Scharf 2 vols Karger,Basle;New York, 1962
Med 36.0134

EUROPEAN SOCIETY OF HEMATOLOGY CONGRESS 7th Proceedings London 1959 2,pt.2: papers Edited by E. Neumark and others Karger,Basle;New York,1960
Med 36.0381

EUROPEAN SOUTHERN OBSERVATORY Auxiliary instrumentation for large telescopes :a conference Geneva 1972 May 2-5 Edited by S. Lanstsen and A. Reiz 525p European southern observatory,Geneva,1972
Obs 6.3599

EUROPEAN SOUTHERN OBSERVATORY Large telescope design :a conference Proceedings Geneva 1971 Mar 1-5 Edited by R.M. West 499p European southern observatory,Geneva, 1971
Obs 6.3601

EUROPEAN SPACE RESEARCH ORGANISATION Electromagnetic radiation in space E.S.R.O. summer school in space physics 3rd Proceedings Alpbach 1965 Jul 19-Aug 13 Edited by J.G. Emming Astrophysics and space science library, 9 viii,307p D.Reidel, Dordrecht,1967
A Math 4.1576

EUROPEAN SPACE RESEARCH ORGANISATION The Significance of space research for fundamental physics :a conference Interlaken 1969 Sep 4 Edited by A.F. Moore and U. Hardy ESRO SO-52 175p european space research organisation,Neuilly-sur-Seine,1971
Obs 6.3603

EUROPEAN SPACE RESEARCH ORGANISATION.PREPARATORY COMMISSION Introduction to solar terrestrial relations Summer school in space physics proceedings Alpbach 1963 Jul.15-Aug.10 Edited by J. Ortner and H. Maseland Astrophysics and space science library ix, 506p D.Reidel,Dordrecht-Holland,1965
A Math 4.1162

EUROPEAN SYMPOSIUM ON CHEMICAL ENGINEERING 1st
Papers Chemical reaction engineering Amsterdam 1957 May 7-9 Instituut van ingenieurs Nederlandse chemische vereniging Held during the 12th meeting of the European federation of chemical engineering International series of monographs on chemical engineering, 1 197p Pergamon,London,1957
Chem E 24.1900

EUROPEAN SYMPOSIUM ON CHEMICAL REACTION ENGINEERING 3rd Proceedings
Chemical reaction engineering Amsterdam 1964 Sep 15-17 Instituut van ingenieurs. Afdeling voor chemische techniek and Nederlandse chemische vereniging.Sectie voor chemische technologie Held under the auspices of the European federation of chemical engineering vi,326p Pergamon press,Oxford,1965 Supplement to "Chemical engineering science".
Chem E 24.1269

EUROPEAN SYMPOSIUM ON THE TIME-OF-FLIGHT MASS SPECTROMETRY 2nd Proceedings
Dynamic mass spectrometry Salford 1969 Jul 1 University of Salford Edited by D. Price and J.E. Williams Heyden,London,1970
Chem 18.2669

EUROPEAN SYMPOSIUM ON TIME-OF-FLIGHT MASS SPECTROMETRY 1st Proceedings
Salford 1967 Jul University of Salford Edited by D. Price and J.E. Williams Pergamon,Oxford,1969
Met 25.1706

EUROPEAN SYMPOSIUM ON TIME-OF-FLIGHT MASS SPECTROSCOPY 1ST Time-of-flight mass spectroscopy :based on the first European symposium Proceedings Salford 1967 Jul 3-5 University of Salford Edited by D. Price and J.E. Williams Pergamon press,Oxford,1969
Chem 18.2590

EUROPEES CONGRES TOEGEPASTE ELEKTRONMICROSCOPIE
Proceedings Ghent 1954 Apr 7-10 Edited by G. Van der Meersche Rijksuniversitet,Ghent,1954 Papers in English,Dutch,German,French and Spanish
Radioth 35.1583

EUSEC CONFERENCE ON ENGINEERING EDUCATION 1st
Proceedings London 1953 Jan 12-17 Conference of engineering societies of western Erope and the United States of America Institution of civil engineers,London,1953
Eng 41.1543

EUSEC CONFERENCE ON ENGINEERING EDUCATION 3rd
Proceedings Paris 1957 Sep Conference of engineering societies of western Europe and the United States of America Institution of mechanical engineers,London,1958
Eng 41.1546

EUSEC MEETING ON ENGINEERING EDUCATION AND TRAINING 4th Proceedings London 1962 Jan Conference of engineering societies of western Europe and the United States of America Institution of electrical engineers, London,1963
Eng 41.1547

EVALUATION OF WEAR TESTING :a symposium presented at the 71st annual meeting San Francisco,Calif. 1968 Jun 23-28 American society for testing and materials American society for testing and materials.Special technical publication, 446 ASTM,Philadelphia, Pa.,1969
Eng 41.8327

EVALUATION TESTS FOR STAINLESS STEELS
Symposium on evaluation tests for stainless steels Atlantic City 1949 Jun 30 American society for testing materials A.S.T.M.Special technical publication, 93 American society for testing materials, Philadelphia,Pa.,1950
Met 25.1967

EVANSTON 1957 Liquid scintillating counting :a conference Proceedings Edited by Carlos G. Bell and F.Newton Hayes Sponsored by the National science foundation Pergamon press,New York,1958 reprinted 1962
Inv Med 37.0014

EVANSTON,ILL 1955 Gas dynamics symposium on aerothermochemistry Proceedings Northwestern university and American rocket society Edited by Donald K. Fleming Supported by the United States.Air force illus 284p Northwestern university, Evanston,Ill.,1956
Chem E 24.1390

EVANSTON,ILL. 1924 The National symposium on colloid chemistry 2nd Papers presented Northwestern university Edited by Harry N. Holmes · Colloid symposium monograph, 2 Chemical catalog company,New York,1925
Bioch 33.1470

EVANSTON,ILL. 1961 Magnetohydrodynamics Biennial gas dynamics symposium 4th proceedings American rocket society and Northwestern university Edited by Ali Bulent Cambel and others Through the generosity of Government agencies and industrial sponsors Northwestern university press,Evanston,Ill., 1962
A Math 4.0623

EVANSTON,ILL. 1965 Dynamic stability of structures :an international conference Proceedings Edited by George Herrmann Sponsored by Northwestern university Pergamon press,London,1967
Eng 41.2531

The EVANSTON COLLOQUIUM.LECTURES ON MATHEMATICS DELIVERED FROM AUG 28 TO SEPT 9,1893...AT NORTHWESTERN UNIVERSITY... By Felix Klein vii,109p Macmillan,New York,1894 Lectures delivered before members of the Congress of mathematics held in connection with the World's Columbian exposition.Reported by A.Ziwet
WSM 43.1751

EVIDENCE FOR GRAVITATIONAL THEORIES
Verifiche delle teorie gravitazionali Scuola internazionale di fisica "Enrico Fermi" 20 corso Rendiconti Varenna 1961 Jun 19-Jul 1 Societa Italiana di fisica Edited by G. Moller Bibliog.,Illus. 264p 24cm Academic press,New York,1962
A Math 4.1555

EVOLUTION Behaviour and evolution : conference 2nd Proceedings Princeton,N. J. 1956 Apr 30-May 5 American psychological association Edited by Anne Roe and George Gaylord Simpson Held in collaboration with the Society for the study of evolution 24cm New Haven,1958
Philos 1.1827

EVOLUTION Cytology,genetics and evolution : university of Pennsylvania bicentennial conference By M. Demerec and others University of Pennsylvania.Bicentennial conference University of Pennsylvania press, Philadelphia,Pa.,1941
Bot 42.6465

EVOLUTION Evoluzione e genetica :colloquio internazionale Relazioni e discussione Rome 1959 Apr 8-11 Accademia nazionale dei Lincei Problemi attuali di scienza et di cultura.Quaderno N., 47 Rome,1960 Text in English and Italian.Summaries in English
Bal 39.0927

EVOLUTION Oxford 1952 Jul Society for experimental biology and Genetical society Society for experimental biology.Symposia, 7 Cambridge university press,Cambridge,1953
Phys 20.0957

EVOLUTION symposium Oxford 1952 Jul Society for experimental biology Genetical society Society for experimental biology. Symposia, 7 Cambridge university press, Cambridge,1953
Gen 34.1258

EVOLUTION :a symposium Papers Oxford 1952 Jul Society for experimental biology and Genetical society Society for experimental biology.Symposia, 7 Cambridge university press,Cambridge,1953
Psy 31.0570

EVOLUTION 7th symposium Oxford 1952 Jul. Society for experimental biology Edited by R. Brown and J.F. Danielli Supported by the British council Society for experimental biology.Symposia, 7 xix,448p 25cm Cambridge university press,Cambridge, 1953
Math 3.0773

EVOLUTION:ITS SCIENCE AND DOCTRINE :a symposium presented to the Royal society of Canada 1959 Edited by Thomas W.M. Cameron Royal society of Canada."Studia varia"series, 4 University of Toronto press,Toronto,1960 Title also in French L'evolution:la science et la doctrine
Bal 39.0915

EVOLUTION:ITS SCIENCE AND DOCTRINE symposium presented to the Royal society of Canada in 1959 Edited by Thomas W.M. Cameron Royal society of Canada.Studia varia series, 4 University of Toronto press,Toronto,1960
Gen 34.1270

EVOLUTION AND ENVIRONMENT :a symposium New Haven,Conn. 1966 Oct 26-28 Edited by Ellen T. Drake Silliman foundation.Lectures Illus,maps xvi,478p Yale university press, New Haven;London,1968 Presented on the occasion of the 100th anniversary of the foundation of the Peabody museum of natural history
Bot 42.0317

The EVOLUTION OF LIVING ORGANISMS :a symposium to mark the centenary of Darwin's 'Origin of species' and of the Royal society of Victoria Melbourne 1959 Dec Edited by G.W. Leeper Melbourne university press, Melbourne,1962
Bal 39.0923

The EVOLUTION OF LIVING ORGANISMS :a symposium to mark the centenary of Darwin's 'Origin of species' and of the Royal society of Victoria Melbourne 1959 Dec 8-11 Edited by G.W. Leiper Melbourne university press,Melbourne,1962
Gen 34.1277

The EVOLUTION OF LIVING ORGANISMS :a symposium to mark the centenary of Darwin's 'Origin of species'and of the Royal society of Victoria Melbourne 1959 Dec Edited by G.W. Leeper illus,maps xi,459p 25cm Melbourne university press,Melbourne,1962
Sco 14.0621

The EVOLUTION OF LIVING ORGANISMS :a symposium to mark the centenary of Darwin's'Origin of species'and of the Royal society of Victoria... Melbourne 1959 Dec 8-11 Edited by G.W. Leeper illus vii, 459p Melbourne university press,Melbourne, 1962
Bot 42.0991

EVOLUTION OF NERVOUS CONTROL from primitive organisms to man:a symposium New York 1956 Dec 29-30 American association for the advancement of science.Section on medical sciences Edited by Allan D. Bass American association for the advancement of science. Publications, 52 Washington,D.C.,1954
Phys 20.2015

EVOLUTION OF NERVOUS CONTROL FROM PRIMITIVE ORGANISMS TO MAN :a symposium New York 1956 Dec 29-30 Edited by Allan O. Bass Organized by the American association for the advancement of science.Section on medical sciences American association for the advancement of science.Publications, 52 American association for the advancement of science,Washington,D.C.,1959
An 32.4357

EVOLUTION OF NERVOUS CONTROL FROM PRIMITIVE ORGANISMS TO MAN :a symposium Proceedings New York 1956 Dec 29-30 American association for the advancement of science. Section on medical sciences Edited by Allan D. Bass American association for the advancement of science.Publication, 52 Washington,D.C.,1959 Arranged by Bernard B.Brodie and Allan D.Bass
Bal 39.1470

EVOLUTION OF THE ATHEROSCLEROTIC PLAQUE Presentations :an international symposium Chicago,Ill. 1963 Mar 28-29 Chicago heart association and American heart association.Council on arteriosclerosis Edited by Richard J. Jones University of Chicago press,Chicago,Ill.;London,1963
Path 30.2350

EVOLUTION OF THE FOREBRAIN:PHYLOGENESIS AND ONTOLOGY OF THE FOREBRAIN :a symposium Lectures Frankfurt 1965 Aug 15-19 and Sprendlingen World federation of neurology Max Planck gesellschaft zur forderung der wissenschaften Edited by R. Hassler and H. Stephan G.Thieme,Stuttgart,1966
Psy 31.3370

EVOLUTION STELLAIRE AVANT LA SEQUENCE PRINCIPALE Liege 1969 Jun 30-Jul 2 Societe des sciences de Liege.Memoires.5s, 19,no 1 377p Liege,1970 Papers in English and French
FA 15.0685

EVOLUTIONARY ASPECTS OF ANIMAL COMMUNICATION; IMPRINTING AND EARLY LEARNING :a symposium Proceedings London 1961 Nov 8-9 Association for the study of animal behaviour Zoological society of London Zoological society of London.Symposia, 8 Zoological society of London,London,1962
An 32.4384

EVOLUTIONARY BIOCHEMISTRY International congress of biochemistry 5th Proceedings Moscow 1961 Aug 10-16 International union of biochemistry I.U.B.symposium series, 23 Pergamon;PWN-Polish scientific publishers,New York;Warsaw,1963
Bioch 33.1332

EVOLUZIONE DELLE STELLE Scuola internazionale di fisica "Enrico Fermi" 28 corso rendiconti Varenna 1962 Aug 20-Sep. 1 Societa Italiana di fisica Edited by L. Gratton 488p Academic press,New York; London,1963
Obs 6.0685

EVOLUZIONE DELLE STELLE Scuola internazionale di fisica'Enrico Fermi' 28 corso Rendiconti Varenna 1962 Aug 20-Sep 1 Societa Italiana di fisica Edited by L. Gratton Academic press,New York;London,1963
A Math 4.1156

EVOLUZIONE E GENETICA :colloquio internazionale Relazioni e discussione Rome 1959 Apr 8-11 Accademia nazionale dei Lincei Problemi attuali di scienza et di cultura.Quaderno N., 47 Rome,1960 Text in English and Italian.Summaries in English
Bal 39.0927

EVOLVING GENES AND PROTEINS New Brunswick,N.J. 1964 Sep 17-18 Edited by Vernon Bryson and Henry J. Vogel With support from the National science foundation xvii,617p Academic press,New York;London,1965
Bot 42.0995

EVOLVING GENES AND PROTEINS :a symposium Rutgers,N.J. 1964 Sep 17-18 Institute of microbiology of Rutgers Edited by V. Bryson and H.J. Vogel With support from the National science foundation Academic press, New York;London,1965
Gen 34.0665

EVOLVING GENES AND PROTEINS :a symposium Proceedings New Brunswick,N.J. 1964 Sep 17-18 Edited by Vernon Bryson and Henry J. Vogel Held at Rutgers university.Institute of microbiology Academic press,New York; London,1965
Bioch 33.0569

EXCERPTA MEDICA.INTERNATIONAL CONGRESS SERIES The Foeto-placental unit an international symposium Proceedings Milan 1968 Sep 4-6 Edited by A. Pecile and C. Finzi Excerpta medica,Amsterdam,1969
Inv Med 37.0260

EXCERPTA MEDICA.INTERNATIONAL CONGRESS SERIES, 25 Congres de la societe europeenne d'opthalmologie 1e Resumees des rapports et des communications Athens 1959 Apr 18-22 European opthalmological society Excerpta medica,Amsterdam;New York,1960 Papers in English,French and German
PGMS 29.0682

EXCERPTA MEDICA.INTERNATIONAL CONGRESS SERIES, 78 Congres international de nephrologie 2 Comptes rendus Prague 1963 Aug 26-30 Societe internationale de nephrologie Societe tchecoslovaque de medecine Edited by J. Vostal and G. Richet Excerpta medica, Amsterdam,1964
Inv Med 37.0220

EXCERPTA MEDICA.INTERNATIONAL CONGRESS SERIES, 83 International congress of endocrinology 2nd Proceedings London 1964 Aug 17-22 Pt 1-2 2 vols Excerpta medica,Amsterdam,1966
Inv Med 37.0114

EXCERPTA MEDICA.INTERNATIONAL CONGRESS SERIES, 101 Androgens in normal and pathological conditions Steroid hormones :a symposium 2nd Proceedings Ghent 1965 Jun 17-19 Rijkauniversiteit te Gent Societe belge d'endrocinologie Edited by A. Vermeulen and D. Exley Excerpta medica,Amsterdam,1966
Inv Med 37.0276

EXCERPTA MEDICA.INTERNATIONAL CONGRESS SERIES, 103 Renal failure and replacement of renal function European dialysis and transport association :a conference 2nd Proceedings Newcastle-upon-Tyne 1965 Vol 2 European dialysis and transplant association Edited by David Kerr Excerpta medica,Amsterdam,1966
Inv Med 37.0130

EXCERPTA MEDICA.INTERNATIONAL CONGRESS SERIES, 112 Endocrinology Pan-American congress of endocrinology 6th Proceedings Mexico City 1965 Oct 10-13 Edited by C. Gual Excerpta medica,Amsterdam,1966
Inv Med 37.0096

EXCERPTA MEDICA.INTERNATIONAL CONGRESS SERIES, 132 Hormonal steroids International congress on hormonal steroids 2nd Proceedings Milan 1966 May 23-28 Edited by L. Martini and others Excerpta medica, Amsterdam,1967
Inv Med 37.0116

EXCERPTA MEDICA.INTERNATIONAL CONGRESS SERIES, 133 Fertility and sterility World congress on fertility and sterility 5th Proceedings Stockholm 1966 Jun 16-22 International fertility association Swedish society of obstetrics and gynaecology Edited by Bjorn Westin and Nils Wiqvist Excerpta medica,Amsterdam,1967
Inv Med 37.0230

EXCERPTA MEDICA.INTERNATIONAL CONGRESS SERIES, 136 Electrotherapeutic sleep and electroanaesthesia :international symposium 1st Proceedings Graz 1966 Sep 12-17 Edited by F.M. Wageneder and others Excerpta medica,Amsterdam,1967
Inv Med 37.0221

EXCERPTA MEDICA.INTERNATIONAL CONGRESS SERIES, 159 Parathyroid hormone and thyrocalcitonin (calcitonin) Parathyroid conference 3rd Proceedings Montreal 1967 Oct 16-20 Edited by Roy V. Talmadge and others Excerpta medica,Amsterdam,1968
Inv Med 37.0211

EXCERPTA MEDICA.INTERNATIONAL CONGRESS SERIES, 161 Protein and polypeptide hormones International symposium on protein and polypeptide hormones Proceedings Liege 1968 May 19-25 Edited by M. Margoulies Excerpta medica,Amsterdam,1968
Inv Med 37.0117

EXCERPTA MEDICA.INTERNATIONAL CONGRESS SERIES, 166 International congress on cell biology 12th Summaries of reports and communications Brussels 1968 Aug 25-31 Edited by P. Dustin Excerpta medica, Amsterdam,1968
Inv Med 37.0250

EXCERPTA MEDICA.INTERNATIONAL CONGRESS SERIES, 184 Progress in endocrinology International congress of endocrinology Proceedings Mexico City 1968 Jun 30-Jul 5 Edited by Carlos Gual and F.J.C. Ebling Excerpta medica,Amsterdam,1969
Inv Med 37.0095

EXCERPTA MEDICA.INTERNATIONAL CONGRESS SERIES, 210 International congress on hormonal steroids 3rd Abstracts of papers Hamburg 1970 Sep 7-12 Edited by V.H.T. James Excerpta medica,Amsterdam,1970
Inv Med 37.0115

EXCERPTA MEDICA.INTERNATIONAL CONGRESS SERIES, 212 Electrotherapeutic sleep and electroanesthesia :international symposium 2nd Proceedings Graz 1969 Sep 8-13 Edited by F.M. Wageneder and others Excerpta medica,Amsterdam,1970
Inv Med 37.0222

EXCERPTA MEDICA,INTERNATIONAL CONGRESS SERIES, 200 Progress in anaesthesiology World congress of anaesthesiologists 4th Proceedings London 1968 Sep 9-13 By T.B. Boulton and others Excerpta Medica Foundation,Amsterdam,1910
PGMS 29.0559

EXCHANGE OF GENETIC MATERIAL :mechanisms and consequences.A symposium Papers Cold Spring Harbor 1958 Cold Spring Harbor biological laboratory Cold Spring Harbor symposia on quantitative biology, 23 Long Island biological association,Cold Spring Harbor,1958
Bioch 33.1279

EXCITATION PHENOMENA Cold Spring Harbor symposia on quantitative biology Papers Cold Spring Harbor 1936 Vol 4 Cold Spring Harbor biological laboratory Long Island biological association,Cold Spring Harbor,1936 Later referred to in vol.9 as "Excitation phenomena"
Bioch 33.1260

EXCITATIONS IN SOLIDS Elementary excitations in solids Cortina lectures and four lectures from the Conference on localized excitations Proceedings Milan 1966 Jul 25-26 Edited by A.A. Maradudin and G.F. Nardelli Bibliog 526p 20cm Plenum press,New York,1969
Cav 7.2677

EXCITONS,MAGNONS AND PHONONS IN MOLECULAR CRYSTALS :an international symposium Proceedings Beirut 1968 Jun 15-18 Edited by A.B. Zahlan Held at the American university of Beirut Cambridge university press,Cambridge,1968
Radioth 35.1689

EXERPTA MEDICA.INTERNATIONAL CONGRESS SERIES, 83 International congress of endocrinology 2nd Proceedings London 1964 Aug 17-22 Pt 1-2 Edited by S. Taylor 2 vols Excerpta medica foundation,London,1965
PGMS 29.0157

EXETER 1955 Recent work on naturally occuring nitrogen heterocyclic compounds :a symposium Report Chemical society Edited by K. Schofield Chemical society.Special publication,3 Chemical society,London,1955
Chem 18.1329

EXETER 1955 Recent work on naturally occurring nitrogen heterocyclic compounds :a symposium Report Chemical society Edited by K. Schofield Chemical society.Special publication, 3 London,1955
Bioch 33.2228

EXETER 1961 Some aspects of the Variscan fold belt Inter-university geological congress 9th ten lectures Edited by Kenneth Coe illus. map Manchester university press,Manchester,1962
Geol 8.1656

EXETER 1962 The International conference on the physics of semiconductors Report Institute of physics and Physical society Edited by A.C. Stickland Under the auspices of the International union of pure and applied physics 909p Institute of physics,London, 1962
Cav 7.0750

EXETER 1964 Limitations of detection in spectrochemical analysis :a conference Papers Organised by the Institute of physics and the physical society.Spectroscopy group Hilger and Watts,London,1964
Radioth 35.1657

EXHIBITION OF SOUTH AFRICAN WILD FLOWERS London 1933 Oct Under the auspices of the Royal horticultural society illus. London, 1934
BG 38.2129

EXISTENCE THEOREMS The Princeton colloquium Princeton 1909 Part 1: fundamental existence theorems By Gilbert Ames Bliss American mathematical society.Colloquium publications, 3 ii,107p 23cm American mathematical society,New York,1913 Reprinted in 1934
P. Math 2.0822

The EXOCRINE PANCREAS :Ciba foundation symposium Ciba foundation Edited by A.V.S. De Reuck and M.P. Cameron Churchill, London,1962
Phys 20.1340

EXPERIMENTAL DIABETES AND ITS RELATION TO THE CLINICAL DISEASE :a symposium Leiden 1952 Jul 14-18 Council for international organizations of medical sciences Edited by J.F. Delafresnaye and G.Howard Smith Blackwell,Oxford,1954
Bioch 33.0448

EXPERIMENTAL HEPATOMAS Biological and biochemical evaluation of malignancy in experimental hepatomas The U.S.-Japan joint conference on biological and biochemical evaluation of malignancy in experimental hepatomas Proceedings Kyoto 1965 Nov 4-5 Held under the auspices of the United States-Japan scientific cooperation program G.A.N.N. Monograph, 1 Japanese cancer association; Japanese foundation for cancer research,Tokyo, 1966
Bioch 33.0952

EXPERIMENTAL HYPERTENSION being the results of a conference Results New York 1945 Feb 9-10 By William Goldring and others New York academy of sciences.Section of biology New York academy of sciences.Special publications, 3 New York academy of sciences,New York,1946
Psy 28.0096

EXPERIMENTAL MECHANICS International congress on experimental mechanics 1st Proceedings New York 1961 Nov 1-3 Society for experimental stress analysis Edited by B.E. Rossi Pergamon press,London, 1963
Eng 41.2920

EXPERIMENTAL MECHANICS International congress on experimental mechanics 2nd Proceedings Washington,D.C. 1965 Sep 28-Oct 1 Society for experimental stress analysis Edited by B.E. Rossi Society for experimental stress analysis,Westport,Conn., 1966
Eng 41.2921

EXPERIMENTAL MEDICINE AND SURGERY IN PRIMATES a conference New York 1967 Sep 27-30 New York academy of sciences New York academy of sciences.Annals, 162,art.1 New York academy of sciences,New York,1969
An 32.5171

EXPERIMENTAL MEDICINE AND SURGERY IN PRIMATES : a conference Papers New York 1967 Sep 27-30 New York academy of sciences Edited by Edward I. Goldsmith and J. Moor-Jankowski New York academy of sciences.Annals, 162,art 1 New York academy of sciences,New York,1969
Path 30.2445

EXPERIMENTAL PEDOLOGY Easter school in agricultural science 11th Proceedings Nottingham 1964 University of Nottingham Edited by E.G. Hallsworth and D.V. Crawford xi,413p Butterworths,London,1965
Bot 42.2090

EXPERIMENTAL PSYCHOPATHOLOGY :forty-fifth annual meeting Proceedings New York 1955 Jun American psychopathological association Edited by Paul H. Hoch and Joseph Zubin Grune and Stratton,New York; London,1957
Psy 28.0197

EXPERIMENTAL STUDIES AND CLINICAL EXPERIENCE - THE ASSESSMENT OF RISK European society for the study of drug toxicity :meeting 6th Proceedings Stockholm 1965 Jun Edited by D.G. Davey European society for the study of drug toxicity.Proceedings,6 International congress series,97 Excerpta medica,Amsterdam, 1965
Pha 16.0331

EXPERIMENTAL TECHNIQUES IN SHOCK AND VIBRATION a colloquium Papers New York 1962 Nov 27 Edited by W.J. Worley Sponsored by the American society of mechanical engineers American society of mechanical engineers,New York,1962
Eng 41.6354

EXPERIMENTS IN GRADUATE TRAINING :a symposium Proceedings London 1962 Mar 22-23 Institution of mechanical engineers Institution of mechanical engineers,London, 1962
Eng 41.1499

EXPERT WORKING GROUP ON SOCIAL ASPECTS OF ECONOMIC DEVELOPMENT IN LATIN AMERICA Social aspects of economic development in Latin America Mexico City 1960 Dec 12-21 Vol 1: papers submitted to the expert working group... Edited by Egbert de Vries and Jose Medina Echavarria Technology and society series Unesco,Paris,1963
Geog 13.5096

EXPLODING WIRES :a conference Papers Boston,Mass. 1959 Apr 2-3 Air force Cambridge research center.Geophysics research directorate Edited by William G. Chace and Howard K. Moore With the cooperation of Lowell technological institute research foundation Plenum press,New York,1959
Chem 18.0191

The EXPLOITATION OF NATURAL ANIMAL POPULATIONS a symposium Durham 1960 Mar 28-31 British ecological society Edited by E.D. Le Cren and M.W. Holdgate British ecological society.Symposia, 2 Oxford,1962
Bal 39.1665

The EXPLOITATION OF NATURAL ANIMAL POPULATIONS 2 a symposium Durham 1960 Mar 28-31 British ecological society Edited by E.David Le Cren and Martin Wyatt Holdgate tables, diagrs,maps x,399p Blackwell scientific publications,Oxford,1962
Sco 14.0721

EXPLORATION IN GROUP RELATIONS :a conference Report Leicester 1957 Sep By E.L. Trist and C. Sofer University of Leicester Tavistock institute of human relations Leicester university press,Leicester,1959
Eng 41.0706

An EXPOSITION OF ADAPTIVE CONTROL :a symposium Proceedings London 1961 Edited by J.H. Westcott Held at the Imperial college of science and technology Pergamon press,London, 1962
Eng 41.5884

EXPOSITION UNIVERSELLE DE 1900 Congres international d'aviculture et de peche Memoires et comptes-rendus Paris 1900 France.Ministere de l'industrie,des postes et des telegraphes Edited by Joseph Perard and Maire 24cm Challamel,Paris,1901
Bal 44.1241

EXPOSITION UNIVERSELLE ET INTERNATIONALE DE SAN FRANCISCO :la science francaise San Francisco France.Ministere de l'instruction publique et des beaux-arts 2 vols Paris, 1915
Philos 1.0013

EXTENSION OF GRAVITY ANOMALIES TO UNSURVEYED AREAS :a symposium Papers Gravity anomalies:unsurveyed areas Columbus,Ohio 1964 Nov 18-20 Ohio state university and American geophysical union Edited by Hyman Orlin Sponsored by the International union of geodesy and geophysics Geophysical monograph, 9 National research council. Publication,1357 American geophysical union, Washington,D.C.,1966
Geod 9.0031

EXTENSION THEORY Contributions to extension theory :a symposium Proceedings Berlin 1967 Aug 14-19 Edited by Jungen Flaschmeyer 279p 24cm VEB Deutscher verlag der wissenschaften,Berlin,1969
P Math 2.3743

EXTERNAL GALAXIES AND QUASI-STELLAR OBJECTS Upsala 1970 Aug 10-14 International astronomical union Edited by D.S. Evans International astronomical union.Symposium, 44 549p Reidel,Dordrecht,1972
TA 15.0658

EXTERNALLY PRESSURIZED BEARINGS :a joint conference Proceedings London 1971 Nov 17-18 Institution of mechanical engineers.Tribology group Institution of production engineers IME,London,1972
Eng 41.8651

EXTRA-GALACTIC RESEARCH Problems of extra-galactic research :a symposium Santa Barbara, Calif. 1961 Aug.10-12 International astronomical union Edited by G.C. McVittie International astronomical union.Symposium, 15 450p Macmillan,New York,1962
A Math 4.1154

EXTRA-GALACTIC RESEARCH Symposium on problems of extra-galactic research Proceedings Santa Barbara,Calif. 1961 Aug 10-12 International astronomical union Edited by G.C. McVittie International astronomical union.Symposium, 15 McMillan, New York,1962
TA 15.0035

EXTRA-GALACTIC RESEARCH The Symposium on problems of extra-galactic research 15th Santa Barbara,Calif. 1961 Aug.10-12 International astronomical union Edited by G. C. McVittie International astronomical union. Symposium, 15 450p Macmillan,New York, 1962
Obs 6.1124

EXTRACTION AND REFINING OF THE RARER METALS A Symposium on extraction metallurgy of some of the less common metals London 1956 Mar 22-23 Institution of mining and metallurgy Institution of mining and metallurgy,London, 1957
Met 25.0500

EXTRACTIVE AND PHYSICAL METALLURGY OF PLUTINIUM based on a symposium San Francisco 1959 Feb 16-17 American institute of mining, metallurgical and petroleum engineers Edited by W.D. Wilkinson Bibliog Interscience,New York;London,1960 Introduction and annotated bibliography by the editor
Met 25.2364

EXTRACTIVE AND PHYSICAL METALLURGY OF PLUTONIUM AND ITS ALLOYS based on a symposium Plutonium and its alloys San Francisco 1959 Feb 16-17 American institute of mining, metallurgical and petroleum engineers.Nuclear metallurgy committee Edited by W.D. Wilkinson Interscience,New York;London,1960 With an introduction and an annotated bibliography by the editor
Met 25.1568

EXTRACTIVE METALLURGY Advances in extractive metallurgy :a symposium Proceedings London 1967 Apr 17-20 Institution of mining and metallurgy Institution of mining and metallurgy,London,1968
Met 25.0494

EXTRACTIVE METALLURGY IN AUSTRALIA:NON-FERROUS METALLURGY Empire mining and metallurgical congress 5th Publications Melbourne 1953 Apr 14-Jun 12 4B Australasian institute of mining and metallurgy Edited by Frank A. Green and others Office of the Congress,Melbourne,1953
Met 25.0498

EXTRACTIVE METALLURGY OF ALUMINUM 1st : international symposium New York 1962 Feb 18-22 Vol 1-2: alumina;aluminum American institute of mining,metallurgical and petroleum engineers.Extractive metallurgy division Edited by Gary Gerard and P.T. Stroup 2 vols John Wiley,New York,1963
Met 25.0578

EXTRACTIVE METALLURGY OF COPPER,NICKEL AND COBALT based on an international symposium New York 1960 Feb 15-18 American institute of mining,metallurgical and petroleum engineers.Extractive metallurgy division Edited by Paul Quenau Bibliog Interscience, New York;London,1961
Met 25.0510

EXTRAPYRAMIDAL SYSTEM The So-called extrapyramidal system :symposium from the sixteenth congress of Scandinavian neurobiologists Oslo 1962 Scandinavian neurological society and Norwegian neurological association Edited by Sigvald Refsum and others Universitetsforlaget,Oslo, 1963
An 32.4655

EXTRATERRESTRIAL CIVILIZATIONS All-Union conference on extraterrestrial civilizations and interstellar communications 1st Proceedings Byurakan 1964 May 20-23 Byurakanskaya astrofizicheskaya observatoriya Edited by S.M. Tovmasyan Translated by Z. Lerman from the Russian Israel program for scientific translations,Jerusalem,1967
TA 15.0322

EYE PERFORMANCE Performance of the eye at low luminances :colloquium Proceedings Delft 1965 Institute for perception RVO-TNO N.V.Optische industrie "De oude Delft". Scientific fund Scientiae serviens Edited by M.A. Bouman and J.J. Vos International congress series, 125 Excerpta medica, Amsterdam,1966
Psy 31.2862

FABRI J. and HAWTHORNE W.R. ed. Selected combustion problems Combustion colloquium Cambridge 1953 Dec 7-11 Agard Butterworths,London,1954
A Math 4.0969

FABRICS Studies in modern fabrics :a conference Papers London 1970 May 26-29 Textile institute Textile institute.Annual conference, 55 Textile institute,Manchester, 1970
Eng 41.8287

FACHKONFERENCE RADIOASTRONOMIE :jahrestagung der deutschen akademie der wissenschaften Berlin 1955 Mar 28- Apr 2 Deutsche akademie der wissenschaften zu Berlin Deutsche akademie der wissenschaften zu Berlin. Abhandlungen.Klasse fur mathematik,physik und technik, 3 135p Akademie-verlag,Berlin, 1958
Obs 6.2016

FACIAL STRUCTURE OF COMPACT CONVEX SETS NATO advanced study institute on facial structure of compact sets and applications Swansea 1972 Jul 2-25 North Atlantic treaty organization Held at University college of Swansea 147p 21cm University college of Swansea,Swansea,1972
P Math 2.4281

FACTOR ANALYSIS Symposium on psychological factor analysis Uppsala 1953 Mar 17-19 Universitet i Uppsala.Institute of statistics Edited by P. Whittle Nordisk psykologi. Monograph series, 3 91p 23cm Almqvist and Wiksell,Stockholm,1953
Math 3.0777

FACTORS AFFECTING THE ABSORBTION AND DISTRIBUTION OF DRUGS Absorption and distribution of drugs :based on a symposium London 1963 Association of medical advisers in the pharmaceutical industry Edited by T.B. Binns E.and S. Livingstone,Edinburgh;London,1964
PGMS 29.0370

FACTORS DETERMINING HUMAN BEHAVIOR Harvard tercentenary conference of arts and sciences Papers Cambridge,Mass. 1936 Aug 31-Sep 12 Harvard university Harvard tercentenary publications Harvard university press, Cambridge,Mass.,1937
Psy 31.0770

FACTORS DETERMINING THE BEHAVIOR OF PLANT PATHOGENS IN SOIL :an international symposium Ecology of soil-borne plant pathogens.Prelude to biological control Berkeley,Calif. 1963 Apr 7-13 Edited by Kenneth F. Baker and William C. Snyder Sponsored by the National research council illus. Murray,London,1965
Mol 45.0575

FACTORS IN THE TRANSFER OF TECHNOLOGY Massachusetts institute of technology conference on the human factor in the transfer of technology Cambridge,Mass. 1966 May 18-20 Edited by W.H. Gruber and D.G. Marquis M.I.T.press,Cambridge,Mass.,1969
Eng 41.1687

FACULTE DES SCIENCES DE PARIS ET DE PROVINCE Low energy nuclear interactions and nuclear structure International conference on nuclear physics Proceedings Paris 1958 Jul 7-12 Under the auspices of International union of pure and applied physics Crosby Lockwood,London,1959
Cav 7.1759

FAILURE AND DEFECTS OF BRIDGES AND STRUCTURES : a symposium Proceedings Tokyo 1957 Sep 15 Japan society of civil engineers Architectural institute of Japan Japan society for the promotion of science,Tokyo, 1959
Eng 41.3015

The FAILURE OF METALS BY FATIGUE :a symposium Proceedings Melbourne 1946 Dec 2-6 University of Melbourne.Faculty of engineering Melbourne university press,Melbourne,1947
Met 25.0909

FAIRBANKS 1964 Review of research on military problems in cold regions The Alaskan science conference 15th Papers By Charles R. Kolb and Fritz M.G. Holmstrom University of Alaska Under the auspices of the American association for the advancement of science 1964 Privately printed
Sco 14.0155

FAIRCHILD TROPICAL GARDENS Biology of the hydra and of some other coelenterates The Physiology and ultrastructure of hydra and of some other coelenterates :a symposium Coral Gables,Fla. 1961 Mar 29-31 Edited by Howard M. Lenhoff and F.Farnsworth Loomis University of Miami press,Coral Gables,Fla., 1961
Bal 39.2183

FALMOUTH,MASS. 1958 Sulfur in proteins :a symposium Proceedings Edited by Reinhold Benesch and others Academic press,New York; London,1959
Radioth 35.0075

FALMOUTH,MASS. 1958 Sulphur in proteins :a symposium Proceedings Edited by Reinhold Benesch and others Academic press,New York; London,1959
Bioch 33.0555

The FAMILY AND ITS FUTURE :a Ciba foundation symposium Proceedings London 1970 Mar 10-12 Edited by Katherine Elliott Churchill,London,1970
Inv Med 37.0075

The FAMILY HEALTH MAINTENANCE DEMONSTRATION : a controlled,long term investigation of family health;proceedings of a round table at the 1953 annual conference of the fund Milbank memorial fund Milbank memorial fund, New York,1954
HE 27.0128

FARADAY SOCIETY A Discussion of the Faraday society 5th Papers Bristol 1949 Apr 12-14 Faraday society.Discussions, 5,1949 Butterworth,London,1959
Cav 7.0760

FARADAY SOCIETY Bonding in metallo-organic compounds Cambridge 1969 Mar 25-27 Faraday society.Discussions, 47 208p 20cm Faraday society,London,1969
Cav 7.2717

FARADAY SOCIETY Catalysis ,with special reference to newer theories of chemical action a general discussion London 1921 Sep 28 Pt 1-2 London,1922 Reprinted from the 'Transactions' of the society,vol 17,pt 3. Chairman of the discussion A.W.Porter
Bot 42.6477

FARADAY SOCIETY Crystal growth :a general discussion Bristol 1949 Apr 12-14 Faraday society.Discussions, 5 Gurney and Jackson,London,1949
Eng 41.4424

FARADAY SOCIETY Crystal structure and chemical constitution :a general discussion London 1929 May 14 London,1929
Bot 42.6478

FARADAY SOCIETY Electrode processes :a general discussion Manchester 1947 Apr 9 Faraday society.Discussions, 1 Butterworths,London,1961
Met 25.2396

FARADAY SOCIETY Electrode reactions of organic compounds Newcastle-upon-Tyne 1968 Faraday society.Discussions, 45 graphs, photos,diagrms 281p 26cm Faraday society, London,1968
Cav 7.3124

FARADAY SOCIETY Energy transfer with special reference to biological systems :a general discussion Nottingham 1959 Apr 14-16 Faraday society.Discussions, 27 Faraday society,London,1959
Radioth 35.1406

FARADAY SOCIETY Energy transfer with special reference to biological systems :a general discussion Nottingham 1959 Apr 14-16 Faraday society.Discussions, 27 The Faraday society,London,1960
Radioth 35.0249

FARADAY SOCIETY Energy transfer with special reference to biological systems :a general discussion Papers Nottingham 1959 Apr 14-16 Faraday society.Discussions, 27 Aberdeen university press,Aberdeen,1960
Bioch 33.0314

FARADAY SOCIETY Flow properties of blood and other biological systems :an informal discussion... Proceedings Oxford 1959 Sep 23-24 Edited by A.L. Copley and G. Stainsby Pergamon press,Oxford,1960
An 32.3471

FARADAY SOCIETY Fundamental processes in radiation chemistry :a general discussion Notre Dame,Ind. 1963 Sep 2-4 Faraday society.Discussions, 36 Butterworths,London, 1963
Radioth 35.1772

FARADAY SOCIETY Homogeneous catalysis with special reference to hydrogenation and oxidation Liverpool 1968 Faraday society. Discussions, 46 diagrms,graphs 227p 24cm Faraday society,London,1968
Cav 7.3125

FARADAY SOCIETY Lipo-proteins :a general discussion Papers Birmingham 1949 Aug 29-31 Faraday society.Discussions, 6 Aberdeen,1949
Bioch 33.0508

FARADAY SOCIETY Liquid crystals and anisotropic malts :a general discussion London 1933 Apr 24-25 held in the Royal institution Faraday society.Discussions, 58 Aberdeen,1933
Bioch 33.1769

FARADAY SOCIETY Mechanism and chemical kinetics of organic reactions in liquid systems :a general discussion London 1941 Sep Gurney and Jackson,London;Edinburgh,1941 Reprint from the society's 'Transactions', vol 37,pt 12
Bot 42.6479

FARADAY SOCIETY Membrane phenomena :a general discussion Nottingham 1956 Apr 10-12 Faraday society.Discussions, 21 Aberdeen university press,Aberdeen,1956
Bioch 33.0947

FARADAY SOCIETY Motions in molecular crystals Oxford 1969 Sep 16-18 Faraday society.Discussions, 48 Bibliog 224p 20cm Faraday society,London,1969
Cav 7.2718

FARADAY SOCIETY Optical studies of adsorbed layers at interfaces London 1970 Dec 14-15 Faraday society.Symposia,1970,4 212p 26cm Faraday society,London,1971
Chem 18.2880

FARADAY SOCIETY Spectroscopy and molecular structure,and optical methods of investigating cell structure :a general discussion Cambridge 1950 Sep 25-28 Faraday society. Discussions, 9 Aberdeen,1951
Bal 39.0230

FARADAY SOCIETY Spectroscopy and molecular structure and optical methods of investigating cell structure :general discussion Cambridge 1950 Sep 25-28 Faraday society.Discussions, 9 Faraday society,Aberdeen,1951
Radioth 35.0453

FARADAY SOCIETY Surface chemistry :a discussion at a joint meeting of the Societe de chimie physique and the Faraday society Papers Bordeaux 1947 Oct 5-9 Illus Butterworths scientific publications,London, 1949 Papers in English and FRENCH
Radioth 35.0342

FARADAY SOCIETY The Cell nucleus :an informal meeting Proceedings Cambridge 1959 Aug 31-Sep 1 Edited by J.S. Mitchell Butterworth,London,1960
An 32.3234

FARADAY SOCIETY The Cell nucleus :an informal meeting Proceedings Cambridge 1959 Aug 31-Sep 1 Held at the University of Cambridge.Department of radiotherapeutics Butterworths,London,1960 Chairman:J.S. Mitchell
Bal 39.0290

FARADAY SOCIETY The Cell nucleus :an informal meeting Proceedings Cambridge 1959 Aug 31-Sep 1 held at the University of Cambridge.Department of radio-therapeutics Butterworths,London,1960
Bioch 33.0946

FARADAY SOCIETY The Cell nucleus :an informal meeting held at the Department of radiotherapeutics,University of Cambridge...by the Faraday society Proceedings Cambridge 1959 Aug 31 - Sep 1 Illus Butterworths,London,1960 Chairman: Professor J.S.Mitchell
Radioth 35.0501

FARADAY SOCIETY The Cell nucleus informal meeting Proceedings Cambridge 1959 Aug 31-Sep 1 Edited by J.S. Mitchell Butterworths,London,1960
Gen 34.0839

FARADAY SOCIETY The Kinetics of proton transfer processes :a general discussion Newcastle-upon-Tyne 1965 Apr 12-14 Faraday society.Discussions, 39 Aberdeen university press,Aberdeen,1966
Bioch 33.0645

FARADAY SOCIETY The Microscope:its design, construction and applications ;a symposium and general discussion London 1920 Jan 14 Edited by F.S. Spiers pls,illus 260p Griffin,London,1920 Faraday society in cooperation with the Technical optics committee of the British science guild. Includes reports of adjourned discussions held in Sheffield,Feb 24 and in London April 21, 1920
WSM 43.5151

FARADAY SOCIETY The Physical chemistry of enzymes :a discussion Papers Oxford 1955 Aug 10-12 Faraday society.Discussions, 20 Aberdeen,1956
Bioch 33.1047

FARADAY SOCIETY The Physical chemistry of proteins :a general discussion Cambridge 1952 Aug 6-8 Faraday society.Discussions, 13 Faraday society,Aberdeen,1953
Radioth 35.0336

FARADAY SOCIETY The Physical chemistry of proteins :a general discussion Papers Cambridge 1952 Aug 6-8 Faraday society. Discussions, 13 Aberdeen university press, Aberdeen,1953
Bioch 33.0510

FARADAY SOCIETY The Physics and chemistry of colloids and their bearing on industrial questions :a general discussion Report London 1920 Oct 25 H.M.S.O.,London,1921
Bioch 33.1522

FARADAY SOCIETY The Properties and functions of membranes,natural and artificial :a general discussion London 1937 Apr By August Krogh and others Faraday society. Transactions, 33, pt 8 London,1937
Bal 39.0358

FARADAY SOCIETY The Size and shape factor in colloidal systems :a discussion Leamington Spa 1951 Jul 18-20 Faraday society. Discussions, 11 Faraday society,Aberdeen, 1952
Bioch 33.1529

FARADAY SOCIETY and SOCIETE DE CHIMIE PHYSIQUE Surface chemistry discussion at a joint meeting...in honour of professor Henri Devaux Papers Bordeaux 1947 Oct 5-9 Butterworths,London,1949 Published as a special supplement to 'Research'
Chem 18.0546

FARADAY SOCIETY.COLLOID AND BIOPHYSICS COMMITTEE Flow properties of blood and other biological systems :an informal discussion Proceedings Oxford 1959 Sep 23-24 Edited by A.L. Copley and G. Stainsby Pergamon, London,1960
PGMS 29.0269

FARADAY SOCIETY.COLLOID AND BIOPHYSICS COMMITTEE Nature and structure of collagen :a discussion Papers London 1953 Mar 26-27 Edited by J.T. Randall Butterworths,London, 1953
Radioth 35.0186

FARADAY SOCIETY.COLLOID AND BIOPHYSICS COMMITTEE Nature and structure of collagen :a discussion Papers London 1953 Mar 26-27 Edited by J.T. Randall and Sylvia Fitton Jackson Butterworths,London,1953
Bioch 33.2346

FARADAY SOCIETY.COLLOID AND BIOPHYSICS COMMITTEE Nature and structure of collagen :a discussion Papers presented London 1953 Mar 26-27 Edited by J.T. Randall and Sylvia Fitton Jackson Butterworths,London,1953
Bal 39.1202

FARADAY SOCIETY.DISCUSSIONS, 5 Crystal growth :a general discussion Bristol 1949 Apr 12-14 Faraday society Gurney and Jackson,London,1949
Eng 41.4424

FARADAY SOCIETY.DISCUSSIONS, 6 Lipo-proteins :a general discussion Papers Birmingham 1949 Aug 29-31 Faraday society Aberdeen,1949
Bioch 33.0508

FARADAY SOCIETY.DISCUSSIONS, 9 Spectroscopy and molecular structure,and optical methods of investigating cell structure :a general discussion Cambridge 1950 Sep 25-28 Faraday society Aberdeen, 1951
Bal 39.0230

FARADAY SOCIETY.DISCUSSIONS, 9 Spectroscopy and molecular structure and optical methods of investigating cell structure :general discussion Cambridge 1950 Sep 25-28 Faraday society Faraday society,Aberdeen,1951
Radioth 35.0453

FARADAY SOCIETY.DISCUSSIONS, 11 The Size and shape factor in colloidal systems :a discussion Leamington Spa 1951 Jul 18-20 Faraday society Faraday society,Aberdeen, 1952
Bioch 33.1529

FARADAY SOCIETY.DISCUSSIONS, 13 The Physical chemistry of proteins :a general discussion Cambridge 1952 Aug 6-8 Faraday society Faraday society,Aberdeen, 1953
Radioth 35.0336

FARADAY SOCIETY.DISCUSSIONS, 13 The Physical chemistry of proteins :a general discussion Papers Cambridge 1952 Aug 6-8 Faraday society Aberdeen university press, Aberdeen,1953
Bioch 33.0510

FARADAY SOCIETY.DISCUSSIONS, 20 The Physical chemistry of enzymes :a discussion Papers Oxford 1955 Aug 10-12 Faraday society Aberdeen,1956
Bioch 33.1047

FARADAY SOCIETY.DISCUSSIONS, 21 Membrane phenomena :a general discussion Nottingham 1956 Apr 10-12 Faraday society Aberdeen university press,Aberdeen,1956
Bioch 33.0947

FARADAY SOCIETY.DISCUSSIONS, 27 Energy transfer with special reference to biological systems :a general discussion Nottingham 1959 Apr 14-16 Faraday society The Faraday society,London,1960
Radioth 35.0249

FARADAY SOCIETY.DISCUSSIONS, 27 Energy transfer with special reference to biological systems :a general discussion Papers Nottingham 1959 Apr 14-16 Faraday society Aberdeen university press,Aberdeen,1960
Bioch 33.0314

FARADAY SOCIETY.DISCUSSIONS, 36 Fundamental processes in radiation chemistry :a general discussion Notre Dame,Ind. 1963 Sep 2-4 Faraday society Butterworths,London,1963
Radioth 35.1772

FARADAY SOCIETY.DISCUSSIONS, 38 Dislocations in solids :a general discussion Gottingen 1964 Sep 15-17 Faraday society Faraday society,London,1964
Eng 41.3532

FARADAY SOCIETY.DISCUSSIONS, 39 The Kinetics of proton transfer processes :a general discussion Newcastle-upon-Tyne 1965 Apr 12-14 Faraday society Aberdeen university press,Aberdeen,1966
Bioch. 33.0645

FARADAY SOCIETY.DISCUSSIONS, 47 Bonding in metallo-organic compounds Cambridge 1969 Mar 25-27 Faraday society 208p 20cm Faraday society,London,1969
Cav 7.2717

FARADAY SOCIETY.DISCUSSIONS, 48 Motions in molecular crystals Oxford 1969 Sep 16-18 Faraday society Bibliog 224p 20cm Faraday society,London,1969
Cav 7.2718

FARADAY SOCIETY.SYMPOSIA,1970,4 Optical studies of adsorbed layers at interfaces London 1970 Dec 14-15 Faraday society 212p 26cm Faraday society,London,1971
Chem 18.2880

FARKAS (L) MEMORIAL SYMPOSIUM 19TH Radiation chemistry of aqueous systems :proceedings of the 19th L.Farkas memorial symposium held at the Hebrew university Jerusalem 1967 Dec 22-29 Hebrew university of Jerusalem Edited by Gabriel Stein Weizmann science press of Israel in co-operation with Interscience,Jerusalem,1968
Chem 18.2485

FARKAS MEMORIAL SYMPOSIUM 19th Proceedings Radiation chemistry of aqueous systems Jerusalem 1967 Dec 27-29 Edited by Gabriel Stein Weizmann science press of Israel, Jerusalem,1968
Radioth 35.1768

FASANO 1969 Electron transport and energy conservation Round table discussion Edited by J.M. Tager and others Adriatica editrice, Bari,1970
Bioch 33.2288

FAST REACTIONS AND PRIMARY PROCESSES IN CHEMICAL KINETICS Nobel symposium 5th Proceedings Sodergarn 1967 Aug 28 Sep 2 Nobel foundation Edited by Stig Claesson Wiley,New York;London,1967
Bioch 33.0328

FAT AS A TISSUE :international research conference Proceedings Philadelphia 1962 Nov 2-3 Edited by Kaare Rodahl and others Held at the Lankenau hospital McGraw-Hill,New York,1963
An 32.5318

FATIGUE Colloquium on fatigue Proceedings Stockholm 1955 May 25-27 International union of theoretical and applied mechanics Edited by Waloddi Weibull and Folke K.G. Odqvist Springer,Berlin,1956
Eng 41.3754

FATIGUE Symposium on fatigue Contributions Cranfield 1952 Mar Ergonomics research society Edited by W.F. Floyd and A.T. Welford Lewis,London,1953
Psy 31.1027

FATIGUE Symposium on fatigue Papers and discussions Prague 1960 Sep 8-10 Czechoslovak scientific-technical society Ceskoslovenska vedecko-technicka spolecnost, 1961 Papers in English,French,German and Russian
Met 25.0936

FATIGUE - AN INTERDISCIPLINARY APPROACH Sagamore army materials research conference 10th Proceedings Raquette Lake,N.Y. 1963 Aug 13-16 Edited by J.J. Burke and others Sponsored by United States.Army materials research agency Syracuse university press, New York,1964
Eng 41.3803

FATIGUE AND FRACTURE OF METALS :a symposium Cambridge,Mass. 1950 Jun 19-22 Edited by William M. Murray Held at the Massachusetts institute of technology Technology press books M.I.T.press;Wiley, Cambridge,Mass.;New York,1952
Eng 41.3740

FATIGUE AND FRACTURE OF METALS :a symposium
Cambridge,Mass. 1950 Jun 19-22
Massachusetts institute of technology Edited by W.M. Murray Technology press;Wiley,New York,1952
Met 25.0911

FATIGUE AT HIGH TEMPERATURE :a symposium
presented at the 71st annual meeting...
San Francisco,Calif. 1968 Jun 23-28
American society for testing and materials
American society for testing and materials. Special technical publication, 459 ASTM, Philadelphia,Pa.,1969
Eng 41.8293

FATIGUE AT HIGH TEMPERATURE :a symposium
presented at the 71st annual meeting
Proceedings San Francisco,Calif. 1968 Jun 23-28 American society for testing materials American society for testing materials.Technical publication, 459 ASTM, Philadelphia,Pa.,1969
Met 25.2583

FATIGUE CRACK PROPAGATION :a symposium
Atlantic City,N.J. 1966 Jun 26-Jul 1
American society for testing and materials
American society for testing and materials. Special technical publication, 415
Eng 41.3814

FATIGUE CRACK PROPAGATION :a symposium
Atlantic City,N.J. 1966 Jun 26-Jul 1
American society for testing materials A.S.T.M.Special technical publication, 415 A.S.T.M. Philadelphia,Pa.,1967
Met 25.0938

FATIGUE IN AIRCRAFT STRUCTURES :an international conference New York 1956 Jan 30-Feb 1 United States.Air research and development command Guggenheim institute of flight structures Edited by Alfred M. Freudenthal Academic press,New York,1956
Eng 41.2772

FATIGUE IN AIRCRAFT STRUCTURES international conference Proceedings New York 1956 Jan 30 -Feb 1 United States.Air force.Solid state sciences division and Guggenheim institute of flight structures Edited by Alfred M. Freudenthal Academic press,New York,1956
Met 25.0921

FATIGUE IN ROLLING CONTACT :symposium
Proceedings London 1963 Mar 28
Arranged by the Institution of mechanical engineers Institution of mechanical engineers,London,1963
Eng 41.3786

FATIGUE OF AIRCRAFT STRUCTURES Los Angeles, Calif. 1956 Sep 16-21 American society for testing materials American society for testing materials.Special technical publication, 203 American society for testing materials,Philadelphia,Pa.,1957
Eng 41.2776

FATIGUE OF AIRCRAFT STRUCTURES Symposium on fatigue tests of aircraft structure:low-cycle, full-scale, and helicopters Los Angeles 1962 Oct 1-3 American society for testing materials A.S.T.M.Special technical publication, 338 A.S.T.M.,Philadelphia,Pa., 1963
Met 25.0935

FATIGUE OF AIRCRAFT STRUCTURES :a symposium
Proceedings Paris 1961 May 16-18
International committee on aeronautical fatigue Edited by W. Barrois and E.L. Ripley International series of monographs in aeronautics and astronautics.Symposia, 12,Div. 9 Pergamon press,Oxford,1963
Met 25.0931

FATIGUE OF METALS Colloquium on fatigue
Proceedings Stockholm 1955 May 25-27
International union of theoretical and applied mechanics Edited by Waloddi Weibull and Folke K.G. Odqvist Springer-verlag,Gottingen; Heidelberg,1956 Lectures in English, French and German
Met 25.0917

FATIGUE OF METALS Conference on the fatigue of metals 2nd Papers Institut metallurgii imini A.A.Baykov Translated by Leo Ronson from the Russian N.L.L.S.T., Boston Spa,1964 Translation edited by P.G. Forrest
Met 25.2505

FATIGUE OF METALS International conference on fatigue of metals Proceedings London 1956 Sep 10-14 New York 1956 Nov 28-30 Institution of mechanical engineers and American society of mechanical engineers Institution of mechanical engineers,London, 1956
Met 25.0919

FATIGUE OF METALS International conference on fatigue of metals Proceedings London 1956 Sep 10-14 and New York 1956 Nov 28-30 Institution of mechanical engineers American society of mechanical engineers Institution of mechanical engineers,London, 1956
Eng 41.3759

FATIGUE OF METALS The Failure of metals by fatigue :a symposium Proceedings Melbourne 1946 Dec 2-6 University of Melbourne.Faculty of engineering Melbourne university press, Melbourne,1947
Met 25.0909

FATIGUE RESISTANCE OF MATERIALS AND METAL STRUCTURAL PARTS international conference Proceedings Warsaw 1960 May Institute of precision mechanics,Warsaw Edited by Alfred Buch Pergamon press;Polish scientific publishers,Oxford;Warsaw,1964
Met 25.0933

FATIGUE-AN INTERDISCIPLINARY APPROACH
Sagamore army materials research conference 10th Raquette Lake,New York 1963 Aug 13-16 United States.Army materials research agency and Syracuse university Edited by John J. Burke and others Syracuse university press,Syracuse,1964
Met 25.2330

FATTY ACID METABOLISM Federation of European biochemical societies meeting 4th Proceedings Oslo 1967 Jul 5 Vol 4: cellular compartmentalization and control of fatty acid metabolism Federation of European biochemical societies Edited by F.C. Gran Organized by the Norwegian biochemical society Universitetsforlaget;Academic press,Oslo; London,1968 Symposium organizer J. Bremer
Bioch 33.1399

FEDERATION BELGE DES GEOGRAPHES PROFESSEURS DE L'ENSEIGNEMENT MOYEN,NORMAL ET TECHNIQUE
Semaine internationale de geographie : international geographical week Report Brussels 1958 Aug 3-10 Ghent,c1958
Geog 13.5602

FEDERATION INTERNATIONALE DE LA PRECONTRAINTE
1st-5th congres Comptes-rendus 1953-66 5 vols Federation internationale de la precontrainte,Paris,1955-66
Eng 41.2928

FEDERATION INTERNATIONALE DES INSTITUTS ET SOCIETES POUR L'ETUDE DE LA RENAISSANCE
Soleil a la renaissance:sciences et mythes : colloque international Brussels 1963 Apr Universite libre de Bruxelles.Institut pour l'etude de la renaissance et de l'humanisme. Publication, 2 584p Presses universitaires de Bruxelles,Brussels;Paris, 1965 Contains Hoskin,M.and Jones,C. 'Problems in late renaissance astronomy'
WSM 43.0046

FEDERATION NATIONALE DES INDUSTRIES ELECTRONIQUES
International congress on microwave tubes 5th Proceedings Paris 1964 Societe francaise des electroniciens et radioelectriciens Societe francaise des ingenieurs et techniciens du vide Academic press,New York,1965
Eng 41.5574

FEDERATION OF BRITISH INDUSTRIES Pattern of research in British industry :a conference Report Eastbourne 1962 Apr 5-7 Federation of British industries,London,1962
Eng 41.1671

FEDERATION OF EUROPEAN BIOCHEMICAL SCIENCES MEETING 2nd Proceedings Vienna 1965 Apr 21-24 symposium on ribonucleic acid structure and function Federation of European biochemical societies Edited by H. Tuppy Pergamon press,Oxofrd,1966
Gen 34.0688

FEDERATION OF EUROPEAN BIOCHEMICAL SOCIETIES
Abstracts of communications presented at the meeting of the Federation of European biochemical societies 1st-2nd,4th London,1964-
Bioch 33.2255

FEDERATION OF EUROPEAN BIOCHEMICAL SOCIETIES
Abstracts of communications presented at the meeting of the federation Wiener medizinischen akademie,Vienna,1965
Radioth 35.0137

FEDERATION OF EUROPEAN BIOCHEMICAL SOCIETIES
Biochemistry of blood platelets :colloquium held on the occasion of the Third meeting of the Federation of European biochemical societies Warsaw 1966 Apr 4-7 Edited by E. Kowalski and S. Niewiarowski Organized by the Polish biochemical society Academic press,New York,1967
Radioth 35.0148

FEDERATION OF EUROPEAN BIOCHEMICAL SOCIETIES
Federation of European biochemical sciences meeting 2nd Proceedings Vienna 1965 Apr 21-24 symposium on ribonucleic acid structure and function Edited by H. Tuppy Pergamon press,Oxofrd,1966
Gen 34.0688

FEDERATION OF EUROPEAN BIOCHEMICAL SOCIETIES
Federation of European biochemical societies meeting 4th Proceedings Oslo 1967 Jul 4 Vol 2: biochemistry of virus replication Edited by S.G. Laland and L.O. Froholm Organized by the Norwegian biochemical society Universitetsforlaget;Academic press,Oslo; London,1968 Symposium organizer A.P. Nygaard
Bioch 33.1397

FEDERATION OF EUROPEAN BIOCHEMICAL SOCIETIES
Federation of European biochemical societies meeting 4th Proceedings Oslo 1967 Jul 5 Vol 4: cellular compartmentalization and control of fatty acid metabolism Edited by F. C. Gran Organized by the Norwegian biochemical society Universitetsforlaget; Academic press,Oslo;London,1968 Symposium organizer J. Bremer
Bioch 33.1399

FEDERATION OF EUROPEAN BIOCHEMICAL SOCIETIES
Federation of European biochemical societies meeting 4th Proceedings Oslo 1967 Jul 7 Vol 5: control of glycogen metabolism Edited by W.J. Whelan Organized by the Norwegian biochemical society Universitetsforlaget;Academic press,Oslo; London,1968 Symposium organizer W.J. Whelan
Bioch 33.1400

FEDERATION OF EUROPEAN BIOCHEMICAL SOCIETIES
Federation of European biochemical societies meeting 5th Prague 1968 Jul enzymes and isoenzymes:structure,properties and function Federation of European biochemical societies.Publications, 18 Academic press,London,1970
Bioch 33.1888

FEDERATION OF EUROPEAN BIOCHEMICAL SOCIETIES
Federation of European biochemical societies meeting 5th Proceedings Prague 1968 Jul Vol 3: mitochondria:structure and function Edited by L. Ernster and Z. Drahota Federation of European biochemical societies. Publications, 17 Academic press,London;New York,1969
Bioch 33.1403

FEDERATION OF EUROPEAN BIOCHEMICAL SOCIETIES
Federation of European biochemical societies meeting 5th Proceedings Prague 1968 Jul Vol 3: mitochondria:structure and function Edited by L. Ernster and Z. Drahota Federation of European biochemical societies. Publications, 17 Academic press,London;New York,1969
Radioth 35.0149

FEDERATION OF EUROPEAN BIOCHEMICAL SOCIETIES
Federation of European biochemical societies meeting 5th Proceedings Prague 1968 Jul gamma globulins;structure and biosynthesis Edited by F. Franek and D. Shugar Federation of European biochemical societies.Publications, 15 Academic press, London;New York,1969
Bioch 33.1402

FEDERATION OF EUROPEAN BIOCHEMICAL SOCIETIES
Federation of European biochemical societies meeting 6th Proceedings Madrid 1969 membranes:structure and function Edited by A. Villanueva and F. Ponz Federation of European biochemical societies.Publications, 20 Academic press,London;New York,1969
Radioth 35.0150

FEDERATION OF EUROPEAN BIOCHEMICAL SOCIETIES
Federation of European biochemical societies meeting 6th Proceedings Madrid 1969 Apr Vol 2: membranes,structure and function Edited by J.R. Villanueva and F. Ponz Federation of European biochemical societies.Publications, 20 Academic press, London,1970
Bioch 33.2321

FEDERATION OF EUROPEAN BIOCHEMICAL SOCIETIES
Federation of European biochemical societies meeting 6th Proceedings Madrid 1969 Apr Vol 3: macromolecules,biosynthesis and function Edited by S. Ochoa and others Federation of European biochemical societies. Publications, 21 Academic press,London,1970
Bioch 33.2322

FEDERATION OF EUROPEAN BIOCHEMICAL SOCIETIES
Federation of European biochemical societies meeting 6th Proceedings Madrid 1969 Apr metabolic regulation and enzyme action Edited by A. Sols and S. Grisolia Federation of European biochemical societies.Publications, 19 Academic press,London,1970
Bioch 33.1889

FEDERATION OF EUROPEAN BIOCHEMICAL SOCIETIES
Federation of European biochemical societies meeting 7th Proceedings Varna 1971 Sep functional units in protein biosynthesis Edited by R.A. Cox and A.A. Hadjiolov Federation of European biochemical societies. Publications, 23 Academic press,London,1972
Bioch 33.2374

FEDERATION OF EUROPEAN BIOCHEMICAL SOCIETIES
Federation of European biochemical societies meeting 7th Proceedings Varna 1971 Sep virus-cell interactions and viral antimetabolites Edited by D. Shugar Federation of European biochemical societies. Publications, 22 Academic press,London,1972
Bioch 33.2323

FEDERATION OF EUROPEAN BIOCHEMICAL SOCIETIES
Structure and activity of enzymes :a symposium Proceedings London 1964 Mar 24 Edited by T.W. Goodwin and others Federation of European biochemical societies.Symposium, 1 Academic press,London;New York,1964
Bioch 33.1404

FEDERATION OF EUROPEAN BIOCHEMICAL SOCIETIES
Structure and activity of enzymes :a symposium 1 London 1964 Mar 24 Edited by T.W. Goodwin and others Academic press,London;New York,1964
Chem 18.1248

FEDERATION OF EUROPEAN BIOCHEMICAL SOCIETIES
Structure and function of transfer RNA and 58-RNA Federation of European biochemical societies meeting 4th Proceedings Oslo 1967 Jul 3-7 Vol 3: structure and function of transfer RNA and 58-RNA Edited by L.O. Froholm and S.H. Laland Academic press, London,1968
Chem 18.2491

FEDERATION OF EUROPEAN BIOCHEMICAL SOCIETIES
The Federation of European biochemical societies meeting 2nd Proceedings Vienna 1965 Apr 21-24 Vol 2: cyclitols and phosphoinositides Edited by H. Kindl Pergamon press,Oxford,1966
Bioch 33.1393

FEDERATION OF EUROPEAN BIOCHEMICAL SOCIETIES
The Federation of European biochemical societies meeting 3rd Warsaw 1966 Apr 4-7 biochemistry of mitochondria; colloquium Edited by E.C. Slater and others Organized by the Polish biochemical society Academic press,London,1967
Bioch 33.1394

FEDERATION OF EUROPEAN BIOCHEMICAL SOCIETIES
The Federation of European biochemical societies meeting 3rd Warsaw 1966 Apr 4-7 genetic elements:properties and function;symposium Edited by D. Shugar Organised by the Polish biochemical society Academic press,London,1967
Bioch 33.1395

FEDERATION OF EUROPEAN BIOCHEMICAL SOCIETIES
The Federation of European biochemical societies meeting 4th Proceedings Oslo 1967 Jul 5 Vol 3: structure and function of transfer RNA and 5 S-RNA Edited by L.O. Froholm and S.G. Laland Organized by the Norwegian biochemical society Universitetsforlaget;Academic press,Oslo; London,1968 Symposium organizer H.G. Zachau
Bioch 33.1398

FEDERATION OF EUROPEAN BIOCHEMICAL SOCIETIES
The Federation of European biochemical societies meeting Oslo 1967 Jul 7 Vol 6: structure and function of the endoplasmic reticulum in animal cells Edited by F.C. Gran Organized by the Norwegian biochemical society Universitetsforlaget;Academic press,Oslo; London,1968 Symposium organizer P.N. Campbell
Bioch 33.1401

FEDERATION OF EUROPEAN BIOCHEMICAL SOCIETIES
The Federation of European biochemical societies meeting 4th Proceedings Oslo 1976 Jul Vol 1: regulation of enzyme activity and allosteric interactions Edited by E. Kvamme and A. Phil Organized by the Norwegian biochemical society Universitetsforlaget;Academic press,Oslo; London,1968 Symposium organizer J.P. Changeux
Bioch 33.1396

FEDERATION OF EUROPEAN BIOCHEMICAL SOCIETIES 2nd meeting Proceedings Antibodies to biologically active molecules Vienna 1965 Apr 21-24 Vol 1 Edited by Bernhard Cinader Pergamon,Oxford,1967
Pha 16.0190

FEDERATION OF EUROPEAN BIOCHEMICAL SOCIETIES. SYMPOSIUM, 1 Structure and activity of enzymes :a symposium Proceedings London 1964 Mar 24 Federation of European biochemical societies Edited by T.W. Goodwin and others Academic press,London;New York, 1964
Bioch 33.1404

FEDERATION OF EUROPEAN BIOCHEMICAL SOCIETIES. SYMPOSIUM,1 Structure and activity of enzymes :symposium Edited by T.W. Goodwin and others Academic press,London,1964
Pha 16.0080

FEDERATION OF EUROPEAN BIOCHEMICAL SOCIETIES MEETING 2nd Proceedings Vienna 1965 Apr 21-24 Vol 1: antibodies to biologically active molecules Edited by B. Cinader Pergamon press,Oxford,1967
Phys 20.1552

The FEDERATION OF EUROPEAN BIOCHEMICAL SOCIETIES MEETING 2nd Proceedings Vienna 1965 Apr 21-24 Vol 2: cyclitols and phosphoinositides Federation of European biochemical societies Edited by H. Kindl Pergamon press,Oxford,1966
Bioch 33.1393

The FEDERATION OF EUROPEAN BIOCHEMICAL SOCIETIES MEETING 3rd Warsaw 1966 Apr 4-7 biochemistry of mitochondria;colloquium Federation of European biochemical societies Edited by E.C. Slater and others Organized by the Polish biochemical society Academic press,London,1967
Bioch 33.1394

The FEDERATION OF EUROPEAN BIOCHEMICAL SOCIETIES MEETING 3rd Warsaw 1966 Apr 4-7 genetic elements:properties and function; symposium Federation of European biochemical societies Edited by D. Shugar Organised by the Polish biochemical society Academic press,London,1967
Bioch 33.1395

FEDERATION OF EUROPEAN BIOCHEMICAL SOCIETIES MEETING 4th Proceedings Oslo 1967 Jul 4 Vol 2: biochemistry of virus replication Federation of European biochemical societies Edited by S.G. Laland and L.O. Froholm Organized by the Norwegian biochemical society Universitetsforlaget; Academic press,Oslo;London,1968 Symposium organizer A.P.Nygaard
Bioch 33.1397

FEDERATION OF EUROPEAN BIOCHEMICAL SOCIETIES MEETING 4th Proceedings Oslo 1967 Jul 5 Vol 4: cellular compartmentalization and control of fatty acid metabolism Federation of European biochemical societies Edited by F.C. Gran Organized by the Norwegian biochemical society Universitetsforlaget;Academic press,Oslo; London,1968 Symposium organizer J. Bremer
Bioch 33.1399

FEDERATION OF EUROPEAN BIOCHEMICAL SOCIETIES MEETING 4th Proceedings Oslo 1967 Jul 7 Vol 5: control of glycogen metabolism Federation of European biochemical societies Edited by W.J. Whelan Organized by the Norwegian biochemical society Universitetsforlaget;Academic press,Oslo; London,1968 Symposium organizer W.J. Whelan
Bioch 33.1400

The FEDERATION OF EUROPEAN BIOCHEMICAL SOCIETIES MEETING 4th Proceedings Oslo 1967 Jul 5 Vol 3: structure and function of transfer RNA and 5 S-RNA Federation of European biochemical societies Edited by L.O. Froholm and S.G. Laland Organized by the Norwegian biochemical society Universitetsforlaget;Academic press,Oslo; London,1968 Symposium organizer H.G. Zachau
Bioch 33.1398

The FEDERATION OF EUROPEAN BIOCHEMICAL SOCIETIES MEETING 4th Proceedings Oslo 1967 Jul 7 Vol 6: structure and function of the endoplasmic reticulum in animal cells Federation of European biochemical societies Edited by F.C. Gran Organized by the Norwegian biochemical society Universitetsforlaget;Academic press,Oslo; London,1968 Symposium organizer P.N. Campbell
Bioch 33.1401

The FEDERATION OF EUROPEAN BIOCHEMICAL SOCIETIES MEETING 4th Proceedings Oslo 1976 Jul Vol 1: regulation of enzyme activity and allosteric interactions Federation of European biochemical societies Edited by E. Kvamme and A. Phil Organized by the Norwegian biochemical society Universitetsforlaget;Academic press,Oslo; London,1968 Symposium organizer J.P. Changeux
Bioch 33.1396

FEDERATION OF EUROPEAN BIOCHEMICAL SOCIETIES MEETING 5th Prague 1968 Jul enzymes and isoenzymes:structure,properties and function Federation of European biochemical societies Federation of European biochemical societies.Publications, 18 Academic press,London,1970
Bioch 33.1888

FEDERATION OF EUROPEAN BIOCHEMICAL SOCIETIES MEETING 5th Proceedings Prague 1968 Jul Vol 3: mitochondria:structure and function Federation of European biochemical societies Edited by L. Ernster and Z. Drahota Federation of European biochemical societies.Publications, 17 Academic press, London;New York,1969
Bioch 33.1403

FEDERATION OF EUROPEAN BIOCHEMICAL SOCIETIES MEETING 5th Proceedings Prague 1968 Jul Vol 3: mitochondria:structure and function Federation of European biochemical societies Edited by L. Ernster and Z. Drahota Federation of European biochemical societies.Publications, 17 Academic press, London;New York,1969
Radioth 35.0149

FEDERATION OF EUROPEAN BIOCHEMICAL SOCIETIES MEETING 5th Proceedings Prague 1968 Jul gamma globulins;structure and biosynthesis Federation of European biochemical societies Edited by F. Franek and D. Shugar Federation of European biochemical societies.Publications, 15 Academic press,London;New York,1969
Bioch 33.1402

FEDERATION OF EUROPEAN BIOCHEMICAL SOCIETIES MEETING 6th Proceedings Madrid 1969 membranes:structure and function Federation of European biochemical societies Edited by A. Villanueva and F. Ponz Federation of European biochemical societies. Publications, 20 Academic press,London;New York,1969
Radioth 35.0150

FEDERATION OF EUROPEAN BIOCHEMICAL SOCIETIES MEETING 6th Proceedings Madrid 1969 Apr Vol 2: membranes,structure and function Federation of European biochemical societies Edited by J.R. Villanueva and F. Ponz Federation of European biochemical societies.Publications, 20 Academic press, London,1970
Bioch 33.2321

FEDERATION OF EUROPEAN BIOCHEMICAL SOCIETIES MEETING 6th Proceedings Madrid 1969 Apr Vol 3: macromolecules, biosynthesis and function Federation of European biochemical societies Edited by S. Ochoa and others Federation of European biochemical societies.Publications, 21 Academic press,London,1970
Bioch 33.2322

FEDERATION OF EUROPEAN BIOCHEMICAL SOCIETIES MEETING 6th Proceedings Madrid 1969 Apr metabolic regulation and enzyme action Federation of European biochemical societies Edited by A. Sols and S. Grisolia Federation of European biochemical societies. Publications, 19 Academic press,London,1970
Bioch 33.1889

FEDERATION OF EUROPEAN BIOCHEMICAL SOCIETIES MEETING 7th Proceedings Varna 1971 Sep functional units in protein biosynthesis Federation of European biochemical societies Edited by R.A. Cox and A.A. Hadjiolov Federation of European biochemical societies.Publications, 23 Academic press,London,1972
Bioch 33.2374

FEDERATION OF EUROPEAN BIOCHEMICAL SOCIETIES MEETING 7th Proceedings Varna 1971 Sep virus-cell interactions and viral antimetabolites Federation of European biochemical societies Edited by D. Shugar Federation of European biochemical societies. Publications, 22 Academic press,London,1972
Bioch 33.2323

FEDERATION OF EUROPEAN BIOCHEMICAL SOCIETIES MEETING 4th Proceedings Structure and function of transfer RNA and 5S-RNA Oslo 1967 Jul 3-7 Vol 3: structure and function of transfer RNA and 5S-RNA Federation of European biochemical societies Edited by L.O. Froholm and S.E. Laland Academic press, London,1968
Chem 18.2491

FEDERATION OF EUROPEAN BIOCHEMICAL SOCIETIES SYMPOSIUM 1st Structure and activity of enzymes London 1914 Mar 24 Edited by T.W. Goodwin and others Academic press, London;New York,1964
Radioth 35.0119

FELDAFING 1964 Circadian clocks :summer school Proceedings Edited by Jurgen Aschoff Sponsored by the North Atlantic treaty organization.Scientific affairs division North-Holland,Amsterdam,1965
An 32.5375

The FELDAFING SUMMER SCHOOL Proceedings Circadian clocks 1964 Sep 7-18 Edited by Jurgen Aschoff Sponsored by the North Atlantic treaty organization.Scientific affairs division North Holland,Amsterdam, 1965
Bal 39.1766

The FELDAFING SUMMER SCHOOL Proceedings Circadian clocks 1964 Sep 7-18 Edited by Jurgen Aschoff Sponsored by the North Atlantic treaty organization North-Holland, Amsterdam,1965
Bot 42.1365

The FERMI SURFACE :an international conference Proceedings Cooperstown,N.Y. 1960 Aug 22-24 General electric research laboratory Edited by W.A. Harrison and M.B. Webb Sponsored by the United States.Air force. Office of scientific research John Wiley,New York,1960
Cav 7.0541

The FERMI SURFACE :an international conference Proceedings Cooperstown,N.Y. 1960 Aug 22-24 United States.Air force.Office of scientific research,and,General electric research laboratory J.Wiley,New York;London, 1960
Chem 18.2382

FERNS Origin and evolution of the ferns :a symposium Papers Edited by Murray F. Buell Torrey botanical club.Memoirs, 21 pt.5 86p Seeman,Durham,N.C.,1964
Bot 42.4272

FERRANTI LTD Linear programming conference mimeograph London 1954 May 17p 1954
Math L 5.1336

FERRANTI LTD Manchester university computer : inaugural conference Manchester 1951 Jul.9-12 Victoria university of Manchester pl. 40p 27cm Manchester,1957
Math L 5.1320

FERROELECTRICITY International meeting on ferroelectricity 2nd Kyoto 1969 Sep 4-9 Science council of Japan Sponsored by the International union of pure and applied physics Physical society of Japan.Journal, 28,suppt Physical society of Japan,1970
Met 25.2865

FERTILITY AND STERILITY World congress on fertility and sterility 5th Proceedings Stockholm 1966 Jun 16-22 Edited by Bjorn Westin and Nils Wiquist Under the supervision of the International fertility association International congress series, 133 Excerpta medica foundation,Amsterdam, 1967
An 32.3899

FERTILITY AND STERILITY World congress on fertility and sterility 5th Proceedings Stockholm 1966 Jun 16-22 International fertility association Swedish society of obstetrics and gynaecology Edited by Bjorn Westin and Nils Wiqvist Excerpta medica. International congress series, 133 Excerpta medica,Amsterdam,1967
Inv Med 37.0230

FERTILIZATION The Biochemistry of fertilization and the gametes :a symposium London 1951 Feb 17 Biochemical society Edited by R.T. Williams Held at the London school of hygiene and tropical medicine Biochemical society.Symposia, 7 Cambridge university press,Cambridge,1951
Bioch 33.1369

FESTBAND ZUM 70.GEBURTSTAG VON ROLF NEVANLINNA 2nd kolloquium vortrage Zurich 1965 Nov.4-6 Edited by H.P. Kunzi and A. Pfluger 149p 24cm Springer-Verlag,Berlin, 1966
P. Math 2.2555

FETAL HOMEOSTASIS Conference on fetal homeostasis 1st Proceedings Edited by Ralph M. Wynn Sponsored by the New York academy of sciences New York academy of sciences,New York,1965
An 32.2519

FETAL HOMEOSTASIS Conference on fetal homeostasis 2nd Proceedings Princeton,N. J. 1966 Apr 3-6 Edited by Ralph M. Wynn Sponsored by the New York academy of sciences New York academy of sciences,New York,1967
An 32.2520

FETAL HOMEOSTASIS 2nd :a conference Proceedings Princeton,N.J. 1966 Apr 3-6 Vol 2 New York academy of sciences Edited by Ralph M. Wynn Sponsored by the William H.Donner foundation New York academy of sciences,New York,1967
Inv Med 37.0263

FIBER COMPOSITE MATERIALS :a seminar Papers Metals Park,Ohio 1964 Oct 17-18 American society for metals American society for metals,Metals Park,Ohio,1965
Eng 41.3697

FIBER SPINNING AND DRAWING :symposium New York 1966 Sep 13-15 American chemical society Edited by Myron J. Coplan Applied polymer symposia, 6 181p Interscience publishers,New York,1967
Chem E 24.0423

FIBER-STRENGTHENED METALLIC COMPOSITES a symposium presented at the 1966 metals congress Chicago 1966 Nov 2-3 American society for testing materials A.S.T. M.Special technical publication, 427 A.S.T.M. Philadelphia,1967
Met 25.2338

FIBROUS PROTEINS Symposium on fibrous proteins Papers Canberra 1967 Aug 14-18 Edited by W.G. Crewther Sponsored by the Australian academy of science Butterworths, Sydney,1968
Bioch 33.0512

FIBROUS PROTEINS Symposium on fibrous proteins Proceedings Leeds 1946 May 23-25 Society of dyers and colourists Edited by C.L. Bird Leeds,1946
Bioch 33.0503

FIBROUS PROTEINS :a symposium Proceedings Leeds 1946 May 23-25 Society of dyers and colourists Edited by C.L. Bird figs,pls 27cm n.p.,c1946
Bal 44.4561

FIBROUS PROTEINS AND THEIR BIOLOGICAL SIGNIFICANCE :a symposium Leeds 1954 Sep Society for experimental biology Society for experimental biology.Symposia, 9 Cambridge university press,Cambridge,1955
Bioch 33.1354

FIBROUS PROTEINS AND THEIR BIOLOGICAL SIGNIFICANCE :a symposium Papers Leeds 1954 Sep Society for experimental biology Edited by R. Brown and J.F. Danielli Society for experimental biology.Symposia, 9 Cambridge university press,Cambridge,1955
An 32.5465

FIELD-ION MICROSCOPY Applications of field-ion microscopy in physical metallurgy and corrosion Conference in physical metallurgy and corrosion Proceedings Atlanta,Ga. 1965 May 15-17 Georgia institute of technology Edited by R.F. Hochman and others Georgia institute of technology.Advanced research projects agency,Atlanta,Ga.,1968
Met 25.2631

FIELD THEORY Lectures on field theory and the many-body problem :international spring school 1st Naples 1961 Universita di Napoli.Instituto di fisica teorica Edited by E.R. Caianiello With support of North Atlantic treaty organization Academic press, New York,1961
Cav 7.0423

FIELD THEORY Theorie quantique des champs Conseil de physique 12eme Rapports Brussels 1961 Oct 9-14 Institut international de physique Solvay and Universite libre de Bruxelles Edited by F. Stoops Solvay conference,1961,12 Interscience,New York,c 1961 13th Solvay conference issued as Conference on physics,q.v.
Cav 7.2077

FIELDS International theoretical physics conference on particles and fields Proceedings Rochester,N.Y. 1967 Aug 28-Sep 1 Edited by C.R. Hagen and others 708p 26cm Interscience,New York,1967 Dedicated to Robert Oppenheimer
Cav 7.2773

FIFTY YEARS OF DARWINISM :modern aspects of evolution Baltimore,Md. 1909 Jan 1 American association for the advancement of science Holt,New York,1909 Centennial addresses in honor of Charles Darwin
Bot 42.6666

FILAMENT WINDING CONFERENCE Pasadena 1961 Mar 28-30 Society of aerospace material and process engineers SAMPE,Azuza,Calif.,n.d. Mimeographed
Met 25.0097

FILARIASIS Chemotherapy of filariasis :a conference New York 1948 May 25 By L.L. Ashburn and others New York academy of sciences Edited by Roy Waldo Miner New York academy of sciences.Annals, 50,art.2 New York,1948
Mol 45.0271

The FINAL FORMING AND SHAPING OF WROUGHT NON-FERROUS METALS :a symposium London 1956 Apr 12 Institute of metals Institute of metals.Monograph and report series, 20 Institute of metals,London,1956
Met 25.0709

FINE STRUCTURE OF CELLS :a symposium held at the 8th congress of cell biology Leiden 1954 International union of biological sciences With financial support of Unesco International union of biological sciences. Series B., 21 Noordhoff,Groningen,1955
Radioth 35.0467

FINE STRUCTURE OF CELLS :symposium held at the VIIIth congress of cell biology Leiden 1954 With financial support of Unesco International union of biological sciences.Series B, 21 Noordhoff,Groningen, 1955
Phys 20.1483

FINE STRUCTURE OF CELLS :symposium held at the 8th congress of cell biology Leiden 1954 With the financial support of Unesco International union of biological sciences. Publications.Series B, 21 Noordhoff, Groningen,1955
An 32.3219

FINE STRUCTURE OF CELLS :a symposium held at the VIIIth congress of cell biology Papers Leiden 1954 International union of biological sciences With financial support of Unesco International union of biological sciences.Publications.Series B, 21 Noordhoff,Groningen,1955
Path 30.2410

FINE STRUCTURE OF CELLS;SYMPOSIUM The Congress of cell biology 8th Papers Leiden 1954 International union of biological sciences With financial support of Unesco International union of biological sciences.Ser.B, 21 Noordhoff,Groningen,1955
Bal 39.0283

The FINE STRUCTURE OF THE SOLAR ATMOSPHERE; EXTENDED ABSTRACTS :a colloquium Anacapri 1966 Jun.6-8 Edited by K.O. Kiepenheuer Sponsored by the Notgemeinschaft der Deutschen wissenschaft Notgemeinschaft der deutschen wissenschaft.Bericht, 12 149p Franz Steiner,Wiesbaden,1966 German title: Die Feinstruktur der sonenatmosphare
Obs 6.0945

The FINISHING OF ALUMINUM :symposium Boston,Mass. 1961 Jun 18-23 American electroplaters' society Edited by G.H. Kissin Reinhold,New York,1963
Met 25.2796

FINITE GROUPS Representation theory of finite groups and related topics :symposium Madison,Wisc. 1970 Apr 14-16 American mathematical society Edited by Irving Reiner American mathematical society.Proceedings of symposia in pure mathematics, 21 v,178p 26cm AMS,Providence,R.I.,1971
P Math 2.4432

FINITE GROUPS Symposium on the theory of finite groups Lectures Urbana,Ill. 1967 Nov 24 American mathematical society Bibliog. 94p 25cm American mathematical society,Providence,R.I.,1969
P Math 2.3581

FINITE GROUPS symposium New York 1959 Apr 23-24 American mathematical society Edited by A.Adrian Albert and Irving Kaplansky Cosponsored by the Institute for defense analysis.Communications division American mathematical society.Proceedings of symposia in pure mathematics, 1 viii,110p 23cm American mathematical society,Providence,R.I., 1959
P. Math 2.2029

FINITE GROUPS institute report Pasadena, Calif. 1960 Aug 1-28 American mathematical society Edited by Marshall Hall Supported by the National science foundation American mathematical society.Proceedings of symposia in pure mathematics, 6 114p 26cm American mathematical society,Providence,R.I., 1962
P. Math 2.2031

FINITE SIMPLE GROUPS :an instructional conference Proceedings Oxford 1969 Sep London mathematical society Edited by M.B. Powell and G. Higman xi,327p 24cm Academic press,London;New York,1971
P Math 2.4431

FINSEN MEMORIAL CONGRESS International congress on photobiology 3rd Proceedings Copenhagen 1960 Edited by B.Chr. Christensen and B. Buchmann illus. Elsevier,Amsterdam,1961
Bot 42.1347

FIRST INTERNATIONAL SYMPOSIUM ON ARCTIC GEOLOGY Geology of the Arctic Calgary 1960 Jan 11-30 Vol 1-2 Alberta society of petroleum geologists Edited by Gilbert O. Raasch illus,maps 26cm 2 vols University of Toronto press,Toronto,1961
Sco 14.2403

FISH The Biochemistry of fish ;a symposium Liverpool 1950 Sep 22 Biochemical society Edited by R.T. Williams Held at the University of Liverpool.Derby Hall Biochemical society.Symposia, 6 Cambridge university press,Cambridge,1961
Bioch 33.1368

FISH BEHAVIOR The Central nervous system and fish behavior :a conference Papers Chicago, Ill. 1967 Apr United States.Air force. Office of scientific research American zoological society.Comparative physiology division Edited by David Ingle Jointly sponsored by Perspectives in biology and medicine illus University of Chicago press, Chicago,Ill.;London,1968
Psy 31.2816

FISH BEHAVIOR The Central nervous system and fish behavior :symposium Papers presented Chicago 1967 Apr Edited by D. Ingle at the University of Chicago University of Chicago press,Chicago;London,1968 Sponsored jointly by the United States Air force office of scientific research, comparative physiology division of the American zoological society,and the academic journal Perspectives in biology and medicine
Bal 39.3266

FISHERIES Symposium on African hydrobiology and inland fisheries 2nd Brazzaville 1956 Jul 3-11 Scientific council for Africa south of the Sahara Scientific council for Africa south of the Sahara.Publications,25 Commission for technical co-operation in Africa,London,1957
Geog 13.2726

FITOTERAPIA rivista di studi ed applicazioni delle piante medicinali Vol 22-39 18 vols Inverni and Della Beffa,Milan,1951-68 Some parts missing
Chem 26.0285

FLAGSTAFF,ARIZ. 1953 Astronomical photoelectric conference Proceedings Lowell observatory Edited by John B. Irwin Sponsored by the National science foundation Indiana university,Bloomington,Ind.,1953
Obs 6.2883

FLAT ROLLED PRODUCTS:ROLLING AND TREATMENT Technical conference 1st Proceedings Chicago 1959 Jan 21 American institute of mining,metallurgical and petroleum engineers. Mechanical working committee Edited by T.E. Dancy and E.L. Robinson Metallurgical society conferences, 1 Interscience,New York;London,1959
Met 25.0731

FLAT ROLLED PRODUCTS III Technical conference 4th Proceedings Chicago,Ill. 1962 Jan 17 American institute of mining, metallurgical and petroleum engineers. Mechanical working committee Metallurgical society conferences, 16 Interscience,New York;London,n.d.
Met 25.0703

FLAVEINS International symposium on flaveins and flavoproteins 3rd Proceedings Durham N.C. 1970 International union of biochemistry International union of biochemistry.Symposium, 39 University park press;Butterworths,Baltimore,Md.;London,1971
Bioch 33.2373

FLAVINS AND FLAVOPROTEINS The Conference on flavins and flavoproteins 2nd Proceedings Nagoya 1967 Aug 14-17 Edited by Kunio Yagi University of Tokyo press,Tokyo,1968
Bioch 33.1076

FLAVINS AND FLAVOPROTEINS :a symposium Proceedings Amsterdam 1965 Jun 10-15 International union of biochemistry Edited by E.C. Slater organized under the auspices of the Nederlandsche akademie van wetenschappen B.B.A. library, 8 illus Elsevier,Amsterdam,1966
Bioch 33.1099

FLAVOPROTEINS International symposium on flaveins and flavoproteins 3rd Proceedings Durham N.C. 1970 International union of biochemistry International union of biochemistry.Symposium, 39 University park press;Butterworths, Baltimore,Md.;London,1971
Bioch 33.2373

FLOODS Conference on the North Sea floods of 31 January-1 February,1953 Papers London 1953 Dec . Institution of civil engineers Institution of civil engineers,London,1954
Geog 13.0706

FLOODS The North sea floods of January 31-February 1,1953 :a conference London 1953 Dec Institution of civil engineers Institution of civil engineers,London,1954
Eng 41.3342

FLORA EUROPAEA SYMPOSIUM 2nd Proceedings Genoa 1961 May 21-28 Edited by V.H. Heywood and R.E.G. Pichi-Sermolli Webbia. Baccolta di scritte botanici, 18 Florence, 1963
BG 38.0895

FLORENCE Endocrinologic and morphologic correlations of the ovary :the Florentine conference Edited by W. Ingiulla and Robert B. Greenblatt
An 32.5225

FLORENCE 1956 and MILAN Congres international d'histoire des sciences 8e Actes Vol 1-3 International union of the history and philosophy of science Academie internationale d'histoire des sciences.Collection de travaux, 9 3 vols Gruppe italiano di storia delle scienze; Hermann,Vinci (Florence);Paris,1958
WSM 43.0057

FLORENCE 1960 Teoria dei gruppi finiti e applicazioni :convegno internazionale Unione matematica Italiana Universita di Firenze Supported by Consiglio nazionale della ricerche vii,156p 24cm Edizioni Cremonese,Rome,1960
P Math 2.2858

FLORENCE 1963 International symposium on iris 1st Report Societa italiana dell'iris Edited by Gian Luigi Sani and others Societa italiana dell'iris,Florence, 1963
BG 38.3303

FLORENCE 1964 Centenario della nascita di Galileo Galilei 1564-1964 4th Atti Vol 2,tomo 1: atti del convegno sulla relativita generale,problemi di energia e onde gravitazionali.Proceedings of the meeting on general electricity problems and gravitational waves Galileo Galilei.Comitato nazionale per le manifestazioni celebrative Galileo Galilei.Comitato nazionale per le manifestazioni celebrative.Publicazioni 267p Barbera,Florence,1965
WSM 43.4484

FLORENCE 1964 Centenario della nascita di Galileo Galilei 1564-1964 4th Atti Vol 2,tomo 2: atti del convegno sulle machie solari.Proceedings of the meeting on sunspots Galileo Galilei.Comitato nazionale per le manifestazioni celebrative Galileo Galilei. Comitato nazionale per le manifestazioni celebrative.Publicazioni port,figs 266p Barbera,Florence,1966
WSM 43.4470

FLORENCE 1964 Convegno sulla relativita: problemi di energia e onde gravitazionali; meeting on general relativity:problems of energy and gravitational waves Atti; proceedings Quatro centenario della nascita di Galileo Galilei,1564-1964.Comitato nazionale per le manifestazioni celebrative Edited by Giorgio Sestini Quatro centenario della nascita di Galileo Galilei,1564-1964. Comitato nazionale per le manifestazioni celebrative.Pubblicazioni, 2,Atti dei convegni, 1 267p Florence,1966
Obs 6.3276

FLORENCE 1965 Hypotensive peptides :an international symposium Proceedings Edited by Ervin C. Erdos and others Springer,Berlin, 1966
Inv Med 37.0252

FLORENCE 1965 Hypotensive peptides : international symposium Proceedings Edited by E.G. Erdos and others Springer,New York, 1966
Pha 16.0118

FLORIDA STATE UNIVERSITY Origins of prebiological systems and of their molecular matrices A Conference on the origins of prebiological systems and of their molecular matrices Proceedings Wakulla Springs,Fla 1963 Oct 27-30 Edited by Sidney W. Fox Bibliog.,illus. xx,482p Academic press,New York;London,1965
Bot 42.1364

FLORIDA STATE UNIVERSITY The Origins of prebiological systems and of their molecular matrices :a conference Proceedings Wakulla springs,Fla. 1963 Oct 27-30 Edited by Sidney W. Fox Under the auspices of the Institute for space biosciences Academic press,New York;London,1965
Gen 34.0456

FLOW MEASUREMENT SYMPOSIUM :presented at Fluid meters golden anniversary Papers Pittsburgh,Pa. 1966 Sep 26-28 American society of mechanical engineers.Fluid measurement research committee American society of mechanical engineers,New York,1966
Eng 41.6717

FLOW PROPERTIES OF BLOOD AND OTHER BIOLOGICAL SYSTEMS :an informal discussion Proceedings Oxford 1959 Sep 23-24 Faraday society.Colloid and biophysics committee British society of rheology Edited by A.L. Copley and G. Stainsby Pergamon,London,1960
PGMS 29.0269

FLOW PROPERTIES OF BLOOD AND OTHER BIOLOGICAL SYSTEMS :an informal discussion... Proceedings Oxford 1959 Sep 23-24 Faraday society and British society of rheology Edited by A.L. Copley and G. Stainsby Pergamon press,Oxford,1960
An 32.3471

FLOW RESEARCH ON BLADING :a symposium Proceedings Brown Boveri symposium 1st Baden,Switzerland 1969 Mar 10-11 Brown, Boveri and company Edited by Lang S. Dzung Elsevier,Amsterdam,1970
Eng 41.6828

FLUCTUATION,RELAXATION AND RESONANCE IN MAGNETIC SYSTEMS Scottish universities' summer school 2nd Lectures Newbattle Abbey 1961 Scottish universities' summer school in physics Edited by D.ter Haar With financial support from the North Atlantic treaty organization Oliver and Boyd, Edinburgh;London,1962
Chem 18.2619

FLUID AMPLIFICATION SYMPOSIUM Proceedings 1965 Oct 26-28 Vol 1 Sponsored by the Harry Diamond laboratories Harry Diamond laboratories,Washington,D.C.,1965
Eng 41.6970

The FLUID DYNAMIC ASPECTS OF SPACE FLIGHT AGARD-NATO specialists' meeting Marseille 1964 Apr 20-24 Vol 1-2 Agard AGARDograph, 87 2 vols Gordon and Breach, New York,1966
Eng 41.6887

FLUID DYNAMICS Fluid dynamics transactions Symposium on fluid dynamics 5th Jablonna 1961 26 Aug.-2 Sep. Vol. 1 Polska akademia nauk Edited by W. Fiszdon Pergamon press,London,1964
A Math 4.0527

FLUID DYNAMICS Fluid dynamics transations,4 Symposium on advanced problems and methods in fluid dynamics 8th Proceedings Tarda, Poland 1967 Sep 18-25 Polska akademia nauk Edited by W. Fiszdon and others bibliog., illus. x,812p 24cm PWN,Warsaw,1969
A Math 4.1680

FLUID DYNAMICS Symposium on geophysical fluid dynamics Woods Hole,Mass. 1969 Sep Woods Hole oceanographic institution Edited by Willem V.R. Malkus Bibliog,illus,port 28cm 2 vols Woods Hole oceanographic institution,Woods Hole,Mass.,1969
A Math 4.1733

FLUID DYNAMICS symposium College park, Md. 1951 Jun 22-23 American mathematical society Edited by M.H. Martin Co-sponsored by the Naval ordnance laboratory American mathematical society.Proceedings of symposia in applied mathematics, 4 McGraw-Hill,New York,1953
A Math 4.0512

FLUID DYNAMICS manuscript By Richard von Mises and Kurt O. Friedrichs Brown university.Advanced instruction and research in mechanics,1941 Providence,R.I.,1942 Containing'The Hodograph method in the theory of compressible fluid'by Stefan Bergman
A Math 4.1292

FLUID DYNAMICS AND APPLIED MATHEMATICS :a symposium Proceedings College Park 1961 Apr 28-29 Edited by J.B. Diaz and S.I. Pai Sponsored by the Institute for fluid dynamics and applied mathematics 207p Gordon and Breach,New York;London,1962
Chem E 24.0371

FLUID DYNAMICS TRANSACTIONS 5th-7th : symposiums Vol 1-3 Edited by W. Fiszden 3 vols Pergamon press,London,1964-67
Eng 41.6532

FLUID DYNAMICS TRANSATIONS,4 Symposium on advanced problems and methods in fluid dynamics 8th Proceedings Tarda,Poland 1967 Sep 18-25 Polska akademia nauk Edited by W. Fiszdon and others bibliog.,illus. x, 812p 24cm PWN,Warsaw,1969
A Math 4.1680

FLUID FLOW Some aspects of fluid flow... :a conference Papers Leamington Spa 1950 Oct 25-28 Institute of physics Arnold, London,1951
Eng 41.6954

FLUID FLOW Two-phase fluid flow :a symposium Proceedings 1962 Feb Institution of mechanical engineers Institution of mechanical engineers,London,1962
Eng 41.7171

FLUID LOGIC AND AMPLIFICATION International conference on fluid logic and amplification 1st Proceedings Cranfield 1965 Sep Edited by C.J. Charnley and H.S. Stephens Organized by the British hydromechanics research association British hydromechanics research association,Cranfield,1965
Eng 41.5895

FLUID MECHANICS Australasian conference on hydraulics and fluid mechanics 1st Proceedings Perth 1962 Dec 6-13 Edited by Richard Silvester Pergamon press,London, 1964
Eng 41.6968

FLUID MECHANICS Biomedical fluid mechanics symposium :presented at Fluids engineering conference Denver,Col. 1966 Apr 25-27 Sponsored by the American society of chemical engineers.Fluids engineering division American society of mechanical engineers,New York,1966
Chem E 24.1797

FLUID MECHANICS Developments in mechanics Midwestern mechanics conference 6th Proceedings Cleveland,Ohio 1963 Apr 1-3 Vol 2,pt 1: fluid mechanics Edited by Simon Ostrach and Robert H. Scanlan Pergamon, London,1965 Combining the 6th Midwestern conference on soil mechanics and the 8th Midwestern conference on fluid mechanics
Eng 41.6959

FLUID MECHANICS Fundamental topics in relativistic fluid mechanics and magnetohydrodynamics :a symposium Proceedings East Lansing,Mich. 1962 Oct Edited by Robert Wasserman and Charles P. Wells Held at Michigan state university Academic press,New York,1963
Eng 41.4547

FLUID MECHANICS Heat transfer and fluid mechanics institute 1958 preprints Berkeley,Calif. 1958 Jun 19-21 University of California Stanford university press, Stanford,Calif.,1958
A Math 4.0517

FLUID MECHANICS Heat transfer and fluid mechanics institute 1961 proceedings Los Angeles 1961 Jun. 19-21 Edited by R.C. Binder and others Stanford university press, Stanford,Calif.,1961
A Math 4.0519

FLUID MECHANICS Heat transfer and fluid mechanics institute 1972 Proceedings Northridge,Calif. 1972 Jun Edited by R.B. Landis and G.J. Hordemann illus xii,430p 25cm Stanford university press,Stanford, Calif.,1972
A Math 4.1876

FLUID MECHANICS Midwestern conference on fluid mechanics 2nd Proceedings Columbus,Ohio 1952 Mar 17-19 Sponsored by Ohio state university.Graduate school Ohio state university.Engineering experiment station.Bulletin, 149 Ohio state university studies:engineering series, 21,no 3 v,529p Ohio state university,Columbus,Ohio,1952
Chem E 24.0402

FLUID MECHANICS Midwestern conference on fluid mechanics 3rd Proceedings Minneapolis 1953 Mar 23-25 Held at the University of Minnesota 783p Minneapolis, Minn.,1953
Chem E 24.0403

FLUID MECHANICS Midwestern conference on fluid mechanics 3rd Proceedings Minneapolis,Minn. 1953 Mar 23-25 University of Minnesota.Institute of technology University of Minnesota, Minneapolis,Minn.,1953
Eng 41.6953

FLUID MECHANICS Midwestern conference on fluid mechanics 4th Proceedings Lafayette,Ind. 1955 Sep 8-9 Held at Purdue university Purdue university.Engineering experiment station.Research series, 128 Purdue university,Lafayette,Ind.,1955
Eng 41.6957

FLUID MECHANICS Midwestern conference on fluid mechanics 4th Proceedings Lafayette,Ind. 1955 Sep 8-9 Held at Purdue university Purdue university.Engineering experiment station.Research series, 128 vii, 370p Purdue university,Lafayette,Ind.,1955
Chem E 24.0404

FLUID MECHANICS Midwestern conference on fluid mechanics 5th Proceedings Ann Arbor,Mich. 1957 Apr 1-2 Held at University of Michigan University of Michigan press,Ann Arbor,Mich.,1957
Eng 41.6958

FLUID MECHANICS Midwestern conference on fluid mechanics 5th Proceedings Ann Arbor,Mich. 1957 Apr 1-2 Held at the University of Michigan University of Michigan.Engineering research institute publications viii,388p University of Michigan press,Ann Arbor,Mich.,1957
Chem E 24.0405

FLUID MECHANICS The Heat transfer and fluid mechanics institute 1960 proceedings Stanford,Calif. 1960 Jun. 15-17 Edited by D.M. Mason and others Stanford university press,Stanford,Calif.,1960
A Math 4.0518

FLUID MECHANICS The Heat transfer and fluid mechanics institute 1962 proceedings Seattle,Wash. 1962 Jun.13-15 Edited by F. Edward Ehlers and others Stanford university press,Stanford,Calif.,1962
A Math 4.0520

FLUID MECHANICS The Heat transfer and fluid mechanics institute 1963 proceedings Pasadena,Calif. 1963 Jun.12-14 Edited by Anatol Roshko and others Stanford university press,Stanford,Calif,1963
A Math 4.0521

FLUID MECHANICS The Heat transfer and fluid mechanics institute 1964 proceedings Berkeley,Calif 1964 Jun.10-12 Edited by Warren H. Gledt and Salomon Levy Stanford university press,Stanford,Calif,1964
A Math 4.0522

FLUID MECHANICS The Heat transfer and fluid mechanics institute 1965 proceedings Los Angeles,Calif. 1965 Jun. 21-23 Edited by Andrew F. Charwat Stanford university press, Stanford,Calif.,1965
A Math 4.0523

FLUID MECHANICS The Heat transfer and fluid mechanics institute 1966 proceedings Santa Clara,Calif. 1966 Jun. 22-24 Edited by Michel A. Saad and James A. Miller Stanford university press,Stanford,Calif.,1966
A Math 4.0524

FLUID MECHANICS The Midwestern conference on fluid mechanics 5th proceedings Ann Arbor 1957 Apr 1-2 Edited by R.C.F. Bartels Under the auspices of the University of Michigan.Engineering research institute University of Michigan.Engineering research institute.Publication University of Michigan press,Ann Arbor,1957
A Math 4.0516

FLUID MECHANICS AND MAGNETOHYDRODYNAMICS Symposium on fundamental topics in fluid mechanics and magnetohydrodynamics Proceedings Michigan State university 1962 Oct 10-11 Michigan State university Edited by Robert Wassermann and Charles P. Wells Academic press,New York,1963
Cav 7.1569

FLUID METERS GOLDEN ANNIVERSARY Flow measurement symposium :presented at Fluid meters golden anniversary Papers Pittsburgh,Pa. 1966 Sep 26-28 American society of mechanical engineers.Fluid measurement research committee American society of mechanical engineers,New York,1966
Eng 41.6717

FLUID MODELS IN GEOPHYSICS Symposium on the use of models in geophysical fluid dynamics 1st Proceedings Baltimore,Md. 1953 Sep 1-4 Edited by Robert R. Long Sponsored by United States.Office of naval research illus, diagrms v,162p U.S.Government printing office,Washington,D.C.,1956
Chem E 24.0202

FLUIDICS Cranfield fluidics conference 2nd Proceedings Cambridge 1967 Turin 1968 Organized by the British hydromechanics research association British hydromechanics research association,Cranfield, 1967-68
Eng 41.5906

FLUIDICS Cranfield fluidics conference 3rd Proceedings Turin 1968 May 8-10 Vol 1-3 Organised by the British hydromechanics research association 3 vols British hydromechanics research association, Cranfield,1968
Eng 41.5907

FLUIDICS I.F.A.C. symposium on fluidics Proceedings London 1968 Nov 4-8 International federation of automatic control Organised by the Royal aeronautical society 1968
Eng 41.5905

FLUIDICS SYMPOSIUM Advances in fluidics Chicago,Ill. 1967 May 9-11 American society of mechanical engineers Edited by Forbes T. Brown American society of mechanical engineers,New York,1967
Eng 41.5898

FLUIDISATION :a symposium Papers London 1963 Mar 11-12 Society of chemical industry. Chemical engineering group and Institution of chemical engineers Sponsored also by the Institute of petroleum 110p Society of chemical industry,London,1964
Chem E 24.1452

FLUIDIZATION International symposium on fluidization Proceedings Eindhoven 1967 Jun 6-9 Edited by A.A.H. Drinkenburg illus 829p Netherlands university press,Amsterdam, 1967
Chem E 24.1918

FLUIDS ENGINEERING CONFERENCE Biomedical fluid mechanics symposium Papers Denver, Colo. 1966 Apr 25-27 American society of mechanical engineers.Fluid mechanics commission Co-sponsored by the Rheology society of America American society of mechanical engineers,New York,1966 Papers presented by the Fluid mechanics commission of the ASME at the Fluids engineering conference, co-sponsored by the Rheology society of America
Eng 41.6540

FLUIDS ENGINEERING CONFERENCE Biomedical fluid mechanics symposium :presented at Fluids engineering conference Denver,Col. 1966 Apr 25-27 Sponsored by the American society of chemical engineers.Fluids engineering division American society of mechanical engineers,New York,1966
Chem E 24.1797

FLUORESCENCE,THEORY AND PRACTICE :a symposium Papers Miami Beach,Fla. 1967 Apr Edited by George G. Guilbault Organixed by the American chemical society.Division of analytical chemistry Arnold;Dekker,London; New York,1967
Col S 12.0091

FOETAL AUTONOMY :a Ciba foundation symposium London 1968 Dec 3-5 Ciba foundation Edited by G.E.W. Wolstenholme and Maeve O'Connor Churchill,London,1969
An 32.2543

FOETAL AUTONOMY :a Ciba foundation symposium London 1969 Ciba foundation Edited by G.E.W. Wolstenholme and Maeve O'Connor Churchill,London,1969
Phys 20.2262

FOETAL AUTONOMY :a Ciba foundation symposium Proceedings London 1968 Dec 3-5 Ciba foundation Edited by G.E.W. Wolstenholme and Maeve O'Connor Churchill,London,1969
Inv Med 37.0237

FOETAL ENDOCRINOLOGY Probleme der fetalen endokrinologie symposium Bonn 1955 Mar 4-5 Edited by H. Nowakowski Deutschen gesellschaft fur endokrinologie.Symposia, 3 Springer,Berlin,1956
An 32.3750

The FOETO-PLACENTAL UNIT an international symposium Proceedings Milan 1968 Sep 4-6 Edited by A. Pecile and C. Finzi Excerpta medica.International congress series Excerpta medica,Amsterdam,1969
Inv Med 37.0260

FOETUS The Mammalian fetus :physiological aspects of development.A symposium Papers Cold Spring Harbor 1954 Cold Spring Harbor biological laboratory Cold Spring Harbor symposia on quantitative biology, 19 Long Island biological association,Cold Spring Harbor,1954
Bioch 33.1275

FOLIA MEDICA NEERLANDICA, 11,5-6 The Living bone :symposium on bone structure and metabolism Edited by R. Steendikk and others Amsterdam,1968
PGMS 29.0177

FOLKESTONE 1957 Anglo-American aeronautical conference 6th Papers Royal aeronautical society Institute of the aeronautical sciences Edited by Joan Bradbrooke and U.A. Libby Royal aeronautical society,London,1959
Eng 41.6962

La FONCTION SPERMATOGENETIQUE DU TESTICULE HUMAIN colloque Paris 1958 Jul 10-12 Societe nationale pour l'etude de la sterilite et de la fecondite Edited by H. Bayle and C. Gouygou Masson,Paris,1958
An 32.3872

Les FONCTIONS DE NIDATION UTERINE ET LEURS TROUBLES :colloque Papers Brussels 1960 Jun 24-26 Societe nationale pour l'etude de la sterilite et de la fecondite Edited by J. Ferin and M. Gaudefroy Masson, Paris,1960
An 32.3834

Les FONCTIONS DU COL UTERIN :colloque Papers Basle 1964 Jul 3-5 Societe nationale pour l'etude de la sterilite et de la fecondite Edited by R. Moricard and R. Wenner Masson,Paris,1964
An 32.3833

FONDATION FRANCAISE D'ETUDES NORDIQUES Congres international de l'industrie morutiere dans l'atlantique-nord :tradition et avenir 1st Rouen 1966 Jan 27-29 Fecamp 1966 Jan 27-29 Tome 1 Fondation francaise d'etudes nordiques.Actes et documents,2 Rouen;Fecamp,1966
Sco 14.1037

FONDATION SINGER-POLIGNAC L' Instinct dans la comportement des animaux et de l'homme : colloque international actes 1954 Jun By M. Autori and others Masson,Paris,1956 Organised by Le Conseil de la Fondation Singer-Polignac
Bal 39.1594

FONDATION SINGER-POLIGNAC L' Instinct dans le comportement des animaux et des hommes : colloque international Actes 1954 Jun By M. Autuori and others Masson,Paris,1956
Psy 31.0644

FONDATION SINGER-POLIGNAC Les Hormones sexuelles colloque international Comptes-rendus Paris 1937 Jun 10-19 Edited by L. Brouha Conferences du college de France Hermann,Paris,1938
Gen 34.1603

FONDATION SINGER-POLIGNAC Les Concepts de Claude Bernard sur le milieu interieur : colloque international Actes Paris 1965 Jun 29-Jul 2 Masson,Paris,1967
Phys 20.1343

FOOD AND AGRICULTURE ORGANIZATION Effects of low doses of radiation on crop plants :a panel Report Rome 1964 Jun 1 International atomic energy agency.Technical reports series, 64 International atomic energy agency, Vienna,1966
Radioth 35.1151

FOOD AND AGRICULTURE ORGANIZATION Mutations in plant breeding Panel on co-ordination of research on the production and use of induced mutations in plant breeding Proceedings Vienna 1966 Jan 17-21 International atomic energy agency.Panel proceedings series 2 vols International atomic energy agency, Vienna,1966-68
Radioth 35.0261

FOOD AND AGRICULTURE ORGANIZATION Mutations in plant breeding II :a research co-ordination meeting on the use of induced mutations in plant breeding Proceedings Vienna 1967 Sep 11-15 International atomic energy agency. Panel proceedings series I.A.E.A.,Vienna, 1968
Radioth 35.0262

FOOD AND AGRICULTURE ORGANIZATION Rice breeding with induced mutations :an FAO-IAEA research co-ordination meeting on the use of induced mutations in rice breeding Report Taipei 1967 Jun 5-9 International atomic energy agency.Technical report series, 86 International atomic energy agency,Vienna,1968
Radioth 35.0271

FOOD SCIENCE :a symposium on quality and preservation of foods Edited by E.C. Bate-Smith and T.N. Morris xvii,319p Cambridge university press,Cambridge,1952 Lectures given at a Summer course in food science arranged by the Board of extra-mural studies of the University of Cambridge,1948
Chem E 24.1922

FOOD SUPPLY AND NATURE CONSERVATION :a symposium Cambridge 1963 Nov 2 Cambridgeshire college of arts and technology Edited by R.I.B. Cooper Cambridge,1964
Bal 39.2963

FORBIDDEN TRANSITIONS Transitions interdites dans les spectres des astres Colloque international d'astrophysique 15ieme Communications Liege 1968 Jun 24-26 Societe royale des sciences de Liege.Memoires, 5s, 17 410p Institut d'astrophysique, Liege,1969
TA 15.0576

FORD FOUNDATION Conference on immuno-reproduction Proceedings La Jolla,Calif. 1962 Sep 9-11 Population council,New York, c1962 Organized by A.Tyler and K.A. Laurence and edited by the latter
Path 30.2340

FORD FOUNDATION Decision processes seminar proceedings Santa Monica,Calif. 1952 summer University of Michigan Edited by R. M. Thrall and others viii,332p 22cm John Wiley and sons,New York,1954
Math 3.0771

FOREBRAIN Evolution of the forebrain : phylogenesis and ontogenesis of the forebrain Edited by R. Hassler and H. Stephan Thieme, Stuttgart,1966 Lectures given at a symposium on "Phylogenesis and ontogenesis of the forebrain",Frankfurt and Sprendlingen,Aug. 1965
An 32.4410

FORENINGEN NORRBOTTENS FRAMJANDE Jorden, skogen,malmen och vattenkraften i morgondagens Norrbotten :a conference Lectures Lulea 1956 Mar 19-20 Edited by Folke Thunberg illus,maps 271p 21cm Foreningen Norrbottens framjande,Lulea,1956
Sco 14.5794

FOREST HYDROLOGY International symposium on forest hydrology Proceedings University Park,Pa. 1965 Aug 29-Sep 10 National science foundation Edited by William E. Sapper and Howard W. Lull Pergamon press, Oxford,1967 A National Science Foundation advanced science seminar
Geog 13.0795

FOREST TREES Physiology of forest trees An International symposium of forest tree physiology 1st Harvard forest 1957 Apr 8-12 Edited by Kenneth Thimann and others Under the auspices of the Maria Moors Cabot foundation Ronald press,New York,1958
Bot 42.1335

FORESTRY International Forestry Congress 3rd programme Helsinki 1940 Jul 1-5 International institute of agriculture Helsinki,1939
Philos 1.2032

FORGEMASTERS' MEETING 1954 Proceedings London 1954 Oct 11-15 Iron and steel institute and National forgemasters' association Joint meeting with the Chambre syndicate de la grosse forge francaise Iron and steel institute.Special report, 60 Iron and steel institute,London,1957
Met 25.0715

FORM AND FUNCTION IN THE HUMAN LUNG :a symposium Proceedings Birmingham 1967 Apr 5-7 Edited by Gordon Cumming and L.B. Hunt Sponsored by the Lederle laboratories Livingstone,Edinburgh,1968
An 32.3805

FORM DISCRIMINATION AS RELATED TO MILITARY PROBLEMS :a symposium Proceedings Medford,Mass. 1957 Apr 4-5 Armed forces-National research council committee on vision Edited by Joseph W. Wulfeck and John H. Taylor National research council.Publication, 561 National research council,Washington,D.C.,1957
Psy 31.0942

FORMAL LANGUAGE DESCRIPTION LANGUAGES FOR COMPUTER PROGRAMMING IFIP working conference on formal language description languages proceedings Vienna 1964 Sep 15-18 International federation for information processing Edited by T.B. Steel bibliog 330p 22cm North-Holland publishing, Amsterdam,1966
Math L 5.1033

FORMAL REPRESENTATION OF HUMAN JUDGEMENT Symposium on cognition 3rd Papers Pittsburgh,Pa. 1967 Apr 13-14 Edited by Benjamin Kleinmuntz Sponsored by the Carnegie-Mellon university Wiley,New York; London,1968
Eng 41.0653

FORMAL SYSTEMS AND RECURSIVE FUNCTIONS The Logic colloquium 8th proceedings Oxford 1963 Jul Edited by J.N. Crossley and M.A.E. Dummett 320p 23cm North-Holland, Amsterdam,1965
P. Math 2.0207

FORSCHUNGSINSTITUT,DAVOS The Immune response and its suppression :an international symposium Davos 1968 Mar 25-28 Edited by E. Sorkin Antibiotica et chematherapia, 15 illus,tables Karger,Basle;New York,1969
Path 30.2545

FORSVARETS FORSKNINGSANSTALT,STOCKHOLM The Conference on structure and reactions of DFP sensitive enzymes Proceedings Stockholm 1966 Sep 5-7 Edited by Edith Heilbronn Research institute of national defence, Stockholm,1967
Bioch 33.1065

FORT COLLINS,COLO 1965 Heritage from Mendel Mendel centennial symposium Proceedings Edited by R. Alexander and E. Derek Styles Sponsored by the Genetics society of America Wisconsin university press,Madison,Wis.,1967
Bot 42.0880

FORT COLLINS,COLO. 1957 Reproduction and infertility :a symposium 3rd Papers Colorado state university.College of veterinary medicine and Colorado state university.Agricultural experiment station Edited by F.X. Gassner and others Pergamon press,London,1958
Chem 26.0313

FORT COLLINS,COLO. 1962 Effects of ionizing radiation on the reproductive system : international symposium Proceedings Colorado state university and American institute of biological sciences Edited by William D. Carlson and F.X. Gassner Pergamon press,Oxford,1964
An 32.0153

FORT COLLINS,COLO. 1965 Heritage from Mendel Mendel centennial symposium Proceedings Genetics society of America Edited by R.Alexander Brink and E.Derek Styles University of Wisconsin press,Madison,Wis., 1967
Gen 34.0994

FORT COLLINS,COLO. 1967 Topological dynamics :an international symposium Edited by Joseph Auslander and Walter H. Gottschalk Held at the University of Colorado Bibliog., Illus. xv,522p 23cm Benjamin,New York; Amsterdam,1968 Dedicated to Frank J.Hahn, 1929-1968
P Math 2.3143

FORT WAINWRIGHT,ALASKA 1960 Symposia on Arctic biology and medicine 1st Proceedings 1: Neural aspects of temperature regulation Edited by John P. Hannon and Eleanor Viereck Under the auspices of the University of Alaska.Geophysical institute illus vii,399p 23cm Arctic aeromedical laboratory,Fort Wainwright,Alaska,1961
Sco 14.0850

FORT WAINWRIGHT,ALASKA 1961 Symposia on Arctic biology and medicine Proceedings 2: Comparative physiology of temperature regulation Edited by John P. Hannon and Eleanor G. Viereck illus viii,455p 23cm Arctic aeromedical laboratory,Fort Wainwright, Alaska,1961
Sco 14.0851

FORT WAINWRIGHT,ALASKA 1962 Symposia on arctic biology and medicine Proceedings influence of cold on host-parasite interactions By Robert I. McClaughry and others Edited by Eleanor G. Viereck Held under the auspices of the University of Alaska. Geophysical institute illus 455p 23cm Fort Wainwright,Alaska,1963
Sco 14.0882

FORT WAINWRIGHT,ALASKA 1963 Symposia on arctic biology and medicine Proceedings 4: frostbite Edited by Eleanor G. Viereck Held under the auspices of the University of Alaska.Geophysical institute illus vi,457p 23cm Arctic aeromedical laboratory,Fort Wainwright,Alaska,1964
Sco 14.0883

FORTRAN A Guide to FORTRAN 4 programming By D.D. McCracken Wiley,New York,1965
Met 25.2208

FORTRAN A Guide to Fortran programming By Daniel D. McCracken J.Wiley,New York,1961
TA 15.0110

FORTRAN Introduction to FORTRAN:a program for self-instruction By S.C. Plumb International business machines corporation vii,203p 23cm McGraw-Hill,New York,1961
Math L 5.0985

FOSFORILIROVANIE I FUNKTSIYA :sympozium Leningrad 1958 Jun 13-18 Akademiya meditsinskikh nauk S.S.S.R. Akademiya meditsinskikh nauk S.S.S.R.Trudy instituta eksperimentalnoi meditsiny Leningrad,1960 Short summaries in English at the end of each chapter
Bioch 33.0561

FOUNDATION ENGINEERING Asian regional conference on soil mechanics and foundation engineering 1st Proceedings New Delhi 1960 Feb International society of soil mechanics and foundation engineering Indian national society of soil mechanics and foundation engineering New Delhi,1960
Eng 41.3172

FOUNDATION ENGINEERING Asian regional conference on soil mechanics and foundation engineering 2nd Proceedings Tokyo 1963 May 1-4 Vol 1-2 International society of soil mechanics and foundation engineering 2 vols 1963
Eng 41.3173

FOUNDATION ENGINEERING Australia-New Zealand conference on soil mechanics and foundation engineering 1st Proceedings Melbourne 1952 Jul University of Melbourne Institution of engineers,Australia 1952
Eng 41.3166

FOUNDATION ENGINEERING Australia-New Zealand conference on soil mechanics and foundation engineering 4th Proceedings Adelaide 1963 Aug 19-23 University of Adelaide 1963
Eng 41.3169

FOUNDATION ENGINEERING Budapest conference on soil mechanics and foundation engineering 3rd Proceedings Budapest 1968 Oct 15-18 Magyar tudomanyos akademia Akademiai kiado, Budapest,1968
Eng 41.3137

FOUNDATION ENGINEERING Congreso panamericano de mecanica de suelos y cimentaciones 1st Proceedings Mexico,D.F. 1959 Sep 7-12 Vol 1-3 Sociedad mexicana de mecanica de suelos 3 vols Mexico,D.F.,1960
Eng 41.3178

FOUNDATION ENGINEERING International conference on soil mechanics and foundation engineering 2nd Proceedings Rotterdam 1948 Jun 21-30 Vol 1-7 7 vols Rotterdam,1948
Eng 41.3139

FOUNDATION ENGINEERING International conference on soil mechanics and foundation engineering 3rd Proceedings Zurich 1953 Aug 16-27 Vol 1-3 3 vols Zurich,1953
Eng 41.3140

FOUNDATION ENGINEERING International conference on soil mechanics and foundation engineering 4th Proceedings London 1957 Aug 12-14 3 vols Butterworths,London, 1957
Eng 41.3141

FOUNDATION ENGINEERING International conference on soil mechanics and foundation engineering 5th Proceedings Paris 1961 Jul 17-22 3 vols Dunod,Paris,1961
Eng 41.3142

FOUNDATION ENGINEERING International conference on soil mechanics and foundation engineering 6th Proceedings Montreal 1965 Sep 8-15 3 vols University of Toronto press,Toronto,1965
Eng 41.3143

FOUNDATION ENGINEERING International conference on soil mechanics and foundation engineering 7th Proceedings Mexico,D.F. 1969 Vol 1-3 3 vols 1969
Eng 41.3144

FOUNDATION ENGINEERING International conference on soils mechanics and foundation engineering Proceedings Cambridge,Mass. 1936 Jun 22-26 Vol 1-3 3 vols Cambridge,Mass.,1936
Eng 41.3138

FOUNDATION ENGINEERING Problems of settlements and compressibility of soils European conference on soil mechanics and foundation engineering Proceedings Wiesbaden 1963 Vol 1-2 2 vols 1963
Eng 41.3175

FOUNDATION ENGINEERING Soil mechanics and foundation engineering Regional conference for Africa 4th Proceedings South African institution of civil engineers 3 vols Balkema,Cape Town,1967
Eng 41.3177

FOUNDATION ENGINEERING Texas conference on soil mechanics and foundation engineering 8th Proceedings Austin,Texas 1956 Sep 14-15 University of Texas.Bureau of engineering research University of Texas,Austin,Texas, 1956
Eng 41.3164

FOUNDATION FOR RESEARCH ON HUMAN BEHAVIOR Communication in organizations :some new research findings.A meeting 1958 Oct Foundation for research on human behavior,Ann Arbor,Mich.,1959 One of four publications of the Foundation bound together
Eng 41.0729

FOUNDATION FOR RESEARCH ON HUMAN BEHAVIOR Modern organization theory :a symposium of the Foundation for research on human behavior Ann Arbor,Mich. 1959 Feb Edited by Mason Haire Wiley,New York,1959 reprinted 1965
Eng 41.0572

FOUNDATION FOR RESEARCH ON HUMAN BEHAVIOR Organization theory in industrial practice :a symposium of the Foundation for research on human behavior 1959 Feb Edited by Mason Haire Wiley,New York;London,1962
Eng 41.0558

FOUNDATION FOR RESEARCH ON HUMAN BEHAVIOR Performance appraisals :effects on employees and their performance.A seminar Ann Arbor, Mich. 1963 Mar Foundation for research on human behavior,Ann Arbor,Mich.,1964
Eng 41.0784

FOUNDATIONS Performance of deep foundations : a symposium San Francisco,Calif. 1968 Jun 23-28 American society for testing materials American society for testing materials.Special technical publication, 444 American society for testing materials,Philadelphia,Pa.,1969
Eng 41.3159

FOUNDATIONS FUND FOR RESEARCH IN PSYCHIATRY The Transmission of schizophrenia:research conference 2nd Proceedings Dorado, Puerto Rico 1967 Jun 26-Jul 1 Edited by David Rosenthal and,Seymour S. Kety Pergamon press,Oxford,1968 The second in a series of three psychiatric research conferences
Psy 28.0084

FOUNDATIONS' FUND FOR RESEARCH IN PSYCHIATRY Transmission of schizophrenia :research conference Proceedings Dorado,Puerto Rico 1967 Jun 26-Jul 1 Edited by Dvid Rosenthal and Seymour S. Kety Foundations' fund for research in psychiatry.Research conference, 2 Pergamon press,Oxford,1968 reprinted 1969
PGMS 29.0566

FOUNDATIONS' FUND FOR RESEARCH IN PSYCHIATRY. RESEARCH CONFERENCE, 2 Transmission of schizophrenia :research conference Proceedings Dorado,Puerto Rico 1967 Jun 26-Jul 1 Foundations' fund for research in psychiatry Edited by Dvid Rosenthal and Seymour S. Kety Pergamon press,Oxford,1968 reprinted 1969
PGMS 29.0566

The FOUNDATIONS OF MATHEMATICS,MATHEMATICAL MACHINES AND THEIR APPLICATIONS colloquium Tihany 1962 Sep 11-15 Magyar tudomanyos akademia.Section of mathematics and physics Edited by Laszlo Kalmar In cooperation with the Bolyai Janos matematikai tarsulat 317p 25cm Akademiai kiado,Budapest,1965
P. Math 2.2359

FOUNDATIONS OF STATISTICAL INFERENCE Joint statistics seminar :a discussion London 1959 Jul 27-28 By Leonard J. Savage and others Birkbeck college Imperial college of science and technology 112p Methuen; Wiley,London;New York,1964
WSM 43.1769

FOUNDATIONS OF STATISTICAL INFERENCE :a symposium Proceedings Waterloo,Ont. 1970 Mar 31-Apr 9 University of Waterloo. Department of statistics Edited by V.P. Godambe and D.A. Sprott viii,519p 25cm Holt,Toronto,1971
Math S 3.1848

FOURIER SERIES Series de Fourier aleatoires mimeograph Montreal 1963 By Jean-Pierre Kahane North Atlantic treaty organization Societe mathematique du Canada Montreal. University.Seminaire de mathematiques superieures, 4 vii,174p 28cm Universite de Montreal,Montreal,1963
P. Math 2.2147

FOX R.H. and others ed. Algebraic geometry and topology :a symposium in honor of S. Lefschetz Princeton,N.J. 1954 Apr 8-10 Princeton mathematical series, 12 viii,399p 24cm Princeton university press,Princeton,N. J.,1957
P. Math 2.0574

FRACTURE An International conference on the atomic mechanisms of fracture Proceedings Swampscott,Mass 1959 Apr 12-16 National research council Edited by Benjamin Lewis Averbach and others 646p Technology press; John Wiley,Cambridge,Mass;New York,1959
Cav 7.0102

FRACTURE An International conference on the atomic mechanisms of fracture Proceedings Swampscott,Mass. 1959 Apr 12-16 Edited by B.L. Averbach and others Sponsored by the National science foundation M.I.T.press; Wiley,Cambridge,Mass.;New York,1959
Eng 41.3767

FRACTURE Atomic mechanisms of fracture :an international conference Proceedings Swampscott,Mass. 1959 Apr 12-16 National research council.Committee for the conference on fracture Edited by B.L. Averback and others Technology press books in science and engineering Technology press,John Wiley; Chapman and Hall,New York;London,1959
Met 25.2323

FRACTURE International conference on fracture 1st Proceedings Sendai 1965 Sep 12-17 Vol 1-3 Edited by T. Yokobori and others 3 vols Japanese society for strength and fracture of materials, Sendai,1966
Eng 41.3807

FRACTURE International conference on fracture 1st Proceedings Sendai 1965 Sep 12-17 Vol 1-3 Organizing committee on fracture Edited by T. Yokobori and others 3 vols Japanese society for strength and fracture of materials,1966
Met 25.0872

FRACTURE International conference on fracture 2nd Proceedings Brighton 1969 Apr Edited by P.L. Pratt and others Chapman and Hall,London,1969
Eng 41.3808

FRACTURE Tewkesbury symposium on fracture 2nd Proceedings Melbourne 1969 Edited by C.J. Osborn and others Held by the University of Melbourne.Fscuity of engineering Butterworths,Sydney,1969 Financial support from the Pearson Tewkesbury bequest
Eng 41.8202

FRACTURE Tewksbury symposium on fracture 1st Proceedings Melbourne 1963 Aug 26-30 Edited by C.J. Osborn University of Melbourne,Melbourne,1965
Eng 41.3802

FRACTURE The Tewksbury symposium 1st Proceedings Melbourne 1963 Aug 26-30 University of Melbourne.Faculty of engineering Edited by C.J. Osborn University of Melbourne,Melbourne,1965
Met 25.0870

FRACTURE The Tewskbury symposium 2nd Proceedings Melbourne 1969 University of Melbourne.Faculty of engineering Edited by C. J. Osborn and others Butterworth,Sydney,1969
Met 25.2579

FRACTURE MECHANICS International symposium on fracture mechanics Kiruna 1967 Aug 7-12 Swedish national committee for mechanics Wolters-Noordhoff,Groningen,1968
Met 25.0874

FRACTURE OF ENGINEERING MATERIALS :a conference Papers New York 1959 Aug Sponsored by the American society for metals American society for metals,Metals Park,Ohio, 1964
Eng 41.3796

FRACTURE OF ENGINEERING MATERIALS a conference Papers New York 1959 Aug American society for metals.Eastern New York chapter A.S.M.,New York,1959
Met 25.0864

FRACTURE OF SOLIDS :an international conference Proceedings Washington,D.C. 1962 Aug 21-24 American institute of mining, metallurgical and petroleum engineers. Institute of metals division Edited by D.C. Drucker and J.J. Gilman Metallurgical society conferences, 20 Interscience,New York;London,1963
Met 25.2247

FRACTURE TOUGHNESS TESTING AND ITS APPLICATIONS a symposium Chicago 1964 Jun 21-25 American society for testing materials American society for testing materials, Philadelphia,Pa.,1965
Met 25.0868

FRACTURE 1969 International conference on fracture 2nd Proceedings Brighton 1969 Apr Edited by P.L. Pratt and others Chapman and Hall,London,1969
Met 25.2327

FRACTURING OF METALS :a seminar...held during the 29th National metal congress and exposition...Co-ordinated by...the metallurgical staff of the Case institute of technology under the direction of George Sachs Chicago,Ill. 1947 Oct 18-24 Sponsored by the American society for metals American society for metals,Cleveland,Ohio,1948
Eng 41.3732

FRACTURING OF METALS a seminar Chicago 1947 Oct 18-24 American society for metals and Case institute of technology American society for metals,Cleveland,Ohio,n.d. A seminar held during the 29th National metal congress and exposition
Met 25.0861

FRANCE.MINISTERE DE L'INDUSTRIE,DES POSTES ET DES TELEGRAPHES Congres international d'aviculture et de peche Memoires et comptes-rendus Paris 1900 Edited by Joseph Perard and Maire 24cm Challamel,Paris,1901
Bal 44.1241

FRANCE.SERVICE GEOGRAPHIQUE DE L'ARMEE Carte du monde au millionieme :Conference internationale 2nd Comptes rendus Paris 1913 Dec Service geographique de l'armee, Paris,1914
Geog 13.5472

FRANCONIA,N.H. 1949 Recent progress in hormone research Laurentian hormone conference Proceedings Vol 5 Edited by Gregory Pincus Academic press,New York, 1950
Bal 39.1376

FRANKFURT 1955 Passivierend filme und deckschichten anlaufschichten mechanismus ihrer entstehung und ihre schutzwirkung gegen korrosion Deutsche gesellschaft fur metallkunde Edited by H. Fischer and K. Wiederholt Springer-verlag,Berlin,1956
Met 25.1944

FRANKFURT 1957 Biophysik,strahlenbiologie und strahlentherapie :vortrage auf der 39 tagung der Deutschen rontgengesellschaft Proceedings Deutsche rontgengesellschaft Edited by B. Rajewsky Urban and Schwarzenberg,Munich,1958
Radioth 35.1617

FRANKFURT 1965 and SPRENDLINGEN Evolution of the forebrain:phylogenesis and ontology of the forebrain :a symposium Lectures World federation of neurology Max Planck gesellschaft zur forderung der wissenschaften Edited by R. Hassler and H. Stephan G.Thieme,Stuttgart,1966
Psy 31.3370

FRANKFURT AM MAIN 1952 Congres international de chimie industrielle und der Dechema-hauptversammlung 1952 Vortrage Dechema-monographie, 21,p245-268 Verlag chemie GMBH,Weinheim,1952
Chem E 24.1906

FRANKLIN Benjamin Atmospheric exploration The Benjamin Franklin memorial symposium Papers American academy of arts and sciences 125p M.I.T.;Wiley,New York,1958
Nap 11.0960

FRANKLIN INSTITUTE Symposium on progress in astrophysics Papers 1939 Feb 17 By Donald H. Menzel and others 1939 From the Proceedings of the American philosophical society,vol 81,no 2.Lacks title page
Obs 6.3230

FRASCATI 1967 Membrane models and the formation of biological membranes International conference on biological membranes 2nd Proceedings Edited by Liana Bolis and B.A. Pethica With the financial support of the North Atlantic treaty organization Illus. xv,337p North-Holland,Amsterdam,1968 A NATO advanced study institute
Pha 16.0016

FRASCATI 1967 Membrane models and the formation of biological membranes The International conference on biological membranes 2nd Proceedings Nato advanced study institute Edited by Liana Bolis and B. A. Perthica Supported by the North Atlantic treaty organization North Holland,Amsterdam, 1968
Inv Med 37.0025

FRASCATI 1971 Cosmic plasma physics :a conference Edited by K. Schindler 369p Plenum press,New York,1972
Obs 6.3625

FREE RADICALS International symposium on free radicals 5th Preprint of papers Uppsala 1961 Jul 6-7 United States.Army. European research office and United States. Air force.Office of scientific research 1961
Chem 18.2625

FREE RADICALS International symposium on free radicals 6th Papers Cambridge 1963 Jul 2-5 University of Cambridge Sponsored in part by Great Britain.Air force. Office of scientific research,OAR Cambridge, 1963 Mimeographed
Chem 18.0414

FREE RADICALS IN BIOLOGICAL SYSTEMS Symposium on free radicals in biological systems Proceedings Stanford,Calif. 1960 Mar 21-23 Stanford university Edited by M. S. Blois Academic press,New York;London,1961
Bioch 33.0309

FREE RADICALS IN BIOLOGICAL SYSTEMS :symposium Proceedings Stanford,Calif. 1960 Mar Edited by M.S. Blois and others illus.,port. Academic press,New York;London,1961
Radioth 35.1407

FREEZE-DRYING Recent research in freezing and drying International symposium on freeze-drying 2nd Papers London 1958 Edited by A.S. Parkes and Audrey U. Smith Organised by the Institute of biology diagrs vii, 320p 25cm Blackwell scientific publications,Oxford,1960
Sco 14.1237

FREEZING AND DRYING :a symposium Papers London 1951 Jun Edited by R.J.C. Harris Held by the Institute of biology Institute of biology,London,1951
Radioth 35.0178

FREEZING AND DRYING :a symposium Report London 1951 Jun Institute of biology Edited by R.J.C. Harris 205p Institute of biology,London,1951
Sco 14.1236

FREEZING AND DRYING OF BIOLOGICAL MATERIALS : a conference New York 1960 Apr 13 New York academy of sciences Edited by Harold T. Meryman and others New York academy of sciences.Annals, 85,2.p501-734 New York, 1960
Bal 39.1259

FREIBURG 1955 Europaischen gesellschaft fur hamatologie congress :colloquium uber aktuelle probleme des transfusionswesens und der immunhamatologie 5th European society of haematology Edited by Herbert Begemann Springer-verlag,Berlin;Gottingen, 1956
Med 36.0022

FREIBURG 1957 Boundary layer research (Grenzschichtforschung) :a symposium Proceedings International union of theoretical and applied mechanics Edited by H. Gortler Springer,Berlin,1958
Eng 41.6961

FREIBURG 1958 Selected topics on ballistics Cranz centenary colloquium. University of Freiburg Agard Edited by Wilbur C. Nelson NATO-AGARD wind tunnel and model testing panel.Publication AGARDograph, 32 illus,port Pergamon press,London,1959
Eng 41.6772

FREIBURG 1960 Neurophysiologie und psychophysik des visuellen systems (the visual systems:neurophysiology and psychophysics)a symposium Edited by Richard Jung and Hans Kornhuber Springer,Berlin,1961 Papers in English,French and German
Psy 31.0355

FREIBURG 1960 The Visual system: neurophysiology and psychophysics :symposium Edited by Richard Jung and Hans Kornhuber Springer,Berlin,1961 Papers in English and German
An 32.4774

FREIBURG 1962 Radio-isotope in der hamatologie :internationales symposion Edited by W. Kiederling and G. Hoffman Nuclear-medizin, 7,supp.1 Schattauer, Stuttgart,1963
Radioth 35.1367

FREIBURG 1962 Zyto- und histochemie in der hamatologie Freiburger symposium :zugleich symposium der gesellschaft deutscher hamatologen 9th Edited by Hans Merker Springer;Blackwell,Berlin;Oxford,1963
Med 36.0394

FREIBURG 1962 Zyto-und histo-chemie in der hamatologie Freiburger symposium :zugleich symposium der gesellschaft deutscher hamatologen 9th Edited by Hans Merker Springer,Berlin,1963
An 32.3186

FREIBURGER SYMPOSIUM 9th :zugleich symposium der gesellschaft deutscher hamatologen Zyto- und histochemie in der hamatologie Freiburg 1962 Oct 25-27 Edited by Hans Merker Springer;Blackwell, Berlin;Oxford,1963
Med 36.0394

FREIBURGER SYMPOSIUM 9th :zugleich symposium der gesellschaft deutscher hamatologen Zyto-und histo-chemie in der hamatologie Freiburg 1962 Oct 25-27 Edited by Hans Merker Springer,Berlin,1963
An 32.3186

FRENCH LICK,IND. 1954 Teaching of pathology,microbiology,immunology,genetics Association of American medical colleges teaching institute 2nd Report Edited by Lura Street Jackson Association of American medical colleges,Chicago,1955
An 32.0993

FRENCH LICK,IND. 1965 Refractory metals and alloys 4 research and development:based on a technical conference Proceedings Vol 1-2 American institute of mining, metallurgical and petroleum engineers. Refractory metals committee Edited by R.I. Jaffee and others Metallurgical society conferences, 41 2 vols Gordon and Breach, New York,1967
Met 25.2250

FRETTING CORROSION Symposium on fretting corrosion New York 1952 Jun 25 American society for testing materials A.S.T.M. Special technical publication, 144 American society for testing materials,Philadelphia,Pa., 1953 Fiftieth anniversary publication
Met 25.1953

FREUDENSTADT 1968 Progress in photosynthesis research The International congress of photosynthesis research Proceedings Vol 1-3 Edited by Helmut Metzner Sponsored by the International union of biological sciences 3 vols International union of biological sciences, Tubingen,1969
Bioch 33.1907

FREUDENSTADT 1968 Progress in photosynthesis research The International congress of photosynthesis research Proceedings Vol 1-3 Edited by Helmut Metzner Sponsored by the International union of biological sciences 3 vols Tubingen, 1969
Bot 42.1905

FRICTION AND WEAR Interdisciplinary approach to friction and wear :a symposium Proceedings 1967 National aeronautics and space administration Washington,D.C.,1968
Eng 41.6272

FRICTION AND WEAR Symposium on friction and wear Proceedings Detroit,Mich. 1957 Jun 10-11 Edited by Robert Davies port Elsevier,Amsterdam,1959
Eng 41.6297

FRIDAY HARBOR 1957 Marine boring and fouling organisms Friday Harbor symposium Edited by D.L. Ray held at the Friday Harbor laboratories Seattle,1959
Bal 39.1950

FRIDAY HARBOR 1960 Nervous inhibition The Friday Harbor symposium 2nd Proceedings Edited by Ernst Florey Sponsored by the National science foundation. Division of regulatory biology Pergamon press,Oxford,1961 Conference also called International symposium on nervous inhibition
Bal 39.1475

FRIDAY HARBOR LABORATORIES Marine boring and fouling organisms Friday Harbor symposium Friday Harbor 1957 Sep Edited by D.L. Ray Seattle,1959
Bal 39.1950

FRIDAY HARBOR SYMPOSIUM Marine boring and fouling organisms Friday Harbor 1957 Sep Edited by D.L. Ray held at the Friday Harbor laboratories Seattle,1959
Bal 39.1950

The FRIDAY HARBOR SYMPOSIUM 2nd Proceedings Nervous inhibition Friday Harbor 1960 May 31-Jun 4 Edited by Ernst Florey Sponsored by the National science foundation. Division of regulatory biology Pergamon press,Oxford,1961 Conference also called International symposium on nervous inhibition
Bal 39.1475

FROBISHER BAY,N.W.T. 1963 Conference of the Arctic co-operatives 1st Minutes Canada.Department of northern affairs and national resources 36cm Ottawa,1963 Typescript
Sco 14.3712

The FRONTAL GRANULAR CORTEX AND BEHAVIOR University Park,Pa. 1962 Aug 8-10 Pennsylvania state university Edited by K. Akert and J.M. Warren McGraw-Hill,New York; London,1964
Psy 31.0654

The FRONTAL GRANULAR CORTEX AND BEHAVIOUR :a symposium University park,Pa. 1962 Aug 8-10 Edited by J.M. Warren and K. Akert Supported by grants from the National science foundation McGraw-Hill,New York,1963
An 32.4656

The FRONTAL LOBES Proceedings New York 1947 Dec 12-13 Association for research in nervous and mental disease Association for research in nervous and mental disease. Research publications, 27 Baltimore,Md., 1948
Phys 20.1992

The FRONTAL LOBES Proceedings of the Association New York 1947 Dec 12-13 Association for research in nervous and mental disease Edited by Mary P. Wheeler Association for research in nervous and mental disease.Research publications, 27 Hafner, New York,1948 reprinted 1966
An 32.5330

FRONTIERS IN CYTOCHEMISTRY:THE PHYSICAL AND CHEMICAL ORGANIZATION OF THE CYTOPLASM :a symposium Chicago,Ill. 1942 Nov 13 Edited by Normand L. Hoerr Biological symposia, 10 Cattell press,Lancaster,Pa., 1943 Held in honour of the 75th birthday of Dr. R.R.Bensley
An 32.3370

FRONTIERS OF NUMERICAL MATHEMATICS : symposium Madison,Wis. 1959 Oct 30-31 United States.Army.Mathematics research center National bureau of standards United States army.Mathematics research center.Publications, 4 xi,132p 24cm University of Wisconsin, Madison,Wis.,1960
Math L 5.3232

FROST ACTION IN SOILS 30th :a symposium Annual meeting Washington 1951 Jan 9-12 Highway research board National research council.Publication,213 Highway research board.Special report 2 illus,tables,diagrs x,385p 25cm Highway research board, Washington,D.C.,1952
Sco 14.0395

Le FROTTEMENT INTERIEUR DES METAUX Comptes rendus Saint-Germain-en-Laye 1960 Oct 13-14 Institut de recherches de la siderurgie francaise and Centre national de recherches metallurgiques Edited by C. Crussard and others Editions metaux,n.p., c1961
Met 25.1026

The FROZEN CELL :a Ciba foundation symposium London 1969 May 20-22 Ciba foundation Edited by G.E.W. Wolstenholme and Maeve O'Connor Churchill,London,1970
An 32.5444

The FROZEN CELL :a Ciba foundation symposium Proceedings London 1969 May 20-22 Ciba foundation Edited by G.E.W. Wolstenholme and Maeve O'Connor illus. ix, 316p Churchill,London,1970
Bot 42.1383

FUEL AND THE FUTURE :a conference Proceedings London 1946 Oct 8-10 Vol 1-3 Great Britain.Ministry of fuel and power Edited by Harry Mitchell Spiers 3 vols H.M.S.O.,London,1948
Chem E 24.1370

FUEL CELLS :a symposium Atlantic City 1959 Sep American chemical society.Gas and fuel division Edited by G.J. Young Reinhold;Chapman and Hall,New York;London,1960
Met 25.1838

FUEL CELLS a symposium Proceedings Atlantic City 1959 Sep American chemical society.Gas and fuel division Edited by G.J. Young v,154p Reinhold;Chapman and Hall,New York;London,1960 Held during the 136th national meeting of the American chemical society
Chem E 24.1400

FUEL ELEMENT FABRICATION with special emphasis on cladding materials:a symposium Proceedings Vienna 1960 May 10-13 Vol 1-2 International atomic energy agency 2 vols Academic press,London;New York,1961
Met 25.1547

FULLY SEPARATED FLOWS Symposium on fully separated flows :a conference Papers Philadelphia,Pa. 1964 May 18-20 American society of mechanical engineers Edited by Arthur G. Hansen American society of mechanical engineers,New York,1964
Eng 41.6514

FUME ARRESTMENT Iron and steel institute annual general meeting,1963 Proceedings London 1963 Nov 26-27 Iron and steel institute Iron and steel institute.Special report, 85 Iron and steel institute,London, 1964
Met 25.0420

FUNCTION ALGEBRAS Summer gathering on function algebras Papers Aarhus 1969 Jul Aarhus universitet.Matematisk institut Aarhus universitet.Matematisk institut.Various publications series, 9 Bibliog. 78p 29cm University of AArhus,AArhus,August,1969
P Math 2.3451

FUNCTION AND STRUCTURE IN MICRO-ORGANISMS : symposium London 1965 Apr Society for general microbiology Society for general microbiology.Symposia, 15 Cambridge university press,Cambridge,1965
Gen 34.0702

FUNCTION AND STRUCTURE IN MICRO-ORGANISMS :a symposium Papers London 1965 Apr Society for general microbiology Held at the Middlesex hospital Society for general microbiology.Symposia, 15 Cambridge university press,Cambridge,1965
Bioch 33.1171

FUNCTION AND STRUCTURE IN MICRO-ORGANISMS London 1965 Apr Society for general microbiology Edited by M.R. Pollock and M.H. Richmond Society for general microbiology. Symposium,15 Cambridge university press, Cambridge,1965
Pha 16.0162

FUNCTION AND TAXONOMIC IMPORTANCE Systematics association symposium 3rd papers Oxford 1957 Apr 7-9 Systematics association Edited by A.J. Cain Systematics association.Publication,3 Systematics association,London,1959
Geol 8.0941

FUNCTION AND TAXONOMIC IMPORTANCE :a symposium Oxford 1957 Apr 7-9 By A.J. Cain Systematics association Systematics association.Publication, 3 Systematics association,London,1959
Gen 34.1623

FUNCTION AND TRAINING OF THE CHEMICAL ENGINEER international conference London 1955 Mar 21-23 Organisation for European economic co-operation 85p O.E.E.C.,Paris,1955
Chem E 24.1760

FUNCTION SPACE Analysis in function space Conference on the theory and applications of analysis in function space Proceedings Dedham,Mass. 1963 Jun 9-13 Edited by William T. Martin and Irving Sega vi,218p 24cm MIT press,Cambridge,Mass.,1964
Math S 3.1659

FUNCTION SPACE Analysis in function space Conference on the theory and applications of analysis in function space proceedings Dedham,Mass. 1963 Jun 9-13 Massachusetts institute of technology Edited by William Ted Martin and Irving Segal Financially supported by United States.Air force.Office of scientific research vi,218p 24cm M.I.T. press,Cambridge,Mass.,1964
P. Math 2.1181

FUNCTION THEORY Contributions to function theory international colloquium papers presented Bombay 1960 Jan 12-19 Tata institute of fundamental research In joint sponsorship with the Sir Dorabji Tata trust xii,231p 24cm Tata institute of fundamental research,Bombay,1960
P. Math 2.1584

FUNCTIONAL ANALYSIS Colloque sur l'analyse fonctionelle 2nd Liege 1964 May 4-6 Centre Belge de recherches mathematiques 166p 25cm Librairie universitaire,Louvain, 1964
P. Math 2.1182

FUNCTIONAL ANALYSIS Colloque sur l'analyse fonctionnelle Louvain 1960 May 25-28 Centre Belge de recherches mathematiques 91p 25cm Libraire universitaire,Louvain,1961
P. Math 2.1184

FUNCTIONAL ANALYSIS International conference on functional analysis and related topics Proceedings Tokyo 1969 Apr 1-8 International mathematical union Science council of Japan Mathematical society of Japan xxxiii,425p 26cm University of Tokyo press,Tokyo,1970
P Math 2.4279

FUNCTIONAL ANALYSIS :a symposium Proceedings Monterey,Calif. 1969 Oct Edited by Carroll O. Wilde bibliog. vii,162p 24cm Academic press,New York;London,1970
P Math 2.3866

FUNCTIONAL ANALYSIS AND RELATED FIELDS :a conference held in honour of Professor Marshall Harvey Stone Proceedings Chicago,Ill. 1968 May 20-24 Edited by Felix E. Browder Supported by United States. Air Force.Office of scientific research port. 241p 24cm Springer,Berlin,1970
P Math 2.3863

FUNCTIONAL BIOCHEMISTRY OF CELL STRUCTURES International congress of biochemistry 5th Proceedings Moscow 1961 Aug 10-16 International union of biochemistry Edited by O. Lindberg I.U.B.symposium series, 22 Pergamon;PWN-Polish scientific publishers,New York;Warsaw,1963
Bioch 33.1331

FUNCTIONAL EQUATIONS United States-Japan seminar on differential and functional equations Proceedings Minneapolis,Ma. 1967 Jun 26-30 Edited by William A. Harris and Yasutaka Sibuya Illus. xvi,585p 24cm Benjamin,New York;Amsterdam,1967
P Math 2.3164

FUNCTIONAL EQUATIONS AND INEQUALITIES Centro internazionale matematico estivo 3 ciclo La Mendola,Trento 1970 Aug 20-28 Centro internazionale matematico estivo 426p 29cm Edizioni cremonese,Rome,1971 Conference coordinator:B.Forte
P Math 2.4213

FUNCTIONAL GROUPS IN ENZYMES Chemical reactivity and biological role of functional groups in enzymes Oxford 1970 Jan Biochemical society Edited by R.M.S. Smellie Biochemical society.Symposium, 31 Academic press,London,1970
Bioch 33.2315

FUNCTIONAL UNITS IN PROTEIN BIOSYNTHESIS Federation of European biochemical societies meeting 7th Proceedings Varna 1971 Sep Federation of European biochemical societies Edited by R.A. Cox and A.A. Hadjiolov Federation of European biochemical societies.Publications, 23 Academic press, London,1972
Bioch 33.2374

FUNCTIONS OF THE CORPUS CALLOSUM London 1964 Sep 4 Ciba foundation Edited by E.G. Ettlinger Ciba foundation study group, 20 Churchill,London,1965
An 32.4657

FUNCTIONS OF THE CORPUS CALLOSUM London 1964 Sep 4 Ciba foundation Edited by E.G. Ettlinger Ciba foundation study groups, 20 Churchill,London,1965
Phys 20.1996

FUNCTIONS OF THE CORPUS CALLOSUM :a symposium Proceedings London 1964 Sep 4 Ciba foundation Edited by E.George Ettlinger Ciba foundation study group, 20 J.and A. Churchill,London,1965 Proceedings dedicated to the Rt.hon.lord Adrian,O.M.,on his retirement
Psy 31.0193

FUNDAMENTAL ASPECTS OF RADIOSENSITIVITY report of a symposium Upton,N.Y. 1961 Jun 5-7 Brookhaven national laboratory. Biology department Brookhaven symposia in biology, 14 BNL 675 Brookhaven national laboratory,Upton,N.Y.,1961
Gen 34.1115

FUNDAMENTAL ISSUES IN ASSOCIATIVE LEARNING :a symposium Proceedings Halifax 1968 Jun Dalhousie university Edited by N.J. Mackintosh and W.K. Honig Dalhousie university press,Halifax,1969
Psy 31.3017

FUNDAMENTAL MECHANISMS OF PHOTOGRAPHIC SENSITIVITY :a symposium Proceedings Bristol 1950 Mar Butterworths,London,1951
Radioth 35.1506

FUNDAMENTAL PARTICLES The International conference on fundamental particles and low temperature Report Cambridge 1946 Jul 22-27 Vol 1: fundamental particles Physical society and Cavendish laboratory Physical society,London,1947
Cav 7.0460

FUNDAMENTAL PHENOMENA IN THE MATERIAL SCIENCES
Symposium on fundamental phenomena in the material sciences 1st Proceedings Boston,Mass. 1963 Feb 1 Vol 1: sintering and plastic deformation Ilikon corporation, Natick Edited by L.J. Bonis and H.H. Hausner Plenum press,New York,1964
Met 25.2280

FUNDAMENTAL PHENOMENA IN THE MATERIALS SCIENCES
2nd,3rd :a symposium Boston,Mass. 1964,1965 Vol 2-3: surface phenomena Ilikon corporation,Natick Edited by L.J. Bonis and others 2 vols Plenum press,New York,1966
Met 25.1616

FUNDAMENTAL PHENOMENA IN THE MATERIALS SCIENCES
4th :a symposium Proceedings Boston,Mass. 1966 Jan 31-Feb 1 Vol 4: fracture of metals,polymers,and glasses Ilikon corporation,Natick Edited by L.J. Bonis and others Plenum press,New York,1967
Met 25.0873

FUNDAMENTAL PROBLEMS IN TURBULENCE AND THEIR RELATION TO GEOPHYSICS SYMPOSIUM proceedings
Marseilles 1961 Sep.4-9 Vol 67 International union of geodesy and geophysics and International union of theoretical and applied mechanics Edited by Francois N. Frenkel Journal of geophysical research,Vol 67,no.8 American geophysical union, Washington D.C.,1962
A Math 4.0577

FUNDAMENTAL PROCESSES IN RADIATION CHEMISTRY
a general discussion Notre Dame,Ind. 1963 Sep 2-4 Faraday society Faraday society.Discussions, 36 Butterworths,London, 1963
Radioth 35.1772

FUNDAMENTAL TOPICS IN RELATIVISTIC FLUID MECHANICS AND MAGNETOHYDRODYNAMICS :a symposium Proceedings East Lansing,Mich. 1962 Oct Edited by Robert Wasserman and Charles P. Wells Held at Michigan state university Academic press,New York,1963
Eng 41.4547

FUNDAMENTALS OF DEFORMATION PROCESSING
Sagamore army materials conference 9th Raquette Lake,New York 1962 Aug 28-31 By Walter A. Backofen and others United States. Army materials research agency and Syracuse university Co-organized and directed by the National research council Syracuse university press,Syracuse,New York, 1964
Met 25.0741

FUNDAMENTALS OF GAS-SURFACE INTERACTIONS :a symposium Proceedings San Diego,Calif. 1966 Dec 14-16 United States.Air force. Office of scientific research General dynamics corporation.General atomic division Edited by Howard M. Saltsburg and others Academic press,New York,1967
Chem 18.2696

FUNDAMENTALS OF KERATINIZATION :a symposium presented at the New York meeting of the New York meeting of the American association for the advancement of science New York 1960 Dec 30 American association for the advancement of science Edited by Earl O. Butcher and Reidar F. Sognnaes American association for the advancement of science. Publications, 70 American association for the advancement of science,Washington,D.C., 1962
An 32.3280

FUNFZIG JAHRE RELATIVITATSTHEORIE...JUBILEE OF RELATIVITY THEORY Verhandlungen Bern 1955 Jul 11-16 Edited by A. Mercier and M. Kervaire Helvetica physica acta.Supplementum, 5 Birkhauser,Basel,1956 In French, German,English
Radioth 35.1599

FUNGI Incompatibility in fungi :a symposium Proceedings Edinburgh 1964 Aug Edited by Karl Esser and John R. Raper viii,124p 24cm Springer,Berlin,1965 Symposium held at the 10th International botanical congress
Bot 42.4713

FUNGUS SPORE :a symposium Proceedings Bristol 1966 Mar 28-Apr 1 Colston research society Edited by M.F. Madelin Colston research society.Symposia, 18 Colston papers, 18 Butterworths,London,1966
Bot 42.4006

FUNKTIONALANALYSIS APPROXIMATIONSTHEORIE NUMERISCHE MATHEMATIK :vortragsauszuge der tagung uber numerische probleme in der approximationstheorie vom 22.bis 25.Juni 1965, und der tagung uber funktionalanalytische methoden in der numerischen mathematik vom 15. bis 20 November,1965 Oberwolfach 1965 Edited by L. Collatz and others International series of numerical mathematics, 7 Bibliog. 232p 24cm Birkhauser Verlag,Basel,1967
P Math 2.3448

FUNKTIONALANALYTISCHE METHODEN DER NUMERISCHEN MATHEMATIK :ein tagung Vortragsauszuge Oberwolfach 1967 Nov 19-25 Edited by L. Collatz and H. Unger International series of numerical mathematics, 12 Bibliog. 143p 24cm Birkhauser Verlag,Basel,1969
P Math 2.3657

FUNKTIONELLE UND MORPHOLOGISCHORGANISATION DER ZELLE Rottach-Egern 1962 Aug Gesellschaft deutscher naturforscher und arzte Springer,Berlin,1963
Gen 34.0830

FUTURE ENVIRONMENTS OF NORTH AMERICA
Warrenton,Va. 1965 Apr Edited by F.Fraser Darling and John P. Milton Natural history press,Garden City,N.Y.,1966
Geog 13.4873

FUTURE OF ARID LANDS International arid land meetings Papers and recommendations University of New Mexico 1955 Apr 26-May 4 Edited by Gilbert F. White American association for the advancement of science. Publications,43 American association for the advancement of science,Washington,D.C.,1956
Geog 13.1467

FUTURE OF IRONMAKING IN THE BLAST FURNACE autumn general meeting... Papers and discussions London 1961 Nov Iron and steel institute Iron and steel institute. Special report, 72 Iron and steel institute, London,1962
Met 25.0368

The FUTURE OF STATISTICS :a conference Proceedings Madison,Wis. 1967 Jun Edited by Donald G. Watts xvi,316p 23cm Academic press,New York;London,1968
Math S 3.1458

GAINESVILLE 1963 Deformation twinning :a conference Proceedings American institute of mining,metallurgical and petroleum engineers and University of Florida. College of engineering In co-operation with Florida institute for continuing university studies Metallurgical society conferences, 25 Gordon and Breach,New York;London,1964
Met 25.1257

GAINESVILLE,FLA. 1954 Monte Carlo methods symposium mimeograph University of Florida. Statistical laboratory Edited by Herbert A. Meyer Sponsored by the Wright air development center Wiley publications in statistics 382p 29cm John Wiley and sons, New York,1956
P. Math 2.2239

GAINESVILLE,FLA. 1954 Symposium on Monte Carlo methods University of Florida. Statistical laboratory Edited by Herbert A. Meyer Sponsored by Wright air development center Wiley publications in statistics bibliog xvi,381p 28cm John Wiley and sons,New York,1956
Math L 5.0746

GAINESVILLE,FLA. 1954 Symposium on Monte Carlo methods University of Florida. Statistical laboratory Edited by Herbert A. Meyer Sponsored by the Wright air development center Wiley publications in statistics Bibliog. xvi,382p 28cm John Wiley and sons,New York,1956
Math 3.0208

GAINESVILLE,FLA. 1962 Caribbean:Venezuelan development,a case history Conference on the Caribbean 13th Papers Edited by A.Curtis Wilgus University of Florida.School of Inter-American studies.Publications.Series 1,13 University of Florida press,Gainesville,Fla., 1963
Geog 13.5242

GAINESVILLE,FLA. 1963 Caribbean:Mexico today Conference on the Caribbean 14th Papers University of Florida.Center for Latin American studies Edited by A.Curtis Wilgus University of Florida.Center for Latin American studies.Publications.Series 1, 14 University of Florida press,Gainesville, Fla.,1964
Geog 13.5202

GAITHERSBURG,MD. 1967 Technical meeting concerning wind loads on buildings and structures Proceedings National bureau of standards Edited by R.D. Marshall and H.C.S. Thom National bureau of standards.Building science series, 30 US government printing office,Washington,D.C.,1970
Eng 41.8274

GALACTIC ASTRONOMY Stony Brook,N.Y. 1968 Jun 19-Jul 17 Edited by Hong-Yee Chiu and A. Muriel 2 vols Gordon and Breach,New York, 1970
TA 15.0656

GALACTIC RESEARCH Co-ordination of galactic research :conference 2nd Saltsjobaden 1957 Jun.17-22 International astronomical union Edited by A. Blaauw and others International astronomical union.Symposium, 7 93p Cambridge university press,Cambridge, 1959
Obs 6.0192

GALACTIC RESEARCH Co-ordination of galactic research a symposium International astronomical union Edited by A. Blaauw International astronomical union.Symposium, 1 59p Cambridge university press,Cambridge, 1955
Obs 6.0191

GALACTIC STRUCTURE Comparison of the large-scale structure of the galactic system with that of other stellar systems :a symposium papers Dublin 1955 Sep 2 International astronomical union Edited by N.G. Roman International astronomical union. Symposium, 5 illus 72p Cambridge university press, Cambridge,1958
Obs 6.1466

GALACTIC STRUCTURE Observational aspects of galactic structure :a summer school typescript Lagonissi 1964 Sep.9-23 Edited by A. Blaauw and L.N. Mavridis Under the auspices of the North Atlantic treaty organization.Science committee 370p Athens, 1965
Obs 6.0193

GALACTIC SYSTEM Present problems concerning the structure and evolution of the galactic system NUFFIC international summer course in science Lecture notes Breukelen 1960 Jul. 28- Aug. 16 Netherlands universities foundation for international cooperation Edited by J.H. Oort and H.G. Quik Supported by North atlantic treaty organization The Hague,1960
Obs 6.2067

GALACTIC SYSTEM Radio astronomy and the galactic system Symposium on radio astronomy and the galactic system Noordwijk 1966 Aug 25-Sep 1 International astronomical union Edited by Hugo van Woerden Organised in co-operation with International scientific radio union International astronomical union. Symposium, 31 xviii,501p Academic press, London;New York,1967
A Math 4.1574

GALAPAGOS The Symposia of the Galapagos international scientific project Berkeley, Calif. 1964 Jan 7-18 Edited by Robert I. Bowman illus,tables xvii,318p California university press,Berkeley;Los Angeles,1966 Conference held aboard ship en route to the Islands from Jan 11-18
Bot 42.2246

The GALAPAGOS :proceedings of the symposia of the Galapagos international scientific project Proceedings Berkeley,Calif. 1964 Jan 7-8 and 1964 Jan 11-18 University of California Edited by Robert I. Bowman illus,pls 27cm University of California press,Berkeley;Los Angeles,1966
Bal 44.4211

GALAPAGOS INTERNATIONAL SCIENTIFIC PROJECT The Galapagos :proceedings of the symposia of the Galapagos international scientific project Proceedings Berkeley,Calif. 1964 Jan 7-8 and 1964 Jan 11-18 University of California Edited by Robert I. Bowman illus,pls 27cm University of California press,Berkeley;Los Angeles,1966
Bal 44.4211

GALAXIES Distribution and motion of interstellar matter in galaxies :a conference Proceedings Princeton,N.J. 1961 Apr 10-20 Institute for advanced study Edited by L. Woltjer 330p W.A.Benjamin,New York,1962
Obs 6.3253

GALAXIES External galaxies and quasi-stellar objects Upsala 1970 Aug 10-14 International astronomical union Edited by D. S. Evans International astronomical union. Symposium, 44 549p Reidel,Dordrecht,1972
TA 15.0658

GALAXIES Non-stable phenomena in galaxies Nestasionarnye yavleniya v galaktikakh Byurakan 1966 May 4-12 International astronomical union International astronomical union.Symposium, 29 480p Akademiya nauk Armyanskey S.S.R.,Erevan,1968
Obs 6.3269

GALAXIES Structure and evolution of galaxies Conference on physics 13th Proceedings Brussels 1964 Sep Universite libre de Bruxelles and Institut international de physique Solvay 174p Interscience,London, 1965
Obs 6.2134

GALAXIES Structure and evolution of galaxies The Conference on physics 13th Proceedings Brussels 1964 Sep Institut international de physique Solvay and Universite libre de Bruxelles Solvay conference,1964,13 Interscience,London,1965 Other Solvay conferences issued as Conseil de physique,q.v.
Cav 7.2078

GALAXIES Structure et evolution de l'univers Conseil de physique 11e Rapports Brussels 1958 Jun 9-13 Universite libre de Bruxelles Institut de physique Solvay Edited by R. Stoops 309p Brussels,1958
Obs 6.3255

GALAXIES Structure et evolution de l'univers Conseil de physique 11eme Rapports Brussels 1958 Jun 9-13 Institut international de physique Solvay and Universite libre de Bruxelles Edited by R. Stoops Solvay conference,1958,11 Brussels, 1958
Cav 7.2084

GALAXIES Study week on nuclei of galaxies Vatican City 1970 Apr 13-18 Edited by D.J. K. O'Connell Accademia pontifica dei nuovi lincei.Scripta varia, 35 795p North-Holland,Amsterdam,1971
Obs 6.3642

GALAXIES Study week on nuclei of galaxies Vatican City 1971 Apr 13-18 Accademia pontifica dei nuovi lincei Edited by D.J.K. O'Connell Accademia pontifica dei nuovi lincei.Scripta varia, 35 795p North-Holland,Amsterdam,1971
TA 15.0661

GALAXIES Vetlesen symposium Galaxies and the universe :the Vetlesen symposium Invited lectures New York 1966 Oct 19 Edited by L. Woltjer 112p Columbia university press, New York,1968
TA 15.0498

GALAXIES AND THE UNIVERSE The Vetlesen symposium New York 1966 Oct 19 Edited by Lodewijk Woltjer 112p Columbia university press,New York,1968
Obs 6.3452

GALAXY The Spiral structure of our galaxy Basle 1969 Aug 29-Sep 4 International astronomical union Edited by W. Becker and G. Contopoulos International astronomical union.Symposium, 38 478p Reidel,Dordrecht, 1970
Obs 6.3320

GALAXY The Spiral structure of our galaxy : symposium Proceedings Basle 1969 Aug 29-Sep 4 International astronomical union Edited by W. Becker and G. Contopoulos International astronomical union.Symposium, 38 478p Reidel,Dordrecht,1970
TA 15.0489

The GALAXY AND THE MAGELLANIC CLOUDS :a symposium Canberra 1963 Mar.18-28 International astronomical union and International scientific radio union Edited by F.J. Kerr and A.W. Rodgers International astronomical union.Symposium, 20 393p Australian academy of science,Canberra,1964 Dedicated to Joseph Lade Pawsey
A Math 4.1158

The GALAXY AND THE MAGELLANIC CLOUDS :a symposium Proceedings Canberra 1963 Mar.18-28 International astronomical union and International scientific radio union Edited by F.J. Kerr and A.W. Rodgers International astronomical union.Symposium, 20 393p Australian academy of science, Canberra,1964 Publication dedicated to memory of Joseph Lade Pawsey
Obs 6.0941

GALILEI Galileo Centenario della nascita di Galileo Galilei 1564-1964 4th Atti Florence 1964 Sep 9-12 Vol 2,tomo 1: atti del convegno sulla relativita generale, problemi di energia e onde gravitazionali. Proceedings of the meeting on general electricity problems and gravitational waves Galileo Galilei.Comitato nazionale per le manifestazioni celebrative Galileo Galilei. Comitato nazionale per le manifestazioni celebrative.Publicazioni 267p Barbera, Florence,1965
WSM 43.4484

GALILEI Galileo Centenario della nascita di Galileo Galilei 1564-1964 4th Atti Florence 1964 Sep 9-12 Vol 2,tomo 2: atti del convegno sulle machie solari. Proceedings of the meeting on sunspots Galileo Galilei.Comitato nazionale per le manifestazioni celebrative Galileo Galilei. Comitato nazionale per le manifestazioni celebrative.Publicazioni port,figs 266p Barbera,Florence,1966
WSM 43.4470

GALILEI Galileo Centenario della nascita di Galileo Galilei 1564-1964 4th Atti Rome 1964 Sep 14-16 Vol 2,tomo 4: atti del convegno sui campi magnetici solari e la spettroscopia ad alta resolutione.Proceedings of the meeting on solar magnetic fields and high resolution spectroscopy Galileo Galilei. Comitato nazionale per le manifestazioni celebrative Galileo Galilei.Comitato nazionale per le manifestazioni celebrative. Publicazione figs 341p Barbera,Florence, 1966
WSM 43.4471

GALILEI Galileo Convegno sulla cosmologia; meeting on cosmology Atti;Proceedings Padua 1964 Sep 14-16 and Venice Quatro centenario della nascita di Galileo Galilei 1564-1964.Comitato nazionale per le manifestazioni celebrative Edited by L. Rosino Organized by Universita di Padua Quatro centenario della nascita di Galileo Galilei.Comitato nazionale per le manifestazioni celebrative.Pubblicazioni, 2, Atti dei convegni, 3 206p Florence,1966
Obs 6.3451

GALILEI Galileo Convegno sulla relativita: problemi di energia e onde gravitazionali; meeting on general relativity:problems of energy and gravitational waves Atti; proceedings Florence 1964 Sep 9-12 Quatro centenario della nascita di Galileo Galilei,1564-1964.Comitato nazionale per le manifestazioni celebrative Edited by Giorgio Sestini Quatro centenario della nascita di Galileo Galilei,1564-1964.Comitato nazionale per le manifestazioni celebrative. Pubblicazioni, 2,Atti dei convegni, 1 267p Florence,1966
Obs 6.3276

GALILEO GALILEI.COMITATO NAZIONALE PER LE MANIFESTAZIONI CELEBRATIVE Centenario della nascita di Galileo Galilei 1564-1964 4th Atti Florence 1964 Sep 9-12 Vol 2,tomo 1: atti del convegno sulla relativita generale,problemi di energia e onde gravitazionali.Proceedings of the meeting on general electricity problems and gravitational waves Galileo Galilei.Comitato nazionale per le manifestazioni celebrative.Publicazioni 267p Barbera,Florence,1965
WSM 43.4484

GALILEO GALILEI.COMITATO NAZIONALE PER LE MANIFESTAZIONI CELEBRATIVE Centenario della nascita di Galileo Galilei 1564-1964 4th Atti Florence 1964 Sep 9-12 Vol 2,tomo 2: atti del convegno sulle machie solari.Proceedings of the meeting on sunspots Galileo Galilei.Comitato nazionale per le manifestazioni celebrative.Publicazioni port, figs 266p Barbera,Florence,1966
WSM 43.4470

GALILEO GALILEI.COMITATO NAZIONALE PER LE MANIFESTAZIONI CELEBRATIVE Centenario della nascita di Galileo Galilei 1564-1964 4th Atti Rome 1964 Sep 14-16 Vol 2, tomo 4: atti del convegno sui campi magnetici solari e la spettroscopia ad alta resolutione. Proceedings of the meeting on solar magnetic fields and high resolution spectroscopy Galileo Galilei.Comitato nazionale per le manifestazioni celebrative.Publicazione figs 341p Barbera,Florence,1966
WSM 43.4471

GALILEO QUADRICENTENNIAL Homage to Galileo Papers Rochester,N.Y. 1964 Oct 8-9 University of Rochester Edited by Morton F. Kaplon xii,139p MIT press,Cambridge,Mass.; London,1965
WSM 43.1985

GAMAGORI 1966 The Dynamic role of molecular constituents in plant-parasite interaction :a conference Proceedings American phytopathological society Edited by Chester J. Mirocha and Ikuzo Uritani American phytopathological society,St.Paul, Minn.,1967
Bot 42.3761

GAMAGORI,JAPAN 1963 Economic development with special reference to east Asia International economic association.Round table conference Proceedings Edited by Kenneth Berrill Macmillan,London,1964
Geog 13.2407

GAME THEORY Theory of games:techniques and applications conference proceedings Toulon 1964 Jun.29-Jul.3 North Atlantic treaty organization.Scientific affairs committee Edited by A. Mensch xi,490p 23cm English universities press,London,1966
Math 3.0241

GAMETES The Biochemistry of fertilization and the gametes :a symposium London 1951 Feb 17 Biochemical society Edited by R.T. Williams Held at the London school of hygiene and tropical medicine Biochemical society.Symposia, 7 Cambridge university press,Cambridge,1951
Bioch 33.1369

GAMMA GLOBULINS;STRUCTURE AND BIOSYNTHESIS Federation of European biochemical societies meeting 5th Proceedings Prague 1968 Jul Federation of European biochemical societies Edited by F. Franek and D. Shugar Federation of European biochemical societies. Publications, 15 Academic press,London;New York,1969
Bioch 33.1402

GAMMA SYSTEM The Role of the gamma system in movement and posture By Ian A. Boyd Association for the aid of crippled children, New York,1964 Record of a colloquium held during the 16th annual meeting of the American academy for cerebral palsy
An 32.4005

GARDEN CITY,N.Y. 1960 Adaptive control systems :a symposium Proceedings Edited by Felix P. Caruthers and Harold Levenstein illus viii,290p 23cm Pergamon press, Oxford,1963
Math 3.1234

GARMISCH 1968 Software engineering :a conference Report Edited by Peter Naur and Brian Randell Sponsored by the North Atlantic treaty organization.Science committee 231p Nato,Brussels,1969
Math L 5.3490

GARMISCH-PARTENKIRCHEN 1956
Internationales colloquium uber halbleiter und phosphore Edited by M. Schon and H. Welker Sponsored by the International union of pure and applied physics Vieweg,Brunswick,1958
Radioth 35.1615

GAS CHROMATOGRAPHIC DETERMINATION OF HORMONAL STEROIDS :a meeting Proceedings Edited by Filippo Polvani and others Organized under the auspices of Accademia nazionale del lincei Academic press,New York, 1967
Inv Med 37.0183

GAS CHROMATOGRAPHY 2nd :a symposium Proceedings Amsterdam 1958 May 19-23 Institute of petroleum.Gas chromatography discussion group Edited by D.H. Desty Also under the auspices of the Nederlandse chemische vereniging xiii,383p Butterworths,London,1958
Chem E 24.0849

GAS CHROMATOGRAPHY 2nd :international symposium Proceedings Ann Arbor 1959 Jun 10-13 Edited by Henry J. Noebels and others Held under the auspices of the Instrument society of America xvi,463p Academic press,New York;London,1961
Chem E 24.0851

GAS CHROMATOGRAPHY 3rd :a symposium Proceedings Edinburgh 1960 Jun 8-10 Institute of petroleum.Gas chromatography discussion group Edited by R.P.W. Scott Organized also by the Society for physical chemistry xvii,466p Butterworths,London, 1960
Chem E 24.0850

GAS CHROMATOGRAPHY DISCUSSION GROUP Gas chromatography 1958 :a symposium 2nd Proceedings Amsterdam 1958 May 19-23 Institute of petroleum.Hydrocarbon research group,and,Nederlandse chemische vereniging Edited by D.H. Desty Butterworths,London, 1958
Chem 18.0566

GAS CHROMATOGRAPHY 1958 2nd :a symposium Proceedings Amsterdam 1958 May 19-23 Institute of petroleum.Hydrocarbon research group,and,Nederlandse chemische vereniging Edited by D.H. Desty Organised by the Gas chromatography discussion group Butterworths, London,1958
Chem 18.0566

GAS CHROMATOGRAPHY 1960 3rd :a symposium Proceedings Edinburgh 1960 Jun 8-10 Society for analytical chemistry,and,Institute of petroleum.Gas chromatography discussion group Edited by R.P.W. Scott Butterworths, London,1960
Chem 18.0567

GAS CHROMATOGRAPHY 1964 5th :a symposium Proceedings Brighton 1964 Sep 8-10 Institute of petroleum.Gas chromatography discussion group Edited by A. Goldop Institute of petroleum,London,1965
Chem 18.0568

GAS CHROMATOGRAPHY 1966 6th :an international symposium on gas chromatography and associated techniques Proceedings Rome 1966 Sep 20-23 Institute of petroleum. Gas chromatography discussion group Edited by A.B. Littlewood Institute of petroleum, London,1967
Chem 18.0569

GAS CHROMATOGRAPHY 1970 International symposium on gas chromatography 8th Proceedings Dublin 1970 Sep 28-Oct 2 Institute of petroleum.Gas chromatography discussion group Edited by R. Stock viii, 445p 25cm Institute of petroleum,London, 1971
Chem 18.2697

GAS CHROMOTOGRAPHY IN BIOLOGY AND MEDICINE : a Ciba foundation symposium London 1969 Feb 5-6 Ciba foundation Edited by Ruth Porter Churchill,London,1969
Phys 20.2256

GAS-COOLED POWER REACTOR Technology of the gas-cooled power reactor and related subjects United Nations conference on the peaceful uses of atomic energy 2nd Papers United Kingdom atomic energy authority United Kingdom atomic energy authority,Harwell,1958
Eng 41.7494

GAS DISCHARGES Conference on gas discharges 1972 Institution of electrical engineers Institute of electrical and electronics engineers Institution of electrical engineers.Conference publication, 90 IEE, London,1972
Eng 41.8570

GAS DISCHARGES Gas discharges and the electricity supply industry :a conference Proceedings Leatherhead,Surrey 1962 May 7-11 Central electricity research laboratories Edited by J.S. Forrest and others 677p Butterworths,London,1962
Nap 11.0123

GAS DISCHARGES AND THE ELECTRICITY SUPPLY INDUSTRY International conference on gas discharges and the electricity supply industry Leatherhead 1962 May 7-11 Edited by J.S. Forrest and others Held at the Central electricity research laboratories Butterworths,London,1962
Eng 41.5415

GAS DISCHARGES AND THE ELECTRICITY SUPPLY INDUSTRY :a conference Proceedings Leatherhead,Surrey 1962 May 7-11 Central electricity research laboratories Edited by J.S. Forrest and others 677p Butterworths, London,1962
Nap 11.0123

GAS DYNAMICS Interstellar gas dynamics Symposium on cosmical gas dynamics 6th Proceedings Yalta 1969 Sep 8-18 International astronomical union International union of theoretical and applied mechanics International astronomical union. Symposium, 39 388p Reidel,Dordrecht,1970
Obs 6.3323

GAS DYNAMICS Magnetohydrodynamics Biennial gas dynamics symposium 4th proceedings Evanston,Ill. 1961 Aug 23-25 American rocket society and Northwestern university Edited by Ali Bulent Cambel and others Through the generosity of Government agencies and industrial sponsors Northwestern university press,Evanston,Ill.,1962
A Math 4.0623

GAS DYNAMICS Rarefied gas dynamics The International symposium on rarefied gas dynamics 2nd Proceedings Berkeley,Calif. 1960 Aug 3-6 University of California Edited by L. Talbot Under the sponsorship of International union of theoretical and applied mechanics Advances in applied mechanics. Supplement, 1 748p Academic press,New York,1961 Co-sponsored by National aeronautics and space administration
Cav 7.2254

GAS DYNAMICS OF COSMIC CLOUDS Symposium on cosmical gas dynamics 2nd Cambridge 1953 Jul 6-11 International union of theoretical and applied mechanics International astronomical union International astronomical union.Symposium series, 2 North-Holland,Amsterdam,1955
Eng 41.4468

GAS DYNAMICS OF COSMIC CLOUDS The Symposium on cosmical gas dynamics Proceedings Cambridge 1953 Jul 6-11 International astronomical union and International union of theoretical and applied mechanics Edited by J.M. Burgess and H.C.van de Hulst International astronomical union.Symposium, 2 North Holland,Amsterdam,1955
TA 15.0217

GAS DYNAMICS SYMPOSIUM ON AEROTHERMOCHEMISTRY Proceedings Evanston,Ill 1955 Aug 22-24 Northwestern university and American rocket society Edited by Donald K. Fleming Supported by the United States.Air force illus 284p Northwestern university, Evanston,Ill.,1956
Chem E 24.1390

The GAS LIQUID CHROMATOGRAPHY OF STEROIDS :a symposium Proceedings Glasgow 1966 Apr 4-6 Society for endocrinology Edited by J.K. Grant Society for endocrinology. Memoirs, 16 Cambridge university press, Cambridge,1967
Gen 34.0556

GAS PHASE MOLECULAR STRUCTURE Austin symposium on gas phase molecular structure Symposium on machine interpretations of Patterson functions and alternative direct approaches and the Austin symposium on gas phase molecular structure Proceedings Austin,Texas 1966 Feb 28-Mar 2 Edited by W. F. Bradley and Harold P. Hansen American crystallographic association.Transactions, 2, 1966 Polycrystal book services,Pittsburgh,Pa. 1966
Chem 18.2633

GAS-SURFACE INTERACTIONS Fundamentals of gas-surface interactions :a symposium Proceedings San Diego,Calif. 1966 Dec 14-16 United States.Air force.Office of scientific research General dynamics corporation.General atomic division Edited by Howard M. Saltsburg and others Academic press,New York,1967
Chem 18.2696

GAS TURBINES Symposium on materials for gas turbines :presented at the 49th annual meeting A.S.T.M. Buffalo,N.Y. 1946 Jun 24-28 American society for testing materials American society for testing materials, Philadelphia,Pa.,1946
Eng 41.3752

GASEOUS NEBULAE The Menzel symposium on solar physics,atomic spectra and gaseous nebulae Cambridge,Mass. 1971 Apr 29 Edited by K.B. Gebbie National bureau of standards.Special publication, 353 203p National bureau of standards,Washington,D.C., 1971
Obs 6.3623

GASTRIC SECRETION:MECHANISMS AND CONTROL : international symposium Proceedings Edmonton 1965 Edited by T.K. Shnitka and others Pergamon press,Oxford,1967
Phys 20.1428

GASTROENTEROLOGY Congres international de gastro-enterologie 7e Brussels 1964 Jun 1-6 1-3 Association des societes nationales europeennes et mediterranneenes de gastro-enterologie 3 vols A.S.N.E.M.G.E., Brussels,c1964
PGMS 29.0357

GATING Symposium on principles of gating presented at the 55th annual meeting Buffalo 1951 Apr 24 American foundrymen's society American foundrymen's society,Chicago,Ill., 1951
Met 25.0659

GATLINBURG 1957 Photoperiodism and related phenomena in plants and animals Conference on photoperiodism American association for the advancement of science Edited by Robert B. Withrow and others American association for the advancement of science.Publication, 55 Washington,D.C.,1959
Bot 42.1826

GATLINBURG 1964 Diffusion in body-centered cubic metals International conference on body-centred cubic materials Papers American society for metals.Oak Ridge chapter and Oak Ridge national laboratory American society for metals,Metals Park,Ohio,1965
Met 25.1283

GATLINBURG 1965 Electron spin resonance and the effects of radiation on biological systems :a conference Edited by Wallace Snipes Organised by the National research council.Subcommittee on radiobiology National research council.Nuclear science series.Report, 43 National research council, Washington,D.C.,1966
Radioth 35.1748

GATLINBURG 1967 Symposium on operating system principles Association for computing machinery ACM,New York,1967
Math L 5.3488

GATLINBURG,TN. 1955 Research in photosynthesis :a conference Papers and discussions National research council. Committee on photobiology Edited by H. Gaffron and others Supported by the National science foundation Interscience,New York; London,1957
Chem 18.2138

GATLINBURG,TENN 1955 Research in photosynthesis :a conference Papers and discussions Edited by H. Gaffron and others Sponsored by the National research council xiv,524p Interscience,London,1957
Bot 42.1869

GATLINBURG,TENN 1963 Symposium on macromolecular aspects of the cell cycle : given at research conference for biology and medicine of the atomic energy commission Sponsored by Oak Ridge national laboratory. Biology division Oak Ridge national laboratory.Symposia, 16 Journal of cellular and comparative physiology, 62,supp.1 Wistar institute of anatomy and biology, Philadelphia,Pa.,1963
PGMS 29.0414

GATLINBURG,TENN. 1955 Research in photosynthesis a conference Papers and discussions Edited by H. Gaffron and others Sponsored by the National research council. Committee on photobiology Interscience,New York,1957
Bioch 33.0758

GATLINBURG,TENN. 1956 Symposium on information theory in biology Edited by H.P. Yockey and others Sponsored by the Oak Ridge national laboratory London,1958 Consists of articles representing authors' results and opinions and additional papers not given at the symposium "The conference was entitled A symposium on information theory in health physics and radiobiology"
Bal 39.1006

GATLINBURG,TENN. 1958 Analytical chemistry in nuclear reactor technology:particle-size analysis Conference on nuclear reactor technology 2nd Papers United States atomic energy commission v,101p Oakridge national laboratory,Oakridge,Tenn.,1959
Chem E 24.1436

GATLINBURG,TENN. 1958 Symposium on genetic approaches to somatic cell variation Sponsored by Oak Ridge national laboratory. Biology division Oak Ridge national laboratory.Symposia, 11 Journal of cellular and comparative physiology, 52,supp.1 Wistar institute of anatomy and biology, Philadelphia,Pa.,1958
Gen 34.1094

GATLINBURG,TENN. 1959 Congenital anomalies of the face and associated structures :an international symposium Proceedings Edited by Samuel Pruzansky Thomas,Springfield,Ill., 1961
An 32.2054

GATLINBURG,TENN. 1959 Symposium on enzyme reaction mechanisms Research conference for biology and medicine 12th Sponsored by the Oak Ridge national laboratory.Biology division Journal of cellular and comparative physiology, 54,suppl.1 Wistar institute of anatomy and biology,Philadelphia,Pa.,1959
Bioch 33.1046

GATLINBURG,TENN. 1960 Symposium on mammalian genetics and reproduction Sponsored by the Oak Ridge national laboratory. Biology division Oak Ridge national laboratory.Symposia, 13 Journal of cellular and comparative physiology, 56,supp.1 Wistar institute of anatomy and biology, Philadelphia,Pa.,1960
Gen 34.1782

GATLINBURG,TENN. 1961 Symposium on recovery of cells from injury :given at research conference for biology and medicine of the atomic energy commission Sponsored by the Oak Ridge national laboratory.Biology division Journal of cellular and comparative physiology, 58,supp.1 Oak Ridge national laboratory.Symposia, 14 Wistar institute of anatomy and biology,Philadelphia,Pa.,1961
Gen 34.1114

GATLINBURG,TENN. 1963 Macromolecular aspects of the cell cycle :a symposium Oak Ridge national laboratory Wistar institute press,Philadelphia,Pa.,1963
An 32.3297

GATLINBURG,TENN. 1963 Symposium on macromolecular aspects of the cell cycle : given at research conference for biology and medicine of the atomic energy commission Sponsored by Oak Ridge national laboratory. Biology division Oak Ridge national laboratory.Symposia, 16 Journal of cellular and comparative physiology, 62,no.2 Wistar institute of anatomy and biology,Philadelphia, Pa.,1963
Gen 34.0842

GATLINBURG,TENN. 1964 Symposium on molecular action of mutagenic and carcinogenic agents Sponsored by the Oak Ridge national laboratory.Biology division Oak Ridge national laboratory.Symposia Journal of cellular and comparative physiology, 64,suppl. 1 Wistar institute of anatomy and biology, Philadelphia,Pa.,1964
Gen 34.1098

GATLINBURG,TENN. 1965 Symposium on hormonal control of protein biosynthesis Oak Ridge national laboratory.Biology division Oak Ridge national laboratory.Annual research conference, 18 Wistar institute of anatomy and biology,Philadelphia,Pa.,1965
Gen 34.0550

GATLINBURG,TENN. 1965 Symposium on hormonal control of protein biosynthesis Research conference for biology and medicine 18th Sponsored by the Oak Ridge national laboratory.Biology division Journal of cellular and comparative physiology, 66,suppl. 1 Wistar institute of anatomy and biology, Philadelphia,Pa.,1965
Bioch 33.0520

GATLINBURG,TENN. 1966 Symposium on differentiation and growth of haemoglobin-and immunoglobin-synthesising cells :given at research conference for biology and medicine of the Atomic energy commission Sponsored by the Oak Ridge national laboratory.Biology division Oak Ridge national laboratory. Biology division.Symposia, 19 Journal of cellular physiology, 67,supp.1 Wistar institute of anatomy and biology,Philadelphia, Pa.,1966
Gen 34.0504

GATLINBURG,TENN. 1967 Symposium on chromosome mechanics at the molecular level : given at research conference for biology and medicine of the atomic energy commission Sponsored by the Oak Ridge national laboratory. Biology division Oak Ridge national laboratory.Symposia, 20 Journal of cellular physiology, 70,suppl Wistar institute of anatomy and biology,Philadelphia,Pa.,1967
Gen 34.0846

GATLINBURG,TENN. 1968 Symposium on molecular aspects of differentiation :given at research conference for biology and medicine of the Atomic energy commission Sponsored by the Oak Ridge national laboratory.Biology division Oak Ridge national laboratory. Biology division.Symposia, 21 Journal of cellular physiology, 72,supp.1 Wistar institute of anatomy and biology,Philadelphia, Pa.,1968
Gen 34.0505

GEARING International conference on gearing Proceedings London 1958 Institution of mechanical engineers illus. Institution of mechanical engineers,London,1960
Eng 41.6224

GELATIN Recent advances in gelatin and glue research :a conference Proceedings Cambridge 1957 Jul 1-5 Edited by G. Stainsby Sponsored by the British gelatine and glue research association Pergamon, London,1958
Bioch 33.2247

GELEEN 1953 and LIEGE 1955 International committee for coal petrology : meeting 1st-2nd proceedings No 1-2 International committee for coal petrology International committee for coal petrology. Proceedings,1,2 Ernest van Aelst,Maastricht, 1954-56
Geol 8.3136

GEMEINSCHAFTSKONFERENZ DER REIHE 'METALL' 6 die kristallisation von metallen aus dem schmelzfluss,der gasphase und durch elektrolytische abscheidung Vortrage Kristallisation Berlin 1968 Mar 28-29 Deutsche akademie der wissenschaften zu Berlin. Sektion der physik.Unterkommission metallphysik VEB verlag fur grundstoffindustrie,Leipzig,1969
Met 25.2622

GENE ACTION IN MICRO-ORGANISMS :conference Papers St.Louis,Mo. Missouri botanical garden Missouri botanical garden.Annals, 32, no.2 Galesburg,Ill.,1945
Gen 34.0741

GENERAL DISCUSSION ON HEAT TRANSFER Proceedings London 1951 Sep 11-12 Institution of mechanical engineers and American society of mechanical engineers xiii,496p Institution of mechanical engineers,London,1951
Chem E 24.0538

GENERAL DISCUSSION ON HEAT TRANSFER Proceedings London 1951 Sep 11-13 Institution of mechanical engineers American society of mechanical engineers Institution of mechanical engineers,London,1952
Eng 41.7146

GENERAL DISCUSSION ON LUBRICATION AND LUBRICANTS Proceedings London 1937 Oct 13-15 Vol 1-2 Institution of mechanical engineers 2 vols Institution of mechanical engineers, London,1937
Met 25.2152

GENERAL DISCUSSION ON LUBRICATION AND LUBRICANTS Proceedings London 1937 Oct 13-15 Vol 1-2 Institution of mechanical engineers 2 vols Institution of mechanical engineers, London,1937
Eng 41.6291

GENERAL DYNAMIC CORPORATION.CONVAIR DIVISION Vistas in astronautics Annual astronautics symposium 1st United States.Air force. Office of scientific research Edited by Morton Alperin and others Pergamon press, London,1958
Eng 41.6880

GENERAL DYNAMICS CORPORATION.CONVAIR DIVISION Vistas in astronautics Astronautics symposium 1st proceedings San Diego 1957 Feb United States.Air force.Office of scientific research Edited by Morton Alperin and Marvin Stern Vistas in astronautics, 1 International series of monographs on aeronautical sciences and space flight. Astronautics division, 1 329p Pergamon press,London,1958
Nap 11.0022

GENERAL DYNAMICS CORPORATION.GENERAL ATOMIC DIVISION Fundamentals of gas-surface interactions :a symposium Proceedings San Diego,Calif. 1966 Dec 14-16 Edited by Howard M. Saltsburg and others Academic press,New York,1967
Chem 18.2696

GENERAL ELECTRIC COMPANY Growth and perfection of crystals International conference on crystal growth Proceedings Cooperstown,N.Y. 1958 Aug 27-29 Edited by R.H. Doremus and others Co-sponsored by United States.Air force.Office of scientific research Wiley;Chapman and Hall,New York; London,1958
Eng 41.4438

GENERAL ELECTRIC COMPANY Structure and properties of thin films An International conference on structure and properties of thin films Proceedings Bolton Landing,N.Y. 1959 Sep 9-11 Edited by C.A. Neugebauer and others John Wiley,New York,1959
Cav 7.0265

GENERAL ELECTRIC COMPANY RESEARCH LABORATORY Progress in very high pressure research :an international conference Proceedings Bolton Landing,N.Y. 1960 Jun 13-14 Edited by F.P. Bundy and others 314p New York, 1961
Geod 9.0254

GENERAL ELECTRIC RESEARCH AND DEVELOPMENT CENTER International symposium on the reactivity of solids 6th Proceedings Schenectady,N. Y. 1968 Aug 25-30 Edited by J.W. Mitchell and others Wiley,New York;London,1969
Met 25.2446

GENERAL ELECTRIC RESEARCH LABORATORY
Dislocations and mechanical properties of crystals :international conference Proceedings Lake Placid,N.Y. 1956 Sep 6-8 Edited by J.C. Fisher and others John Wiley, New York,1957
Cav 7.0804

GENERAL ELECTRIC RESEARCH LABORATORY
Dislocations and mechanical properties of crystals an international conference Lake Placid 1956 Sep 6-8 Edited by J.C. Fisher and others Sponsored by United States.Air research and development command Wiley; Chapman and Hall,New York;London,1957
Met 25.1232

GENERAL ELECTRIC RESEARCH LABORATORY Growth and perfection of crystals International conference on crystal growth Proceedings Cooperstown,New York 1958 Aug 27-29 Wiley; Chapman and Hall,New York;London,1958
Met 25.1507

GENERAL ELECTRIC RESEARCH LABORATORY
Progress in very high pressure research : international conference Proceedings Bolton Landing,New York 1960 Jun 13-14 Edited by F.P. Bundy and others Wiley,New York;London,1961
Met 25.1638

GENERAL ELECTRIC RESEARCH LABORATORY
Structure and properties of thin films :an international conference Proceedings New York 1959 Sep 9-11 Edited by C.A. Neugebauer and others Sponsored by United States.Air force.Office of scientific research Wiley,New York;London,n.d.
Met 25.1204

GENERAL ELECTRIC RESEARCH LABORATORY The Fermi surface :an international conference Proceedings Cooperstown,N.Y. 1960 Aug 22-24 Edited by W.A. Harrison and M.B. Webb Sponsored by the United States.Air force. Office of scientific research John Wiley,New York,1960
Cav 7.0541

GENERAL ELECTRIC RESEARCH LABORATORY The International conference on the electronic properties of metals at low temperatures Report Geneva,N.Y. 1958 Aug 25-29 Hobart college and International union of pure and applied physics 244p c1958 Privately printed report to the sponsors
Cav 7.0895

GENERAL ELECTRIC RESEARCH LABORATORY and UNITED STATES.AIR FORCE.OFFICE OF SCIENTIFIC RESEARCH The Fermi surface :an international conference Proceedings Cooperstown,N.Y. 1960 Aug 22-24 J.Wiley, New York;London,1960
Chem 18.2382

GENERAL MOTORS CORPORATION.RESEARCH LABORATORIES
A Symposium on non-equilibrium thermodynamics, variational techniques and stability Chicago 1965 May 17-19 Edited by Russell J. Donnelly and I. Prigogine University of Chicago,Chicago,1966
A Math 4.0712

GENERAL MOTORS CORPORATION.RESEARCH LABORATORIES
Approximation of functions Symposium on approximation of functions 8th proceedings Warren,Mich. 1964 Aug 31-Sep 2 Edited by Henry L. Garabedian viii,220p 25cm Elsevier publishing co.,Amsterdam,1965
P. Math 2.0978

GENERAL MOTORS CORPORATION.RESEARCH LABORATORIES
Internal stresses and fatigue in metals the symposium Proceedings Detroit 1958 Warren,Mich. 1958 Edited by Gerald M. Rasswiler and William L. Grube Elsevier, Amsterdam,1959
Met 25.0965

GENERAL MOTORS CORPORATION.RESEARCH LABORATORIES
Symposium on the approximation of functions 8th proceedings Warren,Mich. 1964 Aug. 31-Sep.2 Edited by Henry L. Garabedian bibliog. viii,220p 24cm Elsevier, Amsterdam,1965
Math L 5.0348

GENERAL MOTORS CORPORATION.RESEARCH LABORATORIES
The Theory of road traffic flow international symposium 2nd proceedings London 1963 Jun.25-27 Road research laboratory Edited by Joyce Almond ix,406p 24cm Organisation for economic co-operation and development,Paris,1965
Math 3.0922

GENERAL MOTORS CORPORATION.RESEARCH LABORATORIES
Theory of traffic flow symposium 3rd proceedings Warren,Mich. 1959 Dec.7-8 Edited by Robert Herman Elsevier,Amsterdam, 1961
A Math 4.0949

GENERAL MOTORS RESEARCH LABORATORIES
Approximation of functions Symposium on approximation of functions 8th Proceedings Warren,Mich. 1964 Edited by Henry L. Garabedian 220p Elsevier, Amsterdam,1965
TA 15.0097

GENERAL MOTORS RESEARCH LABORATORIES
Cavitation in real liquids :a symposium Proceedings Warren,Mich. 1962 Edited by Robert Davies Elsevier,Amsterdam,1964
Eng 41.6512

GENERAL MOTORS RESEARCH LABORATORIES
Cavitation in real liquids... :a symposium 6th Proceedings Warren,Mich. 1962 Sep 24-25 Edited by Robert Davies port vii,189p Elsevier,Amsterdam,1964
Chem E 24.0240

GENERAL MOTORS RESEARCH LABORATORIES Liquids: structure,properties,solid interactions :the symposium Proceedings Warren,Mich. 1963 Edited by Thomas J. Hughel Elsevier, Amsterdam,1965
Met 25.1514

GENERAL MOTORS RESEARCH LABORATORIES
Symposium on the fluid mechanics of internal flow Warren,Mich. 1965 Edited by Gino Sovran Elsevier,Amsterdam,1967
Eng 41.6553

GENERAL MOTORS RESEARCH LABORATORY.SYMPOSIA, 7
Liquids:structure,properties,solid interactions Symposium on liquids:structure, properties,solid interactions proceedings Warren,Mich. 1963 Sep.5-6 Edited by Thomas J. Hughel Elsevier,Amsterdam,1965
A Math 4.0951

GENERAL RELATIVITY AND COSMOLOGY Scuola internazionale di fisica 'Enrico Fermi' 47 corso Rendiconti Varenna 1969 Jun 30-Jul 12 Societa italiana di fisica Edited by Rainer K. Sachs Academic press New Yorkp 1971cm
A Math 4.1788

GENERAL SESSION ON POWDER METALLURGY Detroit 1959 Apr 20-22 Metal powder industries federation Metal powder industries federation,New York,1959
Met 25.0764

GENERAL SYSTEMS Views on general systems theory The Second systems symposium proceedings Cleveland,Ohio 1963 Apr Case institute of technology Edited by Mihajlo D. Mesarovic xvii,178p 24cm John Wiley and sons,New York;London,1964
Math 3.0880

GENERAL TOPOLOGY AND ITS RELATIONS TO MODERN ANALYSIS AND ALGEBRA Prague topological symposium 2nd Proceedings Prague 1966 Aug 30-Sep 4 Ceskoslovenska akademie ved 365p 24cm Academic press;Academia,New York; Prague,1967
P Math 2.3346

GENERAL TOPOLOGY AND ITS RELATIONS TO MODERN ANALYSIS AND ALGEBRA :conference Proceedings Kanpur 1968 Oct 3-12 illus 332p 24cm Academia,Prague,1971
P Math 2.4109

GENERAL TOPOLOGY AND ITS RELATIONS TO MODERN ANALYSIS AND ALGEBRA :symposium proceedings Prague 1961 Sep 1-8 Ceskoslovenska akademie ved Edited by J. Novak jointly with International mathematical union 363p 25cm Academic press,New York,1962
P. Math 2.0291

GENES Evolving genes and proteins :a symposium Proceedings New Brunswick,N.J. 1964 Sep 17-18 Edited by Vernon Bryson and Henry J. Vogel Held at Rutgers university. Institute of microbiology Academic press,New York;London,1965
Bioch 33.0569

GENES International symposium on genes and chromosomes:structure and function Proceedings Buenos Aires 1964 Nov 30-Dec 4 National cancer institute Edited by Juan I. Valencia and Rhoda F. Grell National cancer institute.Monograph, 18 National cancer institute,Bethesda,Md.,1965
Bioch 33.0987

GENES AND CHROMOSOMES:STRUCTURE AND FUNCTION International symposium on genes and chromosomes structure and function Proceedings Buenos Aires 1964 Nov 30-Dec 4 Edited by Juan I. Valencia and Rhoda F. Grell National cancer institute.Monograph, 18 illus 354p U.S.Dept. of health,education and welfare,Bethesda,Md.,1965
Bot 42.0792

GENES AND CHROMOSOMES:STRUCTURE AND FUNCTION International symposium on genes and chromosomes structure and function Proceedings Buenos Aires 1964 Nov 30-Dec 4 Edited by Juan I. Valencia and others National cancer institute.Monographs, 18 U. S.Department of health,education and welfare, Washington,D.C.,1965
Gen 34.0833

GENES AND CHROMOSONES :structure and organization.A symposium Papers Cold Spring Harbor 1941 Cold Spring Harbor biological laboratory Cold Spring Harbor symposia on quantitative biology, 9 Long Island biological association,Cold Spring Harbor,1941
Bioch 33.1265

GENES AND MUTATION :a symposium Papers Cold Spring Harbor 1951 Cold Spring Harbor biological laboratory Cold Spring Harbor symposia on quantitative biology, 16 Long Island biological association,Cold Spring Harbor,1951
Bioch 33.1272

GENESE DES ROCHES FILONIENNES,FASC.6 International geological congress 19th Proceedings,section 6 Algiers 1952 Algiers,1953
Min 10.0351

The GENESIS OF LANGUAGE:A PSYCHOLINGUISTIC APPROACH conference Proceedings Language development in children Old Point Comfort,Va. 1965 Apr 25-28 National institute of child health and human development.Human communication program Edited by Frank Smith and George A. Miller M. I.T.press,Cambridge,Mass.;London,1966
Psy 31.2746

GENETIC AND ENVIRONMENTAL FACTORS IN HUMAN ABILITY :Eugenics society symposium 1965 Sep-Oct Eugenics society Edited by G. E. Meade and A.S. Parkes Eugenics society. Symposia, 2 Oliver and Boyd,Edinburgh,1966
Gen 34.2042

GENETIC AND ENVIRONMENTAL INFLUENCES ON BEHAVIOUR a symposium 1967 Sep Eugenics society Edited by J.M. Thoday and Alan S. Parkes Eugenics society.Symposium, 4 bibliog.,illus. x,27p 22cm Oliver and Boyd,Edinburgh,1968
Math S 3.1468

GENETIC APPROACHES TO SOMATIC CELL VARIATION Symposium on genetic approaches to somatic cell variation Gatlinburg,Tenn. 1958 Apr 2-5 Sponsored by Oak Ridge national laboratory. Biology division Oak Ridge national laboratory.Symposia, 11 Journal of cellular and comparative physiology, 52,supp.1 Wistar institute of anatomy and biology, Philadelphia,Pa.,1958
Gen 34.1094

The GENETIC CODE :a symposium Papers Cold Spring Harbor 1966 Cold Spring Harbor laboratory of quantitative biology Cold Spring Harbor symposia on quantitative biology, 31 Cold Spring Harbor,1966
Bioch 33.1287

GENETIC CONTROL OF DIFFERENTIATION :report of a symposium Upton,N.Y. 1965 Jun 7-9 Brookhaven national laboratory.Biology department Brookhaven symposia in biology, 18 Brookhaven national laboratory,Upton,N.Y., 1965
Gen 34.0499

GENETIC CONTROL OF DIFFERENTIATION :symposium Report Upton,N.Y. 1965 Jun 7-9 Brookhaven national laboratory.Biology department Brookhaven symposia in biology, 18 BNL931(C-44) Brookhaven national laboratory.Biology department,Upton,N.Y.,1965
Bioch 33.1296

GENETIC COUNSELLING :a report Geneva 1968 Sep 24-30 World health organization World health organization.Expert committee on human genetics.Report, 3 World health organization.Technical report series, 416 Geneva,1969
Gen 34.1970

GENETIC DIVERSITY AND HUMAN BEHAVIOUR The Burg Wartenstein symposium 27th Papers Burg 1964 Sep 16-28 behavioral consequences of genetic differences in man Social science research council.Committee on genetics and behavior Wenner-Gren foundation for anthropological research Edited by G.N. Spuhler Viking fund publications in anthropology, 45 Aldine,Chicago,Ill.,1967
Gen 34.2032

GENETIC EFFECT OF RADIATION Symposium on genetic effect of radiation Proceedings Misima,Japan 1960 Nov 7-9 Under the auspices of the National institute of genetics, Shizvoka Japanese journal of genetics, 36, supp. Genetics society of Japan,1961
Gen 34.1117

GENETIC ELEMENTS Structure and function of genetic elements :symposium Report Upton,N. Y. 1959 Jun 1-3 Brookhaven national laboratory.Biology department Brookhaven symposia in biology, 12 BNL558(C-29) Brookhaven national laboratory.Biology department,Upton,N.Y.,1959
Bioch 33.1291

GENETIC ELEMENTS:PROPERTIES AND FUNCTION; SYMPOSIUM The Federation of European biochemical societies meeting 3rd Warsaw 1966 Apr 4-7 Federation of European biochemical societies Edited by D. Shugar Organised by the Polish biochemical society Academic press,London,1967
Bioch 33.1395

GENETIC INFORMATION Genetics conference : genetic information and the control of protein structure and function 1st Transactions Princeton,N.J. 1959 Oct 19-22 Edited by H. Eldon Sutton Sponsored by the Josiah Macy jr. foundation Josiah Macy jr.foundation,New York,1960
Gen 34.0650

GENETIC MATERIAL Transcription of genetic material :a symposium Papers Cold Spring Harbor 1970 Jun 4-11 Cold Spring Harbor laboratory of quantitative biology Cold Spring Harbor symposia on quantitative biology, 35 Cold Spring Harbor,1971
Bioch 33.2335

GENETIC MECHANISMS :structure and function.A symposium Papers Cold Spring Harbor 1956 Cold Spring Harbor biological laboratory Cold Spring Harbor symposia on quantitative biology, 21 Long Island biological association,Cold Spring Harbor,1956
Bioch 33.1277

GENETIC NEUROLOGY International conference on the development,growth and regeneration of the nervous system Chicago,Ill. 1949 Mar 21-25 Edited by Paul Weiss Sponsored by the International union of biological sciences University of Chicago,Ill.,Chicago,Ill.,1950
An 32.4329

GENETIC NEUROLOGY :problems of the development, growth,and regeneration of the nervous system and of its functions:an international conference Essays Chicago,Ill. 1949 Mar 21-25 International union of biological sciences Edited by Paul Weiss Subsidized by Unesco University of Chicago press, Chicago,Ill.,1950
Psy 31.0170

GENETIC POLYMORPHISMS AND GEOGRAPHIC VARIATIONS IN DISEASE :a conference Proceedings Bethesda,Md. 1960 Feb 23-25 National institute of arthritis and metabolic diseases National heart institute National institutes of health Edited by Barach S. Blumberg Grune and Stratton,New York;London,1961
Gen 34.1984

GENETIC RADIATION DAMAGE Repair from genetic radiation damage and differential radiosensitivity in germ cells :an international symposium Proceedings Leiden 1962 Aug 15-19 By F.H. Sobels Pergamon press,London,1963
Radioth 35.1157

GENETIC RECOMBINATION Symposium on genetic recombination given at research conference for biology and medicine of the atomic energy commission Oak Ridge,Tenn. 1954 Apr 19-21 Sponsored by the Oak Ridge national laboratory. Biology division Oak Ridge national laboratory.Symposia, 7 Journal of cellular and comparative physiology, 45,supp.2 Wistar institute of anatomy and biology, Philadelphia,Pa.,1955
Gen 34.1042

GENETIC RECOMBINATION Symposium on genetic recombination given at the Research conference for biology and medicine of the Atomic Energy commission Oak Ridge,Tenn. 1954 Apr 19-21 Sponsored by Oak Ridge national laboratory. Biology division Wistar institute of anatomy and biology,Philadelphia,Pa.,1955
Reprinted from the Journal of cellular and comparative physiology,Vol.45, suppt.2,May 1955
An 32.0346

GENETIC SELECTION IN MAN Conference on genetics 3rd Princeton,N.J. 1961 Oct 15-18 Edited by William J. Schull Sponsored by the Josiah Macy jr.foundation University of Michigan press,Ann Arbor,Mich., 1963
An 32.0362

GENETIC SELECTION IN MAN Macy conference on genetics 3rd Proceedings Princeton,N.J. 1961 Oct 15-18 Josiah Macy jr. foundation Edited by William J. Schull University of Michigan press,Michigan,1963
Gen 34.1867

GENETICAL ASPECTS OF RADIOSENSITIVITY : mechanisms of repair Vienna 1966 Apr 18-22 International atomic energy agency International atomic energy agency.Proceedings series illus 175p 24cm International atomic energy agency,Vienna,1966
Bot 42.0797

GENETICAL ASPECTS OF RADIOSENSITIVITY:MECHANISMS OF REPAIR :a panel Proceedings Vienna 1966 Apr 18-22 International atomic energy agency International atomic energy agency.Panel proceedings series International atomic energy agency,Vienna,1966
Radioth 35.1194

GENETICAL SOCIETY Evolution Oxford 1952 Jul Society for experimental biology. Symposia, 7 Cambridge university press, Cambridge,1953
Phys 20.0957

GENETICAL SOCIETY Evolution :a symposium Papers Oxford 1952 Jul Society for experimental biology.Symposia, 7 Cambridge university press,Cambridge,1953
Psy 31.0570

GENETICAL SOCIETY Evolution symposium Oxford 1952 Jul Society for experimental biology.Symposia, 7 Cambridge university press,Cambridge,1953
Gen 34.1258

GENETICAL SOCIETY International genetical congress 7th Proceedings Edinburgh 1939 Aug 23-30 Edited by Reginald Crundall Punnett Journal of genetics.Supplement Cambridge university press,Cambridge,1941
Gen 34.2219

GENETICAL VARIATION IN HUMAN POPULATION symposium 1961 Society for the study of human biology Edited by G.A. Harrison Society for the study of human biology. Symposia, 4 Pergamon press,London,1961
Gen 34.1896

GENETICAL VARIATIONS IN HUMAN POPULATIONS :a symposium London 1960 Edited by G.A. Harrison Society for the study of human biology.Symposia, 4 Pergamon press,Oxford, 1961
An 32.2882

GENETICS Biochemical aspects of genetics :a symposium London 1949 Feb 12 Biochemical society Edited by R.T. Williams Held at the Westminster hospital medical school Biochemical society.Symposia, 4 Cambridge university press,Cambridge,1950
Bioch 33.1366

GENETICS Clinical aspects of genetics :a conference Proceedings London 1961 Mar 17-18 Royal college of physicians of London Edited by F.Avery Jones Pitman,London,1961
PGMS 29.0255

GENETICS Conference international de genetique 4e Comptes rendus Paris 1911 Sep 18-23 Societe nationale d'horticulture de France Edited by Ph.de Vilmorin 2 vols Masson,Paris,1913
Gen 34.1000

GENETICS Cytology,genetics and evolution : university of Pennsylvania bicentennial conference By M. Demerec and others University of Pennsylvania.Bicentennial conference University of Pennsylvania press, Philadelphia,Pa.,1941
Bot 42.6465

GENETICS Evoluzione e genetica :colloquio internazionale Relazioni e discussione Rome 1959 Apr 8-11 Accademia nazionale dei Lincei Problemi attuali di scienza et di cultura.Quaderno N., 47 Rome,1960 Text in English and Italian.Summaries in English
Bal 39.0927

GENETICS Exchange of genetic material : mechanisms and consequences.A symposium Papers Cold Spring Harbor 1958 Cold Spring Harbor biological laboratory Cold Spring Harbor symposia on quantitative biology, 23 Long Island biological association,Cold Spring Harbor,1958
Bioch 33.1279

GENETICS Genetic mechanisms :structure and function.A symposium Papers Cold Spring Harbor 1956 Cold Spring Harbor biological laboratory Cold Spring Harbor symposia on quantitative biology, 21 Long Island biological association,Cold Spring Harbor,1956
Bioch 33.1277

GENETICS Genetic selection in man Conference on genetics 3rd Princeton,N. J. 1961 Oct 15-18 Edited by William J. Schull Sponsored by the Josiah Macy jr. foundation University of Michigan press,Ann Arbor,Mich.,1963
An 32.0362

GENETICS Genetics:genetic information and the control of protein structure and function Conference on genetics 1st Transactions Princeton,N.J. 1959 Oct 19-22 Edited by H. Eldon Sutton Sponsored by the Josiah Macy jr. foundation Macy Jr.foundation,New York,1960
An 32.0360

GENETICS Genetics in the 20th century : essays on the progress of genetics during its first fifty years Genetics society of America Edited by L.C. Dunn 2 vols Macmillan,New York,1951 "A compilation of invitation papers presented at the...Golden jubilee of genetics at Ohio State university, Columbus,Ohio,September 11-14,1950"-foreword
An 32.0342

GENETICS Golden jubilee of genetics Genetics in the 20th century :essays on the progress of genetics during its first 50 years Columbus,Ohio 1950 Sep 11-14 Genetics society of America Edited by L.C. Dunn Sponsored by the American institute of biological sciences port Macmillan,New York,1951
Bot 42.0751

GENETICS Human genetics :a symposium Papers Cold Spring Harbor 1964 Cold Spring Harbor laboratory of quantitative biology Cold Spring Harbor symposia on quantitative biology, 29 Cold Spring Harbor, 1964
Bioch 33.1285

GENETICS Hybrid conference report 1900 on hybridisation (the cross-breeding of species) and on the cross-breeding of varieties London 1899 Jul 11-12 Royal horticultural society Royal horticultural society.Journal, 24 Spottiswoode,London,1900 Conference also entitled the 'International congress of genetics' and the 'International conference on hybridization'
Gen 34.0998

GENETICS International conference on genetics 3rd Report London 1906 Jul 30-Aug 3 Royal horticultural society Edited by W. Wilkes Spottiswoode,London,1907
BG 38.0523

GENETICS International congress of biochemistry 6th Abstracts New York 1964 Jul 26-Aug 1 Vol 3: biochemical genetics Scheduled under the auspices of the International union of biochemistry Washington,D.C.,1964
Bioch 33.1340

GENETICS International congress of genetics 6th Proceedings Ithaca,N.Y. 1932 Vol 1-2 Edited by Donald F. Jones Brooklyn botanic garden,Brooklyn,N.Y.,1932 Bound together
Gen 34.1002

GENETICS International congress of genetics 8th Proceedings Stockholm 1948 Jul 7-14 Edited by Gert Bonnier and Robert Larsson Berlingska boktrycheriet,Lund,1949
Gen 34.1004

GENETICS International congress of genetics 9th Proceedings Bellagio 1953 Aug 24-31 Pt 1-2 Edited by G. Montalenti and A. Chiarugi Caryologia, 6,supp. 2 vols Florence,1954
Gen 34.1005

GENETICS International congress of genetics 10th Proceedings Montreal 1958 Aug 20-27 Vol 1-2 University of Toronto press, Toronto,1959 Bound together
Gen 34.1006

GENETICS International congress of genetics 10th Proceedings Montreal 1958 Aug 20-27 Vol 1-2 2 vols University of Toronto press,Toronto,1959
Bot 42.4702

GENETICS International congress of genetics 12th Proceedings Tokyo 1968 Aug 19-28 Vol 1-3 Science council of Japan Under the auspices of the International union of biological sciences 3 vols Science council of Japan,Tokyo,1969
Gen 34.1008

GENETICS International congress of genetics hybridization (the cross-breeding of genera or species)the cross-breeding of varieties... 3rd London 1906 Jul 30-Aug 3 Royal horticultural society Edited by W. Wilks Spottiswoode,London,1907 Conference also entitled 'International conference on hybridisation'
Gen 34.0999

GENETICS International congress of human genetics 1st Proceedings Copenhagen 1956 Aug 1-6 Edited by Tage Kemp and others Karger,Basle,1957
An 32.0355

GENETICS International congress on human genetics 3rd Proceedings Chicago,Ill. 1966 Sep 5-10 plenary sessions and symposia Edited by James F. Crow and James V. Neel Johns Hopkins press,Baltimore,Md.,1967
An 32.0057

GENETICS International genetical congress 7th Proceedings Edinburgh 1939 Aug 23-30 Genetical society Edited by Reginald Crundall Punnett Journal of genetics. Supplement Cambridge university press, Cambridge,1941
Gen 34.2219

GENETICS International genetical congress 7th Proceedings Edinburgh 1939 Aug 23-30 International genetical congress.Permanent international committee Edited by R.C. Punnett Cambridge university press,Cambridge, 1939 Issued as a supplementary volume to the 'Journal of genetics'
Bot 42.6483

GENETICS Internationaler kongress fur vererbungswissenschaft 5 Verhandlungen Berlin 1927 Band 1-2 Edited by Hans Nachtsheim Zeitschrift fur induktive abstammungs- und vererbungslehre. Supplementband, 1 2 vols Borntraeger, Leipzig,1928
Gen 34.1001

GENETICS Mutations Conference on genetics 2nd Princeton,N.J. 1960 Oct 16-19 Edited by William J. Schull Sponsored by the Josiah Macy jr.foundation University of Michigan press,Ann Arbor,Mich.,1962
An 32.0361

GENETICS Population genetics :the nature and causes of genetic variability in populations.A symposium Papers Cold Spring Harbor 1955 Cold Spring Harbor biological laboratory Cold Spring Harbor symposia on quantitative biology, 20 Long Island biological association,Cold Spring Harbor,1955 Dedicated to Vannevar Bush,President of the Carnegie Institute of Washington, 1939-1955
Bioch 33.1276

GENETICS Simposio internacional de genetica symposium Sao Paulo 1966 Jul Edited by F. G. Brieger Held under the auspices of the International union of biological sciences Ciencia e cultura, 19,no.1 Sao Paulo,1967
Gen 34.0995

GENETICS,AND THE INHERITANCE OF INTEGRATED NEUROLOGICAL AND PSYCHIATRIC PATTERNS Proceedings of the Association New York 1953 Dec 11-12 Association for research in nervous and mental disease Edited by Davenport Hooker and Clarence C. Hare Association for research in nervous and mental disease.Research publications, 33 Williams and Wilkins,Baltimore,Md.,1954
An 32.4157

GENETICS,RADIOBIOLOGY AND RADIOLOGY :Mid-western conference Proceedings 1958 May 2 Edited by Wendell G. Scott and Titus Evans Blackwells,Oxford,1959
Radioth 35.1723

GENETICS,RADIOBIOLOGY AND RADIOLOGY :mid-western conference Proceedings 1958 May 2 Edited by Wendell G. Scott and Titus Evans Blackwell,Oxford,1959
An 32.5254

GENETICS,RADIOBIOLOGY AND RADIOLOGY PROCEEDINGS mid-western conference 1958 May National research council.Committee on the biological effects of atomic radiation Edited by Wendell G. Scott and Titus Evans Blackwell,Oxford,1959
Gen 34.1109

GENETICS AND CANCER Annual symposium on fundamental cancer research 13th Papers 1959 Anderson hospital and tumor institute Edited by Russell W. Cumley and others Owen, London,1961
Gen 34.2232

GENETICS AND DENTAL HEALTH :an international symposium Proceedings Bethesda,Md. 1961 Apr 4-6 American dental association. Council on dental research Edited by Carl J. Witkop Symposium on genetics related to dental health, 1 McGraw-Hill,New York,1962
Gen 34.2024

GENETICS AND THE FUTURE OF MAN :a discussion at the Nobel conference St.Peter,Minn. 1965 Jan 7-8 Edited by John D. Roslansky Organized by Gustavus Adolphus college North-Holland,Amsterdam,1966
Gen 34.0062

GENETICS AND TWENTIETH CENTURY DARWINISM :a symposium Cold Spring Harbor 1959 Jun 3-10 Edited by Clara Wooldridge Cold Spring Harbor symposia on quantitative biology, 24 Biological laboratory,Cold Spring Harbor,L.I., 1959
An 32.0043

GENETICS AND TWENTIETH CENTURY DARWINISM :a symposium Papers Cold Spring Harbor 1959 Cold Spring Harbor biological laboratory Cold Spring Harbor symposia on quantitative biology, 24 Long Island biological association,Cold Spring Harbor,1959
Bioch 33.1280

GENETICS CONFERENCE 1st :genetic information and the control of protein structure and function Transactions Princeton,N.J. 1959 Oct 19-22 Edited by H. Eldon Sutton Sponsored by the Josiah Macy jr. foundation Josiah Macy jr.foundation,New York,1960
Gen 34.0650

GENETICS IN PLANT BREEDING report of symposium Upton,N.Y. 1956 May 21-23 Brookhaven national laboratory.Biology department Brookhaven symposia in biology, 9 Brookhaven national laboratory,Upton,N.Y., 1956
Gen 34.1152

GENETICS IN THE 20TH CENTURY :essays on the progress of genetics during its first 50 years Columbus,Ohio 1950 Sep 11-14 Genetics society of America Edited by L.C. Dunn Sponsored by the American institute of biological sciences Macmillan,New York,1951
Gen 34.0991

GENETICS IN THE 20TH CENTURY :essays on the progress of genetics during its first 50 years Golden jubilee of genetics Columbus,Ohio 1950 Sep 11-14 Genetics society of America Edited by L.C. Dunn Sponsored by the American institute of biological sciences port Macmillan,New York,1951
Bot 42.0751

The GENETICS OF COLONIZING SPECIES symposium Proceedings International union of biological sciences Edited by H.G. Baker and G.Ledyard Stebbins International union of biological sciences.Symposium, 1 Academic press,New York;London,1965
Gen 34.1297

GENETICS OF MIGRANT AND ISOLATE POPULATION Conference on human population genetics in Israel Proceedings Jerusalem 1961 Edited by Elisabeth Goldschmidt Published for the Association for the aid of crippled children Williams and Wilkins,New York,1963
Gen 34.1881

GENETICS OF STREPTOMYCES AND OTHER ANTIBIOTIC-PRODUCING MICRO ORGANISMS conference New York,N.Y. 1959 Jan 6 New York academy of sciences Annals of the New York academy of sciences, 81,art 4 New York,1959
Gen 34.1494

GENETICS OF THE IMMUNE RESPONSE :report of a WHO scientific group Geneva 1967 Oct 2-7 World health organization World health organization.Technical report series, 402 World health organization,Geneva,1968
Gen 34.2012

GENETICS SOCIETY OF AMERICA Genetics in the 20th century :essays on the progress of genetics during its first fifty years Edited by L.C. Dunn 2 vols Macmillan,New York, 1951 "A compilation of invitation papers presented at the...Golden jubilee of genetics at Ohio State university,Columbus,Ohio, September 11-14,1950"-foreword
An 32.0342

GENETICS SOCIETY OF AMERICA Genetics in the 20th century :essays on the progress of genetics during its first 50 years Columbus, Ohio 1950 Sep 11-14 Edited by L.C. Dunn Sponsored by the American institute of biological sciences Macmillan,New York,1951
Gen 34.0991

GENETICS SOCIETY OF AMERICA Heritage from Mendel Mendel centennial symposium Proceedings Fort Collins,Colo 1965 Edited by R. Alexander and R.Derek Styles Wisconsin university press,Madison,Wis.,1967
Bot 42.0880

GENETICS SOCIETY OF AMERICA Heritage from Mendel Mendel centennial symposium Proceedings Fort Collins,Colo. 1965 Spe 7-11 Edited by R.Alexander Brink and E.Derek Styles University of Wisconsin press,Madison, Wis.,1967
Gen 34.0994

GENETICS TODAY International congress of genetics 11th Proceedings The Hague 1963 Sep Vol 1-3 Edited by S.J. Geerts 2 vols Pergamon press,Oxford,1963-65
Bot 42.0780

GENETICS TODAY International congress of genetics 11th Proceedings The Hague 1963 Sep Vol 1-3 Edited by S.J. Geerts 3 vols Pergamon press,Oxford,1963-65 Two copies each of vol 2 and 3
Gen 34.1007

GENEVA 1905 Congres federatif international d'anatomie 1e bulletin synthetique Kundig,Geneva,1907
An 32.5266

GENEVA 1927 The World population conference Proceedings Edited by Margaret Sanger Arnold,London,1927
Bal 39.1748

GENEVA 1927 World population conference Proceedings Edited by Margaret Sanger port. Arnold,London,1927
Geog 13.1526

GENEVA 1949 International biometric conference 2nd Proceedings Biometric society 1950 Reproduced from 'Biometrics' vol 6,no.1
Gen 34.2146

GENEVA 1955 Deistvie oblucheniya na organizm :doklady sovetskoi delegatsii na Mezhdunarodnoi konferentsii po mirnomu ispolzovaniyu atomnoi energii Doklady By A. A. Letavet and others Izdatelstvo Akademii nauk SSSR,Moscow,1955
Chem 18.0379

GENEVA 1955 Exploration for nuclear raw materials Edited by Robert D. Nininger Geneva series on the peaceful uses of atomic energy bibliog.,illus. Macmillan and co, London,1956 Data submitted to the International conference on the peaceful uses of atomic energy
Geol 8.1661

GENEVA 1955 International conference on the peaceful uses of atomic energy Proceedings 10-14 United nations 5 vols United Nations,New York,1955-56
Radioth 35.1009

GENEVA 1955 International conference on the peaceful uses of atomic energy Proceedings Vol 8: production technology of the materials used for nuclear energy United nations 627p United nations,New York,1956
Chem E 24.1650

GENEVA 1955 International conference on the peaceful uses of atomic energy Proceedings Vol 9: reactor technology and chemical processing United nations 771p United nations,New York,1956
Chem E 24.1651

GENEVA 1955 International conference on the peaceful uses of atomic energy Proceedings Vol 10: radioactive isotopes and nuclear radiations in medicine United nations United nations,New York,1956
Phys 20.2093

GENEVA 1955 International conference on the peaceful uses of atomic energy Proceedings Vol 12: radioactive isotopes and ionizing radiations in agriculture, physiology and biochemistry United Nations United Nations,New York,1956
Phys 20.2244

GENEVA 1955 International conference on the peaceful uses of atomic energy 1st Proceedings Vol 6: geology of uranium and thorium United Nations United Nations,New York,1956
Min 10.0715

GENEVA 1955 International conference on the peaceful uses of atomic energy 1st Proceedings Vol 8: production technology of the materials used for nuclear nergy United Nations United Nations,New York,1956
Min 10.0718

GENEVA 1955 Metallurgy and fuels Vol 1 Edited by H.M. Finniston and J.P. Howe Progress in nuclear energy, 5 Pergamon press,London,1956 Includes papers presented at the United Nations conference on the peaceful uses of atomic energy
Met 25.1560

GENEVA 1955 Process chemistry Vol 1-2 Edited by F.R. Bruce and others Progress in nuclear energy, 3 Pergamon press,London, 1956 Includes papers presented at the United Nations conference on the peaceful uses of atomic energy
Met 25.1558

GENEVA 1955 Sessiya Akademii nauk S.S.S.R. po mirnomu ispolzovaniyu atomnoi energii... : zasedaniya otdeleniya biologicheskikh nauk Akademiya nauk S.S.S.R.Otdelenie biologicheskikh nauk Izdatelstvo Akademii nauk SSSR,Moscow,1955 Five volumes in box
Chem 18.0878

GENEVA 1955 Technology and engineering Vol 1: reactor coolants,moderators,heat transfer,reactor chemistry and corrosion of reactor materials Edited by R. Hurst and S. McLain Progress in nuclear energy, 4 Pergamon,London,1956 Includes papers presented at the United Nations conference on the peaceful uses of atomic energy
Met 25.1565

GENEVA 1955 The International conference on the peaceful uses of atomic energy Vol 8: production technology of the materials used for nuclear energy United Nations United Nations,New York,1956
Met 25.1564

GENEVA 1956 Discussions on child development :consideration of the biological, psychological and cultural approaches to the understanding of human development and behaviour :meeting Vol 4 By Jean Piaget and others World health organization. Study group on the psychobiological development of the child Edited by J.M. Tanner and Barbel Inhelder Tavistock publications,London,1960
Psy 28.0224

GENEVA 1957 Mental health aspects of the peaceful uses of atomic energy Study group on mental health aspects of the peaceful uses of atomic energy Report World health organization World health organization. Technical report series, 151 World health organization,Geneva,1958
Radioth 35.1057

GENEVA 1957 Strahlenwirkung auf menschliche erbanlagen Papers World health organization Schriften reihe strahlenschutz, 3 Gersbach,Brunswick,1958
Radioth 35.1083

GENEVA 1957 The International symposium on aldosterone Papers and discussions Edited by Alex F. Muller and Cecilia M. O'Connor Organized by the Universite de Geneve.Clinique therapeutique J.and A.Churchill,London,1958
Bioch 33.0416

GENEVA 1958 International conference on the peaceful uses of atomic energy 2nd Vol 29: chemical effects of radiation United Nations United Nations,Geneva,1958
Met 25.1567

GENEVA 1958 International conference on the peaceful uses of atomic energy 2nd Proceedings 18,21-26,29 United Nations 8 vols United Nations,Geneva,1958
Radioth 35.1079

GENEVA 1958 International conference on the peaceful uses of atomic energy 2nd Proceedings Vol 31-32 United Nations 2 vols United Nations,Geneva,1958
Eng 41.4466

GENEVA 1958 Technology,engineering and safety International conference on the peaceful uses of atomic energy 2nd Edited proceedings Vol 2 United nations Edited by R. Hurst and others Progress in nuclear energy, 4 Pergamon press,Oxford, 1960
Met 25.1555

GENEVA 1958 United Nations international conference on the peaceful uses of atomic energy 2nd Proceedings Vol 5-7 United Nations 3 vols United Nations, Geneva,1958
Met 25.1544

GENEVA 1959 United Nations conference on the peaceful uses of atomic energy Proceedings Vol 1-15 8 vols United Nations,New York,1956 Lacks vols 6-7.10-14
Eng 41.7540

GENEVA 1960 The Use of vital and health statistics for genetic and radiation studies seminar Proceedings United nations World health organization United Nations publications, 61,XVII.8 United Nations,New York,1962
Gen 34.1120

GENEVA 1960 Evian 1960 International congress of nephrology 1st Proceedings Edited by G. Richet Organised by Societe de nephrology Karger,Basle;New York,1961 In English and French
Phys 20.1325

GENEVA 1961 Structural interdependence and economic development International conference on input-output techniques 3rd Proceedings United Nations.Secretariat Edited by Tibor Barna and others Sponsored jointly by the Harvard economic research project Macmillan,London,1963
Geog 13.2417

GENEVA 1962 International conference on high-energy physics at CERN 11th Proceedings European organization for nuclear research Edited by J. Prentki Sponsored by the International union of pure and applied physics Bibliog.,Illus. xxiv, 949p 28cm CERN,Geneva,1962
A Math 4.1544

GENEVA 1963 Krebsmetastasierung auf dem blutwege :symposium Schweizerische nationaliga fur krebsekampfung und krebsforschung Schweizerische akademie der medizinischen wissenschaften Schweizerische hamatologe gesellschaft Schweizerische akademie der medizinischen wissenschaften. Bulletin, 20,fasc 1-3 Schwabe,Basel,1964
Radioth 35.0835

GENEVA 1963 Science and technology for development United Nations conference on the application of science and technology for the benefit of the less developed areas Report Vol 1-4 4 vols United Nations,New York, 1963
Geog 13.2389

GENEVA 1963 Tumoren im kindesalter : jahresversammlung der Schweizenschen gesellschaft fUR Padiatrie Schweizerische gesellschaft fur padiatrie Padiatrische fortbildungskurse fur die praxis, 13 Illus Schwabe,Basel,1964
Radioth 35.0836

GENEVA 1964 Planning of radiotherapy facilities Joint IAEA-WHO meeting on planning of radiotherapy facilities Report International atomic energy agency World health organization World health organization.Technical report series, 328 World health organization,Geneva,1966
Radioth 35.1060

GENEVA 1965 Tumors of the alimentary tract in Africans :symposium Proceedings National cancer institute Edited by J.F. Murray Sponsored by the International union against cancer and the Calouste Gulbenkian foundation National cancer institute. Monograph, 25 National cancer institute, Bethesda,Md.,1967
Bioch 33.0990

GENEVA 1966 Polar meteorology W.M.O.-S.C. A.R.-I.C.P.M.Symposium on polar meteorology Proceedings World meteorological organization,and,Scientific committee on Antarctic research Also organized by the International commission on polar meteorology illus vi,540p 28cm World meteorological organization,Geneva,1967
Sco 14.7839

GENEVA 1967 Genetics of the immune response :report of a WHO scientific group World health organization World health organization.Technical report series, 402 World health organization,Geneva,1968
Gen 34.2012

GENEVA 1967 Medical radiation physics Joint IAEA-WHO expert committee on medical radiation physics Report International atomic energy agency World health organization World health organization. Technical report series, 390 World health organization,Geneva,1968
Radioth 35.1381

GENEVA 1968 Genetic counselling :a report World health organization World health organization.Expert committee on human genetics.Report, 3 World health organization.Technical report series, 416 Geneva,1969
Gen 34.1970

GENEVA 1968 Immunology and reproduction Symposium of the International co-ordination committee for the immunology of reproduction 1st Proceedings International co-ordination committee for the immunology of reproduction Edited by R.G. Edwards International planned parenthood association, London,1969
Phys 20.2265

GENEVA 1968 Multiples fonctions d'un jardin botanique :symposium international de Geneve Actes Jardin botanique de Geneve Edited by Jacques Miege Boissiera, 14 Geneva,1969 Symposium held to celebrate the 150th anniversary of the foundation of the botanic garden at Geneva
BG 38.3115

GENEVA 1969 Evian 1960 Sep 1-4 Congres international de nephrologie International congress of nephrology 1st Proceedings Edited by E. Richet S.Karger, Basle;New York,1961
Inv Med 37.0187

GENEVA 1970 Steroid assay by protein binding :a symposium 2nd Transactions Karolinska institutet World health organization Edited by E. Dicfalusy and A. Dicfalusy Karolinska symposia on research methods in reproductive endocrinology, 2 Stockholm,1970
PGMS 29.0589

GENEVA 1971 In vitro methods in reproductive cell biology... symposium 3rd Transactions Karolinska institutet World health organization Edited by E. Dicfalusy and A. Dicfalusy Karolinska symposia on research methods in reproductive endocrinology, 3 Acta endocrinologica.Supplement, 153 Stockholm,1971
PGMS 29.0650

GENEVA 1971 Large telescope design :a conference Proceedings European southern observatory CERN Edited by R.M. West 499p European southern observatory,Geneva, 1971
Obs 6.3601

GENEVA 1972 Auxiliary instrumentation for large telescopes :a conference European southern observatory CERN Edited by S. Lanstsen and A. Reiz 525p European southern observatory,Geneva,1972
Obs 6.3599

GENEVA;VILLARS 1966 Phase stability in metals and alloys Battelle materials science colloquium 1st Proceedings Battelle memorial institute Edited by Peter S. Rudman and others McGraw-Hill series in materials science and engineering McGraw-Hill,New York, 1967
Met 25.1194

GENEVA,N.Y. 1958 The International conference on the electronic properties of metals at low temperatures Report Hobart college and International union of pure and applied physics Co-sponsored by General electric research laboratory 244p c1958 Privately printed report to the sponsors
Cav 7.0895

GENEVA,SWITZERLAND 1953 Discussions on child development ;a consideration of the biological,psychological,and cultural approaches to the understanding of human development and behaviour Vol 1: proceedings World health organization.Study group on the psychobiological development of the child Edited by J.M. Tanner and Barbel Inhelder Tavistock publications,London,1956
Psy 31.1178

GENOA 1892 Congresso botanico internazionale Atti Edited by Ottone Penzig Genoa,1893
Bot 42.0649

GENOA 1961 Flora europaea symposium 2nd Proceedings Edited by V.H. Heywood and R.E.G. Pichi-Sermolli Webbia.Raccolta di scritte botanici, 18 Florence,1963
BG 38.0895

GENOOTSCHAP VOOR GESCHIEDENIS DER GENEESKUNDE, WISKUNDE EN NATUURWETEENSCHAPPEN TE LEIDEN Congres international d'histoire des sciences et congres de la Societe internationale d'histoire de la medecine Actes Amsterdam 1950 Aug 14-21 Academie internationale d'histoire des sciences.Collection de travaux, 6 2 vols Academie d'histoire des sciences;Hermann,Paris,1951-53 The 12th congress of the Societe internationale d'histoire de la medecine is section 4 of the 6th Congres international d'histoire des sciences
WSM 43.0067

GEOCHEMICAL INVESTIGATIONS Symposium de exploracion geoquimica,tom.1-3 International geological congress 20th papers Mexico City 1956 Edited by A. Garcia Rojas and others 2cm Mexico City,1958-60 Text in English,French,Russian and Spanish
Geol 8.3048

GEOCHEMICAL SOCIETY.ORGANIC GEOCHEMISTRY GROUP, EUROPEAN BRANCH   Advances in organic chemistry international meeting proceedings Milan 1962 Sep 10-12 Edited by Umberto Colombo and G.D. Hobson International series of monographs on earth sciences,15 illus. Pergamon press,Oxford,1964
Geol   8.1666

GEOCHIMICA ET COSMOCHIMICA ACTA, 34,supplement 1 Apollo 11 lunar science conference Proceedings Houston,Texas 1970 Jun 5-8 1-3 National aeronautics and space administration Edited by A.A. Levinson 3 vols Pergamon,New York,1970
Min   10.1548

GEODESY   Geodesy in the space age :a symposium Papers Columbus,Ohio 1961 Feb 6-8 Ohio state university Edited by Simo Laurila and others Ohio state university. Institute of geodesy,photogrammetry,and cartography.Publication, 15 Columbus,Ohio, 1961 Privately printed
Geod   9.0032

GEODESY IN :a symposium   the space age Columbus,Ohio 1961 Feb 6-8 Ohio state university Edited by Simo Laurila and others Ohio state university.Institute of geodesy, photogrammetry,and cartography.Publication, 15 Columbus,Ohio,1961 Privately printed
Geod   9.0032

GEOGRAFICHESKOE OBSHCHESTVO S.S.S.R.SEVERNYI FILIAL   Priroda i khozyaistvo severa : materialy pervoi nauchnoi konferentsii Kolskogo otdela geograficheskoe obshchestva S. S.S.R. 1967 Dec Vyp 1 Edited by I.L. Freydin and others maps 314p 29cm Akademiya nauk SSSR,Apatity,1969
Sco   14.8284

GEOGRAPHERS   Congres des geographes allemands notice Halle 1882 Apr 12-14 By Alexis de Tillo Halle,1882 One of fifteen tracts bound together
Philos   1.0875

GEOGRAPHY   Taxonomy and geography Systematics association symposium 4th papers Oxford 1959 Mar 31-Apr 2 Edited by David Nichols Systematics association. Publication,4 Systematics association,London, 1962
Geol   8.0942

GEOLOGIA Y DEPOSITOS DE CARBON DE LA REGION DE SABINAS,ESTADO DE CHIHUAHUA   International geological congress 20th papers Mexico City 1956 By Raymond C. Robeck and others United States geological survey Mexico City, 1956 Bound with two other publications of the congress
Geol   8.3041

GEOLOGICAL SOCIETY OF AMERICA   Crust of the earth :a symposium Proceedings New York 1954 Oct 14-16 Columbia university Edited by Arie Poldervaart Geological society of America.Special paper, 62 New York,1955
Geod   9.0100

GEOLOGICAL SOCIETY OF AMERICA.COMMITTEE ON ROCK MECHANICS   State of stress in the earth's crust :an international conference Proceedings Santa Monica,Calif. 1963 Jun 13-14 Edited by William R. Judd New York, 1964
Geod   9.0143

GEOLOGICAL SOCIETY OF LONDON   The Fossil record a symposium with documentation Swansea 1965 Dec 20-21 Edited by W.B. Harland and others bibliog. Geological society of London,London,1967 The papers in parts 1 and 3 were presented at a joint meeting of the Geological society of London and the Palaeontological association,Swansea, Dec 1965
Geol   8.4440

GEOLOGICAL SOCIETY OF LONDON   The Phanerozoic time-scale :a symposium dedicated to Professor Arthur Holmes Edited by W.B. Harland and others Geological society of London. Quarterly journal,120 S Bibliog viii,458p Geological society of London,London,1964 Based on material presented at the Geological society's symposium on the Phanerozoic time-scale,held in Glasgow 14 February,1964.A supplement to the Quarterly journal of the Geological society of London
Sco   14.0249

GEOLOGICAL SOCIETY OF LONDON   The Phanerozoic time-scale :a symposium dedicated to Professor Arthur Holmes Glasgow 1964 Feb 14 Edited by W.B. Harland and others Geological society of London.Quarterly journal, 120 S Bibliog viii,458p Geological society of London,London,1964 Based on material presented at the Geological society's symposium on the Phanerozoic time-scale,held in Glasgow 14 February,1964.A supplement to the Quarterly journal of the Geological society of London
Bot   42.3194

GEOLOGICAL SOCIETY OF LONDON and INSTITUTE OF PETROLEUM, Salt basins around Africa Institute of petroleum and Geological society of London :joint meeting proceedings London 1965 Mar 3 Institute of petroleum, London,1965
Geol   8.3097

GEOLOGICAL SOCIETY OF SOUTH AFRICA   Symposium on the Bushveld igneous complex and other layered intrusions Proceedings Pretoria 1969 Jul 7-14 Edited by D.J.L. Visser and G. von Gruenewaldt Geological society of South Africa.Special paper, 1 Geological society of South Africa,Johannesburg,1970
Min   10.1476

GEOLOGICAL SURVEY OF BRITISH GUIANA   Inter-Guiana geological conference 5th proceedings Georgetown 1959 Oct.28-Nov.6 Edited by R.B. McConnell Geological survey department,British Guiana,Georgetown,1962
Geol   8.2426

GEOLOGIE DE LA MEDITERRANEE OCCIDENTALE International geological congress 14th etudes et observations Madrid 1926 Vol 1-2 Association pour l'etude geologique de la Mediterranee occidentale 2 vols Liege,1926-35
Geol   8.2954

GEOLOGISCHE CONFERENTIE 1st   report Paramaribo 1950 Sep.23-Oct.3 1950 Mimeographed
Geol   8.2428

GEOLOGISCHE CONFERENTIE see also CONFERENCE GEOLOGIQUE DES GUYANES

GEOLOGISCHE CONFERENTIE see also CONFERENCE GEOLOGIQUE DES TROIS GUYANES

GEOLOGISCHE CONFERENTIE see also INTER-GUIANA GEOLOGICAL CONFERENCE

GEOLOGY C.C.T.A.East-central regional committee for geology :meeting 1st proceedings Dar-es-Salaam 1956 Commission for technical co-operation in Africa south of the Sahara London,1957 Bound with proceedings of three other regional meetings
Geol 8.3115

GEOLOGY C.C.T.A.Southern regional committee for geology :meeting 1st proceedings Salisbury 1955 Sep 19-24 Commission for technical co-operation in Africa south of the Sahara London,1956 Bound with proceedings of three other regional meetings
Geol 8.3114

GEOLOGY Genese des roches filoniennes,fasc.6 International geological congress 19th Proceedings,section 6 Algiers 1952 Algiers,1953
Min 10.0351

GEOLOGY Geologie de la Mediterranee occidentale International geological congress 14th etudes et observations Madrid 1926 Vol 1-2 Association pour l'etude geologique de la Mediterranee occidentale 2 vols Liege,1926-35
Geol 8.2954

GEOLOGY International geological congress 1st- proceedings Paris 1878- Paris, 1886-
Geol 8.4145

GEOLOGY International geological congress 20th papers Mexico City 1956 Gacetas historicas Edited by Manuel Carrera Stampa Mexico City,1956 Bound with two other publications of the congress Text in English, French and Spanish
Geol 8.3030

GEOLOGY International geological congress 20th papers Mexico City 1956 Seccion 4: geohidrologia de regiones aridas y subaridas Edited by A. Garcia Rojas and others Mexico City,1957 Text in English,French,Russian and Spanish
Geol 8.3045

GEOLOGY International geological congress 20th resumes of papers presented Mexico City 1956 Edited by Hans E. Thalmann Mexico City,1956 Text in English,French, German and Spanish
Geol 8.3031

GEOLOGY International geological congress 21st reports Copenhagen 1960 1-22 Edited by Theodor Sorgenfrei Copenhagen,1960
Geol 8.3123

GEOLOGY International geological congress 6th guide-book Zurich 1894 livret-guide geologique dans le Jura et les Alpes de la Suisse Payot,Lausanne,1894
Geol 8.3076

GEOLOGY Paleovolcanologie et ses rapports avec la tectonique,fasc.17 International geological congress 19th Proceedings, section 15 Algiers 1952 Algiers,1954
Min 10.0352

GEOLOGY AND ARCHAEOLOGY OF NORTHERN CYRENAICA, LIBYA Petroleum exploration society of Libya :annual field conference 10th guidebook Cyrenaica 1968 Edited by E.T. Barr Amsterdam,1968
Geol 8.4487

GEOLOGY OF AUSTRALIAN ORE DEPOSITS Empire mining and metallurgical congress 5th Publications Melbourne 1953 Apr 14 Jun 12 1 Australasian institute of mining and metallurgy Edited by A.B. Edwards Office of the congress,and of the Australasian institute of mining and metallurgy,Melbourne, 1953
Met 25.0187

GEOLOGY OF CANADIAN INDUSTRIAL MINERAL DEPOSITS Commonwealth mining and metallurgical congress 6th Papers Montreal 1957 Sep 8-Oct 9 Canadian institute of mining and metallurgy Montreal,1957
Min 10.0716

GEOLOGY OF SHELF SEAS Inter-university geological congress 14th Proceedings Edited by D.T. Donovan Maps Oliver and Boyd,Edinburgh;London,1968
Geog 13.0744

GEOLOGY OF SHELF SEAS Inter-university geological congress 14th Proceedings Hull 1967 Edited by D.T. Donovan illus. viii,160p Oliver and Boyd,Edinburgh;London, 1968
Bot 42.3193

GEOLOGY OF SHELF SEAS Inter-university geological congress 14th proceedings Huel 1967 Jan Edited by D.T. Donovan Oliver Boyd,Edinburgh;London,1968
Geol 8.4480

GEOLOGY OF THE ARCTIC First international symposium on Arctic geology Calgary 1960 Jan 11-30 Vol 1-2 Alberta society of petroleum geologists Edited by Gilbert O. Raasch illus,maps 26cm 2 vols University of Toronto press,Toronto,1961
Sco 14.2403

GEOLOGY OF THE ARCTIC International symposium on Arctic geology 1st Proceedings Calgary 1960 Jan 11-13 Vol 1-3 Edited by Gilbert O. Raasch Under the auspices of the Alberta society of petroleum geologists 3 vols University of Toronto press,Toronto,1961 Vol.3 consists of maps
Geog 13.5390

GEOLOGY OF THE ARCTIC International symposium on arctic geology 1st Proceedings Calgary,Alta. 1960 Jan 11-13 Vol 1-2 Edited by Gilbert O. Raasch Under the auspices of the Alberta society of petroleum geologists 2 vols Toronto,1961 Additional maps in a separate container
Geod 9.0436

GEOLOGY OF THE ARCTIC International symposium on arctic geology 1st Proceedings Calgary,Alberta 1960 Jan 11-13 Edited by Gilbert O. Raasch Under the auspices of Alberta society of petroleum geologists maps 3 vols University of Toronto press,Toronto,1961
Geol 8.2938

GEOMAGNETIC DATA 1958 indices K and C By J. Bartels and others Association of geomagnetism and aeronomy I.A.G.A.Bulletin, 12 m 1 tables viii,113p 24cm I.U.G.G., Paris,1962
Sco 14.7931

GEOMETRICESKOGO SEMINARA trudy Moscow 1963 Tom 1 Organised by the Akademiya nauk S.S.S.R.Institut nauchnoi informatsii. Otdel matematiki 456p 22cm VINITI,Moscow, 1966
P. Math 2.0699

GEOMETRY Summer school in mathematics, geometry and topology proceedings Dundee 1961 Jul 10-22 Queen's college,Dundee 33cm Queen's College,Dundee,1961
P. Math 2.0292

GEOMETRY OF NUMBERS Seminar on geometry of numbers Lectures,mimeograph Princeton,N.J. 1949 Institute for advanced study 90p 26cm Institute for advanced study,Princeton, N.J.,1949
P Math 2.2778

GEOPHYSICAL FLUID DYNAMICS Fluid models in geophysics Symposium on the use of models in geophysical fluid dynamics 1st Proceedings Baltimore,Md. 1953 Sep 1-4 Edited by Robert R. Long Sponsored by United States.Office of naval research illus, diagrms v,162p U.S.Government printing office,Washington,D.C.,1956
Chem E 24.0202

GEOPHYSICAL SCIENCES Mathematical problems in the geophysical sciences Troy,N.Y. 1970 Vol 1-2: geophysical fluid dynamics;inverse problems,dynamo theory,and tides American mathematical society Edited by W.H. Reid Lectures in applied mathematics, 13-14 illus,maps 24cm 2 vols American mathematical society,Providence,R.I.,1971
A Math 4.1742

GEOPHYSICS International geological congress 20th papers Mexico City 1956 Seccion 9: geofisica aplicada,tom.1-2 Edited by A. Garcia Rojas and others Mexico City,1958 Text in English,French,German,Russian and Spanish
Geol 8.3044

GEOPHYSICS Research in geophysics :an international symposium Papers Los Angeles,Calif. 1963 Aug 12-16 Vol 1: sun, upper atmosphere,and space University of California Edited by Hugh Odishaw and others Sponsored by National academy of sciences Massachusetts institute of technology, Cambridge,Mass.,1964
Geod 9.0130

GEOPHYSICS Rock deformation University of California,Los Angeles.Institute of geophysics annual technical conference 1956 Symposium Los Angeles,Calif. 1956 Nov Edited by David Griggs and John Handin Geological society of America.Memoirs,79 New York,1960
Min 10.0348

GEOPHYSICS Solar variations,climatic change, and related geophysical problems :a conference Papers New York 1961 Jan 24-28 New York academy of sciences Edited by W.Franklin N. Furness and others Supported conjointly by the American meteorological society New York. Academy of sciences.Annals, 95,1 740p New York,1961 Conference editor:R.W. Fairbridge
Nap 11.0210

GEOPHYSICS Submarine geology and geophysics : a symposium 17th Proceedings Bristol 1965 Apr 5-9 Colston research society Edited by W.F. Whittard and R. Bradshaw Colston research society.Colston papers, 17 London,1965
Geod 9.0580

GEOPHYSICS AND THE IGY :proceedings of the symposium at the opening of the international geophysical year Proceedings United States national committee for the international geophysical year Edited by Hugh Odishaw and Stanley Ruttenberg National research council.Publication,590 American geophysical union.Geophysical monograph,2 illus vi,210p 26cm Washington,1958
Sco 14.0511

GEORGETOWN Electrical studies on the unanesthetized brain :a symposium Edited by Estelle R. Ramey and Desmond S. O'Doherty Hoeber,New York,1960
VA 19.0238

GEORGETOWN 1959 Inter-Guiana geological conference 5th proceedings Geological survey of British Guiana Edited by R.B. McConnell Geological survey department, British Guiana,Georgetown,1962
Geol 8.2426

GEORGIA INSTITUTE OF TECHNOLOGY Applications of field-ion microscopy in physical metallurgy and corrosion Conference in physical metallurgy and corrosion Proceedings Atlanta,Ga. 1965 May 15-17 Edited by R.F. Hochman and others Georgia institute of technology.Advanced research projects agency, Atlanta,Ga.,1968
Met 25.2631

GEOSCIENCE SOCIETY OF ICELAND Iceland and mid-ocean ridges :a symposium Report Reykjavik 1967 Feb 27-Mar 8 Edited by Sveinbjorn Bjornsson Societas scientiarum islandica.Publication, 38 Reykjavik,1967
Geod 9.0383

GEOTECHNICAL CONFERENCE Shear strength properties of natural soils and rocks :a conference Proceedings Oslo 1967 Vol 1-2 Norsk geoteknisk forening 2 vols Norwegian geotechnical institute,Oslo,1967
Eng 41.3135

GERM CELLS Symposium on the germ cells and earliest stages of development Pallanza 1960 Sep 14-20 International institute of embryology A.Baselli,Milan,1961 At head of title:Istituto Lombardo accademia di scienze e lettere
Bal 39.0459

GERMANY'S CONTRIBUTION TO EUROPEAN ECONOMIC LIFE
Conference on some aspects of the German problem Baarn 1947 Oct 6-11 and Scheveningen 1948 Apr 11-17 By Percy W. Bidwell and others Riviere,Paris,1949 Papers in English and French
Geog 13.2112

GESELLSCHAFT DEUTSCHER NATURFORSCHER UND AERZTE
Versammlung der Gesellschaft deutscher naturforscher und aerzte... 20th Amtlicher bericht Mainz 1842 Sep 19-26 Edited by Johann Groser and Friedrich Karl Bruch 2 pls 28cm Kupferberg,Mainz,1843
Bal 44.6348

GESELLSCHAFT DEUTSCHER NATURFORSCHER UND ARZTE
Funktionelle und morphologischorganisation der zelle Rottach-Egern 1962 Aug Springer, Berlin,1963
Gen 34.0830

GESELLSCHAFT FUR ANGEWANDTE MATHEMATIK UND MECHANIK Electronic digital computers and information processing :conference Proceedings Darmstadt 1956 Oct 25-27 Edited by A. Walther and W. Hoffmann Nachrichtentechnische fachberichte, 4 illus. viii,229p 30cm Friedr.Vieweg und sohn,Brunswick,1956
Math L 5.3247

GESELLSCHAFT FUR ANGEWANDTE MATHEMATIK UND MECHANIK.FACHAUSSCHUSS REGELUNGSMATHEMATIK
Regelungstechnik:moderne theorien und ihre verwendbarkeit :ein tagung Bericht Heidelberg 1956 Sep 25-29 Oldenbourg, Munich,1957
Eng 41.5842

GESELLSCHAFT FUR BIOLOGISCHE CHEMIE
Biochemie des sauerstoffs :colloquium Mosbach 1968 Apr 24-27 Edited by B. Hess and Hj. Staudinger Gesellschaft fur biologische chemie.Colloquia, 19 illus Springer,Berlin,1968 In English and German
Bioch 33.1074

GESELLSCHAFT FUR BIOLOGISCHE CHEMIE
Inhibitions in cell research :colloquium Mosbach 1969 Apr 14-16 Edited by T. Bucher and H. Sies Gesellschaft fur biologische chemie.Colloquia, 20 Springer,Berlin,1969
Bioch 33.1908

GESELLSCHAFT FUR BIOLOGISCHE CHEMIE
Inhibitions in cell research :colloquium Mosbach 1969 Apr 14-16 Edited by T. Bucher and H. Sies Gesellschaft fur biologische chemie.Colloquia, 20 Springer,Berlin,1969
Bot 42.1758

GESELLSCHAFT FUR BIOLOGISCHE CHEMIE
Mammalian reproduction :colloquium Mosbach 1970 Apr 9-11 Edited by H. Gibian and E.J. Plotz Gesellschaft fur biologische chemie. Colloquia, 21 Springer,Heidelburg;New York, 1970
Inv Med 37.0088

GESELLSCHAFT FUR BIOLOGISCHE CHEMIE Neuere ergebnisse aus chemie und stoffwechsel der kohlenhydrate colloquium Mosbach,Baden 1957 May 2-4 Gesellschaft fur biologische chemie.Colloquia, 8 illus Springer, Berlin,1958
Bioch 33.2347

GESELLSCHAFT FUR BIOLOGISCHE CHEMIE
Regulation of gluconeogenesis Conference of the Gesellschaft fur biologische chemie 9th Proceedings Gottingen 1971 Jan 15 Edited by Hans-Dieter Soling and Berend Willms Thieme;Academic press,Stuttgart;London,1971
Bioch 33.2210

GESELLSCHAFT FUR BIOLOGISCHE CHEMIE.COLLOQUIA, 8 Neuere ergebnisse aus chemie und stoffwechsel der kohlenhydrate colloquium Mosbach,Baden 1957 May 2-4 Gesellschaft fur biologische chemie illus Springer, Berlin,1958
Bioch 33.2347

GESELLSCHAFT FUR BIOLOGISCHE CHEMIE.COLLOQUIA, 19 Biochemie des sauerstoffs :colloquium Mosbach 1968 Apr 24-27 Gesellschaft fur biologische chemie Edited by B. Hess and Hj. Staudinger illus Springer,Berlin,1968 In English and German
Bioch 33.1074

GESELLSCHAFT FUR BIOLOGISCHE CHEMIE.COLLOQUIA, 20 Inhibitions in cell research : colloquium Mosbach 1969 Apr 14-16 Gesellschaft fur biologische chemie Edited by T. Bucher and H. Sies Springer,Berlin, 1969
Bioch 33.1908

GESELLSCHAFT FUR BIOLOGISCHE CHEMIE.COLLOQUIA, 20 Inhibitions in cell research : colloquium Mosbach 1969 Apr 14-16 Gesellschaft fur biologische chemie Edited by T. Bucher and H. Sies Springer,Berlin, 1969
Bot 42.1758

GESELLSCHAFT FUR BIOLOGISCHE CHEMIE.COLLOQUIA, 21 Mammalian reproduction :colloquium Mosbach 1970 Apr 9-11 Gesellschaft fur biologische chemie Edited by H. Gibian and E. J. Motz Springer,Berlin,1970
An 32.5461

GESELLSCHAFT FUR BIOLOGISCHE CHEMIE.COLLOQUIA, 21 Mammalian reproduction :colloquium Mosbach 1970 Apr 9-11 Gesellschaft fur biologische chemie Edited by H. Gibian and E. J. Plotz Springer,Heidelburg;New York,1970
Inv Med 37.0088

GESELLSCHAFT FUR PHYSIOLOGISCHE CHEMIE
Induktion und morphogenese Mosbach 1962 May 3-5 Gesellschaft fur physiologische chemie.Colloquium, 13 Springer,Berlin,1963
Gen 34.0857

GESELLSCHAFT FUR PHYSIOLOGISCHE CHEMIE
Chemie und stoffwechsel von binde-und knochengewebe :colloquium Mosbach Baden 1956 Apr 12-14 Gesellschaft fur physiologische chemie.Colloquium, 7 Springer,Berlin,1956
An 32.3221

GESELLSCHAFT FUR PHYSIOLOGISCHE CHEMIE.COLLOQUIUM, 7 Chemie und stoffwechsel von binde- und knochengewebe :colloquium Mosbach Baden 1956 Apr 12-14 Gesellschaft fur physiologische chemie Springer,Berlin,1956
An 32.3221

GESELLSCHAFT FUR PHYSIOLOGISCHE CHEMIE.COLLOQUIUM, 13 Induktion und morphogenese Mosbach 1962 May 3-5 Gesellschaft fur physiologische chemie Springer,Berlin,1963
Gen 34.0857

GESELLSCHAFT FUR PHYSIOLOGISCHE CHEMIE.COLLOQUIUM, 15 Immunchemie :colloquium Mosach 1964 Apr 22-25 Springer,Berlin,1965
Radioth 35.0145

GESELLSCHAFT FUR PHYSIOLOGISCHE CHEMIE.COLLOQUIUM, 17 Molekulare biologie des malignen wachstums :colloquium Mosbach 1966 Apr 21-23 Edited by H. Holzer and A.W. Holldorf Springer,Berlin,1966
Radioth 35.0142

GESTATION Conference on gestation 1st Transactions Princeton,N.J. 1954 Mar 9-11 Edited by Louis B. Flexner Sponsored by the Josiah Macy jr.foundation Josiah Macy jr. foundation,New York,1955
An 32.3815

GESTATION Conference on gestation 2nd-5th Transactions Princeton,N.J. 1955-58 Mar Edited by Claude A. Villee Sponsored by the Josiah Macy jr.foundation 3 vols Josiah Macy jr. foundation,New York,1956-58
An 32.3816

GHENT 1862 Congress international pomologie Federation des societes d'horticulture de Belgique Federation des societes d'horticulture de Belgique.Bulletin, 5 Ghent,1865
Bot 42.0644

GHENT 1950 Jaarlijks symposium over phytopharmacie symposium 2nd Papers Rijkslandbouwhogeschool,Ghent Landbouwhogeschool en de opzoekingsstations, Ghent.Mededelingen, 15,no 1 Rijkslandbouwhogeschool,Ghent,1950 Papers and summaries in English,Flemish and French
Chem 26.0105

GHENT 1953 Jaarlijks symposium over phytopharmacie symposium 5th Verhandelingen Rijkslandbouwhogeschool,Ghent Edited by J.van den Brande Landbouwhogeschool en de opzoekingsstation, Ghent.Mededelingen, 18,no 2 Rijkslandbouwhogeschool,Ghent,1953 Papers and summaries in English,Flemish and French
Chem 26.0106

GHENT 1954 Europees congres toegepaste elektronmicroscopie Proceedings Edited by G. Van der Meersche Rijksuniversitet,Ghent, 1954 Papers in English,Dutch,German, French and Spanish
Radioth 35.1583

GHENT 1955 Biochemical problems of lipids : an international conference 2nd Proceedings Vlaamse chemische vereniging Edited by G. Popjak and E. Le Breton Butterworths,London,1956 Conference president:Professor R.Ruyssen
Chem 18.1500

GHENT 1965 Androgens in normal and pathological conditions Steroid hormones :a symposium 2nd Proceedings Rijkauniversiteit te Gent Societe belge d'endrocinologie Edited by A. Vermeulen and D. Exley Excerpta medica.International congress series, 101 Excerpta medica, Amsterdam,1966
Inv Med 37.0276

GIBBERELLINS Photo-thermoperiodism,the action of different radiations on gibberellins and other active substances Colloque international sur le photothermoperiodisme action des diverses radiations on gibberellines et de quelques autres substances Parma 1957 Jun International union of biological sciences International union of biological sciences.Series B, 34 U.I.S.B., Paris,1958
Bot 42.1420

GIBBERELLINS ...The symposium...presented before the Division of agricultural and food chemistry of the 138th national meeting of the American chemical society Papers New York 1960 Sep American chemical society. Division of agricultural and food chemistry Advances in chemistry series, 28 American chemical society,Washington,1961
Bioch 33.0738

GIF SUR YVETTE 1963 Regulateurs naturels de la croissance vegetales Colloque international sur les substances de croissance vegetales 5e Actes Centre national de la recherche scientifique Centre national de la recherche scientifique.Colloques internationaux, 123 Paris,1964 Text in English and French
Bioch 33.0747

GIF-SUR-YVETTE 1955 Microphysiologie comparee des elements excitables :colloque international Centre national de la recherche scientifique Centre national de la recherche scientifique.Colloques internationaux, 67
Phys 20.2022

GIRTON 1969 Computer science :a symposium Proceedings Institute of mathematics and its applications.Bulletin, 6,no.1 I.M.A., Southend,1970 Held as part of the centenary celebrations of Girton college, Cambridge
Math L 5.3553

GLACIERS Variations of the regime of existing glaciers International association of scientific hydrology :a symposium Proceedings Obergurgl 1962 Sep 10-18 Edited by W. Ward International association of scientific hydrology.Publication,58 312p Gentbrugge,1962
Sco 14.0137

GLACIOLOGY Physics of the movement of the ice The International association of scientific hydrology :a symposium Papers Chamonix 1958 Sep 16-24 International association of scientific hydrology. Publication,47 illus. 393p 22cm Gentbrugge,1958 Papers in English and French
Sco 14.0127

GLASGOW 1875 Navigation: a lecture ... By William Thomson Under the auspices of Glasgow science lecture association 17cm London,Glasgow,1876
Philos 1.0283

GLASGOW 1901 Discussion on the teaching of mathematics... to which is added a report of the British association committee drawn up by the chairman,professor Forsyth British association for the advancement of science Edited by John Perry 2nd edition Macmillan, London;New York,1902
Psy 31.3384

GLASGOW 1901 International engineering congress :report of the proceedings and abstracts of the papers read Edited by J.D. Cormack Asher,Glasgow,1902
Eng 41.1563

GLASGOW 1901 Railways :international engineering congress.Section one of a meeting held at the University of Glasgow Proceedings Clowes,London,1902
Eng 41.1553

GLASGOW 1901 Waterways and maritime works International engineering congress on waterways and maritime works Proceedings Clowes,London,1902
Eng 41.3317

GLASGOW 1954 Conference on nuclear and meson physics Proceedings Edited by E.H. Bellamy and R.G. Moorhouse Sponsored by the International union of pure and applied physics Pergamon press,London;New York,1955
Radioth 35.1563

GLASGOW 1954 The International conference on nuclear and meson physics Proceedings Edited by E.H. Bellamy and R.G. Moorhouse Sponsored by the International union of pure and applied physics 352p Pergamon press, London,1955
Cav 7.0908

GLASGOW 1956 Pump design,testing and operation :a symposium Proceedings National engineering laboratory H.M.S.O., Edinburgh,1966
Eng 41.6824

GLASGOW 1957 Endocrine aspects of breast cancer :a conference Proceedings Edited by Alastair R. Currie Held under the auspices of the University of Glasgow Livingstone, Edinburgh;London,1958
PGMS 29.0455

GLASGOW 1957 Endocrine aspects of breast cancer :a conference Proceedings University of Glasgow Edited by Alastair R. Currie Livingstone,Edinburgh;London,1958
Path 30.2082

GLASGOW 1960 Computing methods and the phase problem in x-ray crystal analysis : conference 2nd report Edited by Ray Pepinsky and others International tracts in computer science and technology and their application,4 port. Pergamon press,Oxford, 1961 The first conference was held in 1950
Min 10.1207

GLASGOW 1960 Computing methods and the phase problem in x-ray crystal analysis conference report Edited by Ray Pepinsky and others International tracts in computer science and technology and their application, 4 viii,326p 23cm Pergamon press,Oxford, 1961
Math L 5.0303

GLASGOW 1960 The Human adrenal cortex :a conference Proceedings Edited by Alastair R. Currie and others Livingstone,Edinburgh; London,1962
An 32.3758

GLASGOW 1964 Activation analysis principles and applications NATO advanced study institute Proceedings North Atlantic treaty organization Edited by J.M.A. Lenihan and S.J. Thomson Academic press,London;New York,1965
Met 25.1689

GLASGOW 1964 Hyperbaric oxygenation International congress on clinical application of hyperbaric oxygen 2nd Proceedings Edited by Iain McA. Ledingham Livingstone, Edinburgh;London,1965
Radioth 35.0845

GLASGOW 1964 Mass spectrometry :Nato advanced study institute on theory,design and applications North Atlantic treaty organization Edited by R.I. Reed Academic press,London;New York,1965
Met 25.1694

GLASGOW 1964 The Phanerozoic time-scale :a symposium dedicated to Professor Arthur Holmes Geological society of London . Edited by W.B. Harland and others Geological society of London.Quarterly journal, 120 S Bibliog viii,458p Geological society of London, London,1964 Based on material presented at the Geological society's symposium on the Phanerozoic time-scale,held in Glasgow 14 February,1964.A supplement to the Quarterly journal of the Geological society of London
Bot 42.3194

GLASGOW 1966 The Gas liquid chromatography of steroids :a symposium Proceedings Society for endocrinology Edited by J.K. Grant Society for endocrinology.Memoirs, 16 Cambridge university press,Cambridge,1967
Gen 34.0556

GLASGOW 1967 Blood flow through organs and tissues :an international conference Proceedings Edited by William H. Bain and A. Murray Harper Livingstone,Edinburgh;London, 1968
Inv Med 37.0009

GLASGOW 1967 Power systems conference 2nd Proceedings Science research council Glasgow,1967
Eng 41.4988

GLASGOW 1968 The Engineering uses of holography :a symposium Proceedings University of Strathclyde Edited by E.R. Robertson and J.M. Harvey Organized in association with the National physical laboratory Cambridge university press, Cambridge,1970
Eng 41.4290

GLASGOW 1968 The Engineering uses of holography :a symposium Proceedings University of Strathclyde National physical laboratory Edited by Elliot R. Robertson and James M. Harvey Cambridge university press, Cambridge,1970
An 32.5364

GLASGOW 1968 The Ocular circulation in health and disease;William Mackenzie centenary symposium Proceedings Edited by E.R. Brown and C.V. Moore Progress in hematology, 6 Heinemann,London,1969
Med 36.0054

GLASGOW 1969 Diffusion processes Thomas Graham memorial symposium Proceedings University of Strathclyde Edited by John N. Sherwood and others 2 vols Gordon and Breach,London,1971
Met 25.2597

GLASS Advances in glass technology International congress on glass 4th Technical papers Washington,D.C. 1962 Jul 8-14 American ceramic society Plenum press, New York,1962
Met 25.0116

GLASS Nucleation and crystallization in glasses and melts :a symposium Toronto 1961 Apr American ceramic society Edited by Margie K. Reser and others American ceramic society,Columbus,Ohio,1962
Met 25.0127

GLAVNAYA ASTRONOMICHESKAYA OBSERVATORIYA,PULKOVO Optical instability of the earth's atmosphere All-union conference 3rd Proceedings Kiev 1962 July Translated by Z. Lerman from the Russian viii,174p 25cm Israel program for scientific translations, Jerusalem,1966
Obs 6.3275

GLAVNAYA ASTRONOMICHESKAYA OBSERVATORIYA,PULKOVO The Moon :a symposium Pulkovo 1960 Dec. International astronomical union Edited by Zdenek Kopal and Zdenka Kadla Mikhailov International astronomical union.Symposium, 14 571p Academic press,London;New York, 1962
A Math 4.1153

GLOBAL ANALYSIS :a symposium Proceedings Berkeley,Calif. 1968 Jul 1-26 American mathematical society Edited by Shing-Shen Chern and Stephen Smale American mathematical society.Proceedings of symposia in pure mathematics, 14,16 bibliog. 25cm American mathematical society,Providence,R.I., 1970
P Math 2.3875

GLOBAL CIRCULATION OF THE ATMOSPHERE :a conference Proceedings London 1969 Aug 25-29 Royal meteorological society American meteorological society Edited by G. A. Corby bibliog.,illus.,port. 257p 26cm Royal meteorological society,London,1969
A Math 4.1651

GLOBAL DIFFERENTIABLE DYNAMICS Conference on global differentiable dynamics Proceedings Cleveland,Ohio 1969 Jun 2-6 Case western reserve university Edited by O. Hajek and others Lecture notes in mathematics, 235 x, 140p 25cm Springer,Berlin,1971
P Math 2.4317

GLOBAL EFFECTS OF ENVIRONMENTAL POLLUTION :a symposium Dallas,Texas 1968 Dec American association for the advancement of science 218p Reidel,Dordrecht,1970
Nap 11.0970

GLUCIDES Conference de l'Union internationale de chimie :rapports sur les hydrates de carbone (Glucides) 10th Rapports Liege 1930 Sep 14-20 Union internationale de chimie Union internationale de chimie,Paris,1930
Chem 18.0066

GLUCONEOGENESIS Regulation of gluconeogenesis Conference of the Gesellschaft fur biologische chemie 9th Proceedings Gottingen 1971 Jan 15 Gesellschaft fur biologische chemie Edited by Hans-Dieter Soling and Berend Willms Thieme;Academic press,Stuttgart;London,1971
Bioch 33.2210

GLUE Recent advances in gelatin and glue research :a conference Proceedings Cambridge 1957 Jul 1-5 Edited by G. Stainsby Sponsored by the British gelatine and glue research association Pergamon, London,1958
Bioch 33.2247

GLUTATHIONE :a symposium London 1958 Feb 15 Biochemical society Edited by E.M. Crook Held at the University of London. Senate House Biochemical society.Symposia, 17 Cambridge university press,Cambridge,1959
Bioch 33.1379

GLUTATHIONE :a symposium London 1958 Feb 15 Biochemical society Edited by E.M. Crook Held at the University of London. Senate house Biochemical society.Symposia, 17 Cambridge university press,Cambridge,1959
Radioth 35.0078

GLUTATHIONE :a symposium Proceedings Ridgefield,Conn. 1953 Nov Edited by S. Colowick and others Academic press,New York, 1954
Bioch 33.0604

GLYCOGEN METABOLISM Control of glycogen metabolism :a Ciba foundation symposium Proceedings London 1963 Jul 23-25 Ciba foundation Edited by W.J. Whelan and Margaret P. Cameron illus Churchill,London, 1964
Bioch 33.0576

GLYCOGEN METABOLISM Federation of European biochemical societies meeting 4th Proceedings Oslo 1967 Jul 7 Vol 5: control of glycogen metabolism Federation of European biochemical societies Edited by W.J. Whelan Organized by the Norwegian biochemical society Universitetsforlaget; Academic press,Oslo;London,1968 Symposium organizer W.J. Whelan
Bioch 33.1400

GODDARD INSTITUTE FOR SPACE STUDIES Earth-moon system The Dynamics of the earth-moon system conference Proceedings New York 1964 Jan.20-21 Edited by B.G. Marsden and A. G.W. Cameron 288p Plenum press,New York, 1966
Obs 6.1169

GODDARD INSTITUTE FOR SPACE STUDIES Nucleosynthesis :a conference Proceedings New York 1965 Jan 25-26 Edited by W.David Arnett and others Gordon and Breach,New York, 1968
TA 15.0306

GODDARD INSTITUTE FOR SPACE STUDIES
Nucleosynthesis :conference Proceedings Greenbelt,Md. 1965 Jan 25-29 Edited by W.D. Arnett and others 273p Gordon and Breach, New York,1968
Obs 6.3363

GODDARD INSTITUTE FOR SPACE STUDIES Origin of the solar system :a conference Proceedings Edited by Robert Jastrow and A.G. W. Cameron 176p Academic press,New York, 1963
Cav 7.2870

GODDARD INSTITUTE FOR SPACE STUDIES Origin of the solar system :a conference Proceedings New York 1962 Jan 23-24 Edited by Robert Jastrow and A.G.W. Cameron Academic press,New York,1963
TA 15.0200

GODDARD INSTITUTE FOR SPACE STUDIES Origin of the solar system conference proceedings New York 1962 Jan.23-24 Edited by Robert Jastrow and A.G.W. Cameron 176p Academic press,New York,1963
Obs 6.0883

GODDARD INSTITUTE FOR SPACE STUDIES Stellar evolution :a symposium proceedings New York 1963 Nov.13-15 Edited by R.F. Stein and A.G.W. Cameron 464p Plenum press,New York,1966
Obs 6.1677

GODDARD INSTITUTE FOR SPACE STUDIES Stellar evolution :international conference Proceedings New York 1963 Nov 13-15 Edited by R.F. Stein and A.G.W. Cameron Plenum press,New York,1966
TA 15.0181

GODDARD INSTITUTE FOR SPACE STUDIES Supernovae and their remnants Conference on supernovae Proceedings Greenbelt,Md. 1967 Edited by P.J. Brancazio and A.G.W. Cameron 240p Gordon and Breach,New York, 1969
Obs 6.3361

GODDARD INSTITUTE FOR SPACE STUDIES The Atmospheres of Venus and Mars :a symposium Tucson,Ariz. 1967 Feb 28-Mar 2 Edited by John C. Brandt and Michael B. McElroy Gordon and Breach,New York,1968
TA 15.0415

GODDARD INSTITUTE FOR SPACE STUDIES The Origin and evolution of atmospheres and oceans a conference New York 1963 Apr 8-9 Edited by Peter J. Brancazio and A.G.W. Cameron Wiley,New York,1964
TA 15.0427

GODDARD INSTITUTE FOR SPACE STUDIES,NEW YORK Origin of the solar system :a conference Proceedings New York 1962 Jan 23-24 Edited by Robert Jastrow and A.G.W. Cameron Academic press,New York,1963
Geod 9.0007

GODDARD SPACE FLIGHT CENTER Gum nebula and related problems Greenbelt,Md. 1971 May 18 Edited by S.P. Maran and others Preliminary edition X-683-71-375 228p goddard space flight center,Greenbelt,Md.,1971
Obs 6.3624

GODDARD SPACE FLIGHT CENTER IAU-NASA symposium The Conference on the nature of the surface of the moon Proceedings Greenbelt,Md. 1965 Apr 15-16 Edited by Wilmot N. Hess and others Sponsored bt the National aeronautics and space administration 320p Johns Hoplins press,Baltimore,1966
TA 15.0203

GODDARD SPACE FLIGHT CENTER Nucleosynthesis : a conference Proceedings New York 1965 Jan 25-26 Edited by W.David Arnett and others Gordon and Breach,New York,1968
TA 15.0306

GOITRE International goitre conference 4th Abstracts of papers presented London 1960 Jul 5-8 American goiter association and London thyroid club Edited by O.de Vaal and others Excerpta medica. International congress series, 26 Excerpta medica foundation,Amsterdam,1960
PGMS 29.0158

GOLDEN GATE METALS CONFERENCE Proceedings High-strength steels for the missile industry San Francisco,Calif. 1960 Feb 4 American society for metals Edited by H.T. Sumsion American society for metals,Novelty,Ohio,1961
Met 25.0487

GOLDEN JUBILEE OF GENETICS Genetics in the 20th century :essays on the progress of genetics during its first fifty years Genetics society of America Edited by L.C. Dunn 2 vols Macmillan,New York,1951 "A compilation of invitation papers presented at the...Golden jubilee of genetics at Ohio State university,Columbus,Ohio,September 11-14, 1950"-foreword
An 32.0342

GOLDEN JUBILEE OF GENSTICS Genetics in the 20th century :essays on the progress of genetics during its first 50 years Columbus, Ohio 1950 Sep 11-14 Genetics society of America Edited by L.C. Dunn Sponsored by the American institute of biological sciences Macmillan,New York,1951
Gen 34.0991

GOLDEN JUBILEE OF GENETICS Genetics in the 20th century :essays on the progress of genetics during its first 50 years Columbus, Ohio 1950 Sep 11-14 Genetics society of America Edited by L.C. Dunn Sponsored by the American institute of biological sciences port Macmillan,New York,1951
Bot 42.0751

GONADOTROPHIC HORMONES Recent research on gonadotrophic hormones Gonadotrophin club meeting 5th Proceedings Edinburgh 1966 Sep 27-29 Gonadotrophin club Edited by E.Trevor Bell and John A. Loraine Livingstone,Edinburgh;London,1967
Inv Med 37.0015

GONADOTROPHIN Developments in the pharmacology and clinical uses of human gonadotrophins :a private scientific meeting Proceedings London 1968 Mar 15-16 Sponsored by Searle (G.D.) and company Searle,High Wycombe,1970
Inv Med 37.0261

GONADOTROPHIN CLUB MEETING 5th Proceedings
Recent research on gonadotrophic hormones
Edinburgh 1966 Sep 27-29 Gonadotrophin
club Edited by E.Trevor Bell and John A.
Loraine Livingstone,Edinburgh;London,1967
Inv Med 37.0015

GONADOTROPHINS Immunoassay of gonadotrophins
symposium Transactions Stockholm 1969
Sep 23-25 Karolinska institutet and
World health organization Edited by E.
Dicfalusy and A. Dicfalusy Karolinska
symposia on research methods in reproductive
endocrinology, 1 Stockholm,1969
PGMS 29.0182

GONADOTROPINS :physiochemical and
immunological properties Ciba foundation
Edited by G.E.W. Wolstenholme and Julie Knight
Ciba foundation study group, 22 Churchill,
London,1965 In honour of Dr.C.Hamburger
Inv Med 37.0243

GONADOTROPINS:PHYSIOCOCHEMICAL AND IMMUNOLOGICAL
PROPERTIES London 1965 Feb 5 Ciba
foundation Edited by G.E.W. Wolstenholme and
Julie Knight Ciba foundation study group,
22 Churchill,London,1965
Phys 20.1409

GONDWANA Symposium sur les series de
Gondwana International geological congress
19th papers Algiers 1952 Edited by
Curt Teichert Algiers,1952 Text in
English,French and German
Geol 8.3012

GORWTH IN RELATION TO DIFFERENTIATION AND
MORPHOGENESIS Cambridge 1947 Jul
Society for experimental biology Edited by J.
F. Danielli and R. Brown Society for
experimental biology, 2 Cambridge
university press,Cambridge,1948
An 32.5434

GOTEBORG 1966 Naturphilosophie bei
Aristotels und Theophrast Symposium
Aristotelicum 4th Verhandlungen Edited
by Ingemar During 292p Stiehm,Heidelberg,
1969
WSM 43.0192

GOTEBORGS UNIVERSITET International
conference on solvent extraction chemistry
Proceedings Gothenburg 1966 Aug 27-Sep 1
By D. Dyrssen and others North-Holland,
Amsterdam,1967
Met 25.2292

GOTHENBURG 1952 International symposium on
the reactivity of solids 2nd Proceedings
Under the auspices of
Ingeniorsvetenskapsakademien and Chalmers
tekniska hogskola 2 vols Gothenburg,1954
Min 10.0614

GOTHENBURG 1966 International conference
on solvent extraction chemistry Proceedings
By D. Dyrssen and others Chalmers university
of technology and Goteborgs universitet
North-Holland,Amsterdam,1967
Met 25.2292

GOTHENBURG 1966 International mineral
processing congress Proceedings
Institution of mining and metallurgy London,
1960
Met 25.2504

GOTTINGEN 1953 Vortrage uber rechenanlagen
Deutsche forschungsgemeinschaft Edited by L.
Biermann 145p 21cm Max-Planck institut
fur physik,Gottingen,1953
Math L 5.1087

GOTTINGEN 1954 International congress of
sedimentology 4th Proceedings Edited by
R. Brinkmann and others Geologische
rundschau,43,heft 2 Enke,Stuttgart,1955
In English,French and German
Geog 13.0285

GOTTINGEN 1964 Dislocations in solids :a
general discussion Faraday society Faraday
society.Discussions, 38 Faraday society,
London,1964
Eng 41.3532

GOTTINGEN 1965 Arabidopsis research :an
international symposium Report Universitat
Gottingen.Arabidopsis information service
Edited by Gerhard Robbelen Gottingen,1965
Supplement to 'Arabidopsis information
service' 1965
Gen 34.1498

GOTTINGEN 1971 Regulation of
gluconeogenesis Conference of the
Gesellschaft fur biologische chemie 9th
Proceedings Gesellschaft fur biologische
chemie Edited by Hans-Dieter Soling and
Berend Willms Thieme;Academic press,
Stuttgart;London,1971
Bioch 33.2210

GRADUATE TRAINING Experiments in graduate
training :a symposium Proceedings London
1962 Mar 22-23 Institution of mechanical
engineers Institution of mechanical
engineers,London,1962
Eng 41.1499

GRAFTING Biological problems of grafting :a
symposium Proceedings Liege 1959 Mar 18-
21 Commission administrative du patrimoine
universitaire de Liege and Council for
international organizations of medical
sciences Edited by F. Albert and G. Lejeune-
Ledant Blackwell,Oxford,1959 Symposium
chairman P.B.Medawar
PGMS 29.0245

GRAFTING Biological problems of grafting :a
symposium Proceedings Liege 1959 Mar 18-
21 Commission administrative du patrimoine
universitaire de Liege Council for
international organizations of medical
sciences Edited by F. Albert and G. Lejeune-
Ledant Blackwell,Oxford,1959
Bal 39.0458

GRAIN-SIZE SYMPOSIUM held during the 16th
annual convention New York 1934 Oct 1
American society for metals American society
for metals,Cleveland,c 1934 Damaged copy.
No title page
Met 25.1264

GRAMMEL Richard The International congress
of applied mechanics 11th Proceedings
Munich 1964 International committee for
the congresses of applied mechanics Edited
by Henry Gortler and Peter Sorger 1184p
Springer-verlag,Berlin,1966 Dedicated to
Richard Grammel
A Math 4.1279

Les GRANDES ACTIVITES DU LOBE TEMPORAL : semaine neurophysiologique:conferences du colloque Paris 1953 Hopital de la Salpetriere.Clinique des malades du systeme nerveux Edited by Th. Alajouanine Actualites neuro-physiologiques Masson,Paris, 1955
Psy 31.0464

Les GRANDES EXPEDITIONS SCIENTIFIQUES ET LA CERATION DES LABORATOIRES MARITIMES Colloque international sur l'histoire de la biologie marine Banyuls-sur-Mer 1963 Sep 2-6 International union of history and philosophy of science Vie et milieu. Supplement, 19 370p Laboratoire Arago; Masson,Banyuls-sur-Mer;Paris,1965
WSM 43.2877

GRAPH THEORY Beitrage zur graphentheorie : internationale kolloquium Vortragen Manebach 1967 May 9-12 Technische hochschule Ilmenau.Mathematische institut Mathematische gesellschaft der D.D.R. Edited by Horst Sachs and others Bibliog.,Illus. 394p 23cm Teubner,Leipzig,1968
P Math 2.3352

GRAPH THEORY Louisiana conference on combinatorics,graph theory and computing Proceedings Baton Rouge,La. 1970 Mar 1-5 Edited by R.C. Mullin and others Held at Louisiana state university vi,464p 25cm Louisiana state university,Baton Rouge,La., 1970
P Math 2.4503

GRAPH THEORY Many facets of graph theory Conference on graph theory Proceedings Kalamazoo 1969 Oct 31-Nov 2 Edited by G. Chartrand and S.F. Kapoor Lecture notes in mathematics, 110 Illus viii,290p 24cm Springer-Verlag,Berlin,1969
P Math 2.3347

GRAPH THEORY Proof techniques in graph theory Ann Arbor graph theory conference 2nd Proceedings Ann Arbor,Mich. 1968 Feb Edited by Frank Harary Bibliog.,Illus. xv, 330p 24cm Academic press,New York;London, 1969
P Math 2.3349

GRAPH THEORY Proof techniques in graph theory Ann Arbor graph theory conference 2nd Proceedings Ann Arbor,Mich. 1968 Feb University of Michigan Edited by Frank Harary xv,330p 24cm Academic press,New York;London,1969
Math S 3.1658

GRAPH THEORY Recent trends in graph theory New York city graph theory conference 1st Proceedings New York 1970 Jun 11-13 Edited by M. Capobianco and others Lecture notes in mathematics, 186 vi,219p 25cm Springer,Berlin,1971
P Math 2.4113

GRAPH THEORY AND APPLICATIONS :conference Proceedings Kalamazoo,Mich. 1972 May 10-13 Edited by Y. Alavi and D.R. Lick Held at Western Michigan university Lecture notes in mathematics, 303 viii,329p 25cm Springer,Berlin,1972
P Math 2.4111

GRAPH THEORY AND ITS APPLICATIONS :an advanced seminar Madison,Wis. 1969 Oct 13-15 United States.Army.Mathematics research center bibliog.,illus. viii,262p 23cm Academic press,New York;London,1970
P Math 2.3821

GRAPH THEORY AND THEORETICAL PHYSICS Edited by H.Frank Harary Bibliog.,Illus. xv,358p 24cm Academic press,New York;London,1967 Based on a NATO advanced study institute held in Paris,1963
P Math 2.3348

GRAPH THEORY AND THEORETICAL PHYSICS :N.A.T.O. advanced study institute Paris 1963 Edited by Frank Harary With the financial support of the North Atlantic treaty organization.Scientific affairs division bibliog.,illus. xv,358p 24cm Academic press,New York;London,1967
Math S 3.1408

The GRAPHIC SYMPOSIUM 3rd Papers Computer graphics:utility,production,art Los Angeles,Calif 1966 University of California,Los Angeles and Informatics inc. Edited by Fred Gruenberger 225p Thompson book co.;Academic press,Washingtom,D. C.;London,1967
TA 15.0098

GRAPHICS SYMPOSIUM Computer graphics:utility, production,art Los Angeles n.d. University of California Edited by Fred Gruenberger Jointly sponsored by Informatics incorporated 225p 22cm Academic press, London,1967
Math L 5.3273

GRAPHICS SYMPOSIUM Computer graphics:utility, production,art Los Angeles,Calif. 1967 University of California Edited by Fred Gruenberger Jointly sponsored by Informatics incorporated Academic press,London,1967
Eng 41.2210

GRASSES Growth of cereals and grasses Easter school in agricultural science 12th Proceedings Nottingham University of Nottingham Edited by F.L. Milthorpe and J.D. Ivins Bibliog.,illus,diagrs,tables xii, 359p Butterworths,London,1966
Bot 42.2035

GRASSLAND International grassland congress 8th Proceedings Reading 1960 Jul 11-21 British grassland society Grassland research institute,Hurley,1961
BG 38.3114

GRATZ 1843 Amtlicher bericht uber die einundzwanzigste versammlung deutscher naturforscher und aerzte 21st Proceedings Deutsche naturforscher und aerzte Edited by L. Langer and A. Schrotter A.Leykam,Gratz, 1844
Hunt B 22.0463

GRAVITATION Relativistic theories of gravitation A Conference on gravitation Proceedings Jablonna 1962 Jul 25-31 Polska akademia nauk Edited by Leopold Infeld With financial support of the International union of pure and applied physics Pergamon press,Oxford,1964 Binder's title in French
TA 15.0085

GRAVITATION Relativistic theories of gravitation Conference on relativistic theories of gravitation proceedings Jablonna 1962 Jul 25-31 Edited by Leopold Infeld Organised and sponsored by Polska akademia nauk Pergamon press,Oxford,1964
A Math 4.1029

GRAVITATIONAL N-BODY PROBLEM Proceedings Cambridge 1970 Aug 12-15 International astronomical union Edited by M. Lecar International astronomical union.Colloquium, 10 441p Reidel,Dordrecht,1972
TA 15.0667

GRAVITATIONAL THEORIES Verifiche delle teorie gravitazionali Scuola internazionale di fisica 'Enrico Fermi' 20 corso Rendiconti Varenna 1961 Jun 19-Jul 1 Societa italiana di fisica Edited by C. Moller Academic press,New York,1962
TA 15.0084

GRAVITATIONAL WAVES Convegno sulla relativita:problemi di energia e onde gravitazionali;meeting on general relativity: problems of energy and gravitational waves Atti;proceedings Florence 1964 Sep 9-12 Quatro centenario della nascita di Galileo Galilei,1564-1964.Comitato nazionale per le manifestazioni celebrative Edited by Giorgio Sestini Quatro centenario della nascita di Galileo Galilei,1564-1964.Comitato nazionale per le manifestazioni celebrative. Pubblicazioni, 2,Atti dei convegni, 1 267p Florence,1966
Obs 6.3276

GRAVITY ANOMALIES:UNSURVEYED AREAS Extension of gravity anomalies to unsurveyed areas :a symposium Papers Columbus,Ohio 1964 Nov 18-20 Ohio state university and American geophysical union Edited by Hyman Orlin Sponsored by the International union of geodesy and geophysics Geophysical monograph, 9 National research council.Publication, 1357 American geophysical union,Washington,D. C.,1966
Geod 9.0031

GRAVITY WAVES NBS semicentennial symposium on gravity waves 3rd Proceedings Washington,D.C. 1951 Jun 18-20 National bureau of standards National bureau of standards.Circular, 521 illus iv,287p U. S.Government printing office,Washington,D.C., 1952
Chem E 24.0215

GRAVITY WAVES NBS semicentennial symposium on gravity waves Proceedings Washington,D. C. 1951 Jun 18-20 National bureau of standards National bureau of standards. Circular, 521 U.S.government printing office, Washington,D.C.,1952
Eng 41.6664

GRAY (L.H.) CONFERENCE 2nd Proceedings Charged particle tracks in solids and liquids Cambridge 1969 Apr Institute of physics and the physical society.Conference series Institute of physics,London,1970
Radioth 35.1702

GRAZ 1966 Electrotherapeutic sleep and electroanaesthesia :international symposium 1st Proceedings Edited by F.M. Wageneder and others Excerpta medica.International congress series, 136 Excerpta medica, Amsterdam,1967
Inv Med 37.0221

GRAZ 1969 Electrotherapeutic sleep and electroanesthesia :international symposium 2nd Proceedings Edited by F.M. Wageneder and others Excerpta medica.International congress series, 212 Excerpta medica, Amsterdam,1970
Inv Med 37.0222

GRAZING IN TERRESTRIAL AND MARINE ENVIRONMENTS a symposium Bangor 1962 Apr 11-14 British ecological society Edited by D.J. Crisp British ecological society.Symposia xvi,322p Blackwell scientific,Oxford,1964
Bot 42.1984

GRAZING IN TERRESTRIAL AND MARINE ENVIRONMENTS a symposium Bangor 1962 Apr 11-14 British ecological society Edited by D.J. Crisp British ecological society.Symposia, 4 Blackwell,Oxford,1964
Bal 39.1671

The GREAT ALASKA EARTHQUAKE OF 1964 hydrology Pt A-B National research council.Committee on the Alaska earthquake National academy of sciences.Publication,1603 illus,maps xvii,441p 29cm 2 vols National academy of sciences,Washington,1968 Part B is a portfolio containing 7 charts
Sco 14.7951

GREAT BRITAIN.ADMIRALTY ADVISORY COMMITTEE ON STRUCTURAL STEEL Brittle fracture in steel :conference Proceedings Cambridge 1959 Sep 28-30 P.3 H.M.S.O.,London,1962
Eng 41.3775

GREAT BRITAIN.AIR FORCE.OFFICE OF SCIENTIFIC RESEARCH,OAR International symposium on free radicals 6th Papers Cambridge 1963 Jul 2-5 University of Cambridge Cambridge,1963 Mimeographed
Chem 18.0414

GREAT BRITAIN.DEPARTMENT OF SCIENTIFIC AND INDUSTRIAL RESEARCH Automatic and manual control Conference on automatic control papers Cranfield 1951 Jul 18-21 xi,584p 25cm Butterworth,London,1951
Math L 5.0021

GREAT BRITAIN.DEPARTMENT OF SCIENTIFIC AND INDUSTRIAL RESEARCH Automatic and manual control Papers Cranfield 1951 Jul 16-21 Institution of electrical engineers Institution of mechanical engineers Edited by A. Tustin Butterworths,London,1952
Psy 31.3398

GREAT BRITAIN.DEPARTMENT OF SCIENTIFIC AND INDUSTRIAL RESEARCH Automatic and manual control :a conference Papers Cranfield 1951 Jul 16-21 Edited by A. Tustin xi,584p Butterworths,London,1952
Chem E 24.1142

GREAT BRITAIN.DEPARTMENT OF SCIENTIFIC AND INDUSTRIAL RESEARCH Automatic and manual control... Conference on automatic control Papers Cranfield 1951 Jul 16-21 Edited by A. Tustin Butterworths,London,1952
Eng 41.6006

GREAT BRITAIN.DEPARTMENT OF SCIENTIFIC AND INDUSTRIAL RESEARCH International congress on high-speed photography 3rd Proceedings London 1956 Sep 10-15 Edited by R.B. Collins xvi,417p Butterworths scientific publications,London,1957
Chem E 24.0452

GREAT BRITAIN.DEPARTMENT OF SCIENTIFIC AND INDUSTRIAL RESEARCH International congress on high-speed photography 3rd proceedings London 1956 Sep 10-15 Edited by R.B. Collins Held under the auspices of the Department of scientific and industrial research Butterworths,London,1957
Eng 41.0439

GREAT BRITAIN.DEPARTMENT OF SCIENTIFIC AND INDUSTRIAL RESEARCH Soils,concrete and bituminous materials :a record of a course dealing with airfield construction... Harmondsworth 1943 Jul-Aug H.M.S.O.,London, 1946
Eng 41.2998

GREAT BRITAIN.DEPARTMENT OF SCIENTIFIC AND INDUSTRIAL RESEARCH The Conference on automatic control Papers Cranfield 1951 Jul 16-21 Edited by A. Tustin 584p Butterworths,London,1952
Cav 7.2360

GREAT BRITAIN.DEPARTMENT OF SCIENTIFIC AND INDUSTRIAL RESEARCH Wind effects on buildings and structures :a conference Proceedings Teddington 1963 Jun 26-28 Vol 1-2 Great Britain.Department of scientific and industrial research.Symposium, 16 2 vols H.M.S.O.,London,1965
Eng 41.2707

GREAT BRITAIN.DEPARTMENT OF SCIENTIFIC AND INDUSTRIAL RESEARCH Wind effects on buildings and structures :a conference Proceedings Teddington 1963 Jun 26-28 Vol 1-2 National physical laboratory. Symposium, 16 H.M.S.O.,London,1963
Eng 41.6904

GREAT BRITAIN.DEPARTMENT OF SCIENTIFIC AND INDUSTRIAL RESEARCH.BUILDING RESEARCH STATION Mechanical properties of non-metallic brittle materials :a conference Proceedings London 1958 Apr Edited by W.H. Walton Butterworths,London,1958
Met 25.0852

GREAT BRITAIN.DEPARTMENT OF SCIENTIFIC AND INDUSTRIAL RESEARCH.SYMPOSIUM, 16 Wind effects on buildings and structures :a conference Proceedings Teddington 1963 Jun 26-28 Vol 1-2 Great Britain. Department of scientific and industrial research 2 vols H.M.S.O.,London,1965
Eng 41.2707

GREAT BRITAIN.LOCAL GOVERNMENT BOARD National conference on the prevention of destitution 1st Proceedings London 1911 May 30-Jun 2 King,London,1911 President of conference:the Rt.Hon.the Lord Mayor of London
Path 30.1285

GREAT BRITAIN.LOCAL GOVERNMENT BOARD National conference on the prevention of destitution 1st Proceedings of the Public health section London 1911 May 30-Jun 2 King,London,1911 President of the Public health section of the Local government board:T. C.Allbutt
Path 30.1284

GREAT BRITAIN.MINISTRY OF FUEL AND POWER Fuel and the future :a conference Proceedings London 1946 Oct 8-10 Vol 1-3 Edited by Harry Mitchell Spiers 3 vols H.M.S.O.,London,1948
Chem E 24.1370

GREAT BRITAIN.MINISTRY OF HEALTH Mathematics and computer science in biology and medicine : conference Proceedings Oxford 1964 July Medical research council H.M.S.O.,London, 1965
HE 27.0125

GREAT BRITAIN.MINISTRY OF HEALTH Mathematics and computer science in biology and medicine conference proceedings Oxford 1964 Jul Co-sponsored by Great Britain.Scottish home and health department Bibliog. xi, 317p 25cm H.M.S.O.,London,1965
Math 3.0775

GREAT BRITAIN.MINISTRY OF SUPPLY Communication theory Applications of communication theory :symposium Papers London 1952 Sep 22-26 Edited by Willis Jackson Supported by the British broadcasting corporation xii,532p 25cm Butterworths,London,1953 Also supported by the Ministry of supply
Math L 5.3249

GREAT BRITAIN.MINISTRY OF SUPPLY Communication theory Applications of communication theory symposium papers London 1952 Sep 22-26 Edited by Willis Jackson· xii,532p Butterworths scientific publications,London,1953
Math 3.0735

GREAT BRITAIN.MINISTRY OF SUPPLY Conference on high-speed automatic calculating machines Report Cambridge 1949 Jun 22-25 University mathematical laboratory,Cambridge, 1950
Eng 41.2175

GREAT BRITAIN.MINISTRY OF SUPPLY Conference on high-speed automatic calculating machines Report Cambridge 1949 Jun 22-25 University mathematical laboratory,Cambridge, 1950
Math L 5.3236

GREAT BRITAIN.MINISTRY OF SUPPLY Isotope techniques conference Proceedings Oxford 1951 Jul Vol 2: industrial and allied research applications Sponsored by the Atomic energy research establishment H.M.S.O. London,1952
Chem 18.2467

GREAT BRITAIN.MINISTRY OF SUPPLY Symposium on information theory 1st report of proceedings, mimeograph London 1950 Sep 26-29 Edited by Willis Jackson 206p 33cm Ministry of supply,London,1950
Math 3.0774

GREAT BRITAIN.MINISTRY OF TECHNOLOGY Adhesion:fundamentals and practice ; international conference Report Nottingham 1966 Sep 20-22 Elsevier,Amsterdam,1969
Met 25.2629

GREAT BRITAIN.MINISTRY OF WORKS Welded structures :a conference Proceedings London 1953 Nov 23-26 H.M.S.O.,London,1954
Eng 41.3021

GREAT BRITAIN.NATIONAL COAL BOARD.MINING RESEARCH ESTABLISHMENT Mechanical properties of non-metallic brittle materials :a conference Proceedings London 1958 Apr Edited by W. H. Walton Butterworths,London,1958
Met 25.0852

GREAT BRITAIN.SCOTTISH HOME AND HEALTH DEPARTMENT Mathematics and computer science in biology and medicine conference proceedings Oxford 1964 Jul Medical research council Great Britain.Ministry of health Bibliog. xi,317p 25cm H.M.S.O.,London,1965
Math 3.0775

GREAT LAKES BASIN Syposium Chicago,Ill. 1959 Dec 29-30 American association for the advancement of science Edited by Howard J. Pincus American association for the advancement of science.Publications,71 American association for the advancement of science,Washington,D.C.,1962
Geog 13.5045

GREENBELT,MA. 1965 Autoionization, astrophysical,theoretical and laboratory experimental aspects Symposium on atomic interactions and space physics Goddard space flight center Edited by Aaron Temkin 155p Mono book corp.,Baltimore,Ma.,1966
A Math 4.0911

GREENBELT,MD 1965 Berkeley,Calif 1965 Dec 29 Magnetic and related stars A.A.S.-N. A.S.A. symposium on the magnetic and other peculiar and metallic-line A stars Proceedings American astronomical society and National aeronautics and space administration Edited by Robert C. Cameron xi,596p Mono book corporation,Baltimore,1967
A Math 4.1575

GREENBELT,MD. 1965 IAU-NASA symposium The Conference on the nature of the surface of the moon Proceedings Goddard space flight center and International astronomical union Edited by Wilmot N. Hess and others Sponsored bt the National aeronautics and space administration 320p Johns Hoplins press,Baltimore,1966
TA 15.0203

GREENBELT,MD. 1965 Magnetic and related stars The Symposium on the magnetic and other peculiar and metallic-line A stars American astronomical society and National aeronautics and space administration Edited by Robert C. Cameron Mono book corporation, Baltimore,Md.,1967
TA 15.0310

GREENBELT,MD. 1965 Nature of the lunar surface IAU-NASA symposium International astronomical union National aeronautics and space administration Edited by W.N. Hess and others Held at the Goddard space flight center Johns Hopkins press,Baltimore,Md., 1966
Min 10.1445

GREENBELT,MD. 1965 Nucleosynthesis : conference Proceedings Goddard institute for space studies Edited by W.D. Arnett and others 273p Gordon and Breach,New York, 1968
Obs 6.3363

GREENBELT,MD. 1965 Symposium on atomic interactions and space physics Autoionization: astrophysical,theoretical and laboratory experimental aspects Papers Edited by Aaron Temkin Mono books,Baltimore,Md.,1966
TA 15.0479

GREENBELT,MD. 1965 and BERKELEY,CALIF. 1965 AAS-NASA magnetic star symposium The Symposium on the magnetic and other peculiar and metallic-line A stars :magnetic and related stars 1st Proceedings American astronomical society National aeronautics and space administration Edited by Robert C. Cameron 596p Mono book corporation,Baltimore,1967 Includes the Helen B.Warner prize lecture of the American astronomical society,by George W.Preston
Obs 6.3432

GREENBELT,MD. 1966 Conference on shock metamorphosis of natural materials 1st Proceedings National aeronautics and space administration.Goddard space flight center Edited by Bevan M. French and Nicholas M. Short Held at the Goddard space flight center Mono book corporation,Baltimore,Md., 1968
Min 10.1443

GREENBELT,MD. 1967 Supernovae and their remnants Conference on supernovae Proceedings Goddard institute for space studies Edited by P.J. Brancazio and A.G.W. Cameron 240p Gordon and Breach,New York, 1969
Obs 6.3361

GREENBELT,MD. 1971 Gum nebula and related problems Goddard space flight center Edited by S.P. Maran and others Preliminary edition X-683-71-375 228p goddard space flight center,Greenbelt,Md.,1971
Obs 6.3624

GRENOBLE 1954 International institute of refrigeration :commissions 1 et 2 International institute of refrigeration. Bulletin.Annexe,1955-2 183p 24cm Paris, 1955 In English and French
Sco 14.0122

GRENOBLE 1960 Congres national de l'Association francaise de calcul 1st mimeograph Association francaise de calcul Association francaise de calcul,1960
Math L 5.0869

GRENOBLE 1960 Plutonium 1960 International conference on plutonium metallurgy 2nd Proceedings Societe Francaise de metallurgie and Commissariat a l'energie atomique Edited by Emmanuel Grison and others Cleaver-Hume,London,1961 Papers in English and French
Met 25.1549

GRENOBLE 1964 Rheology and soil mechanics : a symposium International union of theoretical and applied mechanics Edited by J. Kravtchenko and P.M. Sirieys Springer, Berlin,1966
Eng 41.3097

GRENOBLE 1970 Microscopie electronique 1970 Congres international de microscopie electronique 7e Resumes des communications International federation of societies for electron microscopy Edited by Pierre Favard Societe francaise de microscopie electronique,Paris,1970
Met 25.2604

GRONINGEN 1926 International congress of psychology 8th Proceedings and papers Noordhoff,Groningen,1927
Psy 31.3360

GRONINGEN 1951 International congress of sedimentology 3rd proceedings Edited by Tj.van Andel map Martinus Nijhoff,The Hague,1951
Geol 8.1645

GRONINGEN 1951 Wageningen 1951 Jul 5-12 International congress of sedimentology 3rd Proceedings Edited by Tjeerd Hendrik van Andel Geologisch instituut,Groningen, 1951 In English,French and German
Geog 13.0284

GRONINGEN 1955 Progress in neurobiology International meeting of neurobiologists 1st Proceedings Edited by J.Ariens Kapper Elsevier,Amsterdam,1956
An 32.4351

GRONINGEN 1964 The Nutricia symposium on the adaption of the newborn infant to extra uterine life Edited by J.H.P. Jonxis and others Kroese,Leiden,1964
An 32.1516

GRONLAND 1939 Schaffhausen 1931 Mar.11-12 Naturforschenden gesellschaft Schaffhausen Naturforschenden gesellschaft Schaffhausen.Mitteilungen,Bd.16 1939
Geol 8.2714

GRONLAND 1939 :tagung der Naturforschenden gesellschaft Schaffhausen Schaffhausen 1939 Mar 11-12 Naturforschenden gesellschaft Schaffhausen Naturforschenden gesellschaft Schaffhausen.Mitteilungen,16 illus,maps,4 plates 231p 1939
Sco 14.7861

La GROSSESSE EXTRA-UTERINE :colloque Rapport Paris 1961 Jun 2-4 Societe nationale pour l'etude de la sterilite et de la fecondite Edited by Paul Funck-Brentano Masson,Paris,1961
An 32.3824

GROUND LEVEL CLIMATOLOGY American association for the advancement of science : annual meeting 132nd Symposium on ground level climatology Berkeley,Calif. 1965 Dec Edited by Robert H. Shaw American association for the advancement of science. Publications,86 Washington,D.C.,1967
Geog 13.1034

GROUND WATER International geological congress 20th papers Mexico City 1956 Seccion 4: geohidrologia de regiones aridas y subaridas Edited by A. Garcia Rojas and others Mexico City,1957 Text in English, French,Russian and Spanish
Geol 8.3045

GROUP PROCESSES 2nd :conference Transactions Princeton,N.J. 1955 Oct 9-12 Edited by Bertram Schaffner Sponsored by the Josiah Macy,jr.Foundation 255p 24cm New York,1956
Sco 14.0705

GROUP THEORY Colloque sur la theorie des groupes algebriques 20th colloque Brussells 1962 Jun 5-7 Centre Belge de recherches mathematiques 150p 25cm Librairie universitaire,Louvain,1962
P. Math 2.0573

GROUP THEORY Symposium on group theory abstracts of lectures Cambridge,Mass. 1963 Apr 1-3 Harvard university 59p 23cm Harvard university,Cambridge,Mass.,1963
P Math 2.2792

GROUP THEORY CONFERENCE The Theory of groups international conference proceedings Canberra 1965 Aug 10-20 Australian national university Edited by L.G. Kovacs and B.H. Neumann Bibliog xviii,397p 24cm Gordon and Breach,New York,1967
P Math 2.2793

GROUPEMENT DES MATHEMATICIENS D'EXPRESSION LATINE 3rd reunion comptes rendus Namur 1965 Sep.20-23 Centre Belge de recherches mathematiques 153p 25cm Librairie universitaire,Louvain,1966
P. Math 2.2611

GROUPEMENT POUR L'ADVANCEMENT DES METHODES D'ANALYSE SPECTROGRAPHIQUE DES PRODUITS METALLURGIQUES Congress G.A.M.S. 4th Paris 1951 June 20-22 24cm Paris,1951
Philos 1.0678

GROUTS AND DRILLING MUDS IN ENGINEERING PRACTICE a symposium London 1963 May 22-24 International society of soil mechanics and foundation engineering.British national committee Butterworths,London,1963
Eng 41.3080

GROWTH Aspects of synthesis and order in growth :a symposium Hanover,N.H. 1954 Jun 23-26 Edited by Dorothea Rudnick Society for the study of development and growth. Symposia, 13 Princeton university press, Princeton,N.J.,1954
An 32.2482

GROWTH Dynamics of growth processes :a symposium Williamstown,Mass. 1952 Jun Edited by Edgar J. Boell Society for the study of development and growth.Symposia, 11 Princeton university press,Princeton,N.J.,1954
An 32.2467

GROWTH Problemi di sviluppo :un simposio Milan 1952 Sep Unione zoologie italiana Casa editrice ambrosiana,Milan,1954
An 32.2477

GROWTH :in relation to differentiation and morphogenesis Cambridge 1947 Jul Society for experimental biology Society for experimental biology.Symposia, 2 Cambridge university press,Cambridge,1958
Phys 20.0943

GROWTH in relation to differentiation and morphogenesis.A symposium Cambridge 1947 Jul Society for experimental biology Society for experimental biology.Symposia, 2 Cambridge university press,Cambridge,1948
Bioch 33.1349

GROWTH,IN RELATION TO DIFFERENTIATION AND MORPHOGENESIS Cambridge 1947 Jul Society for experimental biology Society for experimental biology.Symposia, 2 Cambridge university press,Cambridge,1948
Chem 18.2652

The GROWTH,REPLACEMENT AND TYPES OF HAIR : conference Papers New York 1950 Feb 10-11 Edited by J.B. Hamilton New York academy of sciences.Annals, 53,art.3
An 32.4025

GROWTH AND DEVELOPMENT OF MAMMALS Easter school in agricultural science 14th Proceedings Nottingham 1967 University of Nottingham.School of agriculture Edited by G.A. Lodge and G.E. Lamming Butterworths, London,1968
Gen 34.1592

GROWTH AND MATURATION OF THE BRAIN Edited by Dominick P. Purpura and J.P. Schade Progress in brain research, 4 Elsevier,Amsterdam, 1964 Lectures delivered during an interdisciplinary workshop on "Growth and maturation of the brain",1962
An 32.4224

GROWTH AND PERFECTION OF CRYSTALS International conference on crystal growth Proceedings Cooperstown,N.Y. 1958 Aug 27-29 General electric company Edited by R.H. Doremus and others Co-sponsored by United States.Air force.Office of scientific research Wiley;Chapman and Hall,New York;London,1958
Eng 41.4438

GROWTH AND PERFECTION OF CRYSTALS International conference on crystal growth Proceedings Cooperstown,New York 1958 Aug 27-29 United States.Air research and development command and General electric research laboratory Wiley;Chapman and Hall, New York;London,1958
Met 25.1507

GROWTH FACTORS International congress of microbiology 6th Rome 1953 Vol 3: symposium - nutrition and growth factors Edited by W.H. Schopfer and D.D. Woods Fondazione Emanuele Paterno,Rome,1953 Title also in Italian;text in English and French
Bioch 33.1802

GROWTH HORMONE International symposium on growth hormone 1st Proceedings Milan 1967 Sep 11-13 Edited by A. Pecile and E.E. Muller International congress series, 158 Excerpta medica,Amsterdam,1968
Bioch 33.0457

GROWTH HORMONE :a conference Papers New York 1966 Oct 27-28 By Martin Sonenberg New York academy of sciences New York academy of sciences.Annals, 148,p.289-571 New York,1968
Bioch 33.0421

GROWTH INHIBITION AND CHEMOTHERAPY International congress of microbiology 6th Rome 1953 Vol 4: growth inhibition and chemotherapy Edited by H. Eagle and E.B. Chain Fondazione Emanuele Paterno,Rome,1953 Title also in Italian;text in English and French
Bioch 33.1803

GROWTH OF CEREALS AND GRASSES Easter school in agricultural science 12th Proceedings Nottingham University of Nottingham Edited by F.L. Milthorpe and J.D. Ivins Bibliog., illus,diagrs,tables xii,359p Butterworths, London,1966
Bot 42.2035

GROWTH OF CRYSTALS Conference on crystal growth 1st- Proceedings Moscow 1956- 1- Institut kristallografii,Moscow Edited by A.V. Shubnikov and N.N. Sheftal Translated by Consultants' bureau from the Russian Rost kristallov, 1- Consultants bureau,inc.,New York,1958- Includes interim reports published between conferences. Held as part of the international crystallographic congresses
Met 25.1508

GROWTH OF LEAVES Easter school in agricultural science 3rd Proceedings Nottingham 1956 University of Nottingham Edited by F.L. Milthorpe illus. x,223p Butterworths,London,1956
Bot 42.1825

GROWTH OF PROTOZOA New York 1953 Oct 14 By R.W. Miner New York academy of sciences Edited by S.H. Hutner New York academy of sciences.Annals, 56,5 New York,1953
Bot 42.3886

GROWTH OF PROTOZOA :a conference New York 1953 Oct 14 By R.P. Hall and others New York academy of sciences Edited by S.H. Hunter and others New York academy of sciences.Annals, 56,p.815-1095 New York, 1953
Bal 39.2132

GROWTH OF THE NERVOUS SYSTEM :Ciba foundation symposium Ciba foundation Edited by G.E.W. Wolstenholme and M. O'Connor Churchill,London,1968
Phys 20.2046

GROWTH REGULATING SUBSTANCES FOR ANIMAL CELLS IN CULTURE :a symposium Philadelphia 1967 Mar 16 Edited by Vittorio Defendi and Michael Stoker Held at the Wistar institute of anatomy and biology Wistar institute symposium monographs, 7 Wistar institute press,Philadelphia,1967
Radioth 35.0566

GROWTH REGULATORS Plant growth regulators : symposium Papers London 1968 Jan 8-9 Society of chemical industry.Pesticides group Phytochemical society Society of chemical industry.Monograph, 31 Society of chemical industry,London,1968
Bioch 33.2230

GROWTH SUBSTANCES The Biological action of growth substances Cambridge 1956 Sep Society for experimental biology Society for experimental biology.Symposia, 11 Cambridge university press,Cambridge,1956
Bioch 33.1356

GUGGENHEIM INSTITUTE OF FLIGHT STRUCTURES
Fatigue in aircraft structures :an international conference New York 1956 Jan 30-Feb 1 Edited by Alfred M. Freudenthal Academic press,New York,1956
Eng 41.2772

GUGGENHEIM INSTITUTE OF FLIGHT STRUCTURES
Fatigue in aircraft structures international conference Proceedings New York 1956 Jan 30 -Feb 1 Edited by Alfred M. Freudenthal Academic press,New York,1956
Met 25.0921

GUIDE TO THE GEOLOGY AND CULTURE OF GREECE
Petroleum exploration society of Libya :annual field conference 7th guidebook Athens 1965 Apr 9-13 Edited by A.E. Hotz and Peter Norton Amsterdam,1965
Geol 8.4481

GUIDEBOOK TO THE GEOLOGY AND HISTORY OF TUNISIA
Petroleum exploration society of Libya : annual field conference 9th guidebook Tunisia 1967 Edited by Lewis Martin and others Amsterdam,1967
Geol 8.4489

GUILDFORD 1960 International conference on space structures University of Surrey Edited by R.M. Davies Blackwell,Oxford,1967
Eng 41.2739

A GUINNESS SYMPOSIUM held at the Research laboratory St.James's Gate, Dublin
Discussions Aspects of yeast metabolism Dublin 1967 Edited by A.K. Mills and Hans Krebs Guinness symposia Blackwell,Oxford; Edinburgh,1968
Bioch 33.1140

A GUINNESS SYMPOSIUM held at the research laboratory St.James's Gate Dublin
Discussions Aspects of yeast metabolism Dublin 1967 Edited by A.K. Mills and Hans Krebs Guinness symposia Blackwell,Oxford, 1968
Bot 42.1751

GUM NEBULA AND RELATED PROBLEMS Greenbelt,Md. 1971 May 18 Goddard space flight center Edited by S.P. Maran and others Preliminary edition X-683-71-375 228p goddard space flight center,Greenbelt,Md.,1971
Obs 6.3624

GUNMA SYMPOSIUM ON ENDOCRINOLOGY 3rd
Physiological control of iodine metabolism Ikaho 1965 Sep 10-11 Gunma university. Institute of endocrinology Gunma university. Institute of endocrinology.Annual report, 3 Institute of endocrinology,Gunma university, Maebashi,1965
Phys 20.1427

GUNMA SYMPOSIUM ON ENDOCRINOLOGY 1st
Neurosecretion and neural control of internal secretion Minakami 1963 Sep 5-6 Gunma university.Institute of endocrinology Gunma university.Institute of endocrinology.Annual report, 1 Institute of endocrinology,Gunma university,Maebashi,1964
Phys 20.2230

GUNMA SYMPOSIUM ON ENDOCRINOLOGY 4th Sex and reproduction Minakami 1966 Sep 22-23 Gunma university.Institute of endocrinology Gunma university.Institute of endocrinology. Annual report, 4 Institute of endocrinology,Gunma university,Maebashi,1967
Phys 20.1434

GUNMA SYMPOSIUM ON ENDOCRINOLOGY 5th
Molecular basis of endocrinology Kusatsu 1967 Aug 18-19 Gunma university.Institute of endocrinology Gunma university.Institute of endocrinology.Annual report, 5 Institute of endocrinology,Gunma university,Maebashi, 1967
Phys 20.1435

GUNMA UNIVERSITY.INSTITUTE OF ENDOCRINOLOGY
Molecular basis of endocrinology Gunma symposium on endocrinology 5th Kusatsu 1967 Aug 18-19 Gunma university.Institute of endocrinology.Annual report, 5 Institute of endocrinology,Gunma university,Maebashi, 1967
Phys 20.1435

GUNMA UNIVERSITY.INSTITUTE OF ENDOCRINOLOGY
Neurosecretion and neural control of internal secretion Gunma symposium on endocrinology 1st Minakami 1963 Sep 5-6 Gunma university.Institute of endocrinology.Annual report, 1 Institute of endocrinology,Gunma university,Maebashi,1964
Phys 20.2230

GUNMA UNIVERSITY.INSTITUTE OF ENDOCRINOLOGY
Physiological control of iodine metabolism Gunma symposium on endocrinology 3rd Ikaho 1965 Sep 10-11 Gunma university.Institute of endocrinology.Annual report, 3 Institute of endocrinology,Gunma university,Maebashi, 1965
Phys 20.1427

GUNMA UNIVERSITY.INSTITUTE OF ENDOCRINOLOGY
Sex and reproduction Gunma symposium on endocrinology 4th Minakami 1966 Sep 22-23 Gunma university.Institute of endocrinology.Annual report, 4 Institute of endocrinology,Gunma university,Maebashi, 1967
Phys 20.1434

GUSTAVUS ADOLPHUS COLLEGE Genetics and the future of man :a discussion at the Nobel conference St.Peter,Minn. 1965 Jan 7-8 Edited by John D. Roslansky North-Holland, Amsterdam,1966
Gen 34.0062

GYRODYNAMICS symposium Celerina 1962 Aug 20-23 International union of theoretical and applied mechanics Edited by Hans Ziegler Springer-Verlag,Berlin,1963
A Math 4.0314

H-SPACES Reunion de Neuchatel Actes Neuchatel 1970 Aug Edited by Francois Sigrist Lecture notes in mathematics, 196 156p 25cm Springer,Berlin,1971
P Math 2.4121

H 2 REGIONS Interstellar ionized hydrogen :a symposium Charlottesville,Va. 1967 Dec National radio astronomy observatory Edited by Yervant Jerzian Co-sponsored by the Arecibo ionospheric observatory 774p Benjamin,New York,1968
Obs 6.3447

HAEMATIN ENZYMES :a symposium Papers and discussions Canberra 1959 Aug 31-Sep 4 Pt 1 International union of biochemistry Edited by J.E. Falk and others Organized by the Australian academy of science I.U.B.Symposium series, 19 Pergamon press, Oxford,1961
Bioch 33.1079

HAEMATOLOGY Congress of the European society of haematology 7th Proceedings London 1959 2,pt 1: papers European society of haematology Edited by E. Neumark and others Karger,Basle,1960
Med 36.0132

HAEMATOLOGY Congresso nazionale della societa italiana di ematologia 11th Atti Rome 1953 May 10-11 Societa italiana di ematologia Edited by L. Villa and others Edizioni mediche e scientifiche,Rome,1953
Med 36.0341

HAEMATOLOGY Congresso nazionale della societa italiana di ematologia 13th Atti Rome 1955 Oct 6-7 Societa italiana di ematologia Edited by G. Dominici and others Edizioni mediche e scientifiche,Rome,1956
Med 36.0380

HAEMATOLOGY European society of haematology. Congress 6th Transactions Copenhagen 1957 Vol 1: principal papers European society of haematology Edited by Aage Videbaek and others S.Karger,Basle;New York, 1958
Med 36.0128

HAEMATOLOGY European society of haematology congress 6th Transactions Copenhagen 1957 Vol 2: short papers - communications - kurzreferate European society of haematology Edited by Aage Videbaek and others 3 vols Karger,Basle;New York,1958
Med 36.0129

HAEMATOLOGY European society of haematology congress 6th Transactions Copenhagen 1957 Vol 3: panel discussion on "Efficiency and limitations of anticoagulant therapy in arterial thrombosis European society of haematology Edited by T. Astrup Karger,Basle;New York,1958
Med 36.0130

HAEMATOLOGY European society of haematology congress 7th Papers and proceedings London 1959 1-2 European society of haematology Edited by E. Neumark and others 2 vols Karger,Basle;New York,1960
Med 36.0131

HAEMATOLOGY European society of haematology congress 8th Proceedings Vienna 1961 1-2 European society of haematology Edited by Adolf Scharf 2 vols Karger,Basle; New York,1962
Med 36.0134

HAEMATOLOGY European society of hematology congress 7th Proceedings London 1959 2,pt.2: papers Edited by E. Neumark and others Karger,Basle;New York,1960
Med 36.0381

HAEMATOLOGY International congress of haematology 11th Proceedings Sydney 1966 International society of hematology Blackwell,Oxford,1966
Med 36.0203

HAEMATOLOGY International congress of hematology 3rd Proceedings Cambridge 1950 Aug 21-25 International society of hematology Edited by Carl V. Moore Grune and Stratton,New York,1951
Med 36.0256

HAEMATOLOGY International congress of hematology 4th Proceedings Mar del Plata 1952 Sep 20-27 International society of hematology Edited by F. Jimenez de Asua and others Grune and Stratton,New York,1954
Med 36.0013

HAEMATOLOGY International congress of hematology 6th Proceedings Boston 1956 Aug 27-Sep 1 International society of hematology Edited by A. Richardson Jones Grune and Stratton,New York,1958
Med 36.0202

HAEMATOLOGY International congress of hematology :reports,communications,symposia 7th Proceedings Rome 1958 Sep 7-13 3 International society of hematology Edited by Giovanni Di Guglielmo Grune and Stratton,New York,1960
Med 36.0166

HAEMATOLOGY Zyto- und histochemie in der hamatologie Freiburger symposium :zugleich symposium der gesellschaft deutscher hamatologen 9th Freiburg 1962 Oct 25-27 Edited by Hans Merker Springer; Blackwell,Berlin;Oxford,1963
Med 36.0394

HAEMATOLOGY Zyto-und histo-chemie in der hamatologie Freiburger symposium :zugleich symposium der gesellschaft deutscher hamatologen 9th Freiburg 1962 Oct 25-27 Edited by Hans Merker Springer,Berlin, 1963
An 32.3186

HAEMOCYANINS Physiology and biochemistry of haemocyanins :a symposium Naples 1966 Aug 30-Sep 1 Edited by F. Ghiretti held at the Naples zoological station Academic press, London;New York,1968 Sponsored by the Italian national research council,the U.S. national science foundation and the American institute of biological sciences
Bal 39.0522

HAEMOGLOBIN Cambridge 1948 Jun 15-17 Edited by F.J.W. Roughton and J.C. Kendrew Butterworths,London,1949
Col S 12.0049

HAEMOGLOBIN :a symposium Cambridge 1948 Jun Edited by F.J.W. Roughton and J.C. Kendrew port. Butterworths,London,1949 In memory of Sir Joseph Barcroft
Phys 20.0881

HAEMOGLOBIN :a symposium based on a conference. in memory of Sir Joseph Barcroft Proceedings Cambridge 1948 June Edited by F.J.W. Roughton and J.C. Kendrew Butterworths,London,1949
Med 36.0309

HAEMOGLOBIN in memory of Sir Joseph Barcroft a symposium based on a conference held Cambridge 1948 Jun Edited by Francis John Worsley Roughton and John Cowdery Kendrew Butterworths,London,1949
Bioch 33.0490

HAEMOPHILIA International symposium on hemophilia and hemophilioid diseases Proceedings New York 1956 Aug 24-25 National hemophilia foundation Edited by K.M. Brinkhous North Carolina university press, Chapel Hill,1957
Med 36.0050

HAEMOPHILIA International symposium on hemophilia and hemorrhagic states Proceedings Rome 1958 Sep 12 International society of hematology Edited by K.M. Brinkhous North Carolina university press,Chapel Hill,1959 Held in connection with the 7th Congress of the International society of hematology
Med 36.0051

HAEMOPOIESIS Ciba foundation symposium on haemopoiesis :cell production and its regulation London 1960 Feb 2-4 Ciba foundation Edited by G.E.W. Wolstenholme and Maeve O'Connor Churchill,London,1960
An 32.3166

HAEMOPOIESIS Symposium on haemopoiesis cell production and its regulation London 1960 Feb 2-4 Ciba foundation Edited by G.E.W. Wolstenholme and Maeve O'Connor Ciba foundation.Symposia Churchill,London,1960
PGMS 29.0273

HAGUE 1955 International horticultural congress 14th Report Vol 1-2 International society for horticultural science Veenman and Zonen,Wageningen,1955
BG 38.3125

HAHN-MEITNER-INSTITUT FUR KERNFORSCHUNG,BERLIN Der Einsatz elektronischer rechengerate fur forschungs- und entwicklungs- aufgaben conference report Berlin 1960 Mai 9-11 Hahn-Meitner-Institut fur kernforschung Berlin. Berichte, 13 var.p 30cm Hahn-Meitner-Institut,Berlin,1960
Math L 5.0870

HAHNEMANN MEDICAL COLLEGE AND HOSPITAL Epithelial-mesenchymal interactions The Hahnemann symposium 18th Philadelphia 1968 Edited by Raul Fleischmajor and Rupert E. Billingham Williams and Wilkins,Baltimore, 1968 Dedicated to Dr. Johannes Holtfreter
Bal 39.0491

HAHNEMANN MEDICAL COLLEGE AND HOSPITAL Metal-binding in medicine :a symposium Proceedings Edited by Marvin J. Seven and L.Audrey Johnson Lippincott,Philadelphia;Montreal,1960
Bioch 33.0679

HAHNEMANN MEDICAL COLLEGE AND HOSPITAL Sexual function and dysfunction :a symposium 1967 Nov 17 Davis,Philadelphia,1969
Inv Med 37.0083

HAHNEMANN SYMPOSIA, 15 Cancer chemotherapy : basic and clinical application Edited by I. Brodsky and S.B. Kahn Grune and Stratton,New York,1967
Radioth 35.0899

HAHNEMANN SYMPOSIUM 13th Stomach, including related areas in the esophagus and duodinum Edited by Charles Thompson and others Grune and Stratton,New York;London, 1967
An 32.3796

The HAHNEMANN SYMPOSIUM 18th Epithelial-mesenchymal interactions Philadelphia 1968 Hahnemann medical college and hospital Edited by Raul Fleischmajor and Rupert E. Billingham Williams and Wilkins,Baltimore, 1968 Dedicated to Dr. Johannes Holtfreter
Bal 39.0491

HAIFA 1962 Second-order effects in elasticity,plasticity and fluid dynamics International symposium on second-order effects in elasticity,plasticity and fluid dynamics symposium International union of theoretical and applied mechanics Edited by Markus Reiner and David Abir Jerusalem academic press;Pergamon press,Jerusalem;London, 1964
Eng 41.2483

HAIFA 1967 Symposium on computer control of natural resources and public utilities Preprints Sponsored by International federation of automatic control.Committee on applications Haifa,1967
Eng 41.5866

HAKONE 1967 Comparative biochemistry and biophysics of photosynthesis :a conference Papers Edited by K. Shibata and others Sponsored by the Japan-U.S.cooperative science program illus viii,445p University of Tokyo press;University Park press,Tokyo; University Park,Pa.,1968
Bot 42.1903

HAKONE,JAPAN 1967 Comparative biochemistry and biophysics of photosynthesis :a conference Papers presented Edited by K. Shibata and others Sponsored by the Japan-United States cooperative science program University of Tokyo press;University park press,State college,Tokyo;University Park,Pa.,1968
Bioch 33.2349

HALDANE CENTENARY SYMPOSIUM Proceedings Regulation of human respiration Oxford 1961 Jul Edited by D.J.C. Cunningham and B.B. Lloyd bibliog.,port. Blackwells,Oxford, 1963
Phys 20.0857

HALIFAX 1968 Fundamental issues in associative learning :a symposium Proceedings Dalhousie university Edited by N.J. Mackintosh and W.K. Honig Dalhousie university press,Halifax,1969
Psy 31.3017

HALIFAX,NOVA SCOTIA 1971 Toposes,algebraic geometry and logic Conference on connections between category theory and algebraic geometry and intuitionistic logic Partial report Edited by F.W. Lawvere Lecture notes in mathematics, 274 189p 25cm Springer, Berlin,1972
P Math 2.4171

HALLE 1882 Congres des geographes allemands notice By Alexis de Tillo Halle,1882 One of fifteen tracts bound together
Philos 1.0875

HALLE 1959 Purkyne symposium Vortrage und diskussionsbeitrage Deutsche akademie der naturforscher Leopoldina Ceskoslovenska akademie ved Edited by Rudolph Zaunick Nova acta leopoldina.Neue folge, 24 230p Barth,Leipzig,1961
WSM 43.2940

HAMBURG 1932 Hydromechanische probleme des schiffsantriebs Edited by G. Kempf and E. Foerster 1932
Eng 41.6656

HAMBURG 1954 Kapillaren und interstitium: morphologie-funktion-klinik :Hamburger symposium Edited by H. Bartelheimer and H. Kuchmeister Thieme,Stuttgart,1955
An 32.3460

HAMBURG 1964 New aspects in the history and philosophy of astronomy Joint symposium of the International astronomical union and the Union internationale d'histoire et de philosophie des sciences 1st International astronomical union International union of history and philosophy of science Edited by Arthur Beer Vistas in astronomy, 9 illus,ports,maps 318p 25cm Pergamon press,Oxford,1967
Cav 7.2715

HAMBURG 1964 New aspects in the history and philosophy of astronomy Joint symposium of the International astronomical union and the Union internationale d'histoire et de philosophie des sciences 1st International astronomical union International union of history and philosophy of science Edited by Arthur Beer Vistas in astronomy, 9 xvi,317p Pergamon,Oxford, 1967
WSM 43.2439

HAMBURG 1964 Reports on astronomy International astronomical union 12th assembly Edited by Jean-Claude Pecker International astronomical union.Transactions, 12a Academic press,London;New York,1965
A Math 4.1114

HAMBURG 1965 Biophysikalische probleme der strahlen wirkung :jahrestagung der Deutschen gesellschaft fur biophysik Tagungsbericht Deutsche gesellschaft fur biophysik Edited by H. Muth Thieme,Stuttgart,1966
Radioth 35.1420

HAMBURG 1965 Digital simulation in operational research :a conference Proceedings Edited by S.H. Hollingdale Under the aegis of the North Atlantic Treaty Organization.Scientific affairs division English universities press,London,1967
Eng 41.1192

HAMBURG 1970 International congress on hormonal steroids 3rd Abstracts of papers Edited by V.H.T. James Excerpta medica. International congress series, 210 Excerpta medica,Amsterdam,1970
Inv Med 37.0115

HAMBURG 1972 The Role of Schmidt telescopes in astronomy Hamburger sternwarte Edited by U. Hang 160p Hamburg observatory, Hamburg,1972
Obs 6.3716

HAMBURGER STERNWARTE The Role of Schmidt telescopes in astronomy Hamburg 1972 Mar 21-22 Edited by U. Hang 160p Hamburg observatory,Hamburg,1972
Obs 6.3716

HAMILTON,ONT. 1968 Probability and information theory :international symposium Proceedings Edited by M. Behara Lecture notes in mathematics, 89 249p 28cm Springer,Berlin,1969
Math S 3.1439

HAMILTON,ONT. 1968 Probability and information theory :international symposium Proceedings Edited by M. Behara and others Lecture notes in mathematics, 89 iv,555p 28cm Springer-Verlag,Berlin,1969
P Math 2.3519

HANOVER 1966 Contributions to mathematical logic Logic colloquium Proceedings Edited by H.Arnold Schmidt and others Studies in logic and the foundations of mathematics Bibliog. ix,298p 23cm North Holland,Amsterdam,1968
P Math 2.3312

HANOVER 1966 Molekulare struktur und strahlenwirkung :jahrestagung der Deutschen gesellschaft fur biophysik Tagungsbericht Deutsche gesellschaft fur biophysik Edited by H. Glubrecht illus. Thieme,Stuttgart, 1968
Radioth 35.1423

HANOVER,N.H. 1954 Aspects of synthesis and order in growth :a symposium Edited by Dorothea Rudnick Society for the study of development and growth.Symposia, 13 Princeton university press,Princeton,N.J.,1954
An 32.2482

HANOVER,N.H. 1955 Tissue elasticity :a conference Papers Edited by John W. Remington held at Dartmouth college American physiological society,Washington,D.C., 1957
Bal 39.1204

HANOVER,N.H. 1961 New directions in mathematics conference Dartmouth college. Department of mathematics Edited by Robert W. Ritchie Supported by the Alfred P.Sloan foundation iv,124p 22cm Prentice-Hall, Englewood Cliffs,N.J.,1964
P. Math 2.2558

HANOVER,N.H. 1966 Comparative aspects of reproductive failure :an international conference Edited by Kurt Benirschke Springer,Berlin,1967
Inv Med 37.0017

HARDWARE,SOFTWARE,FIRMWARE TRADE-OFFS IEEE-International computer society conference 5th Proceedings Boston,Mass. 1971 Sep 22-24 Institute of electrical and electronics engineers International computer society illus 204p IEEE,New York,1971
Math L 5.3628

HARMONDSWORTH 1943 Soils,concrete and bituminous materials :a record of a course dealing with airfield construction... Great Britain.Department of scientific and industrial research H.M.S.O.,London,1946
Eng 41.2998

HARMONIC ANALYSIS International conference on harmonic analysis Proceedings College Park,Md. 1971 Nov 8-12 Edited by Denny Gulick and Ronald L. Lipsman Held at University of Maryland Lecture notes in mathematics, 266 vi,323p 25cm Springer, Berlin,1972
P Math 2.4283

HARPENDEN 1948 Commonwealth conference on tropical and sub-tropical soils 1st proceedings Commonwealth bureau of soil science Commonwealth bureau of soil science. Technical communications,46 Commonwealth bureau of soil science,Harpenden,1949
Geog 13.1093

HARPER HOSPITAL Ovary Symposium on the physiology and pathology of human reproduction 2nd annual Proceedings Edited by Harold C. Mack Thomas,Springfield,Ill.,1968
An 32.5224

HARRIMAN,N.Y. 1952 Contributions to the theory of partial differential equations Conference on partial differential equations typescript By S. Bergman and others Edited by Lipman Bers and others Sponsored by National research council Annals of mathematics studies, 33 Bibliog 257p 26cm Princeton university press,Princeton,N. J.,1954
P. Math 2.1228

HARRIMAN,N.Y. 1952 Symposium on microseisms Proceedings United States office of naval research and United States air force. Geophysical research directorate Edited by James T. Wilson and Frank Press National research council.Publication, 306 Washington,D.C.,1953
Geod 9.0376

HARROGATE 1954 Conference on work study Papers British institute of management British institute of management,London,1954
Eng 41.1544

HARROGATE 1959 Energy requirements in the iron and steel industry a conference British iron and steel research association BISRA, London,1959
Met 25.0409

HARROGATE 1961 Hydrogen in steel Conference on hydrogen in steel papers and discussions Iron and steel institute Iron and steel institute.Special report, 73 Iron and steel institute,London,1962 No title page
Met 25.1538

HARROGATE 1962 Conference on high-strength steels... Papers and discussions Iron and steel institute Iron and steel institute. Special report, 76 Iron and steel institute, London,1962
Met 25.0486

HARROGATE 1962 International congress of radiation research 2nd Abstracts of papers 1962
Radioth 35.1165

HARROGATE 1962 Radiation effects in physics,chemistry and biology International congress of radiation research 2nd Proceedings Edited by Michael Ebert and Alma Howard North-Holland,Amsterdam,1963
Radioth 35.1141

HARROGATE 1963 Metallurgical developments in carbon steels :a conference Iron and steel institute Sponsored by the British iron and steel research association Iron and steel institute.Special report, 81 Iron and steel institute,London,1963
Met 25.0417

HARROGATE 1963 Steam-plant engineering: present status and future trends ;a convention Proceedings Institution of mechanical engineers Institution of mechanical engineers,London,1964
Eng 41.7276

HARROGATE 1963 Symposium on catalysis in practice Proceedings Edited by J.M. Pirie The Yorkshire meeting of the Institution of chemical engineers 83p Institution of chemical engineers,London,1963
Chem E 24.0803

HARROGATE 1965 Aspects of medical physics International conference on medical physics Review papers Edited by J. Rotblat Convened by the International organization of medical physics Taylor and Francis,London, 1966
Radioth 35.1671

HARRY DIAMOND LABORATORIES Fluid amplification symposium Proceedings 1965 Oct 26-28 Vol 1 Harry Diamond laboratories,Washington,D.C.,1965
Eng 41.6970

HARTFORD,CONN. 1968 National conference on statistical meteorology 1st Proceedings Sponsored by the American meteorological society American meteorlogical society, Boston,Mass.,1968 Co-Sponsor:A.M.S. committee on statistical meteorology
Math S 3.1583

HARVARD BUSINESS SCHOOL Technological planning on the corporate level a conference Proceedings Boston,Mass. 1961 Sep 8-9 Edited by James R. Bright Harvard university, Boston,1962
Met 25.2169

HARVARD COLLEGE OBSERVATORY Contemporary geodesy :a conference Proceedings Cambridge,Mass. 1958 Dec 1-2 Edited by Charles A. Whitten and Kenneth H. Drummond Co-sponsored by the American geophysical union Geophysical monograph, 4 National research council.Publication, 708 American geophysical union,Washington,D.C.,1959
Geod 9.0028

HARVARD ECONOMIC RESEARCH PROJECT Structural interdependence and economic development International conference on input-output techniques 3rd Proceedings Geneva 1961 Sep United Nations.Secretariat Edited by Tibor Barna and others Macmillan,London,1963
Geog 13.2417

HARVARD FOREST 1957 Physiology of forest trees An International symposium of forest tree physiology 1st Edited by Kenneth Thimann and others Under the auspices of the Maria Moors Cabot foundation Ronald press, New York,1958
Bot 42.1335

HARVARD MEDICAL SCHOOL Sensory deprivation : a symposium held at Harvard medical school Cambridge,Mass. 1958 Jun 20-21 Edited by Philip Solomon and others Co-sponsored by the United States.Office of naval research. Physiological psychology branch Harvard university press,Cambridge,Mass.,1965
Psy 31.0128

HARVARD-SMITHSONIAN CONFERENCE ON STELLAR ATMOSPHERES 2nd Proceedings Formation of spectrum lines Cambridge,Mass. 1965 Jan 20-22 Smithsonian institution. Astrophysical laboratory and Harvard university.Astronomical observatory Smithsonian institution.Astrophysical observatory,Research in space. Special report, 174 illus 461p Cambridge,Mass.,1965
Obs 6.2121

HARVARD-SMITHSONIAN CONFERENCE ON STELLAR ATMOSPHERES 3rd Proceedings Theory and observation of normal stellar atmospheres Cambridge,Mass. 1968 Harvard university. Astronomical observatory Smithsonian institution.Astrophysical observatory Edited by O. Gingerich 472p Massachusetts institute of technology press,Cambridge,Mass., 1969
Obs 6.3354

HARVARD SYMPOSIUM ON DIGITAL COMPUTERS AND THEIR APPLICATIONS proceedings Cambridge, Mass. 1961 Apr 3-6 Harvard university. Computation laboratory Harvard university. Computation laboratory.Annals, 31 xiv,332p 27cm Harvard university press,Cambridge,Mass. 1962
Math L 5.0687

HARVARD TERCENTENARY CONFERENCE OF ARTS AND SCIENCES Papers Factors determining human behavior Cambridge,Mass. 1936 Aug 31-Sep 12 Harvard university Harvard tercentenary publications Harvard university press,Cambridge,Mass.,1937
Psy 31.0770

HARVARD UNIVERSITY Factors determining human behavior Harvard tercentenary conference of arts and sciences Papers Cambridge,Mass. 1936 Aug 31-Sep 12 Harvard tercentenary publications Harvard university press, Cambridge,Mass.,1937
Psy 31.0770

HARVARD UNIVERSITY International congress for applied mechanics 5th proceedings Cambridge,Mass 1938 Sep.12-16 Edited by J. P. Den Hartog and H. Peters John Wiley,New York,1939
A Math 4.0325

HARVARD UNIVERSITY Symposium on group theory abstracts of lectures Cambridge,Mass. 1963 Apr 1-3 59p 23cm Harvard university, Cambridge,Mass.,1963
P Math 2.2792

HARVARD UNIVERSITY 1926 International congress of philosophy 6th Proceedings International congress of philosophy. Organizing committee Edited by Edgar Sheffield Brightman Longmans,Green,New York, 1927
Psy 31.2520

HARVARD UNIVERSITY.ASTRONOMICAL OBSERVATORY Theory and observation of normal stellar atmospheres Harvard-Smithsonian conference on stellar atmospheres 3rd Proceedings Cambridge,Mass. 1968 Edited by O. Gingerich 472p Massachusetts institute of technology press,Cambridge,Mass.,1969
Obs 6.3354

HARVARD UNIVERSITY.COMPUTATION LABORATORY Mathematical linguistics and automatic translation Report no.NSF 7-15 var.p 28cm 9 vols Harvard university,Cambridge, Mass.,1961-1965
Math L 5.0932

HARVARD UNIVERSITY.COMPUTATION LABORATORY Symposium on large-scale digital calculating machinery 2nd proceedings Cambridge, Mass. 1949 Sep 13-16 Jointly sponsored by United States.Bureau of ordnance Harvard university.Computation laboratory Annals, 26 illus. xxxviii,393p 27cm Harvard university press,Cambridge,Mass.,1951
Math L 5.0684

HARVARD UNIVERSITY.COMPUTATION LABORATORY Symposium on large-scale digital calculating machinery proceedings Cambridge,Mass. 1947 Jan 7-10 Jointly sponsored by United States.Bureau of ordnance Harvard university. Computation laboratory.Annals, 16 illus. xxix,302p 27cm Harvard university press, Cambridge,Mass.,1948
Math L 5.0681

HARVARD UNIVERSITY.COMPUTATION LABORATORY The International symposium on the theory of switching proceedings Cambridge,Mass. 1957 Apr 2-5 Part 1-2 Harvard university.Computation laboratory.Annals, 29 27cm 2 vols Harvard university press, Cambridge,Mass.,1959
Math L 5.0686

HARVARD UNIVERSITY.GRADUATE SCHOOL OF BUSINESS ADMINISTRATION Automatic data processing conference proceedings Boston,Mass. 1955 Sep 8-9 Edited by Robert N. Anthony illus. viii,194p 21cm Harvard university,Boston, Mass.,1956
Math L 5.0752

HARVARD UNIVERSITY.MUSEUM OF COMPARATIVE ZOOLOGY Phylogeny and evolution of crustacea :a conference Proceedings Cambridge,Mass. 1962 Mar 6-8 Edited by H.B. Whittington and W.D.I. Rolfe Supported by the National science foundation Museum of comparative zoology,Cambridge,Mass.,1963
Bal 39.2305

HARVESTING THE SUN:PHOTOSYNTHESIS IN PLANT LIFE a symposium Chicago 1966 Oct 5-7 Edited by A. San Pietro and others Sponsored by the International minerals and chemical corporation Academic press,New York;London, 1967
Bioch 33.0789

HARVESTING THE SUN:PHOTOSYNTHESIS IN PLANT LIFE a symposium Papers Chicago,Ill. 1966 Oct 5-7 Edited by Anthony San Pietro and others Sponsored by the International minerals and chemical corporation illus. ix,342p Academic press,New York;London,1967
Bot 42.1899

HARWELL 1961 Notes on the Harwell FORTRAN seminar By A.R. Curtis and I.C. Pyle Atlas paper, 19 29p 1961
Math L 5.2520

HARWELL 1962 Neutron dosimetry Symposium on neytron detection,dosimetry and standardization Proceedings 1-2 Held by the International atomic energy agency International atomic energy agency.Proceedings series 2 vols International atomic energy agency,Vienna,1963
Radioth 35.1695

HARWELL 1966 Nature of small defect clusters Consultants symposium 2nd Report Vol 1-2 United Kingdom atomic energy authority.Research group Edited by M. J. Makin 2 vols Atomic energy research establishment,Harwell,1966
Met 25.1243

HARWELL 1968 The Interactions between dislocations and point defects :a symposium Proceedings 1-3 Atomic energy research establishment.Metallurgy division United Kingdom atomic energy authority. Research group.Report AERE-R 5944 3 vols harwell,1968
Met 25.2415

HARWELL 1968 Thermal neutron diffraction International summer school on the accurate determination of neutron intensities and structure factors Proceedings Edited by B.T.M. Willis Harwell postgraduate series Bibliog,illus xiv,229p 24cm Oxford university press,London,1970
Min 10.1524

HARWELL,BERKS 1917 Vacancies and other point defects in metals and alloys a symposium Institute of metals Institute of metals. Monograph and report series, 23 Institute of metals,London,1958
Met 25.1233

HASHISH:ITS CHEMISTRY AND PHARMACOLOGY London 1965 Ciba foundation Churchill, London,1965
Pha 16.0431

HAVERFORD,PA. 1966 Major problems in developmental biology :a symposium Society for developmental biology Edited by Michael Locke Society for developmental biology. Symposia, 25 Academic press,New York;London, 1966
Bioch 33.0912

HAVERFORD,PENN. 1966 Current status of some major problems in developmental biology : symposium Society for developmental biology Edited by Michael Locke Society for developmental biology.Symposia, 25 Academic press,New York;London,1966 Formerly the Society for the study of development and growth
Gen 34.0487

HAWTHORNE W.R. and FABRI J. ed. Selected combustion problems Combustion colloquium Cambridge 1953 Dec 7-11 Agard Butterworths,London,1954
A Math 4.0969

HEALTH OF MANKIND :a Ciba foundation symposium London 1967 Mar 8-10 Ciba foundation Edited by G.E.W. Wolstenholme and Maeve O'Connor Churchill,London,1967 Ciba foundation 100th symposium
Inv Med 37.0267

HEALTH PHYSICS IN NUCLEAR INSTALLATIONS :a symposium Riso 1959 May 25-28 Organization for European economic cooperation European nuclear energy agency Held at the Danish atomic energy centre European nuclear energy agency,Paris,1959
Radioth 35.1110

HEARING MECHANISMS IN VERTEBRATES Ciba foundation symposium...on sensory function 4th Proceedings London 1967 Sep 26-28 Ciba foundation Edited by A.V.S.de Reuck and Julie Knight J.and A.Churchill,London,1968
Psy 31.3372

HEARING MECHANISMS IN VERTEBRATES :a Ciba foundation symposium London Ciba foundation Edited by A.V.S.de Reuck and Julie Knight London,1968
Bal 44.6455

HEARING MECHANISMS IN VERTEBRATES :a Ciba foundation symposium London 1967 Ciba foundation Edited by A.V.S. De Reuck and Julie Knight Churchill,London,1968
Phys 20.1050

HEART ASSOCIATION OF SOUTHEASTERN PENNSYLVANIA International symposium on the myocardial cell structure,function and modification by cardiac drugs 3rd Edited by Stanley A. Briller and Hardley L. Conn University of Pennsylvania press,Philadelphia,Pa.,1966
Phys 20.2221

HEART ASSOCIATION OF SOUTHEASTERN PENNSYLVANIA The Myocardial cell:structure,function and modification by cardiac drugs :international symposium 3rd Edited by Stanley A. Briller and Hardley L. Conn University of Pennsylvania press,Philadelphia,Pa.,1966
Pha 16.0027

HEART MUSCLES Struktur und stoffwechsel des herzmuskels :symposium Munster 1958 Sep 26-27 Edited by W.H. Hauss and H. Losse Thieme,Stuttgart,1959
An 32.3467

HEAT AND MASS TRANSFER BY COMBINED FORCED AND NATURAL CONVECTION :a symposium Proceedings London 1971 Sep 15 Institution of mechanical engineers. Thermodynamics and fluid mechanics group IME, London,1972
Eng 41.8647

HEAT AND MASS TRANSFER IN PROCESS METALLURGY : a symposium Proceedings London 1966 Apr 19-20 Edited by A.W.D. Hills Held by the John Percy research group in process metallurgy ix,252p Institution of mining and metallurgy,London,1967
Chem E 24.0593

HEAT AND MASS TRANSFER IN PROCESS METALLURGY : a symposium Proceedings London 1966 Apr 19-20 John Percy research group in process metallurgy Edited by A.W.D. Hills Institution of mining and metallurgy,London, 1967
Met 25.0317

HEAT TOLERANT METALS FOR AERODYNAMIC APPLICATIONS a conference Proceedings Metals for supersonic aircraft and missiles Alburquerque,New Mexico 1957 Jan 28-29 American society for metals and University of New Mexico Edited by D.W. Grobecker American society for metals,Cleveland,Ohio, 1958
Met 25.1003

HEAT TRANSFER Condensing heat transfer within horizontal tubes National heat transfer conference 2nd Preprints Chicago,Ill. 1958 Aug 18-21 American society of mechanical engineers American society of mechanical engineers,Philadelphia, Pa.,1958
Eng 41.7155

HEAT TRANSFER General discussion on heat transfer Proceedings London 1951 Sep 11-12 Institution of mechanical engineers and American society of mechanical engineers xiii,496p Institution of mechanical engineers,London,1951
Chem E 24.0538

HEAT TRANSFER General discussion on heat transfer Proceedings London 1951 Sep 11-13 Institution of mechanical engineers American society of mechanical engineers Institution of mechanical engineers,London, 1952
Eng 41.7146

HEAT TRANSFER International developments in heat transfer Heat transfer conference 1961-62 Proceedings Boulder,Colo. 1961 Aug 28-Sep 1 and London 1962 Jan 8-12 American society of mechanical engineers American society of mechanical engineers,New York,1963
Eng 41.7168

HEAT TRANSFER International developments in heat transfer International heat transfer conference Papers Boulder,Col. 1961 Aug 28-Sep 1 London 1962 Jan 8-12 Pt 1-5 Arranged by the American society of mechanical engineers 5 vols A.S.M.E.,New York,1961 Co-sponsored by the American institute of chemical engineers,the Institution of mechanical engineers,and the Institution of chemical engineers
Chem E 24.0595

HEAT TRANSFER The Heat transfer and fluid mechanics institute Proceedings Seattle, Wash. 1962 Jun 13-15 University of Washington Edited by Edward Ehlers and others Co-sponsored by Stanford university 293p Stanford university press,Stanford, Calif.,1962
Cav 7.2422

HEAT TRANSFER Theory and fundamental research in heat transfer American society of mechanical engineers :annual meeting Proceedings New York 1960 Nov Edited by J.A. Clark ix,220p Pergamon press,Oxford, 1963
Chem E 24.0582

HEAT TRANSFER Use of secondary surfaces for heat transfer with clean gases :a symposium Proceedings 1960 Institute of mechanical engineers London,1961
Eng 41.7167

HEAT TRANSFER :a symposium Ann Arbor 1952 University of Michigan Edited by B.A. Uhlendorf University of Michigan.Engineering research institute special publications: symposia vii,286p Engineering research institute,Michigan,1953
Chem E 24.0544

HEAT TRANSFER AND FLUID MECHANICS INSTITUTE Heat transfer and fluid mechanics institute Preprints of papers,Proceedings 1957-59,196; 1960- 3 vols Stanford,Calif.,1957- Preprints continued in 1960 as Proceedings
Eng 41.7157

The HEAT TRANSFER AND FLUID MECHANICS INSTITUTE Proceedings Seattle,Wash. 1962 Jun 13-15 University of Washington Edited by Edward Ehlers and others Co-sponsored by Stanford university 293p Stanford university press,Stanford,Calif.,1962
Cav 7.2422

HEAT TRANSFER AND FLUID MECHANICS INSTITUTE 1st- Preprints of papers,proceedings 1949- 1949-
Chem E 24.0594

HEAT TRANSFER AND FLUID MECHANICS INSTITUTE 1958 preprints Berkeley,Calif. 1958 Jun 19-21 University of California Stanford university press,Stanford,Calif.,1958
A Math 4.0517

The HEAT TRANSFER AND FLUID MECHANICS INSTITUTE 1960 proceedings Stanford,Calif. 1960 Jun. 15-17 Edited by D.M. Mason and others Stanford university press,Stanford, Calif.,1960
A Math 4.0518

HEAT TRANSFER AND FLUID MECHANICS INSTITUTE 1961 proceedings Los Angeles 1961 Jun. 19-21 Edited by R.C. Binder and others Stanford university press,Stanford,Calif.,1961
A Math 4.0519

The HEAT TRANSFER AND FLUID MECHANICS INSTITUTE 1962 proceedings Seattle,Wash. 1962 Jun.13-15 Edited by F.Edward Ehlers and others Stanford university press,Stanford, Calif.,1962
A Math 4.0520

The HEAT TRANSFER AND FLUID MECHANICS INSTITUTE 1963 proceedings Pasadena,Calif. 1963 Jun.12-14 Edited by Anatol Roshko and others Stanford university press,Stanford, Calif,1963
A Math 4.0521

The HEAT TRANSFER AND FLUID MECHANICS INSTITUTE 1964 proceedings Berkeley,Calif 1964 Jun.10-12 Edited by Warren H. Gledt and Salomon Levy Stanford university press, Stanford,Calif,1964
A Math 4.0522

The HEAT TRANSFER AND FLUID MECHANICS INSTITUTE 1965 proceedings Los Angeles,Calif. 1965 Jun. 21-23 Edited by Andrew F. Charwat Stanford university press,Stanford,Calif.,1965
A Math 4.0523

The HEAT TRANSFER AND FLUID MECHANICS INSTITUTE 1966 proceedings Santa Clara,Calif. 1966 Jun. 22-24 Edited by Michel A. Saad and James A. Miller Stanford university press, Stanford,Calif.,1966
A Math 4.0524

HEAT TRANSFER AND FLUID MECHANICS INSTITUTE 1968 Proceedings Seattle,Wash. 1968 Jun 17-18 Edited by A.F. Emery and C.A. Depew bibliog.,illus. ix,272p 26cm Stanford university press,Stanford,Calif.,1968
A Math 4.1617

HEAT TRANSFER AND FLUID MECHANICS INSTITUTE 1972 Proceedings Northridge,Calif. 1972 Jun Edited by R.B. Landis and G.J. Hordemann illus xii,430p 25cm Stanford university press,Stanford,Calif.,1972
A Math 4.1876

HEAT TRANSFER AND FLUID MECHANICS INSTITUTE 1970 Proceedings Monterey,Calif. 1970 Jun 10-12 Edited by T. Sarpkaya bibliog.,illus. 370p 26cm Stanford university press, Stanford,1970
A Math 4.1649

HEAT TRANSFER CONFERENCE 1961-62 Proceedings International developments in heat transfer Boulder,Colo. 1961 Aug 28-Sep 1 and London 1962 Jan 8-12 American society of mechanical engineers American society of mechanical engineers,New York,1963
Eng 41.7168

HEAT TREATMENT JOINT COMMITTEE :inaugural meeting Proceedings Heat treatment of metals London 1965 Dec 14-15 Iron and steel institute Institute of metals Iron and steel institute.Special report, 95 London,1966
Met 25.2565

HEAT TREATMENT OF METALS Heat treatment joint committee :inaugural meeting Proceedings London 1965 Dec 14-15 Iron and steel institute Institute of metals Iron and steel institute.Special report, 95 London,1966
Met 25.2565

HEBER,UTAH 1961 Some aspects of internal irradiation :a symposium Proceedings Edited by Thomas F. Dougherty and others Pergamon press,London,1962
Radioth 35.1169

HEBREW UNIVERSITY OF JERUSALEM Paramagnetic resonance :international conference 1st Proceedings Jerusalem 1962 Jul 16-20 Vol 1-2 International union of pure and applied physics Edited by W. Low 2 vols Academic press,New York,1963
Cav 7.1226

HEBREW UNIVERSITY OF JERUSALEM Radiation chemistry of aqueous systems :proceedings of the 19th L.Farkas memorial symposium held at the Hebrew university Jerusalem 1967 Dec 22-29 Edited by Gabriel Stein Weizmann science press of Israel in co-operation with Interscience,Jerusalem,1968
Chem 18.2485

HEERLEN 1927 Congres pour l'avancement des etudes de stratigraphie carbonifere 1er compte rendu Edited by W.J. Jongmans Published by the geologisch-mijnbouwkundig genootschap voor Nederland en kolonien Liege, 1928 Text in English,French and German
Geol 8.2984

HEERLEN 1935 Congres pour l'avancement des etudes de stratigraphie carbonifere 2e compte rendu Tom 1-3 Edited by W.J. Jongmans 3 vols Maestricht,1937-38 Text in English,French and German
Geol 8.2985

HEERLEN 1951 Congres pour l'avancement des etudes de stratigraphie et de geologie du carbonifere 3e compte rendu Tom 1-2 Edited by Geologisch bureau Heerlen Maestricht,1952 Text in English,French and German
Geol 8.2986

HEERLEN 1958 Congres pour l'avancement des etudes de stratigraphie et de geologie du carbonifere 4e compte rendu Tom 1-3 3 vols Maestricht,1960-62 Text in English, French and German
Geol 8.2987

HEERLEN 1958 International congress on coal petrology 1st proceedings International committee for coal petrology. Proceedings,3 Ernest van Aelst,Maastricht, 1960
Geol 8.3137

HEIDELBERG 1956 Regelungstechnik:moderne theorien und ihre verwendbarkeit :ein tagung Bericht Gesellschaft fur angewandte mathematik und mechanik.Fachausschuss regelungsmathematik Verein deutscher ingenieure.Fachgruppe regelungstechnik Oldenbourg,Munich,1957
Eng 41.5842

HEIDELBERG 1958 Kolloquium uber bildwandler und bildspeicherrohren vortrage und diskussionen Edited by Heinrich Siedentopf Heidelberger akademie der wissenschaften.Sitzungenberichte.Mathematisch-naturwissenschaftliche klasse,1959,abh.5 31 plates, 4 tabs 78p Springer-verlag, Heidelberg,1959
Obs 6.1627

HEIDELBERG 1962 Vacuum technique Meeting of the German society for vacuum technique Proceedings Deutsche gesellschaft fur vakuumtechnik Macmillan;Pergamon press, London;Oxford,1965
Cav 7.1062

HEIDELBERG 1964 Assessment of radioactivity in man Symposium on the assessment of radioactive body burdens in man Proceedings 1-2 International atomic energy agency International labour organization World health organization International atomic energy agency.Proceedings series International atomic energy agency, Vienna,1964
Radioth 35.1373

HEIDELBERG 1966 Blanketing effect :I.A.U. colloquium Proceedings International astronomical union Edited by K-H. Bohm International astronomical union.Colloquium Journal of quantitative spectroscopy and radiative transfer, 6,no 5 Pergamon,London, 1966
TA 15.0414

HEIDELBERG 1966 Colloquium on the blanketing effect proceedings Edited by Karl-Heinz Bohm Journal of quantitative spectroscopy and radiative transfer, 6,no.5 Pergamon press,Oxford;London,1966
Obs 6.0207

HEIDELBERG 1967 European conference on elementary particles 4th Proceedings Ruprecht-Karl-universitat Heidelberg and European organization for nuclear research Edited by H. Filthuth viii,550p North-Holland,Amsterdam,1968
A Math 4.1348

HELIUM International conference on low temperature physics 10th Proceedings Moscow 1966 Aug 31-Sep 6 Vol 1: properties of helium International union of pure and applied physics Akademiya nauk S.S.S.R. Edited by M.P. Malkov 552p 22cm Moscow,1967
Cav 7.3030

HELSINKI 1902 Versammlung nordischer naturforscher und artze Verhandlungen der sektion fur anatomie,physiologie und medizinische chemie 1902
Phys 20.0160

HELSINKI 1940 International Forestry Congress 3rd programme International institute of agriculture Helsinki,1939
Philos 1.2032

HELSINKI 1960 International association of scientific hydrology general assembly 12th Papers snow and ice commission International association of scientific hydrology illus,map 588p Gentbrugge,1961 Papers in English,French and German
Sco 14.0131

HELSINKI 1960 International latitude service L' Avenir du service international des latitudes :a symposium proceedings Edited by Paul Melchior Organized by International union of geodesy and geophysics International astronomical union.Symposium, 13 97p Paris,1961 Reprinted from Bulletin geodesique no.59,March 1961
Obs 6.1198

HELSINKI 1960 Symposium on Antarctic glaciology International association of scientific hydrology general assembly Papers International association of scientific hydrology.Publication,55 illus 162p 24cm Gentbrugge,1961 Papers in English and French
Sco 14.0130

HELSINKI 1960 Symposium on radiant energy in the sea Papers International union of geodesy and geophysics Edited by N.G. Jerlov International union of geodesy and geophysics. Monograph,10 illus 115p Paris,1961 Symposium held during the I.U.G.A.assembly at Helsinki
Sco 14.0135

HELSINKI 1960 Symposium on the July 1959 events and associated phenomena International union of geodesy and geophysics International union of geodesy and geophysics. Monograph,7 tables,diagrs 157p 24cm I.U.G.G.,Paris,1960
Sco 14.0246

HELSINKI 1964 Current corrosion research in Scandinavia Scandinavian corrosion congress (NKM) :lectures 4th Teknillisten tieteiden akatemia and Central chemical association Edited by Jori Larinkari and others Organized by the Scandinavian council for applied research Kemian keskusliton julkaisuja, 24 Sanoma Osakeyhtio,Helsinki,1965
Met 25.1901

HELSINKI 1965 Control of cellular growth in adult organisms A Sigrid Juselius foundation symposium Sigrid Juselius foundation Edited by Harald Teir and Tapio Rytomaa Academic press,London;New York,1967
Bioch 33.1007

HELSINKI 1965 Control of cellular growth in adult organisms :a Sigrid Juselius foundation symposium Edited by Harald Teir and Tapio Rytomaa Academic press,London;New York,1967
Pha 16.0177

HELSINKI 1965 Control of cellular growth in adult organisms :a Sigrid Juselius foundation symposium Edited by Harald Tier and Tapio Rytomaa Academic press,London;New York,1967
Phys 20.2232

HELSINKI 1967 Regulatory functions of biological membranes The Sigrid Juselius foundation symposium 2nd Proceedings Sigrid Juselius foundation Edited by Johan Jarnefelt B.B.A.library, 11 illus. viii, 311p Elsevier,Amsterdam,1968
Bot 42.1517

HELSINKI 1967 Regulatory functions of biological membranes The Sigrid Juselius foundation symposium 2nd Proceedings Sigrid Juselius foundation Edited by Johan Jarnefelt BBA library, 11 Elsevier, Amsterdam,1968
Bioch 33.0968

HELSINKI 1967 Regulatory functions of biological membranes :a Sigrid Juselius foundation symposium 2nd Papers Edited by Johan Jarnefelt B.B.A. library,11 Illus. viii,311p Elsevier,Amsterdam,1968
Pha 16.0017

HELSINKI SYMPOSIUM,1965 Aspects of inductive logic Edited by Jaakko Hintikka and Patrick Suppes Studies in logic and the foundations of mathematics vii,320p 23cm North-Holland,Amsterdam,1966 Includes papers read at the International symposium on confirmation and information,Helsinki,1965
P Math 2.2678

HEMATOPOIETIC NEOPLASMS Comparative morphology of hematopoietic neoplasms :a symposium Proceedings Washington,D.C. 1968 Mar 11-12 National cancer institute Edited by Carolyn H. Lingeman and F.M. Garner held at United States.Armed forces institute of pathology National cancer institute. Monograph, 32 National cancer institute, Bethesda,Md.,1969
Bioch 33.0995

HERCEG-NOVI 1968 International symposium on topology and its applications Proceedings Edited by D.R. Kurepa 356p 25cm Savez matematicara,Belgrade,1969
P Math 2.4117

HEREDITARY,DEVELOPMENTAL AND IMMUNOLOGIC ASPECTS OF KIDNEY DISEASE Annual conference on the kidney 13th Proceedings Princeton 1961 Oct 12-13 National kidney disease foundation Edited by Jack Metcoff Northwestern university press,Evanston,1962
Inv Med 37.0157

HEREDITARY,DEVELOPMENTAL AND IMMUNOLOGIC ASPECTS OF KIDNEY DISEASE Annual conference on the kidney 13th Proceedings Princeton,N. J. 1961 Oct 12-13 National kidney disease foundation Edited by Jack Metcoff Northwestern university press,Evanston,Ill., 1962
An 32.2075

HEREDITY A Symposium on the chemical basis of heredity Baltimore,Md. 1956 Jun 19-22 McCollum-Pratt institute Edited by William D. McElroy and Bentley Glass Johns Hopkins press,Baltimore,Md.,1957
Bot 42.4714

HEREDITY A Symposium on the chemical basis of heredity Papers and informal discussions Baltimore,Md. 1956 Jun 19-22 Edited by William D. McElroy and Bentley Glass Sponsored by the McCollum-Pratt institute McCollum-Pratt institute.Contribution, 153 Johns Hopkins press,Baltimore,Md.,1957
Bal 44.6451

HEREDITY Chemical basis of heredity A Symposium on the chemical basis of heredity Papers and informal discussions Baltimore 1956 Jun 19-22 Edited by William D. McElroy and Bentley Glass Sponsored by the McCollum-Pratt institute McCollum-Pratt institute. Contribution, 153 Johns Hopkins press, Baltimore,1957
Bioch 33.0830

HEREDITY The Chemical basis of heredity :a symposium Baltimore,Md. 1956 Jun 19-22 McCollum-Pratt institute Edited by William D. Bentley McElroy and Bentley Glass With support from the Atomic energy commission McCollum-Pratt institute.Contribution,153 Johns Hopkins press,Baltimore,Md.,1957
Chem 18.1555

HEREDITY AND DISEASE The Influence of heredity on disease,with special reference to tuberculosis, cancer and diseases of the nervous system :a discussion London 1908 Nov Royal society of medicine illus. Longmans,London,1909
Gen 34.1979

HEREDITY AND VARIATION IN MICROORGANISMS :a symposium Papers Cold Spring Harbor 1946 Cold Spring Harbor biological laboratory Cold Spring Harbor symposia on quantitative biology, 11 Long Island biological association,Cold Spring Harbor,1946
Bioch 33.1267

HEREDITY COUNSELING :a symposium New York American eugenics society Edited by Helen G. Hammons Hoeber-Harper,New York,1959
PGMS 29.0257

HERING-BREUR CENTENARY SYMPOSIUM Proceedings London 1969 Jul 8-10 Ciba foundation Edited by Ruth Porter Ciba foundation. Symposia Churchill,London,1970
Inv Med 37.0184

HERITAGE FROM MENDEL Mendel centennial symposium Proceedings Fort Collins,Colo 1965 Edited by R. Alexander and E.Derek Styles Sponsored by the Genetics society of America Wisconsin university press,Madison, Wis.,1967
Bot 42.0880

HERITAGE FROM MENDEL Mendel centennial symposium Proceedings Fort Collins,Colo. 1965 Spe 7-11 Genetics society of America Edited by R.Alexander Brink and E.Derek Styles University of Wisconsin press,Madison,Wis., 1967
Gen 34.0994

HERRENCHIEMSEE 1959 Diurese und diuretica : ein internationales symposium Edited by H. Schwiegk and others Sponsored by Ciba foundation Springer,Berlin,1959 t.p. also in English "Diuresis and diuretics"
Inv Med 37.0193

HERSHEY,PA. 1962 Arctic Basin symposium Proceedings Arctic institute of North America,and,United States.Office of naval research illus,maps 313p 23cm Washington,D.C.,1963
Sco 14.5904

HERSTMONCEUX 1962 Transfer of radiation in stellar atmospheres Colloquium on the theory of stellar atmospheres 3rd International astronomical union Edited by C. de Jager and A.B. Underhill 201p Pergamon press,London,1963 Vol 3,no.2 of Journal of quantitative spectroscopy and radiative transfer. AprJun.1963
Obs 6.0875

The HERTZSPRUNG-RUSSELL DIAGRAM :a symposium Proceedings Moscow 1958 Aug.15-16 International astronomical union Edited by Jesse L. Greenstein International astronomical union.Symposium, 10 128p Paris,1959 Reprinted from Annales d'astrophysique 1959,supplements.Fascicule.No. 8
Obs 6.0687

HETEROCYCLIC COMPOUNDS Recent work on naturally occuring nitrogen heterocyclic compounds :a symposium Report Exeter 1955 Jul 13-15 Chemical society Edited by K. Schofield Chemical society.Special publication,3 Chemical society,London,1955
Chem 18.1329

HETEROGENEOUS COMBUSTION :a conference Palm Beach,Fla. 1963 Dec 11-13 American institute of aeronautics and astronautics Edited by Hans G. Wolfhard and others Progress in astronautics and aeronautics, 15 Academic press,New York,1964
Eng 41.7241

HETEROGENEOUS KINETICS AT ELEVATED TEMPERATURES International conference on metallurgy and materials science Proceedings Philadelphia, Pa. 1969 Sep 8-10 Edited by G.R. Belten and W.L. Worrell Held at the University of Pennsylvania Plenum press,New York,1970
Met 25.2783

HEURISTIC PROGRAMMING Artificial intelligence and heuristic programming N.A.T.O. advanced study institute on artificial intelligence and heuristic programming 1st Proceedings Menaggio 1970 Aug North Atlantic treaty organization Edited by N.V. Findler and Bernard Meltzer illus 327p Edinburgh university press,Edinburgh,1971
Math L 5.3853

HIGH ENERGY ASTROPHYSICS Astrofisica delle alte energie Scuola internazionale di fisica 'Enrico Fermi' 35 corso Rendiconti Varenna 1965 Jul 12-24 Societa italiana di fisica Edited by L. Gratton Academic press, New York;London,1966
TA 15.0314

HIGH-ENERGY ASTROPHYSICS Astrofisica delle alte energie Scuola internazionale di fisica "Enrico Fermi" 35 corso rendiconti Varenna 1965 Jul.12-24 Societa Italiana di Fisica Edited by L. Gratton 463p Academic press,New York,1966
A Math 4.1169

HIGH ENERGY ASTROPHYSICS :lectures Les Houches 1966 Vol 3 Universite de Grenoble.Ecole d'ete de physique theorique,Les Houches Edited by C. Dewitt and others Supported by NATO Universite de Grenoble. Ecole d'ete de physique theorique,Les Houches. Lectures,1966 xii,449p Gordon and Breach, New York,1967
A Math 4.1562

HIGH ENERGY ASTROPHYSICS Lectures Les Houches 1966 Vol 1-3 Universite de Grenoble.Ecole d'etude physique theorique,Les Houches Edited by C. DeWitt and others Universite de Grenoble.Ecole d'ete de physique theorique,Les Houches.Lectures,1966 Gordon and Breach,New York,1967 Lectures delivered at Les Houches,1966
TA 15.0312

HIGH-ENERGY ELECTRONS Symposium on high-energy electrons Proceedings Montreux 1964 Sep 7-11 Edited by A. Zuppinger and G. Poretti Springer,Berlin,1965
Radioth 35.1662

HIGH ENERGY NUCLEAR PHYSICS Annual Rochester conference on high energy nuclear physics 3rd Proceedings Rochester,N.Y. 1952 Dec 18-20 Edited by H.Pierre Noyes and others Sponsored by the University of Rochester 110p Interscience,New York,c.1953 Co-sponsored by the National science foundation
Cav 7.1986

HIGH ENERGY NUCLEAR PHYSICS Annual Rochester conference on high energy nuclear physics 4th Proceedings Rochester,N.Y. 1954 Jan 25-27 Edited by H.Pierre Noyes and others Sponsored by the University of Rochester 159p Rochester,N.Y.,1954 Co-sponsored by the National science foundation.Typescript
Cav 7.1987

HIGH ENERGY NUCLEAR PHYSICS Annual Rochester conference on high energy nuclear physics 5th Proceedings Rochester,N.Y. 1955 Jan 31-Feb 2 University of Rochester Edited by H.Pierre Noyes and others Sponsored by the International union of pure and applied physics 196p Interscience,New York,1955 Co-sponsored by the National science foundation in cooperation with the Atomic energy commission and the United States office of naval research
Cav 7.1988

HIGH ENERGY NUCLEAR PHYSICS Annual Rochester conference on high energy nuclear physics 6th Proceedings Rochester,N.Y. 1956 Apr 3-7 Edited by J. Ballam and S.B. Trieman 362p Interscience,New York,1956
Cav 7.1989

HIGH-ENERGY NUCLEAR REACTIONS High-energy nuclear reactions in astrophysics Symposium on high-energy nuclear reactions in astrophysics Invted papers Philadelphia,Pa. 1967 Jan Edited by D.S.P. Shen Sponsored by American astronomical union 281p Benjamin,New York,1967 Four of the ten papers were invited after the symposium to complete the coverage
TA 15.0578

HIGH-ENERGY NUCLEAR REACTIONS IN ASTROPHYSICS Symposium on high-energy nuclear reactions in astrophysics Invted papers Philadelphia,Pa. 1967 Jan Edited by D.S.P. Shen Sponsored by American astronomical union 281p Benjamin,New York,1967 Four of the ten papers were invited after the symposium to complete the coverage
TA 15.0578

HIGH ENERGY PHYSICS Coral Gables conferences on symmetry principles at high energy 4 Proceedings Coral Gables,Fla. 1967 Jan 25-27 University of Miami.Center for theoretical studies Edited by Arnold Perlmutter and Behram Kursunoglu Sponsored jointly by the United States atomic energy commission 253p Freeman,San Francisco; London,1967
A Math 4.1347

HIGH ENERGY PHYSICS International conference on high-energy physics at CERN 11th Proceedings Geneva 1962 Jul 4-11 European organization for nuclear research Edited by J. Prentki Sponsored by the International union of pure and applied physics Bibliog.,Illus. xxiv,949p 28cm CERN,Geneva,1962
A Math 4.1544

HIGH ENERGY PHYSICS Particle interactions at high energies Scottish universities summer school 7th proceedings Newbattle Abbey, Edinburgh 1966 Aug 1-20 Edited by T.W. Preist and L.L.J. Vick 406p Oliver and Boyd,Edinburgh,1967
A Math 4.0875

HIGH ENERGY PHYSICS The International conference on high energy physics 13th Proceedings Berkeley,Calif. 1966 Aug 31-Sep 7 University of California and United States atomic energy commission Edited by Margaret Alston-Garnjost Sponsored by the International union of pure and applied physics Berkeley,Calif.,1967
Cav 7.0189

HIGH-ENERGY PHYSICS International conference on high-energy physics 10th Proceedings Rochester,N.Y. 1960 Aug 25-Sep 1 University of Rochester International union of pure and applied physics Edited by E.C.G. Sudarshan and others Co-sponsored by the United States.Office of naval research illus xxv,890p University of Rochester,Rochester,N. Y.,1960
Cav 7.2891

HIGH-ENERGY PHYSICS International conference on high-energy physics 13th proceedings Berkeley,Calif. 1966 Aug.31-Sep.7 Sponsored by International union of pure and applied physics University of California press,Berkeley;Los Angeles,1967
A Math 4.0891

HIGH-ENERGY PHYSICS AND ELEMENTARY PARTICLES Seminar on elementary particles lectures Trieste 1965 May 3-Jun.30 International atomic energy agency International atomic energy agency.Proceedings series STIPUB117 Vienna,1965
A Math 4.0888

HIGH MAGNETIC FIELDS International conference on high magnetic fields Proceedings Cambridge,Mass. 1961 Nov 1-4 Edited by Henry Kolm Sponsored by the United States.Air force.Office of scientific research. Solid state sciences division M.I.T.press; Wiley,Cambridge,Mass.;New York,1962
Eng 41.5331

HIGH POLYMERS International congress of biochemistry 4th Proceedings Vienna 1958 Sep 1-6 Vol 9: symposium 9 - physical chemistry of high polymers of biological interest International union of biochemistry Edited by O. Kratky Pergamon press,London, 1960
Radioth 35.1941

HIGH PRESSURE RESEARCH Progress in very high pressure research :an international conference Proceedings Bolton Landing,N.Y. 1960 Jun 13-14 Wright air development division. Materials central and General electric company research laboratory Edited by F.P. Bundy and others 314p New York,1961
Geod 9.0254

HIGH PRESSURES The International conference on the physics of solids at high pressures 1st Proceedings Tucson,Ariz. 1965 Apr 20-23 University of Arizona Edited by C.T. Tumizuka and R.M. Emrick Jointly sponsored by the United States.Air force.Office of scientific research Academic press,New York, 1965
Cav 7.2354

HIGH PURITY WATER CORROSION OF METALS seven papers presented at symposia San Francisco 1959 Mar 17-21 Dallas 1960 Mar 14-18 National association of corrosion engineers.Technical committee 3-F on corrosion by high-purity water National association of corrosion engineers,Houston,Texas,1968
Met 25.1933

HIGH SPEED COMPUTER CONFERENCE,1955 Automatic coding for digital computers :a talk presented at the High speed computer conference,Louisiana state university, 16 February 1955 By Grace Murray Hopper 5p 1955
Math L 5.1768

HIGH-SPEED PHOTOGRAPHY International congress on high-speed photography 3rd proceedings London 1956 Sep 10-15 Edited by R.B. Collins Held under the auspices of the Department of scientific and industrial research Butterworths,London,1957
Eng 41.0439

HIGH SPEED TESTING Vol 1: a symposium held at Boston,Massachusetts,December 8,1958 Sponsored by Plas-tech equipment corporation Interscience,New York,1968
Eng 41.3845

HIGH SPEED TESTING Vol 3: third annual symposium held at Boston,Massachusetts,October 26 and 27,1961 Sponsored by Plas-tech equipment corporation Interscience,New York, 1962
Eng 41.3848

HIGH SPEED TESTING Vol 5: 5th international symposium held at Boston, Massachusetts,March 8 and 9,1965 Sponsored by Plas-tech equipment corporation Applied polymer symposia, 1 Interscience,New York, 1965
Eng 41.3849

HIGH STRENGTH BOLTS Institution of structural engineers jubilee symposium on high strength bolts Proceedings London 1959 Jun 10-11 Institution of structural engineers Institution of structural engineers,London,1959
Eng 41.2696

HIGH STRENGTH STEELS FOR AIRCRAFT Dallas, Texas 1958 May 12-16 American society for metals American society for metals,Cleveland, Ohio,1958
Met 25.1002

HIGH TEMPERATURE Kinetics of high-temperature processes :conference Cambridge, Mass. 1958 Jun 23-27 Edited by W.D. Kingery Supported in part by United States air force.Aeronautical research laboratory Technology press books in science and engineering Massachusetts institute of technology press,Cambridge,Mass.,1959
Min 10.1127

HIGH TEMPERATURE MATERIALS Plansee seminar Hochtemperatur werkstoffe 6th Papers Reutte,Austria 1966 Jun 24-26 Metallwerk Plansee A.G. Edited by F. Benesovsky Springer,Vienna;New York,1969 Papers in English,French and German
Met 25.2336

HIGH TEMPERATURE MATERIALS :a conference Cleveland,Ohio 1957 Apr 16-17 American institute of mining,metallurgical and petroleum engineers.High temperature alloys committee Edited by R.F. Hehemann and G. Mervin Ault John Wiley;Chapman and Hall,New York;London,1959
Met 25.1004

HIGH TEMPERATURE MATERIALS 11 a technical conference Cleveland,Ohio 1961 Apr 26-27 American institute of mining, metallurgical and petroleum engineers.High temperature alloys committee Metallurgical society conferences, 18 Interscience,New York;London,1963
Met 25.1006

HIGH TEMPERATURE PHENOMENA Combustion and propulsion:high temperature phenomena Agard colloquium 5th Papers Brunswick 1962 Apr 9-13 Agard Edited by R.P. Hagerty and others Pergamon press,London,1963
Eng 41.7245

HIGH-TEMPERATURE PROPERTIES OF STEELS :joint conference Proceedings Eastbourne 1966 Apr 4-7 British iron and steel research association Iron and steel institute Iron and steel institute.Publications, 97 Iron and steel institute,London,1967
Eng 41.3827

HIGH TEMPERATURE REFRACTORY METALS :a symposium New York 1964 Feb 16-20 1965 Feb Pt 1-2 American institute of mining,metallurgical and petroleum engineers.Refractory metals committee Metallurgical society conferences, 34 2 vols Gordon and Breach,New York,1966-68 Conference pt 1 1965,pt 2 1964
Met 25.1016

HIGH TEMPERATURE STRUCTURES AND MATERIALS Symposium on naval structural mechanics 3rd Proceedings New York 1963 Jan 23-25 Edited by A.M. Freudenthal and others Sponsored by the United States.Office of naval research United States.Office of naval research.Structural mechanics series Pergamon press,London,1964
Eng 41.3806

HIGH TEMPERATURE STRUCTURES AND MATERIALS Symposium on naval structural mechanics Proceedings New York 1963 Jan 23-25 United States.Office of naval research Edited by A.M. Freudenthal and others Pergamon,Oxford,1964
Met 25.1009

HIGH TEMPERATURE TECHNOLOGY An International symposium on high temperature technology 3rd Proceedings Pacific Grove,Calif. 1963 Sep 8-11 International union of pure and applied chemistry.Commission on high temperatures and refractories Organized and directed by Stanford research institute Butterworths,London,1964
Met 25.0992

HIGH TEMPERATURES Conference on extremely high temperatures Boston,Mass. 1958 Mar 18-19 Air force Cambridge research center Edited by Heinz Fischer and Lawrence C. Mansur Wiley;Chapman and Hall,New York;London,1958
Met 25.0249

HIGH VIEW,N.Y. 1959 Quantum electronics Conference on quantum electronics - resonance phenomena Edited by Charles H. Townes Columbia university press,New York,1960
Eng 41.5411

HIGH-ALLOY STEELS Metallurgical developments in high-alloy steels A Joint conference on high-alloy steels Proceedings Scarborough 1964 Jun 2-4 Iron and steel institute and British iron and steel research association Iron and steel institute.Special report, 86 Iron and steel institute,London,1964
Met 25.0488

HIGH-ENERGY ELECTRONS Symposium on high-energy electrons Proceedings Montreux 1964 Sep 7-11 European association of radiologists Edited by A. Zuppinger and G. Poretti Springer,Berlin,1965
Chem 18.0123

HIGH-MELTING METALS Plansee seminar "De re metallica" 3rd Papers Reutte,Austria 1948 Jun 22-26 Metallwerk Plansee AG Edited by F. Benesovsky Pergamon press; Metallwerk Plansee AG;,,London;Reutte,1959
Met 25.1010

HIGH-PURITY WATER CORROSION Symposium on high-purity water corrosion Atlantic City,N. J. 1955 Jun 28 American society for testing materials A.S.T.M.Special technical publication, 179 American society for testing materials,Philadelphia,Pa.,1955
Met 25.1927

HIGH-SPEED PHOTOGRAPHY International congress on high-speed photography 3rd Proceedings London 1956 Sep 10-15 Edited by R.B. Collins Held under the auspices of Great Britain.Department of scientific and industrial research xvi,417p Butterworths scientific publications,London,1957
Chem E 24.0452

HIGH-STRENGTH MATERIALS Berkeley international materials conference :high strength materials- present status and anticipated developments 2nd Proceedings Berkeley 1964 Jun 15-18 Edited by Victor F. Zackay Inorganic materials research division series John Wiley,New York,1965
Met 25.1019

HIGH-STRENGTH STEELS Conference on high-strength steels... Papers and discussions Harrogate 1962 May 23-24 Iron and steel institute Iron and steel institute.Special report, 76 Iron and steel institute,London, 1962
Met 25.0486

HIGH-STRENGTH STEELS FOR THE MISSILE INDUSTRY Golden Gate metals conference Proceedings San Francisco,Calif. 1960 Feb 4 American society for metals Edited by H.T. Sumsion American society for metals,Novelty,Ohio,1961
Met 25.0487

HIGH-TEMPERATURE HIGH-RESOLUTION METALLOGRAPHY a symposium Proceedings Chicago,Ill. 1965 Feb 15 American institute of mining, metallurgical and petroleum engineers.Ferrous metallurgy committee Edited by Hubert I. Aaronson and George S. Ansell Metallurgical society conferences, 38 Gordon and Breach, New York,1967
Met 25.2361

HIGH-TEMPERATURE PROPERTIES OF STEELS the joint conference Proceedings Eastbourne 1966 Apr 4-7 British iron and steel research association and Iron and steel institute Iron and steel institute.Publication, 97 Eyre and Spottiswoode,Margate,1967
Met 25.0429

HIGH-TEMPERATURE STEELS AND ALLOYS FOR GAS TURBINES :a symposium London 1951 Feb 21-22 Iron and steel institute Iron and steel institute.Special report, 43 Iron and steel institute,London,1952
Met 25.0996

HIGH-TEMPERATURE TECHNOLOGY International symposium on high temperature technology 2nd Proceedings Asilomar,Calif 1959 Oct 6-9 Stanford research institute McGraw-Hill,New York,1960
Met 25.2535

HIGHLAND PARK,ILL. 1953 Basic mechanisms in radiobiology.II.Physical and chemical aspects :an informal conference Proceedings Edited by John L. Magee and others Organized by the National research council.Subcommittee on radiobiology National research council. Publications, 305 National academy of science,Washington,D.C.,1953
Radioth 35.1546

HIGHLAND PARK,ILL. 1957 Research in radiology :an informal conference Proceedings Edited by Henry S. Kaplan Organized by the National research council. Subcommittee on radiobiology National research council.Nuclear science series.Report, 22 National research council.Publication, 571 National academy of science,Washington,D. C.,1958
Radioth 35.1116

HIGHLAND PARK,ILL. 1964 Clinical electroretinography :international symposium 3rd Proceedings International society for clinical electroretinography Edited by Hermann M. Burian and Jerry Hart Jacobsen Pergamon press.Symposium publication division, Oxford,1966 Supplement to 'Visual research'
Psy 31.0383

HIGHLAND PARK,ILL. 1964 Clinical electroretinography international symposium 3rd Proceedings International society for clinical electroretinography Edited by Hermann M. Burian and Jerry Hart Jacobson Vision research.Supplement Pergamon,Oxford, 1966
PGMS 29.0340

HIGHLIGHTS OF ASTRONOMY IAU general assembly 13 Invited discources and discussions,and special meetings Prague 1967 International astronomical union Edited by Lubos Perck 548p Reidel,Dordrecht,1968
Obs 6.3528

HIGHLIGHTS OF ASTRONOMY International astronomical union 13th general assembly Prague 1967 Aug Edited by L. Perek illus x,548p 25cm Reidel,Doldrecht,1968
A Math 4.1899

HIGHLIGHTS OF ASTRONOMY International astronomical union 14th general assembly Brighton 1970 Aug Vol 2 International astronomical union Edited by C. de Jager 793p Reidel,Dordrecht,1971
Obs 6.3688

HIGHLIGHTS OF ASTRONOMY International astronomical union.General assembly 14th Invited discourses Brighton 1970 Aug Vol 2 International astronomical union Edited by C.de Jager 793p Reidel,Dordrecht, 1971
TA 15.0655

HIGHWAY NEEDS OF GREAT BRITAIN Conference on the highway needs of Great Britain Proceedings London 1957 Nov 13-15 Institution of civil engineers Institution of civil engineers,London,1958
Eng 41.3260

HIGHWAY RESEARCH BOARD Frost action in soils a symposium 30th Annual meeting Washington 1951 Jan 9-12 National research council.Publication,213 Highway research board.Special report 2 illus,tables,diagrs x,385p 25cm Highway research board, Washington,D.C.,1952
Sco 14.0395

HIGHWAY RESEARCH BOARD Pavement design in frost areas :Annual meeting 42nd eleven reports Washington,D.C. 1963 Jan 7-11 2: design considerations By James F. Haley and others National research council.Publication, 1153,Highway research record,33 illus 270p 25cm Washington,D.C.,1963
Sco 14.0935

HII REGIONS Interstellar ionized hydrogen :a symposium Charlottesville,Va. 1967 Dec National radio astronomy observatory Edited by Yervant Terzian Jointly sponsored by Arecibo ionospheric observatory 774p Benjamin,New York,1968
TA 15.0437

HISTAMINE London 1955 Apr 6-7 Ciba foundation Edited by G.E.W. Wolstenholme and Cecilia M. OConnor Churchill,London,1956
Pha 16.0076

HISTOCHEMISTRY International congress of histochemistry and cytochemistry 3rd Summary reports New York 1968 Aug 18-22 Springer,New York,1968
Radioth 35.0547

HISTOCHEMISTRY International congress of histochemistry and cytochemistry 3rd summary reports New York 1968 Aug 18-22 United States.Office of the secretariat 317p Springer-verlag,New York,1968
Bot 42.0703

HISTOCHEMISTRY International congress of histochemistry and cytochemistry 3rd Summary reports New York 1968 Aug 18-22 Springer,New York,1968
An 32.0438

HISTOCHEMISTRY AND CYTOCHEMISTRY International congress of histochemistry and cytochemistry 1st Proceedings Paris 1960 Edited by R. Wegmann Pergamon,London, 1963
An 32.3391

HISTOCOMPATIBILITY TESTING 1967 Conference and workshop on histocompatibility testing 3r Report Turin 1967 Jun 14-24 St. Vincent Edited by E.S. Curtoni and others Sponsored by the Consiglio nazionale delle ricerche Munksgaard,Copenhagen,1967
Surg 23.0032

HISTONE BIOLOGY Nucleohistones The World conference on histone biology and chemistry 1st Papers 1963 Apr 29-May 2 Edited by James Bonner and Paul Ts'o Holden-Day,San Francisco,1964
Bioch 33.0808

HISTONE BIOLOGY AND CHEMISTRY Nucleohistones The World conference on histone biology and chemistry 1st 1963 Apr 29 - May 3 Edited by James Bonney and Paul Ts'o Holden Day,San Francisco,Calif.,1964
Radioth 35.0121

HISTONE BIOLOGY AND CHEMISTRY Nucleohistones World conference on histone biology and chemistry 1st 1963 Apr 29-May 3 Edited by James Bonner and Paul Ts'o Holden-Day,San Francisco,Calif.,1964
Gen 34.0684

HISTONES :their role in the transfer of genetic information London 1965 Dec 3 Ciba foundation Edited by A.V.S. De Reuck and Julie Knight Ciba foundation study group, 24 Little,Brown and company,Boston,1966 In honour of W.A.Engelhardt
Bioch 33.0888

HISTORICAL ASPECTS OF MICROSCOPY Royal microscopical society conference Papers Oxford 1966 Mar 18 Royal microscopical society Edited by S. Bradbury and G.L'E. Turner viii,227p Cambridge,1967
WSM 43.2520

HISTORICAL ASPECTS OF MICROSCOPY Papers Oxford 1966 Mar 18 Royal microscopical society Edited by S. Bradbury and G.L'E. Turner W.Heffer,Cambridge,1967
Bal 39.0241

The HISTORICAL DEVELOPMENT OF PHYSIOLOGICAL THOUGHT :a symposium New York 1956-57 Edited by Chandler McC. Brooks and Paul F. Cranefield Hafner,New York,1959
Phys 20.0844

HISTORICAL SCIENCES Anglo-Romanian conference on mathematics in the archaeological and historical sciences Proceedings Royal society of London Academia republicii socialiste romania Edited by F.R. Hodson and others viii,565p 24cm Edinburgh university press,Edinburgh, 1971
Math S 3.1826

The HISTORY AND PHILOSOPHY OF KNOWLEDGE OF THE BRAIN AND ITS FUNCTIONS :an Anglo-American symposium London 1957 Jul 15-17 Edited by F.N.L. Poynter Sponsored by Wellcome historical medical library Blackwell,Oxford,1958
An 32.1649

HISTORY AND PHILOSOPHY OF SCIENCE Concept and the role of the model in mathematics and natural and social sciences International union of history and philosophy of science colloquium Proceedings Utrecht 1960 International union of history and philosophy of science.Division of philosophy of science Synthese library 194p Reidel,Dordrecht, 1961 Colloquium organized by H. Freudenthal
WSM 43.0785

HISTORY OF GERMAN GUIDED MISSILES DEVELOPMENT AGARD seminar Munich 1956 Apr Agard Edited by T. Benecke and A.W. Quick AGARDograph, 20 1957
Eng 41.6877

HISTORY OF METALLURGY The Sorby centennial symposium on the history of metallurgy :a symposium Cleveland 1963 Oct 22-23 American society for metals and American institute of mining,metallurgical and petroleum engineers.Metallurgical society Edited by Cyril Stanley Smith Sponsored by the Society for the history of technology Metallurgical society conferences, 27 Gordon and Breach,New York,1965
Met 25.2141

HISTORY OF SCIENCE Congres international d'histoire des sciences 7e Actes Jerusalem 1953 Aug 4-12 International union of history and philosophy of science Edited by F.S. Bodenheimer Academie internationale d'histoire des sciences. Collection de travaux, 8 xii,662p Academie internationale d'histoire des sciences;Hermann,Paris,c1954
WSM 43.0058

HISTORY OF SCIENCE Congres international d'histoire des sciences 8e Actes Florence 1956 Sep 3-9 and Milan Vol 1-3 International union of the history and philosophy of science Academie internationale d'histoire des sciences. Collection de travaux, 9 3 vols Gruppe italiano di storia delle scienze;Hermann,Vinci (Florence);Paris,1958
WSM 43.0057

HISTORY OF SCIENCE Congres international d'histoire des sciences 9e Actes Barcelona 1959 Sep 1-7 and Madrid Vol 1 Union internationale d'histoire des sciences Academie internationale d'histoire des sciences.Collection de travaux, 12 733p Asociacion para la historia de ciencia espanola;Hermann,Barcelona;Paris,1960
WSM 43.0061

HISTORY OF SCIENCE Congres international d'histoire des sciences 11e Actes Warsaw 1965 Aug 24-31 Cracow 1-6 Polska akademia nauk.Comite et institut d'histoire de la science et de la technique 6 vols Ossolineum.Maison d'edition de l'Academie polonaise des sciences,Wroclaw,1968
WSM 43.0079

HISTORY OF SCIENCE Congres international d'histoire des sciences 12e Actes Paris 1968 Tome 1-12 Union internationale d'histoire et de philosophie des sciences Blanchard,Paris,1970-71 President:J. Rostand
WSM 43.5592

HISTORY OF SCIENCE Congres international d'histoire des sciences 5e Actes Lausanne 1947 Sep 30-Oct 6 Union internationale d'histoire des sciences Academie internationale d'histoire des sciences.Collections de travaux, 2 288p Academie internationale d'histoire des sciences;Hermann,Paris,1948
WSM 43.0060

HISTORY OF SCIENCE Congres international d'histoire des sciences et congres de la Societe internationale d'histoire de la medecine Actes Amsterdam 1950 Aug 14-21 Union internationale d'histoire des sciences Genootschap voor geschiedenis der geneeskunde, wiskunde en natuurweteenschappen te Leiden Academie internationale d'histoire des sciences.Collection de travaux, 6 2 vols Academie d'histoire des sciences;Hermann,Paris, 1951-53 The 12th congress of the Societe internationale d'histoire de la medecine is section 4 of the 6th Congres international d'histoire des sciences
WSM 43.0067

HISTORY OF SCIENCE International congress of the history of science 10th Proceedings Ithaca,N.Y. 1962 Aug 8-Sep 2 Vol 1-2 Academie internationale d'histoire des sciences History of science society Co-sponsored by the United States national academy 2 vols Hermann,Paris Congress chairman:H.Guerlac
WSM 43.0030

HISTORY OF SCIENCE Scientific change Symposium on the history of science Papers Oxford 1961 Jul 9-15 Edited by A.C. Crombie Under the auspices of the International union of the history and philosophy of science.Division of history of science Heinemann,London,1963
Geog 13.0053

HISTORY OF SCIENCE Scientific change... Historical studies in the intellectual,social and technical conditions for scientific discovery... :symposium on the history of science Oxford 1961 Jul 9-15 International union of history and philosophy of science.Division of history of science Edited by A.C. Crombie xii,896p Heinemann, London,1963
WSM 43.0029

HISTORY OF SCIENCE Symposium international d'histoire des sciences Actes Turin 1961 Jul 28-30 Comite national des celebrations du 1er centenaire de l'unite d'Italie With the participation of the Union internationale d'histoire et philosophie des sciences Academie internationale d'histoire des sciences.Collections de travaux, 14 189p Gruppe italiano di storia delle science,Vinci (Florence),1964
WSM 43.0059

HISTORY OF SCIENCE Symposium international d'histoire des sciences Actes Vinci (Florence) 1960 Oct 8-10 International union of of the history and philosophy of science Academie internationale d'histoire des sciences.Collection de travaux, 13 235p Gruppe italiano di storia delle scienze,Vinci (Florence),1962
WSM 43.0063

HISTORY OF SCIENCE SOCIETY International congress of the history of science 10th Proceedings Ithaca,N.Y. 1962 Aug 8-Sep 2 Vol 1-2 Co-sponsored by the United States national academy 2 vols Hermann, Paris Congress chairman:H.Guerlac
WSM 43.0030

HIXON FUND COMMITTEE Cerebral mechanisms in behaviour :the Hixon symposium Pasadena, Calif. 1948 Sep Edited by Lloyd A. Jeffress Wiley;Chapman and Hall,New York; London,1951
An 32.4627

HIXON SYMPOSIUM Cerebral mechanisms in behaviour :the Hixon symposium Report Pasadena,Calif. 1948 Sep 20-25 California institute of technology.Hixon committee Edited by Lloyd A. Jeffress Wiley;Chapman and Hall,New York;London,1951
Psy 31.0522

HOBART 1956 Continental drift University of Tasmania.Geology department :symposium 2nd Papers Edited by S.Warren Carey University of Tasmania,Hobart,1958
Min 10.0561

HOBART 1957 Dolerite University of Tasmania.Geology department :symposium 4th Hobart,1958 Typescript
Min 10.0334

HOBART,TASMANIA 1956 Continental drift :a symposium Papers University of Tasmania Edited by S.Warren Carey University of Tasmania.Geology department.Symposia, 2 Hobart,1958
Geod 9.0083

HOBART COLLEGE The International conference on the electronic properties of metals at low temperatures Report Geneva,N.Y. 1958 Aug 25-29 Co-sponsored by General electric research laboratory 244p c1958 Privately printed report to the sponsors
Cav 7.0895

HOBOKEN,N.J. 1957 The Many-body problem :a symposium Proceedings Stevens institute of technology Edited by Jerome K. Percus Interscience,New York;London,1963
Chem 18.2362

HOFFMAN-LAROCHE AND COMPANY,BASLE Chemotherapy of cancer International symposium on the chemotherapy of cancer Proceedings Lugano 1964 Apr 28-May 1 Edited by Placidus Plattner Organised by the Swiss academy of medical sciences Elsevier, Amsterdam,1964 Sponsored by F.Hoffman-Laroche and co.ltd.
Radioth 35.0837

HOFFMANN-LA ROCHE AND COMPANY Challenge of life:biomedical progress and human values Roche anniversary symposium Proceedings Basle 1971 Aug 31-Sep 3 Edited by Robert M. Kunz and Hans Fehr Experientia.Supplementum, 17 456p 24cm Birkhauser,Basle,1972 Symposium held on the occasion of the firm's 75th anniversary.Chairman:Lord Todd
Chem 18.2902

HOFFMANN-LA ROCHE AND CO LTD. International symposium on chemotherapy of cancer Proceedings Lugano 1964 Apr 28-May 1 Swiss academy of medical sciences Edited by Placidus A. Plattner Elsevier,Amsterdam,1964
Pha 16.0335

HOKKAIDO UNIVERSITY.INSTITUTE OF LOW TEMPERATURE SCIENCE Cellular injury and resistance in freezing organisms International conference on low temperature science 2nd Proceedings Sapporo 1966 Aug 14-19 Vol 2: conference on cryobiology Edited by Eizo Asahina illus Hokkaido university. Institute of low temperature science,Sapporo, 1967 Title page in English and Japanese Conference held in commemoration of the 25th anniversary of the establishment of the Institute
Path 30.2625

HOLOGRAPHY The Engineering uses of holography :a symposium Proceedings Glasgow 1968 Sep 17-20 University of Strathclyde Edited by E.R. Robertson and J.M. Harvey Organized in association with the National physical laboratory Cambridge university press,Cambridge,1970
Eng 41.4290

HOLOGRAPHY The Engineering uses of holography :a symposium Proceedings Glasgow 1968 Sep 17-20 University of Strathclyde National physical laboratory Edited by Elliot R. Robertson and James M. Harvey Cambridge university press,Cambridge, 1970
An 32.5364

HOLOMORPHIC MAPPINGS AND MINIMAL SURFACES Carolina conference Proceedings Chapel Hill,N.C. 1970 Mar 30-Apr 3 University of North Carolina.Department of mathematics Edited by Donald A. Eisenman and Laird E. Taylor 107p 28cm University of North Carolina,Chapel Hill,N.C.,1970
P Math 2.4356

HOMEOSTASIS AND FEEDBACK MECHANISMS Cambridge 1963 Sep Society for experimental biology Society for experimental biology.Symposia, 18 Cambridge university press,Cambridge,1964
Phys 20.0962

HOMEOSTATIC REGULATORS :a Ciba foundation symposium held jointly with the Wellcome trust London 1969 Jan 28-30 Ciba foundation and Wellcome trust Edited by G.E.W. Wolstenholme and Julie Knight Churchill, London,1969
Phys 20.2257

HOMEOSTATIC REGULATORS :a Ciba foundation symposium Proceedings London 1969 Jan 28-30 Ciba foundation Wellcome trust Edited by G.E.W. Wolstenholme and Julie Knight Churchill,London,1969
Bioch 33.0934

HOMOGENEOUS CATALYSIS with special reference to hydrogenation and oxidation Liverpool 1968 Faraday society Faraday society. Discussions, 46 diagrms,graphs 227p 24cm Faraday society,London,1968
Cav 7.3125

HOMOTOPIE ET COHOMOLOGIE manuscript Montreal 1964 By Beno Eckmann North Atlantic treaty organization Societe mathematique du Canada University of Montreal.Seminaire de mathematiques superieures, 11 129p 28cm Universite de Montreal,Montreal,1965
P. Math 2.0297

HOMOTRANSPLANTATION:KIDNEY AND OTHER TISSUES Annual conference on the kidney 16th Proceedings 1964 National kidney disease foundation Edited by Jack Metcoff National kidney disease foundation,New York,1966
Inv Med 37.0160

HONEY HABOR,ONT. 1954,1958 Canadian cancer conference Canadian cancer research conference 1st,3rd Proceedings Edited by R.W. Begg illus. 2 vols Academic press,New York,1955,1959
Radioth 35.1958

HONEY HARBOR,ONT. 1962 Canadian cancer conference :Canadian cancer research conference 5th Proceedings National cancer institute of Canada Academic press, New York;London,1963
Gen 34.1993

HONG KONG 1961 Symposium on muscle receptors Proceedings Edited by David Barker Held as part of the Golden jubilee congress of the University of Hong Kong Hong Kong university press,Hong Kong,1962
An 32.4002

HONG KONG 1961 Symposium on muscle receptors :a meeting held...as part of the Golden jubilee congress of the University of Hong Kong Proceedings University of Hong Kong Edited by David Barker Hong Kong university press,Hong Kong,1962
Phys 20.1559

HONG KONG 1961 Symposium on muscle receptors :a meeting held...as part of the golden jubilee congress of the University of Hong Kong Proceedings Edited by David Barker Financial support provided by the University of Hong Kong Hong Kong university press,Hong Kong,1962
Bal 39.1519

HONG KONG 1961 Symposium on the design of high buildings Proceedings of a meeting... held as part of the Golden jubilee congress of the University of Hong Kong University of Hong Kong Edited by Sean Mackey Hong Kong university press,Hong Kong,1962
Eng 41.2685

HONOLULU 1957 World orchid conference 2nd Proceedings Sponsored by the University of Hawaii American orchid society, Cambridge,Mass.,1958
BG 38.3117

HONOLULU 1961 Man's place in the island ecosystem Pacific science congress 10th Symposium Edited by F.R. Fosberg Bishop museum press,Honolulu,1963
Geog 13.1311

HONOLULU 1961 Pacific basin biogeography Pacific science congress 10th Symposium Edited by J.Linsley Gressitt and others Sponsored by the National academy of science Honolulu,1963
Geog 13.1161

HONOLULU 1961 Pacific basin biogeography Pacific science congress :a symposium 10th Proceedings Edited by J.Linsley Gressitt illus,maps x,563p 26cm Bishop museum press,Honolulu,1963
Sco 14.2205

HONOLULU 1961 Pacific island terraces Pacific science congress 10th Pacific science association Edited by Richard J. Russell Annals of geomorphology.New series, 3 Borntraeger,Berlin,1961
Bot 42.3310

HONOLULU 1961 Pacific island terraces: eustatic? Pacific science congress 10th Symposium Edited by Richard J. Russell Annals of geomorphology.Ns.Supplementband,3 Borntraeger,Berlin,1961
Geog 13.0372

HONOLULU 1961 Pacific science congress 10th Abstracts of papers Pacific science association 487p 27cm Honolulu,1961
Sco 14.0136

HONOLULU 1961 Pacific science congress 10th papers the crust of the Pacific basin Edited by Gordon A. Macdonald and Hisashi Kuno Geophysical monographs,6 American geophysical union,Washington,D.C.,1962
Min 10.0421

HONOLULU 1961 Pacific science congress 10th The Matthew Fontaine Maury memorial symposium :papers presented to the tenth Pacific science congress of the Pacific science association 10th Papers Pacific science association,and others Edited by Harry Wexler and others Geophysical monograph,7 American geophysical union. Publication,1036 illus,maps x,228p 25cm American geophysical union,Washington,D.C., 1962
Sco 14.6079

HONOLULU 1961 Plants and the migrations of Pacific peoples :a symposium Bishop (Bernice P) museum,and,University of Hawaii Edited by J. Barrau Sponsored by the National science foundation Bishop museum press,Honolulu,1963 Symposium convened at the 10th Pacific science congress
Geog 13.7202

HONOLULU 1961 The Crust of the Pacific basin :two symposia,presented at the 10th Pacific Edited by Gordon A. MacDonald and Hisashi Kuno Geophysical monograph, 6 National research council.Publication,1035 Washington,D.C.,1961 Based on the papers presented at the symposia
Geod 9.0423

HONOLULU 1961 Venomous and poisonous animals and noxious plants of the Pacific region :a collection of papers based on a symposium in the Public health and medical science division at the tenth Pacific science congress Papers Pacific science association and University of Hawaii Edited by Hugh L. Keegan and W.V. MacFarlane Co-sponsored by the United States.National academy of sciences Pergamon press.Symposium publications division,Oxford,1963
Chem 26.0448

HONOLULU 1967 The Zodiacal light and the interplanetary medium :a conference Papers National aeronautics and space administration. Office of technology utilization NASA SP-150 430p n.a.s.a.,washington,D.C.,1967
Obs 6.3425

HONOLULU 1968 Methodologies of pattern recognition International conference on methodologies of pattern recognition Proceedings University of Hawaii Sponsored by the United States.Air force. Office of scientific research bibliog.,illus. Academic press,New York;London,1969
Math S 3.1521

HONOLULU 1968 Methodologies of pattern recognition International conference on methodologies of pattern recognition Proceedings Edited by Satosi Watanabe Academic press,New York;London,1969
An 32.4412

HOPITAL DE LA SALPETRIERE.CLINIQUE DES MALADES DU SYSTEME NERVEUX Les Grandes activites du lobe temporal :semaine neurophysiologique: conferences du colloque Paris 1953 Edited by Th. Alajouanine Actualites neuro-physiologiques Masson,Paris,1955
Psy 31.0464

HOPITAL DE LA SALPETRIERE.CLINIQUE DES MALADIES DU SYSTEME NERVEUX La Douleur et les douleurs :a symposium Paris 1955 Edited by Th. Alajouanine and others Masson,Paris, 1957
Psy 31.0214

HOPKINS MARINE STATION Contributions to marine biology :lectures and symposia given at the Hopkins marine station...at The midwinter meeting of the Western society of naturalists Pacific Grove,Calif. 1929 Dec 20-21 Western society of naturalists Stanford university press,Stanford,Calif.,1930
Bal 39.1933

HORMONAL CONTROL Symposium on hormonal control of protein biosynthesis Research conference for biology and medicine 18th Gatlinburg,Tenn. 1965 Apr 5-8 Sponsored by the Oak Ridge national laboratory.Biology division Journal of cellular and comparative physiology, 66,suppl.1 Wistar institute of anatomy and biology,Philadelphia,Pa.,1965
Bioch 33.0520

HORMONAL CONTROL OF PROTEIN BIOSYNTHESIS Symposium on hormonal control of protein biosynthesis Gatlinburg,Tenn. 1965 Apr 5-8 Oak Ridge national laboratory.Biology division Oak Ridge national laboratory.Annual research conference, 18 Wistar institute of anatomy and biology,Philadelphia,Pa.,1965
Gen 34.0550

HORMONAL CONTROL SYSTEM :a symposium Proceedings Rancho Santa Fe,Calif. 1967 Oct 20-22 Edited by Edwin B. Stear and Arnold H. Kadish Mathematical biosciences. Supplement, 1 illus. American Elsevier, New York,1969
Math S 3.1543

HORMONAL FACTORS IN CARBOHYDRATE METABOLISM :a colloquium Proceedings London 1952 Jun 30-Jul 3 Ciba foundation Edited by G.E. W. Wolstenholme Ciba foundation colloquia on endocrinology, 6 illus Churchill,London, 1953
Bioch 33.0436

HORMONAL INFLUENCES IN WATER METABOLISM
Anterior pituitary secretion and Hormonal influences in water metabolism :two colloquia Proceedings London 1951 Jul 9-13 and London 1951 Jan 8-10 Ciba foundation Edited by G.E.W. Wolstenholme and Margaret P. Cameron Ciba foundation colloquia on endocrinology, 4 illus Churchill,London, 1952
Bioch 33.0434

HORMONAL REGULATION OF ENERGY METABOLISM :a conference Proceedings 1956 Feb 5 Edited by Laurance W. Kinsell Charles C. Thomas,Springfield,Ill.,1957
Bioch 33.0461

HORMONAL STEROIDS International congress of hormonal steroids :biochemistry, pharmacology and therapeutics 1st Proceedings Milan 1964 Vol 1-2 Edited by L. Martini and A. Pecile 2 vols Academic press,New York; London,1964-65
Gen 34.0546

HORMONAL STEROIDS International congress on hormonal steroids 2nd Proceedings Milan 1966 May 23-28 Edited by L. Martini and others Excerpta medica.International congress series, 132 Excerpta medica, Amsterdam,1967
Inv Med 37.0116

HORMONAL STEROIDS International congress on hormonal steroids 3rd Abstracts of papers Hamburg 1970 Sep 7-12 Edited by V.H.T. James Excerpta medica.International congress series, 210 Excerpta medica,Amsterdam,1970
Inv Med 37.0115

HORMONAL STEROIDS:BIOCHEMISTRY,PHARMACOLOGY AND THERAPEUTICS International congress on hormonal steroids 1st Proceedings Vol 1-2 Edited by L. Martini and A. Pecile 2 vols Academic press,New York,1964-65
Inv Med 37.0154

HORMONE ACTION Mechanisms of hormone action : a Nato advanced study institute Discussions Meersburg 1964 North Atlantic treaty organization Edited by P. Karlson Academic press,New York,1965
Bioch 33.0469

HORMONE ACTION The Biochemistry of steroid hormone action London 1970 Mar Biochemical society Edited by R.M.S. Smellie Biochemical society.Symposium, 32 Academic press,London,1971
Bioch 33.2316

HORMONE PRODUCTION IN ENDOCRINE TUMOURS :a colloquium Proceedings Ciba foundation Edited by G.E.W. Wolstenholme and Maeve O'Connor Ciba foundation colloquia on endocrinology, 12 Little,Brown,Boston,1958
Inv Med 37.0242

HORMONE RESEARCH Recent progress in hormone research Laurentian hormone conference Proceedings Franconia,N.H. 1949 Sep Vol 5 Edited by Gregory Pincus Academic press,New York,1950
Bal 39.1376

HORMONES The Laurentian hormone conference 27th Proceedings Mount Tremblant 1969 Aug 24-29 Edited by E.B. Astwood Recent progress in hormone research, 26 Academic press,New York;London,1970
Inv Med 37.0265

HORMONES The Relation of hormones to development :a symposium Papers Cold Spring Harbor 1942 Cold Spring Harbor biological laboratory Cold Spring Harbor symposia on quantitative biology, 10 Long Island biological association,Cold Spring Harbor,1942
Bioch 33.1266

HORMONES,BRAIN FUNCTION AND BEHAVIOUR :a conference on neuroendocrinology Proceedings New York 1956 May Edited by Hudson Hoagland Academic press,New York,1957
An 32.3748

HORMONES,PSYCHOLOGY AND BEHAVIOUR,AND STEROID HORMONE ADMINISTRATION :two colloquia Proceedings Ciba foundation Edited by G. E.W. Wolstenholme Ciba foundation colloquia on endocrinology,3 Illus. Churchill,London, 1952
VA 19.0225

HORMONES,PSYCHOLOGY AND BEHAVIOUR,AND STEROID HORMONE ADMINISTRATION :two colloquia Proceedings London 1951 Apr 9-12 and London 1950 Feb 23-24 Ciba foundation Edited by G.E.W. Wolstenholme and Margaret P. Cameron Ciba foundation colloquia on endocrinology, 3 illus Churchill,London, 1952
Bioch 33.0433

HORMONES AND ATHEROSCLEROSIS :a conference Proceedings Brighton,Utah 1958 Mar 11-14 National institutes of health. Endocrinology study section Edited by Gregory Pincus Academic press,New York,1959
Bioch 33.0477

HORMONES AND THE AGING PROCESS :a conference Proceedings New York 1955 May 30-31 Edited by Earl T. Engle and Gregory Pincus Academic press,New York,1956
An 32.3738

HORMONES AND THE IMMUNE RESPONSE :Ciba foundation study group London 1970 May 1 Ciba foundation Edited by G.E.W. Wolstenholme and Julie Knight Ciba foundation study group, 36 Churchill,London, 1970
An 32.5453

HORMONES AND THE KIDNEY Society for endocrinology :a meeting 89th Proceedings Cambridge 1962 Sep Society for endocrinology Edited by Peter C. Williams Society for endocrinology.Memoirs, 13 Academic press,London;New York,1963
Inv Med 37.0233

HORMONES AND THE KIDNEY :89th meeting Proceedings Cambridge 1962 Sep Society for endocrinology Edited by Peter C. Williams Society for endocrinology.Memoirs, 13 Academic press,London;New York,1963
Gen 34.0557

HORMONES IN BLOOD :a colloquium Proceedings London 1950 Ciba foundation Edited by G.E.W. Wolstenholme and Elaine C.P. Millar Ciba foundation colloquia on endocrinology, 11 Churchill,London,1957
Inv Med 37.0241

HORMONES IN FISH :a symposium Proceedings London 1959 Oct 13 Zoological society of London Edited by I.Chester Jones Zoological society of London.Symposia, 1 Zoological society of London,London,1960
An 32.3763

Les HORMONES SEXUELLES colloque international Comptes-rendus Paris 1937 Jun 10-19 Edited by L. Brouha Under the auspices of Fondation Singer-Polignac Conferences du college de France Hermann,Paris,1938
Gen 34.1603

HORNSEY COLLEGE OF ART Conference on the teaching of engineering design Papers and summary Scarborough 1964 Apr 1-4 Organised by the Enfield college of technology Illus Institution of engineering designers, London,1964 Co-organized by the Institution of engineering designers and the Hornsey College of art
Eng 41.0309

HORTICULTURE International horticultural congress 9th Reports,Proceedings,Program London 1930 Aug 7-15 Royal horticultural society Under the auspices of the International committee for horticultural congresses London,1930
BG 38.3121

HORTICULTURE International horticultural congress 13th Report London 1952 Sep 8-15 Vol 1-2 Royal horticultural society Royal horticultural society,London, 1953
BG 38.3124

HORTICULTURE International horticultural congress 14th Report Hague 1955 Aug 29-Sep 6 Vol 1-2 International society for horticultural science Veenman and Zonen, Wageningen,1955
BG 38.3125

HORTICULTURE International horticultural congress 15th Proceedings Nice 1958 Vol 1-3 International committee for horticultural congresses Edited by Jean-Claude Garnaud 3 vols Pergamon press, London,1961-62
BG 38.3133

HORTICULTURE International horticultural congress 16th Brussels 1962 Aug 31-Sep 8 Vol 1-5 International society for horticultural sciences 5 vols Duculot, Gembloux,1962
BG 38.3126

HORTICULTURE International horticultural congress 17th Proceedings College Park, Md. 1966 Aug 15-20 Vol 1-4 International society for horticultural sciences Sponsored by the American horticultural society 4 vols 1966-67
BG 38.3134

HORTICULTURE International horticultural congress 18th Proceedings Tel-Aviv 1970 Mar 17-25 International society for horticultural sciences 1970
BG 38.3135

HORTICULTURE International horticultural exhibition and botanical congress :report of proceedings London 1866 May 22-31 London, 1867
Bot 42.4705

HORTICULTURE International tuinbouw-congres 7th Amsterdam 1923 Sep 17-23 Nederlandsche mantschappij voor tuinbouw en plantkunde c1923
Gen 34.1139

HORTUS BELGICUS :catalogue de l'exposition Brussels 1962 Aug-Sep Bibliotheque Albert I,Belgium Edited by Jan Balis Bibliotheque royale de Belgique.Catalogues des expositions, 10 Brussels,1962 Exhibition organised to coincide with the 16th International horticultural congress Book wrongly calls it Botanical congress
BG 38.1857

HOST-PARASITE RELATIONS IN PLANT PATHOLOGY : a symposium Budapest 1964 Oct 19-22 Hungarian academy of sciences.Research institute for plant protection Edited by Z. Kiraly and G. Ubrizsy Sponsored by the Ministry of agriculture 257p Budapest, c1965
Bot 42.4628

HOST SPECIFICITY AMONG PARASITES OF VERTEBRATES Symposium sur la specificite parasitaire des parasites de vertebres 1 er Neuchatel 1957 Universite de Neuchatel. Institut de zoologie Edited by J.G. Baer Paul Attinger,Neuchatel,1957 Title also in English.Text in English and French
Bal 39.1727

HOT CORROSION PROBLEMS ASSOCIATED WITH GAS TURBINES :symposium Proceedings Atlantic City,N.J. 1966 Jun 26-Jul 1 American society for testing and materials American society for testing and materials. Special technical publication, 421 Philadelphia,Pa.,1967 Symposium presented at the annual meeting
Met 25.2788

HOT GASES NBS semicentennial symposium on energy transfer in hot gases Proceedings 1951 Sep 17-18 National bureau of standards National bureau of standards.Circular, 523 U. S.government printing office,Washington,D.C., 1954
Eng 41.7220

HOT PLASMAS Physics of hot plasmas Scottish universities' summer school 9th Proceedings 1968 Jul 28-Aug 16 Edited by B. J. Rye and J.C. Taylor illus xv,455p 25cm Oliver and Boyd,Edinburgh,1970
A Math 4.1886

HOUSE STRUCTURES Symposium on full-scale tests on house structures Los Angeles,Calif. 1956 Sep 18 American society for testing materials American society for testing materials.Special technical publication, 210 American society for testing materials, Philadelphia,Pa.,1956
Eng 41.3018

HOUSTON 1969 Leukemia-Lymphoma Annual clinical conference on cancer 14th Papers Anderson hospital and tumor institute Edited by C.C. Shullenberger and R.W. Cumley Yearbook medical publishers,Chicago,1971
PGMS 29.0678

HOUSTON,TEX. 1953 Finding ancient shorelines :a symposium papers Edited by Jack L. Hough and Henry W. Menard sponsored by Society of economic paleontologists and mineralogists Society of economic paleontologists and mineralogists.Special publication,3 illus. map Society of economic paleontologists and mineralogists, Tulsa,Oka.,1956
Geol 8.1487

HOUSTON,TEX. 1957 Biology of normal and atypical pigment cell growth conference 4th Proceedings New York zoological society Anderson hospital and tumor institute Damon Runyon memorial fund for cancer research Edited by Myron Gordon Academic press,New York,1959
Gen 34.0601

HOUSTON,TEX. 1957 Pigment cell biology Conference on the biology of normal and atypical pigment cell growth 4th Proceedings Edited by Myron Gordon Academic press,New York,1959 For 5th and further conferences see:International pigment cell conference
An 32.3388

HOUSTON,TEX. 1957 Pigment cell biology The Conference on biology of normal and atypical pigment cell growth 4th Proceedings Edited by Myron Gordon Held at the Anderson hospital and tumor institute Academic press,New York,1959 Sponsored jointly by the New York zoological society,the M.D.Anderson Hospital and tumor institute and the Damon Runyon memorial fund for cancer research, inc.
Bioch 33.0937

HOUSTON,TEX. 1958 Radiation biology and cancer Symposium on fundamental cancer research 12th Papers Anderson hospital and tumor institute University of Texas press,Austin,Tex.,1958
Phys 20.0988

HOUSTON,TEX. 1960 Carcinoma of the uterine cervix,endometrium and ovary Annual clinical conference on cancer 5th Papers Anderson hospital and tumor institute Year book medical,Chicago,Ill.,1962
Radioth 35.0828

HOUSTON,TEX. 1960 The Parathyroids :a symposium on advances in parathyroid research Proceedings Edited by Roy O. Greep and Roy V. Talmage Thomas,Springfield,Ill.,1961
An 32.3771

HOUSTON,TEX. 1961 Molecular basis of neoplasia Annual symposium on fundamental cancer research 15th Papers Anderson hospital and tumor institute University of Texas press,Austin,Tex.,1961
Radioth 35.1942

HOUSTON,TEX. 1962 Conceptual advances in immunology and oncology Symposium on fundamental cancer research 16th Papers Harper,New York,1962
Pha 16.0185

HOUSTON,TEX. 1963 Delayed implantation : symposium Rice university Edited by Allen C. Enders Rice university semicentennial series University of Chicago press,Chicago, Ill.,1963
An 32.3890

HOUSTON,TEX. 1963 Viruses,nucleic acids and cancer Annual symposium on fundamental cancer research,1963 17th Papers Anderson hospital and tumor institute Williams and Wilkins,Baltimore,Md.,1963
Radioth 35.0822

HOUSTON,TEX. 1964 Symposium on creep of concrete American concrete institute American concrete institute.Publication,SP-9 American concrete institute,Detroit,Mich., 1964
Eng 41.2932

HOUSTON,TEX. 1967 Complex analysis Conference on complex analysis Proceedings Rice university Edited by H.L. Resnikoff and R.O. Wells Rice university studies, 54,no.4 Bibliog. 84p 23cm William Marsh Rice University,Houston,Tex.,1968
P Math 2.3684

HOUSTON,TEX. 1967 Number theory :a symposium Proceedings American mathematical society Edited by William J. Leveque and Ernst G. Straus American mathematical society.Proceedings of symposia in pure mathematics, 12 Bibliog. v,98p 26cm American mathematical society, Providence,R.I.,1969 Special session on number theory at the 73rd annual meeting of the A.M.S.
P Math 2.3699

HOUSTON,TEX. 1968 Breast cancer:early and late Annual clinical conference on cancer 13th Papers Anderson hospital and tumor institute Yearbook,Chicago,Ill.,1970
Radioth 35.1914

HOUSTON,TEXAS 1956 Brain mechanisms and drug action :a symposium:fourth annual general meeting of the Houston neurological society Papers Houston neurological society Edited by William S. Fields C.C.Thomas,Springfield, Ill.,1957
Psy 31.3371

HOUSTON,TEXAS 1969 Complex analysis 1969 : conference Proceedings Rice university Edited by H.L. Resnikoff and R.O. Wells Rice university studies, 56,no 2 vi,222p 23cm Rice university,Houston,Texas,1971
P Math 2.4357

HOUSTON,TEXAS 1969 Conference on chemical research Proceedings 13: the transuranium elements - the Mendeleev centennial Robert A. Welch foundation Edited by W.O. Milligan 494p R.A.Welch foundation,Houston,Texas,1970
TA 15.0699

HOUSTON,TEXAS 1969 Conference on chemical research Proceedings Robert A.Welsh foundation Robert A.Welch foundation. Conference, 13 photos,graphs,diagrms xii, 494p 23cm Robert A.Welch foundation, Houston,Texas,1970
Cav 7.3126

HOUSTON,TEXAS 1970 Apollo 11 lunar science conference 1st Proceedings Edited by A. A. Levinson Geochimica et cosmochimica acta. Supplement, 1 3 vols Pergamon,New York, 1970
Obs 6.3615

HOUSTON,TEXAS 1970 Apollo 11 lunar science conference Papers National aeronautics and space administration American association for the advancement of science.Miscellaneous publication, 70-1 American association for the advancement of science,Washington,D.C., 1970
Min 10.1474

HOUSTON,TEXAS 1970 Apollo 11 lunar science conference Proceedings 1-3 National aeronautics and space administration Edited by A.A. Levinson Geochimica et cosmochimica acta, 34,supplement 1 3 vols Pergamon,New York,1970
Min 10.1548

HOUSTON,TEXAS 1971 Lunar science conference 2nd Proceedings Edited by A. A. Levinson Geochimica et cosmochimica acta. Supplement, 2 3 vols MIT press,Cambridge, Mass.,1971
Obs 6.3614

HOUSTON NEUROLOGICAL SOCIETY Brain mechanisms and drug action :a symposium:fourth annual general meeting of the Houston neurological society Papers Houston,Texas 1956 Mar 16 Edited by William S. Fields C. C.Thomas,Springfield,Ill.,1957
Psy 31.3371

HOUSTON NEUROLOGICAL SOCIETY.ANNUAL MEETING 4TH Brain mechanisms and drug action :a symposium:fourth annual general meeting of the Houston neurological society Papers Houston,Texas 1956 Mar 16 Houston neurological society Edited by William S. Fields C.C.Thomas,Springfield,Ill.,1957
Psy 31.3371

HOUSTON NEUROLOGICAL SOCIETY.SYMPOSIA, 6 Disorders of the developing nervous system Houston neurological society:annual scientific meeting 8th Houston Tex. 1960 Mar Edited by Williams S. Fields and Murdina M. Desmond Thomas,Springfield,Ill.,1961
An 32.4187

HOUSTON NEUROLOGICAL SOCIETY:ANNUAL SCIENTIFIC MEETING 8th Disorders of the developing nervous system Houston Tex. 1960 Mar Edited by Williams S. Fields and Murdina M. Desmond Houston neurological society.Symposia, 6 Thomas,Springfield,Ill. 1961
An 32.4187

HOUSTON TEX. 1960 Disorders of the developing nervous system Houston neurological society:annual scientific meeting 8th Edited by Williams S. Fields and Murdina M. Desmond Houston neurological society.Symposia, 6 Thomas,Springfield,Ill. 1961
An 32.4187

HUEL 1967 Geology of shelf seas Inter-university geological congress 14th proceedings Edited by D.T. Donovan Oliver Boyd,Edinburgh;London,1968
Geol 8.4480

HULL 1958 Steric effects in conjugated systems :a symposium Chemical society Edited by G.W. Gray Butterworths,London,1958
Chem 18.1687

HULL 1967 Geology of shelf seas Inter-university geological congress 14th Proceedings Edited by D.T. Donovan illus. viii,160p Oliver and Boyd,Edinburgh;London, 1968
Bot 42.3193

The HUMAN ADRENAL CORTEX :a colloquium Proceedings Ciba foundation Edited by G. E.W. Wolstenholme and others Ciba foundation colloquia on endocrinology, 8 Churchill, London,1955
Inv Med 37.0239

The HUMAN ADRENAL CORTEX :a conference Proceedings Glasgow 1960 Jul 11-14 Edited by Alastair R. Currie and others Livingstone,Edinburgh;London,1962
An 32.3758

HUMAN ADRENAL GLAND AND ITS RELATION TO BREAST CANCER Tenovus workshop 1st Proceedings Cardiff 1969 Jun 26-27 Tenovus institute for cancer research Edited by K. Griffiths and E.H.D. Cameron Alpha omega alpha,Cardiff,1969
Bioch 33.2198

HUMAN ADRENAL GLAND AND ITS RELATION TO BREAST CANCER Tenovus workshop 1st Proceedings Cardiff 1969 Jun 26-27 Tenovus institute for cancer research Edited by K. Griffiths and E.H.D. Cameron Alpha omega alpa,Cardiff,1969
Inv Med 37.0270

The HUMAN ADRENAL GLAND AND ITS RELATION TO BREAST CANCER Cardiff 1969 Jun 26-27 Tenovus institute for cancer research Edited by K. Griffiths and E.H.D. Cameron Tenovus workshops, 1 Alpha omega alpha publishing, Cardiff,1969
Radioth 35.0940

HUMAN BEHAVIOR Factors determining human behavior Harvard tercentenary conference of arts and sciences Papers Cambridge,Mass. 1936 Aug 31-Sep 12 Harvard university Harvard tercentenary publications Harvard university press,Cambridge,Mass.,1937
Psy 31.0770

HUMAN CHROMOSOMAL ABNORMALITIES conference Proceedings London 1959 Sep 18-19 King's college hospital medical school Edited by William M. Davidson and D.Robertson Smith illus. Staples press,London,1961
Gen 34.0878

HUMAN DISPLACEMENTS:MEASUREMENT,METHODOLOGICAL ASPECTS Entretiens de Monaco en sciences humaines 1ere session Monaco 1962 May 24-29 Centre international d'etude des problemes humains Edited by Jean Sutter Hachette,Paris,1962 Text in English and French
Geog 13.1566

HUMAN DISPLACEMENTS;LES DEPLACEMENTS HUMAINS 1962 May 24-29 Centre international d'etude des problemes humains Edited by Jean Sutter Entretiens de Monaco en sciences humaines, 1 Hachette,1963 Papers and title page in English and French
Gen 34.1883

HUMAN ECOLOGY FUND Brain and behavior Brain function conference 3rd Proceedings Los Angeles 1963 Vol 3: brain and gonadal function Brain research institute American institute of biological sciences Edited by Roger A. Gorski and Richard E. Whalen UCLA forum in medical sciences, 3 University of California press,Berkeley;Los Angeles,1966
Psy 31.0621

HUMAN FACTORS DIVISION.AIR FORCE DIRECTORATE OF RESEARCH AND DEVELOPMENT Symposium on atherosclerosis Washington,D.C. 1954 Mar 22-23 National research council.Division of medical sciences National research council. Publication, 338 National research council, Washington,D.C.,1955
Med 36.0266

HUMAN FACTORS RESEARCH Vigilance :a symposium Proceedings Santa Barbara,Calif. 1961 Jan United States.Office of naval research Edited by Donald N. Buckner and James J. McGrath McGraw-Hill series in psychology McGraw-Hill,New York,1963
Psy 31.1008

HUMAN GENETICS Biochemistry of human genetics :a symposium Naples 1959 May 13-16 Ciba foundation and International union of biological sciences Edited by G.E.W. Wolstenholme and C.M. O'Connor Ciba Foundation.Symposia Churchill,London,1959
PGMS 29.0251

HUMAN GENETICS International congress of human genetics 2nd Proceedings Rome 1961 Sep 6-12 Vol 1-3 3 vols Instituto G.Mendel,Rome,1963 In English, French,German and Italian
An 32.0155

HUMAN GENETICS International congress of human genetics :plenary sessions and symposia 3rd Proceedings Chicago,Ill. 1966 Sep 5-10 Edited by James F. Crow and James V. Neel John Hopkins press,Baltimore,Md.,1967
PGMS 29.0264

HUMAN GENETICS The International congress of human genetics 2nd Proceedings Rome 1961 Sep 6-12 Vol 1-3 3 vols Institute G.Mendel,Rome,1963
Gen 34.1857

HUMAN GENETICS The International congress of human genetics 3rd Plenary sessions and symposia Chicago,Ill. 1966 Sep 5-10 Edited by James F. Crow and James V. Neel Hopkins,Baltimore,Md.,1967
Gen 34.1858

HUMAN GENETICS :a symposium Cold Spring Harbor 1964 Jun 5-11 Edited by Leonora Frisch Cold Spring Harbor symposia on quantitative biology, 29 Cold Spring Harbor laboratory on quantitative biology,Cold Spring Harbor, L.I.,1964
An 32.0051

HUMAN GENETICS :a symposium Papers Cold Spring Harbor 1964 Cold Spring Harbor laboratory of quantitative biology Cold Spring Harbor symposia on quantitative biology, 29 Cold Spring Harbor,1964
Bioch 33.1285

HUMAN GROWTH A symposium 1960 Society for the study of human biology Edited by G.M. Tanner Society for the study of human biology.Symposia, 3 Pergamon press,London, 1960
Gen 34.1917

HUMAN GROWTH :a symposium Papers London 1960 Edited by J.M. Tanner Society for the study of human biology.Symposia, 3 Pergamon press,Oxford,1960
An 32.2879

HUMAN OPERATOR IN COMPLEX SYSTEMS The Conference on the human operator in complex systems Proceedings Birmingham 1966 Edited by W.T. Singleton and others Bibliog Taylor and Francis,London,1967
Eng 41.0678

HUMAN OVULATION :a symposium Brookline, Mass. 1962 Edited by Chester S. Keefer Sponsored by the Lowell M.Palmer foundation Churchill,London,1965
An 32.3886

HUMAN PALAEOPATHOLOGY Symposium on human palaeopathology Proceedings Washington,D.C. 1965 Jan 14 National academy of sciences. Subcommittee on geographic pathology Edited by Saul Jarcho Yale university press,New Haven,Conn.,1966
An 32.3129

HUMAN PITUITARY HORMONES :a colloquium in honour of B.A.Houssay Proceedings Buenos Aires 1959 Aug 6-8 Ciba foundation Edited by G.E.W. Wolstenholme and Cecilia M. O'Connor Ciba foundation colloquia on endocrinology, 13 illus Churchill,London, 1960
Bioch 33.0440

HUMAN RADIATION CYTOGENETICS :an international symposium Proceedings Edinburgh 1961 Oct 12-15 United Kingdom atomic energy authority Medical research council University of Aberdeen.Department of genetics Edited by H.J. Evans and others North-Holland,Amsterdam,1967
Gen 34.1127

HUMAN RADIATION CYTOGENETICS :an international symposium Proceedings Edinburgh 1966 Oct 12-15 Edited by H.J. Evans and others Sponsored by the Medical research council North-Holland,Amsterdam,1967
Radioth 35.0540

HUMAN TESTIS Workshop conference :a Serono foundation symposium Proceedings Positano 1970 Apr 23-25 Edited by Eugenia Rosemberg and C.Alvin Paulsen Advances in experimental medicine and biology, 10 Plenum,New York; London,1970
Inv Med 37.0190

HUMUS AND PLANT :a symposium Studies about humus Prague 1961 Sep 28-Oct 6 and Brno Sep 28-Oct 6 Edited by S. Prat and V. Rypacek illus 364p Czechoslovak academy of sciences,Prague,1962
Bot 42.2108

HUNGARIAN ACADEMY OF SCIENCES Conference on dimensioning and strength calculations 3rd Proceedings Budapest 1968 Nov Edited by E. Czoboly Akademiai kiado,Budapest,1968
Met 25.2407

HYBRID CONFERENCE REPORT 1900 on hybridisation (the cross-breeding of species) and on the cross-breeding of varieties London 1899 Jul 11-12 Royal horticultural society Royal horticultural society.Journal, 24 Spottiswoode,London,1900 Conference also entitled the 'International congress of genetics' and the 'International conference on hybridization'
Gen 34.0998

HYBRIDISATION International congress of genetics hybridization (the cross-breeding of genera or species)the cross-breeding of varieties... 3rd London 1906 Jul 30-Aug 3 Royal horticultural society Edited by W. Wilks Spottiswoode,London,1907 Conference also entitled 'International conference on hybridisation'
Gen 34.0999

HYDRA Biology of the hydra and of some other coelenterates The Physiology and ultrastructure of hydra and of some other coelenterates :a symposium Coral Gables,Fla. 1961 Mar 29-31 Edited by Howard M. Lenhoff and F.Farnsworth Loomis held at the Fairchild tropical gardens University of Miami press,Coral Gables,Fla.,1961
Bal 39.2183

HYDRATES OF CARBON Conference de l'Union internationale de chimie :rapports sur les hydrates de carbone (Glucides) 10th Rapports Liege 1930 Sep 14-20 Union internationale de chimie Union internationale de chimie,Paris,1930
Chem 18.0066

HYDRAULIC MECHANISMS Conference on hydraulic mechanisms Proceedings London 1954 Mar 26 Institution of mechanical engineers Institution of mechanical engineers,London, 1954
Eng 41.6686

HYDRAULIC SERVO-MECHANISMS Conference on hydraulic servo-mechanisms Proceedings London 1953 Feb 13 Institution of mechanical engineers Institution of mechanical engineers,London,1963
Eng 41.5891

HYDRAULICS Australasian conference on hydraulics and fluid mechanics 1st Proceedings Perth 1962 Dec 6-13 Edited by Richard Silvester Pergamon press,London, 1964
Eng 41.6968

HYDRAULICS Engineering hydraulics 4th conference Proceedings Iowa city,Iowa 1949 Jun 12-15 Iowa institute of hydraulic research Edited by Hunter Rouse Wiley; Chapman and Hall,New York;London,1950
Eng 41.6662

HYDRAULICS AND FLUID MECHANICS Australasian conference on hydraulics and fluid mechanics 2nd proceedings Auckland N.Z. 1965 Dec. 6-11 University of Auckland Edited by C.M. Segedin Sponsored jointly with the New Zealand institution of engineers Conference organising committee,Auckland,N.Z.,1966
A Math 4.0531

HYDRAULICS AND FLUID MECHANICS 1st : Australasian conference Proceedings Nedlands 1962 Dec 6-13 Edited by Richard Silvester Sponsored by the University of Western Australia.Faculty of engineering illus viii,503p Pergamon press,Oxford,1964
Chem E 24.0239

HYDRAULICS CONFERENCE 4th proceedings Engineering hydraulics Iowa city 1949 Jun 12-15 Iowa institute of hydraulic research Edited by Hunter Rouse John Wiley and sons, New York;London,1950
A Math 4.0511

HYDRAULICS CONFERENCE 6th Proceedings Iowa City 1955 Jun 13-15 Edited by L. Landweber and P.G. Hubbard Arranged by the Iowa institute of hydraulic research illus, diagrms 276p Iowa institute of hydraulic research,Iowa City,1956
Chem E 24.0305

HYDROBIOLOGY A Symposium on hydrobiology Madison,Wis. 1940 Sep 4-6 By James G. Needham and others Held at the University of Wisconsin University of Wisconsin,Madison, Wis.,1941 Held under the auspices of the University of Wisconsin and of the Wisconsin alumni research foundation
Bal 44.6459

HYDROBIOLOGY Symposium on African hydrobiology and inland fisheries 2nd Brazzaville 1956 Jul 3-11 Scientific council for Africa south of the Sahara Scientific council for Africa south of the Sahara.Publications,25 Commission for technical co-operation in Africa,London,1957
Geog 13.2726

HYDROGEN BONDING :papers held at the symposium on hydrogen bonding Papers Ljubljana 1957 Jul 29-Aug 3 International union of pure and applied chemistry,and,Union of the chemical societies Edited by D. Hadzi and H. W. Thompson Pergamon press,London,1959
Chem 18.0955

HYDROGEN IN METALS L' Hydrogene dans les metaux :colloque Valduc 1967 Sep 27-28 Centre d'etudes de Bruyeres-le-Chatel Paris, 1969
Met 25.2624

HYDROGEN IN STEEL Conference on hydrogen in steel papers and discussions Harrogate 1961 Oct Iron and steel institute Iron and steel institute.Special report, 73 Iron and steel institute,London,1962 No title page
Met 25.1538

HYDROGEN-BONDED SYSTEMS :symposium on equilibria and reaction kinetics... Proceedings Newcastle-upon-Tyne 1968 Jan 10-12 University of Newcastle-upon-Tyne Edited by A.K. Covington and P. Jones Taylor and Francis,London,1968
Chem 18.0298

L' HYDROGENE DANS LES METAUX :colloque Valduc 1967 Sep 27-28 Centre d'etudes de Bruyeres-le-Chatel Paris,1969
Met 25.2624

HYDROLOGY Application of isotope techniques in hydrology :a comprehensive report Vienna 1961 Nov 6-9 International atomic energy agency International atomic energy agency. Technical reports series, 11 Vienna,1962
Bot 42.3317

HYDROLOGY Conference hydrologique des etats baltiques 4th Compte-rendu Leningrad 1933 Sep 6-22 Union of Soviet socialist republics.Service hydro-meteorologique Leningrad,1934
Geog 13.0767

HYDROLOGY Interterritorial conference on hydrology and water resources 1st Nairobi 1950 Nov Held under the auspices of the East Africa high commission Nairobi,1950
Geog 13.5670

HYDROLOGY Isotope techniques for hydrology Vienna 1962 Dec 17-21 International atomic energy agency International atomic energy agency.Technical reports series, 23 Vienna, 1964
Bot 42.3318

HYDROLYSIS Advances in enzymic hydrolysis of cellular and related materials :a symposium Proceedings Washington,D.C. 1962 Mar Edited by Elwyn T. Reese Sponsored by the American chemical society Pergamon,Oxford, 1963 Including a bibliography for the years 1950-61
Bot 42.1728

HYDROMECHANISCHE PROBLEME DES SCHIFFSANTRIEBS Hamburg 1932 May 1819 Edited by G. Kempf and E. Foerster 1932
Eng 41.6656

HYDROPHOBIC COLLOIDS Utrecht 1937 Nov 5-6 Under the auspices of Nederlandsche chemische vereeniging.Colloid chemistry section Amsterdam,1938 Reprinted from "Chemisch Weekblad" 1938
Col S 12.0100

HYPERBARIC OXYGENATION International congress on clinical application of hyperbaric oxygen 2nd Proceedings Glasgow 1964 Sep Edited by Iain McA. Ledingham Livingstone,Edinburgh;London,1965
Radioth 35.0845

HYPERBOLIC EQUATIONS AND WAVES Battelle Seattle 1968 rencontres Seattle 1968 Edited by M. Froissart Bibliog.,Illus. viii,393p 23cm Springer-Verlag,Berlin,1970
A Math 4.1469

HYPERBOLIC OXYGEN AND RADIATION THERAPY OF CANCER Annual San Francisco cancer symposium 1st Proceedings San Francisco,Calif. 1965 Nov 12-13 Frontiers of radiation therapy and oncology, 1 Karger,Basel;New York,1968
Radioth 35.1970

HYPERSONIC FLOW 11th symposium Bristol 1959 Apr 6-8 Colston research society Edited by A.R. Collar and J. Tinkler Colston papers, 11 Butterworths,London,1960
Eng 41.6759

HYPERSONIC FLOW RESEARCH :a selection of technical papers based mainly a symposium of the American rocket society,held at the Massachusetts institute of technology... Cambridge,Mass. 1961 Aug 16-18 American rocket society Edited by Frederick R. Riddell Progress in astronautics and rocketry, 7 Academic press,NewYork,1962
Eng 41.6771

HYPERSONIC FLOW RESEARCH :a symposium technical papers Cambridge,Mass. 1961 Aug 16-18 American rocket society Edited by Frederick R. Riddell Sponsored jointly with the United States.Air force.Office of scientific research Progress in astronautics and rocketry, 7 Academic press,New York, London,1962
A Math 4.0533

HYPERTENSION Catapres in hypertension :a symposium Proceedings London 1969 Mar Royal college of surgeons Edited by Matthew E. Conolly Butterworths,London,1970
PGMS 29.0608

HYPERTENSION Experimental hypertension being the results of a conference Results New York 1945 Feb 9-10 By William Goldring and others New York academy of sciences.Section of biology New York academy of sciences. Special publications, 3 New York academy of sciences,New York,1946
Psy 28.0096

HYPERTENSION Renal disease and hypertension : tentn annual conference... Proceedings Denver 1971 Mar Atlantic City 1971 May United States.Public health service Edited by Richard H. Thurm Proceedings of the annual conferences of the U.S.Public health service co-operative study (antihypertensive agents), 8 U.S.Dept. of health,education and welfare,Bethesda,1971
PGMS 29.0659

HYPERTENSIVE DRUGS Hypotensive drugs and control of vascular tone in hypertension :a symposium Proceedings London 1956 Apr 5-6 Biological council.Co-ordinating committee for symposia on drug action Edited by M. Harington Pergamon press,London,1956
Med 36.0172

The HYPOPHYSEAL GROWTH HORMONE;NATURE AND ACTIONS an international symposium Proceedings Detroit 1954 Oct 27-29 By Richmond W. Smith and others Sponsored by the Henry Ford Hospital and Edsel B.Ford institute for medical research New York,1955
Bioch 33.0483

The HYPOPHYSEAL GROWTH HORMONE,NATURE AND ACTIONS international symposium Detroit,Mich. 1954 Oct 27-29 Edited by Richmond W. Smith and others Sponsored by the Henry Ford hospital McGraw-Hill,New York,1955
An 32.3736

HYPOPLASIE ET MALFORMATIONS DE L'APPAREIL GENITAL INTERNE DE LA FEMME Papers Strasbourg 1964 May 15-18 Societe francaise de gynecologie Edited by P. Muller Masson, Paris,1964
An 32.3832

HYPOTENSIVE DRUGS AND CONTROL OF VASCULAR TONE IN HYPERTENSION :a symposium Proceedings London 1956 Apr 5-6 Biological council.Co-ordinating committee for symposia on drug action Edited by M. Harington Pergamon press,London,1956
Med 36.0172

HYPOTENSIVE PEPTIDES :an international symposium Florence 1965 Oct 25-29 Edited by Ervin C. Erdos and others Springer, Berlin,1966
Inv Med 37.0252

HYPOTENSIVE PEPTIDES :international symposium Proceedings Florence 1965 Oct 25-29 Edited by E.G. Erdos and others Springer,New York,1966
Pha 16.0118

The HYPOTHALAMUS AND CENTRAL LEVELS OF AUTONOMIC FUNCTION Proceedings of the Association New York 1939 Dec 20-21 Association for research in nervous and mental disease Edited by John F. Fulton and others Association for research in nervous and mental disease.Research publications, 22 Baltimore, Md.,1940
Phys 20.1083

The HYPOTHALAMUS AND CENTRAL LEVELS OF AUTONOMIC FUNCTION Proceedings of the Association New York 1939 Dec 20-21 Association for research in nervous and mental disease Edited by John F. Fulton and others Association for research in nervous and mental disease.Research publications,20 Williams and Wilkins,Baltimore,Md.,1940
Pha 16.0282

The HYPOTHALAMUS AND CENTRAL LEVELS OF AUTONOMIC FUNCTION Proceedings of the association New York 1939 Dec 20-21 Association for research in nervous and mental disease Edited by John F. Fulton and others Association for research in nervous and mental disease.Research publications, 20 Williams and Wilkins,Baltimore,Md.,1940
An 32.4510

HYPOTHERMIA The Physiology of induced hypothermia :a symposium Proceedings Washington,D.C. 1955 Oct 28-29 National research council.Division of medical sciences Edited by Robert D. Dripps National research council.Publications, 451 Washington,D.C., 1956
Phys 20.0987

I.A.U. SYMPOSIUM 32nd Papers Continental drift,secular motion of the Pole and rotation of the earth Stresa 1967 Mar 21-25 International astronomical union and International union of geodesy and geophysics Edited by William Markowitz and B. Guinot illus 107p 25cm D.Riedel,Dordrecht,1968
Sco 14.8118

I.C.N.A.F. ENVIRONMENTAL SYMPOSIUM Contributions Rome 1964 International commission for the Northwest Atlantic fisheries International commission for the Northwest Atlantic fisheries.Special publication,6 figures,tables,graphs 914p ICNAF,Dartmouth,Nova Scotia,1965
Sco 14.5958

I.E.E.CONFERENCE PUBLICATION, 30 Integrated circuits Conference on integrated circuits Eastbourne 1967 May 2-4 Sponsored by the Institution of electrical engineers. Electronics division Institution of electrical engineers,London,1967 Joint sponsors;Institution of electronic and radio engineers and the Institute of electrical and electronics engineers
Eng 41.5461

I.E.E.E. INTERNATIONAL CONVENTION convention record New York 1963 Mar.25-28 Part 4: electronic computers,information theory, human factors in electronics Institute of electrical and electronics engineers 152p Institute of electrical and electronics engineers,New York,1963
Math L 5.2086

I.E.E.E.INTERNATIONAL CONVENTION convention record New York 1965 Mar.22-26 Part 3: computers Institute of electrical and electronics engineers 281p Institute of electrical and electronis engineers,New York, 1965
Math L 5.2087

I.E.E.E.MACHINE TOOLS INDUSTRY CONFERENCE,1964 Graphical language:a step towards computer-aided design By C.A. Lang and R.B. Polansky 15p 1964 Presented at the IEEE machine tools industry conference,1964.
Math L 5.2873

I.F.A.C. SYMPOSIUM ON FLUIDICS Proceedings London 1968 Nov 4-8 International federation of automatic control Organised by the Royal aeronautical society 1968
Eng 41.5905

I.F.A.C. SYMPOSIUM ON IDENTIFICATION IN AUTOMATIC CONTROL SYSTEMS Preprints Identification in automatic control systems Prague 1967 Jun 12-17 Pt 1-2 Ceskoslovenska akademie ved.Institute of information theory and automation Sponsored by International federation of automatic control 2 vols Academia,Prague,1967
Eng 41.6019

I.F.A.C. SYMPOSIUM ON PULSE-RATE AND PULSE-NUMBER SIGNALS IN AUTOMATIC CONTROL Proceedings Budapest 1968 Apr 8-10 Sponsored by the International federation of automatic control International federation of automatic control, Debrecen,1968
Eng 41.6020

I.F.A.C. SYMPOSIUM ON SYSTEM SENSITIVITY AND ADAPTIVITY 2nd Preprints Dubrovnik 1968 Aug 26-31 International federation of automatic control Organised by the Yugoslav committee for electronics and automation I.F. A.C.,Dubrovnik,1968
Eng 41.5889

I.F.A.C.SYMPOSIUM ON THE THEORY OF SELF-ADAPTIVE CONTROL SYSTEMS 2nd Proceedings Teddington 1965 Sep 14-17 International federation of automatic control Edited by P. H. Hammond Plenum press,New York,1966
Eng 41.5888

I.G.U.SYMPOSIUM IN URBAN GEOGRAPHY Proceedings Lund 1960 Aug 15-19 Edited by Knut Norberg Lund studies in geography.Ser.B. Human geography,24 Lunds universitet,Lund, 1962 Held at the 19th International geographical congress
Geog 13.1944

I.N.Q.U.A. Pleistocene extinctions:the search for a cause International association for quaternary research.Congress 7th Proceedings Boulder,Colo. 1965 Aug Edited by P.S. Martin and H.E. Wright Sponsored by the National research council Yale university press,New Haven,Conn.;London, 1967
Geog 13.1320

I.R.E. INTERNATIONAL CONVENTION convention record New York 1960 Mar.21-24 Part 2: circuit theory,electronic computers Institute of radio engineers 203p Institute of radio engineers,New York,1960
Math L 5.2074

I.R.E. INTERNATIONAL CONVENTION convention record New York 1960 Mar.21-24 Part 4: automatic control,information theory Institute of radio engineers 222p Institute of radio engineers,New York,1960
Math L 5.2075

I.R.E. INTERNATIONAL CONVENTION convention record New York 1961 Mar.20-23 Part 2: audio;electronic computers Institute of radio engineers 275p Institute of radio engineers,New York,1961
Math L 5.2077

I.R.E. INTERNATIONAL CONVENTION convention record New York 1961 Mar.20-23 Part 4: automatic control,circuit theory, information theory Institute of radio engineers 286p Institute of radio engineers,New York,1961
Math L 5.2078

I.R.E. INTERNATIONAL CONVENTION convention record New York 1962 Mar.26-29 Part 4: electronic computers,information theory Institute of radio engineers 199p Institute of radio engineers,New York,1962
Math L 5.2084

I.R.E. NATIONAL CONVENTION convention record New York 1953 Mar.23-26 Part 7: electronic computers Institute of radio engineers 71p Institute of radio engineers, New York,1953
Math L 5.2064

I.R.E. NATIONAL CONVENTION convention record New York 1954 Mar.22-25 Part 4: electronic computers and information theory Institute of radio engineers 143p Institute of radio engineers,New York,1954
Math L 5.2065

I.R.E. NATIONAL CONVENTION convention record New York 1955 Mar.21-24 Part 4: computers,information theory,automatic control Institute of radio engineers 208p Institute of radio engineers,New York,1955
Math L 5.2066

I.R.E. NATIONAL CONVENTION convention record New York 1956 Mar.19-22 Part 4: computers,information theory,automatic control Institute of radio engineers 175p Institute of radio engineers,New York,1956
Math L 5.2067

I.R.E. NATIONAL CONVENTION convention record New York 1957 Mar.18-21 Part 2: circuit theory,information theory Institute of radio engineers 212p Institute of radio engineers,New York,1957
Math L 5.2068

I.R.E. NATIONAL CONVENTION convention record New York 1957 Mar.18-21 Part 4: automatic control,electronic computers,medical electronics Institute of radio engineers 179p Institute of radio engineers,New York, 1957
Math L 5.2069

I.R.E. NATIONAL CONVENTION convention record New York 1959 Mar.23-26 Part 4: automatic control,electronic computers, information theory Institute of radio engineers 201p Institute of radio engineers,New York,1959
Math L 5.2072

I.R.E. WESTERN ELECTRONIC SHOW AND CONVENTION convention record Los Angeles 1958 Aug. 19-22 Institute of radio engineers 291p Institute of radio engineers,New York,1958
Math L 5.2070

I.R.E. WESTERN ELECTRONIC SHOW AND CONVENTION convention record Los Angeles 1960 Aug. 23-26 Part 4: computers,man-machine systems Institute of radio engineers 202p Institute of radio engineers,New York,1960
Math L 5.2076

I.R.E.WESTERN ELECTRONIC SHOW AND CONVENTION,1953, AUG.19-21 Transactions of the I.R.E. professional group on circuit theory No. 2: December,1953 Institute of radio engineers 106p I.R.E.,New York,1953 Papers presented at the circuit theory sessions of the Western electronic show and convention.San Francisco,August 1953.
Math L 5.2083

I.U.B.-I.U.B.S. INTERNATIONAL SYMPOSIUM 1st Proceedings Biological structure and function Stockholm 1960 Sep 12-17 Vol 1 International union of biological sciences and International union of biochemistry Edited by T.W. Goodwin and O. Lindberg Academic press,London,1961
An 32.3276

I.U.B.-I.U.B.S. INTERNATIONAL SYMPOSIUM 1st Proceedings Biological structure and function Stockholm 1960 Sep 12-17 Vol 1-2 International union of biological sciences International union of biochemistry Edited by T.W. Goodwin and O. Lindberg illus 2 vols Academic press,London;New York,1961
Bal 44.6450

The I.U.B.-I.U.B.S. INTERNATIONAL SYMPOSIUM 1st Proceedings Biological structure and function Stockholm 1960 Sep 12-17 Vol 1-2 International union of biochemistry and International union of biological sciences Edited by T.W. Goodwin and O. Lindberg illus 2 vols Academic press, London;New York,1961
Bioch 33.0584

The I.U.B.-I.U.B.S.INTERNATIONAL SYMPOSIUM 1st Proceedings Biological struggle and function Stockholm 1960 Sep 12-17 Vol 1-2 International union of biochemistry International union of biological sciences Edited by T.W. Goodwin and O. Lindberg illus. 2 vols Academic press,London;New York, 1961
Bot 42.1344

I.U.B.S.SYMPOSIUM, 2 Towards a theoretical biology :a symposium Papers Lake Como 1967 Aug 3-12 Vol 2: sketches International union of biological sciences Edited by C.H. Waddington Edinburgh university press,Edinburgh,1969
Bal 39.0053

I.U.B.S.SYMPOSIUM SERIES, 8 International congress of biochemistry 4th Proceedings Vienna 1958 Sep 1-6 Vol 6: symposium 6 - biochemistry of morphogenesis International union of biochemistry Edited by W.J. Nickerson Pergamon press,London,1959
Gen 34.0482

I.U.B.SYMPOSIUM SERIES, 1 Origin of life on the earth The International symposium on the origin of life on the earth 1st Proceedings Moscow 1957 Aug 19-24 Akademiya nauk S.S.S.R. Edited by A.I. Oparin and others Organized under the auspices of the International union of biochemistry Pergamon,London,1959 "English-French-German edition edited for the International union of biochemistry by F.Clark and R.L.M.Synge".
Bioch 33.0727

I.U.B.SYMPOSIUM SERIES, 2 The International symposium on enzyme chemistry Proceedings Tokyo 1957 Oct 15-23 and Kyoto 1957 Oct 15-23 International union of biochemistry Organized by Science council of Japan Pergamon,London,1958
Bioch 33.1053

I.U.B.SYMPOSIUM SERIES, 3 International congress of biochemistry 4th Vol 1: symposium 1- carbohydrate chemistry of substances of biological interest International union of biochemistry Edited by M.L. Wolfrom Pergamon,London,1959 Added title page in French and German.Text in English,French and German
Bioch 33.1315

I.U.B.SYMPOSIUM SERIES, 4 International congress of biochemistry 4th Proceedings Vienna 1958 Sep 1-6 Vol 2: symposium 2 - biochemistry of wood International union of biochemistry Edited by K. Kratzi and G. Billek Pergamon,London,1959 Added title page in French and German.Text in English, French and German
Bioch 33.1316

I.U.B.SYMPOSIUM SERIES, 5 International congress of biochemistry 4th Proceedings Vienna 1958 Sep 1-6 Vol 3: symposium 3: biochemistry of the central nervous system International union of biochemistry Edited by F. Brucke Pergamon,London,1959 Added t.p.in French and German
Bot 42.1715

I.U.B.SYMPOSIUM SERIES, 5 International congress of biochemistry 4th Proceedings Vienna 1958 Sep 1-6 Vol 3: symposium 3 - biochemistry of the central nervous system International union of biochemistry Edited by Brucke Pergamon,London,1959 Added title page in French and German.Text in English,French and German
Bioch 33.1317

I.U.B.SYMPOSIUM SERIES, 6 International congress of biochemistry 4th Proceedings Vienna 1958 Sep 1-6 Vol 4: symposium 4 biochemistry of steroids International union of biochemistry Edited by E. Mosettig Pergamon,London,1959 Added title page in French and German.Text in English,French and German.
Bioch 33.1318

I.U.B.SYMPOSIUM SERIES, 7 International congress of biochemistry 4th Proceedings Vienna 1958 Sep 1-6 Vol 5: symposium 5 - biochemistry of antibiotics International union of biochemistry Edited by K.H. Spitzy and R. Brunner Pergamon,London,1959 Added title page in French and German.Text in English,French and German.
Bioch 33.1319

I.U.B.SYMPOSIUM SERIES, 8 International congress of biochemistry 4th Proceedings Vienna 1958 Sep 1-6 Vol 6: symposium 6 - biochemistry of morphogenesis International union of biochemistry Edited by W.J. Nickerson Pergamon,London,1959 Added title page in French and German
Bioch 33.1320

I.U.B.SYMPOSIUM SERIES, 9 International congress of biochemistry 4th Proceedings Vienna 1958 Sep 1-6 Vol 7: symposium 7 - biochemistry of viruses International union of biochemistry Edited by E. Broda and W. Frisch-Niggemeyer Pergamon,London,1959 Added title page in French and German.Text in German and English
Bioch 33.1321

I.U.B.SYMPOSIUM SERIES, 10 International congress of biochemistry 4th Proceedings Vienna 1958 Sep 1-6 Vol 8: symposium 8 - proteins International union of biochemistry Edited by H. Neurath and H. Tuppy Pergamon, London,1960 Added title page in French and German.Text in English and French
Bioch 33.1322

I.U.B.SYMPOSIUM SERIES, 11 International congress of biochemistry 4th Proceedings Vienna 1958 Sep 1-6 Vol 9: symposium 9 - physical chemistry of high polymers of biological interest International union of biochemistry Edited by O. Kratky Pergamon, London,1959 Added title page in French and German.Text in English,French and German.
Bioch 33.1323

I.U.B.SYMPOSIUM SERIES, 12 International congress of biochemistry 4th Proceedings Vienna 1958 Sep 1-6 Vol 10: symposium 10 - blood clotting factors Edited by E. Deutsch Pergamon,London,1959 Added title page in French and German.Text in English and German
Bioch 33.1324

I.U.B.SYMPOSIUM SERIES, 13 International congress of biochemistry 4th Proceedings Vienna 1958 Sep 1-6 Vol 11: symposium 11 - vitamin metabolism International union of biochemistry Edited by W. Umbreit and H. Molitor Pergamon,London,1960 Added title page in French and German
Bioch 33.1325

I.U.B.SYMPOSIUM SERIES, 14 International congress of biochemistry 4th Proceedings Vienna 1958 Sep 1-6 Vol 12: symposium 12 - biochemistry of insects International union of biochemistry Edited by L. Levenbook Pergamon,London,1959 Added title page in French and German.Text in English and German
Bioch 33.1326

I.U.B.SYMPOSIUM SERIES, 14 International congress of biochemistry 4th Proceedings Vienna 1958 Sep 1-6 Vol 14: transactions of the plenary sessions International union of biochemistry Edited by W. Auerswald and O. Hoffmann-Ostenhof Pergamon,London,1959 Added title page in French and German.Text in English,French and German.
Bioch 33.1328

I.U.B.SYMPOSIUM SERIES, 15 International congress of biochemistry 4th Proceedings Vienna 1958 Sep 1-6 Vol 12: colloquia International union of biochemistry Edited by H. Chantrenne and others Pergamon,London, 1959 Added title page in French and German.Text in English,French and German
Bioch 33.1327

I.U.B.SYMPOSIUM SERIES, 15 The International congress of biochemistry 4th Proceedings Vienna 1958 Sep 1-6 13: colloquia International union of biochemistry Edited by O. Hoffmann-Osterhoff Pergamon press,London,1959 Added title page in French and German
Bot 42.1696

I.U.B.SYMPOSIUM SERIES, 17 International congress of biochemistry 4th Vienna 1958 Sep 1-6 Vol 15: biochemistry:abstracts of sectional papers and index to symposia and colloquia International union of biochemistry Pergamon,Oxford,1960 Added title page in French and German.Text in English,French and German
Bioch 33.1329

I.U.B.SYMPOSIUM SERIES, 19 Haematin enzymes a symposium Papers and discussions Canberra 1959 Aug 31-Sep 4 Pt 1 International union of biochemistry Edited by J.E. Falk and others Organized by the Australian academy of science Pergamon press, Oxford,1961
Bioch 33.1079

I.U.B.SYMPOSIUM SERIES, 20 Report of the Commission on enzymes of the International union of biochemistry 1961 International union of biochemistry.Commission on enzymes Pergamon press,Oxford,1961 "Report presented to the Council of the Union, in session during the Fifth International congress of biochemistry at Moscow in August 1961"
Bioch 33.1054

I.U.B.SYMPOSIUM SERIES, 20 Report of the commission on enzymes of the International union of biochemistry 1961 International union of biochemistry.Commission on enzymes Pergamon,Oxford,1961 Report presented to the Council of the Union,in session during the fifth International congress of biochemistry at Moscow,August,1961
Radioth 35.1845

I.U.B.SYMPOSIUM SERIES, 21 International congress of biochemistry 5th Proceedings Moscow 1961 Aug 10-16 Vol 1: biological structure and function at the molecular level International union of biochemistry Edited by V.A. Engelhardt Pergamon;PWN-Polish scientific publishers,London;Warsaw,1963
Bioch 33.1330

I.U.B.SYMPOSIUM SERIES, 22 International congress of biochemistry 5th Proceedings Moscow 1961 Aug 10-16 Vol 2: functional biochemistry of cell structures International union of biochemistry Edited by O. Lindberg Pergamon;PWN-Polish scientific publishers,New York;Warsaw,1963
Bioch 33.1331

I.U.B.SYMPOSIUM SERIES, 23 International congress of biochemistry 5th Proceedings Moscow 1961 Aug 10-16 Vol 3: evolutionary biochemistry International union of biochemistry Pergamon;PWN-Polish scientific publishers,New York;Warsaw,1963
Bioch 33.1332

I.U.B.SYMPOSIUM SERIES, 24 International congress of biochemistry 5th Proceedings Moscow 1961 Aug 10-16 Vol 4: molecular basis of enzyme action and inhibition International union of biochemistry Edited by P.A.E. Desnuelle Pergamon;PWN-Polish scientific publishers,Oxford;Warsaw,1963
Bioch 33.1333

I.U.B.SYMPOSIUM SERIES, 25 International congress of biochemistry 5th Proceedings Moscow 1961 Aug 10-16 Vol 5: intracellular respiration:phosphorylating and non-phosphorylating oxidation reactions International union of biochemistry Edited by E.C. Slater Pergamon;PWN-Polish scientific publishers,Oxford;Warsaw,1963
Bioch 33.1334

I.U.B.SYMPOSIUM SERIES, 26 International congress of biochemistry 5th Proceedings Moscow 1961 Aug 10-16 Vol 6: mechanism of photosynthesis International union of biochemistry Edited by H. Tamiya Pergamon; PWN-Polish scientific publishers,Oxford;Warsaw, 1963
Bioch 33.1335

I.U.B.SYMPOSIUM SERIES, 27 International congress of biochemistry 5th Proceedings Moscow 1961 Aug 10-16 Vol 7: biosynthesis of lipids International union of biochemistry Edited by G. Popjak Pergamon;PWN-Polish scientific publishers, Oxford;Warsaw,1963
Bioch 33.1336

I.U.B.SYMPOSIUM SERIES, 28 International congress of biochemistry 5th Proceedings Moscow 1961 Aug 10-16 Vol 8: biochemical principles of the food industry International union of biochemistry Edited by V.L. Kretovich and E. Pijanowski Pergamon; PWN-Polish scientific publishers,Oxford;Warsaw, 1963
Bioch 33.1337

I.U.B.SYMPOSIUM SERIES, 29 International congress of biochemistry 5th Proceedings Moscow 1961 Aug 10-16 Vol 9: plenary sessions and abstracts of papers International union of biochemistry Pergamon; PWN-Polish scientific publishers,New York; Warsaw,1963
Bioch 33.1338

I.U.B.SYMPOSIUM SERIES, 31 Oxygen in the animal organism :a symposium Proceedings London 1963 Sep 1-5 International union of biochemistry and International union of physiological sciences Edited by Frank Dickens and Eric Neil held at Bedford college Pergamon press,Oxford,1964
Bioch 33.1080

I.U.B.SYMPOSIUM SERIES, 35 Pyridoxal catalysis:enzymes and model systems The International symposium on chemical and biological aspects of pyridoxical catalysis 2nd Proceedings Moscow 1966 Sep 15-21 Edited by E.E. Snell and others sponsored by the International union of biochemistry Interscience,New York,1968 First International symposium on chemical and biological aspects of pyridoxal catalysis published under title chemical and biological aspects of pyridoxical catalysis
Bioch 33.1038

I.U.G.S. see INTERNATIONAL UNION OF GEOLOGICAL SCIENCES

IAU GENERAL ASSEMBLY 11th News bulletin Berkeley,Calif. 1961 Aug. 1-8 International astronomical union Berkeley, 1961
Obs 6.1945

IAU GENERAL ASSEMBLY 13 Invited discources and discussions,and special meetings Highlights of astronomy Prague 1967 International astronomical union Edited by Lubos Perck 548p Reidel,Dordrecht,1968
Obs 6.3528

IAU-NASA SYMPOSIUM Nature of the lunar surface Greenbelt,Md. 1965 Apr 15-16 International astronomical union National aeronautics and space administration Edited by W.N. Hess and others Held at the Goddard space flight center Johns Hopkins press, Baltimore,Md.,1966
Min 10.1445

IAU-NASA SYMPOSIUM The Conference on the nature of the surface of the moon Proceedings Greenbelt,Md. 1965 Apr 15-16 Goddard space flight center and International astronomical union Edited by Wilmot N. Hess and others Sponsored bt the National aeronautics and space administration 320p Johns Hoplins press,Baltimore,1966
TA 15.0203

IBADAN 1964 New elites of tropical Africa International African seminar 6th Studies International African institute Edited by P. C. Lloyd Oxford university press,London,1966
Geog 13.4666

IBADAN 1966 The Population of tropical Africa conference 1st Population council Edited by John C. Caldwell and Chukuka Okonjo Sponsored by the University of Ibadan Longmans,London,1968
Geog 13.6814

IBADAN 1966 West African micropaleontological colloquium 2nd Proceedings Edited by J.E.van Hinte J.E. Brill,Leiden,1966
Geol 8.0344

IBADAN 1967 Systemic mycoses :a Ciba foundation symposium Ciba foundation Edited by G.E.W. Wolstenholme and Ruth Porter Churchill,London,1968
Phys 20.0998

IBADAN 1967 Systemic mycoses :a Ciba foundation symposium in commemoration of William Balfour Baikie Ciba foundation Edited by G.E.W. Wolstenholme and Ruth Porter illus,port Churchill,London,1968
Path 30.2632

ICE AND SNOW:PROPERTIES,PROCESSES,AND APPLICATION a conference held at the Massachusetts Institute of technology Proceedings Cambridge,Mass. 1962 Feb 12-16 Air force Cambridge research laboratories.Terrestrial sciences laboratory Edited by W.D. Kingery Sponsored also by the United States naval civil engineering laboratory illus xv,684p 24cm M.I.T.press,Cambridge,Mass.,1963
Sco 14.0269

ICE AND SNOW:PROPERTIES,PROCESSES AND APPLICATIONS :conference Proceedings Cambridge,Mass. 1962 Feb 12-16 Edited by W. D. Kingery Massachusetts institute of technology press,Cambridge,Mass.,1963
Geog 13.0652

ICE SYMPOSIUM Papers Pittsburgh 1966 Mar American chemical society.Division of colloid and surface chemistry Journal of colloid and interface science,25 no 2 illus 131-294p Academic,New York;London,1967 Symposium chairman H.H.G.Jellinek
Sco 14.7934

ICELAND AND MID-OCEAN RIDGES :a symposium Report Reykjavik 1967 Feb 27-Mar 8 Geoscience society of Iceland Edited by Sveinbjorn Bjornsson Societas scientiarum islandica.Publication, 38 Reykjavik,1967
Geod 9.0383

ICELAND MUSEUM OF NATURAL HISTORY North Atlantic biota and their history :a symposium Reykjavik 1962 Jul Sponsored by the Nato advanced institutes program pl. 430p 24cm Pergamon press,Oxford,1963
Sco 14.8334

ICONOSCOPES Kolloquium uber bildwandler und bildspeicherrohren vortrage und diskussionen Heidelberg 1958 Apr.28-29 Edited by Heinrich Siedentopf Heidelberger akademie der wissenschaften.Sitzungenberichte. Mathematisch- naturwissenschaftliche klasse, 1959,abh.5 31 plates, 4 tabs 78p Springer-verlag,Heidelberg,1959
Obs 6.1627

IDEAS IN MODERN BIOLOGY International congress of zoology 16th Proceedings Washington,D.C. 1963 Aug 20-27 Edited by J. A. Moore Doubleday,NEW York,1965
Radioth 35.0259

IDENTIFICATION AND CLASSICIFICATION OF SOILS Symposium on the identification and classification of soils Atlantic City,N.J. 1950 Jun 29 American society for testing materials American society for testing materials.Special technical publication, 113 American society for testing materials, Philadelphia,Pa.,1951
Eng 41.3146

IDENTIFICATION IN AUTOMATIC CONTROL SYSTEMS I.F.A.C. symposium on identification in automatic control systems Preprints Prague 1967 Jun 12-17 Pt 1-2 Ceskoslovenska akademie ved.Institute of information theory and automation Sponsored by International federation of automatic control 2 vols Academia,Prague,1967
Eng 41.6019

IEEE Comtech symposium on advances in data communications PAPERS New York 1970 Mar 27 Institute of electrical and electronics engineers.Communication technology group.Data communication committee illus 24p IEEE,New York,1970
Math L 5.3626

IEEE-INTERNATIONAL COMPUTER SOCIETY CONFERENCE 5th Proceedings Hardware,software, firmware trade-offs Boston,Mass. 1971 Sep 22-24 Institute of electrical and electronics engineers International computer society illus 204p IEEE,New York,1971
Math L 5.3628

IFAC-IFIP INTERNATIONAL CONFERENCE 2nd Proceedings Digital computer applications to process control,2 Menton 1967 Jun 5-9 International federation of automatic control International federation for information processing Edited by W.E. Miller Sponsored also by the Association francaise de regulation d'automatisme Instrument society of America,Pittsburgh,Pa.,1969
Eng 41.8206

IFAC SYMPOSIUM ON AUTOMATIC CONTROL IN THE PEACEFUL USES OF SPACE Peaceful uses of automation in outer space Stavanger 1965 Jun 21-24 International federation of automatic control Edited by John A. Aseltine Plenum,New York,1966
Eng 41.6002

IFAC SYMPOSIUM ON THE CONTROL OF DISTRIBUTED PARAMETER SYSTEMS Proceedings Banff 1971 Jun 21-23 International federation for automatic control 2 vols IFAC technical committee on theory,Banff,1971
Eng 41.8593

IFIP CONGRESS 1968 invited papers Edinburgh 1968 Aug 5-10 International federation for information processing 1600p 30cm North-Holland,Amsterdam,1968
Math L 5.3123

IFIP CONGRESS 62 proceedings Information processing 1962 Munich 1962 Aug 27-Sep 1 International federation for information processing Edited by Cicely M. Popplewell xvi,780p 30cm North-Holland publishing company,Amsterdam,1962
Math L 5.0912

IFIP CONGRESS 65 Proceedings Information processing 1965 New York 1965 May 24-25 Vol 1-2 Edited by Wayne A. Kalenich Organized by the International federation for information processing 2 vols Spartan; Macmillan,Washington,D.C.;London,1965-66
Eng 41.6001

IFIP CONGRESS 65 proceedings Information processing 1965 New York 1965 May 24-29 Vol. 1 International federation for information processing xv,304p 29cm Spartan books,Washington,D.C.,1965
Math L 5.1045

IFIP CONGRESS 71 Proceedings Information processing 1971 Ljubljana 1971 Aug 23-28 International federation for information processing Edited by C.V. Freiman illus 1621p 2 vols North-Holland,Amsterdam,1972
Math L 5.3841

IFIP WORKING CONFERENCE ON ALGOL AND ITS IMPLEMENTATION Proceedings ALGOL 68 implementation Munich 1970 Jul 20-24 International federation for information processing Edited by J.E.L. Peck illus 375p North-Holland,Amsterdam,1971
Math L 5.3876

IFIP WORKING CONFERENCE ON FORMAL LANGUAGE DESCRIPTION LANGUAGES proceedings Formal language description languages for computer programming Vienna 1964 Sep 15-18 International federation for information processing Edited by T.B. Steel bibliog 330p 22cm North-Holland publishing, Amsterdam,1966
Math L 5.1033

IFIP WORKING CONFERENCE ON SYMBOL MANIPULATION LANGUAGES Proceedings Symbol manipulation languages and techniques Pisa 1966 Sep 5-9 International federation for information processes Edited by Daniel G. Bobrow illus 487p North-Holland, Amsterdam,1968
Math L 5.3822

IGNEOUS INTRUSION Symposium on mechanism of igneous intrusion Proceedings Liverpool 1969 Jan 9-11 Liverpool geological society Edited by Geoffrey Newall and Nicholas Rast Geological journal.Special issue, 2 Liverpool geological society,Liverpool,1970
Min 10.1427

IGU REGIONAL CONFERENCE IN JAPAN Proceedings Tokyo 1957 Aug 28-Sep 3 Nara Science council of Japan,and,International geographical union Edited by Ryuziro Isida and others illus,maps vii,609p 26cm Tokyo,1959
Sco 14.6870

IJMUIDEN 1967 DECUS proceedings 1967 : European seminar Digital equipment computer users'society iii,85p Digital equipment computer users' society,Ijmuiden, 1967
Math L 5.3103

IKAHO 1965 Physiological control of iodine metabolism Gunma symposium on endocrinology 3rd Gunma university.Institute of endocrinology Gunma university.Institute of endocrinology.Annual report, 3 Institute of endocrinology,Gunma university,Maebashi, 1965
Phys 20.1427

ILE DE PORQUEROLLES 1951 Vibrations non lineaires Colloque international des vibrations non lineaires Actes Union internationale de radio-science Organised by the Union internationale de mecanique theorique et appliquee France.Ministere de l'air.Publications scientifiques et techniques, 281 Information technique de l'aeronautique, Paris,1953
Eng 41.6372

ILIKON CORPORATION,NATICK Fundamental phenomena in the materials sciences :a symposium 2nd,3rd Boston,Mass. 1964, 1965 Vol 2-3: surface phenomena Edited by L.J. Bonis and others 2 vols Plenum press,New York,1966
Met 25.1616

ILIKON CORPORATION,NATICK Fundamental phenomena in the materials sciences :a symposium 4th Proceedings Boston,Mass. 1966 Jan 31-Feb 1 Vol 4: fracture of metals,polymers,and glasses Edited by L.J. Bonis and others Plenum press,New York,1967
Met 25.0873

ILIKON CORPORATION,NATICK Symposium on fundamental phenomena in the material sciences 1st Proceedings Boston,Mass. 1963 Feb 1 Vol 1: sintering and plastic deformation Edited by L.J. Bonis and H.H. Hausner Plenum press,New York,1964
Met 25.2280

ILLINOIS INSTITUTE OF TECHNOLOGY.RESEARCH INSTITUTE Computer applications - 1962 Computer applications symposium 9th proceedings Edited by Milton M. Gutterman and Robert S. Hollitch illus. 224p 23cm Spartan books,Baltimore,Md.,1964 The IIT research institute was formerly called the Armour research foundation
Math L 5.0855

ILLINOIS INSTITUTE OF TECHNOLOGY.RESEARCH INSTITUTE Scanning electron microscope 1969 Annual scanning electron microscope symposium 2nd Chicago,Ill. 1969 Apr 29-May 1 Chicago,c1969
Met 25.2870

ILLINOIS INSTITUTE OF TECHNOLOGY.RESEARCH INSTITUTE Scanning electron microscopy symposia 1st-2nd Proceedings Chicago, Ill. 1968-69 2cm I.I.T.,Chicago,Ill., 1968-69
Eng 41.5623

ILLINOIS INSTITUTE OF TECHNOLOGY.RESEARCH INSTITUTE Scanning electron microscopy 1968 Scanning electron microscope :the instrument and its applications Chicago 1968 Apr 30-May 1 Chicago,1968
Met 25.2869

ILLINOIS INSTITUTE OF TECHNOLOGY.RESEARCH INSTITUTE Scanning electron microscopy 1970 Scanning electron microscope symposium 3rd Proceedings Chicago,Ill. 1970 Apr 28-30 I.I.T.research institute,Chicago,Ill., 1970
Eng 41.8204

IMAGE TUBES Photo-electronic image devices Symposium on image tubes and related devices held at the Imperial college 1st Proceedings London 1958 Sep 3-5 Edited by J.D. McGee and W.L. Wilcock Advances in electronics and electron physics, 12 Academic press,New York,1960
Cav 7.1456

IMMEDIATE AND LOW LEVEL EFFECTS OF IONIZING RADIATIONS :symposium Proceedings Venice 1959 Jun 22-26 International atomic energy agency Comitato nazionale per le riccerche nucleari Under the auspices of Unesco International journal of radiation biology.Supplement Taylor and Francis,London, 1960
Gen 34.2221

IMMUNCHEMIE :colloquium Mosach 1964 Apr 22-25 Gesellschaft fur physiologische chemie.Colloquium, 15 Springer,Berlin,1965
Radioth 35.0145

The IMMUNE RESPONSE AND ITS SUPPRESSION :an international symposium Davos 1968 Mar 25-28 Forschungsinstitut,Davos Edited by E. Sorkin Antibiotica et chematherapia, 15 illus,tables Karger,Basle;New York,1969
Path 30.2545

IMMUNITY,CANCER,AND CHEMOTHERAPY 1st :basic relationships on the cellular level Report Buffalo,N.Y. Universita di Milano. Istituto di farmacologia and State university of New York.School of pharmacy Edited by Enrico Mihich Academic press,New York;London,1967
Path 30.2744

IMMUNITY TO PROTOZOA :a symposium British society for immunology Edited by P.C. C. Garnham Blackwell,Oxford,1963
Mol 45.0165

IMMUNITY TO PROTOZOA :a symposium Papers British society for immunology Edited by P.C. C. Garnham and others Blackwell,Oxford,1963
Path 30.2335

IMMUNIZATION IN CHILDHOOD A Symposium on immunization in childhood held in the Wellcome building Proceedings London 1959 May 4-6 Livingstone,London,1960
PGMS 29.0283

IMMUNO-REPRODUCTION Conference on immuno-reproduction Proceedings La Jolla,Calif. 1962 Sep 9-11 Population council and Ford foundation Population council,New York, c1962 Organized by A.Tyler and K.A. Laurence and edited by the latter
Path 30.2340

IMPERIAL BUREAU OF SOIL SCIENCE Conference on soil science problems 1st proceedings Rothamsted 1930 Sep 16-18 Imperial bureau of soil science.Technical communications,17 H.M.S.O.,London,1931
Geog 13.1071

IMPERIAL CANCER RESEARCH FUND Sub-cellular components :preparation and fractionation London 1967 Nov Edited by G.D. Birnie 2nd edition revised,expanded Butterworth, London,1972 Based on a symposium held in 1967
Bioch 33.2275

IMPERIAL CANCER RESEARCH FUND Subcellular components;preparation and fractionation :a symposium London 1967 Nov Edited by G.D. Birnie and Sylvia M. Fox Butterworths,London, 1969
Bioch 33.1016

IMPERIAL CANCER RESEARCH FUND SYMPOSIUM 1st Proceedings Clinical evaluation in breast cancer London 1965 Oct 21-23 Edited by J. L. Haywood and R.D. Bulbrook illus. Academic press,London;New York,1966
Radioth 35.0870

IMPERIAL CANCER RESEARCH FUND SYMPOSIUM 2nd Proceedings Thyroid neoplasia London 1967 Apr Edited by Stretton Young and D.R. Inman Academic press,New York,1968
An 32.3780

IMPERIAL CANCER RESEARCH FUND SYMPOSIUM 2nd Proceedings Thyroid neoplasia London 1967 Apr Imperial cancer research fund Edited by Stretton Young and D.R. Inman Academic press,London;New York,1968
PGMS 29.0180

IMPERIAL CHEMICAL INDUSTRIES.AGRICULTURAL DIVISION Materials technology in steam reforming processes Proceedings 1964 Oct 21-22 Edited by C. Edeleanu Pergamon, Oxford,1966
Met 25.0104

IMPERIAL CHEMICAL INDUSTRIES.PHARMACEUTICALS DIVISION A Symposium on the evaluation of drug toxicity Macclesfield 1957 Oct 1 Edited by A.L. Walpole and A. Spinks Churchill,London,1958
Pha 16.0327

IMPERIAL COLLEGE OF SCIENCE AND TECHNOLOGY Airborne microbes :a symposium Papers London 1967 Apr Society for general microbiology Edited by P.H. Gregory and J.L. Monteith Society for general microbiology. Symposia, 17 Cambridge university press, Cambridge,1967
Bioch 33.1167

IMPERIAL COLLEGE OF SCIENCE AND TECHNOLOGY Foundations of statistical inference Joint statistics seminar :a discussion London 1959 Jul 27-28 By Leonard J. Savage and others 112p Methuen;Wiley,London;New York, 1964
WSM 43.1769

IMPERIAL COLLEGE OF SCIENCE AND TECHNOLOGY Molecular biology of viruses :a symposium Papers London 1968 Apr Society for general microbiology Edited by R.V. Crawford and M.G.P. Stoker Society for general microbiology.Symposia, 18 Cambridge university press,Cambridge,1968
Bioch 33.1178

IMPERIAL COLLEGE OF SCIENCE AND TECHNOLOGY New approaches in cell biology;a symposium The International congress of zoology 15th Proceedings London 1958 Jul Edited by P. M.B. Walker Academic press,London,1960 A section of the Proceedings of the 15th International congress of zoology
Bal 39.0289

IMPERIAL COLLEGE OF SCIENCE AND TECHNOLOGY Organization and control in prokaryotic and eukaryotic cells :a symposium Papers Society for general microbiology Edited by H. P. Charles and B.C.J.G. Knight Society for general microbiology.Symposia, 20 Cambridge university press,Cambridge,1970
Bioch 33.1919

IMPERIAL COLLEGE OF SCIENCE AND TECHNOLOGY Symposium on information theory 3rd Proceedings London 1955 Sep 12-16 Edited by Colin Cherry Butterworths,London, 1956
Psy 31.2003

IMPERIAL COLLEGE OF SCIENCE AND TECHNOLOGY The Foundations of statistical inference :a discussion opened by Professor L.J. Savage at a meeting of the Joint statistics seminar, Birkbeck and Imperial colleges,in the University of London Methuen's monographs on applied probability and statistics Methuen, London,1962
HE 27.0051

IMPERIAL COLLEGE OF SCIENCE AND TECHNOLOGY The Foundations of statistical inference :a discussion opened by Professor L.J.Savage at a meeting of the Joint statistics seminar, Birkbeck and Imperial colleges,in the University of London Methuen's monographs on applied probability and statistics 112p 20cm Methuen,London,1962
Math 3.1311

IMPERIAL COLLEGE OF SCIENCE AND TECHNOLOGY. ELECTRICAL ENGINEERING DEPARTMENT Symposium on information theory 4th Proceedings London 1960 Aug 29-Sep 2 Edited by Colin Cherry Butterworths,London, 1961
Psy 31.3386

IMPERIAL COLLEGE OF SCIENCE AND TECHNOLOGY. ELECTRICAL ENGINEERING DEPARTMENT Symposium on information theory Report of proceedings London 1950 Sep 26-29 IRE professional group on information theory. Publications committee,n.p.,c1950 Foreword by W.Jackson. Duplicated
Psy 31.3221

IMPERIAL COLLEGE OF SCIENCE AND TECHNOLOGY. TECHNICAL OPTICS SECTION Optical design with digital computers symposium proceedings London 1956 June 5-7 viii,93p 26cm Imperial college of science and technology, London,1956
Math L 5.0754

IMPERIAL COLLEGE OF SCIENCE AND TECHNOLOGY. TECHNICAL OPTICS SECTION Optical image assessment using frequency response techniques a summer school Proceedings London 1958 Jul.7-10 Edited by H.H. Hopkins 134p London,1958
Obs 6.1974

IMPERIAL COLLEGE OF SCIENCE AND TECHNOLOGY,LONDON Underwater acoustics :an institute conducted at the Imperial college Proceedings London 1961 Jul 31-Aug 11 Edited by Vernon M. Albers New York,1963
Geod 9.0238

IMPERIAL MALARIA CONFERENCE Proceedings Simla 1909 Oct 12-18 India.General malaria committee Govt.central branch press,Simla, 1910
Mol 45.0490

IMPLANTATION OF OVA :a conference Proceedings London 1957 Nov 27 Society for endocrinology Edited by P. Eckstein Society for endocrinology.Memoirs, 6 Cambridge university press,Cambridge,1959
Phys 20.1383

IMPLANTATION OF OVA :a conference Proceedings London 1957 Nov 27 Society for endocrinology Edited by P. Eckstein held at the Ciba foundation Society for endocrinology.Memoirs, 6 Cambridge university press,Cambridge,1959
Bal 39.1401

IMPROVING EXPERIMENTAL DESIGN AND STATISTICAL ANALYSIS Phi Delta Kappa annual symposium on educational research 7th Discussions Madison,Wisc. 1966 Phi Delta Kappa and University of Wisconsin.Phi Delta Kappa chapter Rand-McNally education series Rand-McNally,Chicago,Ill.,1967
Psy 31.1123

IMPURITIES AND IMPERFECTIONS a seminar held during the 36th national metal congress and exposition American society for metals American society for metals,Cleveland,1955
Met 25.1230

IN VITRO PROCEDURES WITH RADIOISOTOPES IN MEDICINE :symposium Proceedings Vienna 1969 Sep 8-12 International atomic energy agency World health organization IAEA,Vienna,1970
Bioch 33.2184

INCOMPATIBILITY IN FUNGI :a symposium Proceedings Edinburgh 1964 Aug Edited by Karl Esser and John R. Raper viii,124p 24cm Springer,Berlin,1965 Symposium held at the 10th International botanical congress
Bot 42.4713

INDIA 1968 International geographical congress 21st Abstracts of papers Calcutta,1968
Geog 13.6841

INDIA.DEPARTMENT OF ATOMIC ENERGY.COSMIC RAY COMMITTEE Symposium on cosmic rays, elementary particle physics and astrophysics 10th Proceedings Aligarh 1967 Dec 12-16 701p Bombay,1968
TA 15.0553

INDIA.DEPARTMENT OF ATOMIC ENERGY.SPECTROSCOPY DIVISION International conference on spectroscopy Bombay 1967 Jan 8-18 290p International council of scientific unions, Bombay,1968
TA 15.0636

INDIA.GENERAL MALARIA COMMITTEE General malaria committee meeting 3rd Proceedings Madras 1912 Nov 18-20 Govt.Central branch press,Simla,1913
Mol 45.0491

INDIA.GENERAL MALARIA COMMITTEE Imperial malaria conference Proceedings Simla 1909 Oct 12-18 Govt.central branch press, Simla,1910
Mol 45.0490

INDIA.NATIONAL AERONAUTICAL LABORATORY Seminar on aeronautical sciences proceedings Bangalore 1961 Nov 27-Dec.2 Vol 1 National aeronautical laboratory,Bangalore, 1961
A Math 4.0526

INDIAN METEOROLOGICAL SOCIETY Symposium on meteorology in relation to high level aviation over India and surrounding areas Papers New Delhi 1957 Dec 7 Meteorological department,India Edited by U.K. Bose 159p India press,Delhi,1962
Nap 11.0420

INDIAN NATIONAL SOCIETY OF SOIL MECHANICS AND FOUNDATION ENGINEERING Asian regional conference on soil mechanics and foundation engineering 1st Proceedings New Delhi 1960 Feb International society of soil mechanics and foundation engineering New Delhi,1960
Eng 41.3172

INDIAN SCIENCE CONGRESS ASSOCIATION,25TH SESSION The Second city of the empire Calcutta 1938 Edited by P.C. Bagchi illus. Calcutta,1938
Geog 13.4484

INDIANA UNIVERSITY Nonparametric techniques in statistical inference International symposium on nonparametric techniques in statistical inference 1st Proceedings Bloomington,Ind. 1969 Jun 1-6 Edited by Madan Lal Puri bibliog. xiv,623p 24cm Cambridge university press,Cambridge,1970
Math S 3.1616

INDIANA UNIVERSITY SCHOOL OF MEDICINE Symposium on regulation of enzyme activity and synthesis in normal and neoplastic liver 1st Proceedings Indianapolis,Ind. 1962 Oct 1-2 Edited by George Weber Advances in enzyme regulation, 1 Pergamon,Oxford,1963 Technical editor Catherine E.Forrest Weber
Bot 42.1738

INDIANA UNIVERSITY SCHOOL OF MEDICINE Symposium on regulation of enzyme activity and synthesis in normal and neoplastic liver 2nd Proceedings Indianapolis,Ind. 1963 Sep 30-Oct 1 Edited by George Weber Advances in enzyme regulation, 2 Pergamon, Oxford,1964 Technical editor Catherine E. Forrest,Weber
Bot 42.1739

INDIANAPOLIS 1963 Symposium on regulation of enzyme activity and synthesis in normal and neoplastic liver 2nd Proceedings Edited by George Weber Advances in enzyme regulation, 2 Pergamon,Oxford,1964
Radioth 35.0124

INDIANAPOLIS 1964 Regulation of enzyme activity and synthesis in normal and neoplastic tissues :a symposium 3rd Edited by George Weber Advances in enzyme regulation,3 Pergamon,Oxford,1965
Pha 16.0102

INDIANAPOLIS 1965 Regulation of enzyme activity and synthesis in normal and neoplastic tissues :a symposium 4th Edited by George Weber Advances in enzyme regulation,4 Pergamon,Oxford,1966
Pha 16.0103

INDIANAPOLIS 1966 Regulation of enzyme activity and synthesis in normal and neoplastic tissues :a symposium 5th Edited by George Weber Advances in enzyme regulation,5 Pergamon,Oxford,1967
Pha 16.0104

INDIANAPOLIS,IND. 1957 Low level irradiation :a symposium American association for the advancement of science United States atomic energy commission Edited by Austin M. Brues Co-sponsored by the Argonne national laboratory.Division of biological and medical research American association for the advancement of science. Publication, 59 American association for the advancement of science,Washington,D.C., 1959
Radioth 35.1109

INDIANAPOLIS,IND. 1962 Symposium on regulation of enzyme activity and synthesis in normal and neoplastic liver 1st Proceedings Edited by George Weber Held at Indiana university school of medicine Advances in enzyme regulation, 1 Pergamon, Oxford,1963 Technical editor Catherine E. Forrest Weber
Bot 42.1738

INDIANAPOLIS,IND. 1963 Autoregulation of blood flow :international symposium Proceedings Edited by Paul C. Johnson American heart association monographs, 8 American heart association,New York,1964
Phys 20.0898

INDIANAPOLIS,IND. 1963 Symposium on regulation of enzyme activity and synthesis in normal and neoplastic liver 2nd Proceedings Edited by George Weber Held at Indiana university school of medicine Advances in enzyme regulation, 2 Pergamon, Oxford,1964 Technical editor Catherine E. Forrest,Weber
Bot 42.1739

INDIA'S URBAN FUTURE Seminar on urbanization in India Berkeley,Calif. 1960 Jun 26-Jul 2 Edited by Roy Turner University of California press,Berkeley,Calif.;Los Angeles, Calif.,1962 Conference sponsored by Kingsley Davis and others
Geog 13.1947

INDUCTION,PHYSICS,AND ETHICS Colloquium in the philosophy of science Proceedings and discussions Salzburg 1968 Aug 28-31 International union of history and philosophy of science.Division of logic,methodology and philosophy of science Institut fur wissenschaftstheorie,Salzburg Edited by Paul Weingartner and Gerhard Zecha Synthese library x,382p Reidel,Dordrecht,1970
WSM 43.0992

INDUCTION OF MUTATIONS AND THE MUTATION PROCESS a symposium Proceedings Prague 1963 Sep 26-28 Edited by Jiri Veleminsky and Tomas Gichner Nakladatelstvi Ceskosloveske akademie ved,Prague,1965
Gen 34.1099

INDUKTION UND MORPHOGENESE Mosbach 1962 May 3-5 Gesellschaft fur physiologische chemie Gesellschaft fur physiologische chemie.Colloquium, 13 Springer,Berlin,1963
Gen 34.0857

INDUSTRIAL AND BUSINESS PSYCHOLOGY International congress of applied psychology 14th Proceedings Copenhagen 1961 Aug 13-19 International association of applied psychology and Danish psychological association Edited by Gerhard S. Nielsen Munksgaard,Copenhagen,1962
Psy 31.2252

INDUSTRIAL CARBON AND GRAPHITE papers read at the conference... London 1957 Sep 24-26 Society of chemical industry Society of chemical industry,London,1958
Met 25.0155

INDUSTRIAL COMMUNITY RELATIONSHIPS Triennial congress...on the subject of fundamental relationships between all sections of the industrial community 1st Report Cambridge 1928 Jun 27-Jul 3 International association for the study and improvement of human relations and conditions in industry Secretariat I.R.I.,The Hague,1928
Psy 31.2516

INDUSTRIAL COMPUTATION SEMINAR proceedings Endicott,N.Y. 1950 Sept 25-29 Sponsored by International business machines corporation 103p 28cm International business machines corporation,New York,1951
Math L 5.0701

INDUSTRIAL EXPERIMENTS Experimental designs in industry Symposium on design of industrial experiments proceedings Raleigh, N.C. 1956 Nov.5-9 North Carolina state college.Institute of statistics Edited by Victor Chew Supported by the United States. Air force.Office of scientific research Wiley publications in statistics Bibliog. xi,268p 28cm John Wiley and sons,New York, 1958 Being parts A and B of Symposium on design of industrial experiments, proceedings, edited by Victor Chew,1957
Math 3.0583

INDUSTRIAL EXPERIMENTS Symposium on design of industrial experiments proceedings Raleigh,N.C. 1956 Nov.5-9 North Carolina state college.Institute of statistics Edited by Victor Chew Supported by the United States.Air force.Office of scientific research Bibliog. viii,374p 27cm Institute of statistics,University of North Carolina, Raleigh,N.C.,1957
Math 3.0584

INDUSTRIAL MEASUREMENT TECHNIQUES Conference on industrial measurement techniques for on-line computers London 1968 Jun 11-13 Institution of electrical engineers Institution of electrical engineers.Conference publications, 43 I.E.E.,London,1968
Eng 41.2220

INDUSTRIAL PROCESS DESIGN FOR POLLUTION CONTROL a workshop Proceedings New York 1967 Feb 9-10 Vol 1 American institute of chemical engineers.Water committee iv,59p A.I.Ch.E.,New York,1967
Chem E 24.1575

INDUSTRIAL PULMONARY DISEASES :a symposium London 1957 Sep 18-20 London 1958 Mar 25-27 Edited by E.J. King and C.M. Fletcher Illus Churchill,London,1960
HE 27.0030

INDUSTRIALIZATION AND SOCIETY Conference on social implications of industrialization and technical change Papers Chicago,Ill. 1960 Sep 15-22 Edited by Bert F. Hoselitz and Wilbert E. Moore Sponsored by Unesco Unesco,Mouton,1963
Eng 41.1052

INDUSTRIALIZATION AND SOCIETY Conference on the social implications of industrialization and technological change Papers Chicago, Ill. 1960 Sep 15-22 University of Chicago Edited by Bert F. Hoselitz and Wilbert E. Moore Sponsored jointly by Unesco Mouton, The Hague,1963
Geog 13.1608

INDUSTRIALIZED BUILDING AND THE STRUCTURAL ENGINEERS :a symposium London 1966 May 17-19 Institution of structural engineers Institution of structural engineers,London,1966
Eng 41.3381

INDUSTRY FIGHTS CORROSION Corrosion convention Proceedings London 1937 Oct 15-16 Sponsored by Corrosion technology Corrosion technology,London,1957
Eng 41.3719

INDUSTRY FIGHTS CORROSION The Corrosion convention Proceedings London 1957 Oct 15-16 Corrosion technology Corrosion technology,London,c1957
Met 25.1890

INELASTIC SCATTERING OF NEUTRONS Symposium on inelastic scattering of neutrons Bombay 1964 Dec 15-19 Vol 1-2 International atomic energy agency Edited by M.M. Brown International atomic energy agency,Vienna,1965
Cav 7.0267

INELASTIC SCATTERING OF NEUTRONS IN SOLIDS AND LIQUIDS :a symposium Proceedings Vienna 1960 Oct 11-14 International atomic energy agency Vienna,1961
Cav 7.2392

INEQUALITIES :a symposium Proceedings Ohio 1965 Aug 19-27 Edited by Oved Shisha Sponsored by Aerospace research laboratories Bibliog xiv,360p 24cm Academic press, London;New York,1967
P Math 2.2925

INEQUALITIES II Symposium on inequalities 2nd Proceedings 1967 Aug 14-22 Edited by Oved Shisha Held at the United States air force academy xvi,439p 24cm Academic press,New York;London,1970
P Math 2.3847

INEQUALITIES 3 Symposium on inequalities 3rd Proceedings Los Angeles 1969 Sep 1-9 University of California,Los Angeles Edited by Oved Shisha xxii,368p 24cm Academic press,New York;London,1972
P Math 2.4214

INFANT BEHAVIOUR Determinants of infant behaviour :proceedings of a Tavistock study group on mother-infant interaction London 1959 Sep Tavistock clinic.Child development research unit Edited by B.M. Foss Methuen; Wiley,London;New York,1961
Psy 31.1203

INFANT BEHAVIOUR Determinants of infant behaviour 2 Tavistock study group on mother-infant interaction 2nd Proceedings London 1961 Sep Tavistock clinic.Child development research unit Edited by B.M. Foss Methuen;Wiley,London;New York,1963
Psy 31.1208

INFANT BEHAVIOUR Determinants of infant behaviour 3 Tavistock study group on mother-infant interaction 3rd Proceedings London 1963 Sep Tavistock clinic.Child development research unit Edited by B.M. Foss Methuen;Wiley,London;New York,1965
Psy 31.1205

INFANTILE MORTALITY National conference on infantile mortality Report of the proceedings Westminster 1906 Jun 13-14 King,Westminster,1906 President of the conference:Rt.hon.J.Burns,M.P.Chairman:E. Spicer,J.P.
Path 30.0540

INFINITISTIC METHODS Symposium on foundations of mathematics proceedings Warsaw 1959 Sep 2-9 Polska akademia nauk. Instytut matematyczny Under the auspices of the International mathematical union 362p 26cm Pergamon press,Oxford,1961
P. Math 2.0211

The INFLUENCE OF HEREDITY ON DISEASE,WITH SPECIAL REFERENCE TO TUBERCULOSIS, CANCER AND DISEASES OF THE NERVOUS SYSTEM :a discussion London 1908 Nov Royal society of medicine illus. Longmans,London,1909
Gen 34.1979

The INFLUENCE OF HORMONES ON ENZYMES :a conference Papers New York 1951 Jun 5-6 By R.I. Dorfman and others New York academy of sciences New York academy of sciences.Annals, 54,p.531-728 New York,1951
Bioch 33.0450

INFORMAL CONFERENCE ON ALGOL 68 IMPLEMENTATIONS Proceedings Vancouver,B.C. 1969 Aug 29-30 University of British Columbia.Department of computer science Edited by J.E.L. Peck illus 119p University of British Columbia. Department of computer science,Vancouver,B.C., 1969
Math L 5.3845

INFORMATICS INC. Computer graphics:utility, production,art The Graphic symposium 3rd Papers Los Angeles,Calif 1966 Edited by Fred Gruenberger 225p Thompson book co.; Academic press,Washingtom,D.C.;London,1967
TA 15.0098

INFORMATICS INCORPORATED Computer graphics: utility,production,art Graphics symposium Los Angeles n.d. University of California Edited by Fred Gruenberger 225p 22cm Academic press,London,1967
Math L 5.3273

INFORMATICS INCORPORATED Computer graphics: utility,production,art Graphics symposium Los Angeles,Calif. 1967 University of California Edited by Fred Gruenberger Academic press,London,1967
Eng 41.2210

INFORMATICS INCORPORATED Computers and communications - toward a computer utility Computer communications symposium Los Angeles 1967 Mar 20-22 Edited by Fred Gruenberger 219p 23cm Prentice-Hall, Englewood Cliffs,N.J.,1968
Math L 5.3117

INFORMATICS INCORPORATED On-line computing systems symposium proceedings Los Angeles, Calif. 1965 Feb 2-4 University of California Edited by Eric Burgess Data processing library series 152p 28cm American data processing,Detroit,Mich.,1965
Math L 5.1031

INFORMATION Aspects of inductive logic Edited by Jaakko Hintikka and Patrick Suppes Studies in logic and the foundations of mathematics vii,320p 23cm North-Holland, Amsterdam,1966 Includes papers read at the International symposium on confirmation and information,Helsinki,1965
P Math 2.2678

INFORMATION AND CONTROL PROCESSES IN LIVING SYSTEMS :first interdisciplinary conference Proceedings Princeton,N.J. 1965 Feb 28-Mar 3 New York academy of sciences Edited by Diane M. Ramsey New York academy of sciences interdisciplinary communications program,New York,1967
Psy 31.0142

INFORMATION AND DECISION PROCESSES symposium papers Lafayette,Ind. 1959 Apr. Purdue university Edited by Robert E. Machol xi,185p 23cm McGraw-Hill,New York,1960
Math 3.0683

INFORMATION AND PREDICTION IN SCIENCE Symposium de l'Academie internationale de philosophie des sciences Proceedings Brussels 1962 Sep 3-8 Academie internationale de philosophie des sciences Edited by S. Dockx and P. Bernays xi,272p Academic press,London;New York,1965
WSM 43.1136

INFORMATION PROCESSING Information processing 1962 IFIP congress 62 proceedings Munich 1962 Aug 27-Sep 1 International federation for information processing Edited by Cicely M. Popplewell xvi,780p 30cm North-Holland publishing company,Amsterdam,1962
Math L 5.0912

INFORMATION PROCESSING Information processing 1965 IFIP congress 65 proceedings New York 1965 May 24-29 Vol. 1 International federation for information processing xv,304p 29cm Spartan books,Washington,D.C.,1965
Math L 5.1045

INFORMATION PROCESSING International conference on information processing Proceedings Paris 1959 Jun15-20 Unesco Unesco,Paris,1960
Eng 41.5999

INFORMATION PROCESSING International conference on information processing proceedings Paris 1959 June 15-20 Unesco 520p 29cm UNESCO,Paris,1960
Math L 5.0809

INFORMATION PROCESSING IN SIGHT SENSORY SYSTEMS Proceedings of the symposium on information processing in sight sensory systems Proceedings Pasadena,Calif. 1965 Nov 1-3 National institutes of health and California institute of technology Edited by P.W. Nye California institute of technology, Pasadena,Calif.,1965
Psy 31.0419

INFORMATION PROCESSING IN THE NERVOUS SYSTEM International congress of physiological sciences 2nd Proceedings Leiden 1962 Sep 10-17 Edited by R.W. Gerard and J.W. Duyff Under the auspices of the International union of physiological sciences International union of physiological sciences. Proceedings, 3 International congress series, 49 Excerpta medica foundation, Amsterdam,1964
An 32.4392

INFORMATION PROCESSING IN THE NERVOUS SYSTEM : a symposium Proceedings Buffalo,N.Y. 1968 Oct 21-24 Edited by K.N. Leibovic held at the State university of New York Springer,Berlin,1969
Bal 39.1019

INFORMATION PROCESSING IN THE NERVOUS SYSTEM : symposium Proceedings International union of physiological sciences Edited by R. W. Gerard and J.W. Duyff International congress series,49 Excerpta medica,Amsterdam, 1960
Pha 16.0433

INFORMATION PROCESSING MACHINES symposium proceedings Prague 1964 Sep 7-9 Ceskoslovenska akademie ved Edited by Vladimir Bubenik Czechoslovak academy of sciences,Prague,1965
Math L 5.1011

INFORMATION PROCESSING 1965 IFIP congress 65 Proceedings New York 1965 May 24-25 Vol 1-2 Edited by Wayne A. Kalenich Organized by the International federation for information processing 2 vols Spartan; Macmillan,Washington,D.C.;London,1965-66
Eng 41.6001

INFORMATION PROCESSING 1971 IFIP congress 71 Proceedings Ljubljana 1971 Aug 23-28 International federation for information processing Edited by C.V. Freiman illus 1621p 2 vols North-Holland,Amsterdam,1972
Math L 5.3841

INFORMATION RETRIEVAL:A CRITICAL VIEW Annual colloquium on information retrieval 3rd Philadelphia,Pa. 1966 May 12-13 Edited by George Schecter Sponsored by the Association for computing machinery.Special interest group on information retrieval Academic press,New York,1967
Eng 41.7692

INFORMATION SCIENCES Software engineering Symposium on computer and information sciences 3rd Proceedings Miami,Fla. 1969 Dec Vol 1-2 Edited by Julius T. Tou COINS, 3 illus 2 vols Academic press,New York,1970
Math L 5.3830

INFORMATION THEORY A Symposium on information theory Papers read London 1955 Sep 12-16 Edited by Colin Cherry Held at the Royal institution Butterworths,London, 1956
Bal 39.1005

INFORMATION THEORY Colloquium on information theory Proceedings Debrecen 1967 Sep 19-24 Bolyai Janos matematikai tarsulat bibliog.,illus. 24cm 2 vols Janos Bolyai mathematical society,Budapest,1968
Math S 3.1457

INFORMATION THEORY Current trends in information theory conference 7th lectures Pittsburgh,Pa. 1953 Feb.20-21 By Brockway McMillan and others University of Pittsburgh.Department of psychology Current trends in psychology series xii,188p 22cm University of Pittsburgh press, Pittsburgh,Pa.,1953 7th annual conference on Current trends in psychology
Math 3.0767

INFORMATION THEORY International symposium on circuit and information theory transactions Los Angeles 1959 Jun.16-18 Institute of radio engineers Co-sponsored by the International scientific radio union Institute of radio engineers.Transactions on information theory,IT, 5,special supplement xiv,298p Institute of radio engineers,New York,1959 Papers published in advance of the Symposium
Math L 5.2071

INFORMATION THEORY London symposium on information theory 4th papers London 1960 Aug 29 - Sep 2 Seven typescript papers
Math L 5.0858

INFORMATION THEORY Symposium on information theory 1st report of proceedings, mimeograph London 1950 Sep 26-29 Edited by Willis Jackson Supported by Great Britain. Ministry of supply 206p 33cm Ministry of supply,London,1950
Math 3.0774

INFORMATION THEORY Symposium on information theory 3rd papers London 1955 Sep 12-16 Edited by Colin Cherry xii,401p 25cm Butterworths,London,1956
Math 3.0761

INFORMATION THEORY Symposium on information theory 4th Proceedings London 1960 Aug 29-Sep 2 Imperial college of science and technology.Electrical engineering department Edited by Colin Cherry Butterworths,London, 1961
Psy 31.3386

INFORMATION THEORY Symposium on information theory 4th papers London 1960 Aug.29-Sep.2 Edited by Colin Cherry xi,476p 25cm Butterworths,London,1961
Math 3.0762

INFORMATION THEORY Symposium on information theory Report of proceedings London 1950 Sep 26-29 Imperial college of science and technology.Electrical engineering department Institute of radio engineers IRE professional group on information theory. Publications committee,n.p.,c1950 Foreword by W.Jackson. Duplicated
Psy 31.3221

INFORMATION THEORY Symposium on information theory proceedings London 1950 Sep 26-29 Institute of radio engineers Institute of radio engineers.Professional group on information theory, 1 208p Institute of radio engineers,New York,1953
Math L 5.2079

INFORMATION THEORY Symposium on information theory symposium papers London 1955 Sep 12-16 Edited by Colin Cherry xii,401p 25cm Butterworths,London,1956
Math L 5.0760

INFORMATION THEORY Symposium on information theory 3rd Proceedings London 1955 Sep 12-16 Imperial college of science and technology Edited by Colin Cherry Butterworths,London,1956
Psy 31.2003

INFORMATION THEORY IN BIOLOGY Symposium on information theory in biology Gatlinburg, Tenn. 1956 Oct 29-31 Edited by H.P. Yockey and others Sponsored by the Oak Ridge national laboratory London,1958 Consists of articles representing authors' results and opinions and additional papers not given at the symposium "The conference was entitled A symposium on information theory in health physics and radiobiology"
Bal 39.1006

INFORMATIONAL MACROMOLECULES :a symposium Rutgers,N.J. 1962 Sep 5-7 Institute of microbiology of Rutgers Edited by H.J. Vogel and others With support from the National science foundation Academic press,New York, 1963
Gen 34.0664

INFORMATIONAL MACROMOLECULES :a symposium Proceedings New Brunswick,N.J. 1962 Sep 5-7 Rutgers university.Institute of microbiology Edited by Henry J. Vogel and others With support from the National science foundation Academic press,New York; London,1963
Bioch 33.0822

INFORMATIVE MOLECULES IN BIOLOGICAL SYSTEMS
International symposium on uptake of informative molecules by living cells Proceedings Mol,Belgium 1970 Aug 31-Sep 2 North Atlantic treaty organization.Scientific affairs division Edited by L. Ledoux North-Holland,Amsterdam,1971
Bioch 33.2240

INFRARED ASTRONOMY A Discussion on infrared astronomy London 1966 May 1-2 Edited by H. Massey and J. Ring Philosophical transactions,A., 264,p 107-320 Royal society, London,1969
Obs 6.3411

INHIBITION IN THE NERVOUS SYSTEM AND GAMMA-AMINOBUTYRIC ACID :an international symposium Proceedings Duarte,Calif. 1959 May 22-24 Edited by Eugene Roberts Sponsored by the United States.Air force. Office of scientific research Pergamon press, London,1960
An 32.4362

INHIBITION IN THE NERVOUS SYSTEM AND GAMMA-AMINOBUTYRIC ACID :an international symposium held at the City of Hope medical center Proceedings Duarte,Calif. 1959 May 22-24 Edited by Eugene Roberts and others Sponsored by the United States.Air force.Office of scientific research Pergamon press,Oxford,1960
Bal 39.1471

INHIBITIONS IN CELL RESEARCH :colloquium Mosbach 1969 Apr 14-16 Gesellschaft fur biologische chemie Edited by T. Bucher and H. Sies Gesellschaft fur biologische chemie. Colloquia, 20 Springer,Berlin,1969
Bioch 33.1908

INHIBITIONS IN CELL RESEARCH :colloquium Mosbach 1969 Apr 14-16 Gesellschaft fur biologische chemie Edited by T. Bucher and H. Sies Gesellschaft fur biologische chemie. Colloquia, 20 Springer,Berlin,1969
Bot 42.1758

INITIAL EFFECTS OF IONIZING RADIATIONS ON CELLS
International symposium on primary and initial effects of ionizing radiations on living cells Papers:discussions Moscow 1960 Oct Edited by R.J.C. Harris Sponsored by the Akademiya nauk S.S.S.R. Academic press,London;New York,1961
An 32.3277

INITIAL EFFECTS OF IONIZING RADIATIONS ON CELLS
The International symposium on primary and initial effects of ionizing radiations on living cells Papers and discussions Moscow 1960 Oct Edited by R.J.C. Harris Academic press,London;New York,1961 A symposium supported by Unesco and the IAEA and sponsored by the Academy of sciences of the U.S.S.R.
Bioch 33.0961

The INITIAL EFFECTS OF IONIZING RADIATIONS ON CELLS :a symposium Moscow 1960 Oct Unesco International atomic energy agency Edited by R.J.C. Harris Sponsored by the Akademiya nauk S.S.S.R. Academic press, London;New York,1961
Gen 34.2222

The INITIAL EFFECTS OF IONIZING RADIATIONS ON CELLS :a symposium Moscow 1960 Oct Unesco International atomic energy agency Edited by R.J.C. Harris Sponsored by the Akademiya nauk S.S.S.R. Academic press, London;New York,1961
Radioth 35.1127

INNSBRUCK 1931 Commission de l'annee polaire 1932-33 proces-verbaux des seances appendice H Organisation meteorologique internationale E.Ijdo,Leyden,1932
Sco 14.0498

INNSBRUCK 1931 Rapport de la commission internationale de l'annee polaire 1932-33 Vol 2: compte-rendu des travaux de la commission pendant sa deuxieme annee Organisation meteorologique internationale illus,diagrs 188p Ijdo,Leyden,1932
Sco 14.0492

INNSBRUCK 1931 Locarno 1931 Sep Comite meteorologique internationale :seances.. et rapport de 3 commissions et d'une sous-commission Proces-verbaux Comite meteorologique internationale,10 385p 24cm Leyden,1932 In English,French and German
Sco 14.0113

INORGANIC CHEMISTRY OF NITROGEN Recent aspects of the inorganic chemistry of nitrogen a symposium Report Cambridge 1957 Apr 10-12 Chemical society Chemical society. Special publication,10 Chemical society, London,1957 Organized by Dr.Maddock and Dr.Sharpe
Chem 18.1219

INORGANIC CHEMISTRY OF SULPHUR Developments in aromatic chemistry.Applications of electron and nuclear resonance in chemistry.Recent work on the inorganic chemistry of sulphur :a symposium Bristol 1958 Chemical society Chemical society.Special publication,12 Chemical society,London,1958
Chem 18.1229

INORGANIC MATERIALS International course on materials science 1st Essays Tremezzo 1968 Sep 8-20 Accademia nazionale dei lincei. Donegani foundation Edited by Alan W. Searcy and others illus xxv,715p 24cm Interscience,New York;Chichester,1970 Being the 11th course organized by the Donegani foundation
Met 25.2593

INORGANIC MATERIALS RESEARCH DIVISION SERIES, 4
International materials symposium :the structure and chemistry of solid surfaces 4th Proceedings Berkeley,Calif 1968 Jun 17-21 Lawrence radiation laboratory. Inorganic materials research division and University of California.College of chemistry Edited by Gabor A. Somorjai Wiley,New York, 1969
Met 25.2476

INORGANIC NITROGEN METABOLISM A Symposium on inorganic nitrogen metabolism :function of metallo-flavoproteins Papers and informal discussions presented Baltimore,Md. 1955 Jun 21-23 Edited by William D. McElroy and H.Bentley Glass Sponsored by the McCollum-Pratt institute McCollum-Pratt institute.Contribution, 125 Johns Hopkins press,Baltimore,Md.,1956
Bioch 33.0598

INORGANIC NITROGEN METABOLISM A Symposium on inorganic nitrogen metabolism:function of metalloflavoproteins Baltimore,Md. 1955 Jun 21-23 Edited by William D. McElroy and Bentley Glass Sponsored by the McCollum-Pratt institute McCollum-Pratt institute. Contributions, 125 Johns Hopkins press, Baltimore,Md.,1956
Gen 34.0659

INORGANIC NITROGEN METABOLISM:FUNCTION OF METALLO-FLAVOPROTEINS :a symposium Baltimore, Md. 1955 Jun 21-23 Edited by William D. McElroy and Bentley Glass Sponsored by the McCollum-Pratt institute McCollum-Pratt institute.Contributions, 125 Johns Hopkins press,Baltimore,Md.,1956
Mol 45.0279

INORGANIC POLYMERS :an international conference Lectures Nottingham 1961 Jul 18-21 University of Nottingham,and, Chemical society Chemical society.Special publication,15 Chemical society,London,1961
Chem 18.1221

INPUT-OUTPUT Structural interdependence and economic development International conference on input-output techniques 3rd Proceedings Geneva 1961 Sep United Nations.Secretariat Edited by Tibor Barna and others Sponsored jointly by the Harvard economic research project Macmillan,London, 1963
Geog 13.2417

INQUA-NCAR SYMPOSIUM ON CAUSES OF CLIMATIC CHANGE Causes of climatic change :a collection of papers derived from the INQUA-NCAR symposium on causes of climatic change,August 30-31, Boulder,Colorado American meteorological society Edited by J.Murray Mitchell American meteorological society.Meteorological monographs, 8,no 30 Boston,Mass.,1968
Bot 42.6236

INQUA see INTERNATIONAL ASSOCIATION FOR QUATERNARY RESEARCH

INQUA see INTERNATIONAL ASSOCIATION FOR QUATERNARY RESEARCH

INQUA-NCAR SYMPOSIUM ON CAUSES OF CLIMATIC CHANGE International union for quaternary research congress :a collection of papers derived from the INQUA-NCAR symposium 7th Proceedings Boulder,Colo. 1965 Aug 14-Sep 19 5: causes of climatic change International union for quaternary research,and,National center for atmospheric research Edited by J.Murray Mitchell Sponsored by the National research council Meteorological monographs,8,no 30 illus iv,159p 29cm American meteorological association,Boston,Mass.,1968 For other volumes of these proceedings,see also 'International association for quaternary research'
Sco 14.7750

INSECT BEHAVIOUR London 1965 Sep 23-24 Royal entomological society of London Edited by P.T. Haskell Royal entomological society of London.Symposia, 3 Royal entomological society,London,1966
Gen 34.1650

INSECT BIOCHEMISTRY Aspects of insect biochemistry :a symposium London 1965 Apr 1 Biochemical society Edited by T.W. Goodwin Biochemical society.Symposia, 25 Academic press,London;New York,1965
Bioch 33.1387

INSECT PATHOLOGY International conference of insect pathology and biological control 1st Transactions Prague 1958 Aug 13-18 24cm Prague,1958
Philos 1.1598

INSECT POLYMORPHISM symposium Papers and discussions London 1961 Sep 21-22 Royal entomological society of London Edited by J.S. Kennedy Royal entomological society of London.Symposia, 1 Royal entomological society,London,1961
Gen 34.1640

INSECT REPRODUCTION :symposium London 1963 Sep 19-20 Royal entomological society of London Edited by K.C. Highnam Royal entomological society of London.Symposia, 2 Royal entomological society,London,1964
Gen 34.1647

INSECTS International congress of biochemistry 4th Proceedings Vienna 1958 Sep 1-6 Vol 12: symposium 12 - biochemistry of insects International union of biochemistry Edited by L. Levenbook I.U. B.symposium series, 14 Pergamon,London,1959 Added title page in French and German.Text in English and German
Bioch 33.1326

INSTABILITE GRAVITATIONNELLE ET FORMATION DES ETOILES, DES GALAXIES ET DE LEURS STRUCTURES CARACTERISTIQUES Colloque international d'astrophysique 4e Communications Liege 1966 Jun 20-22 Universite de Liege.Institut d'astrophysique Institut d'astrophysique, Cointe-Sclessin,1967 Papers in English
A Math 4.1170

L' INSTINCT DANS LA COMPORTEMENT DES ANIMAUX ET DE L'HOMME :colloque international actes 1954 Jun By M. Autori and others Fondation Singer-Polignac Masson,Paris,1956 Organised by Le Conseil de la Fondation Singer-Polignac
Bal 39.1594

L' INSTINCT DANS LE COMPORTEMENT DES ANIMAUX ET DES HOMMES :colloque international Actes 1954 Jun By M. Autuori and others Fondation Singer-Polignac Masson,Paris,1956
Psy 31.0644

INSTITUT DE PHYSIQUE SOLVAY Structure et evolution de l'univers Conseil de physique 11e Rapports Brussels 1958 Jun 9-13 Edited by R. Stoops 309p Brussels,1958
Obs 6.3255

INSTITUT DE RECHERCHE D'INFORMATIQUE ET D'AUTOMATIQUE Symposium on automatic demonstration Versailles 1968 Dec Edited by M. Laudet and others Lecture notes in mathematics, 125 Bibliog 310p 25cm Springer-Verlag,Berlin,1970
P Math 2.3614

INSTITUT DE RECHERCHES DE LA SIDERURGIE FRANCAISE
Le Frottement interieur des metaux
Comptes rendus Saint-Germain-en-Laye 1960 Oct 13-14 Edited by C. Crussard and others
Editions metaux,n.p.,c1961
Met 25.1026

INSTITUT D'OPTIQUE THEORIQUE ET APPLIQUEE La Theorie des images optiques Memoires Paris 1946 Oct 21-26 Edited by Pierre Fleury
Financed by the Rockefeller foundation
Colloques internationaux du centre de la recherche scientifique Editions de la revue d'optique,Paris,1949
Cav 7.1758

INSTITUT D'OPTIQUE THEORIQUE ET APPLIQUEE
Optical properties and electronic structure of metals and alloys :International colloquium
Proceedings Paris 1965 Sep 13-16 Edited by F. Abeles Co-sponsored by the United States.Air force.European office of aerospace research North Holland,Amsterdam,1966
Cav 7.0004

INSTITUT FUR METALLKUNDE UND METALLPHYSIK DER TECHNISCHEN UNIVERSITAT CLAUSTHAL
Texturen in forschung und praxis international symposium Proceedings Clausthal-Zellerfeld 1968 Oct 2-5 Edited by J. Grewen and G. Wassermann Springer-Verlag,Berlin;Heidelberg, 1969
Met 25.2363

INSTITUT FUR METEOROLOGIE UND GEOPHYSIK
Stratospheric and mesospheric circulation
The International symposium on stratospheric and mesospheric circulation Proceedings Berlin 1962 Aug 20-31 Edited by Richard Scherhag and Gunter Warnecke Institut fur meteorologie und geophysik.Meteorologische abhandlungen,36 illus vi,644p 31cm
Dietrich Reimer,Berlin,1963
Sco 14.0527

INSTITUT FUR PHILOSOPHIE Naturwissenschaft und philosophie Internationales symposium uber naturwissenschaft und philosophie...
Beitrage Leipzig 1959 Oct 8-11 Edited by Gerhard Harig and Josef Schleifstein 437p
Akademie verlag,Berlin,1960 Symposium held to commemorate the 550th anniversary of the Karl-Marx-universitat
WSM 43.0976

INSTITUT FUR WISSENSCHAFTSTHEORIE,SALZBURG
Induction,physics,and ethics Colloquium in the philosophy of science Proceedings and discussions Salzburg 1968 Aug 28-31
Edited by Paul Weingartner and Gerhard Zecha
Synthese library x,382p Reidel,Dordrecht, 1970
WSM 43.0992

INSTITUT GEOGRAFII SIBIRI I DALNEGO VOSTOKA
Problemy geografii Sibiri i Dalnego Vostoka : itogi pervogo nauchnogo sovetshchaniya geografov Sibiri i Dalnego Vostoka 1st 1960 Edited by I.P. Gerosimov and others
136p Irkutskoe knizhnoe izdatelstvo,Irkutsk, 1960
Sco 14.4733

INSTITUT GEOLOGII I GEOFIZIKI,NOVOSIBIRSK
Problemy okhrany prirody Sibiri i Dalnego Vostoka 4th Novosibirsk 1961 Sep 15-19
Edited by G.V. Krylov illus 244p 27cm
Izdatelstvo Sibirskogo otdelenie AN SSSR, Novosibirsk,1963
Sco 14.7681

INSTITUT HENRI POINCARE.SEMINAIRE DE THEORIES PHYSIQUES La Cybernetique:theorie du signal et de l'information :reunions d'etudes et de mises au point Paris 1951 Apr.
Edited by Louis de Broglie vi,317p 22cm
Revue d'optique theorique et instrumentale, Paris,1951
Math L 5.0169

INSTITUT INTERNATIONAL D'ANTHROPOLOGIE:SESSION 2e Prague 1924 Sep 14-21 Nourry, Paris,1926
An 32.2639

INSTITUT INTERNATIONAL DE CHIMIE SOLVAY
Nucleoproteines Conseil de chimie 11eme Brussels 1959 Jun 1-6 Edited by R. Stoops
Interscience;Stoops,New York;Brussels,1960
Radioth 35.0100

INSTITUT INTERNATIONAL DE CHIMIE SOLVAY
Proteines Conseil de chimie 9e Rapports et discussions Brussels 1953 Apr 6-14
Edited by R. Stoops Brussels,1953
Bioch 33.0522

INSTITUT INTERNATIONAL DE CHIMIE SOLVAY
Transfert d'energie dans les gaz 12th Brussels 1962 Nov 5-10 Solvay conference on chemistry,1962 Interscience;R.Stoops,New York;Brussels,c1962
Chem 18.0514

INSTITUT INTERNATIONAL DE CHIMIE SOLVAY.COMITE SCIENTIFIQUE Quelques problemes de chimie minerale :conseil de chimie 10e Rapports et discussions Brussels 1956 May 22-26
Solvay conference on chemistry,1956 R.Stoops, Brussels,1956
Chem 18.1004

INSTITUT INTERNATIONAL DE COOPERATION INTELLECTUELLE Conference internationale de psychotechnique 4th Comptes rendues Paris 1927 Oct 10-14 Alcan,Paris,1929
Papers in English,French,German and Italian
Psy 31.2517

INSTITUT INTERNATIONAL DE COOPERATION INTELLECTUELLE Le Magnetisme Travaux Strasbourg 1939 May 21-25 Supported by Centre national de la recherche scientifique
Collection scientifique,Paris,1940
Cav 7.2241

INSTITUT INTERNATIONAL DE PHILOSOPHIE
Congres international de philosophie des sciences Actes Paris 1949 Oct 17-22
Tome 1- Actualites scientifiques et industrielles,1126,1134,1137,1146,1153,1155-56, 1166-67- Hermann,Paris,1951-
WSM 43.0981

INSTITUT INTERNATIONAL DE PHILOSOPHIE
Demonstration,verification,justification : entretiens de l'Institut international de philosophie Liege 1967 Sep Edited by Philippe Devaux xii,352p Nauwelaerts, Louvain;Paris,1968
WSM 43.0967

INSTITUT INTERNATIONAL DE PHYSIQUE SOLVAY
Atomes et electrons Conseil de physique 3eme Rapports Brussels 1921 Apr 1-6
Solvay conference,1921,3 Paris,1923
Cav 7.2080

INSTITUT INTERNATIONAL DE PHYSIQUE SOLVAY
Electrons et photons Conseil de physique 5eme Rapports Brussels 1927 Oct 24-27 Solvay conference,1927,5 Paris,1928
Cav 7.2081

INSTITUT INTERNATIONAL DE PHYSIQUE SOLVAY
Etat solide Conseil de physique 9 eme Rapports et discussions Brussels 1951 Sep 25-29 R.Stoops,Brussels,1952
Met 25.1135

INSTITUT INTERNATIONAL DE PHYSIQUE SOLVAY
Structure and evolution of galaxies Conference on physics 13th Proceedings Brussels 1964 Sep 174p Interscience, London,1965
Obs 6.2134

INSTITUT INTERNATIONAL DE PHYSIQUE SOLVAY
Structure and evolution of galaxies The Conference on physics 13th Proceedings Brussels 1964 Sep Solvay conference,1964, 13 Interscience,London,1965 Other Solvay conferences issued as Conseil de physique,q.v.
Cav 7.2078

INSTITUT INTERNATIONAL DE PHYSIQUE SOLVAY
Structure et evolution de l'univers Conseil de physique 11eme Rapports Brussels 1958 Jun 9-13 Edited by R. Stoops Solvay conference,1958,11 Brussels,1958
Cav 7.2084

INSTITUT INTERNATIONAL DE PHYSIQUE SOLVAY
Structure et proprietes des noyaux atomiques Conseil de physique 7eme Rapports Brussels 1933 Oct 22-29 Solvay conference, 1933,7 Paris,1934
Cav 7.2083

INSTITUT INTERNATIONAL DE PHYSIQUE SOLVAY
Theorie quantique des champs Conseil de physique 12eme Rapports Brussels 1961 Oct 9-14 Edited by R. Stoops Solvay conference,1961,12 Interscience,New York,c 1961 13th Solvay conference issued as Conference on physics,q.v.
Cav 7.2077

INSTITUT INTERNATIONAL DE STATISTIQUE
Bulletin de l'Institut international de statistique:compte rendus des sessions de l'Institut 19e 1930- Vol 25- v.p., 1931- Wants Vol 31:25th session
Math 3.1329

INSTITUT INTERNATIONAL DU FROID see INTERNATIONAL INSTITUTE OF REFRIGERATION

INSTITUT KRISTALLOGRAFII,MOSCOW Growth of crystals Conference on crystal growth 1st- Proceedings Moscow 1956- 1- Edited by A.V. Shubnikov and N.N. Sheftal Translated by Consultants' bureau from the Russian Rost kristallov, 1- Consultants bureau,inc.,New York,1958- Includes interim reports published between conferences. Held as part of the international crystallographic congresses
Met 25.1508

INSTITUT METALLURGII IMINI A.A.BAYKOV
Conference on the fatigue of metals 2nd Papers Translated by Leo Ronson from the Russian N.L.L.S.T.,Boston Spa,1964 Translation edited by P.G.Forrest
Met 25.2505

INSTITUT NATIONAL DES SCIENCES ET TECHNIQUES NUCLEAIRES Colloque international sur la retention et la migration des ions radioactifs dans les sols Saclay 1962 Oct 16-18 Centre d'etudes nucleaires de Saclay;Presses universitaires de France,Gif-sur-Yvette;Paris, 1963
Eng 41.3174

INSTITUUT VAN INGENIEURS.STEAM PLANT GROUP
Modern steam plant practice :a convention Proceedings The Hague 1971 Apr 28-30 IME, London,1971
Eng 41.8643

INSTITUTE FOR ADVANCED LEARNING IN THE MEDICAL SCIENCES Amino acid pools :distribution and function of free amino acids.Symposium Proceedings Duarte,Calif. 1961 May 19-22 City of hope medical center Edited by Joseph T. Holden illus Elsevier,Amsterdam,1962 Preface:"The committee undertook the organization of the symposium under the title Conference on free amino acids
Bioch 33.0519

INSTITUTE FOR ADVANCED STUDY Analytic functions conference papers Princeton,N.J. 1957 Sep 2-14 By Rolf Nevanlinna and others With the United States.Air force.Office of scientific research Princeton mathematical series, 24 vii,197p 24cm Princeton university press,Princeton,N.J.,1960
P. Math 2.1492

INSTITUTE FOR ADVANCED STUDY Contributions to the theory of Riemann surfaces conference typescript Princeton,N.J. 1951 Dec 14-15 Edited by Lars V. Ahlfors and others In joint sponsorship with Princeton university Annals of mathematics studies, 30 264p 26cm Princeton university press,Princeton,N. J.,1953 Centennial celebration of Riemann's dissertation
P. Math 2.1434

INSTITUTE FOR ADVANCED STUDY Distribution and motion of interstellar matter in galaxies The Conference on problems of the distribution and motion of interstellar matter in galaxies Proceedings Princeton,N.J. 1961 Apr 10-20 Edited by L. Woltjer 330p Benjamin,New York,1962
TA 15.0213

INSTITUTE FOR ADVANCED STUDY Distribution and motion of interstellar matter in galaxies : a conference Proceedings Princeton,N.J. 1961 Apr 10-20 Edited by L. Woltjer 330p W.A.Benjamin,New York,1962
Obs 6.3253

INSTITUTE FOR ADVANCED STUDY Dynamics of climate Conference on the application of numerical integration techniques to the problem of the general circulation proceedings 1955 Oct 26-28 Edited by Richard L. Pfeffer Sponsored jointly by Air force Cambridge research center Pergamon press,Oxford;London,1960
A Math 4.0560

INSTITUTE FOR ADVANCED STUDY Seminar on complex multiplication Princeton,N.J. 1957-58 Lecture notes in mathematics, 21 28cm Springer-verlag,Berlin,1966
P Math 2.2857

INSTITUTE FOR ADVANCED STUDY Seminar on geometry of numbers Lectures,mimeograph Princeton,N.J. 1949 90p 26cm institute for advanced study,Princeton,N.J.,1949
P Math 2.2778

INSTITUTE FOR ADVANCED STUDY Seminars on analytic functions conference typescript Princeton,N.J. 1957 Sep 2-14 Vols. 1-2 Edited by Marston Morse and others Under contract with the United States.Air force. Office of scientific research Bibliog. 25cm 2 vols Princeton,N.J.,1957
P. Math 2.1583

INSTITUTE FOR COMPARATIVE RESEARCH IN HUMAN CULTURE,OSLO Lapps and Norsemen in olden times :a conference Report Oslo 1964 Nov 19-21 Instituttet for sammenlignende kulturforskning.Serie A.Forelesninger,26 Bibliog,maps 168p Universitetsforlaget, Oslo,1967
Sco 14.5603

INSTITUTE FOR DEFENSE ANALYSIS.COMMUNICATIONS DIVISION Finite groups symposium New York 1959 Apr 23-24 American mathematical society Edited by A.Adrian Albert and Irving Kaplansky American mathematical society. Proceedings of symposia in pure mathematics, 1 viii,110p 23cm American mathematical society,Providence,R.I.,1959
P. Math 2.2029

INSTITUTE FOR FLUID DYNAMICS AND APPLIED MATHEMATICS Fluid dynamics and applied mathematics :a symposium Proceedings College Park 1961 Apr 28-29 Edited by J.B. Diaz and S.I. Pai 207p Gordon and Breach, New York;London,1962
Chem E 24.0371

INSTITUTE FOR PERCEPTION RESEARCH IPO Attention and performance 2 Donders centenary symposium on reaction time Proceedings Eindhoven 1968 Jul 29-Aug 2 Edited by W.G. Koster North-Holland, Amsterdam;London,1969
Psy 31.2948

INSTITUTE FOR PERCEPTION RVO-TNO Performance of the eye at low luminances :colloquium Proceedings Delft 1965 Edited by M.A. Bouman and J.J. Vos International congress series, 125 Excerpta medica,Amsterdam,1966
Psy 31.2862

INSTITUTE FOR PERCEPTION RVO-TNO,SOESTERBERG Attention and performance :a symposium Proceedings Driebergen 1966 Aug 17-20 Edited by A.F. Sanders North-Holland, Amsterdam,1967
Psy 31.0959

INSTITUTE FOR SPACE BIOSCIENCES Origins of prebiological systems and of their molecular matrices A Conference on the origins of prebiological systems and of their molecular matrices Proceedings Wakulla Springs,Fla 1963 Oct 27-30 Edited by Sidney W. Fox Bibliog.,illus. xx,482p Academic press,New York;London,1965
Bot 42.1364

INSTITUTE FOR SPACE BIOSCIENCES The Origins of prebiological systems and of their molecular matrices :a conference Proceedings Wakulla Springs 1963 Oct 27-30 Edited by Sidney W. Fox Academic press,New York;London, 1965 Conducted under the auspices of the Institute for space biosciences,the Florida state university and the National aeronautics and space administration
Bal 39.0992

INSTITUTE FOR SPACE BIOSCIENCES The Origins of prebiological systems and of their molecular matrices :a conference Proceedings Wakulla springs,Fla. 1963 Oct 27-30 Florida state university National aeronautics and space administration Edited by Sidney W. Fox Academic press,New York; London,1965
Gen 34.0456

INSTITUTE IN THE THEORY OF NUMBERS report Boulder,Colo. 1959 Jun 21-Jul 17 American mathematical society Edited by Donald C.B. Marsh and James H. Jordan With the support of the National science foundation 350p 27cm University of Colorado,Boulder,Colo., 1959
P. Math 2.1773

INSTITUTE OF ADVANCED STUDY ON THE CHEMISTRY OF EXTENDED DEFECTS IN NON-METALLIC SOLIDS The Chemistry of extended defects in non-metallic solids Proceedings Scotsdale,Ariz. 1969 Apr 16-26 Edited by LeRoy Eyring and Michael O'Keefe North-Holland,Amsterdam,1970
Min 10.1545

INSTITUTE OF AERONAUTICAL SCIENCES Vistas in astronautics Astronautics symposium 2nd Proceedings Denver,Col. 1958 Apr.28 United States.Air force.Office of scientific research Edited by Morton Alperin and Hollingsworth F. Gregory Vistas in astronautics, 2 International series of monographs on aeronautical sciences and space flight.Astronautics division, 2 Pergamon press,London,1959
Nap 11.0023

INSTITUTE OF ARCHAEOLOGY Research seminar on archaeology and related subjects day meeting on statistics and archaeology London 1964 May 30 London,1964 Unbound typescript papers
Math 3.1115

INSTITUTE OF BIOLOGY Biological hazards of atomic energy :a conference Papers 1950 Oct Edited by A. Haddow pl. Clarendon press,Oxford,1952
Radioth 35.0180

INSTITUTE OF BIOLOGY Biological hazards of atomic energy :a conference papers London 1950 Oct Clarendon press,Oxford,1952
HE 27.0097

INSTITUTE OF BIOLOGY Biological problems arising from the control of pests and diseases a symposium Proceedings London 1959 Oct Edited by R.K.S. Wood Institute of biology. Symposia, 9 22cm London,1960
Bal 39.2962

INSTITUTE OF BIOLOGY Biology and ethics :a symposium Proceedings London 1968 Sep 26-27 Edited by F.J. Ebling Institute of biology.Symposia, 18 xxix,145p Academic press,London;New York,1969
Bal 39.0055

INSTITUTE OF BIOLOGY Biology and ethics :a symposium Proceedings London 1968 Sep 26-27 Edited by F.J. Ebling held at the Royal geographical society Institute of biology. Symposia, 18 Academic press,London;New York, 1969
Bal 39.0018

INSTITUTE OF BIOLOGY Biology of deserts Biology of hot and cold deserts :a symposium Proceedings London 1952 Sep 25-27 Edited by J.L. Cloudsley-Thompson Institute of biology,London,1954
Geog 13.1205

INSTITUTE OF BIOLOGY Freezing and drying :a symposium Papers London 1951 Jun Edited by R.J.C. Harris Institute of biology, London,1951
Radioth 35.0178

INSTITUTE OF BIOLOGY Freezing and drying :a symposium Report London 1951 Jun Edited by R.J.C. Harris 205p Institute of biology,London,1951
Sco 14.1236

INSTITUTE OF BIOLOGY Man-made lakes :a symposium Proceedings London 1965 Sep 30-Oct 1 Edited by R. Lowe-McConnell held at the Royal geographical society Institute of biology.Symposia, 15 Academic press,London; New York,1966
Bal 39.1898

INSTITUTE OF BIOLOGY Recent research in freezing and drying International symposium on freeze-drying 2nd Papers London 1958 Edited by A.S. Parkes and Audrey U. Smith diagrs vii,320p 25cm Blackwell scientific publications,Oxford,1960
Sco 14.1237

INSTITUTE OF BIOLOGY Recent research in freezing and drying International symposium on freezing and drying 2nd Proceedings London 1958 Edited by A.S. Parkes and Audrey U. Smith Blackwell,Oxford,1960
An 32.3233

INSTITUTE OF BIOLOGY The Better use of the world's fauna for food London 1962 Edited by J.D. Ovington Institute of biology. Symposia, 11 21cm London,1963
Bal 44.4059

INSTITUTE OF BIOLOGY The Biological significance of climatic changes in Britain Proceedings London 1964 Oct 29-30 Edited by C.G. Johnson and L.P. Smith Held at the Royal geographical society Institute of biology.Symposium, 14 illus x,222p Academic press,London;New York,1965
Bot 42.6678

INSTITUTE OF BIOLOGY The Biological significance of climatic changes in Britain :a symposium held at the Royal geographical society Proceedings London 1964 Oct 29-30 Edited by C.G. Johnson and L.P. Smith Institute of biology.Symposia, 14 Academic press,London;New York,1965
Bal 39.1687

INSTITUTE OF BIOLOGY The Biology of hot and cold deserts... :a symposium Proceedings London 1952 Sep 25-27 Edited by J.L. Cloudsley-Thompson Assisted by a grant from Unesco Institute of biology,London,1954
Bal 39.1858

INSTITUTE OF BIOLOGY The Natural history of aggression :a symposium Proceedings London 1963 Oct 21-22 Edited by J.D. Carthy and F.J. Ebling Institute of biology.Symposia, 13 Academic press,London;New York,1964
Psy 31.0596

INSTITUTE OF BIOLOGY The Natural history of aggression :a symposium held at the British museum(natural history) Proceedings London 1963 Oct 21-22 Edited by J.D. Carthy and F.J. Ebling Institute of biology.Symposia, 13 Academic press,London,1960
Bal 39.1619

INSTITUTE OF BIOLOGY 15th :symposium Man-made lakes London 1965 Sep 30-Oct 1 Edited by R.H. Lowe-McConnell Academic press, London,1966
Geog 13.0836

INSTITUTE OF BIOLOGY.SYMPOSIA, 9 Biological problems arising from the control of pests and diseases :a symposium Proceedings London 1959 Oct Institute of biology Edited by R.K.S. Wood 22cm London,1960
Bal 39.2962

INSTITUTE OF BIOLOGY.SYMPOSIA, 11 The Better use of the world's fauna for food London 1962 Institute of biology Edited by J.D. Ovington 21cm London,1963
Bal 44.4059

INSTITUTE OF BIOLOGY.SYMPOSIA, 13 The Natural history of aggression :a symposium Proceedings London 1963 Oct 21-22 Institute of biology Edited by J.D. Carthy and F.J. Ebling Academic press,London;New York,1964
Psy 31.0596

INSTITUTE OF BIOLOGY.SYMPOSIA, 13 The Natural history of aggression :a symposium held at the British museum(natural history) Proceedings London 1963 Oct 21-22 Institute of biology Edited by J.D. Carthy and F.J. Ebling Academic press,London,1960
Bal 39.1619

INSTITUTE OF BIOLOGY.SYMPOSIA, 14 The Biological significance of climatic changes in Britain :a symposium held at the Royal geographical society Proceedings London 1964 Oct 29-30 Institute of biology Edited by C.G. Johnson and L.P. Smith Academic press,London;New York,1965
Bal 39.1687

INSTITUTE OF BIOLOGY.SYMPOSIA, 15 Man-made lakes :a symposium Proceedings London 1965 Sep 30-Oct 1 Institute of biology Edited by R. Lowe-McConnell held at the Royal geographical society Academic press, London;New York,1966
Bal 39.1898

INSTITUTE OF BIOLOGY.SYMPOSIA, 18 Biology and ethics :a symposium Proceedings London 1968 Sep 26-27 Institute of biology Edited by F.J. Ebling held at the Royal geographical society Academic press,London; New York,1969
Bal 39.0018

INSTITUTE OF BIOLOGY.SYMPOSIA, 18 Biology and ethics :a symposium Proceedings London 1968 Sep 26-27 Institute of biology Edited by F.J. Ebling xxix,145p Academic press, London;New York,1969
Bal 39.0055

INSTITUTE OF BIOLOGY.SYMPOSIA, 19 The Optimum population for Britain :symposium held at the Royal geographical society Proceedings London 1969 Sep 25-26 Edited by L.R. Taylor Academic press,London,1970
Gen 34.2202

INSTITUTE OF BIOLOGY.SYMPOSIUM, 14 The Biological significance of climatic changes in Britain Proceedings London 1964 Oct 29-30 Institute of biology Edited by C.G. Johnson and L.P. Smith Held at the Royal geographical society illus x,222p Academic press,London;New York,1965
Bot 42.6678

The INSTITUTE OF BIOLOGY SYMPOSIUM 11th Better use of the worlds fauna for food London 1961 Edited by J.D. Ovington Institute of biology,London,1963
Geog 13.2660

INSTITUTE OF BIOLOGY SYMPOSIUM 14th Proceedings Biological significance of climatic changes in Britain London 1964 Oct 29-30 Edited by C.G. Johnson and L.P. Smith Academic press,London,1965
Geog 13.1208

INSTITUTE OF BRITISH GEOGRAPHERS and POLSKA AKADEMIA NAUK.INSTYTUT GEOGRAFII Problems of applied geography,2 Anglo-Polish seminar 2nd Proceedings Keele 1962 Sep 9-20 Geographia polonica,3 Panstwowe wydawnictwo naukowe,Warsaw,1964
Geog 13.5616

INSTITUTE OF CHEMISTRY OF IRELAND and UNIVERSITY COLLEGE,DUBLIN Symposium of recent advances in the chemistry of naturally occurring pyrones and related compounds Proceedings Dublin 1955 Jul 12-14 Royal Dublin society.Scientific proceedings,27,no 6 Royal Dublin society,Dublin,1956
Chem 18.1494

INSTITUTE OF ELECTRICAL AND ELECTRONIC ENGINEERS Integrated circuits Conference on integrated circuits Eastbourne 1967 May 2-4 Sponsored by the Institution of electrical engineers.Electronics division I.E.E. Conference publication, 30 Institution of electrical engineers,London,1967 Joint sponsors;Institution of electronic and radio engineers and the Institute of electrical and electronics engineers
Eng 41.5461

INSTITUTE OF ELECTRICAL AND ELECTRONICS ENGINEERS ACM-IEEE symposium on problems in the optimization of data communications systems 2nd Proceedings Palo Alto,Calif. 1971 Oct 20-22 illus 198p ACM;IEEE,New York, 1971
Math L 5.3629

INSTITUTE OF ELECTRICAL AND ELECTRONICS ENGINEERS Conference on gas discharges 1972 Institution of electrical engineers.Conference publication, 90 IEE,London,1972
Eng 41.8570

INSTITUTE OF ELECTRICAL AND ELECTRONICS ENGINEERS Electron and laser beam symposium 8th Proceedings Ann Arbor,Mich 1966 Apr 6-8 Edited by G.I. Haddad University of Michigan, Ann Arbor,Mich.,1966 8th in the series of annual symposia on Electron beam technology initiated in 1959 by Alloyd and held under their sponsorship for seven years.The titles of the symposia vary,therefore those catalogued have been grouped together under the keyword"Electron beams"
Eng 41.5618

INSTITUTE OF ELECTRICAL AND ELECTRONICS ENGINEERS Hardware,software,firmware trade-offs IEEE-International computer society conference 5th Proceedings Boston,Mass. 1971 Sep 22-24 illus 204p IEEE,New York,1971
Math L 5.3628

INSTITUTE OF ELECTRICAL AND ELECTRONICS ENGINEERS I.E.E.E. international convention convention record New York 1963 Mar.25-28 Part 4: electronic computers,information theory,human factors in electronics 152p Institute of electrical and electronics engineers,New York,1963
Math L 5.2086

INSTITUTE OF ELECTRICAL AND ELECTRONICS ENGINEERS I.E.E.E. international convention convention records New York 1964 Mar.23-26 Part 1: circuit theory,computers,control systems,systems science 313p Institute of electrical and electronics engineers,New York, 1964
Math L 5.2085

INSTITUTE OF ELECTRICAL AND ELECTRONICS ENGINEERS I.E.E.E.international convention convention record New York 1965 Mar.22-26 Part 3: computers 281p Institute of electrical and electronis engineers,New York, 1965
Math L 5.2087

INSTITUTE OF ELECTRICAL AND ELECTRONICS ENGINEERS I.E.E.E.international convention digest : synopses of papers presented at the 1971 I.E.E.E. international convention New York 1971 Mar 22-25 bibliog.,illus. 643p Institute of electrical and electronics engineers,New York,1971
Math L 5.3565

INSTITUTE OF ELECTRICAL AND ELECTRONICS ENGINEERS Joint conference on digital methods of measurement Proceedings Canterbury 1969 Jul 23-25 Institution of electronic and radio engineers.Conference proceedings, 15 IERE,London,1969
Eng 41.8305

INSTITUTE OF ELECTRICAL AND ELECTRONICS ENGINEERS. COMMUNICATION TECHNOLOGY GROUP.DATA COMMUNICATION COMMITTEE New York 1970 Mar 27 illus 24p IEEE,New York,1970
Math L 5.3626

INSTITUTE OF ELECTRICAL AND ELECTRONICS ENGINEERS. COMPUTER SOCIETY Annual workshop on microprogramming 5th Preprints Urbana, Ill. 1972 Sep 25-26 Bibliog,diagrs 98p n.p.,1972 Includes 'An annotated bibliography on microprogramming,late 1969-early 1972'
Math L 5.3636

INSTITUTE OF ELECTRICAL AND ELECTRONICS ENGINEERS. GROUP ON AUTOMATIC CONTROL Symposium on nonlinear estimation theory and its applications Proceedings San Diego,Calif. 1970 Sep 21-23 vii,297p 28cm IEEE,New York,1970
Math S 3.1698

INSTITUTE OF ELECTRICAL AND ELECTRONICS ENGINEERS. MAGNETICS GROUP Intermag conference International conference on magnetics 3rd Proceedings Washington,D.C. 1965 Apr 21-23 Institute of electrical and electronics engineers,New York,1965
Eng 41.5345

INSTITUTE OF ELECTRICAL AND ELECTRONICS ENGINEERS. MAGNETICS GROUP 1966 Intermag International magnetics conference 4th Stuttgart 1966 Apr 20-22 I.E.E.E. transactions on magnetics,Mag 2,no 3 I.E.E.E.,New York,1966
Eng 41.5347

INSTITUTE OF ELECTRICAL AND ELECTRONICS ENGINEERS. PROFESSIONAL TECHNICAL GROUP IN AUTOMATIC CONTROL Joint automatic control conference 5th Preprints of conference papers Stanford,Calif. 1964 Jun 24-26 Stanford,Calif.,1964
Eng 41.6014

INSTITUTE OF ELECTRICAL AND ELECTRONICS ENGINEERS. SYSTEMS SCIENCE AND CYBERNETICS GROUP Recent advances in optimization techniques :a symposium Pittsburgh,Pa. 1965 Apr 21-23 Edited by A. Levi and T.P. Vogl Wiley,New York,1966
Eng 41.6003

INSTITUTE OF ELECTRICAL AND ELECTRONICS ENGINEERS. SYSTEMS SCIENCE AND CYBERNETICS GROUP Recent advances in optimization techniques symposium proceedings Pittsburgh,Pa. 1965 Apr.21-23 Edited by Abrahim Lavi and Thomas P. Vogl Jointly sponsored by the Optical society of America Bibliog. xiii, 656p 23cm John Wiley and sons,New York, 1966
Math 3.0240

INSTITUTE OF ELECTRICAL ENGINEERS Conference on machine perception of patterns and pictures Proceedings Teddington 1972 Apr 12-14 Institute of physics.Conference series, 13 illus ix,362p Institute of physics,London, 1972
Math L 5.3882

INSTITUTE OF HORMONE BIOLOGY Steroids in modern medicine :symposium Oxford 1971 Sep 3-4 Pt 1-2 Edited by George A. Christie and Miriam Moore-Robinson Excerpta medica,Amsterdam,1972
Bioch 33.2199

INSTITUTE OF HORMONE BIOLOGY The Treatment of carcinoma of the breast :a symposium Proceedings Cambridge 1967 Sep 9 Edited by Antony S. Jarrett Excerpta medica foundation,1968
PGMS 29.0460

INSTITUTE OF HORMONE BIOLOGY The Treatment of carcinoma of the breat :a symposium Proceedings Cambridge 1967 Sep 7 Syntex pharmaceuticals Edited by Antony S. Jarrett Excerpta medica,Amsterdam,1967
Radioth 35.0926

INSTITUTE OF LOW TEMPERATURE SCIENCE International conference on low temperature science Papers Sapporo,Japan 1966 Aug 14-19 conference on cryobiology.Cellular injury and resistance in freezing organisms Edited by Eizo Asahina illus xxiii,257p 27cm Sapporo,1967
Sco 14.0144

INSTITUTE OF LOW TEMPERATURE SCIENCE Physics of snow and ice International conference on low temperature science 1st Proceedings Sapporo 1966 Aug 14-19 1 pt 2: conference on physics of snow and ice (part 2) Edited by Hirobumi Oura illus 1967p 27cm Hokkaido university,Sapporo,1957
Sco 14.7411

INSTITUTE OF MATHEMATICAL STATISTICS Joint European conference of the Institute of mathematical statistics,the International association for statistics in physical sciences,the Biometric society Berne 1964 Sep 14-18 Berne,1964 13 unbound typescript papers
Math 3.1124

INSTITUTE OF MATHEMATICAL STATISTICS Meeting 26th Ottawa 1963 Aug 29 1963 Typescript papers in folder with 34th session of the International statistical institute, 1963
Math 3.1111

INSTITUTE OF MATHEMATICS AND ITS APPLICATIONS Large sparse sets of linear equations Conference of the Institute of mathematics and its applications Proceedings Oxford 1970 Apr Edited by J.K. Reid Illus 284p Academic press,London,1971
Math L 5.3804

INSTITUTE OF MATHEMATICS AND ITS APPLICATIONS Large sparse sets of linear equations Oxford conference of the Institute of mathematics and its applications Proceedings Oxford 1970 Apr 5-8 Edited by J.K. Reid x,284p 24cm Academic press,London;New York,1971
P Math 2.4464

INSTITUTE OF MATHEMATICS AND ITS APPLICATIONS Large sparse sets of linear equations :a conference Proceedings Oxford 1970 Edited by J.K. Reid Academic press,London, 1971
Eng 41.8474

INSTITUTE OF MATHEMATICS AND ITS APPLICATIONS Numerical analysis :an introduction Birmingham 1965 Edited by Joan Walsh 212p Academic press,London;New York,1966 Based on a symposium held in Birmingham in 1965
A Math 4.1219

INSTITUTE OF MATHEMATICS AND ITS APPLICATIONS
Numerical analysis:an introduction Symposium on numerical analysis Birmingham 1965 Academic press,London,1966
Radioth 35.0631

INSTITUTE OF MATHEMATICS AND ITS APPLICATIONS
Numerical analysis:an introduction Symposium on numerical analysis Birmingham 1965 Edited by J. Walsh bibliog.,diags. xiv, 212p 24cm Academic press,London,1966
Math L 5.0357

INSTITUTE OF MATHEMATICS AND ITS APPLICATIONS
Numerical methods for unconstrained optimization Joint institute of mathematics and its application-National physical laboratory conference Proceedings Teddington 1971 Jan 7-8 Edited by W. Murray xi,144p 24cm Academic press, London;New York,1972
Math S 3.1729

INSTITUTE OF MATHEMATICS AND ITS APPLICATIONS
Optimization Keele conference on optimization and nonlinear programming Keele 1968 Edited by R. Fletcher Academic press, London,1969
Math L 5.3607

INSTITUTE OF MATHEMATICS AND ITS APPLICATIONS
Optimization Keele conference on optimization and nonlinear programming Keele 1968 Edited by R. Fletcher Academic press, London,1969
P Math 2.3755

INSTITUTE OF MATHEMATICS AND ITS APPLICATIONS
Optimization :symposium Keele 1968 May 25-29 Academic press,London,1969
Eng 41.2342

INSTITUTE OF MECHANICAL ENGINEERS Road vehicle aerodynamics Symposium on road vehicle aerodynamics 1st Proceedings London 1969 Nov 6-7 Edited by A.J. Scibor-Rylski Organised by the City university. Department of aeronautics City university, London,1970 Organised in cooperation with the National physical laboratory,the Motor industry research association and the Institution of mechanical engineers
Eng 41.8336

INSTITUTE OF MECHANICAL ENGINEERS Use of secondary surfaces for heat transfer with clean gases :a symposium Proceedings 1960 London,1961
Eng 41.7167

INSTITUTE OF METAL FINISHING.LONDON BRANCH
Control in electroplating :a one-day symposium London 1958 Nov 19 Draper,Teddington,1959
Met 25.2670

INSTITUTE OF METAL FINISHING.LONDON BRANCH
Nickel-chromium plating :a one-day symposium London 1960 Nov 16 Robert Draper, Teddington,1961
Met 25.2039

INSTITUTE OF METALS Equipment for the thermal treatment of non-ferrous metals and alloys a symposium on the metallurgical aspects of the subject London 1952 Mar 26 Institute of metals.Monograph and report series, 14 Institute of metals,London,1953
Met 25.2306

INSTITUTE OF METALS Heat treatment of metals Heat treatment joint committee :inaugural meeting Proceedings London 1965 Dec 14-15 Iron and steel institute.Special report, 95 London,1966
Met 25.2565

INSTITUTE OF METALS Internal stresses in metals and alloys Symposium on internal stresses in metals and alloys London 1947 Oct 15-16 Institute of metals.Monograph and report series, 5 Institute of metals, London,1948
Met 25.0960

INSTITUTE OF METALS Properties of metallic surfaces :a symposium London 1952 Nov 19 Institute of metals.Monograph and report series, 13 Institute of metals,London,1953
Eng 41.3471

INSTITUTE OF METALS Properties of metallic surfaces :a symposium London 1952 Nov 19 Institute of metals.Monograph and report series, 13 Institute of metals,London,1953
Met 25.1593

INSTITUTE OF METALS Science,technology and application of titanium International conference on titanium Proceedings London 1968 May 21-24 Edited by R.I. Jaffee and N.E. Promisel Pergamon,Oxford,1970
Met 25.2852

INSTITUTE OF METALS Structural processes in creep :a symposium Report London 1961 May 3-4 Iron and steel institute.Special report, 70 Iron and steel institute,London, 1961
Met 25.2334

INSTITUTE OF METALS Symposium on internal stresses in metals and alloys Advance copies of papers London 1947 Oct 15-16 Institute of metals,London,1947
Eng 41.3768

INSTITUTE OF METALS Symposium on internal stresses in metals and alloys Proceedings London 1947 Oct 15-16 Monographs and report series, 5 vii,485p Institute of metals,London,1948
Cav 7.1454

INSTITUTE OF METALS The Control of quality in the production of wrought non-ferrous metals and alloys :a symposium London 1954 Apr 28 2: the control of quality in working operations Institute of metals. Monograph and report series, 16 Institute of metals,London,1954
Met 25.0711

INSTITUTE OF METALS The Control of quality in the production of wrought non-ferrous metals and alloys a symposium held... on the occasion of the annual general meeting of the Institute of metals London 1955 Mar 31 Institute of metals.Monograph and report series, 17 Institute of metals,London,1955
Met 25.0504

INSTITUTE OF METALS The Determination of gases in metals :a symposium Report Iron and steel institute and Society for analytical chemistry Iron and steel institute.Special report, 68 Iron and steel institute,London,1960
Met 25.1747

INSTITUTE OF METALS The Final forming and shaping of wrought non-ferrous metals :a symposium London 1956 Apr 12 Institute of metals.Monograph and report series, 20 Institute of metals,London,1956
Met 25.0709

INSTITUTE OF METALS The Mechanism of phase transformations in crystalline solids :an international symposium Proceedings Manchester 1968 Jul 3-5 Institute of metals.Monograph and report series, 33 illus,diagrs 324p 25cm London,1969
Cav 7.2666

INSTITUTE OF METALS The Mechanism of phase transformations in metals :a symposium London 1955 Nov 9 Institute of metals. Monograph and report series, 18 Institute of metals,London,1956
Met 25.1251

INSTITUTE OF METALS The Metallurgy of beryllium :an international conference Proceedings London 1961 Oct 16-18 Institute of metals.Monograph and report series, 28 Chapman and Hall,London,1963
Met 25.0605

INSTITUTE OF METALS The Solidification of metals :the conference Proceedings Brighton 1967 Dec 4-7 Jointly organized with the Institute of physics and the physical society and the Institution of metallurgy Iron and steel institute.Publication, 110 Iron and steel institute,London,1968
Met 25.1529

INSTITUTE OF METALS Thermal and high-strain fatigue :an international conference Proceedings London 1967 Jun 6-7 Metals and metallurgy trust Metals and metallurgy trust.Monograph and report series Metals and metallurgy trust,London,1967
Eng 41.3810

INSTITUTE OF METALS Thermal and high-strain fatigue an international conference Proceedings London 1967 Jun 6-7 Metals and metallurgy trust.Monograph and report series, 32 Metals and metallurgy trust, London,1967
Met 25.0942

INSTITUTE OF METALS Uranium and graphite :a symposium Proceedings London 1962 Mar 20-21 Institute of metals.Monograph, 27 Institute of metals,London,1962
Met 25.1554

INSTITUTE OF METALS Vacancies and other point defects in metals and alloys a symposium Harwell,Berks 1917 Dec 10 Institute of metals.Monograph and report series, 23 Institute of metals,London,1958
Met 25.1233

INSTITUTE OF MICROBIOLOGY OF RUTGERS Evolving genes and proteins :a symposium Rutgers,N.J. 1964 Sep 17-18 Edited by V. Bryson and H.J. Vogel With support from the National science foundation Academic press, New York;London,1965
Gen 34.0665

INSTITUTE OF MICROBIOLOGY OF RUTGERS Informational macromolecules :a symposium Rutgers,N.J. 1962 Sep 5-7 Edited by H.J. Vogel and others With support from the National science foundation Academic press, New York,1963
Gen 34.0664

INSTITUTE OF PACIFIC RELATIONS 5th conference Papers Canadian papers 1933 Banff 1933 Aug 14-28 Canadian institute of international affairs Canadian institute of international affairs,Toronto,1933
Geog 13.4898

INSTITUTE OF PACIFIC RELATIONS CONFERENCE 10th Report Problems of economic reconstruction in the Far East Stratford-on-Avon 1947 Sep 5-20 Institute of Pacific relations,New York,1949
Geog 13.2189

INSTITUTE OF PETROLEUM Corrosion problems of the petroleum industry a joint symosium Papers London 1959 Nov 26-27 Joint symposium with the Corrosion group S.C.I. Monograph, 10 Society of chemical industry, London,1960
Met 25.1934

INSTITUTE OF PETROLEUM Fluidisation :a symposium Papers London 1963 Mar 11-12 Society of chemical industry.Chemical engineering group and Institution of chemical engineers 110p Society of chemical industry,London,1964
Chem E 24.1452

INSTITUTE OF PETROLEUM The Organization of chemical engineering projects :joint symposium Proceedings London 1958 Jun 24-26 Institution of chemical engineers,London,1958 17th meeting of the European federation of chemical engineering
Chem E 24.1543

INSTITUTE OF PETROLEUM The Post-war expansion of the U.K.petroleum industry : summer meeting Report Buxton 1953 Jun 24-27 Edited by George Sell vii,220,xxxvp Institute of petroleum,London,1954
Chem E 24.1705

INSTITUTE OF PETROLEUM.GAS CHROMATOGRAPHY DISCUSSION GROUP Gas chromatography :a symposium 2nd Proceedings Amsterdam 1958 May 19-23 Edited by D.H. Desty Also under the auspices of the Nederlandse chemische vereniging xiii,383p Butterworths,London,1958
Chem E 24.0849

INSTITUTE OF PETROLEUM.GAS CHROMATOGRAPHY DISCUSSION GROUP Gas chromatography :a symposium 3rd Proceedings Edinburgh 1960 Jun 8-10 Edited by R.P.W. Scott Organized also by the Society for physical chemistry xvii,466p Butterworths,London, 1960
Chem E 24.0850

INSTITUTE OF PETROLEUM.GAS CHROMATOGRAPHY DISCUSSION GROUP Gas chromatography 1964 : a symposium 5th Proceedings Brighton 1964 Sep 8-10 Edited by A. Goldop Institute of petroleum,London,1965
Chem 18.0568

INSTITUTE OF PETROLEUM.GAS CHROMATOGRAPHY DISCUSSION GROUP Gas chromatography 1966 : an international symposium on gas chromatography and associated techniques 6th Proceedings Rome 1966 Sep 20-23 Edited by A.B. Littlewood Institute of petroleum, London,1967
Chem 18.0569

INSTITUTE OF PETROLEUM.GAS CHROMATOGRAPHY DISCUSSION GROUP Gas chromatography 1970 International symposium on gas chromatography 8th Proceedings Dublin 1970 Sep 28-Oct 2 Edited by R. Stock viii,445p 25cm Institute of petroleum,London,1971
Chem 18.2697

INSTITUTE OF PETROLEUM.GAS CHROMATOGRAPHY DISCUSSION GROUP and SOCIETY FOR ANALYTICAL CHEMISTRY Gas chromatography 1960 :a symposium 3rd Proceedings Edinburgh 1960 Jun 8-10 Edited by R.P.W. Scott Butterworths,London,1960
Chem 18.0567

INSTITUTE OF PETROLEUM.HYDROCARBON RESEARCH GROUP Spectroscopy :a conference Report London 1962 Mar 28-31 Edited by M.J. Wells Institute of petroleum,London,1962
Chem 18.0193

INSTITUTE OF PETROLEUM.HYDROCARBON RESEARCH GROUP Vapour phase chromatography :a symposium Proceedings London 1956 May 30-Jun 1 Edited by D.H. Desty and C.L.A. Harbourn Butterworths,London,1957
Chem 18.1450

INSTITUTE OF PETROLEUM.HYDROCARBON RESEARCH GROUP Vapour phase chromatography :a symposium Proceedings London 1956 May 30-Jun 1 Edited by D.H. Desty and C.L.A. Harbourn xv, 435p Butterworths,London,1957
Chem E 24.0845

INSTITUTE OF PETROLEUM.HYDROCARBON RESEARCH GROUP and AMERICAN SOCIETY FOR TESTING MATERIALS. COMMITTEE ON MASS SPECTROMETRY Advances in mass spectrometry :a conference Proceedings London 1958 Sep 24-26 Edited by J.D. Waldron Pergamon press.Symposia publications division,London,1959
Chem 18.0197

INSTITUTE OF PETROLEUM.HYDROCARBON RESEARCH GROUP and NEDERLANDSE CHEMISCHE VERENIGING Gas chromatography 1958 :a symposium 2nd Proceedings Amsterdam 1958 May 19-23 Edited by D.H. Desty Organised by the Gas chromatography discussion group Butterworths, London,1958
Chem 18.0566

INSTITUTE OF PETROLEUM.MASS SPECTROMETRY PANEL Mass spectrometry :a conference Report Manchester 1950 Apr 20-21 Institute of petroleum,London,1952
Bioch 33.0282

INSTITUTE OF PETROLEUM.MASS SPECTROSCOPY PANEL Mass spectroscopy :a conference Report Manchester 1950 Apr 20-21 Institute of petroleum,London,1952
Chem 18.0110

INSTITUTE OF PETROLEUM,and GEOLOGICAL SOCIETY OF LONDON Salt basins around Africa Institute of petroleum and Geological society of London :joint meeting proceedings London 1965 Mar 3 Institute of petroleum, London,1965
Geol 8.3097

INSTITUTE OF PETROLEUM AND GEOLOGICAL SOCIETY OF LONDON :joint meeting proceedings Salt basins around Africa London 1965 Mar 3 Institute of petroleum,and,Geological society of London Institute of petroleum, London,1965
Geol 8.3097

INSTITUTE OF PHYSICS Conference on machine perception of patterns and pictures Proceedings Teddington 1972 Apr 12-14 Institute of physics.Conference series, 13 illus ix,362p Institute of physics,London, 1972
Math L 5.3882

INSTITUTE OF PHYSICS Conference on the defects in crystalline solids :held at the H.H. Wills physical laboratory Report Bristol 1954 Jul By N. Bloembergen and others With co-operation of the International union of pure and applied physics.Commission on the solid state 429p Physical society,London, 1955
Cav 7.0345

INSTITUTE OF PHYSICS Electron microscopy and analysis Institute of physics.Electron microscopy and analysis group 25th anniversary meeting Proceedings Cambridge 1971 Jun 29-Jul 1 Edited by W.C. Nixon Institute of physics.Conference series, 10 London,1971
Met 25.2753

INSTITUTE OF PHYSICS European congress on electron microscopy 5th Proceedings Manchester 1972 Institute of physics. Conference series, 14 illus xxviii,683p 26cm Institute of physics,London,1972
Cav 7.3182

INSTITUTE OF PHYSICS Physics of nuclear reactors :a conference Papers London 1956 Jul 3-6 British journal of applied physics.Supplement, 5 Institute of physics, London,1956
Radioth 35.1596

INSTITUTE OF PHYSICS Some aspects of fluid flow :a conference Papers,reports Leamington Spa 1950 Oct 25-28 Edited by H. R. Lang viii,292p Edward Arnold,London, 1951
Chem E 24.0224

INSTITUTE OF PHYSICS Some aspects of fluid flow... :a conference Papers Leamington Spa 1950 Oct 25-28 Arnold,London,1951
Eng 41.6954

INSTITUTE OF PHYSICS The Acceleration of particles to high energies :based on a session arranged by the Electronics group Papers London 1949 May Edited by H.R. Lang Physics in industry Institute of physics, London,1950
Cav 7.1166

INSTITUTE OF PHYSICS The Acceleration of particles to high energies :based on a session arranged by the Electronics group at the Institute of physics convention in May,1949 Physics in industry Institute of physics, London,1950
Eng 41.5499

INSTITUTE OF PHYSICS The International conference on the physics of semiconductors Report Exeter 1962 Jul 16-20 Edited by A. C. Stickland Under the auspices of the International union of pure and applied physics 909p Institute of physics,London, 1962
Cav 7.0750

INSTITUTE OF PHYSICS The Measurement of stress and strain in solids :a conference Proceedings Manchester 1946 Jul 11-13 Institute of physics,London,1948
Eng 41.2898

INSTITUTE OF PHYSICS The Physics of particle analysis :a conference Papers Nottingham 1954 Apr 6-9 British journal of applied physics.Supplement, 3 Institute of physics, London,1954
Geod 9.0289

INSTITUTE OF PHYSICS.CONFERENCE SERIES, 10 Electron microscopy and analysis Institute of physics.Electron microscopy and analysis group 25th anniversary meeting Proceedings Cambridge 1971 Jun 29-Jul 1 Institute of physics Edited by W.C. Nixon London,1971
Met 25.2753

INSTITUTE OF PHYSICS.CONFERENCE SERIES, 13 Conference on machine perception of patterns and pictures Proceedings Teddington 1972 Apr 12-14 Institute of physics National physical laboratory Institute of electrical engineers illus ix,362p Institute of physics,London,1972
Math L 5.3882

INSTITUTE OF PHYSICS.CONFERENCE SERIES, 14 Electron microscopy 1972 European congress on electron microscopy 5th Proceedings Manchester 1972 International federation of societies for electron microscopy Institute of physics,London,1972
Eng 41.8576

INSTITUTE OF PHYSICS.ELECTRON MICROSCOPY AND ANALYSIS GROUP 25th anniversary meeting Proceedings Electron microscopy and analysis Cambridge 1971 Jun 29-Jul 1 Institute of physics Edited by W.C. Nixon Institute of physics.Conference series, 10 London,1971
Met 25.2753

INSTITUTE OF PHYSICS.LONDON AND HOME COUNTIES BRANCH The Design of physics research laboratories :a symposium Proceedings London 1957 Nov 27 bibliog. Chapman and Hall;Reinhold,London;New York,1959
Radioth 35.1634

INSTITUTE OF PHYSICS.MANCHESTER AND DISTRICT BRANCH The Measurement of stress and strain in solids :a conference Manchester 1946 Jul 11-13 Edited by F.A. Vick Physics in industry Institute of physics,London,1948 Book based on the proceedings of the conference
Cav 7.2397

INSTITUTE OF PHYSICS.MANCHESTER AND DISTRICT BRANCH The Measurement of stress and strain in solids :a conference Proceedings Manchester 1946 Jul 11-13 Edited by F.A. Vick Physics in industry Institute of physics,London,1948
Met 25.0880

INSTITUTE OF PHYSICS.MIDLAND BRANCH Vacuum physics :a symposium Birmingham 1950 Jun 27-28 Journal of scientific instruments. Supplement, 1 80,viiip Institute of physics,London,1951
Chem E 24.1353

INSTITUTE OF PHYSICS AND THE PHYSICAL SOCIETY International conference on magnetism Proceedings Nottingham 1964 Sep 6-11 With the support of the International union of pure and applied physics illus 898p London,c1965
Cav 7.2882

INSTITUTE OF PHYSICS AND THE PHYSICAL SOCIETY. CONFERENCE SERIES Charged particle tracks in solids and liquids Gray (L.H.) conference 2nd Proceedings Cambridge 1969 Apr Institute of physics,London,1970
Radioth 35.1702

INSTITUTE OF PHYSICS AND THE PHYSICAL SOCIETY. CONFERENCE SERIES, 5-6 International vacuum congress 4th Manchester 1968 Apr 17-20 Pt 1-2 Joint British committee for vacuum science and technology International union for vacuum science, technique and applications 2 vols Institute of physics,London,c1968
Met 25.2834

INSTITUTE OF PHYSICS AND THE PHYSICAL SOCIETY. ELECTRONICS GROUP Conference on physical aspects of noise in electronic devices Papers Nottingham 1968 Sep 11-13 iv,248p 31cm Peregrinus,Stevenage,1968
Cav 7.2840

INSTITUTE OF PHYSICS AND THE PHYSICAL SOCIETY. SPECTROSCOPY GROUP Limitations of detection in spectrochemical analysis :a conference Papers Exeter 1964 Jul 23 Hilger and Watts,London,1964
Radioth 35.1657

INSTITUTE OF PHYSICS AND THE PHYSICAL SOCIETY. SPECTROSCOPY GROUP Spectroscopy in the metallurgical industry the symposium 2nd Papers Buxton 1962 Jul Edited by L. Bovey Hilger and Watts,London,1963
Met 25.1752

INSTITUTE OF PHYSICS AND THE PHYSICAL SOCIETY. STATIC ELECTRIFICATION GROUP Static electrification The Institute of physics and the Physical society conference on static electrification 2nd Proceedings London 1967 May Institute of physics and the Physical society.Conference series, 4 25cm London,1967
Cav 7.2822

INSTITUTE OF PHYSICS AND THE PHYSICAL SOCIETY. STRESS ANALYSIS GROUP COMMITTEE Physical basis of yield and fracture conference Proceedings Oxford 1966 Sep Edited by A. C. Stickland and R.A. Cook Institute of physics and the physical society.Conference series, 1 Inst.of Physics and the Physical society,London,c1966
Met 25.0869

The INSTITUTE OF PHYSICS AND THE PHYSICAL SOCIETY CONFERENCE ON STATIC ELECTRIFICATION 2nd Proceedings Static electrification London 1967 May Institute of physics and the Physical society.Static electrification group Institute of physics and the Physical society.Conference series, 4 25cm London, 1967
Cav 7.2822

INSTITUTE OF PRECISION MECHANICS,WARSAW Fatigue resistance of materials and metal structural parts international conference Proceedings Warsaw 1960 May Edited by Alfred Buch Pergamon press;Polish scientific publishers,Oxford;Warsaw,1964
Met 25.0933

INSTITUTE OF PRINTING Computer typesetting conference preprints and summaries Brighton 1966 July 15-16 29cm 2 vols Institute of printing,1966
Math L 5.1023

INSTITUTE OF RACE RELATIONS Man,race and Darwin :a conference Papers London 1959 Jan 7-9 Edited by Philip Mason vii,151p Oxford university press,London,1960
WSM 43.3622

INSTITUTE OF RADIO ENGINEERS I.R.E. Western electronic show and convention convention record Los Angeles 1960 Aug.23-26 Part 4: computers,man-machine systems 202p Institute of radio engineers,New York,1960
Math L 5.2076

INSTITUTE OF RADIO ENGINEERS I.R.E. international convention convention record New York 1960 Mar.21-24 Part 2: circuit theory,electronic computers 203p Institute of radio engineers,New York,1960
Math L 5.2074

INSTITUTE OF RADIO ENGINEERS I.R.E. international convention convention record New York 1960 Mar.21-24 Part 4: automatic control,information theory 222p Institute of radio engineers,New York,1960
Math L 5.2075

INSTITUTE OF RADIO ENGINEERS I.R.E. international convention convention record New York 1961 Mar.20-23 Part 2: audio; electronic computers 275p Institute of radio engineers,New York,1961
Math L 5.2077

INSTITUTE OF RADIO ENGINEERS I.R.E. international convention convention record New York 1961 Mar.20-23 Part 4: automatic control,circuit theory,information theory 286p Institute of radio engineers, New York,1961
Math L 5.2078

INSTITUTE OF RADIO ENGINEERS I.R.E. international convention convention record New York 1962 Mar.26-29 Part 4: electronic computers,information theory 199p Institute of radio engineers,New York,1962
Math L 5.2084

INSTITUTE OF RADIO ENGINEERS I.R.E. national convention convention record New York 1953 Mar.23-26 Part 7: electronic computers 71p Institute of radio engineers, New York,1953
Math L 5.2064

INSTITUTE OF RADIO ENGINEERS I.R.E. national convention convention record New York 1954 Mar.22-25 Part 4: electronic computers and information theory 143p Institute of radio engineers,New York,1954
Math L 5.2065

INSTITUTE OF RADIO ENGINEERS I.R.E. national convention convention record New York 1955 Mar.21-24 Part 4: computers, information theory,automatic control 208p Institute of radio engineers,New York,1955
Math L 5.2066

INSTITUTE OF RADIO ENGINEERS I.R.E. national convention convention record New York 1956 Mar.19-22 Part 4: computers, information theory,automatic control 175p Institute of radio engineers,New York,1956
Math L 5.2067

INSTITUTE OF RADIO ENGINEERS I.R.E. national convention convention record New York 1957 Mar.18-21 Part 2: circuit theory, information theory 212p Institute of radio engineers,New York,1957
Math L 5.2068

INSTITUTE OF RADIO ENGINEERS I.R.E. national convention convention record New York 1957 Mar.18-21 Part 4: automatic control, electronic computers,medical electronics 179p Institute of radio engineers,New York, 1957
Math L 5.2069

INSTITUTE OF RADIO ENGINEERS I.R.E. national convention convention record New York 1959 Mar.23-26 Part 4: automatic control, electronic computers,information theory 201p Institute of radio engineers,New York,1959
Math L 5.2072

INSTITUTE OF RADIO ENGINEERS I.R.E. western electronic show and convention convention record Los Angeles 1958 Aug.19-22 291p Institute of radio engineers,New York,1958
Math L 5.2070

INSTITUTE OF RADIO ENGINEERS I.R.E. western electronic show and convention convention record San Francisco 1959 Aug.18-21 Part 4: automatic control,electronic computers, information theory 201p Institute of radio engineers,New York,1959
Math L 5.2073

INSTITUTE OF RADIO ENGINEERS International symposium on circuit and information theory transactions Los Angeles 1959 Jun.16-18 Co-sponsored by the International scientific radio union Institute of radio engineers. Transactions on information theory,IT, 5, special supplement xiv,298p Institute of radio engineers,New York,1959 Papers published in advance of the Symposium
Math L 5.2071

INSTITUTE OF RADIO ENGINEERS Millimeter waves :a symposium Proceedings New York 1959 Mr 31-Apr 2 Co-sponsored by United States.Office of naval research Polytechnic institute of Brooklyn.Microwave research institute symposia series, 9 Polytechnic press,Brooklyn,N.Y.,1960
Cav 7.1715

INSTITUTE OF RADIO ENGINEERS Quasi-optics :a symposium Proceedings New York 1964 Jun 8-10 Co-sponsored by United States.Office of naval research Polytechnic institute of Brooklyn.Microwave research institute symposia series, 14 Polytechnic press,Brooklyn,N.Y., 1964
Cav 7.1721

INSTITUTE OF RADIO ENGINEERS Report of the round-table discussion on fast circuits of I.R. E. committee 4.10 in Tucson,Ariz. mimeograph Tucson,Ariz. 1958 May By Wolfgang J. Poppelbaum University of Illinois.Digital computer laboratory 24p 1958
Math L 5.1678

INSTITUTE OF RADIO ENGINEERS Report on the I. R.E. subcommittee 4.10 round table conference in Austin,Texas mimeograph Austin,Texas 1959 Apr. 30-May 1 By Wolfgang J. Poppelbaum University of Illinois.Digital computer laboratory 20p 1959
Math L 5.1676

INSTITUTE OF RADIO ENGINEERS Solid-state circuit conference digest of technical papers Philadelphia,Pa. 1959 Feb.12-13 Sponsored by American institute of electrical engineers 102p Lewis Winner,1959
Math L 5.2082

INSTITUTE OF RADIO ENGINEERS Symposium on information theory Report of proceedings London 1950 Sep 26-29 IRE professional group on information theory.Publications committee,n.p.,c1950 Foreword by W. Jackson. Duplicated
Psy 31.3221

INSTITUTE OF RADIO ENGINEERS Symposium on information theory proceedings London 1950 Sep 26-29 Institute of radio engineers. Professional group on information theory, 1 208p Institute of radio engineers,New York, 1953
Math L 5.2079

INSTITUTE OF RADIO ENGINEERS Symposium on printed circuits proceedings Philadelphia, Pa. 1955 Jan 20-21 Radio-Electronics-Television manufacturers association illus. 122p 28cm Engineering publishers,New York, 1955
Math L 5.0736

INSTITUTE OF SOCIOLOGY The Social sciences: their relations in theory and in teaching : conference Report London 1935 Sep 22-29 Sponsored jointly by the International student service Le Play House press,London,1936
Geog 13.1613

INSTITUTE OF STATISTICAL MATHEMATICS,TOKYO Seminar on modern methods in number theory Tokyo 1971 Aug 30-Sep 4 29cm Institute of statistical mathematics,Tokyo,1971
P Math 2.4380

INSTITUTE OF THE AERONAUTICAL SCIENCES Anglo-American aeronautical conference 6th Papers Folkestone 1957 Sep 9-12 Edited by Joan Bradbrooke and U.A. Libby Royal aeronautical society,London,1959
Eng 41.6962

INSTITUTE OF WELDING Notch bar testing and its relation to welded construction :symposium London 1951 Dec 5 Institute of welding, London,1953
Eng 41.3834

INSTITUTE OF WELDING Physics of the welding arc :a symposium London 1962 Oct 29-Nov 2 Institute of welding,London,1966
Met 25.0807

INSTITUTE OF WELDING Welding in shipbuilding Joint symposium on welding in shipbuilding London 1961 Oct 30-Nov 3 Institute of welding,London,1962 Jointly organised by Institute of welding,Royal institution of naval architects,Institute of marine engineers, Institution of engineers and shipbuilders in Scotland and North East Coast institution of engineering and shipbuilders
Met 25.0799

INSTITUTE OF WELDING.AUTUMN MEETING 1961 Welding in shipbuilding Joint symposium on welding in shipbuilding London 1961 Oct 30-Nov 3 Institute of welding Institute of welding,London,1962 Jointly organised by Institute of welding,Royal institution of naval architects,Institute of marine engineers, Institution of engineers and shipbuilders in Scotland and North East Coast institution of engineering and shipbuilders
Met 25.0799

INSTITUTION OF CHEMICAL ENGINEERS A.I.Ch.E.-Chem.E. joint meeting Proceedings London 1965 Jun 13-17 4: application of mathematical models in chemical engineering research design and production 1965 A.I.Ch.E. I.Chem.E.Symposium series, 4 I.Chem.E., London,1965
Eng 41.5861

INSTITUTION OF CHEMICAL ENGINEERS A.I.Ch.E.-I.Chem.E. joint meeting Proceedings London 1965 Jun 14 10: mixing - theory related to practice 1965 A.I.Ch.E.-I;Chem.E.symposium series, 10 Institution of chemical engineers,London,1965
Chem E 24.1920

INSTITUTION OF CHEMICAL ENGINEERS A.I.Ch.E.-I.Chem.E.joint meeting Proceeding London 1965 Jun 13-17 1: advances in separation techniques 1965 A.I.Ch.E.-I.Chem.E.Symposium series, 1 Institution of chemical engineers,London,1965
Chem E 24.1506

INSTITUTION OF CHEMICAL ENGINEERS A.I.Ch.E.-I.Chem.E.joint meeting Proceedings London 1965 Jun 13-17 2: chemical engineering under extreme conditions 1965 A.I.Ch.E.-I. Chem.E.Symposium series, 2 Institution of chemical engineers,London,1965
Chem E 24.1507

INSTITUTION OF CHEMICAL ENGINEERS A.I.Ch.E.-I.Chem.E.joint meeting Proceedings London 1965 Jun 13-17 4: application of mathematical models in chemical engineering research, design and production 1965 A.I.Ch. E.-I.Chem.E.Symposium, 4 Institution of chemical engineers,London,1965
Chem E 24.1509

INSTITUTION OF CHEMICAL ENGINEERS A.I.Ch.E.-I.Chem.E.joint meeting Proceedings London 1965 Jun 14 3: reaction kinetics in product and process design 1965 A.I.Ch.E.-I. Chem.E.Symposium series, 3 Institution of chemical engineers,London,1965
Chem E 24.1508

INSTITUTION OF CHEMICAL ENGINEERS A.I.Ch.E.-I.Chem.E.joint meeting Proceedings London 1965 Jun 14 6: transport phenomena 1965 A.I.Ch.E.-I.Chem.E.Symposium series, 6 Institution of chemical engineers,London,1965
Chem E 24.1511

INSTITUTION OF CHEMICAL ENGINEERS A.I.Ch.E.-I.Chem.E.joint meeting Proceedings London 1965 Jun 14 7: management oriented topics 1965 A.I.Ch.E.-I.Chem;E.Symposium series, 7 Institution of chemical engineers,London,1965
Chem E 24.1512

INSTITUTION OF CHEMICAL ENGINEERS A.I.Ch.E.-I.Chem.E.joint meeting Proceedings London 1965 Jun 14 9: new chemical engineering problems in the utilisation of water 1965 A. I.Ch.E.-I.Chem.E.Symposium series, 9 Institution of chemical engineers,London,1965
Chem E 24.1513

INSTITUTION OF CHEMICAL ENGINEERS A.I.Ch.E.-I.ChemE.joint meeting London 1965 Jun 14 5: materials associated with direct energy conversion 1965 A.I.Ch.E.-I.Chem.E.Symposium series, 5 Institution of chemical engineers,London,1965
Chem E 24.1510

INSTITUTION OF CHEMICAL ENGINEERS Chemical engineering in the metallurgical industries a symposium Proceedings Edinburgh 1963 Sep 25-26 Institution of chemical engineers, London,1963
Met 25.0316

INSTITUTION OF CHEMICAL ENGINEERS
Fluidisation :a symposium Papers London 1963 Mar 11-12 Sponsored also by the Institute of petroleum 110p Society of chemical industry,London,1964
Chem E 24.1452

INSTITUTION OF CHEMICAL ENGINEERS I.Chem.E., VTG-VDI joint meeting on the engineering of gas-solid reactions Preprints Brighton 1968 Apr 24-26 characteristics of solid particles which are relevant to gas-solid reactions 43p Institution of chemical engineers,London,1968 In English and German
Chem E 24.1516

INSTITUTION OF CHEMICAL ENGINEERS I.Chem.E., VTG-VDI joint meeting on the engineering of gas-solid reactions Preprints Brighton 1968 Apr 24-26 experiences with the operation of industrial process equipment 51p Institution of chemical engineers,London, 1968 In English and German
Chem E 24.1519

INSTITUTION OF CHEMICAL ENGINEERS I.Chem.E., VTG-VDI joint meeting on the engineering of gas-solid reactions Preprints Brighton 1968 Apr 24-26 physical behaviour of gas-solid systems 69p Institution of chemical engineers,London,1968 In English and German
Chem E 24.1517

INSTITUTION OF CHEMICAL ENGINEERS I.Chem.E., VTG-VDI joint meeting on the engineering of gas-solid reactions Preprints Brighton 1968 Apr 24-26 the technical design and performance of industrial process equipment 52p Institution of chemical engineers,London, 1968 In English and German
Chem E 24.1518

INSTITUTION OF CHEMICAL ENGINEERS
International developments in heat transfer International heat transfer conference Papers Boulder,Col. 1961 Aug 28-Sep 1 London 1962 Jan 8-12 Pt 1-5 Arranged by the American society of mechanical engineers 5 vols A.S.M.E.,New York,1961 Co-sponsored by the American institute of chemical engineers,the Institution of mechanical engineers,and the Institution of chemical engineers
Chem E 24.0595

INSTITUTION OF CHEMICAL ENGINEERS Joint symposium on instrumentation and computation in process development and plant design Papers London 1959 May 11-12 British conference on automation and computation Institution of chemical engineers,London,1959 Sponsored also by the Society of instrument technology,and the British computer society
Math L 5.3290

INSTITUTION OF CHEMICAL ENGINEERS Joint symposium on instrumentation and computation in process development and plant design Proceedings London 1959 May 11-13 Institution of chemical engineers,London,1959
Eng 41.6011

INSTITUTION OF CHEMICAL ENGINEERS Nucleation and crystal growth including precipitation By D.A. Blackadder Industrial research fellow report, 3 Institution of chemical engineers,London,1964
Met 25.1523

INSTITUTION OF CHEMICAL ENGINEERS
Productivity in research :a symposium Proceedings London 1963 Dec 11-12 Institution of chemical engineers,London,1964
Eng 41.1685

INSTITUTION OF CHEMICAL ENGINEERS Symposium on catalysis in practice Proceedings Harrogate 1963 Jun 20-21 Edited by J.M. Pirie 83p Institution of chemical engineers,London,1963
Chem E 24.0803

INSTITUTION OF CHEMICAL ENGINEERS Symposium on chemical engineering education Discussion on papers presented Birmingham 1957 Apr 9-11 1957
Chem E 24.1758

INSTITUTION OF CHEMICAL ENGINEERS Symposium on chemical engineering education Papers Birmingham 1957 Apr 9-11 61p 1957
Chem E 24.1757

INSTITUTION OF CHEMICAL ENGINEERS Symposium on chemical engineering in the metallurgical industries Proceedings Edinburgh 1963 Sep 25-29 147p Institution of chemical engineers,London,1963
Chem E 24.1131

INSTITUTION OF CHEMICAL ENGINEERS Symposium on particle size analysis London 1947 Feb 4 145p Institution of chemical engineers, London,1947 Supplement to Transactions, Institution of chemical engineers,vol.25,1947
Chem E 24.1431

INSTITUTION OF CHEMICAL ENGINEERS Symposium on the handling of solids Proceedings London 1962 Jun 25 Edited by P.A. Rottenburg Institution of chemical engineers, London,1962 3rd congress of the European federation of chemical engineering
Chem E 24.1446

INSTITUTION OF CHEMICAL ENGINEERS Symposium on the interaction between fluids and particles Proceedings London 1962 Jun 20-22 illus 351p Institution of chemical engineers,London,n.d. Third congress on the European federation of chemical engineering
Chem E 24.1447

INSTITUTION OF CHEMICAL ENGINEERS The Engineering of gas-solid reactions :I.Chem.E.-VTG-VDI joint meeting Proceedings Brighton 1968 Apr 24-26 Edited by J.M. Pirie Institution of chemical engineers.Symposium series, 27 illus 255p Institution of chemical engineers,London,1968
Chem E 24.1873

INSTITUTION OF CHEMICAL ENGINEERS The Organization of chemical engineering projects : joint symposium Proceedings London 1958 Jun 24-26 Institution of chemical engineers, London,1958 17th meeting of the European federation of chemical engineering
Chem E 24.1543

INSTITUTION OF CHEMICAL ENGINEERS The Scaling-up of chemical plant and processes : joint symposium Papers London 1957 May 28-29 By R.Edgeworth Johnstone and others Institution of chemical engineers,London,1957 Held jointly with the K.Instituut van ingenieurs and the K.Nederlandse chemische vereniging
Chem E 24.1234

INSTITUTION OF CHEMICAL ENGINEERS.MIDLANDS BRANCH Efficient computer methods for the practising chemical engineer :a symposium Proceedings Nottingham 1967 Apr 19 Edited by J.M. Pine Institution of chemical engineers.Symposium series, 23 illus 194p I.Chem.E.,London,1967
Chem E 24.1809

INSTITUTION OF CHEMICAL ENGINEERS.MIDLANDS BRANCH Efficient computer methods for the practising chemical engineer :a symposium Proceedings Nottingham 1967 Apr 19 Institution of chemical engineers.Symposium series, 23 Institution of chemical engineers,London,1967
Eng 41.5860

INSTITUTION OF CHEMICAL ENGINEERS.MIDLANDS BRANCH The Symposium on the less common means of separation Proceedings Birmingham 1963 Apr 24-26 Edited by J.M. Pirie 127p I. Chem.E.,London,1964
Chem E 24.1915

INSTITUTION OF CHEMICAL ENGINEERS.SYMPOSIUM SERIES, 23 Efficient computer methods for the practising chemical engineer :a symposium Proceedings Nottingham 1967 Apr 19 Edited by J.M. Pine Organised by the Institution of chemical engineers.Midlands branch illus 194p I.Chem.E.,London,1967
Chem E 24.1809

INSTITUTION OF CHEMICAL ENGINEERS.SYMPOSIUM SERIES, 23 Efficient computer methods for the practising chemical engineer :a symposium Proceedings Nottingham 1967 Apr 19 Organised by the Institution of chemical engineers.Midlands branch Institution of chemical engineers,London,1967
Eng 41.5860

INSTITUTION OF CHEMICAL ENGINEERS.SYMPOSIUM SERIES, 27 The Engineering of gas-solid reactions :I.Chem.E.-VTG-VDI joint meeting Proceedings Brighton 1968 Apr 24-26 Institution of chemical engineers Verein deutscher ingenieure Edited by J.M. Pirie illus 255p Institution of chemical engineers,London,1968
Chem E 24.1873

INSTITUTION OF CIVIL ENGINEERS Conference on biology and civil engineering Proceedings London 1948 Sep Institution of civil engineers,London,1949
Geog 13.2688

INSTITUTION OF CIVIL ENGINEERS Conference on coastal engineering 11th Proceedings London 1968 Sep Vol 1-2 2 vols American society of civil engineers,New York, 1969
Eng 41.3347

INSTITUTION OF CIVIL ENGINEERS Conference on prestressed concrete pressure vessels London 1967 Mar 13-17 Institution of civil engineers,London,1968
Eng 41.2891

INSTITUTION OF CIVIL ENGINEERS Conference on the North Sea floods of 31 January-1 February, 1953 Papers London 1953 Dec Institution of civil engineers,London,1954
Geog 13.0706

INSTITUTION OF CIVIL ENGINEERS Conference on the highway needs of Great Britain Proceedings London 1957 Nov 13-15 Institution of civil engineers,London,1958
Eng 41.3260

INSTITUTION OF CIVIL ENGINEERS Conservation of water resources in the United Kingdom :a symposium Proceedings London 1962 Oct 30-Nov 1 Institution of civil engineers,London, 1963
Geog 13.0786

INSTITUTION OF CIVIL ENGINEERS Correlation between calculated and observed stresses and displacements in structures :a conference London 1955 Sep 21-22 2 vols Institution of civl engineers,London,1955
Eng 41.3020

INSTITUTION OF CIVIL ENGINEERS Joint engineering conference Proceedings London 1951 Jun 4-15 Institution of civil engineers, London,1951
Eng 41.1761

INSTITUTION OF CIVIL ENGINEERS Natural draught cooling towers - Ferrybridge and after a conference Proceedings London 1967 Jun 12 Institution of civil engineers,London, 1967
Eng 41.2889

INSTITUTION OF CIVIL ENGINEERS Prestressed concrete :a conference Proceedings London 1949 Feb Institution of civil engineers, London,1949
Eng 41.3019

INSTITUTION OF CIVIL ENGINEERS The Conservation of natural resources :lectures Institution of civil engineers,London,1957
Eng 41.3386

INSTITUTION OF CIVIL ENGINEERS The North sea floods of January 31-February 1,1953 :a conference London 1953 Dec Institution of civil engineers,London,1954
Eng 41.3342

INSTITUTION OF ELECTRICAL ENGINEERS Applications of microelectronics :a symposium Southampton 1965 Sep 21-23 Institution of electronic and radio engineers University of Southampton.Department of electronics Institution of electrical engineers.Conference publication, 14 I.E.E.,London,1965 Reproduced from typescript
Eng 41.5576

INSTITUTION OF ELECTRICAL ENGINEERS Automatic and manual control Papers Cranfield 1951 Jul 16-21 Edited by A. Tustin Under the auspices of Great Britain. Department of scientific and industrial research Butterworths,London,1952
Psy 31.3398

INSTITUTION OF ELECTRICAL ENGINEERS Autotrophic micro-organism :a symposium Papers London 1954 Apr Society for general microbiology Edited by B.A. Fry and J.L. Peel Society for general microbiology. Symposia, 4 Cambridge university press, Cambridge,1954
Bioch 33.1168

INSTITUTION OF ELECTRICAL ENGINEERS Communication theory Applications of communication theory symposium Papers London 1952 Sep 22-26 Edited by Willis Jackson Butterworths,London,1953
Psy 31.2016

INSTITUTION OF ELECTRICAL ENGINEERS Conference on electronic telephone exchanges papers London 1960 Nov 22-24 Collection of typescript papers and reprints
Math L 5.0873

INSTITUTION OF ELECTRICAL ENGINEERS Conference on electronics design Cambridge 1968 Sep 23-27 Institution of electrical engineers.Conference publication, 45 IEE, London,1968
Math L 5.3468

INSTITUTION OF ELECTRICAL ENGINEERS Conference on gas discharges 1972 Institution of electrical engineers.Conference publication, 90 IEE,London,1972
Eng 41.8570

INSTITUTION OF ELECTRICAL ENGINEERS Conference on industrial measurement techniques for on-line computers London 1968 Jun 11-13 Institution of electrical engineers.Conference publications, 43 I.E.E. London,1968
Eng 41.2220

INSTITUTION OF ELECTRICAL ENGINEERS Conference on optimisation techniques in circuit and control applications London 1970 Jun 29-30 Institution of electrical engineers.Conference publication, 66 IEE, London,1970 Organised in association with the Institute of electrical and electronics engineers and the Institute of physics and the Physical society
Eng 41.8587

INSTITUTION OF ELECTRICAL ENGINEERS Conference on power application of controllable semiconductor devices London 1965 Pt 1 Institution of electrical engineers.Conference publications, 17 I.E.E. London,1965
Eng 41.4868

INSTITUTION OF ELECTRICAL ENGINEERS Conference on switching techniques for telecommunications networks London 1969 21-25 Apr Institution of electrical engineers. Conference publications, 52 Institution of electrical engineers,London,1969
Math L 5.3531

INSTITUTION OF ELECTRICAL ENGINEERS Convention on digital-computer techniques proceedings London 1956 Apr 9-14 Edited by W.K. Brasher Institution of electrical engineers.Proceedings, 103,Part B,Suppts.1-3 illus 342p 27cm Institution of electrical engineers,London,1956
Math L 5.0762

INSTITUTION OF ELECTRICAL ENGINEERS Joint conference on digital methods of measurement Proceedings Canterbury 1969 Jul 23-25 Institution of electronic and radio engineers. Conference proceedings, 15 IERE,London,1969
Eng 41.8305

INSTITUTION OF ELECTRICAL ENGINEERS Joint conference on digital processing of signals in communications Proceedings 1972 Institution of electronic and radio engineers. Conference proceedings, 23 IERE,London,1972
Eng 41.8565

INSTITUTION OF ELECTRICAL ENGINEERS Joint engineering conference Proceedings London 1951 Jun 4-15 Institution of civil engineers, London,1951
Eng 41.1761

INSTITUTION OF ELECTRICAL ENGINEERS Magnetic materials and their applications :a conference London 1967 Sep 26-28 Institution of electrical engineers.Conference publication, 13 Institution of electrical engineers, London,1967
Eng 41.5348

INSTITUTION OF ELECTRICAL ENGINEERS Microwave and optical generation and amplification International conference on microwave tubes 6th Cambridge 1966 Institution of electrical engineers.Conference publication, 27 I.E.E.,London,1967
Eng 41.5588

INSTITUTION OF ELECTRICAL ENGINEERS
Multivariable control system design and application U.K.A.C. control convention 4th Manchester 1971 Sep 1-3 United Kingdom automation council Institution of electrical engineers.Conference publication, 78 IEE,London,1971
Eng 41.8590

INSTITUTION OF ELECTRICAL ENGINEERS
Symposium digest Symposium on electronic equipment reliability 2nd London 1962 Oct 24-26 illus. 47p 28cm Institution of electrical engineers,London, 1962
Math L 5.0926

INSTITUTION OF ELECTRICAL ENGINEERS
Symposium on applications of micro-electronics Birmingham 1968 Mar 27 Institution of electrical engineers.Conference publication, 49 Institution of electrical engineers, London,1968
Eng 41.5266

INSTITUTION OF ELECTRICAL ENGINEERS The Impact of users' needs on the design of data processing systems : conference Edinburgh 1964 Mar 31-Apr 3 British computer society Held under the aegis of the United Kingdom automation council 1964 Organized jointly by the British computer society,the British institution of radio engineers and the Institution of electrical engineers
Math L 5.3265

INSTITUTION OF ELECTRICAL ENGINEERS The Interconnection of peripheral equipment and computers colloquium report London 1964 Jan Edited by H.McG. Ross 380p Institution of electrical engineers,London, 1964
Math L 5.1272

INSTITUTION OF ELECTRICAL ENGINEERS The Reliability and maintenance of digital computer systems:managerial and engineering aspects :discussion meetings London 1960 Jan 20-21 70p 28cm British computer society;Institution of electrical engineers, London,1960
Math L 5.3255

INSTITUTION OF ELECTRICAL ENGINEERS.CONFERENCE PUBLICATION, 13 Magnetic materials and their applications :a conference London 1967 Sep 26-28 Sponsored by the Institution of electrical engineers Institution of electrical engineers,London,1967
Eng 41.5348

INSTITUTION OF ELECTRICAL ENGINEERS.CONFERENCE PUBLICATION, 13 Microwave behaviour of ferrimagnetics and plasmas International conference on the microwave behaviour of ferrimagnetics and plasmas London 1965 Sep 13-17 Sponsored by Institution of electrical engineers.Electronics division Institution of electrical engineers,London,1965
Eng 41.5713

INSTITUTION OF ELECTRICAL ENGINEERS.CONFERENCE PUBLICATION, 14 Applications of microelectronics :a symposium Southampton 1965 Sep 21-23 Institution of electronic and radio engineers University of Southampton. Department of electronics Sponsored also by the Institution of electrical engineers I.E. E.,London,1965 Reproduced from typescript
Eng 41.5576

INSTITUTION OF ELECTRICAL ENGINEERS.CONFERENCE PUBLICATION, 27 Microwave and optical generation and amplification International conference on microwave tubes 6th Cambridge 1966 Institution of electrical engineers I.E.E.,London,1967
Eng 41.5588

INSTITUTION OF ELECTRICAL ENGINEERS.CONFERENCE PUBLICATION, 29 Advances in computer control United Kingdom automation council. Control convention 2nd Bristol 1967 Apr 11-14 Organized by the Institution of electrical engineers.Control and automation division I.E.E.,London,1967
Eng 41.5862

INSTITUTION OF ELECTRICAL ENGINEERS.CONFERENCE PUBLICATION, 37 Conference on servocomponents London 1967 Nov 21-23 Sponsored by Institution of electrical engineers.Professional group on measuring and control equipment I.E.E.,London,1967
Eng 41.5879

INSTITUTION OF ELECTRICAL ENGINEERS.CONFERENCE PUBLICATION, 42 Conference on pattern recognition Teddington 1968 Jul 29-31 Organised by Institution of electrical engineers.Control and automation division Institution of electrical engineers,London, 1968
Eng 41.5311

INSTITUTION OF ELECTRICAL ENGINEERS.CONFERENCE PUBLICATION, 49 Symposium on applications of micro-electronics Birmingham 1968 Mar 27 Organized by the Institution of electrical engineers Institution of electrical engineers,London,1968
Eng 41.5266

INSTITUTION OF ELECTRICAL ENGINEERS.CONFERENCE PUBLICATION, 66 Conference on optimisation techniques in circuit and control applications London 1970 Jun 29-30 Institution of electrical engineers IEE, London,1970 Organised in association with the Institute of electrical and electronics engineers and the Institute of physics and the Physical society
Eng 41.8587

INSTITUTION OF ELECTRICAL ENGINEERS.CONFERENCE PUBLICATION, 78 Multivariable control system design and application U.K.A.C. control convention 4th Manchester 1971 Sep 1-3 United Kingdom automation council Organized by the Institution of electrical engineers IEE,London,1971
Eng 41.8590

INSTITUTION OF ELECTRICAL ENGINEERS.CONFERENCE PUBLICATION, 90 Conference on gas discharges 1972 Institution of electrical engineers Institute of electrical and electronics engineers IEE,London,1972
Eng 41.8570

INSTITUTION OF ELECTRICAL ENGINEERS.CONFERENCE PUBLICATIONS, 17 Conference on power application of controllable semiconductor devices London 1965 Pt 1 Institution of electrical engineers I.E.E., London,1965
Eng 41.4868

INSTITUTION OF ELECTRICAL ENGINEERS.CONFERENCE PUBLICATIONS, 43 Conference on industrial measurement techniques for on-line computers London 1968 Jun 11-13 Institution of electrical engineers I.E.E., London,1968
Eng 41.2220

INSTITUTION OF ELECTRICAL ENGINEERS.CONFERENCE REPORT SERIES, 4 Magnetoplasmadynamic electrical power generation :a symposium Report Newcastle-upon-Tyne 1962 Sep 6-8 Institution of electrical engineers Institution of electrical engineers,London, 1963
Eng 41.4521

INSTITUTION OF ELECTRICAL ENGINEERS.CONTROL AND AUTOMATION DIVISION Advances in computer control United Kingdom automation council. Control convention 2nd Bristol 1967 Apr 11-14 Institution of electrical engineers.Conference publication, 29 I.E.E., London,1967
Eng 41.5862

INSTITUTION OF ELECTRICAL ENGINEERS.CONTROL AND AUTOMATION DIVISION Conference on pattern recognition Teddington 1968 Jul 29-31 Institution of electrical engineers.Conference publication, 42 Institution of electrical engineers,London,1968
Eng 41.5311

INSTITUTION OF ELECTRICAL ENGINEERS.ELECTRONICS DIVISION Integrated circuits Conference on integrated circuits Eastbourne 1967 May 2-4 I.E.E.Conference publication, 30 Institution of electrical engineers,London, 1967 Joint sponsors;Institution of electronic and radio engineers and the Institute of electrical and electronics engineers
Eng 41.5461

INSTITUTION OF ELECTRICAL ENGINEERS.ELECTRONICS DIVISION Microwave behaviour of ferrimagnetics and plasmas International conference on the microwave behaviour of ferrimagnetics and plasmas London 1965 Sep 13-17 Institution of electrical engineers. Conference publication, 13 Institution of electrical engineers,London,1965
Eng 41.5713

INSTITUTION OF ELECTRICAL ENGINEERS.ELECTRONICS DIVISION.COMPUTER DESIGN PROFESSIONAL GROUP Conference on computer science and technology Manchester 1969 Jun 30-Jul 3 Institution of electrical engineers.Conference publication, 55 The institution of electrical engineers, London,1969
Math L 5.3503

INSTITUTION OF ELECTRICAL ENGINEERS.PROFESSIONAL GROUP ON MEASURING AND CONTROL EQUIPMENT Conference on servocomponents London 1967 Nov 21-23 Institution of electrical engineers.Conference publication, 37 I.E.E., London,1967
Eng 41.5879

INSTITUTION OF ELECTRONIC AND RADIO ENGINEERS Applications of microelectronics :a symposium Southampton 1965 Sep 21-23 Sponsored also by the Institution of electrical engineers Institution of electrical engineers.Conference publication, 14 I.E.E.,London,1965 Reproduced from typescript
Eng 41.5576

INSTITUTION OF ELECTRONIC AND RADIO ENGINEERS Integrated circuits Conference on integrated circuits Eastbourne 1967 May 2-4 Sponsored by the Institution of electrical engineers.Electronics division I.E.E. Conference publication, 30 Institution of electrical engineers,London,1967 Joint sponsors;Institution of electronic and radio engineers and the Institute of electrical and electronics engineers
Eng 41.5461

INSTITUTION OF ELECTRONIC AND RADIO ENGINEERS Joint conference on digital methods of measurement Proceedings Canterbury 1969 Jul 23-25 Institution of electronic and radio engineers.Conference proceedings, 15 IERE,London,1969
Eng 41.8305

INSTITUTION OF ELECTRONIC AND RADIO ENGINEERS Joint conference on digital processing of signals in communications Proceedings 1972 Institution of electronic and radio engineers. Conference proceedings, 23 IERE,London,1972
Eng 41.8565

INSTITUTION OF ELECTRONIC AND RADIO ENGINEERS. CONFERENCE PROCEEDINGS, 15 Joint conference on digital methods of measurement Proceedings Canterbury 1969 Jul 23-25 Institution of electronic and radio engineers Institution of electrical engineers Institute of electrical and electronics engineers IERE,London,1969
Eng 41.8305

INSTITUTION OF ELECTRONIC AND RADIO ENGINEERS. CONFERENCE PROCEEDINGS, 23 Joint conference on digital processing of signals in communications Proceedings 1972 Institution of electronic and radio engineers Institution of electrical engineers IERE, London,1972
Eng 41.8565

INSTITUTION OF EMCHANICAL ENGINEERS.AUTOMOBILE DIVISION Vibration and noise in motor vehicles :a symposium Proceedings London 1971 Jul 6-7 IME,London,1972
Eng 41.8646

INSTITUTION OF ENGINEERING DESIGNERS Conference on the teaching of engineering design Papers and summary Scarborough 1964 Apr 1-4 Organised by the Enfield college of technology Illus Institution of engineering designers,London,1964 Co-organized by the Institution of engineering designers and the Hornsey College of art
Eng 41.0309

INSTITUTION OF ENGINEERS,AUSTRALIA Australia-New Zealand conference on soil mechanics and foundation engineering 1st Proceedings Melbourne 1952 Jul 1952
Eng 41.3166

INSTITUTION OF MECHANICAL ENGINEERS Automatic and manual control Papers Cranfield 1951 Jul 16-21 Edited by A. Tustin Under the auspices of Great Britain. Department of scientific and industrial research Butterworths,London,1952
Psy 31.3398

INSTITUTION OF MECHANICAL ENGINEERS
Conference on hydraulic mechanisms
Proceedings London 1954 Mar 26
Institution of mechanical engineers,London, 1954
Eng 41.6686

INSTITUTION OF MECHANICAL ENGINEERS
Conference on hydraulic servo-mechanisms
Proceedings London 1953 Feb 13
Institution of mechanical engineers,London, 1963
Eng 41.5891

INSTITUTION OF MECHANICAL ENGINEERS
Conference on lubrication and wear
Proceedings London 1957 Oct 1-3
Institution of mechanical engineers,London, 1957
Eng 41.6295

INSTITUTION OF MECHANICAL ENGINEERS
Conference on steam turbine research and development Proceedings London 1953 Mar 6 Institution of mechanical engineers,London, 1953
Eng 41.7296

INSTITUTION OF MECHANICAL ENGINEERS
Conference on technology of engineering manufacture Proceedings London 1958 Mar 25-27 Institution of mechanical engineers, London,1958
Eng 41.3870

INSTITUTION OF MECHANICAL ENGINEERS
Discussion on critical path analysis planning techniques for engineering management
Proceedings London 1963 Apr 23
Institution of mechanical engineers,London, 1964
Eng 41.0499

INSTITUTION OF MECHANICAL ENGINEERS
Experiments in graduate training :a symposium
Proceedings London 1962 Mar 22-23
Institution of mechanical engineers,London, 1962
Eng 41.1499

INSTITUTION OF MECHANICAL ENGINEERS Fatigue in rolling contact :symposium Proceedings London 1963 Mar 28 Institution of mechanical engineers,London,1963
Eng 41.3786

INSTITUTION OF MECHANICAL ENGINEERS Fatigue of metals International conference on fatigue of metals Proceedings London 1956 Sep 10-14 and New York 1956 Nov 28-30 Institution of mechanical engineers, London,1956
Eng 41.3759

INSTITUTION OF MECHANICAL ENGINEERS General discussion on heat transfer Proceedings London 1951 Sep 11-12 xiii,496p
Institution of mechanical engineers,London, 1951
Chem E 24.0538

INSTITUTION OF MECHANICAL ENGINEERS General discussion on heat transfer Proceedings London 1951 Sep 11-13 Institution of mechanical engineers,London,1952
Eng 41.7146

INSTITUTION OF MECHANICAL ENGINEERS General discussion on lubrication and lubricants Proceedings London 1937 Oct 13-15 Vol 1-2 2 vols Institution of mechanical engineers,London,1937
Met 25.2152

INSTITUTION OF MECHANICAL ENGINEERS General discussion on lubrication and lubricants Proceedings London 1937 Oct 13-15 Vol 1-2 2 vols Institution of mechanical engineers,London,1937
Eng 41.6291

INSTITUTION OF MECHANICAL ENGINEERS
International conference on fatigue of metals Proceedings London 1956 Sep 10-14 New York 1956 Nov 28-30 Institution of mechanical engineers,London,1956
Met 25.0919

INSTITUTION OF MECHANICAL ENGINEERS
International conference on gearing
Proceedings London 1958 illus.
Institution of mechanical engineers,London, 1960
Eng 41.6224

INSTITUTION OF MECHANICAL ENGINEERS
International conference on tribology in iron and steel works 1st Proceedings London 1969 Sep 22-25 Iron and steel institute. Publication, 125 Iron and steel institute, London,1970
Met 25.2562

INSTITUTION OF MECHANICAL ENGINEERS
International developments in heat transfer
International heat transfer conference
Papers Boulder,Col. 1961 Aug 28-Sep 1
London 1962 Jan 8-12 Pt 1-5
Arranged by the American society of mechanical engineers 5 vols A.S.M.E.,New York,1961
Co-sponsored by the American institute of chemical engineers,the Institution of mechanical engineers,and the Institution of chemical engineers
Chem E 24.0595

INSTITUTION OF MECHANICAL ENGINEERS Joint conference on combustion Proceedings Boston,Mass. 1955 Jun 15-17 London 1955 Oct 25-27 viii,457p Institution of mechanical engineers,London,1955
Chem E 24.1376

INSTITUTION OF MECHANICAL ENGINEERS Joint conference on combustion Proceedings Boston,Mass. 1955 Jun 15-17 and London 1955 Oct 25-27 Institute of mechanical engineers,London,1956
Eng 41.7231

INSTITUTION OF MECHANICAL ENGINEERS Joint engineering conference Proceedings London 1951 Jun 4-15 Institution of civil engineers, London,1951
Eng 41.1761

INSTITUTION OF MECHANICAL ENGINEERS Joint international conference on creep session 1-7 New York 1963 Aug 25-29 London 1963 Sep 30-Oct 4 book 1-5 American society of mechanical engineers and American society for testing materials I.M.E. London,c1963 5 books bound together
Met 25.0953

INSTITUTION OF MECHANICAL ENGINEERS Machines for materials and environmental testing Symposium on techniques and equipment for environmental testing; symposium on developments in materials testing machine design :joint conference Manchester 1965 Sep 6-10 Under the aegis of the British national committee on materials Institution of mechanical engineers.Proceedings, 180,pt 3a I.M.E.,London,1966
Met 25.0891

INSTITUTION OF MECHANICAL ENGINEERS Pressure vessel research towards better design :a symposium Proceedings London 1961 Jan Institution of mechanical engineers,London, 1962
Eng 41.2875

INSTITUTION OF MECHANICAL ENGINEERS Steam-plant engineering:present status and future trends ;a convention Proceedings Harrogate 1963 May 2-4 Institution of mechanical engineers,London,1964
Eng 41.7276

INSTITUTION OF MECHANICAL ENGINEERS Symposium on biomechanics Papers London 1959 Apr 17 London,1959
Bal 39.0061

INSTITUTION OF MECHANICAL ENGINEERS Symposium on recent mechanical engineering developments in automatic control Proceedings London 1960 Jan 5-7 Institution of mechanical engineers,London, 1960
Eng 41.6012

INSTITUTION OF MECHANICAL ENGINEERS Symposium on two-phase fluid flow Proceedings London 1962 Feb 7 116p Institution of mechanical engineers,London, 1962
Chem E 24.0372

INSTITUTION OF MECHANICAL ENGINEERS The Properties of materials at high rates of strain :conference Proceedings London 1957 Apr 30-May 2 Institution of mechanical engineers,London,1957
Eng 41.3763

INSTITUTION OF MECHANICAL ENGINEERS The Properties of materials at high rates of strain conference Proceedings London 1957 Apr 30-May 2 Institution of mechanical engineers,London,1957
Met 25.0904

INSTITUTION OF MECHANICAL ENGINEERS The Teaching of automatic control :a discussion Proceedings London 1962 Jan 29 Institution of mechanical engineers,London, 1962
Eng 41.5814

INSTITUTION OF MECHANICAL ENGINEERS The Technology of engineering manufacture a conference Proceedings London 1958 Mar 25-27 Institution of mechanical engineers, London,1958
Met 25.0676

INSTITUTION OF MECHANICAL ENGINEERS Thermodynamic and transport properties of fluids :joint conference Proceedings London 1957 Jul 10-12 viii,217p Institution of mechanical engineers,London, 1958
Chem E 24.0981

INSTITUTION OF MECHANICAL ENGINEERS Two-phase fluid flow :a symposium Proceedings 1962 Feb Institution of mechanical engineers, London,1962
Eng 41.7171

INSTITUTION OF MECHANICAL ENGINEERS.APPLIED MECHANICS GROUP Periodic inspection of pressure vessels :a conference Proceedings London 1972 May 9-11 IME,London,1972
Eng 41.8648

INSTITUTION OF MECHANICAL ENGINEERS.APPLIED MECHANICS GROUP Practical application of fracture mechanics to pressure-vessel technology :a conference Proceedings London 1971 May 3-5 IME,London,1971
Eng 41.8644

INSTITUTION OF MECHANICAL ENGINEERS.AUTOMOBILE DIVISION Air pollution control in transport engines :a symposium Proceedings London 1971 Nov 9-11 IME,London,1972
Eng 41.8645

INSTITUTION OF MECHANICAL ENGINEERS.AUTOMOBILE DIVISION Corrosion and its prevention in motor vehicles a symposium Proceedings London 1968 Mar 28-29 Institution of mechanical engineers.Proceedings, 182 pt 3J Inst.of mechanical engineers,London,1968
Met 25.1943

INSTITUTION OF MECHANICAL ENGINEERS.EDUCATION AND TRAINING GROUP The Conference on the education and training of engineering technicians Proceedings London 1963 Mar 21-22 Institution of mechanical engineers, London,1963
Eng 41.1503

INSTITUTION OF MECHANICAL ENGINEERS.HYDRAULIC PLANT AND MACHINERY GROUP A Conference on oil hydraulic power transmission London 1961 Nov 29-30 Institution of mechanical engineers,London,1962
Eng 41.6685

INSTITUTION OF MECHANICAL ENGINEERS.LUBRICATION AND WEAR GROUP Lubrication and wear convention Proceedings 1963 May 23-25 Institution of mechanical engineers,London, 1964
Eng 41.6305

INSTITUTION OF MECHANICAL ENGINEERS.LUBRICATION AND WEAR GROUP Lubrication and wear in living and artificial human joints :a symposium London 1967 Apr 7 Institution of mechanical engineers.Proceedings, 181, pt.3 Institution of mechanical engineers,London, 1967
Met 25.1626

INSTITUTION OF MECHANICAL ENGINEERS.LUBRICATION AND WEAR GROUP Non-conventional lubricants and bearing materials such as used in nuclear engineering Manchester 1962 Apr 12 Institution of mechanical engineers, London,1962
Eng 41.6303

INSTITUTION OF MECHANICAL ENGINEERS.PROCEEDINGS, 19 Tribology convention Douglas,I.o.M. 1971 May 12-15 Institution of mechanical engineers.Tribology group IME,London,1971
Eng 41.8642

INSTITUTION OF MECHANICAL ENGINEERS.PROCESS ENGINEERING GROUP New engineering materials :a conference Birmingham 1965 Oct 13-14 Institution of mechanical engineers.Proceedings, 180 pt.3D I.Mech.E., London,1965
Met 25.0105

INSTITUTION OF MECHANICAL ENGINEERS.RAILWAY DIVISION Passenger environment :a conference Proceedings London 1972 Mar 23-24 IME,London,1972
Eng 41.8649

INSTITUTION OF MECHANICAL ENGINEERS.STEAM PLANT GROUP Modern steam plant practice :a convention Proceedings The Hague 1971 Apr 28-30 IME,London,1971
Eng 41.8643

INSTITUTION OF MECHANICAL ENGINEERS. THERMODYNAMICS AND FLUID MECHANICS GROUP Heat and mass transfer by combined forced and natural convection :a symposium Proceedings London 1971 Sep 15 IME,London,1972
Eng 41.8647

INSTITUTION OF MECHANICAL ENGINEERS.TRIBOLOGY GROUP Elastohydrodynamic lubrication :a symposium Proceedings Leeds 1972 Apr 11-13 IME,London,1972
Eng 41.8650

INSTITUTION OF MECHANICAL ENGINEERS.TRIBOLOGY GROUP Externally pressurized bearings :a joint conference Proceedings London 1971 Nov 17-18 IME,London,1972
Eng 41.8651

INSTITUTION OF MECHANICAL ENGINEERS.TRIBOLOGY GROUP Tribology convention Douglas,I.o.M. 1971 May 12-15 Institution of mechanical engineers.Proceedings, 19 IME, London,1971
Eng 41.8642

INSTITUTION OF MINING AND METALLURGY Advances in extractive metallurgy :a symposium Proceedings London 1967 Apr 17-20 Institution of mining and metallurgy,London, 1968
Met 25.0494

INSTITUTION OF MINING AND METALLURGY Extraction and refining of the rarer metals A Symposium on extraction metallurgy of some of the less common metals London 1956 Mar 22-23 Institution of mining and metallurgy, London,1957
Met 25.0500

INSTITUTION OF MINING AND METALLURGY International mineral processing congress Proceedings Gothenburg 1966 Aug 27-Sep 1 London,1960
Met 25.2504

INSTITUTION OF MINING AND METALLURGY Mineral resources policy :a symposium Proceedings London 1955 Sep 22 Institution of mining and metallurgy,London,1956
Met 25.0191

INSTITUTION OF MINING AND METALLURGY Recent developments in mineral dressing :a symposium London 1952 Sep 23-25 Institution of mining and metallurgy,London,1953
Met 25.0185

INSTITUTION OF PETROLEUM TECHNOLOGISTS World petroleum congress Proceedings London 1933 Jul 19-25 Vol 2: refining,chemical and testing section Edited by A.E. Dunstan xxvi,956p World petroleum congress,London, 1934
Chem E 24.1708

INSTITUTION OF PRODUCTION ENGINEERS Automation '65 British automation conference 1965 Eastbourne 1965 Nov 7-10 Confederation of British industry Trades union congress United Kingdom automation council Institution of production engineers, London,1965
Eng 41.6018

INSTITUTION OF PRODUCTION ENGINEERS Externally pressurized bearings :a joint conference Proceedings London 1971 Nov 17-18 IME,London,1972
Eng 41.8651

INSTITUTION OF PRODUCTION ENGINEERS The Automatic factory - What does it mean? conference report Margate 1955 June 16-19 illus. 227p 28cm Institution of production engineers,London,1955
Math L 5.0734

INSTITUTION OF PRODUCTION ENGINEERS The Automatic factory - what does it mean? : conference Margate 1955 Jun 16-19 Institution of production engineers,London, 1955
Eng 41.5974

INSTITUTION OF STRUCTURAL ENGINEERS Aluminium in structural engineering :a symposium Proceedings London 1963 Jun 11-12 Aluminium federation,London,1964
Eng 41.2694

INSTITUTION OF STRUCTURAL ENGINEERS Industrialized building and the structural engineers :a symposium London 1966 May 17-19 Institution of structural engineers, London,1966
Eng 41.3381

INSTITUTION OF STRUCTURAL ENGINEERS Institution of structural engineers jubilee symposium on high strength bolts Proceedings London 1959 Jun 10-11 Institution of structural engineers,London,1959
Eng 41.2696

INSTITUTION OF STRUCTURAL ENGINEERS Report on the structural use of aluminium Institution of structural engineers,London, 1962
Eng 41.2671

INSTITUTION OF STRUCTURAL ENGINEERS Thin walled steel structures :a symposium Swansea 1967 Sep 11-14 Edited by K.C. Rockey and H.V. Hill Crosby Lockwood,London,1969
Eng 41.2747

INSTITUTION OF STRUCTURAL ENGINEERS : fiftieth anniversary conference London 1958 Oct 7-10 Institution of structural engineers,London,1958
Eng 41.3023

INSTITUTION OF STRUCTURAL ENGINEERS.SHEAR STUDY GROUP The Shear strength of reinforced concrete beams Report Institution of structural engineers.Series, 49 Institution of structural engineers,London,1969
Eng 41.2955

INSTITUTION OF STRUCTURAL ENGINEERS JUBILEE SYMPOSIUM ON HIGH STRENGTH BOLTS Proceedings London 1959 Jun 10-11 Institution of structural engineers Institution of structural engineers,London, 1959
Eng 41.2696

INSTITUTO GULBENKIAN DE CIENCIA.CENTRO DE CALCULO CIENTIFICO Theory of distributions :an international summer institute Proceedings Lisbon 1964 Sep Sponsored by the North Atlantic treaty organization.Science committee 390p Instituto Gulbenkian de ciencia,Lisbon, 1964
A Math 4.1265

INSTITUTO NACIONAL DE LA INVESTIGACION CIENTIFICA, MEXICO Symposium internacional de topologia algebraica Mexico City 1956 Aug Universidad nacional de Mexico.Instituto de matematicas Sociedad matematica mexicana xii,334p 26cm UNESCO,Mexico City,1958 In memory of Witold Hurewicz
P. Math 2.0290

INSTITUTO NACIONAL PARA LA INVESTIGACION DE RECURSOS MINERALES DE MEXICO Tectonica de la Sierra Madre oriental de Mexico,entre Torreon y Monterrey Mexico City 1956 By Zoltan de Cserna Mexico City,1956 Bound with two other publications of the congress
Geol 8.3040

INSTITUTO TORCUATO DI TELLA Biology of neuroglia Lectures and discussions Buenos Aires 1963 Oct 17-18 Edited by E.D.P.de Robertis and R. Carea Progress in brain research, 15 illus Elsevier,Amsterdam, 1965 Symposium held as part of the 10th Latin-American congress of neurosurgery
Path 30.2466

INSTITUTUL DE FIZICA,BUCHAREST International conference on phenomena in ionized gases 9th Proceedings Bucharest 1969 Institute of physics,Bucharest,1969
Eng 41.8553

INSTITUUT VAN INGENIEURS Chemical reaction engineering European symposium on chemical engineering 1st Papers Amsterdam 1957 May 7-9 Held during the 12th meeting of the European federation of chemical engineering International series of monographs on chemical engineering, 1 197p Pergamon,London,1957
Chem E 24.1900

INSTITUUT VAN INGENIEURS The Scaling-up of chemical plant and processes :joint symposium Papers London 1957 May 28-29 By R. Edgeworth Johnstone and others Society of chemical industry.Chemical engineering group and Institution of chemical engineers Institution of chemical engineers,London,1957 Held jointly with the K.Instituut van ingenieurs and the K.Nederlandse chemische vereniging
Chem E 24.1234

INSTITUUT VAN INGENIEURS.AFDELING VOOR CHEMISCHE TECHNIEK Chemical reaction engineering European symposium on chemical reaction engineering 3rd Proceedings Amsterdam 1964 Sep 15-17 Held under the auspices of the European federation of chemical engineering vi,326p Pergamon press,Oxford, 1965 Supplement to "Chemical engineering science".
Chem E 24.1269

INSTRUCTIONAL CONFERENCE ON DIFFERENTIAL EQUATIONS Edinburgh 1967 Sep 18-22 notes for the introductory lectures on Hilbert spaces and ordinary differential equations Illus. 28,35cm 2 vols Edinburgh,1967 Typescript
P Math 2.3476

INSTRUCTIONAL CONFERENCE ON DIFFERENTIAL EQUATIONS Lectures Edinburgh 1967 Sep 18-22 Edinburgh,1967 Typescript
P Math 2.3477

INSTRUMENT SOCIETY OF AMERICA Gas chromatography :international symposium 2nd Proceedings Ann Arbor 1959 Jun 10-13 Edited by Henry J. Noebels and others xvi, 463p Academic press,New York;London,1961
Chem E 24.0851

INSTRUMENT SOCIETY OF AMERICA Instrumentation requirements for traffic control systems Traffic control :theory and instrumentation Papers New York 1963 Dec 16-17 Edited by Thomas R. Horton Plenum press,New York,1965
Eng 41.3310

INSTRUMENT SOCIETY OF AMERICA Temperature: its measurement and control in science and industry :a symposium 4th Proceedings Columbus,Ohio 1961 Mar 27-31 Vol 3,pt 1: basic concepts,standards and methods Edited by C.M. Herzfeld and Ferdinand Graft Brickwedde Co-sponsored by National bureau of standards Reinhold;Chapman and Hall,New York;London,1962
Cav 7.0048

INSTRUMENT SOCIETY OF AMERICA Temperature: its measurement and control in science and industry :a symposium 4th Proceedings Columbus,Ohio 1961 Mar 27-31 Vol 3,pt2: applied methods and instruments Edited by A. I. Dahl and C.M. Herzfeld Co-sponsored by National bureau of standards 1094p Reinhold publishing,New York,1962
Cav 7.0047

INSTRUMENT SOCIETY OF AMERICA and AMERICAN INSTITUTE OF PHYSICS Temperature,its measurement and control in science and industry :a symposium 4th Proceedings Columbus,Ohio 1961 Mar 27-31 Vol 3,1: basic concepts,standards and methods Edited by Ferdinand Graft Brickwedde With the cooperation of the National bureau of standards xvi,848p Reinhold,New York,1962 General editor:Herzfeld,Charles
Sco 14.0231

INSTRUMENT SOCIETY OF AMERICA and AMERICAN INSTITUTE OF PHYSICS Temperature,its measurement and control in science and industry :a symposium 4th Proceedings Columbus,Ohio 1961 Mar 27-31 Vol 3,2: applied methods and instruments Edited by A. I. Dahl Co-sponsored by the National bureau of standards xiv,1094p Reinhold,New York, 1962 General editor:Herzfeld,Charles
Sco 14.0233

INSTRUMENT SOCIETY OF AMERICA and AMERICAN INSTITUTE OF PHYSICS Temperature,its measurement and control in science and industry :a symposium 4th Proceedings Columbus,Ohio 1961 Mar 27-31 Vol 3,3: Biology and medicine Edited by James D. Hardy With the cooperation of the National bureau of standards xii,683p Reinhold,New York,1962 General editor:Herzfeld,Charles
Sco 14.0232

INSTRUMENTATION Conference Konferentsiya puti razvitiya sovetskogo matematicheskogo mashchinostroeniya i priborostroeniya : plenarnye zasedaniya Moscow 1956 Mar 12-17 illus. 131p 22cm Moscow,1956
Math L 5.0774

INSTRUMENTATION Joint symposium on instrumentation and computation in process development and plant design Papers London 1959 May 11-12 British conference on automation and computation Sponsored by the Institution of chemical engineers Institution of chemical engineers,London,1959 Sponsored also by the Society of instrument technology,and the British computer society
Math L 5.3290

INSTRUMENTATION IN BIOCHEMISTRY London 1966 Apr Biochemical society Edited by T.W. Goodwin Biochemical society.Symposia, 26 Academic press,London;New York,1966
Bioch 33.1388

INSTRUMENTATION IN BIOCHEMISTRY :a symposium London 1966 Apr Biochemical society Edited by T.W. Goodwin Biochemical society. Symposia, 26 Academic press,London,1966
Radioth 35.0135

INSTRUMENTATION REQUIREMENTS FOR TRAFFIC CONTROL SYSTEMS Traffic control :theory and instrumentation Papers New York 1963 Dec 16-17 Instrument society of America Edited by Thomas R. Horton Plenum press,New York, 1965
Eng 41.3310

INSTRUMENTS AND APPARATUS FOR SOIL AND ROCK MECHANICS :a symposium Lafayette,Ind. 1965 Jun 13-18 American society for testing materials American society for testing materials.Special technical publication, 392 American society for testing materials,Philadelphia,Pa.,1965
Eng 41.3158

INSTRUMENTS AND MEASUREMENTS CONFERENCE Transactions Stockholm 1952 Svenska teknologtoreningen,Stockholm,1953
Eng 41.4142

INSULATION OF HIGH VOLTAGES IN VACUUM : international symposium 1964 Massachusetts institute of technology Massachusetts institute of technology, Cambridge,1964
Eng 41.4984

INSULIN Impact of insulin on metabolic pathways International symposium commemorating the 50th anniversary of insulin Jerusalem 1971 Oct 24-29 Vol 1-2: lectures;panel discussions and communications Edited by Eleazar Shafrir Israel journal of medical sciences, 8,no 3,6 2 vols Israel medical association,Jerusalem,1972
Bioch 33.2193

INSULIN Structure and function of polypeptide hormones;insulin Symposium on insulin Proceedings Upton,N.Y. 1965 Nov 8-10 Edited by P.G. Katsoyannis and I.L. Schwartz Sponsored by the Brookhaven national laboratory.Medical department American journal of medicine, 40, no. 5 Yorke medical group.Publication New York, 1966
Bioch 33.0557

INSULIN The Mechanism of action of insulin : a symposium London 1958 Sep 9-10 British insulin manufacturers Edited by F.G. Young and others Blackwell scientific publication, Oxford,1960
PGMS 29.0378

INSULIN The Mechanism of action of insulin a symposium Discussions London 1958 Sep 9-10 Edited by F.G. Young Organized by the British insulin manufacturers Blackwell, Oxford,1960
Bioch 33.0474

INSULIN ACTION Symposium on insulin action Proceedings Toronto 1971 Oct 25-27 Edited by Irving B. Fritz Academic press,New York,1972 Held in honour of the 50th anniversary of the discovery of insulin
Bioch 33.2192

INTEGRATED CIRCUITS Conference on integrated circuits Eastbourne 1967 May 2-4 Sponsored by the Institution of electrical engineers.Electronics division I.E.E. Conference publication, 30 Institution of electrical engineers,London,1967 Joint sponsors;Institution of electronic and radio engineers and the Institute of electrical and electronics engineers
Eng 41.5461

INTEGRATING THE APPROACHES TO MENTAL DISEASE : two conferences Papers and discussions New York New York academy of medicine. Committee on public health Edited by H.D. Kruse Hoeber-Harper,New York,1957
Psy 28.0061

INTEGRATION DANS LES GROUPES TOPOLOGIQUES Montreal 1964 By Geoffrey Fox North Atlantic treaty organization Societe mathematique du Canada Montreal.University. Seminaire de mathematiques superieures, 12 357p 28cm Presses de l'universite de Montreal,Montreal,1966
P. Math 2.2385

INTER-STATE ASTRONOMICAL AND METEOROLOGICAL CONFERENCE Report Adelaide 1905 May 10-16 By Charles Todd 11p South Australian government,Adelaide,1905
Obs 6.3577

INTER-UNION COMMISSION ON RADIO ASTRONOMY
Colloquium on spectra of meteorological variables Proceedings Stockholm 1969 Jun 9-19 Vol 3: radio science illus 28cm American geophysical union,Washington,D.C., 1969
A Math 4.1745

INTER-UNIVERSITY GEOLOGICAL CONGRESS 14th
Proceedings Geology of shelf seas Hull 1967 Edited by D.T. Donovan illus. viii, 160p Oliver and Boyd,Edinburgh;London,1968
Bot 42.3193

INTER-DEPARTMENTAL COMMITTEE ON BATTERIES
Batteries 2 International symposium on batteries :research and development in non-mechanical electrical power sources 4th Brighton 1964 Sep Edited by D.H. Collins Pergamon Press,Oxford,1965
Met 25.1841

INTER-GUIANA GEOLOGICAL CONFERENCE 5th
proceedings Georgetown 1959 Oct.28-Nov. 6 Geological survey of British Guiana Edited by R.B. McConnell Geological survey department,British Guiana,Georgetown,1962
Geol 8.2426

INTER-GUIANA GEOLOGICAL CONFERENCE see also CONFERENCE GEOLOGIQUE DES GUYANES

INTER-GUIANA GEOLOGICAL CONFERENCE see also CONFERENCE GEOLOGIQUE DES TROIS GUYANES

INTER-GUIANA GEOLOGICAL CONFERENCE see also GEOLOGISCHE CONFERENTIE

INTER-NORDIC LAPP COUNCIL see NORDIC LAPP COUNCIL

INTER-UNIVERSITY GEOLOGICAL CONGRESS 14th
Proceedings Geology of shelf seas Edited by D.T. Donovan Maps Oliver and Boyd,Edinburgh;London,1968
Geog 13.0744

INTER-UNIVERSITY GEOLOGICAL CONGRESS 14th
proceedings Geology of shelf seas Huel 1967 Jan Edited by D.T. Donovan Oliver Boyd,Edinburgh;London,1968
Geol 8.4480

INTER-UNIVERSITY GEOLOGICAL CONGRESS 9th ten lectures Some aspects of the Variscan fold belt Exeter 1961 Jan 2-4 Edited by Kenneth Coe illus. map Manchester university press,Manchester,1962
Geol 8.1656

The INTERACTION OF DRUGS AND SUBCELLULAR COMPONENTS IN ANIMAL CELLS :a symposium London 1967 Apr 10-11 Biological council.Co-ordinating committee for symposia on drug action Edited by P.N. Campbell Held in the Middlesex hospital medical school Churchill,London,1968
Bioch 33.0923

INTERACTION OF RADIATION WITH SOLIDS Cairo solid state conference Cairo 1966 Sep 3-8 American university in Cairo Edited by Adli Bishay Plenum press,New York,1967
TA 15.0429

INTERACTION OF RADIATION WITH SOLIDS
International summer school on solid state physics Proceedings Mol 1963 Aug 12-31 Edited by R. Strumane and others North-Holland,Amsterdam,1964
Met 25.1581

INTERACTION OF VIRUSES AND CELLS
International congress of microbiology 6th Rome 1953 Vol 6: symposium - interaction of viruses and cells Edited by F. C. Bawden and G. Penso illus Fondazione Emanuele Paterno,Rome,1953 Title also in Italian;text in English and French
Bioch 33.1804

The INTERACTIONS BETWEEN DISLOCATIONS AND POINT DEFECTS :a symposium Proceedings Harwell 1968 Jul 4-12 1-3 Atomic energy research establishment.Metallurgy division United Kingdom atomic energy authority.Research group.Report AERE-R 5944 3 vols harwell,1968
Met 25.2415

INTERACTIONS DES ELECTRONS,PHONONS ET MAGNONS AVEC LES SURFACES CRISTALLINES :colloque Exposes et communications presentes Lille 1969 Sep 17-19 Societe francaise de physique Journal de physique, 31,no 4,supplement, Colloque,C- 1 Paris,1970
Met 25.2627

INTERACTIVE SYSTEMS FOR EXPERIMENTAL APPLIED MATHEMATICS symposium proceedings Washington,D.C. 1967 Aug Association for computing machinery Edited by M. Klerer and J. Reinfelds Academic press,New York,1968
Math L 5.3495

INTERCOLONIAL MEDICAL CONGRESS OF AUSTRALASIA
4th session Transactions Dunedin 1896 Feb Otago daily times and witness newspapers,Dunedin,1897
An 32.1142

INTERCOLONIAL MEDICAL CONGRESS OF AUSTRALASIA
5th session Transactions Brisbane 1899 Sep Edited by Wilton Love E.Gregory, Brisbane,1901
An 32.1143

The INTERCONNECTION OF PERIPHERAL EQUIPMENT AND COMPUTERS colloquium report London 1964 Jan Institution of electrical engineers Edited by H.McG. Ross 380p Institution of electrical engineers,London,1964
Math L 5.1272

INTERDISCIPLINARY APPROACH TO FRICTION AND WEAR
a symposium Proceedings 1967 National aeronautics and space administration Washington,D.C.,1968
Eng 41.6272

INTERFACES CONFERENCE Papers and abstracts Melbourne 1969 Aug Australian institute of metals Edited by R.C. Gifkins Butterworth, Sydney,1969
Met 25.2596

INTERFACES IN COMPOSITES :a symposium Papers San Francisco,Calif. 1968 Jun 23-28 American society for testing and materials A. S.T.M.Special technical publication, 452 ASTM,Philadelphia,Pa.,1969 Presented at the 71st annual meeting
Met 25.2749

INTERFACIAL STUDIES Clean surfaces their preparation and characterization for interfacial studies :based on a symposium Raleigh,N.C. 1968 Apr 8-10 North Carolina State university Edited by George Goldfinger Dekker,New York,1970
Met 25.2828

INTERFERON :a Ciba foundation symposium London 1967 Apr 19 Ciba foundation Edited by G.E.W. Wolstenholme and Maeve O'Connor Churchill,London,1968
Phys 20.0999

INTERFERON a Ciba foundation symposium dedicated to Alick Isaacs Proceedings London 1967 Apr 19-21 Ciba foundation Edited by G.E.W. Wolstenholme and Maeve O'Connor illus Churchill,London,1968
Bioch 33.1119

INTERHEMISPHERIC RELATIONS AND CEREBRAL DOMINANCE a conference Papers Baltimore,Md. 1961 Apr 23-25 Edited by Vernon B. Mountcastle Johns Hopkins press,Baltimore,Md. 1962
An 32.4652

INTERLAKEN 1969 The Significance of space research for fundamental physics :a conference European space research organisation Edited by A.F. Moore and U. Hardy ESRO SO-52 175p european space research organisation,Neuilly-sur-Seine,1971
Obs 6.3603

INTERMAG 1966 Intermag International magnetics conference 4th Stuttgart 1966 Apr 20-22 Sponsored by the Institute of electrical and electronics engineers.Magnetics group I.E.E.E.transactions on magnetics,Mag 2,no 3 I.E.E.E.,New York,1966
Eng 41.5347

INTERMAG CONFERENCE International conference on magnetics 3rd Proceedings Washington, D.C. 1965 Apr 21-23 Spsonsored by the Institute of electrical and electronics engineers.Magnetics group Institute of electrical and electronics engineers,New York, 1965
Eng 41.5345

INTERNAL AERODYNAMICS (TURBOMACHINERY) :a conference Cambridge 1967 Jul 19-21 Royal society of London Institution of mechanical engineers,London,1970
Eng 41.6829

INTERNAL FLOW Symposium on the fluid mechanics of internal flow Warren,Mich. 1965 General motors research laboratories Edited by Gino Sovran Elsevier,Amsterdam, 1967
Eng 41.6553

INTERNAL FRICTION,DAMPING,AND CYCLIC PLASTICITY a symposium presented before the 67th annual meeting American society for testing materials A.S.T.M.Special technical publication, 378 A.S.T.M.,Philadelphia,1965
Met 25.2340

INTERNAL FRICTION,DAMPING AND CYCLIC PLASTICITY... a symposium presented at the 67th annual meeting,A.S.T.M. Chicago,Ill. 1964 Jun 22 American society for testing and materials American society for testing and materials.Special technical publication, 378
Eng 41.3795

INTERNAL SECRETION OF THE PANCREAS : colloquium London 1955 Jun 21-23 Ciba foundation Edited by G.E.W. Wolstenholme and Cecilia M. O'Connor Ciba foundation colloquia on endocrinology, 9 Churchill, London,1956
An 32.3739

INTERNAL SECRETIONS Cold Spring Harbor symposia on quantitative biology Papers Cold Spring Harbor 1937 Vol 5 Cold Spring Harbor biological laboratory Long Island biological association,Cold Spring Harbor,1937 Later referred to in vol. 9 as "Internal secretions"
Bioch 33.1261

INTERNAL SECRETIONS OF THE PANCREAS The Nature and actions of the internal secretions of the pancreas :a colloquium Proceedings London 1955 Jun 21-23 Ciba foundation Edited by G.E.W. Wolstenholme and Cecilia M. O'Connor Ciba foundation colloquia on endocrinology, 9 illus Churchill,London, 1956
Bioch 33.0438

INTERNAL STRESSES AND FATIGUE IN METALS the symposium Proceedings Detroit 1958 Warren,Mich. 1958 General motors corporation.Research laboratories Edited by Gerald M. Rasswiler and William L. Grube Elsevier,Amsterdam,1959
Met 25.0965

INTERNAL STRESSES IN METALS Symposium on internal stresses in metals and alloys Proceedings London 1947 Oct 15-16 Institute of metals Monographs and report series, 5 vii,485p Institute of metals, London,1948
Cav 7.1454

INTERNAL STRESSES IN METALS AND ALLOYS Symposium on internal stresses in metals and alloys Advance copies of papers London 1947 Oct 15-16 Institute of metals Institute of metals,London,1947
Eng 41.3768

INTERNAL STRESSES IN METALS AND ALLOYS Symposium on internal stresses in metals and alloys London 1947 Oct 15-16 Institute of metals Institute of metals.Monograph and report series, 5 Institute of metals, London,1948
Met 25.0960

INTERNATION UNION OF PURE AND APPLIED CHEMISTRY International colloquium on macromolecules Proceedings Amsterdam 1949 Sep 2-5 D.B. Ceuten's Uitgerers,Amsterdam,1949
Radioth 35.0014

INTERNATIONAL ACADEMY OF ASTRONAUTICS Applied sciences research and utilization of lunar resources Lunar international laboratory (LIL symposium) 4th Proceedings New York 1968 Oct 17 Bibliog, illus 200p 20cm Pergamon press,Oxford, 1970 LIL symposium organized by the International academy of astronautics at the 19th International astronautical congress
Cav 7.2711

INTERNATIONAL ACADEMY OF ASTRONAUTICS Life sciences research and lunar medicine Lunar international laboratory (LIL)symposium 2nd Proceedings Madrid 1966 Oct 13 Edited by Frank J. Malina Pergamon press,Oxford;London, 1967 Held at the 17th International astronautical congress
PGMS 29.0445

INTERNATIONAL ADVANCED SUMMER INSTITUTE ON MICROPROGRAMMING St.Raphael 1971 Aug 30-Sep 10 North Atlantic treaty organization. Scientific affairs division Edited by Guy G. Boulaye and Jean Mermet Actualites scientifiques et industrielles illus 418p Hermann,Paris,1972
Math L 5.3634

INTERNATIONAL AFRICAN INSTITUTE African agrarian systems International African seminar 2nd Studies Leopoldville 1960 Jan Edited by Daniel Biebuyck Oxford university press,London,1963
Geog 13.4631

INTERNATIONAL AFRICAN INSTITUTE New elites of tropical Africa International African seminar 6th Studies Ibadan 1964 Jul Edited by P.C. Lloyd Oxford university press, London,1966
Geog 13.4666

INTERNATIONAL AFRICAN INSTITUTE Social change in modern Africa International African seminar 1st Studies Kampala 1959 Jan Edited by Aidan Southall Oxford university press,London,1965
Geog 13.4662

INTERNATIONAL AFRICAN SEMINAR 2nd Studies African agrarian systems Leopoldville 1960 Jan International African institute Edited by Daniel Biebuyck Oxford university press, London,1963
Geog 13.4631

INTERNATIONAL AFRICAN SEMINAR 6th Studies New elites of tropical Africa Ibadan 1964 Jul International African institute Edited by P.C. Lloyd Oxford university press,London, 1966
Geog 13.4666

INTERNATIONAL AIR CONGRESS Report London 1923 Edited by W.Lockwood Marsh International air congress,London,1923 Being the 2nd air congress although this is not indicated on the title page
Eng 41.6956

INTERNATIONAL ANALOGUE COMPUTATION MEETING 1st, 3rd,4th Proceedings Brussels 1955- Brussels,1956-
Eng 41.2242

INTERNATIONAL ANTI-TUBERCULOSIS ASSOCIATION International tuberculosis conference 7th Report Philadelphia,Pa. 1908 Sep 24-26 Edited by Gotthold Pannwitz port Internationale vereinigung gegen die tuberkulose,Berlin-Charlottenburg,1909 Text in English,French,German;title also in French and German
Path 30.1023

INTERNATIONAL ANTI-TUBERCULOSIS ASSOCIATION International tuberculosis conference 8th Report Stockholm 1909 Jul 8-10 Edited by Gotthold Pannwitz Berlin-Charlottenburg,1910 Text in English,French,German;title also in French and German
Path 30.1024

INTERNATIONAL ANTI-TUBERCULOSIS ASSOCIATION International tuberculosis conference 9th Report Brussels 1910 Oct 6-8 Edited by Gotthold Pannwitz port Berlin-Charlottenburg,1911 Text in English, French,German;title also in French and German
Path 30.1026

INTERNATIONAL ARID LAND MEETINGS Papers and recommendations Future of arid lands University of New Mexico 1955 Apr 26-May 4 Edited by Gilbert F. White American association for the advancement of science. Publications,43 American association for the advancement of science,Washington,D.C.,1956
Geog 13.1467

INTERNATIONAL ASSOCIATION FOR BRIDGE AND STRUCTURAL ENGINEERING 1st-8th :congress Preliminary and final reports 1932-1967 International association for bridge and structural engineering,Zurich,1932-67
Eng 41.2927

INTERNATIONAL ASSOCIATION FOR EARTHQUAKE ENGINEERING Vibration in civil engineering :a symposium Proceedings London 1965 Apr Edited by B.O. Skipp Butterworths,London,1966
Eng 41.6363

INTERNATIONAL ASSOCIATION FOR HYDRAULIC RESEARCH 10th congress Proceedings London 1963 Sep 1-5 Vol 1-5 International association for hydraulic research.British national committee London,1963
Eng 41.6647

INTERNATIONAL ASSOCIATION FOR HYDRAULIC RESEARCH. BRITISH NATIONAL COMMITTEE International association for hydraulic research 10th congress Proceedings London 1963 Sep 1-5 Vol 1-5 London,1963
Eng 41.6647

INTERNATIONAL ASSOCIATION FOR HYDRAULIC RESEARCH SYMPOSIUM Cavitation and hydraulic machinery Proceedings Sendai 1962 Sep 3-5 Edited by F. Numachi Tohoku university, Sendai,163
Eng 41.6518

INTERNATIONAL ASSOCIATION FOR QUATERNARY RESEARCH International association for quaternary research congress 7th Proceedings Boulder,Colo. 1965 Aug 14-Sep 19 Vol 10: arctic and alpine environments Edited by W.H. Osburn and H.E. Wright Sponsored by the National research council illus,maps xi, 308p Indiana university press,Bloomington, Ind.;London,1968 For other volumes of these proceedings,see also International union for quaternary research
Bot 42.2006

INTERNATIONAL ASSOCIATION FOR QUATERNARY RESEARCH International association for quaternary research congress 7th Proceedings Boulder,Colo. 1965 Aug 14-Sep 19 10: arctic and alpine environments Edited by W.H. Osburn and H.E. Wright Sponsored by the National research council illus,maps xi, 308p 25cm Indiana university press, Bloomington,Ind.;London,1968 For other volumes of these proceedings,see also International union for quaternary research
Sco 14.7841

INTERNATIONAL ASSOCIATION FOR QUATERNARY RESEARCH
The Quaternary of the United States :review volume for the 7th congress of the International association for quaternary research Boulder,Colo. 1965 Aug 30-Sep 5 Edited by H.E. Wright and David C. Frey bibliog. Princeton university press, Princeton,N.J.,1965
Geol 8.2527

INTERNATIONAL ASSOCIATION FOR QUATERNARY RESEARCH
7th congress Boulder,Colo. 1965 guidebook for field conference F.Central and south central Alaska Edited by C.B. Schultz and H.T.U. Smith illus 141p 20cm Lincoln,Neb.,1965
Sco 14.0143

INTERNATIONAL ASSOCIATION FOR QUATERNARY RESEARCH
7th :congress Papers International studies on the quaternary Boulder,Colo. 1965 Edited by H.E. Wright and David G. Frey Sponsored by the National research council Geological society of America.Special papers, 84 Geological society of America,New York, 1965
Geog 13.0309

INTERNATIONAL ASSOCIATION FOR QUATERNARY RESEARCH.
CONGRESS 7th Proceedings Pleistocene extinctions:the search for a cause Boulder, Colo. 1965 Aug I.N.Q.U.A. Edited by P.S. Martin and H.E. Wright Sponsored by the National research council Yale university press,New Haven,Conn.;London,1967
Geog 13.1320

INTERNATIONAL ASSOCIATION FOR QUATERNARY RESEARCH.
CONGRESS 7th Proceedings Quaternary paleoecology Boulder,Colo. 1965 Aug-Sep Vol 7 Edited by E.J. Cushing and H.E. Wright Sponsored by the National research council Yale university press,New Haven,Conn.;London, 1967
Geog 13.1318

INTERNATIONAL ASSOCIATION FOR QUATERNARY RESEARCH,
CONGRESS 7TH The Quaternary of the United States :a review volume for the 7th congress of the international association for quaternary research Edited by H.E. Wright and David G. Frey Princeton university press, Princeton,N.J.,1965
Geog 13.4858

INTERNATIONAL ASSOCIATION FOR QUATERNARY RESEARCH
CONGRESS 4th Actes Rome 1953 Aug-Sep and Pisa 1953 Aug-Sep Vol 1-2 Edited by Gian Alberto Blanc illus. Instituto italiano di paleontologia umana,Rome, 1956
Bot 42.4701

INTERNATIONAL ASSOCIATION FOR QUATERNARY RESEARCH
CONGRESS 7th Proceedings Bering land bridge Boulder,Colo. 1965 Aug 30-Sep 5 Edited by D.M. Hopkins Stanford university press,Stanford,Calif.,1967
Bot 42.3247

INTERNATIONAL ASSOCIATION FOR QUATERNARY RESEARCH
CONGRESS 7th Proceedings Boulder, Colo. 1965 Aug 14-Sep 19 Vol 10: arctic and alpine environments International association for quaternary research Edited by W.H. Osburn and H.E. Wright Sponsored by the National research council illus,maps xi,308p Indiana university press,Bloomington, Ind.;London,1968 For other volumes of these proceedings,see also International union for quaternary research
Bot 42.2006

INTERNATIONAL ASSOCIATION FOR QUATERNARY RESEARCH
CONGRESS 7th Proceedings Means of correlation of quaternary successions Boulder,Colo. 1965 Aug 30-Sep 5 Edited by R.B. Morrison and H.E. Wright Utah university press,Salt Lake City,Utah,1968
Bot 42.3246

INTERNATIONAL ASSOCIATION FOR QUATERNARY RESEARCH
CONGRESS 7th Proceedings Pleistocene extinctions:the search for a cause Boulder,Colo. 1965 Aug 30-Sep 5 Vol 6 Edited by P.S. Martin and H.E. Wright Sponsored by the National research council Yale university press,New Haven,Conn.;London, 1967
Bot 42.3244

INTERNATIONAL ASSOCIATION FOR QUATERNARY RESEARCH
CONGRESS 7th Proceedings Progress in oceanography Boulder,Colo. 1965 Aug 30-Sep 5 Vol 4 Edited by M. Sears Pergamon,Oxford,1967
Bot 42.3248

INTERNATIONAL ASSOCIATION FOR QUATERNARY RESEARCH
CONGRESS 7th Proceedings Quaternary paleoecology Boulder,Colo. 1965 Aug 30-Sep 5 Vol 7 Edited by E.J. Cushing and H. E. Wright Sponsored by the National research council Yale university press,New Haven,Conn. London,1967
Bot 42.3245

INTERNATIONAL ASSOCIATION FOR QUATERNARY RESEARCH
CONGRESS 8th Proceedings United States contributions to quaternary research Paris 1969 Edited by S.A. Schumm and W.C. Bradley Geological society of America. Special papers, 123 Geological society of America,Boulder,Colo.,1969
Bot 42.3249

INTERNATIONAL ASSOCIATION FOR QUATERNARY RESEARCH
CONGRESS 7th Proceedings Boulder, Colo. 1965 Aug 14-Sep 19 10: arctic and alpine environments International association for quaternary research Edited by W.H. Osburn and H.E. Wright Sponsored by the National research council illus,maps xi,308p 25cm Indiana university press, Bloomington,Ind.;London,1968 For other volumes of these proceedings,see also International union for quaternary research
Sco 14.7841

INTERNATIONAL ASSOCIATION FOR QUATERNARY RESEARCH
CONGRESS,7TH The Bering land bridge Edited by David M. Hopkins illus,maps xiii, 495p 25cm Stanford university press, Stanford,1967 Includes papers presented at Symposium held at 7th congress of International association for Quaternary research
Sco 14.7504

INTERNATIONAL ASSOCIATION FOR QUATERNARY RESEARCH CONGRESS 7TH International studies on the quaternary Boulder,Colo. 1965 Edited by H.E. Wright and David G. Frey Sponsored by the National research council Geological society of America.Special papers, 84 Geological society of America,New York,1965
Bot 42.3133

INTERNATIONAL ASSOCIATION FOR QUATERNARY RESEARCH CONGRESS 7TH The Quaternary of the United States :review volume for the 7th Congress of the International association for quaternary research Boulder,Colo. 1965 Aug 30-Sep 5 Edited by H.E. Wright and David G. Frey Bibliog. Princeton university press, Princeton,N.J.,1965
Bot 42.6691

INTERNATIONAL ASSOCIATION FOR RADIATION RESEARCH Radiation research International congress of radiation research 3rd Proceedings Cortina d'Ampezzo 1966 Jun 26-Jul 2 Edited by G. Silini North-Holland,Amsterdam,1967
Radioth 35.1173

INTERNATIONAL ASSOCIATION FOR SHELL STRUCTURES Non-classical shell problems :a symposium Proceedings Warsaw 1963 Sep 2-5 Edited by W. Olszak and A. Sawczuk North-Holland; PWN,Amsterdam;Warsaw,1964
Eng 41.2876

INTERNATIONAL ASSOCIATION FOR SHELL STRUCTURES Shell research :a symposium Proceedings Delft 1961 Aug 30-Sep 2 Edited by A.M. Haas and A.L. Bouma North-Holland,Amsterdam, 1961
Eng 41.2860

INTERNATIONAL ASSOCIATION FOR SHELL STRUCTURES Simplified calculation methods of shell structures :colloquium Proceedings Brussels 1961 Sep 4-6 Edited by A. Paduart and R. Dutron North-Holland,Amsterdam,1962
Eng 41.2864

INTERNATIONAL ASSOCIATION FOR SHELL STRUCTURES World conference on shell structures Proceedings San Francisco,Calif. 1962 Oct 1-4 National research council.Publication, 1187 National academy of sciences,Washington, D.C.,1064
Eng 41.3025

INTERNATIONAL ASSOCIATION FOR STATISTICS IN PHYSICAL SCIENCES Joint European conference of the Institute of mathematical statistics,the International association for statistics in physical sciences,the Biometric society Berne 1964 Sep 14-18 Berne,1964 13 unbound typescript papers
Math 3.1124

INTERNATIONAL ASSOCIATION FOR STATISTICS IN PHYSICAL SCIENCES Statistics in the physical sciences :a conference Proceedings Belgrade 1965 Sep 17 International association for statistics in physical sciences.Proceedings, 2 Bibliog 24cm International statistitcal institute,Belgrade, c1966 Reprinted from the Bulletin of the International statistical institute. Proceedings of the 35th session,1965
Math 3.1157

INTERNATIONAL ASSOCIATION FOR THE SCIENTIFIC STUDY OF MENTAL DEFICIENCY Congress of the International association for the scientific study of mental defiency 1st Proceedings Montpellier 1967 Sep 12-20 Edited by B.W. Richards Jackson,Reigate,1968 Title page and papers in English and French
Psy 28.0269

INTERNATIONAL ASSOCIATION FOR THE STUDY AND IMPROVEMENT OF HUMAN RELATIONS AND CONDITIONS IN INDUSTRY Triennial congress...on the subject of fundamental relationships between all sections of the industrial community 1st Report Cambridge 1928 Jun 27-Jul 3 Secretariat I.R.I.,The Hague,1928
Psy 31.2516

INTERNATIONAL ASSOCIATION OF APPLIED PSYCHOLOGY Child and education International congress of applied psychology 14th Proceedings Copenhagen 1961 Aug 13-19 Vol 3 Edited by Gerhard S. Nielsen Munksgaard,Copenhagen,1962
Psy 31.1198

INTERNATIONAL ASSOCIATION OF APPLIED PSYCHOLOGY International congress of applied psychology 14th Proceedings Copenhagen 1961 Aug 13-19 Vol 2: personality research Edited by Stanley Coopersmith Munksgaard, Copenhagen,1952
Psy 31.1968

INTERNATIONAL ASSOCIATION OF APPLIED PSYCHOLOGY International congress of applied psychology 14th Proceedings Copenhagen 1961 Aug 13-19 Vol 5: industrial and business psychology Edited by Gerhard S. Nielsen Munksgaard,Copenhagen,1962
Psy 31.2252

INTERNATIONAL ASSOCIATION OF GEOCHEMISTRY AND COSMOCHEMISTRY Origin and distribution of the elements :a symposium Proceedings Paris 1967 May 8-11 Edited by L.H. Ahrens International series of monographs on earth sciences, 30 1178p Pergamon,Oxford, 1968
Obs 6.3444

INTERNATIONAL ASSOCIATION OF GEODESY Electromagnetic distance measurement :a symposium Oxford 1965 Sep 6-11 Hilger and Watts,London,1967
Geog 13.5530

INTERNATIONAL ASSOCIATION OF GEODESY International symposium on the use of artificial satellites for geodesy 1st Proceedings Washington,D.C. 1962 Apr 26-28 Edited by George Veis Co-sponsored by the International union of geodesy and geophysics North-Holland,Amsterdam,1963
Geod 9.0055

INTERNATIONAL ASSOCIATION OF GEODESY.SPECIAL DUTY GROUP NO 19 Electromagnetic distance measurement :a symposium Oxford 1965 Sep 6-11 Hilger and Watts,London,1967
Eng 41.3278

INTERNATIONAL ASSOCIATION OF GEOMAGNETISM AND AERONOMY and INTERNATIONAL UNION OF GEODESY AND GEOPHYSICS The Toronto meeting Transactions Toronto 1957 Sep 3-14 Edited by T.R. Kaiser and V. Laursen I. A.G.A.bulletin,16 x,412p 25cm North Holland,Copenhagen,1960
Sco 14.0244

INTERNATIONAL ASSOCIATION OF GERONTOLOGY
Aging around the world International congress of gerontology 5th Proceedings San Francisco 1960 Aug Edited by Herman T. Blumenthal Columbia university press,New York;London,1962
An 32.0088

INTERNATIONAL ASSOCIATION OF METEOROLOGY
Scientific proceedings of the International association of meteorology Proceedings Rome 1954 Sep 15 AIM publication, 10,c 594p Butterworths,London,1956 Meeting at the 10th general assembly of International union of geodesy and geophysics
Nap 11.0286

INTERNATIONAL ASSOCIATION OF METEOROLOGY AND ATMOSPHERIC PHYSICS Noctilucent clouds : international symposium Tallinn 1966 Edited by I.A. Khvostikov and G. Witt Co-sponsored by the World meteorological organization illus 235p 27cm Academy of sciences of the USSR.Soviet geophysical committee,Moscow,1967 Symposium papers in English with Russian abstracts
Sco 14.8230

INTERNATIONAL ASSOCIATION OF METEOROLOGY AND ATMOSPHERIC PHYSICS Symposium on the general circulation of the oceans and the atmosphere International geodetic and geophysical union :general assembly 9th Symposium proceedings Brussels 1951 Aug Unesco,Brussels,1951
Geog 13.0960

INTERNATIONAL ASSOCIATION OF METEOROLOGY AND ATMOSPHERIC PHYSICS The General assembly of the International association of meteorology Proceedings Berkeley,Calif 1963 Edited by W.L. Godson 199p Toronto, 1963 Part of the 13th general assembly of International union of geodesy and geophysics
Nap 11.0223

INTERNATIONAL ASSOCIATION OF MICROBIOLOGICAL SOCIETIES.PERMANENT SECTION OF MICROBIOLOGICAL STANDARDIZATION International symposium on immunological methods of biological standardization 15th Proceedings Royaumont 1965 Oct 26-28 Edited by R.H. Regamey and others International association of microbiological societies.Symposia series in immunological standardization, 4 illus, tables Karger,Basle;New York,1967
Path 30.2746

INTERNATIONAL ASSOCIATION OF MICROBIOLOGICAL SOCIETIES.SYMPOSIA SERIES IN IMMUNOLOGICAL STANDARDIZATION, 4 International symposium on immunological methods of biological standardization 15th Proceedings Royaumont 1965 Oct 26-28 International association of microbiological societies.Permanent section of microbiological standardization Edited by R.H. Regamey and others illus,tables Karger,Basle;New York, 1967
Path 30.2746

INTERNATIONAL ASSOCIATION OF PHYSICAL OCEANOGRAPHY SCAR,SCOR,IAPO,IUBS symposium on Antarctic oceanography Papers Santiago de Chile 1966 Sep 13-16 Held by invitation of the Chilean national committee for antarctic research illus,maps xi,268p 23cm Scott Polar research institute for SCAR, Cambridge,1968
Sco 14.8300

INTERNATIONAL ASSOCIATION OF SCIENTIFIC HYDROLOGY
Abstracts of the reports submitted to the XI general assembly of the International union of geodesy and geophysics 101p Moscow,1957 In English,French and Russian;includes also t.-p.in Russian
Sco 14.0126

INTERNATIONAL ASSOCIATION OF SCIENTIFIC HYDROLOGY
Association internationale d'hydrologie scientifique :assemblee generale 6me comtes-rendus Edinburgh 1936 Sep 14-16,18 meetings of the International commissions of snow and of glaciers International association of scientific hydrology.Bulletin, 23 Riga,1938
Sco 14.0116

INTERNATIONAL ASSOCIATION OF SCIENTIFIC HYDROLOGY
Association internationale d'hydrologie scientifique :assemblee generale 8th Proces-verbaux Oslo 1948 Aug 19-28 Tome 2: travaux de la commission de la neige et des glaciers illus 407p Louvain,1948
Sco 14.0118

INTERNATIONAL ASSOCIATION OF SCIENTIFIC HYDROLOGY
Association internationale d'hydrologie scientifique :assemblee generale 9th Rapports et comtes-rendus Brussels 1951 Aug 20-22 Tome 1: 1...2...3:rapports et comtes-rendus des seances de la commission de neige et des glaciers International association of scientific hydrology. Publication,32 Louvain,1951 Papers in English,French and German
Sco 14.0120

INTERNATIONAL ASSOCIATION OF SCIENTIFIC HYDROLOGY
Association internationale d'hydrologie scientifique :reunion Comptes rendus et rapports Washington,D.C. 1939 Sep 4-18 Tome 1-2 2 vols Washington,D.C. Vol 1 imperfect:contains only papers read at meetings of the commissions of potamology and of limnology.In English,French,German and Italian
Sco 14.0117

INTERNATIONAL ASSOCIATION OF SCIENTIFIC HYDROLOGY
Association internationale d'hydrologie scientifiques :assemblee generale de Rome 10th comptes-rendus Rome 1954 Tome 4: commission des neiges et de glaces International association of scientific hydrology.Publication,39 522p 24cm Louvain,1956 Papers in English,French and German
Sco 14.0124

INTERNATIONAL ASSOCIATION OF SCIENTIFIC HYDROLOGY
Soobshcheniye o nauchnykh rabotakh po girdrologii 1963-1966 Bibliography Akademiya nauk S.S.S.R.Sovetskii geografischekii komitet 157p 26cm Moscow, 1967 Introduction and index in English
Sco 14.7518

INTERNATIONAL ASSOCIATION OF SCIENTIFIC HYDROLOGY
The International symposium on scientific aspects of snow and ice avalanches Rapports et discussions Davos 1965 Apr 5-10 International association of scientific hydrology.Publication,69 illus 417p Gentbrugge,1966 In English and French
Sco 14.0142

INTERNATIONAL ASSOCIATION OF SCIENTIFIC HYDROLOGY and WORLD METEOROLOGICAL ASSOCIATION Surface waters :symposium Proceedings Berkeley,Calif. 1963 Aug 19-31 International association of scientific hydrology.Publication,63 illus 615p Gentbrugge,1964 Papers in English,French, German
Sco 14.0138

INTERNATIONAL ASSOCIATION OF SCIENTIFIC HYDROLOGY a symposium Proceedings Variations of the regime of existing glaciers Obergurgl 1962 Sep 10-18 Edited by W. Ward International association of scientific hydrology.Publication,58 312p Gentbrugge, 1962
Sco 14.0137

INTERNATIONAL ASSOCIATION OF SCIENTIFIC HYDROLOGY general assembly Papers Symposium on Antarctic glaciology Helsinki 1960 Jul 25-Aug 6 International association of scientific hydrology.Publication,55 illus 162p 24cm Gentbrugge,1961 Papers in English and French
Sco 14.0130

INTERNATIONAL ASSOCIATION OF SCIENTIFIC HYDROLOGY 11th :general assembly Proceedings Toronto 1957 Sep 3-14 Vol 4: commission on snow and ice Edited by L.J. Tison International association of scientific hydrology.Publication,46 563p 24cm Gentbrugge,1958 Papers in English,French, German,Russian
Sco 14.0125

INTERNATIONAL ASSOCIATION OF SCIENTIFIC HYDROLOGY 12th general assembly Papers Helsinki 1960 Jul 25-Aug 6-8 snow and ice commission International association of scientific hydrology illus,map 588p Gentbrugge,1961 Papers in English,French and German
Sco 14.0131

INTERNATIONAL ASSOCIATION OF SCIENTIFIC HYDROLOGY 13th :general assembly Proceedings Berkeley,Calif. 1963 Aug 19-31 Commission of snow and ice International association of scientific hydrology.Publication,61 286p Gentbrugge,1963
Sco 14.0139

INTERNATIONAL ASSOCIATION OF TERRESTRIAL MAGNETISM AND ELECTRICITY a meeting Transactions of the Rome meeting Rome 1954 Sep 14-25 Edited by V. Laursen IAGA Bulletin, 15 406p Horsholm,Copenhagen, 1957 The meeting took place during the 10th general assembly of the international union of geodesy and geophysics
Nap 11.0287

INTERNATIONAL ASSOCIATION OF THEORETICAL AND APPLIED LIMNOLOGY Symposium on the classification of brackish waters Venice 1958 Apr 8-14 By G.Sven Segerstrale Archivio di oceanografia e limnologia, 11, suppl. Centro nazionale di studi talassografafici del consiglio nazionale delle ricerche,Venice,1959 Title also in Italian:"Simposio sulla classificazione della acque salmastre" Text in English and Italian.
Bal 39.1962

INTERNATIONAL ASTRONAUTICAL CONGRESS.PROCEEDINGS, 8 International astronautical congress 8th proceedings Barcelona 1957 Oct.6-12 International astronautical federation Edited by F. Hecht 607p Springer,Vienna, 1958
Obs 6.0761

INTERNATIONAL ASTRONAUTICAL CONGRESS 17TH Life sciences research and lunar medicine Lunar international laboratory (LIL)symposium 2nd Proceedings Madrid 1966 Oct 13 Edited by Frank J. Malina Organized by International academy of astronautics Pergamon press,Oxford;London,1967 Held at the 17th International astronautical congress
PGMS 29.0445

INTERNATIONAL ASTRONAUTICAL CONGRESS 19TH Applied sciences research and utilization of lunar resources Lunar international laboratory (LIL symposium) 4th Proceedings New York 1968 Oct 17 International academy of astronautics Bibliog,illus 200p 20cm Pergamon press, Oxford,1970 LIL symposium organized by the International academy of astronautics at the 19th International astronautical congress
Cav 7.2711

INTERNATIONAL ASTRONAUTICAL FEDERATION International astronautical congress 8th proceedings Barcelona 1957 Oct.6-12 Edited by F. Hecht International astronautical congress.Proceedings, 8 607p Springer,Vienna,1958
Obs 6.0761

INTERNATIONAL ASTRONOMICAL UNION Abundance determinations in stellar spectra :a symposium Utrecht 1964 Aug.10-14 Edited by Hans Hubenet International astronomical union. Symposium, 26 374p London;New York,1966
Obs 6.0843

INTERNATIONAL ASTRONOMICAL UNION Abundance determinations in stellar spectra :a symposium Proceedings Utrecht 1964 Aug 10-14 International astronomical union.Symposium, 26 Academic press,London,1966
TA 15.0034

INTERNATIONAL ASTRONOMICAL UNION Abundance determinations in stellar spectra :a symposium Utrecht 1964 Aug.10-14 Edited by Hans Hubenet International astronomical union. Symposium, 26 374p Academic press,London; New York,1966
A Math 4.1168

INTERNATIONAL ASTRONOMICAL UNION Aerodynamic phenomena in stellar atmospheres Symposium on cosmical gas dynamics held at the International school of physics "Enrico Fermi" 4th proceedings Varenna 1960 Aug.18-30 Edited by R.N. Thomas and others Under the auspices of Societa Italiana di fisica International astronomical union.Symposium, 12 515p Bologna,c1961 Reprinted from Supplemento Nuovo Cimento,22,no.1,1961
A Math 4.1152

INTERNATIONAL ASTRONOMICAL UNION Aerodynamic phenomena in stellar atmospheres Symposium on cosmical gas dynamics held at the International school of physics "Enrico Fermi" 4th proceedings Varenna 1960 Aug 18-30 Edited by R.N. Thomas and others under the auspices of the Societa italiana di fisica International astronomical union symposium, 12 515p Zanichelli,Bologna,c1961 Reprinted from the 'Supplemento del nuovo cimento',vol.22,no.1,1961
Obs 6.1736

INTERNATIONAL ASTRONOMICAL UNION Aerodynamical phenomena in stellar atmosphwre Nice 1965 Sep Edited by Richard N. Thomas International astronomical union.Symposium, 28 Academic press,London,1967
TA 15.0411

INTERNATIONAL ASTRONOMICAL UNION Blanketing effect :I.A.U. colloquium Proceedings Heidelberg 1966 Mar 17-19 Edited by K-H. Bohm International astronomical union. Colloquium Journal of quantitative spectroscopy and radiative transfer, 6,no 5 Pergamon,London,1966
TA 15.0414

INTERNATIONAL ASTRONOMICAL UNION Co-ordination of galactic research :conference 2nd Saltsjobaden 1957 Jun.17-22 Edited by A. Blaauw and others International astronomical union.Symposium, 7 93p Cambridge university press,Cambridge,1959
Obs 6.0192

INTERNATIONAL ASTRONOMICAL UNION Co-ordination of galactic research a symposium Edited by A. Blaauw International astronomical union.Symposium, 1 59p Cambridge university press,Cambridge,1955
Obs 6.0191

INTERNATIONAL ASTRONOMICAL UNION Comparison of the large-scale structure of the galactic system with that of other stellar systems :a symposium papers Dublin 1955 Sep 2 Edited by N.G. Roman International astronomical union. Symposium, 5 illus 72p Cambridge university press,Cambridge, 1958
Obs 6.1466

INTERNATIONAL ASTRONOMICAL UNION Continental drift,secular motion of the Pole and rotation of the earth I.A.U. symposium 32nd Papers Stresa 1967 Mar 21-25 Edited by William Markowitz and B. Guinot illus 107p 25cm D.Riedel,Dordrecht,1968
Sco 14.8118

INTERNATIONAL ASTRONOMICAL UNION Continental drift,secular motion of the pole,and rotation of the earth Stresa 1967 Mar 21-25 Edited by Wm. Markowitz and B. Guinot International astronomical union.Symposium, 32 Reidel,Dordrecht,1968
TA 15.0424

INTERNATIONAL ASTRONOMICAL UNION Continental drift,secular motion of the pole,and rotation of the earth :a symposium Proceedings Stresa 1967 Mar Edited by W. Markowitz and B. Guinot International astronomical union. Symposium, 32 108p Reidel,Dordrecht,1968
Obs 6.3268

INTERNATIONAL ASTRONOMICAL UNION Determination of radial velocities and their applications Toronto 1966 Jun 20-24 Edited by A.H. Batten and J.F. Heard International astronomical union.Symposium, 30 Academic press,London;New York,1967
TA 15.0307

INTERNATIONAL ASTRONOMICAL UNION Determination of radial velocities and their applications :a symposium Toronto 1966 Jun. 20-24 Edited by A.H. Batten and J.F. Heard International astronomical union.Symposium, 30 262p academic press,London,1967 publication dedicated to the memory of R. Methven Petrie
Obs 6.0147

INTERNATIONAL ASTRONOMICAL UNION Electromagnetic phemomena in cosmical physics a symposium Proceedings Stockholm 1956 Aug.27-31 Edited by B. Lehnert In co-operation with the International union of theoretical and applied mechanics International astronomical union.Symposium, 6 544p Cambridge university press, Cambridge,1958
Obs 6.1054

INTERNATIONAL ASTRONOMICAL UNION Electromagnetic phenomena in cosmical physics Stockholm 1956 Aug 27-31 Edited by B. Lehnert In co-operation with the International union of theoretical and applied mechanics International astronomical union. Symposium, 6 13,544p Cambridge university press,Cambridge,1958
Nap 11.0916

INTERNATIONAL ASTRONOMICAL UNION Electromagnetic phenomena in cosmical physics : a symposium proceedings Stockholm 1956 Aug.27-31 Edited by B. Lehnert International astronomical union.Symposia, 6 Cambridge university press,Cambridge,1958
A Math 4.0615

INTERNATIONAL ASTRONOMICAL UNION External galaxies and quasi-stellar objects Upsala 1970 Aug 10-14 Edited by D.S. Evans International astronomical union.Symposium, 44 549p Reidel,Dordrecht,1972
TA 15.0658

INTERNATIONAL ASTRONOMICAL UNION Extra-galactic research The Symposium on problems of extra-galactic research 15th Santa Barbara,Calif. 1961 Aug.10-12 Edited by G. C. McVittie International astronomical union. Symposium, 15 450p Macmillan,New York, 1962
Obs 6.1124

INTERNATIONAL ASTRONOMICAL UNION Gas dynamics of cosmic clouds Symposium on cosmical gas dynamics 2nd Cambridge 1953 Jul 6-11 International astronomical union.Symposium series, 2 North-Holland, Amsterdam,1955
Eng 41.4468

INTERNATIONAL ASTRONOMICAL UNION Gas dynamics of cosmic clouds The Symposium on cosmical gas dynamics Proceedings Cambridge 1953 Jul 6-11 Edited by J.M. Burgess and H.C.van de Hulst International astronomical union.Symposium, 2 North Holland,Amsterdam,1955
TA 15.0217

INTERNATIONAL ASTRONOMICAL UNION Gas dynamics of cosmic clouds :a symposium Cambridge 1953 Jul.6-11 Edited by J.M. Burgers and H.C.van de Hulst International astronomical union.Symposium, 2 247p North-Holland,Amsterdam,1955
Obs 6.0308

INTERNATIONAL ASTRONOMICAL UNION Gravitational N-body problem Proceedings Cambridge 1970 Aug 12-15 Edited by M. Lecar International astronomical union. Colloquium, 10 441p Reidel,Dordrecht,1972
TA 15.0667

INTERNATIONAL ASTRONOMICAL UNION Highlights of astronomy IAU general assembly 13 Invited discources and discussions,and special meetings Prague 1967 Edited by Lubos Perck 548p Reidel,Dordrecht,1968
Obs 6.3528

INTERNATIONAL ASTRONOMICAL UNION Highlights of astronomy International astronomical union 14th general assembly Brighton 1970 Aug Vol 2 Edited by C.de Jager 793p Reidel,Dordrecht,1971
Obs 6.3688

INTERNATIONAL ASTRONOMICAL UNION Highlights of astronomy International astronomical union.General assembly 14th Invited discourses Brighton 1970 Aug Vol 2 Edited by C.de Jager 793p Reidel,Dordrecht, 1971
TA 15.0655

INTERNATIONAL ASTRONOMICAL UNION IAU general assembly 11th News bulletin Berkeley, Calif. 1961 Aug. 1-8 Berkeley,1961
Obs 6.1945

INTERNATIONAL ASTRONOMICAL UNION IAU-NASA symposium The Conference on the nature of the surface of the moon Proceedings Greenbelt,Md. 1965 Apr 15-16 Edited by Wilmot N. Hess and others Sponsored bt the National aeronautics and space administration 320p Johns Hoplins press,Baltimore,1966
TA 15.0203

INTERNATIONAL ASTRONOMICAL UNION Interstellar gas dynamics Symposium on cosmical gas dynamics 6th Proceedings Yalta 1969 Sep 8-18 Edited by H.J. Habing International astronomical union.Symposium, 39 diagrs,photos,grapns xii,388p 24cm Reidel,Dordrecht,1970
Cav 7.2710

INTERNATIONAL ASTRONOMICAL UNION Interstellar gas dynamics Symposium on cosmical gas dynamics 6th Proceedings Yalta 1969 Sep 8-18 International astronomical union.Symposium, 39 388p Reidel,Dordrecht,1970
TA 15.0541

INTERNATIONAL ASTRONOMICAL UNION Interstellar gas dynamics Symposium on cosmical gas dynamics 6th Proceedings Yalta 1969 Sep 8-18 International astronomical union.Symposium, 39 388p Reidel,Dordrecht,1970
Obs 6.3323

INTERNATIONAL ASTRONOMICAL UNION Mass loss and evolution in close binaries Elsinore 1969 Edited by K. Gyldenkerne and R.M. West International astronomical union.Colloquium, 6 238p Copenhagen university observatory, Copenhagen,1970
Obs 6.3358

INTERNATIONAL ASTRONOMICAL UNION Mass loss from stars Trieste colloquium on astrophysics 2nd Proceedings Trieste 1968 Sep 12-17 Edited by Margherita Hack Astrophysics and space science library, 13 345p Reidel,Dordrecht,1969
TA 15.0620

INTERNATIONAL ASTRONOMICAL UNION Mass loss from stars Trieste colloquium on astrophysics 2nd Proceedings Trieste 1968 Sep 12-17 Edited by Margherita Hack Astrophysics and space science library, 13 345p Reidel,Dordrecht,1969
Obs 6.3628

INTERNATIONAL ASTRONOMICAL UNION Moon Symposium on the moon proceedings Pulkovo 1960 Dec.5-11 Edited by Zdenek Kopal and Zdenka Kadla Mikhailov International astronomical union.Symposia, 14 map 517p Academic press,London,1962
Obs 6.0972

INTERNATIONAL ASTRONOMICAL UNION Moon and planets International space science : symposium 7th Papers Vienna 1966 May 10-18 Edited by A. Dollfus Co-sponsored by the Committee on space research 313p North-Holland,Amsterdam,1967 Papers of one session of the symposium
Obs 6.3206

INTERNATIONAL ASTRONOMICAL UNION Nature of the lunar surface IAU-NASA symposium Greenbelt,Md. 1965 Apr 15-16 Edited by W.N. Hess and others Held at the Goddard space flight center Johns Hopkins press,Baltimore, Md.,1966
Min 10.1445

INTERNATIONAL ASTRONOMICAL UNION New aspects in the history and philosophy of astronomy Joint symposium of the International astronomical union and the Union internationale d'histoire et de philosophie des sciences 1st Hamburg 1964 Aug 22-24 Edited by Arthur Beer Vistas in astronomy, 9 illus,ports,maps 318p 25cm Pergamon press,Oxford,1967
Cav 7.2715

INTERNATIONAL ASTRONOMICAL UNION New aspects in the history and philosophy of astronomy Joint symposium of the International astronomical union and the Union internationale d'histoire et de philosophie des sciences 1st Hamburg 1964 Aug 22-24 Edited by Arthur Beer Vistas in astronomy, 9 xvi,317p Pergamon,Oxford, 1967
WSM 43.2439

INTERNATIONAL ASTRONOMICAL UNION New techniques in space astronomy Munich 1970 Aug 10-14 Edited by F. Labuhn and R. Lust International astronomical union.Symposium, 41 418p Reidel,Dordrecht,1971
Obs 6.3590

INTERNATIONAL ASTRONOMICAL UNION Non-solar X-and gamma-ray astronomy Rome 1969 May 8-10 Edited by L. Gratton International astronomical union.Symposium, 37 425p Reidel,Dordrecht,190
Obs 6.3322

INTERNATIONAL ASTRONOMICAL UNION Non-solar X-and gamma-ray astronomy :symposium Papers Rome 1969 May 8-10 Edited by L. Gratton International astronomical union.Symposium, 37 425p Reidel,Dordrecht,1970
TA 15.0492

INTERNATIONAL ASTRONOMICAL UNION Non-stable phenomena in galaxies Nestasionarnye yavleniya v galaktikakh Byurakan 1966 May 4-12 International astronomical union. Symposium, 29 480p Akademiya nauk Armyanskey S.S.R.,Erevan,1968
Obs 6.3269

INTERNATIONAL ASTRONOMICAL UNION Non-stable stars :a symposium Proceedings Dublin 1955 Sep 1 Edited by George H. Herbig International astronomical union.Symposium, 3 Cambridge university press,Cambridge,1957
TA 15.0033

INTERNATIONAL ASTRONOMICAL UNION Paris symposium on radio astronomy Paris 1958 Jul.30-Aug.6 Edited by Ronald N. Bracewell International astronomical union.Symposium, 9 International scientific radio union. Symposium, 1 612p Stanford university press,Stanford,Calif.,1959
Obs 6.0246

INTERNATIONAL ASTRONOMICAL UNION Paris symposium on radio astronomy Paris 1958 Jul.30-Aug.6 Edited by Ronald N. Bracewell International astronomical union.Symposium, 9 International scientific radio union. Symposium, 1 612p Stanford university press,Stanford,Calif,1959
A Math 4.1151

INTERNATIONAL ASTRONOMICAL UNION Paris symposium on radio astronomy Paris 1958 Jul 30-Aug 6 Edited by Ronald N. Bracewell International astronomical union.Symposium, 9 International scientific radio union. Symposium, 1 612p Stanford university press,Stanford,Calif.,1959
TA 15.0638

INTERNATIONAL ASTRONOMICAL UNION Paris symposium on radio astronomy Paris 1958 Jul 30-Aug 6 Edited by Ronald N. Bracewell International astronomical union.Symposium, 9 International scientific radio union. Symposium, 1 612p Stanford university press,Stanford,Calif,1959
Cav 7.0301

INTERNATIONAL ASTRONOMICAL UNION Physics and dynamics of meteors Tatranska Lomnica 1967 Sep 4-7 Edited by Lubor Kresak and Peter M. Millman International astronomical union. Symposium, 33 Reidel,Dordrecht,1968
TA 15.0422

INTERNATIONAL ASTRONOMICAL UNION Physics and dynamics of meteors :a symposium Proceedings Tatranska Lomnica 1967 Edited by Lubor Kresak and Peter M. Millman International astronomical union.Symposium, 33 525p Reidel,Dordrecht,1968
Obs 6.3264

INTERNATIONAL ASTRONOMICAL UNION Planetary atmospheres Marfa,Texas 1969 Oct 26-31 Edited by C. Sagan and others International astronomical union.Symposium, 40 408p Reidel,Dordrecht,1971
Obs 6.3586

INTERNATIONAL ASTRONOMICAL UNION Planetary nebulae :a symposium Proceedings Tatranska Lomnica 1967 Edited by D.E. Osterbrock and C.R. O'Dell International astronomical union. Symposium, 34 469p Reidel,Dordrecht,1968
Obs 6.3265

INTERNATIONAL ASTRONOMICAL UNION Planetary nebulae :a symposium Tatranska Lomnica 1967 Edited by D.E. Osterbrock and C.R. O'Dell International astronomical union. Symposium, 34 Reidel,Dordrecht,1968
TA 15.0320

INTERNATIONAL ASTRONOMICAL UNION Problems of cosmical aerodynamics Symposium on the motion of gaseous masses of cosmical dimensions Proceedings Paris 1949 Aug 16-19 Edited by J.M. Burgess and H.C.van de Hulst 237p Central air documents office, Dayton,Ohio,1951
Obs 6.3550

INTERNATIONAL ASTRONOMICAL UNION Problems of extra-galactic research :a symposium Santa Barbara,Calif. 1961 Aug.10-12 Edited by G. C. McVittie International astronomical union. Symposium, 15 450p Macmillan,New York, 1962
A Math 4.1154

INTERNATIONAL ASTRONOMICAL UNION Radio astronomy :a symposium Proceedings Jodrell Bank 1955 Aug 25-27 Edited by H.C.van de Hulst International astronomical union. Symposium, 4 xi,409p Cambridge university press,Cambridge,1957
Nap 11.0275

INTERNATIONAL ASTRONOMICAL UNION Radio astronomy :a symposium Proceedings Manchester 1955 Aug 25-27 Edited by H.C. van de Hulst International astronomical union.Symposium, 4 Cambridge,1957
TA 15.0037

INTERNATIONAL ASTRONOMICAL UNION Radio astronomy :a symposium proceedings Manchester 1955 Aug.25-27 Edited by H.C. van de Hulst With financial assistance from Unesco International astronomical union. Symposium, 4 409p Cambridge university press,Cambridge,1957
Obs 6.0847

INTERNATIONAL ASTRONOMICAL UNION Radio astronomy and the galactic system Symposium on radio astronomy and the galactic system Noordwijk 1966 Aug 25-Sep 1 Edited by Hugo van Woerden Organised in co-operation with International scientific radio union International astronomical union.Symposium, 31 xviii,501p Academic press,London;New York,1967
A Math 4.1574

INTERNATIONAL ASTRONOMICAL UNION Solar flares and space research Proceedings of the symposium held on the occasion of the 11th plenary meeting of the Committee on space research Tokyo 1968 May 9-11 Edited by C. de Jager and Z. Svestka 419p North-Holland, Amsterdam,1969 Co-sponsored by International astronomical union,International union of geodesy and geophysics and International union of radio scientists
Obs 6.3335

INTERNATIONAL ASTRONOMICAL UNION Solar magnetic fields Paris 1970 Aug 31-Sep 4 Edited by R. Howard International astronomical union.Symposium, 43 782p Reidel,Dordrecht,1971
Obs 6.3589

INTERNATIONAL ASTRONOMICAL UNION Solar magnetic fields Paris 1970 Aug 31-Sep 4 Edited by R. Howard International astronomical union.Symposium, 43 782p Reidel,Dordrecht,1971
TA 15.0688

INTERNATIONAL ASTRONOMICAL UNION Special classification and multicolour photometry Saltsjobaden 1964 Aug 17-21 Edited by K. Loden International astronomical union. Symposium, 24 Academic press,London;New York,1966
TA 15.0319

INTERNATIONAL ASTRONOMICAL UNION Spectral classification and multicolour photometry Symposium on the spectral classification and multicolour photometry Proceedings Saltsjobaden 1964 Aug.17-21 Edited by K. Loden and others International astronomical union.Symposium, 24 383p Academic press, London;New York,1966
Obs 6.1092

INTERNATIONAL ASTRONOMICAL UNION Spectral classification and multicolour photometry :a symposium Saltsjobaden,Sweden 1964 Aug.17-21 Edited by K. Loden and others International astronomical union.Symposium, 24 383p Academic press,London;New York, 1966
A Math 4.1166

INTERNATIONAL ASTRONOMICAL UNION Stellar and solar magnetic fields Symposium on stellar and solar magnetic fields Munich 1963 Sep. 3-10 Edited by R. Lust International astronomical union.Symposium, 22 460p North Holland,Amsterdam,1965
Obs 6.1148

INTERNATIONAL ASTRONOMICAL UNION Stellar and solar magnetic fields :a symposium Rottach-Egern 1963 Sep.3-10 Edited by R. Lust International astronomical union.Symposium, 22 460p North-Holland,Amsterdam,1965
A Math 4.1160

INTERNATIONAL ASTRONOMICAL UNION Stresa symposium on continental drift,secular motion of the pole and rotation of the earth Stresa, Italy 1967 Mar 21-25 Edited by W. Markowitz and B. Guinot International astronomical union.Symposium, 32 Reidel, Dordrecht,1968
A Math 4.1308

INTERNATIONAL ASTRONOMICAL UNION Structure and development of solar active regions Budapest 1967 Sep 4-8 Edited by K.O. Kiepenheuer International astronomical union. Symposium, 35 Reidel,Dordrecht,1968
TA 15.0420

INTERNATIONAL ASTRONOMICAL UNION Structure and development of solar active regions :a symposium Budapest 1967 Sep 4-8 Edited by K.O. Kiepenheuer International astronomical union.Symposia, 35 Bibliog., Illus. xvii,608p 25cm D.Reidel,Dordrecht, 1968
A Math 4.1602

INTERNATIONAL ASTRONOMICAL UNION Structure and development of solar active regions :a symposium Proceedings Budapest 1967 Sep Edited by K.O. Kiepenheuer International astronomical union.Symposium, 35 608p Reidel,Dordrecht,1968
Obs 6.3266

INTERNATIONAL ASTRONOMICAL UNION Symposium on cosmical gas dynamics 3rd Proceedings Cambridge,Mass. 1957 Jun 24-29 Edited by J. M. Burgers and R.N. Thomas Reviews of modern physics, 30,no 3 American physical society, New York,1958
Eng 41.4467

INTERNATIONAL ASTRONOMICAL UNION Symposium on cosmical gas dynamics 3rd proceedings Cambridge,Mass. 1957 Jun.24-29 Edited by J. M. Burgers and others Financial grant from the Smithsonian institution International astronomical union.Symposium, 8 Smithsonian Institution.Smithsonian symposium publications, 1 American physical society, New York,1958 Reviews of modern physics. Vol.30,no.3,July 1958
Obs 6.0306

INTERNATIONAL ASTRONOMICAL UNION Symposium on problems of extra-galactic research Proceedings Santa Barbara,Calif. 1961 Aug 10-12 Edited by G.C. McVittie International astronomical union.Symposium, 15 McMillan,New York,1962
TA 15.0035

INTERNATIONAL ASTRONOMICAL UNION The Construction of large telescopes :a symposium Edited by D.L. Crawford International astronomical union.Symposium, 27 Academic press,New York,1966
TA 15.0311

INTERNATIONAL ASTRONOMICAL UNION The Construction of large telescopes :a symposium Tucson,Ariz. 1965 Apr 5-12 Pasadena,Calif. Edited by D.L. Crawford International astronomical union.Symposium, 27 234p Academic press,New York;London,1966
Obs 6.3203

INTERNATIONAL ASTRONOMICAL UNION The Crab nebula Jodrell Bank 1970 Aug 5-7 Edited by R.D. Davies and F.G. Smith International astronomical union.Symposium, 46 470p Reidel,Dordrecht,1971
Obs 6.3588

INTERNATIONAL ASTRONOMICAL UNION The Crab nebula Jodrell Bank 1970 Aug 5-7 Edited by R.D. Davies and F.G. Smith International astronomical union.Symposium, 46 470p Reidel,Dordrecht,1971
TA 15.0689

INTERNATIONAL ASTRONOMICAL UNION The Galaxy and the magellanic clouds :a symposium Canberra 1963 Mar.18-28 Edited by F.J. Kerr and A.W. Rodgers International astronomical union.Symposium, 20 393p Australian academy of science,Canberra,1964 Dedicated to Joseph Lade Pawsey
A Math 4.1158

INTERNATIONAL ASTRONOMICAL UNION The Galaxy and the magellanic clouds :a symposium Proceedings Canberra 1963 Mar.18-28 Edited by F.J. Kerr and A.W. Rodgers International astronomical union.Symposium, 20 393p Australian academy of science, Canberra,1964 Publication dedicated to memory of Joseph Lade Pawsey
Obs 6.0941

INTERNATIONAL ASTRONOMICAL UNION The Hertzsprung-Russell diagram :a symposium Proceedings Moscow 1958 Aug.15-16 Edited by Jesse L. Greenstein International astronomical union.Symposium, 10 128p Paris,1959 Reprinted from Annales d'astrophysique 1959,supplements.Fascicule.No. 8
Obs 6.0687

INTERNATIONAL ASTRONOMICAL UNION The International conference on cosmic rays and the earth storm :combined meeting of the International symposium on the earth storm and the International conference on cosmic rays Proceedings Kyoto 1961 Sep 4-15 Vol 2-3 Edited by Ken-ichi Maeda and Osamu Minakawa Co-sponsored by the International scientific radio union Physical society of Japan.Journal, 17,suppl.A-2,3 2 vols physical society of Japan,Tokyo,1962
TA 15.0038

INTERNATIONAL ASTRONOMICAL UNION The Moon :a symposium Pulkovo 1960 Dec. Edited by Zdenek Kopal and Zdenka Kadla Mikhailov Held at Glavnaya astronomicheskaya observatoriya, Pulkovo International astronomical union. Symposium, 14 571p Academic press,London; New York,1962
A Math 4.1153

INTERNATIONAL ASTRONOMICAL UNION The Motion, evolution of orbits,and origin of comets Leningrad 1970 Aug 4-11 Edited by G.A. Chebotarev and others International astronomical union.Symposium, 45 521p Reidel,Dordrecht,1972
TA 15.0687

INTERNATIONAL ASTRONOMICAL UNION The Motion, evolution of orbits,and origin of comets Leningrad 1970 Aug 4-11 Edited by G.A. Chebotarev and others International astronomical union.Symposium, 45 521p Reidel,Dordrecht,1972
Obs 6.3591

INTERNATIONAL ASTRONOMICAL UNION The Rotation of the earth and atomic time standards :a symposium Moscow 1958 Aug Edited by Dirk Brouwer International astronomical union.Symposium, 11 1959 Reprinted from the Astronomical journal,vol 64, no 1268
Obs 6.3199

INTERNATIONAL ASTRONOMICAL UNION The Solar corona :a symposium proceedings Cloudcroft, N.M. 1961 Aug.28-30 Edited by John W. Evans Sponsored by the Air Force Cambridge research laboratories International astronomical union.Symposium, 16 344p Academic press,New York;London,1963
Obs 6.0564

INTERNATIONAL ASTRONOMICAL UNION The Spiral structure of our galaxy Basle 1969 Aug 29-Sep 4 Edited by W. Becker and G. Contopoulos International astronomical union.Symposium, 38 478p Reidel,Dordrecht,1970
Obs 6.3320

INTERNATIONAL ASTRONOMICAL UNION The Spiral structure of our galaxy :symposium Proceedings Basle 1969 Aug 29-Sep 4 Edited by W. Becker and G. Contopoulos International astronomical union.Symposium, 38 478p Reidel,Dordrecht,1970
TA 15.0489

INTERNATIONAL ASTRONOMICAL UNION The System of astronomical constants :a symposium proceedings Paris 1963 May 27-31 Edited by J. Kovalevsky International astronomical union.Symposium, 21 330p Gauthier-Villars, Paris,1965 From the Bulletin astronomique de l'observatoire de Paris,vol.25,fasc.1-3
Obs 6.0986

INTERNATIONAL ASTRONOMICAL UNION The Theory of orbits in the solar system and in stellar systems Thessalonika 1964 Aug 17-22 Edited by George Contopoulos International astronomical union.Symposium, 25 Academic press,London;New York,1966
TA 15.0332

INTERNATIONAL ASTRONOMICAL UNION The Theory of orbits in the solar system and in stellar systems a symposium Thessaloniki 1964 Aug. 17-22 Edited by George Contopoulos Organized in co-operation with Committee on space research International Astronomical Union.Symposium, 25 380p Academic press, London;New York,1966
Obs 6.0409

INTERNATIONAL ASTRONOMICAL UNION The Theory of orbits in the solar system and in stellar systems :a symposium Thessaloniki 1964 Aug. 17-22 Edited by George Contopoulos Organized in co-operation with Committee on space research International astronomical union.Symposium, 25 380p Academic press, London;New York,1966
A Math 4.1167

INTERNATIONAL ASTRONOMICAL UNION Theoretical interpretation of upper atmosphere emissions : a symposium Paris 1962 Jun.25-29 Edited by D.R. Bates International astronomical union.Symposium, 18 264p Pergamon press, New York,1963 Reprinted from'Planetary and space science',Vol.10 1963
Obs 6.0142

INTERNATIONAL ASTRONOMICAL UNION Transfer of radiation in stellar atmospheres Colloquium on the theory of stellar atmospheres 3rd Herstmonceux 1962 Aug.15-16 Edited by C.de Jager and A.B. Underhill 201p Pergamon press,London,1963 Vol 3,no.2 of Journal of quantitative spectroscopy and radiative transfer. AprJun.1963
Obs 6.0875

INTERNATIONAL ASTRONOMICAL UNION Ultraviolet stellar spectra and related ground-based observations Lunteren 1969 Jun 24-27 Edited by L. Houziaux and H.E. Butler International astronomical union. Symposium, 36 361p Reidel, Dordrecht, 1970
Obs 6.3321

INTERNATIONAL ASTRONOMICAL UNION 12th assembly Reports on astronomy Hamburg 1964 Aug. 25-Sep.3 Edited by Jean-Claude Pecker International astronomical union. Transactions, 12a Academic press, London; New York, 1965
A Math 4.1114

INTERNATIONAL ASTRONOMICAL UNION 13th general assembly Highlights of astronomy Prague 1967 Aug Edited by L. Perek illus x,548p 25cm Reidel, Doldrecht, 1968
A Math 4.1899

INTERNATIONAL ASTRONOMICAL UNION 14th general assembly Highlights of astronomy Brighton 1970 Aug Vol 2 International astronomical union Edited by C. de Jager 793p Reidel, Dordrecht, 1971
Obs 6.3688

INTERNATIONAL ASTRONOMICAL UNION. SYMPOSIUM, 5 Comparison of the large-scale structure of the galactic system with that of other stellar systems :a symposium papers Dublin 1955 Sep 2 International astronomical union Edited by N.G. Roman illus 72p Cambridge university press, Cambridge, 1958
Obs 6.1466

INTERNATIONAL ASTRONOMICAL UNION.COLLOQUIUM, 4 Colloquium on variable stars Non-periodic phenomena in variable stars Budapest 1968 Sep 5-9 International astronomical union. Commissions 27 and 42 490p Academic press, Budapest, 1969
Obs 6.3360

INTERNATIONAL ASTRONOMICAL UNION.COLLOQUIUM, 6 Mass loss and evolution in close binaries Elsinore 1969 International astronomical union Edited by K. Gyldenkerne and R.M. West 238p Copenhagen university observatory, Copenhagen, 1970
Obs 6.3358

INTERNATIONAL ASTRONOMICAL UNION.COLLOQUIUM, 10 Gravitational N-body problem Proceedings Cambridge 1970 Aug 12-15 International astronomical union Edited by M. Lecar 441p Reidel, Dordrecht, 1972
TA 15.0667

INTERNATIONAL ASTRONOMICAL UNION.COMMISSION 17 Lunar nomenclature Report of the working group of I.A.U. commission 17:the moon: proposed names for craters on the moon's far side...presented at the 14th assembly... Brighton 1970 43p Harvard college observatory, Cambridge, Mass., 1970 Duplicated
Obs 6.3342

INTERNATIONAL ASTRONOMICAL UNION.COMMISSION 36 Spectrum formation in stars with steady-state extended atmospheres Munich 1969 Apr 16-19 Edited by H.G. Grotl and P. Wellmann International astronomical union.Commission 36. Colloquium, 2 National bureau of standards. Special publication, 332 332p National bureau of standards, Washington, D.C., 1970
Obs 6.3633

INTERNATIONAL ASTRONOMICAL UNION.COMMISSION 36. COLLOQUIUM, 2 Spectrum formation in stars with steady-state extended atmospheres Munich 1969 Apr 16-19 International astronomical union.Commission 36 Edited by H. G. Grotl and P. Wellmann National bureau of standards.Special publication, 332 332p National bureau of standards, Washington, D.C., 1970
Obs 6.3633

INTERNATIONAL ASTRONOMICAL UNION.COMMISSIONS 27 AND 42 Colloquium on variable stars Non-periodic phenomena in variable stars Budapest 1968 Sep 5-9 International astronomical union. Colloquium, 4 490p Academic press, Budapest, 1969
Obs 6.3360

INTERNATIONAL ASTRONOMICAL UNION.COMMISSIONS 27 AND 42 Position of variable stars in the Hertzsprung-Russell diagram Colloquium on variable stars 3rd Bamberg 1965 Aug 11-14 Remeis-sternwarte, Bamberg. Kleine veroffentlichungen, 4,no 40 port 304p Astronomisches institut der universitat Erlangen-Nurnberg, Bamberg, 1965 I.A.U. combined colloquium of commissions 27 and 42 with the theme 'The position of stars'
TA 15.0540

INTERNATIONAL ASTRONOMICAL UNION.GENERAL ASSEMBLY 14th Invited discourses Highlights of astronomy Brighton 1970 Aug Vol 2 International astronomical union Edited by C.de Jager 793p Reidel, Dordrecht, 1971
TA 15.0655

INTERNATIONAL ASTRONOMICAL UNION.SYMPOSIA, 14 Moon Symposium on the moon proceedings Pulkovo 1960 Dec.5-11 International astronomical union Edited by Zdenek Kopal and Zdenka Kadla Mikhailov map 517p Academic press, London, 1962
Obs 6.0972

INTERNATIONAL ASTRONOMICAL UNION.SYMPOSIA, 35 Structure and development of solar active regions :a symposium Budapest 1967 Sep 4-8 International astronomical union Edited by K. O. Kiepenheuer Bibliog., Illus. xvii,608p 25cm D.Reidel, Dordrecht, 1968
A Math 4.1602

INTERNATIONAL ASTRONOMICAL UNION.SYMPOSIUM, 1 Co-ordination of galactic research a symposium International astronomical union Edited by A. Blaauw 59p Cambridge university press, Cambridge, 1955
Obs 6.0191

INTERNATIONAL ASTRONOMICAL UNION.SYMPOSIUM, 2 Gas dynamics of cosmic clouds :a symposium Cambridge 1953 Jul.6-11 International union of theoretical and applied mechanics and International astronomical union Edited by J.M. Burgers and H.C.van de Hulst 247p North-Holland, Amsterdam, 1955
Obs 6.0308

INTERNATIONAL ASTRONOMICAL UNION.SYMPOSIUM, 3 Non-stable stars :a symposium Proceedings Dublin 1955 Sep 1 International astronomical union Edited by George H. Herbig Cambridge university press, Cambridge, 1957
TA 15.0033

INTERNATIONAL ASTRONOMICAL UNION.SYMPOSIUM, 4
Radio astronomy :a symposium Proceedings Manchester 1955 Aug 25-27 International astronomical union Edited by H.C.van de Hulst Cambridge,1957
TA 15.0037

INTERNATIONAL ASTRONOMICAL UNION.SYMPOSIUM, 4
Radio astronomy :a symposium proceedings Manchester 1955 Aug.25-27 International astronomical union Edited by H.C.van de Hulst With financial assistance from Unesco 409p Cambridge university press,Cambridge, 1957
Obs 6.0847

INTERNATIONAL ASTRONOMICAL UNION.SYMPOSIUM, 6
Electromagnetic phemomena in cosmical physics a symposium Proceedings Stockholm 1956 Aug.27-31 International astronomical union Edited by B. Lehnert In co-operation with the International union of theoretical and applied mechanics 544p Cambridge university press,Cambridge,1958
Obs 6.1054

INTERNATIONAL ASTRONOMICAL UNION.SYMPOSIUM, 6
Electromagnetic phenomena in cosmical physics Stockholm 1956 Aug Edited by B. Lehnert Cambridge university press,Cambridge, 1958
Eng 41.4475

INTERNATIONAL ASTRONOMICAL UNION.SYMPOSIUM, 6
Electromagnetic phenomena in cosmical physics Stockholm 1956 Aug 27-31 International astronomical union Edited by B. Lehnert In co-operation with the International union of theoretical and applied mechanics 13,544p Cambridge university press,Cambridge,1958
Nap 11.0916

INTERNATIONAL ASTRONOMICAL UNION.SYMPOSIUM, 7
Co-ordination of galactic research : conference 2nd Saltsjobaden 1957 Jun. 17-22 International astronomical union Edited by A. Blaauw and others 93p Cambridge university press,Cambridge,1959
Obs 6.0192

INTERNATIONAL ASTRONOMICAL UNION.SYMPOSIUM, 8
Symposium on cosmical gas dynamics 3rd proceedings Cambridge,Mass. 1957 Jun.24-29 International union of theoretical and applied mechanics and International astronomical union Edited by J.M. Burgers and others Financial grant from the Smithsonian institution Smithsonian Institution. Smithsonian symposium publications, 1 American physical society,New York,1958 Reviews of modern physics.Vol.30,no.3,July 1958
Obs 6.0306

INTERNATIONAL ASTRONOMICAL UNION.SYMPOSIUM, 9
Paris symposium on radio astronomy Paris 1958 Jul.30-Aug.6 International astronomical union and International scientific radio union Edited by Ronald N. Bracewell International scientific radio union.Symposium, 1 612p Stanford university press, Stanford,Calif.,1959
Obs 6.0246

INTERNATIONAL ASTRONOMICAL UNION.SYMPOSIUM, 9
Paris symposium on radio astronomy Paris 1958 Jul.30-Aug.6 International astronomical union and International scientific radio union Edited by Ronald N. Bracewell International scientific radio union.Symposium, 1 612p Stanford university press, Stanford,Calif,1959
A Math 4.1151

INTERNATIONAL ASTRONOMICAL UNION.SYMPOSIUM, 9
Paris symposium on radio astronomy Paris 1958 Jul 30-Aug 6 International astronomical union International scientific radio union Edited by Ronald N. Bracewell International scientific radio union.Symposium, 1 612p Stanford university press,Stanford,Calif.,1959
TA 15.0638

INTERNATIONAL ASTRONOMICAL UNION.SYMPOSIUM, 9
Paris symposium on radio astronomy Paris 1958 Jul 30-Aug 6 International astronomical union and International scientific radio union Edited by Ronald N. Bracewell International scientific radio union.Symposium, 1 612p Stanford university press, Stanford,Calif,1959
Cav 7.0301

INTERNATIONAL ASTRONOMICAL UNION.SYMPOSIUM, 10
The Hertzsprung-Russell diagram :a symposium Proceedings Moscow 1958 Aug.15-16 International astronomical union Edited by Jesse L. Greenstein 128p Paris,1959 Reprinted from Annales d'astrophysique 1959, supplements.Fascicule.No.8
Obs 6.0687

INTERNATIONAL ASTRONOMICAL UNION.SYMPOSIUM, 11
The Rotation of the earth and atomic time standards :a symposium Moscow 1958 Aug International astronomical union Edited by Dirk Brouwer 1959 Reprinted from the Astronomical journal,vol 64,no 1268
Obs 6.3199

INTERNATIONAL ASTRONOMICAL UNION.SYMPOSIUM, 12
Aerodynamic phenomena in stellar atmospheres Symposium on cosmical gas dynamics held at the International school of physics "Enrico Fermi" 4th proceedings Varenna 1960 Aug.18-30 International astronomical union and International union of theoretical and applied mechanics Edited by R.N. Thomas and others Under the auspices of Societa Italiana di fisica 515p Bologna, c1961 Reprinted from Supplemento Nuovo Cimento,22,no.1,1961
A Math 4.1152

INTERNATIONAL ASTRONOMICAL UNION.SYMPOSIUM, 13
International latitude service L' Avenir du service international des latitudes :a symposium proceedings Helsinki 1960 Jul. 26-Aug.7 Edited by Paul Melchior Organized by International union of geodesy and geophysics 97p Paris,1961 Reprinted from Bulletin geodesique no.59,March 1961
Obs 6.1198

INTERNATIONAL ASTRONOMICAL UNION.SYMPOSIUM, 14
The Moon :a symposium Pulkovo 1960 Dec. International astronomical union Edited by Zdenek Kopal and Zdenka Kadla Mikhailov Held at Glavnaya astronomicheskaya observatoriya, Pulkovo 571p Academic press,London;New York,1962
A Math 4.1153

INTERNATIONAL ASTRONOMICAL UNION.SYMPOSIUM, 15
Extra-galactic research The Symposium on problems of extra-galactic research 15th Santa Barbara,Calif. 1961 Aug.10-12 International astronomical union Edited by G. C. McVittie 450p Macmillan,New York,1962
Obs 6.1124

INTERNATIONAL ASTRONOMICAL UNION.SYMPOSIUM, 15
Problems of extra-galactic research :a symposium Santa Barbara,Calif. 1961 Aug.10-12 International astronomical union Edited by G.C. McVittie 450p Macmillan,New York, 1962
A Math 4.1154

INTERNATIONAL ASTRONOMICAL UNION.SYMPOSIUM, 15
Symposium on problems of extra-galactic research Proceedings Santa Barbara,Calif. 1961 Aug 10-12 International astronomical union Edited by G.C. McVittie McMillan,New York,1962
TA 15.0035

INTERNATIONAL ASTRONOMICAL UNION.SYMPOSIUM, 16
The Solar corona :a symposium proceedings Cloudcroft,N.M. 1961 Aug.28-30 International astronomical union Edited by John W. Evans Sponsored by the Air Force Cambridge research laboratories 344p Academic press,New York;London,1963
Obs 6.0564

INTERNATIONAL ASTRONOMICAL UNION.SYMPOSIUM, 18
Theoretical interpretation of upper atmosphere emissions :a symposium Paris 1962 Jun.25-29 International astronomical union and International union of geodesy and geophysics Edited by D.R. Bates 264p Pergamon press,New York,1963 Reprinted from'Planetary and space science',Vol.10 1963
Obs 6.0142

INTERNATIONAL ASTRONOMICAL UNION.SYMPOSIUM, 19
Le Choix des sites d'observatoires astronomiques (site testing) :a symposium proceedings Rome 1962 Oct 1-2 Edited by J. Rosch and others 382p Gauthier-Villars, Paris,1964 Reprinted from 'Bulletin astronomique de l'observatoire de Paris, vol 24, fasc.2 et 3, 1963
Obs 6.1467

INTERNATIONAL ASTRONOMICAL UNION.SYMPOSIUM, 20
The Galaxy and the magellanic clouds :a symposium Canberra 1963 Mar.18-28 International astronomical union and International scientific radio union Edited by F.J. Kerr and A.W. Rodgers 393p Australian academy of science,Canberra,1964 Dedicated to Joseph Lade Pawsey
A Math 4.1158

INTERNATIONAL ASTRONOMICAL UNION.SYMPOSIUM, 20
The Galaxy and the magellanic clouds :a symposium Proceedings Canberra 1963 Mar. 18-28 International astronomical union and International scientific radio union Edited by F.J. Kerr and A.W. Rodgers 393p Australian academy of science,Canberra,1964 Publication dedicated to memory of Joseph Lade Pawsey
Obs 6.0941

INTERNATIONAL ASTRONOMICAL UNION.SYMPOSIUM, 21
The System of astronomical constants :a symposium proceedings Paris 1963 May 27-31 International astronomical union Edited by J. Kovalevsky 330p Gauthier-Villars, Paris,1965 From the Bulletin astronomique de l'observatoire de Paris,vol.25,fasc.1-3
Obs 6.0986

INTERNATIONAL ASTRONOMICAL UNION.SYMPOSIUM, 22
Stellar and solar magnetic fields Symposium on stellar and solar magnetic fields Munich 1963 Sep.3-10 International astronomical union Edited by R. Lust 460p North Holland,Amsterdam,1965
Obs 6.1148

INTERNATIONAL ASTRONOMICAL UNION.SYMPOSIUM, 22
Stellar and solar magnetic fields :a symposium Rottach-Egern 1963 Sep.3-10 International astronomical union Edited by R. Lust 460p North-Holland,Amsterdam,1965
A Math 4.1160

INTERNATIONAL ASTRONOMICAL UNION.SYMPOSIUM, 24
Special classification and multicolour photometry Saltsjobaden 1964 Aug 17-21 International astronomical union Edited by K. Loden Academic press,London;New York,1966
TA 15.0319

INTERNATIONAL ASTRONOMICAL UNION.SYMPOSIUM, 24
Spectral classification and multicolour photometry Symposium on the spectral classification and multicolour photometry Proceedings Saltsjobaden 1964 Aug.17-21 International astronomical union Edited by K. Loden and others 383p Academic press, London;New York,1966
Obs 6.1092

INTERNATIONAL ASTRONOMICAL UNION.SYMPOSIUM, 24
Spectral classification and multicolour photometry :a symposium Saltsjobaden,Sweden 1964 Aug.17-21 International astronomical union Edited by K. Loden and others 383p Academic press,London;New York,1966
A Math 4.1166

INTERNATIONAL ASTRONOMICAL UNION.SYMPOSIUM, 25
The Theory of orbits in the solar system and in stellar systems Thessalonika 1964 Aug 17-22 International astronomical union Edited by George Contopoulos Academic press, London;New York,1966
TA 15.0332

INTERNATIONAL ASTRONOMICAL UNION.SYMPOSIUM, 25
The Theory of orbits in the solar system and in stellar systems a symposium Thessaloniki 1964 Aug.17-22 International astronomical union Edited by George Contopoulos Organized in co-operation with Committee on space research 380p Academic press,London;New York,1966
Obs 6.0409

INTERNATIONAL ASTRONOMICAL UNION.SYMPOSIUM, 25
The Theory of orbits in the solar system and in stellar systems :a symposium Thessaloniki 1964 Aug.17-22 International astronomical union Edited by George Contopoulos Organized in co-operation with Committee on space research 380p Academic press,London;New York,1966
A Math 4.1167

INTERNATIONAL ASTRONOMICAL UNION.SYMPOSIUM, 26
Abundance determinations in stellar spectra a symposium Utrecht 1964 Aug.10-14 International astronomical union Edited by Hans Hubenet 374p London;New York,1966
Obs 6.0843

INTERNATIONAL ASTRONOMICAL UNION.SYMPOSIUM, 26
Abundance determinations in stellar spectra a symposium Proceedings Utrecht 1964 Aug 10-14 International astronomical union Academic press,London,1966
TA 15.0034

INTERNATIONAL ASTRONOMICAL UNION.SYMPOSIUM, 26
Abundance determinations in stellar spectra a symposium Utrecht 1964 Aug.10-14 International astronomical union Edited by Hans Hubenet 374p Academic press,London; New York,1966
A Math 4.1168

INTERNATIONAL ASTRONOMICAL UNION.SYMPOSIUM, 27
The Construction of large telescopes :a symposium International astronomical union Edited by D.L. Crawford Academic press,New York,1966
TA 15.0311

INTERNATIONAL ASTRONOMICAL UNION.SYMPOSIUM, 27
The Construction of large telescopes :a symposium Tucson,Ariz. 1965 Apr 5-12 Pasadena,Calif. International astronomical union Edited by D.L. Crawford 234p Academic press,New York;London,1966
Obs 6.3203

INTERNATIONAL ASTRONOMICAL UNION.SYMPOSIUM, 28
Aerodynamical phenomena in stellar atmosphwre Nice 1965 Sep International astronomical union Edited by Richard N. Thomas Academic press,London,1967
TA 15.0411

INTERNATIONAL ASTRONOMICAL UNION.SYMPOSIUM, 29
Non-stable phenomena in galaxies Nestasionarnye yavleniya v galaktikakh Byurakan 1966 May 4-12 International astronomical union 480p Akademiya nauk Armyanskey S.S.R.,Erevan,1968
Obs 6.3269

INTERNATIONAL ASTRONOMICAL UNION.SYMPOSIUM, 30
Determination of radial velocities and their applications Toronto 1966 Jun 20-24 International astronomical union and International council of scientific unions Edited by A.H. Batten and J.F. Heard Academic press,London;New York,1967
TA 15.0307

INTERNATIONAL ASTRONOMICAL UNION.SYMPOSIUM, 30
Determination of radial velocities and their applications :a symposium Toronto 1966 Jun.20-24 International astronomical union Edited by A.H. Batten and J.F. Heard 262p academic press,London,1967 publication dedicated to the memory of R. Methven Petrie
Obs 6.0147

INTERNATIONAL ASTRONOMICAL UNION.SYMPOSIUM, 31
Radio astronomy and the galactic system Symposium on radio astronomy and the galactic system Noordwijk 1966 Aug 25-Sep 1 International astronomical union Edited by Hugo van Woerden Organised in co-operation with International scientific radio union xviii,501p Academic press,London;New York, 1967
A Math 4.1574

INTERNATIONAL ASTRONOMICAL UNION.SYMPOSIUM, 32
Continental drift,secular motion of the pole,and rotation of the earth Stresa 1967 Mar 21-25 International astronomical union and International union of geodesy and geophysics Edited by Wm. Markowitz and B. Guinot Reidel,Dordrecht,1968
TA 15.0424

INTERNATIONAL ASTRONOMICAL UNION.SYMPOSIUM, 32
Continental drift,secular motion of the pole,and rotation of the earth :a symposium Proceedings Stresa 1967 Mar International astronomical union International union of geodesy and geophysics Edited by W. Markowitz and B. Guinot 108p Reidel,Dordrecht,1968
Obs 6.3268

INTERNATIONAL ASTRONOMICAL UNION.SYMPOSIUM, 32
Stresa symposium on continental drift, secular motion of the pole and rotation of the earth Stresa,Italy 1967 Mar 21-25 International astronomical union and International union of geodesy and geophysics Edited by W. Markowitz and B. Guinot Reidel, Dordrecht,1968
A Math 4.1308

INTERNATIONAL ASTRONOMICAL UNION.SYMPOSIUM, 33
Physics and dynamics of meteors Tatranska Lomnica 1967 Sep 4-7 International astronomical union Edited by Lubor Kresak and Peter M. Millman Reidel,Dordrecht,1968
TA 15.0422

INTERNATIONAL ASTRONOMICAL UNION.SYMPOSIUM, 33
Physics and dynamics of meteors :a symposium Proceedings Tatranska Lomnica 1967 International astronomical union Edited by Lubor Kresak and Peter M. Millman 525p Reidel,Dordrecht,1968
Obs 6.3264

INTERNATIONAL ASTRONOMICAL UNION.SYMPOSIUM, 34
Planetary nebulae :a symposium Proceedings Tatranska Lomnica 1967 International astronomical union Edited by D. E. Osterbrock and C.R. O'Dell 469p Reidel, Dordrecht,1968
Obs 6.3265

INTERNATIONAL ASTRONOMICAL UNION.SYMPOSIUM, 34
Planetary nebulae :a symposium Tatranska Lomnica 1967 International astronomical union Edited by D.E. Osterbrock and C.R. O'Dell Reidel,Dordrecht,1968
TA 15.0320

INTERNATIONAL ASTRONOMICAL UNION.SYMPOSIUM, 35
Structure and development of solar active regions Budapest 1967 Sep 4-8 International astronomical union Edited by K. O. Kiepenheuer Reidel,Dordrecht,1968
TA 15.0420

INTERNATIONAL ASTRONOMICAL UNION.SYMPOSIUM, 35
Structure and development of solar active regions :a symposium Proceedings Budapest 1967 Sep International astronomical union Edited by K.O. Kiepenheuer 608p Reidel, Dordrecht,1968
Obs 6.3266

INTERNATIONAL ASTRONOMICAL UNION.SYMPOSIUM, 36
Ultraviolet stellar spectra and related ground-based observations Lunteren 1969 Jun 24-27 International astronomical union Edited by L. Houziaux and H.E. Butler 361p Reidel,Dordrecht,1970
Obs 6.3321

INTERNATIONAL ASTRONOMICAL UNION.SYMPOSIUM, 37
Non-solar X- and gamma-ray astronomy Rome 1969 May 8-10 International astronomical union Edited by L. Gratton 425p Reidel, Dordrecht,190
Obs 6.3322

INTERNATIONAL ASTRONOMICAL UNION.SYMPOSIUM, 37
Non-solar X- and gamma-ray astronomy : symposium Papers Rome 1969 May 8-10 International astronomical union Edited by L. Gratton 425p Reidel,Dordrecht,1970
TA 15.0492

INTERNATIONAL ASTRONOMICAL UNION.SYMPOSIUM, 38
The Spiral structure of our galaxy Basle 1969 Aug 29-Sep 4 International astronomical union Edited by W. Becker and G. Contopoulos 478p Reidel,Dordrecht,1970
Obs 6.3320

INTERNATIONAL ASTRONOMICAL UNION.SYMPOSIUM, 38
The Spiral structure of our galaxy : symposium Proceedings Basle 1969 Aug 29-Sep 4 International astronomical union Edited by W. Becker and G. Contopoulos 478p Reidel,Dordrecht,1970
TA 15.0489

INTERNATIONAL ASTRONOMICAL UNION.SYMPOSIUM, 39
Interstellar gas dynamics Symposium on cosmical gas dynamics 6th Proceedings Yalta 1969 Sep 8-18 International astronomical union International union of theoretical and applied mechanics 388p Reidel,Dordrecht,1970
TA 15.0541

INTERNATIONAL ASTRONOMICAL UNION.SYMPOSIUM, 39
Interstellar gas dynamics Symposium on cosmical gas dynamics 6th Proceedings Yalta 1969 Sep 8-18 International astronomical union International union of theoretical and applied mechanics 388p Reidel,Dordrecht,1970
Obs 6.3323

INTERNATIONAL ASTRONOMICAL UNION.SYMPOSIUM, 39
Interstellar gas dynamics Symposium on cosmical gas dynamics 6th Proceedings Yalta 1969 Sep 8-18 International union of theoretical and applied mechanics and International astronomical union Edited by H. J. Habing diagrs,photos,graphs xii,388p 24cm Reidel,Dordrecht,1970
Cav 7.2710

INTERNATIONAL ASTRONOMICAL UNION.SYMPOSIUM, 40
Planetary atmospheres Marfa,Texas 1969 Oct 26-31 International astronomical union Edited by C. Sagan and others 408p Reidel, Dordrecht,1971
Obs 6.3586

INTERNATIONAL ASTRONOMICAL UNION.SYMPOSIUM, 41
New techniques in space astronomy Munich 1970 Aug 10-14 International astronomical union Edited by F. Labuhn and R. Lust 418p Reidel,Dordrecht,1971
Obs 6.3590

INTERNATIONAL ASTRONOMICAL UNION.SYMPOSIUM, 42
White dwarfs St.Andrews 1970 Aug 11-13 Edited by W.J. Luyten 164p Reidel, Dordrecht,1971
TA 15.0573

INTERNATIONAL ASTRONOMICAL UNION.SYMPOSIUM, 43
Solar magnetic fields Paris 1970 Aug 31-Sep 4 International astronomical union Edited by R. Howard 782p Reidel,Dordrecht, 1971
Obs 6.3589

INTERNATIONAL ASTRONOMICAL UNION.SYMPOSIUM, 43
Solar magnetic fields Paris 1970 Aug 31-Sep 4 International astronomical union Edited by R. Howard 782p Reidel,Dordrecht, 1971
TA 15.0688

INTERNATIONAL ASTRONOMICAL UNION.SYMPOSIUM, 44
External galaxies and quasi-stellar objects Upsala 1970 Aug 10-14 International astronomical union Edited by D.S. Evans 549p Reidel,Dordrecht,1972
TA 15.0658

INTERNATIONAL ASTRONOMICAL UNION.SYMPOSIUM, 45
The Motion,evolution of orbits,and origin of comets Leningrad 1970 Aug 4-11 International astronomical union Edited by G. A. Chebotarev and others 521p Reidel, Dordrecht,1972
TA 15.0687

INTERNATIONAL ASTRONOMICAL UNION.SYMPOSIUM, 45
The Motion,evolution of orbits,and origin of comets Leningrad 1970 Aug 4-11 International astronomical union Edited by G. A. Chebotarev and others 521p Reidel, Dordrecht,1972
Obs 6.3591

INTERNATIONAL ASTRONOMICAL UNION.SYMPOSIUM, 46
The Crab nebula Jodrell Bank 1970 Aug 5-7 International astronomical union Edited by R.D. Davies and F.G. Smith 470p Reidel, Dordrecht,1971
Obs 6.3588

INTERNATIONAL ASTRONOMICAL UNION.SYMPOSIUM, 46
The Crab nebula Jodrell Bank 1970 Aug 5-7 International astronomical union Edited by R.D. Davies and F.G. Smith 470p Reidel, Dordrecht,1971
TA 15.0689

INTERNATIONAL ASTRONOMICAL UNION.SYMPOSIUM SERIES, 2
Gas dynamics of cosmic clouds Symposium on cosmical gas dynamics 2nd Cambridge 1953 Jul 6-11 International union of theoretical and applied mechanics International astronomical union North-Holland,Amsterdam,1955
Eng 41.4468

INTERNATIONAL ASTRONOMICAL UNION 14TH ASSEMBLY
Lunar nomenclature Report of the working group of I.A.U. commission 17:the moon: proposed names for craters on the moon's far side...presented at the 14th assembly... Brighton 1970 International astronomical union.Commission 17 43p Harvard college observatory,Cambridge,Mass.,1970
Duplicated
Obs 6.3342

INTERNATIONAL ATOMIC ENERGY AGENCY
Analytical control of radiopharmaceutics :a panel Proceedings Vienna 1969 International atomic energy agency.Panel proceedings series International atomic energy agency,Vienna,1970
Radioth 35.1385

INTERNATIONAL ATOMIC ENERGY AGENCY
Assessment of radioactivity in man Symposium on the assessment of radioactive body burdens in man Proceedings Heidelberg 1964 May 11-16 1-2 International atomic energy agency.Proceedings series International atomic energy agency,Vienna,1964
Radioth 35.1373

INTERNATIONAL ATOMIC ENERGY AGENCY
Biological effects of ionizing radiation at the molecular level Symposium on the biological effects of ionizing radiation at the molecular level Proceedings Brno 1962 Jul 2-6 International atomic energy agency.Proceedings series I.A.E.A.,Vienna, 1962
Radioth 35.1976

INTERNATIONAL ATOMIC ENERGY AGENCY
Biological effects of neutron and protein irradiations Symposium on biological effects of neutron and proton irradiations Proceedings Upton,N.Y. 1963 Oct 7-11 1-2 Held at Brookhaven national laboratory International atomic energy agency.Proceedings series 2 vols International atomic energy agency,Vienna,1964
Radioth 35.1168

INTERNATIONAL ATOMIC ENERGY AGENCY
Biophysical aspects of radiation quality Panel on biophysical aspects of radiation quality 2nd Report Vienna 1967 Apr 10-14 International atomic energy agency.Panel proceedings series International atomic energy agency,Vienna,1968
Radioth 35.1422

INTERNATIONAL ATOMIC ENERGY AGENCY
Biophysical aspects of radiation quality Panel on biophysical aspects of radiation quality Report Vienna 1965 Mar 29-Apr 2 International atomic energy agency. Technical report series, 58 International atomic energy agency,Vienna,1966
Radioth 35.1421

INTERNATIONAL ATOMIC ENERGY AGENCY Chemical effects of nuclear transformations :a symposium Proceedings Prague 1960 Oct 24-27 Vol 1-2 International atomic energy agency.Proceedings series STI-PUB 34 2 vols international atomic energy agency, Vienna,1961
Chem 18.1150

INTERNATIONAL ATOMIC ENERGY AGENCY Chemical effects of nuclear transformations :symposium on chemical effects associated with nuclear reactions and radioactive transformations Proceedings Vienna 1964 Dec 7-11 Vol 1-2 International atomic energy agency.Proceedings series STI-PUB 91 International atomic energy agency,Vienna,1965
Chem 18.1151

INTERNATIONAL ATOMIC ENERGY AGENCY Clinical uses of whole-body counting Panel on clinical uses of whole-body counting Proceedings Vienna 1965 Jun 28-Jul 2 International atomic energy agency.Panel proceedings series International atomic energy agency,Vienna,1966
Radioth 35.1383

INTERNATIONAL ATOMIC ENERGY AGENCY Corrosion of reactor materials the conference Proceedings Salzburg 1962 Jun 4-8 Vol 1-2 2 vols International atomic energy agency,Vienna,1962
Met 25.1937

INTERNATIONAL ATOMIC ENERGY AGENCY Diagnosis and treatment of radioactive poisoning Scientific meeting on the diagnosis and treatment of radioactive poisoning Proceedings Vienna 1962 Oct 15-18 International atomic energy agency.Proceedings series International atomic energy agency, Vienna,1963
Radioth 35.1368

INTERNATIONAL ATOMIC ENERGY AGENCY Economics in managing radioactive wastes :report resulting from two panels of experts on the economics of radioactive waste management International atomic energy agency.Technical reports series, 83 International atomic energy agency,Vienna,1968 Panels convened on 13-17 December 1965 and 17-21 October 1966
Radioth 35.1776

INTERNATIONAL ATOMIC ENERGY AGENCY Effects of ionizing radiation on the nervous system Symposium on ionizing radiation on the nervous system Proceedings Vienna 1961 Jun 5-9 International atomic energy agency.Proceedings series International atomic energy agency, Vienna,1962
Radioth 35.1130

INTERNATIONAL ATOMIC ENERGY AGENCY Effects of ionizing radiation on the nervous system Symposium on the effects of ionizing radiation on the nervous system Vienna 1961 Jun 5-9 International atomic energy agency,Vienna,1962
An 32.4190

INTERNATIONAL ATOMIC ENERGY AGENCY Effects of ionizing radiations on the haematopoietic tissue Panel on the effects of ionizing radiations from different sources on haematopoietic tissue Proceedings Vienna 1966 May 17-20 International atomic energy agency.Panel proceedings series International atomic energy agency,Vienna,1967
Radioth 35.1200

INTERNATIONAL ATOMIC ENERGY AGENCY Effects of radiation on cellular proliferation and differentiation Symposium on the effects of radiation on cellular proliferation and differentiation Proceedings Monaco 1968 Apr 1-5 Organized in co-operation with the Joint commission on applied radioactivity International atomic energy agency,Vienna,1968
Radioth 35.1760

INTERNATIONAL ATOMIC ENERGY AGENCY Effects of radiation on meiotic systems Study group on the effects of radiation on meiotic systems Vienna 1967 May 8-11 International atomic energy agency.Panel proceedings series International atomic energy agency,Vienna,1968
Radioth 35.1201

INTERNATIONAL ATOMIC ENERGY AGENCY Elementary particle theories Internationale universitatswochen fur Kernphysik 5th proceedings Schladming 1966 Feb.24-Mar.9 Edited by Paul Urban Acta physica austriaca. Supplementum, 3 Springer-verlag,Vienna;New York,1966
A Math 4.0889

INTERNATIONAL ATOMIC ENERGY AGENCY Fuel element fabrication with special emphasis on cladding materials:a symposium Proceedings Vienna 1960 May 10-13 Vol 1-2 2 vols Academic press,London;New York,1961
Met 25.1547

INTERNATIONAL ATOMIC ENERGY AGENCY Genetical aspects of radiosensitivity :mechanisms of repair Vienna 1966 Apr 18-22 International atomic energy agency.Proceedings series illus 175p 24cm International atomic energy agency,Vienna,1966
Bot 42.0797

INTERNATIONAL ATOMIC ENERGY AGENCY Genetical aspects of radiosensitivity:mechanisms of repair :a panel Proceedings Vienna 1966 Apr 18-22 International atomic energy agency. Panel proceedings series International atomic energy agency,Vienna,1966
Radioth 35.1194

INTERNATIONAL ATOMIC ENERGY AGENCY High-energy physics and elementary particles Seminar on elementary particles lectures Trieste 1965 May 3-Jun.30 International atomic energy agency.Proceedings series STIPUB117 Vienna,1965
A Math 4.0888

INTERNATIONAL ATOMIC ENERGY AGENCY Immediate and low level effects of ionizing radiations : symposium Proceedings Venice 1959 Jun 22-26 Under the auspices of Unesco International journal of radiation biology. Supplement Taylor and Francis,London,1960
Gen 34.2221

INTERNATIONAL ATOMIC ENERGY AGENCY In vitro procedures with radioisotopes in medicine : symposium Proceedings Vienna 1969 Sep 8-12 IAEA,Vienna,1970
Bioch 33.2184

INTERNATIONAL ATOMIC ENERGY AGENCY Inelastic scattering of neutrons Symposium on inelastic scattering of neutrons Bombay 1964 Dec 15-19 Vol 1-2 Edited by M.M. Brown International atomic energy agency, Vienna,1965
Cav 7.0267

INTERNATIONAL ATOMIC ENERGY AGENCY Inelastic scattering of neutrons in solids and liquids : a symposium Proceedings Vienna 1960 Oct 11-14 Vienna,1961
Cav 7.2392

INTERNATIONAL ATOMIC ENERGY AGENCY Lattice dynamics The International conference on lattice dynamics Proceedings Copenhagen 1963 Aug 5-9 International union of pure and applied physics and University of Pennsylvania Edited by R.F. Wallis 730p Pergamon press,Oxford,1965
Cav 7.0542

INTERNATIONAL ATOMIC ENERGY AGENCY Medical radiation physics Joint IAEA-WHO expert committee on medical radiation physics Report Geneva 1967 Dec 12-18 World health organization.Technical report series, 390 World health organization,Geneva,1968
Radioth 35.1381

INTERNATIONAL ATOMIC ENERGY AGENCY Medical radioisotope scanning Symposium on medical radioisotope scanning Proceedings Athens 1964 Apr 20-24 International atomic energy agency.Proceedings series International atomic energy agency,Vienna,1964
Radioth 35.1376

INTERNATIONAL ATOMIC ENERGY AGENCY Medical radioisotope scanning :a seminar Proceedings Vienna 1959 Feb 25-27 International atomic energy agency,Vienna,1959
Radioth 35.1375

INTERNATIONAL ATOMIC ENERGY AGENCY Medical radioisótope scintigraphy :a symposium Proceedings Salzburg 1968 Aug 6-15 Vol 1-2 Edited by G.R. Stevenson International atomic energy agency.Proceedings series 2 vols International atomic energy agency,Vienna,1969
Radioth 35.1392

INTERNATIONAL ATOMIC ENERGY AGENCY Medical uses of CA47 :a panel Papers 1961 Dec 6-8 International atomic energy agency.Technical report series, 10 International atomic energy agency,Vienna,1962
Radioth 35.1363

INTERNATIONAL ATOMIC ENERGY AGENCY Meteorite research Symposium on meteorite research Proceedings Vienna 1968 Aug 7-13 Edited by Peter M. Millmann Astrophysics and space science library Reidel,Dordrecht,1969
Min 10.1442

INTERNATIONAL ATOMIC ENERGY AGENCY Mutations in plant breeding Panel on co-ordination of research on the production and use of induced mutations in plant breeding Proceedings Vienna 1966 Jan 17-21 International atomic energy agency.Panel proceedings series 2 vols International atomic energy agency, Vienna,1966-68
Radioth 35.0261

INTERNATIONAL ATOMIC ENERGY AGENCY Mutations in plant breeding II :a research co-ordination meeting on the use of induced mutations in plant breeding Proceedings Vienna 1967 Sep 11-15 International atomic energy agency. Panel proceedings series I.A.E.A.,Vienna, 1968
Radioth 35.0262

INTERNATIONAL ATOMIC ENERGY AGENCY Neutron dosimetry Symposium on neytron detection, dosimetry and standardization Proceedings Harwell 1962 Dec 10-14 1-2 International atomic energy agency.Proceedings series 2 vols International atomic energy agency,Vienna,1963
Radioth 35.1695

INTERNATIONAL ATOMIC ENERGY AGENCY Neutron monitoring Symposium on neutron monitoring for radiological protection Proceedings Vienna 1966 Aug 29-Sep 2 International atomic energy agency.Proceedings series International atomic energy agency,Vienna,1967
Radioth 35.1672

INTERNATIONAL ATOMIC ENERGY AGENCY Nuclear accident dosimetry systems Panel on nuclear accident dosimetry systems Proceedings Vienna 1969 Feb 17-21 International atomic energy agency.Panel proceedings series International atomic energy agency,Vienna,1970
Radioth 35.1384

INTERNATIONAL ATOMIC ENERGY AGENCY Nuclear science teaching Panel on nuclear science teaching Report Bangkok 1968 Jul 15-23 International atomic energy agency.Technical reports series, 94 International atomic energy agency,Vienna,1968
Radioth 35.1777

INTERNATIONAL ATOMIC ENERGY AGENCY Nuclear theory International summer school on selected topics in nuclear theory lectures Low Tatra mountains,Czechoslovakia 1962 Aug 20-Sep 8 By N. Austern and others Ceskoslovenska akademie ved.Ustav jaderneho vyzkumu Edited by F. Janouch 452p International atomic energy agency,Vienna,1963
A Math 4.0885

INTERNATIONAL ATOMIC ENERGY AGENCY Planning of radiotherapy facilities Joint IAEA-WHO meeting on planning of radiotherapy facilities Report Geneva 1964 Dec 15-19 World health organization.Technical report series, 328 World health organization,Geneva,1966
Radioth 35.1060

INTERNATIONAL ATOMIC ENERGY AGENCY Production and use of short-lived radioisotopes from reactors Seminar on the practical applications of short-lived radioisotopes produced in small research reactors Proceedings Vienna 1962 Nov 5-9 1-2 International atomic energy agency. Proceedings series 2 vols International atomic energy agency,Vienna,1963
Radioth 35.1369

INTERNATIONAL ATOMIC ENERGY AGENCY Radiation and the control of immune response Panel on radiation and the control of immune response Proceedings Paris 1967 Jun 22-24 International atomic energy agency.Panel proceedings series International atomic energy agency,Vienna,1968
Radioth 35.1202

INTERNATIONAL ATOMIC ENERGY AGENCY Radiation damage and sulphydryl compounds Panel on radiation damage to the biological molecular information system :with special regard to the role of SH-groups Proceedings Vienna 1968 Oct 21-25 International atomic energy agency.Panel proceedings series International atomic energy agency,Vienna,1969
Radioth 35.1764

INTERNATIONAL ATOMIC ENERGY AGENCY Radiation damage in bone Conference on the relation of radiation damage to radiation dose in bone Report and summarized papers Oxford 1960 Apr 10-14 International atomic energy agency, Vienna,1960
Radioth 35.1115

INTERNATIONAL ATOMIC ENERGY AGENCY Radiation damage in solids and reactor materials :a symposium Proceedings Venice 1962 May 7-11 Vol 1-3 3 vols International atomic energy agency,Vienna,1962
Cav 7.2393

INTERNATIONAL ATOMIC ENERGY AGENCY Radiation sensitivity of toxins and animal poisons Panel on the radiation sensitivity of toxins and animal poisons Proceedings Bangkok 1969 May 19-22 International atomic energy agency.Panel proceedings series International atomic energy agency,Vienna,1970
Radioth 35.1912

INTERNATIONAL ATOMIC ENERGY AGENCY Radiation sensitivity of toxins and animal poisons Panel on the radiation sensitivity of toxins and animal poisons Proceedings Bangkok 1969 May 19-22 International atomic energy agency.Panel proceedings series International atomic energy agency,Vienna,1970
Radioth 35.1765

INTERNATIONAL ATOMIC ENERGY AGENCY Radiation-induced cancer :a symposium Proceedings Athens 1969 Apr 28-May 2 International atomic energy agency,Vienna,1969
Radioth 35.1917

INTERNATIONAL ATOMIC ENERGY AGENCY Radioactive dating :symposium proceedings Athens 1962 Nov 19-23 International atomic energy agency,Vienna,1963
Geol 8.4477

INTERNATIONAL ATOMIC ENERGY AGENCY Radioactive dating and methods of low level counting :symposium proceedings Monaco 1967 Mar 2-10 International atomic energy agency,Vienna,1967
Geol 8.4478

INTERNATIONAL ATOMIC ENERGY AGENCY Radioactive dating and methods of low-level counting :a symposium Proceedings Monaco 1967 Mar 2-10 International atomic energy agency.Proceedings series International atomic energy agency,Vienna,1967
Radioth 35.1683

INTERNATIONAL ATOMIC ENERGY AGENCY Radioisotope sample measurement techniques in medicine and biology Symposium on radioisotope sample measurement techniques in medicine and biology Proceedings Vienna 1965 May 24-25 International atomic energy agency.Proceedings series International atomic energy agency,Vienna,1965
Radioth 35.1377

INTERNATIONAL ATOMIC ENERGY AGENCY Radioisotope techniques in the study of protein metabolism Radioisotope techniques in the study of protein metabolism :a panel Vienna 1964 Jun 1-5 International atomic energy agency.Technical reports series, 45 International atomic energy agency,Vienna,1965
Radioth 35.1150

INTERNATIONAL ATOMIC ENERGY AGENCY
Radioisotopes and radiation in entomology :a symposium Proceedings Bombay 1960 Dec 5-9 By H.J. Bhabha and others International atomic energy agency,Vienna,1962
Bal 39.2390

INTERNATIONAL ATOMIC ENERGY AGENCY
Radioisotopes in soil plant nutrition studies The Symposium on the use of radioisotopes in soil-plant nutrition studies Proceedings Bombay 1962 Feb 26-Mar 2 International atomic energy agency,Vienna,1962
Bot 42.1368

INTERNATIONAL ATOMIC ENERGY AGENCY
Radioisotopes in tropical medicine Symposium on the use of radioisotopes in the study of endemic and tropical diseases Proceedings Bangkok 1960 Dec 12-16 International atomic energy agency.Proceedings series International atomic energy agency,Vienna,1962
Radioth 35.1365

INTERNATIONAL ATOMIC ENERGY AGENCY
Radiological health and safety in mining and milling of nuclear materials Symposium on radiological health and safety in mining and milling of nuclear materials :a symposium Proceedings Vienna 1963 Aug 26-31 1-2 In co-operation with the World health organization International atomic energy agency.Proceedings series International atomic energy agency,Vienna,1964
Radioth 35.1164

INTERNATIONAL ATOMIC ENERGY AGENCY Research applications of nuclear pulsed systems Panel on research applications of repetitively pulsed reactors and boosters Proceedings Dubna,U.S.S.R. 1966 Jul 18-22 International atomic energy agency.Panel proceedings series International atomic energy agency,Vienna,1967
Radioth 35.1682

INTERNATIONAL ATOMIC ENERGY AGENCY Rice breeding with induced mutations :an FAO-IAEA research co-ordination meeting on the use of induced mutations in rice breeding Report Taipei 1967 Jun 5-9 International atomic energy agency.Technical report series, 86 International atomic energy agency,Vienna,1968
Radioth 35.0271

INTERNATIONAL ATOMIC ENERGY AGENCY Role of computers in radiotherapy Panel on the role of computers in radiotherapy Report Vienna 1967 Jul 10-14 International atomic energy agency.Panel proceedings series bibliog. International atomic energy agency,Vienna,1968
Radioth 35.1199

INTERNATIONAL ATOMIC ENERGY AGENCY Selected topics in radiation dosimetry Symposium on selected topics in radiation dosimetry Proceedings Vienna 1960 Jun 7-11 International atomic energy agency.Proceedings series International atomic energy agency, Vienna,1961
Radioth 35.1694

INTERNATIONAL ATOMIC ENERGY AGENCY Solid state and chemical radiation dosimetry in medicine and biology :a symposium Proceedings Vienna 1966 Oct 3-7 Edited by M.R.J. Young International atomic energy agency.Proceedings series International atomic energy agency,Vienna,1967
Radioth 35.1673

INTERNATIONAL ATOMIC ENERGY AGENCY
Standardization of radioactive waste categories :report of a panel Vienna 1967 Nov 6-10 International atomic energy agency. Technical reports series, 101 IAEA,Vienna, 1970
Chem 18.2735

INTERNATIONAL ATOMIC ENERGY AGENCY
Sterilisation and preservation of biological tissues by ionizing radiations Panel on radiation sterilisation of biological tissues for transplantation Budapest 1969 Jun 16-20 International atomic energy agency.Panel proceedings series International atomic energy agency,Vienna,1970
Radioth 35.1915

INTERNATIONAL ATOMIC ENERGY AGENCY The Detection and use of tritium in the physical and biological sciences symposium Proceedings Vienna 1961 May 3-10 Vol 1-2 Joint commission on applied radioactivity International atomic energy agency.Proceedings series 2 vols International atomic energy agency,Vienna,1962 Papers in English and Russian.Summaries in English,French,Russian and Spanish
Gen 34.0470

INTERNATIONAL ATOMIC ENERGY AGENCY The Initial effects of ionizing radiations on cells :a symposium Moscow 1960 Oct Edited by R.J.C. Harris Sponsored by the Akademiya nauk S.S.S.R. Academic press, London;New York,1961
Gen 34.2222

INTERNATIONAL ATOMIC ENERGY AGENCY The Initial effects of ionizing radiations on cells :a symposium Moscow 1960 Oct Edited by R.J.C. Harris Sponsored by the Akademiya nauk S.S.S.R. Academic press, London;New York,1961
Radioth 35.1127

INTERNATIONAL ATOMIC ENERGY AGENCY The Seminar on theoretical physics held in Miramare Papers Trieste 1962 Jul 16-Aug 25 Edited by A. Salam STI-PUB-61 637p international atomic energy agency,Vienna,1963
Cav 7.2346

INTERNATIONAL ATOMIC ENERGY AGENCY
Theoretical physics Seminar on theoretical physics Lectures Trieste 1962 Jul 16-25 International atomic energy agency,Vienna,1963
Radioth 35.1655

INTERNATIONAL ATOMIC ENERGY AGENCY
Theoretical physics Seminar on theoretical physics lectures Trieste 1962 Jul.16-Aug. 25 Vienna,1963 Companion volume to International summer school on nuclear theory, 1962
A Math 4.0886

INTERNATIONAL ATOMIC ENERGY AGENCY Tritium in the physical and biological sciences Symposium on the detection and use of tritium in the physical and biological sciences Proceedings Vienna 1961 May 3-10 Vols 1-2 2 vols International atomic energy agency,Vienna,1962 Co-sponsored by the Joint commission on applied radioactivity
Radioth 35.0104

INTERNATIONAL ATOMIC ENERGY AGENCY Use of radioisotopes and supervoltage radiation in radioteletherapy; present status and recommendation Report and background information for a study group convened by the IAEA and WHO Vienna 1959 Aug 3-5 International atomic energy agency, Vienna, 1960
Radioth 35.1146

INTERNATIONAL ATOMIC ENERGY AGENCY.DIVISION OF ATOMIC ENERGY IN AGRICULTURE Effects of low doses of radiation on crop plants :a panel Report Rome 1964 Jun 1 International atomic energy agency.Technical reports series, 64 International atomic energy agency, Vienna, 1966
Radioth 35.1151

INTERNATIONAL ATOMIC ENERGY AGENCY.INTERNATIONAL CENTRE FOR THEORETICAL PHYSICS Plasma physics The International seminar on plasma physics proceedings Trieste 1964 Oct 5-31 International atomic energy agency, Vienna, 1965
A Math 4.0624

INTERNATIONAL ATOMIC ENERGY AGENCY.PANEL PROCEEDINGS SERIES Analytical control of radiopharmaceutics :a panel Proceedings Vienna 1969 Organized by the International atomic energy agency International atomic energy agency, Vienna, 1970
Radioth 35.1385

INTERNATIONAL ATOMIC ENERGY AGENCY.PANEL PROCEEDINGS SERIES Biophysical aspects of radiation quality Panel on biophysical aspects of radiation quality 2nd Report Vienna 1967 Apr 10-14 International atomic energy agency International atomic energy agency, Vienna, 1968
Radioth 35.1422

INTERNATIONAL ATOMIC ENERGY AGENCY.PANEL PROCEEDINGS SERIES Clinical uses of whole-body counting Panel on clinical uses of whole-body counting Proceedings Vienna 1965 Jun 28-Jul 2 International atomic energy agency International atomic energy agency, Vienna, 1966
Radioth 35.1383

INTERNATIONAL ATOMIC ENERGY AGENCY.PANEL PROCEEDINGS SERIES Control of livestock insect pests by the sterile-male technique Panel on the control of livestock insect pests by the sterile-male technique Proceedings Vienna 1967 Jan 23-27 Organized bythe Joint FAO-IAEA division of atomic energy in food and agriculture International atomic energy agency, Vienna, 1968
Radioth 35.0263

INTERNATIONAL ATOMIC ENERGY AGENCY.PANEL PROCEEDINGS SERIES Effects of ionizing radiations on the haematopoietic tissue Panel on the effects of ionizing radiations from different sources on haematopoietic tissue Proceedings Vienna 1966 May 17-20 International atomic energy agency International atomic energy agency, Vienna, 1967
Radioth 35.1200

INTERNATIONAL ATOMIC ENERGY AGENCY.PANEL PROCEEDINGS SERIES Elimination of harmful organisms from food and feed by irradiation Panel on elimination of harmful organisms from food and feed by irradiation Report Zeist 1967 Jun 12-16 Organized by the Joint FAO-IAEA division of atomic energy in food and agriculture International atomic energy agency, Vienna, 1968
Radioth 35.1761

INTERNATIONAL ATOMIC ENERGY AGENCY.PANEL PROCEEDINGS SERIES Enzymological aspects of food irradiation Panel on enzymological aspects of the application of ionizing radiation to food preservation Proceedings Vienna 1968 Apr 8-12 Organized by the Joint FAO-IAEA division of atomic energy in food and agriculture International atomic energy agency, Vienna, 1969
Radioth 35.1762

INTERNATIONAL ATOMIC ENERGY AGENCY.PANEL PROCEEDINGS SERIES Mutations in plant breeding Panel on co-ordination of research on the production and use of induced mutations in plant breeding Proceedings Vienna 1966 Jan 17-21 International atomic energy agency Food and agriculture organization 2 vols International atomic energy agency, Vienna, 1966-68
Radioth 35.0261

INTERNATIONAL ATOMIC ENERGY AGENCY.PANEL PROCEEDINGS SERIES Mutations in plant breeding II :a research co-ordination meeting on the use of induced mutations in plant breeding Proceedings Vienna 1967 Sep 11-15 International atomic energy agency Food and agriculture organization I.A.E.A., Vienna, 1968
Radioth 35.0262

INTERNATIONAL ATOMIC ENERGY AGENCY.PANEL PROCEEDINGS SERIES Nuclear accident dosimetry systems Panel on nuclear accident dosimetry systems Proceedings Vienna 1969 Feb 17-21 Organized by the International atomic energy agency International atomic energy agency, Vienna, 1970
Radioth 35.1384

INTERNATIONAL ATOMIC ENERGY AGENCY.PANEL PROCEEDINGS SERIES Preservation of fruit and vegetables by radiation Panel on preservation of fruit and vegetables by radiation :especially in the tropics Proceedings Vienna 1966 Aug 1-5 Organised by the Joint FAO-IAEA division of atomic energy in food and agriculture International atomic energy agency, Vienna, 1968
Radioth 35.1758

INTERNATIONAL ATOMIC ENERGY AGENCY.PANEL PROCEEDINGS SERIES Radiation and the control of immune response Panel on radiation and the control of immune response Proceedings Paris 1967 Jun 22-24 Organized by the International atomic energy agency International atomic energy agency, Vienna, 1968
Radioth 35.1202

INTERNATIONAL ATOMIC ENERGY AGENCY.PANEL PROCEEDINGS SERIES Radiation damage and sulphydryl compounds Panel on radiation damage to the biological molecular information system :with special regard to the role of SH-groups Proceedings Vienna 1968 Oct 21-25 Organized by the International atomic energy agency International atomic energy agency, Vienna,1969
Radioth 35.1764

INTERNATIONAL ATOMIC ENERGY AGENCY.PANEL PROCEEDINGS SERIES Radiation sensitivity of toxins and animal poisons Panel on the radiation sensitivity of toxins and animal poisons Proceedings Bangkok 1969 May 19-22 Organized by the International atomic energy agency International atomic energy agency,Vienna,1970
Radioth 35.1765

INTERNATIONAL ATOMIC ENERGY AGENCY.PANEL PROCEEDINGS SERIES Research applications of nuclear pulsed systems Panel on research applications of repetitively pulsed reactors and boosters Proceedings Dubna,U.S.S.R. 1966 Jul 18-22 International atomic energy agency International atomic energy agency, Vienna,1967
Radioth 35.1682

INTERNATIONAL ATOMIC ENERGY AGENCY.PANEL PROCEEDINGS SERIES Role of computers in radiotherapy Panel on the role of computers in radiotherapy Report Vienna 1967 Jul 10-14 Organized by the International atomic energy agency bibliog. International atomic energy agency,Vienna,1968
Radioth 35.1199

INTERNATIONAL ATOMIC ENERGY AGENCY.PROCEEDINGS SERIES Assessment of radioactivity in man Symposium on the assessment of radioactive body burdens in man Proceedings Heidelberg 1964 May 11-16 1-2 International atomic energy agency International labour organization World health organization International atomic energy agency,Vienna,1964
Radioth 35.1373

INTERNATIONAL ATOMIC ENERGY AGENCY.PROCEEDINGS SERIES Biological effects of ionizing radiation at the molecular level Symposium on the biological effects of ionizing radiation at the molecular level Proceedings Brno 1962 Jul 2-6 Held by the International atomic energy agency I.A.E.A., Vienna,1962
Radioth 35.1976

INTERNATIONAL ATOMIC ENERGY AGENCY.PROCEEDINGS SERIES Diagnosis and treatment of radioactive poisoning Scientific meeting on the diagnosis and treatment of radioactive poisoning Proceedings Vienna 1962 Oct 15-18 World health organization International atomic energy agency International atomic energy agency,Vienna,1963
Radioth 35.1368

INTERNATIONAL ATOMIC ENERGY AGENCY.PROCEEDINGS SERIES Effects of ionizing radiation on the nervous system Symposium on ionizing radiation on the nervous system Proceedings Vienna 1961 Jun 5-9 Sponsored by the International atomic energy agency International atomic energy agency,Vienna,1962
Radioth 35.1130

INTERNATIONAL ATOMIC ENERGY AGENCY.PROCEEDINGS SERIES Medical radioisotope scanning Symposium on medical radioisotope scanning Proceedings Athens 1964 Apr 20-24 Held by the International atomic energy agency International atomic energy agency,Vienna,1964
Radioth 35.1376

INTERNATIONAL ATOMIC ENERGY AGENCY.PROCEEDINGS SERIES Medical radioisotope scintigraphy : a symposium Proceedings Salzburg 1968 Aug 6-15 Vol 1-2 International atomic energy agency Edited by G.R. Stevenson 2 vols International atomic energy agency, Vienna,1969
Radioth 35.1392

INTERNATIONAL ATOMIC ENERGY AGENCY.PROCEEDINGS SERIES Neutron dosimetry Symposium on neytron detection,dosimetry and standardization Proceedings Harwell 1962 Dec 10-14 1-2 Held by the International atomic energy agency 2 vols International atomic energy agency,Vienna,1963
Radioth 35.1695

INTERNATIONAL ATOMIC ENERGY AGENCY.PROCEEDINGS SERIES Production and use of short-lived radioisotopes from reactors Seminar on the practical applications of short-lived radioisotopes produced in small research reactors Proceedings Vienna 1962 Nov 5-9 1-2 Held by the International atomic energy agency 2 vols International atomic energy agency,Vienna,1963
Radioth 35.1369

INTERNATIONAL ATOMIC ENERGY AGENCY.PROCEEDINGS SERIES Radioactive dating and methods of low-level counting :a symposium Proceedings Monaco 1967 Mar 2-10 International atomic energy agency Joint commission on applied radioactivity International atomic energy agency,Vienna,1967
Radioth 35.1683

INTERNATIONAL ATOMIC ENERGY AGENCY.PROCEEDINGS SERIES Radioisotope sample measurement techniques in medicine and biology Symposium on radioisotope sample measurement techniques in medicine and biology Proceedings Vienna 1965 May 24-25 Held by the International atomic energy agency International atomic energy agency,Vienna,1965
Radioth 35.1377

INTERNATIONAL ATOMIC ENERGY AGENCY.PROCEEDINGS SERIES Radioisotopes in tropical medicine Symposium on the use of radioisotopes in the study of endemic and tropical diseases Proceedings Bangkok 1960 Dec 12-16 International atomic energy agency World health organization International atomic energy agency,Vienna,1962
Radioth 35.1365

INTERNATIONAL ATOMIC ENERGY AGENCY.PROCEEDINGS SERIES Radiological health and safety in mining and milling of nuclear materials Symposium on radiological health and safety in mining and milling of nuclear materials :a symposium Proceedings Vienna 1963 Aug 26-31 1-2 International atomic energy agency International labour organization In co-operation with the World health organization International atomic energy agency,Vienna,1964
Radioth 35.1164

INTERNATIONAL ATOMIC ENERGY AGENCY.PROCEEDINGS SERIES Selected topics in radiation dosimetry Symposium on selected topics in radiation dosimetry Proceedings Vienna 1960 Jun 7-11 Sponsored by the International atomic energy agency International atomic energy agency,Vienna,1961
Radioth 35.1694

INTERNATIONAL ATOMIC ENERGY AGENCY.PROCEEDINGS SERIES Whole-body counting Symposium on whole-body counting Proceedings Vienna 1961 Jun 12-16 International atomic energy agency,Vienna,1962
Radioth 35.1361

INTERNATIONAL AUTOMATION CONGRESS Proceedings Paris 1956 Jul 18-24 Presses academiques europeenes,Brussels,1959
Eng 41.6008

INTERNATIONAL AUTOMATION CONGRESS proceedings Madrid 1958 Oct 13-18 Consejo superior de investigaciones cientificas.Instituto de electricidad y automatica illus. 411p 30cm Instituto de electricidad y automatica, Madrid,1958
Math L 5.0894

INTERNATIONAL BARLEY GENETICS SYMPOSIUM 1st Proceedings Barley genetics 1 Wageningen 1963 Aug 26-31 Centre for agricultural publications and documentation, Wageningen,1964
Gen 34.1560

The INTERNATIONAL BIOCLIMATOLOGICAL CONGRESS 2nd Proceedings Biometeorology London 1960 Sep 4-10 Edited by S.W. Tromp illus xxii,687p 25cm Pergamon,Oxford,1962
Sco 14.0585

INTERNATIONAL BIODETERIORATION SYMPOSIUM 1st Proceedings Biodeterioration of materials: microbiological and allied aspects Southampton 1968 Sep 9-14 Edited by A. Harry Walters and John J. Elphick Elsevier, Amsterdam,1968
Met 25.2483

INTERNATIONAL BIODETERIORATION SYMPOSIUM 1st Proceedings Biodeterioration of materials: microbiological and allied aspects Southampton 1968 Sep 9-14 Edited by Arthur H. Walters and John J. Elphick Bibliog., illus. x,740p Elsevier,Amsterdam,1968
Bot 42.3773

INTERNATIONAL BIOMETEOROLOGICAL CONGRESS 3rd Proceedings Biometeorology Pau 1963 Sep 1-7 Vol 2,pt 1-2 International society of biometeorology Edited by S.W. Tromp and W. H. Weihe 2 vols Pergamon press,Oxford,1967
Geog 13.1212

INTERNATIONAL BIOMETRIC CONFERENCE 1st Woods Hole,Mass. 1947 Sep 5-6 Biometric society 1948 Reprinted from 'Biometrics' December 1947 and March 1948
Gen 34.2145

INTERNATIONAL BIOMETRIC CONFERENCE 2nd Proceedings Geneva 1949 Aug 30-Sep 2 Biometric society 1950 Reproduced from 'Biometrics' vol 6,no.1
Gen 34.2146

INTERNATIONAL BIOMETRIC CONFERENCE 3rd Proceedings Bellagio,Italy 1953 Sep 1-5 Biometric society 1953 Reprinted from 'Biometrica' vol 9,no.4,1953
Gen 34.2147

INTERNATIONAL BIOMETRIC CONFERENCE 5th Cambridge 1963 Sep 10-13 Cambridge,1963 8 unbound typescript papers
Math 3.1127

INTERNATIONAL BIOPHYSICS CONGRESS Stockholm 1961 Jul 31-Aug 4 abstracts of contributed papers Tryckeri aktiebolaget thule,Stockholm, 1961
Radioth 35.1647

INTERNATIONAL BIOPHYSICS CONGRESS :abstracts of contributed papers Stockholm 1961 Jul 31-Aug 4 Tryckeri Aktiebolaget,Stockholm, 1961
Radioth 35.1426

INTERNATIONAL BOTANIC CONGRESS 1905 Regles internationale de la nomenclature botanique adoptees par le congreS international de botanique de Vienne,1905.Deuxieme edition mise au point d'apres les decisions du congres international de botanique de Bruxelles,1910 By John Briquet 2nd edition Fischer,Jena, 1912
Bot 42.6738

INTERNATIONAL BOTANICAL CONGRESS International botanical congress 5th Abstracts Cambridge 1930 Aug 16-23 Cambridge university press,Cambridge,1930 President of congress:prof.Seward.Bound with complete set of documents relating to this congress
Bot 42.5914

INTERNATIONAL BOTANICAL CONGRESS International botanical congress... :minutes of the executive countil 5th Cambridge 1930 Cambridge,1927-35
Bot 42.5961

INTERNATIONAL BOTANICAL CONGRESS Verhandlungen des internationalen botanischen Vienna 1905 Edited by R.von Wettstein and others Fischer,Jena,1906 Title also in French 'Actes du congres international de botanique'
Bot 42.0647

INTERNATIONAL BOTANICAL CONGRESS 5th :a report Proceedings Cambridge 1930 Aug 16-23 Edited by F.T. Brooks and T.F. Chipp port Cambridge university press,Cambridge, 1931
Bot 42.0655

INTERNATIONAL BOTANICAL CONGRESS 5th Abstracts Cambridge 1930 Aug 16-23 International botanical congress Cambridge university press,Cambridge,1930 President of congress:prof.Seward.Bound with complete set of documents relating to this congress
Bot 42.5914

INTERNATIONAL BOTANICAL CONGRESS 5th Abstracts,programs Cambridge 1930 Aug 16-22 Cambridge university press,Cambridge, 1930
BG 38.3127

INTERNATIONAL BOTANICAL CONGRESS 7th
Abstracts Stockholm 1950 section for palaeobotany c1950 Duplicated. Papers in English, French and German. In folder
Bot 42.6225

INTERNATIONAL BOTANICAL CONGRESS 7th
Proceedings Stockholm 1950 Jul 12-20
Edited by Hugo Osvald and Ewert Aberg
Almqvist and Wiksell, Uppsala, 1953
Bot 42.4696

INTERNATIONAL BOTANICAL CONGRESS 7th
Proposals, Abstracts Stockholm 1950
International union of biological sciences
International union of biological sciences, Utrecht, 1950
BG 38.3128

INTERNATIONAL BOTANICAL CONGRESS 8th
Comptes rendus des seances et rapports et communications Paris 1954 Sections 3-27 9 vols Paris, 1955-57
Bot 42.4699

INTERNATIONAL BOTANICAL CONGRESS 8th
Rapports et communications Paris 1954
Section 2: 2.serie Utrecht, 1955
Reprinted from Taxon, vol.4 no.6, pp.121-177
Bot 42.4698

INTERNATIONAL BOTANICAL CONGRESS 8th
Rapports et communications Paris 1954
Sections 2;4-27 9 vols Paris, 1954
Bot 42.4697

INTERNATIONAL BOTANICAL CONGRESS 8th
Reports of communications Paris 1954
1954
BG 38.3129

INTERNATIONAL BOTANICAL CONGRESS 9th
Proceedings Montreal 1959 Aug 19-29
1959
BG 38.3130

INTERNATIONAL BOTANICAL CONGRESS 9th
Proceedings Montreal 1959 Aug 19-29
Vol 1-3 University of Toronto press, Toronto, 1959
Bot 42.4703

INTERNATIONAL BOTANICAL CONGRESS 10th
Proceedings Edinburgh 1964 Aug 1964
BG 38.3131

INTERNATIONAL BOTANICAL CONGRESS 11th
Abstracts Seattle, Wash. 1969 Aug 24-Sep 2 1969 Sep 2-4 1969
BG 38.3253

INTERNATIONAL BOTANICAL CONGRESS 11th
Proceedings Seattle, Wash. 1969 Aug 24-Sep 2 International botanical congress, Washington, D.C., 1970
BG 38.3132

INTERNATIONAL BOTANICAL CONGRESS... 5th :
minutes of the executive countil
Cambridge 1930 International botanical congress Cambridge, 1927-35
Bot 42.5961

INTERNATIONAL BOTANICAL CONGRESS, 5TH
International rules of botanical nomenclature revised by the International botanical congress, Cambridge 1930, compiled from the report of J. Briquet Cambridge 1930
Fischer, Jena, 1935
Bot 42.5221

INTERNATIONAL BOTANICAL CONGRESS 10TH
Chromosome manipulations and plant genetics :a symposium held during the 10th international botanical congress Edinburgh 1964 Aug
Edited by Ralph Riley and K.R. Lewis Oliver and Boyd, Edinburgh, 1966
Bot 42.0789

INTERNATIONAL BOTANICAL CONGRESS 10TH
Chromosome manipulations and plant genetics :a symposium held during the 10th international botanical congress Edinburgh 1964 Aug
Edited by Ralph Riley and K.R. Lewis Oliver and Boyd, Edinburgh; London, 1966
Gen 34.0894

INTERNATIONAL BOTANICAL CONGRESS 10TH
Incompatibility in fungi :a symposium
Proceedings Edinburgh 1964 Aug Edited by Karl Esser and John R. Raper viii,124p 24cm Springer, Berlin, 1965 Symposium held at the 10th International botanical congress
Bot 42.4713

INTERNATIONAL BOTANICAL CONGRESS 10TH
Introduction to the flora of Cambridgeshire... By F.H. Perring and others Published in association with the Cambridge natural history society Cambridge university press, Cambridge, 1964 Reprinted from 'A flora of Cambridgeshire', by Perring, Sell, Walters and Whitehouse, for the use of participants in the Cambridgeshire excursions arranged before the Xth International botanical congress, Edinburgh 1964
Bot 42.5835

INTERNATIONAL BOTANICAL CONGRESS 11TH A
Short history of botany in the United States
Seattle, Wash. 1969 Edited by Joseph Ewan
Hafner, New York; London, 1969 Published on the occasion of the 11th International botanical congress
BG 38.3234

INTERNATIONAL BOTANICAL CONGRESS 11TH
Selection of 20th century botanical art and illustrations International exhibition of twentieth century botanical art 2nd
Catalogue Edited by George H.M. Lawrence port 168p Carnegie-Mellon university, Pittsburgh, Pa., 1969 Presented at the 11th International botanical congress, August 1969
Bot 42.0634

INTERNATIONAL BOTANICAL CONGRESS 7TH A Short history of botany in Sweden Stockholm 1950 Jul By R.E. Fries and others Almqvist and Wiksell, Uppsala, 1950 Published on the occasion of the 7th International botanical congress
BG 38.3233

INTERNATIONAL BRAIN RESEARCH ORGANISATION
Brain mechanisms :international colloquium... on specific and unspecific mechanisms of sensory motor integration Papers Pisa 1961 Edited by Giuseppe Moruzzi and others
Progress in brain research, 1 Elsevier, Amsterdam, 1963
An 32.5230

INTERNATIONAL BRAIN RESEARCH ORGANIZATION
Biology of neuroglia Lectures and discussions Buenos Aires 1963 Oct 17-18 Edited by E.D.P.de Robertis and R. Carea Progress in brain research, 15 illus Elsevier,Amsterdam,1965 Symposium held as part of the 10th Latin-American congress of neurosurgery
Path 30.2466

INTERNATIONAL BRAIN RESEARCH ORGANIZATION
Brain reflexes :the international conference dedicated to the centenary celebration of the publication of I.M.Sechenov's book "Brain reflexes" Moscow Edited by E.A. Asratyan Sponsored by Akademiya nauk S.S.S.R. Progress in brain research, 22 Elsevier, Amsterdam,1968 Sponsored also by the International brain research organization
An 32.4243

INTERNATIONAL BUSINESS MACHINES CORPORATION
Computation seminar proceedings Endicott,N. Y. 1949 Dec 5-9 173p 28cm International business machines corporation, New York,1951
Math L 5.0700

INTERNATIONAL BUSINESS MACHINES CORPORATION
Computation seminar proceedings Endicott,N. Y. 1951 Aug 13-17 148p 28cm International business machines corporation, New York,1951
Math L 5.0702

INTERNATIONAL BUSINESS MACHINES CORPORATION
Control and simulation language :introductory manual 39p 28cm Esso;I.B.M.,London,1963
Math L 5.3262

INTERNATIONAL BUSINESS MACHINES CORPORATION
Control and simulation language reference manual 95p 28cm Esso;I.B.M.,London,1963
Math L 5.3261

INTERNATIONAL BUSINESS MACHINES CORPORATION
Industrial computation seminar proceedings Endicott,N.Y. 1950 Sept 25-29 103p 28cm International business machines corporation, New York,1951
Math L 5.0701

INTERNATIONAL BUSINESS MACHINES CORPORATION
Summer institute on symbolic mathematical computation Proceedings Boston,Mass. 1968 Jun 24-Aug 16 Edited by Robert G. Tobey illus 324p IBM Boston programming center, Cambridge,Mass.,1969
Math L 5.3809

INTERNATIONAL BUSINESS MACHINES CORPORATION
Symposium on digital computing in the aircraft industry proceedings New York 1957 Jan 31-Feb 1 New York university illus. 399p 28cm International business machines corporation,New York,1957
Math L 5.0777

INTERNATIONAL BUSINESS MACHINES CORPORATION
Teaching of computer design Joint IBM-University of Newcastle-upon-Tyne seminar Proceedings Newcastle upon Tyne 1971 Sep 7-10 illus 121p University of Newcastle upon Tyne.Computing laboratory,Newcastle upon Tyne,1972
Math L 5.3723

INTERNATIONAL BUSINESS MACHINES CORPORATION
Teaching of programming at university level Joint IBM-University of Newcastle upon Tyne seminar 3rd Proceedings Cambridge 1970 Sep 8-11 Edited by B. Shaw illus. 153p University of Newcastle,Newcastle,1971
Math L 5.3590

INTERNATIONAL BUSINESS MACHINES CORPORATION
Tests with a phase reversal equipment over the European network I.B.M.France,1960?
Math L 5.0865

INTERNATIONAL BUSINESS MACHINES CORPORATION. RESEARCH CENTER Stochastic point processes :statistical analysis,theory and applications Papers Yorktown Heights,N.Y. 1971 Aug 2-6 Edited by Peter A.W. Lewis xxii,894p 24cm Interscience,New York,1972
Math S 3.1803

INTERNATIONAL CANCER CONGRESS 7th London 1958 Jul 6-12 Abstract of papers Under the auspices of the International union against cancer 1958
Radioth 35.0777

INTERNATIONAL CANCER CONGRESS 8th abstracts of papers Moscow 1962 Jul 22-28 Moscow,1962
Radioth 35.0785

INTERNATIONAL CO-ORDINATION COMMITTEE FOR THE IMMUNOLOGY OF REPRODUCTION Immunology and reproduction Symposium of the International co-ordination committee for the immunology of reproduction 1st Proceedings Geneva 1968 Sep Edited by R.G. Edwards International planned parenthood association, London,1969
Phys 20.2265

INTERNATIONAL COLLEGIUM FOR NEURO-PSYCHOPHARMACY
Neuro-psychopharmacology International congress of neuropharmacology 1st Proceedings Rome 1958 Sep Edited by P.B. Bradley and others Elsevier,Amsterdam,1959
Psy 31.0524

INTERNATIONAL COLLOQUIUM IN THE PHILOSOPHY OF SCIENCE Proceedings Criticism and the growth of knowledge London 1965 Jul 11-17 Edited by Imre Lakatos and Alan Musgrave Held at Bedford college vii,282p 23cm Cambridge university press,Cambridge,1970
Math S 3.1850

INTERNATIONAL COLLOQUIUM IN THE PHILOSOPHY OF SCIENCE Proceedings London 1965 Vol 2: the problem of inductive logic Edited by Imre Lakatos Studies in logic and the foundations of mathematics viii,417p North-Holland,Amsterdam,1968
WSM 43.1218

INTERNATIONAL COLLOQUIUM IN THE PHILOSOPHY OF SCIENCE Proceedings London 1965 Jul 11-17 Vol 4: criticism and the growth of knowledge Edited by Imre Lakatos and Alan Musgrave Cambridge university press, Cambridge,1970
Eng 41.1688

INTERNATIONAL COLLOQUIUM IN THE PHILOSOPHY OF SCIENCE Proceedings Problems in the philosophy of mathematics London 1965 Jul 11-17 Vol 1 British society for the philosophy of science London school of economics and political science Edited by Imre Lakatos Under the auspices of the International union of history and philosophy of science.Division of logic,methodology and philosophy of science Studies in logic and the foundations of mathematics xv,241p North-Holland,Amsterdam,1967
WSM 43.1813

INTERNATIONAL COLLOQUIUM IN THE PHILOSOPHY OF SCIENCE Proceedings Problems in the philosophy of science London 1965 Jul 11-17 Vol 3 British society for the philosophy of science London school of economics and political science Edited by Imre Lakatos and Alan Musgrave Under the auspices of the International union of history and philosophy of science.Division of logic, methodology and philosophy of science Studies in logic and the foundations of mathematics ix,448p North-Holland, Amsterdam,1968
WSM 43.0931

INTERNATIONAL COLLOQUIUM ON DIFFERENTIAL ANALYSIS the Bombay colloquium papers presented Differential analysis Bombay 1964 Jan 7-14 By M.F. Atiyah and others Tata institute of fundamental research Jointly sponsored by the International mathematical union Tata institute of fundamental research. Studies in analysis, 2 viii,253p 25cm Oxford university press,Oxford,1964
P. Math 2.0696

INTERNATIONAL COLLOQUIUM ON LOGIC,PHYSICAL REALITY AND HISTORY Physics,logic and history Denver,Colo. 1966 May 16-20 Edited by Wolfgang Yourgrau and Allen D. Breck xiv,336p Plenum press,New York;London,1970 Based on the 1st colloquium
WSM 43.1116

INTERNATIONAL COLLOQUIUM ON MACROMOLECULES Proceedings Amsterdam 1949 Sep 2-5 Internation union of pure and applied chemistry D.B.Ceuten's Uitgerers,Amsterdam, 1949
Radioth 35.0014

INTERNATIONAL COLLOQUIUM ON MACROMOLECULES Proceedings Amsterdam 1949 Sep 2-5 Under the auspices of the International union of pure and applied chemistry Amsterdam,1950
Col S 12.0101

INTERNATIONAL COLLOQUIUM ON MATHEMATICAL LOGIC AND FOUNDATIONS OF SET THEORY Proceedings Mathematical logic and foundations of set theory Jerusalem 1908 Nov 11-14 Edited by Yehoshua Bar-Hillel Under the auspices of the Israel academy of science and humanities Studies in logic and the foundations of mathematics bibliog. 145p 24cm North-Holland,Amsterdam;London,1970
P Math 2.3795

INTERNATIONAL COLLOQUIUM ON RAPID MIXING AND SAMPLING TECHNIQUES APPLICABLE TO THE STUDY OF BIOCHEMICAL REACTIONS 1st Proceedings Rapid mixing and sampling techniques in biochemistry Philadelphia 1964 Jul 23-24 International union of biochemistry Edited by Britton Chance and others Academic press, New York;London,1964
Bioch 33.2320

INTERNATIONAL COLLOQUIUM ON THE MODERNIZATION OF MATHEMATICS TEACHING IN EUROPEAN COUNTRIES Proceedings Bucharest 1968 Sep 23-Oct 2 Unesco 572p 25cm Editions didactiques et pedagogiques,Bucharest,1968
P Math 2.4020

INTERNATIONAL COMMISSION FOR OPTICS Conference on optical instruments and techniques Proceedings London 1961 Jul 11-14 Edited by K.J. Habell Chapman and Hall,London,1962
Obs 6.3215

INTERNATIONAL COMMISSION FOR OPTICS Optics in metrology Colloquia of the International commission for optics Papers Brussels 1958 May 6-9 Edited by Pol Mollet Pergamon press,Oxford,1960 Papers presented in English,French or German
Cav 7.1631

INTERNATIONAL COMMISSION FOR OPTICS The Conference on photographic and spectrographic optics Proceedings Tokyo 1964 Sep 1-5 and Kyoto 1964 Sep 1-5-,7-8 Under the auspices of the Science council of Japan Japanese journal of applied optics, 4,suppl. 1 669p Tokyo,1965
Obs 6.2050

INTERNATIONAL COMMISSION FOR THE NORTHWEST ATLANTIC FISHERIES I.C.N.A.F. environmental symposium Contributions Rome 1964 International commission for the Northwest Atlantic fisheries.Special publication,6 figures,tables,graphs 914p ICNAF,Dartmouth,Nova Scotia,1965
Sco 14.5958

INTERNATIONAL COMMISSION FOR THE NORTHWEST ATLANTIC FISHERIES North Atlantic fish marking symposium Contributions Woods Hole, Mass. 1961 May Edited by H.W. Graham and others International commission for the Northwest Atlantic fisheries.Special publication,4 370p ICNAF,Dartmouth,Nova Scotia,1963
Sco 14.5957

INTERNATIONAL COMMISSION FOR THE NORTHWEST ATLANTIC FISHERIES North Atlantic fish marking symposium Woods Hole,Mass. 1961 May International commission for the northwest Atlantic fisheries.Special publication,4 Dartmouth,N.S.,1961
Sco 14.7833

INTERNATIONAL COMMISSION FOR THE NORTHWEST ATLANTIC FISHERIES.SPECIAL PUBLICATION,4 North Atlantic fish marking symposium Woods Hole,Mass. 1961 May International commission for the northwest Atlantic fisheries Dartmouth,N.S.,1961
Sco 14.7833

INTERNATIONAL COMMISSION ON LARGE DAMS
International congress on large dams 5th Transactions Paris 1955 May 31-Jun 4 Vol 1-5 5 vols Paris,1956
Eng 41.3334

INTERNATIONAL COMMISSION ON POLAR METEOROLOGY
Polar meteorology W.M.O.-S.C.A.R.-I.C.P.M. Symposium on polar meteorology Proceedings Geneva 1966 Sep 5-9 World meteorological organization,and,Scientific committee on Antarctic research illus vi,540p 28cm World meteorological organization,Geneva,1967
Sco 14.7839

INTERNATIONAL COMMISSION ON RADIOLOGICAL PROTECTION Radiation protection Recommendation of the international commission on radiological protection (adopted September 9,1958) Pergamon press,London,1959
Radioth 35.1138

INTERNATIONAL COMMITTEE FOR COAL PETROLOGY 1st-2nd :meeting proceedings Geleen 1953 Jun 9-11 and Liege 1955 May 23-25 No 1-2 International committee for coal petrology International committee for coal petrology.Proceedings,1,2 Ernest van Aelst, Maastricht,1954-56
Geol 8.3136

INTERNATIONAL COMMITTEE FOR HORTICULTURAL CONGRESSES International horticultural congress 9th Reports,Proceedings,Program London 1930 Aug 7-15 Royal horticultural society London,1930
BG 38.3121

INTERNATIONAL COMMITTEE FOR HORTICULTURAL CONGRESSES International horticultural congress 15th Proceedings Nice 1958 Vol 1-3 Edited by Jean-Claude Garnaud 3 vols Pergamon press,London,1961-62
BG 38.3133

INTERNATIONAL COMMITTEE FOR THE CONGRESSES OF APPLIED MECHANICS International congress on theoretical and applied mechanics 8th proceedings Istanbul 1952 Aug.20-28 Vols 1-2 Istanbul universitesi.Faculty of science Edited by Kerim Erim 2 vols University faculty of science,Istanbul,1953
A Math 4.0326

INTERNATIONAL COMMITTEE FOR THE CONGRESSES OF APPLIED MECHANICS The International congress of applied mechanics 11th Proceedings Munich 1964 Edited by Henry Gortler and Peter Sorger 1184p Springer-verlag,Berlin,1966 Dedicated to Richard Grammel
A Math 4.1279

INTERNATIONAL COMMITTEE OF ELECTROCHEMICAL THERMODYNAMICS AND KINETICS C.I.T.C.E. meeting 6th-9th Proceedings 1954-57 4 vols Butterworths,London,1955-59 Papers in English,French and German
Met 25.1833

INTERNATIONAL COMMITTEE OF FOUNDRY TECHNICAL ASSOCIATIONS International foundry congress 28th Congress papers Vienna 1961 Jun 19-24 Verein Oesterreichischer Giesserfachleute,Vienna,1961 Papers in English,French and German
Met 25.0681

INTERNATIONAL COMMITTEE ON AERONAUTICAL FATIGUE
Fatigue of aircraft structures :a symposium Proceedings Paris 1961 May 16-18 Edited by W. Barrois and E.L. Ripley International series of monographs in aeronautics and astronautics.Symposia, 12,Div.9 Pergamon press,Oxford,1963
Met 25.0931

INTERNATIONAL COMMITTEE ON IMMUNOLOGY
Mechanism of cell and tissue damage produced by immune reactions International symposium on immunopathology 2nd Papers Brook Lodge,Mich. 1961 Edited by Pierre Grabar and Peter Miescher Schwabe,Basle;Stuttgart, 1962
Path 30.2336

INTERNATIONAL COMPUTATION CENTRE Economics of automatic data processing :international symposium papers Rome 1965 Oct 19-22 Edited by A.B. Frielink pl. xiii,384p 22cm North-Holland,Amsterdam,1965
Math L 5.1022

INTERNATIONAL COMPUTATION CENTRE Theory of graphs :international sympoaium Rome 1966 Jul Bibliog.,Illus. xviii,416p 25cm Gordon and Breach;Dunod,New York;Paris,1967
P Math 2.3128

INTERNATIONAL COMPUTER SOCIETY Hardware, software,firmware trade-offs IEEE-International computer society conference 5th Proceedings Boston,Mass. 1971 Sep 22-24 illus 204p IEEE,New York,1971
Math L 5.3628

INTERNATIONAL COMPUTING SYMPOSIUM Proceedings Venice 1972 Apr 12-14 Consiglio nazionale delle ricerche.Cybernetics group illus 634p Venice,1972
Math L 5.3862

INTERNATIONAL CONFERENCE FOR THE SEVENTH REVISION OF THE INTERNATIONAL LISTS OF DISEASES AND CAUSES OF DEATH Manual of the international statistical classification of diseases,injuries,and causes of death based on the seventh revision conference Paris 1955 Feb 21-26 1-2 World health organization 2 vols World health organization,Geneva,1957
Gen 34.1937

INTERNATIONAL CONFERENCE FOR THE SIXTH REVISION OF THE INTERNATIONAL LISTS OF DISEASES AND CAUSES OF DEATH Manual of the international statistical classification of diseases,injuries and causes of death :adapted 1948 by the international conference for the sixth revision of the international lists of diseases and causes of death Paris 1948 1-2 World health organization World health organization.Bulletin.Supplement, 1 2 vols Columbia university press,New York, 1948-49
Med 36.0385

INTERNATIONAL CONFERENCE OF INSECT PATHOLOGY AND BIOLOGICAL CONTROL 1st Transactions Prague 1958 Aug 13-18 24cm Prague,1958
Philos 1.1598

INTERNATIONAL CONFERENCE OF PHYTOPATHOLOGY AND ECONOMIC ENTOMOLOGY Report Wageningen 1923 Jun 24-Jul 1 Edited by T.A.C. Schoevers Committee of management,Wageningen,1923
Bot 42.0650

INTERNATIONAL CONFERENCE OF PURE AND APPLIED CHEMISTRY main lectures 16th Paris 1957 Jul 18-24 International union of pure and applied chemistry Birkhauser,Basle; Stuttgart,1957 Papers in English,French and German
Chem 18.1738

INTERNATIONAL CONFERENCE OF THE ASSOCIATION FOR THE STUDY OF THE EUROPEAN QUATERNARY 2nd trudy Mezhdunarodnaya konferentsiya assotsiatsii po izucheniyu chetvertichnogo perioda Evropy Leningrad 1932 Sep 1-7 Vypusk 1-5 Association for the study of the European quaternary,and,S.S.S.R.- N.K.T.P. Vsesoyuznoe geologo-razvedochnoe obedinenie Edited by D.A. Petrovskii and others Gosudarstvennoe nauchno-tekhnicheskoe geologo-razvedochnoe izdatelstvo,Leningrad;Moscow,1932-34
Geol 8.2982

INTERNATIONAL CONFERENCE OF THE INTERNATIONAL UNION AGAINST TUBERCULOSIS 2nd Transactions London 1921 Jul 26-28 International union against tuberculosis Under the auspices of the British national association for the prevention of tuberculosis Adlard and Newman,London,1921
An 32.2990

The INTERNATIONAL CONFERENCE OF THE WORLD ASSOCIATION FOR THE ADVANCEMENT OF VETERINARY PARASITOLOGY 2nd Proceedings Biology of parasites;emphasis on veterinary parasites Philadelphia 1965 Sep 7-9 World association for the advancement of veterinary parasitology Edited by E.J.L. Soulsby Held at the University of Pennsylvania Academic press,New York;London, 1966 Held...in conjunction with Bicentennial celebrations of medical education in the United States
Bal 39.1737

INTERNATIONAL CONFERENCE ON ATOMIC PHYSICS 1st Proceedings Atomic physics New York City 1968 Jun 3-7 Edited by B. Bederson and others Bibliog.,Illus. xiii,620p 26cm Plenum press,New York,1969
A Math 4.1601

INTERNATIONAL CONFERENCE ON AVIAN TUMOR VIRUSES Proceedings Durham,N.C. 1964 Mar 31-Apr 3 National cancer institute Duke university.School of medicine Edited by Joseph W. Beard National cancer institute. Monograph, 17 National cancer institute, Bethesda,Md.,1964
Bioch 33.1154

INTERNATIONAL CONFERENCE ON AVIAN TUMOUR VIRUSES Proceedings Durham,N.C. 1964 Mar 31-Apr 3 National cancer institute and Duke university.School of medicine Edited by Joseph W. Beard National cancer institute. Monograph, 17 US government printing office, Washington,D.C.,1964
Path 30.2570

The INTERNATIONAL CONFERENCE ON BIOLOGICAL MEMBRANES 2nd Proceedings Membrane models and the formation of biological membranes Frascati 1967·Jun Nato advanced study institute Edited by Liana Bolis and B.A. Perthica Supported by the North Atlantic treaty organization North Holland,Amsterdam,1968
Inv Med 37.0025

INTERNATIONAL CONFERENCE ON BIOLOGICAL MEMBRANES 2nd Proceedings Membrane models and the formation of biological membranes Frascati 1967 Jun Edited by Liana Bolis and B.A. Pethica With the financial support of the North Atlantic treaty organization Illus. xv,337p North-Holland,Amsterdam, 1968 A NATO advanced study institute
Pha 16.0016

INTERNATIONAL CONFERENCE ON BODY-CENTRED CUBIC MATERIALS Papers Diffusion in body-centered cubic metals Gatlinburg 1964 Sep 16-18 American society for metals.Oak Ridge chapter and Oak Ridge national laboratory American society for metals,Metals Park,Ohio, 1965
Met 25.1283

INTERNATIONAL CONFERENCE ON CLOUD PHYSICS Proceedings Toronto 1968 American meteorological society 873p American meteorological society,Boston,Mass.,1969
Nap 11.0973

INTERNATIONAL CONFERENCE ON COMBINATORIAL MATHEMATICS Papers New York 1970 Apr 1-4 New York academy of sciences Edited by Allan Gewirtz and Louis Quintas New York academy of sciences.Annals, 175,art 1 412p 23cm New York academy of sciences,New York, 1970
P Math 2.4510

INTERNATIONAL CONFERENCE ON COMBINATORIAL STRUCTURES Proceedings Calgary,Alta. 1969 Jun Edited by Richard Guy and others xvi,508p 24cm Gordon and Breach,New York, 1970
P Math 2.4511

INTERNATIONAL CONFERENCE ON COMPUTING METHODS IN OPTIMIZATION PROBLEMS 2nd Papers Computing methods in optimization problems San Remo 1968 Sep 9-13 Sponsored by the Society for industrial and applied mathematics Lecture notes in operations research and mathematical economics, 14 bibliog.,illus. Springer,Berlin,1968
Math S 3.1516

INTERNATIONAL CONFERENCE ON COMPUTING METHODS IN OPTIMIZATION PROBLEMS 2nd Papers Computing methods in optimization problems-2 San Remo 1968 Sep 9-13 Edited by Lofti A. Zadeh and others Sponsored by the Society for industrial and applied mathematics Academic press,New York,1969
Eng 41.5865

INTERNATIONAL CONFERENCE ON CONGENITAL MALFORMATIONS 2nd Papers New York 1963 Jul 14-19 International medical congress Edited by Morris Fishbein International medical congress,New York,1864
An 32.2091

INTERNATIONAL CONFERENCE ON COORDINATION CHEMISTRY 6th Advances in the chemistry of coordination compounds Proceedings Detroit,Mich. 1961 Aug 27-Sep 1 American chemical society,and,Wayne state university Edited by Stanley Kirschner Macmillan,New York,1961 Co-sponsored by the United States air force,and the International union of pure and applied physics
Chem 18.0991

INTERNATIONAL CONFERENCE ON COORDINATION CHEMISTRY 8th Proceedings Vienna 1964 Sep 7-11 Edited by V. Gutmann c1964 Damaged copy,title page missing
Chem 18.2001

INTERNATIONAL CONFERENCE ON COORDINATION COMPOUNDS Proceedings Amsterdam 1955 Apr 28-May 3 Nederlandse chemische vereniging Royal Netherlands chemical society,n.p.,c1955
Chem 18.0998

INTERNATIONAL CONFERENCE ON COSMIC RAYS 8th proceedings Jaipur 1963 Dec.2-14 Vols. 1-6 International union of pure and applied physics.Cosmic ray commission Edited by R.R. Daniel and others Sponsored by India. Department of atomic energy 6 vols Bombay, 1964
A Math 4.1159

INTERNATIONAL CONFERENCE ON COSMIC RAYS 11th Invited papers and rapporteur talks Budapest 1969 Aug 25-Sep 4 International union of pure and applied physics Magyar tudomanyos akademia 612p Central research institute of physics,Budapest,1969
TA 15.0551

INTERNATIONAL CONFERENCE ON COSMIC RAYS 11th Proceedings 3: high energy interactions extensive air showers Edited by A. Somogyi Hungarica acta physica.Supplement, 29 766p Akademiai kiado,Budapest,1970
TA 15.0634

INTERNATIONAL CONFERENCE ON COSMIC RAYS 11th Proceedings Budapest 1969 Aug 25-Sep 4 4: muons and neutrons techniques Edited by A. Somogyi Hungarica acta physica.Supplement, 29 Akademiai kiado,Budapest,1970
TA 15.0635

The INTERNATIONAL CONFERENCE ON COSMIC RAYS AND THE EARTH STORM :combined meeting of the International symposium on the earth storm and the International conference on cosmic rays Proceedings Kyoto 1961 Sep 4-15 Vol 2-3 International union of geodesy and geophysics and International astronomical union Edited by Ken-ichi Maeda and Osamu Minakawa Co-sponsored by the International scientific radio union Physical society of Japan.Journal, 17,suppl.A-2,3 2 vols physical society of Japan,Tokyo,1962
TA 15.0038

INTERNATIONAL CONFERENCE ON CRYSTAL GROWTH Proceedings Crystal growth Boston,Mass. 1966 Jun 20-24 International union of pure and applied physics and Air force Cambridge research laboratories Edited by H. Steffen Peiser Physics and chemistry of solids.Supplement Pergamon press,Oxford,1967
Met 25.1520

INTERNATIONAL CONFERENCE ON CRYSTAL GROWTH Proceedings Growth and perfection of crystals Cooperstown,N.Y. 1958 Aug 27-29 General electric company Edited by R.H. Doremus and others Co-sponsored by United States.Air force.Office of scientific research Wiley;Chapman and Hall,New York;London,1958
Eng 41.4438

INTERNATIONAL CONFERENCE ON CRYSTAL GROWTH Proceedings Growth and perfection of crystals Cooperstown,New York 1958 Aug 27-29 United States.Air research and development command and General electric research laboratory Wiley;Chapman and Hall, New York;London,1958
Met 25.1507

INTERNATIONAL CONFERENCE ON CRYSTAL GROWTH 2nd Proceedings Birmingham 1969 Jul 15-19 Edited by F.C. Frank and others Sponsored by International union of pure and applied physics xix,842p 28cm North-Holland,Amsterdam,1968 Reprinted from Journal of crystal growth,3-4,1968
Cav 7.2950

INTERNATIONAL CONFERENCE ON CRYSTAL GROWTH 2nd Proceedings Crystal growth 1968 Birmingham 1968 Jul 15-19 Comite international de croissance crystalline Edited by F.C. Frank and others Jounal of crystal growth,1968,no 3-4 North-Holland, Amsterdam,1968
Min 10.1507

INTERNATIONAL CONFERENCE ON CRYSTAL GROWTH AND EPITAXY FROM THE VAPOUR PHASE 1st Proceedings Crystal growth and epitaxy from the vapour phase Zurich 1970 Sep 23-26 Edited by E. Kaldis and M. Schieber North-Holland,Amsterdam,1971
Met 25.2765

The INTERNATIONAL CONFERENCE ON CRYSTAL LATTICE DEFECTS Proceedings Tokyo 1962 Sep 3-4 Kyoto 1962 Sep 7-12 Pt 1-3 Physical society of Japan Under the auspices of the International union of pure and applied physics Physical society of Japan.Journal, 18; Suppl.1-3 3 vols Physical society of Japan,Tokyo,1963
Cav 7.1343

INTERNATIONAL CONFERENCE ON ELECTRON AND ION BEAM SCIENCE AND TECHNOLOGY 1st Toronto 1964 May 3-7 American institute of mining, metallurgical and petroleum engineers. and Electrochemical society.Electrothermics and metallurgy divisions Edited by Robert Bakish Electrochemical society series Wiley,New York,1965
Met 25.1678

INTERNATIONAL CONFERENCE ON ELECTRON AND ION BEAM SCIENCE AND TECHNOLOGY 1st Toronto 1964 May 3-7 Edited by Robert Bakish Sponsored by the Electrochemical society. Electrothermics and metallurgy divisions Electrochemical society series Wiley,New York,1965
Eng 41.5564

The INTERNATIONAL CONFERENCE ON ELECTRON MICROSCOPY 3rd Proceedings London 1954 Jul 15-21 Edited by R. Ross Royal microscopical society,London,1956
Radioth 35.1582

The INTERNATIONAL CONFERENCE ON ELECTRON MICROSCOPY 3rd Proceedings London 1954 Jul 15-21 International council of scientific unions.Joint commission on electron microscopy Edited by V.E. Cosslett and others Royal microscopical society,London, 1956
Cav 7.1455

INTERNATIONAL CONFERENCE ON ELECTRON MICROSCOPY 4th Berlin 1958 Sep 10-17 Bd 2: biologisch-medizinischer teil Edited by W. Bargmann and others Springer,Berlin,1960
An 32.3165

INTERNATIONAL CONFERENCE ON ELECTRON MICROSCOPY 4TH Intionationaler kongress fur elektronmikroskopie 4th Proceedings Berlin 1958 Sep 10-17 Bd 1: physikalisch-technischer teil International federation of electron microscope societies Edited by G. Mollenstedt and others Springer,Berlin,1960 General editors W.Bargmann and others. Papers in English,French and German
Met 25.2493

INTERNATIONAL CONFERENCE ON ELECTRON SPECTROSCOPY Proceedings Pacific Grove,Calif. 1971 Sep 7-10 Edited by D.A. Shirley xii, 916p 23cm North-Holland,London,1972
Chem 18.2871

INTERNATIONAL CONFERENCE ON ELEMENTARY PARTICLES proceedings Oxford 1965 Sep.19-25 Rutherford high energy laboratory Sponsored by the International union of pure and applied physics Rutherford high energy laboratory, Oxford,1965
A Math 4.0887

The INTERNATIONAL CONFERENCE ON ELEMENTARY PARTICLES Proceedings Aix-en-Provence 1961 Sep 14-20 Vol 1-2 Edited by E. Cremieu-Alcan and others Sponsored by Commissariat a l'energie atomique 2 vols C. E.N.Saclay,Gif-sur-Yvette,1962
Cav 7.0029

INTERNATIONAL CONFERENCE ON ENERGETICS Engineering developments in energy conversion Rochester,N.Y. 1965 Aug 18-20 American society of mechanical engineers American society of mechanical engineers,New York,1965
Eng 41.4865

INTERNATIONAL CONFERENCE ON FATIGUE OF METALS Proceedings Fatigue of metals London 1956 Sep 10-14 and New York 1956 Nov 28-30 Institution of mechanical engineers American society of mechanical engineers Institution of mechanical engineers,London, 1956
Eng 41.3759

INTERNATIONAL CONFERENCE ON FATIGUE OF METALS Proceedings London 1956 Sep 10-14 New York 1956 Nov 28-30 Institution of mechanical engineers and American society of mechanical engineers Institution of mechanical engineers,London,1956
Met 25.0919

INTERNATIONAL CONFERENCE ON FLUID LOGIC AND AMPLIFICATION 1st Proceedings Cranfield 1965 Sep Edited by C.J. Charnley and H.S. Stephens Organized by the British hydromechanics research association British hydromechanics research association,Cranfield, 1965
Eng 41.5895

INTERNATIONAL CONFERENCE ON FRACTURE 1st Proceedings Fracture Sendai 1965 Sep 12-17 Vol 1-3 Organizing committee on fracture Edited by T. Yokobori and others 3 vols Japanese society for strength and fracture of materials,1966
Met 25.0872

INTERNATIONAL CONFERENCE ON FRACTURE 1st Proceedings Sendai 1965 Sep 12-17 Vol 1-3 Edited by T. Yokobori and others 3 vols Japanese society for strength and fracture of materials,Sendai,1966
Eng 41.3807

INTERNATIONAL CONFERENCE ON FRACTURE 2nd Proceedings Brighton 1969 Apr Edited by P.L. Pratt and others Chapman and Hall, London,1969
Met 25.2327

INTERNATIONAL CONFERENCE ON FRACTURE 2nd Proceedings Fracture Brighton 1969 Apr Edited by P.L. Pratt and others Chapman and Hall,London,1969
Eng 41.3808

INTERNATIONAL CONFERENCE ON FUNCTIONAL ANALYSIS AND RELATED TOPICS Proceedings Tokyo 1969 Apr 1-8 International mathematical union Science council of Japan Mathematical society of Japan xxxiii,425p 26cm University of Tokyo press,Tokyo,1970
P Math 2.4279

The INTERNATIONAL CONFERENCE ON FUNDAMENTAL PARTICLES AND LOW TEMPERATURE Report Cambridge 1946 Jul 22-27 Vol 1: fundamental particles Physical society and Cavendish laboratory Physical society,London, 1947
Cav 7.0460

The INTERNATIONAL CONFERENCE ON FUNDAMENTAL PARTICLES AND LOW TEMPERATURES Report Cambridge 1946 Jul 22-27 Vol 2: low temperatures Physical society and Cavendish laboratory Physical society,London, 1947
Cav 7.0458

INTERNATIONAL CONFERENCE ON GAS DISCHARGES AND THE ELECTRICITY SUPPLY INDUSTRY Gas discharges and the electricity supply industry Leatherhead 1962 May 7-11 Edited by J.S. Forrest and others Held at the Central electricity research laboratories Butterworths,London,1962
Eng 41.5415

INTERNATIONAL CONFERENCE ON GEARING Proceedings London 1958 Institution of mechanical engineers illus. Institution of mechanical engineers,London,1960
Eng 41.6224

INTERNATIONAL CONFERENCE ON GENETICS 3rd Report Genetics London 1906 Jul 30-Aug 3 Royal horticultural society Edited by W. Wilkes Spottiswoode,London,1907
BG 38.0523

INTERNATIONAL CONFERENCE ON HARMONIC ANALYSIS Proceedings College Park,Md. 1971 Nov 8-12 Edited by Denny Gulick and Ronald L. Lipsman Held at University of Maryland Lecture notes in mathematics, 266 vi,323p 25cm Springer,Berlin,1972
P Math 2.4283

INTERNATIONAL CONFERENCE ON HIGH-ENERGY PHYSICS 10th Proceedings Rochester,N.Y. 1960 Aug 25-Sep 1 University of Rochester International union of pure and applied physics Edited by E.C.G. Sudarshan and others Co-sponsored by the United States. Office of naval research illus xxv,890p University of Rochester,Rochester,N.Y.,1960
Cav 7.2891

INTERNATIONAL CONFERENCE ON HIGH-ENERGY PHYSICS 13th proceedings Berkeley,Calif. 1966 Aug.31-Sep.7 Sponsored by International union of pure and applied physics University of California press,Berkeley;Los Angeles,1967
A Math 4.0891

The INTERNATIONAL CONFERENCE ON HIGH ENERGY PHYSICS 13th Proceedings Berkeley, Calif. 1966 Aug 31-Sep 7 University of California and United States atomic energy commission Edited by Margaret Alston-Garnjost Sponsored by the International union of pure and applied physics Berkeley, Calif.,1967
Cav 7.0189

INTERNATIONAL CONFERENCE ON HIGH-ENERGY PHYSICS AT CERN 11th Proceedings Geneva 1962 Jul 4-11 European organization for nuclear research Edited by J. Prentki Sponsored by the International union of pure and applied physics Bibliog.,Illus. xxiv, 949p 28cm CERN,Geneva,1962
A Math 4.1544

INTERNATIONAL CONFERENCE ON HIGH MAGNETIC FIELDS Proceedings High magnetic fields Cambridge,Mass. 1961 Nov 1-4 Edited by Henry Kolm Sponsored by the United States. Air force.Office of scientific research.Solid state sciences division M.I.T.press;Wiley, Cambridge,Mass.;New York,1962
Eng 41.5331

INTERNATIONAL CONFERENCE ON HIGH RESOLUTION AUTORADIOGRAPHY OF DIFFUSABLE SUBSTANCES Autoradiography of diffusible substances : based on...lectures presented at an International conference on high resolution autoradiography of diffusible substances,held at the University of Chicago,June 2-4,1968... Edited by Lloyd J. Roth and Walter E. Stumpf Academic press,New York;London,1969
Radioth 35.0207

INTERNATIONAL CONFERENCE ON II-VI SEMICONDUCTING COMPOUNDS Proceedings Providence,R.I. 1967 Sep 6-8 American physical society International union of pure and applied physics Edited by D.G. Thomas 1490p 20cm Benjamin,New York,1967
Cav 7.2681

INTERNATIONAL CONFERENCE ON INFORMATION PROCESSING Proceedings Paris 1959 Jun15-20 Unesco Unesco,Paris,1960
Eng 41.5999

INTERNATIONAL CONFERENCE ON INFORMATION PROCESSING proceedings Paris 1959 June 15-20 Unesco 520p 29cm UNESCO, Paris,1960
Math L 5.0809

INTERNATIONAL CONFERENCE ON INPUT-OUTPUT TECHNIQUES 3rd Proceedings Structural interdependence and economic development Geneva 1961 Sep United Nations.Secretariat Edited by Tibor Barna and others Sponsored jointly by the Harvard economic research project Macmillan,London, 1963
Geog 13.2417

INTERNATIONAL CONFERENCE ON IONIZATION PHENOMENA IN GASES 5th proceedings Munich 1961 Aug 28-Sep 1 Vol. 1 Edited by H. Maecker Organised by Verband Deutscher physikalischer gesellschaften Series in physics North Holland,Amsterdam,1962
A Math 4.0620

INTERNATIONAL CONFERENCE ON LABORATORY ASTROPHYSICS Selection of papers Lunteren 1968 Sep 2-7 Edited by J. Rosenberg Organised under the auspices of the Nederlandse natuurkundig vereniging Physica, 41,no 1 223p North-Holland, Amsterdam,1969
TA 15.0580

INTERNATIONAL CONFERENCE ON LABORATORY ASTROPHYSICS Selection of papers Lunteren 1968 Sep 2-7 Nederlandse natuurkundige vereniging Edited by J. Rosenberg Physica, 41,no 1 223p North-Holland,Amsterdam,1969
Obs 6.3347

The INTERNATIONAL CONFERENCE ON LATTICE DYNAMICS Proceedings Lattice dynamics Copenhagen 1963 Aug 5-9 International union of pure and applied physics and University of Pennsylvania Edited by R.F. Wallis With the support of the International atomic energy agency 730p Pergamon press,Oxford,1965
Cav 7.0542

INTERNATIONAL CONFERENCE ON LEUKEMIA LYMPHOMA Proceedings Ann Arbor 1967 University of Michigan Edited by Chris J.D. Zarafonetis Lea and Febiger;Kimpton,Philadelphia;London, 1968
Med 36.0376

The INTERNATIONAL CONFERENCE ON LOCALIZED EXCITATIONS IN SOLIDS 1st Localized excitations in solids Irvine 1967 Sep 18-22 Edited by R.F. Wallis Plenum press,New York,1968
TA 15.0440

INTERNATIONAL CONFERENCE ON LOCALIZED EXCITATIONS IN SOLIDS 1st Proceedings Localized excitations in solids Irvine 1967 Sep 18-22 International union of pure and applied physics National science foundation Edited by R.F. Wallis xvi,782p 26cm Plenum press,New York,1968
Cav 7.2857

The INTERNATIONAL CONFERENCE ON LOW TEMPERATURE PHYSICS 7th Proceedings Toronto 1960 Aug 29-Sep 3 International union of pure and applied physics Edited by J.M. Graham and A.C. Hollis-Hallett With the co-operation of the Canadian association of physicists University of Toronto press;North-Holland,Toronto;Amsterdam,1961
Cav 7.2330

INTERNATIONAL CONFERENCE ON LOW TEMPERATURE PHYSICS 8th Proceedings London 1962 Sep 16-22 Queen Mary college,London Edited by R.O. Davies Butterworths,London, 1963 Includes Fritz London memorial lecture delivered by Bardeen,J.
Cav 7.1459

INTERNATIONAL CONFERENCE ON LOW TEMPERATURE PHYSICS 8th Proceedings London 1962 Sep 16-23 Edited by R.O. Davies Held at Queen Mary college Butterworths,London, 1963
Eng 41.7396

INTERNATIONAL CONFERENCE ON LOW TEMPERATURE PHYSICS 10th Proceedings Moscow 1966 Aug 31-Sep 6 Vol 1: properties of helium International union of pure and applied physics Akademiya nauk S.S.S.R. Edited by M.P. Malkov 552p 22cm Moscow, 1967
Cav 7.3030

INTERNATIONAL CONFERENCE ON LOW TEMPERATURE PHYSICS 10th Proceedings Moscow 1966 Aug 31-Sep 6 Vol 3: the electronic properties of metals International union of pure and applied physics Akademiya nauk S.S. S.R. 404p 22cm moscow,1967
Cav 7.3032

INTERNATIONAL CONFERENCE ON LOW TEMPERATURE PHYSICS 10th Proceedings Moscow 1966 Aug 31-Sep 6 Vol 4: antiferromagnetism International union of pure and applied physics Akademiya nauk S.S. S.R. Edited by M.P. Malkov 361p 22cm Moscow,1967
Cav 7.3033

INTERNATIONAL CONFERENCE ON LOW TEMPERATURE PHYSICS 10th Proceedings Vol 2 a-b: superconductivity International union of pure and applied physics Akademiya nauk S.S. S.R. Edited by M.P. Malkov 22cm 2 vols
Cav 7.3031

INTERNATIONAL CONFERENCE ON LOW TEMPERATURE PHYSICS 12th Proceedings Kyoto 1970 Sep 4-10 International union of pure and applied physics Physical society of Japan Edited by Eizo Kanda Organized by the Science council of Japan photos,diagrms 895p 26cm Academic press of Japan,Tokyo, 1971
Cav 7.3028

INTERNATIONAL CONFERENCE ON LOW TEMPERATURE SCIENCE Papers Sapporo,Japan 1966 Aug 14-19 conference on cryobiology.Cellular injury and resistance in freezing organisms Edited by Eizo Asahina Sponsored by the Institute of low temperature science illus xxiii,257p 27cm Sapporo,1967
Sco 14.0144

INTERNATIONAL CONFERENCE ON LOW TEMPERATURE SCIENCE 2nd Proceedings Cellular injury and resistance in freezing organisms Sapporo 1966 Aug 14-19 Vol 2: conference on cryobiology Hokkaido university.Institute of low temperature science Edited by Eizo Asahina illus Hokkaido university. Institute of low temperature science,Sapporo, 1967 Title page in English and Japanese Conference held in commemoration of the 25th anniversary of the establishment of the Institute
Path 30.2625

INTERNATIONAL CONFERENCE ON LOW TEMPERATURE SCIENCE Proceedings 1st Physics of snow and ice Sapporo 1966 Aug 14-19 1 pt 2: conference on physics of snow and ice (part 2) Institute of low temperature science Edited by Hirobumi Oura illus 1967p 27cm Hokkaido university,Sapporo,1957
Sco 14.7411

INTERNATIONAL CONFERENCE ON LOW-TEMPERATURE PHYSICS 8th Proceedings Low temperature physics London 1962 Sep 16-22 Edited by R.O. Davies Butterworths,London, 1963
Met 25.1180

INTERNATIONAL CONFERENCE ON LUMINESCENCE Proceedings Budapest 1966 International union of pure and applied physics Magyar tudomanyos akademia Edited by G. Szigeti diagrms 24cm 2 vols Akademiai kiado,Budapest,1968
Cav 7.3097

INTERNATIONAL CONFERENCE ON LUMINESCENCE Proceedings Newark,Del 1969 Aug 25-29 Edited by F. Williams graphs,diagrs xvi, 959p 24cm North-Holland,Amsterdam,1970 Reprinted from 'Journal of luminescence',vol 1-2
Cav 7.2697

INTERNATIONAL CONFERENCE ON MACHINE TRANSLATION OF LANGUAGES AND APPLIED LANGUAGE ANALYSIS proceedings Teddington 1961 Sep 5-8 Vol. 1-2 National physical laboratory National physical laboratory.Proceedings of symposia, 13 23cm 2 vols H.M.S.O., London,1962
Math L 5.0960

INTERNATIONAL CONFERENCE ON MAGNETICS 3rd Proceedings Intermag conference Washington,D.C. 1965 Apr 21-23 Spsonsored by the Institute of electrical and electronics engineers.Magnetics group Institute of electrical and electronics engineers,New York, 1965
Eng 41.5345

INTERNATIONAL CONFERENCE ON MAGNETISM Proceedings Nottingham 1964 Sep Institute of physics and the Physical society London,c1964
Met 25.2354

INTERNATIONAL CONFERENCE ON MAGNETISM Proceedings Nottingham 1964 Sep 6-11 Institute of physics and the Physical society With the support of the International union of pure and applied physics illus 898p London,c1965
Cav 7.2882

INTERNATIONAL CONFERENCE ON MAGNETISM AND CRYSTALLOGRAPHY Proceedings Kyoto 1961 Sep 25-30 Pt 1: magnetism Science council of Japan and Crystallographic society of Japan Edited by Kei Josida and others Held under the auspices of International union of pure and applied physics Physical society of Japan.Journal, 17,B 1 xi,718p Physical society of Japan, Tokyo,1962
Cav 7.1337

INTERNATIONAL CONFERENCE ON MAGNETISM AND CRYSTALLOGRAPHY Proceeding Kyoto 1961 Sep 25-30 Pt 2: electron and neutron diffraction Science council of Japan and Crystallographic society of Japan Edited by Shizuo Miyake and others Sponsored by the International union of pure and applied physics Physical society of Japan.Journal, 17,B 2 396p physical society of Japan, Tokyo,1962
Cav 7.1338

INTERNATIONAL CONFERENCE ON MAGNETISM AND CRYSTALLOGRAPHY Proceeding Kyoto 1961 Sep 25-30 Pt 3: neutron diffraction: study of magnetic materials Science council of Japan and Crystallographic society of Japan Edited by Takeo Nagamiya and Shizno Miyake Under the auspices of International union of pure and applied physics Physical society of Japan.Journal, 17,B 3 71p physical society of Japan,Tokyo,1962
Cav 7.1339

INTERNATIONAL CONFERENCE ON MECHANICAL BEHAVIOR OF MATERIALS Proceedings Kyoto 1971 Aug 15-20 Society of materials science 6 vols Society of materials science,Kyoto,1971
Eng 41.8524

INTERNATIONAL CONFERENCE ON 'MECHANICAL BEHAVIOUR OF MATERIALS' Proceedings Mechanical behaviour of materials Kyoto 1971 Aug 15-20 Society of materials science,Japan Society of materials science,Kyoto,1972
Met 25.2737

INTERNATIONAL CONFERENCE ON MECHANISMS OF SALIVARY SECRETION AND THEIR REGULATION Proceedings Secretory mechanisms of salivary glands Birmingham,Ala. 1966 Aug 9-11 Edited by Leon H. Schneyer and Charlotte A. Schneyer Held at the University of Alabama medical center Academic press,New York;London,1967
Bal 39.1265

INTERNATIONAL CONFERENCE ON MEDICAL ELECTRONICS 2nd Proceedings Medical electronics Paris 1959 Jun 24-27 International federation for medical electronics Edited by C.N. Smyth Iliffe,London,1960
Psy 31.3062

INTERNATIONAL CONFERENCE ON MEDICAL ELECTRONICS 2nd Proceedings Medical electronics Paris 1959 Jun 24-27 Edited by C.N. Smyth Iliffe,London,1960
VA 19.0363

INTERNATIONAL CONFERENCE ON MEDICAL PHYSICS Review papers Aspects of medical physics Harrogate 1965 Sep 8-10 Edited by J. Rotblat Convened by the International organization of medical physics Taylor and Francis,London,1966
Radioth 35.1671

INTERNATIONAL CONFERENCE ON METALLURGY AND MATERIALS SCIENCE Proceedings Heterogeneous kinetics at elevated temperatures Philadelphia,Pa. 1969 Sep 8-10 Edited by G.R. Belten and W.L. Worrell Held at the University of Pennsylvania Plenum press,New York,1970
Met 25.2783

INTERNATIONAL CONFERENCE ON METHODOLOGIES OF PATTERN RECOGNITION Proceedings Methodologies of pattern recognition Honolulu 1968 Jan 24-26 Edited by Satosi Watanabe Academic press,New York;London,1969
An 32.4412

INTERNATIONAL CONFERENCE ON METHODOLOGIES OF PATTERN RECOGNITION Proceedings Methodologies of pattern recognition Honolulu 1968 Jan 24-26 University of Hawaii Sponsored by the United States.Air force.Office of scientific research bibliog., illus. Academic press,New York;London,1969
Math S 3.1521

INTERNATIONAL CONFERENCE ON MICROWAVE TUBES 6th Microwave and optical generation and amplification Cambridge 1966 Institution of electrical engineers Institution of electrical engineers.Conference publication, 27 I.E.E.,London,1967
Eng 41.5588

INTERNATIONAL CONFERENCE ON NON-DESTRUCTIVE TESTING 4th Proceedings London 1963 Sep 9-13 Sponsored by the British national committee for non-destructive testing Butterworths,London,1964
Eng 41.3847

INTERNATIONAL CONFERENCE ON NONLINEAR OSCILLATIONS 5th Abstracts of papers Kiev 1969 Aug 25-Sep 5 267p 20cm Academy of sciences of the USSR,Kiev,1969 Typescript
P Math 2.3734

The INTERNATIONAL CONFERENCE ON NUCLEAR AND MESON PHYSICS Proceedings Glasgow 1954 Edited by E.H. Bellamy and R.G. Moorhouse Sponsored by the International union of pure and applied physics 352p Pergamon press, London,1955
Cav 7.0908

INTERNATIONAL CONFERENCE ON NUCLEAR PHYSICS Proceeding Low energy nuclear interactions and nuclear structure Paris 1958 Jul 7-12 Faculte des sciences de Paris et de province and Centre national de la recherche scientifique Under the auspices of International union of pure and applied physics Crosby Lockwood,London,1959
Cav 7.1759

INTERNATIONAL CONFERENCE ON NUCLEAR STRUCTURE Tokyo 1967 Sep Physical society of Japan illus vi,755p 26cm Physical society of Japan,Tokyo,1968
Cav 7.3196

INTERNATIONAL CONFERENCE ON OPERATIONAL RESEARCH 3rd proceedings Oslo 1963 Jul.1-5 Operational research society,Oslo Edited by G. Kreweras and G. Morlat Sponsored by International federation of operational research societies xii,959p 25cm Dunod; English universities press,Paris;London,1964
Math 3.0915

INTERNATIONAL CONFERENCE ON OPERATIONAL RESEARCH 4th Proceedings Boston,Mass. 1966 Aug 29-Sep 2 Edited by David B. Hertz and Jacques Melese Organised by the International foundation of operational research societies Operations research society of America.Publications in operations research, 15 Wiley-Interscience,New York, 1966
Eng 41.1219

INTERNATIONAL CONFERENCE ON ORGANOGENESIS Organogenesis Baltimore,Md. 1964 Sep 6-12 International institute of embryology Edited by Robert L. DeHaan and Heinrich Ursprung Holt,Rinehart and Winston,New York,1965
Bal 39.0490

INTERNATIONAL CONFERENCE ON PHENOMENA IN IONIZED GASES 9th Proceedings Bucharest 1969 Institutul de fizica,Bucharest Institute of physics,Bucharest,1969
Eng 41.8553

INTERNATIONAL CONFERENCE ON PHOTOCONDUCTIVITY 3rd Proceedings Stanford,Calif. 1969 Aug 12-15 International union of pure and applied physics American physical society United States.Office of naval research Edited by E.M. Pell Bibliog, diagrms,graphs xi,410p 26cm Pergamon, Oxford,1971
Cav 7.3073

INTERNATIONAL CONFERENCE ON PHYSICS Reports on symbols,units and nomenclature approved by the general assembly of the Union at its meeting in London on October 5th,1934 By R.T. Glazebrook International union of pure and applied physics and Physical society 40p The Physical society,London,1935
Nap 11.0481

INTERNATIONAL CONFERENCE ON PHYSICS :reports on symbols,units and nomenclature approved by the general assembly of the Union London 1934 Oct.5 International union of pure and applied physics 40p Physical society, London,1935
Obs 6.1963

INTERNATIONAL CONFERENCE ON PHYSICS Papers London 1934 Vol 1: nuclear physics International union of pure and applied physics and Physical society Physical society,London,1935
Cav 7.1452

INTERNATIONAL CONFERENCE ON PHYSICS Papers and discussions London 1934 Vol 1-2: nuclear physics:solid state of matter International union of pure and applied physics and the Physical society 2 vols Physical society,London,1935
Met 25.2343

INTERNATIONAL CONFERENCE ON PHYSICS papers and discussions London 1934 Vol 2: the solid state of matter International union of pure and applied physics,and,Physical society Physical society,London,1937
Min 10.1078

INTERNATIONAL CONFERENCE ON PLANT GROWTH REGULATION 4th Plant growth regulation Yonkers 1959 Aug 10-14 Sponsored by the Boyce Thompson institute for plant research Iowa state university press, Ames,Iowa,1961
Bot 42.1830

INTERNATIONAL CONFERENCE ON PLANT GROWTH SUBSTANCES 6th Biochemistry and physiology of plant growth substances Ottawa 1967 Jul 24-29 Edited by F. Wightman and G. Setterfield Held at Carleton University Runge press,Ottawa,1968
Bioch 33.0779

INTERNATIONAL CONFERENCE ON PLANT GROWTH SUBSTANCES 6th Biochemistry and physiology of plant growth substances Ottawa 1967 Jul 24-29 Edited by F. Wightman and G. Setterfield Held at the Carleton university xii,1642p Runge press,Ottawa,1968 On the occasion of the centennial of Canadian confederation
Bot 42.1833

INTERNATIONAL CONFERENCE ON PLANT GROWTH SUBSTANCES 6th Proceedings Biochemistry and physiology of plant growth substances Ottawa 1967 Jul 24-29 National research council of Canada Carleton university Edited by F. Wightman and G. Setterfield Runge press,Ottawa,1968
Gen 34.1408

INTERNATIONAL CONFERENCE ON PLANT GROWTH SUBSTANCES 7th Proceedings Canberra 1970 Dec 7-11 Edited by Denis J. Carr Springer,Berlin,1972
Bioch 33.2231

INTERNATIONAL CONFERENCE ON PLANT GROWTH SUBSTANCES 5TH Regulateurs naturels de la croissance vegetales Colloque international sur les substances de croissance vegetales 5e Actes Gif sur Yvette 1963 Jul 15-20 Centre national de la recherche scientifique Centre national de la recherche scientifique. Colloques internationaux, 123 Paris,1964 Text in English and French
Bioch 33.0747

INTERNATIONAL CONFERENCE ON PLUTONIUM METALLURGY 2nd Proceedings Plutonium 1960 Grenoble 1960 Apr 19-22 Societe Francaise de metallurgie and Commissariat a l'energie atomique Edited by Emmanuel Grison and others Cleaver-Hume,London,1961 Papers in English and French
Met 25.1549

INTERNATIONAL CONFERENCE ON RADIATION BIOLOGY AND CANCER Proceedings Kyoto 1966 Nov 1-2 Edited by Tsutomu Sugahara Under the auspices of Nippon societas radiologica. Biological subgroup Radiation society of Japan,1967
Radioth 35.1757

INTERNATIONAL CONFERENCE ON RADIOBIOLOGY 4th Proceedings Progress in radiobiology Cambridge 1955 Aug 14-17 Edited by Joseph S. Mitchell and others port. Oliver and Boyd,Edinburgh;London,1956
Radioth 35.1707

INTERNATIONAL CONFERENCE ON RADIOBIOLOGY 4th Proceedings Progress in radiobiology Edited by Joseph S. Mitchell and others Oliver and Boyd,Edinburgh;London,1956
Phys 20.0993

INTERNATIONAL CONFERENCE ON RADIOBIOLOGY 5th Proceedings Advances in radiobiology Stockholm 1956 Aug 15-19 Edited by George Carl de Hevesy and others port. Oliver and Boyd,Edinburgh;London,1957
Radioth 35.1708

INTERNATIONAL CONFERENCE ON RADIOISOTOPES IN SCIENTIFIC RESEARCH 1st Proceedings Paris 1957 Sep Vol 1: research with radioisotopes in physics and industry Unesco Edited by R.C. Extermann Pergamon,London, 1958
Met 25.1634

INTERNATIONAL CONFERENCE ON RADIOISOTOPES IN SCIENTIFIC RESEARCH 1st Proceedings Radioisotpoes in scientific research Paris 1957 Sep 4-20 1-4 Edited by R.C. Extermann Held under the auspices of Unesco illus. 4 vols Pergamon press,London,1958
Radioth 35.1070

INTERNATIONAL CONFERENCE ON RHEOLOGY 2nd proceedings Oxford 1953 Jul 26-31 Edited by C.G.W. Harrison Butterworths, London,1954
A Math 4.0367

INTERNATIONAL CONFERENCE ON SARCOIDOSIS Proceedings 5th Prague 1969 Jun 16-21 Universita Karlova Edited by Ladislav Levinsky and Frantisek Macholda Univerzita Karlova,Prague,1971
PGMS 29.0666

INTERNATIONAL CONFERENCE ON SINGLE CRYSTAL FILMS Proceedings Single crystal films Blue Bell,Pa. 1963 Edited by M.H. Francombe and Hiroshi Sato Pergamon press,Oxford,1964
Eng 41.4458

INTERNATIONAL CONFERENCE ON SINGLE-CRYSTAL FILMS Proceedings Blue Bell,Pa. 1963 May Philco scientific laboratories Pergamon press,Oxford,1964
Cav 7.0244

INTERNATIONAL CONFERENCE ON SINTERING AND RELATED PHENOMENA Cleveland 1965 Jun 21-23 University of Notre Dame Edited by G.C. Kuczynski and others Gordon and Breach,New York,1967
Met 25.1285

INTERNATIONAL CONFERENCE ON SOIL MECHANICS AND FOUNDATION ENGINEERING 2nd Proceedings Rotterdam 1948 Jun 21-30 Vol 1-7 7 vols Rotterdam,1948
Eng 41.3139

INTERNATIONAL CONFERENCE ON SOIL MECHANICS AND FOUNDATION ENGINEERING 3rd Proceedings Zurich 1953 Aug 16-27 Vol 1-3 3 vols Zurich,1953
Eng 41.3140

INTERNATIONAL CONFERENCE ON SOIL MECHANICS AND FOUNDATION ENGINEERING 4th Proceedings London 1957 Aug 12-14 3 vols Butterworths,London,1957
Eng 41.3141

INTERNATIONAL CONFERENCE ON SOIL MECHANICS AND FOUNDATION ENGINEERING 5th Proceedings Paris 1961 Jul 17-22 3 vols Dunod, Paris,1961
Eng 41.3142

INTERNATIONAL CONFERENCE ON SOIL MECHANICS AND FOUNDATION ENGINEERING 6th Proceedings Montreal 1965 Sep 8-15 3 vols University of Toronto press,Toronto,1965
Eng 41.3143

INTERNATIONAL CONFERENCE ON SOIL MECHANICS AND FOUNDATION ENGINEERING 7th Proceedings Mexico,D.F. 1969 Vol 1-3 3 vols 1969
Eng 41.3144

INTERNATIONAL CONFERENCE ON SOILS MECHANICS AND FOUNDATION ENGINEERING Proceedings Cambridge,Mass. 1936 Jun 22-26 Vol 1-3 3 vols Cambridge,Mass.,1936
Eng 41.3138

INTERNATIONAL CONFERENCE ON SOLVENT EXTRACTION CHEMISTRY Proceedings Gothenburg 1966 Aug 27-Sep 1 By D. Dyrssen and others Chalmers university of technology and Goteborgs universitet North-Holland, Amsterdam,1967
Met 25.2292

INTERNATIONAL CONFERENCE ON SPACE STRUCTURES Guildford 1960 Sep University of Surrey Edited by R.M. Davies Blackwell,Oxford,1967
Eng 41.2739

INTERNATIONAL CONFERENCE ON SPECTROSCOPY Bombay 1967 Jan 8-18 India.Department of atomic energy.Spectroscopy division 290p International council of scientific unions, Bombay,1968
TA 15.0636

INTERNATIONAL CONFERENCE ON SPECTROSCOPY AT RADIO-FREQUENCIES :a conference Proceedings Amsterdam 1950 Sep Netherlands physical society Physica, 17,3-4 169-484p 1951
Cav 7.0055

An INTERNATIONAL CONFERENCE ON STRUCTURE AND PROPERTIES OF THIN FILMS Proceedings Structure and properties of thin films Bolton Landing,N.Y. 1959 Sep 9-11 General electric company Edited by C.A. Neugebauer and others John Wiley,New York,1959
Cav 7.0265

INTERNATIONAL CONFERENCE ON TAXONOMIC BIOCHEMISTRY,PHYSIOLOGY AND SEROLOGY Lawrence,Kan. 1962 Sep Edited by Charles A. Leone xi,728p Ronald press,New York,1964
Bot 42.3437

INTERNATIONAL CONFERENCE ON THE APPLICATIONS OF MODEL THEORY TO ALGEBRA ANALYSIS AND PROBABILITY Proceedings Applications of model theory to algebra,analysis,and probability Pasadena 1967 May 23-26 Edited by W.A.J. Luxemburg Held at the California institute of technology Bibliog. v,307p 23cm Holt,Rinehart and Winston,New York,1968
P Math 2.3241

An INTERNATIONAL CONFERENCE ON THE ATOMIC MECHANISMS OF FRACTURE Proceedings Fracture Swampscott,Mass. 1959 Apr 12-16 Edited by B.L. Averbach and others Sponsored by the National science foundation M.I.T. press;Wiley,Cambridge,Mass.;New York,1959
Eng 41.3767

INTERNATIONAL CONFERENCE ON THE BIOCHEMICAL PROBLEMS OF LIPIDS 5th Proceedings Biochemistry of lipids Vienna 1958 Sep Edited by G. Popjak Pergamon press,Oxford, 1960 Section 18 of the 4th international congress of biochemistry
Col S 12.0102

The INTERNATIONAL CONFERENCE ON THE BIOLOGY OF CUTANEOUS CANCER 1st Philadelphia,Pa. 1962 Apr 6-11 Temple university.School of medicine and International union against cancer Edited by Frederick Urbach With financial support from Merck,Sharp and Dohme. Postgraduate program National cancer institute.Monograph, 10 US government printing office,Washington,D.C.,1963
Path 30.2423

The INTERNATIONAL CONFERENCE ON THE BIOLOGY OF CUTANEOUS CANCER 1st Proceedings Philadelphia 1962 Apr 6-11 National cancer institute Edited by Frederick Urbach National cancer institute.Monograph, 10 U.S. Department of health,education,and welfare, Bethesda,Md.,1963 Sponsored by the skin and cancer hospital,Dept.of dermatology,Temple university school of medicine and The Committee on geographie pathology,Unio internationalis contra cancrum.A Merck Sharp and Dohme medical research conference
Bioch 33.0981

INTERNATIONAL CONFERENCE ON THE CHEMISTRY OF NATURAL PRODUCTS 7th Abstracts Riga 1970 Jun 21-25 International union of pure and applied chemistry Edited by M.N. Kolosov Zinatne,Riga,1970
Bioch 33.2318

INTERNATIONAL CONFERENCE ON THE DEFORMATION OF CRYSTALLINE SOLIDS Ottawa 1966 Aug 22-26 National research council of Canada Canadian journal of physics, 2,pt 2-3 Ottawa,1967
Met 25.2491

INTERNATIONAL CONFERENCE ON THE DEVELOPMENT, GROWTH AND REGENERATION OF THE NERVOUS SYSTEM Genetic neurology Chicago,Ill. 1949 Mar 21-25 Edited by Paul Weiss Sponsored by the International union of biological sciences University of Chicago,Ill.,Chicago,Ill.,1950
An 32.4329

INTERNATIONAL CONFERENCE ON THE EARTH SCIENCES Papers Advances in earth science Cambridge 1964 Sep 30-Oct 2 Edited by P.M. Hurley Massachusetts institute of technology press,Cambridge,Mass.,1966
Min 10.0576

INTERNATIONAL CONFERENCE ON THE EARTH SCIENCES contributions Advances in earth sciences Cambridge,Mass. 1964 Sep 30-Oct 2 Massachusetts institute of technology Edited by P.M. Hurley M.I.T. press,Cambridge,Mass., 1966
Geol 8.4429

The INTERNATIONAL CONFERENCE ON THE ELECTRONIC ASPECTS OF BIOCHEMISTRY Electronic aspects of biochemistry New York 1967 Dec 11-13 By Bernard Pullman and others New York academy of sciences Edited by Philip Feigelson New York academy of sciences. Annals, 158,p.1-438 New York,1969
Bioch 33.0313

The INTERNATIONAL CONFERENCE ON THE ELECTRONIC PROPERTIES OF METALS AT LOW TEMPERATURES Report Geneva,N.Y. 1958 Aug 25-29 Hobart college and International union of pure and applied physics Co-sponsored by General electric research laboratory 244p c1958 Privately printed report to the sponsors
Cav 7.0895

INTERNATIONAL CONFERENCE ON THE MICROWAVE BEHAVIOUR OF FERRIMAGNETICS AND PLASMAS Microwave behaviour of ferrimagnetics and plasmas London 1965 Sep 13-17 Sponsored by Institution of electrical engineers. Electronics division Institution of electrical engineers.Conference publication, 13 Institution of electrical engineers, London,1965
Eng 41.5713

INTERNATIONAL CONFERENCE ON THE ORIGIN OF LIFE Proceedings Chemical evolution and the origin of life Vol 1: molecular evolution Edited by Rene Buvet and C. Ponnamperuma 560p North-Holland,Amsterdam,1971
TA 15.0690

INTERNATIONAL CONFERENCE ON THE PEACEFUL USES OF ATOMIC ENERGY Exploration for nuclear raw materials Geneva 1955 Aug Edited by Robert D. Nininger Geneva series on the peaceful uses of atomic energy bibliog., illus. Macmillan and co,London,1956 Data submitted to the International conference on the peaceful uses of atomic energy
Geol 8.1661

INTERNATIONAL CONFERENCE ON THE PEACEFUL USES OF ATOMIC ENERGY Metallurgy and fuels Geneva 1955 Vol 1 Edited by H.M. Finniston and J.P. Howe Progress in nuclear energy, 5 Pergamon press,London,1956 Includes papers presented at the United Nations conference on the peaceful uses of atomic energy
Met 25.1560

INTERNATIONAL CONFERENCE ON THE PEACEFUL USES OF ATOMIC ENERGY Primenenie radioaktivnykh izotopov v promyshlennosti meditsine i selskom khozyaistve illus. Izdatelstvo akademii nauk S.S.S.R.,Moscow,1956 Papers presented at the International conference on the peaceful uses of atomic energy, Geneva, 1955
Bot 42.0386

INTERNATIONAL CONFERENCE ON THE PEACEFUL USES OF ATOMIC ENERGY Process chemistry Geneva 1955 Vol 1-2 Edited by F.R. Bruce and others Progress in nuclear energy, 3 Pergamon press,London,1956 Includes papers presented at the United Nations conference on the peaceful uses of atomic energy
Met 25.1558

INTERNATIONAL CONFERENCE ON THE PEACEFUL USES OF ATOMIC ENERGY Technology and engineering Geneva 1955 Vol 1: reactor coolants, moderators,heat transfer,reactor chemistry and corrosion of reactor materials Edited by R. Hurst and S. McLain Progress in nuclear energy, 4 Pergamon,London,1956 Includes papers presented at the United Nations conference on the peaceful uses of atomic energy
Met 25.1565

The INTERNATIONAL CONFERENCE ON THE PEACEFUL USES OF ATOMIC ENERGY Geneva 1955 Aug 8-20 Vol 8: production technology of the materials used for nuclear energy United Nations United Nations,New York,1956
Met 25.1564

INTERNATIONAL CONFERENCE ON THE PEACEFUL USES OF ATOMIC ENERGY Proceedings Geneva 1955 Aug 8-20 10-14 United nations 5 vols United Nations,New York,1955-56
Radioth 35.1009

INTERNATIONAL CONFERENCE ON THE PEACEFUL USES OF ATOMIC ENERGY Proceedings Geneva 1955 Aug 8-20 Vol 8: production technology of the materials used for nuclear energy United nations 627p United nations,New York,1956
Chem E 24.1650

INTERNATIONAL CONFERENCE ON THE PEACEFUL USES OF ATOMIC ENERGY Proceedings Geneva 1955 Aug 8-20 Vol 9: reactor technology and chemical processing United nations 771p United nations,New York,1956
Chem E 24.1651

INTERNATIONAL CONFERENCE ON THE PEACEFUL USES OF ATOMIC ENERGY Proceedings Geneva 1955 Aug 8-20 Vol 10: radioactive isotopes and nuclear radiations in medicine United nations United nations,New York,1956
Phys 20.2093

INTERNATIONAL CONFERENCE ON THE PEACEFUL USES OF ATOMIC ENERGY Proceedings Geneva 1955 Aug 8-20 Vol 12: radioactive isotopes and ionizing radiations in agriculture, physiology and biochemistry United Nations United Nations,New York,1956
Phys 20.2244

INTERNATIONAL CONFERENCE ON THE PEACEFUL USES OF ATOMIC ENERGY 2nd Proceedings Geneva 1958 Sep 1-13 18,21-26,29 United Nations 8 vols United Nations, Geneva,1958
Radioth 35.1079

INTERNATIONAL CONFERENCE ON THE PEACEFUL USES OF ATOMIC ENERGY 2nd Proceedings Geneva 1958 Sep 1-13 Vol 31-32 United Nations 2 vols United Nations, Geneva,1958
Eng 41.4466

INTERNATIONAL CONFERENCE ON THE PEACEFUL USES OF ATOMIC ENERGY 1st Proceedings Geneva 1955 Aug 8-20 Vol 6: geology of uranium and thorium United Nations United Nations,New York,1956
Min 10.0715

INTERNATIONAL CONFERENCE ON THE PEACEFUL USES OF ATOMIC ENERGY 1st Proceedings Geneva 1955 Aug 8-20 Vol 8: production technology of the materials used for nuclear nergy United Nations United Nations,New York,1956
Min 10.0718

INTERNATIONAL CONFERENCE ON THE PEACEFUL USES OF ATOMIC ENERGY,2ND Higher order differences in the numerical solution of two-dimensional neutron diffusion equations By J. A. Nohel and W.P. Timlake bibliog. 8p 1958 A paper read at the 2nd United Nations International conference on the peaceful uses of atomic energy
Math L 5.1799

INTERNATIONAL CONFERENCE ON THE PHYSICS OF ELECTRONIC AND ATOMIC COLLISIONS 7th Physics of electronics and atomic collisions Amsterdam 1971 Jul 26-30 Edited by T.R. Govers and F.J. De Heer xi,496p 23cm North Holland,London,1972
Chem 18.2958

The INTERNATIONAL CONFERENCE ON THE PHYSICS OF ELECTRONIC AND ATOMIC COLLISIONS 3rd Proceedings Atomic collision processes London 1963 Jul 22-23 University college, London Edited by M.R.C. McDowell North-Holland,Amsterdam,1964
Cav 7.1461

INTERNATIONAL CONFERENCE ON THE PHYSICS OF ELECTRONIC AND ATOMIC COLLISIONS 5th Invited papers Physics of electronic and atomic collisions Leningrad 1967 Jul 17-23 Edited by Lewis M. Branscomb Sponsored by the International union of pure and applied physics Bibliog,illus xvi,200p 23cm University of Colorado press,Boulder,Colo., 1968
A Math 4.1906

INTERNATIONAL CONFERENCE ON THE PHYSICS OF ELECTRONIC AND ATOMIC COLLISIONS 7th Abstracts of papers Amsterdam 1971 Jul 26-30 Vol 1-2 Edited by L.M. Branscomb and others 23cm 2 vols North-Holland,Amsterdam,1971
Chem 18.2959

The INTERNATIONAL CONFERENCE ON THE PHYSICS OF SEMICONDUCTORS Report Exeter 1962 Jul 16-20 Institute of physics and Physical society Edited by A.C. Stickland Under the auspices of the International union of pure and applied physics 909p Institute of physics,London,1962
Cav 7.0750

INTERNATIONAL CONFERENCE ON THE PHYSICS OF SEMICONDUCTORS 7th Paris 1964 Jul 19-24 Vol 2: plasma effects in solids Edited by J. Bok Academic press;Dunod,New York;Paris,1965
Eng 41.5432

INTERNATIONAL CONFERENCE ON THE PHYSICS OF SEMICONDUCTORS 7th Plasma effects in solids Paris 1964 Vol 2: plasma effects in solids International union of pure and applied physics photos,diagrms, graphs 221p 24cm Dunod,Paris,1965 Co-sponsored by the French physical society, the French ministry for scientific research and Unesco
Cav 7.3047

INTERNATIONAL CONFERENCE ON THE PHYSICS OF SEMICONDUCTORS 8th Proceedings Kyoto 1966 Sep 8-13 Physical society of Japan Under the auspices of the International union of pure and applied physics Physical society of Japan.Journal, 21,suppl. 774p Physical society of Japan, Tokyo,1966
Cav 7.1341

INTERNATIONAL CONFERENCE ON THE PHYSICS OF SEMICONDUCTORS 9th Proceedings Moscow 1968 Jul 23-29 Vol 1-2 Akademiya nauk S.S.S.R. English edition 634p 27cm 2 vols Publishing house 'Nauka',Leningrad,1968
Cav 7.2955

The INTERNATIONAL CONFERENCE ON THE PHYSICS OF SOLIDS AT HIGH PRESSURES 1st Proceedings Tucson,Ariz. 1965 Apr 20-23 University of Arizona Edited by C.T. Tumizuka and R.M. Emrick Jointly sponsored by the United States.Air force.Office of scientific research Academic press,New York,1965
Cav 7.2354

INTERNATIONAL CONFERENCE ON THE STRENGTH OF METALS AND ALLOYS Proceedings Tokyo 1967 Sep 4-8 Japan institute of metals xlvi,1049p 31cm Japan institute of metals, Sendai,1968 In commemoration of the 30th anniversary of the institute
Cav 7.3081

INTERNATIONAL CONFERENCE ON THE UNIVERSAL ASPECTS OF ATMOSPHERIC ELECTRICITY 4th Proceedings Planetary electrodynamics Tokyo 1968 Vol 1-2 Edited by Samuel C. Coroniti and James Hughes 2 vols Gordon and Breach,New York;London,1969
Nap 11.0974

INTERNATIONAL CONFERENCE ON THE USE OF COMPUTERS IN THERAPEUTIC RADIOLOGY Summary report Cambridge 1966 Jun 14-17 British institute of radiology,London,1967
Radioth 35.1198

INTERNATIONAL CONFERENCE ON THE USES OF ISOTOPICALLY LABELED DRUGS IN EXPERIMENTAL PHARMACOLOGY Lectures Isotopes in experimental pharmacology Chicago,Ill. 1964 Jun 7-9 Edited by Lloyd J. Roth University of Chicago press,Chicago,Ill.,1965
Pha 16.0155

INTERNATIONAL CONFERENCE ON THEORETICAL PHYSICS Proceedings Kyoto 1953 Sep 14-24 Tokyo Science council of Japan and Kyoto university Edited by I. Imai Under the auspices of the International union of pure and applied physics 942p Tokyo,1954
Cav 7.1342

INTERNATIONAL CONFERENCE ON THROMBOSIS AND EMBOLISM 1 Proceedings Basle 1954 Edited by Theo Koller and Willy R. Merz Benno Schwabe,Basle,1955
Med 36.0220

INTERNATIONAL CONFERENCE ON TITANIUM Proceedings Science,technology and application of titanium London 1968 May 21-24 Institute of metals American institute of mining,metallurgical and petroleum engineers American society for metals Japan institute of metals Akademiya nauk S.S.S.R. Edited by R.I. Jaffee and N.E. Promisel Pergamon,Oxford,1970
Met 25.2852

INTERNATIONAL CONFERENCE ON TRIBOLOGY IN IRON AND STEEL WORKS 1st Proceedings London 1969 Sep 22-25 Iron and steel institute Institution of mechanical engineers Iron and steel institute.Publication, 125 Iron and steel institute,London,1970
Met 25.2562

INTERNATIONAL CONFERENCE ON TUBERCULOSIS 1st Report Berlin 1902 Oct 22-26 Central international bureau for the prevention of consumption Edited by Gotthold Pannwitz Berlin,1903 Text in English,French,German; title also in French and German
Path 30.1025

INTERNATIONAL CONFERENCE ON VACANCIES AND INTERSTITIALS IN METALS Julich 1968 Sep 23-28 Kernforschungslange Julich Arbeitsgemeinschaft metallphysik Edited by A. Seeger and others North-Holland,Amsterdam, 1970
Met 25.2414

INTERNATIONAL CONFERENCE ON VENOMS 1st Papers Berkeley,Calif. 1954 Dec 27-30 American association for the advancement of science Edited by E.E. Buckley and N. Porges American association for the advancement of science.Publication, 44 des plantes et animaux indesirables
Bal 44.6436

INTERNATIONAL CONFERENCE ON WIND EFFECTS ON BUILDINGS AND STRUCTURES 3rd Proceedings Tokyo 1971 Science council of Japan Saikon,Tokyo,1971
Eng 41.8602

INTERNATIONAL CONFERENCE ON"ELECTRON DIFFRACTION" AND "THE NATURE OF DEFECTS IN CRYSTALS" Abstracts of papers Melbourne 1965 Aug 16-21 Australian academy of science and International union of crystallography Co-sponsored by the International union of pure and applied physics Pergamon press,Oxford, 1966
Met 25.1463

INTERNATIONAL CONGRESS BIOCHEMISTRY 2nd Paris 1952 Jul 21-27 Vol 7: symposium sur la biochemie des steroides Council for international organizations of medical sciences Societe d'edition d'enseignement superieur,Paris,1952 Text in English and French
Bioch 33.1310

INTERNATIONAL CONGRESS FOR APPLIED MECHANICS 1st- Proceedings illus Waltman,1925- Lacks 2nd and 6th
Eng 41.1757

INTERNATIONAL CONGRESS FOR APPLIED MECHANICS 5th proceedings Cambridge,Mass 1938 Sep. 12-16 Harvard university and Massachusetts institute of technology Edited by J.P. Den Hartog and H. Peters John Wiley, New York,1939
A Math 4.0325

INTERNATIONAL CONGRESS FOR ELECTRON MICROSCOPY 5th Electron microscopy Philadelphia 1962 Aug 29-Sep 5 Vol 1: non-biology International federation of electron microscope societies and Electron microscope society of America Edited by Sydney S. Breese Academic press,New York; London,1962
Met 25.1368

INTERNATIONAL CONGRESS FOR ELECTRON MICROSCOPY 5th Electron microscopy Philadelphia, Pa. 1962 Aug 29-Sep 5 Vol 2: biology Edited by Sidney S. Breese Academic press, New York,1962
An 32.3183

INTERNATIONAL CONGRESS FOR ELECTRON MICROSCOPY 6th Electron microscopy 1966 Kyoto 1966 Aug 28-Sep 4 Vol 1: non-biology International federation of electron microscope societies Edited by Ryozi Uyeda Maruzen,Tokyo,1966
Met 25.1388

INTERNATIONAL CONGRESS FOR ELECTRON MICROSCOPY 6th Papers and extended abstracts Electron microscopy 1966 Kyoto 1966 Aug 28-Sep 4 International federation of electron microscope societies Edited by Ryozi Uyeda illus Maruzen,Tokyo,1966
Path 30.2615

The INTERNATIONAL CONGRESS FOR ELECTRON MICROSCOPY 6th Extended abstracts Electron microscopy 1966 Kyoto Aug 28-Sep 4 Vol 1-2 International federation of societies for electron microscopy Edited by Ryozi Uyeda 2 vols Maruzen company,Tokyo, 1966
Cav 7.1340

INTERNATIONAL CONGRESS FOR LOGIC,METHODOLOGY AND PHILOSOPHY OF SCIENCE Proceedings Logic,methodology and philosophy of science Amsterdam 1967 Aug 25-Sep 2 Nederlandse vereniging voor logica en wijsbegeerte der exacte wetenschappen International union for logic,methodology and philosophy of science. Division of logic,methodology and philosophy of science Edited by B.van Rootselaar and J. F. Staal xiii,554p North-Holland,Amsterdam, 1968
WSM 43.0830

INTERNATIONAL CONGRESS FOR LOGIC,METHODOLOGY AND PHILOSOPHY OF SCIENCE Proceedings Logic,methodology and philosophy of science Jerusalem 1964 Aug 26-Sep 2 Israel academy of sciences and humanities International union of history and philosophy of science. Division of logic,methodology and philosophy of science Edited by Yehoshua Bar-Hillel viii,440p North-Holland,Amsterdam,1965
WSM 43.0829

INTERNATIONAL CONGRESS FOR LOGIC,METHODOLOGY AND PHILOSOPHY OF SCIENCE Proceedings Logic,methodology and philosophy of science Stanford,Calif. 1960 Aug 24-Sep 2 International union of history and philosophy of science.Division of logic,methodology and philosophy of science Edited by Ernest Nagel and others ix,661p Stanford university press,Stanford,Calif.,1962
WSM 43.0828

INTERNATIONAL CONGRESS FOR LOGIC,METHODOLOGY AND PHILOSOPHY OF SCIENCE Proceedings Stanford,Calif. 1960 Aug 24-Sep 2 International union of history and philosophy of science.Division of logic,methodology and philosophy of science National academy of sciences Edited by Ernest Nagel and others ix,661p 25cm Stanford university press, Stanford,Calif.,1962
Math S 3.1853

INTERNATIONAL CONGRESS FOR MICROBIOLOGY 8TH Recent progress in microbiology :symposia presented at the 8th international congress for microbiology Montreal 1962 Aug 20-24 International union of microbiological societies Canadian society of microbiologists Edited by N.E. Gibbons University of Toronto press,Toronto,1963
Gen 34.0701

INTERNATIONAL CONGRESS FOR PSYCHIATRY 2nd Report Zurich 1957 Sep 1-7 Vol 1-4 International organization of world congresses for psyciatry.Swiss organizing committee Edited by W.A. Stoll 4cm Fussli arts graphiques,Zurich,1959 Congress languages English,French,German, Italian and Spanish
Psy 28.0407

INTERNATIONAL CONGRESS FOR STEREOLOGY 2nd Proceedings Stereology Chicago,Ill. 1967 Apr 5-13 International society for stereology Edited by Hans Elias Sponsored by the National science foundation Springer, Berlin,1967
An 32.0107

INTERNATIONAL CONGRESS FOR STUDIES ON POPULATION 3rd Bevolkerungsfragen Berlin 1935 Aug 26-Sep 1 Edited by Hans Harmsen and Franz Lohse illus.,maps Lehmann,Munich, 1936
Geog 13.1489

INTERNATIONAL CONGRESS FOR STUDIES ON POPULATION 4th Congres international de la population Paris 1937 July 29-Aug 1 Under the auspices of l Union internationale pour letude scientifique des problemes de la population 8 vols Hermann,Paris,1938
Geog 13.1494

INTERNATIONAL CONGRESS FOR STUDIES ON POPULATION 2ND Bevolkerungsfragen International kongress fur bevolkerungswissenschaft bericht 2nd Berlin 1935 Aug 26-Sep 1 Internationale vereinigung fur bevolkerungswissenschaft.Deutsche ausschluss Edited by Hans Harmson and Franz Lohse Lehmanns,Munich,1936
Gen 34.2069

INTERNATIONAL CONGRESS FOR THE AERONAUTICAL SCIENCES 1st Proceedings Advances in aeronautical sciences Madrid 1958 Sep 8-13 Vol 1-2 Edited by Theodore von Karman International series on aeronautical sciences and space flight.Division 9:symposia 2 vols Pergamon press,London,1959
Eng 41.6963

INTERNATIONAL CONGRESS FOR THE AERONAUTICAL SCIENCES 2nd Proceedings Advances in aeronautical sciences Zurich 1960 Sep 12-16 Vol 3-4 Edited by Theodore von Karman International series in aeronautics and astronautics.Division 9:symposia, 7-8 2 vols Pergamon press,London,1962
Eng 41.6964

INTERNATIONAL CONGRESS OF AMERICANISTS 32nd Proceedings Copenhagen 1956 Aug 8-14 Published with the support of the International council of philosophy and human sciences illus,maps 743p 26cm Munksgaard,Copenhagen,1958
Sco 14.2361

INTERNATIONAL CONGRESS OF ANATOMISTS 8th Wiesbaden 1965 Aug 8-13 abstracts of papers,scientific demonstrations and films 1965 In English,French and German
An 32.1194

INTERNATIONAL CONGRESS OF ANATOMISTS,7TH The Structure of the eye :the symposium held... during the 7th International congress of anatomists New York 1960 Apr 11-13 Edited by George K. Smelser Academic press, New York,1961
An 32.4773

INTERNATIONAL CONGRESS OF ANATOMISTS 8TH Quantitative methods in morphology Symposium on quantitative methods in morphology held... during the eighth International congress of anatomists Proceedings Wiesbaden 1965 Aug 10 Edited by Ewald R. Weibel and Hans Elias Organised by the International society for stereology Illus. Springer,Berlin,1967
An 32.0105

INTERNATIONAL CONGRESS OF APPLIED MECHANICS 10th Proceedings Stresa 1960 Aug 31-Sep 7 Edited by F. Rolla and W.T. Koiter Elsevier, Amsterdam;New York,1962
A Math 4.1280

The INTERNATIONAL CONGRESS OF APPLIED MECHANICS 11th Proceedings Munich 1964 International committee for the congresses of applied mechanics Edited by Henry Gortler and Peter Sorger 1184p Springer-verlag, Berlin,1966 Dedicated to Richard Grammel
A Math 4.1279

INTERNATIONAL CONGRESS OF APPLIED MECHANICS 12th Proceedings Applied mechanics Stanford,Calif. 1968 Aug 26-31 International union of theoretical and applied mechanics Edited by H. Hetenyi and W.G. Vincenti Bibliog,illus xxiv,420p 291cm Springer,Berlin,1969
A Math 4.1722

INTERNATIONAL CONGRESS OF APPLIED PSYCHOLOGY 14th Proceedings Child and education Copenhagen 1961 Aug 13-19 Vol 3 International association of applied psychology and Danish psychological association Edited by Gerhard S. Nielsen Munksgaard,Copenhagen,1962
Psy 31.1198

INTERNATIONAL CONGRESS OF APPLIED PSYCHOLOGY 14th Proceedings Copenhagen 1961 Aug 13-19 Vol 2: personality research International association of applied psychology and Danish psychological association Edited by Stanley Coopersmith Munksgaard,Copenhagen,1952
Psy 31.1968

INTERNATIONAL CONGRESS OF APPLIED PSYCHOLOGY 14th Proceedings Copenhagen 1961 Aug 13-19 Vol 5: industrial and business psychology International association of applied psychology and Danish psychological association Edited by Gerhard S. Nielsen Munksgaard,Copenhagen,1962
Psy 31.2252

INTERNATIONAL CONGRESS OF BIO-CHEMISTRY 1st Cambridge 1949 Aug 19-25 abstracts of communications 1949
Phys 20.1522

INTERNATIONAL CONGRESS OF BIOCHEMISTRY 1st Abstracts of communication Cambridge 1949 Aug 19-25 1949 Text in English and French
Bioch 33.1300

INTERNATIONAL CONGRESS OF BIOCHEMISTRY 1st Report of opening and concluding sessions and three lectures Cambridge 1949 Aug 19-25 Printed for the Biochemical society Glasgow university,Biochemistry department,Glasgow, 1950 Unbound
Bioch 33.1301

INTERNATIONAL CONGRESS OF BIOCHEMISTRY 2nd Paris 1952 Jul 21-27 Vol 1: symposium sur la biochimie de l'hematopoiese Council for international organizations of medical sciences Societe d'edition d'enseignement superieur,Paris,1952 Text in English and French
Bioch 33.1304

INTERNATIONAL CONGRESS OF BIOCHEMISTRY 2nd Paris 1952 Jul 21-27 Vol 2: symposium sur la biogenese des proteines Council for international organizations of medical sciences Societe d'edition d'enseignement superieur,Paris,1952 Text in English and French
Bioch 33.1305

INTERNATIONAL CONGRESS OF BIOCHEMISTRY 2nd Paris 1952 Jul 21-27 Vol 3: symposium sur le cycle tricarboxylique Council for international organizations of medical sciences Societe d'edition d'enseignement superieur,Paris,1952 Text in English, French and German
Bioch 33.1306

INTERNATIONAL CONGRESS OF BIOCHEMISTRY 2nd Paris 1952 Jul 21-27 Vol 4: symposium sur les hormones proteiques et derivees des proteines Council for international organizations of medical sciences Societe d'edition d'enseignement superieur,Paris,1952 Text in English and French
Bioch 33.1307

INTERNATIONAL CONGRESS OF BIOCHEMISTRY 2nd Paris 1952 Jul 21-27 Vol 5: symposium sur le metabolisme microbien Council for international organizations of medical sciences Societe d'edition d'enseignement superieur,Paris,1952 Text in English and French
Bioch 33.1308

INTERNATIONAL CONGRESS OF BIOCHEMISTRY 2nd Paris 1952 Jul 21-27 Vol 6: symposium sur le mode d'action des antibiotiques Council for international organizations of medical sciences Societe d'edition d'enseignement superieur,Paris,1952 Text in English and French
Bioch 33.1309

INTERNATIONAL CONGRESS OF BIOCHEMISTRY 2nd Compte rendu Paris 1952 Jul 21-27 Council for international organizations of medical sciences Societe de chimie biologique.Bulletin, 35,no 1-2 Masson,Paris, 1953 Supplement.Text in English and French
Bioch 33.1302

INTERNATIONAL CONGRESS OF BIOCHEMISTRY 2nd Proceedings Paris 1952 Jul 21-27 Vol 3: symposium sur le cycle tricarboxylique Council for international organizations of medical science Edited by Michel Polonovski Congres international de biochemie, 2 Societe d'edition d'enseignement superieur, Paris,1952
Bot 42.1697

INTERNATIONAL CONGRESS OF BIOCHEMISTRY 2nd
Resumes des communication Paris 1952 Jul 21-27 Council for international organizations of medical sciences Masson, Paris,1952 Text in English and French
Bioch 33.1303

INTERNATIONAL CONGRESS OF BIOCHEMISTRY 3rd
Proceedings Brussels 1955 Aug 1-6 Edited by Claude Liebecq Academic press,New York,1956 Text in English,French and German
Bioch 33.1311

INTERNATIONAL CONGRESS OF BIOCHEMISTRY 3rd
Programme et liste des membres Brussels 1955 Aug 1-6 Council for international organizations of medical sciences Secretariat general;Agence de voyage,Liege; Brussels,c1956 Bound with 3 eme Congres international de biochimie.Resumes des communications
Bioch 33.1312

INTERNATIONAL CONGRESS OF BIOCHEMISTRY 3rd
Rapports (volume provisoire) Brussels 1955 Aug 1-6 Council for international organizations of medical sciences Imprimerie Vaillant-Carmanne,Liege,1955
Bioch 33.1314

INTERNATIONAL CONGRESS OF BIOCHEMISTRY 3rd
Resumes des communications Brussels 1955 Aug 1-6 Council for international organizations of medical sciences c1956 Text in English,French and German.Bound with 3 eme Congres international de biochimie. Programme et liste des membres
Bioch 33.1313

INTERNATIONAL CONGRESS OF BIOCHEMISTRY 4th
Vienna 1958 Sep 1-6 Vol 15: biochemistry: abstracts of sectional papers and index to symposia and colloquia International union of biochemistry I.U.B.symposium series, 17 Pergamon,Oxford,1960 Added title page in French and German.Text in English,French and German
Bioch 33.1329

INTERNATIONAL CONGRESS OF BIOCHEMISTRY 4th
Vol 1: symposium 1- carbohydrate chemistry of substances of biological interest International union of biochemistry Edited by M.L. Wolfrom I.U.B.symposium series, 3 Pergamon,London,1959 Added title page in French and German.Text in English,French and German
Bioch 33.1315

INTERNATIONAL CONGRESS OF BIOCHEMISTRY 4th
Proceedings Vienna 1958 Sep 1-6 Vol 2: symposium 2 - biochemistry of wood International union of biochemistry Edited by K. Kratzi and G. Billek I.U.B.symposium series, 4 Pergamon,London,1959 Added title page in French and German.Text in English,French and German
Bioch 33.1316

INTERNATIONAL CONGRESS OF BIOCHEMISTRY 4th
Proceedings Vienna 1958 Sep 1-6 Vol 3: symposium 3:biochemistry of the central nervous system International union of biochemistry Edited by F. Brucke I.U.B. symposium series, 5 Pergamon,London,1959 Added t.p.in French and German
Bot 42.1715

INTERNATIONAL CONGRESS OF BIOCHEMISTRY 4th
Proceedings Vienna 1958 Sep 1-6 Vol 3: symposium 3 - biochemistry of the central nervous system International union of biochemistry Edited by Brucke I.U.B. symposium series, 5 Pergamon,London,1959 Added title page in French and German.Text in English,French and German
Bioch 33.1317

INTERNATIONAL CONGRESS OF BIOCHEMISTRY 4th
Proceedings Vienna 1958 Sep 1-6 Vol 4: symposium 4 biochemistry of steroids International union of biochemistry Edited by E. Mosettig I.U.B.symposium series, 6 Pergamon,London,1959 Added title page in French and German.Text in English,French and German.
Bioch 33.1318

INTERNATIONAL CONGRESS OF BIOCHEMISTRY 4th
Proceedings Vienna 1958 Sep 1-6 Vol 5: symposium 5 - biochemistry of antibiotics International union of biochemistry Edited by K.H. Spitzy and R. Brunner I.U.B. symposium series, 7 Pergamon,London,1959 Added title page in French and German.Text in English,French and German.
Bioch 33.1319

INTERNATIONAL CONGRESS OF BIOCHEMISTRY 4th
Proceedings Vienna 1958 Sep 1-6 Vol 6: symposium 6 - biochemistry of morphogenesis International union of biochemistry Edited by W.J. Nickerson I.U.B.S.symposium series, 8 Pergamon press,London,1959
Gen 34.0482

INTERNATIONAL CONGRESS OF BIOCHEMISTRY 4th
Proceedings Vienna 1958 Sep 1-6 Vol 6: symposium 6 - biochemistry of morphogenesis International union of biochemistry Edited by W.J. Nickerson I.U.B.symposium series, 8 Pergamon,London,1959 Added title page in French and German
Bioch 33.1320

INTERNATIONAL CONGRESS OF BIOCHEMISTRY 4th
Proceedings Vienna 1958 Sep 1-6 Vol 7: symposium 7 - biochemistry of viruses International union of biochemistry Edited by E. Broda and W. Frisch-Niggemeyer I.U.B. symposium series, 9 Pergamon,London,1959 Added title page in French and German.Text in German and English
Bioch 33.1321

INTERNATIONAL CONGRESS OF BIOCHEMISTRY 4th
Proceedings Vienna 1958 Sep 1-6 Vol 7: symposium 7 - biochemistry of viruses International union of biochemistry Edited by E. Broda and W. Frisch-Niggemeyer Pergamon press,London,1960
Radioth 35.0071

INTERNATIONAL CONGRESS OF BIOCHEMISTRY 4th
Proceedings Vienna 1958 Sep 1-6 Vol 8: symposium 8 - proteins International union of biochemistry Edited by H. Neurath and H. Tuppy I.U.B.symposium series, 10 Pergamon,London,1960 Added title page in French and German.Text in English and French
Bioch 33.1322

INTERNATIONAL CONGRESS OF BIOCHEMISTRY 4th
Proceedings Vienna 1958 Sep 1-6 Vol 8: symposium 8 - proteins International union of biochemistry Edited by H. Neurath and H. Tuppy Pergamon,London,1960
Radioth 35.0072

INTERNATIONAL CONGRESS OF BIOCHEMISTRY 4th
Proceedings Vienna 1958 Sep 1-6 Vol 9: symposium 9 - physical chemistry of high polymers of biological interest International union of biochemistry Edited by O. Kratky I.U.B.symposium series, 11 Pergamon,London,1959 Added title page in French and German.Text in English,French and German.
Bioch 33.1323

INTERNATIONAL CONGRESS OF BIOCHEMISTRY 4th
Proceedings Vienna 1958 Sep 1-6 Vol 9: symposium 9 - physical chemistry of high polymers of biological interest International union of biochemistry Edited by O. Kratky Pergamon press,London,1960
Radioth 35.1941

INTERNATIONAL CONGRESS OF BIOCHEMISTRY 4th
Proceedings Vienna 1958 Sep 1-6 Vol 10: blood clotting factors By E. Deutsch Pergamon press,Oxford,1959
Med 36.0115

INTERNATIONAL CONGRESS OF BIOCHEMISTRY 4th
Proceedings Vienna 1958 Sep 1-6 Vol 10: symposium 10 - blood clotting factors Edited by E. Deutsch I.U.B.symposium series, 12 Pergamon,London,1959 Added title page in French and German.Text in English and German
Bioch 33.1324

INTERNATIONAL CONGRESS OF BIOCHEMISTRY 4th
Proceedings Vienna 1958 Sep 1-6 Vol 11: symposium 11 - vitamin metabolism International union of biochemistry Edited by W. Umbreit and H. Molitor I.U.B.symposium series, 13 Pergamon,London,1960 Added title page in French and German
Bioch 33.1325

INTERNATIONAL CONGRESS OF BIOCHEMISTRY 4th
Proceedings Vienna 1958 Sep 1-6 Vol 12: colloquia International union of biochemistry Edited by H. Chantrenne and others I.U.B.symposium series, 15 Pergamon,London,1959 Added title page in French and German.Text in English,French and German
Bioch 33.1327

INTERNATIONAL CONGRESS OF BIOCHEMISTRY 4th
Proceedings Vienna 1958 Sep 1-6 Vol 12: symposium 12 - biochemistry of insects International union of biochemistry Edited by L. Levenbook I.U.B.symposium series, 14 Pergamon,London,1959 Added title page in French and German.Text in English and German
Bioch 33.1326

INTERNATIONAL CONGRESS OF BIOCHEMISTRY 4th
Proceedings Vienna 1958 Sep 1-6 Vol 14: transactions of the plenary sessions International union of biochemistry Edited by W. Auerswald and O. Hoffmann-Ostenhof I.U.B.symposium series, 14 Pergamon,London,1959 Added title page in French and German.Text in English,French and German.
Bioch 33.1328

The INTERNATIONAL CONGRESS OF BIOCHEMISTRY 4th
Proceedings Vienna 1958 Sep 1-6 13: colloquia International union of biochemistry Edited by O. Hoffmann-Osterhoff I.U.B.Symposium series, 15 Pergamon press, London,1959 Added title page in French and German
Bot 42.1696

INTERNATIONAL CONGRESS OF BIOCHEMISTRY 5th
Proceedings Moscow 1961 Aug 10-16 Vol 1: biological structure and function at the molecular level International union of biochemistry Edited by V.A. Engelhardt I.U.B.symposium series, 21 Pergamon;PWN-Polish scientific publishers,London;Warsaw,1963
Bioch 33.1330

INTERNATIONAL CONGRESS OF BIOCHEMISTRY 5th
Proceedings Moscow 1961 Aug 10-16 Vol 2: functional biochemistry of cell structures International union of biochemistry Edited by O. Lindberg I.U.B. symposium series, 22 Pergamon;PWN-Polish scientific publishers,New York;Warsaw,1963
Bioch 33.1331

INTERNATIONAL CONGRESS OF BIOCHEMISTRY 5th
Proceedings Moscow 1961 Aug 10-16 Vol 3: evolutionary biochemistry International union of biochemistry I.U.B.symposium series, 23 Pergamon;PWN-Polish scientific publishers,New York;Warsaw,1963
Bioch 33.1332

INTERNATIONAL CONGRESS OF BIOCHEMISTRY 5th
Proceedings Moscow 1961 Aug 10-16 Vol 4: molecular basis of enzyme action and inhibition International union of biochemistry Edited by P.A.E. Desnuelle I.U.B.symposium series, 24 Pergamon;PWN-Polish scientific publishers,Oxford;Warsaw, 1963
Bioch 33.1333

INTERNATIONAL CONGRESS OF BIOCHEMISTRY 5th
Proceedings Moscow 1961 Aug 10-16 Vol 5: intracellular respiration:phosphorylating and non-phosphorylating oxidation reactions International union of biochemistry Edited by E.C. Slater I.U.B.symposium series, 25 Pergamon;PWN-Polish scientific publishers, Oxford;Warsaw,1963
Bioch 33.1334

INTERNATIONAL CONGRESS OF BIOCHEMISTRY 5th
Proceedings Moscow 1961 Aug 10-16 Vol 6: mechanism of photosynthesis International union of biochemistry Edited by H. Tamiya I.U.B.symposium series, 26 Pergamon;PWN-Polish scientific publishers, Oxford;Warsaw,1963
Bioch 33.1335

INTERNATIONAL CONGRESS OF BIOCHEMISTRY 5th
Proceedings Moscow 1961 Aug 10-16 Vol 7: biosynthesis of lipids International union of biochemistry Edited by G. Popjak I.U.B.symposium series, 27 Pergamon;PWN-Polish scientific publishers,Oxford;Warsaw, 1963
Bioch 33.1336

INTERNATIONAL CONGRESS OF BIOCHEMISTRY 5th
Proceedings Moscow 1961 Aug 10-16 Vol 8: biochemical principles of the food industry International union of biochemistry Edited by V.L. Kretovich and E. Pijanowski I.U.B.symposium series, 28 Pergamon;PWN-Polish scientific publishers,Oxford;Warsaw, 1963
Bioch 33.1337

INTERNATIONAL CONGRESS OF BIOCHEMISTRY 5th
Proceedings Moscow 1961 Aug 10-16 Vol 9: plenary sessions and abstracts of papers International union of biochemistry I.U.B. symposium series, 29 Pergamon;PWN-Polish scientific publishers,New York;Warsaw,1963
Bioch 33.1338

INTERNATIONAL CONGRESS OF BIOCHEMISTRY 6th
Abstracts New York 1964 Jul 26-Aug 1 Scheduled under the auspices of the International union of biochemistry Washington,D.C.,1964
Bioch 33.1342

INTERNATIONAL CONGRESS OF BIOCHEMISTRY 6th
Abstracts New York 1964 Jul 26-Aug 1 Scheduled under the auspices of the International union of biochemistry Washington,1964
Bioch 33.1343

INTERNATIONAL CONGRESS OF BIOCHEMISTRY 6th
Abstracts New York 1964 Jul 26-Aug 1 Scheduled under the auspices of the International union of biochemistry Washington,1964
Bioch 33.1345

INTERNATIONAL CONGRESS OF BIOCHEMISTRY 6th
Abstracts New York 1964 Jul 26-Aug 1 Vol 2: proteins,peptides,and amino acids Scheduled under the auspices of the International union of biochemistry Washington,D.C.,1964 Title also in French.
Bioch 33.1339

INTERNATIONAL CONGRESS OF BIOCHEMISTRY 6th
Abstracts New York 1964 Jul 26-Aug 1 Vol 3: biochemical genetics Scheduled under the auspices of the International union of biochemistry Washington,D.C.,1964
Bioch 33.1340

INTERNATIONAL CONGRESS OF BIOCHEMISTRY 6th
Abstracts New York 1964 Jul 26-Aug 1 Vol 5: special topics in biochemistry Scheduled under the auspices of the International union of biochemistry Washington,1964
Bioch 33.1341

INTERNATIONAL CONGRESS OF BIOCHEMISTRY 6th
Abstracts New York 1964 Jul 26-Aug 1 Vol 8: cellular organization Scheduled under the auspices of the International union of biochemistry Washington,1964
Bioch 33.1344

INTERNATIONAL CONGRESS OF BIOCHEMISTRY 6th
Abstracts New York 1964 Jul 26-Aug 1 Vol 10: bioenergetics Scheduled under the auspices of the International union of biochemistry Washington,1964
Bioch 33.1346

INTERNATIONAL CONGRESS OF BIOCHEMISTRY 6th
Abstracts New York 1964 Jul 26-Aug 1 Vol 11: author index to abstracts Scheduled under the auspices of the International union of biochemistry Washington,1964
Bioch 33.1347

INTERNATIONAL CONGRESS OF BIOCHEMISTRY 8th : held at Interlaken,Lucerne and Montreux
Abstracts Switzerland 1970 Sep 3-9 International union of biochemistry Edited by J.G. Gregory Staples press,Rochester,1970
Bioch 33.2338

INTERNATIONAL CONGRESS OF BIOCHEMISTRY,4th,papers: section 18 Biochemistry of lipids
International conference on the biochemical problems of lipids 5th Proceedings Vienna 1958 Sep Edited by G. Popjak Pergamon press,Oxford,1960 Section 18 of the 4th international congress of biochemistry
Col S 12.0102

INTERNATIONAL CONGRESS OF BIOCHEMISTRY 5TH
Report of the Commission on enzymes of the International union of biochemistry 1961 International union of biochemistry.Commission on enzymes I.U.B.Symposium series, 20 Pergamon press,Oxford,1961 "Report presented to the Council of the Union,in session during the Fifth International congress of biochemistry at Moscow in August 1961"
Bioch 33.1054

INTERNATIONAL CONGRESS OF BIOCHEMISTRY 5TH
Report of the commission on enzymes of the International union of biochemistry 1961 International union of biochemistry.Commission on enzymes I.U.B.symposium series, 20 Pergamon,Oxford,1961 Report presented to the Council of the Union,in session during the fifth International congress of biochemistry at Moscow,August,1961
Radioth 35.1845

INTERNATIONAL CONGRESS OF BLOOD TRANSFUSION 5th
Proceedings Paris 1954 Sep International society of blood transfusion 1955
Med 36.0388

INTERNATIONAL CONGRESS OF BLOOD TRANSFUSION 7th
Proceedings Rome 1958 Sep 2-6 International society of blood transfusion Edited by Ludwig Hollander Karger,Basle,1959
Med 36.0200

INTERNATIONAL CONGRESS OF BLOOD TRANSFUSION 8th
Programme and abstracts Tokyo 1960 Sep International society of blood transfusion Edited by Ludwig Hollander 1960
Med 36.0201

INTERNATIONAL CONGRESS OF CELL BIOLOGY 12th
Summaries of reports and communications Brussels 1968 Aug 25-31 Edited by P. Dustin and others International congress series, 166 Excerpta medica,Amsterdam,1968
Bioch 33.0963

INTERNATIONAL CONGRESS OF CELL BIOLOGY 12th
Summaries of reports and communications Brussels 1968 Aug 25-31 Edited by P. Dustin and others Under the auspices of the International society for cell biology International congress series, 166 Excerpta medica,Amsterdam,1968
Bot 42.1515

INTERNATIONAL CONGRESS OF CHEMICAL ENGINEERING, CHEMICAL EQUIPMENT AND AUTOMATION 2nd
Lecture summaries Marienbad 1965 Sep 12-19 Czechoslovak scientific-technical society Czechoslovak chemical society Edited by Tomas Misek and Pavel Mitschka 1965
Chem E 24.1907

INTERNATIONAL CONGRESS OF CHEMICAL ENGINEERING, CHEMICAL EQUIPMENT CONSTRUCTION AND AUTOMATION 3rd Lecture summaries Marienbad 1969 Sep Pt A-H Scientific committee Chisa '69,1969 Cover title:Chisa '69
Chem E 24.1872

INTERNATIONAL CONGRESS OF COMPARATIVE PATHOLOGY 3rd Athens 1936 Apr 15-18 Vol 1: reports.2.Section of plant pathology:immunity of plants Editions 'Flamma',Athens,1936
Bot 42.6540

INTERNATIONAL CONGRESS OF ELETROENCEPHALOGRAPHY AND CLINICAL NEUROPHYSIOLOGY 6th Proceedings Recent advances in clinical neurophysiology Vienna 1965 Sep 5-10 functions of the spinal cord; neurophysiological investigations of brain diseases;EEG in stress Edited by L. Widen Electroencephalography and clinical neurophysiology.Supplement,25 Elsevier, Amsterdam,1967
Pha 16.0304

INTERNATIONAL CONGRESS OF ENDOCRINOLOGY Proceedings Progress in endocrinology Mexico City 1968 Jun 30-Jul 5 Edited by Carlos Gual and F.J.C. Ebling Excerpta medica.International congress series, 184 Excerpta medica,Amsterdam,1969
Inv Med 37.0095

INTERNATIONAL CONGRESS OF ENDOCRINOLOGY 2nd Proceedings London 1964 Aug 17-22 Pt 1-2 Edited by S. Taylor Excerpta medica. International congress series, 83 2 vols Excerpta medica foundation,London,1965
PGMS 29.0157

INTERNATIONAL CONGRESS OF ENDOCRINOLOGY 2nd Proceedings London 1964 Aug 17-22 Pt 1-2 Excerpta medica.International congress series, 83 2 vols Excerpta medica,Amsterdam,1966
Inv Med 37.0114

INTERNATIONAL CONGRESS OF ENDOCRINOLOGY 2nd Proceedings London 1964 Aug 17-22 Pt 1-2 International congress series, 83 2 vols Excerpta medica,Amsterdam,1964
Phys 20.1396

INTERNATIONAL CONGRESS OF ENTOMOLOGY 12th Papers Physiology of the insect central nervous system London 1964 Jul Edited by J.E. Treherne and J.W.L. Beament Academic press,London,1965
Pha 16.0276

INTERNATIONAL CONGRESS OF ENTOMOLOGY 12th Papers Symposium on the physiology of the insect central nervous system London 1964 Edited by J.E. Treherne and J.W.L. Beaument London,1965
Bal 39.2458

INTERNATIONAL CONGRESS OF EUGENICS 2nd Papers Eugenics,genetics and the family New York 1921 Sep 22-28 Vol 1 Port 2 vols Williams and Wilkins,Baltimore,Md., 1923
An 32.0031

INTERNATIONAL CONGRESS OF EUGENICS 2nd Papers Eugenics,genetics and the family New York 1921 Sep 22-28 Vol 1-2 port. 2 vols Williams and Wilkins, Baltimore,Md.,1923
Gen 34.1926

The INTERNATIONAL CONGRESS OF EUGENICS 2nd Scientific papers Eugenics in race and state New York 1921 Sep 22-28 Vol 2 Port Williams and Wilkins,Baltimore,1923
An 32.0032

INTERNATIONAL CONGRESS OF EUGENICS 3rd Papers Decade of progress in eugenics New York 1932 Aug 21-23 port.pl. Williams and Wilkins,Baltimore,Md.,1934
Gen 34.1927

INTERNATIONAL CONGRESS OF EUGENICS 1ST Problems in eugenics International eugenics conference 1st Papers London 1912 Jul 24-30 Eugenics education society 2 vols Eugenics education society,London,1913 Later known as the 1st international congress of eugenics
Gen 34.1925

INTERNATIONAL CONGRESS OF GENETICS 3rd hybridization (the cross-breeding of genera or species) the cross-breeding of varieties... London 1906 Jul 30-Aug 3 Royal horticultural society Edited by W. Wilks Spottiswoode,London,1907 Conference also entitled 'International conference on hybridisation'
Gen 34.0999

INTERNATIONAL CONGRESS OF GENETICS 6th Proceedings Ithaca,N.Y. 1932 Vol 1-2 Edited by Donald F. Jones Brooklyn botanic garden,Brooklyn,N.Y.,1932 Bound together
Gen 34.1002

INTERNATIONAL CONGRESS OF GENETICS 8th Proceedings Stockholm 1948 Jul 7-14 Edited by Gert Bonnier and Robert Larsson Berlingska boktrycheriet,Lund,1949
Gen 34.1004

INTERNATIONAL CONGRESS OF GENETICS 9th Proceedings Bellagio 1953 Aug 24-31 Pt 1-2 Edited by G. Montalenti and A. Chiarugi Caryologia, 6,supp. 2 vols Florence,1954
Gen 34.1005

INTERNATIONAL CONGRESS OF GENETICS 10th Proceedings Montreal 1958 Aug 20-27 Vol 1-2 University of Toronto press, Toronto,1959 Bound together
Gen 34.1006

INTERNATIONAL CONGRESS OF GENETICS 10th Proceedings Montreal 1958 Aug 20-27 Vol 1-2 2 vols University of Toronto press,Toronto,1959
Bot 42.4702

INTERNATIONAL CONGRESS OF GENETICS 11th Proceedings Genetics today The Hague 1963 Sep Vol 1-3 Edited by S.J. Geerts 2 vols Pergamon press,Oxford,1963-65
Bot 42.0780

INTERNATIONAL CONGRESS OF GENETICS 11th Proceedings Genetics today The Hague 1963 Sep Vol 1-3 Edited by S.J. Geerts 3 vols Pergamon press,Oxford,1963-65 Two copies each of vol 2 and 3
Gen 34.1007

INTERNATIONAL CONGRESS OF GENETICS 12th
Proceedings Tokyo 1968 Aug 19-28 Vol 1-3 Science council of Japan Under the auspices of the International union of biological sciences 3 vols Science council of Japan,Tokyo,1969
Gen 34.1008

INTERNATIONAL CONGRESS OF GENETICS 1ST
Hybrid conference report 1900 on hybridisation (the cross-breeding of species) and on the cross-breeding of varieties London 1899 Jul 11-12 Royal horticultural society Royal horticultural society.Journal, 24 Spottiswoode,London,1900 Conference also entitled the 'International congress of genetics' and the 'International conference on hybridization'
Gen 34.0998

INTERNATIONAL CONGRESS OF GENETICS 4TH
Conference international de genetique 4e Comptes rendus Paris 1911 Sep 18-23 Societe nationale d'horticulture de France Edited by Ph.de Vilmorin 2 vols Masson, Paris,1913
Gen 34.1000

INTERNATIONAL CONGRESS OF GENETICS 5TH
Internationaler kongress fur vererbungswissenschaft 5 Verhandlungen Berlin 1927 Band 1-2 Edited by Hans Nachtsheim Zeitschrift fur induktive abstammungs- und vererbungslehre. Supplementband, 1 2 vols Borntraeger, Leipzig,1928
Gen 34.1001

INTERNATIONAL CONGRESS OF GENETICS 7TH
International genetical congress 7th Proceedings Edinburgh 1939 Aug 23-30 Genetical society Edited by Reginald Crundall Punnett Journal of genetics. Supplement Cambridge university press, Cambridge,1941
Gen 34.2219

INTERNATIONAL CONGRESS OF GERONTOLOGY 5th
Proceedings Aging around the world San Francisco 1960 Aug International association of gerontology Edited by Herman T. Blumenthal Columbia university press,New York;London,1962
An 32.0088

INTERNATIONAL CONGRESS OF GYNAECOLOGY AND OBSTETRICS 5TH Caesarean section in Great Britain and Ireland... being a report read at the Vme congres international d'obstetrique et de gynecologie at St.Petersburg in September 1910 By Amand Routh tables Sharratt and Hughes,London,1911
Path 30.2982

INTERNATIONAL CONGRESS OF HAEMATOLOGY 11th
Proceedings Sydney 1966 International society of hematology Blackwell,Oxford,1966
Med 36.0203

INTERNATIONAL CONGRESS OF HEMATOLOGY 3rd
Proceedings Cambridge 1950 Aug 21-25 International society of hematology Edited by Carl V. Moore Grune and Stratton,New York, 1951
Med 36.0256

INTERNATIONAL CONGRESS OF HEMATOLOGY 6th
Proceedings Boston 1956 Aug 27-Sep 1 International society of hematology Edited by A. Richardson Jones Grune and Stratton, New York,1958
Med 36.0202

INTERNATIONAL CONGRESS OF HEMATOLOGY 7th : reports,communications,symposia Proceedings Rome 1958 Sep 7-13 3 International society of hematology Edited by Giovanni Di Guglielmo Grune and Stratton, New York,1960
Med 36.0166

INTERNATIONAL CONGRESS OF HEPATIC INSUFFICIENCY AND LIVER DISEASES Vichy 1937 Sep 16-18 2: comptes rendus,discussions et communications diverses Edited by J. Aimard Wallon,Paris,1937 Conference president: professor M.Loeper
Path 30.0948

INTERNATIONAL CONGRESS OF HISTOCHEMISTRY AND CYTOCHEMISTRY 1st Proceedings Histochemistry and cytochemistry Paris 1960 Edited by R. Wegmann Pergamon,London, 1963
An 32.3391

INTERNATIONAL CONGRESS OF HISTOCHEMISTRY AND CYTOCHEMISTRY 3rd Summary reports New York 1968 Aug 18-22 Springer,New York, 1968
Radioth 35.0547

INTERNATIONAL CONGRESS OF HISTOCHEMISTRY AND CYTOCHEMISTRY 3rd summary reports New York 1968 Aug 18-22 United States. Office of the secretariat 317p Springer-verlag,New York,1968
Bot 42.0703

INTERNATIONAL CONGRESS OF HISTOCHEMISTRY AND CYTOCHEMISTRY 3rd Summary reports New York 1968 Aug 18-22 Springer,New York,1968
An 32.0438

INTERNATIONAL CONGRESS OF HORMONAL STEROIDS 1st biochemistry, pharmacology and therapeutics Proceedings Milan 1964 Vol 1-2 Edited by L. Martini and A. Pecile 2 vols Academic press,New York; London,1964-65
Gen 34.0546

INTERNATIONAL CONGRESS OF HORTICULTURE 16TH
Hortus belgicus :catalogue de l'exposition Brussels 1962 Aug-Sep Bibliotheque Albert I,Belgium Edited by Jan Balis Bibliotheque royale de Belgique.Catalogues des expositions, 10 Brussels,1962 Exhibition organised to coincide with the 16th International horticultural congress Book wrongly calls it Botanical congress
BG 38.1857

INTERNATIONAL CONGRESS OF HUMAN GENETICS 1st
Proceedings Copenhagen 1956 Aug 1-6 Edited by Tage Kemp and others Karger,Basle, 1957
An 32.0355

INTERNATIONAL CONGRESS OF HUMAN GENETICS 2nd
Proceedings Rome 1961 Sep 6-12 Vol 1-3 3 vols Instituto G.Mendel,Rome,1963 In English,French,German and Italian
An 32.0155

The INTERNATIONAL CONGRESS OF HUMAN GENETICS 2nd Proceedings Rome 1961 Sep 6-12 Vol 1-3 3 vols Institute G.Mendel,Rome, 1963
Gen 34.1857

INTERNATIONAL CONGRESS OF HUMAN GENETICS 3rd plenary sessions and symposia Proceedings Chicago,Ill. 1966 Sep 5-10 Edited by James F. Crow and James V. Neel John Hopkins press,Baltimore,Md.,1967
PGMS 29.0264

The INTERNATIONAL CONGRESS OF HUMAN GENETICS 3rd Plenary sessions and symposia Chicago, Ill. 1966 Sep 5-10 Edited by James F. Crow and James V. Neel Hopkins,Baltimore,Md.,1967
Gen 34.1858

INTERNATIONAL CONGRESS OF HYGIENE AND DEMOGRAPHY 14TH Hygienischer fuhrer durch Berlin Institut fur infektionskrankheiten and Universitat Berlin.Hygienisches institut Hirschwald,Berlin,1907 Compilation of scientific institutions of Berlin published in connection with the 14th congress In English, French and German
Path 30.1232

INTERNATIONAL CONGRESS OF HYGIENE AND DEMOGRAPHY 14TH Medizinische anstalten auf dem gebiete der volksgesundheitspflege in Preussen festschrift... Prussia.Ministerium der geistlichen,unterrichts- und medizinal-angelegenheiten illus Fischer,Jena,1907 Work issued in connection with the congress held in Berlin,1907
Path 30.1209

INTERNATIONAL CONGRESS OF HYGIENE AND DEMOGRAPHY 7TH Denmark:its medical organization, hygiene and demography Edited by Julius Lehmann and others map Gjellerup, Copenhagen,1891 Presented to the seventh International congress of hygiene and demography,London 1891
Path 30.1192

INTERNATIONAL CONGRESS OF MATHEMATICIANS Outlines of one-hour and half-hour addresses and translations of Russian abstracts of short communications Stockholm 1962 varp 29cm Stockholm,1962
P Math 2.3005

INTERNATIONAL CONGRESS OF MATHEMATICIANS abstracts of brief scientific communications and reports Moscow 1966 Aug nos. 1-15 International mathematical union Moscow,1966
Math 3.1125

INTERNATIONAL CONGRESS OF MATHEMATICIANS short communications abstracts Stockholm 1962 Aug 15-22 International mathematical union 221p 24cm Stockholm,1962
P. Math 2.2602

INTERNATIONAL CONGRESS OF MATHEMATICIANS 1st Proceedings Moscow 1966 Edited by I.G. Petrovsky Bibliog,illus 726p 22cm M.I. R.,Moscow,1968
P Math 2.3006

INTERNATIONAL CONGRESS OF MATHEMATICIANS 5th proceedings Cambridge 1912 Aug 22-28 Vol. 2.Communications to sections 2-4 International mathematical union Edited by E. W. Hobson and A.E.H. Love 657p 26cm Cambridge university press,Cambridge,1913
Math L 5.0751

INTERNATIONAL CONGRESS OF MATHEMATICIANS 11th Proceedings Cambridge,Mass. 1950 Aug 13-Sep 6 Vol 1-2 International mathematical union and American mathematical society 25cm 2 vols American mathematical society,Providence,R.I., 1952
Math L 5.3246

INTERNATIONAL CONGRESS OF MATHEMATICIANS 14 proceedings Stockholm 1962 Aug 15-22 International mathematical union Edited by V. Stenstrom Bibliog. 1,595p 25cm Institut Mittag-Leffler,Djursholm,Sweden,1963
P. Math 2.2601

INTERNATIONAL CONGRESS OF MATHEMATICIANS 15th Proceedings Amsterdam 1954 Sep 2-9 Vol 1-3 Edited by Johan C.H. Gerretson and Johannes de Groot Bibliog 25cm 3 vols North Holland,Amsterdam,1954-57
P Math 2.2875

INTERNATIONAL CONGRESS OF MATHEMATICIANS 16TH Congres international des mathematiciens 1970 Actes Nice 1970 Sep 1-10 Vol 1-3 25cm 3 vols Gauthier-Villars,Paris,1971
P Math 2.3995

INTERNATIONAL CONGRESS OF MEDICINE 17th London 1913 catalogue of the museum Edited by H.W. Armit London,n.d.
An 32.1175

INTERNATIONAL CONGRESS OF MEDICINE 17th London 1913 Aug 6-13 Section 1: anatomy and embryology,pt.2 Frowde;Hodder and Stoughton,London,1914
An 32.1163

INTERNATIONAL CONGRESS OF MEDICINE 17th London 1913 Aug 6-13 Section 3: general pathology and pathological anatomy,pt.2 Frowde;Hodder and Stoughton,London,1914
An 32.1164

INTERNATIONAL CONGRESS OF MEDICINE 17th London 1913 Aug 6-13 general volume Frowde;Hodder and Stoughton,London,1914
An 32.1162

INTERNATIONAL CONGRESS OF MEDICINE 17th : section XII:psychiatry Transactions London 1913 Aug 7-13 Oxford university press;Hodder and Stoughton,London,1913-14 President of congress:J.Crichton-Browne Conference languages:English,French and German
Psy 31.2530

INTERNATIONAL CONGRESS OF MICROBIOLOGY 2nd : a report Proceedings London 1936 Jul 1-Aug 1 International society for microbiology Edited by R. St.John-Brooks port London,1937
Bot 42.0660

INTERNATIONAL CONGRESS OF MICROBIOLOGY 4th Report of proceedings Copenhagen 1947 Jul 20-26 Edited by Mogens Bjorneboe Rosenkilde and Bagger,Copenhagen,1949
Bot 42.0663

INTERNATIONAL CONGRESS OF MICROBIOLOGY 6th Rome 1953 Vol 1: symposium - bacterial cytology Edited by S. Mudd and G. Penso illus Fondazione Emanuele Paterno,Rome,1953 Title also in Italian;text in English and French
Bioch 33.1800

INTERNATIONAL CONGRESS OF MICROBIOLOGY 6th Rome 1953 Vol 2: symposium - microbial metabolism Edited by E.B. Chain Fondazione Emanuele Paterno,Rome,1953 Title also in Italian;text in English and French
Bioch 33.1801

INTERNATIONAL CONGRESS OF MICROBIOLOGY 6th Rome 1953 Vol 3: symposium - nutrition and growth factors Edited by W.H. Schopfer and D.D. Woods Fondazione Emanuele Paterno, Rome,1953 Title also in Italian;text in English and French
Bioch 33.1802

INTERNATIONAL CONGRESS OF MICROBIOLOGY 6th Rome 1953 Vol 4: growth inhibition and chemotherapy Edited by H. Eagle and E.B. Chain Fondazione Emanuele Paterno,Rome,1953 Title also in Italian;text in English and French
Bioch 33.1803

INTERNATIONAL CONGRESS OF MICROBIOLOGY 6th Rome 1953 Vol 6: symposium - interaction of viruses and cells Edited by F.C. Bawden and G. Penso illus Fondazione Emanuele Paterno,Rome,1953 Title also in Italian; text in English and French
Bioch 33.1804

INTERNATIONAL CONGRESS OF MICROBIOLOGY, 6,1 Aetinomyeetales :morphology,biology and systematics Edited by E. Baldacci and P. Redaelli Rome,1953
Bot 42.0664

INTERNATIONAL CONGRESS OF MILITARY MEDICINE AND PHARMACY 5th Report London 1929 May 6-11 By William Seaman Bainbridge Banta,Menasha,Wis.,1929
Path 30.0210

INTERNATIONAL CONGRESS OF NEPHROLOGY 1st Proceedings Congres international de nephrologie Geneva 1969 Sep 1-4 Evian 1960 Sep 1-4 Edited by E. Richet S. Karger,Basle;New York,1961
Inv Med 37.0187

INTERNATIONAL CONGRESS OF NEPHROLOGY 1st Proceedings Geneva 1960 Sep 1-4 Evian 1960 Edited by G. Richet Organised by Societe de nephrology Karger,Basle;New York, 1961 In English and French
Phys 20.1325

INTERNATIONAL CONGRESS OF NEPHROLOGY 4TH Regulation of body fluid volumes by the kidney A Symposium on natriuretic hormone Smolenice castle 1969 Jun Edited by J.H. Cort and B. Lichardus Under the auspices of Ceskoslovenska akademie ved Karger,Basle, 1970
Inv Med 37.0268

INTERNATIONAL CONGRESS OF NEURO-PHARMACOLOGY 1st Proceedings Neuro-psychopharmacology Rome 1958 Sep Collegium for neuro-psychopharmacology Edited by P.B. Bradley and others Elsevier,London,1959
PGMS 29.0192

INTERNATIONAL CONGRESS OF NEUROPATHOLOGY 2nd Proceedings London 1955 Sep 12 Pt 1-2 Edited by W.H. McMenemey Excerpta Medica,Amsterdam,1955
An 32.4163

INTERNATIONAL CONGRESS OF NEUROPATHOLOGY 4th Proceedings Munich 1961 Sep 4-8 Vol 1-3 Edited by H. Jacob 3 vols Thieme, Stuttgart,1962 Papers in English,French, German and Spanish
An 32.4193

INTERNATIONAL CONGRESS OF NEUROPHARMACOLOGY 1st Proceedings Neuro-psychopharmacology Rome 1958 Sep International collegium for neuro-psychopharmacy Edited by P.B. Bradley and others Elsevier,Amsterdam,1959
Psy 31.0524

INTERNATIONAL CONGRESS OF PATHOLOGY Problemi attuali di scienza e di cultura Congresso internazionale di patologia 7 Atti Milan 1968 Sep 5-11 Edited by Alfonso Giordano Accademia nazionale dei Lincei,Rome, 1970
PGMS 29.0618

INTERNATIONAL CONGRESS OF PHILOSOPHY 6th Proceedings Harvard university 1926 Sep 13-17 International congress of philosophy. Organizing committee Edited by Edgar Sheffield Brightman Longmans,Green,New York, 1927
Psy 31.2520

INTERNATIONAL CONGRESS OF PHILOSOPHY.ORGANIZING COMMITTEE International congress of philosophy 6th Proceedings Harvard university 1926 Sep 13-17 Edited by Edgar Sheffield Brightman Longmans,Green,New York, 1927
Psy 31.2520

INTERNATIONAL CONGRESS OF PHOTOBIOLOGY 4th Proceedings Recent progress in photobiology Oxford 1964 Jul Edited by E. J. Bowen Under the auspices of the Comite international de photobiologie Bibliog., illus. vii,400p Blackwell scientific publications,Oxford,1965
Bot 42.1362

INTERNATIONAL CONGRESS OF PHOTOGRAPHY 7th Proceedings London 1928 Jul 9-14 Edited by W. Clark and others port W. Heffer,Cambridge,1928
Chem 18.0247

The INTERNATIONAL CONGRESS OF PHOTOSYNTHESIS RESEARCH Proceedings Progress in photosynthesis research Freudenstadt 1968 Jun 4-8 Vol 1-3 Edited by Helmut Metzner Sponsored by the International union of biological sciences 3 vols Tubingen, 1969
Bot 42.1905

The INTERNATIONAL CONGRESS OF PHOTOSYNTHESIS RESEARCH Proceedings Progress in photosynthesis research Freudenstadt 1968 Jun 4-8 Vol 1-3 Edited by Helmut Metzner Sponsored by the International union of biological sciences 3 vols International union of biological sciences, Tubingen,1969
Bioch 33.1907

INTERNATIONAL CONGRESS OF PHYSICAL MEDICINE 3rd
Proceedings Washington,D.C. 1960
American congress of physical medicine and rehabilitation and American academy of physical medicine and rehabilitation Chicago, Ill.,1960
Phys 20.0990

INTERNATIONAL CONGRESS OF PHYSIOLOGICAL SCIENCES
2nd Proceedings Leiden 1962 Sep 10-17 Vol 3: information processing in the nervous system Edited by R.W. Gerard and J.W. Duyff Under the auspices of the International union of physiological sciences International union of physiological sciences. Proceedings, 3 International congress series, 49 Excerpta medica foundation, Amsterdam,1964
An 32.4392

INTERNATIONAL CONGRESS OF PHYSIOLOGICAL SCIENCES
21st Buenos Aires 1959 Aug 9-15
abstracts of communications 1959
Phys 20.2105

INTERNATIONAL CONGRESS OF PHYSIOLOGICAL SCIENCES
21st Buenos Aires 1959 Aug 9-15
symposia and special lectures 1959
Phys 20.2106

INTERNATIONAL CONGRESS OF PHYSIOLOGICAL SCIENCES
22nd Leiden 1962 Sep 10-17 Vol 3: information processing in the nervous system Edited by R.W. Gerard and J.W. Duyff Under the auspices of the International union of physiological sciences International union of physiological sciences.Proceedings, 3 International congress series, 49 Excerpta medica,Amsterdam,1962
Phys 20.2114

INTERNATIONAL CONGRESS OF PHYSIOLOGICAL SCIENCES
22nd Leiden 1962 Sep 10-17
abstracts of free communications;films and demonstrations Under the auspices of the International union of physiological sciences International union of physiological sciences. Proceedings, 2 International congress series, 48 Excerpta medica,Amsterdam,1962
Phys 20.2113

INTERNATIONAL CONGRESS OF PHYSIOLOGICAL SCIENCES
22nd Leiden 1962 Sep 10-17
lectures and symposia Under the auspices of the International union of physiological sciences International union of physiological sciences.Proceedings, 1,pt 1-2 International congress series, 47 2 vols Excerpta medica,Amsterdam,1962
Phys 20.2112

INTERNATIONAL CONGRESS OF PHYSIOLOGICAL SCIENCES
23rd Tokyo 1965 Sep 1-9 lectures and symposia Under the auspices of the International union of physiological sciences International union of physiological sciences. Proceedings, 4 International congress series, 87 Excerpta medica,Amsterdam,1965
Phys 20.2126

INTERNATIONAL CONGRESS OF PHYSIOLOGICAL SCIENCES
24th Washington,D.C. 1968
abstracts of lectures and symposia International union of physiological sciences International union of physiological sciences. Proceedings, 6 Bethesda,Md.,1968
Phys 20.2154

INTERNATIONAL CONGRESS OF PHYSIOLOGICAL SCIENCES
24th Washington,D.C. 1968
abstracts of volunteer papers and films International union of physiological sciences International union of physiological sciences. Proceedings, 7 Bethesda,Md.,1968
Phys 20.2155

INTERNATIONAL CONGRESS OF PHYSIOLOGICAL SCIENCES,
23RD Japanese physiology:past and present By C.McC. Brooks and K. Koizumi Tokyo,1965 Prepared for the XXIII International congress of physiological sciences,1965
Phys 20.2118

INTERNATIONAL CONGRESS OF PLANT PATHOLOGY 1st
Proceedings Plant pathologist's pocketbook London 1968 Jul Commonwealth mycological institute Commonwealth mycological institute,Kew,1968
BG 38.1757

INTERNATIONAL CONGRESS OF PLANT PATHOLOGY,1ST
Plant pathologist's pocket-book Commonwealth mycological institute Commonwealth mycological institute,Kew,1968
BG 38.1797

INTERNATIONAL CONGRESS OF PREHISTORIC ARCHAEOLOGY
3rd Transactions Norwich 1868 Aug 20-28 and London 1868 Longmans,London, 1869
An 32.2635

INTERNATIONAL CONGRESS OF PSYCHOLOGY
International congress of psychology 9th Proceedings and papers New Haven,Conn. 1929 Sep 1-7 Psychological review co., Princeton,N.J.,1930 Under the presidency of J.M.Cattell
Psy . 31.3393

INTERNATIONAL CONGRESS OF PSYCHOLOGY 3rd
Munich 1896 Aug 4-7 Congress fur psychologie.Internationales organisations-comite Lehmann,Munich,1897 Papers in English,French,German and Italian
Psy 31.3391

INTERNATIONAL CONGRESS OF PSYCHOLOGY 4th
Compte-rendu Paris 1900 Aug 20-26
Edited by Pierre Janet Alcan,Paris,1901
Psy 31.3394

INTERNATIONAL CONGRESS OF PSYCHOLOGY 7th
Proceedings and papers Oxford 1923 Jul 26-Aug 2 International congress of psychology.British organizing committee Edited by Charles S. Myers Cambridge university press,Cambridge,1924 Edited by the president of the congress.Papers in English,French,German and Italian
Psy 31.2510

INTERNATIONAL CONGRESS OF PSYCHOLOGY 8th
Proceedings and papers Groningen 1926
Noordhoff,Groningen,1927
Psy 31.3360

INTERNATIONAL CONGRESS OF PSYCHOLOGY 9th
Proceedings and papers New Haven,Conn. 1929 Sep 1-7 International congress of psychology Psychological review co., Princeton,N.J.,1930 Under the presidency of J.M.Cattell
Psy 31.3393

INTERNATIONAL CONGRESS OF PSYCHOLOGY 10th
Report Copenhagen 1932 Aug 22-27
Levin and Munksgaard,Copenhagen,1935
Under the presidency of professor Rubin
Psy 31.2537

INTERNATIONAL CONGRESS OF PSYCHOLOGY 12th
Proceedings and papers Edinburgh 1948 Jul 23-29 Edited by Mary Collins Oliver and Boyd,Edinburgh;London,1950 Under the presidency of J.Drever
Psy 31.3397

INTERNATIONAL CONGRESS OF PSYCHOLOGY 13th
Proceedings and papers Stockholm 1951 Jul 16-21 International union of scientific psychology Stockholm,1952 Under the presidency of professor Katz Conference languages:English,French and German
Psy 31.2536

INTERNATIONAL CONGRESS OF PSYCHOLOGY 14th
Proceedings Montreal 1954 Jun North-Holland,Amsterdam,1955
Psy 31.3359

INTERNATIONAL CONGRESS OF PSYCHOLOGY 16th
Berichte.Proceedings Bonn 1960 Jul 31-Aug 5 International union of scientific psychology Deutsche gesellschaft fur psychologie Acta psychologica, 19 North-Holland,Amsterdam,1961 President of congress:professor Metzger.Conference languages:English,French and German
Psy 31.3396

INTERNATIONAL CONGRESS OF PSYCHOLOGY.BRITISH ORGANIZING COMMITTEE International congress of psychology 7th Proceedings and papers Oxford 1923 Jul 26-Aug 2 Edited by Charles S. Myers Cambridge university press,Cambridge,1924 Edited by the president of the congress.Papers in English,French,German and Italian
Psy 31.2510

INTERNATIONAL CONGRESS OF PSYCHOLOGY 1ST
Congres international de psychologie physiologique :premiere session 1st Compte-rendu Paris 1889 Societe de psychologie physiologique de Paris Bureau des revues,Paris,1890 Congresses continued as 'International congress of experimental psychology' q.v.
Psy 31.3390

INTERNATIONAL CONGRESS OF PURE AND APPLIED CHEMISTRY 14th main congress lectures and lectures in the sections Lectures Zurich 1955 Jul 21-27 International union of pure and applied chemistry Birkhauser, Basle;Stuttgart,1955 Papers in English, French and German
Chem 18.1737

INTERNATIONAL CONGRESS OF PURE AND APPLIED CHEMISTRY 17th main lectures Lectures Munich 1959 Aug 30-Sep 6 Vol 1-2: inorganic chemistry;biochemistry and applied chemistry International union of pure and applied chemistry 2 vols Butterworths, London,1960 Papers in English,French and German
Chem 18.1739

INTERNATIONAL CONGRESS OF PURE AND APPLIED CHEMISTRY 20th Lectures Moscow 1965 Jul 12-18 International union of pure and applied chemistry and Akademiya nauk S. S.S.R. Butterworths,London,1965
Chem E 24.1274

INTERNATIONAL CONGRESS OF PURE AND APPLIED CHEMISTRY 14TH Congress handbook Zurich 1955 Jul 21-27 International union of pure and applied chemistry I.U.P.A.C. organizing committee,Zurich,c1955 Abstracts in English,French,German and Italian
Chem 18.1736

INTERNATIONAL CONGRESS OF RADIATION RESEARCH 2nd
Abstracts of papers Harrogate 1962 Aug 5-11 1962
Radioth 35.1165

INTERNATIONAL CONGRESS OF RADIATION RESEARCH 2nd
Proceedings Radiation effects in physics,chemistry and biology Harrogate 1962 Aug 5-11 Edited by Michael Ebert and Alma Howard North-Holland,Amsterdam,1963
Radioth 35.1141

INTERNATIONAL CONGRESS OF RADIATION RESEARCH 3rd
Cortina d'Ampezzo 1966 Jun 26-Jul 2 book of abstracts Cortina d'Ampezzo,1966
Radioth 35.1969

INTERNATIONAL CONGRESS OF RADIATION RESEARCH 3rd
Proceedings Radiation research Cortina d'Ampezzo 1966 Jun 26-Jul 2 Edited by G. Silini Under the auspices of the International association for radiation research North-Holland,Amsterdam,1967
Radioth 35.1173

INTERNATIONAL CONGRESS OF RADIOLOGY 7th
Invited papers Copenhagen 1953 Jul 19-24 Acta radiologica.Supplementum, 116 Stockholm,1954
Radioth 35.1311

INTERNATIONAL CONGRESS OF SEDIMENTOLOGY
Proceedings 6th Deltaic and shallow marine deposits Amsterdam 1963 May 29-30 Antwerp 1963 May 31-Jun 1 Edited by L.M. J.U.van Straaten Developments in sedimentology,1 Elsevier,Amsterdam,1964
Geog 13.0304

INTERNATIONAL CONGRESS OF SEDIMENTOLOGY 3rd
Proceedings Groningen 1951 Jul 5-12 Wageningen 1951 Jul 5-12 Edited by Tjeerd Hendrik van Andel Geologisch instituut,Groningen,1951 In English, French and German
Geog 13.0284

INTERNATIONAL CONGRESS OF SEDIMENTOLOGY 3rd
proceedings Groningen 1951 Jul.5-12 Edited by Tj.van Andel map Martinus Nijhoff,The Hague,1951
Geol 8.1645

INTERNATIONAL CONGRESS OF SEDIMENTOLOGY 4th
Proceedings Gottingen 1954 Edited by R. Brinkmann and others Geologische rundschau,43,heft 2 Enke,Stuttgart,1955 In English,French and German
Geog 13.0285

INTERNATIONAL CONGRESS OF SEDIMENTOLOGY 6th
Proceedings Deltaic and shallow marine deposits International congress of sedimentology 6th Proceedings Amsterdam 1963 May 29-30 Antwerp 1963 May 31-Jun 1 Edited by L.M.J.U.van Straaten Developments in sedimentology,1 Elsevier,Amsterdam,1964
Geog 13.0304

INTERNATIONAL CONGRESS OF SOIL SCIENCE 3rd
Oxford 1935 guidebook for the excursion round Britain Clarendon press,Oxford,1935
Geog 13.1078

INTERNATIONAL CONGRESS OF SURFACE ACTIVITY 2nd
Proceedings London 1957 Apr 8-13 Vol 1: gas-liquid and liquid-liquid interface Edited by J.H. Schulman ix,529p Butterworths scientific publications,London, 1957
Chem E 24.0935

INTERNATIONAL CONGRESS OF SURFACE ACTIVITY 2nd
Proceedings London 1957 Apr 8-13 Edited by J.H. Schulman 4 vols Butterworths,London,1957
Col S 12.0278

INTERNATIONAL CONGRESS OF SURFACE ACTIVITY 3rd
Proceedings Cologne 1960 Sep 12-17 Vol 1-4 Universitaetsdruckerei,Mainz,1961
Col S 12.0279

INTERNATIONAL CONGRESS OF THE COLLEGIUM INTERNATIONALE NEURO-PSYCHOPHARMACOLOGICUM 6th Proceedings Present status of psychotropic drugs,pharmacological and clinical aspects Tarragona 1968 Apr 24-27 Collegium internationale neuro-psychopharmacologicum Edited by A. Cerletti and F.J. Bove Excerpta medica foundation. International congress series, 180 Excerpta medica foundation,Amsterdam,1969
Psy 31.2902

INTERNATIONAL CONGRESS OF THE HISTORY OF MEDICINE
10TH Spanish influence on the progress of medical science :with an account of the Wellcome research foundation...founded by sir Henry Wellcome.Commemorating the tenth international congress of the history of medicine... Madrid 1935 Wellcome foundation Bibliog,illus,ports Wellcome foundation,London,1935
Path 30.1591

INTERNATIONAL CONGRESS OF THE HISTORY OF SCIENCE
10th Proceedings Ithaca,N.Y. 1962 Aug 8-Sep 2 Vol 1-2 Academie internationale d'histoire des sciences History of science society Co-sponsored by the United States national academy 2 vols Hermann,Paris Congress chairman:H.Guerlac
WSM 43.0030

INTERNATIONAL CONGRESS OF THE TRANSPLANTATION SOCIETY 1st Paris 1967 Jun 26-30 abstracts Transplantation society Paris, 1967
Surg 23.0036

INTERNATIONAL CONGRESS OF THE TRANSPLANTATION SOCIETY 1st Proceedings Advance in transplantation Paris 1967 Jun 27-30 Transplantation society Edited by J. Dausset and others Munksgaard,Copenhagen,1968
Surg 23.0021

INTERNATIONAL CONGRESS OF ZOOLOGY 2nd-6th,8th-9th Proceedings 1892-1913 8 vols
An 32.0447

The INTERNATIONAL CONGRESS OF ZOOLOGY 15th
Proceedings New approaches in cell biology;a symposium London 1958 Jul Edited by P.M.B. Walker Held at Imperial college of science and technology Academic press,London,1960 A section of the Proceedings of the 15th International congress of zoology
Bal 39.0289

INTERNATIONAL CONGRESS OF ZOOLOGY 16th
Proceedings Ideas in modern biology Washington,D.C. 1963 Aug 20-27 Edited by J. A. Moore Doubleday,NEW York,1965
Radioth 35.0259

INTERNATIONAL CONGRESS OF ZOOLOGY,14TH
Zoological nomenclature Copenhagen decisions on zoological nomenclature additions to...les regles internationales de la nomenclature zoologique Copenhagen 1953 Aug International trust for zoological nomenclature Edited by Francis Hemming International trust for zoological nomenclature,London,1953
BG 38.3107

INTERNATIONAL CONGRESS OF ZOOLOGY,15TH New approaches in cell biology :a symposium Proceedings London 1958 Jul Edited by P. M.B. Walker Academic press,London,1960 Held during the fifteenth international congress of zoology
An 32.3236

INTERNATIONAL CONGRESS OF ZOOLOGY 14TH
Copenhagen decisions on zoological nomenclature additions to,and modifications of, the Regles internationales de la nomenclature Zoological approved and adopted by 14th international congress of zoology,1953 International trust for zoological nomenclature Edited by Francis Hemming International trust for zoological nomenclature,London,1953
Bal 39.4033

INTERNATIONAL CONGRESS OF ZOOLOGY 15TH
Evolution by natural selection :with a foreword by Sir Gavin de Beer By Charles Darwin and Alfred Russel Wallace Published for the Linnean society of London Cambridge university press,Cambridge,1958 Papers originally published in 1858,and reprinted for the 15th international congress of zoology
Gen 34.1309

INTERNATIONAL CONGRESS OF ZOOLOGY 15TH
International code of zoological nomenclature adopted by the fifteenth international congress of zoology International commission on zoological nomenclature London,1961
Bal 39.4042

INTERNATIONAL CONGRESS OF ZOOLOGY 15TH New approaches in cell biology :a symposium Proceedings London 1958 Jul Edited by P. M.B. Walker Academic press,London;New York, 1966 Symposium presented at the fifteenth international congress of zoology
Gen 34.0851

INTERNATIONAL CONGRESS OF ZOOLOGY 4TH Ueber unsere gegenwartige kenntniss vom ursprung des menschen :vortrag gehalten auf dem vierten Internationalen zoologen-congress im Cambridge am 26 August 1898 By Ernst Haeckel 23cm Strauss,Bonn,1898 Bound with three other works by the same author
Bal 44.6531

INTERNATIONAL CONGRESS ON CATALYSIS Philadelphia 1956 Sep Edited by Adalbert Farkas Advances in chemistry and related subjects, 9 xviii,847p Academic press, New York,1957
Chem E 24.0800

INTERNATIONAL CONGRESS ON CATALYSIS 4TH Symposium on electronic phenomena in chemisorption and catalysis on semiconductors Moscow 1968 Jul 2-4 Edited by K. Hauffe and Th. Wolkenstein English edition De Gruyter,Berlin,1969 Held during the fourth international congress on catalysis
Met 25.2475

INTERNATIONAL CONGRESS ON CELL BIOLOGY 12th Summaries of reports and communications Brussels 1968 Aug 25-31 Edited by P. Dustin Excerpta medica.International congress series, 166 Excerpta medica, Amsterdam,1968
Inv Med 37.0250

INTERNATIONAL CONGRESS ON CLINICAL APPLICATION OF HYPERBARIC OXYGEN 2nd Proceedings Hyperbaric oxygenation Glasgow 1964 Sep Edited by Iain McA. Ledingham Livingstone, Edinburgh;London,1965
Radioth 35.0845

INTERNATIONAL CONGRESS ON CLINICAL CHEMISTRY 4th Proceedings Edinburgh 1960 Aug 14-19 Organised by the Association of clinical biochemists E.and S.Livingstone,Edinburgh; London,1961
Bioch 33.2264

INTERNATIONAL CONGRESS ON COAL PETROLOGY 1st proceedings Heerlen 1958 Sep 10-13 International committee for coal petrology. Proceedings,3 Ernest van Aelst,Maastricht, 1960
Geol 8.3137

INTERNATIONAL CONGRESS ON ELECTRON MICROSCOPY 4th Verhandlungen Berlin 1968 Sep 10-17 Band 2: biologisch-medizinischer teil Edited by D. Bargmann and others Springer,Berlin,1969 In English,French and German
Phys 20.1509

INTERNATIONAL CONGRESS ON EXPERIMENTAL MECHANICS 1st Proceedings New York 1961 Nov 1-3 Society for experimental stress analysis Edited by B.E. Rossi Pergamon press,London, 1963
Eng 41.2920

INTERNATIONAL CONGRESS ON EXPERIMENTAL MECHANICS 2nd Proceedings Washington,D.C. 1965 Sep 28-Oct 1 Society for experimental stress analysis Edited by B.E. Rossi Society for experimental stress analysis, Westport,Conn.,1966
Eng 41.2921

INTERNATIONAL CONGRESS ON GLASS 4th Technical papers Advances in glass technology Washington,D.C. 1962 Jul 8-14 American ceramic society Plenum press,New York,1962
Met 25.0116

INTERNATIONAL CONGRESS ON HIGH-SPEED PHOTOGRAPHY 3rd proceedings London 1956 Sep 10-15 Edited by R.B. Collins Held under the auspices of the Department of scientific and industrial research Butterworths,London,1957
Eng 41.0439

INTERNATIONAL CONGRESS ON HIGH-SPEED PHOTOGRAPHY 3rd Proceedings London 1956 Sep 10-15 Edited by R.B. Collins Held under the auspices of Great Britain.Department of scientific and industrial research xvi,417p Butterworths scientific publications,London, 1957
Chem E 24.0452

INTERNATIONAL CONGRESS ON HORMONAL STEROIDS 1st Proceedings Hormonal steroids: biochemistry,pharmacology and therapeutics Vol 1-2 Edited by L. Martini and A. Pecile 2 vols Academic press,New York,1964-65
Inv Med 37.0154

INTERNATIONAL CONGRESS ON HORMONAL STEROIDS 2nd Proceedings Hormonal steroids Milan 1966 May 23-28 Edited by L. Martini and others Excerpta medica.International congress series, 132 Excerpta medica, Amsterdam,1967
Inv Med 37.0116

INTERNATIONAL CONGRESS ON HORMONAL STEROIDS 3rd Abstracts of papers Hamburg 1970 Sep 7-12 Edited by V.H.T. James Excerpta medica.International congress series, 210 Excerpta medica,Amsterdam,1970
Inv Med 37.0115

INTERNATIONAL CONGRESS ON HUMAN FACTORS IN ELECTRONICS :visual distortion from the engineering point of view Papers Stability and distortion of visual space Long Beach,Calif. 1962 May 3-4 By Richard L. Gregory University of Cambridge,Cambridge, 1962
Psy 31.0382

INTERNATIONAL CONGRESS ON HUMAN GENETICS 3rd Proceedings Chicago,Ill. 1966 Sep 5-10 plenary sessions and symposia Edited by James F. Crow and James V. Neel Johns Hopkins press,Baltimore,Md.,1967
An 32.0057

INTERNATIONAL CONGRESS ON LARGE DAMS 5th Transactions Paris 1955 May 31-Jun 4 Vol 1-5 International commission on large dams 5 vols Paris,1956
Eng 41.3334

INTERNATIONAL CONGRESS ON MENTAL RETARDATION 2nd Proceedings Vienna 1961 Aug 14-19 Pt 1: organic bases and biochemical aspects of imbecility Edited by Otto Stur Karger, Basle,1963
An 32.4198

INTERNATIONAL CONGRESS ON METALLIC CORROSION 1st London 1961 Apr 10-15 Society of chemical industry.Corrosion group Butterworths,London,1962
Met 25.1893

INTERNATIONAL CONGRESS ON METALLIC CORROSION 3rd Proceedings Moscow 1966 Vol 1-4 Edited by Ya.M. Kolotyrkin and others 4 vols Mir,Moscow,1969
Met 25.2662

INTERNATIONAL CONGRESS ON MICROWAVE TUBES 3rd Proceedings Munich 1960 Jun 7-11 Edited by J. Wosnik Vieweg,Brunswick,1961
Eng 41.5547

INTERNATIONAL CONGRESS ON MICROWAVE TUBES 4th Proceedings Scheveningen 1962 Sep 3-7 Centrex;Cleaver-Hume,Eindoven;London,1963
Eng 41.5557

INTERNATIONAL CONGRESS ON MICROWAVE TUBES 5th Proceedings Paris 1964 Societe francaise des electroniciens et radioelectriciens Societe francaise des ingenieurs et techniciens du vide Sponsored by the Federation nationale des industries electroniques Academic press,New York,1965
Eng 41.5574

INTERNATIONAL CONGRESS ON PHOTOBIOLOGY 3rd Proceedings Finsen memorial congress Copenhagen 1960 Edited by B.Chr. Christensen and B. Buchmann illus. Elsevier,Amsterdam,1961
Bot 42.1347

INTERNATIONAL CONGRESS ON PROJECT PLANNING BY NETWORK ANALYSIS 2nd Proceedings Amsterdam 1969 Oct 6-10 North-Holland, Amsterdam,1969
Eng 41.1240

INTERNATIONAL CONGRESS ON PROTOZOOLOGY 3rd Abstracts of papers and communications Progress in protozoology Leningrad 1969 Jul 2-10 International commission on protozoology Publishing house 'Nauka', Leningrad,1969
Mol 45.0140

INTERNATIONAL CONGRESS ON RHEOLOGY 2nd Proceedings Oxford 1953 Jul 26-31 Edited by V.G.W. Harrison ix,451p Butterworth scientific publications,London, 1954
Chem E 24.0446

INTERNATIONAL CONGRESS ON RHEOLOGY 1st Proceedings Scheveningen 1948 Sep 21-24 North-Holland,Amsterdam,1949
Col S 12.0103

INTERNATIONAL CONGRESS ON THE APPLICATION OF THE THEORY OF PROBABILITY IN TELEPHONE ENGINEERING AND ADMINISTRATION 1st proceedings Copenhagen 1955 Jun.20-23 Kobnhavns telefon aktieselskab Edited by Arne Jensen 130p 30cm Copenhagen,1957 Published simultaneously as vol.1,1957,no.1 of "Teleteknik"
Math 3.0881

INTERNATIONAL CONGRESS ON THE QUATERNARY (INQUA) 8TH Chetvertichnye otlozheniya Shpitsbergena k VIII kongressu INQUA Parizh 1969 By Yu.A. Lavrushin Kommissiya po izycheniyu chetvertichnogo perioda illus 181p 22cm Izdatelstvo 'Nauka',Moscow,1969
Sco 14.8250

INTERNATIONAL CONGRESS ON THE QUATERNARY (INQUA) 8TH Golotsen k VIII kongressu INQUA Parizh 1969 By M.I. Neishtadt Institut geografii,Moscow illus,maps 228p 27cm Izdatelstvo 'Nauka',Moscow,1969
Sco 14.8260

INTERNATIONAL CONGRESS ON THE QUATERNARY (INQUA) 8TH Poslednii lednikovyi pokrov na severo-zapade evropaiskoi chasti S.S.S.R. k VIII kongressu INQUA,Parizh 1969 Institut geografii,Moscow Edited by I.P. Gerasimov maps 322p 27cm Izdatelstvo 'Nauka', Moscow,1969
Sco 14.8258

INTERNATIONAL CONGRESS ON THE QUATERNARY (INQUA) 8TH Problemy kriolitologii k VIII mezhdunarodnomu kongressu assotsiatsii po izucheniyu chetvertichnogo perioda (INQUA), Parizh 1969 Vyp 1 Moskovskii gosudarstvennii universitet im. M.V.Lomonosova. Geografischekii fakultet.Kafedra kriolitologii i glatsiologii Edited by A.I. Popov Bibliog,illus 176p 23cm Izdatelstvo Moskovskogo universiteta,Moscow,1969
Sco 14.8242

INTERNATIONAL CONGRESS ON THEORETICAL AND APPLIED MECHANICS 8th proceedings Istanbul 1952 Aug.20-28 Vols 1-2 Istanbul universitesi.Faculty of science Edited by Kerim Erim Under the auspices of the International committee for the congresses of applied mechanics 2 vols University faculty of science,Istanbul,1953
A Math 4.0326

INTERNATIONAL CONGRESS ON TROPICAL MEDICINE AND MALARIA 4th Proceedings Tropical medicine and malaria Washington,D.C. 1948 May 10-18 Vol 1-2 United States. Department of State United States.Department of State.International organization and conference series, 1,5 Govt. printing office,Washington,D.C.,1948
Mol 45.0657

INTERNATIONAL CONGRESS ON VACUUM SCIENCE AND TECHNOLOGY 2ND National vacuum symposium 8th Transactions Washington,D.C. 1961 Oct 16-19 Vol 1-2 American vacuum society and International organization for vacuum science and technology Edited by Luther E. Preuss 2 vols Pergamon press, Oxford,1962 Combined with the International congress on vacuum science and technology,2nd
Cav 7.2376

INTERNATIONAL CONGRESSES ON HIGH SPEED PHOTOGRAPHY 1952-70 Index to the international congresses on high speed photography 1952-1970 Edited by J. Wadsworth and M.W. Glover Royal aircraft establishment, Farnborough,c1970
Eng 41.8373

INTERNATIONAL COUNCIL FOR PHILOSOPHY AND HUMANISTIC STUDIES Marx and contemporary scientific thought The Role of Karl Marx in the development of contemporary scientific thought :a symposium Papers Paris 1968 May 8-10 Under the auspices of Unesco International social science council. Publications, 13 xi,612p Mouton,The Hague; Paris,1969
WSM 43.3826

INTERNATIONAL COUNCIL FOR THE EXPLORATION OF THE SEA.DISTANT NORTHERN SEAS COMMITTEE Symposium on the ecology of pelagic fish species in Arctic waters and adjacent seas Papers Charlottenlund 1966 Sep 30-Oct 1 Edited by R.W. Blacker Conseil permanent international pour l'exploration de la mer. Rapports et proces-verbaux des reunions, 158 illus,maps Host,Copenhagen,1968
Sco 14.8291

INTERNATIONAL COUNCIL FOR THE EXPLORATION OF THE SEA.GADOID FISH COMMITTEE Symposium on the ecology of pelagic fish species in Arctic waters and adjacent seas Papers Charlottenlund 1966 Sep 30-Oct 1 Edited by R.W. Blacker Conseil permanent international pour l'exploration de la mer.Rapports et proces-verbaux des reunions, 158 illus,maps Host,Copenhagen,1968
Sco 14.8291

INTERNATIONAL COUNCIL FOR THE EXPLORATION OF THE SEA see also CONSEIL PERMANENT INTERNATIONAL POUR L'EXPLORATION DE LA MER

INTERNATIONAL COUNCIL OF SCIENTIFIC UNIONS Determination of radial velocities and their applications Toronto 1966 Jun 20-24 Edited by A.H. Batten and J.F. Heard International astronomical union.Symposium, 30 Academic press,London;New York,1967
TA 15.0307

INTERNATIONAL COUNCIL OF SCIENTIFIC UNIONS Mechanisms of colour discrimination :an international symposium on the fundamental mechanisms of the chromatic discrimination in animals and man Proceedings Paris 1958 Jul 25-29 Pergamon press:symposium publications division,London,1960
Psy 31.0384

INTERNATIONAL COUNCIL OF SCIENTIFIC UNIONS Mixed commission on the ionosphere meeting 3rd Proceedings Canberra 1952 Aug. 24-26 International scientific radio union Edited by W.J.G. Beynon 194p General secretariat, Brussels,1953
Obs 6.2579

INTERNATIONAL COUNCIL OF SCIENTIFIC UNIONS Solar eclipses and the ionosphere :a symposium Proceedings London 1955 Aug 22-24 Mixed commission on the ionosphere Edited by W.J.G. Beynon and G.H. Brown Journal of atmospheric and terrestrial physics.Special supplement, 6 ix,330p Pergamon press, London;New York,1956
Nap 11.0437

INTERNATIONAL COUNCIL OF SCIENTIFIC UNIONS phenomenes solaires et terrestres Commission mixte pour l'etude des relations entre les phenomenes solaires et terrestres Proceedings Rome 1952 Sep. 3 35p Paris, 1952
Obs 6.2544

INTERNATIONAL COUNCIL OF SCIENTIFIC UNIONS 11th general assembly Record Bombay 1966 Jan 6-11 30cm Rome,1966 Mimeograph
Sco 14.0095

INTERNATIONAL COUNCIL OF SCIENTIFIC UNIONS.JOINT COMMISSION ON ELECTRON MICROSCOPY The International conference on electron microscopy 3rd Proceedings London 1954 Jul 15-21 Edited by V.E. Cosslett and others Royal microscopical society,London, 1956
Cav 7.1455

INTERNATIONAL COUNCIL OF SCIENTIFIC UNIONS. SCIENTIFIC COMMITTEE ON ANTARCTIC RESEARCH SCAR,SCOR,IAPO,IUBS symposium on Antarctic oceanography Papers Santiago de Chile 1966 Sep 13-16 Held by invitation of the Chilean national committee for antarctic research illus,maps xi,268p 23cm Scott Polar research institute for SCAR,Cambridge, 1968
Sco 14.8300

INTERNATIONAL COUNCIL OF SCIENTIFIC UNIONS. SCIENTIFIC COMMITTEE ON OCEANIC RESEARCH SCAR,SCOR,IAPO,IUBS symposium on Antarctic oceanography Papers Santiago de Chile 1966 Sep 13-16 Held by invitation of the Chilean national committee for antarctic research illus,maps xi,268p 23cm Scott Polar research institute for SCAR,Cambridge, 1968
Sco 14.8300

INTERNATIONAL COUNCIL OF THE AERONAUTICAL SCIENCES 3rd congress Proceedings Stockholm 1962 Aug 27-31 American institute of aeronautics and astronautics Spartan;Macmillan,Washington,D.C.;London,1964
Eng 41.6965

INTERNATIONAL COUNCIL OF THE AERONAUTICAL SCIENCES 4th congress Proceedings Paris 1964 Aug 24-28 Edited by Robert R. Dexter Spartan;Macmillan,Washington,D.C.; London,1965
Eng 41.6966

INTERNATIONAL COURSE ON MATERIALS SCIENCE 1st Essays Tremezzo 1968 Sep 8-20 Accademia nazionale dei lincei.Donegani foundation Edited by Alan W. Searcy and others illus xxv,715p 24cm Interscience,New York;Chichester,1970 Being the 11th course organized by the Donegani foundation
Met 25.2593

INTERNATIONAL CRYSTALLOGRAPHIC CONGRESS 1ST- Growth of crystals Conference on crystal growth 1st- Proceedings Moscow 1956- 1- Institut kristallografii,Moscow Edited by A.V. Shubnikov and N.N. Sheftal Translated by Consultants' bureau from the Russian Rost kristallov, 1- Consultants bureau,inc.,New York,1958- Includes interim reports published between conferences. Held as part of the international crystallographic congresses
Met 25.1508

INTERNATIONAL DEVELOPMENTS IN HEAT TRANSFER Heat transfer conference 1961-62 Proceedings Boulder,Colo. 1961 Aug 28-Sep 1 and London 1962 Jan 8-12 American society of mechanical engineers American society of mechanical engineers,New York,1963
Eng 41.7168

INTERNATIONAL DEVELOPMENTS IN HEAT TRANSFER International heat transfer conference Papers Boulder,Col. 1961 Aug 28-Sep 1 London 1962 Jan 8-12 Pt 1-5 Arranged by the American society of mechanical engineers 5 vols A.S.M.E.,New York,1961 Co-sponsored by the American institute of chemical engineers,the Institution of mechanical engineers,and the Institution of chemical engineers
Chem E 24.0595

INTERNATIONAL DIABETES FEDERATION Diabetes International diabetes federation congress 6th Proceedings Stockholm 1967 Jul 30-Aug 4 Edited by J. Ostman and R.D.G. Milner Excerpta medica,Amsterdam,1969
Bioch 33.0462

INTERNATIONAL DIABETES FEDERATION CONGRESS 6th Proceedings Diabetes Stockholm 1967 Jul 30-Aug 4 International diabetes federation Edited by J. Ostman and R.D.G. Milner Excerpta medica,Amsterdam,1969
Bioch 33.0462

INTERNATIONAL ECONOMIC ASSOCIATION Problems in economic development Conference on problems in economic development Proceedings Edited by E.A.G. Robinson Macmillan,London, 1965
Geog 13.2413

INTERNATIONAL ECONOMIC ASSOCIATION 3rd : regional conference Proceedings Economic development for Africa south of the Sahara Addis Ababa 1961 Edited by E.A.G. Robinson Macmillan,London,1965
Geog 13.4646

INTERNATIONAL ECONOMIC ASSOCIATION.ROUND TABLE CONFERENCE Proceedings Economic development with special reference to east Asia Gamagori,Japan 1963 Edited by Kenneth Berrill Macmillan,London,1964
Geog 13.2407

INTERNATIONAL ECONOMIC ASSOCIATION CONFERENCE Proceedings Economic consequences of the size of nations Lisbon 1957 Edited by E. A.G. Robinson Macmillan,London,1960
Geog 13.2036

INTERNATIONAL ECONOMIC ASSOCIATION CONFERENCE Proceedings Economics of take-off into sustained growth Konstanz 1960 Edited by W.W. Rostow Macmillan,London,1963
Geog 13.2064

INTERNATIONAL ENGINEERING CONGRESS :report of the proceedings and abstracts of the papers read Glasgow 1901 Sep 2-3 Edited by J. D. Cormack Asher,Glasgow,1902
Eng 41.1563

INTERNATIONAL ENGINEERING CONGRESS ON WATERWAYS AND MARITIME WORKS Proceedings Waterways and maritime works Glasgow 1901 Sep 3-5 Clowes,London,1902
Eng 41.3317

INTERNATIONAL EPIDEMIOLOGICAL ASSOCIATION 5th Proceedings Primosten 1968 Aug 25-31 Edited by Milutin Srdic and others Savremena administracija,Belgrade,1970
PGMS 29.0635

INTERNATIONAL EUGENICS CONFERENCE 1st Papers Problems in eugenics London 1912 Jul 24-30 Eugenics education society 2 vols Eugenics education society,London, 1913 Later known as the 1st international congress of eugenics
Gen 34.1925

INTERNATIONAL EUGENICS CONGRESS 1st Papers Problems in eugenics London 1912 Jul 24-20 Eugenics education society Eugenics education society,London,1912
Bal 44.6430

INTERNATIONAL EUGENICS CONGRESS 1st Papers Problems in eugenics London 1912 Jul 24-30 Eugenics education society Held at the University of London Eugenics education society,London,1912
Bot 42.6705

INTERNATIONAL EXHIBITION OF BOTANICAL ART AND ILLUSTRATION 2nd Catalogue Pittsburgh,Pa. 1968 Oct 2 and Pittsburgh, Pa. 1969 Apr 15 Hunt botanical library Edited by George H.M. Lawrence Hunt botanical library,Pittsburgh,Pa.,1968
BG 38.0911

INTERNATIONAL EXHIBITION OF TWENTIETH CENTURY BOTANICAL ART Catalogue 2nd Selection of 20th century botanical art and illustrations Edited by George H.M. Lawrence port 168p Carnegie-Mellon university, Pittsburgh,Pa.,1969 Presented at the 11th International botanical congress,August 1969
Bot 42.0634

INTERNATIONAL FEDERATION FOR AUTOMATIC CONTROL IFAC symposium on the control of distributed parameter systems Proceedings Banff 1971 Jun 21-23 2 vols IFAC technical committee on theory,Banff,1971
Eng 41.8593

INTERNATIONAL FEDERATION FOR INFORMATION PROCESSES Symbol manipulation languages and techniques IFIP working conference on symbol manipulation languages Proceedings Pisa 1966 Sep 5-9 Edited by Daniel G. Bobrow illus 487p North-Holland, Amsterdam,1968
Math L 5.3822

INTERNATIONAL FEDERATION FOR INFORMATION PROCESSING ALGOL 68 implementation IFIP working conference on ALGOL and its implementation Proceedings Munich 1970 Jul 20-24 Edited by J.E.L. Peck illus 375p North-Holland,Amsterdam,1971
Math L 5.3876

INTERNATIONAL FEDERATION FOR INFORMATION PROCESSING Formal language description languages for computer programming IFIP working conference on formal language description languages proceedings Vienna 1964 Sep 15-18 Edited by T.B. Steel bibliog 330p 22cm North-Holland publishing,Amsterdam,1966
Math L 5.1033

INTERNATIONAL FEDERATION FOR INFORMATION PROCESSING IFIP congress 1968 invited papers Edinburgh 1968 Aug 5-10 1600p 30cm North-Holland,Amsterdam,1968
Math L 5.3123

INTERNATIONAL FEDERATION FOR INFORMATION PROCESSING Information processing 1962 IFIP congress 62 proceedings Munich 1962 Aug 27-Sep 1 Edited by Cicely M. Popplewell xvi,780p 30cm North-Holland publishing company,Amsterdam,1962
Math L 5.0912

INTERNATIONAL FEDERATION FOR INFORMATION PROCESSING Information processing 1965 IFIP congress 65 Proceedings New York 1965 May 24-25 Vol 1-2 Edited by Wayne A. Kalenich 2 vols Spartan;Macmillan, Washington,D.C.;London,1965-66
Eng 41.6001

INTERNATIONAL FEDERATION FOR INFORMATION PROCESSING Information processing 1965 IFIP congress 65 proceedings New York 1965 May 24-29 Vol. 1 xv,304p 29cm Spartan books,Washington,D.C.,1965
Math L 5.1045

INTERNATIONAL FEDERATION FOR INFORMATION PROCESSING Information processing 1971 IFIP congress 71 Proceedings Ljubljana 1971 Aug 23-28 Edited by C.V. Freiman illus 1621p 2 vols North-Holland, Amsterdam,1972
Math L 5.3841

INTERNATIONAL FEDERATION FOR INFORMATION PROCESSING Microminiaturization in automatic control equipment and in digital computers :IFAC-IFIP symposium Proceedings Munich 1965 Oct 12-23 Edited by J. Berghammer Oldenbourg,Munich,1966
Eng 41.5580

INTERNATIONAL FEDERATION FOR INFORMATION PROCESSING Working conference on symbol manipulation language Papers Pisa 1966 Sep 5-9 1966 Typescript papers in loose-leaf file
Math L 5.3571

INTERNATIONAL FEDERATION FOR MEDICAL ELECTRONICS Medical electronics International conference on medical electronics 2nd Proceedings Paris 1959 Jun 24-27 Edited by C.N. Smyth Iliffe,London,1960
Psy 31.3062

INTERNATIONAL FEDERATION OF AUTOMATIC CONTROL Automatic and remote control Congress of the International federation of automatic control 1st-3rd Proceedings 1960,1963,1966 Butterworths,London,1961-66
Eng 41.8207

INTERNATIONAL FEDERATION OF AUTOMATIC CONTROL I.F.A.C. symposium on fluidics Proceedings London 1968 Nov 4-8 Organised by the Royal aeronautical society 1968
Eng 41.5905

INTERNATIONAL FEDERATION OF AUTOMATIC CONTROL I.F.A.C. symposium on pulse-rate and pulse-number signals in automatic control Proceedings Budapest 1968 Apr 8-10 International federation of automatic control, Debrecen,1968
Eng 41.6020

INTERNATIONAL FEDERATION OF AUTOMATIC CONTROL I.F.A.C. symposium on system sensitivity and adaptivity 2nd Preprints Dubrovnik 1968 Aug 26-31 Organised by the Yugoslav committee for electronics and automation I.F. A.C.,Dubrovnik,1968
Eng 41.5889

INTERNATIONAL FEDERATION OF AUTOMATIC CONTROL I.F.A.C.symposium on the theory of self-adaptive control systems 2nd Proceedings Teddington 1965 Sep 14-17 Edited by P.H. Hammond Plenum press,New York,1966
Eng 41.5888

INTERNATIONAL FEDERATION OF AUTOMATIC CONTROL Identification in automatic control systems I.F.A.C. symposium on identification in automatic control systems Preprints Prague 1967 Jun 12-17 Pt 1-2 Ceskoslovenska akademie ved.Institute of information theory and automation 2 vols Academia,Prague,1967
Eng 41.6019

INTERNATIONAL FEDERATION OF AUTOMATIC CONTROL Microminiaturization in automatic control equipment and in digital computers :IFAC-IFIP symposium Proceedings Munich 1965 Oct 12-23 Edited by J. Berghammer Oldenbourg, Munich,1966
Eng 41.5580

INTERNATIONAL FEDERATION OF AUTOMATIC CONTROL Peaceful uses of automation in outer space IFAC symposium on automatic control in the peaceful uses of space Stavanger 1965 Jun 21-24 Edited by John A. Aseltine Plenum, New York,1966
Eng 41.6002

INTERNATIONAL FEDERATION OF AUTOMATIC CONTROL. COMMITTEE ON APPLICATIONS Symposium on computer control of natural resources and public utilities Preprints Haifa 1967 Sep 11-14 Haifa,1967
Eng 41.5866

INTERNATIONAL FEDERATION OF AUTOMATIC CONTROL. THEORY COMMITTEE International symposium on optimizing and adaptive control 1st Proceedings Rome 1962 Apr 26-28 Edited by Loren E. Bollinger and Emil J. Minnar Instrument society of America,Pittsburgh,Pa., 1962
Eng 41.6017

INTERNATIONAL FEDERATION OF AUTOMATIC CONTROL. THEORY COMMITTEE Sensitivity methods in control theory International symposium on sensitivity analysis Dubrovnik 1964 Aug 31-Sep 5 Yugoslav committee for electronics and automation Edited by L. Radanovic Pergamon press,Oxford,1965
Eng 41.5835

INTERNATIONAL FEDERATION OF ELECTRON MICROSCOPE SOCIETIES Electron microscopy International congress for electron microscopy 5th Philadelphia 1962 Aug 29-Sep 5 Vol 1: non-biology Edited by Sydney S. Breese Academic press,New York;London,1962
Met 25.1368

INTERNATIONAL FEDERATION OF ELECTRON MICROSCOPE SOCIETIES Electron microscopy 1964 European regional conference on electron microscopy 3rd Proceedings Prague 1964 Aug 26-Sep 3 Vol A: non-biology Edited by M. Titlbach Czech.academy of sciences,Prague,1964
Met 25.2494

INTERNATIONAL FEDERATION OF ELECTRON MICROSCOPE SOCIETIES Electron microscopy 1966 International congress for electron microscopy 6th Kyoto 1966 Aug 28-Sep 4 Vol 1: non-biology Edited by Ryozi Uyeda Maruzen, Tokyo,1966
Met 25.1388

INTERNATIONAL FEDERATION OF ELECTRON MICROSCOPE SOCIETIES Electron microscopy 1966 International congress for electron microscopy 6th Papers and extended abstracts Kyoto 1966 Aug 28-Sep 4 Edited by Ryozi Uyeda illus Maruzen,Tokyo,1966
Path 30.2615

INTERNATIONAL FEDERATION OF ELECTRON MICROSCOPE SOCIETIES Intionationaler kongress fur elektronmikroskopie 4th Proceedings Berlin 1958 Sep 10-17 Bd 1: physikalisch-technischer teil Edited by G. Mollenstedt and others Springer,Berlin,1960 General editors W.Bargmann and others.Papers in English,French and German
Met 25.2493

INTERNATIONAL FEDERATION OF OPERATIONAL RESEARCH SOCIETIES International conference on operational research 3rd proceedings Oslo 1963 Jul.1-5 Operational research society,Oslo Edited by G. Kreweras and G. Morlat xii,959p 25cm Dunod;English universities press,Paris;London,1964
Math 3.0915

INTERNATIONAL FEDERATION OF SOCIETIES FOR ELECTRON MICROSCOPY Electron microscopy 1966 The International congress for electron microscopy 6th Extended abstracts Kyoto Aug 28-Sep 4 Vol 1-2 Edited by Ryozi Uyeda 2 vols Maruzen company,Tokyo,1966
Cav 7.1340

INTERNATIONAL FEDERATION OF SOCIETIES FOR ELECTRON MICROSCOPY Electron microscopy 1972 European congress on electron microscopy 5th Proceedings Manchester 1972 Institute of physics.Conference series, 14 Institute of physics,London,1972
Eng 41.8576

INTERNATIONAL FEDFRATION OF SOCIETIES FOR ELECTRON MICROSCOPY Microscopie electronique 1970 Congres international de microscopie electronique 7e Resumes des communications Grenoble 1970 Aug 30-Sep 3 Edited by Pierre Favard Societe francaise de microscopie electronique,Paris,1970
Met 25.2604

INTERNATIONAL FERTILITY ASSOCIATION Fertility and sterility World congress on fertility and sterility 5th Proceedings Stockholm 1966 Jun 16-22 Edited by Bjorn Westin and Nils Wiquist International congress series, 133 Excerpta medica foundation,Amsterdam,1967
An 32.3899

INTERNATIONAL FERTILITY ASSOCIATION Fertility and sterility World congress on fertility and sterility 5th Proceedings Stockholm 1966 Jun 16-22 Edited by Bjorn Westin and Nils Wiqvist Excerpta medica. International congress series, 133 Excerpta medica,Amsterdam,1967
Inv Med 37.0230

INTERNATIONAL FORESTRY CONGRESS 3rd programme Helsinki 1940 Jul 1-5 International institute of agriculture Helsinki,1939
Philos 1.2032

INTERNATIONAL FOUNDATION OF OPERATIONAL RESEARCH SOCIETIES International conference on operational research 4th Proceedings Boston,Mass. 1966 Aug 29-Sep 2 Edited by David B. Hertz and Jacques Melese Operations research society of America.Publications in operations research, 15 Wiley-Interscience, New York,1966
Eng 41.1219

INTERNATIONAL FOUNDRY CONGRESS 28th Congress papers Vienna 1961 Jun 19-24 International committee of foundry technical associations Verein Oesterreichischer Giesserfachleute,Vienna,1961 Papers in English,French and German
Met 25.0681

INTERNATIONAL GENETICAL CONGRESS 7th Proceedings Edinburgh 1939 Aug 23-30 Edited by R.C. Punnett Cambridge university press,Cambridge,1941
Bot 42.0661

INTERNATIONAL GENETICAL CONGRESS 7th Proceedings Edinburgh 1939 Aug 23-30 Genetical society Edited by Reginald Crundall Punnett Journal of genetics. Supplement Cambridge university press, Cambridge,1941
Gen 34.2219

INTERNATIONAL GENETICAL CONGRESS 7th Proceedings Genetics Edinburgh 1939 Aug 23-30 International genetical congress. Permanent international committee Edited by R.C. Punnett Cambridge university press, Cambridge,1939 Issued as a supplementary volume to the 'Journal of genetics'
Bot 42.6483

INTERNATIONAL GENETICAL CONGRESS.PERMANENT INTERNATIONAL COMMITTEE Genetics International genetical congress 7th Proceedings Edinburgh 1939 Aug 23-30 Edited by R.C. Punnett Cambridge university press,Cambridge,1939 Issued as a supplementary volume to the 'Journal of genetics'
Bot 42.6483

INTERNATIONAL GEODETIC AND GEOPHYSICAL UNION 9th general assembly Symposium proceedings Symposium on the general circulation of the oceans and the atmosphere Brussels 1951 Aug International association of meteorology and atmospheric physics Unesco,Brussels,1951
Geog 13.0960

INTERNATIONAL GEOGRAPHICAL CONGRESS
Esplorazione scientifica del Mar Rosso : comunicazione al congresso geografice internazionale Cambridge 1928 Jul Italy. Navy.Hydrographic institute Genoa,1928 Bound with English translation
Geog 13.0854

INTERNATIONAL GEOGRAPHICAL CONGRESS
Guidebook Denmark :contributions to problems discussed in symposia and excursions Edited by Niels Kingo Jacobsen maps xi,372p 24cm Copenhagen,1960
Sco 14.0132

INTERNATIONAL GEOGRAPHICAL CONGRESS Excursion guide Congres international de geographie Paris 1931 Excursion A2: le sud-est du Massif central Edited by Henri Baulig Colin,Paris,1931
Geog 13.0525

INTERNATIONAL GEOGRAPHICAL CONGRESS Guidebooks Congres international de geographie Paris 1931 Colin,Paris,1931
Geog 13.4158

INTERNATIONAL GEOGRAPHICAL CONGRESS Guidebooks to excursions Lisbon 1949 International geographical union 2 vols Lisbon,1949
Geog 13.4139

INTERNATIONAL GEOGRAPHICAL CONGRESS Report Travaux topographiques et cartographiques Amsterdam 1938 Jul International geographical union Amsterdam,1938
Geog 13.5708

INTERNATIONAL GEOGRAPHICAL CONGRESS 11th
Compte rendu Cairo 1925 Apr International geographical union 2 vols Societe royale de geographie dEgypte,Cairo, 1925
Geog 13.5593

INTERNATIONAL GEOGRAPHICAL CONGRESS 12th
Report of the proceedings Cambridge 1928 Jul Cambridge university press, Cambridge,1930
Geog 13.5594

INTERNATIONAL GEOGRAPHICAL CONGRESS 15th
Comptes rendus;excursions Amsterdam 1938 Tom 1-2 International geographical union 17 vols Brill,Leiden,1938 Partly in English
Geog 13.5595

INTERNATIONAL GEOGRAPHICAL CONGRESS 16th
Comptes rendus Lisbon 1949 Tom 1-4 International geographical union 5 vols Lisbon,1950-52 Partly in English
Geog 13.5596

INTERNATIONAL GEOGRAPHICAL CONGRESS 17th
Proceedings International geographical union.General assembly,8th Washington,D.C. 1952 Aug 8-15 National research council, Washington,D.C.,1952
Geog 13.5597

INTERNATIONAL GEOGRAPHICAL CONGRESS 18th
Excursion guides Brazil 1956 1-9 International geographical union.Brazilian national committee 3 vols International geographical union,Rio de Janeiro,1956
Geog 13.5349

INTERNATIONAL GEOGRAPHICAL CONGRESS 19th
Physical geography of Greenland Copenhagen 1960 Edited by Niels Kingo Jacobsen Arranged at the University of Copenhagen. Geographical institute Folia geographica Danica, 9 illus,maps 234p C.A.Reitzel, Copenhagen,1961
Sco 14.8357

INTERNATIONAL GEOGRAPHICAL CONGRESS 19th
Final report,bibliography Stockholm 1960 Aug 6-13 Edited by Staffan Helmfrid illus 138p Stockholm,1963 Held in Norden,ie.Denmark,Finland,Iceland,Norway and Sweden
Sco 14.0133

INTERNATIONAL GEOGRAPHICAL CONGRESS 19th
Papers of the Siljan symposium Advance and retreat of rural settlement Norden 1960 Edited by Gerd Enequist and Gunnar Norling Uppsala universitet.Geografiska institution.Meddelanden.Ser A,160 Geografiska annaler,42,no 4 Uppsala,1960
Geog 13.1898

INTERNATIONAL GEOGRAPHICAL CONGRESS 20th
Guide to London excursions London 1964 Edited by K.M. Clayton London,1964
Geog 13.5695

INTERNATIONAL GEOGRAPHICAL CONGRESS 20th
Proceedings London 1964 Edited by J. Wreford Watson Nelson,London,1967
Geog 13.5622

INTERNATIONAL GEOGRAPHICAL CONGRESS 21st
Abstracts of papers India 1968 Calcutta,1968
Geog 13.6841

INTERNATIONAL GEOGRAPHICAL CONGRESS 6th
Report London 1895 Vol 1-2 Murray, London,1896
Geog 13.5591

INTERNATIONAL GEOGRAPHICAL CONGRESS 6th
Report London 1895 Jul 26-Aug 3 Edited by Hugh Robert Mill illus,maps xxxvi,996p 25cm John Murray,London,1896 Includes papers in French,German and Italian
Sco 14.0104

INTERNATIONAL GEOGRAPHICAL CONGRESS 8th
Report United States 1904 U.S. government printing office,Washington,D.C., 1905
Geog 13.5592

INTERNATIONAL GEOGRAPHICAL CONGRESS,18TH
Essais de geographie :recueil des articles pour le XVIIIe congres international geographique Geografischeskoe obshchestvo soyuza S.S.R. Izdatelstvo akademiya nauk, Moscow;Leningrad,1956
Geog 13.5587

INTERNATIONAL GEOGRAPHICAL CONGRESS,19TH
Finland :guidebooks prepared for the three Finnish excursions Norden 1960 Edited by Helmer Smeds Fennia,84 Helsinki,1960
Geog 13.4337

INTERNATIONAL GEOGRAPHICAL CONGRESS,19TH
Guidebook Denmark :contributions to problems discussed in symposia and excursions Norden 1960 Jul-Aug Edited by Niels Kingo Jacobson Kobenhavns universitets geografiske institut, Copenhagen,1960
Geog 13.4323

INTERNATIONAL GEOGRAPHICAL CONGRESS,19TH I.G.U.symposium in urban geography Proceedings Lund 1960 Aug 15-19 Edited by Knut Norberg Lund studies in geography.Ser.B.Human geography,24 Lunds universitet,Lund,1962 Held at the 19th International geographical congress
Geog 13.1944

INTERNATIONAL GEOGRAPHICAL CONGRESS,19TH Stockholm:structure and development By W. William-Olsson bibliog.,illus. Almqvist and Wiksells,Uppsala,1960
Geog 13.1934

INTERNATIONAL GEOGRAPHICAL CONGRESS,1928 Report of the commission on types of rural settlement International geographical union. Commission on types of rural settlement Newtown;Llanllwchaiarn,1928
Geog 13.1870

INTERNATIONAL GEOGRAPHICAL CONGRESS,20TH Field studies in the British isles Edited by J.A. Steers Nelson,London,1964
Geog 13.3960

INTERNATIONAL GEOGRAPHICAL CONGRESS,20TH The British Isles:a systematic geography Edited by J.Wreford Watson and J.B. Sissons Nelson, London,1964
Geog 13.3962

INTERNATIONAL GEOGRAPHICAL CONGRESS 12TH, CAMBRIDGE Great Britain:essays in regional geography... Edited by Alan G. Ogilvie Cambridge university press,Cambridge, 1928 Published on the occasion of the Twelfth International geographical congress, Cambridge
Geol 8.1767

INTERNATIONAL GEOGRAPHICAL CONGRESS 1938 Rapport de la commission pour la cartographie des surfaces d'aplanissement tertiaires International geographical union.Commission pour la cartographie des surfaces d'aplanissement tertiaires Paris,1938 Report to the International geographical congress held at Amsterdam,1938
Geog 13.0421

INTERNATIONAL GEOGRAPHICAL UNION International geographical congress Guidebooks to excursions Lisbon 1949 2 vols Lisbon,1949
Geog 13.4139

INTERNATIONAL GEOGRAPHICAL UNION International geographical congress 11th Compte rendu Cairo 1925 Apr 2 vols Societe royale de geographie d'Egypte,Cairo, 1925
Geog 13.5593

INTERNATIONAL GEOGRAPHICAL UNION International geographical congress 15th Comptes rendus;excursions Amsterdam 1938 Tom 1-2 17 vols Brill,Leiden,1938 Partly in English
Geog 13.5595

INTERNATIONAL GEOGRAPHICAL UNION International geographical congress 16th Comptes rendus Lisbon 1949 Tom 1-4 5 vols Lisbon,1950-52 Partly in English
Geog 13.5596

INTERNATIONAL GEOGRAPHICAL UNION Natural resources,food and population in inter-tropical Africa :a symposium Report Kampala 1955 Sep 10-17 Edited by L.Dudley Stamp Geographical publications,London,1956
Geog 13.4626

INTERNATIONAL GEOGRAPHICAL UNION Travaux topographiques et cartographiques International geographical congress Report Amsterdam 1938 Jul Amsterdam,1938
Geog 13.5708

INTERNATIONAL GEOGRAPHICAL UNION and SCIENCE COUNCIL OF JAPAN IGU regional conference in Japan Proceedings Tokyo 1957 Aug 28-Sep 3 Nara Edited by Ryuziro Isida and others illus,maps vii,609p 26cm Tokyo, 1959
Sco 14.6870

INTERNATIONAL GEOGRAPHICAL UNION and SCIENCE COUNCIL OF JAPAN Regional geography of Japan :I.G.U.regional conference in Japan Guidebook Japan 1957 1-7 maps Tokyo, 1957
Geog 13.4507

INTERNATIONAL GEOGRAPHICAL UNION.BRAZILIAN NATIONAL COMMITTEE International geographical congress 18th Excursion guides Brazil 1956 1-9 3 vols International geographical union,Rio de Janeiro,1956
Geog 13.5349

INTERNATIONAL GEOGRAPHICAL UNION.COMMISSION POUR LA CARTOGRAPHIE DES SURFACES D'APLANISSEMENT TERTIAIRES Rapport de la commission pour la cartographie des surfaces d'aplanissement tertiaires Paris,1938 Report to the International geographical congress held at Amsterdam,1938
Geog 13.0421

INTERNATIONAL GEOGRAPHICAL UNION.GENERAL ASSEMBLY, 8TH International geographical congress 17th Proceedings Washington,D.C. 1952 Aug 8-15 National research council, Washington,D.C.,1952
Geog 13.5597

INTERNATIONAL GEOGRAPHICAL UNION REGIONAL CONFERENCE Proceedings Tokyo 1957 Aug 28-Sep 3 Science council of Japan,Tokyo, 1959
Geog 13.5598

INTERNATIONAL GEOLOGICAL CONGRESS Congres geologique international 4me London 1888 explication des excursions Edited by W. Topley London,1888
Geol 8.1397

INTERNATIONAL GEOLOGICAL CONGRESS 1st-proceedings Paris 1878- Paris,1886-
Geol 8.4145

INTERNATIONAL GEOLOGICAL CONGRESS 13th comptes rendus Congres geologique international Brussels 1922 Aug Fasc 1-3 3 vols Liege,1924-26
Geog 13.0200

INTERNATIONAL GEOLOGICAL CONGRESS 14th
etudes et observations Geologie de la Mediterranee occidentale Madrid 1926 Vol 1-2 Association pour letude geologique de la Mediterranee occidentale 2 vols Liege,1926-35
Geol 8.2954

INTERNATIONAL GEOLOGICAL CONGRESS 16th
papers Copper resources of the world Washington,D.C. 1935 Vol 1-2 2 vols Washington,D.C.,1935
Min 10.0202

INTERNATIONAL GEOLOGICAL CONGRESS 18th
report London 1948 Part 2: proceedings of section A:problems of geochemistry Edited by C.E. Tilley and S.R. Nockolds London,1950
Min 10.0520

INTERNATIONAL GEOLOGICAL CONGRESS 19th
Symposium sur les gisements de fer du monde Algiers 1952 Tomes 1-2 Edited by F. Blondel and L. Marvier Algiers,1952
Geol 8.3011

INTERNATIONAL GEOLOGICAL CONGRESS 19th
Proceedings,section 15 Paleovolcanologie et ses rapports avec la tectonique,fasc.17 Algiers 1952 Algiers,1954
Min 10.0352

INTERNATIONAL GEOLOGICAL CONGRESS 19th
Proceedings,section 6 Genese des roches filoniennes,fasc.6 Algiers 1952 Algiers, 1953
Min 10.0351

INTERNATIONAL GEOLOGICAL CONGRESS 19th
papers Symposium sur les series de Gondwana Algiers 1952 Edited by Curt Teichert Algiers,1952 Text in English, French and German
Geol 8.3012

INTERNATIONAL GEOLOGICAL CONGRESS 20th
Trabajos Mexico City 1956 43 vols Mexico City,1956-60 Text in English, French,German,Russian and Spanish
Geog 13.7180

INTERNATIONAL GEOLOGICAL CONGRESS 20th
papers El jurasico inferior de Mexico y sus amonitas Mexico City 1956 By Heinrich Karl Erben Translated by U. Hungsberg from the German Mexico City,1956
Geol 8.3033

INTERNATIONAL GEOLOGICAL CONGRESS 20th
papers El jurasico medio y el calloviano de Mexico Mexico City 1956 By Heinrich Karl Erben maps Mexico City,1956
Geol 8.3032

INTERNATIONAL GEOLOGICAL CONGRESS 20th
papers Estudio sobre algunos de los conglomerados rojos del terciario inferior del centro de Mexico Mexico City 1956 By John D. Edwards Translated by Salvador Ulloa from the English Mexico City,1956 Bound with two other publications of the congress
Geol 8.3042

INTERNATIONAL GEOLOGICAL CONGRESS 20th
papers Geologia y depositos de carbon de la region de Sabinas,estado de Chihuahua Mexico City 1956 By Raymond C. Robeck and others United States geological survey Mexico City,1956 Bound with two other publications of the congress
Geol 8.3041

INTERNATIONAL GEOLOGICAL CONGRESS 20th
papers Memoria geologico-minera del estado de Chihuahua Mexico City 1956 By Jenaro Gonzalez Reyna Mexico City,1956 Bound with two other publications of the congress
Geol 8.3029

INTERNATIONAL GEOLOGICAL CONGRESS 20th
papers Mexico City 1956 Gacetas historicas Edited by Manuel Carrera Stampa Mexico City,1956 Bound with two other publications of the congress Text in English, French and Spanish
Geol 8.3030

INTERNATIONAL GEOLOGICAL CONGRESS 20th
papers Mexico City 1956 Seccion 1: vulcanologia del cenozoico,tom 1-2 Edited by A. Garcia Rojas and others Mexico City,1957 Text in English,French,Russian and Spanish
Geol 8.3036

INTERNATIONAL GEOLOGICAL CONGRESS 20th
papers Mexico City 1956 Seccion 11A: petrologia y mineralogia Edited by A. Garcia Rojas Mexico City,1959 Text in English, French,Russian and Spanish
Geol 8.3047

INTERNATIONAL GEOLOGICAL CONGRESS 20th
papers Mexico City 1956 Seccion 13: geologia aplicada a la ingenieria y a la mineria Edited by A. Garcia Rojas and others Mexico City,1959 Text in English,French, German and Spanish
Geol 8.3039

INTERNATIONAL GEOLOGICAL CONGRESS 20th
papers Mexico City 1956 Seccion 2: el mesozoico del hemisferio occidental y sus correlaciones mundiales Edited by A. Garcia Rojas and others Mexico City,1957 Bound with papers from section 3. Text in English, French and Spanish
Geol 8.3034

INTERNATIONAL GEOLOGICAL CONGRESS 20th
papers Mexico City 1956 Seccion 3: geologia del petroleo Edited by A. Garcia Rojas and others Mexico City,1956 Bound with papers from section 2. Text in English, French,German and Russian
Geol 8.3035

INTERNATIONAL GEOLOGICAL CONGRESS 20th
papers Mexico City 1956 Seccion 4: geohidrologia de regiones aridas y subaridas Edited by A. Garcia Rojas and others Mexico City,1957 Text in English,French,Russian and Spanish
Geol 8.3045

INTERNATIONAL GEOLOGICAL CONGRESS 20th
papers Mexico City 1956 Seccion 5: relaciones entre la tectonica y la sedimentacion,tom.1-2 Edited by A. Garcia Rojas Mexico City,1957 Text in English, French,German,Russian and Spanish
Geol 8.3046

INTERNATIONAL GEOLOGICAL CONGRESS 20th papers Mexico City 1956 Seccion 7: paleontologia,taxonomia y evolucion Edited by A. Garcia Rojas and others Mexico City, 1958 Text in English,French,German, Russian and Spanish
Geol 8.3043

INTERNATIONAL GEOLOGICAL CONGRESS 20th papers Mexico City 1956 Seccion 9: geofisica aplicada,tom.1-2 Edited by A. Garcia Rojas and others Mexico City,1958 Text in English,French,German,Russian and Spanish
Geol 8.3044

INTERNATIONAL GEOLOGICAL CONGRESS 20th papers Riqueza minera y yacimientos minerales de Mexico Mexico city 1956 Edited by Jenaro Gonzalez Reyna 3rd edition Mexico city,1956
Geol 8.3025

INTERNATIONAL GEOLOGICAL CONGRESS 20th papers Symposium de exploracion geoquimica,tom.1-3 Mexico City 1956 Edited by A. Garcia Rojas and others 2cm Mexico City,1958-60 Text in English, French,Russian and Spanish
Geol 8.3048

INTERNATIONAL GEOLOGICAL CONGRESS 20th papers Symposium sobre yacimientos de manganeso Mexico city 1956 Tom 1-5 Edited by Jenaro Gonzalez Reyna 2 vols Mexico city,1956 Text in English,French, German and Spanish
Geol 8.3027

INTERNATIONAL GEOLOGICAL CONGRESS 20th papers Symposium sobre yacimientos de petroleo y gas Mexico city 1956 Tom 1-5 Edited by Eduardo J. Guzman 2 vols Mexico city,1956 Text in English,French and Spanish
Geol 8.3026

INTERNATIONAL GEOLOGICAL CONGRESS 20th papers Tamabra limestone of the Poza Rica oilfield Mexico city 1956 By A. Barnetche and L.V. Illing Mexico city,1956 Bound with two other publications of the congress
Geol 8.3028

INTERNATIONAL GEOLOGICAL CONGRESS 20th papers Zonificacion microfaunistica de las calizas cretacicas del este de Mexico By Federico Bonet Associacion mexicana de geologos petroleros Mexico City,1956
Geol 8.3050

INTERNATIONAL GEOLOGICAL CONGRESS 20th papers of a symposium Sistema cambrico,su paleografia y el problema de su base Mexico city 1956 Pt 1-2 Mexico city,1956 Text in English,French and Spanish
Geol 8.3024

INTERNATIONAL GEOLOGICAL CONGRESS 20th resumes of papers presented Mexico City 1956 Edited by Hans E. Thalmann Mexico City,1956 Text in English,French,German and Spanish
Geol 8.3031

INTERNATIONAL GEOLOGICAL CONGRESS 21st Copenhagen 1960 Norden-Denmark,Finland, Iceland,Norway,Sweden,Volume of abstracts Edited by Theodor Sorgenfrei 252p 24cm Copenhagen,1960
Sco 14.0134

INTERNATIONAL GEOLOGICAL CONGRESS 21st Guides to excursions Norden 1960 Oslo, 1960 Unbound
Geog 13.0553

INTERNATIONAL GEOLOGICAL CONGRESS 21st excursion guides:Norway Norden 1960 Oslo,1960
Min 10.0168

INTERNATIONAL GEOLOGICAL CONGRESS 21st excursion guides:Sweden and Finland Norden 1960 Stockholm;Helsinki,1960
Min 10.0169

INTERNATIONAL GEOLOGICAL CONGRESS 21st reports Copenhagen 1960 1-22 Edited by Theodor Sorgenfrei Copenhagen,1960
Geol 8.3123

INTERNATIONAL GEOLOGICAL CONGRESS 4th Etudes sur les schistes cristallins London 1888 London,1888
Geol 8.2141

INTERNATIONAL GEOLOGICAL CONGRESS 4th papers Etudes sur les schistes cristallins London 1888 London,1888
Min 10.0237

INTERNATIONAL GEOLOGICAL CONGRESS 6th guide-book Zurich 1894 livret-guide geologique dans le Jura et les Alpes de la Suisse Payot,Lausanne,1894
Geol 8.3076

INTERNATIONAL GEOLOGICAL CONGRESS 7th-excursion guide books St Petersburg,1897-
Geol 8.4146

INTERNATIONAL GEOLOGICAL CONGRESS.COMMISSION ON THE GEOLOGICAL MAP OF AFRICA,and ASSOCIATION OF AFRICAN GEOLOGICAL SURVEYS Actas y trabajos de las reuniones celebradas en Mexico Association of African geological surveys and commission on the geological map of Africa : joint meeting papers Mexico City 1956 Sep 4-10 Edited by A. Garcia Rojas Mexico City,1959 Text in English,French and Spanish
Geol 8.3038

INTERNATIONAL GEOLOGICAL CONGRESS.EXECUTIVE COMMITTEE Veranderungen des klimas seit dem maximum der letzten eiszeit Internationaler geologen kongress :eine sammlung von berichten unter mitwirking von fachenosson in verschliedenen landern 11 Verlag von generalstabens litografiska anstalt, Stockholm,1910
Bot 42.3774

INTERNATIONAL GEOLOGICAL CONGRESS.INTERNATIONAL COMMISSION ON STRATIGRAPHY Sistema cretacico Un Symposium sobre el cretacico en el hemisferio occidental y su correlacio mundial papers Mexico City 1956 Tom 1-2 Edited by Lewis B. Kellum 2 vols Mexico City,1959 Prepared under the auspices of the International commission on stratigraphy and the International palaeontological union for the 20th International geological congress
Geol 8.3049

INTERNATIONAL GEOLOGICAL CONGRESS.INTERNATIONAL COMMISSION ON STRATIGRAPHY,and INTERNATIONAL PALAEONTOLOGICAL UNION Colloque du jurassique Luxembourg 1962 Aug.1- Ministere des arts et des sciences,Luxembourg, 1964
Geol 8.2510

INTERNATIONAL GEOLOGICAL CONGRESS,19TH La Geologie et les problemes de l'eau en Algerie Tome 1-2 By Georges Drouhin and,Marcel Gautier International geological congress, Algiers,1952
Geol 8.2399

INTERNATIONAL GEOLOGICAL CONGRESS,20TH Actas y trabajos de las reuniones celebradas en Mexico Association of African geological surveys and commission on the geological map of Africa :joint meeting papers Mexico City 1956 Sep 4-10 International geological congress.Commission on the geological map of Africa,and,Association of African geological surveys Edited by A. Garcia Rojas Mexico City,1959 Text in English,French and Spanish
Geol 8.3038

INTERNATIONAL GEOLOGICAL CONGRESS 20 Sistema cretacico Un Symposium sobre el cretacico en el hemisferio occidental y su correlacio mundial papers Mexico City 1956 Tom 1-2 Edited by Lewis B. Kellum 2 vols Mexico City,1959 Prepared under the auspices of the International commission on stratigraphy and the International palaeontological union for the 20th International geological congress
Geol 8.3049

INTERNATIONAL GEOPHYSICAL YEAR Antarctica in the International geophysical year Antarctic symposium Proceedings Northfield,Minn. 1956 Apr.26-27 Co-sponsored by the American geophysical union American geophysical union. Publication, 462 Geophysical monograph map 133p Washington,D.C.,1956
Nap 11.0020

INTERNATIONAL GEOPHYSICAL YEAR Some ionospheric results obtained during the international geophysical year :a symposium Proceedings Brussels 1959 Sep. International scientific radio union.Special commission for the International Geophysical Year Edited by W.J.G. Beynon xi,399p Elsevier,Amsterdam,1960
Nap 11.0053

INTERNATIONAL GEOPHYSICAL YEAR,1957-58 Geophysics and the IGY :proceedings of the symposium at the opening of the international geophysical year Proceedings United States national committee for the international geophysical year Edited by Hugh Odishaw and Stanley Ruttenberg National research council. Publication,590 American geophysical union. Geophysical monograph,2 illus vi,210p 26cm Washington,1958
Sco 14.0511

INTERNATIONAL GEOPHYSICAL YEAR,1957-58 Sbornik materialov rasshirennogo soveshchaniya rabochei gruppy po glyatsiologii Sovetskogo mezhduvedomstvennogo komiteta mezhdunarodnogo geofizicheskogo goda Moscow 1958 May 20-24 Edited by G.A. Avsyuk 165p 20cm Moscow, 1959
Sco 14.0514

INTERNATIONAL GOITER CONFERENCE 4th Abstracts of papers presented London 1960 Jul 5-8 American goiter association and London thyroid club Edited by O.de Vaal and others Excerpta medica. International congress series, 26 Excerpta medica foundation,Amsterdam,1960
PGMS 29.0158

INTERNATIONAL GRASSLAND CONGRESS 8th Proceedings Reading 1960 Jul 11-21 British grassland society Grassland research institute,Hurley,1961
BG 38.3114

INTERNATIONAL HEAT TRANSFER CONFERENCE Papers International developments in heat transfer Boulder,Col. 1961 Aug 28-Sep 1 London 1962 Jan 8-12 Pt 1-5 Arranged by the American society of mechanical engineers 5 vols A.S.M.E.,New York,1961 Co-sponsored by the American institute of chemical engineers,the Institution of mechanical engineers,and the Institution of chemical engineers
Chem E 24.0595

INTERNATIONAL HORTICULTURAL CONGRESS 9th Reports,Proceedings,Program London 1930 Aug 7-15 Royal horticultural society Under the auspices of the International committee for horticultural congresses London,1930
BG 38.3121

INTERNATIONAL HORTICULTURAL CONGRESS 13th Report London 1952 Sep 8-15 Vol 1-2 Royal horticultural society Royal horticultural society,London,1953
BG 38.3124

INTERNATIONAL HORTICULTURAL CONGRESS 14th Report Hague 1955 Aug 29-Sep 6 Vol 1-2 International society for horticultural science Veenman and Zonen, Wageningen,1955
BG 38.3125

INTERNATIONAL HORTICULTURAL CONGRESS 15th Proceedings Nice 1958 Vol 1-3 International committee for horticultural congresses Edited by Jean-Claude Garnaud 3 vols Pergamon press,London,1961-62
BG 38.3133

INTERNATIONAL HORTICULTURAL CONGRESS 16th Brussels 1962 Aug 31-Sep 8 Vol 1-5 International society for horticultural sciences 5 vols Duculot,Gembloux,1962
BG 38.3126

INTERNATIONAL HORTICULTURAL CONGRESS 17th
Proceedings College Park,Md. 1966 Aug 15-20 Vol 1-4 International society for horticultural sciences Sponsored by the American horticultural society 4 vols 1966-67
BG 38.3134

INTERNATIONAL HORTICULTURAL CONGRESS 18th
Proceedings Tel-Aviv 1970 Mar 17-25 International society for horticultural sciences 1970
BG 38.3135

INTERNATIONAL HORTICULTURAL CONGRESS 7TH
International tuinbouw-congres 7th Amsterdam 1923 Sep 17-23 Nederlandsche mantschappij voor tuinbouw en plantkunde c1923
Gen 34.1139

INTERNATIONAL HORTICULTURAL EXHIBITION AND BOTANICAL CONGRESS :report of proceedings London 1866 May 22-31 London,1867
Bot 42.4705

The INTERNATIONAL HORTICULTURAL EXHIBITION AND BOTANICAL CONGRESS Proceedings London 1866 May 22-31 Truscott,London,1866
BG 38.1551

INTERNATIONAL INSTITUTE OF AGRICULTURE
International Forestry Congress 3rd programme Helsinki 1940 Jul 1-5 Helsinki, 1939
Philos 1.2032

INTERNATIONAL INSTITUTE OF CHEMISTRY,and UNIVERSITE LIBRE DE BRUXELLES Reactivity of photo-excited organic molecula Solvay conference on chemistry 13th Proceedings Brussels 1965 Oct 25-30 Interscience, London,1967
Col S 12.0568

INTERNATIONAL INSTITUTE OF EMBRYOLOGY
Organogenesis International conference on organogenesis Baltimore,Md. 1964 Sep 6-12 Edited by Robert L. DeHaan and Heinrich Ursprung Holt,Rinehart and Winston,New York, 1965
Bal 39.0490

INTERNATIONAL INSTITUTE OF EMBRYOLOGY
Symposium on the germ cells and earliest stages of development Pallanza 1960 Sep 14-20 A.Baselli,Milan,1961 At head of title:Istituto Lombardo accademia di scienze e lettere
Bal 39.0459

INTERNATIONAL INSTITUTE OF REFRIGERATION :
commissions 1 et 2 Grenoble 1954 Sep 24-26 International institute of refrigeration. Bulletin.Annexe,1955-2 183p 24cm Paris, 1955 In English and French
Sco 14.0122

INTERNATIONAL INSTITUTE OF REFRIGERATION
Commissions internationales 1 et 2:meetings Proceedings Zurich 1953 Sep 10-12 problems relating to liquefaction and separation of gases... International institute of refrigeration.Bulletin.Annexe, 1954-2 illus. 140p Paris,1954 In English and French
Sco 14.0121

INTERNATIONAL INSTITUTE OF WELDING Welding and allied processes in maintenance and repair work Opatija 1959 Elsevier,Amsterdam; London,1961 Papers in English and French
Met 25.0797

INTERNATIONAL INTERDISCIPLINARY CONFERENCE ON LEARNING,REMEMBERING AND FORGETTING 2nd
Proceedings Organization of recall Princeton,N.J. 1964 Sep 27-30 New York academy of sciences and National institute of child health and human development Edited by Daniel P. Kimble With the support of United States.Office of naval research New York academy of sciences.Interdisciplinary communication program,New York,1967
Psy 31.1012

INTERNATIONAL JOINT CONFERENCE ON ARTIFICIAL INTELLIGENCE 2nd Advance papers London 1971 Sep 1-3 British computer society illus 658p British computer society,London,1971
Math L 5.3848

INTERNATIONAL KONGRESS FUR BEVOLKERUNGSWISSENSCHAFT 2nd bericht Bevolkerungsfragen Berlin 1935 Aug 26-Sep 1 Internationale vereinigung fur bevolkerungswissenschaft.Deutsche ausschluss Edited by Hans Harmson and Franz Lohse Lehmanns,Munich,1936
Gen 34.2069

INTERNATIONAL LABOUR ORGANIZATION Assessment of radioactivity in man Symposium on the assessment of radioactive body burdens in man Proceedings Heidelberg 1964 May 11-16 1-2 International atomic energy agency. Proceedings series International atomic energy agency,Vienna,1964
Radioth 35.1373

INTERNATIONAL LABOUR ORGANIZATION
Radiological health and safety in mining and milling of nuclear materials Symposium on radiological health and safety in mining and milling of nuclear materials :a symposium Proceedings Vienna 1963 Aug 26-31 1-2 In co-operation with the World health organization International atomic energy agency.Proceedings series International atomic energy agency,Vienna,1964
Radioth 35.1164

INTERNATIONAL LATITUDE SERVICE L' Avenir du service international des latitudes :a symposium proceedings Helsinki 1960 Jul. 26-Aug.7 Edited by Paul Melchior Organized by International union of geodesy and geophysics International astronomical union. Symposium, 13 97p Paris,1961
Reprinted from Bulletin geodesique no.59,March 1961
Obs 6.1198

INTERNATIONAL M.T.D.R. CONFERENCE 3rd-9th
Proceedings Advances in machine tool design and research 1963-68 Edited by S. A. Tobias and F. Koenigsberger 9 vols Pergamon press,London,1963-69
Eng 41.3924

INTERNATIONAL MAGNETICS CONFERENCE 4th
1966 Intermag Stuttgart 1966 Apr 20-22 Sponsored by the Institute of electrical and electronics engineers.Magnetics group I.E.E. E.transactions on magnetics,Mag 2,no 3 I. E.E.E.,New York,1966
Eng 41.5347

INTERNATIONAL MAP COMMITTEE.CONFERENCE,2ND
Carte du monde au millionieme :Conference internationale 2nd Comptes rendus Paris 1913 Dec France.Service geographique de l'armee Service geographique de l'armee, Paris,1914
Geog 13.5472

The INTERNATIONAL MATERIALS CONFERENCE 1st
Proceedings Impact of transmission electron microscopy on theories of the strength of crystals Berkeley,Calif 1961 Jul 5-8 Edited by Gareth Thomas and Jack Washburn xi,1022p Interscience,John Wiley, New York,1963
Cav 7.0191

INTERNATIONAL MATERIALS SYMPOSIUM 3rd
Proceedings Ceramic microstructures their analysis,significance and production Berkeley 1966 Jun 13-16 Edited by Richard M. Fulrath and Joseph A. Pask Wiley,New York, 1968
Met 25.0292

INTERNATIONAL MATERIALS SYMPOSIUM 4th :the structure and chemistry of solid surfaces Proceedings Berkeley,Calif 1968 Jun 17-21 Lawrence radiation laboratory.Inorganic materials research division and University of California.College of chemistry Edited by Gabor A. Somorjai Inorganic materials research division series, 4 Wiley,New York, 1969
Met 25.2476

INTERNATIONAL MATHEMATICAL CONFERENCE
Proceedings Several complex variables College Park,Md. 1970 Apr 6-17 1 University of Maryland Edited by John Horvath Lecture notes in mathematics, 155 bibliog. 214p 25cm Springer,Berlin,1970
P Math 2.3890

INTERNATIONAL MATHEMATICAL CONFERENCE
Proceedings Several complex variables College Park,Md. 1970 Apr 6-17 2 University of Maryland Edited by John Horvath Lecture notes in mathematics, 185 214p 25cm Springer,Berlin,1971
P Math 2.3891

INTERNATIONAL MATHEMATICAL UNION Algebraic number theory conference proceedings Brighton 1965 Sep 1-17 London mathematical society Edited by John W.S. Cassels and A. Frohlich xviii,366p 24cm Academic press, New York;London,1967
P Math 2.2773

INTERNATIONAL MATHEMATICAL UNION Coloquio international sobre geometria algebraica Actas Madrid 1965 Sep Edited by Pedro Abellanas Bibliog. 190p 24cm Madrid, 1966 Papers mainly in English and Spanish
P Math 2.3137

INTERNATIONAL MATHEMATICAL UNION
Differential analysis International colloquium on differential analysis :the Bombay colloquium papers presented Bombay 1964 Jan 7-14 By M.F. Atiyah and others Tata institute of fundamental research Tata institute of fundamental research.Studies in analysis, 2 viii,253p 25cm Oxford university press,Oxford,1964
P. Math 2.0696

INTERNATIONAL MATHEMATICAL UNION
Differentialgeometrie und topologie Internationales kolloquium Zurich 1960 Jun 20-25 Schweizerische mathematische gesellschaft Enseignement mathematique. Monographies, 11 159p 24cm L'enseignement mathematique universite,Geneva, 1962
P. Math 2.0697

INTERNATIONAL MATHEMATICAL UNION General topology and its relations to modern analysis and algebra :symposium proceedings Prague 1961 Sep 1-8 Ceskoslovenska akademie ved Edited by J. Novak 363p 25cm Academic press,New York,1962
P. Math 2.0291

INTERNATIONAL MATHEMATICAL UNION
Infinitistic methods Symposium on foundations of mathematics proceedings Warsaw 1959 Sep 2-9 Polska akademia nauk. Instytut matematyczny 362p 26cm Pergamon press,Oxford,1961
P. Math 2.0211

INTERNATIONAL MATHEMATICAL UNION
International conference on functional analysis and related topics Proceedings Tokyo 1969 Apr 1-8 xxxiii,425p 26cm University of Tokyo press,Tokyo,1970
P Math 2.4279

INTERNATIONAL MATHEMATICAL UNION
International congress of mathematicians 5th proceedings Cambridge 1912 Aug 22-28 Vol. 2.Communications to sections 2-4 Edited by E.W. Hobson and A.E.H. Love 657p 26cm Cambridge university press,Cambridge, 1913
Math L 5.0751

INTERNATIONAL MATHEMATICAL UNION
International congress of mathematicians 11th Proceedings Cambridge,Mass. 1950 Aug 13-Sep 6 Vol 1-2 25cm 2 vols American mathematical society,Providence,R.I., 1952
Math L 5.3246

INTERNATIONAL MATHEMATICAL UNION
International congress of mathematicians 14 proceedings Stockholm 1962 Aug 15-22 Edited by V. Stenstrom Bibliog. 1,595p 25cm Institut Mittag-Leffler,Djursholm, Sweden,1963
P. Math 2.2601

INTERNATIONAL MATHEMATICAL UNION
International congress of mathematicians abstracts of brief scientific communications and reports Moscow 1966 Aug nos. 1-15 Moscow,1966
Math 3.1125

INTERNATIONAL MATHEMATICAL UNION
International congress of mathematicians short communications abstracts Stockholm 1962 Aug 15-22 221p 24cm Stockholm,1962
P. Math 2.2602

INTERNATIONAL MATHEMATICAL UNION
International symposium on algebraic number theory proceedings Tokyo 1955 Sep 8-13 Nikko Science council of Japan Edited by Shokichi Iyanaga and Yukiyosi Kawada xx, 265p 26cm Science council of Japan,Tokyo, 1956
P. Math 2.1771

INTERNATIONAL MEDICAL CONGRESS International conference on congenital malformations 2nd Papers New York 1963 Jul 14-19 Edited by Morris Fishbein International medical congress,New York,1864
An 32.2091

INTERNATIONAL MEDICAL CONGRESS 7th London 1881 Revue scientifique de la France et de l'etranger.3e ser,1.annee,2. semestre,tome 28,,no 8 Paris,1881 Includes "Des virus vaccins" by M.Pasteur;: Rapports des sciences biologiques avec la medecine" by M.T.H.Huxley;"Histoire de la physiologie en Angleterre" by M.Michael Foster
Bioch 33.1657

INTERNATIONAL MEDICAL CONGRESS 7th session Transactions London 1881 Aug 2-9 anatomy Kolckmann,London,1881
An 32.1165

INTERNATIONAL MEDICAL CONGRESS 10 Das Arztliche vereinswesen in Deutschland und der deutsche arztvereinsband :festschrift dem X internationalen medizinischen kongress By Eduard Graf Vogel,Leipzig,1890
Path 30.0173

INTERNATIONAL MEDICAL CONGRESS 10 Die Offentliche gesundheits- und krankenpflege der stadt Berlin festschrift stadtischen behorden den mitgliedern des X internationalen medizinischen kongress Edited by Rudolf Ludwig Carl Virchow Hirschwald,Berlin,1890
Path 30.0158

INTERNATIONAL MEDICAL CONGRESS 10TH Anstalten und einrichtungen des offentlichen gesundheitswesen in Preussen :festschrift zum X internationalen medizinischen kongress... Edited by M. Pistor Springer,Berlin,1890
Path 30.1479

INTERNATIONAL MEETING OF MEDICAL CYBERNETICS,2ND Nerve,brain and memory models Edited by N. Niener and J.P. Schade Progress in brain research, 2 Elsevier,Amsterdam,1963 Lectures delivered during a symposium on "Cybernetics of the nervous system",held during the Second International meeting of medical cybernetics,Amsterdam,1962
An 32.4222

INTERNATIONAL MEETING OF MEDICAL CYBERNETICS 2ND Nerve,brain and memory models ..a series of lectures delivered during a symposium on cybernetics of the nervous system...held as part of the second international meeting of medical cybernetics at the Royal academy of sciences of Amsterdam... Lectures Amsterdam 1962 Apr 16-18 Edited by N. Wiener and J.P. Schade Progress in brain research, 2 Elsevier,Amsterdam,1963
Psy 31.0248

INTERNATIONAL MEETING OF NEUROBIOLOGISTS 1st Proceedings Progress in neurobiology Groningen 1955 Aug 3-7 Edited by J.Ariens Kapper Elsevier,Amsterdam,1956
An 32.4351

INTERNATIONAL MEETING OF NEUROBIOLOGISTS 2nd Proceedings Structure and function of the cerebral cortex Amsterdam 1959 Sep 22-25 Edited by D.B. Tower and J.P. Schade Elsevier,Amsterdam,1960
An 32.4646

The INTERNATIONAL MEETING OF NEUROBIOLOGISTS 2nd Proceedings Structure and function of the cerebral cortex Amsterdam 1959 Sep 22-25 Edited by D.B. Tower and J.F. Schade Elsevier,Amsterdam,1960
Bal 39.1494

INTERNATIONAL MEETING OF NEUROBIOLOGISTS 4th Proceedings Structure and function of inhibitory neuronal mechanisms Stockholm 1966 Sep Edited by C.von Euler and others Wenner-Gren center.International symposium series, 10 Pergamon press,Oxford,1968
An 32.5336

INTERNATIONAL MEETING OF NEUROBIOLOGISTS 4th Proceedings Structure and function of inhibitory neural mechanisms Stockholm 1966 Sep Edited by C.von Euler and others Illus. ix,563p Pergamon,Oxford,1968
Pha 16.0312

INTERNATIONAL MEETING OF NEUROBIOLOGISTS,3RD Lectures on the dienecephalon Edited by W. Bargmann and J.P. Schade Progress in brain research, 5 Elsevier,Amsterdam,1964 Lectures delivered during a symposium on the "Structure and function of the dienecephalon held as part of the Third International meeting of neurobiologists,Kiel,1962
An 32.4225

INTERNATIONAL MEETING OF NEUROBIOLOGISTS,3RD The Rhinencephalon and related structures Edited by W. Bargmann and J.D. Schade Progress in brain research, 3 Elsevier, Amsterdam,1963 Lectures delivered during a symposium on the rhinencephalon held as part of the Third international meeting of neurobiologists,Kiel,1962
An 32.4223

INTERNATIONAL MEETING OF NEUROBIOLOGISTS,3RD Topics in basic neurology Edited by W. Bargmann and J.P. Schade Progress in brain research, 6 Elsevier,Amsterdam,1964 Lectures delivered during a symposium on "Topics in basic neurology"held as part of the Third International meeting of neurobiologists, Kiel,1962
An 32.4226

INTERNATIONAL MEETING ON FERROELECTRICITY 2nd Kyoto 1969 Sep 4-9 Science council of Japan Sponsored by the International union of pure and applied physics Physical society of Japan.Journal, 28,suppt Physical society of Japan,1970
Met 25.2865

INTERNATIONAL MERIDIAN CONFERENCE International conference for the purpose of fixing a prime meridian and a universal day 1st Proceedings Washington,D.C. 1884 Oct 1-Nov 1 United States.Department of state 212p Gibson,Washington,D.C.,1884 With five sheets of notes by J.C.Adams in a separate envelope
Obs 6.3169

INTERNATIONAL MINERAL DRESSING CONGRESS Transactions Progress in mineral dressing Stockholm 1957 Sep 18-21 Svenska Gruvforeningen and Jernkontoret Almqvist and Wiksell,Stockholm,1958
Met 25.2286

INTERNATIONAL MINERAL PROCESSING CONGRESS
Proceedings Gothenburg 1966 Aug 27-Sep 1 Institution of mining and metallurgy London,1960
Met 25.2504

INTERNATIONAL MINERAL PROCESSING CONGRESS 7th
Technical papers New York 1964 Sep 20-24 Vol 1 American institute of mining, metallurgical and petroleum engineers and Columbia university Edited by Nathaniel Arbiter Gordon and Breach,New York,1965 Held in conjunction with the centennial of the Henry Krumb school of mines
Met 25.0214

INTERNATIONAL MINERALOGICAL ASSOCIATION
International mineralogical association 4th general meeting Papers and proceedings New Delhi 1964 Dec 15-22 Edited by P.R.J. Naidu and M.N. Viswanathiah Mineralogical society of India,Mysore,1966
Min 10.1550

INTERNATIONAL MINERALOGICAL ASSOCIATION 4th
general meeting Papers and proceedings New Delhi 1964 Dec 15-22 International mineralogical association Mineralogical society of India Edited by P.R.J. Naidu and M.N. Viswanathiah Mineralogical society of India,Mysore,1966
Min 10.1550

INTERNATIONAL MINERALS AND CHEMICAL CORPORATION
Harvesting the sun:photosynthesis in plant life :a symposium Chicago 1966 Oct 5-7 Edited by A. San Pietro and others Academic press,New York;London,1967
Bioch 33.0789

INTERNATIONAL MINERALS AND CHEMICAL CORPORATION
Harvesting the sun:photosynthesis in plant life :a symposium Papers Chicago,Ill. 1966 Oct 5-7 Edited by Anthony San Pietro and others illus. ix,342p Academic press, New York;London,1967
Bot 42.1899

INTERNATIONAL NAVIGATION CONGRESS 19th
Section 1:inland navigation London 1957 Communication 3 S.1-C.3: influence of ice on navigable waterways and on sea and inland ports Permanent international association of navigation congresses illus 260p Permanent international association of navigation congresses,Brussels,1957
Sco 14.0363

INTERNATIONAL NEUROCHEMICAL SYMPOSIUM 1st
Proceedings Biochemistry of the developing nervous system Oxford 1954 Jul 13-17 Edited by Heinrich Waelsch Academic press,New York,1955
An 32.4333

INTERNATIONAL NEUROCHEMICAL SYMPOSIUM 1st
Proceedings Biochemistry of the developing nervous system Oxford 1954 Jul 13-17 Edited by Heinrich Waelsch held at Magdalen college,Oxford Academic press,New York,1955
Bal 39.1468

The INTERNATIONAL NEUROCHEMICAL SYMPOSIUM 1st
Proceedings Biochemistry of the developing nervous system Oxford 1954 Jul 13-17 Edited by Heinrich Waelsch Academic press,New York,1955
Bioch 33.2233

INTERNATIONAL NEUROCHEMICAL SYMPOSIUM 2nd
Proceedings Metabolism of the nervous system Aarhus 1956 Jul Aarhus universitet Edited by Derek Richter Pergamon press,London,1957
Bioch 33.0592

INTERNATIONAL NEUROCHEMICAL SYMPOSIUM 3rd
Proceedings Chemical pathology of the nervous system Strasbourg 1958 Aug Edited by Jordi Folch-Pi Pergamon press, London,1961
An 32.4186

INTERNATIONAL NEUROCHEMICAL SYMPOSIUM 2nd
Proceedings Metabolism of the nervous system Aarhus 1956 Jul Edited by Derek Richter Pergamon,London,1957
Pha 16.0037

INTERNATIONAL NUCLEAR MEDICINE SYMPOSIUM
Proceedings Radioactive isotopes in the localization of tumours London 1967 Edited by N.G. Trott Heinemann medical, London,1969
Radioth 35.1397

INTERNATIONAL OCEANOGRAPHIC CONFERENCE Invited lectures Oceanography New York 1959 Aug 31-Sep 12 Edited by Mary Sears American association for the advancement of science.Publications,67 American association for the advancement of science,Washington,D.C., 1961
Geog 13.0883

INTERNATIONAL OCEANOGRAPHIC CONGRESS Invited lectures Oceanography New York 1959 Aug 31-Sep 12 American association for the advancement of science Edited by Mary Sears American association for the advancement of science.Publication, 67 Washington,D.C., 1961
Geod 9.0576

INTERNATIONAL OCEANOGRAPHIC CONGRESS invited lectures Oceanography New York 1959 Aug 31-Sep 12 Edited by Mary Sears American association for the advancement of science.Publications,67 American association for the advancement of science,Washington,D.C., 1961
Geol 8.1631

The INTERNATIONAL OCEANOGRAPHIC CONGRESS
Lectures Oceanography New York 1959 Aug 31-Sep 12 American association for the advancement of science Edited by Mary Sears American association for the advancement of science.Publication, 67 American association for the advancement of science, Washington,1961
Bal 39.1951

INTERNATIONAL OCEANOGRAPHIC CONGRESS 2nd
Abstracts of papers Moscow 1966 May 30-Jun 9 Academy of sciences,U.S.S.R. Edited by A.P. Vinogradov Moscow,1966
Geod 9.0612

INTERNATIONAL OCEANOGRAPHIC CONGRESS 1st
Preprints of abstracts of papers Washington,D.C. 1959 Aug 31-Sep 12 American association for the advancement of science Edited by Mary Sears illus,maps 1022p 23cm Washington,D.C.,1959
Sco 14.0129

INTERNATIONAL ORGANIZATION FOR VACUUM SCIENCE AND TECHNOLOGY National vacuum symposium 8th Transactions Washington,D.C. 1961 Oct 16-19 Vol 1-2 Edited by Luther E. Preuss 2 vols Pergamon press,Oxford,1962 Combined with the International congress on vacuum science and technology,2nd
Cav 7.2376

INTERNATIONAL ORGANIZATION OF MEDICAL PHYSICS Aspects of medical physics International conference on medical physics Review papers Harrogate 1965 Sep 8-10 Edited by J. Rotblat Taylor and Francis,London,1966
Radioth 35.1671

INTERNATIONAL ORGANIZATION OF WORLD CONGRESSES FOR PSYCIATRY.SWISS ORGANIZING COMMITTEE International congress for psychiatry 2nd Report Zurich 1957 Sep 1-7 Vol 1-4 Edited by W.A. Stoll 4cm Fussli arts graphiques,Zurich,1959 Congress languages English,French,German,Italian and Spanish
Psy 28.0407

INTERNATIONAL ORNITHOLOGICAL CONGRESS.HUNGARIAN COMMITTEE A Review of recent attempts to classify birds :an address delivered before the second ornithological congress on the 18th of May 1891 By R.Bowdler Sharpe pls 25cm International ornithological congress. Hungarian committee,Budapest,1891 Two copies
Bal 44.1537

INTERNATIONAL ORNITHOLOGICAL CONGRESS 2ND A Review of recent attempts to classify birds : an address delivered before the second ornithological congress on the 18th of May 1891 By R.Bowdler Sharpe International ornithological congress.Hungarian committee pls 25cm International ornithological congress.Hungarian committee,Budapest,1891 Two copies
Bal 44.1537

INTERNATIONAL ORNITHOLOGICAL CONGRESS 2ND Ausstellungen der ungarischen vogelfauna... Budapest 1891 Edited by Gyula Madarasz Budapest,1891 One of three papers bound together under spine title 'Vom ornithologi congresse in Budapest 1890'
Bal 44.2249

INTERNATIONAL ORNITHOLOGICAL CONGRESS 2ND Aves Hungariae... :enumeratio systematica avium Hungariae cum notis... Budapest 1891 By Janos Frivaldszky 22cm Franklin, Budapest,1891 One of three papers bound together under spine title 'Vom ornithologi congresse in Budapest 1890'
Bal 44.2247

INTERNATIONAL ORNITHOLOGICAL CONGRESS 2ND Die Vogelsammlung des Bosnisch-Herzegovinischen landesmuseum in Sarajevo Budapest 1891 By C.O. Reiser Budapest, 1891 One of three papers bound together under spine title 'Vom ornithologi congresse in Budapest 1890'
Bal 44.2248

INTERNATIONAL ORNITHOLOGICAL CONGRESS 2ND Die Vogelsammlung des bosnisch-hercegovinischen landesmuseums in Sarajevo Budapest 1891 Edited by Custos O. Reiser 2 pls 22cm Budapest,1891 Paper read at the conference
Bal 44.2369

INTERNATIONAL ORNITHOLOGICAL CONGRESS 2ND Fugleliv i det arctiske Norge ;et populaert foredrag By Robert Collett illus 4,45p 23cm Alb.Cammermeyer,Christiania;Copenhagen, 1892 Papers written for the 2nd International ornithological congress,held in Budapest,May 1891
Sco 14.5891

INTERNATIONAL PALAEONTOLOGICAL UNION Sistema cretacico Un Symposium sobre el cretacico en el hemisferio occidental y su correlacio mundial papers Mexico City 1956 Tom 1-2 Edited by Lewis B. Kellum 2 vols Mexico City,1959 Prepared under the auspices of the International commission on stratigraphy and the International palaeontological union for the 20th International geological congress
Geol 8.3049

INTERNATIONAL PALAEONTOLOGICAL UNION and INTERNATIONAL GEOLOGICAL CONGRESS. INTERNATIONAL COMMISSION ON STRATIGRAPHY, Colloque du jurassique Luxembourg 1962 Aug. 1- Ministere des arts et des sciences, Luxembourg,1964
Geol 8.2510

INTERNATIONAL PHARMACOLOGICAL MEETING Sao Paulo 1966 Jul 24-30 Vol 5: control of growth processes by chemical agents Edited by A.D. Welch Pergamon,Oxford,1968
Pha 16.0257

INTERNATIONAL PHARMACOLOGICAL MEETING 3rd Proceedings Sao Paulo 1966 Jul 24-30 Vol 2: pharmacology of reproduction Edited by E. Diczfalusy Pergamon,Oxford,1968
Pha 16.0256

INTERNATIONAL PHOTOBIOLOGY CONGRESS 4th Proceedings Recent progress in photobiology Oxford 1964 Jul 26-30 Edited by E.J. Bowen Under the auspices of the Comite internationale de photobiologie Blackwell,Oxford,1965
Radioth 35.0253

INTERNATIONAL PHYSIOLOGICAL CONGRESS 11th Abstracts Edinburgh 1923 Jul 23-27 abstracts of communications 1923 Proof copy
Phys 20.2067

INTERNATIONAL PHYSIOLOGICAL CONGRESS 15th Report Moscow 1935 and Leningrad 1935 Edited by B.E. Zbarskii and V.M. Kaganov Moscow;Leningrad,1936
Phys 20.2243

INTERNATIONAL PHYSIOLOGICAL CONGRESS 17th Oxford 1947 Jul 21-25 Abstracts of communications 1947
Phys 20.2074

INTERNATIONAL PHYSIOLOGICAL CONGRESS 18th Copenhagen 1950 Aug 15-18 abstracts of communications 1950
Phys 20.2086

INTERNATIONAL PHYSIOLOGICAL CONGRESS 19th Montreal 1953 Aug 31-Sep 4 abstracts of communications 1953
Phys 20.2089

INTERNATIONAL PHYSIOLOGICAL CONGRESS 20th Brussels 1956 Jul 30-Aug 4 abstracts of communications 1956
Phys 20.2101

INTERNATIONAL PHYSIOLOGICAL CONGRESS 20th Brussels 1956 Jul 30-Aug 4 abstracts of reviews 1956
Phys 20.2102

INTERNATIONAL PHYSIOLOGICAL CONGRESS,20TH Some founders of physiology Edited by Chauncey D. Leake Washington,D.C.,1956
Phys 20.0842

INTERNATIONAL PHYSIOLOGICAL CONGRESS,21st see INTERNATIONAL CONGRESS OF PHYSIOLOGICAL SCIENCES

INTERNATIONAL PHYSIOLOGICAL CONGRESS 20TH Some founders of physiology :contributors to the growth of functional biology Edited by Chauncey D. Leake x,122p Washington,D.C., 1956 Compiled for the 20th international physiological congress held at Brussels,1956
WSM 43.2726

INTERNATIONAL PIGMENT CELL CONFERENCE 5th Papers Pigment cell:molecular,biological and clinical aspects New York 1961 Oct 11-14 Edited by Harold F. Whipple New York academy of sciences.Annals, 100 New York academy of sciences,New York,1963 1st - 4 th conferences called:Conference on the biology of normal and atypical pigment cell growth,q.v.
An 32.3389

INTERNATIONAL PIGMENT CELL CONFERENCE 6th Sofia 1965 May 25-29 International union against cancer Edited by G. Della Porta and O. Muhlbock Springer-verlag,Berlin,1966
Bioch 33.0967

INTERNATIONAL PIGMENT CELL CONFERENCE 6th Structure and control of the melanocyte Sofia 1965 May 25-29 Edited by G. Della Porta and O. Muhlbock Sponsored by the International union against cancer Springer, Berlin,1966
An 32.3425

INTERNATIONAL PIGMENT CELL CONFERENCE 6th The Cell By Carl P. Swanson Edited by Benjamin W. Zweifach and others 2nd edition Foundations of modern biology series Prentice-Hall,Englewood Cliffs,N.J.,1964
An 32.3424

INTERNATIONAL PLASTICS CONVENTION Papers and discussions Plastics progress,1961 London 1961 Jun Edited by Phillip Morgan xi,181p Iliffe;Macmillan,London;New York, 1962
Chem E 24.1014

INTERNATIONAL POLAR COMMISSION Commission polaire internationale :session Proces-verbaux des seances Brussels 1908 May 29-30 Edited by Georges Lecointe 110,clxiip 25cm Brussels,1908 In English,French, German,Russian
Sco 14.0110

INTERNATIONAL QUIET SUN YEAR Solar-terrestrial physics:solar aspects Proceedings of joint I.Q.S.Y.-Cospar symposium London 1967 Jul 17-22 Pt 1 Edited by A.C. Stickland International quiet sun year.Annals, 4 414p MIT press,Cambridge, Mass.,1969
Obs 6.3341

INTERNATIONAL QUIET SUN YEAR.SPECIAL COMMITTEE see SPECIAL COMMITTEE FOR THE INTERNATIONAL QUIET SUN YEAR

INTERNATIONAL RESEARCH CONFERENCE ON PROTEINASE INHIBITORS 1st Proceedings Munich 1970 Nov 4-5 Edited by H. Fritz and H. Tschesche De Gruyter,Berlin,1971
Bioch 33.2205

INTERNATIONAL RESEARCH SEMINAR proceedings Bernoulli 1713:Bayes 1763:Laplace 1818: anniversary volume Berkeley,Calif. 1963 University of California.Statistical laboratory Edited by Jerzy Neyman and Lucien M. LeCam Supported by the National science foundation ix,262p 23cm Springer-verlag, Berlwn,1965
Math 3.0708

INTERNATIONAL RICE RESEARCH INSTITUTE Symposium on rice genetics and cytogenetics Proceedings Los Banos 1963 Feb 4-8 Elsevier,Amsterdam,1964
Gen 34.1559

INTERNATIONAL ROCK GARDEN PLANT CONFERENCE 3rd Report London 1961 Apr 18-28 Edinburgh 1961 Alpine garden society Edited by R.C. Elliott and J.L. Mowat Organized by the Scottish rock garden club Alpine garden society;Scottish rock garden club,London;Penicuik,1961
BG 38.0011

INTERNATIONAL SCHOOL OF PHYSICS 6th course Theory and phenomonology in particle physica "Ettore Majorana" Erice 1968 Jul 13-28 Part A-B European organization for nuclear research Edited by A. Zichichi Co-sponsored by the Weizmann institute of science Bibliog.,Illus. 23cm 2 vols Academic press,New York,1969
A Math 4.1541

INTERNATIONAL SCHOOL OF PHYSICS 'ETTORE MAJORANA' 1st course proceedings Strong, electromagnetic,and weak interactions Erice 1963 May-June Edited by A. Zichichi Sponsored by the European organization for nuclear research W.A.Benjamin,New York; Amsterdam,1964
A Math 4.0846

INTERNATIONAL SCHOOL OF PHYSICS 'ETTORE MAJORANA' 2nd course proceedings Symmetries in elementary particle physics Erice 1964 Aug.-Sep. Edited by A. Zichichi Sponsored by the European organization for nuclear research 429p Academic press,New York; London,1965
A Math 4.0847

INTERNATIONAL SCHOOL OF PHYSICS 'ETTORE MAJORANA 3rd course proceedings Recent developments in particle symmetries Erice 1965 Sep.-Oct. Edited by A. Zichichi Sponsored by the European organization for nuclear research 460p Academic press,New York,1966
A Math 4.0848

INTERNATIONAL SCHOOL OF PHYSICS 'ETTORE MAJORANA' 4th course proceedings strong and weak interactions,present problems Erice 1966 Jun.19-Jul.4 Edited by A. Zichichi Sponsored by the European organization for nuclear research 859p Academic press,New York,1966
A Math 4.0849

INTERNATIONAL SCIENTIFIC CONFERENCE ON LEPROSY
Mittheilungen und verhandlungen Berlin 1897 Oct 11-16 Bd 1-3 2 vols Hirschwald,Berlin,1897-98 Papers in French,English,German,Spanish.Bd 2-3 bound together
Path 30.0701

INTERNATIONAL SCIENTIFIC CONGRESSES HELD SINCE 1930,OR ANNOUNCED FOR 1940 OR LATER
National research council.Library Washington, D.C.,1940 Mimeographed
Bal 39.3923

INTERNATIONAL SCIENTIFIC RADIO UNION
International symposium on circuit and information theory transactions Los Angeles 1959 Jun.16-18 Institute of radio engineers Institute of radio engineers. Transactions on information theory,IT, 5, special supplement xiv,298p Institute of radio engineers,New York,1959 Papers published in advance of the Symposium
Math L 5.2071

INTERNATIONAL SCIENTIFIC RADIO UNION Mixed commission on the ionosphere meeting 3rd Proceedings Canberra 1952 Aug. 24-26 Edited by W.J.G. Beynon Organized by the International council of scientific unions 194p General secretariat,Brussels,1953
Obs 6.2579

INTERNATIONAL SCIENTIFIC RADIO UNION
Monograph on ionospheric radio URSI general assembly 13th Papers of a session London 1960 Sep. Edited by W.J.G. Beynon URSI monographs series 264p Elsevier,Amsterdam; New York,1962
Nap 11.0052

INTERNATIONAL SCIENTIFIC RADIO UNION
Monograph on radio noise of terrestrial origin URSI general assembly 13th Papers London 1960 Sep 6 Commission 4 on radio noise of terrestrial origin Edited by F. Horner URSI monographs series 202p Elsevier, Amsterdam,1962
Nap 11.0268

INTERNATIONAL SCIENTIFIC RADIO UNION
Monograph on radio-wave propagation in the troposphere U.R.S.I. general assembly 13th Papers London 1960 Sep commission 2 on radio and troposphere Edited by J.A. Saxton U.R.S.I. monographs series illus 199p Elsevier,Amsterdam,1962
Nap 11.0533

INTERNATIONAL SCIENTIFIC RADIO UNION Paris symposium on radio astronomy Paris 1958 Jul.30-Aug.6 Edited by Ronald N. Bracewell International astronomical union.Symposium, 9 International scientific radio union. Symposium, 1 612p Stanford university press,Stanford,Calif.,1959
Obs 6.0246

INTERNATIONAL SCIENTIFIC RADIO UNION Paris symposium on radio astronomy Paris 1958 Jul.30-Aug.6 Edited by Ronald N. Bracewell International astronomical union.Symposium, 9 International scientific radio union. Symposium, 1 612p Stanford university press,Stanford,Calif,1959
A Math 4.1151

INTERNATIONAL SCIENTIFIC RADIO UNION Paris symposium on radio astronomy Paris 1958 Jul 30-Aug 6 Edited by Ronald N. Bracewell International astronomical union.Symposium, 9 International scientific radio union. Symposium, 1 612p Stanford university press,Stanford,Calif.,1959
TA 15.0638

INTERNATIONAL SCIENTIFIC RADIO UNION Paris symposium on radio astronomy Paris 1958 Jul 30-Aug 6 Edited by Ronald N. Bracewell International astronomical union.Symposium, 9 International scientific radio union. Symposium, 1 612p Stanford university press,Stanford,Calif,1959
Cav 7.0301

INTERNATIONAL SCIENTIFIC RADIO UNION Radio astronomy and the galactic system Symposium on radio astronomy and the galactic system Noordwijk 1966 Aug 25-Sep 1 International astronomical union Edited by Hugo van Woerden International astronomical union. Symposium, 31 xviii,501p Academic press, London;New York,1967
A Math 4.1574

INTERNATIONAL SCIENTIFIC RADIO UNION Solar-terrestrial physics :a symposium Review papers Belgrade 1966 Aug.29-Sep.2 Edited by J.W. King and W.S. Newman xii,390p Academic press,London;New York,1967
A Math 4.1171

INTERNATIONAL SCIENTIFIC RADIO UNION The Galaxy and the magellanic clouds :a symposium Canberra 1963 Mar.18-28 Edited by F.J. Kerr and A.W. Rodgers International astronomical union.Symposium, 20 393p Australian academy of science,Canberra,1964 Dedicated to Joseph Lade Pawsey
A Math 4.1158

INTERNATIONAL SCIENTIFIC RADIO UNION The Galaxy and the magellanic clouds :a symposium Proceedings Canberra 1963 Mar.18-28 Edited by F.J. Kerr and A.W. Rodgers International astronomical union.Symposium, 20 393p Australian academy of science, Canberra,1964 Publication dedicated to memory of Joseph Lade Pawsey
Obs 6.0941

INTERNATIONAL SCIENTIFIC RADIO UNION The International conference on cosmic rays and the earth storm :combined meeting of the International symposium on the earth storm and the International conference on cosmic rays Proceedings Kyoto 1961 Sep 4-15 Vol 2-3 International union of geodesy and geophysics and International astronomical union Edited by Ken-ichi Maeda and Osamu Minakawa Physical society of Japan.Journal, 17,suppl.A-2,3 2 vols physical society of Japan,Tokyo,1962
TA 15.0038

INTERNATIONAL SCIENTIFIC RADIO UNION URSI general assembly 7th-11th Proceedings Vol 6-10 1946-54
Nap 11.0288

INTERNATIONAL SCIENTIFIC RADIO UNION.COMMISSION ON THE IONOSPHERE Progress in radio science 1960-63 URSI general assembly 14th Proceedings Tokyo 1963 Sep 9-20 Vol 3: the ionosphere Edited by G.M. Brown 196p Elsevier,Amsterdam;London,1965
Nap 11.0086

INTERNATIONAL SCIENTIFIC RADIO UNION.SPECIAL COMMISSION FOR THE INTERNATIONAL GEOPHYSICAL YEAR Some ionospheric results obtained during the international geophysical year :a symposium Proceedings Brussels 1959 Sep. Edited by W.J.G. Beynon xi,399p Elsevier, Amsterdam,1960
Nap 11.0053

INTERNATIONAL SCIENTIFIC RADIO UNION.SYMPOSIUM, 1 Paris symposium on radio astronomy Paris 1958 Jul.30-Aug.6 International astronomical union and International scientific radio union Edited by Ronald N. Bracewell International astronomical union. Symposium, 9 612p Stanford university press,Stanford,Calif.,1959
Obs 6.0246

INTERNATIONAL SCIENTIFIC RADIO UNION.SYMPOSIUM, 1 Paris symposium on radio astronomy Paris 1958 Jul.30-Aug.6 International astronomical union and International scientific radio union Edited by Ronald N. Bracewell International astronomical union. Symposium, 9 612p Stanford university press,Stanford,Calif,1959
A Math 4.1151

INTERNATIONAL SCIENTIFIC RADIO UNION.SYMPOSIUM, 1 Paris symposium on radio astronomy Paris 1958 Jul 30-Aug 6 International astronomical union International scientific radio union Edited by Ronald N. Bracewell International astronomical union.Symposium, 9 612p Stanford university press,Stanford, Calif.,1959
TA 15.0638

INTERNATIONAL SCIENTIFIC RADIO UNION.SYMPOSIUM, 1 Paris symposium on radio astronomy Paris 1958 Jul 30-Aug 6 International astronomical union and International scientific radio union Edited by Ronald N. Bracewell International astronomical union. Symposium, 9 612p Stanford university press,Stanford,Calif,1959
Cav 7.0301

INTERNATIONAL SEAWEED SYMPOSIUM 2nd Proceedings Trondheim 1955 Jul Edited by Trygve Braarud and Nils Andreas Sorensen illus. xiv,220p Pergamon,London,1956
Bot 42.4700

INTERNATIONAL SEMINAR ON CHROMOSOME :its structure and function Proceedings Calcutta 1968 Aug 11-13 University of Calcutta Ramakrisha mission institute of culture University grants commission Edited by Arunkumar Sharma and Archana Sharma Nucleus,1968,Supplementary volume Nucleus, Calcutta,1968
Gen 34.2173

INTERNATIONAL SEMINAR ON OPERATING SYSTEMS PRINCIPLES Submitted papers Belfast 1971 Aug 30-Sep 3 Queen's university of Belfast illus 2 vols Queen's university of Belfast,Belfast,1971
Math L 5.3846

The INTERNATIONAL SEMINAR ON PLASMA PHYSICS proceedings Plasma physics Trieste 1964 Oct 5-31 International atomic energy agency.International centre for theoretical physics International atomic energy agency, Vienna,1965
A Math 4.0624

INTERNATIONAL SIGNIFICAL SUMMER CONFERENCE 9th Conference d'ete internationale de linguistique psychologique Proceedings Amersfoort 1953 Aug 10-15 International society for significs Synthese, 9,issue 3, no 3-4 Kroonder,Bussum,c1954
Psy 31.2514

INTERNATIONAL SOCIAL SCIENCE COUNCIL Marx and contemporary scientific thought The Role of Karl Marx in the development of contemporary scientific thought :a symposium Papers Paris 1968 May 8-10 Under the auspices of Unesco International social science council.Publications, 13 xi,612p Mouton,The Hague;Paris,1969
WSM 43.3826

INTERNATIONAL SOCIETY FOR CELL BIOLOGY Cell growth and cell division Report Liege 1962 May 19-24 Edited by R.J.C. Harris held at the Institute of histology,Liege International society for cell biology. Symposia, 2 illus Academic press,New York;London,1963
Bioch 33.0965

INTERNATIONAL SOCIETY FOR CELL BIOLOGY Cell growth and cell division :a symposium Liege 1962 May 19-24 Edited by R.J.C. Harris International society for cell biology. Symposia, 2 illus. Academic press,New York;London,1963
Gen 34.0863

INTERNATIONAL SOCIETY FOR CELL BIOLOGY Cytogenetics of cells in culture symposium Pasadena,Calif. 1964 Jun 17 Edited by R.J. C. Harris International society for cell biology.Symposia, 3 Academic press,New York,1964
Gen 34.0892

INTERNATIONAL SOCIETY FOR CELL BIOLOGY Cytogentics of cells in culture :a symposium Papers Pasadena,Calif. 1964 Jul 17 Edited by R.J.C. Harris International society for cell biology.Symposia, 3 illus, port ix,307p Academic press,New York; London,1964
Bot 42.0784

INTERNATIONAL SOCIETY FOR CELL BIOLOGY Differentiation and immunology Edited by Katherine Brehme Warren International society for cell biology.Symposia, 7 Academic press,New York;London,1968
Gen 34.0503

INTERNATIONAL SOCIETY FOR CELL BIOLOGY International congress of cell biology 12th Summaries of reports and communications Brussels 1968 Aug 25-31 Edited by P. Dustin and others International congress series, 166 Excerpta medica,Amsterdam,1968
Bot 42.1515

INTERNATIONAL SOCIETY FOR CELL BIOLOGY Intracellular transport Edited by Katherine B. Warren International society for cell biology.Symposia, 5 illus. xvii,325p Academic press,New York;London,1966
Bot 42.1894

INTERNATIONAL SOCIETY FOR CELL BIOLOGY The Interpretation of ultrastructure :symposium Bern 1961 Edited by R.J.C. Harris International society for cell biology. Symposia, 1 illus Academic press,New York;London,1962
Bioch 33.0964

INTERNATIONAL SOCIETY FOR CELL BIOLOGY The Interpretation of ultrastructure :symposium Bern 1961 Edited by R.J.C. Harris International society for cell biology. Symposia, 1 illus. x,438p Academic press,New York;London,1962
Bot 42.1080

INTERNATIONAL SOCIETY FOR CELL BIOLOGY The Relationship between nucleus and cytoplasm :a symposium Proceedings Brussels 1958 Jan 9-13 Experimental cell research.Supplement, 6 Academic press,New York,1959
An 32.2486

INTERNATIONAL SOCIETY FOR CELL BIOLOGY The Relationship between nucleus and cytoplasm : symposium Proceedings Brussels 1958 Jun 9-13 Experimental cell research,Supp. 6, 1959 illus. Academic press,New York,1959
Gen 34.0837

INTERNATIONAL SOCIETY FOR CELL BIOLOGY The Use of radioautography in investigating protein synthesis :a symposium Montreal 1965 Edited by C.P. Leblond and Katherine Brehme Warren International society for cell biology.Symposia, 4 Academic press,New York;London,1965
Bioch 33.0286

INTERNATIONAL SOCIETY FOR CELL BIOLOGY.SYMPOSIA, 1 Interpretation of ultrastructure :a symposium Bern 1961 Edited by R.J.C. Harris Academic press,New York;London,1962
Radioth 35.0517

INTERNATIONAL SOCIETY FOR CELL BIOLOGY.SYMPOSIA, 1 The Interpretation of the ultrastructure Edited by R.J.C. Harris Academic press,London,1962 Based on a symposium held in Berne,September,1961
An 32.3287

INTERNATIONAL SOCIETY FOR CELL BIOLOGY.SYMPOSIA, 1 The Interpretation of the ultrastructure International society for cell biology Edited by R.J.C. Harris Academic press,New York,1962 Based on a symposium held in Berne in September 1961.
Med 36.0177

INTERNATIONAL SOCIETY FOR CELL BIOLOGY.SYMPOSIA, 1 The Interpretation of ultrastructure Edited by R.J.C. Harris Academic press,New York,1962 Based on a symposium held in Berne,September 1961
Phys 20.1461

INTERNATIONAL SOCIETY FOR CELL BIOLOGY.SYMPOSIA, 1 The Interpretation of ultrastructure : symposium Bern 1961 International society for cell biology Edited by R.J.C. Harris illus Academic press,New York; London,1962
Bioch 33.0964

INTERNATIONAL SOCIETY FOR CELL BIOLOGY.SYMPOSIA, 1 The Interpretation of ultrastructure : symposium Bern 1961 International society for cell biology Edited by R.J.C. Harris illus. x,438p Academic press,New York;London,1962
Bot 42.1080

INTERNATIONAL SOCIETY FOR CELL BIOLOGY.SYMPOSIA, 2 Cell growth and cell division Report Liege 1962 May 19-24 International society for cell biology Edited by R.J.C. Harris held at the Institute of histology,Liege illus Academic press,New York;London,1963
Bioch 33.0965

INTERNATIONAL SOCIETY FOR CELL BIOLOGY.SYMPOSIA, 2 Cell growth and cell division :a symposium Liege 1962 May 19-24 Edited by R.J.C. Harris Academic press,New York,1963
An 32.3288

INTERNATIONAL SOCIETY FOR CELL BIOLOGY.SYMPOSIA, 2 Cell growth and cell division :a symposium Liege 1962 May 19-24 International society for cell biology Edited by R.J.C. Harris illus. Academic press,New York;London,1963
Gen 34.0863

INTERNATIONAL SOCIETY FOR CELL BIOLOGY.SYMPOSIA, 2 Cell growth and cell division : symposium Liege 1962 May 19-24 Edited by R.J.C. Harris Illus Academic press,New York;London,1963
Radioth 35.0518

INTERNATIONAL SOCIETY FOR CELL BIOLOGY.SYMPOSIA, 2 Cell growth and cell division;a symposium Report Liege 1962 May 19-24 International society for cell biology Edited by R.J.C. Harris Academic press,New York,1963
Med 36.0178

INTERNATIONAL SOCIETY FOR CELL BIOLOGY.SYMPOSIA, 3 Cytogenetics of cells in culture symposium Pasadena,Calif. 1964 Jun 17 International society for cell biology Edited by R.J.C. Harris Academic press,New York,1964
Gen 34.0892

INTERNATIONAL SOCIETY FOR CELL BIOLOGY.SYMPOSIA, 3 Cytogentics of cells in culture :a symposium Papers Pasadena,Calif. 1964 Jul 17 International society for cell biology Edited by R.J.C. Harris illus,port ix,307p Academic press,New York;London,1964
Bot 42.0784

INTERNATIONAL SOCIETY FOR CELL BIOLOGY.SYMPOSIA, 4 The Use of radioautography in investigating protein synthesis :a symposium Montreal 1965 Edited by C.P. Leblond and Katherine Brehme Warren Sponsored by the International society for cell biology Academic press,New York;London,1965
Bioch 33.0286

INTERNATIONAL SOCIETY FOR CELL BIOLOGY.SYMPOSIA, 5 Intracellular transport :a symposium Edited by Katherine Brehme Warren Academic press,New York,1966
An 32.3435

INTERNATIONAL SOCIETY FOR CELL BIOLOGY.SYMPOSIA, 7 Differentiation and immunology International society for cell biology Edited by Katherine Brehme Warren Academic press,New York;London,1968
Gen 34.0503

INTERNATIONAL SOCIETY FOR CELL BIOLOGY.SYMPOSIA, 8 Cellular dynamics of the neuron Edited by Samuel H. Barondes Academic press, London,1969
An 32.5238

INTERNATIONAL SOCIETY FOR CLINICAL ELECTRORETINOGRAPHY Clinical electroretinography :international symposium 3rd Proceedings Highland Park,Ill. 1964 Oct 14-16 Edited by Hermann M. Burian and Jerry Hart Jacobsen Pergamon press.Symposium publication division,Oxford,1966 Supplement to 'Visual research'
Psy 31.0383

INTERNATIONAL SOCIETY FOR CLINICAL ELECTRORETINOGRAPHY Clinical electroretinography international symposium 3rd Proceedings Highland Park,Ill. 1964 Oct 14-16 Edited by Hermann M. Burian and Jerry Hart Jacobson Vision research. Supplement Pergamon,Oxford,1966
PGMS 29.0340

INTERNATIONAL SOCIETY FOR HORTICULTURAL SCIENCE International horticultural congress 14th Report Hague 1955 Aug 29-Sep 6 Vol 1-2 Veenman and Zonen,Wageningen,1955
BG 38.3125

INTERNATIONAL SOCIETY FOR HORTICULTURAL SCIENCES International horticultural congress 16th Brussels 1962 Aug 31-Sep 8 Vol 1-5 5 vols Duculot,Gembloux,1962
BG 38.3126

INTERNATIONAL SOCIETY FOR HORTICULTURAL SCIENCES International horticultural congress 17th Proceedings College Park,Md. 1966 Aug 15-20 Vol 1-4 Sponsored by the American horticultural society 4 vols 1966-67
BG 38.3134

INTERNATIONAL SOCIETY FOR HORTICULTURAL SCIENCES International horticultural congress 18th Proceedings Tel-Aviv 1970 Mar 17-25 1970
BG 38.3135

INTERNATIONAL SOCIETY FOR MICROBIOLOGY International congress of microbiology :a report 2nd Proceedings London 1936 Jul 1-Aug 1 Edited by R. St.John-Brooks port London,1937
Bot 42.0660

INTERNATIONAL SOCIETY FOR PLANT GEOGRAPHY AND ECOLOGY Pflanzensoziologie und palynologie :international symposium Bericht Edited by Reinhold Tuxen illus. xvii,275p Junk,The Hague,1967
Bot 42.4772

INTERNATIONAL SOCIETY FOR RESEARCH ON THE RETICULOENDOTHELIAL SYSTEM.SYMPOSIA, 3 Reticuloendothelial structure and function : international symposium Rapallo 1958 Aug Edited by John H. Heller Ronald press,New York,1960
An 32.3169

INTERNATIONAL SOCIETY FOR ROCK MECHANICS Large permanent underground openings : international symposium Proceedings Oslo 1969 Sep 23-25 Edited by T.L. Brekke and Finn Jorstrad Universitetsforlaget,Oslo,1970
Eng 41.3133

INTERNATIONAL SOCIETY FOR SIGNIFICS International significal summer conference Conference d'ete internationale de linguistique psychologique 9th Proceedings Amersfoort 1953 Aug 10-15 Synthese, 9,issue 3,no 3-4 Kroonder,Bussum, c1954
Psy 31.2514

INTERNATIONAL SOCIETY FOR STEREOLOGY Quantitative methods in morphology Symposium on quantitative methods in morphology held... during the eighth International congress of anatomists Proceedings Wiesbaden 1965 Aug 10 Edited by Ewald R. Weibel and Hans Elias Illus. Springer,Berlin,1967
An 32.0105

INTERNATIONAL SOCIETY FOR STEREOLOGY Stereology International congress for stereology 2nd Proceedings Chicago,Ill. 1967 Apr 5-13 Edited by Hans Elias Sponsored by the National science foundation Springer,Berlin,1967
An 32.0107

INTERNATIONAL SOCIETY FOR TROPICAL ECOLOGY Recent advances in tropical ecology :a symposium Proceedings Banaras 1967 Jan 16-21 Edited by R. Misra and B. Gopal 2 vols Vararasi,1968
Bot 42.2001

INTERNATIONAL SOCIETY OF BIOMETEOROLOGY Biometeorology International biometeorological congress 3rd Proceedings Pau 1963 Sep 1-7 Vol 2,pt 1-2 Edited by S.W. Tromp and W.H. Weihe 2 vols Pergamon press,Oxford,1967
Geog 13.1212

INTERNATIONAL SOCIETY OF BLOOD TRANSFUSION Congres internationale de la societe internationale de transfusion sanguine 4th Rapports et communications Paris 1951 Jul 21-29 Edited by J. Julliard and others L'expansion scientifique francaise,Paris,1952
Med 36.0389

INTERNATIONAL SOCIETY OF BLOOD TRANSFUSION International congress of blood transfusion 5th Proceedings Paris 1954 Sep 1955
Med 36.0388

INTERNATIONAL SOCIETY OF BLOOD TRANSFUSION International congress of blood transfusion 7th Proceedings Rome 1958 Sep 2-6 Edited by Ludwig Hollander Karger,Basle,1959
Med 36.0200

INTERNATIONAL SOCIETY OF BLOOD TRANSFUSION International congress of blood transfusion 8th Programme and abstracts Tokyo 1960 Sep Edited by Ludwig Hollander 1960
Med 36.0201

INTERNATIONAL SOCIETY OF HEMATOLOGY International congress of haematology 11th Proceedings Sydney 1966 Blackwell,Oxford, 1966
Med 36.0203

INTERNATIONAL SOCIETY OF HEMATOLOGY
International congress of hematology 3rd Proceedings Cambridge 1950 Aug 21-25 Edited by Carl V. Moore Grune and Stratton, New York,1951
Med 36.0256

INTERNATIONAL SOCIETY OF HEMATOLOGY
International congress of hematology 4th Proceedings Mar del Plata 1952 Sep 20-27 Edited by F. Jimenez de Asua and others Grune and Stratton,New York,1954
Med 36.0013

INTERNATIONAL SOCIETY OF HEMATOLOGY
International congress of hematology 6th Proceedings Boston 1956 Aug 27-Sep 1 Edited by A. Richardson Jones Grune and Stratton,New York,1958
Med 36.0202

INTERNATIONAL SOCIETY OF HEMATOLOGY
International congress of hematology :reports, communications,symposia 7th Proceedings Rome 1958 Sep 7-13 3 Edited by Giovanni Di Guglielmo Grune and Stratton,New York,1960
Med 36.0166

INTERNATIONAL SOCIETY OF HEMATOLOGY
International symposium on hemophilia and hemorrhagic states Proceedings Rome 1958 Sep 12 Edited by K.M. Brinkhous North Carolina university press,Chapel Hill,1959 Held in connection with the 7th Congress of the International society of hematology
Med 36.0051

INTERNATIONAL SOCIETY OF LEATHER TRADES' CHEMISTS
The Swelling of proteins,and allied phenomena London 1932 Dec 2 London,1933
Bioch 33.1486

INTERNATIONAL SOCIETY OF ROCK MECHANICS
Congress of the International society of rock mechanics 1st Proceeedings Lisbon 1966 Sep 25-Oct 1 Vol 1-3 3 vols Laboratorio nacional de engenharia civil, Lisbon,1966-67
Eng 41.3176

INTERNATIONAL SOCIETY OF SOIL MECHANICS AND FOUNDATION ENGINEERING
Asian regional conference on soil mechanics and foundation engineering 1st Proceedings New Delhi 1960 Feb Indian national society of soil mechanics and foundation engineering New Delhi,1960
Eng 41.3172

INTERNATIONAL SOCIETY OF SOIL MECHANICS AND FOUNDATION ENGINEERING
Brussels conference on earth pressure problems Proceedings Brussels 1958 Vol 1-3 Brussels,1958
Eng 41.3165

INTERNATIONAL SOCIETY OF SOIL MECHANICS AND FOUNDATION ENGINEERING.BRITISH NATIONAL COMMITTEE
Grouts and drilling muds in engineering practice :a symposium London 1963 May 22-24 Butterworths,London,1963
Eng 41.3080

INTERNATIONAL SOCIETY OF SOIL MECHANICS AND FOUNDATION ENGINEERING.BRITISH NATIONAL COMMITTEE
Pore pressure and suction in soils :a conference London 1960 Mar 30-31 Butterworths,London,1961
Eng 41.3170

INTERNATIONAL SOCIETY OF SOIL SCIENCE.SOIL ZOOLOGY COMMITTEE
Progress in soil zoology Research methods in soil zoology colloquium 1st Papers Rothamsted 1958 Jul 10-14 Edited by P.W. Murphy Bibliog., illus. Butterworths,London,1962
Bot 42.2085

INTERNATIONAL SPACE SCIENCE 7th :symposium
Papers Moon and planets Vienna 1966 May 10-18 International astronomical union Edited by A. Dollfus Co-sponsored by the Committee on space research 313p North-Holland,Amsterdam,1967 Papers of one session of the symposium
Obs 6.3206

INTERNATIONAL SPORE CONFERENCE 4th
Spores 4 Urbana,Ill. 1968 Oct 4-6 American society for microbiology Edited by L.Leon Campbell Under the auspices of the University of Illinois American society for microbiology,1967
Gen 34.0711

INTERNATIONAL STATISTICAL CLASSIFICATION OF DISEASES,INJURIES AND CAUSES OF DEATH
Manual of the international statistical classification of diseases,injuries,and causes of death based on the seventh revision conference Paris 1955 Feb 21-26 1-2 World health organization 2 vols World health organization,Geneva,1957
Gen 34.1937

INTERNATIONAL STATISTICAL CLASSIFICATION OF DISEASES,INJURIES AND CAUSES OF DEATH
Manual of the international statistical classification of diseases,injuries and causes of death :adapted 1948 by the international conference for the sixth revision of the international lists of diseases and causes of death Paris 1948 1-2 World health organization World health organization. Bulletin.Supplement, 1 2 vols Columbia university press,New York,1948-49
Med 36.0385

INTERNATIONAL STATISTICAL INSTITUTE
Session 34th Ottawa 1963 Aug 21-29 1963 Unbound typescript papers
Math 3.1112

INTERNATIONAL STATISTICAL INSTITUTE
Statistics in the physical sciences :a conference Proceedings Belgrade 1965 Sep 17 Under the auspices of the International association for statistics in physical sciences International association for statistics in physical sciences.Proceedings, 2 Bibliog 24cm International statisticcal institute,Belgrade,c1966 Reprinted from the Bulletin of the International statistical institute. Proceedings of the 35th session,1965
Math 3.1157

INTERNATIONAL STUDENT SERVICE
The Social sciences:their relations in theory and in teaching :conference Report London 1935 Sep 22-29 Institute of sociology Le Play House press,London,1936
Geog 13.1613

INTERNATIONAL STUDIES CONFERENCE
The University teaching of international relations international meeting Windsor 1950 Mar 16-20 Edited by Geoffrey L. Goodwin Blackwell, Oxford,1951
Geog 13.0171

INTERNATIONAL STUDIES ON THE QUATERNARY
Boulder,Colo. 1965 Edited by H.E. Wright and David G. Frey Sponsored by the National research council Geological society of America.Special papers, 84 Geological society of America,New York,1965
Bot 42.3133

INTERNATIONAL STUDIES ON THE QUATERNARY
International association for quaternary research :congress 7th Papers Boulder, Colo. 1965 Edited by H.E. Wright and David G. Frey Sponsored by the National research council Geological society of America. Special papers,84 Geological society of America,New York,1965
Geog 13.0309

INTERNATIONAL SUMMER COURSE ON MATERIAL SCIENCE
Proceedings Modern diffraction and imaging techniques in material science Antwerp 1969 Jul 28-Aug 8 North Atlantic treaty organization.Scientific affairs division Edited by S. Amelinckx and others North-Holland,Amsterdam,1970
Met 25.2608

INTERNATIONAL SUMMER COURSE ON PROBABILISTIC METHODS IN ANALYSIS Loutraki 1966 Loutraki,1966 Mimeograph summaries of lectures
Math 3.1116

INTERNATIONAL SUMMER SCHOOL IN RADIO-ASTRONOMY, MANCHESTER Radio astronomy today Papers Jodrell Bank 1962 Edited by H.P. Palmer and others 242p Manchester university press,Manchester,1963 With a foreword by Sir Bernard Lovell
Obs 6.1326

INTERNATIONAL SUMMER SCHOOL OF BRAIN RESEARCH,1ST
Mechanisms of neural regeneration :a workshop...held as part of the first International summer school of brain research.. Amsterdam 1963 Jul 15-26 Edited by M. Singer and J.P. Schade Progress in brain research, 13 Elsevier,Amsterdam,1964
An 32.4233

INTERNATIONAL SUMMER SCHOOL OF BRAIN RESEARCH,1ST
Organization of the spinal cord Edited by J.C. Eccles and J.P. Schade Progress in brain research, 11 Elsevier,Amsterdam,1964 Lectures delivered during a workshop on "Physiology of spinal neurons",held as part of the first International summer school of brain research,Amsterdam,1963
An 32.4231

INTERNATIONAL SUMMER SCHOOL OF BRAIN RESEARCH,1ST
Physiology of spinal neurons :a workshop ... held as part of the first International summer school of brain research... Amsterdam 1963 Jul 15-26 Edited by J.C. Eccles and J.P. Schade Progress in brain research, 12 Elsevier,Amsterdam,1964
An 32.4232

INTERNATIONAL SUMMER SCHOOL OF MOLECULAR BIOPHYSICS Proceedings Molecular biophysics Squaw Valley,Calif. 1964 Aug 17-28 Edited by B. Pullman and M. Weissbluth Sponsored by the North atlantic treaty organization Academic press,New York,1965
Radioth 35.1418

INTERNATIONAL SUMMER SCHOOL OF MOLECULAR BIOPHYSICS Proceedings Molecular biophysics Squaw Valley,Calif. 1964 Aug 17-28 Edited by Bernard Pullman and Mitchel Weissbluth Sponsored jointly by the North Atlantic treaty organization and United States. Office of naval research Academic press,New York;London,1965
Bioch 33.2360

INTERNATIONAL SUMMER SCHOOL ON CRYSTALLOGRAPHIC COMPUTATION Proceedings Ottawa 1969 Aug 4-11 International union of crystallography.Commission on crystallographic computation Edited by F.R. Ahmed Munksgaard,Copenhagen,1970
Min 10.1527

INTERNATIONAL SUMMER SCHOOL ON MATHEMATICAL MODELS OF ACTION AND REACTION Proceedings Differential games and related topics Varenna 1970 Jun 15-27 Consiglio nazionale delle richerche Ente per gli studi monetari, bancari et finanziari 'Luigi Einaudi' Edited by H.W. Kuhn and G.P. Szego x,489p 23cm North-Holland,Amsterdam,1971
P Math 2.4574

INTERNATIONAL SUMMER SCHOOL ON MATHEMATICAL MODELS OF ACTION AND REACTION Proceedings Differential games and related topics Varenna 1970 Jun 15-27 Edited by H.W. Kuhn and G.P. Szego x,489p 23cm North-Holland, Amsterdam;London,1971
Math S 3.1721

INTERNATIONAL SUMMER SCHOOL ON SELECTED TOPICS IN NUCLEAR THEORY lectures Nuclear theory Low Tatra mountains,Czechoslovakia 1962 Aug 20-Sep 8 By N. Austern and others Ceskoslovenska akademie ved.Ustav jaderneho vyzkumu Edited by F. Janouch With the co-operation of the International atomic energy agency 452p International atomic energy agency,Vienna,1963
A Math 4.0885

INTERNATIONAL SUMMER SCHOOL ON SOLID STATE PHYSICS Proceedings Interaction of radiation with solids Mol 1963 Aug 12-31 Edited by R. Strumane and others North-Holland,Amsterdam,1964
Met 25.1581

INTERNATIONAL SUMMER SCHOOL ON THE ACCURATE DETERMINATION OF NEUTRON INTENSITIES AND STRUCTURE FACTORS Proceedings Thermal neutron diffraction Harwell 1968 Jul 1-5 Edited by B.T.M. Willis Harwell postgraduate series Bibliog,illus xiv,229p 24cm Oxford university press,London,1970
Min 10.1524

INTERNATIONAL SWEDENBORG CONGRESS Transactions London 1910 Jul 4-8 Swedenborg society 2nd edition Swedenborg society,London,1911
Phys 20.0723

INTERNATIONAL SYMPOSIA ON COMBUSTION 1st-2nd Combustion institute Combustion institute, Pittsburgh,Pa.,1965
Eng 41.7257

INTERNATIONAL SYMPOSIA ON COMBUSTION 4th-12th Combustion institute 9 vols Combustion institute,Pittsburgh,Pa.,1953-69
Eng 41.7258

INTERNATIONAL SYMPOSIUM 2nd Proceedings X-ray microscopy and x-ray microanalysis Stockholm 1960 Edited by A. Engstrom and others Elsevier,Amsterdam,1960
Met 25.1375

INTERNATIONAL SYMPOSIUM.ENZYMATIC ASPECTS OF METABOLIC REGULATION Proceedings Mexico City 1966 Nov 28-Dec 1 National cancer institute Edited by M.P. Stulberg National cancer institute.Monograph, 27 National cancer institute,Betheseda,Md.,1967
Bioch 33.1062

INTERNATIONAL SYMPOSIUM COMMEMORATING THE 50TH ANNIVERSARY OF INSULIN Impact of insulin on metabolic pathways Jerusalem 1971 Oct 24-29 Vol 1-2: lectures;panel discussions and communications Edited by Eleazar Shafrir Israel journal of medical sciences, 8,no 3,6 2 vols Israel medical association,Jerusalem, 1972
Bioch 33.2193

The INTERNATIONAL SYMPOSIUM FOR CELLULAR CHEMISTRY 1st Proceedings Intracellular membraneous structure Ohtsu 1963 Mar 27-31 Japan society for cell biology Edited by S. Seno and E.V. Cowdry Society for cellular chemistry.Symposia, 14, Suppl. Okayama,1964 In memory of...Dr. Seizo Katsunuma.Japan society for cell biology formerly Japan society for cellular chemistry
Bioch 33.0966

INTERNATIONAL SYMPOSIUM FOR CELLULAR CHEMISTRY 2nd Proceedings Nucleic acid metabolism,cell differentiation and cancer growth Ohtsu 1966 Oct 17-21 Edited by E. V. Cowdry and S. Seno Organised by the Japan society for cell biology Pergamon,London, 1969
Radioth 35.0561

INTERNATIONAL SYMPOSIUM IN THE CHEMISTRY OF CEMENT 4th Proceedings Chemistry of cement Washington,D.C. 1960 Oct 2-7 Vol 1-2 National bureau of standards National bureau of standards.Monograph, 43, vol 1-2 2 vols U.S.government printing office,Washington,D.C.,1962
Eng 41.3026

INTERNATIONAL SYMPOSIUM NEURONTOGENETICUM 5th Proceedings Ontogenesis of the brain:the biochemical,functional and structural development of the nervous system Prague 1967 Universita Karlova and Lekarska spolecnost J.E.Purknye Edited by Lubor Jilek and Stanislav Trojan Universita Karlova, Prague,1968
PGMS 29.0202

An INTERNATIONAL SYMPOSIUM OF FOREST TREE PHYSIOLOGY 1st Physiology of forest trees Harvard forest 1957 Apr 8-12 Edited by Kenneth Thimann and others Under the auspices of the Maria Moors Cabot foundation Ronald press,New York,1958
Bot 42.1335

INTERNATIONAL SYMPOSIUM ON ACETABULARIA 1st Proceedings Brussels 1969 Jun 18-20 Universite libre de Bruxelles.Centre d'etude de l'energie nucleaire Edited by Jean Brachet and Silvano Bonotto illus. xv,300p Academic press,New York;London,1970
Bot 42.3941

INTERNATIONAL SYMPOSIUM ON AGGLOMERATION Agglomeration :based on an international symposium... Philadelphia,Penn. 1961 Apr 12-14 American institute of mining, metallurgical and petroleum engineers Edited by William A. Knepper Interscience,New York; London,1962
Met 25.2288

INTERNATIONAL SYMPOSIUM ON ALDOSTERONE Proceedings Prague 1963 Council for international organization of medical sciences Edited by Etienne Emile Baulier and P. Robel Blackwell,Oxford,1964
Inv Med 37.0013

The INTERNATIONAL SYMPOSIUM ON ALDOSTERONE Papers and discussions Geneva 1957 Jun 7-8 Edited by Alex F. Muller and Cecilia M. O'Connor Organized by the Universite de Geneve.Clinique therapeutique J.and A. Churchill,London,1958
Bioch 33.0416

INTERNATIONAL SYMPOSIUM ON ALGEBRAIC NUMBER THEORY proceedings Tokyo 1955 Sep 8-13 Nikko Science council of Japan Edited by Shokichi Iyanaga and Yukiyosi Kawada Jointly sponsored by the International mathematical union xx,265p 26cm Science council of Japan,Tokyo,1956
P. Math 2.1771

INTERNATIONAL SYMPOSIUM ON ARCTIC GEOLOGY 1st Proceedings Geology of the Arctic Calgary,Alta. 1960 Jan 11-13 Vol 1-2 Edited by Gilbert O. Raasch Under the auspices of the Alberta society of petroleum geologists 2 vols Toronto,1961 Additional maps in a separate container
Geod 9.0436

INTERNATIONAL SYMPOSIUM ON ARCTIC GEOLOGY 1st Proceedings Geology of the arctic Calgary 1960 Jan 11-13 Vol 1-3 Edited by Gilbert O. Raasch Under the auspices of the Alberta society of petroleum geologists 3 vols University of Toronto press,Toronto, 1961 Vol.3 consists of maps
Geog 13.5390

INTERNATIONAL SYMPOSIUM ON ARCTIC GEOLOGY 1st proceedings Geology of the Arctic Calgary,Alberta 1960 Jan 11-13 Edited by Gilbert O. Raasch Under the auspices of Alberta society of petroleum geologists maps 3 vols University of Toronto press,Toronto, 1961
Geol 8.2938

INTERNATIONAL SYMPOSIUM ON AUTOMATIC DIGITAL COMPUTATION 3rd proceedings Teddington 1953 Ot 25-28 National physical laboratory illus vi,296p 26cm National physical laboratory,Teddington,1954
ATH L 5.0724

INTERNATIONAL SYMPOSIUM ON BATTERIES :research and development in non-mechanical electrical power sources 4th Batteries 2 Brighton 1964 Sep Edited by D.H. Collins Sponsored by the Inter-departmental committee on batteries Pergamon Press,Oxford,1965
Met 25.1841

INTERNATIONAL SYMPOSIUM ON BIOSIMULATION 1st Proceedings Aspects of the theory of artificial intelligence Locarno 1960 Jun 29-Jul 5 Edited by C.A. Muses Plenum,New York,1962
Eng 41.5988

INTERNATIONAL SYMPOSIUM ON BLOOD OXYGENATION Proceedings Blood oxygenation Cincinnati,Ohio 1969 Dec 1-3 Edited by Daniel Hershey Plenum press,New York;London, 1970
An 32.5457

The INTERNATIONAL SYMPOSIUM ON BORON Paris 1964 Jul 17-18 Vol 2: preparation, properties,and applications Edited by Gerhart K. Gaule Plenum press,New York,1965
Met 25.0159

The INTERNATIONAL SYMPOSIUM ON CHEMICAL AND BIOLOGICAL ASPECTS OF PYRIDOXICAL CATALYSIS 2nd Proceedings Pyridoxal catalysis: enzymes and model systems Moscow 1966 Sep 15-21 Edited by E.E. Snell and others sponsored by the International union of biochemistry I.U.B.Symposium series, 35 Interscience,New York,1968 First International symposium on chemical and biological aspects of pyridoxal catalysis published under title chemical and biological aspects of pyridoxical catalysis
Bioch 33.1038

INTERNATIONAL SYMPOSIUM ON CHEMOTHERAPY OF CANCER Proceedings Lugano 1964 Apr 28-May 1 Swiss academy of medical sciences Edited by Placidus A. Plattner Sponsored by Hoffmann-La Roche and co ltd. Elsevier,Amsterdam,1964
Pha 16.0335

INTERNATIONAL SYMPOSIUM ON CIRCUIT AND INFORMATION THEORY transaction Los Angeles 1959 Jun.16-18 Institute of radio engineers Co-sponsored by the International scientific radio union Institute of radio engineers.Transactions on information theory, IT, 5,special supplement xiv,298p Institute of radio engineers,New York,1959 Papers published in advance of the Symposium
Math L 5.2071

INTERNATIONAL SYMPOSIUM ON CLASSICAL AND CONTAGIOUS DISCRETE DISTRIBUTIONS proceedings Montreal 1963 Aug 15-20 Canadian mathematical congress Edited by Ganapati P. Patil Supported by the National research council of Canada Bibliog. xiv, 552p 27cm Statistical publishing society, Calctta,1965
Math 3.0709

INTERNATIONAL SYMPOSIUM ON COLD ACCLIMATION Proceedings Buenos Aires 1959 Aug 5-7 Edited by Robert E. Smith and others Federation of American societies for experimental biology.Federation proceedings,19, 4,pt 2,Supplement,5 illus viii,165p Buenos Aires,1966
Sco 14.0862

An INTERNATIONAL SYMPOSIUM ON COMMUNICATION AND SOCIAL INTERACTIONS IN PRIMATES Social communication among primates Montreal 1964 Dec 27-31 Edited by Stuart A. Altmann University of Chicago press,Chicago,1967
Bal 39.1627

INTERNATIONAL SYMPOSIUM ON COMPARATIVE ENDOCRINOLOGY 3rd Proceedings Progress in comparative endocrinology Oiso, Japan 1961 Jun 5-11 Edited by Kiyoshi Takewaki Sponsored by the Zoological society of Japan General and comparative endocrinology.Suppl., 1,1962 Academic press,New York,1962
Bal 39.1381

INTERNATIONAL SYMPOSIUM ON CONDENSATION AND EVAPORATION OF SOLIDS Proceedings Condensation and evaporation of solids Dayton,Ohio 1962 Sep 12-14 United States. Air force.Directorate of materials and processes Edited by Emile Rutner and others Gordon and Breach,New York,1964
Met 25.1518

INTERNATIONAL SYMPOSIUM ON CONFIRMATION AND INFORMATION,1965 Aspects of inductive logic Edited by Jaakko Hintikka and Patrick Suppes Studies in logic and the foundations of mathematics vii,320p 23cm North-Holland,Amsterdam,1966 Includes papers read at the International symposium on confirmation and information,Helsinki,1965
P Math 2.2678

INTERNATIONAL SYMPOSIUM ON CONFIRMATION AND INFORMATION HELSINKI 1965 Aspects of inductive logic Edited by Jaakko Hintikka and Patrick Suppes Studies in logic and the foundations of mathematics vii,320p North-Holland,Amsterdam,1966 Eight of the fourteen papers based on those read at an international symposium on confirmation and information held at Helsinki,Sept.30-Oct.2, 1965
WSM 43.1205

An INTERNATIONAL SYMPOSIUM ON CONFORMATION OF BIOPOLYMERS Madras 1967 Jan 18-21 Vol 1 Edited by G.N. Ramachandran Academic press,London,1967
Col S 12.0132

INTERNATIONAL SYMPOSIUM ON CONFORMATION OF BIOPOLYMERS Papers read Conformation of biopolymers Madras 1967 Jan 18-21 Vol 1-2 University of Madras.Centre of advanced study in biophysics Edited by G.N. Ramachandran Sponsored by the International union of pure and applied physics 2 vols Academic press,London;New York,1967
Bioch 33.0331

INTERNATIONAL SYMPOSIUM ON CONFORMATIONAL ANALYSIS Papers Conformational analysis:scope and present limitations Brussels 1969 Sep Edited by Gregoire Chiurdoglu Organic chemistry, 21
Chem 18.2788

INTERNATIONAL SYMPOSIUM ON CURARE AND CURARE-LIKE AGENTS Proceedings Rio de Janeiro 1957 Aug 5-12 Academia brasileira de ciencias.Conselho nacional de pesquisas and Universidade do Brasil Edited by D. Bovet and others Under the auspices of Unesco illus Elsevier,Amsterdam,1959 Also sponsored by the 'Istituto di sanita',of Rome. Papers in English and French
Chem 26.0103

INTERNATIONAL SYMPOSIUM ON CURARE AND CURARE-LIKE AGENTS Proceedings Curare and curare-like agents Rio de Janeiro 1957 Aug 5-12 Edited by D. Bovet and others Organized under the auspices of Unesco Elsevier, Amsterdam,1959
Pha 16.0243

INTERNATIONAL SYMPOSIUM ON DATA PROCESSING MACHINES Papers Prague 1964 Sep 1964 Collection of typescript papers in folder
Math L 5.3266

INTERNATIONAL SYMPOSIUM ON DATA TRANSMISSION transactions Delft 1960 Sep 19-21 Institute of radio engineers.Benelux section Institute of radio engineers.Transactions on communications systems, 9,no.1 99p I.R.E. professional group on communications systems, New York,1961
Math L 5.0868

INTERNATIONAL SYMPOSIUM ON DISTILLATION Brighton 1960 May 4-6 European federation of chemical engineering 281p Institution of chemical engineers,London,1960 24th meeting of the European Federation of Chemical Engineering
Chem E 24.1303

INTERNATIONAL SYMPOSIUM ON DRUGS AFFECTING LIPID METABOLISM 2nd Papers Drugs affecting lipid metabolism Milan 1965 Pt 1-2 Edited by D. Kritchevsky and others organised by the European society for biochemical pharmacology Progress in biochemical pharmacology, 2-3 illus 2 vols Karger,Basle;New York,1967
Bioch 33.1250

INTERNATIONAL SYMPOSIUM ON EARLY MAN Papers Early man... Philadelphia,Pa. 1937 Mar 17-20 Edited by George Grant MacCurdy illus. Lippincott,Philadelphia,Pa.,1937
Geog 13.1760

INTERNATIONAL SYMPOSIUM ON ELECTRON AND PHOTON INTERACTIONS AT HIGH ENERGIES 4th Proceedings Liverpool 1969 Sep 14-20 Danesbury nuclear physics laboratory International union of pure and applied physics University of Liverpool illus 30cm Danesbury nuclear physics laboratory, Danesbury,1969
A Math 4.1760

INTERNATIONAL SYMPOSIUM ON END RESULTS OF CANCER THERAPY Sandefjord 1963 Sep 16-20 National cancer institute Edited by Sidney J. Cutler National cancer institute.Monograph, 15 U.S.Department of health,education and welfare,Washington,D.C.,1964
Radioth 35.0844

INTERNATIONAL SYMPOSIUM ON END RESULTS OF CANCER THERAPY Proceedings End results of cancer therapy Sandefjord 1963 Sep 16-20 National cancer institute Edited by Sidney J. Cutler National cancer institute.Monograph, 15 National cancer institute,Bethesda,Md., 1964
Bioch 33.0986

INTERNATIONAL SYMPOSIUM ON ENERGY TRANSDUCTION IN RESPIRATION AND PHOTOSYNTHESIS Pugnochiuso 1970 Sep 11-14 International union of biochemistry Edited by E. Quagliariello and others International union of biochemistry.Symposia, 43 Adriatic editrice,Bari,1971
Bioch 33.2289

The INTERNATIONAL SYMPOSIUM ON ENZYME CHEMISTRY Proceedings Tokyo 1957 Oct 15-23 and Kyoto 1957 Oct 15-23 International union of biochemistry Organized by Science council of Japan I.U.B.Symposium series, 2 Pergamon,London,1958
Bioch 33.1053

INTERNATIONAL SYMPOSIUM ON FLAVINS AND FLAVOPROTEINS 3rd Proceedings Durham N.C. 1970 International union of biochemistry International union of biochemistry.Symposium, 39 University park press;Butterworths,Baltimore,Md.;London,1971
Bioch 33.2373

INTERNATIONAL SYMPOSIUM ON FLUIDIZATION Proceedings Eindhoven 1967 Jun 6-9 Edited by A.A.H. Drinkenburg illus 829p Netherlands university press,Amsterdam,1967
Chem E 24.1918

INTERNATIONAL SYMPOSIUM ON FOREST HYDROLOGY Proceedings University Park,Pa. 1965 Aug 29-Sep 10 National science foundation Edited by William E. Sapper and Howard W. Lull Pergamon press,Oxford,1967 A National Science Foundation advanced science seminar
Geog 13.0795

INTERNATIONAL SYMPOSIUM ON FRACTURE MECHANICS Kiruna 1967 Aug 7-12 Swedish national committee for mechanics Wolters-Noordhoff, Groningen,1968
Met 25.0874

INTERNATIONAL SYMPOSIUM ON FREE RADICALS 5th Preprint of papers Uppsala 1961 Jul 6-7 United States.Army.European research office and United States.Air force.Office of scientific research 1961
Chem 18.2625

INTERNATIONAL SYMPOSIUM ON FREE RADICALS 6th Papers Cambridge 1963 Jul 2-5 University of Cambridge Sponsored in part by Great Britain.Air force.Office of scientific research,OAR Cambridge,1963 Mimeographed
Chem 18.0414

INTERNATIONAL SYMPOSIUM ON FREEZE-DRYING 2nd Papers Recent research in freezing and drying London 1958 Edited by A.S. Parkes and Audrey U. Smith Organised by the Institute of biology diagrs vii,320p 25cm Blackwell scientific publications, Oxford,1960
Sco 14.1237

INTERNATIONAL SYMPOSIUM ON FREEZING AND DRYING 2nd Proceedings Recent research in freezing and drying London 1958 Edited by A.S. Parkes and Audrey U. Smith Organized by the Institute of biology Blackwell,Oxford, 1960
An 32.3233

INTERNATIONAL SYMPOSIUM ON GAS CHROMATOGRAPHY 8th Proceedings Gas chromatography 1970 Dublin 1970 Sep 28-Oct 2 Institute of petroleum.Gas chromatography discussion group Edited by R. Stock viii,445p 25cm Institute of petroleum,London,1971
Chem 18.2697

INTERNATIONAL SYMPOSIUM ON GENES AND CHROMOSOMES: STRUCTURE AND FUNCTION Proceedings Buenos Aires 1964 Nov 30-Dec 4 National cancer institute Edited by Juan I. Valencia and Rhoda F. Grell National cancer institute. Monograph, 18 National cancer institute, Bethesda,Md.,1965
Bioch 33.0987

INTERNATIONAL SYMPOSIUM ON GENES AND CHROMOSOMES STRUCTURE AND FUNCTION Proceedings Buenos Aires 1964 Nov 30-Dec 4 Edited by Juan I. Valencia and Rhoda F. Grell National cancer institute.Monograph, 18 illus 354p U.S.Dept. of health,education and welfare, Bethesda,Md.,1965
Bot 42.0792

INTERNATIONAL SYMPOSIUM ON GENES AND CHROMOSOMES STRUCTURE AND FUNCTION Proceedings Buenos Aires 1964 Nov 30-Dec 4 Edited by Juan I. Valencia and others National cancer institute.Monographs, 18 U.S.Department of health,education and welfare,Washington,D.C., 1965
Gen 34.0833

INTERNATIONAL SYMPOSIUM ON GEOPHYSICAL THEORY AND COMPUTERS 3rd Proceedings Cambridge 1966 Jun 27-Jul 5 International upper mantle committee Sponsored by the International union of geodesy and geophysics Geophysical journal, 13;1-3 375p Blackwell,Oxford, 1967
Geod 9.0694

INTERNATIONAL SYMPOSIUM ON GROWTH HORMONE 1st Proceedings Growth hormone Milan 1967 Sep 11-13 Edited by A. Pecile and E.E. Muller International congress series, 158 Excerpta medica,Amsterdam,1968
Bioch 33.0457

INTERNATIONAL SYMPOSIUM ON HEMOPHILIA AND HEMOPHILIOID DISEASES Proceedings New York 1956 Aug 24-25 National hemophilia foundation Edited by K.M. Brinkhous North Carolina university press,Chapel Hill,1957
Med 36.0050

INTERNATIONAL SYMPOSIUM ON HEMOPHILIA AND HEMORRHAGIC STATES Proceedings Rome 1958 Sep 12 International society of hematology Edited by K.M. Brinkhous North Carolina university press,Chapel Hill,1959 Held in connection with the 7th Congress of the International society of hematology
Med 36.0051

INTERNATIONAL SYMPOSIUM ON HIGH SPEED TESTING 7th Proceedings Rheology of solids Boston,Mass. 1969 Mar 17-18 Edited by R.D. Andrews and F.R. Eirich Applied polymer symposia, 12,High speed testing, 7 Interscience,New York,1969
Met 25.2581

INTERNATIONAL SYMPOSIUM ON HIGH TEMPERATURE TECHNOLOGY 2nd Proceedings Asilomar, Calif 1959 Oct 6-9 Stanford research institute McGraw-Hill,New York,1960
Met 25.2535

An INTERNATIONAL SYMPOSIUM ON HIGH TEMPERATURE TECHNOLOGY 3rd Proceedings High temperature technology Pacific Grove,Calif. 1963 Sep 8-11 International union of pure and applied chemistry.Commission on high temperatures and refractories Organized and directed by Stanford research institute Butterworths,London,1964
Met 25.0992

INTERNATIONAL SYMPOSIUM ON IMMUNOLOGICAL METHODS OF BIOLOGICAL STANDARDIZATION 15th Proceedings Royaumont 1965 Oct 26-28 International association of microbiological societies.Permanent section of microbiological standardization Edited by R.H. Regamey and others International association of microbiological societies.Symposia series in immunological standardization, 4 illus, tables Karger,Basle;New York,1967
Path 30.2746

INTERNATIONAL SYMPOSIUM ON IMMUNOPATHOLOGY 2nd Papers Mechanism of cell and tissue damage produced by immune reactions Brook Lodge,Mich. 1961 International committee on immunology and Upjohn company Edited by Pierre Grabar and Peter Miescher Schwabe, Basle;Stuttgart,1962
Path 30.2226

INTERNATIONAL SYMPOSIUM ON IRIS 1st Report Florence 1963 May 14-18 Societa italiana dell'iris Edited by Gian Luigi Sani and others Societa italiana dell'iris, Florence,1963
BG 38.3303

INTERNATIONAL SYMPOSIUM ON ISOTOPE SEPARATION 1st Proceedings Amsterdam 1957 Apr 23-27 Netherlands physical society and International union of pure and applied physics Edited by J. Kistemaker and others xx,704p North-Holland,Amsterdam,1958
Chem E 24.1654

INTERNATIONAL SYMPOSIUM ON LYMPHOLOGY Proceedings Progress in lymphology Zurich 1966 Jul Thieme,Stuttgart,1967
Radioth 35.0922

INTERNATIONAL SYMPOSIUM ON MAGNETO-FLUID DYNAMICS proceedings Magneto-fluid dynamics Williamsburg,Va. 1960 Jan and Washington D.C. Edited by F.N. Frenkel and W.R. Sears Sponsored by the International union of theoretical and applied mechanics National research council.Publication, 829 National research council,Washington,1960
A Math 4.0619

INTERNATIONAL SYMPOSIUM ON MAMMARY CANCER 2nd Proceedings Perugia 1957 Jul 24-29 Edited by Lucio Severi Division of cancer research,Perugia,1958
Radioth 35.0779

INTERNATIONAL SYMPOSIUM ON METABOLISM AND MEMBRANE PERMEABILITY OF ERYTHROCYTES AND THROMBOCYTES 1st Metabolism and membrane permeability of erythrocytes and thrombocytes Vienna 1968 Jun 17-20 Edited by Erwin Deutsch and others Thieme, Stuttgart,1968 In English and German
Radioth 35.0932

1149

INTERNATIONAL SYMPOSIUM ON MOLECULAR STRUCTURE AND SPECTROSCOPY Tokyo 1962 Sep 10-14 International union of pure and applied chemistry Science council of Japan,Tokyo, 1962
Chem 18.2639

INTERNATIONAL SYMPOSIUM ON MULTIVARIATE ANALYSIS Proceedings Dayton,Ohio 1965 Jun 14-19 Edited by P.R. Krishnaiah Sponsored by United States.Air force.Aerospace research laboratories Academic press,New York,1966
Eng 41.2206

INTERNATIONAL SYMPOSIUM ON MULTIVARIATE ANALYSIS Proceedings Dayton,Ohio 1968 Jun 17-22 Edited by Parachuri R. Krishnaiah Held at the Wright state university xx,696p 24cm Academic press,New York;London,1969
Math S 3.1802

INTERNATIONAL SYMPOSIUM ON MYCOTOXINS IN FOODSTUFFS Mycotoxins in foodstuffs :a symposium Proceedings Cambridge,Mass. 1964 Mar 18-19 Edited by Gerald N. Wogan Held at the Massachusetts institute of technology M.I.T.press,Cambridge,Mass.,1965 "This volume comprises the proceedings of an International symposium on mycotoxins in foodstuffs".
Bioch 33.0708

INTERNATIONAL SYMPOSIUM ON NATURAL MAMMALIAN HIBERNATION 3rd Proceedings Mammalian hibernation 3 Toronto 1965 Sep 13-16 Edited by Kenneth C. Fisher and others illus xiv,534p 25cm Oliver and Boyd, Edinburgh;London,1967
Sco 14.8231

INTERNATIONAL SYMPOSIUM ON NERVOUS INHIBITION Nervous inhibition The Friday Harbor symposium 2nd Proceedings Friday Harbor 1960 May 31-Jun 4 Edited by Ernst Florey Sponsored by the National science foundation. Division of regulatory biology Pergamon press,Oxford,1961 Conference also called International symposium on nervous inhibition
Bal 39.1475

An INTERNATIONAL SYMPOSIUM ON NEUROSECRETION Proceedings Society for endocrinology Edited by H. Heller and R.B. Clark Society for endocrinology.Memoirs, 12 Academic press,London;New York,1962
Inv Med 37.0107

INTERNATIONAL SYMPOSIUM ON NEUROSECRETION 4th Strasbourg 1966 Jul 25-27 Edited by F. Stutinsky Springer,Berlin,1967
An 32.3779

INTERNATIONAL SYMPOSIUM ON NONDESTRUCTIVE TESTING OF MATERIALS AND STRUCTURES Vol 1-2 International union of testing and research laboratories for materials and structures 2 vols Paris,1956
Eng 41.2833

INTERNATIONAL SYMPOSIUM ON NONLINEAR DIFFERENTIAL EQUATIONS AND NONLINEAR MECHANICS Colorado Springs,Colo. 1961 Jul 31-Aug 4 Edited by Joseph P. LaSalle and Solomon Lefschetz Sponsored by United States.Air force.Office of scientific research Academic press,New York;London,1963
Eng 41.6385

1150

INTERNATIONAL SYMPOSIUM ON NONPARAMETRIC TECHNIQUES IN STATISTICAL INFERENCE 1st Proceedings Nonparametric techniques in statistical inference Bloomington,Ind. 1969 Jun 1-6 Edited by Madan Lal Puri Held at Indiana university bibliog. xiv,623p 24cm Cambridge university press,Cambridge, 1970
Math S 3.1616

INTERNATIONAL SYMPOSIUM ON NUCLEAR MEDICINE Proceedings 1st Pulmonary and cardiac function bone isotope diagnosis Karlovy Vary 1969 May 13-16 Lekarska spolecnost J.E.Purkyne Edited by O. Andrysek and J. Mestan Universita Karlova,Prague,1970
PGMS 29.0680

INTERNATIONAL SYMPOSIUM ON NUCLEOLUS:ITS STRUCTURE AND FUNCTION Proceedings Montevideo 1965 Dec 5-10 National cancer institute Edited by W.S. Vincent and others National cancer institute.Monograph, 23 National cancer institute,Bethesda,Md.,1966
Bioch 33.0988

INTERNATIONAL SYMPOSIUM ON OLFACTION AND TASTE 1st Proceedings Olfaction and taste Stockholm 1962 Sep Wenner-Gren center Edited by Y. Zotterman Wenner-Gren center. International symposium.Series, 1 Pergamon press,Oxford,1963
An 32.4380

INTERNATIONAL SYMPOSIUM ON OLFACTION AND TASTE 2nd Proceedings Olfaction and taste II Tokyo 1965 Sep Wenner-Gren center Edited by T. Hayashi Wenner-Gren center. International symposium series, 8 Pergamon press,Oxford,1967
An 32.4411

INTERNATIONAL SYMPOSIUM ON OPTIMIZING AND ADAPTIVE CONTROL 1st Proceedings Rome 1962 Apr 26-28 Edited by Loren E. Bollinger and Emil J. Minnar Sponsored by the International federation of automatic control.Theory committee Instrument society of America,Pittsburgh,Pa.,1962
Eng 41.6017

INTERNATIONAL SYMPOSIUM ON PAIN Proceedings Pain Paris 1967 Apr 11-13 Edited by A. Soulairac and others Academic press,London, 1968
Pha 16.0263

The INTERNATIONAL SYMPOSIUM ON POLARIZATION PHENOMENA OF NUCLEONS Basel 1960 Jul 4-8 Universitat Basel Edited by P. Huber and K.P. Meyer Under the sponsorship of the International union of pure and applied physics Helvetica physica acta.Supplementum, 6 Birkhauser,Basel,1961
Cav 7.0135

An INTERNATIONAL SYMPOSIUM ON POLLEN PHYSIOLOGY AND FERTILIZATION 1st Papers Pollen physiology and fertilization Nijmegen 1963 Aug 29-31 Edited by H.F. Linskens North Holland,Amsterdam,1964
Bot 42.1426

INTERNATIONAL SYMPOSIUM ON PRIMARY AND INITIAL EFFECTS OF IONIZING RADIATIONS ON LIVING CELLS Papers:discussions Initial effects of ionizing radiations on cells Moscow 1960 Oct Edited by R.J.C. Harris Sponsored by the Akademiya nauk S.S.S.R. Academic press, London;New York,1961
An 32.3277

The INTERNATIONAL SYMPOSIUM ON PRIMARY AND INITIAL EFFECTS OF IONIZING RADIATIONS ON LIVING CELLS Papers and discussions Initial effects of ionizing radiations on cells Moscow 1960 Oct Edited by R.J.C. Harris Academic press,London;New York,1961 A symposium supported by Unesco and the IAEA and sponsored by the Academy of sciences of the U.S.S.R.
Bioch 33.0961

INTERNATIONAL SYMPOSIUM ON PROTEIN AND POLYPEPTIDE HORMONES Proceedings Protein and polypeptide hormones Liege 1968 May 19-25 Edited by M. Margoulies Excerpta medica.International congress series, 161 Excerpta medica,Amsterdam,1968
Inv Med 37.0117

The INTERNATIONAL SYMPOSIUM ON PROTEIN BIOSYNTHESIS Proceedings Protein biosynthesis Wassenaar 1960 Aug 29-Sep 2 Edited by R.J.C. Harris Under the auspices of Unesco and the Council for international organizations of medical sciences Academic press,London;New York,1961
Bioch 33.0516

INTERNATIONAL SYMPOSIUM ON PROTEIN STRUCTURE AND CRYSTALLOGRAPHY Aspects of protein structure The Symposium on protein structure Proceedings Madras 1963 Jan 14-18 University of Madras.Department of physics Edited by G.N. Ramachandran Academic press, London;New York,1963 The Symposium on protein structure formed part of an International symposium on protein structure and crystallography organized by the University of Madras
Bioch 33.0542

INTERNATIONAL SYMPOSIUM ON PROTEIN STRUCTURE AND CRYSTALLOGRAPHY Crystallography and crystal perfection :a symposium Proceedings Madras 1963 Jan 14-18 Edited by G.N. Ramachandran Academic press,London;New York, 1963 Papers of a symposium which formed part of the International symposium on protein structure and crystallography
Met 25.1448

The INTERNATIONAL SYMPOSIUM ON PROTEIN STRUCTURE AND CRYSTALLOGRAPHY Proceedings Aspects of protein structure Madras 1963 Jan 14-18 University of Madras.Department of physics Edited by G.N. Ramachandran Academic press,London,1963
Cav 7.1495

The INTERNATIONAL SYMPOSIUM ON PROTEIN STRUCTURE AND CRYSTALLOGRAPHY Proceedings Crystallography and crystal perfection Madras 1963 Jan 14-18 session on crystallography and crystal perfection University of Madras.Department of physics Edited by G.N. Ramachandran Academic press, London,1963
Cav 7.1494

INTERNATIONAL SYMPOSIUM ON PULSATILE BLOOD FLOW 1st Proceedings Pulsatile blood flow Philadelphia,Pa. 1963 Apr 11-13 Edited by E.O. Attinger McGraw-Hill,New York,1964
An 32.3528

The INTERNATIONAL SYMPOSIUM ON QUANTUM FLUIDS Quantum fluids Brighton 1965 Aug 16-20 University of Sussex Edited by D.F. Brewer viii,360p 23cm North-Holland,Amsterdam, 1966
Cav 7.2935

INTERNATIONAL SYMPOSIUM ON RADIOSENSITIZING AND RADIOPROTECTIVE DRUGS 2nd Rome 1969 May 6-8 book of abstracts Istituto superiore de sanita,Rome,1969
Radioth 35.1213

INTERNATIONAL SYMPOSIUM ON RADIOSENSITIZING AND RADIOPROTECTIVE DRUGS 2nd Proceedings Radiation protection and sensitization Rome 1969 May 6-8 Edited by Holta Whitehouse Taylor and Francis,London,1970
Radioth 35.1219

The INTERNATIONAL SYMPOSIUM ON RAREFIED GAS DYNAMICS 2nd Proceedings Rarefied gas dynamics Berkeley,Calif. 1960 Aug 3-6 University of California Edited by L. Talbot Under the sponsorship of International union of theoretical and applied mechanics Advances in applied mechanics.Supplement, 1 748p Academic press,New York,1961 Co-sponsored by National aeronautics and space administration
Cav 7.2254

INTERNATIONAL SYMPOSIUM ON RAREFIED GAS DYNAMICS 3rd Proceedings Paris 1962 Jun Vol 1-2 Edited by J.A. Laurmann Advances in applied mechanics.Supplement, 2 2 vols Academic press,New York;London,1963
Eng 41.6774

INTERNATIONAL SYMPOSIUM ON RAREFIED GAS DYNAMICS 4th Proceedings Toronto 1964 Jul Vol 1-2 United States.Air force.Office of scientific research United States.Office of naval research National aeronautics and space administration.Institute for aerospace studies Edited by J.H.de Leeuw Advances in applied mechanics.Supplement, 3 2 vols Academic press,New York;London,1965
Eng 41.6788

INTERNATIONAL SYMPOSIUM ON RAREFIED GAS DYNAMICS 4th proceedings Toronto 1964 Jul. 14-17 Vol 1-2 University of Toronto. Institute for aerospace studies Edited by J. H.de Leeuw Advances in applied mechanics. Supplement, 3 2 vols Academic press,New York,London,1965-66
A Math 4.0532

INTERNATIONAL SYMPOSIUM ON RAREFIED GAS DYNAMICS 5th Proceedings Oxford 1966 Supplement 4 By C.L. Brundin New York; London,1967
A Math 4.1302

INTERNATIONAL SYMPOSIUM ON RAREFIED GAS DYNAMICS 6th Proceedings Cambridge,Mass. 1968 Jul Vol 1-2 United States.Air force.Office of scientific research National aeronautics and space administration United States.Office of naval research Edited by Leon Trilling and Harold Y. Wachman Advances in applied mechanics.Supplement, 5 2 vols Academic press,New York,1969
Eng 41.6797

INTERNATIONAL SYMPOSIUM ON RAREFIED GAS DYNAMICS Proceedings 2nd Berkeley,Calif. 1966 University of California Edited by L. Talbot Advances in applied mechanics. Supplement, 1 Academic press,New York; London,1961
Eng 41.6764

An INTERNATIONAL SYMPOSIUM ON REGULATORY MECHANISMS IN NUCLEIC ACID AND PROTEIN BIOSYNTHESIS Proceedings Regulation of nucleic acid and protein biosynthesis Lunteren 1966 Jun 5-10 Edited by V.V. Konigsberger and L. Bosch B.B.A.library, 10 Elsevier,Amsterdam,1967
Bot 42.1511

An INTERNATIONAL SYMPOSIUM ON REGULATORY MECHANISMS IN NUCLEIC ACID AND PROTEIN BIOSYNTHESIS Proceedings Regulation of nucleic acid and protein biosynthesis Lunteren 1966 Jun 5-10 Edited by V.V. Koningsberger and L. Bosch B.B.A. library, 10 Elsevier,Amsterdam,1967
Bioch 33.0826

INTERNATIONAL SYMPOSIUM ON REPAIR FROM GENETIC RADIATION DAMAGE Proceedings Repair from genetic radiation damage and differential radiosensitivity in germ cells Leiden 1962 Aug 15-19 Riksuniversiteit te Leiden Edited by F.H. Sobels Pergamon press,Oxford, 1963
Gen 34.1121

INTERNATIONAL SYMPOSIUM ON RESEARCH IN CO-CURRENT GAS-LIQUID FLOW Preprints Waterloo, Ontario 1968 Sep 18-19 Vol 1-2 National research council of Canada and Canadian society for chemical engineering Held at the University of Waterloo 2 vols University of Waterloo,Waterloo,Ontario,1968 Mimeographed
Chem E 24.1520

INTERNATIONAL SYMPOSIUM ON RESEARCH IN COCURRENT GAS-LIQUID FLOW Proceedings Cocurrent gas-liquid flow Waterloo,Ont. 1968 Sep 18-19 Edited by Edward Rhodes and Donald S. Scott Canadian society for chemical engineering.Symposium series, 1 illus ix, 698p Plenum press,New York,1969
Chem E 24.1842

INTERNATIONAL SYMPOSIUM ON RESIDUAL GASES IN ELECTRON TUBES AND RELATED VACUUM SYSTEMS 3rd Papers Rome 1967 Mar 14-17 Section 2: sorption-desorption phenomena in high vacuum Italian society of physics Nuovo cimento.Supplement, 5,2 Editrice composori, Bologna,1967
Met 25.2366

INTERNATIONAL SYMPOSIUM ON ROCKET AND SATELLITE METEOROLOGY Washington,D.C. 1962 Apr 23-25 Edited by H. Wexler and J.E. Caskey,jnr Sponsored by Committee on space research ix, 440p North-Holland,Amsterdam,1963
Sco 14.0516

The INTERNATIONAL SYMPOSIUM ON SCIENTIFIC ASPECTS OF SNOW AND ICE AVALANCHES Rapports et discussions Davos 1965 Apr 5-10 International association of scientific hydrology International association of scientific hydrology.Publication,69 illus 417p Gentbrugge,1966 In English and French
Sco 14.0142

INTERNATIONAL SYMPOSIUM ON SECOND-ORDER EFFECTS IN ELASTICITY,PLASTICITY AND FLUID DYNAMICS SYMPOSIUM Second-order effects in elasticity,plasticity and fluid dynamics Haifa 1962 Apr 23-27 International union of theoretical and applied mechanics Edited by Markus Reiner and David Abir Jerusalem academic press;Pergamon press,Jerusalem;London, 1964
Eng 41.2483

INTERNATIONAL SYMPOSIUM ON SENSITIVITY ANALYSIS Sensitivity methods in control theory Dubrovnik 1964 Aug 31-Sep 5 Yugoslav committee for electronics and automation Edited by L. Radanovic Sponsored by the International federation of automatic control. Theory committee Pergamon press,Oxford,1965
Eng 41.5835

INTERNATIONAL SYMPOSIUM ON SOME ASPECTS OF STRONTIUM METABOLISM Proceedings Strontium metabolism Chapelcross 1966 May 5-7 Edited by J.M.A. Lenihan and others Academic press,London;New York,1967
Radioth 35.1719

INTERNATIONAL SYMPOSIUM ON SPECIAL TOPICS IN CERAMICS Proceedings Kinetics of reactions in ionic systems Alfred,N.Y. 1967 Jun 18-23 Edited by T.J. Gray and V.D. Frechette Held at Alfred university Materials science research, 4 Plenum press, New York,1969
Met 25.2750

INTERNATIONAL SYMPOSIUM ON STEREOENCEPHALOTOMY 1st (stereotaxic surgery) Proceedings Philadephia,Pa. 1961 Oct 11-12 Edited by E.A. Spiegel and H.T. Wycis Karger,Basle, 1962
An 32.4200

The INTERNATIONAL SYMPOSIUM ON STRATOSPHERIC AND MESOSPHERIC CIRCULATION Proceedings Stratospheric and mesospheric circulation Berlin 1962 Aug 20-31 Institut fur meteorologie und geophysik Edited by Richard Scherhag and Gunter Warnecke Institut fur meteorologie und geophysik.Meteorologische abhandlungen,36 illus vi,644p 31cm Dietrich Reimer,Berlin,1963
Sco 14.0527

INTERNATIONAL SYMPOSIUM ON STRESS WAVE PROPAGATION IN MATERIALS Pennsylvania state college 1959 Jun 30-Jul 2 United States.Army office of ordinance research Edited by Norman Davids Conducted by the Pennsylvania State university.Department of engineering mechanics New York,1960
Geod 9.0267

INTERNATIONAL SYMPOSIUM ON STRESS WAVE PROPAGATION IN MATERIALS University Park, Pa. 1959 Jun 30-Jul 12 Pennsylvania state university.Department of engineering mechanics Edited by Norman Davids Sponsored by the United States.Army.Office of ordnance research Interscience,New York,1960
Eng 41.3770

INTERNATIONAL SYMPOSIUM ON STRESS WAVE PROPAGATION IN MATERIALS proceedings University Park,Pa 1959 Jun 30-Jul 2 University of Pennsylvania.Department of engineering mechanics Edited by Norman Davids Sponsored by the United States.Army. Office of ordnance research Interscience publishers,New York,1960
A Math 4.0370

INTERNATIONAL SYMPOSIUM ON THE BIOLOGY OF ACETABULARIA 1st Proceedings Brussels 1969 Jun 18-20 and Mol 1969 Jun 18-20 Universite libre de Bruxelles Centre d'etude de l'energie nucleaire Edited by Jean Brachet and Silvano Bonotto Academic press,New York,1970
Bioch 33.2261

INTERNATIONAL SYMPOSIUM ON THE BIOLOGY OF AGGRESSIVE BEHAVIOUR Aggressive behaviour Milan 1968 May 2-4 Edited by S. Garattini and E.B. Sigg Excerpta medica foundation, Amsterdam,1969
Inv Med 37.0086

An INTERNATIONAL SYMPOSIUM ON THE CARDIOVASCULAR AND NEURAL ACTIONS OF BRADYKININ AND RELATED KININS Bradykinin and related kinins; cardiovascular,biochemical,and neural actions Castel di Poggio 1969 Jul 21-25 Edited by F. Sicuteri and others Advances in experimental medicine and biology, 8 Plenum,New York;London,1970
Inv Med 37.0196

INTERNATIONAL SYMPOSIUM ON THE CELL NUCLEUS: METABOLISM AND RADIOSENSITIVITY Proceedings Rijswijk 1966 May 9-12 Taylor and Francis,London,1966
Radioth 35.1751

INTERNATIONAL SYMPOSIUM ON THE CHEMISTRY OF CEMENT 2nd Proceedings Stockholm 1938 July 6-8 Under the auspices of the ingeniorsvetenskapsakademien and Svenska cementforeningen Ingeniorsvetenskapsakademien,Stockholm,1939
Min 10.0598

INTERNATIONAL SYMPOSIUM ON THE CHEMISTRY OF CEMENT 3rd Proceedings London 1952 Under the auspices of the Great Britain. Building research station...and ...Cement and concrete association Cement and concrete association,London,1954
Min 10.0615

INTERNATIONAL SYMPOSIUM ON THE CHEMISTRY OF CEMENT 4th Proceedings Chemistry of cement Washington,D.C. 1960 Oct 2-7 Vol 1-2 National bureau of standards National bureau of standards.Monographs,43 2 vols Washington,D.C.,1962
Min 10.0647

The INTERNATIONAL SYMPOSIUM ON THE CHEMISTRY OF NATURAL PRODUCTS 5th Plenary lectures Chemistry of natural products London 1968 Jul 8-13 International union of pure and applied chemistry.Division of organic chemistry and Chemical society Butterworths,London,1968 "The contents of this book appear in 'Pure and applied chemistry,vol.17, nos.3-4 (196819".Added title page in French
Bioch 33.1392

INTERNATIONAL SYMPOSIUM ON THE CHEMOTHERAPY OF CANCER Proceedings Chemotherapy of cancer Lugano 1964 Apr 28-May 1 Edited by Placidus Plattner Organised by the Swiss academy of medical sciences Elsevier, Amsterdam,1964 Sponsored by F.Hoffman-Laroche and co.ltd.
Radioth 35.0837

An INTERNATIONAL SYMPOSIUM ON THE COMPOSITION, PROPERTIES AND FUNDAMENTAL STRUCTURE OF TOOTH ENAMEL held at the London hospital medical college Report of the proceedings Tooth enamel;its composition,properties and fundamental structure London 1964 Apr 6-7 Edited by Maurice V. Stack and Ronald W. Fearnhead Wright,Bristol,1965 Sponsored by the Wellcome trust,London;the Procter and Gamble Co., Cincinnati;Colgate-Palmolive Ltd., Great Britain;Unilever Ltd.,London
Bal 39.1298

INTERNATIONAL SYMPOSIUM ON THE CONTROL OF CELL DIVISION AND THE INDUCTION OF CANCER Proceedings Control of cell division and the induction of cancer Lima 1963 Jul 1-6 and Cali,Columbia 1963 Jul 1-6 National cancer institute National cancer institute. Monograph, 14 illus National cancer institute,Bethesda,Md.,1964 Introduction, abstracts of each paper and summary in English and Spanish
Bioch 33.0985

INTERNATIONAL SYMPOSIUM ON THE ELECTROPHYSIOLOGY OF THE HEART Proceedings Milan 1963 Edited by B. Taccardi and G. Marchetti Macmillan;Pergamon,New York;Oxford,1965
Pha 16.0286

INTERNATIONAL SYMPOSIUM ON THE LEUKEMIAS : etiology,pathophysiology and treatment Proceedings Detroit 1956 Mar 8-10 Edited by John W. Rebuck and others Sponsored by Henry Ford hospital Academic press,New York,1957
Med 36.0299

INTERNATIONAL SYMPOSIUM ON THE MYOCARDIAL CELL 3rd :structure,function and modification by cardiac drugs Heart association of southeastern Pennsylvania Edited by Stanley A. Briller and Hardley L. Conn University of Pennsylvania press,Philadelphia,Pa.,1966
Phys 20.2221

INTERNATIONAL SYMPOSIUM ON THE NUCLEUS;ITS STRUCTURE AND FUNCTION Montevideo 1965 Dec 5-10 Edited by W.S. Vincent and O.L. Miller National concer institute.Monographs, 23 U.S.government printing office, Washington,D.C.,1966
An 32.3439

The INTERNATIONAL SYMPOSIUM ON THE ORIGIN OF LIFE ON THE EARTH 1st Proceedings Origin of life on the earth Moscow 1957 Aug 19-24 Akademiya nauk S.S.S.R. Edited by A.I. Oparin and others Organized under the auspices of the International union of biochemistry I.U.B.symposium series, 1 Pergamon,London,1959 "English-French-German edition edited for the International union of biochemistry by F.Clark and R.L.M. Synge".
Bioch 33.0727

INTERNATIONAL SYMPOSIUM ON THE PHYSIOLOGY OF DIGESTION IN THE RUMINANT 2ND Papers Physiology of digestion in the ruminant Ames, Iowa 1964 Aug Edited by R.W. Dougherty Butterworths,London,1965
Phys 20.1337

INTERNATIONAL SYMPOSIUM ON THE REACTIVITY OF SOLIDS 4th Amsterdam 1960 May 30th-Jun 4th Nederlandse chemische vereniging and Nederlandse naturkundige vereniging Edited by J.H.de Boer and others Sponsored by the International union of pure and applied physics Elsevier,Amsterdam,1961 Papers in English,French and German
Met 25.1848

INTERNATIONAL SYMPOSIUM ON THE REACTIVITY OF SOLIDS 4th Proceedings Reactivity of solids Amsterdam 1960 May 30-Jun 4 Nederlandse chemische vereniging and International union of pure and applied physics Edited by J.H. DeBoer and others Sponsored also by the International union of pure and applied chemistry illus x,762p
Chem E 24.0882

INTERNATIONAL SYMPOSIUM ON THE REACTIVITY OF SOLIDS 5th Munich 1964 Aug 2-8 Deutsche Bunsen-gesellschaft fur physikalische chemie Edited by G.M. Schwab Sponsored by the International union of pure and applied chemistry Elsevier,Amsterdam,1965 Papers in English,French and German
Met 25.1851

INTERNATIONAL SYMPOSIUM ON THE REACTIVITY OF SOLIDS Proceedings 6th Schenectady,N. Y. 1968 Aug 25-30 United States.Air force. Office of scientific research General electric research and development center International union of pure and applied chemistry. Physical chemistry center Edited by J.W. Mitchell and others Wiley,New York; London,1969
Met 25.2446

INTERNATIONAL SYMPOSIUM ON THE REACTIVITY OF SOLIDS 2nd Proceedings Gothenburg 1952 Jun 9-13 Under the auspices of Ingeniorsvetenskapsakademien and Chalmers tekniska hogskola 2 vols Gothenburg,1954
Min 10.0614

INTERNATIONAL SYMPOSIUM ON THE REACTIVITY OF SOLIDS 4th Proceedings Amsterdam 1960 May 30-Jun 4 Edited by J.H.de Boer and others Elsevier,Amsterdam,1961
Min 10.0648

INTERNATIONAL SYMPOSIUM ON THE REACTIVITY OF SOLIDS 5th Plenary lectures Munich 1964 Aug 2-8 International union of pure and applied chemistry,and,Deutsche bunsengesellschaft fur physikalische chemie Butterworths,London,1965
Min 10.0662

INTERNATIONAL SYMPOSIUM ON THE REACTIVITY OF SOLIDS 5th papers Munich 1964 Aug 2-8 Edited by G.-M. Schwab Elsevier, Amsterdam,1965
Min 10.1278

INTERNATIONAL SYMPOSIUM ON THE THEORY OF MODELS proceedings Theory of models Berkeley, Calif. 1963 Jun 25-Jul 11 Edited by J.W. Addison and others xv,494p 23cm North-Holland,Amsterdam,1965
P. Math 2.0205

INTERNATIONAL SYMPOSIUM ON THE THEORY OF ROAD TRAFFIC FLOW London 1963 Jun 25-27 London,1963 Unbound mimeograph papers
Math 3.1113

INTERNATIONAL SYMPOSIUM ON THE THEORY OF ROAD TRAFFIC FLOW London 1963 Jun 25-27 Road research laboratory Edited by Joyce Almond Organization for economic cooperation and development,Paris,1965
Eng 41.3311

The INTERNATIONAL SYMPOSIUM ON THE THEORY OF SWITCHING proceedings Cambridge,Mass. 1957 Apr 2-5 Part 1-2 Harvard university.Computation laboratory Harvard university.Computation laboratory.Annals, 29 27cm 2 vols Harvard university press, Cambridge,Mass.,1959
Math L 5.0686

INTERNATIONAL SYMPOSIUM ON THE THEORY OF TRAFFIC FLOW 3rd Proceedings Vehicular traffic science New York 1965 Jun Edited by Leslie C. Edie and others Under the auspices of the Operations research society of America.Transportation science section illus x,373p 24cm American Elsevier,New York, 1967
Math 3.1137

INTERNATIONAL SYMPOSIUM ON THE USE OF ARTIFICIAL SATELLITES FOR GEODESY 1st Proceedings Washington,D.C. 1962 Apr 26-28 Committee on space research and International association of geodesy Edited by George Veis Co-sponsored by the International union of geodesy and geophysics North-Holland,Amsterdam,1963
Geod 9.0055

INTERNATIONAL SYMPOSIUM ON TISSUE TRANSPLANTATION Proceedings Santiago 1961 Aug 30-Sep 2 Edited by Alberto P. Cristoffanini and Gustavo Hoecker Sponsored by the Universidad de Chile Universidad de Chile,Santiago,Chile, 1962
An 32.3286

The INTERNATIONAL SYMPOSIUM ON TISSUE TRANSPLANTATION Proceedings Santiago 1961 Aug 30-Sep 2 Edited by Alberto P. Cristoffanini and Gustavo Hoecker Sponsored by Universidad de Chile Universidad de Chile, Santiago,1962
PGMS 29.0244

INTERNATIONAL SYMPOSIUM ON TOPOLOGY AND ITS APPLICATIONS Proceedings Herceg-Novi 1968 Aug 28-31 Edited by D.R. Kurepa 356p 25cm Savez matematicara,Belgrade,1969
P Math 2.4117

INTERNATIONAL SYMPOSIUM ON TUMOR VIRUSES 1st
Subviral carcinogenesis 1966 Edited by Yohei Ito Sponsored by the Aichi cancer center.Research institute and the Japanese cancer association Aichi cancer center, Nagoya,1967
Bioch 33.1191

INTERNATIONAL SYMPOSIUM ON UPTAKE OF INFORMATIVE MOLECULES BY LIVING CELLS Proceedings
Informative molecules in biological systems Mol,Belgium 1970 Aug 31-Sep 2 North Atlantic treaty organization.Scientific affairs division Edited by L. Ledoux North-Holland,Amsterdam,1971
Bioch 33.2240

INTERNATIONAL SYMPOSIUM ON WORLD CLIMATE 8000 TO NOUGHT B.C. Proceedings World climate from 8000 to nought B.C. London 1966 Apr 18-19 Royal meteorological society London, 1966
Sco 14.0581

INTERNATIONAL SYMPOSIUM ON X-RAY MICROSCOPY AND X-RAY MICROANALYSIS 2nd Proceedings
Stockholm 1960 Feb Edited by Arne Engstrom and others illus x,542p Elsevier, Amsterdam,1960
Cav 7.2234

INTERNATIONAL SYMPOSIUM ON X-RAY OPTICS AND X-RAY MICROANALYSIS 3rd X-ray optics and x-ray microanalysis Stanford,Calif. 1962 Aug 22-24 Edited by H.H. Pattee Sponsored by Stanford university.Biophysics laboratory Academic press,New York,1963
Eng 41.4462

INTERNATIONAL SYMPOSIUM ON X-RAY OPTICS AND X-RAY MICROANALYSIS 3rd Proceedings
Stanford,Calif. 1962 Aug 22-24 Edited by H. H. Pattee,jr. and others Sponsored by Stanford university.Biophysics laboratory illus xvii,622p Academic press,New York, 1963
Cav 7.2189

INTERNATIONAL SYMPOSIUM ON X-RAY MICROTECHNIQUES 3rd X-ray optics and X-ray microanalysis Stanford,Calif. 1962 Aug 22-24 Stanford university.Biophysics laboratory Edited by H.H. Pattee and others With the financial support of the United States.Office of naval research and the National science foundation Academic press,New York;London, 1963
Met 25.1379

INTERNATIONAL SYMPOSIUM UBER NEUROSEKRETION 2nd
Proceedings Lund 1957 Jul 1-6
Edited by W. Bargmann and others Springer, Berlin,1958 Trilingual,English,French and German
An 32.3745

INTERNATIONAL TECHNICAL CONFERENCE ON THE PROTECTION OF NATURE Preparatory documents
Lake Success 1949 Aug 22-29
International union for the protection of nature.Secretariat Unesco,Paris,1949
Text in English and French
Bal 39.3922

INTERNATIONAL TELETRAFFIC CONGRESS 4th
London 1964 Jul 15-21 London,1964
Unbound typescript papers
Math 3.1126

INTERNATIONAL THEORETICAL PHYSICS CONFERENCE ON PARTICLES AND FIELDS Proceedings
Rochester,N.Y. 1967 Aug 28-Sep 1 Edited by C.R. Hagen and others 708p 26cm Interscience,New York,1967 Dedicated to Robert Oppenheimer
Cav 7.2773

INTERNATIONAL TRUST FOR ZOOLOGICAL NOMENCLATURE
Zoological nomenclature Copenhagen decisions on zoological nomenclature additions to...les regles internationales de la nomenclature zoologique Copenhagen 1953 Aug Edited by Francis Hemming International trust for zoological nomenclature,London,1953
BG 38.3107

INTERNATIONAL TUBERCULOSIS CONFERENCE 7th
Report Philadelphia,Pa. 1908 Sep 24-26
International anti-tuberculosis association
Edited by Gotthold Pannwitz port
Internationale vereinigung gegen die tuberkulose,Berlin-Charlottenburg,1909
Text in English,French,German;title also in French and German
Path 30.1023

INTERNATIONAL TUBERCULOSIS CONFERENCE 8th
Report Stockholm 1909 Jul 8-10
International anti-tuberculosis association
Edited by Gotthold Pannwitz Berlin-Charlottenburg,1910 Text in English, French,German;title also in French and German
Path 30.1024

INTERNATIONAL TUBERCULOSIS CONFERENCE 9th
Report Brussels 1910 Oct 6-8
International anti-tuberculosis association
Edited by Gotthold Pannwitz port Berlin-Charlottenburg,1911 Text in English, French,German;title also in French and German
Path 30.1026

INTERNATIONAL TUINBOUW-CONGRES 7th
Amsterdam 1923 Sep 17-23 Nederlandsche mantschappij voor tuinbouw en plantkunde c1923
Gen 34.1139

INTERNATIONAL UNION AGAINST CANCER Cancer control :Latin American regional conference
Papers Santiago 1967 Nov 25-28 Sponsored by the Committee on national cancer control programmes of the commission on cqncer control International union against cancer.Technical report series, 1 International union against cancer,Geneva,1968
Radioth 35.0888

INTERNATIONAL UNION AGAINST CANCER Cellular control mechanisms and cancer :a conference Proceedings Amsterdam 1963 Sep 9-13
Edited by P. Emmelot and O. Muhlbock
Elsevier,Amsterdam,1964 Conference held on the occasion of the 50th anniversary of the Netherlands cancer institute
An 32.2504

INTERNATIONAL UNION AGAINST CANCER
Choriocarcinoma :a conference Transactions Baguio 1965 Jan Edited by James F. Holland and Myroslaw M. Hreshchyshya International union against cancer.Monograph series, 3 Springer,Berlin,1967
Radioth 35.0886

INTERNATIONAL UNION AGAINST CANCER
Conference on cellular control mechanisms and cancer Proceedings Amsterdam 1963 Sep 9-13 Edited by P. Emmelot and O. Muhlbock Elsevier,Amsterdam,1964 Organized to commemorate the 50th anniversary of the Netherlands cancer institute
Path 30.2459

INTERNATIONAL UNION AGAINST CANCER European conference on tumour biology Warsaw 1961 May 22-27 International union against cancer. Acta, 18,no 1-2 U.I.C;C.,Louvain,1962
Radioth 35.0817

INTERNATIONAL UNION AGAINST CANCER
International cancer congress 7th London 1958 Jul 6-12 Abstract of papers 1958
Radioth 35.0777

INTERNATIONAL UNION AGAINST CANCER
International pigment cell conference 6th Sofia 1965 May 25-29 Edited by G. Della Porta and O. Muhlbock Springer-verlag,Berlin, 1966
Bioch 33.0967

INTERNATIONAL UNION AGAINST CANCER Specific tumour antigens :a symposium Sukhumi 1965 May Edited by R.J.C. Harris International union against cancer.Monograph series, 2 Munksgaard,Copenhagen,1967
Radioth 35.0901

INTERNATIONAL UNION AGAINST CANCER Structure and control of the melanocyte International pigment cell conference 6th Sofia 1965 May 25-29 Edited by G. Della Porta and O. Muhlbock Springer,Berlin,1966
An 32.3425

INTERNATIONAL UNION AGAINST CANCER The International conference on the biology of cutaneous cancer 1st Philadelphia,Pa. 1962 Apr 6-11 Edited by Frederick Urbach With financial support from Merck,Sharp and Dohme.Postgraduate program National cancer institute.Monograph, 10 US government printing office,Washington,D.C.,1963
Path 30.2423

INTERNATIONAL UNION AGAINST CANCER.CANCER RESEARCH COMMISION Cellular control mechanisms and cancer :conference Proceedings Amsterdam 1963 Sep 9-13 Edited by P. Emmelot and O. Muhlbock Illus Elsevier,Amsterdam,1964
Radioth 35.0530

INTERNATIONAL UNION AGAINST TUBERCULOSIS
International conference of the International union against tuberculosis 2nd Transactions London 1921 Jul 26-28 Under the auspices of the British national association for the prevention of tuberculosis Adlard and Newman,London,1921
An 32.2990

INTERNATIONAL UNION FOR LOGIC,METHODOLOGY AND PHILOSOPHY OF SCIENCE.DIVISION OF LOGIC, METHODOLOGY AND PHILOSOPHY OF SCIENCE
Logic,methodology and philosophy of science International congress for logic,methodology and philosophy of science Proceedings Amsterdam 1967 Aug 25-Sep 2 Edited by B. van Rootselaar and J.F. Staal xiii,554p North-Holland,Amsterdam,1968
WSM 43.0830

INTERNATIONAL UNION FOR QUATERNARY RESEARCH
Chetvertichnye otlozheniya Shpitsbergena k VIII kongressu INQUA Parizh 1969 By Yu.A. Lavrushin Kommissiya po izycheniyu chetvertichnogo perioda illus 181p 22cm Izdatelstvo 'Nauka',Moscow,1969
Sco 14.8250

INTERNATIONAL UNION FOR QUATERNARY RESEARCH
Golotsen k VIII kongressu INQUA Parizh 1969 By M.I. Neishtadt Institut geografii,Moscow illus,maps 228p 27cm Izdatelstvo 'Nauka', Moscow,1969
Sco 14.8260

INTERNATIONAL UNION FOR QUATERNARY RESEARCH
Poslednii lednikovyi pokrov na severo-zapade evropaiskoi chasti S.S.S.R. k VIII kongressu INQUA,Parizh 1969 Institut geografii,Moscow Edited by I.P. Gerasimov maps 322p 27cm Izdatelstvo 'Nauka',Moscow,1969
Sco 14.8258

INTERNATIONAL UNION FOR QUATERNARY RESEARCH
Problemy kriolitologii k VIII mezhdunarodnomu kongressu assotsiatsii po izucheniyu chetvertichnogo perioda (INQUA),Parizh 1969 Vyp 1 Moskovskii gosudarstvennii universitet im. M.V.Lomonosova.Geografischekii fakultet.Kafedra kriolitologii i glatsiologii Edited by A.I. Popov Bibliog,illus 176p 23cm Izdatelstvo Moskovskogo universiteta, Moscow,1969
Sco 14.8242

INTERNATIONAL UNION FOR QUATERNARY RESEARCH and NATIONAL CENTER FOR ATMOSPHERIC RESEARCH
International union for quaternary research congress :a collection of papers derived from the INQUA-NCAR symposium 7th Proceedings Boulder,Colo. 1965 Aug 14-Sep 19 5: causes of climatic change Edited by J.Murray Mitchell Sponsored by the National research council Meteorological monographs,8,no 30 illus iv,159p 29cm American meteorological association,Boston,Mass.,1968 For other volumes of these proceedings,see also 'International association for quaternary research'
Sco 14.7750

INTERNATIONAL UNION FOR QUATERNARY RESEARCH CONGRESS 7th :a collection of papers derived from the INQUA-NCAR symposium Proceedings Boulder,Colo. 1965 Aug 14-Sep 19 5: causes of climatic change International union for quaternary research, and,National center for atmospheric research Edited by J.Murray Mitchell Sponsored by the National research council Meteorological monographs,8,no 30 illus iv,159p 29cm American meteorological association,Boston, Mass.,1968 For other volumes of these proceedings,see also 'International association for quaternary research'
Sco 14.7750

INTERNATIONAL UNION FOR THE CONSERVATION OF NATURE AND NATURAL RESOURCES 6th : technical meeting Proceedings and papers Edinburgh 1956 Jun Society for the promotion of nature reserves,London,1956 Held simultaneously with the Unions 5th general assembly
Geog 13.2698

INTERNATIONAL UNION FOR THE CONSERVATION OF NATURE AND NATURAL RESOURCES. TECHNICAL MEETING 8th Ecological effects of biological and chemical control of undesirable plants and animal symposium Warsaw 1960 Jul 15-24 Edited by D.J. Kuenen Leiden, 1961 Title also in French:Effets ecologiques du controle biologique et chimique des plantes et animaux undesirables
Bal 44.6435

INTERNATIONAL UNION FOR THE PROTECTION OF NATURE. SECRETARIAT International technical conference on the protection of nature Preparatory documents Lake Success 1949 Aug 22-29 Unesco,Paris,1949 Text in English and French
Bal 39.3922

INTERNATIONAL UNION FOR THE SCIENTIFIC INVESTIGATION OF POPULATION PROBLEMS.GENERAL ASSEMBLY 2nd :report Proceedings Problems of population London 1931 Jun 15-18 Edited by G.H.L.F. Pitt-Rivers Allen and Unwin,London,1932
Gen 34.2068

INTERNATIONAL UNION FOR VACUUM SCIENCE.GERMAN NATIONAL COMMITTEE International vacuum congress 3rd Transactions Stuttgart 1965 Jun 28-Jul 2 1: invited papers Edited by H. Adam Pergamon,Oxford,1966 Papers mainly in English,with summaries in French and German.Title pages in English, French and German
Met 25.1691

INTERNATIONAL UNION FOR VACUUM SCIENCE,TECHNIQUE AND APPLICATIONS International vacuum congress 3rd Transactions Stuttgart 1965 Jun 28-Jul 2 Vol 1: invited papers Edited by H. Adam 153p Pergamon press, Oxford,1966 Cover title'Advances in vacuum science and technology'
Cav 7.0015

INTERNATIONAL UNION FOR VACUUM SCIENCE,TECHNIQUE AND APPLICATIONS International vacuum congress 4th Manchester 1968 Apr 17-20 Pt 1-2 Institute of physics and the Physical society.Conference series, 5-6 2 vols Institute of physics,London,c1968
Met 25.2834

INTERNATIONAL UNION OF BIOCHEMISTRY Biological structure and function I.U.B.-I.U.B.S. international symposium 1st Proceedings Stockholm 1960 Sep 12-17 Vol 1 Edited by T.W. Goodwin and O. Lindberg Academic press,London,1961
An 32.3276

INTERNATIONAL UNION OF BIOCHEMISTRY Biological structure and function I.U.B.-I.U.B.S. international symposium 1st Proceedings Stockholm 1960 Sep 12-17 Vol 1-2 Edited by T.W. Goodwin and O. Lindberg illus 2 vols Academic press, London;New York,1961
Bal 44.6450

INTERNATIONAL UNION OF BIOCHEMISTRY Biological structure and function IUB-IUBS international symposium 1st Proceedings Stockholm 1960 Sep 12-17 Vol 1-2 Edited by T.W. Goodwin and O. Lindberg 2 vols Academic press,London,1961
Gen 34.0826

INTERNATIONAL UNION OF BIOCHEMISTRY Biological structure and function The I.U.B.-I.U.B.S. international symposium 1st Proceedings Stockholm 1960 Sep 12-17 Vol 1-2 Edited by T.W. Goodwin and O. Lindberg illus 2 vols Academic press, London;New York,1961
Bioch 33.0584

INTERNATIONAL UNION OF BIOCHEMISTRY Biological structure and function : international symposium 1st Proceedings Stockholm 1960 Sep 12-17 Vol 1-2 Edited by T.W. Goodwin and D. Lindberg Academic press,London;New York,1961
Phys 20.1539

INTERNATIONAL UNION OF BIOCHEMISTRY Biological struggle and function The I.U.B.-I.U.B.S.international symposium 1st Proceedings Stockholm 1960 Sep 12-17 Vol 1-2 Edited by T.W. Goodwin and O. Lindberg illus. 2 vols Academic press, London;New York,1961
Bot 42.1344

INTERNATIONAL UNION OF BIOCHEMISTRY Flavins and flavoproteins :a symposium Proceedings Amsterdam 1965 Jun 10-15 Edited by E.C. Slater organized under the auspices of the Nederlandsche akademie van wetenschappen B.B.A. library, 8 illus Elsevier,Amsterdam, 1966
Bioch 33.1099

INTERNATIONAL UNION OF BIOCHEMISTRY Haematin enzymes :a symposium Papers and discussions Canberra 1959 Aug 31-Sep 4 Pt 1 Edited by J.E. Falk and others Organized by the Australian academy of science I.U.B. Symposium series, 19 Pergamon press,Oxford, 1961
Bioch 33.1079

INTERNATIONAL UNION OF BIOCHEMISTRY International congress of biochemistry 4th Vienna 1958 Sep 1-6 Vol 15: biochemistry:abstracts of sectional papers and index to symposia and colloquia I.U.B. symposium series, 17 Pergamon,Oxford,1960 Added title page in French and German.Text in English,French and German
Bioch 33.1329

INTERNATIONAL UNION OF BIOCHEMISTRY International congress of biochemistry 4th Vol 1: symposium 1- carbohydrate chemistry of substances of biological interest Edited by M.L. Wolfrom I.U.B.symposium series, 3 Pergamon,London,1959 Added title page in French and German.Text in English,French and German
Bioch 33.1315

INTERNATIONAL UNION OF BIOCHEMISTRY International congress of biochemistry 4th Proceedings Vienna 1958 Sep 1-6 Vol 2: symposium 2 - biochemistry of wood Edited by K. Kratzi and G. Billek I.U.B.symposium series, 4 Pergamon,London,1959 Added title page in French and German.Text in English,French and German
Bioch 33.1316

INTERNATIONAL UNION OF BIOCHEMISTRY
International congress of biochemistry 4th Proceedings Vienna 1958 Sep 1-6 Vol 3: symposium 3:biochemistry of the central nervous system Edited by F. Brucke I.U.B. symposium series, 5 Pergamon,London,1959 Added t.p.in French and German
Bot 42.1715

INTERNATIONAL UNION OF BIOCHEMISTRY
International congress of biochemistry 4th Proceedings Vienna 1958 Sep 1-6 Vol 3: symposium 3 - biochemistry of the central nervous system Edited by Brucke I.U.B. symposium series, 5 Pergamon,London,1959 Added title page in French and German.Text in English,French and German
Bioch 33.1317

INTERNATIONAL UNION OF BIOCHEMISTRY
International congress of biochemistry 4th Proceedings Vienna 1958 Sep 1-6 Vol 4: symposium 4 biochemistry of steroids Edited by E. Mosettig I.U.B.symposium series, 6 Pergamon,London,1959 Added title page in French and German.Text in English,French and German.
Bioch 33.1318

INTERNATIONAL UNION OF BIOCHEMISTRY
International congress of biochemistry 4th Proceedings Vienna 1958 Sep 1-6 Vol 5: symposium 5 - biochemistry of antibiotics Edited by K.H. Spitzy and R. Brunner I.U.B. symposium series, 7 Pergamon,London,1959 Added title page in French and German.Text in English,French and German.
Bioch 33.1319

INTERNATIONAL UNION OF BIOCHEMISTRY
International congress of biochemistry 4th Proceedings Vienna 1958 Sep 1-6 Vol 6: symposium 6 - biochemistry of morphogenesis Edited by W.J. Nickerson I.U.B.S.symposium series, 8 Pergamon press,London,1959
Gen 34.0482

INTERNATIONAL UNION OF BIOCHEMISTRY
International congress of biochemistry 4th Proceedings Vienna 1958 Sep 1-6 Vol 6: symposium 6 - biochemistry of morphogenesis Edited by W.J. Nickerson I.U.B.symposium series, 8 Pergamon,London,1959 Added title page in French and German
Bioch 33.1320

INTERNATIONAL UNION OF BIOCHEMISTRY
International congress of biochemistry 4th Proceedings Vienna 1958 Sep 1-6 Vol 7: symposium 7 - biochemistry of viruses Edited by E. Broda and W. Frisch-Niggemeyer I.U.B. symposium series, 9 Pergamon,London,1959 Added title page in French and German.Text in German and English
Bioch 33.1321

INTERNATIONAL UNION OF BIOCHEMISTRY
International congress of biochemistry 4th Proceedings Vienna 1958 Sep 1-6 Vol 7: symposium 7 - biochemistry of viruses Edited by E. Broda and W. Frisch-Niggemeyer Pergamon press,London,1960
Radioth 35.0071

INTERNATIONAL UNION OF BIOCHEMISTRY
International congress of biochemistry 4th Proceedings Vienna 1958 Sep 1-6 Vol 8: symposium 8 - proteins Edited by H. Neurath and H. Tuppy I.U.B.symposium series, 10 Pergamon,London,1960 Added title page in French and German.Text in English and French
Bioch 33.1322

INTERNATIONAL UNION OF BIOCHEMISTRY
International congress of biochemistry 4th Proceedings Vienna 1958 Sep 1-6 Vol 8: symposium 8 - proteins Edited by H. Neurath and H. Tuppy Pergamon,London,1960
Radioth 35.0072

INTERNATIONAL UNION OF BIOCHEMISTRY
International congress of biochemistry 4th Proceedings Vienna 1958 Sep 1-6 Vol 9: symposium 9 - physical chemistry of high polymers of biological interest Edited by O. Kratky I.U.B.symposium series, 11 Pergamon,London,1959 Added title page in French and German.Text in English,French and German.
Bioch 33.1323

INTERNATIONAL UNION OF BIOCHEMISTRY
International congress of biochemistry 4th Proceedings Vienna 1958 Sep 1-6 Vol 9: symposium 9 - physical chemistry of high polymers of biological interest Edited by O. Kratky Pergamon press,London,1960
Radioth 35.1941

INTERNATIONAL UNION OF BIOCHEMISTRY
International congress of biochemistry 4th Proceedings Vienna 1958 Sep 1-6 Vol 11: symposium 11 - vitamin metabolism Edited by W. Umbreit and H. Molitor I.U.B.symposium series, 13 Pergamon,London,1960 Added title page in French and German
Bioch 33.1325

INTERNATIONAL UNION OF BIOCHEMISTRY
International congress of biochemistry 4th Proceedings Vienna 1958 Sep 1-6 Vol 12: colloquia Edited by H. Chantrenne and others I.U.B.symposium series, 15 Pergamon,London, 1959 Added title page in French and German.Text in English,French and German
Bioch 33.1327

INTERNATIONAL UNION OF BIOCHEMISTRY
International congress of biochemistry 4th Proceedings Vienna 1958 Sep 1-6 Vol 12: symposium 12 - biochemistry of insects Edited by L. Levenbook I.U.B.symposium series, 14 Pergamon,London,1959 Added title page in French and German.Text in English and German
Bioch 33.1326

INTERNATIONAL UNION OF BIOCHEMISTRY
International congress of biochemistry 4th Proceedings Vienna 1958 Sep 1-6 Vol 14: transactions of the plenary sessions Edited by W. Auerswald and O. Hoffmann-Ostenhof I.U. B.symposium series, 14 Pergamon,London,1959 Added title page in French and German.Text in English,French and German.
Bioch 33.1328

INTERNATIONAL UNION OF BIOCHEMISTRY
International congress of biochemistry 5th Proceedings Moscow 1961 Aug 10-16 Vol 1: biological structure and function at the molecular level Edited by V.A. Engelhardt I.U.B.symposium series, 21 Pergamon;PWN-Polish scientific publishers,London;Warsaw, 1963
Bioch 33.1330

INTERNATIONAL UNION OF BIOCHEMISTRY
International congress of biochemistry 5th Proceedings Moscow 1961 Aug 10-16 Vol 2: functional biochemistry of cell structures Edited by O. Lindberg I.U.B.symposium series, 22 Pergamon;PWN-Polish scientific publishers,New York;Warsaw,1963
Bioch 33.1331

INTERNATIONAL UNION OF BIOCHEMISTRY
International congress of biochemistry 5th Proceedings Moscow 1961 Aug 10-16 Vol 3: evolutionary biochemistry I.U.B.symposium series, 23 Pergamon;PWN-Polish scientific publishers,New York;Warsaw,1963
Bioch 33.1332

INTERNATIONAL UNION OF BIOCHEMISTRY
International congress of biochemistry 5th Proceedings Moscow 1961 Aug 10-16 Vol 4: molecular basis of enzyme action and inhibition Edited by P.A.E. Desnuelle I.U.B.symposium series, 24 Pergamon;PWN-Polish scientific publishers,Oxford;Warsaw,1963
Bioch 33.1333

INTERNATIONAL UNION OF BIOCHEMISTRY
International congress of biochemistry 5th Proceedings Moscow 1961 Aug 10-16 Vol 5: intracellular respiration:phosphorylating and non-phosphorylating oxidation reactions Edited by E.C. Slater I.U.B.symposium series, 25 Pergamon;PWN-Polish scientific publishers,Oxford;Warsaw,1963
Bioch 33.1334

INTERNATIONAL UNION OF BIOCHEMISTRY
International congress of biochemistry 5th Proceedings Moscow 1961 Aug 10-16 Vol 6: mechanism of photosynthesis Edited by H. Tamiya I.U.B.symposium series, 26 Pergamon;PWN-Polish scientific publishers, Oxford;Warsaw,1963
Bioch 33.1335

INTERNATIONAL UNION OF BIOCHEMISTRY
International congress of biochemistry 5th Proceedings Moscow 1961 Aug 10-16 Vol 7: biosynthesis of lipids Edited by G. Popjak I.U.B.symposium series, 27 Pergamon;PWN-Polish scientific publishers, Oxford;Warsaw,1963
Bioch 33.1336

INTERNATIONAL UNION OF BIOCHEMISTRY
International congress of biochemistry 5th Proceedings Moscow 1961 Aug 10-16 Vol 8: biochemical principles of the food industry Edited by V.L. Kretovich and E. Pijanowski I.U.B.symposium series, 28 Pergamon;PWN-Polish scientific publishers,Oxford;Warsaw, 1963
Bioch 33.1337

INTERNATIONAL UNION OF BIOCHEMISTRY
International congress of biochemistry 5th Proceedings Moscow 1961 Aug 10-16 Vol 9: plenary sessions and abstracts of papers I.U.B.symposium series, 29 Pergamon;PWN-Polish scientific publishers,New York;Warsaw, 1963
Bioch 33.1338

INTERNATIONAL UNION OF BIOCHEMISTRY
International congress of biochemistry 6th Abstracts New York 1964 Jul 26-Aug 1 Vol 2: proteins,peptides,and amino acids Washington,D.C.,1964 Title also in French.
Bioch 33.1339

INTERNATIONAL UNION OF BIOCHEMISTRY
International congress of biochemistry 6th Abstracts New York 1964 Jul 26-Aug 1 Vol 3: biochemical genetics Washington,D.C., 1964
Bioch 33.1340

INTERNATIONAL UNION OF BIOCHEMISTRY
International congress of biochemistry 6th Abstracts New York 1964 Jul 26-Aug 1 Vol 5: special topics in biochemistry Washington,1964
Bioch 33.1341

INTERNATIONAL UNION OF BIOCHEMISTRY
International congress of biochemistry 6th Abstracts New York 1964 Jul 26-Aug 1 Vol 8: cellular organization Washington,1964
Bioch 33.1344

INTERNATIONAL UNION OF BIOCHEMISTRY
International congress of biochemistry 6th Abstracts New York 1964 Jul 26-Aug 1 Vol 10: bioenergetics Washington,1964
Bioch 33.1346

INTERNATIONAL UNION OF BIOCHEMISTRY
International congress of biochemistry 6th Abstracts New York 1964 Jul 26-Aug 1 Vol 11: author index to abstracts Washington, 1964
Bioch 33.1347

INTERNATIONAL UNION OF BIOCHEMISTRY
International congress of biochemistry 6th Abstracts New York 1964 Jul 26-Aug 1 Washington,D.C.,1964
Bioch 33.1342

INTERNATIONAL UNION OF BIOCHEMISTRY
International congress of biochemistry 6th Abstracts New York 1964 Jul 26-Aug 1 Washington,1964
Bioch 33.1343

INTERNATIONAL UNION OF BIOCHEMISTRY
International congress of biochemistry 6th Abstracts New York 1964 Jul 26-Aug 1 Washington,1964
Bioch 33.1345

INTERNATIONAL UNION OF BIOCHEMISTRY
International congress of biochemistry :held at Interlaken,Lucerne and Montreux 8th Abstracts Switzerland 1970 Sep 3-9 Edited by J.G. Gregory Staples press, Rochester,1970
Bioch 33.2338

INTERNATIONAL UNION OF BIOCHEMISTRY
International symposium on energy transduction in respiration and photosynthesis Pugnochiuso 1970 Sep 11-14 Edited by E. Quagliariello and others International union of biochemistry.Symposia, 43 Adriatic editrice,Bari,1971
Bioch 33.2289

INTERNATIONAL UNION OF BIOCHEMISTRY
International symposium on flaveins and flavoproteins 3rd Proceedings Durham N. C. 1970 International union of biochemistry.Symposium, 39 University park press;Butterworths,Baltimore,Md.;London,1971
Bioch 33.2373

INTERNATIONAL UNION OF BIOCHEMISTRY Origin of life on the earth The International symposium on the origin of life on the earth 1st Proceedings Moscow 1957 Aug 19-24 Akademiya nauk S.S.S.R. Edited by A.I. Oparin and others I.U.B.symposium series, 1 Pergamon,London,1959 "English-French-German edition edited for the International union of biochemistry by F.Clark and R.L.M. Synge".
Bioch 33.0727

INTERNATIONAL UNION OF BIOCHEMISTRY Oxygen in the animal organism :a symposium Proceedings London 1963 Sep 1-5 Edited by Frank Dickens and Eric Neil Pergamon press,Oxford,1964
PGMS 29.0407

INTERNATIONAL UNION OF BIOCHEMISTRY Oxygen in the animal organism :a symposium Proceedings London 1963 Sep 1-5 Edited by Frank Dickens and Eric Neil held at Bedford college I.U.B.Symposium series, 31 Pergamon press,Oxford,1964
Bioch 33.1080

INTERNATIONAL UNION OF BIOCHEMISTRY Problems of atmospheric circulation The Space science symposium 6th Papers of a session Mar del Plata,Argentina 1965 May 11-19 Edited by R.V. Garcia and T.F. Malone Co-sponsored by the International union of physiological sciences 186p Spartan books,Washington,D.C. 1966
Nap 11.0212

INTERNATIONAL UNION OF BIOCHEMISTRY
Pyridoxal catalysis:enzymes and model systems The International symposium on chemical and biological aspects of pyridoxical catalysis 2nd Proceedings Moscow 1966 Sep 15-21 Edited by E.E. Snell and others I.U.B. Symposium series, 35 Interscience,New York, 1968 First International symposium on chemical and biological aspects of pyridoxal catalysis published under title chemical and biological aspects of pyridoxical catalysis
Bioch 33.1038

INTERNATIONAL UNION OF BIOCHEMISTRY Rapid mixing and sampling techniques in biochemistry International colloquium on rapid mixing and sampling techniques applicable to the study of biochemical reactions 1st Proceedings Philadelphia 1964 Jul 23-24 Edited by Britton Chance and others Academic press,New York;London,1964
Bioch 33.2320

INTERNATIONAL UNION OF BIOCHEMISTRY Rapid mixing and sampling techniques in biochemistry nternational coloquium 1st Proceedings Philadelphia,Pa. 1964 Jul 23-24 Edited by Britton Chance and others Academic press,New York,1964
Pha 16.0141

INTERNATIONAL UNION OF BIOCHEMISTRY The International congress of biochemistry 4th Proceedings Vienna 1958 Sep 1-6 13: colloquia Edited by O. Hoffmann-Osterhoff I.U.B.Symposium series, 15 Pergamon press, London,1959 Added title page in French and German
Bot 42.1696

INTERNATIONAL UNION OF BIOCHEMISTRY The International symposium on enzyme chemistry Proceedings Tokyo 1957 Oct 15-23 and Kyoto 1957 Oct 15-23 Organized by Science council of Japan I.U.B.Symposium series, 2 Pergamon,London,1958
Bioch 33.1053

INTERNATIONAL UNION OF BIOCHEMISTRY.SYMPOSIA, 43 International symposium on energy transduction in respiration and photosynthesis Pugnochiuso 1970 Sep 11-14 International union of biochemistry Edited by E. Quagliariello and others Adriatic editrice, Bari,1971
Bioch 33.2289

INTERNATIONAL UNION OF BIOCHEMISTRY.SYMPOSIUM, 39 International symposium on flaveins and flavoproteins 3rd Proceedings Durham N.C. 1970 International union of biochemistry University park press; Butterworths,Baltimore,Md.;London,1971
Bioch 33.2373

INTERNATIONAL UNION OF BIOCHEMISTRY.SYMPOSIUM SERIES, 30 Chemical and biological aspects of pyridoxal catalysis :a symposium of the International union of biochemistry Proceedings Rome 1960 Oct Edited by E.E. Snell Pergamon,Oxford,1963
Radioth 35.0131

INTERNATIONAL UNION OF BIOLOGICAL SCIENCES
Biochemistry of human genetics symposium Ciba foundation Edited by G.E.W. Wolstenholme and Cecilia M. O'Connor Ciba foundation.Symposia Churchill,London,1959
Gen 34.0602

INTERNATIONAL UNION OF BIOLOGICAL SCIENCES
Biological control Scientific basis of an international biological control organization Les Bases scientifiques d'une organisation internationale pour la lutte biologique Stockholm 1948 Aug 5-7 International union of biological sciences.Serie B.Colloques, 5 Secretairiat general de I.U.I.S.B.,Paris,1949 Title also in English: The scientific bases of an international biological control organization.Text in English and French
Bal 39.3920

INTERNATIONAL UNION OF BIOLOGICAL SCIENCES
Biological structure and function I.U.B.-I.U.B.S. international symposium 1st Proceedings Stockholm 1960 Sep 12-17 Vol 1 Edited by T.W. Goodwin and O. Lindberg Academic press,London,1961
An 32.3276

INTERNATIONAL UNION OF BIOLOGICAL SCIENCES
Biological structure and function I.U.B.-I.U.B.S. international symposium 1st Proceedings Stockholm 1960 Sep 12-17 Vol 1-2 Edited by T.W. Goodwin and O. Lindberg illus 2 vols Academic press, London;New York,1961
Bal 44.6450

INTERNATIONAL UNION OF BIOLOGICAL SCIENCES
Biological structure and function IUB-IUBS international symposium 1st Proceedings Stockholm 1960 Sep 12-17 Vol 1-2 Edited by T.W. Goodwin and O. Lindberg 2 vols Academic press,London,1961
Gen 34.0826

INTERNATIONAL UNION OF BIOLOGICAL SCIENCES
Biological structure and function The I.U.B.-I.U.B.S. international symposium 1st Proceedings Stockholm 1960 Sep 12-17 Vol 1-2 Edited by T.W. Goodwin and O. Lindberg illus 2 vols Academic press, London;New York,1961
Bioch 33.0584

INTERNATIONAL UNION OF BIOLOGICAL SCIENCES
Biological structure and function : international symposium 1st Proceedings Stockholm 1960 Sep 12-17 Vol 1-2 Edited by T.W. Goodwin and D. Lindberg Academic press,London;New York,1961
Phys 20.1539

INTERNATIONAL UNION OF BIOLOGICAL SCIENCES
Biological struggle and function The I.U.B.-I.U.B.S.international symposium 1st Proceedings Stockholm 1960 Sep 12-17 Vol 1-2 Edited by T.W. Goodwin and O. Lindberg illus. 2 vols Academic press, London;New York,1961
Bot 42.1344

INTERNATIONAL UNION OF BIOLOGICAL SCIENCES
Biometrical genetics :international symposium Proceedings Ottawa 1958 Aug Edited by Oscar Kempthorne Sponsored by the Biometrics society International series of monographs on biometry International union of biological sciences.Series B, 38 Pergamon press,London,1960
Gen 34.0299

INTERNATIONAL UNION OF BIOLOGICAL SCIENCES
Ciba foundation symposium on biochemistry of human genetics Proceedings Naples 1959 May 13-16 Edited by G.E.W. Wolstenholme and Cecilia M. O'Connor illus Churchill,London, 1959
Bioch 33.0890

INTERNATIONAL UNION OF BIOLOGICAL SCIENCES
Fine structure of cells :a symposium held at the VIIIth congress of cell biology Papers Leiden 1954 With financial support of Unesco International union of biological sciences.Publications.Series B, 21 Noordhoff,Groningen,1955
Path 30.2410

INTERNATIONAL UNION OF BIOLOGICAL SCIENCES
Fine structure of cells :a symposium held at the 8th congress of cell biology Leiden 1954 With financial support of Unesco International union of biological sciences. Series B., 21 Noordhoff,Groningen,1955
Radioth 35.0467

INTERNATIONAL UNION OF BIOLOGICAL SCIENCES
Fine structure of cells;symposium The Congress of cell biology 8th Papers Leiden 1954 With financial support of Unesco International union of biological sciences.Ser.B, 21 Noordhoff,Groningen,1955
Bal 39.0283

INTERNATIONAL UNION OF BIOLOGICAL SCIENCES
Genetic neurology International conference on the development,growth and regeneration of the nervous system Chicago,Ill. 1949 Mar 21-25 Edited by Paul Weiss University of Chicago,Ill.,Chicago,Ill.,1950
An 32.4329

INTERNATIONAL UNION OF BIOLOGICAL SCIENCES
Genetic neurology :problems of the development, growth,and regeneration of the nervous system and of its functions:an international conference Essays Chicago,Ill. 1949 Mar 21-25 Edited by Paul Weiss Subsidized by Unesco University of Chicago press,Chicago, Ill.,1950
Psy 31.0170

INTERNATIONAL UNION OF BIOLOGICAL SCIENCES
International botanical congress 7th Proposals,Abstracts Stockholm 1950 International union of biological sciences, Utrecht,1950
BG 38.3128

INTERNATIONAL UNION OF BIOLOGICAL SCIENCES
International congress of genetics 12th Proceedings Tokyo 1968 Aug 19-28 Vol 1-3 Science council of Japan 3 vols Science council of Japan,Tokyo,1969
Gen 34.1008

INTERNATIONAL UNION OF BIOLOGICAL SCIENCES
Photo-thermoperiodism,the action of different radiations on gibberellins and other active substances Colloque international sur le photothermoperiodisme action des diverses radiations on gibberellines et de quelques autres substances Parma 1957 Jun International union of biological sciences. Series B, 34 U.I.S.B.,Paris,1958
Bot 42.1420

INTERNATIONAL UNION OF BIOLOGICAL SCIENCES
Progress in photosynthesis research The International congress of photosynthesis research Proceedings Freudenstadt 1968 Jun 4-8 Vol 1-3 Edited by Helmut Metzner 3 vols Tubingen,1969
Bot 42.1905

INTERNATIONAL UNION OF BIOLOGICAL SCIENCES
Progress in photosynthesis research The International congress of photosynthesis research Proceedings Freudenstadt 1968 Jun 4-8 Vol 1-3 Edited by Helmut Metzner 3 vols International union of biological sciences,Tubingen,1969
Bioch 33.1907

INTERNATIONAL UNION OF BIOLOGICAL SCIENCES
Role des anaerobies dans la nature Congres international des microbiologistes de langue francaise Brussels 1949 May 23-27 International union of biological sciences. Series B.Colloques, 7 Secreteriat general de l'U.I.S.B.,Paris,1949
Bal 39.3919

INTERNATIONAL UNION OF BIOLOGICAL SCIENCES
SCAR,SCOR,IAPO,IUBS symposium on Antarctic oceanography Papers Santiago de Chile 1966 Sep 13-16 Held by invitation of the Chilean national committee for antarctic research illus,maps xi,268p 23cm Scott Polar research institute for SCAR,Cambridge, 1968
Sco 14.8300

INTERNATIONAL UNION OF BIOLOGICAL SCIENCES
Simposio internacional de genetica symposium Sao Paulo 1966 Jul Edited by F.G. Brieger Ciencia e cultura, 19,no.1 Sao Paulo,1967
Gen 34.0995

INTERNATIONAL UNION OF BIOLOGICAL SCIENCES
The Genetics of colonizing species :symposium Proceedings Asilomar,Calif. 1964 Feb 12-16 Edited by H.G. Baker and G.Ledyard Stebbins International union of biological sciences. Symposia, 1 Academic press,New York;London, 1965
Bot 42.0869

INTERNATIONAL UNION OF BIOLOGICAL SCIENCES
The Genetics of colonizing species symposium Proceedings Edited by H.G. Baker and G. Ledyard Stebbins International union of biological sciences.Symposium, 1 Academic press,New York;London,1965
Gen 34.1297

INTERNATIONAL UNION OF BIOLOGICAL SCIENCES
Towards a theoretical biology 1: prolegomena.An I.U.B.S.symposium Edited by C. H. Waddington viii,234p 26cm Edinburgh university press,Edinburgh,1968 Essays written after,and in the light of,the 1st IUBS symposium held at Bellagio,Aug.28-Sep.3,1966
Math 3.1361

INTERNATIONAL UNION OF BIOLOGICAL SCIENCES
Towards a theoretical biology 2: sketches. An I.U.B.S. symposium Edited by C.H. Waddington illus.,port. v,351p 26cm Edinburgh university press,Edinburgh,1968 Essays written after,and in the light of the 1st I.U.B.S.symposium,held at Bellagio,Aug 28-Sep 3,1966
Math S 3.1469

INTERNATIONAL UNION OF BIOLOGICAL SCIENCES
Towards a theoretical biology 4: essays:an IUBS symposium Edited by C.H. Waddington 299p 26cm Edinburgh university press, Edinburgh,1972
Math S 3.1866

INTERNATIONAL UNION OF BIOLOGICAL SCIENCES
Towards a theoretical biology :a collection of essays written as a result of a symposium... Bellagio 1966 Aug 28-Sep 3 Edited by C.H. Waddington 253p 26cm Edinburgh university press,Edinburgh,1970
Math S 3.1819

INTERNATIONAL UNION OF BIOLOGICAL SCIENCES
Towards a theoretical biology :a symposium Papers Lake Como 1967 Aug 3-12 Vol 2: sketches Edited by C.H. Waddington I.U.B.S. symposium, 2 Edinburgh university press, Edinburgh,1969
Bal 39.0053

INTERNATIONAL UNION OF BIOLOGICAL SCIENCES
Towards a theoretical biology :an IUBS symposium Lake Como 1967 Aug 3-12 2: sketches Edited by C.H. Waddington Edinburgh university press,Edinburgh,1968
Bal 39.0017

INTERNATIONAL UNION OF BIOLOGICAL SCIENCES
Towards a theoretical biology 1.Prolegomena:an I.U.B.S. symposium Essays Bellagio 1966 Aug 28-Sep 3 Edited by C.H. Waddington 234p Edinburgh university press,Edinburgh, 1968 Essays written after and in the light of the symposium
WSM 43.2897

INTERNATIONAL UNION OF BIOLOGICAL SCIENCES 10th general assembly Stockholm 1950 Jul Under the auspices of Unesco International union of biological sciences.Series A, 10 International union of biological sciences, Paris,1951
BG 38.3108

INTERNATIONAL UNION OF BIOLOGICAL SCIENCES 15th general assembly Prague 1964 Under the auspices of Unesco International union of biological sciences.Series A, 15 International union of biological sciences, Utrecht,1964
BG 38.3109

INTERNATIONAL UNION OF BIOLOGICAL SCIENCES.SERIE B.COLLOQUES, 5 Biological control Scientific basis of an international biological control organization Les Bases scientifiques d'une organisation internationale pour la lutte biologique Stockholm 1948 Aug 5-7 International union of biological sciences Secretairiat general de I.U.I.S.B.,Paris,1949 Title also in English: The scientific bases of an international biological control organization. Text in English and French
Bal 39.3920

INTERNATIONAL UNION OF BIOLOGICAL SCIENCES.SERIES B.COLLOQUES, 7 Role des anaerobies dans la nature Congres international des microbiologistes de langue francaise Brussels 1949 May 23-27 International union of biological sciences Secreteriat general de l'U.I.S.B.,Paris,1949
Bal 39.3919

INTERNATIONAL UNION OF BIOLOGICAL SCIENCES. SYMPOSIA, 1 The Genetics of colonizing species :symposium Proceedings Asilomar, Calif. 1964 Feb 12-16 International union of biological sciences Edited by H.G. Baker and G.Ledyard Stebbins Academic press,New York;London,1965
Bot 42.0869

INTERNATIONAL UNION OF BIOLOGICAL SCIENCES. SYMPOSIUM, 1 The Genetics of colonizing species symposium Proceedings International union of biological sciences Edited by H.G. Baker and G.Ledyard Stebbins Academic press,New York;London,1965
Gen 34.1297

INTERNATIONAL UNION OF CRYSTALLOGRAPHERS. COMMISSION ON CRYSTALLOGRAPHIC COMPUTING
Crystallographic computing :international summer school Proceedings Ottawa 1969 Aug 4-11 Edited by F.R. Ahmed and others Munksgaard,Copenhagen,1970
Chem 18.2733

INTERNATIONAL UNION OF CRYSTALLOGRAPHY
International conference on 'Electron diffraction' and 'The nature of defects in crystals' Abstracts of papers Melbourne 1965 Aug 16-21 With the support of the International union of pure and applied physics. Commission on the solid state Pergamon press,Oxford,1966 Extended abstracts of papers to be presented at the conference
Cav 7.1584

INTERNATIONAL UNION OF CRYSTALLOGRAPHY
International conference on"electron diffraction" and "the nature of defects in crystals" Abstracts of papers Melbourne 1965 Aug 16-21 Co-sponsored by the International union of pure and applied physics Pergamon press,Oxford,1966
Met 25.1463

INTERNATIONAL UNION OF CRYSTALLOGRAPHY.COMMISSION ON CRYSTALLOGRAPHIC COMPUTATION
International summer school on crystallographic computation Proceedings Ottawa 1969 Aug 4-11 Edited by F.R. Ahmed Munksgaard,Copenhagen,1970
Min 10.1527

INTERNATIONAL UNION OF CRYSTALLOGRAPHY.COMMISSION ON CRYSTALLOGRAPHIC COMPUTING
Crystallographic computing :a conference Proceedings Ottawa 1969 Aug 4-11 Edited by F.R. Ahmed and others Bibliog 384p 20cm Munksgaard,Copenhagen,1970
Cav 7.2655

INTERNATIONAL UNION OF GEODESY AND GEOPHYSICS
A Symposium on atmospheric diffusion and air pollution proceedings Oxford 1958 Aug 24-29 Edited by F.N. Frenkel and P.A. Sheppard Advances in geophysics, 6 Academic press, New York;London,1959
A Math 4.0576

INTERNATIONAL UNION OF GEODESY AND GEOPHYSICS
Abstracts of the reports submitted to the XI general assembly of the International union of geodesy and geophysics International association of scientific hydrology 101p Moscow,1957 In English,French and Russian; includes also t.-p.in Russian
Sco 14.0126

INTERNATIONAL UNION OF GEODESY AND GEOPHYSICS
Antarctic symposium Papers Buenos Aires 1959 Nov 17-25 International union of geodesy and geophysics.Monograph,5 95p Paris,1960
Sco 14.0128

INTERNATIONAL UNION OF GEODESY AND GEOPHYSICS
Atmospheric diffusion and air pollution :a symposium Proceedings Oxford 1958 Aug 24-29 Edited by F.N. Frenkel and P.A. Sheppard Advances in physics, 6 Academic press,New York,1959
Eng 41.6946

INTERNATIONAL UNION OF GEODESY AND GEOPHYSICS
Atmospheric diffusion and air pollution :a symposium Proceedings Oxford 1958 Aug 24-29 Edited by F.N. Frenkiel and P.A. Sheppard New York,1959
Geod 9.0521

INTERNATIONAL UNION OF GEODESY AND GEOPHYSICS
Continental drift,secular motion of the Pole and rotation of the earth I.A.U. symposium 32nd Papers Stresa 1967 Mar 21-25 Edited by William Markowitz and B. Guinot illus 107p 25cm D.Reidel,Dordrecht,1968
Sco 14.8118

INTERNATIONAL UNION OF GEODESY AND GEOPHYSICS
Continental drift,secular motion of the pole, and rotation of the earth Stresa 1967 Mar 21-25 Edited by Wm. Markowitz and B. Guinot International astronomical union.Symposium, 32 Reidel,Dordrecht,1968
TA 15.0424

INTERNATIONAL UNION OF GEODESY AND GEOPHYSICS
Continental drift,secular motion of the pole, and rotation of the earth :a symposium Proceedings Stresa 1967 Mar Edited by W. Markowitz and B. Guinot International astronomical union.Symposium, 32 108p Reidel,Dordrecht,1968
Obs 6.3268

INTERNATIONAL UNION OF GEODESY AND GEOPHYSICS
Fundamental problems in turbulence and their relation to geophysics symposium proceedings Marseilles 1961 Sep.4-9 Vol 67 Edited by Francois N. Frenkel Journal of geophysical research,Vol 67,no.8 American geophysical union,Washington D.C.,1962
A Math 4.0577

INTERNATIONAL UNION OF GEODESY AND GEOPHYSICS
Gravity anomalies:unsurveyed areas Extension of gravity anomalies to unsurveyed areas :a symposium Papers Columbus,Ohio 1964 Nov 18-20 Ohio state university and American geophysical union Edited by Hyman Orlin Geophysical monograph, 9 National research council.Publication,1357 American geophysical union,Washington,D.C.,1966
Geod 9.0031

INTERNATIONAL UNION OF GEODESY AND GEOPHYSICS
International latitude service L' Avenir du service international des latitudes :a symposium proceedings Helsinki 1960 Jul. 26-Aug.7 Edited by Paul Melchior International astronomical union.Symposium, 13 97p Paris,1961 Reprinted from Bulletin geodesique no.59,March 1961
Obs 6.1198

INTERNATIONAL UNION OF GEODESY AND GEOPHYSICS
International symposium on the use of artificial satellites for geodesy 1st Proceedings Washington,D.C. 1962 Apr 26-28 Committee on space research and International association of geodesy Edited by George Veis North-Holland,Amsterdam,1963
Geod 9.0055

INTERNATIONAL UNION OF GEODESY AND GEOPHYSICS
Stresa symposium on continental drift,secular motion of the pole and rotation of the earth Stresa,Italy 1967 Mar 21-25 Edited by W. Markowitz and B. Guinot International astronomical union.Symposium, 32 Reidel, Dordrecht,1968
A Math 4.1308

INTERNATIONAL UNION OF GEODESY AND GEOPHYSICS
Symposium on radiant energy in the sea Papers Helsinki 1960 Aug 4-5 Edited by N. G. Jerlov International union of geodesy and geophysics.Monograph,10 illus 115p Paris, 1961 Symposium held during the I.U.G.A. assembly at Helsinki
Sco 14.0135

INTERNATIONAL UNION OF GEODESY AND GEOPHYSICS
Symposium on the July 1959 events and associated phenomena Helsinki 1960 Jul 28 International union of geodesy and geophysics. Monograph,7 tables,diagrs 157p 24cm I. U.G.G.,Paris,1960
Sco 14.0246

INTERNATIONAL UNION OF GEODESY AND GEOPHYSICS
Symposium on tropical meteorology Proceedings Rotorua,N.Z. 1963 Nov 5-13 World meteorological organization Edited by J.W. Hutchings illus 751p New Zealand meteorological service,Wellington,N.Z.,1964
Nap 11.0650

INTERNATIONAL UNION OF GEODESY AND GEOPHYSICS
The Development of geodesy and geophysics in Switzerland :commemorative book presented to the participants in the XIVth general assembly of the International union of geodesy and geophysics Swiss academy of natural sciences Edited by J.C. Thams illus,maps 98p Zurich,1967 In English and French, including also French t.-p.
Sco 14.0145

INTERNATIONAL UNION OF GEODESY AND GEOPHYSICS
The East African rift system :a seminar Papers Nairobi 1965 Apr 12-17 International upper mantle committee Nairobi, 1965
Geod 9.0451

INTERNATIONAL UNION OF GEODESY AND GEOPHYSICS
The International conference on cosmic rays and the earth storm :combined meeting of the International symposium on the earth storm and the International conference on cosmic rays Proceedings Kyoto 1961 Sep 4-15 Vol 2-3 Edited by Ken-ichi Maeda and Osamu Minakawa Co-sponsored by the International scientific radio union Physical society of Japan.Journal, 17,suppl.A-2,3 2 vols physical society of Japan,Tokyo,1962
TA 15.0038

INTERNATIONAL UNION OF GEODESY AND GEOPHYSICS
The Symposium on numerical weather prediction Proceedings Tokyo 1960 Nov 7-13 Meteorological agency,Japan Science council of Japan 655p Meteorological society of Japan,Tokyo,1962
Nap 11.0923

INTERNATIONAL UNION OF GEODESY AND GEOPHYSICS
The World rift systems :a symposium Report Ottawa 1965 Sep 4-5 National research council,Canada and International upper mantle committee Edited by T.N. Irvine Geological survey of Canada.Paper, 66-14 Queen's printer,Ottawa,1967
Geod 9.0138

INTERNATIONAL UNION OF GEODESY AND GEOPHYSICS
Theoretical interpretation of upper atmosphere emissions :a symposium Paris 1962 Jun.25-29 Edited by D.R. Bates International astronomical union.Symposium, 18 264p Pergamon press,New York,1963 Reprinted from'Planetary and space science',Vol.10 1963
Obs 6.0142

INTERNATIONAL UNION OF GEODESY AND GEOPHYSICS
Union geodsique et geophysique internationale : assemblee generale 11th Comptes rendus Toronto 1957 Sep 3-14 284p Paris,1958 Papers in English and French
Sco 14.7407

INTERNATIONAL UNION OF GEODESY AND GEOPHYSICS and INTERNATIONAL ASSOCIATION OF GEOMAGNETISM AND AERONOMY The Toronto meeting Transactions Toronto 1957 Sep 3-14 Edited by T.R. Kaiser and V. Laursen I. A.G.A.bulletin,16 x,412p 25cm North Holland,Copenhagen,1960
Sco 14.0244

INTERNATIONAL UNION OF GEODESY AND GEOPHYSICS and WORLD METEOROLOGICAL ORGANIZATION
Symposium on tropical meteorology Proceedings Rotorua,N.Z. 1963 Nov 5-13 Edited by J.W. Hutchings New Zealand meteorological service,Wellington,N.Z.,1964
Geog 13.0990

INTERNATIONAL UNION OF GEODESY AND GEOPHYSICS
13th :general assembly comtes-rendus Berkeley,Calif. 1963 Aug 19-31 Edited by G. R. Laclavere and G.P. Garland 301p 24cm Paris,1965.
Sco 14.0140

INTERNATIONAL UNION OF GEODESY AND GEOPHYSICS
14th :general assembly Abstracts of papers Zurich 1967 Sep 25-Oct 7 Vol 6: International association of scientific hydrology 1967
Sco 14.7409

INTERNATIONAL UNION OF GEODESY AND GEOPHYSICS 6me assemblee generale Edinburgh 1936 Sep 14-26 184p 25cm Camelot press,London, 1937 In English and French
Sco 14.0114

INTERNATIONAL UNION OF GEODESY AND GEOPHYSICS.
GENERAL ASSEMBLY 14TH Soobshcheniye o nauchnykh rabotakh po girdrologii 1963-1966 Bibliography Akademiya nauk S.S.S.R. Sovetskii geografischekii komitet Presented to the International association of scientific hydrology 157p 26cm Moscow,1967 Introduction and index in English
Sco 14.7518

INTERNATIONAL UNION OF GEODESY AND GEOPHYSICS.
INTERNATIONAL ASSOCIATION OF SCIENTIFIC HYDROLOGY see INTERNATIONAL ASSOCIATION OF SCIENTIFIC HYDROLOGY

INTERNATIONAL UNION OF GEODESY AND GEOPHYSICS,
GENERAL ASSEMBLY, 10TH Scientific proceedings of the International association of meteorology Proceedings Rome 1954 Sep 15 International association of meteorology AIM publication, 10,c 594p Butterworths, London,1956 Meeting at the 10th general assembly of International union of geodesy and geophysics
Nap 11.0286

INTERNATIONAL UNION OF GEODESY AND GEOPHYSICS, GENERAL ASSEMBLY, 13TH,1963 The General assembly of the International association of meteorology Proceedings Berkeley,Calif 1963 International association of meteorology and atmospheric physics Edited by W.L. Godson 199p Toronto,1963 Part of the 13th general assembly of International union of geodesy and geophysics
Nap 11.0223

INTERNATIONAL UNION OF GEODESY AND GEOPHYSICS, GENERAL ASSEMBLY,10TH Transactions of the Rome meeting International association of terrestrial magnetism and electricity a meeting Rome 1954 Sep 14-25 Edited by V. Laursen IAGA Bulletin, 15 406p Horsholm, Copenhagen,1957 The meeting took place during the 10th general assembly of the international union of geodesy and geophysics
Nap 11.0287

INTERNATIONAL UNION OF GEOLOGICAL SCIENCES Origin and distribution of the elements Symposium on the origin and distribution of the elements Paris 1967 Edited by L.H. Ahrens International series of monographs on earth sciences, 30 xvii,1178p 24cm Pergamon press,Oxford,1968
TA 15.0565

INTERNATIONAL UNION OF GEOLOGICAL SCIENCES. COMMISSION ON STRATIGRAPHY. COMMITTEE ON MEDITERRANEAN NEOGENE STRATIGRAPHY see COMMITTEE ON MEDITERRANEAN NEOGENE STRATIGRAPHY

INTERNATIONAL UNION OF GEOLOGICAL SCIENCES, International upper mantle committee The Upper mantle symposium papers New Delhi 1964 Dec. Edited by Charles H. Smith and Theodor Sorgenfrei Det Berlingske Bogtrykkeri,Copenhagen,1965
Geol 8.2191

INTERNATIONAL UNION OF HISTORY AND PHILOSOPHY OF SCIENCE Colloque international sur l'histoire de la biologie marine Les Grandes expeditions scientifiques et la creation des laboratoires maritimes Banyuls-sur-Mer 1963 Sep 2-6 Vie et milieu.Supplement, 19 370p Laboratoire Arago;Masson,Banyuls-sur-Mer;Paris,1965
WSM 43.2877

INTERNATIONAL UNION OF HISTORY AND PHILOSOPHY OF SCIENCE Congres international d'histoire des sciences 7e Actes Jerusalem 1953 Aug 4-12 Edited by F.S. Bodenheimer Academie internationale d'histoire des sciences.Collection de travaux, 8 xii,662p Academie internationale d'histoire des sciences;Hermann,Paris,c1954
WSM 43.0058

INTERNATIONAL UNION OF HISTORY AND PHILOSOPHY OF SCIENCE La Science au seizieme siecle : colloque international de Royaumont Royaumont 1957 Jul 1-4 Histoire de la pensee, 2 Hermann,Paris,1960
WSM 43.0911

INTERNATIONAL UNION OF HISTORY AND PHILOSOPHY OF SCIENCE New aspects in the history and philosophy of astronomy Joint symposium of the International astronomical union and the Union internationale d'histoire et de philosophie des sciences 1st Hamburg 1964 Aug 22-24 Edited by Arthur Beer Vistas in astronomy, 9 illus,ports,maps 318p 25cm Pergamon press,Oxford,1967
Cav 7.2715

INTERNATIONAL UNION OF HISTORY AND PHILOSOPHY OF SCIENCE New aspects in the history and philosophy of astronomy Joint symposium of the International astronomical union and the Union internationale d'histoire et de philosophie des sciences 1st Hamburg 1964 Aug 22-24 Edited by Arthur Beer Vistas in astronomy, 9 xvi,317p Pergamon, Oxford,1967
WSM 43.2439

INTERNATIONAL UNION OF HISTORY AND PHILOSOPHY OF SCIENCE.DIVISION OF HISTORY OF SCIENCE Scientific change...Historical studies in the intellectual,social and technical conditions for scientific discovery... :symposium on the history of science Oxford 1961 Jul 9-15 Edited by A.C. Crombie xii,896p Heinemann, London,1963
WSM 43.0029

INTERNATIONAL UNION OF HISTORY AND PHILOSOPHY OF SCIENCE.DIVISION OF LOGIC,METHODOLOGY AND PHILOSOPHY OF SCIENCE Induction,physics, and ethics Colloquium in the philosophy of science Proceedings and discussions Salzburg 1968 Aug 28-31 Edited by Paul Weingartner and Gerhard Zecha Synthese library x,382p Reidel,Dordrecht,1970
WSM 43.0992

INTERNATIONAL UNION OF HISTORY AND PHILOSOPHY OF SCIENCE.DIVISION OF LOGIC,METHODOLOGY AND PHILOSOPHY OF SCIENCE International congress for logic,methodology and philosophy of science Proceedings Stanford,Calif. 1960 Aug 24-Sep 2 Edited by Ernest Nagel and others ix,661p 25cm Stanford university press,Stanford,Calif.,1962
Math S 3.1853

INTERNATIONAL UNION OF HISTORY AND PHILOSOPHY OF SCIENCE.DIVISION OF LOGIC,METHODOLOGY AND PHILOSOPHY OF SCIENCE Logic,methodology and philosophy of science International congress for logic,methodology and philosophy of science Proceedings Jerusalem 1964 Aug 26-Sep 2 Edited by Yehoshua Bar-Hillel viii,440p North-Holland,Amsterdam,1965
WSM 43.0829

INTERNATIONAL UNION OF HISTORY AND PHILOSOPHY OF SCIENCE.DIVISION OF LOGIC,METHODOLOGY AND PHILOSOPHY OF SCIENCE Logic,methodology and philosophy of science International congress for logic,methodology and philosophy of science Proceedings Stanford,Calif. 1960 Aug 24-Sep 2 Edited by Ernest Nagel and others ix,661p Stanford university press, Stanford,Calif.,1962
WSM 43.0828

INTERNATIONAL UNION OF PURE AND APPLIED PHYSICS
International conference on physics : reports on symbols,units and nomenclature approved by the general assembly of the Union London 1934 Oct.5 40p Physical society, London,1935
Obs 6.1963

INTERNATIONAL UNION OF PURE AND APPLIED PHYSICS
International conference on the physics of semiconductors 8th Proceedings Kyoto 1966 Sep 8-13 Physical society of Japan Physical society of Japan.Journal, 21,suppl. 774p Physical society of Japan,Tokyo,1966
Cav 7.1341

INTERNATIONAL UNION OF PURE AND APPLIED PHYSICS
International conference on theoretical physics Proceedings Kyoto 1953 Sep 14-24 Tokyo Science council of Japan and Kyoto university Edited by I. Imai 942p Tokyo,1954
Cav 7.1342

INTERNATIONAL UNION OF PURE AND APPLIED PHYSICS
International conference on"electron diffraction" and "the nature of defects in crystals" Abstracts of papers Melbourne 1965 Aug 16-21 Australian academy of science and International union of crystallography Pergamon press,Oxford,1966
Met 25.1463

INTERNATIONAL UNION OF PURE AND APPLIED PHYSICS
International meeting on ferroelectricity 2nd Kyoto 1969 Sep 4-9 Science council of Japan Physical society of Japan. Journal, 28,suppt Physical society of Japan, 1970
Met 25.2865

INTERNATIONAL UNION OF PURE AND APPLIED PHYSICS
International symposium on electron and photon interactions at high energies 4th Proceedings Liverpool 1969 Sep 14-20 illus 30cm Danesbury nuclear physics laboratory,Danesbury,1969
A Math 4.1760

INTERNATIONAL UNION OF PURE AND APPLIED PHYSICS
International symposium on the reactivity of solids 4th Amsterdam 1960 May 30th-Jun 4th Nederlandse chemische vereniging and Nederlandse naturkundige vereniging Edited by J.H.de Boer and others Elsevier,Amsterdam,1961 Papers in English, French and German
Met 25.1848

INTERNATIONAL UNION OF PURE AND APPLIED PHYSICS
Internationales colloquium uber halbleiter und phosphore Garmisch-Partenkirchen 1956 Aug 28-Sep 15 Edited by M. Schon and H. Welker Vieweg,Brunswick,1958
Radioth 35.1615

INTERNATIONAL UNION OF PURE AND APPLIED PHYSICS
Lattice dynamics The International conference on lattice dynamics Proceedings Copenhagen 1963 Aug 5-9 Edited by R.F. Wallis With the support of the International atomic energy agency 730p Pergamon press, Oxford,1965
Cav 7.0542

INTERNATIONAL UNION OF PURE AND APPLIED PHYSICS
Localized excitations in solids International conference on localized excitations in solids 1st Proceedings Irvine 1967 Sep 18-22 Edited by R.F. Wallis xvi,782p 26cm Plenum press,New York,1968
Cav 7.2857

INTERNATIONAL UNION OF PURE AND APPLIED PHYSICS
Low energy nuclear interactions and nuclear structure International conference on nuclear physics Proceedings Paris 1958 Jul 7-12 Faculte des sciences de Paris et de province and Centre national de la recherche scientifique Crosby Lockwood, London,1959
Cav 7.1759

INTERNATIONAL UNION OF PURE AND APPLIED PHYSICS
Magnetic and electric resonance and relaxation Colloque Ampere :international conference 11th Proceedings Eindhoven 1962 Jul 2-7 Netherlands physical society Edited by J. Smidt 789p North-Holland, Amsterdam,1963
Cav 7.0712

INTERNATIONAL UNION OF PURE AND APPLIED PHYSICS
Paramagnetic resonance :international conference 1st Proceedings Jerusalem 1962 Jul 16-20 Vol 1-2 Edited by W. Low Sponsored by the Hebrew university of Jerusalem 2 vols Academic press,New York, 1963
Cav 7.1226

INTERNATIONAL UNION OF PURE AND APPLIED PHYSICS
Physics of electronic and atomic collisions International conference on the physics of electronic and atomic collisions 5th Invited papers Leningrad 1967 Jul 17-23 Edited by Lewis M. Branscomb Bibliog,illus xvi,200p 23cm University of Colorado press, Boulder,Colo.,1968
A Math 4.1906

INTERNATIONAL UNION OF PURE AND APPLIED PHYSICS
Plasma effects in solids International conference on the physics of semiconductors 7th Paris 1964 Vol 2: plasma effects in solids photos,diagrms,graphs 221p 24cm Dunod,Paris,1965 Co-sponsored by the French physical society,the French ministry for scientific research and Unesco
Cav 7.3047

INTERNATIONAL UNION OF PURE AND APPLIED PHYSICS
Relativistic theories of gravitation A Conference on gravitation Proceedings Jablonna 1962 Jul 25-31 Polska akademia nauk Edited by Leopold Infeld Pergamon press,Oxford,1964 Binder's title in French
TA 15.0085

INTERNATIONAL UNION OF PURE AND APPLIED PHYSICS
Rutherford jubilee international conference Proceedings Manchester 1961 Sep 4-8 Victoria university of Manchester Edited by J.B. Birks 856p Haywood,London,1961 Includes commemorative session 'Rutherford at Manchester'
Cav 7.1952

INTERNATIONAL UNION OF PURE AND APPLIED PHYSICS
Rutherford jubilee international conference proceedings Manchester 1961 Sep.4-8 Victoria university of Manchester Edited by J.B. Birks Heywood,London,1961 To commemorate the discoveries of Rutherford at Manchester
A Math 4.0826

INTERNATIONAL UNION OF PURE AND APPLIED PHYSICS
Semi-conducting materials :a conference Proceedings Reading 1950 Jul 10-15 Edited by H.K. Henisch Butterworths,London, 1951
Eng 41.5373

INTERNATIONAL UNION OF PURE AND APPLIED PHYSICS
Semi-conducting materials :a conference Proceedings Reading 1950 Jul 10-15 Royal society of London Edited by H.K. Henisch illus xv,281p Butterworths,London,1951
Cav 7.1925

INTERNATIONAL UNION OF PURE AND APPLIED PHYSICS
Semi-conducting materials :a conference Proceedings Reading 1950 Jul 10-15 With the cooperation of the Royal society of London London,1951
Geod 9.0286

INTERNATIONAL UNION OF PURE AND APPLIED PHYSICS
Solid state physics in electronics and telecommunications :international conference Proceedings Brussels 1958 Jun 2-7 Vol 3,pt 1: magnetic and optical properties Edited by M. Desirant and J.L. Michiels Academic press,London,1960
Cav 7.0387

INTERNATIONAL UNION OF PURE AND APPLIED PHYSICS
Solid state physics in electronics and telecommunications :international conference Proceedings Brussels 1958 Jun 2-7 Vol 1-2,pt 1-2: semiconductors Edited by M. Desirant and J.L. Michiels 2 vols Academic press,London,1960
Cav 7.0388

INTERNATIONAL UNION OF PURE AND APPLIED PHYSICS
Statistical mechanics of equilibrum and non-equilibrium The Symposium on statistical mechanics and thermodynamics Proceedings Aachen 1964 Jun 15-20 Edited by J. Meixner 274p North-Holland,Amsterdam,1965
Cav 7.0001

INTERNATIONAL UNION OF PURE AND APPLIED PHYSICS
The International conference on crystal lattice defects Proceedings Tokyo 1962 Sep 3-4 Kyoto 1962 Sep 7-12 Pt 1-3 Physical society of Japan Physical society of Japan.Journal, 18; Suppl.1-3 3 vols Physical society of Japan,Tokyo,1963
Cav 7.1343

INTERNATIONAL UNION OF PURE AND APPLIED PHYSICS
The International conference on high energy physics 13th Proceedings Berkeley,Calif. 1966 Aug 31-Sep 7 University of California and United States atomic energy commission Edited by Margaret Alston-Garnjost Berkeley, Calif.,1967
Cav 7.0189

INTERNATIONAL UNION OF PURE AND APPLIED PHYSICS
The International conference on low temperature physics 7th Proceedings Toronto 1960 Aug 29-Sep 3 Edited by J.M. Graham and A.C. Hollis-Hallett With the co-operation of the Canadian association of physicists University of Toronto press;North-Holland,Toronto;Amsterdam,1961
Cav 7.2330

INTERNATIONAL UNION OF PURE AND APPLIED PHYSICS
The International conference on nuclear and meson physics Proceedings Glasgow 1954 Edited by E.H. Bellamy and R.G. Moorhouse 352p Pergamon press,London,1955
Cav 7.0908

INTERNATIONAL UNION OF PURE AND APPLIED PHYSICS
The International conference on the electronic properties of metals at low temperatures Report Geneva,N.Y. 1958 Aug 25-29 Co-sponsored by General electric research laboratory 244p c1958 Privately printed report to the sponsors
Cav 7.0895

INTERNATIONAL UNION OF PURE AND APPLIED PHYSICS
The International conference on the physics of semiconductors Report Exeter 1962 Jul 16-20 Institute of physics and Physical society Edited by A.C. Stickland 909p Institute of physics,London,1962
Cav 7.0750

INTERNATIONAL UNION OF PURE AND APPLIED PHYSICS
The International symposium on polarization phenomena of nucleons Basel 1960 Jul 4-8 Universitat Basel Edited by P. Huber and K.P. Meyer Helvetica physica acta.Supplementum, 6 Birkhauser,Basel,1961
Cav 7.0135

INTERNATIONAL UNION OF PURE AND APPLIED PHYSICS
X-ray microscopy and microradiography Symposium on x-ray microscopy and microradiography Proceedings Cambridge 1956 Edited by V.E. Coslett and others Academic press,New York,1957
Radioth 35.1600

INTERNATIONAL UNION OF PURE AND APPLIED PHYSICS
X-ray microscopy and microradiography The Symposium on X-ray microscopy and microradiography Proceedings Cambridge 1956 Aug 16-25 Cavendish laboratory Edited by V.E. Cosslett and others Academic press, New York,1957
Cav 7.0461

INTERNATIONAL UNION OF PURE AND APPLIED PHYSICS and INTERNATIONAL UNION OF PURE AND APPLIED CHEMISTRY Reactivity of solids : international symposium 4th Proceedings Amsterdam 1960 May 30-Jun 4 Edited by J.H. de Boer and others Elsevier,Amsterdam,1961
Chem 18.0596

INTERNATIONAL UNION OF PURE AND APPLIED PHYSICS
meeting Proceedings Conference on statistical mechanics and thermodynamics. Foundations and applications Copenhagen 1966 Edited by Thor A. Bak table 582p 25cm Benjamin,New York,1967
Cav 7.3022

INTERNATIONAL UNION OF PURE AND APPLIED PHYSICS. COMMISSION ON THE SOLID STATE International conference on 'Electron diffraction' and 'The nature of defects in crystals' Abstracts of papers Melbourne 1965 Aug 16-21 Australian academy of science and International union of crystallography Pergamon press,Oxford,1966 Extended abstracts of papers to be presented at the conference
Cav 7.1584

INTERNATIONAL UNION OF PURE AND APPLIED PHYSICS. COMMISSION ON THE SOLID STATE Conference on the defects in crystalline solids :held at the H.H.Wills physical laboratory Report Bristol 1954 Jul By N. Bloembergen and others University of Bristol and Institute of physics 429p Physical society, London,1955
Cav 7.0345

INTERNATIONAL UNION OF PURE AND APPLIED PHYSICS. COSMIC RAY COMMISSION International conference on cosmic rays 8th proceedings Jaipur 1963 Dec.2-14 Vols. 1-6 Edited by R.R. Daniel and others Sponsored by India.Department of atomic energy 6 vols Bombay,1964
A Math 4.1159

INTERNATIONAL UNION OF PURE AND APPLIED PHYSICS. SEMICONDUCTOR COMMISSION Amorphous and liquid semiconductors :a conference Proceedings Cambridge 1969 Sep 24-27 Edited by Nevill Francis Mott Bibliog,diagrs, figs 626p 20cm North-Holland,Amsterdam, 1970
Cav 7.2678

INTERNATIONAL UNION OF PURE AND APPLIED PHYSICS, and PHYSICAL SOCIETY International conference on physics papers and discussions London 1934 Vol 2: the solid state of matter Physical society,London,1937
Min 10.1078

INTERNATIONAL UNION OF PURE AND APPLIED PHYSICS AND THE PHYSICAL SOCIETY International conference on physics Papers and discussions London 1934 Vol 1-2: nuclear physics: solid state of matter 2 vols Physical society,London,1935
Met 25.2343

INTERNATIONAL UNION OF RADIO SCIENCE.SWEDISH NATIONAL COMMITTEE Colloquium on spectra of meteorological variables Proceedings Stockholm 1969 Jun 9-19 Vol 3: radio science illus 28cm American geophysical union,Washington,D.C.,1969
A Math 4.1745

INTERNATIONAL UNION OF SCIENTIFIC PSYCHOLOGY International congress of psychology 13th Proceedings and papers Stockholm 1951 Jul 16-21 Stockholm,1952 Under the presidency of professor Katz Conference languages:English,French and German
Psy 31.2536

INTERNATIONAL UNION OF SCIENTIFIC PSYCHOLOGY International congress of psychology 16th Berichte.Proceedings Bonn 1960 Jul 31-Aug 5 Acta psychologica, 19 North-Holland, Amsterdam,1961 President of congress: professor Metzger.Conference languages:English, French and German
Psy 31.3396

INTERNATIONAL UNION OF TESTING AND RESEARCH LABORATORIES FOR MATERIALS AND STRUCTURE Shell research :a symposium Proceedings Delft 1961 Aug 30-Sep 2 Edited by A.M. Haas and A.L. Bouma North-Holland,Amsterdam, 1961
Eng 41.2860

INTERNATIONAL UNION OF TESTING AND RESEARCH LABORATORIES FOR MATERIALS AND STRUCTURES International symposium on nondestructive testing of materials and atructures Vol 1-2 2 vols Paris,1956
Eng 41.2833

INTERNATIONAL UNION OF THE HISTORY AND PHILOSOPHY OF SCIENCE Congres international d'histoire des sciences 8e Actes Florence 1956 Sep 3-9 and Milan Vol 1-3 Academie internationale d'histoire des sciences.Collection de travaux, 9 3 vols Gruppe italiano di storia delle scienze; Hermann,Vinci (Florence);Paris,1958
WSM 43.0057

INTERNATIONAL UNION OF THE HISTORY AND PHILOSOPHY OF SCIENCE.DIVISION OF HISTORY OF SCIENCE Scientific change Symposium on the history of science Papers Oxford 1961 Jul 9-15 Edited by A.C. Crombie Heinemann,London,1963
Geog 13.0053

INTERNATIONAL UNION OF THEORETICAL AND APPLIED MECHANICS A Symposium on atmospheric diffusion and air pollution proceedings Oxford 1958 Aug 24-29 Edited by F.N. Frenkel and P.A. Sheppard Advances in geophysics, 6 Academic press,New York; London,1959
A Math 4.0576

INTERNATIONAL UNION OF THEORETICAL AND APPLIED MECHANICS Aerodynamic phenomena in stellar atmospheres Symposium on cosmical gas dynamics held at the International school of physics "Enrico Fermi" 4th proceedings Varenna 1960 Aug.18-30 Edited by R.N. Thomas and others Under the auspices of Societa Italiana di fisica International astronomical union.Symposium, 12 515p Bologna,c1961 Reprinted from Supplemento Nuovo Cimento,22,no.1,1961
A Math 4.1152

INTERNATIONAL UNION OF THEORETICAL AND APPLIED MECHANICS Aerodynamic phenomena in stellar atmospheres Symposium on cosmical gas dynamics held at the International school of physics "Enrico Fermi" 4th proceedings Varenna 1960 Aug 18-30 Edited by R.N. Thomas and others under the auspices of the Societa italiana di fisica International astronomical union symposium, 12 515p Zanichelli,Bologna,c1961 Reprinted from the 'Supplemento del nuovo cimento',vol.22,no. 1,1961
Obs 6.1736

INTERNATIONAL UNION OF THEORETICAL AND APPLIED MECHANICS Applied mechanics International congress of applied mechanics 12th Proceedings Stanford,Calif. 1968 Aug 26-31 Edited by H. Hetenyi and W.G. Vincenti Bibliog,illus xxiv,420p 291cm Springer,Berlin,1969
A Math 4.1722

INTERNATIONAL UNION OF THEORETICAL AND APPLIED MECHANICS Atmospheric diffusion and air pollution :a symposium Proceedings Oxford 1958 Aug 24-29 Edited by F.N. Frenkel and P. A. Sheppard Advances in physics, 6 Academic press,New York,1959
Eng 41.6946

INTERNATIONAL UNION OF THEORETICAL AND APPLIED MECHANICS Atmospheric diffusion and air pollution :a symposium Proceedings Oxford 1958 Aug 24-29 Edited by F.N. Frenkiel and P. A. Sheppard New York,1959
Geod 9.0521

INTERNATIONAL UNION OF THEORETICAL AND APPLIED MECHANICS Boundary layer research (Grenzschichtforschung) :a symposium Proceedings Freiburg 1957 Aug 26-29 Edited by H. Gortler Springer,Berlin,1958
Eng 41.6961

INTERNATIONAL UNION OF THEORETICAL AND APPLIED MECHANICS Colloquium on fatigue Proceedings Stockholm 1955 May 25-27 Edited by Waloddi Weibull and Folke K.G. Odqvist Springer,Berlin,1956
Eng 41.3754

INTERNATIONAL UNION OF THEORETICAL AND APPLIED MECHANICS Colloquium on fatigue Proceedings Stockholm 1955 May 25-27 Edited by Waloddi Weibull and Folke K.G. Odqvist Springer-verlag,Gottingen;Heidelberg, 1956 Lectures in English,French and German
Met 25.0917

INTERNATIONAL UNION OF THEORETICAL AND APPLIED MECHANICS Creep in structures :a colloquium Stanford,Calif. 1960 Jul 11-15 Edited by Nicholas J. Hoff Springer,Berlin, 1962
Eng 41.2529

INTERNATIONAL UNION OF THEORETICAL AND APPLIED MECHANICS Creep in structures colloquium Stanford,Calif. 1960 Jul 11-15 Edited by Nicholas J. Hoff Springer-verlag,Berlin,1962
Met 25.0959

INTERNATIONAL UNION OF THEORETICAL AND APPLIED MECHANICS Dynamics of satellites :a symposium Paris 1962 May 28-30 Edited by Maurice Roy Springer,Berlin,1963
Eng 41.6890

INTERNATIONAL UNION OF THEORETICAL AND APPLIED MECHANICS Electromagnetic phemomena in cosmical physics a symposium Proceedings Stockholm 1956 Aug.27-31 International astronomical union Edited by B. Lehnert International astronomical union.Symposium, 6 544p Cambridge university press, Cambridge,1958
Obs 6.1054

INTERNATIONAL UNION OF THEORETICAL AND APPLIED MECHANICS Electromagnetic phenomena in cosmical physics Stockholm 1956 Aug 27-31 International astronomical union Edited by B. Lehnert International astronomical union. Symposium, 6 13,544p Cambridge university press,Cambridge,1958
Nap 11.0916

INTERNATIONAL UNION OF THEORETICAL AND APPLIED MECHANICS Fundamental problems in turbulence and their relation to geophysics symposium proceedings Marseilles 1961 Sep.4-9 Vol 67 Edited by Francois N. Frenkel Journal of geophysical research,Vol 67,no.8 American geophysical union, Washington D.C.,1962
A Math 4.0577

INTERNATIONAL UNION OF THEORETICAL AND APPLIED MECHANICS Gas dynamics of cosmic clouds Symposium on cosmical gas dynamics 2nd Cambridge 1953 Jul 6-11 International astronomical union.Symposium series, 2 North-Holland,Amsterdam,1955
Eng 41.4468

INTERNATIONAL UNION OF THEORETICAL AND APPLIED MECHANICS Gas dynamics of cosmic clouds The Symposium on cosmical gas dynamics Proceedings Cambridge 1953 Jul 6-11 Edited by J.M. Burgess and H.C.van de Hulst International astronomical union.Symposium, 2 North Holland,Amsterdam,1955
TA 15.0217

INTERNATIONAL UNION OF THEORETICAL AND APPLIED MECHANICS Gas dynamics of cosmic clouds : a symposium Cambridge 1953 Jul.6-11 Edited by J.M. Burgers and H.C.van de Hulst International astronomical union.Symposium, 2 247p North-Holland,Amsterdam,1955
Obs 6.0308

INTERNATIONAL UNION OF THEORETICAL AND APPLIED MECHANICS Gyrodynamics symposium Celerina 1962 Aug 20-23 Edited by Hans Ziegler Springer-Verlag,Berlin,1963
A Math 4.0314

INTERNATIONAL UNION OF THEORETICAL AND APPLIED MECHANICS Interstellar gas dynamics Symposium on cosmical gas dynamics 6th Proceedings Yalta 1969 Sep 8-18 Edited by H.J. Habing International astronomical union.Symposium, 39 diagrs,photos,graphs xii,388p 24cm Reidel,Dordrecht,1970
Cav 7.2710

INTERNATIONAL UNION OF THEORETICAL AND APPLIED MECHANICS Interstellar gas dynamics Symposium on cosmical gas dynamics 6th Proceedings Yalta 1969 Sep 8-18 International astronomical union.Symposium, 39 388p Reidel,Dordrecht,1970
TA 15.0541

INTERNATIONAL UNION OF THEORETICAL AND APPLIED MECHANICS Interstellar gas dynamics Symposium on cosmical gas dynamics 6th Proceedings Yalta 1969 Sep 8-18 International astronomical union.Symposium, 39 388p Reidel,Dordrecht,1970
Obs 6.3323

INTERNATIONAL UNION OF THEORETICAL AND APPLIED MECHANICS Kreisel probleme : internationale symposium Vortrage Celerina, Switzerland 1962 Aug 20-23 Edited by Hans Ziegler Berlin,1963 Papers in English, French and German
Geod 9.0236

INTERNATIONAL UNION OF HISTORY AND PHILOSOPHY OF SCIENCE.DIVISION OF LOGIC,METHODOLOGY AND PHILOSOPHY OF SCIENCE Problems in the philosophy of science International colloquium in the philosophy of science Proceedings London 1965 Jul 11-17 Vol 3 British society for the philosophy of science London school of economics and political science Edited by Imre Lakatos and Alan Musgrave Studies in logic and the foundations of mathematics ix,448p North-Holland,Amsterdam,1968
WSM 43.0931

INTERNATIONAL UNION OF HISTORY AND PHILOSOPHY OF SCIENCE.DIVISION OF PHILOSOPHY OF SCIENCE Concept and the role of the model in mathematics and natural and social sciences International union of history and philosophy of science colloquium Proceedings Utrecht 1960 Synthese library 194p Reidel, Dordrecht,1961 Colloquium organized by H. Freudenthal
WSM 43.0785

INTERNATIONAL UNION OF HISTORY AND PHILOSOPHY OF SCIENCE.DIVISION OF PHILOSOPHY OF SCIENCE The Concept and the role of the model in mathematics and natural and social sciences :a colloquium Proceedings Utrecht 1960 Jan Edited by B.H. Kazemier and D. Vuysje 164p Reidel,Dordrecht,1961
WSM 43.1797

INTERNATIONAL UNION OF HISTORY AND PHILOSOPHY OF SCIENCE COLLOQUIUM Proceedings Concept and the role of the model in mathematics and natural and social sciences Utrecht 1960 International union of history and philosophy of science.Division of philosophy of science Synthese library 194p Reidel,Dordrecht, 1961 Colloquium organized by H. Freudenthal
WSM 43.0785

INTERNATIONAL UNION OF MICROBIOLOGICAL SOCIETIES Recent progress in microbiology :symposia presented at the 8th international congress for microbiology Montreal 1962 Aug 20-24 Edited by N.E. Gibbons University of Toronto press,Toronto,1963
Gen 34.0701

INTERNATIONAL UNION OF OF THE HISTORY AND PHILOSOPHY OF SCIENCE Symposium international d'histoire des sciences Actes Vinci (Florence) 1960 Oct 8-10 Academie internationale d'histoire des sciences. Collection de travaux, 13 235p Gruppe italiano di storia delle scienze,Vinci (Florence),1962
WSM 43.0063

INTERNATIONAL UNION OF PHYSIOLOGICAL SCIENCES Information processing in the nervous system : symposium Proceedings Edited by R.W. Gerard and J.W. Duyff International congress series,49 Excerpta medica,Amsterdam,1960
Pha 16.0433

INTERNATIONAL UNION OF PHYSIOLOGICAL SCIENCES International congress of physiological sciences 2nd Proceedings Leiden 1962 Sep 10-17 Vol 3: information processing in the nervous system Edited by R.W. Gerard and J.W. Duyff International union of physiological sciences.Proceedings, 3 International congress series, 49 Excerpta medica foundation,Amsterdam,1964
An 32.4392

INTERNATIONAL UNION OF PHYSIOLOGICAL SCIENCES International congress of physiological sciences 22nd Leiden 1962 Sep 10-17 Vol 3: information processing in the nervous system Edited by R.W. Gerard and J.W. Duyff International union of physiological sciences. Proceedings, 3 International congress series, 49 Excerpta medica,Amsterdam,1962
Phys 20.2114

INTERNATIONAL UNION OF PHYSIOLOGICAL SCIENCES International congress of physiological sciences 22nd Leiden 1962 Sep 10-17 abstracts of free communications;films and demonstrations International union of physiological sciences.Proceedings, 2 International congress series, 48 Excerpta medica,Amsterdam,1962
Phys 20.2113

INTERNATIONAL UNION OF PHYSIOLOGICAL SCIENCES International congress of physiological sciences 22nd Leiden 1962 Sep 10-17 lectures and symposia International union of physiological sciences.Proceedings, 1,pt 1-2 International congress series, 47 2 vols Excerpta medica,Amsterdam,1962
Phys 20.2112

INTERNATIONAL UNION OF PHYSIOLOGICAL SCIENCES International congress of physiological sciences 23rd Tokyo 1965 Sep 1-9 lectures and symposia International union of physiological sciences.Proceedings, 4 International congress series, 87 Excerpta medica,Amsterdam,1965
Phys 20.2126

INTERNATIONAL UNION OF PHYSIOLOGICAL SCIENCES International congress of physiological sciences 24th Washington,D.C. 1968 abstracts of lectures and symposia International union of physiological sciences. Proceedings, 6 Bethesda,Md.,1968
Phys 20.2154

INTERNATIONAL UNION OF PHYSIOLOGICAL SCIENCES International congress of physiological sciences 24th Washington,D.C. 1968 abstracts of volunteer papers and films International union of physiological sciences. Proceedings, 7 Bethesda,Md.,1968
Phys 20.2155

INTERNATIONAL UNION OF PHYSIOLOGICAL SCIENCES Oxygen in the animal organism :a symposium Proceedings London 1963 Sep 1-5 Edited by Frank Dickens and Eric Neil Pergamon press,Oxford,1964
PGMS 29.0407

INTERNATIONAL UNION OF PHYSIOLOGICAL SCIENCES Oxygen in the animal organism :a symposium Proceedings London 1963 Sep 1-5 Edited by Frank Dickens and Eric Neil held at Bedford college I.U.B.Symposium series, 31 Pergamon press,Oxford,1964
Bioch 33.1080

INTERNATIONAL UNION OF PHYSIOLOGICAL SCIENCES Problems of atmospheric circulation The Space science symposium 6th Papers of a session Mar del Plata,Argentina 1965 May 11-19 Committee on space research and International union of biochemistry Edited by R.V. Garcia and T.F. Malone 186p Spartan books,Washington,D.C.,1966
Nap 11.0212

INTERNATIONAL UNION OF PURE AND APPLIED CHEMISTRY
Congress handbook Zurich 1955 Jul 21-27 I.U.P.A.C. organizing committee,Zurich,c1955 Abstracts in English,French,German and Italian
Chem 18.1736

INTERNATIONAL UNION OF PURE AND APPLIED CHEMISTRY
International colloquium on macromolecules Proceedings Amsterdam 1949 Sep 2-5 Amsterdam,1950
Col S 12.0101

INTERNATIONAL UNION OF PURE AND APPLIED CHEMISTRY
International conference of pure and applied chemistry main lectures 16th Lectures Paris 1957 Jul 18-24 Birkhauser, Basle;Stuttgart,1957 Papers in English, French and German
Chem 18.1738

INTERNATIONAL UNION OF PURE AND APPLIED CHEMISTRY
International conference on the chemistry of natural products 7th Abstracts Riga 1970 Jun 21-25 Edited by M.N. Kolosov Zinatne,Riga,1970
Bioch 33.2318

INTERNATIONAL UNION OF PURE AND APPLIED CHEMISTRY
International congress of pure and applied chemistry main congress lectures and lectures in the sections 14th Lectures Zurich 1955 Jul 21-27 Birkhauser,Basle;Stuttgart, 1955 Papers in English,French and German
Chem 18.1737

INTERNATIONAL UNION OF PURE AND APPLIED CHEMISTRY
International congress of pure and applied chemistry main lectures 17th Lectures Munich 1959 Aug 30-Sep 6 Vol 1-2: inorganic chemistry;biochemistry and applied chemistry 2 vols Butterworths,London,1960 Papers in English,French and German
Chem 18.1739

INTERNATIONAL UNION OF PURE AND APPLIED CHEMISTRY
International symposium on molecular structure and spectroscopy Tokyo 1962 Sep 10-14 Science council of Japan,Tokyo,1962
Chem 18.2639

INTERNATIONAL UNION OF PURE AND APPLIED CHEMISTRY
International symposium on the reactivity of solids 5th Munich 1964 Aug 2-8 Deutsche Bunsen-gesellschaft fur physikalische chemie Edited by G.M. Schwab Elsevier, Amsterdam,1965 Papers in English,French and German
Met 25.1851

INTERNATIONAL UNION OF PURE AND APPLIED CHEMISTRY
Isotope mass effects in chemistry and biology Symposium on isotope mass effects in chemistry and biology Proceedings Vienna 1963 Dec 9-13 Edited by N. Grell Pure and applied chemistry, 8,no.3-4 Butterworths, London,1964
Radioth 35.0421

INTERNATIONAL UNION OF PURE AND APPLIED CHEMISTRY
Modern chemistry in industry :a symposium Proceedings Eastbourne 1968 Mar 11-14 Edited by J.G. Gregory 311p Society of chemical industry,London,1968
Chem E 24.1551

INTERNATIONAL UNION OF PURE AND APPLIED CHEMISTRY
Modern chemistry in industry :an International union of pure and applied chemistry symposium Proceedings Eastbourne 1968 Mar 11-14 Edited by J.G. Gregory illus 311p Society of chemical industry, London,1968
Chem E 24.1874

INTERNATIONAL UNION OF PURE AND APPLIED CHEMISTRY
Reactivity of solids International symposium on the reactivity of solids 4th Proceedings Amsterdam 1960 May 30-Jun 4 Nederlandse chemische vereniging and International union of pure and applied physics Edited by J.H. DeBoer and others illus x,762p
Chem E 24.0882

INTERNATIONAL UNION OF PURE AND APPLIED CHEMISTRY
The Chemistry and biochemistry of fungi and yeasts :a symposium Proceedings Dublin 1963 Jul 18-20 sponsored by the International union of pure and applied chemistry and Royal Irish academy.Irish national committee for chemistry Butterworths,London,1963 Reprinted from 'Pure and applied chemistry',vol.7,no.,4.Added title-page in French:Chimie et biochimie des champignons inferieurs et des levures
Bioch 33.1146

INTERNATIONAL UNION OF PURE AND APPLIED CHEMISTRY
The Chemistry and biochemistry of fungi and yeasts :symposium Proceedings Dublin 1963 Jul 18-20 Butterworths,London,1963 Reprinted from 'Pure and applied chemistry' vol.7,no.4
Gen 34.0729

INTERNATIONAL UNION OF PURE AND APPLIED CHEMISTRY
Thermodynamic and transport properties of fluids :joint conference Proceedings London 1957 Jul 10-12 viii,217p Institution of mechanical engineers,London, 1958
Chem E 24.0981

INTERNATIONAL UNION OF PURE AND APPLIED CHEMISTRY
Thermodynamics and thermochemistry Symposium on thermodynamics and thermochemistry Plenary lectures Lund 1963 Jul 18-23 Butterworths,London,1964
Chem E 24.0988

INTERNATIONAL UNION OF PURE AND APPLIED CHEMISTRY and CHEMICAL SOCIETY International conference on co-ordination chemistry Lectures delivered and abstracts of papers London 1959 Apr 6-11 Chemical society. Special publication,13 Chemical society, London,1959
Chem 18.1220

INTERNATIONAL UNION OF PURE AND APPLIED CHEMISTRY and INTERNATIONAL UNION OF PURE AND APPLIED PHYSICS Reactivity of solids : international symposium 4th Proceedings Amsterdam 1960 May 30-Jun 4 Edited by J.H. de Boer and others Elsevier,Amsterdam,1961
Chem 18.0596

INTERNATIONAL UNION OF PURE AND APPLIED CHEMISTRY and UNION OF THE CHEMICAL SOCIETIES Hydrogen bonding :papers held at the symposium on hydrogen bonding Papers Ljubljana 1957 Jul 29-Aug 3 Edited by D. Hadzi and H.W. Thompson Pergamon press,London,1959
Chem 18.0955

INTERNATIONAL UNION OF PURE AND APPLIED CHEMISTRY. PHYSICAL CHEMISTRY CENTER International symposium on the reactivity of solids 6th Proceedings Schenectady,N.Y. 1968 Aug 25-30 Edited by J.W. Mitchell and others Wiley,New York;London,1969
Met 25.2446

INTERNATIONAL UNION OF PURE AND APPLIED CHEMISTRY. APPLIED CHEMISTRY SECTION The Wood chemistry symposium Proceedings Montreal 1961 Aug 9-11 Pure and applied chemistry, 5,nos 1-2 illus Butterworths,London,1962 Added title page in French "Colloque sur la chimie du bois".
Bioch 33.1212

INTERNATIONAL UNION OF PURE AND APPLIED CHEMISTRY. COMMISSION ON HIGH TEMPERATURES AND REFRACTORIES High temperature technology An International symposium on high temperature technology 3rd Proceedings Pacific Grove,Calif. 1963 Sep 8-11 Organized and directed by Stanford research institute Butterworths,London,1964
Met 25.0992

INTERNATIONAL UNION OF PURE AND APPLIED CHEMISTRY. DIVISION FOR PHYSICAL CHEMISTRY Reactivity of solids :international symposium 5th Proceedings Munich 1964 Aug 2-8 Edited by G.-M. Schwab Organised by the Deutsche Bunsen-gesellschaft fur physikalische chemie Elsevier,Amsterdam,1965
Chem 18.0608

INTERNATIONAL UNION OF PURE AND APPLIED CHEMISTRY. DIVISION OF ORGANIC CHEMISTRY Chemistry of natural products The International symposium on the chemistry of natural products 5th Plenary lectures London 1968 Jul 8-13 Butterworths,London,1968 "The contents of this book appear in 'Pure and applied chemistry,vol.17, nos.3-4 (196819". Added title page in French
Bioch 33.1392

INTERNATIONAL UNION OF PURE AND APPLIED CHEMISTRY. MACROMOLECULAR DIVISION Structure of organic solids Microsymposium 'Structure of organic solids' 2nd Lectures Prague 1968 Sep 16-19 In conjunction with Ceskoslovenska akademie ved illus 550p 25cm Butterworths,London,1969 Simultaneously published in 'Pure and applied chemistry',vol 18,no 4,1969
Cav 7.2659

INTERNATIONAL UNION OF PURE AND APPLIED CHEMISTRY. PHYSICAL CHEMISTRY SECTION and ROYAL NETHERLANDS CHEMICAL SOCIETY Molecular spectroscopy :general and introductory lectures presented at the fifth European congress on molecular spectroscopy 5th Amsterdam 1961 May 29-Jun 3 Co-sponsored by the Netherlands physical society Butterworths,London,1962 Reprinted from 'Pure and applied chemistry',vol 4,no 1
Chem 18.2315

INTERNATIONAL UNION OF PURE AND APPLIED CHEMISTRY. PROTEIN COMMISSION Symposium on protein structure Papers Paris 1957 Jul 25-29 Edited by Albert Neuberger Methuen;Wiley, London;New York,1958
Bioch 33.0535

INTERNATIONAL UNION OF PURE AND APPLIED CHEMISTRY. PROTEIN COMMISSION Symposium on protein structure Paris 1957 Jul 25-29 Edited by Albert Neuberger Methuen;Wiley,London New York,1958
Gen 34.0667

INTERNATIONAL UNION OF PURE AND APPLIED CHEMISTRY. SECTION OF ORGANIC CHEMISTRY Theoretical organic chemistry :Kekule symposium Papers London 1958 Sep 15-17 Organized by the Chemical society Butterworths,London,1959
Chem 18.2165

INTERNATIONAL UNION OF PURE AND APPLIED CHEMISTRY. SECTION OF ORGANIC CHEMISTRY and CHEMICAL SOCIETY Theoretical organic chemistry : Kekule symposium Papers London 1958 Sep 15-17 Butterworths,London,1959
Chem 18.1604

INTERNATIONAL UNION OF PURE AND APPLIED CHEMISTRY, and DEUTSCHE BUNSENGESELLSCHAFT FUR PHYSIKALISCHE CHEMIE International symposium on the reactivity of solids 5th Plenary lectures Munich 1964 Aug 2-8 Butterworths,London,1965
Min 10.0662

An INTERNATIONAL UNION OF PURE AND APPLIED CHEMISTRY SYMPOSIUM Proceedings Modern chemistry in industry Eastbourne 1968 Mar 11-14 Society of chemical industry Edited by J.G. Gregory London,1968
Bioch 33.1874

INTERNATIONAL UNION OF PURE AND APPLIED PHYSICS Conference on nuclear and meson physics Proceedings Glasgow 1954 Jul 13-17 Edited by E.H. Bellamy and R.G. Moorhouse Pergamon press,London;New York,1955
Radioth 35.1563

INTERNATIONAL UNION OF PURE AND APPLIED PHYSICS Conference on the electron capture and higher processes in nuclear decays Proceedings Debrecen 1968 Jul 15-18 Vol 1-3 diagrs 20cm 3 vols Eotvos Lorand physical society,Budapest,1968
Cav 7.2700

INTERNATIONAL UNION OF PURE AND APPLIED PHYSICS Conformation of biopolymers International symposium on conformation of biopolymers Papers read Madras 1967 Jan 18-21 Vol 1-2 University of Madras.Centre of advanced study in biophysics Edited by G.N. Ramachandran 2 vols Academic press,London; New York,1967
Bioch 33.0331

INTERNATIONAL UNION OF PURE AND APPLIED PHYSICS Crystal growth International conference on crystal growth Proceedings Boston,Mass. 1966 Jun 20-24 Edited by H.Steffen Peiser Physics and chemistry of solids.Supplement Pergamon press,Oxford,1967
Met 25.1520

INTERNATIONAL UNION OF PURE AND APPLIED PHYSICS High energy nuclear physics Annual Rochester conference on high energy nuclear physics 5th Proceedings Rochester,N.Y. 1955 Jan 31-Feb 2 University of Rochester Edited by H.Pierre Noyes and others 196p Interscience,New York,1955 Co-sponsored by the National science foundation in cooperation with the Atomic energy commission and the United States office of naval research
Cav 7.1988

INTERNATIONAL UNION OF PURE AND APPLIED PHYSICS
International conference on II-VI semiconducting compounds Proceedings Providence,R.I. 1967 Sep 6-8 Edited by D.G. Thomas 1490p 20cm Benjamin,New York, 1967
Cav 7.2681

INTERNATIONAL UNION OF PURE AND APPLIED PHYSICS
International conference on cosmic rays 11th Invited papers and rapporteur talks Budapest 1969 Aug 25-Sep 4 612p Central research institute of physics,Budapest,1969
TA 15.0551

INTERNATIONAL UNION OF PURE AND APPLIED PHYSICS
International conference on crystal growth 2nd Proceedings Birmingham 1969 Jul 15-19 Edited by F.C. Frank and others xix, 842p 28cm North-Holland,Amsterdam,1968 Reprinted from Journal of crystal growth,3-4, 1968
Cav 7.2950

INTERNATIONAL UNION OF PURE AND APPLIED PHYSICS
International conference on elementary particles proceedings Oxford 1965 Sep.19-25 Rutherford high energy laboratory Rutherford high energy laboratory,Oxford,1965
A Math 4.0887

INTERNATIONAL UNION OF PURE AND APPLIED PHYSICS
International conference on high-energy physics 10th Proceedings Rochester,N.Y. 1960 Aug 25-Sep 1 Edited by E.C.G. Sudarshan and others Co-sponsored by the United States. Office of naval research illus xxv,890p University of Rochester,Rochester,N.Y.,1960
Cav 7.2891

INTERNATIONAL UNION OF PURE AND APPLIED PHYSICS
International conference on high-energy physics 13th proceedings Berkeley,Calif. 1966 Aug.31-Sep.7 University of California press,Berkeley;Los Angeles,1967
A Math 4.0891

INTERNATIONAL UNION OF PURE AND APPLIED PHYSICS
International conference on high-energy physics at CERN 11th Proceedings Geneva 1962 Jul 4-11 European organization for nuclear research Edited by J. Prentki Bibliog.,Illus. xxiv,949p 28cm CERN, Geneva,1962
A Math 4.1544

INTERNATIONAL UNION OF PURE AND APPLIED PHYSICS
International conference on low temperature physics 10th Proceedings Moscow 1966 Aug 31-Sep 6 Vol 1: properties of helium Edited by M.P. Malkov 552p 22cm Moscow, 1967
Cav 7.3030

INTERNATIONAL UNION OF PURE AND APPLIED PHYSICS
International conference on low temperature physics 10th Proceedings Moscow 1966 Aug 31-Sep 6 Vol 3: the electronic properties of metals 404p 22cm Moscow, 1967
Cav 7.3032

INTERNATIONAL UNION OF PURE AND APPLIED PHYSICS
International conference on low temperature physics 10th Proceedings Moscow 1966 Aug 31-Sep 6 Vol 4: antiferromagnetism Edited by M.P. Malkov 361p 22cm Moscow, 1967
Cav 7.3033

INTERNATIONAL UNION OF PURE AND APPLIED PHYSICS
International conference on low temperature physics 10th Proceedings Vol 2 a-b: superconductivity Edited by M.P. Malkov 22cm 2 vols
Cav 7.3031

INTERNATIONAL UNION OF PURE AND APPLIED PHYSICS
International conference on low temperature physics 12th Proceedings Kyoto 1970 Sep 4-10 Edited by Eizo Kanda Organized by the Science council of Japan photos,diagrms 895p 26cm Academic press of Japan,Tokyo, 1971
Cav 7.3028

INTERNATIONAL UNION OF PURE AND APPLIED PHYSICS
International conference on luminescence Proceedings Budapest 1966 Edited by G. Szigeti diagrms 24cm 2 vols Akademiai kiado,Budapest,1968
Cav 7.3097

INTERNATIONAL UNION OF PURE AND APPLIED PHYSICS
International conference on magnetism Proceedings Nottingham 1964 Sep 6-11 Institute of physics and the Physical society illus 898p London,c1965
Cav 7.2882

INTERNATIONAL UNION OF PURE AND APPLIED PHYSICS
International conference on magnetism and crystallography Proceedings Kyoto 1961 Sep 25-30 Pt 2: electron and neutron diffraction Science council of Japan and Crystallographic society of Japan Edited by Shizuo Miyake and others Physical society of Japan.Journal, 17,B 2 396p physical society of Japan,Tokyo,1962
Cav 7.1338

INTERNATIONAL UNION OF PURE AND APPLIED PHYSICS
International conference on magnetism and crystallography Proceedings Kyoto 1961 Sep 25-30 Pt 3: neutron diffraction:study of magnetic materials Science council of Japan and Crystallographic society of Japan Edited by Takeo Nagamiya and Shizno Miyake Physical society of Japan.Journal, 17,B 3 71p physical society of Japan,Tokyo, 1962
Cav 7.1339

INTERNATIONAL UNION OF PURE AND APPLIED PHYSICS
International conference on magnetism and crystallography Proceedings Kyoto 1961 Sep 25-30 Pt 1: magnetism Science council of Japan and Crystallographic society of Japan Edited by Kei Josida and others Physical society of Japan.Journal, 17,B 1 xi,718p Physical society of Japan, Tokyo,1962
Cav 7.1337

INTERNATIONAL UNION OF PURE AND APPLIED PHYSICS
International conference on photoconductivity 3rd Proceedings Stanford,Calif. 1969 Aug 12-15 Edited by E. M. Pell Bibliog,diagrms,graphs xi,410p 26cm Pergamon,Oxford,1971
Cav 7.3073

INTERNATIONAL UNION OF PURE AND APPLIED PHYSICS
International conference on physics Papers London 1934 Vol 1: nuclear physics Physical society,London,1935
Cav 7.1452

INTERSTELLAR MATTER Distribution and motion of interstellar matter in galaxies :a conference Proceedings Princeton,N.J. 1961 Apr 10-20 Institute for advanced study Edited by L. Woltjer 330p W.A.Benjamin,New York,1962
Obs 6.3253

INTERTERRITORIAL CONFERENCE ON HYDROLOGY AND WATER RESOURCES 1st Nairobi 1950 Nov Held under the auspices of the East Africa high commission Nairobi,1950
Geog 13.5670

INTIONATIONALER KONGRESS FUR ELEKTRONMIKROSKOPIE 4th Proceedings Berlin 1958 Sep 10-17 Bd 1: physikalisch-technischer teil International federation of electron microscope societies Edited by G. Mollenstedt and others Springer,Berlin,1960 General editors W.Bargmann and others. Papers in English,French and German
Met 25.2493

INTRACELLULAR MEMBRANEOUS STRUCTURE The International symposium for cellular chemistry 1st Proceedings Ohtsu 1963 Mar 27-31 Japan society for cell biology Edited by S. Seno and E.V. Cowdry Society for cellular chemistry.Symposia, 14,Suppl. Okayama,1964 In memory of...Dr. Seizo Katsunuma.Japan society for cell biology formerly Japan society for cellular chemistry
Bioch 33.0966

INTRACELLULAR RESPIRATION:PHOSPHORYLATING AND NON-PHOSPHORYLATING OXIDATION REACTIONS International congress of biochemistry 5th Proceedings Moscow 1961 Aug 10-16 International union of biochemistry Edited by E.C. Slater I.U.B.symposium series, 25 Pergamon;PWN-Polish scientific publishers, Oxford;Warsaw,1963
Bioch 33.1334

INTRACELLULAR TRANSPORT :a symposium Edited by Katherine Brehme Warren International society for cell biology. Symposia, 5 Academic press,New York,1966
An 32.3435

The INTRACRANIAL PRESSURE IN HEALTH AND DISEASE an investigation of the most recent advances Proceedings of the Association New York 1927 Dec 28-29 Association for research in nervous and mental disease Edited by Charles A. Elsberg and others Association for research in nervous and mental disease. Research publications, 8 Williams and Wilkins,Baltimore,Md.,1929
An 32.4155

INTRODUCTION TO ELECTRON BEAM TECHNOLOGY Edited by Robert Bakish John Wiley,New York; London,1962
Met 25.1670

INTRODUCTION TO MODERNITY :a symposium on eighteenth century thought University of Texas.Department of Germanic languages Edited by Robert Mollenauer 175p University of Texas press,Austin,Texas,1965 Essays presented in 1962 as the 4th in a series of yearly symposia
WSM 43.0649

INTRODUCTION TO SOLAR TERRESTRIAL RELATIONS The Summer school in space physics Proceedings Alpbach 1963 Jul 15-Aug 10 European preparatory commission for space research Edited by J. Ortner and H. Maseland Astrophysics and space science library Reidel,Dordrecht,1965
TA 15.0425

INTRODUCTION TO SYSTEM PROGRAMMING symposium proceedings London 1962 July Edited by Peter Wegner A.P.I.C.studies in data processing, 4 x,316p 23cm Academic press,London,1964
Math L 5.0947

INTRODUCTORY PHYSICAL METALLURGY Chicago 1947 Oct 18-24 By Clyde W. Mason American society for metals A.S.M.,Ohio,1947 A series of lectures...presented to members of the A.S.M.,during the twenty-ninth National metal congress and exposition,Chicago October 18 to 24,1
Met 25.0030

INTUITIONISM AND PROOF THEORY Conference on intuitionism and proof theory Proceedings Buffalo,N.Y. 1968 Aug Edited by A. Kino Studies in logic and the foundations of mathematics Bibliog. viii,516p 23cm North Holland,Amsterdam;London,1970
P Math 2.3613

INTUITIONISM AND PROOF THEORY :summer conference Proceedings Buffalo,N.Y. 1968 State university of New York Edited by A. Kino and others Studies in logic and the foundations of mathematics viii,516p North-Holland,Amsterdam;London,1970
WSM 43.1239

INVARIANT IMBEDDING :summer workshop Proceedings Los Angeles,Calif. 1970 Jun-Aug Edited by R.E. Bellman and E.D. Denman Held at University of southern California Lecture notes in operations research and mathematical systems, 52 148p 25cm Springer,Berlin,1971
P Math 2.4296

The INVENTORY PAPERS :international symposium on anthropology Anthropology today:an encyclopedic inventory Edited by A. L. Kroeber xv,958p University of Chicago press,Chicago,Ill;London,1965 Pre-symposium papers originally prepared as a basis for the symposium held in 1952
WSM 43.3613

INVERTEBRATE NERVOUS SYSTEMS The Conference on invertebrate nervous systems :their significance for mammalian neurophysiology Papers Pasadena,Calif. 1966 Jan 10-12 Edited by C.A.G. Wiersma Sponsored by the National institute of health University of Chicago press,Chicago,Ill.;London,1967 Dedicated to the memory of George Howard Parker
Bal 44.6453

INVERTEBRATE PHYSIOLOGY Recent advances in invertebrate physiology :a symposium Eugene 1955 Sep Edited by B.T. Scheer and others held at the University of Oregon University of Oregon publications,Eugene,Oregon,1957 Sponsored by the National science foundation, Tektronix foundation,the University of Oregon
Bal 39.2107

The INVESTIGATION OF HYPOTHALAMIC-PITUITARY-ADRENAL FUNCTION :a symposium Proceedings London 1967 Feb 22-23 Society for endocrinology Edited by V.H.T. James and J. Landon Society for endocrinology.Memoirs, 17 Cambridge university press,Cambridge,1968
Inv Med 37.0120

The INVESTIGATION OF HYPOTHALAMIC-PITUITARY-ADRENAL FUNCTION :a symposium Proceedings London 1967 Feb 22-23 Society for endocrinology Royal society of medicine. Endocrine section Edited by V.H.T. James and J. Landon Sponsored by Ciba,Horsham Society for endocrinology.Memoirs, 17 Cambridge university press,Cambridge,1968
Gen 34.0558

INVITATIONAL CONFERENCE ON PHARMACY MANPOWER Proceedings Challenge to pharmacy in the seventies San Francisco 1970 Sep 10-12 Edited by Joe B. Graber and Donald C. Brodie Sponsored by the University of California. School of pharmacy U.S.Department of health, education and welfare,Washington,D.C.,1972
PGMS 29.0674

ION EXCHANGE Symposium on ion exchange and chromatography in analytical chemistry presented at the fifty-ninth annual meeting Atlantic City,N.J. 1956 Jun 18 American society for testing materials A.S.T.M. Special technical publication, 195 A.S.T.M., Philadelphia,Pa.,1958
Met 25.1750

ION TRANSPORT ACROSS MEMBRANES :a symposium Papers New York 1953 Oct 2-3 Edited by Hans T. Clarke and David Nachmansohn held at Columbia university.College of physicians and surgeons Academic press,New York,1954 Incorporating papers presented at a symposium on the role of proteins in ion transport across membranes together with six invited articles on allied topics
Bal 39.0560

IONIZATION IN HIGH-TEMPERATURE GASES :a selection of technical papers based mainly on the American rocket society conference on ions in flames and rocket exhausts Palm Springs,Calif. 1962 Oct 10-12 Edited by Kurt E. Shuler Progress in astronautics and aeronautics, 12 Academic press,New York, 1963
Eng 41.4543

IONIZATION PHENOMENA IN GASES International conference on ionization phenomena in gases 5th proceedings Munich 1961 Aug 28-Sep 1 Vol. 1 Edited by H. Maecker Organised by Verband Deutscher physikalischer gesellschaften Series in physics North Holland,Amsterdam,1962
A Math 4.0620

IONIZED GASES International conference on phenomena in ionized gases 9th Proceedings Bucharest 1969 Institutul de fizica,Bucharest Institute of physics, Bucharest,1969
Eng 41.8553

IONIZED GASES The Physics of ionized gases : a conference proceedings,typescript London 1953 Apr University college,London Edited by J.B. Hasted Under the auspices of the Royal society of London.Warren research fund London,1953
A Math 4.0614

IONIZING RADIATION Effects of ionizing radiation on the nervous system Symposium on the effects of ionizing radiation on the nervous system Vienna 1961 Jun 5-9 Sponsored by the International atomic energy agency International atomic energy agency, Vienna,1962
An 32.4190

IONIZING RADIATIONS Ciba foundation symposium on ionizing radiations and cell metabolism Proceedings London 1956 Mar 6-9 Ciba foundation Edited by G.E.W. Wolstenholme and C.M. O'Connor illus. Churchill,London,1956
Radioth 35.1355

IONIZING RADIATIONS ON CELLS Initial effects of ionizing radiations on cells The International symposium on primary and initial effects of ionizing radiations on living cells Papers and discussions Moscow 1960 Oct Edited by R.J.C. Harris Academic press, London;New York,1961 A symposium supported by Unesco and the IAEA and sponsored by the Academy of sciences of the U.S.S.R.
Bioch 33.0961

IONOSPHERE Conference on the physics of the ionosphere Report Cambridge 1954 Sep 6-9 Physical society Cavendish laboratory 406p Physical society,London,1955
Nap 11.0926

IONOSPHERE Mixed commission on the ionosphere meeting 3rd Proceedings Canberra 1952 Aug. 24-26 International scientific radio union Edited by W.J.G. Beynon Organized by the International council of scientific unions 194p General secretariat,Brussels,1953
Obs 6.2579

IONOSPHERE N.A.T.O. Advanced study institute Electron density profiles in the ionosphere and exosphere :a symposium Proceedings Skeikampen,Norway 1961 Apr 17-26 Norwegian defence research establishment Edited by B. Maehlum Sponsored by N.A.T.O. science committee N.A.T.O. conference series, 2 428p Pergamon press,Oxford,1962
Nap 11.0391

IONOSPHERE Progress in radio science 1960-63 URSI general assembly 14th Proceedings Tokyo 1963 Sep 9-20 Vol 3: the ionosphere International scientific radio union.Commission on the ionosphere Edited by G.M. Brown 196p Elsevier,Amsterdam;London, 1965
Nap 11.0086

IONOSPHERE Solar eclipses and the ionosphere a symposium London 1955 Aug.22-24 Edited by W.J.G. Beynon and G.M. Brown Under the auspices of the Mixed commission on the ionosphere Pergamon press,London,1956 Vol.6 of special supplements to Journal of atmospheric and terrestrial physics
Obs 6.0176

IONOSPHERE Some ionospheric results obtained during the international geophysical year :a symposium Proceedings Brussels 1959 Sep. International scientific radio union.Special commission for the International Geophysical Year Edited by W.J.G. Beynon xi,399p Elsevier,Amsterdam,1960
Nap 11.0053

IONOSPHERIC RADIO Monograph on ionospheric radio URSI general assembly 13th Papers of a session London 1960 Sep. International scientific radio union Edited by W.J.G. Beynon URSI monographs series 264p Elsevier,Amsterdam;New York,1962
Nap 11.0052

IOWA CITY 1949 Engineering hydraulics Hydraulics conference 4th proceedings Iowa institute of hydraulic research Edited by Hunter Rouse John Wiley and sons,New York; London,1950
A Math 4.0511

IOWA CITY 1955 Hydraulics conference 6th Proceedings Edited by L. Landweber and P.G. Hubbard Arranged by the Iowa institute of hydraulic research illus,diagrms 276p Iowa institute of hydraulic research,Iowa City, 1956
Chem E 24.0305

IOWA CITY 1962 A Review of space research : symposium Report National aeronautics and space administration Under the auspices of the Space science board National research council.Publication,1079 Washington,D.C., 1962
Obs 6.3259

IOWA CITY,IOWA 1949 Engineering hydraulics 4th conference Proceedings Iowa institute of hydraulic research Edited by Hunter Rouse Wiley;Chapman and Hall,New York;London,1950
Eng 41.6662

IOWA INSTITUTE OF HYDRAULIC RESEARCH Engineering hydraulics 4th conference Proceedings Iowa city,Iowa 1949 Jun 12-15 Edited by Hunter Rouse Wiley;Chapman and Hall,New York;London,1950
Eng 41.6662

IOWA INSTITUTE OF HYDRAULIC RESEARCH Engineering hydraulics Hydraulics conference 4th proceedings Iowa city 1949 Jun 12-15 Edited by Hunter Rouse John Wiley and sons,New York;London,1950
A Math 4.0511

IOWA INSTITUTE OF HYDRAULIC RESEARCH Hydraulics conference 6th Proceedings Iowa City 1955 Jun 13-15 Edited by L. Landweber and P.G. Hubbard illus,diagrms 276p Iowa institute of hydraulic research, Iowa City,1956
Chem E 24.0305

IOWA STATE COLLEGE Statistics and mathematics in biology conference Ames,Iowa 1952 Jun-Jul Edited by Oscar Kempthorne and others Co-sponsored by Biometrics society xi,632p 23cm Iowa state college press,Ames, Iowa,1954
Math 3.0764

IRIS International symposium on iris 1st Report Florence 1963 May 14-18 Societa italiana dell'iris Edited by Gian Luigi Sani and others Societa italiana dell'iris, Florence,1963
BG 38.3303

IRIS CULTIVEES Conference internationale des iris 1ere Actes et comptes-rendus Paris 1922 Societe nationale d'horticulture de France Societe nationale d'horticulture, Paris,1923
BG 38.2574

IRISH NATIONAL COMMITTEE FOR CHEMISTRY The Chemistry and biochemistry of fungi and yeasts symposium Proceedings Dublin 1963 Jul 18-20 Butterworths,London,1963 Reprinted from 'Pure and applied chemistry' vol.7,no.4
Gen 34.0729

IRKUTSK 1966 Prikladnaya geografiya : materially Irkutskogo Soveshchaniya Edited by V.V. Vorob'yev 171p 20cm Irkutsk,1966 Abstracts of papers at conference in 1966
Sco 14.0405

IRON Symposium sur les gisements de fer du monde International geological congress 19th Algiers 1952 Tomes 1-2 Edited by F. Blondel and L. Marvier Algiers,1952
Geol 8.3011

IRON AND ITS DILUTE SOLID SOLUTIONS :a conference Proceedings Detroit 1961 Oct 23 American institute of mining, metallurgical and petroleum engineers Edited by C.W. Spencer and F.E. Werner Interscience; Wiley,New York;London,1963
Met 25.1305

The IRON AND STEEL ENGINEERS GROUP MEETING 43rd Report Effect of the various steelmaking processes on the energy balances of integrated iron and steelworks London 1961 Dec 1 Iron and steel institute. Engineers group London,1962
Met 25.2522

IRON AND STEEL ENGINEERS GROUP MEETING 47th Papers and discussions Ore mining and materials handling Metz 1963 Jun 10-14 Iron and steel institute.Engineers group Iron and steel institute.Special report, 82 Iron and steel institute,London,1963
Met 25.0208

IRON AND STEEL INSTITUTE Air and water pollution in the iron and steel industry proceedings of the air pollution meeting 25th and 26th Sept. and the water pollution meeting 11th and 12th Dec Proceedings London 1957 Sep 25-26 London 1957 Dec 11-12 Iron and steel institute.Special report, 61 Iron and steel institute,London,1958
Met 25.0359

IRON AND STEEL INSTITUTE Challenge for steel The Conference on 'The challenge for steel' :I. S.I. autumn general meeting Proceedings 1968 Nov Iron and steel institute. Publication, 119 Iron and steel institute, London,1969
Met 25.2241

IRON AND STEEL INSTITUTE Clean steel Iron and steel institute annual general meeting, 1962 Papers and discussions London 1962 Nov 28-29 Iron and steel institute.Special report, 77 Iron and steel institute,London, 1963
Met 25.0412

IRON AND STEEL INSTITUTE Conference on high-strength steels... Papers and discussions Harrogate 1962 May 23-24 Iron and steel institute.Special report, 76 Iron and steel institute,London,1962
Met 25.0486

IRON AND STEEL INSTITUTE Continuous casting of steel Iron and steel institute :autumn general meeting Proceedings London 1964 Nov 25-26 Iron and steel institute. Special report, 89 London, 1965
Met 25.2569

IRON AND STEEL INSTITUTE Control of composition in steelmaking :annual general meeting Papers and discussions London 1966 May 4-5 Iron and steel institute. Special publication, 99 London, 1967
Met 25.2557

IRON AND STEEL INSTITUTE Forgemasters' meeting 1954 Proceedings London 1954 Oct 11-15 Joint meeting with the Chambre syndicate de la grosse forge francaise Iron and steel institute. Special report, 60 Iron and steel institute, London, 1957
Met 25.0715

IRON AND STEEL INSTITUTE Fume arrestment Iron and steel institute annual general meeting, 1963 Proceedings London 1963 Nov 26-27 Iron and steel institute. Special report, 85 Iron and steel institute, London, 1964
Met 25.0420

IRON AND STEEL INSTITUTE Future of ironmaking in the blast furnace autumn general meeting... Papers and discussions London 1961 Nov Iron and steel institute. Special report, 72 Iron and steel institute, London, 1962
Met 25.0368

IRON AND STEEL INSTITUTE Heat treatment of metals Heat treatment joint committee : inaugural meeting Proceedings London 1965 Dec 14-15 Iron and steel institute. Special report, 95 London, 1966
Met 25.2565

IRON AND STEEL INSTITUTE High-temperature properties of steels :joint conference Proceedings Eastbourne 1966 Apr 4-7 Iron and steel institute. Publications, 97 Iron and steel institute, London, 1967
Eng 41.3827

IRON AND STEEL INSTITUTE High-temperature properties of steels the joint conference Proceedings Eastbourne 1966 Apr 4-7 Iron and steel institute. Publication, 97 Eyre and Spottiswoode, Margate, 1967
Met 25.0429

IRON AND STEEL INSTITUTE High-temperature steels and alloys for gas turbines :a symposium London 1951 Feb 21-22 Iron and steel institute. Special report, 43 Iron and steel institute, London, 1952
Met 25.0996

IRON AND STEEL INSTITUTE Hydrogen in steel Conference on hydrogen in steel papers and discussions Harrogate 1961 Oct Iron and steel institute. Special report, 73 Iron and steel institute, London, 1962 No title page
Met 25.1538

IRON AND STEEL INSTITUTE International conference on tribology in iron and steel works 1st Proceedings London 1969 Sep 22-25 Iron and steel institute. Publication, 125 Iron and steel institute, London, 1970
Met 25.2562

IRON AND STEEL INSTITUTE Ironmaking tomorrow conference Proceedings London 1966 Nov 22-23 Iron and steel institute. Publication, 102 Iron and steel institute, London, 1967
Met 25.0380

IRON AND STEEL INSTITUTE Joint conference on low-alloy steels organized by BISRA and the Iron and steel institute Proceedings Scarborough 1965 Apr 2-4 Iron and steel institute. Publication, 114 illus 269p London, 1969
Met 25.2566

IRON AND STEEL INSTITUTE Machinability :a conference jointly organized by the iron and steel institute the Institute of metals, the Institution of production engineers and the Institution of metallurgists Proceedings London 1965 Oct 4-6 Iron and steel institute. Special report, 94 Iron and steel institute, London, 1967
Met 25.0739

IRON AND STEEL INSTITUTE Metallurgical developments in carbon steels :a conference Harrogate 1963 May 15-16 Sponsored by the British iron and steel research association Iron and steel institute. Special report, 81 Iron and steel institute, London, 1963
Met 25.0417

IRON AND STEEL INSTITUTE Metallurgical developments in high-alloy steels A Joint conference on high-alloy steels Proceedings Scarborough 1964 Jun 2-4 Iron and steel institute. Special report, 86 Iron and steel institute, London, 1964
Met 25.0488

IRON AND STEEL INSTITUTE Non-destructive examination in the steel industry Conference on non-destructive examination applied to process control in the steel industry Papers and discussions Swansea 1967 Jan 3-5 Iron and steel institute. Publication, 103 illus 193p London, 1967
Met 25.2601

IRON AND STEEL INSTITUTE Optimization of steel product yield Conference on optimization of steel product yield Proceedings London 1967 May 3-4 Iron and steel institute. Publication, 107 London, 1967 Conference held at the Institute's annual general meeting
Met 25.2561

IRON AND STEEL INSTITUTE Pilot plants in the iron and steel industry autumn general meeting 1965 Papers and discussions London 1965 Nov 30-Dec 1 Iron and steel institute. Special report, 96 Iron and steel institute, London, 1966
Met 25.0367

IRON AND STEEL INSTITUTE Production of wide steel strip :a symposium London 1960 May 3-5 Iron and steel institute. Special report, 67 Iron and steel institute, London, 1960
Met 25.0726

IRON AND STEEL INSTITUTE Productivity in the iron and steel industry Iron and steel institute annual general meeting, 1962 Papers and discussions London 1962 May Iron and steel institute. Special report, 75 Iron and steel institute, London, 1962
Met 25.0411

INTERNATIONAL UNION OF THEORETICAL AND APPLIED MECHANICS Magneto-fluid dynamics International symposium on magneto-fluid dynamics proceedings Williamsburg,Va. 1960 Jan and Washington D.C. Edited by F. N. Frenkel and W.R. Sears National research council.Publication, 829 National research council,Washington,1960
A Math 4.0619

INTERNATIONAL UNION OF THEORETICAL AND APPLIED MECHANICS Magneto-fluid dynamics :a symposium Proceedings Williamsburgh,Va, 1960 Jan and Washington,D.C. 1960 Jan Edited by F.N. Frankiel and W.R. Sears National academy of sciences.Publication, 829 American physical society,New York,1960
Eng 41.4488

INTERNATIONAL UNION OF THEORETICAL AND APPLIED MECHANICS Non-homogeneity in elasticity and plasticity :a symposium Warsaw 1958 Sep 2-9 Edited by W. Olszak Pergamon press, London,1959
Eng 41.2488

INTERNATIONAL UNION OF THEORETICAL AND APPLIED MECHANICS Problems of cosmical aerodynamics Symposium on the motion of gaseous masses of cosmical dimensions Proceedings Paris 1949 Aug 16-19 Edited by J.M. Burgess and H.C.van de Hulst 237p Central air documents office,Dayton,Ohio,1951
Obs 6.3550

INTERNATIONAL UNION OF THEORETICAL AND APPLIED MECHANICS Rarefied gas dynamics The International symposium on rarefied gas dynamics 2nd Proceedings Berkeley,Calif. 1960 Aug 3-6 University of California Edited by L. Talbot Advances in applied mechanics.Supplement, 1 748p Academic press,New York,1961 Co-sponsored by National aeronautics and space administration
Cav 7.2254

INTERNATIONAL UNION OF THEORETICAL AND APPLIED MECHANICS Rheology and soil mechanics :a symposium Grenoble 1964 Apr 1-8 Edited by J. Kravtchenko and P.M. Sirieys Springer, Berlin,1966
Eng 41.3097

INTERNATIONAL UNION OF THEORETICAL AND APPLIED MECHANICS Second-order effects in elasticity,plasticity and fluid dynamics International symposium on second-order effects in elasticity,plasticity and fluid dynamics symposium Haifa 1962 Apr 23-27 Edited by Markus Reiner and David Abir Jerusalem academic press;Pergamon press, Jerusalem;London,1964
Eng 41.2483

INTERNATIONAL UNION OF THEORETICAL AND APPLIED MECHANICS Stress waves in anelastic solids :symposium Providence,R.I. 1963 Apr 3-5 Edited by Herbert Kolsky and William Prager Springer,Berlin,1964
Eng 41.3805

INTERNATIONAL UNION OF THEORETICAL AND APPLIED MECHANICS Stress waves in anelastic solids IUTAM symposium Proceedings Providence R.I. 1963 April 3-5 Edited by Herbert Kolsky and William Prager Supported by the National science foundation Springer-verlag,Berlin,1964
A Math 4.0369

INTERNATIONAL UNION OF THEORETICAL AND APPLIED MECHANICS Symposium on cosmical gas dynamics 3rd Proceedings Cambridge,Mass. 1957 Jun 24-29 Edited by J.M. Burgers and R.N. Thomas Reviews of modern physics, 30, no 3 American physical society,New York,1958
Eng 41.4467

INTERNATIONAL UNION OF THEORETICAL AND APPLIED MECHANICS Symposium on cosmical gas dynamics 3rd proceedings Cambridge,Mass. 1957 Jun.24-29 Edited by J.M. Burgers and others Financial grant from the Smithsonian institution International astronomical union. Symposium, 8 Smithsonian Institution. Smithsonian symposium publications, 1 American physical society,New York,1958 Reviews of modern physics.Vol.30,no.3,July 1958
Obs 6.0306

INTERNATIONAL UNION OF THEORETICAL AND APPLIED MECHANICS Symposium on the theory of thin elastic shells proceedings Delft 1959 Aug 24-28 Edited by W.T. Koiter North-Holland,Amsterdam,1960
A Math 4.0366

INTERNATIONAL UNION OF THEORETICAL AND APPLIED MECHANICS The Theory of thin elastic shells :a symposium Proceedings Delft 1959 Aug 24-28 North-Holland,Amsterdam,1960
Eng 41.2854

INTERNATIONAL UNION OF THEORETICAL AND APPLIED MECHANICS Theory of thin shells :a symposium 2nd Proceedings Copenhagen 1967 Sep 5-9 Edited by F.I. Niordson Springer,Berlin,1969
Eng 41.2892

INTERNATIONAL UNION OF THEORETICAL AND APPLIED MECHANICS Verformung und fliessen des festkorpers Deformation and flow of solids : colloquium Madrid 1955 Sep 26-30 Edited by R. Grammel Springer,Berlin,1956 Papers in English,French and German
Met 25.1203

INTERNATIONAL UPPER MANTLE COMMITTEE International symposium on geophysical theory and computers 3rd Proceedings Cambridge 1966 Jun 27-Jul 5 Sponsored by the International union of geodesy and geophysics Geophysical journal, 13;1-3 375p Blackwell,Oxford,1967
Geod 9.0694

INTERNATIONAL UPPER MANTLE COMMITTEE Phase transformations and the earth's interior :a conference Canberra 1969 Jan 6-10 Edited by A.E. Ringwood and D.H. Green North-Holland,Amsterdam,1970
Min 10.1459

INTERNATIONAL UPPER MANTLE COMMITTEE The East African rift system :a seminar Papers Nairobi 1965 Apr 12-17 Sponsored by the International union of geodesy and geophysics Nairobi,1965
Geod 9.0451

INTERNATIONAL UPPER MANTLE COMMITTEE The Symposium on non-elastic processes in the mantle Proceedings Newcastle-upon-Tyne 1966 Feb 21-25 Edited by D.C. Tozer With financial assistance of the International union of geodesy and geophysics Geophysical journal, 14;1-4 Blackwell,Oxford,1967
Geod 9.0695

INTERNATIONAL UPPER MANTLE COMMITTEE The World rift system symposium report Ottawa 1965 Sep 4-5 sponsored by Canada.Department of mines and technical surveys Geological survey of Canada.Paper,66-14 Ottawa,1967
Geol 8.4451

INTERNATIONAL UPPER MANTLE COMMITTEE The World rift systems :a symposium Report Ottawa 1965 Sep 4-5 Edited by T.N. Irvine Sponsored by the International union of geodesy and geophysics Geological survey of Canada.Paper, 66-14 Queen's printer,Ottawa, 1967
Geod 9.0138

INTERNATIONAL VACUUM CONGRESS 3rd Transactions Stuttgart 1965 Jun 28-Jul 2 1: invited papers International union for vacuum science.German national committee Edited by H. Adam Pergamon,Oxford,1966 Papers mainly in English,with summaries in French and German.Title pages in English, French and German
Met 25.1691

INTERNATIONAL VACUUM CONGRESS 3rd Transactions Stuttgart 1965 Jun 28-Jul 2 Vol 1: invited papers International union for vacuum science,technique and applications Edited by H. Adam 153p Pergamon press,Oxford,1966 Cover title'Advances in vacuum science and technology'
Cav 7.0015

INTERNATIONAL VACUUM CONGRESS 4th Manchester 1968 Apr 17-20 Pt 1-2 Joint British committee for vacuum science and technology International union for vacuum science,technique and applications Institute of physics and the Physical society.Conference series, 5-6 2 vols Institute of physics, London,c1968
Met 25.2834

INTERNATIONAL WOOD CHEMISTRY SYMPOSIUM International botanical congress 11th Abstracts Seattle,Wash. 1969 Aug 24-Sep 2 1969 Sep 2-4 1969
BG 38.3253

The INTERNATIONAL WOOL TEXTILE RESEARCH CONFERENCE Proceedings Sydney 1955 Aug 22-Sep 9 Vol A-F 7 vols commonwealth scientific and industrial research organization,Melbourne,1956 The conference took the form of a series of sessions...at Sydney,Geelong and Melbourne
Bioch 33.1481

INTERNATIONAL WORKING MEETING ON SOIL MICROMORPHOLOGY 2nd Proceedings Soil micromorphology Arnhem 1964 Sep 22-25 Edited by A. Jongerius Elsevier,Amsterdam, 1964
Eng 41.3115

INTERNATIONAL WORKING SESSION ON THE TEACHING OF SCHOOL BIOLOGY :report of an OECD working session Biology today:its role in education Organization for economic cooperation and development Edited by Paul Duvigneaud English edition prepared by L.S. Comber New thinking in school science OECD,Paris,1966
Gen 34.2234

INTERNATIONAL WORKING-MEETING ON SOIL MICROMORPHOLOGY 2nd proceedings Soil micromorphology Arnhem 1964 Sep.22-25 Edited by A. Jongerius Elsevier,Amsterdam, 1964
Geol 8.1511

INTERNATIONALE ARBEITSTAGUNG UBER DIE SILUR DEVON-GRENZE UND DIE STRATIGRAPHIE VON SILUR UND DEVON 2nd symposiums-band Bonn 1960 Brussels 1960 Edited by H.K. Erben Stuttgart,1962
Geol 8.2417

INTERNATIONALE GESELLSCHAFT ZUR ERFORSCHUNG DER ARKTIS MIT LUFTFAHRZEUGEN 2te :ordentl. versammlung verhandlungen Leningrad 1928 Jun 18-23 Edited by A. Berson and L. Breitfuss Petermann's mitteilungen. Erganzungsheft,201 maps 76p 28cm J. Perthes,Gotha,1929
Sco 14.0105

INTERNATIONALE GESELLSCHAFT ZUR ERFORSCHUNG DER ARKTIS MIT LUFTFAHRZEUGEN 2te :polar konferenz Verhandlungen Leningrad 1928 Jun 18-23 Edited by P. Wittenberg Russian edition illus,maps xxii,194p 26cm Leningrad,1930 Includes also t.-p. in Russian:"Trudy vtoroi polyarnoi konferentsii". Explanatory introduction in English
Sco 14.0107

INTERNATIONALE GESELLSCHAFT ZUR ERFORSCHUNG DER ARKTIS MIT LUFTFAHRZEUGEN see also INTERNATIONALE STUDIENGESELLSCHAFT ZUR ERFORSCHUNG DER ARKTIS MIT DEM LUFTSCHIFF

INTERNATIONALE JAHRESTAGUNG DES ARBEITSAUSSCHUSS MEDIZIN 10th Verhandlungsbericht Krebs - dokumentation und statistik maligner tumoren Berlin 1965 Oct 25-28 Edited by Gustav Wagner Schattauer,Stuttgart,1966
Radioth 35.0877

INTERNATIONALE POLAR-KONFERENZ :verhandlungen und ergebnisse 3te St.Petersburg 1881 Aug 1-6 Akademie der wissenschaften,St. Petersburg,1881 Parallel French-German texts.One of 8 pamphlets bound together
Philos 1.2259

INTERNATIONALE STUDIENGESELLSCHAFT ZUR ERFORSCHUNG DER ARKTIS MIT DEM LUFTSCHIFF 1ste ordentl.versammlung Verhandlungen Berlin 1926 Nov 9-13 Edited by Leonid Breitfuss Petermann's mitteilungen. Erganzungsheft,191 illus,maps 115p 28cm Justus Perthes,Gotha,1927
Sco 14.0106

INTERNATIONALE STUDIENGESELLSCHAFT ZUR ERFORSCHUNG DER ARKTIS MIT DEM LUFTSCHIFF (AEROARCTIC) 1 :versammlung Verhandlungen Berlin 1926 Nov 9-13 Edited by Leonid Breitfuss map 115p J. Perthus,Gotha,1927 Contains a paper by Shaw,W.N. 'The influence of the North Polar region upon the meteorology of the northern hemisphere'
Nap 11.0895

INTERNATIONALE STUDIENGESELLSCHAFT ZUR ERFORSCHUNG DER ARKTIS MIT DEM LUFTSCHIFF see also INTERNATIONALE GESELLSCHAFT ZUR ERFORSCHUNG DER ARKTIS MIT LUFTFAHRZEUGEN

INTERNATIONALE TAGUNG FUR ALPINE METEOROLOGIE 9. Wissenschaftliche abhandlungen Brig 1966 Sep14-17 Zermatt 1966 Sep 14-17 Edited by Karin Schram and J.C. Thams Schweizerische meteorologische zentralanstalt. Veroffentlichungen,4 illus 366p City-druck a.g.,Zurich,1967 In French,German, and Italian
Sco 14.7410

INTERNATIONALE TAGUNG UBER MATHEMATISCHE STATISTIK UND IHRE ANWENDUNGEN Berlin 1962 Sep.4-8 Edited by Erna Weber Deutsche akademie der wissenschaften zu Berlin. Abhandlungen,Klasse fur mathematik,physik und technik,4 126p 30cm Akademie-verlag, Berlin,1964
Math 3.0701

INTERNATIONALE UNIVERSITATSWOCHEN FUR KERNPHYSIK 5th proceedings Elementary particle theories Schladming 1966 Feb.24-Mar.9 Edited by Paul Urban Sponsored by the International Atomic Energy Agency Acta physica austriaca.Supplementum, 3 Springer-verlag,Vienna;New York,1966
A Math 4.0889

INTERNATIONALE VEREINIGUNG FUR BEVOLKERUNGSWISSENSCHAFT.DEUTSCHE AUSSCHLUSS Bevolkerungsfragen International kongress fur bevolkerungswissenschaft bericht 2nd Berlin 1935 Aug 26-Sep 1 Edited by Hans Harmson and Franz Lohse Lehmanns,Munich,1936
Gen 34.2069

INTERNATIONALE VEREINIGUNG FUR VEGETATIONSKUNDE Bericht Anthropogene vegetation Stolzenau 1961 Edited by Reinhold Tuxen illus. Junk,The Hague,1966
Bot 42.4761

INTERNATIONALE VEREINIGUNG FUR VEGETATIONSKUNDE Bericht Biosoziologie Stolzenau 1960 Edited by Reinhold Tuxen illus. Junk,The Hague,1965
Bot 42.4762

INTERNATIONALE VEREINIGUNG FUR VEGETATIONSKUNDE Bericht Pflanzensoziologie und landschaftsokologie Stolzenau 1963 Edited by Reinhold Tuxen illus. Junk,The Hague,1968
Bot 42.4765

INTERNATIONALE VEREINIGUNG FUR VEGETATIONSKUNDE Bericht Pflanzensoziologische systematik Stolzenau 1964 Edited by Reinhold Tuxen illus. Junk,The Hague,1968
Bot 42.4764

INTERNATIONALER CONGRESS FUR PSYCHOLOGIE 3rd Munich 1896 Aug 4-7 Congress fur psychologie.Internationales organisations-comite Lehmann,Munich,1897 Papers in English,French,German and Italian
Psy 31.2518

INTERNATIONALER GEOLOGEN KONGRESS 11 :eine sammlung von berichten unter mitwirking von fachenosson in verschliedenen landern Veranderungen des klimas seit dem maximum der letzten eiszeit International geological congress.Executive committee Verlag von generalstabens litografiska anstalt,Stockholm, 1910
Bot 42.3774

INTERNATIONALER KONGRESS FUR ELEKTRONENMIKROSKOPIE 4th Verhandlungen Berlin 1958 Sep 10-17 Band 1: physikalisch-technischer teil Edited by G. Mollenstedt and others Springer,Berlin,1960
Eng 41.5532

INTERNATIONALER KONGRESS FUR MEDIZINISCHE PHOTOGRAPHIE UND KINEMATOGRAPHIE 1st Abhandlungen Dusseldorf 1960 Sep 27-30 Edited by Heinz Orbach Thieme,Stuttgart,1962
An 32.0311

INTERNATIONALER KONGRESS FUR VERERBUNGSWISSENSCHAFT 5 Verhandlungen Berlin 1927 Band 1-2 Edited by Hans Nachtsheim Zeitschrift fur induktive abstammungs- und vererbungslehre. Supplementband, 1 2 vols Borntraeger, Leipzig,1928
Gen 34.1001

INTERNATIONALER KONGRESS UBER KERNPHYSIK UND QUANTENELEKTRODYNAMIK Basel 1948 Sep 5-9 Helvetica physica acta, 23,supp.3 Basel, 1950
Radioth 35.1271

INTERNATIONALES COLLOQUIUM UBER HALBLEITER UND PHOSPHORE Garmisch-Partenkirchen 1956 Aug 28-Sep 15 Edited by M. Schon and H. Welker Sponsored by the International union of pure and applied physics Vieweg,Brunswick, 1958
Radioth 35.1615

INTERNATIONALES KOLLOQUIUM Differentialgeometrie und topologie Zurich 1960 Jun 20-25 Schweizerische mathematische gesellschaft Sponsored by the International mathematical union Enseignement mathematique. Monographies, 11 159p 24cm L'enseignement mathematique universite,Geneva, 1962
P. Math 2.0697

INTERNATIONALES PSUCHOANALYTISCHES KONGRESS 5TH Zur psychoanalyse der kriegsneurosen : diskussion gehalten auf dem V internationalen psychoanalytischen kongress Beitrage Budapest 1918 Sep 28-29 By Sigmund Freud and others Internationale psychoanalytische bbibliothek, 1 Internationaler psychoanelytischer verlag,Leipzig;Vienna,1919
Psy 28.0171

INTERNATIONALES SYMPOSIUM UBER NATURWISSENSCHAFT UND PHILOSOPHIE... Beitrage Naturwissenschaft und philosophie Leipzig 1959 Oct 8-11 Institut fur philosophie Karl-Sudhoff institut fur geschichte der medizin und naturwissenschaften Edited by Gerhard Harig and Josef Schleifstein 437p Akademie verlag,Berlin,1960 Symposium held to commemorate the 550th anniversary of the Karl-Marx-universitat
WSM 43.0976

INTERNATIONL GEOGRAPHICAL CONGRESS 20th Abstracts of papers United Kingdom 1964 Edited by F.E.Ian Hamilton Nelson,London, 1964
Geog 13.5599

The INTERNEURON :a conference Proceedings Los Angeles,Calif. 1967 Sep Edited by Mary A.B. Brazier Sponsored by the Brain research institute UCLA forum in medical sciences, 11 University of California press,Berkeley; Los Angeles,Calif.,1969
An 32.5473

INTERPLANETARY MEDIUM The Zodiacal light and the interplanetary medium :a conference Papers Honolulu 1967 Jan 30-Feb 3 National aeronautics and space administration. Office of technology utilization NASA SP-150 430p n.a.s.a.,Washington,D.C.,1967
Obs 6.3425

INTERPLAS 1961 see INTERNATIONAL PLASTICS CONVENTION

INTERPRETATION OF TESTS AND CORRELATION WITH SERVICE Cleveland 1950 Oct 23-27 By M. F. Garwood and others American society for metals A.S.M.,1950 Lectures presented during the 32nd National metal congress and exposition
Met 25.0883

INTERPRETATION OF TESTS AND CORRELATION WITH SERVICE :a series of four educational lectures...presented during the thirty-second National metal congress and exposition Chicago,Ill. 1950 Oct 23-27 By M.F. Garwood and others American society for metals American society for metals,Cleveland, Ohio,1951
Eng 41.3837

The INTERPRETATION OF THE ULTRASTRUCTURE International society for cell biology Edited by R.J.C. Harris International society for cell biology.Symposia, 1 Academic press,New York,1962 Based on a symposium held in Berne in September 1961.
Med 36.0177

INTERPRETATION OF ULTRASTRUCTURE :a symposium Bern 1961 Edited by R.J.C. Harris International society for cell biology.Symposia, 1 Academic press,New York;London,1962
Radioth 35.0517

The INTERPRETATION OF ULTRASTRUCTURE : symposium Bern 1961 International society for cell biology Edited by R.J.C. Harris International society for cell biology.Symposia, 1 illus Academic press, New York;London,1962
Bioch 33.0964

The INTERPRETATION OF ULTRASTRUCTURE : symposium Bern 1961 International society for cell biology Edited by R.J.C. Harris International society for cell biology.Symposia, 1 illus. x,438p Academic press,New York;London,1962
Bot 42.1080

INTERSOCIETY COMMITTEE FOR RESEARCH POTENTIAL IN PATHOLOGY,INC. Quantitive electron microscopy :a symposium Proceedings Washington,D.C. 1964 Mar 30 - Apr 3 Edited by Gunter F. Bahr and Elmar H. Zeitler Laboratory investigation, 14 no.6 illus. viii,605p Baltimore,Md.,1965
Bot 42.4731

INTERSPECIFIC HYBRIDIZATION IN PLANTS Symposium on interspecific hybridization in plants Sofia 1964 Nov 10-12 Edited by R. Georgieva and D. Tsikov Izdatelstvo Bolgarskoi akademii nauk,Sofia,1965 In Russian,English summaries
Gen 34.1025

INTERSTELLAR COMMUNICATION Extraterrestrial civilizations All-Union conference on extraterrestrial civilizations and interstellar communications 1st Proceedings Byurakan 1964 May 20-23 Byurakanskaya astrofizicheskaya observatoriya Edited by S.M. Tovmasyan Translated by Z. Lerman from the Russian Israel program for scientific translations,Jerusalem,1967
TA 15.0322

INTERSTELLAR GAS DYNAMICS Symposium on cosmical gas dynamics 6th Proceedings Yalta 1969 Sep 8-18 International astronomical union International union of theoretical and applied mechanics International astronomical union.Symposium, 39 388p Reidel,Dordrecht,1970
TA 15.0541

INTERSTELLAR GAS DYNAMICS Symposium on cosmical gas dynamics 6th Proceedings Yalta 1969 Sep 8-18 International astronomical union International union of theoretical and applied mechanics International astronomical union.Symposium, 39 388p Reidel,Dordrecht,1970
Obs 6.3323

INTERSTELLAR GAS DYNAMICS Symposium on cosmical gas dynamics 6th Proceedings Yalta 1969 Sep 8-18 International union of theoretical and applied mechanics and International astronomical union Edited by H. J. Habing International astronomical union. Symposium, 39 diagrs,photos,graphs xii, 388p 24cm Reidel,Dordrecht,1970
Cav 7.2710

INTERSTELLAR GRAINS :an international colloquium Proceedings Troy,N.Y. 1965 Aug 24-26 Rensselaer polytechnic institute Edited by J.Mayo Greenberg and T.P. Roark NASA SP-140 National aeronautics and space administration,Washington,D.C.,1967
TA 15.0445

INTERSTELLAR IONIZED HYDROGEN :a symposium Charlottesville,Va. 1967 Dec National radio astronomy observatory Edited by Yervant Jerzian Co-sponsored by the Arecibo ionospheric observatory 774p Benjamin,New York,1968
Obs 6.3447

INTERSTELLAR IONIZED HYDROGEN :a symposium Charlottesville,Va. 1967 Dec National radio astronomy observatory Edited by Yervant Terzian Jointly sponsored by Arecibo ionospheric observatory 774p Benjamin,New York,1968
TA 15.0437

INTERSTELLAR MATTER Distribution and motion of interstellar matter in galaxies The Conference on problems of the distribution and motion of interstellar matter in galaxies Proceedings Princeton,N.J. 1961 Apr 10-20 Institute for advanced study Edited by L. Woltjer 330p Benjamin,New York,1962
TA 15.0213

IRON AND STEEL INSTITUTE Recent developments in annealing London 1963 May 2 Iron and steel institute.Special report, 79 Iron and steel institute,London,1963
Met 25.0418

IRON AND STEEL INSTITUTE Refractories for oxygen steelmaking technical sessions of the annual general meeting... Papers and discussions London 1962 May 3 Iron and steel institute.Special report, 74 Iron and steel institute,London,1962
Met 25.0273

IRON AND STEEL INSTITUTE Stainless steels Conference on stainless steels for the fabricator and user Proceedings Birmingham 1968 Sep 10-12 Iron and steel institute, London,1969
Met 25.2242

IRON AND STEEL INSTITUTE Steelmaking in the basic arc furnace Iron and steel institute conference :technical sessions Papers and discussions Sheffield 1964 Jul 6-10 Iron and steel institute.Special report, 87 London,1964
Met 25.2558

IRON AND STEEL INSTITUTE Steels for reactor pressure circuits :a symposium London 1960 Nov 30-Dec 2 Iron and steel institute. Special report, 69 Iron and steel institute, London,1961
Met 25.0414

IRON AND STEEL INSTITUTE Strong tough structural steels joint conference... Proceedings Scarborough 1967 Apr 4-6 Iron and steel institute.Publication, 104 Eyre and Spottiswoode,Margate,1967
Met 25.0428

IRON AND STEEL INSTITUTE Structural processes in creep :a symposium Report London 1961 May 3-4 Iron and steel institute.Special report, 70 Iron and steel institute,London,1961
Met 25.2334

IRON AND STEEL INSTITUTE Symposium on powder metallurgy Iron and steel institute.Special report, 38 Iron and steel institute,London, 1947
Met 25.0753

IRON AND STEEL INSTITUTE Symposium on powder metallurgy London 1954 Dec 1-2 Iron and steel institute.Special report, 58 Iron and steel institute,London,1956
Met 25.0761

IRON AND STEEL INSTITUTE Symposium on powder metallurgy London 1954 Dec 1-2 Iron and steel institute.Special report, 58 illus. Iron and steel institute,London,1956
Eng 41.3477

IRON AND STEEL INSTITUTE Symposium on sinter London 1953 Nov Iron and steel institute. Special report, 53 Iron and steel institute, London,1955
Met 25.0321

IRON AND STEEL INSTITUTE Symposium on steel making (acid and basic open-hearth practice) Iron and steel institute.Special report, 22 Iron and steel institute,London,1938
Met 25.0324

IRON AND STEEL INSTITUTE Symposium on the welding of iron and steel London 1935 May 2-3 Vol 1 2 vols Iron and steel institute,London,1935
Eng 41.3947

IRON AND STEEL INSTITUTE Symposium on the welding of iron and steel London 1935 May 2-3 Vol 1-2 2 vols Iron and steel institute,London,1935
Met 25.0783

IRON AND STEEL INSTITUTE The Determination of gases in metals :a symposium Report Organised in conjunction with the Institute of metals Iron and steel institute.Special report, 68 Iron and steel institute,London, 1960
Met 25.1747

IRON AND STEEL INSTITUTE The Solidification of metals :the conference Proceedings Brighton 1967 Dec 4-7 Jointly organized with the Institute of physics and the physical society and the Institution of metallurgy Iron and steel institute.Publication, 110 Iron and steel institute,London,1968
Met 25.1529

IRON AND STEEL INSTITUTE Thermal and high-strain fatigue an international conference Proceedings London 1967 Jun 6-7 Metals and metallurgy trust.Monograph and report series, 32 Metals and metallurgy trust, London,1967
Met 25.0942

IRON AND STEEL INSTITUTE Vacuum degassing of steel Iron and steel institute annual general meeting 1965 Papers and discussions London 1965 May 5-6 Iron and steel institute. Special report, 92 Iron and steel institute, London,1965
Met 25.0372

IRON AND STEEL INSTITUTE.ANNUAL GENERAL MEETING 1962 Refractories for oxygen steelmaking technical sessions of the annual general meeting... Papers and discussions London 1962 May 3 Iron and steel institute Iron and steel institute.Special report, 74 Iron and steel institute,London,1962
Met 25.0273

IRON AND STEEL INSTITUTE.AUTUMN GENERAL MEETING 1961 Future of ironmaking in the blast furnace autumn general meeting... Papers and discussions London 1961 Nov Iron and steel institute Iron and steel institute. Special report, 72 Iron and steel institute, London,1962
Met 25.0368

IRON AND STEEL INSTITUTE.AUTUMN GENERAL MEETING 1965 Pilot plants in the iron and steel industry autumn general meeting 1965 Papers and discussions London 1965 Nov 30-Dec 1 Iron and steel institute Iron and steel institute.Special report, 96 Iron and steel institute,London,1966
Met 25.0367

IRON AND STEEL INSTITUTE.ENGINEERS GROUP Effect of the various steelmaking processes on the energy balances of integrated iron and steelworks The Iron and steel engineers group meeting 43rd Report London 1961 Dec 1 London,1962
Met 25.2522

IRON AND STEEL INSTITUTE.ENGINEERS GROUP Ore mining and materials handling Iron and steel engineers group meeting 47th Papers and discussions Metz 1963 Jun 10-14 Iron and steel institute.Special report, 82 Iron and steel institute,London,1963
Met 25.0208

IRON AND STEEL INSTITUTE.WELDING COMMITTEE Report on the symposium on the welding of iron and steel,May 1935 Iron and steel institute, London,1936
Met 25.0782

IRON AND STEEL INSTITUTE CONFERENCE :technical sessions Papers and discussions Steelmaking in the basic arc furnace Sheffield 1964 Jul 6-10 Iron and steel institute Iron and steel institute.Special report, 87 London,1964
Met 25.2558

IRON AND STEEL WORKS International conference on tribology in iron and steel works 1st Proceedings London 1969 Sep 22-25 Iron and steel institute Institution of mechanical engineers Iron and steel institute.Publication, 125 Iron and steel institute,London,1970
Met 25.2562

IRON DEFICIENCY Iron deficiency;pathogenesis, clinical aspects and therapy Clinical symposium on iron deficiency Proceedings Basle 1969 Edited by L. Hallberg and others Colloquia Geigy Academic press, London,1970
Med 36.0383

IRON DEFICIENCY;PATHOGENESIS,CLINICAL ASPECTS AND THERAPY Clinical symposium on iron deficiency Proceedings Basle 1969 Edited by L. Hallberg and others Colloquia Geigy Academic press,London,1970
Med 36.0383

IRON PROTEINS Non-heme iron proteins;role in energy conversion :a symposium Proceedings Yellow Springs,Ohio 1965 Mar 22-24 Edited by Anthony San Pietro Sponsored by the Charles F.Kettering research laboratory Charles F.Kettering research laboratory. Contribution, 201 Antioch press,Yellow Springs,Ohio,1965
Bioch 33.0546

IRONMAKING TOMORROW :conference Proceedings London 1966 Nov 22-23 Iron and steel institute Iron and steel institute. Publication, 102 Iron and steel institute, London,1967
Met 25.0380

IRVINE 1967 Localized excitations in solids International conference on localized excitations in solids 1st Proceedings International union of pure and applied physics National science foundation Edited by R.F. Wallis xvi,782p 26cm Plenum press,New York,1968
Cav 7.2857

IRVINE 1967 Localized excitations in solids The International conference on localized excitations in solids 1st Edited by R.F. Wallis Plenum press,New York, 1968
TA 15.0440

ISOANTIGENS AND CELL INTERACTIONS :a symposium Philadelphia,Pa. 1965 Mar 17 Wistar institute of anatomy and biology Edited by Joy Palm Wistar institute symposium monograph, 3 Wistar institute press,Philadelphia,Pa.,1965
Path 30.2416

ISOTOPE MASS EFFECTS IN CHEMISTRY AND BIOLOGY Symposium on isotope mass effects in chemistry and biology Proceedings Vienna 1963 Dec 9-13 Edited by N. Grell Held by the International union of pure and applied chemistry Pure and applied chemistry, 8,no. 3-4 Butterworths,London,1964
Radioth 35.0421

ISOTOPE SEPARATION International symposium on isotope separation 1st Proceedings Amsterdam 1957 Apr 23-27 Netherlands physical society and International union of pure and applied physics Edited by J. Kistemaker and others xx,704p North-Holland,Amsterdam,1958
Chem E 24.1654

ISOTOPE TECHNIQUES CONFERENCE Proceedings Oxford 1951 Jul Vol 2: industrial and allied research applications Great Britain. Ministry of supply Sponsored by the Atomic energy research establishment H.M.S.O., London,1952
Chem 18.2467

ISOTOPE TECHNIQUES CONFERENCE Proceedings Radioisotope techniques Oxford 1951 Jul 16-20 Vol 1: medical and physiological applications Sponsored by the Atomic energy research establishment H.M.S.O.,London,1953 Subsequent conference called Radioisotope conference,q.v.
Radioth 35.1348

ISOTOPE TECHNIQUES FOR HYDROLOGY Vienna 1962 Dec 17-21 International atomic energy agency International atomic energy agency. Technical reports series, 23 Vienna,1964
Bot 42.3318

ISOTOPES A Symposium on the use of isotopes in biology and medicine Addresses Madison, Wisc. 1947 Sep 10-15 Wisconsin alumni research foundation University of Wisconsin press,Madison,Wisc.,1948
Chem 18.1259

ISOTOPES A Symposium on the use of isotopes in biology and medicine Madison,Wis. 1947 Sep 10-13 By Hans T. Clarke and others University of Wisconsin University of Wisconsin press,Madison,1948
Bioch 33.0283

ISOTOPES Application of isotope techniques in hydrology :a comprehensive report Vienna 1961 Nov 6-9 International atomic energy agency International atomic energy agency. Technical reports series, 11 Vienna,1962
Bot 42.3317

ISOTOPES Conference on applications of isotopes in scientific research Proceedings Melbourne 1950 Aug 14-17 Commonwealth scientific and industrial research organization University of Melbourne. Chemistry department University of Melbourne, Melbourne,1951
Radioth 35.1495

ISOTOPES Isotope techniques for hydrology Vienna 1962 Dec 17-21 International atomic energy agency International atomic energy agency.Technical reports series, 23 Vienna, 1964
Bot 42.3318

ISOTOPES IN BIOCHEMISTRY Ciba foundation conference on isotopes in biochemistry London 1951 Mar 12-15 Ciba foundation Edited by G.E.W. Wolstenholme Illus J.and A.Churchill,London,1951
Radioth 35.0023

ISOTOPES IN BIOCHEMISTRY Ciba foundation conference on isotopes in biochemistry Papers London 1951 Mar 12-15 Ciba foundation Edited by J.N. Davidson Churchill,London,1951
Bioch 33.0274

ISOTOPES IN EXPERIMENTAL PHARMACOLOGY International conference on the uses of isotopically labeled drugs in experimental pharmacology Lectures Chicago,Ill. 1964 Jun 7-9 Edited by Lloyd J. Roth University of Chicago press,Chicago,Ill.,1965
Pha 16.0155

ISPRA,VARESE 1960 Radiation damage nei solidi Scuola internazionale di fisica "Enrico Fermi" 28 corso Rendiconti Societa Italiana di fisica Edited by D.S. Billington Academic press,New York;London, 1962
Met 25.1579

ISPRA (VARESE) 1960 Radiation damage nei solidi Scuola internazionale di fisica 'Enrico Fermi' 18 corso Rendiconti Societa italiana di fisica Edited by D.S. Billington Academic press,New York;London, 1962 Contributions in English and French
Chem 18.1144

ISRAEL ACADEMY OF SCIENCE AND HUMANITIES Mathematical logic and foundations of set theory International colloquium on mathematical logic and foundations of set theory Proceedings Jerusalem 1908 Nov 11-14 Edited by Yehoshua Bar-Hillel Studies in logic and the foundations of mathematics bibliog. 145p 24cm North-Holland, Amsterdam;London,1970
P Math 2.3795

ISRAEL ACADEMY OF SCIENCES AND HUMANITIES Logic,methodology and philosophy of science International congress for logic,methodology and philosophy of science Proceedings Jerusalem 1964 Aug 26-Sep 2 Edited by Yehoshua Bar-Hillel viii,440p North-Holland,Amsterdam,1965
WSM 43.0829

ISTANBUL 1952 International congress on theoretical and applied mechanics 8th proceedings Vols 1-2 Istanbul universitesi.Faculty of science Edited by Kerim Erim Under the auspices of the International committee for the congresses of applied mechanics 2 vols University faculty of science,Istanbul,1953
A Math 4.0326

ISTANBUL 1960 Navigation systems for aircraft and space vehicles :papers presented at the AGARD avionics panel meeting Agard Edited by T.G. Thorne AGARDograph, 55 Pergamon press,London,1962
Eng 41.5706

ISTANBUL 1964 Modern quantum chemistry : Istanbul international school of quantum chemistry Lectures Pt 3: action of light and organic crystals North Atlantic treaty organization.Pure science bureau Edited by Oktay Sinanoglu Organized with the cooperattion of Orta Dogu teknik universitesi Istanbul lectures:modern quantum chemistry,3 Academic press,New York;London,1965
Chem 18.2238

ISTANBUL 1970 Advanced radar systems 19th Avionics panel meeting Agard. Avionics panel AGARD conference proceedings, 66 Agard,Neuilly-sur-Seine,1970
Eng 41.8322

ISTANBUL LECTURES:MODERN QUANTUM CHEMISTRY,3 Modern quantum chemistry :Istanbul international school of quantum chemistry Lectures Istanbul 1964 Aug 16-Sep 5 Pt 3: action of light and organic crystals North Atlantic treaty organization.Pure science bureau Edited by Oktay Sinanoglu Organized with the cooperattion of Orta Dogu teknik universitesi Academic press,New York;London, 1965
Chem 18.2238

ISTANBUL UNIVERSITESI.FACULTY OF SCIENCE International congress on theoretical and applied mechanics 8th proceedings Istanbul 1952 Aug.20-28 Vols 1-2 Edited by Kerim Erim Under the auspices of the International committee for the congresses of applied mechanics 2 vols University faculty of science,Istanbul,1953
A Math 4.0326

ISTITUTO DE RICERCHE FARMACOLOGICHE "MARIO NEGRI" Antidepressant drugs :international symposium 1st Proceedings Milan 1966 Apr 25-27 Edited by S. Garattini and M.N.G. Dukes Excerpta medica international congress series,122 Excerpta medica,Amsterdam,1967
Pha 16.0046

ISTITUTO DI GIACIMENTI MINERALI UNIVERSITARIA Remobilization of ores and minerals Convegno sulle rimobilizzazione del minerali metallici e non metallici Rendiconti minerali metallici e non metallici Papers Cagliari 1969 Aug Cagliari,c1969
Min 10.1489

ISTITUTO DI RICERCHE FARMACOLOGICHE "MARIO NEGRI" Symposium on catecholamines 2nd Milan 1965 Jul 4-9 American society for pharmacology and experimental therapeutics Edited by George H. Acheson Williams and Wilkins company,Baltimore,1966 Reprinted from the Pharmacological reviews,18,no.1
Bioch 33.0430

ISTITUTO GREGORIO MENDEL Symposium internationale geneticae medicae 1st Atti Rome 1953 Sep 6-7 1 Edited by Luigi Gedda Analecta genetica, 1 Edizioni dell' istituto Gregorio Mendel,Rome,1954
Med 36.0148

ISTITUTO GREGORIO MENDEL Symposium internazionale geneticae medicae 1 Atti Rome 1953 Sep 6-7 4: il parto indolore Edited by Luigi Gedda Analecta genetica, 1 Edizioni dell' istituto Gregorio Mendel,Rome, 1956
Med 36.0151

ISTITUTO GREGORIO MENDEL Symposium internazionale geneticae medicae 1st Atti Rome 1953 Sep 6-7 2: krampfbereitschaft Edited by Luigi Gedda Analecta genetica, 1 Edizioni dell' istituto Gregorio Mendel,Rome, 1954
Med 36.0149

ISTITUTO GREGORIO MENDEL Symposium internazionale geneticae medicae Rome 1953 Sep 6-7 3: chondrodysplasie Edited by Luigi Gedda Analecta genetica, 1 Edizioni dell' istituto Gregorio Mendel,Rome, 1954
Med 36.0150

ISTITUTO LOMBARDO ACCADEMIA DI SCIENZE E LETTERE Symposium on the germ cells and earliest stages of development Pallanza 1960 Sep 14-20 International institute of embryology A. Baselli,Milan,1961 At head of title: Istituto Lombardo accademia di scienze e lettere
Bal 39.0459

ISTITUTO NAZIONALE DE ALTA MATEMATICA Symposia mathematica Rome 1967-68 1,3 bibliog. 25cm 2 vols Academic press, New York;London,1969
P Math 2.3778

ISTITUTO NAZIONALE DI ALTA MATEMATICA Meccanica non lineare e stabilita Symposia mathematica Rome 1970 Feb 23-26 Istituto nazionale di alta matematica.Pubblicazione, 6 395p 25cm Academic press,London;New York,1971
P Math 2.3989

ISTITUTO NAZIONALE DI ALTA MATEMATICA Questioni di elasticita non linearizzata Edited by Antonio Signorini Cremonese,Rome, 1960 Extract of conference held 1959-60 published in 'Rendiconti di matematica e delle sue applicazioni,'serie 5
Eng 41.2484

ISTITUTO PER LO STUDIO SPERIMENTALE DI PROBLEMI SOCIALI CON TECNICHE FILMOLOGICHE Conferenza internazionale di informazione visiva :modalita dell'informazione visiva: ricerca scientifica e azione politica 1mo 1-2 2 vols Istituto per lo studio sperimentale di problemi sociali con tecniche filmologiche,Milan,1961 Papers in English, French and Italian
Psy 31.2545

ITALIAN SOCIETY OF PHYSICS International symposium on residual gases in electron tubes and related vacuum systems 3rd Papers Rome 1967 Mar 14-17 Section 2: sorption-desorption phenomena in high vacuum Nuovo cimento.Supplement, 5,2 Editrice composori, Bologna,1967
Met 25.2366

ITALY.NAVY.HYDROGRAPHIC INSTITUTE International geographical congress Esplorazione scientifica del Mar Rosso : comunicazione al congresso geografice internazionale Cambridge 1928 Jul Genoa, 1928 Bound with English translation
Geog 13.0854

ITHACA,N.Y. May 13-16 The Transfer of calcium and strontium across biological membranes :a conference Proceedings Edited by R.H. Wasserman Academic press,New York; London,1963
Inv Med 37.0227

ITHACA,N.Y. 1926 International congress of plant sciences Proceedings Vol 1-2 Edited by B.M. Duggar 2 vols Banta,Menasha, Wis.,1929
Bot 42.0653

ITHACA,N.Y. 1932 International congress of genetics 6th Proceedings Vol 1-2 Edited by Donald F. Jones Brooklyn botanic garden,Brooklyn,N.Y.,1932 Bound together
Gen 34.1002

ITHACA,N.Y. 1935 The Symposium on colloid chemistry 12th Papers presented Edited by Harry Boyer Weiser Colloid symposium monograph, 12 Williams and Wilkins, Baltimore,1936
Bioch 33.1479

ITHACA,N.Y. 1946 Preparation and characteristics of solid luminescent materials a conference American physical society. Division of electron optics Edited by Gorton R. Fonda and Frederick Seitz Published under the auspices of the National research council graphs xv,459p 22cm Wiley,New York,1948
Cav 7.3061

ITHACA,N.Y. 1946 Preparation and characteristics of solid luminescent materials a symposium Sponsored by the American physical society.Division of electron optics Wiley;Chapman and Hall,New York;London,1948
Eng 41.5491

ITHACA,N.Y. 1948 Phase transformations in solids :a symposium Papers Edited by R. Smoluchowski and others Sponsored by the National research council.Committee on solids Wiley,New York,1951
Min 10.0640

ITHACA,N.Y. 1962 International congress of the history of science 10th Proceedings Vol 1-2 Academie internationale d'histoire des sciences History of science society Co-sponsored by the United States national academy 2 vols Hermann,Paris Congress chairman:H.Guerlac
WSM 43.0030

ITHACA,N.Y. 1962 Symposium (international) on combustion 9th Papers Combustion institute Academic press,New York;London, 1963
Chem 18.0439

ITHACA,N.Y. 1962 The Transfer of calcium and strontium across biological membranes :a conference Proceedings Edited by R.H. Wasserman held at Cornell university Academic press,New York;London,1963
Bioch 33.1021

Libraries and collections

A.Biol.	Department of Applied Biology.
A.Math.	Department of Applied Mathematics and Theoretical Physics.
An.	Department of Anatomy.
B.G.	Cory Library, Botanic Garden.
Bal.	Balfour and Newton Libraries, Department of Zoology.
Bioch.	Colman Library, Department of Biochemistry.
Bot.	Department of Botany.
Cav.	Rayleigh Library, Cavendish Laboratory, Department of Physics.
Chem.	Joint Chemical Library.
Chem.E.	Department of Chemical Engineering.
Col.S.	Oliver Gatty Library, Colloid Science Laboratory.
Eng.	Department of Engineering.
Gen.	Department of Genetics.
Geod.	Department of Geodesy and Geophysics.
Geog.	Department of Geography.
Geol.	Sedgwick Geology Library, Sedgwick Museum.
H.E.	Human Ecology Library (now incorporated in the Department of Medicine).
Hunt.	Hunterian collection, University Library (formerly in the Department of Pathology).
Inv.Med.	Department of Investigative Medicine.
Math.	Wishart Library, Statistical Laboratory.
Math.L.	University Computer Laboratory (formerly the Mathematical Laboratory).

Math.S.	Wishart Library, Statistical Laboratory.
Med.	University Medical Library (formerly the Department of Medicine Library).
Met.	Department of Metallurgy and Materials Science.
Min.	Department of Mineralogy and Petrology.
Mol.	Nuttall Library (Parasitology), Molteno Institute.
Nap.	Napier Shaw Library (Meteorology), Department of Physics.
Obs.	Observatories Library, Institute of Astronomy.
P.G.M.S.	University Medical Library (formerly Post-Graduate Medical Library).
P.Math.	Department of Pure Mathematics and Mathematical Statistics.
Path.	Kanthack Library, Department of Pathology.
Philos.	Cambridge Philosophical Society book collection, Scientific Periodicals Library.
Phys.	Department of Physiology.
Psy.	Department of Experimental Psychology.
Radioth.	University Medical Library (formerly Radiotherapeutics Library).
Sco.	Scott Polar Research Institute.
Surg.	Department of Surgery.
T.A.	Institute of Astronomy.
V.A.	Sub-Department of Veterinary Anatomy.
W.S.M.	Whipple Science Museum.

Union Catalogue of Scientific Libraries in the University of Cambridge

Scientific Conference Proceedings
1644-1972

II

Compiled at the Scientific Periodicals Library,
University of Cambridge

MANSELL
1975

Mansell Information/Publishing Limited
3 Bloomsbury Place, London WC1A 2QA

International Standard Book Number: 0 7201 0531 5

Text reproduced from computer printout supplied by the compilers, printed by photolithography and bound at The Scolar Press Limited, Ilkley, Yorkshire

ITHACA,N.Y. 1962 The Transfer of calcium and strontium across biological membranes :a conference Proceedings Edited by R.H. Wassermann Academic press,New York,1963
Pha 16.0200

ITHACA,N.Y. 1963 The Nature of time :a meeting convened by T.Gold and H.Bondi Cornell university Edited by T. Gold xiv, 248p Cornell university press,Ithaca,N.Y., 1967
WSM 43.1104

ITHACA,N.Y. 1963 The Nature of time : report of a meeting Cornell university Edited by T. Gold and D.L. Schumacher ix, 248p 23cm Cornell university,Ithaca,N.Y., 1967
Cav 7.2764

ITHACA,N.Y. 1964 Metric geometry over affine spaces Cooperative summer seminar 1st By Ernst Snapper Mathematical association of America Held at Cornell university bibliog. iv,166p 28cm Mathematical association of America,Buffalo,N. Y.,1964 Duplicated notes
P Math 2.3829

ITHACA,N.Y. 1965 Relativity theory and astrophysics 1:relativity and cosmology Lectures By W. Bonner and others American mathematical society.Summer seminar in applied mathematics Edited by Jurgen Ehlers Lectures in applied mathematics, 8 xvi, 292p American mathematical society, Providence,R.I.,1967
TA 15.0350

ITHACA,N.Y. 1965 Relativity theory and astrophysics 2:galactic structure Lectures By E. Burbidge and others American mathematical society.Summer seminar in applied mathematics Edited by Jurgen Ehlers viii, 220p American mathematical society, Providence,R.I.,1967
TA 15.0351

ITHACA,N.Y. 1968 The Emergence of order in developing systems Society for developmental biology Edited by Michael Locke Society for developmental biology.Symposia, 27 350p Academic press,New York,1968
Bot 42.1097

ITHACA,NEW YORK 1962 Symposium on combustion 9th proceedings Organized by the Combustion Institute Academic press,New York,1963
A Math 4.0970

IUB-IUBS INTERNATIONAL SYMPOSIUM 1st Proceedings Biological structure and function Stockholm 1960 Sep 12-17 Vol 1-2 International union of biochemistry International union of biological sciences Edited by T.W. Goodwin and O. Lindberg 2 vols Academic press,London,1961
Gen 34.0826

IUTAM COLLOQUIUM 1955 Verformung und fliessen des festkorpers Deformation and flow of solids :colloquium Madrid 1955 Sep 26-30 International union of theoretical and applied mechanics Edited by R. Grammel Springer,Berlin,1956 Papers in English, French and German
Met 25.1203

JAARLIJKS SYMPOSIUM OVER PHYTOPHARMACIE 2nd symposium Papers Ghent 1950 Apr 23 Rijkslandbouwhogeschool,Ghent Landbouwhogeschool en de opzoekingsstations, Ghent.Mededelingen, 15,no 1 Rijkslandbouwhogeschool,Ghent,1950 Papers and summaries in English,Flemish and French
Chem 26.0105

JAARLIJKS SYMPOSIUM OVER PHYTOPHARMACIE 5th symposium Verhandelingen Ghent 1953 May 5 Rijkslandbouwhogeschool,Ghent Edited by J.van den Brande Landbouwhogeschool en de opzoekingsstation,Ghent.Mededelingen, 18,no 2 Rijkslandbouwhogeschool,Ghent,1953 Papers and summaries in English,Flemish and French
Chem 26.0106

JABLONNA 1961 Fluid dynamics transactions Symposium on fluid dynamics 5th Vol. 1 Polska akademia nauk Edited by W. Fiszdon Pergamon press,London,1964
A Math 4.0527

JABLONNA 1962 Relativistic theories of gravitation A Conference on gravitation Proceedings Polska akademia nauk Edited by Leopold Infeld With financial support of the International union of pure and applied physics Pergamon press,Oxford,1964 Binder's title in French
TA 15.0085

JABLONNA 1962 Relativistic theories of gravitation Conference on relativistic theories of gravitation proceedings Edited by Leopold Infeld Organised and sponsored by Polska.akademia nauk Pergamon press,Oxford, 1964
A Math 4.1029

JAIPUR 1963 International conference on cosmic rays 8th proceedings Vols. 1-6 International union of pure and applied physics.Cosmic ray commission Edited by R.R. Daniel and others Sponsored by India. Department of atomic energy 6 vols Bombay, 1964
A Math 4.1159

JAPAN 1957 Regional geography of Japan :I. G.U.regional conference in Japan Guidebook 1-7 International geographical union,and, Science council of Japan maps Tokyo,1957
Geog 13.4507

JAPAN CONGRESS ON TESTING MATERIALS 6th Kyoto 1962 Oct 10111 Science council of Japan Japan society for testing materials, Kyoto,1963
Met 25.2860

JAPAN ENDOCRINOLOGICAL SOCIETY Annual meeting of Japan endocrinological society 42nd Proceedings Maebashi 1969 May 15-17 1969
PGMS 29.0587

JAPAN ENDOCRINOLOGICAL SOCIETY Annual meeting of Japan endocrinological society 43rd Proceedings Osaka 1970 Mar 10-12 Kyoto,1970
PGMS 29.0588

JAPAN INSTITUTE OF METALS International conference on the strength of metals and alloys Proceedings Tokyo 1967 Sep 4-8 xlvi,1049p 31cm Japan institute of metals, Sendai,1968 In commemoration of the 30th anniversary of the institute
Cav 7.3081

JAPAN INSTITUTE OF METALS Science,technology and application of titanium International conference on titanium Proceedings London 1968 May 21-24 Edited by R.I. Jaffee and N.E. Promisel Pergamon,Oxford,1970
Met 25.2852

JAPAN INSTITUTE OF METALS The Strength of metals and alloys the international conference in commemoration of the 30th anniversary of the Japan institute of metals Proceedings Tokyo 1967 Sep 4-8 Japan institute of metals.Transactions, 9 suppl. Japan institute of metals,Tokyo,1968
Met 25.0829

JAPAN NATIONAL COMMITTEE FOR THEORETICAL AND APPLIED MECHANICS Japan national congress for applied mechanics 1-16th Proceedings 1952-67 16 vols Science council of Japan, Tokyo,1952-67
Eng 41.1758

JAPAN NATIONAL COMMITTEE FOR THEORETICAL AND APPLIED MECHANICS Japan national congress for applied mechanics 18th Proceedings Tokyo 1968 Nov 8-9 Bibliog,illus 188p 26cm Central scientific publishers,Tokyo, 1970
A Math 4.1725

JAPAN NATIONAL CONGRESS FOR APPLIED MECHANICS 1-16th Proceedings 1952-67 Sponsored by Japan national committee for theoretical and applied mechanics 16 vols Science council of Japan,Tokyo,1952-67
Eng 41.1758

JAPAN NATIONAL CONGRESS FOR APPLIED MECHANICS 18th Proceedings Tokyo 1968 Nov 8-9 Science council of Japan Japan national committee for theoretical and applied mechanics Bibliog,illus 188p 26cm Central scientific publishers,Tokyo,1970
A Math 4.1725

JAPAN SOCIETY FOR CELL BIOLOGY Intracellular membraneous structure The International symposium for cellular chemistry 1st Proceedings Ohtsu 1963 Mar 27-31 Edited by S. Seno and E.V. Cowdry Society for cellular chemistry.Symposia, 14,Suppl. Okayama,1964 In memory of...Dr. Seizo Katsunuma.Japan society for cell biology formerly Japan society for cellular chemistry
Bioch 33.0966

JAPAN SOCIETY FOR CELL BIOLOGY Nucleic acid metabolism,cell differentiation and cancer growth International symposium for cellular chemistry 2nd Proceedings Ohtsu 1966 Oct 17-21 Edited by E.V. Cowdry and S. Seno Pergamon,London,1969
Radioth 35.0561

JAPAN SOCIETY FOR THE PROMOTION OF SCIENCE and UNESCO Unesco symposium on physical oceanography Proceedings Tokyo 1955 Oct 19-22 Unesco and Japan society for the promotion of science,Tokyo,1957
Geog 13.0875

JAPAN SOCIETY OF CIVIL ENGINEERS Failure and defects of bridges and structures :a symposium Proceedings Tokyo 1957 Sep 15 Japan society for the promotion of science,Tokyo, 1959
Eng 41.3015

JAPAN SOCIETY OF CIVIL ENGINEERS Prestressed structures :a symposium Proceedings Tokyo 1959 Sep 14 Japan society for the promotion of science,Tokyo,1960
Eng 41.3016

JAPAN SOCIETY OF CIVIL ENGINEERS Safety of structures symposium 2nd Proceedings Kyoto 1955 Sep 6 Science council of Japan, Kyoto,1956
Eng 41.3014

JAPAN SOCIETY OF CIVIL ENGINEERS World conference on earthquake engineering 2nd Proceedings Tokyo 1960 Jul 11-18 and Kyoto 1960 Jul 11-18 Vol 1-3 Organised by the Science council of Japan 3 vols Science council of Japan,Tokyo,1960
Eng 41.6395

JAPAN-U.S.COOPERATIVE SCIENCE PROGRAM Comparative biochemistry and biophysics of photosynthesis :a conference Papers Hakone 1967 Aug 12-15 Edited by K. Shibata and others illus viii,445p University of Tokyo press;University Park press,Tokyo; University Park,Pa.,1968
Bot 42.1903

JAPAN-UNITED STATES COOPERATIVE SCIENCE PROGRAM Comparative biochemistry and biophysics of photosynthesis :a conference Papers presented Hakone,Japan 1967 Aug 12-15 Edited by K. Shibata and others University of Tokyo press;University park press,State college,Tokyo;University Park,Pa.,1968
Bioch 33.2349

JAPAN-UNITED STATES SEMINAR ON ORDINARY DIFFERENTIAL FUNCTIONAL EQUATIONS Proceedings Kyoto 1971 Sep 6-11 Edited by Minoru Urabe Lecture notes in mathematics, 243 viii,332p 25cm Springer, Berlin,1971
P Math 2.4318

JAPANESE CONGRESS ON TESTING MATERIALS 1st-8th Proceedings 1957-65 8 vols Japan society for testing materials,Kyoto,1958-65
Eng 41.3844

JAPANESE NATIONAL COMMISSION FOR UNESCO Symposium on typhoons Proceedings Tokyo 1954 Nov 9-12 illus 257p Japanese national commission for Unesco,Tokyo,1955
Nap 11.0612

JAPANESE NATIONAL COMMISSION FOR UNESCO Unesco symposium on typhoons Proceedings Tokyo 1954 Nov 9-12 Japanese national commission for Unesco,Tokyo,1955
Geog 13.0962

JARDIN BOTANIQUE DE GENEVE Multiples fonctions d'un jardin botanique :symposium international de Geneve Actes Geneve 1968 Jul 29-Aug 3 Edited by Jacques Miege Boissiera, 14 Geneva,1969 Symposium held to celebrate the 150th anniversary of the foundation of the botanic garden at Geneva
BG 38.3115

JASPER,ALTA. 1955 Jurassic and carboniferous of western Canada,with related papers :a symposium papers Edited by A.J. Goodman sponsored by American association of petroleum geologists American association of petroleum geologists,Tulsa,Oka.,1958 John Andrew Allan memorial volume
Geol 8.2712

JEKYLL ISLAND,GA 1964 Estuaries :a symposium Papers University of Georgia. Marine institute Edited by George H. Lauff American association for the advance of science.Publication, 83 Washington,D.C., 1967
Bot 42.3345

JEKYLL ISLAND,GA. 1964 Estuaries Conference on estuaries Papers University of Georgia.Marine institute Edited by George H. Lauff American association for the advancement of science.Publications,83 New York,1967
Geog 13.0743

JELLINEK H.H.G. Ice symposium Papers Pittsburgh 1966 Mar American chemical society.Division of colloid and surface chemistry Journal of colloid and interface science,25 no 2 illus 131-294p Academic, New York;London,1967 Symposium chairman H. H.G.Jellinek
Sco 14.7934

JERNKONTORET Progress in mineral dressing International mineral dressing congress Transactions Stockholm 1957 Sep 18-21 Almqvist and Wiksell,Stockholm,1958
Met 25.2286

JERUSALEM 1908 Mathematical logic and foundations of set theory International colloquium on mathematical logic and foundations of set theory Proceedings Edited by Yehoshua Bar-Hillel Under the auspices of the Israel academy of science and humanities Studies in logic and the foundations of mathematics bibliog. 145p 24cm North-Holland,Amsterdam;London,1970
P Math 2.3795

JERUSALEM 1952 Desert research : international symposium Proceedings Jointly sponsored by the Research council of Israel Research council of Israel.Special publications,2 Research council of Israel, Jerusalem,1953
Geog 13.1458

JERUSALEM 1953 Congres international d'histoire des sciences 7e Actes International union of history and philosophy of science Edited by F.S. Bodenheimer Academie internationale d'histoire des sciences.Collection de travaux, 8 xii,662p Academie internationale d'histoire des sciences;Hermann,Paris,c1954
WSM 43.0058

JERUSALEM 1961 Genetics of migrant and isolate population Conference on human population genetics in Israel Proceedings Edited by Elisabeth Goldschmidt Published for the Association for the aid of crippled children Williams and Wilkins,New York,1963
Gen 34.1881

JERUSALEM 1962 Paramagnetic resonance : international conference 1st Proceedings Vol 1-2 International union of pure and applied physics Edited by W. Low Sponsored by the Hebrew university of Jerusalem 2 vols Academic press,New York,1963
Cav 7.1226

JERUSALEM 1964 Logic,methodology and philosophy of science International congress for logic,methodology and philosophy of science Proceedings Israel academy of sciences and humanities International union of history and philosophy of science.Division of logic,methodology and philosophy of science Edited by Yehoshua Bar-Hillel viii,440p North-Holland,Amsterdam,1965
WSM 43.0829

JERUSALEM 1967 Radiation chemistry of aqueous systems Farkas memorial symposium 19th Proceedings Edited by Gabriel Stein Weizmann science press of Israel,Jerusalem, 1968
Radiotb 35.1768

JERUSALEM 1967 Radiation chemistry of aqueous systems :proceedings of the 19th L. Farkas memorial symposium held at the Hebrew university Hebrew university of Jerusalem Edited by Gabriel Stein Weizmann science press of Israel in co-operation with Interscience,Jerusalem,1968
Chem 18.2485

JERUSALEM 1971 Impact of insulin on metabolic pathways International symposium commemorating the 50th anniversary of insulin Vol 1-2: lectures;panel discussions and communications Edited by Eleazar Shafrir Israel journal of medical sciences, 8,no 3,6 2 vols Israel medical association,Jerusalem, 1972
Bioch 33.2193

JESSE P.GREENSTEIN MEMORIAL SYMPOSIUM Papers Amino acids,proteins and cancer biochemistry New York 1959 Sep 16 Edited by John T. Edsall Port Academic press,New York,1960
Radioth 35.0083

The JESSE P.GREENSTEIN MEMORIAL SYMPOSIUM Papers Amino acids,proteins and cancer biochemistry Washington,D.C. 1959 Sep 16 American chemical society.Division of biological chemistry Edited by John T. Edsall Academic press,New York;London,1960 With a biographical article on Dr. Greenstein and a bibliography of his writings
Bioch 33.2345

JODRELL BANK 1955 Radio astronomy :a symposium Proceedings International astronomical union Edited by H.C.van de Hulst International astronomical union. Symposium, 4 xi,409p Cambridge university press,Cambridge,1957
Nap 11.0275

JODRELL BANK 1962 Radio astronomy today Papers International summer school in radio-astronomy,Manchester Edited by H.P. Palmer and others 242p Manchester university press,Manchester,1963 With a foreword by Sir Bernard Lovell
Obs 6.1326

JODRELL BANK 1962 Radio astronomy today Summer school in radio astronomy Papers Victoria university of Manchester Edited by H.P. Palmer and others 242p Manchester university press,Manchester,1963
TA 15.0226

JODRELL BANK 1962 Radio astronomy today : international summer school Papers Victoria university of Manchester Edited by H.P. Palmer and others Manchester university press,Manchester,1963 Foreword by Sir Bernard Lovell
Cav 7.1751

JODRELL BANK 1970 The Crab nebula International astronomical union Edited by R. D. Davies and F.G. Smith International astronomical union.Symposium, 46 470p Reidel,Dordrecht,1971
Obs 6.3588

JODRELL BANK 1970 Aug 5-7 The Crab nebula International astronomical union Edited by R. D. Davies and F.G. Smith International astronomical union.Symposium, 46 470p Reidel,Dordrecht,1971
TA 15.0689

JODRELL BANK EXPERIMENTAL STATION Meteors Symposium on meteor physics Proceedings Manchester 1954 Jul. Edited by T.R. Kaiser Journal of atmospheric and terrestrial physics. Special supplement, 2 204p Pergamon press,London,1955
Obs 6.0923

JOHANNESBURG 1962 Karoo symposium South African association for the advancement of science :annual congress papers South African journal of science,59,no.5 Johannesburg,1963
Geol 8.2941

JOHANNESBURG 1963 South African institution of civil engineers diamond jubilee convention Proceedings Johannesburg,1963
Eng 41.3389

JOHN DALTON AND THE PROGRESS OF SCIENCE : papers presented at a conference of historians of science Manchester 1966 Sep 19-24 Manchester literary and philosophical society xxii,352p Manchester university press;Barnes and Noble,Manchester;New York,1968 Conference held to mark the bicentenary of Dalton's birth.Co-sponsored by the Royal society,Chemical society and Society of chemical industry
WSM 43.2304

JOHN INNES HORTICULTURAL INSTITUTION Symposium on chromosome breakage Contributions London 1952 Jun 9-11 By C. D. Darlington and others Heredity, 5, supplement Oliver and Boyd,London;Edinburgh, 1953
Path 30.0219

JOHN PERCY RESEARCH GROUP IN PROCESS METALLURGY Heat and mass transfer in process metallurgy :a symposium Proceedings London 1966 Apr 19-20 Edited by A.W.D. Hills Institution of mining and metallurgy,London, 1967
Met 25.0317

JOHN PERCY RESEARCH GROUP IN PROCESS METALLURGY Heat and mass transfer in process metallurgy :a symposium Proceedings London 1966 Apr 19-20 Edited by A.W.D. Hills ix, 252p Institution of mining and metallurgy, London,1967
Chem M 24.0593

JOHNS HOPKINS CONFERENCE ON RESEARCH NEEDS IN DYSLEXIA AND RELATED APHASIC DISORDERS Papers Reading disability:progress and research needs in dyslexia Baltimore,Md. 1961 Nov 15-17 Edited by John Money Johns Hopkins press,Baltimore,Md.,1962
An 32.4203

JOHNS HOPKINS UNIVERSITY Drugs in our society :a conference Baltimore,Md. 1963 Nov Edited by Paul Talalay Johns Hopkins press,Baltimore,Md.,1964
Pha 16.0237

JOHNS HOPKINS UNIVERSITY The Symposium on colloid chemistry 7th Papers presented Baltimore 1929 Jun 20-22 Edited by Harry Boyer Weiser Colloid symposium annual, 7 Wiley;Chapman and Hall,New York;London,1939 Colloid symposium annual known also as Colloid symposium monograph
Bioch 33.1475

JOHNS HOPKINS UNIVERSITY.OPERATION RESEARCH OFFICE The Delayed effects of whole body radiation :a symposium Bethesda,Md. 1959 Oct 29 Edited by Bernard B. Watson Jointly sponsored by the Walter Reid army institute of research Johns Hopkins press,Baltimore,Md., 1960
Radioth 35.1968

JOHNSON RESEARCH FOUNDATION Control of energy metabolism A Colloquium on metabolic control and a Symposium on control of energy metabolism Proceedings Philadelphia 1965 May 20-21 Edited by Britton Chance and others Johnson research foundation.Colloquia Academic press,New York;London,1965 "In celebration of the bicentennial of the University of Pennsylvania school of medicine"
Bioch 33.1077

JOHNSON RESEARCH FOUNDATION Control of energy metabolism Colloquium on metabolic control :and symposium on control of metabolism Philadelphia,Pa. 1965 May 20-21 Edited by Britton Chance and others Academic press,New York;London,1965
Gen 34.0614

JOHNSON RESEARCH FOUNDATION Energy-linked functions of mitochondria :colloquium Papers presented Philadelphia 1963 Apr 13 Edited by Britton Chance Johnson research foundation.Colloquia, 1 Academic press,New York;London,1963
Bioch 33.1081

JOHNSON RESEARCH FOUNDATION Energy-linked functions of mitochondria :first colloquium of the Johnson research foundation of the University of Pennsylvania Papers Philadelphia 1963 Apr 13 Edited by Britton Chance Academic press,NEw York,1963
Radioth 35.0412

JOHNSON RESEARCH FOUNDATION Hemes and hemoproteins :colloquium Proceedings Philadelphia 1966 Apr 16-17 Edited by Britton Chance and others Johnson research foundation.Colloquia, 3 Academic press,New York,1966
Bioch 33.0507

JOHNSON RESEARCH FOUNDATION Probes of structure and function of macromolecules and membranes Colloquium of the Johnson research foundation 5th Proceedings Philadelphia, Pa. 1969 Apr 19-21 Vol 1-2: probes and membrane function;probes of enzymes and hemoproteins Edited by Britton Chance and others Academic press,New York,1971
Bioch 33.2190

JOHNSON RESEARCH FOUNDATION.COLLOQUIA, 1 Energy-linked functions of mitochondria : colloquium Papers presented Philadelphia 1963 Apr 13 Johnson research foundation Edited by Britton Chance Academic press,New York;London,1963
Bioch 33.1081

JOHNSON RESEARCH FOUNDATION.COLLOQUIA, 3 Hemes and hemoproteins :colloquium Proceedings Philadelphia 1966 Apr 16-17 Johnson research foundation Edited by Britton Chance and others Academic press,New York,1966
Bioch 33.0507

JOHNSON RESEARCH FOUNDATION COLLOQUIUM 3RD Control of energy metabolism Colloquium on metabolic control :and symposium on control of metabolism Philadelphia,Pa. 1965 May 20-21 Johnson research foundation Edited by Britton Chance and others Academic press,New York;London,1965
Gen 34.0614

JOINT AUTOMATIC CONTROL CONFERENCE 4th Preprints of technical papers Minneapolis 1963 Jul 19-21 American institute of chemical engineers xiv,680p American institute of chemical engineers,New York,1963
Chem A 24.1172

JOINT AUTOMATIC CONTROL CONFERENCE 5th Preprints of conference papers Stanford, Calif. 1964 Jun 24-26 Under the sponsorship of the Institute of electrical and electronics engineers.Professional technical group in automatic control Stanford,Calif., 1964
Eng 41.6014

JOINT AUTOMATIC CONTROL CONFERENCE 6th Preprints of conference papers New York 1965 Jun 22-25 Sponsored by American society of mechanical engineers.Automatic control division New York,1965
Eng 41.6015

JOINT AUTOMATIC CONTROL CONFERENCE 7th Preprints of conference papers Washington, D.C. 1966 Aug 17-19 Sponsored by the American institute of aeronautics and astronautics Washington,D.C.,1966
Eng 41.6016

JOINT AUTOMATIC CONTROL CONFERENCE 8th-9th Preprints,Technical papers University of Pennsylvania University of Michigan 1967-68
Eng 41.8208

JOINT BRITISH COMMITTEE FOR VACUUM SCIENCE AND TECHNOLOGY International vacuum congress 4th Manchester 1968 Apr 17-20 Pt 1-2 Institute of physics and the Physical society.Conference series, 5-6 2 vols Institute of physics,London,c1968
Met 25.2834

JOINT COMMISSION ON APPLIED RADIOACTIVITY Effects of radiation on cellular proliferation and differentiation Symposium on the effects of radiation on cellular proliferation and differentiation Proceedings Monaco 1968 Apr 1-5 International atomic energy agency International atomic energy agency,Vienna,1968
Radioth 35.1760

JOINT COMMISSION ON APPLIED RADIOACTIVITY Radioactive dating and methods of low-level counting :a symposium Proceedings Monaco 1967 Mar 2-10 International atomic energy agency.Proceedings series International atomic energy agency,Vienna,1967
Radioth 35.1683

JOINT COMMISSION ON APPLIED RADIOACTIVITY The Detection and use of tritium in the physical and biological sciences symposium Proceedings Vienna 1961 May 3-10 Vol 1-2 Sponsored by the International atomic energy agency International atomic energy agency.Proceedings series 2 vols International atomic energy agency,Vienna,1962 Papers in English and Russian.Summaries in English,French,Russian and Spanish
Gen 34.0470

JOINT COMMISSION ON APPLIED RADIOACTIVITY Tritium in the physical and biological sciences Symposium on the detection and use of tritium in the physical and biological sciences Proceedings Vienna 1961 May 3-10 Vols 1-2 Sponsored by the International atomic energy agency 2 vols International atomic energy agency,Vienna,1962 Co-sponsored by the Joint commission on applied radioactivity
Radioth 35.0104

JOINT COMMITTEE ON LATIN AMERICAN STUDIES Social science in Latin America Conference on Latin American studies Papers Rio de Janeiro 1965 Mar 29-31 Edited by Manuel Diegues and Bryce Wood Jointly sponsored by the Latin American center for research in the social sciences,Rio de Janeiro Columbia university press,New York,1967
Geog 13.5108

JOINT COMPUTER CONFERENCE Report on the Washington Joint AIEE-IRE-ACM computer conference,1953 By D.W. Davies 16p 1954
Math L 5.1138

JOINT COMPUTER CONFERENCE Report on visit to New York for the Joint AIEE-IRE-ACM computer conference,1952 mimeograph By D.W. Davies 19p 1953
Math L 5.1137

JOINT COMPUTER CONFERENCE 1st- proceedings 1951- 1952- Conferences 1-19 sponsored by the American institute of electrical engineers,the Institute of radio engineers and the Association for computing machinery.Conference 20 onwards sponsored by the American federation of information processing societies
Math L 5.0704

JOINT CONFERENCE ON COMBUSTION Proceedings Boston,Mass. 1955 Jun 15-17 London 1955 Oct 25-27 Institution of mechanical engineers and American society of mechanical engineers viii,457p Institution of mechanical engineers,London,1955
Chem E 24.1376

JOINT CONFERENCE ON COMBUSTION Proceedings Boston,Mass. 1955 Jun 15-17 and London 1955 Oct 25-27 Institution of mechanical engineers American society of mechanical engineers Institute of mechanical engineers, London,1956
Eng 41.7231

JOINT CONFERENCE ON DIGITAL METHODS OF MEASUREMENT Proceedings Canterbury 1969 Jul 23-25 Institution of electronic and radio engineers Institution of electrical engineers Institute of electrical and electronics engineers Institution of electronic and radio engineers.Conference proceedings, 15 IERE,London,1969
Eng 41.8305

JOINT CONFERENCE ON DIGITAL PROCESSING OF SIGNALS IN COMMUNICATIONS Proceedings 1972 Institution of electronic and radio engineers Institution of electrical engineers Institution of electronic and radio engineers. Conference proceedings, 23 IERE,London,1972
Eng 41.8565

A JOINT CONFERENCE ON HIGH-ALLOY STEELS Proceedings Metallurgical developments in high-alloy steels Scarborough 1964 Jun 2-4 Iron and steel institute and British iron and steel research association Iron and steel institute.Special report, 86 Iron and steel institute,London,1964
Met 25.0488

JOINT CONFERENCE ON LOW-ALLOY STEELS organized by BISRA and the Iron and steel institute Proceedings Scarborough 1965 Apr 2-4 British iron and steel research association Iron and steel institute Iron and steel institute.Publication, 114 illus 269p London,1969
Met 25.2566

A JOINT DISCUSSION ON VISION held at the Imperial college of science London 1932 Jun 3 Physical society Optical society Physical society,London,1932 Bound with 'Report of a discussion on audition' by the Physical society,q.v.
Psy 31.2610

A JOINT DISCUSSION ON VISION Report London 1932 Jun.3 Physical society and Optical society 327p Physical society, London,c1932
Obs 6.1962

JOINT ENGINEERING CONFERENCE Proceedings London 1951 Jun 4-15 Institution of civil engineers Institution of mechanical engineers Institution of electrical engineers Institution of civil engineers, London,1951
Eng 41.1761

JOINT EUROPEAN CONFERENCE OF THE INSTITUTE OF MATHEMATICAL STATISTICS,THE INTERNATIONAL ASSOCIATION FOR STATISTICS IN PHYSICAL SCIENCES,THE BIOMETRIC SOCIETY Berne 1964 Sep 14-18 Berne,1964 13 unbound typescript papers
Math 3.1124

JOINT FAO-IAEA DIVISION OF ATOMIC ENERGY IN FOOD AND AGRICULTURE Control of livestock insect pests by the sterile-male technique Panel on the control of livestock insect pests by the sterile-male technique Proceedings Vienna 1967 Jan 23-27 International atomic energy agency.Panel proceedings series International atomic energy agency,Vienna,1968
Radioth 35.0263

JOINT FAO-IAEA DIVISION OF ATOMIC ENERGY IN FOOD AND AGRICULTURE Elimination of harmful organisms from food and feed by irradiation Panel on elimination of harmful organisms from food and feed by irradiation Report Zeist 1967 Jun 12-16 International atomic energy agency.Panel proceedings series International atomic energy agency,Vienna,1968
Radioth 35.1761

JOINT FAO-IAEA DIVISION OF ATOMIC ENERGY IN FOOD AND AGRICULTURE Enzymological aspects of food irradiation Panel on enzymological aspects of the application of ionizing radiation to food preservation Proceedings Vienna 1968 Apr 8-12 International atomic energy agency.Panel proceedings series International atomic energy agency,Vienna,1969
Radioth 35.1762

JOINT FAO-IAEA DIVISION OF ATOMIC ENERGY IN FOOD AND AGRICULTURE Preservation of fruit and vegetables by radiation Panel on preservation of fruit and vegetables by radiation :especially in the tropics Proceedings Vienna 1966 Aug 1-5 International atomic energy agency.Panel proceedings series International atomic energy agency,Vienna,1968
Radioth 35.1758

JOINT IAEA-WHO EXPERT COMMITTEE ON MEDICAL RADIATION PHYSICS Report Medical radiation physics Geneva 1967 Dec 12-18 International atomic energy agency World health organization World health organization.Technical report series, 390 World health organization,Geneva,1968
Radioth 35.1381

JOINT IAEA-WHO MEETING ON PLANNING OF RADIOTHERAPY FACILITIES Report Planning of radiotherapy facilities Geneva 1964 Dec 15-19 International atomic energy agency World health organization World health organization.Technical report series, 328 World health organization,Geneva,1966
Radioth 35.1060

JOINT IBM-UNIVERSITY OF NEWCASTLE-UPON-TYNE SEMINAR Proceedings Newcastle upon Tyne 1969 Sep 8-12 University of Newcastle upon Tyne computer laboratory International business machines corporation Edited by N.S. M. Cox University of Newcastle upon Tyne, Newcastle upon Tyne,1970
Math L 5.3354

JOINT IBM-UNIVERSITY OF NEWCASTLE-UPON-TYNE SEMINAR Proceedings Teaching of computer design Newcastle upon Tyne 1971 Sep 7-10 International business machines corporation University of Newcastle-upon-Tyne.Computing laboratory illus 121p University of Newcastle upon Tyne.Computing laboratory,Newcastle upon Tyne,1972
Math L 5.3723

JOINT IBM-UNIVERSITY OF NEWCASTLE UPON TYNE SEMINAR 3rd Proceedings Teaching of programming at university level Cambridge 1970 Sep 8-11 University of Newcastle upon Tyne.Computing laboratory International business machines corporation Edited by B. Shaw illus. 153p University of Newcastle, Newcastle,1971
Math L 5.3590

JOINT INSTITUTE FOR LABORATORY ASTROPHYSICS Wolf Rayet stars :a symposium Proceedings Boulder,Colo. 1968 Jun 10-14 Edited by K.B. Gebbie and R.N. Thomas National bureau of standards.Special publication, 307 277p National bureau of standards,Washington,D.C., 1968
TA 15.0568

JOINT INSTITUTE FOR LABORATORY ASTROPHYSICS Wolf-Rayet stars :a symposium Proceedings Boulder,Colo. 1958 Jun 10-14 Edited by K.B. Gebbie and R.N. Thomas National bureau of standards.Special publication, 307 277p National bureau of standards,Washington,D.C., 1968
Obs 6.3438

JOINT INSTITUTE OF MATHEMATICS AND ITS APPLICATION-NATIONAL PHYSICAL LABORATORY CONFERENCE Proceedings Numerical methods for unconstrained optimization Teddington 1971 Jan 7-8 National physical laboratory Institute of mathematics and its applications Edited by W. Murray xi,144p 24cm Academic press,London;New York,1972
Math S 3.1729

JOINT INTERNATIONAL CONFERENCE ON CREEP session 1-7 New York 1963 Aug 25-29 London 1963 Sep 30-Oct 4 book 1-5 American society of mechanical engineers and American society for testing materials Jointly sponsored by Institution of mechanical engineers I.M.E.,London,c1963 5 books bound together
Met 25.0953

JOINT STATISTICS SEMINAR :a discussion Foundations of statistical inference London 1959 Jul 27-28 By Leonard J. Savage and others Birkbeck college Imperial college of science and technology 112p Methuen; Wiley,London;New York,1964
WSM 43.1769

JOINT STATISTICS SEMINAR,1959 The Foundations of statistical inference :a discussion opened by Professor L.J. Savage at a meeting of the Joint statistics seminar, Birkbeck and Imperial colleges,in the University of London Birkbeck college and Imperial college of science and technology Methuen's monographs on applied probability and statistics Methuen,London,1962
HE 27.0051

JOINT STATISTICS SEMINAR,1959 The Foundations of statistical inference :a discussion opened by Professor L.J.Savage at a meeting of the Joint statistics seminar, Birkbeck and Imperial colleges,in the University of London Birkbeck college and Imperial college of science and technology Methuen's monographs on applied probability and statistics 112p 20cm Methuen,London, 1962
Math 3.1311

JOINT SYMPOSIUM OF THE INTERNATIONAL ASTRONOMICAL UNION AND THE UNION INTERNATIONALE D'HISTOIRE ET DE PHILOSOPHIE DES SCIENCES 1st New aspects in the history and philosophy of astronomy Hamburg 1964 Aug 22-24 International astronomical union International union of history and philosophy of science Edited by Arthur Beer Vistas in astronomy, 9 illus,ports,maps 318p 25cm Pergamon press,Oxford,1967
Cav 7.2715

JOINT SYMPOSIUM ON INSTRUMENTATION AND COMPUTATION IN PROCESS DEVELOPMENT AND PLANT DESIGN Papers London 1959 May 11-12 British conference on automation and computation Sponsored by the Institution of chemical engineers Institution of chemical engineers,London,1959 Sponsored also by the Society of instrument technology,and the British computer society
Math L 5.3290

JOINT SYMPOSIUM ON INSTRUMENTATION AND COMPUTATION IN PROCESS DEVELOPMENT AND PLANT DESIGN Proceedings London 1959 May 11-13 Institution of chemical engineers Society of instrument technology British computer society Institution of chemical engineers,London,1959
Eng 41.6011

JOINT SYMPOSIUM ON WELDING IN SHIPBUILDING Welding in shipbuilding London 1961 Oct 30-Nov 3 Institute of welding Institute of welding,London,1962 Jointly organised by Institute of welding,Royal institution of naval architects,Institute of marine engineers, Institution of engineers and shipbuilders in Scotland and North East Coast institution of engineering and shipbuilders
Met 25.0799

JOKKMOKK,SWEDEN 1953 Samiid dilit : foredrag vid den nordiska Samekonferensen Edited by Kalle Nickul and others illus 344p 21cm Merkur,Oslo,1957
Sco 14.5602

JOKKMOKK,SWEDEN 1953 1956 The Lapps today in Finland,Norway and Sweden :en annexe: le departement de Laponie,notes documentaires Proceedings 1 Nordic Lapp council Edited by Rowland G.P. Hill Ecole pratique des hautes etudes.Sorbonne.Bibliotheque arctique et antarctique,1 maps 227p 24cm Mouton, Paris;The Hague,1960 Annex in French
Sco 14.5586

JORDEN,SKOGEN,MALMEN OCH VATTENKRAFTEN I MORGONDAGENS NORRBOTTEN :a conference Lectures Lulea 1956 Mar 19-20 Edited by Folke Thunberg Sponsored by Foreningen Norrbottens framjande illus,maps 271p 21cm Foreningen Norrbottens framjande,Lulea, 1956
Sco 14.5794

JOSIAH MACY,JR.FOUNDATION Group processes : conference 2nd Transactions Princeton,N. J. 1955 Oct 9-12 Edited by Bertram Schaffner 255p 24cm New York,1956
Sco 14.0705

JOSIAH MACY FOUNDATION Conference on cold injury 1st Transactions 1-6 248p 23cm 6 vols New York,1952-60
Sco 14.0875

JOSIAH MACY JR. FOUNDATION Biological antioxidants :a conference 1st Transactions New York 1946 Oct 10-11 Edited by Cosmo G. Mackenzie New York,1946
Chem 18.2646

JOSIAH MACY JR. FOUNDATION Genetic selection in man Macy conference on genetics 3rd Proceedings Princeton,N.J. 1961 Oct 15-18 Edited by William J. Schull University of Michigan press,Michigan,1963
Gen 34.1867

JOSIAH MACY JR. FOUNDATION Oxygen supply to the human foetus :a symposium Princeton,N.J. 1957 Dec Edited by James Walker and Alec C. Turnbull Blackwell,Oxford,1959
Phys 20.1003

JOSIAH MACY JR. FOUNDATION Physiology of prematurity conference 4th Transactions Princeton,N.J. 1959 Mar 25-26 Edited by Jonathan T. Lanham New York,1960
PGMS 29.0288

JOSIAH MACY JR.FOUNDATION Adrenal cortex Conference on adrenal cortex 1st-5th Transactions Edited by Elaine P. Ralli 5 vols Josiah Macy foundation,New York,1950-54
An 32.3730

JOSIAH MACY JR.FOUNDATION Adrenal cortex :a conference 1st Transactions New York 1949 Nov 21-22 Edited by Elaine P. Ralli and others Josiah Macy Jr.foundation,New York, 1950
Bioch 33.0464

JOSIAH MACY JR.FOUNDATION Adrenal cortex :a conference 2nd Transactions New York 1950 Nov 16-17 Edited by Elaine P. Ralli Josiah Macy Jr.foundation,New York,1951
Bioch 33.0465

JOSIAH MACY JR.FOUNDATION Adrenal cortex :a conference 3rd Transactions New York 1951 Nov 15-16 Edited by Elaine P. Ralli Josiah Macy Jr.foundation,New York,1952
Bioch 33.0466

JOSIAH MACY JR.FOUNDATION Adrenal cortex :a conference 4th Transactions New York 1952 Nov 12-14 Edited by Elaine P. Ralli Josiah Macy Jr.foundation,New York,1953
Bioch 33.0467

JOSIAH MACY JR.FOUNDATION Adrenal cortex :a conference 5th Transactions Princeton,N. J. 1953 Nov 4-5 Edited by Elaine P. Ralli Josiah Macy Jr.foundation,New York,1954
Bioch 33.0468

JOSIAH MACY JR.FOUNDATION Central nervous system and behaviour Conference on the central nervous system and behaviour 1st Transactions 1958 Feb 23-26 Edited by Mary A.B. Brazier New York,1959
Phys 20.2239

JOSIAH MACY JR.FOUNDATION Connective tissues Conference on connective tissues 1st Transactions New York 1950 Apr 24-25 Edited by Charles Ragan Josiaph Macy jr. foundation,New York,1951
Path 30.2487

JOSIAH MACY JR.FOUNDATION Connective tissues Conference on connective tissues 1st-5th Transactions New York 1950 Princeton,N.J. 1953-54 Edited by Charles Ragan Josiah Macy,jr.foundation,New York,1951-54
An 32.3218

JOSIAH MACY JR.FOUNDATION Cybernetics : circular causal and feedback mechanisms in biological and social systems Transactions New York 1951 Mar 15-16 Edited by Heinz von Foerster Josiah Macy,jr. foundation,New York,1952
Psy 31.2046

JOSIAH MACY JR.FOUNDATION Cybernetics: circular causal and feedback mechanisms in biological and social systems Conference on cybernetics 7th Transactions New York 1950 Mar 23-24 Edited by Heinz von Foerster Josiah Macy jr. foundation,New York,1951
Psy 31.1064

JOSIAH MACY JR.FOUNDATION Cybernetics: circular causal and feedback mechanisms in biological and social systems Conference on cybernetics 9th Transactions New York 1952 Mar 20-21 Edited by Heinz von Foerster Josiah Macy jr. foundation,New York,1953
Psy 31.1063

JOSIAH MACY JR.FOUNDATION Genetic selection in man Conference on genetics 3rd Princeton,N.J. 1961 Oct 15-18 Edited by William J. Schull University of Michigan press,Ann Arbor,Mich.,1963
An 32.0362

JOSIAH MACY JR.FOUNDATION Genetics:genetic information and the control of protein structure and function Conference on genetics 1st Transactions Princeton,N.J. 1959 Oct 19-22 Edited by H.Eldon Sutton Macy Jr.foundation,New York,1960
An 32.0360

JOSIAH MACY JR.FOUNDATION Genetics conference :genetic information and the control of protein structure and function 1st Transactions Princeton,N.J. 1959 Oct 19-22 Edited by H.Eldon Sutton Josiah Macy jr.foundation,New York,1960
Gen 34.0650

JOSIAH MACY JR.FOUNDATION Gestation Conference on gestation 1st Transactions Princeton,N.J. 1954 Mar 9-11 Edited by Louis B. Flexner Josiah Macy jr. foundation, New York,1955
An 32.3815

JOSIAH MACY JR.FOUNDATION Gestation Conference on gestation 2nd-5th Transactions Princeton,N.J. 1955-58 Mar Edited by Claude A. Villee 3 vols Josiah Macy jr. foundation,New York,1956-58
An 32.3816

JOSIAH MACY JR.FOUNDATION Mutations
Conference on genetics 2nd Princeton,N.J. 1960 Oct 16-19 Edited by William J. Schull University of Michigan press,Ann Arbor,Mich.,1962
An 32.0361

JOSIAH MACY JR.FOUNDATION Nerve impulse
Conference on nerve impulse 3rd Transactions New York 1952 Mar 3-4 Edited by H.Houston Merritt Josiah Macy jr. foundation,New York,1952
An 32.4315

JOSIAH MACY JR.FOUNDATION Physiology of prematurity Conference on physiology of prematurity 1st-4th Transactions Princeton,N.J. 1956-59 Mar Edited by Jonathan T. Lanman 4 vols Josiah Macy jr. foundation,New York,1957-60
An 32.3819

JOSIAH MACY JR.FOUNDATION Polysaccarides in biology Conference on polysaccharides in biology 1st-5th Transactions Princeton, N.J. 1955-1959 Edited by Georg F. Springer 5 vols Josiah Macy,Jr.foundation,New York, 1955-59
An 32.1493

JOSIAH MACY JR.FOUNDATION Somatic cell genetics Macy conference on genetics 4th Princeton,N.J. 1962 Oct 15-17 Edited by Robert S. Krooth illus. University of Michigan press,Ann Arbor,Mich.,1964
Gen 34.0891

JOSIAH MACY JR.FOUNDATION The Central nervous system and behavior :a conference 3rd Transactions Princeton,N.J. 1960 Feb 21-24 Edited by Mary A.B. Brazier With the cooperation of the National science foundation Josiah Macy,jr.foundation,New York,1960
Psy 31.0063

JOURNAL OF ATMOSPHERIC AND TERRESTIAL PHYSICS. SPECIAL SUPPLEMENT,5 The Airglow and the aurorae :a symposium Belfast 1955 Sep Edited by E.B. Armstrong and A. Dalgarno illus x,420p 25cm Pergamon press,London; New York,1956
Sco 14.0598

JOURNAL OF GEOPHYSICAL RESEARCH,Vol 67,no.8 Fundamental problems in turbulence and their relation to geophysics symposium proceedings Marseilles 1961 Sep.4-9 Vol 67 International union of geodesy and geophysics and International union of theoretical and applied mechanics Edited by Francois N. Frenkel American geophysical union, Washington D.C.,1962
A Math 4.0577

JULICH 1968 International conference on vacancies and interstitials in metals Kernforschungslange Julich Arbeitsgemeinschaft metallphysik Edited by A. Seeger and others North-Holland,Amsterdam, 1970
Met 25.2414

JULICH 1968 Vacancies and interstitials in metals :conference Proceedings Edited by A. Seeger and others Bibliog 1074p 22cm North-Holland,Amsterdam,1970
Cav 7.2686

JURASSIC El jurasico inferior de Mexico y sus amonitas International geological congress 20th papers Mexico City 1956 By Heinrich Karl Erben Translated by U. Hungsberg from the German Mexico City,1956
Geol 8.3033

JURASSIC El jurasico medio y el calloviano de Mexico International geological congress 20th papers Mexico City 1956 By Heinrich Karl Erben maps Mexico City,1956
Geol 8.3032

JURASSIC AND CARBONIFEROUS OF WESTERN CANADA,WITH RELATED PAPERS :a symposium papers Jasper,Alta. 1955 Sep.15-16 Edited by A.J. Goodman sponsored by American association of petroleum geologists American association of petroleum geologists,Tulsa,Oka.,1958 John Andrew Allan memorial volume
Geol 8.2712

KAISER FOUNDATION RESEARCH INSTITUTE
Comparative biochemistry of photoreactive systems Symposium on comparative biology 1st Papers Richmond,Calif. Edited by Mary Belle Allen Kaiser foundation research institute.Symposia on comparative biology, 1 Academic press,New York;London,1960
Bioch 33.0334

KAISER FOUNDATION RESEARCH INSTITUTE
Comparative biochemistry of photoreactive systems :annual symposium on comparative biology of the Kaiser foundation research institute 1st Papers Edited by Mary Belle Allen Kaiser foundation research institute.Symposia on comparative biology,1 Academic press,New York;London,1960
Chem 18.2139

KAISER FOUNDATION RESEARCH INSTITUTE.SYMPOSIA ON COMPARATIVE BIOLOGY, 1 Comparative biochemistry of photoreactive systems Symposium on comparative biology 1st Papers Richmond,Calif. Kaiser foundation research institute Edited by Mary Belle Allen Academic press,New York;London,1960
Bioch 33.0334

KALAMAZOO 1969 Many facets of graph theory Conference on graph theory Proceedings Edited by G. Chartrand and S.F. Kapoor Lecture notes in mathematics, 110 Illus viii,290p 24cm Springer-Verlag,Berlin,1969
P Math 2.3347

KALAMAZOO,MICH. 1966 Microbial protoplasts, spheroplasts and L-forms :a conference Proceedings Edited by Lucien B. Guze sponsored by the Upjohn company Williams and Wilkins,Baltimore,1968
Bioch 33.1142

KALAMAZOO,MICH. 1970 Analytic theory of differentiel equations :a conference Proceedings Edited by P.F. Hsieh and A.W.J. Stoddart Held at Western Michigan university Lecture notes in mathematics, 183 bibliog. vi,225p 25cm Springer,Berlin,1971
P Math 2.3876

KALAMAZOO,MICH. 1971 The Theory of arithmetic functions :conference Proceedings Edited by Anthony A. Gioia and Donald L. Goldsmith Held at Western Michigan university Lecture notes in mathematics, 251 287p 25cm Springer,Berlin,1972
P Math 2.4382

KALAMAZOO,MICH. 1972 Graph theory and applications :conference Proceedings Edited by Y. Alavi and D.R. Lick Held at Western Michigan university Lecture notes in mathematics, 303 viii,329p 25cm Springer, Berlin,1972
P Math 2.4111

KAMPALA 1955 Natural resources,food and population in inter-tropical Africa :a symposium Report Edited by L.Dudley Stamp Held under the auspices of the International geographical union Geographical publications, London,1956
Geog 13.4626

KAMPALA 1959 Social change in modern Africa International African seminar 1st Studies International African institute Edited by Aidan Southall Oxford university press,London,1965
Geog 13.4662

KANDY 1956 Study of tropical vegetation Kandy symposium Proceedings Unesco. International advisory committee for humid tropics research illus. Unesco,Paris,1958
BG 38.2155

KANDY,CEYLON 1956 Study of tropical vegetation :a symposium Proceedings Unesco Jointly organized by the Government of Ceylon Humid tropics research Unesco,Paris,1958
Geog 13.1261

KANDY SYMPOSIUM Proceedings Study of tropical vegetation Kandy 1956 Mar 19-21 Unesco.International advisory committee for humid tropics research illus. Unesco,Paris, 1958
BG 38.2155

KANPUR 1968 General topology and its relations to modern analysis and algebra : conference Proceedings illus 332p 24cm Academia,Prague,1971
P Math 2.4109

KANSAS CITY 1965 Communication theory and research:international symposium 1st Proceedings University of Missouri and National society for the study of communication Edited by Lee Thayer Thomas, Springfield,Ill.,1967
Psy 31.1068

KAPILLAREN UND INTERSTITIUM:MORPHOLOGIE-FUNKTION-KLINIK :Hamburger symposium Hamburg 1954 Oct 29-31 Edited by H. Bartelheimer and H. Kuchmeister Thieme,Stuttgart,1955
An 32.3460

KARL-MARX-UNIVERSITAT Naturwissenschaft und philosophie Internationales symposium uber naturwissenschaft und philosophie... Beitrage Leipzig 1959 Oct 8-11 Institut fur philosophie Karl-Sudhoff institut fur geschichte der medizin und naturwissenschaften Edited by Gerhard Harig and Josef Schleifstein 437p Akademie verlag,Berlin,1960 Symposium held to commemorate the 550th anniversary of the Karl-Marx-universitat
WSM 43.0976

KARL-SUDHOFF INSTITUT FUR GESCHICHTE DER MEDIZIN UND NATURWISSENSCHAFTEN Naturwissenschaft und philosophie Internationales symposium uber naturwissenschaft und philosophie... Beitrage Leipzig 1959 Oct 8-11 Edited by Gerhard Harig and Josef Schleifstein 437p Akademie verlag,Berlin,1960 Symposium held to commemorate the 550th anniversary of the Karl-Marx-universitat
WSM 43.0976

KARLOVY VARY 1969 Pulmonary and cardiac function bone isotope diagnosis International symposium on nuclear medicine 1st Proceedings Lekarska spolecnost J.E. Purkyne Edited by O. Andrysek and J. Mestan Universita Karlova,Prague,1970
PGMS 29.0680

KAROLINSKA INSTITUTET Immunoassay of gonadotrophins ...symposium Transactions Stockholm 1969 Sep 23-25 Edited by E. Dicfalusy and A. Dicfalusy Karolinska symposia on research methods in reproductive endocrinology, 1 Stockholm,1969
PGMS 29.0182

KAROLINSKA INSTITUTET In vitro methods in reproductive cell biology... symposium 3rd Transactions Geneva 1971 Jan 25-27 Edited by E. Dicfalusy and A. Dicfalusy Karolinska symposia on research methods in reproductive endocrinology, 3 Acta endocrinologica.Supplement, 153 Stockholm, 1971
PGMS 29.0650

KAROLINSKA INSTITUTET Steroid assay by protein binding :a symposium 2nd Transactions Geneva 1970 Mar 23-25 Edited by E. Dicfalusy and A. Dicfalusy Karolinska symposia on research methods in reproductive endocrinology, 2 Stockholm, 1970
PGMS 29.0589

KAROLINSKA SYMPOSIA ON RESEARCH METHODS IN REPRODUCTIVE ENDOCRINOLOGY, 1 Immunoassay of gonadotrophins ...symposium Transactions Stockholm 1969 Sep 23-25 Karolinska institutet and World health organization Edited by E. Dicfalusy and A. Dicfalusy Stockholm,1969
PGMS 29.0182

KAROLINSKA SYMPOSIA ON RESEARCH METHODS IN REPRODUCTIVE ENDOCRINOLOGY, 2 Steroid assay by protein binding :a symposium 2nd Transactions Geneva 1970 Mar 23-25 Karolinska institutet World health organization Edited by E. Dicfalusy and A. Dicfalusy Stockholm,1970
PGMS 29.0589

KAROLINSKA SYMPOSIA ON RESEARCH METHODS IN REPRODUCTIVE ENDOCRINOLOGY, 3 In vitro methods in reproductive cell biology... symposium 3rd Transactions Geneva 1971 Jan 25-27 Karolinska institutet World health organization Edited by E. Dicfalusy and A. Dicfalusy Acta endocrinologica. Supplement, 153 Stockholm,1971
PGMS 29.0650

KAROO SYMPOSIUM South African association for the advancement of science :annual congress papers Johannesburg 1962 South African journal of science,59,no.5 Johannesburg,1963
Geol 8.2941

KATATA 1966 Katata conference on the theory of partial differential equations and on the theory of complex manifolds Proceedings Bibliog.,Illus. vi,107p 25cm Research institute for mathematical science, Kyoto university,Kyoto,1966
P Math 2.3188

KATATA CONFERENCE ON THE THEORY OF PARTIAL DIFFERENTIAL EQUATIONS AND ON THE THEORY OF COMPLEX MANIFOLDS Proceedings Katata 1966 Sep 18-22 Bibliog.,Illus. vi,107p 25cm Research institute for mathematical science,Kyoto university,Kyoto,1966
P Math 2.3188

KEELE 1956 Chemisorption :a symposium Proceedings Chemical society Edited by W.E. Garner Butterworths,London,1957
Chem 18.0618

KEELE 1956 Chemisorption :symposium Proceedings Chemical society Edited by W.E. Garner Butterworths,London,1957
Col S 12.0266

KEELE 1962 Problems of applied geography,2 Anglo-Polish seminar 2nd Proceedings Polska akademia nauk.Instytut geografii,and, Institute of British geographers Geographia polonica,3 Panstwowe wydawnictwo naukowe, Warsaw,1964
Geog 13.5616

KEELE 1968 Optimization Keele conference on optimization and nonlinear programming Institute of mathematics and its applications Edited by R. Fletcher Academic press,London, 1969
Math L 5.3607

KEELE 1968 Optimization Keele conference on optimization and nonlinear programming Institute of mathematics and its applications Edited by R. Fletcher Academic press,London, 1969
P Math 2.3755

KEELE 1968 Optimization :symposium Institute of mathematics and its applications Academic press,London,1969
Eng 41.2342

KEELE,STAFFS 1966 Aurora and airglow proceedings of the Nato advanced study institute Proceedings Nato advanced study institute Edited by Billy M. McCormac illus vii,689p 24cm Reinhold,New York, 1967
Sco 14.7473

KEELE CONFERENCE ON OPTIMIZATION AND NONLINEAR PROGRAMMING Optimization Keele 1968 Institute of mathematics and its applications Edited by R. Fletcher Academic press,London, 1969
Math L 5.3607

KEELE CONFERENCE ON OPTIMIZATION AND NONLINEAR PROGRAMMING Optimization Keele 1968 Institute of mathematics and its applications Edited by R. Fletcher Academic press,London, 1969
P Math 2.3755

KENDALL AWARD SYMPOSIUM Solid surfaces and the gas-solid interface :papers presented at the Kendall award symposium honoring Stephen Brunauer Papers St.Louis,Mo. 1961 Mar 27-29 American chemical society.Division of colloid and surface chemistry Edited by Lewellyn E. Copeland and others American chemical society.Advances in chemistry series, 33 American chemical society,Washington,D.C., 1961
Chem 18.0616

KENTUCKY SYMPOSIUM Learning theory, personality theory,and clinical research :the Kentucky symposium Papers Lexington,Ky. 1953 Mar 13-14 University of Kentucky. College of arts and sciences.Department of psychology Wiley;Chapman and Hall,New York; London,1954
Psy 31.1839

KERATINIZATION Fundamentals of keratinization :a symposium presented at the New York meeting of the New York meeting of the American association for the advancement of science New York 1960 Dec 30 American association for the advancement of science Edited by Earl O. Butcher and Reidar F. Sognnaes American association for the advancement of science.Publications, 70 American association for the advancement of science,Washington,D.C.,1962
An 32.3280

KERNFORSCHUNGSLANGE JULICH International conference on vacancies and interstitials in metals Julich 1968 Sep 23-28 Edited by A. Seeger and others North-Holland,Amsterdam, 1970
Met 25.2414

KERNICTERUS AND ITS IMPORTANCE IN CEREBRAL PALSY American academy for cerebral palsy:annual meeting 11th Papers New Orleans Thomas, Springfield,Ill.,1961
An 32.4184

KHARTOUM 1965 Agricultural development in the Sudan Philosophical society of the Sudan annual conference 13th Proceedings and papers Vol 1-2 Edited by D.J. Shaw Co-sponsored by the Sudan agricultural society maps 2 vols Philosophical society of the Sudan,Khartoum,1966 Typescript
Geog 13.4700

KIEV 1962 Optical instability of the earth's atmosphere All-union conference 3rd Proceedings Akademiya nauk S.S.S.R. Astronomical council Glavnaya astronomicheskaya observatoriya,Pulkovo Translated by Z. Lerman from the Russian viii,174p 25cm Israel program for scientific translations,Jerusalem,1966
Obs 6.3275

KIEV 1969 International conference on nonlinear oscillations 5th Abstracts of papers 267p 20cm Academy of sciences of the USSR,Kiev,1969 Typescript
P Math 2.3734

KIEV 1969 Qualitative methods in the theory of nonlinear oscillations 5th conference Proceedings Vol 2-4 Edited by Yu.A. Mitropolski Bibliog,illus 26cm 3 vols Ukrainian academy of sciences. Institute of mathematics,Kiev,1970
A Math 4.1724

KINETICS Heterogeneous kinetics at elevated temperatures International conference on metallurgy and materials science Proceedings Philadelphia,Pa. 1969 Sep 8-10 Edited by G. R. Belten and W.L. Worrell Held at the University of Pennsylvania Plenum press,New York,1970
Met 25.2783

KINETICS,EQUILIBRIA AND PERFORMANCE OF HIGH TEMPERATURE SYSTEMS 1st :a conference Proceedings Los Angeles 1959 Nov 2-5 Combustion institute.Western states section Edited by Gilbert S. Bahn and Edward E. Zukoski x,255p Combustion institute,Los Angeles,1960
Chem E 24.1394

The KINETICS AND MECHANISM OF INORGANIC REACTION IN SOLUTION;A SURVEY OF RECENT WORK :a symposium Report London 1954 Feb 4 Chemical society Edited by K.W. Sykes Chemical society.Special publication, 1 London,1954
Bioch 33.0631

KINETICS OF CELLULAR PROLIFERATION Conference on fundamental problems and technics for the study of the kinetics of cellular proliferation Proceedings Salt Lake City,Utah 1959 Jan 19-21 Edited by Frederick Stohlman Grune and Stratton,New York;London,1959
An 32.3235

KINETICS OF HIGH-TEMPERATURE PROCESSES Cambridge,Mass. 1958 Jun 23-27 Edited by W. D. Kingery Supported in part by United States air force.Aeronautical research laboratory Technology press books in science and engineering Massachusetts institute of technology press,Cambridge,Mass.,1959
Min 10.1127

The KINETICS OF PROTON TRANSFER PROCESSES :a general discussion Newcastle-upon-Tyne 1965 Apr 12-14 Faraday society Faraday society.Discussions, 39 Aberdeen university press,Aberdeen,1966
Bioch 33.0645

KINETICS OF REACTIONS IN IONIC SYSTEMS International symposium on special topics in ceramics Proceedings Alfred,N.Y. 1967 Jun 18-23 Edited by T.J. Gray and V.D. Frechette Held at Alfred university Materials science research, 4 Plenum press, New York,1969
Met 25.2750

KING'S COLLEGE HOSPITAL MEDICAL SCHOOL Human chromosomal abnormalities conference Proceedings London 1959 Sep 18-19 Edited by William M. Davidson and D.Robertson Smith illus. Staples press,London,1961
Gen 34.0878

KING'S COLLEGE HOSPITAL MEDICAL SCHOOL Symposium on nuclear sex London 1957 Sep Edited by D.Robertson Smith and William M. Davidson Heinemann,London,1958
Gen 34.0879

KINGSTON,JAMAICA 1962 Caribbean geological conference 3rd transactions Edited by Edward Robinson Jamaica geological survey department,Kingston,Jamaica,1966
Geol 8.2425

KINGSTON,ONT 1969 Conference on Universal algebra Proceedings Held at Queen's university,Kingston,Ont. Queen's papers in pure and applied mathematics, 25 Bibliog. 273p 27cm Queen's university,Kingston,Ont., 1970
P Math 2.3712

KINGSTON,R.I. 1957 Developmental cytology Society for the study of development and growth Edited by Dorothea Rudnick Society for the study of development and growth. Symposia, 16 Ronald press,New York,1957
Gen 34.0849

KINGSTON COLLEGE OF TECHNOLOGY Computer aided circuit design Conference on CACD Proceedings 1968 Mar 26-28 Jointly sponsored by Design electronics Design electronics,London,1968
Eng 41.2226

KININ HORMONES Bradykinin and related kinins; cardiovascular,biochemical,and neural actions An International symposium on the cardiovascular and neural actions of bradykinin and related kinins Castel di Poggio 1969 Jul 21-25 Edited by F. Sicuteri and others Advances in experimental medicine and biology, 8 Plenum,New York; London,1970
Inv Med 37.0196

KIRUNA 1967 International symposium on fracture mechanics Swedish national committee for mechanics Wolters-Noordhoff, Groningen,1968
Met 25.0874

KITT PEAK NATIONAL OBSERVATORY.SPACE DIVISION The Atmospheres of Venus and Mars :a symposium Tucson,Ariz. 1967 Feb 28-Mar 2 Edited by John C. Brandt and Michael B. McElroy Gordon and Breach,New York,1968
TA 15.0415

KNOKKE 1959 The Transparancy of the cornea a symposium Council for international organizations of medical sciences Edited by William Stewart Duke-Elder and E.S. Perkins Blackwell,Oxford,1960
An 32.4772

KNOXSVILLE 1955 Symposium on corrosion fundamentals :a series of lectures... University of Tennessee.Chemical engineering department and National association of corrosion engineers.Southeast region Edited by Anton De S. Brasunas and N.E. Stansbury University of Tennessee press,Knoxville,Tenn., 1956
Met 25.1884

KNOXVILLE,TENN. 1959 Lectures in biological sciences :a symposium By Norman G. Anderson and others Sigma Xi.Tennessee chapter Oak Ridge national laboratory. Biology division Edited by J.Ives Townsend 110p University of Tennessee press,Knoxville, Tenn.,1963
Bot 42.4744

KOBNHAVNS TELEFON AKTIESELSKAB International congress on the application of the theory of probability in telephone engineering and administration 1st proceedings Copenhagen 1955 Jun.20-23 Edited by Arne Jensen 130p 30cm Copenhagen,1957 Published simultaneously as vol.1,1957,no.1 of "Teleteknik"
Math 3.0881

KOLLOQUIUM UBER BILDWANDLER UND BILDSPEICHERROHREN vortrage und diskussionen Heidelberg 1958 Apr.28-29 Edited by Heinrich Siedentopf Heidelberger akademie der wissenschaften.Sitzungenberichte. Mathematisch- naturwissenschaftliche klasse, 1959,abh.5 31 plates, 4 tabs 78p Springer-verlag,Heidelberg,1959
Obs 6.1627

KOLLOQUIUM UBER CYTOSTATICA Chemotherapeutische probleme maligner tumoren Hemer 1958 Jul 28 Edited by F. Meythaler Symposien aktueller therapeutischer probleme, 1 Enke,Stuttgart,1958
Radioth 35.1082

KOLNER ANTHROPOLOGISCHE GESELLSCHAFT Deutsche anthropologische gesellschaft: versammlung 49th Bericht Cologne 1927 Sep 11-17 Edited by Walter Venn-Koln Kabitzsch,Leipzig,1928
An 32.2563

KOMITET VSESOYUZNIKH GIDROLOGICHESKIKH S'EZDOV Russian hydrological congress 1st Proceedings Leningrad 1924 May 7-14 tables,maps 623p 26cm Leningrad,1925
Sco 14.0111

KONECHNYE GRUPPY :a seminar Papers Minsk 1965 Edited by Ya.G. Berkovich and others Bibliog,port 191p 21cm Minsk, 1966 In honour of S.A.Cunihin
P Math 2.3070

KONFERENCJA POSWIENCONEJ OMOWIENIU WSTEPNYCH WYNIKOW POLSKIEJ WYPRAWY NA SPITSBERGEN Warsaw 1958 Przeland geofizyczny Rocznik,3, pt.2 Warsaw,1960
Sco 14.2969

KONFERENTSIYA PO PROBLEME 'VSAIMODEISTVIE ATMOSFERY I GIDROSFERY V SEVERNOI CHASTI ATLANTICHESKOGO OKEANA' Conference on the problem 'Interactions between the atmosphere and the hydrosphere in the North Atlantic 2nd Papers Leningrad 1961 May 25-30 Leningradskii gidrometeorologicheskii institut 281p Izdatelstvo Leningradskogo universiteta, Leningrad,1964
Sco 14.5941

KONFERENTSIYA PUTI RAZVITIYA SOVETSKOGO MATEMATICHESKOGO MASHCHINOSTROENIYA I PRIBOROSTROENIYA :plenarnye zasedaniya Conference Moscow 1956 Mar 12-17 illus. 131p 22cm Moscow,1956
Math L 5.0774

KONGRESS ZUR BEKAMPFUNG DER TUBERKULOSE ALS VOLKSKRANKHEIT Bericht Berlin 1899 May 24-27 Deutsches central-komite zur errichtung von heilstatten fur lungenkranke Edited by Gotthold Pannwitz Berlin,1899
Path 30.0770

KONSTANZ 1960 Economics of take-off into sustained growth International economic association conference Proceedings Edited by W.W. Rostow Macmillan,London,1963
Geog 13.2064

KONSTANZ 1969 Pyridine-nucleotic-dependent dehydrogenases :advanced study institute Proceedings Universitat Konstanz Edited by Horst Sund Springer,Berlin,1970
Bioch 33.2292

KREBS - DOKUMENTATION UND STATISTIK MALIGNER TUMOREN Internationale jahrestagung des Arbeitsausschuss medizin 10th Verhandlungsbericht Berlin 1965 Oct 25-28 Edited by Gustav Wagner Schattauer,Stuttgart, 1966
Radioth 35.0877

KREBSMETASTASIERUNG AUF DEM BLUTWEGE : symposium Geneva 1963 Jun 27-29 Schweizerische nationaliga fur krebsekampfung und krebsforschung Schweizerische akademie der medizinischen wissenschaften Schweizerische hamatologe gesellschaft Schweizerische akademie der medizinischen wissenschaften.Bulletin, 20,fasc 1-3 Schwabe,Basel,1964
Radioth 35.0835

KREFELD 1959 Pliozan und pleistozan :am mittel - und niederrhein Edited by R. Teichmuller and G.v.d. Brelie Fortschritte in der geologie von Rheinland und Westfalen, 4 illus. 412p Krefeld,1959
Bot 42.3327

KREISEL PROBLEME :internationale symposium Vortrage Celerina,Switzerland 1962 Aug 20-23 International union of theoretical and applied mechanics and Swiss federal institute of technology Edited by Hans Ziegler Berlin,1963 Papers in English, French and German
Geod 9.0236

KRISTALLISATION Gemeinschaftskonferenz der reihe 'Metall' :die kristallisation von metallen aus dem schmelzfluss,der gasphase und durch elektrolytische abscheidung 6 Vortrage Berlin 1968 Mar 28-29 Deutsche akademie der wissenschaften zu Berlin.Sektion der physik.Unterkommission metallphysik VEB verlag fur grundstoffindustrie,Leipzig,1969
Met 25.2622

KUCHING 1963 Ecological research in humid tropical vegetation :a symposium Unesco. Science cooperation office for south-east Asia x,376p Unesco,Sarawak,1965
Bot 42.4771

KURAMOCHI BOUNDARIES OF RIEMANN SURFACES :a symposium Papers Kyoto 1965 Oct Edited by F-Y. Maeda and M. Ohtsuka Lecture notes in mathematics, 58 Bibliog 102p 28cm Springer-verlag,Berlin,1968
P Math 2.3051

KUSATSU 1967 Molecular basis of endocrinology Gunma symposium on endocrinology 5th Gunma university. Institute of endocrinology Gunma university. Institute of endocrinology.Annual report, 5 Institute of endocrinology,Gunma university, Maebashi,1967
Phys 20.1435

KYOTO Aug 28-Sep 4 Electron microscopy 1966 The International congress for electron microscopy 6th Extended abstracts Vol 1-2 International federation of societies for electron microscopy Edited by Ryozi Uyeda 2 vols Maruzen company,Tokyo,1966
Cav 7.1340

KYOTO 1953 Tokyo International conference on theoretical physics Proceedings Science council of Japan and Kyoto university Edited by I. Imai Under the auspices of the International union of pure and applied physics 942p Tokyo,1954
Cav 7.1342

KYOTO 1954 Prestressed concrete and composite beams :a symposium Proceedings Science council of Japan Science council of Japan,Kyoto,1955
Eng 41.3013

KYOTO 1955 Safety of structures symposium 2nd Proceedings Science council of Japan Japan society of civil engineers Science council of Japan,Kyoto,1956
Eng 41.3014

KYOTO 1961 International conference on magnetism and crystallography Proceedings Pt 2: electron and neutron diffraction Science council of Japan and Crystallographic society of Japan Edited by Shizuo Miyake and others Sponsored by the International union of pure and applied physics Physical society of Japan.Journal, 17,B 2 396p physical society of Japan, Tokyo,1962
Cav 7.1338

KYOTO 1961 International conference on magnetism and crystallography Proceedings Pt 3: neutron diffraction:study of magnetic materials Science council of Japan and Crystallographic society of Japan Edited by Takeo Nagamiya and Shizno Miyake Under the auspices of International union of pure and applied physics Physical society of Japan. Journal, 17,B 3 71p physical society of Japan,Tokyo,1962
Cav 7.1339

KYOTO 1961 International conference on magnetism and crystallography Proceedings Pt 1: magnetism Science council of Japan and Crystallographic society of Japan Edited by Kei Josida and others Held under the auspices of International union of pure and applied physics Physical society of Japan.Journal, 17,B 1 xi,718p Physical society of Japan,Tokyo,1962
Cav 7.1337

KYOTO 1961 The International conference on cosmic rays and the earth storm :combined meeting of the International symposium on the earth storm and the International conference on cosmic rays Proceedings Vol 2-3 International union of geodesy and geophysics and International astronomical union Edited by Ken-ichi Maeda and Osamu Minakawa Co-sponsored by the International scientific radio union Physical society of Japan. Journal, 17,suppl.A-2,3 2 vols physical society of Japan,Tokyo,1962
TA 15.0038

KYOTO 1962 Japan congress on testing materials 6th Science council of Japan Japan society for testing materials,Kyoto,1963
Met 25.2860

KYOTO 1965 Biological and biochemical evaluation of malignancy in experimental hepatomas The U.S.-Japan joint conference on biological and biochemical evaluation of malignancy in experimental hepatomas Proceedings Held under the auspices of the United States-Japan scientific cooperation program G.A.N.N.Monograph, 1 Japanese cancer association;Japanese foundation for cancer research,Tokyo,1966
Bioch 33.0952

KYOTO 1965 Kuramochi boundaries of Riemann surfaces :a symposium Papers Edited by F-Y. Maeda and M. Ohtsuka Lecture notes in mathematics, 58 Bibliog 102p 28cm Springer-verlag,Berlin,1968
P Math 2.3051

KYOTO 1966 Biological and chemical aspects of oxygenases The United States-Japan symposium on oxygenases Proceedings Edited by Konrad Bloch and Osamu Hayaishi held under the auspices of the United States-Japan committee on scientific co-operation Maruzen company,Tokyo,1966
Bioch 33.1102

KYOTO 1966 Electron microscopy 1966 International congress for electron microscopy 6th Vol 1: non-biology International federation of electron microscope societies Edited by Ryozi Uyeda Maruzen,Tokyo,1966
Met 25.1388

KYOTO 1966 Electron microscopy 1966 International congress for electron microscopy 6th Papers and extended abstracts International federation of electron microscope societies Edited by Ryozi Uyeda illus Maruzen,Tokyo,1966
Path 30.2615

KYOTO 1966 International conference on radiation biology and cancer Proceedings Edited by Tsutomu Sugahara Under the auspices of Nippon societas radiologica. Biological subgroup Radiation society of Japan,1967
Radioth 35.1757

KYOTO 1966 International conference on the physics of semiconductors 8th Proceedings Physical society of Japan Under the auspices of the International union of pure and applied physics Physical society of Japan.Journal, 21,suppl. 774p Physical society of Japan, Tokyo,1966
Cav 7.1341

KYOTO 1968 Comparative cellular and species radiosensitivity :international seminar Proceedings Edited by Victor P. Bond and Tsutomu Suyahara Iyaku Shoin,Tokyo, 1969
Radioth 35.0554

KYOTO 1969 International meeting on ferroelectricity 2nd Science council of Japan Sponsored by the International union of pure and applied physics Physical society of Japan.Journal, 28,suppt Physical society of Japan,1970
Met 25.2865

KYOTO 1969 Seminar on contact manifolds Lectures By Koichi Ogiue and others Kyoto university.Research institute for mathematical sciences Okayama university.Study group of geometry.Publication, 4 ii,83p 26cm Okayama university.Department of mathematics, Okayama,1970
P Math 2.4119

KYOTO 1970 International conference on low temperature physics 12th Proceedings International union of pure and applied physics Physical society of Japan Edited by Eizo Kanda Organized by the Science council of Japan photos,diagrms 895p 26cm Academic press of Japan,Tokyo,1971
Cav 7.3028

KYOTO 1971 International conference on mechanical behavior of materials Proceedings Society of materials science 6 vols Society of materials science,Kyoto,1971
Eng 41.8524

KYOTO 1971 Japan-United States seminar on ordinary differential functional equations Proceedings Edited by Minoru Urabe Lecture notes in mathematics, 243 viii,332p 25cm Springer,Berlin,1971
P Math 2.4318

KYOTO 1971 Mechanical behaviour of materials International conference on 'mechanical behaviour of materials' Proceedings Society of materials science, Japan Society of materials science,Kyoto, 1972
Met 25.2737

KYOTO UNIVERSITY International conference on theoretical physics Proceedings Kyoto 1953 Sep 14-24 Tokyo Edited by I. Imai Under the auspices of the International union of pure and applied physics 942p Tokyo, 1954
Cav 7.1342

KYOTO UNIVERSITY.RESEARCH INSTITUTE FOR MATHEMATICAL SCIENCES Seminar on contact manifolds Lectures Kyoto 1969 Jun 23-25 By Koichi Ogiue and others Okayama university.Study group of geometry.Publication, 4 ii,83p 26cm Okayama university. Department of mathematics,Okayama,1970
P Math 2.4119

LA CORUNA 1968 Perspectivas de Galicia ante el segundo plan de desarrollo Instituto "Jose Cornide' de estudios coruneses La Coruna,1968
Geog 13.6798

LA JOLLA 1966 Entire functions and related parts of analysis :summer institute Lecture notes American mathematical society 28cm 2 vols 1966 Mimeographed
P Math 2.3682

LA JOLLA 1966 Entire functions and related parts of analysis :symposium Proceedings American mathematical society Edited by Jacob Korevaar and others American mathematical society.Proceedings of Symposia in pure mathematics, 11 illus. vi,554p 26cm American mathematical society, Providence,R.I.,1968
P Math 2.3199

LA JOLLA,CALIF 1967 Control mechanisms in developmental processes :a symposium Society for developmental biology Edited by Michael Locke Society for developmental biology. Symposia, 26 Developmental biology. Supplement, 1 Academic press,New York,1967
Bioch 33.0911

LA JOLLA,CALIF. 1962 Conference on immuno-reproduction Proceedings Population council and Ford foundation Population council,New York,c1962 Organized by A. Tyler and K.A.Laurence and edited by the latter
Path 30.2340

LA JOLLA,CALIF. 1965 Advances in tactical rocket propulsion :a colloquium Agard Edited by S.S. Penner AGARD conference proceedings, 1 Technivision,Slough,1968
Eng 41.6896

LA JOLLA,CALIF. 1965 Categorical algebra conference proceedings Edited by S. Eilenberg and others Supported by the United States.Air force.Office of scientific research 562p 24cm Springer-Verlag,Berlin,1966
P. Math 2.2027

LA JOLLA,CALIF. 1967 Control mechanisms in developmental processes :a symposium Edited by Michael Locke Society for developmental biology.Symposia, 26,supp.1 Academic press, New York,1967 Formerly the Society for the study of development and growth
An 32.2535

LA MENDOLA,TRENTO 1970 Functional equations and inequalities Centro internazionale matematico estivo 3 ciclo Centro internazionale matematico estivo 426p 29cm Edizioni cremonese,Rome,1971 Conference coordinator:B.Forte
P Math 2.4213

LA PLATA 1960 Symposium on stellar evolution papers Universidad nacional de La Plata.Observatorio astronomico Edited by J. Sahade 309p Astronomical observatory,La Plata,1962 Scientific meeting organized to celebrate the sesquicentennial of the May revolution
Obs 6.1505

LA TOUR DE PEILZ,SWITZERLAND 1962 New thinking in school biology The Seminar on the reform of biology teaching :report on the OECD seminar Organization for economic co-operation and development New thinking in school science Paris,1963
Gen 34.2062

LABELLED PROTEINS IN TRACER STUDIES Conference on problems connected with the preparation and use of labelled proteins in tracer studies Proceedings Pisa 1966 Jan 17-19 Edited by L. Donato and others Sponsored by the Universita di Pisa.Nuclear medicine center EUR 2950 d,f,e European atomic energy community,Brussels,1966
Radioth 35.1752

LABORATORY SHEAR TESTING OF SOILS :a symposium Ottawa 1963 Sep 9-11 National research council of Canada American society for testing materials.Special technical publication, 361 American society for testing materials,Philadelphia,Pa.,1964
Eng 41.3157

LACTIC ACIDS Chemistry and metabolism of L- and D- lactic acids :a conference Papers New York 1964 Nov 12-14 New York academy of sciences Edited by Harold E. Whipple New York academy of sciences.Annals, 119,p.851-1165 New York,1965
Bioch 33.0563

LACTOSE OPERON :conference Proceedings Cold Spring Harbor 1969 Sep Cold Spring Harbor laboratory of quantitative biology Cold Spring harbor symposia on quantitative biology Cold Spring Harbor,L.I.,N.Y.,1970
Bioch 33.2251

LAFAYETTE 1965 Determination of nonmetallic compounds in steel a symposium presented at the sixty-eighth annual meeting American society for testing materials A.S.T.M.Special technical publication, 393 A.S.T.M. Philadelphia,1966
Met 25.2303

LAFAYETTE,IND 1959 Thermodynamic and transport properties of gases liquids and solids Symposium on thermal properties 3rd Papers Sponsored by the American society of mechanical engineers:Heat transfer division x,472p American society of mechanical engineers;McGraw-Hill,New York,1959
Chem E 24.0984

LAFAYETTE,IND. 1940 Purdue conference on soil mechanics and its application Proceedings Purdue university Purdue university,Lafayette,Ind.,1940
Eng 41.3163

LAFAYETTE,IND. 1955 Midwestern conference on fluid mechanics 4th Proceedings Held at Purdue university Purdue university. Engineering experiment station.Research series, 128 Purdue university,Lafayette,Ind.,1955
Eng 41.6957

LAFAYETTE,IND. 1955 Midwestern conference on fluid mechanics 4th Proceedings Held at Purdue university Purdue university. Engineering experiment station.Research series, 128 vii,370p Purdue university,Lafayette, Ind.,1955
Chem E 24.0404

LAFAYETTE,IND. 1957 Recent advances in the engineering sciences:their impact on engineering education Conference on science and technology for deans of engineering Proceedings Purdue research foundation and Purdue university viii,257p McGraw-Hill, New York,1958
Chem E 24.1074

LAFAYETTE,IND. 1959 Information and decision processes symposium papers Purdue university Edited by Robert E. Machol xi, 185p 23cm McGraw-Hill,New York,1960
Math 3.0683

LAFAYETTE,IND. 1963 Arithmetical algebraic geometry conference proceedings Purdue university.Division of mathematical sciences Edited by Otto F.G. Schilling Harper's series in modern mathematics Bibliog. vii, 200p 22cm Harper and Row,New York,1965
P. Math 2.1770

LAFAYETTE,IND. 1965 Advances in electron metallography a symposium presented at the sixty-eighth annual meeting Vol 6 American society for testing materials A.S.T.M.Special technical publication, 396 A.S.T.M. Philadelphia,1966
Met 25.1399

LAFAYETTE,IND. 1965 Advances in thermophysical properties at extreme temperatures and pressures Symposium on thermophysical properties 3rd Papers Edited by Serge Gratch Sponsored by the American society of mechanical engineers American society of mechanical engineers,New York,1965
Eng 41.7121

LAFAYETTE,IND. 1965 Behaviour of materials at cryogenic temperatures a symposium American society for testing materials A.S.T.M.Special technical publication, 387 American society for testing materials, Philadelphia,Pa.,1916
Met 25.0976

LAFAYETTE,IND. 1965 Instruments and apparatus for soil and rock mechanics :a symposium American society for testing materials American society for testing materials.Special technical publication, 392 American society for testing materials, Philadelphia,Pa.,1965
Eng 41.3158

LAFAYETTE,IND. 1970 Mathematical software symposium Mathematical software based on the proceedings of the Mathematical software symposium Edited by John R. Rive Held at Purdue university Association for computing machinery.Monograph series 515p Academic press,New York,1971
Math L 5.3806

LAFAYETTE,INDIANA 1963 Permafrost international conference 1st Proceedings National research council Presented by the Building research advisory board National research council.Publication,1287 DLC,GB641. 16 1963 illus,maps,tables,graphs 563p National academy of sciences,Washington,1966
Sco 14.0390

LAGONISSI 1964 Sep.9-23 Observational aspects of galactic structure :a summer school typescript Edited by A. Blaauw and L.N. Mavridis Under the auspices of the North Atlantic treaty organization.Science committee 370p Athens,1965
Obs 6.0193

LAGONISSI,ATHENS 1965 Solar physics NATO advanced study institute on solar physics Proceedings North Atlantic treaty organization.Division of scientific affairs Interscience,London,1967
Obs 6.3280

LAKE ARROWHEAD 1966 Proceedings of the symposium on the arctic heat budget and atmospheric circulation University of California,and,Rand corporation Edited by J. O. Fletcher Prepared for the National science foundation Memorandum RM-5233-NSF illus,diagrs 566p The Rand corporation, Santa Monica,1966
Sco 14.2407

LAKE COMO 1966 and LAKE COMO 1967 Towards a theoretical biology :an I.U.B.S. symposium Vol 1-3 International union of biological sciences Edited by C.H. Waddington 3 vols Edinburgh university press,Edinburgh,1970
Bot 42.0320

LAKE COMO 1967 Towards a theoretical biology :a symposium Papers Vol 2: sketches International union of biological sciences Edited by C.H. Waddington I.U.B.S. symposium, 2 Edinburgh university press, Edinburgh,1969
Bal 39.0053

LAKE COMO 1967 Towards a theoretical biology :an IUBS symposium 2: sketches International union of biological sciences Edited by C.H. Waddington Edinburgh university press,Edinburgh,1968
Bal 39.0017

LAKE GENEVA,WIS. 1952 Structure and properties of solid surfaces :conference Papers Edited by Robert Gomer and Cyril Stanley Smith Organized by the National research council.Committee on solids University of Chicago press,Chicago,Ill.,1953
Col S 12.0275

LAKE GENEVA,WISC. 1952 Structure and properties of solid surfaces :a conference Edited by Robert Gomer and Cyril Stanley Smith Arranged by the National research council illus 491p University of Chicago press, Chicago,Ill.,1953
Cav 7.1344

LAKE GENEVA,WISC. 1952 Structure and properties of solid surfaces :a conference National research council Edited by Robert Gomer and Cyril Stanley Smith University of Chicago press,Chicago,1953
Met 25.1594

LAKE PLACID 1956 Dislocations and mechanical properties of crystals an international conference United States.Air force.Office of scientific research and General electric research laboratory Edited by J.C. Fisher and others Sponsored by United States.Air research and development command Wiley;Chapman and Hall,New York; London,1957
Met 25.1232

LAKE PLACID,N.Y. 1956 Dislocations and mechanical properties of crystals : international conference Proceedings United States.Air force.Office of scientific research and General electric research laboratory Edited by J.C. Fisher and others John Wiley,New York,1957
Cav 7.0804

LAKE SUCCESS 1949 International technical conference on the protection of nature Preparatory documents International union for the protection of nature.Secretariat Unesco,Paris,1949 Text in English and French
Bal 39.3922

LAKES Man-made lakes Institute of biology : symposium 15th London 1965 Sep 30-Oct 1 Edited by R.H. Lowe-McConnell Academic press, London,1966
Geog 13.0836

LANCASTER 1969 Approximation theory :a symposium Proceedings Edited by A. Talbot Academic press,London,1970
Eng 41.2344

LANCASTER 1969 Approximation theory :a symposium Proceedings Edited by A. Talbot viii,356p 25cm Academic press,New York; London,1970
P Math 2.3850

LANCASTER,PA. 1958 Viscoelasticity - phenomenological aspects :a symposium Papers Edited by J.T. Bergen Sponsored by the Armstrong Cork company x,150p Academic press,New York;London,1960
Chem E 24.0413

LANDBOUWHOGESCHOOL,WAGENINGEN Cell differentiation and morphogenesis : international lecture course Wageningen 1965 Apr 26-29 Edited by W. Beermann and others North-Holland,Amsterdam,1969
An 32.5439

LANDBOUWHOGESCHOOL,WAGENINGEN Cell differentiation and morphogenesis... : international lecture course Wageningen 1965 Apr 26-29 By W. Beerman North-Holland, Amsterdam,1966
Gen 34.2211

LANDBOUWHOOGESCHOOL,WAGENINGEN Cell differentiation and morphogenesis : international lecture course Wageningen 1965 Apr 26-29 By W. Beermann and others Organized by the Agricultural university of Wageningen North Holland,Amsterdam,1966 The fourth international symposium organized by the Agricultural university of Wageningen
Bot 42.1098

LANDBOUWHOOGESCHOOL,WAGENINGEN Cell differentiation and morphogenesis : international lecture course Wageningen 1965 Apr 26-29 Edited by W. Beermann and others Organized by the Agricultural university of Wageningen North-Holland, Amsterdam,1966 "The fourth international symposium organized by the Agricultural university of Wageningen"
Bioch 33.1020

LANGUAGE Cognition and the development of language :annual symposium 4th Papers Pittsburgh,Pa. 1968 Apr 11-12 Carnegie-Mellon university Edited by John R. Hayes Wiley,New York,1970
Psy 31.3113

LANGUAGE AND THOUGHT IN SCHIZOPHRENIA : collected papers presented at a meeting of the American psychiatric association...and brought up to date Papers Chicago,Ill. 1939 May 12 American psychiatric association Edited by J.S. Kasanin University of California press,Berkeley;Los Angeles,1954 Preface by N.D.C.Lewis
Psy 28.0082

LANGUAGE DEVELOPMENT IN CHILDREN The Genesis of language:a psycholinguistic approach conference Proceedings Old Point Comfort, Va. 1965 Apr 25-28 National institute of child health and human development.Human communication program Edited by Frank Smith and George A. Miller M.I.T.press,Cambridge, Mass.;London,1966
Psy 31.2746

LANGUAGE MACHINES Picture language machines : a conference Proceedings Canberra 1969 Feb 24-28 Commonwealth scientific and industrial research organisation Australian national university.Research school of physical sciences Edited by S. Kaneff Academic press,London,1970
Psy 31.3083

LANKENAU CONFERENCE 5th Proceedings Mind as a tissue Philadelphia,Pa. Lankenau hospital Edited by Charles Rupp Harper and Row,Hoeber medical division,New York,1968
Psy 31.2898

LANKENAU HOSPITAL Fat as a tissue : international research conference Proceedings Philadelphia 1962 Nov 2-3 Edited by Kaare Rodahl and others McGraw-Hill,New York,1963
An 32.5318

LANKENAU HOSPITAL Mind as a tissue Lankenau conference 5th Proceedings Philadelphia,Pa. Edited by Charles Rupp Harper and Row,Hoeber medical division,New York,1968
Psy 31.2898

LANKENAU HOSPITAL Muscle as a tissue :a conference Proceedings Philadelphia,Pa. 1960 Nov 3-4 Edited by Kaare Rodahl and Steven M. Horvath McGraw-Hill,New York,1961
An 32.3999

LANKENAU HOSPITAL Nerve as a tissue :a conference 4th Proceedings Philadelphia, Pa. 1964 Nov 12-13 Edited by Kaare Rodahl and Bela Issekutz Harper,New York,1966
Pha 16.0299

LANKENAU HOSPITAL Nerve as a tissue : conference Proceedings Philadelphia,Pa. 1964 Nov 12-13 Edited by Kaare Rodahl and Bela Issekutz Harper,London,1966
An 32.3430

LARGE-CAPACITY MEMORY TECHNIQUES FOR COMPUTING SYSTEMS :a symposium Washington,D.C. 1961 May 23-25 Edited by Marshall C. Yovits Sponsored by the Association for computing machinery.Information systems branch Association for computing machinery.Monograph series Macmillan,New York;London,1962
Eng 41.2196

LARGE DAMS International congress on large dams 5th Transactions Paris 1955 May 31-Jun 4 Vol 1-5 International commission on large dams 5 vols Paris,1956
Eng 41.3334

LARGE FATIGUE TESTING MACHINES AND THEIR RESULTS a symposium held at the 1957 annual meeting of the society Atlantic City,N.J. 1957 Jun 18 American society for testing materials American society for testing materials,Cleveland,Ohio,1957
Met 25.0922

LARGE PERMANENT UNDERGROUND OPENINGS : international symposium Proceedings Oslo 1969 Sep 23-25 International society for rock mechanics Edited by T.L. Brekke and Finn Jorstrad Universitetsforlaget,Oslo,1970
Eng 41.3133

LARGE SPARSE SETS OF LINEAR EQUATIONS Conference of the Institute of mathematics and its applications Proceedings Oxford 1970 Apr Institute of mathematics and its applications Edited by J.K. Reid Illus 284p Academic press,London,1971
Math L 5.3804

LARGE SPARSE SETS OF LINEAR EQUATIONS Oxford conference of the Institute of mathematics and its applications Proceedings Oxford 1970 Apr 5-8 Institute of mathematics and its applications Edited by J.K. Reid x, 284p 24cm Academic press,London;New York, 1971
P Math 2.4464

LARGE SPARSE SETS OF LINEAR EQUATIONS :a conference Proceedings Oxford 1970 Institute of mathematics and its applications Edited by J.K. Reid Academic press,London, 1971
Eng 41.8474

LARGE TELESCOPE DESIGN :a conference Proceedings Geneva 1971 Mar 1-5 European southern observatory CERN Edited by R.M. West 499p European southern observatory,Geneva,1971
Obs 6.3601

LAS CRUCES,N.M. 1972 ACM conference on proving assertions about programs Proceedings Association for computing machinery.SIGPLAN Association for computing machinery.SIGACT illus 211p ACM,New York, 1972 Published as SIGPLAN notices,vol 7, no 1,Jan 1972 and SIGACT news no 14,Jan 1972
Math L 5.3859

LASER BEAMS Electron and laser beam symposium 7th Proceedings University Park,Pa. 1965 Mar 31-Apr 2 Pennsylvania state university Sponsored also by the Alloyd general corporation Pennsylvania state university,University Park,Pa.,1965
Eng 41.5617

LASER BEAMS Electron and laser beam symposium 8th Proceedings Ann Arbor, Mich 1966 Apr 6-8 University of Michigan Institute of electrical and electronics engineers Edited by G.I. Haddad University of Michigan,Ann Arbor,Mich.,1966 8th in the series of annual symposia on Electron beam technology initiated in 1959 by Alloyd and held under their sponsorship for seven years. The titles of the symposia vary,therefore those catalogued have been grouped together under the keyword"Electron beams"
Eng 41.5618

LATE EFFECTS OF RADIATION :a colloquium Proceedings Chicago,Ill. 1969 May Edited by R.J.M. Fry and others Taylor and Francis,London,1970
Radioth 35.1221

LATE-TYPE STARS Colloquium on late-type stars proceedings Trieste 1966 Jun.13-17 Osservatorio astronomico di Trieste Edited by Margherita Hack 465p Trieste,1967
Obs 6.0712

LATIN AMERICAN CENTER FOR RESEARCH IN THE SOCIAL SCIENCES,RIO DE JANEIRO Social science in Latin America Conference on Latin American studies Papers Rio de Janeiro 1965 Mar 29-31 Joint committee on Latin American studies Edited by Manuel Diegues and Bryce Wood Columbia university press,New York,1967
Geog 13.5108

LATIN-AMERICAN CONGRESS OF NEURO-SURGERY 10TH Biology of neuroglia Lectures and discussions Buenos Aires 1963 Oct 17-18 Instituto Torcuato di Tella and International brain research organization Edited by E.D.P.de Robertis and R. Carea Progress in brain research, 15 illus Elsevier,Amsterdam,1965 Symposium held as part of the 10th Latin-American congress of neurosurgery
Path 30.2466

LATIN-AMERICAN CONGRESS OF NEUROSURGERY,10TH Biology of neuroglia :a symposium... held as part of the 10th Latin-American congress of neurosurgery Buenos Aires 1963 Oct 17-18 Edited by E.D.P.de Robertis and R. Carrea Progress in brain research, 15 Elsevier, Amsterdam,1965
An 32.4235

LATIN-AMERICAN GEOLOGY conference proceedings Austin,Tex. 1954 Mar.29-30 University of Texas.Department of geology,and, University of Texas Institute of Latin-American studies Edited by Fred M. Bullard map University of Texas,Austin,Tex.,1955
Geol 8.2706

LATITUDE SERVICE International latitude service L' Avenir du service international des latitudes :a symposium proceedings Helsinki 1960 Jul.26-Aug.7 Edited by Paul Melchior Organized by International union of geodesy and geophysics International astronomical union.Symposium, 13 97p Paris,1961 Reprinted from Bulletin geodesique no.59,March 1961
Obs 6.1198

LATTICE DEFECTS IN QUENCHED METALS :an international conference Chicago 1964 Jun 15-17 Edited by R.M.J. Cotterill and others Sponsored by the United States atomic energy commission Academic press,New York; London,1965
Met 25.1240

LATTICE DYNAMICS The International conference on lattice dynamics Proceedings Copenhagen 1963 Aug 5-9 International union of pure and applied physics and University of Pennsylvania Edited by R.F. Wallis With the support of the International atomic energy agency 730p Pergamon press, Oxford,1965
Cav 7.0542

LATTICE IMPERFECTIONS Imperfections in nearly perfect crystals :a symposium Papers Pocono Manor,Pa. 1950 Oct 12-14 Edited by William Shockley and others Sponsored by the National research council.Committee on solids John Wiley,New York,1952
Cav 7.2045

LATTICE THEORY Trends in lattice theory By Garrett Birkhoff and others Edited by J.C. Abbott Van Nostrand mathematical studies, 31 Bibliog. ix,215p 20cm Van Nostrand Reinhold,New York,1970 Written versions of four talks given to a symposium on "Trends in Lattice theory" held at the U.S.Naval academy,May 1966
P Math 2.3710

LATTICE THEORY symposium Monterey,Calif. 1959 Apr 16-18 American mathematical society Edited by R.P. Dilworth With the financial support of the National science foundation American mathematical society. Proceedings of symposia in pure mathematics, 2 viii,208p 26cm American mathematical society,Providence,R.I.,1961
P. Math 2.2030

LATTICES Phonons in perfect lattices and in lattices with point imperfections N.A.T.O. advanced study institute Papers Aberdeen 1965 Aug 9-27 Scottish universities' summer school in physics Edited by R.W.H. Stevenson With financial support of North Atlantic treaty organization Scottish universities' summer school in physics, 6 448p Oliver and Boyd,Edinburgh,1966
TA 15.0212

LAURENTIAN HORMONE CONFERENCE Proceedings Recent progress in hormone research Franconia,N.H. 1949 Sep Vol 5 Edited by Gregory Pincus Academic press,New York,1950
Bal 39.1376

The LAURENTIAN HORMONE CONFERENCE 27th Proceedings Mount Tremblant 1969 Aug 24-29 Edited by E.B. Astwood Recent progress in hormone research, 26 Academic press,New York;London,1970
Inv Med 37.0265

LAURENTIAN SYMPOSIUM Brain mechanisms and consciousness :a symposium Council for international organizations of medical sciences Edited by J.F. Delafresnaye Blackwell,Oxford,1954
Psy 31.0519

LAUSANNE 1947 Congres international d'histoire des sciences 5e Actes Union internationale d'histoire des sciences Academie internationale d'histoire des sciences.Collections de travaux, 2 288p Academie internationale d'histoire des sciences;Hermann,Paris,1948
WSM 43.0060

LAUSANNE 1958 Symposium on noxious effects of low level radiation :physical elements and biological aspects Schweizerische akademie der medizinischen wissenschaften Schwabe, Basle;Stuttgart,1958
Radioth 35.1078

LAWRENCE 1964 Bibliography and natural history :a conference Essays University of Kansas.Linda Hall library of science and technology Edited by Thomas R. Buckman University of Kansas.Publications.Library series, 27 ix,148p University of Kansas libraries,Lawrence,Kansas,1966
WSM 43.3212

LAWRENCE,KAN 1964 Mechanisms of inorganic reactions :symposium Papers Edited by R. Kent Murmann and others Sponsored by the American chemical society.Inorganic chemistry division American chemical society.Advances in chemistry series, 49 vii,266p American chemical society,Washington,D.C.,1965
Chem E 24.0629

LAWRENCE,KAN. 1962 International conference on taxonomic biochemistry, physiology and serology Edited by Charles A. Leone xi,728p Ronald press,New York,1964
Bot 42.3437

LAWRENCE,KAN. 1964 Mechanisms of inorganic reactions American chemical society.Division of inorganic chemistry Edited by R.Kent Murmann and others American chemical society. Advances in chemistry series,49 American chemical society,Washington,D.C.,1965 Symposium chairman:J.Kleinberg
Chem 18.1002

LAWRENCE,KANSAS 1968 Symposium on the structure of low-medium mass nuclei 3rd Aerospace research laboratories Edited by J. P. Davidson vi,294p 24cm University press of Kansas,Lawrence,Kansas,1968
Cav 7.3099

LAWRENCE RADIATION LABORATORY.INORGANIC MATERIALS DIVISION Electron microscopy and strength of crystals Berkeley international materials conference :the impact of transmission electron microscopy on theories of the strength of crystals 1st Proceedings Berkeley 1961 Jul 5-8 Edited by Gareth Thomas and Jack Washburn Interscience,New York;London,1963
Met 25.1211

LAWRENCE RADIATION LABORATORY.INORGANIC MATERIALS RESEARCH DIVISION International materials symposium :the structure and chemistry of solid surfaces 4th Proceedings Berkeley, Calif 1968 Jun 17-21 Edited by Gabor A. Somorjai Inorganic materials research division series, 4 Wiley,New York,1969
Met 25.2476

LAYER SILICATES A.G.I. short course on layer silicates :lecture notes New Orleans,La. 1967 Nov 17-19 American geological institute American geological institute,Washington,D.C., 1967 Typescript
Min 10.1287

LAYERED INTRUSIONS Symposium on the Bushveld igneous complex and other layered intrusions Proceedings Pretoria 1969 Jul 7-14 Geological society of South Africa Edited by D.J.L. Visser and G.von Gruenewaldt Geological society of South Africa.Special paper, 1 Geological society of South Africa,Johannesburg,1970
Min 10.1476

LEAMINGTON SPA 1950 Some aspects of fluid flow :a conference Papers,reports Edited by H.R. Lang Organized by the Institute of physics viii,292p Edward Arnold,London, 1951
Chem E 24.0224

LEAMINGTON SPA 1950 Some aspects of fluid flow... :a conference Papers Institute of physics Arnold,London,1951
Eng 41.6954

LEAMINGTON SPA 1950 Wire ropes in mines :a conference Proceedings Arranged by the Institution of mining and metallurgy Institution of mining and metallurgy,London, 1951
Eng 41.8248

LEAMINGTON SPA 1951 The Size and shape factor in colloidal systems :a discussion Faraday society Faraday society.Discussions, 11 Faraday society,Aberdeen,1952
Bioch 33.1529

LEAMINGTON-SPA 1951 The All-basic open-hearth furnace being also the report of the 36th steelmaking conference British ceramic research association and British iron and steel research association Iron and steel institute.Special report, 46 Iron and steel institute,London,1952
Met 25.0330

LEARNING,REMEMBERING AND FORGETTING Organization of recall International interdisciplinary conference on learning, remembering and forgetting 2nd Proceedings Princeton,N.J. 1964 Sep 27-30 New York academy of sciences and National institute of child health and human development Edited by Daniel P. Kimble With the support of United States.Office of naval research New York academy of sciences. Interdisciplinary communication program,New York,1967
Psy 31.1012

LEARNING,REMEMBERING AND FORGETTING Readiness to remember Conference on learning, remembering and forgetting 3rd Proceedings Princeton,N.J. 1965 Vol 1-2 New York academy of sciences Edited by Daniel P. Kimble Gordon and Breach,New York,1969
Psy 31.3281

LEARNING THEORY,PERSONALITY THEORY,AND CLINICAL RESEARCH :the Kentucky symposium Papers Kentucky symposium Lexington,Ky. 1953 Mar 13-14 University of Kentucky.College of arts and sciences.Department of psychology Wiley;Chapman and Hall,New York;London,1954
Psy 31.1839

LEATHERHEAD 1960 Aerodynamic capture of particles :a conference Proceedings Edited by E.G. Richardson Pergamon press,London, 1960
Eng 41.6967

LEATHERHEAD 1960 Aerodynamic capture of particles :a conference Proceedings Edited by E.G. Richardson viii,200p Pergamon press,Oxford,1960
Chem E 24.0223

LEATHERHEAD 1962 Gas discharges and the electricity supply industry International conference on gas discharges and the electricity supply industry Edited by J.S. Forrest and others Held at the Central electricity research laboratories Butterworths,London,1962
Eng 41.5415

LEATHERHEAD,SURREY 1962 Gas discharges and the electricity supply industry :a conference Proceedings Central electricity research laboratories Edited by J.S. Forrest and others 677p Butterworths,London,1962
Nap 11.0123

LEBANESE SOCIETY OF THE FRIENDS OF THE TREES
Colloque sur la protection et la conservation de la nature dans le Proche-orient :symposium on the protection and conservation of nature in the Near East Beirut 1954 Jun 3-8 Beirut,1954 Text in English and French
Bal 39.1787

LECTURE NOTES PREPARED IN CONNECTION WITH THE SUMMER INSTITUTE ON ALGEBRAIC GEOMETRY
typescript Woods Hole,Mass. 1964 Jul 6-31 By Shreeram S. Abhyankar and others American mathematical society 29cm American mathematical society,1964
P. Math 2.0575

LECTURES AND LABORATORY EXERCISES ON STRAIN GAGE TECHNIQUES Los Angeles,Calif. 1958 Aug 18-29 By William M. Murray and Peter K. Stein University of California Los Angeles Society for experimental stress analysis University of California Los Angeles,Los Angeles,Calif.,1958
Eng 41.2906

LECTURES IN APPLIED MATHEMATICS :summer seminar proceedings Boulder,Colo. 1957 Jun.23-Jul.19 Vol. 1: probability and related topics in physical sciences By Mark Kac American mathematical society Supported by United States.Armed services xiii,266p 24cm Interscience publishers, London;New York,1959 With special lectures by G.E.Uhlenbeck,A.R.Hibbs,and Balth. van der Pol
Math 3.0844

LECTURES IN APPLIED MATHEMATICS:PROCEEDINGS OF THE SUMMER SEMINAR,BOULDER, COLO. 1957, 2
Lectures on fluid mechanics By Sydney Goldstein xvi,307p Interscience,New York, 1960 Seminar held under the joint sponsorship of the University of Colorado, and the American mathematical society.
Chem E 24.0355

LECTURES IN APPLIED MATHEMATICS, 3 Summer seminar in applied mathematics 1st Proceedings Boulder,Col. 1957 Jun 23-Jul 21 Vol 3: partial differential equations By Lipman Bers and others Edited by Alton S. Householder and others Sponsored by the American Mathematical Society xiii,343p Interscience,New York,1964 Dedicated to Arthur Norman Milgram
A Math 4.1266

LECTURES IN APPLIED MATHEMATICS, 8
Relativity theory and astrophysics 1: relativity and cosmology Lectures Ithaca,N. Y. 1965 Jul 25-Aug 20 By W. Bonner and others American mathematical society.Summer seminar in applied mathematics Edited by Jurgen Ehlers xvi,292p American mathematical society,Providence,R.I.,1967
TA 15.0350

LECTURES IN BIOLOGICAL SCIENCES :a symposium Knoxville,Tenn. 1959 Dec 3-5 By Norman G. Anderson and others Sigma Xi.Tennessee chapter Oak Ridge national laboratory. Biology division Edited by J.Ives Townsend 110p University of Tennessee press,Knoxville, Tenn.,1963
Bot 42.4744

LECTURES IN THEORETICAL PHYSICS, 1-5
Lectures in theoretical physics Boulder,Col. 1958-62 Vol 1-5 University of Colorado.Summer institute for theoretical physics Edited by Wesley E. Brittin and others Interscience,New York,1959-63
Cav 7.0292

LECTURES IN THEORETICAL PHYSICS, 11a,pt 1-2
Elementary particle physics Boulder conference on particle physics held during the first part of the 11th Boulder summer institute for theoretical physics 5th Proceedings Boulder,Colo. 1968 Jun 17-Aug 23 Vol 1-2 University of Colorado. Department of physics and astrophysics Edited by Kalyana T. Mahanthappa and others Bibliog,diagrms xxi,629p 23cm 2 vols Gordon and Breach,New York,1969 Dedicated to George Gamow
Cav 7.3008

LECTURES IN THEORETICAL PHYSICS, 11b
Quantum fluids and nuclear matter Boulder summer institute for theoretical physics 11th Proceedings Boulder,Colo. 1968 Jun 17-Aug 23 University of Colorado.Department of physics and astrophysics Edited by Kalyana T. Mahanthappa and Wesley E. Brittin diagrms xiv,428p 23cm Gordon and Breach, New York,1969 Dedicated to George Gamow
Cav 7.3009

LECTURES IN THEORETICAL PHYSICS, 11c
Symposium on atomic collision processes Boulder summer institute for theoretical physics 11th Proceedings Boulder,Colo. 1968 Jun 17-Aug 23 University of Colorado. Department of physics and astrophysics Edited by Kalyana T. Mahanthappa and others diagrms xiii,337p 23cm Gordon and Breach, London,1969 Dedicated to George Gamow
Cav 7.3010

LECTURES IN THEORETICAL PHYSICS, 11d
Mathematical methods in theoretical physics Boulder summer institute for theoretical physics 11th Proceedings Boulder,Colo. 1968 Jun 17-Aug 23 University of Colorado. Department of physics and astrophysics Edited by Kalyana T. Mahanthappa and Wesley E. Brittin diagrms xiv,648p 23cm Gordon and Breach,New York,1969
Cav 7.3011

LECTURES IN THEORETICAL PHYSICS,BOULDER.,1963,6
Lectures in theoretical physics 6th Boulder 1963 Vol. 6 University of Colorado.Summer institute for theoretical physics Edited by W.E. Brittin and W.R. Chappell Sponsored by the University of Colorado.Department of physics and astrophysics Colorado university press, Boulder,Colo.,1964
A Math 4.0834

LECTURES IN THEORETICAL PHYSICS,BOULDER,1964,7a
Lectures in theoretical physics 7th Boulder,Col. 1964 7a: Lorentz group University of Colorado.Summer institute for theoretical physics Edited by Wesley E. Britten and Asim O. Barut Sponsored by the University of Colorado.Department of physics and astrophysics University of Colorado press,Boulder,Col.,1965 The Lorentz group symposium volume is dedicated to Professor E. Wigner
A Math 4.1335

LECTURES IN THEORETICAL PHYSICS,BOULDER,1964,7c
Lectures in theoretical physics 7th Boulder,Col. 1964 Vol. 7c: statistical physics,weak interactions,field theory University of Colorado.Summer institute for theoretical physics Edited by Wesley E. Brittin Sponsored by the University of Colorado.Department of physics and astrophysics University of Colorado press, Boulder,Col.,1965
A Math 4.0835

LECTURES IN THEORETICAL PHYSICS,1959
Theoretical physics: lecture notes Waltham, Mass. 1959 By F.E. Low Brandeis university.Summer institute in theoretical physics Edited by Wayne A. Mills Brandeis university,Waltham,Mass.,1959
A Math 4.0828

LECTURES IN THEORETICAL PHYSICS,1960
Theoretical physics :lecture notes Typescript Waltham,Mass. 1960 By C. Moller and others Brandeis university.Summer institute in theoretical physics Edited by J. Stachel and others Brandeis university, Waltham,Mass.,1960
Cav 7.0314

LECTURES IN THEORETICAL PHYSICS,1960
Theoretical physics :lecture notes Waltham, Mass. 1960 By C. Moller and others Brandeis university.Summer institute in theoretical physics Edited by J. Stachel and others Brandeis university,Waltham,Mass, c1961
A Math 4.0829

LECTURES IN THEORETICAL PHYSICS,1961,1
Lectures in theoretical physics Waltham,Mass. 1961 Pt 1 By R.J. Eden Brandeis university.Summer institute in theoretical physics Benjamin,New York,1962
Cav 7.0315

LECTURES IN THEORETICAL PHYSICS,1961,2
Lectures in theoretical physics Waltham,Mass 1961 Vol. 2 By M.E. Rose and E.C.G. Sudarshan Brandeis university.Summer institute in theoretical physics Benjamin, New York,1962
A Math 4.0830

LECTURES IN THEORETICAL PHYSICS,1961,2
Lectures in theoretical physics Waltham,Mass. 1961 Pt 2 By M.E. Rose and E.C.G. Sudarshan Brandeis university.Summer institute in theoretical physics Benjamin, New York,1962
Cav 7.0316

LECTURES IN THEORETICAL PHYSICS,1962,1
Lectures on particle and field theory Waltham,Mass 1962 Vol 1: elementary particle physics and field theory By I. Fulton and others Brandeis university.Summer institute in theoretical physics Benjamin, New York,1963
Cav 7.0317

LECTURES IN THEORETICAL PHYSICS,1962,1
Lectures on particles and field theory Waltham,Mass. 1962 Vol. 1: elementary particle physics and field theory By I. Fulton and others Brandeis university.Summer institute in theoretical physics Benjamin, New York,1963
A Math 4.0831

LECTURES IN THEORETICAL PHYSICS,1962,2
Lectures in theoretical physics Waltham,Mass. 1962 Vol 2: astrophysics and the many-body problem By E.N. Parker and J.S. Goldstein Brandeis university.Summer institute in theoretical physics Benjamin, New York,1963
Cav 7.0318

LECTURES IN THEORETICAL PHYSICS,1963,1
Lectures on strong and electromagnetic interactions Waltham,Mass. 1963 By P.T. Matthews and others Brandeis university. Summer institute in theoretical physics Brandeis University,Waltham,Mass.,1964
A Math 4.0036

LECTURES IN THEORETICAL PHYSICS,1963,2
Lectures on astrophysics and weak interactions Waltham,Mass. 1963 By S. Hayakawa and others Brandeis university.Summer institute in theoretical physics Brandeis university, Waltham,Mass.,1963
A Math 4.1087

LECTURES IN THEORETICAL PHYSICS,1964,1
Lectures on general relativity Waltham,Mass. 1964 By A. Trautman Brandeis university. Summer institute in theoretical physics Prentice-Hall,Englewood Cliffs,N.J.,1965
A Math 4.1025

LECTURES IN THEORETICAL PHYSICS,1964,1
Lectures on general relativity Waltham,Mass. 1964 By A. Trautman and others Brandeis university.Summer institute in theoretical physics Prentice-Hall,Englewood Cliffs,N.J., 1965
Cav 7.0320

LECTURES IN THEORETICAL PHYSICS,1964,2
Lectures on particles and field theory Waltham,Mass. 1964 By K. Johnson and others Brandeis university.Summer institute in theoretical physics Prentice-Hall, Englewood Cliffs,N.J.,1965
Cav 7.0321

LECTURES IN THEORETICAL PHYSICS,1965,1
Lectures on particle symmetries and axiomatic field theory Papers Waltham,Mass. 1965 1: axiomatic field theory Brandeis university.Summer institute in theoretical physics Edited by M. Chretien and S. Deser Gordon and Breach,New York,1966
A Math 4.1334

LECTURES IN TOPOLOGY :conference Ann Arbor 1940 Jun.24-Jul.6 University of Michigan Edited by Raymond L. Wilder and William L. Ayres vii,316p 23cm University of Michigan press,Ann Arbor, Mich., 1941
P. Math 2.0288

LECTURES ON FUNCTIONS OF A COMPLEX VARIABLE : conference proceedings Ann Arbor, Mich. 1953 Jun 17-Jul1 University of Michigan Edited by Wilfred Kaplan and others Bibliog 435p 24cm University of Michigan,Ann Arbor, Mich.,1955
P. Math 2.1533

LECTURES ON THEORETICAL PHYSICS,1964,2
Lectures on particles and field theory Waltham,Mass 1964 Vol. 2 By K. Johnson and others Brandeis university. Summer institute in theoretical physics Prentice-Hall,New York,1965
A Math 4.0833

LEDERLE LABORATORIES Form and function in the human lung :a symposium Proceedings Birmingham 1967 Apr 5-7 Edited by Gordon Cumming and L.B. Hunt Livingstone,Edinburgh, 1968
An 32.3805

LEE,MASS. 1955 Concepts of biology Conference report National research council. Division of biology and agriculture.Biology council Edited by R.W. Gerard and R.B. Stevens National research council. Publication, 560 Washington,D.C.,1958
Bal 39.0989

LEE WILSON ENGINEERING COMPANY,INC. The Annealing of low carbon steel :the international symposium Proceedings Cleveland 1957 Oct 29-30 Lee Wilson engineering company,Cleveland,Ohio,1958
Met 25.0447

LEEDS 1946 Fibrous proteins Symposium on fibrous proteins Proceedings Society of dyers and colourists Edited by C.L. Bird Leeds,1946
Bioch 33.0503

LEEDS 1946 Fibrous proteins :a symposium Proceedings Society of dyers and colourists Edited by C.L. Bird figs,pls 27cm n.p., c1946
Bal 44.4561

LEEDS 1954 Fibrous proteins and their biological significance :a symposium Papers Society for experimental biology Edited by R. Brown and J.F. Danielli Society for experimental biology.Symposia, 9 Cambridge university press,Cambridge,1955
An 32.5465

LEEDS 1954 Fibrous proteins and their biological significance :a symposium Society for experimental biology Society for experimental biology.Symposia, 9 Cambridge university press,Cambridge,1955
Bioch 33.1354

LEEDS 1956 Metals and enzyme activity :a symposium Biochemical society Edited by E. M. Crook Held at the University of Leeds Biochemical society.Symposia, 15 Cambridge university press,Cambridge,1958
Bioch 33.1377

LEEDS 1956 Metals and enzyme activity :a symposium 15 Biochemical society Edited by E.M. Crook Biochemical society symposia, 15 Cambridge university press,Cambridge,1958
Chem 18.1269

LEEDS 1957 Symposium on coal preparation 2nd Papers University of Leeds.Department of mining xiii,513p University of Leeds, Leeds,1957
Chem E 24.1392

LEEDS 1964 Biosynthetic pathways in higher plants Plant phenolics group symposium Proceedings Plant phenolics group Edited by J.B. Pridham and T. Swain Academic press, London;New York,1965
Bot 42.1883

LEEDS 1964 Biosynthetic pathways in higher plants Plant phenolics group symposium Proceedings Plant phenolics group Edited by J.B. Pridham and T. Swain Academic press, London;New York,1965 For later publications of this body see Phytochemical society
Gen 34.1406

LEEDS 1964 Biosynthetic pathways in higher plants The Plant phenolics group symposium Proceedings Plant phenolics group Edited by J.B. Pridham and T. Swain Academic press, London;New York,1965
Bioch 33.0783

LEEDS 1967 Summer school in logic :N.A.T.O. advanced study institute:meeting of the association for symbolic logic Proceedings North Atlantic treaty organization Association for symbolic logic Edited by M.H. Lob Lecture notes in mathematics, 70 Bibliog. 331p 28cm Springer,Berlin,1968
P Math 2.3120

LEEDS 1967 Symposium on tetanus in Great Britain Leeds general infirmary.Tetanus unit Edited by M. Ellis c1967 Typescript
PGMS 29.0042

LEEDS 1968 The Renal stone research symposium Proceedings Medical research council and University of Leeds Edited by A. Hodgkinson and B.E.C. Nordin Churchill, London,1969
PGMS 29.0240

LEEDS 1972 Elastohydrodynamic lubrication : a symposium Proceedings Institution of mechanical engineers.Tribology group IME, London,1972
Eng 41.8650

LEEDS GENERAL INFIRMARY.TETANUS UNIT Symposium on tetanus in Great Britain Leeds 1967 Apr 7-8 Edited by M. Ellis c1967 Typescript
PGMS 29.0042

LEGUMES Nutrition of the legumes Easter school in agricultural science 5th Proceedings Nottingham 1958 University of Nottingham Edited by E.G. Hallsworth Butterworths,London,1958
Bioch 33.0776

LEGUMES Nutrition of the legumes Easter school in agricultural sciences 5th Proceedings Nottingham 1958 Edited by E. G. Hallsworth Butterworths,London,1958
Bot 42.1781

LEICESTER 1957 Exploration in group relations :a conference Report By E.L. Trist and C. Sofer University of Leicester Tavistock institute of human relations Leicester university press,Leicester,1959
Eng 41.0706

LEICESTER 1963 People in the countryside United Nations European study group Report Edited by John Higgs NCSS 694 National council of social service,London,1966
Geog 13.1901

LEICESTER 1963 The Civitas capitals of Roman Britain :a conference Papers Edited by J.S. Wacher Bibliog,illus 124p Leicester university press,Leicester,1966
Geog 13.3213

LEICESTER 1965 Sets models and recursion theory :summer school in mathematical logic and Tenth logic colloquium Proceedings Edited by John N. Crossley Studies in logic and the foundations of mathematics Bibliog. vii,329p 23cm North Holland,Amsterdam,1967
P Math 2.3118

LEICESTER 1966 Aspects of Tethyan biogeography Systematics association symposium 7th papers Systematics association,and,Palaeontological association Edited by C.G. Adams and D.V. Ager Systematics association.Publication,7 Systematics association,London,1967
Geol 8.4466

LEICESTER 1966 Aspects of Tethyan biogeography :a symposium Systematics association Edited by C.G. Adams and D.V. Ager Systematics association.Publication, 7 illus. vi,336p British museum,London, 1967
Bot 42.3273

LEICESTER 1966 The Study of urban history : international round-table conference Proceedings Urban history group Edited by H.J. Dyos Arnold,London,1968
Geog 13.6762

LEICESTERSHIRE PROGRAMMED LEARNING GROUP Aspects of educational technology :the proceedings of the programmed learning conference Proceedings Loughborough 1966 Apr 15-18 Edited by Derick Unwin and John Leedham Methuen,London,1967
Psy 28.0233

LEIDEN 1952 Experimental diabetes and its relation to the clinical disease :a symposium Council for international organizations of medical sciences Edited by J.F. Delafresnaye and G.Howard Smith Blackwell,Oxford,1954
Bioch 33.0448

LEIDEN 1954 Fine structure of cells :a symposium held at the VIIIth congress of cell biology Papers International union of biological sciences With financial support of Unesco International union of biological sciences.Publications.Series B, 21 Noordhoff,Groningen,1955
Path 30.2410

LEIDEN 1954 Fine structure of cells :a symposium held at the 8th congress of cell biology International union of biological sciences With financial support of Unesco International union of biological sciences. Series B., 21 Noordhoff,Groningen,1955
Radioth 35.0467

LEIDEN 1954 Fine structure of cells : symposium held at the VIIIth congress of cell biology With financial support of Unesco International union of biological sciences. Series B, 21 Noordhoff,Groningen,1955
Phys 20.1483

LEIDEN 1954 Fine structure of cells : symposium held at the 8th congress of cell biology With the financial support of Unesco International union of biological sciences. Publications.Series B, 21 Noordhoff, Groningen,1955
An 32.3219

LEIDEN 1954 Fine structure of cells: symposium The Congress of cell biology 8th Papers International union of biological sciences With financial support of Unesco International union of biological sciences.Ser.B, 21 Noordhoff,Groningen,1955
Bal 39.0283

LEIDEN 1960 Quantitative methods in pharmacology :a symposium Proceedings Edited by H.de Jonge North-Holland,Amsterdam, 1961
HE 27.0109

LEIDEN 1962 International congress of physiological sciences 2nd Proceedings Vol 3: information processing in the nervous system Edited by R.W. Gerard and J.W. Duyff Under the auspices of the International union of physiological sciences International union of physiological sciences.Proceedings, 3 International congress series, 49 Excerpta medica foundation,Amsterdam,1964
An 32.4392

LEIDEN 1962 International congress of physiological sciences 22nd Vol 3: information processing in the nervous system Edited by R.W. Gerard and J.W. Duyff Under the auspices of the International union of physiological sciences International union of physiological sciences.Proceedings, 3 International congress series, 49 Excerpta medica,Amsterdam,1962
Phys 20.2114

LEIDEN 1962 International congress of physiological sciences 22nd abstracts of free communications;films and demonstrations Under the auspices of the International union of physiological sciences International union of physiological sciences. Proceedings, 2 International congress series, 48 Excerpta medica,Amsterdam,1962
Phys 20.2113

LEIDEN 1962 International congress of physiological sciences 22nd lectures and symposia Under the auspices of the International union of physiological sciences International union of physiological sciences. Proceedings, 1,pt 1-2 International congress series, 47 2 vols Excerpta medica,Amsterdam,1962
Phys 20.2112

LEIDEN 1962 Repair from genetic radiation damage and differential radiosensitivity in germ cells International symposium on repair from genetic radiation damage Proceedings Riksuniversiteit te Leiden Edited by F.H. Sobels Pergamon press,Oxford,1963
Gen 34.1121

LEIDEN 1962 Repair from genetic radiation damage and differential radiosensitivity in germ cells :an international symposium Proceedings By F.H. Sobels Pergamon press, London,1963
Radioth 35.1157

LEIDEN 1969 Mononuclear phagocytes Conference on mononuclear phagocytes Edited by Ralph van Furth Blackwell scientific, Oxford;Edinburgh,1970
An 32.5450

LEIPZIG 1956 Plasmas in physik und astronomie Probleme des plasmas in physik und astronomie :eine tagung berichte Physikalische gesellschaft in der deutschen demokratischen republik 223p Akademie-verlag,Berlin,1958
Obs 6.1366

LEIPZIG 1959 Naturwissenschaft und philosophie Internationales symposium uber naturwissenschaft und philosophie... Beitrage Institut fur philosophie Karl-Sudhoff institut fur geschichte der medizin und naturwissenschaften Edited by Gerhard Harig and Josef Schleifstein 437p Akademie verlag,Berlin,1960 Symposium held to commemorate the 550th anniversary of the Karl-Marx-universitat
WSM 43.0976

LEKARSKA SPOLECNOST J.E.PURKNYE Ontogenesis of the brain:the biochemical,functional and structural development of the nervous system International symposium neurontogeneticum 5th Proceedings Prague 1967 Edited by Lubor Jilek and Stanislav Trojan Universita Karlova,Prague,1968
PGMS 29.0202

LEKARSKA SPOLECNOST J.E.PURKYNE Pulmonary and cardiac function bone isotope diagnosis International symposium on nuclear medicine 1st Proceedings Karlovy Vary 1969 May 13-16 Edited by O. Andrysek and J. Mestan Universita Karlova,Prague,1970
PGMS 29.0680

LEMONT,ILL. 1953 Symposium on large scale digital computing machinery proceedings Edited by J.C. Chu Sponsored by Argonne national laboratory illus. 295p 25cm Argonne national laboratory,Lemont,Ill.,1953 Photocopy
Math L 5.0737

LENINGRAD 1924 Russian hydrological congress 1st Proceedings Komitet vsesoyuznikh gidrologicheskikh S'ezdov tables,maps 623p 26cm Leningrad,1925
Sco 14.0111

LENINGRAD 1928 Internationale gesellschaft zur erforschung der Arktis mit luftfahrzeugen : ordentl.versammlung 2te verhandlungen Edited by A. Berson and L. Breitfuss Petermann's mitteilungen.Erganzungsheft,201 maps 76p 28cm J.Perthes,Gotha,1929
Sco 14.0105

LENINGRAD 1928 Internationale gesellschaft zur erforschung der Arktis mit luftfahrzeugen : polar konferenz 2te Verhandlungen Edited by P. Wittenberg Russian edition illus, maps xxii,194p 26cm Leningrad,1930 Includes also t.-p.in Russian:"Trudy vtoroi polyarnoi konferentsii". Explanatory introduction in English
Sco 14.0107

LENINGRAD 1929 U.S.S.R.congress of genetics plant and animal breeding.Proceedings Trudi vsesoyuznogo sveztsda po lenetike selektsii semenovodstvy i plemennomy zhivothovodstvy Edited by N.I. Wulff and others 5 vols Izdanie Redaktsionnoi kollegii ceda,Leningrad,1930
Bot 42.0654

LENINGRAD 1930 Rapport de la commission internationale de l'annee polaire 1932-33 Vol 1: compte-rendu des travaux de la commission pendant sa premiere annee de travail... Organization meteorologique internationale illus,diagrms 152p Ijdo, Leyden,1933
Sco 14.0493

LENINGRAD 1932 Mezhdunarodnaya konferentsiya assotsiatsii po izucheniyu chetvertichnogo perioda Evropy International conference of the Association for the study of the European quaternary 2nd trudy Vypusk 1-5 Association for the study of the European quaternary,and,S.S.S.R.- N.K.T.P. Vsesoyuznoe geologo-razvedochnoe obedinenie Edited by D.A. Petrovskii and others Gosudarstvennoe nauchno-tekhnicheskoe geologo-razvedochnoe izdatelstvo,Leningrad;Moscow,1932-34
Geol 8.2982

LENINGRAD 1933 Conference hydrologique des etats baltiques 4th Compte-rendu Union of Soviet socialist republics.Service hydro-meteorologique Leningrad,1934
Geog 13.0767

LENINGRAD 1958 Fosforilirovanie i funktsiya :sympozium Akademiya meditsinskikh nauk S.S.S.R. Akademiya meditsinskikh nauk S. S.S.R.Trudy instituta eksperimentalnoi meditsiny Leningrad,1960 Short summaries in English at the end of each chapter
Bioch 33.0561

LENINGRAD 1961 Konferentsiya po probleme 'Vsaimodeistvie atmosfery i gidrosfery v severnoi chasti Atlanticheskogo okeana' Conference on the problem 'Interactions between the atmosphere and the hydrosphere in the North Atlantic 2nd Papers Leningradskii gidrometeorologicheskii institut 281p Izdatelstvo Leningradskogo universiteta, Leningrad,1964
Sco 14.5941

LENINGRAD 1961 Trudy chetvertogo vsesoyuznogo matematicheskogo sezda Tom 1-2 Bibliog. 27cm 2 vols Leningrad, 1963-64
P. Math 2.2606

LENINGRAD 1967 Physics of electronic and atomic collisions International conference on the physics of electronic and atomic collisions 5th Invited papers Edited by Lewis M. Branscomb Sponsored by the International union of pure and applied physics Bibliog,illus xvi,200p 23cm University of Colorado press,Boulder,Colo., 1968
A Math 4.1906

LENINGRAD 1968 Kainozoiskaya istoriya polyarnogo basseina i ee vliyanie na razvitie landshaftov severnykh territorii Edited by N. A. Belov and others 155p 20cm Geografischeskoe obshchestvo S.S.S.R., Leningrad,1968
Sco 14.8290

LENINGRAD 1969 Progress in protozoology International congress on protozoology 3rd Abstracts of papers and communications International commission on protozoology Publishing house 'Nauka',Leningrad,1969
Mol 45.0140

LENINGRAD 1970 The Motion,evolution of orbits,and origin of comets International astronomical union Edited by G.A. Chebotarev and others International astronomical union. Symposium, 45 521p Reidel,Dordrecht,1972
Obs 6.3591

LENINGRAD 1970 The Motion,evolution of orbits,and origin of comets International astronomical union Edited by G.A. Chebotarev and others International astronomical union. Symposium, 45 521p Reidel,Dordrecht,1972
TA 15.0687

LENINGRADSKII GIDROMETEOROLOGICHESKII INSTITUT Konferentsiya po probleme 'Vsaimodeistvie atmosfery i gidrosfery v severnoi chasti Atlanticheskogo okeana' Conference on the problem 'Interactions between the atmosphere and the hydrosphere in the North Atlantic 2nd Papers Leningrad 1961 May 25-30 281p Izdatelstvo Leningradskogo universiteta, Leningrad,1964
Sco 14.5941

LEONARD DE VINCI ET L'EXPERIENCE SCIENTIFIQUE AU XVIE SIECLE :colloque international Paris 1952 Jul 4-7 Centre national de la recherche scientifique.Sciences humaines viii,273p Presses universitaires de France, Paris,1953
WSM 43.1938

LEOPOLDVILLE 1960 African agrarian systems International African seminar 2nd Studies International African institute Edited by Daniel Biebuyck Oxford university press, London,1963
Geog 13.4631

LEOPOLDVILLE 1964 Etudes sur les termites africains :colloque international Comptes rendus Universite de Louvain Edited by A. Bouillon Under the auspices of Unesco figs, pls,tables 27cm Masson,Paris,1964
Bal 44.4805

LEPETIT COLLOQUIUM ON BIOLOGY AND MEDICINE 3rd Proceedings Cell interactions London 1971 Nov Edited by Luigi G. Silvestri North-Holland,Amsterdam,1972
Bioch 33.2267

LEPETIT COLLOQUIUM ON THE BIOLOGY OF ONCOGENIC VIRUSES 2nd Proceedings Paris 1970 Nov Edited by Luigi G. Silvestri North-Holland,Amsterdam,1971
Bioch 33.2299

LEPROSY International scientific conference on leprosy Mittheilungen und verhandlungen Berlin 1897 Oct 11-16 Bd 1-3 2 vols Hirschwald,Berlin,1897-98 Papers in French,English,German,Spanish.Bd 2-3 bound together
Path 30.0701

LERNER MARINE LABORATORY Marine bio-acoustics :symposium 1st,2nd Proceedings Bimini,Bahamas 1963 Apr 11-13 New York 1966 Apr 13-15 Vol 1-2 By William N. Tavolga 2 vols Pergamon press,Oxford,1964-67 Vol.1 held at the Lerner marine laboratory Bimini;vol.2 held at the American museum of natural history,New York
Bal 39.1534

LES HOUCHES 1960 Dispersion relations and elementary particles By M.L. Goldberger Universite de Grenoble.Ecole d'ete de physique theorique,Les Houches Edited by C.de Witt and R Omnes Universite de Grenoble.Ecole d'ete de physique theorique,Les Houches,1960 670p Hermann,Paris,1960
A Math 4.0839

LES HOUCHES 1962 Geophysics:the earths environment lectures Universite de Grenoble.Ecole d'ete de physique theorique,Les Houches Edited by Cecile De witt Universite de Grenoble.Ecole d'ete de physique theorique,Les Houches.Lectures,1963 Gordon and Breach,New York;London,1963
A Math 4.0578

LES HOUCHES 1962 Geophysique exterieure Papers Universite de Grenoble.Ecole d'ete de physique theorique,Les Houches Edited by Cecile De Witt and others Universite de Grenoble.Ecole d'ete de physique theorique.Les Houches. Lectures,1962 illus xiv,624p Gordon and Breach,New York,1963 Papers in English or French
Nap 11.0163

LES HOUCHES 1963 Relativity,groups and topology Papers Universite de Grenoble. Ecole d'ete de physique theorique,Les Houches Edited by C. DeWitt and B. DeWitt Universite de Grenoble.Ecole d'eete de physique theorique, Les Houches. Lectures,1963 929p Gordon, Breach,London,1964
TA 15.0343

LES HOUCHES 1963 Relativity,groups and topology lectures Universite de Grenoble. Ecole d'ete de physique theorique,Les Houches Edited by C. DeWitt and B. DeWitt Universite de Grenoble.Ecole d'ete de physique theorique, Les Houches.Lectures,1963 Gordon and Breach, New York;London,1963
A Math 4.1030

LES HOUCHES 1965 High energy physics Universite de Grenoble.Ecole d'ete de physique theorique,Les Houches Edited by C. DeWitt and Maurice Jacob Universite de Grenoble. Ecole d'ete de physique theorique,Les Houches, 1965 509p Gordon and Breach,New York,1966
A Math 4.0864

LES HOUCHES 1966 High energy astrophysics Lectures Vol 1-3 Universite de Grenoble.Ecole d'etude physique theorique,Les Houches Edited by C. DeWitt and others Universite de Grenoble.Ecole d'ete de physique theorique,Les Houches.Lectures,1966 Gordon and Breach,New York,1967 Lectures delivered at Les Houches,1966
TA 15.0312

LES HOUCHES 1966 High energy astrophysics : lectures Vol 3 Universite de Grenoble. Ecole d'ete de physique theorique,Les Houches Edited by C. Dewitt and others Supported by NATO Universite de Grenoble.Ecole d'ete de physique theorique,Les Houches. Lectures,1966 xii,449p Gordon and Breach,New York,1967
A Math 4.1562

LETAVET A.A. and others Deistvie oblucheniya na organizm :doklady sovetskoi delegatsii na Mezhdunarodnoi konferentsii po mirnomu ispolzovaniyu atomnoi energii Doklady Geneva 1955 Jul 1-5 Izdatelstvo Akademii nauk SSSR,Moscow,1955
Chem 18.0879

LEUKAEMIA Ciba foundation symposium on leukaemia research London 1953 Nov 16-19 Ciba foundation Edited by G.E.W. Wolstenholme and Margaret P. Cameron Ciba foundation.Symposia Churchill,London,1954
Radioth 35.0800

LEUKAEMIA RESEARCH :a Ciba foundation symposium Ciba foundation Edited by G.E. W. Wolstenholme Churchill,London,1954
Med 36.0366

LEUKAEMIA SOCIETY Perspectives in leukaemia : a symposium Proceedings New Orleans 1966 By William Dameshek and R.M. Dutcher Grune and Stratton,New York,1968
Med 36.0103

LEUKEMIA Conference on murine leukemia Papers Philadelphia,Pa. 1965 Oct 13-15 National cancer institute.Virology research branch and Albert Einstein medical center Edited by Marvin A. Rich and John B. Moloney National cancer institute.Monograph, 22 US government printing office,Washington,D.C., 1966
Path 30.2460

LEUKEMIA International conference on leukemia lymphoma Proceedings Ann Arbor 1967 University of Michigan Edited by Chris J.D. Zarafonetis Lea and Febiger; Kimpton,Philadelphia;London,1968
Med 36.0376

LEUKEMIA-LYMPHOMA Annual clinical conference on cancer 14th Papers Houston 1969 Anderson hospital and tumor institute Edited by C.C. Shullenberger and R.W. Cumley Yearbook medical publishers,Chicago,1971
PGMS 29.0678

LEUKEMIAS International symposium on the leukemias :etiology,pathophysiology and treatment Proceedings Detroit 1956 Mar 8-10 Edited by John W. Rebuck and others Sponsored by Henry Ford hospital Academic press,New York,1957
Med 36.0299

LEUSCHNER OBSERVATORY The Conference for instructors of astronomy :lecture notes Berkeley,Calif. 1954 Aug 12-Sep 8 Edited by Robert Fleischer and others Co-sponsored by National science foundation 265p Berkeley,Calif.,1954 Typescript
Obs 6.2054

LEXINGTON,KY. 1953 Kentucky symposium Learning theory,personality theory,and clinical research :the Kentucky symposium Papers University of Kentucky.College of arts and sciences.Department of psychology Wiley;Chapman and Hall,New York;London,1954
Psy 31.1839

LEXINGTON,KY. 1953 Learning theory, personality theory,and clinical research :the Kentucky symposium Lectures University of Kentucky.Department of psychology Wiley; Chapman and Hall,New York;London,1954
Psy 28.0195

LEXINGTON,KY. 1955 Collectivization of agriculture in eastern Europe Conference on collectivization in eastern Europe Papers Edited by Irwin T. Sanders maps University of Kentucky press,Lexington,Ky.,1958
Geog 13.2582

LEXINGTON,KY. 1971 Symposium on topological dynamics and ergodic theory Proceedings University of Kentucky. Department of mathematics 80p 28cm University of Kentucky,Lexington,Ky.,1971 Duplicated notes
P Math 2.4189

LEYDEN 1960 Quantitative methods in pharmacology :a symposium Edited by H.de Jonge North-Holland,Amsterdam,1961
Pha 16.0415

LEYDEN 1960 Quantitative methods in pharmacology :a symposium Proceedings Edited by H. de Jonge Bibliog,illus xx, 391p 23cm North-Holland,Amsterdam,1961
Math 3.1168

LEYDEN 1962 Protein metabolism :influence of growth hormone,anabolic steroids,and nutrition in health and disease:an international symposium Edited by F. Gross Sponsored by Ciba foundation Springer,Berlin, 1967
Inv Med 37.0255

LEYDEN 1962 Protein metabolism :influence of growth hormone anabolic steroids, and nutrition in health and disease :an international symposium Edited by F. Gross Sponsored by Ciba Springer,Berlin;Gottingen, 1967 Chairman A.Querido
PGMS 29.0406

LHISTOIRE SOCIALE.SOURCES ET METHODES Saint-Cloud 1965 May 15-16 Ecole normale superieure de Saint-Cloud Presses universitaires de France,Paris,1967
Geog 13.1610

LIBLICE 1961 Symposium on mechanisms of immunological tolerance Proceedings Ceskoslovenska akademie ved Edited by M. Hasek and others illus. Academic press,New York,1963
BG 23.0056

LIBLICE 1965 Ecology of aphidophagous insects :symposium Proceedings Ceskoslovenska akademie ved Junk;Academia, The Hague;Prague,1966
Bal 39.2844

LIBLICE 1969 Symposium of pediatric group working physiology 2nd Proceedings Universita Karlova Edited by M. Macek Universita Karlova,Prague,1970
PGMS 29.0599

LIBLICE,CZECH. 1962 Prague conference on information theory,statistical decision functions, random processes 3rd transactions Ceskoslovenska akademie ved Edited by Jaroslav Kozesnik Organized by the Institute of information theory and automation 84625 Czechoslovak academy of sciences, Prague,1964 Dedicated to the memory of Antonin Spacek
Math 3.0697

LIBRARY SERVICES Brasenose conference on the automation of libraries Anglo-American conference on the mechanization of library services Proceedings Oxford 1966 Jun 30-Jul 3 Edited by John Harrison and Peter Laslett Sponsored by the Old Dominion foundation of New York Mansell,Chicago,Ill., 1967
Eng 41.7672

LIBRARY SERVICES Research into library services in higher education :papers presented at a conference London 1967 Nov 3 Society for research into higher education Society for research into higher education, London,1968
Eng 41.7669

LIBYA South-central Libya and northern Chad Petroleum exploration society of Libya :annual field conference 8th Guidebook Libya 1966 Edited by James J. Williams Amsterdam, 1966
Geog 13.0598

LIBYA 1966 South-central Libya and northern Chad Petroleum exploration society of Libya :annual field conference 8th Guidebook Edited by James J. Williams Amsterdam,1966
Geog 13.0598

LIBYA 1966 South-central Libya and northern Chad:a guidebook Petroleum exploration society of Libya :annual field conference 8th guidebook Edited by James J. Williams Amsterdam,1966
Geol 8.4490

LIDINGO 1965 Muscular afferent and motor control Nobel symposium 1st Proceedings Nobel foundation Edited by Ragnar Granit Wiley;Almqvist and Wiksell,New York;Stockholm, 1966
An 32.4008

LIEBIG AND AFTER LIEBIG:A CENTURY OF PROGRESS IN AGRICULTURAL CHEMISTRY A Symposium of papers presented before the sections of chemistry and agriculture of the American association for the advancement of science... in commemoration of the 100th anniversary of the publication of Liebig's 'Organic chemistry in its application to agriculture and physiology' Philadelphia,Pa. 1940 Dec 30 American association for the advancement of science Edited by Forest Ray Moulton American association for the advancement of science.Publication, 16 111p American association for the advancement of science, Washington,D.C.,1942
WSM 43.2238

LIEGE 1930 Conference de l'Union internationale de chimie :rapports sur les hydrates de carbone (Glucides) 10th Rapports Union internationale de chimie Union internationale de chimie,Paris,1930
Chem 18.0066

LIEGE 1954 Radiobiology symposium proceedings Edited by Z.M. Bacq and P. Alexandra London,1955
Philos 1.1413

LIEGE 1954 Radiobiology symposium 1954 Proceedings Edited by Z.M. Bacq and Peter Alexander Butterworths,London,1955
Radioth 35.1712

LIEGE 1957 Sensitivity reactions to drugs : a symposium Council for international organizations of medical sciences Edited by M.L. Rosenheim and R. Moulton Blackwell, Oxford,1958
Pha 16.0187

LIEGE 1959 Biological problems of grafting a symposium Proceedings Commission administrative du patrimoine universitaire de Liege and Council for international organizations of medical sciences Edited by F. Albert and G. Lejeune-Ledant Blackwell, Oxford,1959 Symposium chairman P.B. Medawar
PGMS 29.0245

LIEGE 1959 Biological problems of grafting a symposium Proceedings Commission administrative du patrimoine universitaire de Liege Council for international organizations of medical sciences Edited by F. Albert and G. Lejeune-Ledant Blackwell, Oxford,1959
Bal 39.0458

LIEGE 1959 Selected combustion problems II: transport phenomena;ignition;altitude behaviour and scaling of aeroengines Combustion colloquium Agard Butterworths, London,1956
Eng 41.7228

LIEGE 1962 Cell growth and cell division;a symposium Report International society for cell biology Edited by R.J.C. Harris International society for cell biology. Symposia, 2 Academic press,New York,1963
Med 36.0178

LIEGE 1962 Cell growth and cell division Report International society for cell biology Edited by R.J.C. Harris held at the Institute of histology,Liege International society for cell biology. Symposia, 2 illus Academic press,New York;London,1963
Bioch 33.0965

LIEGE 1962 Cell growth and cell division : a symposium Edited by R.J.C. Harris International society for cell biology. Symposia, 2 Academic press,New York,1963
An 32.3288

LIEGE 1962 Cell growth and cell division : a symposium International society for cell biology Edited by R.J.C. Harris International society for cell biology. Symposia, 2 illus. Academic press,New York;London,1963
Gen 34.0863

LIEGE 1962 Cell growth and cell division : symposium Edited by R.J.C. Harris International society for cell biology. Symposia, 2 Illus Academic press,New York;London,1963
Radioth 35.0518

LIEGE 1964 Colloque sur l'analyse fonctionelle 2nd Centre Belge de recherches mathematiques 166p 25cm Librairie universitaire,Louvain,1964
P. Math 2.1182

LIEGE 1966 Instabilite gravitationnelle et formation des etoiles, des galaxies et de leurs structures caracteristiques Colloque international d'astrophysique 4e Communications Universite de Liege.Institut d'astrophysique Institut d'astrophysique, Cointe-Sclessin,1967 Papers in English
A Math 4.1170

LIEGE 1967 Cellular and molecular aspects of floral induction :a symposium Proceedings Edited by Georges Bernier Bibliog.,illus, plates xv,492p Longman,London,1970
Bot 42.1440

LIEGE 1967 Colloque international de geographie appliquee comptes rendus 3rd Commission de geographie appliquee Christans, Liege,c1967
Geog 13.7167

LIEGE 1967 Demonstration,verification, justification :entretiens de l'Institut international de philosophie Institut international de philosophie Edited by Philippe Devaux xii,352p Nauwelaerts, Louvain;Paris,1968
WSM 43.0967

LIEGE 1967 Performance forecast of selected static energy conversion devices : 29th meeting of the AGARD propulsion and energetics panel Edited by G.W. Sherman and L. Devol AGARD,Brussels,1967
Eng 41.4869

LIEGE 1968 Protein and polypeptide hormones International symposium on protein and polypeptide hormones Proceedings Edited by M. Margoulies Excerpta medica. International congress series, 161 Excerpta medica,Amsterdam,1968
Inv Med 37.0117

LIEGE 1968 Transitions interdites dans les spectres des astres Colloque international d'astrophysique 15ieme Communications Societe royale des sciences de Liege.Memoires, 5s, 17 410p Institut d'astrophysique, Liege,1969
TA 15.0576

LIEGE 1969 Evolution stellaire avant la sequence principale Societe des sciences de Liege.Memoires.5s, 19,no 1 377p Liege, 1970 Papers in English and French
TA 15.0685

LIFE Chemical evolution and the origin of life International conference on the origin of life Proceedings Vol 1: molecular evolution Edited by Rene Buvet and C. Ponnamperuma 560p North-Holland,Amsterdam, 1971
TA 15.0690

LIFE,LAND AND WATER Environmental studies of the Glacial Lake Agassiz region :a conference Proceedings 1966 Edited by William J. Mayer-Oakes University of Manitoba. Department of anthropology.Occasional papers, 1 maps xvi,414p University of Manitoba press,Winnipeg,1967
Bot 42.2101

LIFE SCIENCES The Symposium on advancing frontiers of life sciences Proceedings New Delhi 1960-61 Dec 30-Jan 1 National institute of sciences of India National institute of sciences of India.Bulletin, 19 New Delhi,1962 Symposium held on the occasion of the Silver Jubilee of the National institute of sciences of India
Bal 39.0009

LIFE SCIENCES RESEARCH AND LUNAR MEDICINE Lunar international laboratory (LIL)symposium 2nd Proceedings Madrid 1966 Oct 13 Edited by Frank J. Malina Organized by International academy of astronautics Pergamon press,Oxford;London,1967 Held at the 17th International astronautical congress
PGMS 29.0445

LIFE STRESS AND BODILY DISEASE Proceedings of the association New York 1949 Dec 2-3 Association for research in nervous and mental disease Edited by Harold G. Wolff Association for research in nervous and mental disease.Research publications, 29 Hafner, New York,1950 reprinted 1968
An 32.4852

LIGHT AND LIFE A Symposium on light and life McCollum-Pratt institute Edited by William D. McElroy and Bentley Glass With support from the National science foundation McCollum-Pratt institute.Contribution,302 Johns Hopkins press,Baltimore,Md.,1961
Chem 18.0168

LIGHT AND LIFE A Symposium on light and life Papers and discussions presented Baltmore,Md. 1960 Mar 28-31 Edited by William D. McElroy and H.Bentley Glass Sponsored by the McCollum-Pratt institute McCollum-Pratt institute.Contribution, 302 John Hopkins press,Baltimore,Md.,1961
Bioch 33.0600

LIGHT AND LIFE A Symposium on light and life Papers and informal discussions Baltimore,Md. 1960 Mar 28-31 Edited by William D. McElroy and Bentley Glass Sponsored by the McCollum-Pratt institute McCollum-Pratt institute.Contribution, 302 Johns Hopkins university press,Baltimore,Md.,1961
Bal 44.6447

LIGHT AS A ECOLOGICAL FACTOR :a symposium Cambridge 1965 Mar 30-Apr 1 British ecological society Edited by Richard Bainbridge and G. Evans British ecological society.Symposia, 6 Blackwell,Oxford,1966
Bal 39.1683

LIGHT AS AN ECOLOGICAL FACTOR British ecological society :a symposium 6th Proceedings Cambridge 1965 Mar 30-Apr 1 British ecological society Edited by Richard Bainbridge and others British ecological society.Symposia, 6 Bibliog.,tables,diagrs 452p Blackwell,Oxford,1966
Bot 42.2093

LIGHT AS AN ECOLOGICAL FACTOR British ecological society symposium 6th Cambridge 1965 Mar 30-Apr 1 Edited by Richard Bainbridge and others Blackwell,Oxford,1966
Geog 13.1217

LIGHT MICROSCOPY Symposium on light microscopy American society for testing materials 55th annual meeting Papers New York 1952 Jun 25 American society for testing materials American society for testing materials.Special publication, 143, Fiftieth anniversary publication Philadelphia,Pa.,1953
Min 10.1571

LIGHTNING ROD CONFERENCE Report Meteorological society Physical society Edited by G.J. Symons Bibliog,illus x,261p E.and F.N.Spon,London,1882 With appendices 1-L.Appendix G contains catalogue of works upon lightning conductors,pp.143-174
Nap 11.0940

LILLE 1955 Convention on electric traction by single-phase current at industrial frequency Documents Lille,1955
Eng 41.4842

LILLE 1969 Interactions des electrons, phonons et magnons avec les surfaces cristallines :colloque Exposes et communications presentes Societe francaise de physique Journal de physique, 31,no 4, supplement,Colloque,C- 1 Paris,1970
Met 25.2627

LIMA 1963 and CALI,COLUMBIA 1963 Control of cell division and the induction of cancer International symposium on the control of cell division and the induction of cancer Proceedings National cancer institute National cancer institute. Monograph, 14 illus National cancer institute,Bethesda,Md.,1964 Introduction, abstracts of each paper and summary in English and Spanish
Bioch 33.0985

LIMA 1963 and CALI,COLUMBIA 1963 The Control of cell division and the induction of cancer :international symposium Proceedings Edited by C.C. Congdon and Pablo Mori-Chavez National cancer institute.Monographs, 14 Illus U.S.department of health,education and welfare;National cancer institute,Bethesda,Md., 1964 In Spanish and English
Radioth 35.0532

LIMITATIONS OF DETECTION IN SPECTROCHEMICAL ANALYSIS :a conference Papers Exeter 1964 Jul 23 Organised by the Institute of physics and the physical society.Spectroscopy group Hilger and Watts,London,1964
Radioth 35.1657

LINCOLN,NEBRASKA 1953 MAR 26-27 Current theory and research in motivation :a symposium Papers By Judson S. Brown and others University of Nebraska.Department of psychology University of Nebraska press, Lincoln,Nebraska,1953
Psy 31.1857

LINCOLN,NEBRASKA 1962- Nebraska symposium on motivation 1962- University of Nebraska.Department of psychology University of Nebraska.Current theory and research in motivation symposium, 10- University of Nebraska press,Lincoln,Nebraska,1962-
Psy 31.1858

LINCOLN LABORATORY Semiconductor surface physics Conference on the physics of semiconductor surfaces Proceedings Philadelphia,Pa. 1956 Jun 4-6 Edited by R. H. Kingston Sponsored by the United States. Office of naval research,the University of Pennsylvania and the Lincoln laboratory University of Pennsylvania,Philadelphia,Pa., 1957
Eng 41.5399

LIND BICENTENARY SYMPOSIUM :a conference on scurvy and vitamin C in honour of James Lind Proceedings Edinburgh 1953 May 22-23 Nutrition society Nutrition society. Proceedings,12,3 Cambridge university press, London,1953
Sco 14.0848

LINDAU 1966 Winds and turbulence in stratosphere,mesosphere and ionosphere North Atlantic treaty organization.Advanced study institute Proceedings North Atlantic treaty organization.Science committee Edited by K. Rawer 421p North Holland,Amsterdam, 1968
Nap 11.0955

LINEAR ALGEBRAS conference report Shelter Island,N.Y. 1956 Jun 6-8 By A. Adrian Albert National research council. Division of mathematics National research council.Publications, 502 v,59p 28cm National academy of sciences.National research council,Washington,D.C.,1957
P. Math 2.2033

LINEAR EQUATIONS Large sparse sets of linear equations Conference of the Institute of mathematics and its applications Proceedings Oxford 1970 Apr Institute of mathematics and its applications Edited by J. K. Reid Illus 284p Academic press,London, 1971
Math L 5.3804

LINEAR EQUATIONS Large sparse sets of linear equations :a conference Proceedings Oxford 1970 Institute of mathematics and its applications Edited by J.K. Reid Academic press,London,1971
Eng 41.8474

LINEAR PROGRAMMING CONFERENCE mimeograph London 1954 May Ferranti ltd 17p 1954
Math L 5.1336

LINEAR PROGRAMMING SYMPOSIUM 2nd Proceedings Washington,D.C. 1955 Jan 27-29 Vol 1-2 National bureau of standards United States.Air force. Directorate of management analysis Sponsored by the United States.Air force.Office of scientific research 24cm 2 vols National bureau of standards,Washington,D.C.,1955
Math L 5.3248

LINNEAN SOCIETY OF LONDON Evolution by natural selection :with a foreword by Sir Gavin de Beer By Charles Darwin and Alfred Russel Wallace Cambridge university press, Cambridge,1958 Papers originally published in 1858,and reprinted for the 15th international congress of zoology
Gen 34.1309

LINNEAN SOCIETY OF LONDON Speciation in tropical environments symposium Papers London 1968 Oct 31-Nov 1 Linnean society of London.Biological journal, 1,no.1-2 Academic press,London,1969
Gen 34.1298

LINNEAN SOCIETY OF LONDON The Darwin-Wallace celebration London 1908 Jul 1 Linnean society,London,1908
BG 38.1428

LIPID METABOLISM Drugs affecting lipid matabolism The Symposium on drugs affecting lipid metabolism Proceedings Milan 1960 Edited by S. Garattini and R. Paoletti illus Elsevier,Amsterdam,1961
Bioch 33.1237

LIPID METABOLISM Drugs affecting lipid metabolism International symposium on drugs affecting lipid metabolism 2nd Papers Milan 1965 Pt 1-2 Edited by D. Kritchevsky and others organised by the European society for biochemical pharmacology Progress in biochemical pharmacology, 2-3 illus 2 vols Karger,Basle;New York,1967
Bioch 33.1250

LIPID METABOLISM The Control of lipid metabolism :a symposium Oxford 1963 Jul 19 Biochemical society Edited by J.K. Grant Biochemical society.Symposia, 24 Academic press,London;New York,1963 Organized by G. Popjak
Bioch 33.1386

LIPID METABOLISM :a symposium London 1952 Feb 16 Biochemical society Edited by R.T. Williams Held at the London school of hygiene and tropical medicine Biochemical society.Symposia, 9 Cambridge university press,Cambridge,1952
Bioch 33.1371

LIPIDS Biochemical problems of lipids :an international conference 2nd Proceedings Ghent 1955 Jul 27-30 Vlaamse chemische vereniging Edited by G. Popjak and E. Le Breton Butterworths,London,1956 Conference president:Professor R.Ruyssen
Chem 18.1500

LIPIDS Biochemistry of lipids International conference on the biochemical problems of lipids 5th Proceedings Vienna 1958 Sep Edited by G. Popjak Pergamon press,Oxford,1960 Section 18 of the 4th international congress of biochemistry
Col S 12.0102

LIPIDS International congress of biochemistry 5th Proceedings Moscow 1961 Aug 10-16 Vol 7: biosynthesis of lipids International union of biochemistry Edited by G. Popjak I.U.B.symposium series, 27 Pergamon;PWN-Polish scientific publishers, Oxford;Warsaw,1963
Bioch 33.1336

LIPIDS Metabolism and physiological significance of lipids :advanced study course Proceedings Cambridge 1963 Sep Edited by R.M.C. Dawson and D.N. Rhodes Wiley,London, 1964
Pha 16.0127

LIPIDS Metabolism and physiological significance of lipids :advanced study course Proceedings Cambridge 1963 Sep 16-21 Edited by R.M.C. Dawson and Douglas N. Rhodes Wiley,London,1964
Bioch 33.1235

LIPO-PROTEINS :a general discussion Papers Birmingham 1949 Aug 29-31 Faraday society Faraday society.Discussions, 6 Aberdeen,1949
Bioch 33.0508

LIPOPROTEINS Plasma lipoproteins London 1971 Apr Biochemical society Edited by R.M. S. Smellie Biochemical society.Symposium, 33 Academic press,London,1971
Bioch 33.2317

LIQUID CRYSTALS AND ANISOTROPIC MALTS :a general discussion London 1933 Apr 24-25 Faraday society held in the Royal institution Faraday society.Discussions, 58 Aberdeen,1933
Bioch 33.1769

LIQUID METALS Corrosion by liquid metals Proceedings of the sessions on corrosion by liquid metals of the 1969 fall meeting of the Metallurgical society of AIME Philadelphia, Pa. 1969 Oct 13-16 American institution of mining engineers.Metallurgical society Edited by J.E. Draley and J.R. Weeks Metallurgical society of AIME.Publications Plenum press,New York,1970
Met 25.2790

LIQUID METALS AND SOLIDIFICATION a seminar held during the 39th national metal congress and exposition Chicago 1957 Nov 2-8 American society for metals A.S.M.,Cleveland, Ohio,1958
Met 25.1510

LIQUID SCINTILLATING COUNTING :a conference Proceedings Evanston 1957 Aug 20-22 Edited by Carlos G. Bell and F.Newton Hayes Sponsored by the National science foundation Pergamon press,New York,1958 reprinted 1962
Inv Med 37.0014

LIQUID SCINTILLATION COUNTING Simple preparation techniques for liquid scintillation counting The Beckman summer school 1st Proceedings London 1966 Aug 15-18 Beckman instruments London,c1966
Bioch 33.2300

LIQUID SCINTILLATION COUNTING Techniques for liquid scintillation counting The Beckman summer school 2nd Proceedings 1967 Beckman instruments Glenrothes,c1967
Bioch 33.2301

LIQUIDS:STRUCTURE,PROPERTIES,SOLID INTERACTIONS Symposium on liquids:structure,properties, solid interactions proceedings Warren,Mich. 1963 Sep.5-6 Edited by Thomas J. Hughel General motors research laboratory.Symposia, 7 Elsevier,Amsterdam,1965
A Math 4.0951

LIQUIDS:STRUCTURE,PROPERTIES,SOLID INTERACTIONS the symposium Proceedings Warren,Mich. 1963 General motors research laboratories Edited by Thomas J. Hughel Elsevier, Amsterdam,1965
Met 25.1514

LISBON 1880 Congres international d'anthropologie et d'archeologie prehistoriques 9th Compte rendu Academie royale des sciences,Lisbon,1884
An 32.2637

LISBON 1949 International geographical congress Guidebooks to excursions International geographical union 2 vols Lisbon,1949
Geog 13.4139

LISBON 1949 International geographical congress 16th Comptes rendus Tom 1-4 International geographical union 5 vols Lisbon,1950-52 Partly in English
Geog 13.5596

LISBON 1957 Economic consequences of the size of nations International economic association conference Proceedings Edited by E.A.G. Robinson Macmillan,London,1960
Geog 13.2036

LISBON 1960 Cerebral localization and organization :a symposium Proceedings World federation of neurology Edited by Georges Schattenbrand and Clinton N. Woolsey University of Wisconsin press,Madison,Wis.; Milwaukee,Wis.,1964
Psy 31.0254

LISBON 1960 Reunion pleniere de l'association pour l'etude taxonique de la flore d'Afrique tropicale 4 Comptes rendus Edited by A. Fernandes Junta de investigacoes do ultramar,Lisbon,1962
Bot 42.2646

LISBON 1964 Theory of distributions :an international summer institute Proceedings Instituto Gulbenkian de ciencia.Centro de calculo cientifico Sponsored by the North Atlantic treaty organization.Science committee 390p Instituto Gulbenkian de ciencia,Lisbon, 1964
A Math 4.1265

LISBON 1964 Theory of distributions : international summer institute Proceedings Centro de calculo cientifico Sponsored by the North Atlantic treaty organization.Science committee Illus. xviii,390p 25cm Centro de calculo cientifico,Lisbon,1964 In English and French
P Math 2.3180

LISBON 1965 Queuing theory:recent developments and applications conference proceedings North Atlantic treaty organization.Science committee Edited by R. Cruon 224p 23cm English universities press,London,1967
Math 3.0703

LISBON 1966 Congress of the International society of rock mechanics 1st Proceeedings Vol 1-3 International society of rock mechanics 3 vols Laboratorio nacional de engenharia civil, Lisbon,1966-67
Eng 41.3176

LISTERIC INFECTION Symposium on listeric infection 2nd Papers and discussions Bozeman,Montana 1962 Aug 29-31 Montana state college Edited by M.L. Gray With the support of the National institute of allergy and infectious diseases Montana state college,Bozeman,Montana,1963
Path 30.2590

LITERATURE OF THE COMBUSTION OF PETROLEUM :a symposium Papers American chemical society national meeting 129th Dallas,Texas 1956 Apr American chemical society Advances in chemistry series, 20 Washington, D.C.,1958
Eng 41.7239

LITERATURE OF THE COMBUSTION OF PETROLEUM :a symposium held...at the 129th national meeting of the American chemical society Papers Dallas,Texas 1956 Apr American chemical society American chemical society.Advances in chemistry series,20 American chemistry society,Washington,D.C.,1958
Chem 18.0449

The LIVER :...symposium Proceedings Bristol 1967 Apr 3-7 Colston research society Edited by A.E. Read Colston research society.Symposium, 19 Colston papers, 19 Butterworths,London,1967
PGMS 29.0367

LIVER DISEASE :a Ciba foundation symposium Proceedings Ciba foundation Edited by Sheila Sherlock Churchill,London,1955
Med 36.0318

LIVER INJURY Conference on liver injury 10th Transactions New York 1951 May 21-22 Edited by F.W. Hoffbauer Churchill, London,1951
Med 36.0189

LIVERPOOL 1950 The Biochemistry of fish ;a symposium Biochemical society Edited by R. T. Williams Held at the University of Liverpool.Derby Hall Biochemical society. Symposia, 6 Cambridge university press, Cambridge,1961
Bioch 33.1368

LIVERPOOL 1954 The Comparative endocrinology of vertebrates :a conference Proceedings Part 1: the comparative physiology of reproduction and the effects of sex hormones in vertebrates Society for endocrinology Edited by I.Chester Jones and P. Eckstein Society for endocrinology. Memoirs, 4 Cambridge university press, Cambridge,1955
Phys 20.1378

LIVERPOOL 1954 The Comparative endocrinology of vertebrates :a conference Proceedings Part 2: the hormonal control of water and salt electrolyte metabolism in vertebrates Society for endocrinology Edited by I.Chester Jones and P. Eckstein Society for endocrinology.Memoirs, 5 Cambridge university press,Cambridge,1955
Phys 20.1379

LIVERPOOL 1954 The Comparative endocrinology of vertebrates :a conference Proceedings Pt 1: the comparative physiology of reproduction and the effects of sex hormones in vertebrates Society for endocrinology Edited by I.Chester Jones and P. Eckstein held at the University of Liverpool.Department of zoology Society for endocrinology.Memoirs, 4 Cambridge university press,Cambridge,1955
Bal 39.1400

LIVERPOOL 1962 Enzyme chemistry of phenolic compounds :a symposium Proceedings Plant phenolics group Edited by J.B. Pridham Pergamon press.Symposium publications division, Oxford,1963
Chem 18.1247

LIVERPOOL 1964 Controls of metamorphism :a symposium Edited by Wallace S. Pitcher and Glenys W. Flinn Under the auspices of the Liverpool geological society Liverpool and Manchester geological journal,special issue, 1965 Oliver and Boyd,Edinburgh;London,1965
Geol 8.3168

LIVERPOOL 1964 Phenetic and phylogenetic classification :a symposium Edited by Vernon H. Heywood and J. McNeill Systematics association.Publication, 6 xi,164p Systematics assoc.,London,1964
Bot 42.3439

LIVERPOOL 1966 The Role of the adsorbed state in heterogenous catalysis :a general discussion Faraday society Faraday society. Discussions,41 Faraday society,London,1966
Min 10.0623

LIVERPOOL 1968 Homogeneous catalysis with special reference to hydrogenation and oxidation Faraday society Faraday society. Discussions, 46 diagrms,graphs 227p 24cm Faraday society,London,1968
Cav 7.3125

LIVERPOOL 1969 International symposium on electron and photon interactions at high energies 4th Proceedings Danesbury nuclear physics laboratory International union of pure and applied physics University of Liverpool illus 30cm Danesbury nuclear physics laboratory,Danesbury,1969
A Math 4.1760

LIVERPOOL 1969 Natural substances formed biologically from mevalonic acid :a symposium Biochemical society Biochemical society. Symposia, 29 Academic press,London,1970
Chem 18.2550

LIVERPOOL 1969 Natural substances formed biologically from mevalonic acid :a symposium Biochemical society Edited by T.W. Goodwin Biochemical society.Symposia, 29 Academic press,London;New York,1970
Bioch 33.1391

LIVERPOOL 1969 Symposium on mechanism of igneous intrusion Proceedings Liverpool geological society Edited by Geoffrey Newall and Nicholas Rast Geological journal.Special issue, 2 Liverpool geological society, Liverpool,1970
Min 10.1427

LIVERPOOL 1970 Liverpool singularities symposium II Symposium on singularities of smooth manifolds and maps Proceedings University of Liverpool.Department of pure mathematics Edited by C.T.C. Wall Lecture notes in mathematics, 209 280p 25cm Springer,Berlin,1971
P Math 2.4123

LIVERPOOL 1970 Symposium on singularities of smooth manifolds and maps Liverpool singularities symposium I Proceedings University of Liverpool.Department of pure mathematics Edited by C.T.C. Wall Lecture notes in mathematics, 192 318p 25cm Springer,Berlin,1971
P Math 2.4122

LIVERPOOL GEOLOGICAL SOCIETY Controls of metamorphism :a symposium Liverpool 1964 Jan Edited by Wallace S. Pitcher and Glenys W. Flinn Liverpool and Manchester geological journal,special issue,1965 Oliver and Boyd, Edinburgh;London,1965
Geol 8.3168

LIVERPOOL GEOLOGICAL SOCIETY Symposium on mechanism of igneous intrusion Proceedings Liverpool 1969 Jan 9-11 Edited by Geoffrey Newall and Nicholas Rast Geological journal. Special issue, 2 Liverpool geological society,Liverpool,1970
Min 10.1427

LIVERPOOL SINGULARITIES SYMPOSIUM I Proceedings Symposium on singularities of smooth manifolds and maps Liverpool 1970 Aug -Sep University of Liverpool.Department of pure mathematics Edited by C.T.C. Wall Lecture notes in mathematics, 192 318p 25cm Springer,Berlin,1971
P Math 2.4122

LIVERPOOL SINGULARITIES SYMPOSIUM II Symposium on singularities of smooth manifolds and maps Proceedings Liverpool 1970 Aug-Sep University of Liverpool.Department of pure mathematics Edited by C.T.C. Wall Lecture notes in mathematics, 209 280p 25cm Springer,Berlin,1971
P Math 2.4123

The LIVING BONE :symposium on bone structure and metabolism Edited by R. Steendikk and others Folia medica neerlandica, 11,5-6 Amsterdam,1968
PGMS 29.0177

LJUBLJANA 1957 Hydrogen bonding :papers held at the symposium on hydrogen bonding Papers International union of pure and applied chemistry,and,Union of the chemical societies Edited by D. Hadzi and H.W. Thompson Pergamon press,London,1959
Chem 18.0955

LJUBLJANA 1971 Information processing 1971 IFIP congress 71 Proceedings International federation for information processing Edited by C.V. Freiman illus 1621p 2 vols North-Holland,Amsterdam,1972
Math L 5.3841

LOAD-BEARING BRICKWORK :a conference Proceedings London 1964 Nov 18-19 British ceramic society British ceramic society.Proceedings, 4 British ceramic society,Stoke-on-Trent,1965
Eng 41.2941

LOAD TESTS OF BEARING CAPACITY OF SOILS Symposium on load tests of bearing capacity of soils Atlantic City,N.J. 1947 Jun 16-20 American society for testing materials American society for testing materials.Special technical publication, 79 American society for testing materials,Philadelphia,Pa.,1948
Eng 41.3145

LOCAL ATOMIC ARRANGEMENTS STUDIED BY X-RAY DIFFRACTION :a symposium Proceedings Chicago,Ill. 1965 Feb 15 American institute of mining,metallurgical and petroleum engineers Edited by J.B. Cohen and J.E. Hilliard Metallurgical society conferences, 36 Gordon and Breach,New York, 1966
Met 25.1461

LOCAL FIELDS summer school proceedings Driebergen 1966 Jul 25-Aug 6 Netherlands universities foundation for international co-operation Edited by T.A. Springer viii, 214p 24cm Springer,Berlin,1967
P Math 2.2774

LOCAL FLORAS :a conference Proceedings London 1961 Nov 24-25 Botanical society of the British Isles Edited by P.J. Wanstall Botanical society of the British Isles.B.S.B.I. conference reports, 7 London,1963
Bot 42.4712

LOCALIZATION OF FUNCTION IN THE CEREBRAL CORTEX an investigation of the most recent advances Proceedings of the Association New York 1932 Dec 28-29 Association for research in nervous and mental disease Edited by Samuel T. Orton and others Association for research in nervous and mental disease.Research publications, 13 Williams and Wilkins, Baltimore,Md.,1934
An 32.4632

LOCALIZED EXCITATIONS IN SOLIDS International conference on localized excitations in solids 1st Proceedings Irvine 1967 Sep 18-22 International union of pure and applied physics National science foundation Edited by R.F. Wallis xvi,782p 26cm Plenum press,New York,1968
Cav 7.2857

LOCALIZED EXCITATIONS IN SOLIDS The International conference on localized excitations in solids 1st Irvine 1967 Sep 18-22 Edited by R.F. Wallis Plenum press,New York,1968
TA 15.0440

LOCARNO 1960 Aspects of the theory of artificial intelligence International symposium on biosimulation 1st Proceedings Edited by C.A. Muses Plenum, New York,1962
Eng 41.5988

LOCKHEED AIRCRAFT CORPORATION.LOCKHEED MISSILES AND SPACE DIVISION Magnetodynamics of conducting fluids Symposium on magnetohydrodynamics 3rd Palo Alto, Calif. 1958 Nov 21-22 Edited by Daniel Bershader viii,145p 24cm Stanford university press,Stanford,Calif.,1959
Cav 7.2767

LOCKHEED AIRCRAFT CORPORATION.LOCKHEED MISSILES AND SPACE DIVISION Radiation and waves in plasmas Symposium on magnetohydrodynamics 5th Palo Alto,Calif. 1960 Edited by Morton Mitchner 156p Stanford university press,Stanford,Calif.,1961
Cav 7.2881

LOCKHEED AIRCRAFT CORPORATION.MISSILE SYSTEMS DIVISION Magnetohydrodynamics :a symposium Palo Alto,Calif. 1956 Dec 29 Edited by Rolf K.M. Landshoff Stanford university press,Stanford,Calif.,1957 The 1st symposium on magnetohydrodynamics sponsored by Lockheed
Eng 41.4472

LOCKHEED AIRCRAFT CORPORATION.MISSILE SYSTEMS DIVISION Magnetohydrodynamics of conducting fluids Symposium on magnetohydrodynamics 3rd Palo Alto, Calif. 1958 Nov 21-22 Edited by Daniel Bershader Stanford university press,Stanford, Calif.,1959 1st symposium called 'Magnetohydrodynamics';2nd symposium called 'Plasma in a magnetic field',q.v.
Eng 41.4474

LOCKHEED AIRCRAFT CORPORATION.MISSILE SYSTEMS DIVISION Plasma hydromagnetics Symposium on magnetohydrodynamics 6th Proceedings Palo Alto,Calif. 1961 Dec 15-16 Edited by Daniel Bershader Stanford university press,Stanford,Calif.,1962
Eng 41.4510

LOCKHEED AIRCRAFT CORPORATION.MISSILE SYSTEMS DIVISION Radiation and waves Symposium on magnetohydrodynamics 5th Palo Alto, Calif. 1960 Dec 15-16 Edited by Morton Mitchener Stanford university press,Stanford, Calif.,1961
Eng 41.4496

LOCKHEED AIRCRAFT CORPORATION.MISSILE SYSTEMS DIVISION The Plasma in a magnetic field : a symposium on magnetohydrodynamics Palo Alto,Calif. 1957 Dec 16 Edited by Rolf K.M. Landshoff Stanford university press, Stanford,Calif.,1958 The 2nd symposium on magnetohydrodynamics sponsored by Lockheed.For later symposia see 'Symposium on magnetohydrodynamics'
Eng 41.4473

LOCKHEED AIRCRAFT CORPORATION.MISSILES AND SPACE DIVISION Magnetodynamics of conducting fluids Lockheed symposium on magnetohydrodynamics 3rd proceedings Palo Alto 1958 Nov.21-22 Edited by Daniel Bershader Stanford university press,Stanford, Calif.,1959
A Math 4.0616

LOCKHEED AIRCRAFT CORPORATION.MISSILES AND SPACE DIVISION Plasma acceleration Lockheed symposium on magnetohydrodynamics 4th proceedings Palo Alto 1959 Dec.2 Edited by Sidney W. Kash Stanford university press, Stanford,Calif.,1960
A Math 4.0618

LOCKHEED AIRCRAFT CORPORATION.MISSILES AND SPACE DIVISION Plasma acceleration Symposium on magnetohydrodynamics 4th Palo Alto, Calif. 1959 Dec 2 Edited by Sidney W. Kash Stanford university press,Stanford,Calif.,1960
Eng 41.4489

LOCKHEED AIRCRAFT CORPORATION.MISSILES AND SPACE DIVISION Propagation and instabilities in plasmas Symposium on magnetohydrodynamics 7th Proceedings Palo Alto,Calif. 1962 Dec 14-15 Edited by Walter I. Futterman Stanford university press,Stanford,Calif.,1963
Eng 41.4523

LOCKHEED AIRCRAFT CORPORATION.MISSILES AND SPACE DIVISION Radiation and waves in plasmas Lockheed symposium on magnetohydrodynamics 5th proceedings Palo Alto 1960 Dec 15-16 Edited by Morton Mitchner Stanford university press,Stanford,Calif.,1961
A Math 4.0621

LOCKHEED MISSILES AND SPACE COMPANY Weld imperfections a symposium Proceedings Palo Alto 1966 Sep 19-21 Edited by A.R. Pfluger and R.E. Lewis Co-sponsored by the United States.Air force.Materials laboratory and the United States office of naval research Addison-Wesley,Reading,Mass.,1968
Met 25.2319

LOCKHEED SYMPOSIUM ON MAGNETOHYDRODYNAMICS 3rd proceedings Magnetodynamics of conducting fluids Palo Alto 1958 Nov.21-22 Edited by Daniel Bershader Sponsored by Lockheed aircraft corporation.Missiles and space division Stanford university press, Stanford,Calif.,1959
A Math 4.0616

LOCKHEED SYMPOSIUM ON MAGNETOHYDRODYNAMICS 4th proceedings Plasma acceleration Palo Alto 1959 Dec.2 Edited by Sidney W. Kash Sponsored by Lockheed aircraft corporation. Missiles and space division Stanford university press,Stanford,Calif.,1960
A Math 4.0618

LOCKHEED SYMPOSIUM ON MAGNETOHYDRODYNAMICS 5th proceedings Radiation and waves in plasmas Palo Alto 1960 Dec 15-16 Edited by Morton Mitchner Sponsored by Lockheed aircraft corporation.Missiles and space division Stanford university press,Stanford. Calif.,1961
A Math 4.0621

LODZ 1966 Conference on analytic functions Proceedings Polska akademia nauk.Instytut matematyczny Bibliog 36p 24cm Lodz, 1966
P Math 2.3052

LOGIC International congress for logic, methodology and philosophy of science proceedings Jerusalem 1964 Aug 26-Sep 2 Edited by Yehoshua Bar-Hillel and others Studies in logic and the foundations of mathematics viii,440p 23cm North-Holland, Amsterdam,1965
P. Math 2.0206

LOGIC International congress for logic, methodology and philosophy of science proceedings Stanford,Calif. 1960 Aug.24-Sep.2 Edited by Ernest Nagel and others ix, 661p 26cm Stanford university press, Stanford, Calif.,1962
P. Math 2.0209

LOGIC Scandinavian logic symposium 2nd Proceedings Oslo 1970 Jun 18-20 Universitet i Oslo Edited by J.E. Fenstad Studies in logic and the foundations of mathematics, 63 vi,405p 23cm North-Holland,Amsterdam,1971
P Math 2.4056

LOGIC Summer school in logic :N.A.T.O. advanced study institute:meeting of the association for symbolic logic Proceedings Leeds 1967 Aug 7-23 North Atlantic treaty organization Association for symbolic logic Edited by M.H. Lob Lecture notes in mathematics, 70 Bibliog. 331p 28cm Springer,Berlin,1968
P Math 2.3120

LOGIC,METHODOLOGY AND PHILOSOPHY OF SCIENCE International congress for logic,methodology and philosophy of science Proceedings Amsterdam 1967 Aug 25-Sep 2 Nederlandse vereniging voor logica en wijsbegeerte der exacte wetenschappen International union for logic,methodology and philosophy of science. Division of logic,methodology and philosophy of science Edited by B.van Rootselaar and J. F. Staal xiii,554p North-Holland,Amsterdam, 1968
WSM 43.0830

LOGIC,METHODOLOGY AND PHILOSOPHY OF SCIENCE International congress for logic,methodology and philosophy of science Proceedings Jerusalem 1964 Aug 26-Sep 2 Israel academy of sciences and humanities International union of history and philosophy of science. Division of logic,methodology and philosophy of science Edited by Yehoshua Bar-Hillel viii,440p North-Holland,Amsterdam,1965
WSM 43.0829

LOGIC,METHODOLOGY AND PHILOSOPHY OF SCIENCE International congress for logic,methodology and philosophy of science Proceedings Stanford,Calif. 1960 Aug 24-Sep 2 International union of history and philosophy of science.Division of logic,methodology and philosophy of science Edited by Ernest Nagel and others ix,661p Stanford university press,Stanford,Calif.,1962
WSM 43.0828

LOGIC AND FOUNDATIONS OF SCIENCE Evert W. Beth memorial colloquium :logic and foundations of science Actes Paris 1964 May 19-21 Institut Henri Poincare Edited by Jean-Louis Destouches viii,137p Reidel,Dordrecht,1967 Organized by J-L. Destouches
WSM 43.1216

LOGIC COLLOQUIUM Proceedings Contributions to mathematical logic Hanover 1966 Aug Edited by H.Arnold Schmidt and others Studies in logic and the foundations of mathematics Bibliog. ix,298p 23cm North Holland,Amsterdam,1968
P Math 2.3312

The LOGIC COLLOQUIUM 8th proceedings Formal systems and recursive functions Oxford 1963 Jul Edited by J.N. Crossley and M.A.E. Dummett 320p 23cm North-Holland,Amsterdam,1965
P. Math 2.0207

LOGIC COLLOQUIUM 1969 Summer school and colloquium in mathematical logic Proceedings Manchester 1969 Aug 3-23 Victoria university of Manchester Edited by R.O. Gandy and C.M.E. Yates Studies in logic and the foundations of mathematics, 61 xiv,451p 23cm North-Holland,Amsterdam,1971
P Math 2.4057

LOGICAL SYSTEMS CONTAINING ONLY A FINITE NUMBER OF SYMBOLS Montreal 1966 By Leon Henkin North Atlantic treaty organization Societe mathematique du Canada Universite de Montreal.Seminaire de mathematiques superieures, 21 illus. 48p 28cm Presses de l'Universite de Montreal,Montreal, 1967
P Math 2.3740

LONDON Adrenergic neurotransmission :Ciba foundation study group Ciba foundation Edited by G.E.W. Wolstenholme and Maeve O'Connor Ciba foundation study group, 33 Churchill,London,1968
Phys 20.2050

LONDON Adrenergic neurotransmission :Ciba foundation study group 33rd Ciba foundation Edited by G.E.W. Wolstenholme and Maeve O'Connor Churchill,London,1968
Pha 16.0075

LONDON Colour vision :physiology and experimental psychology Ciba foundation Edited by A.V.S.de Reuck and Julie Knight Churchill,London,1965
Psy 31.3373

LONDON Enzymes and drug action :Ciba foundation symposium 8th Ciba foundation, and,Biological council.Co-ordinating committee for symposia on drug action Edited by J.L. Mongar and A.V.S. De Reuck Churchill,London, 1962
Pha 16.0105

LONDON Physiology and experimental psychology :a Ciba foundation symposium Ciba foundation Edited by A.V.S. De Reuck and Julie Knight Churchill,London,1965
Phys 20.2226

LONDON Quality control in metal finishing based on a symposium at the Borough Polytechnic Borough Polytechnic,London Edited by G. Isserlis Columbine press, Manchester;London,1967
Met 25.1983

LONDON Apr Symbiotic associations :a symposium Proceedings Society for general microbiology Society for general microbiology.Symposia, 13 Cambridge university press,Cambridge,1963
Gen 34.1355

LONDON 1866 International horticultural exhibition and botanical congress :report of proceedings London,1867
Bot 42.4705

LONDON 1866 The International horticultural exhibition and botanical congress Proceedings Truscott,London,1866
BG 38.1551

LONDON 1876 Conferences held in connection with special loan collection of scientific apparatus :physics and mechanics South Kensington museum 420p Chapman and Hall, London,1876
Obs 6.1966

LONDON 1881 International medical congress 7th Revue scientifique de la France et de l'etranger.3e ser,1.annee,2.semestre,tome 28,,no 8 Paris,1881 Includes "Des virus vaccins" by M.Pasteur;:Rapports des sciences biologiques avec la medecine" by M.T. H.Huxley;"Histoire de la physiologie en Angleterre" by M.Michael Foster
Bioch 33.1657

LONDON 1881 International medical congress 7th session Transactions anatomy Kolckmann,London,1881
An 32.1165

LONDON 1888 Congres geologique international 4me explication des excursions International geological congress Edited by W. Topley London,1888
Geol 8.1397

LONDON 1888 Etudes sur les schistes cristallins International geological congress 4th London,1888
Geol 8.2141

LONDON 1888 Etudes sur les schistes cristallins International geological congress 4th papers London,1888
Min 10.0237

LONDON 1891 Conifer conference Report Edited by W. Wilks and John Weathers Royal horticultural society.Journal, 14 Spottiswoode,London,1892
BG 38.2423

LONDON 1895 International geographical congress 6th Report Edited by Hugh Robert Mill illus,maps xxxvi,996p 25cm John Murray,London,1896 Includes papers in French,German and Italian
Sco 14.0104

LONDON 1895 International geographical congress 6th Report Vol 1-2 Murray, London,1896
Geog 13.5591

LONDON 1899 Hybrid conference report 1900 on hybridisation (the cross-breeding of species)and on the cross-breeding of varieties Royal horticultural society Royal horticultural society.Journal, 24 Spottiswoode,London,1900 Conference also entitled the 'International congress of genetics' and the 'International conference on hybridization'
Gen 34.0998

LONDON 1905 Optical convention 1st Proceedings Bibliog,illus 247p Norgate and Williams,London,1905
WSM 43.5433

LONDON 1906 Genetics International conference on genetics 3rd Report Royal horticultural society Edited by W. Wilkes Spottiswoode,London,1907
BG 38.0523

LONDON 1906 International congress of genetics hybridization (the cross-breeding of genera or species)the cross-breeding of varieties... 3rd Royal horticultural society Edited by W. Wilks Spottiswoode, London,1907 Conference also entitled 'International conference on hybridisation'
Gen 34.0999

LONDON 1908 The Darwin-Wallace celebration Linnean society of London Linnean society, London,1908
BG 38.1428

LONDON 1908 The Darwin-Wallace celebration Linnean society of London London,1908
Philos 1.0757

LONDON 1908 The Influence of heredity on disease,with special reference to tuberculosis, cancer and diseases of the nervous system :a discussion Royal society of medicine illus. Longmans,London,1909
Gen 34.1979

LONDON 1910 International Swedenborg congress Transactions Swedenborg society 2nd edition Swedenborg society,London,1911
Phys 20.0723

LONDON 1911 National conference on the prevention of destitution 1st Proceedings Great Britain.Local government board King, London,1911 President of conference:the Rt.Hon.the Lord Mayor of London
Path 30.1285

LONDON 1911 National conference on the prevention of destitution 1st Proceedings of the Public health section Great Britain. Local government board King,London,1911 President of the Public health section of the Local government board:T.C.Allbutt
Path 30.1284

LONDON 1912 Optical convention 2nd Proceedings tables,illus 359p Hodder and Stoughton for University of London press, London,1912
WSM 43.5434

LONDON 1912 Problems in eugenics International eugenics conference 1st Papers Eugenics education society 2 vols Eugenics education society,London,1913 Later known as the 1st international congress of eugenics
Gen 34.1925

LONDON 1912 Problems in eugenics International eugenics congress 1st Papers Eugenics education society Eugenics education society,London,1912
Bal 44.6430

LONDON 1912 Problems in eugenics International eugenics congress 1st Papers Eugenics education society Held at the University of London Eugenics education society,London,1912
Bot 42.6705

LONDON 1913 International congress of medicine 17th Section 1: anatomy and embryology,pt.2 Frowde;Hodder and Stoughton, London,1914
An 32.1163

LONDON 1913 International congress of medicine 17th Section 3: general pathology and pathological anatomy,pt.2 Frowde;Hodder and Stoughton,London,1914
An 32.1164

LONDON 1913 International congress of medicine 17th catalogue of the museum Edited by H.W. Armit London,n.d.
An 32.1175

LONDON 1913 International congress of medicine 17th general volume Frowde; Hodder and Stoughton,London,1914
An 32.1162

LONDON 1913 International congress of medicine :section XII:psychiatry 17th Transactions Oxford university press;Hodder and Stoughton,London,1913-14 President of congress:J.Crichton-Browne Conference languages:English,French and German
Psy 31.2530

LONDON 1914 Structure and activity of enzymes Federation of European biochemical societies symposium 1st Edited by T.W. Goodwin and others Academic press,London;New York,1964
Radioth 35.0119

LONDON 1920 The Microscope:its design, construction and applications ;a symposium and general discussion Faraday society Edited by F.S. Spiers pls,illus 260p Griffin, London,1920 Faraday society in cooperation with the Technical optics committee of the British science guild. Includes reports of adjourned discussions held in Sheffield,Feb 24 and in London April 21, 1920
WSM 43.5151

LONDON 1920 The Physics and chemistry of colloids and their bearing on industrial questions :a general discussion Report Faraday society Physical society of London H.M.S.O.,London,1921
Bioch 33.1522

LONDON 1921 Catalysis ,with special reference to newer theories of chemical action a general discussion Pt 1-2 Faraday society London,1922 Reprinted from the 'Transactions' of the society,vol 17,pt 3. Chairman of the discussion A.W.Porter
Bot 42.6477

LONDON 1921 International conference of the International union against tuberculosis 2nd Transactions International union against tuberculosis Under the auspices of the British national association for the prevention of tuberculosis Adlard and Newman, London,1921
An 32.2990

LONDON 1923 International air congress Report Edited by W.Lockwood Marsh International air congress,London,1923 Being the 2nd air congress although this is not indicated on the title page
Eng 41.6956

LONDON 1924 Imperial botanical conference : a report Proceedings Edited by F.T. Brooks port Cambridge university press,Cambridge, 1925
Bot 42.0651

LONDON 1926 Optical convention 3rd Proceedings Pt 1-2 2 vols Optical convention,London,1926
WSM 43.5435

LONDON 1928 International congress of photography 7th Proceedings Edited by W. Clark and others port W.Heffer,Cambridge, 1928
Chem 18.0247

LONDON 1929 Crystal structure and chemical constitution :a general discussion Faraday society London,1929
Bot 42.6478

LONDON 1929 International congress of military medicine and pharmacy 5th Report By William Seaman Bainbridge Banta,Menasha, Wis.,1929
Path 30.0210

LONDON 1929 Sexual reform congress 3rd Proceedings World league for sexual reform Edited by Norman Haire World league for sexual reform.Congress, 3rd Kegan Paul, Trench,Trubner,London,1930
Psy 28.0356

LONDON 1930 International horticultural congress 9th Reports,Proceedings,Program Royal horticultural society Under the auspices of the International committee for horticultural congresses London,1930
BG 38.3121

LONDON 1931 A Discussion held...at the Imperial college of science Report Physical society Physical society,London, 1931 Bound with 'Report of a joint discussion on vision...'by the Physical and optical society
Psy 31.3361

LONDON 1931 Conifers in cultivation Conifer conference Royal horticultural society Edited by F.J. Chittenden Royal horticultural society,London,1932
BG 38.2424

LONDON 1931 Problems of population International union for the scientific investigation of population problems.General assembly :report 2nd Proceedings Edited by G.H.L.F. Pitt-Rivers Allen and Unwin, London,1932
Gen 34.2068

LONDON 1932 A Joint discussion on vision Report Physical society and Optical society 327p Physical society,London,c1932
Obs 6.1962

LONDON 1932 A Joint discussion on vision held at the Imperial college of science Physical society Optical society Physical society,London,1932 Bound with 'Report of a discussion on audition' by the Physical society,q.v.
Psy 31.2610

LONDON 1932 Report of a joint discussion on vision held...at the Imperial college of science Report Physical society and Optical society Physical society,London,1932
Psy 31.0411

LONDON 1932 The Swelling of proteins,and allied phenomena International society of leather trades' chemists London,1933
Bioch 33.1486

LONDON 1933 Exhibition of South African wild flowers Under the auspices of the Royal horticultural society illus. London,1934
BG 38.2129

LONDON 1933 Liquid crystals and anisotropic malts :a general discussion Faraday society held in the Royal institution Faraday society.Discussions, 58 Aberdeen,1933
Bioch 33.1769

LONDON 1933 World petroleum congress Proceedings Vol 2: refining,chemical and testing section Edited by A.E. Dunstan Organised by the Institution of petroleum technologists xxvi,956p World petroleum congress,London,1934
Chem E 24.1708

LONDON 1934 International conference on physics Papers Vol 1: nuclear physics International union of pure and applied physics and Physical society Physical society,London,1935
Cav 7.1452

LONDON 1934 International conference on physics Papers and discussions Vol 1-2: nuclear physics:solid state of matter International union of pure and applied physics and the Physical society 2 vols Physical society,London,1935
Met 25.2343

LONDON 1934 International conference on physics papers and discussions Vol 2: the solid state of matter International union of pure and applied physics,and,Physical society Physical society,London,1937
Min 10.1078

LONDON 1934 International conference on physics :reports on symbols,units and nomenclature approved by the general assembly of the Union International union of pure and applied physics 40p Physical society, London,1935
Obs 6.1963

LONDON 1935 Symposium on the welding of iron and steel Vol 1 Iron and steel institute 2 vols Iron and steel institute, London,1935
Eng 41.3947

LONDON 1935 Symposium on the welding of iron and steel Vol 1-2 Iron and steel institute 2 vols Iron and steel institute, London,1935
Met 25.0783

LONDON 1935 The Social sciences:their relations in theory and in teaching : conference Report Institute of sociology Sponsored jointly by the International student service Le Play House press,London,1936
Geog 13.1613

LONDON 1936 International congress of microbiology :a report 2nd Proceedings International society for microbiology Edited by R. St.John-Brooks port London, 1937
Bot 42.0660

LONDON 1936 Rock gardens and rock plants conference Reports Royal horticultural society Alpine garden society Edited by F. J. Chittenden Royal horticultural society, London,1936
BG 38.3310

LONDON 1937 General discussion on lubrication and lubricants Proceedings Vol 1-2 Institution of mechanical engineers 2 vols Institution of mechanical engineers, London,1937
Eng 41.6291

LONDON 1937 General discussion on lubrication and lubricants Proceedings Vol 1-2 Institution of mechanical engineers 2 vols Institution of mechanical engineers, London,1937
Met 25.2152

LONDON 1937 Industry fights corrosion Corrosion convention Proceedings Sponsored by Corrosion technology Corrosion technology, London,1957
Eng 41.3719

LONDON 1937 The Properties and functions of membranes,natural and artificial :a general discussion By August Krogh and others Faraday society Faraday society.Transactions, 33, pt 8 London,1937
Bal 39.0358

LONDON 1938 Ornamental flowering trees and shrubs conference Report Royal horticultural society Edited by F.J. Chittenden illus. Royal horticultural society,London,1940
BG 38.2822

LONDON 1941 Mechanism and chemical kinetics of organic reactions in liquid systems :a general discussion Faraday society Gurney and Jackson,London;Edinburgh, 1941 Reprint from the society's 'Transactions',vol 37,pt 12
Bot 42.6479

LONDON 1943 Conference on the ultra-fine structure of coal and cokes held at the Royal Institution Proceedings British coal utilisation research association 23cm London,1944
Philos 1.0687

LONDON 1943 Empire mining and metallurgical congress 4th Proceedings Pt 1-2 Edited by F. Higham 2 vols London, 1950
Min 10.0726

LONDON 1946 Empire scientific conference Report Vol 1-2 Royal society of London 2 vols London,1948
Bioch 33.1683

LONDON 1946 Empire scientific conference, 1946 Vol 1-2 Royal society of London 2 vols London,1948
Bal 39.3917

LONDON 1946 Fuel and the future :a conference Proceedings Vol 1-3 Great Britain.Ministry of fuel and power Edited by Harry Mitchell Spiers 3 vols H.M. S.O.,London,1948
Chem E 24.1370

LONDON 1946 Meteorological factors in radio-wave propagation :a conference Report Physical society Royal meteorological society 325p Physical society,London,1946
Nap 11.0925

LONDON 1946 The Royal society empire scientific conference 1st report London, 1948
Col S 12.0199

LONDON 1947 Internal stresses in metals and alloys Symposium on internal stresses in metals and alloys Institute of metals Institute of metals.Monograph and report series, 5 Institute of metals,London,1948
Met 25.0960

LONDON 1947 Symposium on internal stresses in metals and alloys Advance copies of papers Institute of metals Institute of metals,London,1947
Eng 41.3768

LONDON 1947 Symposium on internal stresses in metals and alloys Proceedings Institute of metals Monographs and report series, 5 vii,485p Institute of metals,London,1948
Cav 7.1454

LONDON 1947 Symposium on particle size analysis Institution of chemical engineers and Society of chemical industry. Road and building materials group 145p Institution of chemical engineers,London,1947 Supplement to Transactions,Institution of chemical engineers,vol.25,1947
Chem E 24.1431

LONDON 1947 The Biochemical reactions of chemical warfare agents :a symposium Biochemical society Edited by R.T. Williams Held at the University college hospital medical school Biochemical society.Symposia, 2 Cambridge university press,Cambridge, 1948
Bioch 33.1364

LONDON 1947 The Biochemical reactions of chemical warfare agents :a symposium Edited by R.T. Williams Biochemical society. Symposia, 2 Cambridge university press, Cambridge,1948
Phys 20.1524

LONDON 1947 The Conference on emission spectra of the night sky and aurorae Papers Royal society of London.Gassiot committee 139p Physical society,London,1948
Cav 7.1453

LONDON 1947 The Conference on emission spectra of the night sky and aurorae Papers Royal society of London.Gassiot committee 139p Physical society,London,1948
TA 15.0039

LONDON 1947 The Emission spectra of the night sky and aurorae :a conference Papers Royal society of London.Gassiot committee 139p London,1948
Obs 6.1964

LONDON 1947 The Relation of optical form to biological activity in amino-acid series :a symposium held at University college hospital medical school Edited by R.T. Williams Biochemical society.Symposia, 1 Cambridge university press,Cambridge,1948
Radioth 35.0291

LONDON 1947 The Relation of optical form to biological activity in the amino-acid series :a symposium Biochemical society Edited by R.T. Williams Held at the University college hospital medical school Biochemical society.Symposia, 1 Cambridge university press,Cambridge,1948
Bioch 33.1363

LONDON 1947 The Relation of optical form to biological activity in the amino-acid series :a symposium Edited by R.T. Williams Biochemical society.Symposia, 1 Cambridge university press,Cambridge,1948
Phys 20.1523

LONDON 1947 The Relation of optical form to biological activity in the amino-acid series :a symposium 1 Biochemical society Edited by R.T. Williams Biochemical society symposia,1 Cambridge university press, Cambridge,1948
Chem 18.1275

LONDON 1947 Tuberculosis in the commonwealth 1947 Commonwealth and empire health and tuberculosis conference Transactions National association for the prevention of tuberculosis N.A.P.T.,London, 1947
HE 27.0132

LONDON 1948 British flowering plants and modern systematic methods The Conference on the study of critical British groups Report Botanical society of the British Isles Edited by A.J. Wilmott Botanical society of the British Isles.B.S.B.I.conference reports, 1 London,1949
Bot 42.4706

LONDON 1948 British medical association annual meeting Proceedings Butterworth's, London,1949
Radioth 35.0666

LONDON 1948 Conference on biology and civil engineering Proceedings Institution of civil engineers Institution of civil engineers,London,1949
Geog 13.2688

LONDON 1948 Detergents,wetting and emulsifying agents :a symposium Papers Organized by the Society of chemical industry. London section 48p Society of chemical industry,London,1950
Chem E 24.0923

LONDON 1948 International geological congress 18th report Part 2: proceedings of section A:problems of geochemistry Edited by C.E. Tilley and S.R. Nockolds London,1950
Min 10.0520

LONDON 1948 Partition chromatography :a symposium Biochemical society Edited by R. T. Williams and R.L.M. Synge Biochemical society.Symposia, 3 Cambridge university press,Cambridge,1949
Gen 34.0622

LONDON 1948 Partition chromatography :a symposium Biochemical society Edited by R. T. Williams and R.L.M. Synge Held at the London school of hygiene and tropical medicine Biochemical society.Symposia, 3 Cambridge university press,Cambridge,1949
Bioch 33.1365

LONDON 1948 Partition chromatography :a symposium Edited by R.T. Williams and R.L.M. Synge Biochemical society.Symposia, 3 Cambridge university press,Cambridge,1951
Phys 20.1525

LONDON 1948 Partition chromatography :a symposium 3 Biochemical society Edited by R.T. Williams and R.L.M. Synge Biochemical society symposia,3 Cambridge university press,Cambridge,1950
Chem 18.1276

LONDON 1948 Scientific information conference Reports and papers submitted Royal society of London London,1948
Bioch 33.1749

LONDON 1948 Scientific information conference :report and papers submitted Royal society of London Royal society,London, 1948
Bot 42.0662

LONDON 1948 The Royal society scientific information conference Reports and papers Royal society of London Royal society,London, 1948
Chem 18.2006

LONDON 1948 The Royal society scientific information conference :report and papers submitted Royal society of London Royal society,London,1948
Bal 39.3921

LONDON 1948 The Scientific information conference report and papers submitted Royal society of London Royal society,London,1948
Gen 34.0054

LONDON 1949 Biochemical aspects of genetics :a symposium Biochemical society Edited by R.T. Williams Biochemical society. Symposia, 4 Cambridge university press, Cambridge,1950
Gen 34.0733

LONDON 1949 Biochemical aspects of genetics :a symposium Biochemical society Edited by R.T. Williams Held at the Westminster hospital medical school Biochemical society.Symposia, 4 Cambridge university press,Cambridge,1950
Bioch 33.1366

LONDON 1949 Biochemical aspects of genetics :a symposium Edited by R.T. Williams Biochemical society.Symposia, 4 Cambridge university press,Cambridge,1950
Phys 20.1526

LONDON 1949 Biological oxidation of aromatic rings :a symposium Biochemical society Edited by R.T. Williams Held at the London school of hygiene and tropical medicine Biochemical society.Symposia, 5 Cambridge university press,Cambridge,1950
Bioch 33.1367

LONDON 1949 Biological oxidation of aromatic rings :a symposium Edited by R.T. Williams Biochemical society.Symposia, 5 Cambridge university press,Cambridge,1950
Phys 20.1527

LONDON 1949 Nature of the bacterial surface Society for general microbiology : symposium 1st Proceedings Edited by A.A. Miles and N.W. Pirie Blackwell,Oxford,1949
Col S 12.0036

LONDON 1949 Prestressed concrete :a conference Proceedings Institution of civil engineers Institution of civil engineers,London,1949
Eng 41.3019

LONDON 1949 The Acceleration of particles to high energies :based on a session arranged by the Electronics group Papers Institute of physics Edited by H.R. Lang Physics in industry Institute of physics,London,1950
Cav 7.1166

LONDON 1949 The Nature of the bacterial surface :a symposium Society for general microbiology Edited by A.A. Miles and N.W. Pirie illus. Blackwell,Oxford,1949
Gen 34.0697

LONDON 1950 Biological hazards of atomic energy :a conference papers Institute of biology and Atomic scientists' association Clarendon press,Oxford,1952
HE 27.0097

LONDON 1950 Camellias and magnolias conference Report Royal horticultural society illus. Royal horticultural society, London,1950
BG 38.2668

LONDON 1950 Hormones in blood :a colloquium Proceedings Ciba foundation Edited by G.E.W. Wolstenholme and Elaine C.P. Millar Ciba foundation colloquia on endocrinology, 11 Churchill,London,1957
Inv Med 37.0241

LONDON 1950 Steroid hormones and tumour growth,and Steroid hormones and enzymes :two colloquia Proceedings Ciba foundation Edited by G.E.W. Wolstenholme and Margaret P. Cameron Ciba foundation colloquia on endocrinology, 1 Churchill,London,1952
Phys 20.1366

LONDON 1950 Steroid metabolism and estimation :a colloquium Proceedings Ciba foundation Edited by G.E.W. Wolstenholme Ciba foundation colloquia on endocrinology, 2 illus Churchill,London,1952
Bioch 33.0432

LONDON 1950 Symposium on information theory 1st report of proceedings, mimeograph Edited by Willis Jackson Supported by Great Britain.Ministry of supply 206p 33cm Ministry of supply,London,1950
Math 3.0774

LONDON 1950 Symposium on information theory Report of proceedings Imperial college of science and technology.Electrical engineering department Institute of radio engineers IRE professional group on information theory.Publications committee,n.p., c1950 Foreword by W.Jackson. Duplicated
Psy 31.3221

LONDON 1950 Symposium on information theory proceedings Institute of radio engineers Institute of radio engineers. Professional group on information theory, 1 208p Institute of radio engineers,New York, 1953
Math L 5.2079

LONDON 1950 Synthesis and metabolism of adrenocortical steroids :a colloquium Proceedings Ciba foundation Edited by W. Klyne Ciba foundation colloquia on endocrinology, 7 Churchill,London,1953
Inv Med 37.0137

LONDON 1950 The Application of scientific methods to industrial and service medicine :a conference Proceedings Medical research council H.M.S.O.,London,1951
Bioch 33.1542

LONDON 1950 World power conference 4th Transactions Vol 1-5 Edited by W.O. Skeat 5 vols Lund,Humphries,London,1952
Eng 41.7439

LONDON 1950 London 1950 Aug 9-10 Steroid metabolism and estimation :two colloquia Ciba foundation Edited by G.E.W. Wolstenholme and Margaret Cameron Ciba foundation colloquia on endocrinology, 2 Churchill,London,1952
Phys 20.1367

LONDON 1950 and LONDON 1950 Steroid hormones and tumour growth and steroid hormones and enzymes :two colloquia Proceedings Ciba foundation Edited by G.E. W. Wolstenholme Ciba foundation colloquia on endocrinology, 1 illus Churchill,London, 1952
Bioch 33.0431

LONDON 1951 Building research congress 1951 Papers presented in division 1-3 London,1951
Eng 41.3372

LONDON 1951 Ciba foundation conference on isotopes in biochemistry Ciba foundation Edited by G.E.W. Wolstenholme Illus J.and A.Churchill,London,1951
Radioth 35.0023

LONDON 1951 Ciba foundation conference on isotopes in biochemistry Papers Ciba foundation Edited by J.N. Davidson Churchill,London,1951
Bioch 33.0274

LONDON 1951 Freezing and drying :a symposium Papers Edited by R.J.C. Harris Held by the Institute of biology Institute of biology,London,1951
Radioth 35.0178

LONDON 1951 Freezing and drying :a symposium Report Institute of biology Edited by R.J.C. Harris 205p Institute of biology,London,1951
Sco 14.1236

LONDON 1951 General discussion on heat transfer Proceedings Institution of mechanical engineers American society of mechanical engineers Institution of mechanical engineers,London,1952
Eng 41.7146

LONDON 1951 General discussion on heat transfer Proceedings Institution of mechanical engineers and American society of mechanical engineers xiii,496p Institution of mechanical engineers,London, 1951
Chem E 24.0538

LONDON 1951 High-temperature steels and alloys for gas turbines :a symposium Iron and steel institute Iron and steel institute. Special report, 43 Iron and steel institute, London,1952
Met 25.0996

LONDON 1951 Joint engineering conference Proceedings Institution of civil engineers Institution of mechanical engineers Institution of electrical engineers Institution of civil engineers,London,1951
Eng 41.1761

LONDON 1951 Metabolism and functions in nervous tissue :a symposium Biochemical society Edited by R.T. Williams Held at the London school of hygiene and tropical medicine Biochemical society.Symposia, 8 Cambridge university press,Cambridge,1952
Bioch 33.1370

LONDON 1951 Notch bar testing and its relation to welded construction :symposium Organized by the Institute of welding Institute of welding,London,1953
Eng 41.3834

LONDON 1951 Prestressed concrete statically indeterminate structures :a symposium Cement and concrete association Cement and concrete association,London,1953
Eng 41.3006

LONDON 1951 The Biochemistry of fertilization and the gametes :a symposium Biochemical society Edited by R.T. Williams Held at the London school of hygiene and tropical medicine Biochemical society. Symposia, 7 Cambridge university press, Cambridge,1951
Pal 39.0388

LONDON 1951 The Biochemistry of fertilization and the gametes :a symposium Biochemical society Edited by R.T. Williams Held at the London school of hygiene and tropical medicine Biochemical society. Symposia, 7 Cambridge university press, Cambridge,1951
Bioch 33.1369

LONDON 1951 Welding and riveting larger aluminium structures :a symposium Proceedings Aluminium development association Aluminium development association,London,1952
Eng 41.3941

LONDON 1951 London 1951 Jan 8-10 Anterior pituitary secretion and hormonal influences in water metabolism :two colloquia Ciba foundation Edited by G.E.W. Wolstenholme and Margaret P. Cameron Ciba foundation colloquia on endocrinology, 4 Churhcill,London,1952
Phys 20.1365

LONDON 1951 and LONDON 1950 Hormones, psychology and behaviour,and steroid hormone administration :two colloquia Proceedings Ciba foundation Edited by G.E.W. Wolstenholme and Margaret P. Cameron Ciba foundation colloquia on endocrinology, 3 illus Churchill,London,1952
Bioch 33.0433

LONDON 1951 and LONDON 1951 Anterior pituitary secretion and Hormonal influences in water metabolism :two colloquia Proceedings Ciba foundation Edited by G.E.W. Wolstenholme and Margaret P. Cameron Ciba foundation colloquia on endocrinology, 4 illus Churchill,London,1952
Bioch 33.0434

LONDON 1952 Bioassay of anterior pituitary and adrenocortical hormones :a colloquium proceedings Ciba foundation Edited by G.E. W. Wolstenholme Ciba foundation colloquia on endocrinology, 5 illus Churchill,London, 1953
Bioch 33.0435

LONDON 1952 Biology of deserts Biology of hot and cold deserts :a symposium Proceedings Edited by J.L. Cloudsley-Thompson Organized by the Institute of biology Institute of biology,London,1954
Geog 13.1205

LONDON 1952 Commission for maritime meteorology abridged final report of the first session Report Commission for maritime meteorology 108p 28cm World meteorological organization,Geneva,1952
Sco 14.0349

LONDON 1952 Communication theory Applications of communication theory : symposium Papers Edited by Willis Jackson Supported by the British broadcasting corporation xii,532p 25cm Butterworths, London,1953 Also supported by the Ministry of supply
Math L 5.3249

LONDON 1952 Communication theory Applications of communication theory symposium papers British broadcasting corporation Great Britain.Ministry of supply Edited by Willis Jackson xii,532p Butterworths scientific publications,London,1953
Math 3.0735

LONDON 1952 Communication theory Applications of communication theory symposium Papers Institution of electrical engineers Edited by Willis Jackson Butterworths,London, 1953
Psy 31.2016

LONDON 1952 Concrete shell roof construction :a symposium Proceedings Cement and concrete association Cement and concrete association,London,1954
Eng 41.3008

LONDON 1952 Equipment for the thermal treatment of non-ferrous metals and alloys a symposium on the metallurgical aspects of the subject Institute of metals Institute of metals.Monograph and report series, 14 Institute of metals,London,1953
Met 25.2306

LONDON 1952 Hormonal factors in carbohydrate metabolism :a colloquium Proceedings Ciba foundation Edited by G.E. W. Wolstenholme Ciba foundation colloquia on endocrinology, 6 illus Churchill,London, 1953
Bioch 33.0436

LONDON 1952 Immunochemistry :a symposium Biochemical society Edited by R.T. Williams Held at the London school of hygiene and tropical medicine Biochemical society. Symposia, 10 Cambridge university press, Cambridge,1953
Bioch 33.1372

LONDON 1952 International horticultural congress 13th Report Vol 1-2 Royal horticultural society Royal horticultural society,London,1953
BG 38.3124

LONDON 1952 International symposium on the chemistry of cement 3rd Proceedings Under the auspices of the Great Britain. Building research station...and ...Cement and concrete association Cement and concrete association,London,1954
Min 10.0615

LONDON 1952 Lipid metabolism :a symposium Biochemical society Edited by R.T. Williams Held at the London school of hygiene and tropical medicine Biochemical society. Symposia, 9 Cambridge university press, Cambridge,1952
Bioch 33.1371

LONDON 1952 Mammalian germ cells :Ciba foundation symposium Ciba foundation Edited by G.E.W. Wolstenholme and others Churchill,London,1953
Phys 20.1478

LONDON 1952 Physical chemistry of melts Symposium on the nature of molten slags and salts Nuffield research group in extraction metallurgy Institution of mining and metallurgy,London,1953
Met 25.1787

LONDON 1952 Properties of metallic surfaces :a symposium Institute of metals Institute of metals.Monograph and report series, 13 Institute of metals,London,1953
Met 25.1593

LONDON 1952 Properties of metallic surfaces :a symposium Organised by the Institute of metals Institute of metals. Monograph and report series, 13 Institute of metals,London,1953
Eng 41.3471

LONDON 1952 Recent developments in mineral dressing :a symposium Institution of mining and metallurgy Institution of mining and metallurgy,London,1953
Met 25.0185

LONDON 1952 Soft magnetic materials for telecommunications :a symposium Post Office. Engineering research station Edited by C.E. Richards and A.C. Lynch Pergamon press, London,1953
Eng 41.5085

LONDON 1952 Symposium on chromosome breakage Contributions By C.D. Darlington and others John Innes horticultural institution Heredity, 5,supplement Oliver and Boyd,London;Edinburgh,1953
Path 30.0219

LONDON 1952 Synthesis and metabolism of adrenocortical steroids :a colloquium Proceedings Ciba foundation Edited by W. Klyne and G.E.W. Wolstenholme Ciba foundation colloquia on endocrinology, 7 illus Churchill,London,1953
Bioch 33.0437

LONDON 1952 The Biology of hot and cold deserts... :a symposium Proceedings Institute of biology Edited by J.L. Cloudsley-Thompson Assisted by a grant from Unesco Institute of biology,London,1954
Bal 39.1858

LONDON 1952 The Chemical structure of proteins :a Ciba foundation symposium Proceedings Ciba foundation Edited by G.E. W. Wolstenholme and Margaret P. Cameron illus Churchill,London,1953
Bioch 33.0500

LONDON 1952 The Spinal cord :a Ciba foundation symposium Ciba foundation Edited by G.E.W. Wolstenholme Churchill, London,1953
An 32.4129

LONDON 1953 Adaptation in micro-organisms : symposium Society for general microbiology Society for general microbiology.Symposia, 3 Cambridge university press,Cambridge,1953
Gen 34.0694

LONDON 1953 Adaption in micro-organisms :a symposium Papers Society for general microbiology Edited by R. Davies and E.F. Gale Held at the Royal institution Society for general microbiology.Symposia, 3 Cambridge university press,Cambridge,1953
Bioch 33.1166

LONDON 1953 Biological transformation of starch and cellulose Biochemical society Edited by R.T. Williams Biochemical society. Symposia, 11 Cambridge university press, Cambridge,1953
Bot 42.1690

LONDON 1953 Biological transformations of starch and cellulose :a symposium Biochemical society Held at the London school of hygiene and tropical medicine Biochemical society.Symposia, 11 Cambridge university press,Cambridge,1953
Bioch 33.1373

LONDON 1953 Ciba foundation symposium on leukaemia research Ciba foundation Edited by G.E.W. Wolstenholme and Margaret P. Cameron Ciba foundation.Symposia Churchill,London, 1954
Radioth 35.0800

LONDON 1953 Conference on hydraulic servo-mechanisms Proceedings Institution of mechanical engineers Institution of mechanical engineers,London,1963
Eng 41.5891

LONDON 1953 Conference on steam turbine research and development Proceedings Institution of mechanical engineers Institution of mechanical engineers,London, 1953
Eng 41.7296

LONDON 1953 Conference on the North Sea floods of 31 January-1 February,1953 Papers Institution of civil engineers Institution of civil engineers,London,1954
Geog 13.0706

LONDON 1953 EUSEC conference on engineering education 1st Proceedings Conference of engineering societies of western Erope and the United States of America Institution of civil engineers,London,1953
Eng 41.1543

LONDON 1953 Nature and structure of collagen :a discussion Papers Edited by J. T. Randall Convened by the Faraday society. Colloid and biophysics committee Butterworths,London,1953
Radioth 35.0186

LONDON 1953 Nature and structure of collagen :a discussion Papers Faraday society.Colloid and biophysics committee Edited by J.T. Randall and Sylvia Fitton Jackson Butterworths,London,1953
An 32.5311

LONDON 1953 Nature and structure of collagen :a discussion Papers Faraday society.Colloid and biophysics committee Edited by J.T. Randall and Sylvia Fitton Jackson Butterworths,London,1953
Bioch 33.2346

LONDON 1953 Nature and structure of collagen :a discussion Papers presented Faraday society.Colloid and biophysics committee Edited by J.T. Randall and Sylvia Fitton Jackson Butterworths,London,1953
Bal 39.1202

LONDON 1953 Peripheral circulation in man : Ciba foundation symposium Ciba foundation Edited by G.E.W. Wolstenholme and Jessie S. Freeman Churchill,London,1954
An 32.3459

LONDON 1953 Preservation and transplantation of normal tissues :Ciba foundation symposium Ciba foundation Edited by G.E.W. Wolstenholme and Margaret P. Cameron Churchill,London,1954
An 32.3211

LONDON 1953 Symposium on sinter Iron and steel institute Iron and steel institute. Special report, 53 Iron and steel institute, London,1955
Met 25.0321

LONDON 1953 The Determination of adrenocortical steroids and their metabolites Proceedings Society for endocrinology Edited by P. Eckstein and S. Zuckermann Society for endocrinology.Memoirs Dobson, London,1953
Med 36.0126

LONDON 1953 The Determination of adrenocortical steroids and their metabolites : a conference Proceedings Society for endocrinology Royal society of medicine. Endocrinological section Edited by P. Eckstein and S. Zuckerman Society for endocrinology.Memoirs, 2 Cambridge university press,Cambridge,1955
Radioth 35.1940

LONDON 1953 The North sea floods of January 31-February 1,1953 :a conference Institution of civil engineers Institution of civil engineers,London,1954
Eng 41.3342

LONDON 1953 The Physics of ionized gases : a conference proceedings,typescript University college,London Edited by J.B. Hasted Under the auspices of the Royal society of London.Warren research fund London,1953
A Math 4.0614

LONDON 1953 The Thyroid gland :a symposium Proceedings Society for endocrinology and Royal society of medicine Edited by P. Eckstein and S. Zuckerman Society for endocrinology.Memoirs, 1 1953
Radioth 35.0713

LONDON 1953 Welded structures :a conference Proceedings Great Britain. Ministry of works H.M.S.O.,London,1954
Eng 41.3021

LONDON 1953 World conference on medical education 1st Proceedings World medical association Oxford university press,London, 1953
Med 36.0370

LONDON 1953 World conference on medical education 1st proceedings Edited by Hugh Clegg Held under the auspices of the World medical association Oxford university press,London,1954
An 32.0991

LONDON 1954 Autotrophic micro-organism :a symposium Papers Society for general microbiology Edited by B.A. Fry and J.L. Peel Held at the Institution of electrical engineers Society for general microbiology. Symposia, 4 Cambridge university press, Cambridge,1954
Bioch 33.1168

LONDON 1954 Ciba foundation symposium on chemistry and biology of pteridines Papers Ciba foundation Edited by G.E.W. Wolstenholme and Margaret P. Cameron Ciba foundation symposia J. and A. Churchill, London,1954
Chem 18.1327

LONDON 1954 Ciba foundation symposium on the chemistry and biology of purines Edited by G.E.W. Wolstenholme and M. Cameron Illus J.and A.Churchill,London,1954
Radioth 35.0058

LONDON 1954 Commonwealth entomological conference... 6th Report Commonwealth institute of entomology London,1954
Bal 39.2874

LONDON 1954 Conference on hydraulic mechanisms Proceedings Institution of mechanical engineers Institution of mechanical engineers,London,1954
Eng 41.6686

LONDON 1954 Forgemasters' meeting 1954 Proceedings Iron and steel institute and National forgemasters' association Joint meeting with the Chambre syndicate de la grosse forge francaise Iron and steel institute.Special report, 60 Iron and steel institute,London,1957
Met 25.0715

LONDON 1954 Linear programming conference mimeograph Ferranti ltd 17p 1954
Math L 5.1336

LONDON 1954 Mix design and quality control of concrete :a symposium Proceedings Cement and concrete association Cement and concrete association,London,1955
Eng 41.2929

LONDON 1954 Species concept in paleontology Systematics association symposium 2nd papers Systematics association Edited by Peter C. Sylvester-Bradley Systematics association.Publication, 2 Systematics association,London,1956
Geol 8.0940

LONDON 1954 Species studies in the British flora Species concept in its relation to the British flora :a conference Report Botanical society of the British Isles Edited by J.E. Lousley B.S.B.I.conference reports,1954 Botanical society of the British Isles,London,1955
BG 38.1259

LONDON 1954 Symposium on powder metallurgy Iron and steel institute Iron and steel institute.Special report, 58 Iron and steel institute,London,1956
Met 25.0761

LONDON 1954 Symposium on powder metallurgy Iron and steel institute Iron and steel institute.Special report, 58 illus. Iron and steel institute,London,1956
Eng 41.3477

LONDON 1954 The Chemical pathology of animal pigments :a symposium Biochemical society Edited by R.T. Williams Held at the London school of hygiene and tropical medicine Biochemical society.Symposia, 12 Cambridge university press,Cambridge,1954
Bioch 33.1374

LONDON 1954 The Control of quality in the production of wrought non-ferrous metals and alloys :a symposium 2: the control of quality in working operations Institute of metals Institute of metals.Monograph and report series, 16 Institute of metals, London,1954
Met 25.0711

LONDON 1954 The Human adrenal cortex : colloqium Ciba foundation Edited by G.E.W. Wolstenholme and Margaret Cameron Ciba foundation colloquia on endocrinology, 14 Churchill,London,1962
Phys 20.1380

LONDON 1954 The International conference on electron microscopy 3rd Proceedings Edited by R. Ross Royal microscopical society,London,1956
Radioth 35.1582

LONDON 1954 The International conference on electron microscopy 3rd Proceedings International council of scientific unions. Joint commission on electron microscopy Edited by V.E. Cosslett and others Royal microscopical society,London,1956
Cav 7.1455

LONDON 1954 The Kinetics and mechanism of inorganic reaction in solution;a survey of recent work :a symposium Report Chemical society Edited by K.W. Sykes Chemical society.Special publication, 1 London,1954
Bioch 33.0631

LONDON 1954 The Support of medical research :a symposium Council for international organisations of medical sciences Edited by H. Himsworth and J.F. Delafresnaye Blackwell,Oxford,1956
PGMS 29.0496

LONDON 1954 The Technique and significance of oestrogen :a conference proceedings Society for endocrinology Royal society of medicine.Endocrinological section Edited by P. Eckstein and S. Zuckerman Society for endocrinology.Memoirs, 3 Cambridge university press,Cambridge,1955
Radioth 35.0045

LONDON 1954 The Technique and significance of oestrogen determinations :a conference Proceedings Society for endocrinology and Royal society of medicine.Endocrinological section Edited by P. Eckstein and S. Zuckerman Society for endocrinology.Memoirs, 3 Cambridge university press,Cambridge, 1955
Phys 20.1384

LONDON 1955 Ciba foundation symposium on paper electrophoresis Ciba foundation Edited by G.E. Wolstenholme and E.C.P. Millar Ciba foundation.Symposia Churchill,London, 1958
PGMS 29.0516

LONDON 1955 Ciba foundation symposium on paper electrophoresis Proceedings Ciba foundation Edited by G.E.W. Wolstenholme and Elaine C.P. Millar Ciba foundation symposia Churchill,London,1956
Chem 18.1445

LONDON 1955 Ciba foundation symposium on paper electrophoresis proceedings Ciba foundation Edited by G.E.W. Wolstenholme and Elaine C.P. Miller Illus Churchill,London, 1956
Radioth 35.0361

LONDON 1955 Ciba foundation symposium on porphyrin biosynthesis and metabolism Ciba foundation Edited by G.E.W. Wolstenholme and E. Miller Churchill,London,1955
Radioth 35.0088

LONDON 1955 Ciba foundation symposium on porphyrin biosynthesis and metabolism Papers Ciba foundation Churchill,London,1955
Chem 18.2637

LONDON 1955 Ciba foundation symposium on porphyrin biosynthesis and metabolism Proceedings Ciba foundation Edited by G.E. W. Wolstenholme and Elaine C.P. Millar illus Churchill,London,1955
Bioch 33.0572

LONDON 1955 Correlation between calculated and observed stresses and displacements in structures :a conference Institution of civil engineers 2 vols Institution of civl engineers,London,1955
Eng 41.3020

LONDON 1955 Function and training of the chemical engineer :international conference Organisation for European economic co-operation 85p O.E.E.C.,Paris,1955
Chem E 24.1760

LONDON 1955 Histamine :Ciba foundation symposium Ciba foundation Edited by G.E.W. Wolstenholme and Cecilia M. O'Connor Churchill,London,1956
Pha 16.0076

LONDON 1955 Information theory A Symposium on information theory Papers read Edited by Colin Cherry Held at the Royal institution Butterworths,London,1956
Bal 39.1005

LONDON 1955 Information theory Symposium on information theory 3rd papers Edited by Colin Cherry xii,401p 25cm Butterworths,London,1956
Math 3.0761

LONDON 1955 Information theory Symposium on information theory symposium papers Edited by Colin Cherry xii,401p 25cm Butterworths,London,1956
Math L 5.0760

LONDON 1955 Internal secretion of the pancreas :colloquium Ciba foundation Edited by G.E.W. Wolstenholme and Cecilia M. O'Connor Ciba foundation colloquia on endocrinology, 9 Churchill,London,1956
An 32.3739

LONDON 1955 Internal secretions of the pancreas The Nature and actions of the internal secretions of the pancreas :a colloquium Proceedings Ciba foundation Edited by G.E.W. Wolstenholme and Cecilia M. O'Connor Ciba foundation colloquia on endocrinology, 9 illus Churchill,London, 1956
Bioch 33.0438

LONDON 1955 International congress of neuropathology 2nd Proceedings Pt 1-2 Edited by W.H. McMenemey Excerpta Medica, Amsterdam,1955
An 32.4163

LONDON 1955 Mechanisms of microbial pathogenicity :a symposium Papers Society for general microbiology Edited by J.W. Howie and A.J. O'Hea Held at the Royal institution Society for general microbiology. Symposia, 5 Cambridge university press, Cambridge,1955
Bioch 33.1172

LONDON 1955 Mechanisms of microbial pathogenicity :a symposium Papers Society for general microbiology Edited by J.W. Howie and A.J. O'Hea Society for general microbiology.Symposia, 5 Cambridge university press,Cambridge,1955
Path 30.2608

LONDON 1955 Mineral resources policy :a symposium Proceedings Institution of mining and metallurgy Institution of mining and metallurgy,London,1956
Met 25.0191

LONDON 1955 Paper electrophoresis Ciba foundation symposium Edited by G.E.W. Wolstenholme and Elaine C.S. Millar Churchill,London,1958
Col S 12.0070

LONDON 1955 Peptide chemistry :a symposium Report Chemical society Edited by A.D. Jenkins and D.F. Elliott Chemical society. Special publication, 2 London,1955 Organised by D.H.Hey
Bioch 33.0498

LONDON 1955 Peptide chemistry :a symposium Report Chemical society Edited by A.D. Jenkins and D.F. Elliott Chemical society. Special publication,2 Chemical society, London,1955 Symposium organized by professor Hey
Chem 18.1524

LONDON 1955 Plant growth substances The Chemistry and mode of action of plant growth substances :a symposium Proceedings Edited by R.L. Wain and F. Wightman Butterworths, London,1956
BG 38.0165

LONDON 1955 Solar eclipses and the ionosphere :a symposium Edited by W.J.G. Beynon and G.M. Brown Under the auspices of the Mixed commission on the ionosphere Pergamon press,London,1956 Vol.6 of special supplements to Journal of atmospheric and terrestrial physics
Obs 6.0176

LONDON 1955 Solar eclipses and the ionosphere :a symposium Proceedings Mixed commission on the ionosphere Edited by W.J.G. Beynon and G.H. Brown Under the auspices of the International council of scientific unions Journal of atmospheric and terrestrial physics. Special supplement, 6 ix,330p Pergamon press,London;New York,1956
Nap 11.0437

LONDON 1955 Symposium on information theory 3rd Proceedings Imperial college of science and technology Edited by Colin Cherry Butterworths,London,1956
Psy 31.2003

LONDON 1955 The Biochemistry of vitamin B12 :a symposium Biochemical society Edited by R.T. Williams Held at the London school of hygiene and tropical medicine Biochemical society.Symposia, 13 Cambridge university press,Cambridge,1955
Bioch 33.1375

LONDON 1955 The Chemistry and mode of action of plant growth substances :a symposium Edited by R.L. Wain and F. Wightman Butterworth scientific,London,1956
Radioth 35.0053

LONDON 1955 The Control of quality in the production of wrought non-ferrous metals and alloys a symposium held... on the occasion of the annual general meeting of the Institute of metals Institute of metals Institute of metals.Monograph and report series, 17 Institute of metals,London,1955
Met 25.0504

LONDON 1955 The Mechanism of phase transformations in metals :a symposium Institute of metals Institute of metals. Monograph and report series, 18 Institute of metals,London,1956
Met 25.1251

LONDON 1956 Bacterial anatomy Society for general microbiology :symposium 6th Papers Edited by E.T.C. Spooner and B.A.D. Stocker Cambridge university press,Cambridge, 1956
Col S 12.0037

LONDON 1956 Bacterial anatomy :a symposium Papers Society for general microbiology Edited by E.T.C. Spooner and B.A.D. Stocker Held at the Royal institution Society for general microbiology.Symposia, 6 Cambridge university press,Cambridge,1956
Bioch 33.1169

LONDON 1956 Ciba foundation symposium on ionizing radiations and cell metabolism Proceedings Ciba foundation Edited by G.E. W. Wolstenholme and C.M. O'Connor illus. Churchill,London,1956
Radioth 35.1355

LONDON 1956 Ciba foundation symposium on the chemistry and biology of purines Edited by G.E.W. Wolstenholme and C.M. O'Connor Illus Churchill,London,1957
Radioth 35.0059

LONDON 1956 Ciba foundation symposium on the chemistry and biology of purines Papers Ciba foundation Edited by G.E.W. Wolstenholme and Cecilia M. O'Connor Ciba foundation symposia J. and A. Churchill, London,1957
Chem 18.1326

LONDON 1956 Ciba foundation symposium on the chemistry and biology of purines Papers Ciba foundation Edited by G.E.W. Wolstenholme and Cecilia M. O'Connor illus Churchill,London,1957
Bioch 33.0814

LONDON 1956 Connective tissue :a symposium Papers and discussions Edited by R.E. Tunbridge and others Organized by the Council for international organizations of medical sciences Blackwell,Oxford,1957
Path 30.2490

LONDON 1956 Convention on digital-computer techniques proceedings Institution of electrical engineers Edited by W.K. Brasher Institution of electrical engineers. Proceedings, 103,Part B,Suppts.1-3 illus 342p 27cm Institution of electrical engineers,London,1956
Math L 5.0762

LONDON 1956 Extraction and refining of the rarer metals A Symposium on extraction metallurgy of some of the less common metals Institution of mining and metallurgy Institution of mining and metallurgy,London, 1957
Met 25.0500

LONDON 1956 Hypotensive drugs and control of vascular tone in hypertension :a symposium Proceedings Biological council.Co-ordinating committee for symposia on drug action Edited by M. Harington Pergamon press,London,1956
Med 36.0172

LONDON 1956 International congress on high-speed photography 3rd Proceedings Edited by R.B. Collins Held under the auspices of Great Britain.Department of scientific and industrial research xvi,417p Butterworths scientific publications,London, 1957
Chem E 24.0452

LONDON 1956 International congress on high-speed photography 3rd proceedings Edited by R.B. Collins Held under the auspices of the Department of scientific and industrial research Butterworths,London,1957
Eng 41.0439

LONDON 1956 Optical design with digital computers symposium proceedings Imperial college of science and technology.Technical optics section viii,93p 26cm Imperial college of science and technology,London,1956
Math L 5.0754

LONDON 1956 Physics of nuclear reactors :a conference Papers Arranged by the Institute of physics British journal of applied physics.Supplement, 5 Institute of physics,London,1956
Radioth 35.1596

LONDON 1956 Progress in the study of the British flora :a conference Report Botanical society of the British Isles Edited by J.E. Lousley Botanical society of the British Isles.B.S.B.I.conference reports, 5 London,1957
Bot 42.4710

LONDON · 1956 Recent advances in the chemistry of colouring matter :a symposium Report Chemical society Chemical society. Special publication,4 Chemical society, London,1956 Organized by professor Braude
Chem 18.1214

LONDON 1956 Regulation and mode of action of thyroid hormones :a colloquium Proceedings Ciba foundation Edited by G.E. W. Wolstenholme and Elaine C.P. Millar Ciba foundation colloquia on endocrinology, 10 Churchill,London,1957
Inv Med 37.0240

LONDON 1956 Regulation and mode of action of thyroid hormones :a colloquium Proceedings Ciba foundation Edited by G.E. W. Wolstenholme and Elaine C.P. Millar Ciba foundation colloquia on endocrinology, 10 illus Churchill,London,1957
Bioch 33.0439

LONDON 1956 The Ciba foundation symposium on the nature of viruses Ciba foundation Edited by G.E.W. Wolstenholme and E.C.P. Millar Ciba foundation.Symposia Churchill, London,1957
Radioth 35.0752

LONDON 1956 The Final forming and shaping of wrought non-ferrous metals :a symposium Institute of metals Institute of metals. Monograph and report series, 20 Institute of metals,London,1956
Met 25.0709

LONDON 1956 The Strength of concrete structures :a symposium Proceedings Cement and concrete association Cement and concrete association,London,1958
Eng 41.2930

LONDON 1956 The Structure of nucleic acids and their role in protein synthesis :a symposium Biochemical society Edited by E. M. Crook Held at the London school of hygiene and tropical medicine Biochemical society.Symposia, 14 Cambridge university press,Cambridge,1957
Bioch 33.1376

LONDON 1956 Vapour phase chromatography :a symposium Proceedings Edited by D.H. Desty and C.L.A. Harbourn Sponsored by the Institute of petroleum.Hydrocarbon research group xv,435p Butterworths,London,1957
Chem E 24.0845

LONDON 1956 Vapour phase chromatography :a symposium Proceedings Institute of petroleum.Hydrocarbon research group Edited by D.H. Desty and C.L.A. Harbourn Butterworths,London,1957
Chem 18.1450

LONDON 1956 New York 1956 Nov 28-30 International conference on fatigue of metals Proceedings Institution of mechanical engineers and American society of mechanical engineers Institution of mechanical engineers,London,1956
Met 25.0919

LONDON 1956 and NEW YORK 1956 Fatigue of metals International conference on fatigue of metals Proceedings Institution of mechanical engineers American society of mechanical engineers Institution of mechanical engineers,London,1956
Eng 41.3759

LONDON 1957 Ciba foundation ssymposium on the chemistry and biology of mucopolysaccharides Proceedings Ciba foundation Edited by G.E.W. Wolstenholme and Maeve O'Connor Churchill,London,1958
Bioch 33.1202

LONDON 1957 Ciba foundation symposium on the chemistry and biology of mucopolysaccarides Edited by G.E.W. Wolstenholme and M. O'Connor Illus J.and A. Churchill,London,1958
Radioth 35.0057

LONDON 1957 Ciba foundation symposium on the chemistry and biology of mucopolysaccharides Ciba foundation Edited by G.E.W. Wolstenholme and Maeve O'Connor Churchill,London,1958
An 32.1490

LONDON 1957 Ciba foundation symposium on the neurological basis of behaviour :in commemoration of Sir Charles Sherrington,O.M., G.B.E.,F.R.S.,1857-1952 Papers Ciba foundation Edited by G.E.W. Wolstenholme and Cecilia M. O'Connor Churchill,London,1958
Psy 31.0512

LONDON 1957 Conference on lubrication and wear Proceedings Institution of mechanical engineers Institution of mechanical engineers,London,1957
Eng 41.6295

LONDON 1957 Conference on the highway needs of Great Britain Proceedings Institution of civil engineers Institution of civil engineers,London,1958
Eng 41.3260

LONDON 1957 Implantation of ova :a conference Proceedings Society for endocrinology Edited by P. Eckstein Society for endocrinology.Memoirs, 6 Cambridge university press,Cambridge,1959
Phys 20.1383

LONDON 1957 Implantation of ova :a conference Proceedings Society for endocrinology Edited by P. Eckstein held at the Ciba foundation Society for endocrinology.Memoirs, 6 Cambridge university press,Cambridge,1959
Bal 39.1401

LONDON 1957 Industrial carbon and graphite papers read at the conference... Society of chemical industry Society of chemical industry,London,1958
Met 25.0155

LONDON 1957 Industry fights corrosion The Corrosion convention Proceedings Corrosion technology Corrosion technology, London,c1957
Met 25.1890

LONDON 1957 International conference on soil mechanics and foundation engineering 4th Proceedings 3 vols Butterworths, London,1957
Eng 41.3141

LONDON 1957 International congress of surface activity 2nd Proceedings Vol 1: gas-liquid and liquid-liquid interface Edited by J.H. Schulman ix,529p Butterworths scientific publications,London, 1957
Chem E 24.0935

LONDON 1957 International congress of surface activity 2nd Proceedings Edited by J.H. Schulman 4 vols Butterworths, London,1957
Col S 12.0278

LONDON 1957 International navigation congress Section 1:inland navigation 19th Communication 3 S.1-C.3: influence of ice on navigable waterways and on sea and inland ports Permanent international association of navigation congresses illus 260p Permanent international association of navigation congresses,Brussels,1957
Sco 14.0363

LONDON 1957 Microbial ecology :a symposium Papers Society for general microbiology Edited by R.E.O. Williams and C.C. Spicer Held at the Royal institution Society for general microbiology.Symposia, 7 Cambridge university press,Cambridge,1957
Bioch 33.1175

LONDON 1957 Microbial ecology :a symposium Papers Society for general microbiology Edited by R.E.O. Williams and C.C. Spicer Society for general microbiology.Symposia, 7 Cambridge,1957
Path 30.2585

LONDON 1957 Structure and function of subcellular components :a symposium Biochemical society Edited by E.M. Crook Biochemical society.Symposia, 16 Pl. Cambridge university press,Cambridge,1959
Radioth 35.0069

LONDON 1957 Structure and function of subcellular components :a symposium Biochemical society Edited by E.M. Crook Held at the University of London.Senate House Biochemical society.Symposia, 16 Cambridge university press,Cambridge,1959
Bot 42.1493

LONDON 1957 Symposium on nuclear sex 1st Proceedings Edited by D.Robertson Smith and William M. Davidson Heinemann, London,1958
Inv Med 37.0197

LONDON 1957 Symposium on nuclear sex King's college hospital medical school Edited by D.Robertson Smith and William M. Davidson Heinemann,London,1958
Gen 34.0879

LONDON 1957 The Biology of hair growth :a conference Papers Edited by William Montagna and Richard A. Ellis Sponsored by the British society for research on ageing Academic press,New York,1958
An 32.4034

LONDON 1957 The Ciba foundation symposium on the cerebrospinal fluid :production, circulation and absorption Edited by G.E.W. Wolstenholme and Cecilia M. O'Connor Churchill,London,1958
An 32.4167

LONDON 1957 The Design of physics research laboratories :a symposium Proceedings Held by the Institute of physics.London and home counties branch bibliog. Chapman and Hall; Reinhold,London;New York,1959
Radioth 35.1634

LONDON 1957 The History and philosophy of knowledge of the brain and its functions :an Anglo-American symposium Edited by F.N.L. Poynter Sponsored by Wellcome historical medical library Blackwell,Oxford,1958
An 32.1649

LONDON 1957 The Properties of materials at high rates of strain :conference Proceedings Institution of mechanical engineers Institution of mechanical engineers,London, 1957
Eng 41.3763

LONDON 1957 The Properties of materials at high rates of strain conference Proceedings Institution of mechanical engineers Institution of mechanical engineers,London, 1957
Met 25.0904

LONDON 1957 The Scaling-up of chemical plant and processes :joint symposium Papers By R.Edgeworth Johnstone and others Society of chemical industry.Chemical engineering group and Institution of chemical engineers Institution of chemical engineers, London,1957 Held jointly with the K. Instituut van ingenieurs and the K.Nederlandse chemische vereniging
Chem E 24.1234

LONDON 1957 The Scope of physical anthropology and its place in academic studies a symposium Edited by D.F. Roberts and J.S. Weiner Society for the study of human biology.Symposia, 1 Wenner-Gren foundation for anthropological research,London,1958
An 32.2880

LONDON 1957 The Scope of physical anthropology and its place in academic studies a symposium held at the Ciba foundation Papers Society for the study of human biology Edited by D.F. Roberts and J.S. Weiner Wenner-Gren foundation for anthropological research,London,1958
HE 27.0143

LONDON 1957 The Structure and function of subcellular components :a symposium Biochemical society Edited by E.M. Crook Held at the University of London.Senate House Biochemical society.Symposia, 16 Cambridge university press,Cambridge,1959
Bioch 33.1378

LONDON 1957 Thermodynamic and transport properties of fluids :joint conference Proceedings International union of pure and applied chemistry and Institution of mechanical engineers viii,217p Institution of mechanical engineers,London,1958
Chem E 24.0981

LONDON 1957 5-Hydroxytryptamine :a symposium Proceedings Edited by G.P. Lewis Pergamon press,London,1958
An 32.1489

LONDON 1957 5-Hydroxytryptamine :a symposium Proceedings Edited by G.P. Lewis Pergamon press,London,1958
Radioth 35.0788

LONDON 1957 London 1957 DEC 11-12 Air and water pollution in the iron and steel industry proceedings of the air pollution meeting 25th and 26th Sept. and the water pollution meeting 11th and 12th Dec Proceedings Iron and steel institute Iron and steel institute.Special report, 61 Iron and steel institute,London,1958
Met 25.0359

LONDON 1957 London 1958 Mar 25-27 Industrial pulmonary diseases :a symposium Edited by E.J. King and C.M. Fletcher Illus Churchill,London,1960
HE 27.0030

LONDON 1958 Advances in mass spectrometry : a conference Proceedings Institute of petroleum.Hydrocarbon research group,and, American society for testing materials. Committee on mass spectrometry Edited by J.D. Waldron Pergamon press.Symposia publications division,London,1959
Chem 18.0197

LONDON 1958 Ciba foundation symposium on amino acids and peptides with antimetabolic activity Ciba foundation Edited by G.E.W. Wolstenholme and Cecilia M. O'Connor Ciba foundation symposia Churchill,London,1958
Chem 18.1523

LONDON 1958 Ciba foundation symposium on amino acids and peptides with antimetabolic activity Proceedings Ciba foundation Edited by G.E.W. Wolstenholme and Cecilia M. O'Connor illus Churchill,London,1958
Bioch 33.0570

LONDON 1958 Ciba foundation symposium on carcinogenesis:mechanisms of action Ciba foundation Edited by G.E.W. Wolstenholme and Maeve O'Connor Illus Churchill,London,1959
Radioth 35.0107

LONDON 1958 Ciba foundation symposium on the biosynthesis of terpenes and sterols Papers Ciba foundation Edited by G.E.W. Wolstenholme and Maeve O'Connor Ciba foundation symposia Churchill,London,1959
Chem 18.1511

LONDON 1958 Ciba foundation symposium on the biosynthesis of terpenes and sterols Proceedings Ciba foundation Edited by G.E. W. Wolstenholme and Maeve O'Connor illus Churchill,London,1959
Bioch 33.0571

LONDON 1958 Ciba foundation symposium on the regulation of cell metabolism Ciba foundation Edited by G.E.W. Wolstenholme and Cecilia M. O'Connor Illus J.and A. Churchill,London,1959
Radioth 35.0493

LONDON 1958 Ciba foundation symposium on the regulation of cell metabolism Proceedings Ciba foundation Edited by G.E. W. Wolstenholme and Cecilia M. O'Connor illus Churchill,London,1959
Bioch 33.0575

LONDON 1958 Conference on technology of engineering manufacture Proceedings Institution of mechanical engineers Institution of mechanical engineers,London, 1958
Eng 41.3870

LONDON 1958 Control in electroplating :a one-day symposium Institute of metal finishing.London branch Draper,Teddington, 1959
Met 25.2670

LONDON 1958 Glutathione :a symposium Biochemical society Edited by E.M. Crook Held at the University of London.Senate House Biochemical society.Symposia, 17 Cambridge university press,Cambridge,1959
Bioch 33.1379

LONDON 1958 Glutathione :a symposium Biochemical society Edited by E.M. Crook Held at the University of London.Senate house Biochemical society.Symposia, 17 Cambridge university press,Cambridge,1959
Radioth 35.0078

LONDON 1958 Institution of structural engineers :fiftieth anniversary conference Institution of structural engineers,London, 1958
Eng 41.3023

LONDON 1958 International cancer congress 7th Abstract of papers Under the auspices of the International union against cancer 1958
Radioth 35.0777

LONDON 1958 International conference on gearing Proceedings Institution of mechanical engineers illus. Institution of mechanical engineers,London,1960
Eng 41.6224

LONDON 1958 Mechanical properties of non-metallic brittle materials :a conference Proceedings Edited by W.H. Walton Organized by the National coal board.Mining research establishment Butterworths,London, 1958
Eng 41.3645

LONDON 1958 Mechanical properties of non-metallic brittle materials :a conference Proceedings Great Britain.National coal board.Mining research establishment and Great Britain.Department of scientific and industrial research.Building research station Edited by W.H. Walton Butterworths,London, 1958
Met 25.0852

LONDON 1958 Medical biology and etruscan origins :a Ciba foundation symposium Ciba foundation Edited by G.E.W. Wolstenholme and Cecilia M. O'Connor Ciba foundation.Symposia, 50 illus. Churchill,London,1959
Gen 34.1877

LONDON 1958 Natural selection in human populations Society for the study of human biology Edited by D.F. Roberts and G.A. Harrison Society for the study of human biology.Symposia, 2 Pergamon press,London, 1959
Gen 34.1863

LONDON 1958 Natural selection in human populations :a symposium Edited by D.F. Roberts and G.A. Harrison Society for the study of human biology.Symposia, 2 Pergamon press,Oxford,1959
An 32.2881

LONDON 1958 New approaches in cell biology a symposium Proceedings Edited by P.M.B. Walker Academic press,London;New York,1966 Symposium presented at the fifteenth international congress of zoology
Gen 34.0851

LONDON 1958 New approaches in cell biology a symposium Proceedings Edited by P.M.B. Walker Academic press,London,1960 Held during the fifteenth international congress of zoology
An 32.3236

LONDON 1958 New approaches in cell biology a symposium Proceedings Edited by P.M.B. Walker Illus Academic press,London;New York,1960 Held during the 15th International congress of zoology
Radioth 35.0506

LONDON 1958 New approaches in cell biology; a symposium The International congress of zoology 15th Proceedings Edited by P.M.B. Walker Held at Imperial college of science and technology Academic press,London,1960 A section of the Proceedings of the 15th International congress of zoology
Bal 39.0289

LONDON 1958 Optical image assessment using frequency response techniques :a summer school Proceedings Imperial college of science and technology.Technical optics section Edited by H.H. Hopkins 134p London,1958
Obs 6.1974

LONDON 1958 Quantitative methods in human pharmacology and therapeutics :a symposium Proceedings Biological council.Co-ordinating committee for symposia on drug action Edited by D.R. Laurence Biological council.Co-ordinating committee for symposia on drug action.Symposium on drug action series,3 Pergamon,London,1959
Pha 16.0419

LONDON 1958 Quantitative paper chromatography of steroids :a conference Proceedings Society for endocrinology Edited by D. Abelson and R.V. Brooks Society for endocrinology.Memoirs, 8 Cambridge university press,Cambridge,1960
Bioch 33.2355

LONDON 1958 Recent research in freezing and drying International symposium on freeze-drying 2nd Papers Edited by A.S. Parkes and Audrey U. Smith Organised by the Institute of biology diagrs vii,320p 25cm Blackwell scientific publications, Oxford,1960
Sco 14.1237

LONDON 1958 Recent research in freezing and drying International symposium on freezing and drying 2nd Proceedings Edited by A.S. Parkes and Audrey U. Smith Organized by the Institute of biology Blackwell,Oxford,1960
An 32.3233

LONDON 1958 Sex differentiation and development :a symposium Proceedings Society for endocrinology Edited by C.R. Austin Society for endocrinology.Memoirs, 7 Cambridge university press,Cambridge,1960
Gen 34.0583

LONDON 1958 Sex differentiation and development :a symposium Proceedings Society for endocrinology Edited by C.R. Austin held at the Royal society of medicine Society for endocrinology.Memoirs, 7 Cambridge university press,Cambridge,1960
Bal 39.1402

LONDON 1958 Strategy of chemotherapy :a symposium Papers Society for general microbiology Edited by S.T. Cowan and Elizabeth Rowatt Held at the Royal institution Society for general microbiology. Symposia, 8 Cambridge university press, Cambridge,1958
Bioch 33.1180

LONDON 1958 Symposium on clinical trials Royal society of medicine Edited by Edward Charles Dodds Pfizer,Tonbridge,1958
Med 36.0116

LONDON 1958 The Business computer symposium Papers No 1-22 Electronic engineering association and Office appliance and business equipment Collection of 22 papers and other mimeographed material
Math L 5.3252

LONDON 1958 The Mechanism of action of insulin :a symposium British insulin manufacturers Edited by F.G. Young and others Blackwell scientific publication, Oxford,1960
PGMS 29.0378

LONDON 1958 The Mechanism of action of insulin a symposium Discussions Edited by F.G. Young Organized by the British insulin manufacturers Blackwell,Oxford,1960
Bioch 33.0474

LONDON 1958 The Organization of chemical engineering projects :joint symposium Proceedings Institute of petroleum and Institution of chemical engineers Institution of chemical engineers,London,1958 17th meeting of the European federation of chemical engineering
Chem E 24.1543

LONDON 1958 The Physical chemistry of metallic solutions and intermetallic compounds a symposium Proceedings Vol 1-2 National physical laboratory National physical laboratory.Symposium, 9 2 vols H.M.S.O.,London,1959
Met 25.1779

LONDON 1958 The Physical properties of polymers :symposium Papers Organised by the Society of chemical industry.Plastics and polymers group Society of chemical industry. Monograph, 5 293p Society of chemical industry,London,1959
Chem E 24.1019

LONDON 1958 The Protection of motor vehicles from corrosion :a symposium Corrosion group S.C.I.monograph, 4 Society of chemical industry,London,1958
Met 25.1940

LONDON 1958 The Strategy of chemotherapy : a symposium Proceedings By S.T. Cowan and Elizabeth Rowatt Society for general microbiology Society for general microbiology.Symposia, 8 Cambridge university press,Cambridge,1958
Radioth 35.0225

LONDON 1958 The Strategy of chemotherapy : symposium Society for general microbiology Edited by S.T. Cowan and Elisabeth Rowatt Society for general microbiology.Symposium,8 Cambridge university press,Cambridge,1958
Pha 16.0262

LONDON 1958 The Technology of engineering manufacture a conference Proceedings Institution of mechanical engineers Institution of mechanical engineers,London, 1958
Met 25.0676

LONDON 1958 Theoretical organic chemistry : Kekule symposium Papers International union of pure and applied chemistry.Section of organic chemistry Organized by the Chemical society Butterworths,London,1959
Chem 18.2165

LONDON 1958 Theoretical organic chemistry : Kekule symposium Papers International union of pure and applied chemistry.Section of organic chemistry,and,Chemical society Butterworths,London,1959
Chem 18.1604

LONDON 1958 Tools of biological research symposium 1st Proceedings Surgical research society Edited by Hedley J.B. Atkins Blackwell,Oxford,1959
Gen 34.0465

LONDON 1958 Virus growth and variation :a symposium Society for general microbiology Edited by A. Isaacs and B.W. Lacey Society for general microbiology.Symposia, 9 Cambridge university press for the Society of general microbiology,Cambridge,1959
Path 30.2743

LONDON 1958 Oxford Symposium (international) on combustion 7th Papers Combustion institute Butterworths scientific publications,London,1959 Earlier symposia entitled 'Symposium (international) on combustion,flame and explosion phenomena'
Chem 18.0437

LONDON 1958 3-5 Photo-electronic image devices Symposium on image tubes and related devices held at the Imperial college 1st Proceedings Edited by J.D. McGee and W.L. Wilcock Advances in electronics and electron physics, 12 Academic press,New York,1960
Cav 7.1456

LONDON 1959 A Darwin centenary :a conference Report Botanical society of the British Isles Edited by P.J. Wanstall Botanical society of the British Isles.B.S.B.I. conference reports, 6 port. London,1961 Centenary of the publication of the Origin of species
Bot 42.4711

LONDON 1959 A Symposium on immunization in childhood held in the Wellcome building Proceedings Livingstone,London,1960
PGMS 29.0283

LONDON 1959 Biochemistry of human genetics Ciba foundation symposium Ciba foundation Edited by G.E.W. Wolstenholme and Cecilia M. O'Connor Churchill,London,1959
An 32.0371

LONDON 1959 Biological problems arising from the control of pests and diseases :a symposium Proceedings Institute of biology Edited by R.K.S. Wood Institute of biology. Symposia, 9 22cm London,1960
Bal 39.2962

LONDON 1959 Cancer of the cervix : diagnosis of early forms Ciba foundation Edited by G.E.W. Wolstenholme and Maeve O'Connor Ciba foundation study group, 3 Churchill,London,1959
Radioth 35.1960

LONDON 1959 Congress of the European society of haematology 7th Proceedings 2,pt 1: papers European society of haematology Edited by E. Neumark and others Karger,Basle,1960
Med 36.0132

LONDON 1959 Corrosion problems of the petroleum industry a joint symosium Papers Institute of petroleum and Chemical engineering group Joint symposium with the Corrosion group S.C.I.Monograph, 10 Society of chemical industry,London,1960
Met 25.1934

LONDON 1959 Cyclical activity in endocrine systems :a symposium Proceedings Zoological society of London Edited by E.J.W. Barrington Zoological society of London. Symposia, 2 Zoological society of London, London,1960
An 32.3764

LONDON 1959 DEUCE users' colloquium on partial differential equations papers English electric company ltd DEUCE news, 45 91p English electric,Nelson,Staffs., 1959
Math L 5.2408

LONDON 1959 Determinants of infant behaviour :proceedings of a Tavistock study group on mother-infant interaction Tavistock clinic.Child development research unit Edited by B.M. Foss Methuen;Wiley,London;New York,1961
Psy 31.1203

LONDON 1959 Electron microscopy in anatomy A Symposium on the ultrastructure of cells Proceedings Anatomical society of Great Britain and Ireland Edited by J.D. Boyd and others Edward Arnold,London,1961
Bioch 33.2297

LONDON 1959 Electron microscopy in anatomy a symposium Proceedings Anatomical society of Great Britain and Ireland Arnold, London,1961
Phys 20.1491

LONDON 1959 Electron microscopy in anatomy a symposium papers Anatomical society of Great Britain and Ireland Illus. Arnold, London,1961
VA 19.0356

LONDON 1959 European society of haematology congress 7th Papers and proceedings 1-2 European society of haematology Edited by E. Neumark and others 2 vols Karger,Basle;New York,1960
Med 36.0131

LONDON 1959 European society of hematology congress 7th Proceedings 2,pt.2: papers Edited by E. Neumark and others Karger,Basle;New York,1960
Med 36.0381

LONDON 1959 Foundations of statistical inference Joint statistics seminar :a discussion By Leonard J. Savage and others Birkbeck college Imperial college of science and technology 112p Methuen;Wiley,London; New York,1964
WSM 43.1769

LONDON 1959 Hormones in fish :a symposium Proceedings Zoological society of London Edited by I.Chester Jones Zoological society of London.Symposia, 1 Zoological society of London,London,1960
An 32.3763

LONDON 1959 Human chromosomal abnormalities conference Proceedings King's college hospital medical school Edited by William M. Davidson and D.Robertson Smith illus. Staples press,London,1961
Gen 34.0878

LONDON 1959 Institution of structural engineers jubilee symposium on high strength bolts Proceedings Institution of structural engineers Institution of structural engineers,London,1959
Eng 41.2696

LONDON 1959 International conference on co-ordination chemistry Lectures delivered and abstracts of papers Chemical society,and, International union of pure and applied chemistry Chemical society.Special publication,13 Chemical society,London,1959
Chem 18.1220

LONDON 1959 Joint symposium on instrumentation and computation in process development and plant design Papers British conference on automation and computation Sponsored by the Institution of chemical engineers Institution of chemical engineers,London,1959 Sponsored also by the Society of instrument technology,and the British computer society
Math L 5.3290

LONDON 1959 Joint symposium on instrumentation and computation in process development and plant design Proceedings Institution of chemical engineers Society of instrument technology British computer society Institution of chemical engineers, London,1959
Eng 41.6011

LONDON 1959 Man,race and Darwin :a conference Papers Royal anthropological institute of Great Britain and Ireland Institute of race relations Edited by Philip Mason vii,151p Oxford university press, London,1960
WSM 43.3622

LONDON 1959 Methods of learning and techniques of teaching Annual conference of the association for the study of medical education 2nd Proceedings Association for the study of medical education Edited by J.R. Ellis Pitman medical publishing company, London,c1960
PGMS 29.0493

LONDON 1959 Pain and itch :nervous mechanisms Ciba foundation Edited by G.E.W. Wolstenholme and Maeve O'Connor Ciba foundation study groups, 1 Churchill, London,1959
Phys 20.2017

LONDON 1959 Pain and itch nervous mechanisms Ciba foundation Edited by G.E.W. Wolstenholme and M. O'Connor Ciba foundation study group, 1 Churchill,London, 1959
PGMS 29.0539

LONDON 1959 Pain and itch:nervous mechanisms Ciba foundation Edited by G.E.W. Wolstenholme and Maeve O'Connor Ciba foundation study group, 1 Churchill,London, 1959
An 32.4535

LONDON 1959 Pathogenesis and treatment of occlusive arterial disease :a conference Proceedings Royal college of physicians of London Edited by L. McDonald Pitman,London, 1960
PGMS 29.0081

LONDON 1959 Pathogenesis and treatment of occlusive arterial disease :a conference Proceedings Royal college of physicians of London Edited by Lawson McDonald Pitman, London,1960
Path 30.2430

LONDON 1959 Ploypeptides which affect smooth muscles and blood vessels... :a symposium Proceedings Edited by M. Schachter Biological council.Co-ordinating committee for symposia on drug action. Symposium on drug action series,4 Pergamon, Oxford,1960
Pha 16.0439

LONDON 1959 Polypeptides which affect smooth muscles and blood vessels :a symposium Proceedings Biological council.Coordinating committee for symposia on drug action Edited by M. Schachter Biological council. Coordinating committee for symposia on drug action. Series, 4 Pergamon press,London, 1960
PGMS 29.0380

LONDON 1959 Steric aspects of the chemistry and biochemistry of natural products Biochemical society Edited by J.K. Grant and W. Klyne Biochemical society.Symposia, 19 Cambridge university press,Cambridge,1960
Chem 18.2613

LONDON 1959 Steric aspects of the chemistry and biochemistry of natural products a symposium Biochemical society Edited by J.K. Grant and W. Klyne Held at the University of London.Senate House Biochemical society.Symposia, 19 Cambridge university press,Cambridge,1960
Bioch 33.1381

LONDON 1959 Steric course of microbiological reactions Ciba foundation Edited by G.E.W. Wolstenholme and C.M. O'Connor Ciba foundation study group, 2 Churchill,London,1959
PGMS 29.0540

LONDON 1959 Steric course of microbiological reactions :in honour of Dr.V. Prelog Ciba foundation Edited by G.E.W. Wolstenholme and Cecilia M. O'Connor Ciba foundation study group, 2 illus Churchill,London,1959
Bioch 33.1120

LONDON 1959 Symposium on biomechanics Papers Institution of mechanical engineers London,1959
Bal 39.0061

LONDON 1959 The Biosynthesis and secretion of adrenocortical steroids :a symposium Biochemical society Edited by F. Clark and J. K. Grant Held at the University of London. Senate House Biochemical society.Symposia, 18 Cambridge university press,Cambridge,1960
Bioch 33.1380

LONDON 1959 The Use of computers in production control conference proceedings Central London productivity association 69p 32cm Central London productivity association, London,1959
Math L 5.0843

LONDON 1959 Virus growth and variation :a symposium Papers Society for general microbiology Edited by A. Isaacs and B.W. Lacey Held at the University of London Society for general microbiology.Symposia, 9 Cambridge university press,Cambridge,1959
Bioch 33.1182

LONDON 1959 Virus virulence and pathogenicity Ciba foundation Edited by G. E.W. Wolstenholme and C.M. O'Connor Ciba foundation study group, 4 Churchill,London, 1960
PGMS 29.0541

LONDON 1960 Adrenergic mechanisms Ciba foundation Committee for symposia on drug action Edited by J.R. Vane and others Ciba foundation.Symposia Churchill,London,1960
PGMS 29.0379

LONDON 1960 Adrenergic mechanisms :Ciba foundation symposium Ciba foundation Edited by J.R. Vane and others Churchill, London,1960
Pha 16.0070

LONDON 1960 Adrenergic mechanisms :a Ciba foundation symposium Ciba foundation Edited by J.R. Vane and others Churchill, London,1960
Phys 20.0972

LONDON 1960 Biochemistry of mucopolysaccharides of connective tissue :a symposium By J.N. Davidson Biochemical society Edited by F. Clark and J.K. Grant 4th edition Biochemical society.Symposia, 20 Bibliog Cambridge university press, Cambridge,1961
Radioth 35.0098

LONDON 1960 Biometeorology The International bioclimatological congress 2nd Proceedings Edited by S.W. Tromp illus xxii,687p 25cm Pergamon,Oxford,1962
Sco 14.0585

LONDON 1960 Cell mechanism in hormone production and action :a symposium Proceedings Society for endocrinology Edited by P.C. Williams and C.R. Austin Society for endocrinology.Memoirs, 11 Cambridge university press,Cambridge,1961
Phys 20.1363

LONDON 1960 Cell mechanisms in hormone production and action :a symposium Proceedings Society for endocrinology Edited by P.C. Williams and C.R. Austin Society for endocrinology.Memoirs,11 Cambridge university press,Cambridge,1961
Pha 16.0091

LONDON 1960 Chemical pathology in relation to clinical medicine The Adrenal cortex :a symposium Proceedings Association of clinical pathologists Edited by G.K. McGowan and M. Sandler Pitman,London,1961
An 32.3751

LONDON 1960 Ciba foundation symposium on haemopoiesis :cell production and its regulation Ciba foundation Edited by G.E.W. Wolstenholme and Maeve O'Connor Churchill, London,1960
An 32.3166

LONDON 1960 Ciba foundation symposium on quinones in electron transport Ciba foundation Edited by G.E.W. Wolstenholme and Cecilia M. O'Connor Churchill,London,1961
Radioth 35.0099

LONDON 1960 Ciba foundation symposium on quinones in electron transport Papers Ciba foundation Edited by G.E.W. Wolstenholme and Cecilia M. O'Connor Churchill,London,1961
Chem 18.1489

LONDON 1960 Ciba foundation symposium on the nature of sleep Ciba foundation Edited by G.E.W. Wolstenholme and Maeve O'Connor Churchill,London,1961 Under the chairmanship of sir John Eccles
Psy 31.3387

LONDON 1960 Conference on electronic telephone exchanges papers Institution of electrical engineers Collection of typescript papers and reprints
Math L 5.0873

LONDON 1960 Congenital malformations :Ciba foundation symposium Ciba foundation Edited by G.E.W. Wolstenholme and Cecilia M. O'Connor Churchill,London,1960
An 32.2079

LONDON 1960 Congential malformations : symposium Ciba foundation Edited by G.E.W. Wolstenholme and Cecilia M. O'Connor Ciba foundation.Symposia Churchill,London,1960
Gen 34.0517

LONDON 1960 DEUCE users' colloquium on Monte Carlo methods papers English electric company DEUCE news, 52 21p English electric,Nelson,Staffs.,1960 Typescript
Math L 5.2410

LONDON 1960 DEUCE users' colloquium on linear algebra papers English electric company DEUCE news, 57 83p English electric,Nelson,Staffs.,1960
Math L 5.2409

LONDON 1960 Genetical variations in human populations :a symposium Edited by G.A. Harrison Society for the study of human biology.Symposia, 4 Pergamon press,Oxford, 1961
An 32.2882

LONDON 1960 Human growth :a symposium Papers Edited by J.M. Tanner Society for the study of human biology.Symposia, 3 Pergamon press,Oxford,1960
An 32.2879

LONDON 1960 Information theory Symposium on information theory 4th papers Edited by Colin Cherry xi,476p 25cm Butterworths,London,1961
Math 3.0762

LONDON 1960 International goitre conference 4th Abstracts of papers presented American goiter association and London thyroid club Edited by O.de Vaal and others Excerpta medica.International congress series, 26 Excerpta medica foundation,Amsterdam,1960
PGMS 29.0158

LONDON 1960 London symposium on information theory 4th papers Seven typescript papers
Math L 5.0858

LONDON 1960 Metabolic effects of adrenal hormones Ciba foundation Edited by G.E.W. Wolstenholme and Maeve O'Connor Ciba foundation study group, 6 Churchill,London, 1960
Gen 34.0540

LONDON 1960 Metabolic effects of adrenal hormones in honour of G.W.Thorn Ciba foundation Edited by G.E.W. Wolstenholme and Maeve O'Connor Ciba foundation study group, 6 Churchill,London,1960
Bioch 33.0442

LONDON 1960 Microbial genetics Society for general microbiology Society for general microbiology.Symposia, 10 Cambridge university press,Cambridge,1960 Two copies
Gen 34.0748

LONDON 1960 Microbial genetics :a symposium Papers Society for general microbiology Edited by W. Hayes and R.C. Clowes Held at the Royal institution Society for general microbiology.Symposia, 10 Cambridge university press,Cambridge,1960
Bioch 33.1176

LONDON 1960 Microbial genetics :a symposium Papers Society for general microbiology Edited by W. Hayes and R.C. Clowes Society for general microbiology. Symposia, 10 Cambridge university press, Cambridge,1960
Path 30.2605

LONDON 1960 Models for decision :a conference British computer society Operational research society Under the auspices of the United Kingdom automation council English universities press,London, 1965
Eng 41.1270

LONDON 1960 Monograph on ionospheric radio URSI general assembly 13th Papers of a session International scientific radio union Edited by W.J.G. Beynon URSI monographs series 264p Elsevier,Amsterdam;New York, 1962
Nap 11.0052

LONDON 1960 Monograph on radio noise of terrestrial origin URSI general assembly 13th Papers Commission 4 on radio noise of terrestrial origin International scientific radio union Edited by F. Horner URSI monographs series 202p Elsevier,Amsterdam, 1962
Nap 11.0268

LONDON 1960 Monograph on radio-wave propagation in the troposphere U.R.S.I. general assembly 13th Papers commission 2 on radio and troposphere International scientific radio union Edited by J.A. Saxton U.R.S.I. monographs series illus 199p Elsevier,Amsterdam,1962
Nap 11.0533

LONDON 1960 Nickel-chromium plating :a one-day symposium Institute of metal finishing. London branch Robert Draper,Teddington,1961
Met 25.2039

LONDON 1960 Pore pressure and suction in soils :a conference International society of soil mechanics and foundation engineering. British national committee Butterworths, London,1961
Eng 41.3170

LONDON 1960 Powders in industry :a symposium Papers Society of chemical industry.Surface activity group Society of chemical industry.Monograph v,447p Society of chemical industry,London,1961
Chem E 24.1445

LONDON 1960 Production of wide steel strip a symposium Iron and steel institute Iron and steel institute.Special report, 67 Iron and steel institute,London,1960
Met 25.0726

LONDON 1960 Railway electrification at industrial frequency British railways electrification conference Proceedings British transport commission Illus British transport commission,London,1960
Eng 41.1605

LONDON 1960 Recent advances in renal diseases :a conference Proceedings Royal college of physicians of London Edited by M. D. Milne Pitman,London,1960
PGMS 29.0224

LONDON 1960 Sensory specialization in responses to environmental demands :a symposium Proceedings Zoological society of London Edited by O.E. Lowenstein Zoological society of London.Symposia, 3 Zoological society of London,London,1960
An 32.4541

LONDON 1960 Steels for reactor pressure circuits :a symposium Iron and steel institute Iron and steel institute.Special report, 69 Iron and steel institute,London, 1961
Met 25.0414

LONDON 1960 Symposium on haemopoiesis cell production and its regulation Ciba foundation Edited by G.E.W. Wolstenholme and Maeve O'Connor Ciba foundation.Symposia Churchill,London,1960
PGMS 29.0273

LONDON 1960 Symposium on information theory 4th Proceedings Imperial college of science and technology.Electrical engineering department Edited by Colin Cherry Butterworths,London,1961
Psy 31.3386

LONDON 1960 Symposium on recent mechanical engineering developments in automatic control Proceedings Institution of mechanical engineers Institution of mechanical engineers,London,1960
Eng 41.6012

LONDON 1960 The Biochemistry of mucopolysaccharides of connective tissue :a symposium Biochemical society Edited by F. Clark and J.K. Grant Held at the Royal college of surgeons of England Biochemical society.Symposia, 20 Cambridge university press,Cambridge,1961
Bioch 33.1382

LONDON 1960 The Nature of sleep :a Ciba foundation symposium Proceedings Ciba foundation Edited by G.E.W. Wolstenholme and Maeve O'Connor Churchill,London,1961
An 32.4378

LONDON 1960 The Reliability and maintenance of digital computer systems: managerial and engineering aspects :discussion meetings British computer society and Institution of electrical engineers 70p 28cm British computer society;Institution of electrical engineers,London,1960
Math L 5.3255

LONDON 1960 Tools of biological research symposium 3rd Proceedings Third series Surgical research society Edited by Hedley J.B. Atkins Blackwell,Oxford,1961
Gen 34.0467

LONDON 1960 Vertebrate locomotion :a symposium Proceedings Zoological society of London Edited by J.E. Harris Zoological society of London.Symposia, 5 Zoological society of London,London,1961
An 32.0459

LONDON 1960 Vertebrate locomotion :a symposium Proceedings Zoological society of London Edited by J.E. Harris Zoological society of London.Symposia,5 Illus. Zoological society,London,1961
VA 19.0086

LONDON 1960 Vertebrate locomotion :a symposium 5 Proceedings Zoological society of London Edited by J.E. Harris illus,1 plate vi,132p 26cm London,1961
Sco 14.0730

LONDON 1960 World orchid conference 3rd Proceedings Sponsored by the American orchid society Royal horticultural society,London, 1960
BG 38.3118

LONDON 1960 11-13 Ciba foundation symposium on quinones in electron transport Proceedings Ciba foundation Edited by G.E. W. Wolstenholme and Cecilia M. O'Connor illus Churchill,London,1961
Bioch 33.0573

LONDON 1961 A Conference on oil hydraulic power transmission Institution of mechanical engineers.Hydraulic plant and machinery group Institution of mechanical engineers,London, 1962
Eng 41.6685

LONDON 1961 An Exposition of adaptive control :a symposium Proceedings Edited by J.H. Westcott Held at the Imperial college of science and technology Pergamon press, London,1962
Eng 41.5884

LONDON 1961 Better use of the world's fauna for food The Institute of biology symposium 11th Edited by J.D. Ovington Institute of biology,London,1963
Geog 13.2660

LONDON 1961 Biological acoustics Proceedings Zoological society of London Edited by P.T. Haskell and F.C. Fraser Zoological society of London.Symposia, 7 London,1962
An 32.4826

LONDON 1961 Ciba foundation symposium on the exocrine pancreas :normal and abnormal functions Edited by A.V.S. De Reuck and Margaret P. Cameron Churchill,London,1962
An 32.3659

LONDON 1961 Ciba foundation symposium on transplantation Edited by G.E.W. Wolstenholme and Margaret P. Cameron Churchill,London,1962
An 32.3281

LONDON 1961 Clinical aspects of genetics Royal college of physicians scientific conference 5th Proceedings Royal college of physicians Edited by F.Avery Jones Pitman medical,London,1961
Gen 34.1981

LONDON 1961 Clinical aspects of genetics : a conference Proceedings Royal college of physicians of London Edited by F.Avery Jones Pitman,London,1961
PGMS 29.0255

LONDON 1961 Conference on optical instruments and techniques Proceedings Edited by K.J. Habell Chapman and Hall, London,1962 Held at Imperial college
Cav 7.1457

LONDON 1961 Conference on optical instruments and techniques Proceedings Edited by K.J. Habell Under the auspices of the International commission for optics Chapman and Hall,London,1962
Obs 6.3215

LONDON 1961 Curare and curare-like agents : Ciba foundation study group 12th Ciba foundation Edited by A.V.S. De Reuck Churchill,London,1962
Pha 16.0044

LONDON 1961 Determinants of infant behaviour 2 Tavistock study group on mother-infant interaction 2nd Proceedings Tavistock clinic.Child development research unit Edited by B.M. Foss Methuen;Wiley, London;New York,1963
Psy 31.1208

LONDON 1961 Effect of the various steelmaking processes on the energy balances of integrated iron and steelworks The Iron and steel engineers group meeting 43rd Report Iron and steel institute.Engineers group London,1962
Met 25.2522

LONDON 1961 Enzymes and drug action Ciba foundation and Co-ordinating committee for symposia on drug action Edited by J.L. Mongar and A.V.S. De Reuck Churchill,London, 1962
Bioch 33.1872

LONDON 1961 Evolutionary aspects of animal communication;imprinting and early learning :a symposium Proceedings Association for the study of animal behaviour Zoological society of London Zoological society of London. Symposia, 8 Zoological society of London, London,1962
An 32.4384

LONDON 1961 Future of ironmaking in the blast furnace autumn general meeting... Papers and discussions Iron and steel institute Iron and steel institute.Special report, 72 Iron and steel institute,London, 1962
Met 25.0368

LONDON 1961 Immunoassay of hormones Ciba foundation Edited by G.E.W. Wolstenholme and Margaret P. Cameron Ciba foundation colloquia on endocrinology, 14 Churchill, London,1962
Gen 34.0654

LONDON 1961 Immunoassay of hormones :a colloquium Proceedings Ciba foundation Edited by G.E.W. Wolstenholme and Margaret P. Cameron Ciba foundation colloquia on endocrinology, 14 Churchill,London,1962
Bioch 33.0441

LONDON 1961 Immunoassay of hormones : colloquium Ciba foundation Ciba foundation colloquia on endocrinology, 14 Churchill, London,1962
Phys 20.1400

LONDON 1961 Insect polymorphism symposium Papers and discussions Royal entomological society of London Edited by J.S. Kennedy Royal entomological society of London.Symposia, 1 Royal entomological society,London,1961
Gen 34.1640

LONDON 1961 International congress on metallic corrosion 1st Society of chemical industry.Corrosion group Butterworths,London,1962
Met 25.1893

LONDON 1961 Local floras :a conference Proceedings Botanical society of the British Isles Edited by P.J. Wanstall Botanical society of the British Isles.B.S.B.I. conference reports, 7 London,1963
Bot 42.4712

LONDON 1961 Microbial reaction to environment :a symposium Papers Society for general microbiology Edited by G.C. Meynell and H. Gooder Held at the Royal institution Society for general microbiology. Symposia, 11 Cambridge university press, Cambridge,1961
Bioch 33.1177

LONDON 1961 Microbial reaction to environment :a symposium Papers Society for general microbiology Edited by G.G. Meynell and H. Gooder Society for general microbiology.Symposia, 11 pls Cambridge university press,Cambridge,1961
Path 30.2607

LONDON 1961 Photo-electronic image devices A Symposium on photo-electronic image devices held at Imperial college 2nd Proceedings Edited by J.D. McGee and others Advances in electronics and electron physics, 16 Academic press,New York,1962
Cav 7.1458

LONDON 1961 Photo-electronic image devices symposium 2nd Proceedings Edited by J. D. McGee and others Held at the Imperial college of science and technology Advances in electronics and electron physics, 16 Academic press,New York;London,1962
Met 25.2638

LONDON 1961 Plastics progress,1961 International plastics convention Papers and discussions Edited by Phillip Morgan xi, 181p Iliffe;Macmillan,London;New York,1962
Chem E 24.1014

LONDON 1961 Pressure vessel research towards better design :a symposium Proceedings Institution of mechanical engineers British welding research association Institution of mechanical engineers,London,1962
Eng 41.2875

LONDON 1961 Progesterone and the defence mechanism of pregnancy Ciba foundation Edited by G.E.W. Wolstenholme and Margaret P. Cameron Ciba foundation study group, 9 Churchill,London,1961
An 32.3753

LONDON 1961 Progesterone and the defence mechanism of pregnancy 9th Ciba foundation Edited by G.E.W. Wolstenholme and Margaret Cameron Ciba foundation study group, 9 Churchill,London,1961
Phys 20.1388

LONDON 1961 Structural processes in creep : a symposium Report Iron and steel institute and Institute of metals Iron and steel institute.Special report, 70 Iron and steel institute,London,1961
Met 25.2334

LONDON 1961 Symposium on drug action and modern therapy Anglo-German medical society Anglo-German medical review, 1,no.4 Schattauer,Stuttgart,1962 Text in English and German
Gen 34.1973

LONDON 1961 The Mechanism of action of water soluble vitamins :a meeting 11 Ciba foundation Edited by A.V.S. de Reuck and Maeve O'Connor Ciba foundation study group, 11 J.and A.Churchill,London,1961
Chem 18.1283

LONDON 1961 The Metallurgy of beryllium : an international conference Proceedings Institute of metals Institute of metals. Monograph and report series, 28 Chapman and Hall,London,1963
Met 25.0605

LONDON 1961 The Structure and biosynthesis of macro-molecules Biochemical society Edited by D.J. Bell and J.K. Grant Held at University of London.Senate house Biochemical society.Symposia, 21 Cambridge, 1962 In commemoration of the 50th anniversary of the Biochemical society
Bal 39.0515

LONDON 1961 The Structure and biosynthesis of macromolecules :a symposium Biochemical society Edited by D.J. Bell and J.K. Grant Held at the University of London.Senate House Biochemical society.Symposia, 21 Cambridge university press,Cambridge,1962 In commemoration of the 50th anniversary of the Biochemical society
Bioch 33.1383

LONDON 1961 The Structure and biosynthesis of macromolecules :a symposium Biochemical society Edited by J.K. Grant and D.J. Bell Biochemical society symposia,21 Cambridge university press,Cambridge,1962
Chem 18.1277

LONDON 1961 The Water relations of plants : a symposium British ecological society Edited by A.J. Rutter and F.H. Whitehead British ecological society, 3 illus. x, 349p Blackwell, London, 1963
Bot 42.1379

LONDON 1961 Underwater acoustics :an institute conducted at the Imperial college Proceedings Imperial college of science and technology, London and Pennsylvania state university Edited by Vernon M. Albers New York, 1963
Geod 9.0238

LONDON 1961 Welding in shipbuilding Joint symposium on welding in shipbuilding Institute of welding Institute of welding, London, 1962 Jointly organised by Institute of welding, Royal institution of naval architects, Institute of marine engineers, Institution of engineers and shipbuilders in Scotland and North East Coast institution of engineering and shipbuilders
Met 25.0799

LONDON 1961 Edinburgh 1961 International rock garden plant conference 3rd Report Alpine garden society Edited by R.C. Elliott and J.L. Mowat Organized by the Scottish rock garden club Alpine garden society; Scottish rock garden club, London; Penicuik, 1961
BG 38.0011

LONDON 1962 Cellular basis and aetiology of late somatic effects of ionizing radiation : a symposium Edited by R.J.C. Harris Under the auspices of Unesco Academic press, London; New York, 1963
Radioth 35.1750

LONDON 1962 Clean steel Iron and steel institute annual general meeting, 1962 Papers and discussions Iron and steel institute Iron and steel institute. Special report, 77 Iron and steel institute, London, 1963
Met 25.0412

LONDON 1962 Clinical trials :a symposium Report Pharmaceutical society of Great Britain Pharmaceutical press, London, 1962
Pha 16.0130

LONDON 1962 Comparative aspects of neurohypophysical morphology and function :a symposium Proceedings Zoological society of London Edited by H. Heller Zoological society of London. Symposia, 9 Zoological society of London, 1963
An 32.3765

LONDON 1962 Conference on classification proceedings Aslib Aslib proceedings, 14, no.8 215-266p 25cm Aslib, London, 1962
Math 3.0776

LONDON 1962 Conference on design methods : conference on systematic and intuitive methods in engineering industrial design, architecture and communications Papers Edited by J. Christopher Jones and D.G. Thornley Pergamon press, London, 1963
Eng 41.0304

LONDON 1962 Conservation of water resources in the United Kingdom :a symposium Proceedings Institution of civil engineers Institution of civil engineers, London, 1963
Geog 13.0786

LONDON 1962 Dental anthropology :a symposium Edited by D.R. Brothwell Society for the study of human biology. Symposia, 5 Pergamon press, Oxford, 1963
An 32.2883

LONDON 1962 EUSEC meeting on engineering education and training 4th Proceedings Conference of engineering societies of western Europe and the United States of America Institution of electrical engineers, London, 1963
Eng 41.1547

LONDON 1962 Experiments in graduate training :a symposium Proceedings Institution of mechanical engineers Institution of mechanical engineers, London, 1962
Eng 41.1499

LONDON 1962 International conference on low temperature physics 8th Proceedings Edited by R.O. Davies Held at Queen Mary college Butterworths, London, 1963
Eng 41.7396

LONDON 1962 International conference on low temperature physics 8th Proceedings Queen Mary college, London Edited by R.O. Davies Butterworths, London, 1963 Includes Fritz London memorial lecture delivered by Bardeen, J.
Cav 7.1459

LONDON 1962 Introduction to system programming symposium proceedings Edited by Peter Wegner A.P.I.C. studies in data processing, 4 x, 316p 23cm Academic press, London, 1964
Math L 5.0947

LONDON 1962 Low temperature physics International conference on low-temperature physics 8th Proceedings Edited by R.O. Davies Butterworths, London, 1963
Met 25.1180

LONDON 1962 Microbial classification :a symposium Papers Society for general microbiology Edited by G.C. Ainsworth and P. H.A. Sneath Held at the Royal institution Society for general microbiology. Symposia, 12 Cambridge university press, Cambridge, 1962
Bioch 33.1174

LONDON 1962 Microbial classification : symposium Society for general microbiology Society for general microbiology. Symposia, 12 Cambridge university press, Cambridge, 1962
Gen 34.0714

LONDON 1962 Physics of the welding arc :a symposium Institute of welding Institute of welding, London, 1966
Met 25.0807

LONDON 1962 Productivity in the iron and steel industry Iron and steel institute annual general meeting, 1962 Papers and discussions Iron and steel institute Iron and steel institute. Special report, 75 Iron and steel institute, London, 1962
Met 25.0411

LONDON 1962 Refractories for oxygen steelmaking technical sessions of the annual general meeting... Papers and discussions Iron and steel institute Iron and steel institute.Special report, 74 Iron and steel institute,London,1962
Met 25.0273

LONDON 1962 Spectroscopy :a conference Report Institute of petroleum.Hydrocarbon research group Edited by M.J. Wells Institute of petroleum,London,1962
Chem 18.0193

LONDON 1962 Symposium digest Symposium on electronic equipment reliability 2nd Institution of electrical engineers illus. 47p 28cm Institution of electrical engineers,London,1962
Math L 5.0926

LONDON 1962 Symposium on process optimisation Proceedings Institution of chemical engineers Edited by J.M. Pirie 70p Institution of chemical engineers,London, 1962 3rd congress of the European federation of chemical engineering
Chem E 24.1630

LONDON 1962 Symposium on the handling of solids Proceedings Institution of chemical engineers Edited by P.A. Rottenburg Institution of chemical engineers,London,1962 3rd congress of the European federation of chemical engineering
Chem E 24.1446

LONDON 1962 Symposium on the interaction between fluids and particles Proceedings Institution of chemical engineers illus 351p Institution of chemical engineers, London,n.d. Third congress on the European federation of chemical engineering
Chem E 24.1447

LONDON 1962 Symposium on two-phase fluid flow Proceedings Arranged by the Institution of mechanical engineers 116p Institution of mechanical engineers,London, 1962
Chem E 24.0372

LONDON 1962 The Accuracy of Industrial measurement of length and diameter :the conference National physical laboratory National physical laboratory.Conference, 14 H.M.S.O.,London,1963
Met 25.2166

LONDON 1962 The Better use of the world's fauna for food Institute of biology Edited by J.D. Ovington Institute of biology. Symposia, 11 21cm London,1963
Bal 44.4059

LONDON 1962 The Changing flora of Britain : a conference Report Botanical society of the British Isles Edited by J.E. Lousley Botanical society of the British Isles.B.S.B.I. conference reports, 3 London,1953
Bot 42.4708

LONDON 1962 The Primates :a symposium Proceedings Edited by John Napier and N.A. Barnicot Zoological society of London. Symposia, 10 Zoological society,London,1963
An 32.0580

LONDON 1962 The Structure and function of the membranes and surfaces of cells :a symposium Biochemical society Edited by D. J. Bell and J.K. Grant Biochemical society. Symposia, 22 Cambridge university press, Cambridge,1963
Bioch 33.1384

LONDON 1962 The Structure and function of the membranes and surfaces of cells :a symposium Biochemical society Edited by D. J. Bell and J.K. Grant Held at the Middlesex hospital medical school Biochemical society. Symposia, 22 Cambridge university press, Cambridge,1963 Organized by J.K.Grant
Bal 39.0294

LONDON 1962 The Teaching of automatic control :a discussion Proceedings Institution of mechanical engineers Institution of mechanical engineers,London, 1962
Eng 41.5814

LONDON 1962 Ultra sound as a diagnostic and surgical tool :based on the international symposium held at the Royal college of surgeons Edited by Douglas Gordon port. Livingstone,Edinburgh;London,1964
Radioth 35.1412

LONDON 1962 Uranium and graphite :a symposium Proceedings Institute of metals Institute of metals.Monograph, 27 Institute of metals,London,1962
Met 25.1554

LONDON 1963 Absorption and distribution of drugs :a symposium Association of medical advisers in the pharmaceutical industry Edited by T.B. Binns Livingstone,Edinburgh; London,1964
Pha 16.0239

LONDON 1963 Absorption and distribution of drugs :based on a symposium Association of medical advisers in the pharmaceutical industry Edited by T.B. Binns E.and S. Livingstone,Edinburgh;London,1964
PGMS 29.0370

LONDON 1963 Acute renal failure Symposium on acute renal failure Papers Edited by Stanley Shaldon and G.C. Cook Blackwell,Oxford,1964
Inv Med 37.0195

LONDON 1963 Acute renal failure :a symposium Edited by S. Shaldon and G.C. Cook Held at the Royal free hospital Blackwell, Oxford,1964
PGMS 29.0225

LONDON 1963 Aetiology of diabetes mellitus and its complications :colloquium Ciba foundation Edited by Margaret P. Cameron and Maeve O'Connor Ciba foundation colloquia on endocrinology, 15 Churchill,London,1964
Phys 20.1406

LONDON 1963 Aluminium in structural engineering :a symposium Proceedings Institution of structural engineers Aluminium federation Aluminium federation, London,1964
Eng 41.2694

LONDON 1963 Animal behaviour and drug action :annual meeting Proceedings Ciba foundation and Biological council.Co-ordinating committee for symposia on drug action Edited by A.V.S.de Reuck and Julie Knight Biological council.Co-ordinating committee for symposia on drug action.Annual meeting, 10 Ciba foundation symposia Churchill,London,1964
Psy 31.0526

LONDON 1963 Animal behaviour and drug action :symposium Ciba foundation,and, Biological council.Co-ordinating committee for symposia on drug action Edited by Hannah Steinberg and others Churchill,London,1964
Pha 16.0030

LONDON 1963 Atomic collision processes The International conference on the physics of electronic and atomic collisions 3rd Proceedings University college,London Edited by M.R.C. McDowell North-Holland, Amsterdam,1964
Cav 7.1461

LONDON 1963 British ecological society jubilee symposium Edited by A. Macfadyen and P.J. Newbound Blackwell,Oxford,1964 Supplement to "Journal of ecology",vol 52 and "Journal of animal ecology",vol 33
Geog 13.1309

LONDON 1963 Cell electrophoresis :a symposium British biophysical society Edited by E.J. Ambrose held at the Chester Beatty research institute illus Churchill, London,1965
Bal 39.0034

LONDON 1963 Cell electrophoresis :a symposium Edited by E.J. Ambrose Convened by the British biophysical society Churchill, London,1965
Radioth 35.1414

LONDON 1963 Cellular injury :a Ciba foundation symposium Papers Ciba foundation Edited by A.V.S. De Reuck and Julie Knight Churchill,London,1964
An 32.5317

LONDON 1963 Control of glycogen metabolism a Ciba foundation symposium Proceedings Ciba foundation Edited by W.J. Whelan and Margaret P. Cameron illus Churchill,London, 1964
Bioch 33.0576

LONDON 1963 Determinants of infant behaviour 3 Tavistock study group on mother-infant interaction 3rd Proceedings Tavistock clinic.Child development research unit Edited by B.M. Foss Methuen;Wiley, London;New York,1965
Psy 31.1205

LONDON 1963 Discussion on critical path analysis planning techniques for engineering management Proceedings Institution of mechanical engineers Institution of mechanical engineers,London,1964
Eng 41.0499

LONDON 1963 Fatigue in rolling contact : symposium Proceedings Arranged by the Institution of mechanical engineers Institution of mechanical engineers,London, 1963
Eng 41.3786

LONDON 1963 Fluidisation :a symposium Papers Society of chemical industry.Chemical engineering group and Institution of chemical engineers Sponsored also by the Institute of petroleum 110p Society of chemical industry,London,1964
Chem E 24.1452

LONDON 1963 Fume arrestment Iron and steel institute annual general meeting,1963 Proceedings Iron and steel institute Iron and steel institute.Special report, 85 Iron and steel institute,London,1964
Met 25.0420

LONDON 1963 Grouts and drilling muds in engineering practice :a symposium International society of soil mechanics and foundation engineering.British national committee Butterworths,London,1963
Eng 41.3080

LONDON 1963 Insect reproduction :symposium Royal entomological society of London Edited by K.C. Highnam Royal entomological society of London.Symposia, 2 Royal entomological society,London,1964
Gen 34.1647

LONDON 1963 International association for hydraulic research 10th congress Proceedings Vol 1-5 International association for hydraulic research.British national committee London,1963
Eng 41.6647

LONDON 1963 International conference on non-destructive testing 4th Proceedings Sponsored by the British national committee for non-destructive testing Butterworths, London,1964
Eng 41.3847

LONDON 1963 International symposium on the theory of road traffic flow London,1963 Unbound mimeograph papers
Math 3.1113

LONDON 1963 International symposium on the theory of road traffic flow Road research laboratory Edited by Joyce Almond Organization for economic cooperation and development,Paris,1965
Eng 41.3311

LONDON 1963 Lysosmes :a Ciba foundation symposium Ciba foundation Edited by A.V.S. De Reuck and Margaret P. Cameron Churchill, London,1963
An 32.5316

LONDON 1963 Lysosomes :a Ciba foundation symposium Papers Ciba foundation Edited by A.V.S. De Reuck and Margaret P. Cameron Ciba foundation.Symposia Churchill,London, 1963
Bal 39.0298

LONDON 1963 Lysosomes :a Ciba foundation symposium Papers Ciba foundation Edited by A.V.S. De Reuck and Margaret P. Cameron illus Churchill,London,1963
Bioch 33.0935

LONDON 1963 Oxygen in the animal organism : a symposium Proceedings International union of biochemistry and International union of physiological sciences Edited by Frank Dickens and Eric Neil Pergamon press, Oxford,1964
PGMS 29.0407

LONDON 1963 Oxygen in the animal organism : a symposium Proceedings International union of biochemistry and International union of physiological sciences Edited by Frank Dickens and Eric Neil held at Bedford college I.U.B.Symposium series, 31 Pergamon press,Oxford,1964
Bioch 33.1080

LONDON 1963 Prevention of hospital infection :the personal factor;report of a conference Royal society of health Edited by George Godber and others Royal society of health,London,1963
PGMS 29.0584

LONDON 1963 Productivity in research :a symposium Preprint of papers Institution of chemical engineers 89p Institution of chemical engineers,London,1963
Chem E 24.1794

LONDON 1963 Productivity in research :a symposium Proceedings Institution of chemical engineers Institution of chemical engineers,London,1964
Eng 41.1685

LONDON 1963 Recent developments in annealing Iron and steel institute Iron and steel institute.Special report, 79 Iron and steel institute,London,1963
Met 25.0418

LONDON 1963 Rheomacrodex low molecular weight dextrain :a symposium Proceedings Edited by H.G. Melrose Pharmacia limited, London,1964
PGMS 29.0609

LONDON 1963 Small animal anaesthesia :a symposium Proceedings British small animals veterinary association Universities federation for animal welfare Edited by Oliver Graham-Jones Pergamon press,London, 1964
Radioth 35.0270

LONDON 1963 Small animal anaesthesia :a symposium Proceedings British small animals veterinary association,and, Universities federation for animal welfare Edited by Oliver Graham-Jones Illus. Pergamon press,London,1964
VA 19.0264

LONDON 1963 Symbiotic associations :a symposium Papers Society for general microbiology Edited by P.S. Nutman and Barbara Mosse Society for general microbiology.Symposia, 13 Cambridge university press,Cambridge,1963
Bioch 33.1181

LONDON 1963 The Conference on the education and training of engineering technicians Proceedings Sponsored by Institution of mechanical engineers.Education and training group Institution of mechanical engineers,London,1963
Eng 41.1503

LONDON 1963 The Natural history of aggression :a symposium Proceedings Institute of biology Edited by J.D. Carthy and F.J. Ebling Institute of biology. Symposia, 13 Academic press,London;New York, 1964
Psy 31.0596

LONDON 1963 The Natural history of aggression :a symposium held at the British museum(natural history) Proceedings Institute of biology Edited by J.D. Carthy and F.J. Ebling Institute of biology. Symposia, 13 Academic press,London,1960
Bal 39.1619

LONDON 1963 The Theory of road traffic flow international symposium 2nd proceedings Road research laboratory Edited by Joyce Almond With financial support from General motors corporation. Research laboratories ix,406p 24cm Organisation for economic co-operation and development,Paris,1965
Math 3.0922

LONDON 1963 The Thyroid and its diseases : a conference Proceedings Royal college of physicians of London Edited by A.Stuart Mason Pitman,London,1963
PGMS 29.0159

LONDON 1964 A Symposium on Continental drift papers By P.M.S. Blackett and others Royal society of London Royal society of London.Philosophical transactions.Series A,no. 258,no.1088 map Royal society of London, London,1965
Geol 8.1262

LONDON 1964 A Symposium on agents affecting fertility Biological council.Co-ordinating committee for symposia on drug action Edited by C.R. Austin and J.S. Perry Churchill,London,1965
An 32.3896

LONDON 1964 A Symposium on continental drift Papers By P.M.S. Blackett and others Organised for the Royal society of London Royal society of London.Philosophical transactions.Series A,258,no.1088 vi,323p Royal society of London,London,1965
Sco 14.0260

LONDON 1964 A Symposium on continental drift Papers Royal society of London Royal society of London.Philosophical transactions.Series A, 258 London,1965 Symposium organized for the Royal society by P. M.S.Blackett,Sir Edward Bullard, and S.K. Runcorn
Geod 9.0129

LONDON 1964 A Symposium on continental drift Royal society Edited by P.M.S. Blackett and others Royal society. Philosophical transactions,1088 Royal society,London,1965
Geog 13.5647

LONDON 1964 Biological aspects of social problems :a symposium Eugenics society Edited by J.E. Meade and A.S. Parkes Eugenics society.Symposia,1 Oliver and Boyd, Edinburgh;London,1965
Geog 13.1550

LONDON 1964 Biological significance of climatic changes in Britain Institute of biology symposium 14th Proceedings Edited by C.G. Johnson and L.P. Smith Academic press,London,1965
Geog 13.1208

LONDON 1964 Cellular biology of myxovirus infections Ciba foundation Edited by G.E.W. Wolstenholme and Julie Knight Ciba foundation.Symposia Churchill,London,1964
Gen 34.0727

LONDON 1964 Comparative biology of reproduction in mammals :international symposium Proceedings Society for the study of fertility,and,Zoological society of London Edited by I.W. Rowlands Zoological society of London.Symposia,15 Academic press, London,1966
VA 19.0088

LONDON 1964 Comparative biology of reproduction in mammals :symposium Proceedings Society for the study of fertility Zoological society of London Edited by I.W. Rowlands Zoological society of London.Symposia, 15 Academic press, London,1966
An 32.3887

LONDON 1964 Complement :a Ciba foundation symposium Ciba foundation Edited by G.E.W. Wolstenholme and Julie Knight Churchill, London,1965
Pha 16.0189

LONDON 1964 Conference on dust cooling Proceedings Vol 1-2 Queen Mary college.Nuclear engineering department 2 vols London,1964 Mimeogra?h
Chem E 24.1464

LONDON 1964 Continuous casting of steel Iron and steel institute :autumn general meeting Proceedings Iron and steel institute Iron and steel institute.Special report, 89 London,1965
Met 25.2569

LONDON 1964 Functions of the corpus callosum Ciba foundation Edited by E.G. Ettlinger Ciba foundation study group, 20 Churchill,London,1965
An 32.4657

LONDON 1964 Functions of the corpus callosum Ciba foundation Edited by E.G. Ettlinger Ciba foundation study groups, 20 Churchill,London,1965
Phys 20.1996

LONDON 1964 Functions of the corpus callosum :a symposium Proceedings Ciba foundation Edited by E.George Ettlinger Ciba foundation study group, 20 J.and A. Churchill,London,1965 Proceedings dedicated to the Rt.hon.lord Adrian,O.M.,on his retirement
Psy 31.0193

LONDON 1964 International congress of endocrinology 2nd Proceedings Pt 1-2 Edited by S. Taylor Exerpta medica. International congress series, 83 2 vols Excerpta medica foundation,London,1965
PGMS 29.0157

LONDON 1964 International congress of endocrinology 2nd Proceedings Pt 1-2 Excerpta medica.International congress series, 83 2 vols Excerpta medica, Amsterdam,1966
Inv Med 37.0114

LONDON 1964 International congress of endocrinology 2nd Proceedings Pt 1-2 International congress series, 83 2 vols Excerpta medica,Amsterdam,1964
Phys 20.1396

LONDON 1964 International geographical congress 20th Guide to London excursions Edited by K.M. Clayton London,1964
Geog 13.5695

LONDON 1964 International geographical congress 20th Proceedings Edited by J. Wreford Watson Nelson,London,1967
Geog 13.5622

LONDON 1964 International teletraffic congress 4th London,1964 Unbound typescript papers
Math 3.1126

LONDON 1964 Load-bearing brickwork :a conference Proceedings British ceramic society British ceramic society.Proceedings, 4 British ceramic society,Stoke-on-Trent, 1965
Eng 41.2941

LONDON 1964 Microbial behaviour 'in vivo' and 'in vitro' :a symposium Papers Society for general microbiology Edited by H. Smith and Joan Taylor Held at the Royal institution Society for general microbiology. Symposia, 14 Cambridge university press, Cambridge,1964
Bioch 33.1173

LONDON 1964 Microbial behaviour 'in vivo' and 'in vitro' :a symposium Papers Society for general microbiology Edited by H. Smith and Joan Taylor Society for general microbiology.Symposia, 14 figs Cambridge university press,Cambridge,1964
Path 30.2610

LONDON 1964 Microbial behaviour in vivo and in vitro Society for general microbiology Society for general microbiology.Symposia, 14 Cambridge university press,Cambridge,1964
Gen 34.0700

LONDON 1964 Model testing :one-day meeting Proceedings Cement and concrete association Cement and concrete association,London,1964
Eng 41.2934

LONDON 1964 Physiology of the insect central nervous system International congress of entomology 12th Papers Edited by J.E. Treherne and J.W.L. Beament Academic press,London,1965
Pha 16.0276

LONDON 1964 Psychosocial aspects of drug taking :conference Proceedings Edited by R. G. Andry Pergamon,Oxford,1965 Summarised by Derrick Sington
Pha 16.0060

LONDON 1964 Radar techniques for detection tracking and navigation :8th symposium of the AGARD avionics panel Proceedings Agard Edited by W.T. Blackband AGARDograph, 100 Gordon and Breach,New York,1966
Eng 41.5717

LONDON 1964 Recent advances in selenium physics A Symposium on solid and liquid state selenium physics... held at the Chemical society Proceedings European selenium-tellurium committee Edited by Sven Walden and others Pergamon press,Oxford,1965
Cav 7.1462

LONDON 1964 Research seminar on archaeology and related subjects day meeting on statistics and archaeology Institute of archaeology London,1964 Unbound typescript papers
Math 3.1115

LONDON 1964 Species study in the British flora The Conference on the species concept in its relation to the British flora Report Botanical society of the British Isles Edited by J.E. Lousley Botanical society of the British Isles.B.S.B.I.conference reports, 4 London,1955
Bot 42.4709

LONDON 1964 Structure and activity of enzymes :a symposium Proceedings Federation of European biochemical societies Edited by T.W. Goodwin and others Federation of European biochemical societies.Symposium, 1 Academic press,London;New York,1964
Bioch 33.1404

LONDON 1964 Structure and activity of enzymes :a symposium 1 Federation of European biochemical societies Edited by T.W. Goodwin and others Academic press,London; New York,1964
Chem 18.1248

LONDON 1964 Symposium on advanced medicine a conference Proceedings Royal college of physicians of London Edited by Nigel Compston Pitman,London,1965
PGMS 29.0057

LONDON 1964 Symposium on the physiology of the insect central nervous system International congress of entomology 12th Papers Edited by J.E. Treherne and J.W.L. Beaument London,1965
Bal 39.2458

LONDON 1964 Teaching and research in human biology :a symposium Edited by G.Ainsworth Harrison Society for the study of human biology.Symposia, 6 Pergamon press,Oxford, 1964
An 32.2884

LONDON 1964 The Biological significance of climatic changes in Britain Proceedings Institute of biology Edited by C.G. Johnson and L.P. Smith Held at the Royal geographical society Institute of biology. Symposium, 14 illus x,222p Academic press,London;New York,1965
Bot 42.6678

LONDON 1964 The Biological significance of climatic changes in Britain :a symposium held at the Royal geographical society Proceedings Institute of biology Edited by C.G. Johnson and L.P. Smith Institute of biology.Symposia, 14 Academic press,London; New York,1965
Bal 39.1687

LONDON 1964 The Biology of survival :a symposium 13th Proceedings Edited by O.G. Edholm Organized by the Physiological society,the society for the study of human biology,and the Zoological society of London Zoological society of London.Symposia, 13 Academic press,London,1964
Phys 20.2132

LONDON 1964 The Interconnection of peripheral equipment and computers colloquium report Institution of electrical engineers Edited by H.McG. Ross 380p Institution of electrical engineers,London,1964
Math L 5.1272

LONDON 1964 The Scientific basis of drug therapy in psychiatry :a symposium Proceedings Edited by John Marks and C.M.B. Pare Pergamon,Oxford,1965
Pha 16.0028

LONDON 1964 Tooth enamel;its composition, properties and fundamental structure An International symposium on the composition, properties and fundamental structure of tooth enamel held at the London hospital medical college Report of the proceedings Edited by Maurice V. Stack and Ronald W. Fearnhead Wright,Bristol,1965 Sponsored by the Wellcome trust,London;the Procter and Gamble Co., Cincinnati;Colgate-Palmolive Ltd.,Great Britain;Unilever Ltd.,London
Bal 39.1298

LONDON 1965 A Discussion on the physics of the moon and its environment :a meeting Papers Royal society of London Royal society of London.Proceedings.Series A. Mathematical and physical sciences, 296,1446 Royal society,London,1967
Geod 9.0015

LONDON 1965 A.I.Ch.E.-Chem.E. joint meeting Proceedings 4: application of mathematical models in chemical engineering research design and production American institute of chemical engineers Institution of chemical engineers 1965 A.I.Ch.E.-I.Chem. E.Symposium series, 4 I.Chem.E.,London, 1965
Eng 41.5861

LONDON 1965 A.I.Ch.E.-I.Chem.E. joint meeting Proceedings 10: mixing - theory related to practice American institute of chemical engineers Institution of chemical engineers 1965 A.I.Ch.E.-I;Chem.E.symposium series, 10 Institution of chemical engineers,London,1965
Chem E 24.1920

LONDON 1965 A.I.Ch.E.-I.Chem.E.joint meeting Proceeding 1: advances in separation techniques American institute of chemical engineers and Institution of chemical engineers 1965 A.I.Ch.E.-I.Chem.E. Symposium series, 1 Institution of chemical engineers,London,1965
Chem E 24.1506

LONDON 1965 A.I.Ch.E.-I.Chem.E.joint meeting Proceedings 2: chemical engineering under extreme conditions American institute of chemical engineers and Institution of chemical engineers 1965 A.I. Ch.E.-I.Chem.E.Symposium series, 2 Institution of chemical engineers,London,1965
Chem E 24.1507

LONDON 1965 A.I.Ch.E.-I.Chem.E.joint meeting Proceedings 3: reaction kinetics in product and process design American institute of chemical engineers and Institution of chemical engineers 1965 A.I. Ch.E.-I.Chem.E.Symposium series, 3 Institution of chemical engineers,London,1965
Chem E 24.1508

LONDON 1965 A.I.Ch.E.-I.Chem.E.joint meeting Proceedings 4: application of mathematical models in chemical engineering research, design and production American institute of chemical engineers and Institution of chemical engineers 1965 A.I. Ch.E.-I.Chem.E.Symposium, 4 Institution of chemical engineers,London,1965
Chem E 24.1509

LONDON 1965 A.I.Ch.E.-I.Chem.E.joint meeting Proceedings 6: transport phenomena American institute of chemical engineers and Institution of chemical engineers 1965 A.I.Ch.E.-I.Chem.E.Symposium series, 6 Institution of chemical engineers,London,1965
Chem E 24.1511

LONDON 1965 A.I.Ch.E.-I.Chem.E.joint meeting Proceedings 7: management oriented topics American institute of chemical engineers and Institution of chemical engineers 1965 A.I.Ch.E.-I.Chem;E. Symposium series, 7 Institution of chemical engineers,London,1965
Chem E 24.1512

LONDON 1965 A.I.Ch.E.-I.Chem.E.joint meeting Proceedings 9: new chemical engineering problems in the utilisation of water American institute of chemical engineers and Institution of chemical engineers 1965 A.I.Ch.E.-I.Chem.E.Symposium series, 9 Institution of chemical engineers,London,1965
Chem E 24.1513

LONDON 1965 A.I.Ch.E.-I.ChemE.joint meeting 5: materials associated with direct energy conversion American institute of chemical engineers and Institution of chemical engineers 1965 A.I.Ch.E.-I.Chem.E. Symposium series, 5 Institution of chemical engineers,London,1965
Chem E 24.1510

LONDON 1965 Advanced medicine A Symposium on advanced medicine 2nd Proceedings Edited by J.R. Trounce Held at the Royal college of physicians Pitman medical,London,1966
Inv Med 37.0213

LONDON 1965 Aspects of insect biochemistry a symposium Biochemical society Edited by T.W. Goodwin Biochemical society.Symposia, 25 Academic press,London;New York,1965
Bioch 33.1387

LONDON 1965 Aspects of insect biochemistry symposium Biochemical society Edited by T. W. Goodwin Biochemical society.Symposia, 25 Academic press,London;New York,1965
Gen 34.1646

LONDON 1965 Ciba foundation symposium...on sensory function :touch,heat and pain 2nd Proceedings Ciba foundation Edited by A.V. S. de Reuck and Julie Knight J.and A. Churchill,London,1966
Psy 31.0286

LONDON 1965 Clinical evaluation in breast cancer Imperial cancer research fund symposium 1st Proceedings Edited by J.L. Haywood and R.D. Bulbrook illus. Academic press,London;New York,1966
Radioth 35.0870

LONDON 1965 Conference on power application of controllable semiconductor devices Pt 1 Institution of electrical engineers Institution of electrical engineers.Conference publications, 17 I.E.E.,London,1965
Eng 41.4868

LONDON 1965 Criticism and the growth of knowledge International colloquium in the philosophy of science Proceedings Edited by Imre Lakatos and Alan Musgrave Held at Bedford college vii,282p 23cm Cambridge university press,Cambridge,1970
Math S 3.1850

LONDON 1965 Development of the lung :a Ciba foundation symposium Ciba foundation Edited by A.V.S. De Reuck and Ruth Porter Churchill,London,1967
Phys 20.2220

LONDON 1965 Development of the lung :a Ciba foundation symposium Ciba foundation Edited by A.V.S. Reuck and Ruth Porter Churchill,London,1967
An 32.3803

LONDON 1965 Egg implantation Ciba foundation Edited by G.E.W. Wolstenholme and Maeve O'Connor Ciba foundation study group, 23 Churchill,London,1966
An 32.3889

LONDON 1965 Egg implantation Ciba foundation Edited by G.E.W. Wolstenholme and Maeve O'Connor Ciba foundation study group, 23 Churchill,London,1966
Phys 20.1408

LONDON 1965 Embryopathic activity of drugs a symposium Biological council.Co-ordinating committee for symposia on drug action Edited by J.M. Robson and others Churchill,London,1965
Pha 16.0320

LONDON 1965 Function and structure in micro-organisms :a symposium Papers Society for general microbiology Held at the Middlesex hospital Society for general microbiology.Symposia, 15 Cambridge university press,Cambridge,1965
Bioch 33.1171

LONDON 1965 Function and structure in micro-organisms :symposium Society for general microbiology Edited by M.R. Pollock and M.H. Richmond Society for general microbiology.Symposium,15 Cambridge university press,Cambridge,1965
Pha 16.0162

LONDON 1965 Function and structure in micro-organisms :symposium Society for general microbiology Society for general microbiology.Symposia, 15 Cambridge university press,Cambridge,1965
Gen 34.0702

LONDON 1965 Gonadotropins:physiocochemical and immunological properties Ciba foundation Edited by G.E.W. Wolstenholme and Julie Knight Ciba foundation study group, 22 Churchill, London,1965
Phys 20.1409

LONDON 1965 Hashish:its chemistry and pharmacology :a Ciba foundation study group 21st Ciba foundation Churchill,London,1965
Pha 16.0431

LONDON 1965 Heat treatment of metals Heat treatment joint committee :inaugural meeting Proceedings Iron and steel institute Institute of metals Iron and steel institute.Special report, 95 London, 1966
Met 25.2565

LONDON 1965 Histones :their role in the transfer of genetic information Ciba foundation Edited by A.V.S. De Reuck and Julie Knight Ciba foundation study group, 24 Little,Brown and company,Boston,1966 In honour of W.A.Engelhardt
Bioch 33.0888

LONDON 1965 Insect behaviour Royal entomological society of London Edited by P. T. Haskell Royal entomological society of London.Symposia, 3 Royal entomological society,London,1966
Gen 34.1650

LONDON 1965 International colloquium in the philosophy of science Proceedings Vol 2: the problem of inductive logic Edited by Imre Lakatos Studies in logic and the foundations of mathematics viii,417p North-Holland,Amsterdam,1968
WSM 43.1218

LONDON 1965 International colloquium in the philosophy of science Proceedings Vol 4: criticism and the growth of knowledge Edited by Imre Lakatos and Alan Musgrave Cambridge university press,Cambridge,1970
Eng 41.1688

LONDON 1965 Machinability :a conference jointly organized by the iron and steel institute the Institute of metals,the Institution of production engineers and the Institution of metallurgists Proceedings Iron and steel institute Iron and steel institute.Special report, 94 Iron and steel institute,London,1967
Met 25.0739

LONDON 1965 Man-made lakes Institute of biology :symposium 15th Edited by R.H. Lowe-McConnell Academic press,London,1966
Geog 13.0836

LONDON 1965 Man-made lakes :a symposium Proceedings Institute of biology Edited by R. Lowe-McConnell held at the Royal geographical society Institute of biology. Symposia, 15 Academic press,London;New York, 1966
Bal 39.1898

LONDON 1965 Microwave behaviour of ferrimagnetics and plasmas International conference on the microwave behaviour of ferrimagnetics and plasmas Sponsored by Institution of electrical engineers. Electronics division Institution of electrical engineers.Conference publication, 13 Institution of electrical engineers, London,1965
Eng 41.5713

LONDON 1965 Nerves and hormonal mechanisms of integration :a symposium Society for experimental biology Society for experimental biology.Symposia, 20 Cambridge university press,Cambridge,1966
An 32.3785

LONDON 1965 Photo-electronic image devices symposium 3rd Proceedings Edited by J. D. McGee and others Advances in electronics and electron physics, 22 a-b 2 vols Academic press,New York;London,1966
Met 25.2639

LONDON 1965 Pilot plants in the iron and steel industry autumn general meeting 1965 Papers and discussions Iron and steel institute Iron and steel institute.Special report, 96 Iron and steel institute,London, 1966
Met 25.0367

LONDON 1965 Plastics in building structures :a conference Proceedings Organised by the Plastics institute Plastics institute.Transactions and journal.Conference supplement, 1 Pergamon press,London,1966
Eng 41.3670

LONDON 1965 Preimplantation stages of pregnancy :Ciba foundation symposium Ciba foundation Edited by G.E.W. Wolstenholme and M. O'Connor Churchill,London,1965
Phys 20.1407

LONDON 1965 Preimplantation stages of pregnancy :Ciba foundation symposium Ciba foundation Edited by G.E.W. Wolstenholme and M. O'Connor London,1965
Bal 39.3558

LONDON 1965 Preimplantation stages of pregnancy :a Ciba foundation symposium Ciba foundation Edited by G.E.W. Wolstenholme and Maeve O'Connor Churchill,London,1965
An 32.2517

LONDON 1965 Principles of biomolecular organisation :a CIBA foundation symposium Ciba foundation Edited by G.E.W. Wolstenholme and Maeve O'Connor Churchill, London,1966
An 32.1523

LONDON 1965 Principles of biomolecular organization :Ciba foundation symposium Ciba foundation Edited by G.E.W. Wolstenholme and Maeve O'Connor Churchill,London,1966
Phys 20.1505

LONDON 1965 Principles of biomolecular organization :a Ciba foundation symposium Proceedings Ciba foundation Edited by G.E. W. Wolstenholme and Maeve O'Connor Churchill, London,1966
Bioch 33.0311

LONDON 1965 Principles of biomolecular organization :a Ciba foundation symposium Proceedings Ciba foundation Edited by G.E. W. Wolstenholme and Maeve O'Connor Churchill, London,1966
Bot 42.1503

LONDON 1965 Problems in the philosophy of mathematics International colloquium in the philosophy of science Proceedings Vol 1 British society for the philosophy of science London school of economics and political science Edited by Imre Lakatos Under the auspices of the International union of history and philosophy of science.Division of logic,methodology and philosophy of science Studies in logic and the foundations of mathematics xv,241p North-Holland, Amsterdam,1967
WSM 43.1813

LONDON 1965 Problems in the philosophy of science International colloquium in the philosophy of science Proceedings Vol 3 British society for the philosophy of science London school of economics and political science Edited by Imre Lakatos and Alan Musgrave Under the auspices of the International union of history and philosophy of science.Division of logic,methodology and philosophy of science Studies in logic and the foundations of mathematics ix,448p North-Holland,Amsterdam,1968
WSM 43.0931

LONDON 1965 Salt basins around Africa Institute of petroleum and Geological society of London :joint meeting proceedings Institute of petroleum,and,Geological society of London Institute of petroleum,London,1965
Geol 8.3097

LONDON 1965 The Countryside in 1970 : conference 2nd Proceedings Council for nature,and,Royal society of arts Jointly sponsored by the Nature conservancy Royal society of arts;Nature conservancy,London,1966
Geog 13.1899

LONDON 1965 The Skeletal biology of earlier human populations :a symposium Edited by D.R. Brothwell Society for the study of human biology.Symposia, 8 Pergamon press,Oxford,1968
An 32.2885

LONDON 1965 The Soil resources of tropical Africa :a symposium African studies association of the United Kingdom Edited by R.P. Moss Cambridge university press, Cambridge,1968
Geog 13.4644

LONDON 1965 The Structure of concrete and its behaviour under load :an international conference Proceedings Cement and concrete association Cement and concrete association, London,1968
Eng 41.3007

LONDON 1965 The Use of thin films in physical investigations :a NATO advanced study institute held at the Imperial college of science and technology Edited by J.C. Anderson Sponsored by the North Atlantic treaty organization Academic press,London, 1966
Eng 41.4379

LONDON 1965 Touch,heat and pain :Ciba foundation symposium Proceedings Ciba foundation Edited by A.V.S. De Reuck and Julie Knight Churchill,London,1966
An 32.5335

LONDON 1965 Vacuum degassing of steel Iron and steel institute annual general meeting 1965 Papers and discussions Iron and steel institute Iron and steel institute.Special report, 92 Iron and steel institute,London, 1965
Met 25.0372

LONDON 1965 Vibration in civil engineering a symposium Proceedings Edited by B.O. Skipp Organised by the International association for earthquake engineering Butterworths,London,1966
Eng 41.6363

LONDON 1966 A Discussion on infrared astronomy Edited by H. Massey and J. Ring Philosophical transactions,A., 264,p 107-320 Royal society,London,1969
Obs 6.3411

LONDON 1966 A.C.T.P.Summer school on systems programming Lectures 1969 Collection of typescript copies of lectures
Math L 5.3420

LONDON 1966 Biochemical studies of antimicrobial drugs :a symposium Papers Society for general microbiology Edited by B. A. Newton and P.E. Reynolds Held at the Royal institution Society for general microbiology.Symposia, 16 Cambridge university press,Cambridge,1966
Bioch 33.1170

LONDON 1966 Biochemical studies of antimicrobial drugs :symposium Society for general microbiology Society for general microbiology.Symposia, 16 Cambridge university press,Cambridge,1966
Gen 34.0705

LONDON 1966 Biochemistry studies of antimicrobial drugs :symposium 16th Society for general microbiology Edited by B. A. Newton and P.E. Reynolds Society for general microbiology.Symposium,15 Cambridge university press,Cambridge,1966
Pha 16.0336

LONDON 1966 Control of composition in steelmaking :annual general meeting Papers and discussions Iron and steel institute Iron and steel institute.Special publication, 99 London,1967
Met 25.2557

LONDON 1966 Design guide for medical group practice centres :a study...incorporating papers given at a symposium College of general practitioners National building agency College of general practitioners, London,c1966
PGMS 29.0555

LONDON 1966 Drug responses in man :a symposium Papers Ciba foundation Edited by Gordon Wolstenholme and Ruth Porter Churchill,London,1967
Pha 16.0253

LONDON 1966 Effects of external stimuli on reproduction :a Ciba foundation study group Edited by G.E.W. Wolstenholme and Maeve O'Connor Ciba foundation study group, 26 Churchill,London,1967
An 32.3895

LONDON 1966 Endocrinology of the testis : colloquium Ciba foundation Ciba foundation colloquia on endocrinology, 16 Churchill, London,1967
Phys 20.1412

LONDON 1966 Endocrinology of the testis : colloquium Ciba foundation Edited by G.E.W. Wolstenholme and Maeve O'Connor Ciba foundation colloquia on endocrinology, 16 Churchill,London,1967
An 32.3893

LONDON 1966 Heat and mass transfer in process metallurgy :a symposium Proceedings Edited by A.W.D. Hills Held by the John Percy research group in process metallurgy ix,252p Institution of mining and metallurgy, London,1967
Chem E 24.0593

LONDON 1966 Heat and mass transfer in process metallurgy :a symposium Proceedings John Percy research group in process metallurgy Edited by A.W.D. Hills Institution of mining and metallurgy,London, 1967
Met 25.0317

LONDON 1966 Industrialized building and the structural engineers :a symposium Institution of structural engineers Institution of structural engineers,London, 1966
Eng 41.3381

LONDON 1966 Instrumentation in biochemistry Biochemical society Edited by T.W. Goodwin Biochemical society.Symposia, 26 Academic press,London;New York,1966
Bioch 33.1388

LONDON 1966 Instrumentation in biochemistry :a symposium Biochemical society Edited by T.W. Goodwin Biochemical society.Symposia, 26 Academic press,London, 1966
Radioth 35.0135

LONDON 1966 International symposium on world climate 8000 to 0 b.c. proceedings Royal meteorological society Edited by J.S. Sawyer Royal meteorological society,London, 1966
A Math 4.0579

LONDON 1966 Ironmaking tomorrow : conference Proceedings Iron and steel institute Iron and steel institute. Publication, 102 Iron and steel institute, London,1967
Met 25.0380

LONDON 1966 Medicine and culture :a historical symposium Proceedings Wellcome institute of the history of medicine Wenner-Gren foundation for anthropological research Wellcome institute of the history of medicine. Publications.New series, 15 vi,321p Wellcome institute of the history of medicine, London,1969
WSM 43.3330

LONDON 1966 Pathology of laboratory rats and mice Nuffield foundation.Advisory committee Edited by Ernest Cotchin and Francis J.C. Roe Held at the Royal veterinary field station Blackwell,Oxford, 1967
Path 30.2438

LONDON 1966 Sample preparation techniques for liquid scintillation counting :1966 summer school Proceedings Beckman instruments limited Beckman instruments limited, Glenrothes,n.d.
Radioth 35.1778

LONDON 1966 Simple preparation techniques for liquid scintillation counting The Beckman summer school 1st Proceedings Beckman instruments London,c1966
Bioch 33.2300

LONDON 1966 Structural steelwork :a conference Proceedings British constructional steelwork association British constructional steelwork association,London, 1966
Eng 41.3024

LONDON 1966 The Teaching of ecology :a symposium British ecological society Edited by J.M. Lambert British ecological society.Symposia, 7 illus. xi,294p Blackwell scientific,Oxford,1967
Bot 42.1988

LONDON 1966 Thiamine deficiency : biochemical lesions and their clinical sihnificance Ciba foundation Edited by G.E. W. Wolstenholme and Maeve O'Connor Ciba foundation study group, 28 Churchill,London, 1967 In honour of Sir Rudolph Peters
Bioch 33.0574

LONDON 1966 World climate from 8000 to nought B.C. International symposium on world climate 8000 to nought B.C. Proceedings Royal meteorological society London,1966
Sco 14.0581

LONDON 1967 A Symposium on carbenoxolone sodium Edited by J.M. Robson and F.M. Sullivan Sponsored by Berk pharmaceuticals limited Butterworths,London,1968
Surg 23.0013

LONDON 1967 A Symposium on the interaction of drugs and subcellular components in animal cells Biological council.Co-ordinating committee for symposia on drug action Edited by P.N. Campbell Pl. viii,355p 21cm Churchill,London,1968
Pha 16.0252

LONDON 1967 A Symposium on the interaction of drugs and subcellular components in animal cells Co-ordinating committee for symposia on drug action Edited by P.N. Campbell Illus Churchill,London,1968
Radioth 35.0273

LONDON 1967 Advances in extractive metallurgy :a symposium Proceedings Institution of mining and metallurgy Institution of mining and metallurgy,London, 1968
Met 25.0494

LONDON 1967 Airborne microbes :a symposium Papers Society for general microbiology Edited by P.H. Gregory and J.L. Monteith Held at the Imperial college of science and technology Society for general microbiology. Symposia, 17 Cambridge university press, Cambridge,1967
Bioch 33.1167

LONDON 1967 Calcitonin Symposium on thyrocalcitonin and the C cells Proceedings Edited by Selwyn Taylor Heinemann,London, 1968
An 32.3778

LONDON 1967 Cell differentiation :a Ciba foundation symposium Ciba foundation Edited by A.V.S. De Reuck and Julie Knight Churchill,London,1967
An 32.2531

LONDON 1967 Cell differentiation :a Ciba foundation symposium Ciba foundation Edited by A.V.S. De Reuck and Julie Knight Churchill,London,1967
Phys 20.1511

LONDON 1967 Cell differentiation :a Ciba foundation symposium Ciba foundation Edited by A.V.S. De Reuck and Julie Knight Ciba foundation.Symposia Churchill,London, 1967
Bal 39.0317

LONDON 1967 Cell differentiation :a Ciba foundation symposium Proceedings Ciba foundation Edited by A.V.S. De Reuck and Julie Knight Churchill,London,1967
Bioch 33.0889

LONDON 1967 Chemistry and biology of mucopolysaccharides Ciba foundation Edited by G.E.W. Wolstenholme and Maeve O'Connor Ciba foundation.Symposia Churchill,London, 1958
PGMS 29.0401

LONDON 1967 Conference on prestressed concrete pressure vessels Institution of civil engineers Institution of civil engineers,London,1968
Eng 41.2891

LONDON 1967 Conference on servocomponents Sponsored by Institution of electrical engineers.Professional group on measuring and control equipment Institution of electrical engineers.Conference publication, 37 I.E.E., London,1967
Eng 41.5879

LONDON 1967 Diazepam in anaesthesia :a symposium Proceedings Edited by Peter F. Knight and C.G. Burgess Sponsored by Roche Products Wright,Bristol,1968
PGMS 29.0039

LONDON 1967 Growth of the nervous system : a Ciba foundation symposia Ciba foundation Edited by G.E.W. Wolstenholme and Maeve O'Connor Churchill,London,1968
An 32.2536

LONDON 1967 Health of mankind :a Ciba foundation symposium Ciba foundation Edited by G.E.W. Wolstenholme and Maeve O'Connor Churchill,London,1967 Ciba foundation 100th symposium
Inv Med 37.0267

LONDON 1967 Hearing mechanisms in vertebrates Ciba foundation symposium...on sensory function 4th Proceedings Ciba foundation Edited by A.V.S.de Reuck and Julie Knight J.and A.Churchill,London,1968
Psy 31.3372

LONDON 1967 Hearing mechanisms in vertebrates :a Ciba foundation symposium Ciba foundation Edited by A.V.S. De Reuck and Julie Knight Churchill,London,1968
Phys 20.1050

LONDON 1967 Interferon :a Ciba foundation symposium Ciba foundation Edited by G.E.W. Wolstenholme and Maeve O'Connor Churchill, London,1968
Phys 20.0999

LONDON 1967 Interferon a Ciba foundation symposium dedicated to Alick Isaacs Proceedings Ciba foundation Edited by G.E. W. Wolstenholme and Maeve O'Connor illus Churchill,London,1968
Bioch 33.1119

LONDON 1967 Lubrication and wear in living and artificial human joints :a symposium Institution of mechanical engineers. Lubrication and wear group and British orthopaedic association Institution of mechanical engineers.Proceedings, 181, pt.3 Institution of mechanical engineers,London, 1967
Met 25.1626

LONDON 1967 Magnetic materials and their applications :a conference Sponsored by the Institution of electrical engineers Institution of electrical engineers.Conference publication, 13 Institution of electrical engineers,London,1967
Eng 41.5348

LONDON 1967 Mathematical model building in economics and industry :a conference Papers Corporation for economic and industrial research vii,165p 23cm Griffin,London, 1968
Math S 3.1407

LONDON 1967 Mathematical model building in economics and industry :a conference Papers Organized by the Corporation for economic and industrial research Griffin,London,1968
Eng 41.1223

LONDON 1967 Moon and planets,2 Cospar plenary meeting :joint open meeting of working groups 1,2 and 5 10th Committee on space research Royal society of London Edited by A. Dollfus 196p North-Holland, Amsterdam,1968
Obs 6.3419

LONDON 1967 Natural draught cooling towers Ferrybridge and after :a conference Proceedings Institution of civil engineers Institution of civil engineers,London,1967
Eng 41.2889

LONDON 1967 Optimization of steel product yield Conference on optimization of steel product yield Proceedings Iron and steel institute Iron and steel institute. Publication, 107 London,1967 Conference held at the Institute's annual general meeting
Met 25.2561

LONDON 1967 Plant cell organelles The Phytochemical group symposium Proceedings Phytochemical group Edited by J.B. Pridham Academic press,London,1968
Bioch 33.0782

LONDON 1967 Plant cell organelles :a symposium Proceedings Phytochemical group Edited by J.B. Pridham xiii,261p Academic press,London,1968
Bot 42.4737

LONDON 1967 Radioactive isotopes in the localization of tumours International nuclear medicine symposium Proceedings Edited by N.G. Trott Heinemann medical, London,1969
Radioth 35.1397

LONDON 1967 Research into library services in higher education :papers presented at a conference Society for research into higher education Society for research into higher education,London,1968
Eng 41.7669

LONDON 1967 Solar-terrestrial physics: solar aspects Proceedings of joint I.Q.S.Y.-Cospar symposium Pt 1 International quiet sun year Committee on space research Edited by A.C. Stickland International quiet sun year.Annals, 4 414p MIT press, Cambridge,Mass.,1969
Obs 6.3341

LONDON 1967 Static electrification The Institute of physics and the Physical society conference on static electrification 2nd Proceedings Institute of physics and the Physical society.Static electrification group Institute of physics and the Physical society. Conference series, 4 25cm London,1967
Cav 7.2822

LONDON 1967 Sub-cellular components : preparation and fractionation Imperial cancer research fund Edited by G.D. Birnie 2nd edition revised,expanded Butterworth, London,1972 Based on a symposium held in 1967
Bioch 33.2275

LONDON 1967 Subcellular components; preparation and fractionation :a symposium Imperial cancer research fund Edited by G.D. Birnie and Sylvia M. Fox Butterworths,London, 1969
Bioch 33.1016

LONDON 1967 The Interaction of drugs and subcellular components in animal cells :a symposium Biological council.Co-ordinating committee for symposia on drug action Edited by P.N. Campbell Held in the Middlesex hospital medical school Churchill,London, 1968
Bioch 33.0923

LONDON 1967 The Investigation of hypothalamic-pituitary-adrenal function :a symposium Proceedings Society for endocrinology Edited by V.H.T. James and J. Landon Society for endocrinology.Memoirs, 17 Cambridge university press,Cambridge,1968
Inv Med 37.0120

LONDON 1967 The Investigation of hypothalamic-pituitary-adrenal function :a symposium Proceedings Society for endocrinology Royal society of medicine. Endocrine section Edited by V.H.T. James and J. Landon Sponsored by Ciba,Horsham Society for endocrinology.Memoirs, 17 Cambridge university press,Cambridge,1968
Gen 34.0558

LONDON 1967 The Investigation of hypothalamic-pituitary-adrenal function :a symposium Proceedings Society for endocrinology and Ciba laboratories ltd Edited by V.H.T. James and J. Landon Sponsored also by the Royal society of medicine.Endocrine section Society for endocrinology.Memoirs, 17 Cambridge university press,Cambridge,1968
Phys 20.1421

LONDON 1967 Thermal and high-strain fatigue :an international conference Proceedings Metals and metallurgy trust Organized by the Institute of metals Metals and metallurgy trust.Monograph and report series Metals and metallurgy trust,London, 1967
Eng 41.3810

LONDON 1967 Thermal and high-strain fatigue an international conference Proceedings Institute of metals and Iron and steel institute Metals and metallurgy trust.Monograph and report series, 32 Metals and metallurgy trust,London,1967
Met 25.0942

LONDON 1967 Thyroid neoplasia Imperial cancer research fund symposium 2nd Proceedings Edited by Stretton Young and D.R. Inman Academic press,New York,1968
An 32.3780

LONDON 1967 Thyroid neoplasia Imperial cancer research fund symposium 2nd Proceedings Imperial cancer research fund Edited by Stretton Young and D.R. Inman Academic press,London;New York,1968
PGMS 29.0180

LONDON 1968 Bacterial episomes and plasmids :a Ciba foundation symposium Proceedings Ciba foundation Edited by G.E. W. Wolstenholme and M. O'Connor Churchill, London,1969
Bioch 33.1118

LONDON 1968 Biology and ethics :a symposium Proceedings Institute of biology Edited by F.J. Ebling Institute of biology. Symposia, 18 xxix,145p Academic press, London;New York,1969
Bal 39.0055

LONDON 1968 Biology and ethics :a symposium Proceedings Institute of biology Edited by F.J. Ebling held at the Royal geographical society Institute of biology. Symposia, 18 Academic press,London;New York, 1969
Bal 39.0018

LONDON 1968 Chemistry of natural products The International symposium on the chemistry of natural products 5th Plenary lectures International union of pure and applied chemistry.Division of organic chemistry and Chemical society Butterworths,London,1968 "The contents of this book appear in 'Pure and applied chemistry,vol.17, nos.3-4 (196819". Added title page in French
Bioch 33.1392

LONDON 1968 Circulatory and respiratory mass transport :a Ciba foundation symposium Ciba foundation Edited by G.E.W. Wolstenholme and Julie Knight Churchill, London,1969
An 32.5222

LONDON 1968 Circulatory and respiratory mass transport :a Ciba foundation symposium Ciba foundation Edited by G.E.W. Wolstenholme and Julie Knight Churchill, London,1969
Inv Med 37.0236

LONDON 1968 Circulatory and respiratory mass transport :a Ciba foundation symposium Ciba foundation Edited by G.E.W. Wolstenholme and Julie Knight illus.,pl. x, 310p Churchill,London,1969
Phys 20.1969

LONDON 1968 Conference on coastal engineering 11th Proceedings Vol 1-2 American society of civil engineers Institution of civil engineers 2 vols American society of civil engineers,New York, 1969
Eng 41.3347

LONDON 1968 Conference on industrial measurement techniques for on-line computers Institution of electrical engineers Institution of electrical engineers.Conference publications, 43 I.E.E.,London,1968
Eng 41.2220

LONDON 1968 Corrosion and its prevention in motor vehicles a symposium Proceedings Institution of mechanical engineers.Automobile division and British joint corrosion group Institution of mechanical engineers. Proceedings, 182 pt 3J Inst.of mechanical engineers,London,1968
Met 25.1943

LONDON 1968 Developments in the pharmacology and clinical uses of human gonadotrophins :a private scientific meeting Proceedings Sponsored by Searle (G.D.) and company Searle,High Wycombe,1970
Inv Med 37.0261

LONDON 1968 Disorders of carbohydrate metabolism :a symposium Proceedings Association of clinical pathologists.Chemical pathology sub-committee Edited by G.K. McGowan and G. Walters held at the Royal society of medicine Chemical pathology in relation to clinical medicine, 5 Journal of clinical pathology.Supplement, 2 British medical association,London,1969
Bioch 33.1205

LONDON 1968 Foetal autonomy :a Ciba foundation symposium Ciba foundation Edited by G.E.W. Wolstenholme and Maeve O'Connor Churchill,London,1969
An 32.2543

LONDON 1968 Foetal autonomy :a Ciba foundation symposium Proceedings Ciba foundation Edited by G.E.W. Wolstenholme and Maeve O'Connor Churchill,London,1969
Inv Med 37.0237

LONDON 1968 I.F.A.C. symposium on fluidics Proceedings International federation of automatic control Organised by the Royal aeronautical society 1968
Eng 41.5905

LONDON 1968 Mechanisms of motor skill development Centre for advanced study in the developmental sciences study group on 'Mechanisms of motor skill development' Proceedings Ciba foundation Edited by Kevin Connolly Academic press,London;New York,1970 Being the 4th study group in a C.A.S.D.S. programme on 'The origins of human behaviour' held jointly with the Ciba foundation
Psy 31.3101

LONDON 1968 Molecular biology of viruses : a symposium Papers Society for general microbiology Edited by R.V. Crawford and M.G. P. Stoker Held at the Imperial college of science and technology Society for general microbiology.Symposia, 18 Cambridge university press,Cambridge,1968
Bioch 33.1178

LONDON 1968 Molecular biology of viruses : symposium Society for general microbiology Society for general microbiology.Symposia, 18 Cambridge university press,Cambridge,1968
Gen 34.0709

LONDON 1968 National symposium on urinary tract infection 1st Proceedings Edited by Francis O'Grady and William Brumfitt Sponsored by the Beecham research laboratories Oxford university press,London,1968
PGMS 29.0238

LONDON 1968 Obesity.Medical and scientific aspects :a symposium Proceedings Obesity association of Great Britain Edited by I. McLean Baird and Alan N. Howard Obesity association of Great Britain.Symposium, 1 Livingstone,Edinburgh;London,1969
Path 30.2446

LONDON 1968 Obesity; medical and scientific aspects :a symposium 1st Proceedings Obesity association of Great Britain Edited by I.McLean Baird and Alan N. Howard Livingstone,Edinburgh;London,1969
Inv Med 37.0010

LONDON 1968 Photo-electronic image devices symposium 4th Proceedings Edited by J. D. McGee and others Advances in electronics and electron physics, 28a 552p Academic press,London,1969
Obs 6.3413

LONDON 1968 Plant growth regulators : symposium Papers Society of chemical industry.Pesticides group Phytochemical society Society of chemical industry. Monograph, 31 Society of chemical industry, London,1968
Bioch 33.2230

LONDON 1968 Plant pathologist's pocketbook International congress of plant pathology 1st Proceedings Commonwealth mycological institute Commonwealth mycological institute, Kew,1968
BG 38.1757

LONDON 1968 Porphyrins and related compounds Biochemical society Edited by T. W. Goodwin Biochemical society.Symposia, 28 Academic press,London;New York,1968
Bioch 33.1390

LONDON 1968 Porphyrins and related compounds Biochemical society Edited by T. W. Goodwin Biochemical society.Symposia, 28 Academic press,London;New York,1968
Bot 42.1755

LONDON 1968 Porphyrins and related compounds :a symposium Biochemical society Edited by T.W. Goodwin Biochemical society. Symposia, 28 Academic press,London,1968
Chem 18.2761

LONDON 1968 Production and hazards of a hyperbaric oxygen environment :a symposium Proceedings Edited by G.S. Innes Pergamon press,Oxford,1970
Radioth 35.1400

LONDON 1968 Progress in anaesthesiology World congress of anaesthesiologists 4th Proceedings By T.B. Boulton and others Excerpta Medica,International congress series, 200 Excerpta Medica Foundation,Amsterdam, 1970
PGMS 29.0559

LONDON 1968 Science,technology and application of titanium International conference on titanium Proceedings Institute of metals American institute of mining,metallurgical and petroleum engineers American society for metals Japan institute of metals Akademiya nauk S.S.S.R. Edited by R.I. Jaffee and N.E. Promisel Pergamon, Oxford,1970
Met 25.2852

LONDON 1968 Scientific basis of drug dependence a symposium Biological council.Co-ordinating committee for symposia on drug action Edited by Hannah Steinberg J.and A. Churchill,London,1969
PGMS 29.0398

LONDON 1968 Speciation in tropical environments symposium Papers Linnean society of London British ecological society. Tropical group Linnean society of London. Biological journal, 1,no.1-2 Academic press,London,1969
Gen 34.1298

LONDON 1968 The Molecular biology of viruses :a symposium Papers Society for general microbiology Edited by L.V. Crawford and M.G.P. Stoker Society for general microbiology.Symposia, 18 Cambridge university press for the Society of general microbiology,Cambridge,1968
Path 30.2742

LONDON 1968 The Molecular biology of viruses :a symposium Proceedings Society for general microbiology Edited by L.V. Crawford and M.G.P. Stoker Society for general microbiology.Symposia, 18 Cambridge university press,Cambridge,1968
Radioth 35.0268

LONDON 1968 The Molecular biology of viruses :a symposium Society for general microbiology Edited by L.V. Crawford and M.G. P. Stoker Society for general microbiology. Symposia, 18 Cambridge university press, Cambridge,1968
An 32.5380

LONDON 1968 The Molecular biology of viruses :a symposium Society for general microbiology Society for general microbiology.Symposia, 18 Cambridge university press,Cambridge,1968
PGMS 29.0416

LONDON 1968 The Structures and functions of proteolytic enzymes a discussion Royal society of London Edited by D.C. Phillips and others Royal society of London. Philosophical transactions.Ser.B, 257,p.63-266 Royal society,London,1970
Bioch 33.1917

LONDON 1968 Urinary tract infection National symposium on urinary tract infection 1st Proceedings Edited by Francis O'Grady and William Brumfitt Oxford medical publications Oxford university press,London, 1968
Surg 23.0010

LONDON 1969 Automation and data processing in pathology :a symposium College of pathologists Edited by T.P. Whitehead Journal of clinical pathology,London,1969
HE 27.0161

LONDON 1969 Bacterial endocarditis : national symposium Proceedings Beecham research laboratories Edited by P.B. Beeson and M. Ridley London,c1970
PGMS 29.0600

LONDON 1969 British biochemistry past and present Essays Biochemical society Edited by T.W. Goodwin Biochemical society. Symposium, 30 Academic press,London,1970
Bioch 33.2314

LONDON 1969 Catapres in hypertension :a symposium Proceedings Royal college of surgeons Edited by Matthew E. Conolly Butterworths,London,1970
PGMS 29.0608

LONDON 1969 Conference on steel in architecture Proceedings Organized by the British constructional steelwork association British constructional steelwork association, London,1970
Eng 41.0269

LONDON 1969 Conference on switching techniques for telecommunications networks Sponsored by the Institution of electrical engineers Institution of electrical engineers.Conference publications, 52 Institution of electrical engineers,London, 1969
Math L 5.3531

LONDON 1969 Control of organelle development :a symposium Proceedings Society for experimental biology Edited by P. L. Miller Society for experimental biology. Symposia, 24 Cambridge university press, Cambridge,1970
Gen 34.2205

LONDON 1969 Control of organelle development :a symposium Society for experimental biology Society for experimental biology.Symposia, 24 Cambridge university press,Cambridge,1970
Bioch 33.2340

LONDON 1969 Domiciliary care of the patient with cancer :symposium Marie Curie foundation fund Edited by R.W. Raven Heinemann,London,1970
PGMS 29.0561

LONDON 1969 Foetal autonomy :a Ciba foundation symposium Ciba foundation Edited by G.E.W. Wolstenholme and Maeve O'Connor Churchill,London,1969
Phys 20.2262

LONDON 1969 Gas chromotography in biology and medicine :a Ciba foundation symposium Ciba foundation Edited by Ruth Porter Churchill,London,1969
Phys 20.2256

LONDON 1969 Global circulation of the atmosphere :a conference Proceedings Royal meteorological society American meteorological society Edited by G.A. Corby bibliog.,illus.,port. 257p 26cm Royal meteorological society,London,1969
A Math 4.1651

LONDON 1969 Hering-Breur centenary symposium Proceedings Ciba foundation Edited by Ruth Porter Ciba foundation. Symposia Churchill,London,1970
Inv Med 37.0184

LONDON 1969 Homeostatic regulators :a Ciba foundation symposium Proceedings Ciba foundation Wellcome trust Edited by G.E.W. Wolstenholme and Julie Knight Churchill, London,1969
Bioch 33.0934

LONDON 1969 Homeostatic regulators :a Ciba foundation symposium held jointly with the Wellcome trust Ciba foundation and Wellcome trust Edited by G.E.W. Wolstenholme and Julie Knight Churchill,London,1969
Phys 20.2257

LONDON 1969 International conference on tribology in iron and steel works 1st Proceedings Iron and steel institute Institution of mechanical engineers Iron and steel institute.Publication, 125 Iron and steel institute,London,1970
Met 25.2562

LONDON 1969 Microbial growth Papers Society for general microbiology Society for general microbiology.Symposia, 19 Cambridge university press,Cambridge,1969
Path 30.2597

LONDON 1969 Microbial growth :a symposium Society for general microbiology Edited by Pauline M. Meadow and S.J. Pirt Society for general microbiology.Symposia, 19 Cambridge university press,Cambridge,1969
An 32.5379

LONDON 1969 Microbial growth :symposium Society for general microbiology Society for general microbiology.Symposia, 19 Cambridge university press,Cambridge,1969
Gen 34.0710

LONDON 1969 Mutation as cellular process : a Ciba foundation symposium Ciba foundation Edited by G.E.W. Wolstenholme and M. O'Connor illus Chuchill,London,1969
Bioch 33.0891

LONDON 1969 Road vehicle aerodynamics Symposium on road vehicle aerodynamics 1st Proceedings Edited by A.J. Scibor-Rylski Organised by the City university.Department of aeronautics City university,London,1970 Organised in cooperation with the National physical laboratory,the Motor industry research association and the Institution of mechanical engineers
Eng 41.8336

LONDON 1969 Symposium on calcium and cellular function Essays Biological council.Co-ordinating committee for symposia on drug action Edited by A.W. Cuthbert Macmillan,London,1969
Bioch 33.2264

LONDON 1969 Symposium on variation in mammalian populations Proceedings Zoological society of London Edited by Robert J. Berry Zoological society of London. Symposia, 26 Academic press,London,1970
Gen 34.2196

LONDON 1969 The Commonwealth mining and metallurgical congress 9th Proceedings Vol 4: physical and fabrication metallurgy Commonwealth council of mining and metallurgical institutions Edited by M.J. Jones Institution of mining and metallurgy, London,1970
Met 25.2847

LONDON 1969 The Frozen cell :a Ciba foundation symposium Ciba foundation Edited by G.E.W. Wolstenholme and Maeve O'Connor Churchill,London,1970
An 32.5444

LONDON 1969 The Frozen cell :a Ciba foundation symposium Proceedings Ciba foundation Edited by G.E.W. Wolstenholme and Maeve O'Connor illus. ix,316p Churchill, London,1970
Bot 42.1383

LONDON 1969 The Optimum population for Britain :symposium held at the Royal geographical society Proceedings Edited by L.R. Taylor Institute of biology.Symposia, 19 Academic press,London,1970
Gen 34.2202

LONDON 1969 Water requirements for kidney dialysis Organised by the Elga group of companies Elga Publications,Lane End,Bucks, 1969
PGMS 29.0239

LONDON 1970 A Discussion on solar studies with special reference to space observations British national committee on space research Philosophical transactions, 270A,1202 195p royal society of London,London,1971
Obs 6.3606

LONDON 1970 British computer society. CODASYL data base group illus 59p British computer society,London,1971
Math L 5.3861

LONDON 1970 Conference in mathematical logic Proceedings Edited by Wilfrid Hodges Held at Bedford college Lecture notes in mathematics, 255 vii,351p 25cm Springer, Berlin,1972
P Math 2.4059

LONDON 1970 Conference on optimisation techniques in circuit and control applications Institution of electrical engineers Institution of electrical engineers.Conference publication, 66 IEE,London,1970 Organised in association with the Institute of electrical and electronics engineers and the Institute of physics and the Physical society
Eng 41.8587

LONDON 1970 Hormones and the immune response :Ciba foundation study group Ciba foundation Edited by G.E.W. Wolstenholme and Julie Knight Ciba foundation study group, 36 Churchill,London,1970
An 32.5453

LONDON 1970 Molecular properties of drug receptors :a Ciba foundation symposium Ciba foundation Edited by Ruth Pater and Maeve O'Connor Churchill,London,1970
An 32.5477

LONDON 1970 New research in plant anatomy Edited by N.K.B. Robson and D.F. Cutler Linnean society of London.Botany.Supplement, 63,1 Academic press,London,1970 Symposium arranged by the Plant anatomy group of the Linnean society
Bot 42.1118

LONDON 1970 Occupational health for the undergraduate medical student Society of occupational medicine Edited by John R. Glover Society of occupational medicine, London,c1971
PGMS 29.0667

LONDON 1970 Optical studies of adsorbed layers at interfaces Faraday society Faraday society.Symposia,1970,4 212p 26cm Faraday society,London,1971
Chem 18.2880

LONDON 1970 Organization and control in prokaryotic and eukaryotic cells :a symposium Society for general microbiology Edited by H. P. Charles and B.C.J.G. Knight Society for general microbiology.Symposia, 20 Cambridge university press,Cambridge,1970
An 32.5447

LONDON 1970 Studies in modern fabrics :a conference Papers Textile institute Textile institute.Annual conference, 55 Textile institute,Manchester,1970
Eng 41.8287

LONDON 1970 Symposium on the adverse effects of drugs 1st Proceedings Edited by Gillian C. Hanson Sponsored by Beecham research laboratories Beecham research laboratories,London,1971
PGMS 29.0673

LONDON 1970 The Biochemistry of steroid hormone action Biochemical society Edited by R.M.S. Smellie Biochemical society. Symposium, 32 Academic press,London,1971
Bioch 33.2316

LONDON 1970 The Family and its future :a Ciba foundation symposium Proceedings Edited by Katherine Elliott Churchill,London, 1970
Inv Med 37.0075

LONDON 1970 The Modern design of wind-sensitive structures :a seminar Proceedings Cement industry research and information association CIRIA,London,1971
Eng 41.8276

LONDON 1970 The Pineal gland :a Ciba foundation symposium Ciba foundation Edited by G.E.W. Wolstenholme and Julie Knight Churchill;Livingstone,Edinburgh;London,1971
An 32.5456

LONDON 1970 The Social impact of modern biology :an international conference Papers and discussions British society for social responsibility in science Edited by Watson Fuller 256p Routledge and Kegan Paul, London,1971
WSM 43.0589

LONDON 1971 Air pollution control in transport engines :a symposium Proceedings Institution of mechanical engineers.Automobile division Advanced school engineers. Combustion engines group IME,London,1972
Eng 41.8645

LONDON 1971 BCS October 71 conference on April 71 report Collected papers British computer society.CODASYL data base task group illus 221p British computer society,London, 1972
Math L 5.3865

LONDON 1971 Biological oxydation of nitrogen in organic molecules :symposium Proceedings Chelsea college Edited by J.W. Bridges and others Taylor and Francis,London, 1972
Bioch 33.2286

LONDON 1971 Cell interactions Lepetit colloquium on biology and medicine 3rd Proceedings Edited by Luigi G. Silvestri North-Holland,Amsterdam,1972
Bioch 33.2267

LONDON 1971 Ciba foundation symposium on carbon fluorine compounds :chemistry biochemistry and biological Ciba foundation Edited by Katherine Elliott and John Birch Elsevier,Amsterdam,1972
Bioch 33.2244

LONDON 1971 Ciba foundation symposium on growth control in cell culture Papers Ciba foundation Edited by G.E.W. Wolstenholme and Julie Knight Churchill;Livingstone,Edinburgh, 1971
Bioch 33.2263

LONDON 1971 Externally pressurized bearings :a joint conference Proceedings Institution of mechanical engineers.Tribology group Institution of production engineers IME,London,1972
Eng 41.8651

LONDON 1971 Heat and mass transfer by combined forced and natural convection :a symposium Proceedings Institution of mechanical engineers.Thermodynamics and fluid mechanics group IME,London,1972
Eng 41.8647

LONDON 1971 International joint conference on artificial intelligence 2nd Advance papers British computer society illus 658p British computer society,London,1971
Math L 5.3848

LONDON 1971 Mathematics of contemporary physics :instructional conference Proceedings London mathematical society North Atlantic treaty organization Edited by R.F. Streater xi,274p 24cm Academic press,London,1972
A Math 4.1864

LONDON 1971 Microbes and biological productivity :a symposium Papers Society for general microbiology Edited by David E. Hughes and Anthony H. Rose Society for general microbiology.Symposia, 21 Cambridge university press,Cambridge,1971
Bioch 33.2302

LONDON 1971 Plasma lipoproteins Biochemical society Edited by R.M.S. Smellie Biochemical society.Symposium, 33 Academic press,London,1971
Bioch 33.2317

LONDON 1971 Practical application of fracture mechanics to pressure-vessel technology :a conference Proceedings Institution of mechanical engineers.Applied mechanics group IME,London,1971
Eng 41.8644

LONDON 1971 The Babbage memorial meeting British computer society Royal statistical society illus 40p British computer society,London,1971
Math L 5.3612

LONDON 1971 The Babbage memorial meeting Report of the proceedings Royal statistical society British computer society iii,39p 30cm British computer society,London,1971
Math S 3.1858

LONDON 1971 Vibration and noise in motor vehicles :a symposium Proceedings Institution of emchanical engineers.Automobile division Advanced school of automotive studies IME,London,1972
Eng 41.8646

LONDON 1972 Ciba foundation symposium on pathogenic mycoplasma Proceeings Ciba foundation Edited by Katherine Elliott and Joan Birch Associated scientific publishers, Amsterdam,1972
Bioch 33.2296

LONDON 1972 Minicomputers in instrumentation and control :proceedings of a three-day short course Polytechnic of North London Edited by Y. Paker and others Polytechnic of North London,London,1972
Eng 41.8589

LONDON 1972 Passenger environment :a conference Proceedings Institution of mechanical engineers.Railway division IME, London,1972
Eng 41.8649

LONDON 1972 Periodic inspection of pressure vessels :a conference Proceedings Institution of mechanical engineers.Applied mechanics group IME,London,1972
Eng 41.8648

LONDON;CAMBRIDGE;OXFORD 1946 The Empire scientific conference Reports Vol 1-2 Royal society of London 2 vols Royal society,London,1948
Gen 34.0053

LONDON,ONT. 1965 Systems and computer science Conference on systems and computer science :held at the University of Western Ontario Proceedings Edited by John F. Hart and Satoru Takasu Bibliog.,Illus. xiv,249p 25cm Toronto university press,Toronto,1967
P Math 2.3494

LONDON,ONTARIO 1951 Auroral physics Conference on auroral physics 1st Proceedings Edited by N.C. Gersen and others Sponsored jointly by the University of Western Ontario.Department of physics. Air force Cambridge research center.Geophysical research papers,30 illus xxvi,462p Cambridge,Mass. 1954
Sco 14.0613

LONDON MATHEMATICAL SOCIETY Algebraic number theory conference proceedings Brighton 1965 Sep 1-17 Edited by John W.S. Cassels and A. Frohlich With the support of the International mathematical union xviii,366p 24cm Academic press,New York;London,1967
P Math 2.2773

LONDON MATHEMATICAL SOCIETY Finite simple groups :an instructional conference Proceedings Oxford 1969 Sep Edited by M. B. Powell and G. Higman xi,327p 24cm Academic press,London;New York,1971
P Math 2.4431

LONDON MATHEMATICAL SOCIETY Mathematics of contemporary physics :instructional conference Proceedings London 1971 Aug 23-Sep 11 Edited by R.F. Streater xi,274p 24cm Academic press,London,1972
A Math 4.1864

LONDON SCHOOL OF ECONOMICS AND POLITICAL SCIENCE Problems in the philosophy of mathematics International colloquium in the philosophy of science Proceedings London 1965 Jul 11-17 Vol 1 Edited by Imre Lakatos Under the auspices of the International union of history and philosophy of science.Division of logic,methodology and philosophy of science Studies in logic and the foundations of mathematics xv,241p North-Holland, Amsterdam,1967
WSM 43.1813

LONDON SCHOOL OF ECONOMICS AND POLITICAL SCIENCE
Problems in the philosophy of science International colloquium in the philosophy of science Proceedings London 1965 Jul 11-17 Vol 3 Edited by Imre Lakatos and Alan Musgrave Under the auspices of the International union of history and philosophy of science.Division of logic,methodology and philosophy of science Studies in logic and the foundations of mathematics ix,448p North-Holland,Amsterdam,1968
WSM 43.0931

LONDON SCHOOL OF HYGIENE AND TROPICAL MEDICINE
Biological oxidation of aromatic rings :a symposium London 1949 Nov 12 Biochemical society Edited by R.T. Williams Biochemical society.Symposia, 5 Cambridge university press,Cambridge,1950
Bioch 33.1367

LONDON SCHOOL OF HYGIENE AND TROPICAL MEDICINE
Biological transformations of starch and cellulose :a symposium London 1953 Feb 21 Biochemical society Biochemical society. Symposia, 11 Cambridge university press, Cambridge,1953
Bioch 33.1373

LONDON SCHOOL OF HYGIENE AND TROPICAL MEDICINE
Immunochemistry :a symposium London 1952 Nov 15 Biochemical society Edited by R.T. Williams Biochemical society.Symposia, 10 Cambridge university press,Cambridge,1953
Bioch 33.1372

LONDON SCHOOL OF HYGIENE AND TROPICAL MEDICINE
Lipid metabolism :a symposium London 1952 Feb 16 Biochemical society Edited by R.T. Williams Biochemical society.Symposia, 9 Cambridge university press,Cambridge,1952
Bioch 33.1371

LONDON SCHOOL OF HYGIENE AND TROPICAL MEDICINE
Metabolism and functions in nervous tissue :a symposium London 1951 Nov 10 Biochemical society Edited by R.T. Williams Biochemical society.Symposia, 8 Cambridge university press,Cambridge,1952
Bioch 33.1370

LONDON SCHOOL OF HYGIENE AND TROPICAL MEDICINE
Partition chromatography :a symposium London 1948 Oct 30 Biochemical society Edited by R.T. Williams and R.L.M. Synge Biochemical society.Symposia, 3 Cambridge university press,Cambridge,1949
Bioch 33.1365

LONDON SCHOOL OF HYGIENE AND TROPICAL MEDICINE
The Biochemistry of fertilization and the gametes :a symposium London 1951 Feb 17 Biochemical society Edited by R.T. Williams Biochemical society.Symposia, 7 Cambridge university press,Cambridge,1951
Bal 39.0388

LONDON SCHOOL OF HYGIENE AND TROPICAL MEDICINE
The Biochemistry of fertilization and the gametes :a symposium London 1951 Feb 17 Biochemical society Edited by R.T. Williams Biochemical society.Symposia, 7 Cambridge university press,Cambridge,1951
Bioch 33.1369

LONDON SCHOOL OF HYGIENE AND TROPICAL MEDICINE
The Biochemistry of vitamin B12 :a symposium London 1955 Feb 19 Biochemical society Edited by R.T. Williams Biochemical society. Symposia, 13 Cambridge university press, Cambridge,1955
Bioch 33.1375

LONDON SCHOOL OF HYGIENE AND TROPICAL MEDICINE
The Chemical pathology of animal pigments :a symposium London 1954 Feb 20 Biochemical society Edited by R.T. Williams Biochemical society.Symposia, 12 Cambridge university press,Cambridge,1954
Bioch 33.1374

LONDON SCHOOL OF HYGIENE AND TROPICAL MEDICINE
The Structure of nucleic acids and their role in protein synthesis :a symposium London 1956 Feb 18 Biochemical society Edited by E.M. Crook Biochemical society.Symposia, 14 Cambridge university press,Cambridge,1957
Bioch 33.1376

LONDON SYMPOSIUM ON INFORMATION THEORY 4th
papers London 1960 Aug 29 - Sep 2 Seven typescript papers
Math L 5.0858

LONG ASTON RESEARCH STATION Nitrogen metabolism in plants Recent aspects of nitrogen metabolism in plants :a symposium Proceedings Bristol 1967 Apr 18-19 Edited by E.J. Hewitt and C.V. Cutting Long Ashton research station.Symposia, 1 Academic press,London;New York,1968
Bot 42.1893

LONG BEACH,CALIF. 1962 Stability and distortion of visual space International congress on human factors in electronics : visual distortion from the engineering point of view Papers By Richard L. Gregory University of Cambridge,Cambridge,1962
Psy 31.0382

LONG BEACH,CALIF. 1966 World orchid conference 5th Proceedings Edited by Lloyd R.de Garmo Fifth world orchid conference,Long beach,Calif.,1966
BG 38.3120

LONG ISLAND BIOLOGICAL ASSOCIATION.BIOLOGICAL LABORATORY Cold Spring Harbor symposia on quantitative biology Papers Cold Spring Harbor,L.I.,N.Y. 1935 Summer Vol 3 Long Island biological association,Cold Spring Harbor,L.I.,N.Y.,1935
Chem 18.2002

LOS ANGELES n.d. Computer graphics:utility, production,art Graphics symposium University of California Edited by Fred Gruenberger Jointly sponsored by Informatics incorporated 225p 22cm Academic press, London,1967
Math L 5.3273

LOS ANGELES 1941 Alloy constructional steels By Herbert J. French American society for metals,Cleveland,Ohio,1942 With a section on corrosion in collaboration with Francis L.Laque.A series of lectures presented at the 4th Western metal congress and exposition
Met 25.0480

LOS ANGELES 1949 Monte Carlo method :a symposium Proceedings National bureau of standards Edited by A.S. Householder and others Sponsored by the Rand corporation National bureau of standards.Applied mathematics series, 12 United States government printing office,Washington,D.C., 1951
Radioth 35.0586

LOS ANGELES 1951 Simultaneous linear equations and the determination of Eigenvalues symposium Proceedings National bureau of standards Edited by L.J. Paige and Olga Taussky jointly sponsored by United States. Office of naval research United States. National bureau of standards. Applied mathematics series, 29 bibliog. 125p 26cm National bureau of standards,Washington, D.C.,1953
Math L 5.0131

LOS ANGELES 1953 Zirconium and zirconium alloys American society for metals American society for metals,Cleveland,Ohio, 1953 A symposium on zirconium and zirconium alloys presented to members of the A.S.M. during the eighth Western metal congress and exposition,Los Angeles, march 23 to 27, 1953
Met 25.0556

LOS ANGELES 1956 Metals :presented at the 2nd Pacific area national meeting American society for testing materials A.S.T.M. Special technical publication, 196 American society for testing materials,Philadelphia,Pa., 1957
Met 25.1036

LOS ANGELES 1956 Symposium on nondestructive testing presented to the 2nd Pacific area national meeting American society for testing materials A.S.T.M. Special technical publication, 213 A.S.T.M., Philadelphia,1957
Met 25.1323

LOS ANGELES 1956 Symposium on radioisotopes presented at the second Pacific area national meeting American society for testing materials A.S.T.M.Special technical publication, 215 A.S.T.M.,Philadelphia,1958
Met 25.1559

LOS ANGELES 1956 Symposium on titanium held at the second Pacific area national meeting American society for testing materials A.S.T.M.Special technical publication, 204 American society for testing materials,Philadelphia,Pa.,1957
Met 25.0571

LOS ANGELES 1957 New computers - a report from the manufacturers symposium Sponsored by Association for computing machinery illus. 132p 28cm Association for computing machinery,Los Angeles,1957
Math L 5.0792

LOS ANGELES 1957 Short-time high-temperature testing Short-time elevated temperature testing of metals :a symposium Proceedings American society for metals American society for metals,Cleveland,Ohio, 1958
Met 25.0983

LOS ANGELES 1958 I.R.E. western electronic show and convention convention record Institute of radio engineers 291p Institute of radio engineers,New York,1958
Math L 5.2070

LOS ANGELES 1958 Let the computer decide : a symposium papers Association for computing machinery.Los Angeles chapter Association for computing machinery,Los Angeles,1958
Math L 5.2373

LOS ANGELES 1959 International symposium on circuit and information theory transactions Institute of radio engineers Co-sponsored by the International scientific radio union Institute of radio engineers. Transactions on information theory,IT, 5, special supplement xiv,298p Institute of radio engineers,New York,1959 Papers published in advance of the Symposium
Math L 5.2071

LOS ANGELES 1959 Kinetics,equilibria and performance of high temperature systems :a conference 1st Proceedings Combustion institute.Western states section Edited by Gilbert S. Bahn and Edward E. Zukoski x,255p Combustion institute,Los Angeles,1960
Chem E 24.1394

LOS ANGELES 1960 I.R.E. Western electronic show and convention convention record Part 4: computers,man-machine systems Institute of radio engineers 202p Institute of radio engineers,New York,1960
Math L 5.2076

LOS ANGELES 1961 Association for computing machinery :national meeting 16th Proceedings 22cm ACM,New York,1961
Math L 5.3452

LOS ANGELES 1961 Heat transfer and fluid mechanics institute 1961 proceedings Edited by R.C. Binder and others Stanford university press,Stanford,Calif.,1961
A Math 4.0519

LOS ANGELES 1962 Brain function conference 2nd Proceedings Vol 2: RNA and brain function memory and learning Brain research institute and American institute of biological sciences Edited by Mary A.B. Brazier With the support of the United States.Air force.Office of scientific research U.C.L.A. forum in medical sciences, 2 University of California press,Berkeley;Los Angeles,1964
Psy 31.0246

LOS ANGELES 1962 Symposium on fatigue tests of aircraft structure:low-cycle,full-scale, and helicopters American society for testing materials A.S.T.M.Special technical publication, 338 A.S.T.M.,Philadelphia,Pa., 1963
Met 25.0935

LOS ANGELES 1963 Brain and behavior Brain function conference 3rd Proceedings Vol 3: brain and gonadal function Brain research institute American institute of biological sciences Edited by Roger A. Gorski and Richard E. Whalen With the support of the Human ecology fund UCLA forum in medical sciences, 3 University of California press,Berkeley;Los Angeles,1966
Psy 31.0621

LOS ANGELES 1963 Technical conference on refractory metals and alloys 3rd Proceedings American institute of mining, metallurgical and petroleum engineers. Refractory metals committee Metallurgical society conferences, 30 Gordon and Breach, New York,1966
Met 25.1017

LOS ANGELES 1964 Computing methods in optimization problems :a conference Proceedings Edited by A.V. Balakrishnan and Lucien W. Neustadt illus x,327p 24cm Academic press,New York;London,1964
Math 3.1205

LOS ANGELES 1967 Applications related phenomena in titanium alloys :a symposium American society for testing materials A.S.T. M.Special technical publication, 432 A.S.T.M. Philadelphia,Pa.,1968
Met 25.0609

LOS ANGELES 1967 Computers and communications - toward a computer utility Computer communications symposium University of California Informatics incorporated Edited by Fred Gruenberger 219p 23cm Prentice-Hall,Englewood Cliffs,N.J.,1968
Math L 5.3117

LOS ANGELES 1967 Energy bands in metals and alloys based on a symposium American institute of mining,metallurgical and petroleum engineers.Committee on alloy phases Edited by L.H. Bennett and J.T. Waber Metallurgical society conferences, 45 Gordon and Breach,New York,1968
Met 25.1060

LOS ANGELES 1969 Inequalities 3 Symposium on inequalities 3rd Proceedings University of California,Los Angeles Edited by Oved Shisha xxii,368p 24cm Academic press,New York;London,1972
P Math 2.4214

LOS ANGELES,CALIF 1966 Computer graphics: utility,production,art The Graphic symposium 3rd Papers University of California,Los Angeles and Informatics inc. Edited by Fred Gruenberger 225p Thompson book co.; Academic press,Washingtom,D.C.;London,1967
TA 15.0098

LOS ANGELES,CALIF. 1956 Cement and concrete Technical papers American society for testing materials American society for testing materials.Special technical publication, 205 American society for testing materials,Philadelphia,Pa.,1958
Eng 41.2845

LOS ANGELES,CALIF. 1956 Designs and tests of building structures .Seismic and shock loading...Glued laminated and other constructions American society for testing materials American society for testing materials.Special technial publication, 209 American society for testing materials, Philadelphia,Pa.,1957
Eng 41.3017

LOS ANGELES,CALIF. 1956 Fatigue of aircraft structures American society for testing materials American society for testing materials.Special technical publication, 203 American society for testing materials,Philadelphia,Pa.,1957
Eng 41.2776

LOS ANGELES,CALIF. 1956 Rock deformation University of California,Los Angeles.Institute of geophysics :annual technical conference 1956 Symposium Edited by David Griggs and John Handin Geological society of America. Memoirs,79 New York,1960
Min 10.0348

LOS ANGELES,CALIF. 1956 Symposium on full-scale tests on house structures American society for testing materials American society for testing materials.Special technical publication, 210 American society for testing materials,Philadelphia,Pa.,1956
Eng 41.3018

LOS ANGELES,CALIF. 1956 Symposium on titanium :presented at the second Pacific area national meeting American society for testing materials American society for testing materials.Special technical publication, 204 American society for testing materials,Philadelphia,Pa.,1957
Eng 41.3598

LOS ANGELES,CALIF. 1956,1957 Radiation effects on materials :a symposium Vol 1-2 American society for testing materials and Atomic industrial forum A.S.T.M. Special technical publication, 208,220 2 vols American society for testing materials, Philadelphia,Pa.,1957
Met 25.1571

LOS ANGELES,CALIF. 1957 Advanced propulsion systems symposium Proceedings Edited by Morton Alperin and George P. Sutton International series on aeronautical sciences and space flight.Division 9:symposia, 2 Pergamon press,London,1959
Eng 41.7370

LOS ANGELES,CALIF. 1958 Lectures and laboratory exercises on strain gage techniques By William M. Murray and Peter K. Stein University of California Los Angeles Society for experimental stress analysis University of California Los Angeles,Los Angeles,Calif., 1958
Eng 41.2906

LOS ANGELES,CALIF. 1960 Vistas in astronautics Astronautics symposium 3rd Proceedings United States.Air force.Office of scientific research jointly sponsored Society of automative engineers Vistas in astronautics, 3 illus 266p Society of automotive engineers,New York,1960
Nap 11.0024

LOS ANGELES,CALIF. 1961 Brain and behavior, vol.1 Conference on brain and behavior 1st Proceedings Vol 1 Edited by Mary A.B. Brazier American institute of biological sciences,Washington,D.C.,1961
An 32.4382

LOS ANGELES,CALIF. 1961 Brain function : conference 1st Proceedings Vol 1: cortical excitability and steady potentials; relations of basic research to space biology Brain research institute and American institute of biological sciences Edited by Mary A.B. Brazier U.C.L.A. forum in medical sciences, 1 University of California press, Berkeley;Los Angeles,1963
Psy 31.0255

LOS ANGELES,CALIF. 1961 Metallurgy of semiconductor materials a technical conference American institute of mining,metallurgical and petroleum engineers.Semiconductors committee Edited by John B. Schroeder Metallurgical society conferences, 15 Interscience,New York;London,1962
Met 25.1158

LOS ANGELES,CALIF. 1962 Brain and behavior, vol.2:the internal environment and alimentary behavior Conference on brain and behavior 2nd Proceedings Edited by Mary A.B. Brazier American institute of biological sciences,Washington,D.C.,1963
An 32.4383

LOS ANGELES,CALIF. 1962 Symposium on properties of surfaces :presented at the fourth Pacific area national meeting American society for testing materials American society for testing materials.Special technical publication, 340 Philadelphia,1963
Met 25.2829

LOS ANGELES,CALIF. 1962 Symposium on radiation effects on metals and neutron dosimetry American society for testing materials A.S.T.M.Special technical publication, 341 American society for testing and materials,Philadelphia,Pa.,1963
Met 25.1585

LOS ANGELES,CALIF. 1963 Brain function :a conference 3rd Proceedings Vol 3: speech,language and communication Brain research institute Edited by Edward C. Carterette With the support of the United States.Air force.Office of scientific research UCLA forum in medical sciences, 4 University of California press,Berkeley;Los Angeles,1966
Psy 31.0458

LOS ANGELES,CALIF. 1964 Computing problems in optimization problems :a conference Proceedings Edited by A.V. Balakrishnan Academic press,New York,1964
Eng 41.5857

LOS ANGELES,CALIF. 1965 On-line computing systems symposium proceedings University of California Edited by Eric Burgess Jointly sponsored by Informatics incorporated Data processing library series 152p 28cm American data processing,Detroit,Mich.,1965
Math L 5.1031

LOS ANGELES,CALIF. 1965 The Heat transfer and fluid mechanics institute 1965 proceedings Edited by Andrew F. Charwat Stanford university press,Stanford,Calif.,1965
A Math 4.0523

LOS ANGELES,CALIF. 1967 Axiomatic set theory Symposium in pure mathematics 14th Proceedings Edited by Dana S. Scott Held at the University of California Los Angeles American mathematical society.Proceedings of symposia in pure mathematics, 13,pt 1 v, 474p 26cm American mathematical society, Providence,R.I.,1971
P Math 2.4061

LOS ANGELES,CALIF. 1967 Computer graphics: utility,production,art Graphics symposium University of California Edited by Fred Gruenberger Jointly sponsored by Informatics incorporated Academic press,London,1967
Eng 41.2210

LOS ANGELES,CALIF. 1967 Mathematical theory of control :a conference Proceedings Edited by A.V. Balakrishnan and Lucian W. Neustadt bibliog. xv,459p 24cm Academic press,London,1967
Math S 3.1402

LOS ANGELES,CALIF. 1967 Mathematical theory of control :a conference Proceedings Edited by A.V. Balakrishnan and Lucien W. Neustadt Academic press,New York;London,1967
Eng 41.5763

LOS ANGELES,CALIF. 1967 The Interneuron :a conference Proceedings Edited by Mary A.B. Brazier Sponsored by the Brain research institute UCLA forum in medical sciences, 11 University of California press,Berkeley; Los Angeles,Calif.,1969
An 32.5473

LOS ANGELES,CALIF. 1968 Combinatorics American mathematical society Edited by Theodore S. Motzkin American mathematical society.Proceedings of symposia in pure mathematics, 19 viii,255p 26cm AMS, Providence,R.I.,1971
P Math 2.4513

LOS ANGELES,CALIF. 1970 Invariant imbedding :summer workshop Proceedings Edited by R.E. Bellman and E.D. Denman Held at University of southern California Lecture notes in operations research and mathematical systems, 52 148p 25cm Springer,Berlin, 1971
P Math 2.4296

LOS BANOS 1963 Symposium on rice genetics and cytogenetics Proceedings Sponsored by the International rice research institute Elsevier,Amsterdam,1964
Gen 34.1559

LOUGHBOROUGH 1966 Aspects of educational technology :the proceedings of the programmed learning conference Proceedings Association for programmed learning and Leicestershire programmed learning group Edited by Derick Unwin and John Leedham Methuen,London,1967
Psy 28.0233

LOUGHBOROUGH 1968 Wind effects on buildings and structures :a symposium Proceedings Vol 1-2 Loughborough university of technology Edited by D.J. Johns 2 vols Loughborough,1968
Eng 41.2762

LOUGHBOROUGH UNIVERSITY OF TECHNOLOGY Wind effects on buildings and structures :a symposium Proceedings Loughborough 1968 Apr 2-4 Vol 1-2 Edited by D.J. Johns 2 vols Loughborough,1968
Eng 41.2762

LOUISIANA CONFERENCE ON COMBINATORICS,GRAPH THEORY AND COMPUTING Proceedings Baton Rouge,La. 1970 Mar 1-5 Edited by R.C. Mullin and others Held at Louisiana state university vi,464p 25cm Louisiana state university,Baton Rouge,La.,1970
P Math 2.4503

LOUISIANA STATE DEPARTMENT OF HOSPITALS
Serological fractions in schizophrenia :a research symposium New Orleans 1961 Jun 13-14 Edited by Robert G. Heath Harper and Row for the Commonwealth fund,New York,1963
PGMS 29.0324

LOUISIANA STATE UNIVERSITY Louisiana conference on combinatorics,graph theory and computing Proceedings Baton Rouge,La. 1970 Mar 1-5 Edited by R.C. Mullin and others vi,464p 25cm Louisiana state university,Baton Rouge,La.,1970
P Math 2.4503

LOUISIANA STATE UNIVERSITY Serological fractions in schizophrenia :a research symposium New Orleans 1961 Jun 13-14 Edited by Robert G. Heath Harper and Row for the Commonwealth fund,New York,1963
PGMS 29.0324

LOUISIANA STATE UNIVERSITY 1959 Coastal geography conference 2nd By Richard J. Russell Sponsored by the Office of naval research.Geography branch,Washington,D.C.,1959
Geog 13.0712

LOUISVILLE,KY. 1962 Algae and man Edited by Daniel F. Jackson illus. x,434p Plenum press,New York,1964 Based on lectures presented at the NATO advanced study institute
Bot 42.3911

LOUTRAKI 1966 International summer course on probabilistic methods in analysis Loutraki,1966 Mimeograph summaries of lectures
Math 3.1116

LOUTRAKI,GREECE 1966 Probability methods in analysis symposium lectures Edited by D. A. Kappos Supported by North Atlantic treaty organization.Scientific affairs division Lecture notes in mathematics, 31 329p 28cm Springer-Verlag,Berlin,1967
P. Math 2.2425

LOUTRAKI,GREECE 1966 Probability methods in anlysis symposium lectures North Atlantic treaty organization.Scientific affairs division Edited by D.A. Kappos Lecture notes in mathematics, 31 329p 28cm Springer-verlag,Berlin,1967
Math 3.0711

LOUVAIN 1953 Symposium de l'Association de psychologie scientifique de langue francaise : la perception 2eme Textes et communications By Henri Pieron and others Association de psychologie scientifique de langue francaise Presses universitaires de France,Paris,1955
Psy 31.0887

LOUVAIN 1960 Biological approaches to cancer chemotherapy :a symposium Edited by R. J.C. Hams Under the auspices of Unesco Academic press,London;New York,1961
Pha 16.0337

LOUVAIN 1960 Biological approaches to cancer chemotherapy :a symposium Edited by R. J.C. Harris Under the auspices of Unesco Academic press,London;New York,1960
Radioth 35.0238

LOUVAIN 1960 Biological approaches to cancer chemotherapy :a symposium Edited by R. J.C. Harris Under the auspices of Unesco and the World health organization Academic press, London;New York,1961
Bioch 33.0960

LOUVAIN 1962 Methods of separation of subcellular structural components :a symposium Biochemical society Edited by J.K. Grant Held jointly with the Societe belge de biochimie Biochemical society.Symposia, 23 Cambridge university press,Cambridge,1963
Radioth 35.0136

LOUVAIN 1962 Methods of separation of subcellular structural components :a symposium Biochemical society and Societe belge de biochimie Edited by J.K. Grant Held in the University of Louvain.Physiological institute Biochemical society.Symposia, 23 Cambridge university press,Cambridge,1963 Organized by C.de Duve
Bioch 33.1385

LOVELACE FOUNDATION FOR MEDICAL EDUCATION AND RESEARCH Physics and medicine of the upper atmosphere Symposium on the physics and medicine of the upper atmosphere Proceedings San Antonio,Tex. 1951 Nov 6-9 Edited by Clayton S. White and Otis O. Benson Sponsored by School of aviation medicine. Randolph Field illus xxix,611p University of New Mexico press,Albuquerque,N.M. 1952
Nap 11.0635

LOW ENERGY ELECTRON DIFFRACTION Symposium on low energy electron diffraction Tucson,Ariz. 1968 Feb 4-7 American crystallographic association Edited by David H. Templeton and Gabor A. Somorjai American crystallographic association.Transactions, 4 American crystallographic association,Pittsburgh,1968
Met 25.1476

LOW ENERGY NUCLEAR INTERACTIONS AND NUCLEAR STRUCTURE International conference on nuclear physics Proceedings Paris 1958 Jul 7-12 Faculte des sciences de Paris et de province and Centre national de la recherche scientifique Under the auspices of International union of pure and applied physics Crosby Lockwood,London,1959
Cav 7.1759

LOW LEVEL IRRADIATION :a symposium Indianapolis,Ind. 1957 Dec 30 American association for the advancement of science United States atomic energy commission Edited by Austin M. Brues Co-sponsored by the Argonne national laboratory.Division of biological and medical research American association for the advancement of science. Publication, 59 American association for the advancement of science,Washington,D.C., 1959
Radioth 35.1109

LOW-LUMINOSITY STARS Symposium on low-luminosity stars Proceedings Charlottesville,Va. 1968 University of Virginia Edited by Shiv S. Kumar 542p Gordon and Breach,New York,1969
Obs 6.3627

LOW-MEDIUM MASS NUCLEI Symposium on the structure of low-medium mass nuclei 3rd Lawrence,Kansas 1968 Apr 18-20 Aerospace research laboratories Edited by J.P. Davidson vi,294p 24cm University press of Kansas,Lawrence,Kansas,1968
Cav 7.3099

LOW TATRA MOUNTAINS,CZECHOSLOVAKIA 1962 Nuclear theory International summer school on selected topics in nuclear theory lectures By N. Austern and others Ceskoslovenska akademie ved.Ustav jaderneho vyzkumu Edited by F. Janouch With the co-operation of the International atomic energy agency 452p International atomic energy agency,Vienna,1963
A Math 4.0885

LOW TEMPERATURE PHYSICS International conference on low temperature physics 8th Proceedings London 1962 Sep 16-22 Queen Mary college,London Edited by R.O. Davies Butterworths,London,1963 Includes Fritz London memorial lecture delivered by Bardeen,J.
Cav 7.1459

LOW TEMPERATURE PHYSICS International conference on low temperature physics 8th Proceedings London 1962 Sep 16-23 Edited by R.O. Davies Held at Queen Mary college Butterworths,London,1963
Eng 41.7396

LOW TEMPERATURE PHYSICS International conference on low temperature physics 10th Proceedings Moscow 1966 Aug 31-Sep 6 Vol 1: properties of helium International union of pure and applied physics Akademiya nauk S.S.S.R. Edited by M.P. Malkov 552p 22cm Moscow,1967
Cav 7.3030

LOW TEMPERATURE PHYSICS International conference on low temperature physics 10th Proceedings Moscow 1966 Aug 31-Sep 6 Vol 3: the electronic properties of metals International union of pure and applied physics Akademiya nauk S.S.S.R. 404p 22cm moscow,1967
Cav 7.3032

LOW TEMPERATURE PHYSICS International conference on low temperature physics 10th Proceedings Moscow 1966 Aug 31-Sep 6 Vol 4: antiferromagnetism International union of pure and applied physics Akademiya nauk S. S.S.R. Edited by M.P. Malkov 361p 22cm Moscow,1967
Cav 7.3033

LOW TEMPERATURE PHYSICS International conference on low temperature physics 10th Proceedings Vol 2 a-b: superconductivity International union of pure and applied physics Akademiya nauk S.S.S.R. Edited by M.P. Malkov 22cm 2 vols
Cav 7.3031

LOW TEMPERATURE PHYSICS International conference on low temperature physics 12th Proceedings Kyoto 1970 Sep 4-10 International union of pure and applied physics Physical society of Japan Edited by Eizo Kanda Organized by the Science council of Japan photos,diagrms 895p 26cm Academic press of Japan,Tokyo,1971
Cav 7.3028

LOW TEMPERATURE PHYSICS International conference on low-temperature physics 8th Proceedings London 1962 Sep 16-22 Edited by R.O. Davies Butterworths,London, 1963
Met 25.1180

LOW TEMPERATURE PHYSICS The International conference on low temperature physics 7th Proceedings Toronto 1960 Aug 29-Sep 3 International union of pure and applied physics Edited by J.M. Graham and A.C. Hollis-Hallett With the co-operation of the Canadian association of physicists University of Toronto press;North-Holland, Toronto;Amsterdam,1961
Cav 7.2330

LOW TEMPERATURE SCIENCE Physics of snow and ice International conference on low temperature science 1st Proceedings Sapporo 1966 Aug 14-19 1 pt 2: conference on physics of snow and ice (part 2) Institute of low temperature science Edited by Hirobumi Oura illus 1967p 27cm Hokkaido university,Sapporo,1957
Sco 14.7411

LOW TEMPERATURE TEST METHODS AND STANDARDS FOR CONTAINERS :a symposium Chicago 1953 Dec 10 National research council Edited by Earl C. Myers and Norbert J. Leiner Sponsored by the Quartermaster food and container institute illus vi,126p 23cm Washington,1954
Sco 14.0889

LOW TEMPERATURES Symposium on effects of low temperatures on the properties of materials : presented at a meeting of the Philadelphia district,A.S.T.M. Philadelphia,Pa. 1946 Mar 19 American society for testing materials American society for testing materials.Special technical publication, 78 American society for testing materials, Philadelphia,Pa.,1950
Eng 41.3778

LOW TEMPERATURES The International conference on fundamental particles and low temperatures Report Cambridge 1946 Jul 22-27 Vol 2: low temperatures Physical society and Cavendish laboratory Physical society,London,1947
Cav 7.0458

LOWELL M.PALMER FOUNDATION Human ovulation : a symposium Brookline,Mass. 1962 Edited by Chester S. Keefer Churchill,London,1965
An 32.3886

LOWELL OBSERVATORY Astronomical photoelectric conference Proceedings Flagstaff,Ariz. 1953 Aug.31 - Sep. 1 Edited by John B. Irwin Sponsored by the National science foundation Indiana university,Bloomington,Ind.,1953
Obs 6.2883

LOWER VERTEBRATE PHYLOGENY Current problems of lower vertebrate phylogeny Nobel symposium 4th Proceedings Stockholm 1967 Jun Edited by Tor Orvig Wiley,New York,1968
Bal 39.2054

LUBRICANTS Symposium on lubricants Chicago, Ill. 1937 Mar 3 American society for testing materials American society for testing materials,Philadelphia,Pa.,1937
Eng 41.6265

LUBRICATION Elastohydrodynamic lubrication : a symposium Proceedings Leeds 1972 Apr 11-13 Institution of mechanical engineers. Tribology group IME,London,1972
Eng 41.8650

LUBRICATION AND LUBRICANTS General discussion on lubrication and lubricants Proceedings London 1937 Oct 13-15 Vol 1-2 Institution of mechanical engineers 2 vols Institution of mechanical engineers, London,1937
Met 25.2152

LUBRICATION AND LUBRICANTS General discussion on lubrication and lubricants Proceedings London 1937 Oct 13-15 Vol 1-2 Institution of mechanical engineers 2 vols Institution of mechanical engineers, London,1937
Eng 41.6291

LUBRICATION AND WEAR Conference on lubrication and wear Proceedings London 1957 Oct 1-3 Institution of mechanical engineers Institution of mechanical engineers,London,1957
Eng 41.6295

LUBRICATION AND WEAR Conference on lubrication and wear 1957 Dec Scientific lubrication Scientific publications,Broseley, 1957
Eng 41.6294

LUBRICATION AND WEAR CONVENTION Proceedings 1963 May 23-25 Institution of mechanical engineers.Lubrication and wear group Institution of mechanical engineers,London, 1964
Eng 41.6305

LUBRICATION AND WEAR GROUP ANNUAL MEETING 1ST Non-conventional lubricants and bearing materials such as used in nuclear engineering Manchester 1962 Apr 12 Institution of mechanical engineers.Lubrication and wear group Institution of mechanical engineers, London,1962
Eng 41.6303

LUBRICATION AND WEAR IN LIVING AND ARTIFICIAL HUMAN JOINTS :a symposium London 1967 Apr 7 Institution of mechanical engineers.Lubrication and wear group and British orthopaedic association Institution of mechanical engineers.Proceedings, 181, pt.3 Institution of mechanical engineers,London, 1967
Met 25.1626

LUBRICATION (TRIBOLOGY) EDUCATION AND RESEARCH a report on the present position and industry's needs Great Britain.Department of education and science H.M.S.O.,London, 1966
Met 25.1618

LUCCA 1843 Congresso scientifico italiano 5th Atti Unione degli scienzati italiani 30cm Lucca,1844 Secretary general of conference:prof.Pacini
Bal 44.6331

LUGANO 1964 Chemotherapy of cancer International symposium on the chemotherapy of cancer Proceedings Edited by Placidus Plattner Organised by the Swiss academy of medical sciences Elsevier,Amsterdam,1964 Sponsored by F.Hoffman-Laroche and co.ltd.
Radioth 35.0837

LUGANO 1964 International symposium on chemotherapy of cancer Proceedings Swiss academy of medical sciences Edited by Placidus A. Plattner Sponsored by Hoffmann-La Roche and co ltd. Elsevier,Amsterdam,1964
Pha 16.0335

LULEA 1956 Jorden,skogen,malmen och vattenkraften i morgondagens Norrbotten :a conference Lectures Edited by Folke Thunberg Sponsored by Foreningen Norrbottens framjande illus,maps 271p 21cm Foreningen Norrbottens framjande,Lulea,1956
Sco 14.5794

LUMINESCENCE International conference on luminescence Proceedings Budapest 1966 International union of pure and applied physics Magyar tudomanyos akademia Edited by G. Szigeti diagrms 24cm 2 vols Akademiai kiado,Budapest,1968
Cav 7.3097

LUMINESCENCE International conference on luminescence Proceedings Newark,Del 1969 Aug 25-29 Edited by F. Williams graphs, diagrs xvi,959p 24cm North-Holland, Amsterdam,1970 Reprinted from 'Journal of luminescence',vol 1-2
Cav 7.2697

LUMINESCENCE International symposium on luminescence Papers Balatonvilagos,Hungary 1961 Jun 7-10 Roland Eotvos physical society Budapest,1962 Papers in English,German, Russian,reprinted from 'Acta physica Academiae scientiarum hungaricae',tom 14,fasc 2-3
Chem 18.0474

The LUMINESCENCE OF BIOLOGICAL SYSTEMS :a conference Proceedings Asilomar,Calif. 1954 Mar 28-Apr 2 National research council. Committee on photobiology Edited by Frank H. Johnson American assoc.for the advancement of science,Washington,D.C.,1955
Bal 39.0080

The LUMINESCENCE OF BIOLOGICAL SYSTEMS : conference Proceedings Pacific Grove, Calif. 1954 Mar28-Apr 2 Edited by Frank H. Johnson Sponsored by National academy of sciences.Committee on photobiology American association for the advancement of science, Washington D.C.,1955
An 32.1502

LUMINESCENCE OF ORGANIC AND INORGANIC MATERIALS international conference New York 1962 New York university Edited by Hartmut P. Kallman and Grace Marmor Spruch Sponsored by the United States.Air force.Aeronautical research laboratory xxiv,664p Wiley,New York,1962
Cav 7.1253

LUMINESCENT MATERIALS Preparation and characteristics of solid luminescent materials a conference Ithaca,N.Y. 1946 Oct 24-26 American physical society.Division of electron optics Edited by Gorton R. Fonda and Frederick Seitz Published under the auspices of the National research council graphs xv, 459p 22cm Wiley,New York,1948
Cav 7.3061

LUNAR INTERNATIONAL LABORATORY (LIL SYMPOSIUM) 4th Proceedings Applied sciences research and utilization of lunar resources New York 1968 Oct 17 International academy of astronautics Bibliog,illus 200p 20cm Pergamon press,Oxford,1970 LIL symposium organized by the International academy of astronautics at the 19th International astronautical congress
Cav 7.2711

LUNAR INTERNATIONAL LABORATORY (LIL) SYMPOSIUM 2nd Proceedings Life sciences research and lunar medicine Madrid 1966 Oct 13 Edited by Frank J. Malina Organized by International academy of astronautics Pergamon press,Oxford;London,1967 Held at the 17th International astronautical congress
PGMS 29.0445

LUNAR NOMENCLATURE Report of the working group of I.A.U. commission 17:the moon: proposed names for craters on the moon's far side...presented at the 14th assembly... Brighton 1970 International astronomical union.Commission 17 43p Harvard college observatory,Cambridge,Mass.,1970 Duplicated
Obs 6.3342

LUNAR RESOURCES Applied sciences research and utilization of lunar resources Lunar international laboratory (LIL symposium) 4th Proceedings New York 1968 Oct 17 International academy of astronautics Bibliog,illus 200p 20cm Pergamon press, Oxford,1970 LIL symposium organized by the International academy of astronautics at the 19th International astronautical congress
Cav 7.2711

LUNAR SCIENCE Apollo 11 lunar science conference 1st Proceedings Houston, Texas 1970 Jan 5-8 Edited by A.A. Levinson Geochimica et cosmochimica acta.Supplement, 1 3 vols Pergamon,New York,1970
Obs 6.3615

LUNAR SCIENCE Apollo 11 lunar science conference Papers Houston,Texas 1970 Jan 5-8 National aeronautics and space administration American association for the advancement of science.Miscellaneous publication, 70-1 American association for the advancement of science,Washington,D.C., 1970
Min 10.1474

LUNAR SCIENCE Apollo 11 lunar science conference Proceedings Houston,Texas 1970 Jun 5-8 1-3 National aeronautics and space administration Edited by A.A. Levinson Geochimica et cosmochimica acta, 34,supplement 1 3 vols Pergamon,New York, 1970
Min 10.1548

LUNAR SCIENCE CONFERENCE 2nd Proceedings Houston,Texas 1971 Jan 11-14 Edited by A.A. Levinson Geochimica et cosmochimica acta. Supplement, 2 3 vols MIT press,Cambridge, Mass.,1971
Obs 6.3614

LUNAR SURFACE Nature of the lunar surface IAU-NASA symposium Greenbelt,Md. 1965 Apr 15-16 International astronomical union National aeronautics and space administration Edited by W.N. Hess and others Held at the Goddard space flight center Johns Hopkins press,Baltimore,Md.,1966
Min 10.1445

The LUNAR SURFACE LAYER MATERIALS AND CHARACTERISTICS :a conference proceedings Boston 1963 May Edited by John W. Salisbury and Peter E. Glaser Financially supported by Air Force Cambridge Research Laboratories 532p Academic Press,New York, 1964
Obs 6.1507

LUND 1953 On the machine calculation of characters of the symmetric group :comptes rendus du douzieme Conges des Mathematiciens Scandinaves By Stig Comet Swedish board for computing machinery,Stockholm,1954
Math L 5.1123

LUND 1954 Rydberg centennial conference on atomic spectroscopy Proceedings Edited by Bengt Edlen Arranged by the Swedish national committee for physics Physiographisk sallskaps i Lund.Forhandlingar.N.F., 65,21 108p c.k.w.gleerup,Lund,1954
Obs 6.3285

LUND 1957 International symposium uber neurosekretion 2nd Proceedings Edited by W. Bargmann and others Springer,Berlin, 1958 Trilingual,English,French and German
An 32.3745

LUND 1960 I.G.U.symposium in urban geography Proceedings Edited by Knut Norberg Lund studies in geography.Ser.B. Human geography,24 Lunds universitet,Lund, 1962 Held at the 19th International geographical congress
Geog 13.1944

LUND 1963 Thermodynamics and thermochemistry Symposium on thermodynamics and thermochemistry Plenary lectures International union of pure and applied chemistry and Swedish chemical society Butterworths,London,1964
Chem E 24.0988

LUND 1969 Lund international conference on elementary particles Proceedings Edited by G.von Dardel Bibliog. 452p 30cm Berlingska Boktryckeriet,Lund,1969
A Math 4.1600

LUND INTERNATIONAL CONFERENCE ON ELEMENTARY PARTICLES Proceedings Lund 1969 Jun 25-Jul 1 Edited by G.von Dardel Bibliog. 452p 30cm Berlingska Boktryckeriet,Lund, 1969
A Math 4.1600

LUNG Form and function in the human lung :a symposium Proceedings Birmingham 1967 Apr 5-7 Edited by Gordon Cumming and L.B. Hunt Sponsored by the Lederle laboratories Livingstone,Edinburgh,1968
An 32.3805

LUNTEREN 1966 Regulation of nucleic acid and protein biosynthesis An International symposium on regulatory mechanisms in nucleic acid and protein biosynthesis Proceedings Edited by V.V. Konigsberger and L. Bosch B.B. A.library, 10 Elsevier,Amsterdam,1967
Bot 42.1511

LUNTEREN 1966 Regulation of nucleic acid and protein biosynthesis An International symposium on regulatory mechanisms in nucleic acid and protein biosynthesis Proceedings Edited by V.V. Koningsberger and L. Bosch B. B.A. library, 10 Elsevier,Amsterdam,1967
Bioch 33.0826

LUNTEREN 1966 Regulation of nucleic acid and protein biosynthesis :international synposium Edited by V.V. Koningsberger and L. Bosch B.B.A.library, 10 Illus Elsevier, Amsterdam,1967
Radioth 35.0140

LUNTEREN 1968 International conference on laboratory astrophysics Selection of papers Edited by J. Rosenberg Organised under the auspices of the Nederlandse natuurkundig vereniging Physica, 41,no 1 223p North-Holland,Amsterdam,1969
TA 15.0580

LUNTEREN 1968 International conference on laboratory astrophysics Selection of papers Nederlandse natuurkundige vereniging Edited by J. Rosenberg Physica, 41,no 1 223p North-Holland,Amsterdam,1969
Obs 6.3347

LUNTEREN 1969 Ultraviolet stellar spectra and related ground-based observations International astronomical union Edited by L. Houziaux and H.E. Butler International astronomical union.Symposium, 36 361p Reidel,Dordrecht,1970
Obs 6.3321

LUXEMBOURG 1962 Colloque du jurassique International geological congress. International commission on stratigraphy,and, International palaeontological union Ministere des arts et des sciences,Luxembourg, 1964
Geol 8.2510

LVOV 1962 Vsesoyuznaya ornitologicheskyaya konferentsiya 11-17 Sentyabrya,1962 3rd Materialy 20cm Lvov,1962
Philos 1.1626

LYMPH :a conference New York 1946 Sep 16 By Philip D. McMaster and others New York academy of sciences New York academy of sciences.Annals, 46,p679-882 New York,1946
Bal 39.1342

LYMPH :a conference Papers New York 1945 Apr 13-14 New York academy of sciences New York academy of sciences.Annals, 46,art.8 New York,1946
Phys 20.0883

LYMPHOLOGY Progress in lymphology International symposium on lymphology Proceedings Zurich 1966 Jul Thieme, Stuttgart,1967
Radioth 35.0922

LYON 1947 Relations entre les phenomenes solaires et geophysiques Proceedings Centre national de la recherche scientifique Edited by Jean Dufay Supported by the Rockefeller foundation Centre national de la recherche scientifique.Colloques internationaux, 9 312p Centre national de la recherche scientifique,Lyon,1947
Obs 6.2010

LYON 1948 Le Calcul de probabilites et ses applications colloque international proceedings Centre national de la recherche scientifique Edited by Maurice Frechet Centre national de la recherche scientifique. Colloques internationaux 130p 27cm Centre national de la recherche scientifique, Paris,1949
Math 3.0678

LYSOSMES :a Ciba foundation symposium London 1963 Feb 12-14 Ciba foundation Edited by A.V.S. De Reuck and Margaret P. Cameron Churchill,London,1963
An 32.5316

LYSOSOMES :a Ciba foundation symposium Papers London 1963 Feb 12-14 Ciba foundation Edited by A.V.S. De Reuck and Margaret P. Cameron Ciba foundation.Symposia Churchill,London,1963
Bal 39.0298

LYSOSOMES :a Ciba foundation symposium Papers London 1963 Feb 12-14 Ciba foundation Edited by A.V.S. De Reuck and Margaret P. Cameron illus Churchill,London, 1963
Bioch 33.0935

MACCLESFIELD 1957 A Symposium on the evaluation of drug toxicity Imperial chemical industries.Pharmaceuticals division Edited by A.L. Walpole and A. Spinks Churchill,London,1958
Pha 16.0327

MCCOLLUM-PRATT INSTITUTE A Symposium on amino acid metabolism Papers and discussions presented Baltimore,Md. 1954 Jun 14-17 Edited by William D. McElroy and H.Bentley Glass McCollum-Pratt institute.Contribution, 105 Johns Hopkins press,Baltimore,1955
Bioch 33.0599

MCCOLLUM-PRATT INSTITUTE A Symposium on inorganic nitrogen metabolism :function of metallo-flavoproteins Papers and informal discussions presented Baltimore,Md. 1955 Jun 21-23 Edited by William D. McElroy and H.Bentley Glass McCollum-Pratt institute. Contribution, 125 Johns Hopkins press, Baltimore,Md.,1956
Bioch 33.0598

MCCOLLUM-PRATT INSTITUTE A Symposium on light and life Papers and discussions presented Baltmore,Md. 1960 Mar 28-31 Edited by William D. McElroy and H.Bentley Glass McCollum-Pratt institute.Contribution, 302 John Hopkins press,Baltimore,Md.,1961
Bioch 33.0600

MCCOLLUM-PRATT INSTITUTE A Symposium on light and life Papers and informal discussions Baltimore,Md. 1960 Mar 28-31 Edited by William D. McElroy and Bentley Glass McCollum-Pratt institute.Contribution, 302 Johns Hopkins university press,Baltimore,Md., 1961
Bal 44.6447

MCCOLLUM-PRATT INSTITUTE A Symposium on the chemical basis of development Baltimore,Md 1958 Mar 24-27 Edited by William D. McElroy and Bentley Glass Johns Hopkins press, Baltimore,Md.,1958
Bal 39.0454

MCCOLLUM-PRATT INSTITUTE A Symposium on the chemical basis of development Baltimore,Md. 1958 Mar 24-27 Edited by William D. McElroy and Bentley Glass McCollum-Pratt institute. Symposia illus. Johns Hopkins press, Baltimore,Md.,1958
Gen 34.0852

MCCOLLUM-PRATT INSTITUTE A Symposium on the chemical basis of heredity Baltimore,Md. 1956 Jun 19-22 Edited by William D. McElroy and Bentley Glass Johns Hopkins press, Baltimore,Md.,1957
Bot 42.4714

MCCOLLUM-PRATT INSTITUTE A Symposium on the chemical basis of heredity Baltimore,Md. 1956 Jun 19-22 Edited by William D. McElroy and Bentley Glass McCollum-Pratt institute. Contribution, 153 Johns Hopkins press, Baltimore,Md.,1957
Gen 34.0853

MCCOLLUM-PRATT INSTITUTE A Symposium on the chemical basis of heredity Papers and informal discussions Baltimore,Md. 1956 Jun 19-22 Edited by William D. McElroy and Bentley Glass McCollum-Pratt institute. Contribution, 153 Johns Hopkins press, Baltimore,Md.,1957
Bal 44.6451

MCCOLLUM-PRATT INSTITUTE A Symposium on the mechanism of enzyme action Edited by W.D. McElroy and Bentley Glass McCollum-Pratt institute,Contributions, 70 Johns Hopkins press,Baltimore,Md.,1954
Radioth 35.0034

MCCOLLUM-PRATT INSTITUTE A Symposium on the mechanism of enzyme action Papers and discussions Baltimore 1953 Jun 16-19 Edited by William D. McElroy and Bentley Glass McCollum-Pratt institute.Contribution, 70 Johns Hopkins press,Baltimore,Md.,1954
Bioch 33.1059

MCCOLLUM-PRATT INSTITUTE Amino acid metabolism :a symposium Papers and discussion Baltimore,Md. 1954 Jun 14-17 Edited by William D. McElroy and H.Bentley Glass McCollum-Pratt institute.Contribution, 108 Johns Hopkins press,Baltimore,Md.,1955
Mol 45.0292

MCCOLLUM-PRATT INSTITUTE Chemical basis of development A Symposium on the chemical basis of development Papers and informal discussions Baltimore 1958 Mar 24-27 Edited by William D. McElroy and Bentley Glass McCollum-Pratt institute.Contribution, 234 Johns Hopkins press,Baltimore,1958
Bioch 33.0829

MCCOLLUM-PRATT INSTITUTE Chemical basis of heredity A Symposium on the chemical basis of heredity Papers and informal discussions Baltimore 1956 Jun 19-22 Edited by William D. McElroy and Bentley Glass McCollum-Pratt institute.Contribution, 153 Johns Hopkins press,Baltimore,1957
Bioch 33.0830

MCCOLLUM-PRATT INSTITUTE Inorganic nitrogen metabolism:function of metallo-flavoproteins : a symposium Baltimore,Md. 1955 Jun 21-23 Edited by William D. McElroy and Bentley Glass McCollum-Pratt institute.Contributions, 125 Johns Hopkins press,Baltimore,Md.,1956
Mol 45.0279

MCCOLLUM-PRATT INSTITUTE Phosphorus metabolism A Symposium on the role of phosphorous in the metabolism of plants and animals Papers and discussions Baltimore, Md. 1951 Jun 18-21 and Baltimore,Md. 1952 Jun 16-19 Vol 1-2 Edited by William D. McElroy and H.Bentley Glass McCollum-Pratt institute.Contribution, 23,36 2 vols Johns Hopkins press,Baltimore,Md., 1951-52
Bot 42.1670

MCCOLLUM-PRATT INSTITUTE Phosphorus metabolism A Symposium on the role of phosphorus in the metabolism of plants and animals Baltimore,Md. 1951 Jun 18-21 Vol 1 Edited by W.D. McElroy and B. Glass McCollum-Pratt institute.Contributions, 23 John Hopkins press,Baltimore,Md.,1951
Radioth 35.0028

MCCOLLUM-PRATT INSTITUTE Phosphorus metabolism A Symposium on the role of phosphorus in the metabolism of plants and animals Baltimore,Md. 1951 Jun 18-21 Vol 2 Edited by W.D. McElroy and B. Glass McCollum-Pratt institute.Contributions, 36 John Hopkins press,Baltimore,Md.,1952
Radioth 35.0029

MCCOLLUM-PRATT INSTITUTE Phosphorus metabolism A Symposium on the role of phosphorus in the metabolism of plants and animals Papers and discussions presented Baltimore,Md. 1951 Jun 18-21 and Baltimore,Md. 1952 Jun 16-19 Vol 1-2 Edited by William D. McElroy and H.Bentley Glass McCollum-Pratt institute.Contribution, 23,36 2 vols Johns Hopkins press, Baltimore,Md.,1951-52
Bioch 33.0597

MCCOLLUM-PRATT INSTITUTE The Chemical basis of development :a symposium Baltimore,Md. 1958 Mar 24-27 Edited by William D. McElroy and Bentley Glass Johns Hopkins press, Baltimore,Md.,1958
An 32.2488

MCCOLLUM-PRATT INSTITUTE The Chemical basis of development :a symposium Baltimore,Md. 1958 Mar 24-27 Edited by William D. McElroy and Bentley Glass port Johns Hopkins, Baltimore,Md.,1958
Mol 45.0257

MCCOLLUM-PRATT INSTITUTE The Chemical basis of heredity :a symposium Baltimore,Md. 1956 Jun 19-22 Edited by W.D. McElroy and Bentley Glass McCollum-Pratt institute. Contributions, 153 John Hopkins press, Baltimore,Md.,1957
Radioth 35.0219

MCCOLLUM-PRATT INSTITUTE The Chemical basis of heredity :a symposium Baltimore,Md. 1956 Jun 19-22 Edited by William D. McElroy and Bentley Glass McCollum-Pratt institute. Contributions, 153 illus. Johns Hopkins press,Baltimore,Md.,1957
Mol 45.0193

MCCOLLUM-PRATT INSTITUTE A Symposium on amino-acid metabolism Papers Baltimore,Md. 1954 Jun 14-17 Edited by William D. McElroy and H.Bentley Glass McCollum-Pratt institute. Contribution,105 Johns Hopkins,Baltimore,Md., 1955
Chem 18.1532

MCCOLLUM-PRATT INSTITUTE A Symposium on light and life Edited by William D. McElroy and Bentley Glass With support from the National science foundation McCollum-Pratt institute.Contribution,302 Johns Hopkins press,Baltimore,Md.,1961
Chem 18.0168

MCCOLLUM-PRATT INSTITUTE A symposium on the mechanism of enzyme activity Edited by William D. McElroy McCollum-Pratt institute. Contribution,70 Johns Hopkins press, Baltimore,Md.,1954
Chem 18.1270

MCCOLLUM-PRATT INSTITUTE The Chemical basis of heredity :a symposium Baltimore,Md. 1956 Jun 19-22 Edited by William D.Bentley McElroy and Bentley Glass With support from the Atomic energy commission McCollum-Pratt institute.Contribution,153 Johns Hopkins press,Baltimore,Md.,1957
Chem 18.1555

MCCOLLUM-PRATT INSTITUTE The Mechanism of enzyme action :a symposium Baltimore,Md. 1953 Jun 16-19 Edited by William D. McElroy and Bentley Glass McCollum-Pratt institute. Contributions,70 Johns Hopkins press, Baltimore,Md.,1954
Col S 12.0113

MCGILL UNIVERSITY The World congress of psychiatry 3rd Proceedings Montreal 1961 Jun 4-10 2 vols University of Toronto press;McGill university press,Toronto;Montreal, 1961 Conference languages:English,French, German and Spanish
Psy 31.2540

MCGILL UNIVERSITY World congress of psychiatry Proceedings Montreal 1961 Jun 4-10 Vol 1-2 Edited by R.A. Cleghorn and others 2 vols University of Toronto press;McGill university press,Toronto;Montreal, 1961 Papers in English,French,German and Spanish
PGMS 29.0322

MCGILL UNIVERSITY.DEPARTMENT OF PSYCHIATRY McGill university conference on depression and allied states Papers Montreal 1959 Mar 19-21 Canadian psychiatric association journal, 4,Special supplement, 1 Canadian psychiatric association journal, Ottawa,1959
Psy 28.0086

MCGILL UNIVERSITY CONFERENCE ON DEPRESSION AND ALLIED STATES Papers Montreal 1959 Mar 19-21 McGill university.Department of psychiatry Canadian psychiatric association journal, 4,Special supplement, 1 Canadian psychiatric association journal, Ottawa,1959
Psy 28.0086

MACHINABILITY :a conference jointly organized by the iron and steel institute the Institute of metals,the Institution of production engineers and the Institution of metallurgists Proceedings London 1965 Oct 4-6 Iron and steel institute Iron and steel institute. Special report, 94 Iron and steel institute, London,1967
Met 25.0739

MACHINE INTELLIGENCE WORKSHOP 1st- Essays Edinburgh 1965- University of Edinburgh. Experimental programming unit Edited by Donald Michie and others Oliver and Boyd; Edinburgh university press,Edinburgh,1967- The 5th contains a previously unpublished report by Alan Turing
Math S 3.1818

MACHINE INTELLIGENCE WORKSHOP 3rd;4th Proceedings 3-4 University of Edinburgh Edited by Donald Michie and Bernard Meltzer Edinburgh university press, Edinburgh,1968-69
Psy 31.3066

MACHINE PERCEPTION OF PATTERNS AND PICTURES Conference on machine perception of patterns and pictures Proceedings Teddington 1972 Apr 12-14 Institute of physics National physical laboratory Institute of electrical engineers Institute of physics.Conference series, 13 illus ix,362p Institute of physics,London,1972
Math L 5.3882

MACHINE TOOL DESIGN AND RESEARCH Advances in machine tool design and research International M.T.D.R. conference 3rd-9th Proceedings 1963-68 Edited by S.A. Tobias and F. Koenigsberger 9 vols Pergamon press,London,1963-69
Eng 41.3924

MACHINE TRANSLATION OF LANGUAGES International conference on machine translation of languages and applied language analysis proceedings Teddington 1961 Sep 5-8 Vol. 1-2 National physical laboratory National physical laboratory. Proceedings of symposia, 13 23cm 2 vols H.M.S.O.,London,1962
Math L 5.0960

Les MACHINES A CALCULER ET LA PENSEE HUMAINE colloque international Paris 1951 Jan 8-13 Centre national de la recherche scientifique Centre national de la recherche scientifique.,Colloques internationaux,37 illus. xix,570p 24cm Centre national de la recherche scientifique,Paris,1953
Math L 5.0716

MACHINES FOR MATERIALS AND ENVIRONMENTAL TESTING Symposium on techniques and equipment for environmental testing; symposium on developments in materials testing machine design :joint conference Manchester 1965 Sep 6-10 Institution of mechanical engineers and Society of environmental engineers Under the aegis of the British national committee on materials Institution of mechanical engineers.Proceedings, 180,pt 3a I.M.E.,London,1966
Met 25.0891

MACHINING-THEORY AND PRACTICE By Hans Ernst and others American society for metals, Cleveland,Ohio,1950 A series of thirteen educational lectures on machining-theory and practice presented to members of the A.S.M. during the thirty-first National metal congress and exposition,Cleveland.
Met 25.0717

MACKENZIE William The Ocular circulation in health and disease;William Mackenzie centenary symposium Proceedings Glasgow 1968 Sep 23-23 Edited by E.R. Brown and C.V. Moore Progress in hematology, 6 Heinemann,London, 1969
Med 36.0054

MACROMOLECULAR ASPECTS OF THE CELL CYCLE Symposium on macromolecular aspects of the cell cycle :given at research conference for biology and medicine of the atomic energy commission Gatlinburg,Tenn 1963 Apr 8-11 Sponsored by Oak Ridge national laboratory. Biology division Oak Ridge national laboratory.Symposia, 16 Journal of cellular and comparative physiology, 62,supp.1 Wistar institute of anatomy and biology, Philadelphia,Pa.,1963
PGMS 29.0414

MACROMOLECULAR ASPECTS OF THE CELL CYCLE Symposium on macromolecular aspects of the cell cycle :given at research conference for biology and medicine of the atomic energy commission Gatlinburg,Tenn. 1963 Apr 8-11 Sponsored by Oak Ridge national laboratory. Biology division Oak Ridge national laboratory.Symposia, 16 Journal of cellular and comparative physiology, 62,no.2 Wistar institute of anatomy and biology,Philadelphia, Pa.,1963
Gen 34.0842

MACROMOLECULAR ASPECTS OF THE CELL CYCLE :a symposium Gatlinburg,Tenn. 1963 Apr 8-11 Oak Ridge national laboratory Wistar institute press,Philadelphia,Pa.,1963
An 32.3297

MACROMOLECULAR COMPLEXES :a symposium Urbana,Ill. 1959 Sep Edited by M.V. Edds Society of general physiologists.Annual symposia, 6 illus. Ronald press,New York, 1961
Radioth 35.1408

MACROMOLECULAR SYNTHESIS Cytodifferentiation and macromolecular synthesis :a symposium Asilomar,Calif. 1962 Jun Society for the study of development and growth Edited by Michael Locke Society for the study of development and growth.Symposia, 21 Academic press,New York;London,1963
Bioch 33.1009

MACROMOLECULAR SYNTHESIS Cytodifferentiation and macromolecular synthesis :a symposium Papers Asilomar,Calif. 1962 Jun Society for the study of development and growth Edited by Michael Locke Society for the study of development and growth.Symposia, 21 Academic press,New York;London,1963
Bot 42.1496

MACROMOLECULES Informational macromolecules : a symposium Proceedings New Brunswick,N.J. 1962 Sep 5-7 Rutgers university.Institute of microbiology Edited by Henry J. Vogel and others With support from the National science foundation Academic press,New York; London,1963
Bioch 33.0822

MACROMOLECULES International colloquium on macromolecules Proceedings Amsterdam 1949 Sep 2-5 Internation union of pure and applied chemistry D.B.Ceuten's Uitgerers, Amsterdam,1949
Radioth 35.0014

MACROMOLECULES International colloquium on macromolecules Proceedings Amsterdam 1949 Sep 2-5 Under the auspices of the International union of pure and applied chemistry Amsterdam,1950
Col S 12.0101

MACROMOLECULES Synthesis and structure of macromolecules :a symposium Papers Cold Spring Harbor 1963 Cold Spring Harbor laboratory of quantitative biology Cold Spring Harbor symposia on quantitative biology, 28 Cold Spring Harbor,1963
Bioch 33.1284

MACROMOLECULES The Biological replication of macromolecules :a symposium Society for experimental biology Society for experimental biology.Symposia, 12 Cambridge university press,Cambridge,1958
Bioch 33.1357

MACROMOLECULES The Structure and biosynthesis of macro-molecules London 1961 Mar 27-28 Biochemical society Edited by D.J. Bell and J.K. Grant Held at University of London.Senate house Biochemical society.Symposia, 21 Cambridge, 1962 In commemoration of the 50th anniversary of the Biochemical society
Bal 39.0515

MACROMOLECULES The Structure and biosynthesis of macromolecules :a symposium London 1961 Mar 27-28 Biochemical society Edited by D.J. Bell and J.K. Grant Held at the University of London.Senate House Biochemical society.Symposia, 21 Cambridge university press,Cambridge,1962 In commemoration of the 50th anniversary of the Biochemical society
Bioch 33.1383

MACROMOLECULES,BIOSYNTHESIS AND FUNCTION Federation of European biochemical societies meeting 6th Proceedings Madrid 1969 Apr Federation of European biochemical societies Edited by S. Ochoa and others Federation of European biochemical societies. Publications, 21 Academic press,London,1970
Bioch 33.2322

MACROMOLECULES AND BEHAVIOUR Manhattan,Kan 1964 Apr 20-22 Edited by John Gaito North Holland;Meredith,Amsterdam;New York,1966 Revised and current versions of presentations at a Conference held at Kansas state university on April 20 to 22,1964 under the sponsorship of the Office of naval research to discuss the role of macromolecules in complex behavior.
Bal 39.1622

MACROMOLECULES AND MEMBRANES Probes of structure and function of macromolecules and membranes Colloquium of the Johnson research foundation 5th Proceedings Philadelphia, Pa. 1969 Apr 19-21 Vol 1-2: probes and membrane function;probes of enzymes and hemoproteins Johnson research foundation Edited by Britton Chance and others Academic press,New York,1971
Bioch 33.2190

MACROMOLECULES HELICOIDALES EN SOLUTION Societe de chimie physique.Reunion 17e Paris 1967 May 2-5 Societe de chimie physique Journal de chimie physique et de physico-chimie biologique, 65,no 1 Paris, 1968
Bioch 33.0326

MACY CONFERENCE ON GENETICS 3rd Proceedings Genetic selection in man Princeton,N.J. 1961 Oct 15-18 Josiah Macy jr. foundation Edited by William J. Schull University of Michigan press,Michigan,1963
Gen 34.1867

MACY CONFERENCE ON GENETICS 4th Somatic cell genetics Princeton,N.J. 1962 Oct 15-17 Josiah Macy jr.foundation Edited by Robert S. Krooth illus. University of Michigan press,Ann Arbor,Mich.,1964
Gen 34.0891

MACY CONFERENCE 3RD The Central nervous system and behavior :translations from the Russian literature collected for the participants of the third Macy conference on the central nervous system and behavior... under the sponsorship of the Josiah Macy jr. foundation,the National science foundation, Princeton,N.J.,February 21-24,1960 National institutes of health.Russian scientific translation program US department of health, education and welfare,Washington,D.C.,1960 Duplicated
Psy 31.2903

MADISON 1968 Error correcting codes :a symposium Proceedings United States.Army. Mathematics research center Edited by Henry B. Mann United States.Army.Mathematics research center.Publications, 22 Wiley,New York,1968
Math L 5.3514

MADISON,WIS 1964 Asymptotic solutions of differential equations and their applications a symposium proceedings Edited by C.H. Wilcox United States Army.Mathematics Research Center.Publicatons, 13 John Wiley and sons,New York,1964
I Math 4.0246

MADISON,WIS. A Symposium on respiratory enzymes Lectures and discussions By Otto Meyerhof and others Given at the University of Wisconsin University of Wisconsin,Madison, Wis.,1942 reprinted 1948 Sponsored jointly by the universities of Wisconsin and Chicago
Bal 39.1344

MADISON,WIS. 1923 The National symposium on colloid chemistry 1st Papers and discussions University of Wisconsin Edited by J.Howard Mathews Colloid symposium monograph, 1 Department of chemistry, university of Wisconsin,Madison,Wis.,1923
Bioch 33.1469

MADISON,WIS. 1934 The Symposium on colloid chemistry 11th Papers presented Edited by Harry Boyer Weiser Colloid symposium monograph, 11 Williams and Wilkins, Baltimore,1935
Bioch 33.1478

MADISON,WIS. 1940 A Symposium on hydrobiology By James G. Needham and others Held at the University of Wisconsin University of Wisconsin,Madison,Wis.,1941 Held under the auspices of the University of Wisconsin and of the Wisconsin alumni research foundation
Bal 44.6459

MADISON,WIS. 1947 A Symposium on the use of isotopes in biology and medicine By Hans T. Clarke and others University of Wisconsin University of Wisconsin press,Madison,1948
Bioch 33.0283

MADISON,WIS. 1948 Symposium on combustion and flame and explosion phenomena 3rd Standing committee on combustion symposia xiii,748p Williams and Wilkins,Baltimore,Md., 1949
Chem E 24.1404

MADISON,WIS. 1950 A Symposium on steroid hormones Papers University of Wisconsin Edited by Edgar S. Gordon University of Wisconsin press,1950 Published in celebration of the hundredth anniversary of the founding of the University of Wisconsin
Bioch 33.0455

MADISON,WIS. 1955 The Computing laboratory in the university conference papers Edited by Preston C. Hammer pl. xv,236p 23cm University of Wisconsin press,Madison,Wis., 1957
Math L 5.0784

MADISON,WIS. 1958 On numerical approximation symposium proceedings United States.Army.Mathematics research center Edited by Rudolph E. Langer United States army.Mathematics research center.Publications, 1 x,462p 24cm University of Wisconsin press,Madison,Wis.,1959
Math L 5.0267

MADISON,WIS. 1959 Boundary problems in differential equations symposium proceedings United States.Army.Mathematics research center Edited by Rudolph E. Langer United States army.Mathematics research center.Publications, 2 x,320p 23cm University of Wisconsin press,Madison,Wis.,1960
Math L 5.0281

MADISON,WIS. 1959 Developing cell systems and their control :a symposium Society for the study of development and growth Edited by Dorothea Rudnick at the University of Wisconsin Society for the study of development and growth.Symposia, 18 Ronald press,New York,1960
Bioch 33.1010

MADISON,WIS. 1959 Frontiers of numerical mathematics :symposium United States.Army. Mathematics research center National bureau of standards United States army.Mathematics research center.Publications, 4 xi,132p 24cm University of Wisconsin,Madison,Wis., 1960
Math L 5.3232

MADISON,WIS. 1960 Partial differential equations and continuum mechanics :an international conference Proceedings United States.Army.Mathematics research center Edited by R.E. Langer bibliog.,illus. xv, 397p 24cm University of Wisconsin press, Madison,Wis.,1961
A Math 4.1614

MADISON,WIS. 1960 The Conference on research on the radiotherapy of cancer Proceedings American cancer society,New York, 1961
Radioth 35.1133

MADISON,WIS. 1961 Electromagnetic waves :a symposium proceedings United States.Army. Mathematics research center Edited by Rudolph E. Langer 396p University of Wisconsin press,Madison,Wis.,1962
Obs 6.1032

MADISON,WIS. 1961 Electromagnetic waves symposium proceedings United States.Army. Mathematics research center Edited by Rudolph E. Langer University of Wisconsin press,Madison,Wis.,1962 photolith
A Math 4.0684

MADISON,WIS. 1962 Nonlinear problems :a symposium Proceedings United States.Army. Mathematics research center Edited by R.E. Langer bibliog.,illus. xiii,321p 24cm University of Wisconsin press,Madison,Wis., 1963
A Math 4.1615

MADISON,WIS. 1962 Statistical theory of reliability advanced seminar proceedings United States.Army.Mathematics research center Edited by Marvin Zelen United States army. Mathematics research center.Publications, 9 xvii,166p 23cm University of Wisconsin press,Madison,Wis.,1963
Math 3.0700

MADISON,WIS. 1963 Recent advances in matrix theory :advanced seminar 2nd proceedings United States army.Mathematics research center. Edited by Hans Schneider United States army.Mathematics research center. Publication, 12 xi,142p 25cm University of Wisconsin press,Madison,Wis.,1964
Math L 5.0328

MADISON,WIS. 1963 Recent advances in matrix theory advanced seminar 2nd proceedings United States.Army.Mathematics research center Edited by Hans Schneider United States army.Mathematics research center. Publications, 12 xi,142p 24cm University of Wisconsin press,Madison; Milwaukee,Wis.,1964
Math 3.0164

MADISON,WIS. 1963 Recent advances in matrix theory seminar 2nd proceedings United States.Army.Mathematics research center Edited by Hans Schneider United States army. Mathematics research center.Publications, 12 xi,142p 24cm University of Wisconsin press, Madison,Wis.,1964
P. Math 2.2087

MADISON,WIS. 1963 Stochastic models in medicine and biology :a symposium Proceedings Edited by John Gurland Conducted by United States.Army.Mathematics research center United States.Army. Mathematics research center.Publications, 10 University of Wisconsin press,Madison,Wis., 1964
Gen 34.0419

MADISON,WIS. 1964 Asymptotic solutions of differential equations and their applications symposium proceedings United States.Army. Mathematics research center Edited by Calvin H. Wilcox United States.Army.Mathematics research center.Publications, 13 x,249p 22cm John Wiley and sons,New York,1964 Dedicated to Professor Rudolph E.Langer
Math 3.0291

MADISON,WIS. 1964 Error in digital computation advanced seminar proceedings Vol. 1 United States.Army.Mathematics research center Edited by Louis B. Rall United States army.Mathematics research center. Publications, 14 Bibliog. ix,324p 24cm John Wiley and sons,New York,1965
Math 3.0242

MADISON,WIS. 1965 Error in digital computation symposium proceedings Vol. 2 United States.Army.Mathematics research center Edited by Louis B. Rall United States army.Mathematics research center. Publications, 15 x,288p 24cm John Wiley and sons,New York,1965
Math 3.0243

MADISON,WIS. 1965 Perturbation theory and its application in quantum mechanics seminar proceedings United States army.Mathematics research center Edited by Calvin H. Wilcox Organized jointly with University of Wisconsin. Theoretical chemistry institute 428p John Wiley and sons,New York;London,1966
A Math 4.0809

MADISON,WIS. 1966 Numerical solutions of nonlinear differential equations :advanced symposium proceedings United States.Army. Mathematics research center Edited by Donald Greenspan United States.Army.Mathematics research center.Publications, 17 John Wiley, New York,1966
A Math 4.1220

MADISON,WIS. 1966 Numerical solutions of nonlinear differential equations advanced seminar proceedings United States.Army. Mathematics research center Edited by Donald Greenspan United States army.Mathematics research center.Publication, 17 x,343p 23cm John Wiley and sons,New York,1966
Math L 5.0358

MADISON,WIS. 1966 Spectral analysis of time series :advanced seminar Proceedings United States.Army.Mathematics research center and University of Wisconsin.Statistics department Edited by Bernard Harris United States.Army.Mathematics research center. Publications, 18 illus. x,319p 24cm Wiley,New York,1967
Math S 3.1456

MADISON,WIS. 1967 Markov processes and potential theory :a symposium Proceedings Edited by Joshua Chover Sponsored by the United States.Army.Mathematics research center United States army.Mathematics research center. Publications, 19 Bibliog x,235p 24cm John Wiley,New York,1967
Math 3.1189

MADISON,WIS. 1967 Stochastic optimization and control :advanced seminar Proceedings United States.Army.Mathematics research center Edited by Herman Karreman United States.Army. Mathematics research center.Publications, 20 bibliog.,illus. xii,217p 23cm Wiley,New York,1968
Math S 3.1403

MADISON,WIS. 1967 The Future of statistics a conference Proceedings Edited by Donald G. Watts xvi,316p 23cm Academic press, New York;London,1968
Math S 3.1458

MADISON,WIS. 1968 Error correcting codes United States.Army.Mathematics research center Edited by Henry B. Mann United States.Army. Mathematics research center, 21 Bibliog., Illus. xii,231p 23cm Wiley,New York,1968
P Math 2.3700

MADISON,WIS. 1968 Theory and applications of spline functions :a symposium Proceedings United States.Army.Mathematics research center United States.Army.Mathematics research center Bibliog.,Illus. xi,212p 24cm Academic press,New York;London,1969
P Math 2.3487

MADISON,WIS. 1968 Theory and applications of spline functions :an advanced seminar Proceedings United States.Army.Mathematics research center Edited by T.N.E. Greville Held at the University of Wisconsin United States.Army.Mathematics research center. Publications, 22 xi,212p 24cm Academic press,New York;London,1969
Math S 3.1674

MADISON,WIS. 1968 Theory and applications of spline functions :an advanced seminar Proceedings United States.Army.Mathematics research center Edited by T.N.E. Greville United States.Army.Mathematics research center. Publications, 22 Academic press,New York; London,1969
Math L 5.3443

MADISON,WIS. 1969 Approximations with special emphasis on spline functions :a symposium Proceedings United States.Army. Mathematics research center Edited by I.J. Schoenberg bibliog. Academic press,New York;London,1969
Math S 2.1504

MADISON,WIS. 1969 Graph theory and its applications :an advanced seminar United States.Army.Mathematics research center bibliog.,illus. viii,262p 23cm Academic press,New York;London,1970
P Math 2.3821

MADISON,WIS. 1969 Statistical computation : a conference Proceedings Edited by Roy C. Milton and John A. Nelder Organized by the University of Wisconsin bibliog. Academic press,New York;london,1969
Math S 3.1503

MADISON,WIS. 1970 Approximations with special emphasis on spline functions :a symposium United States.Army. Mathematics research center Edited by I.J. Schoenberg Held at the University of Wisconsin United States.Army.Mathematics research center.Publications, 23 xi,488p 24cm Academic press,London;New York,1971
Math S 3.1675

MADISON,WIS. 1971 Waves on beaches and resulting sediment transport Proceedings Edited by R.E. Meyer illus vii,462p 23cm Academic press,New York,1972
A Math 4.1885

MADISON,WIS, 1941 A Symposium on respiratory enzymes Addresses By Otto Meyerhof and others held at the University of Wisconsin University of Wisconsin press, Madison,Wis.,1942
Bioch 33.1661

MADISON,WISC. 1941 A Symposium on respiratory enzymes Addresses University of Wisconsin,and,University of Chicago University of Wisconsin press,Madison,Wisc., 1942
Chem 18.1249

MADISON,WISC. 1947 A Symposium on the use of isotopes in biology and medicine Addresses Wisconsin alumni research foundation University of Wisconsin press, Madison,Wisc.,1948
Chem 18.1259

MADISON,WISC. 1948 Symposium on combustion, flame and explosion phenomena 3rd Papers Standing committee on combustion symposia Williams and Wilkins,Baltimore,Md.,1949 Chairman of committee:B.Lewis.Later volumes entitled 'Symposium (international) on combustion,flame and explosion phenomena'
Chem 18.0433

MADISON,WISC. 1948 Symposium on combustion, flame and explosion phenomena abstracts of papers 3rd Standing committee on combustion symposia University of Wisconsin, Madison,Wisc.,1948
Chem 18.0432

MADISON,WISC. 1960 Partial differential equations and continuum mechanics :an international conference Proceedings United States army.Mathematics research center Edited by Rudolph E. Langer United States army.Mathematics research center.Publication, 5 Madison,Wisc.,1961
Geod 9.0202

MADISON,WISC. 1965 International symposium on organometallic chemistry 2nd Abstracts of proceedings National science foundation, and,Wisconsin alumni research foundation Co-sponsored by the United States.Army research office,Durham Madison,Wisc.,1965 Mimeographed
Chem 18.1589

MADISON,WISC. 1965 Perturbations theory and its applications in quantum mechanics an advanced seminar Papers United States.Army. Mathematics research center,and,University of Wisconsin.Theoretical chemistry institute Edited by Calvin H. Wilcox J.Wiley,New York, 1966
Chem 18.0692

MADISON,WISC. 1966 Improving experimental design and statistical analysis Phi Delta Kappa annual symposium on educational research 7th Discussions Phi Delta Kappa and University of Wisconsin.Phi Delta Kappa chapter Rand-McNally education series Rand-McNally,Chicago,Ill.,1967
Psy 31.1123

MADISON,WISC. 1970 Nonlinear functional analysis and applications Mathematics research center advanced seminar Proceedings University of Wisconsin.Mathematics research center vii,586p 24cm Academic press,New York;London,1971
P Math 2.4286

MADISON,WISC. 1970 Representation theory of finite groups and related topics :symposium American mathematical society Edited by Irving Reiner American mathematical society. Proceedings of symposia in pure mathematics, 21 v,178p 26cm AMS,Providence,R.I.,1971
P Math 2.4432

MADISON,WISC. 1971 Mathematics research center symposium on contributions to nonlinear functional analysis Proceedings University of Wisconsin.Mathematics research center xii, 672p 24cm Academic press,New York;London, 1971
P Math 2.4288

MADRAS 1912 General malaria committee meeting 3rd Proceedings India.General malaria committee Govt.Central branch press, Simla,1913
Mol 45.0491

MADRAS 1963 Aspects of protein structure The International symposium on protein structure and crystallography Proceedings University of Madras.Department of physics Edited by G.N. Ramachandran Academic press, London,1963
Cav 7.1495

MADRAS 1963 Aspects of protein structure The Symposium on protein structure Proceedings University of Madras.Department of physics Edited by G.N. Ramachandran Academic press,London;New York,1963 The Symposium on protein structure formed part of an International symposium on protein structure and crystallography organized by the University of Madras
Bioch 33.0542

MADRAS 1963 Aspects of protein structure : a symposium Proceedings Edited by G.N. Ramachandran Organized by the University of Madras Academic press,London;New York,1963
Gen 34.0658

MADRAS 1963 Crystallography and crystal perfection The International symposium on protein structure and crystallography Proceedings session on crystallography and crystal perfection University of Madras. Department of physics Edited by G.N. Ramachandran Academic press,London,1963
Cav 7.1494

MADRAS 1963 Crystallography and crystal perfection :a symposium Proceedings Edited by G.N. Ramachandran Academic press,London; New York,1963 Papers of a symposium which formed part of the International symposium on protein structure and crystallography
Met 25.1448

MADRAS 1963 Crystallography and crystal perfection :symposium Edited by G.N. Ramachandran Organized by University of Madras Academic press,London;New York,1963
Min 10.1193

MADRAS 1967 An International symposium on conformation of biopolymers Vol 1 Edited by G.N. Ramachandran Academic press,London, 1967
Col S 12.0132

MADRAS 1967 Conformation of biopolymers International symposium on conformation of biopolymers Papers read Vol 1-2 University of Madras.Centre of advanced study in biophysics Edited by G.N. Ramachandran Sponsored by the International union of pure and applied physics 2 vols Academic press, London;New York,1967
Bioch 33.0331

MADRID 1926 Geologie de la Mediterranee occidentale International geological congress 14th etudes et observations Vol 1-2 Association pour l'etude geologique de la Mediterranee occidentale 2 vols Liege, 1926-35
Geol 8.2954

MADRID 1955 Verformung und fliessen des festkorpers Deformation and flow of solids : colloquium International union of theoretical and applied mechanics Edited by R. Grammel Springer,Berlin,1956 Papers in English,French and German
Met 25.1203

MADRID 1958 Advances in aeronautical sciences International congress for the aeronautical sciences 1st Proceedings Vol 1-2 Edited by Theodore von Karman International series on aeronautical sciences and space flight.Division 9:symposia 2 vols Pergamon press,London,1959
Eng 41.6963

MADRID 1958 International automation congress proceedings Consejo superior de investigaciones cientificas.Instituto de electricidad y automatica illus. 411p 30cm Instituto de electricidad y automatica, Madrid,1958
Math L 5.0894

MADRID 1964 Coloquio sobre quimica fisica de processos en superficies solidas Trabajos Consejo superior de investigaciones cientificas Consejo superior de investigaciones cientificas,Madrid,1965 Conference celebrating the 25th anniversary of C.S.I.C.
Chem 18.0609

MADRID 1965 Coloquio international sobre geometria algebraica Actas Consejo superior de investigaciones cientificas. Instituto "Jorge Juan" de matematicas and International mathematical union Edited by Pedro Abellanas Bibliog. 190p 24cm Madrid,1966 Papers mainly in English and Spanish
P Math 2.3137

MADRID 1966 Life sciences research and lunar medicine Lunar international laboratory (LIL)symposium 2nd Proceedings Edited by Frank J. Malina Organized by International academy of astronautics Pergamon press,Oxford;London,1967 Held at the 17th International astronautical congress
PGMS 29.0445

MADRID 1967 Cerebral circulation European congress of neurosurgery 3rd Edited by W. Luyendijk Progress in brain research, 30 Elsevier,Amsterdam,1968
An 32.4248

MADRID 1969 Colloquium spectroscopicum internationale 15th Vol 1-2: plenary conferences;summary of papers 22cm 2 vols Hilger,London,1969-70
Chem 18.2847

MADRID 1969 Federation of European biochemical societies meeting 6th Proceedings Vol 2: membranes,structure and function Federation of European biochemical societies Edited by J.R. Villanueva and F. Ponz Federation of European biochemical societies.Publications, 20 Academic press, London,1970
Bioch 33.2321

MADRID 1969 Federation of European biochemical societies meeting 6th Proceedings Vol 3: macromolecules, biosynthesis and function Federation of European biochemical societies Edited by S. Ochoa and others Federation of European biochemical societies.Publications, 21 Academic press,London,1970
Bioch 33.2322

MADRID 1969 Federation of European biochemical societies meeting 6th Proceedings membranes:structure and function Federation of European biochemical societies Edited by A. Villanueva and F. Ponz Federation of European biochemical societies. Publications, 20 Academic press,London;New York,1969
Radioth 35.0150

MADRID 1969 Federation of European biochemical societies meeting 6th Proceedings metabolic regulation and enzyme action Federation of European biochemical societies Edited by A. Sols and S. Grisolia Federation of European biochemical societies. Publications, 19 Academic press,London,1970
Bioch 33.1889

MAEBASHI 1969 Annual meeting of Japan endocrinological society 42nd Proceedings Japan endocrinological society 1969
PGMS 29.0587

MAGADAN 1962 Bogatstva zemli na sluzhbu rodine :materialy soveshchaniya rabotnikov selskogo khozyaistva oblastei i avlovonykh kraynego Severa Materialy Ministerstvo selskogo khozyaistva,R.S.F.R. 159p 20cm izdatelstvo ministerstva selskogo khozyaistva RSFSR,Moscow,1962
Sco 14.4924

MAGELLANIC CLOUDS The Galaxy and the magellanic clouds :a symposium Canberra 1963 Mar.18-28 International astronomical union and International scientific radio union Edited by F.J. Kerr and A.W. Rodgers International astronomical union.Symposium, 20 393p Australian academy of science, Canberra,1964 Dedicated to Joseph Lade Pawsey
A Math 4.1158

MAGELLANIC CLOUDS The Galaxy and the magellanic clouds :a symposium Proceedings Canberra 1963 Mar.18-28 International astronomical union and International scientific radio union Edited by F.J. Kerr and A.W. Rodgers International astronomical union.Symposium, 20 393p Australian academy of science,Canberra,1964 Publication dedicated to memory of Joseph Lade Pawsey
Obs 6.0941

MAGNETIC AND ELECTRIC RESONANCE AND RELAXATION Colloque Ampere :international conference 11th Proceedings Eindhoven 1962 Jul 2-7 Netherlands physical society Edited by J. Smidt Under the auspices of the International union of pure and applied physics 789p North-Holland,Amsterdam,1963
Cav 7.0712

MAGNETIC AND RELATED STARS A.A.S.-N.A.S.A. symposium on the magnetic and other peculiar and metallic-line A stars Proceedings Greenbelt,Md 1965 Nov 8-10 Berkeley,Calif 1965 Dec 29 American astronomical society and National aeronautics and space administration Edited by Robert C. Cameron xi,596p Mono book corporation,Baltimore,1967
A Math 4.1575

MAGNETIC AND RELATED STARS The Symposium on the magnetic and other peculiar and metallic-line A stars Greenbelt,Md. 1965 Nov 8-10 American astronomical society and National aeronautics and space administration Edited by Robert C. Cameron Mono book corporation, Baltimore,Md.,1967
TA 15.0310

MAGNETIC MATERIALS AND THEIR APPLICATIONS :a conference London 1967 Sep 26-28 Sponsored by the Institution of electrical engineers Institution of electrical engineers.Conference publication, 13 Institution of electrical engineers,London, 1967
Eng 41.5348

MAGNETIC PROPERTIES OF METALS AND ALLOYS Cleveland 1958 Oct 25-26 By R.M. Bozorth and others American society for metals American society for metals,Cleveland,Ohio, 1959 A seminar presented to members of A. S.M.during the National metal congress and exposition
Met 25.2353

MAGNETIC RESONANCE Magnetic and electric resonance and relaxation Colloque Ampere : international conference 11th Proceedings Eindhoven 1962 Jul 2-7 Netherlands physical society Edited by J. Smidt Under the auspices of the International union of pure and applied physics 789p North-Holland,Amsterdam,1963
Cav 7.0712

MAGNETIC STARS AAS-NASA magnetic star symposium The Symposium on the magnetic and other peculiar and metallic-line A stars : magnetic and related stars 1st Proceedings Greenbelt,Md. 1965 Nov 8-10 and Berkeley,Calif. 1965 Dec 29 American astronomical society National aeronautics and space administration Edited by Robert C. Cameron 596p Mono book corporation, Baltimore,1967 Includes the Helen B. Warner prize lecture of the American astronomical society,by George W.Preston
Obs 6.3432

MAGNETICS Intermag conference International conference on magnetics 3rd Proceedings Washington,D.C. 1965 Apr 21-23 Spsonsored by the Institute of electrical and electronics engineers.Magnetics group Institute of electrical and electronics engineers,New York,1965
Eng 41.5345

MAGNETICS 1966 Intermag International magnetics conference 4th Stuttgart 1966 Apr 20-22 Sponsored by the Institute of electrical and electronics engineers.Magnetics group I.E.E.E.transactions on magnetics,Mag 2,no 3 I.E.E.E.,New York,1966
Eng 41.5347

MAGNETISM International conference on magnetism Proceedings Nottingham 1964 Sep 6-11 Institute of physics and the Physical society With the support of the International union of pure and applied physics illus 898p London,c1965
Cav 7.2882

MAGNETISM International conference on magnetism Proceedings Nottingham 1964 Sep Institute of physics and the Physical society London,c1964
Met 25.2354

MAGNETISM International conference on magnetism and crystallography Proceedings Kyoto 1961 Sep 25-30 Pt 2: electron and neutron diffraction Science council of Japan and Crystallographic society of Japan Edited by Shizuo Miyake and others Sponsored by the International union of pure and applied physics Physical society of Japan.Journal, 17,B 2 396p physical society of Japan, Tokyo,1962
Cav 7.1338

MAGNETISM International conference on magnetism and crystallography Proceedings Kyoto 1961 Sep 25-30 Pt 3: neutron diffraction:study of magnetic materials Science council of Japan and Crystallographic society of Japan Edited by Takeo Nagamiya and Shizno Miyake Under the auspices of International union of pure and applied physics Physical society of Japan. Journal, 17,B 3 71p physical society of Japan,Tokyo,1962
Cav 7.1339

MAGNETISM International conference on magnetism and crystallography Proceedings Kyoto 1961 Sep 25-30 Pt 1: magnetism Science council of Japan and Crystallographic society of Japan Edited by Kei Josida and others Held under the auspices of International union of pure and applied physics Physical society of Japan. Journal, 17,B 1 xi,718p Physical society of Japan,Tokyo,1962
Cav 7.1337

MAGNETISM Le Magnetisme Travaux Strasbourg 1939 May 21-25 Universite de Strasbourg.Institut de physique and Institut international de cooperation intellectuelle Supported by Centre national de la recherche scientifique Collection scientifique,Paris,1940
Cav 7.2241

MAGNETISM Magnetism and the cosmos N.A.T.O. advanced study institute on planetary and stellar magnetism Newcastle-upon-Tyne 1965 Apr North Atlantic treaty organisation Edited by W.R. Hindmarsh and others Oliver and Boyd,Edinburgh;London,1967
Geol 8.4428

MAGNETISM Magnetism and the cosmos NATO Advanced study institute on planetary and stellar magnetism Newcastle-upon-Tyne 1965 Apr North Atlantic treaty organization and University of Newcastle upon Tyne Edinburgh, 1967
Geod 9.0567

MAGNETISM Teoria del magnetismo nei metalli di transizione Scuola internazionale di fisica 'Enrico Fermi' 37 corso Rendiconti Varenna 1966 Jun 6-25 Societa italiana di fisica Edited by W. Marshall diagrs 455p 24cm Academic press,London,1967
Cav 7.2612

MAGNETISM AND THE COSMOS North Atlantic treaty organization advanced study institute on planetary and stellar magnetism Newcastle-upon-Tyne 1965 North Atlantic treaty organization Edited by W.R. Hindmarsh and others 436p Oliver and Boyd,Edinburgh,1967
Obs 6.3350

MAGNETO-FLUID DYNAMICS International symposium on magneto-fluid dynamics proceedings Williamsburg,Va. 1960 Jan and Washington D.C. Edited by F.N. Frenkel and W.R. Sears Sponsored by the International union of theoretical and applied mechanics National research council. Publication, 829 National research council, Washington,1960
A Math 4.0619

MAGNETO-FLUID DYNAMICS :a symposium Proceedings Williamsburgh,Va, 1960 Jan and Washington,D.C. 1960 Jan Edited by F. N. Frankiel and W.R. Sears Sponsored by the International union of theoretical and applied mechanics National academy of sciences. Publication, 829 American physical society, New York,1960
Eng 41.4488

MAGNETODYNAMICS OF CONDUCTING FLUIDS Lockheed symposium on magnetohydrodynamics 3rd proceedings Palo Alto 1958 Nov.21-22 Edited by Daniel Bershader Sponsored by Lockheed aircraft corporation.Missiles and space division Stanford university press, Stanford,Calif.,1959
A Math 4.0616

MAGNETODYNAMICS OF CONDUCTING FLUIDS Symposium on magnetohydrodynamics 3rd Palo Alto,Calif. 1958 Nov 21-22 Lockheed aircraft corporation.Lockheed missiles and space division Edited by Daniel Bershader viii,145p 24cm Stanford university press, Stanford,Calif.,1959
Cav 7.2767

MAGNETOHYDRODYNAMICS Advances in magnetohydrodynamics :a colloquium Proceedings Sheffield 1961 Oct Edited by I.A. McGrath and others Organized by the University of Sheffield.Department of fuel technology and chemical engineering Pergamon press,London,1963
Eng 41.4504

MAGNETOHYDRODYNAMICS Advances in magnetohydrodynamics :a colloquium proceedings Sheffield 1961 Oct. Edited by I.A. McGrath and others Organized by University of Sheffield.Department of fuel technology and chemical engineering Pergamon press,London,1963
A Math 4.0622

MAGNETOHYDRODYNAMICS Biennial gas dynamics symposium 4th proceedings Evanston,Ill. 1961 Aug 23-25 American rocket society and Northwestern university Edited by Ali Bulent Cambel and others Through the generosity of Government agencies and industrial sponsors Northwestern university press,Evanston,Ill., 1962
A Math 4.0623

MAGNETOHYDRODYNAMICS Engineering aspects of magnetohydrodynamics Symposium on the engineering aspects of magnetohydrodynamics 3rd Proceedings Rochester,N.Y. 1962 Mar 28-29 Edited by Norman W. Mather and George W. Sutton Gordon and Breach,New York,1964
Eng 41.4546

MAGNETOHYDRODYNAMICS Fundamental topics in relativistic fluid dynamics and magnetohydrodynamics :a symposium proceedings Ann Arbor 1962 Oct.10-11 Edited by Robert Wasserman and Charles P. Wells Academic press,New York,London,1963
A Math 4.0528

MAGNETOHYDRODYNAMICS Fundamental topics in relativistic fluid mechanics and magnetohydrodynamics :a symposium Proceedings East Lansing,Mich. 1962 Oct Edited by Robert Wasserman and Charles P. Wells Held at Michigan state university Academic press,New York,1963
Eng 41.4547

MAGNETOHYDRODYNAMICS Magnetohydrodynamics of conducting fluids Symposium on magnetohydrodynamics 3rd Palo Alto, Calif. 1958 Nov 21-22 Edited by Daniel Bershader Sponsored by Lockheed aircraft corporation.Missile systems division Stanford university press,Stanford,Calif.,1959 1st symposium called 'Magnetohydrodynamics'; 2nd symposium called 'Plasma in a magnetic field',q.v.
Eng 41.4474

MAGNETOHYDRODYNAMICS Plasma acceleration Symposium on magnetohydrodynamics 4th Palo Alto,Calif. 1959 Dec 2 Edited by Sidney W. Kash Sponsored by the Lockheed aircraft corporation.Missiles and space division Stanford university press,Stanford, Calif.,1960
Eng 41.4489

MAGNETOHYDRODYNAMICS Plasma hydromagnetics Symposium on magnetohydrodynamics 6th Proceedings Palo Alto,Calif. 1961 Dec 15-16 Edited by Daniel Bershader Sponsored by Lockheed aircraft corporation.Missile systems division Stanford university press,Stanford, Calif.,1962
Eng 41.4510

MAGNETOHYDRODYNAMICS Propagation and instabilities in plasmas Symposium on magnetohydrodynamics 7th Proceedings Palo Alto,Calif. 1962 Dec 14-15 Edited by Walter I. Futterman Sponsored by Lockheed aircraft corporation.Missiles and space division Stanford university press,Stanford, Calif.,1963
Eng 41.4523

MAGNETOHYDRODYNAMICS Radiation and waves Symposium on magnetohydrodynamics 5th Palo Alto,Calif. 1960 Dec 15-16 Edited by Morton Mitchener Sponsored by Lockheed aircraft corporation.Missile systems division Stanford university press,Stanford,Calif.,1961
Eng 41.4496

MAGNETOHYDRODYNAMICS Radiation and waves in plasmas Symposium on magnetohydrodynamics 5th Palo Alto,Calif. 1960 Lockheed aircraft corporation.Lockheed missiles and space division Edited by Morton Mitchner 156p Stanford university press,Stanford, Calif.,1961
Cav 7.2881

MAGNETOHYDRODYNAMICS Symposium on the engineering aspects of magnetohydrodynamics 2nd Proceedings Philadelphia,Pa. 1961 Mar 9-10 Edited by Clifford Mannal and Norman W. Mather Sponsored by the American institute of electrical engineers Columbia university press,New York;London,1962
Eng 41.4497

MAGNETOHYDRODYNAMICS The Plasma in a magnetic field :a symposium on magnetohydrodynamics Palo Alto,Calif. 1957 Dec 16 Edited by Rolf K.M. Landshoff Sponsored by the Lockheed aircraft corporation. Missile systems division Stanford university press,Stanford,Calif.,1958 The 2nd symposium on magnetohydrodynamics sponsored by Lockheed.For later symposia see 'Symposium on magnetohydrodynamics'
Eng 41.4473

MAGNETOHYDRODYNAMICS :a symposium Palo Alto,Calif. 1956 Dec 29 Edited by Rolf K.M. Landshoff Sponsored by the Lockheed aircraft corporation.Missile systems division Stanford university press,Stanford,Calif.,1957 The 1st symposium on magnetohydrodynamics sponsored by Lockheed
Eng 41.4472

MAGNETOHYDRODYNAMICS OF CONDUCTING FLUIDS Symposium on magnetohydrodynamics 3rd Palo Alto,Calif. 1958 Nov 21-22 Edited by Daniel Bershader Sponsored by Lockheed aircraft corporation.Missile systems division Stanford university press,Stanford,Calif.,1959 1st symposium called 'Magnetohydrodynamics'; 2nd symposium called 'Plasma in a magnetic field',q.v.
Eng 41.4474

MAGNETOPLASMADYNAMIC ELECTRICAL POWER GENERATION a symposium Report Newcastle-upon-Tyne 1962 Sep 6-8 Institution of electrical engineers Institution of electrical engineers.Conference report series, 4 Institution of electrical engineers,London, 1963
Eng 41.4521

MAGNETOSPHERE Natural electromagnetic phenomena below 30 kilocycles NATO advanced study institute Proceedings Bad Homburg 1963 Jul 22-Aug 2 United States.Office of naval research and Naval ordnance laboratory Edited by D.F. Bleil vii,470p Plenum press,New York,1964
Nap 11.0067

MAGYAR MATEMATIKAI KONGRESSZUS 2 abstracts of lectures,mimeograph Budapest 1960 Aug 24-31 Vols. 1-2 and suppt. 29cm 3 vols Budapest,1960
P. Math 2.2607

MAGYAR TUDOMANYOS AKADEMIA Abelian groups colloquium proceedings Tihany,Hungary 1963 Sep 2-7 Edited by L. Fuchs and E.T. Schmidt 162p 24cm Akademiai kiado, Budapest,1964
P. Math 2.2028

MAGYAR TUDOMANYOS AKADEMIA International conference on luminescence Proceedings Budapest 1966 Edited by G. Szigeti diagrms 24cm 2 vols Akademiai kiado, Budapest,1968
Cav 7.3097

MAGYAR TUDOMANYOS AKADEMIA Budapest conference on soil mechanics and foundation engineering 3rd Proceedings Budapest 1968 Oct 15-18 Akademiai kiado,Budapest,1968
Eng 41.3137

MAGYAR TUDOMANYOS AKADEMIA International conference on cosmic rays 11th Invited papers and rapporteur talks Budapest 1969 Aug 25-Sep 4 612p Central research institute of physics,Budapest,1969
TA 15.0551

MAGYAR TUDOMANYOS AKADEMIA.INSTITUTE OF NUCLEAR RESEARCH Conference on the electron capture and higher processes in nuclear decays Proceedings Debrecen 1968 Jul 15-18 Vol 1-3 diagrs 20cm 3 vols Eotvos Lorand physical society,Budapest,1968
Cav 7.2700

MAGYAR TUDOMANYOS AKADEMIA.SECTION OF MATHEMATICS AND PHYSICS The Foundations of mathematics,mathematical machines and their applications colloquium Tihany 1962 Sep 11-15 Edited by Laszlo Kalmar In cooperation with the Bolyai Janos matematikai tarsulat 317p 25cm Akademiai kiado,Budapest,1965
P. Math 2.2359

MAGYAR TUDOMANYOS AKADEMIA Regeneration and wound healing Proceedings Budapest 1960 Nov 8-9 Symposia biologica hungarica, 3 Akademiai kiado,Budapest,1960
Phys 20.1001

MAINZ 1842 Versammlung der Gesellschaft deutscher naturforscher und aerzte... 20th Amtlicher bericht Gesellschaft deutscher naturforscher und aerzte Edited by Johann Groser and Friedrich Karl Bruch 2 pls 28cm Kupferberg,Mainz,1843
Bal 44.6348

MAJOR PROBLEMS IN DEVELOPMENT BIOLOGY :a symposium Edited by Michael Locke Society for the study of development and growth.Symposia, 25 Academic press,London; New York,1966
An 32.2529

MAJOR PROBLEMS IN DEVELOPMENTAL BIOLOGY :a symposium Haverford,Pa. 1966 Jun Society for developmental biology Edited by Michael Locke Society for developmental biology.Symposia, 25 Academic press,New York;London,1966
Bioch 33.0912

MALARIA General malaria committee meeting 3rd Proceedings Madras 1912 Nov 18-20 India.General malaria committee Govt.Central branch press,Simla,1913
Mol 45.0491

MALARIA Imperial malaria conference Proceedings Simla 1909 Oct 12-18 India. General malaria committee Govt.central branch press,Simla,1910
Mol 45.0490

MALIGNANCY Biological and biochemical evaluation of malignancy in experimental hepatomas The U.S.-Japan joint conference on biological and biochemical evaluation of malignancy in experimental hepatomas Proceedings Kyoto 1965 Nov 4-5 Held under the auspices of the United States-Japan scientific cooperation program G.A.N.N. Monograph, 1 Japanese cancer association; Japanese foundation for cancer research,Tokyo, 1966
Bioch 33.0952

MALIGNANT TRANSFORMATION BY VIRUSES :a symposium Chicago,Ill. 1966 Feb 26-27 Edited by W.H. Kirsten Recent advances in cancer research, 6 Springer,Berlin,1966
Radioth 35.0856

MALVERN 1961 Microminiaturisation and its effect on computer elements discussion meetings proceedings Royal radar establishment.Radio components research and development sub-committee 15:computer elements 15p Royal radar establishment,Malvern,1961
Math L 5.1848

The MAMMALIAN FETUS :physiological aspects of development.A symposium Papers Cold Spring Harbor 1954 Cold Spring Harbor biological laboratory Cold Spring Harbor symposia on quantitative biology, 19 Long Island biological association,Cold Spring Harbor,1954
Bioch 33.1275

The MAMMALIAN FETUS:PHYSIOLOGICAL ASPECTS OF DEVELOPMENT :symposium Cold Spring Harbor 1954 Jun 7-14 Cold Spring Harbor symposia on quantitative biology, 19 Long Island biological association,New York,1954
Phys 20.1376

MAMMALIAN GENETICS AND REPRODUCTION Symposium on mammalian genetics and reproduction Gatlinburg,Tenn. 1960 Apr 4-7 Sponsored by the Oak Ridge national laboratory. Biology division Oak Ridge national laboratory.Symposia, 13 Journal of cellular and comparative physiology, 56,supp.1 Wistar institute of anatomy and biology, Philadelphia,Pa.,1960
Gen 34.1782

MAMMALIAN GERM CELLS :Ciba foundation symposium London 1952 Jun 17-20 Ciba foundation Edited by G.E.W. Wolstenholme and others Churchill,London,1953
Phys 20.1478

MAMMALIAN HIBERNATION 3 International symposium on natural mammalian hibernation 3rd Proceedings Toronto 1965 Sep 13-16 Edited by Kenneth C. Fisher and others illus xiv,534p 25cm Oliver and Boyd,Edinburgh; London,1967
Sco 14.8231

MAMMALIAN REPRODUCTION :colloquium Mosbach 1970 Apr 9-11 Gesellschaft fur biologische chemie Edited by H. Gibian and E. J. Motz Gesellschaft fur biologische chemie. Colloquia, 21 Springer,Berlin,1970
An 32.5461

MAMMALIAN REPRODUCTION :colloquium Mosbach 1970 Apr 9-11 Gesellschaft fur biologische chemie Edited by H. Gibian and E. J. Plotz Gesellschaft fur biologische chemie. Colloquia, 21 Springer,Heidelburg;New York, 1970
Inv Med 37.0088

MAMMARY CANCER International symposium on mammary cancer 2nd Proceedings Perugia 1957 Jul 24-29 Edited by Lucio Severi Division of cancer research,Perugia,1958
Radioth 35.0779

MAMMARY TUMORS IN MICE A Symposium on mammary tumors in mice National cancer institute Edited by Forest Ray Moulton American association for the advancement of science.Publication, 22 American association for the advancement of science, Washington,D.C.,1945
Gen 34.1813

MAN,CULTURE AND ANIMALS American association for the advancement of science :annual meeting 128th Symposium Denver,Colo. 1961 Dec 30 Edited by Anthony Leeds and Andrew P. Vayda American association for the advancement of science.Publications,78 Washington,D.C.,1963
Geog 13.1313

MAN,RACE AND DARWIN :a conference London 1959 Jan 7-9 Royal anthropological institute of Great Britain and Ireland Institute of race relations Edited by Philip Mason vii, 151p Oxford university press,London,1960
WSM 43.3622

MAN-MADE LAKES :a symposium Proceedings London 1965 Sep 30-Oct 1 Institute of biology Edited by R. Lowe-McConnell held at the Royal geographical society Institute of biology.Symposia, 15 Academic press, London;New York,1966
Bal 39.1898

MAN-MADE LAKES Institute of biology : symposium 15th London 1965 Sep 30-Oct 1 Edited by R.H. Lowe-McConnell Academic press, London,1966
Geog 13.0836

MANAGERIAL ASPECTS OF DIGITAL COMPUTER INSTALLATIONS symposium papers Washington,D.C. 1953 Mar.30 Organized by United States.Navy mathematical computing advisory panel 36p Office of naval research,Washington,D.C.,1953
Math L 5.1792

MANCHESTER 1946 The Measurement of stress and strain in solids :a conference Institute of physics.Manchester and district branch Edited by F.A. Vick Physics in industry Institute of physics,London,1948 Book based on the proceedings of the conference
Cav 7.2397

MANCHESTER 1946 The Measurement of stress and strain in solids :a conference Proceedings Institute of physics Institute of physics,London,1948
Eng 41.2898

MANCHESTER 1946 The Measurement of stress and strain in solids :a conference Proceedings Institute of physics.Manchester and district branch Edited by F.A. Vick Physics in industry Institute of physics, London,1948
Met 25.0880

MANCHESTER 1947 Electrode processes :a general discussion Faraday society Faraday society.Discussions, 1 Butterworths,London, 1961
Met 25.2396

MANCHESTER 1950 Mass spectrometry :a conference Report Organized by the Institute of petroleum.Mass spectrometry panel Institute of petroleum,London,1952
Bioch 33.0282

MANCHESTER 1950 Mass spectroscopy :a conference Report Institute of petroleum. Mass spectroscopy panel Institute of petroleum,London,1952
Chem 18.0110

MANCHESTER 1951 Manchester university computer :inaugural conference Victoria university of Manchester in co-operation with Ferranti ltd pl. 40p 27cm Manchester,1957
Math L 5.1320

MANCHESTER 1954 Meteors Symposium on meteor physics Proceedings Jodrell Bank experimental station Edited by T.R. Kaiser Journal of atmospheric and terrestrial physics. Special supplement, 2 204p Pergamon press,London,1955
Obs 6.0923

MANCHESTER 1955 Astronomical optics and related subjects A Symposium on astronomical optics and related subjects proceedings Edited by Zdenek Kopal Under the auspices of Victoria university of Manchester.Astronomy department 428p North-Holland,Amsterdam, 1956 Dedicated to the memory of Peter Yorke Millns
Obs 6.0975

MANCHESTER 1955 Radio astronomy :a symposium Proceedings International astronomical union Edited by H.C.van de Hulst International astronomical union. Symposium, 4 Cambridge,1957
TA 15.0037

MANCHESTER 1955 Radio astronomy :a symposium proceedings International astronomical union Edited by H.C.van de Hulst With financial assistance from Unesco International astronomical union.Symposium, 4 409p Cambridge university press, Cambridge,1957
Obs 6.0847

MANCHESTER 1960 Symposium on chemical process hazards with special reference to plant design Proceedings Edited by D.M. Pirie 117p Institution of chemical engineers,London,1960
Chem E 24.1395

MANCHESTER 1961 Rutherford jubilee international conference Proceedings Victoria university of Manchester Edited by J.B. Birks Under the sponsorship of International union of pure and applied physics 856p Haywood,London,1961 Includes commemorative session 'Rutherford at Manchester'
Cav 7.1952

MANCHESTER 1961 Rutherford jubilee international conference proceedings Victoria university of Manchester Edited by J.B. Birks Co-sponsored by the International union of pure and applied physics Heywood, London,1961 To commemorate the discoveries of Rutherford at Manchester
A Math 4.0826

MANCHESTER 1962 Non-conventional lubricants and bearing materials such as used in nuclear engineering Institution of mechanical engineers.Lubrication and wear group Institution of mechanical engineers, London,1962
Eng 41.6303

MANCHESTER 1965 Machines for materials and environmental testing Symposium on techniques and equipment for environmental testing; symposium on developments in materials testing machine design :joint conference Institution of mechanical engineers and Society of environmental engineers Under the aegis of the British national committee on materials Institution of mechanical engineers.Proceedings, 180,pt 3a I.M.E.,London,1966
Met 25.0891

MANCHESTER 1965 Pulse radiolysis : international symposium Proceedings Edited by M. Ebert and others Academic press,London, 1965
Radioth 35.1773

MANCHESTER 1966 John Dalton and the progress of science :papers presented at a conference of historians of science Manchester literary and philosophical society xxii,352p Manchester university press;Barnes and Noble,Manchester;New York,1968 Conference held to mark the bicentenary of Dalton's birth.Co-sponsored by the Royal society,Chemical society and Society of chemical industry
WSM 43.2304

MANCHESTER 1967 Nutrition in renal disease a conference Proceedings Edited by Geoffrey Merton Berlyne Livingstone;Williams, Edinburgh;Baltimore,1968
Inv Med 37.0021

MANCHESTER 1968 International vacuum congress 4th Pt 1-2 Joint British committee for vacuum science and technology International union for vacuum science,technique and applications Institute of physics and the Physical society.Conference series, 5-6 2 vols Institute of physics, London,c1968
Met 25.2834

MANCHESTER 1968 Osborne Reynolds and engineering science today Osborne Reynolds centenary symposium Papers By Jack Allen and others Edited by D.M. McDowell and J.D. Jackson Manchester university press;Barnes and Noble,Manchester;New York,1970
Eng 41.1755

MANCHESTER 1968 The Mechanism of phase transformations in crystalline solids :an international symposium Proceedings Institute of metals Institute of metals. Monograph and report series, 33 illus, diagrs 324p 25cm London,1969
Cav 7.2666

MANCHESTER 1969 Conference on computer science and technology Sponsored by the institution of electrical engineers. Electronics division.Computer design professional group Institution of electrical engineers.Conference publication, 55 The institution of electrical engineers,London, 1969
Math L 5.3503

MANCHESTER 1969 Logic colloquium 1969 Summer school and colloquium in mathematical logic Proceedings Victoria university of Manchester Edited by R.O. Gandy and C.M.E. Yates Studies in logic and the foundations of mathematics, 61 xiv,451p 23cm North-Holland,Amsterdam,1971
P Math 2.4057

MANCHESTER 1971 Multivariable control system design and application U.K.A.C. control convention 4th United Kingdom automation council Organized by the Institution of electrical engineers Institution of electrical engineers.Conference publication, 78 IEE,London,1971
Eng 41.8590

MANCHESTER 1972 Electron microscopy 1972 European congress on electron microscopy 5th Proceedings International federation of societies for electron microscopy Institute of physics.Conference series, 14 Institute of physics,London,1972
Eng 41.8576

MANCHESTER 1972 European congress on electron microscopy 5th Proceedings Institute of physics Institute of physics. Conference series, 14 illus xxviii,683p 26cm Institute of physics,London,1972
Cav 7.3182

MANCHESTER COLLEGE OF SCIENCE AND TECHNOLOGY Modern yarn production from man-made fibres (and their conversion into fabrics) :a symposium Edited by G.R. Wray illus. Columbine press,Manchester;London,1960 Lectures given at Manchester college of science and technology
Eng 41.3661

MANCHESTER LITERARY AND PHILOSOPHICAL SOCIETY John Dalton and the progress of science : papers presented at a conference of historians of science Manchester 1966 Sep 19-24 xxii,352p Manchester university press;Barnes and Noble,Manchester;New York,1968 Conference held to mark the bicentenary of Dalton's birth.Co-sponsored by the Royal society,Chemical society and Society of chemical industry
WSM 43.2304

MANCHESTER UNIVERSITY COMPUTER :inaugural conference Manchester 1951 Jul.9-12 Victoria university of Manchester in co-operation with Ferranti ltd pl. 40p 27cm Manchester,1957
Math L 5.1320

MANEBACH 1967 Beitrage zur graphentheorie : internationale kolloquium Vortragen Technische hochschule Ilmenau.Mathematische institut Mathematische gesellschaft der D.D.R. Edited by Horst Sachs and others Bibliog.,Illus. 394p 23cm Teubner, Leipzig,1968
P Math 2.3352

MANGANESE Symposium sobre yacimientos de manganeso International geological congress 20th papers Mexico city 1956 Tom 1-5 Edited by Jenaro Gonzalez Reyna 2 vols Mexico city,1956 Text in English,French, German and Spanish
Geol 8.3027

MANHATTAN,KAN 1964 Macromolecules and behaviour Edited by John Gaito North Holland;Meredith,Amsterdam;New York,1966 Revised and current versions of presentations at a Conference held at Kansas state university on April 20 to 22,1964 under the sponsorship of the Office of naval research to discuss the role of macromolecules in complex behavior.
Bal 39.1622

MANIFOLD AND MAPS Liverpool singularities symposium II Symposium on singularities of smooth manifolds and maps Proceedings Liverpool 1970 Aug-Sep University of Liverpool.Department of pure mathematics Edited by C.T.C. Wall Lecture notes in mathematics, 209 280p 25cm Springer, Berlin,1971
P Math 2.4123

MANIFOLDS Seminar on contact manifolds Lectures Kyoto 1969 Jun 23-25 By Koichi Ogiue and others Kyoto university.Research institute for mathematical sciences Okayama university.Study group of geometry.Publication, 4 ii,83p 26cm Okayama university. Department of mathematics,Okayama,1970
P Math 2.4119

MANIFOLDS Topology of 3-manifolds and related topics :a symposium proceedings Athens, Ga. 1961 Aug 14 - Sep 8 University of Georgia.Topology institute Edited by M.K. Fort supported by United States.Office of naval research and National science foundation viii,256p 24cm Englewood Cliffs, N.J.,1962
P. Math 2.0296

MANIFOLDS - AMSTERDAM 1970 NUFFIC summer school on manifolds Proceedings Amsterdam 1970 Aug 17-29 Edited by N.H. Kuiper Lecture notes in mathematics, 197 230p 25cm Springer,Berlin,1971
P Math 2.4116

MANIFOLDS AND MAPS Symposium on singularities of smooth manifolds and maps Liverpool singularities symposium I Proceedings Liverpool 1970 Aug -Sep University of Liverpool.Department of pure mathematics Edited by C.T.C. Wall Lecture notes in mathematics, 192 318p 25cm Springer,Berlin,1971
P Math 2.4122

MANILA 1964 United nations regional cartographic conference for Asia and the Far East 4th 2: proceedings of the conference and technical papers United nations. Department of economic and social affairs United nations,New York,1966
Geog 13.7226

MAN'S PLACE IN THE ISLAND ECOSYSTEM Pacific science congress 10th Symposium Honolulu 1961 Aug 21-Sep 6 Edited by F.R. Fosberg Bishop museum press,Honolulu,1963
Geog 13.1311

MANS ROLE IN CHANGING THE FACE OF THE EARTH Princeton,N.J. 1955 Jun 16-22 Wenner-Gren foundation for anthropological research Edited by William L. Thomas and others Sponsored jointly by the National science foundation University of Chicago press, Chicago,Ill.;London,1956
Geog 13.1427

MANTLES OF THE EARTH AND TERRESTRIAL PLANETS N.A.T.O. advanced study institute :a conference Proceedings Newcastle upon Tyne 1966 Mar 30-Apr 7 University of Newcastle upon Tyne.School of physics Edited by S.K. Runcorn Sponsored by North Atlantic treaty organization.Science office Interscience, London,1967
TA 15.0202

MANTLES OF THE EARTH AND TERRESTRIAL PLANETS : Nato advanced study institute Proceedings University of Newcastle-upon-Tyne.School of physics Edited by S.K. Runcorn Interscience,London,1967
Geod 9.0016

MANY-BODY PROBLEM Lectures on field theory and the many-body problem :international spring school 1st Naples 1961 Universita di Napoli.Instituto di fisica teorica Edited by E.R. Caianiello With support of North Atlantic treaty organization Academic press,New York,1961
Cav 7.0423

MANY-BODY PROBLEM Lectures on the many-body problem :international spring school 2nd Naples 1962 Ravello 1963 Vol 1-2 Universita di Napoli.Instituto di fisica teorica Edited by E.R. Caianiello With support of North Atlantic treaty organization 2 vols Academic press,New York,1962-64
Cav 7.0424

The MANY-BODY PROBLEM :le probleme a N corps: cours donnes a l'ecole d'ete de physique theorique
Chem 18.2620

The MANY-BODY PROBLEM :a symposium Proceedings Hoboken,N.J. 1957 Jan 28-29 Stevens institute of technology Edited by Jerome K. Percus Interscience,New York; London,1963
Chem 18.2362

MAPPINGS Conference on monotone mappings and open mappings 1st Proceedings Binghamton,N.Y. 1970 Oct 8-11 State university of New York,Binghampton Edited by Louis F. McAuley xxii,422p 25cm State university of New York,Binghamton,N.Y.,1971 Proceedings dedicated to the memory of G.T. Whyburn
P Math 2.4118

MAR DEL PLATA 1952 International congress of hematology 4th Proceedings International society of hematology Edited by F. Jimenez de Asua and others Grune and Stratton,New York,1954
Med 36.0013

MAR DEL PLATA,ARGENTINA 1965 Problems of atmospheric circulation The Space science symposium 6th Papers of a session Committee on space research and International union of biochemistry Edited by R.V. Garcia and T.F. Malone Co-sponsored by the International union of physiological sciences 186p Spartan books,Washington,D.C. 1966
Nap 11.0212

MARCHWOOD,SOUTHAMPTON 1963 The Mechanism of corrosion by fuel impurities :international conference Marchwood engineering laboratories Edited by H.R. Johnson and D.J. Littler Butterworths,London,1963
Met 25.1939

MARCHWOOD ENGINEERING LABORATORIES The Mechanism of corrosion by fuel impurities : international conference Marchwood, Southampton 1963 May 20-24 Edited by H.R. Johnson and D.J. Littler Butterworths,London, 1963
Met 25.1939

MARFA,TEXAS 1969 Planetary atmospheres International astronomical union Edited by C. Sagan and others International astronomical union.Symposium, 40 408p Reidel,Dordrecht, 1971
Obs. 6.3586

MARGATE 1955 The Automatic factory - What does it mean? conference report Institution of production engineers illus. 227p 28cm Institution of production engineers,London,1955
Math L 5.0734

MARGATE 1955 The Automatic factory - what does it mean? :conference Institution of production engineers Institution of production engineers,London,1955
Eng 41.5974

MARIA MOORS CABOT FOUNDATION Physiology of forest trees An International symposium of forest tree physiology 1st Harvard forest 1957 Apr 8-12 Edited by Kenneth Thimann and others Ronald press,New York, 1958
Bot 42.1335

MARIE CURIE FOUNDATION FUND Domiciliary care of the patient with cancer :symposium London 1969 Edited by R.W. Raven Heinemann,London, 1970
PGMS 29.0561

MARIENBAD 1965 International congress of chemical engineering,chemical equipment and automation 2nd Lecture summaries Czechoslovak scientific-technical society Czechoslovak chemical society Edited by Tomas Misek and Pavel Mitschka 1965
Chem E 24.1907

MARIENBAD 1969 International congress of chemical engineering,chemical equipment construction and automation 3rd Lecture summaries Pt A-H Scientific committee Chisa '69,1969 Cover title:Chisa '69
Chem E 24.1872

MARINE BIO-ACOUSTICS 1st,2nd :symposium Proceedings Bimini,Bahamas 1963 Apr 11-13 New York 1966 Apr 13-15 Vol 1-2 By William N. Tavolga 2 vols Pergamon press,Oxford,1964-67 Vol.1 held at the Lerner marine laboratory Bimini;vol.2 held at the American museum of natural history,New York
Bal 39.1534

MARINE BIOLOGICAL LABORATORY Subcellular particles :a symposium held during the meeting of the Society of general physiologists at the Marine biological laboratory Woods Hole,Mass. 1958 Jun 9-11 Society of general physiologists Edited by Teru Hayashi Published for the American physiological society Ronald press,New York,1959
Bioch 33.1014

MARINE BIOLOGICAL LABORATORY The Control of nuclear activity :a symposium held under the auspices of the Society of general physiologists at its annual meeting at the Marine biological laboratory Woods Hole,Mass. 1966 Aug 31-Sep 3 Society of general physiologists Edited by Lester Goldstein Prentice-Hall,Englewood Cliffs,N.J.,1967
Bioch 33.1013

MARINE BIOLOGY Colloque international sur l'histoire de la biologie marine Les Grandes expeditions scientifiques et la creation des laboratoires maritimes Banyuls-sur-Mer 1963 Sep 2-6 International union of history and philosophy of science Vie et milieu. Supplement, 19 370p Laboratoire Arago; Masson,Banyuls-sur-Mer;Paris,1965
WSM 43.2877

MARINE BIOLOGY Contributions to marine biology :lectures and symposia given at the Hopkins marine station...at The midwinter meeting of the Western society of naturalists Pacific Grove,Calif. 1929 Dec 20-21 Western society of naturalists Stanford university press,Stanford,Calif.,1930
Bal 39.1933

MARINE BORING AND FOULING ORGANISMS Friday Harbor symposium Friday Harbor 1957 Sep Edited by D.L. Ray held at the Friday Harbor laboratories Seattle,1959
Bal 39.1950

MARINE ORGANISMS Biochemistry and pharmacology of compounds derived from marine organisms :a conference Papers New York 1960 Apr 6-8 New York academy of sciences and American institute of biological sciences Edited by Ross F. Nigrelli New York academy of sciences.Annals, 90,article 3 New York,1960 Co-sponsored by the New York zoological society and the United States navy,Office of naval research
Chem 26.0302

MARKETS AND MARKETING AS FACTORS OF DEVELOPMENT IN THE MEDITERRANEAN BASIN Mediterranean social sciences research council.Assembly 2nd Papers Cairo 1962 Dec 1-4 Edited by C.A.O. van Nieuwenhuijze Institute of social studies,The Hague.Publications.Series maior,11 Mouton,The Hague,1963
Geog 13.4117

MARKOV PROCESSES AND POTENTIAL THEORY :a symposium Proceedings Madison,Wis. 1967 May 1-3 Edited by Joshua Chover Sponsored by the United States.Army. Mathematics research center United States army.Mathematics research center.Publications, 19 Bibliog x,235p 24cm John Wiley,New York,1967
Math 3.1189

MARS The Atmospheres of Venus and Mars :a symposium Tucson,Ariz. 1967 Feb 28-Mar 2 Goddard institute for space studies and Kitt Peak national observatory.Space division Edited by John C. Brandt and Michael B. McElroy Gordon and Breach,New York,1968
TA 15.0415

MARSEILLE 1949 Les Proprietes optiques des lames minces solides Proceedings Centre national de la recherche scientifique Supported by the Rockefeller foundation Centre national de la recherche scientifique. Colloques internationaux, 23 176p Centre national de la recherche scientifique,Paris, 1950
Obs 6.2009

MARSEILLE 1961 Mecanique de la turbulence : colloque international Centre national de la recherche scientifique Centre national de la recherche scientifique.Colloques internationaux, 108 470p Editions du centre national de la recherche scientifique, Paris,1962
Chem E 24.0370

MARSEILLE 1961 Mecanique de la turbulence colloque international Mecanique de la turbulence Centre national de la recherche scientifique Centre national de la recherche scientifique.Colloques internationaux, 108 Centre national de la recherche scientifique, Paris,1962
A Math 4.0525

MARSEILLE 1964 AGARD-NATO specialists' meeting The Fluid dynamic aspects of space flight Vol 1-2 Agard AGARDograph, 87 2 vols Gordon and Breach,New York,1966
Eng 41.6887

MARSEILLE 1969 Archeologie et calculateurs: problemes semiologiques et mathematiques : colloque international Centre national de la recherche scientifique Centre national de la recherche scientifique.Colloques internationaux bibliog.,illus. 371p 28cm Centre nationale de la recherche scientifique, Paris,1970
Math S 3.1631

MARSEILLES 1961 Fundamental problems in turbulence and their relation to geophysics symposium proceedings Vol 67 International union of geodesy and geophysics and International union of theoretical and applied mechanics Edited by Francois N. Frenkel Journal of geophysical research, Vol 67, no.8 American geophysical union, Washington D.C., 1962
A Math 4.0577

MARSEILLES 1963 Mecanismes de regulation des activites cellulaires chez les microorganismes Colloque international Centre national de la recherche scientifique Centre national de la recherche scientifique. Colloques internationaux, 124 Paris, 1965
Gen 34.0707

MARTIN COMPANY RESEARCH INSTITUTE FOR ADVANCED STUDIES Environment-sensitive mechanical behaviour :a conference Proceedings Baltimore 1965 Jun 7-8 Pt 1-2 American institute of mining, metallurgical and petroleum engineers. Physical metallurgy committee and United States. Army research office, Durham Edited by A.R.C. Westwood and N.S. Stoloff Metallurgical society conferences, 35 2 vols Gordon and Breach, New York, 1966
Met 25.1963

MARTIN SCHEERER MEMORIAL MEETINGS ON COGNITIVE PSYCHOLOGY Cognition, theory, promise : papers read at the Martin Scheerer memorial meetings on cognitive psychology Papers University of Kansas 1962 May 7-9 Edited by Constance Scheerer Harper and Row, New York, 1964
Psy 31.1919

MARTINGALES :a meeting Report Oberwolfach 1970 May 17-23 Mathematisches forschungsinstitut, Oberwolfach Edited by Hermann Dinges Lecture notes in mathematics, 190 75p 25cm Springer, Berlin, 1971
P Math 2.4559

MARX AND CONTEMPORARY SCIENTIFIC THOUGHT The Role of Karl Marx in the development of contemporary scientific thought :a symposium Papers Paris 1968 May 8-10 International council for philosophy and humanistic studies International social science council Under the auspices of Unesco International social science council. Publications, 13 xi,612p Mouton, The Hague; Paris, 1969
WSM 43.3826

MASS LOSS AND EVOLUTION IN CLOSE BINARIES Elsinore 1969 International astronomical union Edited by K. Gyldenkerne and R.M. West International astronomical union. Colloquium, 6 238p Copenhagen university observatory, Copenhagen, 1970
Obs 6.3358

MASS LOSS FROM STARS Trieste colloquium on astrophysics 2nd Proceedings Trieste 1968 Sep 12-17 Edited by Margherita Hack Sponsored by the International astronomical union Astrophysics and space science library, 13 345p Reidel, Dordrecht, 1969
Obs 6.3628

MASS LOSS FROM STARS Trieste colloquium on astrophysics 2nd Proceedings Trieste 1968 Sep 12-17 Edited by Margherita Hack Sponsored by the International astronomical union Astrophysics and space science library, 13 345p Reidel, Dordrecht, 1969
TA 15.0620

MASS MOTIONS IN SOLAR FLARES AND RELATED PHENOMENA Nobel symposium 9th Proceedings Anacapri 1968 Jun 10-12 Nobel foundation Edited by Yngve Ohman 245p 24cm Wiley, New York, 1968
Cav 7.2971

MASS SPECTROMETRY :Nato advanced study institute on theory, design and applications Glasgow 1964 Aug North Atlantic treaty organization Edited by R.I. Reed Academic press, London; New York, 1965
Met 25.1694

MASS SPECTROMETRY :a conference Report Manchester 1950 Apr 20-21 Organized by the Institute of petroleum. Mass spectrometry panel Institute of petroleum, London, 1952
Bioch 33.0282

MASS SPECTROSCOPY Advances in mass spectrometry :a conference Proceedings London 1958 Sep 24-26 Institute of petroleum. Hydrocarbon research group, and, American society for testing materials. Committee on mass spectrometry Edited by J.D. Waldron Pergamon press. Symposia publications division, London, 1959
Chem 18.0197

MASS SPECTROSCOPY :a conference Report Manchester 1950 Apr 20-21 Institute of petroleum. Mass spectroscopy panel Institute of petroleum, London, 1952
Chem 18.0110

MASSACHUSETTS 1952 Dynamics of growth processes :a symposium Papers Society for the study of development and growth Edited by Edgar J. Boell Society for the study of development and growth. Symposia, 11 Princeton university press, Princeton, N.J., 1954
Radioth 35.0197

MASSACHUSETTS 1968 Symposium of plasma dynamics proceedings National academy of sciences Edited by Frances H. Clauser Sponsored by the United States. Air force. Office of scientific research Pergamon press, London, 1959
A Math 4.0617

MASSACHUSETTS INSTITUTE OF TECHNOLOGY Advances in earth sciences International conference on the earth sciences contributions Cambridge, Mass. 1964 Sep 30-Oct 2 Edited by P.M. Hurley M.I.T. press, Cambridge, Mass., 1966
Geol 8.4429

MASSACHUSETTS INSTITUTE OF TECHNOLOGY Analysis in function space Conference on the theory and applications of analysis in function space proceedings Dedham, Mass. 1963 Jun 9-13 Edited by William Ted Martin and Irving Segal Financially supported by United States. Air force. Office of scientific research vi,218p 24cm M.I.T. press, Cambridge, Mass., 1964
P. Math 2.1181

MASSACHUSETTS INSTITUTE OF TECHNOLOGY
Conference on soil stabilization Proceedings Cambridge,Mass. 1952 Jun 18-20 M.I.T.press, Cambridge,Mass.,1952
Eng 41.3134

MASSACHUSETTS INSTITUTE OF TECHNOLOGY
Conference on the theory and applications of analysis in function space Proceedings Dedham,Mass. 1963 Jun 9-13 Edited by William T. Martin and Irving Segal vi,218p 24cm MIT press,Cambridge,Mass.,1964
Cav 7.3154

MASSACHUSETTS INSTITUTE OF TECHNOLOGY
Fatigue and fracture of metals :a symposium Cambridge,Mass. 1950 Jun 19-22 Edited by W. M. Murray Technology press;Wiley,New York, 1952
Met 25.0911

MASSACHUSETTS INSTITUTE OF TECHNOLOGY
Fatigue and fracture of metals :a symposium Cambridge,Mass. 1950 Jun 19-22 Edited by William M. Murray Technology press books M. I.T.press;Wiley,Cambridge,Mass.;New York,1952
Eng 41.3740

MASSACHUSETTS INSTITUTE OF TECHNOLOGY
Hypersonic flow research :a selection of technical papers based mainly a symposium of the American rocket society,held at the Massachusetts institute of technology... Cambridge,Mass. 1961 Aug 16-18 American rocket society Edited by Frederick R. Riddell Progress in astronautics and rocketry, 7 Academic press,NewYork,1962
Eng 41.6771

MASSACHUSETTS INSTITUTE OF TECHNOLOGY
Insulation of high voltages in vacuum : international symposium 1964 Massachusetts institute of technology,Cambridge,1964
Eng 41.4984

MASSACHUSETTS INSTITUTE OF TECHNOLOGY
International congress for applied mechanics 5th proceedings Cambridge,Mass 1938 Sep. 12-16 Edited by J.P. Den Hartog and H. Peters John Wiley,New York,1939
A Math 4.0325

MASSACHUSETTS INSTITUTE OF TECHNOLOGY
Mechanical wear :a summer conference Proceedings Cambridge,Mass. 1948 Jun Edited by John T. Burwell Jointly sponsored by the General motors corp. and the Chrysler corp. American society for metals,1950
Met 25.1597

MASSACHUSETTS INSTITUTE OF TECHNOLOGY
Microsomal particles and protein synthesis :a symposium Papers presented Cambridge,Mass. 1958 Feb 5-8 Biophysical society Edited by Richard B. Roberts Biophysical society. Symposia, 1 Pergamon,London,1958
Bioch 33.0924

MASSACHUSETTS INSTITUTE OF TECHNOLOGY Mid-century Convocation on the social implications of scientific progress Verbatim account Cambridge,Mass. 1949 Mar 31-Apr 2 Edited by John Ely Burchard xx,549p Technology press of the Massachusetts institute of technology,Cambridge,Mass.,1950
Cav 7.0406

MASSACHUSETTS INSTITUTE OF TECHNOLOGY
Mycotoxins in foodstuffs :a symposium Proceedings Cambridge,Mass. 1964 Mar 18-19 Edited by Gerald N. Wogan M.I.T.press, Cambridge,Mass.,1965 "This volume comprises the proceedings of an International symposium on mycotoxins in foodstuffs".
Bioch 33.0708

MASSACHUSETTS INSTITUTE OF TECHNOLOGY
Neurosciences research symposium summaries :an anthology from the 'Neurosciences research program bulletin' Cambridge,Mass. 1969 Neurosciences research program.Intensive study program Edited by Francis O. Schmitt and others Rockefeller university press;MIT press,New York;Cambridge,Mass.,1970
Bioch 33.2164

MASSACHUSETTS INSTITUTE OF TECHNOLOGY Recent research on carnitine:its relation to lipid metabolism :a symposium Papers presented Cambridge,Mass. 1964 Jul 24-25 Edited by George Wolf M.I.T. press,Cambridge,Mass., 1965
Bioch 33.1249

MASSACHUSETTS INSTITUTE OF TECHNOLOGY
Sensory communication The Symposium on principles of sensory communication Contributions Dedham,Mass. 1959 Jul 19-Aug 1 Edited by W.A. Rosenblith M.I.T.press; Wiley,New York;London,1961 Held at M.I.T. 's Endicott House,1961
Bal 39.1518

MASSACHUSETTS INSTITUTE OF TECHNOLOGY The National symposium on colloid chemistry 4th Papers presented Cambridge,Mass. 1926 Jun 23-25 Edited by Harry Boyer Weiser Colloid symposium monograph, 4 Chemical catalog company,New York,1926
Bioch 33.1472

MASSACHUSETTS INSTITUTE OF TECHNOLOGY United States conference on prestressed concrete 1st Proceedings Cambridge,Mass. 1951 Aug 14-16 Cambridge,Mass.,1951
Eng 41.3012

MASSACHUSETTS INSTITUTE OF TECHNOLOGY.DEPARTMENT OF ELECTRICAL ENGINEERING Problems in multiple access computer system design : special summer session Cambridge,Mass 1968 Jul 8-19 1968 Administrative material, class notes and other course material in loose-leaf binder
Math L 5.3474

MASSACHUSETTS INSTITUTE OF TECHNOLOGY.DEPARTMENT OF METALLURGY The Physical chemistry of steelmaking :a conference Proceedings Dedham,Mass. 1956 May 28-Jun 3 Edited by John F. Elliott Wiley;Chapman and Hall,New York;London,1958
Met 25.1789

MASSACHUSETTS INSTITUTE OF TECHNOLOGY.PROJECT MAC
Project MAC conference on concurrent systems and parallel computation Record Woods Hole,Mass. 1970 Jun 2-5 bibliog., illus. 199p Association for computing machinery,New York,1970
Math L 5.3600

MASSACHUSETTS INSTITUTE OF TECHNOLOGY CONFERENCE ON THE HUMAN FACTOR IN THE TRANSFER OF TECHNOLOGY Factors in the transfer of technology Cambridge,Mass. 1966 May 18-20 Edited by W.H. Gruber and D.G. Marquis M.I.T. press,Cambridge,Mass.,1969
Eng 41.1687

MATERIAL SCIENCE Modern diffraction and imaging techniques in material science International summer course on material science Proceedings Antwerp 1969 Jul 28-Aug 8 North Atlantic treaty organization. Scientific affairs division Edited by S. Amelinckx and others North-Holland,Amsterdam, 1970
Met 25.2608

MATERIALS International conference on mechanical behavior of materials Proceedings Kyoto 1971 Aug 15-20 Society of materials science 6 vols Society of materials science,Kyoto,1971
Eng 41.8524

MATERIALS ENGINEERING EXPOSITION AND CONGRESS Symposium on applications of modern metallographic techniques Philadelphia,Pa. 1969 Oct 13-16 American society for testing and materials American society for metals American society for testing and materials. Special technical publication, 480 ASTM, Philadelphia,Pa.,1970 Symposium presented at the exposition and congress
Met 25.2607

MATERIALS RESEARCH Strengthening mechanisms, metals and ceramics Sagamore army materials research conference 12th Proceedings Raquette Lake,N.Y. 1965 Aug 24-27 Army materials research agency Syracuse university Syracuse university press, Syracuse,N.Y.,1966
Met 25.2746

MATERIALS RESEARCH Ultrafine-grain metals Sagamore army materials research conference 16th Proceedings Raquette Lake,N.Y. 1969 Aug 19-22 Army materials research agency Syracuse university Syracuse university press,Syracuse,N.Y.,1970
Met 25.2747

MATERIALS SCIENCE Dislocation dynamics Battelle materials science colloquium 2nd Seattle,Wash. 1967 May 1-16 Harrison Hot Springs,B.C. Battelle memorial institute Edited by Alan R. Rosenfield and others McGraw-Hill series in materials science McGraw-Hill,New York,1968
Met 25.2253

MATERIALS SCIENCE Heterogeneous kinetics at elevated temperatures International conference on metallurgy and materials science Proceedings Philadelphia,Pa. 1969 Sep 8-10 Edited by G.R. Belten and W.L. Worrell Held at the University of Pennsylvania Plenum press,New York,1970
Met 25.2783

MATERIALS SCIENCE Mechanical behaviour of materials International conference on 'mechanical behaviour of materials' Proceedings Kyoto 1971 Aug 15-20 Society of materials science,Japan Society of materials science,Kyoto,1972
Met 25.2737

MATERIALS SCIENCE RESEARCH :a conference Proceedings Raleigh 1964 Nov 16-18 Vol 3: role of grain boundaries and surfaces in ceramics North Carolina state university Edited by W.Wurth Kriegel and Hayne Palmour Plenum press,New York,1966
Met 25.0288

MATERIALS SCIENCE RESEARCH, 1 Research conference on structure and properties of engineering materials Proceedings Raleigh, N.C. 1962 Mar 12-13 North Carolina state college and United States.Army research office,Durham Plenum press,New York,1963
Met 25.2281

MATERIALS TECHNOLOGY IN STEAM REFORMING PROCESSES Proceedings 1964 Oct 21-22 Imperial chemical industries.Agricultural division Edited by C. Edeleanu Pergamon,Oxford,1966
Met 25.0104

MATERIALS TESTING MACHINE DESIGN Machines for materials and environmental testing Symposium on techniques and equipment for environmental testing; symposium on developments in materials testing machine design :joint conference Manchester 1965 Sep 6-10 Institution of mechanical engineers and Society of environmental engineers Under the aegis of the British national committee on materials Institution of mechanical engineers.Proceedings, 180,pt 3a I.M.E.,London,1966
Met 25.0891

MATERIALY PALEONTOLOGICHESKOGO SOVESCHANIYA PO PALEOZOYU Moscow 1951 May 14-17 Edited by T.G. Sarycheva Akademiya nauk S.S. S.R.,Moscow,1953
Geol 8.0705

MATERIALY SOVESHCHANIYA PO PSIKHOLOGII 1955 Jul 1-6 Akademiya pedagogicheskikh nauk R.S. F.S.R.Institut psikhologii Edited by G.N. Voskresenskii and others Izdatekstvo Akademii pedagogicheskikh nauk RSFSR,Moscow, 1957
Psy 31.2576

MATERIALY VIII VSESOYUZNOGO MEZHDUVEDOMSTVENNOGO SOVESHCHANIYA PO GEOKRIOLOGII MERZLOTOVEDENIYU Yakutsk 1966 Vyp 6: geomorfologicheskaya sektsiya Edited by S.P. Kachurin and G.F. Gravis 151p 22cm Knizhnoe izdatelstvo,Yakutsk,1966
Sco 14.7714

MATHEMATICAL ASSOCIATION OF AMERICA Metric geometry over affine spaces Cooperative summer seminar 1st Ithaca,N.Y. 1964 By Ernst Snapper Held at Cornell university bibliog. iv,166p 28cm Mathematical association of America,Buffalo,N.Y.,1964 Duplicated notes
P Math 2.3829

MATHEMATICAL EDUCATION OF ENGINEERS :report of the O.E.C.D. seminar Paris 1965 Organization for economic co-operation and development OECD,Paris,1965
Eng 41.1838

MATHEMATICAL INSTITUTE,OXFORD Combinatorial mathematics and its application :conference Proceedings Oxford 1969 Jul 7-10 Edited by J.J.A. Welsh x,364p 24cm Academic press,New York;London,1971
Math S 3.1661

MATHEMATICAL LOGIC Conference in mathematical logic Proceedings London 1970 Aug 24-28 Edited by Wilfrid Hodges Held at Bedford college Lecture notes in mathematics, 255 vii,351p 25cm Springer, Berlin,1972
P Math 2.4059

MATHEMATICAL LOGIC AND FOUNDATIONS OF SET THEORY International colloquium on mathematical logic and foundations of set theory Proceedings Jerusalem 1908 Nov 11-14 Edited by Yehoshua Bar-Hillel Under the auspices of the Israel academy of science and humanities Studies in logic and the foundations of mathematics bibliog. 145p 24cm North-Holland,Amsterdam;London,1970
P Math 2.3795

MATHEMATICAL METHODS IN SOLID STATE AND SUPERFLUID THEORY Scottish universities' summer school in physics 8th Papers St. Andrews 1967 Jul 31-Aug 19 Edited by R.C. Clark and G.H. Derrick Scottish universities' summer school in physics,1967 xvi,400p 25cm Oliver and Boyd,Edinburgh, 1969
Cav 7.2759

MATHEMATICAL METHODS IN THE SOCIAL SCIENCES,1959 1st Stanford symposium proceedings Stanford,Calif. 1959 Jun 15-24 Stanford university Edited by Kenneth J. Arrow and others Financed by the United States.Office of naval research Stanford university. Stanford mathematical studies in the social sciences, 4 vii,365p 24cm Stanford university press,Stanford,Calif.,1960
P. Math 2.2259

MATHEMATICAL METHODS IN THEORETICAL PHYSICS Boulder summer institute for theoretical physics 11th Proceedings Boulder,Colo. 1968 Jun 17-Aug 23 University of Colorado. Department of physics and astrophysics Edited by Kalyana T. Mahanthappa and Wesley E. Brittin Lectures in theoretical physics, 11d diagrms xiv,648p 23cm Gordon and Breach,New York,1969
Cav 7.3011

MATHEMATICAL MODEL BUILDING IN ECONOMICS AND INDUSTRY :a conference Papers London 1967 Jul 4-6 Corporation for economic and industrial research vii,165p 23cm Griffin,London,1968
Math S 3.1407

MATHEMATICAL MODEL BUILDING IN ECONOMICS AND INDUSTRY :a conference Papers London 1967 Jul 4-6 Organized by the Corporation for economic and industrial research Griffin, London,1968
Eng 41.1223

MATHEMATICAL MODELS OF ACTION AND REACTION Differential games and related topics International summer school on mathematical models of action and reaction Proceedings Varenna 1970 Jun 15-27 Edited by H.W. Kuhn and G.P. Szego x,489p 23cm North-Holland, Amsterdam;London,1971
Math S 3.1721

MATHEMATICAL OPTIMIZATION TECHNIQUES symposium Santa Monica,Calif. 1960 Oct 18-20 University of California Edited by Richard Bellman xii,346p 24cm University of California press,Berkeley,Calif.,1963
P. Math 2.2288

MATHEMATICAL OPTIMIZATION TECHNIQUES symposium papers Santa Monica,Calif. 1960 Oct.18-20 University of California Edited by Richard Bellman With assistance from the National science foundation xii,346p 24cm University of California press,Berkeley,Calif., 1963
Math 3.0238

MATHEMATICAL PROBABILITY AND ITS APPLICATIONS symposium Applied probability New York 1955 Apr.14-15 American mathematical society Edited by L.A. MacColl Cosponsored by the United States.Army.Office of ordnance research American mathematical society.Proceedings of symposia in applied mathematics, 7 v,104p 26cm McGraw-Hill,New York,1957
Math 3.0682

MATHEMATICAL PROBLEMS IN THE GEOPHYSICAL SCIENCES Troy,N.Y. 1970 Vol 1-2: geophysical fluid dynamics;inverse problems,dynamo theory, and tides American mathematical society Edited by W.H. Reid Lectures in applied mathematics, 13-14 illus,maps 24cm 2 vols American mathematical society, Providence,R.I.,1971
A Math 4.1742

MATHEMATICAL PROGRAMMING Princeton symposium on mathematical programming Princeton,N.J. 1967 Aug 14-18 Princeton university Edited by Harold W. Kuhn vi,620p 24cm Princeton university press,Princeton,N.J.,1970
Math S 3.1687

MATHEMATICAL PROGRAMMING TECHNIQUES Applications of mathematical programming techniques :a conference Proceedings Cambridge 1968 Jun 24-28 North Atlantic treaty organization.Science committee Edited by E.M.L. Beale ix,451p 24cm English universities press,London,1970
Math S 3.1686

MATHEMATICAL SOCIETY OF JAPAN International conference on functional analysis and related topics Proceedings Tokyo 1969 Apr 1-8 xxxiii,425p 26cm University of Tokyo press, Tokyo,1970
P Math 2.4279

MATHEMATICAL SOFTWARE SYMPOSIUM Mathematical software based on the proceedings of the Mathematical software symposium Lafayette, Ind. 1970 Apr 1-3 Edited by John R. Rice Held at Purdue university Association for computing machinery.Monograph series 515p Academic press,New York,1971
Math L 5.3806

MATHEMATICAL STATISTICS AND PROBABILITY Berkeley symposium on mathematical statistics and probability 2nd Proceedings Berkeley,Calif. 1950 Jul 31-Aug 12 University of California.Statistical laboratory Edited by Jerzy Neyman University of California press,Berkeley,Calif.; Los Angeles,Calif.,1951
Radioth 35.0599

MATHEMATICAL STATISTICS AND PROBABILITY Berkeley symposium on mathematical statistics and probability 3rd Proceedings Berkeley,Calif. 1954 Dec 26-31 Berkeley, Calif. 1955 Jul,Aug Vol 4: contributions to biology and problems of health University of California.Statistical laboratory Edited by Jerzy Neyman University of California press,Berkeley,Calif.; Los Angeles,Calif.,1956
Radioth 35.0606

MATHEMATICAL THEORY OF AUTOMATA Symposium on mathematical theory of automata proceedings New York 1962 Apr 24-26 Microwave research institute Partly sponsored by American institute of electrical engineers Polytechnic institute of Brooklyn.Microwave research institute.Symposia series, 12 xix, 640p 22cm Polytechnic press,Brooklyn,1962
Math L 5.0996

MATHEMATICAL THEORY OF CONTROL :a conference Proceedings Los Angeles,Calif. 1967 Jan 30-Feb 1 Edited by A.V. Balakrishnan and Lucian W. Neustadt bibliog. xv,459p 24cm Academic press,London,1967
Math S 3.1402

MATHEMATICAL THEORY OF CONTROL :a conference Proceedings Los Angeles,Calif. 1967 Jan 30-Feb 1 Edited by A.V. Balakrishnan and Lucien W. Neustadt Academic press,New York; London,1967
Eng 41.5763

MATHEMATICS Congress of Scandinavian mathematicians 15th Proceedings Oslo 1968 Aug 12-16 Edited by K.E. Aubert and W. Ljunggren Lecture notes in mathematics, 118 Bibliog. 162p 25cm Springer-verlag, Berlin,1970
P Math 2.3240

MATHEMATICS Congres international des mathematiciens :les 265 communications individuelles Nice 1970 Sep vii,290p 24cm Gauthier-Villars,Paris,1970
P Math 2.3775

MATHEMATICS Elasticity American mathematical society.Symposium in applied mathematics 3rd Proceedings Ann Arbor, Mich. 1949 Jun 14-16 Edited by R.V. Churchill and others Cosponsored by the American society of mechanical engineers. Applied mechanics division McGraw-Hill,New York,1950
Col S 12.0335

MATHEMATICS International congress of mathematicians 1st Proceedings Moscow 1966 Edited by I.G. Petrovsky Bibliog, illus 726p 22cm M.I.R.,Moscow,1968
P Math 2.3006

MATHEMATICS International congress of mathematicians 5th proceedings Cambridge 1912 Aug 22-28 Vol. 2. Communications to sections 2-4 International mathematical union Edited by E. W. Hobson and A.E.H. Love 657p 26cm Cambridge university press,Cambridge,1913
Math L 5.0751

MATHEMATICS International congress of mathematicians 11th Proceedings Cambridge,Mass. 1950 Aug 13-Sep 6 Vol 1-2 International mathematical union and American mathematical society 25cm 2 vols American mathematical society,Providence,R.I., 1952
Math L 5.3246

MATHEMATICS International congress of mathematicians 14 proceedings Stockholm 1962 Aug 15-22 International mathematical union Edited by V. Stenstrom Bibliog. 1, 595p 25cm Institut Mittag-Leffler, Djursholm,Sweden,1963
P. Math 2.2601

MATHEMATICS International congress of mathematicians 15th Proceedings Amsterdam 1954 Sep 2-9 Vol 1-3 Edited by Johan C.H. Gerretson and Johannes de Groot Bibliog 25cm 3 vols North Holland,Amsterdam,1954-57
P Math 2.2875

MATHEMATICS International congress of mathematicians abstracts of brief scientific communications and reports Moscow 1966 Aug nos. 1-15 International mathematical union Moscow,1966
Math 3.1125

MATHEMATICS International congress of mathematicians short communications abstracts Stockholm 1962 Aug 15-22 International mathematical union 221p 24cm Stockholm,1962
P. Math 2.2602

MATHEMATICS Mezhdunarodne matematicheskie olimpiady 1st-7th Zadachi,resheniya, itogi By E.A. Morozova and I.S. Petrakov Bibliog.,Illus. 175p 20cm Moscow,1967
P Math 2.3273

MATHEMATICS Outlines of one-hour and half-hour addresses and translations of Russian abstracts of short communications Stockholm 1962 International congress of mathematicians varp 29cm Stockholm,1962
P Math 2.3005

MATHEMATICS Problems in the philosophy of mathematics International colloquium in the philosophy of science Proceedings London 1965 Jul 11-17 Vol 1 British society for the philosophy of science London school of economics and political science Edited by Imre Lakatos Under the auspices of the International union of history and philosophy of science.Division of logic,methodology and philosophy of science Studies in logic and the foundations of mathematics xv,241p North-Holland,Amsterdam,1967
WSM 43.1813

MATHEMATICS Rocnik matematicke Olympiady 5th-8th,10th-16th Illus.port. 17cm 10 vols Statni Pedagogicke Nakladatelstvi, Prague,1957-68
P Math 2.3275

MATHEMATICS Summer school in mathematics, geometry and topology proceedings Dundee 1961 Jul 10-22 Queen's college,Dundee 33cm Queen's College,Dundee,1961
P. Math 2.0292

MATHEMATICS Symposia mathematica Rome 1967-68 1,3 Istituto nazionale de alta matematica bibliog. 25cm 2 vols Academic press,New York;London,1969
P Math 2.3778

MATHEMATICS Symposium international des sciences physiques et mathematiques dans la premiere moitie du XVII siecle Actes Pisa 1958 Jun 16-18 Vinci (Florence) Union internationale d'histoire et philosophie des sciences Academie internationale d'histoire des sciences.Collection de travaux, 11 x, 278p Gruppe italiano di storia delle scienze: Hermann,Vinci (Florence);Paris,c1959
WSM 43.0066

MATHEMATICS The Ditchley mathematical conference :a report by the British co-chairman (Prof.B.Thwaites) of the Anglo-American conference on mathematical education Ditchley Park 1966 Sep 9-12 School mathematics project Cambridge conference on school mathematics S.M.P.,London,1966
Eng 41.1829

MATHEMATICS AND COMPUTER SCIENCE IN BIOLOGY AND MEDICINE Conference on mathematics and computer science in biology and medicine Proceedings Oxford 1964 Jul Medical research council 317p H.M.S.O.,London,1965
Bot 42.0512

MATHEMATICS AND COMPUTER SCIENCE IN BIOLOGY AND MEDICINE :conference Proceedings Oxford 1964 July Medical research council Co-sponsored by Great Britain.Ministry of health H.M.S.O.,London,1965
HE 27.0125

MATHEMATICS AND COMPUTER SCIENCE IN BIOLOGY AND MEDICINE proceedings conference Oxford 1964 Jul Medical research council Great Britain.Ministry of health Co-sponsored by Great Britain.Scottish home and health department Bibliog. xi,317p 25cm H.M.S. O.,London,1965
Math 3.0775

MATHEMATICS OF CONTEMPORARY PHYSICS : instructional conference Proceedings London 1971 Aug 23-Sep 11 London mathematical society North Atlantic treaty organization Edited by R.F. Streater xi, 274p 24cm Academic press,London,1972
A Math 4.1864

MATHEMATICS RESEARCH CENTER ADVANCED SEMINAR Proceedings Nonlinear functional analysis and applications Madison,Wisc. 1970 Oct 12-14 University of Wisconsin.Mathematics research center vii,586p 24cm Academic press,New York;London,1971
P Math 2.4286

MATHEMATICS RESEARCH CENTER SYMPOSIUM ON CONTRIBUTIONS TO NONLINEAR FUNCTIONAL ANALYSIS Proceedings Madison,Wisc. 1971 Apr 12-14 University of Wisconsin.Mathematics research center xii,672p 24cm Academic press,New York;London,1971
P Math 2.4288

MATHEMATICS TEACHING International colloquium on the modernization of mathematics teaching in European countries Proceedings Bucharest 1968 Sep 23-Oct 2 Unesco 572p 25cm Editions didactiques et pedagogiques, Bucharest,1968
P Math 2.4020

MATHEMATICS TEACHING Symposium on modernization of mathematics teaching in European countries Proceedings Bucharest 1968 Sep 23-Oct 2 Unesco Bibliog 573p 24cm Editions didactiques et pedagogiques, Bucharest,1969
A Math 4.1688

MATHEMATISCH CENTRUM,AMSTERDAM MC-25 information symposium Amsterdam By J.W.de Bakker and others Mathematical centre tracts, 37 illus varp Mathematisch centrum, Amsterdam,1971 Dedicated to Prof.van Wijngaarden on the occasion of the centre's 25th anniversary
Math L 5.3857

MATHEMATISCHE GESELLSCHAFT DER D.D.R. Beitrage zur graphentheorie :internationale kolloquium Vortragen Manebach 1967 May 9-12 Edited by Horst Sachs and others Bibliog.,Illus. 394p 23cm Teubner, Leipzig,1968
P Math 2.3352

MATHEMATISCHE STATISTIK UND WAHRSCHEINLICHKEITSTHEORIE conference Oberwolfach 1966 Apr 17-24 Mathematisches forschungsinstitut,Oberwolfach Oberwolfach, 1966 Mimeograph resumes of papers
Math 3.1120

MATHEMATISCHES FORSCHUNGSINSTITUT,OBERWOLFACH Algebraische zahlentheorie conference an account Oberwolfach 1964 Sep 6-12 Edited by Helmut Hasse and Peter Roquette Mathematisches forschungsinstitut,Oberwolfach, 2 264p 19cm Bibliographisches institut, Mannheim,1966
P. Math 2.1769

MATHEMATISCHES FORSCHUNGSINSTITUT,OBERWOLFACH Martingales Report Oberwolfach 1970 May 17-23 Lecture notes in mathematics, 190 75p 25cm Springer,Berlin,1971
Math S 3.1781

MATHEMATISCHES FORSCHUNGSINSTITUT,OBERWOLFACH Martingales :a meeting Report Oberwolfach 1970 May 17-23 Edited by Hermann Dinges Lecture notes in mathematics, 190 75p 25cm Springer,Berlin,1971
P Math 2.4559

MATHEMATISCHES FORSCHUNGSINSTITUT,OBERWOLFACH Mathematische statistik und wahrscheinlichkeitstheorie conference Oberwolfach 1966 Apr 17-24 Oberwolfach, 1966 Mimeograph resumes of papers
Math 3.1120

MATRIX METHODS IN STRUCTURAL MECHANICS :a conference Proceedings Dayton,Ohio 1965 Oct 26-28 United States.Air force systems command United States.Air university 1966
Eng 41.2732

MATRIX THEORY Recent advances in matrix theory :advanced seminar 2nd proceedings Madison,Wis. 1963 Oct.14-16 United States army.Mathematics research center. Edited by Hans Schneider United States army. Mathematics research center.Publication, 12 xi,142p 25cm University of Wisconsin press, Madison,Wis.,1964
Math L 5.0328

MATRIX THEORY Recent advances in matrix theory seminar 2nd proceedings Madison, Wis. 1963 Oct 14-16 United States.Army. Mathematics research center Edited by Hans Schneider United States army.Mathematics research center.Publications, 12 xi,142p 24cm University of Wisconsin press,Madison, Wis.,1964
P. Math 2.2087

MATROIDS Theorie des matroides Rencontre Franco-britannique Proceedings Brest 1970 May 14-15 Edited by C.P. Bruter Lecture notes in mathematics, 211 108p 25cm Springer,Berlin,1970
P Math 2.4508

The MATTHEW FONTAINE MAURY MEMORIAL SYMPOSIUM 10th :papers presented to the tenth Pacific science congress of the Pacific science association Papers Pacific science congress 10th Honolulu 1961 Aug 21-Sep 6 Pacific science association,and others Edited by Harry Wexler and others Geophysical monograph,7 American geophysical union.Publication,1036 illus,maps x,228p 25cm American geophysical union,Washington,D. C.,1962
Sco 14.6079

MAX PLANCK GESELLSCHAFT ZUR FORDERUNG DER WISSENSCHAFTEN Evolution of the forebrain: phylogenesis and ontology of the forebrain :a symposium Lectures Frankfurt 1965 Aug 15-19 and Sprendlingen Edited by R. Hassler and H. Stephan G.Thieme,Stuttgart,1966
Psy 31.3370

MAYAGUEZ 1959 Caribbean geological conference 2nd transactions Edited by John D. Weaver University of Puerto Rico, Mayaguez,1960
Geol 8.2424

MAYAGUEZ,PUERTO RICO 1965 Differential equations and dynamical systems :an international symposium Proceedings Edited by Jack K. Hale and Joseph P. LaSalle Bibliog,illus,port xvii,544p 24cm Academic press,New York;London,1967 Dedicated to Solomon Lefschetz
P Math 2.2957

MC-25 INFORMATION SYMPOSIUM Amsterdam By J. W.de Bakker and others Mathematisch centrum, Amsterdam Mathematical centre tracts, 37 illus varp Mathematisch centrum,Amsterdam, 1971 Dedicated to Prof.van Wijngaarden on the occasion of the centre's 25th anniversary
Math L 5.3857

MEANS OF CORRELATION OF QUATERNARY SUCCESSIONS International association for quaternary research congress 7th Proceedings Boulder,Colo. 1965 Aug 30-Sep 5 Edited by R.B. Morrison and H.E. Wright Utah university press,Salt Lake City,Utah,1968
Bot 42.3246

MEASUREMENT Joint conference on digital methods of measurement Proceedings Canterbury 1969 Jul 23-25 Institution of electronic and radio engineers Institution of electrical engineers Institute of electrical and electronics engineers Institution of electronic and radio engineers. Conference proceedings, 15 IERE,London,1969
Eng 41.8305

MEASUREMENT IN PERSONALITY AND COGNITION :a conference Princeton,N.J. 1960 Oct Educational testing service Edited by Samuel Messick and John Ross Wiley,New York;London, 1962
Psy 31.2152

MEASUREMENT OF GRASSLAND PRODUCTIVITY Easter school in agricultural science 6th Proceedings Nottingham 1959 University of Nottingham Edited by J.D. Ivins Butterworths,London,1959
Bot 42.1951

The MEASUREMENT OF STRESS AND STRAIN IN SOLIDS a conference Manchester 1946 Jul 11-13 Institute of physics.Manchester and district branch Edited by F.A. Vick Physics in industry Institute of physics,London,1948 Book based on the proceedings of the conference
Cav 7.2397

The MEASUREMENT OF STRESS AND STRAIN IN SOLIDS a conference Proceedings Manchester 1946 Jul 11-13 Institute of physics Institute of physics,London,1948
Eng 41.2898

The MEASUREMENT OF STRESS AND STRAIN IN SOLIDS a conference Proceedings Manchester 1946 Jul 11-13 Institute of physics. Manchester and district branch Edited by F.A. Vick Physics in industry Institute of physics,London,1948
Met 25.0880

MEASUREMENT OF VISUAL FUNCTION Symposium on the measurement of visual function : proceedings of Spring meeting,1965 Proceedings 1965 Armed forces-National research council committee on vision Edited by Milton A. Whitcomb and William Benson National research council,Washington,D.C.,1968
Psy 31.0427

MEASUREMENTS OF EXOCRINE AND ENDOCRINE FUNCTIONS OF THE PANCREAS... 2nd :applied seminar Proceedings Association of clinical scientists Edited by F.W. Sunderman and F.W. Sunderman,Jr. Lippincott,Philadelphia,Pa., 1961
Pha 16.0094

MECANIQUE DE LA TURBULENCE :colloque international Marseille 1961 Aug 28-Sep 2 Centre national de la recherche scientifique Centre national de la recherche scientifique.Colloques internationaux, 108 470p Editions du centre national de la recherche scientifique,Paris,1962
Chem E 24.0370

MECANIQUE DE LA TURBULENCE colloque international Marseille 1961 Aug 28-Sep 2 Mecanique de la turbulence Centre national de la recherche scientifique Centre national de la recherche scientifique. Colloques internationaux, 108 Centre national de la recherche scientifique,Paris, 1962
A Math 4.0525

MECANISME PHYSIOLOGIQUE DE LA SECRETION LACTEE colloque international Strasbourg 1950 Aug 22-29 Centre national de la recherche scientifique Centre national de la recherche scientifique.Colloques internationaux, 32 C. N.R.S.,Paris,1950
Phys 20.2212

MECANISMES DE REGULATION DES ACTIVITES CELLULAIRES CHEZ LES MICROORGANISMES Colloque international Marseilles 1963 Jul 23-27 Centre national de la recherche scientifique Centre national de la recherche scientifique.Colloques internationaux, 124 Paris,1965
Gen 34.0707

MECCANICA NON LINEARE E STABILITA Symposia mathematica Rome 1970 Feb 23-26 Istituto nazionale di alta matematica Istituto nazionale di alta matematica.Pubblicazione, 6 395p 25cm Academic press,London;New York,1971
P Math 2.3989

MECHANICAL BEHAVIOR OF CRYSTALLINE SOLIDS a symposium Proceedings New York 1962 Apr 28-29 Ceramic education council and National bureau of standards Sponsored by the Edward Orton junior ceramic foundation National bureau of standards monograph, 59 U.S.Government printing office,Washington,1963 Lithographed
Met 25.1213

MECHANICAL BEHAVIOR OF MATERIALS International conference on mechanical behavior of materials Proceedings Kyoto 1971 Aug 15-20 Society of materials science 6 vols Society of materials science,Kyoto, 1971
Eng 41.8524

MECHANICAL BEHAVIOUR OF MATERIALS International conference on 'mechanical behaviour of materials' Proceedings Kyoto 1971 Aug 15-20 Society of materials science, Japan Society of materials science,Kyoto, 1972
Met 25.2737

MECHANICAL LANGUAGE STRUCTURES Working conference on mechanical language structures New York 1963 Aug 14-16 Association for computing machinery Collection of 8 mimeograph abstracts of papers
Math L 5.0928

MECHANICAL PROPERTIES OF ENGINEERING CERAMICS : a conference Proceedings Raleigh,N.C. 1960 North Carolina state college and United States.Army.Office of ordnance research Edited by W.Wurth Kriegel and Hayne Palmour Interscience,New York;London,1961
Met 25.0853

MECHANICAL PROPERTIES OF INTERMETALLIC COMPOUNDS a symposium held during the 115th meeting of the Electrochemical society Philadelphia 1959 May 3-7 Electrochemical society Edited by J.H. Westbrook John Wiley,London,1960
Met 25.2277

MECHANICAL PROPERTIES OF METALS AT LOW TEMPERATURE the NBS semicentennial symposium Proceedings National bureau of standards National bureau of standards. Circular, 520 United States printing office, Washington,D.C.,1952
Met 25.0972

MECHANICAL PROPERTIES OF METALS AT LOW TEMPERATURES :NBS semi-centennial symposium Washington,D.C. 1951 May 14-15 By N.P. Allen and others United States. National bureau of standards United States. National bureau of standards.Circular,520 illus,tables,diagrs iv,206p 24cm Washington,D.C.,1952
Sco 14.1244

MECHANICAL PROPERTIES OF NON-METALLIC BRITTLE MATERIALS :a conference Proceedings London 1958 Apr Edited by W.H. Walton Organized by the National coal board.Mining research establishment Butterworths,London, 1958
Eng 41.3645

MECHANICAL PROPERTIES OF NON-METALLIC BRITTLE MATERIALS :a conference Proceedings London 1958 Apr Great Britain.National coal board.Mining research establishment and Great Britain.Department of scientific and industrial research.Building research station Edited by W.H. Walton Butterworths,London, 1958
Met 25.0852

MECHANICAL WEAR :a summer conference Proceedings Cambridge,Mass. 1948 Jun Massachusetts institute of technology and American society of mechanical engineers Edited by John T. Burwell Jointly sponsored by the General motors corp. and the Chrysler corp. American society for metals,1950
Met 25.1597

MECHANICAL WORKING AND STEEL PROCESSING COMMITTEE 2ND Operating metallurgy conference 9th Proceedings Philadelphia,Pa. 1966 Dec 5-9 American institute of mining, metallurgical and petroleum engineers Metallurgical society of the AIME.Conference series, 50 Gordon and Breach,New York,1968
Met 25.2572

MECHANICAL WORKING OF STEEL 1 Technical conference 5th Proceedings Pittsburgh 1963 Jan 15-16 American institute of mining, metallurgical and petroleum engineers. Mechanical working committee Metallurgical society conferences, 21 Gordon and Breach, New York,1964
Met 25.0728

MECHANICAL WORKING OF STEEL 2 The Technical conference 6th Proceedings Chicago,Ill. 1964 Jan 30-31 American institute of mining, metallurgical and petroleum engineers. Mechanical working and steel processing committee Edited by T.G. Bradbury Sponsored by the Chicago section Metallurgical society conferences, 26 Gordon and Breach,New York,1965
Met 25.2244

MECHANICS Developments in mechanics Midwestern mechanics conference 6th Proceedings Cleveland,Ohio 1963 Apr 1-3 Vol 2,pt 1: fluid mechanics Edited by Simon Ostrach and Robert H. Scanlan Pergamon, London,1965 Combining the 6th Midwestern conference on soil mechanics and the 8th Midwestern conference on fluid mechanics
Eng 41.6959

MECHANICS Developments in mechanics Midwestern mechanics conference 7th proceedings Ann Arbor,Mich 1961 Sep. 6-8 Vol 1 Edited by J.F. Lay and L.E. Malvern North Holland,Amsterdam,1961
A Math 4.0313

MECHANICS Developments in theoretical and applied mechanics Southeastern conference on theoretical and applied technics 4th Proceedings New Orleans,La. 1968 Feb 29-Mar 1 Edited by D. Frederick Sponsored by Tulane university illus 637p 26cm Pergamon,Oxford,1970
A Math 4.1903

MECHANICS National congress of theoretical and applied mechanics 1st Proceedings Varna 1969 Nov 3-6 Bulgarsko knizhovno druzhestvo.National committee on theoretical and applied mechanics Bibliog,illus 632p 24cm Bulgarsko knizhovno druzhestvo,Sofia, 1971 Papers in English,French and Russian
A Math 4.1723

MECHANISATION OF THOUGHT PROCESSES :a symposium Proceedings Teddington 1958 Nov 24-27 Vol 1-2 National physical laboratory National physical laboratory. Symposia, 10 2 vols H.M.S.O.,London,1959
Eng 41.6010

MECHANISATION OF THOUGHT PROCESSES symposium proceedings Teddington 1958 Nov 24-27 National physical laboratory National physical laboratory.Proceedings of symposia, 10 23cm 2 vols H.M.S.O.,London,1959
Math L 5.0820

MECHANISM AND CHEMICAL KINETICS OF ORGANIC REACTIONS IN LIQUID SYSTEMS :a general discussion London 1941 Sep Faraday society Gurney and Jackson,London;Edinburgh, 1941 Reprint from the society's 'Transactions',vol 37,pt 12
Bot 42.6479

The MECHANISM OF ACTION OF INSULIN :a symposium London 1958 Sep 9-10 British insulin manufacturers Edited by F.G. Young and others Blackwell scientific publication,Oxford,1960
PGMS 29.0378

The MECHANISM OF ACTION OF INSULIN a symposium Discussions London 1958 Sep 9-10 Edited by F.G. Young Organized by the British insulin manufacturers Blackwell, Oxford,1960
Bioch 33.0474

MECHANISM OF ACTION OF STEROID HORMONES :a conference Proceedings Dedham,Mass. 1960 Edited by Claude A. Villee and Lewis L. Engel International series of monographs on pure and applied biology Pergamon,Oxford, 1961
Inv Med 37.0219

MECHANISM OF ACTION OF STEROID HORMONES : conference Proceedings Dedham,Mass. 1960 Edited by Claude A. Villee and Lewis L. Engel International series of monographs on pure and applied biology.Symposium division,1 Pergamon,Oxford,1961
Pha 16.0088

The MECHANISM OF ACTION OF WATER SOLUBLE VITAMINS 11 :a meeting London 1961 Apr 28 Ciba foundation Edited by A.V.S. de Reuck and Maeve O'Connor Ciba foundation study group,11 J.and A.Churchill,London,1961
Chem 18.1283

MECHANISM OF CELL AND TISSUE DAMAGE PRODUCED BY IMMUNE REACTIONS International symposium on immunopathology 2nd Papers Brook Lodge,Mich. 1961 International committee on immunology and Upjohn company Edited by Pierre Grabar and Peter Miescher Schwabe, Basle;Stuttgart,1962
Path 30.2336

MECHANISM OF CYTOPLASMIC STREAMING,CELL MOVEMENT, AND THE SALTATORY MOTION OF SUBCELLULAR PARTICLES :a symposium Proceedings Primitive motile systems in cell biology Princeton,N.J. 1963 Apr 2-5 Edited by Robert D. Allen and Noburo Kamiya illus. Academic press,New York;London,1964
Bot 42.3760

MECHANISM OF ENZYME ACTION A Symposium on the mechanism of enzyme action Papers and discussions Baltimore 1953 Jun 16-19 Edited by William D. McElroy and Bentley Glass Sponsored by the McCollum-Pratt institute McCollum-Pratt institute.Contribution, 70 Johns Hopkins press,Baltimore,Md.,1954
Bioch 33.1059

The MECHANISM OF ENZYME ACTION Baltimore, Md. 1953 Jun 16-19 Edited by William D. McElroy and Bentley Glass Sponsored by McCollum-Pratt institute McCollum-Pratt institute.Contributions,70 Johns Hopkins press,Baltimore,Md.,1954
Col S 12.0113

The MECHANISM OF HETEROGENEOUS CATALYSIS :a symposium Proceedings Amsterdam 1959 Nov 12-13 Royal Netherlands chemical society. Section for inorganic and physical chemistry Edited by J.H. de Boer and others Elsevier monographs.Chemistry series Elsevier, Amsterdam,1960
Chem 18.0455

The MECHANISM OF HETEROGENEOUS CATALYSIS Proceedings Amsterdam 1959 Nov 12-13 Edited by J.H. De Boer and others vii,179p Elsevier,Amsterdam,1960
Chem E 24.0797

The MECHANISM OF PHASE TRANSFORMATIONS IN CRYSTALLINE SOLIDS :an international symposium Proceedings Manchester 1968 Jul 3-5 Institute of metals Institute of metals.Monograph and report series, 33 illus,diagrs 324p 25cm London,1969
Cav 7.2666

The MECHANISM OF PHASE TRANSFORMATIONS IN METALS a symposium London 1955 Nov 9 Institute of metals Institute of metals. Monograph and report series, 18 Institute of metals,London,1956
Met 25.1251

MECHANISM OF PHOTOSYNTHESIS International congress of biochemistry 5th Proceedings Moscow 1961 Aug 10-16 International union of biochemistry Edited by H. Tamiya I.U.B. symposium series, 26 Pergamon;PWN-Polish scientific publishers,Oxford;Warsaw,1963
Bioch 33.1335

The MECHANISM OF PROTEIN SYNTHESIS :a symposium Proceedings Cold Spring Harbor 1969 Cold Spring Harbor laboratory of quantitative biology Cold Spring Harbor symposia on quantitative biology, 34 Cold Spring Harbor,1969
Bioch 33.1887

MECHANISM OF RELEASE OF BIOGENIC AMINES : international symposium Proceedings Stockholm 1965 Feb Edited by U.S. Von Euler and others Wenner-Gren center. International symposium series,5 Pergamon, Oxford,1965
Pha 16.0071

MECHANISMS IN BIOLOGICAL COMPETITION symposium Southampton 1960 Sep Society for experimental biology Society for experimental biology.Symposia, 15 Cambridge university press,Cambridge,1961
Gen 34.1348

MECHANISMS OF CELL DIVISION Conference on the mechanisms of cell divisions 2nd Proceedings New York 1960 Oct 7 By M.J. Kopac and others New York academy of sciences Edited by Paul R. Gross New York academy of sciences.Annals, 90,p.345-613 New York,1960
Bal 39.0291

MECHANISMS OF COLOUR DISCRIMINATION :an international symposium on the fundamental mechanisms of the chromatic discrimination in animals and man Proceedings Paris 1958 Jul 25-29 International council of scientific unions Pergamon press:symposium publications division,London,1960
Psy 31.0384

MECHANISMS OF CONGENITAL MALFORMATION :second scientific conference Proceedings New York 1954 Jun 15-16 Association for the aid of crippled children Edited by Harold Wolff Association for the aid of crippled children,New York,1954
An 32.2034

MECHANISMS OF EMBRYONIC DEVELOPMENT Recent studies in the mechanisms of embryonic development :a conference New York 1947 Jan 10-11 New York academy of sciences. Section of biology Edited by Robert Rugh New York academy of sciences.Annals, 49,p 661-866 New York,1948
Bal 39.0438

MECHANISMS OF HARD TISSUE DESTRUCTION :a symposium presented at the Philadelphia meeting of the AAAS... Philadelphia,Pa. 1962 Dec 29-30 Edited by Reidar F. Sognnaes American association for the advancement of science.Publications, 75 American association for the advancement of science, Washington,D.C.,1963
An 32.3661

MECHANISMS OF HORMONE ACTION Meersburg 1964 Edited by Peter Karlson Supported by the North Atlantic treaty organization Academic press,New York,1965 Some papers are in German
Pha 16.0092

MECHANISMS OF HORMONE ACTION a NATO advanced study institute Meersburg 1964 North Atlantic treaty organization Edited by P. Karlson Academic press;Thieme,New York; Stuttgart,1965
Gen 34.0549

MECHANISMS OF HORMONE ACTION :a Nato advanced study institute Discussions Meersburg 1964 North Atlantic treaty organization Edited by P. Karlson Academic press,New York, 1965
Bioch 33.0469

MECHANISMS OF IMMUNOLOGICAL TOLERANCE Symposium on mechanisms of immunological tolerance Proceedings Liblice 1961 Nov 8-10 Ceskoslovenska akademie ved Edited by M. Hasek and others illus. Academic press,New York,1963
BG 23.0056

MECHANISMS OF INORGANIC REACTIONS Lawrence, Kan. 1964 Jun 21-24 American chemical society.Division of inorganic chemistry Edited by R.Kent Murmann and others American chemical society.Advances in chemistry series, 49 American chemical society,Washington,D.C., 1965 Symposium chairman:J.Kleinberg
Chem 18.1002

MECHANISMS OF INORGANIC REACTIONS :symposium Papers Lawrence,Kan 1964 Jun 21-24 Edited by R.Kent Murmann and others Sponsored by the American chemical society. Inorganic chemistry division American chemical society.Advances in chemistry series, 49 vii,266p American chemical society, Washington,D.C.,1965
Chem E 24.0629

MECHANISMS OF MICROBIAL PATHOGENICITY :a symposium Papers London 1955 Apr Society for general microbiology Edited by J. W. Howie and A.J. O'Hea Held at the Royal institution Society for general microbiology. Symposia, 5 Cambridge university press, Cambridge,1955
Bioch 33.1172

MECHANISMS OF MICROBIAL PATHOGENICITY :a symposium Papers London 1955 Apr Society for general microbiology Edited by J. W. Howie and A.J. O'Hea Society for general microbiology.Symposia, 5 Cambridge university press,Cambridge,1955
Path 30.2608

MECHANISMS OF MOTOR SKILL DEVELOPMENT Centre for advanced study in the developmental sciences study group on 'Mechanisms of motor skill development' Proceedings London 1968 Nov Ciba foundation Edited by Kevin Connolly Academic press,London;New York,1970 Being the 4th study group in a C.A.S.D.S. programme on 'The origins of human behaviour' held jointly with the Ciba foundation
Psy 31.3101

MECHANISMS OF NEURAL REGENERATION :a workshop...held as part of the first International summer school of brain research.. Amsterdam 1963 Jul 15-26 Edited by M. Singer and J.P. Schade Progress in brain research, 13 Elsevier,Amsterdam,1964
An 32.4233

MECHANIZATION OF THOUGHT PROCESSES National physical laboratory symposium 10th Proceedings Teddington 1958 Nov 24-27 Vol 1-2 National physical laboratory 2 vols H.M.S.O.,London,1959
Psy 31.2019

MEDFORD,MASS. 1957 Form discrimination as related to military problems :a symposium Proceedings Armed forces-National research council committee on vision Edited by Joseph W. Wulfeck and John H. Taylor National research council.Publication, 561 National research council,Washington,D.C.,1957
Psy 31.0942

MEDICAL BIOLOGY AND ETRUSCAN ORIGINS :a Ciba foundation symposium London 1958 Apr Ciba foundation Edited by G.E.W. Wolstenholme and Cecilia M. O'Connor Ciba foundation.Symposia, 50 illus. Churchill, London,1959
Gen 34.1877

MEDICAL EDUCATION Medicine,a life-long study World conference on medical education 2 Proceedings Chicago 1959 World medical association British medical journal,London, 1961
Med 36.0371

MEDICAL EDUCATION World conference on medical education 1st Proceedings London 1953 World medical association Oxford university press,London,1953
Med 36.0370

MEDICAL EDUCATION World conference on medical education 1st proceedings London 1953 Edited by Hugh Clegg Held under the auspices of the World medical association Oxford university press,London, 1954
An 32.0991

MEDICAL ELECTRONICS International conference on medical electronics 2nd Proceedings Paris 1959 Jun 24-27 International federation for medical electronics Edited by C.N. Smyth Iliffe,London,1960
Psy 31.3062

MEDICAL ELECTRONICS International conference on medical electronics 2nd Proceedings Paris 1959 Jun 24-27 Edited by C.N. Smyth Iliffe,London,1960
VA 19.0363

MEDICAL GENETICS Symposium internationale geneticae medicae 1st Atti Rome 1953 Sep 6-7 1 Istituto Gregorio Mendel Edited by Luigi Gedda Analecta genetica, 1 Edizioni dell' istituto Gregorio Mendel,Rome, 1954
Med 36.0148

MEDICAL GENETICS Symposium internazionale geneticae medicae 1st Atti Rome 1953 Sep 6-7 2: krampfbereitschaft Istituto Gregorio Mendel Edited by Luigi Gedda Analecta genetica, 1 Edizioni dell' istituto Gregorio Mendel,Rome,1954
Med 36.0149

MEDICAL GENETICS Symposium internazionale geneticae medicae Rome 1953 Sep 6-7 3: chondrodysplasie Istituto Gregorio Mendel Edited by Luigi Gedda Analecta genetica, 1 Edizioni dell' istituto Gregorio Mendel,Rome, 1954
Med 36.0150

MEDICAL PHOTOGRAPHY Internationaler kongress fur medizinische photographie und Kinematographie 1st Abhandlungen Dusseldorf 1960 Sep 27-30 Edited by Heinz Orbach Thieme,Stuttgart,1962
An 32.0311

MEDICAL PHYSICS Aspects of medical physics International conference on medical physics Review papers Harrogate 1965 Sep 8-10 Edited by J. Rotblat Convened by the International organization of medical physics Taylor and Francis,London,1966
Radioth 35.1671

MEDICAL RADIATION PHYSICS Joint IAEA-WHO expert committee on medical radiation physics Report Geneva 1967 Dec 12-18 International atomic energy agency World health organization World health organization.Technical report series, 390 World health organization,Geneva,1968
Radioth 35.1381

MEDICAL RADIOISOTOPE SCANNING Symposium on medical radioisotope scanning Proceedings Athens 1964 Apr 20-24 Held by the International atomic energy agency International atomic energy agency.Proceedings series International atomic energy agency, Vienna,1964
Radioth 35.1376

MEDICAL RADIOISOTOPE SCANNING :a seminar Proceedings Vienna 1959 Feb 25-27 International atomic energy agency World health organization International atomic energy agency,Vienna,1959
Radioth 35.1375

MEDICAL RADIOISOTOPE SCINTIGRAPHY :a symposium Proceedings Salzburg 1968 Aug 6-15 Vol 1-2 International atomic energy agency Edited by G.R. Stevenson International atomic energy agency.Proceedings series 2 vols International atomic energy agency,Vienna,1969
Radioth 35.1392

MEDICAL RESEARCH COUNCIL Human radiation cytogenetics :an international symposium Proceedings Edinburgh 1961 Oct 12-15 Edited by H.J. Evans and others North-Holland,Amsterdam,1967
Gen 34.1127

MEDICAL RESEARCH COUNCIL Human radiation cytogenetics :an international symposium Proceedings Edinburgh 1966 Oct 12-15 Edited by H.J. Evans and others North-Holland,Amsterdam,1967
Radioth 35.0540

MEDICAL RESEARCH COUNCIL Mathematics and computer science in biology and medicine Conference on mathematics and computer science in biology and medicine Proceedings Oxford 1964 Jul 317p H.M.S.O.,London,1965
Bot 42.0512

MEDICAL RESEARCH COUNCIL Mathematics and computer science in biology and medicine : conference Proceedings Oxford 1964 July Co-sponsored by Great Britain. Ministry of health H.M.S.O.,London,1965
HF 27.0125

MEDICAL RESEARCH COUNCIL Mathematics and computer science in biology and medicine conference proceedings Oxford 1964 Jul Co-sponsored by Great Britain. Scottish home and health department Bibliog. xi, 317p 25cm H.M.S.O.,London,1965
Math 3.0775

MEDICAL RESEARCH COUNCIL Mathematics and computer science in biology and medicine conference proceedings Oxford 1964 Jul Co-sponsored by Ministry of health,bibliog, illus.,pl. vii,316p 24cm H.M.S.O.,London, 1965
Math L 5.1007

MEDICAL RESEARCH COUNCIL The Application of scientific methods to industrial and service medicine :a conference Proceedings London 1950 Mar 29-31 H.M.S.O.,London,1951
Bioch 33.1542

MEDICAL RESEARCH COUNCIL The Renal stone research symposium Proceedings Leeds 1968 Apr 17-20 Edited by A. Hodgkinson and B. E.C. Nordin Churchill,London,1969
PGMS 29.0240

MEDICAL RESEARCH COUNCIL CONFERENCE Oxford 1964 Oxford,1964 Resumes of papers given at the conference
Math 3.1121

MEDICAL SOCIETY OF THE COUNTY OF NEW YORK.SPECIAL COMMITTEE ON INFANT MORTALITY Symposium on the placenta :its form and functions with particular reference to the prevention of birth defects and fetal deaths New York 1964 Mar 6 Edited by Daniel Bergsma and others Sponsored by the National Foundation-March of dimes.Great New York Chapter Birth defects original article series, 1,no.1 National foundation,New York,1965
PGMS 29.0212

MEDICAL USES OF CA47 :a panel Papers 1961 Dec 6-8 Sponsored by the International atomic energy agency International atomic energy agency.Technical report series, 10 International atomic energy agency,Vienna,1962
Radioth 35.1363

MEDICINAL PLANTS Symposium on medicinal plants :8th Pacific science congress Quezon city 1953 Nov 16-28 Pacific science association Under the auspices of the National research council of the Philippines Pacific science congress.Proceedings, 8th,4a National research council of the Philippines, Quezon city,1954
BG 38.3111

MEDICINE International congress of medicine 17th London 1913 Aug 6-13 Section 1: anatomy and embryology,pt.2 Frowde;Hodder and Stoughton,London,1914
An 32.1163

MEDICINE International congress of medicine 17th London 1913 Aug 6-13 Section 3: general pathology and pathological anatomy,pt. 2 Frowde;Hodder and Stoughton,London,1914
An 32.1164

MEDICINE International congress of medicine 17th London 1913 Aug 6-13 general volume Frowde;Hodder and Stoughton,London, 1914
An 32.1162

MEDICINE International medical congress 7th session Transactions London 1881 Aug 2-9 anatomy Kolckmann,London,1881
An 32.1165

MEDICINE,A LIFE-LONG STUDY World conference on medical education 2 Proceedings Chicago 1959 World medical association British medical journal,London,1961
Med 36.0371

MEDICINE AND CULTURE :a historical symposium Proceedings London 1966 Sep 27-20 Wellcome institute of the history of medicine Wenner-Gren foundation for anthropological research Wellcome institute of the history of medicine.Publications.New series, 15 vi, 321p Wellcome institute of the history of medicine,London,1969
WSM 43.3330

MEDIKAMENTOSE PATHOGENESE FETALER MISSBILDUNGEN symposium Edited by Th. Koller and H. Erb Acta genetica et statistica medica S. Karger,Basle,1964
An 32.2090

MEDITERRANEAN SOCIAL SCIENCES RESEARCH COUNCIL. ASSEMBLY 2nd Papers Markets and marketing as factors of development in the Mediterranean basin Cairo 1962 Dec 1-4 Edited by C.A.O.van Nieuwenhuijze Institute of social studies,The Hague.Publications. Series maior,11 Mouton,The Hague,1963
Geog 13.4117

MEERSBURG 1964 Mechanisms of hormone action :a NATO advanced study institute Edited by Peter Karlson Supported by the North Atlantic treaty organization Academic press,New York,1965 Some papers are in German
Pha 16.0092

MEERSBURG 1964 Mechanisms of hormone action :a Nato advanced study institute Discussions North Atlantic treaty organization Edited by P. Karlson Academic press,New York,1965
Bioch 33.0469

MEERSBURG 1964 Mechanisms of hormone action a NATO advanced study institute North Atlantic treaty organization Edited by P. Karlson Academic press;Thieme,New York; Stuttgart,1965
Gen 34.0549

MELANOMA Biology of melanomas Conference on the biology of normal and atypical pigment cell growth New York 1946 Nov 15-16 New York academy of sciences.Section of biology New York academy of sciences.Special publications, 4 New York,1948
Phys 20.0952

MELBOURNE Antarctic meteorology Proceedings Australia.Special committee for the International geophysical year,and,Special committee for Antarctic research Under the auspices of the Australian academy of science illus,maps,tables,diagrs xviii,483p 25cm Pergamon press,London,1960
Sco 14.6145

MELBOURNE 1908 Australasian medical congress 8th session Transactions Vol 1-3 Edited by Alan Lewers 3 vols Kemp, Melbourne,1909
An 32.1144

MELBOURNE 1946 The Failure of metals by fatigue :a symposium Proceedings University of Melbourne.Faculty of engineering Melbourne university press,Melbourne,1947
Met 25.0909

MELBOURNE 1950 Conference on applications of isotopes in scientific research Proceedings Commonwealth scientific and industrial research organization University of Melbourne.Chemistry department University of Melbourne,Melbourne,1951
Radioth 35.1495

MELBOURNE 1952 Australia-New Zealand conference on soil mechanics and foundation engineering 1st Proceedings University of Melbourne Institution of engineers, Australia 1952
Eng 41.3166

MELBOURNE 1953 Empire mining and metallurgical congress 5th Proceedings Vol 4: extractive metallurgy in Australia Australasian institute of mining and metallurgy Edited by J.C. Richards Melbourne,1953
Met 25.2524

MELBOURNE 1953 Extractive metallurgy in Australia:non-ferrous metallurgy Empire mining and metallurgical congress 5th Publications 4B Australasian institute of mining and metallurgy Edited by Frank A. Green and others Office of the Congress,Melbourne,1953
Met 25.0498

MELBOURNE 1953 Geology of Australian ore deposits Empire mining and metallurgical congress 5th Publications 1 Australasian institute of mining and metallurgy Edited by A.B. Edwards Office of the congress,and of the Australasian institute of mining and metallurgy,Melbourne, 1953
Met 25.0187

MELBOURNE 1953 Ore dressing methods in Australia and adjacent territories Empire mining and metallurgical congress 5th Publications 3 Australasian institute of mining and metallurgy Edited by H.H. Dunkin Office of the congress,and of the Australasian institute of mining and metallurgy,Melbourne,1953
Met 25.0205

MELBOURNE 1956 World congress on physical education Report Australian physical education association Sponsored by the Commonwealth national fitness council Australian physical education association, Melbourne,1956
HE 27.0219

MELBOURNE 1958 Radiation biology Australasian conference on radiation biology 2nd Proceedings Australian radiation society Edited by J.H. Martin Butterworths, London,1959
Gen 34.1110

MELBOURNE 1959 Antarctic meteorology :a symposium Proceedings Commonwealth bureau of meteorology and International geophysical year Under the auspices of the Australian academy of science 483p Pergamon press,Oxford;London,1960
Nap 11.0026

MELBOURNE 1959 Antarctic meteorology :a symposium Proceedings Under the auspices of Australian academy of science Pergamon press,Oxford,1960
Geog 13.0989

MELBOURNE 1959 The Evolution of living organisms :a symposium to mark the centenary of Darwin's 'Origin of species' and of the Royal society of Victoria Edited by G.W. Leeper Melbourne university press,Melbourne, 1962
Bal 39.0923

MELBOURNE 1959 The Evolution of living organisms :a symposium to mark the centenary of Darwin's 'Origin of species' and of the Royal society of Victoria Edited by G.W. Leiper Melbourne university press,Melbourne, 1962
Gen 34.1277

MELBOURNE 1959 The Evolution of living organisms :a symposium to mark the centenary of Darwin's 'Origin of species'and of the Royal society of Victoria Edited by G.W. Leeper illus,maps xi,459p 25cm Melbourne university press,Melbourne,1962
Sco 14.0621

MELBOURNE 1959 The Evolution of living organisms :a symposium to mark the centenary of Darwin's'Origin of species'and of the Royal society of Victoria... Edited by G.W. Leeper illus vii,459p Melbourne university press, Melbourne,1962
Bot 42.0991

MELBOURNE 1963 Fracture Tewksbury symposium on fracture 1st Proceedings Edited by C.J. Osborn University of Melbourne,Melbourne,1965
Eng 41.3802

MELBOURNE 1963 Fracture The Tewksbury symposium 1st Proceedings University of Melbourne.Faculty of engineering Edited by C. J. Osborn University of Melbourne,Melbourne, 1965
Met 25.0870

MELBOURNE 1965 International conference on 'Electron diffraction' and 'The nature of defects in crystals' Abstracts of papers Australian academy of science and International union of crystallography With the support of the International union of pure and applied physics. Commission on the solid state Pergamon press,Oxford,1966
Extended abstracts of papers to be presented at the conference
Cav 7.1584

MELBOURNE 1965 International conference on"electron diffraction" and "the nature of defects in crystals" Abstracts of papers Australian academy of science and International union of crystallography Co-sponsored by the International union of pure and applied physics Pergamon press,Oxford, 1966
Met 25.1463

MELBOURNE 1965 The Thymus :experimental and clinical studies:a Ciba foundation symposium Ciba foundation Edited by G.E.W. Wolstenholme and Ruth Porter Churchill, London,1966
Phys 20.0905

MELBOURNE 1965 The Thymus:experimental and clinical studies :a Ciba foundation symposium Ciba foundation Edited by G.E.W. Wolstenholme and Ruth Porter Churchill, London,1966
An 32.3703

MELBOURNE 1965 The Thymus:experimental and clinical studies Ciba foundation symposium Ciba foundation and Walter and Eliza Hall institute of medical research Edited by G.E. W. Wolstenholme and Ruth Porter illus Churchill,London,1966 In honour of sir Macfarlane Burnet,chairman of the conference
Path 30.2461

MELBOURNE 1969 Fracture Tewkesbury symposium on fracture 2nd Proceedings Edited by C.J. Osborn and others Held by the University of Melbourne.Fsculty of engineering Butterworths,Sydney,1969 Financial support from the Pearson Tewkesbury bequest
Eng 41.8202

MELBOURNE 1969 Fracture The Tewskbury symposium 2nd Proceedings University of Melbourne.Faculty of engineering Edited by C. J. Osborn and others Butterworth,Sydney,1969
Met 25.2579

MELBOURNE 1969 Interfaces conference Papers and abstracts Australian institute of metals Edited by R.C. Gifkins Butterworth, Sydney,1969
Met 25.2596

MELLON INSTITUTE Association for computing machinery :meeting Proceedings Pittsburgh 1952 May 2-3 illus. 305p 28cm Rimbach, Pittsburgh,1952
Math L 5.3239

MELTS Physical chemistry of melts Symposium on the nature of molten slags and salts London 1952 Feb 20 Nuffield research group in extraction metallurgy Institution of mining and metallurgy,London, 1953
Met 25.1787

MEMBRANE MODELS AND THE FORMATION OF BIOLOGICAL MEMBRANES International conference on biological membranes 2nd Proceedings Frascati 1967 Jun Edited by Liana Bolis and B.A. Pethica With the financial support of the North Atlantic treaty organization Illus. xv,337p North-Holland,Amsterdam, 1968 A NATO advanced study institute
Pha 16.0016

MEMBRANE MODELS AND THE FORMATION OF BIOLOGICAL MEMBRANES The International conference on biological membranes 2nd Proceedings Frascati 1967 Jun Nato advanced study institute Edited by Liana Bolis and B.A. Perthica Supported by the North Atlantic treaty organization North Holland,Amsterdam, 1968
Inv Med 37.0025

MEMBRANE PHENOMENA :a general discussion Nottingham 1956 Apr 10-12 Faraday society Faraday society.Discussions, 21 Aberdeen university press,Aberdeen,1956
Bioch 33.0947

MEMBRANE TRANSPORT AND METABOLISM :a symposium Proceedings Prague 1960 Edited by A. Kleinzeller and A. Kotyk Academic press,New York,1961
Phys 20.1326

MEMBRANE TRANSPORT AND METABOLISM :a symposium Proceedings Prague 1960 Aug 22-27 Ceskoslovenska akademie ved.Institute of biology Edited by A. Kleinzeller and A. Kotyk Academic press,London;New York,1961
Bioch 33.0972

MEMBRANES The Properties and functions of membranes,natural and artificial :a general discussion London 1937 Apr By August Krogh and others Faraday society Faraday society.Transactions, 33, pt 8 London,1937
Bal 39.0358

MEMBRANES:STRUCTURE AND FUNCTION Federation of European biochemical societies meeting 6th Proceedings Madrid 1969 Federation of European biochemical societies Edited by A. Villanueva and F. Ponz Federation of European biochemical societies.Publications, 20 Academic press,London;New York,1969
Radioth 35.0150

MEMBRANES,STRUCTURE AND FUNCTION Federation of European biochemical societies meeting 6th Proceedings Madrid 1969 Apr Federation of European biochemical societies Edited by J.R. Villanueva and F. Ponz Federation of European biochemical societies. Publications, 20 Academic press,London,1970
Bioch 33.2321

MEMBRANES AND SURFACES OF CELLS The Structure and function of the membranes and surfaces of cells :a symposium London 1962 Mar 9 Biochemical society Edited by D.J. Bell and J.K. Grant Biochemical society. Symposia, 22 Cambridge university press, Cambridge,1963
Bioch 33.1384

MEMBRANES AND SURFACES OF CELLS The Structure and function of the membranes and surfaces of cells :a symposium London 1962 Mar 9 Biochemical society Edited by D.J. Bell and J.K. Grant Held at the Middlesex hospital medical school Biochemical society. Symposia, 22 Cambridge university press, Cambridge,1963 Organized by J.K.Grant
Bal 39.0294

MEMORIA GEOLOGICO-MINERA DEL ESTADO DE CHIHUAHUA International geological congress 20th papers Mexico City 1956 By Jenaro Gonzalez Reyna Mexico City,1956 Bound with two other publications of the congress
Geol 8.3029

MEMORIAL LECTURE MEETING ON THE 10TH ANNIVERSARY OF THE FOUNDATION OF NATIONAL RESEARCH INSTITUTE FOR METALS Proceedings Tokyo 1966 Jun 28-30 National research institute for metals,Tokyo National research institute for metals,Tokyo,1966
Met 25.0113

MEMORY TECHNIQUES Large-capacity memory techniques for computing systems :a symposium Washington,D.C. 1961 May 23-25 Edited by Marshall C. Yovits Sponsored by the Association for computing machinery. Information systems branch Association for computing machinery.Monograph series Macmillan,New York;London,1962
Eng 41.2196

MENAGGIO 1970 Artificial intelligence and heuristic programming N.A.T.O. advanced study institute on artificial intelligence and heuristic programming 1st Proceedings North Atlantic treaty organization Edited by N.V. Findler and Bernard Meltzer illus 327p Edinburgh university press,Edinburgh, 1971
Math L 5.3853

MENDEL CENTENNIAL SYMPOSIUM Proceedings Heritage from Mendel Fort Collins,Colo 1965 Edited by R. Alexander and E.Derek Styles Sponsored by the Genetics society of America Wisconsin university press,Madison, Wis.,1967
Bot 42.0880

MENDEL CENTENNIAL SYMPOSIUM Proceedings Heritage from Mendel Fort Collins,Colo. 1965 Spe 7-11 Genetics society of America Edited by R.Alexander Brink and E.Derek Styles University of Wisconsin press,Madison,Wis., 1967
Gen 34.0994

MENDEL MEMORIAL SYMPOSIUM Symposia Csav The Physiology of gene and mutation expression symposium on the mutational process Proceedings Brno 1965 Aug 4-7 Prague 1965 Aug 9-11 Edited by Margita Kohoutova and J. Hubacek Organized by the Ceskoslovenska akademie ved.Institute of microbiology Academia,Prague,1967
Gen 34.0770

MENDELISM By Reginald Crundall Punnett American edition Wilshire,New York,1909
Gen 34.1026

MENSTRUATION AND ITS DISORDERS :conference Proceedings Edited by Earl T. Engle Held under the auspices of the National committee on maternal health Thomas, Springfield,Ill.,1950
An 32.3903

MENTAL DISEASE Integrating the approaches to mental disease :two conferences Papers and discussions New York New York academy of medicine.Committee on public health Edited by H.D. Kruse Hoeber-Harper,New York,1957
Psy 28.0061

MENTAL HEALTH ASPECTS OF THE PEACEFUL USES OF ATOMIC ENERGY Study group on mental health aspects of the peaceful uses of atomic energy Report Geneva 1957 Oct 21-26 World health organization World health organization.Technical report series, 151 World health organization,Geneva,1958
Radioth 35.1057

MENTAL HEALTH IN LATER MATURITY :a conference Papers Washington,D.C. 1941 May 23-24 United States.Public health service United States.Public health service.Public health reports.Supplement, 168 US government printing office,Washington,D.C.,1942
Psy 31.3388

MENTAL HEALTH RESEARCH FUND Prospects in psychiatric research :the proceedings of the Oxford conference of the Mental health research fund Proceedings Oxford 1952 Mar Edited by J.M. Tanner Blackwell,Oxford, 1953
Psy 28.0036

MENTAL RETARDATION Congress of the International association for the scientific study of mental defiency 1st Proceedings Montpellier 1967 Sep 12-20 International association for the scientific study of mental deficiency Edited by B.W. Richards Jackson, Reigate,1968 Title page and papers in English and French
Psy 28.0269

MENTAL RETARDATION International congress on mental retardation 2nd Proceedings Vienna 1961 Aug 14-19 Pt 1: organic bases and biochemical aspects of imbecility Edited by Otto Stur Karger,Basle,1963
An 32.4198

MENTAL RETARDATION Proceedings of the Association New York 1959 Dec 11-12 Association for research in nervous and mental disease Edited by Lawrence C. Kolb and others Association for research in nervous and mental disease.Research publications, 39 Williams and Wilkins,Baltimore,Md.,1962
An 32.4201

MENTHOL AND MENTHOL-CONTAINING EXTERNAL REMEDIES use,mode of effect and tolerance in children: international symposium Proceedings Paris 1966 Apr 27-28 Vick international Edited by F.H. Dost and B. Leiber Georg Thieme verlag,Stuttgart,1967
PGMS 29.0391

MENTON 1967 Digital computer applications to process control,2 IFAC-IFIP international conference 2nd Proceedings International federation of automatic control International federation for information processing Edited by W.E. Miller Sponsored also by the Association francaise de regulation d'automatisme Instrument society of America,Pittsburgh,Pa.,1969
Eng 41.8206

The MENZEL SYMPOSIUM ON SOLAR PHYSICS,ATOMIC SPECTRA AND GASEOUS NEBULAE Cambridge, Mass. 1971 Apr 29 Edited by K.B. Gebbie National bureau of standards.Special publication, 353 203p National bureau of standards,Washington,D.C.,1971
Obs 6.3623

MERCK,SHARP AND DOHME.POSTGRADUATE PROGRAM The International conference on the biology of cutaneous cancer 1st Philadelphia,Pa. 1962 Apr 6-11 Temple university.School of medicine and International union against cancer Edited by Frederick Urbach National cancer institute.Monograph, 10 US government printing office,Washington,D.C., 1963
Path 30.2423

MERISTEMS AND DIFFERENTIATION report of symposium Upton,N.Y. 1963 Jan 3-5 Brookhaven national laboratory Brookhaven symposia in biology, 16 BNL 805 Brookhaven national laboratory,Upton,N.Y.,1964
Gen 34.1414

MERISTEMS AND DIFFERENTIATION :symposium Report Upton,N.Y. 1963 Jun 3-5 Brookhaven national laboratory.Biology department Brookhaven symposia in biology, 16 BNL805(C-38) Brookhaven national laboratory.Biology department,Upton,N.Y.,1964
Bioch 33.1294

MESON PHYSICS The International conference on nuclear and meson physics Proceedings Glasgow 1954 Edited by E.H. Bellamy and R. G. Moorhouse Sponsored by the International union of pure and applied physics 352p Pergamon press,London,1955
Cav 7.0908

MESOZOIC International geological congress 20th papers Mexico City 1956 Seccion 2: el mesozoico del hemisferio occidental y sus correlaciones mundiales Edited by A. Garcia Rojas and others Mexico City,1957 Bound with papers from section 3. Text in English, French and Spanish
Geol 8.3034

METABOLIC AND TOXIC DISEASES OF THE NERVOUS SYSTEM Proceedings of the Association New York 1952 Dec 12-13 Association for research in nervous and mental disease Edited by H.Houston Merritt and Clarence C. Hare Association for research in nervous and mental disease.Research publications, 32 Williams and Wilkins,Baltimore,Md.,1953
An 32.4156

METABOLIC CONTROL Control of energy metabolism Colloquium on metabolic control : and symposium on control of metabolism Philadelphia,Pa. 1965 May 20-21 Johnson research foundation Edited by Britton Chance and others Academic press,New York;London, 1965
Gen 34.0614

METABOLIC CONTROL MECHANISMS IN ANIMAL CELLS : a symposium Proceedings Boston,Mass. 1963 May 28-30 National cancer institute Edited by William J. Rutter Sponsored by the Tissue culture association and National cancer institute National cancer institute. Monograph, 13 illus National cancer institute,Bethesda,Md.,1964
Bioch 33.0984

METABOLIC EFFECTS OF ADRENAL HORMONES London 1960 Jul 18 Ciba foundation Edited by G.E. W. Wolstenholme and Maeve O'Connor Ciba foundation study group, 6 Churchill,London, 1960
Gen 34.0540

METABOLIC EFFECTS OF ADRENAL HORMONES in honour of G.W.Thorn London 1960 Jul 15 Ciba foundation Edited by G.E.W. Wolstenholme and Maeve O'Connor Ciba foundation study group, 6 Churchill,London, 1960
Bioch 33.0442

METABOLIC REGULATION AND ENZYME ACTION Federation of European biochemical societies meeting 6th Proceedings Madrid 1969 Apr Federation of European biochemical societies Edited by A. Sols and S. Grisolia Federation of European biochemical societies. Publications, 19 Academic press,London,1970
Bioch 33.1889

The METABOLIC ROLES OF CITRATE :a symposium in honour of...Sir Hans Krebs Oxford 1967 Jul Biochemical society Edited by T.W. Goodwin Biochemical society.Symposia, 27 Academic press,London;New York,1968
Bioch 33.1389

METABOLISM Endogenous metabolism with special reference to bacteria New York 1963 Jan 21 New York academy of sciences Edited by Carl Lamanna New York academy of sciences.Annals, 102,3 New York,1963
Bot 42.3726

METABOLISM International congress of microbiology 6th Rome 1953 Vol 2: symposium - microbial metabolism Edited by E.B. Chain Fondazione Emanuele Paterno,Rome, 1953 Title also in Italian;text in English and French
Bioch 33.1801

METABOLISM Symposium on salt and water metabolism Papers New York 1959 Dec 11-12 Edited by Alfred P. Fishman Sponsored by the New York heart association Circulation, 21,no 5,pt 2 American heart association,New York,1959
Phys 20.1327

METABOLISM AND FUNCTIONS IN NERVOUS TISSUE : a symposium London 1951 Nov 10 Biochemical society Edited by R.T. Williams Held at the London school of hygiene and tropical medicine Biochemical society. Symposia, 8 Cambridge university press, Cambridge,1952
Bioch 33.1370

METABOLISM AND MEMBRANE PERMEABILITY OF ERYTHROCYTES AND THROMBOCYTES International symposium on metabolism and membrane permeability of erythrocytes and thrombocytes 1st Vienna 1968 Jun 17-20 Edited by Erwin Deutsch and others Thieme,Stuttgart,1968 In English and German
Radioth 35.0932

METABOLISM AND PHYSIOLOGICAL SIGNIFICANCE OF LIPIDS :advanced study course Proceedings Cambridge 1963 Sep Edited by R.M.C. Dawson and D.N. Rhodes Wiley,London,1964
Pha 16.0127

METABOLISM AND PHYSIOLOGICAL SIGNIFICANCE OF LIPIDS :advanced study course Proceedings Cambridge 1963 Sep 16-21 Edited by R.M. C. Dawson and Douglas N. Rhodes Wiley,London, 1964
Bioch 33.1235

METABOLISM OF AMINES IN THE BRAIN :a symposium Proceedings Edinburgh 1968 Jul 11 British pharmacological society Scandinavian pharmacological society Edited by G. Hooper Macmillan,London,1969
An 32.5040

METABOLISM OF THE NERVOUS SYSTEM International neurochemical symposium 2nd Proceedings Aarhus 1956 Jul Aarhus universitet Edited by Derek Richter Pergamon press,London,1957
Bioch 33.0592

METABOLISM OF THE NERVOUS SYSTEM
International neurochemical symposium 2nd Proceedings Aarhus 1956 Jul Edited by Derek Richter Pergamon,London,1957
Pha 16.0037

METAL-BINDING IN MEDICINE :a symposium Proceedings Edited by Marvin J. Seven and L.Audrey Johnson Sponsored by Hahnemann medical college and hospital Lippincott, Philadelphia;Montreal,1960
Bioch 33.0679

METAL CONGRESS AND EXPOSITION 10TH Short-time high-temperature testing Short-time elevated temperature testing of metals :a symposium Proceedings Los Angeles 1957 Mar 25-29 American society for metals American society for metals,Cleveland,Ohio, 1958
Met 25.0983

METAL CORROSION IN THE ATMOSPHERE :symposium Proceedings Boston,Mass. 1967 Jun 25-30 American society for testing and materials American society for testing and materials. Special technical publication, 435 Philadelphia,Pa.,1968 Symposium presented at the annual meeting
Met 25.2792

METAL INTERFACES a seminar held during the thirty-third national metal congress and exposition Detroit 1951 Oct 13-19 American society for metals American society for metals,Cleveland,Ohio,1952
Met 25.1261

METAL MATRIX COMPOSITES a symposium presented at the 70th annual meeting... Boston 1967 Jun 25-30 American society for testing materials A.S.T.M.Special technical publication, 438 A.S.T.M.,Philadelphia,1968
Met 25.2337

METAL MOLYBDENUM The Technology of molybdenum and its alloys :a symposium Proceedings Washington,D.C. 1956 Sep 18-19 United States.Office of naval research Edited by Julius J. Harwood American society for metals,Cleveland,Ohio,1958
Met 25.0546

METAL POWDER INDUSTRIES FEDERATION General session on powder metallurgy Detroit 1959 Apr 20-22 Metal powder industries federation, New York,1959
Met 25.0764

METAL POWDER INDUSTRIES FEDERATION Progress in powder metallurgy Annual technical meeting 16th Proceedings Chicago 1960 Apr 25-27 Metal powder industries federation, Chicago,1960
Met 25.0769

METAL POWDER INDUSTRIES FEDERATION Progress in powder metallurgy Annual technical meeting 17th Proceedings Cleveland 1961 Apr 24-26 Metal powder industries federation,Cleveland,Ohio,1961
Met 25.0770

METAL POWDER INDUSTRIES FEDERATION.ANNUAL MEETING, 15TH General session on powder metallurgy Detroit 1959 Apr 20-22 Metal powder industries federation Metal powder industries federation,New York,1959
Met 25.0764

METAL POWDER PRODUCTS Symposium on testing metal powders and metal powder products Papers Cleveland 1952 Mar 5 American society for testing materials American society for testing materials.Special technical publication, 140 87p American society for testing materials,Philadelphia,Pa., 1953
Chem E 24.1121

METAL POWDERS Symposium on testing metal powders and metal powder products Papers Cleveland 1952 Mar 5 American society for testing materials American society for testing materials.Special technical publication, 140 87p American society for testing materials,Philadelphia,Pa.,1953
Chem E 24.1121

METAL THORIUM The Conference on thorium Proceedings Cleveland,Ohio 1956 Oct 11 American society for metals and United States atomic energy commission Edited by Harley A. Wilhelm American society for metals,Cleveland,Ohio,1958
Met 25.1561

METAL TRANSFORMATIONS Buhl international conference on materials 2nd Informal proceedings Pittsburgh,Pa. 1966 Mar 16-17 Buhl foundation Edited by William W. Mullins and Milton C. Shaw Gordon and Breach,London; New York,1968
Met 25.2514

METALLIC CORROSION International congress on metallic corrosion 1st London 1961 Apr 10-15 Society of chemical industry. Corrosion group Butterworths,London,1962
Met 25.1893

METALLIC CORROSION International congress on metallic corrosion 3rd Proceedings Moscow 1966 Vol 1-4 Edited by Ya.M. Kolotyrkin and others 4 vols Mir,Moscow, 1969
Met 25.2662

METALLIC MATERIALS AT LOW TEMPERATURES
Symposium on effect of temperature on the brittle behaviour of metals with particular reference to low temperatures Atlantic City 1953 Jun 28-30 American society for testing materials A.S.T.M.Special technical publication, 158 American society for testing materials,Philadelphia,Pa.,1954
Met 25.0850

METALLIC MATERIALS FOR LOW-TEMPERATURE SERVICE
Symposium on evaluation of metallic materials in design for low-temperature service presented at the 64th annual meeting Atlantic City,N.J. 1961 Jun 27-28 American society for testing materials A.S.T.M. Special technical publication, 302 A.S.T.M., Philadelphia,1962
Met 25.2335

METALLIC SOLID SOLUTIONS:THEIR ELECTRONIC AND ATOMIC STRUCTURE :a symposium Proceedings Orsay 1962 Jul Universite de Paris a la Sorbonne W.A.Benjamin,New York,1963
Cav 7.1737

METALLO-FLAVOPROTEINS A Symposium on inorganic nitrogen metabolism :function of metallo-flavoproteins Papers and informal discussions presented Baltimore,Md. 1955 Jun 21-23 Edited by William D. McElroy and H.Bentley Glass Sponsored by the McCollum-Pratt institute McCollum-Pratt institute.Contribution, 125 Johns Hopkins press,Baltimore,Md.,1956
Bioch 33.0598

METALLO-ORGANIC COMPOUNDS Bonding in metallo-organic compounds Cambridge 1969 Mar 25-27 Faraday society Faraday society.Discussions, 47 208p 20cm Faraday society,London, 1969
Cav 7.2717

METALLOFLAVOPROTEINS A Symposium on inorganic nitrogen metabolism:function of metalloflavoproteins Baltimore,Md. 1955 Jun 21-23 Edited by William D. McElroy and Bentley Glass Sponsored by the McCollum-Pratt institute McCollum-Pratt institute. Contributions, 125 Johns Hopkins press, Baltimore,Md.,1956
Gen 34.0659

METALLOGRAPHIC SPECIMEN PREPARATION Symposium on methods of metallographic specimen preparation Atlantic City 1960 Jun 28 American society for testing materials A.S.T.M.Special technical publication, 285 American society for testing materials,Philadelphia,Pa.,1960
Met 25.1381

METALLOGRAPHIC TECHNIQUES Materials engineering exposition and congress Symposium on applications of modern metallographic techniques Philadelphia,Pa. 1969 Oct 13-16 American society for testing and materials American society for metals American society for testing and materials. Special technical publication, 480 ASTM, Philadelphia,Pa.,1970 Symposium presented at the exposition and congress
Met 25.2607

METALLOGRAPHY IN COLOR Symposium on metallography in color (1948) presented to the 51st annual meeting Detroit 1948 Jun 21 American society for testing materials A.S.T.M.Special technical publication, 86 American society for testing materials, Philadelphia,Pa.,1949
Met 25.1342

METALLOGRAPHY 1963 Sorby centenary conference Papers and discussions Sheffield 1963 May 9 Iron and steel institute Sponsored by the Company of Cutlers in Hallamshire Iron and steel institute.Special report, 80 Iron and steel institute,London,1964
Met 25.2360

METALLURGICAL ACHIEVEMENTS Birmingham metallurgical society Edited by W.O. Alexander Pergamon press,Oxford,1965 Selection of papers presented at the Birmingham metallurgical society's Diamond Jubilee session, 1963-64
Met 25.0080

METALLURGICAL ACHIEVEMENTS :selection of papers presented at the Birmingham metallurgical society's diamond jubilee session 1963-64 Birmingham 1963-64 Birmingham metallurgical society Edited by W. O. Alexander Pergamon press,London,1965
Eng 41.3485

METALLURGICAL DEVELOPMENTS IN CARBON STEELS : a conference Harrogate 1963 May 15-16 Iron and steel institute Sponsored by the British iron and steel research association Iron and steel institute.Special report, 81 Iron and steel institute,London,1963
Met 25.0417

METALLURGICAL DEVELOPMENTS IN HIGH-ALLOY STEELS A Joint conference on high-alloy steels Proceedings Scarborough 1964 Jun 2-4 Iron and steel institute and British iron and steel research association Iron and steel institute.Special report, 86 Iron and steel institute,London,1964
Met 25.0488

METALLURGICAL SOCIETY CONFERENCES Precipitation from iron-base alloys :a symposium Cleveland 1963 Oct 21 American institute of mining,metallurgical and petroleum engineers.Ferrous metallurgy committee Edited by Gilbert R. Speich and John B. Clark Gordon and Breach,New York; London,1965
Met 25.2525

METALLURGICAL SOCIETY CONFERENCES, 1 Flat rolled products:rolling and treatment Technical conference 1st Proceedings Chicago 1959 Jan 21 American institute of mining,metallurgical and petroleum engineers. Mechanical working committee Edited by T.E. Dancy and E.L. Robinson Interscience,New York;London,1959
Met 25.0731

METALLURGICAL SOCIETY CONFERENCES, 3 Quality requirements of super-duty steels a technical conference Proceedings Pittsburgh 1958 May 5-6 American institute of mining,metallurgical and petroleum engineers.Committee on physical chemistry of steelmaking Edited by R.W. Lindsay Interscience,New York;London,1959
Met 25.0482

METALLURGICAL SOCIETY CONFERENCES, 4 Physical metallurgy of stress-corrosion fracture Pittsburgh 1959 Apr 2-3 American institute of mining,metallurgical and petroleum engineers Edited by Thor N. Rhodin Interscience,New York;London,1959
Met 25.2503

METALLURGICAL SOCIETY CONFERENCES, 5 Properties of elemental and compound semiconductors a technical conference Boston, Mass. 1959 Aug 31-Sep 2 American institute of mining,metallurgical and petroleum engineers Edited by Harry C. Gatos Interscience,New York;London,1960
Met 25.1160

METALLURGICAL SOCIETY CONFERENCES, 7,8 Physical chemistry of process metallurgy :an international symposium Pittsburgh,Pa. 1959 Apr 27-May 1 American institute of mining,metallurgical and petroleum engineers Edited by George R. St.Pierre Interscience, New York;London,1961
Met 25.0491

METALLURGICAL SOCIETY CONFERENCES, 9 Response of metals to high velocity deformation Technical conference Estes Park,Col. 1960 Jul 11-12 American institute of mining,metallurgical and petroleum engineers Interscience,New York, 1961
Met 25.0905

METALLURGICAL SOCIETY CONFERENCES, 10 Columbium metallurgy :a symposium Bolting Landing,N.Y. 1960 Jun 9-10 American institute of mining,metallurgical and petroleum engineers Edited by D.L. Douglass and F.W. Kunz Interscience,New York;London, 1958
Met 25.2533

METALLURGICAL SOCIETY CONFERENCES, 11 Refractory metals and alloys :a technical conference Detroit 1960 May 25-26 American institute of mining,metallurgical and petroleum engineers.Refractory metals committee Edited by M. Semchyshen and J.J. Harwood Interscience,New York;London,1961
Met 25.2532

METALLURGICAL SOCIETY CONFERENCES, 12 Metallurgy of elemental and compound semiconductors a technical conference Proceedings Boston,Mass. 1960,Aug. 29-31 American institute of mining,metallurgical and petroleum engineers Edited by Ralph O. Grubel Interscience,New York;London,1961
Met 25.1146

METALLURGICAL SOCIETY CONFERENCES, 13 Bar and allied products Technical conference 3rd Proceedings Pittsburgh 1961 Jan 18 American institute of mining,metallurgical and petroleum engineers.Mechanical working committee Interscience,New York;London,1961
Met 25.0729

METALLURGICAL SOCIETY CONFERENCES, 15 Metallurgy of semiconductor materials a technical conference Los Angeles,Calif. 1961 Aug 30-Sep 1 American institute of mining,metallurgical and petroleum engineers. Semiconductors committee Edited by John B. Schroeder Interscience,New York;London,1962
Met 25.1158

METALLURGICAL SOCIETY CONFERENCES, 16 Flat rolled products III Technical conference 4th Proceedings Chicago,Ill. 1962 Jan 17 American institute of mining,metallurgical and petroleum engineers.Mechanical working committee Interscience,New York;London,n.d.
Met 25.0703

METALLURGICAL SOCIETY CONFERENCES, 18 High temperature materials 11 a technical conference Cleveland,Ohio 1961 Apr 26-27 American institute of mining,metallurgical and petroleum engineers.High temperature alloys committee Interscience,New York;London,1963
Met 25.1006

METALLURGICAL SOCIETY CONFERENCES, 19 Metallurgy of advanced electronic materials a technical conference Philadelphia,Pa. 1962 Aug 27-29 American institute of mining, metallurgical and petroleum engineers Edited by Geoffrey E. Brock Interscience,New York; London,1963
Met 25.1155

METALLURGICAL SOCIETY CONFERENCES, 20 Fracture of solids :an international conference Proceedings Washington,D.C. 1962 Aug 21-24 American institute of mining, metallurgical and petroleum engineers. Institute of metals division Edited by D.C. Drucker and J.J. Gilman Interscience,New York;London,1963
Met 25.2247

METALLURGICAL SOCIETY CONFERENCES, 21 Mechanical working of steel 1 Technical conference 5th Proceedings Pittsburgh 1963 Jan 15-16 American institute of mining, metallurgical and petroleum engineers. Mechanical working committee Gordon and Breach,New York,1964
Met 25.0728

METALLURGICAL SOCIETY CONFERENCES, 22 Metallurgy at high pressures and high temperatures :a symposium Proceedings Dallas,Tex. 1963 Feb 25-26 American institute of mining,metallurgical and petroleum engineers.Physical chemistry of steelmaking committee Edited by K.A. Gschneidner and others Gordon and Breach,New York;London,1964
Met 25.1639

METALLURGICAL SOCIETY CONFERENCES, 23 New types of metal powders :a symposium Proceedings Cleveland,Ohio 1963 Oct 24 American institute of mining,metallurgical and petroleum engineers.Powder metallurgy committee Edited by Henry H. Hausner Gordon and Breach,New York;London,1964
Met 25.2509

METALLURGICAL SOCIETY CONFERENCES, 23 New types of metal powders :a symposium Proceedings Cleveland,Ohio 1963 Oct 24 American institute of mining,metallurgical and petroleum engineers.Powder metallurgy committee Edited by Henry H. Hausner Gordon and Breach,New York;London,1964
Met 25.0766

METALLURGICAL SOCIETY CONFERENCES, 24 Unit processes in hydrometallurgy based on an international symposium Dallas 1963 Feb 24-28 American institute of mining, metallurgical and petroleum engineers. Extractive metallurgy division and Society of mining engineers.Minerals benefication division Edited by Milton E. Wadsworth and Franklin T. Davis Gordon and Breach,New York; London,1964
Met 25.2289

METALLURGICAL SOCIETY CONFERENCES, 25 Deformation twinning :a conference Proceedings Gainesville 1963 Mar 21-22 American institute of mining,metallurgical and petroleum engineers and University of Florida.College of engineering In co-operation with Florida institute for continuing university studies Gordon and Breach,New York;London,1964
Met 25.1257

METALLURGICAL SOCIETY CONFERENCES, 26
Mechanical working of steel 2 The Technical conference 6th Proceedings Chicago,Ill. 1964 Jan 30-31 American institute of mining, metallurgical and petroleum engineers. Mechanical working and steel processing committee Edited by T.G. Bradbury Sponsored by the Chicago section Gordon and Breach,New York,1965
Met 25.2244

METALLURGICAL SOCIETY CONFERENCES, 27 The Sorby centennial symposium on the history of metallurgy :a symposium Cleveland 1963 Oct 22-23 American society for metals and American institute of mining,metallurgical and petroleum engineers.Metallurgical society Edited by Cyril Stanley Smith Sponsored by the Society for the history of technology Gordon and Breach,New York,1965
Met 25.2141

METALLURGICAL SOCIETY CONFERENCES, 29
Alloying behaviour and effects in concentrated solid solutions :based on a symposium Cleveland 1963 Oct 21 American institute of mining,metallurgical and petroleum engineers Edited by T.B. Massalski Gordon and Breach,Ohio,1963
Met 25.1242

METALLURGICAL SOCIETY CONFERENCES, 30
Technical conference on refractory metals and alloys 3rd Proceedings Los Angeles 1963 Dec 9-10 American institute of mining, metallurgical and petroleum engineers. Refractory metals committee Gordon and Breach,New York,1966
Met 25.1017

METALLURGICAL SOCIETY CONFERENCES, 31
Application of fracture toughness parameters to structural metals :a symposium 1964 Oct 20 American institute of mining, metallurgical and petroleum engineers. Structural materials technical committee Edited by Hermann D. Greenberg Gordon and Breach,New York;London,1966
Met 25.0867

METALLURGICAL SOCIETY CONFERENCES, 32
Process simulation and control in iron and steelmaking :a symposium Proceedings New York 1964 Feb 17-18 American institute of mining,metallurgical and petroleum engineers Edited by J.M. Uys and H.L. Bishop Gordon and Breach,New York,1966
Met 25.0376

METALLURGICAL SOCIETY CONFERENCES, 34 High temperature refractory metals :a symposium New York 1964 Feb 16-20 1965 Feb Pt 1-2 American institute of mining, metallurgical and petroleum engineers. Refractory metals committee 2 vols Gordon and Breach,New York,1966-68 Conference pt 1 1965,pt 2 1964
Met 25.1016

METALLURGICAL SOCIETY CONFERENCES, 35
Environment-sensitive mechanical behaviour :a conference Proceedings Baltimore 1965 Jun 7-8 Pt 1-2 American institute of mining,metallurgical and petroleum engineers. Physical metallurgy committee and United States.Army research office,Durham Edited by A.R.C. Westwood and N.S. Stoloff Sponsored by the Martin company research institute for advanced studies 2 vols Gordon and Breach, New York,1966
Met 25.1963

METALLURGICAL SOCIETY CONFERENCES, 36 Local atomic arrangements studied by x-ray diffraction :a symposium Proceedings Chicago,Ill. 1965 Feb 15 American institute of mining,metallurgical and petroleum engineers Edited by J.B. Cohen and J.E. Hilliard Gordon and Breach,New York, 1966
Met 25.1461

METALLURGICAL SOCIETY CONFERENCES, 38 High-temperature high-resolution metallography :a symposium Proceedings Chicago,Ill. 1965 Feb 15 American institute of mining, metallurgical and petroleum engineers.Ferrous metallurgy committee Edited by Hubert I. Aaronson and George S. Ansell Gordon and Breach,New York,1967
Met 25.2361

METALLURGICAL SOCIETY CONFERENCES, 39
Pyrometallurgical processes in nonferrous metallurgy :a symposium Pittsburgh 1965 Nov 29-Dec 1 American institute of mining, metallurgical and mining engineers. Extractive metallurgy division Edited by J.N. Anderson and P.E. Queneau Gordon and Breach,New York, 1967
Met 25.2530

METALLURGICAL SOCIETY CONFERENCES, 40
Methods of materials selection :a symposium American institute of mining,metallurgical and petroleum engineers and University of Florida Gordon and Breach,New York;London, 1968
Met 25.2502

METALLURGICAL SOCIETY CONFERENCES, 41
Refractory metals and alloys 4 research and development:based on a technical conference Proceedings French Lick,Ind. 1965 Oct 3-5 Vol 1-2 American institute of mining, metallurgical and petroleum engineers. Refractory metals committee Edited by R.I. Jaffee and others 2 vols Gordon and Breach, New York,1967
Met 25.2250

METALLURGICAL SOCIETY CONFERENCES, 45
Energy bands in metals and alloys based on a symposium Los Angeles 1967 American institute of mining,metallurgical and petroleum engineers.Committee on alloy phases Edited by L.H. Bennett and J.T. Waber Gordon and Breach,New York,1968
Met 25.1060

METALLURGICAL SOCIETY CONFERENCES, 46 Work hardening :based on a symposium Chicago,Ill. 1966 Nov American institute of mining, metallurgical and petroleum engineers. Institute of metals division.Chemistry and physics committee Edited by J.P. Hirth and J. Weertman Gordon and Breach,New York,1968
Met 25.2866

METALLURGICAL SOCIETY CONFERENCES, 47
Bolton Landing conference oxide dispersion strengthening:a symposium 2nd Bolton Landing,New York 1966 Jun 27-29 American institute of mining,metallurgical and petroleum engineers Edited by G.S. Ansell Gordon and Breach,New York,1968
Met 25.2339

METALLURGICAL SOCIETY CONFERENCES, 49
Continuous processing and process control :a symposium Proceedings Philadelphia,Penn. 1966 Dec 5-8 American institute of mining, metallurgical and petroleum engineers Edited by Thomas R. Ingraham Gordon and Breach,New York,1968 Symposium sponsored by the Electric furnace conference,the Mechanical working and steel processing conference and the Physical chemistry of steelmaking conference of the Metallurgical society
Met 25.2846

METALLURGICAL SOCIETY OF AIME.PUBLICATIONS
Corrosion by liquid metals Proceedings of the sessions on corrosion by liquid metals of the 1969 fall meeting of the Metallurgical society of AIME Philadelphia,Pa. 1969 Oct 13-16 American institution of mining engineers.Metallurgical society Edited by J. E. Draley and J.R. Weeks Plenum press,New York,1970
Met 25.2790

METALLURGY Congress G.A.M.S. 4th
Paris 1951 June 20-22 Groupement pour l'advancement des methodes d'analyse spectrographique des produits metallurgiques 24cm Paris,1951
Philos 1.0678

METALLURGY Empire mining and metallurgical congress 4th Proceedings London 1949 Jul 9-23 Pt 1-2 Edited by F. Higham 2 vols London,1950
Min 10.0726

METALLURGY Research in chemical and extraction metallurgy Symposium on 'recent progress in research in chemical and extraction metallurgy' Proceedings Sydney 1965 Feb 24-26 Australasian institute of mining and metallurgy Edited by J.T. Woodcock and others Australasian institute of mining and metallurgy.Monograph series, 2 Melbourne,1965 reprinted 1967
Met 25.2780

METALLURGY World metallurgical congress 1st
Proceedings Detroit,Mich. 1951 Oct 14-19 American society for metals Edited by William Marsh Baldwin 833p American society for metals,Cleveland,Ohio,1952
Met A 61 25.0044

METALLURGY World metallurgical congress 2nd
Story of 2nd world metallurgical congress held in the United States of America 1957 Oct-Nov American society for metals Edited by Kingsley W. Given 225p A.S.M.,Cleveland, Ohio,1958
Met 25.0045

METALLURGY AND FUELS Geneva 1955 Vol 1
Edited by H.M. Finniston and J.P. Howe Progress in nuclear energy, 5 Pergamon press,London,1956 Includes papers presented at the United Nations conference on the peaceful uses of atomic energy
Met 25.1560

METALLURGY AT HIGH PRESSURES AND HIGH TEMPERATURES :a symposium Proceedings Dallas,Tex. 1963 Feb 25-26 American institute of mining,metallurgical and petroleum engineers.Physical chemistry of steelmaking committee Edited by K.A. Gschneidner and others Metallurgical society conferences, 22 Gordon and Breach,New York; London,1964
Met 25.1639

METALLURGY INFORMATION MEETING 5th
Proceedings Oak Ridge 1955 Apr 11-13 Pt 1-2 United States atomic energy commission 2 vols Washington,1960
Met 25.1552

METALLURGY OF ADVANCED ELECTRONIC MATERIALS
a technical conference Philadelphia,Pa. 1962 Aug 27-29 American institute of mining, metallurgical and petroleum engineers Edited by Geoffrey E. Brock Metallurgical society conferences, 19 Interscience,New York; London,1963
Met 25.1155

The METALLURGY OF BERYLLIUM :an international conference Proceedings London 1961 Oct 16-18 Institute of metals Institute of metals.Monograph and report series, 28 Chapman and Hall,London,1963
Met 25.0605

METALLURGY OF ELEMENTAL AND COMPOUND SEMICONDUCTORS a technical conference Proceedings Boston,Mass. 1960,Aug. 29-31 American institute of mining, metallurgical and petroleum engineers Edited by Ralph O. Grubel Metallurgical society conferences, 12 Interscience,New York; London,1961
Met 25.1146

METALLURGY OF SEMICONDUCTOR MATERIALS a technical conference Los Angeles,Calif. 1961 Aug 30-Sep 1 American institute of mining,metallurgical and petroleum engineers. Semiconductors committee Edited by John B. Schroeder Metallurgical society conferences, 15 Interscience,New York;London,1962
Met 25.1158

METALLWERK PLANSEE A.G. High temperature materials Plansee seminar Hochtemperatur werkstoffe 6th Papers Reutte,Austria 1966 Jun 24-26 Edited by F. Benesovsky Springer,Vienna;New York,1969 Papers in English,French and German
Met 25.2336

METALLWERK PLANSEE A.G. Metals for the space age Plansee seminar "De re metallica" 5th Papers Reutte,Austria 1964 Jun 22-26 Edited by F. Benesovsky Metallwerk Plansee A G,Reutte,1965 Papers in English,French and German
Met 25.0091

METALLWERK PLANSEE AG High-melting metals Plansee seminar "De re metallica" 3rd Papers Reutte,Austria 1948 Jun 22-26 Edited by F. Benesovsky Pergamon press; Metallwerk Plansee AG;,,London;Reutte,1959
Met 25.1010

METALLWERK PLANSEE AG Powder metallurgy in the nuclear age Plansee seminar 'De re metallica' 4th Proceedings Reutte, Austria 1961 Jun 20-24 Edited by F. Benesorsky Metallwerk Plansee AG,Reutte,1962 Papers in English,French and German
Met 25.2724

METALS Heat treatment of metals Heat treatment joint committee :inaugural meeting Proceedings London 1965 Dec 14-15 Iron and steel institute Institute of metals Iron and steel institute.Special report, 95 London,1966
Met 25.2565

METALS Symposium on internal stresses in metals and alloys Advance copies of papers London 1947 Oct 15-16 Institute of metals Institute of metals,London,1947
Eng 41.3768

METALS Vacancies and interstitials in metals conference Proceedings Julich 1968 Sep 23-28 Edited by A. Seeger and others Bibliog 1074p 22cm North-Holland, Amsterdam,1970
Cav 7.2686

METALS :presented at the 2nd Pacific area national meeting Los Angeles 1956 Sep 17-21 American society for testing materials A.S.T.M.Special technical publication, 196 American society for testing materials, Philadelphia,Pa.,1957
Met 25.1036

METALS AND ENZYME ACTIVITY :a symposium Leeds 1956 Jul 13 Biochemical society Edited by E.M. Crook Held at the University of Leeds Biochemical society.Symposia, 15 Cambridge university press,Cambridge,1958
Bioch 33.1377

METALS AND ENZYME ACTIVITY 15 :a symposium Leeds 1956 Jul 13 Biochemical society Edited by E.M. Crook Biochemical society symposia,15 Cambridge university press, Cambridge,1958
Chem 18.1269

METALS AND METALLURGY TRUST Thermal and high-strain fatigue :an international conference Proceedings London 1967 Jun 6-7 Organized by the Institute of metals Metals and metallurgy trust.Monograph and report series Metals and metallurgy trust,London, 1967
Eng 41.3810

METALS CONGRESS 1966 Fiber-strengthened metallic composites a symposium presented at the 1966 metals congress Chicago 1966 Nov 2-3 American society for testing materials A.S.T.M.Special technical publication, 427 A. S.T.M.,Philadelphia,1967
Met 25.2338

METALS FOR SUPERSONIC AIRCRAFT AND MISSILES Heat tolerant metals for aerodynamic applications :a conference Proceedings Alburquerque,New Mexico 1957 Jan 28-29 American society for metals and University of New Mexico Edited by D.W. Grobecker American society for metals,Cleveland,Ohio, 1958
Met 25.1003

METALS FOR THE SPACE AGE Plansee seminar "De re metallica" 5th Papers Reutte,Austria 1964 Jun 22-26 Metallwerk Plansee A.G. Edited by F. Benesovsky Metallwerk Plansee A G,Reutte,1965 Papers in English,French and German
Met 25.0091

METALS PARK,OHIO 1961 Ultra high purity metals :a seminar Papers American society for metals Edited by R.L. Smith 264p Metals Park,Ohio,1962
Cav 7.0051

METALS PARK,OHIO 1964 Fiber composite materials :a seminar Papers American society for metals American society for metals,Metals Park,Ohio,1965
Eng 41.3697

METEORITE RESEARCH Symposium on meteorite research Proceedings Vienna 1968 Aug 7-13 International atomic energy agency Unesco Edited by Peter M. Millmann Astrophysics and space science library Reidel,Dordrecht,1969
Min 10.1442

METEORITES Researches on meteorites :a symposian Proceedings Tempe,Ariz. 1961 Mar.10 University of Arizona 227p John Wiley,New York;London,1962
Obs 6.1243

METEORITES Researches on meteorites :a symposium Papers Arizona state university 1961 Mar 10 Edited by Carleton B. Moore Wiley,New York;London,1962
Min 10.0499

METEOROLOGICAL AGENCY,JAPAN The Symposium on numerical weather prediction Proceedings Tokyo 1960 Nov 7-13 Co-sponsored by the International union of geodesy and geophysics 655p Meteorological society of Japan,Tokyo, 1962
Nap 11.0923

METEOROLOGICAL DEPARTMENT,INDIA Symposium on meteorology in relation to high level aviation over India and surrounding areas Papers New Delhi 1957 Dec 7 Edited by U.K. Bose Under the auspices of Indian meteorological society 159p India press,Delhi,1962
Nap 11.0420

METEOROLOGICAL FACTORS IN RADIO-WAVE PROPAGATION a conference Report London 1946 Apr 8 Physical society Royal meteorological society 325p Physical society,London,1946
Nap 11.0925

METEOROLOGICAL SOCIETY Lightning rod conference Report Edited by G.J. Symons Bibliog,illus x,261p E.and F.N.Spon,London, 1882 With appendices 1-L.Appendix G contains catalogue of works upon lightning conductors,pp.143-174
Nap 11.0940

METEOROLOGICAL VARIABLES Colloquium on spectra of meteorological variables Proceedings Stockholm 1969 Jun 9-19 Vol 3: radio science Inter-union commission on radio astronomy International union of radio science.Swedish national committee illus 28cm American geophysical union,Washington,D. C.,1969
A Math 4.1745

METEOROLOGY Antarctic meteorology :a symposium Proceedings Melbourne 1959 Feb. 18-25 Commonwealth bureau of meteorology and International geophysical year Under the auspices of the Australian academy of science 483p Pergamon press,Oxford;London, 1960
Nap 11.0026

METEOROLOGY La Meteorologie... Numero special consacre a la meteorologie Alpine 1957,Janvier-Juin Societe meteorologigne de France,Paris,1957 Ce numero contient les communications presentees au 1Ve. Congres international de meteorologie Alpine
Sco 14.0288

METEOROLOGY Scientific proceedings of the International association of meteorology Proceedings Rome 1954 Sep 15 International association of meteorology AIM publication, 10,c 594p Butterworths, London,1956 Meeting at the 10th general assembly of International union of geodesy and geophysics
Nap 11.0286

METEORS Physics and dynamics of meteors Tatranska Lomnica 1967 Sep 4-7 International astronomical union Edited by Lubor Kresak and Peter M. Millman International astronomical union.Symposium, 33 Reidel,Dordrecht,1968
TA 15.0422

METEORS Physics and dynamics of meteors :a symposium Proceedings Tatranska Lomnica 1967 International astronomical union Edited by Lubor Kresak and Peter M. Millman International astronomical union.Symposium, 33 525p Reidel,Dordrecht,1968
Obs 6.3264

METEORS Symposium on meteor physics Proceedings Manchester 1954 Jul. Jodrell Bank experimental station Edited by T.R. Kaiser Journal of atmospheric and terrestrial physics.Special supplement, 2 204p Pergamon press,London,1955
Obs 6.0923

METHIONINE BIOSYNTHESIS Symposium on transmethylation and methionine biosynthesis Argonne 1964 Edited by Stanley K. Shapiro and Fritz Schlenk Sponsored by the Argonne national laboratory.Division of biological and medical research Chicago university press, Chicago,1965
Bioch 33.0607

METHODES NOUVELLES DE SPECTROSCOPIE INSTRUMENTALE Exposes Orsay 1966 Apr 25-29 Centre national de la recherche scientifique 2nd edition Centre national de la recherche scientifique.Colloques internationaux, 161 344p CNRS,Paris,1969
Obs 6.3376

METHODOLOGIES New methods of thought and procedure Symposium on methodologies Contributions Pasadena,Calif. 1967 May 22-24 California institute of technology.Office for industrial association Society for morphological research Edited by F. Zwicky and A.G. Wilson viii,338p Springer,Berlin, 1967
WSM 43.2639

METHODOLOGIES New methods of thought and procedure Symposium on methodologies Contributions Pasadena,Calif. 1967 May 22-24 Edited by F. Zwicky and A.G. Wilson illus. viii,338p 24cm Springer,Berlin, 1967
Math S 3.1389

METHODOLOGIES OF PATTERN RECOGNITION International conference on methodologies of pattern recognition Proceedings Honolulu 1968 Jan 24-26 University of Hawaii Sponsored by the United States.Air force.Office of scientific research bibliog., illus. Academic press,New York;London,1969
Math S 3.1521

METHODOLOGIES OF PATTERN RECOGNITION International conference on methodologies of pattern recognition Proceedings Honolulu 1968 Jan 24-26 Edited by Satosi Watanabe Academic press,New York;London,1969
An 32.4412

METHODOLOGY International congress for logic, methodology and philosophy of science proceedings Jerusalem 1964 Aug 26-Sep 2 Edited by Yehoshua Bar-Hillel and others Studies in logic and the foundations of mathematics viii,440p 23cm North-Holland, Amsterdam,1965
P. Math 2.0206

METHODOLOGY International congress for logic, methodology and philosophy of science proceedings Stanford,Calif. 1960 Aug.24-Sep.2 Edited by Ernest Nagel and others ix, 661p 26cm Stanford university press, Stanford, Calif.,1962
P. Math 2.0209

METHODOLOGY IN BASIC GENETICS National institutes of health.Genetics study section Edited by Walter J. Burdette Sponsored by the University of Texas Holden-Day,San Francisco,Calif.,1963 Three copies
Gen 34.1012

METHODOLOGY IN HUMAN GENETICS National institutes of health.Genetics study section Edited by Walter J. Burdette Sponsored by the Roscoe B.Jackson memorial laboratory Holden-Day,San Francisco,Calif.,1962
Gen 34.1010

METHODOLOGY IN MAMMALIAN GENETICS National institutes of health.Genetics study section Edited by Walter J. Burdette Sponsored by the Roscoe B.Jackson memorial laboratory Holden-Day,San Francisco,Calif.,1963
Gen 34.1011

The METHODOLOGY OF PLANT ECO-PHYSIOLOGY Montpellier 1962 Edited by F.E. Eckardt Unesco.Arid zone research series, 25 Bibliog.,illus. 531p Unesco;H.M.S.O.,Paris; London,1965
Bot 42.1981

METHODS IN PALAEOMAGNETISM NATO Advanced study institute on procedures in palaeomagnetism Proceedings Newcastle-upon-Tyne 1964 Apr 1-10 North Atlantic treaty organization and University of Newcastle upon Tyne Edited by D.W. Collinson and others Developments in solid earth geophysics, 3 Amsterdam,1967
Geod 9.0568

METHODS OF LEARNING AND TECHNIQUES OF TEACHING Annual conference of the association for the study of medical education 2nd Proceedings London 1959 Oct 15-16 Association for the study of medical education Edited by J.R. Ellis Pitman medical publishing company, London,c1960
PGMS 29.0493

METHODS OF LOCAL AND GLOBAL DIFFERENTIAL GEOMETRY IN GENERAL RELATIVITY :conference Pittsburgh,Pa. 1970 Jul 13-17 Edited by D. Farnsworth and others Lecture notes in physics illus 190p 25cm Springer, Heidelberg,1972
A Math 4.1789

METHODS OF MATERIALS SELECTION :a symposium American institute of mining,metallurgical and petroleum engineers and University of Florida Metallurgical society conferences, 40 Gordon and Breach,New York;London,1968
Met 25.2502

METHODS OF NUMERICAL APPROXIMATION lectures delivered at a Summer school... Oxford 1965 Sep Edited by D.C. Handscomb Bibliog., Illus. ix,218p 23cm Pergamon press, Oxford,1966
A Math 4.1578

METHODS OF SEPARATION OF SUBCELLULAR STRUCTURAL COMPONENTS :a symposium Louvain 1962 May 11 Biochemical society Edited by J.K. Grant Held jointly with the Societe belge de biochimie Biochemical society. Symposia, 23 Cambridge university press, Cambridge,1963
Radioth 35.0136

METHODS OF SEPARATION OF SUBCELLULAR STRUCTURAL COMPONENTS :a symposium Louvain 1962 May 11 Biochemical society and Societe belge de biochimie Edited by J.K. Grant Held in the University of Louvain. Physiological institute Biochemical society. Symposia, 23 Cambridge university press, Cambridge,1963 Organized by C.de Duve
Bioch 33.1385

METRIC GEOMETRY OVER AFFINE SPACES Cooperative summer seminar 1st Ithaca, N.Y. 1964 By Ernst Snapper Mathematical association of America Held at Cornell university bibliog. iv,166p 28cm Mathematical association of America,Buffalo,N. Y.,1964 Duplicated notes
P Math 2.3829

METROLOGY Engineering dimensional metrology: a symposium proceedings Teddington 1953 October 21-24 Vols 1-2 National physical laboratory 26cm London,1955
Philos 1.0402

METZ 1963 Ore mining and materials handling Iron and steel engineers group meeting 47th Papers and discussions Iron and steel institute.Engineers group Iron and steel institute.Special report, 82 Iron and steel institute,London,1963
Met 25.0208

MEVALONIC ACID Natural substances formed biologically from mevalonic acid :a symposium Liverpool 1969 Apr Biochemical society Edited by T.W. Goodwin Biochemical society. Symposia, 29 Academic press,London;New York, 1970
Bioch 33.1391

MEXICO El jurasico medio y el calloviano de Mexico International geological congress 20th papers Mexico City 1956 By Heinrich Karl Erben maps Mexico City,1956
Geol 8.3032

MEXICO,D.F. 1959 Congreso panamericano de mecanica de suelos y cimentaciones 1st Proceedings Vol 1-3 Sociedad mexicana de mecanica de suelos 3 vols Mexico,D.F., 1960
Eng 41.3178

MEXICO,D.F. 1969 International conference on soil mechanics and foundation engineering 7th Proceedings Vol 1-3 3 vols 1969
Eng 41.3144

MEXICO CITY 1956 Actas y trabajos de las reuniones celebradas en Mexico Association of African geological surveys and commission on the geological map of Africa :joint meeting papers International geological congress. Commission on the geological map of Africa,and, Association of African geological surveys Edited by A. Garcia Rojas Mexico City,1959 Text in English,French and Spanish
Geol 8.3038

MEXICO CITY 1956 El jurasico inferior de Mexico y sus amonitas International geological congress 20th papers By Heinrich Karl Erben Translated by U. Hungsberg from the German Mexico City,1956
Geol 8.3033

MEXICO CITY 1956 El jurasico medio y el calloviano de Mexico International geological congress 20th papers By Heinrich Karl Erben maps Mexico City,1956
Geol 8.3032

MEXICO CITY 1956 Estudio sobre algunos de los conglomerados rojos del terciario inferior del centro de Mexico International geological congress 20th papers By John D. Edwards Translated by Salvador Ulloa from the English Mexico City,1956 Bound with two other publications of the congress
Geol 8.3042

MEXICO CITY 1956 Geologia y depositos de carbon de la region de Sabinas,estado de Chihuahua International geological congress 20th papers By Raymond C. Robeck and others United States geological survey Mexico City,1956 Bound with two other publications of the congress
Geol 8.3041

MEXICO CITY 1956 International geological congress 20th Trabajos 43 vols Mexico City,1956-60 Text in English,French, German,Russian and Spanish
Geog 13.7180

MEXICO CITY 1956 International geological congress 20th papers Gacetas historicas Edited by Manuel Carrera Stampa Mexico City, 1956 Bound with two other publications of the congress Text in English,French and Spanish
Geol 8.3030

MEXICO CITY 1956 International geological congress 20th papers Seccion 1: vulcanologia del cenozoico,tom 1-2 Edited by A. Garcia Rojas and others Mexico City,1957 Text in English,French,Russian and Spanish
Geol 8.3036

MEXICO CITY 1956 International geological congress 20th papers Seccion 11A: petrologia y mineralogia Edited by A. Garcia Rojas Mexico City,1959 Text in English, French,Russian and Spanish
Geol 8.3047

MEXICO CITY 1956 International geological congress 20th papers Seccion 13: geologia aplicada a la ingenieria y a la mineria Edited by A. Garcia Rojas and others Mexico City,1959 Text in English,French, German and Spanish
Geol 8.3039

MEXICO CITY 1956 International geological congress 20th papers Seccion 2: el mesozoico del hemisferio occidental y sus correlaciones mundiales Edited by A. Garcia Rojas and others Mexico City,1957 Bound with papers from section 3. Text in English, French and Spanish
Geol 8.3034

MEXICO CITY 1956 International geological congress 20th papers Seccion 3: geologia del petroleo Edited by A. Garcia Rojas and others Mexico City,1956 Bound with papers from section 2. Text in English,French, German and Russian
Geol 8.3035

MEXICO CITY 1956 International geological congress 20th papers Seccion 4: geohidrologia de regiones aridas y subaridas Edited by A. Garcia Rojas and others Mexico City,1957 Text in English,French,Russian and Spanish
Geol 8.3045

MEXICO CITY 1956 International geological congress 20th papers Seccion 5: relaciones entre la tectonica y la sedimentacion,tom.1-2 Edited by A. Garcia Rojas Mexico City,1957 Text in English, French,German,Russian and Spanish
Geol 8.3046

MEXICO CITY 1956 International geological congress 20th papers Seccion 7: paleontologia,taxonomia y evolucion Edited by A. Garcia Rojas and others Mexico City, 1958 Text in English,French,German, Russian and Spanish
Geol 8.3043

MEXICO CITY 1956 International geological congress 20th papers Seccion 9: geofisica aplicada,tom.1-2 Edited by A. Garcia Rojas and others Mexico City,1958 Text in English,French,German,Russian and Spanish
Geol 8.3044

MEXICO CITY 1956 International geological congress 20th resumes of papers presented Edited by Hans E. Thalmann Mexico City,1956 Text in English,French,German and Spanish
Geol 8.3031

MEXICO CITY 1956 Memoria geologico-minera del estado de Chihuahua International geological congress 20th papers By Jenaro Gonzalez Reyna Mexico City,1956 Bound with two other publications of the congress
Geol 8.3029

MEXICO CITY 1956 Riqueza minera y yacimientos minerales de Mexico International geological congress 20th papers Edited by Jenaro Gonzalez Reyna 3rd edition Mexico city,1956
Geol 8.3025

MEXICO CITY 1956 Sistema cambrico,su paleografia y el problema de su base International geological congress 20th papers of a symposium Pt 1-2 Mexico city, 1956 Text in English,French and Spanish
Geol 8.3024

MEXICO CITY 1956 Sistema cretacico Un Symposium sobre el cretacico en el hemisferio occidental y su correlacio mundial papers Tom 1-2 Edited by Lewis B. Kellum 2 vols Mexico City,1959 Prepared under the auspices of the International commission on stratigraphy and the International palaeontological union for the 20th International geological congress
Geol 8.3049

MEXICO CITY 1956 Symposium de exploracion geoquimica,tom.1-3 International geological congress 20th papers Edited by A. Garcia Rojas and others 2cm Mexico City,1958-60 Text in English,French,Russian and Spanish
Geol 8.3048

MEXICO CITY 1956 Symposium internacional de topologia algebraica Universidad nacional de Mexico.Instituto de matematicas Instituto nacional de la investigacion cientifica,Mexico Sociedad matematica mexicana xii,334p 26cm UNESCO,Mexico City,1958 In memory of Witold Hurewicz
P. Math 2.0290

MEXICO CITY 1956 Symposium sobre yacimientos de manganeso International geological congress 20th papers Tom 1-5 Edited by Jenaro Gonzalez Reyna 2 vols Mexico city,1956 Text in English,French, German and Spanish
Geol 8.3027

MEXICO CITY 1956 Symposium sobre yacimientos de petroleo y gas International geological congress 20th papers Tom 1-5 Edited by Eduardo J. Guzman 2 vols Mexico city,1956 Text in English,French and Spanish
Geol 8.3026

MEXICO CITY 1956 Tamabra limestone of the Poza Rica oilfield International geological congress 20th papers By A. Barnetche and L.V. Illing Mexico city,1956 Bound with two other publications of the congress
Geol 8.3028

MEXICO CITY 1956 Tectonica de la Sierra Madre oriental de Mexico,entre Torreon y Monterrey By Zoltan de Cserna Instituto nacional para la investigacion de recursos minerales de Mexico Mexico City,1956 Bound with two other publications of the congress
Geol 8.3040

MEXICO CITY 1960 Social aspects of economic development in Latin America Vol 1: papers submitted to the expert working group... Expert working group on social aspects of economic development in Latin America Edited by Egbert de Vries and Jose Medina Echavarria Technology and society series Unesco,Paris, 1963
Geog 13.5096

MEXICO CITY 1962 Social research and rural life in central America,Mexico and the Caribbean region :seminar Proceedings Unesco,and,United Nations.Economic commission for Latin America Edited by Egbert de Vries Unesco,Paris,1966
Geog 13.5161

MEXICO CITY 1965 Endocrinology Pan-American congress of endocrinology 6th Proceedings Edited by C. Gual Excerpta medica.International congress series, 112 Excerpta medica,Amsterdam,1966
Inv Med 37.0096

MEXICO CITY 1965 Pan-American congress of endocrinology 6th Proceedings Edited by C. Gual Excerpta medica.International congress series, 112 Excerpta medica foundation,Amsterdam,1966
Inv Med 37.0254

MEXICO CITY 1965 Pan-American congress of endocrinology 6th Proceedings Edited by C. Gual Excerpta medica.International congress series, 112 Excerpta medica foundation,New York,c1965
PGMS 29.0167

MEXICO CITY 1966 International symposium. Enzymatic aspects of metabolic regulation Proceedings National cancer institute Edited by M.P. Stulberg National cancer institute.Monograph, 27 National cancer institute,Betheseda,Md.,1967
Bioch 33.1062

MEXICO CITY 1967 World petroleum congress 7th proceedings Vol 2: origin of oil, geology and geophysics bibliogs,illus Elsevier publishing co ltd,Barking,1967
Geol 8.4427

MEXICO CITY 1968 Progress in endocrinology International congress of endocrinology Proceedings Edited by Carlos Gual and F.J.C. Ebling Excerpta medica.International congress series, 184 Excerpta medica, Amsterdam,1969
Inv Med 37.0095

MEZHDUNARODNAYA ASSOTSIATSIYA NAUCHNOI GIDROLOGII see INTERNATIONAL ASSOCIATION OF SCIENTIFIC HYDROLOGY

MEZHDUNARODNAYA KONFERENTSIYA ASSOTSIATSII PO IZUCHENIYU CHETVERTICHNOGO PERIODA EVROPY International conference of the Association for the study of the European quaternary 2nd trudy Leningrad 1932 Sep 1-7 Vypusk 1-5 Association for the study of the European quaternary,and,S.S.S.R.- N.K.T.P.Vsesoyuznoe geologo-razvedochnoe obedinenie Edited by D. A. Petrovskii and others Gosudarstvennoe nauchno-tekhnicheskoe geologo-razvedochnoe izdatelstvo,Leningrad;Moscow,1932-34
Geol 8.2982

MEZHDUNARODNAYA KONFERENTSIYA PO TEORII ANALITICHESKIKH FUNKTSII Sovremennye problemy teorii analiticheskikh funktsii Erevan 1965 Sep 6-14 Edited by M.A. Lavrentev and others 361p 27cm Moscow, 1966
P. Math 2.2408

MEZHDUNARODNE MATEMATICHESKIE OLIMPIADY 1st-7th Zadachi,resheniya,itogi By E.A. Morozova and I.S. Petrakov Bibliog.,Illus. 175p 20cm Moscow,1967
P Math 2.3273

MEZHDUNARODNOE OBSHCHESTVO IZUCHENIYA POLYARNYKH STRAN POSREDSTVOM VOZDUKHOPLAVATELNYKH APPARATOV see INTERNATIONALE GESELLSCHAFT ZUR ERFORSCHUNG DER ARKTIS MIT LUFTFAHRZEUGEN

MEZHDUNARODNYI PHYSIOLOGICHESKII KONGRESS see INTERNATIONAL PHYSIOLOGICAL CONGRESS

MEZHDUVEDOMSTVENNAYA KOMMISSIYA PO IZUCHENIYU ANTARKTIKI Osnovnye itogi izucheniya Antarktiki za 10 let :doklady vsesoyuznogo soveshchaniya po izucheniyu Antarktiki 1966 god Moscow 1966 Apr 19-20 map Izdatelstvo 'Nauka',Moscow,1967
Sco 14.6068

MIAMI 1965 The Chromosome;structural and functional aspects :a symposium Vol 1 Tissue culture association Edited by Clyde Johnson Dawe Williams and Wilkins,Baltimore, Md.,1966 Organized by George Yerganian
Bal 39.0339

MIAMI,FLA. 1961 Advances in neuroendocrinology Symposium on neuroendocrinology Edited by Andrew V. Nalbandov Supported by a grant from the National institute of neurological diseases and blindness University of Illinois press, Urbana,Ill.,1963
An 32.3766

MIAMI,FLA. 1965 Chromosome:structural and functional aspects Tissue culture association symposia 5th Edited by George Yerganian illus. xiii,107p 26cm Tissue cultural association,n.p.,1965
Bot 42.4734

MIAMI,FLA. 1965 The Chromosome:structural and functional aspects :a symposium Edited by Clyde J. Dawe Tissue culture association. Publications, 1 Tissue culture association, 1965
An 32.3302

MIAMI,FLA. 1965 The Chromosome:structural and functional aspects :a symposium Tissue culture association Edited by Clyde Johnson-Dawe Williams and Wilkins,Baltimore,Md.,1965 Organized by George Yerganian
Gen 34.0783

MIAMI,FLA. 1969 Microcirculation,perfusion, and transplantation of organs Conference on microcirculation,perfusion and transplantation of organs Proceedings University of Miami school of medicine United States.Office of naval research Edited by Theodore I. Malinin and others Academic press,London;New York, 1970
An 32.5458

MIAMI,FLA. 1969 Software engineering Symposium on computer and information sciences 3rd Proceedings Vol 1-2 Edited by Julius T. Tou COINS, 3 illus 2 vols Academic press,New York,1970
Math L 5.3830

MIAMI,FLA. 1971 Nucleic acid-protein interactions and nucleic acid synthesis in viral infection :symposium University of Miami.Department of biochemistry Papanicolaou cancer research institute Miami winter symposia, 2 North-Holland,Amsterdam, 1971
Bioch 33.2241

MIAMI,FLO. 1961 Advances in neuroendocrinology Symposium on neuroendocrinology Edited by A.V. Nalbandov Sponsored by the National institute of neurological diseases and blindness University of Illinois press,Urbana,Ill.,1963
Phys 20.1404

MIAMI BEACH,FLA. 1967 Fluorescence,theory and practice :a symposium Papers Edited by George G. Guilbault Organixed by the American chemical society.Division of analytical chemistry Arnold;Dekker,London; New York,1967
Col S 12.0091

MIAMI WINTER SYMPOSIA, 2 Nucleic acid-protein interactions and nucleic acid synthesis in viral infection :symposium Miami,Fla. 1971 Jan 18-22 University of Miami.Department of biochemistry Papanicolaou cancer research institute North-Holland,Amsterdam,1971
Bioch 33.2241

MICHELSON (CHR.) INSTITUTT FOR VIDENSKAP OG AANDSFRIHET Radiation trapped in the earth's magnetic field :the advanced study institute Proceedings Bergen 1965 Aug 16-Sep 3 Edited by Billy M. McCormac Astrophysics and space science library, 5 901p Reidel,Dordrecht,1966
Obs 6.3282

MICHIGAN.DEPARTMENT OF SOCIAL WELFARE.DIVISION OF SERVICES FOR THE BLIND Psycholgical diagnosis and counseling of the adult blind : selected papers from the proceedings of the University of Michigan conference for the blind Ann Arbor,Mich. 1947 Edited by Wilma Donahue and Donald Dabelstein Co-sponsored by the University of Michigan. Institute for human adjustment American foundation for the blind,New York,1950
Psy 31.2118

MICHIGAN CONFERENCE ON GENETICS Approaches to the genetic analysis of mammalian cells : Michigan conference on genetics Lectures University of Michigan Edited by Donald J. Merchant and James V. Neel University of Michigan press,Ann Arbor,Mich.,1962
Gen 34.0911

MICHIGAN STATE COLLEGE Biology of phosphorus A Symposium in general biology Papers presented East Lansing,Mich. 1951 Apr 25-27 By G.Evelyn Hutchinson and others Edited by Lester F. Wolterink Michigan state college press,1952
Bioch 33.0590

MICHIGAN STATE UNIVERSITY Fundamental topics in relativistic fluid mechanics and magnetohydrodynamics :a symposium Proceedings East Lansing,Mich. 1962 Oct Edited by Robert Wasserman and Charles P. Wells Academic press,New York,1963
Eng 41.4547

MICHIGAN STATE UNIVERSITY Symposium on fundamental topics in fluid mechanics and magnetohydrodynamics Proceedings Michigan State university 1962 Oct 10-11 Edited by Robert Wassermann and Charles P. Wells Academic press,New York,1963
Cav 7.1569

MICHIGAN STATE UNIVERSITY Topology of manifolds :a conference Proceedings East Lansing,Mich. 1967 Mar 15-17 Edited by John G. Hocking Bibliog. viii,161p 21cm Prindle,Weber and Schmidt,Boston,Mass.,1968
P Math 2.3351

MICHIGAN STATE UNIVERSITY 1962 Symposium on fundamental topics in fluid mechanics and magnetohydrodynamics Proceedings Michigan State university Edited by Robert Wassermann and Charles P. Wells Academic press,New York, 1963
Cav 7.1569

MICHIGAN STATE UNIVERSITY.COLLEGE OF ENGINEERING Seminar in systems Proceedings Edited by G.D.S. MacLellan and Samuel Mercer Michigan state university,East Lansing,Mich., 1958
Eng 41.6009

MICHIGAN SYMPOSIUM ON ASTROPHYSICS Proceedings Ann Arbor,Mich. 1953 Jun.29-Jul.24 University of Michigan.Observatory Ann Arbor, 1953 Typescript
Obs 6.1978

MICRO-ELECTRONICS Symposium on applications of micro-electronics Birmingham 1968 Mar 27 Organized by the Institution of electrical engineers Institution of electrical engineers.Conference publication, 49 Institution of electrical engineers, London,1968
Eng 41.5266

MICRO-ORGANISMS Adaption in micro-organisms : a symposium Papers London 1953 Apr Society for general microbiology Edited by R. Davies and E.F. Gale Held at the Royal institution Society for general microbiology. Symposia, 3 Cambridge university press, Cambridge,1953
Bioch 33.1166

MICRO-ORGANISMS Autotrophic micro-organism : a symposium Papers London 1954 Apr Society for general microbiology Edited by B. A. Fry and J.L. Peel Held at the Institution of electrical engineers Society for general microbiology.Symposia, 4 Cambridge university press,Cambridge,1954
Bioch 33.1168

MICRO-ORGANISMS Function and structure in micro-organisms :a symposium Papers London 1965 Apr Society for general microbiology Held at the Middlesex hospital Society for general microbiology.Symposia, 15 Cambridge university press,Cambridge,1965
Bioch 33.1171

MICROBES Airborne microbes :a symposium Papers London 1967 Apr Society for general microbiology Edited by P.H. Gregory and J.L. Monteith Held at the Imperial college of science and technology Society for general microbiology.Symposia, 17 Cambridge university press,Cambridge,1967
Bioch 33.1167

MICROBES Microbes and biological productivity :a symposium Papers London 1971 Apr Society for general microbiology Edited by David E. Hughes and Anthony H. Rose Society for general microbiology.Symposia, 21 Cambridge university press,Cambridge,1971
Bioch 33.2302

MICROBES AND BIOLOGICAL PRODUCTIVITY :a symposium Papers London 1971 Apr Society for general microbiology Edited by David E. Hughes and Anthony H. Rose Society for general microbiology.Symposia, 21 Cambridge university press,Cambridge,1971
Bioch 33.2302

MICROBIAL BEHAVIOUR IN VIVO AND IN VITRO London 1964 Apr Society for general microbiology Society for general microbiology.Symposia, 14 Cambridge university press,Cambridge,1964
Gen 34.0700

MICROBIAL BEHAVIOUR 'IN VIVO' AND 'IN VITRO' : a symposium Papers London 1964 Apr Society for general microbiology Edited by H. Smith and Joan Taylor Held at the Royal institution Society for general microbiology. Symposia, 14 Cambridge university press, Cambridge,1964
Bioch 33.1173

MICROBIAL BEHAVIOUR 'IN VIVO' AND 'IN VITRO' : a symposium Papers London 1964 Apr Society for general microbiology Edited by H. Smith and Joan Taylor Society for general microbiology.Symposia, 14 figs Cambridge university press,Cambridge,1964
Path 30.2610

MICROBIAL CLASSIFICATION :symposium London 1962 Apr Society for general microbiology Society for general microbiology.Symposia, 12 Cambridge university press,Cambridge,1962
Gen 34.0714

MICROBIAL CLASSIFICATION :a symposium Papers London 1962 Apr Society for general microbiology Edited by G.C. Ainsworth and P. H.A. Sneath Held at the Royal institution Society for general microbiology.Symposia, 12 Cambridge university press,Cambridge,1962
Bioch 33.1174

MICROBIAL ECOLOGY :a symposium Papers London 1957 Apr Society for general microbiology Edited by R.E.O. Williams and C. C. Spicer Held at the Royal institution Society for general microbiology.Symposia, 7 Cambridge university press,Cambridge,1957
Bioch 33.1175

MICROBIAL ECOLOGY :a symposium Papers London 1957 Apr Society for general microbiology Edited by R.E.O. Williams and C. C. Spicer Society for general microbiology. Symposia, 7 Cambridge,1957
Path 30.2585

MICROBIAL GENETICS London 1960 Apr Society for general microbiology Society for general microbiology.Symposia, 10 Cambridge university press,Cambridge,1960 Two copies
Gen 34.0748

MICROBIAL GENETICS :a symposium Papers London 1960 Apr Society for general microbiology Edited by W. Hayes and R.C. Clowes Held at the Royal institution Society for general microbiology.Symposia, 10 Cambridge university press,Cambridge,1960
Bioch 33.1176

MICROBIAL GENETICS :a symposium Papers London 1960 Apr Society for general microbiology Edited by W. Hayes and R.C. Clowes Society for general microbiology. Symposia, 10 Cambridge university press, Cambridge,1960
Path 30.2605

MICROBIAL GROWTH :a symposium London 1969 Apr Society for general microbiology Edited by Pauline M. Meadow and S.J. Pirt Society for general microbiology.Symposia, 19 Cambridge university press,Cambridge,1969
An 32.5379

MICROBIAL GROWTH :symposium London 1969 Apr Society for general microbiology Society for general microbiology.Symposia, 19 Cambridge university press,Cambridge,1969
Gen 34.0710

MICROBIAL GROWTH Papers London 1969 Apr Society for general microbiology Society for general microbiology.Symposia, 19 Cambridge university press,Cambridge,1969
Path 30.2597

MICROBIAL METABOLISM International congress of biochemistry 2nd Paris 1952 Jul 21-27 Vol 5: symposium sur le metabolisme microbien Council for international organizations of medical sciences Societe d'edition d'enseignement superieur,Paris,1952 Text in English and French
Bioch 33.1308

MICROBIAL METABOLISM International congress of microbiology 6th Rome 1953 Vol 2: symposium - microbial metabolism Edited by E.B. Chain Fondazione Emanuele Paterno, Rome,1953 Title also in Italian;text in English and French
Bioch 33.1801

MICROBIAL PATHOGENICITY Mechanisms of microbial pathogenicity :a symposium Papers London 1955 Apr Society for general microbiology Edited by J.W. Howie and A.J. O'Hea Held at the Royal institution Society for general microbiology.Symposia, 5 Cambridge university press,Cambridge,1955
Bioch 33.1172

MICROBIAL PATHOGENICITY Mechanisms of microbial pathogenicity :a symposium Papers London 1955 Apr Society for general microbiology Edited by J.W. Howie and A.J. O'Hea Society for general microbiology. Symposia, 5 Cambridge university press, Cambridge,1955
Path 30.2608

MICROBIAL PROTOPLASTS,SPHEROPLASTS AND L-FORMS a conference Proceedings Kalamazoo, Mich. 1966 Nov 10-11 Edited by Lucien B. Guze sponsored by the Upjohn company Williams and Wilkins,Baltimore,1968
Bioch 33.1142

MICROBIAL REACTION TO ENVIRONMENT :a symposium Papers London 1961 Apr Society for general microbiology Edited by G.C. Meynell and H. Gooder Held at the Royal institution Society for general microbiology.Symposia, 11 Cambridge university press,Cambridge,1961
Bioch 33.1177

MICROBIAL REACTION TO ENVIRONMENT :a symposium Papers London 1961 Apr Society for general microbiology Edited by G.G. Meynell and H. Gooder Society for general microbiology.Symposia, 11 pls Cambridge university press,Cambridge,1961
Path 30.2607

MICROBIOLOGY Biodeterioration of materials: microbiological and allied aspects International biodeterioration symposium 1st Proceedings Southampton 1968 Sep 9-14 Edited by Arthur H. Walters and John J. Elphick Bibliog.,illus. x,740p Elsevier, Amsterdam,1968
Bot 42.3773

MICROBIOLOGY International congress of microbiology :a report 2nd Proceedings London 1936 Jul 1-Aug 1 International society for microbiology Edited by R. St. John-Brooks port London,1937
Bot 42.0660

MICROBIOLOGY Recent progress in microbiology symposia presented at the 8th international congress for microbiology Montreal 1962 Aug 20-24 International union of microbiological societies Canadian society of microbiologists Edited by N.E. Gibbons University of Toronto press,Toronto,1963
Gen 34.0701

MICROBIOLOGY YESTERDAY AND TODAY :a symposium New Brunswick,N.J. 1958 Jun 5 Rutgers university.Institute of microbiology Edited by Vernon Bryson Rutgers university press,New Brunswick,N.J.,1959 Held in honour of the 70th birthday of Selman A. Waksman
Bot 42.3720

MICROCIRCULATION Symposium on techniques used in the study of microcirculation Papers Oxford 1968 Sep British microcirculation society Edited by D.R. Chambers and P.A.G. Monro British microcirculation society, Cambridge,1969
An 32.5223

The MICROCIRCULATION :symposium on factors influencing exchange of substances across capillary wall Conference on microcirculatory physiology and pathology,5th Buffalo,N.Y. 1958 Apr 1 Edited by S.R.M. Reynolds and Benjamin W. Zweifach University of Illinois press,Urbana,Ill.,1959
Phys 20.0888

MICROCIRCULATION:SYMPOSIUM ON FACTORS INFLUENCING EXCHANGE OF SUBSTANCES ACROSS CAPILLARY WALL A Sterotaxic atlas of the dog's brain Buffalo,N.Y. 1958 Apr 1 By Robert K.S. Lim and others American association of anatomists Edited by S.R.M. Reynolds and Benjamin W. Zweifach Thomas,Springfield,Ill., 1960
VA 19.0362

MICROCIRCULATION,PERFUSION,AND TRANSPLANTATION OF ORGANS Conference on microcirculation, perfusion and transplantation of organs Proceedings Miami,Fla. 1969 Oct 30 University of Miami school of medicine United States.Office of naval research Edited by Theodore I. Malinin and others Academic press,London;New York,1970
An 32.5458

MICROCIRCULATORY PHYSIOLOGY Vascular patterns as related to functions Conference on microcirculatory physiology and pathology 2nd Philadelphia,Pa. 1955 Apr 5 Williams and Wilkins,Baltimore,Md.,1955
An 32.3461

MICROMINIATURIZATION IN AUTOMATIC CONTROL EQUIPMENT AND IN DIGITAL COMPUTERS :IFAC-IFIP symposium Proceedings Munich 1965 Oct 12-23 International federation of automatic control International federation for information processing Edited by J. Berghammer Oldenbourg,Munich,1966
Eng 41.5580

MICROMINIATURIZATION OF ELECTRONIC ASSEMBLIES Symposium on microminiaturization of electronic assemblies Proceedings Washington,D.C. 1958 Sep 30-Oct 1 Diamond ordnance fuze laboratories Edited by Eleonor F. Horsey and Laurence D. Shergalis Hayden, New York,1960
Eng 41.5541

MICROORGANISMS Heredity and variation in microorganisms :a symposium Papers Cold Spring Harbor 1946 Cold Spring Harbor biological laboratory Cold Spring Harbor symposia on quantitative biology, 11 Long Island biological association,Cold Spring Harbor,1946
Bioch 33.1267

MICROPALEONTOLOGY West African micropaleontological colloquium 2nd Proceedings Ibadan 1966 Jun 18-Jul 1 Edited by J.E.van Hinte J.E.Brill,Leiden, 1966
Geol 8.0344

MICROPHYSIOLOGIE COMPAREE DES ELEMENTS EXCITABLES colloque international Gif-sur-Yvette 1955 Jul 19-23 Centre national de la recherche scientifique Centre national de la recherche scientifique.Colloques internationaux, 67
Phys 20.2022

MICROPLASTICITY Edited by Charles J. McMahon Advances in materials research, 2 Interscience,New York,1968
Met 25.1038

MICROPROGRAMMING Annual workshop on microprogramming 4th Proceedings Santa Cruz,Calif. 1971 Sep 13-14 Association for computing machinery Association for computing machinery,New York,1971
Math L 5.3631

MICROPROGRAMMING Annual workshop on microprogramming 5th Preprints Urbana, Ill. 1972 Sep 25-26 Association for computing machinery Institute of electrical and electronics engineers.Computer society Bibliog,diagrs 98p n.p.,1972 Includes 'An annotated bibliography on microprogramming, late 1969-early 1972'
Math L 5.3636

MICROPROGRAMMING International advanced summer institute on microprogramming St. Raphael 1971 Aug 30-Sep 10 North Atlantic treaty organization.Scientific affairs division Edited by Guy G. Boulaye and Jean Mermet Actualites scientifiques et industrielles illus 418p Hermann,Paris, 1972
Math L 5.3634

MICROPROGRAMMING Workshop on microprogramming 3rd Preprints Buffalo, N.Y. 1970 OCT Association for computing machinery bibliog.,illus. Association for computing machinery,New York,1970
Math L 5.3574

The MICROSCOPE:ITS DESIGN,CONSTRUCTION AND APPLICATIONS ;a symposium and general discussion London 1920 Jan 14 Faraday society Edited by F.S. Spiers pls,illus 260p Griffin,London,1920 Faraday society in cooperation with the Technical optics committee of the British science guild. Includes reports of adjourned discussions held in Sheffield,Feb 24 and in London April 21, 1920
WSM 43.5151

MICROSCOPIE ELECTRONIQUE 1970 Congres international de microscopie electronique 7e Resumes des communications Grenoble 1970 Aug 30-Sep 3 International federation of societies for electron microscopy Edited by Pierre Favard Societe francaise de microscopie electronique,Paris,1970
Met 25.2604

MICROSCOPY Historical aspects of microscopy Papers Oxford 1966 Mar 18 Royal microscopical society Edited by S. Bradbury and G.L'E. Turner W.Heffer,Cambridge,1967
Bal 39.0241

MICROSCOPY Historical aspects of microscopy Royal microscopical society conference Papers Oxford 1966 Mar 18 Royal microscopical society Edited by S. Bradbury and G.L'E. Turner viii,227p Cambridge,1967
WSM 43.2520

MICROSCOPY Symposium on microscopy American society for testing materials :annual meeting 62nd papers Atlantic City,N.J. 1959 Jun 25-26 American society for testing materials.Special technical publications,257 American society for testing materials, Philadelphia,Pa.,1959
Min 10.1046

MICROSEISMS Symposium on microseisms Proceedings Harriman,N.Y. 1952 Sep 4-6 United States office of naval research and United States air force. Geophysical research directorate Edited by James T. Wilson and Frank Press National research council. Publication, 306 Washington,D.C.,1953
Geod 9.0376

MICROSOMAL PARTICLES AND PROTEIN BIOSYNTHESIS : symposium Papers Cambridge,Mass. 1958 Feb 5-8 Biophysical society Edited by Richard B. Roberts Biophysical society. Symposia, 1 Washington academy of sciences, Washington,D.C.;London,1958
Gen 34.0657

MICROSOMAL PARTICLES AND PROTEIN SYNTHESIS Biophysical society :symposium 1st Papers Cambridge,Mass. 1958 Feb 5-8 Edited by Richard B. Roberts Pergamon press,London, 1958
Col S 12.0053

MICROSOMAL PARTICLES AND PROTEIN SYNTHESIS : a symposium Cambridge,Mass. 1958 Feb 5-8 Biophysical society Edited by Richard B. Roberts Biophysical society.Symposia, 1 Pergamon press,London,1958
Radioth 35.0077

MICROSOMAL PARTICLES AND PROTEIN SYNTHESIS :a symposium Papers Cambridge,Mass 1958 Feb 5-8 Biophysical society Edited by Richard B. Roberts Biophysical society. Symposia, 1 Pergamon,London;New York,1958
Bot 42.1686

MICROSOMAL PARTICLES AND PROTEIN SYNTHESIS :a symposium Papers presented Cambridge, Mass. 1958 Feb 5-8 Biophysical society Edited by Richard B. Roberts At the Massachusetts institute of technology Biophysical society.Symposia, 1 Pergamon, London,1958
Bioch 33.0924

MICROSOMES AND DRUG OXIDATIONS :a symposium Bethesda,Md. 1968 Feb 16-17 Edited by J. R. Gillette and others Academic press,New York;London,1969
Bioch 33.1089

MICROSTRUCTURE OF CERAMIC MATERIALS :a symposium Proceedings Pittsburgh 1963 Apr 27-28 Ceramic education council and National bureau of standards Sponsored by Edward Orton junior ceramic foundation National bureau of standards.Miscellaneous publication, 257 U.S.dept of commerce, Washington D.C.,1964
Met 25.0285

MICROSYMPOSIUM 'STRUCTURE OF ORGANIC SOLIDS' 2nd Lectures Structure of organic solids Prague 1968 Sep 16-19 International union of pure and applied chemistry.Macromolecular division Czechoslovak chemical society In conjunction with Ceskoslovenska akademie ved illus 550p 25cm Butterworths,London,1969 Simultaneously published in 'Pure and applied chemistry',vol 18,no 4,1969
Cav 7.2659

MICROWAVE AND OPTICAL GENERATION AND AMPLIFICATION International conference on microwave tubes 6th Cambridge 1966 Institution of electrical engineers Institution of electrical engineers.Conference publication, 27 I.E.E.,London,1967
Eng 41.5588

MICROWAVE BEHAVIOUR OF FERRIMAGNETICS AND PLASMAS International conference on the microwave behaviour of ferrimagnetics and plasmas London 1965 Sep 13-17 Sponsored by Institution of electrical engineers. Electronics division Institution of electrical engineers.Conference publication, 13 Institution of electrical engineers, London,1965
Eng 41.5713

MICROWAVE RESEARCH INSTITUTE Computer processing in communications Symposium on computer processing in communications Proceedings New York 1969 Edited by Jerome Fox Microwave research institute. Symposia series, 19 illus. 850p Polytechnic institute of Brooklyn,Brooklyn,N.Y. 1970
Math L 5.3558

MICROWAVE RESEARCH INSTITUTE Electronic waveguides :symposium Proceedings New York 1958 Apr 8-10 Polytechnic institute of Brooklyn Microwave research institute. Symposia series, 8 Polytechnic press, Brooklyn,N.Y.,1958
Eng 41.5649

MICROWAVE RESEARCH INSTITUTE Millimeter waves :a symposium Proceedings New York 1959 Mr 31-Apr 2 Co-sponsored by United States.Office of naval research Polytechnic institute of Brooklyn.Microwave research institute symposia series, 9 Polytechnic press,Brooklyn,N.Y.,1960
Cav 7.1715

MICROWAVE RESEARCH INSTITUTE Modern network synthesis :a symposium Proceedings New York 1955 Apr 13-15 Polytechnic institute of Brooklyn Microwave research institute. Symposia series Polytechnic institute of Brooklyn,New York,1956
Eng 41.5117

MICROWAVE RESEARCH INSTITUTE Modern network synthesis (audio to microwaves) :a symposium Proceedings New York 1952 Apr 16-18 Polytechnic institute of Brooklyn Microwave research institute.Symposia series Polytechnic institute of Brooklyn,New York, 1956
Eng 41.5116

MICROWAVE RESEARCH INSTITUTE Optical maser : a symposium Proceedings New York 1963 Apr 16-19 Polytechnic institute of Brooklyn Edited by Jerome Fox Microwave research institute.Symposia series, 13 Polytechnic press,Brooklyn,N.Y.,1963
Eng 41.5417

MICROWAVE RESEARCH INSTITUTE Quasi-optics :a symposium Proceedings New York 1964 Jun 8-10 Co-sponsored by United States.Office of naval research Polytechnic institute of Brooklyn.Microwave research institute symposia series, 14 Polytechnic press,Brooklyn,N.Y., 1964
Cav 7.1721

MICROWAVE RESEARCH INSTITUTE Symposium on computer processing in communications New York 1969 Microwave research institute. Symposium series, 19 Polytechnic press, Brooklyn,N.Y.,1970
Eng 41.8313

MICROWAVE RESEARCH INSTITUTE Symposium on computers and automata Proceedings New York 1971 Apr 13-15 Edited by Jerome Fox Microwave research institute.Symposia series, 21 illus 653p Polytechnic press,Brooklyn, N.Y.,1971
Math L 5.3635

MICROWAVE RESEARCH INSTITUTE Symposium on mathematical theory of automata proceedings New York 1962 Apr 24-26 Partly sponsored by American institute of electrical engineers Polytechnic institute of Brooklyn.Microwave research institute.Symposia series, 12 xix, 640p 22cm Polytechnic press,Brooklyn,1962
Math L 5.0996

MICROWAVE RESEARCH INSTITUTE Symposium on nonlinear circuit analysis Proceedings New York 1953 Apr 23-24 Polytechnic institute of Brooklyn Edited by Jerome Fox Microwave research institute.Network symposia series, 2 Polytechnic institute of Brooklyn,New York, 1953
Eng 41.5920

MICROWAVE RESEARCH INSTITUTE Symposium on nonlinear circuit analysis Proceedings New York 1956 Apr 25-27 Polytechnic institute of Brooklyn Edited by Jerome Fox Microwave research institute.Network symposia series, 6 Polytechnic institute of Brooklyn,New York, 1957
Eng 41.5922

MICROWAVE RESEARCH INSTITUTE Symposium on turbulence of fluids and plasmas Proceedings New York 1968 Apr 16-18 Polytechnic institute of Brooklyn Edited by Jerome Fox Microwave research institute.Symposia series, 18 512p 17cm Polytechnic press,Brooklyn, N.Y.,1969
Cav 7.2920

MICROWAVE RESEARCH INSTITUTE Turbulence of fluids and plasmas Symposium on turbulence of fluids and plasmas Proceedings New York 1968 Apr 16-18 Edited by Jerome Fox Microwave research institute.Symposia series, 18 bibliog.,illus. 512p 22cm Polytechnic press,Brooklyn,N.Y.,1969
A Math 4.1650

MICROWAVE RESEARCH INSTITUTE Turbulence of fluids and plasmas Symposium on turbulence of fluids and plasmas Proceedings New York 1968 Apr 16-18 Polytechnic institute of Brooklyn Edited by Jerome Fox Microwave research institute.Symposia series, 18 Bibliog.,Illus. xxix,511p 23cm Polytechnic press,Brooklyn N.Y.,1969
A Math 4.1508

MICROWAVE RESEARCH INSTITUTE.NETWORK SYMPOSIA SERIES, 2 Symposium on nonlinear circuit analysis Proceedings New York 1953 Apr 23-24 Polytechnic institute of Brooklyn Edited by Jerome Fox Sponsored by the Microwave research institute Polytechnic institute of Brooklyn,New York,1953
Eng 41.5920

MICROWAVE RESEARCH INSTITUTE.NETWORK SYMPOSIA SERIES, 6 Symposium on nonlinear circuit analysis Proceedings New York 1956 Apr 25-27 Polytechnic institute of Brooklyn Edited by Jerome Fox Sponsored by the Microwave research institute Polytechnic institute of Brooklyn,New York,1957
Eng 41.5922

MICROWAVE RESEARCH INSTITUTE.SYMPOSIA SERIES
Modern network synthesis :a symposium Proceedings New York 1955 Apr 13-15 Polytechnic institute of Brooklyn Sponsored also by the Microwave research institute Polytechnic institute of Brooklyn,New York, 1956
Eng 41.5117

MICROWAVE RESEARCH INSTITUTE.SYMPOSIA SERIES
Modern network synthesis (audio to microwaves) a symposium Proceedings New York 1952 Apr 16-18 Polytechnic institute of Brooklyn Sponsored also by the Microwave research institute Polytechnic institute of Brooklyn, New York,1956
Eng 41.5116

MICROWAVE RESEARCH INSTITUTE.SYMPOSIA SERIES, 8
Electronic waveguides :symposium Proceedings New York 1958 Apr 8-10 Polytechnic institute of Brooklyn Microwave research institute Polytechnic press, Brooklyn,N.Y.,1958
Eng 41.5649

MICROWAVE RESEARCH INSTITUTE.SYMPOSIA SERIES, 13
Optical maser :a symposium Proceedings New York 1963 Apr 16-19 Polytechnic institute of Brooklyn Edited by Jerome Fox Sponsored also by the Microwave research institute Polytechnic press,Brooklyn,N.Y., 1963
Eng 41.5417

MICROWAVE RESEARCH INSTITUTE.SYMPOSIA SERIES, 17
Modern optics The Symposium on modern optics Proceedings New York 1967 Mar 22-24 xxxv,804p 23cm Polytechnic press, Brooklyn,N.Y.,1967
Cav 7.2815

MICROWAVE RESEARCH INSTITUTE.SYMPOSIA SERIES, 21
Symposium on computers and automata Proceedings New York 1971 Apr 13-15 Microwave research institute Edited by Jerome Fox illus 653p Polytechnic press, Brooklyn,N.Y.,1971
Math L 5.3635

MICROWAVE RESEARCH INSTITUTE.SYMPOSIUM SERIES, 19
Symposium on computer processing in communications New York 1969 Microwave research institute Polytechnic press, Brooklyn,N.Y.,1970
Eng 41.8313

MICROWAVE TUBES International congress on microwave tubes 3rd Proceedings Munich 1960 Jun 7-11 Edited by J. Wosnik Vieweg, Brunswick,1961
Eng 41.5547

MICROWAVE TUBES International congress on microwave tubes 4th Proceedings Scheveningen 1962 Sep 3-7 Centrex;Cleaver-Hume,Eindoven;London,1963
Eng 41.5557

MICROWAVE TUBES International congress on microwave tubes 5th Proceedings Paris 1964 Societe francaise des electroniciens et radioelectriciens Societe francaise des ingenieurs et techniciens du vide Sponsored by the Federation nationale des industries electroniques Academic press,New York,1965
Eng 41.5574

MICROWAVE TUBES Microwave and optical generation and amplification International conference on microwave tubes 6th Cambridge 1966 Institution of electrical engineers Institution of electrical engineers.Conference publication, 27 I.E.E., London,1967
Eng 41.5588

MID-CENTURY Convocation on the social implications of scientific progress Verbatim account Cambridge,Mass. 1949 Mar 31-Apr 2 Massachusetts institute of technology Edited by John Ely Burchard xx,549p Technology press of the Massachusetts institute of technology,Cambridge,Mass.,1950
Cav 7.0406

MIDDLESEX HOSPITAL Function and structure in micro-organisms :a symposium Papers London 1965 Apr Society for general microbiology Society for general microbiology.Symposia, 15 Cambridge university press,Cambridge,1965
Bioch 33.1171

MIDDLESEX HOSPITAL MEDICAL SCHOOL The Interaction of drugs and subcellular components in animal cells :a symposium London 1967 Apr 10-11 Biological council. Co-ordinating committee for symposia on drug action Edited by P.N. Campbell Churchill, London,1968
Bioch 33.0923

MIDDLESEX HOSPITAL MEDICAL SCHOOL The Structure and function of the membranes and surfaces of cells :a symposium London 1962 Mar 9 Biochemical society Edited by D.J. Bell and J.K. Grant Biochemical society. Symposia, 22 Cambridge university press, Cambridge,1963 Organized by J.K.Grant
Bal 39.0294

MIDWEST CATEGORY SEMINAR 5 Reports By M. Andre and others Edited by J.W. Gray and S. Mac Lane Lecture notes in mathematics, 195 255p 25cm Springer,Berlin,1971
P Math 2.4425

MIDWESTERN CONFERENCE ON ERGODIC THEORY 1st Proceedings Contributions to ergodic theory and probability Columbus,Ohio 1970 Mar 27-30 Held at Ohio state university Lecture notes in mathematics, 160 bibliog. vii,278p 25cm Springer,Berlin,1970
P Math 2.3939

MIDWESTERN CONFERENCE ON ERGODIC THEORY AND PROBABILITY 1st Proceedings Contributions to ergodic theory and probability Columbus,Ohio 1970 Mar 27-30 Held at the Ohio state university Lecture notes in mathematics, 160 vii,278p 26cm Springer,Berlin,1970
Math S 3.1779

MIDWESTERN CONFERENCE ON FLUID MECHANICS 2nd Proceedings Columbus,Ohio 1952 Mar 17-19 Sponsored by Ohio state university. Graduate school Ohio state university. Engineering experiment station.Bulletin, 149 Ohio state university studies:engineering series, 21,no 3 v,529p Ohio state university,Columbus,Ohio,1952
Chem E 24.0402

MIDWESTERN CONFERENCE ON FLUID MECHANICS 3rd
Proceedings Minneapolis 1953 Mar 23-25
Held at the University of Minnesota 783p
Minneapolis,Minn.,1953
Chem E 24.0403

MIDWESTERN CONFERENCE ON FLUID MECHANICS 3rd
Proceedings Minneapolis,Minn. 1953 Mar
23-25 University of Minnesota.Institute of
technology University of Minnesota,
Minneapolis,Minn.,1953
Eng 41.6953

MIDWESTERN CONFERENCE ON FLUID MECHANICS 4th
Proceedings Lafayette,Ind. 1955 Sep 8-9
Held at Purdue university Purdue university.
Engineering experiment station.Research series,
128 Purdue university,Lafayette,Ind.,1955
Eng 41.6957

MIDWESTERN CONFERENCE ON FLUID MECHANICS 4th
Proceedings Lafayette,Ind. 1955 Sep 8-9
Held at Purdue university Purdue university.
Engineering experiment station.Research series,
128 vii,370p Purdue university,Lafayette,
Ind.,1955
Chem E 24.0404

MIDWESTERN CONFERENCE ON FLUID MECHANICS 5th
Proceedings Ann Arbor,Mich. 1957 Apr 1-
2 Held at University of Michigan
University of Michigan press,Ann Arbor,Mich.,
1957
Eng 41.6958

MIDWESTERN CONFERENCE ON FLUID MECHANICS 5th
Proceedings Ann Arbor,Mich. 1957 Apr 1-
2 Held at the University of Michigan
University of Michigan.Engineering research
institute publications viii,388p
University of Michigan press,Ann Arbor,Mich.,
1957
Chem E 24.0405

The MIDWESTERN CONFERENCE ON FLUID MECHANICS 5th
proceedings Ann Arbor 1957 Apr 1-2
Edited by R.C.F. Bartels Under the auspices
of the University of Michigan.Engineering
research institute University of Michigan.
Engineering research institute.Publication
University of Michigan press,Ann Arbor,1957
A Math 4.0516

MIDWESTERN CONFERENCE ON SOLID MECHANICS 4th
proceedings Austin,Tex. 1959 Sept 9-11
University of Texas,Austin,Tex,1959
A Math 4.0368

MIDWESTERN MECHANICS CONFERENCE 6th
Proceedings Developments in mechanics
Cleveland,Ohio 1963 Apr 1-3 Vol 2,pt 1:
fluid mechanics Edited by Simon Ostrach and
Robert H. Scanlan Pergamon,London,1965
Combining the 6th Midwestern conference on
soil mechanics and the 8th Midwestern
conference on fluid mechanics
Eng 41.6959

MIDWESTERN MECHANICS CONFERENCE 7th
proceedings Developments in mechanics
Ann Arbor,Mich 1961 Sep. 6-8 Vol 1
Edited by J.E. Lay and L.E. Malvern North
Holland,Amsterdam,1961
A Math 4.0313

MIGRATION Postwar problems of migration
Milbank memorial fund annual conference
Papers New York 1946 Oct 29-30 Milbank
memorial fund Milbank memorial fund,New York,
1947
Geog 13.1564

MILAN 1952 Problemi di sviluppo :un
simposio Unione zoologie italiana Casa
editrice ambrosiana,Milan,1954
An 32.2477

MILAN 1957 Psychotropic drugs :an
international symposium Proceedings Edited
by S. Garattini and V. Ghetti illus
Elsevier,Amsterdam,1957 Chairman of
conference:professor E.Trabucchi
Chem 26.0307

MILAN 1957 Psychotropic drugs :
international symposium Proceedings Edited
by S. Garattini and V. Ghetti Elsevier,
Amsterdam,1957
Psy 31.0523

MILAN 1957 Psychotropic drugs :
international symposium Proceedings Edited
by S. Garrantini and V. Ghetti Elsevier,
Amsterdam,1957
Pha 16.0029

MILAN 1960 Atti del convegno per le
celebrazioni del cinquantenario della morte di
G.V.Schiaparelli Osservatorio astronomico di
Brera Milan,1960
Obs 6.0263

MILAN 1960 Drugs affecting lipid
matabolism The Symposium on drugs affecting
lipid metabolism Proceedings Edited by S.
Garattini and R. Paoletti illus Elsevier,
Amsterdam,1961
Bioch 33.1237

MILAN 1960 Drugs affecting lipid
metabolism :symposium Proceedings Edited
by S. Garattini and R. Paoletti Elsevier,
Amsterdam,1961
Pha 16.0128

MILAN 1962 Advances in organic chemistry
international meeting proceedings
Geochemical society.Organic geochemistry group,
European branch Edited by Umberto Colombo
and G.D. Hobson International series of
monographs on earth sciences,15 illus.
Pergamon press,Oxford,1964
Geol 8.1666

MILAN 1963 International symposium on the
electrophysiology of the heart Proceedings
Edited by B. Taccardi and G. Marchetti
Macmillan;Pergamon,New York;Oxford,1965
Pha 16.0286

MILAN 1964 International congress of
hormonal steroids :biochemistry, pharmacology
and therapeutics 1st Proceedings Vol
1-2 Edited by L. Martini and A. Pecile
2 vols Academic press,New York;London,1964-
65
Gen 34.0546

MILAN 1964 Symposium on radiosensitizers
and radioprotective drugs 1st Edited
by R. Paoletti and R. Vertus Organized by
the European society for biochemical
pharmacology Progress in biochemical
pharmacology, 1 Karger,Basel,1965
Radioth 35.1943

MILAN 1965 Drugs affecting lipid metabolism International symposium on drugs affecting lipid metabolism 2nd Papers Pt 1-2 Edited by D. Kritchevsky and others organised by the European society for biochemical pharmacology Progress in biochemical pharmacology, 2-3 illus 2 vols Karger,Basle;New York,1967
Bioch 33.1250

MILAN 1965 Symposium on catecholamines 2nd American society for pharmacology and experimental therapeutics Edited by George H. Acheson Held at the Istituto di ricerche farmacologiche "Mario Negri" Williams and Wilkins company,Baltimore,1966 Reprinted from the Pharmacological reviews,18,no.1
Bioch 33.0430

MILAN 1965 Symposium on catecholamines 2nd Edited by George H. Acheson Pharmacological reviews, 18,no.1 Williams and Wilkins,Baltimore,Md.,1966 Reprint
An 32.3777

MILAN 1966 Antidepressant drugs : international symposium 1st Proceedings Istituto de ricerche farmacologiche "Mario Negri" Edited by S. Garattini and M.N.G. Dukes Excerpta medica international congress series,122 Excerpta medica,Amsterdam,1967
Pha 16.0046

MILAN 1966 Elementary excitations in solids Cortina lectures and four lectures from the Conference on localized excitations Proceedings Edited by A.A. Maradudin and G.F. Nardelli Bibliog 526p 20cm Plenum press,New York,1969
Cav 7.2677

MILAN 1966 Hormonal steroids International congress on hormonal steroids 2nd Proceedings Edited by L. Martini and others Excerpta medica.International congress series, 132 Excerpta medica, Amsterdam,1967
Inv Med 37.0116

MILAN 1967 Advanced navigational techniques :symposium Proceedings AGARD Edited by W.T. Blackband AGARD.Conference proceedings, 28 Technivision,Maidenhead, 1970 "...the record of the proceedings of the 14th symposium of the Avionics panel..."- foreward
Eng 41.5726

MILAN 1967 Growth hormone International symposium on growth hormone 1st Proceedings Edited by A. Pecile and E.E. Muller International congress series, 158 Excerpta medica,Amsterdam,1968
Bioch 33.0457

MILAN 1968 Aggressive behaviour International symposium on the biology of aggressive behaviour Edited by S. Garattini and E.B. Sigg Excerpta medica foundation, Amsterdam,1969
Inv Med 37.0086

MILAN 1968 Problemi attuali di scienza e di cultura Congresso internazionale di patologia 7 Atti International congress of pathology Edited by Alfonso Giordano Accademia nazionale dei Lincei,Rome,1970
PGMS 29.0618

MILAN 1968 The Foeto-placental unit an international symposium Proceedings Edited by A. Pecile and C. Finzi Excerpta medica. International congress series Excerpta medica,Amsterdam,1969
Inv Med 37.0260

MILBANK MEMORIAL FUND The Family health maintenance demonstration :a controlled,long term investigation of family health; proceedings of a round table at the 1953 annual conference of the fund Milbank memorial fund,New York,1954
HE 27.0128

MILBANK MEMORIAL FUND ANNUAL CONFERENCE Papers Postwar problems of migration New York 1946 Oct 29-30 Milbank memorial fund Milbank memorial fund,New York,1947
Geog 13.1564

MILITARY MEDICINE International congress of military medicine and pharmacy 5th Report London 1929 May 6-11 By William Seaman Bainbridge Banta,Menasha,Wis.,1929
Path 30.0210

MILLIMETER WAVES :a symposium Proceedings New York 1959 Mr 31-Apr 2 Microwave research institute and Institute of radio engineers Co-sponsored by United States. Office of naval research Polytechnic institute of Brooklyn.Microwave research institute symposia series, 9 Polytechnic press,Brooklyn,N.Y.,1960
Cav 7.1715

MINAKAMI 1963 Neurosecretion and neural control of internal secretion Gunma symposium on endocrinology 1st Gunma university.Institute of endocrinology Gunma university.Institute of endocrinology.Annual report, 1 Institute of endocrinology,Gunma university,Maebashi,1964
Phys 20.2230

MINAKAMI 1966 Sex and reproduction Gunma symposium on endocrinology 4th Gunma university.Institute of endocrinology Gunma university.Institute of endocrinology.Annual report, 4 Institute of endocrinology,Gunma university,Maebashi,1967
Phys 20.1434

MIND AS A TISSUE Lankenau conference 5th Proceedings Philadelphia,Pa. Lankenau hospital Edited by Charles Rupp Harper and Row,Hoeber medical division,New York,1968
Psy 31.2898

MINERAL CHEMISTRY Quelques problemes de chimie minerale :conseil de chimie 10e Rapports et discussions Brussels 1956 May 22-26 Institut international de chimie Solvay.Comite scientifique Solvay conference on chemistry,1956 R.Stoops,Brussels,1956
Chem 18.1004

MINERAL DEPOSITS Geology of Canadian industrial mineral deposits Commonwealth mining and metallurgical congress 6th Papers Montreal 1957 Sep 8-Oct 9 Canadian institute of mining and metallurgy Montreal,1957
Min 10.0716

MINERAL DRESSING Progress in mineral dressing International mineral dressing congress Transactions Stockholm 1957 Sep 18-21 Svenska Gruvforeningen and Jernkontoret Almqvist and Wiksell,Stockholm, 1958
Met 25.2286

MINERAL PROCESSING International mineral processing congress 7th Technical papers New York 1964 Sep 20-24 Vol 1 American institute of mining,metallurgical and petroleum engineers and Columbia university Edited by Nathaniel Arbiter Gordon and Breach,New York,1965 Held in conjunction with the centennial of the Henry Krumb school of mines
Met 25.0214

MINERAL PROCESSING International mineral processing congress Proceedings Gothenburg 1966 Aug 27-Sep 1 Institution of mining and metallurgy London,1960
Met 25.2504

MINERAL RESOURCES POLICY :a symposium Proceedings London 1955 Sep 22 Institution of mining and metallurgy Institution of mining and metallurgy,London, 1956
Met 25.0191

MINERALOGICAL SOCIETY OF INDIA International mineralogical association 4th general meeting Papers and proceedings New Delhi 1964 Dec 15-22 Edited by P.R.J. Naidu and M. N. Viswanathiah Mineralogical society of India,Mysore,1966
Min 10.1550

MINERALOGY International geological congress 20th papers Mexico City 1956 Seccion 11A: petrologia y mineralogia Edited by A. Garcia Rojas Mexico City,1959 Text in English,French,Russian and Spanish
Geol 8.3047

MINERALS Ecological aspects of the mineral nutrition of plants :a symposium Sheffield 1968 Apr 1-5 British ecological society Edited by Ian H. Rorison and others British ecological society.Symposia, 9 illus, plates xxi,484p Blackwell scientific, Oxford,1969
Bot 42.2107

MINERALS Riqueza minera y yacimientos minerales de Mexico International geological congress 20th papers Mexico city 1956 Edited by Jenaro Gonzalez Reyna 3rd edition Mexico city,1956
Geol 8.3025

MINICOMPUTERS IN INSTRUMENTATION AND CONTROL proceedings of a three-day short course London 1972 Polytechnic of North London Edited by Y. Paker and others Polytechnic of North London,London,1972
Eng 41.8589

MINING Empire mining and metallurgical congress 4th Proceedings London 1949 Jul 9-23 Pt 1-2 Edited by F. Higham 2 vols London,1950
Min 10.0726

MINING AND METALLURGY The Commonwealth mining and metallurgical congress 9th Proceedings London 1969 May 3-24 Vol 4: physical and fabrication metallurgy Commonwealth council of mining and metallurgical institutions Edited by M.J. Jones Institution of mining and metallurgy, London,1970
Met 25.2847

MINISTERSTVO SELSKOGO KHOZYAISTVA,R.S.F.R. Bogatstva zemli na sluzhbu rodine :materialy soveshchaniya rabotnikov selskogo khozyaistva oblastei i avlovonykh kraynego Severa Materialy Magadan 1962 Feb 22-24 159p 20cm Izdatelstvo ministerstva selskogo khozyaistva RSFSR,Moscow,1962
Sco 14.4924

MINISTRY OF HEALTH, Mathematics and computer science in biology and medicine conference proceedings Oxford 1964 Jul Medical research council bibliog,illus.,pl. vii, 316p 24cm H.M.S.O.,London,1965
Math L 5.1007

MINNEAPOLIS 1925 The National symposium on colloid chemistry 3rd Papers presented University of Minnesota Edited by Harry N. Holmes and Harry B. Weiser Colloid symposium monograph, 3 Chemical catalog company,New York,1925
Bioch 33.1471

MINNEAPOLIS 1953 Midwestern conference on fluid mechanics 3rd Proceedings Held at the University of Minnesota 783p Minneapolis,Minn.,1953
Chem E 24.0403

MINNEAPOLIS 1963 Joint automatic control conference 4th Preprints of technical papers American institute of chemical engineers xiv,680p American institute of chemical engineers,New York,1963
Chem E 24.1172

MINNEAPOLIS,MA. 1967 United States-Japan seminar on differential and functional equations Proceedings Edited by William A. Harris and Yasutaka Sibuya Illus. xvi,585p 24cm Benjamin,New York;Amsterdam,1967
P Math 2.3164

MINNEAPOLIS,MINN. 1953 Midwestern conference on fluid mechanics 3rd Proceedings University of Minnesota. Institute of technology University of Minnesota,Minneapolis,Minn.,1953
Eng 41.6953

MINNEAPOLIS,MINN. 1962 Conference on the thymus;the thymus in immunology :structure, function and role in disease Proceedings Edited by Robert A. Good and Ann E. Gabrielson Harper and Row,New York,1964
Med 36.0382

MINNEAPOLIS,MINN. 1964 Conference on complex analysis proceedings United States. Air force.Office of scientific research Edited by A. Aeppli and others Bibliog. vii,308p 24cm Springer-Verlag,Berlin, Heidelberg,1965
P. Math 2.1582

MINSK 1965 Konechnye gruppy :a seminar Papers Edited by Ya.G. Berkovich and others Bibliog,port 191p 21cm Minsk,1966 In honour of S.A.Cunihin
P Math 2.3070

MISIMA,JAPAN 1960 Symposium on genetic effect of radiation Proceedings Under the auspices of the National institute of genetics, Shizvoka Japanese journal of genetics, 36, supp. Genetics society of Japan,1961
Gen 34.1117

MISSOURI BOTANICAL GARDEN Gene action in micro-organisms :conference Papers St. Louis,Mo. Missouri botanical garden.Annals, 32,no.2 Galesburg,Ill.,1945
Gen 34.0741

MITOCHONDRIA Biochemical aspects of the biogenesis of mitochondria The Round table discussion on the biochemical aspects of the biogenesis of mitochondria Proceedings Polignane a Mare 1967 May 15-18 Edited by E.C. Slater and others Adriatica editrice, Bari,1968
Bioch 33.1073

MITOCHONDRIA Energy level and metabolic control in mitochondria The Round table discussion on the energy level and metabolic control in mitochondria Proceedings Polignano a Mare 1968 May 13-16 Edited by S. Papa and others Organized by the University of Bari.Department of biochemistry and the University of Amsterdam.Laboratory of biochemistry Adriatica editrice,Bari,1969
Bioch 33.1075

MITOCHONDRIA Energy-linked functions of mitochondria :colloquium Papers presented Philadelphia 1963 Apr 13 Johnson research foundation Edited by Britton Chance Johnson research foundation.Colloquia, 1 Academic press,New York;London,1963
Bioch 33.1081

MITOCHONDRIA Regulation of metabolic processes in mitochondria :a symposium Proceedings Bari 1965 Apr 26-May 1 Edited by J.M. Tager and others organized by the University of Bari,Dept.of Biochemistry and the University of Amsterdam,Laboratory of Biochemistry B.B.A. library, 7 Elsevier, Amsterdam,1966
Bioch 33.1101

MITOCHONDRIA The Federation of European biochemical societies meeting 3rd Warsaw 1966 Apr 4-7 biochemistry of mitochondria;colloquium Federation of European biochemical societies Edited by E.C. Slater and others Organized by the Polish biochemical society Academic press,London, 1967
Bioch 33.1394

MITOCHONDRIA:STRUCTURE AND FUNCTION Federation of European biochemical societies meeting 5th Proceedings Prague 1968 Jul Federation of European biochemical societies Edited by L. Ernster and Z. Drahota Federation of European biochemical societies.Publications, 17 Academic press, London;New York,1969
Bioch 33.1403

MITOCHONDRIA:STRUCTURE AND FUNCTION Federation of European biochemical societies meeting 5th Proceedings Prague 1968 Jul Federation of European biochemical societies Edited by L. Ernster and Z. Drahota Federation of European biochemical societies.Publications, 17 Academic press, London;New York,1969
Radioth 35.0149

MITOCHONDRIA AND OTHER CYTOPLASMIC INCLUSIONS a symposium Oxford 1955 Sep Society for experimental biology Society for experimental biology.Symposia, 10 Cambridge university press,Cambridge,1957
Bioch 33.1355

MITOCHONDRIA AND OTHER CYTOPLASMIC INCLUSIONS : a symposium Papers Oxford 1955 Sep Society for experimental biology Society for experimental biology.Symposia, 10 Cambridge university press,Cambridge,1957
Radioth 35.0472

MITOCHONDRIA AND OTHER CYTOPLASMIC INCLUSIONS : a symposium Oxford 1955 Sep Society for experimental biology.Symposia, 10 Cambridge university press,Cambridge,1957
An 32.3392

MITOCHONDRIAL STRUCTURE AND COMPARTMENTATION The Round table discussion on mitochondrial structure and compartmentation Proceedings Polignano a Mare 1966 May 23-25 Edited by E. Quagliariello and others organized by the University of Bari,Department of biochemistry and the University of Amsterdam.Laboratory of biochemistry Adriatica editrice,Bari,1967
Bioch 33.1093

MIX DESIGN AND QUALITY CONTROL OF CONCRETE :a symposium Proceedings London 1954 May Cement and concrete association Cement and concrete association,London,1955
Eng 41.2929

MIXED COMMISSION ON THE IONOSPHERE Solar eclipses and the ionosphere :a symposium London 1955 Aug.22-24 Edited by W.J.G. Beynon and G.M. Brown Pergamon press,London, 1956 Vol.6 of special supplements to Journal of atmospheric and terrestrial physics
Obs 6.0176

MIXED COMMISSION ON THE IONOSPHERE Solar eclipses and the ionosphere :a symposium Proceedings London 1955 Aug 22-24 Edited by W.J.G. Beynon and G.H. Brown Under the auspices of the International council of scientific unions Journal of atmospheric and terrestrial physics.Special supplement, 6 ix,330p Pergamon press,London;New York,1956
Nap 11.0437

MIXED COMMISSION ON THE IONOSPHERE 3rd meeting Proceedings Canberra 1952 Aug. 24-26 International scientific radio union Edited by W.J.G. Beynon Organized by the International council of scientific unions 194p General secretariat,Brussels,1953
Obs 6.2579

MODEL TESTING :one-day meeting Proceedings London 1964 Mar 17 Cement and concrete association Cement and concrete association, London,1964
Eng 41.2934

MODEL THEORY Theory of models
International symposium on the theory of models proceedings Berkeley, Calif. 1963 Jun 25-Jul 11 Edited by J.W. Addison and others xv,494p 23cm North-Holland, Amsterdam,1965
P. Math 2.0205

MODELS AND ANALOGUES IN BIOLOGY Bristol 1960 Sep 6-12 Society for experimental biology Edited by J.W.L. Beament Society for experimental biology.Symposia, 14 Cambridge university press,Cambridge,1960
Phys 20.0939

MODELS AND ANALOGUES IN BIOLOGY Bristol 1960 Sep 6-12 Society for experimental biology Society for experimental biology symposia, 14 vii,255p Cambridge university press,Cambridge,1960
WSM 43.2826

MODELS AND ANALOGUES IN BIOLOGY :a symposium Bristol 1960 Sep 6-12 Society for experimental biology.Symposia, 14 Cambridge university press,Cambridge,1960
An 32.0058

MODELS FOR DECISION :a conference London 1960 Oct 13-144 British computer society Operational research society Under the auspices of the United Kingdom automation council English universities press,London, 1965
Eng 41.1270

MODELS FOR THE PERCEPTION OF SPEECH AND VISUAL FORM :a symposium Proceedings Boston, Mass. 1964 Nov 11-14 Air force Cambridge research laboratories.Data science laboratory Edited by Weiant Wathen-Dunn MIT press, Cambridge,Mass.,1967
Psy 31.2951

MODELS IN MATHEMATICS AND NATURAL AND SOCIAL SCIENCES The Concept and the role of the model in mathematics and natural and social sciences :a colloquium Proceedings Utrecht 1960 Jan International union of history and philosophy of science.Division of philosophy of science Edited by B.H. Kazemier and D. Vuysje 164p Reidel,Dordrecht,1961
WSM 43.1797

MODERN CHEMISTRY IN INDUSTRY An International union of pure and applied chemistry symposium Proceedings Eastbourne 1968 Mar 11-14 Society of chemical industry Edited by J.G. Gregory London,1968
Bioch 33.1874

MODERN CHEMISTRY IN INDUSTRY :a symposium Proceedings Eastbourne 1968 Mar 11-14 International union of pure and applied chemistry Edited by J.G. Gregory 311p Society of chemical industry,London,1968
Chem & 24.1551

MODERN CHEMISTRY IN INDUSTRY :an International union of pure and applied chemistry symposium Proceedings Eastbourne 1968 Mar 11-14 International union of pure and applied chemistry Edited by J.G. Gregory illus 311p Society of chemical industry,London, 1968
Chem E 24.1874

The MODERN DESIGN OF WIND-SENSITIVE STRUCTURES a seminar Proceedings London 1970 Jun 18 Cement industry research and information association CIRIA,London,1971
Eng 41.8276

MODERN DIFFRACTION AND IMAGING TECHNIQUES IN MATERIAL SCIENCE International summer course on material science Proceedings Antwerp 1969 Jul 28-Aug 8 North Atlantic treaty organization.Scientific affairs division Edited by S. Amelinckx and others North-Holland,Amsterdam,1970
Met 25.2608

MODERN NETWORK SYNTHESIS :a symposium Proceedings New York 1955 Apr 13-15 Polytechnic institute of Brooklyn Sponsored also by the Microwave research institute Microwave research institute.Symposia series Polytechnic institute of Brooklyn,New York, 1956
Eng 41.5117

MODERN NETWORK SYNTHESIS (AUDIO TO MICROWAVES) a symposium Proceedings New York 1952 Apr 16-18 Polytechnic institute of Brooklyn Sponsored also by the Microwave research institute Microwave research institute.Symposia series Polytechnic institute of Brooklyn,New York,1956
Eng 41.5116

MODERN OPTICS The Symposium on modern optics Proceedings New York 1967 Mar 22-24 Microwave research institute.Symposia series, 17 xxxv,804p 23cm Polytechnic press, Brooklyn,N.Y.,1967
Cav 7.2815

MODERN ORGANIZATION THEORY :a symposium of the Foundation for research on human behavior Ann Arbor,Mich. 1959 Feb Foundation for research on human behavior Edited by Mason Haire Wiley,New York,1959 reprinted 1965
Eng 41.0572

MODERN PSYCHOLOGY University of Pennsylvania bicentennial conference By Charles S. Myers and others University of Pennsylvania press, Philadelphia,Pa.,1941
Eng 41.1542

MODERN QUANTUM CHEMISTRY :Istanbul international school of quantum chemistry Lectures Istanbul 1964 Aug 16-Sep 5 Pt 3: action of light and organic crystals North Atlantic treaty organization.Pure science bureau Edited by Oktay Sinanoglu Organized with the cooperattion of Orta Dogu teknik universitesi Istanbul lectures:modern quantum chemistry,3 Academic press,New York; London,1965
Chem 18.2238

MODERN RESEARCH TECHNIQUES IN PHYSICAL METALLURGY a seminar... held during the thirty-fourth national metal congress and exposition Philadelphia 1952 Oct 18-24 American society for metals American society for metals,Cleveland,Ohio,1953
Met 25.1640

MODERN STEAM PLANT PRACTICE :a convention Proceedings The Hague 1971 Apr 28-30 Institution of mechanical engineers.Steam plant group Institut van ingenieurs.Steam plant group IME,London,1971
Eng 41.8643

MODERN YARN PRODUCTION FROM MAN-MADE FIBRES (AND THEIR CONVERSION INTO FABRICS) :a symposium Manchester college of science and technology Edited by G.R. Wray illus. Columbine press,Manchester;London,1960 Lectures given at Manchester college of science and technology
Eng 41.3661

MODERNITY Introduction to modernity :a symposium on eighteenth century thought University of Texas.Department of Germanic languages Edited by Robert Mollenauer 175p University of Texas press,Austin,Texas,1965 Essays presented in 1962 as the 4th in a series of yearly symposia
WSM 43.0649

MOL 1963 Interaction of radiation with solids International summer school on solid state physics Proceedings Edited by R. Strumane and others North-Holland,Amsterdam, 1964
Met 25.1581

MOL 1965 Chemical stability of tritiated pryimidine nucleosides and some problems raised by their use in biology :a colloquium Proceedings Edited by R. Goutier and others EUR 3290 f-e Euratom,Brussels,1967
Radioth 35.1754

MOL,BELGIUM 1970 Informative molecules in biological systems International symposium on uptake of informative molecules by living cells Proceedings North Atlantic treaty organization.Scientific affairs division Edited by L. Ledoux North-Holland,Amsterdam, 1971
Bioch 33.2240

MOLDING MACHINES Symposium on molding machines American foundrymen's society.Plant and plant equipment committee American foundrymen's society,n.p.,n.d.
Met 25.0648

MOLECULAR ACTION OF MUTAGENIC AND CARCINOGENIC AGENTS Symposium on molecular action of mutagenic and carcinogenic agents Gatlinburg, Tenn. 1964 Apr 6-9 Sponsored by the Oak Ridge national laboratory.Biology division Oak Ridge national laboratory.Symposia Journal of cellular and comparative physiology, 64,suppl.1 Wistar institute of anatomy and biology,Philadelphia,Pa.,1964
Gen 34.1098

MOLECULAR AND CELLULAR BASIS OF ANTIBODY FORMATION :a symposium Proceedings Prague 1964 Jun 1-5 Ceskoslovenska akademie ved.Institute of microbiology Edited by J. Sterzl Czechoslovak academy of sciences;Academic press,Prague;New York,1965
An 32.3701

MOLECULAR AND CELLULAR BASIS OF ANTIBODY FORMATION a symposium Proceedings Prague 1964 Jun 1-5 Edited by J. Sterzl Organized by the Ceskoslovenska akademie ved. Institute of microbiology. Immunological department Czechoslovak academy of sciences; Academic press,Prague;New York,1965
Bioch 33.0554

MOLECULAR AND CELLULAR STRUCTURE Synthesis of molecular and cellular structure :a symposium Papers Waltham,Mass. 1960 Society for the study of development and growth Edited by Dorothea Rudnick Held at Brandeis university Society for the study of development and growth.Symposia, 19 Ronald press,New York,1961
Bioch 33.1012

MOLECULAR AND RADIATION BIOLOGY Conference on molecular and radiation biology Ardsley-on-Hudson,N.Y. 1959 Dec 2-4 Edited by R.A. Deering Organised by the National research council.Subcommittee on radiobiology National research council.Nuclear science series.Report, 31 NAS-NRC 823 National research council,Washington,D.C.,1961
Radioth 35.1735

MOLECULAR ARCHITECTURE IN CELL PHYSIOLOGY :a symposium Woods Hole,Mass. 1964 Sep 8-11 Edited by Tern Hayashi and Andrew G. Szent-Gyorgyi Under the auspices of the Society of general physiologists Prentice-Hall,Englewood Cliffs,N.J.,1966 Held at the 11th annual meeting of the society
Gen 34.0668

MOLECULAR ASPECTS OF DIFFERENTIATION Symposium on molecular aspects of differentiation :given at research conference for biology and medicine of the Atomic energy commission Gatlinburg,Tenn. 1968 Apr 8-11 Sponsored by the Oak Ridge national laboratory. Biology division Oak Ridge national laboratory.Biology division.Symposia, 21 Journal of cellular physiology, 72,supp.1 Wistar institute of anatomy and biology, Philadelphia,Pa.,1968
Gen 34.0505

MOLECULAR BASIS OF ENDOCRINOLOGY Gunma symposium on endocrinology 5th Kusatsu 1967 Aug 18-19 Gunma university.Institute of endocrinology Gunma university.Institute of endocrinology.Annual report, 5 Institute of endocrinology,Gunma university,Maebashi, 1967
Phys 20.1435

MOLECULAR BASIS OF ENZYME ACTION AND INHIBITION International congress of biochemistry 5th Proceedings Moscow 1961 Aug 10-16 International union of biochemistry Edited by P.A.E. Desnuelle I.U.B.symposium series, 24 Pergamon;PWN-Polish scientific publishers, Oxford;Warsaw,1963
Bioch 33.1333

MOLECULAR BASIS OF NEOPLASIA Annual symposium on fundamental cancer research 15th Papers Houston,Tex. 1961 Anderson hospital and tumor institute University of Texas press,Austin,Tex.,1961
Radioth 35.1942

MOLECULAR BASIS OF SOME ASPECTS OF MENTAL ACTIVITY :a NATO advanced study institute Proceedings Drammen 1965 Aug 2-14 North Atlantic treaty organization Edited by Otto Walaas 2 vols Academic press,London, 1966-67
Pha 16.0053

MOLECULAR BEAMS AND REACTION KINETICS Scuola internazionale di fisica 'Enrico Fermi' 44 corso Rendiconti Varenna 1968 Jul 29-Aug 10 Societa italiana di fisica Edited by C. Schlier Academic press,New York,1970
Chem 18.2814

MOLECULAR BIOLOGY A Symposium on molecular biology Chicago,Ill. 1956-57 Nov-Mar Edited by Raymond E. Zirkle Held under the auspices of the Chicago-Frankfurt inter-university program illus. University of Chicago press,Chicago,Ill.,1959
Gen 34.0647

MOLECULAR BIOLOGY A Symposium on molecular biology :a seminar series and a symposium Lectures Chicago 1956-57 Nov-Mar University of Chicago Edited by Raymond E. Zirkle Under the auspices of the Chicago-Frankfurt inter-university program University of Chicago press,Chicago,1959
Bioch 33.0839

MOLECULAR BIOLOGY A Symposium on molecular biology :a seminar series and a symposium lectures Chicago 1956-57 Nov-Mar University of Chicago Edited by Raymond E. Zirkle Under the auspices of the Chicago-Frankfurt inter-university program University of Chicago press,Chicago,1959
Bal 39.0617

MOLECULAR BIOLOGY :elementary processes of nerve conduction and muscle contraction.A symposium Papers New York 1958 Sep 25-30 Rockefeller institute Edited by David Nachmansohn Sponsored by Columbia university and the Rockefeller institute Academic press, New York;London,1960
Bioch 33.0831

MOLECULAR BIOLOGY OF VIRUSES Symposium of the molecular biology of viruses Proceedings Edmonton,Alberta 1966 Jun 27-30 University of Alberta.Faculty of medicine Edited by John S. Colter and William Paranchych Academic press,New York;London,1967
Path 30.2745

MOLECULAR BIOLOGY OF VIRUSES :symposium London 1968 Apr Society for general microbiology Society for general microbiology.Symposia, 18 Cambridge university press,Cambridge,1968
Gen 34.0709

MOLECULAR BIOLOGY OF VIRUSES :a symposium Papers London 1968 Apr Society for general microbiology Edited by R.V. Crawford and M.G.P. Stoker Held at the Imperial college of science and technology Society for general microbiology.Symposia, 18 Cambridge university press,Cambridge,1968
Bioch 33.1178

The MOLECULAR BIOLOGY OF VIRUSES :a symposium London 1968 Apr Society for general microbiology Edited by L.V. Crawford and M.G.P. Stoker Society for general microbiology.Symposia, 18 Cambridge university press,Cambridge,1968
An 32.5380

The MOLECULAR BIOLOGY OF VIRUSES :a symposium London 1968 Apr Society for general microbiology Society for general microbiology.Symposia, 18 Cambridge university press,Cambridge,1968
PGMS 29.0416

The MOLECULAR BIOLOGY OF VIRUSES :a symposium Papers London 1968 Apr Society for general microbiology Edited by L.V. Crawford and M.G.P. Stoker Society for general microbiology.Symposia, 18 Cambridge university press for the Society of general microbiology,Cambridge,1968
Path 30.2742

The MOLECULAR BIOLOGY OF VIRUSES :a symposium Proceedings London 1968 Apr Society for general microbiology Edited by L.V. Crawford and M.G.P. Stoker Society for general microbiology.Symposia, 18 Cambridge university press,Cambridge,1968
Radioth 35.0268

MOLECULAR BIOPHYSICS International summer school of molecular biophysics Proceedings Squaw Valley,Calif. 1964 Aug 17-28 Edited by B. Pullman and M. Weissbluth Sponsored by the North atlantic treaty organization Academic press,New York,1965
Radioth 35.1418

MOLECULAR BIOPHYSICS International summer school of molecular biophysics Proceedings Squaw Valley,Calif. 1964 Aug 17-28 Edited by Bernard Pullman and Mitchel Weissbluth Sponsored jointly by the North Atlantic treaty organization and United States.Office of naval research Academic press,New York;London,1965
Bioch 33.2360

MOLECULAR BIOPHYSICS :international summer school Proceedings Squaw Valley,Calif. 1964 Aug 17-28 North Atlantic treaty organization,and,United States.Office of naval research Edited by Bernard Pullmann and Mitchel Weissbluth Academic press,New York; London,1965
Chem 18.2249

MOLECULAR CRYSTALS Motions in molecular crystals Oxford 1969 Sep 16-18 Faraday society Faraday society.Discussions, 48 Bibliog 224p 20cm Faraday society,London, 1969
Cav 7.2718

MOLECULAR ENZYMOLOGY AND PROTEIN GROUP Chemical reactivity and biological role of functional groups in enzymes :a symposium... Proceedings Oxford 1970 Jan Edited by R. M.S. Smellie Biochemical society.Symposia, 31 Academic press,London,1970 Molecular enzymology and protein group of the Biochemical society and the Chemical society
Chem 18.2745

MOLECULAR MECHANISMS OF TEMPERATURE ADAPTATION a symposium Berkeley,Calif. 1965 Dec 27-29 American association for the advancement of science Edited by C.Ladd Prosser American association for the advancement of science.Publication, 84 American association for the advancement of science,Washington,D.C.,1967
Gen 34.0538

MOLECULAR MODIFICATION IN DRUG DESIGN :a symposium at the 145th meeting of the American chemical society New York 1963 Sep 9-10 American chemical society Edited by Robert F. Gould American chemical society.Advances in chemistry series, 45 American chemical society,Washington,D.C.,1964
Radioth 35.0429

MOLECULAR PROPERTIES OF DRUG RECEPTORS :a Ciba foundation symposium London 1970 Jan 27-29 Ciba foundation Edited by Ruth Pater and Maeve O'Connor Churchill,London, 1970
An 32.5477

MOLECULAR RELAXATION PROCESSES Symposium on relaxation methods in relation to molecular structure Lectures and papers Aberystwyth 1965 Jul 7-9 Chemical society Academic press,London,1966
Pha 16.0369

MOLECULAR RELAXATION PROCESSES :a symposium Lectures and synopses of papers Aberystwyth 1965 Jul 7-9 Chemical society Chemical society.Special publication,20 Chemical society;Academic press,London;New York,1966
Chem 18.1224

MOLECULAR SPECTROSCOPY 5th :general and introductory lectures presented at the fifth European congress on molecular spectroscopy Amsterdam 1961 May 29-Jun 3 International union of pure and applied chemistry.Physical chemistry section,and,Royal Netherlands chemical society Co-sponsored by the Netherlands physical society Butterworths, London,1962 Reprinted from 'Pure and applied chemistry',vol 4,no 1
Chem 18.2315

MOLECULAR STRUCTURE Spectroscopy and molecular structure,and optical methods of investigating cell structure :a general discussion Cambridge 1950 Sep 25-28 Faraday society Faraday society.Discussions, 9 Aberdeen,1951
Bal 39.0230

MOLECULAR STRUCTURE AND BIOLOGICAL SPECIFICITY a symposium Washington,D.C. 1955 Oct 28-29 American institute of biological sciences Edited by Linus Pauling and Harvey A. Hana Sponsored by the United States. Office of naval research American institute of biological sciences.Publications, 2 American institute of biological sciences, Washington,D.C.,1957
An 32.3156

MOLECULAR STRUCTURE AND BIOLOGICAL SPECIFICITY a symposium Papers Washington,D.C. 1955 Oct 28-29 United States.Office of naval research Edited by Linus Pauling and Harvey A. Itano Arranged by the American institute of biological sciences American institute of biological sciences.Publication,2 American institute of biological sciences,Washington,D. C.,1957
Chem 18.2137

MOLECULAR STRUCTURE AND FUNCTIONAL ACTIVITY OF NERVE CELLS :a symposium Washington, D.C. 1955 Jun 3-4 American institute of biological sciences Edited by Robert G. Grenell and L.J. Mullins Sponsored by the United States.Office of naval research American institute of biological sciences. Publications, 1 American institute of biological sciences,Washington,D.C.,1956
An 32.4348

MOLECULAR STRUCTURE AND SPECTROSCOPY International symposium on molecular structure and spectroscopy Tokyo 1962 Sep 10-14 International union of pure and applied chemistry Science council of Japan,Tokyo, 1962
Chem 18.2639

MOLEKULARE BIOLOGIE DES MALIGNEN WACHSTUMS : colloquium Mosbach 1966 Apr 21-23 Edited by H. Holzer and A.W. Holldorf Gesellschaft fur physiologische chemie. Colloquium, 17 Springer,Berlin,1966
Radioth 35.0142

MOLEKULARE STRUKTUR UND STRAHLENWIRKUNG : jahrestagung der Deutschen gesellschaft fur biophysik Tagungsbericht Hanover 1966 Jan 2-3 Deutsche gesellschaft fur biophysik Edited by H. Glubrecht illus. Thieme, Stuttgart,1968
Radioth 35.1423

MOLYBDENUM Metal molybdenum The Technology of molybdenum and its alloys :a symposium Proceedings Washington,D.C. 1956 Sep 18-19 United States.Office of naval research Edited by Julius J. Harwood American society for metals,Cleveland,Ohio,1958
Met 25.0546

MONACO 1906 Congres international d'anthropologie et d'archaeologie prehistoriques 13th Compte rendu Tome 1-2 2 vols Imprimerie de Monaco,Monaco, 1907-08
An 32.2638

MONACO 1920 Congres de l'alpinisme Comptes rendus Tomes 1-2 22cm 2 vols Paris,1921
Sco 14.1300

MONACO 1962 Human displacements: measurement,methodological aspects Entretiens de Monaco en sciences humaines 1ere session Centre international d'etude des problemes humains Edited by Jean Sutter Hachette,Paris,1962 Text in English and French
Geog 13.1566

MONACO 1967 Radioactive dating and methods of low level counting :symposium proceedings organized by the International atomic energy agency International atomic energy agency, Vienna,1967
Geol 8.4478

MONACO 1967 Radioactive dating and methods of low-level counting :a symposium Proceedings International atomic energy agency Joint commission on applied radioactivity International atomic energy agency.Proceedings series International atomic energy agency,Vienna,1967
Radioth 35.1683

MONACO 1968 Effects of radiation on cellular proliferation and differentiation Symposium on the effects of radiation on cellular proliferation and differentiation Proceedings International atomic energy agency Organized in co-operation with the Joint commission on applied radioactivity International atomic energy agency,Vienna,1968
Radioth 35.1760

MONKS WOOD 1965 Pesticides in the environment and their effects on wildlife :an advanced study Proceedings North Atlantic treaty organization Edited by N.W. Moore Journal of applied ecology.Supplement, 3 311p Blackwells,Oxford,1966
Bot 42.1983

MONKS WOOD 1965 Pesticides in the environment and their effects on wildlife :an advanced study institute sponsored by the North Atlantic treaty organization,Monks Wood experimental station,England Proceedings Edited by N.W. Moore Journal of applied ecology, 3,suppl. Blackwell,Oxford,1965
Bal 39.1676

MONOMOLECULAR LAYERS :a symposium Philadelphia 1951 Dec 27 American association for the advancement of science Edited by Harry Sobotka vii,207p American association for the advancement of science, Washington, D.C.,1954
Chem E 24.0932

MONONUCLEAR PHAGOCYTES Conference on mononuclear phagocytes Leiden 1969 Sep 2-5 Edited by Ralph van Furth Blackwell scientific,Oxford;Edinburgh,1970
An 32.5450

MONSOONS Symposium on monsoons of the world Proceedings New Delhi 1958 Feb 19-21 Meteorological office,India Under the auspices of the World meteorological organization x,270p Manager of publications,Delhi,1960
Nap 11.0284

MONSOONS OF THE WORLD New Delhi 1958 Feb 19-21 India.Meteorological department Delhi,1960
Geog 13.6752

MONTANA STATE COLLEGE Symposium on listeric infection 2nd Papers and discussions Bozeman,Montana 1962 Aug 29-31 Edited by M. L. Gray With the support of the National institute of allergy and infectious diseases Montana state college,Bozeman,Montana,1963
Path 30.2590

MONTE CARLO METHOD :a symposium Proceedings Los Angeles 1949 Jun 29-Jul 1 National bureau of standards Edited by A.S. Householder and others Sponsored by the Rand corporation National bureau of standards. Applied mathematics series, 12 United States government printing office,Washington,D. C.,1951
Radioth 35.0586

MONTE CARLO METHODS DEUCE users' colloquium on Monte Carlo methods papers London 1960 May 17 English electric company DEUCE news, 52 21p English electric,Nelson, Staffs.,1960 Typescript
Math L 5.2410

MONTE CARLO METHODS Symposium on Monte Carlo methods Gainesville,Fla. 1954 Mar.16-17 University of Florida.Statistical laboratory Edited by Herbert A. Meyer Sponsored by the Wright air development center Wiley publications in statistics Bibliog. xvi, 382p 28cm John Wiley and sons,New York, 1956
Math 3.0208

MONTE CARLO METHODS Symposium on Monte Carlo methods Gainesville,Fla. 1954 Mar 16-17 University of Florida.Statistical laboratory Edited by Herbert A. Meyer Sponsored by Wright air development center Wiley publications in statistics bibliog xvi, 381p 28cm John Wiley and sons,New York, 1956
Math L 5.0746

MONTE CARLO METHODS symposium mimeograph Gainesville,Fla. 1954 Mar 16-17 University of Florida.Statistical laboratory Edited by Herbert A. Meyer Sponsored by the Wright air development center Wiley publications in statistics 382p 29cm John Wiley and sons, New York,1956
P. Math 2.2239

MONTEREY,CALIF. 1959 Lattice theory symposium American mathematical society Edited by R.P. Dilworth With the financial support of the National science foundation American mathematical society.Proceedings of symposia in pure mathematics, 2 viii,208p 26cm American mathematical society, Providence,R.I.,1961
P. Math 2.2030

MONTEREY,CALIF. 1960 Electrostatic propulsion... a selection of technical papers based mainly on a symposium of the American rocket society... American rocket society Edited by David B. Langmuir and others Progress in astronautics and rocketry, 5 Academic press,New York;London,1961
Eng 41.6888

MONTEREY,CALIF. 1969 Functional analysis : a symposium Proceedings Edited by Carroll O. Wilde bibliog. vii,162p 24cm Academic press,New York;London,1970
P Math 2.3866

MONTEREY,CALIF. 1970 Heat transfer and fluid mechanics institute 1970 Proceedings Edited by T. Sarpkaya bibliog.,illus. 370p 26cm Stanford university press,Stanford,1970
A Math 4.1649

MONTEVIDEO 1959 Brain mechanism and learning :a symposium Council for international organizations of medical sciences Edited by J.F. Delafresnaye Under the auspices of Unesco Blackwell,Oxford,1960 Under the joint auspices of Unesco and WHO
Psy 31.0608

MONTEVIDEO 1959 Brain mechanism and learning :a symposium Council for international organizations of medical sciences Under the joint auspices of Unesco World health organization Blackwell,Oxford, 1961
Bal 39.1604

MONTEVIDEO 1959 Brain mechanisms and learning :a symposium By J.F. Delafresnaye Council for international organizations of medical sciences Blackwell,Oxford,1961
An 32.4647

MONTEVIDEO 1959 Brain mechanisms and learnings :a symposium Edited by J.F. Pelafresnaye Organised by Council for international organisations of medical sciences Blackwell,Oxford,1961
VA 19.0236

MONTEVIDEO 1959 Oxytocin :an international symposium Proceedings Universidad de Uraguay.Facultad de medicina.Servicio de fisiologia obstetrica Edited by R. Caldeyro-Barcia and H. Heller Pergamon press,London, 1961
VA 19.0237

MONTEVIDEO 1959 Oxytocin :an international symposium Proceedings Universidad de Uruguay.Facultad de medicina.Servicio de fisiologia obstetrica Edited by R. Caldeyro-Barcia and H. Heller Pergamon,Oxford,1961
Pha 16.0082

MONTEVIDEO 1959 Oxytocin :an international symposium Proceedings Universidad de Uruguay.Facultad de medicina.Servicio de fisiologia obstetrica Edited by R. Caldeyro-Barcia and H. Heller Pergamon press,Oxford, 1961
Phys 20.1387

MONTEVIDEO 1959 Oxytocin :an international symposium Proceedings Universidad de Uruguay.Facultad de medicina.Servicio de fisiologica obstetrica Edited by R. Caldeyro-Barcia and H. Heller Pergamon,Oxford,1961
An 32.5322

MONTEVIDEO 1965 International symposium on nucleolus:its structure and function Proceedings National cancer institute Edited by W.S. Vincent and others National cancer institute.Monograph, 23 National cancer institute,Bethesda,Md.,1966
Bioch 33.0988

MONTEVIDEO 1965 International symposium on the nucleus;its structure and function Edited by W.S. Vincent and O.L. Miller National concer institute.Monographs, 23 U. S.government printing office,Washington,D.C., 1966
An 32.3439

MONTEVIDEO,URUGUAY 1965 The Nucleolus:its structure and function :international symposium Proceedings Edited by W.S. Vincent and O.L. Miller National cancer institute.Monographs, 23 National cancer institute,Bethesda,Md.,1966 In Spanish and English
Radioth 35.0533

MONTICELLO,ILL. 1955 Symposium on circuit analysis Proceedings University of Illinois 1955
Eng 41.5160

MONTICELLO,ILL. 1955 Ultrasound in biology and medicine :a symposium Edited by Elizabeth Kelly Sponsored by University of Illinois.Bioacoustics laboratory American institute of biological sciences.Publications, 3 American institute of biological sciences,Washington,D.C.,1957
An 32.0393

MONTICELLO,ILL. 1963-68 Allerton conference on circuit and system theory 1st-6th Proceedings Sponsored by the University of Illinois 6 vols University of Illinois,Evanston,Ill.,1963-68
Eng 41.5240

MONTPELLIER 1955 Colloque sur la diffusion Actes By J. Salvinien and others France. Ministere de l'air.Publications scientifiques et techniques de l'air.Notes techniques, 59 96p Service de documentation et d'information technique de l'aeronautique, Paris,1956
Chem E 24.0760

MONTPELLIER 1962 The Methodology of plant eco-physiology Edited by F.E. Eckardt Unesco.Arid zone research series, 25 Bibliog.,illus. 531p Unesco;H.M.S.O.,Paris; London,1965
Bot 42.1981

MONTPELLIER 1967 Congress of the International association for the scientific study of mental defiency 1st Proceedings International association for the scientific study of mental deficiency Edited by B.W. Richards Jackson,Reigate,1968 Title page and papers in English and French
Psy 28.0269

MONTPELLIER 1967 Studies on Abelian groups Symposium on the theory of Abelian groups Edited by B. Charles Bibliog. x,356p 25cm Dunod,Paris,1968
P Math 2.3507

MONTREAL 1953 International physiological congress 19th abstracts of communications 1953
Phys 20.2089

MONTREAL 1954 International congress of psychology 14th Proceedings North-Holland,Amsterdam,1955
Psy 31.3359

MONTREAL 1957 Geology of Canadian industrial mineral deposits Commonwealth mining and metallurgical congress 6th Papers Canadian institute of mining and metallurgy Montreal,1957
Min 10.0716

MONTREAL 1958 International congress of genetics 10th Proceedings Vol 1-2 University of Toronto press,Toronto,1959 Bound together
Gen 34.1006

MONTREAL 1958 International congress of genetics 10th Proceedings Vol 1-2 2 vols University of Toronto press,Toronto, 1959
Bot 42.4702

MONTREAL 1959 International botanical congress 9th Proceedings Vol 1-3 University of Toronto press,Toronto,1959
Bot 42.4703

MONTREAL 1959 International botanical congress 9th Proceedings 1959
BG 38.3130

MONTREAL 1959 McGill university conference on depression and allied states Papers McGill university.Department of psychiatry Canadian psychiatric association journal, 4, Special supplement, 1 Canadian psychiatric association journal,Ottawa,1959
Psy 28.0086

MONTREAL 1961 The Wood chemistry symposium Proceedings International union of pure and applied chemistry.Applied chemistry section Pure and applied chemistry, 5,nos 1-2 illus Butterworths,London,1962 Added title page in French "Colloque sur la chimie du bois".
Bioch 33.1212

MONTREAL 1961 The World congress of psychiatry 3rd Proceedings Canadian psychiatric association and McGill university 2 vols University of Toronto press;McGill university press,Toronto;Montreal, 1961 Conference languages:English,French, German and Spanish
Psy 31.2540

MONTREAL 1961 World congress of psychiatry Proceedings Vol 1-2 Canadian psychiatric association McGill university Edited by R.A. Cleghorn and others 2 vols University of Toronto press;McGill university press,Toronto;Montreal,1961 Papers in English,French,German and Spanish
PGMS 29.0322

MONTREAL 1962 Problemes aux limites dans les equations aux derivees partielles By Jacques L. Lions Societe mathematique du Canada Montreal.University.Seminaire de mathematiques superieures, 1 176p 28cm Universite de Montreal,Montreal,1965
P. Math 2.1283

MONTREAL 1962 Recent progress in microbiology :symposia presented at the 8th international congress for microbiology International union of microbiological societies Canadian society of microbiologists Edited by N.E. Gibbons University of Toronto press,Toronto,1963
Gen 34.0701

MONTREAL 1963 International symposium on classical and contagious discrete distributions proceedings Canadian mathematical congress Edited by Ganapati P. Patil Supported by the National research council of Canada Bibliog. xiv,552p 27cm Statistical publishing society,Calctta,1965
Math 3.0709

MONTREAL 1963 Promenades aleatoires et mouvement brownien :introduction a la theorie des probabilites mimeograph By Anatole Joffe North Atlantic treaty organization Societe mathematique du Canada Montreal. University.Seminaire de mathematiques superieures, 7 vii,143p 28cm Universite de Montreal,Montreal,1964
P. Math 2.2145

MONTREAL 1963 Series de Fourier aleatoires mimeograph By Jean-Pierre Kahane North Atlantic treaty organization Societe mathematique du Canada Montreal.University. Seminaire de mathematiques superieures, 4 vii,174p 28cm Universite de Montreal, Montreal,1963
P. Math 2.2147

MONTREAL 1964 Categories non-abeliennes By Peter Hilton North Atlantic treaty organization Societe mathematique du Canada Montreal.University.Seminaire de mathematiques superieures, 10 138p 28cm Universite de Montreal,Montreal,1964
P. Math 2.0265

MONTREAL 1964 Homotopie et cohomologie manuscript By Beno Eckmann North Atlantic treaty organization Societe mathematique du Canada University of Montreal.Seminaire de mathematiques superieures, 11 129p 28cm Universite de Montreal,Montreal,1965
P. Math 2.0297

MONTREAL 1964 Integration dans les groupes topologiques By Geoffrey Fox North Atlantic treaty organization Societe mathematique du Canada Montreal.University. Seminaire de mathematiques superieures, 12 357p 28cm Presses de l'universite de Montreal,Montreal,1966
P. Math 2.2385

MONTREAL 1964 Social communication among primates An International symposium on communication and social interactions in primates Edited by Stuart A. Altmann University of Chicago press,Chicago,1967
Bal 39.1627

MONTREAL 1964 Theorie des valuations mimeograph By Paulo Ribenboim North Atlantic treaty organization Societe mathematique du Canada 313p 28cm Presses de l'universite de Montreal,Montreal,1964
P. Math 2.1974

MONTREAL 1965 International conference on soil mechanics and foundation engineering 6th Proceedings 3 vols University of Toronto press,Toronto,1965
Eng 41.3143

MONTREAL 1965 The Use of radioautography in investigating protein synthesis :a symposium Edited by C.P. Leblond and Katherine Brehme Warren Sponsored by the International society for cell biology International society for cell biology. Symposia, 4 Academic press,New York;London, 1965
Bioch 33.0286

MONTREAL 1966 Logical systems containing only a finite number of symbols By Leon Henkin North Atlantic treaty organization Societe mathematique du Canada Universite de Montreal.Seminaire de mathematiques superieures, 21 illus. 48p 28cm Presses de l'Universite de Montreal,Montreal, 1967
P Math 2.3740

MONTREAL 1967 Parathyroid hormone and thyrocalcitonin (calcitonin) Parathyroid conference 3rd Proceedings Edited by Roy V. Talmadge and others Excerpta medica. International congress series, 159 Excerpta medica,Amsterdam,1968
Inv Med 37.0211

MONTREAL 1967 Parathyroid hormone and thyrocalcitonin (calcitonin) :parathyroid conference 3rd Proceedings Edited by Roy V. Talmage and others Excerpta medica international congress series,159 Illus. viii,535p Excerpta medica foundation, Amsterdam,1968
Pha 16.0099

MONTREAL 1967 The Physics of selenium and tellurium :international symposium Proceedings Selenium-Tellurium development association Edited by W.Charles Cooper Bibliog,photos,diagrs ix,380p 23cm Pergamon press,Oxford,1969
Cav 7.2670

MONTREAL.UNIVERSITY.SEMINAIRE DE MATHEMATIQUES SUPERIEURES, 1 Problemes aux limites dans les equations aux derivees partielles Montreal 1962 By Jacques L. Lions Societe mathematique du Canada 176p 28cm Universite de Montreal,Montreal,1965
P. Math 2.1283

MONTREAL.UNIVERSITY.SEMINAIRE DE MATHEMATIQUES SUPERIEURES, 4 Series de Fourier aleatoires mimeograph Montreal 1963 By Jean-Pierre Kahane North Atlantic treaty organization Societe mathematique du Canada vii,174p 28cm Universite de Montreal, Montreal,1963
P. Math 2.2147

MONTREAL.UNIVERSITY.SEMINAIRE DE MATHEMATIQUES SUPERIEURES, 7 Promenades aleatoires et mouvement brownien :introduction a la theorie des probabilites mimeograph Montreal 1963 By Anatole Joffe North Atlantic treaty organization Societe mathematique du Canada vii,143p 28cm Universite de Montreal,Montreal,1964
P. Math 2.2145

MONTREAL.UNIVERSITY.SEMINAIRE DE MATHEMATIQUES SUPERIEURES, 10 Categories non-abeliennes Montreal 1964 By Peter Hilton North Atlantic treaty organization Societe mathematique du Canada 138p 28cm Universite de Montreal,Montreal,1964
P. Math 2.0265

MONTREUX 1964 Symposium on high-energy electrons Proceedings Edited by A. Zuppinger and G. Poretti Springer,Berlin, 1965
Radioth 35.1662

MONTREUX 1964 Symposium on high-energy electrons Proceedings European association of radiologists Edited by A. Zuppinger and G. Poretti Springer,Berlin,1965
Chem 18.0123

MOON Symposium on the moon proceedings Pulkovo 1960 Dec.5-11 International astronomical union Edited by Zdenek Kopal and Zdenka Kadla Mikhailov International astronomical union.Symposia, 14 map 517p Academic press,London,1962
Obs 6.0972

The MOON :a symposium Pulkovo 1960 Dec. International astronomical union Edited by Zdenek Kopal and Zdenka Kadla Mikhailov Held at Glavnaya astronomicheskaya observatoriya,Pulkovo International astronomical union.Symposium, 14 571p Academic press,London;New York,1962
A Math 4.1153

MOON AND PLANETS International space science symposium 7th Papers Vienna 1966 May 10-18 International astronomical union Edited by A. Dollfus Co-sponsored by the Committee on space research 313p North-Holland,Amsterdam,1967 Papers of one session of the symposium
Obs 6.3206

MOON AND PLANETS,2 Cospar plenary meeting : joint open meeting of working groups 1,2 and 5 10th London 1967 Jul 26-27 Committee on space research Royal society of London Edited by A. Dollfus 196p North-Holland, Amsterdam,1968
Obs 6.3419

MOORE Eliakim Hastings and others The New Haven mathematical colloquium lectures New Haven 1906 September 3-8 American mathematical society.13th summer meeting Under the auspices of Yale university Americian mathematical society.Colloquium, 5 Yale university press,New Haven,1910
Philos 1.0198

MORE MINOR HORRORS By Arthur E. Shipley illus. London,1916
Mol 45.0345

MORPHOGENESE :colloque international Strasbourg 1949 Jul 4-11 Centre national de la recherche scientifique Centre national de la recherche scientifique.Colloques internationaux, 28 C.N.R.S.,Paris,1951
An 32.2463

MORPHOGENESIS Growth,in relation to differentiation and morphogenesis Cambridge 1947 Jul Society for experimental biology Society for experimental biology.Symposia, 2 Cambridge university press,Cambridge,1948
Chem 18.2652

MORPHOGENESIS International congress of biochemistry 4th Proceedings Vienna 1958 Sep 1-6 Vol 6: symposium 6 - biochemistry of morphogenesis International union of biochemistry Edited by W.J. Nickerson I.U.B.S.symposium series, 8 Pergamon press,London,1959
Gen 34.0482

MORPHOGENESIS International congress of biochemistry 4th Proceedings Vienna 1958 Sep 1-6 Vol 6: symposium 6 - biochemistry of morphogenesis International union of biochemistry Edited by W.J. Nickerson I.U.B.symposium series, 8 Pergamon,London,1959 Added title page in French and German
Bioch 33.1320

MORSKIE MLEKOPITAYUSHCHIE Sbornik uklyuchaet materialy tretego vsesoyuznogo soveshchaniya po morskim mlekopitayushchim Vladivostok 1966 Akademiya nauk S.S.S.R.Ministerstvo po rybnogo khozyaistva S.S.S.R.Ikhtiologicheskaya komissiya Edited by A.V.A. Arsenev and others 342p 27cm Izdatelstvo 'Nauka', Moscow,1969
Sco 14.8244

MOSACH 1964 Immunchemie :colloquium Gesellschaft fur physiologische chemie. Colloquium, 15 Springer,Berlin,1965
Radioth 35.0145

MOSBACH 1962 Induktion und morphogenese Gesellschaft fur physiologische chemie Gesellschaft fur physiologische chemie. Colloquium, 13 Springer,Berlin,1963
Gen 34.0857

MOSBACH 1966 Molekulare biologie des malignen wachstums :colloquium Edited by H. Holzer and A.W. Holldorf Gesellschaft fur physiologische chemie.Colloquium, 17 Springer,Berlin,1966
Radioth 35.0142

MOSBACH 1968 Biochemie des sauerstoffs : colloquium Gesellschaft fur biologische chemie Edited by B. Hess and Hj. Staudinger Gesellschaft fur biologische chemie.Colloquia, 19 illus Springer,Berlin,1968 In English and German
Bioch 33.1074

MOSBACH 1969 Inhibitions in cell research : colloquium Gesellschaft fur biologische chemie Edited by T. Bucher and H. Sies Gesellschaft fur biologische chemie.Colloquia, 20 Springer,Berlin,1969
Bioch 33.1908

MOSBACH 1969 Inhibitions in cell research : colloquium Gesellschaft fur biologische chemie Edited by T. Bucher and H. Sies Gesellschaft fur biologische chemie.Colloquia, 20 Springer,Berlin,1969
Bot 42.1758

MOSBACH 1970 Mammalian reproduction : colloquium Gesellschaft fur biologische chemie Edited by H. Gibian and E.J. Motz Gesellschaft fur biologische chemie.Colloquia, 21 Springer,Berlin,1970
An 32.5461

MOSBACH 1970 Mammalian reproduction : colloquium Gesellschaft fur biologische chemie Edited by H. Gibian and E.J. Plotz Gesellschaft fur biologische chemie.Colloquia, 21 Springer,Heidelberg;New York,1970
Inv Med 37.0088

MOSBACH,BADEN 1957 Neuere ergebnisse aus chemie und stoffwechsel der kohlenhydrate colloquium Gesellschaft fur biologische chemie Gesellschaft fur biologische chemie. Colloquia, 8 illus Springer,Berlin,1958
Bioch 33.2347

MOSBACH BADEN 1956 Chemie und stoffwechsel von binde-und knochengewebe :colloquium Gesellschaft fur physiologische chemie Gesellschaft fur physiologische chemie. Colloquium, 7 Springer,Berlin,1956
An 32.3221

MOSCOW Brain reflexes :the international conference dedicated to the centenary celebration of the publication of I.M. Sechenov's book "Brain reflexes" Edited by E. A. Asratyan Sponsored by Akademiya nauk S.S. S.R. Progress in brain research, 22 Elsevier,Amsterdam,1968 Sponsored also by the International brain research organization
An 32.4243

MOSCOW 1892 Congres internationale d'anthropologie et d'archeologie prehistorique et de zoologie 1e-2e partie: materiaux... concernant les expositions,les excursions et les rapports sur des questions touchant les congres illus.,pl. Moscow,1893 2 volumes bound together
Phys 20.0376

MOSCOW 1935 and LENINGRAD 1935 International physiological congress 15th Report Edited by B.E. Zbarskii and V.M. Kaganov Moscow;Leningrad,1936
Phys 20.2243

MOSCOW 1943 Obshchee sobranie... Akademiya nauk S.S.S.R. Akademiya nauk S.S.S. R,Moscow,1944
Philos 1.0031

MOSCOW 1947 Obshchee sobranie... posvyashchennie trudtsatiletiyu velikoi oktyabr'skoi sotsialistichesk revolyutsii Akademiya nauk S.S.S.R Akademiya nauk S.S.S. R,Moscow,Leningrad,1948
Philos 1.0032

MOSCOW 1948 The Situation in biological science :proceedings of the Lenin academy of agricultural sciences of the U.S.S.R., session. verbatim report By Trefim Denisovich Lysenko and others Vsesoyuznaya akademiya selsko-khozyaistvennuikh nauk imeni Lenina Foreign languages publishing house,Moscow,1949
Bal 39.3918

MOSCOW 1948 The Situation in biological science verbatim report Proceedings Vsesoyuznaya akademiya selskokozyaistennyk nauk imeni V.I.Lenina Foreign languages publishing house,Moscow,1949 Translated from the Russian
Gen 34.1210

MOSCOW 1951 Materialy paleontologicheskogo soveschaniya po paleozoyu :ochorednye zadachi paleontologii v dele pomoshchi praktike Edited by T.G. Sarycheva Akademiya nauk S.S. S.R.,Moscow,1953
Geol 8.0705

MOSCOW 1953 Soveshchaniya po voprosam kosmogonii, proiskhozhdenie kosmicheskikh luchei 3 trudy Edited by Viktor Amazaspovich Ambartsumyan 319p Akademia Nauk SSSR,Moscow,1954
Obs 6.0060

MOSCOW 1956 Conference Konferentsiya puti razvitiya sovetskogo matematicheskogo mashchinostroeniya i priborostroeniya : plenarnye zasedaniya illus. 131p 22cm Moscow,1956
Math L 5.0774

MOSCOW 1956 Materially po obshchemu merzlotovedeniyu :mezhduvedomstvennoe coveshchanie po merzlotovedeniyu 7 Papers Akademiya nauk S.S.S.R.Institut merzlotovedeniya im V.A.Obrucheva Edited by V.F. Zhukov and I.Ya. Baranov 271p 26cm Moscow,1959 Papers read at the 7th interdepartmental conference on permafrost studies, Moscow,1956
Sco 14.0399

MOSCOW 1956 Materialy po fizike i mekhanike merzlykh gruntov : mezhduvedomstvennoe soveshchanie po merzlotovedeniyu 7 Papers Akademiya nauk S.S.S.R.Institut merzlotovedeniya im V.A. Obrucheva Edited by N.A. Tsytovich 107p 26cm Moscow,1959
Sco 14.0398

MOSCOW 1956 Sessiya Akademii nauk SSSR po nauchnym problemam avtomatizatsii proizvodstva plenarnye zasedaniya Akademiya nauk S.S.S.R. Edited by D.Ya. Libenson and others 271p 22cm Akademiya nauk SSSR,Moscow,1957
Math L 5.0775

MOSCOW 1956 Soviet electrochemistry Conference on electrochemistry 4th Proceedings Vol 1-3 Translated by Consultants bureau from the Russian New York, 1961
Met 25.1821

MOSCOW 1956- Growth of crystals Conference on crystal growth 1st- Proceedings 1- Institut kristallografii,Moscow Edited by A.V. Shubnikov and N.N. Sheftal Translated by Consultants' bureau from the Russian Rost kristallov, 1- Consultants bureau,inc.,New York,1958- Includes interim reports published between conferences.Held as part of the international crystallographic congresses
Met 25.1508

MOSCOW 1957 Fonctions d'une variable complexe:problemes contemporaines Colloque de l'U.R.S.S. 3rd selection of papers presented Edited by A.I. Markushevitch Translated by L. Nicolas from the Russian Monographies internationales de mathematiques modernes, 1 271p 24cm Gauthier-Villars, Paris,1962
P. Math 2.1557

MOSCOW 1957 Materialy vsesoyuznogo soveshchaniya po izuchenoyu chetvertichnogo perioda Tom 3: chetvertichnye otlozheniya aziatskoi chasti S.S.S.R. Akademiya nauk,S.S. S.R.Kommisiya po izucheniyu chetvertichnogo perioda Edited by G.F. Lungergauzen and others illus,maps 442p 27cm Izdatelstvo Akademii nauk SSSR,Moscow,1961
Sco 14.4765

MOSCOW 1957 Origin of life on the earth The International symposium on the origin of life on the earth 1st Proceedings Akademiya nauk S.S.S.R. Edited by A.I. Oparin and others Organized under the auspices of the International union of biochemistry I.U.B.symposium series, 1 Pergamon,London,1959 "English-French-German edition edited for the International union of biochemistry by F.Clark and R.L.M. Synge".
Bioch 33.0727

MOSCOW 1957 Radiobiology All-union scientific and technical conference on the application of radioactive isotopes :a portion of the proceedings...in English translation Consultants bureau,New York,1959
Radioth 35.1731

MOSCOW 1957 The Origin of life on the earth Reports on the international symposium Edited by A.I. Oparin and others Publishing house of the Academy of sciences of the U.S.S. R.,n.p.,c1957
Bal 39.0990

MOSCOW 1958 Sbornik materialov rasshirennogo soveshchaniya rabochei gruppy po glyatsiologii Sovetskogo mezhduvedomstvennogo komiteta mezhdunarodnogo geofizicheskogo goda Edited by G.A. Avsyuk 165p 20cm Moscow, 1959
Sco 14.0514

MOSCOW 1958 Speleologiya i karstovedenie Soveshchanii po speleologiv i karstovedeni Moskovskoe obshchestvo ispytatelei prirody Moscow,1959
Philos 1.1382

MOSCOW 1958 The Hertzsprung-Russell diagram :a symposium Proceedings International astronomical union Edited by Jesse L. Greenstein International astronomical union.Symposium, 10 128p Paris,1959 Reprinted from Annales d'astrophysique 1959,supplements.Fascicule.No. 8
Obs 6.0687

MOSCOW 1958 The Rotation of the earth and atomic time standards :a symposium International astronomical union Edited by Dirk Brouwer International astronomical union.Symposium, 11 1959 Reprinted from the Astronomical journal,vol 64,no 1268
Obs 6.3199

MOSCOW 1960 Initial effects of ionizing radiations on cells International symposium on primary and initial effects of ionizing radiations on living cells Papers: discussions Edited by R.J.C. Harris Sponsored by the Akademiya nauk S.S.S.R. Academic press,London;New York,1961
An 32.3277

MOSCOW 1960 Initial effects of ionizing radiations on cells The International symposium on primary and initial effects of ionizing radiations on living cells Papers and discussions Edited by R.J.C. Harris Academic press,London;New York,1961 A symposium supported by Unesco and the IAEA and sponsored by the Academy of sciences of the U. S.S.R.
Bioch 33.0961

MOSCOW 1960 The Initial effects of ionizing radiations on cells :a symposium Unesco International atomic energy agency Edited by R.J.C. Harris Sponsored by the Akademiya nauk S.S.S.R. Academic press, London;New York,1961
Gen 34.2222

MOSCOW 1960 The Initial effects of ionizing radiations on cells :a symposium Unesco International atomic energy agency Edited by R.J.C. Harris Sponsored by the Akademiya nauk S.S.S.R. Academic press, London;New York,1961
Radioth 35.1127

MOSCOW 1961 International congress of biochemistry 5th Proceedings Vol 1: biological structure and function at the molecular level International union of biochemistry Edited by V.A. Engelhardt I.U. B.symposium series, 21 Pergamon;PWN-Polish scientific publishers,London;Warsaw,1963
Bioch 33.1330

MOSCOW 1961 International congress of biochemistry 5th Proceedings Vol 2: functional biochemistry of cell structures International union of biochemistry Edited by O. Lindberg I.U.B.symposium series, 22 Pergamon;PWN-Polish scientific publishers,New York;Warsaw,1963
Bioch 33.1331

MOSCOW 1961 International congress of biochemistry 5th Proceedings Vol 3: evolutionary biochemistry International union of biochemistry I.U.B.symposium series, 23 Pergamon;PWN-Polish scientific publishers,New York;Warsaw,1963
Bioch 33.1332

MOSCOW 1961 International congress of biochemistry 5th Proceedings Vol 4: molecular basis of enzyme action and inhibition International union of biochemistry Edited by P.A.E. Desnuelle I. U.B.symposium series, 24 Pergamon;PWN-Polish scientific publishers,Oxford;Warsaw, 1963
Bioch 33.1333

MOSCOW 1961 International congress of biochemistry 5th Proceedings Vol 5: intracellular respiration:phosphorylating and non-phosphorylating oxidation reactions International union of biochemistry Edited by E.C. Slater I.U.B.symposium series, 25 Pergamon;PWN-Polish scientific publishers, Oxford;Warsaw,1963
Bioch 33.1334

MOSCOW 1961 International congress of biochemistry 5th Proceedings Vol 6: mechanism of photosynthesis International union of biochemistry Edited by H. Tamiya I.U.B.symposium series, 26 Pergamon;PWN-Polish scientific publishers,Oxford;Warsaw, 1963
Bioch 33.1335

MOSCOW 1961 International congress of biochemistry 5th Proceedings Vol 7: biosynthesis of lipids International union of biochemistry Edited by G. Popjak I.U.B. symposium series, 27 Pergamon;PWN-Polish scientific publishers,Oxford;Warsaw,1963
Bioch 33.1336

MOSCOW 1961 International congress of biochemistry 5th Proceedings Vol 8: biochemical principles of the food industry International union of biochemistry Edited by V.L. Kretovich and E. Pijanowski I.U.B. symposium series, 28 Pergamon;PWN-Polish scientific publishers,Oxford;Warsaw,1963
Bioch 33.1337

MOSCOW 1961 International congress of biochemistry 5th Proceedings Vol 9: plenary sessions and abstracts of papers International union of biochemistry I.U.B. symposium series, 29 Pergamon;PWN-Polish scientific publishers,New York;Warsaw,1963
Bioch 33.1338

MOSCOW 1961 and KAZAN 1964 New techniques in astronomy :conferences Proceedings Edited by H.C. Ingrao 446p Gordon and Breach,New York,1971
Obs 6.3598

MOSCOW 1962 International cancer congress 8th abstracts of papers Moscow,1962
Radioth 35.0785

MOSCOW 1963 Geometriceskogo seminara trudy Tom 1 Organised by the Akademiya nauk S.S.S.R.Institut nauchnoi informatsii.Otdel matematiki 456p 22cm VINITI,Moscow,1966
P. Math 2.0699

MOSCOW 1963 Stroenie i razvitie zemnoii kory conference proceedings Edited by P.N. Kropokkim Akademiya nauk SSSR,Mowcow,1964 Includes English title-page,Structure and evolution of the crust.(Proceedings of a conference on the problems of tectonics in Moscow)
Geol 8.1367

MOSCOW 1963 Voprosy sravnitelnoi tektoniki drevnikh platform conference proceedings Edited by A.A. Bogdanov and others Akademiya nauk SSSR,Moscow,1964 Includes English title-page:Problems of comparative tectonics of old platforms (Proceedings of the conference on the problems of tectonics in Moscow)
Geol 8.1366

MOSCOW 1965 Atmospheric turbulence and radio wave propagation Atmosfernaya turbulentnostui i rasprostranenie radioboli : an international colloquium Proceedings Edited by A.M. Yaglom and V.I. Tatarsky 370p Moscow,1967
A Math 4.1307

MOSCOW 1965 International congress of pure and applied chemistry 20th Lectures International union of pure and applied chemistry and Akademiya nauk S.S.S.R. Butterworths,London,1965
Chem E 24.1274

MOSCOW 1966 International conference on low temperature physics 10th Proceedings Vol 1: properties of helium International union of pure and applied physics Akademiya nauk S.S.S.R. Edited by M.P. Malkov 552p 22cm Moscow,1967
Cav 7.3030

MOSCOW 1966 International conference on low temperature physics 10th Proceedings Vol 3: the electronic properties of metals International union of pure and applied physics Akademiya nauk S.S.S.R. 404p 22cm moscow,1967
Cav 7.3032

MOSCOW 1966 International conference on low temperature physics 10th Proceedings Vol 4: antiferromagnetism International union of pure and applied physics Akademiya nauk S.S.S.R. Edited by M.P. Malkov 361p 22cm Moscow,1967
Cav 7.3033

MOSCOW 1966 International congress of mathematicians 1st Proceedings Edited by I.G. Petrovsky Bibliog,illus 726p 22cm M.I.R.,Moscow,1968
P Math 2.3006

MOSCOW 1966 International congress of mathematicians abstracts of brief scientific communications and reports nos. 1-15 International mathematical union Moscow,1966
Math 3.1125

MOSCOW 1966 International congress on metallic corrosion 3rd Proceedings Vol 1-4 Edited by Ya.M. Kolotyrkin and others 4 vols Mir,Moscow,1969
Met 25.2662

MOSCOW 1966 International oceanographic congress 2nd Abstracts of papers Academy of sciences,U.S.S.R. Edited by A.P. Vinogradov Moscow,1966
Geod 9.0612

MOSCOW 1966 Osnovnye itogi izucheniya Antarktiki za 10 let :doklady vsesoyuznogo soveshchaniya po izucheniyu Antarktiki 1966 god Mezhduvedomstvennaya kommissiya po izucheniyu Antarktiki map Izdatelstvo 'Nauka',Moscow,1967
Sco 14.6068

MOSCOW 1966 Pyridoxal catalysis:enzymes and model systems The International symposium on chemical and biological aspects of pyridoxical catalysis 2nd Proceedings Edited by E.E. Snell and others sponsored by the International union of biochemistry I.U. B.Symposium series, 35 Interscience,New York,1968 First International symposium on chemical and biological aspects of pyridoxal catalysis published under title chemical and biological aspects of pyridoxical catalysis
Bioch 33.1038

MOSCOW 1968 International conference on the physics of semiconductors 9th Proceedings Vol 1-2 Akademiya nauk S. S.S.R. English edition 634p 27cm 2 vols Publishing house 'Nauka',Leningrad,1968
Cav 7.2955

MOSCOW 1968 Symposium on electronic phenomena in chemisorption and catalysis on semiconductors Edited by K. Hauffe and T. Wolkenstein De Gruyter,Berlin,1969
Chem 18.2698

MOSCOW 1968 Symposium on electronic phenomena in chemisorption and catalysis on semiconductors Edited by K. Hauffe and Th. Wolkenstein English edition De Gruyter, Berlin,1969 Held during the fourth international congress on catalysis
Met 25.2475

MOSKOVSKII KHIMIKO-TEKHNOLOGICHESKII INSTITUT IMENI D.I.MENDELEYEVA Electrometallurgy of chloride solutions All-union seminar on applied electrochemistry 5th Reports Dnepropetrovsk 1962 Oct 17-19 Edited by V. V. Stender Translated by Consultants bureau from the Russian Consultants bureau,New York, 1965
Met 25.1843

MOSKOVSKOE OBSHCHESTVO ISPYTATELEI PRIRODY Speleologiya i karstovedenie Soveshchanii po speleologiv i karstovedeni Moscow 1958 Dec. 17-18 Moscow,1959
Philos 1.1382

MOSSBAUER EFFECT Applications of the Mossbauer effect in chemistry and solid-state physics :report of a panel Vienna 1965 Apr 26-30 International atomic energy agency International atomic energy agency.Technical reports series,50 International atomic energy agency,Vienna,1966
Min 10.1131

The MOSSBAUER EFFECT AND ITS APPLICATION IN CHEMISTRY New York 1966 Sep 12 American chemical society Edited by E. Gould American chemical society.Advances in chemistry series,68 American chemical society,Washington,D.C.,1967 Symposium organized by C.W.Seidel.Chairman:R.H.Herber
Chem 18.1190

MOSSBAUER EFFECT METHODOLOGY Symposium on Mossbauer effect methodology 7th Proceedings New York 1971 Jan 31 New England nuclear corporation Edited by Irwin J. Gruverman xii,308p 23cm Plenum press, London,1971
Chem 18.2961

MOSSBAUER EFFECT METHODOLOGY Symposium on Mossbauer effect methodology 1st Proceedings New York 1965 Feb 2-5 New England nuclear corporation Edited by I.J. Gruverman Plenum press,New York,1965
Chem 18.2826

MOSSBAUER EFFECT METHODOLOGY Symposium on Mossbauer effect methodology 6th Proceedings New York 1970 Jan 25 New England nuclear corporation Edited by Irwin J. Gruverman viii,237p 23cm Plenum press, New York,1971
Chem 18.2805

The MOTION,EVOLUTION OF ORBITS,AND ORIGIN OF COMETS Leningrad 1970 Aug 4-11 International astronomical union Edited by G. A. Chebotarev and others International astronomical union.Symposium, 45 521p Reidel,Dordrecht,1972
TA 15.0687

The MOTION,EVOLUTION OF ORBITS,AND ORIGIN OF COMETS Leningrad 1970 Aug 4-11 International astronomical union Edited by G. A. Chebotarev and others International astronomical union.Symposium, 45 521p Reidel,Dordrecht,1972
Obs 6.3591

MOTIONS IN MOLECULAR CRYSTALS Oxford 1969 Sep 16-18 Faraday society Faraday society. Discussions, 48 Bibliog 224p 20cm Faraday society,London,1969
Cav 7.2718

MOTIVATION Current theory and research in motivation :a symposium Papers Lincoln, Nebraska 1953 Jan 15-16 Mar 26-27 By Judson S. Brown and others University of Nebraska.Department of psychology University of Nebraska press,Lincoln,Nebraska,1953
Psy 31.1857

MOTIVATION Nebraska symposium on motivation Lincoln,Nebraska 1962- 1962- University of Nebraska.Department of psychology University of Nebraska.Current theory and research in motivation symposium, 10- University of Nebraska press,Lincoln, Nebraska,1962-
Psy 31.1858

MOTOR INDUSTRY RESEARCH ASSOCIATION Road vehicle aerodynamics Symposium on road vehicle aerodynamics 1st Proceedings London 1969 Nov 6-7 Edited by A.J. Scibor-Rylski Organised by the City university. Department of aeronautics City university, London,1970 Organised in cooperation with the National physical laboratory,the Motor industry research association and the Institution of mechanical engineers
Eng 41.8336

MOTOR LUBRICANTS Symposium on motor lubricants New York 1933 Mar 8 American society for testing materials Society for automotive engineers American society for testing materials,Philadelphia,Pa.,1933
Eng 41.6264

MOTOR SKILLS Mechanisms of motor skill development Centre for advanced study in the developmental sciences study group on 'Mechanisms of motor skill development' Proceedings London 1968 Nov Ciba foundation Edited by Kevin Connolly Academic press,London;New York,1970 Being the 4th study group in a C.A.S.D.S. programme on 'The origins of human behaviour' held jointly with the Ciba foundation
Psy 31.3101

MOTOR VEHICLES Air pollution control in transport engines :a symposium Proceedings London 1971 Nov 9-11 Institution of mechanical engineers.Automobile division Advanced school engineers.Combustion engines group IME,London,1972
Eng 41.8645

MOTOR VEHICLES Vibration and noise in motor vehicles :a symposium Proceedings London 1971 Jul 6-7 Institution of emchanical engineers.Automobile division Advanced school of automotive studies IME,London,1972
Eng 41.8646

MOUNT HOLYOKE COLLEGE Cell,organism and milieu :a symposium South Hadley,Mass. 1958 Jun 9-11 Society for the study of development and growth Edited by Dorothea Rudnick Society for the study of development and growth.Symposia, 17 Ronald press,New York,1959
Bioch 33.1008

MOUNT TREMBLANT 1969 The Laurentian hormone conference 27th Proceedings Edited by E.B. Astwood Recent progress in hormone research, 26 Academic press,New York;London,1970
Inv Med 37.0265

MUCOPOLYSACCARIDES Ciba foundation symposium on the chemistry and biology of mucopolysaccarides London 1957 Apr 23-25 Edited by G.E.W. Wolstenholme and M. O'Connor Illus J.and A.Churchill,London,1958
Radioth 35.0057

MUCOPOLYSACCHARIDES Ciba foundation ssymposium on the chemistry and biology of mucopolysaccharides Proceedings London 1957 Apr 23-25 Ciba foundation Edited by G. E.W. Wolstenholme and Maeve O'Connor Churchill,London,1958
Bioch 33.1202

MUCOPOLYSACHARIDES The Biochemistry of mucopolysaccharides of connective tissue :a symposium London 1960 Feb 13 Biochemical society Edited by F. Clark and J.K. Grant Held at the Royal college of surgeons of England Biochemical society.Symposia, 20 Cambridge university press,Cambridge,1961
Bioch 33.1382

MULTI-FLOW SYMPOSIUM Reports and proceedings Philadelphia,Pa. 1963 Nov 17-22 American society of mechanical engineers. Hydraulics and heat-transfer division American society of mechanical engineers,New York,1963
Eng 41.6513

MULTI-PHASE FLOW SYMPOSIUM Papers Philadelphia,Pa. 1963 Nov 17-22 Edited by Norman J. Lipstein Presented at the winter annual meeting of the American society of mechanical engineers illus,diagrms 99p American society of mechanical engineers,New York,1963
Chem E 24.0241

MULTICOLOUR PHOTOMETRY Spectral classification and multicolour photometry Symposium on the spectral classification and multicolour photometry Proceedings Saltsjobaden 1964 Aug.17-21 International astronomical union Edited by K. Loden and others International astronomical union. Symposium, 24 383p Academic press,London; New York,1966
Obs 6.1092

MULTIPLE MOLECULAR FORMS OF ENZYMES 1st :a conference Papers New York 1961 Feb 1-3 New York academy of sciences Edited by Felix Wroblewski and others New York academy of sciences.Annals, 94,p 655-1030 New York, 1961
Bioch 33.1028

MULTIPLE MOLECULAR FORMS OF ENZYMES 2nd :a conference Papers New York 1966 Dec 1-3 New York academy of sciences Edited by Edward M. Weyer New York academy of sciences. Annals, 151,p 1-689 New York,1968
Bioch 33.1029

MULTIPLES FONCTIONS D'UN JARDIN BOTANIQUE : symposium international de Geneve Actes Geneva 1968 Jul 29-Aug 3 Jardin botanique de Geneve Edited by Jacques Miege Boissiera, 14 Geneva,1969 Symposium held to celebrate the 150th anniversary of the foundation of the botanic garden at Geneva
BG 38.3115

MULTIVARIABLE CONTROL SYSTEM DESIGN AND APPLICATION U.K.A.C. control convention 4th Manchester 1971 Sep 1-3 United Kingdom automation council Organized by the Institution of electrical engineers Institution of electrical engineers.Conference publication, 78 IEE,London,1971
Eng 41.8590

MULTIVARIATE ANALYSIS International symposium on multivariate analysis Proceedings Dayton,Ohio 1965 Jun 14-19 Edited by P.R. Krishnaiah Sponsored by United States.Air force.Aerospace research laboratories Academic press,New York,1966
Eng 41.2206

MULTIVARIATE ANALYSIS International symposium on multivariate analysis Proceedings Dayton,Ohio 1968 Jun 17-22 Edited by Parachuri R. Krishnaiah Held at the Wright state university xx,696p 24cm Academic press,New York;London,1969
Math S 3.1802

MULTIVARIATE ANALYSIS :international symposium Proceedings Dayton,Ohio 1965 Jun 14-19 Edited by Paruchuri R. Krishnaiah Sponsored by Aerospace research laboratories Bibliog, port xix,592p 23cm Academic press,New York,1966
Math 3.1159

MUNICH 1896 International congress of psychology 3rd Congress fur psychologie.Internationales organisations-comite Lehmann,Munich,1897 Papers in English,French,German and Italian
Psy 31.3391

MUNICH 1896 Internationaler congress fur psychologie 3rd Congress fur psychologie.Internationales organisations-comite Lehmann,Munich,1897 Papers in English,French,German and Italian
Psy 31.2518

MUNICH 1910-11 Die Abstammungslehre :zwolf gemeinverstandliche vortrage uber die deszendtheorie im licht der neuren forschung gehalten im Winter- semester 1910-11 im Munchner verein fur naturkunde By O. Abel and others Fischer,Jena,1911
An 32.0010

MUNICH 1956 History of German guided missiles development :AGARD seminar Agard Edited by T. Benecke and A.W. Quick AGARDograph, 20 1957
Eng 41.6877

MUNICH 1959 International congress of pure and applied chemistry main lectures 17th Lectures Vol 1-2: inorganic chemistry; biochemistry and applied chemistry International union of pure and applied chemistry 2 vols Butterworths,London,1960 Papers in English,French and German
Chem 18.1739

MUNICH 1960 International congress on microwave tubes 3rd Proceedings Edited by J. Wosnik Vieweg,Brunswick,1961
Eng 41.5547

MUNICH 1961 International conference on ionization phenomena in gases 5th proceedings Vol. 1 Edited by H. Maecker Organised by Verband Deutscher physikalischer gesellschaften Series in physics North Holland,Amsterdam,1962
A Math 4.0620

MUNICH 1961 International congress of neuropathology 4th Proceedings Vol 1-3 Edited by H. Jacob 3 vols Thieme, Stuttgart,1962 Papers in English,French, German and Spanish
An 32.4193

MUNICH 1962 Information processing 1962 IFIP congress 62 proceedings International federation for information processing Edited by Cicely M. Popplewell xvi,780p 30cm North-Holland publishing company,Amsterdam, 1962
Math L 5.0912

MUNICH 1963 Stellar and solar magnetic fields Symposium on stellar and solar magnetic fields International astronomical union Edited by R. Lust International astronomical union.Symposium, 22 460p North Holland,Amsterdam,1965
Obs 6.1148

MUNICH 1964 International symposium on the reactivity of solids 5th Deutsche Bunsen-gesellschaft fur physikalische chemie Edited by G.M. Schwab Sponsored by the International union of pure and applied chemistry Elsevier,Amsterdam,1965 Papers in English,French and German
Met 25.1851

MUNICH 1964 International symposium on the reactivity of solids 5th Plenary lectures International union of pure and applied chemistry,and,Deutsche bunsengesellschaft fur physikalische chemie Butterworths,London, 1965
Min 10.0662

MUNICH 1964 International symposium on the reactivity of solids 5th papers Edited by G.-M. Schwab Elsevier,Amsterdam,1965
Min 10.1278

MUNICH 1964 Reactivity of solids : international symposium 5th Proceedings International union of pure and applied chemistry.Division for physical chemistry Edited by G.-M. Schwab Organised by the Deutsche Bunsen-gesellschaft fur physikalische chemie Elsevier,Amsterdam,1965
Chem 18.0608

MUNICH 1964 The International congress of applied mechanics 11th Proceedings International committee for the congresses of applied mechanics Edited by Henry Gortler and Peter Sorger 1184p Springer-verlag, Berlin,1966 Dedicated to Richard Grammel
A Math 4.1279

MUNICH 1965 Microminiaturization in automatic control equipment and in digital computers :IFAC-IFIP symposium Proceedings International federation of automatic control International federation for information processing Edited by J. Berghammer Oldenbourg,Munich,1966
Eng 41.5580

MUNICH 1966 Displays for command and control centers :a conference Proceedings AGARD Edited by I.J. Gabelman AGARD. Conference proceedings, 23 Technivision services,Slough,1969 Proceedings of the 11th technical meeting of the AGARD avionics panel
Eng 41.5609

MUNICH 1967 New experimental techniques in propulsion and energetics research :a technical meeting Proceedings Agard Edited by David Andrews and Jean Surugue Agard conference proceedings, 38 Technivision,Slough,1970
Eng 41.8359

MUNICH 1969 Spectrum formation in stars with steady-state extended atmospheres International astronomical union.Commission 36 Edited by H.G. Grotl and P. Wellmann International astronomical union.Commission 36. Colloquium, 2 National bureau of standards. Special publication, 332 332p National bureau of standards,Washington,D.C.,1970
Obs 6.3633

MUNICH 1970 ALGOL 68 implementation IFIP working conference on ALGOL and its implementation Proceedings International federation for information processing Edited by J.E.L. Peck illus 375p North-Holland,Amsterdam,1971
Math L 5.3876

MUNICH 1970 International research conference on proteinase inhibitors 1st Proceedings Edited by H. Fritz and H. Tschesche De Gruyter,Berlin,1971
Bioch 33.2205

MUNICH 1970 New techniques in space astronomy International astronomical union Edited by F. Labuhn and R. Lust International astronomical union.Symposium, 41 418p Reidel,Dordrecht,1971
Obs 6.3590

MUNSTER 1958 Struktur und stoffwechsel des herzmuskels :symposium Edited by W.H. Hauss and H. Losse Thieme,Stuttgart,1959
An 32.3467

MURINE LEUKEMIA Conference on murine leukemia Papers Philadelphia,Pa. 1965 Oct 13-15 National cancer institute.Virology research branch and Albert Einstein medical center Edited by Marvin A. Rich and John B. Moloney National cancer institute. Monograph, 22 US government printing office, Washington,D.C.,1966
Path 30.2460

MUSCLE Symposium on muscle Budapest 1966 Sep 12-16 Edited by E. Ernst and F.B. Straub Symposia biologica hungarica, 8 Akademia Kaido,Budapest,1968
An 32.4011

Le MUSCLE :etude de biologie et de pathologie. Colloque Compte rendu Royaumont 1950 Aug 31-Sep 6 Council for international organizations of medical sciences L'expansion scientifique,1952 In English and French
Bioch 33.2245

MUSCLE :a symposium Proceedings Edmonton,Alberta 1964 Jun 1-4 University of Alberta.Faculty of medicine Edited by W.M. Paul and others Pergamon press,Oxford,1965 First symposium of the University of Alberta medical school held as part of the celebrations of the 50th year of this medical faculty
Bioch 33.2244

MUSCLE :symposium Proceedings Edmonton, Alta. 1964 Jun 1-4 University of Alberta. Faculty of medicine Edited by W.M. Paul and others Pergamon press,Oxford,1965 50th anniversary of the University of Alberta medical school
An 32.4007

Le MUSCLE:ETUDE DE BIOLOGIE ET DE PATHOLOGIE : colloque Compte rendu Royaumont 1950 Aug 31-Sep 6 Organise par la Conseil pour la co-ordination des congres internationaux des sciences medicales L'Expansion scientifique francaise,1950 In French and English
Phys 20.1557

MUSCLE AS A TISSUE :a conference Proceedings Philadelphia,Pa. 1960 Nov 3-4 Edited by Kaare Rodahl and Steven M. Horvath Held at the Lankenau hospital McGraw-Hill,New York, 1961
An 32.3999

MUSCLE CONTRACTION Biochemistry of muscle contraction :a conference Proceedings Dedham,Mass. 1962 May Edited by John Gergely Retina foundation.Institute of biological and medical sciences.Monographs and conferences, 2 Churchill,London,1964
Bioch 33.2231

MUSCLE RECEPTORS Symposium on muscle receptors :a meeting held...as part of the golden jubilee congress of the University of Hong Kong Proceedings Hong Kong 1961 Sep Edited by David Barker Financial support provided by the University of Hong Kong Hong Kong university press,Hong Kong,1962
Bal 39.1519

MUSCULAR AFFERANTS AND MOTOR CONTROL Nobel symposium 1st Proceedings Sodergarn, Sweden 1965 Jun Nobel institute for neurophysiology Edited by Ragnar Granit Wiley;Almqvist and Wiksell,New York;Stockholm, 1966
Psy 31.0250

MUSCULAR AFFERENT AND MOTOR CONTROL Nobel symposium 1st Proceedings Lidingo 1965 Jun Nobel foundation Edited by Ragnar Granit Wiley;Almqvist and Wiksell,New York; Stockholm,1966
An 32.4008

MUSCULAR AFFERENTS AND MOTOR CONTROL Nobel symposium 1st Proceedings Sodergarn 1965 Jun Nobel foundation Almqvist and Wiksell;Wiley,Stockholm;New York,1966
Phys 20.2242

MUSCULAR CONTRACTION :a conference New York 1947 May 30 By Alexander Sandow and others New York academy of sciences New York academy of sciences.Annals, 47 p665-930 New York,1947
Bal 39.1422

MUSEUM OF NATURAL HISTORY,REYKJAVIK and UNIVERSITY OF ICELAND North atlantic biota and their history :a symposium Reykjavik 1962 Jul 12-25 Edited by Askell Love and Doris Love Sponsored by NATO advanced study institutes program illus. Pergamon press,Oxford,1963
Geog 13.1211

MUTATION AS CELLULAR PROCESS :a Ciba foundation symposium London 1969 Feb 11-13 Ciba foundation Edited by G.E.W. Wolstenholme and M. O'Connor illus Chuchill,London,1969
Bioch 33.0891

MUTATIONS Conference on genetics 2nd Princeton,N.J. 1960 Oct 16-19 Edited by William J. Schull Sponsored by the Josiah Macy jr.foundation University of Michigan press,Ann Arbor,Mich.,1962
An 32.0361

MUTATIONS IN PLANT BREEDING Panel on co-ordination of research on the production and use of induced mutations in plant breeding Proceedings Vienna 1966 Jan 17-21 International atomic energy agency Food and agriculture organization International atomic energy agency.Panel proceedings series 2 vols International atomic energy agency, Vienna,1966-68
Radioth 35.0261

MUTATIONS IN PLANT BREEDING II :a research co-ordination meeting on the use of induced mutations in plant breeding Proceedings Vienna 1967 Sep 11-15 International atomic energy agency Food and agriculture organization International atomic energy agency.Panel proceedings series I.A.E.A., Vienna,1968
Radioth 35.0262

MYCOSIS Systemic mycoses :a Ciba foundation symposium Ibadan 1967 Mar 29-31 Ciba foundation Edited by G.E.W. Wolstenholme and Ruth Porter Churchill,London,1968
Phys 20.0998

MYCOTOXINS IN FOODSTUFFS :a symposium Proceedings Cambridge,Mass. 1964 Mar 18-19 Edited by Gerald N. Wogan Held at the Massachusetts institute of technology M.I.T. press,Cambridge,Mass.,1965 "This volume comprises the proceedings of an International symposium on mycotoxins in foodstuffs".
Bioch 33.0708

MYOCARDIAL CELL International symposium on the myocardial cell :structure,function and modification by cardiac drugs 3rd Heart association of southeastern Pennsylvania Edited by Stanley A. Briller and Hardley L. Conn University of Pennsylvania press, Philadelphia,Pa.,1966
Phys 20.2221

MYOCARDIAL CELL The Myocardial cell: structure,function and modification by cardiac drugs :international symposium 3rd Heart association of southeastern Pennsylvania Edited by Stanley A. Briller and Hardley L. Conn University of Pennsylvania press, Philadelphia,Pa.,1966
Pha 16.0027

N.A.T.O. ADVANCED STUDY INSTITUTE Electron density profiles in the ionosphere and exosphere :a symposium Proceedings Skeikampen,Norway 1961 Apr 17-26 Norwegian defence research establishment Edited by B. Maehlum Sponsored by N.A.T.O. science committee N.A.T.O. conference series, 2 428p Pergamon press,Oxford,1962
Nap 11.0391

N.A.T.O. ADVANCED STUDY INSTITUTE :a conference Proceedings Mantles of the earth and terrestrial planets Newcastle upon Tyne 1966 Mar 30-Apr 7 University of Newcastle upon Tyne.School of physics Edited by S.K. Runcorn Sponsored by North Atlantic treaty organization.Science office Interscience,London,1967
TA 15.0202

N.A.T.O. ADVANCED STUDY INSTITUTE.SUMMER SCHOOL Programming languages Villard de Lans 1966 North Atlantic treaty organization. Science committee Edited by F. Genuys 395p Academic press,London,1968
TA 15.0589

N.A.T.O. ADVANCED STUDY INSTITUTE ON ARTIFICIAL INTELLIGENCE AND HEURISTIC PROGRAMMING 1st Proceedings Artificial intelligence and heuristic programming Menaggio 1970 Aug North Atlantic treaty organization Edited by N.V. Findler and Bernard Meltzer illus 327p Edinburgh university press,Edinburgh, 1971
Math L 5.3853

N.A.T.O. ADVANCED STUDY INSTITUTE ON PLANETARY AND STELLAR MAGNETISM Magnetism and the cosmos Newcastle-upon-Tyne 1965 Apr North Atlantic treaty organisation Edited by W.R. Hindmarsh and others Oliver and Boyd, Edinburgh;London,1967
Geol 8.4428

N.A.T.O. SCIENCE COMMITTEE N.A.T.O. Advanced study institute Electron density profiles in the ionosphere and exosphere :a symposium Proceedings Skeikampen,Norway 1961 Apr 17-26 Norwegian defence research establishment Edited by B. Maehlum N.A.T.O. conference series, 2 428p Pergamon press,Oxford, 1962
Nap 11.0391

N.A.T.O.ADVANCED RESEARCH INSTITUTE ON APPLICATIONS OF THERMOLUMINESCENCE TO GEOLOGICAL PROBLEMS Proceedings Thermoluminescence of geological materials Spoleto 1966 Sep 5-16 North Atlantic treaty organization Edited by D.J. McDougall Academic press,London,1968
Min 10.0628

N.A.T.O.ADVANCED STUDY INSTITUTE Papers Phonons in perfect lattices and in lattices with point imperfections Aberdeen 1965 Aug 9-27 Scottish universities' summer school in physics Edited by R.W.H. Stevenson With financial support of North Atlantic treaty organization Scottish universities' summer school in physics, 6 448p Oliver and Boyd,Edinburgh,1966
TA 15.0212

N.V.OPTISCHE INDUSTRIE "DE OUDE DELFT'.SCIENTIFIC FUND SCIENTIAE SERVIENS Performance of the eye at low luminances :colloquium Proceedings Delft 1965 Edited by M.A. Bouman and J.J. Vos International congress series, 125 Excerpta medica,Amsterdam,1966
Psy 31.2862

N-BODY-PROBLEM Gravitational N-body problem Proceedings Cambridge 1970 Aug 12-15 International astronomical union Edited by M. Lecar International astronomical union. Colloquium, 10 441p Reidel,Dordrecht,1972
TA 15.0667

NACHRICHTENTECHNISCHE GESELLSCHAFT Electronic digital computers and information processing :conference Proceedings Darmstadt 1956 Oct 25-27 Edited by A. Walther and W. Hoffmann Nachrichtentechnische fachberichte, 4 illus. viii,229p 30cm Friedr.Vieweg und sohn,Brunswick,1956
Math L 5.3247

NAGOYA 1967 Flavins and flavoproteins The Conference on flavins and flavoproteins 2nd Proceedings Edited by Kunio Yagi University of Tokyo press,Tokyo,1968
Bioch 33.1076

NAIROBI 1950 Interterritorial conference on hydrology and water resources 1st Held under the auspices of the East Africa high commission Nairobi,1950
Geog 13.5670

NAIROBI 1965 The East African rift system : a seminar Papers International upper mantle committee Sponsored by the International union of geodesy and geophysics Nairobi,1965
Geod 9.0451

NAIROBI,KENYA 1968 Biology of reproduction in mammals :a symposium Proceedings Society for the study of fertility Edited by J.S. Perry and I.W. Rowlands Journal of reproduction and fertility.Supplement, 6 Blackwell,Oxford;Edinburgh,1969
Bal 39.1404

NAMUR 1965 Groupement des mathematiciens d'expression latine 3rd reunion comptes rendus Centre Belge de recherches mathematiques 153p 25cm Librairie universitaire,Louvain,1966
P. Math 2.2611

NANCY 1955 Echanges de matieres au cours de la genese des roches grenues acides et basiques Centre national de la recherche scientifique :colloque international 68th Proceedings Edited by M. Roubault Sciences de la terre,no.hors serie Ecole superieure de geologie,Nancy,1955
Min 10.0713

NANCY 1965 Adsorption et croissance cristalline :colloque international Centre national de la recherche scientifique Centre national de la recherche scientifique. Colloques internationaux, 152 Centre national de la recherche scientifique,Paris, 1965
Met 25.1621

NANCY 1968 Dosage des elements a l'etat de traces dans les roches et autres substances : colloque Comptes-rendus Centre national de la recherche scientifique Centre national de la recherche scientifique.Colloques nationaux, 923 CNRS,Paris,1970
Min 10.1475

NANCY 1969 Seminaire su les algebres completes Edited by J.P. Ferrier Lecture notes in mathematics, 164 68p 25cm Springer,Berlin,1970
P Math 2.3865

NANSEN Fridtjof Stauferland-expedition,1959-60 Vortrage des Fridtjof-Nansen-gedachtnis-symposions uber Spitzbergen in Nansens 100. geburtsjahr (geb 10.10.1861) Wurzburg 1961 Apr 3-11 Edited by Julius Budel and Alfred Wirthmann Ergebnisse der Stauferland-expedition,3 illus,7 plates 86p 25cm Franz Steiner,Wiesbaden,1965
Sco 14.2974

NAPLES 1948 Embriologia e genetica congress Consiglio nazionale per la ricerca Per iniziativa del Centre di studio di biologia,di citologia genetica e di enzimologia Stazione zoologica di Napoli. Publicazioni, 21 Naples,1949
Gen 34.0525

NAPLES 1959 Biochemistry of human genetics a symposium Ciba foundation and International union of biological sciences Edited by G.E.W. Wolstenholme and C.M. O'Connor Ciba Foundation.Symposia Churchill,London,1959
PGMS 29.0251

NAPLES 1959 Ciba foundation symposium on biochemistry of human genetics Proceedings Ciba foundation International union of biological sciences Edited by G.E.W. Wolstenholme and Cecilia M. O'Connor illus Churchill,London,1959
Bioch 33.0890

NAPLES 1959 Congresso dell'Unione matematica italiana 6to Atti Unione matematica italiana Bibliog,illus 497p 24cm Edizioni Cremonese,Rome,1960
P Math 2.3008

NAPLES 1961 Lectures on field theory and the many-body problem :international spring school 1st Universita di Napoli. Instituto di fisica teorica Edited by E.R. Caianiello With support of North Atlantic treaty organization Academic press,New York, 1961
Cav 7.0423

NAPLES 1962 Problemes du developpement economique dans les pays mediterraneens colloque international Actes Ecole pratique des hautes etudes,Paris.Centre de sociologie europeenne Edited by Jean Cuisenier Maison des sciences de l'homme. Recherches mediterraneennes,5 Mouton,The Hague,1963
Geog 13.4116

NAPLES 1962 1963 Lectures on the many-body problem :international spring school 2nd Vol 1-2 Universita di Napoli. Instituto di fisica teorica Edited by E.R. Caianiello With support of North Atlantic treaty organization 2 vols Academic press, New York,1962-64
Cav 7.0424

NAPLES 1966 Physiology and biochemistry of haemocyanins :a symposium Edited by F. Ghiretti held at the Naples zoological station Academic press,London;New York,1968 Sponsored by the Italian national research council,the U.S. national science foundation and the American institute of biological sciences
Bal 39.0522

NASA-UNIVERSITY CONFERENCE ON THE SCIENCE AND TECHNOLOGY OF SPACE EXPLORATION proceedings Chicago 1962 Nov.1-3 Vol 1-2 United States.National aeronautics and space administration.NASA SP, 11 NASA SP 11 2 vols office of scientific and technical information,Washington,D.C.,1962
A Math 4.1157

NATICK,MASS. 1956 Protection and functioning of the hands in cold climates :a conference Proceedings Edited by Frank R. Fisher Sponsored by the United States.Army. Headquarters Quartermaster research and development command illus vi,176p 23cm National academy of sciences,Washington,1957
Sco 14.0858

NATIONAL ACADEMY OF SCIENCE Pacific basin biogeography Pacific science congress 10th Symposium Honolulu 1961 Aug 21-Sep 6 Edited by J.Linsley Gressitt and others Honolulu,1963
Geog 13.1161

NATIONAL ACADEMY OF SCIENCES Conference on oceanwave spectra 1961 Englewood Cliffs,N. J.,1963
Eng 41.6696

NATIONAL ACADEMY OF SCIENCES International congress for logic,methodology and philosophy of science Proceedings Stanford,Calif. 1960 Aug 24-Sep 2 Edited by Ernest Nagel and others ix,661p 25cm Stanford university press,Stanford,Calif.,1962
Math S 3.1853

NATIONAL ACADEMY OF SCIENCES Quantum mechanical methods in valence theory :a conference Papers Shelter Island,L.I.,N.Y. 1951 Sep 8-10 With the support of the United States.Office of naval research United States.Office of naval research.Physics branch. Publication,4 U.S.government printing office, Washington,D.C.,1952
Chem 18.2237

NATIONAL ACADEMY OF SCIENCES Symposium of plasma dynamics proceedings Massachusetts 1968 Jun.9-13 Edited by Frances H. Clauser Sponsored by the United States.Air force. Office of scientific research Pergamon press, London,1959
A Math 4.0617

NATIONAL ACADEMY OF SCIENCES.COMMITTEE ON PHOTOBIOLOGY The Luminescence of biological systems :conference Proceedings Pacific Grove,Calif. 1954 Mar28-Apr 2 Edited by Frank H. Johnson American association for the advancement of science, Washington D.C.,1955
An 32.1502

NATIONAL ACADEMY OF SCIENCES.COMMITTEE ON POLAR RESEARCH Symposium on Antarctic logistics Boulder,Colo. 1962 Aug 13-17 Under the auspices of the Scientific committee on Antarctic research.WOrking party on logistics illus xvl,778p 28cm Washington,D.C.,1963
Sco 14.6199

NATIONAL ACADEMY OF SCIENCES.SUBCOMMITTEE ON GEOGRAPHIC PATHOLOGY Human palaeopathology Symposium on human palaeopathology Proceedings Washington,D.C. 1965 Jan 14 Edited by Saul Jarcho Yale university press,New Haven,Conn.,1966
An 32.3129

NATIONAL ADVISORY NEUROLOGICAL DISEASES AND BLINDNESS COUNCIL The Process of aging in the nervous system :a conference Proceedings Bethesda,Md. 1957 Jan 30-Feb 1 Edited by James E. Birren and others Symposia in neuroanatomical sciences, 5 Blackwell, Oxford,1959
An 32.4171

NATIONAL AERONAUTICAL LABORATORY,BANGALORE Seminar on aeronautical sciences Proceedings 1961 Nov 27-Dec 2 Vol 1 National aeronautical laboratory,Bangalore,1961
Eng 41.6969

NATIONAL AERONAUTICS AND SPACE ADMINISTRATION A Review of space research :symposium Report Iowa City 1962 Jun 17-Aug 10 Under the auspices of the Space science board National research council.Publication,1079 Washington, D.C.,1962
Obs 6.3259

NATIONAL AERONAUTICS AND SPACE ADMINISTRATION AAS-NASA magnetic star symposium The Symposium on the magnetic and other peculiar and metallic-line A stars :magnetic and related stars 1st Proceedings Greenbelt, Md. 1965 Nov 8-10 and Berkeley,Calif. 1965 Dec 29 Edited by Robert C. Cameron 596p Mono book corporation,Baltimore,1967 Includes the Helen B.Warner prize lecture of the American astronomical society,by George W. Preston
Obs 6.3432

NATIONAL AERONAUTICS AND SPACE ADMINISTRATION Apollo 11 lunar science conference Papers Houston,Texas 1970 Jan 5-8 American association for the advancement of science. Miscellaneous publication, 70-1 American association for the advancement of science, Washington,D.C.,1970
Min 10.1474

NATIONAL AERONAUTICS AND SPACE ADMINISTRATION Apollo 11 lunar science conference Proceedings Houston,Texas 1970 Jun 5-8 1-3 Edited by A.A. Levinson Geochimica et cosmochimica acta, 34,supplement 1 3 vols Pergamon,New York,1970
Min 10.1548

NATIONAL AERONAUTICS AND SPACE ADMINISTRATION Basic aerodynamic noise research :a conference Washington,D.C. 1969 Jul 14-15 Edited by Ira R. Schwartz NASA SP-207 N.A.S.A., Washington,D.C.,1969
Eng 41.4224

NATIONAL AERONAUTICS AND SPACE ADMINISTRATION IAU-NASA symposium The Conference on the nature of the surface of the moon Proceedings Greenbelt,Md. 1965 Apr 15-16 Goddard space flight center and International astronomical union Edited by Wilmot N. Hess and others 320p Johns Hoplins press,Baltimore,1966
TA 15.0203

NATIONAL AERONAUTICS AND SPACE ADMINISTRATION Interdisciplinary approach to friction and wear :a symposium Proceedings 1967 Washington,D.C.,1968
Eng 41.6272

NATIONAL AERONAUTICS AND SPACE ADMINISTRATION International symposium on rarefied gas dynamics 6th Proceedings Cambridge,Mass. 1968 Jul Vol 1-2 Edited by Leon Trilling and Harold Y. Wachman Advances in applied mechanics.Supplement, 5 2 vols Academic press,New York,1969
Eng 41.6797

NATIONAL AERONAUTICS AND SPACE ADMINISTRATION
Magnetic and related stars A.A.S.-N.A.S.A. symposium on the magnetic and other peculiar and metallic-line A stars Proceedings Greenbelt,Md 1965 Nov 8-10 Berkeley,Calif 1965 Dec 29 Edited by Robert C. Cameron xi,596p Mono book corporation,Baltimore,1967
A Math 4.1575

NATIONAL AERONAUTICS AND SPACE ADMINISTRATION
Magnetic and related stars The Symposium on the magnetic and other peculiar and metallic-line A stars Greenbelt,Md. 1965 Nov 8-10 Edited by Robert C. Cameron Mono book corporation,Baltimore,Md.,1967
TA 15.0310

NATIONAL AERONAUTICS AND SPACE ADMINISTRATION
Nature of the lunar surface IAU-NASA symposium Greenbelt,Md. 1965 Apr 15-16 Edited by W.N. Hess and others Held at the Goddard space flight center Johns Hopkins press,Baltimore,Md.,1966
Min 10.1445

NATIONAL AERONAUTICS AND SPACE ADMINISTRATION
Oceanography from space :proceedings of conference on the feasibility of conducting oceanographic exploratiions from aircraft, manned orbital and lunar laboratories Proceedings Woods Hole 1964 Aug 24-28 Woods Hole oceanographic institution Woods Hole oceanographic institution,Woods Hole,Mass. 1965
Sco 14.0355

NATIONAL AERONAUTICS AND SPACE ADMINISTRATION
Origins of prebiological systems and of their molecular matrices A Conference on the origins of prebiological systems and of their molecular matrices Proceedings Wakulla Springs,Fla 1963 Oct 27-30 Edited by Sidney W. Fox Bibliog.,illus. xx,482p Academic press,New York;London,1965
Bot 42.1364

NATIONAL AERONAUTICS AND SPACE ADMINISTRATION
Symposium on viscous drag reduction Proceedings Dallas,Texas 1968 Sep 24-25 Edited by C.S. Wells illus xi,500p 25cm Plenum press,New York,1969
A Math 4.1736

NATIONAL AERONAUTICS AND SPACE ADMINISTRATION
The Origins of prebiological systems and of their molecular matrices :a conference Proceedings Wakulla springs,Fla. 1963 Oct 27-30 Edited by Sidney W. Fox Under the auspices of the Institute for space biosciences Academic press,New York;London, 1965
Gen 34.0456

NATIONAL AERONAUTICS AND SPACE ADMINISTRATION
The Plasma space science symposium proceedings Washington D.C. 1963 Jun 11-14 Catholic university of America Edited by C.C. Chang and S.S. Huang Astrophysics and space science library, 3 D.Reidel,Dordrecht,1965
A Math 4.0625

NATIONAL AERONAUTICS AND SPACE ADMINISTRATION
The Scientific results from the orbiting astronomical observatory (OAO-2) Amherst, Mass. 1971 Aug 23-24 Edited by A.D. Code NASA SP-310 590p nasa,washington,D.C.,1972
Obs 6.3626

NATIONAL AERONAUTICS AND SPACE ADMINISTRATION. GODDARD INSTITUTE FOR SPACE STUDIES
Stellar evolution :a symposium proceedings New York 1963 Nov. Edited by R.F. Stein and A.G.W. Cameron 464p Plenum press,New York,1966
A Math 4.1164

NATIONAL AERONAUTICS AND SPACE ADMINISTRATION. GODDARD SPACE FLIGHT CENTER Conference on shock metamorphosis of natural materials 1st Proceedings Greenbelt,Md. 1966 Apr 14-16 Edited by Bevan M. French and Nicholas M. Short Held at the Goddard space flight center Mono book corporation,Baltimore,Md., 1968
Min 10.1443

NATIONAL AERONAUTICS AND SPACE ADMINISTRATION. INSTITUTE FOR AEROSPACE STUDIES
International symposium on rarefied gas dynamics 4th Proceedings Toronto 1964 Jul Vol 1-2 Edited by J.H.de Leeuw Advances in applied mechanics.Supplement, 3 2 vols Academic press,New York;London,1965
Eng 41.6788

NATIONAL AERONAUTICS AND SPACE ADMINISTRATION. OFFICE OF TECHNOLOGY UTILIZATION The Zodiacal light and the interplanetary medium : a conference Papers Honolulu 1967 Jan 30-Feb 3 NASA SP-150 430p n.a.s.a., washington,D.C.,1967
Obs 6.3425

NATIONAL AEROSPACE ELECTRONICS CONFERENCE 1971
Structure of a multiprocessor using microprogrammable building blocks presented at the National aerospace electronics conference 1971 By R,L. Davis and S. Zucker Burroughs corporation.Defense,space and special systems group illus 15p Burroughs corporation, Paoli,Pa.,1971
Math L 5.3766

NATIONAL AIR POLLUTION CONTROL ADMINISTRATION
Environmental protection Symposium on multiple-source urban diffusion models Proceedings 1970 Air pollution control office.Publication,AP- 86 US government printing office,Washington,D.C.,1970
Eng 41.8616

NATIONAL ASSOCIATION FOR THE PREVENTION OF TUBERCULOSIS Tuberculosis in the commonwealth 1947 Commonwealth and empire health and tuberculosis conference Transactions London 1947 N.A.P.T.,London, 1947
HE 27.0132

NATIONAL ASSOCIATION OF CORROSION ENGINEERS. SOUTHEAST REGION Symposium on corrosion fundamentals :a series of lectures... Knoxsville 1955 Mar 1-3 Edited by Anton De S. Brasunas and E.E. Stansbury University of Tennessee press,Knoxville,Tenn.,1956
Met 25.1884

NATIONAL ASSOCIATION OF CORROSION ENGINEERS. TECHNICAL COMMITTEE 3-F ON CORROSION BY HIGH-PURITY WATER High purity water corrosion of metals seven papers presented at symposia San Francisco 1959 Mar 17-21 Dallas 1960 Mar 14-18 National association of corrosion engineers,Houston,Texas,1968
Met 25.1933

NATIONAL ASSOCIATION OF CORROSION ENGINEERS. TECHNICAL GROUP COMMITTEE T-5 Short course on process industry corrosion Proceedings Ohio 1960 Sep 12-16 Edited by V.D. Miller National association of corrosion engineers,n.p.,1960
Met 25.1935

NATIONAL BIOPHYSICS CONFERENCE 1st Proceedings Columbus,Ohio 1957 Mar 4-6 Edited by H. Quastler and H.J. Morowitz Yale university press,NEW Haven,Conn.,1959
Radioth 35.0231

The NATIONAL BIOPHYSICS CONFERENCE 1st Proceedings Biophysics conference Columbus,Ohio 1957 Mar 4-6 Edited by Henry Quastler and Harold J. Morowitz With the support of the United States.Air force.Office of scientific research Yale university press, New Haven,1959
Bioch 33.0308

NATIONAL BUILDING AGENCY Design guide for medical group practice centres :a study... incorporating papers given at a symposium London 1966 Jun 3 College of general practitioners,London,c1966
PGMS 29.0555

NATIONAL BUREAU OF STANDARDS Advances in cryogenic engineering Cryogenic engineering conference 6th Proceedings Boulder,Colo. 1960 Aug 23-26 Edited by K.D. Timmerhaus Advances in cryogenic engineering, 6 Plenum press,New York,1961
Eng 41.7398

NATIONAL BUREAU OF STANDARDS Characteristics and applications of resistance strain gages :a symposium Proceedings Washington,D.C. 1951 Nov 8-9 National bureau of standards. Circular, 528 Washington,D.C.,1954
Eng 41.2902

NATIONAL BUREAU OF STANDARDS Chemistry of cement International symposium in the chemistry of cement 4th Proceedings Washington,D.C. 1960 Oct 2-7 Vol 1-2 National bureau of standards.Monograph, 43, vol 1-2 2 vols U.S.government printing office,Washington,D.C.,1962
Eng 41.3026

NATIONAL BUREAU OF STANDARDS Chemistry of cement International symposium on the chemistry of cement 4th Proceedings Washington,D.C. 1960 Oct 2-7 Vol 1-2 National bureau of standards.Monographs,43 2 vols Washington,D.C.,1962
Min 10.0647

NATIONAL BUREAU OF STANDARDS Critical phenomena Conference on phenomena in the neighbourhood of critical points Proceedings Washington,D.C. 1965 Apr 5-8 Edited by M.S. Green and J.V. Sengers National bureau of standards.Miscellaneous publication, 273 xiv, 242p 27cm National bureau of standards, Washington,D.C.,1966
Cav 7.2931

NATIONAL BUREAU OF STANDARDS Electrochemical constants NBS semicentennial symposium 7th Proceedings Washington,D.C. 1951 Sep 19-21 National bureau of standards.Circular, 524 U.S.Government printing office, Washington,D.C.,1953 One of 12 semicentennial symposia
Met 25.2401

NATIONAL BUREAU OF STANDARDS Ellipsometry in the measurement of surfaces and thin films symposium Proceedings Washington,D.C. 1963 Sep 5-6 Edited by E. Passaglia and others National bureau of standards. Miscellaneous publication, 256 U.S. Government printing office,Washington,D.C., 1964
Met 25.1687

NATIONAL BUREAU OF STANDARDS Frontiers of numerical mathematics :symposium Madison,Wis. 1959 Oct 30-31 United States army. Mathematics research center.Publications, 4 xi,132p 24cm University of Wisconsin, Madison,Wis.,1960
Math L 5.3232

NATIONAL BUREAU OF STANDARDS Gravity waves NBS semicentennial symposium on gravity waves 3rd Proceedings Washington,D.C. 1951 Jun 18-20 National bureau of standards. Circular, 521 illus iv,287p U.S. Government printing office,Washington,D.C., 1952
Chem E 24.0215

NATIONAL BUREAU OF STANDARDS Gravity waves NBS semicentennial symposium on gravity waves Proceedings Washington,D.C. 1951 Jun 18-20 National bureau of standards.Circular, 521 U. S.government printing office,Washington,D.C., 1952
Eng 41.6664

NATIONAL BUREAU OF STANDARDS Linear programming symposium 2nd Proceedings Washington,D.C. 1955 Jan 27-29 Vol 1-2 Sponsored by the United States.Air force. Office of scientific research 24cm 2 vols National bureau of standards,Washington,D.C., 1955
Math L 5.3248

NATIONAL BUREAU OF STANDARDS Mechanical behavior of crystalline solids a symposium Proceedings New York 1962 Apr 28-29 Sponsored by the Edward Orton junior ceramic foundation National bureau of standards monograph, 59 U.S.Government printing office,Washington,1963 Lithographed
Met 25.1213

NATIONAL BUREAU OF STANDARDS Mechanical properties of metals at low temperature the NBS semicentennial symposium Proceedings National bureau of standards.Circular, 520 United States printing office,Washington,D.C., 1952
Met 25.0972

NATIONAL BUREAU OF STANDARDS Microstructure of ceramic materials :a symposium Proceedings Pittsburgh 1963 Apr 27-28 Sponsored by Edward Orton junior ceramic foundation National bureau of standards. Miscellaneous publication, 257 U.S.dept of commerce,Washington D.C.,1964
Met 25.0285

NATIONAL BUREAU OF STANDARDS Miscellaneous physical tables :Planck's radiation functions and electronic functions National bureau of standards,Washington,D.C.,1941
Radioth 35.1504

NATIONAL BUREAU OF STANDARDS Monte Carlo method :a symposium Proceedings Los Angeles 1949 Jun 29-Jul 1 Edited by A.S. Householder and others Sponsored by the Rand corporation National bureau of standards. Applied mathematics series, 12 United States government printing office,Washington,D.C.,1951
Radioth 35.0586

NATIONAL BUREAU OF STANDARDS NBS semicentennial symposium on energy transfer in hot gases Proceedings 1951 Sep 17-18 National bureau of standards.Circular, 523 U.S.government printing office,Washington,D.C.,1954
Eng 41.7220

NATIONAL BUREAU OF STANDARDS Numerical analysis :symposium 6th Proceedings Santa Monica,Calif. 1953 Aug 26-28 American mathematical society Edited by John H. Curtiss American mathematical society. Proceedings of symposia in applied mathematics, 6 vi,303p McGraw-Hill,New York,1956
Math L 5.3245

NATIONAL BUREAU OF STANDARDS Pyrometry The Papers and discussion of a symposium on pyrometry Chicago,Ill. 1919 Sep illus 701p New York,1920
WSM 43.4920

NATIONAL BUREAU OF STANDARDS Pyrometry :a symposium Chicago 1919 Sep American institute of mining and metallurgical engineers,New York City,1920
Met 25.0294

NATIONAL BUREAU OF STANDARDS Simultaneous linear equations and the determination of Eigenvalues symposium Proceedings Los Angeles 1951 Edited by L.J. Paige and Olga Taussky jointly sponsored by United States. Office of naval research United States. National bureau of standards. Applied mathematics series, 29 bibliog. 125p 26cm National bureau of standards,Washington, D.C.,1953
Math L 5.0131

NATIONAL BUREAU OF STANDARDS Technical meeting concerning wind loads on buildings and structures Proceedings Gaithersburg,Md. 1967 Jan 27-28 Edited by R.D. Marshall and H. C.S. Thom National bureau of standards. Building science series, 30 US government printing office,Washington,D.C.,1970
Eng 41.8274

NATIONAL BUREAU OF STANDARDS Temperature:its measurement and control in science and industry :a symposium 4th Proceedings Columbus,Ohio 1961 Mar 27-31 Vol 3,pt 1: basic concepts,standards and methods American institute of physics and Instrument society of America Edited by C.M. Herzfeld and Ferdinand Graft Brickwedde Reinhold;Chapman and Hall,New York;London,1962
Cav 7.0048

NATIONAL BUREAU OF STANDARDS Temperature:its measurement and control in science and industry :a symposium 4th Proceedings Columbus,Ohio 1961 Mar 27-31 Vol 3,pt2: applied methods and instruments American institute of physics and Instrument society of America Edited by A.I. Dahl and C. M. Herzfeld 1094p Reinhold publishing,New York,1962
Cav 7.0047

NATIONAL BUREAU OF STANDARDS Temperature:its measurement and control in science and industry :a symposium New York 1939 Nov 2-4 Edited by C.O. Fairchild 1362p Reinhold publishing,New York,1941
Cav 7.0046

NATIONAL BUREAU OF STANDARDS Temperature,its measurement and control in science and industry :a symposium Records New York City 1939 Nov 2-4 Edited by C.O. Fairchild Under the auspices of the American institute of physics xiii,1362p 23cm Reinhold publishing,New York,1941
Sco 14.0234

NATIONAL BUREAU OF STANDARDS Temperature,its measurement and control in science and industry :a symposium 4th Proceedings Columbus,Ohio 1961 Mar 27-31 Vol 3,1: basic concepts,standards and methods American institute of physics,and,Instrument society of America Edited by Ferdinand Graft Brickwedde xvi,848p Reinhold,New York,1962 General editor:Herzfeld,Charles
Sco 14.0231

NATIONAL BUREAU OF STANDARDS Temperature,its measurement and control in science and industry :a symposium 4th Proceedings Columbus,Ohio 1961 Mar 27-31 Vol 3,2: applied methods and instruments American institute of physics,and,Instrument society of America Edited by A.I. Dahl xiv,1094p Reinhold,New York,1962 General editor: Herzfeld,Charles
Sco 14.0233

NATIONAL BUREAU OF STANDARDS Temperature,its measurement and control in science and industry :a symposium 4th Proceedings Columbus,Ohio 1961 Mar 27-31 Vol 3,3: Biology and medicine American institute of physics,and,Instrument society of America Edited by James D. Hardy xii,683p Reinhold, New York,1962 General editor:Herzfeld, Charles
Sco 14.0232

NATIONAL BUREAU OF STANDARDS Temperature - its measurement and control in science and industry :a symposium 2nd Papers New York 1939 Nov 2-4 Reinhold,New York,1941
Met 25.0298

NATIONAL BUREAU OF STANDARDS.SEMICENTENNIAL SYMPOSIUM Characteristics and applications of resistance strain gages :a symposium Proceedings Washington,D.C. 1951 Nov 8-9 National bureau of standards National bureau of standards.Circular, 528 Washington,D.C.,1954
Eng 41.2902

NATIONAL CANCER CONFERENCE 1st Proceedings Bethesda,Md 1949 Feb 25-27 American cancer society American cancer society,1949
Med 36.0003

NATIONAL CANCER INSTITUTE A Symposium on mammary tumors in mice Edited by Forest Ray Moulton American association for the advancement of science.Publication, 22 American association for the advancement of science,Washington,D.C.,1945
Gen 34.1813

NATIONAL CANCER INSTITUTE Analysis of carcinogenic air pollutants :symposium Proceedings Cincinnati,Ohio 1961 Aug 29-31 Edited by Eugene Sawicki and Kenneth Cassel National cancer institute.Monograph, 9 illus U.S.Department of health,education and welfare,Bethesda,Md.,1962 Sponsored by the Laboratory of engineering and physical sciences of the Division of air pollution,U.S. Department of health education,and welfare, Public health service.Robert A.Taft sanitary engineering center
Bioch 33.0980

NATIONAL CANCER INSTITUTE Analytic cell culture Syverton memorial symposium Proceedings Detroit 1961 Jun 6-7 Edited by Robert E. Stevenson Sponsored by Tissue culture association and Cell culture collection committee of the viruses and cancer board,National cancer institute National cancer institute.Monograph, 7 U.S. Department of health,education and welfare, Bethesda,Md.,1962
Bioch 33.0979

NATIONAL CANCER INSTITUTE Cell tissue and organ culture Decennial review conference on cell tissue and organ culture 2nd Proceedings Bedford,Pa. 1966 Sep 11-15 Edited by Benton B. Westfall National cancer institute.Monograph, 26 National cancer institute,Bethesda,Md.,1967
Bioch 33.0991

NATIONAL CANCER INSTITUTE Comparative morphology of hematopoietic neoplasms :a symposium Proceedings Washington,D.C. 1968 Mar 11-12 Edited by Carolyn H. Lingeman and F.M. Garner held at United States.Armed forces institute of pathology National cancer institute.Monograph, 32 National cancer institute,Bethesda,Md.,1969
Bioch 33.0995

NATIONAL CANCER INSTITUTE Conference on experimental clinical cancer chemotherapy Washington,D.C. 1959 Nov 11-12 Edited by B. H. Morrison National cancer institute. Monograph, 3 U.S.Department of health, education and welfare,Washington,D.C.,1960
Radioth 35.0809

NATIONAL CANCER INSTITUTE Conference on radiobiology and radiotherapy Proceedings Colorado Springs 1965 Nov 1-3 Edited by J. A. Del Regato National cancer institute. Monograph, 24 National cancer institute, Bethesda,Md.,1967
Bioch 33.0989

NATIONAL CANCER INSTITUTE Control of cell division and the induction of cancer International symposium on the control of cell division and the induction of cancer Proceedings Lima 1963 Jul 1-6 and Cali, Columbia 1963 Jul 1-6 National cancer institute.Monograph, 14 illus National cancer institute,Bethesda,Md.,1964 Introduction,abstracts of each paper and summary in English and Spanish
Bioch 33.0985

NATIONAL CANCER INSTITUTE End results of cancer therapy International symposium on end results of cancer therapy Proceedings Sandefjord 1963 Sep 16-20 Edited by Sidney J. Cutler National cancer institute. Monograph, 15 National cancer institute, Bethesda,Md.,1964
Bioch 33.0986

NATIONAL CANCER INSTITUTE International conference on Avian tumor viruses Proceedings Durham,N.C. 1964 Mar 31-Apr 3 Edited by Joseph W. Beard National cancer institute.Monograph, 17 National cancer institute,Bethesda,Md.,1964
Bioch 33.1154

NATIONAL CANCER INSTITUTE International conference on avian tumour viruses Proceedings Durham,N.C. 1964 Mar 31-Apr 3 Edited by Joseph W. Beard National cancer institute.Monograph, 17 US government printing office,Washington,D.C.,1964
Path 30.2570

NATIONAL CANCER INSTITUTE International symposium.Enzymatic aspects of metabolic regulation Proceedings Mexico City 1966 Nov 28-Dec 1 Edited by M.P. Stulberg National cancer institute.Monograph, 27 National cancer institute,Betheseda,Md.,1967
Bioch 33.1062

NATIONAL CANCER INSTITUTE International symposium on end results of cancer therapy Sandefjord 1963 Sep 16-20 Edited by Sidney J. Cutler National cancer institute. Monograph, 15 U.S.Department of health, education and welfare,Washington,D.C.,1964
Radioth 35.0844

NATIONAL CANCER INSTITUTE International symposium on genes and chromosomes:structure and function Proceedings Buenos Aires 1964 Nov 30-Dec 4 Edited by Juan I. Valencia and Rhoda F. Grell National cancer institute. Monograph, 18 National cancer institute, Bethesda,Md.,1965
Bioch 33.0987

NATIONAL CANCER INSTITUTE International symposium on nucleolus:its structure and function Proceedings Montevideo 1965 Dec 5-10 Edited by W.S. Vincent and others National cancer institute.Monograph, 23 National cancer institute,Bethesda,Md.,1966
Bioch 33.0988

NATIONAL CANCER INSTITUTE Metabolic control mechanisms in animal cells :a symposium Proceedings Boston,Mass. 1963 May 28-30 Edited by William J. Rutter Sponsored by the Tissue culture association and National cancer institute National cancer institute. Monograph, 13 illus National cancer institute,Bethesda,Md.,1964
Bioch 33.0984

NATIONAL CANCER INSTITUTE Neoplasms and related disorders of invertebrate and lower vertebrate animals :a symposium Proceedings Washington,D.C. 1968 Jun 19-21 Edited by Clyde J. Dawe and John C. Harshbarger held at the Smithsonian institution National cancer institute.Monograph, 31 National cancer institute,Bethesda,Md.,1969
Bioch 33.0994

NATIONAL CANCER INSTITUTE Symposium on organ culture :studies of development,function,and disease Washington,D.C. 1962 May 29-31 Edited by Clyde J. Dawe Sponsored by Tissue culture association and National cancer institute National cancer institute. Monograph, 11 U.S.Department of health, education,and welfare,Washington,D.C.,1963
Bioch 33.2367

NATIONAL CANCER INSTITUTE Symposium on some problems of normal and abnormal differentiation and development Proceedings Bar Harbor,Maine 1959 Jun 15-17 Edited by Nathan Kaliss National cancer institute. Monograph, 2 U.S.Department of health, education and welfare,Washington,D.C.,1960
Radioth 35.0808

NATIONAL CANCER INSTITUTE The International conference on the biology of cutaneous cancer 1st Proceedings Philadelphia 1962 Apr 6-11 Edited by Frederick Urbach National cancer institute.Monograph, 10 U.S. Department of health,education,and welfare, Bethesda,Md.,1963 Sponsored by the skin and cancer hospital,Dept.of dermatology,Temple university school of medicine and The Committee on geographie pathology,Unio internationalis contra cancrum.A Merck Sharp and Dohme medical research conference
Bioch 33.0981

NATIONAL CANCER INSTITUTE Toward a less harmful cigarette The World conference on smoking and health :a workshop Proceedings 1967 Sep 11-13 Edited by Ernest L. Wynder and Dietrich Hoffmann National cancer institute.Monograph, 28 National cancer institute,Bethesda,Md.,1968
Bioch 33.0992

NATIONAL CANCER INSTITUTE Tumors of the alimentary tract in Africans :symposium Proceedings Geneva 1965 Nov 29-Dec 5 Edited by J.F. Murray Sponsored by the International union against cancer and the Calouste Gulbenkian foundation National cancer institute.Monograph, 25 National cancer institute,Bethesda,Md.,1967
Bioch 33.0990

NATIONAL CANCER INSTITUTE.MONOGRAPH, 18 International symposium on genes and chromosomes structure and function Proceedings Buenos Aires 1964 Nov 30-Dec 4 Edited by Juan I. Valencia and Rhoda F. Grell illus 354p U.S.Dept. of health,education and welfare,Bethesda,Md.,1965
Bot 42.0792

NATIONAL CANCER INSTITUTE.MONOGRAPHS, 18 International symposium on genes and chromosomes structure and function Proceedings Buenos Aires 1964 Nov 30-Dec 4 Edited by Juan I. Valencia and others U.S. Department of health,education and welfare, Washington,D.C.,1965
Gen 34.0833

NATIONAL CANCER INSTITUTE.VIROLOGY RESEARCH BRANCH Conference on murine leukemia Papers Philadelphia,Pa. 1965 Oct 13-15 Edited by Marvin A. Rich and John B. Moloney National cancer institute.Monograph, 22 US government printing office,Washington,D.C., 1966
Path 30.2460

NATIONAL CANCER INSTITUTE OF CANADA Canadian cancer conference :Canadian cancer research conference 5th Proceedings Honey harbor, Ont. 1962 Jun 10-14 Academic press,New York;London,1963
Gen 34.1993

NATIONAL CENTER FOR ATMOSPHERIC RESEARCH and INTERNATIONAL UNION FOR QUATERNARY RESEARCH International union for quaternary research congress :a collection of papers derived from the INQUA-NCAR symposium 7th Proceedings Boulder,Colo. 1965 Aug 14-Sep 19 5: causes of climatic change Edited by J.Murray Mitchell Sponsored by the National research council Meteorological monographs,8,no 30 illus iv,159p 29cm American meteorological association,Boston,Mass.,1968 For other volumes of these proceedings,see also International association for quaternary research
Sco 14.7750

NATIONAL CENTER FOR ATMOSPHERIC RESEARCH.HIGH ALTITUDE OBSERVATORY Resonance lines in astrophysics :manuscripts presented at a conference... Boulder,Colo. 1968 Sep 9-11 480p National center for atmospheric research,Boulder,Colo.,1968
Obs 6.3349

NATIONAL COAL BOARD Chemical engineering in the coal industry :an international conference Cheltenham 1956 June Edited by Forbes W. Sharpley 141p Pergamon press,London,n.d.
Chem E 24.1603

NATIONAL COAL BOARD.MINING RESEARCH ESTABLISHMENT Mechanical properties of non-metallic brittle materials :a conference Proceedings London 1958 Apr Edited by W.H. Walton Butterworths,London,1958
Eng 41.3645

NATIONAL COMMITTEE ON MATERNAL HEALTH Menstruation and its disorders :conference Proceedings Edited by Earl T. Engle Thomas, Springfield,Ill.,1950
An 32.3903

NATIONAL CONFERENCE OF THE OFFICE MANAGEMENT ASSOCIATION papers The Scope for electronic computers in the office Brighton 1955 May Office management association illus. 102p 26cm Office management association,London,1955
Math L 5.0727

NATIONAL CONFERENCE ON INFANTILE MORTALITY Report of the proceedings Westminster 1906 Jun 13-14 King,Westminster,1906 President of the conference:Rt.hon.J.Burns,M.P. Chairman:E.Spicer,J.P.
Path 30.0540

NATIONAL CONFERENCE ON STATISTICAL METEOROLOGY 1st Proceedings Hartford,Conn. 1968 May 27-29 Sponsored by the American meteorological society American meteorlogical society,Boston,Mass.,1968 Co-Sponsor:A.M.S.committee on statistical meteorology
Math S 3.1583

NATIONAL CONFERENCE ON STRATA reports Science publishing house,Peking,1964 12 pts in box
Geol 8.2613

NATIONAL CONFERENCE ON THE PREVENTION OF DESTITUTION 1st Proceedings London 1911 May 30-Jun 2 Great Britain.Local government board King,London,1911 President of conference:the Rt.Hon.the Lord Mayor of London
Path 30.1285

NATIONAL CONFERENCE ON THE PREVENTION OF DESTITUTION 1st Proceedings of the Public health section London 1911 May 30-Jun 2 Great Britain.Local government board King,London,1911 President of the Public health section of the Local government board:T.C.Allbutt
Path 30.1284

NATIONAL CONFERENCE ON TUBE TECHNIQUES 3rd Proceedings New York 1956 Sep 12-14 Sponsored by the Advisory group on electron tubes.Working group on semiconductor devices New York university press,New York,1958
Eng 41.5619

NATIONAL CONFERENCE ON TUBE TECHNIQUES 4th Proceedings New York 1958 Sep 10-12 Sponsored by the Advisory group on electron tubes.Working group on tube techniques New York university press,New York,1959
Eng 41.5620

NATIONAL CONFERENCE ON TUBE TECHNIQUES 5th Proceedings Advances in electron tube techniques New York 1960 Sep Edited by David Slater Sponsored by the Advisory group on electron tubes.Working group on tube techniques Pergamon press,Oxford,1961
Eng 41.5621

NATIONAL CONFERENCE ON WEATHER MODIFICATION 1st Proceedings Albany,N.Y. 1968 Apr 28-May 1 American meteorological society State university of New York 532p American meteorological society,Boston,Mass.,1968
Nap 11.0975

NATIONAL CONFERENCE ON WEATHER MODIFICATION 2nd Proceedings Santa Barbara,Calif. 1970 Apr 6-9 National science foundation. Atmospheric sciences section American meteorological society 436p American meteorological society,Boston,Mass.,1970
Nap 11.0977

NATIONAL CONGRESS OF APPLIED MECHANICS 1st proceedings Chicago 1951 June 11-16 Edited by Eli Sternberg United States national committee on theoretical and applied mechanics American society of mechanical engineers,New York,1952
A Math 4.0327

NATIONAL CONGRESS OF THEORETICAL AND APPLIED MECHANICS 1st Proceedings Varna 1969 Nov 3-6 Bulgarsko knizhovno druzhestvo. National committee on theoretical and applied mechanics Bibliog,illus 632p 24cm Bulgarsko knizhovno druzhestvo,Sofia,1971 Papers in English,French and Russian
A Math 4.1723

NATIONAL DIE CASTING EXPOSITION AND CONGRESS 1st Technical papers Detroit 1960 Nov 8-11 Society of die casting engineers Society of die casting engineers,Detroit,1960
Met 25.0630

NATIONAL ENGINEERING LABORATORY Pump design, testing and operation :a symposium Proceedings Glasgow 1956 Apr 12-14 H.M.S. O.,Edinburgh,1966
Eng 41.6824

NATIONAL FORGEMASTERS' ASSOCIATION Forgemasters' meeting 1954 Proceedings London 1954 Oct 11-15 Joint meeting with the Chambre syndicate de la grosse forge francaise Iron and steel institute.Special report, 60 Iron and steel institute,London, 1957
Met 25.0715

NATIONAL FOUNDATION-MARCH OF DIMES Chicago conference Standardization in human cytogenetics Chicago,Ill. 1966 Sep 3-4,10 Birth defects:original article series, 2,no. 2 National foundation -March of dimes,New York,1966
Gen 34.0895

NATIONAL FOUNDATION-MARCH OF DIMES.GREAT NEW YORK CHAPTER Symposium on the placenta :its form and functions with particular reference to the prevention of birth defects and fetal deaths New York 1964 Mar 6 Medical society of the county of New York.Special committee on infant mortality Edited by Daniel Bergsma and others Birth defects original article series, 1,no.1 National foundation,New York,1965
PGMS 29.0212

NATIONAL HEART INSTITUTE Genetic polymorphisms and geographic variations in disease :a conference Proceedings Bethesda, Md. 1960 Feb 23-25 Edited by Barach S. Blumberg Grune and Stratton,New York;London, 1961
Gen 34.1984

NATIONAL HEAT TRANSFER CONFERENCE 2nd Preprints Condensing heat transfer within horizontal tubes Chicago,Ill. 1958 Aug 18-21 American society of mechanical engineers American society of mechanical engineers, Philadelphia,Pa.,1958
Eng 41.7155

NATIONAL HEMOPHILIA FOUNDATION International symposium on hemophilia and hemophilioid diseases Proceedings New York 1956 Aug 24-25 Edited by K.M. Brinkhous North Carolina university press,Chapel Hill,1957
Med 36.0050

NATIONAL INSTITUTE OF ALLERGY AND INFECTIOUS DISEASES Conference on newer respiratory disease viruses Bethesda,Md 1962 Oct 3-5 Edited by Clayton G. Loosli Sponsored by University of Southern California.School of medicine American review of respiratory diseases, 88,3 pt 2 American thoracic society,1963
PGMS 29.0321

NATIONAL INSTITUTE OF ALLERGY AND INFECTIOUS DISEASES Symposium on listeric infection 2nd Papers and discussions Bozeman, Montana 1962 Aug 29-31 Montana state college Edited by M.L. Gray Montana state college,Bozeman,Montana,1963
Path 30.2590

NATIONAL INSTITUTE OF ARTHRITIS AND METABOLIC DISEASES Genetic polymorphisms and geographic variations in disease :a conference Proceedings Bethesda,Md. 1960 Feb 23-25 Edited by Barach S. Blumberg Grune and Stratton,New York;London,1961
Gen 34.1984

NATIONAL INSTITUTE OF CHILD HEALTH AND HUMAN DEVELOPMENT Organization of recall International interdisciplinary conference on learning,remembering and forgetting 2nd Proceedings Princeton,N.J. 1964 Sep 27-30 Edited by Daniel P. Kimble With the support of United States.Office of naval research New York academy of sciences.Interdisciplinary communication program,New York,1967
Psy 31.1012

NATIONAL INSTITUTE OF CHILD HEALTH AND HUMAN DEVELOPMENT.HUMAN COMMUNICATION PROGRAM Language development in children The Genesis of language:a psycholinguistic approach conference Proceedings Old Point Comfort, Va. 1965 Apr 25-28 Edited by Frank Smith and George A. Miller M.I.T.press,Cambridge, Mass.;London,1966
Psy 31.2746

NATIONAL INSTITUTE OF GENETICS,SHIZVOKA Symposium on genetic effect of radiation Proceedings Misima,Japan 1960 Nov 7-9 Japanese journal of genetics, 36,supp. Genetics society of Japan,1961
Gen 34.1117

NATIONAL INSTITUTE OF HEALTH Invertebrate nervous systems The Conference on invertebrate nervous systems :their significance for mammalian neurophysiology Papers Pasadena,Calif. 1966 Jan 10-12 Edited by C.A.G. Wiersma University of Chicago press,Chicago,Ill.;London,1967 Dedicated to the memory of George Howard Parker
Bal 44.6453

NATIONAL INSTITUTE OF MENTAL HEALTH Ethnopharmacologic search for psychoactive drugs :a symposium Proceedings San Francisco 1967 Jan 28-30 Edited by Daniel H. Efron and others National institute of mental health.Pharmacology section.Workshop series,2 United States.Public health service publication,1645 U.S.Dept.of public health, Washington,D.C.,1967
Pha 16.0065

NATIONAL INSTITUTE OF NEUROLOGICAL DISEASES AND BLINDNESS Advances in neuroendocrinology Symposium on neuroendocrinology Miami,Fla. 1961 Dec 6-8 Edited by Andrew V. Nalbandov University of Illinois press,Urbana,Ill.,1963
An 32.3766

NATIONAL INSTITUTE OF NEUROLOGICAL DISEASES AND BLINDNESS Advances in neuroendocrinology Symposium on neuroendocrinology Miami,Flo. 1961 Dec 6-8 Edited by A.V. Nalbandov University of Illinois press,Urbana,Ill.,1963
Phys 20.1404

NATIONAL INSTITUTE OF NEUROLOGICAL DISEASES AND BLINDNESS Neural mechanisms of the auditory and vertibular systems :a conference Proceedings Bethesda,Md. 1959 Jun 11-13 Edited by G.L. Rasmussen and W.F. Windle Symposia in neuroanatomical sciences, 6 illus. xiv,422p Thomas,Springfield,Ill., 1960
Phys 20.1975

NATIONAL INSTITUTE OF SCIENCES OF INDIA Plant and animal viruses :a symposium Proceedings Cuttack 1961-62 Dec 31-Jan 1 National institute of sciences of India. Bulletin, 24 illus. vi,248p National institute of sciences of India,New Delhi,1963
Bot 42.3739

NATIONAL INSTITUTE OF SCIENCES OF INDIA Symposium on newer trends in taxonomy Proceedings New Delhi 1966 Jan 28-30 National institute of sciences of India. Bulletin, 34 New Delhi,1967
Bal 39.2079

NATIONAL INSTITUTE OF SCIENCES OF INDIA The Symposium on advancing frontiers of life sciences Proceedings New Delhi 1960-61 Dec 30-Jan 1 National institute of sciences of India.Bulletin, 19 New Delhi,1962 Symposium held on the occasion of the Silver Jubilee of the National institute of sciences of India
Bal 39.0009

NATIONAL INSTITUTE OF SCIENCES OF INDIA The Symposium on plant and animal viruses Proceedings Cuttack 1961-62 Dec 31-Jan 1 National institute of sciences of India. Bulletin, 24,1963 National institute of sciences of India,New Delhi,1963 Held under the joint auspices of the National institute and the Council of scientific and industrial research
Bal 39.1071

NATIONAL INSTITUTES OF HEALTH Conference on newer respiratory disease viruses Bethesda, Md 1962 Oct 3-5 Edited by Clayton G. Loosli Sponsored by University of Southern California.School of medicine American review of respiratory diseases, 88,3 pt 2 American thoracic society,1963
PGMS 29.0321

NATIONAL INSTITUTES OF HEALTH Current topics in biochemistry Bethseda,Md. 1971 Sep Edited by C.B. Anfinsen National institutes of health.Lectures in biomedical sciences Academic press,New York,1972
Bioch 33.2168

NATIONAL INSTITUTES OF HEALTH Genetic polymorphisms and geographic variations in disease :a conference Proceedings Bethesda, Md. 1960 Feb 23-25 Edited by Barach S. Blumberg Grune and Stratton,New York;London, 1961
Gen 34.1984

NATIONAL INSTITUTES OF HEALTH Proceedings of the symposium on information processing in sight sensory systems Proceedings Pasadena, Calif. 1965 Nov 1-3 Edited by P.W. Nye California institute of technology,Pasadena, Calif.,1965
Psy 31.0419

NATIONAL INSTITUTES OF HEALTH.BIOPHYSICS AND BIOPHYSICAL CHEMISTRY STUDY SECTION
Biophysical science :a study program Boulder, Colo. 1958 Jul 20 Aug 16 Edited by J.L. Oncley and others Wiley,New York,1959
Bioch 33.0329

NATIONAL INSTITUTES OF HEALTH.BIOPHYSICS AND BIOPHYSICAL CHEMISTRY STUDY SECTION
Biophysical science :a study program Boulder, Colo. 1959 Jul 20-Aug 16 Edited by J.L. Oncley and others Wiley,New York,1959 Published also in the Reviews of modern physics,Jan.and Apr.1959
Bal 39.0026

NATIONAL INSTITUTES OF HEALTH.ENDOCRINE STUDY SECTION Thyrotropin :a conference Proceedings 1962 Edited by Sidney C. Werner Thomas,Springfield,Ill.,1963 Introduction:"This conference is the second on tropic hormones of the pituitary gland initiated by the Endocrine study section...
An 32.3762

NATIONAL INSTITUTES OF HEALTH.ENDOCRINOLOGY STUDY SECTION Hormones and atherosclerosis :a conference Proceedings Brighton,Utah 1958 Mar 11-14 Edited by Gregory Pincus Academic press,New York,1959
Bioch 33.0477

NATIONAL INSTITUTES OF HEALTH.GENETICS STUDY SECTION Methodology in basic genetics Edited by Walter J. Burdette Sponsored by the University of Texas Holden-Day,San Francisco,Calif.,1963 Three copies
Gen 34.1012

NATIONAL INSTITUTES OF HEALTH.GENETICS STUDY SECTION Methodology in human genetics Edited by Walter J. Burdette Sponsored by the Roscoe B.Jackson memorial laboratory Holden-Day,San Francisco,Calif.,1962
Gen 34.1010

NATIONAL INSTITUTES OF HEALTH.GENETICS STUDY SECTION Methodology in mammalian genetics Edited by Walter J. Burdette Sponsored by the Roscoe B.Jackson memorial laboratory Holden-Day,San Francisco,Calif.,1963
Gen 34.1011

NATIONAL KIDNEY DISEASE FOUNDATION
Angiotensin systems and experimental renal diseases Annual conference on the kidney 14th Proceedings Edited by Jack Metcoff Churchill,London,1963 By 24 authors and other participants
Inv Med 37.0159

NATIONAL KIDNEY DISEASE FOUNDATION
Hereditary,developmental and immunologic aspects of kidney disease Annual conference on the kidney 13th Proceedings Princeton 1961 Oct 12-13 Edited by Jack Metcoff Northwestern university press,Evanston,1962
Inv Med 37.0157

NATIONAL KIDNEY DISEASE FOUNDATION
Hereditary,developmental and immunologic aspects of kidney disease Annual conference on the kidney 13th Proceedings Princeton, N.J. 1961 Oct 12-13 Edited by Jack Metcoff Northwestern university press,Evanston,Ill., 1962
An 32.2075

NATIONAL KIDNEY DISEASE FOUNDATION
Homotransplantation:kidney and other tissues Annual conference on the kidney 16th Proceedings 1964 Edited by Jack Metcoff National kidney disease foundation,New York, 1966
Inv Med 37.0160

NATIONAL KIDNEY DISEASE FOUNDATION Renal metabolism and epidemiology of some renal diseases Annual conference on the kidney Proceedings Princeton 1961 Oct 12-13 Edited by Jack Metcoff National kidney disease foundation,New York,1964
Inv Med 37.0158

NATIONAL METAL CONGRESS AND EXPOSITION
Magnetic properties of metals and alloys Cleveland 1958 Oct 25-26 By R.M. Bozorth and others American society for metals American society for metals,Cleveland,Ohio, 1959 A seminar presented to members of A. S.M.during the National metal congress and exposition
Met 25.2353

NATIONAL METAL CONGRESS AND EXPOSITION,1952
Behaviour of metals at low temperatures :a series of three educational lectures on behaviour of metals at low temperatures presented to members of the ASM during the thirtyfourth National metal congress and exposition,Philadelphia,October 20 t 24,1952 Lecures By R.M. Brick and others American society for metals diagrs iv,112p 24cm Cleveland,Ohio,1953
Sco 14.1245

NATIONAL METAL CONGRESS AND EXPOSITION 17TH
Principles of heat treatment Chicago 1935 By M.A. Grossman American society for metals, Cleveland,Ohio,1953,reprinted 1955 A series of educational lectures...first presented to members of the ASM during the 17th National Metal Congress and Exposition, Chicago, 1935, and later extended to include the more recent developments
Met 25.0434

NATIONAL METAL CONGRESS AND EXPOSITION 17TH
Principles of heat treatment Chicago 1935 By M.A. Grossman 3rd edition American society for metals,Cleveland,Ohio,1940 Lectures first presented to members of the A.S. M. during the 17th National Metal Congress and exposition,Chicago, 1935 and later extended to include the more recent developments
Met 25.0433

NATIONAL METAL CONGRESS AND EXPOSITION 19TH
Metallographic technique for steel By J.R. Vilella American society for metals, Philadelphia,Pa.,1938 A series of three educational lectures presented during the 19th National metal congress and exposition
Met 25.1347

NATIONAL METAL CONGRESS AND EXPOSITION 19TH
Open-hearth steel making Atlantic City 1937 Oct 18-22 By Earnshaw Cook A series of five educational lectures presented to members of the A.S.M. during the nineteenth National metal congress and exposition, Atlantic City, N.J. October 18 to 22,1937
Met 25.0329

NATIONAL METAL CONGRESS AND EXPOSITION 20TH
The Pyrometry of solids and surfaces Detroit 1938 Oct 17-21 By Robert B. Sosman American society for metals,Cleveland,1940 A series of three educational lectures presented to members of the American society for metals during the twentieth National metal congress and exposition Detroit,Michigan,Oct. 17 to 21.1938
Met 25.0297

NATIONAL METAL CONGRESS AND EXPOSITION 21ST
Visual examination of steel By George M. Enos American society for metals,Cleveland, Ohio,1940 A series of three educational lectures presented during the 21st National metal congress and exposition
Met 25.1315

NATIONAL METAL CONGRESS AND EXPOSITION 22ND
Strength of metals under combined stresses Cleveland 1940 Oct 21-25 By Maxwell Gensamer American society for metals, Cleveland,Ohio,1940 A series of educational lectures presented to members of the A.S.M. during the twenty-second National metal congress and exposition Cleveland, Ohio, October 2-25,1940
Met 25.0815

NATIONAL METAL CONGRESS AND EXPOSITION 22ND
Strength of metals under combined stresses :a series of educational lectures...presented... during the twenty-second National metal congress and exhibition Cleveland,Ohio 1940 Oct 21-25 By Maxwell Gensamer American society for metals American society for metals,Cleveland,Ohio,1941
Eng 41.3742

NATIONAL METAL CONGRESS AND EXPOSITION 23RD
The Flow of heat in metals Philadelphia 1941 Oct 20-25 By J.B. Austin American society for metals,Cleveland,Ohio,1942 A series of five educational lectures presented to members of the American society for metals at the twenty-third National metal congress and exposition Philadelphia Pa. Oct. 20-25, 1941
Met 25.0304

NATIONAL METAL CONGRESS AND EXPOSITION 27TH
Induction heating Cleveland 1946 Feb 4-8 By H.B. Osborn and others American society for metals A.S.M.,Cleveland,Ohio,1946
Met 25.0237

NATIONAL METAL CONGRESS AND EXPOSITION 27TH
Magnesium Cleveland 1948 Feb 4-8 By L.M. Pidgeon and others A series of five educational lectures on magnesium presented to members of the A.S.M. during the twenty-seventh National metal congress and exposition, Cleveland,February 4 to 8.1946
Met 25.0542

NATIONAL METAL CONGRESS AND EXPOSITION 27TH
Surface stressing of metals :a series of five educational lectures...presented...during the twenty-seventh National metal congress and exposition Cleveland,Ohio 1946 Feb 4-8 By H.F. Moore and others American society for metals American society for metals, Cleveland,Ohio,1947
Eng 41.3731

NATIONAL METAL CONGRESS AND EXPOSITION 28TH
Electronic methods of inspection of metals By H.F. Hamburg and others American society for metals,Cleveland,Ohio,1947 A series of seven educational lectures presented during the 28th National metal congress and exposition
Met 25.1313

NATIONAL METAL CONGRESS AND EXPOSITION 28TH
Physical metallurgy of aluminum alloys By W. L. Fink American society for metals, Cleveland,Ohio,1949 A series of five educational lectures presented during the 28th National metal congress and exposition
Met 25.1308

NATIONAL METAL CONGRESS AND EXPOSITION 28TH
The Structure of cast iron Atlantic City 1946 Nov 18-22 By Alfred Boyles American society for metals,Cleveland,Ohio,1947, reprinted 1949 A series of three educational lectures...presented to members of the ASM during the twenty-eighth National metal congress and exposition, Atlantic City, November 18 to 22,1946
Met 25.0401

NATIONAL METAL CONGRESS AND EXPOSITION 29TH
Copper and copper alloys Chicago 1947 Oct 18-24 By Owen W. Ellis American society for metals,Cleveland,Ohio,1948 A series of lectures on copper and copper alloys presented to members of the A.S.M. during the twenty-ninth National Metal Congress and Exposition, Chicago,October 18 to 24,1947
Met 25.0532

NATIONAL METAL CONGRESS AND EXPOSITION 29TH
Fracturing of metals :a seminar...held during the 29th National metal congress and exposition...Co-ordinated by...the metallurgical staff of the Case institute of technology under the direction of George Sachs Chicago,Ill. 1947 Oct 18-24 Sponsored by the American society for metals American society for metals,Cleveland,Ohio,1948
Eng 41.3732

NATIONAL METAL CONGRESS AND EXPOSITION 29TH
Fracturing of metals a seminar Chicago 1947 Oct 18-24 American society for metals and Case institute of technology American society for metals,Cleveland,Ohio,n.d. A seminar held during the 29th National metal congress and exposition
Met 25.0861

NATIONAL METAL CONGRESS AND EXPOSITION 29TH
Introductory physical metallurgy Chicago 1947 Oct 18-24 By Clyde W. Mason American society for metals A.S.M.,Ohio,1947 A series of lectures...presented to members of the A.S.M.,during the twenty-ninth National metal congress and exposition,Chicago October 18 to 24,1
Met 25.0030

NATIONAL METAL CONGRESS AND EXPOSITION 30TH
Cold working of metals Philadelphia 1948 Oct 25-29 American society for metals American society for metals,Cleveland,Ohio, 1949
Met 25.1199

NATIONAL METAL CONGRESS AND EXPOSITION 30TH
Grain control in industrial metallurgy By J. E. Burke American society for metals, Cleveland,1949 A series of four educational lectured presented to members of the A.S.M. during the 30th National metal congress and exposition Philadelphia, Oct. 23-29,1948
Met 25.1265

NATIONAL METAL CONGRESS AND EXPOSITION 30TH
Metallurgy and magnetism Philadelphia 1948 Oct 23-29 By James K. Stanley American society for metals,Cleveland,Ohio,1949 Lectures presented to members of the A.S.M. during the 30th National metal congress and exposition
Met 25.1104

NATIONAL METAL CONGRESS AND EXPOSITION 30TH
Properties of metals in materials engineering Philadelphia 1948 Oct 23-29 By R.L. Templin and others American society for metals,Cleveland,Ohio,1949 A series of eight educational lectures presented to members of the A.S.M. during the thirtieth National metal congress and exposition
Met 25.0826

NATIONAL METAL CONGRESS AND EXPOSITION 31ST
Machining-theory and practice By Hans Ernst and others American society for metals, Cleveland,Ohio,1950 A series of thirteen educational lectures on machining-theory and practice presented to members of the A.S.M. during the thirty-first National metal congress and exposition,Cleveland.
Met 25.0717

NATIONAL METAL CONGRESS AND EXPOSITION 31ST
Machining-theory and practice... :a series of thirteen educational lectures...presented... during the thirty first National metal congress and exposition By Hans Ernst and others American society for metals American society for metals,Cleveland,Ohio, 1950
Eng 41.3909

NATIONAL METAL CONGRESS AND EXPOSITION 31ST
Thermodynamics in physical metallurgy :a seminar held during the 31st national metal congress and exposition Cleveland,Ohio 1949 Oct 15-21 American society for metals American society for metals,Cleveland,Ohio, 1952
Met 25.2372

NATIONAL METAL CONGRESS AND EXPOSITION 32ND
Atom movements Chicago 1950 Oct 21-27 American society for metals American society for metals,Cleveland,1951 A seminar held during the 32nd national metal congress and exposition
Met 25.1272

NATIONAL METAL CONGRESS AND EXPOSITION 32ND
High temperature properties of metals Chicago 1950 Oct 23-27 By E.R. Parker and others American society for metals,Cleveland, Ohio,1951 A series of five educational lectures presented to members of the A.S.M. during the 32nd National metal congress and exposition
Met 25.0978

NATIONAL METAL CONGRESS AND EXPOSITION 32ND
Interpretation of tests and correlation with service Chicago 1950 Oct 23-27 By M.F. Garwood and others American society for metals,Cleveland,Ohio,1951 Lectures presented during the thirty-second National metal congress and exposition.
Met 25.0811

NATIONAL METAL CONGRESS AND EXPOSITION 32ND
Interpretation of tests and correlation with service Cleveland 1950 Oct 23-27 By M.F. Garwood and others American society for metals A.S.M.,1950 Lectures presented during the 32nd National metal congress and exposition
Met 25.0883

NATIONAL METAL CONGRESS AND EXPOSITION 32ND
Interpretation of tests and correlation with service :a series of four educational lectures. presented during the thirty-second National metal congress and exposition Chicago,Ill. 1950 Oct 23-27 By M.F. Garwood and others American society for metals American society for metals,Cleveland,Ohio,1951
Eng 41.3837

NATIONAL METAL CONGRESS AND EXPOSITION 33RD
Metal interfaces a seminar held during the thirty-third national metal congress and exposition Detroit 1951 Oct 13-19 American society for metals American society for metals,Cleveland,Ohio,1952
Met 25.1261

NATIONAL METAL CONGRESS AND EXPOSITION 33RD
Residual stress measurements Detroit 1951 Oct 15-19 By R.G. Treuting and others A series of four educational lectures presented to members of the A.S.M. during the 33rd National metal congress and exposition
Met 25.0962

NATIONAL METAL CONGRESS AND EXPOSITION 34TH
Behaviour of metals at low temperatures By R. M. Brick and others American society for metals,Cleveland,Ohio,1953 A series of lectures presented at the 34th National metal congress and exposition
Met 25.0973

NATIONAL METAL CONGRESS AND EXPOSITION 34TH
Modern research techniques in physical metallurgy a seminar... held during the thirty-fourth national metal congress and exposition Philadelphia 1952 Oct 18-24 American society for metals American society for metals,Cleveland,Ohio,1953
Met 25.1640

NATIONAL METAL CONGRESS AND EXPOSITION 35TH
Fatigue By Thomas J. Dolan and others American society for metals,Cleveland,Ohio, 1954 A series of lectures presented during the 35th metal congress and exposition
Met 25.0914

NATIONAL METAL CONGRESS AND EXPOSITION 35TH
Relation of properties to microstructure Cleveland 1953 Oct 17-23 American society for metals American society for metals, Cleveland,Ohio,1954 A symposium presented at the 35th National metal congress and exposition
Met 25.1201

NATIONAL METAL CONGRESS AND EXPOSITION 35TH
Relation of properties to microstructure :a seminar held during the thirty-fifth National metal congress and exposition Cleveland,Ohio 1953 Oct 17-23 Sponsored by the American society for metals American society for metals,Cleveland,Ohio,1954
Eng 41.3501

NATIONAL METAL CONGRESS AND EXPOSITION 35TH
Surface protection against wear and corrosion : two series of educational lectures...presented. during the thirty-fifth National metal congress and exposition Cleveland,Ohio 1953 Oct 19-23 By H.S. Avery and others American society for metals American society for metals,Cleveland,Ohio,1954
Eng 41.3715

NATIONAL METAL CONGRESS AND EXPOSITION 35TH
Surface protection against wear and corrosion.. educational lectures... presented during the thirty-fifth national metal congress and exposition Cleveland 1953 Oct 19-23 By H. S. Avery and others American society for metals American society for metals,Cleveland, Ohio,1954
Met 25.1977

NATIONAL METAL CONGRESS AND EXPOSITION 36TH
Impurities and imperfections a seminar held during the 36th national metal congress and exposition American society for metals American society for metals,Cleveland,1955
Met 25.1230

NATIONAL METAL CONGRESS AND EXPOSITION 37TH
Theory of alloy phases a seminar Philadelphia 1955 Oct 15-21 American society for metals American society for metals,Cleveland,Ohio,1956 A seminar held during the 37th National metal congress and exposition
Met 25.1051

NATIONAL METAL CONGRESS AND EXPOSITION 38TH
Creep and recovery a seminar on creep and recovery of metals Cleveland 1956 Oct 6-12 American society for metals American society for metals,Cleveland,1957 Held during the 38th National metal congress and exposition
Met 25.0950

NATIONAL METAL CONGRESS AND EXPOSITION 38TH
Fatigue durability of carburized steel By J. B. Bidwell and others General motors corporation American society for metals, Cleveland,Ohio,1957 A series of educational lectures presented by members of the General motors corporation research staff to the members of the A.S.M. during the 38th National metal congress and exposition
Met 25.0920

NATIONAL METAL CONGRESS AND EXPOSITION 39TH
Liquid metals and solidification a seminar held during the 39th national metal congress and exposition Chicago 1957 Nov 2-8 American society for metals A.S.M.,Cleveland, Ohio,1958
Met 25.1510

NATIONAL METAL CONGRESS 40TH Quenching of steels :a compilation of papers presented at the National metal congress Cleveland,Ohio 1948 Oct 27-31 American society for metals American society for metals,Cleveland,Ohio, 1951
Met 25.0446

NATIONAL MULTIPLE SCLEROSIS SOCIETY New research techniques of neuroanatomy :a symposium Edited by William F. Windle Thomas,Springfield,Ill.,1957
An 32.4352

NATIONAL NORTHERN DEVELOPMENT CONFERENCE 3rd
Proceedings Edmonton,Alberta 1964 Oct 21-23 Alberta and northwest chamber of mines and resources,and,Edmonton chamber of commerce illus 170p 28cm Alberta,c1965
Sco 14.7424

NATIONAL OCEANOGRAPHIC COUNCIL The Commonwealth oceanographic conference Proceedings Wormley,Surrey 1954 Oct 18-22 vi,93p 25cm Cambridge university press, Cambridge,1955
Sco 14.0123

NATIONAL PHYSICAL LABORATORY Attention in neurophysiology :an international conference Proceedings Teddington 1967 Oct Edited by C.R. Evans and T.B. Mulholland port Butterworths,London,1970
Psy 31.2849

NATIONAL PHYSICAL LABORATORY Boundary layer effects in aerodynamics :a symposium 5th Proceedings Teddington 1955 Mar 31-Apr 1 illus,diagrms H.M.S.O.,London,1955 Mimeograph
Chem E 24.0229

NATIONAL PHYSICAL LABORATORY Boundary layer effects in aerodynamics :a symposium 1955 London,1955
Eng 41.6960

NATIONAL PHYSICAL LABORATORY Boundary layer effects in aerodynamics: a symposium proceedings Teddington 1955 March 31 28cm London,1955
Philos 1.0437

NATIONAL PHYSICAL LABORATORY Commission internationale de l'eclairage en succession a la Commission internationale de photometrie 8e session Recueil des travaux et compte rendu Cambridge 1931 Sep Cambridge university press,Cambridge,1932
Psy 31.2562

NATIONAL PHYSICAL LABORATORY Conference on machine perception of patterns and pictures Proceedings Teddington 1972 Apr 12-14 Institute of physics.Conference series, 13 illus ix,362p Institute of physics,London, 1972
Math L 5.3882

NATIONAL PHYSICAL LABORATORY Creep and fracture of metals at high temperature :a symposium Proceedings Teddington 1954 May 31-Jun 2 National physical laboratory, Teddington,Middx,1955
Met 25.0949

NATIONAL PHYSICAL LABORATORY Engineering dimensional metrology :a symposium Proceedings Teddington 1953 Oct 21-24 Vol 1-2 2 vols H.M.S.O.,London,1955
Eng 41.3927

NATIONAL PHYSICAL LABORATORY Engineering dimensional metrology:a symposium proceedings Teddington 1953 October 21-24 Vols 1-2 26cm London,1955
Philos 1.0402

NATIONAL PHYSICAL LABORATORY International conference on machine translation of languages and applied language analysis proceedings Teddington 1961 Sep 5-8 Vol. 1-2 National physical laboratory.Proceedings of symposia, 13 23cm 2 vols H.M.S.O., London,1962
Math L 5.0960

NATIONAL PHYSICAL LABORATORY International symposium on automatic digital computation 3rd proceedings Teddington 1953 Ot 25-28 illus vi,296p 26cm National physical laboratory,Teddington,1954
ATH L 5.0724

NATIONAL PHYSICAL LABORATORY Mechanisation of thought processes :a symposium Proceedings Teddington 1958 Nov 24-27 Vol 1-2 National physical laboratory. Symposia, 10 2 vols H.M.S.O.,London,1959
Eng 41.6010

NATIONAL PHYSICAL LABORATORY Mechanisation of thought processes symposium proceedings Teddington 1958 Nov 24-27 National physical laboratory.Proceedings of symposia, 10 23cm 2 vols H.M.S.O.,London,1959
Math L 5.0820

NATIONAL PHYSICAL LABORATORY Mechanization of thought processes National physical laboratory symposium 10th Proceedings Teddington 1958 Nov 24-27 Vol 1-2 2 vols H.M.S.O.,London,1959
Psy 31.2019

NATIONAL PHYSICAL LABORATORY Numerical methods for unconstrained optimization Joint institute of mathematics and its application-National physical laboratory conference Proceedings Teddington 1971 Jan 7-8 Edited by W. Murray xi,144p 24cm Academic press,London;New York,1972
Math S 3.1729

NATIONAL PHYSICAL LABORATORY Relation between the structure and mechanical properties of metals National physical laboratory symposium 15th Proceedings Teddington 1963 Jan 7-9 Vol 1-2 2 vols H.M.S.O.,London,1963
Met 25.1214

NATIONAL PHYSICAL LABORATORY Road vehicle aerodynamics Symposium on road vehicle aerodynamics 1st Proceedings London 1969 Nov 6-7 Edited by A.J. Scibor-Rylski Organised by the City university.Department of aeronautics City university,London,1970 Organised in cooperation with the National physical laboratory,the Motor industry research association and the Institution of mechanical engineers
Eng 41.8336

NATIONAL PHYSICAL LABORATORY Sea-going qualities of ships :a seminar Teddington 1961 Oct 17-18 H.M.S.O.,London,1963
Eng 41.6932

NATIONAL PHYSICAL LABORATORY The Accuracy of industrial measurement of length and diameter : conference Proceedings Teddington 1962 Apr 17-18 National physical laboratory. Conference, 14 H.M.S.O.,London,1963
Eng 41.3933

NATIONAL PHYSICAL LABORATORY The Control of noise :a conference Proceedings Teddington 1961 Jun 26-28 National physical laboratory. Symposia, 12 H.M.S.O.,London,1962
Eng 41.4218

NATIONAL PHYSICAL LABORATORY The Engineering uses of holography :a symposium Proceedings Glasgow 1968 Sep 17-20 Edited by Elliot R. Robertson and James M. Harvey Cambridge university press,Cambridge,1970
An 32.5364

NATIONAL PHYSICAL LABORATORY The Engineering uses of holography :a symposium Proceedings Glasgow 1968 Sep 17-20 University of Strathclyde Edited by E.R. Robertson and J.M. Harvey Cambridge university press,Cambridge, 1970
Eng 41.4290

NATIONAL PHYSICAL LABORATORY The Physical chemistry of metallic solutions and intermetallic compounds :a symposium Proceedings London 1958 Jun 4-6 Vol 1-2 National physical laboratory.Symposium, 9 2 vols H.M.S.O.,London,1959
Met 25.1779

NATIONAL PHYSICAL LABORATORY Visual problems of colour :a symposium 6 Proceedings Teddington 1957 Sep 23-25 Vol 1-2 2 vols H.M.S.O.,London,1958
Psy 31.0353

NATIONAL PHYSICAL LABORATORY Visual problems of colour :a symposium Teddington 1957 Sep 23-25 Vol 1-2 National physical laboratory.Symposia, 8 2 vols H.M.S.O., London,1958
Phys 20.1033

NATIONAL PHYSICAL LABORATORY Wind effects on buildings and structures :a conference Proceedings Teddington 1963 Jun 26-28 Vol 1-2 National physical laboratory. Symposium, 16 H.M.S.O.,London,1963
Eng 41.6904

NATIONAL PHYSICAL LABORATORY.CONFERENCE, 14 The Accuracy of Industrial measurement of length and diameter :the conference London 1962 Apr 17-18 National physical laboratory H.M.S.O.,London,1963
Met 25.2166

NATIONAL PHYSICAL LABORATORY.CONFERENCE, 14 The Accuracy of industrial measurement of length and diameter :conference Proceedings Teddington 1962 Apr 17-18 National physical laboratory H.M.S.O.,London,1963
Eng 41.3933

NATIONAL PHYSICAL LABORATORY.SYMPOSIA, 10 Mechanisation of thought processes :a symposium Proceedings Teddington 1958 Nov 24-27 Vol 1-2 National physical laboratory 2 vols H.M.S.O.,London,1959
Eng 41.6010

NATIONAL PHYSICAL LABORATORY.SYMPOSIA, 12 The Control of noise :a conference Proceedings Teddington 1961 Jun 26-28 National physical laboratory H.M.S.O.,London, 1962
Eng 41.4218

NATIONAL PHYSICAL LABORATORY.SYMPOSIUM, 15
The Relation between the structure and mechanical properties of metals :a conference Proceedings Teddington 1963 Jan 7-9 Vol 1-2 2 vols H.M.S.O.,London,1963
Eng 41.3530

NATIONAL PHYSICAL LABORATORY.SYMPOSIUM, 16
Wind effects on buildings and structures :a conference Proceedings Teddington 1963 Jun 26-28 Vol 1-2 Great Britain. Department of scientific and industrial research National physical laboratory H.M. S.O.,London,1963
Eng 41.6904

NATIONAL PHYSICAL LABORATORY SYMPOSIUM 10th
Mechanization of thought processes Teddington 1958 Nov 24-27 Vol 1-2 National physical laboratory 2 vols H.M.S. O.,London,1959
Psy 31.2019

NATIONAL PHYSICAL LABORATORY SYMPOSIUM 15th
Proceedings Relation between the structure and mechanical properties of metals Teddington 1963 Jan 7-9 Vol 1-2 National physical laboratory 2 vols H.M.S. O.,London,1963
Met 25.1214

NATIONAL PRESTRESSED CONCRETE SHORT COURSE 2nd
Papers Advanced prestressed concrete design Dayton beach,Fla. 1958 Jan University of Florida.College of engineering. Bulletin series, 76 University of Florida, Gainesville,Fla.,1958
Eng 41.2847

NATIONAL RADIO ASTRONOMY OBSERVATORY
Interstellar ionized hydrogen :a symposium Charlottesville,Va. 1967 Dec Edited by Yervant Jerzian Co-sponsored by the Arecibo ionospheric observatory 774p Benjamin,New York,1968
Obs 6.3447

NATIONAL RADIO ASTRONOMY OBSERVATORY
Interstellar ionized hydrogen :a symposium Charlottesville,Va. 1967 Dec Edited by Yervant Terzian Jointly sponsored by Arecibo ionospheric observatory 774p Benjamin,New York,1968
TA 15.0437

NATIONAL RESEARCH COUNCIL Conference on non-crystalline solids Alfred,N.Y. 1958 Sep 3-5 Edited by V.D. Frechette John Wiley,New York;London,1960
Met 25.0126

NATIONAL RESEARCH COUNCIL Conference on nuclear processes in geologic settings Proceedings Williamsbay 1953 Sep,21-23 University of Chicago Williams Bay,Wisconsin, 1953
Philos 1.0514

NATIONAL RESEARCH COUNCIL Conference on training in applied mathematics proceedings New York 1953 Oct 22-24 American mathematical society 97p 27cm Columbia university,New York,1953
Math L 5.0709

NATIONAL RESEARCH COUNCIL Contributions to the theory of partial differential equations Conference on partial differential equations typescript Harriman,N.Y. 1952 Oct By S. Bergman and others Edited by Lipman Bers and others Annals of mathematics studies, 33 Bibliog 257p 26cm Princeton university press,Princeton,N.J.,1954
P. Math 2.1228

NATIONAL RESEARCH COUNCIL Ecology of soil-borne plant pathogens.Prelude to biological control Factors determining the behavior of plant pathogens in soil :an international symposium Berkeley,Calif. 1963 Apr 7-13 Edited by Kenneth F. Baker and William C. Snyder illus. Murray,London,1965
Mol 45.0575

NATIONAL RESEARCH COUNCIL Endocrines in development Shelter Island,N.Y. 1956 Sep 11-13 Edited by Ray L. Watterson Developmental biology conference series,1956 University of Chicago.Committee on publications in biology and medicine University of Chicago press,Chicago,Ill.,1959
Gen 34.0553

NATIONAL RESEARCH COUNCIL Fracture An International conference on the atomic mechanisms of fracture Proceedings Swampscott,Mass 1959 Apr 12-16 Edited by Benjamin Lewis Averbach and others 646p Technology press;John Wiley,Cambridge,Mass;New York,1959
Cav 7.0102

NATIONAL RESEARCH COUNCIL Fundamentals of deformation processing Sagamore army materials conference 9th Raquette Lake,New York 1962 Aug 28-31 By Walter A. Backofen and others United States.Army materials research agency and Syracuse university Syracuse university press,Syracuse,New York, 1964
Met 25.0741

NATIONAL RESEARCH COUNCIL International association for quaternary research congress 7th Proceedings Boulder,Colo. 1965 Aug 14-Sep 19 Vol 10: arctic and alpine environments International association for quaternary research Edited by W.H. Osburn and H.E. Wright illus,maps xi,308p Indiana university press,Bloomington,Ind.; London,1968 For other volumes of these proceedings,see also International union for quaternary research
Bot 42.2006

NATIONAL RESEARCH COUNCIL International association for quaternary research congress 7th Proceedings Boulder,Colo. 1965 Aug 14-Sep 19 10: arctic and alpine environments International association for quaternary research Edited by W.H. Osburn and H.E. Wright illus,maps xi,308p 25cm Indiana university press,Bloomington,Ind.;London,1968 For other volumes of these proceedings,see also International union for quaternary research
Sco 14.7841

NATIONAL RESEARCH COUNCIL International studies on the quaternary Boulder,Colo. 1965 Edited by H.E. Wright and David G. Frey Geological society of America.Special papers, 84 Geological society of America,New York, 1965
Bot 42.3133

NATIONAL RESEARCH COUNCIL International studies on the quaternary International association for quaternary research :congress 7th Papers Boulder,Colo. 1965 Edited by H.E. Wright and David G. Frey Geological society of America.Special papers,84 Geological society of America,New York,1965
Geog 13.0309

NATIONAL RESEARCH COUNCIL International union for quaternary research congress :a collection of papers derived from the INQUA-NCAR symposium 7th Proceedings Boulder, Colo. 1965 Aug 14-Sep 19 5: causes of climatic change International union for quaternary research,and,National center for atmospheric research Edited by J.Murray Mitchell Meteorological monographs,8,no 30 illus iv,159p 29cm American meteorological association,Boston,Mass.,1968 For other volumes of these proceedings,see also 'International association for quaternary research'
Sco 14.7750

NATIONAL RESEARCH COUNCIL Permafrost international conference 1st Proceedings Lafayette,Indiana 1963 Nov 11-15 Presented by the Building research advisory board National research council.Publication,1287 DLC,GB641.16 1963 illus,maps,tables,graphs 563p National academy of sciences,Washington, 1966
Sco 14.0390

NATIONAL RESEARCH COUNCIL Pleistocene extinctions:the search for a cause International association for quaternary research.Congress 7th Proceedings Boulder,Colo. 1965 Aug I.N.Q.U.A. Edited by P.S. Martin and H.E. Wright Yale university press,New Haven,Conn.;London,1967
Geog 13.1320

NATIONAL RESEARCH COUNCIL Pleistocene extinctions:the search for a cause International association for quaternary research congress 7th Proceedings Boulder,Colo. 1965 Aug 30-Sep 5 Vol 6 Edited by P.S. Martin and H.E. Wright Yale university press,New Haven,Conn.;London,1967
Bot 42.3244

NATIONAL RESEARCH COUNCIL Pyrometry The Papers and discussion of a symposium on pyrometry Chicago,Ill. 1919 Sep illus 701p New York,1920
WSM 43.4920

NATIONAL RESEARCH COUNCIL Quaternary paleoecology International association for quaternary research.Congress 7th Proceedings Boulder,Colo. 1965 Aug-Sep Vol 7 Edited by E.J. Cushing and H.E. Wright Yale university press,New Haven,Conn.;London, 1967
Geog 13.1318

NATIONAL RESEARCH COUNCIL Quaternary paleoecology International association for quaternary research congress 7th Proceedings Boulder,Colo. 1965 Aug 30-Sep 5 Vol 7 Edited by E.J. Cushing and H. E. Wright Yale university press,New Haven, Conn.;London,1967
Bot 42.3245

NATIONAL RESEARCH COUNCIL Radiation-induced chromosome aberrations Conference on biochemical and biophysical mechanisms in the production of radiation- induced chromosome aberrations San Juan,Puerto Rico 1961 Nov 16-18 Edited by Sheldon Wolff Columbia university press,New York;London,1963
Gen 34.1122

NATIONAL RESEARCH COUNCIL Radiation-induced chromosome aberrations :report of a conference on biochemical and biophysical mechanisms in the production of radiation-induced chromosome aberrations San Juan,P.R. 1961 Nov 16-18 Edited by Sheldon Wolff illus. xii,304p Columbia university press,New York;London,1963
Bot 42.4730

NATIONAL RESEARCH COUNCIL Research in photosynthesis :a conference Papers and discussions Gatlinburg,Tenn 1955 Oct 25-29 Edited by H. Gaffron and others xiv,524p Interscience,London,1957
Bot 42.1869

NATIONAL RESEARCH COUNCIL Structure and properties of solid surfaces :a conference Lake Geneva,Wisc. 1952 Sep Edited by Robert Gomer and Cyril Stanley Smith University of Chicago press,Chicago,1953
Met 25.1594

NATIONAL RESEARCH COUNCIL Structure and properties of solid surfaces :a conference Lake Geneva,Wisc. 1952 Sep Edited by Robert Gomer and Cyril Stanley Smith illus 491p University of Chicago press,Chicago,Ill. 1953
Cav 7.1344

NATIONAL RESEARCH COUNCIL Symposium on naval hydrodynamics proceedings Washington D.C. 1956 Sep.24-28 United states.Office of naval research Edited by F.S. Sherman National research council.Publication, 515 National research council,Washington D.C.,1957
A Math 4.0530

NATIONAL RESEARCH COUNCIL Symposium on radiobiology :the basic aspects of radiation effects on living systems Oberlin,Ohio 1950 Jun 14-18 Edited by James J. Nickson Wiley;Chapman and Hall,New York;London,1952
Radioth 35.1717

NATIONAL RESEARCH COUNCIL Symposium on stress Army medical services,Washington,D.C., 1953
Med 36.0265

NATIONAL RESEARCH COUNCIL.COMMITTEE FOR THE CONFERENCE ON FRACTURE Atomic mechanisms of fracture :an international conference Proceedings Swampscott,Mass. 1959 Apr 12-16 Edited by B.L. Averback and others Technology press books in science and engineering Technology press,John Wiley; Chapman and Hall,New York;London,1959
Met 25.2323

NATIONAL RESEARCH COUNCIL.COMMITTEE ON ANAESTHESIA Uptake and distribution of anaesthetic agents :a conference New York 1962 Apr 23-24 Edited by E.M. Papper and Richard J. Kitz McGraw-Hill,New York,1963
Pha 16.0043

NATIONAL RESEARCH COUNCIL.COMMITTEE ON PHOTOBIOLOGY Photosynthetic mechanisms of green plants :a symposium Papers Warrenton, Va. 1963 Oct 14-18 National research council.Publication,1145 Washington,1963
Bioch 33.0774

NATIONAL RESEARCH COUNCIL.COMMITTEE ON PHOTOBIOLOGY Research in photosynthesis : a conference Papers and discussions Gatlinburg,Te. 1955 Oct 25-29 Edited by H. Gaffron and others Supported by the National science foundation Interscience,New York; London,1957
Chem 18.2138

NATIONAL RESEARCH COUNCIL.COMMITTEE ON PHOTOBIOLOGY Research in photosynthesis a conference Papers and discussions Gatlinburg,Tenn. 1955 Oct 25-29 Edited by H. Gaffron and others Interscience,New York, 1957
Bioch 33.0758

NATIONAL RESEARCH COUNCIL.COMMITTEE ON PHOTOBIOLOGY The Luminescence of biological systems :a conference Proceedings Asilomar,Calif. 1954 Mar 28-Apr 2 Edited by Frank H. Johnson American assoc.for the advancement of science,Washington,D.C.,1955
Bal 39.0080

NATIONAL RESEARCH COUNCIL.COMMITTEE ON SHIP STRUCTURE DESIGN An Evaluation of current knowledge of the mechanics of brittle fracture including papers and discussions presented at Conference on brittle fracture mechanics held.. at Massachusetts institute of technology Cambridge,Mass. 1953 Oct 13-16 By D.C. Drucker SSC 69 N.A.S.-N.R.C.Division of engineering and industrial research,Washington, D.C.,1954
Eng 41.3743

NATIONAL RESEARCH COUNCIL.COMMITTEE ON SOLIDS Imperfections in nearly perfect crystals :a symposium Papers Pocono Manor,Pa. 1950 Oct 12-14 Edited by William Shockley and others John Wiley,New York,1952
Cav 7.2045

NATIONAL RESEARCH COUNCIL.COMMITTEE ON SOLIDS Imperfections in nearly perfect crystals :a symposium Pocono Manor 1950 Oct 12-14 Edited by W. Shockley and others Wiley; Chapman and Hall,New York;London,1952
Met 25.1226

NATIONAL RESEARCH COUNCIL.COMMITTEE ON SOLIDS Imperfections in nearly perfect crystals :a symposium Pocono Manor,Pa. 1950 Oct 12-14 Edited by W. Shockley and others Wiley; Chapman and Hall,New York;London,1952
Eng 41.4427

NATIONAL RESEARCH COUNCIL.COMMITTEE ON SOLIDS Phase transformations in solids :symposium New York 1948 Aug 23-26 Edited by R. Smoluchowski and others Wiley;Chapman and Hall,New York;London,1951
Met 25.1250

NATIONAL RESEARCH COUNCIL.COMMITTEE ON SOLIDS Structure and properties of solid surfaces : conference Papers Lake Geneva,Wis. 1952 Sep Edited by Robert Gomer and Cyril Stanley Smith University of Chicago press,Chicago, Ill.,1953
Col S 12.0275

NATIONAL RESEARCH COUNCIL.COMMITTEE ON THE BIOLOGICAL EFFECTS OF ATOMIC RADIATION Genetics,radiobiology and radiology proceedings mid-western conference 1958 May Edited by Wendell G. Scott and Titus Evans Blackwell,Oxford,1959
Gen 34.1109

NATIONAL RESEARCH COUNCIL.COMMITTEE ON TISSUE TRANSPLANTATION Cell-bound antibodies :a conference Proceedings Washington,D.C. 1963 May 10 Edited by Bernard Amos and Hilary Koprowski Wistar institute press, Philadelphia,1963
Bioch 33.0534

NATIONAL RESEARCH COUNCIL.COMMITTEE ON TISSUE TRANSPLANTATION Cell-bound antibodies : conference Proceedings Washington,D.C. 1963 May 10 Edited by Bernard Amos and Hilary Koprowski Wistar institute press, Philadelphia,Pa.,1963
An 32.3296

NATIONAL RESEARCH COUNCIL.COMMITTEE ON TISSUE TRANSPLANTATION Cross-reacting antigens and neoantigens (with implications for autoimmunity and cancer immunity):a conference Proceedings Edited by John J. Trentin illus,tables Williams and Wilkins,Baltimore, Md.,1967
Path 30.2544

NATIONAL RESEARCH COUNCIL.DIVISION OF BIOLOGY AND AGRICULTURE.BIOLOGY COUNCIL Concepts of biology Conference report Lee,Mass. 1955 Oct 12-14 Edited by R.W. Gerard and R.B. Stevens National research council. Publication, 560 Washington,D.C.,1958
Bal 39.0989

NATIONAL RESEARCH COUNCIL.DIVISION OF MATHEMATICS Linear algebras conference report Shelter Island,N.Y. 1956 Jun 6-8 By A. Adrian Albert National research council. Publications, 502 v,59p 28cm National academy of sciences.National research council, Washington,D.C.,1957
P. Math 2.2033

NATIONAL RESEARCH COUNCIL.DIVISION OF MEDICAL SCIENCES Symposium on atherosclerosis Papers 1954 Mar 22-23 Held at the request of the United States.Air force.Directorate of research and development.Human factors division National research council. Publication, 338 National research council, Washington,D.C.,1955
Path 30.2349

NATIONAL RESEARCH COUNCIL.DIVISION OF MEDICAL SCIENCES Symposium on atherosclerosis Washington,D.C. 1954 Mar 22-23 Sponsored by Human factors division.Air force directorate of research and development National research council.Publication, 338 National research council,Washington,D.C.,1955
Med 36.0266

NATIONAL RESEARCH COUNCIL.DIVISION OF MEDICAL SCIENCES Symposium on preventive and social psychiatry Washington,D.C. 1957 Apr 15-17 US government printing office, Washington,D.C.,c1957
Psy 28.0058

NATIONAL RESEARCH COUNCIL.DIVISION OF MEDICAL SCIENCES The Physiology of induced hypothermia :a symposium Proceedings Washington,D.C. 1955 Oct 28-29 Edited by Robert D. Dripps National research council. Publications, 451 Washington,D.C.,1956
Phys 20.0987

NATIONAL RESEARCH COUNCIL.DIVISION OF PHYSICAL SCIENCES. COMMITTEE ON SOLIDS Phase transformations in solids :a symposium Papers Cornell university 1948 Aug 23-26 Edited by R. Smoluchowski and others J.Wiley; Chapman and Hall,New York;London,1951
Chem 18.2149

NATIONAL RESEARCH COUNCIL.PUBLICATION, 338 Symposium on atherosclerosis Washington,D.C. 1954 Mar 22-23 National research council. Division of medical sciences Sponsored by Human factors division.Air force directorate of research and development National research council,Washington,D.C.,1955
Med 36.0266

NATIONAL RESEARCH COUNCIL.PUBLICATION, 829 Magneto-fluid dynamics International symposium on magneto-fluid dynamics proceedings Williamsburg,Va. 1960 Jan and Washington D.C. Edited by F.N. Frenkel and W.R. Sears Sponsored by the International union of theoretical and applied mechanics National research council, Washington,1960
A Math 4.0619

NATIONAL RESEARCH COUNCIL.SUBCOMMITTEE ON RADIOBIOLOGY Basic mechanisms in radiobiology.II.Physical and chemical aspects : an informal conference Proceedings Highland Park,Ill. 1953 May 7-9 Edited by John L. Magee and others National research council.Publications, 305 National academy of science,Washington,D.C.,1953
Radioth 35.1546

NATIONAL RESEARCH COUNCIL.SUBCOMMITTEE ON RADIOBIOLOGY Cellular aspects of basic mechanisms in radiobiology :an informal conference Proceedings Bear Mountain,N.Y. 1955 May 12-14 Edited by Harvey M. Patt and E.L. Powers National research council. Nuclear science series.Report, 18 NAS-NRC 450 National research council,Washington,D.C. 1956
Radioth 35.1716

NATIONAL RESEARCH COUNCIL.SUBCOMMITTEE ON RADIOBIOLOGY Conference on molecular and radiation biology Ardsley-on-Hudson,N.Y. 1959 Dec 2-4 Edited by R.A. Deering National research council.Nuclear science series.Report, 31 NAS-NRC 823 National research council,Washington,D.C.,1961
Radioth 35.1735

NATIONAL RESEARCH COUNCIL.SUBCOMMITTEE ON RADIOBIOLOGY Electron spin resonance and the effects of radiation on biological systems a conference Gatlinburg 1965 May 10-12 Edited by Wallace Snipes National research council.Nuclear science series.Report, 43 National research council,Washington,D.C.,1966
Radioth 35.1748

NATIONAL RESEARCH COUNCIL.SUBCOMMITTEE ON RADIOBIOLOGY Research in radiology :an informal conference Proceedings Highland park,Ill. 1957 May 10-12 Edited by Henry S. Kaplan National research council.Nuclear science series.Report, 22 National research council.Publication, 571 National academy of science,Washington,D.C.,1958
Radioth 35.1116

NATIONAL RESEARCH COUNCIL,CANADA The World rift systems :a symposium Report Ottawa 1965 Sep 4-5 Edited by T.N. Irvine Sponsored by the International union of geodesy and geophysics Geological survey of Canada.Paper, 66-14 Queen's printer,Ottawa, 1967
Geod 9.0138

NATIONAL RESEARCH COUNCIL OF CANADA Biochemistry and physiology of plant growth substances International conference on plant growth substances 6th Proceedings Ottawa 1967 Jul 24-29 Edited by F. Wightman and G. Setterfield Runge press, Ottawa,1968
Gen 34.1408

NATIONAL RESEARCH COUNCIL OF CANADA International conference on the deformation of crystalline solids Ottawa 1966 Aug 22-26 Canadian journal of physics, 2,pt 2-3 Ottawa,1967
Met 25.2491

NATIONAL RESEARCH COUNCIL OF CANADA International symposium on classical and contagious discrete distributions proceedings Montreal 1963 Aug 15-20 Canadian mathematical congress Edited by Ganapati P. Patil Bibliog. xiv,552p 27cm Statistical publishing society,Calctta,1965
Math 3.0709

NATIONAL RESEARCH COUNCIL OF CANADA Laboratory shear testing of soils :a symposium Ottawa 1963 Sep 9-11 American society for testing materials.Special technical publication, 361 American society for testing materials,Philadelphia,Pa.,1964
Eng 41.3157

NATIONAL RESEARCH COUNCIL OF CANADA Wind effects on buildings and structures : international research seminar Proceedings Ottawa 1967 Sep 11-15 Vol 1-2 2 vols University of Toronto press,Toronto, 1968
Eng 41.2746

NATIONAL RESEARCH COUNCIL OF THE PHILIPPINES Pacific science congress 8th proceedings Quezon City 1953 Nov 16-28 Vol 2,2A: geology and geophysics meteorology Pacific science association 2 vols National research council of the Philippines,Quezon City,1956
Geol 8.3070

NATIONAL RESEARCH INSTITUTE FOR METALS,TOKYO Memorial lecture meeting on the 10th anniversary of the foundation of National research institute for metals Proceedings Tokyo 1966 Jun 28-30 National research institute for metals,Tokyo,1966
Met 25.0113

NATIONAL SCIENCE FOUNCATION Plants and the migrations of Pacific peoples :a symposium Honolulu 1961 Aug 21-Sep 6 Bishop (Bernice P) museum,and,University of Hawaii Edited by J. Barrau Bishop museum press,Honolulu,1963 Symposium convened at the 10th Pacific science congress
Geog 13.7202

NATIONAL SCIENCE FOUNDATION Algebraic groups and discontinuous subgroups Summer mathematical institute 12th Boulder, Colo. 1965 Jul 5-Aug 6 American mathematical society Edited by Armand Borel and George D. Mostow American mathematical society.Proceedings of symposia in pure mathematics, 9 vii,426p 26cm American mathematical society,Providence,R.I.,1966
P Math 2.2719

NATIONAL SCIENCE FOUNDATION Antarctica in the International geophysical year based on a symposium on the Antarctic American geophysical union and United States national committee for the International geophysical year Geophysical monograph, 1 American geophysical union.Publication, 462 illus,maps v,133p 26cm American geophysical union,Washington,D.C.,1956
Sco 14.8201

NATIONAL SCIENCE FOUNDATION Astronomical photoelectric conference Proceedings Flagstaff,Ariz. 1953 Aug.31 - Sep. 1 Lowell observatory Edited by John B. Irwin Indiana university,Bloomington,Ind.,1953
Obs 6.2883

NATIONAL SCIENCE FOUNDATION Bacterial endotoxins :a symposium Proceedings New Brunswick,N.J. 1963 Sep 4-6 Rutgers university.Institute of microbiology Edited by Maurice Landy and Werner Braun Rutgers university.Institute of microbiology,New Brunswick,N.J.,c1963
Path 30.2611

NATIONAL SCIENCE FOUNDATION Bernoulli 1713: Bayes 1763:Laplace 1818:anniversary volume International research seminar proceedings Berkeley,Calif. 1963 University of California.Statistical laboratory Edited by Jerzy Neyman and Lucien M. LeCam ix,262p 23cm Springer-verlag,Berlwn,1965
Math 3.0708

NATIONAL SCIENCE FOUNDATION Conference on carbon 3rd Proceedings Buffalo,N.Y. 1957 Jun 17-21 By S. Morzowski and others University of Buffalo and United States. Office of naval research Pergamon press, London,1959
Met 25.0156

NATIONAL SCIENCE FOUNDATION Differential geometry Tucson, Ariz. 1960 Feb 18-19 American mathematical society Edited by Carl B. Allendoerfer American mathematical society.Proceedings of symposia in pure mathematics, 3 vii,200p 26cm American mathematical society,Providence,R.I.,1961
P. Math 2.0698

NATIONAL SCIENCE FOUNDATION Ergodic theory international symposium papers New Orleans, La. 1961 Oct.23-27 Tulane university Edited by Fred B. Wright xii,316p Academic press,New York;London,1963
Math 3.0044

NATIONAL SCIENCE FOUNDATION Ergodic theory international symposium proceedings New Orleans,La. 1961 Oct 23-27 Edited by Fred B. Wright xii,316p 24cm Academic press, New York,London,1963
P. Math 2.2188

NATIONAL SCIENCE FOUNDATION Evolving genes and proteins New Brunswick,N.J. 1964 Sep 17-18 Edited by Vernon Bryson and Henry J. Vogel xvii,617p Academic press,New York; London,1965
Bot 42.0995

NATIONAL SCIENCE FOUNDATION Finite groups institute report Pasadena,Calif. 1960 Aug 1-28 American mathematical society Edited by Marshall Hall American mathematical society.Proceedings of symposia in pure mathematics, 6 114p 26cm American mathematical society,Providence,R.I., 1962
P. Math 2.2031

NATIONAL SCIENCE FOUNDATION Fracture An International conference on the atomic mechanisms of fracture Proceedings Swampscott,Mass. 1959 Apr 12-16 Edited by B.L. Averbach and others M.I.T.press;Wiley, Cambridge,Mass.;New York,1959
Eng 41.3767

NATIONAL SCIENCE FOUNDATION Informational macromolecules :a symposium Proceedings New Brunswick,N.J. 1962 Sep 5-7 Rutgers university.Institute of microbiology Edited by Henry J. Vogel and others Academic press, New York;London,1963
Bioch 33.0822

NATIONAL SCIENCE FOUNDATION Institute in the theory of numbers report Boulder,Colo. 1959 Jun 21-Jul 17 American mathematical society Edited by Donald C.B. Marsh and James H. Jordan 350p 27cm University of Colorado,Boulder,Colo.,1959
P. Math 2.1773

NATIONAL SCIENCE FOUNDATION International symposium on forest hydrology Proceedings University Park,Pa. 1965 Aug 29-Sep 10 Edited by William E. Sapper and Howard W. Lull Pergamon press,Oxford,1967 A National Science Foundation advanced science seminar
Geog 13.0795

NATIONAL SCIENCE FOUNDATION Lattice theory symposium Monterey,Calif. 1959 Apr 16-18 American mathematical society Edited by R.P. Dilworth American mathematical society. Proceedings of symposia in pure mathematics, 2 viii,208p 26cm American mathematical society,Providence,R.I.,1961
P. Math 2.2030

NATIONAL SCIENCE FOUNDATION Localized excitations in solids International conference on localized excitations in solids 1st Proceedings Irvine 1967 Sep 18-22 Edited by R.F. Wallis xvi,782p 26cm Plenum press,New York,1968
Cav 7.2857

NATIONAL SCIENCE FOUNDATION Man's role in changing the face of the earth :international symposium Princeton,N.J. 1955 Jun 16-22 Wenner-Gren foundation for anthropological research Edited by William L. Thomas and others University of Chicago press,Chicago, Ill.;London,1956
Geog 13.1427

NATIONAL SCIENCE FOUNDATION Mathematical optimization techniques symposium papers Santa Monica,Calif. 1960 Oct.18-20 University of California Edited by Richard Bellman xii,346p 24cm University of California press,Berkeley,Calif.,1963
Math 3.0238

NATIONAL SCIENCE FOUNDATION Nonlinear partial differential equations :a symposium on methods of solution Newark,Del. 1965 Dec 27-29 Edited by W.F. Ames xv,316p Academic press,New York;London,1967
Chem E 24.0140

NATIONAL SCIENCE FOUNDATION Number theory conference proceedings Boulder,Colo. 1963 Aug 5-24 University of Colorado 121p 27cm University of Colorado,Boulder,Colo., 1963
P. Math 2.1772

NATIONAL SCIENCE FOUNDATION Organizational biosynthesis :a symposium New Brunswick,N.J. 1966 Sep 8-10 By Henry J. Vogel and others illus. xx,549p Academic press,New York; London,1967
Bot 42.1512

NATIONAL SCIENCE FOUNDATION Organizational biosynthesis :a symposium Proceedings New Brunswick,N.J. 1966 Sep 8-10 Rutgers university.Institute of microbiology Edited by Henry J. Vogel and others Academic press, New York;London,1967
Bioch 33.0832

NATIONAL SCIENCE FOUNDATION Phylogeny and evolution of crustacea :a conference Proceedings Cambridge,Mass. 1962 Mar 6-8 Harvard university.Museum of comparative zoology Edited by H.B. Whittington and W.D.I. Rolfe Museum of comparative zoology, Cambridge,Mass.,1963
Bal 39.2305

NATIONAL SCIENCE FOUNDATION Proceedings of the symposium on the arctic heat budget and atmospheric circulation Lake Arrowhead 1966 Jan-Feb University of California,and, Rand corporation Edited by J.O. Fletcher Memorandum RM-5233-NSF illus,diagrs 566p The Rand corporation,Santa Monica,1966
Sco 14.2407

NATIONAL SCIENCE FOUNDATION Semigroups Symposium on semigroups Proceedings Detroit,Mich. 1968 Jun 27th-29th Wayne state university Edited by Karl W. Folley bibliog. xi,277p 24cm Academic press,New York;London,1969
P Math 2.3927

NATIONAL SCIENCE FOUNDATION Stellar atmospheres :a conference proceedings Bloomington,Ind. 1954 Oct 1-2 Edited by Marshal H. Wrubel 183p Indiana university, Bloomington,Ind.,1955
Obs 6.1900

NATIONAL SCIENCE FOUNDATION Stereology International congress for stereology 2nd Proceedings Chicago,Ill. 1967 Apr 5-13 International society for stereology Edited by Hans Elias Springer,Berlin,1967
An 32.0107

NATIONAL SCIENCE FOUNDATION Symposium on engineering applications of random function theory and probability 1st Proceedings Edited by J.L. Bogdanoff and F. Kozin Wiley, New York,1963
Eng 41.5949

NATIONAL SCIENCE FOUNDATION Symposium on engineering applications of random function theory and probability 1st proceedings Purdue university Edited by John L. Bogdanoff and Frank Kozin x,421p 24cm John Wiley and sons,New York,1963
Math 3.0879

NATIONAL SCIENCE FOUNDATION The Astrometric conference 2nd Cincinnati 1959 May 17-21 University of Cincinnati Edited by S. Aa. Strand and O.G. Franz 58p Astronomical journal,1960 Reprinted from the Astronomical journal,vol 65,no 4,May 1960
Obs 6.3243

NATIONAL SCIENCE FOUNDATION The Central nervous system and behavior :a conference 3rd Transactions Princeton,N.J. 1960 Feb 21-24 Josiah Macy jr.foundation Edited by Mary A.B. Brazier Josiah Macy,jr.foundation, New York,1960
Psy 31.0063

NATIONAL SCIENCE FOUNDATION The Conference for instructors of astronomy :lecture notes Berkeley,Calif. 1954 Aug 12-Sep 8 Leuschner observatory Edited by Robert Fleischer and others 265p Berkeley,Calif., 1954 Typescript
Obs 6.2054

NATIONAL SCIENCE FOUNDATION The Frontal granular cortex and behaviour :a symposium University park,Pa. 1962 Aug 8-10 Edited by J.M. Warren and K. Akert McGraw-Hill,New York,1963
An 32.4656

NATIONAL SCIENCE FOUNDATION Theory of numbers symposium Pasadena,Calif. 1963 Nov 21-22 American mathematical society Edited by Albert Leon Whiteman American mathematical society.Proceedings of symposia in pure mathematics, 8 vii,214p 26cm American mathematical society,Providence,R.I., 1965
P. Math 2.1774

NATIONAL SCIENCE FOUNDATION and AMERICAN GEOPHYSICAL UNION.CLOUD PHYSICS COMMITTEE Physics of precipitation Cloud physics conference 2nd Proceedings Woods Hole, Mass. 1959 Jun 3-5 Edited by Helmut Weickmann Geophysical monograph,5 National research council.Publication,746 illus,map xii,435p 25cm American geophysical union, Washington,1960
Sco 14.0536

NATIONAL SCIENCE FOUNDATION and WISCONSIN ALUMNI RESEARCH FOUNDATION International symposium on organometallic chemistry 2nd Abstracts of proceedings Madison,Wisc. 1965 Aug 30-Sep 3 Co-sponsored by the United States.Army research office,Durham Madison, Wisc.,1965 Mimeographed
Chem 18.1589

NATIONAL SCIENCE FOUNDATION.ADVANCED SCIENCE SEMINAR PROJECTS Conference on transformation groups Proceedings New Orleans 1967 May 8-Jun 2 Edited by Paul S. Mostert Bibliog.,Illus. xiii,456p 24cm Springer-Verlag,Berlin,1968
P Math 2.3395

NATIONAL SCIENCE FOUNDATION.ATMOSPHERIC SCIENCES SECTION National conference on weather modification 2nd Proceedings Santa Barbara,Calif. 1970 Apr 6-9 436p American meteorological society,Boston,Mass., 1970
Nap 11.0977

NATIONAL SCIENCE FOUNDATION.DIVISION OF REGULATORY BIOLOGY Nervous inhibition The Friday Harbor symposium 2nd Proceedings Friday Harbor 1960 May 31-Jun 4 Edited by Ernst Florey Pergamon press, Oxford,1961 Conference also called International symposium on nervous inhibition
Bal 39.1475

NATIONAL SOCIETY FOR THE STUDY OF COMMUNICATION Communication theory and research: international symposium 1st Proceedings Kansas City 1965 Mar 24-27 Edited by Lee Thayer Thomas,Springfield,Ill.,1967
Psy 31.1068

The NATIONAL SYMPOSIUM ON COLLOID CHEMISTRY 1st Papers and discussions Madison,Wis. 1923 Jun 12-15 University of Wisconsin Edited by J.Howard Mathews Colloid symposium monograph, 1 Department of chemistry, university of Wisconsin,Madison,Wis.,1923
Bioch 33.1469

The NATIONAL SYMPOSIUM ON COLLOID CHEMISTRY 2nd Papers presented Evanston,Ill. 1924 Jun 18-21 Northwestern university Edited by Harry N. Holmes Colloid symposium monograph, 2 Chemical catalog company,New York,1925
Bioch 33.1470

The NATIONAL SYMPOSIUM ON COLLOID CHEMISTRY 3rd Papers presented Minneapolis 1925 Jun 17-19 University of Minnesota Edited by Harry N. Holmes and Harry B. Weiser Colloid symposium monograph, 3 Chemical catalog company,New York,1925
Bioch 33.1471

The NATIONAL SYMPOSIUM ON COLLOID CHEMISTRY 4th Papers presented Cambridge,Mass. 1926 Jun 23-25 Massachusetts institute of technology Edited by Harry Boyer Weiser Colloid symposium monograph, 4 Chemical catalog company,New York,1926
Bioch 33.1472

The NATIONAL SYMPOSIUM ON COLLOID CHEMISTRY 5th Papers presented Ann Arbor,Mich. 1927 Jun 22-24 University of Michigan Edited by Harry Boyer Weiser Colloid symposium monograph, 5 Chemical catalog company,New York,1928
Bioch 33.1473

NATIONAL SYMPOSIUM ON URINARY TRACT INFECTION 1st Proceedings London 1968 April Edited by Francis O'Grady and William Brumfitt Sponsored by the Beecham research laboratories Oxford university press,London,1968
PGMS 29.0238

NATIONAL SYMPOSIUM ON URINARY TRACT INFECTION 1st Proceedings Urinary tract infection London 1968 Apr Edited by Francis O'Grady and William Brumfitt Oxford medical publications Oxford university press, London,1968
Surg 23.0010

NATIONAL SYMPOSIUM ON VACUUM TECHNOLOGY Transactions Chicago,Ill 1956 Oct 10-11 Committee on vacuum techniques Edited by Edmond S. Perry and John H. Durant Pergamon press,London,1957 Later symposia published by the American vacuum society
Met 25.1629

NATIONAL SYMPOSIUM ON VACUUM TECHNOLOGY 7th Transactions Cleveland,Ohio 1960 Oct 12-14 American vacuum society Edited by C. Robert Meissner Pergamon press,Oxford,1961 Earlier symposia published by the Committee on vacuum techniques
Met 25.1669

The NATIONAL SYMPOSIUM ON VACUUM TECHNOLOGY 7th Transactions Cleveland,Ohio 1960 Oct 12-14 American vacuum society Edited by C. Robert Meissner 427p Pergamon press,Oxford, 1961 8th symposium called 'National vacuum symposium' q.v.
Cav 7.2375

NATIONAL TELEMETERING CONFERENCE,CHICAGO 1961 Nanosecond data acquisition system By Marcus R. McCraven and Richard C. Epps Lawrence radiation laboratory UCRL-6255 30p 1961 a paper read at the National telemetering conference,Chicago 1961
Math L 5.2266

NATIONAL VACUUM SYMPOSIUM 8th Transactions Washington,D.C. 1961 Oct 16-19 Vol 1-2 American vacuum society and International organization for vacuum science and technology Edited by Luther E. Preuss 2 vols Pergamon press,Oxford,1962 Combined with the International congress on vacuum science and technology,2nd
Cav 7.2376

NATIONAL VITAMIN FOUNDATION Rat quality:a consideration of heredity,diet and disease :a symposium Proceedings New York 1952 Jan 31 By W.E. Heston and others National vitamin foundation,New York,1953
Radioth 35.0218

NATO ADVANCED STUDY INSTITUTE Aurora and airglow proceedings of the Nato advanced study institute Proceedings Keele,Staffs 1966 Aug 15-26 Edited by Billy M. McCormac illus vii,689p 24cm Reinhold,New York, 1967
Sco 14.7473

NATO ADVANCED STUDY INSTITUTE Membrane models and the formation of biological membranes The International conference on biological membranes 2nd Proceedings Frascati 1967 Jun Edited by Liana Bolis and B.A. Perthica Supported by the North Atlantic treaty organization North Holland, Amsterdam,1968
Inv Med 37.0025

NATO ADVANCED STUDY INSTITUTE Papers Meteorological and astronomical influences on radio wave propagation Corfu 1961 North Atlantic treaty organization.Division of scientific affairs NATO conference series, 3 318p Pergamon press,Oxford,1963
Nap 11.0915

NATO ADVANCED STUDY INSTITUTE Proceedings Natural electromagnetic phenomena below 30 kilocycles Bad Homburg 1963 Jul 22-Aug 2 United States.Office of naval research and Naval ordnance laboratory Edited by D.F. Bleil vii,470p Plenum press,New York,1964
Nap 11.0067

NATO ADVANCED STUDY INSTITUTE MEETING ON PALAEOGEOPHYSICS Proceedings Palaeogeophysics Newcastle-upon-Tyne 1968 North Atlantic treaty organization Edited by Stanley K. Runcorn Bibliog,illus,charts,maps xv,518p 24cm Academic press,London;New York,1970
Min 10.1458

NATO ADVANCED STUDY INSTITUTE ON FACIAL STRUCTURE OF COMPACT SETS AND APPLICATIONS Swansea 1972 Jul 2-25 North Atlantic treaty organization Held at University college of Swansea 147p 21cm University college of Swansea,Swansea,1972
P Math 2.4281

NATO ADVANCED STUDY INSTITUTE ON MASS SPECTROMETRY Mass spectrometry :Nato advanced study institute on theory,design and applications Glasgow 1964 Aug North Atlantic treaty organization Edited by R.I. Reed Academic press,London;New York,1965
Met 25.1694

NATO ADVANCED STUDY INSTITUTE ON PHYSICS OF THE SOLAR CORONA Proceedings Physics of the solar corona Athens 1970 Sep 6-17 North Atlantic treaty organization Edited by C.J. Macris Astrophysics and space science library, 27 illus xii,345p 24cm Reidel,Dordrecht,1971
A Math 4.1803

NATO ADVANCED STUDY INSTITUTE ON PLANETARY AND STELLAR MAGNETISM Magnetism and the cosmos Newcastle-upon-Tyne 1965 Apr North Atlantic treaty organization and University of Newcastle upon Tyne Edinburgh, 1967
Geod 9.0567

NATO ADVANCED STUDY INSTITUTE ON PROCEDURES IN PALAEOMAGNETISM Proceedings Methods in palaeomagnetism Newcastle-upon-Tyne 1964 Apr 1-10 North Atlantic treaty organization and University of Newcastle upon Tyne Edited by D.W. Collinson and others Developments in solid earth geophysics, 3 Amsterdam,1967
Geod 9.0568

NATO ADVANCED STUDY INSTITUTE ON SOLAR PHYSICS Proceedings Solar physics Lagonissi, Athens 1965 Sep North Atlantic treaty organization.Division of scientific affairs Interscience,London,1967
Obs 6.3280

NATO ADVANCED STUDY INSTITUTE ON THE PHYSICS OF THE SOLAR CORONA Proceedings Physics of the solar corona Athens 1970 Sep 6-17 North Atlantic treaty organization Edited by C.J. Macris Astrophysics and space science library, 27 345p Reidel,Dordrecht,1971
Obs 6.3607

NATO CONFERENCE SERIES, 3 Meteorological and astronomical influences on radio wave propagation NATO Advanced study institute Papers Corfu 1961 North Atlantic treaty organization.Division of scientific affairs 318p Pergamon press,Oxford,1963
Nap 11.0915

NATO PALAEOCLIMATES CONFERENCE :a NATO advanced study institute proceedings Problems in palaeoclimatology Newcastle-upon-Tyne 1963 Jan.7-12 Edited by A.E.M. Nairn sponsored by the North Atlantic treaty organisation.Science office John Wiley, London,1964
Geol 8.1353

NATO SUMMER SCHOOL Proceedings Critical rotatory dispersion and circular dichroism in organic chemistry Bonn 1965 Sep 24-Oct 10 North Atlantic treaty organization Edited by G. Snatzke Heyden,London,1967
Chem 18.2791

NATO SYMPOSIUM ON COMMUNICATION PROCESSES Proceedings Communication processes Washington,D.C. 1963 North Atlantic treaty organization.Advisory group on human factors Edited by Frank A. Geldard and others NATO conference series, 4 298p Pergamon press, Oxford,1965
Cav 7.2423

NATURA ARTIS MAGISTRA TE AMSTERDAM.ZOOLOGISCH GENOOTSCHAP Untersuchungen zur morphologie und systematik der vogel By Max Furbringer Bijdragentot de dierkunde, 15 Holkema;Fischer,Amsterdam;Jena,1888
Gen 34.2134

NATURAL DRAUGHT COOLING TOWERS - FERRYBRIDGE AND AFTER :a conference Proceedings London 1967 Jun 12 Institution of civil engineers Institution of civil engineers, London,1967
Eng 41.2889

NATURAL ELECTROMAGNETIC PHENOMENA BELOW 30 KILOCYCLES NATO advanced study institute Proceedings Bad Homburg 1963 Jul 22-Aug 2 United States.Office of naval research and Naval ordnance laboratory Edited by D.F. Bleil vii,470p Plenum press,New York,1964
Nap 11.0067

NATURAL HISTORY Bibliography and natural history :a conference Essays Lawrence 1964 Jun 25-27 University of Kansas.Linda Hall library of science and technology Edited by Thomas R. Buckman University of Kansas.Publications.Library series, 27 ix, 148p University of Kansas libraries,Lawrence, Kansas,1966
WSM 43.3212

The NATURAL HISTORY OF AGGRESSION :a symposium Proceedings London 1963 Oct 21-22 Institute of biology Edited by J.D. Carthy and F.J. Ebling Institute of biology. Symposia, 13 Academic press,London;New York, 1964
Psy 31.0596

The NATURAL HISTORY OF AGGRESSION :a symposium held at the British museum(natural history) Proceedings London 1963 Oct 21-22 Institute of biology Edited by J.D. Carthy and F.J. Ebling Institute of biology. Symposia, 13 Academic press,London,1960
Bal 39.1619

NATURAL HISTORY SOCIETY Climate :an inquiry into the causes of its differences,and its influence on vegetable life Torquay 1863 Feb By C. Daubeny Bohn,Oxford,1863 Being the substance of four lectures delivered before the Natural history society
Bot 42.2042

NATURAL MATERIALS Conference on shock metamorphosis of natural materials 1st Proceedings Greenbelt,Md. 1966 Apr 14-16 National aeronautics and space administration. Goddard space flight center Edited by Bevan M. French and Nicholas M. Short Held at the Goddard space flight center Mono book corporation,Baltimore,Md.,1968
Min 10.1443

NATURAL POLYMERS Solution properties of natural polymers :an international symposium Proceedings Edinburgh 1967 Jul 25-28 Chemical society Chemical society.Special publication, 23 Chemical society,London, 1968
Bioch 33.0632

NATURAL PRODUCTS International conference on the chemistry of natural products 7th Abstracts Riga 1970 Jun 21-25 International union of pure and applied chemistry Edited by M.N. Kolosov Zinatne, Riga,1970
Bioch 33.2318

NATURAL PRODUCTS Sapropeli ikh ispolzovanie : po matrialam konferentsii 1956 Akademiya nauk Belorusskoi S.S.R.Institut torfa Izdatelstvo akademii nauk B.S.S.R.,Minsk,1958
Bot 42.6733

NATURAL RESOURCES,FOOD AND POPULATION IN INTER-TROPICAL AFRICA :a symposium Report Kampala 1955 Sep 10-17 Edited by L.Dudley Stamp Held under the auspices of the International geographical union Geographical publications,London,1956
Geog 13.4626

NATURAL RESOURCES COMMITTEE Low dams United States.Department of the interior. Natural resources committee U.S.government printing office,Washington,D.C.,1939
Eng 41.3319

NATURAL RESOURCES IN SCOTLAND :symposium Edinburgh 1960 Oct 31 - Nov 2 Royal society of Edinburgh 25cm Edinburgh,1961
Philos 1.1867

The NATURAL RUBBER PRODUCERS' RESEARCH ASSOCIATION Jubillee conference Proceedings Cambridge 1964 Natural rubber producers' research association Edited by L. Mullins Bibliog.,illus,plate viii,250p McLaren,London,1965
Bot 42.1427

NATURAL RUBBER PRODUCERS' RESEARCH ASSOCIATION JUBILEE CONFERENCE Proceedings Cambridge 1964 Natural rubber producers' research association Edited by L. Mullins Maclaren,London,1965
Bioch 33.1740

NATURAL SELECTION AND ADAPTATION :annual general meeting Papers Philadelphia,Pa. 1949 Apr American philosophical society American philosophical society.Proceedings, 93,no.6 Philadelphia,Pa.,1949
Gen 34.1305

NATURAL SELECTION IN HUMAN POPULATIONS London 1958 Nov 8 Society for the study of human biology Edited by D.F. Roberts and G.A. Harrison Society for the study of human biology.Symposia, 2 Pergamon press,London, 1959
Gen 34.1863

NATURAL SELECTION IN HUMAN POPULATIONS :a symposium London 1958 Nov 8 Edited by D.F. Roberts and G.A. Harrison Society for the study of human biology.Symposia, 2 Pergamon press,Oxford,1959
An 32.2881

NATURAL SUBSTANCES FORMED BIOLOGICALLY FROM MEVALONIC ACID :a symposium Liverpool 1969 Apr Biochemical society Biochemical society.Symposia, 29 Academic press,London,1970
Chem 18.2550

NATURAL SUBSTANCES FORMED BIOLOGICALLY FROM MEVALONIC ACID :a symposium Liverpool 1969 Apr Biochemical society Edited by T.W. Goodwin Biochemical society. Symposia, 29 Academic press,London;New York, 1970
Bioch 33.1391

The NATURE AND ACTIONS OF THE INTERNAL SECRETIONS OF THE PANCREAS :a colloquium Proceedings Internal secretions of the pancreas London 1955 Jun 21-23 Ciba foundation Edited by G.E.W. Wolstenholme and Cecilia M. O'Connor Ciba foundation colloquia on endocrinology, 9 illus Churchill,London, 1956
Bioch 33.0438

The NATURE AND FUNCTION OF PEROXISOMES (MICROBODIES,GLYOXOSOMES) :a conference Papers New York 1969 May 16-17 New York academy of sciences Edited by James Hogg New York academy of sciences.Annals, 168 p. 209-381 New York,1969
Bot 42.4758

NATURE AND ORIGIN OF STRENGTH OF MATERIALS Properties of crystalline solids :...symposium on recent progress in materials sciences; symposium on nature and origin of strength of materials Philadelphia 1960 Jun 27 A.S.T.M.Special technical publication Philadelphia, 1961 Includes an introductory paper by W. O.Baker on 'The national role of materials research and development'
Met 25.2490

The NATURE AND SIGNIFICANCE OF THE ANTIBODY RESPONSE :a symposium Papers and discussions New York 1951 Mar 21-22 New York academy of medicine.Section on microbiology Edited by A.M. Pappenheimer New York academy of medicine.Section on microbiology.Symposia, 5 Columbia university press,New York,1953
Bioch 33.1156

The NATURE AND SIGNIFICANCE OF THE ANTIBODY RESPONSE Papers and discussions New York 1951 Mar 21-22 New York academy of medicine.Section of microbiology Edited by A. M. Pappenheimer New York academy of sciences. Section of microbiology.Symposia, 5 Columbia university press,New York,1953
Path 30.2532

The NATURE AND SIGNIFICANCE OF THE ANTIBODY RESPONSE;SYMPOSIUM... New York 1951 Mar 21-22 New York academy of medicine.Section on microbiology Edited by Alwin M. Pappenheimer Held at the New York academy of medicine New York academy of medicine. Section on microbiology.Symposia, 5 Columbia university press,New York,1953
Med 36.0272

NATURE AND STRUCTURE OF COLLAGEN :a discussion Papers London 1953 Mar 26-27 Edited by J.T. Randall Convened by the Faraday society.Colloid and biophysics committee Butterworths,London,1953
Radioth 35.0186

NATURE AND STRUCTURE OF COLLAGEN :a discussion Papers London 1953 Mar 26-27 Faraday society.Colloid and biophysics committee Edited by J.T. Randall and Sylvia Fitton Jackson Butterworths,London,1953
Bioch 33.2346

NATURE AND STRUCTURE OF COLLAGEN :a discussion Papers London 1953 Mar 26-27 Faraday society.Colloid and biophysics committee Edited by J.T. Randall and Sylvia Fitton Jackson Butterworths,London,1953
An 32.5311

NATURE AND STRUCTURE OF COLLAGEN :a discussion Papers presented London 1953 Mar 26-27 Faraday society.Colloid and biophysics committee Edited by J.T. Randall and Sylvia Fitton Jackson Butterworths,London,1953
Bal 39.1202

NATURE CONSERVANCY The Countryside in 1970 : conference 2nd Proceedings London 1965 Nov 10-12 Council for nature,and,Royal society of arts Royal society of arts;Nature conservancy,London,1966
Geog 13.1899

NATURE CONSERVANCY.CONSULTATIVE COMMITTEE ON GREY SEALS A Seals symposium Proceedings Cambridge 1964 Sep 9 Edited by E.A. Smith illus 101p 28cm Edinburgh,1965
Sco 14.1172

NATURE CONSERVANCY,EDINBURGH.CONSULTATIVE COMMITTEE ON GREY SEALS AND FISHERIES A Seals symposium Proceedings Cambridge 1964 Sep 9 Edited by E.A. Smith maps, tables 28cm Edinburgh,1965 Number 84 of 100 mimeographed copies.In folder
Bal 44.5093

NATURE OF DEFECTS IN CRYSTALS International conference on 'Electron diffraction' and 'The nature of defects in crystals' Abstracts of papers Melbourne 1965 Aug 16-21 Australian academy of science and International union of crystallography With the support of the International union of pure and applied physics. Commission on the solid state Pergamon press,Oxford,1966 Extended abstracts of papers to be presented at the conference
Cav 7.1584

NATURE OF DEFECTS IN CRYSTALS International conference on"electron diffraction" and "the nature of defects in crystals" Abstracts of papers Melbourne 1965 Aug 16-21 Australian academy of science and International union of crystallography Co-sponsored by the International union of pure and applied physics Pergamon press,Oxford, 1966
Met 25.1463

The NATURE OF SLEEP :a Ciba foundation symposium Proceedings London 1960 Jun 27-29 Ciba foundation Edited by G.E.W. Wolstenholme and Maeve O'Connor Churchill, London,1961
An 32.4378

NATURE OF SMALL DEFECT CLUSTERS Consultants symposium 2nd Report Harwell 1966 Jul 4-6 Vol 1-2 United Kingdom atomic energy authority.Research group Edited by M. J. Makin 2 vols Atomic energy research establishment,Harwell,1966
Met 25.1243

NATURE OF THE BACTERIAL SURFACE Society for general microbiology :symposium 1st Proceedings London 1949 Apr 20 Edited by A.A. Miles and N.W. Pirie Blackwell,Oxford, 1949
Col S 12.0036

The NATURE OF THE BACTERIAL SURFACE :a symposium London 1949 Apr 20 Society for general microbiology Edited by A.A. Miles and N.W. Pirie illus. Blackwell, Oxford,1949
Gen 34.0697

NATURE OF THE LUNAR SURFACE IAU-NASA symposium Greenbelt,Md. 1965 Apr 15-16 International astronomical union National aeronautics and space administration Edited by W.N. Hess and others Held at the Goddard space flight center Johns Hopkins press, Baltimore,Md.,1966
Min 10.1445

The NATURE OF TIME :a meeting convened by T. Gold and H.Bondi Ithaca,N.Y. 1963 May 30-Jun 1 Cornell university Edited by T. Gold xiv,248p Cornell university press, Ithaca,N.Y.,1967
WSM 43.1104

The NATURE OF TIME :report of a meeting Ithaca,N.Y. 1963 May 30-Jun 1 Cornell university Edited by T. Gold and D.L. Schumacher ix,248p 23cm Cornell university,Ithaca,N.Y.,1967
Cav 7.2764

The NATURE OF VIRUS MULTIPLICATION :a symposium Papers Oxford 1952 Apr Society for general microbiology Edited by Paul Fildes and W.E. Van Heyningen Held at the University of Oxford Society for general microbiology.Symposia, 2 Cambridge university press,Cambridge,1953
Bioch 33.1179

The NATURE OF VIRUS MULTIPLICATION :a symposium Proceedings Oxford 1952 Apr 16-17 Society for general microbiology Edited by Paul G. Fildes and W.E. Van Heyningen Society for general microbiology. Symposia, 2 Cambridge university press, Cambridge,1953
Radioth 35.0183

NATURFORSCHENDEN GESELLSCHAFT SCHAFFHAUSEN Gronland 1939 :tagung der Naturforschenden gesellschaft Schaffhausen Schaffhausen 1931 Mar.11-12 Naturforschenden gesellschaft Schaffhausen.Mitteilungen,Bd.16 1939
Geol 8.2714

NATURFORSCHENDEN GESELLSCHAFT SCHAFFHAUSEN Gronland 1939 :tagung der Naturforschenden gesellschaft Schaffhausen Schaffhausen 1939 Mar 11-12 Naturforschenden gesellschaft Schaffhausen.Mitteilungen,16 illus,maps,4 plates 231p 1939
Sco 14.7861

NATURPHILOSOPHIE BEI ARISTOTELS UND THEOPHRAST Symposium Aristotelicum 4th Verhandlungen Goteborg 1966 Aug Edited by Ingemar During 292p Stiehm,Heidelberg,1969
WSM 43.0192

NATURWISSENSCHAFT UND PHILOSOPHIE Internationales symposium uber naturwissenschaft und philosophie... Beitrage Leipzig 1959 Oct 8-11 Institut fur philosophie Karl-Sudhoff institut fur geschichte der medizin und naturwissenschaften Edited by Gerhard Harig and Josef Schleifstein 437p Akademie verlag,Berlin,1960 Symposium held to commemorate the 550th anniversary of the Karl-Marx-universitat
WSM 43.0976

NAUCHNO-ISSLEDOVATELSKII INSTITUT POLYARNOGO ZEMLEDELIYA,ZHIVOTNOVODSTVA I PROMYSLOVOGO KHOZYAISTVA Doklady 6 rasshirennoi sessii uchenogo soveta instituta 1953 Mar 2-9 Vyp 2-3 Leningrad,1953-54
Sco 14.4902

NAVAL HYDRODYNAMICS Symposium on naval hydrodynamics proceedings Washington D.C. 1956 Sep.24-28 United states.Office of naval research Edited by F.S. Sherman Sponsored by the National research council National research council.Publication, 515 National research council,Washington D.C.,1957
A Math 4.0530

NAVAL HYDRODYNAMICS Symposium on naval hydrodynamics :ship motions and drag reduction 5th Bergen 1964 Sep 10-12 Edited by J.K. Lunde and S.W. Doroff Sponsored by the United States.Office of naval research,and Office of naval research-Department of the navy,Washington,D.C.,1965
A Math 4.1303

NAVAL ORDNANCE LABORATORY Fluid dynamics symposium College park,Md. 1951 Jun 22-23 American mathematical society Edited by M.H. Martin American mathematical society. Proceedings of symposia in applied mathematics, 4 McGraw-Hill,New York,1953
A Math 4.0512

NAVAL ORDNANCE LABORATORY Natural electromagnetic phenomena below 30 kilocycles NATO advanced study institute Proceedings Bad Homburg 1963 Jul 22-Aug 2 Edited by D. F. Bleil vii,470p Plenum press,New York, 1964
Nap 11.0067

NAVAL STRUCTURAL MECHANICS Symposium on naval structural mechanics 1st Proceedings Stanford,Calif. 1958 Aug 11-14 United States.Office of naval research and Stanford university Edited by J.Norman Goodier and Nicholas J. Hoff International series on aeronautical sciences and space flight Pergamon,Oxford,1960
Met 25.1300

NAVAL STRUCTURAL MECHANICS 2nd :a symposium Proceedings Plasticity Rhode Island 1960 Apr 5-7 United States.Office of naval research and Brown university Edited by E.H. Lee and P.S. Symonds Pergamon press,New York,1960
Met 25.0845

NAVAL STRUCTURAL MECHANICS SYMPOSIUM 2nd Proceedings Plasticity Providence,R.I. 1960 Apr 5-7 United States.Office of naval research Brown university Edited by E.H. Lee and P.S. Symonds United States.Office of naval research.Structural mechanics series Pergamon press,London,1960
Eng 41.2479

NAVIGATION SYSTEMS FOR AIRCRAFT AND SPACE VEHICLES :papers presented at the AGARD avionics panel meeting Istanbul 1960 Oct 3-8 Agard Edited by T.G. Thorne AGARDograph, 55 Pergamon press,London,1962
Eng 41.5706

NAVY-INDUSTRY ZINC SYMPOSIUM ON CATHODIC PROTECTION development,application and specification Zinc as a galvanic anode 1955 Apr 21-22 United States.Department of the navy.Bureau of ships and American zinc institute U.S.Navy dept.,Washington,D.C., 1955
Met 25.2500

NBS SEMICENTENNIAL SYMPOSIUM 7th Proceedings Electrochemical constants Washington,D.C. 1951 Sep 19-21 National bureau of standards National bureau of standards.Circular, 524 U.S.Government printing office,Washington,D.C.,1953 One of 12 semicentennial symposia
Met 25.2401

NBS SEMICENTENNIAL SYMPOSIUM ON ENERGY TRANSFER IN HOT GASES Proceedings 1951 Sep 17-18 National bureau of standards National bureau of standards.Circular, 523 U.S. government printing office,Washington,D.C., 1954
Eng 41.7220

NBS SEMICENTENNIAL SYMPOSIUM ON GRAVITY WAVES
Proceedings Gravity waves Washington,D.C. 1951 Jun 18-20 National bureau of standards National bureau of standards. Circular, 521 U.S.government printing office, Washington,D.C.,1952
Eng 41.6664

NEAR EAST Colloque sur la protection et la conservation de la nature dans le Proche-orient :symposium on the protection and conservation of nature in the Near East Beirut 1954 Jun 3-8 Lebanese society of the friends of the trees Unesco-Middle East science cooperation office Beirut,1954 Text in English and French
Bal 39.1787

NEAR EASTERN CULTURE AND SOCIETY
Princeton,N.J. 1947 Mar Edited by T.Cuyler Young Princeton oriental studies,15 illus. Princeton university press,Princeton,N.J.,1951
Geog 13.1673

NEBRASKA SYMPOSIUM ON MOTIVATION Lincoln, Nebraska 1962- 1962- University of Nebraska.Department of psychology University of Nebraska.Current theory and research in motivation symposium, 10- University of Nebraska press,Lincoln,Nebraska,1962-
Psy 31.1858

NEDERLANDSCHE AKADEMIE VAN WETENSCHAPPEN
Flavins and flavoproteins :a symposium Proceedings Amsterdam 1965 Jun 10-15 International union of biochemistry Edited by E.C. Slater B.B.A. library, 8 illus Elsevier,Amsterdam,1966
Bioch 33.1099

NEDERLANDSCHE CHEMISCHE VEREENIGING.COLLOID CHEMISTRY SECTION Hydrophobic colloids : symposium on the dynamics of hydrophobic suspensions and emulsions Utrecht 1937 Nov 5-6 Amsterdam,1938 Reprinted from "Chemisch Weekblad" 1938
Col S 12.0100

NEDERLANDSCHE MANTSCHAPPIJ VOOR TUINBOUW EN PLANTKUNDE International tuinbouw-congres 7th Amsterdam 1923 Sep 17-23 c1923
Gen 34.1139

NEDERLANDSE CHEMISCHE VERENIGING Chemical reaction engineering European symposium on chemical engineering 1st Papers Amsterdam 1957 May 7-9 Held during the 12th meeting of the European federation of chemical engineering International series of monographs on chemical engineering, 1 197p Pergamon,London,1957
Chem E 24.1900

NEDERLANDSE CHEMISCHE VERENIGING Gas chromatography :a symposium 2nd Proceedings Amsterdam 1958 May 19-23 Institute of petroleum.Gas chromatography discussion group Edited by D.H. Desty xiii, 383p Butterworths,London,1958
Chem E 24.0849

NEDERLANDSE CHEMISCHE VERENIGING
International conference on coordination compounds Proceedings Amsterdam 1955 Apr 28-May 3 Royal Netherlands chemical society, n.p.,c1955
Chem 18.0998

NEDERLANDSE CHEMISCHE VERENIGING
International symposium on the reactivity of solids 4th Amsterdam 1960 May 30th-Jun 4th Edited by J.H.de Boer and others Sponsored by the International union of pure and applied physics Elsevier,Amsterdam,1961 Papers in English,French and German
Met 25.1848

NEDERLANDSE CHEMISCHE VERENIGING The Scaling-up of chemical plant and processes :joint symposium Papers London 1957 May 28-29 By R.Edgeworth Johnstone and others Society of chemical industry.Chemical engineering group and Institution of chemical engineers Institution of chemical engineers, London,1957 Held jointly with the K. Instituut van ingenieurs and the K.Nederlandse chemische vereniging
Chem E 24.1234

NEDERLANDSE CHEMISCHE VERENIGING and INSTITUTE OF PETROLEUM.HYDROCARBON RESEARCH GROUP Gas chromatography 1958 :a symposium 2nd Proceedings Amsterdam 1958 May 19-23 Edited by D.H. Desty Organised by the Gas chromatography discussion group Butterworths,London,1958
Chem 18.0566

NEDERLANDSE CHEMISCHE VERENIGING.SECTIE VOOR CHEMISCHE TECHNOLOGIE Chemical reaction engineering European symposium on chemical reaction engineering 3rd Proceedings Amsterdam 1964 Sep 15-17 Held under the auspices of the European federation of chemical engineering vi,326p Pergamon press,Oxford,1965 Supplement to "Chemical engineering science".
Chem E 24.1269

NEDERLANDSE KERAMISCHE VERENIGING Science of ceramics :a conference 3rd Proceedings Bristol 1965 Jul 5-8 Edited by G.H. Stewart Under the auspices of the European ceramic association Academic press,London; New York,1967
Met 25.0286

NEDERLANDSE NATURKUNDIGE VERENIGING
International symposium on the reactivity of solids 4th Amsterdam 1960 May 30th-Jun 4th Edited by J.H.de Boer and others Sponsored by the International union of pure and applied physics Elsevier,Amsterdam,1961 Papers in English,French and German
Met 25.1848

NEDERLANDSE NATUURKUNDIG VERENIGING
International conference on laboratory astrophysics Selection of papers Lunteren 1968 Sep 2-7 Edited by J. Rosenberg Physica, 41,no 1 223p North-Holland, Amsterdam,1969
TA 15.0580

NEDERLANDSE NATUURKUNDIGE VERENIGING
International conference on laboratory astrophysics Selection of papers Lunteren 1968 Sep 2-7 Edited by J. Rosenberg Physica, 41,no 1 223p North-Holland, Amsterdam,1969
Obs 6.3347

NEDERLANDSE VERENIGING VOOR LOGICA EN WIJSBEGEERTE DER EXACTE WETENSCHAPPEN Logic,methodology and philosophy of science International congress for logic,methodology and philosophy of science Proceedings Amsterdam 1967 Aug 25-Sep 2 Edited by B. van Rootselaar and J.F. Staal xiii,554p North-Holland,Amsterdam,1968
WSM 43.0830

NEDERLANSKE GEOLOGISCH-MIJNBOUW KUNDIG GENOOTSCHAP.GEOLOGISCHE SECTIE Quaternary changes in level,especially in the Netherlands a symposium Utrecht 1954 Mar 5-6 Geologie en mijnbouw,ns,16,no.6 The Hague, 1954
Geog 13.0371

NEDLANDS 1962 Hydraulics and fluid mechanics :Australasian conference 1st Proceedings Edited by Richard Silvester Sponsored by the University of Western Australia.Faculty of engineering illus viii,503p Pergamon press,Oxford,1964
Chem E 24.0239

NEOPLASIA Molecular basis of neoplasia Annual symposium on fundamental cancer research 15th Papers Houston,Tex. 1961 Anderson hospital and tumor institute University of Texas press,Austin,Tex.,1961
Radioth 35.1942

NEOPLASMS AND RELATED DISORDERS OF INVERTEBRATE AND LOWER VERTEBRATE ANIMALS :a symposium Proceedings Washington,D.C. 1968 Jun 19-21 National cancer institute Edited by Clyde J. Dawe and John C. Harshbarger held at the Smithsonian institution National cancer institute.Monograph, 31 National cancer institute,Bethesda,Md.,1969
Bioch 33.0994

The NEOTROPICAL BOTANY CONFERENCE Proceedings Trinidad 1962 Jul 2-6 Association of tropical biology.Bulletin, 1 Trinidad,1962
BG 38.3113

NEPHROLOGY Congres international de nephrologie International congress of nephrology 1st Proceedings Geneva 1969 Sep 1-4 Evian 1960 Sep 1-4 Edited by G. Richet S.Karger,Basle;New York,1961
Inv Med 37.0187

NEPHROLOGY Congres international de nephrologie 2 Comptes rendus Prague 1963 Aug 26-30 Societe internationale de nephrologie Societe tchecoslovaque de medecine Edited by J. Vostal and G. Richet Excerpta medica.International congress series, 78 Excerpta medica,Amsterdam,1964
Inv Med 37.0220

NEPHROLOGY International congress of nephrology 1st Proceedings Geneva 1960 Sep 1-4 Evian 1960 Edited by G. Richet Organised by Societe de nephrology Karger, Basle;New York,1961 In English and French
Phys 20.1325

NEPTUNE Reunions scientifiques...pour le centenaire de la decouverte de Neptune reports Paris 1946 Oct.22-24 Comite national Francais d'astronomie Edited by P. Couderc illus. 164p Imprimerie nouvelle, Orleans,1948
Obs 6.0418

NERVE,BRAIN AND MEMORY MODELS ..a series of lectures delivered during a symposium on cybernetics of the nervous system...held as part of the second international meeting of medical cybernetics at the Royal academy of sciences of Amsterdam... Lectures Amsterdam 1962 Apr 16-18 Edited by N. Wiener and J.P. Schade Progress in brain research, 2 Elsevier,Amsterdam,1963
Psy 31.0248

NERVE AS A TISSUE :conference Proceedings Philadelphia,Pa. 1964 Nov 12-13 Edited by Kaare Rodahl and Bela Issekutz Held at the Lankenau hospital Harper,London,1966
An 32.3430

NERVE AS A TISSUE 4th :a conference Proceedings Philadelphia,Pa. 1964 Nov 12-13 Edited by Kaare Rodahl and Bela Issekutz Sponsored by the Lankenau hospital Harper,New York,1966
Pha 16.0299

NERVE IMPULSE Conference on nerve impulse 3rd Transactions New York 1952 Mar 3-4 Edited by H.Houston Merritt Sponsored by the Josiah Macy jr.foundation Josiah Macy jr. foundation,New York,1952
An 32.4315

NERVE IMPULSE 1st-3rd :conferences Transactions New York 1950-1952 3 vols Macy foundation,New York,1950-52
Phys 20.2002

NERVES AND HORMONAL MECHANISMS OF INTEGRATION a symposium London 1965 Society for experimental biology Society for experimental biology.Symposia, 20 Cambridge university press,Cambridge,1966
An 32.3785

NERVOUS AND HORMONAL MECHANISMS OF INTEGRATION St.Andrews 1965 Sep Society for experimental biology and American society of zoologists Society for experimental biology.Symposia, 20 Cambridge university press,Cambridge,1966
Psy 31.0137

NERVOUS AND HORMONAL MECHANISMS OF INTEGRATION symposium Proceedings Society of experimental biology Edited by G.M. Hughes Society of experimental biology.Symposia,20 Cambridge university press,Cambridge,1966
Pha 16.0048

NERVOUS CONTROL Evolution of nervous control from primitive organisms to man :a symposium Proceedings New York 1956 Dec 29-30 American association for the advancement of science.Section on medical sciences Edited by Allan D. Bass American association for the advancement of science.Publication, 52 Washington,D.C.,1959 Arranged by Bernard B.Brodie and Allan D.Bass
Bal 39.1470

NERVOUS CONTROL OF THE HEART :symposium Atlantic City,N.J. 1963 Apr 16-20 Edited by Walter C. Randall Williams and Wilkins, Baltimore,Md.,1965 Held during the annual spring meetings of the American physiological society and the Federation of American societies for experimental biology
An 32.3811

NERVOUS INHIBITION The Friday Harbor symposium 2nd Proceedings Friday Harbor 1960 May 31-Jun 4 Edited by Ernst Florey Sponsored by the National science foundation. Division of regulatory biology Pergamon press,Oxford,1961 Conference also called International symposium on nervous inhibition
Bal 39.1475

NERVOUS SYSTEM Biochemistry of the developing nervous system International neurochemical symposium 1st Proceedings Oxford 1954 Jul 13-17 Edited by Heinrich Waelsch held at Magdalen college,Oxford Academic press,New York,1955
Bal 39.1468

NERVOUS SYSTEM Biochemistry of the developing nervous system The International neurochemical symposium 1st Proceedings Oxford 1954 Jul 13-17 Edited by Heinrich Waelsch Academic press,New York,1955
Bioch 33.2233

NERVOUS SYSTEM Genetic neurology International conference on the development, growth and regeneration of the nervous system Chicago,Ill. 1949 Mar 21-25 Edited by Paul Weiss Sponsored by the International union of biological sciences University of Chicago, Ill.,Chicago,Ill.,1950
An 32.4329

NERVOUS SYSTEM Inhibition in the nervous system and gamma-aminobutyric acid :an international symposium held at the City of Hope medical center Proceedings Duarte, Calif. 1959 May 22-24 Edited by Eugene Roberts and others Sponsored by the United States.Air force.Office of scientific research Pergamon press,Oxford,1960
Bal 39.1471

NERVOUS SYSTEM Metabolism of the nervous system International neurochemical symposium 2nd Proceedings Aarhus 1956 Jul Aarhus universitet Edited by Derek Richter Pergamon press,London,1957
Bioch 33.0592

NERVOUS TISSUE Metabolism and functions in nervous tissue :a symposium London 1951 Nov 10 Biochemical society Edited by R.T. Williams Held at the London school of hygiene and tropical medicine Biochemical society.Symposia, 8 Cambridge university press,Cambridge,1952
Bioch 33.1370

NESTASIONARNYE YAVLENIYA V GALAKTIKAKH Non-stable phenomena in galaxies Byurakan 1966 May 4-12 International astronomical union International astronomical union.Symposium, 29 480p Akademiya nauk Armyanskey S.S.R., Erevan,1968
Obs 6.3269

NETHERLANDS CANCER INSTITUTE Cellular control mechanisms and cancer :a conference Proceedings Amsterdam 1963 Sep 9-13 Edited by P. Emmelot and O. Muhlbock Under the auspices of the International union against cancer Elsevier,Amsterdam,1964 Conference held on the occasion of the 50th anniversary of the Netherlands cancer institute
An 32.2504

NETHERLANDS CANCER INSTITUTE Conference on cellular control mechanisms and cancer Proceedings Amsterdam 1963 Sep 9-13 Edited by P. Emmelot and O. Muhlbock Elsevier,Amsterdam,1964 Organized to commemorate the 50th anniversary of the Netherlands cancer institute
Path 30.2459

NETHERLANDS PHYSICAL SOCIETY International conference on spectroscopy at radio-frequencies :a conference Proceedings Amsterdam 1950 Sep Physica, 17,3-4 169-484p 1951
Cav 7.0055

NETHERLANDS PHYSICAL SOCIETY Magnetic and electric resonance and relaxation Colloque Ampere :international conference 11th Proceedings Eindhoven 1962 Jul 2-7 Edited by J. Smidt Under the auspices of the International union of pure and applied physics 789p North-Holland,Amsterdam,1963
Cav 7.0712

NETHERLANDS PHYSICAL SOCIETY Molecular spectroscopy :general and introductory lectures presented at the fifth European congress on molecular spectroscopy 5th Amsterdam 1961 May 29-Jun 3 International union of pure and applied chemistry.Physical chemistry section,and,Royal Netherlands chemical society Butterworths,London,1962 Reprinted from 'Pure and applied chemistry', vol 4,no 1
Chem 18.2315

NETHERLANDS RADIOBIOLOGICAL SOCIETY Program of the joint meeting 1969
Radioth 35.1763

NETHERLANDS UNIVERSITIES FOUNDATION FOR INTERNATIONAL CO-OPERATION Local fields summer school proceedings Driebergen 1966 Jul 25-Aug 6 Edited by T.A. Springer viii,214p 24cm Springer,Berlin,1967
P Math 2.2774

NETHERLANDS UNIVERSITIES FOUNDATION FOR INTERNATIONAL COOPERATION Asymptotic distribution modulo 1 Nuffic international summer session in science papers Breukelen 1962 Aug 1-11 Edited by J.F. Koksma and L. Kuipers With the financial support of North Atlantic treaty organization Bibliog. 203p 24cm Noordhoff,Groningen,1964
P. Math 2.1826

NETHERLANDS UNIVERSITIES FOUNDATION FOR INTERNATIONAL COOPERATION Present problems concerning the structure and evolution of the galactic system NUFFIC international summer course in science Lecture notes Breukelen 1960 Jul. 28- Aug. 16 Edited by J.H. Oort and H.G. Quik Supported by North atlantic treaty organization The Hague,1960
Obs 6.2067

NETHERLANDS UNIVERSITIES FOUNDATION FOR INTERNATIONAL COOPERATION and DUTCH PHYSICAL SOCIETY Selected topics in nuclear spectroscopy :N.U.F.F.I.C. international summer course in science, Nijenrode Castle... Proceedings Breukelen, Netherlands 1963 Jul 30-Aug 17 Edited by B. J. Verhaar With financial support from the North Atlantic treaty organization North-Holland,Amsterdam,1964
Chem 18.2279

NETWORK ANALYSIS International congress on project planning by network analysis 2nd Proceedings Amsterdam 1969 Oct 6-10 North-Holland,Amsterdam,1969
Eng 41.1240

NETWORK THEORY Recent developments in network theory a symposium Proceedings Cranfield 1961 Sep College of aeronautics, Cranfield.Department of electrical and control engineering Edited by S.R. Deards Pergamon press,Oxford,1963
Cav 7.0582

NEUCHATEL 1957 Symposium sur la specificite parasitaire des parasites de vertebres 1 er Universite de Neuchatel. Institut de zoologie Edited by J.G. Baer Paul Attinger,Neuchatel,1957 Title also in English.Text in English and French
Bal 39.1727

NEUCHATEL 1970 H-spaces Reunion de Neuchatel Actes Edited by Francois Sigrist Lecture notes in mathematics, 196 156p 25cm Springer,Berlin,1971
P Math 2.4121

NEUERE ERGEBNISSE AUS CHEMIE UND STOFFWECHSEL DER KOHLENHYDRATE colloquium Mosbach, Baden 1957 May 2-4 Gesellschaft fur biologische chemie Gesellschaft fur biologische chemie.Colloquia, 8 illus Springer,Berlin,1958
Bioch 33.2347

NEUORANATOMY New research techniques of neuroanatomy :a symposium Edited by William F. Windle Sponsored by the National multiple sclerosis society Thomas,Springfield,Ill., 1957
An 32.4352

NEURAL MECHANISMS OF THE AUDITORY AND VERTIBULAR SYSTEMS :a conference Proceedings Bethesda,Md. 1959 Jun 11-13 Edited by G.L. Rasmussen and W.F. Windle Sponsored by the National institute of neurological diseases and blindness Symposia in neuroanatomical sciences, 6 illus. xiv,422p Thomas, Springfield,Ill.,1960
Phys 20.1975

NEURAL MECHANISMS OF THE AUDITORY AND VESTIBULAR SYSTEMS :a conference Bethesda,Md. 1959 Jun 11-13 Edited by Grant L. Rasmussen and William F. Windle Symposia in neuroanatomical sciences, 6 Thomas, Springfield,Ill.,1960
An 32.4821

NEURAL NETWORKS School on neural networks Proceedings Edited by E.R. Caianiello With the assistance of North Atlantic treaty organization.Advanced study institute programme Springer,Berlin,1968
Psy 31.3067

NEURO-PSYCHOPHARMACOLOGY International congress of neuro-pharmacology 1st Proceedings Rome 1958 Sep Collegium for neuro-psychopharmacology Edited by P.B. Bradley and others Elsevier,London,1959
PGMS 29.0192

NEURO-PSYCHOPHARMACOLOGY International congress of neuropharmacology 1st Proceedings Rome 1958 Sep International collegium for neuro-psychopharmacy Edited by P.B. Bradley and others Elsevier,Amsterdam, 1959
Psy 31.0524

NEUROBIOLOGY Progress in neurobiology International meeting of neurobiologists 1st Proceedings Groningen 1955 Aug 3-7 Edited by J.Ariens Kapper Elsevier,Amsterdam, 1956
An 32.4351

NEUROCHEMISTRY Biochemistry of the developing nervous system International neurochemical symposium 1st Proceedings Oxford 1954 Jul 13-17 Edited by Heinrich Waelsch Academic press,New York,1955
An 32.4333

NEUROCHEMISTRY Chemical pathology of the nervous system International neurochemical symposium 3rd Proceedings Strasbourg 1958 Aug Edited by Jordi Folch-Pi Pergamon press,London,1961
An 32.4186

NEUROCHEMISTRY Regional neurochemistry... International neurochemical symposium 4th Proceedings Edited by Seymour S. Kety and Joel Elkes Pergamon press,London,1961
VA 19.0235

NEUROCHEMISTRY OF NUCLEOTIDES AND AMINO ACIDS : a symposium Papers and discussions presented Philadelphia 1958 Apr 24-25 American academy of neurology.Section of neurochemistry Edited.by Roscoe O. Brady and Donald B. Tower Wiley,New York,1960
Bioch 33.2222

NEUROENDOCRINOLOGY Advances in neuroendocrinology Symposium on neuroendocrinology Miami,Fla. 1961 Dec 6-8 Edited by Andrew V. Nalbandov Supported by a grant from the National institute of neurological diseases and blindness University of Illinois press,Urbana,Ill.,1963
An 32.3766

NEUROENDOCRINOLOGY Advances in neuroendocrinology Symposium on neuroendocrinology Miami,Flo. 1961 Dec 6-8 Edited by A.V. Nalbandov Sponsored by the National institute of neurological diseases and blindness University of Illinois press, Urbana,Ill.,1963
Phys 20.1404

NEUROHYPOPHYSIS :symposium Proceedings Bristol 1956 Apr 9-12 Colston research society Edited by H. Heller Colston research society.Symposia, 8 Colston papers, 8 Butterworths,London,1957
Bioch 33.0444

The NEUROHYPOPHYSIS :symposium Proceedings Bristol 1956 Apr 9-12 Colston research society Edited by H. Heller Colston papers, 8 Butterworth,London,1957
An 32.4689

The NEUROHYPOPHYSIS :symposium Proceedings Bristol 1956 Apr 9-12 Colston research society Edited by H. Heller Colston papers, 8 Butterworths,London,1957
VA 19.0224

The NEUROLOGIC AND PSYCHIATRIC ASPECTS OF THE DISORDERS OF AGING Proceedings of the Association New York 1955 Dec 9-10 Association for research in nervous and mental disease Edited by Joseph Earle Moore and others Association for research in nervous and mental disease.Research publications, 35 Williams and Wilkins,Baltimore,Md.,1956
An 32.4166

NEUROLOGICAL BASIS OF BEHAVIOUR Ciba foundation symposium on the neurological basis of behaviour :in commemoration of Sir Charles Sherrington,O.M.,G.B.E.,F.R.S.,1857-1952 Papers London 1957 Jul 2-4 Ciba foundation Edited by G.E.W. Wolstenholme and Cecilia M. O'Connor Churchill,London,1958
Psy 31.0512

NEUROLOGY AND PSYCHIATRY IN CHILDHOOD Proceedings of the Association New York 1954 Dec 10-11 Association for research in nervous and mental disease Edited by Rustin McIntosh and Clarence C. Hare Association for research in nervous and mental disease. Research publications, 34 Williams and Wilkins,Baltimore,Md.,1954
An 32.4161

NEUROMUSCULAR DISORDERS (the motor unit and its disorders) Proceedings of the Association New York 1958 Dec 12-13 Association for research in nervous and mental disease Edited by Raymond D. Adams and others Association for research in nervous and mental disease.Research publications, 38 Williams and Wilkins,Baltimore,Md.,1960
An 32.4185

The NEURON :a symposium Cold Spring Harbor 1952 Jun 6-13 Cold Spring Harbor symposia on quantitative biology, 17 Cold Spring Harbor,N.Y.,1952
Phys 20.0956

The NEURON :a ymposium Cold Spring Harbor 1952 Jun 6-13 Cold Spring Harbor biological laboratory Edited by Katherine Brehme Warren Cold Spring Harbor symposia on quantitative biology, 17 Biological laboratory,Cold Spring Harbor,1952
An 32.4330

The NEURON :a symposium Papers Cold Spring Harbor 1952 Cold Spring Harbor biological laboratory Cold Spring Harbor symposia on quantitative biology, 17 Long Island biological association,Cold Spring Harbor,1952
Bioch 33.1273

NEUROPATHOLOGY International congress of neuropathology 2nd Proceedings London 1955 Sep 12 Pt 1-2 Edited by W.H. McMenemey Excerpta Medica,Amsterdam,1955
An 32.4163

NEUROPATHOLOGY International congress of neuropathology 4th Proceedings Munich 1961 Sep 4-8 Vol 1-3 Edited by H. Jacob 3 vols Thieme,Stuttgart,1962 Papers in English,French,German and Spanish
An 32.4193

NEUROPHYSIOLOGIE UND PSYCHOPHYSIK DES VISUELLEN SYSTEMS (the visual systems: neurophysiology and psychophysics)a symposium Freiburg 1960 Aug 28-Sep 3 Edited by Richard Jung and Hans Kornhuber Springer, Berlin,1961 Papers in English,French and German
Psy 31.0355

NEUROPHYSIOLOGY Attention in neurophysiology an international conference Proceedings Teddington 1967 Oct National physical laboratory Edited by C.R. Evans and T.B. Mulholland port Butterworths,London,1970
Psy 31.2849

NEUROPHYSIOLOGY Conference on neurophysiology in relation to anesthesiology Seattle 1966 May 13-14 Anesthesiology,28, no 1 American society of anesthesiologists, Lancaster,Pa.,1967
Pha 16.0052

NEUROPHYSIOLOGY Recent advances in clinical neurophysiology International congress of eletroencephalography and clinical neurophysiology 6th Proceedings Vienna 1965 Sep 5-10 functions of the spinal cord; neurophysiological investigations of brain diseases;EEG in stress Edited by L. Widen Electroencephalography and clinical neurophysiology.Supplement,25 Elsevier, Amsterdam,1967
Pha 16.0304

NEUROPHYSIOLOGY AND EMOTION :a conference Proceedings New York 1965 Dec 10-11 Russell Sage foundation Rockefeller university Edited by David C. Glass Rockefeller university press;Russell Sage foundation,New York,1967
Psy 31.2900

NEUROSCIENCES :a study program Boulder, Colo. 1966 Jul Neurosciences research program.Intensive study program Edited by Gardner C. Quarton and others Rockefeller university press,New York,1967
Bioch 33.2220

NEUROSCIENCES RESEARCH PROGRAM Neurosciences research symposium summaries :an anthology from the Neurosciences research program bulletin Vol 1-2 Edited by Francis O. Schmitt and others Neurosciences research program bulletin 2 vols M.I.T.Press, Cambridge,Mass.,1966-67 This volume reports 9 of the twenty work sessions held in the last 3 years by the N.R.P. Work session reports are by Albert L.Lehninger and others
Bioch 33.2221

NEUROSCIENCES RESEARCH PROGRAM.INTENSIVE STUDY PROGRAM Neurosciences :a study program Boulder,Colo. 1966 Jul Edited by Gardner C. Quarton and others Rockefeller university press,New York,1967
Bioch 33.2220

NEUROSCIENCES RESEARCH PROGRAM.INTENSIVE STUDY PROGRAM Neurosciences research symposium summaries :an anthology from the 'Neurosciences research program bulletin' Cambridge,Mass. 1969 Edited by Francis O. Schmitt and others Sponsored by the Massachusetts institute of technology Rockefeller university press;MIT press,New York;Cambridge,Mass.,1970
Bioch 33.2164

NEUROSCIENCES RESEARCH SYMPOSIUM SUMMARIES : an anthology from the Neurosciences research program bulletin Vol 1-2 Neurosciences research program Edited by Francis O. Schmitt and others Neurosciences research program bulletin 2 vols M.I.T. Press,Cambridge,Mass.,1966-67 This volume reports 9 of the twenty work sessions held in the last 3 years by the N.R.P. Work session reports are by Albert L.Lehninger and others
Bioch 33.2221

NEUROSCRETION International symposium uber neurosekretion 2nd Proceedings Lund 1957 Jul 1-6 Edited by W. Bargmann and others Springer,Berlin,1958 Trilingual, English,French and German
An 32.3745

NEUROSECRETION An International symposium on neurosecretion Proceedings Society for endocrinology Edited by H. Heller and R.B. Clark Society for endocrinology.Memoirs, 12 Academic press,London;New York,1962
Inv Med 37.0107

NEUROSECRETION International symposium on neurosecretion 4th Strasbourg 1966 Jul 25-27 Edited by F. Stutinsky Springer, Berlin,1967
An 32.3779

NEUROSECRETION AND NEURAL CONTROL OF INTERNAL SECRETION Gunma symposium on endocrinology 1st Minakami 1963 Sep 5-6 Gunma university.Institute of endocrinology Gunma university.Institute of endocrinology.Annual report, 1 Institute of endocrinology,Gunma university,Maebashi, 1964
Phys 20.2230

NEUROTOXICITY OF DRUGS European society for the study of drug toxicity :meeting 8th Proceedings Prague 1966 Jun-Jul European society for the study of drug toxicity. Proceedings,8 International congress series, 118 Excerpta medica,Amsterdam,1967
Pha 16.0332

NEUROVEGETATIVE SYSTEM Symposion uber das neurovegetative system der gesunden und kranken haut des menschen Vienna 1957 May 30-Jun 1 Edited by E. Anderson and others Acta neurovegetativa, 18 Springer,Vienna, 1958 In English,French and German
An 32.4046

NEUTRON CRYSTALLOGRAPHY Thermal neutron diffraction International summer school on the accurate determination of neutron intensities and structure factors Proceedings Harwell 1968 Jul 1-5 Edited by B.T.M. Willis Harwell postgraduate series Bibliog,illus xiv,229p 24cm Oxford university press,London,1970
Min 10.1524

NEUTRON DOSIMETRY Symposium on neytron detection,dosimetry and standardization Proceedings Harwell 1962 Dec 10-14 1-2 Held by the International atomic energy agency International atomic energy agency. Proceedings series 2 vols International atomic energy agency,Vienna,1963
Radioth 35.1695

NEUTRON MONITORING Symposium on neutron monitoring for radiological protection Proceedings Vienna 1966 Aug 29-Sep 2 International atomic energy agency International atomic energy agency.Proceedings series International atomic energy agency, Vienna,1967
Radioth 35.1672

NEUTRON PHYSICS :symposium Proceedings Troy,N.Y. 1961 May 5-6 Rensselaer polytechnic institute Edited by M.L. Yeater Co-sponsored by the American nuclear society Nuclear science and technology:a series of monographs and text books, 2 xiii,303p Academic press,New York,1962
Cav 7.2543

NEUTRONS Inelastic scattering of neutrons Symposium on inelastic scattering of neutrons Bombay 1964 Dec 15-19 Vol 1-2 International atomic energy agency Edited by M.M. Brown International atomic energy agency,Vienna,1965
Cav 7.0267

NEUTRONS Inelastic scattering of neutrons in solids and liquids :a symposium Proceedings Vienna 1960 Oct 11-14 International atomic energy agency Vienna,1961
Cav 7.2392

NEW ADRENERGIC BLOCKING DRUGS.THEIR PHARMACOLOGICAL,BIOCHEMICAL AND CLINICAL ACTIONS :a conference Papers New York 1966 Feb 24-26 By N.C. Moran New York academy of sciences New York academy of sciences.Annals, 139,p.541-1009 New York, 1967
Bioch 33.0422

NEW APPROACHES IN CELL BIOLOGY :a symposium Proceedings London 1958 Jul Edited by P.M.B. Walker Academic press,London;New York, 1966 Symposium presented at the fifteenth international congress of zoology
Gen 34.0851

NEW APPROACHES IN CELL BIOLOGY :a symposium Proceedings London 1958 Jul Edited by P.M.B. Walker Academic press,London,1960 Held during the fifteenth international congress of zoology
An 32.3236

NEW APPROACHES IN CELL BIOLOGY :a symposium Proceedings London 1958 Jul Edited by P.M.B. Walker Illus Academic press,London; New York,1960 Held during the 15th International congress of zoology
Radioth 35.0506

NEW APPROACHES IN CELL BIOLOGY;A SYMPOSIUM The International congress of zoology 15th Proceedings London 1958 Jul Edited by P. M.B. Walker Held at Imperial college of science and technology Academic press,London, 1960 A section of the Proceedings of the 15th International congress of zoology
Bal 39.0289

NEW ASPECTS IN THE HISTORY AND PHILOSOPHY OF ASTRONOMY Joint symposium of the International astronomical union and the Union internationale d'histoire et de philosophie des sciences 1st Hamburg 1964 Aug 22-24 International astronomical union International union of history and philosophy of science Edited by Arthur Beer Vistas in astronomy, 9 illus,ports,maps 318p 25cm Pergamon press,Oxford,1967
Cav 7.2715

NEW ASPECTS IN THE HISTORY AND PHILOSOPHY OF ASTRONOMY Joint symposium of the International astronomical union and the Union internationale d'histoire et de philosophie des sciences 1st Hamburg 1964 Aug 22-24 International astronomical union International union of history and philosophy of science Edited by Arthur Beer Vistas in astronomy, 9 xvi,317p Pergamon,Oxford, 1967
WSM 43.2439

NEW BRUNSWICK,N.J. Nucleic acids in immunology :a symposium Proceedings Edited by O.J. Plescia and W. Braun Held at Rutgers university.Institute of microbiology Springer,NEW York,1968
Radioth 35.0152

NEW BRUNSWICK,N.J. 1953 Some conjugated proteins;a symposium The Annual conference on protein metabolism 9th six lectures Rutgers university.Bureau of biological research Edited by William H. Cole Rutgers university press,New Brunswick,N.J.,1953
Bioch 33.0543

NEW BRUNSWICK,N.J. 1958 Microbiology yesterday and today :a symposium Rutgers university.Institute of microbiology Edited by Vernon Bryson Rutgers university press, New Brunswick,N.J.,1959 Held in honour of the 70th birthday of Selman A.Waksman
Bot 42.3720

NEW BRUNSWICK,N.J. 1958 Serological and biochemical comparisons of proteins Rutgers university.Bureau of biological research. Annual conferences on protein metabolism 14th Papers Edited by William H. Cole Rutgers university press,New Brunswick,N.J., 1958
Col S 12.0073

NEW BRUNSWICK,N.J. 1961 New developments in tissue culture :annual research conference Rutgers university.Bureau of biological research Edited by James W. Green Annual research conference of the bureau of biological research, 17 Rutgers university press,New Brunswick,N.J.,1961
An 32.3177

NEW BRUNSWICK,N.J. 1962 Informational macromolecules :a symposium Proceedings Rutgers university.Institute of microbiology Edited by Henry J. Vogel and others With support from the National science foundation Academic press,New York;London,1963
Bioch 33.0822

NEW BRUNSWICK,N.J. 1963 Bacterial endotoxins :a symposium Proceedings Rutgers university.Institute of microbiology Edited by Maurice Landy and Werner Braun With the support of the National science foundation Rutgers university.Institute of microbiology,New Brunswick,N.J.,c1963
Path 30.2611

NEW BRUNSWICK,N.J. 1964 Evolving genes and proteins Edited by Vernon Bryson and Henry J. Vogel With support from the National science foundation xvii,617p Academic press,New York;London,1965
Bot 42.0995

NEW BRUNSWICK,N.J. 1964 Evolving genes and proteins :a symposium Proceedings Edited by Vernon Bryson and Henry J. Vogel Held at Rutgers university.Institute of microbiology Academic press,New York;London,1965
Bioch 33.0569

NEW BRUNSWICK,N.J. 1966 Organizational biosynthesis :a symposium By Henry J. Vogel and others With support from the National science foundation illus. xx,549p Academic press,New York;London,1967
Bot 42.1512

NEW BRUNSWICK,N.J. 1966 Organizational biosynthesis :a symposium Proceedings Rutgers university.Institute of microbiology Edited by Henry J. Vogel and others With support from the National science foundation Academic press,New York;London,1967
Bioch 33.0832

NEW COMPUTERS - A REPORT FROM THE MANUFACTURERS symposium Los Angeles 1957 Mar.1 Sponsored by Association for computing machinery illus. 132p 28cm Association for computing machinery,Los Angeles,1957
Math L 5.0792

NEW DELHI 1957 Symposium on meteorology in relation to high level aviation over India and surrounding areas Papers Meteorological department,India Edited by U.K. Bose Under the auspices of Indian meteorological society 159p India press,Delhi,1962
Nap 11.0420

NEW DELHI 1958 Monsoons of the world : symposium India.Meteorological department Delhi,1960
Geog 13.6752

NEW DELHI 1958 Symposium on monsoons of the world Proceedings Meteorological office,India Under the auspices of the World meteorological organization x,270p Manager of publications,Delhi,1960
Nap 11.0284

NEW DELHI 1960 Asian regional conference on soil mechanics and foundation engineering 1st Proceedings International society of soil mechanics and foundation engineering Indian national society of soil mechanics and foundation engineering New Delhi,1960
Eng 41.3172

NEW DELHI 1960 Plant embryology :a symposium Council of scientific and industrial research.Biological research committee Council of scientific and industrial research,New Delhi,1962
Bot 42.1240

NEW DELHI 1960 Termites in the humid tropics :symposium Proceedings Unesco Zoological survey of India Paris,1962
Bal 39.4249

NEW DELHI 1960-61 The Symposium on advancing frontiers of life sciences Proceedings National institute of sciences of India National institute of sciences of India.Bulletin, 19 New Delhi,1962 Symposium held on the occasion of the Silver Jubilee of the National institute of sciences of India
Bal 39.0009

NEW DELHI 1964 International mineralogical association 4th general meeting Papers and proceedings International mineralogical association Mineralogical society of India Edited by P.R.J. Naidu and M.N. Viswanathiah Mineralogical society of India,Mysore,1966
Min 10.1550

NEW DELHI 1964 The Upper mantle symposium papers International union of geological sciences,International upper mantle committee Edited by Charles H. Smith and Theodor Sorgenfrei Det Berlingske Bogtrykkeri, Copenhagen,1965
Geol 8.2191

NEW DELHI 1966 Symposium on newer trends in taxonomy Proceedings National institute of sciences of India National institute of sciences of India.Bulletin, 34 New Delhi, 1967
Bal 39.2079

NEW DELHI 1969 Control processes in multicellular organisms :a Ciba foundation symposium Ciba foundation Edited by G.E.W. Wolstenholme and Julie Knight Churchill, London,1970
An 32.5436

NEW DELHI 1969 Control processes in multicellular organisms :a Ciba foundation symposium Ciba foundation Edited by G.E.W. Wolstenholme and Julie Knight Churchill, London,1970
Phys 20.2271

NEW DEVELOPMENTS IN TISSUE CULTURE :annual research conference New Brunswick,N.J. 1961 Rutgers university.Bureau of biological research Edited by James W. Green Annual research conference of the bureau of biological research, 17 Rutgers university press,New Brunswick,N.J.,1961
An 32.3177

NEW DIRECTIONS IN MATHEMATICS conference Hanover,N.H. 1961 Nov.3-4 Dartmouth college.Department of mathematics Edited by Robert W. Ritchie Supported by the Alfred P. Sloan foundation iv,124p 22cm Prentice-Hall,Englewood Cliffs,N.J.,1964
P. Math 2.2558

NEW ELITES OF TROPICAL AFRICA International African seminar 6th Studies Ibadan 1964 Jul International African institute Edited by P.C. Lloyd Oxford university press, London,1966
Geog 13.4666

NEW ENGINEERING MATERIALS :a conference Birmingham 1965 Oct 13-14 Institution of mechanical engineers.Process engineering group Institution of mechanical engineers. Proceedings, 180 pt.3D I.Mech.E.,London,1965
Met 25.0105

NEW ENGLAND NUCLEAR CORPORATION Annual symposium on tracer methodology 5th Proceedings Washington,D.C. 1961 Oct 20 Edited by Seymour Rothschild Advances in tracer methodology, 1 Plenum press,New York,1963 Includes selected papers from the first four annual symposia
Gen 34.0633

NEW ENGLAND NUCLEAR CORPORATION Mossbauer effect methodology Symposium on Mossbauer effect methodology 7th Proceedings New York 1971 Jan 31 Edited by Irwin J. Gruverman xii,308p 23cm Plenum press, London,1971
Chem 18.2961

NEW ENGLAND NUCLEAR CORPORATION Mossbauer effect methodology Symposium on Mossbauer effect methodology 1st Proceedings New York 1965 Feb 2-5 Edited by I.J. Gruverman Plenum press,New York,1965
Chem 18.2826

NEW ENGLAND NUCLEAR CORPORATION Mossbauer effect methodology Symposium on Mossbauer effect methodology 6th Proceedings New York 1970 Jan 25 Edited by Irwin J. Gruverman viii,237p 23cm Plenum press, New York,1971
Chem 18.2805

NEW EXPERIMENTAL TECHNIQUES IN PROPULSION AND ENERGETICS RESEARCH :a technical meeting Proceedings Munich 1967 Sep 11-15 Agard Edited by David Andrews and Jean Surugue Agard conference proceedings, 38 Technivision,Slough,1970
Eng 41.8359

NEW FABRICATION TECHNIQUES Southwestern metal congress Papers American society for metals American society for metals,Cleveland, Ohio,n.d.
Met 25.0684

NEW HAVEN 1906 The New Haven mathematical colloquium lectures By Eliakim Hastings Moore and others American mathematical society.13th summer meeting Under the auspices of Yale university Americian mathematical society.Colloquium, 5 Yale university press,New Haven,1910
Philos 1.0198

NEW HAVEN,CONN. 1929 International congress of psychology 9th Proceedings and papers International congress of psychology Psychological review co., Princeton,N.J.,1930 Under the presidency of J.M.Cattell
Psy 31.3393

NEW HAVEN,CONN. 1956 Symposium (International) on combustion 6th Combustion institute xxv,943p Reinhold; Chapman and Hall,New York;London,1957
Chem E 24.1407

NEW HAVEN,CONN. 1956 Symposium (international) on combustion,flame and explosion phenomena 6th Papers Combustion institute Reinhold;Chapman and Hall,New York;London,1957 Earlier symposia organized by the Standing committee on combustion.Later symposia entitled 'Symposium (international) on combustion'
Chem 18.2602

NEW HAVEN,CONN. 1966 Evolution and environment :a symposium Edited by Ellen T. Drake Silliman foundation.Lectures Illus, maps xvi,478p Yale university press,New Haven;London,1968 Presented on the occasion of the 100th anniversary of the foundation of the Peabody museum of natural history
Bot 42.0317

The NEW HAVEN MATHEMATICAL COLLOQUIUM lectures New Haven 1906 September 3-8 By Eliakim Hastings Moore and others American mathematical society.13th summer meeting Under the auspices of Yale university Americian mathematical society. Colloquium, 5 Yale university press,New Haven,1910
Philos 1.0198

The NEW HAVEN MATHEMATICAL COLLOQUIUM 5th Lectures New Haven,Conn. 1906 Sep 5-8 By Eliakim Hastings Moore and others American mathematical society American mathematical society.Colloquium lectures, 2 x,222p 24cm University microfilms,Ann Arbor,Mich.,1967 Facsimile produced by microfilm-xerography of the original book published in 1910 by Yale university press,New Haven,Conn.
P Math 2.2864

NEW MATERIALS Proceedings of the Royal society Series A:mathematical and physical sciences Vol 282: discussion on new materials Royal society of London Royal society,London,1964 Organized by J.D. Bernal and others
Eng 41.3430

NEW METHODS OF THOUGHT AND PROCEDURE Symposium on methodologies Contributions Pasadena,Calif. 1967 May 22-24 California institute of technology.Office for industrial association Society for morphological research Edited by F. Zwicky and A.G. Wilson viii,338p Springer,Berlin,1967
WSM 43.2639

NEW METHODS OF THOUGHT AND PROCEDURE Symposium on methodologies Contributions Pasadena,Calif. 1967 May 22-24 Edited by F. Zwicky and A.G. Wilson illus. viii,338p 24cm Springer,Berlin,1967
Math S 3.1389

NEW ORLEANS Kernicterus and its importance in cerebral palsy American academy for cerebral palsy:annual meeting 11th Papers Thomas,Springfield,Ill.,1961
An 32.4184

NEW ORLEANS 1961 Serological fractions in schizophrenia :a research symposium Louisiana state department of hospitals Tulane university Louisiana state university Edited by Robert G. Heath Harper and Row for the Commonwealth fund,New York,1963
PGMS 29.0324

NEW ORLEANS 1966 Perspectives in leukaemia a symposium Proceedings By William Dameshek and R.M. Dutcher Leukaemia society Grune and Stratton,New York,1968
Med 36.0103

NEW ORLEANS 1967 Conference on transformation groups Proceedings Edited by Paul S. Mostert Sponsored by the National science foundation.Advanced science seminar projects Bibliog.,Illus. xiii,456p 24cm Springer-Verlag,Berlin,1968
P Math 2.3395

NEW ORLEANS,LA. 1952 Studies in schizophrenia :a multidisciplinary approach to mind-brain relationships Tulane university of Louisiana.Department of psychiatry and neurology Edited by Robert G. Heath,chairman Published for the Commonwealth fund Harvard university press,Cambridge,Mass.,1954 Transcript of a series of invitational meetings
Psy 28.0081

NEW ORLEANS,LA. 1961 Ergodic theory international symposium papers Tulane university Edited by Fred B. Wright Sponsored by the National science foundation xii,316p Academic press,New York;London,1963
Math 3.0044

NEW ORLEANS,LA. 1961 Ergodic theory international symposium proceedings Edited by Fred B. Wright Sponsored by the National science foundation xii,316p 24cm Academic press,New York,London,1963
P. Math 2.2188

NEW ORLEANS,LA. 1965 Acquisition of skills a conference Papers United States.Army. Medical research and development command Edited by Edward A. Bilodeau Academic press, New York;London,1966
Psy 31.1005

NEW ORLEANS,LA. 1967 A.G.I. short course on layer silicates :lecture notes American geological institute American geological institute,Washington,D.C.,1967 Typescript
Min 10.1287

NEW ORLEANS,LA. 1968 Developments in theoretical and applied mechanics Southeastern conference on theoretical and applied technics 4th Proceedings Edited by D. Frederick Sponsored by Tulane university illus 637p 26cm Pergamon, Oxford,1970
A Math 4.1903

NEW ORLEANS,LA. 1969 Composite materials: testing and design :a symposium American society for testing and materials American society for testing and materials.Special technical publication, 460 ASTM,Philadelphia, Pa.,1969
Eng 41.8290

NEW ORLEANS,LA. 1969 Symposium on composite materials:testing and design American society for testing materials. Committee D-30 high modulus fibers and their composites A.S.T.M.Special technical publication, 460 Philadelphia,Pa.,1969 Presented at a meeting of committee D-30
Met 25.2859

NEW PERSPECTIVES IN BIOLOGY :a symposium Proceedings Rehovoth 1963 Jun 10-17 Edited by Michael Sela Held at the Weizmann institute of science B.B.A.Library, 4 xviii,285p Elsevier,Amsterdam,1964 A symposium held on the occasion of the inaugration of the Ullmann institute of life sciences
Bot 42.1502

NEW PERSPECTIVES IN BIOLOGY :a symposium Proceedings Rehovoth 1963 Jun 10-17 Edited by Michael Sela Held at the Weizmann institute of science B.B.A.library, 4 Elsevier,Amsterdam,1964 A symposium held, on the occasion of the inauguration of the Ullmann institute of life sciences
Bioch 33.0606

NEW PERSPECTIVES IN BIOLOGY :a symposium Proceedings Rehovoth,Israel 1963 Jun 10-17 Edited by Michael Sela . B.B.A.library,4 Elsevier,Amsterdam,1964
Pha 16.0273

NEW PERSPECTIVES IN ORGANIZATION RESEARCH : consisting of papers from a conference on research in organizations...and a seminar on the social science of organizations Pittsburgh,Pa. 1962 Jun.22-24 Pittsburgh, Pa. 1962 Jun.10-23 Edited by W.W. Cooper and others Sponsored by United States.Office of naval research,Bibliog. xxii,606p 24cm John Wiley and sons,New York,1964
Math 3.0910

NEW PHYSICAL AND CHEMICAL PROPERTIES OF METALS OF VERY HIGH PURITY :international symposia Paris 1959 Oct 12-14 Centre national de la recherche scientifique Centre national de la recherche scientifique.International symposia, 40 Gordon and Breach,New York, 1965
Met 25.1500

NEW RESEARCH IN PLANT ANATOMY London 1970 Sep 18 Edited by N.K.B. Robson and D.F. Cutler Linnean society of London.Botany. Supplement, 63,1 Academic press,London,1970 Symposium arranged by the Plant anatomy group of the Linnean society
Bot 42.1118

NEW RESEARCH TECHNIQUES OF NEUROANATOMY :a symposium Edited by William F. Windle Sponsored by the National multiple sclerosis society Thomas,Springfield,Ill.,1957
An 32.4352

NEW TECHNIQUES IN ASTRONOMY :conferences Proceedings Moscow 1961 Apr 18-20 and Kazan 1964 May 12-14 Edited by H.C. Ingrao 446p Gordon and Breach,New York,1971
Obs 6.3598

NEW TECHNIQUES IN SPACE ASTRONOMY Munich 1970 Aug 10-14 International astronomical union Edited by F. Labuhn and R. Lust International astronomical union.Symposium, 41 418p Reidel,Dordrecht,1971
Obs 6.3590

NEW THINKING IN SCHOOL BIOLOGY The Seminar on the reform of biology teaching :report on the OECD seminar La Tour de Peilz, Switzerland 1962 Dep 4-14 Organization for economic co-operation and development New thinking in school science Paris,1963
Gen 34.2062

NEW TYPES OF METAL POWDERS :a symposium Proceedings Cleveland,Ohio 1963 Oct 24 American institute of mining,metallurgical and petroleum engineers.Powder metallurgy committee Edited by Henry H. Hausner Metallurgical society conferences, 23 Gordon and Breach,New York;London,1964
Met 25.2509

NEW TYPES OF METAL POWDERS :a symposium Proceedings Cleveland,Ohio 1963 Oct 24 American institute of mining,metallurgical and petroleum engineers.Powder metallurgy committee Edited by Henry H. Hausner Metallurgical society conferences, 23 Gordon and Breach,New York;London,1964
Met 25.0766

NEW YORK Cellular and humoral aspects of the hypersensitive states :symposium Edited by H. Sherwood Lawrence New York academy of medicine.Section microbiology.Symposia,9 Cassell,London,1959
Pha 16.0181

NEW YORK Heredity counseling :a symposium American eugenics society Edited by Helen G. Hammons Hoeber-Harper,New York,1959
PGMS 29.0257

NEW YORK Integrating the approaches to mental disease :two conferences Papers and discussions New York academy of medicine. Committee on public health Edited by H.D. Kruse Hoeber-Harper,New York,1957
Psy 28.0061

NEW YORK 1921 Eugenics,genetics and the family International congress of eugenics 2nd Papers Vol 1 Port 2 vols Williams and Wilkins,Baltimore,Md.,1923
An 32.0031

NEW YORK 1921 Eugenics,genetics and the family International congress of eugenics 2nd Papers Vol 1-2 port. 2 vols Williams and Wilkins,Baltimore,Md.,1923
Gen 34.1926

NEW YORK 1921 Eugenics in race and state The International congress of eugenics 2nd Scientific papers Vol 2 Port Williams and Wilkins,Baltimore,1923
An 32.0032

NEW YORK 1926 The Cerebellum :an investigation of recent advances Proceedings of the Association Association for research in nervous and mental disease Edited by Frederick Tilney and others Association for research in nervous and mental disease. Research publications, 6 Williams and Wilkins,Baltimore,Md.,1929
An 32.4680

NEW YORK 1926 Theory of continental drift : a symposium on the origin and movement of land masses both intercontinental and intra-continental,as proposed by Alfred Wegener By W.A.J.M. Waterschoot van der Gracht American association of petroleum geologists American association of petroleum geologists,Tulsa,Oka., 1928
Geol 8.2130

NEW YORK 1927 The Intracranial pressure in health and disease :an investigation of the most recent advances Proceedings of the Association Association for research in nervous and mental disease Edited by Charles A. Elsberg and others Association for research in nervous and mental disease. Research publications, 8 Williams and Wilkins,Baltimore,Md.,1929
An 32.4155

NEW YORK 1928 The Vegetative nervous system :an investigation of the most recent advances Proceedings of the Association Association for research in nervous and mental disease Edited by Walter Timme and others Association for research in nervous and mental disease.Research publications, 9 Williams and Wilkins,Baltimore,Md.,1930
An 32.4884

NEW YORK 1932 Decade of progress in eugenics International congress of eugenics 3rd Papers port.pl. Williams and Wilkins,Baltimore,Md.,1934
Gen 34.1927

NEW YORK 1932 Localization of function in the cerebral cortex :an investigation of the most recent advances Proceedings of the Association Association for research in nervous and mental disease Edited by Samuel T. Orton and others Association for research in nervous and mental disease.Research publications, 13 Williams and Wilkins, Baltimore,Md.,1934
An 32.4632

NEW YORK 1933 Symposium on motor lubricants American society for testing materials Society for automotive engineers American society for testing materials, Philadelphia,Pa.,1933
Eng 41.6264

NEW YORK 1934 Grain-size symposium held during the 16th annual convention American society for metals American society for metals,Cleveland,c 1934 Damaged copy.No title page
Met 25.1264

NEW YORK 1934 Sensation:its mechanisms and disturbances :an investigation of the most recent advances Proceedings of the association Association for research in nervous and mental disease Edited by Clarence A. Patten and others Association for research in nervous and mental disease. Research publications, 15 Williams and Wilkins,Baltimore,Md.,1935
An 32.4523

NEW YORK 1936 The Pituitary gland :an investigation of the most recent advances Proceedings of the Association Association for research in nervous and mental disease Edited by Walter Timme and others Association for research in nervous and mental disease.Research publications,17 Williams and Wilkins,Baltimore,Md.,1938
Pha 16.0275

NEW YORK 1937 The Circulation of the brain and spinal cord :a symposium on blood supply Proceedings of the Association Association for research in nervous and mental disease Edited by Stanley Cobb and others Association for research in nervous and mental disease.Research publications, 18 Hafner, New York,1966 Facsimile of 1938 edition
An 32.4133

NEW YORK 1939 Temperature :its measurement and control in science and industry :a symposium Papers Under the auspices of the American institute of physics Reinhold,New York,1941
Eng 41.4125

NEW YORK 1939 Temperature :its measurement and control in science and industry : symposium Papers Under the auspices of American institute of physics Reinhold,New York,1941
Min 10.0599

NEW YORK 1939 Temperature:its measurement and control in science and industry :a symposium American institute of physics and National bureau of standards Edited by C.O. Fairchild 1362p Reinhold publishing, New York,1941
Cav 7.0046

NEW YORK 1939 Temperature:its measurement and control in science and industry :a symposium Papers American institute of physics With the cooperation of the National bureau of standards Reinhold,New York,1941
Chem 18.0636

NEW YORK 1939 Temperature - its measurement and control in science and industry :a symposium 2nd Papers American institute of physics and National bureau of standards Reinhold,New York,1941
Met 25.0298

NEW YORK 1939 Temperature - its measurement and control in science and industry :a symposium Papers Under the auspices of the American institute of physics Reinhold,New York,1941
Radioth 35.1431

NEW YORK 1939 Temperature - its measurement and control in science and industry :a symposium Papers Under the auspices of the American institute of physics xiii,1362p Reinhold,New York,1941
Chem 3 24.0478

NEW YORK 1939 The Hypothalamus and central levels of autonomic function Proceedings of the Association Association for research in nervous and mental disease Edited by John F. Fulton and others Association for research in nervous and mental disease.Research publications, 22 Baltimore,Md.,1940
Phys 20.1083

NEW YORK 1939 The Hypothalamus and central levels of autonomic function Proceedings of the Association Association for research in nervous and mental disease Edited by John F. Fulton and others Association for research in nervous and mental disease.Research publications,20 Williams and Wilkins, Baltimore,Md.,1940
Pha 16.0282

NEW YORK 1939 The Hypothalamus and central levels of autonomic function Proceedings of the association Association for research in nervous and mental disease Edited by John F. Fulton and others Association for research in nervous and mental disease.Research publications, 20 Williams and Wilkins, Baltimore,Md.,1940
An 32.4510

NEW YORK 1940 The Diseases of the basal ganglia Proceedings of the association Association for research in nervous and mental disease Edited by Tracy J. Putnam and others Association for research in nervous and mental disease.Research publications, 21 Williams and Wilkins,Baltimore,Md.,1942
An 32.4511

NEW YORK 1941 The Role of nutritional deficiency on nervous and mental disease Proceedings of the Association Association for research in nervous and mental disease Edited by Stanley Cobb and others Association for research in nervous and mental disease.Research publications, 22 Williams and Wilkins,Baltimore,Md.,1943
An 32.4377

NEW YORK 1941 The Ultracentrifuge :a conference By D.A. MacInnes and others New York academy of sciences.Section of physics and chemistry New York academy of sciences. Annals, 43,p 173-252 New York,1942
Bioch 33.1491

NEW YORK 1942 Pain Proceedings Association for research in nervous and mental disease Association for research in nervous and mental disease.Research publications, 23 Baltimore,Md.,1943
Phys 20.2003

NEW YORK 1942 Pain Proceedings of the association Association for research in nervous and mental disease Edited by Harold G. Wolff and others Association for research in nervous and mental disease.Research publications, 23 Williams and Wilkins, Baltimore,Md.,1945
An 32.4520

NEW YORK 1943 Trauma of the central nervous system Proceedings of the Association Association for research in nervous and mental disease Association for research in nervous and mental disease. Research publications, 24 Williams and Wilkins,Baltimore,Md.,1945
An 32.4090

NEW YORK 1943 Trauma of the central nervous system Proceedings of the Association Association for research in nervous and mental disease Association for research in nervous and mental disease. Research publications, 24 Williams and Wilkins,Baltimore,Md.,1945
An 32.5227

NEW YORK 1944 Energy relationships in enzyme reactions :a conference New York academy of sciences.Section of physics and chemistry New York academy of sciences. Annals, 45,p.357-436 New York,1944
Bioch 33.1027

NEW YORK 1945 Amino acid analysis of proteins :a conference Papers By William H. Stein and others New York academy of sciences New York academy of sciences.Annals, 47,p.57-240 New York,1946
Bioch 33.0552

NEW YORK 1945 Experimental hypertension being the results of a conference Results By William Goldring and others New York academy of sciences.Section of biology New York academy of sciences.Special publications, 3 New York academy of sciences,New York, 1946
Psy 28.0096

NEW YORK 1945 Lymph :a conference Papers New York academy of sciences New York academy of sciences.Annals, 46,art.8 New York,1946
Phys 20.0883

NEW YORK 1946 Biological antioxidants :a conference 1st Transactions Josiah Macy jr. foundation Edited by Cosmo G. Mackenzie New York,1946
Chem 18.2646

NEW YORK 1946 Biology of melanomas A Conference on the biology of normal and atypical pigment cell growth New York academy of sciences.Section of biology New York academy of sciences.Special publications, 4 pls New York academy of sciences,New York,1948
Path 30.2492

NEW YORK 1946 Biology of melanomas Conference on the biology of normal and atypical pigment cell growth New York academy of sciences.Section of biology New York academy of sciences.Special publications, 4 New York,1948
Phys 20.0952

NEW YORK 1946 Biology of melanomas Conference on the biology of normal and atypical pigment cell growth Results By Myron Gordon and others New York academy of sciences.Special publications, 6 New York academy of sciences,New York,1948
An 32.3374

NEW YORK 1946 Bioluminescence :a conference By E.Newton Harvey and others New York academy of sciences.Section of biology New York academy of sciences.Annals, 49,p 327-482 New York,1948
Bal 39.0084

NEW YORK 1946 Chromatography :a conference Papers By Harold G, Cassidy and others New York academy of sciences.Section of physics and chemistry New York academy of sciences. Annals, 49,p.141-326 New York,1948
Bioch 33.2362

NEW YORK 1946 Lymph :a conference By Philip D. McMaster and others New York academy of sciences New York academy of sciences.Annals, 46,p679-882 New York,1946
Bal 39.1342

NEW YORK 1946 Postwar problems of migration Milbank memorial fund annual conference Papers Milbank memorial fund Milbank memorial fund,New York,1947
Geog 13.1564

NEW YORK 1946 The Physico-chemical mechanism of nerve activity :a conference By D. Nachmansohn and others New York academy of sciences New York academy of sciences. Annals, 47, p 375-602 New York,1946
Bal 39.1449

NEW YORK 1947 Culture and personality :an interdisciplinary conference Proceedings Viking fund Edited by S.Stansfeld Sargent and Marian W. Smith Viking fund,New York, 1949
Psy 31.1931

NEW YORK 1947 Ecology of health :institute on public health Papers New York academy of medicine Edited by E.H.L. Corwin Commonwealth fund,New York,1949 Issued for the centennial celebration of the New York academy of medicine
An 32.2035

NEW YORK 1947 Muscular contraction :a conference By Alexander Sandow and others New York academy of sciences New York academy of sciences.Annals, 47 p665-930 New York,1947
Bal 39.1422

NEW YORK 1947 Recent studies in the mechanisms of embryonic development :a conference New York academy of sciences. Section of biology Edited by Robert Rugh New York academy of sciences.Annals, 49,p 661-866 New York,1948
Bal 39.0438

NEW YORK 1947 The Frontal lobes Proceedings Association for research in nervous and mental disease Association for research in nervous and mental disease. Research publications, 27 Baltimore,Md., 1948
Phys 20.1992

NEW YORK 1947 The Frontal lobes Proceedings of the Association Association for research in nervous and mental disease Edited by Mary P. Wheeler Association for research in nervous and mental disease. Research publications, 27 Hafner,New York, 1948 reprinted 1966
An 32.5330

NEW YORK 1948 Adrenal cortex :a conference Papers New York academy of sciences Edited by Robert Gaunt New York academy of sciences. Annuals, 50,509-678 New York,1949
Inv Med 37.0274

NEW YORK 1948 Chemotherapy of filariasis : a conference By L.L. Ashburn and others New York academy of sciences Edited by Roy Waldo Miner New York academy of sciences. Annals, 50,art.2 New York,1948
Mol 45.0271

NEW YORK 1948 Ocean surface waves New York academy of sciences.Oceanography and meteorology section Edited by B. Haurwitz New York academy of sciences.Annals, 51,3 New York,1949 Series of papers resulting from a conference on surface waves held by the section of oceanography and meteorology of the New York academy of sciences March 18-19,1948
Geod 9.0602

NEW YORK 1948 Phase transformations in solids :symposium Edited by R. Smoluchowski and others Sponsored by the National research council.Committee on solids Wiley; Chapman and Hall,New York;London,1951
Met 25.1250

NEW YORK 1948 Teleological mechanisms :a conference By Lawrence K. Frank and others New York academy of sciences New York academy of sciences.Annals, 50,p. 187-278 New York,1948
Bal 39.1003

NEW YORK 1948 The Pathogenesis and pathology of viral diseases :a symposium New York academy of medicine.Section on microbiology Edited by John G. Kidd New York academy of medicine.Section on microbiology.Symposia, 3 Columbia university press,New York,1950
Path 30.2554

NEW YORK 1949 Adrenal cortex :a conference 1st Transactions Edited by Elaine P. Ralli and others Sponsored by the Josiah Macy Jr.foundation Josiah Macy Jr.foundation, New York,1950
Bioch 33.0464

NEW YORK 1949 Life stress and bodily disease Proceedings of the association Association for research in nervous and mental disease Edited by Harold G. Wolff Association for research in nervous and mental disease.Research publications, 29 Hafner, New York,1950 reprinted 1968
An 32.4852

NEW YORK 1949 Problem of land connections across the South Atlantic with special reference to the Mesozoic Symposium on the role of the South Atlantic basin in biogeography and evolution Proceedings By Maurice Ewing and others Edited by Ernst Mayr American museum of natural history. Bulletin, 99,3 American museum of natural history,New York,1952
Bot 42.2341

NEW YORK 1949 Renal function :conference 1st Transactions Edited by Stanley E. Bradley Josiah Macy,jr.,foundation,New York, 1950
Pha 16.0284

NEW YORK 1949 The Physics of powder metallurgy :a symposium Sylvania electric products,inc.Metallurgical laboratories Edited by Walter E. Kingston McGraw-Hill,New York,1951
Met 25.0752

NEW YORK 1949 The Problem of land connections across the south Atlantic ,with special reference to the Mesozoic :a symposium Proceedings Edited by Ernst Mayr Held at the 4th annual meeting of the Society for the study of evolution American museum of natural history.Bulletin, 99,art.3 New York, 1952
An 32.2158

NEW YORK 1950 Adrenal cortex :a conference 2nd Transactions Edited by Elaine P. Ralli Sponsored by the Josiah Macy Jr. foundation Josiah Macy Jr.foundation,New York,1951
Bioch 33.0465

NEW YORK 1950 Connective tissues
Conference on connective tissues 1st
Transactions Josiah Macy jr.foundation
Edited by Charles Ragan Josiaph Macy jr.
foundation,New York,1951
Path 30.2487

NEW YORK 1950 Cybernetics:circular causal
and feedback mechanisms in biological and
social systems Conference on cybernetics
7th Transactions Josiah Macy jr.foundation
Edited by Heinz von Foerster Josiah Macy jr.
foundation,New York,1951
Psy 31.1064

NEW YORK 1950 Patterns of organization in
the central nervous system Proceedings of
the association Association for research in
nervous and mental disease Edited by Philip
Bard Association for research in nervous and
mental disease.Research publications, 30
Hafner,New York,1952 reprinted 1968
An 32.4853

NEW YORK 1950 Patterns of recognition in
the central nervous system Proceedings
Association for research in nervous and mental
disease Association for research in nervous
and mental disease.Research publications, 30
Baltimore,Md.,1952
Phys 20.2238

NEW YORK 1950 The Growth,replacement and
types of hair :conference Papers Edited by
J.B. Hamilton New York academy of sciences.
Annals, 53,art.3
An 32.4025

NEW YORK 1950 The Theory of electro-
magnetic waves :a symposium Proceedings
Washington square college of arts and science
New York university.Institute for mathematics
and mechanics Under the auspices of Air
force Cambridge research laboratories.
Geophysics research directorate Interscience,
New York;London,1951
Radioth 35.1507

NEW YORK 1950 Princeton,N.J. 1953-54
Connective tissues Conference on connective
tissues 1st-5th Transactions Edited by
Charles Ragan Sponsored by the Josiah Macy
jr.foundation Josiah Macy,jr.foundation,New
York,1951-54
An 32.3218

NEW YORK 1950-1952 Nerve impulse :
conferences 1st-3rd Transactions 3 vols
Macy foundation,New York,1950-52
Phys 20.2002

NEW YORK 1951 Adrenal cortex :a conference
3rd Transactions Edited by Elaine P.
Ralli Sponsored by the Josiah Macy Jr.
foundation Josiah Macy Jr.foundation,New
York,1952
Bioch 33.0466

NEW YORK 1951 Cybernetics :circular causal
and feedback mechanisms in biological and
social systems Transactions Josiah Macy jr.
foundation Edited by Heinz von Foerster
Josiah Macy,jr. foundation,New York,1952
Psy 31.2046

NEW YORK 1951 Liver injury Conference on
liver injury 10th Transactions Edited by
F.W. Hoffbauer Churchill,London,1951
Med 36.0189

NEW YORK 1951 Pigment cell growth
Conference on the biology of normal and
atypical pigment cell growth 3rd
Proceedings Edited by Myron Gordon
Sponsored by the New York zoological society
Academic press,New York,1953
An 32.3368

NEW YORK 1951 The Chick embryo in
biological research :a conference Papers
Edited by Roy Waldo Miner and D.A. Karnofsky
Held by the New York academy of sciences.
Section of biology New York academy of
sciences.Annals, 55,art.2 New York academy
of sciences,New York,1952
An 32.2265

NEW YORK 1951 The Influence of hormones on
enzymes :a conference Papers By R.I.
Dorfman and others New York academy of
sciences New York academy of sciences.Annals,
54,p.531-728 New York,1951
Bioch 33.0450

NEW YORK 1951 The Nature and significance
of the antibody response Papers and
discussions New York academy of medicine.
Section of microbiology Edited by A.M.
Pappenheimer New York academy of sciences.
Section of microbiology.Symposia, 5
Columbia university press,New York,1953
Path 30.2532

NEW YORK 1951 The Nature and significance
of the antibody response :a symposium Papers
and discussions New York academy of medicine.
Section on microbiology Edited by A.M.
Pappenheimer New York academy of medicine.
Section on microbiology.Symposia, 5
Columbia university press,New York,1953
Bioch 33.1156

NEW YORK 1951 The Nature and significance
of the antibody response;symposium... New
York academy of medicine.Section on
microbiology Edited by Alwin M. Pappenheimer
Held at the New York academy of medicine New
York academy of medicine.Section on
microbiology.Symposia, 5 Columbia
university press,New York,1953
Med 36.0272

NEW YORK 1952 Adrenal cortex :a conference
4th Transactions Edited by Elaine P.
Ralli Sponsored by the Josiah Macy Jr.
foundation Josiah Macy Jr.foundation,New
York,1953
Bioch 33.0467

NEW YORK 1952 Cybernetics:circular causal
and feedback mechanisms in biological and
social systems Conference on cybernetics
9th Transactions Josiah Macy jr.foundation
Edited by Heinz von Foerster Josiah Macy jr.
foundation,New York,1953
Psy 31.1063

NEW YORK 1952 Metabolic and toxic diseases
of the nervous system Proceedings of the
Association Association for research in
nervous and mental disease Edited by H.
Houston Merritt and Clarence C. Hare
Association for research in nervous and mental
disease.Research publications, 32 Williams
and Wilkins,Baltimore,Md.,1953
An 32.4156

NEW YORK 1952 Modern network synthesis (audio to microwaves) :a symposium Proceedings Polytechnic institute of Brooklyn Sponsored also by the Microwave research institute Microwave research institute.Symposia series Polytechnic institute of Brooklyn,New York,1956
Eng 41.5116

NEW YORK 1952 Nerve impulse Conference on nerve impulse 3rd Transactions Edited by H.Houston Merritt Sponsored by the Josiah Macy jr.foundation Josiah Macy jr. foundation,New York,1952
An 32.4315

NEW YORK 1952 Prematurity,congenital malformation,and birth injury :a conference Proceedings Sponsored by the Association for the aid of crippled children Association for the aid of crippled children,New York,1953
An 32.2030

NEW YORK 1952 Rat quality:a consideration of heredity,diet and disease :a symposium Proceedings By W.F. Heston and others National vitamin foundation National vitamin foundation,New York,1953
Radioth 35.0218

NEW YORK 1952 Symposium on direct shear testing of soils American society for testing materials American society for testing materials.Special technical publication, 131 American society for testing materials,Philadelphia,Pa.,1953
Eng 41.3156

NEW YORK 1952 Symposium on fatigue with emphasis on statistical approach Pt 2 American society for testing materials A.S.T. M.Special technical publication American society for testing materials,Philadelphia 3, Pa.,1953
Met 25.0913

NEW YORK 1952 Symposium on fretting corrosion American society for testing materials A.S.T.M.Special technical publication, 144 American society for testing materials,Philadelphia,Pa.,1953 Fiftieth anniversary publication
Met 25.1953

NEW YORK 1952 Symposium on light microscopy American society for testing materials A.S.T.M.Special technical publication, 143 American society for testing materials,Philadelphia,Pa.,1953
Met 25.1333

NEW YORK 1952 Symposium on light microscopy American society for testing materials 55th annual meeting Papers American society for testing materials American society for testing materials.Special publication, 143,Fiftieth anniversary publication Philadelphia,Pa.,1953
Min 10.1571

NEW YORK 1952 Symposium on nondestructive testing presented to the 55th annual meeting... American society for testing materials A. S.T.M.Special technical publication, 145 A.S. T.M.,Philadelphia,1953
Met 25.2359

NEW YORK 1952 Symposium on strength and ductility of metals at elevated temperatures with particular reference to effects of notches and metallurgical changes American society for testing materials A.S.T.M. Special technical publication, 128 A.S.T.M., Philadelphia,Pa.,1953
Met 25.0979

NEW YORK 1952 Symposium on tin American society for testing materials A.S.T.M. Special technical publication, 141 American society for testing materials,Philadelphia,Pa., 1953 Fiftieth anniversary publication. A symposium held at the 55th annual meeting of the American society for testing materials
Met 25.0567

NEW YORK 1953 Conference on training in applied mathematics proceedings American mathematical society Sponsored also by the National research council 97p 27cm Columbia university,New York,1953
Math L 5.0709

NEW YORK 1953 Genetics,and the inheritance of integrated neurological and psychiatric patterns Proceedings of the Association Association for research in nervous and mental disease Edited by Davenport Hooker and Clarence C. Hare Association for research in nervous and mental disease.Research publications, 33 Williams and Wilkins, Baltimore,Md.,1954
An 32.4157

NEW YORK 1953 Growth of protozoa By R.W. Miner New York academy of sciences Edited by S.H. Hutner New York academy of sciences. Annals, 56,5 New York,1953
Bot 42.3886

NEW YORK 1953 Growth of protozoa :a conference By R.P. Hall and others New York academy of sciences Edited by S.H. Hunter and others New York academy of sciences.Annals, 56,p.815-1095 New York, 1953
Bal 39.2132

NEW YORK 1953 I.R.E. national convention convention record Part 7: electronic computers Institute of radio engineers 71p Institute of radio engineers,New York,1953
Math L 5.2064

NEW YORK 1953 Ion transport across membranes :a symposium Papers Edited by Hans T. Clarke and David Nachmansohn held at Columbia university.College of physicians and surgeons Academic press,New York,1954 Incorporating papers presented at a symposium on the role of proteins in ion transport across membranes together with six invited articles on allied topics
Bal 39.0560

NEW YORK 1953 Psychiatry and the law :the proceedings of the forty-third annual meeting of the American psychopathological association Proceedings American psychopathological association Edited by Paul H. Hoch and Joseph Zubin Grune and Stratton,New York; London,1955
Psy 28.0311

NEW YORK 1953 Symposium on nonlinear circuit analysis Proceedings Polytechnic institute of Brooklyn Edited by Jerome Fox Sponsored by the Microwave research institute Microwave research institute.Network symposia series, 2 Polytechnic institute of Brooklyn,New York,1953
Eng 41.5920

NEW YORK 1953 The Surgical clinics of North America New York number: symposium on the surgery of cancer Saunders,Philadelphia, Pa.;London,1953
Radioth 35.0754

NEW YORK 1954 Crust of the earth :a symposium Edited by Arie Poldervaart Organized by the Columbia university. Department of geology Geological society of America.Special papers,62 Geological society of America,New York,1955
Geog 13.0219

NEW YORK 1954 Crust of the earth :a symposium Proceedings Columbia university Edited by Arie Poldervaart Under the auspices of the Geological society of America Geological society of America.Special paper, 62 New York,1955
Geod 9.0100

NEW YORK 1954 I.R.E. national convention convention record Part 4: electronic computers and information theory Institute of radio engineers 143p Institute of radio engineers,New York,1954
Math L 5.2065

NEW YORK 1954 Mechanisms of congenital malformation :second scientific conference Proceedings Association for the aid of crippled children Edited by Harold Wolff Association for the aid of crippled children, New York,1954
An 32.2034

NEW YORK 1954 Neurology and psychiatry in childhood Proceedings of the Association Association for research in nervous and mental disease Edited by Rustin McIntosh and Clarence C. Hare Association for research in nervous and mental disease.Research publications, 34 Williams and Wilkins, Baltimore,Md.,1954
An 32.4161

NEW YORK 1955 Applied probability Mathematical probability and its applications symposium American mathematical society Edited by L.A. MacColl Cosponsored by the United States.Army.Office of ordnance research American mathematical society.Proceedings of symposia in applied mathematics, 7 v,104p 26cm McGraw-Hill,New York,1957
Math 3.0682

NEW YORK 1955 Experimental psychopathology forty-fifth annual meeting Proceedings American psychopathological association Edited by Paul H. Hoch and Joseph Zubin Grune and Stratton,New York;London,1957
Psy 28.0197

NEW YORK 1955 Habitat of oil :symposium papers American association of petroleum geologists Edited by Lewis G. Weeks illus. viii,1384p 26cm American association of petroleum geologists,Tulsa,Okl.,1958 The symposium was held during the 40th annual meeting of the American association of petroleum geologists
Geol 8.1346

NEW YORK 1955 Hormones and the aging process :a conference Proceedings Edited by Earl T. Engle and Gregory Pincus Academic press,New York,1956
An 32.3738

NEW YORK 1955 I.R.E. national convention convention record Part 4: computers, information theory,automatic control Institute of radio engineers 208p Institute of radio engineers,New York,1955
Math L 5.2066

NEW YORK 1955 Modern network synthesis :a symposium Proceedings Polytechnic institute of Brooklyn Sponsored also by the Microwave research institute Microwave research institute.Symposia series Polytechnic institute of Brooklyn,New York, 1956
Eng 41.5117

NEW YORK 1955 Symposium on trace analysis papers Rockefeller foundation and Sloan-Kettering institute Edited by John H. Yoe and Henry J. Koch Wiley;Chapman and Hall,New York;London,1957
Met 25.1720

NEW YORK 1955 The Neurologic and psychiatric aspects of the disorders of aging Proceedings of the Association Association for research in nervous and mental disease Edited by Joseph Earle Moore and others Association for research in nervous and mental disease.Research publications, 35 Williams and Wilkins,Baltimore,Md.,1956
An 32.4166

NEW YORK 1956 Evolution of nervous control from primitive organisms to man:a symposium American association for the advancement of science.Section on medical sciences Edited by Allan D. Bass American association for the advancement of science.Publications, 52 Washington,D.C.,1954
Phys 20.2015

NEW YORK 1956 Evolution of nervous control from primitive organisms to man :a symposium Edited by Allan O. Bass Organized by the American association for the advancement of science.Section on medical sciences American association for the advancement of science. Publications, 52 American association for the advancement of science,Washington,D.C., 1959
An 32.4357

NEW YORK 1956 Evolution of nervous control from primitive organisms to man :a symposium Proceedings American association for the advancement of science.Section on medical sciences Edited by Allan D. Bass American association for the advancement of science. Publication, 52 Washington,D.C.,1959 Arranged by Bernard B.Brodie and Allan D.Bass
Bal 39.1470

NEW YORK 1956 Fatigue in aircraft structures :an international conference United States.Air research and development command Guggenheim institute of flight structures Edited by Alfred M. Freudenthal Academic press,New York,1956
Eng 41.2772

NEW YORK 1956 Fatigue in aircraft structures international conference Proceedings United States.Air force.Solid state sciences division and Guggenheim institute of flight structures Edited by Alfred M. Freudenthal Academic press,New York,1956
Met 25.0921

NEW YORK 1956 Hormones,brain function and behaviour :a conference on neuroendocrinology Proceedings Edited by Hudson Hoagland Academic press,New York,1957
An 32.3748

NEW YORK 1956 I.R.E. national convention convention record Part 4: computers, information theory,automatic control Institute of radio engineers 175p Institute of radio engineers,New York,1956
Math L 5.2067

NEW YORK 1956 International symposium on hemophilia and hemophilioid diseases Proceedings National hemophilia foundation Edited by K.M. Brinkhous North Carolina university press,Chapel Hill,1957
Med 36.0050

NEW YORK 1956 National conference on tube techniques 3rd Proceedings Sponsored by the Advisory group on electron tubes.Working group on semiconductor devices New York university press,New York,1958
Eng 41.5619

NEW YORK 1956 Psychopathology of communication :forty-sixth annual meeting Proceedings American psychopathological association Edited by Paul H. Hoch and Joseph Zubin Grune and Stratton,New York; London,1958
Psy 28.0196

NEW YORK 1956 Symposium on nonlinear circuit analysis Proceedings Polytechnic institute of Brooklyn Edited by Jerome Fox Sponsored by the Microwave research institute Microwave research institute.Network symposia series, 6 Polytechnic institute of Brooklyn,New York,1957
Eng 41.5922

NEW YORK 1956 The Brain and human behavior 1956 meeting Proceedings Association for research in nervous and mental disease Association for research in nervous and mental disease.Research publications, 36 Williams and Wilkins,Baltimore,Md.,1958
Psy 31.0494

NEW YORK 1956 The Brain and human behaviour Proceedings of the Association Association for research in nervous and mental disease Edited by Harry C. Solomon and others Association for research in nervous and mental disease.Research publications, 36 Williams and Wilkins,Baltimore,Md.,1958
An 32.4353

NEW YORK 1956 Wound healing and tissue repair :a symposium Edited by W.Bradford Patterson Developmental biology conferences series,1956 University of Chicago press, Chicago,Ill.,1959
An 32.3161

NEW YORK 1956-57 The Historical development of physiological thought :a symposium Edited by Chandler McC. Brooks and Paul F. Cranefield Hafner,New York,1959
Phys 20.0844

NEW YORK 1957 Cellular biology nucleic acids and viruses :a conference Papers Edited by Thomas M. Rivers Held at the New York academy of sciences New York academy of sciences.Special publications, 5 Port New York academy of sciences,New York,1957
Radioth 35.0496

NEW YORK 1957 Cellular biology nucleic acids and viruses :conference Papers New York academy of sciences Edited by Thomas M. Rivers New York academy of sciences.Special publications, 5 New York academy of sciences,New York,1957
Gen 34.0724

NEW YORK 1957 Determinism and freedom in the age of modern science:a philosophical symposium New York university institute of philosophy :annual symposium 1st Proceedings Edited by Sidney Hook xv,237p New York university press,New York,1965
WSM 43.1662

NEW YORK 1957 I.R.E. national convention convention record Part 2: circuit theory, information theory Institute of radio engineers 212p Institute of radio engineers,New York,1957
Math L 5.2068

NEW YORK 1957 I.R.E. national convention convention record Part 4: automatic control,electronic computers,medical electronics Institute of radio engineers 179p Institute of radio engineers,New York, 1957
Math L 5.2069

NEW YORK 1957 Symposium on digital computing in the aircraft industry proceedings New York university Jointly sponsored by International business machines corporation illus. 399p 28cm International business machines corporation, New York,1957
Math L 5.0777

NEW YORK 1957 The Effect of pharmacologic agents on the nervous system Proceedings of the Association Association for research in nervous and mental disease Edited by Francis J. Braceland Association for research in nervous and mental disease.Research publications, 37 Williams and Wilkins, Baltimore,Md.,1959
An 32.4169

NEW YORK 1957 The Effect of pharmacologic agents on the nervous system :proceedings of the Association... Association for research in nervous and mental disease Edited by Francis J. Braceland Association for research in nervous and mental disease. Research publications,37 Williams and Wilkins,Baltimore,Md.,1959
Pha 16.0036

NEW YORK 1957 Vacuum metallurgy New York university.Department of metallurgical engineering Edited by Rointan F. Bunshah Reinhold.Chapman and Hall,New York;London,1958 Lectures presented during the course on vacuum metallurgy
Met 25.1630

NEW YORK 1958 Chlorpropamide and diabetes mellitus :a conference Papers By Martin G. Goldner and others New York academy of sciences New York academy of sciences.Annals, 74,p.407-1028 New York,1959
Bioch 33.0419

NEW YORK 1958 Electronic waveguides : symposium Proceedings Polytechnic institute of Brooklyn Microwave research institute Microwave research institute. Symposia series, 8 Polytechnic press, Brooklyn,N.Y.,1958
Eng 41.5649

NEW YORK 1958 Molecular biology : elementary processes of nerve conduction and muscle contraction.A symposium Papers Rockefeller institute Edited by David Nachmansohn Sponsored by Columbia university and the Rockefeller institute Academic press, New York;London,1960
Bioch 33.0831

NEW YORK 1958 National conference on tube techniques 4th Proceedings Sponsored by the Advisory group on electron tubes.Working group on tube techniques New York university press,New York,1959
Eng 41.5620

NEW YORK 1958 Neuromuscular disorders (the motor unit and its disorders) Proceedings of the Association Association for research in nervous and mental disease Edited by Raymond D. Adams and others Association for research in nervous and mental disease.Research publications, 38 Williams and Wilkins, Baltimore,Md.,1960
An 32.4185

NEW YORK 1958 Photoreception Papers New York academy of sciences and National council to combat blindness Edited by Jerome J. Wolken New York academy of sciences. Annals, 74,art.2 New York,1958
Phys 20.2225

NEW YORK 1959 Amino acids,proteins and cancer biochemistry Jesse P.Greenstein memorial symposium Papers Edited by John T. Edsall Port Academic press,New York,1960
Radioth 35.0083

NEW YORK 1959 Cellular and humoral aspects of the hypersensitive states :a symposium New York academy of medicine.Section on microbiology Edited by H.Sherwood Lawrence New York academy of medicine.Section of microbiology.Symposia, 9 illus,diagrs Cassell,London,1959
Path 30.2526

NEW YORK 1959 Composite materials and composite structures Sagamore ordnance materials research conference 6th Proceedings United States.Army.Ordnance materials research office and United States.Army.Office of ordnance research Arrangements by Syracuse university research institute New York,1959 Mimeograph
Met 25.0093

NEW YORK 1959 Current trends in research and clinical management in diabetes :a conference Papers By Peter H. Forsham and others New York academy of sciences New York academy of sciences.Annals, 82,p.191-644 New York,1959
Bioch 33.0420

NEW YORK 1959 Finite groups symposium American mathematical society Edited by A. Adrian Albert and Irving Kaplansky Cosponsored by the Institute for defense analysis.Communications division American mathematical society.Proceedings of symposia in pure mathematics, 1 viii,110p 23cm American mathematical society,Providence,R.I., 1959
P. Math 2.2029

NEW YORK 1959 Fracture of engineering materials :a conference Papers Sponsored by the American society for metals American society for metals,Metals Park,Ohio,1964
Eng 41.3796

NEW YORK 1959 Fracture of engineering materials a conference Papers American society for metals.Eastern New York chapter A.S.M.,New York,1959
Met 25.0864

NEW YORK 1959 I.R.E. national convention convention record Part 4: automatic control,electronic computers,information theory Institute of radio engineers 201p Institute of radio engineers,New York,1959
Math L 5.2072

NEW YORK 1959 International oceanographic congress preprints of abstracts of papers Edited by Mary Sears bibliog. American association for the advancement of science, Washington,D.C.,1959
Geol 8.1647

NEW YORK 1959 Mental retardation Proceedings of the Association Association for research in nervous and mental disease Edited by Lawrence C. Kolb and others Association for research in nervous and mental disease.Research publications, 39 Williams and Wilkins,Baltimore,Md.,1962
An 32.4201

NEW YORK 1959 Millimeter waves :a symposium Proceedings Microwave research institute and Institute of radio engineers Co-sponsored by United States.Office of naval research Polytechnic institute of Brooklyn. Microwave research institute symposia series, 9 Polytechnic press,Brooklyn,N.Y.,1960
Cav 7.1715

NEW YORK 1959 Oceanography International oceanographic conference Invited lectures Edited by Mary Sears American association for the advancement of science.Publications,67 American association for the advancement of science,Washington,D.C.,1961
Geog 13.0883

NEW YORK 1959 Oceanography International oceanographic congress Invited lectures American association for the advancement of science Edited by Mary Sears American association for the advancement of science. Publication, 67 Washington,D.C.,1961
Geod 9.0576

NEW YORK 1959 Oceanography International oceanographic congress invited lectures Edited by Mary Sears American association for the advancement of science.Publications,67 American association for the advancement of science,Washington,D.C.,1961
Geol 8.1631

NEW YORK 1959 Oceanography The International oceanographic congress Lectures American association for the advancement of science Edited by Mary Sears American association for the advancement of science.Publication, 67 American association for the advancement of science, Washington,1961
Bal 39.1951

NEW YORK 1959 Quantum electronics Conference on quantum electronics-resonance phenomena Papers Columbia university Edited by Charles H. Townes 601p Columbia university press,New York,1960
Cav 7.2332

NEW YORK 1959 Structure and properties of thin films :an international conference Proceedings United States.Air research and development command and General electric research laboratory Edited by C.A. Neugebauer and others Sponsored by United States.Air force.Office of scientific research Wiley,New York;London,n.d.
Met 25.1204

NEW YORK 1959 Symposium on basic research American association for the advancement of science Edited by Dael Wolfle American association for the advancement of science. Publication, 56 Washington,D.C.,1959
Eng 41.1676

NEW YORK 1959 Symposium on salt and water metabolism Papers Edited by Alfred P. Fishman Sponsored by the New York heart association Circulation, 21,no 5,pt 2 American heart association,New York,1959
Phys 20.1327

NEW YORK 1959 Vacuum metallurgy conference Transactions New York university.Department of metallurgical engineering Edited by Rointan F. Bunshah New York university press, New York,1960
Met 25.2259

NEW YORK 1959 Verbal learning and verbal behavior :a conference Proceedings United States.Office of naval research.Psychological sciences division and New York university. Department of psychology Edited by Charles N. Cofer and Barbara S. Musgrave McGraw-Hill series in psychology McGraw-Hill,New York, 1961
Psy 31.1905

NEW YORK 1960 Advances in electron tube techniques National conference on tube techniques 5th Proceedings Edited by David Slater Sponsored by the Advisory group on electron tubes.Working group on tube techniques Pergamon press,Oxford,1961
Eng 41.5621

NEW YORK 1960 Amino acids,peptides,and proteins :a conference Papers By Karl Folkers and others New York academy of sciences New York academy of sciences.Annals, 88,p.533-770 New York,1960
Bioch 33.0489

NEW YORK 1960 Biochemistry and pharmacology of compounds derived from marine organisms :a conference Papers New York academy of sciences and American institute of biological sciences Edited by Ross F. Nigrelli New York academy of sciences.Annals, 90,article 3 New York,1960 Co-sponsored by the New York zoological society and the United States navy,Office of naval research
Chem 26.0302

NEW YORK 1960 Biophysics of physiological and pharmacological actions The American association for the advancement of science meeting 127th Papers presented American association for the advancement of science Edited by Abraham M. Shanes and others American association for the advancement of science.Publication, 69 Washington,D.C., 1961
Bal 39.1257

NEW YORK 1960 Conference on the mechanisms of cell divisions 2nd Proceedings By M. J. Kopac and others New York academy of sciences Edited by Paul R. Gross New York academy of sciences.Annals, 90,p.345-613 New York,1960
Bal 39.0291

NEW YORK 1960 Extractive metallurgy of copper,nickel and cobalt based on an international symposium American institute of mining,metallurgical and petroleum engineers.Extractive metallurgy division Edited by Paul Quenau Bibliog Interscience, New York;London,1961
Met 25.0510

NEW YORK 1960 Freezing and drying of biological materials :a conference New York academy of sciences Edited by Harold T. Meryman and others New York academy of sciences.Annals, 85,2.p501-734 New York, 1960
Bal 39.1259

NEW YORK 1960 Fundamentals of keratinization :a symposium presented at the New York meeting of the New York meeting of the American association for the advancement of science American association for the advancement of science Edited by Earl O. Butcher and Reidar F. Sognnaes American association for the advancement of science. Publications, 70 American association for the advancement of science,Washington,D.C., 1962
An 32.3280

NEW YORK 1960 Gibberellins ...The symposium...presented before the Division of agricultural and food chemistry of the 138th national meeting of the American chemical society Papers American chemical society. Division of agricultural and food chemistry Advances in chemistry series, 28 American chemical society,Washington,1961
Bioch 33.0738

NEW YORK 1960 I.R.E. international convention convention record Part 2: circuit theory,electronic computers Institute of radio engineers 203p Institute of radio engineers,New York,1960
Math L 5.2074

NEW YORK 1960 I.R.E. international convention convention record Part 4: automatic control,information theory Institute of radio engineers 222p Institute of radio engineers,New York,1960
Math L 5.2075

NEW YORK 1960 Research conference on shear strength of cohesive soils University of Colorado American society of civil engineers Boulder,Colo.,1960
Eng 41.3171

NEW YORK 1960 Spermatozoan motility :a symposium American association for the advancement of science Edited by D.W. Bishop Sponsored by the American society of zoologists and the Society of general physiologists American association for the advancement of science.Publication, 72 Washington,D.C.,1962
Bal 39.0391

NEW YORK 1960 The Structure of the eye : the symposium held...during the 7th International congress of anatomists Edited by George K. Smelser Academic press,New York, 1961
An 32.4773

NEW YORK 1960 Theory and fundamental research in heat transfer American society of mechanical engineers :annual meeting Proceedings Edited by J.A. Clark ix,220p Pergamon press,Oxford,1963
Chem E 24.0582

NEW YORK 1960 Ultrastructure and metabolism of the nervous system Proceedings of the Association Association for research in nervous and mental disease Edited by Saul R. Korey and others Association for research in nervous and mental disease.Research publications, 40 Williams and Wilkins, Baltimore,Md.,1962
An 32.4199

NEW YORK 1961 Cerebrovascular disease Proceedings of the Association Association for research in nervous and mental disease Edited by Clark H. Millikan Association for research in nervous and mental disease. Research publications, 41 Williams and Wilkins,Baltimore,Md.,1966
An 32.4214

NEW YORK 1961 I.R.E. international convention convention record Part 2: audio;electronic computers Institute of radio engineers 275p Institute of radio engineers,New York,1961
Math L 5.2077

NEW YORK 1961 I.R.E. international convention convention record Part 4: automatic control,circuit theory,information theory Institute of radio engineers 286p Institute of radio engineers,New York,1961
Math L 5.2078

NEW YORK 1961 International congress on experimental mechanics 1st Proceedings Society for experimental stress analysis Edited by B.E. Rossi Pergamon press,London, 1963
Eng 41.2920

NEW YORK 1961 Multiple molecular forms of enzymes :a conference 1st Papers New York academy of sciences Edited by Felix Wroblewski and others New York academy of sciences.Annals, 94,p 655-1030 New York, 1961
Bioch 33.1028

NEW YORK 1961 Pigment cell:molecular, biological and clinical aspects International pigment cell conference 5th Papers Edited by Harold E. Whipple New York academy of sciences.Annals, 100 New York academy of sciences,New York,1963 1st - 4 th conferences called:Conference on the biology of normal and atypical pigment cell growth,q.v.
An 32.3389

NEW YORK 1961 Solar variations,climatic change,and related geophysical problems New York academy of sciences New York academy of sciences.Annals, 95(1) New York,1961
Bot 42.3302

NEW YORK 1961 Solar variations,climatic change,and related geophysical problems :a conference Papers New York academy of sciences Edited by W.Franklin N. Furness and others Supported conjointly by the American meteorological society New York.Academy of sciences.Annals, 95,1 740p New York,1961 Conference editor:R.W.Fairbridge
Nap 11.0210

NEW YORK 1961 Solar variations,climatic change and related geophysical problems : conference Papers New York academy of sciences,and,American meteorological society Edited by Rhodes W. Fairbridge New York academy of sciences.Annals,95,art.1 New York academy of sciences,New York,1961
Geog 13.1055

NEW YORK 1962 Blood groups in infrahuman species :a conference By Ray D. Owen and others New York academy of sciences Edited by C. Cohen New York academy of sciences. Annals, 97,p1-328 New York,1962
Bal 39.1341

NEW YORK 1962 Comparative biology of calcified tissue papers New York academy of sciences New York academy of sciences. Annals,109,art.1 New York,1963
Geol 8.0971

NEW YORK 1962 Disorders of communication Proceedings of the Association Association for research in nervous and mental disease Edited by David McK. Rioch and Edwin A. Weinstein Association for research in nervous and mental disease.Research publications, 42 Williams and Wilkins, Baltimore,Md.,1964
An 32.4561

NEW YORK 1962 Electronic structure and alloy chemistry of the transition elements :a symposium American institute of mining, metallurgical and petroleum engineers. Metallurgical society Edited by Paul Adams Beck 251p Interscience,New York,1963
Cav 7.1711

NEW YORK 1962 Experimental techniques in shock and vibration :a colloquium Papers Edited by W.J. Worley Sponsored by the American society of mechanical engineers American society of mechanical engineers,New York,1962
Eng 41.6354

NEW YORK 1962 Extractive metallurgy of aluminum :international symposium 1st Vol 1-2: alumina;aluminum American institute of mining,metallurgical and petroleum engineers.Extractive metallurgy division Edited by Gary Gerard and P.T. Stroup 2 vols John Wiley,New York,1963
Met 25.0578

NEW YORK 1962 I.R.E. international convention convention record Part 4: electronic computers,information theory Institute of radio engineers 199p Institute of radio engineers,New York,1962
Math L 5.2084

NEW YORK 1962 Luminescence of organic and inorganic materials :international conference New York university Edited by Hartmut P. Kallman and Grace Marmor Spruch Sponsored by the United States.Air force.Aeronautical research laboratory xxiv,664p Wiley,New York,1962
Cav 7.1253

NEW YORK 1962 Mechanical behavior of crystalline solids a symposium Proceedings Ceramic education council and National bureau of standards Sponsored by the Edward Orton junior ceramic foundation National bureau of standards monograph, 59 U.S. Government printing office,Washington,1963 Lithographed
Met 25.1213

NEW YORK 1962 Origin of the solar system : a conference Proceedings Goddard institute for space studies Edited by Robert Jastrow and A.G.W. Cameron Academic press,New York, 1963
TA 15.0200

NEW YORK 1962 Origin of the solar system : a conference Proceedings Goddard institute for space studies,New York Edited by Robert Jastrow and A.G.W. Cameron Academic press, New York,1963
Geod 9.0007

NEW YORK 1962 Origin of the solar system conference proceedings Goddard institute for space studies Edited by Robert Jastrow and A.G.W. Cameron 176p Academic press,New York,1963
Obs 6.0883

NEW YORK 1962 Recovery and recrystallization of metals :a symposium Proceedings American institute of mining, metallurgical and petroleum engineers Edited by L. Himmel Interscience,New York;London, 1963
Met 25.1267

NEW YORK 1962 Recovery and recrystallization of metals a symposium Proceedings American institute of mining, metallurgical and petroleum engineers.Physical metallurgy committee Edited by L. Himmel Interscience,New York,1963
Cav 7.1720

NEW YORK 1962 Superconductors technical sessions American institute of mining, metallurgical and petroleum engineers Edited by M. Tanenbaum and W.V. Wright Interscience, New York;London,1962
Met 25.1181

NEW YORK 1962 Symposium on mathematical theory of automata proceedings Microwave research institute Partly sponsored by American institute of electrical engineers Polytechnic institute of Brooklyn.Microwave research institute.Symposia series, 12 xix, 640p 22cm Polytechnic press,Brooklyn,1962
Math L 5.0996

NEW YORK 1962 Symposium on stress-strain-time-temperature relationships in materials American society for testing materials A.S.T.M.Special technical publication, 325,Materials science series, 3 American society for testing and materials,Philadelphia,Pa.,1962
Met 25.1034

NEW YORK 1962 Symposium on stress-strain-time-temperature relationships in materials : presented at the 65th annual meeting A.S.T.M. American society for testing and materials American society for testing and materials. Special technical publication, 325 American society for testing and materials.Materials science series, 3
Eng 41.3789

NEW YORK 1962 Techniques in electron metallography 1963 Symposium on advances in techniques in electron metallography American society for testing materials. Subcommittee 11 on electron microstructure of metals A.S.T.M.Special technical publication, 339 American society for testing and materials,Philadelphia,Pa.,1963
Met 25.1378

NEW YORK 1962 Uptake and distribution of anaesthetic agents :a conference Edited by E. M. Papper and Richard J. Kitz Sponsored by the National research council.Committee on anaesthesia McGraw-Hill,New York,1963
Pha 16.0043

NEW YORK 1963 Endocrines and the central nervous system Proceedings of the Association Association for research in nervous and mental disease Edited by Rachmiel Levine Association for research in nervous and mental disease.Research publications, 43 Williams and Wilkins, Baltimore,Md.,1966
An 32.4403

NEW YORK 1963 Endogenous metabolism with special reference to bacteria New York academy of sciences Edited by Carl Lamanna New York academy of sciences.Annals, 102,3 New York,1963
Bot 42.3726

NEW YORK 1963 High temperature structures and materials Symposium on naval structural mechanics 3rd Proceedings Edited by A.M. Freudenthal and others Sponsored by the United States.Office of naval research United States.Office of naval research. Structural mechanics series Pergamon press, London,1964
Eng 41.3806

NEW YORK 1963 High temperature structures and materials Symposium on naval structural mechanics Proceedings United States.Office of naval research Edited by A.M. Freudenthal and others Pergamon,Oxford,1964
Met 25.1009

NEW YORK 1963 I.E.E.E. international convention convention record Part 4: electronic computers,information theory,human factors in electronics Institute of electrical and electronics engineers 152p Institute of electrical and electronics engineers,New York,1963
Math L 5.2086

NEW YORK 1963 Instrumentation requirements for traffic control systems Traffic control : theory and instrumentation Papers Instrument society of America Edited by Thomas R. Horton Plenum press,New York,1965
Eng 41.3310

NEW YORK 1963 International conference on congenital malformations 2nd Papers International medical congress Edited by Morris Fishbein International medical congress,New York,1864
An 32.2091

NEW YORK 1963 Molecular modification in drug design :a symposium at the 145th meeting of the American chemical society American chemical society Edited by Robert F. Gould American chemical society.Advances in chemistry series, 45 American chemical society,Washington,D.C.,1964
Radioth 35.0429

NEW YORK 1963 Optical maser :a symposium Proceedings Polytechnic institute of Brooklyn Edited by Jerome Fox Sponsored also by the Microwave research institute Microwave research institute.Symposia series, 13 Polytechnic press,Brooklyn,N.Y.,1963
Eng 41.5417

NEW YORK 1963 Psychology of perception : the proceedings of the fifty-third annual meeting of the American psychological association Proceedings American psychopathological association Edited by Paul H. Hoch and Joseph Zubin Grune and Stratton,London,1965 Contains the 1963 Samuel Hamilton award lecture by Heinrich Kluver
Psy 31.0958

NEW YORK 1963 Some biochemical and immunological aspects of host-parasite relationships :a conference By Clark P. Read and others New York academy of sciences Edited by T.C. Cheng New York academy of sciences.Annals, 113 p 1-510 New York,1963
Bal 39.1733

NEW YORK 1963 Stellar evolution :a symposium proceedings Edited by R.F. Stein and A.G.W. Cameron Organized by Goddard institute for space studies 464p Plenum press,New York,1966
Obs 6.1677

NEW YORK 1963 Stellar evolution :a symposium proceedings Edited by R.F. Stein and A.G.W. Cameron Organized by the National aeronautics and space administration.Goddard institute for space studies 464p Plenum press,New York,1966
A Math 4.1164

NEW YORK 1963 Stellar evolution : international conference Proceedings Goddard institute for space studies Edited by R.F. Stein and A.G.W. Cameron Plenum press,New York,1966
TA 15.0181

NEW YORK 1963 Stochastic processes in mathematical physics and engineering symposium 16 proceedings American mathematical society Edited by Richard Bellman Cosponsored by the Society for industrial and applied mathematics American mathematical society.Proceedings of symposia in applied mathematics, 16 viii,318p 25cm American mathematical society,Providence,R.I.,1964
Math 3.0567

NEW YORK 1963 The Origin and evolution of atmospheres and oceans :a conference Goddard institute for space studies Edited by Peter J. Brancazio and A.G.W. Cameron Wiley,New York,1964
TA 15.0427

NEW YORK 1963 The Regulation of respiration :a conference By Harry Grundfest and others New York academy of sciences Edited by G.G. Nahas New York academy of sciences.Annals, 109,p411-948 New York,1963
Bal 39.1355

NEW YORK 1963 Vitamin B 12 coenzymes :a conference Papers New York academy of sciences Edited by Harold E. Whipple New York academy of sciences.Annals, 112,p.547-921 New York,1964
Bioch 33.0565

NEW YORK 1963 Working conference on mechanical language structures Association for computing machinery Collection of 8 mimeograph abstracts of papers
Math L 5.0928

NEW YORK 1963 London 1963 Sep 30-Oct 4 Joint international conference on creep session 1-7 book 1-5 American society of mechanical engineers and American society for testing materials Jointly sponsored by Institution of mechanical engineers I.M.E.,London,c1963 5 books bound together
Met 25.0953

NEW YORK 1964 Adipose tissue metabolism and obesity :a conference Papers New York academy of sciences Edited by Harold E. Whipple New York academy of sciences.Annals, 131,p 1-683 New York,1965
Bioch 33.0562

NEW YORK 1964 Cerebrospinal fluid and the regulation of ventilation :a symposium 6 Proceedings Downstate medical center Edited by Chandler McC. Brooks and others Blackwell,Oxford,1965
Pha 16.0278

NEW YORK 1964 Chemistry and metabolism of L- and D- lactic acids :a conference Papers New York academy of sciences Edited by Harold E. Whipple New York academy of sciences.Annals, 119,p.851-1165 New York, 1965
Bioch 33.0563

NEW YORK 1964 Differentiation and development :a symposium Proceedings New York heart association Journal of experimental zoology, 157,no.1 Wistar institute of anatomy and biology,Philadelphia, Pa.,1964
Gen 34.0530

NEW YORK 1964 Earth-moon system The Dynamics of the earth-moon system conference Proceedings Goddard institute for space studies Edited by B.G. Marsden and A.G.W. Cameron 288p Plenum press,New York,1966
Obs 6.1169

NEW YORK 1964 I.E.E.E. international convention convention records Part 1: circuit theory,computers,control systems, systems science Institute of electrical and electronics engineers 313p Institute of electrical and electronics engineers,New York, 1964
Math L 5.2085

NEW YORK 1964 International congress of biochemistry 6th Abstracts Scheduled under the auspices of the International union of biochemistry Washington,D.C.,1964
Bioch 33.1342

NEW YORK 1964 International congress of biochemistry 6th Abstracts Scheduled under the auspices of the International union of biochemistry Washington,1964
Bioch 33.1343

NEW YORK 1964 International congress of biochemistry 6th Abstracts Scheduled under the auspices of the International union of biochemistry Washington,1964
Bioch 33.1345

NEW YORK 1964 International congress of biochemistry 6th Abstracts Vol 2: proteins,peptides,and amino acids Scheduled under the auspices of the International union of biochemistry Washington,D.C.,1964 Title also in French.
Bioch 33.1339

NEW YORK 1964 International congress of biochemistry 6th Abstracts Vol 3: biochemical genetics Scheduled under the auspices of the International union of biochemistry Washington,D.C.,1964
Bioch 33.1340

NEW YORK 1964 International congress of biochemistry 6th Abstracts Vol 5: special topics in biochemistry Scheduled under the auspices of the International union of biochemistry Washington,1964
Bioch 33.1341

NEW YORK 1964 International congress of biochemistry 6th Abstracts Vol 8: cellular organization Scheduled under the auspices of the International union of biochemistry Washington,1964
Bioch 33.1344

NEW YORK 1964 International congress of biochemistry 6th Abstracts Vol 10: bioenergetics Scheduled under the auspices of the International union of biochemistry Washington,1964
Bioch 33.1346

NEW YORK 1964 International congress of biochemistry 6th Abstracts Vol 11: author index to abstracts Scheduled under the auspices of the International union of biochemistry Washington,1964
Bioch 33.1347

NEW YORK 1964 International mineral processing congress 7th Technical papers Vol 1 American institute of mining, metallurgical and petroleum engineers and Columbia university Edited by Nathaniel Arbiter Gordon and Breach,New York,1965 Held in conjunction with the centennial of the Henry Krumb school of mines
Met 25.0214

NEW YORK 1964 Process simulation and control in iron and steelmaking :a symposium Proceedings American institute of mining, metallurgical and petroleum engineers Edited by J.M. Uys and H.L. Bishop Metallurgical society conferences, 32 Gordon and Breach, New York,1966
Met 25.0376

NEW YORK 1964 Quasi-optics :a symposium Proceedings Microwave research institute and Institute of radio engineers Co-sponsored by United States.Office of naval research Polytechnic institute of Brooklyn. Microwave research institute symposia series, 14 Polytechnic press,Brooklyn,N.Y.,1964
Cav 7.1721

NEW YORK 1964 Symposium on the placenta : its form and functions with particular reference to the prevention of birth defects and fetal deaths Medical society of the county of New York.Special committee on infant mortality Edited by Daniel Bergsma and others Sponsored by the National Foundation-March of dimes.Great New York Chapter Birth defects original article series, 1,no.1 National foundation,New York,1965
PGMS 29.0212

NEW YORK 1964 The Thalamus :international symposium Proceedings Edited by Dominick P. Purpura and Melvin D. Yahr Sponsored by the Parkinson's disease information and research center Columbia university press,New York, 1966
An 32.4661

NEW YORK 1964 The Thalamus :international symposium 1st Proceedings Edited by Dominick P. Purpura and Melvin D. Yahr Sponsored by the Parkinson's disease information and research center Columbia university press,New York;London,1966
Pha 16.0303

NEW YORK 1964 1965 Feb High temperature refractory metals :a symposium Pt 1-2 American institute of mining, metallurgical and petroleum engineers. Refractory metals committee Metallurgical society conferences, 34 2 vols Gordon and Breach,New York,1966-68 Conference pt 1 1965,pt 2 1964
Met 25.1016

NEW YORK 1965 Biochemistry and pharmacology of the basal ganglia :symposium 2nd Proceedings Parkinson's disease information and research center Edited by Erminio Costa and others Raven press,Hewlett, N.Y.,1966
Pha 16.0059

NEW YORK 1965 Biochemistry of copper Symposium on copper in biological systems Proceedings Edited by Jack Peisach and others Academic press,New York;London,1966
Bioch 33.1068

NEW YORK 1965 Biological membranes :recent progress New York academy of sciences New York academy of sciences.Annals, 137,art.2 New York,1966
Phys 20.0964

NEW YORK 1965 Biological membranes:recent progress :a conference New York academy of sciences Edited by Edward M. Weyer and others New York academy of science.Annals, 137,p.403-1048 New York,1966
Bioch 33.0921

NEW YORK 1965 Biological membranes:recent progress :a conference Papers Edited by Edward M. Weyer New York academy of sciences. Annals,137,art 2 NNew York,1966
Pha 16.0014

NEW YORK 1965 Comparative psychopathology animal and human Proceedings American psychopathological association Edited by Joseph Zubin and Howard F. Hunt Grune and Stratton,New York;London,1967 Proceedings of the 55th annual meeting of the Association
Psy 31.1682

NEW YORK 1965 I.E.E.E.international convention convention record Part 3: computers Institute of electrical and electronics engineers 281p Institute of electrical and electronis engineers,New York, 1965
Math L 5.2087

NEW YORK 1965 Information processing 1965 IFIP congress 65 Proceedings Vol 1-2 Edited by Wayne A. Kalenich Organized by the International federation for information processing 2 vols Spartan;Macmillan, Washington,D.C.;London,1965-66
Eng 41.6001

NEW YORK 1965 Information processing 1965 IFIP congress 65 proceedings Vol. 1 International federation for information processing xv,304p 29cm Spartan books, Washington,D.C.,1965
Math L 5.1045

NEW YORK 1965 Joint automatic control conference 6th Preprints of conference papers Sponsored by American society of mechanical engineers.Automatic control division New York,1965
Eng 41.6015

NEW YORK 1965 Mossbauer effect methodology Symposium on Mossbauer effect methodology 1st Proceedings New England nuclear corporation Edited by I.J. Gruverman Plenum press,New York,1965
Chem 18.2826

NEW YORK 1965 Neurophysiology and emotion : a conference Proceedings Russell Sage foundation Rockefeller university Edited by David C. Glass Rockefeller university press;Russell Sage foundation,New York,1967
Psy 31.2900

NEW YORK 1965 Nucleosynthesis :a conference Proceedings Goddard space flight center and Goddard institute for space studies Edited by W.David Arnett and others Gordon and Breach,New York,1968
TA 15.0306

NEW YORK 1965 Vehicular traffic science International symposium on the theory of traffic flow 3rd Proceedings Edited by Leslie C. Edie and others Under the auspices of the Operations research society of America. Transportation science section illus x, 373p 24cm American Elsevier,New York,1967
Math 3.1137

NEW YORK 1966 Biological membranes :recent progress:aconference By Kenneth S. Cole and others New York academy of sciences Edited by Werner R. Loewenstein New York academy of sciences.Annals, 137,2 p 403-1048 New York, 1966
Bot 42.1508

NEW YORK 1966 Biological membranes:recent progress :a conference By Kenneth S. Cole New York academy of sciences Edited by Werner R. Loewenstein New York academy of sciences.Annals, 137,p 403-1048 New York, 1966
Bal 39.0325

NEW YORK 1966 Cholinergic mechanisms :a conference Edited by S. Ehrenpreis New York academy of sciences.Annals, 144,art.2 New York academy of sciences,New York,1967
An 32.5248

NEW YORK 1966 Cholinergic mechanisms :a conference Papers Edited by S. Ehrenpreis New York academy of sciences.Annals,144,art 2 New York academy of sciences,New York,1967
Pha 16.0311

NEW YORK 1966 Cholinergic mechanisms :a conference Papers New York academy of sciences Edited by Edward M. Weyer New York academy of sciences.Annals, 144,p. 383-936 New York,1967
Bioch 33.0564

NEW YORK 1966 Fiber spinning and drawing : symposium American chemical society Edited by Myron J. Coplan Applied polymer symposia, 6 181p Interscience publishers,New York, 1967
Chem E 24.0423

NEW YORK 1966 Galaxies and the universe The Vetlesen symposium Edited by Lodewijk Woltjer 112p Columbia university press,New York,1968
Obs 6.3452

NEW YORK 1966 Growth hormone :a conference Papers By Martin Sonenberg New York academy of sciences New York academy of sciences.Annals, 148,p.289-571 New York,1968
Bioch 33.0421

NEW YORK 1966 Multiple molecular forms of enzymes :a conference 2nd Papers New York academy of sciences Edited by Edward M. Weyer New York academy of sciences.Annals, 151,p 1-689 New York,1968
Bioch 33.1029

NEW YORK 1966 New adrenergic blocking drugs.Their pharmacological,biochemical and clinical actions :a conference Papers By N. C. Moran New York academy of sciences New York academy of sciences.Annals, 139,p.541-1009 New York,1967
Bioch 33.0422

NEW YORK 1966 Plant growth regulators :a conference Papers New York academy of sciences Edited by Edward M. Weyer New York academy of sciences.Annals, 144,p 1-382 New York,1967
Bioch 33.0780

NEW YORK 1966 Safety in air and ammonia plants :a symposium 9 Chemical engineering progress 96p A.I.Ch.E.,New York,1967 A C.E.P.technical manual
Chem E 24.1576

NEW YORK 1966 The Mossbauer effect and its application in chemistry American chemical society Edited by E. Gould American chemical society.Advances in chemistry series, 68 American chemical society,Washington,D.C., 1967 Symposium organized by C.W.Seidel. Chairman:R.H.Herber
Chem 18.1190

NEW YORK 1966 Werner centennial :a symposium Papers American chemical society American chemical society.Advances in chemistry series,62 American chemical society,Washington,D.C.,1967 Symposium chairman:G.B.Kauffman
Chem 18.0990

NEW YORK 1966-67 Bifurcation theory and nonlinear eigenvalue problems :a seminar Lectures Edited by Joseph B. Keller and Stuart Antman Held at the Courant institute of mathematical sciences Bibliog. xiv,434p 23cm Benjamin,New York;Amsterdam,1969
P Math 2.3648

NEW YORK 1967 Biological interfaces:flows and exchanges :a symposium Proceedings Sponsored by the New York heart association Basic science symposia, 9 Little,Brown, Boston,Mass.,1968 Simultaneously published in the Journal of general physiology, vol.52,1968
Phys 20.2247

NEW YORK 1967 Biology and behavior 3 Environmental influences :a conference Proceedings Russell Sage foundation Rockefeller institute Edited by David C. Glass Rockefeller university press;Russell Sage foundation,New York,1968 Third and last in a series of conferences on biology and behavior
Psy 31.3096

NEW YORK 1967 Electronic aspects of biochemistry The International conference on the electronic aspects of biochemistry By Bernard Pullman and others New York academy of sciences Edited by Philip Feigelson New York academy of sciences.Annals, 158,p.1-438 New York,1969
Bioch 33.0313

NEW YORK 1967 Experimental medicine and surgery in primates :a conference New York academy of sciences New York academy of sciences.Annals, 162,art.1 New York academy of sciences,New York,1969
An 32.5171

NEW YORK 1967 Experimental medicine and surgery in primates :a conference Papers New York academy of sciences Edited by Edward I. Goldsmith and J. Moor-Jankowski New York academy of sciences.Annals, 162,art 1 New York academy of sciences,New York,1969
Path 30.2445

NEW YORK 1967 Industrial process design for pollution control :a workshop Proceedings Vol 1 American institute of chemical engineers.Water committee iv,59p A.I.Ch.E.,New York,1967
Chem E 24.1575

NEW YORK 1967 Mathematical aspects of computer science :a symposium Proceedings American mathematical society Edited by J.T. Schwartz American mathematical society. Proceedings of symposia in applied mathematics, 19 Bibliog,illus 224p 26cm American mathematical society,Providence,R.I.,1967
P Math 2.3001

NEW YORK 1967 Modern optics The Symposium on modern optics Proceedings Microwave research institute.Symposia series, 17 xxxv,804p 23cm Polytechnic press, Brooklyn,N.Y.,1967
Cav 7.2815

NEW YORK 1967 Product design through the engineering analysis of its end use :an approach and a case history.Presented at the Design engineering conference Sponsored by the American society of mechanical engineers. Design engineering division American society of mechanical engineers,New York,1967
Eng 41.0326

NEW YORK 1967 University education in computing science :a conference on graduate academic and related research programs in computing science Proceedings Edited by Aaron Finerman Sponsored by New York state science and technology foundation ACM monograph series xvi,237p 23cm Academic press,New York,1968
Math L 5.3107

NEW YORK 1968 Applications of categorical algebra :a symposium Proceedings American mathematical society Edited by Alex Heller American mathematical society.Proceedings of symposia in pure mathematics, 17 Bibliog v,231p 26cm American mathematical society, Providence,R.I.,1970
P Math 2.3508

NEW YORK 1968 Applied sciences research and utilization of lunar resources Lunar international laboratory (LIL symposium) 4th Proceedings International academy of astronautics Bibliog,illus 200p 20cm Pergamon press,Oxford,1970 LIL symposium organized by the International academy of astronautics at the 19th International astronautical congress
Cav 7.2711

NEW YORK 1968 International congress of histochemistry and cytochemistry 3rd Summary reports Springer,New York,1968
Radioth 35.0547

NEW YORK 1968 International congress of histochemistry and cytochemistry 3rd summary reports United States.Office of the secretariat 317p Springer-verlag,New York, 1968
Bot 42.0703

NEW YORK 1968 International congress of histochemistry and cytochemistry 3rd Summary reports Springer,New York,1968
An 32.0438

NEW YORK 1968 Symposium on turbulence of fluids and plasmas Proceedings Polytechnic institute of Brooklyn Edited by Jerome Fox Co-sponsored by the Microwave research institute Microwave research institute. Symposia series, 18 512p 17cm Polytechnic press,Brooklyn,N.Y.,1969
Cav 7.2920

NEW YORK 1968 Turbulence of fluids and plasmas Symposium on turbulence of fluids and plasmas Proceedings Edited by Jerome Fox Sponsored by the Microwave research institute Microwave research institute. Symposia series, 18 bibliog.,illus. 512p 22cm Polytechnic press,Brooklyn,N.Y.,1969
A Math 4.1650

NEW YORK 1968 Turbulence of fluids and plasmas Symposium on turbulence of fluids and plasmas Proceedings Polytechnic institute of Brooklyn Edited by Jerome Fox Co-sponsored by the Microwave research institute Microwave research institute. Symposia series, 18 Bibliog.,Illus. xxix, 511p 23cm Polytechnic press,Brooklyn N.Y., 1969
A Math 4.1508

NEW YORK 1969 Computer processing in communications Symposium on computer processing in communications Proceedings Microwave research institute Edited by Jerome Fox Microwave research institute. Symposia series, 19 illus. 850p Polytechnic institute of Brooklyn,Brooklyn,N.Y. 1970
Math L 5.3558

NEW YORK 1969 Plastic deformation of polymers Symposium on plastic deformation of polymers Selected papers American chemical society Dekker,New York,1971
Met 25.2576

NEW YORK 1969 Symposium on computer processing in communications Microwave research institute Microwave research institute.Symposium series, 19 Polytechnic press,Brooklyn,N.Y.,1970
Eng 41.8313

NEW YORK 1969 The Culture of algae :a symposium Proceedings Edited by J.E. Zajic Sponsored by the American chemical society. Division of microbial chemistry and technology illus,tables ix,154p Plenum press,New York; London,1970
Bot 42.3942

NEW YORK 1969 The Nature and function of peroxisomes (microbodies,glyoxosomes) :a conference Papers New York academy of sciences Edited by James Hogg New York academy of sciences.Annals, 168 p. 209-381 New York,1969
Bot 42.4758

NEW YORK 1969 The Pharmacological and chemotherapeutic properties of niridazole and other antischistosomal compounds :a conference By F.C. Goble and others New York academy of sciences Edited by Francis C. Goble New York academy of sciences.Annals, 160,art.2 New York,1969
Mol 45.0272

NEW YORK 1970 Computer networks Courantter science symposium 3rd Courant institute of mathematical sciences Edited by Randall Rustin diagrs 205p Prentice-Hall,Englewood Cliffs,N.J.,1972
Math L 5.3638

NEW YORK 1970 Computers in algebra and number theory Symposium in applied mathematics Proceedings By Garrett Birkhoff and Marshall Hall American mathematical society Society for industrial and applied mathematics SIAM-AMS proceedings, 4 vii,200p 26cm AMS,Providence,R.I., 1971
P Math 2.4476

NEW YORK 1970 Institute of electrical and electronics engineers.Communication technology group.Data communication committee illus 24p IEEE,New York,1970
Math L 5.3626

NEW YORK 1970 International conference on combinatorial mathematics Papers New York academy of sciences Edited by Allan Gewirtz and Louis Quintas New York academy of sciences.Annals, 175,art 1 412p 23cm New York academy of sciences,New York,1970
P Math 2.4510

NEW YORK 1970 Mossbauer effect methodology Symposium on Mossbauer effect methodology 6th Proceedings New England nuclear corporation Edited by Irwin J. Gruverman viii,237p 23cm Plenum press,New York,1971
Chem 18.2805

NEW YORK 1970 Recent trends in graph theory New York city graph theory conference 1st Proceedings Edited by M. Capobianco and others Lecture notes in mathematics, 186 vi,219p 25cm Springer,Berlin,1971
P Math 2.4113

NEW YORK 1971 I.E.E.E.international convention digest :synopses of papers presented at the 1971 I.E.E.E. international convention Institute of electrical and electronics engineers bibliog.,illus. 643p Institute of electrical and electronics engineers,New York,1971
Math L 5.3565

NEW YORK 1971 Mossbauer effect methodology Symposium on Mossbauer effect methodology 7th Proceedings New England nuclear corporation Edited by Irwin J. Gruverman xii,308p 23cm Plenum press,London,1971
Chem 18.2961

NEW YORK 1971 Symposium on computers and automata Proceedings Microwave research institute Edited by Jerome Fox Microwave research institute.Symposia series, 21 illus 653p Polytechnic press,Brooklyn,N.Y., 1971
Math L 5.3635

NEW YORK,N.Y. 1959 Genetics of streptomyces and other antibiotic-producing micro organisms conference New York academy of sciences Annals of the New York academy of sciences, 81,art 4 New York,1959
Gen 34.1494

NEW YORK ACADEMY OF MEDICINE Ecology of health :institute on public health Papers New York 1947 Apr 1-3 Edited by E.H.L. Corwin Commonwealth fund,New York,1949 Issued for the centennial celebration of the New York academy of medicine
An 32.2035

NEW YORK ACADEMY OF MEDICINE The Nature and significance of the antibody response; symposium... New York 1951 Mar 21-22 New York academy of medicine.Section on microbiology Edited by Alwin M. Pappenheimer New York academy of medicine.Section on microbiology.Symposia, 5 Columbia university press,New York,1953
Med 36.0272

NEW YORK ACADEMY OF MEDICINE.COMMITTEE ON PUBLIC HEALTH Integrating the approaches to mental disease :two conferences Papers and discussions New York Edited by H.D. Kruse Hoeber-Harper,New York,1957
Psy 28.0061

NEW YORK ACADEMY OF MEDICINE.SECTION OF MICROBIOLOGY The Nature and significance of the antibody response Papers and discussions New York 1951 Mar 21-22 Edited by A.M. Pappenheimer New York academy of sciences.Section of microbiology.Symposia, 5 Columbia university press,New York,1953
Path 30.2532

NEW YORK ACADEMY OF MEDICINE.SECTION OF MICROBIOLOGY.SYMPOSIA, 9 Cellular and humoral aspects of the hypersensitive states : a symposium New York 1959 New York academy of medicine.Section on microbiology Edited by H.Sherwood Lawrence illus,diagrs Cassell,London,1959
Path 30.2526

NEW YORK ACADEMY OF MEDICINE.SECTION ON MICROBIOLOGY Cellular and humoral aspects of the hypersensitive states :a symposium New York 1959 Edited by H.Sherwood Lawrence New York academy of medicine. Section of microbiology.Symposia, 9 illus, diagrs Cassell,London,1959
Path 30.2526

NEW YORK ACADEMY OF MEDICINE.SECTION ON MICROBIOLOGY The Nature and significance of the antibody response :a symposium Papers and discussions New York 1951 Mar 21-22 Edited by A.M. Pappenheimer New York academy of medicine.Section on microbiology.Symposia, 5 Columbia university press,New York,1953
Bioch 33.1156

NEW YORK ACADEMY OF MEDICINE.SECTION ON MICROBIOLOGY The Nature and significance of the antibody response;symposium... New York 1951 Mar 21-22 Edited by Alwin M. Pappenheimer Held at the New York academy of medicine New York academy of medicine. Section on microbiology.Symposia, 5 Columbia university press,New York,1953
Med 36.0272

NEW YORK ACADEMY OF MEDICINE.SECTION ON MICROBIOLOGY The Pathogenesis and pathology of viral diseases :a symposium New York 1948 Dec 14-15 Edited by John G. Kidd New York academy of medicine.Section on microbiology.Symposia, 3 Columbia university press,New York,1950
Path 30.2554

NEW YORK ACADEMY OF MEDICINE.SECTION ON MICROBIOLOGY.SYMPOSIA, 3 The Pathogenesis and pathology of viral diseases : a symposium New York 1948 Dec 14-15 New York academy of medicine.Section on microbiology Edited by John G. Kidd Columbia university press,New York,1950
Path 30.2554

NEW YORK ACADEMY OF MEDICINE.SECTION ON MICROBIOLOGY.SYMPOSIA, 5 The Nature and significance of the antibody response :a symposium Papers and discussions New York 1951 Mar 21-22 New York academy of medicine. Section on microbiology Edited by A.M. Pappenheimer Columbia university press,New York,1953
Bioch 33.1156

NEW YORK ACADEMY OF MEDICINE.SECTION ON MICROBIOLOGY.SYMPOSIA, 5 The Nature and significance of the antibody response; symposium... New York 1951 Mar 21-22 New York academy of medicine.Section on microbiology Edited by Alwin M. Pappenheimer Held at the New York academy of medicine Columbia university press,New York,1953
Med 36.0272

NEW YORK ACADEMY OF SCIENCES Adipose tissue metabolism and obesity :a conference Papers New York 1964 Dec 14-17 Edited by Harold E. Whipple New York academy of sciences.Annals, 131,p 1-683 New York,1965
Bioch 33.0562

NEW YORK ACADEMY OF SCIENCES Adrenal cortex : a conference Papers New York 1948 Mar 21-22 Edited by Robert Gaunt New York academy of sciences.Annuals, 50,509-678 New York, 1949
Inv Med 37.0274

NEW YORK ACADEMY OF SCIENCES Amino acid analysis of proteins :a conference Papers New York 1945 Oct 12-13 By William H. Stein and others New York academy of sciences.Annals, 47,p.57-240 New York,1946
Bioch 33.0552

NEW YORK ACADEMY OF SCIENCES Amino acids, peptides,and proteins :a conference Papers New York 1960 Feb 25-26 By Karl Folkers and others New York academy of sciences. Annals, 88,p.533-770 New York,1960
Bioch 33.0489

NEW YORK ACADEMY OF SCIENCES Biochemistry and pharmacology of compounds derived from marine organisms :a conference Papers New York 1960 Apr 6-8 Edited by Ross F. Nigrelli New York academy of sciences.Annals, 90,article 3 New York,1960 Co-sponsored by the New York zoological society and the United States navy,Office of naval research
Chem 26.0302

NEW YORK ACADEMY OF SCIENCES Biological membranes :recent progress New York 1965 Oct 4-7 New York academy of sciences.Annals, 137,art.2 New York,1966
Phys 20.0964

NEW YORK ACADEMY OF SCIENCES Biological membranes :recent progress:aconference New York 1966 Jul 14 By Kenneth S. Cole and others Edited by Werner R. Loewenstein New York academy of sciences.Annals, 137,2 p 403-1048 New York,1966
Bot 42.1508

NEW YORK ACADEMY OF SCIENCES Biological membranes:recent progress :a conference New York 1965 Oct 4-6 Edited by Edward M. Weyer and others New York academy of science. Annals, 137,p.403-1048 New York,1966
Bioch 33.0921

NEW YORK ACADEMY OF SCIENCES Biological membranes:recent progress :a conference New York 1966 Jul 14 By Kenneth S. Cole Edited by Werner R. Loewenstein New York academy of sciences.Annals, 137,p 403-1048 New York,1966
Bal 39.0325

NEW YORK ACADEMY OF SCIENCES Blood groups in infrahuman species :a conference New York 1962 May 3 By Ray D. Owen and others Edited by C. Cohen New York academy of sciences.Annals, 97,p1-328 New York,1962
Bal 39.1341

NEW YORK ACADEMY OF SCIENCES Cellular biology nucleic acids and viruses :a conference Papers New York 1957 Jun 7-9 Edited by Thomas M. Rivers New York academy of sciences.Special publications, 5 Port New York academy of sciences,New York,1957
Radioth 35.0496

NEW YORK ACADEMY OF SCIENCES Cellular biology nucleic acids and viruses :conference Papers New York 1957 Jan 7-9 Edited by Thomas M. Rivers New York academy of sciences.Special publications, 5 New York academy of sciences,New York,1957
Gen 34.0724

NEW YORK ACADEMY OF SCIENCES Chemistry and metabolism of L- and D- lactic acids :a conference Papers New York 1964 Nov 12-14 Edited by Harold E. Whipple New York academy of sciences.Annals, 119,p.851-1165 New York,1965
Bioch 33.0563

NEW YORK ACADEMY OF SCIENCES Chemotherapy of filariasis :a conference New York 1948 May 25 By L.L. Ashburn and others Edited by Roy Waldo Miner New York academy of sciences. Annals, 50,art.2 New York,1948
Mol 45.0271

NEW YORK ACADEMY OF SCIENCES Chlorpropamide and diabetes mellitus :a conference Papers New York 1958 Sep 25-27 By Martin G. Goldner and others New York academy of sciences.Annals, 74,p.407-1028 New York, 1959
Bioch 33.0419

NEW YORK ACADEMY OF SCIENCES Cholinergic mechanisms :a conference Papers New York 1966 May 19-21 Edited by Edward M. Weyer New York academy of sciences.Annals, 144,p. 383-936 New York,1967
Bioch 33.0564

NEW YORK ACADEMY OF SCIENCES Comparative biology of calcified tissue papers New York 1962 Oct 12-13 New York academy of sciences.Annals,109,art.1 New York,1963
Geol 8.0971

NEW YORK ACADEMY OF SCIENCES Conference on physicochemical mechanism of nerve activity and Conference on muscular contraction 2nd Edited by David A. Nachmansohn New York academy of sciences.Annals, 81,p 215-510 New York,1959
Bal 39.1450

NEW YORK ACADEMY OF SCIENCES Conference on the mechanisms of cell divisions 2nd Proceedings New York 1960 Oct 7 By M.J. Kopac and others Edited by Paul R. Gross New York academy of sciences.Annals, 90,p.345-613 New York,1960
Bal 39.0291

NEW YORK ACADEMY OF SCIENCES Current trends in research and clinical management in diabetes :a conference Papers New York 1959 Apr 10-11 By Peter H. Forsham and others New York academy of sciences.Annals, 82,p.191-644 New York,1959
Bioch 33.0420

NEW YORK ACADEMY OF SCIENCES Electronic aspects of biochemistry The International conference on the electronic aspects of biochemistry New York 1967 Dec 11-13 By Bernard Pullman and others Edited by Philip Feigelson New York academy of sciences. Annals, 158,p.1-438 New York,1969
Bioch 33.0313

NEW YORK ACADEMY OF SCIENCES Endogenous metabolism with special reference to bacteria New York 1963 Jan 21 Edited by Carl Lamanna New York academy of sciences.Annals, 102,3 New York,1963
Bot 42.3726

NEW YORK ACADEMY OF SCIENCES Experimental medicine and surgery in primates :a conference New York 1967 Sep 27-30 New York academy of sciences.Annals, 162,art.1 New York academy of sciences,New York,1969
An 32.5171

NEW YORK ACADEMY OF SCIENCES Experimental medicine and surgery in primates :a conference Papers New York 1967 Sep 27-30 Edited by Edward I. Goldsmith and J. Moor-Jankowski New York academy of sciences.Annals, 162,art 1 New York academy of sciences,New York,1969
Path 30.2445

NEW YORK ACADEMY OF SCIENCES Fetal homeostasis Conference on fetal homeostasis 1st Proceedings Edited by Ralph M. Wynn New York academy of sciences,New York,1965
An 32.2519

NEW YORK ACADEMY OF SCIENCES Fetal homeostasis Conference on fetal homeostasis 2nd Proceedings Princeton,N.J. 1966 Apr 3-6 Edited by Ralph M. Wynn New York academy of sciences,New York,1967
An 32.2520

1629

NEW YORK ACADEMY OF SCIENCES Fetal homeostasis :a conference 2nd Proceedings Princeton,N.J. 1966 Apr 3-6 Vol 2 Edited by Ralph M. Wynn Sponsored by the William H.Donner foundation New York academy of sciences,New York,1967
Inv Med 37.0263

NEW YORK ACADEMY OF SCIENCES Freezing and drying of biological materials :a conference New York 1960 Apr 13 Edited by Harold T. Meryman and others New York academy of sciences.Annals, 85,2.p501-734 New York, 1960
Bal 39.1259

NEW YORK ACADEMY OF SCIENCES Genetics of streptomyces and other antibiotic-producing micro organisms conference New York,N.Y. 1959 Jan 6 Annals of the New York academy of sciences, 81,art 4 New York,1959
Gen 34.1494

NEW YORK ACADEMY OF SCIENCES Growth hormone : a conference Papers New York 1966 Oct 27-28 By Martin Sonenberg New York academy of sciences.Annals, 148,p.289-571 New York,1968
Bioch 33.0421

NEW YORK ACADEMY OF SCIENCES Growth of protozoa New York 1953 Oct 14 By R.W. Miner Edited by S.H. Hutner New York academy of sciences.Annals, 56,5 New York, 1953
Bot 42.3886

NEW YORK ACADEMY OF SCIENCES Growth of protozoa :a conference New York 1953 Oct 14 By R.P. Hall and others Edited by S.H. Hunter and others New York academy of sciences.Annals, 56,p.815-1095 New York, 1953
Bal 39.2132

NEW YORK ACADEMY OF SCIENCES Information and control processes in living systems :first interdisciplinary conference Proceedings Princeton,N.J. 1965 Feb 28-Mar 3 Edited by Diane M. Ramsey New York academy of sciences interdisciplinary communications program,New York,1967
Psy 31.0142

NEW YORK ACADEMY OF SCIENCES International conference on combinatorial mathematics Papers New York 1970 Apr 1-4 Edited by Allan Gewirtz and Louis Quintas New York academy of sciences.Annals, 175,art 1 412p 23cm New York academy of sciences,New York, 1970
P Math 2.4510

NEW YORK ACADEMY OF SCIENCES Lymph :a conference New York 1946 Sep 16 By Philip D. McMaster and others New York academy of sciences.Annals, 46,p679-882 New York,1946
Bal 39.1342

NEW YORK ACADEMY OF SCIENCES Lymph :a conference Papers New York 1945 Apr 13-14 New York academy of sciences.Annals, 46, art.8 New York,1946
Phys 20.0883

1630

NEW YORK ACADEMY OF SCIENCES Multiple molecular forms of enzymes :a conference 1st Papers New York 1961 Feb 1-3 Edited by Felix Wroblewski and others New York academy of sciences.Annals, 94,p 655-1030 New York,1961
Bioch 33.1028

NEW YORK ACADEMY OF SCIENCES Multiple molecular forms of enzymes :a conference 2nd Papers New York 1966 Dec 1-3 Edited by Edward M. Weyer New York academy of sciences.Annals, 151,p 1-689 New York, 1968
Bioch 33.1029

NEW YORK ACADEMY OF SCIENCES Muscular contraction :a conference New York 1947 May 30 By Alexander Sandow and others New York academy of sciences.Annals, 47 p665-930 New York,1947
Bal 39.1422

NEW YORK ACADEMY OF SCIENCES New adrenergic blocking drugs.Their pharmacological, biochemical and clinical actions :a conference Papers New York 1966 Feb 24-26 By N.C. Moran New York academy of sciences.Annals, 139,p.541-1009 New York,1967
Bioch 33.0422

NEW YORK ACADEMY OF SCIENCES Nuclear magnetic resonance :a conference Papers 1957 Nov 23 New York academy of sciences. Annals,70,art 4 New York academy of sciences, New York,1958
Chem 18.2305

NEW YORK ACADEMY OF SCIENCES Organization of recall International interdisciplinary conference on learning,remembering and forgetting 2nd Proceedings Princeton,N. J. 1964 Sep 27-30 Edited by Daniel P. Kimble With the support of United States. Office of naval research New York academy of sciences.Interdisciplinary communication program,New York,1967
Psy 31.1012

NEW YORK ACADEMY OF SCIENCES Photoreception Papers New York 1958 Jan 31-Feb 1 Edited by Jerome J. Wolken New York academy of sciences.Annals, 74,art.2 New York,1958
Phys 20.2225

NEW YORK ACADEMY OF SCIENCES Plant growth regulators :a conference Papers New York 1966 May 16-18 Edited by Edward M. Weyer New York academy of sciences.Annals, 144,p 1-382 New York,1967
Bioch 33.0780

NEW YORK ACADEMY OF SCIENCES Readiness to remember Conference on learning,remembering and forgetting 3rd Proceedings Princeton,N.J. 1965 Vol 1-2 Edited by Daniel P. Kimble Gordon and Breach,New York,1969
Psy 31.3281

NEW YORK ACADEMY OF SCIENCES Solar variations,climatic change,and related geophysical problems New York 1961 Jan New York academy of sciences.Annals, 95(1) New York,1961
Bot 42.3302

NEW YORK ACADEMY OF SCIENCES Solar variations,climatic change,and related geophysical problems :a conference Papers New York 1961 Jan 24-28 Edited by W. Franklin N. Furness and others Supported conjointly by the American meteorological society New York.Academy of sciences.Annals, 95,1 740p New York,1961 Conference editor:R.W.Fairbridge
Nap 11.0210

NEW YORK ACADEMY OF SCIENCES Some biochemical and immunological aspects of host-parasite relationships :a conference New York 1963 Apr 23-25 By Clark P. Read and others Edited by T.C. Cheng New York academy of sciences.Annals, 113 p 1-510 New York,1963
Bal 39.1733

NEW YORK ACADEMY OF SCIENCES Teleological mechanisms :a conference New York 1948 Oct 13 By Lawrence K. Frank and others New York academy of sciences.Annals, 50,p. 187-278 New York,1948
Bal 39.1003

NEW YORK ACADEMY OF SCIENCES The Influence of hormones on enzymes :a conference Papers New York 1951 Jun 5-6 By R.I. Dorfman and others New York academy of sciences.Annals, 54,p.531-728 New York,1951
Bioch 33.0450

NEW YORK ACADEMY OF SCIENCES The Nature and function of peroxisomes (microbodies, glyoxosomes) :a conference Papers New York 1969 May 16-17 Edited by James Hogg New York academy of sciences.Annals, 168 p. 209-381 New York,1969
Bot 42.4758

NEW YORK ACADEMY OF SCIENCES The Pharmacological and chemotherapeutic properties of niridazole and other antischistosomal compounds :a conference New York 1969 Oct 6 By F.C. Goble and others Edited by Francis C. Goble New York academy of sciences.Annals, 160,art.2 New York,1969
Mol 45.0272

NEW YORK ACADEMY OF SCIENCES The Physico-chemical mechanism of nerve activity :a conference New York 1946 Feb 8-9 By D. Nachmansohn and others New York academy of sciences.Annals, 47, p 375-602 New York, 1946
Bal 39.1449

NEW YORK ACADEMY OF SCIENCES The Regulation of respiration :a conference New York 1963 Jun 24 By Harry Grundfest and others Edited by G.G. Nahas New York academy of sciences.Annals, 109,p411-948 New York,1963
Bal 39.1355

NEW YORK ACADEMY OF SCIENCES Vitamin B 12 coenzymes :a conference Papers New York 1963 Apr 10-11 Edited by Harold E. Whipple New York academy of sciences.Annals, 112,p.547-921 New York,1964
Bioch 33.0565

NEW YORK ACADEMY OF SCIENCES and AMERICAN METEOROLOGICAL SOCIETY Solar variations, climatic change and related geophysical problems :conference Papers New York 1961 Jan 24-28 Edited by Rhodes W. Fairbridge New York academy of sciences. Annals,95,art.1 New York academy of sciences, New York,1961
Geog 13.1055

NEW YORK ACADEMY OF SCIENCES.OCEANOGRAPHY AND METEOROLOGY SECTION Ocean surface waves New York 1948 Mar 18-19 Edited by B. Haurwitz New York academy of sciences.Annals, 51,3 New York,1949 Series of papers resulting from a conference on surface waves held by the section of oceanography and meteorology of the New York academy of sciences March 18-19,1948
Geod 9.0602

NEW YORK ACADEMY OF SCIENCES.SECTION OF BIOLOGY Biology of melanomas A Conference on the biology of normal and atypical pigment cell growth New York 1946 Nov 15-16 New York academy of sciences.Special publications, 4 pls New York academy of sciences,New York, 1948
Path 30.2492

NEW YORK ACADEMY OF SCIENCES.SECTION OF BIOLOGY Biology of melanomas Conference on the biology of normal and atypical pigment cell growth New York 1946 Nov 15-16 New York academy of sciences.Special publications, 4 New York,1948
Phys 20.0952

NEW YORK ACADEMY OF SCIENCES.SECTION OF BIOLOGY Bioluminescence :a conference New York 1946 Nov 8 By E.Newton Harvey and others New York academy of sciences.Annals, 49,p 327-482 New York,1948
Bal 39.0084

NEW YORK ACADEMY OF SCIENCES.SECTION OF BIOLOGY Experimental hypertension being the results of a conference Results New York 1945 Feb 9-10 By William Goldring and others New York academy of sciences.Special publications, 3 New York academy of sciences,New York,1946
Psy 28.0096

NEW YORK ACADEMY OF SCIENCES.SECTION OF BIOLOGY Recent studies in the mechanisms of embryonic development :a conference New York 1947 Jan 10-11 Edited by Robert Rugh New York academy of sciences.Annals, 49,p 661-866 New York,1948
Bal 39.0438

NEW YORK ACADEMY OF SCIENCES.SECTION OF BIOLOGY The Chick embryo in biological research :a conference Papers New York 1951 Dec 7-8 Edited by Roy Waldo Miner and D.A. Karnofsky New York academy of sciences.Annals, 55,art.2 New York academy of sciences,New York,1952
An 32.2265

NEW YORK ACADEMY OF SCIENCES.SECTION OF MICROBIOLOGY.SYMPOSIA, 5 The Nature and significance of the antibody response Papers and discussions New York 1951 Mar 21-22 New York academy of medicine.Section of microbiology Edited by A.M. Pappenheimer Columbia university press,New York,1953
Path 30.2532

NEW YORK ACADEMY OF SCIENCES.SECTION OF PHYSICS AND CHEMISTRY Chromatography Conference papers 1946 Nov 29-30 By Harold G. Cassidy and others New York academy of sciences. Annals, 49;2 New York academy of sciences, New York,1948
Chem 18.2636

NEW YORK ACADEMY OF SCIENCES.SECTION OF PHYSICS AND CHEMISTRY Chromatography :a conference Papers New York 1946 Nov 29-30 By Harold G, Cassidy and others New York academy of sciences.Annals, 49,p.141-326 New York,1948
Bioch 33.2362

NEW YORK ACADEMY OF SCIENCES.SECTION OF PHYSICS AND CHEMISTRY Energy relationships in enzyme reactions :a conference New York 1944 Feb 11-12 New York academy of sciences. Annals, 45,p.357-436 New York,1944
Bioch 33.1027

NEW YORK ACADEMY OF SCIENCES.SECTION OF PHYSICS AND CHEMISTRY The Ultracentrifuge :a conference New York 1941 Nov 14-15 By D. A. MacInnes and others New York academy of sciences.Annals, 43,p 173-252 New York,1942
Bioch 33.1491

NEW YORK CITY 1939 Temperature,its measurement and control in science and industry :a symposium Records National bureau of standards Edited by C.O. Fairchild Under the auspices of the American institute of physics xiii,1362p 23cm Reinhold publishing,New York,1941
Sco 14.0234

NEW YORK CITY 1960 Symposium on flow visualization :A.S.M.E. annual meeting Presentation summaries By S.J. Kline and others American society of mechanical engineers illus var.p A.S.M.E.,New York, 1960
Chem E 24.0329

NEW YORK CITY 1968 Atomic physics International conference on atomic physics 1st Proceedings Edited by B. Bederson and others Bibliog.,Illus. xiii,620p 26cm Plenum press,New York,1969
A Math 4.1601

NEW YORK CITY GRAPH THEORY CONFERENCE 1st Proceedings Recent trends in graph theory New York 1970 Jun 11-13 Edited by M. Capobianco and others Lecture notes in mathematics, 186 vi,219p 25cm Springer, Berlin,1971
P Math 2.4113

NEW YORK HEART ASSOCIATION Biological interfaces:flows and exchanges :a symposium Proceedings New York 1967 Dec 8-9 Basic science symposia, 9 Little,Brown,Boston, Mass.,1968 Simultaneously published in the Journal of general physiology,vol.52,1968
Phys 20.2247

NEW YORK HEART ASSOCIATION Symposium on salt and water metabolism Papers New York 1959 Dec 11-12 Edited by Alfred P. Fishman Circulation, 21,no 5,pt 2 American heart association,New York,1959
Phys 20.1327

NEW YORK HEART ASSOCIATION.ANNUAL SYMPOSIUM 5TH Differentiation and development :a symposium Proceedings New York 1964 Feb 7-8 New York heart association Journal of experimental zoology, 157,no.1 Wistar institute of anatomy and biology,Philadelphia, Pa.,1964
Gen 34.0530

NEW YORK STATE MUSEUM AND SCIENCE SERVICE. GEOLOGICAL SURVEY Origin of anorthosite and related topics Annual George H.Hudson symposium 2nd Papers Plattsburgh,N.Y. 1966 Edited by Yngvar W. Isachsen New York state museum and science service.Memoir, 18 New York state museum and science service, Albany,N.Y.,1968
Min 10.1428

NEW YORK STATE SCIENCE AND TECHNOLOGY FOUNDATION Population biology and evolution symposium Proceedings Syracuse,N.Y. 1967 Jun 7-9 Edited by R.C. Lewontin Sponsored by Syracuse university Syracuse university press,Syracuse,N.Y.,1968
Gen 34.1070

NEW YORK STATE SCIENCE AND TECHNOLOGY FOUNDATION University education in computing science : a conference on graduate academic and related research programs in computing science Proceedings New York 1967 Jun 5-8 Edited by Aaron Finerman ACM monograph series xvi, 237p 23cm Academic press,New York,1968
Math L 5.3107

NEW YORK UNIVERSITY Luminescence of organic and inorganic materials :international conference New York 1962 Edited by Hartmut P. Kallman and Grace Marmor Spruch Sponsored by the United States.Air force. Aeronautical research laboratory xxiv,664p Wiley,New York,1962
Cav 7.1253

NEW YORK UNIVERSITY Symposium on digital computing in the aircraft industry proceedings New York 1957 Jan 31-Feb 1 Jointly sponsored by International business machines corporation illus. 399p 28cm International business machines corporation, New York,1957
Math L 5.0777

NEW YORK UNIVERSITY Verbal behavior and learning Conference on verbal learning 2nd Proceedings Edited by C.N. Cofer and Barbara S. Musgrave McGraw-Hill,New York, 1963
Psy 31.2737

NEW YORK UNIVERSITY.DEPARTMENT OF METALLURGICAL ENGINEERING Vacuum metallurgy New York 1957 Jun 10-14 Edited by Rointan F. Bunshah Reinhold.Chapman and Hall,New York;London,1958 Lectures presented during the course on vacuum metallurgy
Met 25.1630

NEW YORK UNIVERSITY.DEPARTMENT OF METALLURGICAL ENGINEERING Vacuum metallurgy conference Transactions New York 1959 Jun 1-3 Edited by Rointan F. Bunshah New York university press,New York,1960
Met 25.2259

NEW YORK UNIVERSITY.DEPARTMENT OF PSYCHOLOGY Verbal learning and verbal behavior :a conference Proceedings New York 1959 Edited by Charles N. Cofer and Barbara S. Musgrave McGraw-Hill series in psychology McGraw-Hill,New York,1961
Psy 31.1905

NEW YORK UNIVERSITY.INSTITUTE FOR MATHEMATICS AND MECHANICS The Theory of electro-magnetic waves :a symposium Proceedings New York 1950 Jun 6-8 Under the auspices of Air force Cambridge research laboratories.Geophysics research directorate Interscience,New York; London,1951
Radioth 35.1507

NEW YORK UNIVERSITY INSTITUTE OF PHILOSOPHY 1st annual symposium Proceedings , Determinism and freedom in the age of modern science:a philosophical symposium New York 1957 Feb 9-10 Edited by Sidney Hook xv, 237p New York university press,New York,1965
WSM 43.1662

NEW YORK ZOOLOGICAL SOCIETY Biology of normal and atypical pigment cell growth conference 4th Proceedings Houston,Tex. 1957 Nov 14-16 Edited by Myron Gordon Academic press,New York,1959
Gen 34.0601

NEW YORK ZOOLOGICAL SOCIETY Pigment cell growth Conference on the biology of normal and atypical pigment cell growth 3rd Proceedings New York 1951 Nov 15-17 Edited by Myron Gordon Academic press,New York,1953
An 32.3368

NEW ZEALAND.NATIONAL COMMITTEE FOR THE INTERNATIONAL GEOPHYSICAL YEAR Symposium on Antarctic research Wellington,N.Z. 1958 Feb 18-22 figures,tables varp Department of scientific and industrial research, Wellington,N.Z.,1958
Sco 14.6067

NEW ZEALAND EXHIBITION reports and awards of the jurors Dunedin 1865 New Zealand commissioners,Dunedin,1866
Philos 1.0056

NEW ZEALAND INSTITUTION OF ENGINEERS Australasian conference on hydraulics and fluid mechanics 2nd proceedings Auckland N.Z. 1965 Dec.6-11 University of Auckland Edited by C.M. Segedin Conference organising committee,Auckland,N.Z.,1966
A Math 4.0531

NEWARK,DEL 1969 International conference on luminescence Proceedings Edited by F. Williams graphs,diagrs xvi,959p 24cm North-Holland,Amsterdam,1970 Reprinted from 'Journal of luminescence',vol 1-2
Cav 7.2697

NEWARK,DEL. 1965 Nonlinear partial differential equations :a symposium on methods of solution Edited by W.F. Ames Supported by the National science foundation xv,316p Academic press,New York;London,1967
Chem E 24.0140

NEWBATTLE ABBEY 1961 Fluctuation, relaxation and resonance in magnetic systems Scottish universities' summer school 2nd Lectures Scottish universities' summer school in physics Edited by D.ter Haar With financial support from the North Atlantic treaty organization Oliver and Boyd, Edinburgh;London,1962
Chem 18.2619

NEWBATTLE ABBEY 1968 Physics of hot plasmas Scottish universities' summer school 9th Proceedings 1968 Jul 28-Aug 16 Edited by B.J. Rye and J.C. Taylor illus xv,455p 25cm Oliver and Boyd,Edinburgh, 1970
A Math 4.1886

NEWBATTLE ABBEY,EDINBURGH 1966 Particle interactions at high energies Scottish universities summer school 7th proceedings Edited by T.W. Preist and L.L.J. Vick 406p Oliver and Boyd,Edinburgh,1967
A Math 4.0875

NEWCASTLE DISEASE VIRUS:AN EVOLVING PATHOGEN : international symposium Proceedings 1963 University of Wisconsin American association of avian pathologists Edited by Robert Paul Hanson University of Wisconsin press,Madison,Wis.,1964
Gen 34.0728

NEWCASTLE-UPON-TYNE 1962 Magnetoplasmadynamic electrical power generation :a symposium Report Institution of electrical engineers Institution of electrical engineers.Conference report series, 4 Institution of electrical engineers, London,1963
Eng 41.4521

NEWCASTLE-UPON-TYNE 1964 Methods in palaeomagnetism NATO Advanced study institute on procedures in palaeomagnetism Proceedings North Atlantic treaty organization and University of Newcastle upon Tyne Edited by D.W. Collinson and others Developments in solid earth geophysics, 3 Amsterdam,1967
Geod 9.0568

NEWCASTLE-UPON-TYNE 1965 Magnetism and the cosmos NATO Advanced study institute on planetary and stellar magnetism North Atlantic treaty organization and University of Newcastle upon Tyne Edinburgh, 1967
Geod 9.0567

NEWCASTLE-UPON-TYNE 1965 Magnetism and the cosmos North Atlantic treaty organization advanced study institute on planetary and stellar magnetism North Atlantic treaty organization Edited by W.R. Hindmarsh and others 436p Oliver and Boyd,Edinburgh,1967
Obs 6.3350

NEWCASTLE-UPON-TYNE 1965 Renal failure and replacement of renal function European dialysis and transport association :a conference 2nd Proceedings Vol 2 European dialysis and transplant association Edited by David Kerr Excerpta medica. International congress series, 103 Excerpta medica,Amsterdam,1966
Inv Med 37.0130

NEWCASTLE-UPON-TYNE 1965 The Kinetics of proton transfer processes :a general discussion Faraday society Faraday society. Discussions, 39 Aberdeen university press, Aberdeen,1966
Bioch 33.0645

NEWCASTLE UPON TYNE 1966 Mantles of the earth and terrestrial planets N.A.T.O. advanced study institute :a conference Proceedings University of Newcastle upon Tyne.School of physics Edited by S.K. Runcorn Sponsored by North Atlantic treaty organization.Science office Interscience, London,1967
TA 15.0202

NEWCASTLE-UPON-TYNE 1966 The Symposium on non-elastic processes in the mantle Proceedings International upper mantle committee Edited by D.C. Tozer With financial assistance of the International union of geodesy and geophysics Geophysical journal, 14;1-4 Blackwell,Oxford,1967
Geod 9.0695

NEWCASTLE-UPON-TYNE 1968 Electrode reactions of organic compounds Faraday society Faraday society.Discussions, 45 graphs,photos,diagrms 281p 26cm Faraday society,London,1968
Cav 7.3124

NEWCASTLE-UPON-TYNE 1968 Palaeogeophysics NATO advanced study institute meeting on palaeogeophysics Proceedings North Atlantic treaty organization Edited by Stanley K. Runcorn Bibliog,illus,charts,maps xv,518p 24cm Academic press,London;New York,1970
Min 10.1458

NEWCASTLE UPON TYNE 1969 Joint IBM-University of Newcastle-upon-Tyne seminar Proceedings University of Newcastle upon Tyne computer laboratory International business machines corporation Edited by N.S. M. Cox University of Newcastle upon Tyne, Newcastle upon Tyne,1970
Math L 5.3354

NEWCASTLE-UPON-TYNE 1970 Conference on welding creep-resistant steels Proceedings Welding institute Welding institute,Abington, 1970 Conference director:B.Phelps
Met 25.2725

NEWCASTLE UPON TYNE 1971 Teaching of computer design Joint IBM-University of Newcastle-upon-Tyne seminar Proceedings International business machines corporation University of Newcastle-upon-Tyne.Computing laboratory illus 121p University of Newcastle upon Tyne.Computing laboratory, Newcastle upon Tyne,1972
Math L 5.3723

NEWCASTLE-UPON-TYNE 1968 Hydrogen-bonded systems :symposium on equilibria and reaction kinetics... Proceedings University of Newcastle-upon-Tyne Edited by A.K. Covington and P. Jones Taylor and Francis,London,1968
Chem 18.0298

NEWCASTLE-UPON-TYNE 1963 Problems in palaeoclimatology NATO palaeoclimates conference :a NATO advanced study institute proceedings Edited by A.E.M. Nairn sponsored by the North Atlantic treaty organisation.Science office John Wiley, London,1964
Geol 8.1353

NEWCASTLE-UPON-TYNE 1963 Problems in palaeoclimatology Palaeoclimates conference Proceedings North Atlantic treaty organization Edited by A.E.M. Nairn Interscience,London,1964
Geog 13.1058

NEWCASTLE-UPON-TYNE 1965 Magnetism and the cosmos N.A.T.O. advanced study institute on planetary and stellar magnetism North Atlantic treaty organisation Edited by W.R. Hindmarsh and others Oliver and Boyd, Edinburgh;London,1967
Geol 8.4428

NEWER ENGINEERING MATERIALS :a symposium on recent developments in new materials for use in engineering South Birmingham technical college Edited by R.F. Winters Macmillan, London,1969 Based on a short course held at South Birmingham technical college
Eng 41.3447

NEWER METALS Symposium on newer metals presented at the third Pacific area national meeting... San Francisco,Calif. 1959 Oct 15 American society for testing materials A.S.T.M.Special technical publication, 272 A. S.T.M.,Philadelphia,Pa.,1960
Met 25.0514

NEWER RESPIRATORY DISEASE VIRUSES Conference on newer respiratory disease viruses Bethesda,Md 1962 Oct 3-5 National institutes of health National institute of allergy and infectious diseases Edited by Clayton G. Loosli Sponsored by University of Southern California.School of medicine American review of respiratory diseases, 88,3 pt 2 American thoracic society,1963
PGMS 29.0321

NICE 1958 International horticultural congress 15th Proceedings Vol 1-3 International committee for horticultural congresses Edited by Jean-Claude Garnaud 3 vols Pergamon press,London,1961-62
BG 38.3133

NICE 1958 Topographie et la geologie des profondeurs oceaniques :colloque international Centre national de la recherche scientifique. Colloques internationaux,83 C.N.R.S.,Paris, 1959
Geog 13.0734

NICE 1960 Space research The Space science symposium 1st Proceedings Edited by Hilde Kallmann-Bijl Sponsored by Committee on space research xvi,1195p North-Holland,Amsterdam,1960
Nap 11.0317

NICE 1965 Aerodynamical phenomena in stellar atmosphwre International astronomical union Edited by Richard N. Thomas International astronomical union. Symposium, 28 Academic press,London,1967
TA 15.0411

NICE 1968 La Recherche en enseignement programme:tendances actuelles :colloque international Actes North Atlantic treaty organization Sciences du comportement, 8 Dunod,Paris,1969 Director of the colloquium:A.de Brisson.Papers in English and French
Psy 31.3117

NICE 1969 Physique fondamentale et astrophysique :colloque Compte-rendu Centre national de la recherche scientifique Journal de physique.Supplement, 30,no 11-12 Centre national de la recherche scientifique. Colloques nationaux, 925 162p Centre national de la recherche scientifique,Paris, 1970
TA 15.0499

NICE 1969 Symposium on optimization Edited by A.V. Balakrishnan and others Lecture notes in mathematics, 132 Bibliog., Illus. 350p 26cm Springer-Verlag,Berlin, 1970
P Math 2.3719

NICE 1969 Symposium on optimization Proceedings Edited by A.V. Balakrishnan and others Lecture notes in mathematics, 132 350p 25cm Springer,Berlin,1970
Math S 3.1685

NICE 1970 Congres international des mathematiciens :les 265 communications individuelles vii,290p 24cm Gauthier-Villars,Paris,1970
P Math 2.3775

NICE 1970 Sep 1-10 Congres international des mathematiciens 1970 Actes Vol 1-3 25cm 3 vols Gauthier-Villars,Paris,1971
P Math 2.3995

NICKEL-CHROMIUM PLATING :a one-day symposium London 1960 Nov 16 Institute of metal finishing.London branch Robert Draper, Teddington,1961
Met 25.2039

NIEBOROW 1959 Problems of applied geography Anglo-Polish seminar 1st Proceedings Polska akademia nauk.Instytut geografii Geographical studies,25 Panstwowe wydawnictwo naukowe,Warsaw,1961
Geog 13.5617

NIJMEGEN 1963 Pollen physiology and fertilization An International symposium on pollen physiology and fertilization 1st Papers Edited by H.F. Linskens North Holland,Amsterdam,1964
Bot 42.1426

NILPOTENT GROUPS Canadian mathematical congress :summer seminar Notes of lectures Edmonton,Alta. 1957 Aug 12-30 By Philip Hall 2nd edition iii,76p 25cm Queen Mary college,London,1957
P Math 2.4544

NIOBIUM Technology of columbium (Niobium) :a symposium Washington,D.C. 1958 May 15-16 Electrochemical society.Electrothermics and metallurgy division Wiley;Chapman and Hall, London,1958
Met 25.0587

NIPPON SOCIETAS RADIOLOGICA.BIOLOGICAL SUBGROUP International conference on radiation biology and cancer Proceedings Kyoto 1966 Nov 1-2 Edited by Tsutomu Sugahara Radiation society of Japan,1967
Radioth 35.1757

NIRIDAZOLE The Pharmacological and chemotherapeutic properties of niridazole and other antischistosomal compounds :a conference New York 1969 Oct 6 By F.C. Goble and others New York academy of sciences Edited by Francis C. Goble New York academy of sciences.Annals, 160,art.2 New York,1969
Mol 45.0272

NITRA 1966 Pathological wilt of plants :an international symposium Edited by J. Smolak and E. Haspel-Horvatovic illus. 127p Swets and Zeitlinger,Bratislava;Amsterdam,1970
Bot 42.4634

NITROGEN Biological oxydation of nitrogen in organic molecules :symposium Proceedings London 1971 Dec 19-22 Chelsea college Edited by J.W. Bridges and others Taylor and Francis,London,1972
Bioch 33.2286

NITROGEN Nitrogen in the tropics A Review of nitrogen in the tropics,with particular reference to pastures :a symposium Commonwealth scientific and industrial research organisation.Division of tropical pastures Commonwealth bureau of pastures and field crops.Bulletin, 46 Commonwealth agricultural bureaux,Farnham Royal,1962
Bot 42.1784

NITROGEN Utilization of nitrogen and its compounds by plants :a symposium Reading 1958 Sep 15-19 Society for experimental biology Society for experimental biology. Symposia, 13 Cambridge university press, Cambridge,1959
Bioch 33.1358

NITROGEN HETEROCYCLIC COMPOUNDS Recent work on naturally occurring nitrogen heterocyclic compounds :a symposium Report Exeter 1955 Jul 13-15 Chemical society Edited by K. Schofield Chemical society.Special publication, 3 London,1955
Bioch 33.2228

NITROGEN IN THE TROPICS A Review of nitrogen in the tropics,with particular reference to pastures :a symposium Commonwealth scientific and industrial research organisation.Division of tropical pastures Commonwealth bureau of pastures and field crops.Bulletin, 46 Commonwealth agricultural bureaux,Farnham Royal,1962
Bot 42.1784

NITROGEN METABOLISM IN PLANTS Recent aspects of nitrogen metabolism in plants :a symposium Proceedings Bristol 1967 Apr 18-19 Long Aston research station Edited by E.J. Hewitt and C.V. Cutting Long Ashton research station.Symposia, 1 Academic press,London; New York,1968
Bot 42.1893

NOBEL CONFERENCE 1ST Genetics and the future of man :a discussion at the Nobel conference St.Peter,Minn. 1965 Jan 7-8 Edited by John D. Roslansky Organized by Gustavus Adolphus college North-Holland,Amsterdam,1966
Gen 34.0062

NOBEL FOUNDATION Elementary particle theory: relativistic groups and analyticity Nobel symposium 8th Proceedings Aspenasgarden, Lerum 1968 May 19-25 Edited by Nils Svartholm illus 399p 24cm Wiley,New York,1968
A Math 4.1765

NOBEL FOUNDATION Fast reactions and primary processes in chemical kinetics Nobel symposium 5th Proceedings Sodergarn 1967 Aug 28 Sep 2 Edited by Stig Claesson Wiley,New York;London,1967
Bioch 33.0328

NOBEL FOUNDATION Mass motions in solar flares and related phenomena Nobel symposium 9th Proceedings Anacapri 1968 Jun 10-12 Edited by Yngve Ohman 245p 24cm Wiley, New York,1968
Cav 7.2971

NOBEL FOUNDATION Muscular afferent and motor control Nobel symposium 1st Proceedings Lidingo 1965 Jun Edited by Ragnar Granit Wiley;Almqvist and Wiksell,New York;Stockholm, 1966
An 32.4008

NOBEL FOUNDATION Muscular afferents and motor control Nobel symposium 1st Proceedings Sodergarn 1965 Jun Almqvist and Wiksell;Wiley,Stockholm;New York,1966
Phys 20.2242

NOBEL FOUNDATION Prostaglandins Nobel symposium 2nd Proceedings Stockholm 1966 Jun Edited by Sune Bergstrom and Bengt Samuelsson Wiley;Almqvist and Wiksell,New York;Stockholm,1967
Inv Med 37.0019

NOBEL FOUNDATION Symmetry and function of biological systems at the macromolecular level Nobel symposium 11th Procedings Sodergarn 1968 Aug 26-29 Edited by Arne Engstrom and Bror Strandberg Wiley,New York, 1969
Bioch 33.0327

NOBEL INSTITUTE FOR NEUROPHYSIOLOGY Muscular afferants and motor control Nobel symposium 1st Proceedings Sodergarn,Sweden 1965 Jun Edited by Ragnar Granit Wiley;Almqvist and Wiksell,New York;Stockholm,1966
Psy 31.0250

NOBEL SYMPOSIUM 1st Proceedings Muscular afferants and motor control Sodergarn,Sweden 1965 Jun Nobel institute for neurophysiology Edited by Ragnar Granit Wiley;Almqvist and Wiksell,New York;Stockholm, 1966
Psy 31.0250

NOBEL SYMPOSIUM 1st Proceedings Muscular afferent and motor control Lidingo 1965 Jun Nobel foundation Edited by Ragnar Granit Wiley;Almqvist and Wiksell,New York; Stockholm,1966
An 32.4008

NOBEL SYMPOSIUM 1st Proceedings Muscular afferents and motor control Sodergarn 1965 Jun Nobel foundation Almqvist and Wiksell;Wiley,Stockholm;New York, 1966
Phys 20.2242

NOBEL SYMPOSIUM 2nd Proceedings Prostaglandins Stockholm 1966 Jun Nobel foundation Edited by Sune Bergstrom and Bengt Samuelsson Wiley;Almqvist and Wiksell, New York;Stockholm,1967
Inv Med 37.0019

NOBEL SYMPOSIUM 4th Proceedings Current problems of lower vertebrate phylogeny Stockholm 1967 Jun Edited by Tor Orvig Wiley,New York,1968
Bal 39.2054

NOBEL SYMPOSIUM 5th Proceedings Fast reactions and primary processes in chemical kinetics Sodergarn 1967 Aug 28 Sep 2 Nobel foundation Edited by Stig Claesson Wiley,New York;London,1967
Bioch 33.0328

NOBEL SYMPOSIUM 8th Proceedings Elementary particle theory:relativistic groups and analyticity Aspenasgarden,Lerum 1968 May 19-25 Nobel foundation Edited by Nils Svartholm illus 399p 24cm Wiley,New York,1968
A Math 4.1765

NOBEL SYMPOSIUM 9th Proceedings Mass motions in solar flares and related phenomena Anacapri 1968 Jun 10-12 Nobel foundation Edited by Yngve Ohman 245p 24cm Wiley, New York,1968
Cav 7.2971

NOBEL SYMPOSIUM 11th Procedings Symmetry and function of biological systems at the macromolecular level Sodergarn 1968 Aug 26-29 Nobel foundation Edited by Arne Engstrom and Bror Strandberg Wiley,New York, 1969
Bioch 33.0327

NOCTILUCENT CLOUDS :international symposium Tallinn 1966 Special committee for the International quiet sun year and International association of meteorology and atmospheric physics Edited by I.A. Khvostikov and G. Witt Co-sponsored by the World meteorological organization illus 235p 27cm Academy of sciences of the USSR. Soviet geophysical committee,Moscow,1967 Symposium papers in English with Russian abstracts
Sco 14.8230

NOISE Conference on physical aspects of noise in electronic devices Papers Nottingham 1968 Sep 11-13 Institute of physics and the Physical society.Electronics group iv,248p 31cm Peregrinus,Stevenage, 1968
Cav 7.2840

NOISE The Control of noise :a conference Proceedings Teddington 1961 Jun 26-28 National physical laboratory National physical laboratory.Symposia, 12 H.M.S.O., London,1962
Eng 41.4218

NOISE,SHOCK TUBES,MAGNETIC EFFECTS,INSTABILITY AND MIXING Agard colloquium 3rd Papers Palermo 1958 Mar 17-21 Agard. Combustion and propulsion panel Edited by M. W. Thring and others Pergamon press,London, 1958
Eng 41.7232

NON-ABELIAN CATEGORIES Categories non-abeliennes Montreal 1964 By Peter Hilton North Atlantic treaty organization Societe mathematique du Canada Montreal.University. Seminaire de mathematiques superieures, 10 138p 28cm Universite de Montreal,Montreal, 1964
P. Math 2.0265

NON-CLASSICAL SHELL PROBLEMS :a symposium Proceedings Warsaw 1963 Sep 2-5 International association for shell structures Edited by W. Olszak and A. Sawczuk North-Holland;PWN,Amsterdam;Warsaw,1964
Eng 41.2876

NON-COMPACT GROUPS IN PARTICLE PHYSICS conference proceedings Wisconsin,Me. 1966 May 5-6 University of Wisconsin. Graduate school Edited by Yutze Chow 216p Benjamin,New York,1966
A Math 4.0890

NON-CONVENTIONAL LUBRICANTS AND BEARING MATERIALS SUCH AS USED IN NUCLEAR ENGINEERING Manchester 1962 Apr 12 Institution of mechanical engineers.Lubrication and wear group Institution of mechanical engineers, London,1962
Eng 41.6303

NON-DESTRUCTIVE EXAMINATION IN THE STEEL INDUSTRY Conference on non-destructive examination applied to process control in the steel industry Papers and discussions Swansea 1967 Jan 3-5 Swansea and district metallurgical society Iron and steel institute Society of non-destructive examination Iron and steel institute. Publication, 103 illus 193p London,1967
Met 25.2601

NON-DESTRUCTIVE TESTING International conference on non-destructive testing 4th Proceedings London 1963 Sep 9-13 Sponsored by the British national committee for non-destructive testing Butterworths, London,1964
Eng 41.3847

NON-EQUILIBRIUM THERMODYNAMICS A Symposium on non-equilibrium thermodynamics, variational techniques and stability Chicago 1965 May 17-19 University of Chicago General motors corporation.Research laboratories Universite libre de Bruxelles Edited by Russell J. Donnelly and I. Prigogine University of Chicago,Chicago,1966
A Math 4.0712

NON-EQUILIBRIUM THERMODYNAMICS VARIATIONAL TECHNIQUES AND STABILITY :a symposium Proceedings Chicago,Ill. 1965 May 17-19 Edited by Russell J. Donnelly and others University of Chicago press,Chicago,Ill.; London,1966
Eng 41.7117

NON-HEME IRON PROTEINS;ROLE IN ENERGY CONVERSION a symposium Proceedings Yellow Springs, Ohio 1965 Mar 22-24 Edited by Anthony San Pietro Sponsored by the Charles F.Kettering research laboratory Charles F.Kettering research laboratory.Contribution, 201 Antioch press,Yellow Springs,Ohio,1965
Bioch 33.0546

NON-HOMOGENEITY IN ELASTICITY AND PLASTICITY a symposium Warsaw 1958 Sep 2-9 International union of theoretical and applied mechanics Edited by W. Olszak Pergamon press,London,1959
Eng 41.2488

NON-LINEAR VIBRATIONS Vibrations non lineaires Colloque international des vibrations non lineaires Actes Ile de Porquerolles 1951 Sep 18-21 Union internationale de radio-science Organised by the Union internationale de mecanique theorique et appliquee France.Ministere de l'air.Publications scientifiques et techniques, 281 Information technique de l'aeronautique, Paris,1953
Eng 41.6372

NON-METALLIC SOLIDS The Chemistry of extended defects in non-metallic solids Proceedings Scotsdale,Ariz. 1969 Apr 16-26 Institute of advanced study on the chemistry of extended defects in non-metallic solids Edited by LeRoy Eyring and Michael O'Keefe North-Holland,Amsterdam,1970
Min 10.1545

NON-NUMERICAL COMPUTATION Advances in programming and non-numerical computation summer school proceedings Oxford 1963 University of Oxford.Computing laboratory Edited by L. Fox bibliog viii,218p 23cm Pergamon press,Oxford,1966
Math L 5.1026

NON-NUMERICAL COMPUTATION Advances in programming and non-numerical computation summer school proceedings Oxford 1963 University of Oxford.Computing laboratory Edited by L. Fox viii,218p 24cm Pergamon press,Oxford,1966
Math 3.0239

NON-PERIODIC PHENOMENA IN VARIABLE STARS Colloquium on variable stars Budapest 1968 Sep 5-9 International astronomical union. Commissions 27 and 42 International astronomical union.Colloquium, 4 490p Academic press,Budapest,1969
Obs 6.3360

NON-SOLAR X- AND GAMMA-RAY ASTRONOMY Rome 1969 May 8-10 International astronomical union Edited by L. Gratton International astronomical union.Symposium, 37 425p Reidel,Dordrecht,190
Obs 6.3322

NON-SOLAR X- AND GAMMA-RAY ASTRONOMY : symposium Papers Rome 1969 May 8-10 International astronomical union Edited by L. Gratton International astronomical union. Symposium, 37 425p Reidel,Dordrecht,1970
TA 15.0492

NON-STABLE PHENOMENA IN GALAXIES Nestasionarnye yavleniya v galaktikakh Byurakan 1966 May 4-12 International astronomical union International astronomical union.Symposium, 29 480p Akademiya nauk Armyanskey S.S.R.,Erevan,1968
Obs 6.3269

NON-STABLE STARS :a symposium Proceedings Dublin 1955 Sep 1 International astronomical union Edited by George H. Herbig International astronomical union. Symposium, 3 Cambridge university press, Cambridge,1957
TA 15.0033

NON-STANDARD ANALYSIS Contributions to non-standard analysis :a symposium Papers Oberwolfach 1970 Jul 19-25 Edited by W.A.J. Luxemburg and A. Robinson Studies in logic and the foundations of mathematics, 69 vi, 289p 23cm North-Holland,Amsterdam;London, 1972
P Math 2.4060

NON-CRYSTALLINE SOLIDS Conference on non-crystalline solids Alfred,N.Y. 1958 Sep 3-5 National research council and United States.Air research and development command Edited by V.D. Frechette John Wiley,New York; London,1960
Met 25.0126

NON-DESTRUCTIVE TESTING Papers on non-destructive testing presented at a meeting of committee E-7... 1953 Jul 2 American society for testing materials.Committee E-7 on non-destructive testing American society for testing materials,Philadelphia,Pa.,1954
Met 25.1318

NON-METALLIC BRITTLE MATERIALS Mechanical properties of non-metallic brittle materials : a conference Proceedings London 1958 Apr Great Britain.National coal board.Mining research establishment and Great Britain. Department of scientific and industrial research.Building research station Edited by W.H. Walton Butterworths,London,1958
Met 25.0852

NONDESTRUCTIVE TESTING Symposium on nondestructive testing presented to the 2nd Pacific area national meeting Los Angeles 1956 Sep 17-18 American society for testing materials A.S.T.M.Special technical publication, 213 A.S.T.M.,Philadelphia,1957
Met 25.1323

NONDESTRUCTIVE TESTING Symposium on nondestructive testing presented to the 55th annual meeting... New York 1952 Jun 26 American society for testing materials A.S.T. M.Special technical publication, 145 A.S.T.M. Philadelphia,1953
Met 25.2359

NONLINEAR CIRCUIT ANALYSIS Symposium on nonlinear circuit analysis Proceedings New York 1953 Apr 23-24 Polytechnic institute of Brooklyn Edited by Jerome Fox Sponsored by the Microwave research institute Microwave research institute.Network symposia series, 2 Polytechnic institute of Brooklyn,New York,1953
Eng 41.5920

NONLINEAR CIRCUIT ANALYSIS Symposium on nonlinear circuit analysis Proceedings New York 1956 Apr 25-27 Polytechnic institute of Brooklyn Edited by Jerome Fox Sponsored by the Microwave research institute Microwave research institute.Network symposia series, 6 Polytechnic institute of Brooklyn,New York,1957
Eng 41.5922

NONLINEAR DIFFERENTIAL EQUATIONS International symposium on nonlinear differential equations and nonlinear mechanics Colorado Springs,Colo. 1961 Jul 31-Aug 4 Edited by Joseph P. LaSalle and Solomon Lefschetz Sponsored by United States.Air force.Office of scientific research Academic press,New York;London,1963
Eng 41.6385

NONLINEAR DIFFERENTIAL EQUATIONS International symposium on nonlinear differential equations and nonlinear mechanics proceedings Colorado Springs,Col 1961 Jul 31-Aug 4 Edited by Joseph P. La Salle and Solomon Lefschetz Sponsored by the United States.Air force.Office of scientific research Academic press,New York,1963
A Math 4.0245

NONLINEAR DIFFERENTIAL EQUATIONS Numerical solutions of nonlinear differential equations : advanced symposium proceedings Madison,Wis. 1966 May 9-11 United States.Army. Mathematics research center Edited by Donald Greenspan United States.Army.Mathematics research center.Publications, 17 John Wiley, New York,1966
A Math 4.1220

NONLINEAR ESTIMATION THEORY Symposium on nonlinear estimation theory and its applications Proceedings San Diego,Calif. 1970 Sep 21-23 United States.Air force. Office of scientific research Institute of electrical and electronics engineers.Group on automatic control vii,297p 28cm IEEE,New York,1970
Math S 3.1698

NONLINEAR FUNCTIONAL ANALYSIS Mathematics research center symposium on contributions to nonlinear functional analysis Proceedings Madison,Wisc. 1971 Apr 12-14 University of Wisconsin.Mathematics research center xii, 672p 24cm Academic press,New York;London, 1971
P Math 2.4288

NONLINEAR FUNCTIONAL ANALYSIS :a symposium Proceedings Chicago,Ill. 1968 Apr 16-19 American mathematical society Edited by Felix E. Browder American mathematical society.Proceedings of symposia in pure mathematics, 18,part 1 bibliog. v,296p 26cm American mathematical society, Providence,R.I.,1970
P Math 2.3864

NONLINEAR FUNCTIONAL ANALYSIS AND APPLICATIONS Mathematics research center advanced seminar Proceedings Madison,Wisc. 1970 Oct 12-14 University of Wisconsin.Mathematics research center vii,586p 24cm Academic press,New York;London,1971
P Math 2.4286

NONLINEAR MECHANICS International symposium on nonlinear differential equations and nonlinear mechanics proceedings Colorado Springs,Col 1961 Jul 31-Aug 4 Edited by Joseph P. La Salle and Solomon Lefschetz Sponsored by the United States.Air force. Office of scientific research Academic press, New York,1963
A Math 4.0245

NONLINEAR OSCILLATIONS Conference on nonlinear oscillations 4th Proceedings Prague 1967 Sep 5-9 Edited by Jan Gonda Academia kiado,Prague,1968
Eng 41.6391

NONLINEAR OSCILLATIONS International conference on nonlinear oscillations 5th Abstracts of papers Kiev 1969 Aug 25-Sep 5 267p 20cm Academy of sciences of the USSR, Kiev,1969 Typescript
P Math 2.3734

NONLINEAR OSCILLATIONS Qualitative methods in the theory of nonlinear oscillations 5th conference Proceedings Kiev 1969 Vol 2-4 Edited by Yu.A. Mitropolski Bibliog, illus 26cm 3 vols Ukrainian academy of sciences.Institute of mathematics,Kiev,1970
A Math 4.1724

NONLINEAR PARTIAL DIFFERENTIAL Numerische losung nichtlinear partieller differential- und integrodifferentialgleichungen :conferenz Proceedings Oberwolfach 1971 Nov 28-Dec 4 Edited by R. Ansorge Lecture notes in mathematics, 267 vi,339p 25cm Springer, Berlin,1972
P Math 2.4294

NONLINEAR PARTIAL DIFFERENTIAL EQUATIONS :a symposium on methods of solution Edited by W.F. Ames xv,316p Academic press,New York,1967
A Math 4.1267

NONLINEAR PARTIAL DIFFERENTIAL EQUATIONS :a symposium on methods of solution Newark, Del. 1965 Dec 27-29 Edited by W.F. Ames Supported by the National science foundation xv,316p Academic press,New York;London,1967
Chem E 24.0140

NONLINEAR PROBLEMS :a symposium Proceedings Madison,Wis. 1962 Apr 30-May 2 United States.Army.Mathematics research center Edited by R.E. Langer bibliog.,illus. xiii, 321p 24cm University of Wisconsin press, Madison,Wis.,1963
A Math 4.1615

NONPARAMETRIC TECHNIQUES IN STATISTICAL INFERENCE International symposium on nonparametric techniques in statistical inference 1st Proceedings Bloomington,Ind. 1969 Jun 1-6 Edited by Madan Lal Puri Held at Indiana university bibliog. xiv,623p 24cm Cambridge university press,Cambridge,1970
Math S 3.1616

NOORDWIJK 1964 The Parathyroid glands: ultrastructure,secretion and function :a symposium 2nd Papers Edited by P.J. Gaillard and others University of Chicago press,Chicago,Ill.,1965
Phys 20.1417

NOORDWIJK 1964 The Parathyroid glands; ultrastructure,secretion and function : symposium 2nd Edited by Peter J. Gaillard and others University of Chicago press, Chicago,Ill.,1965
Pha 16.0291

NOORDWIJK 1966 Peptides European peptide symposium 8th Proceedings Edited by H.C. Beyerman and others North-Holland,Amsterdam, 1967
Bioch 33.0539

NOORDWIJK 1966 Radio astronomy and the galactic system Symposium on radio astronomy and the galactic system International astronomical union Edited by Hugo van Woerden Organised in co-operation with International scientific radio union International astronomical union.Symposium, 31 xviii,501p Academic press,London;New York,1967
A Math 4.1574

NOORDWIJK AAN ZEE 1963 Science of ceramics a conference 2nd Proceedings British ceramic society Edited by G.H. Stewart Academic press,London;New York,1965
Met 25.0278

NOORDWIJK AM ZEE 1964 The Parathyroid glands:ultrastructure,secretion and function : a symposium 2nd Papers Edited by Pieter J. Gaillard Bibliog.,Illus. xii,353p University of Chicago press,Chicago,Ill.,1965
Pha 16.0098

NORDEN 1960 Advance and retreat of rural settlement International geographical congress 19th Papers of the Siljan symposium Edited by Gerd Enequist and Gunnar Norling Uppsala universitet.Geografiska institution.Meddelanden.Ser A,160 Geografiska annaler,42,no 4 Uppsala,1960
Geog 13.1898

NORDEN 1960 International geographical congress,19th Finland :guidebooks prepared for the three Finnish excursions Edited by Helmer Smeds Fennia,84 Helsinki,1960
Geog 13.4337

NORDEN 1960 International geographical congress,19th Guidebook Denmark : contributions to problems discussed in symposia and excursions Edited by Niels Kingo Jacobson Kobenhavns universitets geografiske institut,Copenhagen,1960
Geog 13.4323

NORDEN 1960 International geological congress 21st Guides to excursions Oslo, 1960 Unbound
Geog 13.0553

NORDEN 1960 International geological congress 21st excursion guides:Norway Oslo,1960
Min 10.0168

NORDEN 1960 International geological congress 21st excursion guides:Sweden and Finland Stockholm;Helsinki,1960
Min 10.0169

NORDEN 1960 International geographical congress 19th Final report,bibliography Stockholm 1960 Aug 6-13 Edited by Staffan Helmfrid illus 138p Stockholm,1963 Held in Norden,ie.Denmark,Finland,Iceland, Norway and Sweden
Sco 14.0133

NORDEN 1960 International geological congress 21st Copenhagen 1960 Norden-Denmark,Finland,Iceland,Norway,Sweden,Volume of abstracts Edited by Theodor Sorgenfrei 252p 24cm Copenhagen,1960
Sco 14.0134

NORDIC LAPP COUNCIL The Lapps today in Finland,Norway and Sweden :en annexe:le departement de Laponie,notes documentaires Proceedings Jokkmokk,Sweden 1953 Karasjok, Norway 1956 1 Edited by Rowland G.P. Hill Ecole pratique des hautes etudes. Sorbonne.Bibliotheque arctique et antarctique, 1 maps 227p 24cm Mouton,Paris;The Hague,1960 Annex in French
Sco 14.5586

NORDIC RADIATION PROTECTION CONFERENCE 1st Proceedings Edited by Kurt Liden and Erik Lindgren Acta radiologica.Supplementum, 254 Stockholm,1966
Radioth 35.1295

NORDIC-SUMMER SCHOOL IN MATHEMATICS 5th Proceedings Algebraic geometry Oslo 1970 Oslo 1970 Aug 5-25 Edited by F. Oort vii, 332p 25cm Wolters-Noordoff,Groningen,1972
P Math 2.4173

NORDIC SUMMER SCHOOL IN MATHEMATICS 1968 On characteristic classes.Chs 1-6 :lectures given at the Nordic summer school in mathematics, June 16th-July 6th,1968 By Arunas Liulevicius 99p 29cm Aarhus universitet. Matematisk institut,Aarhus,1968
P Math 2.4582

NORDIC SUMMER SCHOOL IN MATHEMATICS 1968 On general cohomology,chs.1-9 :lectures given at the Nordic summer school in mathematics,June 16th-July 6th,1968 By Albrecht Dold 61p 29cm Matematisk institut,Aarhus universitet, Aarhus,1968
P Math 2.3800

NORDIC SUMMER SCHOOL OF MATHEMATICS 1967 Topics in several complex variables By A. Douady and others Enseignement mathematique. Monographies.Serie, 2 bibliog. 119p 24cm L'Enseignement mathematique,Geneva,1968 Notes of lectures given at the Nordic summer school of mathematics.Helsinki,1967
P Math 2.3763

NORDISK KIRURGISK FORENING 27th meeting transactions Oslo 1955 Jun 2-4 Acta chirurgica Scandinavica, 110,fasc.1-2 Stockholm,1955
Radioth 35.1274

NORDISK METEOROLOGMODE 3 Foredragenes resumeer Copenhagen 1963 Danske meteorologiske institut.Meddeleser,17 illus, maps xi,87p Charlottenlund,1964 Mimeographed
Sco 14.0141

NORMAL AND MALIGNANT CELL GROWTH :symposium Proceedings Chicago,Ill. 1968 Feb 24-25 Edited by R.J.M. Fry and others Sponsored by the University of Chicago.Cancer training grant Recent results in cancer research, 17 Springer,New York,1969
Radioth 35.0865

NORMAN,OKLA. 1950 Social psychology at the crossroads :the university of Oklahoma lectures in social psychology Papers University of Oklahoma Edited by John H. Rohrer and Mazafer Sherif Harper,New York, 1951
Psy 31.1363

NORRBOTTEN Jorden,skogen,malmen och vattenkraften i morgondagens Norrbotten :a conference Lectures Lulea 1956 Mar 19-20 Edited by Folke Thunberg Sponsored by Foreningen Norrbottens framjande illus,maps 271p 21cm Foreningen Norrbottens framjande, Lulea,1956
Sco 14.5794

NORSK GEOTEKNISK FORENING Geotechnical conference Shear strength properties of natural soils and rocks :a conference Proceedings Oslo 1967 Vol 1-2 2 vols Norwegian geotechnical institute,Oslo, 1967
Eng 41.3135

NORTH AMERICAN WILDLIFE CONFERENCE Transactions 15th San Francisco 1950 Mar 6-9 Wildlife management institute Edited by Ethel M. Quee 681p 23cm Washington,D.C., 1950
Sco 14.8304

NORTH ATLANTIC BIOTA AND THEIR HISTORY Reykjavik 1962 Jul 12-25 University of Iceland,and,Museum of natural history, Reykjavik Edited by Askell Love and Doris Love Sponsored by NATO advanced study institutes program illus. Pergamon press, Oxford,1963
Geog 13.1211

NORTH ATLANTIC BIOTA AND THEIR HISTORY :a symposium Reykjavik 1962 Jul University of Iceland and Iceland museum of natural history Sponsored by the Nato advanced institutes program pl. 430p 24cm Pergamon press,Oxford,1963
Sco 14.8334

NORTH ATLANTIC FISH MARKING SYMPOSIUM Woods Hole,Mass. 1961 May International commission for the northwest Atlantic fisheries International commission for the northwest Atlantic fisheries.Special publication,4 Dartmouth,N.S.,1961
Sco 14.7833

NORTH ATLANTIC FISH MARKING SYMPOSIUM Contributions Woods Hole,Mass. 1961 May International commission for the Northwest Atlantic fisheries Edited by H.W. Graham and others International commission for the Northwest Atlantic fisheries.Special publication,4 370p ICNAF,Dartmouth,Nova Scotia,1963
Sco 14.5957

NORTH ATLANTIC TREATY ORGANISATION Magnetism and the cosmos N.A.T.O. advanced study institute on planetary and stellar magnetism Newcastle-upon-Tyne 1965 Apr Edited by W.R. Hindmarsh and others Oliver and Boyd, Edinburgh;London,1967
Geol 8.4428

NORTH ATLANTIC TREATY ORGANISATION.ADVISORY GROUP FOR AERONAUTICAL RESEARCH AND DEVELOPMENT Combustion researches and reviews,1955 AGARD combustion panel meetings 6th-7th Invited papers Scheveningen 1954 May Paris 1954 Nov illus xv,187p Butterworths scientific publications,London,1955
Chem E 24.1386

NORTH ATLANTIC TREATY ORGANISATION.ADVISORY GROUP FOR AERONAUTICAL RESEARCH AND DEVELOPMENT
Polar atmosphere :a symposium Oslo 1956 July 2-8 Pt 1: meteorology section Edited by R.C. Sutcliffe AGARDograph 29,1 xii, 341p 26cm Pergamon press,London,1958
Sco 14.2367

NORTH ATLANTIC TREATY ORGANISATION.AGARD IONOSPHERIC RESEARCH COMMITTEE Arctic communications proceedings of the eighth meeting of the Agard ionaspheric research committee 8th Proceedings Athens 1963 Jul Edited by B. Landmark tables,graphs 297p Pergamon,Oxford;New York,1964
Sco 14.2468

NORTH ATLANTIC TREATY ORGANIZATION
Activation analysis principles and applications NATO advanced study institute Proceedings Glasgow 1964 Aug Edited by J. M.A. Lenihan and S.J. Thomson Academic press, London;New York,1965
Met 25.1689

NORTH ATLANTIC TREATY ORGANIZATION
Artificial intelligence and heuristic programming N.A.T.O. advanced study institute on artificial intelligence and heuristic programming 1st Proceedings Menaggio 1970 Aug Edited by N.V. Findler and Bernard Meltzer illus 327p Edinburgh university press,Edinburgh,1971
Math L 5.3853

NORTH ATLANTIC TREATY ORGANIZATION
Asymptotic distribution modulo 1 Nuffic international summer session in science papers Breukelen 1962 Aug 1-11 Netherlands universities foundation for international cooperation Edited by J.F. Koksma and L. Kuipers Bibliog. 203p 24cm Noordhoff,Groningen,1964
P. Math 2.1826

NORTH ATLANTIC TREATY ORGANIZATION
Biochemistry of chloroplasts :a Nato advanced study institute Proceedings Aberystwyth 1965 Aug Vol 1-2 Edited by T.W. Goodwin 2 vols Academic press,London;New York,1966-67
Bot 42.1429

NORTH ATLANTIC TREATY ORGANIZATION
Categories non-abeliennes Montreal 1964 By Peter Hilton Societe mathematique du Canada Montreal.University.Seminaire de mathematiques superieures, 10 138p 28cm Universite de Montreal,Montreal,1964
P. Math 2.0265

NORTH ATLANTIC TREATY ORGANIZATION Circadian clocks The Feldafing summer school Proceedings 1964 Sep 7-18 Edited by Jurgen Aschoff North-Holland,Amsterdam,1965
Bot 42.1365

NORTH ATLANTIC TREATY ORGANIZATION
Colloquium on convexity Proceedings Copenhagen 1965 Aug 12-19 Sponsored under the Nato advanced study institute programme Bibliog,illus xii,325p 25cm Kobenhavens universitets matematiske inst.,Copenhagen,1967 In English,French and German
P Math 2.3016

NORTH ATLANTIC TREATY ORGANIZATION Critical rotatory dispersion and circular dichroism in organic chemistry NATO summer school Proceedings Bonn 1965 Sep 24-Oct 10 Edited by G. Snatzke Heyden,London,1967
Chem 18.2791

NORTH ATLANTIC TREATY ORGANIZATION Decision making and age :N.A.T.O. advanced study institute Papers Thessaloniki 1967 Aug 14-19 Edited by A.T. Welford and J.E. Birren Interdisciplinary topics in gerontology, 4 Karger,Basle;New York,c1968
Psy 31.3208

NORTH ATLANTIC TREATY ORGANIZATION
Fluctuation,relaxation and resonance in magnetic systems Scottish universities' summer school 2nd Lectures Newbattle Abbey 1961 Scottish universities' summer school in physics Edited by D.ter Haar Oliver and Boyd,Edinburgh;London,1962
Chem 18.2619

NORTH ATLANTIC TREATY ORGANIZATION
Fondements de la geometrie algebrique moderne Montreal 1964 By Jean Dieudonne Societe mathematique du Canada University of Montreal.Seminaire de mathematiques superieures, 8 151p 28cm Universite de Montreal,Montreal,1964
P. Math 2.0507

NORTH ATLANTIC TREATY ORGANIZATION Graph theory and theoretical physics Edited by H. Frank Harary Bibliog.,Illus. xv,358p 24cm Academic press,New York;London,1967 Based on a NATO advanced study institute held in Paris,1963
P Math 2.3348

NORTH ATLANTIC TREATY ORGANIZATION Homotopie et cohomologie manuscript Montreal 1964 By Beno Eckmann Societe mathematique du Canada University of Montreal.Seminaire de mathematiques superieures, 11 129p 28cm Universite de Montreal,Montreal,1965
P. Math 2.0297

NORTH ATLANTIC TREATY ORGANIZATION
Integration dans les groupes topologiques Montreal 1964 By Geoffrey Fox Societe mathematique du Canada Montreal.University. Seminaire de mathematiques superieures, 12 357p 28cm Presses de l'universite de Montreal,Montreal,1966
P. Math 2.2385

NORTH ATLANTIC TREATY ORGANIZATION La Recherche en enseignement programme:tendances actuelles :colloque international Actes Nice 1968 May 13-17 Sciences du comportement, 8 Dunod,Paris,1969 Director of the colloquium:A.de Brisson.Papers in English and French
Psy 31.3117

NORTH ATLANTIC TREATY ORGANIZATION Lectures on field theory and the many-body problem : international spring school 1st Naples 1961 Universita di Napoli.Instituto di fisica teorica Edited by E.R. Caianiello Academic press,New York,1961
Cav 7.0423

NORTH ATLANTIC TREATY ORGANIZATION Lectures on the many-body problem :international spring school 2nd Naples 1962 Ravello 1963 Vol 1-2 Universita di Napoli. Instituto di fisica teorica Edited by E.R. Caianiello 2 vols Academic press,New York, 1962-64
Cav 7.0424

NORTH ATLANTIC TREATY ORGANIZATION Logical systems containing only a finite number of symbols Montreal 1966 By Leon Henkin Societe mathematique du Canada Universite de Montreal.Seminaire de mathematiques superieures, 21 illus. 48p 28cm Presses de l'Universite de Montreal,Montreal, 1967
P Math 2.3740

NORTH ATLANTIC TREATY ORGANIZATION Magnetism and the cosmos NATO Advanced study institute on planetary and stellar magnetism Newcastle-upon-Tyne 1965 Apr Edinburgh,1967
Geod 9.0567

NORTH ATLANTIC TREATY ORGANIZATION Magnetism and the cosmos North Atlantic treaty organization advanced study institute on planetary and stellar magnetism Newcastle-upon-Tyne 1965 Edited by W.R. Hindmarsh and others 436p Oliver and Boyd,Edinburgh, 1967
Obs 6.3350

NORTH ATLANTIC TREATY ORGANIZATION Mass spectrometry :Nato advanced study institute on theory,design and applications Glasgow 1964 Aug Edited by R.I. Reed Academic press,London;New York,1965
Met 25.1694

NORTH ATLANTIC TREATY ORGANIZATION Mathematics of contemporary physics : instructional conference Proceedings London 1971 Aug 23-Sep 11 Edited by R.F. Streater xi,274p 24cm Academic press, London,1972
A Math 4.1864

NORTH ATLANTIC TREATY ORGANIZATION Mechanisms of hormone action :a NATO advanced study institute Meersburg 1964 Edited by Peter Karlson Academic press,New York,1965 Some papers are in German
Pha 16.0092

NORTH ATLANTIC TREATY ORGANIZATION Mechanisms of hormone action :a Nato advanced study institute Discussions Meersburg 1964 Edited by P. Karlson Academic press, New York,1965
Bioch 33.0469

NORTH ATLANTIC TREATY ORGANIZATION Mechanisms of hormone action a NATO advanced study institute Meersburg 1964 Edited by P. Karlson Academic press;Thieme,New York; Stuttgart,1965
Gen 34.0549

NORTH ATLANTIC TREATY ORGANIZATION Membrane models and the formation of biological membranes International conference on biological membranes 2nd Proceedings Frascati 1967 Jun Edited by Liana Bolis and B.A. Pethica Illus. xv,337p North-Holland,Amsterdam,1968 A NATO advanced study institute
Pha 16.0016

NORTH ATLANTIC TREATY ORGANIZATION Methods in palaeomagnetism NATO Advanced study institute on procedures in palaeomagnetism Proceedings Newcastle-upon-Tyne 1964 Apr 1-10 Edited by D.W. Collinson and others Developments in solid earth geophysics, 3 Amsterdam,1967
Geod 9.0568

NORTH ATLANTIC TREATY ORGANIZATION Molecular basis of some aspects of mental activity :a NATO advanced study institute Proceedings Drammen 1965 Aug 2-14 Edited by Otto Walaas 2 vols Academic press,London,1966-67
Pha 16.0053

NORTH ATLANTIC TREATY ORGANIZATION Molecular biophysics International summer school of molecular biophysics Proceedings Squaw Valley,Calif. 1964 Aug 17-28 Edited by B. Pullman and M. Weissbluth Academic press,New York,1965
Radioth 35.1418

NORTH ATLANTIC TREATY ORGANIZATION NATO advanced study institute on facial structure of compact sets and applications Swansea 1972 Jul 2-25 Held at University college of Swansea 147p 21cm University college of Swansea,Swansea,1972
P Math 2.4281

NORTH ATLANTIC TREATY ORGANIZATION Palaeogeophysics NATO advanced study institute meeting on palaeogeophysics Proceedings Newcastle-upon-Tyne 1968 Edited by Stanley K. Runcorn Bibliog,illus, charts,maps xv,518p 24cm Academic press, London;New York,1970
Min 10.1458

NORTH ATLANTIC TREATY ORGANIZATION Pesticides in the environment and their effects on wildlife :an advanced study Proceedings Monks Wood 1965 Jul 1-14 Edited by N.W. Moore Journal of applied ecology.Supplement, 3 311p Blackwells, Oxford,1966
Bot 42.1983

NORTH ATLANTIC TREATY ORGANIZATION Pesticides in the environment and their effects on wildlife :an advanced study institute sponsored by the North Atlantic treaty organization,Monks Wood experimental station,England Proceedings Monks Wood 1965 Jul 1-14 Edited by N.W. Moore Journal of applied ecology, 3,suppl. Blackwell, Oxford,1965
Bal 39.1676

NORTH ATLANTIC TREATY ORGANIZATION Phonons in perfect lattices and in lattices with point imperfections Scottish universities' summer school 6th Lectures Aberdeen 1965 Aug 9-27 Scottish universities' summer school in physics Edited by R.W.H. Stevenson Oliver and Boyd,Edinburgh;London,1966
Chem 18.2643

NORTH ATLANTIC TREATY ORGANIZATION Physics of the solar corona NATO advanced study institute on physics of the solar corona Proceedings Athens 1970 Sep 6-17 Edited by C.J. Macris Astrophysics and space science library, 27 illus xii,345p 24cm Reidel,Dordrecht,1971
A Math 4.1803

NORTH ATLANTIC TREATY ORGANIZATION Physics of the solar corona NATO advanced study institute on the physics of the solar corona Proceedings Athens 1970 Sep 6-17 Edited by C.J. Macris Astrophysics and space science library, 27 345p Reidel,Dordrecht, 1971
Obs 6.3607

NORTH ATLANTIC TREATY ORGANIZATION Plasma waves in space and in the laboratory :a NATO advanced study institute Roros,Norway 1968 Apr 17-26 Edited by J.O. Thomas and B.J. Landmark Bibliog.,Illus. 487p 23cm Edinburgh university press,Edinburgh,1969
A Math 4.1505

NORTH ATLANTIC TREATY ORGANIZATION Problems in palaeoclimatology Palaeoclimates conference Proceedings Newcastle-upon-Tyne 1963 Jan 7-12 Edited by A.E.M. Nairn Interscience,London,1964
Geog 13.1058

NORTH ATLANTIC TREATY ORGANIZATION Promenades aleatoires et mouvement brownien : introduction a la theorie des probabilites mimeograph Montreal 1963 By Anatole Joffe Societe mathematique du Canada Montreal.University.Seminaire de mathematiques superieures, 7 vii,143p 28cm Universite de Montreal,Montreal,1964
P. Math 2.2145

NORTH ATLANTIC TREATY ORGANIZATION Selected topics in nuclear spectroscopy :N.U.F.F.I.C. international summer course in science, Nijenrode Castle... Proceedings Breukelen, Netherlands 1963 Jul 30-Aug 17 Netherlands universities foundation for international cooperation,and,Dutch physical society Edited by B.J. Verhaar North-Holland, Amsterdam,1964
Chem 18.2279

NORTH ATLANTIC TREATY ORGANIZATION Series de Fourier aleatoires mimeograph Montreal 1963 By Jean-Pierre Kahane Societe mathematique du Canada Montreal.University. Seminaire de mathematiques superieures, 4 vii,174p 28cm Universite de Montreal, Montreal,1963
P. Math 2.2147

NORTH ATLANTIC TREATY ORGANIZATION Structure and function of connective and skeletal tissue advanced study institute Proceedings St. Andrews 1964 Jun 15-25 Butterworths,London, 1965
An 32.3423

NORTH ATLANTIC TREATY ORGANIZATION Summer school in logic :N.A.T.O. advanced study institute:meeting of the association for symbolic logic Proceedings Leeds 1967 Aug 7-23 Edited by M.H. Lob Lecture notes in mathematics, 70 Bibliog. 331p 28cm Springer,Berlin,1968
P Math 2.3120

NORTH ATLANTIC TREATY ORGANIZATION Techniques in endocrine research :a NATO advanced study institute Proceedings Stratford-upon-Avon 1962 Sep Edited by Peter Eckstein and Francis Knowles Under the auspices of the University of Birmingham Academic press,London,1963
Gen 34.1964

NORTH ATLANTIC TREATY ORGANIZATION Techniques in endocrine research :a NATO advanced study institute Proceedings Stratford-upon-Avon 1962 Sep Edited by Peter Eckstein and Francis Knowles Under the auspices of the University of Birmingham Academic press,London,1963
An 32.3767

NORTH ATLANTIC TREATY ORGANIZATION Techniques in endocrine research :a Nato advanced study institute Proceedings Stratford-upon-Avon 1962 Sep Edited by P. Eckstein and F. Knowles Under the auspices of the University of Birmingham Academic press,New York,1963
Phys 20.1399

NORTH ATLANTIC TREATY ORGANIZATION The Use of thin films in physical investigations :a NATO advanced study institute held at the Imperial college of science and technology London 1965 Jul 19-24 Edited by J.C. Anderson Academic press,London,1966
Eng 41.4379

NORTH ATLANTIC TREATY ORGANIZATION Thermoluminescence of geological materials N. A.T.O.advanced research institute on applications of thermoluminescence to geological problems Proceedings Spoleto 1966 Sep 5-16 Edited by D.J. McDougall Academic press,London,1968
Min 10.0628

NORTH ATLANTIC TREATY ORGANIZATION Theorie des valuations mimeograph Montreal 1964 By Paulo Ribenboim Societe mathematique du Canada 313p 28cm Presses de l'universite de Montreal,Montreal,1964
P. Math 2.1974

NORTH ATLANTIC TREATY ORGANIZATION Tunneling phenomena in solids NATO advanced study institute on tunneling phenomena in solids Proceedings Riso,Denmark 1967 Edited by Elias Burstein and Stig Lundqvist Bibliog 580p 18cm Plenum press,New York,1969
Cav 7.2667

NORTH ATLANTIC TREATY ORGANIZATION Tunneling phenomena in solids :lectures presented at the 1967 NATO advanced study institute at Riso, Denmark Riso 1967 Edited by Elias Burstein and Stig Lundqvist Plenum press,New York,1969
Eng 41.8302

NORTH ATLANTIC TREATY ORGANIZATION and UNITED STATES.OFFICE OF NAVAL RESEARCH Molecular biophysics :international summer school Proceedings Squaw Valley,Calif. 1964 Aug 17-28 Edited by Bernard Pullmann and Mitchel Weissbluth Academic press,New York;London, 1965
Chem 18.2249

NORTH ATLANTIC TREATY ORGANIZATION.ADVANCED STUDY INSTITUTE Mantles of the earth and terrestrial planets :Nato advanced study institute Proceedings University of Newcastle-upon-Tyne.School of physics Edited by S.K. Runcorn Interscience,London,1967
Geod 9.0016

NORTH ATLANTIC TREATY ORGANIZATION.ADVANCED STUDY INSTITUTE Proceedings Winds and turbulence in stratosphere,mesosphere and ionosphere Lindau 1966 Sep 18-Oct 1 North Atlantic treaty organization.Science committee Edited by K. Rawer 421p North Holland,Amsterdam,1968
Nap 11.0955

NORTH ATLANTIC TREATY ORGANIZATION.ADVANCED STUDY INSTITUTE Proceedings Structure and evolution of the galaxy Athens 1969 Sep 8-19 Edited by L.N. Mavridis Astrophysics and space science library, 22 312p D. Reidel,Dordrecht,1971
Obs 6.3456

NORTH ATLANTIC TREATY ORGANIZATION.ADVISORY GROUP FOR AERONAUTICAL RESEARCH AND DEVELOPMENT Advances in materials research in the N.A.T.O. nations :a symposium Proceedings Washington,D.C. 1961 May Edited by H. Brooks and others AGARD 62 Pergamon press, Oxford,1963
Met 25.0096

NORTH ATLANTIC TREATY ORGANIZATION.ADVISORY GROUP FOR AERONAUTICAL RESEARCH AND DEVELOPMENT Polar atmosphere symposium Oslo 1956 Jul 2-8 Pt 2: ionospheric section Agardograph,29, pt.2 diagrs xiii,212p 26cm Pergamon, London,1957
Sco 14.0528

NORTH ATLANTIC TREATY ORGANIZATION.ADVISORY GROUP FOR AERONAUTICAL RESEARCH AND DEVELOPMENT Science and technology of tungsten,tantalum, molybdenum,niobium and their alloys Agard conference on refractory metals Proceedings Oslo 1963 Jun 23-26 AGARDograph 82 Pergamon press,Oxford,1964
Met 25.1011

NORTH ATLANTIC TREATY ORGANIZATION.ADVISORY GROUP FOR AERONAUTICAL RESEARCH AND DEVELOPMENT Selected combustion problems:fundamentals and aeronautical applications Combustion colloquium Proceedings Cambridge 1953 Dec 7-11 viii,534p Butterworths,London, 1954
Chem E 24.1912

NORTH ATLANTIC TREATY ORGANIZATION.ADVISORY GROUP FOR AERONAUTICAL RESEARCH AND DEVELOPMENT. STRUCTURES AND MATERIALS PANEL Stress corrosion cracking in aircraft structural materials :an Agard conference Proceedings Turin 1967 Apr 18-19 AGARD conference proceedings series, 18 Nato,1967
Met 25.2484

NORTH ATLANTIC TREATY ORGANIZATION.ADVISORY GROUP ON HUMAN FACTORS Communication processes NATO symposium on communication processes Proceedings Washington,D.C. 1963 Edited by Frank A. Geldard and others NATO conference series, 4 298p Pergamon press, Oxford,1965
Cav 7.2423

NORTH ATLANTIC TREATY ORGANIZATION.DIVISION OF SCIENTIFIC AFFAIRS Meteorological and astronomical influences on radio wave propagation NATO Advanced study institute Papers Corfu 1961 NATO conference series, 3 318p Pergamon press,Oxford,1963
Nap 11.0915

NORTH ATLANTIC TREATY ORGANIZATION.DIVISION OF SCIENTIFIC AFFAIRS Regeneration in animals and related problems :an international symposium Athens 1964 Apr Edited by V. Kiortsis and H.A.L. Trampusch North-Holland, Amsterdam,1965
An 32.5438

NORTH ATLANTIC TREATY ORGANIZATION.DIVISION OF SCIENTIFIC AFFAIRS Regeneration in animals and related problems :an international symposium Athens 1964 Apr Edited by V. Kiortsis and H.A.L. Trampusch North-Holland, Amsterdam,1965 Text in English and French
Bal 39.1320

NORTH ATLANTIC TREATY ORGANIZATION.DIVISION OF SCIENTIFIC AFFAIRS Solar physics NATO advanced study institute on solar physics Proceedings Lagonissi,Athens 1965 Sep Interscience,London,1967
Obs 6.3280

NORTH ATLANTIC TREATY ORGANIZATION.PURE SCIENCE BUREAU Modern quantum chemistry :Istanbul international school of quantum chemistry Lectures Istanbul 1964 Aug 16-Sep 5 Pt 3: action of light and organic crystals Edited by Oktay Sinanoglu Organized with the cooperattion of Orta Dogu teknik universitesi Istanbul lectures:modern quantum chemistry,3 Academic press,New York;London,1965
Chem 18.2238

NORTH ATLANTIC TREATY ORGANIZATION.SCIENCE COMMITTEE Applications of mathematical programming techniques :a conference Proceedings Cambridge 1968 Jun 24-28 Edited by E.M.L. Beale ix,451p 24cm English universities press,London,1970
Math S 3.1686

NORTH ATLANTIC TREATY ORGANIZATION.SCIENCE COMMITTEE Observational aspects of galactic structure :a summer school typescript Lagonissi 1964 Sep.9-23 Edited by A. Blaauw and L.N. Mavridis 370p Athens,1965
Obs 6.0193

NORTH ATLANTIC TREATY ORGANIZATION.SCIENCE COMMITTEE Programming languages N.A.T.O. advanced study institute.Summer school Villard de Lans 1966 Edited by F. Genuys 395p Academic press,London,1968
TA 15.0589

NORTH ATLANTIC TREATY ORGANIZATION.SCIENCE COMMITTEE Programming languages :N.A.T.O. advanced summer school Villard-de-Lans 1966 Edited by F. Genuys illus. 395p Academic press,London;New York,1968
Math L 5.3577

NORTH ATLANTIC TREATY ORGANIZATION.SCIENCE COMMITTEE Queuing theory:recent developments and applications conference proceedings Lisbon 1965 Sep.27-Oct.1 Edited by R. Cruon 224p 23cm English universities press,London,1967
Math 3.0703

NORTH ATLANTIC TREATY ORGANIZATION.SCIENCE COMMITTEE Software engineering :a conference Report Garmisch 1968 Oct 7-11 Edited by Peter Naur and Brian Randell 231p Nato,Brussels,1969
Math L 5.3490

NORTH ATLANTIC TREATY ORGANIZATION.SCIENCE COMMITTEE Software engineering technique : a conference Report Rome 1969 Oct 27-31 Edited by J.N. Buxton and B. Randall illus. 1964p NATO science committee,Brussels,1970
Math 5.3610

NORTH ATLANTIC TREATY ORGANIZATION.SCIENCE COMMITTEE Software engineering techniques a conference Report Rome 1969 27-31 Oct Edited by J.N. Buxton and B. Randell Illus 164p NATO science committee,Brussels,1970
Math L 5.3545

NORTH ATLANTIC TREATY ORGANIZATION.SCIENCE COMMITTEE Theory of distributions :an international summer institute Proceedings Lisbon 1964 Sep Instituto Gulbenkian de ciencia.Centro de calculo cientifico 390p Instituto Gulbenkian de ciencia,Lisbon,1964
A Math 4.1265

NORTH ATLANTIC TREATY ORGANIZATION.SCIENCE COMMITTEE Theory of distributions : international summer institute Proceedings Lisbon 1964 Sep Centro de calculo cientifico Illus. xviii,390p 25cm Centro de calculo cientifico,Lisbon,1964 In English and French
P Math 2.3180

NORTH ATLANTIC TREATY ORGANIZATION.SCIENCE COMMITTEE Winds and turbulence in stratosphere,mesosphere and ionosphere North Atlantic treaty organization.Advanced study institute Proceedings Lindau 1966 Sep 18-Oct 1 Edited by K. Rawer 421p North Holland,Amsterdam,1968
Nap 11.0955

NORTH ATLANTIC TREATY ORGANIZATION.SCIENCE OFFICE Mantles of the earth and terrestrial planets N.A.T.O. advanced study institute :a conference Proceedings Newcastle upon Tyne 1966 Mar 30-Apr 7 University of Newcastle upon Tyne.School of physics Edited by S.K. Runcorn Interscience,London,1967
TA 15.0202

NORTH ATLANTIC TREATY ORGANIZATION.SCIENTIFIC AFFAIRS COMMITTEE Applications of mathematical programming techniques :a conference Cambridge 1968 Jun 24-28 Edited by E.M.L. Beale EUP,London,1970
Eng 41.8254

NORTH ATLANTIC TREATY ORGANIZATION.SCIENTIFIC AFFAIRS COMMITTEE Theory of games: techniques and applications conference proceedings Toulon 1964 Jun.29-Jul.3 Edited by A. Mensch xi,490p 23cm English universities press,London,1966
Math 3.0241

NORTH ATLANTIC TREATY ORGANIZATION.SCIENTIFIC AFFAIRS DIVISION Circadian clocks The Feldafing summer school Proceedings 1964 Sep 7-18 Edited by Jurgen Aschoff North Holland,Amsterdam,1965
Bal 39.1766

NORTH ATLANTIC TREATY ORGANIZATION.SCIENTIFIC AFFAIRS DIVISION Circadian clocks :summer school Proceedings Feldafing 1964 Sep 7-18 Edited by Jurgen Aschoff North-Holland, Amsterdam,1965
An 32.5375

NORTH ATLANTIC TREATY ORGANIZATION.SCIENTIFIC AFFAIRS DIVISION Colloquium on algebraic topology Aarhus 1962 Aug 1-10 Aarhus universitet.Matematisk institut 113p 29cm Aarhus universitet,1962
P. Math 2.0293

NORTH ATLANTIC TREATY ORGANIZATION.SCIENTIFIC AFFAIRS DIVISION Combinatorial methods in probability theory colloquium lectures, mimeograph Aarhus 1962 Aug 1-10 Aarhus universitet.Matematisk institut 126p 29cm Aarhus universitet,Aarhus,1962
P. Math 2.2189

NORTH ATLANTIC TREATY ORGANIZATION.SCIENTIFIC AFFAIRS DIVISION Combinatorial methods in probability theory colloquium lectures, mimeograph Aarhus 1962 Aug.1-10 Aarhus universitet.Matematisk institut 126p 29cm Aarhus universitet,Aarhus,1962
Math 3.0570

NORTH ATLANTIC TREATY ORGANIZATION.SCIENTIFIC AFFAIRS DIVISION Digital simulation in operational research :a conference Proceedings Hamburg 1965 Sep 6-10 Edited by S.H. Hollingdale English universities press,London,1967
Eng 41.1192

NORTH ATLANTIC TREATY ORGANIZATION.SCIENTIFIC AFFAIRS DIVISION Graph theory and theoretical physics :N.A.T.O. advanced study institute Paris 1963 Edited by Frank Harary bibliog.,illus. xv,358p 24cm Academic press,New York;London,1967
Math S 3.1408

NORTH ATLANTIC TREATY ORGANIZATION.SCIENTIFIC AFFAIRS DIVISION Informative molecules in biological systems International symposium on uptake of informative molecules by living cells Proceedings Mol,Belgium 1970 Aug 31-Sep 2 Edited by L. Ledoux North-Holland, Amsterdam,1971
Bioch 33.2240

NORTH ATLANTIC TREATY ORGANIZATION.SCIENTIFIC AFFAIRS DIVISION International advanced summer institute on microprogramming St. Raphael 1971 Aug 30-Sep 10 Edited by Guy G. Boulaye and Jean Mermet Actualites scientifiques et industrielles illus 418p Hermann,Paris,1972
Math L 5.3634

NORTH ATLANTIC TREATY ORGANIZATION.SCIENTIFIC AFFAIRS DIVISION Modern diffraction and imaging techniques in material science International summer course on material science Proceedings Antwerp 1969 Jul 28-Aug 8 Edited by S. Amelinckx and others North-Holland,Amsterdam,1970
Met 25.2608

NORTH ATLANTIC TREATY ORGANIZATION.SCIENTIFIC AFFAIRS DIVISION Probability methods in anlysis symposium lectures Loutraki,Greece 1966 May 22-Jun 4 Edited by D.A. Kappos Lecture notes in mathematics, 31 329p 28cm Springer-verlag,Berlin,1967
Math 3.0711

NORTH ATLANTIC TREATY ORGANIZATION.SCIENTIFIC AFFAIRS DIVISION Structure and functions of connective and skeletal tissue an advanced study institute Proceedings St.Andrews 1964 Jun 15-25 Butterworths,London,1964
Path 30.2421

NORTH ATLANTIC TREATY ORGANIZATION ADVANCED STUDY INSTITUTE ON PLANETARY AND STELLAR MAGNETISM Magnetism and the cosmos Newcastle-upon-Tyne 1965 North Atlantic treaty organization Edited by W.R. Hindmarsh and others 436p Oliver and Boyd,Edinburgh,1967
Obs 6.3350

NORTH CAROLINA CONSORTIUM ON AIR POLLUTION Environmental protection Symposium on multiple-source urban diffusion models Proceedings 1970 Air pollution control office.Publication,AP- 86 US government printing office,Washington,D.C.,1970
Eng 41.8616

NORTH CAROLINA STATE COLLEGE Mechanical properties of engineering ceramics :a conference Proceedings Raleigh,N.C. 1960 Edited by W.Wurth Kriegel and Hayne Palmour Interscience,New York;London,1961
Met 25.0853

NORTH CAROLINA STATE COLLEGE Research conference on structure and properties of engineering materials Proceedings Raleigh, N.C. 1962 Mar 12-13 Materials science research, 1 Plenum press,New York,1963
Met 25.2281

NORTH CAROLINA STATE COLLEGE.INSTITUTE OF STATISTICS Experimental designs in industry Symposium on design of industrial experiments proceedings Raleigh,N.C. 1956 Nov.5-9 Edited by Victor Chew Supported by the United States.Air force. Office of scientific research Wiley publications in statistics Bibliog. xi, 268p 28cm John Wiley and sons,New York, 1958 Being parts A and B of Symposium on design of industrial experiments, proceedings, edited by Victor Chew,1957
Math 3.0583

NORTH CAROLINA STATE COLLEGE.INSTITUTE OF STATISTICS Symposium on design of industrial experiments proceedings Raleigh, N.C. 1956 Nov.5-9 Edited by Victor Chew Supported by the United States.Air force. Office of scientific research Bibliog. viii,374p 27cm Institute of statistics, University of North Carolina,Raleigh,N.C.,1957
Math 3.0584

NORTH CAROLINA STATE UNIVERSITY Clean surfaces their preparation and characterization for interfacial studies : based on a symposium Raleigh,N.C. 1968 Apr 8-10 Edited by George Goldfinger Dekker, New York,1970
Met 25.2828

NORTH CAROLINA STATE UNIVERSITY Materials science research :a conference Proceedings Raleigh 1964 Nov 16-18 Vol 3: role of grain boundaries and surfaces in ceramics Edited by W.Wurth Kriegel and Hayne Palmour Plenum press,New York,1966
Met 25.0288

NORTH PACIFIC FUR SEAL CONFERENCE Proceedings Washington,D.C. 1955-57 5 vols washington D.C.,1957 Mimeographed
Sco 14.1174

NORTH SEA FLOODS Conference on the North Sea floods of 31 January-1 February,1953 Papers London 1953 Dec Institution of civil engineers Institution of civil engineers, London,1954
Geog 13.0706

The NORTH SEA FLOODS OF JANUARY 31-FEBRUARY 1, 1953 :a conference London 1953 Dec Institution of civil engineers Institution of civil engineers,London,1954
Eng 41.3342

NORTHAMPTON 1964 The Use of redundancy in system design :a symposium Society of instrument technology Society of instrument technology,1964
Eng 41.5963

NORTHFIELD,MINN. 1956 Antarctica in the International geophysical year Antarctic symposium Proceedings International geophysical year Co-sponsored by the American geophysical union American geophysical union.Publication, 462 Geophysical monograph map 133p Washington,D.C.,1956
Nap 11.0020

NORTHRIDGE,CALIF. 1972 Heat transfer and fluid mechanics institute 1972 Proceedings Edited by R.B. Landis and G.J. Hordemann illus xii,430p 25cm Stanford university press,Stanford,Calif.,1972
A Math 4.1876

NORTHWESTERN UNIVERSITY Dynamic stability of structures :an international conference Proceedings Evanston,Ill. 1965 Oct 18-20 Edited by George Herrmann Pergamon press, London,1967
Eng 41.2531

NORTHWESTERN UNIVERSITY Gas dynamics symposium on aerothermochemistry Proceedings Evanston,Ill 1955 Aug 22-24 Edited by Donald K. Fleming Supported by the United States.Air force illus 284p Northwestern university,Evanston,Ill.,1956
Chem E 24.1390

NORTHWESTERN UNIVERSITY Magnetohydrodynamics Biennial gas dynamics symposium 4th proceedings Evanston,Ill. 1961 Aug 23-25 Edited by Ali Bulent Cambel and others Through the generosity of Government agencies and industrial sponsors Northwestern university press,Evanston,Ill.,1962
A Math 4.0623

NORTHWESTERN UNIVERSITY The National symposium on colloid chemistry 2nd Papers presented Evanston,Ill. 1924 Jun 18-21 Edited by Harry N. Holmes Colloid symposium monograph, 2 Chemical catalog company,New York,1925
Bioch 33.1470

NORWEGIAN DEFENCE RESEARCH ESTABLISHMENT N.A. T.O. Advanced study institute Electron density profiles in the ionosphere and exosphere :a symposium Proceedings Skeikampen,Norway 1961 Apr 17-26 Edited by B. Maehlum Sponsored by N.A.T.O. science committee N.A.T.O. conference series, 2 428p Pergamon press,Oxford,1962
Nap 11.0391

NORWEGIAN NEUROLOGICAL ASSOCIATION The So-called extrapyramidal system :symposium from the sixteenth congress of Scandinavian neurobiologists Oslo 1962 Edited by Sigvald Refsum and others Universitetsforlaget,Oslo,1963
An 32.4655

NORWICH 1868 and LONDON 1868 International congress of prehistoric archaeology 3rd Transactions Longmans, London,1869
An 32.2635

NORWICH 1935 British association for the advancement of science :annual meeting 105th Report London,1935
Geog 13.3987

NORWICH 1968 Dormancy and survival :a symposium Papers Society for experimental biology Society for experimental biology. Symposia, 23 Cambridge university press, Cambridge,1969
Gen 34.2224

NORWICH 1968 Dormancy and survival :a symposium Society for experimental biology Society for experimental biology.Symposia, 23 Cambridge university press,Cambridge,1969
Bioch 33.1362

NORWICH 1968 Dormancy and survival : symposium Papers Society for experimental biology Edited by H.W. Woolhouse Society for experimental biology.Symposia, 23 illus vi,598p 23cm Cambridge university press, Cambridge,1969
Sco 14.8232

NORWICH PHARMACAL COMPANY Symposium on pyelonephritis Edinburgh 1966 Livingstone,London,1967
PGMS 29.0236

NOTCH BAR TESTING AND ITS RELATION TO WELDED CONSTRUCTION :symposium London 1951 Dec 5 Organized by the Institute of welding Institute of welding,London,1953
Eng 41.3834

NOTGEMEINSCHAFT DER DEUTSCHEN WISSENSCHAFT The Fine structure of the solar atmosphere; extended abstracts :a colloquium Anacapri 1966 Jun.6-8 Edited by K.O. Kiepenheuer Notgemeinschaft der deutschen wissenschaft. Bericht, 12 149p Franz Steiner,Wiesbaden, 1966 German title:Die Feinstruktur der sonenatmosphare
Obs 6.0945

NOTRE DAME,IND. 1963 Fundamental processes in radiation chemistry :a general discussion Faraday society Faraday society.Discussions, 36 Butterworths,London,1963
Radioth 35.1772

NOTRE DAME CONFERENCE 2ND International conference on sintering and related phenomena Cleveland 1965 Jun 21-23 University of Notre Dame Edited by G.C. Kuczynski and others Gordon and Breach,New York,1967
Met 25.1285

NOTTINGHAM Growth of cereals and grasses Easter school in agricultural science 12th Proceedings University of Nottingham Edited by F.L. Milthorpe and J.D. Ivins Bibliog.,illus,diagrs,tables xii,359p Butterworths,London,1966
Bot 42.2035

NOTTINGHAM 1954 The Physics of particle analysis :a conference Papers Institute of physics British journal of applied physics. Supplement, 3 Institute of physics,London, 1954
Geod 9.0289

NOTTINGHAM 1954 The Physics of particle size analysis :a conference Edited by H.R. Lang British journal of applied physics. Supplement, 3 218,viiip Institute of physics,London,1954
Obs 6.1961

NOTTINGHAM 1955 Soil zoology Easter school in agricultural science 2nd Proceedings University of Nottingham Edited by D.Keith McE. Kevan Butterworths, London,1955
Bal 39.2006

NOTTINGHAM 1955 Soil zoology Easter school in agricultural science 2nd Proceedings University of Nottingham Edited by D.Keith McE. Kevan Butterworths scientific publications illus. xiv,512p Butterworths,London,1955
Bot 42.2070

NOTTINGHAM 1956 Antibiotics and mould metabolites :a symposium Report Chemical society Chemical society.Special publication, 5 Chemical society,London,1956 Organized by professor Johnson
Chem 18.1215

NOTTINGHAM 1956 Growth of leaves Easter school in agricultural science 3rd Proceedings University of Nottingham Edited by F.L. Milthorpe illus. x,223p Butterworths,London,1956
Bot 42.1825

NOTTINGHAM 1956 Membrane phenomena Faraday society Faraday society.Discussions, 21 288p Aberdeen university press, Aberdeen,1956
Bot 42.1565

NOTTINGHAM 1956 Membrane phenomena :a general discussion Faraday society Faraday society.Discussions, 21 Aberdeen university press,Aberdeen,1956
Bioch 33.0947

NOTTINGHAM 1957 Control of the plant environment Easter school in agricultural science 4th Proceedings University of Nottingham Edited by J.P. Hudson Butterworths,London,1957
BG 38.0605

NOTTINGHAM 1957 Control of the plant environment Easter school in agricultural science 4th Proceedings University of Nottingham Edited by J.P. Hudson illus. xvi,240p Butterworths,London,1957
Bot 42.1943

NOTTINGHAM 1958 Nutrition of the legumes Easter school in agricultural science 5th Proceedings University of Nottingham Edited by E.G. Hallsworth Butterworths, London,1958
Bioch 33.0776

NOTTINGHAM 1958 Nutrition of the legumes Easter school in agricultural sciences 5th Proceedings Edited by E.G. Hallsworth Butterworths,London,1958
Bot 42.1781

NOTTINGHAM 1959 Energy transfer with special reference to biological systems :a general discussion Faraday society Faraday society.Discussions, 27 Faraday society, London,1959
Radioth 35.1406

NOTTINGHAM 1959 Energy transfer with special reference to biological systems :a general discussion Faraday society Faraday society.Discussions, 27 The Faraday society, London,1960
Radioth 35.0249

NOTTINGHAM 1959 Energy transfer with special reference to biological systems :a general discussion Papers Faraday society Faraday society.Discussions, 27 Aberdeen university press,Aberdeen,1960
Bioch 33.0314

NOTTINGHAM 1959 Measurement of grassland productivity Easter school in agricultural science 6th Proceedings University of Nottingham Edited by J.D. Ivins Butterworths,London,1959
Bot 42.1951

NOTTINGHAM 1960 Digestive physiology and nutrition of the ruminant Easter school in agricultural sciences 7th Proceedings University of Nottingham Edited by D. Lewis Butterworth,London,1961
VA 19.0296

NOTTINGHAM 1961 Conference on anodising aluminium Proceedings Aluminium development association and University of Nottingham Aluminium development association, London,1961
Met 25.2047

NOTTINGHAM 1961 Conference on anodising aluminium Proceedings University of Nottingham Convened by the Aluminium development association Aluminium development association,London,1962
Eng 41.3602

NOTTINGHAM 1961 Inorganic polymers :an international conference Lectures University of Nottingham,and,Chemical society Chemical society.Special publication,15 Chemical society,London,1961
Chem 18.1221

NOTTINGHAM 1964 Experimental pedology Easter school in agricultural science 11th Proceedings University of Nottingham Edited by E.G. Hallsworth and D.V. Crawford xi,413p Butterworths,London,1965
Bot 42.2090

NOTTINGHAM 1964 International conference on magnetism Proceedings Institute of physics and the Physical society With the support of the International union of pure and applied physics illus 898p London,c1965
Cav 7.2882

NOTTINGHAM 1964 International conference on magnetism Proceedings Institute of physics and the Physical society London, c1964
Met 25.2354

NOTTINGHAM 1964 The Development of industrial societies :papers read at the Nottingham conference of the British sociological association British sociological association Edited by Paul Halmos Sociological review monograph, 8 University of Keele,Keele,1964
Eng 41.0914

NOTTINGHAM 1965 Science in the sixth form : a conference Sponsored by the Schools council Schools council publications.Working paper, 4 H.M.S.O.,London,1966
Eng 41.8193

NOTTINGHAM 1966 Adhesion:fundamentals and practice ;international conference Report Great Britain.Ministry of technology Elsevier,Amsterdam,1969
Met 25.2629

NOTTINGHAM 1966 Reproduction in the female mammals Easter school in agricultural science 13th Proceedings University of Nottingham Edited by G.E. Lamming and E.C. Amoroso Butterworths,London,1967
An 32.5323

NOTTINGHAM 1966 The Alkali metals :an international symposium Chemical society,and, University of Nottingham Chemical society. Special publication,22 Chemical society, London,1966
Chem 18.1226

NOTTINGHAM 1967 Efficient computer methods for the practising chemical engineer :a symposium Proceedings Edited by J.M. Pine Organised by the Institution of chemical engineers.Midlands branch Institution of chemical engineers.Symposium series, 23 illus 194p I.Chem.E.,London,1967
Chem E 24.1809

NOTTINGHAM 1967 Efficient computer methods for the practising chemical engineer :a symposium Proceedings Organised by the Institution of chemical engineers.Midlands branch Institution of chemical engineers. Symposium series, 23 Institution of chemical engineers,London,1967
Eng 41.5860

NOTTINGHAM 1967 Growth and development of mammals Easter school in agricultural science 14th Proceedings University of Nottingham.School of agriculture Edited by G. A. Lodge and G.E. Lamming Butterworths, London,1968
Gen 34.1592

NOTTINGHAM 1968 Conference on physical aspects of noise in electronic devices Papers Institute of physics and the Physical society.Electronics group iv,248p 31cm Peregrinus,Stevenage,1968
Cav 7.2840

NOTTINGHAM 1968 Root growth Easter school in agricultural science 15th Proceedings University of Nottingham Edited by William J. Whittington Bibliog., illus. xi,450p Butterworths,London,1969
Bot 42.1437

NOUMEA 1964 Phytochimie et plantes medicinales des terres du Pacifique :colloque international Centre national de la recherche scientifique Centre national de la recherche scientifique.Colloques internationaux,144 Editions du Centre national de la recherche scientifique,Paris, 1966
Chem 18.1296

NOUMEA,NEW CALEDONIA 1964 Phytochimie et plantes medicinales des terres du Pacifique colloque Proceedings Centre nationale de la recherche scientifique Centre national de la recherche scientifique.Colloques internationaux, 144 Editions du centre national de la recherche scientifique,Paris, 1966
Chem 26.0137

NOUVELLES PROPRIETES PHYSIQUES, MECANIQUES ET CHIMIQUES Colloque international Paris 1966 Sep 26-Oct 1 Centre national de la recherche scientifique Centre national de la recherche scientifique.Colloques internationaux, 167 Editions du centre national de la recherche scientifique,Paris, 1968
Met 25.1503

NOUVELLES PROPRIETES PHYSIQUES ET CHIMIQUES DES METAUX DE TRES HAUT PURETE colloque international Paris 1959 Oct 12-14 Centre national de la recherche scientifique Centre national de la recherche scientifique. Colloques internationaux, 40 Centre national de la recherche scientifique,Paris, 1960 English translation also available
Met 25.1499

NOVAE AND WHITE DWARFS Colloque international d'astrophysique Reports Paris 1939 Jul.17-23 1: observations of novae By Knut Lundmark and others Actualites scientifiques et industrielles, 901 College de France.Conferences, 13 140p Hermann,Paris,1941
Obs 6.1146

NOVAE AND WHITE DWARFS Colloque international d'astrophysique reports Paris 1939 Jul 17-23 2: nova theory supernovae By P. Swings and others Actualites scientifiques et industrielles, 902 College de France.Conferences, 14 267p Hermann,Paris,1941
Obs 6.1720

NOVAK J. ed. General topology and its relations to modern analysis and algebra : symposium proceedings Prague 1961 Sep 1-8 Ceskoslovenska akademie ved jointly with International mathematical union 363p 25cm Academic press,New York,1962
P. Math 2.0291

NOVOSIBIRSK 1961 Problemy okhrany prirody Sibiri i Dalnego Vostoka 4th Institut geologii i geofiziki,Novosibirsk Edited by G. V. Krylov illus 244p 27cm Izdatelstvo Sibirskogo otdelenie AN SSSR,Novosibirsk,1963
Sco 14.7681

NOVOSIBIRSK 1968 Colloquium on methods of optimization Edited by N.N. Moiseev Lecture notes in mathematics, 112 Illus. 293p 25cm Springer-Verlag,Berlin,1970
P Math 2.3531

NOVOSIBIRSK 1968 Colloquium on methods of optimization Edited by N.N. Moiseev Lecture notes in mathematics, 112 293p 26cm Springer,Berlin,1970
Math S 3.1688

NUCLEAR ACCIDENT DOSIMETRY SYSTEMS Panel on nuclear accident dosimetry systems Proceedings Vienna 1969 Feb 17-21 Organized by the International atomic energy agency International atomic energy agency. Panel proceedings series International atomic energy agency,Vienna,1970
Radioth 35.1384

NUCLEAR ACTIVITY The Control of nuclear activity :a symposium Papers Woods Hole, Mass. 1966 Aug 31-Sep 3 Edited by Lester Goldstein Held under the auspices of the Society of general physiologists Prentice-Hall,Englewood Cliffs,N.J.,1967 Original title of the symposium 'Cytoplasmic and environmental influences on nuclear behavior'
Bot 42.4736

NUCLEAR ACTIVITY The Control of nuclear activity :a symposium held under the auspices of the Society of general physiologists at its annual meeting at the Marine biological laboratory Woods Hole,Mass. 1966 Aug 31-Sep 3 Society of general physiologists Edited by Lester Goldstein Prentice-Hall, Englewood Cliffs,N.J.,1967
Bioch 33.1013

NUCLEAR AND MESON PHYSICS Conference on nuclear and meson physics Proceedings Glasgow 1954 Jul 13-17 Edited by E.H. Bellamy and R.G. Moorhouse Sponsored by the International union of pure and applied physics Pergamon press,London;New York,1955
Radioth 35.1563

NUCLEAR AND PARTICLE PHYSICS Summer institute in nuclear and particle physics Proceedings Toronto 1967 Canadian association of physicists.Theoretical physics division Edited by B. Margolis and C.S. Lam 547p 24cm Benjamin,New York,1968
Cav 7.3100

NUCLEAR DECAYS Conference on the electron capture and higher processes in nuclear decays Proceedings Debrecen 1968 Jul 15-18 Vol 1-3 International union of pure and applied physics Magyar tudomanyos akademia. Institute of nuclear research diagrs 20cm 3 vols Eotvos Lorand physical society, Budapest,1968
Cav 7.2700

NUCLEAR MAGNETIC RESONANCE :a conference Papers 1957 Nov 23 New York academy of sciences New York academy of sciences.Annals, 70,art 4 New York academy of sciences,New York,1958
Chem 18.2305

NUCLEAR METALLURGY A Symposium on behavior of materials in reactor environment 1956 Feb 20 By R.C. Dalzell and others American institute of mining and metallurgical engineers.Institute of metals division I.M.D. Special report series, 2 American institute of mining and metallurgical engineers,New York,1956 Lithoprinted
Met 25.1928

NUCLEAR METALLURGY a symposium 1955 Oct 17 American institute of mining and metallurgical engineers.Institute of metals division I.M.D.Special report series A.I.M. E.,New York,1955 Lithoprinted
Met 25.1573

NUCLEAR METHODS FOR MEASURING SOIL DENSITY AND MOISTURE Symposium on nuclear methods for measuring soil density and moisture Atlantic City,N.J. 1960 Jun 27 American society for testing materials American society for testing materials.Special technical publication, 293 American society for testing materials,Philadelphia,Pa.,1961
Eng 41.3152

NUCLEAR PHYSICS Low energy nuclear interactions and nuclear structure International conference on nuclear physics Proceedings Paris 1958 Jul 7-12 Faculte des sciences de Paris et de province and Centre national de la recherche scientifique Under the auspices of International union of pure and applied physics Crosby Lockwood, London,1959
Cav 7.1759

NUCLEAR SCIENCE AND ENGINEERING CORPORATION Radioactivity for pharmaceutical and allied research laboratories :a symposium Uniontown, Pa. Edited by Abraham Edelmann Academic press,New York;London,1960
Radioth 35.1725

NUCLEAR SCIENCE TEACHING Panel on nuclear science teaching Report Bangkok 1968 Jul 15-23 International atomic energy agency Unesco International atomic energy agency. Technical reports series, 94 International atomic energy agency,Vienna,1968
Radioth 35.1777

NUCLEAR SEX Symposium on nuclear sex 1st Proceedings London 1957 Edited by D. Robertson Smith and William M. Davidson Heinemann,London,1958
Inv Med 37.0197

NUCLEAR SPECTROSCOPY Selected topics in nuclear spectroscopy :N.U.F.F.I.C. international summer course in science, Nijenrode Castle... Proceedings Breukelen, Netherlands 1963 Jul 30-Aug 17 Netherlands universities foundation for international cooperation,and,Dutch physical society Edited by B.J. Verhaar With financial support from the North Atlantic treaty organization North-Holland,Amsterdam,1964
Chem 18.2279

NUCLEAR STRUCTURE International conference on nuclear structure Tokyo 1967 Sep Physical society of Japan illus vi,755p 26cm Physical society of Japan,Tokyo,1968
Cav 7.3196

NUCLEAR THEORY International summer school on selected topics in nuclear theory lectures Low Tatra mountains,Czechoslovakia 1962 Aug 20-Sep 8 By N. Austern and others Ceskoslovenska akademie ved.Ustav jaderneho vyzkumu Edited by F. Janouch With the co-operation of the International atomic energy agency 452p International atomic energy agency,Vienna,1963
A Math 4.0885

NUCLEATION AND CRYSTALLIZATION IN GLASSES AND MELTS :a symposium Toronto 1961 Apr American ceramic society Edited by Margie K. Reser and others American ceramic society,Columbus,Ohio,1962
Met 25.0127

NUCLEIC ACID Cambridge 1946 Jul Society for experimental biology Society for experimental biology.Symposia, 1 Cambridge university press,Cambridge,1947
Phys 20.0942

NUCLEIC ACID :a symposium Cambridge 1946 Jul Society for experimental biology Society for experimental biology.Symposia, 1 Cambridge university press,Cambridge,1947
Bioch 33.1348

NUCLEIC ACID :a symposium Cambridge 1946 Jul Society for experimental biology Society for experimental biology.Symposia, 1 Cambridge university press,Cambridge,1947
Bot 42.1722

NUCLEIC ACID :a symposium Cambridge 1946 Jul Society for experimental biology Society for experimental biology.Symposia, 1 Cambridge university press,Cambridge,1947
Radioth 35.0009

NUCLEIC ACID :symposium Cambridge 1946 Jul Society for experimental biology Society for experimental biology.Symposia, 1 Cambridge university press,Cambridge,1947
Gen 34.0683

NUCLEIC ACID 1 :a symposium Papers Cambridge 1946 Jul Society for experimental biology Society for experimental biology.Symposia,1 Cambridge university press,Cambridge,1947
Chem 18.1554

NUCLEIC ACID METABOLISM,CELL DIFFERENTIATION AND CANCER GROWTH International symposium for cellular chemistry 2nd Proceedings Ohtsu 1966 Oct 17-21 Edited by E.V. Cowdry and S. Seno Organised by the Japan society for cell biology Pergamon,London,1969
Radioth 35.0561

NUCLEIC ACID-PROTEIN INTERACTIONS AND NUCLEIC ACID SYNTHESIS IN VIRAL INFECTION : symposium Miami,Fla. 1971 Jan 18-22 University of Miami.Department of biochemistry Papanicolaou cancer research institute Miami winter symposia, 2 North-Holland,Amsterdam, 1971
Bioch 33.2241

NUCLEIC ACID SYMPOSIUM Papers 1st Cambridge 1946 Jul Society for experimental biology Society for experimental biology.Symposia, 1 plates 290p Cambridge university press,Cambridge, 1947
Cav 7.0459

NUCLEIC ACIDS Complexes of biologically active substances with nucleic acid and their mode of action :a research symposium Proceedings Washington,D.C. 1970 Mar 16-19 Walter Reed army institute of research Edited by F.E. Hahn Progress in molecular and subcellular biology, 2 Springer,Berlin, 1971
Bioch 33.2242

NUCLEIC ACIDS The Structure of nucleic acids and their role in protein synthesis :a symposium London 1956 Feb 18 Biochemical society Edited by E.M. Crook Held at the London school of hygiene and tropical medicine Biochemical society.Symposia, 14 Cambridge university press,Cambridge,1957
Bioch 33.1376

NUCLEIC ACIDS AND NUCLEOPROTEINS Cold Spring Harbor 1947 Jun 11-20 Cold Spring Harbor laboratory of quantitative biology Cold Spring Harbor symposia on quantitative biology, 12 Port Biological laboratory,New York, 1947
Radioth 35.0157

NUCLEIC ACIDS AND NUCLEOPROTEINS :a symposium Papers Cold Spring Harbor 1947 Cold Spring Harbor biological laboratory Cold Spring Harbor symposia on quantitative biology, 12 Long Island biological association,Cold Spring Harbor,1947
Bioch 33.1268

NUCLEIC ACIDS IN IMMUNOLOGY :a symposium Proceedings New Brunswick,N.J. Edited by O.J. Plescia and W. Braun Held at Rutgers university.Institute of microbiology Springer,NEW York,1968
Radioth 35.0152

NUCLEOHISTONES The World conference on histone biology and chemistry 1st 1963 Apr 29 - May 3 Edited by James Bonney and Paul Ts'o Holden Day,San Francisco,Calif., 1964
Radioth 35.0121

NUCLEOHISTONES The World conference on histone biology and chemistry 1st Papers 1963 Apr 29-May 2 Edited by James Bonner and Paul Ts'o Holden-Day,San Francisco,1964
Bioch 33.0808

NUCLEOHISTONES World conference on histone biology and chemistry 1st 1963 Apr 29-May 3 Edited by James Bonner and Paul Ts'o Holden-Day,San Francisco,Calif.,1964
Gen 34.0684

NUCLEOLUS International symposium on nucleolus:its structure and function Proceedings Montevideo 1965 Dec 5-10 National cancer institute Edited by W.S. Vincent and others National cancer institute. Monograph, 23 National cancer institute, Bethesda,Md.,1966
Bioch 33.0988

The NUCLEOLUS:ITS STRUCTURE AND FUNCTION : international symposium Proceedings Montevideo,Uruguay 1965 Dec 5-10 Edited by W.S. Vincent and O.L. Miller National cancer institute.Monographs, 23 National cancer institute,Bethesda,Md.,1966 In Spanish and English
Radioth 35.0533

NUCLEOPROTEINES Conseil de chimie 11eme Brussels 1959 Jun 1-6 Institut international de chimie Solvay Edited by R. Stoops Interscience;Stoops,New York;Brussels, 1960
Radioth 35.0100

NUCLEOSYNTHESIS :a conference Proceedings New York 1965 Jan 25-26 Goddard space flight center and Goddard institute for space studies Edited by W.David Arnett and others Gordon and Breach,New York,1968
TA 15.0306

NUCLEOSYNTHESIS :conference Proceedings Greenbelt,Md. 1965 Jan 25-29 Goddard institute for space studies Edited by W.D. Arnett and others 273p Gordon and Breach, New York,1968
Obs 6.3363

NUCLEOTIDES Neurochemistry of nucleotides and amino acids :a symposium Papers and discussions presented Philadelphia 1958 Apr 24-25 American academy of neurology. Section of neurochemistry Edited by Roscoe O. Brady and Donald B. Tower Wiley,New York, 1960
Bioch 33.2222

NUCLEOTIDES Role of nucleotides for the function and conformation of enzymes The Alfred Benzon symposium 1 Proceedings Copenhagen 1968 Sep 9-11 Alfred Benzon foundation Edited by Herman M. Kalckar Held at the premises of the Royal Danish academy of sciences and letters Munksgaard, Copenhagen,1969
.Bioch 33.1913

NUCLEUS International symposium on the nucleus;its structure and function Montevideo 1965 Dec 5-10 Edited by W.S. Vincent and O.L. Miller National concer institute.Monographs, 23 U.S.government printing office,Washington,D.C.,1966
An 32.3439

NUCLEUS The Relationship between nucleus and cytoplasm :a symposium Proceedings Brussels 1958 Jan 9-13 Organised by the International society for cell biology Experimental cell research.Supplement, 6 Academic press,New York,1959
An 32.2486

NUFFIC INTERNATIONAL SUMMER COURSE IN SCIENCE Lecture notes Present problems concerning the structure and evolution of the galactic system Breukelen 1960 Jul. 28- Aug. 16 Netherlands universities foundation for international cooperation Edited by J.H. Oort and H.G. Quik Supported by North atlantic treaty organization The Hague,1960
Obs 6.2067

NUFFIC INTERNATIONAL SUMMER SESSION IN SCIENCE papers Asymptotic distribution modulo 1 Breukelen 1962 Aug 1-11 Netherlands universities foundation for international cooperation Edited by J.F. Koksma and L. Kuipers With the financial support of North Atlantic treaty organization Bibliog. 203p 24cm Noordhoff,Groningen,1964
P. Math 2.1826

NUFFIC SUMMER SCHOOL ON MANIFOLDS Proceedings Manifolds - Amsterdam 1970 Amsterdam 1970 Aug 17-29 Edited by N.H. Kuiper Lecture notes in mathematics, 197 230p 25cm Springer,Berlin,1971
P Math 2.4116

NUFFIELD FOUNDATION.ADVISORY COMMITTEE Pathology of laboratory rats and mice London 1966 Apr 19-21 Edited by Ernest Cotchin and Francis J.C. Roe Held at the Royal veterinary field station Blackwell,Oxford, 1967
Path 30.2438

NUFFIELD PROVINCIAL HOSPITALS TRUST A Balanced teaching hospital :a symposium Birmingham 1963 Sep Edited by Thomas McKeown viii,131p 28cm Oxford university press,London,1965
PGMS 29.0500

NUFFIELD RESEARCH GROUP IN EXTRACTION METALLURGY Physical chemistry of melts Symposium on the nature of molten slags and salts London 1952 Feb 20 Institution of mining and metallurgy,London,1953
Met 25.1787

NUMBER THEORY Computers in number theory Science research council Atlas symposium 2nd Proceedings Oxford 1969 Aug 18-23 Science research council Edited by A.O.L. Atkin and B.J. Birch xvii,433p 24cm Academic press,London;New York,1971
P Math 2.4381

NUMBER THEORY Institute in the theory of numbers report Boulder,Colo. 1959 Jun 21-Jul 17 American mathematical society Edited by Donald C.B. Marsh and James H. Jordan With the support of the National science foundation 350p 27cm University of Colorado,Boulder,Colo.,1959
P. Math 2.1773

NUMBER THEORY Number theory conference proceedings Boulder,Colo. 1963 Aug 5-24 University of Colorado With the support of the National science foundation 121p 27cm University of Colorado,Boulder,Colo.,1963
P. Math 2.1772

NUMBER THEORY Seminar on modern methods in number theory Tokyo 1971 Aug 30-Sep 4 Held at the Institute of statistical mathematics,Tokyo 29cm Institute of statistical mathematics,Tokyo,1971
P Math 2.4380

NUMBER THEORY 1969 number theory institute Summer institute on number theory Stony Brook, N.Y. 1969 Jul 7-Aug 1 American mathematical society Edited by Donald J. Lewis American mathematical society. Proceedings of symposia in pure mathematics, 20 xiii,451p 26cm AMS,Providence,R.I., 1971
P Math 2.4383

NUMBER THEORY :a colloquium held at Debrecen 1968 Apr 4-8 Edited by Paul Turan Colloquia mathematica societatis Janos Bolyai, 2 bibliog. 244p 24cm North-Holland, Amsterdam,1970
P Math 2.3898

NUMBER THEORY :a symposium Proceedings Houston,Tex. 1967 Jan 24-28 American mathematical society Edited by William J. Leveque and Ernst G. Straus American mathematical society.Proceedings of symposia in pure mathematics, 12 Bibliog. v,98p 26cm American mathematical society, Providence,R.I.,1969 Special session on number theory at the 73rd annual meeting of the A.M.S.
P Math 2.3699

NUMBER THEORY CONFERENCE proceedings Boulder,Colo. 1963 Aug 5-24 University of Colorado With the support of the National science foundation 121p 27cm University of Colorado,Boulder,Colo.,1963
P. Math 2.1772

NUMERICAL ANALYSIS Conference on applications of numerical analysis Proceedings Dundee 1971 Mar 23-26 Edited by John L. Morris Lecture notes in mathematics, 228 x,358p 25cm Springer, Berlin,1971
P Math 2.4463

NUMERICAL ANALYSIS Numerical analysis:an introduction Symposium on numerical analysis Birmingham 1965 Institute of mathematics and its applications Edited by J. Walsh bibliog.,diags. xiv,212p 24cm Academic press,London,1966
Math L 5.0357

NUMERICAL ANALYSIS Symposium on questions of numerical analysis proceedings Rome 1958 Jun.30-Jul.1 Provisional international computation centre 79p Rome,1958
Math L 5.2374

NUMERICAL ANALYSIS Symposium on the theory of numerical numbers Proceedings Dundee 1970 Sep 15-23 Edited by John L. Morris Lecture notes in mathematics, 193 vi,152p 25cm Springer,Berlin,1971
P Math 2.4462

NUMERICAL ANALYSIS :an introduction Birmingham 1965 Institute of mathematics and its applications Edited by Joan Walsh 212p Academic press,London;New York,1966 Based on a symposium held in Birmingham in 1965
A Math 4.1219

NUMERICAL ANALYSIS :an introduction Edited by J. Walsh Academic press,London;New York,1966 Based on a symposium held in Birmingham in 1965
Eng 41.2326

NUMERICAL ANALYSIS 6th :symposium Proceedings Santa Monica,Calif. 1953 Aug 26-28 American mathematical society Edited by John H. Curtiss Cosponsored by National bureau of standards American mathematical society.Proceedings of symposia in applied mathematics, 6 vi,303p McGraw-Hill,New York,1956
Math L 5.3245

NUMERICAL APPROXIMATION On numerical approximation symposium proceedings Madison,Wis. 1958 Apr. 21-23 United States. Army.Mathematics research center Edited by Rudolph E. Langer United States army. Mathematics research center.Publications, 1 x,462p 24cm University of Wisconsin press, Madison,Wis.,1959
Math L 5.0267

NUMERICAL MATHEMATICS Frontiers of numerical mathematics :symposium Madison,Wis. 1959 Oct 30-31 United States.Army.Mathematics research center National bureau of standards United States army.Mathematics research center. Publications, 4 xi,132p 24cm University of Wisconsin,Madison,Wis.,1960
Math L 5.3232

NUMERICAL MATHEMATICS Funktionalanalysis approximationstheorie numerische mathematik : vortragsauszuge der tagung uber numerische probleme in der approximationstheorie vom 22. bis 25.Juni 1965,und der tagung uber funktionalanalytische methoden in der numerischen mathematik vom 15.bis 20 November, 1965 Oberwolfach 1965 Edited by L. Collatz and others International series of numerical mathematics, 7 Bibliog. 232p 24cm Birkhauser Verlag,Basel,1967
P Math 2.3448

NUMERICAL METHODS FOR UNCONSTRAINED OPTIMIZATION Joint institute of mathematics and its application-National physical laboratory conference Proceedings Teddington 1971 Jan 7-8 National physical laboratory Institute of mathematics and its applications Edited by W. Murray xi,144p 24cm Academic press,London;New York,1972
Math S 3.1729

NUMERICAL SOLUTION OF ORDINARY AND PARTIAL DIFFERENTIAL EQUATIONS based on a Summer school Oxford 1961 Aug-Sep University of Oxford.Computing laboratory Edited by L. Fox Adiwes international series in the engineering sciences Addison-Wesley series in computer science and information processing Pergamon press;Addison-Wesley,London;Reading, Mass.,1962
TA 15.0378

NUMERICAL SOLUTION OF ORDINARY AND PARTIAL DIFFERENTIAL EQUATIONS based on a summer school held at Oxford August-September 1961 Oxford 1961 Aug-Sep Oxford university. Computing laboratory Oxford,1962
Geod 9.0179

NUMERICAL SOLUTION OF ORDINARY AND PARTIAL DIFFERENTIAL EQUATIONS summer school Oxford 1961 Aug.-Sep. By L. Fox University of Oxford.Computing laboratory bibliog. ix,509p 23cm Pergamon press, Oxford,1962
Math L 5.0308

NUMERICAL SOLUTION OF PARTIAL DIFFERENTIAL EQUATIONS :a symposium Proceedings College Park,Md. 1965 May 3-8 Edited by James H. Bramble Academic press,New York; London,1966
Eng 41.2331

NUMERICAL SOLUTION OF PARTIAL DIFFERENTIAL EQUATIONS :symposium Proceedings College Park,Md. 1965 May 3-8 Edited by James H. Bramble 373p 23cm Academic press,New York,1966
Math L 5.3110

NUMERICAL SOLUTIONS OF NONLINEAR DIFFERENTIAL EQUATIONS :advanced symposium proceedings Madison,Wis. 1966 May 9-11 United States.Army.Mathematics research center Edited by Donald Greenspan United States. Army.Mathematics research center.Publications, 17 John Wiley,New York,1966
A Math 4.1220

NUMERICAL SOLUTIONS OF NONLINEAR DIFFERENTIAL EQUATIONS advanced seminar proceedings Madison,Wis. 1966 May 9-11 United States. Army.Mathematics research center Edited by Donald Greenspan United States army. Mathematics research center.Publication, 17 x,343p 23cm John Wiley and sons,New York, 1966
Math L 5.0358

NUMERICAL TAXONOMY Colloquium in numerical taxonomy Proceedings St.Andrews 1968 Sep Edited by A.J. Cole Academic press,London, 1969
HE 27.0022

NUMERICAL TAXONOMY Colloquium in numerical taxonomy Proceedings St.Andrews 1968 Sep. Edited by A.J. Cole Academic press,London, 1969
Math L 5.3440

NUMERICAL TAXONOMY Colloquium in numerical taxonomy Proceedings St.Andrews 1968 Sep 2-4 Edited by A.J. Cole Academic press, London;New York,1969
Gen 34.2149

NUMERICAL TAXONOMY :colloquium Proceedings St.Andrews 1968 Sep University of St. Andrews Edited by Alfred J. Cole Academic press,London,1969
Bot 42.3452

NUMERICAL WEATHER PREDICTION The Symposium on numerical weather prediction Proceedings Tokyo 1960 Nov 7-13 Meteorological agency,Japan and Science council of Japan Sponsored jointly by International union of geodesy and geophysics 655p Meteorological society of Japan,Tokyo,1962
Nap 11.0421

NUMERISCHE LOSUNG NICHTLINEAR PARTIELLER DIFFERENTIAL- UND INTEGRODIFFERENTIALGLEICHUNGEN :conferenz Proceedings Oberwolfach 1971 Nov 28-Dec 4 Edited by R. Ansorge Lecture notes in mathematics, 267 vi,339p 25cm Springer, Berlin,1972
P Math 2.4294

The NUTRICIA SYMPOSIUM ON THE ADAPTION OF THE NEWBORN INFANT TO EXTRA UTERINE LIFE Groningen 1964 Feb Edited by J.H.P. Jonxis and others Kroese,Leiden,1964
An 32.1516

NUTRITION AND GROWTH FACTORS International congress of microbiology 6th Rome 1953 Vol 3: symposium - nutrition and growth factors Edited by W.H. Schopfer and D. D. Woods Fondazione Emanuele Paterno,Rome, 1953 Title also in Italian;text in English and French
Bioch 33.1802

NUTRITION AND GROWTH FACTORS :a symposium Nutrizione e fattori di crescita Edited by W. H. Schopfer and D.D. Woods International congress of microbiology, 6,2 Rome,1953
Bot 42.0665

NUTRITION IN RENAL DISEASE :a conference Proceedings Manchester 1967 Jun 29-30 Edited by Geoffrey Merton Berlyne Livingstone;Williams,Edinburgh;Baltimore,1968
Inv Med 37.0021

NUTRITION OF THE LEGUMES Easter school in agricultural science 5th Proceedings Nottingham 1958 University of Nottingham Edited by E.G. Hallsworth Butterworths, London,1958
Bioch 33.0776

NUTRITION OF THE LEGUMES Easter school in agricultural sciences 5th Proceedings Nottingham 1958 Edited by E.G. Hallsworth Butterworths,London,1958
Bot 42.1781

NUTRITION SOCIETY Lind bicentenary symposium a conference on scurvy and vitamin C in honour of James Lind Proceedings Edinburgh 1953 May 22-23 Nutrition society.Proceedings, 12,3 Cambridge university press,London,1953
Sco 14.0848

NUTRIZIONE E FATTORI DI CRESCITA Nutrition and growth factors :a symposium Edited by W. H. Schopfer and D.D. Woods International congress of microbiology, 6,2 Rome,1953
Bot 42.0665

NYASALAND ECONOMIC SYMPOSIUM Papers Economic development in Africa Blantyre 1962 Jul 18-28 Edited by E.F. Jackson Blackwell,Oxford,1965
Geog 13.4650

OAK RIDGE 1955 Metallurgy information meeting 5th Proceedings Pt 1-2 United States atomic energy commission 2 vols Washington,1960
Met 25.1552

OAK RIDGE,TENN. 1948 Discussion on the present status of radiation genetics :given at the information meeting for biology and medicine of the Atomic energy commission Sponsored by Oak Ridge national laboratory. Biological division Wistar institute of anatomy and biology,Philadelphia,Pa.,1950 Reprinted from the Journal of cellular and comparative physiology,Vol 35,suppt 1, 1950
An 32.0357

OAK RIDGE,TENN. 1948 Discussion on the present status of radiation genetics information meeting... Sponsored by the Oak Ridge national laboratory Oak Ridge national laboratory.Biology division.Symposia, 1 Journal of cellular and comparative physiology, 35,supp Wistar institute of anatomy and biology,Philadelphia,Pa.,1950
Gen 34.1103

OAK RIDGE,TENN. 1953 Symposium on effects of radiation and other deleterious agents on embryonic development Sponsored by the Oak Ridge national laboratory.Biology division Wistar institute of anatomy and biology, Philadelphia,Pa.,1954 Reprinted from the Journal of cellular and comparative physiology Vol 43,Suppt 1,May 1954
An 32.2027

OAK RIDGE,TENN. 1954 Symposium on genetic recombination given at research conference for biology and medicine of the atomic energy commission Sponsored by the Oak Ridge national laboratory.Biology division Oak Ridge national laboratory.Symposia, 7 Journal of cellular and comparative physiology, 45,supp.2 Wistar institute of anatomy and biology,Philadelphia,Pa.,1955
Gen 34.1042

OAK RIDGE,TENN. 1954 Symposium on genetic recombination given at the Research conference for biology and medicine of the Atomic Energy commission Sponsored by Oak Ridge national laboratory.Biology division Wistar institute of anatomy and biology,Philadelphia,Pa.,1955 Reprinted from the Journal of cellular and comparative physiology,Vol.45, suppt.2,May 1955
An 32.0346

OAK RIDGE,TENN. 1956 Roentdens,rads and riddles :a symposium on supervoltage radiation therapy United States atomic energy commission Edited by Milton Friedman and others Held at the Oak Ridge institute of nuclear studies U.S.atomic energy commission, Washington,D.C.,1959
Radioth 35.1090

OAK RIDGE,TENN. 1963 Dynamic clinical studies with radioisotopes :a symposium held at the Oak Ridge institute of nuclear studies Proceedings Edited by R.M. Kniseley and W.N. Tauke Sponsored by the United States atomic energy commission.Division of technical information U.S.A.E.C.,Washington,D.C.,1964
Radioth 35.1403

OAK RIDGE,TENN. 1966 Compartments pools and spaces in medical physiology :a symposium Edited by Per.Erik E. Bergner and others Held at the Oak Ridge institute of nuclear studies United States atomic energy commission.Symposium series, 11 U.S.atomic energy commission,1967
Radioth 35.0927

OAK RIDGE INSTITUTE OF NUCLEAR STUDIES Compartments pools and spaces in medical physiology :a symposium Oak Ridge,Tenn. 1966 Oct 24-27 Edited by Per.Erik E. Bergner and others United States atomic energy commission.Symposium series, 11 U.S.atomic energy commission,1967
Radioth 35.0927

OAK RIDGE INSTITUTE OF NUCLEAR STUDIES Roentdens,rads and riddles :a symposium on supervoltage radiation therapy Oak Ridge, Tenn. 1956 Jul 15-18 United States atomic energy commission Edited by Milton Friedman and others U.S.atomic energy commission, Washington,D.C.,1959
Radioth 35.1090

OAK RIDGE NATIONAL LABORATORY Diffusion in body-centered cubic metals International conference on body-centred cubic materials Papers Gatlinburg 1964 Sep 16-18 American society for metals, Metals Park, Ohio, 1965
Met 25.1283

OAK RIDGE NATIONAL LABORATORY Macromolecular aspects of the cell cycle :a symposium Gatlinburg, Tenn. 1963 Apr 8-11 Wistar institute press, Philadelphia, Pa., 1963
An 32.3297

OAK RIDGE NATIONAL LABORATORY Symposium on information theory in biology Gatlinburg, Tenn. 1956 Oct 29-31 Edited by H.P. Yockey and others London, 1958 Consists of articles representing authors' results and opinions and additional papers not given at the symposium "The conference was entitled A symposium on information theory in health physics and radiobiology"
Bal 39.1006

OAK RIDGE NATIONAL LABORATORY.ANNUAL RESEARCH CONFERENCE, 18 Symposium on hormonal control of protein biosynthesis Gatlinburg, Tenn. 1965 Apr 5-8 Oak Ridge national laboratory.Biology division Wistar institute of anatomy and biology, Philadelphia, Pa., 1965
Gen 34.0550

OAK RIDGE NATIONAL LABORATORY.BIOLOGY DIVISION Lectures in biological sciences :a symposium Knoxville, Tenn. 1959 Dec 3-5 By Norman G. Anderson and others Edited by J.Ives Townsend 110p University of Tennessee press, Knoxville, Tenn., 1963
Bot 42.4744

OAK RIDGE NATIONAL LABORATORY.BIOLOGY DIVISION Symposium on chromosome mechanics at the molecular level :given at research conference for biology and medicine of the atomic energy commission Gatlinburg, Tenn. 1967 Apr 10-13 Oak Ridge national laboratory.Symposia, 20 Journal of cellular physiology, 70, suppl Wistar institute of anatomy and biology, Philadelphia, Pa., 1967
Gen 34.0846

OAK RIDGE NATIONAL LABORATORY.BIOLOGY DIVISION Symposium on differentiation and growth of haemoglobin-and immunoglobin-synthesising cells :given at research conference for biology and medicine of the Atomic energy commission Gatlinburg, Tenn. 1966 Apr 4-7 Oak Ridge national laboratory.Biology division. Symposia, 19 Journal of cellular physiology, 67, supp.1 Wistar institute of anatomy and biology, Philadelphia, Pa., 1966
Gen 34.0504

OAK RIDGE NATIONAL LABORATORY.BIOLOGY DIVISION Symposium on effects of radiation and other deleterious agents on embryonic development Oak Ridge, Tenn. 1953 Apr 20-21 Wistar institute of anatomy and biology, Philadelphia, Pa., 1954 Reprinted from the Journal of cellular and comparative physiology Vol 43, Suppt 1, May 1954
An 32.2027

OAK RIDGE NATIONAL LABORATORY.BIOLOGY DIVISION Symposium on enzyme reaction mechanisms Research conference for biology and medicine 12th Gatlinburg, Tenn. 1959 Apr 1-4 Journal of cellular and comparative physiology, 54, suppl.1 Wistar institute of anatomy and biology, Philadelphia, Pa., 1959
Bioch 33.1046

OAK RIDGE NATIONAL LABORATORY.BIOLOGY DIVISION Symposium on genetic approaches to somatic cell variation Gatlinburg, Tenn. 1958 Apr 2-5 Oak Ridge national laboratory.Symposia, 11 Journal of cellular and comparative physiology, 52, supp.1 Wistar institute of anatomy and biology, Philadelphia, Pa., 1958
Gen 34.1094

OAK RIDGE NATIONAL LABORATORY.BIOLOGY DIVISION Symposium on genetic recombination given at research conference for biology and medicine of the atomic energy commission Oak Ridge, Tenn. 1954 Apr 19-21 Oak Ridge national laboratory.Symposia, 7 Journal of cellular and comparative physiology, 45, supp.2 Wistar institute of anatomy and biology, Philadelphia, Pa., 1955
Gen 34.1042

OAK RIDGE NATIONAL LABORATORY.BIOLOGY DIVISION Symposium on genetic recombination given at the Research conference for biology and medicine of the Atomic Energy commission Oak Ridge, Tenn. 1954 Apr 19-21 Wistar institute of anatomy and biology, Philadelphia, Pa., 1955 Reprinted from the Journal of cellular and comparative physiology, Vol.45, suppt.2, May 1955
An 32.0346

OAK RIDGE NATIONAL LABORATORY.BIOLOGY DIVISION Symposium on hormonal control of protein biosynthesis Gatlinburg, Tenn. 1965 Apr 5-8 Oak Ridge national laboratory.Annual research conference, 18 Wistar institute of anatomy and biology, Philadelphia, Pa., 1965
Gen 34.0550

OAK RIDGE NATIONAL LABORATORY.BIOLOGY DIVISION Symposium on hormonal control of protein biosynthesis Research conference for biology and medicine 18th Gatlinburg, Tenn. 1965 Apr 5-8 Journal of cellular and comparative physiology, 66, suppl.1 Wistar institute of anatomy and biology, Philadelphia, Pa., 1965
Bioch 33.0520

OAK RIDGE NATIONAL LABORATORY.BIOLOGY DIVISION Symposium on macromolecular aspects of the cell cycle :given at research conference for biology and medicine of the atomic energy commission Gatlinburg, Tenn 1963 Apr 8-11 Oak Ridge national laboratory.Symposia, 16 Journal of cellular and comparative physiology, 62, supp.1 Wistar institute of anatomy and biology, Philadelphia, Pa., 1963
PGMS 29.0414

OAK RIDGE NATIONAL LABORATORY.BIOLOGY DIVISION Symposium on macromolecular aspects of the cell cycle :given at research conference for biology and medicine of the atomic energy commission Gatlinburg, Tenn. 1963 Apr 8-11 Oak Ridge national laboratory.Symposia, 16 Journal of cellular and comparative physiology, 62, no.2 Wistar institute of anatomy and biology, Philadelphia, Pa., 1963
Gen 34.0842

OAK RIDGE NATIONAL LABORATORY.BIOLOGY DIVISION
Symposium on mammalian genetics and reproduction Gatlinburg,Tenn. 1960 Apr 4-7 Oak Ridge national laboratory.Symposia, 13 Journal of cellular and comparative physiology, 56,supp.1 Wistar institute of anatomy and biology,Philadelphia,Pa.,1960
Gen 34.1782

OAK RIDGE NATIONAL LABORATORY.BIOLOGY DIVISION
Symposium on molecular action of mutagenic and carcinogenic agents Gatlinburg,Tenn. 1964 Apr 6-9 Oak Ridge national laboratory. Symposia Journal of cellular and comparative physiology, 64,suppl.1 Wistar institute of anatomy and biology,Philadelphia,Pa.,1964
Gen 34.1098

OAK RIDGE NATIONAL LABORATORY.BIOLOGY DIVISION
Symposium on molecular aspects of differentiation :given at research conference for biology and medicine of the Atomic energy commission Gatlinburg,Tenn. 1968 Apr 8-11 Oak Ridge national laboratory.Biology division. Symposia, 21 Journal of cellular physiology, 72,supp.1 Wistar institute of anatomy and biology,Philadelphia,Pa.,1968
Gen 34.0505

OAK RIDGE NATIONAL LABORATORY.BIOLOGY DIVISION
Symposium on recovery of cells from injury : given at research conference for biology and medicine of the atomic energy commission Gatlinburg,Tenn. 1961 Apt 3-6 Journal of cellular and comparative physiology, 58,supp. 1 Oak Ridge national laboratory.Symposia, 14 Wistar institute of anatomy and biology, Philadelphia,Pa.,1961
Gen 34.1114

OAK RIDGE NATIONAL LABORATORY.BIOLOGY DIVISION. SYMPOSIA, 1 Discussion on the present status of radiation genetics information meeting... Oak Ridge,Tenn. 1948 Mar 26-27 Sponsored by the Oak Ridge national laboratory. Journal of cellular and comparative physiology, 35,supp Wistar institute of anatomy and biology,Philadelphia,Pa.,1950
Gen 34.1103

OAK RIDGE NATIONAL LABORATORY.BIOLOGY DIVISION. SYMPOSIA, 19 Symposium on differentiation and growth of haemoglobin-and immunoglobin-synthesising cells :given at research conference for biology and medicine of the Atomic energy commission Gatlinburg, Tenn. 1966 Apr 4-7 Sponsored by the Oak Ridge national laboratory.Biology division Journal of cellular physiology, 67,supp.1 Wistar institute of anatomy and biology, Philadelphia,Pa.,1966
Gen 34.0504

OAK RIDGE NATIONAL LABORATORY.BIOLOGY DIVISION. SYMPOSIA, 21 Symposium on molecular aspects of differentiation :given at research conference for biology and medicine of the Atomic energy commission Gatlinburg,Tenn. 1968 Apr 8-11 Sponsored by the Oak Ridge national laboratory.Biology division Journal of cellular physiology, 72,supp.1 Wistar institute of anatomy and biology,Philadelphia, Pa.,1968
Gen 34.0505

OAK RIDGE NATIONAL LABORATORY.SYMPOSIA
Symposium on molecular action of mutagenic and carcinogenic agents Gatlinburg,Tenn. 1964 Apr 6-9 Sponsored by the Oak Ridge national laboratory.Biology division Journal of cellular and comparative physiology, 64,suppl. 1 Wistar institute of anatomy and biology, Philadelphia,Pa.,1964
Gen 34.1098

OAK RIDGE NATIONAL LABORATORY.SYMPOSIA, 7
Symposium on genetic recombination given at research conference for biology and medicine of the atomic energy commission Oak Ridge, Tenn. 1954 Apr 19-21 Sponsored by the Oak Ridge national laboratory.Biology division Journal of cellular and comparative physiology, 45,supp.2 Wistar institute of anatomy and biology,Philadelphia,Pa.,1955
Gen 34.1042

OAK RIDGE NATIONAL LABORATORY.SYMPOSIA, 11
Symposium on genetic approaches to somatic cell variation Gatlinburg,Tenn. 1958 Apr 2-5 Sponsored by Oak Ridge national laboratory. Biology division Journal of cellular and comparative physiology, 52,supp.1 Wistar institute of anatomy and biology,Philadelphia, Pa.,1958
Gen 34.1094

OAK RIDGE NATIONAL LABORATORY.SYMPOSIA, 13
Symposium on mammalian genetics and reproduction Gatlinburg,Tenn. 1960 Apr 4-7 Sponsored by the Oak Ridge national laboratory. Biology division Journal of cellular and comparative physiology, 56,supp.1 Wistar institute of anatomy and biology,Philadelphia, Pa.,1960
Gen 34.1782

OAK RIDGE NATIONAL LABORATORY.SYMPOSIA, 14
Symposium on recovery of cells from injury : given at research conference for biology and medicine of the atomic energy commission Gatlinburg,Tenn. 1961 Apt 3-6 Sponsored by the Oak Ridge national laboratory.Biology division Journal of cellular and comparative physiology, 58,supp.1 Wistar institute of anatomy and biology,Philadelphia,Pa.,1961
Gen 34.1114

OAK RIDGE NATIONAL LABORATORY.SYMPOSIA, 16
Symposium on macromolecular aspects of the cell cycle :given at research conference for biology and medicine of the atomic energy commission Gatlinburg,Tenn 1963 Apr 8-11 Sponsored by Oak Ridge national laboratory. Biology division Journal of cellular and comparative physiology, 62,supp.1 Wistar institute of anatomy and biology,Philadelphia, Pa.,1963
PGMS 29.0414

OAK RIDGE NATIONAL LABORATORY.SYMPOSIA, 16
Symposium on macromolecular aspects of the cell cycle :given at research conference for biology and medicine of the atomic energy commission Gatlinburg,Tenn. 1963 Apr 8-11 Sponsored by Oak Ridge national laboratory. Biology division Journal of cellular and comparative physiology, 62,no.2 Wistar institute of anatomy and biology,Philadelphia, Pa.,1963
Gen 34.0842

OAK RIDGE NATIONAL LABORATORY.SYMPOSIA, 20
Symposium on chromosome mechanics at the molecular level :given at research conference for biology and medicine of the atomic energy commission Gatlinburg,Tenn. 1967 Apr 10-13 Sponsored by the Oak Ridge national laboratory. Biology division Journal of cellular physiology, 70,suppl Wistar institute of anatomy and biology,Philadelphia,Pa.,1967
Gen 34.0846

OBERGURGL 1962 Variations of the regime of existing glaciers International association of scientific hydrology :a symposium Proceedings Edited by W. Ward International association of scientific hydrology.Publication,58 312p Gentbrugge, 1962
Sco 14.0137

OBERLIN,OHIO 1950 Symposium on radiobiology :the basic aspects of radiation effects on living systems Edited by James J. Nickson Sponsored by the National research council Wiley;Chapman and Hall,New York; London,1952
Radioth 35.1717

OBERWOLFACH 1964 Algebraische zahlentheorie conference an account Mathematisches forschungsinstitut,Oberwolfach Edited by Helmut Hasse and Peter Roquette Mathematisches forschungsinstitut,Oberwolfach, 2 264p 19cm Bibliographisches institut, Mannheim,1966
P. Math 2.1769

OBERWOLFACH 1966 Mathematische statistik und wahrscheinlichkeitstheorie conference Mathematisches forschungsinstitut,Oberwolfach Oberwolfach,1966 Mimeograph resumes of papers
Math 3.1120

OBERWOLFACH 1967 Funktionalanalytische methoden der numerischen mathematik :ein tagung Vortragsauszuge Edited by L. Collatz and H. Unger International series of numerical mathematics, 12 Bibliog. 143p 24cm Birkhauser Verlag,Basel,1969
P Math 2.3657

OBERWOLFACH 1968 Abstract spaces and approximation :a conference Proceedings Edited by P.L. Butzer and Bela Sz-Nagy International series on numerical mathematics, 10 bibliog.,port. 423p 24cm Birkhauser,Basel,1969
P Math 2.3749

OBERWOLFACH 1970 Contributions to non-standard analysis :a symposium Papers Edited by W.A.J. Luxemburg and A. Robinson Studies in logic and the foundations of mathematics, 69 vi,289p 23cm North-Holland,Amsterdam;London,1972
P Math 2.4060

OBERWOLFACH 1970 Martingales Report Mathematisches forschungsinstitut,Oberwolfach Lecture notes in mathematics, 190 75p 25cm Springer,Berlin,1971
Math S 3.1781

OBERWOLFACH 1970 Martingales :a meeting Report Mathematisches forschungsinstitut, Oberwolfach Edited by Hermann Dinges Lecture notes in mathematics, 190 75p 25cm Springer,Berlin,1971
P Math 2.4559

OBERWOLFACH 1971 Numerische losung nichtlinear partieller differential- und integrodifferentialgleichungen :conferenz Proceedings Edited by R. Ansorge Lecture notes in mathematics, 267 vi,339p 25cm Springer,Berlin,1972
P Math 2.4294

OBERWOLFACH 1965 Funktionalanalysis approximationstheorie numerische mathematik : vortragsauszuge der tagung uber numerische probleme in der approximationstheorie vom 22. bis 25.Juni 1965,und der tagung uber funktionalanalytische methoden in der numerischen mathematik vom 15.bis 20 November, 1965 Edited by L. Collatz and others International series of numerical mathematics, 7 Bibliog. 232p 24cm Birkhauser Verlag,Basel,1967
P Math 2.3448

OBESITY Adipose tissue metabolism and obesity :a conference Papers New York 1964 Dec 14-17 New York academy of sciences Edited by Harold E. Whipple New York academy of sciences.Annals, 131,p 1-683 New York, 1965
Bioch 33.0562

OBESITY.MEDICAL AND SCIENTIFIC ASPECTS :a symposium Proceedings London 1968 Obesity association of Great Britain Edited by I.McLean Baird and Alan N. Howard Obesity association of Great Britain.Symposium, 1 Livingstone,Edinburgh;London,1969
Path 30.2446

OBESITY; MEDICAL AND SCIENTIFIC ASPECTS 1st a symposium Proceedings London 1968 Obesity association of Great Britain Edited by I.McLean Baird and Alan N. Howard Livingstone,Edinburgh;London,1969
Inv Med 37.0010

OBESITY ASSOCIATION OF GREAT BRITAIN Obesity. Medical and scientific aspects :a symposium Proceedings London 1968 Edited by I. McLean Baird and Alan N. Howard Obesity association of Great Britain.Symposium, 1 Livingstone,Edinburgh;London,1969
Path 30.2446

OBESITY ASSOCIATION OF GREAT BRITAIN Obesity; medical and scientific aspects :a symposium 1st Proceedings London 1968 Edited by I.McLean Baird and Alan N. Howard Livingstone,Edinburgh;London,1969
Inv Med 37.0010

OBESITY ASSOCIATION OF GREAT BRITAIN.SYMPOSIUM, 1 Obesity.Medical and scientific aspects : a symposium Proceedings London 1968 Obesity association of Great Britain Edited by I.McLean Baird and Alan N. Howard Livingstone,Edinburgh;London,1969
Path 30.2446

OBLIQUE IONOSPHERIC RADIOWAVE PROPAGATION : at frequencies near the lowest usable high frequency:a symposium Rome 1965 Agard AGARD.Conference proceedings, 13 Technivision,Slough,1969 The 11th annual symposium of the AGARD Electromagnetic wave propagation committee
Eng 41.5683

OBSERVATION AND INTERPRETATION... 9th symposium proceedings Bristol 1957 Apr 1-4 Colston research society Edited by S. Korner Colston papers, 9 xiii,218p 25cm Butterworths,London,1957
Math 3.0972

OBSERVATION AND INTERPRETATION:A SYMPOSIUM OF PHILOSOPHERS AND PHYSICISTS Colston research society symposium 9th Proceedings Bristol 1957 Apr 1-4 Edited by S. Korner xiv,218p Butterworths,London, 1957
WSM 43.1022

OBSERVATION AND INTERPRETATION:A SYMPOSIUM OF PHILOSOPHERS AND PHYSICISTS Proceedings Bristol 1957 Apr Colston research society Edited by S. Korner Colston research society. Symposia, 9 Butterworths,London,1957
Psy 31.2431

OBSERVATIONAL ASPECTS OF GALACTIC STRUCTURE :a summer school typescript Lagonissi 1964 Sep.9-23 Edited by A. Blaauw and L.N. Mavridis Under the auspices of the North Atlantic treaty organization.Science committee 370p Athens,1965
Obs 6.0193

OCCLUSIVE ARTERIAL DISEASE Pathogenesis and treatment of occlusive arterial disease :a conference Proceedings London 1959 Nov 13-14 Royal college of physicians of London Edited by L. McDonald Pitman,London,1960
PGMS 29.0081

OCCLUSIVE ARTERIAL DISEASE Pathogenesis and treatment of occlusive arterial disease :a conference Proceedings London 1959 Nov 13-14 Royal college of physicians of London Edited by Lawson McDonald Pitman,London,1960
Path 30.2430

OCCUPATIONAL HEALTH FOR THE UNDERGRADUATE MEDICAL STUDENT London 1970 May 14 Society of occupational medicine Edited by John R. Glover Society of occupational medicine, London,c1971
PGMS 29.0667

OCEAN SURFACE WAVES New York 1948 Mar 18-19 New York academy of sciences.Oceanography and meteorology section Edited by B. Haurwitz New York academy of sciences.Annals, 51,3 New York,1949 Series of papers resulting from a conference on surface waves held by the section of oceanography and meteorology of the New York academy of sciences March 18-19,1948
Geod 9.0602

OCEANOGRAPHIE GEOLOGIQUE ET GEOPHYSIQUE DE LA MEDITERRANE OCCIDENTALE :colloque international Travaux Villefranche sur Mer 1961 Apr 4-8 Centre national de la recherche scientifique Paris,1962
Geod 9.0581

OCEANOGRAPHY International oceanographic conference Invited lectures New York 1959 Aug 31-Sep 12 Edited by Mary Sears American association for the advancement of science.Publications,67 American association for the advancement of science,Washington,D.C., 1961
Geog 13.0883

OCEANOGRAPHY International oceanographic congress Invited lectures New York 1959 Aug 31-Sep 12 American association for the advancement of science Edited by Mary Sears American association for the advancement of science.Publication, 67 Washington,D.C., 1961
Geod 9.0576

OCEANOGRAPHY International oceanographic congress invited lectures New York 1959 Aug 31-Sep 12 Edited by Mary Sears American association for the advancement of science.Publications,67 American association for the advancement of science,Washington,D.C., 1961
Geol 8.1631

OCEANOGRAPHY International oceanographic congress 1st Preprints of abstracts of papers Washington,D.C. 1959 Aug 31-Sep 12 American association for the advancement of science Edited by Mary Sears illus,maps 1022p 23cm Washington,D.C.,1959
Sco 14.0129

OCEANOGRAPHY Progress in oceanography International association for quaternary research congress 7th Proceedings Boulder,Colo. 1965 Aug 30-Sep 5 Vol 4 Edited by M. Sears Pergamon,Oxford,1967
Bot 42.3248

OCEANOGRAPHY The Commonwealth oceanographic conference Proceedings Wormley,Surrey 1954 Oct 18-22 Arranged by the National oceanographic council vi,93p 25cm Cambridge university press,Cambridge,1955
Sco 14.0123

OCEANOGRAPHY The International oceanographic congress Lectures New York 1959 Aug 31-Sep 12 American association for the advancement of science Edited by Mary Sears American association for the advancement of science.Publication, 67 American association for the advancement of science, Washington,1961
Bal 39.1951

OCEANOGRAPHY Unesco symposium on physical oceanography Proceedings Tokyo 1955 Oct 19-22 Unesco,and,Japan society for the promotion of science Unesco and Japan society for the promotion of science,Tokyo, 1957
Geog 13.0875

OCEANOGRAPHY FROM SPACE :proceedings of conference on the feasibility of conducting oceanographic exploratiions from aircraft, manned orbital and lunar laboratories Proceedings Woods Hole 1964 Aug 24-28 Woods Hole oceanographic institution Sponsored by National aeronautics and space administration Woods Hole oceanographic institution,Woods Hole,Mass.,1965
Sco 14.0355

OCEANS The Origin and evolution of atmospheres and oceans :a conference New York 1963 Apr 8-9 Goddard institute for space studies Edited by Peter J. Brancazio and A.G.W. Cameron Wiley,New York,1964
TA 15.0427

OCEANWAVE SPECTRA Conference on oceanwave spectra 1961 National academy of sciences Englewood Cliffs,N.J.,1963
Eng 41.6696

The OCULAR CIRCULATION IN HEALTH AND DISEASE; WILLIAM MACKENZIE CENTENARY SYMPOSIUM Proceedings Glasgow 1968 Sep 23-23 Edited by E.R. Brown and C.V. Moore Progress in hematology, 6 Heinemann,London,1969
Med 36.0054

OESTROGEN DETERMINATIONS The Technique and significance of oestrogen :a conference proceedings London 1954 Feb 17 Society for endocrinology Royal society of medicine. Endocrinological section Edited by P. Eckstein and S. Zuckerman Society for endocrinology.Memoirs, 3 Cambridge university press,Cambridge,1955
Radioth 35.0045

OFFICE APPLIANCE AND BUSINESS EQUIPMENT The Business computer symposium Papers London 1958 Dec 1-3 No 1-22 Collection of 22 papers and other mimeographed material
Math L 5.3252

OFFICE MANAGEMENT ASSOCIATION The Scope for electronic computers in the office National conference of the Office management association papers Brighton 1955 May illus. 102p 26cm Office management association,London,1955
Math L 5.0727

OFFICE OF NAVAL RESEARCH.GEOGRAPHY BRANCH, Coastal geography conference 2nd Louisiana state university 1959 Apr 6-9 By Richard J. Russell Washington,D.C.,1959
Geog 13.0712

OHIO 1960 Short course on process industry corrosion Proceedings Ohio State university.Department of metallurgical engineering and National association of corrosion engineers.Technical group committee T-5 Edited by V.D. Miller National association of corrosion engineers,n.p.,1960
Met 25.1935

OHIO 1965 Inequalities :a symposium Proceedings Edited by Oved Shisha Sponsored by Aerospace research laboratories Bibliog xiv,360p 24cm Academic press, London;New York,1967
P Math 2.2925

OHIO AGRICULTURAL EXPERIMENT STATION Trace elements :the conference held at Ohio agricultural experiment station Proceedings Wooster,Ohio 1957 Oct 14-16 Edited by C.A. Lamb and others Academic press,New York; London,1958
Bioch 33.0348

OHIO STATE UNIVERSITY Contributions to ergodic theory and probability Midwestern conference on ergodic theory 1st Proceedings Columbus,Ohio 1970 Mar 27-30 Lecture notes in mathematics, 160 bibliog. vii,278p 25cm Springer,Berlin,1970
P Math 2.3939

OHIO STATE UNIVERSITY Contributions to ergodic theory and probability Midwestern conference on ergodic theory and probability 1st Proceedings Columbus,Ohio 1970 Mar 27-30 Lecture notes in mathematics, 160 vii,278p 26cm Springer,Berlin,1970
Math S 3.1779

OHIO STATE UNIVERSITY Gravity anomalies: unsurveyed areas Extension of gravity anomalies to unsurveyed areas :a symposium Papers Columbus,Ohio 1964 Nov 18-20 Edited by Hyman Orlin Sponsored by the International union of geodesy and geophysics Geophysical monograph, 9 National research council.Publication,1357 American geophysical union,Washington,D.C.,1966
Geod 9.0031

OHIO STATE UNIVERSITY The Symposium on colloid chemistry 9th Papers presented Columbus,Ohio 1931 Jun Edited by Harry Boyer Weiser Colloid symposium monograph, 9 Journal of physical chemistry,New York, 1931
Bioch 33.1476

OHIO STATE UNIVERSITY.DEPARTMENT OF METALLURGICAL ENGINEERING Short course on process industry corrosion Proceedings Ohio 1960 Sep 12-16 Edited by V.D. Miller National association of corrosion engineers,n.p.,1960
Met 25.1935

OHIO STATE UNIVERSITY.GRADUATE SCHOOL Midwestern conference on fluid mechanics 2nd Proceedings Columbus,Ohio 1952 Mar 17-19 Ohio state university.Engineering experiment station.Bulletin, 149 Ohio state university studies:engineering series, 21,no 3 v,529p Ohio state university,Columbus, Ohio,1952
Chem E 24.0402

OHTSU 1963 Intracellular membraneous structure The International symposium for cellular chemistry 1st Proceedings Japan society for cell biology Edited by S. Seno and E.V. Cowdry Society for cellular chemistry.Symposia, 14,Suppl. Okayama,1964 In memory of...Dr. Seizo Katsunuma.Japan society for cell biology formerly Japan society for cellular chemistry
Bioch 33.0966

OHTSU 1966 Nucleic acid metabolism,cell differentiation and cancer growth International symposium for cellular chemistry 2nd Proceedings Edited by E.V. Cowdry and S. Seno Organised by the Japan society for cell biology Pergamon,London,1969
Radioth 35.0561

OISO,JAPAN 1961 Progress in comparative endocrinology International symposium on comparative endocrinology 3rd Proceedings Edited by Kiyoshi Takewaki Sponsored by the Zoological society of Japan General and comparative endocrinology.Suppl., 1,1962 Academic press,New York,1962
Bal 39.1381

OKHA 1959 Materialy soveshchaniya po razrabotke unifitsirovannykh stratigraficheskikh skhem Sakhalina,Kamchatki, Kurilskikh i Komandorskikh Ostrovov Vsesoyuznyi nauchno-issledovatelskii geolorazvedochnyi neftyanoi institut, Moscow Edited by T.A. Dementeva illus 340p 27cm Gosudarstvemmoe nauchno-tekhnicheskoe izdatelstvo neftyanoi i gorno-toplivnoi literatury,Moscow,1961
Sco 14.4756

1689

OLD DOMINION FOUNDATION OF NEW YORK
Brasenose conference on the automation of libraries Anglo-American conference on the mechanization of library services Proceedings Oxford 1966 Jun 30-Jul 3 Edited by John Harrison and Peter Laslett Mansell,Chicago,Ill.,1967
Eng 41.7672

OLD POINT COMFORT,VA. 1965 Language development in children The Genesis of language:a psycholinguistic approach conference Proceedings National institute of child health and human development.Human communication program Edited by Frank Smith and George A. Miller M.I.T.press,Cambridge, Mass.;London,1966
Psy 31.2746

OLFACTION AND TASTE International symposium on olfaction and taste 1st Proceedings Stockholm 1962 Sep Wenner-Gren center Edited by Y. Zotterman Wenner-Gren center. International symposium.Series, 1 Pergamon press,Oxford,1963
An 32.4380

OLFACTION AND TASTE II International symposium on olfaction and taste 2nd Proceedings Tokyo 1965 Sep Wenner-Gren center Edited by T. Hayashi Wenner-Gren center.International symposium series, 8 Pergamon press,Oxford,1967
An 32.4411

OLFACTION AND TASTE 2 Wenner-Gren center international symposium 2nd Proceedings Tokyo 1965 Sep Wenner-Gren center Edited by Takashi Hayashi Wenner-Gren center international symposium series, 8 Pergamon press,Oxford,1967
Psy 31.0296

ON-LINE COMPUTING SYSTEMS symposium proceedings Los Angeles,Calif. 1965 Feb 2-4 University of California Edited by Eric Burgess Jointly sponsored by Informatics incorporated Data processing library series 152p 28cm American data processing,Detroit,Mich.,1965
Math L 5.1031

ON TEACHING MATHEMATICS Southampton mathematical conference :a report on some present-day problems in the teaching of mathematics Southampton 1961 Apr Commonwealth library of science,technology and engineering.Mathematics division, 1 Pergamon press,Oxford,1961
Eng 41.1813

ONE-DAY SYMPOSIUM ON STRUCTURAL LIGHTWEIGHT CONCRETE Brighton 1962 Jun 26 Vol 1-2 Reinforced concrete association 2 vols RCA,London,1963
Eng 41.2968

The ONTOGENESIS OF GRAMMAR :a theoretical symposium Edited by D.I. Slobin Academic press,New York;London,1971 Based on a symposium held in December 1965 in Berkeley at the meeting of the American association for the advancement of science
Psy 31.3299

1690

ONTOGENESIS OF THE BRAIN:THE BIOCHEMICAL, FUNCTIONAL AND STRUCTURAL DEVELOPMENT OF THE NERVOUS SYSTEM International symposium neurontogeneticum 5th Proceedings Prague 1967 Universita Karlova and Lekarska spolecnost J.E.Purknye Edited by Lubor Jilek and Stanislav Trojan Universita Karlova,Prague,1968
PGMS 29.0202

ONTOGENY OF INSECTS :a symposium Acta Prague 1959 Ceskoslovenska akademie ved. Sekce biologicko - lekarska Edited by Ivan Hrdy Czechoslovak academy of sciences,Prague, 1960 Text in English,French,German and Russian
Bal 39.2433

OOSTERBECK 1962 Soil organisms Colloquium on soil fauna and soil microflora and their relationships Proceedings Edited by J. Doeksen and Joseph van der Drift North Holland,Amsterdam,1963
Bot 42.4690

OPATIJA 1959 Welding and allied processes in maintenance and repair work International institute of welding Elsevier,Amsterdam; London,1961 Papers in English and French
Met 25.0797

OPERATING METALLURGY CONFERENCE 9th Proceedings Mechanical working and steel processing committee 2nd Philadelphia,Pa. 1966 Dec 5-9 American institute of mining, metallurgical and petroleum engineers Metallurgical society of the AIME.Conference series, 50 Gordon and Breach,New York,1968
Met 25.2572

OPERATING SYSTEMS PRINCIPLES International seminar on operating systems principles Submitted papers Belfast 1971 Aug 30-Sep 3 Queen's university of Belfast illus 2 vols Queen's university of Belfast,Belfast,1971
Math L 5.3846

OPERATING SYSTEMS PRINCIPLES Symposium on operating systems principles 3rd Stanford,Calif. 1971 Oct 18-20 Association for computing machinery illus 163p Association for computing machinery,New York, 1971
Math L 5.3824

OPERATIONAL RESEARCH International conference on operational research 3rd proceedings Oslo 1963 Jul.1-5 Operational research society,Oslo Edited by G. Kreweras and G. Morlat Sponsored by International federation of operational research societies xii,959p 25cm Dunod; English universities press,Paris;London,1964
Math 3.0915

OPERATIONAL RESEARCH International conference on operational research 4th Proceedings Boston,Mass. 1966 Aug 29-Sep 2 Edited by David B. Hertz and Jacques Melese Organised by the International foundation of operational research societies Operations research society of America.Publications in operations research, 15 Wiley-Interscience, New York,1966
Eng 41.1219

OPERATIONAL RESEARCH SOCIETY Models for decision :a conference London 1960 Oct 13-144 Under the auspices of the United Kingdom automation council English universities press,London,1965
Eng 41.1270

OPERATIONAL RESEARCH SOCIETY,OSLO International conference on operational research 3rd proceedings Oslo 1963 Jul.1-5 Edited by G. Kreweras and G. Morlat Sponsored by International federation of operational research societies xii,959p 25cm Dunod;English universities press,Paris; London,1964
Math 3.0915

OPERATIONS RESEARCH IN RESEARCH DEVELOPMENT :a conference Proceedings Cleveland,Ohio 1962 Edited by Barton V. Dean Held at the Case institute of technology Wiley,New York; London,1963
Eng 41.1173

OPERATIONS RESEARCH SOCIETY OF AMERICA. TRANSPORTATION SCIENCE SECTION Vehicular traffic science International symposium on the theory of traffic flow 3rd Proceedings New York 1965 Jun Edited by Leslie C. Edie and others illus x,373p 24cm American Elsevier,New York,1967
Math 3.1137

OPTHALMOLOGY Congres de la societe europeenne d'opthalmologie 1e Resumees des rapports et des communications Athens 1959 Apr 18-22 European opthalmological society Excerpta medica.International congress series, 25 Excerpta medica, Amsterdam;New York,1960 Papers in English, French and German
PGMS 29.0682

OPTICAL CONVENTION 1st Proceedings London 1905 May 30-Jun 3 Bibliog,illus 247p Norgate and Williams,London,1905
WSM 43.5433

OPTICAL CONVENTION 2nd Proceedings London 1912 tables,illus 359p Hodder and Stoughton for University of London press, London,1912
WSM 43.5434

OPTICAL CONVENTION 3rd Proceedings London 1926 Apr 12-17 Pt 1-2 2 vols Optical convention,London,1926
WSM 43.5435

OPTICAL CONVENTION 3RD Catalogue of optical and general scientific instruments illus London,1926 Convention held in 1926
WSM 43.5432

OPTICAL DESIGN WITH DIGITAL COMPUTERS symposium proceedings London 1956 June 5-7 Imperial college of science and technology.Technical optics section viii,93p 26cm Imperial college of science and technology,London,1956
Math L 5.0754

OPTICAL IMAGE ASSESSMENT USING FREQUENCY RESPONSE TECHNIQUES :a summer school Proceedings London 1958 Jul.7-10 Imperial college of science and technology.Technical optics section Edited by H.H. Hopkins 134p London,1958
Obs 6.1974

OPTICAL INSTABILITY OF THE EARTH'S ATMOSPHERE All-union conference 3rd Proceedings Kiev 1962 July Akademiya nauk S.S.S.R. Astronomical council Glavnaya astronomicheskaya observatoriya,Pulkovo Translated by Z. Lerman from the Russian viii,174p 25cm Israel program for scientific translations,Jerusalem,1966
Obs 6.3275

OPTICAL INSTRUMENTS Conference on optical instruments and techniques Proceedings London 1961 Jul 11-14 Edited by K.J. Habell Chapman and Hall,London,1962 Held at Imperial college
Cav 7.1457

OPTICAL INSTRUMENTS Conference on optical instruments and techniques Proceedings London 1961 Jul 11-14 Edited by K.J. Habell Under the auspices of the International commission for optics Chapman and Hall,London,1962
Obs 6.3215

OPTICAL MASER :a symposium Proceedings New York 1963 Apr 16-19 Polytechnic institute of Brooklyn Edited by Jerome Fox Sponsored also by the Microwave research institute Microwave research institute. Symposia series, 13 Polytechnic press, Brooklyn,N.Y.,1963
Eng 41.5417

OPTICAL PROCESSING OF INFORMATION :a symposium Washington,D.C. 1962 Edited by D.K. Pollock and others Cleaver-Hume, Baltimore,Md.,1963
Eng 41.5877

OPTICAL PROPERTIES AND ELECTRONIC STRUCTURE OF METALS AND ALLOYS :International colloquium Proceedings Paris 1965 Sep 13-16 Institut d'optique theorique et appliquee Edited by F. Abeles Co-sponsored by the United States.Air force.European office of aerospace research North Holland,Amsterdam, 1966
Cav 7.0004

OPTICAL SOCIETY A Joint discussion on vision Report London 1932 Jun.3 327p Physical society,London,c1932
Obs 6.1962

OPTICAL SOCIETY A Joint discussion on vision held at the Imperial college of science London 1932 Jun 3 Physical society,London, 1932 Bound with 'Report of a discussion on audition' by the Physical society,q.v.
Psy 31.2610

OPTICAL SOCIETY Photo-electric cells and their applications :a discussion at a joint meeting of the Physical and Optical societies 1930 Edited by John S. Anderson Physical and Optical societis,London,1930
Eng 41.4623

OPTICAL SOCIETY Report of a joint discussion on vision held...at the Imperial college of science Report London 1932 Jun 3 Physical society,London,1932
Psy 31.0411

OPTICAL SOCIETY OF AMERICA Recent advances in optimization techniques symposium proceedings Pittsburgh,Pa. 1965 Apr.21-23 Institute of electrical and electronics engineers.Systems science and cybernetics group Edited by Abrahim Lavi and Thomas P. Vogl Bibliog. xiii,656p 23cm John Wiley and sons,New York,1966
Math 3.0240

OPTICAL STUDIES OF ADSORBED LAYERS AT INTERFACES London 1970 Dec 14-15 Faraday society Faraday society.Symposia,1970,4 212p 26cm Faraday society,London,1971
Chem 18.2880

OPTICS Symposium on recent advances in optics Proceedings Oxford 1959 Mar.20 University of Oxford 138-180p London,1959
Obs 6.1989

OPTICS IN METROLOGY Colloquia of the International commission for optics Papers Brussels 1958 May 6-9 International commission for optics Edited by Pol Mollet Pergamon press,Oxford,1960 Papers presented in English,French or German
Cav 7.1631

OPTIMISATION TECHNIQUES Conference on optimisation techniques in circuit and control applications London 1970 Jun 29-30 Institution of electrical engineers Institution of electrical engineers.Conference publication, 66 IEE,London,1970 Organised in association with the Institute of electrical and electronics engineers and the Institute of physics and the Physical society
Eng 41.8587

OPTIMIZATION Colloquium on methods of optimization Novosibirsk 1968 Jun Edited by N.N. Moiseev Lecture notes in mathematics, 112 Illus. 293p 25cm Springer-Verlag, Berlin,1970
P Math 2.3531

OPTIMIZATION Colloquium on methods of optimization Novosibirsk 1968 Jun Edited by N.N. Moiseev Lecture notes in mathematics, 112 293p 26cm Springer,Berlin,1970
Math S 3.1688

OPTIMIZATION Computing methods in optimization problems International conference on computing methods in optimization problems 2nd Papers San Remo 1968 Sep 9-13 Sponsored by the Society for industrial and applied mathematics Lecture notes in operations research and mathematical economics, 14 bibliog.,illus. Springer,Berlin,1968
Math S 3.1516

OPTIMIZATION Keele conference on optimization and nonlinear programming Keele 1968 Institute of mathematics and its applications Edited by R. Fletcher Academic press,London,1969
Math L 5.3607

OPTIMIZATION Keele conference on optimization and nonlinear programming Keele 1968 Institute of mathematics and its applications Edited by R. Fletcher Academic press,London,1969
P Math 2.3755

OPTIMIZATION Numerical methods for unconstrained optimization Joint institute of mathematics and its application-National physical laboratory conference Proceedings Teddington 1971 Jan 7-8 National physical laboratory Institute of mathematics and its applications Edited by W. Murray xi,144p 24cm Academic press,London;New York,1972
Math S 3.1729

OPTIMIZATION Recent advances in optimization techniques :a symposium Pittsburgh,Pa. 1965 Apr 21-23 Edited by A. Levi and T.P. Vogl Sponsored by the Institute of electrical and electronics engineers.Systems science and cybernetics group Wiley,New York, 1966
Eng 41.6003

OPTIMIZATION Symposium on optimization Nice 1969 Jun 29-Jul 5 Edited by A.V. Balakrishnan and others Lecture notes in mathematics, 132 Bibliog.,Illus. 350p 26cm Springer-Verlag,Berlin,1970
P Math 2.3719

OPTIMIZATION Symposium on optimization Proceedings Nice 1969 Jun 29-Jul 5 Edited by A.V. Balakrishnan and others Lecture notes in mathematics, 132 350p 25cm Springer,Berlin,1970
Math S 3.1685

OPTIMIZATION :symposium Keele 1968 May 25-29 Institute of mathematics and its applications Academic press,London,1969
Eng 41.2342

OPTIMIZATION OF STEEL PRODUCT YIELD Conference on optimization of steel product yield Proceedings London 1967 May 3-4 Iron and steel institute Iron and steel institute.Publication, 107 London,1967 Conference held at the Institute's annual general meeting
Met 25.2561

OPTIMIZATION TECHNIQUES Recent advances in optimization techniques symposium proceedings Pittsburgh,Pa. 1965 Apr.21-23 Institute of electrical and electronics engineers.Systems science and cybernetics group Edited by Abrahim Lavi and Thomas P. Vogl Jointly sponsored by the Optical society of America Bibliog. xiii,656p 23cm John Wiley and sons,New York,1966
Math 3.0240

OPTIMIZING International symposium on optimizing and adaptive control 1st Proceedings Rome 1962 Apr 26-28 Edited by Loren E. Bollinger and Emil J. Minnar Sponsored by the International federation of automatic control.Theory committee Instrument society of America,Pittsburgh,Pa., 1962
Eng 41.6017

The OPTIMUM POPULATION FOR BRITAIN :symposium held at the Royal geographical society Proceedings London 1969 Sep 25-26 Edited by L.R. Taylor Institute of biology. Symposia, 19 Academic press,London,1970
Gen 34.2202

OPTIQUE DES RAYONS X ET MICROANALYSE Congres international sur l'optique des rayons X et la microanalyse 4e Orsay 1965 Sep 7-10 Edited by Raymond Castaing Hermann,Paris, 1966
Met 25.1460

ORBELI L.A. ed. Uspekhi biologicheskikh nauk v S.S.S.R. za 25 let, 1917-42 :symposium Akademiya nauk S.S.S.R.Otdelenie biologicheskikh nauk 356p 26cm Moscow, 1945
Philos 1.1394

ORBITING ASTRONOMICAL OBSERVATORY The Scientific results from the orbiting astronomical observatory (OAO-2) Amherst, Mass. 1971 Aug 23-24 National aeronautics and space administration Edited by A.D. Code NASA SP-310 590p nasa,washington,D.C.,1972
Obs 6.3626

ORBITS The Theory of orbits in the solar system and in stellar systems a symposium Thessaloniki 1964 Aug.17-22 International astronomical union Edited by George Contopoulos Organized in co-operation with Committee on space research International Astronomical Union.Symposium, 25 380p Academic press,London;New York,1966
Obs 6.0409

ORBITS The Theory of orbits in the solar system and in stellar systems :a symposium Thessaloniki 1964 Aug.17-22 International astronomical union Edited by George Contopoulos Organized in co-operation with Committee on space research International astronomical union.Symposium, 25 380p Academic press,London;New York,1966
A Math 4.1167

ORCHIDS European orchid congress 2nd Proceedings Paris 1969 Apr 24-26 Comite national interprofessionnel de l'horticulture et des pepinieres.Monographie, 2 C.N.I.H., Rungis,1969
BG 38.3116

ORCHIDS World orchid conference 2nd Proceedings Honolulu 1957 Sep 19-23 Sponsored by the University of Hawaii American orchid society,Cambridge,Mass.,1958
BG 38.3117

ORCHIDS World orchid conference 3rd Proceedings London 1960 May 30-Jun 22 Sponsored by the American orchid society Royal horticultural society,London,1960
BG 38.3118

ORCHIDS World orchid conference 4th Proceedings Singapore 1963 Oct illus. Straits Times press,Singapore,1964
BG 38.3119

ORCHIDS World orchid conference 5th Proceedings Long beach,Calif. 1966 Apr 11-22 Edited by Lloyd R.de Garmo Fifth world orchid conference,Long beach,Calif.,1966
BG 38.3120

ORCHIDS World orchid conference 6th Proceedings Sydney 1969 Sep Edited by Murray J.G. Corrigan and others 1971
BG 38.1753

ORCHIDS UNDER THE SOUTHERN CROSS World orchid conference 6th Commemorative brochure Sydney 1969 1969
BG 38.1754

ORDER OF THE PROCEEDINGS AT THE DARWIN CELEBRATION... :with a sketch of Darwin's life Cambridge 1909 Jun 22-24 University of Cambridge 23p Cambridge university press,Cambridge,1909
WSM 43.2842

ORDINARY AND PARTIAL DIFFERENTIAL EQUATIONS Numerical solution of ordinary and partial differential equations based on a summer school held at Oxford August-September 1961 Oxford 1961 Aug-Sep Oxford university. Computing laboratory Oxford,1962
Geod 9.0179

ORDINARY DIFFERENTIAL EQUATIONS Symposium on the numerical treatment of ordinary differential equations, integral and integro-differential equations 4th proceedings Rome 1960 Sep.20-24 Provisional international computation centre bibliog. 679p 26cm Birkhauser,Basel,1960
Math L 5.0299

ORDNANCE MATERIALS Composite materials and composite structures Sagamore ordnance materials research conference 6th Proceedings New York 1959 Aug 8-21 United States.Army.Ordnance materials research office and United States.Army.Office of ordnance research Arrangements by Syracuse university research institute New York,1959 Mimeograph
Met 25.0093

ORE DRESSING METHODS IN AUSTRALIA AND ADJACENT TERRITORIES Empire mining and metallurgical congress 5th Publications Melbourne 1953 Apr 14-Jun 12 3 Australasian institute of mining and metallurgy Edited by H.H. Dunkin Office of the congress,and of the Australasian institute of mining and metallurgy,Melbourne,1953
Met 25.0205

ORE MINING AND MATERIALS HANDLING Iron and steel engineers group meeting 47th Papers and discussions Metz 1963 Jun 10-14 Iron and steel institute.Engineers group Iron and steel institute.Special report, 82 Iron and steel institute,London,1963
Met 25.0208

OREGON STATE UNIVERSITY Oregon state university annual biology colloquium arctic biology:ten papers presented at the 1957 and one at the 1965 biology colloquium 18th Proceedings Cornvallis,Ore. 1957 Edited by Henry P. Hansen 2nd edition Illus,maps 318p 24cm Oregon state university press, Cornvallis,Ore.,1967
Sco 14.7840

OREGON STATE UNIVERSITY ANNUAL BIOLOGY COLLOQUIUM 18th arctic biology:ten papers presented at the 1957 and one at the 1965 biology colloquium Proceedings Cornvallis,Ore. 1957 Oregon state university Edited by Henry P. Hansen 2nd edition Illus,maps 318p 24cm Oregon state university press, Cornvallis,Ore.,1967
Sco 14.7840

ORES AND MINERALS Remobilization of ores and minerals Convegno sulle rimobilizzazione del minerali metallici e non metallici Rendiconti minerali metallici e non metallici Papers Cagliari 1969 Aug Associazzione mineralia sarda Istituto di giacimenti minerali universitaria Cagliari,c1969
Min 10.1489

ORGANOMETALLIC CHEMISTRY International symposium on organometallic chemistry 2nd Abstracts of proceedings Madison,Wisc. 1965 Aug 30-Sep 3 National science foundation,and,Wisconsin alumni research foundation Co-sponsored by the United States. Army research office,Durham Madison,Wisc., 1965 Mimeographed
Chem 18.1589

ORGAN CULTURE Symposium on organ culture : studies of development,function,and disease Washington,D.C. 1962 May 29-31 National cancer institute Edited by Clyde J. Dawe Sponsored by Tissue culture association and National cancer institute National cancer institute.Monograph, 11 U.S.Department of health,education,and welfare,Washington,D.C., 1963
Bioch 33.2367

ORGANELLE DEVELOPMENT Control of organelle development :a symposium London 1969 Sep Society for experimental biology Society for experimental biology.Symposia, 24 Cambridge university press,Cambridge,1970
Bioch 33.2340

ORGANELLE DEVELOPMENT Control of organelle development :a symposium Proceedings London 1969 Sep 8-12 Society for experimental biology Edited by P.L. Miller Society for experimental biology.Symposia, 24 Cambridge university press,Cambridge,1970
Gen 34.2205

ORGANIC BASES AND BIOCHEMICAL ASPECTS OF IMBECILITY International congress on mental retardation 2nd Proceedings Vienna 1961 Aug 14-19 Edited by Otto Stur Karger,Basle,1963
An 32.4198

ORGANIC CHEMISTRY Advances in organic chemistry international meeting proceedings Milan 1962 Sep 10-12 Geochemical society. Organic geochemistry group,European branch Edited by Umberto Colombo and G.D. Hobson International series of monographs on earth sciences,15 illus. Pergamon press,Oxford, 1964
Geol 8.1666

ORGANIC REACTION MECHANISMS :an international symposium Lectures Cork,Ireland 1964 Jul 20-25 Chemical society,and,University college,Cork Chemical society.Special publication,19 Chemical society,London,1965
Chem 18.1223

ORGANIC SOLIDS Structure of organic solids Microsymposium 'Structure of organic solids' 2nd Lectures Prague 1968 Sep 16-19 International union of pure and applied chemistry.Macromolecular division Czechoslovak chemical society In conjunction with Ceskoslovenska akademie ved illus 550p 25cm Butterworths,London,1969 Simultaneously published in 'Pure and applied chemistry',vol 18,no 4,1969
Cav 7.2659

ORGANISATION FOR EUROPEAN ECONOMIC CO-OPERATION Function and training of the chemical engineer :international conference London 1955 Mar 21-23 85p O.E.E.C.,Paris,1955
Chem E 24.1760

ORGANISATION METEOROLOGIQUE INTERNATIONALE Commission de l'annee polaire 1932-33 proces-verbaux des seances appendice H Innsbruck 1931 Sept 23-26 E.Ijdo,Leyden,1932
Sco 14.0498

ORGANISATION METEOROLOGIQUE INTERNATIONALE Rapport de la commission internationale de l'annee polaire 1932-33 Copenhagen 1933 May Vol 3: Compte-rendu des travaux de la commission Octobre 1931-Mai 1933 Ijdo,Leyden, 1934
Sco 14.0491

ORGANISATION METEOROLOGIQUE INTERNATIONALE Rapport de la commission internationale de l'annee polaire 1932-33 Innsbruck 1931 Sep Vol 2: compte-rendu des travaux de la commission pendant sa deuxieme annee illus, diagrs 188p Ijdo,Leyden,1932
Sco 14.0492

ORGANIZATION AND CONTROL IN PROKARYOTIC AND EUKARYOTIC CELLS :a symposium London 1970 Apr Society for general microbiology Edited by H.P. Charles and B.C.J.G. Knight Society for general microbiology.Symposia, 20 Cambridge university press,Cambridge,1970
An 32.5447

ORGANIZATION AND CONTROL IN PROKARYOTIC AND EUKARYOTIC CELLS :a symposium Papers Society for general microbiology Edited by H. P. Charles and B.C.J.G. Knight Held at Imperial college of science and technology Society for general microbiology.Symposia, 20 Cambridge university press,Cambridge,1970
Bioch 33.1919

ORGANIZATION FOR ECONOMIC CO-OPERATION AND DEVELOPMENT Mathematical education of engineers :report of the O.E.C.D. seminar Paris 1965 OECD,Paris,1965
Eng 41.1838

ORGANIZATION FOR ECONOMIC CO-OPERATION AND DEVELOPMENT New thinking in school biology The Seminar on the reform of biology teaching :report on the OECD seminar La Tour de Peilz,Switzerland 1962 Dep 4-14 New thinking in school science Paris,1963
Gen 34.2062

ORGANIZATION FOR ECONOMIC COOPERATION AND DEVELOPMENT Biology today:its role in education International working session on the teaching of school biology :report of an OECD working session Edited by Paul Duvigneaud English edition prepared by L.S. Comber New thinking in school science OECD,Paris,1966
Gen 34.2234

ORGANIZATION FOR EUROPEAN ECONOMIC COOPERATION Health physics in nuclear installations :a symposium Riso 1959 May 25-28 Held at the Danish atomic energy centre European nuclear energy agency,Paris,1959
Radioth 35.1110

ORGANIZATION METEOROLOGIQUE INTERNATIONALE Rapport de la commission internationale de l'annee polaire 1932-33 Leningrad 1930 August Vol 1: compte-rendu des travaux de la commission pendant sa premiere annee de travail... illus,diagrms 152p Ijdo, Leyden,1933
Sco 14.0493

The ORGANIZATION OF CHEMICAL ENGINEERING PROJECTS joint symposium Proceedings London 1958 Jun 24-26 Institute of petroleum and Institution of chemical engineers Institution of chemical engineers,London,1958 17th meeting of the European federation of chemical engineering
Chem E 24.1543

ORGANIZATION OF RECALL International interdisciplinary conference on learning, remembering and forgetting 2nd Proceedings Princeton,N.J. 1964 Sep 27-30 New York academy of sciences and National institute of child health and human development Edited by Daniel P. Kimble With the support of United States.Office of naval research New York academy of sciences. Interdisciplinary communication program,New York,1967
Psy 31.1012

ORGANIZATION RESEARCH New perspectives in organization research :consisting of papers from a conference on research in organizations. and a seminar on the social science of organizations Pittsburgh,Pa. 1962 Jun.22-24 Pittsburgh,Pa. 1962 Jun.10-23 Edited by W.W. Cooper and others Sponsored by United States.Office of naval research, Bibliog. xxii,606p 24cm John Wiley and sons,New York,1964
Math 3.0910

ORGANIZATION SCIENTIFIQUE DES JARDINS BOTANIQUES Colloque international de l'union international des sciences biologiques Paris 1953 Jun 4-7 Under the auspices of Unesco International union of biological sciences. Series B:Colloquia, 13 International union of biological sciences,Paris,1953
BG 38.3110

ORGANIZATION THEORY IN INDUSTRIAL PRACTICE : a symposium of the Foundation for research on human behavior 1959 Feb Foundation for research on human behavior Edited by Mason Haire Wiley,New York;London,1962
Eng 41.0558

ORGANIZATIONAL BIOSYNTHESIS :a symposium New Brunswick,N.J. 1966 Sep 8-10 By Henry J. Vogel and others With support from the National science foundation illus. xx,549p Academic press,New York;London,1967
Bot 42.1512

ORGANIZATIONAL BIOSYNTHESIS :a symposium Proceedings New Brunswick,N.J. 1966 Sep 8-10 Rutgers university.Institute of microbiology Edited by Henry J. Vogel and others With support from the National science foundation Academic press,New York; London,1967
Bioch 33.0832

ORGANIZING COMMITTEE ON FRACTURE Fracture International conference on fracture 1st Proceedings Sendai 1965 Sep 12-17 Vol 1-3 Edited by T. Yokobori and others 3 vols Japanese society for strength and fracture of materials,1966
Met 25.0872

ORGANOGENESIS International conference on organogenesis Baltimore,Md. 1964 Sep 6-12 International institute of embryology Edited by Robert L. DeHaan and Heinrich Ursprung Holt,Rinehart and Winston,New York,1965
Bal 39.0490

ORIENTATION EFFECTS IN THE MECHANICAL BEHAVIOR OF ANISOTROPIC STRUCTURAL MATERIALS :a symposium Seattle,Wash. 1965 Oct 31-Nov 5 American society for testing materials American society for testing materials.Special technical publication, 405 American society for testing materials,Philadelphia,Pa.,1966
Eng 41.3701

ORIENTATION EFFECTS IN THE MECHANICAL BEHAVIOUR OF ANISOTROPIC STRUCTURAL MATERIALS a symposium presented at the 5th Pacific area national meeting American society for testing materials A.S.T.M.Special technical publication, 405 A.S.T.M.,Philadelphia,1966
Met 25.2341

ORIGIN AND DISTRIBUTION OF THE ELEMENTS Symposium on the origin and distribution of the elements Paris 1967 International union of geological sciences Edited by L.H. Ahrens International series of monographs on earth sciences, 30 xvii,1178p 24cm Pergamon press,Oxford,1968
TA 15.0565

ORIGIN AND DISTRIBUTION OF THE ELEMENTS :a symposium Proceedings Paris 1967 May 8-11 International association of geochemistry and cosmochemistry Edited by L. H. Ahrens International series of monographs on earth sciences, 30 1178p Pergamon, Oxford,1968
Obs 6.3444

The ORIGIN AND EVOLUTION OF ATMOSPHERES AND OCEANS :a conference New York 1963 Apr 8-9 Goddard institute for space studies Edited by Peter J. Brancazio and A.G.W. Cameron Wiley,New York,1964
TA 15.0427

ORIGIN AND EVOLUTION OF MAN :a symposium Cold Spring Harbor 1950 Jun 9-17 Edited by Katherine Brehme Warren Cold Spring Harbor symposia on quantitative biology, 15 Long Island biological association,Cold Spring Harbor,N.Y.,1950
An 32.2807

ORIGIN AND EVOLUTION OF MAN :a symposium Papers Cold Spring Harbor 1950 Cold Spring Harbor biological laboratory Cold Spring Harbor symposia on quantitative biology, 15 Long Island biological association,Cold Spring Harbor,1950
Bioch 33.1271

ORIGIN AND EVOLUTION OF THE FERNS :a symposium Papers Edited by Murray F. Buell Torrey botanical club.Memoirs, 21 pt.5 86p Seeman,Durham,N.C.,1964
Bot 42.4272

ORIGIN OF ANORTHOSITE AND RELATED TOPICS Annual George H.Hudson symposium 2nd Papers Plattsburgh,N.Y. 1966 New York state museum and science service.Geological survey State university of New York.College at Plattsburgh Edited by Yngvar W. Isachsen New York state museum and science service. Memoir, 18 New York state museum and science service,Albany,N.Y.,1968
Min 10.1428

ORIGIN OF LIFE ON THE EARTH The International symposium on the origin of life on the earth 1st Proceedings Moscow 1957 Aug 19-24 Akademiya nauk S.S.S.R. Edited by A.I. Oparin and others Organized under the auspices of the International union of biochemistry I.U.B.symposium series, 1 Pergamon,London,1959 "English-French-German edition edited for the International union of biochemistry by F.Clark and R.L.M. Synge".
Bioch 33.0727

The ORIGIN OF LIFE ON THE EARTH Reports on the international symposium Moscow 1957 Aug Edited by A.I. Oparin and others Publishing house of the Academy of sciences of the U.S.S.R.,n.p.,c1957
Bal 39.0990

ORIGIN OF THE SOLAR SYSTEM :a conference Proceedings Goddard institute for space studies Edited by Robert Jastrow and A.G.W. Cameron 176p Academic press,New York,1963
Cav 7.2870

ORIGIN OF THE SOLAR SYSTEM conference proceedings New York 1962 Jan.23-24 Goddard institute for space studies Edited by Robert Jastrow and A.G.W. Cameron 176p Academic press,New York,1963
Obs 6.0883

ORIGINS OF HUMAN BEHAVIOUR Mechanisms of motor skill development Centre for advanced study in the developmental sciences study group on 'Mechanisms of motor skill development' Proceedings London 1968 Nov Ciba foundation Edited by Kevin Connolly Academic press,London;New York,1970 Being the 4th study group in a C.A.S.D.S. programme on 'The origins of human behaviour' held jointly with the Ciba foundation
Psy 31.3101

ORIGINS OF PREBIOLOGICAL SYSTEMS AND OF THEIR MOLECULAR MATRICES A Conference on the origins of prebiological systems and of their molecular matrices Proceedings Wakulla Springs,Fla 1963 Oct 27-30 Institute for space biosciences Florida state university National aeronautics and space administration Edited by Sidney W. Fox Bibliog.,illus. xx, 482p Academic press,New York;London,1965
Bot 42.1364

The ORIGINS OF PREBIOLOGICAL SYSTEMS AND OF THEIR MOLECULAR MATRICES :a conference Proceedings Wakulla Springs 1963 Oct 27-30 Edited by Sidney W. Fox Academic press, New York;London,1965 Conducted under the auspices of the Institute for space biosciences,the Florida state university and the National aeronautics and space administration
Bal 39.0992

The ORIGINS OF PREBIOLOGICAL SYSTEMS AND OF THEIR MOLECULAR MATRICES :a conference Proceedings Wakulla springs,Fla. 1963 Oct 27-30 Florida state university National aeronautics and space administration Edited by Sidney W. Fox Under the auspices of the Institute for space biosciences Academic press,New York;London,1965
Gen 34.0456

ORLANDO,FLA. 1959 Space trajectories :a symposium Papers Radiation incorporated, Orlando,Fla. and Advanced research projects agency Co-sponsored by the American astronautical society Academic press,New York,1960
Geod 9.0014

ORLANDO,FLA. 1964 Southern metals-materials conference on advances in aerospace materials Proceedings American society of metals.Orlando chapter Edited by Henry M. Otte and Saul R. Locke Plenum press,New York, 1965
Met 25.2282

ORNAMENTAL FLOWERING TREES AND SHRUBS conference Report London 1938 Apr 26-29 Royal horticultural society Edited by F. J. Chittenden illus. Royal horticultural society,London,1940
BG 38.2822

ORNITHOLOGY Progress and prospects in ornithology being the centenary symposium Proceedings Cambridge 1959 Mar 20-23 British ornithologists' union Ibis, 101,no 3-4,Centenary celebration number 25cm London, 1959
Bal 44.1519

ORNITHOLOGY Vsesoyuznaya ornitologicheskyaya konferentsiya 11-17 Sentyabrya,1962 3rd Materialy Lvov 1962 September 11-17 20cm Lvov,1962
Philos 1.1626

ORSAY 1962 La Structure des solutions solides metalliques Centre national de la recherche scientifique Centre national de la recherche scientifique.Colloques internationaux, 118 Centre national de la recherche scientifique,Paris,1962 Papers in English and French
Met 25.1210

ORSAY 1962 Metallic solid solutions:their electronic and atomic structure :a symposium Proceedings Universite de Paris a la Sorbonne W.A.Benjamin,New York,1963
Cav 7.1737

ORSAY 1964 La Theorie du potentiel : colloque international Centre national de la recherche scientifique Edited by M. Brelot and others Centre national de la recherche scientifique.Colloques internationaux, 146 x, 312p 25cm Centre national de la recherche scientifique,Paris,1965
Math 3.0050

ORSAY 1965 Optique des rayons X et microanalyse Congres international sur l'optique des rayons X et la microanalyse 4e Edited by Raymond Castaing Hermann, Paris,1966
Met 25.1460

ORSAY 1966 Methodes nouvelles de spectroscopie instrumentale Exposes Centre national de la recherche scientifique 2nd edition Centre national de la recherche scientifique.Colloques internationaux, 161 344p CNRS,Paris,1969
Obs 6.3376

ORTA DOGU TEKNIK UNIVERSITESI Modern quantum chemistry :Istanbul international school of quantum chemistry Lectures Istanbul 1964 Aug 16-Sep 5 Pt 3: action of light and organic crystals North Atlantic treaty organization.Pure science bureau Edited by Oktay Sinanoglu Istanbul lectures:modern quantum chemistry,3 Academic press,New York; London,1965
Chem 18.2238

ORTHOGONAL EXPANSIONS AND THEIR CONTINUOUS ANALOGUES :a conference Proceedings Edwardsville 1967 Apr 27-29 Edited by Deborah Tepper Haimo Bibliog. xx,307p 24cm Feffer and Simons Inc.,London;Amsterdam, 1968
P Math 2.3408

OSAKA 1967 Structure and function of cytochromes The Symposium on structural and chemical aspects of cytochromes Proceedings Edited by Kazuo Okunuki and others University of Tokyo press,Tokyo,1968
Bioch 33.1100

OSAKA 1970 Annual meeting of Japan endocrinological society 43rd Proceedings Japan endocrinological society Kyoto,1970
PGMS 29.0588

OSBORNE REYNOLDS AND ENGINEERING SCIENCE TODAY Osborne Reynolds centenary symposium Papers Manchester 1968 Sep By Jack Allen and others Edited by D.M. McDowell and J.D. Jackson Manchester university press;Barnes and Noble,Manchester;New York,1970
Eng 41.1755

OSBORNE REYNOLDS CENTENARY SYMPOSIUM Papers Osborne Reynolds and engineering science today Manchester 1968 Sep By Jack Allen and others Edited by D.M. McDowell and J.D. Jackson Manchester university press;Barnes and Noble,Manchester;New York,1970
Eng 41.1755

OSLO 1948 Association internationale d'hydrologie scientifique :assemblee generale 8th Proces-verbaux Tome 2: travaux de la commission de la neige et des glaciers illus 407p Louvain,1948
Sco 14.0118

OSLO 1955 Nordisk kirurgisk forening 27th meeting transactions Acta chirurgica Scandinavica, 110,fasc.1-2 Stockholm,1955
Radioth 35.1274

OSLO 1956 Polar atmosphere :a symposium Pt 1: meteorology section Edited by R.C. Sutcliffe Published for and on behalf of North Atlantic treaty organisation.Advisory group for aeronautical research and development AGARDograph 29,1 xii,341p 26cm Pergamon press,London,1958
Sco 14.2367

OSLO 1956 Polar atmosphere symposium Pt 2: ionospheric section North Atlantic treaty organization.Advisory group for aeronautical research and development Agardograph,29,pt.2 diagrs xiii,212p 26cm Pergamon,London, 1957
Sco 14.0528

OSLO 1956 Symposium on chemotherapy of cancer International union against cancer International union against cancer.Acta, 13, no 3 U.I.C.C.,Louvain,1957
Radioth 35.0814

OSLO 1957 Concrete shell roof construction symposium 2nd Proceedings Teknisk ukeblad,Norway Teknisk ukeblad,Oslo,1958
Eng 41.3009

OSLO 1962 The So-called extrapyramidal system :symposium from the sixteenth congress of Scandinavian neurobiologists Scandinavian neurological society and Norwegian neurological association Edited by Sigvald Refsum and others Universitetsforlaget,Oslo, 1963
An 32.4655

OSLO 1963 International conference on operational research 3rd proceedings Operational research society,Oslo Edited by G. Kreweras and G. Morlat Sponsored by International federation of operational research societies xii,959p 25cm Dunod; English universities press,Paris;London,1964
Math 3.0915

OSLO 1963 Science and technology of tungsten,tantalum,molybdenum,niobium and their alloys Agard conference on refractory metals Proceedings North Atlantic treaty organization.Advisory group for aeronautical research and development AGARDograph 82 Pergamon press,Oxford,1964
Met 25.1011

OSLO 1964 Lapps and Norsemen in olden times :a conference Report Institute for comparative research in human culture,Oslo Instituttet for sammenlignende kulturforskning. Serie A.Forelesninger,26 Bibliog,maps 168p Universitetsforlaget,Oslo,1967
Sco 14.5603

OSLO 1965 Conference on pulmonary circulation Edited by Carsten Muller Universitetsforlaget,Oslo,1966
An 32.3575

OSLO 1967 Federation of European biochemical societies meeting 4th Proceedings Vol 2: biochemistry of virus replication Federation of European biochemical societies Edited by S.G. Laland and L.O. Froholm Organized by the Norwegian biochemical society Universitetsforlaget; Academic press,Oslo;London,1968 Symposium organizer A.P.Nygaard
Bioch 33.1397

OSLO 1967 Federation of European biochemical societies meeting 4th Proceedings Vol 4: cellular compartmentalization and control of fatty acid metabolism Federation of European biochemical societies Edited by F.C. Gran Organized by the Norwegian biochemical society Universitetsforlaget;Academic press,Oslo; London,1968 Symposium organizer J. Bremer
Bioch 33.1399

OSLO 1967 Federation of European biochemical societies meeting 4th Proceedings Vol 5: control of glycogen metabolism Federation of European biochemical societies Edited by W.J. Whelan Organized by the Norwegian biochemical society Universitetsforlaget;Academic press,Oslo; London,1968 Symposium organizer W.J. Whelan
Bioch 33.1400

OSLO 1967 Geotechnical conference Shear strength properties of natural soils and rocks a conference Proceedings Vol 1-2 Norsk geoteknisk forening 2 vols Norwegian geotechnical institute,Oslo,1967
Eng 41.3135

OSLO 1967 Structure and function of transfer RNA and 58-RNA Federation of European biochemical societies meeting 4th Proceedings Vol 3: structure and function of transfer RNA and 58-RNA Federation of European biochemical societies Edited by L.O. Froholm and S.E. Laland Academic press, London,1968
Chem 18.2491

OSLO 1967 The Federation of European biochemical societies meeting 4th Proceedings Vol 3: structure and function of transfer RNA and 5 S-RNA Federation of European biochemical societies Edited by L.O. Froholm and S.G. Laland Organized by the Norwegian biochemical society Universitetsforlaget;Academic press,Oslo; London,1968 Symposium organizer H.G. Zachau
Bioch 33.1398

OSLO 1967 The Federation of European biochemical societies meeting 4th Proceedings Vol 6: structure and function of the endoplasmic reticulum in animal cells Federation of European biochemical societies Edited by F.C. Gran Organized by the Norwegian biochemical society Universitetsforlaget;Academic press,Oslo; London,1968 Symposium organizer P.N. Campbell
Bioch 33.1401

OSLO 1968 Congress of Scandinavian mathematicians 15th Proceedings Edited by K.E. Aubert and W. Ljunggren Lecture notes in mathematics, 118 Bibliog. 162p 25cm Springer-verlag,Berlin,1970
P Math 2.3240

OSLO 1969 Large permanent underground openings :international symposium Proceedings International society for rock mechanics Edited by T.L. Brekke and Finn Jorstrad Universitetsforlaget,Oslo,1970
Eng 41.3133

OSLO 1970 Algebraic geometry Oslo 1970 Nordic-summer school in mathematics 5th Proceedings Edited by F. Oort vii,332p 25cm Wolters-Noordoff,Groningen,1972
P Math 2.4173

OSLO 1970 Scandinavian logic symposium 2nd Proceedings Universitet i Oslo Edited by J.E. Fenstad Studies in logic and the foundations of mathematics, 63 vi,405p 23cm North-Holland,Amsterdam,1971
P Math 2.4056

OSLO 1976 The Federation of European biochemical societies meeting 4th Proceedings Vol 1: regulation of enzyme activity and allosteric interactions Federation of European biochemical societies Edited by E. Kvamme and A. Phil Organized by the Norwegian biochemical society Universitetsforlaget;Academic press,Oslo; London,1968 Symposium organizer J.P. Changeux
Bioch 33.1396

OSNOVNYE ITOGI IZUCHENIYA ANTARKTIKI ZA 10 LET doklady vsesoyuznogo soveshchaniya po izucheniyu Antarktiki 1966 god Moscow 1966 Apr 19-20 Mezhduvedomstvennaya kommissiya po izucheniyu Antarktiki map Izdatelstvo 'Nauka',Moscow,1967
Sco 14.6068

OSSERVATORIO ASTRONOMICO DI TRIESTE Colloquium on late-type stars Proceedings Trieste 1966 Jun 13-17 Edited by Margherita Hack 465p Trieste,1966
TA 15.0315

OSSERVATORIO ASTRONOMICO DI TRIESTE Colloquium on late-type stars proceedings Trieste 1966 Jun.13-17 Edited by Margherita Hack 465p Trieste,1967
Obs 6.0712

OTTAWA 1932 The Symposium on colloid chemistry 10th Papers presented Edited by Harry Boyer Weiser Colloid symposium monograph, 10 Journal of physical chemistry, New York,1932
Bioch 33.1477

OTTAWA 1957 The Structure and reactivity of electronically-excited species :a symposium Preprint of papers Chemical institute of Canada.Physical chemistry division Part-financed by the National research council of Canada Chemical institute of Canada.Physical chemistry division,1957 Chairman: Professor K.J.Laidler
Chem 18.0227

OTTAWA 1958 Biometrical genetics : international symposium Proceedings International union of biological sciences Edited by Oscar Kempthorne Sponsored by the Biometrics society International series of monographs on biometry International union of biological sciences.Series B, 38 Pergamon press,London,1960
Gen 34.0299

OTTAWA 1959 The Symposium on physical processes in the sun-earth environment Proceedings Canada.Defence research board Edited by C. Collins DRTE no 1025 xii,392p Ottawa,1960
Nap 11.0140

OTTAWA 1963 Centennial of entomology in Canada,1863-1963 :a commemorative joint meeting Proceedings Entomological society of Canada,and,Entomological society of Ontario Canadian entomologist,96,pt.1-2 475p Carelton university,Ottawa,1963
Sco 14.2471

OTTAWA 1963 Laboratory shear testing of soils :a symposium National research council of Canada American society for testing materials.Special technical publication, 361 American society for testing materials, Philadelphia,Pa.,1964
Eng 41.3157

OTTAWA 1963 Meeting 26th Institute of mathematical statistics 1963 Typescript papers in folder with 34th session of the International statistical institute, 1963
Math 3.1111

OTTAWA 1963 Session 34th International statistical institute 1963 Unbound typescript papers
Math 3.1112

OTTAWA 1965 The World rift system symposium report International upper mantle committee sponsored by Canada. Department of mines and technical surveys Geological survey of Canada.Paper,66-14 Ottawa,1967
Geol 8.4451

OTTAWA 1965 The World rift systems :a symposium Report National research council, Canada and International upper mantle committee Edited by T.N. Irvine Sponsored by the International union of geodesy and geophysics Geological survey of Canada.Paper, 66-14 Queen's printer,Ottawa,1967
Geod 9.0138

OTTAWA 1966 International conference on the deformation of crystalline solids National research council of Canada Canadian journal of physics, 2,pt 2-3 Ottawa,1967
Met 25.2491

OTTAWA 1967 Biochemistry and physiology of plant growth substances International conference on plant growth substances 6th Edited by F. Wightman and G. Setterfield Held at Carleton University Runge press, Ottawa,1968
Bioch 33.0779

OTTAWA 1967 Biochemistry and physiology of plant growth substances International conference on plant growth substances 6th Edited by F. Wightman and G. Setterfield Held at the Carleton university xii,1642p Runge press,Ottawa,1968 On the occasion of the centennial of Canadian confederation
Bot 42.1833

OTTAWA 1967 Biochemistry and physiology of plant growth substances International conference on plant growth substances 6th Proceedings National research council of Canada Carleton university Edited by F. Wightman and G. Setterfield Runge press, Ottawa,1968
Gen 34.1408

OTTAWA 1967 Colloquium spectroscopicum internationale 13th Proceedings Canadian association for applied spectroscopy Carleton university Co-sponsored by the Society for applied spectroscopy xix,460p 23cm Hilger,London,1968
Chem 18.2845

OTTAWA 1967 Wind effects on buildings and structures :international research seminar Proceedings Vol 1-2 National research council of Canada 2 vols University of Toronto press,Toronto,1968
Eng 41.2746

OTTAWA 1969 Crystallographic computing :a conference Proceedings International union of crystallography.Commission on crystallographic computing Edited by F.R. Ahmed and others Bibliog 384p 20cm Munksgaard,Copenhagen,1970
Cav 7.2655

OTTAWA 1969 Crystallographic computing : international summer school Proceedings International union of crystallographers. Commission on crystallographic computing Edited by F.R. Ahmed and others Munksgaard, Copenhagen,1970
Chem 18.2733

OTTAWA 1969 International summer school on crystallographic computation Proceedings International union of crystallography. Commission on crystallographic computation Edited by F.R. Ahmed Munksgaard,Copenhagen, 1970
Min 10.1527

OTTICA QUANTISTICA Scuola internazionale di fisica "Enrico Fermi" 42 corso Rendiconti Varenna 1967 Jul 31-Aug 19 Societa Italiana di fisica Edited by R.J. Glauber Bibliog.,Illus.port. xvi,159p 25cm Academic press,New York,1968
A Math 4.1599

OUTLINES OF GEOLOGIC HISTORY WITH ESPECIAL REFERENCE TO NORTH AMERICA American association for the advancement of science. Section E :meetings Papers Baltimore,Md. 1908 Dec Edited by Bailey Willis and Rollin D. Salisbury University of Chicago press, Chicago,Ill.,1910
Geog 13.0599

OVARY Endocrinologic and morphologic correlations of the ovary :the Florentine conference Florence Edited by W. Ingiulla and Robert B. Greenblatt
An 32.5225

OVARY Symposium on the physiology and pathology of human reproduction 2nd annual Proceedings Harper hospital Wayne state university Edited by Harold C. Mack Thomas, Springfield,Ill.,1968
An 32.5224

OVULATION Control of ovulation :a conference Proceedings Dedham,Mass. 1960 Feb 26-28 Edited by Claude A. Villee Sponsored by the Association for the aid of crippled children Pergamon press,Oxford,1961
An 32.3823

OXFORD 1923 International congress of psychology 7th Proceedings and papers International congress of psychology.British organizing committee Edited by Charles S. Myers Cambridge university press,Cambridge, 1924 Edited by the president of the congress.Papers in English,French,German and Italian
Psy 31.2510

OXFORD 1935 International congress of soil science 3rd guidebook for the excursion round Britain Clarendon press,Oxford,1935
Geog 13.1078

OXFORD 1938 Report of the Men of the trees summer school and conference Men of the trees illus. Men of the trees,London,c1938
BG 38.0336

OXFORD 1947 International physiological congress 17th Abstracts of communications 1947
Phys 20.2074

OXFORD 1950 The Study of the distribution of British plants :a conference Report Botanical society of the British Isles Botanical society of the British Isles.B.S.B.I. conference reports, 2 London,1951
Bot 42.4707

OXFORD 1951 Isotope techniques conference Proceedings Vol 2: industrial and allied research applications Great Britain.Ministry of supply Sponsored by the Atomic energy research establishment H.M.S.O.,London,1952
Chem 18.2467

OXFORD 1951 Radioisotope techniques Isotope techniques conference Proceedings Vol 1: medical and physiological applications Sponsored by the Atomic energy research establishment H.M.S.O.,London,1953 Subsequent conference called Radioisotope conference,q.v.
Radioth 35.1348

OXFORD 1952 Evolution Society for experimental biology and Genetical society Society for experimental biology.Symposia, 7 Cambridge university press,Cambridge,1953
Phys 20.0957

OXFORD 1952 Evolution :a symposium Papers Society for experimental biology and Genetical society Society for experimental biology.Symposia, 7 Cambridge university press,Cambridge,1953
Psy 31.0570

OXFORD 1952 Evolution symposium 7th Society for experimental biology Edited by R. Brown and J.F. Danielli Supported by the British council Society for experimental biology.Symposia, 7 xix,448p 25cm Cambridge university press,Cambridge,1953
Math 3.0773

OXFORD 1952 Evolution symposium Society for experimental biology Genetical society Society for experimental biology.Symposia, 7 Cambridge university press,Cambridge,1953
Gen 34.1258

OXFORD 1952 Prospects in psychiatric research :the proceedings of the Oxford conference of the Mental health research fund Proceedings Mental health research fund Edited by J.M. Tanner Blackwell,Oxford,1953
Psy 28.0036

OXFORD 1952 The Nature of virus multiplication :a symposium Papers Society for general microbiology Edited by Paul Fildes and W.E. Van Heyningen Held at the University of Oxford Society for general microbiology.Symposia, 2 Cambridge university press,Cambridge,1953
Bioch 33.1179

OXFORD 1952 The Nature of virus multiplication :a symposium Proceedings Society for general microbiology Edited by Paul G. Fildes and W.E. Van Heyningen Society for general microbiology.Symposia, 2 Cambridge university press,Cambridge,1953
Radioth 35.0183

OXFORD 1953 International conference on rheology 2nd proceedings Edited by C.G. W. Harrison Butterworths,London,1954
A Math 4.0367

OXFORD 1953 International congress on rheology 2nd Proceedings Edited by V.G. W. Harrison ix,451p Butterworth scientific publications,London,1954
Chem E 24.0446

OXFORD 1953 Rocket exploration of the upper atmosphere :a conference papers United States.Upper atmosphere rocket research panel and Royal society of London.Gassiot committee Edited by R.L.F. Boyd and M.J. Seaton Journal of atmospheric and terrestrial physics.Special supplement, 1 376p Pergamon press,London,1954
Nap 11.0073

OXFORD 1954 Biochemistry of the developing nervous system International neurochemical symposium 1st Proceedings Edited by Heinrich Waelsch Academic press,New York, 1955
An 32.4333

OXFORD 1954 Biochemistry of the developing nervous system International neurochemical symposium 1st Proceedings Edited by Heinrich Waelsch held at Magdalen college, Oxford Academic press,New York,1955
Bal 39.1468

OXFORD 1954 Biochemistry of the developing nervous system The International neurochemical symposium 1st Proceedings Edited by Heinrich Waelsch Academic press, New York,1955
Bioch 33.2233

OXFORD 1954 Radioisotope conference 2nd Proceedings 1: medical and physiological applications Edited by J.E. Johnston and others Sponsored by the Atomic energy research establishment Butterworths,London, 1954 First conference called Isotope techniques conference,q.v.
Radioth 35.1349

OXFORD 1954 Radioisotope conference 2nd Proceedings Vol 1: medical and physiological applications Edited by J.R. Johnston and others Sponsored by the Atomic energy research establishment Butterworths, London,1954 Previous conference has title "Isotope techniques conference",Oxford,1951
Bioch 33.0300

OXFORD 1955 Mitochondria and other cytoplasmic inclusions :a symposium Papers Society for experimental biology Society for experimental biology.Symposia, 10 Cambridge university press,Cambridge,1957
Radioth 35.0472

OXFORD 1955 Mitochondria and other cytoplasmic inclusions :a symposium Papers Society for experimental biology.Symposia, 10 Cambridge university press,Cambridge,1957
An 32.3392

OXFORD 1955 Mitochondria and other cytoplasmic inclusions :a symposium Society for experimental biology Society for experimental biology.Symposia, 10 Cambridge university press,Cambridge,1957
Bioch 33.1355

OXFORD 1955 The Physical chemistry of enzymes :a discussion Papers Faraday society Faraday society.Discussions, 20 Aberdeen,1956
Bioch 33.1047

OXFORD 1957 Function and taxonomic importance Systematics association symposium 3rd papers Systematics association Edited by A.J. Cain Systematics association. Publication,3 Systematics association,London, 1959
Geol 8.0941

OXFORD 1957 Function and taxonomic importance :a symposium By A.J. Cain Systematics association Systematics association.Publication, 3 Systematics association,London,1959
Gen 34.1623

OXFORD 1958 A Symposium on atmospheric diffusion and air pollution proceedings International union of theoretical and applied mechanics and International union of geodesy and geophysics Edited by F.N. Frenkel and P.A. Sheppard Advances in geophysics, 6 Academic press,New York; London,1959
A Math 4.0576

OXFORD 1958 Atmospheric diffusion and air pollution :a symposium Proceedings International union of theoretical and applied mechanics International union of geodesy and geophysics Edited by F.N. Frenkel and P.A. Sheppard Advances in physics, 6 Academic press,New York,1959
Eng 41.6946

OXFORD 1958 Atmospheric diffusion and air pollution :a symposium Proceedings International union of theoretical and applied mechanics and International union of geodesy and geophysics Edited by F.N. Frenkiel and P.A. Sheppard New York,1959
Geod 9.0521

OXFORD 1958 The Pharmacology of plant phenolics :a symposium Proceedings Edited by J.W. Fairbairn Organised by the Plant phenolics group Academic press,London;New York,1959
Bioch 33.0752

OXFORD 1958 The Pharmacology of plant phenolics :a symposium Proceedings Plant phenolics group Edited by J.W. Fairbairn Academic press,London;New York,1959
Chem 26.0265

OXFORD 1959 Flow properties of blood and other biological systems :an informal discussion Proceedings Faraday society. Colloid and biophysics committee British society of rheology Edited by A.L. Copley and G. Stainsby Pergamon,London,1960
PGMS 29.0269

OXFORD 1959 Flow properties of blood and other biological systems :an informal discussion... Proceedings Faraday society and British society of rheology Edited by A.L. Copley and G. Stainsby Pergamon press, Oxford,1960
An 32.3471

OXFORD 1959 Symposium on recent advances in optics Proceedings University of Oxford 138-180p London,1959
Obs 6.1989

OXFORD 1959 Taxonomy and geography Systematics association symposium 4th Papers Systematics association Edited by David Nichols Systematics association. Publication,4 Systematics association,London, 1962
Geog 13.1162

OXFORD 1959 Taxonomy and geography Systematics association symposium 4th papers Edited by David Nichols Systematics association.Publication,4 Systematics association,London,1962
Geol 8.0942

OXFORD 1959 The Biology of weeds :a symposium British ecological society Edited by John L. Harper British ecological society.Symposia, 1 Bibliog,maps,2 plates, tables,diagrs xv,256p Blackwell scientific, Oxford,1960
Bot 42.2032

OXFORD 1960 Radiation damage in bone Conference on the relation of radiation damage to radiation dose in bone Report and summarized papers International atomic energy agency International atomic energy agency,Vienna,1960
Radioth 35.1115

OXFORD 1961 Numerical solution of ordinary and partial differential equations based on a Summer school University of Oxford.Computing laboratory Edited by L. Fox Adiwes international series in the engineering sciences Addison-Wesley series in computer science and information processing Pergamon press;Addison-Wesley,London;Reading,Mass.,1962
TA 15.0378

OXFORD 1961 Numerical solution of ordinary and partial differential equations based on a summer school held at Oxford August-September 1961 Oxford university.Computing laboratory Oxford,1962
Geod 9.0179

OXFORD 1961 Numerical solution of ordinary and partial differential equations summer school By L. Fox University of Oxford. Computing laboratory bibliog. ix,509p 23cm Pergamon press,Oxford,1962
Math L 5.0308

OXFORD 1961 Numerical solutions of ordinary and partial differential equations : summer school Lectures University of Oxford.Computing laboratory Edited by L. Fox In collaboration with the University of Oxford. Delegacy for extra-mural studies Pergamon press,Oxford,1962
Cav 7.0830

OXFORD 1961 Regulation of human respiration Haldane centenary symposium Proceedings Edited by D.J.C. Cunningham and B.B. Lloyd bibliog.,port. Blackwells, Oxford,1963
Phys 20.0857

OXFORD 1961 Science of ceramics :a conference 1st Proceedings British ceramic society Edited by G.H. Stewart Academic press,London;New York,1962
Met 25.0274

OXFORD 1961 Scientific change Symposium on the history of science Papers Edited by A.C. Crombie Under the auspices of the International union of the history and philosophy of science.Division of history of science Heinemann,London,1963
Geog 13.0053

OXFORD 1961 Scientific change...Historical studies in the intellectual,social and technical conditions for scientific discovery.. symposium on the history of science International union of history and philosophy of science.Division of history of science Edited by A.C. Crombie xii,896p Heinemann, London,1963
WSM 43.0029

OXFORD 1962 Cytochemical progress in electron microscopy :a symposium Royal microscopical society Royal microscopical society.Journal, 81,pt 3-4 London,1963
Bot 42.4729

OXFORD 1962 Symposium on cytochemical progress in electron microscopy Royal microscopical society Journal of the Royal microscopical society.Ser.3, 81,pt 3-4,p.107-117 London,1963
Bal 39.0235

OXFORD 1962 Symposium on cytochemical progress in electron microscopy Royal microscopical society Journal of the Royal microscopical society,1962,pt.3,4 Royal microscopical society,London,1963
An 32.3182

OXFORD 1963 Advances in programming and non-numerical computation summer school proceedings University of Oxford.Computing laboratory Edited by L. Fox bibliog viii, 218p 23cm Pergamon press,Oxford,1966
Math L 5.1026

OXFORD 1963 Advances in programming and non-numerical computation summer school proceedings University of Oxford.Computing laboratory Edited by L. Fox viii,218p 24cm Pergamon press,Oxford,1966
Math 3.0239

OXFORD 1963 Formal systems and recursive functions The Logic colloquium 8th proceedings Edited by J.N. Crossley and M.A. E. Dummett 320p 23cm North-Holland, Amsterdam,1965
P. Math 2.0207

OXFORD 1963 The Control of lipid metabolism :a symposium Biochemical society Edited by J.K. Grant Biochemical society. Symposia, 24 Academic press,London;New York, 1963 Organized by G.Popjak
Bioch 33.1386

OXFORD 1964 Chromosomes today The Oxford chromosome conference 1st Proceedings Edited by C.D. Darlington and K.R. Lewis Oliver and Boyd,Edinburgh;London,1966
Mol 45.0190

OXFORD 1964 Mathematics and computer science in biology and medicine Conference on mathematics and computer science in biology and medicine Proceedings Medical research council 317p H.M.S.O.,London,1965
Bot 42.0512

OXFORD 1964 Mathematics and computer science in biology and medicine :conference Proceedings Medical research council Co-sponsored by Great Britain.Ministry of health H.M.S.O.,London,1965
HE 27.0125

OXFORD 1964 Mathematics and computer science in biology and medicine conference proceedings Medical research council Great Britain.Ministry of health Co-sponsored by Great Britain.Scottish home and health department Bibliog. xi,317p 25cm H.M.S.O.,London,1965
Math 3.0775

OXFORD 1964 Mathematics and computer science in biology and medicine conference proceedings Medical research council Co-sponsored by Ministry of health,bibliog,illus., pl. vii,316p 24cm H.M.S.O.,London,1965
Math L 5.1007

OXFORD 1964 Medical research council conference Oxford,1964 Resumes of papers given at the conference
Math 3.1121

OXFORD 1964 Recent progress in photobiology International congress of photobiology 4th Proceedings Edited by E.J. Bowen Under the auspices of the Comite international de photobiologie Bibliog., illus. vii,400p Blackwell scientific publications,Oxford,1965
Bot 42.1362

OXFORD 1964 Recent progress in photobiology International photobiology congress 4th Proceedings Edited by E.J. Bowen Under the auspices of the Comite internationale de photobiologie Blackwell, Oxford,1965
Radioth 35.0253

OXFORD 1964 Oxford 1967 Sep 5-8 Chromosomes today :Oxford chromosome conference Proceedings Vol 1-2 Edited by C.D. Darlington and K.R. Lewis illus. 2 vols Oliver and Boyd,Edinburgh. London,1966-68
Gen 34.0893

OXFORD 1964 and OXFORD 1967 Chromosomes today :Oxford chromosome conference Proceedings Vol 1-2 Edited by C.D. Darlington and K.R. Lewis Heredity.Supplement series, 19,24 illus Oliver and Boyd,Edinburgh,1966-68
Bot 42.0788

OXFORD 1965 Electromagnetic distance measurement :a symposium International association of geodesy Hilger and Watts, London,1967
Geog 13.5530

OXFORD 1965 Electromagnetic distance measurement :a symposium International association of geodesy.Special duty group no 19 Hilger and Watts,London,1967
Eng 41.3278

OXFORD 1965 International conference on elementary particles proceedings Rutherford high energy laboratory Sponsored by the International union of pure and applied physics Rutherford high energy laboratory, Oxford,1965
A Math 4.0887

OXFORD 1965 Methods of numerical approximation lectures delivered at a Summer school... Edited by D.C. Handscomb Bibliog. Illus. ix,218p 23cm Pergamon press, Oxford,1966
A Math 4.1578

OXFORD 1966 Brasenose conference on the automation of libraries Anglo-American conference on the mechanization of library services Proceedings Edited by John Harrison and Peter Laslett Sponsored by the Old Dominion foundation of New York Mansell, Chicago,Ill.,1967
Eng 41.7672

OXFORD 1966 Historical aspects of microscopy Papers Royal microscopical society Edited by S. Bradbury and G.L'E. Turner W.Heffer,Cambridge,1967
Bal 39.0241

OXFORD 1966 Historical aspects of microscopy Royal microscopical society conference Papers Royal microscopical society Edited by S. Bradbury and G.L'E. Turner viii,227p Cambridge,1967
WSM 43.2520

OXFORD 1966 International symposium on rarefied gas dynamics 5th Proceedings Supplement 4 By C.L. Brundin New York; London,1967
A Math 4.1302

OXFORD 1966 Physical basis of yield and fracture conference Proceedings Institute of physics and the physical society.Stress analysis group committee Edited by A.C. Stickland and R.A. Cook Institute of physics and the physical society.Conference series, 1 Inst.of Physics and the Physical society, London,c1966
Met 25.0869

OXFORD 1966 Rarefied gas dynamics : international symposium 5th Proceedings Vol 1-2 Edited by C.L. Brundin Advances in applied mechanics.Supplement, 4 2 vols Academic press,New York;London,1967
Chem R 24.0409

OXFORD 1966 The Wates foundation symposium on arterial chemoreceptors Proceedings Edited by R.W. Torrance Blackwell,Oxford, 1968
An 32.3810

OXFORD 1966 The Wates foundation symposium on arterial chemoreceptors Proceedings Edited by R.W. Torrance pl. xiv,402p Blackwell scientific publications,Oxford, Edinburgh,1968
Phys 20.1968

OXFORD 1967 Aspects of cell motility :a symposium Papers Society for experimental biology Society for experimental biology. Symposia, 22 Cambridge university press, Cambridge,1968
Radioth 35.0473

OXFORD 1967 Aspects of cell motility :a symposium Society for experimental biology Society for experimental biology.Symposia, 22 Cambridge university press,Cambridge,1968
An 32.5206

OXFORD 1967 Aspects of cell motility :a symposium Society for experimental biology Society for experimental biology.Symposia, 22 Cambridge university press,Cambridge,1968
Bioch 33.1361

OXFORD 1967 Chromosomes today 2 Oxford chromosome conference 2nd Proceedings Edited by C.D. Darlington and K.R. Lewis Oliver and Boyd,Edinburgh,1969
Gen 34.2179

OXFORD 1967 Computational problems in abstract algebra :a conference Proceedings Science research council Held under the auspices of the Atlas computer laboratory bibliog.,illus. x,402p 23cm Pergamon press,Oxford,1970
P Math 2.3758

OXFORD 1967 The Metabolic roles of citrate a symposium in honour of...Sir Hans Krebs Biochemical society Edited by T.W. Goodwin Biochemical society.Symposia, 27 Academic press,London;New York,1968
Bioch 33.1389

OXFORD 1968 Symposium on techniques used in the study of microcirculation Papers British microcirculation society Edited by D. R. Chambers and P.A.G. Monro British microcirculation society,Cambridge,1969
An 32.5223

OXFORD 1969 Combinatorial mathematics and its application :conference Proceedings Edited by J.J.A. Welsh Held at the Mathematical institute,Oxford x,364p 24cm Academic press,New York;London,1971
Math S 3.1661

OXFORD 1969 Combinatorial mathematics and its applications :conference Proceedings University of Oxford.Mathematics institute Edited by D.J.A. Welsh x,364p 24cm Academic press,London;New York,1971
P Math 2.4514

OXFORD 1969 Computers in number theory Science research council Atlas symposium 2nd Proceedings Science research council Edited by A.O.L. Atkin and B.J. Birch xvii, 433p 24cm Academic press,London;New York, 1971
P Math 2.4381

OXFORD 1969 Finite simple groups :an instructional conference Proceedings London mathematical society Edited by M.B. Powell and G. Higman xi,327p 24cm Academic press,London;New York,1971
P Math 2.4431

OXFORD 1969 Motions in molecular crystals Faraday society Faraday society.Discussions, 48 Bibliog 224p 20cm Faraday society, London,1969
Cav 7.2718

OXFORD 1970 Chemical reactivity and biological role of functional groups in enzymes Biochemical society Edited by R.M. S. Smellie Biochemical society.Symposium, 31 Academic press,London,1970
Bioch 33.2315

OXFORD 1970 Chemical reactivity and biological role of functional groups in enzymes :a symposium... Proceedings Molecular enzymology and protein group Edited by R.M.S. Smellie Biochemical society. Symposia, 31 Academic press,London,1970 Molecular enzymology and protein group of the Biochemical society and the Chemical society
Chem 18.2745

OXFORD 1970 Large sparse sets of linear equations Conference of the Institute of mathematics and its applications Proceedings Institute of mathematics and its applications Edited by J.K. Reid Illus 284p Academic press,London,1971
Math L 5.3804

OXFORD 1970 Large sparse sets of linear equations Oxford conference of the Institute of mathematics and its applications Proceedings Institute of mathematics and its applications Edited by J.K. Reid x, 284p 24cm Academic press,London;New York, 1971
P Math 2.4464

OXFORD 1970 Large sparse sets of linear equations :a conference Proceedings Institute of mathematics and its applications Edited by J.K. Reid Academic press,London, 1971
Eng 41.8474

OXFORD 1971 Steroids in modern medicine : symposium Pt 1-2 Institute of hormone biology Syntex pharmaceuticals Edited by George A. Christie and Miriam Moore-Robinson Excerpta medica,Amsterdam,1972
Bioch 33.2199

The OXFORD CHROMOSOME CONFERENCE 1st Proceedings Chromosomes today Oxford 1964 Jul 28-31 Edited by C.D. Darlington and K.R. Lewis Oliver and Boyd,Edinburgh;London, 1966
Mol 45.0190

OXFORD CHROMOSOME CONFERENCE 2nd Proceedings Chromosomes today 2 Oxford 1967 Sep 5-8 Edited by C.D. Darlington and K. R. Lewis Oliver and Boyd,Edinburgh,1969
Gen 34.2179

OXFORD CHROMOSOME CONFERENCE 1ST,2ND Chromosomes today :Oxford chromosome conference Proceedings Oxford 1964 Jul 28-31 and Oxford 1967 Sep 5-8 Vol 1-2 Edited by C.D. Darlington and K.R. Lewis Heredity.Supplement series, 19,24 illus Oliver and Boyd,Edinburgh,1966-68
Bot 42.0788

OXFORD CHROMOSOME CONFERENCE 1ST,2ND Chromosomes today :Oxford chromosome conference Proceedings Oxford 1964 Jun 28-31 Oxford 1967 Sep 5-8 Vol 1-2 Edited by C.D. Darlington and K.R. Lewis illus. 2 vols Oliver and Boyd,Edinburgh. London,1966-68
Gen 34.0893

OXFORD CONFERENCE OF THE INSTITUTE OF MATHEMATICS AND ITS APPLICATIONS Proceedings Large sparse sets of linear equations Oxford 1970 Apr 5-8 Institute of mathematics and its applications Edited by J.K. Reid x, 284p 24cm Academic press,London;New York, 1971
P Math 2.4464

OXFORD UNIVERSITY.COMPUTING LABORATORY Numerical solution of ordinary and partial differential equations based on a summer school held at Oxford August-September 1961 Oxford 1961 Aug-Sep Oxford,1962
Geod 9.0179

OXIDASES AND RELATED REDOX SYSTEMS :a symposium Proceedings Amherst,Mass. 1964 Jul 15-19 Vol 1-2 Edited by Tsoo E. King and others held at Amherst college 2 vols Wiley,New York,1965
Bioch 33.1085

OXIDATION REACTIONS International congress of biochemistry 5th Proceedings Moscow 1961 Aug 10-16 Vol 5: intracellular respiration:phosphorylating and non-phosphorylating oxidation reactions International union of biochemistry Edited by E.C. Slater I.U.B.symposium series, 25 Pergamon;PWN-Polish scientific publishers, Oxford;Warsaw,1963
Bioch 33.1334

OXIDE DISPERSION STRENGTHENING Bolton Landing conference oxide dispersion strengthening:a symposium 2nd Bolton Landing,New York 1966 Jun 27-29 American institute of mining,metallurgical and petroleum engineers Edited by G.S. Ansell Metallurgical society conferences, 47 Gordon and Breach,New York,1968
Met 25.2339

OXYGEN Biochemie des sauerstoffs :colloquium Mosbach 1968 Apr 24-27 Gesellschaft fur biologische chemie Edited by B. Hess and Hj. Staudinger Gesellschaft fur biologische chemie.Colloquia, 19 illus Springer, Berlin,1968 In English and German
Bioch 33.1074

OXYGEN IN THE ANIMAL ORGANISM :a symposium Proceedings London 1963 Sep 1-5 International union of biochemistry and International union of physiological sciences Edited by Frank Dickens and Eric Neil Pergamon press,Oxford,1964
PGMS 29.0407

OXYGEN IN THE ANIMAL ORGANISM :a symposium Proceedings London 1963 Sep 1-5 International union of biochemistry and International union of physiological sciences Edited by Frank Dickens and Eric Neil held at Bedford college I.U.B.Symposium series, 31 Pergamon press,Oxford,1964
Bioch 33.1080

OXYGEN SUPPLY TO THE HUMAN FOETUS :a symposium Princeton,N.J. 1957 Dec Council for international organizations of medical sciences and Josiah Macy Jr. foundation Edited by James Walker and Alec C. Turnbull Blackwell,Oxford,1959
Phys 20.1003

OXYGEN SUPPLY TO THE HUMAN FOETUS :a symposium Princeton,N.J. 1957 Dec Edited by James Walker and Alec C. Turnbull Organised by the Council for international organizations of medical sciences Blackwell, Oxford,1959
An 32.2514

OXYTOCIN :an international symposium Proceedings Montevideo 1959 Universidad de Uraguay.Facultad de medicina. Servicio de fisiologia obstetrica Edited by R. Caldeyro-Barcia and H. Heller Pergamon press,London,1961
VA 19.0237

OXYTOCIN :an international symposium Proceedings Montevideo 1959 Universidad de Uruguay.Facultad de medicina. Servicio de fisiologia obstetrica Edited by R. Caldeyro-Barcia and H. Heller Pergamon, Oxford,1961
Pha 16.0082

OXYTOCIN :an international symposium Proceedings Montevideo 1959 Universidad de Uruguay.Facultad de medicina. Servicio de fisiologia obstetrica Edited by R. Caldeyro-Barcia and H. Heller Pergamon press,Oxford,1961
Phys 20.1387

OXYTOCIN :an international symposium Proceedings Montevideo 1959 Universidad de Uruguay.Facultad de medicina. Servicio de fisiologica obstetrica Edited by R. Caldeyro-Barcia and H. Heller Pergamon, Oxford,1961
An 32.5322

OZONE CHEMISTRY AND TECHNOLOGY :international ozone conference Proceedings Chicago, Ill. 1956 Nov 28-30 Armour research foundation American chemical society. Advances in chemistry series,21 American chemical society,Washington,D.C.,1959 Edited by the staff of the American chemical society
Chem 18.0031

PACIFIC AREA NATIONAL MEETING 2ND Symposium on nondestructive testing presented to the 2nd Pacific area national meeting Los Angeles 1956 Sep 17-18 American society for testing materials A.S.T.M.Special technical publication, 213 A.S.T.M.,Philadelphia,1957
Met 25.1323

PACIFIC AREA NATIONAL MEETING 2ND Symposium on radioisotopes presented at the second Pacific area national meeting Los Angeles 1956 Sep 21 American society for testing materials A.S.T.M.Special technical publication, 215 A.S.T.M.,Philadelphia,1958
Met 25.1559

PACIFIC AREA NATIONAL MEETING 2ND Symposium on titanium held at the second Pacific area national meeting Los Angeles 1956 Sep 17-18 American society for testing materials A.S.T.M.Special technical publication, 204 American society for testing materials, Philadelphia,Pa.,1957
Met 25.0571

PACIFIC AREA NATIONAL MEETING 3RD Newer metals Symposium on newer metals presented at the third Pacific area national meeting... San Francisco,Calif. 1959 Oct 15 American society for testing materials A.S.T.M. Special technical publication, 272 A.S.T.M., Philadelphia,Pa.,1960
Met 25.0514

PACIFIC BASIN The Crust of the Pacific basin two symposia,presented at the 10th Pacific Honolulu 1961 Aug 21-Sep 6 Edited by Gordon A. MacDonald and Hisashi Kuno Geophysical monograph, 6 National research council.Publication,1035 Washington,D.C., 1961 Based on the papers presented at the symposia
Geod 9.0423

PACIFIC BASIN The Crust of the Pacific basin two symposia,presented at the 10th Pacific Honolulu 1961 Aug 21-Sep 6 Edited by Gordon A. MacDonald and Hisashi Kuno Geophysical monograph, 6 National research council.Publication,1035 Washington,D.C., 1961 Based on the papers presented at the symposia
Geod 9.0423

PACIFIC BASIN BIOGEOGRAPHY Pacific science congress 10th Symposium Honolulu 1961 Aug 21-Sep 6 Edited by J.Linsley Gressitt and others Sponsored by the National academy of science Honolulu,1963
Geog 13.1161

PACIFIC BASIN BIOGEOGRAPHY Pacific science congress :a symposium 10th Proceedings Honolulu 1961 Aug-Sep Edited by J.Linsley Gressitt illus,maps x,563p 26cm Bishop museum press,Honolulu,1963
Sco 14.2205

PACIFIC GROVE,CALIF. 1929 Contributions to marine biology :lectures and symposia given at the Hopkins marine station...at The midwinter meeting of the Western society of naturalists Western society of naturalists Stanford university press,Stanford,Calif.,1930
Bal 39.1933

PACIFIC GROVE,CALIF. 1954 The Luminescence of biological systems :conference Proceedings Edited by Frank H. Johnson Sponsored by National academy of sciences. Committee on photobiology American association for the advancement of science, Washington D.C.,1955
An 32.1502

PACIFIC GROVE,CALIF. 1963 Asilomar conference The Conference on reactions between complex nuclei Edited by A. Ghiorso and others 456p University of California press,Berkeley,Calif.,1963
Cav 7.0090

PACIFIC GROVE,CALIF. 1963 High temperature technology An International symposium on high temperature technology 3rd Proceedings International union of pure and applied chemistry.Commission on high temperatures and refractories Organized and directed by Stanford research institute Butterworths,London,1964
Met 25.0992

PACIFIC GROVE,CALIF. 1971 International conference on electron spectroscopy Proceedings Edited by D.A. Shirley xii, 916p 23cm North-Holland,London,1972
Chem 18.2871

PACIFIC ISLAND TERRACES Pacific science congress 10th Honolulu 1961 Pacific science association Edited by Richard J. Russell Annals of geomorphology.New series, 3 Borntraeger,Berlin,1961
Bot 42.3310

PACIFIC ISLAND TERRACES:EUSTATIC? Pacific science congress 10th Symposium Honolulu 1961 Aug 21-Sep 6 Edited by Richard J. Russell Annals of geomorphology.Ns. Supplementband,3 Borntraeger,Berlin,1961
Geog 13.0372

PACIFIC PALISADES 1965 Aggression and defence:neural mechanisms and social patterns (brain function,vol 5) Conference on brain function 5th Proceedings Edited by Carmine D. Clemente and Donald B. Lindsley UCLA forum in medical sciences, 7 University of California press,Berkeley;Los Angeles,Calif.,1967
An 32.5475

PACIFIC SCIENCE ASSOCIATION Pacific science congress 10th Abstracts of papers Honolulu 1961 487p 27cm Honolulu,1961
Sco 14.0136

PACIFIC SCIENCE ASSOCIATION Pacific science congress 7th proceedings Auckland 1949 Feb 2-Mar 4 and Christchurch,N.Z. 1949 Feb 2-Mar 4 Vol 2: geology Wellington,1953
Geol 8.3069

PACIFIC SCIENCE ASSOCIATION Pacific science congress 8th proceedings Quezon City 1953 Nov 16-28 Vol 2,2A: geology and geophysics meteorology Under the auspices of the National research council of the Philippines 2 vols National research council of the Philippines,Quezon City,1956
Geol 8.3070

PACIFIC SCIENCE ASSOCIATION Symposium on medicinal plants :8th Pacific science congress Quezon city 1953 Nov 16-28 Under the auspices of the National research council of the Philippines Pacific science congress. Proceedings, 8th,4a National research council of the Philippines,Quezon city,1954
BG 38.3111

PACIFIC SCIENCE ASSOCIATION The Pacific science congress 6th Proceedings Berkeley,Calif. 1939 Jul 24-Aug 12 San Francisco Vol 2 illus 451-914p 24cm University of California press,Berkeley,Calif, 1940
Sco 14.0115

PACIFIC SCIENCE ASSOCIATION Venomous and poisonous animals and noxious plants of the Pacific region :a collection of papers based on a symposium in the Public health and medical science division at the tenth Pacific science congress Papers Honolulu 1961 Aug 21-Sep 6 Edited by Hugh L. Keegan and W. V. MacFarlane Co-sponsored by the United States.National academy of sciences Pergamon press.Symposium publications division,Oxford, 1963
Chem 26.0448

PACIFIC SCIENCE ASSOCIATION,and others Pacific science congress 10th The Matthew Fontaine Maury memorial symposium :papers presented to the tenth Pacific science congress of the Pacific science association 10th Papers Honolulu 1961 Aug 21-Sep 6 Edited by Harry Wexler and others Geophysical monograph,7 American geophysical union.Publication,1036 illus,maps x,228p 25cm American geophysical union,Washington,D. C.,1962
Sco 14.6079

PACIFIC SCIENCE CONGRESS 10th Pacific island terraces Honolulu 1961 Pacific science association Edited by Richard J. Russell Annals of geomorphology.New series, 3 Borntraeger,Berlin,1961
Bot 42.3310

PACIFIC SCIENCE CONGRESS 10th :a symposium Proceedings Pacific basin biogeography Honolulu 1961 Aug-Sep Edited by J.Linsley Gressitt illus,maps x,563p 26cm Bishop museum press,Honolulu,1963
Sco 14.2205

PACIFIC SCIENCE CONGRESS 10th Abstracts of papers Honolulu 1961 Pacific science association 487p 27cm Honolulu,1961
Sco 14.0136

PACIFIC SCIENCE CONGRESS 10th Symposium Mans place in the island ecosystem Honolulu 1961 Aug 21-Sep 6 Edited by F.R. Fosberg Bishop museum press,Honolulu,1963
Geog 13.1311

PACIFIC SCIENCE CONGRESS 10th Symposium Pacific basin biogeography Honolulu 1961 Aug 21-Sep 6 Edited by J.Linsley Gressitt and others Sponsored by the National academy of science Honolulu,1963
Geog 13.1161

PACIFIC SCIENCE CONGRESS 10th Symposium Pacific island terraces:eustatic? Honolulu 1961 Aug 21-Sep 6 Edited by Richard J. Russell Annals of geomorphology.Ns. Supplementband,3 Borntraeger,Berlin,1961
Geog 13.0372

PACIFIC SCIENCE CONGRESS 10th papers Honolulu 1961 Aug 21-Sep 6 the crust of the Pacific basin Edited by Gordon A. Macdonald and Hisashi Kuno Geophysical monographs,6 American geophysical union, Washington,D.C.,1962
Min 10.0421

The PACIFIC SCIENCE CONGRESS 6th Proceedings Berkeley,Calif. 1939 Jul 24-Aug 12 San Francisco Vol 2 Pacific science association illus 451-914p 24cm University of California press,Berkeley,Calif, 1940
Sco 14.0115

PACIFIC SCIENCE CONGRESS 7th proceedings Auckland 1949 Feb 2-Mar 4 and Christchurch,N.Z. 1949 Feb 2-Mar 4 Vol 2: geology Pacific science association Wellington,1953
Geol 8.3069

PACIFIC SCIENCE CONGRESS 8th proceedings Quezon City 1953 Nov 16-28 Vol 2,2A: geology and geophysics meteorology Pacific science association Under the auspices of the National research council of the Philippines 2 vols National research council of the Philippines,Quezon City,1956
Geol 8.3070

PACIFIC SCIENCE CONGRESS.PROCEEDINGS, 8th,4a Symposium on medicinal plants :8th Pacific science congress Quezon city 1953 Nov 16-28 Pacific science association Under the auspices of the National research council of the Philippines National research council of the Philippines,Quezon city,1954
BG 38.3111

PACIFIC SCIENCE CONGRESS,10TH The Crust of the Pacific basin :based on papers in two symposia...presented at the Tenth Pacific science congress Edited by Gordon A. Macdonald and Hisashi Kuno Geophysical monographs,6 American geophysical union, Washington,D.C.,1962
Geog 13.0745

PACIFIC SCIENCE CONGRESS,10TH,1961 The Crust of the Pacific basin :two symposia,presented at the 10th Pacific Honolulu 1961 Aug 21-Sep 6 Edited by Gordon A. MacDonald and Hisashi Kuno Geophysical monograph, 6 National research council.Publication,1035 Washington,D.C.,1961 Based on the papers presented at the symposia
Geod 9.0423

PACIFIC SCIENCE CONGRESS,11TH Proceedings of the symposium on Pacific-Antarctic sciences : papers presented at the eleventh Pacific congress 11th Proceedings Tokyo 1966 Aug 23-27 Edited by Takesi Nagata Japanese Antarctic research expedition.Scientific reports.Special issue,1 n.p.,1967
Sco 14.6039

PACIFIC SCIENCE CONGRESS,8TH Symposium on medicinal plants :8th Pacific science congress Quezon city 1953 Nov 16-28 Pacific science association Under the auspices of the National research council of the Philippines Pacific science congress.Proceedings, 8th,4a National research council of the Philippines, Quezon city,1954
BG 38.3111

PACIFIC SCIENCE CONGRESS 10TH Plants and the migrations of Pacific peoples :a symposium Honolulu 1961 Aug 21-Sep 6 Bishop (Bernice P) museum,and,University of Hawaii Edited by J. Barrau Sponsored by the National science foundation Bishop museum press,Honolulu,1963 Symposium convened at the 10th Pacific science congress
Geog 13.7202

PACIFIC SCIENCE CONGRESS 10TH The Crust of the Pacific basin :two symposia,presented at the 10th Pacific Honolulu 1961 Aug 21-Sep 6 Edited by Gordon A. MacDonald and Hisashi Kuno Geophysical monograph, 6 National research council.Publication,1035 Washington, D.C.,1961 Based on the papers presented at the symposia
Geod 9.0423

PACIFIC SCIENCE CONGRESS 10TH The Matthew Fontaine Maury memorial symposium :papers presented to the tenth Pacific science congress of the Pacific science association 10th Papers Honolulu 1961 Aug 21-Sep 6 Pacific science association,and others Edited by Harry Wexler and others Geophysical monograph,7 American geophysical union.Publication,1036 illus,maps x,228p 25cm American geophysical union,Washington,D. C.,1962
Sco 14.6079

PACIFIC SCIENCE CONGRESS 10TH Venomous and poisonous animals and noxious plants of the Pacific region :a collection of papers based on a symposium in the Public health and medical science division at the tenth Pacific science congress Papers Honolulu 1961 Aug 21-Sep 6 Pacific science association and University of Hawaii Edited by Hugh L. Keegan and W.V. MacFarlane Co-sponsored by the United States.National academy of sciences Pergamon press.Symposium publications division, Oxford,1963
Chem 26.0448

PACIFIC-ANTARCTIC SCIENCES Pacific science congress,11th Proceedings of the symposium on Pacific-Antarctic sciences :papers presented at the eleventh Pacific congress 11th Proceedings Tokyo 1966 Aug 23-27 Edited by Takesi Nagata Japanese Antarctic research expedition.Scientific reports.Special issue,1 n.p.,1967
Sco 14.6039

PACKARD INSTRUMENT COMPANY Annual symposium on tracer methodology 5th Proceedings Washington,D.C. 1961 Oct 20 Edited by Seymour Rothschild Advances in tracer methodology, 1 Plenum press,New York,1963 Includes selected papers from the first four annual symposia
Gen 34.0633

PADUA 1964 and VENICE Convegno sulla cosmologia;meeting on cosmology Atti; Proceedings Quatro centenario della nascita di Galileo Galilei 1564-1964.Comitato nazionale per le manifestazioni celebrative Edited by L. Rosino Organized by Universita di Padua Quatro centenario della nascita di Galileo Galilei.Comitato nazionale per le manifestazioni celebrative.Pubblicazioni, 2, Atti dei convegni, 3 206p Florence,1966
Obs 6.3451

PAIN International symposium on pain Proceedings Paris 1967 Apr 11-13 Edited by A. Soulairac and others Academic press, London,1968
Pha 16.0263

PAIN La Douleur et les douleurs :a symposium Paris 1955 Hopital de la Salpetriere. Clinique des maladies du systeme nerveux Edited by Th. Alajouanine and others Masson, Paris,1957
Psy 31.0214

PAIN Proceedings New York 1942 Dec 18-19 Association for research in nervous and mental disease Association for research in nervous and mental disease.Research publications, 23 Baltimore,Md.,1943
Phys 20.2003

PAIN Proceedings of the association New York 1942 Dec 18-19 Association for research in nervous and mental disease Edited by Harold G. Wolff and others Association for research in nervous and mental disease.Research publications, 23 Williams and Wilkins,Baltimore,Md.,1945
An 32.4520

PAIN AND ITCH :nervous mechanisms London 1959 Mar 10 Ciba foundation Edited by G.E.W. Wolstenholme and Maeve O'Connor Ciba foundation study groups, 1 Churchill,London,1959
Phys 20.2017

PAIN AND ITCH nervous mechanisms London 1959 Mar 10 Ciba foundation Edited by G.E. W. Wolstenholme and M. O'Connor Ciba foundation study group, 1 Churchill,London, 1959
PGMS 29.0539

PAKISTAN Scientific problems of the humid tropical zone deltas and their implication :a symposium Proceedings Dacca 1964 Feb 24-Mar 2 Unesco Humid tropics research Unesco,Paris,1966 Partly in French
Geog 13.5396

PALAEOBOTANICAL SOCIETY.SPECIAL SESSION,1964 A Symposium on floristics and stratigraphy of Gondwanaland Proceedings Birbal Sahni institute of palaeobotany 1964 Dec Palaeobotanical society illus. 154p 24cm Birbal Sahni institute of palaeobotany,Lucknow, 1966
Bot 42.3271

PALAEOCLIMATES Problems in palaeoclimatology NATO palaeoclimates conference :a NATO advanced study institute proceedings Newcastle-upon-Tyne 1963 Jan.7-12 Edited by A.E.M. Nairn sponsored by the North Atlantic treaty organisation.Science office John Wiley,London,1964
Geol 8.1353

PALAEOCLIMATES CONFERENCE Proceedings Problems in palaeoclimatology Newcastle-upon-Tyne 1963 Jan 7-12 North Atlantic treaty organization Edited by A.E.M. Nairn Interscience,London,1964
Geog 13.1058

PALAEOCLIMATOLOGY Problems in palaeoclimatology Palaeoclimates conference Proceedings Newcastle-upon-Tyne 1963 Jan 7-12 North Atlantic treaty organization Edited by A.E.M. Nairn Interscience,London, 1964
Geog 13.1058

PALAEOGEOPHYSICS NATO advanced study institute meeting on palaeogeophysics Proceedings Newcastle-upon-Tyne 1968 North Atlantic treaty organization Edited by Stanley K. Runcorn Bibliog,illus,charts,maps xv,518p 24cm Academic press,London;New York,1970
Min 10.1458

PALAEOMAGNETISM Methods in palaeomagnetism NATO Advanced study institute on procedures in palaeomagnetism Proceedings Newcastle-upon-Tyne 1964 Apr 1-10 North Atlantic treaty organization and University of Newcastle upon Tyne Edited by D.W. Collinson and others Developments in solid earth geophysics, 3 Amsterdam,1967
Geod 9.0568

PALAEONTOLOGICAL ASSOCIATION The Fossil record a symposium with documentation Swansea 1965 Dec 20-21 Edited by W.B. Harland and others sponsored by the Geological society of London bibliog. Geological society of London,London,1967 The papers in parts 1 and 3 were presented at a joint meeting of the Geological society of London and the Palaeontological association, Swansea,Dec 1965
Geol 8.4440

PALAEONTOLOGICAL ASSOCIATION nd SYSTEMATICS ASSOCIATION, Aspects of Tethyan biogeography Systematics association symposium 7th papers Leicester 1966 Sep 21-23 Edited by C.G. Adams and D.V. Ager Systematics association.Publication,7 Systematics association,London,1967
Geol 8.4466

PALAEONTOLOGY Problemes actuels de paleontologie (evolution des vertebres) Paris 1966 Jun 6-11 Centre national de la recherche scientifique Centre national de la recherche scientifique.Colloques internationaux.Actes, 163 pls 27cm C.N.R.S.,Paris,1967 Organized by J.P.Lehman. Papers in English,French and German
Bal 44.4640

PALAEONTOLOGY Species concept in paleontology Systematics association symposium 2nd papers London 1954 May 19 Systematics association Edited by Peter C. Sylvester-Bradley Systematics association. Publication,2 Systematics association,London, 1956
Geol 8.0940

PALAEONTOLOGY The Species concept in palaeontology :a symposium Systematics association Edited by Peter C.S. Bradley Systematics association.Publication, 2 Systematics association,London,1956
Bot 42.3421

PALEOECOLOGY Quaternary paleoecology International association for quaternary research.Congress 7th Proceedings Boulder,Colo. 1965 Aug-Sep Vol 7 Edited by E.J. Cushing and H.E. Wright Sponsored by the National research council Yale university press,New Haven,Conn.;London,1967
Geog 13.1318

PALEONTOLOGIE :colloque internationale Paris 1947 Apr Centre national de la recherche scientifique Centre national de la recherche scientifique.Colloques internationaux, 21 C.N.R.S.,Paris,1950
An 32.2180

PALEONTOLOGY International geological congress 20th papers Mexico City 1956 Seccion 7: paleontologia,taxonomia y evolucion Edited by A. Garcia Rojas and others Mexico City,1958 Text in English,French,German, Russian and Spanish
Geol 8.3043

PALEONTOLOGY Karoo symposium South African association for the advancement of science : annual congress papers Johannesburg 1962 South African journal of science,59,no.5 Johannesburg,1963
Geol 8.2941

PALEONTOLOGY Materialy paleontologicheskogo soveschaniya po paleozoyu :ochorednye zadachi paleontologii v dele pomoshchi praktike Moscow 1951 May 14-17 Edited by T.G. Sarycheva Akademiya nauk S.S.S.R.,Moscow, 1953
Geol 8.0705

PALEOVOLCANOLOGIE ET SES RAPPORTS AVEC LA TECTONIQUE,FASC.17 International geological congress 19th Proceedings, section 15 Algiers 1952 Algiers,1954
Min 10.0352

PALEOZOOLOGY Materialy paleontologicheskogo soveschaniya po paleozoyu :ochorednye zadachi paleontologii v dele pomoshchi praktike Moscow 1951 May 14-17 Edited by T.G. Sarycheva Akademiya nauk S.S.S.R.,Moscow, 1953
Geol 8.0705

PALERMO 1958 Noise,shock tubes,magnetic effects,instability and mixing Agard colloquium 3rd Papers Agard.Combustion and propulsion panel Edited by M.W. Thring and others Pergamon press,London,1958
Eng 41.7232

PALERMO 1958 Noise,shock tubes,magnetic effects,instability and mixing Agard colloquium 3rd Papers Palermo 1958 Mar 17-21 Agard.Combustion and propulsion panel Edited by M.W. Thring and others Pergamon press,London,1958
Eng 41.7232

PALLANZA 1960 Symposium on the germ cells and earliest stages of development International institute of embryology A. Baselli,Milan,1961 At head of title: Istituto Lombardo accademia di scienze e lettere
Bal 39.0459

PALM BEACH,FLA. 1963 Heterogeneous combustion :a conference American institute of aeronautics and astronautics Edited by Hans G. Wolfhard and others Progress in astronautics and aeronautics, 15 Academic press,New York,1964
Eng 41.7241

PALM SPRINGS,CALIF. 1962 Ionization in high-temperature gases :a selection of technical papers based mainly on the American rocket society conference on ions in flames and rocket exhausts Edited by Kurt E. Shuler Progress in astronautics and aeronautics, 12 Academic press,New York,1963
Eng 41.4543

PALO ALTO 1958 Magnetodynamics of conducting fluids Lockheed symposium on magnetohydrodynamics 3rd proceedings Edited by Daniel Bershader Sponsored by Lockheed aircraft corporation.Missiles and space division Stanford university press, Stanford,Calif.,1959
A Math 4.0616

PALO ALTO 1959 Plasma acceleration Lockheed symposium on magnetohydrodynamics 4th proceedings Edited by Sidney W. Kash Sponsored by Lockheed aircraft corporation. Missiles and space division Stanford university press,Stanford,Calif.,1960
A Math 4.0618

PALO ALTO 1960 Radiation and waves in plasmas Lockheed symposium on magnetohydrodynamics 5th proceedings Edited by Morton Mitchner Sponsored by Lockheed aircraft corporation.Missiles and space division Stanford university press, Stanford,Calif.,1961
A Math 4.0621

PALO ALTO 1966 Weld imperfections a symposium Proceedings Lockheed missiles and space company Edited by A.R. Pfluger and R.E. Lewis Co-sponsored by the United States. Air force.Materials laboratory and the United States office of naval research Addison-Wesley,Reading,Mass.,1968
Met 25.2319

PALO ALTO,CALIF. 1956 Magnetohydrodynamics a symposium Edited by Rolf K.M. Landshoff Sponsored by the Lockheed aircraft corporation. Missile systems division Stanford university press,Stanford,Calif.,1957 The 1st symposium on magnetohydrodynamics sponsored by Lockheed
Eng 41.4472

PALO ALTO,CALIF. 1957 The Plasma in a magnetic field :a symposium on magnetohydrodynamics Edited by Rolf K.M. Landshoff Sponsored by the Lockheed aircraft corporation.Missile systems division Stanford university press,Stanford,Calif.,1958 The 2nd symposium on magnetohydrodynamics sponsored by Lockheed.For later symposia see 'Symposium on magnetohydrodynamics'
Eng 41.4473

PALO ALTO,CALIF. 1958 Magnetodynamics of conducting fluids Symposium on magnetohydrodynamics 3rd Lockheed aircraft corporation.Lockheed missiles and space division Edited by Daniel Bershader viii,145p 24cm Stanford university press, Stanford,Calif.,1959
Cav 7.2767

PALO ALTO,CALIF. 1958 Magnetohydrodynamics of conducting fluids Symposium on magnetohydrodynamics 3rd Edited by Daniel Bershader Sponsored by Lockheed aircraft corporation.Missile systems division Stanford university press,Stanford,Calif.,1959 1st symposium called 'Magnetohydrodynamics'; 2nd symposium called 'Plasma in a magnetic field',q.v.
Eng 41.4474

PALO ALTO,CALIF. 1959 Plasma acceleration Symposium on magnetohydrodynamics 4th Edited by Sidney W. Kash Sponsored by the Lockheed aircraft corporation.Missiles and space division Stanford university press, Stanford,Calif.,1960
Eng 41.4489

PALO ALTO,CALIF. 1960 Radiation and waves Symposium on magnetohydrodynamics 5th Edited by Morton Mitchener Sponsored by Lockheed aircraft corporation.Missile systems division Stanford university press,Stanford, Calif.,1961
Eng 41.4496

PALO ALTO,CALIF. 1960 Radiation and waves in plasmas Symposium on magnetohydrodynamics 5th Lockheed aircraft corporation. Lockheed missiles and space division Edited by Morton Mitchner 156p Stanford university press,Stanford,Calif.,1961
Cav 7.2881

PALO ALTO,CALIF. 1961 Plasma hydromagnetics Symposium on magnetohydrodynamics 6th Proceedings Edited by Daniel Bershader Sponsored by Lockheed aircraft corporation.Missile systems division Stanford university press,Stanford, Calif.,1962
Eng 41.4510

PALO ALTO,CALIF. 1962 Propagation and instabilities in plasmas Symposium on magnetohydrodynamics 7th Proceedings Edited by Walter I. Futterman Sponsored by Lockheed aircraft corporation.Missiles and space division Stanford university press, Stanford,Calif.,1963
Eng 41.4523

PALO ALTO,CALIF. 1966 A.G.I. short course on chain silicates :lecture notes American geological institute American geological institute,Washington,D.C.,1966 Typescript
Min 10.1286

PALO ALTO,CALIF. 1971 ACM-IEEE symposium on problems in the optimization of data communications systems 2nd Proceedings Association for computing machinery Institute of electrical and electronics engineers illus 198p ACM;IEEE,New York, 1971
Math L 5.3629

PALYNOLOGY IN OIL EXPLORATION Symposium San Francisco,Calif. 1962 Edited by Aureal T. Cross Southwestern association of petroleum geologists.Society of economic paleontologists and mineralogists.Special publication,11 Tulsa,Okla.,1964
Geol 8.0978

PAN-AMERICAN CONGRESS OF ENDOCRINOLOGY 6th Proceedings Endocrinology Mexico City 1965 Oct 10-13 Edited by C. Gual Excerpta medica.International congress series, 112 Excerpta medica,Amsterdam,1966
Inv Med 37.0096

PAN-AMERICAN CONGRESS OF ENDOCRINOLOGY 6th Proceedings Mexico City 1965 Oct 10-15 Edited by C. Gual Excerpta medica. International congress series, 112 Excerpta medica foundation,New York,c1965
PGMS 29.0167

PAN-AMERICAN CONGRESS OF ENDOCRINOLOGY 6th Proceedings Mexico City 1965 Oct 10-15 Edited by C. Gual Excerpta medica. International congress series, 112 Excerpta medica foundation,Amsterdam,1966
Inv Med 37.0254

The PAN AMERICAN SCIENTIFIC CONGRESS 2nd Proceedings Astronomy,meteorology and seismology Washington,D.C. 1915-16 Dec 27-Jan 8 Pan American union Edited by Glen Levin Swiggett illus xiv,847p Government printing office,Washington,D.C.,1917
Nap 11.0838

PAN-AMERICAN SCIENTIFIC CONGRESS.MEXICO CITY,1932 Distribucion y regimen de la humedad atmosferica en la region del norte de Chile By Enrique Taulis illus 170p Chile,1932 Typescript.Paper presented at the 12th Pan-American scientific congress,Mexico City,1932
Nap 11.0888

PAN AMERICAN UNION and RESEARCH INSTITUTE FOR THE STUDY OF MAN Plantation systems : seminar Papers and discussion summaries San Juan,Puerto Rico 1957 Nov 17-23 Pan American union.Social science monographs,7 Washington,D.C.,1959
Geog 13.5120

PANCREAS Ciba foundation symposium on the exocrine pancreas :normal and abnormal functions London 1961 May 30-Jun 1 Edited by A.V.S. De Reuck and Margaret P. Cameron Churchill,London,1962
An 32.3659

PANCREAS Internal secretions of the pancreas The Nature and actions of the internal secretions of the pancreas :a colloquium Proceedings London 1955 Jun 21-23 Ciba foundation Edited by G.E.W. Wolstenholme and Cecilia M. O'Connor Ciba foundation colloquia on endocrinology, 9 illus Churchill,London,1956
Bioch 33.0438

PANCREATIC ISLETS The Structure and metabolism of the pancreatic islets :an international Wenner-Gren symposium Proceedings Uppsala 1963 Aug 29-Sep 2 and Stockholm Wenner-Gren center Edited by S.E. Brolin and others Wenner-Gren center. International symposium series, 3 Pergamon press,Oxford,1964
Bioch 33.0485

PANEL ON BIOPHYSICAL ASPECTS OF RADIATION QUALITY 2nd Report Biophysical aspects of radiation quality Vienna 1967 Apr 10-14 International atomic energy agency International atomic energy agency.Panel proceedings series International atomic energy agency,Vienna,1968
Radioth 35.1422

PANEL ON BIOPHYSICAL ASPECTS OF RADIATION QUALITY Report Biophysical aspects of radiation quality Vienna 1965 Mar 29-Apr 2 International atomic energy agency International atomic energy agency.Technical report series, 58 International atomic energy agency,Vienna,1966
Radioth 35.1421

PANEL ON CLINICAL USES OF WHOLE-BODY COUNTING Proceedings Clinical uses of whole-body counting Vienna 1965 Jun 28-Jul 2 International atomic energy agency International atomic energy agency.Panel proceedings series International atomic energy agency,Vienna,1966
Radioth 35.1383

PANEL ON CO-ORDINATION OF RESEARCH ON THE PRODUCTION AND USE OF INDUCED MUTATIONS IN PLANT BREEDING Proceedings Mutations in plant breeding Vienna 1966 Jan 17-21 International atomic energy agency Food and agriculture organization International atomic energy agency.Panel proceedings series 2 vols International atomic energy agency, Vienna,1966-68
Radioth 35.0261

PANEL ON ELIMINATION OF HARMFUL ORGANISMS FROM FOOD AND FEED BY IRRADIATION Report Elimination of harmful organisms from food and feed by irradiation Zeist 1967 Jun 12-16 Organized by the Joint FAO-IAEA division of atomic energy in food and agriculture International atomic energy agency.Panel proceedings series International atomic energy agency,Vienna,1968
Radioth 35.1761

PANEL ON ENZYMOLOGICAL ASPECTS OF THE APPLICATION OF IONIZING RADIATION TO FOOD PRESERVATION Proceedings Enzymological aspects of food irradiation Vienna 1968 Apr 8-12 Organized by the Joint FAO-IAEA division of atomic energy in food and agriculture International atomic energy agency.Panel proceedings series International atomic energy agency,Vienna,1969
Radioth 35.1762

PANEL ON NUCLEAR ACCIDENT DOSIMETRY SYSTEMS Proceedings Nuclear accident dosimetry systems Vienna 1969 Feb 17-21 Organized by the International atomic energy agency International atomic energy agency.Panel proceedings series International atomic energy agency,Vienna,1970
Radioth 35.1384

PANEL ON NUCLEAR SCIENCE TEACHING Report Nuclear science teaching Bangkok 1968 Jul 15-23 International atomic energy agency Unesco International atomic energy agency. Technical reports series, 94 International atomic energy agency,Vienna,1968
Radioth 35.1777

PANEL ON PRESERVATION OF FRUIT AND VEGETABLES BY RADIATION :especially in the tropics Proceedings Preservation of fruit and vegetables by radiation Vienna 1966 Aug 1-5 Organised by the Joint FAO-IAEA division of atomic energy in food and agriculture International atomic energy agency.Panel proceedings series International atomic energy agency,Vienna,1968
Radioth 35.1758

PANEL ON RADIATION AND THE CONTROL OF IMMUNE RESPONSE Proceedings Radiation and the control of immune response Paris 1967 Jun 22-24 Organized by the International atomic energy agency International atomic energy agency.Panel proceedings series International atomic energy agency,Vienna,1968
Radioth 35.1202

PANEL ON RADIATION DAMAGE TO THE BIOLOGICAL MOLECULAR INFORMATION SYSTEM :with special regard to the role of SH-groups Proceedings Radiation damage and sulphydryl compounds Vienna 1968 Oct 21-25 Organized by the International atomic energy agency International atomic energy agency.Panel proceedings series International atomic energy agency,Vienna,1969
Radioth 35.1764

PANEL ON RADIATION STERILISATION OF BIOLOGICAL TISSUES FOR TRANSPLANTATION Sterilisation and preservation of biological tissues by ionizing radiations Budapest 1969 Jun 16-20 International atomic energy agency International atomic energy agency.Panel proceedings series International atomic energy agency,Vienna,1970
Radioth 35.1915

PANEL ON RESEARCH APPLICATIONS OF REPETITIVELY PULSED REACTORS AND BOOSTERS Proceedings Research applications of nuclear pulsed systems Dubna,U.S.S.R. 1966 Jul 18-22 International atomic energy agency International atomic energy agency.Panel proceedings series International atomic energy agency,Vienna,1967
Radioth 35.1682

PANEL ON THE CONTROL OF LIVESTOCK INSECT PESTS BY THE STERILE-MALE TECHNIQUE Proceedings Control of livestock insect pests by the sterile-male technique Vienna 1967 Jan 23-27 Organized bythe Joint FAO-IAEA division of atomic energy in food and agriculture International atomic energy agency.Panel proceedings series International atomic energy agency,Vienna,1968
Radioth 35.0263

PANEL ON THE EFFECTS OF IONIZING RADIATIONS FROM DIFFERENT SOURCES ON HAEMATOPOIETIC TISSUE Proceedings Effects of ionizing radiations on the haematopoietic tissue Vienna 1966 May 17-20 International atomic energy agency International atomic energy agency.Panel proceedings series International atomic energy agency,Vienna,1967
Radioth 35.1200

PANEL ON THE RADIATION SENSITIVITY OF TOXINS AND ANIMAL POISONS Proceedings Radiation sensitivity of toxins and animal poisons Bangkok 1969 May 19-22 International atomic energy agency International atomic energy agency.Panel proceedings series International atomic energy agency,Vienna,1970
Radioth 35.1912

PANEL ON THE RADIATION SENSITIVITY OF TOXINS AND ANIMAL POISONS Proceedings Radiation sensitivity of toxins and animal poisons Bangkok 1969 May 19-22 Organized by the International atomic energy agency International atomic energy agency.Panel proceedings series International atomic energy agency,Vienna,1970
Radioth 35.1765

PANEL ON THE ROLE OF COMPUTERS IN RADIOTHERAPY Report Role of computers in radiotherapy Vienna 1967 Jul 10-14 Organized by the International atomic energy agency International atomic energy agency.Panel proceedings series bibliog. International atomic energy agency,Vienna,1968
Radioth 35.1199

PAPANICOLAOU CANCER RESEARCH INSTITUTE Nucleic acid-protein interactions and nucleic acid synthesis in viral infection :symposium Miami,Fla. 1971 Jan 18-22 Miami winter symposia, 2 North-Holland,Amsterdam,1971
Bioch 33.2241

PAPER CHROMATOGRAPHY Quantitative paper chromatography of steroids :a conference Proceedings London 1958 Jul 1 Society for endocrinology Edited by D. Abelson and R. V. Brooks Society for endocrinology.Memoirs, 8 Cambridge university press,Cambridge, 1960
Bioch 33.2355

PAPER ELECTROPHORESIS Ciba foundation symposium London 1955 Jul 27-29 Edited by G.E.W. Wolstenholme and Elaine C.S. Millar Churchill,London,1958
Col S 12.0070

PAPER ELECTROPHORESIS Ciba foundation symposium on paper electrophoresis London 1955 Jul 27-29 Ciba foundation Edited by G. E. Wolstenholme and E.C.P. Millar Ciba foundation.Symposia Churchill,London,1958
PGMS 29.0516

PAPER ELECTROPHORESIS Ciba foundation symposium on paper electrophoresis Proceedings London 1955 Jul 27-29 Ciba foundation Edited by G.E.W. Wolstenholme and Elaine C.P. Millar Ciba foundation symposia Churchill,London,1956
Chem 18.1445

PAPERS ON NON-DESTRUCTIVE TESTING presented at a meeting of committee E-7... 1953 Jul 2 American society for testing materials. Committee E-7 on non-destructive testing American society for testing materials, Philadelphia,Pa.,1954
Met 25.1318

PAPERS ON SOILS Symposium on time rates of loading in soil tension Symposium on Atterberg limits,Session on soils and Symposium on soils for engineering purposes Atlantic City,N.J. 1959 Jun 22-23 San Francisco,Calif. 1959 Oct 16 American society for testing materials.Special technical publication, 254 American society for testing materials, Philadelphia,Pa.,1960 Papers presented at various A.S.T.M. meetings during 1959
Eng 41.3151

PARAMAGNETIC RESONANCE 1st :international conference Proceedings Jerusalem 1962 Jul 16-20 Vol 1-2 International union of pure and applied physics Edited by W. Low Sponsored by the Hebrew university of Jerusalem 2 vols Academic press,New York, 1963
Cav 7.1226

PARAMARIBO 1950 Geologische conferentie 1st report 1950 Mimeographed
Geol 8.2428

PARAMETER SYSTEMS IFAC symposium on the control of distributed parameter systems Proceedings Banff 1971 Jun 21-23 International federation for automatic control 2 vols IFAC technical committee on theory, Banff,1971
Eng 41.8593

PARASITES Problems in systematics of parasites :a symposium Proceedings American association for the advancement of science Edited by Gerald D. Schmidt University park press,Baltimore,Md.,1969
Mol 45.0148

PARASITES Symposium sur la specificite parasitaire des parasites de vertebres 1st Universite de Neuchatel.Institut de zoologie 24cm Neuchatel,1957
Philos 1.1588

PARATHYROID CONFERENCE 3rd Proceedings Parathyroid hormone and thyrocalcitonin (calcitonin) Montreal 1967 Oct 16-20 Edited by Roy V. Talmadge and others Excerpta medica.International congress series, 159 Excerpta medica,Amsterdam,1968
Inv Med 37.0211

The PARATHYROID GLANDS:ULTRASTRUCTURE,SECRETION AND FUNCTION 2nd :a symposium Papers Noordwijk 1964 Aug 25-29 Edited by P.J. Gaillard and others University of Chicago press,Chicago,Ill.,1965
Phys 20.1417

The PARATHYROID GLANDS:ULTRASTRUCTURE,SECRETION AND FUNCTION 2nd :a symposium Papers Noordwijk am Zee 1964 Aug 25-29 Edited by Pieter J. Gaillard Bibliog.,Illus. xii, 353p University of Chicago press,Chicago,Ill. 1965
Pha 16.0098

The PARATHYROID GLANDS;ULTRASTRUCTURE,SECRETION AND FUNCTION 2nd :symposium Noordwijk 1964 Aug 25-29 Edited by Peter J. Gaillard and others University of Chicago press,Chicago,Ill.,1965
Pha 16.0291

PARATHYROID HORMONE AND THYROCALCITONIN (CALCITONIN) Parathyroid conference 3rd Proceedings Montreal 1967 Oct 16-20 Edited by Roy V. Talmadge and others Excerpta medica.International congress series, 159 Excerpta medica,Amsterdam,1968
Inv Med 37.0211

PARATHYROID HORMONE AND THYROCALCITONIN CALCITONIN 3rd :parathyroid conference Proceedings Montreal 1967 Oct 16-20 Edited by Roy V. Talmage and others Excerpta medica international congress series,159 Illus. viii,535p Excerpta medica foundation,Amsterdam,1968
Pha 16.0099

The PARATHYROIDS :a symposium on advances in parathyroid research Proceedings Houston,Tex. 1960 Edited by Roy O. Greep and Roy V. Talmage Thomas,Springfield,Ill., 1961
An 32.3771

PARIS Physiopathology of the reticular endothelial system :a symposium Council for international organizations of medical sciences Edited by B.N. Halpern Blackwell, Oxford,1957 Summary of each paper in French
An 32.3154

PARIS 1878- International geological congress 1st- proceedings Paris,1886-
Geol 8.4145

PARIS 1889 Congres international de psychologie physiologique :premiere session 1st Compte-rendu Societe de psychologie physiologique de Paris Bureau des revues, Paris,1890 Congresses continued as 'International congress of experimental psychology' q.v.
Psy 31.3390

PARIS 1896 Conference internationale des etoiles fondamentales Proceedings Sponsored by the Bureau des longitudes,Paris 90p Gauthier-Villars,Paris,1896
Obs 6.2315

PARIS 1900 Congres international d'aviculture et de peche Memoires et comptes-rendus France.Ministere de l'industrie,des postes et des telegraphes Edited by Joseph Perard and Maire 24cm Challamel,Paris,1901
Bal 44.1241

PARIS 1900 Congres international de botanique Actes Edited by Emile Perrot Lons-le-Saunier,1900 On the occasion of 'L'exposition universelle de 1900'
Bot 42.0648

PARIS 1900 International congress of psychology 4th Compte-rendu Edited by Pierre Janet Alcan,Paris,1901
Psy 31.3394

PARIS 1911 Conference international de genetique 4e Comptes rendus Societe nationale d'horticulture de France Edited by Ph.de Vilmorin 2 vols Masson,Paris,1913
Gen 34.1000

PARIS 1911 Congres international des ephemerides astronomiques Rapports Bureau des longitudes,Paris 53p Gauthier-Villars, Paris,1912
Obs 6.2483

PARIS 1913 Carte du monde au millionieme : Conference internationale 2nd Comptes rendus France.Service geographique de l'armee Service geographique de l'armee, Paris,1914
Geog 13.5472

PARIS 1922 Iris cultivees Conference internationale des iris 1ere Actes et comptes-rendus Societe nationale d'horticulture de France Societe nationale d'horticulture,Paris,1923
BG 38.2574

PARIS 1927 Conference internationale de psychotechnique 4th Comptes rendues Conferences internationales de psychotechnique and Institut international de cooperation intellectuelle Alcan,Paris,1929 Papers in English,French,German and Italian
Psy 31.2517

PARIS 1928 Reunion internationale de chimie physique rapports et discussions Edited by R. Audubert and M.L. Claudel Paris, 1929
Philos 1.0668

PARIS 1930 La Constitution des cometes : conference By F. Baldet Actualites scientifiques et industrielles, 16 illus 23p Hermann,Paris,1930
Obs 6.2607

PARIS 1931 Congres international de geographie International geographical congress Excursion guide Excursion A2: le sud-est du Massif central Edited by Henri Baulig Colin,Paris,1931
Geog 13.0525

PARIS 1931 Congres international de geographie International geographical congress Guidebooks Colin,Paris,1931
Geog 13.4158

PARIS 1931 Congres national d'astronomie rapports Comite national francais d'astronomie 203p Paris,1932
Obs 6.2481

PARIS 1937 Congres international de la population International congress for studies on population 4th Under the auspices of l' Union internationale pour l'etude scientifique des problemes de la population 8 vols Hermann,Paris,1938
Geog 13.1494

PARIS 1937 Etudes et recherches sur les phytohormones By Peter Boysen-Jensen and others League of Nations.Committee on intellectual co-operation Organized in collaboration with Union internationale des sciences biologiques League of Nations,Paris, 1937
Bot 42.1815

PARIS 1937 Les Hormones sexuelles colloque international Comptes-rendus Edited by L. Brouha Under the auspices of Fondation Singer-Polignac Conferences du college de France Hermann,Paris,1938
Gen 34.1603

PARIS 1939 Novae and white dwarfs Colloque international d'astrophysique Reports 1: observations of novae By Knut Lundmark and others Actualites scientifiques et industrielles, 901 College de France. Conferences, 13 140p Hermann,Paris,1941
Obs 6.1146

PARIS 1939 Novae and white dwarfs Colloque international d'astrophysique reports 2: nova theory supernovae By P. Swings and others Actualites scientifiques et industrielles, 902 College de France. Conferences, 14 267p Hermann,Paris,1941
Obs 6.1720

PARIS 1939 Novae and white dwarfs Colloque international d'astrophysique reports 3: white dwarfs By G.P. Kuiper and others Actualites scientifiques et industrielles, 903 College de France. Conferences, 15 267p Hermann,Paris,1941
Obs 6.0996

PARIS 1946 La Theorie des images optiques Memoires Centre national de la recherche scientifique and Institut d'optique theorique et appliquee Edited by Pierre Fleury Financed by the Rockefeller foundation Colloques internationaux du centre de la recherche scientifique Editions de la revue d'optique,Paris,1949
Cav 7.1758

PARIS 1946 Reunions scientifiques...pour le centenaire de la decouverte de Neptune reports Comite national Francais d'astronomie Edited by P. Couderc illus. 164p Imprimerie nouvelle,Orleans,1948
Obs 6.0418

PARIS 1947 Endocrinologie des arthropodes Colloque international sur l'endocrinologie des arthropodes Centre national de la recherche scientifique Centre national de la recherche scientifique.Colloques internationaux, 4 Centre national de la recherch scientifique,Paris,1948
Bal 39.2291

PARIS 1947 Paleontologie :colloque internationale Centre national de la recherche scientifique Centre national de la recherche scientifique.Colloques internationaux, 21 C.N.R.S.,Paris,1950
An 32.2180

PARIS 1948 Cinetique et mecanisme des reactions d'inflammation et de combustion en phase gazeuse Centre national de la recherche scientifique With the financial support of the Rockefeller foundation Centre national de la recherche scientifique. Colloques internationaux,16 Centre national de la recherche scientifique,Paris,1949 With summaries in English
Chem 18.0431

PARIS 1948 La Conference generale des poids et mesures 9eme comptes rendus Comite international des poids et mesures 127p Gauthier-Villars,Paris,1949
Obs 6.2388

PARIS 1948 Manual of the international statistical classification of diseases, injuries and causes of death :adapted 1948 by the international conference for the sixth revision of the international lists of diseases and causes of death 1-2 World health organization World health organization.Bulletin.Supplement, 1 2 vols Columbia university press,New York,1948-49
Med 36.0385

PARIS 1948 Unites biologiques douees de continuite genetique colloque international Centre national de la recherche scientifique Centre national de la recherche scientifique. Colloques internationaux, 8 Centre national de la recherche scientifique,Paris, 1949
Gen 34.0988

PARIS 1949 Congres international de philosophie des sciences Actes Tome 1- Institut international de philosophie Actualites scientifiques et industrielles,1126, 1134,1137,1146,1153,1155-56,1166-67- Hermann, Paris,1951-
WSM 43.0981

PARIS 1949 Problems of cosmical aerodynamics Symposium on the motion of gaseous masses of cosmical dimensions Proceedings International union of theoretical and applied mechanics International astronomical union Edited by J. M. Burgess and H.C.van de Hulst 237p Central air documents office,Dayton,Ohio,1951
Obs 6.3550

PARIS 1950 Structure et physiologie des societes animales :colloque international Centre national de la recherche scientifique Centre national de la recherche scientifique. Colloques internationaux, 34 C.N.R.S.,Paris, 1952
Phys 20.0540

PARIS 1951 Congress G.A.M.S. 4th Groupement pour l'advancement des methodes d'analyse spectrographique des produits metallurgiques 24cm Paris,1951
Philos 1.0678

PARIS 1951 Congres internationale de la societe internationale de transfusion sanguine 4th Rapports et communications International society of blood transfusion Edited by J. Julliard and others L'expansion scientifique francaise,Paris,1952
Med 36.0389

PARIS 1951 La Cybernetique:theorie du signal et de l'information :reunions d'etudes et de mises au point Institut Henri Poincare. Seminaire de theories physiques Edited by Louis de Broglie vi,317p 22cm Revue d'optique theorique et instrumentale,Paris, 1951
Math L 5.0169

PARIS 1951 Les Machines a calculer et la pensee humaine colloque international Centre national de la recherche scientifique Centre national de la recherche scientifique., Colloques internationaux,37 illus. xix, 570p 24cm Centre national de la recherche scientifique,Paris,1953
Math L 5.0716

PARIS 1951 Villes et campagnes Semaine sociologique 2nd Compte rendu Centre d'etudes sociologiques Edited by Georges Friedmann Ecole pratique des hautes etudes, Paris.Bibliotheque generale.Section 6 Colin, Paris,1953
Geog 13.4272

PARIS 1952 International congress biochemistry 2nd Vol 7: symposium sur la biochemie des steroides Council for international organizations of medical sciences Societe d'edition d'enseignement superieur,Paris,1952 Text in English and French
Bioch 33.1310

PARIS 1952 International congress of biochemistry 2nd Vol 1: symposium sur la biochimie de l'hematopoiese Council for international organizations of medical sciences Societe d'edition d'enseignement superieur,Paris,1952 Text in English and French
Bioch 33.1304

PARIS 1952 International congress of biochemistry 2nd Vol 2: symposium sur la biogenese des proteines Council for international organizations of medical sciences Societe d'edition d'enseignement superieur,Paris,1952 Text in English and French
Bioch 33.1305

PARIS 1952 International congress of biochemistry 2nd Vol 3: symposium sur le cycle tricarboxylique Council for international organizations of medical sciences Societe d'edition d'enseignement superieur,Paris,1952 Text in English, French and German
Bioch 33.1306

PARIS 1952 International congress of biochemistry 2nd Vol 4: symposium sur les hormones proteiques et derivees des proteines Council for international organizations of medical sciences Societe d'edition d'enseignement superieur,Paris,1952 Text in English and French
Bioch 33.1307

PARIS 1952 International congress of biochemistry 2nd Vol 5: symposium sur le metabolisme microbien Council for international organizations of medical sciences Societe d'edition d'enseignement superieur,Paris,1952 Text in English and French
Bioch 33.1308

PARIS 1952 International congress of biochemistry 2nd Vol 6: symposium sur le mode d'action des antibiotiques Council for international organizations of medical sciences Societe d'edition d'enseignement superieur,Paris,1952 Text in English and French
Bioch 33.1309

PARIS 1952 International congress of biochemistry 2nd Compte rendu Council for international organizations of medical sciences Societe de chimie biologique. Bulletin, 35,no 1-2 Masson,Paris,1953 Supplement.Text in English and French
Bioch 33.1302

PARIS 1952 International congress of biochemistry 2nd Proceedings Vol 3: symposium sur le cycle tricarboxylique Council for international organizations of medical science Edited by Michel Polonovski Congres international de biochemie, 2 Societe d'edition d'enseignement superieur, Paris,1952
Bot 42.1697

PARIS 1952 International congress of biochemistry 2nd Resumes des communication Council for international organizations of medical sciences Masson, Paris,1952 Text in English and French
Bioch 33.1303

PARIS 1952 Leonard de Vinci et l'experience scientifique au XVIe siecle : colloque international Centre national de la recherche scientifique.Sciences humaines viii,273p Presses universitaires de France, Paris,1953
WSM 43.1938

PARIS 1953 Les Grandes activites du lobe temporal :semaine neurophysiologique: conferences du colloque Hopital de la Salpetriere.Clinique des malades du systeme nerveux Edited by Th. Alajouanine Actualites neuro-physiologiques Masson,Paris, 1955
Psy 31.0464

PARIS 1953 Organization scientifique des jardins botaniques Colloque international de l'union international des sciences biologiques Under the auspices of Unesco International union of biological sciences.Series B: Colloquia, 13 International union of biological sciences,Paris,1953
BG 38.3110

PARIS 1953 Principes fondamentaux de classification stellaire Proceedings Centre national de la recherche scientifique With the support of the Rockefeller foundation Centre national de la recherche scientifique. Colloques internationaux, 55 190p Centre national de la recherche scientifique,Paris, 1955
Obs 6.2008

PARIS 1954 International botanical congress 8th Comptes rendus des seances et rapports et communications Sections 3-27 9 vols Paris,1955-57
Bot 42.4699

PARIS 1954 International botanical congress 8th Rapports et communications Section 2: 2.serie Utrecht,1955 Reprinted from Taxon,vol.4 no.6,pp.121-177
Bot 42.4698

PARIS 1954 International botanical congress 8th Rapports et communications Sections 2;4-27 9 vols Paris,1954
Bot 42.4697

PARIS 1954 International botanical congress 8th Reports of communications 1954
BG 38.3129

PARIS 1954 International congress of blood transfusion 5th Proceedings International society of blood transfusion 1955
Med 36.0388

PARIS 1954 Les Divisions ecologiques du monde moyens d'expression,nomenclature cartographie :colloques internationaux Centre national de la recherche scientifique illus xii,235p Centre national de la recherche scientifique,Paris,1955
Bot 42.2071

PARIS 1955 International congress on large dams 5th Transactions Vol 1-5 International commission on large dams 5 vols Paris,1956
Eng 41.3334

PARIS 1955 La Douleur et les douleurs :a symposium Hopital de la Salpetriere.Clinique des maladies du systeme nerveux Edited by Th. Alajouanine and others Masson,Paris,1957
Psy 31.0214

PARIS 1955 Manual of the international statistical classification of diseases, injuries,and causes of death based on the seventh revision conference 1-2 World health organization 2 vols World health organization,Geneva,1957
Gen 34.1937

PARIS 1956 Air intake problems in supersonic propulsion AGARD.Combustion and propulsion panel meeting 11th Agard Edited by J. Fabri AGARDograph, 27 Pergamon press,London,1958
Eng 41.6753

PARIS 1956 International automation congress Proceedings Presses academiques europeenes,Brussels,1959
Eng 41.6008

PARIS 1956 and 1957 Structures algebriques et structures topologiques : conferences a l'Institut Henri Poincare de la Sorbonne By H. Cartan and others Societe mathematique de France en accord avec L' Association des Professeurs de mathematique de l'enseignement public Enseignement mathematique.Monographies, 7 198p Geneva, Paris,1958
P. Math 2.0241

PARIS 1957 EUSEC conference on engineering education 3rd Proceedings Conference of engineering societies of western Europe and the United States of America Institution of mechanical engineers,London,1958
Eng 41.1546

PARIS 1957 Effect du gel dans les materiaux de construction :colloque Compte rendu Pt 1-2 Association francaise de recherches et d'essais sur les materiaux et les constructions Cahiers de la recherche theorique et experimentale sur les materiaux et les structures,8,9 illus,diagrs 81p 27cm Paris,1959
Sco 14.1263

PARIS 1957 International conference of pure and applied chemistry main lectures 16th Lectures International union of pure and applied chemistry Birkhauser,Basle; Stuttgart,1957 Papers in English,French and German
Chem 18.1738

PARIS 1957 International conference on radioisotopes in scientific research 1st Proceedings Vol 1: research with radioisotopes in physics and industry Unesco Edited by R.C. Extermann Pergamon,London, 1958
Met 25.1634

PARIS 1957 Les Peroxydes organiques en radiobiologie :colloque By R. Latarjet and others Actions chimiques et biologiques des radiations.Serie, 4 Masson,Paris,1958
Radioth 35.1720

PARIS 1957 Les Relations entre precambrien et cambrien problemes des series intermediaires colloque international Centre national de la recherche scientifique Centre national de la recherche scientifique. Colloques internationaux,66 bibliog. Centre national de la recherche scientifique, Paris,1958
Geol 8.2716

PARIS 1957 Radioisotpoes in scientific research International conference on radioisotopes in scientific research 1st Proceedings 1-4 Edited by R.C. Extermann Held under the auspices of Unesco illus. 4 vols Pergamon press,London,1958
Radioth 35.1070

PARIS 1957 Symposium on protein structure Edited by Albert Neuberger Sponsored by the International union of pure and applied chemistry.Protein commission Methuen;Wiley, London New York,1958
Gen 34.0667

PARIS 1957 Symposium on protein structure Papers Edited by Albert Neuberger Sponsored by the International union of pure and applied chemistry.Protein commission Methuen;Wiley,London;New York,1958
Bioch 33.0535

PARIS 1958 La Fonction spermatogenetique du testicule humain :colloque Societe nationale pour l'etude de la sterilite et de la fecondite Edited by H. Bayle and C. Gouygou Masson,Paris,1958
An 32.3872

PARIS 1958 Le Calcul des probabilites et ses applications colloque international proceedings Centre national de la recherche scientifique Edited by Georges Darmois Centre national de la recherche scientifique. Colloques internationaux, 87 196p 24cm Centre national de la recherche scientifique, Paris,1959
Math 3.0679

PARIS 1958 Low energy nuclear interactions and nuclear structure International conference on nuclear physics Proceedings Faculte des sciences de Paris et de province and Centre national de la recherche scientifique Under the auspices of International union of pure and applied physics Crosby Lockwood,London,1959
Cav 7.1759

PARIS 1958 Mechanisms of colour discrimination :an international symposium on the fundamental mechanisms of the chromatic discrimination in animals and man Proceedings International council of scientific unions Pergamon press:symposium publications division,London,1960
Psy 31.0384

PARIS 1958 Paris symposium on radio astronomy International astronomical union International scientific radio union Edited by Ronald N. Bracewell International astronomical union.Symposium, 9 International scientific radio union.Symposium, 1 612p Stanford university press, Stanford,Calif.,1959
TA 15.0638

PARIS 1958 Paris symposium on radio astronomy International astronomical union and International scientific radio union Edited by Ronald N. Bracewell International astronomical union.Symposium, 9 International scientific radio union.Symposium, 1 612p Stanford university press, Stanford,Calif.,1959
Obs 6.0246

PARIS 1958 Paris symposium on radio astronomy International astronomical union and International scientific radio union Edited by Ronald N. Bracewell International astronomical union.Symposium, 9 International scientific radio union.Symposium, 1 612p Stanford university press, Stanford,Calif,1959
A Math 4.1151

PARIS 1958 Paris symposium on radio astronomy International astronomical union and International scientific radio union Edited by Ronald N. Bracewell International astronomical union.Symposium, 9 International scientific radio union.Symposium, 1 612p Stanford university press, Stanford,Calif,1959
Cav 7.0301

PARIS 1959 International conference on information processing Proceedings Unesco Unesco,Paris,1960
Eng 41.5999

PARIS 1959 International conference on information processing proceedings Unesco 520p 29cm UNESCO,Paris,1960
Math L 5.0809

PARIS 1959 Medical electronics International conference on medical electronics 2nd Proceedings International federation for medical electronics Edited by C.N. Smyth Iliffe, London,1960
Psy 31.3062

PARIS 1959 Medical electronics International conference on medical electronics 2nd Proceedings Edited by C. N. Smyth Iliffe,London,1960
VA 19.0363

PARIS 1959 New physical and chemical properties of metals of very high purity : international symposia Centre national de la recherche scientifique Centre national de la recherche scientifique.International symposia, 40 Gordon and Breach,New York,1965
Met 25.1500

PARIS 1959 Nouvelles proprietes physiques et chimiques des metaux de tres haut purete colloque international Centre national de la recherche scientifique Centre national de la recherche scientifique.Colloques internationaux, 40 Centre national de la recherche scientifique,Paris,1960 English translation also available
Met 25.1499

PARIS 1959 Thyroide et exploration thyroidienne medullo-surrenale :rapports de la vie reunion des endocrinologistes de langue francaise Endocrinologistes de langue francaise Annales d'endocrinologie, 19,no.4 Masson;Doin,Paris,1959 At head of title: Colloque d'endocrinologie 1959
An 32.3755

PARIS 1960 Histochemistry and cytochemistry International congress of histochemistry and cytochèmistry 1st Proceedings Edited by R. Wegmann Pergamon, London,1963
An 32.3391

PARIS 1961 Colloque international sur la formation des ingenieurs en Europe Comte-rendu Paris,1961
Chem E 24.1779

PARIS 1961 Fatigue of aircraft structures : a symposium Proceedings International committee on aeronautical fatigue Edited by W. Barrois and E.L. Ripley International series of monographs in aeronautics and astronautics.Symposia, 12,Div.9 Pergamon press,Oxford,1963
Met 25.0931

PARIS 1961 International conference on soil mechanics and foundation engineering 5th Proceedings 3 vols Dunod,Paris,1961
Eng 41.3142

PARIS 1961 La Grossesse extra-uterine : colloque Rapport Societe nationale pour l'etude de la sterilite et de la fecondite Edited by Paul Funck-Brentano Masson,Paris, 1961
An 32.3824

PARIS 1961 Problemes actuels de paleontologie (evolution des vertebres) Centre national de la recherche scientifique Centra national de la recherche scientifique. Colloques internationaux.Actes, 104 27cm C. N.R.S.,Paris,1962 Organized by J.P. Lehmann
Bal 44.4639

PARIS 1962 Biologie antarctique :symposium 1er Comptes-rendus Scientific committee for Antarctic research Edited by Robert Carrick and others Hermann,Paris,1964
Bal 39.1669

PARIS 1962 Biologie antarctique :symposium 1er Comptes-rendus Edited by Robert Carrick and others Organized by the Scientific committee on Antarctic research Actualites scientifiques et industrielles,1312 illus,maps 651p 25cm Herman,Paris,1964 Preface by professor P.P.Grasse
Sco 14.6172

PARIS 1962 Dynamics of satellites :a symposium International union of theoretical and applied mechanics Edited by Maurice Roy Springer,Berlin,1963
Eng 41.6890

PARIS 1962 International symposium on rarefied gas dynamics 3rd Proceedings Vol 1-2 Edited by J.A. Laurmann Advances in applied mechanics.Supplement, 2 2 vols Academic press,New York;London,1963
Eng 41.6774

PARIS 1962 Physiologie comportement et ecologie des acridiens en rapport avec la phase :colloque international Actes Centre national de la recherche scientifique Centre national de la recherche scientifique. Colloques internationaux, 114 Paris,1962 Organised by P.O.Albrecht
Bal 39.2555

PARIS 1962 Theoretical interpretation of upper atmosphere emissions :a symposium International astronomical union and International union of geodesy and geophysics Edited by D.R. Bates International astronomical union.Symposium, 18 264p Pergamon press,New York,1963 Reprinted from'Planetary and space science',Vol.10 1963
Obs 6.0142

PARIS 1963 Congres international de stratigraphie et de geologie du carbonifere 5e compte rendu Tom 1-3 3 vols Paris, 1964 Text in English,French and German
Geol 8.2988

PARIS 1963 Cytologie de l'adenohypophyse : colloque international Centre national de la recherche scientifique Edited by Jacques Benoit and Christian Da Lage Centre national de la recherche scientifique.Colloques internationaux, 128 C.N.R.S.,Paris,1963
An 32.3770

PARIS 1963 Graph theory and theoretical physics :N.A.T.O. advanced study institute Edited by Frank Harary With the financial support of the North Atlantic treaty organization.Scientific affairs division bibliog.,illus. xv,358p 24cm Academic press,New York;London,1967
Math S 3.1408

PARIS 1963 Science et loi... semaine internationale de synthese 5e Discussions et conclusions By Abel Rey and others Centre international de synthese vi,228p Alcan,Paris,1934
WSM 43.0874

PARIS 1963 Structure and metabolism of corticosteroids :a symposium Proceedings Universite de Paris.Faculte de medicine. Laboratoire de chimie biologique Edited by Jorge R. Pasqualini and Max F. Jayle Academic press,London;New York,1964
Inv Med 37.0175

PARIS 1963 The System of astronomical constants :a symposium proceedings International astronomical union Edited by J. Kovalevsky International astronomical union. Symposium, 21 330p Gauthier-Villars,Paris, 1965 From the Bulletin astronomique de l'observatoire de Paris,vol.25,fasc.1-3
Obs 6.0986

PARIS 1964 International conference on the physics of semiconductors 7th Vol 2: plasma effects in solids Edited by J. Bok Academic press;Dunod,New York;Paris,1965
Eng 41.5432

PARIS 1964 International congress on microwave tubes 5th Proceedings Societe francaise des electroniciens et radioelectriciens Societe francaise des ingenieurs et techniciens du vide Sponsored by the Federation nationale des industries electroniques Academic press,New York,1965
Eng 41.5574

PARIS 1964 International council of the aeronautical sciences 4th congress Proceedings Edited by Robert R. Dexter Spartan;Macmillan,Washington,D.C.;London,1965
Eng 41.6966

PARIS 1964 Plasma effects in solids International conference on the physics of semiconductors 7th Vol 2: plasma effects in solids International union of pure and applied physics photos,diagrms, graphs 221p 24cm Dunod,Paris,1965 Co-sponsored by the French physical society, the French ministry for scientific research and Unesco
Cav 7.3047

PARIS 1964 The International symposium on Boron Vol 2: preparation,properties,and applications Edited by Gerhart K. Gaule Plenum press,New York,1965
Met 25.0159

PARIS 1965 Antitumoral effects of vinca rosea alkaloids :first symposium of the G.E.C. A. Proceedings European cancer chemotherapy group Edited by S. Garattini and E.M. Sproston International congress series, 106 Excerpta medica foundation, Amsterdam,1966
Radioth 35.0875

PARIS 1965 Les Concepts de Claude Bernard sur le milieu interieur :colloque international Actes Organise par la Fondation Singer-Polignac Masson,Paris,1967
Phys 20.1343

PARIS 1965 Les Problemes agraires des Ameriques latines C.N.R.S.colloques internationaux C.N.R.S.,Paris,1965
Geog 13.6822

PARIS 1965 Mathematical education of engineers :report of the O.E.C.D. seminar Organization for economic co-operation and development OECD,Paris,1965
Eng 41.1838

PARIS 1965 Optical properties and electronic structure of metals and alloys : International colloquium Proceedings Institut d'optique theorique et appliquee Edited by F. Abeles Co-sponsored by the United States.Air force.European office of aerospace research North Holland,Amsterdam, 1966
Cav 7.0004

PARIS 1966 Menthol and menthol-containing external remedies use,mode of effect and tolerance in children:international symposium Proceedings Vick international Edited by F. H. Dost and B. Leiber Georg Thieme verlag, Stuttgart,1967
PGMS 29.0391

PARIS 1966 Nouvelles proprietes physiques, mecaniques et chimiques Colloque international Centre national de la recherche scientifique Centre national de la recherche scientifique. Colloques internationaux, 167 Editions du centre national de la recherche scientifique, Paris,1968
Met 25.1503

PARIS 1966 Problemes actuels de paleontologie (evolution des vertebres) Centre national de la recherche scientifique Centre national de la recherche scientifique. Colloques internationaux.Actes, 163 pls 27cm C.N.R.S.,Paris,1967 Organized by J. P.Lehman.Papers in English,French and German
Bal 44.4640

PARIS 1967 Advance in transplantation International congress of the Transplantation society 1st Proceedings Transplantation society Edited by J. Dausset and others Munksgaard,Copenhagen,1968
Surg 23.0021

PARIS 1967 Dialysis and renal transplantation European dialysis and transplant association conference 4th Proceedings Edited by David S. Kerr and others European dialysis and transplant association.Proceedings, 4 International congress series, 155 Excerpta medica, Amsterdam,1968
Surg 23.0043

PARIS 1967 International congress of the Transplantation society 1st abstracts Transplantation society Paris,1967
Surg 23.0036

PARIS 1967 Macromolecules helicoidales en solution Societe de chimie physique.Reunion 17e Societe de chimie physique Journal de chimie physique et de physico-chimie biologique, 65,no 1 Paris,1968
Bioch 33.0326

PARIS 1967 Origin and distribution of the elements Symposium on the origin and distribution of the elements International union of geological sciences Edited by L.H. Ahrens International series of monographs on earth sciences, 30 xvii,1178p 24cm Pergamon press,Oxford,1968
TA 15.0565

PARIS 1967 Pain International symposium on pain Proceedings Edited by A. Soulairac and others Academic press,London,1968
Pha 16.0263

PARIS 1967 Radiation and the control of immune response Panel on radiation and the control of immune response Proceedings Organized by the International atomic energy agency International atomic energy agency. Panel proceedings series International atomic energy agency,Vienna,1968
Radioth 35.1202

PARIS 1967 8-11 Origin and distribution of the elements :a symposium Proceedings International association of geochemistry and cosmochemistry Edited by L.H. Ahrens International series of monographs on earth sciences, 30 1178p Pergamon,Oxford,1968
Obs 6.3444

PARIS 1968 Marx and contemporary scientific thought The Role of Karl Marx in the development of contemporary scientific thought :a symposium Papers International council for philosophy and humanistic studies International social science council Under the auspices of Unesco International social science council.Publications, 13 xi,612p Mouton,The Hague;Paris,1969
WSM 43.3826

PARIS 1968 Scientific basis of cancer chemotherapy :seminar Edited by Georges Mathe Sponsored by the European organization for research into cancer treatment Recent results in cancer research, 21 Heinemann medical,London,1969
Radioth 35.0866

PARIS 1969 European orchid congress 2nd Proceedings Comite national interprofessionnel de l'horticulture et des pepinieres.Monographie, 2 C.N.I.H.,Rungis, 1969
BG 38.3116

PARIS 1969 Structure et proprietes des surfaces des solides :colloque international Centre national de la recherche scientifique Centre national de la recherche scientifique. Colloques internationaux, 187 Paris,1970
Met 25.2626

PARIS 1969 United States contributions to quaternary research International association for quaternary research congress 8th Proceedings Edited by S.A. Schumm and W.C. Bradley Geological society of America. Special papers, 123 Geological society of America,Boulder,Colo.,1969
Bot 42.3249

PARIS 1970 Lepetit colloquium on the biology of oncogenic viruses 2nd Proceedings Edited by Luigi G. Silvestri North-Holland,Amsterdam,1971
Bioch 33.2299

PARIS 1970 Solar magnetic fields International astronomical union Edited by R. Howard International astronomical union. Symposium, 43 782p Reidel,Dordrecht,1971
Obs 6.3589

PARIS 1970 Aug 31-Sep 4 Solar magnetic fields International astronomical union Edited by R. Howard International astronomical union.Symposium, 43 782p Reidel,Dordrecht,1971
TA 15.0688

PARIS SYMPOSIUM ON RADIO ASTRONOMY Paris 1958 Jul.30-Aug.6 International astronomical union and International scientific radio union Edited by Ronald N. Bracewell International astronomical union.Symposium, 9 International scientific radio union. Symposium, 1 612p Stanford university press,Stanford,Calif.,1959
Obs 6.0246

PARIS SYMPOSIUM ON RADIO ASTRONOMY Paris 1958 Jul.30-Aug.6 International astronomical union and International scientific radio union Edited by Ronald N. Bracewell International astronomical union.Symposium, 9 International scientific radio union. Symposium, 1 612p Stanford university press,Stanford,Calif,1959
A Math 4.1151

PARIS SYMPOSIUM ON RADIO ASTRONOMY Paris 1958 Jul 30-Aug 6 International astronomical union International scientific radio union Edited by Ronald N. Bracewell International astronomical union.Symposium, 9 International scientific radio union.Symposium, 1 612p Stanford university press, Stanford,Calif.,1959
TA 15.0638

PARIS SYMPOSIUM ON RADIO ASTRONOMY Paris 1958 Jul 30-Aug 6 International astronomical union and International scientific radio union Edited by Ronald N. Bracewell International astronomical union.Symposium, 9 International scientific radio union. Symposium, 1 612p Stanford university press,Stanford,Calif,1959
Cav 7.0301

PARK CITY,UTAH 1970 Symposium on several complex variables Proceedings Edited by R. M. Brooks Lecture notes in mathematics, 184 234p 25cm Springer,Berlin,1971
P Math 2.4354

PARKINSON'S DISEASE INFORMATION AND RESEARCH CENTER Biochemistry and pharmacology of the basal ganglia :symposium 2nd Proceedings New York 1965 Nov 29-30 Edited by Erminio Costa and others Raven press,Hewlett,N.Y.,1966
Pha 16.0059

PARKINSON'S DISEASE INFORMATION AND RESEARCH CENTER The Thalamus :international symposium Proceedings New York 1964 Edited by Dominick P. Purpura and Melvin D. Yahr Columbia university press,New York,1966
An 32.4661

PARKINSON'S DISEASE INFORMATION AND RESEARCH CENTER The Thalamus :international symposium 1st Proceedings New York 1964 Edited by Dominick P. Purpura and Melvin D. Yahr Columbia university press,New York;London,1966
Pha 16.0303

PARMA 1957 Photo-thermoperiodism,the action of different radiations on gibberellins and other active substances Colloque international sur le photothermoperiodisme action des diverses radiations on gibberellines et de quelques autres substances International union of biological sciences International union of biological sciences. Series B, 34 U.I.S.B.,Paris,1958
Bot 42.1420

PARTIAL DIFFERENTIAL EQUATIONS Contributions to the theory of partial differential equations Conference on partial differential equations typescript Harriman,N.Y. 1952 Oct By S. Bergman and others Edited by Lipman Bers and others Sponsored by National research council Annals of mathematics studies, 33 Bibliog 257p 26cm Princeton university press,Princeton,N.J.,1954
P. Math 2.1228

PARTIAL DIFFERENTIAL EQUATIONS DEUCE users' colloquium on partial differential equations papers London 1959 Oct 6 English electric company ltd DEUCE news, 45 91p English electric,Nelson,Staffs.,1959
Math L 5.2408

PARTIAL DIFFERENTIAL EQUATIONS Katata conference on the theory of partial differential equations and on the theory of complex manifolds Proceedings Katata 1966 Sep 18-22 Bibliog.,Illus. vi,107p 25cm Research institute for mathematical science,Kyoto university,Kyoto,1966
P Math 2.3188

PARTIAL DIFFERENTIAL EQUATIONS Problemes aux limites dans les equations aux derivees partielles Montreal 1962 By Jacques L. Lions Societe mathematique du Canada Montreal.University.Seminaire de mathematiques superieures, 1 176p 28cm Universite de Montreal,Montreal,1965
P. Math 2.1283

PARTIAL DIFFERENTIAL EQUATIONS :a symposium Proceedings Berkeley,Calif. 1960 Apr 21-22 American mathematical society Edited by Charles B. Morrey American mathematical society.Proceedings of symposia in pure mathematics, 4 Bibliog.,Illus. vi,169p 26cm American mathematical society, Providence,R.I.,1961
A Math 4.1468

PARTIAL DIFFERENTIAL EQUATIONS AND CONTINUUM MECHANICS :an international conference Proceedings Madison,Wis. 1960 Jun 7-15 United States.Army.Mathematics research center Edited by R.E. Langer bibliog.,illus. xv, 397p 24cm University of Wisconsin press, Madison,Wis.,1961
A Math 4.1614

PARTIAL DIFFERENTIAL EQUATIONS AND CONTINUUM MECHANICS :an international conference Proceedings Madison,Wisc. 1960 Jun 7-15 United States army.Mathematics research center Edited by Rudolph E. Langer United States army.Mathematics research center.Publication, 5 Madison,Wisc.,1961
Geod 9.0202

PARTICLE ANALYSIS The Physics of particle analysis :a conference Papers Nottingham 1954 Apr 6-9 Institute of physics British journal of applied physics.Supplement, 3 Institute of physics,London,1954
Geod 9.0289

PARTICLE INTERACTIONS AT HIGH ENERGIES Scottish universities summer school 7th proceedings Newbattle Abbey,Edinburgh 1966 Aug 1-20 Edited by T.W. Preist and L.L.J. Vick 406p Oliver and Boyd,Edinburgh,1967
A Math 4.0875

PARTICLE PHYSICS Argomenti scelti di fisica delle particelle Scuola internazionale di fisica "Enrico Fermi" 41 corso Rendiconti Varenna 1957 Jul 17-29 Societa Italiana di fisica Edited by J. Steinberger Bibliog., Illus. vii,194p 24cm Academic press,New York,1967
A Math 4.1545

PARTICLE PHYSICS Non-compact groups in particle physics conference proceedings Wisconsin,Me. 1966 May 5-6 University of Wisconsin.Graduate school Edited by Yutze Chow 216p Benjamin,New York,1966
A Math 4.0890

PARTICLE PHYSICS Phenomenology in particle physics :conference Proceedings Pasadena, Calif. 1971 Mar 25-26 California institute of technology.Division of physics,mathematics and astronomy Edited by C.B. Chiu and others illus xvi,814p 28cm California institute of technology,Pasadena,Calif.,1971
A Math 4.1766

PARTICLE SIZE ANALYSIS Symposium on particle size analysis London 1947 Feb 4 Institution of chemical engineers and Society of chemical industry. Road and building materials group 145p Institution of chemical engineers,London,1947 Supplement to Transactions,Institution of chemical engineers,vol.25,1947
Chem E 24.1431

PARTICLE SIZE MEASUREMENT Symposium on particle size measurement Papers American society for testing materials American society for testing materials.Special technical publication, 234 v,310p American society for testing materials,Philadelphia,Pa., 1959 Presented at the 61st annaual meeting of the American society for testing materials
Chem E 24.1440

PARTICLES Aerodynamic capture of particles : a conference Proceedings Leatherhead 1960 Edited by E.G. Richardson Pergamon press,London,1960
Eng 41.6967

PARTICLES International theoretical physics conference on particles and fields Proceedings Rochester,N.Y. 1967 Aug 28-Sep 1 Edited by C.R. Hagen and others 708p 26cm Interscience,New York,1967 Dedicated to Robert Oppenheimer
Cav 7.2773

PARTICLES AND FIELDS International conference on particles and fields 1967 Proceedings Rochester,N.Y. 1967 Aug 28-Sep 1 University of Rochester Edited by Carl Richard Hagen and others Interscience,New York,1967
A Math 4.1349

PARTITION CHROMATOGRAPHY :a symposium London 1948 Oct 30 Biochemical society Edited by R.T. Williams and R.L.M. Synge Biochemical society.Symposia, 3 Cambridge university press,Cambridge,1949
Gen 34.0622

PARTITION CHROMATOGRAPHY :a symposium London 1948 Oct 30 Biochemical society Edited by R.T. Williams and R.L.M. Synge Held at the London school of hygiene and tropical medicine Biochemical society. Symposia, 3 Cambridge university press, Cambridge,1949
Bioch 33.1365

PARTITION CHROMATOGRAPHY :a symposium London 1948 Oct 30 Edited by R.T. Williams and R.L.M. Synge Biochemical society. Symposia, 3 Cambridge university press, Cambridge,1951
Phys 20.1525

PARTITION CHROMATOGRAPHY 3 :a symposium London 1948 Oct 30 Biochemical society Edited by R.T. Williams and R.L.M. Synge Biochemical society symposia,3 Cambridge university press,Cambridge,1950
Chem 18.1276

PASADENA 1961 Filament winding conference Society of aerospace material and process engineers SAMPE,Azuza,Calif.,n.d. Mimeographed
Met 25.0097

PASADENA 1967 Applications of model theory to algebra,analysis,and probability International conference on the applications of model theory to algebra analysis and probability Proceedings Edited by W.A.J. Luxemburg Held at the California institute of technology Bibliog. v,307p 23cm Holt,Rinehart and Winston,New York,1968
P Math 2.3241

PASADENA,CALIF. Aug 28-Sep 3 Symposium (international) on combustion 8th Papers Combustion institute Williams and Wilkins, Baltimore,Md.,1962
Chem 18.0438

PASADENA,CALIF. 1948 Cerebral mechanisms in behaviour :the Hixon symposium Edited by Lloyd A. Jeffress Under the auspices of the Hixon fund committee Wiley;Chapman and Hall, New York;London,1951
An 32.4627

PASADENA,CALIF. 1948 Hixon symposium Cerebral mechanisms in behaviour :the Hixon symposium Report California institute of technology.Hixon committee Edited by Lloyd A. Jeffress Wiley;Chapman and Hall,New York; London,1951
Psy 31.0522

PASADENA,CALIF. 1960 Advanced propulsion techniques :Agard combustion and propulsion panel technical meeting Proceedings Agard Edited by S.S. Penner Pergamon press,London, 1961
Eng 41.7373

PASADENA,CALIF. 1960 Finite groups institute report American mathematical society Edited by Marshall Hall Supported by the National science foundation American mathematical society.Proceedings of symposia in pure mathematics, 6 114p 26cm American mathematical society,Providence,R.I., 1962
P. Math 2.2031

PASADENA,CALIF. 1963 The Heat transfer and fluid mechanics institute 1963 proceedings Edited by Anatol Roshko and others Stanford university press,Stanford,Calif,1963
A Math 4.0521

PASADENA,CALIF. 1963 Theory of numbers symposium American mathematical society Edited by Albert Leon Whiteman Under a grant from the National science foundation American mathematical society.Proceedings of symposia in pure mathematics, 8 vii,214p 26cm American mathematical society, Providence,R.I.,1965
P. Math 2.1774

PASADENA,CALIF. 1964 Cytogenetics of cells in culture symposium International society for cell biology Edited by R.J.C. Harris International society for cell biology. Symposia, 3 Academic press,New York,1964
Gen 34.0892

PASADENA,CALIF. 1964 Cytogentics of cells in culture :a symposium Papers International society for cell biology Edited by R.J.C. Harris International society for cell biology.Symposia, 3 illus, port ix,307p Academic press,New York; London,1964
Bot 42.0784

PASADENA,CALIF. 1964 Solar wind The Conference on the solar wind Proceedings California institute of technology.Jet propulsion laboratory Edited by Robert J. Mackin and Marcia Neugebauer JPL TR 32-630 400p pergamon press,Oxford,1966
TA 15.0201

PASADENA,CALIF. 1964 The Solar wind :a conference proceedings Edited by Robert J. Mackin and Marcia Neugebauer Sponsored by the California institute of technology.Jet propulsion laboratory Pergamon press,Oxford, 1966
A Math 4.1165

PASADENA,CALIF. 1965 Proceedings of the symposium on information processing in sight sensory systems Proceedings National institutes of health and California institute of technology Edited by P.W. Nye California institute of technology,Pasadena, Calif.,1965
Psy 31.0419

PASADENA,CALIF. 1966 Invertebrate nervous systems The Conference on invertebrate nervous systems :their significance for mammalian neurophysiology Papers Edited by C.A.G. Wiersma Sponsored by the National institute of health University of Chicago press,Chicago,Ill.;London,1967 Dedicated to the memory of George Howard Parker
Bal 44.6453

PASADENA,CALIF. 1967 New methods of thought and procedure Symposium on methodologies Contributions California institute of technology.Office for industrial association Society for morphological research Edited by F. Zwicky and A.G. Wilson viii,338p Springer,Berlin,1967
WSM 43.2639

PASADENA,CALIF. 1967 New methods of thought and procedure Symposium on methodologies Contributions Edited by F. Zwicky and A.G. Wilson illus. viii,338p 24cm Springer,Berlin,1967
Math S 3.1389

PASADENA,CALIF. 1971 Phenomenology in particle physics :conference Proceedings California institute of technology.Division of physics,mathematics and astronomy Edited by C.B. Chiu and others illus xvi,814p 28cm California institute of technology,Pasadena, Calif.,1971
A Math 4.1766

PASSENGER ENVIRONMENT :a conference Proceedings London 1972 Mar 23-24 Institution of mechanical engineers.Railway division IME,London,1972
Eng 41.8649

PASSIVIEREND FILME UND DECKSCHICHTEN ANLAUFSCHICHTEN mechanismus ihrer entstehung und ihre schutzwirkung gegen korrosion Frankfurt 1955 Oct 13-14 Deutsche gesellschaft fur metallkunde Edited by H. Fischer and K. Wiederholt Springer-verlag,Berlin,1956
Met 25.1944

The PATHOGENESIS AND PATHOLOGY OF VIRAL DISEASES a symposium New York 1948 Dec 14-15 New York academy of medicine.Section on microbiology Edited by John G. Kidd New York academy of medicine.Section on microbiology.Symposia, 3 Columbia university press,New York,1950
Path 30.2554

PATHOGENESIS AND TREATMENT OF OCCLUSIVE ARTERIAL DISEASE :a conference Proceedings London 1959 Nov 13-14 Royal college of physicians of London Edited by L. McDonald Pitman,London,1960
PGMS 29.0081

PATHOGENESIS AND TREATMENT OF OCCLUSIVE ARTERIAL DISEASE :a conference Proceedings London 1959 Nov 13-14 Royal college of physicians of London Edited by Lawson McDonald Pitman,London,1960
Path 30.2430

PATHOGENIC MYCOPLASMA Ciba foundation symposium on pathogenic mycoplasma Proceeings London 1972 Jan 25-27 Ciba foundation Edited by Katherine Elliott and Joan Birch Associated scientific publishers, Amsterdam,1972
Bioch 33.2296

PATHOLOGICAL WILT OF PLANTS :an international symposium Nitra 1966 Sep 1-6 Edited by J. Smolak and E. Haspel-Horvatovic illus. 127p Swets and Zeitlinger,Bratislava;Amsterdam,1970
Bot 42.4634

PATHOLOGY Vascular patterns as related to functions Conference on microcirculatory physiology and pathology 2nd Philadelphia,Pa. 1955 Apr 5 Williams and Wilkins,Baltimore,Md.,1955
An 32.3461

PATHOLOGY AND MICROBIOLOGY Chinese society of pathology and microbiolgy Proceedings Canton 1935 Nov 5-8 Chinese medical journal.Supplement, 1 Chinese medical journal,Peiping,1935
An 32.1220

PATHOLOGY OF LABORATORY RATS AND MICE London 1966 Apr 19-21 Nuffield foundation.Advisory committee Edited by Ernest Cotchin and Francis J.C. Roe Held at the Royal veterinary field station Blackwell,Oxford, 1967
Path 30.2438

PATTERN OF RESEARCH IN BRITISH INDUSTRY :a conference Report Eastbourne 1962 Apr 5-7 Federation of British industries Federation of British industries,London,1962
Eng 41.1671

PATTERN RECOGNITION Conference on pattern recognition Teddington 1968 Jul 29-31 Organised by Institution of electrical engineers.Control and automation division Institution of electrical engineers.Conference publication, 42 Institution of electrical engineers,London,1968
Eng 41.5311

PATTERN RECOGNITION Methodologies of pattern recognition International conference on methodologies of pattern recognition Proceedings Honolulu 1968 Jan 24-26 Edited by Satosi Watanabe Academic press,New York;London,1969
An 32.4412

PATTERNS OF ORGANIZATION IN THE CENTRAL NERVOUS SYSTEM Proceedings of the association New York 1950 Dec 15-16 Association for research in nervous and mental disease Edited by Philip Bard Association for research in nervous and mental disease. Research publications, 30 Hafner,New York, 1952 reprinted 1968
An 32.4853

PEOPLE IN THE COUNTRYSIDE United Nations European study group Report Leicester 1963 Edited by John Higgs NCSS 694 National council of social service,London,1966
Geog 13.1901

PEPTIDE CHEMISTRY :a symposium Report London 1955 Mar 30 Chemical society Edited by A.D. Jenkins and D.F. Elliott Chemical society.Special publication, 2 London,1955 Organised by D.H.Hey
Bioch 33.0498

PEPTIDE CHEMISTRY :a symposium Report London 1955 Mar 30 Chemical society Edited by A.D. Jenkins and D.F. Elliott Chemical society.Special publication,2 Chemical society,London,1955 Symposium organized by professor Hey
Chem 18.1524

PEPTIDES Ciba foundation symposium on amino acids and peptides with antimetabolic activity London 1958 Mar 18-20 Ciba foundation Edited by G.E.W. Wolstenholme and Cecilia M. O'Connor Ciba foundation symposia Churchill,London,1958
Chem 18.1523

PEPTIDES European peptide symposium 8th Proceedings Noordwijk 1966 Sep Edited by H.C. Beyerman and others North-Holland, Amsterdam,1967
Bioch 33.0539

PEPTIDES International congress of biochemistry 6th Abstracts New York 1964 Jul 26-Aug 1 Vol 2: proteins,peptides, and amino acids Scheduled under the auspices of the International union of biochemistry Washington,D.C.,1964 Title also in French.
Bioch 33.1339

PERCEPTION Models for the perception of speech and visual form :a symposium Proceedings Boston,Mass. 1964 Nov 11-14 Air force Cambridge research laboratories.Data science laboratory Edited by Weiant Wathen-Dunn MIT press,Cambridge,Mass.,1967
Psy 31.2951

PERCEPTION Symposium de l'Association de psychologie scientifique de langue francaise : la perception 2eme Textes et communications Louvain 1953 Sep 26-28 By Henri Pieron and others Association de psychologie scientifique de langue francaise Presses universitaires de France,Paris,1955
Psy 31.0887

PERCEPTION AND PERSONALITY :a symposium Papers Personal and social factors in perception Denver,Colo. American psychological association Edited by Jerome S. Bruner and David Krech Duke university publications Duke university press,Durham,N. C.,1950 Papers originally appeared in the 'Journal of personality',vol 18,no 1-2,Sept and Dec. 1949
Psy 31.0945

PERFORMANCE APPRAISALS :effects on employees and their performance.A seminar Ann Arbor, Mich. 1963 Mar Foundation for research on human behavior Foundation for research on human behavior,Ann Arbor,Mich.,1964
Eng 41.0784

PERFORMANCE FORECAST OF SELECTED STATIC ENERGY CONVERSION DEVICES :29th meeting of the AGARD propulsion and energetics panel Liege 1967 Jun 12-16 Edited by G.W. Sherman and L. Devol AGARD,Brussels,1967
Eng 41.4869

PERFORMANCE OF DEEP FOUNDATIONS :a symposium San Francisco,Calif. 1968 Jun 23-28 American society for testing materials American society for testing materials.Special technical publication, 444 American society for testing materials,Philadelphia,Pa.,1969
Eng 41.3159

PERFORMANCE OF THE EYE AT LOW LUMINANCES : colloquium Proceedings Delft 1965 Institute for perception RVO-TNO N.V. Optische industrie "De oude Delft'.Scientific fund Scientiae serviens Edited by M.A. Bouman and J.J. Vos International congress series, 125 Excerpta medica,Amsterdam,1966
Psy 31.2862

PERINATAL MEDICINE European congress of perinatal medicine 1st Berlin 1968 Mar 28-30 Edited by Peter John Hungerford and others Thieme;Academic press,Stuttgart; New York,1969
Phys 20.2246

PERIODIC INSPECTION OF PRESSURE VESSELS :a conference Proceedings London 1972 May 9-11 Institution of mechanical engineers. Applied mechanics group IME,London,1972
Eng 41.8648

PERIPHERAL CIRCULATION IN MAN :Ciba foundation symposium London 1953 May 11-13 Ciba foundation Edited by G.E.W. Wolstenholme and Jessie S. Freeman Churchill, London,1954
An 32.3459

PERMAFROST INTERNATIONAL CONFERENCE 1st Proceedings Lafayette,Indiana 1963 Nov 11-15 National research council Presented by the Building research advisory board National research council.Publication,1287 DLC,GB641.16 1963 illus,maps,tables,graphs 563p National academy of sciences,Washington, 1966
Sco 14.0390

PERMANENT INTERNATIONAL ASSOCIATION OF NAVIGATION CONGRESSES International navigation congress Section 1:inland navigation 19th London 1957 Communication 3 S.1-C.3: influence of ice on navigable waterways and on sea and inland ports illus 260p Permanent international association of navigation congresses,Brussels,1957
Sco 14.0363

PERMEABILITY AND CAPILLARITY OF SOILS :a symposium Atlantic city,N.J. 1966 Jun 26-Jul 1 American society for testing materials American society for testing materials.Special technical publication, 417 American society for testing and materials, Philadelphia,Pa.,1967
Eng 41.3102

PERMEABILITY AND THE NATURE OF CELL MEMBRANES Cold Spring Harbor symposia on quantitative biology Papers Vol 8 Cold Spring Harbor biological laboratory Long Island biological association,Cold Spring Harbor,1940 Later referred to in vol.9 as "Permeability and the nature of cell membranes"
Bioch 33.1264

PEROXISOMES The Nature and function of peroxisomes (microbodies,glyoxosomes) :a conference Papers New York 1969 May 16-17 New York academy of sciences Edited by James Hogg New York academy of sciences. Annals, 168 p. 209-381 New York,1969
Bot 42.4758

Les PEROXYDES ORGANIQUES EN RADIOBIOLOGIE : colloque Paris 1957 Jan 9-10 By R. Latarjet and others Actions chimiques et biologiques des radiations.Serie, 4 Masson, Paris,1958
Radioth 35.1720

PERSONAL AND SOCIAL FACTORS IN PERCEPTION Perception and personality :a symposium Papers Denver,Colo. American psychological association Edited by Jerome S. Bruner and David Krech Duke university publications Duke university press,Durham,N.C.,1950 Papers originally appeared in the 'Journal of personality',vol 18,no 1-2,Sept and Dec. 1949
Psy 31.0945

PERSONALITY RESEARCH International congress of applied psychology 14th Proceedings Copenhagen 1961 Aug 13-19 International association of applied psychology and Danish psychological association Edited by Stanley Coopersmith Munksgaard,Copenhagen, 1952
Psy 31.1968

PERSONNEL TRAINING Conference on training personnel for the computing machine field 1st proceedings Detroit,Mich. 1954 June 22-23 Wayne state university Edited by Arvid W. Jacobson Sponsored also by the Association for computing machinery,the Industrial mathematics society and the Institute of radio engineers 104p 23cm Wayne university press,Detroit,Mich.,1955
Math L 5.0730

PERSPECTIVAS DE GALICIA ANTE EL SEGUNDO PLAN DE DESARROLLO La Coruna 1968 May-Jun Instituto "Jose Cornide de estudios coruneses La Coruna,1968
Geog 13.6798

PERSPECTIVES IN BIOLOGY AND MEDICINE The Central nervous system and fish behavior :a conference Papers Chicago,Ill. 1967 Apr United States.Air force.Office of scientific research American zoological society. Comparative physiology division Edited by David Ingle illus University of Chicago press,Chicago,Ill.:London,1968
Psy 31.2816

PERSPECTIVES IN LEUKAEMIA :a symposium Proceedings New Orleans 1966 By William Dameshek and R.M. Dutcher Leukaemia society Grune and Stratton,New York,1968
Med 36.0103

PERSPECTIVES IN PHYTOCHEMISTRY Phytochemical society symposium Proceedings Cambridge 1968 Apr Phytochemical society Edited by J. B. Harborne and T. Swain Academic press, London;New York,1969 For earlier publications of this body see Plant phenolics group
Gen 34.1407

PERTH 1962 Australasian conference on hydraulics and fluid mechanics 1st Proceedings Edited by Richard Silvester Pergamon press,London,1964
Eng 41.6968

PERTURBATION THEORY AND ITS APPLICATION IN QUANTUM MECHANICS seminar proceedings Madison,Wis. 1965 Oct 4-6 United States army.Mathematics research center Edited by Calvin H. Wilcox Organized jointly with University of Wisconsin.Theoretical chemistry institute 428p John Wiley and sons,New York;London,1966
A Math 4.0809

PERTURBATIONS THEORY AND ITS APPLICATIONS IN QUANTUM MECHANICS an advanced seminar Papers Madison,Wisc. 1965 Oct 4-6 United States.Army.Mathematics research center, and,University of Wisconsin.Theoretical chemistry institute Edited by Calvin H. Wilcox J.Wiley,New York,1966
Chem 18.0692

PERUGIA 1957 International symposium on mammary cancer 2nd Proceedings Edited by Lucio Severi Division of cancer research, Perugia,1958
Radioth 35.0779

PESTICIDES IN THE ENVIRONMENT AND THEIR EFFECTS ON WILDLIFE :an advanced study Proceedings Monks Wood 1965 Jul 1-14 North Atlantic treaty organization Edited by N.W. Moore Journal of applied ecology. Supplement, 3 311p Blackwells,Oxford, 1966
Bot 42.1983

PESTICIDES IN THE ENVIRONMENT AND THEIR EFFECTS ON WILDLIFE :an advanced study institute sponsored by the North Atlantic treaty organization,Monks Wood experimental station, England Proceedings Monks Wood 1965 Jul 1-14 Edited by N.W. Moore Journal of applied ecology, 3,suppl. Blackwell,Oxford, 1965
Bal 39.1676

PETROLEUM International geological congress 20th papers Mexico City 1956 Seccion 3: geologia del petroleo Edited by A. Garcia Rojas and others Mexico City,1956 Bound with papers from section 2. Text in English, French,German and Russian
Geol 8.3035

PETROLEUM Literature of the combustion of petroleum :a symposium held...at the 129th national meeting of the American chemical society Papers Dallas,Texas 1956 Apr American chemical society American chemical society.Advances in chemistry series,20 American chemistry society,Washington,D.C., 1958
Chem 18.0449

PETROLEUM Symposium sobre yacimientos de petroleo y gas International geological congress 20th papers Mexico city 1956 Tom 1-5 Edited by Eduardo J. Guzman 2 vols Mexico city,1956 Text in English,French and Spanish
Geol 8.3026

PETROLEUM World petroleum congress Proceedings London 1933 Jul 19-25 Vol 2: refining,chemical and testing section Edited by A.E. Dunstan Organised by the Institution of petroleum technologists xxvi, 956p World petroleum congress,London,1934
Chem E 24.1708

PETROLEUM EXPLORATION SOCIETY OF LIBYA 10th : annual field conference guidebook Geology and archaeology of northern Cyrenaica, Libya Cyrenaica 1968 Edited by E.T. Barr Amsterdam,1968
Geol 8.4487

PETROLEUM EXPLORATION SOCIETY OF LIBYA 7th : annual field conference guidebook Guide to the geology and culture of Greece Athens 1965 Apr 9-13 Edited by E.E. Hotz and Peter Norton Amsterdam,1965
Geol 8.4481

PETROLEUM EXPLORATION SOCIETY OF LIBYA 8th : annual field conference Guidebook South-central Libya and northern Chad Libya 1966 Edited by James J. Williams Amsterdam,1966
Geog 13.0598

PETROLEUM EXPLORATION SOCIETY OF LIBYA 8th : annual field conference guidebook South-central Libya and northern Chad:a guidebook Libya 1966 Edited by James J. Williams Amsterdam,1966
Geol 8.4490

PETROLEUM EXPLORATION SOCIETY OF LIBYA 9th : annual field conference guidebook Guidebook to the geology and history of Tunisia Tunisia 1967 Edited by Lewis Martin and others Amsterdam,1967
Geol 8.4489

PETROLOGY International geological congress 20th papers Mexico City 1956 Seccion 11A: petrologia y mineralogia Edited by A. Garcia Rojas Mexico City,1959 Text in English,French,Russian and Spanish
Geol 8.3047

PFIZER FOUNDATION Diabetes melitus :the opening symposium of the Pfizer foundation of the post-graduate medical school,University of Edinburgh Edinburgh 1965 Edited by L.J.P. Duncan University of Edinburgh.Pfizer medical monographs Edinburgh university press,Edinburgh,1966
PGMS 29.0179

PFLANZENSOZIOLOGIE UND LANDSCHAFTSOKOLOGIE Internationale vereinigung fur vegetationskunde Bericht Stolzenau 1963 Edited by Reinhold Tuxen illus. Junk,The Hague,1968
Bot 42.4765

PFLANZENSOZIOLOGIE UND PALYNOLOGIE : international symposium Bericht International society for plant geography and ecology Edited by Reinhold Tuxen illus. xvii,275p Junk,The Hague,1967
Bot 42.4772

PFLANZENSOZIOLOGISCHE SYSTEMATIK Internationale vereinigung fur vegetationskunde Bericht Stolzenau 1964 Edited by Reinhold Tuxen illus. Junk,The Hague,1968
Bot 42.4764

PHAGE AND ORIGINS OF MOLECULAR BIOLOGY symposium Cold Spring harbor laboratory of quantitative biology Edited by John Cairns and others Cold Spring harbor symposia, 32 Cold Spring harbor laboratory of quantitative biology,Long Island,N.Y.,1966
Gen 34.0908

The PHANEROZOIC TIME-SCALE :a symposium dedicated to Professor Arthur Holmes Glasgow 1964 Feb 14 Geological society of London Edited by W.B. Harland and others Geological society of London.Quarterly journal, 120 S Bibliog viii,458p Geological society of London,London,1964 Based on material presented at the Geological society's symposium on the Phanerozoic time-scale,held in Glasgow 14 February,1964.A supplement to the Quarterly journal of the Geological society of London
Bot 42.3194

PHANEROZOIC TIME-SCALE The Phanerozoic time-scale :a symposium dedicated to Professor Arthur Holmes Geological society of London Edited by W.B. Harland and others Geological society of London.Quarterly journal,120 S Bibliog viii,458p Geological society of London,London,1964 Based on material presented at the Geological society's symposium on the Phanerozoic time-scale,held in Glasgow 14 February,1964.A supplement to the Quarterly journal of the Geological society of London
Sco 14.0249

PHARMACEUTICAL SOCIETY OF GREAT BRITAIN Clinical trials :a symposium Report London 1962 Apr 5 Pharmaceutical press,London,1962
Pha 16.0130

PHARMACOKINETICS AND MODE OF ACTION OF ORAL HYPOGLYCAEMIC AGENTS 3rd conference Capri 1969 Apr 2-3 Edited by A. Loubatieres and A.E. Renold Acta diabetologica latina, 6 1969
PGMS 29.0400

The PHARMACOLOGICAL AND CHEMOTHERAPEUTIC PROPERTIES OF NIRIDAZOLE AND OTHER ANTISCHISTOSOMAL COMPOUNDS :a conference New York 1969 Oct 6 By F.C. Goble and others New York academy of sciences Edited by Francis C. Goble New York academy of sciences.Annals, 160,art.2 New York,1969
Mol 45.0272

PHARMACOLOGY Quantitative methods in pharmacology :a symposium Proceedings Leyden 1960 May 10-13 Edited by H. de Jonge Bibliog,illus xx,391p 23cm North-Holland,Amsterdam,1961
Math 3.1168

The PHARMACOLOGY OF PLANT PHENOLICS :a symposium Proceedings Oxford 1958 Apr 10-11 Edited by J.W. Fairbairn Organised by the Plant phenolics group Academic press, London;New York,1959
Bioch 33.0752

The PHARMACOLOGY OF PLANT PHENOLICS :a symposium Proceedings Oxford 1958 Apr 10-11 Plant phenolics group Edited by J.W. Fairbairn Academic press,London;New York, 1959
Chem 26.0265

PHASE AND FREQUENCY INSTABILITIES IN ELECTROMAGNETIC WAVE PROPAGATION Agard Agard conference proceedings, 33 Technivision,Slough,1970 Contains the proceedings of the Electromagnetic wave propagation committee's thirteenth symposium
Eng 41.8319

PHASE STABILITY IN METALS AND ALLOYS Battelle materials science colloquium 1st Proceedings Geneva;Villars 1966 Mar 7-12 Battelle memorial institute Edited by Peter S. Rudman and others McGraw-Hill series in materials science and engineering McGraw-Hill,New York,1967
Met 25.1194

PHASE TRANSFORMATIONS Phase transformations in solids :a symposium Papers Ithaca,N.Y. 1948 Aug 23-26 Edited by R. Smoluchowski and others Sponsored by the National research council.Committee on solids Wiley,New York, 1951
Min 10.0640

PHASE TRANSFORMATIONS AND THE EARTH'S INTERIOR a conference Canberra 1969 Jan 6-10 International upper mantle committee Australiann academy of science Edited by A.E. Ringwood and D.H. Green North-Holland, Amsterdam,1970
Min 10.1459

PHASE TRANSFORMATIONS IN METALS The Mechanism of phase transformations in metals : a symposium London 1955 Nov 9 Institute of metals Institute of metals.Monograph and report series, 18 Institute of metals, London,1956
Met 25.1251

PHASE TRANSFORMATIONS IN SOLIDS :symposium New York 1948 Aug 23-26 Edited by R. Smoluchowski and others Sponsored by the National research council.Committee on solids Wiley;Chapman and Hall,New York;London,1951
Met 25.1250

PHASE TRANSFORMATIONS IN SOLIDS :a symposium Papers Cornell university 1948 Aug 23-26 Edited by R. Smoluchowski and others Sponsored by the National research council. Division of physical sciences. Committee on solids J.Wiley;Chapman and Hall,New York; London,1951
Chem 18.2149

PHASE TRANSFORMATIONS IN SOLIDS :a symposium Papers Ithaca,N.Y. 1948 Aug 23-26 Edited by R. Smoluchowski and others Sponsored by the National research council. Committee on solids Wiley,New York,1951
Min 10.0640

PHENETIC AND PHYLOGENETIC CLASSIFICATION :a symposium Liverpool 1964 Apr 9 Edited by Vernon H. Heywood and J. McNeill Systematics association.Publication, 6 xi, 164p Systematics assoc.,London,1964
Bot 42.3439

PHENOLIC COMPOUNDS Recent developments in the chemistry of natural phenolic compounds :a symposium Proceedings 1960 Apr Plant phenolics group Edited by W.D. Ollis Pergamon press,Oxford,1961
Chem 18.1495

PHENOMENES SOLAIRES ET TERRESTRES Commission mixte pour l'etude des relations entre les phenomenes solaires et terrestres Proceedings Rome 1952 Sep. 3 International council of scientific unions 35p Paris,1952
Obs 6.2544

PHENOMENOLOGY IN PARTICLE PHYSICS :conference Proceedings Pasadena,Calif. 1971 Mar 25-26 California institute of technology. Division of physics,mathematics and astronomy Edited by C.B. Chiu and others illus xvi, 814p 28cm California institute of technology,Pasadena,Calif.,1971
A Math 4.1766

PHI DELTA KAPPA Improving experimental design and statistical analysis Phi Delta Kappa annual symposium on educational research 7th Discussions Madison,Wisc. 1966 Rand-McNally education series Rand-McNally, Chicago,Ill.,1967
Psy 31.1123

PHI DELTA KAPPA ANNUAL SYMPOSIUM ON EDUCATIONAL RESEARCH 7th Discussions Improving experimental design and statistical analysis Madison,Wisc. 1966 Phi Delta Kappa and University of Wisconsin.Phi Delta Kappa chapter Rand-McNally education series Rand-McNally,Chicago,Ill.,1967
Psy 31.1123

PHILADELPHIA 1941 The Flow of heat in metals By J.B. Austin American society for metals,Cleveland,Ohio,1942 A series of five educational lectures presented to members of the American society for metals at the twenty-third National metal congress and exposition Philadelphia Pa. Oct. 20-25,1941
Met 25.0304

PHILADELPHIA 1948 Cold working of metals American society for metals American society for metals,Cleveland,Ohio,1949
Met 25.1199

PHILADELPHIA 1948 Metallurgy and magnetism By James K. Stanley American society for metals,Cleveland,Ohio,1949 Lectures presented to members of the A.S.M. during the 30th National metal congress and exposition
Met 25.1104

PHILADELPHIA 1948 Properties of metals in materials engineering By R.L. Templin and others American society for metals,Cleveland, Ohio,1949 A series of eight educational lectures presented to members of the A.S.M. during the thirtieth National metal congress and exposition
Met 25.0826

PHILADELPHIA 1951 Monomolecular layers :a symposium American association for the advancement of science Edited by Harry Sobotka vii,207p American association for the advancement of science,Washington, D.C., 1954
Chem E 24.0932

PHILADELPHIA 1951 Sex in microorganisms :a symposium American association for the advancement of science Edited by D.H. Wenrich illus v,362p American association for the advancement of science, Washington,D.C.,1954
Bot 42.0758

PHILADELPHIA 1952 Modern research techniques in physical metallurgy a seminar... held during the thirty-fourth national metal congress and exposition American society for metals American society for metals,Cleveland, Ohio,1953
Met 25.1640

PHILADELPHIA 1955 Powder metallurgy in nuclear engineering Conference on powder metallurgy in atomic energy with additional papers By Henry H. Hausner United States atomic energy commission and American society for metals American society for metals,Cleveland,Ohio,1958
Met 25.0767

PHILADELPHIA 1955 Theory of alloy phases a seminar American society for metals American society for metals,Cleveland,Ohio, 1956 A seminar held during the 37th National metal congress and exposition
Met 25.1051

PHILADELPHIA 1956 International congress on catalysis Edited by Adalbert Farkas Advances in chemistry and related subjects, 9 xviii,847p Academic press,New York,1957
Chem E 24.0800

PHILADELPHIA 1958 Neurochemistry of nucleotides and amino acids :a symposium Papers and discussions presented American academy of neurology.Section of neurochemistry Edited by Roscoe O. Brady and Donald B. Tower Wiley,New York,1960
Bioch 33.2222

PHILADELPHIA 1959 Mechanical properties of intermetallic compounds a symposium held during the 115th meeting of the Electrochemical society Electrochemical society Edited by J.H. Westbrook John Wiley,London,1960
Met 25.2277

PHILADELPHIA 1959 Symposium on electrode processes Transactions Electrochemical society.Theoretical electrochemistry division Edited by Ernest Yeager Wiley,New York; London,1961
Met 25.1820

PHILADELPHIA 1960 Decomposition of austenite by diffusional processes :a symposium Proceedings American institute of mining,metallurgical and petroleum engineers Edited by V.F. Zackay and H.I. Aaronson Interscience,New York;London,1962
Met 25.1247

PHILADELPHIA 1960 Properties of crystalline solids :...symposium on recent progress in materials sciences;symposium on nature and origin of strength of materials A. S.T.M.Special technical publication Philadelphia,1961 Includes an introductory paper by W.O.Baker on 'The national role of materials research and development'
Met 25.2490

PHILADELPHIA 1962 Electron microscopy International congress for electron microscopy 5th Vol 1: non-biology International federation of electron microscope societies and Electron microscope society of America Edited by Sydney S. Breese Academic press, New York;London,1962
Met 25.1368

PHILADELPHIA 1962 Fat as a tissue : international research conference Proceedings Edited by Kaare Rodahl and others Held at the Lankenau hospital McGraw-Hill,New York,1963
An 32.5318

PHILADELPHIA 1962 The International conference on the biology of cutaneous cancer 1st Proceedings National cancer institute Edited by Frederick Urbach National cancer institute.Monograph, 10 U.S.Department of health,education,and welfare,Bethesda,Md.,1963 Sponsored by the skin and cancer hospital, Dept.of dermatology,Temple university school of medicine and The Committee on geographie pathology,Unio internationalis contra cancrum. A Merck Sharp and Dohme medical research conference
Bioch 33.0981

PHILADELPHIA 1963 Energy-linked functions of mitochondria :colloquium Papers presented Johnson research foundation Edited by Britton Chance Johnson research foundation. Colloquia, 1 Academic press,New York; London,1963
Bioch 33.1081

PHILADELPHIA 1963 Energy-linked functions of mitochondria :first colloquium of the Johnson research foundation of the University of Pennsylvania Papers Johnson research foundation Edited by Britton Chance Academic press,NEW York,1963
Radioth 35.0412

PHILADELPHIA 1964 Rapid mixing and sampling techniques in biochemistry International colloquium on rapid mixing and sampling techniques applicable to the study of biochemical reactions 1st Proceedings International union of biochemistry Edited by Britton Chance and others Academic press, New York;London,1964
Bioch 33.2320

PHILADELPHIA 1964 The Thymus :a symposium Edited by Vittorio Defendi and Donald Metcalf Held at the Wistar institute of anatomy and biology Wistar institute symposium monograph, 2 Wistar institute press,Philadelphia,Pa., 1964
An 32.3702

PHILADELPHIA 1965 Biology of parasites; emphasis on veterinary parasites The International conference of the World association for the advancement of veterinary parasitology 2nd Proceedings World association for the advancement of veterinary parasitology Edited by E.J.L. Soulsby Held at the University of Pennsylvania Academic press,New York;London,1966 Held...in conjunction with Bicentennial celebrations of medical education in the United States
Bal 39.1737

PHILADELPHIA 1965 Control of energy metabolism A Colloquium on metabolic control and a Symposium on control of energy metabolism Proceedings Johnson research foundation Edited by Britton Chance and others Johnson research foundation.Colloquia Academic press,New York;London,1965 "In celebration of the bicentennial of the University of Pennsylvania school of medicine"
Bioch 33.1077

PHILADELPHIA 1966 Hemes and hemoprotein : colloquium 3rd Proceedings University of Pennsylvania.Johnson research foundation Edited by Britton Chance and others Academic press,New York,1966
Phys 20.1551

PHILADELPHIA 1966 Hemes and hemoproteins : colloquium Proceedings Johnson research foundation Edited by Britton Chance and others Johnson research foundation.Colloquia, 3 Academic press,New York,1966
Bioch 33.0507

PHILADELPHIA 1967 Growth regulating substances for animal cells in culture :a symposium Edited by Vittorio Defendi and Michael Stoker Held at the Wistar institute of anatomy and biology Wistar institute symposium monographs, 7 Wistar institute press,Philadelphia,1967
Radioth 35.0566

PHILADELPHIA 1967 Temper embrittlement in steel a symposium American society for testing materials.Committee A-I on steel A.S. T.M.Special technical publication, 407 A.S.T. M.,Philadelphia,1968
Met 25.0455

PHILADELPHIA 1968 Epithelial-mesenchymal interactions The Hahnemann symposium 18th Hahnemann medical college and hospital Edited by Raul Fleischmajor and Rupert E. Billingham Williams and Wilkins,Baltimore, 1968 Dedicated to Dr. Johannes Holtfreter
Bal 39.0491

PHILADELPHIA,PA. Mind as a tissue Lankenau conference 5th Proceedings Lankenau hospital Edited by Charles Rupp Harper and Row,Hoeber medical division,New York,1968
Psy 31.2898

PHILADELPHIA,PA. 1893 Proceedings commemorative of the 150th anniversary of the foundation of the American philosophical society phylogeny of an acquired characteristic By Alpheus Hyatt American philosophical society.Proceedings,32 Philadelphia,Pa.,1894
Geol 8.0856

PHILADELPHIA,PA. 1908 International tuberculosis conference 7th Report International anti-tuberculosis association Edited by Gotthold Pannwitz port Internationale vereinigung gegen die tuberkulose,Berlin-Charlottenburg,1909 Text in English,French,German;title also in French and German
Path 30.1023

PHILADELPHIA,PA. 1937 Early man... International symposium on early man Papers Edited by George Grant MacCurdy illus. Lippincott,Philadelphia,Pa.,1937
Geog 13.1760

PHILADELPHIA,PA. 1940 American polar explorations Centenary celebration of the Wilkes exploring expedition of the United States navy 1838-1842 and symposium on American polar explorations Papers By Edwin G. Conklin and others American philosophical society.Proceedings, 82 519-950p American philosophical society, Philadelphia,Pa.,1940
Sco 14.8094

PHILADELPHIA,PA. 1940 Liebig and after Liebig:a century of progress in agricultural chemistry A Symposium of papers presented before the sections of chemistry and agriculture of the American association for the advancement of science...in commemoration of the 100th anniversary of the publication of Liebig's 'Organic chemistry in its application to agriculture and physiology' American association for the advancement of science Edited by Forest Ray Moulton American association for the advancement of science. Publication, 16 111p American association for the advancement of science,Washington,D.C., 1942
WSM 43.2238

PHILADELPHIA,PA. 1941 Controlled atmospheres :the symposium...presented before the annual convention American society for metals A.S.M.,Cleveland,Ohio,c1941
Met 25.0250

PHILADELPHIA,PA. 1944 Symposium on plastics American society for testing and materials American society for testing materials,Philadelphia,Pa.,1944
Eng 41.3642

PHILADELPHIA,PA. 1944 Symposium on stress-corrosion cracking of metals American institute of mining and metallurgical engineers American society for testing materials American society for testing materials;American institute of mining and metallurgical engineers,Philadelphia,Pa.;New York,1945
Eng 41.3717

PHILADELPHIA,PA. 1944 Symposium on stress-corrosion cracking of metals American society for testing materials and American institute of mining and metallurgical engineers.Institute of metals division A.S.T. M.,New York,1945
Met 25.1952

PHILADELPHIA,PA. 1946 Symposium on effects of low temperatures on the properties of materials :presented at a meeting of the Philadelphia district,A.S.T.M. American society for testing materials American society for testing materials.Special technical publication, 78 American society for testing materials,Philadelphia,Pa.,1950
Eng 41.3778

PHILADELPHIA,PA. 1949 Natural selection and adaptation :annual general meeting Papers American philosophical society American philosophical society.Proceedings, 93,no.6 Philadelphia,Pa.,1949
Gen 34.1305

PHILADELPHIA,PA. 1955 Symposium on printed circuits proceedings Radio-Electronics-Television manufacturers association Jointly sponsored by Institute of radio engineers illus. 122p 28cm Engineering publishers, New York,1955
Math L 5.0736

PHILADELPHIA,PA. 1955 Vascular patterns as related to functions Conference on microcirculatory physiology and pathology 2nd Williams and Wilkins,Baltimore,Md., 1955
An 32.3461

PHILADELPHIA,PA. 1956 Semiconductor surface physics Conference on the physics of semiconductor surfaces Proceedings Edited by R.H. Kingston Sponsored by the United States.Office of naval research,the University of Pennsylvania and the Lincoln laboratory University of Pennsylvania,Philadelphia,Pa., 1957
Eng 41.5399

PHILADELPHIA,PA. 1957 Automatic coding symposium proceedings Edited by Nancy S. Glenn Franklin institute.Journal.Monograph, 3 vii,116p 24cm Journal of the Franklin institute,Philadelphia,Pa.,1957
Math L 5.0781

PHILADELPHIA,PA. 1957 Symposium on brittle failure of rotor forgings presented at special meeting for delegates to world metals congress American society for testing materials A.S.T.M.Special technical publication, 231 American society for testing materials, Philadelphia,Pa,1957
Met 25.1325

PHILADELPHIA,PA. 1957 Symposium on elevated temperature strain gages American society for testing materials Sponsored by United States.Naval air material center A.S.T.M.Special technical publication, 230 American society for testing materials, Philadelphia,Pa,1958
Met 25.0985

PHILADELPHIA,PA. 1959 Commemoration of the centennial of the publication of the origin of the species :annual general meeting American philosophical society American philosophical society.Proceedings, 103,no.2 Philadelphia, Pa.,1959
Gen 34.1265

PHILADELPHIA,PA. 1959 Solid-state circuit conference digest of technical papers Institute of radio engineers Sponsored by American institute of electrical engineers 102p Lewis Winner,1959
Math L 5.2082

PHILADELPHIA,PA. 1959 Symposium on electrode processes Transactions Electrochemical society.Theoretical electrochemistry division and United States.Air force.Office of scientific research Wiley,New York,1961
Chem 18.2624

PHILADELPHIA,PA. 1960 Congenital heart disease... :an international symposium Edited by Dryden P. Morse Sponsored by Deborah hospital Blackwell,Oxford,1962
An 32.2064

PHILADELPHIA,PA. 1960 Muscle as a tissue : a conference Proceedings Edited by Kaare Rodahl and Steven M. Horvath Held at the Lankenau hospital McGraw-Hill,New York,1961
An 32.3999

PHILADELPHIA,PA. 1960 Properties of crystalline solids ...Symposium on recent progress in materials sciences;Symposium on nature and origin of strength of materials. Presented at the 63rd annual meeting A.S.T.M. American society for testing materials American society for testing materials.Special technical publication, 283 American society for testing materials,Philadelphia,Pa.,1961
Eng 41.3790

PHILADELPHIA,PA. 1960 The Symposium on solidification presented at the 64th AFS castings congress and exposition American foundrymen's society American foundrymen's society,Des Plaines,Ill.,1961
Met 25.0644

PHILADELPHIA,PA. 1961 Symposium on the engineering aspects of magnetohydrodynamics 2nd Proceedings Edited by Clifford Mannal and Norman W. Mather Sponsored by the American institute of electrical engineers Columbia university press,New York;London,1962
Eng 41.4497

PHILADELPHIA,PA. 1962 Electron microscopy International congress for electron microscopy 5th Vol 2: biology Edited by Sidney S. Breese Academic press,New York,1962
An 32.3183

PHILADELPHIA,PA. 1962 Mechanisms of hard tissue destruction :a symposium presented at the Philadelphia meeting of the AAAS... Edited by Reidar F. Sognnaes American association for the advancement of science. Publications, 75 American association for the advancement of science,Washington,D.C., 1963
An 32.3661

PHILADELPHIA,PA. 1962 Metallurgy of advanced electronic materials a technical conference American institute of mining, metallurgical and petroleum engineers Edited by Geoffrey E. Brock Metallurgical society conferences, 19 Interscience,New York; London,1963
Met 25.1155

PHILADELPHIA,PA. 1962 The International conference on the biology of cutaneous cancer 1st Temple university.School of medicine and International union against cancer Edited by Frederick Urbach With financial support from Merck,Sharp and Dohme. Postgraduate program National cancer institute.Monograph, 10 US government printing office,Washington,D.C.,1963
Path 30.2423

PHILADELPHIA,PA. 1962 The Physics and chemistry of ceramics :a symposium Proceedings United States.Office of naval research Edited by Cyrus Klingsberg Gordon and Breach,New York;London,1963
Met 25.0277

PHILADELPHIA,PA. 1963 Multi-flow symposium Reports and proceedings American society of mechanical engineers.Hydraulics and heat-transfer division American society of mechanical engineers,New York,1963
Eng 41.6513

PHILADELPHIA,PA. 1963 Multi-phase flow symposium Papers Edited by Norman J. Lipstein Presented at the winter annual meeting of the American society of mechanical engineers illus,diagrms 99p American society of mechanical engineers,New York,1963
Chem E 24.0241

PHILADELPHIA,PA. 1963 Pulsatile blood flow International symposium on pulsatile blood flow 1st Proceedings Edited by E.O. Attinger McGraw-Hill,New York,1964
An 32.3528

PHILADELPHIA,PA. 1963 Pulsatile blood flow international symposium 1st Proceedings Edited by E.O. Attinger McGraw-Hill,New York, 1964
Phys 20.0899

PHILADELPHIA,PA. 1964 Nerve as a tissue :a conference 4th Proceedings Edited by Kaare Rodahl and Bela Issekutz Sponsored by the Lankenau hospital Harper,New York,1966
Pha 16.0299

PHILADELPHIA,PA. 1964 Nerve as a tissue : conference Proceedings Edited by Kaare Rodahl and Bela Issekutz Held at the Lankenau hospital Harper,London,1966
An 32.3430

PHILADELPHIA,PA. 1964 Plasticization and plasticizer processes :a symposium Sponsored by the American chemical society.Division of industrial and engineering chemistry Advances in chemistry series, 48 ix,200p American chemical society,Washington,D.C.,1965 Symposium chairman:Norbert A.J.Platzer
Chem E 24.1045

PHILADELPHIA,PA. 1964 Rapid mixing and sampling techniques in biochemistry : nternational coloquium 1st Proceedings International union of biochemistry Edited by Britton Chance and others Academic press, New York,1964
Pha 16.0141

PHILADELPHIA,PA. 1964 Symposium on fully separated flows :a conference Papers American society of mechanical engineers Edited by Arthur G. Hansen American society of mechanical engineers,New York,1964
Eng 41.6514

PHILADELPHIA,PA. 1965 Commemoration of the publication of Gregor Mendel's pioneer experiments in genetics :annual general meeting Papers American philosophical society American philosophical society. Proceedings, 109,no.4 American philosophical society,Philadelphia,Pa.,1965
Gen 34.2207

PHILADELPHIA,PA. 1965 Conference on murine leukemia Papers National cancer institute. Virology research branch and Albert Einstein medical center Edited by Marvin A. Rich and John B. Moloney National cancer institute.Monograph, 22 US government printing office,Washington,D.C.,1966
Path 30.2460

PHILADELPHIA,PA. 1965 Control of energy metabolism Colloquium on metabolic control : and symposium on control of metabolism Johnson research foundation Edited by Britton Chance and others Academic press,New York;London,1965
Gen 34.0614

PHILADELPHIA,PA. 1965 isoantigens and cell interactions :a symposium Wistar institute of anatomy and biology Edited by Joy Palm Wistar institute symposium monograph, 3 Wistar institute press,Philadelphia,Pa.,1965
Path 30.2416

PHILADELPHIA,PA. 1966 Information retrieval:a critical view Annual colloquium on information retrieval 3rd Edited by George Schecter Sponsored by the Association for computing machinery.Special interest group on information retrieval Academic press,New York,1967
Eng 41.7692

PHILADELPHIA,PA. 1966 Mechanical working and steel processing committee 2nd Operating metallurgy conference 9th Proceedings American institute of mining,metallurgical and petroleum engineers Metallurgical society of the AIME.Conference series, 50 Gordon and Breach,New York,1968
Met 25.2572

PHILADELPHIA,PA. 1967 High-energy nuclear reactions in astrophysics Symposium on high-energy nuclear reactions in astrophysics Invted papers Edited by D.S.P. Shen Sponsored by American astronomical union 281p Benjamin,New York,1967 Four of the ten papers were invited after the symposium to complete the coverage
TA 15.0578

PHILADELPHIA,PA. 1969 Corrosion by liquid metals Proceedings of the sessions on corrosion by liquid metals of the 1969 fall meeting of the Metallurgical society of AIME American institution of mining engineers. Metallurgical society Edited by J.E. Draley and J.R. Weeks Metallurgical society of AIME. Publications Plenum press,New York,1970
Met 25.2790

PHILADELPHIA,PA. 1969 Heterogeneous kinetics at elevated temperatures International conference on metallurgy and materials science Proceedings Edited by G. R. Belten and W.L. Worrell Held at the University of Pennsylvania Plenum press,New York,1970
Met 25.2783

PHILADELPHIA,PA. 1969 Materials engineering exposition and congress Symposium on applications of modern metallographic techniques American society for testing and materials American society for metals American society for testing and materials.Special technical publication, 480 ASTM,Philadelphia,Pa.,1970 Symposium presented at the exposition and congress
Met 25.2607

PHILADELPHIA,PA. 1969 Probes of structure and function of macromolecules and membranes Colloquium of the Johnson research foundation 5th Proceedings Vol 1-2: probes and membrane function;probes of enzymes and hemoproteins Johnson research foundation Edited by Britton Chance and others Academic press,New York,1971
Bioch 33.2190

PHILADELPHIA,PENN. 1961 Agglomeration : based on an international symposium... American institute of mining,metallurgical and petroleum engineers Edited by William A. Knepper Interscience,New York;London,1962
Met 25.2288

PHILADELPHIA,PENN. 1966 Continuous processing and process control :a symposium Proceedings American institute of mining, metallurgical and petroleum engineers Edited by Thomas R. Ingraham Metallurgical society conferences, 49 Gordon and Breach,New York, 1968 Symposium sponsored by the Electric furnace conference,the Mechanical working and steel processing conference and the Physical chemistry of steelmaking conference of the Metallurgical society
Met 25.2846

PHILADEPHIA,PA. 1961 International symposium on stereoencephalotomy (stereotaxic surgery) 1st Proceedings Edited by E.A. Spiegel and H.T. Wycis Karger,Basle,1962
An 32.4200

PHILCO CORPORATION Single-crystal films :a conference Proceedings Blue Bell,Penn. 1963 May United States.Office of naval research and University of Pennsylvania Edited by Maurice H. Francombe and Hiroshi Sato Pergamon press,Oxford,1964
Met 25.0177

PHILCO SCIENTIFIC LABORATORIES International conference on single-crystal films Proceedings Blue Bell,Pa. 1963 May Pergamon press,Oxford,1964
Cav 7.0244

PHILIP,Prince,consort of Elizabeth II,Queen of Great Britain Conference across a continent The Duke of Edinburgh's commonwealth study conference 2nd Canada 1962 May 13-Jun 6 Macmillan,Toronto,1963
Geog 13.5039

PHILOSOPHICAL SOCIETY OF THE SUDAN 13th : annual conference Proceedings and papers Agricultural development in the Sudan Khartoum 1965 Dec 3-6 Vol 1-2 Edited by D.J. Shaw Co-sponsored by the Sudan agricultural society maps 2 vols Philosophical society of the Sudan,Khartoum, 1966 Typescript
Geog 13.4700

PHILOSOPHY Determinism and freedom in the age of modern science:a philosophical symposium New York university institute of philosophy :annual symposium 1st Proceedings New York 1957 Feb 9-10 Edited by Sidney Hook xv,237p New York university press,New York,1965
WSM 43.1662

PHILOSOPHY International congress of philosophy 6th Proceedings Harvard university 1926 Sep 13-17 International congress of philosophy.Organizing committee Edited by Edgar Sheffield Brightman Longmans, Green,New York,1927
Psy 31.2520

PHILOSOPHY OF SCIENCE Boston studies in the philosophy of science Boston colloquium 1961-62 Proceedings Boston,Mass. 1961,1962 Vol 1 Edited by Marx W. Wartofsky viii,212p Reidel,Dordrecht,1963
WSM 43.0747

PHILOSOPHY OF SCIENCE Boston studies in the philosophy of science Boston colloquium 1964-1966 Proceedings Boston,Mass. 1964-66 Vol 3: in memory of N.R.Hanson Edited by R. S. Cohen and Marx W. Wartofsky xlix,489p Humanities press,New York,1968
WSM 43.0749

PHILOSOPHY OF SCIENCE Boston studies in the philosophy of science Boston colloquium 1966-1968 Proceedings Boston,Mass. 1966-68 Vol 4-5 Edited by R.S. Cohen and Marx W. Wartofsky 2 vols Reidel,Dordrecht,1969
WSM 43.0750

PHILOSOPHY OF SCIENCE Congres international de philosophie des sciences Actes Paris 1949 Oct 17-22 Tome 1- Institut international de philosophie Actualites scientifiques et industrielles,1126,1134,1137, 1146,1153,1155-56,1166-67- Hermann,Paris, 1951-
WSM 43.0981

PHILOSOPHY OF SCIENCE Criticism and the growth of knowledge International colloquium in the philosophy of science Proceedings London 1965 Jul 11-17 Edited by Imre Lakatos and Alan Musgrave Held at Bedford college vii,282p 23cm Cambridge university press,Cambridge,1970
Math S 3.1850

PHILOSOPHY OF SCIENCE Current issues in the philosophy of science :symposia of scientists and philosophers Proceedings Chicago,Ill. 1959 Dec 27-30 American association for the advancement of science Edited by Herbert Feigl and Grover Maxwell xi,484p Holt, Rinehart and Winston,New York,1961 Proceedings of section L of the Association
WSM 43.0784

PHILOSOPHY OF SCIENCE Induction,physics,and ethics Colloquium in the philosophy of science Proceedings and discussions Salzburg 1968 Aug 28-31 International union of history and philosophy of science. Division of logic,methodology and philosophy of science Institut fur wissenschaftstheorie, Salzburg Edited by Paul Weingartner and Gerhard Zecha Synthese library x,382p Reidel,Dordrecht,1970
WSM 43.0992

PHILOSOPHY OF SCIENCE International colloquium in the philosophy of science Proceedings London 1965 Vol 2: the problem of inductive logic Edited by Imre Lakatos Studies in logic and the foundations of mathematics viii,417p North-Holland, Amsterdam,1968
WSM 43.1218

PHILOSOPHY OF SCIENCE International congress for logic, methodology and philosophy of science proceedings Jerusalem 1964 Aug 26-Sep 2 Edited by Yehoshua Bar-Hillel and others Studies in logic and the foundations of mathematics viii,440p 23cm North-Holland,Amsterdam,1965
P. Math 2.0206

PHILOSOPHY OF SCIENCE International congress for logic, methodology and philosophy of science proceedings Stanford,Calif. 1960 Aug.24-Sep.2 Edited by Ernest Nagel and others ix,661p 26cm Stanford university press,Stanford, Calif.,1962
P. Math 2.0209

PHILOSOPHY OF SCIENCE International congress for logic,methodology and philosophy of science Proceedings Stanford,Calif. 1960 Aug 24-Sep 2 International union of history and philosophy of science.Division of logic, methodology and philosophy of science National academy of sciences. Edited by Ernest Nagel and others ix,661p 25cm Stanford university press,Stanford,Calif.,1962
Math S 3.1853

PHILOSOPHY OF SCIENCE Logic,methodology and philosophy of science International congress for logic,methodology and philosophy of science Proceedings Amsterdam 1967 Aug 25-Sep 2 Nederlandse vereniging voor logica en wijsbegeerte der exacte wetenschappen International union for logic,methodology and philosophy of science.Division of logic, methodology and philosophy of science Edited by B.van Rootselaar and J.F. Staal xiii,554p North-Holland,Amsterdam,1968
WSM 43.0830

PHILOSOPHY OF SCIENCE Logic,methodology and philosophy of science International congress for logic,methodology and philosophy of science Proceedings Jerusalem 1964 Aug 26-Sep 2 Israel academy of sciences and humanities International union of history and philosophy of science.Division of logic, methodology and philosophy of science Edited by Yehoshua Bar-Hillel viii,440p North-Holland,Amsterdam,1965
WSM 43.0829

PHILOSOPHY OF SCIENCE Problems in the philosophy of science International colloquium in the philosophy of science Proceedings London 1965 Jul 11-17 Vol 3 British society for the philosophy of science London school of economics and political science Edited by Imre Lakatos and Alan Musgrave Under the auspices of the International union of history and philosophy of science.Division of logic,methodology and philosophy of science Studies in logic and the foundations of mathematics ix,448p North-Holland,Amsterdam,1968
WSM 43.0931

PHILOSOPHY OF SCIENCE :the Delaware seminar Delaware seminar in the philosophy of science Vol 1-: 1961- University of Delaware Edited by Bernard Baumrin Interscience,New York;London,1963-
WSM 43.0757

PHONON AND PHONON INTERACTIONS Aarhus 1963 Aug 12-14 Aarhus universitet.Summer school Edited by Thor.A. Bak and others Aarhus summer school lectures,1963 xi,640p W.A. Benjamin,New York;Amsterdam,1964
Cav 7.0002

PHONONS Phonons in perfect lattices and in lattices with point imperfections N.A.T.O. advanced study institute Papers Aberdeen 1965 Aug 9-27 Scottish universities' summer school in physics Edited by R.W.H. Stevenson With financial support of North Atlantic treaty organization Scottish universities' summer school in physics, 6 448p Oliver and Boyd,Edinburgh,1966
TA 15.0212

PHONONS IN PERFECT LATTICES AND IN LATTICES WITH POINT IMPERFECTIONS Scottish universities' summer school 6th Lectures Aberdeen 1965 Aug 9-27 Scottish universities' summer school in physics Edited by R.W.H. Stevenson With financial help from the North Atlantic treaty organization Oliver and Boyd,Edinburgh; London,1966
Chem 18.2643

PHOSPHOINOSITIDES The Federation of European biochemical societies meeting 2nd Proceedings Vienna 1965 Apr 21-24 Vol 2: cyclitols and phosphoinositides Federation of European biochemical societies Edited by H. Kindl Pergamon press,Oxford, 1966
Bioch 33.1393

PHOSPHORIC ESTERS AND RELATED COMPOUNDS :a symposium Report Cambridge 1957 Apr 9-12 Chemical society Chemical society. Special publication,8 Chemical society, London,1957 Organized by Dr.Kenner and Dr. Brown
Chem 18.1217

PHOSPHORIC ESTERS AND RELATED COMPOUNDS :a symposium held at the Chemical society anniversary meeting Report Cambridge 1957 Apr 9-12 Chemical society Chemical society.Special publication, 8 Chemical society,London,1957
Bioch 33.0842

PHOSPHORUS Biology of phosphorus A Symposium in general biology Papers presented East Lansing,Mich. 1951 Apr 25-27 By G.Evelyn Hutchinson and others Edited by Lester F. Wolterink Held at Michigan state college Michigan state college press,1952
Bioch 33.0590

PHOSPHORUS METABOLISM A Symposium on the role of phosphorous in the metabolism of plants and animals Papers and discussions Baltimore,Md. 1951 Jun 18-21 and Baltimore,Md. 1952 Jun 16-19 Vol 1-2 Edited by William D. McElroy and H.Bentley Glass Sponsored by the McCollum-Pratt institute McCollum-Pratt institute. Contribution, 23,36 2 vols Johns Hopkins press,Baltimore,Md.,1951-52
Bot 42.1670

PHOSPHORUS METABOLISM A Symposium on the role of phosphorus in the metabolism of plants and animals Baltimore,Md. 1951 Jun 18-21 Vol 1 Edited by W.D. McElroy and B. Glass Sponsored by McCollum-Pratt institute McCollum-Pratt institute.Contributions, 23 John Hopkins press,Baltimore,Md.,1951
Radioth 35.0028

PHOSPHORUS METABOLISM A Symposium on the role of phosphorus in the metabolism of plants and animals Baltimore,Md. 1951 Jun 18-21 Vol 2 Edited by W.D. McElroy and B. Glass Sponsored by the McCollum-Pratt institute McCollum-Pratt institute. Contributions, 36 John Hopkins press, Baltimore,Md.,1952
Radioth 35.0029

PHOSPHORUS METABOLISM A Symposium on the role of phosphorus in the metabolism of plants and animals Papers and discussions presented Baltimore,Md. 1951 Jun 18-21 and Baltimore,Md. 1952 Jun 16-19 Vol 1-2 Edited by William D. McElroy and H.Bentley Glass Sponsored by the McCollum-Pratt institute McCollum-Pratt institute. Contribution, 23,36 2 vols Johns Hopkins press,Baltimore,Md.,1951-52
Bioch 33.0597

PHOSPHORYLATION Fosforilirovanie i funktsiya sympozium Leningrad 1958 Jun 13-18 Akademiya meditsinskikh nauk S.S.S.R. Akademiya meditsinskikh nauk S.S.S.R.Trudy instituta eksperimentalnoi meditsiny Leningrad,1960 Short summaries in English at the end of each chapter
Bioch 33.0561

PHOTO-ELECTRIC IMAGE DEVICES Symposium on photo-electric image devices 5th Proceedings Advances in electronics and electron physics Academic press,New York, 1972
Eng 41.8579

PHOTO-ELECTRONIC IMAGE DEVICES A Symposium on photo-electronic image devices held at Imperial college 2nd Proceedings London 1961 Sep 5-8 Edited by J.D. McGee and others Advances in electronics and electron physics, 16 Academic press,New York,1962
Cav 7.1458

PHOTO-ELECTRONIC IMAGE DEVICES Photo-electronic image devices Symposium on image tubes and related devices held at the Imperial college 1st Proceedings London 1958 Sep 3-5 Edited by J.D. McGee and W.L. Wilcock Advances in electronics and electron physics, 12 Academic press,New York,1960
Cav 7.1456

PHOTO-ELECTRONIC IMAGE DEVICES 2nd : symposium Proceedings London 1961 Sep 5-8 Edited by J.D. McGee and others Held at the Imperial college of science and technology Advances in electronics and electron physics, 16 Academic press,New York;London,1962
Met 25.2638

PHOTO-ELECTRONIC IMAGE DEVICES 3rd : symposium Proceedings London 1965 Sep 20-24 Edited by J.D. McGee and others Advances in electronics and electron physics, 22 a-b 2 vols Academic press,New York; London,1966
Met 25.2639

PHOTO-ELECTRONIC IMAGE DEVICES 4th : symposium Proceedings London 1968 Edited by J.D. McGee and others Advances in electronics and electron physics, 28a 552p Academic press,London,1969
Obs 6.3413

PHOTO-THERMOPERIODISM,THE ACTION OF DIFFERENT RADIATIONS ON GIBBERELLINS AND OTHER ACTIVE SUBSTANCES Colloque international sur le photothermoperiodisme action des diverses radiations on gibberellines et de quelques autres substances Parma 1957 Jun International union of biological sciences International union of biological sciences. Series B, 34 U.I.S.B.,Paris,1958
Bot 42.1420

PHOTOBIOLOGY Basic mechanisms in photochemistry and photobiology :an international symposium Caracas 1967 Dec 4-8 Edited by J.W. Longworth Photochemistry and photobiology, 7,no 6 Pergamon press, London,1968
Radioth 35.0279

PHOTOBIOLOGY Finsen memorial congress International congress on photobiology 3rd Proceedings Copenhagen 1960 Edited by B. Chr. Christensen and B. Buchmann illus. Elsevier,Amsterdam,1961
Bot 42.1347

PHOTOBIOLOGY Recent progress in photobiology International congress of photobiology 4th Proceedings Oxford 1964 Jul Edited by E. J. Bowen Under the auspices of the Comite international de photobiologie Bibliog., illus. vii,400p Blackwell scientific publications,Oxford,1965
Bot 42.1362

PHOTOBIOLOGY Recent progress in photobiology International photobiology congress 4th Proceedings Oxford 1964 Jul 26-30 Edited by E.J. Bowen Under the auspices of the Comite internationale de photobiologie Blackwell,Oxford,1965
Radioth 35.0253

The PHOTOCHEMICAL APPARATUS;ITS STRUCTURE AND FUNCTION :a symposium Report Upton,N. Y. 1958 Jun 16-18 Brookhaven national laboratory.Biology department Brookhaven symposia in biology, 11 BNL512(C-28) Brookhaven national laboratory.Biology department,Upton,N.Y.,1959
Bioch 33.1290

PHOTOCHEMICAL REACTIONS Cold Spring Harbor symposia on quantitative biology Papers Cold Spring Harbor 1935 Vol 3 Cold Spring Harbor biological laboratory Long Island biological association,Cold Spring Harbor,1935 Later referred to in vol.9 as "Photochemical reactions"
Bioch 33.1259

PHOTOCHEMISTRY Basic mechanisms in photochemistry and photobiology :an international symposium Caracas 1967 Dec 4-8 Edited by J.W. Longworth Photochemistry and photobiology, 7,no 6 Pergamon press, London,1968
Radioth 35.0279

PHOTOCONDUCTIVITY International conference on photoconductivity 3rd Proceedings Stanford,Calif. 1969 Aug 12-15 International union of pure and applied physics American physical society United States.Office of naval research Edited by E. M. Pell Bibliog,diagrms,graphs xi,410p 26cm Pergamon,Oxford,1971
Cav 7.3073

PHOTOCONDUCTIVITY CONFERENCE Atlantic City,N. J. 1954 Nov 4-6 Edited by R.G. Breckenridge Sponsored by the University of Pennsylvania Wiley;Chapman and Hall,New York; London,1956
Eng 41.5395

PHOTOELASTICITY :an international symposium Proceedings Chicago,Ill. 1961 Oct United States.Army.Research office Edited by M.M. Frocht Pergamon press,London,1963
Eng 41.2913

PHOTOGRAPHIC SENSITIVITY Fundamental mechanisms of photographic sensitivity A Conference on fundamental mechanisms of photographic sensitivity :held in the H.H. Wills physical laboratory Proceedings Bristol 1950 March University of Bristol Edited by J.W. Mitchell Butterworths,London, 1951
Cav 7.0347

PHOTOGRAPHIC SENSITIVITY Fundamental mechanisms of photographic sensitivity :a symposium Proceedings Bristol 1950 Mar Butterworths,London,1951
Radioth 35.1506

PHOTOGRAPHY Congres international de photographie scientifique et appliquee 8e Comptes rendus Dresden 1931 Aug 3-7 Edited by L.P. Clerc Revue d'optique theorique et instrumentale,Paris,1933
Chem 18.0248

PHOTOGRAPHY Index to the international congresses on high speed photography 1952-1970 Edited by J. Wadsworth and M.W. Glover Royal aircraft establishment,Farnborough,c1970
Eng 41.8373

PHOTOGRAPHY International congress of photography 7th Proceedings London 1928 Jul 9-14 Edited by W. Clark and others port W.Heffer,Cambridge,1928
Chem 18.0247

PHOTOPERIODISM AND RELATED PHENOMENA IN PLANTS AND ANIMALS Conference on photoperiodism Gatlinburg 1957 American association for the advancement of science Edited by Robert B. Withrow and others American association for the advancement of science.Publication, 55 Washington,D.C.,1959
Bot 42.1826

PHOTOREACTIVE SYSTEMS Comparative biochemistry of photoreactive systems Symposium on comparative biology 1st Papers Richmond,Calif. Kaiser foundation research institute Edited by Mary Belle Allen Kaiser foundation research institute. Symposia on comparative biology, 1 Academic press,New York;London,1960
Bioch 33.0334

PHOTOREACTIVE SYSTEMS Comparative biochemistry of photoreactive systems :annual symposium on comparative biology of the Kaiser foundation research institute 1st Papers Kaiser foundation research institute Edited by Mary Belle Allen Kaiser foundation research institute.Symposia on comparative biology,1 Academic press,New York;London, 1960
Chem 18.2139

PHOTORECEPTION Papers New York 1958 Jan 31-Feb 1 New York academy of sciences and National council to combat blindness Edited by Jerome J. Wolken New York academy of sciences.Annals, 74,art.2 New York,1958
Phys 20.2225

PHOTORESPIRATION Photosynthesis and photorespiration :a conference Canberra 1970 Nov 23-Dec 5 Australian academy of sciences Edited by M.D. Hatch and others Interscience,New York,1971
Bioch 33.2221

PHOTOSYNTHESIS Bacterial photosynthesis :a symposium Yellow Springs,Ohio 1963 Edited by Howard Gest and others Sponsored by the Charles F.Kettering research laboratory Charles F.Kettering research laboratory. Contribution, 112 illus. xvi,523p Antioch press,Yellow Springs,Ohio,1963
Bot 42.3800

PHOTOSYNTHESIS Carbon dioxide fixation and photosynthesis Sheffield 1950 Jul Society for experimental biology Society for experimental biology.Symposia, 5 Cambridge university press,Cambridge,1951
Phys 20.0953

PHOTOSYNTHESIS Carbon dioxide fixation and photosynthesis :a symposium Sheffield 1950 Jul Society for experimental biology Society for experimental biology.Symposia, 5 Cambridge university press,Cambridge,1951
Bioch 33.1351

PHOTOSYNTHESIS Comparative biochemistry and biophysics of photosynthesis :a conference Papers Hakone 1967 Aug 12-15 Edited by K. Shibata and others Sponsored by the Japan-U. S.cooperative science program illus viii, 445p University of Tokyo press;University Park press,Tokyo;University Park,Pa.,1968
Bot 42.1903

PHOTOSYNTHESIS Comparative biochemistry and biophysics of photosynthesis :a conference Papers presented Hakone,Japan 1967 Aug 12-15 Edited by K. Shibata and others Sponsored by the Japan-United States cooperative science program University of Tokyo press;University park press,State college,Tokyo;University Park,Pa.,1968
Bioch 33.2349

PHOTOSYNTHESIS Currents in photosyntheses Western-Europe conference on photosynthesis 2nd Proceedings Woudschoten,Zeist 1965 Sep 6-12 Edited by J.B. Thomas and J.C. Goedheer A.D.Donker,Rotterdam,1966
Bioch 33.0805

PHOTOSYNTHESIS Currents in photosynthesis The Western-Europe conference on photosynthesis 2nd Proceedings Woudschoten,Zeist 1965 Sep Edited by J.B. Thomas and J.C. Goedheer 486,23p Donker, Rotterdam,1966
Bot 42.1890

PHOTOSYNTHESIS International congress of biochemistry 5th Proceedings Moscow 1961 Aug 10-16 Vol 6: mechanism of photosynthesis International union of biochemistry Edited by H. Tamiya I.U.B. symposium series, 26 Pergamon;PWN-Polish scientific publishers,Oxford;Warsaw,1963
Bioch 33.1335

PHOTOSYNTHESIS International symposium on energy transduction in respiration and photosynthesis Pugnochiuso 1970 Sep 11-14 International union of biochemistry Edited by E. Quagliariello and others International union of biochemistry.Symposia, 43 Adriatic editrice,Bari,1971
Bioch 33.2289

PHOTOSYNTHESIS Research in photosynthesis :a conference Papers and discussions Gatlinburg,Te. 1955 Oct 25-29 National research council.Committee on photobiology Edited by H. Gaffron and others Supported by the National science foundation Interscience, New York;London,1957
Chem 18.2138

PHOTOSYNTHESIS Research in photosynthesis :a conference Papers and discussions Gatlinburg,Tenn 1955 Oct 25-29 Edited by H. Gaffron and others Sponsored by the National research council xiv,524p Interscience,London,1957
Bot 42.1869

PHOTOSYNTHESIS Research in photosynthesis a conference Papers and discussions Gatlinburg,Tenn. 1955 Oct 25-29 Edited by H. Gaffron and others Sponsored by the National research council.Committee on photobiology Interscience,New York,1957
Bioch 33.0758

PHOTOSYNTHESIS AND PHOTORESPIRATION :a conference Canberra 1970 Nov 23-Dec 5 Australian academy of sciences Edited by M.D. Hatch and others Interscience,New York,1971
Bioch 33.2221

PHOTOSYNTHESIS IN PLANT LIFE Harvesting the sun:photosynthesis in plant life :a symposium Chicago 1966 Oct 5-7 Edited by A. San Pietro and others Sponsored by the International minerals and chemical corporation Academic press,New York;London, 1967
Bioch 33.0789

PHOTOSYNTHESIS IN PLANT LIFE Harvesting the sun:pHotosynthesis in plant life :a symposium Papers Chicago,Ill. 1966 Oct 5-7 Edited by Anthony San Pietro and others Sponsored by the International minerals and chemical corporation illus. ix,342p Academic press,New York;London,1967
Bot 42.1899

PHOTOSYNTHESIS RESEARCH Progress in photosynthesis research The International congress of photosynthesis research Proceedings Freudenstadt 1968 Jun 4-8 Vol 1-3 Edited by Helmut Metzner Sponsored by the International union of biological sciences 3 vols International union of biological sciences,Tubingen,1969
Bioch 33.1907

PHOTOSYNTHESIS RESEARCH Progress in photosynthesis research The International congress of photosynthesis research Proceedings Freudenstadt 1968 Jun 4-8 Vol 1-3 Edited by Helmut Metzner Sponsored by the International union of biological sciences 3 vols Tubingen,1969
Bot 42.1905

PHOTOSYNTHETIC APPARATUS Energy conversion by the photosynthetic apparatus :symposium Report Upton,N.Y. 1966 Jun 6-9 Brookhaven national laboratory.Biology department Brookhaven symposia in biology, 19 BNL989(C-48) Brookhaven national laboratory.Biology department,Upton,N.Y.,1967
Bioch 33.1297

PHOTOSYNTHETIC MECHANISMS OF GREEN PLANTS :a symposium Papers Warrenton,Va. 1963 Oct 14-18 Sponsored by the National research council.Committee on photobiology National research council.Publication,1145 Washington, 1963
Bioch 33.0774

PHYLOGENESIS AND ONTOGENESIS OF THE FOREBRAIN Evolution of the forebrain:phylogenesis and ontology of the forebrain :a symposium Lectures Frankfurt 1965 Aug 15-19 and Sprendlingen World federation of neurology Max Planck gesellschaft zur forderung der wissenschaften Edited by R. Hassler and H. Stephan G.Thieme,Stuttgart,1966
Psy 31.3370

PHYLOGENY AND EVOLUTION OF CRUSTACEA :a conference Proceedings Cambridge,Mass. 1962 Mar 6-8 Harvard university.Museum of comparative zoology Edited by H.B. Whittington and W.D.I. Rolfe Supported by the National science foundation Museum of comparative zoology,Cambridge,Mass.,1963
Bal 39.2305

PHYSICAL ACTIVITY IN HEALTH AND DISEASE : international symposium Proceedings Beitostolen 1966 Edited by K. Evang and K. Lange Anderson Williams and Wilkins, Baltimore,1967
HE 27.0178

PHYSICAL AND FABRICATION METALLURGY The Commonwealth mining and metallurgical congress 9th Proceedings London 1969 May 3-24 Commonwealth council of mining and metallurgical institutions Edited by M.J. Jones Institution of mining and metallurgy, London,1970
Met 25.2847

PHYSICAL ANTHROPOLOGY The Scope of physical anthropology and its place in academic studies a symposium London 1957 Nov Edited by D. F. Roberts and J.S. Neiner Society for the study of human biology.Symposia, 1 Wenner-Gren foundation for anthropological research, London,1958
An 32.2880

PHYSICAL BASIS OF YIELD AND FRACTURE conference Proceedings Oxford 1966 Sep Institute of physics and the physical society.Stress analysis group committee Edited by A.C. Stickland and R.A. Cook Institute of physics and the physical society. Conference series, 1 Inst.of Physics and the Physical society,London,c1966
Met 25.0869

PHYSICAL CHEMISTRY Reunion internationale de chimie physique rapports et discussions Paris 1928 October 8-12 Edited by R. Audubert and M.L. Claudel Paris,1929
Philos 1.0668

The PHYSICAL CHEMISTRY OF ENZYMES :a discussion Papers Oxford 1955 Aug 10-12 Faraday society Faraday society. Discussions, 20 Aberdeen,1956
Bioch 33.1047

PHYSICAL CHEMISTRY OF HIGH POLYMERS International congress of biochemistry 4th Proceedings Vienna 1958 Sep 1-6 Vol 9: symposium 9 - physical chemistry of high polymers of biological interest International union of biochemistry Edited by O. Kratky Pergamon press,London,1960
Radioth 35.1941

PHYSICAL CHEMISTRY OF HIGH POLYMERS OF BIOLOGICAL INTEREST International congress of biochemistry 4th Proceedings Vienna 1958 Sep 1-6 Vol 9: symposium 9 - physical chemistry of high polymers of biological interest International union of biochemistry Edited by O. Kratky I.U.B.symposium series, 11 Pergamon,London,1959 Added title page in French and German.Text in English, French and German.
Bioch 33.1323

PHYSICAL CHEMISTRY OF MELTS Symposium on the nature of molten slags and salts London 1952 Feb 20 Nuffield research group in extraction metallurgy Institution of mining and metallurgy,London,1953
Met 25.1787

The PHYSICAL CHEMISTRY OF METALLIC SOLUTIONS AND INTERMETALLIC COMPOUNDS :a symposium Proceedings London 1958 Jun 4-6 Vol 1-2 National physical laboratory National physical laboratory.Symposium, 9 2 vols H.M.S.O.,London,1959
Met 25.1779

PHYSICAL CHEMISTRY OF PROCESS METALLURGY :an international symposium Pittsburgh,Pa. 1959 Apr 27-May 1 American institute of mining,metallurgical and petroleum engineers Edited by George R. St.Pierre Metallurgical society conferences, 7,8 Interscience,New York;London,1961
Met 25.0491

The PHYSICAL CHEMISTRY OF PROTEINS :a general discussion Cambridge 1952 Aug 6-8 Faraday society Faraday society. Discussions, 13 Faraday society,Aberdeen, 1953
Radioth 35.0336

The PHYSICAL CHEMISTRY OF PROTEINS :a general discussion Papers Cambridge 1952 Aug 6-8 Faraday society Faraday society. Discussions, 13 Aberdeen university press, Aberdeen,1953
Bioch 33.0510

The PHYSICAL CHEMISTRY OF STEELMAKING :a conference Proceedings Dedham,Mass. 1956 May 28-Jun 3 Edited by John F. Elliott Sponsored by Massachusetts institute of technology.Department of metallurgy Wiley; Chapman and Hall,New York;London,1958
Met 25.1789

PHYSICAL EDUCATION World congress on physical education Report Melbourne 1956 Nov 16-21 Australian physical education association Sponsored by the Commonwealth national fitness council Australian physical education association,Melbourne,1956
HE 27.0219

PHYSICAL ELECTRONICS Conference on physical electronics Report Cambridge,Mass. 1949 Apr 7-9 By E.B. Callick 86p 27cm British joint services mission,Washington,D.C., 1949
Math L 5.3263

PHYSICAL GEOGRAPHY OF GREENLAND International geographical congress 19th Copenhagen 1960 Edited by Niels Kingo Jacobsen Arranged at the University of Copenhagen.Geographical institute Folia geographica Danica, 9 illus,maps 234p C.A.Reitzel,Copenhagen,1961
Sco 14.8357

PHYSICAL MEDICINE International congress of physical medicine 3rd Proceedings Washington,D.C. 1960 American congress of physical medicine and rehabilitation and American academy of physical medicine and rehabilitation Chicago,Ill.,1960
Phys 20.0990

PHYSICAL METALLURGY OF STRESS-CORROSION FRACTURE Pittsburgh 1959 Apr 2-3 American institute of mining,metallurgical and petroleum engineers Edited by Thor N. Rhodin Metallurgical society conferences, 4 Interscience,New York;London,1959
Met 25.2503

The PHYSICAL PROPERTIES OF POLYMERS :symposium Papers London 1958 Apr 15-17 Organised by the Society of chemical industry. Plastics and polymers group Society of chemical industry.Monograph, 5 293p Society of chemical industry,London,1959
Chem E 24.1019

PHYSICAL SCIENCES Symposium international des sciences physiques et mathematiques dans la premiere moitie du XVII siecle Actes Pisa 1958 Jun 16-18 Vinci (Florence) Union internationale d'histoire et philosophie des sciences Academie internationale d'histoire des sciences.Collection de travaux, 11 x,278p Gruppe italiano di storia delle scienze;Hermann,Vinci (Florence);Paris,c1959
WSM 43.0066

PHYSICAL SOCIETY A Discussion held...at the Imperial college of science Report London 1931 Jun 19 Physical society,London,1931 Bound with 'Report of a joint discussion on vision...'by the Physical and optical society
Psy 31.3361

PHYSICAL SOCIETY A Joint discussion on vision Report London 1932 Jun.3 327p Physical society,London,c1932
Obs 6.1962

PHYSICAL SOCIETY A Joint discussion on vision held at the Imperial college of science London 1932 Jun 3 Physical society,London, 1932 Bound with 'Report of a discussion on audition' by the Physical society,q.v.
Psy 31.2610

PHYSICAL SOCIETY Conference on the physics of the ionosphere Report Cambridge 1954 Sep 6-9 406p Physical society,London,1955
Nap 11.0926

PHYSICAL SOCIETY International conference on physics Papers London 1934 Vol 1: nuclear physics Physical society,London,1935
Cav 7.1452

PHYSICAL SOCIETY Lightning rod conference Report Edited by G.J. Symons Bibliog,illus x,261p E.and F.N.Spon,London,1882 With appendices 1-L.Appendix G contains catalogue of works upon lightning conductors,pp.143-174
Nap 11.0940

PHYSICAL SOCIETY Meteorological factors in radio-wave propagation :a conference Report London 1946 Apr 8 325p Physical society, London,1946
Nap 11.0925

PHYSICAL SOCIETY Photo-electric cells and their applications :a discussion at a joint meeting of the Physical and Optical societies 1930 Edited by John S. Anderson Physical and Optical societis,London,1930
Eng 41.4623

PHYSICAL SOCIETY Physical memoirs selected and translated from foreign sources under the direction of the Physical society of London Vol 1,1-3 3 vols Taylor and Francis, London,1888-90
Nap 11.0484

PHYSICAL SOCIETY Report of a joint discussion on vision held...at the Imperial college of science Report London 1932 Jun 3 Physical society,London,1932
Psy 31.0411

PHYSICAL SOCIETY Report of the relativity theory of gravitation By Arthur Stanley Eddington 91p London,1918
Obs 6.0532

PHYSICAL SOCIETY The International conference on fundamental particles and low temperature Report Cambridge 1946 Jul 22-27 Vol 1: fundamental particles Physical society,London,1947
Cav 7.0460

PHYSICAL SOCIETY The International conference on fundamental particles and low temperatures Report Cambridge 1946 Jul 22-27 Vol 2: low temperatures Physical society,London,1947
Cav 7.0458

PHYSICAL SOCIETY The International conference on the physics of semiconductors Report Exeter 1962 Jul 16-20 Edited by A. C. Stickland Under the auspices of the International union of pure and applied physics 909p Institute of physics,London, 1962
Cav 7.0750

PHYSICAL SOCIETY and INTERNATIONAL UNION OF PURE AND APPLIED PHYSICS; International conference on physics papers and discussions London 1934 Vol 2: the solid state of matter Physical society,London,1937
Min 10.1078

PHYSICAL SOCIETY OF JAPAN International conference on low temperature physics 12th Proceedings Kyoto 1970 Sep 4-10 Edited by Eizo Kanda Organized by the Science council of Japan photos,diagrms 895p 26cm Academic press of Japan,Tokyo,1971
Cav 7.3028

PHYSICAL SOCIETY OF JAPAN International conference on nuclear structure Tokyo 1967 Sep illus vi,755p 26cm Physical society of Japan,Tokyo,1968
Cav 7.3196

PHYSICAL SOCIETY OF JAPAN International conference on the physics of semiconductors 8th Proceedings Kyoto 1966 Sep 8-13 Under the auspices of the International union of pure and applied physics Physical society of Japan.Journal, 21,suppl. 774p Physical society of Japan,Tokyo,1966
Cav 7.1341

PHYSICAL SOCIETY OF LONDON The Physics and chemistry of colloids and their bearing on industrial questions :a general discussion Report London 1920 Oct 25 H.M.S.O., London,1921
Bioch 33.1522

The PHYSICO-CHEMICAL MECHANISM OF NERVE ACTIVITY a conference New York 1946 Feb 8-9 By D. Nachmansohn and others New York academy of sciences New York academy of sciences.Annals, 47, p 375-602 New York, 1946
Bal 39.1449

PHYSICOCHEMICAL MECHANISM OF NERVE ACTIVITY Conference on physicochemical mechanism of nerve activity and Conference on muscular contraction 2nd New York academy of sciences Edited by David A. Nachmansohn New York academy of sciences.Annals, 81,p 215-510 New York,1959
Bal 39.1450

PHYSICS International conference on physics Papers London 1934 Vol 1: nuclear physics International union of pure and applied physics and Physical society Physical society,London,1935
Cav 7.1452

PHYSICS International conference on physics Papers and discussions London 1934 Vol 1-2: nuclear physics:solid state of matter International union of pure and applied physics and the Physical society 2 vols Physical society,London,1935
Met 25.2343

PHYSICS Mathematics of contemporary physics : instructional conference Proceedings London 1971 Aug 23-Sep 11 London mathematical society North Atlantic treaty organization Edited by R.F. Streater xi, 274p 24cm Academic press,London,1972
A Math 4.1864

PHYSICS The Significance of space research for fundamental physics :a conference Interlaken 1969 Sep 4 European space research organisation Edited by A.F. Moore and U. Hardy ESRO SO-52 175p european space research organisation,Neuilly-sur-Seine, 1971
Obs 6.3603

PHYSICS,LOGIC AND HISTORY International colloquium on logic,physical reality and history Denver,Colo. 1966 May 16-20 Edited by Wolfgang Yourgrau and Allen D. Breck xiv,336p Plenum press,New York;London,1970 Based on the 1st colloquium
WSM 43.1116

The PHYSICS AND CHEMISTRY OF CERAMICS :a symposium Proceedings Philadelphia,Pa. 1962 May 28-30 United States.Office of naval research Edited by Cyrus Klingsberg Gordon and Breach,New York;London,1963
Met 25.0277

The PHYSICS AND CHEMISTRY OF COLLOIDS AND THEIR BEARING ON INDUSTRIAL QUESTIONS :a general discussion Report London 1920 Oct 25 Faraday society Physical society of London H.M.S.O.,London,1921
Bioch 33.1522

PHYSICS AND DYNAMICS OF METEORS Tatranska Lomnica 1967 Sep 4-7 International astronomical union Edited by Lubor Kresak and Peter M. Millman International astronomical union.Symposium, 33 Reidel, Dordrecht,1968
TA 15.0422

PHYSICS AND DYNAMICS OF METEORS :a symposium Proceedings Tatranska Lomnica 1967 International astronomical union Edited by Lubor Kresak and Peter M. Millman International astronomical union.Symposium, 33 525p Reidel,Dordrecht,1968
Obs 6.3264

PHYSICS AND MEDICINE OF THE UPPER ATMOSPHERE Symposium on the physics and medicine of the upper atmosphere Proceedings San Antonio, Tex. 1951 Nov 6-9 Lovelace foundation for medical education and research Edited by Clayton S. White and Otis O. Benson Sponsored by School of aviation medicine. Randolph Field illus xxix,611p University of New Mexico press,Albuquerque,N.M. 1952
Nap 11.0635

PHYSICS OF ELECTRONIC AND ATOMIC COLLISIONS International conference on the physics of electronic and atomic collisions 5th Invited papers Leningrad 1967 Jul 17-23 Edited by Lewis M. Branscomb Sponsored by the International union of pure and applied physics Bibliog,illus xvi,200p 23cm University of Colorado press,Boulder,Colo., 1968
A Math 4.1906

PHYSICS OF ELECTRONICS AND ATOMIC COLLISIONS International conference on the physics of electronic and atomic collisions 7th Invited papers and progress reports Amsterdam 1971 Jul 26-30 Edited by T.R. Govers and F.J. De Heer xi,496p 23cm North Holland,London,1972
Chem 18.2958

The PHYSICS OF EXPERIMENTAL METHOD By H.J.J. Braddick Chapman and Hall,London,1954
Cav 7.0302

PHYSICS OF HOT PLASMAS Scottish universities' summer school 9th Proceedings 1968 Jul 28-Aug 16 Edited by B. J. Rye and J.C. Taylor illus xv,455p 25cm Oliver and Boyd,Edinburgh,1970
A Math 4.1886

The PHYSICS OF IONIZED GASES :a conference proceedings,typescript London 1953 Apr University college,London Edited by J.B. Hasted Under the auspices of the Royal society of London.Warren research fund London,1953
A Math 4.0614

PHYSICS OF NUCLEAR REACTORS :a conference Papers London 1956 Jul 3-6 Arranged by the Institute of physics British journal of applied physics.Supplement, 5 Institute of physics,London,1956
Radioth 35.1596

The PHYSICS OF PARTICLE SIZE ANALYSIS :a conference Nottingham 1954 Apr.6-9 Edited by H.R. Lang British journal of applied physics.Supplement, 3 218,viiip Institute of physics,London,1954
Obs 6.1961

The PHYSICS OF POWDER METALLURGY :a symposium New York 1949 Aug 24-26 Sylvania electric products,inc.Metallurgical laboratories Edited by Walter E. Kingston McGraw-Hill,New York,1951
Met 25.0752

PHYSICS OF PRECIPITATION Cloud physics conference 2nd Proceedings Woods Hole, Mass. 1959 Jun 3-5 American geophysical union.Cloud physics committee,and,National science foundation Edited by Helmut Weickmann Geophysical monograph,5 National research council.Publication,746 illus,map xii,435p 25cm American geophysical union, Washington,1960
Sco 14.0536

The PHYSICS OF SELENIUM AND TELLURIUM : international symposium Proceedings Montreal 1967 Oct 12-13 Selenium-Tellurium development association Edited by W.Charles Cooper Bibliog,photos,diagrs ix,380p 23cm Pergamon press,Oxford,1969
Cav 7.2670

PHYSICS OF SNOW AND ICE International conference on low temperature science 1st Proceedings Sapporo 1966 Aug 14-19 1 pt 2: conference on physics of snow and ice (part 2) Institute of low temperature science Edited by Hirobumi Oura illus 1967p 27cm Hokkaido university,Sapporo,1957
Sco 14.7411

PHYSICS OF SOLIDS The International conference on the physics of solids at high pressures 1st Proceedings Tucson,Ariz. 1965 Apr 20-23 University of Arizona Edited by C.T. Tumizuka and R.M. Emrick Jointly sponsored by the United States.Air force.Office of scientific research Academic press,New York,1965
Cav 7.2354

PHYSICS OF THE SOLAR CORONA NATO advanced study institute on physics of the solar corona Proceedings Athens 1970 Sep 6-17 North Atlantic treaty organization Edited by C.J. Macris Astrophysics and space science library, 27 illus xii,345p 24cm Reidel,Dordrecht,1971
A Math 4.1803

PHYSICS OF THE SOLAR CORONA NATO advanced study institute on the physics of the solar corona Proceedings Athens 1970 Sep 6-17 North Atlantic treaty organization Edited by C.J. Macris Astrophysics and space science library, 27 345p Reidel,Dordrecht,1971
Obs 6.3607

PHYSICS OF THE SOLAR SYSTEM Conference on physics of the solar system and re-entry dynamics proceedings Blacksburg,Va. 1961 Jul.31-Aug.11 Part 1: physics of the solar system Virginia polytechnic institute. Engineering experiment station. Virginia polytechnic institute.Engineering experiment station series, 149 Virginia polytechnic institute.Bulletin, 55,9 Virginia polytechnic institute,Blacksburg,Va.,1962
A Math 4.1155

PHYSICS OF THE WELDING ARC :a symposium London 1962 Oct 29-Nov 2 Institute of welding Institute of welding,London,1966
Met 25.0807

PHYSICS WITH INTERSECTING STORAGE RINGS Scuola internazionale di fisica 'Enrico Fermi' 46 corso Rendiconti Varenna 1969 Jun 16-26 Societa italiana di fisica Bibliog, photos 24cm Academic press,New York,1971
Cav 7.3012

PHYSIKALISCHE GESELLSCHAFT IN DER DEUTSCHEN DEMOKRATISCHEN REPUBLIK Plasmas in physik und astronomie Probleme des plasmas in physik und astronomie :eine tagung berichte Leipzig 1956 Oct.8-11 223p Akademie-verlag,Berlin,1958
Obs 6.1366

PHYSIOLOGICAL CONTROL OF IODINE METABOLISM Gunma symposium on endocrinology 3rd Ikaho 1965 Sep 10-11 Gunma university.Institute of endocrinology Gunma university.Institute of endocrinology.Annual report, 3 Institute of endocrinology,Gunma university,Maebashi, 1965
Phys 20.1427

PHYSIOLOGICAL MECHANISMS IN ANIMAL BEHAVIOUR Cambridge 1949 Jul Society for experimental biology Edited by J.F. Danielli and R. Brown Society for experimental biology.Symposia, 4 Cambridge university press,Cambridge,1950
An 32.4327

PHYSIOLOGICAL MECHANISMS IN ANIMAL BEHAVIOUR Cambridge 1949 Jul Society for experimental biology Society for experimental biology.Symposia, 4 Cambridge university press,Cambridge,1950
Phys 20.0945

PHYSIOLOGICAL MECHANISMS IN ANIMAL BEHAVIOUR Papers Cambridge 1949 Jul Society for experimental biology Society for experimental biology.Symposia, 4 Cambridge university press,Cambridge,1950
Psy 31.0562

PHYSIOLOGICAL PSYCHOLOGY Congres international de psychologie physiologique : premiere session 1st Compte-rendu Paris 1889 Societe de psychologie physiologique de Paris Bureau des revues,Paris,1890 Congresses continued as 'International congress of experimental psychology' q.v.
Psy 31.3390

PHYSIOLOGICAL SCIENCES International congress of the International union of physiological sciences 12th Proceedings Leiden 1962 Sep 10-17 Vol 1-3 International union of physiological sciences Excerpta medica foundation.International congress series, 47-49 3 vols Excerpta medica foundation,Amsterdam,c1962 Vol 3: information processing in the nervous system
Psy 28.0408

PHYSIOLOGICAL SOCIETY The Biology of survival :a symposium 13th Proceedings London 1964 May 7-8 Edited by O.G. Edholm Organized by the Physiological society,the society for the study of human biology,and the Zoological society of London Zoological society of London.Symposia, 13 Academic press,London,1964
Phys 20.2132

PHYSIOLOGIE (BEWEG"NG) DER SPERMIEN : symposium Budapest 1960 Oct Edited by I. Toro Symposia biologica hungarica, 4 Akademia Kiado,Budapest,1964
An 32.3880

PHYSIOLOGIE COMPORTEMENT ET ECOLOGIE DES ACRIDIENS EN RAPPORT AVEC LA PHASE : colloque international Actes Paris 1962 Apr 9-13 Centre national de la recherche scientifique Centre national de la recherche scientifique.Colloques internationaux, 114 Paris,1962
Organised by F.O.Albrecht
Bal 39.2555

PHYSIOLOGY International congress of physiological sciences 24th Washington, D.C. 1968 abstracts of lectures and symposia International union of physiological sciences International union of physiological sciences.Proceedings, 6 Bethesda,Md.,1968
Phys 20.2154

PHYSIOLOGY International congress of physiological sciences 24th Washington, D.C. 1968 abstracts of volunteer papers and films International union of physiological sciences International union of physiological sciences.Proceedings, 7 Bethesda,Md.,1968
Phys 20.2155

PHYSIOLOGY AND BIOCHEMISTRY OF HAEMOCYANINS : a symposium Naples 1966 Aug 30-Sep 1 Edited by F. Ghiretti held at the Naples zoological station Academic press,London;New York,1968 Sponsored by the Italian national research council,the U.S. national science foundation and the American institute of biological sciences
Bal 39.0522

PHYSIOLOGY AND EXPERIMENTAL PSYCHOLOGY :a Ciba foundation symposium London Ciba foundation Edited by A.V.S. De Reuck and Julie Knight Churchill,London,1965
Phys 20.2226

The PHYSIOLOGY AND ULTRASTRUCTURE OF HYDRA AND OF SOME OTHER COELENTERATES :a symposium Biology of the hydra and of some other coelenterates Coral Gables,Fla. 1961 Mar 29-31 Edited by Howard M. Lenhoff and F. Farnsworth Loomis held at the Fairchild tropical gardens University of Miami press, Coral Gables,Fla.,1961
Bal 39.2183

PHYSIOLOGY OF DIGESTION IN THE RUMINANT International symposium on the physiology of digestion in the ruminant 2nd Papers Ames, Iowa 1964 Aug Edited by R.W. Dougherty Butterworths,London,1965
Phys 20.1337

PHYSIOLOGY OF DIGESTION IN THE RUMINANT 2nd : International symposium Papers Ames, Iowa 1964 Aug Edited by R.W. Dougherty and others Butterworths,Washington,1965
VA 19.0304

PHYSIOLOGY OF FOREST TREES An International symposium of forest tree physiology 1st Harvard forest 1957 Apr 8-12 Edited by Kenneth Thimann and others Under the auspices of the Maria Moors Cabot foundation Ronald press,New York,1958
Bot 42.1335

The PHYSIOLOGY OF GENE AND MUTATION EXPRESSION symposium on the mutational process Proceedings Symposia Csav Brno 1965 Aug 4-7 Prague 1965 Aug 9-11 Edited by Margita Kohoutova and J. Hubacek Organized by the Ceskoslovenska akademie ved.Institute of microbiology Academia,Prague,1967
Gen 34.0770

The PHYSIOLOGY OF INDUCED HYPOTHERMIA :a symposium Proceedings Washington,D.C. 1955 Oct 28-29 National research council. Division of medical sciences Edited by Robert D. Dripps National research council. Publications, 451 Washington,D.C.,1956
Phys 20.0987

PHYSIOLOGY OF MUSCULAR EXERCISE :a symposium Proceedings Dallas 1966 Feb 7-9 American heart association Edited by Carleta B. Chapman Circulation research.Supplement, 20-21 New York,1967
Inv Med 37.0053

PHYSIOLOGY OF PRE-MATURITY Conference on physiology of prematurity 1st-4th Transactions Princeton,N.J. 1956-59 Mar Edited by Jonathan T. Lanman Sponsored by the Josiah Macy jr.foundation 4 vols Josiah Macy jr. foundation,New York,1957-60
An 32.3819

PHYSIOLOGY OF PREMATURITY 4th conference Transactions Princeton,N.J. 1959 Mar 25-26 Edited by Jonathan T. Lanham Sponsored by Josiah Macy Jr. foundation New York,1960
PGMS 29.0288

PHYSIOLOGY OF SPINAL NEURONS :a workshop ... held as part of the first International summer school of brain research... Amsterdam 1963 Jul 15-26 Edited by J.C. Eccles and J.P. Schade Progress in brain research, 12 Elsevier,Amsterdam,1964
An 32.4232

PHYSIOLOGY OF THE DOMESTIC FOWL :a symposium Proceedings Sutton Bonington 1964 Dec 16-18 British Egg marketing board Edited by C. Horton-Smith and E.C. Amoroso British egg marketing board symposium,1 Oliver and Boyd,Edinburgh;London,1966
VA 19.0308

PHYSIOLOGY OF THE INSECT CENTRAL NERVOUS SYSTEM International congress of entomology 12th Papers London 1964 Jul Edited by J.E. Treherne and J.W.L. Beament Academic press, London,1965
Pha 16.0276

PHYSIOLOGY OF THE INSECT CENTRAL NERVOUS SYSTEM Symposium on the physiology of the insect central nervous system International congress of entomology 12th Papers London 1964 Edited by J.E. Treherne and J. W.L. Beament London,1965
Bal 39.2458

PHYSIOPATHOLOGY OF THE RETICULAR ENDOTHELIAL SYSTEM :a symposium Paris Council for international organizations of medical sciences Edited by B.N. Halpern Blackwell, Oxford,1957 Summary of each paper in French
An 32.3154

PHYSIQUE FONDAMENTALE ET ASTROPHYSIQUE : colloque Compte-rendu Nice 1969 Centre national de la recherche scientifique Journal de physique.Supplement, v30,no 11-12 Centre national de la recherche scientifique. Colloques nationaux, 925 162p Centre national de la recherche scientifique,Paris, 1970
TA 15.0499

PHYTOCHEMICAL ECOLOGY Phytochemical society symposium 8th Annual proceedings Englefield Green,Surrey 1971 Apr Phytochemical society Academic press,London, 1972
Bioch 33.2229

PHYTOCHEMICAL GROUP Plant cell organelles The Phytochemical group symposium Proceedings London 1967 Apr 10-12 Edited by J.B. Pridham Academic press,London,1968
Bioch 33.0782

PHYTOCHEMICAL GROUP Plant cell organelles :a symposium Proceedings London 1967 Apr 10-12 Edited by J.B. Pridham xiii,261p Academic press,London,1968
Bot 42.4737

PHYTOCHEMICAL GROUP Terpenoids in plants :a symposium Proceedings Aberystwyth 1966 Apr 12-14 Edited by J.B. Pridham Academic press,London;New York,1967
Chem 18.1510

PHYTOCHEMICAL GROUP ANNUAL GENERAL MEETING 1966 Terpenoids in plants :a symposium Proceedings Aberystwyth 1966 Apr 12-14 Phytochemical group Edited by J.B. Pridham Academic press,London;New York,1967
Chem 18.1510

The PHYTOCHEMICAL GROUP SYMPOSIUM Proceedings Plant cell organelles London 1967 Apr 10-12 Phytochemical group Edited by J.B. Pridham Academic press,London,1968
Bioch 33.0782

PHYTOCHEMICAL PHYLOGENY :a symposium Proceedings Bristol 1969 Mar 31-Apr 2 Phytochemical society Edited by Jeffrey B. Harborne Bibliog.,illus,maps xiii,334p Academic press,London,1970
Bot 42.3453

PHYTOCHEMICAL SOCIETY Perspectives in phytochemistry Phytochemical society symposium Proceedings Cambridge 1968 Apr Edited by J.B. Harborne and T. Swain Academic press,London;New York,1969 For earlier publications of this body see Plant phenolics group
Gen 34.1407

PHYTOCHEMICAL SOCIETY Perspectives in phytochemistry :a symposium Proceedings Cambridge 1968 Apr Edited by J.B. Harborne and T. Swain Bibliog.,illus. Academic press,London;New York,1969
Bot 42.1436

PHYTOCHEMICAL SOCIETY Phytochemical ecology Phytochemical society symposium 8th Annual proceedings Englefield Green,Surrey 1971 Apr Academic press,London,1972
Bioch 33.2229

PHYTOCHEMICAL SOCIETY Phytochemical phylogeny :a symposium Proceedings Bristol 1969 Mar 31-Apr 2 Edited by Jeffrey B. Harborne Bibliog.,illus,maps xiii,334p Academic press,London,1970
Bot 42.3453

PHYTOCHEMICAL SOCIETY Plant growth regulators :symposium Papers London 1968 Jan 8-9 Society of chemical industry. Monograph, 31 Society of chemical industry, London,1968
Bioch 33.2230

PHYTOCHEMICAL SOCIETY SYMPOSIUM Proceedings Perspectives in phytochemistry Cambridge 1968 Apr Phytochemical society Edited by J. B. Harborne and T. Swain Academic press, London;New York,1969 For earlier publications of this body see Plant phenolics group
Gen 34.1407

PHYTOCHEMICAL SOCIETY SYMPOSIUM 8th Annual proceedings Phytochemical ecology Englefield Green,Surrey 1971 Apr Phytochemical society Academic press,London, 1972
Bioch 33.2229

PHYTOCHIMIE ET PLANTES MEDICINALES DES TERRES DU PACIFIQUE :colloque international Noumea 1964 Apr 28-May 5 Centre national de la recherche scientifique Centre national de la recherche scientifique.Colloques internationaux,144 Editions du Centre national de la recherche scientifique,Paris, 1966
Chem 18.1296

PHYTOCHIMIE ET PLANTES MEDICINALES DES TERRES DU PACIFIQUE colloque Proceedings Noumea,New Caledonia 1964 Apr 28-May 5 Centre nationale de la recherche scientifique Centre national de la recherche scientifique. Colloques internationaux, 144 Editions du centre national de la recherche scientifique, Paris,1966
Chem 26.0137

PHYTOPHARMACY Jaarlijks symposium over phytopharmacie symposium 2nd Papers Ghent 1950 Apr 23 Rijkslandbouwhogeschool, Ghent Landbouwhogeschool en de opzoekingsstations,Ghent.Mededelingen, 15,no 1 Rijkslandbouwhogeschool,Ghent,1950 Papers and summaries in English,Flemish and French
Chem 26.0105

PHYTOPHARMACY Jaarlijks symposium over phytopharmacie symposium 5th Verhandelingen Ghent 1953 May 5 Rijkslandbouwhogeschool,Ghent Edited by J. van den Brande Landbouwhogeschool en de opzoekingsstation,Ghent.Mededelingen, 18,no 2 Rijkslandbouwhogeschool,Ghent,1953 Papers and summaries in English,Flemish and French
Chem 26.0106

PICTURE LANGUAGE MACHINES :a conference Proceedings Canberra 1969 Feb 24-28 Australian national university Edited by S. Kaneff illus. 425p Academic press,London, 1970
Math L 5.3579

PICTURE LANGUAGE MACHINES :a conference Proceedings Canberra 1969 Feb 24-28 Commonwealth scientific and industrial research organisation Australian national university.Research school of physical sciences Edited by S. Kaneff Academic press,London,1970
Psy 31.3083

PICTURE LANGUAGE MACHINES :a conference Proceedings Canberra 1969 Feb 24-28 Edited by S. Kaneff Held at the Australian national university Academic press,London; New York,1970
An 32.5428

PIGMENT CELL International pigment cell conference 6th Sofia 1965 May 25-29 International union against cancer Edited by G. Della Porta and O. Muhlbock Springer-verlag,Berlin,1966
Bioch 33.0967

PIGMENT CELL:MOLECULAR,BIOLOGICAL AND CLINICAL ASPECTS International pigment cell conference 5th Papers New York 1961 Oct 11-14 Edited by Harold E. Whipple New York academy of sciences.Annals, 100 New York academy of sciences,New York,1963 1st - 4 th conferences called:Conference on the biology of normal and atypical pigment cell growth,q.v.
An 32.3389

PIGMENT CELL BIOLOGY Biology of normal and atypical pigment cell growth conference 4th Proceedings Houston,Tex. 1957 Nov 14-16 New York zoological society Anderson hospital and tumor institute Damon Runyon memorial fund for cancer research Edited by Myron Gordon Academic press,New York,1959
Gen 34.0601

PIGMENT CELL BIOLOGY Conference on the biology of normal and atypical pigment cell growth 4th Proceedings Houston,Tex. 1957 Nov 14-16 Edited by Myron Gordon Academic press,New York,1959 For 5th and further conferences see:International pigment cell conference
An 32.3388

PIGMENT CELL BIOLOGY The Conference on biology of normal and atypical pigment cell growth 4th Proceedings Houston,Tex. 1957 Nov 14-16 Edited by Myron Gordon Held at the Anderson hospital and tumor institute Academic press,New York,1959 Sponsored jointly by the New York zoological society,the M.D.Anderson Hospital and tumor institute and the Damon Runyon memorial fund for cancer research, inc.
Bioch 33.0937

PIGMENT CELL GROWTH Biology of melanomas A Conference on the biology of normal and atypical pigment cell growth New York 1946 Nov 15-16 New York academy of sciences. Section of biology New York academy of sciences.Special publications, 4 pls New York academy of sciences,New York,1948
Path 30.2492

PIGMENT CELL GROWTH Conference on the biology of normal and atypical pigment cell growth 3rd Proceedings New York 1951 Nov 15-17 Edited by Myron Gordon Sponsored by the New York zoological society Academic press,New York,1953
An 32.3368

PIGMENTARY SYSTEM Symposium on the biology of skin 16th Proceedings Portland,Ore. 1966 Edited by William Montagna and Funan Hu Held at the University of Oregon medical school Advances in biology of skin, 8 Oregon regional primate research center, 183 Pergamon press,Oxford,1967
An 32.4043

PILOT PLANTS IN THE IRON AND STEEL INDUSTRY autumn general meeting 1965 Papers and discussions London 1965 Nov 30-Dec 1 Iron and steel institute Iron and steel institute.Special report, 96 Iron and steel institute,London,1966
Met 25.0367

The PINEAL GLAND :a Ciba foundation symposium London 1970 Jun 30-Jul 2 Ciba foundation Edited by G.E.W. Wolstenholme and Julie Knight Churchill; Livingstone,Edinburgh;London,1971
An 32.5456

PISA 1957 Celebrazione della Accademia del cimento nel tricentenario della fondazione Domus galilaeana Accademia del cimento 49 pls 79p Domus galilaeana,Pisa,1958 Limited edition.With appendices
WSM 43.4472

PISA 1958 Vinci (Florence) Symposium international des sciences physiques et mathematiques dans la premiere moitie du XVII siecle Actes Union internationale d'histoire et philosophie des sciences Academie internationale d'histoire des sciences.Collection de travaux, 11 x,278p Gruppe italiano di storia delle scienze; Hermann,Vinci (Florence);Paris,c1959
WSM 43.0066

PISA 1961 Brain mechanisms :international colloquium...on specific and unspecific mechanisms of sensory motor integration Papers Edited by Giuseppe Moruzzi and others Sponsored by the International brain research organisation Progress in brain research, 1 Elsevier,Amsterdam,1963
An 32.5230

PISA 1966 Labelled proteins in tracer studies Conference on problems connected with the preparation and use of labelled proteins in tracer studies Proceedings Edited by L. Donato and others Sponsored by the Universita di Pisa.Nuclear medicine center EUR 2950 d,f,e European atomic energy community,Brussels,1966
Radioth 35.1752

PISA 1966 Symbol manipulation languages and techniques IFIP working conference on symbol manipulation languages Proceedings International federation for information processes Edited by Daniel G. Bobrow illus 487p North-Holland,Amsterdam,1968
Math L 5.3822

PISA 1966 Working conference on symbol manipulation language Papers International federation for information processing 1966 Typescript papers in loose-leaf file
Math L 5.3571

PITTSBURGH 1946 Atmospheric exposure tests on non-ferrous metals :a symposium American society for testing materials American society for testing materials,Philadelphia,Pa., 1946
Met 25.1924

PITTSBURGH 1952 Association for computing machinery :meeting Proceedings Jointly sponsored by the Mellon institute illus. 305p 28cm Rimbach,Pittsburgh,1952
Math L 5.3239

PITTSBURGH 1958 Quality requirements of super-duty steels a technical conference Proceedings American institute of mining, metallurgical and petroleum engineers. Committee on physical chemistry of steelmaking Edited by R.W. Lindsay Metallurgical society conferences, 3 Interscience,New York; London,1959
Met 25.0482

PITTSBURGH 1959 Physical metallurgy of stress-corrosion fracture American institute of mining,metallurgical and petroleum engineers Edited by Thor N. Rhodin Metallurgical society conferences, 4 Interscience,New York;London,1959
Met 25.2503

PITTSBURGH 1961 Bar and allied products Technical conference 3rd Proceedings American institute of mining,metallurgical and petroleum engineers.Mechanical working committee Metallurgical society conferences, 13 Interscience,New York;London,1961
Met 25.0729

PITTSBURGH 1963 Buhl international conference on materials :transition metal compounds,transport and magnetic properties 1st Informal proceedings Buhl foundation Edited by E.R. Schatz Gordon and Breach,New York,1964
Chem 18.2630

PITTSBURGH 1963 Mechanical working of steel 1 Technical conference 5th Proceedings American institute of mining, metallurgical and petroleum engineers. Mechanical working committee Metallurgical society conferences, 21 Gordon and Breach, New York,1964
Met 25.0728

PITTSBURGH 1963 Microstructure of ceramic materials :a symposium Proceedings Ceramic education council and National bureau of standards Sponsored by Edward Orton junior ceramic foundation National bureau of standards.Miscellaneous publication, 257 U.S. dept of commerce,Washington D.C.,1964
Met 25.0285

PITTSBURGH 1964 Applications of fundamental thermodynamics to metallurgical processes Conference on the thermodynamic properties of materials 1st Proceedings University of Pittsburgh.Center for the study of thermodynamic properties of materials Edited by G.R. Fitterer Gordon and Breach, New York,1967
Met 25.2443

PITTSBURGH 1965 Problem solving:research, method and theory Annual symposium on cognition 1st Papers Carnegie institute of technology Edited by Benjamin Kleinmutz Wiley,New York,1966
Psy 31.1911

PITTSBURGH 1965 Pyrometallurgical processes in nonferrous metallurgy :a symposium American institute of mining, metallurgical and mining engineers.Extractive metallurgy division Edited by J.N. Anderson and P.E. Queneau Metallurgical society conferences, 39 Gordon and Breach,New York, 1967
Met 25.2530

PITTSBURGH 1966 Ice symposium Papers American chemical society.Division of colloid and surface chemistry Journal of colloid and interface science,25 no 2 illus 131-294p Academic,New York;London,1967 Symposium chairman H.H.G.Jellinek
Sco 14.7934

PITTSBURGH,PA 1954 Symposium (international) on combustion 5th combustion in engines and combustion kinetics Combustion institute Under the auspices of the Standing committee on combustion symposia v,802p Reinhold;Chapman and Hall,New York; London,1955
Chem E 24.1914

PITTSBURGH,PA. 1952 Botanical books,prints and drawings from the collection of Mrs Roy Arthur Hunt illus. Carnegie institute, Pittsburgh,Pa.,1951
BG 38.0934

PITTSBURGH,PA. 1953 Current trends in information theory conference 7th lectures By Brockway McMillan and others University of Pittsburgh.Department of psychology Current trends in psychology series xii,188p 22cm University of Pittsburgh press,Pittsburgh,Pa.,1953 7th annual conference on Current trends in psychology
Math 3.0767

PITTSBURGH,PA. 1954 Current trends in psychology and the behavioral sciences... :six lectures Lectures By John T. Wilson and others University of Pittsburgh.Department of psychology University of Pittsburgh press, Pittsburgh,1954
Psy 31.2597

PITTSBURGH,PA. 1954 Symposium (international) on combustion,flame and explosion phenomena :combustion in engines and combustion kinetics 5th Papers Standing committee on combustion symposia Published for the Combustion institute Reinhold; Chapman and Hall,New York;London,1955 Later symposia organized by the Combustion institute
Chem 18.0435

PITTSBURGH,PA. 1959 Physical chemistry of process metallurgy :an international symposium American institute of mining,metallurgical and petroleum engineers Edited by George R. St. Pierre Metallurgical society conferences, 7,8 Interscience,New York;London,1961
Met 25.0491

PITTSBURGH,PA. 1960 Conference on electronic computation 2nd Proceedings American society of civil engineers American society of civil engineers,Kansas City,Miss., 1960
Eng 41.2718

PITTSBURGH,PA. 1962 Pittsburgh,Pa. 1962 Jun.10-23 New perspectives in organization research :consisting of papers from a conference on research in organizations. and a seminar on the social science of organizations Edited by W.W. Cooper and others Sponsored by United States.Office of naval research,Bibliog. xxii,606p 24cm John Wiley and sons,New York,1964
Math 3.0910

PITTSBURGH,PA. 1963 Adanson bicentennial symposium Hunt botanical library Hunt botanical library,Pittsburgh,Pa.,1963
BG 38.0952

PITTSBURGH,PA. 1963 Exhibition of Redouteana Hunt botanical library Hunt botanical library,Pittsburgh,Pa.,1963
BG 38.0935

PITTSBURGH,PA. 1965 Recent advances in optimization techniques :a symposium Edited by A. Levi and T.P. Vogl Sponsored by the Institute of electrical and electronics engineers.Systems science and cybernetics group Wiley,New York,1966
Eng 41.6003

PITTSBURGH,PA. 1965 Recent advances in optimization techniques symposium proceedings Institute of electrical and electronics engineers.Systems science and cybernetics group Edited by Abrahim Lavi and Thomas P. Vogl Jointly sponsored by the Optical society of America Bibliog. xiii, 656p 23cm John Wiley and sons,New York, 1966
Math 3.0240

PITTSBURGH,PA. 1966 Concepts and the structure of memory Symposium on cognition 2nd Papers Edited by Benjamin Kleinmuntz Sponsored by the Carnegie institute of technology Wiley,New York;London,1967
Eng 41.0652

PITTSBURGH,PA. 1966 Flow measurement symposium :presented at Fluid meters golden anniversary Papers American society of mechanical engineers.Fluid measurement research committee American society of mechanical engineers,New York,1966
Eng 41.6717

PITTSBURGH,PA. 1966 Metal transformations Buhl international conference on materials 2nd Informal proceedings Buhl foundation Edited by William W. Mullins and Milton C. Shaw Gordon and Breach,London;New York,1968
Met 25.2514

PITTSBURGH,PA. 1966 Problem solving: research,method and theory Symposium on cognition 1st Edited by Benjamin Kleinmuntz Sponsored by the Carnegie institute of technology Wiley,New York; London,1966
Eng 41.1291

PITTSBURGH,PA. 1967 Formal representation of human judgement Symposium on cognition 3rd Papers Edited by Benjamin Kleinmuntz Sponsored by the Carnegie-Mellon university Wiley,New York;London,1968
Eng 41.0653

PITTSBURGH,PA. 1968 Cognition and the development of language :annual symposium 4th Papers Carnegie-Mellon university Edited by John R. Hayes Wiley,New York,1970
Psy 31.3113

PITTSBURGH,PA. 1968 and PITTSBURGH,PA. 1969 International exhibition of botanical art and illustration 2nd Catalogue Hunt botanical library Edited by George H.M. Lawrence Hunt botanical library, Pittsburgh,Pa.,1968
BG 38.0911

PITTSBURGH,PA. 1970 Methods of local and global differential geometry in general relativity :conference Edited by D. Farnsworth and others Lecture notes in physics illus 190p 25cm Springer, Heidelberg,1972
A Math 4.1789

PITTSBURGH,PENN. 1963 Buhl international conference on materials :transition metal compounds ;transport and magnetic properties 1st Informal proceedings Buhl foundation and Carnegie institute of technology, Pittsburgh Edited by E.R. Schatz Gordon and Breach,New York;London,1964
Met 25.2351

PITUITARY,ADRENAL AND THE BRAIN :an international conference on the pituitary-adrenal axis and the nervous system Proceedings Vierhouten 1969 Jul 22-24 Edited by D. De Wied and J.A.W.M. Weijnen Organized by the Rudolf Magnus institute for pharmacology Progress in brain research, 32 Elsevier,Amsterdam,1970
An 32.5235

The PITUITARY GLAND :an investigation of the most recent advances Proceedings of the Association New York 1936 Dec 28-29 Association for research in nervous and mental disease Edited by Walter Timme and others Association for research in nervous and mental disease.Research publications,17 Williams and Wilkins,Baltimore,Md.,1938
Pha 16.0275

PITUITARY HORMONES Human pituitary hormones : a colloquium in honour of B.A.Houssay Proceedings Buenos Aires 1959 Aug 6-8 Ciba foundation Edited by G.E.W. Wolstenholme and Cecilia M. O'Connor Ciba foundation colloquia on endocrinology, 13 illus Churchill,London,1960
Bioch 33.0440

PLACENTA Symposium on the placenta :its form and functions with particular reference to the prevention of birth defects and fetal deaths New York 1964 Mar 6 Medical society of the county of New York.Special committee on infant mortality Edited by Daniel Bergsma and others Sponsored by the National Foundation-March of dimes.Great New York Chapter Birth defects original article series, 1,no.1 National foundation,New York,1965
PGMS 29.0212

PLANETARY AND STELLAR MAGNETISM Magnetism and the cosmos North Atlantic treaty organization advanced study institute on planetary and stellar magnetism Newcastle-upon-Tyne 1965 North Atlantic treaty organization Edited by W.R. Hindmarsh and others 436p Oliver and Boyd,Edinburgh,1967
Obs 6.3350

PLANETARY ATMOSPHERES Atmosphere of the earth and planets A Symposium on planetary atmospheres Papers Williams Bay,Wisc. 1947 Sep 8-10 University of Chicago Edited by Gerard P. Kuiper 366p Chicago,Ill.,1949 Symposium held in connection with the 50th anniversary of the Yerkes observatory
Geod 9.0523

PLANETARY ATMOSPHERES Marfa,Texas 1969 Oct 26-31 International astronomical union Edited by C. Sagan and others International astronomical union.Symposium, 40 408p Reidel,Dordrecht,1971
Obs 6.3586

PLANETARY ATMOSPHERES The Symposium on planetary atmospheres Yerkes observatory 1947 Sep 8-10 University of Chicago Edited by Gerard P. Kuiper vii,366p Chicago,1949 In connection with the 15th anniversary of the Yerkes observatory
Nap 11.0340

PLANETARY ELECTRODYNAMICS International conference on the universal aspects of atmospheric electricity 4th Proceedings Tokyo 1968 Vol 1-2 Edited by Samuel C. Coroniti and James Hughes 2 vols Gordon and Breach,New York;London,1969
Nap 11.0974

PLANETARY NEBULAE :a symposium Tatranska Lomnica 1967 International astronomical union Edited by D.E. Osterbrock and C.R. O'Dell International astronomical union.Symposium, 34 Reidel,Dordrecht,1968
TA 15.0320

PLANETARY NEBULAE :a symposium Proceedings Tatranska Lomnica 1967 International astronomical union Edited by D.E. Osterbrock and C.R. O'Dell International astronomical union.Symposium, 34 469p Reidel,Dordrecht, 1968
Obs 6.3265

PLANNING OF RADIOTHERAPY FACILITIES Joint IAEA-WHO meeting on planning of radiotherapy facilities Report Geneva 1964 Dec 15-19 International atomic energy agency World health organization World health organization.Technical report series, 328 World health organization,Geneva,1966
Radioth 35.1060

PLANSEE PROCEEDINGS 1948 High-melting metals Plansee seminar "De re metallica" 3rd Papers Reutte,Austria 1948 Jun 22-26 Metallwerk Plansee AG Edited by F. Benesovsky Pergamon press;Metallwerk Plansee AG;,,London;Reutte,1959
Met 25.1010

PLANSEE PROCEEDINGS 1961 Powder metallurgy in the nuclear age Plansee seminar 'De re metallica' 4th Proceedings Reutte, Austria 1961 Jun 20-24 Metallwerk Plansee AG Edited by F. Benesorsky Metallwerk Plansee AG,Reutte,1962 Papers in English, French and German
Met 25.2724

PLANSEE PROCEEDINGS 1964 Metals for the space age Plansee seminar "De re metallica" 5th Papers Reutte,Austria 1964 Jun 22-26 Metallwerk Plansee A.G. Edited by F. Benesovsky Metallwerk Plansee A G,Reutte, 1965 Papers in English,French and German
Met 25.0091

PLANSEE SEMINAR 6th Hochtemperatur werkstoffe Papers High temperature materials Reutte,Austria 1966 Jun 24-26 Metallwerk Plansee A.G. Edited by F. Benesovsky Springer,Vienna;New York,1969 Papers in English,French and German
Met 25.2336

PLANSEE SEMINAR 'DE RE METALLICA' 4th Proceedings Powder metallurgy in the nuclear age Reutte,Austria 1961 Jun 20-24 Metallwerk Plansee AG Edited by F. Benesorsky Metallwerk Plansee AG,Reutte,1962 Papers in English,French and German
Met 25.2724

PLANSEE SEMINAR "DE RE METALLICA" 3rd Papers High-melting metals Reutte, Austria 1948 Jun 22-26 Metallwerk Plansee AG Edited by F. Benesovsky Pergamon press; Metallwerk Plansee AG;,,London;Reutte,1959
Met 25.1010

PLANSEE SEMINAR "DE RE METALLICA" 5th Papers Metals for the space age Reutte, Austria 1964 Jun 22-26 Metallwerk Plansee A.G. Edited by F. Benesovsky Metallwerk Plansee A G,Reutte,1965 Papers in English, French and German
Met 25.0091

PLANT ANATOMY New research in plant anatomy London 1970 Sep 18 Edited by N.K.B. Robson and D.F. Cutler Linnean society of London. Botany.Supplement, 63,1 Academic press, London,1970 Symposium arranged by the Plant anatomy group of the Linnean society
Bot 42.1118

PLANT AND ANIMAL BREEDING U.S.S.R.congress of genetics plant and animal breeding. Proceedings Trudi vsesoyuznogo sveztsda po lenetike selektsii semenovodstvy i plemennomy zhivothovodstvy Leningrad 1929 Jan 10-16 Edited by N.I. Wulff and others 5 vols Izdanie Redaktsionnoi kollegii ceda,Leningrad, 1930
Bot 42.0654

PLANT AND ANIMAL NUTRITION IN RELATION TO SOIL AND CLIMATIC FACTORS British commonwealth scientific official conference.Specialist conference in agriculture 1st Proceedings Adelaide 1949 Aug 22-Sep 15 and Canberra 1949 Aug 22-Sep 15 H.M.S.O.,London,1951
Geog 13.2510

PLANT AND ANIMAL NUTRITION IN RELATION TO SOIL AND CLIMATIC FACTORS The Australian environment handbook prepared for the conference on plant and animal nutrition... 1949 Aug Commonwealth scientific and industrial research organization 2nd edition British commonwealth official scientific specialist conferences,Special agricultural conference Commonwealth scientific and industrial research organization,Melbourne, 1950
Gen 34.1372

PLANT AND ANIMAL VIRUSES The Symposium on plant and animal viruses Proceedings Cuttack 1961-62 Dec 31-Jan 1 National institute of sciences of India National institute of sciences of India.Bulletin, 24, 1963 National institute of sciences of India, New Delhi,1963 Held under the joint auspices of the National institute and the Council of scientific and industrial research
Bal 39.1071

PLANT AND ANIMAL VIRUSES :a symposium Proceedings Cuttack 1961-62 Dec 31-Jan 1 National institute of sciences of India National institute of sciences of India. Bulletin, 24 illus. vi,248p National institute of sciences of India,New Delhi,1963
Bot 42.3739

PLANT AND PROCESS DYNAMIC CHARACTERISTICS :a conference Proceedings Cambridge 1956 Apr 4-6 Society of instrument technicians Academic press;Butterworths,New York;London, 1957
Eng 41.5908

PLANT AND PROCESS DYNAMIC CHARACTERISTICS :a conference Proceedings Cambridge 1956 Apr 4-6 Society of instrument technology xii,246p Butterworths,London,1957
Chem E 24.1157

PLANT BIOLOGY TODAY:ADVANCES AND CHALLENGES : a symposium Denver,Colo. 1961 Dec American association for the advancement of science Botanical society of America Edited by William A. Jensen and Leroy G. Kavaljian illus Macmillan,London,1963
Bot 42.0298

PLANT BREEDING INSTITUTE Cereal rust conferences Cambridge 1964 Jun-Jul Cambridge,1966
Bot 42.4125

PLANT CELL ORGANELLES The Phytochemical group symposium Proceedings London 1967 Apr 10-12 Phytochemical group Edited by J. B. Pridham Academic press,London,1968
Bioch 33.0782

PLANT CELL ORGANELLES :a symposium Proceedings London 1967 Apr 10-12 Phytochemical group Edited by J.B. Pridham xiii,261p Academic press,London,1968
Bot 42.4737

PLANT DESIGN Symposium on chemical process hazards with special reference to plant design Proceedings Manchester 1960 Mar 29-31 Edited by D.M. Pirie 117p Institution of chemical engineers,London,1960
Chem E 24.1395

PLANT DISTRIBUTION The Study of the distribution of British plants :a conference Report Oxford 1950 Mar 31-Apr 2 Botanical society of the British Isles Botanical society of the British Isles.B.S.B.I. conference reports, 2 . London,1951
Bot 42.4707

PLANT EMBRYOLOGY :a symposium New Delhi 1960 Nov 11-14 Council of scientific and industrial research.Biological research committee Council of scientific and industrial research,New Delhi,1962
Bot 42.1240

PLANT ENVIRONMENT Control of the plant environment Easter school in agricultural science 4th Proceedings Nottingham 1957 University of Nottingham Edited by J. P. Hudson Butterworths,London,1957
BG 38.0605

PLANT GROWTH International conference on plant growth substances 7th Proceedings Canberra 1970 Dec 7-11 Edited by Denis J. Carr Springer,Berlin,1972
Bioch 33.2231

PLANT GROWTH REGULATION International conference on plant growth regulation 4th Yonkers 1959 Aug 10-14 Sponsored by the Boyce Thompson institute for plant research Iowa state university press,Ames,Iowa,1961
Bot 42.1830

PLANT GROWTH REGULATORS :a conference Papers New York 1966 May 16-18 New York academy of sciences Edited by Edward M. Weyer New York academy of sciences.Annals, 144,p 1-382 New York,1967
Bioch 33.0780

PLANT GROWTH REGULATORS :symposium Papers London 1968 Jan 8-9 Society of chemical industry.Pesticides group Phytochemical society Society of chemical industry. Monograph, 31 Society of chemical industry, London,1968
Bioch 33.2230

PLANT GROWTH SUBSTANCES Biochemistry and physiology of plant growth substances International conference on plant growth substances 6th Ottawa 1967 Jul 24-29 Edited by F. Wightman and G. Setterfield Held at Carleton University Runge press, Ottawa,1968
Bioch 33.0779

PLANT GROWTH SUBSTANCES Biochemistry and physiology of plant growth substances International conference on plant growth substances 6th Ottawa 1967 Jul 24-29 Edited by F. Wightman and G. Setterfield Held at the Carleton university xii,1642p Runge press,Ottawa,1968 On the occasion of the centennial of Canadian confederation
Bot 42.1833

PLANT GROWTH SUBSTANCES Regulateurs naturels de la croissance vegetales Colloque international sur les substances de croissance vegetales 5e Actes Gif sur Yvette 1963 Jul 15-20 Centre national de la recherche scientifique Centre national de la recherche scientifique.Colloques internationaux, 123 Paris,1964 Text in English and French
Bioch 33.0747

PLANT GROWTH SUBSTANCES The Chemistry and mode of action of plant growth substances :a symposium Proceedings London 1955 Jul Edited by R.L. Wain and F. Wightman Butterworths,London,1956
BG 38.0165

PLANT METABOLISM Recent advances in the study of plant metabolism :a symposium Proceedings Allahabad 1958 Jan 22-24 Sponsored by the University grants commission Allahabad university studies.Botany section. Symposium supplement port. vi,140p Allahabad,1958
Bot 42.4704

PLANT-PARASITE INTERACTION The Dynamic role of molecular constituents in plant-parasite interaction :a conference Proceedings Gamagori 1966 May 15-21 American phytopathological society Edited by Chester J. Mirocha and Ikuzo Uritani American phytopathological society,St.Paul,Minn.,1967
Bot 42.3761

PLANT PATHOLOGIST'S POCKETBOOK International congress of plant pathology 1st Proceedings London 1968 Jul Commonwealth mycological institute Commonwealth mycological institute,Kew,1968
BG 38.1757

PLANT PATHOLOGY Host-parasite relations in plant pathology :a symposium Budapest 1964 Oct 19-22 Hungarian academy of sciences. Research institute for plant protection Edited by Z. Kiraly and G. Ubrizsy Sponsored by the Ministry of agriculture 257p Budapest,c1965
Bot 42.4628

PLANT PHENOLICS The Pharmacology of plant phenolics :a symposium Proceedings Oxford 1958 Apr 10-11 Edited by J.W. Fairbairn Organised by the Plant phenolics group Academic press,London;New York,1959
Bioch 33.0752

PLANT PHENOLICS GROUP Biosynthetic pathways in higher plants Plant phenolics group symposium Proceedings Leeds 1964 Apr 13-15 Edited by J.B. Pridham and T. Swain Academic press,London;New York,1965
Bot 42.1883

PLANT PHENOLICS GROUP Biosynthetic pathways in higher plants Plant phenolics group symposium Proceedings Leeds 1964 Apr 13-15 Edited by J.B. Pridham and T. Swain Academic press,London;New York,1965 For later publications of this body see Phytochemical society
Gen 34.1406

PLANT PHENOLICS GROUP Biosynthetic pathways in higher plants The Plant phenolics group symposium Proceedings Leeds 1964 Apr 13-15 Edited by J.B. Pridham and T. Swain Academic press,London;New York,1965
Bioch 33.0783

PLANT PHENOLICS GROUP Enzyme chemistry of phenolic compounds :a symposium Proceedings Liverpool 1962 Apr 11-12 Edited by J.B. Pridham Pergamon press.Symposium publications division,Oxford,1963
Chem 18.1247

PLANT PHENOLICS GROUP Recent developments in the chemistry of natural phenolic compounds :a symposium Proceedings 1960 Apr Edited by W.D. Ollis Pergamon press,Oxford,1961
Chem 18.1495

PLANT PHENOLICS GROUP The Pharmacology of plant phenolics :a symposium Proceedings Oxford 1958 Apr 10-11 Edited by J.W. Fairbairn Academic press,London;New York, 1959
Bioch 33.0752

PLANT PHENOLICS GROUP The Pharmacology of plant phenolics :a symposium Proceedings Oxford 1958 Apr 10-11 Edited by J.W. Fairbairn Academic press,London;New York, 1959
Chem 26.0265

PLANT PHENOLICS GROUP ANNUAL GENERAL MEETING 3RD Recent developments in the chemistry of natural phenolic compounds :a symposium Proceedings 1960 Apr Plant phenolics group Edited by W.D. Ollis Pergamon press,Oxford, 1961
Chem 18.1495

PLANT PHENOLICS GROUP SYMPOSIUM Proceedings Biosynthetic pathways in higher plants Leeds 1964 Apr 13-15 Plant phenolics group Edited by J.B. Pridham and T. Swain Academic press,London;New York,1965
Bot 42.1883

PLANT PHENOLICS GROUP SYMPOSIUM Proceedings Biosynthetic pathways in higher plants Leeds 1964 Apr 13-15 Plant phenolics group Edited by J.B. Pridham and T. Swain Academic press,London;New York,1965 For later publications of this body see Phytochemical society
Gen 34.1406

The PLANT PHENOLICS GROUP SYMPOSIUM Proceedings Biosynthetic pathways in higher plants Leeds 1964 Apr 13-15 Plant phenolics group Edited by J.B. Pridham and T. Swain Academic press,London;New York,1965
Bioch 33.0783

PLANT SCIENCES International congress of plant sciences Proceedings Ithaca,N.Y. 1926 Aug 16-23 Vol 1-2 Edited by B.M. Duggar 2 vols Banta,Menasha,Wis.,1929
Bot 42.0653

PLANTATION SYSTEMS :seminar Papers and discussion summaries San Juan,Puerto Rico 1957 Nov 17-23 Research institute for the study of man,and,Pan American union Pan American union.Social science monographs,7 Washington,D.C.,1959
Geog 13.5120

PLANTS Les Cultures de tissus de plantes Proceedings Strasburg 1970 Jul 6-10 Centre national de la recherche scientifique Centre national de la recherche scientifique. Colloques internationaux, 193 CNRS,Paris, 1971
Bioch 33.2223

PLANTS AND THE MIGRATIONS OF PACIFIC PEOPLES Honolulu 1961 Aug 21-Sep 6 Bishop Bernice P museum,and,University of Hawaii Edited by J. Barrau Sponsored by the National science foundation Bishop museum press,Honolulu,1963 Symposium convened at the 10th Pacific science congress
Geog 13.7202

PLAS-TECH EQUIPMENT CORPORATION High speed testing Vol 1: a symposium held at Boston, Massachusetts,December 8,1958 Interscience, New York,1968
Eng 41.3845

PLAS-TECH EQUIPMENT CORPORATION High speed testing Vol 3: third annual symposium held at Boston,Massachusetts,October 26 and 27,1961 Interscience,New York,1962
Eng 41.3848

PLAS-TECH EQUIPMENT CORPORATION High speed testing Vol 5: 5th international symposium held at Boston,Massachusetts,March 8 and 9, 1965 Applied polymer symposia, 1 Interscience,New York,1965
Eng 41.3849

PLASMA ACCELERATION Lockheed symposium on magnetohydrodynamics 4th proceedings Palo Alto 1959 Dec.2 Edited by Sidney W. Kash Sponsored by Lockheed aircraft corporation.Missiles and space division Stanford university press,Stanford,Calif.,1960
A Math 4.0618

PLASMA ACCELERATION Symposium on magnetohydrodynamics 4th Palo Alto, Calif. 1959 Dec 2 Edited by Sidney W. Kash Sponsored by the Lockheed aircraft corporation. Missiles and space division Stanford university press,Stanford,Calif.,1960
Eng 41.4489

PLASMA ASTROPHYSICS Astrofisica del plasma Scuola internazionale di fisica 'Enrico Fermi' 39 corso Rendiconti Varenna 1966 Jul 11-30 Societa italiana di fisica Edited by P. A. Sturrock Academic press,New York;London, 1967
TA 15.0328

PLASMA ASTROPHYSICS Astrofisica del plasma Scuola internazionale di fisica "Enrico Fermi" 39 corso Rendiconti Varenna 1966 Jul 11-30 Societa italiana di Fisica Edited by P. A. Sturrock xvi,364p Academic press,New York;London,1967
A Math 4.1577

PLASMA DYNAMICS Symposium of plasma dynamics Woods Hole,Mass. 1958 Jun 9-13 Edited by Francis H. Clauser Sponsored by United States.Air force.Office of scientific research Pergamon press;Addison-Wesley,London;Reading, Mass.,1960
Eng 41.4479

PLASMA DYNAMICS Symposium of plasma dynamics proceedings Massachusetts 1968 Jun.9-13 National academy of sciences Edited by Frances H. Clauser Sponsored by the United States.Air force.Office of scientific research Pergamon press,London,1959
A Math 4.0617

PLASMA EFFECTS IN SOLIDS International conference on the physics of semiconductors 7th Paris 1964 Vol 2: plasma effects in solids International union of pure and applied physics photos,diagrms, graphs 221p 24cm Dunod,Paris,1965 Co-sponsored by the French physical society, the French ministry for scientific research and Unesco
Cav 7.3047

PLASMA EFFECTS IN SOLIDS Plasma effects in solids International conference on the physics of semiconductors 7th Paris 1964 International union of pure and applied physics photos,diagrms,graphs 221p 24cm Dunod,Paris,1965 Co-sponsored by the French physical society,the French ministry for scientific research and Unesco
Cav 7.3047

PLASMA HYDROMAGNETICS Symposium on magnetohydrodynamics 6th Proceedings Palo Alto,Calif. 1961 Dec 15-16 Edited by Daniel Bershader Sponsored by Lockheed aircraft corporation.Missile systems division Stanford university press,Stanford,Calif.,1962
Eng 41.4510

The PLASMA IN A MAGNETIC FIELD :a symposium on magnetohydrodynamics Palo Alto,Calif. 1957 Dec 16 Edited by Rolf K.M. Landshoff Sponsored by the Lockheed aircraft corporation. Missile systems division Stanford university press,Stanford,Calif.,1958 The 2nd symposium on magnetohydrodynamics sponsored by Lockheed.For later symposia see 'Symposium on magnetohydrodynamics'
Eng 41.4473

PLASMA LIPOPROTEINS London 1971 Apr Biochemical society Edited by R.M.S. Smellie Biochemical society.Symposium, 33 Academic press,London,1971
Bioch 33.2317

PLASMA PHYSICS Cosmic plasma physics :a conference Frascati 1971 Sep 20-24 Edited by K. Schindler 369p Plenum press, New York,1972
Obs 6.3625

PLASMA PHYSICS The International seminar on plasma physics proceedings Trieste 1964 Oct 5-31 International atomic energy agency. International centre for theoretical physics International atomic energy agency,Vienna,1965
A Math 4.0624

PLASMA SHEATH Electromagnetic aspects of hypersonic flight Symposium on the plasma sheath :its effect upon reentry communication and detection 2nd Proceedings Boston, Mass. 1962 Apr 10-12 Edited by Walter Rotman and others Sponsored by the Air force Cambridge research laboratories. Electromagnetic radiation laboratory Spartan books;Cleaver-Hume,Baltimore,Md.;London,1964
Eng 41.4544

The PLASMA SPACE SCIENCE SYMPOSIUM proceedings Washington D.C. 1963 Jun 11-14 Catholic university of America Edited by C.C. Chang and S.S. Huang In co-operation with the National aeronautics and space administration Astrophysics and space science library, 3 D.Reidel,Dordrecht,1965
A Math 4.0625

PLASMA THEORY Teoria dei plasma Scuola internazionale di fisica 'Enrico Fermi' 25 corso Rendiconti Varenna 1962 Jul 9-21 Societa italiana di fisica Edited by M.N. Rosenbluth Academic press,New York;London, 1964
Eng 41.4527

PLASMA WAVES IN SPACE AND IN THE LABORATORY a NATO advanced study institute Roros, Norway 1968 Apr 17-26 North Atlantic treaty organization Edited by J.O. Thomas and B.J. Landmark Bibliog.,Illus. 487p 23cm Edinburgh university press,Edinburgh, 1969
A Math 4.1505

PLASMAS Radiation and waves in plasmas Lockheed symposium on magnetohydrodynamics 5th proceedings Palo Alto 1960 Dec 15-16 Edited by Morton Mitchner Sponsored by Lockheed aircraft corporation.Missiles and space division Stanford university press, Stanford,Calif.,1961
A Math 4.0621

PLASMAS IN PHYSIK UND ASTRONOMIE Probleme des plasmas in physik und astronomie :eine tagung berichte Leipzig 1956 Oct.8-11 Physikalische gesellschaft in der deutschen demokratischen republik 223p Akademie-verlag,Berlin,1958
Obs 6.1366

PLASTIC DEFORMATION OF POLYMERS Symposium on plastic deformation of polymers Selected papers New York 1969 Sep 10-11 American chemical society Dekker,New York,1971
Met 25.2576

The PLASTIC WORKING METALS Papers and discussions a symposium Cleveland 1936 Oct 19-23 American society for metals American society for metals,Cleveland,Ohio,1936 The symposium presented before the eighteenth annual convention of the American society for metals.
Met 25.0679

PLASTICITY Naval structural mechanics :a symposium 2nd Proceedings Rhode Island 1960 Apr 5-7 United States.Office of naval research and Brown university Edited by E.H. Lee and P.S. Symonds Pergamon press,New York,1960
Met 25.0845

PLASTICITY Naval structural mechanics symposium 2nd Proceedings Providence,R.I. 1960 Apr 5-7 United States.Office of naval research Brown university Edited by E.H. Lee and P.S. Symonds United States. Office of naval research.Structural mechanics series Pergamon press,London,1960
Eng 41.2479

PLASTICIZATION AND PLASTICIZER PROCESSES :a symposium Philadelphia,Pa. 1964 Apr 6-7 Sponsored by the American chemical society. Division of industrial and engineering chemistry Advances in chemistry series, 48 ix,200p American chemical society,Washington, D.C.,1965 Symposium chairman:Norbert A.J. Platzer
Chem E 24.1045

PLASTICS Plastics progress,1961 International plastics convention Papers and discussions London 1961 Jun Edited by Phillip Morgan xi,181p Iliffe;Macmillan, London;New York,1962
Chem E 24.1014

PLASTICS Symposium on plastics Philadelphia,Pa. 1944 Feb 22-23 American society for testing and materials American society for testing materials,Philadelphia,Pa., 1944
Eng 41.3642

PLASTICS IN BUILDING STRUCTURES :a conference Proceedings London 1965 Jun 14-16 Organised by the Plastics institute Plastics institute.Transactions and journal.Conference supplement, 1 Pergamon press,London,1966
Eng 41.3670

PLASTICS INSTITUTE Plastics in building structures :a conference Proceedings London 1965 Jun 14-16 Plastics institute. Transactions and journal.Conference supplement, 1 Pergamon press,London,1966
Eng 41.3670

PLASTICS INSTITUTE.TRANSACTIONS AND JOURNAL. CONFERENCE SUPPLEMENT, 1 Plastics in building structures :a conference Proceedings London 1965 Jun 14-16 Organised by the Plastics institute Pergamon press,London,1966
Eng 41.3670

PLASTICS PROGRESS,1961 International plastics convention Papers and discussions London 1961 Jun Edited by Phillip Morgan xi,181p Iliffe;Macmillan,London;New York, 1962
Chem E 24.1014

PLATTSBURGH,N.Y. 1966 Origin of anorthosite and related topics Annual George H.Hudson symposium 2nd Papers New York state museum and science service.Geological survey State university of New York.College at Plattsburgh Edited by Yngvar W. Isachsen New York state museum and science service. Memoir, 18 New York state museum and science service,Albany,N.Y.,1968
Min 10.1428

PLEISTOCENE EXTINCTIONS:THE SEARCH FOR A CAUSE International association for quaternary research.Congress 7th Proceedings Boulder,Colo. 1965 Aug I.N.Q.U.A. Edited by P.S. Martin and H.E. Wright Sponsored by the National research council Yale university press,New Haven,Conn.;London,1967
Geog 13.1320

PLEISTOCENE EXTINCTIONS:THE SEARCH FOR A CAUSE International association for quaternary research congress 7th Proceedings Boulder,Colo. 1965 Aug 30-Sep 5 Vol 6 Edited by P.S. Martin and H.E. Wright Sponsored by the National research council Yale university press,New Haven,Conn.;London, 1967
Bot 42.3244

PLIOZAN UND PLEISTOZAN :am mittel - und niederrhein Krefeld 1959 Edited by R. Teichmuller and G.v.d. Brelie Fortschritte in der geologie von Rheinland und Westfalen, 4 illus. 412p Krefeld,1959
Bot 42.3327

PLOYPEPTIDES WHICH AFFECT SMOOTH MUSCLES AND BLOOD VESSELS... :a symposium Proceedings London 1959 Mar 23-24 Edited by M. Schachter Biological council.Co-ordinating committee for symposia on drug action. Symposium on drug action series,4 Pergamon, Oxford,1960
Pha 16.0439

PLUTONIUM AND ITS ALLOYS Extractive and physical metallurgy of plutonium and its alloys based on a symposium San Francisco 1959 Feb 16-17 American institute of mining, metallurgical and petroleum engineers.Nuclear metallurgy committee Edited by W.D. Wilkinson Interscience,New York;London,1960 With an introduction and an annotated bibliography by the editor
Met 25.1568

PLUTONIUM 1960 International conference on plutonium metallurgy 2nd Proceedings Grenoble 1960 Apr 19-22 Societe Francaise de metallurgie and Commissariat a l'energie atomique Edited by Emmanuel Grison and others Cleaver-Hume,London,1961 Papers in English and French
Met 25.1549

PNEUMATIC AND HYDRAULIC CONTROL SYSTEMS : seminar on pneumohydraulic automation 1968 Vol 1-2 Edited by M.A. Aizerman Translated by P. Linnik from the Russian 2 vols Pergamon press,London,1968
Eng 41.5903

POCONO MANOR 1950 Imperfections in nearly perfect crystals :a symposium National research council.Committee on solids Edited by W. Shockley and others Wiley;Chapman and Hall,New York;London,1952
Met 25.1226

POCONO MANOR,PA. 1950 Imperfections in nearly perfect crystals :a symposium Edited by W. Shockley and others Sponsored by the National research council.Committee on solids Wiley;Chapman and Hall,New York;London,1952
Eng 41.4427

POCONO MANOR,PA. 1950 Imperfections in nearly perfect crystals :a symposium Papers Edited by William Shockley and others Sponsored by the National research council. Committee on solids John Wiley,New York,1952
Cav 7.2045

POCONO MANOR,PA. 1950 Imperfections in nearly perfect crystals :symposium Edited by W. Shockley sponsored by,National research council Wiley,New York,1952
Min 10.1091

POINT BARROW 1967 Working party conference for the international biological program human adaptability study of circumpolar populations Report Canadian national committee for the international biological program,and,United States national committee for the international biological program variousp 29cm Naval arctic research laboratory,Point Barrow,1967
Sco 14.7529

POITIERS 1968 Symposium (International) on combustion 12 Abstracts of papers Combustion institute xxx,228p Combustion institute,Pittsburgh,Pa.,1968
Chem E 24.1408

POLAR ATMOSPHERE :a symposium Oslo 1956 July 2-8 Pt 1: meteorology section Edited by R.C. Sutcliffe Published for and on behalf of North Atlantic treaty organisation.Advisory group for aeronautical research and development AGARDograph 29,1 xii,341p 26cm Pergamon press,London,1958
Sco 14.2367

POLAR ATMOSPHERE SYMPOSIUM Oslo 1956 Jul 2-8 Pt 2: ionospheric section North Atlantic treaty organization.Advisory group for aeronautical research and development Agardograph,29,pt.2 diagrs xiii,212p 26cm Pergamon,London,1957
Sco 14.0528

POLAR METEOROLOGY W.M.O.-S.C.A.R.-I.C.P.M. Symposium on polar meteorology Proceedings Geneva 1966 Sep 5-9 World meteorological organization,and,Scientific committee on Antarctic research Also organized by the International commission on polar meteorology illus vi,540p 28cm World meteorological organization,Geneva,1967
Sco 14.7839

POLAR REGIONS Congres international pour l'etude des regions polaires Brussels 1906 Sep 7-11 programme 40p Hayez,Brussels, 1906 In English,French and German
Sco 14.0108

POLAR REGIONS Congres international pour l'etude des regions polaires Rapport d'ensemble Brussels 1906 Sep 7-11 documents preliminaires et comte rendu des seances Hayez,Brussels,1906
Sco 14.0109

POLAR REGIONS Internationale polar-konferenz verhandlungen und ergebnisse 3te Bericht St.Petersburg 1881 Aug 1-6 Akademie der wissenschaften,St.Petersburg,1881 Parallel French-German texts.One of 8 pamphlets bound together
Philos 1.2259

POLARONS AND EXCITONS Excitations in semiconductors,polarons and excitons Scottish universities' summer school in physics 3rd Papers St.Andrews 1962 Jul 30-Aug 18 Edited by C.G. Kuper and G.D. Whitfield Scottish universities' summer school in physics,1962,3 Oliver and Boyd, Edinburgh,1963
Cav 7.2136

POLARONS AND EXCITONS Scottish universities' summer school 3rd Lectures St.Andrews 1962 Jul 30-Aug 18 Scottish universities' summer school in physics Edited by C.G. Kuper and G.D. Whitfield With financial support from the North Atlantic treaty organization.Science committee Oliver and Boyd,Edinburgh,1963
Chem 18.2640

POLIGNANE A MARE 1967 Biochemical aspects of the biogenesis of mitochondria The Round table discussion on the biochemical aspects of the biogenesis of mitochondria Proceedings Edited by E.C. Slater and others Adriatica editrice,Bari,1968
Bioch 33.1073

POLIGNANO A MARE 1966 Mitochondrial structure and compartmentation The Round table discussion on mitochondrial structure and compartmentation Proceedings Edited by E. Quagliariello and others organized by the University of Bari,Department of biochemistry and the University of Amsterdam.Laboratory of biochemistry Adriatica editrice,Bari,1967
Bioch 33.1093

POLIGNANO A MARE 1968 Energy level and metabolic control in mitochondria The Round table discussion on the energy level and metabolic control in mitochondria Proceedings Edited by S. Papa and others Organized by the University of Bari.Department of biochemistry and the University of Amsterdam.Laboratory of biochemistry Adriatica editrice,Bari,1969
Bioch 33.1075

POLISH BIOCHEMICAL SOCIETY Biochemistry of blood platelets :colloquium held on the occasion of the Third meeting of the Federation of European biochemical societies Warsaw 1966 Apr 4-7 Edited by E. Kowalski and S. Niewiarowski Academic press,New York, 1967
Radioth 35.0148

POLLEN PHYSIOLOGY AND FERTILIZATION An International symposium on pollen physiology and fertilization 1st Papers Nijmegen 1963 Aug 29-31 Edited by H.F. Linskens North Holland,Amsterdam,1964
Bot 42.1426

POLLUTION Air and water pollution in the iron and steel industry proceedings of the air pollution meeting 25th and 26th Sept. and the water pollution meeting 11th and 12th Dec Proceedings London 1957 Sep 25-26 London 1957 Dec 11-12 Iron and steel institute Iron and steel institute.Special report, 61 Iron and steel institute,London,1958
Met 25.0359

POLLUTION Global effects of environmental pollution :a symposium Dallas,Texas 1968 Dec American association for the advancement of science 218p Reidel,Dordrecht,1970
Nap 11.0970

POLLUTION Industrial process design for pollution control :a workshop Proceedings New York 1967 Feb 9-10 Vol 1 American institute of chemical engineers.Water committee iv,59p A.I.Ch.E.,New York,1967
Chem E 24.1575

POLSKA AKADEMIA NAUK Fluid dynamics transactions Symposium on fluid dynamics 5th Jablonna 1961 26 Aug.-2 Sep. Vol. 1 Edited by W. Fiszdon Pergamon press, London,1964
A Math 4.0527

POLSKA AKADEMIA NAUK Fluid dynamics transations,4 Symposium on advanced problems and methods in fluid dynamics 8th Proceedings Tarda,Poland 1967 Sep 18-25 Edited by W. Fiszdon and others bibliog., illus. x,812p 24cm PWN,Warsaw,1969
A Math 4.1680

POLSKA AKADEMIA NAUK Relativistic theories of gravitation A Conference on gravitation Proceedings Jablonna 1962 Jul 25-31 Edited by Leopold Infeld With financial support of the International union of pure and applied physics Pergamon press,Oxford,1964 Binder's title in French
TA 15.0085

POLSKA AKADEMIA NAUK Relativistic theories of gravitation Conference on relativistic theories of gravitation proceedings Jablonna 1962 Jul 25-31 Edited by Leopold Infeld Pergamon press,Oxford,1964
A Math 4.1029

POLSKA AKADEMIA NAUK Sesja kopernikowska... Warsaw 1953 Sep 15-16 482p Panstowe wydawnictwo naukowe,Warsaw,1955
WSM 43.2653

POLSKA AKADEMIA NAUK.COMITE ET INSTITUT D'HISTOIRE DE LA SCIENCE ET DE LA TECHNIQUE Congres international d'histoire des sciences 11e Actes Warsaw 1965 Aug 24-31 Cracow 1-6 6 vols Ossolineum.Maison d'edition de l'Academie polonaise des sciences, Wroclaw,1968
WSM 43.0079

POLSKA AKADEMIA NAUK.INSTYTUT GEOGRAFII Problems of applied geography Anglo-Polish seminar 1st Proceedings Nieborow 1959 Sep 15-18 Geographical studies,25 Panstwowe wydawnictwo naukowe,Warsaw,1961
Geog 13.5617

POLSKA AKADEMIA NAUK.INSTYTUT GEOGRAFII and INSTITUTE OF BRITISH GEOGRAPHERS Problems of applied geography,2 Anglo-Polish seminar 2nd Proceedings Keele 1962 Sep 9-20 Geographia polonica,3 Panstwowe wydawnictwo naukowe,Warsaw,1964
Geog 13.5616

POLSKA AKADEMIA NAUK.INSTYTUT MATEMATYCZNY Conference on analytic functions Proceedings Lodz 1966 Sep 1-7 Bibliog 36p 24cm Lodz,1966
P Math 2.3052

POLSKA AKADEMIA NAUK.INSTYTUT MATEMATYCZNY Infinitistic methods Symposium on foundations of mathematics proceedings Warsaw 1959 Sep 2-9 Under the auspices of the International mathematical union 362p 26cm Pergamon press,Oxford,1961
P. Math 2.0211

POLYMER SYMPOSIA, 14 Transitions and relaxations in polymers :symposium Papers Atlantic City,N.J. 1965 Sep 13-14 American chemical society Edited by Raymond F. Boyer Interscience,New York,1966
Met 25.2704

POLYMERS Plastic deformation of polymers Symposium on plastic deformation of polymers Selected papers New York 1969 Sep 10-11 American chemical society Dekker,New York, 1971
Met 25.2576

POLYMERS The Physical properties of polymers symposium Papers London 1958 Apr 15-17 Organised by the Society of chemical industry. Plastics and polymers group Society of chemical industry.Monograph, 5 293p Society of chemical industry,London,1959
Chem E 24.1019

POLYMERS Thermoanalysis of fibers and fiber-forming polymers :symposium Papers Atlantic City,N.J. 1965 Sep 17 American chemical society Edited by Robert F. Schwenker Applied polymer symposia, 2 Interscience,New York,1966
Met 25.2705

POLYMERS Transitions and relaxations in polymers :symposium Papers Atlantic City,N. J. 1965 Sep 13-14 American chemical society Edited by Raymond F. Boyer Polymer symposia, 14 Interscience,New York,1966
Met 25.2704

POLYPEPTIDE HORMONES Structure and function of polypeptide hormones;insulin Symposium on insulin Proceedings Upton,N.Y. 1965 Nov 8-10 Edited by P.G. Katsoyannis and I.L. Schwartz Sponsored by the Brookhaven national laboratory.Medical department American journal of medicine, 40, no. 5 Yorke medical group.Publication New York, 1966
Bioch 33.0557

POLYPEPTIDES WHICH AFFECT SMOOTH MUSCLES AND BLOOD VESSELS :a symposium Proceedings London 1959 Mar 23-24 Biological council. Coordinating committee for symposia on drug action Edited by M. Schachter Biological council.Coordinating committee for symposia on drug action. Series, 4 Pergamon press, London,1960
PGMS 29.0380

POLYSACCARIDES IN BIOLOGY Conference on polysaccharides in biology 1st-5th Transactions Princeton,N.J. 1955-1959 Edited by Georg F. Springer Sponsored by Josiah Macy Jr.foundation 5 vols Josiah Macy,Jr.foundation,New York,1955-59
An 32.1493

POLYTECHNIC INSTITUTE OF BROOKLYN Electronic waveguides :symposium Proceedings New York 1958 Apr 8-10 Microwave research institute Microwave research institute.Symposia series, 8 Polytechnic press,Brooklyn,N.Y.,1958
Eng 41.5649

POLYTECHNIC INSTITUTE OF BROOKLYN Modern network synthesis :a symposium Proceedings New York 1955 Apr 13-15 Sponsored also by the Microwave research institute Microwave research institute.Symposia series Polytechnic institute of Brooklyn,New York, 1956
Eng 41.5117

POLYTECHNIC INSTITUTE OF BROOKLYN Modern network synthesis (audio to microwaves) :a symposium Proceedings New York 1952 Apr 16-18 Sponsored also by the Microwave research institute Microwave research institute.Symposia series Polytechnic institute of Brooklyn,New York,1956
Eng 41.5116

POLYTECHNIC INSTITUTE OF BROOKLYN Optical maser :a symposium Proceedings New York 1963 Apr 16-19 Edited by Jerome Fox Sponsored also by the Microwave research institute Microwave research institute. Symposia series, 13 Polytechnic press, Brooklyn,N.Y.,1963
Eng 41.5417

POLYTECHNIC INSTITUTE OF BROOKLYN Symposium on nonlinear circuit analysis Proceedings New York 1953 Apr 23-24 Edited by Jerome Fox Sponsored by the Microwave research institute Microwave research institute. Network symposia series, 2 Polytechnic institute of Brooklyn,New York,1953
Eng 41.5920

POLYTECHNIC INSTITUTE OF BROOKLYN Symposium on nonlinear circuit analysis Proceedings New York 1956 Apr 25-27 Edited by Jerome Fox Sponsored by the Microwave research institute Microwave research institute. Network symposia series, 6 Polytechnic institute of Brooklyn,New York,1957
Eng 41.5922

POLYTECHNIC INSTITUTE OF BROOKLYN Symposium on turbulence of fluids and plasmas Proceedings New York 1968 Apr 16-18 Edited by Jerome Fox Co-sponsored by the Microwave research institute Microwave research institute.Symposia series, 18 512p 17cm Polytechnic press,Brooklyn,N.Y.,1969
Cav 7.2920

POLYTECHNIC INSTITUTE OF BROOKLYN Turbulence of fluids and plasmas Symposium on turbulence of fluids and plasmas Proceedings New York 1968 Apr 16-18 Edited by Jerome Fox Co-sponsored by the Microwave research institute Microwave research institute. Symposia series, 18 Bibliog.,Illus. xxix, 511p 23cm Polytechnic press,Brooklyn N.Y., 1969
A Math 4.1508

POLYTECHNIC INSTITUTE OF BROOKLYN.DEPARTMENT OF AERONAUTICAL ENGINEERING AND APPLIED MECHANICS Conference on high-speed aeronautics Proceedings Brooklyn,N.Y. 1955 Jan 20-22 Edited by Antonio Ferri and others Polytechnic institute of Brooklyn,Brooklyn,N.Y. 1955
Eng 41.6955

POLYTECHNIC OF NORTH LONDON Minicomputers in instrumentation and control :proceedings of a three-day short course London 1972 Edited by Y. Paker and others Polytechnic of North London,London,1972
Eng 41.8589

PONTIFICIA ACADEMIA SCIENTIARUM Brain and conscious experiemce :study week Vatican 1964 Sep 28-Oct 4 Edited by John Carew Eccles Springer,Berlin,1966
Phys 20.2041

PONTIFICIA ACADEMIA SCIENTIARUM Brain and conscious experience :study week Vatican 1964 Sep 28-Oct 4 Edited by John C. Eccles Springer,Berlin,1966
Pha 16.0309

PONTIFICIA ACADEMIA SCIENTIARUM Stellar populations :a conference Proceedings Vatican observatory 1957 May 20-28 Edited by D.J.K. O'Connell Pontificia academia scientiarum.Scripta varia, 16 Specola vaticana.Ricerche astronomiche, 5 illus 544p North-Holland,Amsterdam,1958
Cav 7.2390

PONTIFICIA ACADEMIA SCIENTIARUM Stellar populations conferences proceedings Vatican 1957 May 20-28 Edited by D.J.K. O'Connell Pontificia academia scientiarum, 16 Specola vaticana.Ricerche astronomiche, 5 North-Holland,Amsterdam,1958
A Math 4.1150

POPULATION Bevolkerungsfragen International congress for studies on population 3rd Berlin 1935 Aug 26-Sep 1 Edited by Hans Harmsen and Franz Lohse illus. maps Lehmann,Munich,1936
Geog 13.1489

POPULATION Congres international de la population International congress for studies on population 4th Paris 1937 July 29-Aug 1 Under the auspices of l' Union internationale pour l'etude scientifique des problemes de la population 8 vols Hermann, Paris,1938
Geog 13.1494

POPULATION Problems of population International union for the scientific investigation of population problems.General assembly :report 2nd Proceedings London 1931 Jun 15-18 Edited by G.H.L.F. Pitt-Rivers Allen and Unwin,London,1932
Gen 34.2068

POPULATION The World population conference Proceedings Geneva 1927 Aug 29-Sep 3 Edited by Margaret Sanger Arnold,London,1927
Bal 39.1748

POPULATION World population conference Proceedings Geneva 1927 Aug 29-Sep 3 Edited by Margaret Sanger port. Arnold, London,1927
Geog 13.1526

POPULATION BIOLOGY AND EVOLUTION :an international symposium Proceedings Syracuse 1967 Jun 7-9 Edited by Richard C. Lewontin Sponsored by Syracuse university, New York state science and technology foundation Syracuse university press,New York,1968
Bal 39.0940

POPULATION BIOLOGY AND EVOLUTION symposium Proceedings Syracuse,N.Y. 1967 Jun 7-9 New York state science and technology foundation Edited by R.C. Lewontin Sponsored by Syracuse university Syracuse university press,Syracuse,N.Y.,1968
Gen 34.1070

POPULATION COUNCIL Conference on immuno-reproduction Proceedings La Jolla,Calif. 1962 Sep 9-11 Population council,New York, c1962 Organized by A.Tyler and K.A. Laurence and edited by the latter
Path 30.2340

POPULATION GENETICS :the nature and causes of genetic variability in populations.A symposium Papers Cold Spring Harbor 1955 Cold Spring Harbor biological laboratory Cold Spring Harbor symposia on quantitative biology, 20 Long Island biological association,Cold Spring Harbor,1955 Dedicated to Vannevar Bush,President of the Carnegie Institute of Washington, 1939-1955
Bioch 33.1276

The POPULATION OF TROPICAL AFRICA 1st conference Ibadan 1966 Jan 3-7 Population council Edited by John C. Caldwell and Chukuka Okonjo Sponsored by the University of Ibadan Longmans,London,1968
Geog 13.6814

POPULATION STUDIES :animal ecology and demography.A symposium Papers Cold Spring Harbor 1957 Cold Spring Harbor biological laboratory Cold Spring Harbor symposia on quantitative biology, 22 Long Island biological association,Cold Spring Harbor,1957
Bioch 33.1278

PORE PRESSURE AND SUCTION IN SOILS :a conference London 1960 Mar 30-31 International society of soil mechanics and foundation engineering.British national committee Butterworths,London,1961
Eng 41.3170

POROUS MATERIALS Structure and properties of porous materials Colston research society : symposium 10th Proceedings Bristol 1958 Mar 24-27 Edited by D.H. Everett and F. S. Stone Butterworths,London,1958
Min 40 77 10.0655

PORPHYRIN BIOSYNTHESIS Ciba foundation symposium on porphyrin biosynthesis and metabolism Papers London 1955 Feb 8-10 Ciba foundation Churchill,London,1955
Chem 18.2637

PORPHYRIN BIOSYNTHESIS AND METABOLISM Ciba foundation symposium on porphyrin biosynthesis and metabolism London 1955 Feb 8-10 Ciba foundation Edited by G.E.W. Wolstenholme and E. Miller Churchill,London,1955
Radioth 35.0088

PORPHYRIN BIOSYNTHESIS AND METABOLISM Ciba foundation symposium on porphyrin biosynthesis and metabolism Proceedings London 1955 Feb 8-10 Ciba foundation Edited by G.E.W. Wolstenholme and Elaine C.P. Millar illus Churchill,London,1955
Bioch 33.0572

PORPHYRINS AND RELATED COMPOUNDS London 1968 Apr Biochemical society Edited by T.W. Goodwin Biochemical society.Symposia, 28 Academic press,London;New York,1968
Bot 42.1755

PORPHYRINS AND RELATED COMPOUNDS London 1968 Apr Biochemical society Edited by T.W. Goodwin Biochemical society.Symposia, 28 Academic press,London;New York,1968
Bioch 33.1390

PORPHYRINS AND RELATED COMPOUNDS :a symposium London 1968 Apr Biochemical society Edited by T.W. Goodwin Biochemical society.Symposia, 28 Academic press,London, 1968
Chem 18.2761

PORTLAND,ORE. 1964 Aging Symposium on the biology of skin 14th Proceedings Edited by William Montagna Held at the University of Oregon medical school Advances in biology of skin, 6 Oregon regional primate research center, 55 Pergamon press, Oxford,1965
An 32.4041

PORTLAND,ORE. 1965 Carcinogenesis Symposium on the biology of skin 15th Proceedings Edited by William Montagna and Richard L. Dobson Held at the University of Oregon medical school Advances in biology of skin, 7 Oregon regional primate research center, 87 Pergamon press,Oxford,1966
An 32.4042

PORTLAND,ORE. 1966 Pigmentary system Symposium on the biology of skin 16th Proceedings Edited by William Montagna and Funan Hu Held at the University of Oregon medical school Advances in biology of skin, 8 Oregon regional primate research center, 183 Pergamon press,Oxford,1967
An 32.4043

PORTSMOUTH,N.H. 1958 Recent advances in atmospheric electricity Conference on atmospheric electricity 2nd Proceedings Edited by L.G. Smith Sponsored by Air force Cambridge research center.Geophysics research directorate.Aerophysics laboratory illus 646p Pergamon,London,1958
Nap 11.0928

PORTSMOUTH,N.H. 1959 Cumulus dynamics Conference on cumulus convection 1st Proceedings Edited by Charles E. Anderson Sponsored by the Air Force Cambridge research center.Geophysics research directorate ix, 211p Pergamon press,Oxford,1960
Nap 11.0012

POSITANO 1970 Human testis Workshop conference :a Serono foundation symposium Proceedings Edited by Eugenia Rosemberg and C.Alvin Paulsen Advances in experimental medicine and biology, 10 Plenum,New York; London,1970
Inv Med 37.0190

POSITION OF VARIABLE STARS IN THE HERTZSPRUNG-RUSSELL DIAGRAM Colloquium on variable stars 3rd Bamberg 1965 Aug 11-14 International astronomical union.Commissions 27 and 42 Remeis-sternwarte,Bamberg.Kleine veroffentlichungen, 4,no 40 port 304p Astronomisches institut der universitat Erlangen-Nurnberg,Bamberg,1965 I.A.U. combined colloquium of commissions 27 and 42 with the theme 'The position of stars'
TA 15.0540

POSITRON ANNIHILATION CONFERENCE :conference Proceedings Detroit,Mich. 1965 Jul 27-29 Wayne state university Edited by A.T. Stewart and L.O. Roellig xvi,438p 24cm Academic press,New York,1967
Cav 7.2855

POST OFFICE.ENGINEERING RESEARCH STATION Soft magnetic materials for telecommunications a symposium London 1952 Apr Edited by C. E. Richards and A.C. Lynch Pergamon press, London,1953
Eng 41.5085

The POST-WAR EXPANSION OF THE U.K.PETROLEUM INDUSTRY :summer meeting Report Buxton 1953 Jun 24-27 Institute of petroleum Edited by George Sell vii,220, xxxvp Institute of petroleum,London,1954
Chem E 24.1705

POSTWAR PROBLEMS OF MIGRATION Milbank memorial fund annual conference Papers New York 1946 Oct 29-30 Milbank memorial fund Milbank memorial fund,New York,1947
Geog 13.1564

POTSDAM,N.Y. 1962 The Conference on electromagnetic scattering :interdisciplinary conference Papers Clarkson college of technology and Air force Cambridge research laboratories Edited by Milton Kerker Co-sponsored by the American chemical society.Division of colloid and surface chemistry International series of monographs on electromagnetic waves, 5 591p Pergamon press,Oxford,1963
TA 15.0256

POVUNGNITUK,QUE. 1966 Conference of the Arctic cooperatives 2nd Minutes 93p 35cm n.p.,1966 Typescript
Sco 14.3711

POWDER METALLURGY General session on powder metallurgy Detroit 1959 Apr 20-22 Metal powder industries federation Metal powder industries federation,New York,1959
Met 25.0764

POWDER METALLURGY New types of metal powders a symposium Proceedings Cleveland,Ohio 1963 Oct 24 American institute of mining, metallurgical and petroleum engineers.Powder metallurgy committee Edited by Henry H. Hausner Metallurgical society conferences, 23 Gordon and Breach,New York;London,1964
Met 25.2509

POWDER METALLURGY New types of metal powders a symposium Proceedings Cleveland,Ohio 1963 Oct 24 American institute of mining, metallurgical and petroleum engineers.Powder metallurgy committee Edited by Henry H. Hausner Metallurgical society conferences, 23 Gordon and Breach,New York;London,1964
Met 25.0766

POWDER METALLURGY Progress in powder metallurgy Annual technical meeting 16th Proceedings Chicago 1960 Apr 25-27 Metal powder industries federation Metal powder industries federation,Chicago,1960
Met 25.0769

POWDER METALLURGY Progress in powder metallurgy Annual technical meeting 17th Proceedings Cleveland 1961 Apr 24-26 Metal powder industries federation Metal powder industries federation,Cleveland,Ohio, 1961
Met 25.0770

POWDER METALLURGY Symposium on powder metallurgy Iron and steel institute Iron and steel institute.Special report, 38 Iron and steel institute,London,1947
Met 25.0753

POWDER METALLURGY Symposium on powder metallurgy London 1954 Dec 1-2 Iron and steel institute Iron and steel institute. Special report, 58 Iron and steel institute, London,1956
Met 25.0761

POWDER METALLURGY Symposium on powder metallurgy London 1954 Dec 1-2 Iron and steel institute Iron and steel institute. Special report, 58 illus. Iron and steel institute,London,1956
Eng 41.3477

POWDER METALLURGY Symposium on powder metallurgy :spring meeting Buffalo,N.Y. 1943 American society for testing and materials American society for testing and materials,Philadelphia,Pa.,1943
Eng 41.3474

POWDER METALLURGY The Physics of powder metallurgy :a symposium New York 1949 Aug 24-26 Sylvania electric products,inc. Metallurgical laboratories Edited by Walter E. Kingston McGraw-Hill,New York,1951
Met 25.0752

POWDER METALLURGY IN NUCLEAR ENGINEERING Conference on powder metallurgy in atomic energy with additional papers Philadelphia 1955 Oct 20 By Henry H. Hausner United States atomic energy commission and American society for metals American society for metals,Cleveland,Ohio,1958
Met 25.0767

POWDER METALLURGY IN THE NUCLEAR AGE Plansee seminar 'De re metallica' 4th Proceedings Reutte,Austria 1961 Jun 20-24 Metallwerk Plansee AG Edited by F. Benesorsky Metallwerk Plansee AG,Reutte,1962 Papers in English,French and German
Met 25.2724

POWDERS IN INDUSTRY :a symposium Papers London 1960 Sep 29-30 Society of chemical industry.Surface activity group Society of chemical industry.Monograph v,447p Society of chemical industry,London,1961
Chem E 24.1445

POWER SYSTEMS COMPUTATION CONFERENCE 2nd Proceedings Stockholm 1966 Jun 22-Jul 1 Pt 1-5 5 vols Stockholm,1966
Eng 41.4985

POWER SYSTEMS CONFERENCE 2nd Proceedings Glasgow 1967 Jan 5-6 Science research council Glasgow,1967
Eng 41.4988

POZA RICA Tamabra limestone of the Poza Rica oilfield International geological congress 20th papers Mexico city 1956 By A. Barnetche and L.V. Illing Mexico city,1956 Bound with two other publications of the congress
Geol 8.3028

PRA-OPERATIVE TUMORBESTRAHLUNG Deutsche rontgenkongress 1970 Vortrage Edited by Otto Hugg Urban and Schwarzenberg,Munich, 1971
Radioth 35.1939

PRACTICAL APPLICATION OF FRACTURE MECHANICS TO PRESSURE-VESSEL TECHNOLOGY :a conference Proceedings London 1971 May 3-5 Institution of mechanical engineers.Applied mechanics group IME,London,1971
Eng 41.8644

PRACTICAL FRACTURE MECHANICS FOR STRUCTURAL STEEL Symposium on fracture toughness concepts for weldable structural steel Proceedings Risley 1969 Apr Edited by M.O. Dobson Published for the United Kingdom atomic energy authority Chapman and Hall,London,1970
Eng 41.3822

PRACTICAL FRACTURE MECHANICS FOR STRUCTURAL STEEL Symposium on fracture toughness concepts for weldable structural steel Proceedings Risley 1969 Apr United Kingdom atomic energy authority Edited by M.O. Dobson U.K. A.E.A. in association with Chapman and Hall, Risley,1969
Met 25.2465

PRAGER ARBEITSTAGUNG UBER DIE STRATIGRAPHIE DES SILURS UND DES DEVONS 1958 :a symposium papers Prague 1958 Aug.30-Sep.6 Ustredni ustav geologicky,Prague Edited by Josef Svoboda Nakladetalstvi ceskoslovenske akademie ved,Prague,1960
Geol 8.2416

PRAGUE 1924 Institut international d'anthropologie:session 2e Nourry, Paris,1926
An 32.2639

PRAGUE 1958 International conference of insect pathology and biological control 1st Transactions 24cm Prague,1958
Philos 1.1598

PRAGUE 1958 Prager arbeitstagung uber die stratigraphie des Silurs und des Devons (1958) a symposium papers Ustredni ustav geologicky,Prague Edited by Josef Svoboda Nakladetalstvi ceskoslovenske akademie ved, Prague,1960
Geol 8.2416

PRAGUE 1959 Ontogeny of insects :a symposium Acta Ceskoslovenska akademie ved. Sekce biologicko - lekarska Edited by Ivan Hrdy Czechoslovak academy of sciences,Prague, 1960 Text in English,French,German and Russian
Bal 39.2433

PRAGUE 1960 Chemical effects of nuclear transformations :a symposium Proceedings Vol 1-2 International atomic energy agency International atomic energy agency.Proceedings series STI-PUB 34 2 vols international atomic energy agency,Vienna,1961
Chem 18.1150

PRAGUE 1960 Membrane transport and metabolism :a symposium Proceedings Ceskoslovenska akademie ved.Institute of biology Edited by A. Kleinzeller and A. Kotyk Academic press,London;New York,1961
Bot 42.1483

PRAGUE 1960 Membrane transport and metabolism :a symposium Proceedings Ceskoslovenska akademie ved.Institute of biology Edited by A. Kleinzeller and A. Kotyk Academic press,London;New York,1961
Bioch 33.0972

PRAGUE 1960 Membrane transport and metabolism :a symposium Proceedings Edited by A. Kleinzeller and A. Kotyk Academic press,New York,1961
Phys 20.1326

PRAGUE 1960 Symposium on fatigue Papers and discussions Czechoslovak scientific-technical society Ceskoslovenska vedecko-technicka spolecnost,1961 Papers in English,French,German and Russian
Met 25.0936

PRAGUE 1961 General topology and its relations to modern analysis and algebra : symposium proceedings Ceskoslovenska akademie ved Edited by J. Novak jointly with International mathematical union 363p 25cm Academic press,New York,1962
P. Math 2.0291

PRAGUE 1961 and BRNO Sep 28-Oct 6 Studies about humus Humus and plant :a symposium Edited by S. Prat and V. Rypacek illus 364p Czechoslovak academy of sciences,Prague,1962
Bot 42.2108

PRAGUE 1962 The Czechoslovak plant virologists conference 5th Proceedings Czechoslovak plant virologists illus 364p Publishing house of the Czechoslovak academy of sciences,Prague,1964
Bot 42.3745

PRAGUE 1963 Aldosterone :a symposium Edited by E.E. Baulieu and P. Robel Organized by the Council for international organizations of medical sciences Blackwells, Oxford,1964
Phys 20.1398

PRAGUE 1963 Congres international de nephrologie 2 Comptes rendus Societe internationale de nephrologie Societe tchecoslovaque de medecine Edited by J. Vostal and G. Richet Excerpta medica. International congress series, 78 Excerpta medica,Amsterdam,1964
Inv Med 37.0220

PRAGUE 1963 Induction of mutations and the mutation process :a symposium Proceedings Edited by Jiri Veleminsky and Tomas Gichner Nakladatelstvi Ceskosloveske akademie ved, Prague,1965
Gen 34.1099

PRAGUE 1963 International symposium on aldosterone Proceedings Council for international organization of medical sciences Edited by Etienne Emile Baulier and P. Robel Blackwell,Oxford,1964
Inv Med 37.0013

PRAGUE 1963 Water stress in plants :a symposium Proceedings Ceskoslovenska akademie ved Edited by Bohdan Slavik 318p Junk,The Hague,1965
Bot 42.1881

PRAGUE 1964 Blood groups of animals European animal blood group conference 9th Proceedings Ceskoslovenska akademie ved and European society for animal blood group research Edited by Josef Matousek Junk,The Hague,1964
Surg 23.0044

PRAGUE 1964 Electron microscopy 1964 European regional conference on electron microscopy 3rd Proceedings Vol A: non-biology International federation of electron microscope societies Edited by M. Titlbach Czech.academy of sciences,Prague,1964
Met 25.2494

PRAGUE 1964 Information processing machines symposium proceedings Ceskoslovenska akademie ved Edited by Vladimir Bubenik Czechoslovak academy of sciences,Prague,1965
Math L 5.1011

PRAGUE 1964 International symposium on data processing machines Papers 1964 Collection of typescript papers in folder
Math L 5.3266

PRAGUE 1964 International union of biological sciences :general assembly 15th Under the auspices of Unesco International union of biological sciences. Series A, 15 International union of biological sciences,Utrecht,1964
BG 38.3109

PRAGUE 1964 Molecular and cellular basis of antibody formation :a symposium Proceedings Ceskoslovenska akademie ved. Institute of microbiology Edited by J. Sterzl Czechoslovak academy of sciences; Academic press,Prague;New York,1965
An 32.3701

PRAGUE 1964 Molecular and cellular basis of antibody formation a symposium Proceedings Edited by J. Sterzl Organized by the Ceskoslovenska akademie ved.Institute of microbiology. Immunological department Czechoslovak academy of sciences;Academic press,Prague;New York,1965
Bioch 33.0554

PRAGUE 1965 Electronics and vacuum physics Czechoslovak conference 3rd Transactions Ceskoslovenska akademie ved Edited by Libor Paty Sponsored by the Universita Karlova Academia,Prague,1967
Eng 41.5453

PRAGUE 1965 Prague conference on information theory,statistical decision functions and random processes 4th Lectures Selected translations in mathematical statistics and probability, 8 bibliog. American mathematical society, Providence,R.I.,1970
Math S 3.1577

PRAGUE 1966 General topology and its relations to modern analysis and algebra Prague topological symposium 2nd Proceedings Ceskoslovenska akademie ved 365p 24cm Academic press;Academia,New York; Prague,1967
P Math 2.3346

PRAGUE 1966 Neurotoxicity of drugs European society for the study of drug toxicity :meeting 8th Proceedings European society for the study of drug toxicity.Proceedings,8 International congress series,118 Excerpta medica, Amsterdam,1967
Pha 16.0332

PRAGUE 1967 Conference on nonlinear oscillations 4th Proceedings Edited by Jan Gonda Academia kiado,Prague,1968
Eng 41.6391

PRAGUE 1967 Highlights of astronomy IAU general assembly 13 Invited discources and discussions,and special meetings International astronomical union Edited by Lubos Perck 548p Reidel,Dordrecht,1968
Obs 6.3528

PRAGUE 1967 Highlights of astronomy International astronomical union 13th general assembly Edited by L. Perek illus x,548p 25cm Reidel,Doldrecht,1968
A Math 4.1899

PRAGUE 1967 Identification in automatic control systems I.F.A.C. symposium on identification in automatic control systems Preprints Pt 1-2 Ceskoslovenska akademie ved.Institute of information theory and automation Sponsored by International federation of automatic control 2 vols Academia,Prague,1967
Eng 41.6019

PRAGUE 1967 Ontogenesis of the brain:the biochemical,functional and structural development of the nervous system International symposium neurontogeneticum 5th Proceedings Universita Karlova and Lekarska spolecnost J.E.Purknye Edited by Lubor Jilek and Stanislav Trojan Universita Karlova,Prague,1968
PGMS 29.0202

PRAGUE 1968 Federation of European biochemical societies meeting 5th enzymes and isoenzymes:structure,properties and function Federation of European biochemical societies Federation of European biochemical societies.Publications, 18 Academic press,London,1970
Bioch 33.1888

PRAGUE 1968 Federation of European biochemical societies meeting 5th Proceedings Vol 3: mitochondria:structure and function Federation of European biochemical societies Edited by L. Ernster and Z. Drahota Federation of European biochemical societies.Publications, 17 Academic press,London;New York,1969
Bioch 33.1403

PRAGUE 1968 Federation of European biochemical societies meeting 5th Proceedings Vol 3: mitochondria:structure and function Federation of European biochemical societies Edited by L. Ernster and Z. Drahota Federation of European biochemical societies.Publications, 17 Academic press,London;New York,1969
Radioth 35.0149

PRAGUE 1968 Federation of European biochemical societies meeting 5th Proceedings gamma globulins;structure and biosynthesis Federation of European biochemical societies Edited by F. Franek and D. Shugar Federation of European biochemical societies.Publications, 15 Academic press,London;New York,1969
Bioch 33.1402

PRAGUE 1968 Structure of organic solids Microsymposium 'Structure of organic solids' 2nd Lectures International union of pure and applied chemistry.Macromolecular division Czechoslovak chemical society In conjunction with Ceskoslovenska akademie ved illus 550p 25cm Butterworths,London,1969 Simultaneously published in 'Pure and applied chemistry',vol 18,no 4,1969
Cav 7.2659

PRAGUE 1969 International conference on sarcoidosis 5th Proceedings Universita Karlova Edited by Ladislav Levinsky and Frantisek Macholda Univerzita Karlova,Prague, 1971
PGMS 29.0666

PRAGUE CONFERENCE ON INFORMATION THEORY, STATISTICAL DECISION FUNCTIONS, RANDOM PROCESSES 3rd transactions Liblice, Czech. 1962 Jun 5-13 Ceskoslovenska akademie ved Edited by Jaroslav Kozesnik Organized by the Institute of information theory and automation 84625 Czechoslovak academy of sciences,Prague,1964 Dedicated to the memory of Antonin Spacek
Math 3.0697

PRAGUE CONFERENCE ON INFORMATION THEORY, STATISTICAL DECISION FUNCTIONS AND RANDOM PROCESSES 4th Lectures Prague 1965 Aug 31-Sep 11 Selected translations in mathematical statistics and probability, 8 bibliog. American mathematical society, Providence,R.I.,1970
Math S 3.1577

PRAGUE TOPOLOGICAL SYMPOSIUM 2nd Proceedings General topology and its relations to modern analysis and algebra Prague 1966 Aug 30-Sep 4 Ceskoslovenska akademie ved 365p 24cm Academic press; Academia,New York;Prague,1967
P Math 2.3346

PREBIOLOGICAL SYSTEMS AND THEIR MOLECULAR MATRICES Origins of prebiological systems and of their molecular matrices A Conference on the origins of prebiological systems and of their molecular matrices Proceedings Wakulla Springs,Fla 1963 Oct 27-30 Institute for space biosciences Florida state university National aeronautics and space administration Edited by Sidney W. Fox Bibliog.,illus. xx,482p Academic press,New York;London,1965
Bot 42.1364

PRECIPITATION FROM IRON-BASE ALLOYS :a symposium Cleveland 1963 Oct 21 American institute of mining,metallurgical and petroleum engineers.Ferrous metallurgy committee Edited by Gilbert R. Speich and John B. Clark Metallurgical society conferences Gordon and Breach,New York; London,1965
Met 25.2525

PRECIPITATION HARDENING Age hardening of metals the symposium on precipitation hardening (age hardening) presented before the twenty-first annual convention Papers and discussions Chicago 1939 Oct 23-27 American society for metals American society for metals,Cleveland,1940
Met 25.1249

PREDICTION IN SCIENCE Information and prediction in science Symposium de l'Academie internationale de philosophie des sciences Proceedings Brussels 1962 Sep 3-8 Academie internationale de philosophie des sciences Edited by S. Dockx and P. Bernays xi,272p Academic press,London;New York,1965
WSM 43.1136

PREGNANCY Preimplantation stages of pregnancy :Ciba foundation symposium London 1965 Apr 13-15 Ciba foundation Edited by G. E.W. Wolstenholme and M. O'Connor London, 1965
Bal 39.3558

PREIMPLANTATION STAGES OF PREGNANCY :Ciba foundation symposium London 1965 Apr 13-15 Ciba foundation Edited by G.E.W. Wolstenholme and M. O'Connor Churchill, London,1965
Phys 20.1407

PREIMPLANTATION STAGES OF PREGNANCY :Ciba foundation symposium London 1965 Apr 13-15 Ciba foundation Edited by G.E.W. Wolstenholme and M. O'Connor London,1965
Bal 39.3558

PREIMPLANTATION STAGES OF PREGNANCY :a Ciba foundation symposium London 1965 Ciba foundation Edited by G.E.W. Wolstenholme and Maeve O'Connor Churchill,London,1965
An 32.2517

PREMATURITY Gestational age,size and maturity :a symposium on the causes and associations of prematurity... Spastics society.Medical education and information unit Edited by Michael Dawkins and W.G. MacGregor Spastics society.Clinics in developmental medicine, 19 Spastics society in association with Heinemann,London,1964 Based on the proceedings of an international study group at Oxford,1964
Psy 28.0278

PREMATURITY Physiology of pre-maturity Conference on physiology of prematurity 1st-4th Transactions Princeton,N.J. 1956-59 Mar Edited by Jonathan T. Lanman Sponsored by the Josiah Macy jr.foundation 4 vols Josiah Macy jr. foundation,New York,1957-60
An 32.3819

PREMATURITY Physiology of prematurity conference 4th Transactions Princeton,N. J. 1959 Mar 25-26 Edited by Jonathan T. Lanham Sponsored by Josiah Macy Jr. foundation New York,1960
PGMS 29.0288

PREMATURITY,CONGENITAL MALFORMATION,AND BIRTH INJURY :a conference Proceedings New York 1952 Jun 5-6 Sponsored by the Association for the aid of crippled children Association for the aid of crippled children, New York,1953
An 32.2030

PREPARATION AND CHARACTERISTICS OF SOLID LUMINESCENT MATERIALS :a conference Ithaca,N.Y. 1946 Oct 24-26 American physical society.Division of electron optics Edited by Gorton R. Fonda and Frederick Seitz Published under the auspices of the National research council graphs xv,459p 22cm Wiley,New York,1948
Cav 7.3061

PREPARATION AND CHARACTERISTICS OF SOLID LUMINESCENT MATERIALS :a symposium Ithaca,N.Y. 1946 Oct 24-26 Sponsored by the American physical society.Division of electron optics Wiley;Chapman and Hall,New York;London,1948
Eng 41.5491

PRESENT PROBLEMS CONCERNING THE STRUCTURE AND EVOLUTION OF THE GALACTIC SYSTEM NUFFIC international summer course in science Lecture notes Breukelen 1960 Jul. 28- Aug. 16 Netherlands universities foundation for international cooperation Edited by J.H. Oort and H.G. Quik Supported by North atlantic treaty organization The Hague,1960
Obs 6.2067

PRESENT STATUS OF PSYCHOTROPIC DRUGS, PHARMACOLOGICAL AND CLINICAL ASPECTS International congress of the Collegium internationale neuro-psychopharmacologicum 6th Proceedings Tarragona 1968 Apr 24-27 Collegium internationale neuro-psychopharmacologicum Edited by A. Cerletti and F.J. Bove Excerpta medica foundation. International congress series, 180 Excerpta medica foundation,Amsterdam,1969
Psy 31.2902

PRESERVATION AND TRANSPLANTATION OF NORMAL TISSUES :Ciba foundation symposium London 1953 Mar 16-18 Ciba foundation Edited by G.E.W. Wolstenholme and Margaret P. Cameron Churchill,London,1954
An 32.3211

PRESERVATION OF FRUIT AND VEGETABLES BY RADIATION Panel on preservation of fruit and vegetables by radiation :especially in the tropics Proceedings Vienna 1966 Aug 1-5 Organised by the Joint FAO-IAEA division of atomic energy in food and agriculture International atomic energy agency.Panel proceedings series International atomic energy agency,Vienna,1968
Radioth 35.1758

PRESSURE VESSEL RESEARCH TOWARDS BETTER DESIGN a symposium Proceedings London 1961 Jan Institution of mechanical engineers British welding research association Institution of mechanical engineers,London, 1962
Eng 41.2875

PRESSURE-VESSEL TECHNOLOGY Practical application of fracture mechanics to pressure-vessel technology :a conference Proceedings London 1971 May 3-5 Institution of mechanical engineers.Applied mechanics group IME,London,1971
Eng 41.8644

PRESSURE VESSELS Periodic inspection of pressure vessels :a conference Proceedings London 1972 May 9-11 Institution of mechanical engineers.Applied mechanics group IME,London,1972
Eng 41.8648

PRESTRESS Federation internationale de la precontrainte 1st-5th congres Comptes-rendus 1953-66 5 vols Federation internationale de la precontrainte,Paris,1955-66
Eng 41.2928

PRESTRESSED CONCRETE Symposium on the application of prestressed concrete to machinery structures Slough 1964 Jan 14-15 Prestressed concrete development group Prestressed concrete development group,London, 1965
Eng 41.2948

PRESTRESSED CONCRETE United States conference on prestressed concrete 1st Proceedings Cambridge,Mass. 1951 Aug 14-16 Massachusetts institute of technology Cambridge,Mass.,1951
Eng 41.3012

PRESTRESSED CONCRETE World conference on prestressed concrete :conference Proceedings San Francisco,Calif. 1957 Jul-Aug University of California San Francisco,Calif. 1957
Eng 41.3011

PRESTRESSED CONCRETE :a conference Proceedings London 1949 Feb Institution of civil engineers Institution of civil engineers,London,1949
Eng 41.3019

PRESTRESSED CONCRETE AND COMPOSITE BEAMS :a symposium Proceedings Kyoto 1954 Sep 4 Science council of Japan Science council of Japan,Kyoto,1955
Eng 41.3013

PRESTRESSED CONCRETE DEVELOPMENT GROUP Symposium on the application of prestressed concrete to machinery structures Slough 1964 Jan 14-15 Prestressed concrete development group,London,1965
Eng 41.2948

PRESTRESSED CONCRETE PRESSURE VESSELS Conference on prestressed concrete pressure vessels London 1967 Mar 13-17 Institution of civil engineers Institution of civil engineers,London,1968
Eng 41.2891

PRESTRESSED CONCRETE STATICALLY INDETERMINATE STRUCTURES :a symposium London 1951 Sep 24-25 Cement and concrete association Cement and concrete association, London,1953
Eng 41.3006

PRESTRESSED STRUCTURES :a symposium Proceedings Tokyo 1959 Sep 14 Japan society of civil engineers Architectural institute of Japan Japan society for the promotion of science,Tokyo,1960
Eng 41.3016

PRETORIA 1969 Symposium on the Bushveld igneous complex and other layered intrusions Proceedings Geological society of South Africa Edited by D.J.L. Visser and G.von Gruenewaldt Geological society of South Africa.Special paper, 1 Geological society of South Africa,Johannesburg,1970
Min 10.1476

The PREVENTION OF ACCIDENTS IN CHILDHOOD : report on a seminar Spa,Belgium 1958 Jul 16-25 World health organization.Regional office for Europe World health organization, Copenhagen,1960
PGMS 29.0280

PREVENTION OF HOSPITAL INFECTION :the personal factor;report of a conference London 1963 Jun 19 Royal society of health Edited by George Godber and others Royal society of health,London,1963
PGMS 29.0584

PREVENTIVE AND SOCIAL PSYCHIATRY :a symposium Proceedings Washington 1957 Apr 15-17 Army institute of research 1957
Med 36.0011

PRIMATES Experimental medicine and surgery in primates :a conference Papers New York 1967 Sep 27-30 New York academy of sciences Edited by Edward I. Goldsmith and J. Moor-Jankowski New York academy of sciences. Annals, 162,art 1 New York academy of sciences,New York,1969
Path 30.2445

PRIMATES :studies in adaptation and variability Wenner-Gren foundation for anthropological research Edited by Phyllis C. Jay Bibliog Holt,Rinehart and Winston,New York,1968 Dedicated to K.R.L.Hall,and including his bibliography.Work based on the results of a symposium on primate social behaviour sponsored by the Wenner-Gren foundation for anthropological research and held at Burg Wartenstein,Austria
Psy 31.2925

The PRIMATES :a symposium Proceedings London 1962 Apr 12-14 Edited by John Napier and N.A. Barnicot Zoological society of London.Symposia, 10 Zoological society, London,1963
An 32.0580

PRIME MERIDIAN Adoption d'un premier meridien unique et d'une heure universelle : conference Proces-verbaux Washington,D.C. 1884 Oct 1-Nov 1 United States.Department of state 216p Washington,D.C.,1884
Obs 6.3168

PRIME MERIDIAN International meridian conference International conference for the purpose of fixing a prime meridian and a universal day 1st Proceedings Washington,D.C. 1884 Oct 1-Nov 1 United States.Department of state 212p Gibson, Washington,D.C.,1884 With five sheets of notes by J.C.Adams in a separate envelope
Obs 6.3169

PRIMENENIE RADIOAKTIVNYKH IZOTOPOV V PROMYSHLENNOSTI MEDITSINE I SELSKOM KHOZYAISTVE illus. Izdatelstvo akademii nauk S.S.S.R.,Moscow,1956 Papers presented at the International conference on the peaceful uses of atomic energy, Geneva, 1955
Bot 42.0386

PRIMITIVE MOTILE SYSTEMS IN CELL BIOLOGY A Symposium on the mechanism of cytoplasmic streaming,cell movement,and the saltatory motion of subcellular particles Princeton,N. J. 1963 Apr 2-5 Edited by Robert D. Allen and Noburo Kamiya Held at Princeton university Academic press,New York;London, 1964
Bal 39.0297

PRIMITIVE MOTILE SYSTEMS IN CELL BIOLOGY Mechanism of cytoplasmic streaming,cell movement,and the saltatory motion of subcellular particles :a symposium Proceedings Princeton,N.J. 1963 Apr 2-5 Edited by Robert D. Allen and Noburo Kamiya illus. Academic press,New York;London,1964
Bot 42.3760

PRIMOSTEN 1968 International epidemiological association 5th Proceedings Edited by Milutin Srdic and others Savremena administracija,Belgrade, 1970
PGMS 29.0635

PRINCETON 1909 The Princeton colloquium Part 1: fundamental existence theorems By Gilbert Ames Bliss American mathematical society.Colloquium publications, 3 ii,107p 23cm American mathematical society,New York, 1913 Reprinted in 1934
P. Math 2.0822

PRINCETON 1961 Hereditary,developmental and immunologic aspects of kidney disease Annual conference on the kidney 13th Proceedings National kidney disease foundation Edited by Jack Metcoff Northwestern university press,Evanston,1962
Inv Med 37.0157

PRINCETON 1961 Renal metabolism and epidemiology of some renal diseases Annual conference on the kidney Proceedings National kidney disease foundation Edited by Jack Metcoff National kidney disease foundation,New York,1964
Inv Med 37.0158

PRINCETON, N.J. 1963 Differential and combinatorial topology :a symposium in honor of Marston Morse Edited by Stewart S. Cairns Supported by United States.Air force.Office of scientific research.Mathematics division vi, 265p 24cm Princeton university press, Princeton, N.J.,1965
P. Math 2.0295

PRINCETON,N.J. 1946 Problems of mathematics conference 2 Princeton university Edited by John W. Tukey Princeton university.Bicentennial conferences, Series 2,2 32p 22cm Princeton university press,Princeton,N.J.,1947
P. Math 2.2557

PRINCETON,N.J. 1947 Near eastern culture and society :a symposium on the meeting of east and west Edited by T.Cuyler Young Princeton oriental studies,15 illus. Princeton university press,Princeton,N.J.,1951
Geog 13.1673

PRINCETON,N.J. 1949 Seminar on geometry of numbers Lectures,mimeograph Institute for advanced study 90p 26cm Institute for advanced study,Princeton,N.J.,1949
P Math 2.2778

PRINCETON,N.J. 1951 Contributions to the theory of Riemann surfaces conference typescript Institute for advanced study Edited by Lars V. Ahlfors and others In joint sponsorship with Princeton university Annals of mathematics studies, 30 264p 26cm Princeton university press,Princeton,N. J.,1953 Centennial celebration of Riemann's dissertation
P. Math 2.1434

PRINCETON,N.J. 1953 Adrenal cortex :a conference 5th Transactions Edited by Elaine P. Ralli Sponsored by the Josiah Macy Jr.foundation Josiah Macy Jr.foundation,New York,1954
Bioch 33.0468

PRINCETON,N.J. 1954 Algebraic geometry and topology :a symposium in honor of S.Lefschetz Edited by R.H. Fox and others Princeton mathematical series, 12 viii,399p 24cm Princeton university press,Princeton,N.J.,1957
P. Math 2.0574

PRINCETON,N.J. 1954 Gestation Conference on gestation 1st Transactions Edited by Louis B. Flexner Sponsored by the Josiah Macy jr.foundation Josiah Macy jr. foundation,New York,1955
An 32.3815

PRINCETON,N.J. 1955 Group processes : conference 2nd Transactions Edited by Bertram Schaffner Sponsored by the Josiah Macy,jr.Foundation 255p 24cm New York, 1956
Sco 14.0705

PRINCETON,N.J. 1955 Man's role in changing the face of the earth :international symposium Wenner-Gren foundation for anthropological research Edited by William L. Thomas and others Sponsored jointly by the National science foundation University of Chicago press,Chicago,Ill.;London,1956
Geog 13.1427

PRINCETON,N.J. 1955-1959 Polysaccarides in biology Conference on polysaccharides in biology 1st-5th Transactions Edited by Georg F. Springer Sponsored by Josiah Macy Jr.foundation 5 vols Josiah Macy,Jr. foundation,New York,1955-59
An 32.1493

PRINCETON,N.J. 1955-58 Gestation Conference on gestation 2nd-5th Transactions Edited by Claude A. Villee Sponsored by the Josiah Macy jr.foundation 3 vols Josiah Macy jr. foundation,New York, 1956-58
An 32.3816

PRINCETON,N.J. 1956 Behaviour and evolution :conference 2nd Proceedings American psychological association Edited by Anne Roe and George Gaylord Simpson Held in collaboration with the Society for the study of evolution 24cm New Haven,1958
Philos 1.1827

PRINCETON,N.J. 1956-59 Physiology of pre- maturity Conference on physiology of prematurity 1st-4th Transactions Edited by Jonathan T. Lanman Sponsored by the Josiah Macy jr.foundation 4 vols Josiah Macy jr. foundation,New York,1957-60
An 32.3819

PRINCETON,N.J. 1957 Analytic functions conference papers By Rolf Nevanlinna and others Institute for advanced study With the United States.Air force.Office of scientific research Princeton mathematical series, 24 vii,197p 24cm Princeton university press,Princeton,N.J.,1960
P. Math 2.1492

PRINCETON,N.J. 1957 Oxygen supply to the human foetus :a symposium Council for international organizations of medical sciences and Josiah Macy Jr. foundation Edited by James Walker and Alec C. Turnbull Blackwell,Oxford,1959
Phys 20.1003

PRINCETON,N.J. 1957 Oxygen supply to the human foetus :a symposium Edited by James Walker and Alec C. Turnbull Organised by the Council for international organizations of medical sciences Blackwell,Oxford,1959
An 32.2514

PRINCETON,N.J. 1957 Seminars on analytic functions conference typescript Vols. 1-2 Institute for advanced study Edited by Marston Morse and others Under contract with the United States.Air force.Office of scientific research Bibliog. 25cm 2 vols Princeton,N.J.,1957
P. Math 2.1583

PRINCETON,N.J. 1957-58 Seminar on complex multiplication Institute for advanced study Lecture notes in mathematics, 21 28cm Springer-verlag,Berlin,1966
P Math 2.2857

PRINCETON,N.J. 1958 Connective tissue, thrombosis and atherosclerosis :a conference Proceedings Edited by Irvine H. Page Academic press,New York,1959
An 32.3232

PRINCETON,N.J. 1959 Genetics:genetic information and the control of protein structure and function Conference on genetics 1st Transactions Edited by H. Eldon Sutton Sponsored by the Josiah Macy jr. foundation Macy Jr.foundation,New York,1960
An 32.0360

PRINCETON,N.J. 1959 Genetics conference : genetic information and the control of protein structure and function 1st Transactions Edited by H.Eldon Sutton Sponsored by the Josiah Macy jr.foundation Josiah Macy jr. foundation,New York,1960
Gen 34.0650

PRINCETON,N.J. 1959 Physiology of prematurity conference 4th Transactions Edited by Jonathan T. Lanham Sponsored by Josiah Macy Jr. foundation New York,1960
PGMS 29.0288

PRINCETON,N.J. 1960 Measurement in personality and cognition :a conference Educational testing service Edited by Samuel Messick and John Ross Wiley,New York;London, 1962
Psy 31.2152

PRINCETON,N.J. 1960 Mutations Conference on genetics 2nd Edited by William J. Schull Sponsored by the Josiah Macy jr. foundation University of Michigan press,Ann Arbor,Mich.,1962
An 32.0361

PRINCETON,N.J. 1960 The Central nervous system and behavior :a conference 3rd Transactions Josiah Macy jr.foundation Edited by Mary A.B. Brazier With the cooperation of the National science foundation Josiah Macy,jr.foundation,New York,1960
Psy 31.0063

PRINCETON,N.J. 1961 Distribution and motion of interstellar matter in galaxies The Conference on problems of the distribution and motion of interstellar matter in galaxies Proceedings Institute for advanced study Edited by L. Woltjer 330p Benjamin,New York,1962
TA 15.0213

PRINCETON,N.J. 1961 Distribution and motion of interstellar matter in galaxies :a conference Proceedings Institute for advanced study Edited by L. Woltjer 330p W.A.Benjamin,New York,1962
Obs 6.3253

PRINCETON,N.J. 1961 Genetic selection in man Conference on genetics 3rd Edited by William J. Schull Sponsored by the Josiah Macy jr.foundation University of Michigan press,Ann Arbor,Mich.,1963
An 32.0362

PRINCETON,N.J. 1961 Genetic selection in man Macy conference on genetics 3rd Proceedings Josiah Macy jr. foundation Edited by William J. Schull University of Michigan press,Michigan,1963
Gen 34.1867

PRINCETON,N.J. 1961 Hereditary, developmental and immunologic aspects of kidney disease Annual conference on the kidney 13th Proceedings National kidney disease foundation Edited by Jack Metcoff Northwestern university press,Evanston,Ill., 1962
An 32.2075

PRINCETON,N.J. 1962 Somatic cell genetics Macy conference on genetics 4th Josiah Macy jr.foundation Edited by Robert S. Krooth illus. University of Michigan press, Ann Arbor,Mich.,1964
Gen 34.0891

PRINCETON,N.J. 1962 Symposium on thermophysical properties progress in international research on thermodynamic and transport properties 2nd Papers American society of mechanical engineers.Standing committee on thermophysical properties American society of mechanical engineers; Academic press,New York,1962
Chem 18.0326

PRINCETON,N.J. 1963 Primitive motile systems in cell biology A Symposium on the mechanism of cytoplasmic streaming,cell movement,and the saltatory motion of subcellular particles Edited by Robert D. Allen and Noburo Kamiya Held at Princeton university Academic press,New York;London, 1964
Bal 39.0297

PRINCETON,N.J. 1963 Primitive motile systems in cell biology Mechanism of cytoplasmic streaming,cell movement,and the saltatory motion of subcellular particles :a symposium Proceedings Edited by Robert D. Allen and Noburo Kamiya illus. Academic press,New York;London,1964
Bot 42.3760

PRINCETON,N.J. 1964 Organization of recall International interdisciplinary conference on learning,remembering and forgetting 2nd Proceedings New York academy of sciences and National institute of child health and human development Edited by Daniel P. Kimble With the support of United States.Office of naval research New York academy of sciences. Interdisciplinary communication program,New York,1967
Psy 31.1012

PRINCETON,N.J. 1965 Brain mechanisms underlying speech and language Edited by F.L. Darley and Millikan Grune and Stratton,New York,1967
Psy 31.0460

PRINCETON,N.J. 1965 Information and control processes in living systems :first interdisciplinary conference Proceedings New York academy of sciences Edited by Diane M. Ramsey New York academy of sciences interdisciplinary communications program,New York,1967
Psy 31.0142

PRINCETON,N.J. 1965 Readiness to remember Conference on learning,remembering and forgetting 3rd Proceedings Vol 1-2 New York academy of sciences Edited by Daniel P. Kimble Gordon and Breach,New York, 1969
Psy 31.3281

PRINCETON,N.J. 1965 Princeton,N.J. 1966 Feb Conferences on cellular dynamics : interdisciplinary conferences 3rd and 4th Proceedings Edited by Lee D. Peachey New York academy of sciences,New York,1967
Phys 20.1513

PRINCETON,N.J. 1966 Fetal homeostasis Conference on fetal homeostasis 2nd Proceedings Edited by Ralph M. Wynn Sponsored by the New York academy of sciences New York academy of sciences,New York,1967
An 32.2520

PRINCETON,N.J. 1966 Fetal homeostasis :a conference 2nd Proceedings Vol 2 New York academy of sciences Edited by Ralph M. Wynn Sponsored by the William H.Donner foundation New York academy of sciences,New York,1967
Inv Med 37.0263

PRINCETON,N.J. 1967 Princeton symposium on mathematical programming Princeton university Edited by Harold W. Kuhn vi, 620p 24cm Princeton university press, Princeton,N.J.,1970
Math S 3.1687

PRINCETON,N.J. 1968-69 Seminar on algebraic groups and related finite groups By A. Borel and others Partially sponsored by Office of scientific research,Office of aerospace research,United States air force Lecture notes in mathematics, 131 Bibliog. not con.no.p 25cm Springer-Verlag,Berlin, 1970
P Math 2.3582

PRINCETON,N.J. 1969 Selected mathematical papers Symposium on problems in analysis Papers Princeton university Princeton mathematical series, 31 Bibliog,port x, 351p 24cm Princeton university press, Princeton,N.J.,1970 Symposium held in honour of S.Bochner
P Math 2.3992

The PRINCETON COLLOQUIUM Princeton 1909 Part 1: fundamental existence theorems By Gilbert Ames Bliss American mathematical society.Colloquium publications, 3 ii,107p 23cm American mathematical society,New York, 1913 Reprinted in 1934
P. Math 2.0822

PRINCETON SYMPOSIUM ON MATHEMATICAL PROGRAMMING Princeton,N.J. 1967 Aug 14-18 Princeton university Edited by Harold W. Kuhn vi, 620p 24cm Princeton university press, Princeton,N.J.,1970
Math S 3.1687

PRINCETON UNIVERSITY Contributions to the theory of Riemann surfaces conference typescript Princeton,N.J. 1951 Dec 14-15 Institute for advanced study Edited by Lars V. Ahlfors and others Annals of mathematics studies, 30 264p 26cm Princeton university press,Princeton,N.J.,1953 Centennial celebration of Riemann's dissertation
P. Math 2.1434

PRINCETON UNIVERSITY Primitive motile systems in cell biology A Symposium on the mechanism of cytoplasmic streaming,cell movement,and the saltatory motion of subcellular particles Princeton,N.J. 1963 Apr 2-5 Edited by Robert D. Allen and Noburo Kamiya Academic press,New York;London,1964
Bal 39.0297

PRINCETON UNIVERSITY Princeton symposium on mathematical programming Princeton,N.J. 1967 Aug 14-18 Edited by Harold W. Kuhn vi, 620p 24cm Princeton university press, Princeton,N.J.,1970
Math S 3.1687

PRINCETON UNIVERSITY Problems of mathematics conference 2 Princeton,N.J. 1946 Dec. 17-19 Edited by John W. Tukey Princeton university.Bicentennial conferences,Series 2,2 32p 22cm Princeton university press, Princeton,N.J.,1947
P. Math 2.2557

PRINCETON UNIVERSITY Selected mathematical papers Symposium on problems in analysis Papers Princeton,N.J. 1969 Apr 1-3 Princeton mathematical series, 31 Bibliog, port x,351p 24cm Princeton university press,Princeton,N.J.,1970 Symposium held in honour of S.Bochner
P Math 2.3992

PRINCIPLES AND METHODS OF COLONIAL ADMINISTRATION Bristol 1950 Apr Colston research society Colston papers Butterworths,London, 1950
Geog 13.3107

PRINCIPLES OF BIOMOLECULAR ORGANISATION :a CIBA foundation symposium London 1965 Jan 9-11 Ciba foundation Edited by G.E.W. Wolstenholme and Maeve O'Connor Churchill, London,1966
An 32.1523

PRINCIPLES OF BIOMOLECULAR ORGANIZATION : Ciba foundation symposium London 1965 Jul 9-11 Ciba foundation Edited by G.E.W. Wolstenholme and Maeve O'Connor Churchill, London,1966
Phys 20.1505

PRINCIPLES OF BIOMOLECULAR ORGANIZATION :a Ciba foundation symposium Proceedings London 1965 Jun 9-11 Ciba foundation Edited by G.E.W. Wolstenholme and Maeve O'Connor Churchill,London,1966
Bioch 33.0311

PRINCIPLES OF BIOMOLECULAR ORGANIZATION :a Ciba foundation symposium Proceedings London 1965 Jun 9-11 Ciba foundation Edited by G.E.W. Wolstenholme and Maeve O'Connor Churchill,London,1966
Bot 42.1503

PRINTED CIRCUITS Symposium on printed circuits proceedings Philadelphia,Pa. 1955 Jan 20-21 Radio-Electronics-Television manufacturers association Jointly sponsored by Institute of radio engineers illus. 122p 28cm Engineering publishers,New York, 1955
Math L 5.0736

PRIRODA I KHOZYAISTVO SEVERA :materialy pervoi nauchnoi konferentsii Kolskogo otdela geograficheskoe obshchestva S.S.S.R. 1967 Dec Vyp 1 Geograficheskoe obshchestvo S.S.S.R.Severnyi filial Edited by I.L. Freydin and others maps 314p 29cm Akademiya nauk SSSR,Apatity,1969
Sco 14.8284

Les PROBABILITES SUR LES STRUCTURES ALGEBRIQUES Clermont-Ferrand 1969 Jun 30-Jul 5 Centre national de la recherche scientifique Centre national de la recherche scientifique. Colloques internationaux, 186 361p 25cm CNRS,Paris,1970
P Math 2.4558

Les PROBABILITES SUR LES STRUCTURES ALGEBRIQUES colloque international Clermont-Ferrand 1969 Jun 30-Jul 5 Centre national de la recherche scientifique Centre national de la recherche scientifique.Colloques internationaux, 186 361p 25cm Editions du CNRS,Paris,1970 Oeganized by A. Badrikian and P.L.Hennequin
Math S 3.1780

PROBABILITY Mathematical statistics and probability Berkeley symposium on mathematical statistics and probability 5th Proceedings Berkeley,Calif. 1965 Jun-Jul Berkeley,Calif. 1966 Dec-Jan Edited by Lucien M.le Cam and Jerzy Neyman bibliog., illus. 26cm 6 vols University of California press,Berkeley,Los Angeles,Calif., 1967
Math S 3.1455

PROBABILITY Mathematical statistics and probability Berkeley symposium on mathematical statistics and probability 3r proceedings Berkeley,Calif. 1954 Dec.26-31 Berkeley,Calif. 1955 Jul-Aug Vols. 1-5 University of California.Statistical laboratory Edited by Jerzy Neyman 26cm 5 vols University of California press,Berkeley; Los Angeles,Calif.,1956
Math 3.0687

PROBABILITY Mathematical statistics and probability Berkeley symposium on mathematical statistics and probability 1st proceedings Berkeley,Calif. 1945 Aug.13-18 Berkeley,Calif. 1946 Jan.27-29 University of California.Statistical laboratory Edited by Jerzy Neyman viii, 501p 26cm University of California press, Berkeley;Los Angeles,Calif.,1949
Math 3.0685

PROBABILITY Symposium on engineering applications of random function theory and probability 1st Proceedings National science foundation Purdue university Edited by J.L. Bogdanoff and F. Kozin Wiley, New York,1963
Eng 41.5949

PROBABILITY Symposium on engineering applications of random function theory and probability 1st proceedings Purdue university Edited by John L. Bogdanoff and Frank Kozin Supported by National science foundation x,421p 24cm John Wiley and sons,New York,1963
Math 3.0879

PROBABILITY Symposium on mathematical statistics and probability Proceedings Berkeley,Calif. 1945 Aug.13-18 and 1946 Jan.27-29 University of California. Department of mathematics Edited by Jerzy Neyman 501p University of California press, Berkeley;Los Angeles,1949
Nap 11.0048

PROBABILITY AND INFORMATION THEORY : international symposium Proceedings Hamilton,Ont. 1968 Apr 4-5 Edited by M. Behara Lecture notes in mathematics, 89 249p 28cm Springer,Berlin,1969
Math S 3.1439

PROBABILITY AND INFORMATION THEORY : international symposium Proceedings Hamilton,Ont. 1968 Apr 4-5 Edited by M. Behara and others Lecture notes in mathematics, 89 iv,555p 28cm Springer-Verlag,Berlin,1969
P Math 2.3519

PROBABILITY METHODS IN ANALYSIS symposium lectures Loutraki,Greece 1966 May 22-Jun.4 Edited by D.A. Kappos Supported by North Atlantic treaty organization.Scientific affairs division Lecture notes in mathematics, 31 329p 28cm Springer-Verlag,Berlin,1967
P. Math 2.2425

PROBABILITY METHODS IN ANLYSIS symposium lectures Loutraki,Greece 1966 May 22-Jun 4 North Atlantic treaty organization. Scientific affairs division Edited by D.A. Kappos Lecture notes in mathematics, 31 329p 28cm Springer-verlag,Berlin,1967
Math 3.0711

PROBABILITY THEORY Combinatorial methods in probability theory colloquium lectures, mimeograph Aarhus 1962 Aug.1-10 Aarhus universitet.Matematisk institut Supported by the North Atlantic treaty organization. Scientific affairs division 126p 29cm Aarhus universitet,Aarhus,1962
Math 3.0570

PROBAILITY Lectures in applied mathematics : summer seminar proceedings Boulder,Colo. 1957 Jun.23-Jul.19 Vol. 1: probability and related topics in physical sciences By Mark Kac American mathematical society Supported by United States.Armed services xiii,266p 24cm Interscience publishers, London;New York,1959 With special lectures by G.E.Uhlenbeck,A.R.Hibbs,and Balth. van der Pol
Math 3.0844

PROBES OF STRUCTURE AND FUNCTION OF MACROMOLECULES AND MEMBRANES Colloquium of the Johnson research foundation 5th Proceedings Philadelphia,Pa. 1969 Apr 19-21 Vol 1-2: probes and membrane function; probes of enzymes and hemoproteins Johnson research foundation Edited by Britton Chance and others Academic press,New York,1971
Bioch 33.2190

The PROBLEM OF LAND CONNECTIONS ACROSS THE SOUTH ATLANTIC ,with special reference to the Mesozoic :a symposium Proceedings New York 1949 Dec 28-29 Edited by Ernst Mayr Held at the 4th annual meeting of the Society for the study of evolution American museum of natural history.Bulletin, 99,art.3 New York,1952
An 32.2158

PROBLEM OF LAND CONNECTIONS ACROSS THE SOUTH ATLANTIC WITH SPECIAL REFERENCE TO THE MESOZOIC Symposium on the role of the South Atlantic basin in biogeography and evolution Proceedings New York 1949 Dec 28-29 By Maurice Ewing and others Edited by Ernst Mayr American museum of natural history.Bulletin, 99,3 American museum of natural history,New York,1952
Bot 42.2341

PROBLEM SOLVING:RESEARCH,METHOD AND THEORY Annual symposium on cognition 1st Papers Pittsburgh 1965 Apr 15-16 Carnegie institute of technology Edited by Benjamin Kleinmutz Wiley,New York,1966
Psy 31.1911

PROBLEM SOLVING:RESEARCH,METHOD AND THEORY Symposium on cognition 1st Pittsburgh, Pa. 1966 Edited by Benjamin Kleinmuntz Sponsored by the Carnegie institute of technology Wiley,New York;London,1966
Eng 41.1291

PROBLEME DER FETALEN ENDOKRINOLOGIE symposium Bonn 1955 Mar 4-5 Edited by H. Nowakowski Deutschen gesellschaft fur endokrinologie.Symposia, 3 Springer,Berlin, 1956
An 32.3750

PROBLEME DES PLASMAS IN PHYSIK UND ASTRONOMIE bericht :eine tagung Plasmas in physik und astronomie Leipzig 1956 Oct.8-11 Physikalische gesellschaft in der deutschen demokratischen republik 223p Akademie-verlag,Berlin,1958
Obs 6.1366

PROBLEMES ACTUELS DE PALEONTOLOGIE EVOLUTION DES VERTEBRES Centre national de la recherche scientifique Centre national de la recherche scientifique.Colloques internationaux,104 C. N.R.S.,Paris,1962
Geol 8.0315

PROBLEMES ACTUELS DE PALEONTOLOGIE (EVOLUTION DES VERTEBRES) Paris 1961 May 29-Jun 3 Centre national de la recherche scientifique Centra national de la recherche scientifique. Colloques internationaux.Actes, 104 27cm C. N.R.S.,Paris,1962 Organized by J.P. Lehmann
Bal 44.4639

PROBLEMES ACTUELS DE PALEONTOLOGIE (EVOLUTION DES VERTEBRES) Paris 1966 Jun 6-11 Centre national de la recherche scientifique Centre national de la recherche scientifique. Colloques internationaux.Actes, 163 pls 27cm C.N.R.S.,Paris,1967 Organized by J. P.Lehman.Papers in English,French and German
Bal 44.4640

Les PROBLEMES AGRAIRES DES AMERIQUES LATINES Paris 1965 Oct 11-16 C.N.R.S.colloques internationaux C.N.R.S.,Paris,1965
Geog 13.6822

PROBLEMES DU DEVELOPPEMENT ECONOMIQUE DANS LES PAYS MEDITERRANEENS colloque international Actes Naples 1962 Oct 28-Nov 2 Ecole pratique des hautes etudes,Paris.Centre de sociologie europeenne Edited by Jean Cuisenier Maison des sciences de lhomme. Recherches mediterraneennes,5 Mouton,The Hague,1963
Geog 13.4116

PROBLEMI ATTUALI DI SCIENZA E DI CULTURA Congresso internazionale di patologia 7 Atti Milan 1968 Sep 5-11 International congress of pathology Edited by Alfonso Giordano Accademia nazionale dei Lincei,Rome, 1970
PGMS 29.0618

PROBLEMI DI SVILUPPO :un simposio Milan 1952 Sep Unione zoologie italiana Casa editrice ambrosiana,Milan,1954
An 32.2477

PROBLEMS IN ECONOMIC DEVELOPMENT Conference on problems in economic development Proceedings International economic association Edited by E.A.G. Robinson Macmillan,London,1965
Geog 13.2413

PROBLEMS IN EUGENICS International eugenics conference 1st Papers London 1912 Jul 24-30 Eugenics education society 2 vols Eugenics education society,London,1913 Later known as the 1st international congress of eugenics
Gen 34.1925

PROBLEMS IN EUGENICS International eugenics congress 1st Papers London 1912 Jul 24-20 Eugenics education society Eugenics education society,London,1912
Bal 44.6430

PROBLEMS IN EUGENICS International eugenics congress 1st Papers London 1912 Jul 24-30 Eugenics education society Held at the University of London Eugenics education society,London,1912
Bot 42.6705

PROBLEMS IN PALAEOCLIMATOLOGY Palaeoclimates conference Proceedings Newcastle-upon-Tyne 1963 Jan 7-12 North Atlantic treaty organization Edited by A.E.M. Nairn Interscience,London,1964
Geog 13.1058

PROBLEMS IN SYSTEMATICS OF PARASITES :a symposium Proceedings American association for the advancement of science Edited by Gerald D. Schmidt University park press,Baltimore,Md.,1969
Mol 45.0148

PROBLEMS IN THE PHILOSOPHY OF MATHEMATICS International colloquium in the philosophy of science Proceedings London 1965 Jul 11-17 Vol 1 British society for the philosophy of science London school of economics and political science Edited by Imre Lakatos Under the auspices of the International union of history and philosophy of science.Division of logic,methodology and philosophy of science Studies in logic and the foundations of mathematics xv,241p North-Holland,Amsterdam,1967
WSM 43.1813

PROBLEMS IN THE PHILOSOPHY OF SCIENCE International colloquium in the philosophy of science Proceedings London 1965 Jul 11-17 Vol 3 British society for the philosophy of science London school of economics and political science Edited by Imre Lakatos and Alan Musgrave Under the auspices of the International union of history and philosophy of science.Division of logic, methodology and philosophy of science Studies in logic and the foundations of mathematics ix,448p North-Holland, Amsterdam,1968
WSM 43.0931

PROBLEMS OF APPLIED GEOGRAPHY Anglo-Polish seminar 1st Proceedings Nieborow 1959 Sep 15-18 Polska akademia nauk.Instytut geografii Geographical studies,25 Panstwowe wydawnictwo naukowe,Warsaw,1961
Geog 13.5617

PROBLEMS OF APPLIED GEOGRAPHY,2 Anglo-Polish seminar 2nd Proceedings Keele 1962 Sep 9-20 Polska akademia nauk.Instytut geografii, and,Institute of British geographers Geographia polonica,3 Panstwowe wydawnictwo naukowe,Warsaw,1964
Geog 13.5616

PROBLEMS OF COSMICAL AERODYNAMICS Symposium on the motion of gaseous masses of cosmical dimensions Proceedings Paris 1949 Aug 16-19 International union of theoretical and applied mechanics International astronomical union Edited by J.M. Burgess and H.C.van de Hulst 237p Central air documents office, Dayton,Ohio,1951
Obs 6.3550

PROBLEMS OF ECONOMIC RECONSTRUCTION IN THE FAR EAST Institute of Pacific relations conference 10th Report Stratford-on-Avon 1947 Sep 5-20 Institute of Pacific relations, New York,1949
Geog 13.2189

PROBLEMS OF EXTRA-GALACTIC RESEARCH :a symposium Santa Barbara,Calif. 1961 Aug. 10-12 International astronomical union Edited by G.C. McVittie International astronomical union.Symposium, 15 450p Macmillan,New York,1962
A Math 4.1154

PROBLEMS OF GROWTH IN INDUSTRIAL UNDERTAKINGS By L. Urwick British institute of management. Winter proceedings 1948-49, 2 British institute of management,London,1949
Eng 41.0531

PROBLEMS OF MATHEMATICS 2 conference Princeton,N.J. 1946 Dec.17-19 Princeton university Edited by John W. Tukey Princeton university.Bicentennial conferences, Series 2,2 32p 22cm Princeton university press,Princeton,N.J.,1947
P. Math 2.2557

PROBLEMS OF POPULATION International union for the scientific investigation of population problems.General assembly :report 2nd Proceedings London 1931 Jun 15-18 Edited by G.H.L.F. Pitt-Rivers Allen and Unwin, London,1932
Gen 34.2068

PROBLEMS OF SETTLEMENTS AND COMPRESSIBILITY OF SOILS European conference on soil mechanics and foundation engineering Proceedings Wiesbaden 1963 Vol 1-2 2 vols 1963
Eng 41.3175

PROBLEMY GEOGRAFII SIBIRI I DALNEGO VOSTOKA 1st itogi pervogo nauchnogo sovetshchaniya geografov Sibiri i Dalnego Vostoka 1960 Institut geografii Sibiri i Dalnego Vostoka Edited by I.P. Gerosimov and others 136p Irkutskoe knizhnoe izdatelstvo,Irkutsk,1960
Sco 14.4733

PROCEDURES FOR TESTING SOILS nomenclatures and definitions,standard methods,suggested methods American society for testing materials American society for testing materials,Philadelphia,Pa.,1958
Eng 41.3154

PROCEEDINGS OF A SYMPOSIUM ON HIGH LATITUDE PARTICLES Proceedings Alpbach 1964 Mar 19-26 Edited by B. Maehlum illus vii, 320p 24cm Logos press,London,1965
Sco 14.7472

PROCEEDINGS OF THE SYMPOSIUM ON INFORMATION PROCESSING IN SIGHT SENSORY SYSTEMS Proceedings Pasadena,Calif. 1965 Nov 1-3 National institutes of health and California institute of technology Edited by P.W. Nye California institute of technology, Pasadena,Calif.,1965
Psy 31.0419

PROCEEDINGS OF THE SYMPOSIUM ON THE ARCTIC HEAT BUDGET AND ATMOSPHERIC CIRCULATION Lake Arrowhead 1966 Jan-Feb University of California,and,Rand corporation Edited by J. O. Fletcher Prepared for the National science foundation Memorandum RM-5233-NSF illus,diagrs 566p The Rand corporation, Santa Monica,1966
Sco 14.2407

PROCESS CHEMISTRY Geneva 1955 Vol 1-2 Edited by F.R. Bruce and others Progress in nuclear energy, 3 Pergamon press,London, 1956 Includes papers presented at the United Nations conference on the peaceful uses of atomic energy
Met 25.1558

PROCESS DEVELOPMENT AND PLANT DESIGN Joint symposium on instrumentation and computation in process development and plant design Proceedings London 1959 May 11-13 Institution of chemical engineers Society of instrument technology British computer society Institution of chemical engineers, London,1959
Eng 41.6011

PROCESS METALLURGY Heat and mass transfer in process metallurgy :a symposium Proceedings London 1966 Apr 19-20 Edited by A.W.D. Hills Held by the John Percy research group in process metallurgy ix,252p Institution of mining and metallurgy,London,1967
Chem E 24.0593

PROCESS METALLURGY Heat and mass transfer in process metallurgy :a symposium Proceedings London 1966 Apr 19-20 John Percy research group in process metallurgy Edited by A.W.D. Hills Institution of mining and metallurgy, London,1967
Met 25.0317

PROCESS METALLURGY Physical chemistry of process metallurgy :an international symposium Pittsburgh,Pa. 1959 Apr 27-May 1 American institute of mining,metallurgical and petroleum engineers Edited by George R. St. Pierre Metallurgical society conferences, 7,8 Interscience,New York;London,1961
Met 25.0491

The PROCESS OF AGING IN THE NERVOUS SYSTEM :a conference Proceedings Bethesda,Md. 1957 Jan 30-Feb 1 Edited by James E. Birren and others Sponsored by the National advisory neurological diseases and blindness council Symposia in neuroanatomical sciences, 5 Blackwell,Oxford,1959
An 32.4171

PROCESS OPTIMISATION Symposium on process optimisation Proceedings London 1962 Jun 26 Institution of chemical engineers Edited by J.M. Pirie 70p Institution of chemical engineers,London,1962 3rd congress of the European federation of chemical engineering
Chem E 24.1630

PROCESS SIMULATION AND CONTROL IN IRON AND STEELMAKING :a symposium Proceedings New York 1964 Feb 17-18 American institute of mining,metallurgical and petroleum engineers Edited by J.M. Uys and H.L. Bishop Metallurgical society conferences, 32 Gordon and Breach,New York,1966
Met 25.0376

PROCESSING OF OPTICAL DATA BY ORGANISMS AND BY MACHINES Elaborazione di dati offici de parte di organism e di macchine Scuola internazionale di fisica "Enrico Fermi" 43 corso Rendaconti Varenna 1968 Jul 15-27 Societa italiana di fisica Edited by W. Reichardt Academic press,New York;London, 1969
An 32.5470

PRODUCT DESIGN THROUGH THE ENGINEERING ANALYSIS OF ITS END USE :an approach and a case history.Presented at the Design engineering conference New York 1967 May 15-18 Sponsored by the American society of mechanical engineers.Design engineering division American society of mechanical engineers,New York,1967
Eng 41.0326

PRODUCTION AND HAZARDS OF A HYPERBARIC OXYGEN ENVIRONMENT :a symposium Proceedings London 1968 Jan Edited by G.S. Innes Pergamon press,Oxford,1970
Radioth 35.1400

PRODUCTION AND USE OF SHORT-LIVED RADIOISOTOPES FROM REACTORS Seminar on the practical applications of short-lived radioisotopes produced in small research reactors Proceedings Vienna 1962 Nov 5-9 1-2 Held by the International atomic energy agency International atomic energy agency.Proceedings series 2 vols International atomic energy agency,Vienna,1963
Radioth 35.1369

PRODUCTION OF WIDE STEEL STRIP :a symposium London 1960 May 3-5 Iron and steel institute Iron and steel institute.Special report, 67 Iron and steel institute,London, 1960
Met 25.0726

PRODUCTION TECHNOLOGY OF NUCLEAR ENERGY MATERIALS The International conference on the peaceful uses of atomic energy Geneva 1955 Aug 8-20 Vol 8: production technology of the materials used for nuclear energy United Nations United Nations,New York,1956
Met 25.1564

PRODUCTIVITY IN RESEARCH :a symposium Preprint of papers London 1963 Dec 11-12 Institution of chemical engineers 89p Institution of chemical engineers,London,1963
Chem E 24.1794

PRODUCTIVITY IN RESEARCH :a symposium Proceedings London 1963 Dec 11-12 Institution of chemical engineers Institution of chemical engineers,London,1964
Eng 41.1685

PRODUCTIVITY IN THE IRON AND STEEL INDUSTRY Iron and steel institute annual general meeting,1962 Papers and discussions London 1962 May Iron and steel institute Iron and steel institute.Special report, 75 Iron and steel institute,London,1962
Met 25.0411

PROGESTERONE :its regulatory effect on the myometrium Ciba foundation Edited by G. E.W. Wolstenholme and Julie Knight Ciba foundation study group, 34 Churchill,London, 1969 In memory of Brenda M.Schofield
Inv Med 37.0249

PROGESTERONE AND THE DEFENCE MECHANISM OF PREGNANCY London 1961 Feb 10 Ciba foundation Edited by G.E.W. Wolstenholme and Margaret P. Cameron Ciba foundation study group, 9 Churchill,London,1961
An 32.3753

PROGESTERONE AND THE DEFENCE MECHANISM OF PREGNANCY 9th London 1961 Feb 10 Ciba foundation Edited by G.E.W. Wolstenholme and Margaret Cameron Ciba foundation study group, 9 Churchill,London, 1961
Phys 20.1388

PROGNOSTIC FACTORS IN BREAST CANCER :symposium Proceedings Cardiff 1967 Apr 12-14 Tenovus institute for cancer research Tenovus symposia, 1 Livingstone,Edinburgh; London,1968
Radioth 35.0935

PROGRAMMED LEARNING Aspects of educational technology :the proceedings of the programmed learning conference Proceedings Loughborough 1966 Apr 15-18 Association for programmed learning and Leicestershire programmed learning group Edited by Derick Unwin and John Leedham Methuen,London,1967
Psy 28.0233

PROGRAMMED LEARNING RESEARCH La Recherche en enseignement programme:tendances actuelles : colloque international Actes Nice 1968 May 13-17 North Atlantic treaty organization Sciences du comportement, 8 Dunod,Paris, 1969 Director of the colloquium:A.de Brisson.Papers in English and French
Psy 31.3117

PROGRAMMING Advances in programming and non-numerical computation summer school proceedings Oxford 1963 University of Oxford.Computing laboratory Edited by L. Fox bibliog viii,218p 23cm Pergamon press, Oxford,1966
Math L 5.1026

PROGRAMMING Advances in programming and non-numerical computation summer school proceedings Oxford 1963 University of Oxford.Computing laboratory Edited by L. Fox viii,218p 24cm Pergamon press,Oxford,1966
Math 3.0239

PROGRAMMING Applications of mathematical programming techniques :a conference Cambridge 1968 Jun 24-28 North Atlantic treaty organization.Scientific affairs committee Edited by E.M.L. Beale EUP, London,1970
Eng 41.8254

PROGRAMMING LANGUAGES N.A.T.O. advanced study institute.Summer school Villard de Lans 1966 North Atlantic treaty organization.Science committee Edited by F. Genuys 395p Academic press,London,1968
TA 15.0589

PROGRAMMING LANGUAGES :N.A.T.O. advanced summer school Villard-de-Lans 1966 Edited by F. Genuys Sponsored by the North Atlantic treaty organization.Science committee illus. 395p Academic press,London;New York, 1968
Math L 5.3577

PROGRESS AND PROSPECTS IN ORNITHOLOGY being the centenary symposium Proceedings Cambridge 1959 Mar 20-23 British ornithologists' union Ibis, 101,no 3-4, Centenary celebration number 25cm London, 1959
Bal 44.1519

PROGRESS IN ANAESTHESIOLOGY World congress of anaesthesiologists 4th Proceedings London 1968 Sep 9-13 By T.B. Boulton and others Excerpta Medica,International congress series, 200 Excerpta Medica Foundation,Amsterdam,1910
PGMS 29.0559

PROGRESS IN ASTRONAUTICS AND ROCKETRY, 3 Energy conversion for space power... based on a symposium Selected technical papers Santa Monica,Calif. 1960 Sep 27-30 American rocket society.Space power systems committee Edited by Nathan W. Snyder Academic press,New York;London,1961
Met 25.2165

PROGRESS IN BRAIN RESEARCH, 2 Nerve,brain and memory models Edited by N. Niener and J. P. Schade Elsevier,Amsterdam,1963 Lectures delivered during a symposium on "Cybernetics of the nervous system",held during the Second International meeting of medical cybernetics,Amsterdam,1962
An 32.4222

PROGRESS IN BRAIN RESEARCH, 3 The Rhinencephalon and related structures Edited by W. Bargmann and J.D. Schade Elsevier, Amsterdam,1963 Lectures delivered during a symposium on the rhinencephalon held as part of the Third international meeting of neurobiologists,Kiel,1962
An 32.4223

PROGRESS IN BRAIN RESEARCH, 4 Growth and maturation of the brain Edited by Dominick P. Purpura and J.P. Schade Elsevier,Amsterdam, 1964 Lectures delivered during an interdisciplinary workshop on "Growth and maturation of the brain",1962
An 32.4224

PROGRESS IN BRAIN RESEARCH, 5 Lectures on the dienecephalon Edited by W. Bargmann and J.P. Schade Elsevier,Amsterdam,1964 Lectures delivered during a symposium on the "Structure and function of the dienecephalon held as part of the Third International meeting of neurobiologists,Kiel,1962
An 32.4225

PROGRESS IN BRAIN RESEARCH, 6 Topics in basic neurology Edited by W. Bargmann and J. P. Schade Elsevier,Amsterdam,1964 Lectures delivered during a symposium on "Topics in basic neurology"held as part of the Third International meeting of neurobiologists, Kiel,1962
An 32.4226

PROGRESS IN BRAIN RESEARCH, 9 The Developing brain Edited by Williamina A. Himwich and Harold E. Himwich Elsevier, Amsterdam,1964 Derived from a Symposium held at Galesburg,Ill.,April 1963
An 32.4229

PROGRESS IN BRAIN RESEARCH, 10 Structure and function of the epiphysis cerebri :an international round-table conference Proceedings Amsterdam 1963 Jul 10-13 Edited by J.Ariens Kappers and J.P. Schade Elsevier,Amsterdam,1964
An 32.4230

PROGRESS IN BRAIN RESEARCH, 11 Organization of the spinal cord Edited by J.C. Eccles and J.P. Schade Elsevier,Amsterdam,1964 Lectures delivered during a workshop on "Physiology of spinal neurons",held as part of the first International summer school of brain research,Amsterdam,1963
An 32.4231

PROGRESS IN BRAIN RESEARCH, 12 Physiology of spinal neurons :a workshop ...held as part of the first International summer school of brain research... Amsterdam 1963 Jul 15-26 Edited by J.C. Eccles and J.P. Schade Elsevier,Amsterdam,1964
An 32.4232

PROGRESS IN BRAIN RESEARCH, 13 Mechanisms of neural regeneration :a workshop...held as part of the first International summer school of brain research... Amsterdam 1963 Jul 15-26 Edited by M. Singer and J.P. Schade Elsevier,Amsterdam,1964
An 32.4233

PROGRESS IN BRAIN RESEARCH, 14 Degeneration patterns in the nervous system :a workshop ... held as part of the first International summer school of brain research... Amsterdam 1963 Jul 15-26 Edited by M. Singer and J.P. Schade Elsevier,Amsterdam,1965
An 32.4234

PROGRESS IN BRAIN RESEARCH, 15 Biology of neuroglia :a symposium... held as part of the 10th Latin-American congress of neurosurgery Buenos Aires 1963 Oct 17-18 Edited by E.D. P.de Robertis and R. Carrea Elsevier, Amsterdam,1965
An 32.4235

PROGRESS IN BRAIN RESEARCH, 18 Sleep mechanisms Symposium on the physiological pharmacological and clinical aspects of sleep Proceedings Zurich 1964 Sep 18-19 Edited by K. Akert and others Elsevier,Amsterdam, 1965
An 32.4238

PROGRESS IN BRAIN RESEARCH, 22 Brain reflexes :the international conference dedicated to the centenary celebration of the publication of I.M.Sechenov's book "Brain reflexes" Moscow Edited by E.A. Asratyan Sponsored by Akademiya nauk S.S.S.R. Elsevier,Amsterdam,1968 Sponsored also by the International brain research organization
An 32.4243

PROGRESS IN BRAIN RESEARCH, 23 Sensory mechanisms :a workshop...held during the second International summer school of brain research... amsterdam 1964 Edited by Y. Zotterman Elsevier,Amsterdam,1967
An 32.4244

PROGRESS IN BRAIN RESEARCH, 25 The Cerebellum Edited by C.A. Fox and R.S. Snider Elsevier,Amsterdam,1967 Derived from a meeting held in the Netherlands,1965
An 32.4246

PROGRESS IN BRAIN RESEARCH, 30 Cerebral circulation European congress of neurosurgery 3rd Madrid 1967 Apr Edited by W. Luyendijk Elsevier,Amsterdam, 1968
An 32.4248

PROGRESS IN COMPARATIVE ENDOCRINOLOGY International symposium on comparative endocrinology 3rd Proceedings Oiso, Japan 1961 Jun 5-11 Edited by Kiyoshi Takewaki Sponsored by the Zoological society of Japan General and comparative endocrinology.Suppl., 1,1962 Academic press,New York,1962
Bal 39.1381

PROGRESS IN ENDOCRINOLOGY International congress of endocrinology Proceedings Mexico City 1968 Jun 30-Jul 5 Edited by Carlos Gual and F.J.C. Ebling Excerpta medica.International congress series, 184 Excerpta medica,Amsterdam,1969
Inv Med 37.0095

PROGRESS IN ENDOCRINOLOGY The Edinburgh meeting on endocrinology Proceedings Edinburgh 1959 Aug 16-20 Pt 1-2 Society for endocrinology Edited by K. Fotherby and others Society for endocrinology.Memoirs, 9-10 2 vols Cambridge university press,Cambridge,1960-1961
Bioch 33.0484

PROGRESS IN ENDOCRINOLOGY Proceedings Edinburgh 1959 Aug 16-20 Vol 1-2 Society for endocrinology Edited by K. Fotherby and others Society for endocrinology.Memoirs, 9-10 2 vols Cambridge university press,Cambridge,1964
Inv Med 37.0082

PROGRESS IN LYMPHOLOGY International symposium on lymphology Proceedings Zurich 1966 Jul Thieme,Stuttgart,1967
Radioth 35.0922

PROGRESS IN MINERAL DRESSING International mineral dressing congress Transactions Stockholm 1957 Sep 18-21 Svenska Gruvforeningen and Jernkontoret Almqvist and Wiksell,Stockholm,1958
Met 25.2286

PROGRESS IN NEUROBIOLOGY International meeting of neurobiologists 1st Proceedings Groningen 1955 Aug 3-7 Edited by J.Ariens Kapper Elsevier,Amsterdam, 1956
An 32.4351

PROGRESS IN NUCLEAR ENERGY, 4 Technology, engineering and safety International conference on the peaceful uses of atomic energy 2nd Edited proceedings Geneva 1958 Vol 2 United nations Edited by R. Hurst and others Pergamon press,Oxford, 1960
Met 25.1555

PROGRESS IN NUCLEAR ENERGY, 4 Technology and engineering Geneva 1955 Vol 1: reactor coolants,moderators,heat transfer, reactor chemistry and corrosion of reactor materials Edited by R. Hurst and S. McLain Pergamon,London,1956 Includes papers presented at the United Nations conference on the peaceful uses of atomic energy
Met 25.1565

PROGRESS IN NUCLEAR ENERGY, 5 Metallurgy and fuels Geneva 1955 Vol 1 Edited by H.M. Finniston and J.P. Howe Pergamon press,London,1956 Includes papers presented at the United Nations conference on the peaceful uses of atomic energy
Met 25.1560

PROGRESS IN OCEANOGRAPHY International association for quaternary research congress 7th Proceedings Boulder,Colo. 1965 Aug 30-Sep 5 Vol 4 Edited by M. Sears Pergamon,Oxford,1967
Bot 42.3248

PROGRESS IN PHOTOBIOLOGY Finsen memorial congress International congress on photobiology 3rd Proceedings Copenhagen 1960 Edited by B.Chr. Christensen and B. Buchmann illus. Elsevier,Amsterdam,1961
Bot 42.1347

PROGRESS IN PHOTOSYNTHESIS RESEARCH The International congress of photosynthesis research Proceedings Freudenstadt 1968 Jun 4-8 Vol 1-3 Edited by Helmut Metzner Sponsored by the International union of biological sciences 3 vols International union of biological sciences, Tubingen,1969
Bioch 33.1907

PROGRESS IN PHOTOSYNTHESIS RESEARCH The International congress of photosynthesis research Proceedings Freudenstadt 1968 Jun 4-8 Vol 1-3 Edited by Helmut Metzner Sponsored by the International union of biological sciences 3 vols Tubingen, 1969
Bot 42.1905

PROGRESS IN POWDER METALLURGY Annual technical meeting 16th Proceedings Chicago 1960 Apr 25-27 Metal powder industries federation Metal powder industries federation,Chicago,1960
Met 25.0769

PROGRESS IN POWDER METALLURGY Annual technical meeting 17th Proceedings Cleveland 1961 Apr 24-26 Metal powder industries federation Metal powder industries federation,Cleveland,Ohio,1961
Met 25.0770

PROGRESS IN PROTOZOOLOGY International congress on protozoology 3rd Abstracts of papers and communications Leningrad 1969 Jul 2-10 International commission on protozoology Publishing house 'Nauka', Leningrad,1969
Mol 45.0140

PROGRESS IN RADIO SCIENCE 1960-63 URSI general assembly 14th Proceedings Tokyo 1963 Sep 9-20 Vol 3: the ionosphere International scientific radio union. Commission on the ionosphere Edited by G.M. Brown 196p Elsevier,Amsterdam;London,1965
Nap 11.0086

PROGRESS IN RADIOBIOLOGY International conference on radiobiology 4th Proceedings Cambridge 1955 Aug 14-17 Edited by Joseph S. Mitchell and others port. Oliver and Boyd,Edinburgh;London,1956
Radioth 35.1707

PROGRESS IN RADIOBIOLOGY International conference on radiobiology 4th Proceedings Edited by Joseph S. Mitchell and others Oliver and Boyd, Edinburgh; London, 1956
Phys 20.0993

PROGRESS IN SOIL ZOOLOGY Research methods in soil zoology colloquium 1st Papers Rothamsted 1958 Jul 10-14 International society of soil science. Soil zoology committee Edited by P.W. Murphy Bibliog., illus. Butterworths, London, 1962
Bot 42.2085

PROGRESS IN THE STUDY OF THE BRITISH FLORA :a conference Report London 1956 Apr 13-15 Botanical society of the British Isles Edited by J.E. Lousley Botanical society of the British Isles. B.S.B.I. conference reports, 5 London, 1957
Bot 42.4710

PROGRESS IN VERY HIGH PRESSURE RESEARCH :an international conference Proceedings Bolton Landing, N.Y. 1960 Jun 13-14 Wright air development division. Materials central and General electric company research laboratory Edited by F.P. Bundy and others 314p New York, 1961
Geod 9.0254

PROGRESS IN VERY HIGH PRESSURE RESEARCH : international conference Proceedings Bolton Landing, New York 1960 Jun 13-14 United States. Air research and development command. Materials central and General electric research laboratory Edited by F.P. Bundy and others Wiley, New York; London, 1961
Met 25.1638

PROJECT MAC CONFERENCE ON CONCURRENT SYSTEMS AND PARALLEL COMPUTATION Record Woods Hole, Mass. 1970 Jun 2-5 Massachusetts institute of technology. Project MAC bibliog., illus. 199p Association for computing machinery, New York, 1970
Math L 5.3600

PROJECT PLANNING International congress on project planning by network analysis 2nd Proceedings Amsterdam 1969 Oct 6-10 North-Holland, Amsterdam, 1969
Eng 41.1240

PROJECTIVE GEOMETRY CONFERENCE Proceedings Chicago 1967 Jun 19-30 Bibliog v, 162p 28cm University of Illinois at Chicago Circle, Chicago, 1967 Mimeographed
P Math 2.3367

PROKARYOTIC AND EUKARYOTIC CELLS Organization and control in prokaryotic and eukaryotic cells :a symposium Papers Society for general microbiology Edited by H. P. Charles and B.C.J.G. Knight Held at Imperial college of science and technology Society for general microbiology. Symposia, 20 Cambridge university press, Cambridge, 1970
Bioch 33.1919

PROLIFERATION AND SPREAD OF NEOPLASTIC CELLS Annual symposium on fundamental cancer research 21st 1967 Anderson hospital and tumor institute Williams and Wilkins, Baltimore, Md., 1968
Radioth 35.0550

PROOF TECHNIQUES IN GRAPH THEORY Ann Arbor graph theory conference 2nd Proceedings Ann Arbor, Mich. 1968 Feb Edited by Frank Harary Bibliog., Illus. xv, 330p 24cm Academic press, New York; London, 1969
P Math 2.3349

PROOF TECHNIQUES IN GRAPH THEORY Ann Arbor graph theory conference 2nd Proceedings Ann Arbor, Mich. 1968 Feb University of Michigan Edited by Frank Harary xv, 330p 24cm Academic press, New York; London, 1969
Math S 3.1658

PROPAGATION AND INSTABILITIES IN PLASMAS Symposium on magnetohydrodynamics 7th Proceedings Palo Alto, Calif. 1962 Dec 14-15 Edited by Walter I. Futterman Sponsored by Lockheed aircraft corporation. Missiles and space division Stanford university press, Stanford, Calif., 1963
Eng 41.4523

PROPELLANT CHEMISTRY Advanced propellant chemistry :a symposium Papers Detroit, Mich. 1965 Apr 6-7 American chemical society. Division of fuel chemistry, and, American institute of aeronautics and astronautics. Propellants and combustion technical committee American chemical society. Advances in chemistry series, 54 American chemical society, Washington, D.C., 1966 Presented at the 149th meeting of the American chemical society. Symposium chairman: Richard T. Holtzmann
Chem 18.0977

The PROPERTIES AND FUNCTIONS OF MEMBRANES, NATURAL AND ARTIFICIAL :a general discussion London 1937 Apr By August Krogh and others Faraday society Faraday society. Transactions, 33, pt 8 London, 1937
Bal 39.0358

PROPERTIES OF CRYSTALLINE SOLIDS ... Symposium on recent progress in materials sciences; Symposium on nature and origin of strength of materials. Presented at the 63rd annual meeting A.S.T.M. Philadelphia, Pa. 1960 Jun 27 American society for testing materials American society for testing materials. Special technical publication, 283 American society for testing materials, Philadelphia, Pa., 1961
Eng 41.3790

PROPERTIES OF CRYSTALLINE SOLIDS :... symposium on recent progress in materials sciences; symposium on nature and origin of strength of materials Philadelphia 1960 Jun 27 A.S.T.M. Special technical publication Philadelphia, 1961 Includes an introductory paper by W.O. Baker on 'The national role of materials research and development'
Met 25.2490

PROPERTIES OF ELEMENTAL AND COMPOUND SEMICONDUCTORS a technical conference Boston, Mass. 1959 Aug 31-Sep 2 American institute of mining, metallurgical and petroleum engineers Edited by Harry C. Gatos Metallurgical society conferences, 5 Interscience, New York; London, 1960
Met 25.1160

PROPERTIES OF GRAIN BOUNDARIES Colloque de metallurgie proprietes des joints de grains 4th Saclay 1960 Jun 27-28 Centre d'etudes nucleaires,Saclay Presses universitaires de France,Paris,1961 Papers in English and French
Met 25.1266

PROPERTIES OF MATERIALS Symposium on major effects of minor constituents on the properties of materials Atlantic City,N.J. 1961 Jun 26 American society for testing materials American society for testing and materials.Materials science series, 2 American society for testing and materials. Special technical publications, 304 American society for testing materials,Philadelphia,Pa., 1962
Eng 41.3429

The PROPERTIES OF MATERIALS AT HIGH RATES OF STRAIN :conference Proceedings London 1957 Apr 30-May 2 Institution of mechanical engineers Institution of mechanical engineers,London,1957
Eng 41.3763

The PROPERTIES OF MATERIALS AT HIGH RATES OF STRAIN conference Proceedings London 1957 Apr 30-May 2 Institution of mechanical engineers Institution of mechanical engineers,London,1957
Met 25.0904

PROPERTIES OF METALLIC SURFACES :a symposium London 1952 Nov 19 Institute of metals Institute of metals.Monograph and report series, 13 Institute of metals,London,1953
Met 25.1593

PROPERTIES OF METALLIC SURFACES :a symposium London 1952 Nov 19 Organised by the Institute of metals Institute of metals. Monograph and report series, 13 Institute of metals,London,1953
Eng 41.3471

PROPERTIES OF REACTOR MATERIALS AND THE EFFECTS OF RADIATION DAMAGE :international conference Proceedings Berkeley Castle 1961 May 30-Jun 2 Berkeley nuclear laboratories,Gloucestershire Edited by D.J. Littler Butterworths,London,1962
Met 25.1574

PROPERTIES OF REACTOR MATERIALS AND THE EFFECTS OF RADIATION DAMAGE :international conference Proceedings Berkeley Castle, Glos. 1961 May 30-Jun 2 Central electricity generating board.Berkeley nuclear laboratories Edited by D.J. Littler xv, 562p Butterworths,London,1962
Cav 7.2864

PROPERTIES OF SURFACES Symposium on properties of surfaces :presented at the fourth Pacific area national meeting Los Angeles,Calif. 1962 Oct 4 American society for testing materials American society for testing materials.Special technical publication, 340 Philadelphia,1963
Met 25.2829

PROPIETA OTTICHE DEI SOLIDI Scuola internazionale di fisica 'Enrico Fermi' 34 corso Rendiconti Varenna 1965 Jun 28-Jul 10 Societa italiana di fisica Edited by J. Tauc Academic press,New York;London,1966
TA 15.0443

PROPRIETES OPTIQUES ET ACOUSTIQUES DES FLUIDES COMPRIMEES ET ACTIONS INTERMOLECULAIRES Bellevue 1957 Jul 1-6 Centre national de la recherche scientifique Edited by B. Vodar Centre national de la recherche scientifique. Colloques internationaux,77 Centre national de la recherche scientifique,Paris,1959
Chem 18.2309

PROPULSION New experimental techniques in propulsion and energetics research :a technical meeting Proceedings Munich 1967 Sep 11-15 Agard Edited by David Andrews and Jean Surugue Agard conference proceedings, 38 Technivision,Slough,1970
Eng 41.8359

PROPULSION SYSTEMS Advanced propulsion systems symposium Proceedings Los Angeles, Calif. 1957 Dec 11-13 Edited by Morton Alperin and George P. Sutton International series on aeronautical sciences and space flight.Division 9:symposia, 2 Pergamon press,London,1959
Eng 41.7370

PROSPECTS IN PSYCHIATRIC RESEARCH :the proceedings of the Oxford conference of the Mental health research fund Proceedings Oxford 1952 Mar Mental health research fund Edited by J.M. Tanner Blackwell, Oxford,1953
Psy 28.0036

PROSTAGLANDIN :a symposium Shrewsbury 1967 Oct 16-17 Worcester foundation for experimental biology Edited by Peter W. Ramwell and Jane E. Shaw Wiley,New York,1968
Inv Med 37.0185

PROSTAGLANDINS Nobel symposium 2nd Proceedings Stockholm 1966 Jun Nobel foundation Edited by Sune Bergstrom and Bengt Samuelsson Wiley;Almqvist and Wiksell, New York;Stockholm,1967
Inv Med 37.0019

PROTECTION AND FUNCTIONING OF THE HANDS IN COLD CLIMATES :a conference Proceedings Natick,Mass. 1956 Apr 23-24 Edited by Frank R. Fisher Sponsored by the United States.Army.Headquarters Quartermaster research and development command illus vi, 176p 23cm National academy of sciences, Washington,1957
Sco 14.0858

The PROTECTION OF MOTOR VEHICLES FROM CORROSION a symposium London 1958 Mar 11-12 Corrosion group S.C.I.monograph, 4 Society of chemical industry,London,1958
Met 25.1940

PROTECTION OF NATURE International technical conference on the protection of nature Preparatory documents Lake Success 1949 Aug 22-29 International union for the protection of nature.Secretariat Unesco, Paris,1949 Text in English and French
Bal 39.3922

PROTEIN AND POLYPEPTIDE HORMONES International symposium on protein and polypeptide hormones Proceedings Liege 1968 May 19-25 Edited by M. Margoulies Excerpta medica.International congress series, 161 Excerpta medica,Amsterdam,1968
Inv Med 37.0117

PROTEIN BIOSYNTHESIS Symposium on hormonal control of protein biosynthesis Research conference for biology and medicine 18th Gatlinburg,Tenn. 1965 Apr 5-8 Sponsored by the Oak Ridge national laboratory.Biology division Journal of cellular and comparative physiology, 66,suppl.1 Wistar institute of anatomy and biology,Philadelphia,Pa.,1965
Bioch 33.0520

PROTEIN BIOSYNTHESIS The International symposium on protein biosynthesis Proceedings Wassenaar 1960 Aug 29-Sep 2 Edited by R.J.C. Harris Under the auspices of Unesco and the Council for international organizations of medical sciences Academic press,London;New York,1961
Bioch 33.0516

PROTEIN BIOSYNTHESIS :a symposium Wassenaar 1960 Aug 29-Sep 2 Edited by R.J. C. Harris Under the auspice of Unesco Academic press,London;New York,1961 Co-sponsored by the Council for international organizations of medical sciences
Radioth 35.0102

PROTEIN BIOSYNTHESIS :a symposium Wassenaar 1960 Aug 29-Sep 2 Unesco Edited by R.J.C. Harris Sponsored also by the Council for international organizations of medical sciences Academic press,London;New York,1961
An 32.1509

PROTEIN BIOSYNTHESIS :a symposium Wassennaar 1960 Aug 29-Sep 2 Unesco Council for international organizations of medical sciences Edited by R.J.C. Harris Academic press,London;New York,1961
Gen 34.0652

PROTEIN CHEMISTRY Cold Spring Harbor symposia on quantitative biology Papers Cold Spring Harbor 1938 Cold Spring Harbor biological laboratory Long Island biological association,Cold Spring Harbor,1938 Later referred to in vol.9 as "Protein chemistry"
Bioch 33.1262

PROTEIN HORMONES International congress of biochemistry 2nd Paris 1952 Jul 21-27 Vol 4: symposium sur les hormones proteiques et derivees des proteines Council for international organizations of medical sciences Societe d'edition d'enseignément superieur,Paris,1952 Text in English and French
Bioch 33.1307

PROTEIN METABOLISM Some conjugated proteins; a symposium The Annual conference on protein metabolism 9th six lectures New Brunswick,N.J. 1953 Jan 30-31 Rutgers university.Bureau of biological research Edited by William H. Cole Rutgers university press,New Brunswick,N.J.,1953
Bioch 33.0543

PROTEIN METABOLISM :influence of growth hormone,anabolic steroids,and nutrition in health and disease:an international symposium Leyden 1962 Jun 25-29 Edited by F. Gross Sponsored by Ciba foundation Springer,Berlin,1967
Inv Med 37.0255

PROTEIN METABOLISM :influence of growth hormone anabolic steroids, and nutrition in health and disease :an international symposium Leyden 1962 Jun 25-29 Edited by F. Gross Sponsored by Ciba Springer,Berlin; Gottingen,1967 Chairman A.Querido
PGMS 29.0406

PROTEIN STRUCTURE Aspects of protein structure The International symposium on protein structure and crystallography Proceedings Madras 1963 Jan 14-18 University of Madras.Department of physics Edited by G.N. Ramachandran Academic press, London,1963
Cav 7.1495

PROTEIN STRUCTURE Aspects of protein structure The Symposium on protein structure Proceedings Madras 1963 Jan 14-18 University of Madras.Department of physics Edited by G.N. Ramachandran Academic press, London;New York,1963 The Symposium on protein structure formed part of an International symposium on protein structure and crystallography organized by the University of Madras
Bioch 33.0542

PROTEIN STRUCTURE Crystallography and crystal perfection The International symposium on protein structure and crystallography Proceedings Madras 1963 Jan 14-18 session on crystallography and crystal perfection University of Madras. Department of physics Edited by G.N. Ramachandran Academic press,London,1963
Cav 7.1494

PROTEIN STRUCTURE Symposium on protein structure Papers Paris 1957 Jul 25-29 Edited by Albert Neuberger Sponsored by the International union of pure and applied chemistry.Protein commission Methuen;Wiley, London;New York,1958
Bioch 33.0535

PROTEIN STRUCTURE Symposium on protein structure Paris 1957 Jul 25-29 Edited by Albert Neuberger Sponsored by the International union of pure and applied chemistry.Protein commission Methuen;Wiley, London New York,1958
Gen 34.0667

PROTEIN STRUCTURE AND FUNCTION Upton,N.Y. 1960 Jun 6-8 Brookhaven national laboratory. Biology department Brookhaven symposia in biology,13 BNL-608 Brookhaven national laboratory,biology dep.,Upton,N.Y.,1960
Pha 16.0120

PROTEIN STRUCTURE AND FUNCTION :a symposium Report Upton,N.Y. 1960 Jun 6-8 Brookhaven national laboratory.Biology department Brookhaven symposia in biology, 13 BNL608(C-30) Brookhaven national laboratory.Biology department,Upton,N.Y.,1960
Bioch 33.1292

PROTEIN SYNTHESIS Microsomal particles and protein synthesis :a symposium Papers Cambridge,Mass 1958 Feb 5-8 Biophysical society Edited by Richard B. Roberts Biophysical society.Symposia, 1 Pergamon, London;New York,1958
Bot 42.1686

PROTEIN SYNTHESIS Microsomal particles and protein synthesis :a symposium Papers presented Cambridge,Mass. 1958 Feb 5-8 Biophysical society Edited by Richard B. Roberts At the Massachusetts institute of technology Biophysical society.Symposia, 1 Pergamon,London,1958
Bioch 33.0924

PROTEIN SYNTHESIS The Mechanism of protein synthesis :a symposium Proceedings Cold Spring Harbor 1969 Cold Spring Harbor laboratory of quantitative biology Cold Spring Harbor symposia on quantitative biology, 34 Cold Spring Harbor,1969
Bioch 33.1887

PROTEIN SYNTHESIS The Use of radioautography in investigating protein synthesis :a symposium Montreal 1965 Edited by C.P. Leblond and Katherine Brehme Warren Sponsored by the International society for cell biology International society for cell biology.Symposia, 4 Academic press,New York;London,1965
Bioch 33.0286

PROTEIN UTILIZATION BY POULTRY :a symposium Proceedings Sutton Bonington 1965 Sep 22-23 British egg marketing board Edited by R.A. Morton and E.C. Amoroso British egg marketing board symposium,2 Oliver and Boyd, Edinburgh;London,1967
VA 19.0309

PROTEINASE INHIBITORS International research conference on proteinase inhibitors 1st Proceedings Munich 1970 Nov 4-5 Edited by H. Fritz and H. Tschesche De Gruyter, Berlin,1971
Bioch 33.2205

PROTEINES Conseil de chimie 9e Rapports et discussions Brussels 1953 Apr 6-14 Institut international de chimie Solvay and Universite libre de Bruxelles Edited by R. Stoops Brussels,1953
Bioch 33.0522

PROTEINS International congress of biochemistry 2nd Paris 1952 Jul 21-27 Vol 2: symposium sur la biogenese des proteines Council for international organizations of medical sciences Societe d'edition d'enseignement superieur,Paris,1952 Text in English and French
Bioch 33.1305

PROTEINS International congress of biochemistry 4th Proceedings Vienna 1958 Sep 1-6 Vol 8: symposium 8 - proteins International union of biochemistry Edited by H. Neurath and H. Tuppy I.U.B.symposium series, 10 Pergamon,London,1960 Added title page in French and German.Text in English and French
Bioch 33.1322

PROTEINS International congress of biochemistry 4th Proceedings Vienna 1958 Sep 1-6 Vol 8: symposium 8 - proteins International union of biochemistry Edited by H. Neurath and H. Tuppy Pergamon,London, 1960
Radioth 35.0072

PROTEINS Microsomal particles and protein synthesis Biophysical society :symposium 1st Papers Cambridge,Mass. 1958 Feb 5-8 Edited by Richard B. Roberts Pergamon press, London,1958
Col S 12.0053

PROTEINS Serological and biochemical comparisons of proteins Rutgers university. Bureau of biological research.Annual conferences on protein metabolism 14th Papers New Brunswick,N.J. 1958 Jan 24-25 Edited by William H. Cole Rutgers university press,New Brunswick,N.J.,1958
Col S 12.0073

PROTEINS Structure,function and evolution in proteins :symposium Report Upton,N.Y. 1968 Jun 3-5 Brookhaven national laboratory. Biology department Brookhaven symposia in biology, 21 BNL50116(C-53) Brookhaven national laboratory.Biology department,Upton,N.Y.,1969
Bioch 33.1299

PROTEINS Structure and function of proteins at the three-dimensional level :a symposium Papers Cold Spring Harbor 1972 Cold Spring Harbor laboratory of quantitative biology Cold Spring Harbor symposia on quantitative biology, 36 Cold Spring Harbor, 1972
Bioch 33.2334

PROTEINS Subunit structure of proteins : biochemical and genetic aspects.Symposium Report Upton,N.Y. 1964 Jun 1-3 Brookhaven national laboratory.Biology department Brookhaven symposia in biology, 17 · BNL869(C-40) Brookhaven national laboratory.Biology department,Upton,N.Y.,1964
Bioch 33.1295

PROTEINS The Swelling of proteins,and allied phenomena London 1932 Dec 2 International society of leather trades' chemists London,1933
Bioch 33.1486

PROTEINS,PEPTIDES,AND AMINO ACIDS International congress of biochemistry 6th Abstracts New York 1964 Jul 26-Aug 1 Scheduled under the auspices of the International union of biochemistry Washington,D.C.,1964 Title also in French.
Bioch 33.1339

PROTEOLYTIC ENZYMES The Structures and functions of proteolytic enzymes a discussion London 1968 Dec 5-6 Royal society of London Edited by D.C. Phillips and others Royal society of London.Philosophical transactions.Ser.B, 257,p.63-266 Royal society,London,1970
Bioch 33.1917

PROTIDES OF THE BIOLOGICAL FLUIDS 11th-13th,15th colloquia Proceedings Bruges 1963-65,67 Edited by H. Peeters Elsevier, Amsterdam,1964-68 In English,French and German
Bioch 33.0538

PROTIDES OF THE BIOLOGICAL FLUIDS 5th : colloquium Proceedings Bruges 1957 Edited by H. Peeters Elsevier,Amsterdam,1958
Col S 12.0126

PROTON TRANSFER PROCESSES The Kinetics of proton transfer processes :a general discussion Newcastle-upon-Tyne 1965 Apr 12-14 Faraday society Faraday society. Discussions, 39 Aberdeen university press, Aberdeen,1966
Bioch 33.0645

PROTOPLASTS Microbial protoplasts, spheroplasts and L-forms :a conference Proceedings Kalamazoo,Mich. 1966 Nov 10-11 Edited by Lucien B. Guze sponsored by the Upjohn company Williams and Wilkins, Baltimore,1968
Bioch 33.1142

PROTOZOA Growth of protozoa New York 1953 Oct 14 By R.W. Miner New York academy of sciences Edited by S.H. Hutner New York academy of sciences.Annals, 56,5 New York, 1953
Bot 42.3886

PROTOZOA Growth of protozoa :a conference New York 1953 Oct 14 By R.P. Hall and others New York academy of sciences Edited by S.H. Hunter and others New York academy of sciences.Annals, 56,p.815-1095 New York, 1953
Bal 39.2132

PROTOZOA Immunity to protozoa :a symposium British society for immunology Edited by P.C. C. Garnham Blackwell,Oxford,1963
Mol 45.0165

PROTOZOA Immunity to protozoa :a symposium Papers British society for immunology Edited by P.C.C. Garnham and others Blackwell,Oxford,1963
Path 30.2335

PROTOZOOLOGY Progress in protozoology International congress on protozoology 3rd Abstracts of papers and communications Leningrad 1969 Jul 2-10 International commission on protozoology Publishing house 'Nauka',Leningrad,1969
Mol 45.0140

PROVIDENCE,R.I. 1959 Cutaneous innervation Brown university symposium on the biology of skin,1959 Proceedings Edited by William Montagna Advances in biology of skin, 1 Pergamon press,Oxford,1960
An 32.4037

PROVIDENCE,R.I. 1959 Cutaneous innervation Brown university symposium Proceedings Brown university Edited by William Montagna Advances in the biology of skin,1 Pergamon press,London,1966
VA 19.0232

PROVIDENCE,R.I. 1960 Blood vessels and circulation Brown university symposium on the biology of skin,1960 Proceedings Edited by William Montagna and Richard A. Ellis Advances in biology of skin, 2 Pergamon press,Oxford,1961
An 32.4038

PROVIDENCE,R.I. 1960 Plasticity Naval structural mechanics symposium 2nd Proceedings United States.Office of naval research Brown university Edited by E.H. Lee and P.S. Symonds United States.Office of naval research.Structural mechanics series Pergamon press,London,1960
Eng 41.2479

PROVIDENCE,R.I. 1961 Eccrine sweat glands and eccrine sweating Brown university symposium on the biology of skin,1961 Proceedings Edited by William Montagna and others Advances in biology of skin, 3 Pergamon press,Oxford,1962
An 32.4039

PROVIDENCE,R.I. 1962 Sebaceous glands Brown university symposium on the biology of skin,1962 Proceedings Edited by William Montagna and others Advances in biology of skin, 4 Pergamon press,Oxford,1963
An 32.4040

PROVIDENCE,R.I. 1962 Symposium on time series analysis proceedings Brown university Edited by Murray Rosenblatt Also sponsored by United States.Office of naval research xiv,497p 25cm John Wiley and sons,New York,1963
Math 3.0695

PROVIDENCE,R.I. 1962 The Symposium on time series analysis Proceedings Brown university United States.Office of naval research Edited by Murray Rosenblatt xiv, 497p 26cm Wiley,New York;London,1963
Cav 7.2744

PROVIDENCE,R.I. 1963 Stress waves in anelastic solids :symposium International union of theoretical and applied mechanics Edited by Herbert Kolsky and William Prager Springer,Berlin,1964
Eng 41.3805

PROVIDENCE,R.I. 1963 Wound healing Brown university symposium on the biology of skin, 1963 Proceedings Edited by William Montagna and Rupert E. Billingham Advances in biology of skin, 5 Pergamon press, Oxford,1964
An 32.4044

PROVIDENCE,R.I. 1967 International conference on II-VI semiconducting compounds Proceedings American physical society International union of pure and applied physics Edited by D.G. Thomas 1490p 20cm Benjamin,New York,1967
Cav 7.2681

PROVIDENCE R.I. 1963 Stress waves in anelastic solids IUTAM symposium Proceedings International union of theoretical and applied mechanics Edited by Herbert Kolsky and William Prager Supported by the National science foundation Springer-verlag,Berlin, 1964
A Math 4.0369

PROVING ASSERTIONS ABOUT PROGRAMS ACM conference on proving assertions about programs Proceedings Las Cruces,N.M. 1972 Jan 6-7 Association for computing machinery.SIGPLAN Association for computing machinery.SIGACT illus 211p ACM,New York, 1972 Published as SIGPLAN notices,vol 7, no 1,Jan 1972 and SIGACT news no 14,Jan 1972
Math L 5.3859

PROVISIONAL INTERNATIONAL COMPUTATION CENTRE Symposium on questions of numerical analysis proceedings Rome 1958 Jun.30-Jul.1 79p Rome,1958
Math L 5.2374

PROVISIONAL INTERNATIONAL COMPUTATION CENTRE Symposium on the numerical treatment of ordinary differential equations, integral and integro-differential equations 4th proceedings Rome 1960 Sep.20-24 bibliog. 679p 26cm Birkhauser,Basel,1960
Math L 5.0299

PROVISIONAL INTERNATIONAL COMPUTATION CENTRE Symposium on the numerical treatment of partial differential equations with real characteristics proceedings Rome 1959 Jan.28-30 158p Birkhauser,Basel,1959
Math L 5.2378

PSYCHIATRIC RESEARCH Prospects in psychiatric research :the proceedings of the Oxford conference of the Mental health research fund Proceedings Oxford 1952 Mar Mental health research fund Edited by J.M. Tanner Blackwell,Oxford,1953
Psy 28.0036

PSYCHIATRY International congress for psychiatry 2nd Report Zurich 1957 Sep 1-7 Vol 1-4 International organization of world congresses for psyciatry. Swiss organizing committee Edited by W.A. Stoll 4cm Fussli arts graphiques,Zurich, 1959 Congress languages English,French, German,Italian and Spanish
Psy 28.0407

PSYCHIATRY International congress of medicine :section XII:psychiatry 17th Transactions London 1913 Aug 7-13 Oxford university press;Hodder and Stoughton,London, 1913-14 President of congress:J.Crichton-Browne Conference languages:English,French and German
Psy 31.2530

PSYCHIATRY Symposium on preventive and social psychiatry Washington,D.C. 1957 Apr 15-17 Walter Reed army medical center. Institute of research and National research council.Division of medical sciences US government printing office,Washington,D.C., c1957
Psy 28.0058

PSYCHIATRY The World congress of psychiatry 3rd Proceedings Montreal 1961 Jun 4-10 Canadian psychiatric association and McGill university 2 vols University of Toronto press;McGill university press,Toronto; Montreal,1961 Conference languages: English,French,German and Spanish
Psy 31.2540

PSYCHIATRY World congress of psychiatry Proceedings Montreal 1961 Jun 4-10 Vol 1-2 Canadian psychiatric association McGill university Edited by R.A. Cleghorn and others 2 vols University of Toronto press;McGill university press,Toronto;Montreal, 1961 Papers in English,French,German and Spanish
PGMS 29.0322

PSYCHIATRY AND THE LAW :the proceedings of the forty-third annual meeting of the American psychopathological association Proceedings New York 1953 Jun American psychopathological association Edited by Paul H. Hoch and Joseph Zubin Grune and Stratton,New York;London,1955
Psy 28.0311

PSYCHOBIOLOGICAL DEVELOPMENT OF THE CHILD Discussions on child development : consideration of the biological,psychological and cultural approaches to the understanding of human development and behaviour :meeting Geneva 1956 Vol 4 By Jean Piaget and others World health organization.Study group on the psychobiological development of the child Edited by J.M. Tanner and Barbel Inhelder Tavistock publications,London,1960
Psy 28.0224

PSYCHOBIOLOGICAL DEVELOPMENT OF THE CHILD Discussions on child development ;a consideration of the biological,psychological, and cultural approaches to the understanding of human development and behaviour Geneva, Switzerland 1953 Vol 1: proceedings World health organization.Study group on the psychobiological development of the child Edited by J.M. Tanner and Barbel Inhelder Tavistock publications,London,1956
Psy 31.1178

PSYCHOLGICAL DIAGNOSIS AND COUNSELING OF THE ADULT BLIND :selected papers from the proceedings of the University of Michigan conference for the blind Ann Arbor,Mich. 1947 United States.Federal security agency. Office of vocational rehabilitation and Michigan.Department of social welfare.Division of services for the blind Edited by Wilma Donahue and Donald Dabelstein Co-sponsored by the University of Michigan.Institute for human adjustment American foundation for the blind,New York,1950
Psy 31.2118

PSYCHOLINGUISTIC PAPERS :a conference Proceedings Edinburgh 1966 Mar 18-20 Edinburgh university.Faculty of social sciences.Research board Edited by J. Lyons and R.J. Wales Edinburgh university press, Edinburgh,1966
Psy 31.2745

PSYCHOLOGY International congress of psychology 3rd Munich 1896 Aug 4-7 Congress fur psychologie.Internationales organisations-comite Lehmann,Munich,1897 Papers in English,French,German and Italian
Psy 31.3391

PSYCHOLOGY International congress of psychology 4th Compte-rendu Paris 1900 Aug 20-26 Edited by Pierre Janet Alcan,Paris,1901
Psy 31.3394

PSYCHOLOGY International congress of psychology 7th Proceedings and papers Oxford 1923 Jul 26-Aug 2 International congress of psychology.British organizing committee Edited by Charles S. Myers Cambridge university press,Cambridge,1924 Edited by the president of the congress.Papers in English,French,German and Italian
Psy 31.2510

PSYCHOLOGY International congress of psychology 8th Proceedings and papers Groningen 1926 Noordhoff,Groningen,1927
Psy 31.3360

PSYCHOLOGY International congress of psychology 9th Proceedings and papers New Haven,Conn. 1929 Sep 1-7 International congress of psychology Psychological review co.,Princeton,N.J.,1930 Under the presidency of J.M.Cattell
Psy 31.3393

PSYCHOLOGY International congress of psychology 10th Report Copenhagen 1932 Aug 22-27 Levin and Munksgaard,Copenhagen, 1935 Under the presidency of professor Rubin
Psy 31.2537

PSYCHOLOGY International congress of psychology 12th Proceedings and papers Edinburgh 1948 Jul 23-29 Edited by Mary Collins Oliver and Boyd,Edinburgh;London, 1950 Under the presidency of J.Drever
Psy 31.3397

PSYCHOLOGY International congress of psychology 13th Proceedings and papers Stockholm 1951 Jul 16-21 International union of scientific psychology Stockholm, 1952 Under the presidency of professor Katz Conference languages:English,French and German
Psy 31.2536

PSYCHOLOGY International congress of psychology 14th Proceedings Montreal 1954 Jun North-Holland,Amsterdam,1955
Psy 31.3359

PSYCHOLOGY International congress of psychology 16th Berichte.Proceedings Bonn 1960 Jul 31-Aug 5 International union of scientific psychology Deutsche gesellschaft fur psychologie Acta psychologica, 19 North-Holland,Amsterdam, 1961 President of congress:professor Metzger.Conference languages:English,French and German
Psy 31.3396

PSYCHOLOGY Internationaler congress fur psychologie 3rd Munich 1896 Aug 4-7 Congress fur psychologie.Internationales organisations-comite Lehmann,Munich,1897 Papers in English,French,German and Italian
Psy 31.2518

PSYCHOLOGY OF HUMAN LEARNING Categories of human learning :symposium on the psychology of human learning Papers Ann Arbor,Mich. 1962 Jan 31-Feb 1 United States.Office of naval research and University of Michigan Edited by Arthur W. Melton Academic press, New York;London,1964
Psy 31.1003

PSYCHOLOGY OF PERCEPTION :the proceedings of the fifty-third annual meeting of the American psychological association Proceedings New York 1963 Feb American psychopathological association Edited by Paul H. Hoch and Joseph Zubin Grune and Stratton,London,1965 Contains the 1963 Samuel Hamilton award lecture by Heinrich Kluver
Psy 31.0958

PSYCHOPATHOLOGY OF COMMUNICATION :forty-sixth annual meeting Proceedings New York 1956 Jun American psychopathological association Edited by Paul H. Hoch and Joseph Zubin Grune and Stratton,New York; London,1958
Psy 28.0196

PSYCHOSOCIAL ASPECTS OF DRUG TAKING : conference Proceedings London 1964 Sep 25 Edited by R.G. Andry Pergamon, Oxford,1965 Summarised by Derrick Sington
Pha 16.0060

PSYCHOTECHNIQUE Conference internationale de psychotechnique 4th Comptes rendues Paris 1927 Oct 10-14 Conferences internationales de psychotechnique and Institut international de cooperation intellectuelle Alcan,Paris,1929 Papers in English,French,German and Italian
Psy 31.2517

PSYCHOTROPIC DRUGS Present status of psychotropic drugs,pharmacological and clinical aspects International congress of the Collegium internationale neuro-psychopharmacologicum 6th Proceedings Tarragona 1968 Apr 24-27 Collegium internationale neuro-psychopharmacologicum Edited by A. Cerletti and F.J. Bove Excerpta medica foundation.International congress series, 180 Excerpta medica foundation, Amsterdam,1969
Psy 31.2902

PSYCHOTROPIC DRUGS :an international symposium Proceedings Milan 1957 May 9-11 Edited by S. Garattini and V. Ghetti illus Elsevier,Amsterdam,1957 Chairman of conference:professor E.Trabucchi
Chem 26.0307

PSYCHOTROPIC DRUGS :international symposium Proceedings Milan 1957 May 9-11 Edited by S. Garattini and V. Ghetti Elsevier,Amsterdam,1957
Psy 31.0523

PSYCHOTROPIC DRUGS :international symposium Proceedings Milan 1957 May 9-11 Edited by S. Garrantini and V. Ghetti Elsevier,Amsterdam,1957
Pha 16.0029

PTERIDINES Ciba foundation symposium on chemistry and biology of pteridines Papers London 1954 Mar 22-26 Ciba foundation Edited by G.E.W. Wolstenholme and Margaret P. Cameron Ciba foundation symposia J. and A. Churchill,London,1954
Chem 18.1327

PTERIDINES Ciba foundation symposium on the chemistry and biology of purines London 1954 Mar 22-26 Edited by G.E.W. Wolstenholme and M. Cameron Illus J.and A.Churchill, London,1954
Radioth 35.0058

PUGNOCHIUSO 1970 International symposium on energy transduction in respiration and photosynthesis International union of biochemistry Edited by E. Quagliariello and others International union of biochemistry. Symposia, 43 Adriatic editrice,Bari,1971
Bioch 33.2289

PULKOVO Vsesoyuznaya astrometricheskaya konferentsiya sostoyavsheicya v Pulokove 10- trudy Vol 10- Edited by M.S. Zverev Leningrad,1954-
Obs 6.1923

PULKOVO 1960 Moon Symposium on the moon proceedings International astronomical union Edited by Zdenek Kopal and Zdenka Kadla Mikhailov International astronomical union. Symposia, 14 map 517p Academic press, London,1962
Obs 6.0972

PULKOVO 1960 The Moon :a symposium International astronomical union Edited by Zdenek Kopal and Zdenka Kadla Mikhailov Held at Glavnaya astronomicheskaya observatoriya, Pulkovo International astronomical union. Symposium, 14 571p Academic press,London; New York,1962
A Math 4.1153

PULLMAN,WASH. 1970 Conference on general topology Proceedings Washington state university.Department of mathematics iv,136p 27cm Washington state university.Department of mathematics,Pullman,Wash.,1970
P Math 2.4108

PULMONARY AND CARDIAC FUNCTION BONE ISOTOPE DIAGNOSIS International symposium on nuclear medicine 1st Proceedings Karlovy Vary 1969 May 13-16 Lekarska spolecnost J.E.Purkyne Edited by O. Andrysek and J. Mestan Universita Karlova,Prague,1970
PGMS 29.0680

PULMONARY CIRCULATION Conference on pulmonary circulation Oslo 1965 Sep 10-11 Edited by Carsten Muller Universitetsforlaget,Oslo,1966
An 32.3575

PULMONARY CIRCULATION AND RESPIRATORY FUNCTION a symposium held at Queen's college Dundee 1955 Sep 15-16 Supported by the University of St.Andrews University of St. Andrews,St.Andrews,1956
An 32.3562

PULSATILE BLOOD FLOW International symposium on pulsatile blood flow 1st Proceedings Philadelphia,Pa. 1963 Apr 11-13 Edited by E.O. Attinger McGraw-Hill,New York,1964
An 32.3528

PULSATILE BLOOD FLOW 1st :international symposium Proceedings Philadelphia,Pa. 1963 Apr 11-13 Edited by E.O. Attinger McGraw-Hill,New York,1964
Phys 20.0899

PULSE RADIOLYSIS :international symposium Proceedings Manchester 1965 Apr Edited by M. Ebert and others Academic press, London,1965
Radioth 35.1773

PULSE-RATE AND PULSE-NUMBER SIGNALS IN AUTOMATIC CONTROL I.F.A.C. symposium on pulse-rate and pulse-number signals in automatic control Proceedings Budapest 1968 Apr 8-10 Sponsored by the International federation of automatic control International federation of automatic control,Debrecen,1968
Eng 41.6020

PULSED-RADIOSOURCES AND HIGH ENERGY ACTIVITY IN SUPERNOVA REMNANTS Proceedings Rome 1969 Dec 18-20 illus 326p 27cm Accademia nazionale dei lincei,Rome,1972
A Math 4.1898

PUMP DESIGN,TESTING AND OPERATION :a symposium Proceedings Glasgow 1956 Apr 12-14 National engineering laboratory H.M.S.O., Edinburgh,1966
Eng 41.6824

PUNTA DEL ESTE 1951 Symposium sobre algunos problemas matematicos que se estan estudiando en Latina America Universidad de la republica,Montevideo.Instituto de matematica y estadistica Bibliog 183p 24cm UNESCO para America Latina,Montevideo, 1952
P Math 2.3007

PURDUE CONFERENCE ON SOIL MECHANICS AND ITS APPLICATION Proceedings Lafayette,Ind. 1940 Sep 2-6 Purdue university Purdue university,Lafayette,Ind.,1940
Eng 41.3163

PURDUE RESEARCH FOUNDATION Recent advances in the engineering sciences:their impact on engineering education Conference on science and technology for deans of engineering Proceedings Lafayette,Ind. 1957 Sep 9-12 viii,257p McGraw-Hill,New York,1958
Chem E 24.1074

PURDUE UNIVERSITY Information and decision processes symposium papers Lafayette,Ind. 1959 Apr. Edited by Robert E. Machol xi, 185p 23cm McGraw-Hill,New York,1960
Math 3.0683

PURDUE UNIVERSITY Midwestern conference on fluid mechanics 4th Proceedings Lafayette,Ind. 1955 Sep 8-9 Purdue university.Engineering experiment station. Research series, 128 Purdue university, Lafayette,Ind.,1955
Eng 41.6957

PURDUE UNIVERSITY Midwestern conference on fluid mechanics 4th Proceedings Lafayette,Ind. 1955 Sep 8-9 Purdue university.Engineering experiment station. Research series, 128 vii,370p Purdue university,Lafayette,Ind.,1955
Chem E 24.0404

PURDUE UNIVERSITY Purdue conference on soil mechanics and its application Proceedings Lafayette,Ind. 1940 Sep 2-6 Purdue university,Lafayette,Ind.,1940
Eng 41.3163

PURDUE UNIVERSITY Recent advances in the engineering sciences:their impact on engineering education Conference on science and technology for deans of engineering Proceedings Lafayette,Ind. 1957 Sep 9-12 viii,257p McGraw-Hill,New York,1958
Chem E 24.1074

PURDUE UNIVERSITY Symposium on engineering applications of random function theory and probability 1st Proceedings Edited by J. L. Bogdanoff and F. Kozin Wiley,New York, 1963
Eng 41.5949

PURDUE UNIVERSITY Symposium on engineering applications of random function theory and probability 1st proceedings Edited by John L. Bogdanoff and Frank Kozin Supported by National science foundation x,421p 24cm John Wiley and sons,New York,1963
Math 3.0879

PURDUE UNIVERSITY.DIVISION OF MATHEMATICAL SCIENCES Arithmetical algebraic geometry conference proceedings Lafayette,Ind. 1963 Dec 5-7 Edited by Otto F.G. Schilling Harper's series in modern mathematics Bibliog. vii,200p 22cm Harper and Row, New York,1965
P. Math 2.1770

PURDUE UNIVERSITY.THERMOPHYSICAL PROPERTIES RESEARCH CENTER Conference on thermal conductivity 8th Proceedings West Lafayette,Ind. 1968 Oct 7-10 Edited by C.Y. Ho and R.E. Taylor xx,1169p 29cm Plenum press,New York,1969
Cav 7.3075

PURINES Ciba foundation symposium on the chemistry and biology of purines London 1956 May 8-10 Edited by G.E.W. Wolstenholme and C.M. O'Connor Illus Churchill,London, 1957
Radioth 35.0059

PURINES Ciba foundation symposium on the chemistry and biology of purines Papers London 1956 May 8-10 Ciba foundation Edited by G.E.W. Wolstenholme and Cecilia M. O'Connor Ciba foundation symposia J. and A. Churchill,London,1957
Chem 18.1326

PURINES Ciba foundation symposium on the chemistry and biology of purines Papers London 1956 May 8-10 Ciba foundation Edited by G.E.W. Wolstenholme and Cecilia M. O'Connor illus Churchill,London,1957
Bioch 33.0814

PURKYNE SYMPOSIUM Vortrage und diskussionsbeitrage Halle 1959 Oct 31-Nov 1 Deutsche akademie der naturforscher Leopoldina Ceskoslovenska akademie ved Edited by Rudolph Zaunick Nova acta leopoldina.Neue folge, 24 230p Barth, Leipzig,1961
WSM 43.2940

PURPOSIVE SYSTEMS American society for cybernetics :annual symposium 1st Proceedings Edited by Heinz von Foerster and others Spartan books,New York;Washington,D.C. 1968
Psy 31.3068

PYELONEPHRITIS Renal disease and hypertension :tentn annual conference... Proceedings Denver 1971 Mar Atlantic City 1971 May United States.Public health service Edited by Richard H. Thurm Proceedings of the annual conferences of the U. S.Public health service co-operative study (antihypertensive agents), 8 U.S.Dept. of health,education and welfare,Bethesda,1971
PGMS 29.0659

PYELONEPHRITIS Symposium on pyelonephritis Edinburgh 1966 Norwich pharmacal company and Eaton Laboratories division Livingstone,London,1967
PGMS 29.0236

PYMATUNING,PA. 1959 The Ecology of algae : a symposium Edited by C.A. Tryon and R.T. Hartman Pymatuning laboratory of field biology.Special publication, 2 Pittsburgh, Pa.,1960
Bot 42.3900

PYRIDINE-NUCLEOTIC-DEPENDENT DEHYDROGENASES : advanced study institute Proceedings Konstanz 1969 Sep 15-20 Universitat Konstanz Edited by Horst Sund Springer, Berlin,1970
Bioch 33.2292

PYRIDOXAL CATALYSIS Chemical and biological aspects of pyridoxal catalysis :a symposium of the International union of biochemistry Proceedings Rome 1960 Oct Edited by E.E. Snell International union of biochemistry. Symposium series, 30 Pergamon,Oxford,1963
Radioth 35.0131

PYRIDOXAL CATALYSIS:ENZYMES AND MODEL SYSTEMS The International symposium on chemical and biological aspects of pyridoxical catalysis 2nd Proceedings Moscow 1966 Sep 15-21 Edited by E.E. Snell and others sponsored by the International union of biochemistry I.U. B.Symposium series, 35 Interscience,New York,1968 First International symposium on chemical and biological aspects of pyridoxal catalysis published under title chemical and biological aspects of pyridoxical catalysis
Bioch 33.1038

PYROMETALLURGICAL PROCESSES IN NONFERROUS METALLURGY :a symposium Pittsburgh 1965 Nov 29-Dec 1 American institute of mining,metallurgical and mining engineers. Extractive metallurgy division Edited by J.N. Anderson and P.E. Queneau Metallurgical society conferences, 39 Gordon and Breach, New York,1967
Met 25.2530

PYROMETRY The Papers and discussion of a symposium on pyrometry Chicago,Ill. 1919 Sep American institute of mining and metallurgical engineers National research council National bureau of standards illus 701p New York,1920
WSM 43.4920

PYROMETRY :a symposium Chicago 1919 Sep American institute of mining and metallurgical engineers and National bureau of standards American institute of mining and metallurgical engineers,New York City,1920
Met 25.0294

PYRONES Symposium of recent advances in the chemistry of naturally occurring pyrones and related compounds Proceedings Dublin 1955 Jul 12-14 University college,Dublin,and, Institute of chemistry of Ireland Royal Dublin society.Scientific proceedings,27,no 6 Royal Dublin society,Dublin,1956
Chem 18.1494

QUALITATIVE METHODS IN THE THEORY OF NONLINEAR OSCILLATIONS 5th conference Proceedings Kiev 1969 Vol 2-4 Edited by Yu.A. Mitropolski Bibliog,illus 26cm 3 vols Ukrainian academy of sciences.Institute of mathematics,Kiev,1970
A Math 4.1724

QUALITY CONTROL IN METAL FINISHING based on a symposium at the Borough Polytechnic London Borough Polytechnic,London Edited by G. Isserlis Columbine press,Manchester; London,1967
Met 25.1983

QUALITY REQUIREMENTS OF SUPER-DUTY STEELS a technical conference Proceedings Pittsburgh 1958 May 5-6 American institute of mining,metallurgical and petroleum engineers.Committee on physical chemistry of steelmaking Edited by R.W. Lindsay Metallurgical society conferences, 3 Interscience,New York;London,1959
Met 25.0482

QUANTITATIVE ELECTRON MICROSCOPY :a symposium Proceedings Washington,D.C. 1964 Mar30-Apr 3 Armed forces institute of pathology Edited by Gunter F. Bahr and Elmar H. Zeitler Laboratory investigation, 14,no 6 Williams and Wilkins,Baltimore,Md.,1965
An 32.0162

QUANTITATIVE INHERITANCE colloquium Papers Edinburgh 1950 Apr 4-6 University of Edinburgh.Institute of animal genetics Edited by E.C.R. Reeve and C.H. Waddington Under the auspices of the Agricultural research council H.M.S.O.,London,1952
Gen 34.1083

QUANTITATIVE METHODS IN HUMAN PHARMACOLOGY AND THERAPEUTICS :a symposium Proceedings London 1958 May 24-25 Biological council. Co-ordinating committee for symposia on drug action Edited by D.R. Laurence Biological council.Co-ordinating committee for symposia on drug action.Symposium on drug action series, 3 Pergamon,London,1959
Pha 16.0419

QUANTITATIVE METHODS IN MORPHOLOGY Symposium on quantitative methods in morphology held... during the eighth International congress of anatomists Proceedings Wiesbaden 1965 Aug 10 Edited by Ewald R. Weibel and Hans Elias Organised by the International society for stereology Illus. Springer,Berlin,1967
An 32.0105

QUANTITATIVE METHODS IN PHARMACOLOGY Leyden 1960 May 10-13 Edited by H.de Jonge North-Holland,Amsterdam,1961
Pha 16.0415

QUANTITATIVE METHODS IN PHARMACOLOGY :a symposium Proceedings Leiden 1960 May 10-13 Edited by H.de Jonge North-Holland, Amsterdam,1961
HE 27.0109

QUANTITATIVE METHODS IN PHARMACOLOGY :a symposium Proceedings Leyden 1960 May 10-13 Edited by H. de Jonge Bibliog,illus xx,391p 23cm North-Holland,Amsterdam,1961
Math 3.1168

QUANTITATIVE PAPER CHROMATOGRAPHY OF STEROIDS : a conference Proceedings London 1958 Jul 1 Society for endocrinology Edited by D. Abelson and R.V. Brooks Society for endocrinology.Memoirs, 8 Cambridge university press,Cambridge,1960
Bioch 33.2355

QUANTITIVE ELECTRON MICROSCOPY :a symposium Proceedings Washington,D.C. 1964 Mar 30 Apr 3 Edited by Gunter F. Bahr and Elmar H. Zeitler Sponsored by the Intersociety committee for research potential in pathology, inc. Laboratory investigation, 14 no.6 illus. viii,605p Baltimore,Md.,1965
Bot 42.4731

QUANTITIVE ELECTRON MICROSCOPY :a symposium Proceedings Washington,D.C. 1964 Mar 30-Apr 3 Edited by Gunter F. Bahr and Elmar H. Zeitler held at the Armed forces institute of pathology Laboratory investigation, 14, no 6 Washington,D.C.,1966 Sponsored jointly by the Intersociety committee for research potential in pathology and the Armed forces institute of pathology
Bal 39.0239

QUANTUM ASPECTS OF POLYPEPTIDES AND POLYNUCLEOTIDES :a symposium Papers Stanford,Calif. 1963 Mar 25-29 Edited by M. Weissbluth Biopolymers.Symposia,1 Interscience,New York,1964
Pha 16.0117

QUANTUM ELECTRONICS Conference on quantum electronics - resonance phenomena High View, N.Y. 1959 Sep 14-16 Edited by Charles H. Townes Columbia university press,New York, 1960
Eng 41.5411

QUANTUM FLUIDS The International symposium on quantum fluids Brighton 1965 Aug 16-20 University of Sussex Edited by D.F. Brewer viii,360p 23cm North-Holland,Amsterdam, 1966
Cav 7.2935

QUANTUM FLUIDS AND NUCLEAR MATTER Boulder summer institute for theoretical physics 11th Proceedings Boulder,Colo. 1968 Jun 17-Aug 23 University of Colorado.Department of physics and astrophysics Edited by Kalyana T. Mahanthappa and Wesley E. Brittin Lectures in theoretical physics, 11b diagrms xiv,428p 23cm Gordon and Breach, New York,1969 Dedicated to George Gamow
Cav 7.3009

QUANTUM MECHANICAL METHODS IN VALENCE THEORY : a conference Papers Shelter Island,L.I., N.Y. 1951 Sep 8-10 National academy of sciences With the support of the United States.Office of naval research United States.Office of naval research.Physics branch. Publication,4 U.S.government printing office, Washington,D.C.,1952
Chem 18.2237

QUANTUM MECHANICS Perturbation theory and its application in quantum mechanics seminar proceedings Madison,Wis. 1965 Oct 4-6 United States army.Mathematics research center Edited by Calvin H. Wilcox Organized jointly with University of Wisconsin.Theoretical chemistry institute 428p John Wiley and sons,New York;London,1966
A Math 4.0809

QUANTUM OPTICS Ottica quantistica Scuola internazionale di fisica "Enrico Fermi" 42 corso Rendiconti Varenna 1967 Jul 31-Aug 19 Societa Italiana di fisica Edited by R. J. Glauber Bibliog.,Illus.port. xvi,159p 25cm Academic press,New York,1968
A Math 4.1599

QUANTUM THEORY AND BEYOND :essays and discussions arising from a colloquium Cambridge 1968 Edited by Ted Bastin ix, 345p Cambridge university press,Cambridge, 1971 Based on papers read at an informal colloquium held in Cambridge 1968
WSM 43.1119

QUANTUM THEORY AND BEYOND :essays and discussions arising from a colloquium Cambridge 1968 Jul Royal society of London Theoria incorporated Carnegie institution of Washington illus viii,345p 24cm Cambridge university press,Cambridge,1971 Based on papers read at an informal colloquium
A Math 4.1759

QUARTERMASTER FOOD AND CONTAINER INSTITUTE Low temperature test methods and standards for containers :a symposium Chicago 1953 Dec 10 National research council Edited by Earl C. Myers and Norbert J. Leiner illus vi,126p 23cm Washington,1954
Sco 14.0889

QUASARS AND HIGH ENERGY ASTRONOMY ;including proceedings of 2nd Texas conference on relativistic astrophysics Edited by K.N. Douglas and others 485p Gordon and Breach, New York,1969 Texas conference held in 1964
TA 15.0490

QUASI-OPTICS :a symposium Proceedings New York 1964 Jun 8-10 Microwave research institute and Institute of radio engineers Co-sponsored by United States.Office of naval research Polytechnic institute of Brooklyn. Microwave research institute symposia series, 14 Polytechnic press,Brooklyn,N.Y.,1964
Cav 7.1721

QUASI-STELLAR OBJECTS External galaxies and quasi-stellar objects Upsala 1970 Aug 10-14 International astronomical union Edited by D.S. Evans International astronomical union.Symposium, 44 549p Reidel,Dordrecht, 1972
TA 15.0658

QUASI STELLAR SOURCES AND GRAVITATIONAL COLLAPSE Texas symposium on relativistic astrophysics 1st Proceedings Dallas, Texas 1963 Dec 16-18 By Ivor Robinson and others Southwest center for advanced studies Sponsored by University of Texas 475p 25cm University of Chicago press,Chicago,Ill.,1965
Cav 7.1985

QUASI-STELLAR SOURCES AND GRAVITATIONAL COLLAPSE Texas symposium on relativistic astrophysics 1st proceedings Dallas, Texas 1963 Dec. 16-18 Southwest center for advanced studies Edited by Ivor Robinson and others sponsored by University of Texas xvii,475p 25cm University of Chicago press, Chicago,1965
Obs 6.1457

QUASI-STELLAR SOURCES AND GRAVITATIONAL COLLAPSE Texas symposium on relativistic astrophysics 1st proceedings Dallas, Texas 1963 Dec.16-18 Southwest center for advanced studies Edited by Ivor Robinson and others Sponsored by University of Texas xvii,475p 25cm University of Chicago press, Chicago,1965
A Math 4.1161

QUASIFORMAL MAPPINGS Romanian-Finnish seminar on Teichmuller spaces and quasiformal mappings Proceedings Brasov 1969 Edited by Cabiria Andreian Cazacu 307p 21cm Publishing house of the Academy of the socialist republic of Romania,Bucharest,1971
P Math 2.4355

QUATERNARY International studies on the quaternary International association for quaternary research :congress 7th Papers Boulder,Colo. 1965 Edited by H.E. Wright and David G. Frey Sponsored by the National research council Geological society of America.Special papers,84 Geological society of America,New York,1965
Geog 13.0309

QUATERNARY CHANGES IN LEVEL,ESPECIALLY IN THE NETHERLANDS Utrecht 1954 Mar 5-6 Organized by the Nederlanske geologisch-mijnbouw kundig genootschap.Geologische sectie Geologie en mijnbouw,ns,16,no.6 The Hague, 1954
Geog 13.0371

QUATERNARY PALEOECOLOGY International association for quaternary research.Congress 7th Proceedings Boulder,Colo. 1965 Aug-Sep Vol 7 Edited by E.J. Cushing and H.E. Wright Sponsored by the National research council Yale university press,New Haven,Conn. London,1967
Geog 13.1318

QUATERNARY PALEOECOLOGY International association for quaternary research congress 7th Proceedings Boulder,Colo. 1965 Aug 30-Sep 5 Vol 7 Edited by E.J. Cushing and H.E. Wright Sponsored by the National research council Yale university press,New Haven,Conn.;London,1967
Bot 42.3245

QUATRIEME CONGRES INTERNATIONAL DE MICROBIOLOGIE International congress of microbiology 4th Report of proceedings Copenhagen 1947 Jul 20-26 Edited by Mogens Bjorneboe Rosenkilde and Bagger,Copenhagen,1949
Bot 42.0663

QUATRO CENTENARIO DELLA NASCITA DI GALILEO GALILEI,1564-1964.COMITATO NAZIONALE PER LE MANIFESTAZIONI CELEBRATIVE Convegno sulla relativita:problemi di energia e onde gravitazionali;meeting on general relativity: problems of energy and gravitational waves Atti;proceedings Florence 1964 Sep 9-12 Edited by Giorgio Sestini Quatro centenario della nascita di Galileo Galilei,1564-1964. Comitato nazionale per le manifestazioni celebrative.Pubblicazioni, 2,Atti dei convegni, 1 267p Florence,1966
Obs 6.3276

QUATRO CENTENARIO DELLA NASCITA DI GALILEO GALILEI 1564-1964.COMITATO NAZIONALE PER LE MANIFESTAZIONI CELEBRATIVE Convegno sulla cosmologia;meeting on cosmology Atti; Proceedings Padua 1964 Sep 14-16 and Venice Edited by L. Rosino Organized by Universita di Padua Quatro centenario della nascita di Galileo Galilei.Comitato nazionale per le manifestazioni celebrative. Pubblicazioni, 2,Atti dei convegni, 3 206p Florence,1966
Obs 6.3451

QUEEN MARY COLLEGE.NUCLEAR ENGINEERING DEPARTMENT Conference on dust cooling Proceedings London 1964 Dec 21-22 Vol 1-2 2 vols London,1964 Mimeogra?h
Chem E 24.1464

QUEEN MARY COLLEGE,LONDON International conference on low temperature physics 8th Proceedings London 1962 Sep 16-22 Edited by R.O. Davies Butterworths,London,1963 Includes Fritz London memorial lecture delivered by Bardeen,J.
Cav 7.1459

QUEEN'S COLLEGE,DUNDEE Summer school in mathematics,geometry and topology proceedings Dundee 1961 Jul 10-22 33cm Queen's College,Dundee,1961
P. Math 2.0292

QUEEN'S UNIVERSITY,KINGSTON,ONT. Conference on Universal algebra Proceedings Kingston, Ont 1969 Oct Queen's papers in pure and applied mathematics, 25 Bibliog. 273p 27cm Queen's university,Kingston,Ont.,1970
P Math 2.3712

QUEEN'S UNIVERSITY OF BELFAST International seminar on operating systems principles Submitted papers Belfast 1971 Aug 30-Sep 3 illus 2 vols Queen's university of Belfast, Belfast,1971
Math L 5.3846

QUENCHING OF STEELS :a compilation of papers presented at the National metal congress Cleveland,Ohio 1948 Oct 27-31 American society for metals American society for metals,Cleveland,Ohio,1951
Met 25.0446

QUESTIONI DI ELASTICITA NON LINEARIZZATA Istituto nazionale di alta matematica Edited by Antonio Signorini Cremonese,Rome,1960 Extract of conference held 1959-60 published in 'Rendiconti di matematica e delle sue applicazioni,'serie 5
Eng 41.2484

QUESTIONS ON ALGEBRAIC VARIETIES Centro matematico estivo 3 ciclo Varenna 1969 Sep 7-17 Centro matematico estivo 343p 28cm Edizioni cremonese,Rome,1970 Conference coordinator:E.Marchionna
P Math 2.4172

QUEUING THEORY:RECENT DEVELOPMENTS AND APPLICATIONS conference proceedings Lisbon 1965 Sep.27-Oct.1 North Atlantic treaty organization.Science committee Edited by R. Cruon 224p 23cm English universities press,London,1967
Math 3.0703

QUEZON CITY 1953 Pacific science congress 8th proceedings Vol 2,2A: geology and geophysics meteorology Pacific science association Under the auspices of the National research council of the Philippines 2 vols National research council of the Philippines,Quezon City,1956
Geol 8.3070

QUEZON CITY 1953 Symposium on medicinal plants :8th Pacific science congress Pacific science association Under the auspices of the National research council of the Philippines Pacific science congress. Proceedings, 8th,4a National research council of the Philippines,Quezon city,1954
BG 38.3111

QUINONES Ciba foundation symposium on quinones in electron transport London 1960 May 11-13 Ciba foundation Edited by G.E.W. Wolstenholme and Cecilia M. O'Connor Churchill,London,1961
Radioth 35.0099

QUINONES Ciba foundation symposium on quinones in electron transport Papers London 1960 May 11-13 Ciba foundation Edited by G.E.W. Wolstenholme and Cecilia M. O'Connor Churchill,London,1961
Chem 18.1489

QUINONES IN ELECTRON TRANSPORT Ciba foundation symposium on quinones in electron transport Proceedings London 1960 May 11-13 Ciba foundation Edited by G.E.W. Wolstenholme and Cecilia M. O'Connor illus Churchill,London,1961
Bioch 33.0573

RADAR SYSTEMS Advanced radar systems 19th Avionics panel meeting Istanbul 1970 May 25-29 Agard.Avionics panel AGARD conference proceedings, 66 Agard,Neuilly-sur-Seine,1970
Eng 41.8322

RADAR TECHNIQUES FOR DETECTION TRACKING AND NAVIGATION :8th symposium of the AGARD avionics panel Proceedings London 1964 Sep 21-25 Agard Edited by W.T. Blackband AGARDograph, 100 Gordon and Breach,New York,1966
Eng 41.5717

RADIAL VELOCITIES Determination of radial velocities and their applications Toronto 1966 Jun 20-24 International astronomical union and International council of scientific unions Edited by A.H. Batten and J.F. Heard International astronomical union. Symposium, 30 Academic press,London;New York,1967
TA 15.0307

RADIAL VELOCITIES Determination of radial velocities and their applications :a symposium Toronto 1966 Jun.20-24 International astronomical union Edited by A.H. Batten and J.F. Heard International astronomical union. Symposium, 30 262p academic press,London, 1967 publication dedicated to the memory of R.Methven Petrie
Obs 6.0147

RADIANT ENERGY Symposium on radiant energy in the sea Papers Helsinki 1960 Aug 4-5 International union of geodesy and geophysics Edited by N.G. Jerlov International union of geodesy and geophysics.Monograph,10 illus 115p Paris,1961 Symposium held during the I.U.G.A.assembly at Helsinki
Sco 14.0135

RADIATION Comparative effects of radiation : a conference Report San Juan 1960 Feb 15-19 Edited by Milton Burton and others Wiley,New York,1960
Radioth 35.0947

RADIATION Initial effects of ionizing radiations on cells International symposium on primary and initial effects of ionizing radiations on living cells Papers: discussions Moscow 1960 Oct Edited by R. J.C. Harris Sponsored by the Akademiya nauk S.S.S.R. Academic press,London;New York,1961
An 32.3277

RADIATION Interaction of radiation with solids Cairo solid state conference Cairo 1966 Sep 3-8 American university in Cairo Edited by Adli Bishay Plenum press,New York, 1967
TA 15.0429

RADIATION Late effects of radiation :a colloquium Proceedings Chicago,Ill. 1969 May Edited by R.J.M. Fry and others Taylor and Francis,London,1970
Radioth 35.1221

RADIATION Symposium on effects of radiation and other deleterious agents on embryonic development Oak Ridge,Tenn. 1953 Apr 20-21 Sponsored by the Oak Ridge national laboratory. Biology division Wistar institute of anatomy and biology,Philadelphia,Pa.,1954
Reprinted from the Journal of cellular and comparative physiology Vol 43,Suppt 1,May 1954
An 32.2027

RADIATION Symposium on noxious effects of low level radiation :physical elements and biological aspects Lausanne 1958 Mar 27-29 Schweizerische akademie der medizinischen wissenschaften Schwabe,Basle;Stuttgart,1958
Radioth 35.1078

RADIATION AND THE CONTROL OF IMMUNE RESPONSE Panel on radiation and the control of immune response Proceedings Paris 1967 Jun 22-24 Organized by the International atomic energy agency International atomic energy agency.Panel proceedings series
International atomic energy agency,Vienna,1968
Radioth 35.1202

RADIATION AND WAVES Symposium on magnetohydrodynamics 5th Palo Alto, Calif. 1960 Dec 15-16 Edited by Morton Mitchener Sponsored by Lockheed aircraft corporation.Missile systems division
Stanford university press,Stanford,Calif.,1961
Eng 41.4496

RADIATION AND WAVES IN PLASMAS Lockheed symposium on magnetohydrodynamics 5th proceedings Palo Alto 1960 Dec 15-16 Edited by Morton Mitchner Sponsored by Lockheed aircraft corporation.Missiles and space division Stanford university press, Stanford,Calif.,1961
A Math 4.0621

RADIATION AND WAVES IN PLASMAS Symposium on magnetohydrodynamics 5th Palo Alto, Calif. 1960 Lockheed aircraft corporation. Lockheed missiles and space division Edited by Morton Mitchner 156p Stanford university press,Stanford,Calif.,1961
Cav 7.2881

RADIATION BIOLOGY Australasian conference on radiation biology 2nd Proceedings Melbourne 1958 Dec 15-18 Australian radiation society Edited by J.H. Martin Butterworths,London,1959
Gen 34.1110

RADIATION BIOLOGY International conference on radiation biology and cancer Proceedings Kyoto 1966 Nov 1-2 Edited by Tsutomu Sugahara Under the auspices of Nippon societas radiologica.Biological subgroup Radiation society of Japan,1967
Radioth 35.1757

RADIATION BIOLOGY AND CANCER Animal symposium on fundamental cancer research 12th Papers 1958 Owen,London,1960
Radioth 35.1101

RADIATION BIOLOGY AND CANCER Symposium on fundamental cancer research 12th Papers Houston,Tex. 1958 Anderson hospital and tumor institute University of Texas press, Austin,Tex.,1958
Phys 20.0988

RADIATION CHEMISTRY OF AQUEOUS SYSTEMS Farkas memorial symposium 19th Proceedings Jerusalem 1967 Dec 27-29 Edited by Gabriel Stein Weizmann science press of Israel, Jerusalem,1968
Radioth 35.1768

RADIATION CHEMISTRY OF AQUEOUS SYSTEMS : proceedings of the 19th L.Farkas memorial symposium held at the Hebrew university Jerusalem 1967 Dec 22-29 Hebrew university of Jerusalem Edited by Gabriel Stein Weizmann science press of Israel in co-operation with Interscience,Jerusalem,1968
Chem 18.2485

RADIATION DAMAGE Properties of reactor materials and the effects of radiation damage : international conference Proceedings Berkeley Castle,Glos. 1961 May 30-Jun 2 Central electricity generating board.Berkeley nuclear laboratories Edited by D.J. Littler xv,562p Butterworths,London,1962
Cav 7.2864

RADIATION DAMAGE AND SULPHYDRYL COMPOUNDS Panel on radiation damage to the biological molecular information system :with special regard to the role of SH-groups Proceedings Vienna 1968 Oct 21-25 Organized by the International atomic energy agency International atomic energy agency.Panel proceedings series International atomic energy agency,Vienna,1969
Radioth 35.1764

RADIATION DAMAGE IN BONE Conference on the relation of radiation damage to radiation dose in bone Report and summarized papers Oxford 1960 Apr 10-14 International atomic energy agency International atomic energy agency,Vienna,1960
Radioth 35.1115

RADIATION DAMAGE IN SOLIDS Radiation damage nei solidi Scuola internazionale di fisica 'Enrico Fermi' 18 corso Rendiconti Ispra (Varese) 1960 Sep 5-24 Societa italiana di fisica Edited by D.S. Billington Academic press,New York;London,1962 Contributions in English and French
Chem 18.1144

RADIATION DAMAGE IN SOLIDS Radiation damage nei solidi Scuola internazionale di fisica "Enrico Fermi" 28 corso Rendiconti Ispra, Varese 1960 Sep 5-24 Societa Italiana di fisica Edited by D.S. Billington Academic press,New York;London,1962
Met 25.1579

RADIATION DAMAGE IN SOLIDS AND REACTOR MATERIALS Proceedings :a symposium Venice 1962 May 7-11 Vol 1-3 International atomic energy agency 3 vols International atomic energy agency,Vienna,1962
Cav 7.2393

RADIATION DAMAGE NEI SOLIDI Scuola internazionale di fisica 'Enrico Fermi' 18 corso Rendiconti Ispra (Varese) 1960 Sep 5-24 Societa italiana di fisica Edited by D.S. Billington Academic press,New York; London,1962 Contributions in English and French
Chem 18.1144

RADIATION DAMAGE NEI SOLIDI Scuola internazionale di fisica "Enrico Fermi" 28 corso Rendiconti Ispra,Varese 1960 Sep 5-24 Societa Italiana di fisica Edited by D. S. Billington Academic press,New York;London, 1962
Met 25.1579

RADIATION DOSIMETRY Selected topics in radiation dosimetry Symposium on selected topics in radiation dosimetry Proceedings Vienna 1960 Jun 7-11 Sponsored by the International atomic energy agency International atomic energy agency.Proceedings series International atomic energy agency, Vienna,1961
Radioth 35.1694

RADIATION EFFECTS IN PHYSICS,CHEMISTRY AND BIOLOGY International congress of radiation research 2nd Proceedings Harrogate 1962 Aug 5-11 Edited by Michael Ebert and Alma Howard North-Holland, Amsterdam,1963
Radioth 35.1141

RADIATION EFFECTS ON MATERIALS :a symposium Los Angeles,Calif. 1956,1957 Vol 1-2 American society for testing materials and Atomic industrial forum A.S.T.M. Special technical publication, 208,220 2 vols American society for testing materials, Philadelphia,Pa.,1957
Met 25.1571

RADIATION GENETICS Discussion on the present status of radiation genetics :given at the information meeting for biology and medicine of the Atomic energy commission Oak Ridge, Tenn. 1948 Mar 26-27 Sponsored by Oak Ridge national laboratory.Biological division Wistar institute of anatomy and biology, Philadelphia,Pa.,1950 Reprinted from the Journal of cellular and comparative physiology, Vol 35,suppt 1, 1950
An 32.0357

RADIATION GENETICS Discussion on the present status of radiation genetics information meeting... Oak Ridge,Tenn. 1948 Mar 26-27 Sponsored by the Oak Ridge national laboratory Oak Ridge national laboratory.Biology division. Symposia, 1 Journal of cellular and comparative physiology, 35,supp Wistar institute of anatomy and biology,Philadelphia, Pa.,1950
Gen 34.1103

RADIATION INCORPORATED,ORLANDO,FLA. Space trajectories :a symposium Papers Orlando, Fla. 1959 Dec 14-15 Co-sponsored by the American astronautical society Academic press,New York,1960
Geod 9.0014

RADIATION-INDUCED CANCER :a symposium Proceedings Athens 1969 Apr 28-May 2 International atomic energy agency World health organization International atomic energy agency,Vienna,1969
Radioth 35.1917

RADIATION-INDUCED CHROMOSOME ABERRATIONS Conference on biochemical and biophysical mechanisms in the production of radiation-induced chromosome aberrations San Juan, Puerto Rico 1961 Nov 16-18 Edited by Sheldon Wolff Sponsored by the National research council Columbia university press, New York;London,1963
Gen 34.1122

RADIATION-INDUCED CHROMOSOME ABERRATIONS : report of a conference on biochemical and biophysical mechanisms in the production of radiation-induced chromosome aberrations San Juan,P.R. 1961 Nov 16-18 Edited by Sheldon Wolff Sponsored by the National research council illus. xii,304p Columbia university press,New York;London,1963
Bot 42.4730

RADIATION PROTECTION Conference on radiation protection in accelerator environments Proceedings Chilton,Berks. 1969 Mar Edited by G.R. Stevenson Sponsored by the Science research council Rutherford laboratory,Chilton,1969 Organized by and held at the Rutherford laboratory
Radioth 35.1393

RADIATION PROTECTION Conference on radiobiology and radiation protection Stockholm 1952 Sep 15-20 Acta radiologica, 41,fasc.1 Stockholm,1954
Radioth 35.1228

RADIATION PROTECTION Nordic radiation protection conference 1st Proceedings Edited by Kurt Liden and Erik Lindgren Acta radiologica.Supplementum, 254 Stockholm,1966
Radioth 35.1295

RADIATION PROTECTION AND SENSITIZATION International symposium on radiosensitizing and radioprotective drugs 2nd Proceedings Rome 1969 May 6-8 Edited by Holta Whitehouse Taylor and Francis,London,1970
Radioth 35.1219

RADIATION RESEARCH International congress of radiation research 2nd Abstracts of papers Harrogate 1962 Aug 5-11 1962
Radioth 35.1165

RADIATION RESEARCH International congress of radiation research 3rd Cortina d'Ampezzo 1966 Jun 26-Jul 2 book of abstracts Cortina d'Ampezzo,1966
Radioth 35.1969

RADIATION RESEARCH International congress of radiation research 3rd Proceedings Cortina d'Ampezzo 1966 Jun 26-Jul 2 Edited by G. Silini Under the auspices of the International association for radiation research North-Holland,Amsterdam,1967
Radioth 35.1173

RADIATION RESEARCH Radiation effects in physics,chemistry and biology International congress of radiation research 2nd Proceedings Harrogate 1962 Aug 5-11 Edited by Michael Ebert and Alma Howard North-Holland,Amsterdam,1963
Radioth 35.1141

RADIATION SENSITIVITY OF TOXINS AND ANIMAL POISONS Panel on the radiation sensitivity of toxins and animal poisons Proceedings Bangkok 1969 May 19-22 International atomic energy agency International atomic energy agency.Panel proceedings series International atomic energy agency,Vienna,1970
Radioth 35.1912

RADIATION SENSITIVITY OF TOXINS AND ANIMAL POISONS Panel on the radiation sensitivity of toxins and animal poisons Proceedings Bangkok 1969 May 19-22 Organized by the International atomic energy agency International atomic energy agency. Panel proceedings series International atomic energy agency,Vienna,1970
Radioth 35.1765

RADIATION TRAPPED IN THE EARTH'S MAGNETIC FIELD the advanced study institute Proceedings Bergen 1965 Aug 16-Sep 3 Michelson (Chr.) institutt for videnskap og aandsfrihet Edited by Billy M. McCormac Astrophysics and space science library, 5 901p Reidel, Dordrecht,1966
Obs 6.3282

RADIO ASTRONOMY Fachkonference radioastronomie :jahrestagung der deutschen akademie der wissenschaften Berlin 1955 Mar 28- Apr 2 Deutsche akademie der wissenschaften zu Berlin Deutsche akademie der wissenschaften zu Berlin.Abhandlungen. Klasse fur mathematik,physik und technik, 3 135p Akademie-verlag,Berlin,1958
Obs 6.2016

RADIO ASTRONOMY Paris symposium on radio astronomy Paris 1958 Jul.30-Aug.6 International astronomical union and International scientific radio union Edited by Ronald N. Bracewell International astronomical union.Symposium, 9 International scientific radio union.Symposium, 1 612p Stanford university press, Stanford,Calif.,1959
Obs 6.0246

RADIO ASTRONOMY Paris symposium on radio astronomy Paris 1958 Jul.30-Aug.6 International astronomical union and International scientific radio union Edited by Ronald N. Bracewell International astronomical union.Symposium, 9 International scientific radio union.Symposium, 1 612p Stanford university press, Stanford,Calif,1959
A Math 4.1151

RADIO ASTRONOMY Paris symposium on radio astronomy Paris 1958 Jul 30-Aug 6 International astronomical union International scientific radio union Edited by Ronald N. Bracewell International astronomical union.Symposium, 9 International scientific radio union.Symposium, 1 612p Stanford university press, Stanford,Calif.,1959
TA 15.0638

RADIO ASTRONOMY Paris symposium on radio astronomy Paris 1958 Jul 30-Aug 6 International astronomical union and International scientific radio union Edited by Ronald N. Bracewell International astronomical union.Symposium, 9 International scientific radio union.Symposium, 1 612p Stanford university press, Stanford,Calif,1959
Cav 7.0301

RADIO ASTRONOMY Radio astronomy today : international summer school Papers Jodrell Bank 1962 Victoria university of Manchester Edited by H.P. Palmer and others Manchester university press,Manchester,1963 Foreword by Sir Bernard Lovell
Cav 7.1751

RADIO-ASTRONOMY Radio astronomy today Papers Jodrell Bank 1962 International summer school in radio-astronomy,Manchester Edited by H.P. Palmer and others 242p Manchester university press,Manchester,1963 With a foreword by Sir Bernard Lovell
Obs 6.1326

RADIO ASTRONOMY :a symposium Proceedings Jodrell Bank 1955 Aug 25-27 International astronomical union Edited by H.C.van de Hulst International astronomical union. Symposium, 4 xi,409p Cambridge university press,Cambridge,1957
Nap 11.0275

RADIO ASTRONOMY :a symposium Proceedings Manchester 1955 Aug 25-27 International astronomical union Edited by H.C.van de Hulst International astronomical union. Symposium, 4 Cambridge,1957
TA 15.0037

RADIO ASTRONOMY :a symposium proceedings Manchester 1955 Aug.25-27 International astronomical union Edited by H.C.van de Hulst With financial assistance from Unesco International astronomical union.Symposium, 4 409p Cambridge university press, Cambridge,1957
Obs 6.0847

RADIO ASTRONOMY AND THE GALACTIC SYSTEM Symposium on radio astronomy and the galactic system Noordwijk 1966 Aug 25-Sep 1 International astronomical union Edited by Hugo van Woerden Organised in co-operation with International scientific radio union International astronomical union.Symposium, 31 xviii,501p Academic press,London;New York,1967
A Math 4.1574

RADIO ASTRONOMY TODAY Summer school in radio astronomy Papers Jodrell Bank 1962 Victoria university of Manchester Edited by H.P. Palmer and others 242p Manchester university press,Manchester,1963
TA 15.0226

RADIO-ELECTRONICS-TELEVISION MANUFACTURERS ASSOCIATION Symposium on printed circuits proceedings Philadelphia,Pa. 1955 Jan 20-21 Jointly sponsored by Institute of radio engineers illus. 122p 28cm Engineering publishers,New York,1955
Math L 5.0736

RADIO-ISOTOPE IN DER HAMATOLOGIE : internationales symposion Freiburg 1962 Mar 1-3 Edited by W. Kiederling and G. Hoffman Nuclear-medizin, 7,supp.1 Schattauer,Stuttgart,1963
Radioth 35.1367

RADIO NOISE OF TERRESTRIAL ORIGIN Monograph on radio noise of terrestrial origin URSI general assembly 13th Papers London 1960 Sep 6 Commission 4 on radio noise of terrestrial origin International scientific radio union Edited by F. Horner URSI monographs series 202p Elsevier,Amsterdam, 1962
Nap 11.0268

RADIO WAVE PROPAGATION Meteorological and astronomical influences on radio wave propagation NATO Advanced study institute Papers Corfu 1961 North Atlantic treaty organization.Division of scientific affairs Published for and on behalf of North Atlantic treaty organization.Science committee. Division of scientific affairs NATO Conference series, 3 318p Pergamon press, Oxford,1963
Nap 11.0349

RADIO WAVE PROPAGATION Meteorological and astronomical influences on radio wave propagation NATO Advanced study institute Papers Corfu 1961 North Atlantic treaty organization.Division of scientific affairs NATO conference series, 3 318p Pergamon press,Oxford,1963
Nap 11.0915

RADIO-WAVE PROPAGATION Meteorological factors in radio-wave propagation :a conference Report London 1946 Apr 8 Physical society Royal meteorological society 325p Physical society,London,1946
Nap 11.0925

RADIO-WAVE PROPAGATION Monograph on radio-wave propagation in the troposphere U.R.S.I. general assembly 13th Papers London 1960 Sep commission 2 on radio and troposphere International scientific radio union Edited by J.A. Saxton U.R.S.I. monographs series illus 199p Elsevier, Amsterdam,1962
Nap 11.0533

RADIOACTIVE DATING :a symposium Athens 1962 Nov 19-13 International atomic energy agency Joint commission on applied radioactivity International atomic energy agency.Proceedings series International atomic energy agency,Vienna,1963
Bot 42.3319

RADIOACTIVE DATING :symposium proceedings Athens 1962 Nov 19-23 sponsored by the International atomic energy agency International atomic energy agency,Vienna,1963
Geol 8.4477

RADIOACTIVE DATING AND METHODS OF LOW LEVEL COUNTING :symposium proceedings Monaco 1967 Mar 2-10 organized by the International atomic energy agency International atomic energy agency,Vienna,1967
Geol 8.4478

RADIOACTIVE DATING AND METHODS OF LOW-LEVEL COUNTING :a symposium Proceedings Monaco 1967 Mar 2-10 International atomic energy agency Joint commission on applied radioactivity International atomic energy agency.Proceedings series International atomic energy agency,Vienna,1967
Radioth 35.1683

RADIOACTIVE ISOTOPES Primenenie radioaktivnykh izotopov v promyshlennosti meditsine i selskom khozyaistve illus. Izdatelstvo akademii nauk S.S.S.R.,Moscow,1956 Papers presented at the International conference on the peaceful uses of atomic energy, Geneva,1955
Bot 42.0386

RADIOACTIVE ISOTOPES Radiobiologiya Trudy vsesoyuznoi nauchno-tekhnicheskoi konferentsii po primeneniyu radioaktivnykh i stabilnykh izotopov i izduchenii v narodnom khozyaistve i nauke 1957 Apr 4-12 Akademiya nauk S.S.S.R. Edited by A.M. Kuzin Izdatelstvo akademii nauk S.S.S.R.,Moscow,1955
Bot 42.0385

RADIOACTIVE ISOTOPES Symposium uber die markierang der proteine mittels radioaktiver isotopen und deren anwendung in biologie und medizin Bern 1964 Nov 14 Schweizerische akademie der medizinischen wissenschaften. Bulletin, 21,fasc.3-4 Schwabe,Basle, Stuttgart,1965 Parallel text in German, French and Italian
Radioth 35.1187

RADIOACTIVE ISOTOPES IN THE LOCALIZATION OF TUMOURS International nuclear medicine symposium Proceedings London 1967 Edited by N.G. Trott Heinemann medical, London,1969
Radioth 35.1397

RADIOACTIVE POISONING Diagnosis and treatment of radioactive poisoning Scientific meeting on the diagnosis and treatment of radioactive poisoning Proceedings Vienna 1962 Oct 15-18 World health organization International atomic energy agency International atomic energy agency.Proceedings series International atomic energy agency,Vienna,1963
Radioth 35.1368

RADIOACTIVE WASTE Standardization of radioactive waste categories :report of a panel Vienna 1967 Nov 6-10 International atomic energy agency International atomic energy agency.Technical reports series, 101 IAEA,Vienna,1970
Chem 18.2735

RADIOACTIVITY Biochemisch nachweisbare strahlenwirkungen und deren beziehungen zur strahlentherapie :symposium der Arbeitsgemeinschaft fur strahlenbiologie Stuttgart 1969 May 11 Arbeitsgemeinschaft fur strahlenbiologie Edited by G.B. Gerber Thieme,Stuttgart,1970
Radioth 35.1222

RADIOACTIVITY FOR PHARMACEUTICAL AND ALLIED RESEARCH LABORATORIES :a symposium Uniontown,Pa. Edited by Abraham Edelmann Sponsored by Nuclear science and engineering corporation Academic press,New York;London, 1960
Radioth 35.1725

RADIOAKTIVE IOSTOPE IN KLINIK UND FORSCHUNG Band 3: vortrage am Gasteiner internationalen symposium 1958 Edited by K. Fellinger and H. Vettel Strahlentherapie. Sonderband, 38 Urban and Schwarzenberg, Munich;Berlin,1958
Radioth 35.1073

RADIOAUTOGRAPHY The Use of radioautography in investigating protein synthesis :a symposium Montreal 1965 Edited by C.P. Leblond and Katherine Brehme Warren Sponsored by the International society for cell biology International society for cell biology.Symposia, 4 Academic press,New York;London,1965
Bioch 33.0286

RADIOBIOLOGIYA Trudy vsesoyuznoi nauchno-tekhnicheskoi konferentsii po primeneniyu radioaktivnykh i stabilnykh izotopov i izduchenii v narodnom khozyaistve i nauke 1957 Apr 4-12 Akademiya nauk S.S.S.R. Edited by A.M. Kuzin Izdatelstvo akademii nauk S.S.S.R.,Moscow,1955
Bot 42.0385

RADIOBIOLOGY Advances in radiobiology International conference on radiobiology 5th Proceedings Stockholm 1956 Aug 15-19 Edited by George Carl de Hevesy and others port. Oliver and Boyd,Edinburgh;London,1957
Radioth 35.1708

RADIOBIOLOGY All-union scientific and technical conference on the application of radioactive isotopes :a portion of the proceedings...in English translation Moscow 1957 Consultants bureau,New York,1959
Radioth 35.1731

RADIOBIOLOGY Australasian conference on radiobiology 3rd Proceedings Sydney 1960 Aug 15-18 Australian radiation society Edited by P.L.T. Ilbery Butterworths,London, 1961
Radioth 35.1727

RADIOBIOLOGY Conference on radiobiology and radiation protection Stockholm 1952 Sep 15-20 Acta radiologica, 41,fasc.1 Stockholm, 1954
Radioth 35.1228

RADIOBIOLOGY Conference on radiobiology and radiotherapy Proceedings Colorado Springs 1965 Nov 1-3 National cancer institute Edited by J.A. Del Regato National cancer institute.Monograph, 24 National cancer institute,Bethesda,Md.,1967
Bioch 33.0989

RADIOBIOLOGY International congress of radiology 7th Invited papers Copenhagen 1953 Jul 19-24 Acta radiologica.Supplementum, 116 Stockholm,1954
Radioth 35.1311

RADIOBIOLOGY Progress in radiobiology International conference on radiobiology 4th Proceedings Cambridge 1955 Aug 14-17 Edited by Joseph S. Mitchell and others port. Oliver and Boyd,Edinburgh;London,1956
Radioth 35.1707

RADIOBIOLOGY Progress in radiobiology International conference on radiobiology 4th Proceedings Edited by Joseph S. Mitchell and others Oliver and Boyd, Edinburgh;London,1956
Phys 20.0993

RADIOBIOLOGY Radiobiology symposium proceedings Liege 1954 Aug-Sep Edited by Z.M. Bacq and P. Alexandra London,1955
Philos 1.1413

RADIOBIOLOGY AT THE INTRA-CELLULAR LEVEL U.C. L.A.conference on radiobiology 1st Proceedings Catalina island 1957 Sep 9-12 University of California Edited by T.G. Hennessy and others ports. Pergamon press, London,1959
Radioth 35.1722

RADIOBIOLOGY SYMPOSIUM proceedings Liege 1954 Aug-Sep Edited by Z.M. Bacq and P. Alexandra London,1955
Philos 1.1413

RADIOBIOLOGY SYMPOSIUM 1954 Proceedings Liege 1954 Aug-Sep Edited by Z.M. Bacq and Peter Alexander Butterworths,London,1955
Radioth 35.1712

RADIOGRAPHY Recovery and repair mechanics in radiography :symposium Report Upton,N.Y. 1967 Jun 5-7 Brookhaven national laboratory. Biology department Brookhaven symposia in biology, 20 BNL50058)C-51) Brookhaven national laboratory.Biology department,Upton,N. Y.,1968
Bioch 33.1298

RADIOGRAPHY Symposium on radiography 1st and 2nd Atlantic City 1936,1942 American society for testing materials A.S.T. M.,Philadelphia,1943
Met 25.1317

RADIOISOTOPE Conference on radioiodine Proceedings Chicago,Ill. 1956 Nov 5-6 Argonne cancer research hospital United States atomic energy commission Edited by D. E. Clark ACRH-100 University of Chicago, Chicago,Ill.,1956
Radioth 35.1076

RADIOISOTOPE CONFERENCE 2nd Proceedings Oxford 1954 Jul 19-23 1: medical and physiological applications Edited by J.E. Johnston and others Sponsored by the Atomic energy research establishment Butterworths, London,1954 First conference called Isotope techniques conference,q.v.
Radioth 35.1349

RADIOISOTOPE CONFERENCE 2nd Proceedings Oxford 1954 Jul 19-23 Vol 1: medical and physiological applications Edited by J.E. Johnston and others Sponsored by the Atomic energy research establishment Butterworths, London,1954 Previous conference has title "Isotope techniques conference",Oxford,1951
Bioch 33.0300

RADIOISOTOPE SAMPLE MEASUREMENT TECHNIQUES IN MEDICINE AND BIOLOGY Symposium on radioisotope sample measurement techniques in medicine and biology Proceedings Vienna 1965 May 24-25 Held by the International atomic energy agency International atomic energy agency.Proceedings series International atomic energy agency,Vienna,1965
Radioth 35.1377

RADIOISOTOPE TECHNIQUES Isotope techniques conference Proceedings Oxford 1951 Jul Vol 2: industrial and allied research applications Great Britain.Ministry of supply Sponsored by the Atomic energy research establishment H.M.S.O.,London,1952
Chem 18.2467

RADIOISOTOPE TECHNIQUES Isotope techniques conference Proceedings Oxford 1951 Jul 16-20 Vol 1: medical and physiological applications Sponsored by the Atomic energy research establishment H.M.S.O.,London,1953 Subsequent conference called Radioisotope conference,q.v.
Radioth 35.1348

RADIOISOTOPE TECHNIQUES IN THE STUDY OF PROTEIN METABOLISM Radioisotope techniques in the study of protein metabolism :a panel Vienna 1964 Jun 1-5 International atomic energy agency International atomic energy agency. Technical reports series, 45 International atomic energy agency,Vienna,1965
Radioth 35.1150

RADIOISOTOPE TECHNIQUES IN THE STUDY OF PROTEIN METABOLISM :a panel Radioisotope techniques in the study of protein metabolism Vienna 1964 Jun 1-5 International atomic energy agency International atomic energy agency.Technical reports series, 45 International atomic energy agency,Vienna,1965
Radioth 35.1150

RADIOISOTOPES In vitro procedures with radioisotopes in medicine :symposium Proceedings Vienna 1969 Sep 8-12 International atomic energy agency World health organization IAEA,Vienna,1970
Bioch 33.2184

RADIOISOTOPES Production and use of short-lived radioisotopes from reactors Seminar on the practical applications of short-lived radioisotopes produced in small research reactors Proceedings Vienna 1962 Nov 5-9 1-2 Held by the International atomic energy agency International atomic energy agency.Proceedings series 2 vols International atomic energy agency,Vienna,1963
Radioth 35.1369

RADIOISOTOPES Radioisotpoes in scientific research International conference on radioisotopes in scientific research 1st Proceedings Paris 1957 Sep 4-20 1-4 Edited by R.C. Extermann Held under the auspices of Unesco illus. 4 vols Pergamon press,London,1958
Radioth 35.1070

RADIOISOTOPES Symposium on radioisotopes presented at the second Pacific area national meeting Los Angeles 1956 Sep 21 American society for testing materials A.S.T.M. Special technical publication, 215 A.S.T.M., Philadelphia,1956
Met 25.1559

RADIOISOTOPES Symposium on radiosotopes in metals analysis and testing presented to the sixty-second annual meeting Atlantic City,N. J. 1959 Hun 22 American society for testing materials A.S.T.M.Special technical publication, 261 A.S.T.M.,Philadelphia,Pa., 1960
Met 25.1741

RADIOISOTOPES Symposium on the use of radioisotopes in soil mechanics Cleveland, Ohio 1952 Mar 5 American society for testing materials American society for testing materials.Special technical publication, 134 American society for testing materials,Philadelphia,Pa.,1953
Eng 41.3147

RADIOISOTOPES AND RADIATION IN ENTOMOLOGY :a symposium Proceedings Bombay 1960 Dec 5-9 By H.J. Bhabha and others Sponsored by the International atomic energy agency International atomic energy agency,Vienna,1962
Bal 39.2390

RADIOISOTOPES IN SCIENTIFIC RESEARCH International conference on radioisotopes in scientific research 1st Proceedings Paris 1957 Sep Vol 1: research with radioisotopes in physics and industry Unesco Edited by R.C. Extermann Pergamon,London, 1958
Met 25.1634

RADIOISOTOPES IN SOIL PLANT NUTRITION STUDIES The Symposium on the use of radioisotopes in soil-plant nutrition studies Proceedings Bombay 1962 Feb 26-Mar 2 International atomic energy agency United Nations.Food and agriculture organization International atomic energy agency,Vienna,1962
Bot 42.1368

RADIOISOTOPES IN TROPICAL MEDICINE Symposium on the use of radioisotopes in the study of endemic and tropical diseases Proceedings Bangkok 1960 Dec 12-16 International atomic energy agency World health organization International atomic energy agency.Proceedings series International atomic energy agency,Vienna,1962
Radioth 35.1365

RADIOISOTPOES IN SCIENTIFIC RESEARCH International conference on radioisotopes in scientific research 1st Proceedings Paris 1957 Sep 4-20 1-4 Edited by R. C. Extermann Held under the auspices of Unesco illus. 4 vols Pergamon press, London,1958
Radioth 35.1070

RADIOLOGICAL HEALTH AND SAFETY IN MINING AND MILLING OF NUCLEAR MATERIALS Symposium on radiological health and safety in mining and milling of nuclear materials :a symposium Proceedings Vienna 1963 Aug 26-31 1-2 International atomic energy agency International labour organization In co-operation with the World health organization International atomic energy agency.Proceedings series International atomic energy agency, Vienna,1964
Radioth 35.1164

RADIOLOGY International congress of radiology 7th Invited papers Copenhagen 1953 Jul 19-24 Acta radiologica.Supplementum, 116 Stockholm,1954
Radioth 35.1311

RADIOLOGY Research in radiology :an informal conference Proceedings Highland park,Ill. 1957 May 10-12 Edited by Henry S. Kaplan Organized by the National research council. Subcommittee on radiobiology National research council.Nuclear science series.Report, 22 National research council.Publication, 571 National academy of science,Washington,D. C.,1958
Radioth 35.1116

RADIOPROTECTIVE DRUGS Symposium on radiosensitizers and radioprotective drugs 1st Milan 1964 Edited by R. Paoletti and R. Vertus Organized by the European society for biochemical pharmacology Progress in biochemical pharmacology, 1 Karger,Basel,1965
Radioth 35.1943

RADIOSENSITIZERS Symposium on radiosensitizers and radioprotective drugs 1st Milan 1964 Edited by R. Paoletti and R. Vertus Organized by the European society for biochemical pharmacology Progress in biochemical pharmacology, 1 Karger,Basel,1965
Radioth 35.1943

RADIOSENSITIZING AND RADIOPROTECTIVE DRUGS International symposium on radiosensitizing and radioprotective drugs 2nd Rome 1969 May 6-8 book of abstracts Istituto superiore de sanita,Rome,1969
Radioth 35.1213

RADIOTHERAPY Biochemisch nachweisbare strahlenwirkungen und deren beziehungen zur strahlentherapie :symposium der Arbeitsgemeinschaft fur strahlenbiologie Stuttgart 1969 May 11 Arbeitsgemeinschaft fur strahlenbiologie Edited by G.B. Gerber Thieme,Stuttgart,1970
Radioth 35.1222

RADIOTHERAPY Conference on radiobiology and radiotherapy Proceedings Colorado Springs 1965 Nov 1-3 National cancer institute Edited by J.A. Del Regato National cancer institute.Monograph, 24 National cancer institute,Bethesda,Md.,1967
Bioch 33.0989

RADIOTHERAPY Planning of radiotherapy facilities Joint IAEA-WHO meeting on planning of radiotherapy facilities Report Geneva 1964 Dec 15-19 International atomic energy agency World health organization World health organization.Technical report series, 328 World health organization,Geneva, 1966
Radioth 35.1060

RAILWAY ELECTRIFICATION AT INDUSTRIAL FREQUENCY British railways electrification conference Proceedings London 1960 Oct 3 British transport commission Illus British transport commission,London,1960
Eng 41.1605

RAILWAYS :international engineering congress. Section one of a meeting held at the University of Glasgow Proceedings Glasgow 1901 Sep 3-5 Clowes,London,1902
Eng 41.1553

RAIN STIMULATION Artificial stimulation of rain Conference on the physics of cloud and precipitation particles 1st Proceedings Woods Hole,Mass. 1955 Sep 7-10 Air force Cambridge research center.Geophysics research directorate Edited by Helmut Weickmann and Waldo Smith Co-sponsored by United States. Office of naval research.Geophysics branch illus 442p Pergamon,London,1957
Nap 11.0931

RALEIGH 1964 Materials science research :a conference Proceedings Vol 3: role of grain boundaries and surfaces in ceramics North Carolina state university Edited by W. Wurth Kriegel and Hayne Palmour Plenum press, New York,1966
Met 25.0288

RALEIGH,N.C. 1956 Experimental designs in industry Symposium on design of industrial experiments proceedings North Carolina state college.Institute of statistics Edited by Victor Chew Supported by the United States.Air force.Office of scientific research Wiley publications in statistics Bibliog. xi,268p 28cm John Wiley and sons,New York, 1958 Being parts A and B of Symposium on design of industrial experiments, proceedings, edited by Victor Chew,1957
Math 3.0583

RALEIGH,N.C. 1956 Symposium on design of industrial experiments proceedings North Carolina state college.Institute of statistics Edited by Victor Chew Supported by the United States.Air force.Office of scientific research Bibliog. viii,374p 27cm Institute of statistics,University of North Carolina,Raleigh,N.C.,1957
Math 3.0584

RALEIGH,N.C. 1960 Mechanical properties of engineering ceramics :a conference Proceedings North Carolina state college and United States.Army.Office of ordnance research Edited by W.Wurth Kriegel and Hayne Palmour Interscience,New York;London,1961
Met 25.0853

RALEIGH,N.C. 1962 Research conference on structure and properties of engineering materials Proceedings North Carolina state college and United States.Army research office,Durham Materials science research, 1 Plenum press,New York,1963
Met 25.2281

RALEIGH,N.C. 1968 Clean surfaces their preparation and characterization for interfacial studies :based on a symposium North Carolina State university Edited by George Goldfinger Dekker,New York,1970
Met 25.2828

RAMAKRISHA MISSION INSTITUTE OF CULTURE International seminar on chromosome :its structure and function Proceedings Calcutta 1968 Aug 11-13 Edited by Arunkumar Sharma and Archana Sharma Nucleus, 1968,Supplementary volume Nucleus,Calcutta, 1968
Gen 34.2173

RANCHO SANTA FE,CALIF. 1967 Hormonal control system :a symposium Proceedings Edited by Edwin B. Stear and Arnold H. Kadish Mathematical biosciences.Supplement, 1 illus. American Elsevier,New York,1969
Math S 3.1543

RAND CORPORATION Monte Carlo method :a symposium Proceedings Los Angeles 1949 Jun 29-Jul 1 National bureau of standards Edited by A.S. Householder and others National bureau of standards.Applied mathematics series, 12 United States government printing office,Washington,D.C., 1951
Radioth 35.0586

RAND CORPORATION State of stress in the earth's crust :an international conference Proceedings Santa Monica,Calif. 1963 Jun 13-14 Edited by William R. Judd New York, 1964
Geod 9.0143

RAND CORPORATION The Rand symposium on mathematical programming:linear programming and recent extensions :a conference proceedings Santa Monica,Calif. 1959 Mar. 16-20 Edited by Philip Wolfe R-351 123p rand corporation,Santa Monica,Calif.,1960
Math L 5.2453

RAND CORPORATION and UNIVERSITY OF CALIFORNIA Proceedings of the symposium on the arctic heat budget and atmospheric circulation Lake Arrowhead 1966 Jan-Feb Edited by J.O. Fletcher Prepared for the National science foundation Memorandum RM-5233-NSF illus, diagrs 566p The Rand corporation,Santa Monica,1966
Sco 14.2407

The RAND SYMPOSIUM ON MATHEMATICAL PROGRAMMING: LINEAR PROGRAMMING AND RECENT EXTENSIONS :a conference proceedings Santa Monica, Calif. 1959 Mar.16-20 Edited by Philip Wolfe sponsored by Rand corporation R-351 123p rand corporation,Santa Monica,Calif., 1960
Math L 5.2453

RANDOM FOURIER SERIES Series de Fourier aleatoires mimeograph Montreal 1963 By Jean-Pierre Kahane North Atlantic treaty organization Societe mathematique du Canada Montreal.University.Seminaire de mathematiques superieures, 4 vii,174p 28cm Universite de Montreal,Montreal,1963
P. Math 2.2147

RANDOM FUNCTION THEORY Symposium on engineering applications of random function theory and probability 1st Proceedings National science foundation Purdue university Edited by J.L. Bogdanoff and F. Kozin Wiley,New York,1963
Eng 41.5949

RANDOM FUNCTION THEORY Symposium on engineering applications of random function theory and probability 1st proceedings Purdue university Edited by John L. Bogdanoff and Frank Kozin Supported by National science foundation x,421p 24cm John Wiley and sons,New York,1963
Math 3.0879

RANDOM WALKS Promenades aleatoires et mouvement brownien :introduction a la theorie des probabilites mimeograph Montreal 1963 By Anatole Joffe North Atlantic treaty organization Societe mathematique du Canada Montreal.University.Seminaire de mathematiques superieures, 7 vii,143p 28cm Universite de Montreal,Montreal,1964
P. Math 2.2145

RAPALLO 1958 Reticuloendothelial structure and function :international symposium Edited by John H. Heller International society for research on the reticuloendothelial system. Symposia, 3 Ronald press,New York,1960
An 32.3169

RAPID MIXING AND SAMPLING TECHNIQUES IN BIOCHEMISTRY International colloquium on rapid mixing and sampling techniques applicable to the study of biochemical reactions 1st Proceedings Philadelphia 1964 Jul 23-24 International union of biochemistry Edited by Britton Chance and others Academic press,New York;London,1964
Bioch 33.2320

RAPID MIXING AND SAMPLING TECHNIQUES IN BIOCHEMISTRY 1st :nternational coloquium Proceedings Philadelphia,Pa. 1964 Jul 23-24 International union of biochemistry Edited by Britton Chance and others Academic press,New York,1964
Pha 16.0141

RAPPORT DE LA COMMISSION INTERNATIONALE DE L'ANNEE POLAIRE 1932-33 Copenhagen 1933 May Vol 3: Compte-rendu des travaux de la commission Octobre 1931-Mai 1933 Organisation meteorologique internationale Ijdo,Leyden,1934
Sco 14.0491

RAPPORT DE LA COMMISSION INTERNATIONALE DE L'ANNEE POLAIRE 1932-33 Innsbruck 1931 Sep Vol 2: compte-rendu des travaux de la commission pendant sa deuxieme annee Organisation meteorologique internationale illus,diagrs 188p Ijdo,Leyden,1932
Sco 14.0492

RAPPORT DE LA COMMISSION INTERNATIONALE DE L'ANNEE POLAIRE 1932-33 Leningrad 1930 August Vol 1: compte-rendu des travaux de la commission pendant sa premiere annee de travail... Organization meteorologique internationale illus,diagrms 152p Ijdo, Leyden,1933
Sco 14.0493

RAQUETTE LAKE,N.Y. 1963 Fatigue - an interdisciplinary approach Sagamore army materials research conference 10th Proceedings Edited by J.J. Burke and others Sponsored by United States.Army materials research agency Syracuse university press, New York,1964
Eng 41.3803

RAQUETTE LAKE,N.Y. 1965 Strengthening mechanisms,metals and ceramics Sagamore army materials research conference 12th Proceedings Army materials research agency Syracuse university Syracuse university press,Syracuse,N.Y.,1966
Met 25.2746

RAQUETTE LAKE,N.Y. 1969 Ultrafine-grain metals Sagamore army materials research conference 16th Proceedings Army materials research agency Syracuse university Syracuse university press, Syracuse,N.Y.,1970
Met 25.2747

RAQUETTE LAKE,NEW YORK 1962 Fundamentals of deformation processing Sagamore army materials conference 9th By Walter A. Backofen and others United States.Army materials research agency and Syracuse university Co-organized and directed by the National research council Syracuse university press,Syracuse,New York,1964
Met 25.0741

RAQUETTE LAKE,NEW YORK 1963 Fatigue-an interdisciplinary approach Sagamore army materials research conference 10th United States.Army materials research agency and Syracuse university Edited by John J. Burke and others Syracuse university press, Syracuse,1964
Met 25.2330

RAQUETTE LAKE,NEW YORK 1966-67 Surfaces and interfaces 1-2 Sagamore army materials research conference 13th-14th Proceedings United States.Army materials research agency and Syracuse university Edited by John J. Burke and others Syracuse university press, New York,1967-68
Met 25.2365

RAREFIED GAS DYNAMICS International symposium on rarefied gas dynamics 3rd Proceedings Paris 1962 Jun Vol 1-2 Edited by J.A. Laurmann Advances in applied mechanics.Supplement, 2 2 vols Academic press,New York;London,1963
Eng 41.6774

RAREFIED GAS DYNAMICS International symposium on rarefied gas dynamics 4th Proceedings Toronto 1964 Jul Vol 1-2 United States.Air force.Office of scientific research United States.Office of naval research National aeronautics and space administration.Institute for aerospace studies Edited by J.H.de Leeuw Advances in applied mechanics.Supplement, 3 2 vols Academic press,New York;London,1965
Eng 41.6788

RAREFIED GAS DYNAMICS International symposium on rarefied gas dynamics 4th proceedings Toronto 1964 Jul.14-17 Vol 1-2 University of Toronto.Institute for aerospace studies Edited by J.H.de Leeuw Advances in applied mechanics.Supplement, 3 2 vols Academic press,New York,London,1965-66
A Math 4.0532

RAREFIED GAS DYNAMICS International symposium on rarefied gas dynamics 5th Proceedings Oxford 1966 Supplement 4 By C.L. Brundin New York;London,1967
A Math 4.1302

RAREFIED GAS DYNAMICS International symposium on rarefied gas dynamics 6th Proceedings Cambridge,Mass. 1968 Jul Vol 1-2 United States.Air force.Office of scientific research National aeronautics and space administration United States.Office of naval research Edited by Leon Trilling and Harold Y. Wachman Advances in applied mechanics.Supplement, 5 2 vols Academic press,New York,1969
Eng 41.6797

RAREFIED GAS DYNAMICS International symposium on rarefied gas dynamics 2nd Proceedings Berkeley,Calif. 1966 University of California Edited by L. Talbot Advances in applied mechanics.Supplement, 1 Academic press,New York;London,1961
Eng 41.6764

RAREFIED GAS DYNAMICS The International symposium on rarefied gas dynamics 2nd proceedings Berkeley,Calif. 1960 Aug.3-6 University of California Edited by L. Talbot Advances in applied mechanics.Supplement, 1 Academic press,New York,London,1961
A Math 4.0529

RAREFIED GAS DYNAMICS 2nd :international symposium Proceedings Berkeley,Calif. 1960 Edited by L. Talbot Held at the University of California Advances in applied mechanics.Supplement, 1 x,748p Academic press,New York;London,1961
Chem E 24.0408

RAREFIED GAS DYNAMICS 5th :international symposium Proceedings Oxford 1966 Jul Vol 1-2 Edited by C.L. Brundin Advances in applied mechanics.Supplement, 4 2 vols Academic press,New York;London,1967
Chem E 24.0409

RAREFIED GAS DYNAMICS 6th :international symposium Proceedings Cambridge,Mass. 1968 Jul Vol 1-2 Edited by Leon Trilling and Harold Y. Wachman Advances in applied mechanics.Supplement, 5 2 vols Academic press,New York;London,1969
Chem E 24.1814

RAT QUALITY:A CONSIDERATION OF HEREDITY,DIET AND DISEASE :a symposium Proceedings New York 1952 Jan 31 By W.E. Heston and others National vitamin foundation National vitamin foundation,New York,1953
Radioth 35.0218

RATS AND MICE Pathology of laboratory rats and mice London 1966 Apr 19-21 Nuffield foundation.Advisory committee Edited by Ernest Cotchin and Francis J.C. Roe Held at the Royal veterinary field station Blackwell, Oxford,1967
Path 30.2438

RAZVITIE PROIZVODITELNYKH SIL VOSTOCHNOI SIBIRI chernaya metallurgiya Trudy 1958 Aug 18-26 Akademiya nauk S.S.S.R.Sovet po izucheniyu proizvoditelnykh sil Edited by G. I. Lyudogovskii maps 275p 27cm Izdatelstvo Akademii nauk SSSR,Moscow,1966
Sco 14.8216

REACTION TIME Attention and performance 2 Donders centenary symposium on reaction time Proceedings Eindhoven 1968 Jul 29-Aug 2 Institute for perception research IPO Edited by W.G. Koster North-Holland,Amsterdam; London,1969
Psy 31.2948

REACTIONS OF COORDINATED LIGANDS AND HOMOGENEOUS CATALYSIS :a symposium Papers Washington,D.C. 1962 Mar 22-24 American chemical society.Division of inorganic chemistry American chemical society.Advances in chemistry series,37 American chemical society,Washington,D.C.,1963 Symposium chairman:D.H.Busch
Chem 18.1010

REACTIONS OF FREE RADICALS IN THE GAS PHASE :a symposium Report Cambridge 1957 Apr 9-12 Chemical society Chemical society. Special publication,9 Chemical society, London,1957 Organized by Dr.Sugden
Chem 18.1218

REACTIVE METALS 3rd annual conference Proceedings Buffalo 1958 May 27-29 Vol 2 American institute of mining, metallurgical and petroleum engineers.Niagara frontier section Edited by W.R. Clough Interscience,New York;London,1959
Met 25.0505

REACTIVITY OF PHOTO-EXCITED ORGANIC MOLECULA Solvay conference on chemistry 13th Proceedings Brussels 1965 Oct 25-30 International institute of chemistry,and, Universite libre de Bruxelles Interscience, London,1967
Col S 12.0568

REACTIVITY OF SOLIDS International symposium on the reactivity of solids 4th Amsterdam 1960 May 30th-Jun 4th Nederlandse chemische vereniging and Nederlandse naturkundige vereniging Edited by J.H.de Boer and others Sponsored by the International union of pure and applied physics Elsevier,Amsterdam,1961 Papers in English,French and German
Met 25.1848

REACTIVITY OF SOLIDS International symposium on the reactivity of solids 4th Proceedings Amsterdam 1960 May 30-Jun 4 Nederlandse chemische vereniging and International union of pure and applied physics Edited by J.H. DeBoer and others Sponsored also by the International union of pure and applied chemistry illus x,762p
Chem M 24.0882

REACTIVITY OF SOLIDS International symposium on the reactivity of solids 5th Munich 1964 Aug 2-8 Deutsche Bunsen-gesellschaft fur physikalische chemie Edited by G.M. Schwab Sponsored by the International union of pure and applied chemistry Elsevier, Amsterdam,1965 Papers in English,French and German
Met 25.1851

REACTIVITY OF SOLIDS International symposium on the reactivity of solids 4th Proceedings Amsterdam 1960 May 30-Jun 4 Edited by J.H.de Boer and others Elsevier, Amsterdam,1961
Min 10.0648

REACTIVITY OF SOLIDS International symposium on the reactivity of solids 5th Plenary lectures Munich 1964 Aug 2-8 International union of pure and applied chemistry,and,Deutsche bunsengesellschaft fur physikalische chemie Butterworths,London, 1965
Min 10.0662

REACTIVITY OF SOLIDS International symposium on the reactivity of solids 5th papers Munich 1964 Aug 2-8 Edited by G.-M. Schwab Elsevier,Amsterdam,1965
Min 10.1278

REACTIVITY OF SOLIDS International symposium on the reactivity of solids 6th Proceedings Schenectady,N.Y. 1968 Aug 25-30 United States.Air force.Office of scientific research General electric research and development center International union of pure and applied chemistry. Physical chemistry center Edited by J.W. Mitchell and others Wiley,New York;London,1969
Met 25.2446

REACTIVITY OF SOLIDS 4th :international symposium Proceedings Amsterdam 1960 May 30-Jun 4 International union of pure and applied physics,and,International union of pure and applied chemistry Edited by J.H. de Boer and others Elsevier,Amsterdam,1961
Chem 18.0596

REACTIVITY OF SOLIDS 5th :international symposium Proceedings Munich 1964 Aug 2-8 International union of pure and applied chemistry.Division for physical chemistry Edited by G.-M. Schwab Organised by the Deutsche Bunsen-gesellschaft fur physikalische chemie Elsevier,Amsterdam,1965
Chem 18.0608

REACTOR MATERIALS Properties of reactor materials and the effects of radiation damage : international conference Proceedings Berkeley Castle,Glos. 1961 May 30-Jun 2 Central electricity generating board.Berkeley nuclear laboratories Edited by D.J. Littler xv,562p Butterworths,London,1962
Cav 7.2864

REACTOR MATERIALS Radiation damage in solids and reactor materials :a symposium Proceedings Venice 1962 May 7-11 Vol 1-3 International atomic energy agency 3 vols International atomic energy agency, Vienna,1962
Cav 7.2393

READINESS TO REMEMBER Conference on learning, remembering and forgetting 3rd Proceedings Princeton,N.J. 1965 Vol 1-2 New York academy of sciences Edited by Daniel P. Kimble Gordon and Breach,New York,1969
Psy 31.3281

READING 1950 Semi-conducting materials :a conference Proceedings Edited by H.K. Henisch Under the auspices of the International union of pure and applied physics Butterworths,London,1951
Eng 41.5373

READING 1950 Semi-conducting materials :a conference Proceedings International union of pure and applied physics and University of Reading With the cooperation of the Royal society of London London,1951
Geod 9.0286

READING 1950 Semi-conducting materials :a conference Proceedings Royal society of London Edited by H.K. Henisch Under the auspices of the International union of pure and applied physics illus xv,281p Butterworths,London,1951
Cav 7.1925

READING 1958 British mathematical colloquium 10 lectures,typescript 84p 33cm University of Reading,Reading,1958
P. Math 2.2604

READING 1958 Utilization of nitrogen and its compounds by plants :a symposium Society for experimental biology Society for experimental biology.Symposia, 13 Cambridge university press,Cambridge,1959
Bioch 33.1358

READING 1960 International grassland congress 8th Proceedings British grassland society Grassland research institute,Hurley,1961
BG 38.3114

READING DISABILITY:PROGRESS AND RESEARCH NEEDS IN DYSLEXIA Johns Hopkins conference on research needs in dyslexia and related aphasic disorders Papers Baltimore,Md. 1961 Nov 15-17 Edited by John Money Johns Hopkins press,Baltimore,Md.,1962
An 32.4203

REAL-TIME CONTROL OF ELECTRIC POWER SYSTEMS Symposium of real-time control of electric power systems Baden 1971 Sep 27-28 Brown, Boveri and company Edited by Edmund Handschin Elsevier,Amsterdam,1972 The second Brown,Boveri symposium
Eng 41.8558

RECENT ADVANCES IN ADHESIVE TECHNOLOGY Adhesion ...Recent developments in adhesion science...Recent advances in adhesives technology...Presented at the 66th annual meeting A.S.T.M. Atlantic city,N.J. 1963 Jun 26 American society for testing materials American society for testing materials.Special technical publication, 360 American society for testing materials, Philadeplphia,Pa.,1964
Eng 41.4027

RECENT ADVANCES IN CLINICAL NEUROPHYSIOLOGY International congress of eletroencephalography and clinical neurophysiology 6th Proceedings Vienna 1965 Sep 5-10 functions of the spinal cord; neurophysiological investigations of brain diseases;EEG in stress Edited by L. Widen Electroencephalography and clinical neurophysiology.Supplement,25 Elsevier, Amsterdam,1967
Pha 16.0304

RECENT ADVANCES IN GELATIN AND GLUE RESEARCH : a conference Proceedings Cambridge 1957 Jul 1-5 Edited by G. Stainsby Sponsored by the British gelatine and glue research association Pergamon,London,1958
Bioch 33.2247

RECENT ADVANCES IN INVERTEBRATE PHYSIOLOGY : a symposium Eugene 1955 Sep Edited by B.T. Scheer and others held at the University of Oregon University of Oregon publications,Eugene,Oregon,1957 Sponsored by the National science foundation,Tektronix foundation,the University of Oregon
Bal 39.2107

RECENT ADVANCES IN MATRIX THEORY 2nd : advanced seminar proceedings Madison, Wis. 1963 Oct.14-16 United States army. Mathematics research center. Edited by Hans Schneider United States army.Mathematics research center.Publication, 12 xi,142p 25cm University of Wisconsin press,Madison, Wis.,1964
Math L 5.0328

RECENT ADVANCES IN MATRIX THEORY 2nd advanced seminar proceedings Madison, Wis. 1963 Oct.14-16 United States.Army. Mathematics research center Edited by Hans Schneider United States army.Mathematics research center.Publications, 12 xi,142p 24cm University of Wisconsin press,Madison; Milwaukee,Wis.,1964
Math 3.0164

RECENT ADVANCES IN MATRIX THEORY 2nd seminar proceedings Madison,Wis. 1963 Oct 14-16 United States.Army.Mathematics research center Edited by Hans Schneider United States army.Mathematics research center. Publications, 12 xi,142p 24cm University of Wisconsin press,Madison,Wis., 1964
P. Math 2.2087

RECENT ADVANCES IN OPTIMIZATION TECHNIQUES : a symposium Pittsburgh,Pa. 1965 Apr 21-23 Edited by A. Levi and T.P. Vogl Sponsored by the Institute of electrical and electronics engineers.Systems science and cybernetics group Wiley,New York,1966
Eng 41.6003

RECENT ADVANCES IN OPTIMIZATION TECHNIQUES symposium proceedings Pittsburgh,Pa. 1965 Apr.21-23 Institute of electrical and electronics engineers.Systems science and cybernetics group Edited by Abrahim Lavi and Thomas P. Vogl Jointly sponsored by the Optical society of America Bibliog. xiii, 656p 23cm John Wiley and sons,New York, 1966
Math 3.0240

RECENT ADVANCES IN RENAL DISEASES :a conference Proceedings London 1960 Jul 22-23 Royal college of physicians of London Edited by M.D. Milne Pitman,London, 1960
PGMS 29.0224

RECENT ADVANCES IN STRESS CORROSION :a symposium Stockholm 1960 Oct 12 Royal Swedish academy of engineering sciences. Corrosion committee Edited by Ake Bresle Almqvist and Wiksell,Stockholm,1961
Met 25.1960

RECENT ADVANCES IN THE STUDY OF PLANT METABOLISM a symposium Proceedings Allahabad 1958 Jan 22-24 Sponsored by the University grants commission Allahabad university studies.Botany section.Symposium supplement port. vi,140p Allahabad,1958
Bot 42.4704

RECENT ADVANCES IN TROPICAL ECOLOGY :a symposium Proceedings Banaras 1967 Jan 16-21 International society for tropical ecology Edited by R. Misra and B. Gopal 2 vols Vararasi,1968
Bot 42.2001

RECENT ASPECTS OF NITROGEN METABOLISM IN PLANTS a symposium Proceedings Nitrogen metabolism in plants Bristol 1967 Apr 18-19 Long Aston research station Edited by E. J. Hewitt and C.V. Cutting Long Ashton research station.Symposia, 1 Academic press,London;New York,1968
Bot 42.1893

RECENT DEVELOPMENTS IN ADHESION SCIENCE
Adhesion ...Recent developments in adhesion science...Recent advances in adhesives technology...Presented at the 66th annual meeting A.S.T.M. Atlantic city,N.J. 1963 Jun 26 American society for testing materials American society for testing materials.Special technical publication, 360 American society for testing materials, Philadeplphia,Pa.,1964
Eng 41.4027

RECENT DEVELOPMENTS IN ANNEALING London 1963 May 2 Iron and steel institute Iron and steel institute.Special report, 79 Iron and steel institute,London,1963
Met 25.0418

RECENT DEVELOPMENTS IN CELL PHYSIOLOGY :a symposium Bristol 1954 Mar 29-Apr 1 Colston research society Edited by J.A. Kitching Held at the University of Bristol Colston papers, 7 London,1954
Bal 39.0332

RECENT DEVELOPMENTS IN CELL PHYSIOLOGY : symposium Proceedings Bristol 1954 Mar 29-Apr 1 Colston research society Edited by J.A. Kitching Colston research society.Symposia, 7 Colston papers, 7 Butterworth,London,1954
Radioth 35.0463

RECENT DEVELOPMENTS IN CELL PHYSIOLOGY : symposium Proceedings Bristol 1954 Mar 29-Apr 1 Colston research society Edited by J.A. Kitching Colston research society.Symposia, 7 Colston papers, 7 Butterworths,London,1954
Bioch 33.0936

RECENT DEVELOPMENTS IN CELL PHYSIOLOGY 7th : symposium Proceedings Bristol 1954 Mar 29-Apr 1 Colston research society Edited by J.A. Kitching Colston papers, 7 Butterworths,London,1954
Phys 20.1481

RECENT DEVELOPMENTS IN MINERAL DRESSING :a symposium London 1952 Sep 23-25 Institution of mining and metallurgy Institution of mining and metallurgy,London, 1953
Met 25.0185

RECENT DEVELOPMENTS IN PARTICLE SYMMETRIES International school of physics 'Ettore Majorana 3rd course proceedings Erice 1965 Sep.-Oct. Edited by A. Zichichi Sponsored by the European organization for nuclear research 460p Academic press,New York,1966
A Math 4.0848

RECENT PROGRESS IN COMBINATORICS Waterloo conference on combinatorics 3rd Proceedings Waterloo,Ont. 1969 May 20-31 Edited by W.T. Tutte Supported by N.A.T.O. and National research council of Canada xiv, 347p 24cm Academic press,New York;London, 1969
P Math 2.3354

RECENT PROGRESS IN COMBINATORICS Waterloo conference on combinatorics 3rd Proceedings Waterloo,Ont. 1969 May 21-30 University of Waterloo xiv,341p 24cm Academic press,New York;London,1969
Math S 3.1660

RECENT PROGRESS IN HORMONE RESEARCH
Laurentian hormone conference Proceedings Franconia,N.H. 1949 Sep Vol 5 Edited by Gregory Pincus Academic press,New York,1950
Bal 39.1376

RECENT PROGRESS IN HORMONE RESEARCH, 26 The Laurentian hormone conference 27th Proceedings Mount Tremblant 1969 Aug 24-29 Edited by E.B. Astwood Academic press,New York;London,1970
Inv Med 37.0265

RECENT PROGRESS IN MATERIALS SCIENCE
Properties of crystalline solids :...symposium on recent progress in materials sciences; symposium on nature and origin of strength of materials Philadelphia 1960 Jun 27 A.S.T.M.Special technical publication Philadelphia, 1961 Includes an introductory paper by W. O.Baker on 'The national role of materials research and development'
Met 25.2490

RECENT PROGRESS IN MICROBIOLOGY :symposia presented at the 8th international congress for microbiology Montreal 1962 Aug 20-24 International union of microbiological societies Canadian society of microbiologists Edited by N.E. Gibbons University of Toronto press,Toronto,1963
Gen 34.0701

RECENT PROGRESS IN PHOTOBIOLOGY
International congress of photobiology 4th Proceedings Oxford 1964 Jul Edited by E. J. Bowen Under the auspices of the Comite international de photobiologie Bibliog., illus. vii,400p Blackwell scientific publications,Oxford,1965
Bot 42.1362

RECENT PROGRESS IN PHOTOBIOLOGY
International photobiology congress 4th Proceedings Oxford 1964 Jul 26-30 Edited by E.J. Bowen Under the auspices of the Comite internationale de photobiologie Blackwell,Oxford,1965
Radioth 35.0253

RECENT PROGRESS IN THE ENDOCRINOLOGY OF REPRODUCTION conference Proceedings Syracuse,N.Y. 1958 Jun 9-12 Edited by Charles W. Lloyd Academic press,New York; London,1959
VA 19.0282

RECENT RESEARCH IN FREEZING AND DRYING
International symposium on freezing and drying 2nd Proceedings London 1958 Edited by A.S. Parkes and Audrey U. Smith Organized by the Institute of biology Blackwell,Oxford, 1960
An 32.3233

RECENT RESEARCH ON CARNITINE:ITS RELATION TO LIPID METABOLISM :a symposium Papers presented Cambridge,Mass. 1964 Jul 24-25 Edited by George Wolf at the Massachusetts institute of technology M.I.T. press,Cambridge,Mass.,1965
Bioch 33.1249

RECENT RESEARCH ON GONADOTROPHIC HORMONES
Gonadotrophin club meeting 5th
Proceedings Edinburgh 1966 Sep 27-29
Gonadotrophin club Edited by E.Trevor Bell
and John A. Loraine Livingstone,Edinburgh;
London,1967
Inv Med 37.0015

RECENT STUDIES IN THE MECHANISMS OF EMBRYONIC
DEVELOPMENT :a conference New York
1947 Jan 10-11 New York academy of sciences.
Section of biology Edited by Robert Rugh
New York academy of sciences.Annals, 49,p 661-
866 New York,1948
Bal 39.0438

RECENT TRENDS IN GRAPH THEORY New York city
graph theory conference 1st Proceedings
New York 1970 Jun 11-13 Edited by M.
Capobianco and others Lecture notes in
mathematics, 186 vi,219p 25cm Springer,
Berlin,1971
P Math 2.4113

RECENT WORK ON NATURALLY OCCURRING NITROGEN
HETEROCYCLIC COMPOUNDS :a symposium
Report Exeter 1955 Jul 13-15 Chemical
society Edited by K. Schofield Chemical
society.Special publication, 3 London,1955
Bioch 33.2228

RECENT WORK ON THE INORGANIC CHEMISTRY OF SULPHUR
Developments in aromatic chemistry.
Applications of electron and nuclear resonance
in chemistry.Recent work on the inorganic
chemistry of sulphur :symposia Bristol
1958 Chemical society Chemical society.
Special publication, 12 Chemical society,
London,1958
Bioch 33.0630

RECHENANLAGEN Vortrage uber rechenanlagen
Gottingen 1953 Mar 19-21 Deutsche
forschungsgemeinschaft Edited by L. Biermann
145p 21cm Max-Planck institut fur physik,
Gottingen,1953
Math L 5.1087

La RECHERCHE EN ENSEIGNEMENT PROGRAMME:TENDANCES
ACTUELLES :colloque international Actes
Nice 1968 May 13-17 North Atlantic
treaty organization Sciences du comportement,
8 Dunod,Paris,1969 Director of the
colloquium:A.de Brisson.Papers in English and
French
Psy 31.3117

RECINIAO BRASILIERA DE GENETICA HUMANA
Parana: 1958: Nov 10-15 Universidade do
Parana Sociedade Brasileira de genetica
Edited by N. Freire-Maice and A. Quelee-
Solgrado port. Universidade do Parana,1959
Gen 34.1895

RECOVERY AND RECRYSTALLIZATION OF METALS :a
symposium Proceedings New York 1962
Feb 20-21 American institute of mining,
metallurgical and petroleum engineers Edited
by L. Himmel Interscience,New York;London,
1963
Met 25.1267

RECOVERY AND REPAIR MECHANICS IN RADIOGRAPHY :
symposium Report Upton,N.Y. 1967 Jun
5-7 Brookhaven national laboratory.Biology
department Brookhaven symposia in biology,
20 BNL50058)C-51) Brookhaven national
laboratory.Biology department,Upton,N.Y.,1968
Bioch 33.1298

RECOVERY AND REPAIR MECHANISMS IN RADIOBIOLOGY
report of symposium Upton,N.Y. 1967 Jun
5-7 Brookhaven national laboratory.Biology
department Brookhaven symposia in biology,
20 BNL 50058 Brookhaven national
laboratory,Upton,N.Y.,1968
Gen 34.1128

RECOVERY OF CELLS FROM INJURY Symposium on
recovery of cells from injury :given at
research conference for biology and medicine
of the atomic energy commission Gatlinburg,
Tenn. 1961 Apt 3-6 Sponsored by the Oak
Ridge national laboratory.Biology division
Journal of cellular and comparative physiology,
58,supp.1 Oak Ridge national laboratory.
Symposia, 14 Wistar institute of anatomy
and biology,Philadelphia,Pa.,1961
Gen 34.1114

RECOVERY OF METALS Recovery and
recrystallization of metals a symposium
Proceedings New York 1962 Feb 20-21
American institute of mining,metallurgical and
petroleum engineers.Physical metallurgy
committee Edited by L. Himmel Interscience,
New York,1963
Cav 7.1720

RECRYSTALLIZATION,GRAIN GROWTH AND TEXTURES a
seminar Papers 1965 Oct 16-17
American society for metals A.S.M.,Metals
Park,Ohio,1966
Met 25.1269

REDOX SYSTEMS Oxidases and related redox
systems :a symposium Proceedings Amherst,
Mass. 1964 Jul 15-19 Vol 1-2 Edited
by Tsoo E. King and others held at Amherst
college 2 vols Wiley,New York,1965
Bioch 33.1085

REDUNDANCY The Use of redundancy in system
design :a symposium Northampton 1964 Feb
14 Society of instrument technology
Society of instrument technology,1964
Eng 41.5963

REFRACTORIES FOR OXYGEN STEELMAKING technical
sessions of the annual general meeting...
Papers and discussions London 1962 May
3 Iron and steel institute Iron and steel
institute.Special report, 74 Iron and steel
institute,London,1962
Met 25.0273

REFRACTORY METALS Science and technology of
tungsten,tantalum,molybdenum,niobium and their
alloys Agard conference on refractory metals
Proceedings Oslo 1963 Jun 23-26 North
Atlantic treaty organization.Advisory group
for aeronautical research and development
AGARDograph 82 Pergamon press,Oxford,1964
Met 25.1011

REFRACTORY METALS AND ALLOYS :a technical
conference Detroit 1960 May 25-26
American institute of mining,metallurgical and
petroleum engineers.Refractory metals
committee Edited by M. Semchyshen and J.J.
Harwood Metallurgical society conferences,
11 Interscience,New York;London,1961
Met 25.2532

REFRACTORY METALS AND ALLOYS 3:APPLIED ASPECTS Technical conference on refractory metals and alloys 3rd Proceedings Los Angeles 1963 Dec 9-10 American institute of mining, metallurgical and petroleum engineers. Refractory metals committee Metallurgical society conferences, 30 Gordon and Breach, New York,1966
Met 25.1017

REFRACTORY METALS AND ALLOYS 4 research and development:based on a technical conference Proceedings French Lick,Ind. 1965 Oct 3-5 Vol 1-2 American institute of mining,metallurgical and petroleum engineers. Refractory metals committee Edited by R.I. Jaffee and others Metallurgical society conferences, 41 2 vols Gordon and Breach, New York,1967
Met 25.2250

REFRESHER COURSE IN RECENT ADVANCES IN CYTOGENETICS AND DEVELOPMENTAL GENETICS Corvallis,Ore. 1962 Aug 27-28 American society of zoologists American zoologist, 3,no.1 Utica,N.Y.,1963
Gen 34.0856

REGELUNGSTECHNIK:MODERNE THEORIEN UND IHRE VERWENDBARKEIT :ein tagung Bericht Heidelberg 1956 Sep 25-29 Gesellschaft fur angewandte mathematik und mechanik. Fachausschuss regelungsmathematik Verein deutscher ingenieure.Fachgruppe regelungstechnik Oldenbourg,Munich,1957
Eng 41.5842

REGENERATION AND WOUND HEALING :symposium Budapest 1960 Nov 8-9 Edited by G. Szanto Symposia biologica Hungarica, 3 Akademiai Kiado,Budapest,1964
An 32.0093

REGENERATION AND WOUND HEALING Proceedings Budapest 1960 Nov 8-9 Magyar tudonanyos akademia Symposia biologica hungarica, 3 Akademiai kiado,Budapest,1960
Phys 20.1001

REGENERATION IN ANIMALS AND RELATED PROBLEMS an international symposium Athens 1964 Apr Edited by V. Kiortsis and H.A.L. Trampusch Sponsored by the North Atlantic treaty organization.Division of scientific affairs North-Holland,Amsterdam,1965
An 32.5438

REGENERATION IN ANIMALS AND RELATED PROBLEMS an international symposium Athens 1964 Apr Edited by V. Kiortsis and H.A.L. Trampusch Sponsored by the North Atlantic treaty organization.Division of scientific affairs North-Holland,Amsterdam,1965 Text in English and French
Bal 39.1320

REGIONAL CONFERENCE FOR AFRICA 4th Proceedings Soil mechanics and foundation engineering South African institution of civil engineers 3 vols Balkema,Cape Town, 1967
Eng 41.3177

REGIONAL GEOGRAPHY OF JAPAN :I.G.U.regional conference in Japan Guidebook Japan 1957 1-7 International geographical union, and,Science council of Japan maps Tokyo, 1957
Geog 13.4507

REGIONAL NEUROCHEMISTRY... International neurochemical symposium 4th Proceedings Edited by Seymour S. Kety and Joel Elkes Pergamon press,London,1961
VA 19.0235

REGIONALISATION ET DEVELOPPEMENT Centre national de la recherche scientifique colloques internationaux sciences humaines Strasbourg 1967 Jun 26-30 Centre national de la recherche scientifique Centre national de la recherche scientifique,Paris,1968 Papers in English and French
Geog 13.7207

REGULATEURS NATURELS DE LA CROISSANCE VEGETALES Colloque international sur les substances de croissance vegetales 5e Actes Gif sur Yvette 1963 Jul 15-20 Centre national de la recherche scientifique Centre national de la recherche scientifique.Colloques internationaux, 123 Paris,1964 Text in English and French
Bioch 33.0747

REGULATION AND MODE OF ACTION OF THYROID HORMONES a colloquium Proceedings London 1956 Jun 20-22 Ciba foundation Edited by G. E.W. Wolstenholme and Elaine C.P. Millar Ciba foundation colloquia on endocrinology, 10 Churchill,London,1957
Inv Med 37.0240

REGULATION AND MODE OF ACTION OF THYROID HORMONES a colloquium Proceedings London 1956 Jun 20-22 Ciba foundation Edited by G. E.W. Wolstenholme and Elaine C.P. Millar Ciba foundation colloquia on endocrinology, 10 illus Churchill,London,1957
Bioch 33.0439

REGULATION OF BODY FLUID VOLUMES BY THE KIDNEY A Symposium on natriuretic hormone Smolenice castle 1969 Jun Edited by J.H. Cort and B. Lichardus Under the auspices of Ceskoslovenska akademie ved Karger,Basle, 1970
Inv Med 37.0268

REGULATION OF CELL METABOLISM Ciba foundation symposium on the regulation of cell metabolism London 1958 Jul 28-30 Ciba foundation Edited by G.E.W. Wolstenholme and Cecilia M. O'Connor Illus J.and A. Churchill,London,1959
Radioth 35.0493

REGULATION OF CELL METABOLISM Ciba foundation symposium on the regulation of cell metabolism Proceedings London 1958 Jul 28-30 Ciba foundation Edited by G.E.W. Wolstenholme and Cecilia M. O'Connor illus Churchill,London,1959
Bioch 33.0575

REGULATION OF ENZYME ACTIVITY AND ALLOSTERIC INTERACTIONS The Federation of European biochemical societies meeting 4th Proceedings Oslo 1976 Jul Federation of European biochemical societies Edited by E. Kvamme and A. Phil Organized by the Norwegian biochemical society Universitetsforlaget;Academic press,Oslo; London,1968 Symposium organizer J.P. Changeux
Bioch 33.1396

REGULATION OF ENZYME ACTIVITY AND SYNTHESIS IN NORMAL AND NEOPLASTIC TISSUES 3rd :a symposium Indianapolis 1964 Oct 5-6 Edited by George Weber Advances in enzyme regulation,3 Pergamon,Oxford,1965
Pha 16.0102

REGULATION OF ENZYME ACTIVITY AND SYNTHESIS IN NORMAL AND NEOPLASTIC TISSUES 4th :a symposium Indianapolis 1965 Oct 4-5 Edited by George Weber Advances in enzyme regulation,4 Pergamon,Oxford,1966
Pha 16.0103

REGULATION OF ENZYME ACTIVITY AND SYNTHESIS IN NORMAL AND NEOPLASTIC TISSUES 5th :a symposium Indianapolis 1966 Oct 3-4 Edited by George Weber Advances in enzyme regulation,5 Pergamon,Oxford,1967
Pha 16.0104

REGULATION OF GLUCONEOGENESIS Conference of the Gesellschaft fur biologische chemie 9th Proceedings Gottingen 1971 Jan 15 Gesellschaft fur biologische chemie Edited by Hans-Dieter Soling and Berend Willms Thieme;Academic press,Stuttgart;London,1971
Bioch 33.2210

REGULATION OF HUMAN RESPIRATION Haldane centenary symposium Proceedings Oxford 1961 Jul Edited by D.J.C. Cunningham and B.B. Lloyd bibliog.,port. Blackwells,Oxford, 1963
Phys 20.0857

REGULATION OF METABOLIC PROCESSES IN MITOCHONDRIA a symposium Proceedings Bari 1965 Apr 26-May 1 Edited by J.M. Tager and others organized by the University of Bari,Dept.of Biochemistry and the University of Amsterdam, Laboratory of Biochemistry B.B.A. library, 7 Elsevier,Amsterdam,1966
Bioch 33.1101

REGULATION OF METABOLIC PROCESSES IN MITOCHONDRIA symposium Proceedings Bari 1965 Apr 26-May 1 Edited by J.M. Tager and others Organized by the Universiteit van Amsterdam and Universita degli studi,Bari B.B.A. library, 7 Elsevier,Amsterdam,1966
Bal 39.0308

REGULATION OF NUCLEIC ACID AND PROTEIN BIOSYNTHESIS An International symposium on regulatory mechanisms in nucleic acid and protein biosynthesis Proceedings Lunteren 1966 Jun 5-10 Edited by V.V. Konigsberger and L. Bosch B.B.A.library, 10 Elsevier, Amsterdam,1967
Bot 42.1511

REGULATION OF NUCLEIC ACID AND PROTEIN BIOSYNTHESIS An International symposium on regulatory mechanisms in nucleic acid and protein biosynthesis Proceedings Lunteren 1966 Jun 5-10 Edited by V.V. Koningsberger and L. Bosch B.B.A. library, 10 Elsevier, Amsterdam,1967
Bioch 33.0826

REGULATION OF NUCLEIC ACID AND PROTEIN BIOSYNTHESIS :international symposium Lunteren 1966 Jun 5-10 Edited by V.V. Koningsberger and L. Bosch B.B.A.library, 10 Illus Elsevier,Amsterdam,1967
Radioth 35.0140

The REGULATION OF RESPIRATION :a conference New York 1963 Jun 24 By Harry Grundfest and others New York academy of sciences Edited by G.G. Nahas New York academy of sciences.Annals, 109,p411-948 New York,1963
Bal 39.1355

REGULATORY FUNCTIONS OF BIOLOGICAL MEMBRANES The Sigrid Juselius foundation symposium 2nd Proceedings Helsinki 1967 Nov 6-9 Sigrid Juselius foundation Edited by Johan Jarnefelt B.B.A.library, 11 illus. viii, 311p Elsevier,Amsterdam,1968
Bot 42.1517

REGULATORY FUNCTIONS OF BIOLOGICAL MEMBRANES The Sigrid Juselius foundation symposium 2nd Proceedings Helsinki 1967 Nov 6-9 Sigrid Juselius foundation Edited by Johan Jarnefelt BBA library, 11 Elsevier, Amsterdam,1968
Bioch 33.0968

REGULATORY FUNCTIONS OF BIOLOGICAL MEMBRANES 2nd a Sigrid Juselius foundation symposium Papers Helsinki 1967 Nov 6-9 Edited by Johan Jarnefelt B.B.A. library,11 Illus. viii,311p Elsevier,Amsterdam,1968
Pha 16.0017

REHOVOTH 1963 New perspectives in biology : a symposium Proceedings Edited by Michael Sela Held at the Weizmann institute of science B.B.A.Library, 4 xviii,285p Elsevier,Amsterdam,1964 A symposium held on the occasion of the inaugration of the Ullmann institute of life sciences
Bot 42.1502

REHOVOTH 1963 New perspectives in biology : a symposium Proceedings Edited by Michael Sela Held at the Weizmann institute of science B.B.A.library, 4 Elsevier, Amsterdam,1964 A symposium held,on the occasion of the inauguration of the Ullmann institute of life sciences
Bioch 33.0606

REHOVOTH,ISRAEL 1963 New perspectives in biology :a symposium Proceedings Edited by Michael Sela B.B.A.library,4 Elsevier, Amsterdam,1964
Pha 16.0273

REINFORCED CONCRETE ASSOCIATION One-day symposium on structural lightweight concrete Brighton 1962 Jun 26 Vol 1-2 2 vols RCA,London,1963
Eng 41.2968

RELATION BETWEEN THE STRUCTURE AND MECHANICAL PROPERTIES OF METALS National physical laboratory symposium 15th Proceedings Teddington 1963 Jan 7-9 Vol 1-2 National physical laboratory 2 vols H.M.S. O.,London,1963
Met 25.1214

The RELATION BETWEEN THE STRUCTURE AND MECHANICAL PROPERTIES OF METALS :a conference Proceedings Teddington 1963 Jan 7-9 Vol 1-2 National physical laboratory. Symposium, 15 2 vols H.M.S.O.,London,1963
Eng 41.3530

The RELATION OF HORMONES TO DEVELOPMENT :a symposium Papers Cold Spring Harbor 1942 Cold Spring Harbor biological laboratory Cold Spring Harbor symposia on quantitative biology, 10 Long Island biological association,Cold Spring Harbor,1942
Bioch 33.1266

The RELATION OF OPTICAL FORM TO BIOLOGICAL ACTIVITY IN AMINO-ACID SERIES :a symposium held at University college hospital medical school London 1947 Feb 15 Edited by R.T. Williams Biochemical society. Symposia, 1 Cambridge university press, Cambridge,1948
Radioth 35.0291

The RELATION OF OPTICAL FORM TO BIOLOGICAL ACTIVITY IN THE AMINO-ACID SERIES :a symposium London 1947 Feb 15 Biochemical society Edited by R.T. Williams Held at the University college hospital medical school Biochemical society.Symposia, 1 Cambridge university press,Cambridge, 1948
Bioch 33.1363

The RELATION OF OPTICAL FORM TO BIOLOGICAL ACTIVITY IN THE AMINO-ACID SERIES :a symposium London 1947 Feb 15 Edited by R.T. Williams Biochemical society. Symposia, 1 Cambridge university press, Cambridge,1948
Phys 20.1523

The RELATION OF OPTICAL FORM TO BIOLOGICAL ACTIVITY IN THE AMINO-ACID SERIES 1 :a symposium London 1947 Feb 15 Biochemical society Edited by R.T. Williams Biochemical society symposia,1 Cambridge university press,Cambridge,1948
Chem 18.1275

RELATION OF PROPERTIES TO MICROSTRUCTURE Cleveland 1953 Oct 17-23 American society for metals American society for metals, Cleveland,Ohio,1954 A symposium presented at the 35th National metal congress and exposition
Met 25.1201

RELATION OF PROPERTIES TO MICROSTRUCTURE :a seminar held during the thirty-fifth National metal congress and exposition Cleveland, Ohio 1953 Oct 17-23 Sponsored by the American society for metals American society for metals,Cleveland,Ohio,1954
Eng 41.3501

Les RELATIONS ENTRE PRECAMBRIEN ET CAMBRIEN PROBLEMES DES SERIES INTERMEDIAIRES Paris 1957 Jun 27- Jul 4 Centre national de la recherche scientifique Centre national de la recherche scientifique.Colloques internationaux,66 bibliog. Centre national de la recherche scientifique,Paris,1958
Geol 8.2716

The RELATIONSHIP BETWEEN NUCLEUS AND CYTOPLASM a symposium Proceedings Brussels 1958 Jan 9-13 Organised by the International society for cell biology Experimental cell research.Supplement, 6 Academic press,New York,1959
An 32.2486

The RELATIONSHIP BETWEEN NUCLEUS AND CYTOPLASM symposium Proceedings Brussels 1958 Jun 9-13 Organized by the International society for cell biology Experimental cell research,Supp. 6,1959 illus. Academic press,New York,1959
Gen 34.0837

RELATIVISTIC ASTROPHYSICS Quasi stellar sources and gravitational collapse Texas symposium on relativistic astrophysics 1st Proceedings Dallas,Texas 1963 Dec 16-18 By Ivor Robinson and others Southwest center for advanced studies Sponsored by University of Texas 475p 25cm University of Chicago press,Chicago,Ill.,1965
Cav 7.1985

RELATIVISTIC ASTROPHYSICS Quasi-stellar sources and gravitational collapse Texas symposium on relativistic astrophysics 1st proceedings Dallas,Texas 1963 Dec. 16-18 Southwest center for advanced studies Edited by Ivor Robinson and others sponsored by University of Texas xvii,475p 25cm University of Chicago press,Chicago,1965
Obs 6.1457

RELATIVISTIC ASTROPHYSICS Quasi-stellar sources and gravitational collapse Texas symposium on relativistic astrophysics 1st proceedings Dallas,Texas 1963 Dec.16-18 Southwest center for advanced studies Edited by Ivor Robinson and others Sponsored by University of Texas xvii,475p 25cm University of Chicago press,Chicago,1965
A Math 4.1161

RELATIVISTIC FLUID MECHANICS Fundamental topics in relativistic fluid dynamics and magnetohydrodynamics :a symposium proceedings Ann Arbor 1962 Oct.10-11 Edited by Robert Wasserman and Charles P. Wells Academic press,New York,London,1963
A Math 4.0528

RELATIVISTIC THEORIES OF GRAVITATION A Conference on gravitation Proceedings Jablonna 1962 Jul 25-31 Polska akademia nauk Edited by Leopold Infeld With financial support of the International union of pure and applied physics Pergamon press, Oxford,1964 Binder's title in French
TA 15.0085

RELATIVISTIC THEORIES OF GRAVITATION Conference on relativistic theories of gravitation proceedings Jablonna 1962 Jul 25-31 Edited by Leopold Infeld Organised and sponsored by Polska akademia nauk Pergamon press,Oxford,1964
A Math 4.1029

RELATIVITY Astrophysics and general relativity Brandeis university.Summer institute in theoretical physics 11th Proceedings Waltham,Mass. 1968 Jun 17-Jul 26 Edited by Max Chretien and others 2 vols Gordon and Breach,New York,1971
TA 15.0657

RELATIVITY Astrophysics and general relativity Brandeis university.Summer institute of theoretical physics 11th Waltham,Mass. 1968 Vol 1 Edited by M. Chretien and others illus 300p 23cm 2 vols Gordon and Breach,New York,1969
A Math 4.1791

RELATIVITY Centenario della nascita di Galileo Galilei 1564-1964 4th Atti Florence 1964 Sep 9-12 Vol 2,tomo 1: atti del convegno sulla relativita generale, problemi di energia e onde gravitazionali. Proceedings of the meeting on general electricity problems and gravitational waves Galileo Galilei.Comitato nazionale per le manifestazioni celebrative Galileo Galilei. Comitato nazionale per le manifestazioni celebrative.Publicazioni 267p Barbera, Florence,1965
WSM 43.4484

RELATIVITY Convegno sulla relativita: problemi di energia e onde gravitazionali; meeting on general relativity:problems of energy and gravitational waves Atti; proceedings Florence 1964 Sep 9-12 Quatro centenario della nascita di Galileo Galilei,1564-1964.Comitato nazionale per le manifestazioni celebrative Edited by Giorgio Sestini Quatro centenario della nascita di Galileo Galilei,1564-1964.Comitato nazionale per le manifestazioni celebrative. Pubblicazioni, 2,Atti dei convegni, 1 267p Florence,1966
Obs 6.3276

RELATIVITY Einstein symposium Entstehung, entwicklung und perspektiven der Einsteinschen gravitationstheorie vortrage und diskussionen Berlin 1965 Nov 2-5 Deutsche akademie der wissenschaften zu Berlin Edited by H-J. Treder 314p Akademie-verlag, Berlin,1966
Obs 6.1765

RELATIVITY Funfzig jahre relativitatstheorie. Jubilee of relativity theory Verhandlungen Bern 1955 Jul 11-16 Edited by A. Mercier and M. Kervaire Helvetica physica acta. Supplementum, 5 Birkhauser,Basel,1956 In French,German,English
Radioth 35.1599

RELATIVITY Lectures on general relativity Waltham,Mass. 1964 By A. Trautman Brandeis university.Summer institute in theoretical physics Lectures in theoretical physics,1964,1 Prentice-Hall,Englewood Cliffs,N.J.,1965
A Math 4.1025

RELATIVITY Methods of local and global differential geometry in general relativity : conference Pittsburgh,Pa. 1970 Jul 13-17 Edited by D. Farnsworth and others Lecture notes in physics illus 190p 25cm Springer,Heidelberg,1972
A Math 4.1789

RELATIVITY Relativity conference in the midwest Proceedings Cincinnati,Ohio 1969 Jun 2-6 Edited by Moshe Carmel and others Sponsored jointly by the Aerospace research laboratories and the University of Cincinnati bibliog.,illus. xii,381p 26cm Plenum,New York,1970
A Math 4.1622

RELATIVITY Relativity conference of the Midwest Proceedings Cincinnati,Ohio 1969 Jun 2-6 Aerospace research laboratories University of Cincinnati Edited by M. Carmeli and others 381p Plenum press,New York,1970
TA 15.0506

RELATIVITY,GROUPS AND TOPOLOGY lectures Les Houches 1963 Universite de Grenoble. Ecole d'ete de physique theorique,Les Houches Edited by C. DeWitt and B. DeWitt Universite de Grenoble.Ecole d'ete de physique theorique, Les Houches.Lectures,1963 Gordon and Breach, New York;London,1963
A Math 4.1030

RELATIVITY,GROUPS AND TOPOLOGY Papers Les Houches 1963 Universite de Grenoble.Ecole d'ete de physique theorique,Les Houches Edited by C. DeWitt and B. DeWitt Universite de Grenoble.Ecole d'eete de physique theorique, Les Houches. Lectures,1963 929p Gordon, Breach,London,1964
TA 15.0343

RELATIVITY CONFERENCE IN THE MIDWEST Proceedings Relativity Cincinnati,Ohio 1969 Jun 2-6 Edited by Moshe Carmel and others Sponsored jointly by the Aerospace research laboratories and the University of Cincinnati bibliog.,illus. xii,381p 26cm Plenum,New York,1970
A Math 4.1622

RELATIVITY CONFERENCE OF THE MIDWEST Proceedings Relativity Cincinnati,Ohio 1969 Jun 2-6 Aerospace research laboratories University of Cincinnati Edited by M. Carmeli and others 381p Plenum press,New York,1970
TA 15.0506

RELATIVITY THEORY AND ASTROPHYSICS 1: relativity and cosmology Lectures Ithaca,N.Y. 1965 Jul 25-Aug 20 By W. Bonner and others American mathematical society.Summer seminar in applied mathematics Edited by Jurgen Ehlers Lectures in applied mathematics, 8 xvi,292p American mathematical society,Providence,R.I.,1967
TA 15.0350

RELATIVITY THEORY AND ASTROPHYSICS 2:galactic structure Lectures Ithaca,N.Y. 1965 Jul 25-Aug 20 By E. Burbidge and others American mathematical society.Summer seminar in applied mathematics Edited by Jurgen Ehlers viii,220p American mathematical society,Providence,R.I.,1967
TA 15.0351

The RELIABILITY AND MAINTENANCE OF DIGITAL COMPUTER SYSTEMS:MANAGERIAL AND ENGINEERING ASPECTS :discussion meetings London 1960 Jan 20-21 British computer society and Institution of electrical engineers 70p 28cm British computer society; Institution of electrical engineers,London, 1960
Math L 5.3255

REMEIS-STERNWARTE,BAMBERG.KLEINE VEROFFENTLICHUNGEN, 4,no 40 Position of variable stars in the Hertzsprung-Russell diagram Colloquium on variable stars 3rd Bamberg 1965 Aug 11-14 International astronomical union.Commissions 27 and 42 port 304p Astronomisches institut der universitat Erlangen-Nurnberg,Bamberg,1965 I.A.U. combined colloquium of commissions 27 and 42 with the theme 'The position of stars'
TA 15.0540

REMOBILIZATION OF ORES AND MINERALS Convegno sulle rimobilizzazione del minerali metallici e non metallici Rendiconti minerali metallici e non metallici Papers Cagliari 1969 Aug Associazzione mineralia sarda Istituto di giacimenti minerali universitaria Cagliari,c1969
Min 10.1489

The REMOTE EFFECTS OF CANCER ON THE NERVOUS SYSTEM :a symposium Proceedings Rochester N.Y. 1964 Sep 30-Oct 1 Edited by Russell Brain and Forbes H. Norris Sponsored by the University of Rochester school of medicine and dentistry. Division of neurology Contemporary neurology symposia, 1 Grune and Stratton,New York;London,1965
PGMS 29.0451

RENAL DISEASE Recent advances in renal diseases :a conference Proceedings London 1960 Jul 22-23 Royal college of physicians of London Edited by M.D. Milne Pitman, London,1960
PGMS 29.0224

RENAL DISEASE AND HYPERTENSION :tentn annual conference... Proceedings Denver 1971 Mar Atlantic City 1971 May United States.Public health service Edited by Richard H. Thurm Proceedings of the annual conferences of the U.S.Public health service co-operative study (antihypertensive agents), 8 U.S.Dept. of health,education and welfare, Bethesda,1971
PGMS 29.0659

RENAL FAILURE Acute renal failure :a symposium London 1963 Sep Edited by S. Shaldon and G.C. Cook Held at the Royal free hospital Blackwell,Oxford,1964
PGMS 29.0225

RENAL FAILURE AND REPLACEMENT OF RENAL FUNCTION European dialysis and transport association a conference 2nd Proceedings Newcastle-upon-Tyne 1965 Vol 2 European dialysis and transplant association Edited by David Kerr Excerpta medica.International congress series, 103 Excerpta medica, Amsterdam,1966
Inv Med 37.0130

RENAL FUNCTION 1st :conference Transactions New York 1949 Oct 20-21 Edited by Stanley E. Bradley Josiah Macy,jr., foundation,New York,1950
Pha 16.0284

RENAL METABOLISM AND EPIDEMIOLOGY OF SOME RENAL DISEASES Annual conference on the kidney Proceedings Princeton 1961 Oct 12-13 National kidney disease foundation Edited by Jack Metcoff National kidney disease foundation,New York,1964
Inv Med 37.0158

The RENAL STONE RESEARCH SYMPOSIUM Proceedings Leeds 1968 Apr 17-20 Medical research council and University of Leeds Edited by A. Hodgkinson and B.E.C. Nordin Churchill, London,1969
PGMS 29.0240

RENCONTRE FRANCO-BRITANNIQUE Proceedings Theorie des matroides Brest 1970 May 14-15 Edited by C.P. Bruter Lecture notes in mathematics, 211 108p 25cm Springer, Berlin,1970
P Math 2.4508

RENSSELAER POLYTECHNIC INSTITUTE Interstellar grains :an international colloquium Proceedings Troy,N.Y. 1965 Aug 24-26 Edited by J.Mayo Greenberg and T.P. Roark NASA SP-140 National aeronautics and space administration,Washington,D.C.,1967
TA 15.0445

RENSSELAER POLYTECHNIC INSTITUTE Neutron physics :symposium Proceedings Troy,N.Y. 1961 May 5-6 Edited by M.L. Yeater Co-sponsored by the American nuclear society Nuclear science and technology:a series of monographs and text books, 2 xiii,303p Academic press,New York,1962
Cav 7.2543

REPAIR FROM GENETIC RADIATION DAMAGE AND DIFFERENTIAL RADIOSENSITIVITY IN GERM CELLS International symposium on repair from genetic radiation damage Proceedings Leiden 1962 Aug 15-19 Riksuniversiteit te Leiden Edited by F.H. Sobels Pergamon press,Oxford, 1963
Gen 34.1121

REPAIR FROM GENETIC RADIATION DAMAGE AND DIFFERENTIAL RADIOSENSITIVITY IN GERM CELLS an international symposium Proceedings Leiden 1962 Aug 15-19 By F.H. Sobels Pergamon press,London,1963
Radioth 35.1157

REPLICATION AND RECOMBINATION OF GENETIC MATERIAL an international conference Papers Edited by W.J. Peacock and R.D. Brock Sponsored by the Australian academy of science Australian academy of science,Canberra,1968
Bot 42.6703

REPLICATION OF DNA :a symposium Abstracts Cold Spring Harbor 1971 Sep Cold Spring Harbor laboratory of quantitative biology Cold Spring Harbor symposia on quantitative biology Cold Spring Harbor,1971
Bioch 33.2333

REPLICATION OF DNA IN MICRO-ORGANISM :a symposium Papers Cold Spring Harbor 1968 Cold Spring Harbor laboratory of quantitative biology Cold Spring Harbor symposia on quantitative biology, 33 Cold Spring Harbor,1968
Bioch 33.1289

REPLICATION OF DNA IN MICRO-ORGANISMS :a symposium Papers Cold Spring Harbor 1968 Jun 5-12 Cold Spring Harbor laboratory of quantitative biology Cold Spring Harbor symposia on quantitative biology, 33 2 vols Cold Spring Harbor,1968
Radioth 35.0246

REPORT OF A JOINT DISCUSSION ON VISION held... at the Imperial college of science Report London 1932 Jun 3 Physical society and Optical society Physical society,London,1932
Psy 31.0411

REPORT ON THE SYMPOSIUM ON THE WELDING OF IRON AND STEEL,MAY 1935 Iron and steel institute.Welding committee Iron and steel institute,London,1936
Met 25.0782

REPORTS ON ASTRONOMY International astronomical union 12th assembly Hamburg 1964 Aug.25-Sep.3 Edited by Jean-Claude Pecker International astronomical union.Transactions, 12a Academic press, London;New York,1965
A Math 4.1114

REPRESENTATION THEORY OF FINITE GROUPS AND RELATED TOPICS :symposium Madison, Wisc. 1970 Apr 14-16 American mathematical society Edited by Irving Reiner American mathematical society.Proceedings of symposia in pure mathematics, 21 v,178p 26cm AMS, Providence,R.I.,1971
P Math 2.4432

REPRODUCTION:MOLECULAR,SUBCELLULAR AND CELLULAR a symposium Edited by Michael Locke Society for the study of development and growth.Symposia, 24 Academic press,New York; London,1965
An 32.2526

REPRODUCTION;MOLECULAR,SUBCELLULAR AND CELLULAR a symposium Carleton,Minn. 1965 Jun Society for developmental biology Edited by Michael Locke Society for developmental biology.Symposia, 24 Academic press,London, 1965 The Society for developmental biology formerly known as the Society for the study of development
Bioch 33.0900

REPRODUCTION AND INFERTILITY 3rd :a symposium Papers Fort Collins,Colo. 1957 Jul 1-4 Colorado state university. College of veterinary medicine and Colorado state university.Agricultural experiment station Edited by F.X. Gassner and others Pergamon press,London,1958
Chem 26.0313

REPRODUCTION IN MAMMALS Biology of reproduction in mammals :a symposium Proceedings Nairobi,Kenya 1968 Apr Society for the study of fertility Edited by J.S. Perry and I.W. Rowlands Journal of reproduction and fertility.Supplement, 6 Blackwell,Oxford;Edinburgh,1969
Bal 39.1404

REPRODUCTION IN THE FEMALE MAMMALS Easter school in agricultural science 13th Proceedings Nottingham 1966 University of Nottingham Edited by G.E. Lamming and E.C. Amoroso Butterworths,London,1967
An 32.5323

REPRODUCTION MOLECULAR,SUBCELLULAR AND CELLULAR symposium Crelton,Minn. 1965 Jun Society for developmental biology Edited by Michael Locke Society for developmental biology.Symposia, 24 Academic press,New York;London,1965
Gen 34.0485

REPRODUCTIVE BIOLOGY AND TAXONOMY OF VASCULAR PLANTS :conference Report Birmingham 1965 Apr 23-24 Botanical society of the British Isles Edited by J.G. Hawkes Botanical society of the British Isles.B.S.B.I. conference reports, 9 183p Pergamon press,Oxford,1966
Bot 42.0307

RESEARCH APPLICATIONS OF NUCLEAR PULSED SYSTEMS Panel on research applications of repetitively pulsed reactors and boosters Proceedings Dubna,U.S.S.R. 1966 Jul 18-22 International atomic energy agency International atomic energy agency.Panel proceedings series International atomic energy agency,Vienna,1967
Radioth 35.1682

RESEARCH CONFERENCE FOR BIOLOGY AND MEDICINE 12th Symposium on enzyme reaction mechanisms Gatlinburg,Tenn. 1959 Apr 1-4 Sponsored by the Oak Ridge national laboratory.Biology division Journal of cellular and comparative physiology, 54,suppl.1 Wistar institute of anatomy and biology,Philadelphia,Pa.,1959
Bioch 33.1046

RESEARCH CONFERENCE FOR BIOLOGY AND MEDICINE 18th Symposium on hormonal control of protein biosynthesis Gatlinburg,Tenn. 1965 Apr 5-8 Sponsored by the Oak Ridge national laboratory. Biology division Journal of cellular and comparative physiology, 66,suppl.1 Wistar institute of anatomy and biology,Philadelphia, Pa.,1965
Bioch 33.0520

RESEARCH CONFERENCE FOR BIOLOGY AND MEDICINE 7TH Symposium on genetic recombination given at the Research conference for biology and medicine of the Atomic Energy commission Oak Ridge,Tenn. 1954 Apr 19-21 Sponsored by Oak Ridge national laboratory.Biology division Wistar institute of anatomy and biology, Philadelphia,Pa.,1955 Reprinted from the Journal of cellular and comparative physiology, Vol.45, suppt.2,May 1955
An 32.0346

RESEARCH CONFERENCE ON SHEAR STRENGTH OF COHESIVE SOILS New York 1960 University of Colorado American society of civil engineers Boulder,Colo.,1960
Eng 41.3171

RESEARCH CONFERENCE ON STRUCTURE AND PROPERTIES OF ENGINEERING MATERIALS Proceedings Raleigh,N.C. 1962 Mar 12-13 North Carolina state college and United States.Army research office,Durham Materials science research, 1 Plenum press,New York,1963
Met 25.2281

RESEARCH COUNCIL OF ISRAEL Desert research : international symposium Proceedings Jerusalem 1952 May 7-14 Research council of Israel.Special publications,2 Research council of Israel,Jerusalem,1953
Geog 13.1458

RESEARCH IN CHEMICAL AND EXTRACTION METALLURGY Symposium on 'recent progress in research in chemical and extraction metallurgy' Proceedings Sydney 1965 Feb 24-26 Australasian institute of mining and metallurgy Edited by J.T. Woodcock and others Australasian institute of mining and metallurgy.Monograph series, 2 Melbourne, 1965 reprinted 1967
Met 25.2780

RESEARCH IN PHOTOSYNTHESIS :a conference Papers and discussions Gatlinburg,Te. 1955 Oct 25-29 National research council. Committee on photobiology Edited by H. Gaffron and others Supported by the National science foundation Interscience,New York; London,1957
Chem 18.2138

RESEARCH IN PHOTOSYNTHESIS :a conference Papers and discussions Gatlinburg,Tenn 1955 Oct 25-29 Edited by H. Gaffron and others Sponsored by the National research council xiv,524p Interscience,London,1957
Bot 42.1869

RESEARCH IN PHOTOSYNTHESIS a conference Papers and discussions Gatlinburg,Tenn. 1955 Oct 25-29 Edited by H. Gaffron and others Sponsored by the National research council.Committee on photobiology Interscience,New York,1957
Bioch 33.0758

RESEARCH IN PSYCHOTHERAPY :a conference Proceedings Washington,D.C. 1958 Apr 9-12 American psychological association. Division of clinical psychology Edited by Eli Rubinstein and Morris B. Parloff National publishing co.,Washington,D.C.,1959
Psy 31.1627

RESEARCH IN RADIOLOGY :an informal conference Proceedings Highland park,Ill. 1957 May 10-12 Edited by Henry S. Kaplan Organized by the National research council.Subcommittee on radiobiology National research council. Nuclear science series.Report, 22 National research council.Publication, 571 National academy of science,Washington,D.C.,1958
Radioth 35.1116

RESEARCH INSTITUTE FOR THE STUDY OF MAN and PAN AMERICAN UNION Plantation systems : seminar Papers and discussion summaries San Juan,Puerto Rico 1957 Nov 17-23 Pan American union.Social science monographs,7 Washington,D.C.,1959
Geog 13.5120

RESEARCH INTO LIBRARY SERVICES IN HIGHER EDUCATION :papers presented at a conference London 1967 Nov 3 Society for research into higher education Society for research into higher education,London,1968
Eng 41.7669

RESEARCH METHODOLOGY AND NEEDS IN PERINATAL STUDIES :conference Proceedings Chapel Hill,N.C. 1963 Sep 9 Edited by Sidney S. Chipman and others Thomas, Springfield,Ill.,1966
An 32.2096

RESEARCH METHODS IN SOIL ZOOLOGY 1st colloquium Papers Progress in soil zoology Rothamsted 1958 Jul 10-14 International society of soil science.Soil zoology committee Edited by P.W. Murphy Bibliog.,illus. Butterworths,London,1962
Bot 42.2085

RESEARCH SEMINAR ON ARCHAEOLOGY AND RELATED SUBJECTS day meeting on statistics and archaeology London 1964 May 30 Institute of archaeology London,1964 Unbound typescript papers
Math 3.1115

RESEARCHES ON METEORITES :a symposian Proceedings Tempe,Ariz. 1961 Mar.10 University of Arizona 227p John Wiley,New York;London,1962
Obs 6.1243

RESEARCHES ON METEORITES :a symposium Papers Arizona state university 1961 Mar 10 Edited by Carleton B. Moore Wiley,New York; London,1962
Min 10.0499

RESIDUAL STRESS MEASUREMENTS Detroit 1951 Oct 15-19 By R.G. Treuting and others A series of four educational lectures presented to members of the A.S.M. during the 33rd National metal congress and exposition
Met 25.0962

RESISTANCE STRAIN GAGES Characteristics and applications of resistance strain gages :a symposium Proceedings Washington,D.C. 1951 Nov 8-9 National bureau of standards National bureau of standards.Circular, 528 Washington,D.C.,1954
Eng 41.2902

RESONANCE LINES IN ASTROPHYSICS :manuscripts presented at a conference... Boulder,Colo. 1968 Sep 9-11 National center for atmospheric research.High altitude observatory Washburn observatory 480p National center for atmospheric research,Boulder,Colo.,1968
Obs 6.3349

RESPIRATION International symposium on energy transduction in respiration and photosynthesis Pugnochiuso 1970 Sep 11-14 International union of biochemistry Edited by E. Quagliariello and others International union of biochemistry.Symposia, 43 Adriatic editrice,Bari,1971
Bioch 33.2289

RESPIRATION Regulation of human respiration Haldane centenary symposium Proceedings Oxford 1961 Jul Edited by D.J.C. Cunningham and B.B. Lloyd bibliog.,port. Blackwells,Oxford,1963
Phys 20.0857

RESPIRATION The Regulation of respiration :a conference New York 1963 Jun 24 By Harry Grundfest and others New York academy of sciences Edited by G.G. Nahas New York academy of sciences.Annals, 109,p411-948 New York,1963
Bal 39.1355

RESPIRATORY ENZYMES A Symposium on respiratory enzymes Addresses Madison,Wis, 1941 Sep 11-17 By Otto Meyerhof and others held at the University of Wisconsin University of Wisconsin press,Madison,Wis., 1942
Bioch 33.1661

RESPIRATORY ENZYMES A Symposium on respiratory enzymes Addresses Madison,Wisc. 1941 Sep 11-17 University of Wisconsin, and,University of Chicago University of Wisconsin press,Madison,Wisc.,1942
Chem 18.1249

RESPIRATORY ENZYMES A Symposium on respiratory enzymes Lectures and discussions Madison,Wis. By Otto Meyerhof and others Given at the University of Wisconsin University of Wisconsin,Madison,Wis.,1942 reprinted 1948 Sponsored jointly by the universities of Wisconsin and Chicago
Bal 39.1344

RESPONSE OF METALS TO HIGH VELOCITY DEFORMATION Technical conference Estes Park,Col. 1960 Jul 11-12 American institute of mining, metallurgical and petroleum engineers Metallurgical society conferences, 9 Interscience,New York,1961
Met 25.0905

RESPONSE OF THE NERVOUS SYSTEM TO IONIZING RADIATION :an international symposium Proceedings Chicago,Ill. 1960 Sep 7-9 Edited by Thomas J. Haley and Ray S. Snider Academic press,New York;London,1962
Radioth 35.1730

RESPONSE OF THE NERVOUS SYSTEM TO IONIZING RADIATION :an international symposium Proceedings Chicago,Ill. 1960 Sep 7-9 Edited by Thomas J. Hayley and Ray S. Snider Academic press,New York,1902
An 32.4191

RESULTAS SCIENTIFIQUES DU CONGRES INTERNATIONAL DE BOTANIQUE : wissenschaftliche ergebnisse des internationalen botanischen kongresses Wien 1905 Vienna 1905 By J. P. Lotsy Edited by R.von Wettstein and others Publications scientifiques de l'association internationale des botanishes, 1 Fischer,Jena,1906
Bot 42.0645

RETARDATION OF EVAPORATION BY MONOLAYERS : transport processes Edited by Victor K. La Mer xx,277p Academic press,New York; London,1962 Papers presented at a symposium sponsored by the Division of colloid and surface chemistry of the American chemical society,Sep.1960
Chem E 24.0945

RETENTION AND MIGRATION OF RADIOACTIVE IONS IN SOILS Colloque international sur la retention et la migration des ions radioactifs dans les sols Saclay 1962 Oct 16-18 Institut national des sciences et techniques nucleaires Centre d'etudes nucleaires de Saclay;Presses universitaires de France,Gif-sur-Yvette;Paris,1963
Eng 41.3174

RETICULAR FORMATION OF THE BRAIN Henry Ford hospital international symposium Contributions Detroit,Mich. 1957 Mar 14-16 Henry Ford hospital Edited by Herbert H. Jasper and others J.and A.Churchill,London, 1957
Psy 31.0230

RETICULOENDOTHELIAL STRUCTURE AND FUNCTION : international symposium Rapallo 1958 Aug Edited by John H. Heller International society for research on the reticuloendothelial system.Symposia, 3 Ronald press,New York,1960
An 32.3169

RETINA FOUNDATION.INSTITUTE OF BIOLOGICAL AND MEDICAL SCIENCES.MONOGRAPHS AND CONFERENCES, 2 Biochemistry of muscle contraction :a conference Proceedings Dedham,Mass. 1962 May Edited by John Gergely Churchill, London,1964
Bioch 33.2231

REUNION DE NEUCHATEL Actes H-spaces Neuchatel 1970 Aug Edited by Francois Sigrist Lecture notes in mathematics, 196 156p 25cm Springer,Berlin,1971
P Math 2.4121

REUNION INTERNATIONALE DE CHIMIE PHYSIQUE rapports et discussions Paris 1928 October 8-12 Edited by R. Audubert and M.L. Claudel Paris,1929
Philos 1.0668

REUNION PLENIERE DE L'ASSOCIATION POUR L'ETUDE TAXONIQUE DE LA FLORE D'AFRIQUE TROPICALE 4 Comptes rendus Lisbon 1960 Sep 16-23 Edited by A. Fernandes Junta de investigacoes do ultramar,Lisbon,1962
Bot 42.2646

REUNIONS SCIENTIFIQUES...POUR LE CENTENAIRE DE LA DECOUVERTE DE NEPTUNE reports Paris 1946 Oct.22-24 Comite national Francais d'astronomie Edited by P. Couderc illus. 164p Imprimerie nouvelle,Orleans,1948
Obs 6.0418

REUTTE,AUSTRIA 1948 High-melting metals Plansee seminar "De re metallica" 3rd Papers Metallwerk Plansee AG Edited by F. Benesovsky Pergamon press;Metallwerk Plansee AG;,,London;Reutte,1959
Met 25.1010

REUTTE,AUSTRIA 1961 Powder metallurgy in the nuclear age Plansee seminar 'De re metallica' 4th Proceedings Metallwerk Plansee AG Edited by F. Benesorsky Metallwerk Plansee AG,Reutte,1962 Papers in English,French and German
Met 25.2724

REUTTE,AUSTRIA 1964 Metals for the space age Plansee seminar "De re metallica" 5th Papers Metallwerk Plansee A.G. Edited by F. Benesovsky Metallwerk Plansee A G,Reutte, 1965 Papers in English,French and German
Met 25.0091

REUTTE,AUSTRIA 1966 High temperature materials Plansee seminar Hochtemperatur werkstoffe 6th Papers Metallwerk Plansee A.G. Edited by F. Benesovsky Springer,Vienna;New York,1969 Papers in English,French and German
Met 25.2336

A REVIEW OF NITROGEN IN THE TROPICS,WITH PARTICULAR REFERENCE TO PASTURES :a symposium Nitrogen in the tropics Commonwealth scientific and industrial research organisation.Division of tropical pastures Commonwealth bureau of pastures and field crops.Bulletin, 46 Commonwealth agricultural bureaux,Farnham Royal,1962
Bot 42.1784

REYKJAVIK 1962 North Atlantic biota and their history :a symposium University of Iceland and Iceland museum of natural history Sponsored by the Nato advanced institutes program pl. 430p 24cm Pergamon press,Oxford,1963
Sco 14.8334

REYKJAVIK 1962 North atlantic biota and their history :a symposium University of Iceland,and,Museum of natural history, Reykjavik Edited by Askell Love and Doris Love Sponsored by NATO advanced study institutes program illus. Pergamon press, Oxford,1963
Geog 13.1211

REYKJAVIK 1967 Iceland and mid-ocean ridges :a symposium Report Geoscience society of Iceland Edited by Sveinbjorn Bjornsson Societas scientiarum islandica. Publication, 38 Reykjavik,1967
Geod 9.0383

REYKJAVIK 1967 Surtsey research conference Proceedings Surtsey reseafch society Sponsored by the American institute of biological sciences Surtsey research society, Reykjavik,c1967
Sco 14.7944

REYNOLDS METALS COMPANY Welding aluminum : welding,brazing,soldering Reynolds metals company,Louisville,Ky.,1953
Eng 41.3943

RHENIUM :a symposium Chicago 1960 May 3-4 Electrochemical society.Electrothermics and metallurgy division Edited by B.W. Gonser Elsevier,Amsterdam;New York,1962
Met 25.0585

RHEOLOGY International conference on rheology 2nd proceedings Oxford 1953 Jul 26-31 Edited by C.G.W. Harrison Butterworths,London,1954
A Math 4.0367

RHEOLOGY International congress on rheology 2nd Proceedings Oxford 1953 Jul 26-31 Edited by V.G.W. Harrison ix,451p Butterworth scientific publications,London, 1954
Chem E 24.0446

RHEOLOGY International congress on rheology 1st Proceedings Scheveningen 1948 Sep 21-24 North-Holland,Amsterdam,1949
Col S 12.0103

RHEOLOGY Symposium on rheology :presented at the ASME applied mechanics and fluids engineering conference Washington,D.C. 1965 Jun 7-9 American society of mechanical engineers Edited by A.W. Marris and J.T.S. Wang American society of mechanical engineers,New York,1965
Eng 41.4418

RHEOLOGY AND SOIL MECHANICS :a symposium Grenoble 1964 Apr 1-8 International union of theoretical and applied mechanics Edited by J. Kravtchenko and P.M. Sirieys Springer, Berlin,1966
Eng 41.3097

RHEOLOGY OF SOLIDS International symposium on high speed testing 7th Proceedings Boston,Mass. 1969 Mar 17-18 Edited by R.D. Andrews and F.R. Eirich Applied polymer symposia, 12,High speed testing, 7 Interscience,New York,1969
Met 25.2581

RHEOLOGY SOCIETY OF AMERICA Biomedical fluid mechanics symposium Papers Denver,Colo. 1966 Apr 25-27 American society of mechanical engineers.Fluid mechanics commission American society of mechanical engineers,New York,1966 Papers presented by the Fluid mechanics commission of the ASME at the Fluids engineering conference,co-sponsored by the Rheology society of America
Eng 41.6540

RHEOMACRODEX :a symposium Proceedings Cardiff 1964 Nov 23 and London 1964 Nov 24 1 Edited by H. Scarborough and H.G. Melrose Pharmacia limited,London,1965
PGMS 29.0610

RHEOMACRODEX LOW MOLECULAR WEIGHT DEXTRAIN :a symposium Proceedings London 1963 Feb 1 Edited by H.G. Melrose Pharmacia limited, London,1964
PGMS 29.0609

RHODE ISLAND 1960 Plasticity Naval structural mechanics :a symposium 2nd Proceedings United States.Office of naval research and Brown university Edited by E.H. Lee and P.S. Symonds Pergamon press,New York,1960
Met 25.0845

RIBONUCLEIC ACID Federation of European biochemical sciences meeting 2nd Proceedings Vienna 1965 Apr 21-24 symposium on ribonucleic acid structure and function Federation of European biochemical societies Edited by H. Tuppy Pergamon press,Oxofrd,1966
Gen 34.0688

RICE BREEDING WITH INDUCED MUTATIONS :an FAO-IAEA research co-ordination meeting on the use of induced mutations in rice breeding Report Taipei 1967 Jun 5-9 Food and agriculture organization International atomic energy agency International atomic energy agency.Technical report series, 86 International atomic energy agency,Vienna,1968
Radioth 35.0271

RICE GENETICS AND CYTOGENETICS Symposium on rice genetics and cytogenetics Proceedings Los Banos 1963 Feb 4-8 Sponsored by the International rice research institute Elsevier,Amsterdam,1964
Gen 34.1559

RICE UNIVERSITY Complex analysis Conference on complex analysis Proceedings Houston,Tex. 1967 Apr 13-15 Edited by H.L. Resnikoff and R.O. Wells Rice university studies, 54,no.4 Bibliog. 84p 23cm William Marsh Rice University,Houston,Tex., 1968
P Math 2.3684

RICE UNIVERSITY Complex analysis 1969 : conference Proceedings Houston,Texas 1969 Mar 26-29 Edited by H.L. Resnikoff and R.O. Wells Rice university studies, 56,no 2 vi,222p 23cm Rice university,Houston,Texas, 1971
P Math 2.4357

RICE UNIVERSITY Delayed implantation : symposium Houston,Tex. 1963 Jan 23-26 Edited by Allen C. Enders Rice university semicentennial series University of Chicago press,Chicago,Ill.,1963
An 32.3890

RICHMOND,CALIF. Comparative biochemistry of photoreactive systems Symposium on comparative biology 1st Papers Kaiser foundation research institute Edited by Mary Belle Allen Kaiser foundation research institute.Symposia on comparative biology, 1 Academic press,New York;London,1960
Bioch 33.0334

RIDGEFIELD,CONN. 1953 Glutathione :a symposium Proceedings Edited by S. Colowick and others Academic press,New York, 1954
Bioch 33.0604

RIEMANN SURFACES Contributions to the theory of Riemann surfaces conference typescript Princeton,N.J. 1951 Dec 14-15 Institute for advanced study Edited by Lars V. Ahlfors and others In joint sponsorship with Princeton university Annals of mathematics studies, 30 264p 26cm Princeton university press,Princeton,N.J.,1953 Centennial celebration of Riemann's dissertation
P. Math 2.1434

RIGA 1970 International conference on the chemistry of natural products 7th Abstracts International union of pure and applied chemistry Edited by M.N. Kolosov Zinatne,Riga,1970
Bioch 33.2318

RIJKAUNIVERSITEIT TE GENT Androgens in normal and pathological conditions Steroid hormones :a symposium 2nd Proceedings Ghent 1965 Jun 17-19 Edited by A. Vermeulen and D. Exley Excerpta medica. International congress series, 101 Excerpta medica,Amsterdam,1966
Inv Med 37.0276

RIJKSLANDBOUWHOGESCHOOL,GHENT Jaarlijks symposium over phytopharmacie symposium 2nd Papers Ghent 1950 Apr 23 Landbouwhogeschool en de opzoekingsstations, Ghent.Mededelingen, 15,no 1 Rijkslandbouwhogeschool,Ghent,1950 Papers and summaries in English,Flemish and French
Chem 26.0105

RIJKSLANDBOUWHOGESCHOOL,GHENT Jaarlijks symposium over phytopharmacie symposium 5th Verhandelingen Ghent 1953 May 5 Edited by J.van den Brande Landbouwhogeschool en de opzoekingsstation,Ghent.Mededelingen, 18,no 2 Rijkslandbouwhogeschool,Ghent,1953 Papers and summaries in English,Flemish and French
Chem 26.0106

RIJKSUNIVERSITEIT TE AMSTERDAM Elementary particles :conference Proceedings Amsterdam 1971 Jun 30-Jul 6 Edited by A.G. Tenner and H.J.G. Vettman illus viii,472p 26cm North-Holland,Amsterdam,1972
A Math 4.1767

RIJKSUNIVERSITEIT TE UTRECHT The Solar spectrum symposium proceedings Utrecht 1963 Aug.26-31 Edited by C.de Jager Astrophysics and space science library D. Reidel,Dordrecht-Holland,1965
A Math 4.1163

RIJSWIJK 1966 International symposium on the cell nucleus:metabolism and radiosensitivity Proceedings Taylor and Francis,London,1966
Radioth 35.1751

RIKSUNIVERSITEIT TE LEIDEN Repair from genetic radiation damage and differential radiosensitivity in germ cells International symposium on repair from genetic radiation damage Proceedings Leiden 1962 Aug 15-19 Edited by F.H. Sobels Pergamon press,Oxford, 1963
Gen 34.1121

RIO DE JANEIRO 1941 Symposium sobre raios cosmicos Academia Brasileira de Ciencias 26cm Rio de Janeiro 1943
Philos 1.0379

RIO DE JANEIRO 1952 Sao Paulo New research techniques in physics :a symposium Proceedings Academia brasileira de ciencias and Unesco.Centro de cooperacion cientifica para America Latina Sponsored by International union of pure and applied physics illus 447p Unesco,Rio de Janeiro, 1954 Under the auspices of Conselho nacional de pesquisas de Brasil and co-sponsored by the university of Sao Paulo
Cav 7.1977

RIO DE JANEIRO 1957 Curare and curare-like agents International symposium on curare and curare-like agents Proceedings Edited by D. Bovet and others Organized under the auspices of Unesco Elsevier,Amsterdam,1959
Pha 16.0243

RIO DE JANEIRO 1957 International symposium on curare and curare-like agents Proceedings Academia brasileira de ciencias. Conselho nacional de pesquisas and Universidade do Brasil Edited by D. Bovet and others Under the auspices of Unesco illus Elsevier,Amsterdam,1959 Also sponsored by the 'Istituto di sanita',of Rome. Papers in English and French
Chem 26.0103

RIO DE JANEIRO 1960 Specialized tissues of the heart Symposium on the specialized tissues of the heart Proceedings Edited by Antonio Paes de Carvalho and others Elsevier, Amsterdam,1961
An 32.3512

RIO DE JANEIRO 1965 Social science in Latin America Conference on Latin American studies Papers Joint committee on Latin American studies Edited by Manuel Diegues and Bryce Wood Jointly sponsored by the Latin American center for research in the social sciences,Rio de Janeiro Columbia university press,New York,1967
Geog 13.5108

RIQUEZA MINERA Y YACIMIENTOS MINERALES DE MEXICO International geological congress 20th papers Mexico city 1956 Edited by Jenaro Gonzalez Reyna 3rd edition Mexico city, 1956
Geol 8.3025

RISLEY 1969 Practical fracture mechanics for structural steel Symposium on fracture toughness concepts for weldable structural steel Proceedings Edited by M.O. Dobson Published for the United Kingdom atomic energy authority Chapman and Hall, London, 1970
Eng 41.3822

RISLEY 1969 Practical fracture mechanics for structural steel Symposium on fracture toughness concepts for weldable structural steel Proceedings United Kingdom atomic energy authority Edited by M.O. Dobson U.K. A.E.A. in association with Chapman and Hall, Risley, 1969
Met 25.2465

RISO 1959 Health physics in nuclear installations :a symposium Organization for European economic cooperation European nuclear energy agency Held at the Danish atomic energy centre European nuclear energy agency, Paris, 1959
Radioth 35.1110

RISO 1967 Tunneling phenomena in solids : lectures presented at the 1967 NATO advanced study institute at Riso, Denmark North Atlantic treaty organization Edited by Elias Burstein and Stig Lundqvist Plenum press, New York, 1969
Eng 41.8302

RISO, DENMARK 1967 Tunneling phenomena in solids NATO advanced study institute on tunneling phenomena in solids Proceedings North Atlantic treaty organization Edited by Elias Burstein and Stig Lundqvist Bibliog 580p 18cm Plenum press, New York, 1969
Cav 7.2667

RNA Structure and function of transfer RNA and 58-RNA Federation of European biochemical societies meeting 4th Proceedings Oslo 1967 Jul 3-7 Vol 3: structure and function of transfer RNA and 58-RNA Federation of European biochemical societies Edited by L.O. Froholm and S.E. Laland Academic press, London, 1968
Chem 18.2491

RNA The Federation of European biochemical societies meeting 4th Proceedings Oslo 1967 Jul 5 Vol 3: structure and function of transfer RNA and 5 S-RNA Federation of European biochemical societies Edited by L.O. Froholm and S.G. Laland Organized by the Norwegian biochemical society Universitetsforlaget; Academic press, Oslo; London, 1968 Symposium organizer H.G. Zachau
Bioch 33.1398

RNA AND BRAIN FUNCTION MEMORY AND LEARNING Brain function conference 2nd Proceedings Los Angeles 1962 Brain research institute and American institute of biological sciences Edited by Mary A.B. Brazier With the support of the United States. Air force. Office of scientific research U.C.L.A. forum in medical sciences, 2 University of California press, Berkeley; Los Angeles, 1964
Psy 31.0246

ROAD RESEARCH LABORATORY The Theory of road traffic flow international symposium 2nd proceedings London 1963 Jun.25-27 Edited by Joyce Almond With financial support from General motors corporation. Research laboratories ix, 406p 24cm Organisation for economic co-operation and development, Paris, 1965
Math 3.0922

ROAD VEHICLE AERODYNAMICS Symposium on road vehicle aerodynamics 1st Proceedings London 1969 Nov 6-7 Edited by A.J. Scibor-Rylski Organised by the City university. Department of aeronautics City university, London, 1970 Organised in cooperation with the National physical laboratory, the Motor industry research association and the Institution of mechanical engineers
Eng 41.8336

ROBERT A. WELCH FOUNDATION Conference on chemical research Proceedings Houston, Texas 1969 Nov 17-19 13: the transuranium elements - the Mendeleev centennial Edited by W.O. Milligan 494p R.A. Welch foundation, Houston, Texas, 1970
TA 15.0699

ROBERT A. WELCH FOUNDATION. CONFERENCE, 13 Conference on chemical research Proceedings Houston, Texas 1969 Nov 17-19 Robert A. Welsh foundation photos, graphs, diagrms xii, 494p 23cm Robert A. Welch foundation, Houston, Texas, 1970
Cav 7.3126

ROBERT A. WELSH FOUNDATION Conference on chemical research Proceedings Houston, Texas 1969 Nov 17-19 Robert A. Welch foundation. Conference, 13 photos, graphs, diagrms xii, 494p 23cm Robert A. Welch foundation, Houston, Texas, 1970
Cav 7.3126

ROBERT JONES AND AGNES HUNT ORTHOPAEDIC HOSPITAL MANAGEMENT COMMITTEE Transplantation antigens and tissue typing symposium Proceedings Charles Salt research centre 1968 Nov 11-13 Edited by M.W. Elves and N.W. Nisbet Oswestry, 1969
Surg 23.0031

ROCHE ANNIVERSARY SYMPOSIUM Proceedings Challenge of life: biomedical progress and human values Basle 1971 Aug 31-Sep 3 Hoffmann-La Roche and company Edited by Robert M. Kunz and Hans Fehr Experientia. Supplementum, 17 456p 24cm Birkhauser, Basle, 1972 Symposium held on the occasion of the firm's 75th anniversary. Chairman: Lord Todd
Chem 18.2902

ROCHE COMPANIES Jubilee volume dedicated to Emil Christoph Barell...on the occasion of the fiftieth anniversary of his association with the house of Roche Roche, Basle, 1946
Radioth 35.0141

ROCHE PRODUCTS Diazepam in anaesthesia :a symposium Proceedings London 1967 Jun 30 Edited by Peter F. Knight and C.G. Burgess Wright, Bristol, 1968
PGMS 29.0039

ROCHESTER 1961 Data acquisition and processing in biology and medicine :a conference Proceedings Edited by K. Enslein Pergamon,Oxford;London,1962
PGMS 29.0509

ROCHESTER,N.Y. 1952 High energy nuclear physics Annual Rochester conference on high energy nuclear physics 3rd Proceedings Edited by H.Pierre Noyes and others Sponsored by the University of Rochester 110p Interscience,New York,c.1953 Co-sponsored by the National science foundation
Cav 7.1986

ROCHESTER,N.Y. 1954 High energy nuclear physics Annual Rochester conference on high energy nuclear physics 4th Proceedings Edited by H.Pierre Noyes and others Sponsored by the University of Rochester 159p Rochester,N.Y.,1954 Co-sponsored by the National science foundation.Typescript
Cav 7.1987

ROCHESTER,N.Y. 1955 High energy nuclear physics Annual Rochester conference on high energy nuclear physics 5th Proceedings University of Rochester Edited by H.Pierre Noyes and others Sponsored by the International union of pure and applied physics 196p Interscience,New York,1955 Co-sponsored by the National science foundation in cooperation with the Atomic energy commission and the United States office of naval research
Cav 7.1988

ROCHESTER,N.Y. 1956 High energy nuclear physics Annual Rochester conference on high energy nuclear physics 6th Proceedings Edited by J. Ballam and S.B. Trieman 362p Interscience,New York,1956
Cav 7.1989

ROCHESTER,N.Y. 1958 Cinefluorography Annual symposium on cinefluorography 1st Proceedings Edited by George H.S. Ramsey and others Sponsored by University of Rochester. Department of radiology Thomas,Springfield, Ill.,1960
VA 19.0381

ROCHESTER,N.Y. 1960 International conference on high-energy physics 10th Proceedings University of Rochester International union of pure and applied physics Edited by E.C.G. Sudarshan and others Co-sponsored by the United States. Office of naval research illus xxv,890p University of Rochester,Rochester,N.Y.,1960
Cav 7.2891

ROCHESTER,N.Y. 1962 Engineering aspects of magnetohydrodynamics Symposium on the engineering aspects of magnetohydrodynamics 3rd Proceedings Edited by Norman W. Mather and George W. Sutton Gordon and Breach,New York,1964
Eng 41.4546

ROCHESTER,N.Y. 1964 Galileo quadricentennial Homage to Galileo Papers University of Rochester Edited by Morton F. Kaplon xii,139p MIT press,Cambridge,Mass.; London,1965
WSM 43.1985

ROCHESTER,N.Y. 1965 Engineering developments in energy conversion International conference on energetics American society of mechanical engineers American society of mechanical engineers,New York,1965
Eng 41.4865

ROCHESTER,N.Y. 1967 International conference on particles and fields 1967 Proceedings University of Rochester Edited by Carl Richard Hagen and others Interscience,New York,1967
A Math 4.1349

ROCHESTER,N.Y. 1967 International theoretical physics conference on particles and fields Proceedings Edited by C.R. Hagen and others 708p 26cm Interscience, New York,1967 Dedicated to Robert Oppenheimer
Cav 7.2773

ROCHESTER N.Y. 1964 The Remote effects of cancer on the nervous system :a symposium Proceedings Edited by Russell Brain and Forbes H. Norris Sponsored by the University of Rochester school of medicine and dentistry. Division of neurology Contemporary neurology symposia, 1 Grune and Stratton,New York; London,1965
PGMS 29.0451

ROCK DEFORMATION University of California, Los Angeles.Institute of geophysics :annual technical conference 1956 Symposium Los Angeles,Calif. 1956 Nov Edited by David Griggs and John Handin Geological society of America.Memoirs,79 New York,1960
Min 10.0348

ROCK FORMATION Genese des roches filoniennes, fasc.6 International geological congress 19th Proceedings,section 6 Algiers 1952 Algiers,1953
Min 10.0351

ROCK GARDEN PLANTS International rock garden plant conference 3rd Report London 1961 Apr 18-28 Edinburgh 1961 Alpine garden society Edited by R.C. Elliott and J. L. Mowat Organized by the Scottish rock garden club Alpine garden society;Scottish rock garden club,London;Penicuik,1961
BG 38.0011

ROCK GARDENS AND ROCK PLANTS conference Reports London 1936 May 5-7 Royal horticultural society Alpine garden society Edited by F.J. Chittenden Royal horticultural society,London,1936
BG 38.3310

ROCK MECHANICS Congress of the International society of rock mechanics 1st Proceeedings Lisbon 1966 Sep 25-Oct 1 Vol 1-3 International society of rock mechanics 3 vols Laboratorio nacional de engenharia civil,Lisbon,1966-67
Eng 41.3176

ROCK MECHANICS Testing techniques for rock mechanics :a symposium Seattle,Wash. 1965 Oct 31-Nov 5 American society for testing and materials American society for testing and materials.Special technical publication, 402 American society for testing materials, Philadelphia,Pa.,1966
Eng 41.3161

ROCKEFELLER FOUNDATION Computing methods and the phase problem in X-ray crystal analysis :a conference Report Pennsylvania state college 1950 Apr 6-8 Pennsylvania state college,State College,Pa.,1952
Cav 7.1789

ROCKEFELLER FOUNDATION Symposium on trace analysis papers New York 1955 Nov 2-4 Edited by John H. Yoe and Henry J. Koch Wiley;Chapman and Hall,New York;London,1957
Met 25.1720

ROCKEFELLER INSTITUTE Biology and behavior 3 Environmental influences :a conference Proceedings New York 1967 Apr 21-22 Edited by David C. Glass Rockefeller university press;Russell Sage foundation,New York,1968 Third and last in a series of conferences on biology and behavior
Psy 31.3096

ROCKEFELLER INSTITUTE Molecular biology : elementary processes of nerve conduction and muscle contraction.A symposium Papers New York 1958 Sep 25-30 Edited by David Nachmansohn Sponsored by Columbia university and the Rockefeller institute Academic press, New York;London,1960
Bioch 33.0831

ROCKEFELLER UNIVERSITY Neurophysiology and emotion :a conference Proceedings New York 1965 Dec 10-11 Edited by David C. Glass Rockefeller university press;Russell Sage foundation,New York,1967
Psy 31.2900

ROCKET AND SATELLITE METEOROLOGY International symposium on rocket and satellite meteorology Washington,D.C. 1962 Apr 23-25 Edited by H. Wexler and J.E. Caskey,jnr Sponsored by Committee on space research ix,440p North-Holland,Amsterdam, 1963
Sco 14.0516

ROCKET EXPLORATION Rocket exploration of the upper atmosphere :a conference papers Oxford 1953 Aug 24-26 United States.Upper atmosphere rocket research panel and Royal society of London.Gassiot committee Edited by R.L.F. Boyd and M.J. Seaton Journal of atmospheric and terrestrial physics.Special supplement, 1 376p Pergamon press,London, 1954
Nap 11.0073

ROCKET EXPLORATION OF THE UPPER ATMOSPHERE :a conference papers Oxford 1953 Aug 24-26 United States.Upper atmosphere rocket research panel and Royal society of London. Gassiot committee Edited by R.L.F. Boyd and M.J. Seaton Journal of atmospheric and terrestrial physics.Special supplement, 1 376p Pergamon press,London,1954
Nap 11.0073

ROCKET PROPULSION Advances in tactical rocket propulsion :a colloquium La Jolla, Calif. 1965 Apr 22-23 Agard Edited by S. S. Penner AGARD conference proceedings, 1 Technivision,Slough,1968
Eng 41.6896

ROCNIK MATEMATICKE OLYMPIADY 5th-8th,10th-16th Illus.port. 17cm 10 vols Statni Pedagogicke Nakladatelstvi,Prague,1957-68
P Math 2.3275

ROE Anne and SIMPSON George Gaylord ed. Behaviour and evolution :conference 2nd Proceedings Princeton,N.J. 1956 Apr 30-May 5 American psychological association Held in collaboration with the Society for the study of evolution 24cm New Haven,1958
Philos 1.1827

ROENTGENS,RADS AND RIDDLES :a symposium on supervoltage radiation therapy Oak Ridge, Tenn. 1956 Jul 15-18 United States atomic energy commission Edited by Milton Friedman and others Held at the Oak Ridge institute of nuclear studies U.S.atomic energy commission,Washington,D.C.,1959
Radioth 35.1090

ROLAND EOTVOS PHYSICAL SOCIETY International symposium on luminescence Papers Balatonvilagos,Hungary 1961 Jun 7-10 Budapest,1962 Papers in English,German, Russian,reprinted from 'Acta physica Academiae scientiarum hungaricae',tom 14,fasc 2-3
Chem 18.0474

ROLE DES ANAEROBIES DANS LA NATURE Congres international des microbiologistes de langue francaise Brussels 1949 May 23-27 International union of biological sciences International union of biological sciences. Series B.Colloques, 7 Secreteriat general de l'U.I.S.B.,Paris,1949
Bal 39.3919

The ROLE OF CHROMOSOMES IN DEVELOPMENT Amherst,Mass. 1964 Jun Society for the study of development and growth Edited by Michael Locke Society for the study of development and growth.Symposia, 23 illus. Academic press,New York;London,1964
Gen 34.0859

The ROLE OF CHROMOSOMES IN DEVELOPMENT :a symposium Amherst,Mass. 1964 Jun Edited by Michael Locke Society for the study of development and growth.Symposia, 23 Academic press,New York;London,1964
An 32.2506

The ROLE OF CHROMOSOMES IN DEVELOPMENT :a symposium Amherst,Mass. 1964 Jun Society for the study of development and growth Edited by Michael Locke Society for the study of development and growth.Symposia, 23 Academic press,New York;London,1964 The Society for the study of development and growth known later as the Society for developmental biology
Bioch 33.1011

ROLE OF COMPUTERS IN RADIOTHERAPY Panel on the role of computers in radiotherapy Report Vienna 1967 Jul 10-14 Organized by the International atomic energy agency International atomic energy agency.Panel proceedings series bibliog. International atomic energy agency,Vienna,1968
Radioth 35.1199

ROLE OF CYCLIC AMP IN CELL FUNCTION : symposium San Diego,Calif. 1970 Feb American college of neuropsychopharmacology Edited by Paul Greengard and Erminio Costa Advances in biochemical psychopharmacology, 3 Raven press,New York,1970
Bioch 33.2211

The ROLE OF KARL MARX IN THE DEVELOPMENT OF CONTEMPORARY SCIENTIFIC THOUGHT :a symposium Papers Marx and contemporary scientific thought Paris 1968 May 8-10 International council for philosophy and humanistic studies International social science council Under the auspices of Unesco International social science council. Publications, 13 xi,612p Mouton,The Hague; Paris,1969
WSM 43.3826

ROLE OF NUCLEOTIDES FOR THE FUNCTION AND CONFORMATION OF ENZYMES The Alfred Benzon symposium 1 Proceedings Copenhagen 1968 Sep 9-11 Alfred Benzon foundation Edited by Herman M. Kalckar Held at the premises of the Royal Danish academy of sciences and letters Munksgaard,Copenhagen, 1969
Bioch 33.1913

The ROLE OF NUTRITIONAL DEFICIENCY ON NERVOUS AND MENTAL DISEASE Proceedings of the Association New York 1941 Dec 19-20 Association for research in nervous and mental disease Edited by Stanley Cobb and others Association for research in nervous and mental disease.Research publications, 22 Williams and Wilkins,Baltimore,Md.,1943
An 32.4377

ROLE OF PROTEINS IN ION TRANSPORT ACROSS MEMBRANES Ion transport across membranes : a symposium Papers New York 1953 Oct 2-3 Edited by Hans T. Clarke and David Nachmansohn held at Columbia university.College of physicians and surgeons Academic press,New York,1954 Incorporating papers presented at a symposium on the role of proteins in ion transport across membranes together with six invited articles on allied topics
Bal 39.0560

The ROLE OF SCHMIDT TELESCOPES IN ASTRONOMY Hamburg 1972 Mar 21-22 Hamburger sternwarte Edited by U. Hang 160p Hamburg observatory,Hamburg,1972
Obs 6.3716

The ROLE OF THE ADSORBED STATE IN HETEROGENOUS CATALYSIS Liverpool 1966 Apr 4-6 Faraday society Faraday society.Discussions, 41 Faraday society,London,1966
Min 10.0623

The ROLE OF THE GAMMA SYSTEM IN MOVEMENT AND POSTURE By Ian A. Boyd Association for the aid of crippled children,New York,1964 Record of a colloquium held during the 16th annual meeting of the American academy for cerebral palsy
An 32.4005

ROMANIAN-FINNISH SEMINAR ON TEICHMULLER SPACES AND QUASIFORMAL MAPPINGS Proceedings Brasov 1969 Edited by Cabiria Andreian Cazacu 307p 21cm Publishing house of the Academy of the socialist republic of Romania, Bucharest,1971
P Math 2.4355

ROME 1952 phenomenes solaires et terrestres Commission mixte pour l'etude des relations entre les phenomenes solaires et terrestres Proceedings International council of scientific unions 35p Paris, 1952
Obs 6.2544

ROME 1953 Congresso internazionale de microbiologia symposia 6e Abstracts Instituto superiore di Sanita,Rome,1953
Bot 42.0667

ROME 1953 Congresso nazionale della societa italiana di ematologia 11th Atti Societa italiana di ematologia Edited by L. Villa and others Edizioni mediche e scientifiche,Rome,1953
Med 36.0341

ROME 1953 International congress of microbiology 6th Vol 1: symposium - bacterial cytology Edited by S. Mudd and G. Penso illus Fondazione Emanuele Paterno, Rome,1953 Title also in Italian;text in English and French
Bioch 33.1800

ROME 1953 International congress of microbiology 6th Vol 2: symposium - microbial metabolism Edited by E.B. Chain Fondazione Emanuele Paterno,Rome,1953 Title also in Italian;text in English and French
Bioch 33.1801

ROME 1953 International congress of microbiology 6th Vol 3: symposium - nutrition and growth factors Edited by W.H. Schopfer and D.D. Woods Fondazione Emanuele Paterno,Rome,1953 Title also in Italian; text in English and French
Bioch 33.1802

ROME 1953 International congress of microbiology 6th Vol 4: growth inhibition and chemotherapy Edited by H. Eagle and E.B. Chain Fondazione Emanuele Paterno,Rome,1953 Title also in Italian; text in English and French
Bioch 33.1803

ROME 1953 International congress of microbiology 6th Vol 6: symposium - interaction of viruses and cells Edited by F. C. Bawden and G. Penso illus Fondazione Emanuele Paterno,Rome,1953 Title also in Italian;text in English and French
Bioch 33.1804

ROME 1953 Symposium internationale geneticae medicae 1st Atti 1 Istituto Gregorio Mendel Edited by Luigi Gedda Analecta genetica, 1 Edizioni dell' istituto Gregorio Mendel,Rome,1954
Med 36.0148

ROME 1953 Symposium internazionale geneticae medicae 3: chondrodysplasie Istituto Gregorio Mendel Edited by Luigi Gedda Analecta genetica, 1 Edizioni dell' istituto Gregorio Mendel,Rome,1954
Med 36.0150

ROME 1953 Symposium internazionale geneticae medicae 1 Atti 4: il parto indolore Istituto Gregorio Mendel Edited by Luigi Gedda Analecta genetica, 1 Edizioni dell' istituto Gregorio Mendel,Rome, 1956
Med 36.0151

ROME 1953 Symposium internazionale geneticae medicae 1st Atti 2: krampfbereitschaft Istituto Gregorio Mendel Edited by Luigi Gedda Analecta genetica, 1 Edizioni dell' istituto Gregorio Mendel,Rome, 1954
Med 36.0149

ROME 1953 and PISA 1953 International association for quaternary research congress 4th Actes Vol 1-2 Edited by Gian Alberto Blanc illus. Instituto italiano di paleontologia umana,Rome, 1956
Bot 42.4701

ROME 1954 Association internationale d'hydrologie scientifiques :assemblee generale de Rome 10th comptes-rendus Tome 4: commission des neiges et de glaces International association of scientific hydrology.Publication,39 522p 24cm Louvain,1956 Papers in English,French and German
Sco 14.0124

ROME 1954 Scientific proceedings of the International association of meteorology Proceedings International association of meteorology AIM publication, 10,c 594p Butterworths,London,1956 Meeting at the 10th general assembly of International union of geodesy and geophysics
Nap 11.0286

ROME 1954 Transactions of the Rome meeting International association of terrestrial magnetism and electricity a meeting Edited by V. Laursen IAGA Bulletin, 15 406p Horsholm,Copenhagen,1957 The meeting took place during the 10th general assembly of the international union of geodesy and geophysics
Nap 11.0287

ROME 1955 Congresso nazionale della societa italiana di ematologia 13th Atti Societa italiana di ematologia Edited by G. Dominici and others Edizioni mediche e scientifiche,Rome,1956
Med 36.0380

ROME 1958 International congress of blood transfusion 7th Proceedings International society of blood transfusion Edited by Ludwig Hollander Karger,Basle,1959
Med 36.0200

ROME 1958 International congress of hematology :reports,communications,symposia 7th Proceedings 3 International society of hematology Edited by Giovanni Di Guglielmo Grune and Stratton,New York,1960
Med 36.0166

ROME 1958 International symposium on hemophilia and hemorrhagic states Proceedings International society of hematology Edited by K.M. Brinkhous North Carolina university press,Chapel Hill,1959 Held in connection with the 7th Congress of the International society of hematology
Med 36.0051

ROME 1958 Neuro-psychopharmacology International congress of neuro-pharmacology 1st Proceedings Collegium for neuro-psychopharmacology Edited by P.B. Bradley and others Elsevier,London,1959
PGMS 29.0192

ROME 1958 Neuro-psychopharmacology International congress of neuropharmacology 1st Proceedings International collegium for neuro-psychopharmacy Edited by P.B. Bradley and others Elsevier,Amsterdam,1959
Psy 31.0524

ROME 1958 Symposium on questions of numerical analysis proceedings Provisional international computation centre 79p Rome, 1958
Math L 5.2374

ROME 1959 Evoluzione e genetica :colloquio internazionale Relazioni e discussione Accademia nazionale dei Lincei Problemi attuali di scienza et di cultura.Quaderno N., 47 Rome,1960 Text in English and Italian.Summaries in English
Bal 39.0927

ROME 1959 Symposium on the numerical treatment of partial differential equations with real characteristics proceedings Provisional international computation centre 158p Birkhauser,Basel,1959
Math L 5.2378

ROME 1960 Chemical and biological aspects of pyridoxal catalysis :a symposium of the International union of biochemistry Proceedings Edited by E.E. Snell International union of biochemistry.Symposium series, 30 Pergamon,Oxford,1963
Radioth 35.0131

ROME 1960 Symposium on the numerical treatment of ordinary differential equations, integral and integro-differential equations 4th proceedings Provisional international computation centre bibliog. 679p 26cm Birkhauser,Basel,1960
Math L 5.0299

ROME 1961 International congress of human genetics 2nd Proceedings Vol 1-3 3 vols Instituto G.Mendel,Rome,1963 In English,French,German and Italian
An 32.0155

ROME 1961 Symposium on solar seeing 2nd Proceedings International electronic and nuclear exhibition,Rome Under the auspices of Consiglio nazionale delle ricerche 158p Rome,1962
Obs 6.2051

ROME 1961 The International congress of human genetics 2nd Proceedings Vol 1-3 3 vols Institute G.Mendel,Rome,1963
Gen 34.1857

ROME 1962 International symposium on optimizing and adaptive control 1st Proceedings Edited by Loren E. Bollinger and Emil J. Minnar Sponsored by the International federation of automatic control. Theory committee Instrument society of America,Pittsburgh,Pa.,1962
Eng 41.6017

ROME 1962 Le Choix des sites d'observatoires astronomiques (site testing) : a symposium proceedings Edited by J. Rosch and others International astronomical union. Symposium, 19 382p Gauthier-Villars,Paris, 1964 Reprinted from 'Bulletin astronomique de l'observatoire de Paris, vol 24, fasc.2 et 3, 1963
Obs 6.1467

ROME 1964 Centenario della nascita di Galileo Galilei 1564-1964 4th Atti Vol 2,tomo 4: atti del convegno sui campi magnetici solari e la spettroscopia ad alta resolutione.Proceedings of the meeting on solar magnetic fields and high resolution spectroscopy Galileo Galilei.Comitato nazionale per le manifestazioni celebrative Galileo Galilei.Comitato nazionale per le manifestazioni celebrative.Publicazione figs 341p Barbera,Florence,1966
WSM 43.4471

ROME 1964 Effects of low doses of radiation on crop plants :a panel Report Food and agriculture organization International atomic energy agency.Division of atomic energy in agriculture International atomic energy agency.Technical reports series, 64 International atomic energy agency, Vienna,1966
Radioth 35.1151

ROME 1964 I.C.N.A.F. environmental symposium Contributions International commission for the Northwest Atlantic fisheries International commission for the Northwest Atlantic fisheries.Special publication,6 figures,tables,graphs 914p ICNAF,Dartmouth,Nova Scotia,1965
Sco 14.5958

ROME 1965 Economics of automatic data processing :international symposium papers International computation centre Edited by A. B. Frielink pl. xiii,384p 22cm North-Holland,Amsterdam,1965
Math L 5.1022

ROME 1965 Oblique ionospheric radiowave propagation :at frequencies near the lowest usable high frequency:a symposium Agard AGARD.Conference proceedings, 13 Technivision,Slough,1969 The 11th annual symposium of the AGARD Electromagnetic wave propagation committee
Eng 41.5683

ROME 1966 Gas chromatography 1966 :an international symposium on gas chromatography and associated techniques 6th Proceedings Institute of petroleum.Gas chromatography discussion group Edited by A.B. Littlewood Institute of petroleum,London,1967
Chem 18.0569

ROME 1966 Theory of graphs :international sympoaium International computation centre Bibliog.,Illus. xviii,416p 25cm Gordon and Breach;Dunod,New York;Paris,1967
P Math 2.3128

ROME 1967 International symposium on residual gases in electron tubes and related vacuum systems 3rd Papers Section 2: sorption-desorption phenomena in high vacuum Italian society of physics Nuovo cimento. Supplement, 5,2 Editrice composori,Bologna, 1967
Met 25.2366

ROME 1967-68 Symposia mathematica 1,3 Istituto nazionale de alta matematica bibliog. 25cm 2 vols Academic press,New York;London,1969
P Math 2.3778

ROME 1968 Electron microscopy,1968 European regional conference on electron microscopy 4th Pre-congress abstracts Edited by Daria Steve Bocciarelli Tipografia poliglotta vaticana,Rome,1968
Met 25.1395

ROME 1968 Electron microscopy 1968 European regional conference on electron microscopy 4th Pre-congress abstracts Vol 2 Edited by D.S. Bocciarelli Tipografia polyglotta vaticana,Rome,1968 Chairman:J.S.Mitchell
Radioth 35.0559

ROME 1969 International symposium on radiosensitizing and radioprotective drugs 2nd book of abstracts Istituto superiore de sanita,Rome,1969
Radioth 35.1213

ROME 1969 Non-solar X- and gamma-ray astronomy International astronomical union Edited by L. Gratton International astronomical union.Symposium, 37 425p Reidel,Dordrecht,190
Obs 6.3322

ROME 1969 Non-solar X- and gamma-ray astronomy :symposium Papers International astronomical union Edited by L. Gratton International astronomical union.Symposium, 37 425p Reidel,Dordrecht,1970
TA 15.0492

ROME 1969 Pulsed-radiosources and high energy activity in supernova remnants Proceedings illus 326p 27cm Accademia nazionale dei lincei,Rome,1972
A Math 4.1898

ROME 1969 Radiation protection and sensitization International symposium on radiosensitizing and radioprotective drugs 2nd Proceedings Edited by Holta Whitehouse Taylor and Francis,London,1970
Radioth 35.1219

ROME 1969 Software engineering technique : a conference Report Edited by J.N. Buxton and B. Randall Sponsored by the North Atlantic treaty organization.Science committee illus. 1964p NATO science committee, Brussels,1970
Math 5.3610

ROME 1969 Software engineering techniques : a conference Report Edited by J.N. Buxton and B. Randell Sponsored by the North Atlantic treaty organization.Science committee Illus 164p NATO science committee,Brussels, 1970
Math L 5.3545

ROME 1970 Meccanica non lineare e stabilita Symposia mathematica Istituto nazionale di alta matematica Istituto nazionale di alta matematica.Pubblicazione, 6 395p 25cm Academic press,London;New York,1971
P Math 2.3989

ROOT GROWTH Easter school in agricultural science 15th Proceedings Nottingham 1968 University of Nottingham Edited by William J. Whittington Bibliog.,illus. xi, 450p Butterworths,London,1969
Bot 42.1437

ROROS,NORWAY 1968 Plasma waves in space and in the laboratory :a NATO advanced study institute North Atlantic treaty organization Edited by J.O. Thomas and B.J. Landmark Bibliog.,Illus. 487p 23cm Edinburgh university press,Edinburgh,1969
A Math 4.1505

ROSCOE B.JACKSON MEMORIAL LABORATORY Methodology in human genetics National institutes of health.Genetics study section Edited by Walter J. Burdette Holden-Day,San Francisco,Calif.,1962
Gen 34.1010

ROSCOE B.JACKSON MEMORIAL LABORATORY Methodology in mammalian genetics National institutes of health.Genetics study section Edited by Walter J. Burdette Holden-Day,San Francisco,Calif.,1963
Gen 34.1011

ROSHKO Anatol and others ed. The Heat transfer and fluid mechanics institute 1963 proceedings Pasadena,Calif. 1963 Jun.12-14 Stanford university press,Stanford,Calif,1963
A Math 4.0521

ROTATION La Rotazione come fenomeno e fattore evolutivo nell'universo Bologna 1971 Oct 8-9 Edited by G. Righini 212p Societa astronomica italiana,Florence,1972
Obs 6.3715

The ROTATION OF THE EARTH AND ATOMIC TIME STANDARDS :a symposium Moscow 1958 Aug International astronomical union Edited by Dirk Brouwer International astronomical union.Symposium, 11 1959 Reprinted from the Astronomical journal,vol 64, no 1268
Obs 6.3199

La ROTAZIONE COME FENOMENO E FATTORE EVOLUTIVO NELL'UNIVERSO Bologna 1971 Oct 8-9 Edited by G. Righini 212p Societa astronomica italiana,Florence,1972
Obs 6.3715

ROTHAMSTED 1930 Conference on soil science problems 1st proceedings Imperial bureau of soil science Imperial bureau of soil science.Technical communications,17 H.M.S.O., London,1931
Geog 13.1071

ROTHAMSTED 1958 Progress in soil zoology Research methods in soil zoology colloquium 1st Papers International society of soil science.Soil zoology committee Edited by P.W. Murphy Bibliog.,illus. Butterworths, London,1962
Bot 42.2085

ROTHSCHILD REPORT Framework for government research and development ;presented to Parliament by the Lord Privy Seal Great Britain.Lord Privy Seal Cmnd 4814 HMSO, London,1971 Contains the Rothschild and Dainton reports
Met 25.2797

ROTORUA,N.Z. 1963 Symposium on tropical meteorology Proceedings World meteorological organization Edited by J.W. Hutchings With the cooperation of the International union of geodesy and geophysics illus 751p New Zealand meteorological service,Wellington,N.Z.,1964
Nap 11.0650

ROTORUA,N.Z. 1963 Symposium on tropical meteorology Proceedings World meteorological organization,and,International union of geodesy and geophysics Edited by J. W. Hutchings New Zealand meteorological service,Wellington,N.Z.,1964
Geog 13.0990

ROTTACH-EGERN 1962 Funktionelle und morphologischorganisation der zelle Gesellschaft deutscher naturforscher und arzte Springer,Berlin,1963
Gen 34.0830

ROTTACH-EGERN 1963 Stellar and solar magnetic fields :a symposium International astronomical union Edited by R. Lust International astronomical union.Symposium, 22 460p North-Holland,Amsterdam,1965
A Math 4.1160

ROTTERDAM 1948 International conference on soil mechanics and foundation engineering 2nd Proceedings Vol 1-7 7 vols Rotterdam,1948
Eng 41.3139

ROUEN 1964 Colloque international sur "Le marche des bois du nord et la region economique de Haute-Normandie Fondation Francaise d'etudes nordiques Edited by Jean Malaurie Fondation Francaise d'etudes nordiques.Actes et documents,1 256p Rouen, 1964 Mimeograph
Sco 14.2469

ROUEN 1966 Fecamp 1966 Jan 27-29 Congres international de l'industrie morutiere dans l'atlantique-nord :tradition et avenir 1st Tome 1 Fondation francaise d'etudes nordiques Fondation francaise d'etudes nordiques.Actes et documents,2 Rouen;Fecamp, 1966
Sco 14.1037

ROUND TABLE DISCUSSION Electron transport and energy conservation Fasano 1969 May 12-15 Edited by J.M. Tager and others Adriatica editrice,Bari,1970
Bioch 33.2288

The ROUND TABLE DISCUSSION ON MITOCHONDRIAL STRUCTURE AND COMPARTMENTATION Proceedings Mitochondrial structure and compartmentation Polignano a Mare 1966 May 23-25 Edited by E. Quagliariello and others organized by the University of Bari,Department of biochemistry and the University of Amsterdam.Laboratory of biochemistry Adriatica editrice,Bari,1967
Bioch 33.1093

The ROUND TABLE DISCUSSION ON THE BIOCHEMICAL ASPECTS OF THE BIOGENESIS OF MITOCHONDRIA Proceedings Biochemical aspects of the biogenesis of mitochondria Polignane a Mare 1967 May 15-18 Edited by E.C. Slater and others Adriatica editrice,Bari,1968
Bioch 33.1073

The ROUND TABLE DISCUSSION ON THE ENERGY LEVEL AND METABOLIC CONTROL IN MITOCHONDRIA Proceedings Energy level and metabolic control in mitochondria Polignano a Mare 1968 May 13-16 Edited by S. Papa and others Organized by the University of Bari.Department of biochemistry and the University of Amsterdam.Laboratory of biochemistry Adriatica editrice,Bari,1969
Bioch 33.1075

ROYAL AERONAUTICAL SOCIETY Anglo-American aeronautical conference 6th Papers Folkestone 1957 Sep 9-12 Edited by Joan Bradbrooke and U.A. Libby Royal aeronautical society,London,1959
Eng 41.6962

ROYAL AERONAUTICAL SOCIETY I.F.A.C. symposium on fluidics Proceedings London 1968 Nov 4-8 International federation of automatic control 1968
Eng 41.5905

ROYAL AERONAUTICAL SOCIETY The Crack propagation symposium Proceedings Cranfield 1961 Sep Vol 1-2 Aeronautical research council 2 vols College of aeronautics,Cranfield,1962
Met 25.0865

ROYAL ANTHROPOLOGICAL INSTITUTE OF GREAT BRITAIN AND IRELAND Man,race and Darwin :a conference Papers London 1959 Jan 7-9 Edited by Philip Mason vii,151p Oxford university press,London,1960
WSM 43.3622

ROYAL COLLEGE OF PHYSICIANS Advanced medicine A Symposium on advanced medicine 2nd Proceedings London 1965 Nov 29-Dec 3 Edited by J.R. Trounce Pitman medical,London, 1966
Inv Med 37.0213

ROYAL COLLEGE OF PHYSICIANS OF LONDON Clinical aspects of genetics :a conference Proceedings London 1961 Mar 17-18 Edited by F.Avery Jones Pitman,London,1961
PGMS 29.0255

ROYAL COLLEGE OF PHYSICIANS OF LONDON Pathogenesis and treatment of occlusive arterial disease :a conference Proceedings London 1959 Nov 13-14 Edited by L. McDonald Pitman,London,1960
PGMS 29.0081

ROYAL COLLEGE OF PHYSICIANS OF LONDON Pathogenesis and treatment of occlusive arterial disease :a conference Proceedings London 1959 Nov 13-14 Edited by Lawson McDonald Pitman,London,1960
Path 30.2430

ROYAL COLLEGE OF PHYSICIANS OF LONDON Recent advances in renal diseases :a conference Proceedings London 1960 Jul 22-23 Edited by M.D. Milne Pitman,London,1960
PGMS 29.0224

ROYAL COLLEGE OF PHYSICIANS OF LONDON Symposium on advanced medicine :a conference Proceedings London 1964 Nov 16-20 Edited by Nigel Compston Pitman,London,1965
PGMS 29.0057

ROYAL COLLEGE OF PHYSICIANS OF LONDON The Thyroid and its diseases :a conference Proceedings London 1963 Mar 15-16 Edited by A.Stuart Mason Pitman,London,1963
PGMS 29.0159

ROYAL COLLEGE OF PHYSICIANS SCIENTIFIC CONFERENCE 5th Proceedings Clinical aspects of genetics London 1961 Mar 17-18 Royal college of physicians Edited by F.Avery Jones Pitman medical,London,1961
Gen 34.1981

ROYAL COLLEGE OF SURGEONS Catapres in hypertension :a symposium Proceedings London 1969 Mar Edited by Matthew E. Conolly Butterworths,London,1970
PGMS 29.0608

ROYAL COLLEGE OF SURGEONS OF ENGLAND The Biochemistry of mucopolysaccharides of connective tissue :a symposium London 1960 Feb 13 Biochemical society Edited by F. Clark and J.K. Grant Biochemical society. Symposia, 20 Cambridge university press, Cambridge,1961
Bioch 33.1382

ROYAL ENTOMOLOGICAL SOCIETY OF LONDON.SYMPOSIA, 1 Insect polymorphism symposium Papers and discussions London 1961 Sep 21-22 Royal entomological society of London Edited by J.S. Kennedy Royal entomological society, London,1961
Gen 34.1640

ROYAL ENTOMOLOGICAL SOCIETY OF LONDON.SYMPOSIA, 2 Insect reproduction :symposium London 1963 Sep 19-20 Royal entomological society of London Edited by K.C. Highnam Royal entomological society,London,1964
Gen 34.1647

ROYAL ENTOMOLOGICAL SOCIETY OF LONDON.SYMPOSIA, 3 Insect behaviour London 1965 Sep 23-24 Royal entomological society of London Edited by P.T. Haskell Royal entomological society,London,1966
Gen 34.1650

ROYAL FREE HOSPITAL Acute renal failure :a symposium London 1963 Sep Edited by S. Shaldon and G.C. Cook Blackwell,Oxford,1964
PGMS 29.0225

ROYAL GEOGRAPHICAL SOCIETY Biology and ethics :a symposium Proceedings London 1968 Sep 26-27 Institute of biology Edited by F.J. Ebling Institute of biology.Symposia, 18 Academic press,London;New York,1969
Bal 39.0018

ROYAL GEOGRAPHICAL SOCIETY Man-made lakes :a symposium Proceedings London 1965 Sep 30-Oct 1 Institute of biology Edited by R. Lowe-McConnell Institute of biology.Symposia, 15 Academic press,London;New York,1966
Bal 39.1898

ROYAL GEOGRAPHICAL SOCIETY The Biological significance of climatic changes in Britain Proceedings London 1964 Oct 29-30 Institute of biology Edited by C.G. Johnson and L.P. Smith Institute of biology. Symposium, 14 illus x,222p Academic press,London;New York,1965
Bot 42.6678

ROYAL GEOGRAPHICAL SOCIETY The Biological significance of climatic changes in Britain :a symposium held at the Royal geographical society Proceedings London 1964 Oct 29-30 Institute of biology Edited by C.G. Johnson and L.P. Smith Institute of biology. Symposia, 14 Academic press,London;New York, 1965
Bal 39.1687

ROYAL HORTICULTURAL SOCIETY Conifers in cultivation Conifer conference London 1931 Nov 10-12 Edited by F.J. Chittenden Royal horticultural society,London,1932
BG 38.2424

ROYAL HORTICULTURAL SOCIETY Hybrid conference report 1900 on hybridisation (the cross-breeding of species)and on the cross-breeding of varieties London 1899 Jul 11-12 Royal horticultural society.Journal, 24 Spottiswoode,London,1900 Conference also entitled the 'International congress of genetics' and the 'International conference on hybridization'
Gen 34.0998

ROYAL HORTICULTURAL SOCIETY International congress of genetics hybridization (the cross-breeding of genera or species)the cross-breeding of varieties... 3rd London 1906 Jul 30-Aug 3 Edited by W. Wilks Spottiswoode,London,1907 Conference also entitled 'International conference on hybridisation'
Gen 34.0999

ROYAL HORTICULTURAL SOCIETY International horticultural congress 13th Report London 1952 Sep 8-15 Vol 1-2 Royal horticultural society,London,1953
BG 38.3124

ROYAL HORTICULTURAL SOCIETY Rock gardens and rock plants conference Reports London 1936 May 5-7 Edited by F.J. Chittenden Royal horticultural society,London,1936
BG 38.3310

ROYAL INSTITUTION Adaption in micro-organisms :a symposium Papers London 1953 Apr Society for general microbiology Edited by R. Davies and E.F. Gale Society for general microbiology.Symposia, 3 Cambridge university press,Cambridge,1953
Bioch 33.1166

ROYAL INSTITUTION Bacterial anatomy :a symposium Papers London 1956 Apr Society for general microbiology Edited by E. T.C. Spooner and B.A.D. Stocker Society for general microbiology.Symposia, 6 Cambridge university press,Cambridge,1956
Bioch 33.1169

ROYAL INSTITUTION Biochemical studies of antimicrobial drugs :a symposium Papers London 1966 Apr Society for general microbiology Edited by B.A. Newton and P.E. Reynolds Society for general microbiology. Symposia, 16 Cambridge university press, Cambridge,1966
Bioch 33.1170

ROYAL INSTITUTION Information theory A Symposium on information theory Papers read London 1955 Sep 12-16 Edited by Colin Cherry Butterworths,London,1956
Bal 39.1005

ROYAL INSTITUTION Mechanisms of microbial pathogenicity :a symposium Papers London 1955 Apr Society for general microbiology Edited by J.W. Howie and A.J. O'Hea Society for general microbiology.Symposia, 5 Cambridge university press,Cambridge,1955
Bioch 33.1172

ROYAL INSTITUTION Microbial behaviour 'in vivo' and 'in vitro' :a symposium Papers London 1964 Apr Society for general microbiology Edited by H. Smith and Joan Taylor Society for general microbiology. Symposia, 14 Cambridge university press, Cambridge,1964
Bioch 33.1173

ROYAL INSTITUTION Microbial classification : a symposium Papers London 1962 Apr Society for general microbiology Edited by G. C. Ainsworth and P.H.A. Sneath Society for general microbiology.Symposia, 12 Cambridge university press,Cambridge,1962
Bioch 33.1174

ROYAL INSTITUTION Microbial ecology :a symposium Papers London 1957 Apr Society for general microbiology Edited by R. E.O. Williams and C.C. Spicer Society for general microbiology.Symposia, 7 Cambridge university press,Cambridge,1957
Bioch 33.1175

ROYAL INSTITUTION Microbial genetics :a symposium Papers London 1960 Apr Society for general microbiology Edited by W. Hayes and R.C. Clowes Society for general microbiology.Symposia, 10 Cambridge university press,Cambridge,1960
Bioch 33.1176

ROYAL INSTITUTION Microbial reaction to environment :a symposium Papers London 1961 Apr Society for general microbiology Edited by G.C. Meynell and H. Gooder Society for general microbiology.Symposia, 11 Cambridge university press,Cambridge,1961
Bioch 33.1177

ROYAL INSTITUTION Strategy of chemotherapy : a symposium Papers London 1958 Apr Society for general icrobiology Edited by S. T. Cowan and Elizabeth Rowatt Society for general microbiology.Symposia, 8 Cambridge university press,Cambridge,1958
Bioch 33.1180

ROYAL IRISH ACADEMY The Chemistry and biochemistry of fungi and yeasts :symposium Proceedings Dublin 1963 Jul 18-20 Butterworths,London,1963 Reprinted from 'Pure and applied chemistry' vol.7,no.4
Gen 34.0729

ROYAL METEOROLOGICAL SOCIETY Global circulation of the atmosphere :a conference Proceedings London 1969 Aug 25-29 Edited by G.A. Corby bibliog.,illus.,port. 257p 26cm Royal meteorological society,London, 1969
A Math 4.1651

ROYAL METEOROLOGICAL SOCIETY International symposium on world climate 8000 to 0 b.c. proceedings London 1966 Apr 18-19 Edited by J.S. Sawyer Royal meteorological society, London,1966
A Math 4.0579

ROYAL METEOROLOGICAL SOCIETY Meteorological factors in radio-wave propagation :a conference Report London 1946 Apr 8 325p Physical society,London,1946
Nap 11.0925

ROYAL METEOROLOGICAL SOCIETY World climate from 8000 to nought B.C. International symposium on world climate 8000 to nought B.C. Proceedings London 1966 Apr 18-19 London, 1966
Sco 14.0581

ROYAL MICROSCOPICAL SOCIETY Cytochemical progress in electron microscopy :a symposium Oxford 1962 Jul 2-4 Royal microscopical society.Journal, 81,pt 3-4 London,1963
Bot 42.4729

ROYAL MICROSCOPICAL SOCIETY Historical aspects of microscopy Papers Oxford 1966 Mar 18 Edited by S. Bradbury and G.L'E. Turner W.Heffer,Cambridge,1967
Bal 39.0241

ROYAL MICROSCOPICAL SOCIETY Historical aspects of microscopy Royal microscopical society conference Papers Oxford 1966 Mar 18 Edited by S. Bradbury and G.L'E. Turner viii,227p Cambridge,1967
WSM 43.2520

ROYAL MICROSCOPICAL SOCIETY Symposium on cytochemical progress in electron microscopy Oxford 1962 Jul 2-4 Journal of the Royal microscopical society.Ser.3, 81,pt 3-4,p.107-117 London,1963
Bal 39.0235

ROYAL MICROSCOPICAL SOCIETY Symposium on cytochemical progress in electron microscopy Oxford 1962 Jul 2-4 Journal of the Royal microscopical society,1962,pt.3,4 Royal microscopical society,London,1963
An 32.3182

ROYAL NETHERLANDS CHEMICAL SOCIETY and INTERNATIONAL UNION OF PURE AND APPLIED CHEMISTRY.PHYSICAL CHEMISTRY SECTION Molecular spectroscopy :general and introductory lectures presented at the fifth European congress on molecular spectroscopy 5th Amsterdam 1961 May 29-Jun 3 Co-sponsored by the Netherlands physical society Butterworths,London,1962 Reprinted from Pure and applied chemistry,vol 4,no 1
Chem 18.2315

ROYAL NETHERLANDS CHEMICAL SOCIETY.SECTION FOR INORGANIC AND PHYSICAL CHEMISTRY The Mechanism of heterogeneous catalysis :a symposium Proceedings Amsterdam 1959 Nov 12-13 Edited by J.H. de Boer and others Elsevier monographs.Chemistry series Elsevier,Amsterdam,1960
Chem 18.0455

ROYAL RADAR ESTABLISHMENT.RADIO COMPONENTS RESEARCH AND DEVELOPMENT SUB-COMMITTEE 15: COMPUTER ELEMENTS Microminiaturisation and its effect on computer elements discussion meetings proceedings Malvern 1961 Aug.30 15p Royal radar establishment,Malvern,1961
Math L 5.1848

ROYAL SOCIETY A Symposium on continental drift London 1964 Mar Edited by P.M.S. Blackett and others Royal society. Philosophical transactions,1088 Royal society,London,1965
Geog 13.5647

ROYAL SOCIETY Soft X-ray band spectra and the electronic structure of metals and materials :a conference Proceedings Strathclyde 1967 Sep 18-21 Edited by Derek J. Fabian Academic press,London;New York, 1968
Met 25.1470

The ROYAL SOCIETY EMPIRE SCIENTIFIC CONFERENCE 1st report London 1946 Jun 17-Jul 8 London,1948
Col S 12.0199

ROYAL SOCIETY OF ARTS and COUNCIL FOR NATURE The Countryside in 1970 :conference 2nd Proceedings London 1965 Nov 10-12 Jointly sponsored by the Nature conservancy Royal society of arts;Nature conservancy, London,1966
Geog 13.1899

ROYAL SOCIETY OF CANADA Evolution:its science and doctrine :a symposium presented to the Royal society of Canada 1959 Edited by Thomas W.M. Cameron Royal society of Canada. "Studia varia"series, 4 University of Toronto press,Toronto,1960 Title also in French L'evolution:la science et la doctrine
Bal 39.0915

ROYAL SOCIETY OF CANADA Evolution:its science and doctrine symposium presented to the Royal society of Canada in 1959 Edited by Thomas W.M. Cameron Royal society of Canada.Studia varia series, 4 University of Toronto press,Toronto,1960
Gen 34.1270

ROYAL SOCIETY OF CANADA The Canadian Northwest:its potentialities :symposium presented to the Royal society of Canada in 1958 Papers Edmonton 1958 Jun 3 Edited by Frank H. Underhill Royal society of Canada.Studia varia.Series 3 map 104p 24cm University of Toronto press,Toronto, 1959
Sco 14.4228

ROYAL SOCIETY OF CANADA The Canadian northwest:its potentialities :symposium Edmonton 1958 Jun Edited by Frank H. Underhill Royal society of Canada."Studia varia" series,3 University of Toronto press, Toronto,1959
Geog 13.5038

ROYAL SOCIETY OF EDINBURGH Natural resources in Scotland :symposium Edinburgh 1960 Oct 31 - Nov 2 25cm Edinburgh,1961
Philos 1.1867

ROYAL SOCIETY OF HEALTH Prevention of hospital infection :the personal factor;report of a conference London 1963 Jun 19 Edited by George Godber and others Royal society of health,London,1963
PGMS 29.0584

ROYAL SOCIETY OF LONDON A Discussion on the physics of the moon and its environment :a meeting Papers London 1965 Jun 3-4 Royal society of London.Proceedings.Series A. Mathematical and physical sciences, 296,1446 Royal society,London,1967
Geod 9.0015

ROYAL SOCIETY OF LONDON A Symposium on Continental drift papers London 1964 Mar 19-20 By P.M.S. Blackett and others Royal society of London.Philosophical transactions. Series A,no.258,no.1088 map Royal society of London,London,1965
Geol 8.1262

ROYAL SOCIETY OF LONDON A Symposium on continental drift Papers London 1964 Mar 19-20 By P.M.S. Blackett and others Royal society of London.Philosophical transactions. Series A,258,no.1088 vi,323p Royal society of London,London,1965
Sco 14.0260

ROYAL SOCIETY OF LONDON A Symposium on continental drift Papers London 1964 Mar 19-20 Royal society of London.Philosophical transactions.Series A, 258 London,1965 Symposium organized for the Royal society by P. M.S.Blackett,Sir Edward Bullard, and S.K. Runcorn
Geod 9.0129

ROYAL SOCIETY OF LONDON Amorphous and liquid semiconductors :a conference Proceedings Cambridge 1969 Sep 24-27 Edited by Nevill Francis Mott Bibliog,diagrs,figs 626p 20cm North-Holland,Amsterdam,1970
Cav 7.2678

ROYAL SOCIETY OF LONDON Anglo-Romanian conference on mathematics in the archaeological and historical sciences Proceedings Edited by F.R. Hodson and others viii,565p 24cm Edinburgh university press, Edinburgh,1971
Math S 3.1826

ROYAL SOCIETY OF LONDON Empire scientific conference Report London 1946 Jun-Jul Vol 1-2 2 vols London,1948
Bioch 33.1683

ROYAL SOCIETY OF LONDON Internal aerodynamics (turbomachinery) :a conference Cambridge 1967 Jul 19-21 Institution of mechanical engineers,London,1970
Eng 41.6829

ROYAL SOCIETY OF LONDON Moon and planets,2 Cospar plenary meeting :joint open meeting of working groups 1,2 and 5 10th London 1967 Jul 26-27 Edited by A. Dollfus 196p North-Holland,Amsterdam,1968
Obs 6.3419

ROYAL SOCIETY OF LONDON Quantum theory and beyond :essays and discussions arising from a colloquium Cambridge 1968 Jul illus viii,345p 24cm Cambridge university press, Cambridge,1971 Based on papers read at an informal colloquium
A Math 4.1759

ROYAL SOCIETY OF LONDON Scientific information conference Reports and papers submitted London 1948 Jun 21 Jul 2 London,1948
Bioch 33.1749

ROYAL SOCIETY OF LONDON Semi-conducting materials :a conference Proceedings Reading 1950 Jul 10-15 Edited by H.K. Henisch Under the auspices of the International union of pure and applied physics illus xv,281p Butterworths, London,1951
Cav 7.1925

ROYAL SOCIETY OF LONDON Semi-conducting materials :a conference Proceedings Reading 1950 Jul 10-15 International union of pure and applied physics and University of Reading London,1951
Geod 9.0286

ROYAL SOCIETY OF LONDON The Empire scientific conference Reports London; Cambridge;Oxford 1946 Jun-Jul Vol 1-2 2 vols Royal society,London,1948
Gen 34.0053

ROYAL SOCIETY OF LONDON The Royal society scientific information conference Reports and papers London 1948 Jun 21-Jul 2 Royal society,London,1948
Chem 18.2006

ROYAL SOCIETY OF LONDON The Royal society scientific information conference :report and papers submitted London 1948 Jun 21-Jul 2 Royal society,London,1948
Bal 39.3921

ROYAL SOCIETY OF LONDON The Scientific information conference report and papers submitted London 1948 Jun 21-Jul 2 Royal society,London,1948
Gen 34.0054

ROYAL SOCIETY OF LONDON The Structures and functions of proteolytic enzymes a discussion London 1968 Dec 5-6 Edited by D.C. Phillips and others Royal society of London. Philosophical transactions.Ser.B, 257,p.63-266 Royal society,London,1970
Bioch 33.1917

ROYAL SOCIETY OF LONDON Tissue proteinases Wates symposium Proceedings Cambridge 1970 Apr 6-7 Edited by A.J. Barrett and J.T. Dingle Held at the Strangeways research laboratory North-Holland,Amsterdam,1971
Bioch 33.2284

ROYAL SOCIETY OF LONDON.GASSIOT COMMITTEE Rocket exploration of the upper atmosphere :a conference papers Oxford 1953 Aug 24-26 Edited by R.L.F. Boyd and M.J. Seaton Journal of atmospheric and terrestrial physics. Special supplement, 1 376p Pergamon press,London,1954
Nap 11.0073

ROYAL SOCIETY OF LONDON.GASSIOT COMMITTEE The Conference on emission spectra of the night sky and aurorae Papers London 1947 Jul 7-10 139p Physical society,London,1948
Cav 7.1453

ROYAL SOCIETY OF LONDON.GASSIOT COMMITTEE The Conference on emission spectra of the night sky and aurorae Papers London 1947 Jul 7-10 139p Physical society,London,1948
TA 15.0039

ROYAL SOCIETY OF LONDON.GASSIOT COMMITTEE The Emission spectra of the night sky and aurorae :a conference Papers London 1947 Jul.7-10 139p London,1948
Obs 6.1964

ROYAL SOCIETY OF LONDON.MATHEMATICAL TABLES COMMITTEE Mathematical tables Vol. 3: Table of binomial coefficients Edited by J.C. P. Miller viii,162p 29cm Cambridge university press,Cambridge,1954
P. Math 2.2337

ROYAL SOCIETY OF LONDON.WARREN RESEARCH FUND The Physics of ionized gases :a conference proceedings,typescript London 1953 Apr University college,London Edited by J.B. Hasted London,1953
A Math 4.0614

ROYAL SOCIETY OF MEDICINE Disorders of carbohydrate metabolism :a symposium Proceedings London 1968 Nov 29-30 Association of clinical pathologists.Chemical pathology sub-committee Edited by G.K. McGowan and G. Walters Chemical pathology in relation to clinical medicine, 5 Journal of clinical pathology.Supplement, 2 British medical association,London,1969
Bioch 33.1205

ROYAL SOCIETY OF MEDICINE Symposium on clinical trials London 1958 Apr Edited by Edward Charles Dodds Pfizer,Tonbridge, 1958
Med 36.0116

ROYAL SOCIETY OF MEDICINE The Influence of heredity on disease,with special reference to tuberculosis, cancer and diseases of the nervous system :a discussion London 1908 Nov illus. Longmans,London,1909
Gen 34.1979

ROYAL SOCIETY OF MEDICINE The Thyroid gland : a symposium Proceedings London 1953 Feb 25 Edited by P. Eckstein and S. Zuckerman Society for endocrinology.Memoirs, 1 1953
Radioth 35.0713

ROYAL SOCIETY OF MEDICINE.ENDOCRINE SECTION The Investigation of hypothalamic-pituitary-adrenal function :a symposium Proceedings London 1967 Feb 22-23 Edited by V.H.T. James and J. Landon Sponsored by Ciba, Horsham Society for endocrinology.Memoirs, 17 Cambridge university press,Cambridge,1968
Gen 34.0558

ROYAL SOCIETY OF MEDICINE.ENDOCRINE SECTION The Investigation of hypothalamic-pituitary-adrenal function :a symposium Proceedings London 1967 Feb 22-23 Society for endocrinology and Ciba laboratories ltd Edited by V.H.T. James and J. Landon Society for endocrinology.Memoirs, 17 Cambridge university press,Cambridge,1968
Phys 20.1421

ROYAL SOCIETY OF MEDICINE.ENDOCRINOLOGICAL SECTION The Determination of adrenocortical steroids and their metabolites : a conference Proceedings London 1953 May 21 Edited by P. Eckstein and S. Zuckerman Society for endocrinology.Memoirs, 2 Cambridge university press,Cambridge,1955
Radioth 35.1940

ROYAL SOCIETY OF MEDICINE.ENDOCRINOLOGICAL SECTION The Technique and significance of oestrogen :a conference proceedings London 1954 Feb 17 Edited by P. Eckstein and S. Zuckerman Society for endocrinology.Memoirs, 3 Cambridge university press,Cambridge, 1955
Radioth 35.0045

ROYAL SOCIETY OF MEDICINE.ENDOCRINOLOGICAL SECTION The Technique and significance of oestrogen determinations :a conference Proceedings London 1954 Feb 17 Edited by P. Eckstein and S. Zuckerman Society for endocrinology.Memoirs, 3 Cambridge university press,Cambridge,1955
Phys 20.1384

ROYAL SOCIETY OF VICTORIA The Evolution of living organisms :a symposium to mark the centenary of Darwin's 'Origin of species' and of the Royal society of Victoria Melbourne 1959 Dec Edited by G.W. Leeper Melbourne university press,Melbourne,1962
Bal 39.0923

ROYAL SOCIETY OF VICTORIA The Evolution of living organisms :a symposium to mark the centenary of Darwin's 'Origin of species' and of the Royal society of Victoria Melbourne 1959 Dec 8-11 Edited by G.W. Leiper Melbourne university press,Melbourne,1962
Gen 34.1277

The ROYAL SOCIETY SCIENTIFIC INFORMATION CONFERENCE :report and papers submitted London 1948 Jun 21-Jul 2 Royal society of London Royal society,London,1948
Bal 39.3921

The ROYAL SOCIETY SCIENTIFIC INFORMATION CONFERENCE Reports and papers London 1948 Jun 21-Jul 2 Royal society of London Royal society,London,1948
Chem 18.2006

ROYAL STATISTICAL SOCIETY Royal statistical society.Research section and industrial applications joint conference Cardiff 1964 Sep 28-Oct 1 Cardiff,1964 Unbound mimeograph papers
Math 3.1114

ROYAL STATISTICAL SOCIETY The Babbage memorial meeting London 1971 Oct 18 illus 40p British computer society,London, 1971
Math L 5.3612

ROYAL STATISTICAL SOCIETY The Babbage memorial meeting Report of the proceedings London 1971 Oct 18 iii,39p 30cm British computer society,London,1971
Math S 3.1858

ROYAL STATISTICAL SOCIETY.INDUSTRIAL APPLICATIONS SECTION Statistical method in industrial production conference proceedings Sheffield 1950 Sep 29-Oct 1 89p 25cm Royal statistical society,London,1951
Math 3.0821

ROYAL SWEDISH ACADEMY OF ENGINEERING SCIENCES. CORROSION COMMITTEE Recent advances in stress corrosion :a symposium Stockholm 1960 Oct 12 Edited by Ake Bresle Almqvist and Wiksell,Stockholm,1961
Met 25.1960

ROYAL VETERINARY FIELD STATION Pathology of laboratory rats and mice London 1966 Apr 19-21 Nuffield foundation.Advisory committee Edited by Ernest Cotchin and Francis J.C. Roe Blackwell,Oxford,1967
Path 30.2438

ROYAUMONT 1950 Le Muscle :etude de biologie et de pathologie.Colloque Compte rendu Council for international organizations of medical sciences L'expansion scientifique,1952 In English and French
Bioch 33.2245

ROYAUMONT 1950 Le Muscle:etude de biologie et de pathologie :colloque Compte rendu Organise par la Conseil pour la co-ordination des congres internationaux des sciences medicales L'Expansion scientifique francaise, 1950 In French and English
Phys 20.1557

ROYAUMONT 1957 La Science au seizieme siecle :colloque international de Royaumont International union of history and philosophy of science Histoire de la pensee, 2 Hermann,Paris,1960
WSM 43.0911

ROYAUMONT 1959 Ciba foundation symposium on cellular aspects of immunity Proceedings Ciba foundation Edited by G.E.W. Wolstenholme and Maeve O'Connor Churchill, London,1960
Bioch 33.1117

ROYAUMONT 1965 International symposium on immunological methods of biological standardization 15th Proceedings International association of microbiological societies.Permanent section of microbiological standardization Edited by R.H. Regamey and others International association of microbiological societies.Symposia series in immunological standardization, 4 illus, tables Karger,Basle;New York,1967
Path 30.2746

RUDOLF MAGNUS INSTITUTE FOR PHARMACOLOGY Pituitary,adrenal and the brain :an international conference on the pituitary-adrenal axis and the nervous system Proceedings Vierhouten 1969 Jul 22-24 Edited by D. De Wied and J.A.W.M. Weijnen Progress in brain research, 32 Elsevier, Amsterdam,1970
An 32.5235

RUPRECHT-KARL-UNIVERSITAT HEIDELBERG European conference on elementary particles 4th Proceedings Heidelberg 1967 Sep 20-27 Edited by H. Filthuth viii,550p North-Holland,Amsterdam,1968
A Math 4.1348

RUSCHLIKON 1967 Constructive aspects of the fundamental theorem of algebra :a symposium Proceedings Edited by Bruno Dejon and Peter Henrici Bibliog.,illus. vii,337p 24cm Wiley,New York,1969
P Math 2.3485

RUSSELL SAGE FOUNDATION Biology and behavior 3 Environmental influences :a conference Proceedings New York 1967 Apr 21-22 Edited by David C. Glass Rockefeller university press;Russell Sage foundation,New York,1968 Third and last in a series of conferences on biology and behavior
Psy 31.3096

RUSSELL SAGE FOUNDATION Neurophysiology and emotion :a conference Proceedings New York 1965 Dec 10-11 Edited by David C. Glass Rockefeller university press;Russell Sage foundation,New York,1967
Psy 31.2900

RUSSIAN HYDROLOGICAL CONGRESS 1st Proceedings Leningrad 1924 May 7-14 Komitet vsesoyuznikh gidrologicheskikh S'ezdov tables,maps 623p 26cm Leningrad,1925
Sco 14.0111

RUSTS Cereal rust conferences Cambridge 1964 Jun-Jul Plant breeding institute Cambridge,1966
Bot 42.4125

RUTGERS,N.J. 1962 Informational macromolecules :a symposium Institute of microbiology of Rutgers Edited by H.J. Vogel and others With support from the National science foundation Academic press,New York, 1963
Gen 34.0664

RUTGERS,N.J. 1964 Evolving genes and proteins :a symposium Institute of microbiology of Rutgers Edited by V. Bryson and H.J. Vogel With support from the National science foundation Academic press, New York;London,1965
Gen 34.0665

RUTGERS INTERDISCIPLINARY RESEARCH CENTER Drugs and youth Rutgers symposium on drug abuse Proceedings Edited by J.R. Wittenborn and others Held at Rutgers university Charles C.Thomas,Springfield,Ill., 1969
Inv Med 37.0235

RUTGERS SYMPOSIUM ON DRUG ABUSE Proceedings Drugs and youth Rutgers interdisciplinary research center Edited by J.R. Wittenborn and others Held at Rutgers university Charles C.Thomas,Springfield,Ill.,1969
Inv Med 37.0235

RUTGERS UNIVERSITY.BUREAU OF BIOLOGICAL RESEARCH New developments in tissue culture :annual research conference New Brunswick,N.J. 1961 Edited by James W. Green Annual research conference of the bureau of biological research, 17 Rutgers university press,New Brunswick,N.J.,1961
An 32.3177

RUTGERS UNIVERSITY.BUREAU OF BIOLOGICAL RESEARCH Some conjugated proteins;a symposium The Annual conference on protein metabolism 9th six lectures New Brunswick,N.J. 1953 Jan 30-31 Edited by William H. Cole Rutgers university press,New Brunswick,N.J.,1953
Bioch 33.0543

RUTGERS UNIVERSITY.BUREAU OF BIOLOGICAL RESEARCH. ANNUAL CONFERENCES ON PROTEIN METABOLISM 14th Papers Serological and biochemical comparisons of proteins New Brunswick,N.J. 1958 Jan 24-25 Edited by William H. Cole Rutgers university press,New Brunswick,N.J., 1958
Col S 12.0073

RUTGERS UNIVERSITY.INSTITUTE OF MICROBIOLOGY Bacterial endotoxins :a symposium Proceedings New Brunswick,N.J. 1963 Sep 4-6 Edited by Maurice Landy and Werner Braun With the support of the National science foundation Rutgers university.Institute of microbiology,New Brunswick,N.J.,c1963
Path 30.2611

RUTGERS UNIVERSITY.INSTITUTE OF MICROBIOLOGY Evolving genes and proteins :a symposium Proceedings New Brunswick,N.J. 1964 Sep 17-18 Edited by Vernon Bryson and Henry J. Vogel Academic press,New York;London,1965
Bioch 33.0569

RUTGERS UNIVERSITY.INSTITUTE OF MICROBIOLOGY
Informational macromolecules :a symposium Proceedings New Brunswick,N.J. 1962 Sep 5-7 Edited by Henry J. Vogel and others With support from the National science foundation Academic press,New York;London,1963
Bioch 33.0822

RUTGERS UNIVERSITY.INSTITUTE OF MICROBIOLOGY
Nucleic acids in immunology :a symposium Proceedings New Brunswick,N.J. Edited by O. J. Plescia and W. Braun Springer,NEW York, 1968
Radioth 35.0152

RUTGERS UNIVERSITY.INSTITUTE OF MICROBIOLOGY
Organizational biosynthesis :a symposium Proceedings New Brunswick,N.J. 1966 Sep 8-10 Edited by Henry J. Vogel and others With support from the National science foundation Academic press,New York;London, 1967
Bioch 33.0832

RUTHERFORD Ernest,baron Rutherford jubilee international conference Proceedings Manchester 1961 Sep 4-8 Victoria university of Manchester Edited by J.B. Birks Under the sponsorship of International union of pure and applied physics 856p Haywood,London,1961 Includes commemorative session 'Rutherford at Manchester'
Cav 7.1952

RUTHERFORD Ernest,baron Rutherford jubilee international conference proceedings Manchester 1961 Sep.4-8 Victoria university of Manchester Edited by J.B. Birks Co-sponsored by the International union of pure and applied physics Heywood, London,1961 To commemorate the discoveries of Rutherford at Manchester
A Math 4.0826

RUTHERFORD HIGH ENERGY LABORATORY
International conference on elementary particles proceedings Oxford 1965 Sep.19-25 Sponsored by the International union of pure and applied physics Rutherford high energy laboratory,Oxford,1965
A Math 4.0887

RUTHERFORD JUBILEE INTERNATIONAL CONFERENCE
Proceedings Manchester 1961 Sep 4-8 Victoria university of Manchester Edited by J.B. Birks Under the sponsorship of International union of pure and applied physics 856p Haywood,London,1961 Includes commemorative session 'Rutherford at Manchester'
Cav 7.1952

RUTHERFORD JUBILEE INTERNATIONAL CONFERENCE
proceedings Manchester 1961 Sep.4-8 Victoria university of Manchester Edited by J.B. Birks Co-sponsored by the International union of pure and applied physics Heywood, London,1961 To commemorate the discoveries of Rutherford at Manchester
A Math 4.0826

RUTHERFORD LABORATORY Conference on radiation protection in accelerator environments Proceedings Chilton,Berks. 1969 Mar Edited by G.R. Stevenson Sponsored by the Science research council Rutherford laboratory,Chilton,1969 Organized by and held at the Rutherford laboratory
Radioth 35.1393

RYDBERG J.R. Rydberg centennial conference on atomic spectroscopy Proceedings Lund 1954 Jul 1-5 Edited by Bengt Edlen Arranged by the Swedish national committee for physics Physiographisk sallskaps i Lund. Forhandlingar.N.F., 65,21 108p c.k.w. gleerup,Lund,1954
Obs 6.3285

S.C.A.R. see SCIENTIFIC COMMITTEE ON ANTARCTIC RESEARCH

S.S.S.R.- N.K.T.P.VSESOYUZNOE GEOLOGO-RAZVEDOCHNOE OBEDINENIE and ASSOCIATION FOR THE STUDY OF THE EUROPEAN QUATERNARY, Mezhdunarodnaya konferentsiya assotsiatsii po izucheniyu chetvertichnogo perioda Evropy International conference of the Association for the study of the European quaternary 2nd trudy Leningrad 1932 Sep 1-7 Vypusk 1-5 Edited by D.A. Petrovskii and others Gosudarstvennoe nauchno-tekhnicheskoe geologo-razvedochnoe izdatelstvo,Leningrad;Moscow,1932-34
Geol 8.2982

SAAS-FEE 1971 Theorie des atmospheres stellaires :conference By D. Mihalas and others 312p Observatoire de Geneve, Sauverny,1971
Obs 6.3631

SACLAY 1957 Symposium de metallurgie speciale Centre d'etudes nucleaires,Saclay Presses universitaires de France,Paris,1958
Met 25.2517

SACLAY 1958 Diffusion a l'etat solide Colloque sur la diffusion a l'etat solide Centre d'etudes nuleaires,Saclay: and Commissariat a l'energie atomique Centre d'etudes nucleaires de Saclay;North Holland, Gif-sur-Yvette;Amsterdam,1959
Met 25.1277

SACLAY 1959 Colloque de metallurgie corrosion,seche et aqueuse 3e Centre d'etudes nucleaires,Saclay Sponsored by the Institut national des sciences et techniques nucleaires Centre d'etudes nucleaires de Saclay;North Holland,Paris;Amsterdam,1960
Met 25.1891

SACLAY 1960 Colloque de metallurgie proprietes des joints de grains 4th Centre d'etudes nucleaires,Saclay Presses universitaires de France,Paris,1961 Papers in English and French
Met 25.1266

SACLAY 1962 Colloque international sur la retention et la migration des ions radioactifs dans les sols Institut national des sciences et techniques nucleaires Centre d'etudes nucleaires de Saclay;Presses universitaires de France,Gif-sur-Yvette;Paris,1963
Eng 41.3174

SAFETY IN AIR AND AMMONIA PLANTS :a symposium New York 1966 9 Chemical engineering progress 96p A.I.Ch.E. New York,1967 A C.E.P.technical manual
Chem E 24.1576

SAFETY OF STRUCTURES SYMPOSIUM 2nd Proceedings Kyoto 1955 Sep 6 Science council of Japan Japan society of civil engineers Science council of Japan,Kyoto, 1956
Eng 41.3014

SAGAMORE ARMY MATERIALS CONFERENCE 9th Fundamentals of deformation processing Raquette Lake,New York 1962 Aug 28-31 By Walter A. Backofen and others United States. Army materials research agency and Syracuse university Co-organized and directed by the National research council Syracuse university press,Syracuse,New York, 1964
Met 25.0741

SAGAMORE ARMY MATERIALS RESEARCH CONFERENCE 12th Strengthening mechanisms,metals and ceramics Raquette Lake,N.Y. 1965 Aug 24-27 Army materials research agency Syracuse university Syracuse university press, Syracuse,N.Y.,1966
Met 25.2746

SAGAMORE ARMY MATERIALS RESEARCH CONFERENCE 10th Fatigue-an interdisciplinary approach Raquette Lake,New York 1963 Aug 13-16 United States.Army materials research agency and Syracuse university Edited by John J. Burke and others Syracuse university press, Syracuse,1964
Met 25.2330

SAGAMORE ARMY MATERIALS RESEARCH CONFERENCE 10th Proceedings Fatigue - an interdisciplinary approach Raquette Lake,N.Y. 1963 Aug 13-16 Edited by J.J. Burke and others Sponsored by United States.Army materials research agency Syracuse university press,New York,1964
Eng 41.3803

SAGAMORE ARMY MATERIALS RESEARCH CONFERENCE 13th-14th Proceedings Surfaces and interfaces 1-2 Raquette Lake,New York 1966-67 United States.Army materials research agency and Syracuse university Edited by John J. Burke and others Syracuse university press,New York,1967-68
Met 25.2365

SAGAMORE ARMY MATERIALS RESEARCH CONFERENCE 16th Proceedings Ultrafine-grain metals Raquette Lake,N.Y. 1969 Aug 19-22 Army materials research agency Syracuse university Syracuse university press, Syracuse,N.Y.,1970
Met 25.2747

SAGAMORE ORDNANCE MATERIALS RESEARCH CONFERENCE Proceedings 6th Composite materials and composite structures New York 1959 Aug 8-21 United States.Army.Ordnance materials research office and United States.Army. Office of ordnance research Arrangements by Syracuse university research institute New York,1959 Mimeograph
Met 25.0093

ST.ANDREWS 1962 Excitations in semiconductors,polarons and excitons Scottish universities' summer school in physics 3rd Papers Edited by C.G. Kuper and G.D. Whitfield Scottish universities' summer school in physics,1962,3 Oliver and Boyd,Edinburgh,1963
Cav 7.2136

ST.ANDREWS 1964 Structure and function of connective and skeletal tissue :advanced study institute Proceedings Organized under the auspices of the North Atlantic treaty organization Butterworths,London,1965
An 32.3423

ST.ANDREWS 1964 Structure and functions of connective and skeletal tissue an advanced study institute Proceedings North Atlantic treaty organization.Scientific affairs division Butterworths,London,1964
Path 30.2421

ST.ANDREWS 1964 Symposium on aspects of the gene symposium Proceedings University of St.Andrews.Biological society Edited by Michael D.B. Burt and Peter W. Barlow Biological journal, 4,supp. University of St.Andrews.Biological society,St.Andrews,1964
Gen 34.0762

ST.ANDREWS 1965 Nervous and hormonal mechanisms of integration Society for experimental biology and American society of zoologists Society for experimental biology.Symposia, 20 Cambridge university press,Cambridge,1966
Psy 31.0137

ST.ANDREWS 1967 Mathematical methods in solid state and superfluid theory Scottish universities' summer school in physics 8th Papers Edited by R.C. Clark and G.H. Derrick Scottish universities' summer school in physics,1967 xvi,400p 25cm Oliver and Boyd,Edinburgh,1969
Cav 7.2759

ST.ANDREWS 1968 Numerical taxonomy Colloquium in numerical taxonomy Proceedings Edited by A.J. Cole Academic press,London; New York,1969
Gen 34.2149

ST.ANDREWS 1968 Numerical taxonomy Colloquium in numerical taxonomy Proceedings Edited by A.J. Cole Academic press,London, 1969
HE 27.0022

ST.ANDREWS 1968 Numerical taxonomy Colloquium in numerical taxonomy Proceedings Edited by A.J. Cole Academic press,London, 1969
Math L 5.3440

ST.ANDREWS 1968 Numerical taxonomy : colloquium Proceedings University of St. Andrews Edited by Alfred J. Cole Academic press,London,1969
Bot 42.3452

ST.ANDREWS 1970 White dwarfs Edited by W. J. Luyten International astronomical union. Symposium, 42 164p Reidel,Dordrecht,1971
TA 15.0573

ST.LOUIS,MISS. 1968 Annual meeting of the Association for academic surgery 2nd Proceedings Association for academic surgery Edited by G.D. Zuidema and D.B. Skinner Current topics in surgical research, 1 Academic press,New York;London,1969
Surg 23.0069

ST.LOUIS,MO. Gene action in micro-organisms : conference Papers Missouri botanical garden Missouri botanical garden.Annals, 32, no.2 Galesburg,Ill.,1945
Gen 34.0741

ST.LOUIS,MO. 1936 The Symposium on colloid chemistry 13th Papers presented Edited by Harry Boyer Weiser Colloid symposium monograph, 13 William and Wilkins,Baltimore, 1937
Bioch 33.1480

ST.LOUIS,MO. 1951 American institute of mining and metallurgical engineers :annual meeting 1951 Symposium:problems of clay and laterite genesis Sponsored by the Karl Eilers Memorial Fund A.I.M.E.,New York,1952
Min 10.0319

ST.LOUIS,MO. 1961 Direct observation of imperfections in crystals :technical conference Proceedings American institute of mining,metallurgical and petroleum engineers.Metallurgical society Interscience, New York,1962
Cav 7.2011

ST.LOUIS,MO. 1961 Kendall award symposium Solid surfaces and the gas-solid interface : papers presented at the Kendall award symposium honoring Stephen Brunauer Papers American chemical society.Division of colloid and surface chemistry Edited by Lewellyn E. Copeland and others American chemical society.Advances in chemistry series,33 American chemical society,Washington,D.C.,1961
Chem 18.0616

ST.LOUIS CONGRESS 1904 The Historical relations between medicine and surgery to the end of the sixteenth century:an address By T. Clifford Allbutt Macmillan,London;New York, 1905 Delivered at the St.Louis congress, 1904
Path 30.1646

ST.PETER,MINN. 1965 Genetics and the future of man :a discussion at the Nobel conference Edited by John D. Roslansky Organized by Gustavus Adolphus college North-Holland,Amsterdam,1966
Gen 34.0062

ST.PETERSBURG 1881 Internationale polar-konferenz :verhandlungen und ergebnisse 3te Bericht Akademie der wissenschaften,St. Petersburg,1881 Parallel French-German texts.One of 8 pamphlets bound together
Philos 1.2259

ST.PETERSBURG 1897 Congres geologique international 7th guide des excursions International geological congress St. Petersburg,1897
Geol 8.2786

ST.RAPHAEL 1971 International advanced summer institute on microprogramming North Atlantic treaty organization.Scientific affairs division Edited by Guy G. Boulaye and Jean Mermet Actualites scientifiques et industrielles illus 418p Hermann,Paris, 1972
Math L 5.3634

SAINT-CLOUD 1965 L'histoire sociale. Sources et methodes :colloque Ecole normale superieure de Saint-Cloud Presses universitaires de France,Paris,1967
Geog 13.1610

SAINT-GERMAIN-EN-LAYE 1960 Le Frottement interieur des metaux Comptes rendus Institut de recherches de la siderurgie francaise and Centre national de recherches metallurgiques Edited by C. Crussard and others Editions metaux,n.p., c1961
Met 25.1026

SALFORD 1967 European symposium on time-of-flight mass spectrometry 1st Proceedings University of Salford Edited by D. Price and J.E. Williams Pergamon,Oxford,1969
Met 25.1706

SALFORD 1967 Time-of-flight mass spectroscopy :based on the first European symposium Proceedings University of Salford Edited by D. Price and J.E. Williams Pergamon press,Oxford,1969
Chem 18.2590

SALFORD 1969 Dynamic mass spectrometry European symposium on the time-of-flight mass spectrometry 2nd Proceedings 1 University of Salford Edited by D. Price and J.E. Williams Heyden,London,1970
Chem 18.2669

SALINE WATER CONVERSION :a symposium Papers American chemical society national meeting 137th Cleveland,Ohio 1960 Apr American chemical society.Division of water and waste chemistry Advances in chemistry series American chemical society,Washington,D.C.,1960
Eng 41.7535

SALISBURY 1955 C.C.T.A.Southern regional committee for geology :meeting 1st proceedings Commission for technical co-operation in Africa south of the Sahara London,1956 Bound with proceedings of three other regional meetings
Geol 8.3114

SALISBURY,S.AUSTRALIA 1957 Data processing and automatic computing machines conference Australia.Department of supply.Weapons research establishment Six typescript papers
Math L 5.0776

SALISBURY,S.AUSTRALIA 1957 Data processing and automatic computing machines conference proceedings Vol. 1-2 Australia. Department of supply.Weapons research establishment illus. 29cm 2 vols Weapons research establishment,Salisbury, Australia,1957
Math L 5.0805

SALIVARY SECRETION Secretory mechanisms of salivary glands International conference on mechanisms of salivary secretion and their regulation Proceedings Birmingham,Ala. 1966 Aug 9-11 Edited by Leon H. Schneyer and Charlotte A. Schneyer Held at the University of Alabama medical center Academic press,New York;London,1967
Bal 39.1265

SALMON Symposium on pink salmon Lectures Vancouver,B.C. 1960 Oct 13-15 University of British Columbia.Institute of fisheries Edited by N.J. Wilimovsky University of British Columbia.Institute of fisheries.H.R. Macmillan lectures in fisheries, 3 27cm Vancouver,B.C.,1962
Bal 44.4988

SALT Sea,salt and plants seminar Proceedings Bhavnagar 1965 Dec 20-23 Central salt and marine chemicals research institute Edited by V. Krishnamurthy illus. xv,372p CSMCRI,Bhavnagar,1967
Bot 42.1991

SALT BASINS Salt basins around Africa Institute of petroleum and Geological society of London :joint meeting proceedings London 1965 Mar 3 Institute of petroleum, and,Geological society of London Institute of petroleum,London,1965
Geol 8.3097

SALT BASINS AROUND AFRICA Institute of petroleum and Geological society of London : joint meeting proceedings London 1965 Mar 3 Institute of petroleum,and,Geological society of London Institute of petroleum, London,1965
Geol 8.3097

SALT LAKE CITY,UTAH 1959 Kinetics of cellular proliferation Conference on fundamental problems and technics for the study of the kinetics of cellular proliferation Proceedings Edited by Frederick Stohlman Grune and Stratton,New York;London,1959
An 32.3235

SALT LAKE CITY,UTAH 1970 Symposium (international) on combustion 13th Proceedings Combustion institute xix,1190p 26cm Combustion institute,Pittsburgh,Pa., 1971
Chem 18.2863

SALT MARSH CONFERENCE Proceedings Sapelo island,Ga. 1958 Mar 25-28 University of Georgia.Marine institute University of Georgia.Marine institute,Athens,Ga.,1959
Geog 13.0713

SALTSJOBADEN 1957 Co-ordination of galactic research :conference 2nd International astronomical union Edited by A. Blaauw and others International astronomical union.Symposium, 7 93p Cambridge university press,Cambridge,1959
Obs 6.0192

SALTSJOBADEN 1964 Spectral classification and multicolour photometry Symposium on the spectral classification and multicolour photometry Proceedings International astronomical union Edited by K. Loden and others International astronomical union. Symposium, 24 383p Academic press,London; New York,1966
Obs 6.1092

SALTSJOBADEN 1964 Special classification and multicolour photometry International astronomical union Edited by K. Loden International astronomical union.Symposium, 24 Academic press,London;New York,1966
TA 15.0319

SALTSJOBADEN 1964 Symposium on virus and cancer Papers Swedish cancer society and Unio nordica contra cancrum Edited by H. Bergstrand and K.E. Hellstrom illus Balder, Stockholm,1965 Symposium arranged by the Swedish cancer society in conjunction with the annual general meeting of the Unio nordica contra cancrum
Path 30.2363

SALTSJOBADEN,SWEDEN 1964 Spectral classification and multicolour photometry :a symposium International astronomical union Edited by K. Loden and others International astronomical union.Symposium, 24 383p Academic press,London;New York,1966
A Math 4.1166

SALZBURG 1962 Corrosion of reactor materials the conference Proceedings Vol 1-2 International atomic energy agency 2 vols International atomic energy agency, Vienna,1962
Met 25.1937

SALZBURG 1968 Induction,physics,and ethics Colloquium in the philosophy of science Proceedings and discussions International union of history and philosophy of science. Division of logic,methodology and philosophy of science Institut fur wissenschaftstheorie, Salzburg Edited by Paul Weingartner and Gerhard Zecha Synthese library x,382p Reidel,Dordrecht,1970
WSM 43.0992

SALZBURG 1968 Medical radioisotope scintigraphy :a symposium Proceedings Vol 1-2 International atomic energy agency Edited by G.R. Stevenson International atomic energy agency.Proceedings series 2 vols International atomic energy agency, Vienna,1969
Radioth 35.1392

SAMPLE PREPARATION TECHNIQUES FOR LIQUID SCINTILLATION COUNTING :1966 summer school Proceedings London 1966 Aug 15-18 Beckman instruments limited Beckman instruments limited,Glenrothes,n.d.
Radioth 35.1778

SAN ANTONIO,TEX. 1951 Physics and medicine of the upper atmosphere Symposium on the physics and medicine of the upper atmosphere Proceedings Lovelace foundation for medical education and research Edited by Clayton S. White and Otis O. Benson Sponsored by School of aviation medicine.Randolph Field illus xxix,611p University of New Mexico press, Albuquerque,N.M.,1952
Nap 11.0635

SAN DIEGO 1957 Vistas in astronautics Astronautics symposium 1st proceedings United States.Air force.Office of scientific research Edited by Morton Alperin and Marvin Stern Co-sponsored by General dynamics corporation.Convair division Vistas in astronautics, 1 International series of monographs on aeronautical sciences and space flight.Astronautics division, 1 329p Pergamon press,London,1958
Nap 11.0022

SAN DIEGO,CALIF. 1966 Fundamentals of gas-surface interactions :a symposium Proceedings United States.Air force.Office of scientific research General dynamics corporation.General atomic division Edited by Howard M. Saltsburg and others Academic press,New York,1967
Chem 18.2696

SAN DIEGO,CALIF. 1970 Role of cyclic AMP in cell function :symposium American college of neuropsychopharmacology Edited by Paul Greengard and Erminio Costa Advances in biochemical psychopharmacology, 3 Raven press,New York,1970
Bioch 33.2211

SAN DIEGO,CALIF. 1970 Symposium on nonlinear estimation theory and its applications Proceedings United States.Air force.Office of scientific research Institute of electrical and electronics engineers.Group on automatic control vii, 297p 28cm IEEE,New York,1970
Math S 3.1698

SAN DIEGO,CALIF. 1971 Statistical models and turbulence :a symposium Proceedings Edited by M. Rosenblatt and C. Van Atta Lecture notes in physics, 12 viii,492p 25cm Springer,Berlin,1972
Math S 3.1832

SAN FRANCISCO Exposition universelle et internationale de San Francisco :la science francaise France.Ministere de l'instruction publique et des beaux-arts 2 vols Paris, 1915
Philos 1.0013

SAN FRANCISCO 1950 North American wildlife conference 15th Transactions Wildlife management institute Edited by Ethel M. Quee 681p 23cm Washington,D.C.,1950
Sco 14.8304

SAN FRANCISCO 1959 Extractive and physical metallurgy of plutinium ...based on a symposium American institute of mining, metallurgical and petroleum engineers Edited by W.D. Wilkinson Bibliog Interscience,New York;London,1960 Introduction and annotated bibliography by the editor
Met 25.2364

SAN FRANCISCO 1959 I.R.E. western electronic show and convention convention record Part 4: automatic control, electronic computers,information theory Institute of radio engineers 201p Institute of radio engineers,New York,1959
Math L 5.2073

SAN FRANCISCO 1959 Plutonium and its alloys Extractive and physical metallurgy of plutonium and its alloys based on a symposium American institute of mining,metallurgical and petroleum engineers.Nuclear metallurgy committee Edited by W.D. Wilkinson Interscience,New York;London,1960 With an introduction and an annotated bibliography by the editor
Met 25.1568

SAN FRANCISCO 1959 Dallas 1960 Mar 14-18 High purity water corrosion of metals seven papers presented at symposia National association of corrosion engineers.Technical committee 3-F on corrosion by high-purity water National association of corrosion engineers,Houston,Texas,1968
Met 25.1933

SAN FRANCISCO 1960 Aging around the world International congress of gerontology 5th Proceedings International association of gerontology Edited by Herman T. Blumenthal Columbia university press,New York;London,1962
An 32.0088

SAN FRANCISCO 1967 Ethnopharmacologic search for psychoactive drugs :a symposium Proceedings Edited by Daniel H. Efron and others Sponsored by the National institute of mental health National institute of mental health.Pharmacology section.Workshop series,2 United States.Public health service publication,1645 U.S.Dept.of public health, Washington,D.C.,1967
Pha 16.0065

SAN FRANCISCO 1970 Challenge to pharmacy in the seventies Invitational conference on pharmacy manpower Proceedings Edited by Joe B. Graber and Donald C. Brodie Sponsored by the University of California.School of pharmacy U.S.Department of health,education and welfare,Washington,D.C.,1972
PGMS 29.0674

SAN FRANCISCO,CALIF. 1956 Arcs in inert atmospheres and vacuum a symposium Electrochemical society.Electrothermics and metallurgy division Edited by W.E. Kuhn Wiley;Chapman and Hall,New York;London,1956
Met 25.0248

SAN FRANCISCO,CALIF. 1957 World conference on prestressed concrete :conference Proceedings University of California San Francisco,Calif.,1957
Eng 41.3011

SAN FRANCISCO,CALIF. 1958 Biological psychiatry :the proceedings of the scientific sessions of the Society... Vol 1 Society of biological psychiatry Edited by J. Masserman Grune and Stratton,New York; London,1959
Psy 31.1624

SAN FRANCISCO,CALIF. 1959 Newer metals Symposium on newer metals presented at the third Pacific area national meeting... American society for testing materials A.S.T.M.Special technical publication, 272 A.S.T.M. Philadelphia,Pa.,1960
Met 25.0514

SAN FRANCISCO,CALIF. 1960 High-strength steels for the missile industry Golden Gate metals conference Proceedings American society for metals Edited by H.T. Sumsion American society for metals,Novelty,Ohio,1961
Met 25.0487

SAN FRANCISCO,CALIF. 1961 Chemical reactions in the lower and upper atmosphere :a symposium Proceedings Stanford research institute 390p Wiley,New York,1961
Nap 11.0929

SAN FRANCISCO,CALIF. 1961 Chemical reactions in the lower and upper atmosphere : an international symposium Proceedings Stanford research institute Interscience,New York;London,1961
Chem 18.0476

SAN FRANCISCO,CALIF. 1962 Palynology in oil exploration Symposium Edited by Aureal T. Cross Southwestern association of petroleum geologists.Society of economic paleontologists and mineralogists.Special publication,11 Tulsa,Okla.,1964
Geol 8.0978

SAN FRANCISCO,CALIF. 1962 World conference on shell structures Proceedings University of California International association for shell structures Building research advisory board National research council.Publication, 1187 National academy of sciences,Washington, D.C.,1064
Eng 41.3025

SAN FRANCISCO,CALIF. 1965 Hyperbolic oxygen and radiation therapy of cancer Annual San Francisco cancer symposium 1st Proceedings Frontiers of radiation therapy and oncology, 1 Karger,Basel;New York,1968
Radioth 35.1970

SAN FRANCISCO,CALIF. 1966 Electron beam therapy Annual San Francisco cancer symposium 2nd Proceedings Frontiers of radiation therapy and oncology, 2 Karger, Basel;New York,1968
Radioth 35.1208

SAN FRANCISCO,CALIF. 1968 Atomic absorption spectroscopy :a symposium presented at the seventy-first annual meeting... Proceedings American society for testing and materials American society for testing and materials.Publication, 443 ASTM,Philadelphia, Pa.,1969
Met 25.2647

SAN FRANCISCO,CALIF. 1968 Evaluation of wear testing :a symposium presented at the 71st annual meeting American society for testing and materials American society for testing and materials.Special technical publication, 446 ASTM,Philadelphia,Pa.,1969
Eng 41.8327

SAN FRANCISCO,CALIF. 1968 Fatigue at high temperature :a symposium presented at the 71st annual meeting Proceedings American society for testing materials American society for testing materials.Technical publication, 459 ASTM,Philadelphia,Pa.,1969
Met 25.2583

SAN FRANCISCO,CALIF. 1968 Fatigue at high temperature :a symposium presented at the 71st annual meeting... American society for testing and materials American society for testing and materials.Special technical publication, 459 ASTM,Philadelphia,Pa.,1969
Eng 41.8293

SAN FRANCISCO,CALIF. 1968 Interfaces in composites :a symposium Papers American society for testing and materials A.S.T.M. Special technical publication, 452 ASTM, Philadelphia,Pa.,1969 Presented at the 71st annual meeting
Met 25.2749

SAN FRANCISCO,CALIF. 1968 Performance of deep foundations :a symposium American society for testing materials American society for testing materials.Special technical publication, 444 American society for testing materials,Philadelphia,Pa.,1969
Eng 41.3159

SAN FRANCSICO,CALIF. 1968 Vibration effects of earthquakes on soils and foundations :a symposium American society for testing and materials American society for testing and materials.Special technical publication, 450 American society for testing materials,Philadelphia,Pa.,1969
Eng 41.3162

SAN JUAN 1960 Anales de las primeras jornadas geologicas Argentinas Tom 1-3 Asociacion geologica Argentina Buenos Aires, 1960-62
Geol 8.3165

SAN JUAN 1960 Comparative effects of radiation :a conference Report Edited by Milton Burton and others Wiley,New York,1960
Radioth 35.0947

SAN JUAN,P.R. 1961 Radiation-induced chromosome aberrations :report of a conference on biochemical and biophysical mechanisms in the production of radiation-induced chromosome aberrations Edited by Sheldon Wolff Sponsored by the National research council illus. xii,304p Columbia university press, New York;London,1963
Bot 42.4730

SAN JUAN,PUERTO RICO 1957 Plantation systems :seminar Papers and discussion summaries Research institute for the study of man,and,Pan American union Pan American union.Social science monographs,7 Washington, D.C.,1959
Geog 13.5120

SAN JUAN,PUERTO RICO 1961 Radiation-induced chromosome aberrations Conference on biochemical and biophysical mechanisms in the production of radiation- induced chromosome aberrations Edited by Sheldon Wolff Sponsored by the National research council Columbia university press,New York;London,1963
Gen 34.1122

SAN REMO 1968 Computing methods in optimization problems International conference on computing methods in optimization problems 2nd Papers Sponsored by the Society for industrial and applied mathematics Lecture notes in operations research and mathematical economics, 14 bibliog.,illus. Springer,Berlin,1968
Math S 3.1516

SAN REMO 1968 Computing methods in optimization problems-2 International conference on computing methods in optimization problems 2nd Papers Edited by Lofti A. Zadeh and others Sponsored by the Society for industrial and applied mathematics Academic press,New York,1969
Eng 41.5865

SANDEFJORD 1963 End results of cancer therapy International symposium on end results of cancer therapy Proceedings National cancer institute Edited by Sidney J. Cutler National cancer institute.Monograph, 15 National cancer institute,Bethesda,Md., 1964
Bioch 33.0986

SANDEFJORD 1963 International symposium on end results of cancer therapy National cancer institute Edited by Sidney J. Cutler National cancer institute.Monograph, 15 U.S. Department of health,education and welfare, Washington,D.C.,1964
Radioth 35.0844

SANTA BARBARA,CALIF. 1961 Extra-galactic research The Symposium on problems of extra-galactic research 15th International astronomical union Edited by G.C. McVittie International astronomical union.Symposium, 15 450p Macmillan,New York,1962
Obs 6.1124

SANTA BARBARA,CALIF. 1961 Problems of extra-galactic research :a symposium International astronomical union Edited by G. C. McVittie International astronomical union. Symposium, 15 450p Macmillan,New York, 1962
A Math 4.1154

SANTA BARBARA,CALIF. 1961 Symposium on problems of extra-galactic research Proceedings International astronomical union Edited by G.C. McVittie International astronomical union.Symposium, 15 McMillan, New York,1962
TA 15.0035

SANTA BARBARA,CALIF. 1961 Vigilance :a symposium Proceedings United States.Office of naval research Edited by Donald N. Buckner and James J. McGrath Conducted by Human factors research McGraw-Hill series in psychology McGraw-Hill,New York,1963
Psy 31.1008

SANTA BARBARA,CALIF. 1965 Coastal engineering :a specialty conference American society of civil engineers American society of civil engineers,New York,1966
Eng 41.3346

SANTA BARBARA,CALIF. 1965 Soviet and east European agriculture Conference on Soviet agricultural affairs 2nd Papers University of California,Berkeley.Center of Slavic and east European studies Edited by Jerzy F. Karcz Russian and east European studies University of California press, Berkeley,Calif.;Los Angeles,Calif.,1967
Geog 13.2584

SANTA BARBARA,CALIF. 1970 National conference on weather modification 2nd Proceedings National science foundation. Atmospheric sciences section American meteorological society 436p American meteorological society,Boston,Mass.,1970
Nap 11.0977

SANTA CLARA,CALIF. 1966 The Heat transfer and fluid mechanics institute 1966 proceedings Edited by Michel A. Saad and James A. Miller Stanford university press, Stanford,Calif.,1966
A Math 4.0524

SANTA CRUZ,CALIF. 1971 Annual workshop on microprogramming 4th Proceedings Association for computing machinery Association for computing machinery,New York, 1971
Math L 5.3631

SANTA MONICA,CALIF. 1952 Decision processes seminar proceedings University of Michigan Edited by R.M. Thrall and others Supported by Ford foundation viii,332p 22cm John Wiley and sons,New York,1954
Math 3.0771

SANTA MONICA,CALIF. 1953 Numerical analysis :symposium 6th Proceedings American mathematical society Edited by John H. Curtiss Cosponsored by National bureau of standards American mathematical society. Proceedings of symposia in applied mathematics, 6 vi,303p McGraw-Hill,New York,1956
Math L 5.3245

SANTA MONICA,CALIF. 1959 The Rand symposium on mathematical programming:linear programming and recent extensions :a conference proceedings Edited by Philip Wolfe sponsored by Rand corporation R-351 123p rand corporation,Santa Monica,Calif., 1960
Math L 5.2453

SANTA MONICA,CALIF. 1960 Energy conversion for space power :a symposium American rocket society Edited by Nathan W. Snyder Progress in astronautics and rocketry, 3 Academic press,New York,1961
Eng 41.4971

SANTA MONICA,CALIF. 1960 Energy conversion for space power... based on a symposium Selected technical papers American rocket society.Space power systems committee Edited by Nathan W. Snyder Progress in astronautics and rocketry, 3 Academic press,New York; London,1961
Met 25.2165

SANTA MONICA,CALIF. 1960 Mathematical optimization techniques symposium University of California Edited by Richard Bellman xii,346p 24cm University of California press,Berkeley,Calif.,1963
P. Math 2.2288

SANTA MONICA,CALIF. 1960 Mathematical optimization techniques symposium papers University of California Edited by Richard Bellman With assistance from the National science foundation xii,346p 24cm University of California press,Berkeley,Calif., 1963
Math 3.0238

SANTA MONICA,CALIF. 1963 State of stress in the earth's crust :an international conference Proceedings Geological society of America.Committee on rock mechanics and Rand corporation Edited by William R. Judd New York,1964
Geod 9.0143

SANTIAGO 1961 International symposium on tissue transplantation Proceedings Edited by Alberto P. Cristoffanini and Gustavo Hoecker Sponsored by the Universidad de Chile Universidad de Chile,Santiago,Chile, 1962
An 32.3286

SANTIAGO 1961 The International symposium on tissue transplantation Proceedings Edited by Alberto P. Cristoffanini and Gustavo Hoecker Sponsored by Universidad de Chile Universidad de Chile,Santiago,1962
PGMS 29.0244

SANTIAGO 1967 Cancer control :Latin American regional conference Papers International union against cancer Sponsored by the Committee on national cancer control programmes of the commission on cqncer control International union against cancer.Technical report series, 1 International union against cancer,Geneva,1968
Radioth 35.0888

SANTIAGO,CHILE 1966 Symposium on antarctic oceanography main review papers Scientific committee on antarctic research Bibliog., Illus. xi,268p 23cm Scott polar research institute,Cambridge,1966
A Math 4.1501

SANTIAGO DE CHILE 1966 SCAR,SCOR,IAPO,IUBS symposium on Antarctic oceanography Papers International council of scientific unions. Scientific committee on Antarctic research International council of scientific unions. Scientific committee on oceanic research International association of physical oceanography International union of biological sciences Held by invitation of the Chilean national committee for antarctic research illus,maps xi,268p 23cm Scott Polar research institute for SCAR,Cambridge, 1968
Sco 14.8300

SANTIAGO DE CHILE 1966 Tratado antartico informe de la quarte reunion consultiva 4th Proceedings Antarctic treaty plate 86p 20cm Santiago de Chile,n.d.
Sco 14.6116

SAO PAULO 1966 International pharmacological meeting Vol 5: control of growth processes by chemical agents Edited by A.D. Welch Pergamon,Oxford,1968
Pha 16.0257

SAO PAULO 1966 International pharmacological meeting 3rd Proceedings Vol 2: pharmacology of reproduction Edited by E. Diczfalusy Pergamon,Oxford,1968
Pha 16.0256

SAO PAULO 1966 Simposio internacional de genetica symposium Edited by F.G. Brieger Held under the auspices of the International union of biological sciences Ciencia e cultura, 19,no.1 Sao Paulo,1967
Gen 34.0995

SAPELO ISLAND,GA. 1958 Salt marsh conference Proceedings University of Georgia.Marine institute University of Georgia.Marine institute,Athens,Ga.,1959
Geog 13.0713

SAPPORO 1966 Cellular injury and resistance in freezing organisms International conference on low temperature science 2nd Proceedings Vol 2: conference on cryobiology Hokkaido university.Institute of low temperature science Edited by Eizo Asahina illus Hokkaido university.Institute of low temperature science,Sapporo,1967 Title page in English and Japanese Conference held in commemoration of the 25th anniversary of the establishment of the Institute
Path 30.2625

SAPPORO 1966 Physics of snow and ice International conference on low temperature science 1st Proceedings 1 pt 2: conference on physics of snow and ice (part 2) Institute of low temperature science Edited by Hirobumi Oura illus 1967p 27cm Hokkaido university,Sapporo,1957
Sco 14.7411

SAPPORO,JAPAN 1966 International conference on low temperature science Papers conference on cryobiology.Cellular injury and resistance in freezing organisms Edited by Eizo Asahina Sponsored by the Institute of low temperature science illus xxiii,257p 27cm Sapporo,1967
Sco 14.0144

SAPROPELI IKH ISPOLZOVANIE :po matrialam konferentsii 1956 Akademiya nauk Belorusskoi S.S.R.Institut torfa Izdatelstvo akademii nauk B.S.S.R.,Minsk,1958
Bot 42.6733

SARANAC INN,N.Y. 1928 Commission internationale de l'eclairage en succession a la Commission internationale de photometrie 7e session Recueil des travaux et compte rendu National physical laboratory, Teddington,c1929
Psy 31.2561

SARCOIDOSIS International conference on sarcoidosis 5th Proceedings Prague 1969 Jun 16-21 Universita Karlova Edited by Ladislav Levinsky and Frantisek Macholda Univerzita Karlova,Prague,1971
PGMS 29.0666

SATELLITES Dynamics of satellites :a symposium Paris 1962 May 28-30 International union of theoretical and applied mechanics Edited by Maurice Roy Springer, Berlin,1963
Eng 41.6890

SBORNIK MATERIALOV RASSHIRENNOGO SOVESHCHANIYA RABOCHEI GRUPPY PO GLYATSIOLOGII SOVETSKOGO MEZHDUVEDOMSTVENNOGO KOMITETA MEZHDUNARODNOGO GEOFIZICHESKOGO GODA Moscow 1958 May 20-24 Edited by G.A. Avsyuk 165p 20cm Moscow,1959
Sco 14.0514

SBORNIK UKLYUCHAET MATERIALY TRETEGO VSESOYUZNOGO SOVESHCHANIYA PO MORSKIM MLEKOPITAYUSHCHIM Morskie mlekopitayushchie Vladivostok 1966 Akademiya nauk S.S.S.R.Ministerstvo po rybnogo khozyaistva S.S.S.R.Ikhtiologicheskaya komissiya Edited by A.V.A. Arsenev and others 342p 27cm Izdatelstvo 'Nauka', Moscow,1969
Sco 14.8244

The SCALING-UP OF CHEMICAL PLANT AND PROCESSES joint symposium Papers London 1957 May 28-29 By R.Edgeworth Johnstone and others Society of chemical industry.Chemical engineering group and Institution of chemical engineers Institution of chemical engineers,London,1957 Held jointly with the K.Instituut van ingenieurs and the K. Nederlandse chemische vereniging
Chem E 24.1234

SCANDINAVIAN CORROSION CONGRESS (NKM) 4th : lectures Current corrosion research in Scandinavia Helsinki 1964 Nov 24-27 Teknillisten tieteiden akatemia and Central chemical association Edited by Jori Larinkari and others Organized by the Scandinavian council for applied research Kemian keskusliton julkaisuja, 24 Sanoma Osakeyhtio,Helsinki,1965
Met 25.1901

SCANDINAVIAN COUNCIL FOR APPLIED RESEARCH Current corrosion research in Scandinavia Scandinavian corrosion congress (NKM) : lectures 4th Helsinki 1964 Nov 24-27 Teknillisten tieteiden akatemia and Central chemical association Edited by Jori Larinkari and others Kemian keskusliton julkaisuja, 24 Sanoma Osakeyhtio,Helsinki, 1965
Met 25.1901

SCANDINAVIAN LOGIC SYMPOSIUM 2nd Proceedings Oslo 1970 Jun 18-20 Universitet i Oslo Edited by J.E. Fenstad Studies in logic and the foundations of mathematics, 63 vi,405p 23cm North-Holland,Amsterdam,1971
P Math 2.4056

SCANDINAVIAN METEOROLOGICAL MEETING see NORDISK METEOROLOGMODE

SCANDINAVIAN NEUROLOGICAL SOCIETY The So-called extrapyramidal system :symposium from the sixteenth congress of Scandinavian neurobiologists Oslo 1962 Edited by Sigvald Refsum and others Universitetsforlaget,Oslo,1963
An 32.4655

SCANDINAVIAN PHARMACOLOGICAL SOCIETY Metabolism of amines in the brain :a symposium Proceedings Edinburgh 1968 Jul 11 Edited by G. Hooper Macmillan,London,1969
An 32.5040

SCANDINAVIAN SOCIETY FOR CLINICAL CHEMISTRY AND CLINICAL PHYSIOLOGY 6th meeting Upsala 1954 Sep 10-11 and Stockholm 1954 Sep 13 Scandinavian journal of clinical and laboratory investigation, 7,supp.20 Upsala,1955
Radioth 35.1971

SCANNING ELECTRON MICROSCOPE SYMPOSIUM 3rd Proceedings Scanning electron microscopy 1970 Chicago,Ill. 1970 Apr 28-30 Sponsored by Illinois institute of technology. Research institute I.I.T.research institute, Chicago,Ill.,1970
Eng 41.8204

SCANNING ELECTRON MICROSCOPE 1969 Annual scanning electron microscope symposium 2nd Chicago,Ill. 1969 Apr 29-May 1 Illinois institute of technology.Research institute Chicago,c1969
Met 25.2870

SCANNING ELECTRON MICROSCOPY SYMPOSIA 1st-2nd Proceedings Chicago,Ill. 1968-69 Sponsored by the Illinois institute of technology.Research institute 2cm I.I.T., Chicago,Ill.,1968-69
Eng 41.5623

SCANNING ELECTRON MICROSCOPY 1968 Scanning electron microscope :the instrument and its applications Chicago 1968 Apr 30-May 1 Illinois institute of technology.Research institute Chicago,1968
Met 25.2869

SCANNING ELECTRON MICROSCOPY 1970 Scanning electron microscope symposium 3rd Proceedings Chicago,Ill. 1970 Apr 28-30 Sponsored by Illinois institute of technology. Research institute I.I.T.research institute, Chicago,Ill.,1970
Eng 41.8204

SCAR,SCOR,IAPO,IUBS SYMPOSIUM ON ANTARCTIC OCEANOGRAPHY Papers Santiago de Chile 1966 Sep 13-16 International council of scientific unions.Scientific committee on Antarctic research International council of scientific unions.Scientific committee on oceanic research International association of physical oceanography International union of biological sciences Held by invitation of the Chilean national committee for antarctic research illus,maps xi,268p 23cm Scott Polar research institute for SCAR,Cambridge, 1968
Sco 14.8300

SCARBOROUGH 1964 Conference on the teaching of engineering design Papers and summary Organised by the Enfield college of technology Illus Institution of engineering designers,London,1964 Co-organized by the Institution of engineering designers and the Hornsey College of art
Eng 41.0309

SCARBOROUGH 1964 Metallurgical developments in high-alloy steels A Joint conference on high-alloy steels Proceedings Iron and steel institute and British iron and steel research association Iron and steel institute.Special report, 86 Iron and steel institute,London,1964
Met 25.0488

SCARBOROUGH 1965 Joint conference on low-alloy steels organized by BISRA and the Iron and steel institute Proceedings British iron and steel research association Iron and steel institute Iron and steel institute. Publication, 114 illus 269p London,1969
Met 25.2566

SCARBOROUGH 1967 Strong tough structural steels joint conference... Proceedings British iron and steel research association and Iron and steel institute Iron and steel institute.Publication, 104 Eyre and Spottiswoode,Margate,1967
Met 25.0428

SCHAFFHAUSEN 1931 Gronland 1939 :tagung der Naturforschenden gesellschaft Schaffhausen Naturforschenden gesellschaft Schaffhausen Naturforschenden gesellschaft Schaffhausen. Mitteilungen,Bd.16 1939
Geol 8.2714

SCHAFFHAUSEN 1939 Gronland 1939 :tagung der Naturforschenden gesellschaft Schaffhausen Naturforschenden gesellschaft Schaffhausen Naturforschenden gesellschaft Schaffhausen. Mitteilungen,16 illus,maps,4 plates 231p 1939
Sco 14.7861

SCHENECTADY,N.Y. 1968 International symposium on the reactivity of solids 6th Proceedings United States.Air force. Office of scientific research General electric research and development center International union of pure and applied chemistry. Physical chemistry center Edited by J.W. Mitchell and others Wiley,New York; London,1969
Met 25.2446

SCHEVENINGEN 1948 International congress on rheology 1st Proceedings North-Holland,Amsterdam,1949
Col S 12.0103

SCHEVENINGEN 1954 Paris 1954 Nov Combustion researches and reviews,1955 AGARD combustion panel meetings 6th-7th Invited papers North Atlantic treaty organisation. Advisory group for aeronautical research and development illus xv,187p Butterworths scientific publications,London,1955
Chem E 24.1386

SCHEVENINGEN 1954 and PARIS 1954 Combustion researches and reviews 1955 AGARD combustion panel meetings 6th and 7th Papers Agard Edited by B.P. Mullins AGARDograph, 9 Butterworths,London,1955
Eng 41.7222

SCHEVENINGEN 1962 International congress on microwave tubes 4th Proceedings Centrex;Cleaver-Hume,Eindoven;London,1963
Eng 41.5557

SCHIAPARELLI Giovanni Virginio Atti del convegno per le celebrazioni del cinquantenario della morte di G.V.Schiaparelli Milan 1960 Oct.1-3 Osservatorio astronomico di Brera Milan,1960
Obs 6.0263

SCHIZOPHRENIA Language and thought in schizophrenia :collected papers presented at a meeting of the American psychiatric association...and brought up to date Papers Chicago,Ill. 1939 May 12 American psychiatric association Edited by J.S. Kasanin University of California press, Berkeley;Los Angeles,1954 Preface by N.D. C.Lewis
Psy 28.0082

SCHIZOPHRENIA Serological fractions in schizophrenia :a research symposium New Orleans 1961 Jun 13-14 Louisiana state department of hospitals Tulane university Louisiana state university Edited by Robert G. Heath Harper and Row for the Commonwealth fund,New York,1963
PGMS 29.0324

SCHIZOPHRENIA Studies in schizophrenia :a multidisciplinary approach to mind-brain relationships New Orleans,La. 1952 Jun 11-13 Tulane university of Louisiana.Department of psychiatry and neurology Edited by Robert G. Heath,chairman Published for the Commonwealth fund Harvard university press, Cambridge,Mass.,1954 Transcript of a series of invitational meetings
Psy 28.0081

SCHIZOPHRENIA The Transmission of schizophrenia:research conference 2nd Proceedings Dorado,Puerto Rico 1967 Jun 26-Jul 1 Foundations fund for research in psychiatry Edited by David Rosenthal and, Seymour S. Kety Pergamon press,Oxford,1968 The second in a series of three psychiatric research conferences
Psy 28.0084

SCHLADMING 1966 Elementary particle theories Internationale universitatswochen fur Kernphysik 5th proceedings Edited by Paul Urban Sponsored by the International Atomic Energy Agency Acta physica austriaca. Supplementum, 3 Springer-verlag,Vienna;New York,1966
A Math 4.0889

SCHLOSS-REINHARDSBRUNN 1968 Stofftransport und stoffverteilung in zellen hoherer pflanzen internationales symposium Edited by Kurt Mothes and others Deutsche akademie der wissenschaften zu Berlin.Klasse fur medizin. Abhandlungen, 4a illus. iv,215p Akademie-verlag,Berlin,1968 Title also in English 'Transport and distribution of matter in cells of higher plants'
Bot 42.1900

SCHMIDT TELESCOPE The Role of Schmidt telescopes in astronomy Hamburg 1972 Mar 21-22 Hamburger sternwarte Edited by U. Haug 160p Hamburg observatory,Hamburg,1972
Obs 6.3716

SCHOOL MATHEMATICS PROJECT The Ditchley mathematical conference :a report by the British co-chairman (Prof.B.Thwaites) of the Anglo-American conference on mathematical education Ditchley Park 1966 Sep 9-12 S. M.P.,London,1966
Eng 41.1829

SCHOOL OF AVIATION MEDICINE.RANDOLPH FIELD Physics and medicine of the upper atmosphere Symposium on the physics and medicine of the upper atmosphere Proceedings San Antonio, Tex. 1951 Nov 6-9 Lovelace foundation for medical education and research Edited by Clayton S. White and Otis O. Benson illus xxix,611p University of New Mexico press, Albuquerque,N.M.,1952
Nap 11.0635

1965

SCHOOL ON NEURAL NETWORKS Proceedings Neural networks Edited by E.R. Caianiello With the assistance of North Atlantic treaty organization. Advanced study institute programme Springer, Berlin, 1968
Psy 31.3067

SCHOOLS COUNCIL Science in the sixth form :a conference Nottingham 1965 Mar Schools council publications. Working paper, 4 H.M. S.O., London, 1966
Eng 41.8193

SCHWEIZERISCHE AKADEMIE DER MEDIZINISCHEN WISSENSCHAFTEN Chemotherapy of cancer International symposium on the chemotherapy of cancer Proceedings Lugano 1964 Apr 28-May 1 Edited by Placidus Plattner Organised by the Swiss academy of medical sciences Elsevier, Amsterdam, 1964 Sponsored by F. Hoffman-Laroche and co.ltd.
Radioth 35.0837

SCHWEIZERISCHE AKADEMIE DER MEDIZINISCHEN WISSENSCHAFTEN Krebsmetastasierung auf dem blutwege :symposium Geneva 1963 Jun 27-29 Schweizerische hamatologe gesellschaft Schweizerische akademie der medizinischen wissenschaften. Bulletin, 20, fasc 1-3 Schwabe, Basel, 1964
Radioth 35.0835

SCHWEIZERISCHE AKADEMIE DER MEDIZINISCHEN WISSENSCHAFTEN Symposium on noxious effects of low level radiation :physical elements and biological aspects Lausanne 1958 Mar 27-29 Schwabe, Basle; Stuttgart, 1958
Radioth 35.1078

SCHWEIZERISCHE AKADEMIE DER MEDIZINISCHEN WISSENSCHAFTEN Szintigraphie und radiokardiographie :symposion Basel 1962 Sep 28-30 Schwabe, Basel; Stuttgart, 1963
Radioth 35.1371

SCHWEIZERISCHE GESELLSCHAFT FUR PADIATRIE Tumoren im kindesalter :jahresversammlung der Schweizenschen gesellschaft fUR Padiatrie Geneva 1963 Padiatrische fortbildungskurse fur die praxis, 13 Illus Schwabe, Basel, 1964
Radioth 35.0836

SCHWEIZERISCHE HAMATOLOGE GESELLSCHAFT Krebsmetastasierung auf dem blutwege : symposium Geneva 1963 Jun 27-29 Schweizerische nationaliga fur krebsekampfung und krebsforschung Schweizerische akademie der medizinischen wissenschaften Schweizerische akademie der medizinischen wissenschaften. Bulletin, 20, fasc 1-3 Schwabe, Basel, 1964
Radioth 35.0835

SCHWEIZERISCHE MATHEMATISCHE GESELLSCHAFT Differentialgeometrie und topologie Internationales kolloquium Zurich 1960 Jun 20-25 Sponsored by the International mathematical union Enseignement mathematique. Monographies, 11 159p 24cm L'enseignement mathematique universite, Geneva, 1962
P. Math 2.0697

1966

SCHWEIZERISCHE NATIONALIGA FUR KREBSEKAMPFUNG UND KREBSFORSCHUNG Krebsmetastasierung auf dem blutwege :symposium Geneva 1963 Jun 27-29 Schweizerische hamatologe gesellschaft Schweizerische akademie der medizinischen wissenschaften. Bulletin, 20, fasc 1-3 Schwabe, Basel, 1964
Radioth 35.0835

SCHWEIZERISCHES INSTITUT FUR HOCHGEBIRGESPHYSIOLOGIE UND TUBERKULOSEFORSCHUNG Verhandlungen der klimatologischen tagung in Davos 1925 Davos 1925 illus vii, 576p Benno Schwabe, Basel, c.1925
Nap 11.0894

SCIENCE, TECHNOLOGY AND APPLICATION OF TITANIUM International conference on titanium Proceedings London 1968 May 21-24 Institute of metals American institute of mining, metallurgical and petroleum engineers American society for metals Japan institute of metals Akademiya nauk S.S.S.R. Edited by R.I. Jaffee and N.E. Promisel Pergamon, Oxford, 1970
Met 25.2852

SCIENCE AND LAW Science et loi... semaine internationale de synthese 5e Discussions et conclusions Paris 1963 May 29-Jun 3 By Abel Rey and others Centre international de synthese vi, 228p Alcan, Paris, 1934
WSM 43.0874

SCIENCE AND TECHNOLOGY FOR DEVELOPMENT United Nations conference on the application of science and technology for the benefit of the less developed areas Report Geneva 1963 Feb Vol 1-4 4 vols United Nations, New York, 1963
Geog 13.2389

SCIENCE AND TECHNOLOGY OF TUNGSTEN, TANTALUM, MOLYBDENUM, NIOBIUM AND THEIR ALLOYS Agard conference on refractory metals Proceedings Oslo 1963 Jun 23-26 North Atlantic treaty organization. Advisory group for aeronautical research and development AGARDograph 82 Pergamon press, Oxford, 1964
Met 25.1011

La SCIENCE AU SEIZIEME SIECLE :colloque international de Royaumont Royaumont 1957 Jul 1-4 International union of history and philosophy of science Histoire de la pensee, 2 Hermann, Paris, 1960
WSM 43.0911

SCIENCE COUNCIL OF JAPAN International conference on functional analysis and related topics Proceedings Tokyo 1969 Apr 1-8 xxxiii, 425p 26cm University of Tokyo press, Tokyo, 1970
P Math 2.4279

SCIENCE COUNCIL OF JAPAN International conference on low temperature physics 12th Proceedings Kyoto 1970 Sep 4-10 International union of pure and applied physics Physical society of Japan Edited by Eizo Kanda photos, diagrms 895p 26cm Academic press of Japan, Tokyo, 1971
Cav 7.3028

SCIENCE COUNCIL OF JAPAN International conference on magnetism and crystallography Proceedings Kyoto 1961 Sep 25-30 Pt 2: electron and neutron diffraction Edited by Shizuo Miyake and others Sponsored by the International union of pure and applied physics Physical society of Japan.Journal, 17,B 2 396p physical society of Japan, Tokyo,1962
Cav 7.1338

SCIENCE COUNCIL OF JAPAN International conference on magnetism and crystallography Proceedings Kyoto 1961 Sep 25-30 Pt 3: neutron diffraction:study of magnetic materials Edited by Takeo Nagamiya and Shizno Miyake Under the auspices of International union of pure and applied physics Physical society of Japan.Journal, 17,B 3 71p physical society of Japan,Tokyo, 1962
Cav 7.1339

SCIENCE COUNCIL OF JAPAN International conference on magnetism and crystallography Proceedings Kyoto 1961 Sep 25-30 Pt 1: magnetism Edited by Kei Josida and others Held under the auspices of International union of pure and applied physics Physical society of Japan.Journal, 17,B 1 xi,718p Physical society of Japan,Tokyo,1962
Cav 7.1337

SCIENCE COUNCIL OF JAPAN International conference on theoretical physics Proceedings Kyoto 1953 Sep 14-24 Tokyo Edited by I. Imai Under the auspices of the International union of pure and applied physics 942p Tokyo,1954
Cav 7.1342

SCIENCE COUNCIL OF JAPAN International conference on wind effects on buildings and structures 3rd Proceedings Tokyo 1971 Saikon,Tokyo,1971
Eng 41.8602

SCIENCE COUNCIL OF JAPAN International congress of genetics 12th Proceedings Tokyo 1968 Aug 19-28 Vol 1-3 Under the auspices of the International union of biological sciences 3 vols Science council of Japan,Tokyo,1969
Gen 34.1008

SCIENCE COUNCIL OF JAPAN International meeting on ferroelectricity 2nd Kyoto 1969 Sep 4-9 Sponsored by the International union of pure and applied physics Physical society of Japan.Journal, 28,suppt Physical society of Japan,1970
Met 25.2865

SCIENCE COUNCIL OF JAPAN International symposium on algebraic number theory proceedings Tokyo 1955 Sep 8-13 Nikko Edited by Shokichi Iyanaga and Yukiyosi Kawada Jointly sponsored by the International mathematical union xx,265p 26cm Science council of Japan,Tokyo,1956
P. Math 2.1771

SCIENCE COUNCIL OF JAPAN Japan congress on testing materials 6th Kyoto 1962 Oct 10111 Japan society for testing materials, Kyoto,1963
Met 25.2860

SCIENCE COUNCIL OF JAPAN Japan national congress for applied mechanics 18th Proceedings Tokyo 1968 Nov 8-9 Bibliog, illus 188p 26cm Central scientific publishers,Tokyo,1970
A Math 4.1725

SCIENCE COUNCIL OF JAPAN Prestressed concrete and composite beams :a symposium Proceedings Kyoto 1954 Sep 4 Science council of Japan,Kyoto,1955
Eng 41.3013

SCIENCE COUNCIL OF JAPAN Safety of structures symposium 2nd Proceedings Kyoto 1955 Sep 6 Science council of Japan, Kyoto,1956
Eng 41.3014

SCIENCE COUNCIL OF JAPAN Symposium on typhoons Proceedings Tokyo 1954 Nov 9-12 illus 257p Japanese national commission for Unesco,Tokyo,1955
Nap 11.0612

SCIENCE COUNCIL OF JAPAN The Conference on photographic and spectrographic optics Proceedings Tokyo 1964 Sep 1-5 and Kyoto 1964 Sep 1-5-,7-8 International commission for optics Japanese journal of applied optics, 4,suppl.1 669p Tokyo, 1965
Obs 6.2050

SCIENCE COUNCIL OF JAPAN The International symposium on enzyme chemistry Proceedings Tokyo 1957 Oct 15-23 and Kyoto 1957 Oct 15-23 International union of biochemistry I.U.B.Symposium series, 2 Pergamon,London,1958
Bioch 33.1053

SCIENCE COUNCIL OF JAPAN The Symposium on numerical weather prediction Proceedings Tokyo 1960 Nov 7-13 Co-sponsored by the International union of geodesy and geophysics 655p Meteorological society of Japan,Tokyo, 1962
Nap 11.0923

SCIENCE COUNCIL OF JAPAN World conference on earthquake engineering 2nd Proceedings Tokyo 1960 Jul 11-18 and Kyoto 1960 Jul 11-18 Vol 1-3 Japan society of civil engineers 3 vols Science council of Japan,Tokyo,1960
Eng 41.6395

SCIENCE COUNCIL OF JAPAN and INTERNATIONAL GEOGRAPHICAL UNION IGU regional conference in Japan Proceedings Tokyo 1957 Aug 28-Sep 3 Nara Edited by Ryuziro Isida and others illus,maps vii, 609p 26cm Tokyo,1959
Sco 14.6870

SCIENCE COUNCIL OF JAPAN and INTERNATIONAL GEOGRAPHICAL UNION Regional geography of Japan :I.G.U.regional conference in Japan Guidebook Japan 1957 1-7 maps Tokyo, 1957
Geog 13.4507

SCIENCE ET LOI... 5e semaine internationale de synthese Discussions et conclusions Paris 1963 May 29-Jun 3 By Abel Rey and others Centre international de synthese vi, 228p Alcan,Paris,1934
WSM 43.0874

SCIENCE IN THE SIXTH FORM :a conference Nottingham 1965 Mar Sponsored by the Schools council Schools council publications. Working paper, 4 H.M.S.O.,London,1966
Eng 41.8193

SCIENCE OF CERAMICS 1st :a conference Proceedings Oxford 1961 Jun 26-30 British ceramic society Edited by G.H. Stewart Academic press,London;New York,1962
Met 25.0274

SCIENCE OF CERAMICS 2nd :a conference Proceedings Noordwijk aan Zee 1963 May 13-17 British ceramic society Edited by G. H. Stewart Academic press,London;New York, 1965
Met 25.0278

SCIENCE OF CERAMICS 3rd :a conference Proceedings Bristol 1965 Jul 5-8 British ceramic society and Nederlandse keramische vereniging Edited by G.H. Stewart Under the auspices of the European ceramic association Academic press,London;New York, 1967
Met 25.0286

SCIENCE RESEARCH COUNCIL Computational problems in abstract algebra :a conference Proceedings Oxford 1967 Aug 29-Sep 2 Held under the auspices of the Atlas computer laboratory bibliog.,illus. x,402p 23cm Pergamon press,Oxford,1970
P Math 2.3758

SCIENCE RESEARCH COUNCIL Computers in number theory Science research council Atlas symposium 2nd Proceedings Oxford 1969 Aug 18-23 Edited by A.O.L. Atkin and B.J. Birch xvii,433p 24cm Academic press, London;New York,1971
P Math 2.4381

SCIENCE RESEARCH COUNCIL Conference on radiation protection in accelerator environments Proceedings Chilton,Berks. 1969 Mar Edited by G.R. Stevenson Rutherford laboratory,Chilton,1969 Organized by and held at the Rutherford laboratory
Radioth 35.1393

SCIENCE RESEARCH COUNCIL Power systems conference 2nd Proceedings Glasgow 1967 Jan 5-6 Glasgow,1967
Eng 41.4988

SCIENCE RESEARCH COUNCIL ATLAS SYMPOSIUM 2nd Proceedings Computers in number theory Oxford 1969 Aug 18-23 Science research council Edited by A.O.L. Atkin and B.J. Birch xvii,433p 24cm Academic press, London;New York,1971
P Math 2.4381

SCIENCE SURVEY 1 Edited by A.W. Haslett and John St.John illus 360p Longacre press, Vista books,London,1960 Foreword by sir George Thomson
Sco 14.6129

SCIENCES AND MYTHS Soleil a la renaissance: sciences et mythes :colloque international Brussels 1963 Apr Federation internationale des instituts et societes pour l'etude de la Renaissance and Belgique. Ministere de l'education et de la culture Universite libre de Bruxelles.Institut pour l'etude de la renaissance et de l'humanisme. Publication, 2 584p Presses universitaires de Bruxelles,Brussels;Paris, 1965 Contains Hoskin,M.and Jones,C. 'Problems in late renaissance astronomy'
WSM 43.0046

SCIENTIFIC BASIS OF CANCER CHEMOTHERAPY : seminar Paris 1968 Mar 22-23 Edited by Georges Mathe Sponsored by the European organization for research into cancer treatment Recent results in cancer research, 21 Heinemann medical,London,1969
Radioth 35.0866

The SCIENTIFIC BASIS OF DRUG THERAPY IN PSYCHIATRY :a symposium Proceedings London 1964 Sep 7-8 Edited by John Marks and C.M.B. Pare Pergamon,Oxford,1965
Pha 16.0028

SCIENTIFIC CHANGE Symposium on the history of science Papers Oxford 1961 Jul 9-15 Edited by A.C. Crombie Under the auspices of the International union of the history and philosophy of science.Division of history of science Heinemann,London,1963
Geog 13.0053

SCIENTIFIC CHANGE...HISTORICAL STUDIES IN THE INTELLECTUAL,SOCIAL AND TECHNICAL CONDITIONS FOR SCIENTIFIC DISCOVERY... :symposium on the history of science Oxford 1961 Jul 9-15 International union of history and philosophy of science.Division of history of science Edited by A.C. Crombie xii,896p Heinemann,London,1963
WSM 43.0029

SCIENTIFIC COMMITTEE FOR ANTARCTIC RESEARCH Biologie antarctique :symposium 1er Comptes-rendus Paris 1962 Sep 2-8 Edited by Robert Carrick and others Hermann,Paris, 1964
Bal 39.1669

SCIENTIFIC COMMITTEE ON ANTARCTIC RESEARCH Antarctic geology :international symposium 1st Proceedings Cape Town 1963 Sep 16-21 Edited by Raymond J. Adie illus xx,758p 27cm North-Holland,Amsterdam,1964
Sco 14.6125

SCIENTIFIC COMMITTEE ON ANTARCTIC RESEARCH Antarctic ice and water masses Proceedings Tokyo 1970 Sep 15 Edited by George Deacon illus x,114p 23cm Scientific committee on antarctic research,1971
A Math 4.1746

SCIENTIFIC COMMITTEE ON ANTARCTIC RESEARCH Biologie antarctique :symposium 1er Comptes-rendus Paris 1962 Sep 2-8 Edited by Robert Carrick and others Actualites scientifiques et industrielles,1312 illus, maps 651p 25cm Herman,Paris,1964 Preface by professor P.P.Grasse
Sco 14.6172

SCIENTIFIC COMMITTEE ON ANTARCTIC RESEARCH Symposium on antarctic oceanography main review papers Santiago,Chile 1966 Sep 13-16 Bibliog.,Illus. xi,268p 23cm Scott polar research institute,Cambridge,1966
A Math 4.1501

SCIENTIFIC COMMITTEE ON ANTARCTIC RESEARCH and WORLD METEOROLOGICAL ORGANIZATION Polar meteorology W.M.O.-S.C.A.R.-I.C.P.M. Symposium on polar meteorology Proceedings Geneva 1966 Sep 5-9 Also organized by the International commission on polar meteorology illus vi,540p 28cm World meteorological organization,Geneva,1967
Sco 14.7839

SCIENTIFIC COMMITTEE ON ANTARCTIC RESEARCH., International union of geological sciences Antarctic geology :international symposium 1st proceedings Capetown 1963 Sep.16-21 Edited by Raymond J. Adie illus. North-Holland,Amsterdam,1964
Geol 8.2509

SCIENTIFIC COUNCIL FOR AFRICA SOUTH OF THE SAHARA Symposium on African hydrobiology and inland fisheries 2nd Brazzaville 1956 Jul 3-11 Published under the auspices of the Commission for technical co-operation in Africa south of the Sahara Scientific council for Africa south of the Sahara. Publication, 25 Brazzaville,1956 Title also in French:Symposium sur l'hydriobiologie et la peche en eaux douces en Afrique;text in English and French
Bal 39.1894

SCIENTIFIC COUNCIL FOR AFRICA SOUTH OF THE SAHARA Symposium on African hydrobiology and inland fisheries 2nd Brazzaville 1956 Jul 3-11 Scientific council for Africa south of the Sahara.Publications,25 Commission for technical co-operation in Africa,London,1957
Geog 13.2726

SCIENTIFIC INFORMATION The Royal society scientific information conference Reports and papers London 1948 Jun 21-Jul 2 Royal society of London Royal society,London, 1948
Chem 18.2006

SCIENTIFIC INFORMATION CONFERENCE :report and papers submitted London 1948 Jun 21-Jul 2 Royal society of London Royal society,London,1948
Bot 42.0662

SCIENTIFIC INFORMATION CONFERENCE Reports and papers submitted London 1948 Jun 21 Jul 2 Royal society of London London,1948
Bioch 33.1749

The SCIENTIFIC INFORMATION CONFERENCE report and papers submitted London 1948 Jun 21-Jul 2 Royal society of London Royal society,London,1948
Gen 34.0054

SCIENTIFIC MEETING ON THE DIAGNOSIS AND TREATMENT OF RADIOACTIVE POISONING Proceedings Diagnosis and treatment of radioactive poisoning Vienna 1962 Oct 15-18 World health organization International atomic energy agency International atomic energy agency.Proceedings series International atomic energy agency,Vienna,1963
Radioth 35.1368

SCIENTIFIC PROBLEMS OF THE HUMID TROPICAL ZONE DELTAS AND THEIR IMPLICATION :a symposium Proceedings Dacca 1964 Feb 24-Mar 2 Unesco Jointly sponsored by the government of Pakistan Humid tropics research Unesco, Paris,1966 Partly in French
Geog 13.5396

The SCIENTIFIC RESULTS FROM THE ORBITING ASTRONOMICAL OBSERVATORY (OAO-2) Amherst, Mass. 1971 Aug 23-24 National aeronautics and space administration Edited by A.D. Code NASA SP-310 590p nasa,washington,D.C.,1972
Obs 6.3626

SCIENTIFIC SOCIETY OF MECHANICAL ENGINEERS Colloquium spectroscopicum internationale 14th Proceedings Debrecen 1967 Aug 7-12 Vol 1-3 1579p 24cm 3 vols Hilger, London,1968
Chem 18.2846

SCIENTIFIC THEORIES Beitrage zur entwicklung der wissenschaftstheorie in 19 jahrhundert : colloquium Vortrage und diskussionen Dusseldorf 1965 Dec 10-11 and Dusseldorf 1966 Dec 9-10 Edited by A. Diemer Studien zur wissenschaftstheorie, 1 234p Hain, Meisenheim am Glan,1968
WSM 43.0983

SCIENTIFIC THEORIES System und klassifikkation in wissenschaft und dokumentation :colloquium Vortrage und diskussionen Dusseldorf 1967 Apr 27-29 Edited by A. Diemer Studien zur wissenschaftstheorie, 2 183p Hain, Meisenheim am Glan,1968
WSM 43.0984

The SCOPE OF PHYSICAL ANTHROPOLOGY AND ITS PLACE IN ACADEMIC STUDIES :a symposium London 1957 Nov Edited by D.F. Roberts and J.S. Neiner Society for the study of human biology.Symposia, 1 Wenner-Gren foundation for anthropological research,London,1958
An 32.2880

The SCOPE OF PHYSICAL ANTHROPOLOGY AND ITS PLACE IN ACADEMIC STUDIES :a symposium held at the Ciba foundation Papers London 1957 Nov 6 Society for the study of human biology Edited by D.F. Roberts and J.S. Weiner Wenner-Gren foundation for anthropological research,London,1958
HE 27.0143

SCOTSDALE,ARIZ. 1969 The Chemistry of extended defects in non-metallic solids Proceedings Institute of advanced study on the chemistry of extended defects in non-metallic solids Edited by LeRoy Eyring and Michael O'Keefe North-Holland,Amsterdam,1970
Min 10.1545

SCOTTISH UNIVERSITIES' SUMMER SCHOOL 2nd Lectures Fluctuation,relaxation and resonance in magnetic systems Newbattle Abbey 1961 Scottish universities' summer school in physics Edited by D.ter Haar With financial support from the North Atlantic treaty organization Oliver and Boyd, Edinburgh;London,1962
Chem 18.2619

SCOTTISH UNIVERSITIES' SUMMER SCHOOL 6th Lectures Phonons in perfect lattices and in lattices with point imperfections Aberdeen 1965 Aug 9-27 Scottish universities' summer school in physics Edited by R.W.H. Stevenson With financial help from the North Atlantic treaty organization Oliver and Boyd,Edinburgh; London,1966
Chem 18.2643

SCOTTISH UNIVERSITIES SUMMER SCHOOL 7th proceedings Particle interactions at high energies Newbattle Abbey,Edinburgh 1966 Aug 1-20 Edited by T.W. Preist and L.L.J. Vick 406p Oliver and Boyd,Edinburgh,1967
A Math 4.0875

SCOTTISH UNIVERSITIES' SUMMER SCHOOL 9th Proceedings Physics of hot plasmas 1968 Jul 28-Aug 16 Edited by B.J. Rye and J.C. Taylor illus xv,455p 25cm Oliver and Boyd,Edinburgh,1970
A Math 4.1886

SCOTTISH UNIVERSITIES' SUMMER SCHOOL IN PHYSICS Fluctuation,relaxation and resonance in magnetic systems Scottish universities' summer school 2nd Lectures Newbattle Abbey 1961 Edited by D.ter Haar With financial support from the North Atlantic treaty organization Oliver and Boyd, Edinburgh;London,1962
Chem 18.2619

SCOTTISH UNIVERSITIES' SUMMER SCHOOL IN PHYSICS Phonons in perfect lattices and in lattices with point imperfections N.A.T.O.advanced study institute Papers Aberdeen 1965 Aug 9-27 Edited by R.W.H. Stevenson With financial support of North Atlantic treaty organization Scottish universities' summer school in physics, 6 448p Oliver and Boyd,Edinburgh,1966
TA 15.0212

SCOTTISH UNIVERSITIES' SUMMER SCHOOL IN PHYSICS Phonons in perfect lattices and in lattices with point imperfections Scottish universities' summer school 6th Lectures Aberdeen 1965 Aug 9-27 Edited by R.W.H. Stevenson With financial help from the North Atlantic treaty organization Oliver and Boyd, Edinburgh;London,1966
Chem 18.2643

SCOTTISH UNIVERSITIES' SUMMER SCHOOL IN PHYSICS 1st Papers Dispersion relations 1960 Edited by G.R. Screaton Scottish universities' summer school in physics,1960,1 Oliver and Boyd,Edinburgh,1961
Cav 7.2134

SCOTTISH UNIVERSITIES' SUMMER SCHOOL IN PHYSICS 3rd Papers Excitations in semiconductors,polarons and excitons St. Andrews 1962 Jul 30-Aug 18 Edited by C.G. Kuper and G.D. Whitfield Scottish universities' summer school in physics,1962,3 Oliver and Boyd,Edinburgh,1963
Cav 7.2136

SCOTTISH UNIVERSITIES' SUMMER SCHOOL IN PHYSICS 8th Papers Mathematical methods in solid state and superfluid theory St.Andrews 1967 Jul 31-Aug 19 Edited by R.C. Clark and G.H. Derrick Scottish universities' summer school in physics,1967 xvi,400p 25cm Oliver and Boyd,Edinburgh,1969
Cav 7.2759

SCUOLA INTERNATIONALE DI FISICA 'ENRICO FERMI' 14 corso Rendiconti Teorie ergodiche Varenna 1960 May 23-31 Societa Italiana di Fisica Edited by P. Caldirola Academic press,New York;London,1961
A Math 4.0950

SCUOLA INTERNATIONALE DI FISICA 'ENRICO FERMI'. 32 corso rendiconti weak interactions and high energy neutrino physics Varenna 1964 Jun.15-27 Societa Italiana di Fisica Edited by T.D. Lee 334p Academic press,New York,1966
A Math 4.0845

SCUOLA INTERNAZIONALE DE FISICA 'ENRICO FERMI' 27 corso Rendiconti Dispersione ed assorbimento del suono Varenna 1962 Aug 6-18 Societa italiana di fisica Edited by D. Sette Academic press,New York;London,1963 Papers in English and French
Chem 18.0517

SCUOLA INTERNAZIONALE DI FISICA 'ENRICO FERMI' Rendiconti Varenna Corso 10-14;16-: 1959- Societa italiana di fisica Nicola Zanichelli,Bologna,1960-
Cav 7.2388

SCUOLA INTERNAZIONALE DI FISICA 'ENRICO FERMI' 18 corso Rendiconti Radiation damage nei solidi Ispra (Varese) 1960 Sep 5-24 Societa italiana di fisica Edited by D.S. Billington Academic press,New York;London, 1962 Contributions in English and French
Chem 18.1144

SCUOLA INTERNAZIONALE DI FISICA 'ENRICO FERMI' 20 corso Rendiconti Verifiche delle teorie gravitazionali Varenna 1961 Jun 19-Jul 1 Societa italiana di fisica Edited by C. Moller Academic press,New York,1962
TA 15.0084

SCUOLA INTERNAZIONALE DI FISICA 'ENRICO FERMI' 25 corso Rendiconti Teoria dei plasma Varenna 1962 Jul 9-21 Societa italiana di fisica Edited by N.M. Rosenbluth Academic press,New York;London,1964
Eng 41.4527

SCUOLA INTERNAZIONALE DI FISICA 'ENRICO FERMI' 34 corso Rendiconti Propieta ottiche dei solidi Varenna 1965 Jun 28-Jul 10 Societa italiana di fisica Edited by J. Tauc Academic press,New York;London,1966
TA 15.0443

SCUOLA INTERNAZIONALE DI FISICA 'ENRICO FERMI' 35 corso Rendiconti Astrofisica delle alte energie Varenna 1965 Jul 12-24 Societa italiana di fisica Edited by L. Gratton Academic press,New York;London,1966
TA 15.0314

SCUOLA INTERNAZIONALE DI FISICA 'ENRICO FERMI' 37 corso Rendiconti Teoria del magnetismo nei metalli di transizione Varenna 1966 Jun 6-25 Societa italiana di fisica Edited by W. Marshall diagrs 455p 24cm Academic press,London,1967
Cav 7.2612

SCUOLA INTERNAZIONALE DI FISICA 'ENRICO FERMI' 39 corso Rendiconti Astrofisica del plasma Varenna 1966 Jul 11-30 Societa italiana di fisica Edited by P.A. Sturrock Academic press,New York;London,1967
TA 15.0328

SCUOLA INTERNAZIONALE DI FISICA 'ENRICO FERMI' 44 corso Rendiconti Molecular beams and reaction kinetics Varenna 1968 Jul 29-Aug 10 Societa italiana di fisica Edited by C. Schlier Academic press,New York,1970
Chem 18.2814

SCUOLA INTERNAZIONALE DI FISICA 'ENRICO FERMI' 46 corso Rendiconti Physics with intersecting storage rings Varenna 1969 Jun 16-26 Societa italiana di fisica Bibliog,photos 24cm Academic press,New York,1971
Cav 7.3012

SCUOLA INTERNAZIONALE DI FISICA 'ENRICO FERMI' 47 corso Rendiconti Varenna 1969 Jun 30-Jul 12 Societa italiana di fisica Edited by Rainer K. Sachs Academic press New Yorkp 1971cm
A Math 4.1788

SCUOLA INTERNAZIONALE DI FISICA "ENRICO FERMI" 20 corso Rendiconti Verifiche delle teorie gravitazionali Varenna 1961 Jun 19-Jul 1 Societa Italiana di fisica Edited by G. Moller Bibliog.,Illus. 264p 24cm Academic press,New York,1962
A Math 4.1555

SCUOLA INTERNAZIONALE DI FISICA "ENRICO FERMI" 28 corso Rendiconti Radiation damage nei solidi Ispra,Varese 1960 Sep 5-24 Societa Italiana di fisica Edited by D.S. Billington Academic press,New York;London, 1962
Met 25.1579

SCUOLA INTERNAZIONALE DI FISICA "ENRICO FERMI" 28 corso rendiconti Evoluzione delle stelle Varenna 1962 Aug 20-Sep.1 Societa Italiana di fisica Edited by L. Gratton 488p Academic press,New York;London,1963
Obs 6.0685

SCUOLA INTERNAZIONALE DI FISICA "ENRICO FERMI" 35 corso rendiconti Astrofisica delle alte energie Varenna 1965 Jul.12-24 Societa Italiana di Fisica Edited by L. Gratton 463p Academic press,New York,1966
A Math 4.1169

SCUOLA INTERNAZIONALE DI FISICA "ENRICO FERMI" 39 corso Rendiconti Astrofisica del plasma Varenna 1966 Jul 11-30 Societa italiana di Fisica Edited by P.A. Sturrock xvi,364p Academic press,New York;London,1967
A Math 4.1577

SCUOLA INTERNAZIONALE DI FISICA "ENRICO FERMI" 41 corso Rendiconti Argomenti scelti di fisica delle particelle Varenna 1957 Jul 17-29 Societa Italiana di fisica Edited by J. Steinberger Bibliog.,Illus. vii,194p 24cm Academic press,New York,1967
A Math 4.1545

SCUOLA INTERNAZIONALE DI FISICA "ENRICO FERMI" 42 corso Rendiconti Ottica quantistica Varenna 1967 Jul 31-Aug 19 Societa Italiana di fisica Edited by R.J. Glauber Bibliog.,Illus.port. xvi,159p 25cm Academic press,New York,1968
A Math 4.1599

SCUOLA INTERNAZIONALE DI FISICA "ENRICO FERMI" 43 corso Rendaconti Elaborazione di dati offici de parte di organism e di macchine Varenna 1968 Jul 15-27 Societa italiana di fisica Edited by W. Reichardt Academic press,New York;London,1969
An 32.5470

SCUOLA INTERNAZIONALE DI FISICA'ENRICO FERMI' 28 corso Rendiconti Evoluzione delle stelle Varenna 1962 Aug 20-Sep 1 Societa Italiana di fisica Edited by L. Gratton Academic press,New York;London,1963
A Math 4.1156

SEA,SALT AND PLANTS seminar Proceedings Bhavnagar 1965 Dec 20-23 Central salt and marine chemicals research institute Edited by V. Krishnamurthy illus. xv,372p CSMCRI,Bhavnagar,1967
Bot 42.1991

SEA-GOING QUALITIES OF SHIPS :a seminar Teddington 1961 Oct 17-18 National physical laboratory H.M.S.O.,London,1963
Eng 41.6932

SEA WATER Conference on physical and chemical properties of sea water Report Easton,Md. 1958 Sep 4-5 United States. Office of naval research and National research council. Committee on oceanography National research council.Publication, 600 Washington,D.C.,1959
Geod 9.0610

A SEALS SYMPOSIUM Proceedings Cambridge 1964 Sep 9 Nature conservancy.Consultative committee on grey seals Edited by E.A. Smith illus 101p 28cm Edinburgh,1965
Sco 14.1172

A SEALS SYMPOSIUM Proceedings Cambridge 1964 Sep 9 Nature conservancy,Edinburgh. Consultative committee on grey seals and fisheries Edited by E.A. Smith maps,tables 28cm Edinburgh,1965 Number 84 of 100 mimeographed copies.In folder
Bal 44.5093

SEARLE (G.D.) AND COMPANY Developments in the pharmacology and clinical uses of human gonadotrophins :a private scientific meeting Proceedings London 1968 Mar 15-16 Searle, High Wycombe,1970
Inv Med 37.0261

SEATTLE 1963 Sudden death in infants Conference on causes of sudden death in infants Proceedings United States. Department of health,education and welfare Edited by Ralph J. Wedgwood and others United States.Public health service publication,1412 U.S.govt.printing office, Washington,D.C.,1965
HE 27.0158

SEATTLE 1965 Stress-corrosion cracking of titanium :a symposium American society for testing materials A.S.T.M.Special technical publication, 397 American society for testing and materials,Philadelphia,Pa.,1966
Met 25.1962

SEATTLE 1966 Conference on neurophysiology in relation to anesthesiology Anesthesiology, 28,no 1 American society of anesthesiologists,Lancaster,Pa.,1967
Pha 16.0052

SEATTLE 1968 Category theory,homology theory and their application :a conference Proceedings Vol 1 Supported by the Battelle memorial institute Lecture notes in mathematics, 86 Bibliog,Illus vi,216p 28cm Springer-Verlag,Berlin,1969
P Math 2.3343

SEATTLE 1968 Category theory,homology theory and their applications Vol 3 Supported by the Battelle memorial institute Lecture notes in mathematics, 99 Bibliog., Illus. iv,489p 28cm Springer-Verlag, Berlin,1969
P Math 2.3345

SEATTLE 1968 Category theory,homology theory and their applications :a conference Vol 2 Supported by the Battelle memorial institute Lecture notes in mathematics, 92 Bibliog.,Illus. vi,216p 28cm Springer-Verlag,Berlin,1969
P Math 2.3344

SEATTLE 1968 Clear air turbulence and its detection A Symposium on clear air turbulence and its detection Proceedings Boeing scientific research laboratories.Flight sciences laboratory Edited by Yih-Ho Pao and Arnold Goldburg 542p 22cm Plenum press, New York,1969
Cav 7.2921

SEATTLE 1968 Hyperbolic equations and waves Battelle Seattle 1968 rencontres Edited by M. Froissart Bibliog.,Illus. viii,393p 23cm Springer-Verlag,Berlin,1970
A Math 4.1469

SEATTLE,WAS. 1969 Molecular orbital studies in chemical pharmacology :a symposium Edited by L.B. Kier Springer,Berlin;New York, 1970
Radioth 35.1427

SEATTLE,WASH. 1959 Biological and economic aspects of fisheries management :a conference Proceedings University of Washington.College of fisheries Edited by James A. Crutchfield figures vi,160p University of Washington, Seattle,Wash.,1959
Sco 14.4563

SEATTLE,WASH. 1961 Convexity symposium American mathematical society Edited by Victor L. Klee American mathematical society. Proceedings of symposia in pure mathematics, 7 xv,516p 26cm American mathematical society,Providence, R.I.,1963
P. Math 2.0487

SEATTLE,WASH. 1962 The Heat transfer and fluid mechanics institute Proceedings University of Washington Edited by Edward Ehlers and others Co-sponsored by Stanford university 293p Stanford university press, Stanford,Calif.,1962
Cav 7.2422

SEATTLE,WASH. 1962 The Heat transfer and fluid mechanics institute 1962 proceedings Edited by F.Edward Ehlers and others Stanford university press,Stanford,Calif.,1962
A Math 4.0520

SEATTLE,WASH. 1965 Orientation effects in the mechanical behavior of anisotropic structural materials :a symposium American society for testing materials American society for testing materials.Special technical publication, 405 American society for testing materials,Philadelphia,Pa.,1966
Eng 41.3701

SEATTLE,WASH. 1965 Testing techniques for rock mechanics :a symposium American society for testing and materials American society for testing and materials.Special technical publication, 402 American society for testing materials,Philadelphia,Pa.,1966
Eng 41.3161

SEATTLE,WASH. 1967 Harrison Hot Springs,B.C. Dislocation dynamics Battelle materials science colloquium 2nd Battelle memorial institute Edited by Alan R. Rosenfield and others McGraw-Hill series in materials science McGraw-Hill,New York,1968
Met 25.2253

SEATTLE,WASH. 1968 Heat transfer and fluid mechanics institute 1968 Proceedings Edited by A.F. Emery and C.A. Depew bibliog., illus. ix,272p 26cm Stanford university press,Stanford,Calif.,1968
A Math 4.1617

SEATTLE,WASH. 1969 A Short history of botany in the United States Edited by Joseph Ewan Hafner,New York;London,1969 Published on the occasion of the 11th International botanical congress
BG 38.3234

SEATTLE,WASH. 1969 International botanical congress 11th Proceedings International botanical congress,Washington,D.C.,1970
BG 38.3132

SEATTLE,WASH. 1969 1969 SEP 2-4 International botanical congress 11th Abstracts 1969
BG 38.3253

SEATTLE,WASH. 1970 Aircraft wake turbulence and its detection :a symposium Proceedings Boeing company.Scientific research laboratories Edited by J.H. Olsen and others Plenum press,New York,1971
Eng 41.8333

SEATTLE,WASH. 1971 Seminar on algebraic topology Proceedings Battelle memorial institute.Seattle research center Edited by Peter J. Hilton Lecture notes in mathematics, 249 vi,110p 25cm Springer,Berlin,1971
P Math 2.4115

SEAWEED International seaweed symposium 2nd Proceedings Trondheim 1955 Jul Edited by Trygve Braarud and Nils Andreas Sorensen illus. xiv,220p Pergamon,London, 1956
Bot 42.4700

SEBACEOUS GLANDS Brown university symposium on the biology of skin,1962 Proceedings Providence,R.I. 1962 Jan 27-28 Edited by William Montagna and others Advances in biology of skin, 4 Pergamon press,Oxford, 1963
An 32.4040

SECOND-ORDER EFFECTS IN ELASTICITY,PLASTICITY AND FLUID DYNAMICS International symposium on second-order effects in elasticity,plasticity and fluid dynamics symposium Haifa 1962 Apr 23-27 International union of theoretical and applied mechanics Edited by Markus Reiner and David Abir Jerusalem academic press;Pergamon press,Jerusalem;London,1964
Eng 41.2483

The SECOND SYSTEMS SYMPOSIUM proceedings Views on general systems theory Cleveland, Ohio 1963 Apr Case institute of technology Edited by Mihajlo D. Mesarovic xvii,178p 24cm John Wiley and sons,New York;London, 1964
Math 3.0880

SECRETORY MECHANISMS OF SALIVARY GLANDS International conference on mechanisms of salivary secretion and their regulation Proceedings Birmingham,Ala. 1966 Aug 9-11 Edited by Leon H. Schneyer and Charlotte A. Schneyer Held at the University of Alabama medical center Academic press,New York; London,1967
Bal 39.1265

SEDIMENT Waves on beaches and resulting sediment transport Proceedings Madison,Wis. 1971 Oct Edited by R.E. Meyer illus vii,462p 23cm Academic press,New York,1972
A Math 4.1885

SEDIMENTATION International geological congress 20th papers Mexico City 1956 Seccion 5: relaciones entre la tectonica y la sedimentacion,tom.1-2 Edited by A. Garcia Rojas Mexico City,1957 Text in English, French,German,Russian and Spanish
Geol 8.3046

SEDIMENTOLOGY Deltaic and shallow marine deposits International congress of sedimentology 6th Proceedings Amsterdam 1963 May 29-30 Antwerp 1963 May 31-Jun 1 Edited by L.M.J.U.van Straaten Developments in sedimentology,1 Elsevier,Amsterdam,1964
Geog 13.0304

SEDIMENTOLOGY International congress of sedimentology 3rd Proceedings Groningen 1951 Jul 5-12 Wageningen 1951 Jul 5-12 Edited by Tjeerd Hendrik van Andel Geologisch instituut,Groningen,1951 In English,French and German
Geog 13.0284

SEDIMENTOLOGY International congress of sedimentology 3rd proceedings Groningen 1951 Jul.5-12 Edited by Tj.van Andel map Martinus Nijhoff,The Hague,1951
Geol 8.1645

SEDIMENTOLOGY International congress of sedimentology 4th Proceedings Gottingen 1954 Edited by R. Brinkmann and others Geologische rundschau,43,heft 2 Enke, Stuttgart,1955 In English,French and German
Geog 13.0285

SEGEDIN C.M. ed. Australasian conference on hydraulics and fluid mechanics 2nd proceedings Auckland N.Z. 1965 Dec.6-11 University of Auckland Sponsored jointly with the New Zealand institution of engineers Conference organising committee,Auckland,N.Z., 1966
A Math 4.0531

SEIDEL Carl W. and HERBER Rolf H. The Mossbauer effect and its application in chemistry New York 1966 Sep 12 American chemical society Edited by E. Gould American chemical society.Advances in chemistry series,68 American chemical society,Washington,D.C.,1967 Symposium organized by C.W.Seidel.Chairman:R.H.Herber
Chem 18.1190

SEISMOLOGY Astronomy,meteorology and seismology The Pan American scientific congress 2nd Proceedings Washington,D.C. 1915-16 Dec 27-Jan 8 Pan American union Edited by Glen Levin Swiggett illus xiv, 847p Government printing office,Washington,D. C.,1917
Nap 11.0838

SELECTED COMBUSTION PROBLEMS Combustion colloquium Cambridge 1953 Dec 7-11 Agard Edited by W.R. Hawthorne and J. Fabri Butterworths,London,1954
A Math 4.0969

SELECTED COMBUSTION PROBLEMS:FUNDAMENTALS AND AERONAUTICAL APPLICATIONS Combustion colloquium Cambridge 1953 Dec 7-11 Agard Butterworths,London,1954
Eng 41.7219

SELECTED COMBUSTION PROBLEMS:FUNDAMENTALS AND AERONAUTICAL APPLICATIONS Combustion colloquium Proceedings Cambridge 1953 Dec 7-11 North Atlantic treaty organization. Advisory group for aeronautical research and development viii,534p Butterworths,London, 1954
Chem E 24.1912

SELECTED COMBUSTION PROBLEMS II:TRANSPORT PHENOMENA;IGNITION;ALTITUDE BEHAVIOUR AND SCALING OF AEROENGINES Combustion colloquium Liege 1959 Dec 5-9 Agard Butterworths,London,1956
Eng 41.7228

SELECTED MATHEMATICAL PAPERS Symposium on problems in analysis Papers Princeton,N.J. 1969 Apr 1-3 Princeton university Princeton mathematical series, 31 Bibliog, port x,351p 24cm Princeton university press,Princeton,N.J.,1970 Symposium held in honour of S.Bochner
P Math 2.3992

SELECTED TOPICS IN RADIATION DOSIMETRY Symposium on selected topics in radiation dosimetry Proceedings Vienna 1960 Jun 7-11 Sponsored by the International atomic energy agency International atomic energy agency.Proceedings series International atomic energy agency,Vienna,1961
Radioth 35.1694

SELECTED TOPICS ON BALLISTICS Cranz centenary colloquium.University of Freiburg Freiburg 1958 Agard Edited by Wilbur C. Nelson NATO-AGARD wind tunnel and model testing panel.Publication AGARDograph, 32 illus,port Pergamon press,London,1959
Eng 41.6772

SELECTION OF 20TH CENTURY BOTANICAL ART AND ILLUSTRATIONS International exhibition of twentieth century botanical art 2nd Catalogue Edited by George H.M. Lawrence port 168p Carnegie-Mellon university, Pittsburgh,Pa.,1969 Presented at the 11th International botanical congress,August 1969
Bot 42.0634

SELECTIVE TOXICITY AND ANTIBIOTICS Edinburgh 1948 Jul Society for experimental biology Edited by J.F. Danielli and R. Brown Society for experimental biology.Symposia, 3 Cambridge university press,Cambridge,1949
Radioth 35.1931

SELECTIVE TOXICITY AND ANTIBIOTICS Edinburgh 1948 Jul Society for experimental biology Society for experimental biology.Symposia, 3 Cambridge university press,Cambridge,1949
Phys 20.2224

SELECTIVE TOXICITY AND ANTIBIOTICS :a symposium Edinburgh 1948 Jul Society for experimental biology Society for experimental biology.Symposia, 3 Cambridge university press,Cambridge,1949
Bioch 33.1350

SELENIUM PHYSICS Recent advances in selenium physics A Symposium on solid and liquid state selenium physics... held at the Chemical society Proceedings London 1964 June European selenium-tellurium committee Edited by Sven Walden and others Pergamon press, Oxford,1965
Cav 7.1462

SELENIUM-TELLURIUM DEVELOPMENT ASSOCIATION The Physics of selenium and tellurium : international symposium Proceedings Montreal 1967 Oct 12-13 Edited by W. Charles Cooper Bibliog,photos,diagrs ix, 380p 23cm Pergamon press,Oxford,1969
Cav 7.2670

SELF-ADAPTIVE CONTROL SYSTEMS I.F.A.C. symposium on the theory of self-adaptive control systems 2nd Proceedings Teddington 1965 Sep 14-17 International federation of automatic control Edited by P. H. Hammond Plenum press,New York,1966
Eng 41.5888

SELF-ORGANIZING SYSTEMS :an interdisciplinary conference Proceedings Chicago,Ill. 1959 May 5-6 Edited by Marshall C. Yovits and Scott Cameron Sponsored by the United States.Office of naval research.Information systems branch International tracts in computer science and technology and their application, 2 Pergamon press,New York, 1960
Eng 41.1162

SELF-ORGANIZING SYSTEMS 1962 conference proceedings Chicago,Ill. 1962 May 22-23 United States.Office of naval research Edited by Marshall C. Yovits and others Jointly sponsored by Armour research foundation ix,563p 24cm Spartan books, Washington,D.C.,1962
Math L 5.0992

SEMAINE INTERNATIONALE DE GEOGRAPHIE : international geographical week Report Brussels 1958 Aug 3-10 Federation Belge des geographes professeurs de lenseignement moyen,normal et technique Ghent,c1958
Geog 13.5602

SEMAINE SOCIOLOGIQUE 2nd Compte rendu Villes et campagnes Paris 1951 Mar Centre detudes sociologiques Edited by Georges Friedmann Ecole pratique des hautes etudes,Paris.Bibliotheque generale.Section 6 Colin,Paris,1953
Geog 13.4272

SEMI-CONDUCTING MATERIALS :a conference Proceedings Reading 1950 Jul 10-15 Edited by H.K. Henisch Under the auspices of the International union of pure and applied physics Butterworths,London,1951
Eng 41.5373

SEMI-CONDUCTING MATERIALS :a conference Proceedings Reading 1950 Jul 10-15 Royal society of London Edited by H.K. Henisch Under the auspices of the International union of pure and applied physics illus xv,281p Butterworths, London,1951
Cav 7.1925

SEMICENTENIAL SYMPOSIUM ON OPTICAL IMAGE EVALUATION Proceedings Optical image evaluation Washington,D.C. 1951 Oct 18 - 20 National bureau of standards National bureau of standards.Circular, 526 289p Washington,D.C.,1954
Obs 6.2129

SEMICONDUCTING COMPOUNDS International conference on II-VI semiconducting compounds Proceedings Providence,R.I. 1967 Sep 6-8 American physical society International union of pure and applied physics Edited by D.G. Thomas 1490p 20cm Benjamin,New York, 1967
Cav 7.2681

SEMICONDUCTOR DEVICES IN POWER ENGINEERING : a symposium Edited by J. Seymour Held at Woolwich polytechnic Pitman,London,1968
Eng 41.4867

SEMICONDUCTOR SURFACE PHYSICS Conference on the physics of semiconductor surfaces Proceedings Philadelphia,Pa. 1956 Jun 4-6 Edited by R.H. Kingston Sponsored by the United States.Office of naval research,the University of Pennsylvania and the Lincoln laboratory University of Pennsylvania, Philadelphia,Pa.,1957
Eng 41.5399

SEMICONDUCTORS Amorphous and liquid semiconductors :a conference Proceedings Cambridge 1969 Sep 24-27 International union of pure and applied physics. Semiconductor commission Royal society of London Edited by Nevill Francis Mott Bibliog,diagrs,figs 626p 20cm North-Holland,Amsterdam,1970
Cav 7.2678

SEMICONDUCTORS Conference on power application of controllable semiconductor devices London 1965 Pt 1 Institution of electrical engineers Institution of electrical engineers.Conference publications, 17 I.E.E.,London,1965
Eng 41.4868

SEMICONDUCTORS Excitations in semiconductors, polarons and excitons Scottish universities' summer school in physics 3rd Papers St. Andrews 1962 Jul 30-Aug 18 Edited by C.G. Kuper and G.D. Whitfield Scottish universities' summer school in physics,1962,3 Oliver and Boyd,Edinburgh,1963
Cav 7.2136

SEMICONDUCTORS International conference on the physics of semiconductors 8th Proceedings Kyoto 1966 Sep 8-13 Physical society of Japan Under the auspices of the International union of pure and applied physics Physical society of Japan.Journal, 21,suppl. 774p Physical society of Japan, Tokyo,1966
Cav 7.1341

SEMICONDUCTORS International conference on the physics of semiconductors 9th Proceedings Moscow 1968 Jul 23-29 Vol 1-2 Akademiya nauk S.S.S.R. English edition 634p 27cm 2 vols Publishing house 'Nauka',Leningrad,1968
Cav 7.2955

SEMICONDUCTORS Symposium on electronic phenomena in chemisorption and catalysis on semiconductors Moscow 1968 Jul 2-4 Edited by K. Hauffe and T. Wolkenstein De Gruyter,Berlin,1969
Chem 18.2698

SEMICONDUCTORS The International conference on the physics of semiconductors Report Exeter 1962 Jul 16-20 Institute of physics and Physical society Edited by A.C. Stickland Under the auspices of the International union of pure and applied physics 909p Institute of physics,London, 1962
Cav 7.0750

SEMICONDUCTORS AND PHOSPHORS Internationales colloquium uber halbleiter und phosphore Garmisch-Partenkirchen 1956 Aug 28-Sep 15 Edited by M. Schon and H. Welker Sponsored by the International union of pure and applied physics Vieweg,Brunswick,1958
Radioth 35.1615

SEMIGROUPS Symposium on semigroups Proceedings Detroit,Mich. 1968 Jun 27th-29th Wayne state university Edited by Karl W. Folley Sponsored by the National science foundation bibliog. xi,277p 24cm Academic press,New York;London,1969
P Math 2.3927

SEMINAIRE DE PROBABILITES 1e Strasbourg 1966-67 Nov-Feb Universite de Strasbourg. Institut de mathematique Lecture notes in mathematics, 39 189p 28cm Springer, Berlin,1967
Math 3.1188

SEMINAIRE DE PROBABILITES 2me Strasbourg 1967 Mar-Oct Universite de Strasbourg. Institut de mathematique Lecture notes in mathematics, 51 Bibliog 199p 28cm Springer,Berlin,1968
P Math 2.2996

SEMINAIRE DE PROBABILITES 2me Strasbourg 1967 Mar-Oct Universite de Strasbourg. Institut de mathematique Lecture notes in mathematics, 51 199p 28cm Springer, Berlin,1968
Math S 3.1438

SEMINAIRE DE PROBABILITES 3me Strasbourg 1967-68 Universite de Strasbourg.Institut de mathematique Lecture notes in mathematics, 88 Bibliog. 229p 28cm Springer,Berlin, 1969
P Math 2.3517

SEMINAIRE DE PROBABILITES 4me Strasbourg 1968-69 Universite de Strasbourg.Institut de mathematique Lecture notes in mathematics, 124 Bibliog. 282p 25cm Springer-Verlag,Berlin,1970
P Math 2.3518

SEMINAIRE SU LES ALGEBRES COMPLETES Nancy 1969 Edited by J.P. Ferrier Lecture notes in mathematics, 164 68p 25cm Springer, Berlin,1970
P Math 2.3865

SEMINAR IN SYSTEMS Proceedings Edited by G.D.S. MacLellan and Samuel Mercer Held at Michigan state university.College of engineering Michigan state university,East Lansing,Mich.,1958
Eng 41.6009

SEMINAR ON AERONAUTICAL SCIENCES Proceedings 1961 Nov 27-Dec 2 Vol 1 National aeronautical laboratory,Bangalore National aeronautical laboratory,Bangalore,1961
Eng 41.6969

SEMINAR ON AERONAUTICAL SCIENCES proceedings Bangalore 1961 Nov 27-Dec.2 Vol 1 India.National aeronautical laboratory National aeronautical laboratory,Bangalore, 1961
A' Math 4.0526

SEMINAR ON ALGEBRAIC GROUPS AND RELATED FINITE GROUPS Princeton,N.J. 1968-69 By A. Borel and others Partially sponsored by Office of scientific research,Office of aerospace research,United States air force Lecture notes in mathematics, 131 Bibliog. not con.no.p 25cm Springer-Verlag,Berlin, 1970
P Math 2.3582

SEMINAR ON ALGEBRAIC TOPOLOGY Proceedings Seattle,Wash. 1971 Feb 22-26 Battelle memorial institute.Seattle research center Edited by Peter J. Hilton Lecture notes in mathematics, 249 vi,110p 25cm Springer, Berlin,1971
P Math 2.4115

SEMINAR ON COMPLEX MULTIPLICATION Princeton, N.J. 1957-58 Institute for advanced study Lecture notes in mathematics, 21 28cm Springer-verlag,Berlin,1966
P Math 2.2857

SEMINAR ON CONTACT MANIFOLDS Lectures Kyoto 1969 Jun 23-25 By Koichi Ogiue and others Kyoto university.Research institute for mathematical sciences Okayama university. Study group of geometry.Publication, 4 ii, 83p 26cm Okayama university.Department of mathematics,Okayama,1970
P Math 2.4119

SEMINAR ON DIFFERENTIAL EQUATIONS AND DYNAMICAL SYSTEMS Lectures College Park,Md. 1967 Aug Edited by G.Stephen Jones Lecture notes in mathematics, 60 Bibliog v,106p 28cm Springer-verlag,Berlin,1968
P Math 2.3046

SEMINAR ON DIFFERENTIAL EQUATIONS AND DYNAMICAL SYSTEMS Lectures 2nd College Park, Md 1970 Jul-Aug Edited by J.A. Yorke Held at the University of Maryland Lecture notes in mathematics, 144 viii,268p 17cm Springer-Verlag,Berlin,1970
P Math 2.3534

SEMINAR ON ELEMENTARY PARTICLES lectures High-energy physics and elementary particles Trieste 1965 May 3-Jun.30 International atomic energy agency International atomic energy agency.Proceedings series STIPUB117 Vienna,1965
A Math 4.0888

SEMINAR ON GEOMETRY OF NUMBERS Lectures, mimeograph Princeton,N.J. 1949 Institute for advanced study 90p 26cm Institute for advanced study,Princeton,N.J., 1949
P Math 2.2778

SEMINAR ON MODERN METHODS IN NUMBER THEORY Tokyo 1971 Aug 30-Sep 4 Held at the Institute of statistical mathematics,Tokyo 29cm Institute of statistical mathematics, Tokyo,1971
P Math 2.4380

The SEMINAR ON THE REFORM OF BIOLOGY TEACHING report on the OECD seminar New thinking in school biology La Tour de Peilz, Switzerland 1962 Dep 4-14 Organization for economic co-operation and development New thinking in school science Paris,1963
Gen 34.2062

SEMINAR ON THEORETICAL PHYSICS Lectures Theoretical physics Trieste 1962 Jul 16-25 Organized by the International atomic energy agency International atomic energy agency, Vienna,1963
Radioth 35.1655

SEMINAR ON THEORETICAL PHYSICS lectures Theoretical physics Trieste 1962 Jul.16-Aug.25 Organized by the International atomic energy agency Vienna,1963 Companion volume to International summer school on nuclear theory,1962
A Math 4.0886

The SEMINAR ON THEORETICAL PHYSICS held in Miramare Papers Trieste 1962 Jul 16-Aug 25 International atomic energy agency Edited by A. Salam STI-PUB-61 637p international atomic energy agency,Vienna,1963
Cav 7.2346

SEMINAR ON URBANIZATION IN INDIA Indias urban future Berkeley,Calif. 1960 Jun 26-Jul 2 Edited by Roy Turner University of California press,Berkeley,Calif.;Los Angeles, Calif.,1962 Conference sponsored by Kingsley Davis and others
Geog 13.1947

SEMINARS ON ANALYTIC FUNCTIONS conference typescript Princeton,N.J. 1957 Sep 2-14 Vols. 1-2 Institute for advanced study Edited by Marston Morse and others Under contract with the United States.Air force. Office of scientific research Bibliog. 25cm 2 vols Princeton,N.J.,1957
P. Math 2.1583

SENDAI 1962 International association for hydraulic research symposium Cavitation and hydraulic machinery Proceedings Edited by F. Numachi Tohoku university,Sendai,163
Eng 41.6518

SENDAI 1965 Fracture International conference on fracture 1st Proceedings Vol 1-3 Organizing committee on fracture Edited by T. Yokobori and others 3 vols Japanese society for strength and fracture of materials,1966
Met 25.0872

SENDAI 1965 International conference on fracture 1st Proceedings Vol 1-3 Edited by T. Yokobori and others 3 vols Japanese society for strength and fracture of materials,Sendai,1966
Eng 41.3807

SENSATION:ITS MECHANISMS AND DISTURBANCES :an investigation of the most recent advances Proceedings of the association New York 1934 Dec 27-28 Association for research in nervous and mental disease Edited by Clarence A. Patten and others Association for research in nervous and mental disease. Research publications, 15 Williams and Wilkins,Baltimore,Md.,1935
An 32.4523

SENSITIVITY METHODS IN CONTROL THEORY International symposium on sensitivity analysis Dubrovnik 1964 Aug 31-Sep 5 Yugoslav committee for electronics and automation Edited by L. Radanovic Sponsored by the International federation of automatic control.Theory committee Pergamon press,Oxford,1965
Eng 41.5835

SENSITIVITY REACTIONS TO DRUGS Liege 1957 Jul 9-12 Council for international organizations of medical sciences Edited by M.L. Rosenheim and R. Moulton Blackwell, Oxford,1958
Pha 16.0187

SENSORY COMMUNICATION Symposium on principles of sensory communication Contributions Boston,Mass. 1961 Jul 19-Aug 7 Edited by W.A. Rosenblith Held at the Massachusetts institute of technology M.I.T. press;Wiley,Boston,Mass.;New York,1961
Phys 20.2108

SENSORY COMMUNICATION Symposium on principles of sensory communication Contributions Cambridge,Mass. 1959 Jul 19-Aug 1 Edited by Walter Rosenblith M.I.T. press;Wiley,New York,1961
An 32.4395

SENSORY COMMUNICATION The Symposium on principles of sensory communication Contributions Dedham,Mass. 1959 Jul 19-Aug 1 Massachusetts institute of technology Edited by W.A. Rosenblith M.I.T.press;Wiley, New York;London,1961 Held at M.I.T.'s Endicott House,1961
Bal 39.1518

SENSORY COMMUNICATION :symposium on principles of sensory communication Contributions Endicott House,M.I.T. 1959 Jul 19-Aug 1 Edited by Walter A. Rosenblith M.I.T.press, Cambridge,Mass.
Psy 31.0282

SENSORY COMMUNICATIONS The Symposium on principles of sensory communications Contributions Dedham,Mass. 1959 Jul 19-Aug 1 United States.Air force.Office of scientific research Edited by Walter A. Rosenblith MIT press,Cambridge,Mass.,1961
Psy 31.2842

SENSORY DEPRIVATION :a symposium held at Harvard medical school Cambridge,Mass. 1958 Jun 20-21 Harvard medical school and Boston city hospital Edited by Philip Solomon and others Co-sponsored by the United States.Office of naval research. Physiological psychology branch Harvard university press,Cambridge,Mass.,1965
Psy 31.0128

SENSORY FUNCTION Ciba foundation symposium... on sensory function :touch,heat and pain 2nd Proceedings London 1965 Sep 21-23 Ciba foundation Edited by A.V.S. de Reuck and Julie Knight J.and A.Churchill,London, 1966
Psy 31.0286

SENSORY FUNCTION Hearing mechanisms in vertebrates Ciba foundation symposium...on sensory function 4th Proceedings London 1967 Sep 26-28 Ciba foundation Edited by A. V.S.de Reuck and Julie Knight J.and A. Churchill,London,1968
Psy 31.3372

SENSORY MECHANISMS :a workshop...held during the second International summer school of brain research... amsterdam 1964 Edited by Y. Zotterman Progress in brain research, 23 Elsevier,Amsterdam,1967
An 32.4244

SENSORY RECEPTORS :a symposium Cold Spring Harbor 1965 Jun 4-11 Cold Spring Harbor symposia on quantitative biology, 30 Cold Spring Harbor,N.Y.,1965
Phys 20.1058

SENSORY RECEPTORS :a symposium Cold Spring Harbor,L.I.,N.Y. 1965 Jun 4-11 Cold Spring Harbor laboratory of quantitative biology Edited by John Cairns Cold Spring Harbor symposia on quantitative biology, 30 Cold Spring Harbor laboratory of quantitative biology,Cold Spring Harbor,L.I.,N.Y.,1965
Psy 31.0298

SENSORY RECEPTORS :a symposium Papers Cold Spring Harbor 1965 Cold Spring Harbor laboratory of quantitative biology Cold Spring Harbor symposia on quantitative biology, 30 Cold Spring Harbor,1965
Bioch 33.1286

SENSORY SPECIALIZATION IN RESPONSES TO ENVIRONMENTAL DEMANDS :a symposium London 1960 Mar 4 Zoological society of London Edited by O.E. Lowenstein Zoological society of London.Symposia, 3 Zoological society of London,London,1960
An 32.4541

SEPARATION The Symposium on the less common means of separation Proceedings Birmingham 1963 Apr 24-26 Institution of chemical engineers.Midlands branch Edited by J.M. Pirie 127p I.Chem.E.,London,1964
Chem E 24.1915

SEROLOGICAL AND BIOCHEMICAL COMPARISONS OF PROTEINS Rutgers university.Bureau of biological research.Annual conferences on protein metabolism 14th Papers New Brunswick,N.J. 1958 Jan 24-25 Edited by William H. Cole Rutgers university press,New Brunswick,N.J.,1958
Col S 12.0073

SEROLOGICAL FRACTIONS IN SCHIZOPHRENIA :a research symposium New Orleans 1961 Jun 13-14 Louisiana state department of hospitals Tulane university Louisiana state university Edited by Robert G. Heath Harper and Row for the Commonwealth fund,New York,1963
PGMS 29.0324

SERVOCOMPONENTS Conference on servocomponents London 1967 Nov 21-23 Sponsored by Institution of electrical engineers.Professional group on measuring and control equipment Institution of electrical engineers.Conference publication, 37 I.E.E., London,1967
Eng 41.5879

SESJA KOPERNIKOWSKA... Warsaw 1953 Sep 15-16 Polska akademia nauk 482p Panstowe wydawnictwo naukowe,Warsaw,1955
WSM 43.2653

SESSIYA AKADEMII NAUK SSSR PO NAUCHNYM PROBLEMAM AVTOMATIZATSII PROIZVODSTVA :plenarnye zasedaniya Moscow 1956 Oct 15-20 Akademiya nauk S.S.S.R. Edited by D.Ya. Libenson and others 271p 22cm Akademiya nauk SSSR,Moscow,1957
Math L 5.0775

SET-VALUED MAPPINGS,SELECTIONS AND TOPOLOGICAL PROPERTIES Proceedings Buffalo,N.Y. 1969 May 8-10 Edited by W.M. Fleischman Held at the State university of New York Lecture notes in mathematics, 171 Springer, Berlin,1970
Math S 3.1657

SETS MODELS AND RECURSION THEORY :summer school in mathematical logic and Tenth logic colloquium Proceedings Leicester 1965 Aug-Sep Edited by John N. Crossley Studies in logic and the foundations of mathematics Bibliog. vii,329p 23cm North Holland, Amsterdam,1967
P Math 2.3118

SEVERAL COMPLEX VARIABLES International mathematical conference Proceedings College Park,Md. 1970 Apr 6-17 1 University of Maryland Edited by John Horvath Lecture notes in mathematics, 155 bibliog. 214p 25cm Springer,Berlin,1970
P Math 2.3890

SEVERAL COMPLEX VARIABLES International mathematical conference Proceedings College Park,Md. 1970 Apr 6-17 2 University of Maryland Edited by John Horvath Lecture notes in mathematics, 185 214p 25cm Springer,Berlin,1971
P Math 2.3891

SEX AND BEHAVIOR :conferences Papers Berkeley,Calif. 1961 Committee for research in problems of sex Edited by Frank A. Beach Wiley,New York,1965 Based on conferences held in 1961 and 1962
Psy 31.2709

SEX AND REPRODUCTION Gunma symposium on endocrinology 4th Minakami 1966 Sep 22-23 Gunma university.Institute of endocrinology Gunma university.Institute of endocrinology.Annual report, 4 Institute of endocrinology,Gunma university,Maebashi, 1967
Phys 20.1434

SEX DIFFERENTIATION AND DEVELOPMENT :a symposium Proceedings London 1958 Apr 10-11 Society for endocrinology Edited by C.R. Austin Society for endocrinology. Memoirs, 7 Cambridge university press, Cambridge,1960
Gen 34.0583

SEX DIFFERENTIATION AND DEVELOPMENT :a symposium Proceedings London 1958 Apr 10-11 Society for endocrinology Edited by C.R. Austin held at the Royal society of medicine Society for endocrinology.Memoirs, 7 Cambridge university press,Cambridge,1960
Bal 39.1402

SEX IN MICROORGANISMS :a symposium Philadelphia 1951 Dec 30 American association for the advancement of science Edited by D.H. Wenrich illus v,362p American association for the advancement of science,Washington,D.C.,1954
Bot 42.0758

SEXUAL FUNCTION AND DYSFUNCTION :a symposium 1967 Nov 17 Hahnemann medical college and hospital Davis,Philadelphia,1969
Inv Med 37.0083

SEXUAL REFORM CONGRESS 3rd Proceedings London 1929 Sep 8-14 World league for sexual reform Edited by Norman Haire World league for sexual reform.Congress, 3rd Kegan Paul,Trench,Trubner,London,1930
Psy 28.0356

SHEAR STRENGTH OF COHESIVE SOILS Research conference on shear strength of cohesive soils New York 1960 University of Colorado American society of civil engineers Boulder, Colo.,1960
Eng 41.3171

SHEAR STRENGTH PROPERTIES OF NATURAL SOILS AND ROCKS :a conference Proceedings Geotechnical conference Oslo 1967 Vol 1-2 Norsk geoteknisk forening 2 vols Norwegian geotechnical institute,Oslo,1967
Eng 41.3135

SHEAR TESTING OF SOILS Laboratory shear testing of soils :a symposium Ottawa 1963 Sep 9-11 National research council of Canada American society for testing materials.Special technical publication, 361 American society for testing materials,Philadelphia,Pa.,1964
Eng 41.3157

SHEAR TESTING OF SOILS Symposium on direct shear testing of soils New York 1952 Jun 26 American society for testing materials American society for testing materials.Special technical publication, 131 American society for testing materials,Philadelphia,Pa.,1953
Eng 41.3156

SHEAR TESTING OF SOILS Symposium on shear testing of soils Atlantic City,N.J. 1939 Jun 28 American society for testing materials American society for testing materials,Philadelphia,Pa.,1939
Eng 41.2988

SHEET MATERIALS FOR HIGH TEMPERATURE SERVICE Papers Dallas 1957 May 12-16 American society for metals American society for metals,Cleveland,1958
Met 25.1001

SHEFFIELD 1950 Carbon dioxide fixation and photosynthesis Society for experimental biology Society for experimental biology. Symposia, 5 Cambridge university press, Cambridge,1951
Phys 20.0953

SHEFFIELD 1950 Carbon dioxide fixation and photosynthesis :a symposium Society for experimental biology Society for experimental biology.Symposia, 5 Cambridge university press,Cambridge,1951
Bioch 33.1351

SHEFFIELD 1950 Statistical method in industrial production conference proceedings Royal statistical society.Industrial applications section 89p 25cm Royal statistical society,London,1951
Math 3.0821

SHEFFIELD 1956 Air pollution :based on papers given at a conference held at the University of Sheffield Edited by M.W. Thring 245p Butterworths,London,1957
Nap 11.0607

SHEFFIELD 1961 Advances in magnetohydrodynamics :a colloquium Proceedings Edited by I.A. McGrath and others Organized by the University of Sheffield.Department of fuel technology and chemical engineering Pergamon press,London, 1963
Eng 41.4504

SHEFFIELD 1961 Advances in magnetohydrodynamics :a colloquium proceedings Edited by I.A. McGrath and others Organized by University of Sheffield. Department of fuel technology and chemical engineering Pergamon press,London,1963
A Math 4.0622

SHEFFIELD 1962 The Transition state :a symposium University of Sheffield,and, Chemical society Chemical society.Special publication,16 Chemical society,London,1962
Chem 18.1222

SHEFFIELD 1963 Sorby centenary conference Papers and discussions Iron and steel institute Sponsored by the Company of Cutlers in Hallamshire Iron and steel institute.Special report, 80 Iron and steel institute,London,1964
Met 25.2360

SHEFFIELD 1964 Steelmaking in the basic arc furnace Iron and steel institute conference :technical sessions Papers and discussions Iron and steel institute Iron and steel institute.Special report, 87 London,1964
Met 25.2558

SHEFFIELD 1966 Aromaticity :an international symposium Chemical society,and, University of Sheffield Chemical society. Special publication,21 Chemical society, London,1967
Chem 18.1225

SHEFFIELD 1966 Aspects of the biology of ageing :a symposium Society for experimental biology Edited by H.W. Woolhouse Society for experimental biology.Symposia, 21 Cambridge university press,Cambridge,1967
An 32.5482

SHEFFIELD 1966 Aspects of the biology of ageing :a symposium Society for experimental biology Society for experimental biology. Symposia, 21 Cambridge university press, Cambridge,1967
Bioch 33.1360

SHEFFIELD 1966 Deformation under hot working conditions :the conference Proceedings University of Sheffield. Department of metallurgy Edited by C.M. Sellars and W.J.McG. Tegart Iron and Steel institute,London,1968
Met 25.2402

SHEFFIELD 1968 Ecological aspects of the mineral nutrition of plants :a symposium British ecological society Edited by Ian H. Rorison and others British ecological society.Symposia, 9 illus,plates xxi, 484p Blackwell scientific,Oxford,1969
Bot 42.2107

SHEFFIELD 1970 Software 70 :conference Proceedings Software world Edited by David J. Evans illus 197p Transcripta books, London,1970
Math L 5.3871

SHEFFIELD UNIVERSITY Computer aided circuit design Conference on CACD Proceedings 1968 Mar 26-28 Jointly sponsored by Design electronics Design electronics,London,1968
Eng 41.2226

SHELL PROBLEMS Non-classical shell problems : a symposium Proceedings Warsaw 1963 Sep 2-5 International association for shell structures Edited by W. Olszak and A. Sawczuk North-Holland;PWN,Amsterdam;Warsaw, 1964
Eng 41.2876

SHELL RESEARCH :a symposium Proceedings Delft 1961 Aug 30-Sep 2 International union of testing and research laboratories for materials and structure International association for shell structures Edited by A. M. Haas and A.L. Bouma North-Holland, Amsterdam,1961
Eng 41.2860

SHELL STRUCTURES World conference on shell structures Proceedings San Francisco,Calif. 1962 Oct 1-4 University of California International association for shell structures Building research advisory board National research council.Publication,1187 National academy of sciences,Washington,D.C.,1064
Eng 41.3025

SHELLS Theory of thin shells :a symposium 2nd Proceedings Copenhagen 1967 Sep 5-9 International union of theoretical and applied mechanics Edited by F.I. Niordson Springer, Berlin,1969
Eng 41.2892

SHELTER ISLAND,L.I.,N.Y. 1951 Quantum mechanical methods in valence theory :a conference Papers National academy of sciences With the support of the United States.Office of naval research United States.Office of naval research.Physics branch. Publication,4 U.S.government printing office, Washington,D.C.,1952
Chem 18.2237

SHELTER ISLAND,N.Y. 1956 Endocrines in development National research council Edited by Ray L. Watterson Developmental biology conference series,1956 University of Chicago.Committee on publications in biology and medicine University of Chicago press, Chicago,Ill.,1959
Gen 34.0553

SHELTER ISLAND,N.Y. 1956 Linear algebras conference report By A.Adrian Albert National research council.Division of mathematics National research council. Publications, 502 v,59p 28cm National academy of sciences.National research council, Washington,D.C.,1957
P. Math 2.2033

SHIPS Sea-going qualities of ships :a seminar Teddington 1961 Oct 17-18 National physical laboratory H.M.S.O.,London, 1963
Eng 41.6932

SHOCK AND VIBRATION Experimental techniques in shock and vibration :a colloquium Papers New York 1962 Nov 27 Edited by W.J. Worley Sponsored by the American society of mechanical engineers American society of mechanical engineers,New York,1962
Eng 41.6354

SHOCK AND VIBRATION INSTRUMENTATION;A REVIEW OF THE LATEST DEVELOPMENTS Urbana,Ill. 1955 Jun 14-16 American society of mechanical engineers.Applied mechanics division New York,1956
Eng 41.6324

SHOCK METAMORPHOSIS Conference on shock metamorphosis of natural materials 1st Proceedings Greenbelt,Md. 1966 Apr 14-16 National aeronautics and space administration. Goddard space flight center Edited by Bevan M. French and Nicholas M. Short Held at the Goddard space flight center Mono book corporation,Baltimore,Md.,1968
Min 10.1443

SHOCK PATHOGENESIS AND THERAPY :an international symposium Stockholm Jun 27-30 Edited by K.D. Bock Sponsored by Ciba Springer-verlag,Berlin,1962
PGMS 29.0080

SHORT COURSE ON PROCESS INDUSTRY CORROSION Proceedings Ohio 1960 Sep 12-16 Ohio State university.Department of metallurgical engineering and National association of corrosion engineers.Technical group committee T-5 Edited by V.D. Miller National association of corrosion engineers,n.p.,1960
Met 25.1935

SHORT-TERM CHANGES IN NEURAL ACTIVITY AND BEHAVIOUR :a conference Cambridge 1969 Jul 15-18 Edited by Gabriel Horn and Robert A. Hinde Cambridge university press, Cambridge,1970
An 32.5239

SHORT-TIME ELEVATED TEMPERATURE TESTING OF METALS a symposium Proceedings Short-time high-temperature testing Los Angeles 1957 Mar 25-29 American society for metals American society for metals,Cleveland,Ohio, 1958
Met 25.0983

SHORT-TIME HIGH-TEMPERATURE TESTING Short-time elevated temperature testing of metals :a symposium Proceedings Los Angeles 1957 Mar 25-29 American society for metals American society for metals,Cleveland,Ohio, 1958
Met 25.0983

SHREWSBURY 1967 Prostaglandin :a symposium Worcester foundation for experimental biology Edited by Peter W. Ramwell and Jane E. Shaw Wiley,New York,1968
Inv Med 37.0185

SIAM-AMS PROCEEDINGS, 4 Computers in algebra and number theory Symposium in applied mathematics Proceedings New York 1970 Mar 25-26 By Garrett Birkhoff and Marshall Hall American mathematical society Society for industrial and applied mathematics vii,200p 26cm AMS,Providence,R.I.,1971
P Math 2.4476

SIC-SAM SYMPOSIUM,1966 PANON-1B:a programming language for symbol manipulation By A.Caracciolo di Forino and others 18p Universita di Pisa,Pisa,1966? Presented at the SIC-SAM Symposium.Washington,D.C.,1966
Math L 5.1136

SIGMA XI.TENNESSEE CHAPTER Lectures in biological sciences :a symposium Knoxville, Tenn. 1959 Dec 3-5 By Norman G. Anderson and others Edited by J.Ives Townsend 110p University of Tennessee press,Knoxville,Tenn., 1963
Bot 42.4744

SIGNALS IN COMMUNICATIONS Joint conference on digital processing of signals in communications Proceedings 1972 Institution of electronic and radio engineers Institution of electrical engineers Institution of electronic and radio engineers. Conference proceedings, 23 IERE,London,1972
Eng 41.8565

The SIGNIFICANCE OF SPACE RESEARCH FOR FUNDAMENTAL PHYSICS :a conference Interlaken 1969 Sep 4 European space research organisation Edited by A.F. Moore and U. Hardy ESRO SO-52 175p european space research organisation,Neuilly-sur-Seine, 1971
Obs 6.3603

SIGNIFICS International significal summer conference Conference d'ete internationale de linguistique psychologique 9th Proceedings Amersfoort 1953 Aug 10-15 International society for significs Synthese, 9,issue 3,no 3-4 Kroonder,Bussum,c1954
Psy 31.2514

SIGRID JUSELIUS FOUNDATION Control of cellular growth in adult organisms A Sigrid Juselius foundation symposium Helsinki 1965 Oct 4-6 Edited by Harald Teir and Tapio Rytomaa Academic press,London;New York,1967
Bioch 33.1007

SIGRID JUSELIUS FOUNDATION Control of cellular growth in adult organisms :a Sigrid Juselius foundation symposium Helsinki 1965 Oct 4-6 Edited by Harald Teir and Tapio Rytomaa Academic press,London;New York,1967
Pha 16.0177

SIGRID JUSELIUS FOUNDATION Control of cellular growth in adult organisms :a Sigrid Juselius foundation symposium Helsinki 1965 Oct 4-6 Edited by Harald Tier and Tapio Rytomaa Academic press,London;New York,1967
Phys 20.2232

SIGRID JUSELIUS FOUNDATION Regulatory functions of biological membranes The Sigrid Juselius foundation symposium 2nd Proceedings Helsinki 1967 Nov 6-9 Edited by Johan Jarnefelt B.B.A.library, 11 illus. viii,311p Elsevier,Amsterdam,1968
Bot 42.1517

SIGRID JUSELIUS FOUNDATION Regulatory functions of biological membranes The Sigrid Juselius foundation symposium 2nd Proceedings Helsinki 1967 Nov 6-9 Edited by Johan Jarnefelt BBA library, 11 Elsevier,Amsterdam,1968
Bioch 33.0968

SIGRID JUSELIUS FOUNDATION Regulatory functions of biological membranes :a Sigrid Juselius foundation symposium 2nd Papers Helsinki 1967 Nov 6-9 Edited by Johan Jarnefelt B.B.A. library,11 Illus. viii, 311p Elsevier,Amsterdam,1968
Pha 16.0017

A SIGRID JUSELIUS FOUNDATION SYMPOSIUM Control of cellular growth in adult organisms Helsinki 1965 Oct 4-6 Sigrid Juselius foundation Edited by Harald Teir and Tapio Rytomaa Academic press,London;New York,1967
Bioch 33.1007

The SIGRID JUSELIUS FOUNDATION SYMPOSIUM 2nd Proceedings Regulatory functions of biological membranes Helsinki 1967 Nov 6-9 Sigrid Juselius foundation Edited by Johan Jarnefelt B.B.A.library, 11 illus. viii, 311p Elsevier,Amsterdam,1968
Bot 42.1517

The SIGRID JUSELIUS FOUNDATION SYMPOSIUM 2nd Proceedings Regulatory functions of biological membranes Helsinki 1967 Nov 6-9 Sigrid Juselius foundation Edited by Johan Jarnefelt BBA library, 11 Elsevier, Amsterdam,1968
Bioch 33.0968

SILICON AND ITS BINARY SYSTEMS By A.S. Berezhnoi Translated by Consultants' bureau from the Russian Consultants bureau,New York, 1960
Met 25.0162

SILJAN SYMPOSIUM Advance and retreat of rural settlement International geographical congress 19th Papers of the Siljan symposium Norden 1960 Edited by Gerd Enequist and Gunnar Norling Uppsala universitet.Geografiska institution. Meddelanden.Ser A,160 Geografiska annaler,42, no 4 Uppsala,1960
Geog 13.1898

SIMLA 1909 Imperial malaria conference Proceedings India.General malaria committee Govt.central branch press,Simla,1910
Mol 45.0490

SIMPLE PREPARATION TECHNIQUES FOR LIQUID SCINTILLATION COUNTING The Beckman summer school 1st Proceedings London 1966 Aug 15-18 Beckman instruments London,c1966
Bioch 33.2300

SIMPLIFIED CALCULATION METHODS OF SHELL STRUCTURES :colloquium Proceedings Brussels 1961 Sep 4-6 International association for shell structures Edited by A. Paduart and R. Dutron North-Holland, Amsterdam,1962
Eng 41.2864

SIMPOSIO INTERNACIONAL DE ACCLIMATION AL FRIO Actas Buenos Aires 1959 Aug 5-7 By Robert E. Smith and others Institute Antartico Argentino.Publicacion,9 illus 414p 1962
Sco 14.0863

SIMPOSIO INTERNACIONAL DE GENETICA symposium Sao Paulo 1966 Jul Edited by F.G. Brieger Held under the auspices of the International union of biological sciences Ciencia e cultura, 19,no.1 Sao Paulo,1967
Gen 34.0995

SIMPSON George Gaylord and ROE Anne ed. Behaviour and evolution :conference 2nd Proceedings Princeton,N.J. 1956 Apr 30-May 5 American psychological association Held in collaboration with the Society for the study of evolution 24cm New Haven,1958
Philos 1.1827

SIMULTANEOUS LINEAR EQUATIONS AND THE DETERMINATION OF EIGENVALUES SYMPOSIUM Proceedings Los Angeles 1951 National bureau of standards Edited by L.J. Paige and Olga Taussky jointly sponsored by United States.Office of naval research United States.National bureau of standards. Applied mathematics series, 29 bibliog. 125p 26cm National bureau of standards, Washington,D.C.,1953
Math L 5.0131

SINGAPORE 1958 Centenary and bicentenary congress of biology :papers delivered in the university of Malaya in commemoration of the works of Darwin,Wallace and Linnaeus Proceedings Edited by R.D. Purchon Bibliog, illus,port,maps,tables 333p University of Malaya press;Oxford university press,Singapore; London,1960 Includes papers on anthropology
Bal 39.3924

SINGAPORE 1963 World orchid conference 4th Proceedings illus. Straits Times press,Singapore,1964
BG 38.3119

SINGLE CRYSTAL FILMS International conference on single crystal films Proceedings Blue Bell,Pa. 1963 Edited by M.H. Francombe and Hiroshi Sato Pergamon press,Oxford,1964
Eng 41.4458

SINGLE-CRYSTAL FILMS International conference on single-crystal films Proceedings Blue Bell,Pa. 1963 May Philco scientific laboratories Pergamon press,Oxford,1964
Cav 7.0244

SINGLE-CRYSTAL FILMS :a conference Proceedings Blue Bell,Penn. 1963 May United States.Office of naval research and University of Pennsylvania Edited by Maurice H. Francombe and Hiroshi Sato Sponsored jointly by Philco corporation Pergamon press, Oxford,1964
Met 25.0177

SINO-AMERICAN COOPERATION COMMITTEE Recent developments in biochemistry colloquium Proceedings Taipei 1967 Aug 17-19 Edited by Hsien-Wen Li and others Under the auspices of the Academia Sinica Academia Sinica,n.p.,1968
Gen 34.0619

SINTER Symposium on sinter London 1953 Nov Iron and steel institute Iron and steel institute.Special report, 53 Iron and steel institute,London,1955
Met 25.0321

SINTERING AND RELATED PHENOMENA ·International conference on sintering and related phenomena Cleveland 1965 Jun 21-23 University of Notre Dame Edited by G.C. Kuczynski and others Gordon and Breach,New York,1967
Met 25.1285

SIR DORABJI TATA TRUST Contributions to function theory international colloquium papers presented Bombay 1960 Jan 12-19 Tata institute of fundamental research xii, 231p 24cm Tata institute of fundamental research,Bombay,1960
P. Math 2.1584

SISTEMA CAMBRICO,SU PALEOGRAFIA Y EL PROBLEMA DE SU BASE International geological congress 20th papers of a symposium Mexico city 1956 Pt 1-2 Mexico city,1956 Text in English,French and Spanish
Geol 8.3024

SISTEMA CRETACICO Un Symposium sobre el cretacico en el hemisferio occidental y su correlacio mundial papers Mexico City 1956 Tom 1-2 Edited by Lewis B. Kellum 2 vols Mexico City,1959 Prepared under the auspices of the International commission on stratigraphy and the International palaeontological union for the 20th International geological congress
Geol 8.3049

SITE TESTING Le Choix des sites d'observatoires astronomiques (site testing) : a symposium proceedings Rome 1962 Oct 1-2 Edited by J. Rosch and others International astronomical union.Symposium, 19 382p Gauthier-Villars,Paris,1964 Reprinted from 'Bulletin astronomique de l'observatoire de Paris, vol 24, fasc.2 et 3, 1963
Obs 6.1467

The SITUATION IN BIOLOGICAL SCIENCE : proceedings of the Lenin academy of agricultural sciences of the U.S.S.R., session. verbatim report Moscow 1948 Jul 31-Aug 7 By Trefim Denisovich Lysenko and others Vsesoyuznaya akademiya selsko-khozyaistvennuikh nauk imeni Lenina Foreign languages publishing house,Moscow,1949
Bal 39.3918

The SITUATION IN BIOLOGICAL SCIENCE verbatim report Proceedings Moscow 1948 Jul 31-Aug 7 Vsesoyuznaya akademiya selskokozyaistennyk nauk imeni V.I.Lenina Foreign languages publishing house,Moscow,1949 Translated from the Russian
Gen 34.1210

The SIZE AND SHAPE FACTOR IN COLLOIDAL SYSTEMS a discussion Leamington Spa 1951 Jul 18-20 Faraday society Faraday society. Discussions, 11 Faraday society,Aberdeen, 1952
Bioch 33.1529

SKANDINAVISKE NATURFORSKERE Skandinaviske naturforskeres syvende mode Forhandlingar Christiania 1856 Jul 12-18 Werner, Christiania,1857 Includes two papers by J. Wolley
Bal 44.1980

SKANDINAVISKE NATURFORSKERES SYVENDE MODE Forhandlingar Christiania 1856 Jul 12-18 Skandinaviske naturforskere Werner, Christiania,1857 Includes two papers by J. Wolley
Bal 44.1980

SKEIKAMPEN,NORWAY 1961 N.A.T.O. Advanced study institute Electron density profiles in the ionosphere and exosphere :a symposium Proceedings Norwegian defence research establishment Edited by B. Maehlum Sponsored by N.A.T.O. science committee N.A.T.O. conference series, 2 428p Pergamon press,Oxford,1962
Nap 11.0391

The SKELETAL BIOLOGY OF EARLIER HUMAN POPULATIONS a symposium London 1965 Nov Edited by D.R. Brothwell Society for the study of human biology.Symposia, 8 Pergamon press,Oxford,1968
An 32.2885

SKIN Aging Symposium on the biology of skin 14th Proceedings Portland,Ore. 1964 May 9-10 Edited by William Montagna Held at the University of Oregon medical school Advances in biology of skin, 6 Oregon regional primate research center, 55 Pergamon press,Oxford,1965
An 32.4041

SKIN Blood vessels and circulation Brown university symposium on the biology of skin, 1960 Proceedings Providence,R.I. 1960 Jan 30-31 Edited by William Montagna and Richard A. Ellis Advances in biology of skin, 2 Pergamon press,Oxford,1961
An 32.4038

SKIN Carcinogenesis Symposium on the biology of skin 15th Proceedings Portland,Ore. 1965 Edited by William Montagna and Richard L. Dobson Held at the University of Oregon medical school Advances in biology of skin, 7 Oregon regional primate research center, 87 Pergamon press, Oxford,1966
An 32.4042

SKIN Cutaneous innervation Brown university symposium on the biology of skin, 1959 Proceedings Providence,R.I. 1959 Jan 24-25 Edited by William Montagna Advances in biology of skin, 1 Pergamon press,Oxford,1960
An 32.4037

SKIN Pigmentary system Symposium on the biology of skin 16th Proceedings Portland,Ore. 1966 Edited by William Montagna and Funan Hu Held at the University of Oregon medical school Advances in biology of skin, 8 Oregon regional primate research center, 183 Pergamon press,Oxford, 1967
An 32.4043

SLEEP Ciba foundation symposium on the nature of sleep London 1960 Jun 27-29 Ciba foundation Edited by G.E.W. Wolstenholme and Maeve O'Connor Churchill, London,1961 Under the chairmanship of sir John Eccles
Psy 31.3387

SLEEP Sleep mechanisms Symposium on the physiological pharmacological and clinical aspects of sleep Proceedings Zurich 1964 Sep 18-19 Edited by K. Akert and others Progress in brain research, 18 Elsevier, Amsterdam,1965
An 32.4238

SLEEP The Nature of sleep :a Ciba foundation symposium Proceedings London 1960 Jun 27-29 Ciba foundation Edited by G.E.W. Wolstenholme and Maeve O'Connor Churchill, London,1961
An 32.4378

SLEEP MECHANISMS Symposium on the physiological pharmacological and clinical aspects of sleep Proceedings Zurich 1964 Sep 18-19 Edited by K. Akert and others Progress in brain research, 18 Elsevier, Amsterdam,1965
An 32.4238

SLOAN-KETTERING INSTITUTE Symposium on trace analysis papers New York 1955 Nov 2-4 Edited by John H. Yoe and Henry J. Koch Wiley;Chapman and Hall,New York;London,1957
Met 25.1720

SLOUGH 1964 Symposium on the application of prestressed concrete to machinery structures Prestressed concrete development group Prestressed concrete development group, London,1965
Eng 41.2948

SMALL ANIMAL ANAESTHESIA :a symposium Proceedings London 1963 Jul British small animals veterinary association Universities federation for animal welfare Edited by Oliver Graham-Jones Pergamon press, London,1964
Radioth 35.0270

SMALL ANIMAL ANAESTHESIA :a symposium Proceedings London 1963 Jul British small animals veterinary association,and, Universities federation for animal welfare Edited by Oliver Graham-Jones Illus. Pergamon press,London,1964
VA 19.0264

SMALL BLOOD VESSEL INVOLVEMENT IN DIABETES MELLITUS :conference Proceedings Warrenton,Va. 1963 Mar 25-27 American institute of biological sciences Edited by Marvin D. Siperstein and others American institute of biological sciences,Washington,D. C.,1964
Bioch 33.0482

SMALL-ANGLE X-RAY SCATTERING :the conference Proceedings Syracuse 1965 Jun 24-26 American crystallographic association and Syracuse university Edited by H. Brumberger Co-sponsored by the United States Army. Research office and the National science foundation Gordon and Breach,New York,1967
Met 25.1466

SMITHSONIAN ASTROPHYSICAL OBSERVATORY Contemporary geodesy :a conference Proceedings Cambridge,Mass. 1958 Dec 1-2 Edited by Charles A. Whitten and Kenneth H. Drummond Co-sponsored by the American geophysical union Geophysical monograph, 4 National research council.Publication, 708 American geophysical union,Washington,D.C., 1959
Geod 9.0028

SMITHSONIAN INSTITUTION Neoplasms and related disorders of invertebrate and lower vertebrate animals :a symposium Proceedings Washington,D.C. 1968 Jun 19-21 National cancer institute Edited by Clyde J. Dawe and John C. Harshbarger National cancer institute.Monograph, 31 National cancer institute,Bethesda,Md.,1969
Bioch 33.0994

SMITHSONIAN INSTITUTION Symposium on cosmical gas dynamics 3rd proceedings Cambridge,Mass. 1957 Jun.24-29 International union of theoretical and applied mechanics and International astronomical union Edited by J.M. Burgers and others International astronomical union.Symposium, 8 Smithsonian Institution.Smithsonian symposium publications, 1 American physical society,New York,1958 Reviews of modern physics.Vol.30,no.3,July 1958
Obs 6.0306

SMITHSONIAN INSTITUTION.ASTROPHYSICAL OBSERVATORY Theory and observation of normal stellar atmospheres Harvard-Smithsonian conference on stellar atmospheres 3rd Proceedings Cambridge,Mass. 1968 Edited by O. Gingerich 472p Massachusetts institute of technology press,Cambridge,Mass.,1969
Obs 6.3354

SMITHSONIAN INSTITUTION.SMITHSONIAN SYMPOSIUM PUBLICATIONS, 1 Symposium on cosmical gas dynamics 3rd proceedings Cambridge, Mass. 1957 Jun.24-29 International union of theoretical and applied mechanics and International astronomical union Edited by J. M. Burgers and others Financial grant from the Smithsonian institution International astronomical union.Symposium, 8 American physical society,New York,1958 Reviews of modern physics.Vol.30,no.3,July 1958
Obs 6.0306

SMOKING AND HEALTH Toward a less harmful cigarette The World conference on smoking and health :a workshop Proceedings 1967 Sep 11-13 National cancer institute Edited by Ernest L. Wynder and Dietrich Hoffmann National cancer institute.Monograph, 28 National cancer institute,Bethesda,Md.,1968
Bioch 33.0992

SMOLENICE CASTLE 1969 Regulation of body fluid volumes by the kidney A Symposium on natriuretic hormone Edited by J.H. Cort and B. Lichardus Under the auspices of Ceskoslovenska akademie ved Karger,Basle, 1970
Inv Med 37.0268

SOCEITY FOR ENDOCRINOLOGY Endocrine genetics a symposium Proceedings Cambridge 1966 Mar 29-31 Edited by S.G. Spicket and J.G.M. Shire Society for endocrinology.Memoirs,15 Cambridge university press,Cambridge,1967
Pha 16.0093

SOCIAL AND GENETIC INFLUENCES ON LIFE AND DEATH a symposium Papers 1966 Sep Eugenics society Edited by Robert Platt and A.S. Parkes Eugenics society.Symposia, 3 Oliver and Boyd,Edinburgh;London,1967
Gen 34.1911

SOCIAL ASPECTS OF ECONOMIC DEVELOPMENT IN LATIN AMERICA Mexico City 1960 Dec 12-21 Vol 1: papers submitted to the expert working group... Expert working group on social aspects of economic development in Latin America Edited by Egbert de Vries and Jose Medina Echavarria Technology and society series Unesco,Paris,1963
Geog 13.5096

SOCIAL CHANGE IN MODERN AFRICA International African seminar 1st Studies Kampala 1959 Jan International African institute Edited by Aidan Southall Oxford university press,London,1965
Geog 13.4662

SOCIAL COMMUNICATION AMONG PRIMATES An International symposium on communication and social interactions in primates Montreal 1964 Dec 27-31 Edited by Stuart A. Altmann University of Chicago press,Chicago,1967
Bal 39.1627

The SOCIAL IMPACT OF MODERN BIOLOGY :an international conference Papers and discussions London 1970 Nov 26-28 British society for social responsibility in science Edited by Watson Fuller 256p Routledge and Kegan Paul,London,1971
WSM 43.0589

SOCIAL PSYCHOLOGY AT THE CROSSROADS :the university of Oklahoma lectures in social psychology Papers Norman,Okla. 1950 Apr 6-11 University of Oklahoma Edited by John H. Rohrer and Mazafer Sherif Harper,New York,1951
Psy 31.1363

SOCIAL RESEARCH AND RURAL LIFE IN CENTRAL AMERICA, MEXICO AND THE CARIBBEAN REGION :seminar Proceedings Mexico City 1962 Oct 17-27 Unesco,and,United Nations.Economic commission for Latin America Edited by Egbert de Vries Unesco,Paris,1966
Geog 13.5161

SOCIAL SCIENCE IN LATIN AMERICA Conference on Latin American studies Papers Rio de Janeiro 1965 Mar 29-31 Joint committee on Latin American studies Edited by Manuel Diegues and Bryce Wood Jointly sponsored by the Latin American center for research in the social sciences,Rio de Janeiro Columbia university press,New York,1967
Geog 13.5108

SOCIAL SCIENCE RESEARCH COUNCIL.COMMITTEE ON GENETICS AND BEHAVIOR Genetic diversity and human behaviour The Burg Wartenstein symposium 27th Papers Burg 1964 Sep 16-28 behavioral consequences of genetic differences in man Edited by G.N. Spuhler Viking fund publications in anthropology, 45 Aldine,Chicago,Ill.,1967
Gen 34.2032

SOCIAL SCIENCE RESEARCH COUNCIL.JOINT COMMITTEE ON LATIN AMERICAN STUDIES see JOINT COMMITTEE ON LATIN AMERICAN STUDIES

The SOCIAL SCIENCES:THEIR RELATIONS IN THEORY AND IN TEACHING :conference Report London 1935 Sep 22-29 Institute of sociology Sponsored jointly by the International student service Le Play House press,London,1936
Geog 13.1613

SOCIEDAD MATEMATICA MEXICANA Symposium internacional de topologia algebraica Mexico City 1956 Aug Universidad nacional de Mexico.Instituto de matematicas Instituto nacional de la investigacion cientifica,Mexico xii,334p 26cm UNESCO,Mexico City,1958 In memory of Witold Hurewicz
P. Math 2.0290

SOCIEDAD MEXICANA DE MECANICA DE SUELOS Congreso panamericano de mecanica de suelos y cimentaciones 1st Proceedings Mexico,D. F. 1959 Sep 7-12 Vol 1-3 3 vols Mexico,D.F.,1960
Eng 41.3178

SOCIEDADE BRASILEIRA DE GENETICA Reciniao Brasiliera de genetica humana Parana: 1958: Nov 10-15 Edited by N. Freire-Maice and A. Quelee-Solgrado port. Universidade do Parana,1959
Gen 34.1895

SOCIETA ITALIANA DELL'IRIS International symposium on iris 1st Report Florence 1963 May 14-18 Edited by Gian Luigi Sani and others Societa italiana dell'iris,Florence, 1963
BG 38.3303

SOCIETA ITALIANA DI EMATOLOGIA Congresso nazionale della societa italiana di ematologia 11th Atti Rome 1953 May 10-11 Edited by L. Villa and others Edizioni mediche e scientifiche,Rome,1953
Med 36.0341

SOCIETA ITALIANA DI EMATOLOGIA Congresso nazionale della societa italiana di ematologia 13th Atti Rome 1955 Oct 6-7 Edited by G. Dominici and others Edizioni mediche e scientifiche,Rome,1956
Med 36.0380

SOCIETA ITALIANA DI FISICA Aerodynamic phenomena in stellar atmospheres Symposium on cosmical gas dynamics held at the International school of physics "Enrico Fermi" 4th proceedings Varenna 1960 Aug 18-30 International astronomical union and International union of theoretical and applied mechanics Edited by R.N. Thomas and others International astronomical union symposium, 12 515p Zanichelli,Bologna,c1961 Reprinted from the 'Supplemento del nuovo cimento',vol.22,no.1,1961
Obs 6.1736

SOCIETA ITALIANA DI FISICA Astrofisica del plasma Scuola internazionale di fisica 'Enrico Fermi' 39 corso Rendiconti Varenna 1966 Jul 11-30 Edited by P.A. Sturrock Academic press,New York;London,1967
TA 15.0328

SOCIETA ITALIANA DI FISICA Astrofisica delle alte energie Scuola internazionale di fisica "Enrico Fermi" 35 corso rendiconti Varenna 1965 Jul.12-24 Edited by L. Gratton 463p Academic press,New York,1966
A Math 4.1169

SOCIETA ITALIANA DI FISICA Dispersione ed assorbimento del suono Scuola internazionale de fisica 'Enrico Fermi' 27 corso Rendiconti Varenna 1962 Aug 6-18 Edited by D. Sette Academic press,New York;London, 1963 Papers in English and French
Chem 18.0517

SOCIETA ITALIANA DI FISICA Elaborazione di dati offici de parte di organism e di macchine Scuola internazionale di fisica "Enrico Fermi" 43 corso Rendaconti Varenna 1968 Jul 15-27 Edited by W. Reichardt Academic press, New York;London,1969
An 32.5470

SOCIETA ITALIANA DI FISICA Evoluzione delle stelle Scuola internazionale di fisica'Enrico Fermi' 28 corso Rendiconti Varenna 1962 Aug 20-Sep 1 Edited by L. Gratton Academic press,New York;London,1963
A Math 4.1156

SOCIETA ITALIANA DI FISICA Molecular beams and reaction kinetics Scuola internazionale di fisica 'Enrico Fermi' 44 corso Rendiconti Varenna 1968 Jul 29-Aug 10 Edited by C. Schlier Academic press,New York, 1970
Chem 18.2814

SOCIETA ITALIANA DI FISICA Ottica quantistica Scuola internazionale di fisica "Enrico Fermi" 42 corso Rendiconti Varenna 1967 Jul 31-Aug 19 Edited by R.J. Glauber Bibliog.,Illus.port. xvi,159p 25cm Academic press,New York,1968
A Math 4.1599

SOCIETA ITALIANA DI FISICA Physics with intersecting storage rings Scuola internazionale di fisica 'Enrico Fermi' 46 corso Rendiconti Varenna 1969 Jun 16-26 Bibliog,photos 24cm Academic press,New York,1971
Cav 7.3012

SOCIETA ITALIANA DI FISICA Radiation damage nei solidi Scuola internazionale di fisica 'Enrico Fermi' 18 corso Rendiconti Ispra (Varese) 1960 Sep 5-24 Edited by D.S. Billington Academic press,New York;London, 1962 Contributions in English and French
Chem 18.1144

SOCIETA ITALIANA DI FISICA Radiation damage nei solidi Scuola internazionale di fisica "Enrico Fermi" 28 corso Rendiconti Ispra, Varese 1960 Sep 5-24 Edited by D.S. Billington Academic press,New York;London, 1962
Met 25.1579

SOCIETA ITALIANA DI FISICA Scuola internazionale di fisica 'Enrico Fermi' Rendiconti Varenna Corso 10-14;16-: 1959- Nicola Zanichelli,Bologna,1960-
Cav 7.2388

SOCIETA ITALIANA DI FISICA Scuola internazionale di fisica 'Enrico Fermi' 47 corso Rendiconti Varenna 1969 Jun 30-Jul 12 Edited by Rainer K. Sachs Academic press New Yorkp 1971cm
A Math 4.1788

SOCIETA ITALIANA DI FISICA Teoria dei plasma Scuola internazionale di fisica 'Enrico Fermi' 25 corso Rendiconti Varenna 1962 Jul 9-21 Edited by N.M. Rosenbluth Academic press,New York;London,1964
Eng 41.4527

SOCIETA ITALIANA DI FISICA Teoria del magnetismo nei metalli di transizione Scuola internazionale di fisica 'Enrico Fermi' 37 corso Rendiconti Varenna 1966 Jun 6-25 Edited by W. Marshall diagrs 455p 24cm Academic press,London,1967
Cav 7.2612

SOCIETA ITALIANA DI FISICA Teorie ergodiche Scuola internationale di fisica 'Enrico Fermi' 14 corso Rendiconti Varenna 1960 May 23-31 Edited by P. Caldirola Academic press, New York;London,1961
A Math 4.0950

SOCIETA ITALIANA DI FISICA Verifiche delle teorie gravitazionali Scuola internazionale di fisica "Enrico Fermi" 20 corso Rendiconti Varenna 1961 Jun 19-Jul 1 Edited by G. Moller Bibliog.,Illus. 264p 24cm Academic press,New York,1962
A Math 4.1555

SOCIETA ITALIANA DI FISICA weak interactions and high energy neutrino physics Scuola internationale di fisica 'Enrico Fermi'. 32 corso rendiconti Varenna 1964 Jun.15-27 Edited by T.D. Lee 334p Academic press,New York,1966
A Math 4.0845

SOCIETA ITALIANA DI FISICA Argomenti scelti di fisica delle particelle Scuola internazionale di fisica "Enrico Fermi" 41 corso Rendiconti Varenna 1957 Jul 17-29 Edited by J. Steinberger Bibliog.,Illus. vii,194p 24cm Academic press,New York,1967
A Math 4.1545

SOCIETA ITALIANA DI FISICA Astrofisica del plasma Scuola internazionale di fisica "Enrico Fermi" 39 corso Rendiconti Varenna 1966 Jul 11-30 Edited by P.A. Sturrock xvi,364p Academic press,New York; London,1967
A Math 4.1577

SOCIETA ITALIANA DI FISICA Astrofisica delle alte energie Scuola internazionale di fisica 'Enrico Fermi' 35 corso Rendiconti Varenna 1965 Jul 12-24 Edited by L. Gratton Academic press,New York;London,1966
TA 15.0314

SOCIETA ITALIANA DI FISICA Evoluzione delle stelle Scuola internazionale di fisica "Enrico Fermi" 28 corso rendiconti Varenna 1962 Aug 20-Sep.1 Edited by L. Gratton 488p Academic press,New York; London,1963
Obs 6.0685

SOCIETA ITALIANA DI FISICA Propieta ottiche dei solidi Scuola internazionale di fisica 'Enrico Fermi' 34 corso Rendiconti Varenna 1965 Jun 28-Jul 10 Edited by J. Tauc Academic press,New York;London,1966
TA 15.0443

SOCIETATIS JANOS BOLYAI.COLLOQUIA MATHEMATICA, 4 Colloquium on combinatorial theory Proceedings Balatonfured 1969 Aug 24 Bolyai Janos matematikai tarsulat Edited by P. Erdos and others 1201p 24cm 3 vols North-Holland,Amsterdam;London,1970
Math S 3.1656

SOCIETE BELGE DE BIOCHIMIE Methods of separation of subcellular structural components :a symposium Louvain 1962 May 11 Biochemical society Edited by J.K. Grant Biochemical society.Symposia, 23 Cambridge university press,Cambridge,1963
Radioth 35.0136

SOCIETE BELGE DE BIOCHIMIE Methods of separation of subcellular structural components :a symposium Louvain 1962 May 11 Edited by J.K. Grant Held in the University of Louvain.Physiological institute Biochemical society.Symposia, 23 Cambridge university press,Cambridge,1963 Organized by C.de Duve
Bioch 33.1385

SOCIETE BELGE D'ENDROCINOLOGIE Androgens in normal and pathological conditions Steroid hormones :a symposium 2nd Proceedings Ghent 1965 Jun 17-19 Edited by A. Vermeulen and D. Exley Excerpta medica. International congress series, 101 Excerpta medica,Amsterdam,1966
Inv Med 37.0276

SOCIETE DE CHIMIE PHYSIQUE Macromolecules helicoidales en solution Societe de chimie physique.Reunion 17e Paris 1967 May 2-5 Journal de chimie physique et de physico-chimie biologique, 65,no 1 Paris,1968
Bioch 33.0326

SOCIETE DE CHIMIE PHYSIQUE Surface chemistry a discussion at a joint meeting of the Societe de chimie physique and the Faraday society Papers Bordeaux 1947 Oct 5-9 Illus Butterworths scientific publications, London,1949 Papers in English and FRENCH
Radioth 35.0342

SOCIETE DE CHIMIE PHYSIQUE and FARADAY SOCIETY Surface chemistry discussion at a joint meeting...in honour of professor Henri Devaux Papers Bordeaux 1947 Oct 5-9 Butterworths,London,1949 Published as a special supplement to 'Research'
Chem 18.0546

SOCIETE DE CHIMIE PHYSIQUE.REUNION 17e Macromolecules helicoidales en solution Paris 1967 May 2-5 Societe de chimie physique Journal de chimie physique et de physico-chimie biologique, 65,no 1 Paris, 1968
Bioch 33.0326

SOCIETE DE NEPHROLOGY International congress of nephrology 1st Proceedings Geneva 1960 Sep 1-4 Evian 1960 Edited by G. Richet Karger,Basle;New York,1961 In English and French
Phys 20.1325

SOCIETE DE NEURO-CHIRURGIE DE LANGUE FRANCAISE Les Cranio-pharyngiomes :rapport presente a la reunion annelle de la Societe de neuro-chirurgie de langue francaise By J. Rougerie and M. Fardeau Masson,Paris,1962
An 32.4204

SOCIETE DE PSYCHOLOGIE PHYSIOLOGIQUE DE PARIS Congres international de psychologie physiologique :premiere session 1st Compte-rendu Paris 1889 Bureau des revues,Paris,1890 Congresses continued as 'International congress of experimental psychology' q.v.
Psy 31.3390

SOCIETE FRANCAISE DE GYNECOLOGIE Hypoplasie et malformations de l'appareil genital interne de la femme Papers Strasbourg 1964 May 15-18 Edited by P. Muller Masson,Paris, 1964
An 32.3832

SOCIETE FRANCAISE DE METALLURGIE Plutonium 1960 International conference on plutonium metallurgy 2nd Proceedings Grenoble 1960 Apr 19-22 Edited by Emmanuel Grison and others Cleaver-Hume,London,1961 Papers in English and French
Met 25.1549

SOCIETE FRANCAISE DE METALLURGIE Semaine d'etude de la physique des metaux 1952 formation du grain dans les metaux par recristallisation Editions metaux,Paris,1952
Met 25.1262

SOCIETE FRANCAISE DE PHYSIQUE Interactions des electrons,phonons et magnons avec les surfaces cristallines :colloque Exposes et communications presentes Lille 1969 Sep 17-19 Journal de physique, 31,no 4,supplement, Colloque,C- 1 Paris,1970
Met 25.2627

SOCIETE FRANCAISE DES ELECTRONICIENS ET RADIOELECTRICIENS International congress on microwave tubes 5th Proceedings Paris 1964 Sponsored by the Federation nationale des industries electroniques Academic press,New York,1965
Eng 41.5574

SOCIETE FRANCAISE DES INGENIEURS ET TECHNICIENS DU VIDE International congress on microwave tubes 5th Proceedings Paris 1964 Sponsored by the Federation nationale des industries electroniques Academic press, New York,1965
Eng 41.5574

SOCIETE INTERNATIONALE D'HISTOIRE DE LA MEDECINE. CONGRE 12E Congres international d'histoire des sciences et congres de la Societe internationale d'histoire de la medecine Actes Amsterdam 1950 Aug 14-21 Union internationale d'histoire des sciences Genootschap voor geschiedenis der geneeskunde, wiskunde en natuurweteenschappen te Leiden Academie internationale d'histoire des sciences.Collection de travaux, 6 2 vols Academie d'histoire des sciences;Hermann,Paris, 1951-53 The 12th congress of the Societe internationale d'histoire de la medecine is section 4 of the 6th Congres international d'histoire des sciences
WSM 43.0067

SOCIETE MATHEMATIQUE DE FRANCE Structures algebriques et structures topologiques : conferences a l'Institut Henri Poincare de la Sorbonne Paris 1956 Feb.9 and 1957 Jun. 6 By H. Cartan and others en accord avec L' Association des Professeurs de mathematique de l'enseignement public Enseignement mathematique.Monographies, 7 198p Geneva, Paris,1958
P. Math 2.0241

SOCIETE MATHEMATIQUE DU CANADA Promenades aleatoires et mouvement brownien :introduction a la theorie des probabilites mimeograph Montreal 1963 By Anatole Joffe North Atlantic treaty organization Montreal. University.Seminaire de mathematiques superieures, 7 vii,143p 28cm Universite de Montreal,Montreal,1964
P. Math 2.2145

SOCIETE MATHEMATIQUE DU CANADA Series de Fourier aleatoires mimeograph Montreal 1963 By Jean-Pierre Kahane North Atlantic treaty organization Montreal.University. Seminaire de mathematiques superieures, 4 vii,174p 28cm Universite de Montreal, Montreal,1963
P. Math 2.2147

SOCIETE MATHEMATIQUE DU CANADA Integration dans les groupes topologiques Montreal 1964 By Geoffrey Fox North Atlantic treaty organization Montreal.University.Seminaire de mathematiques superieures, 12 357p 28cm Presses de l'universite de Montreal, Montreal,1966
P. Math 2.2385

SOCIETE MATHEMATIQUE DU CANADA Categories non-abeliennes Montreal 1964 By Peter Hilton North Atlantic treaty organization Montreal.University.Seminaire de mathematiques superieures, 10 138p 28cm Universite de Montreal,Montreal,1964
P. Math 2.0265

SOCIETE MATHEMATIQUE DU CANADA Fondements de la geometrie algebrique moderne Montreal 1964 By Jean Dieudonne North Atlantic treaty organization University of Montreal. Seminaire de mathematiques superieures, 8 151p 28cm Universite de Montreal, Montreal, 1964
P. Math 2.0507

SOCIETE MATHEMATIQUE DU CANADA Homotopie et cohomologie manuscript Montreal 1964 By Beno Eckmann North Atlantic treaty organization University of Montreal. Seminaire de mathematiques superieures, 11 129p 28cm Universite de Montreal, Montreal, 1965
P. Math 2.0297

SOCIETE MATHEMATIQUE DU CANADA Logical systems containing only a finite number of symbols Montreal 1966 By Leon Henkin North Atlantic treaty organization Universite de Montreal. Seminaire de mathematiques superieures, 21 illus. 48p 28cm Presses de l'Universite de Montreal, Montreal, 1967
P Math 2.3740

SOCIETE MATHEMATIQUE DU CANADA Problemes aux limites dans les equations aux derivees partielles Montreal 1962 By Jacques L. Lions Montreal. University. Seminaire de mathematiques superieures, 1 176p 28cm Universite de Montreal, Montreal, 1965
P. Math 2.1283

SOCIETE MATHEMATIQUE DU CANADA Theorie des valuations mimeograph Montreal 1964 By Paulo Ribenboim North Atlantic treaty organization 313p 28cm Presses de l'universite de Montreal, Montreal, 1964
P. Math 2.1974

SOCIETE NATIONALE D'HORTICULTURE DE FRANCE Conference international de genetique 4e Comptes rendus Paris 1911 Sep 18-23 Edited by Ph. de Vilmorin 2 vols Masson, Paris, 1913
Gen 34.1000

SOCIETE NATIONALE D'HORTICULTURE DE FRANCE Iris cultivees Conference internationale des iris 1ere Actes et comptes-rendus Paris 1922 Societe nationale d'horticulture, Paris, 1923
BG 38.2574

SOCIETE NATIONALE POUR L'ETUDE DE LA STERILITE ET DE LA FECONDITE La Fonction spermatogenetique du testicule humain : colloque Paris 1958 Jul 10-12 Edited by H. Bayle and C. Gouygou Masson, Paris, 1958
An 32.3872

SOCIETE NATIONALE POUR L'ETUDE DE LA STERILITE ET DE LA FECONDITE La Grossesse extra-uterine :colloque Rapport Paris 1961 Jun 2-4 Edited by Paul Funck-Brentano Masson, Paris, 1961
An 32.3824

SOCIETE NATIONALE POUR L'ETUDE DE LA STERILITE ET DE LA FECONDITE Les Fonctions de nidation uterine et leurs troubles :colloque Papers Brussels 1960 Jun 24-26 Edited by J. Ferin and M. Gaudefroy Masson, Paris, 1960
An 32.3834

SOCIETE NATIONALE POUR L'ETUDE DE LA STERILITE ET DE LA FECONDITE Les Fonctions du col uterin :colloque Papers Basle 1964 Jul 3-5 Edited by R. Moricard and R. Wenner Masson, Paris, 1964
An 32.3833

SOCIETY FOR ANALYTICAL CHEMISTRY The Determination of gases in metals :a symposium Report Organised in conjunction with the Institute of metals Iron and steel institute. Special report, 68 Iron and steel institute, London, 1960
Met 25.1747

SOCIETY FOR ANALYTICAL CHEMISTRY and INSTITUTE OF PETROLEUM.GAS CHROMATOGRAPHY DISCUSSION GROUP Gas chromatography 1960 : a symposium 3rd Proceedings Edinburgh 1960 Jun 8-10 Edited by R.P.W. Scott Butterworths, London, 1960
Chem 18.0567

SOCIETY FOR APPLIED SPECTROSCOPY Colloquium spectroscopicum internationale 13th Proceedings Ottawa 1967 Jun 19-23 Canadian association for applied spectroscopy Carleton university xix, 460p 23cm Hilger, London, 1968
Chem 18.2845

SOCIETY FOR AUTOMOTIVE ENGINEERS Symposium on motor lubricants New York 1933 Mar 8 American society for testing materials, Philadelphia, Pa., 1933
Eng 41.6264

SOCIETY FOR CELLULAR CHEMISTRY.SYMPOSIA, 14, Suppl. Intracellular membraneous structure The International symposium for cellular chemistry 1st Proceedings Ohtsu 1963 Mar 27-31 Japan society for cell biology Edited by S. Seno and E.V. Cowdry Okayama, 1964 In memory of...Dr. Seizo Katsunuma.Japan society for cell biology formerly Japan society for cellular chemistry
Bioch 33.0966

SOCIETY FOR DEVELOPMENTAL BIOLOGY Communication in development :a symposium Proceedings Edited by A. Lang Society for developmental biology.Symposia, 28 Developmental biology.Supplement, 3 Academic press, New York, 1969
Gen 34.2155

SOCIETY FOR DEVELOPMENTAL BIOLOGY Communication in developpment :symposium Boulder, Colo. 1969 Jun 16-18 Society for developmental biology.Symposium, 28 Academic press, New York, 1969
Bioch 33.2255

SOCIETY FOR DEVELOPMENTAL BIOLOGY Control mechanisms in developmental processes :a symposium La Jolla, Calif 1967 Jun Edited by Michael Locke Society for developmental biology.Symposia, 26 Developmental biology. Supplement, 1 Academic press, New York, 1967
Bioch 33.0911

SOCIETY FOR DEVELOPMENTAL BIOLOGY Control mechanisms in developmental processes :a symposium Proceedings Edited by Michael Locke Society for developmental biology. Symposia, 26 Developmental biology. Supplement, 1 Academic press, New York; London, 1967
Gen 34.2153

SOCIETY FOR DEVELOPMENTAL BIOLOGY Current status of some major problems in developmental biology :symposium Haverford,Penn. 1966 Jun Edited by Michael Locke Society for developmental biology.Symposia, 25 Academic press,New York;London,1966 Formerly the Society for the study of development and growth
Gen 34.0487

SOCIETY FOR DEVELOPMENTAL BIOLOGY Emergence of order in developing systems :a symposium Proceedings Edited by Michael Locke Society for developmental biology.Symposia, 27 Developmental biology.Supplement, 2 Academic press,New York;London,1968
Gen 34.2154

SOCIETY FOR DEVELOPMENTAL BIOLOGY Major problems in developmental biology :a symposium Haverford,Pa. 1966 Jun Edited by Michael Locke Society for developmental biology. Symposia, 25 Academic press,New York;London, 1966
Bioch 33.0912

SOCIETY FOR DEVELOPMENTAL BIOLOGY Reproduction;molecular,subcellular and cellular :a symposium Carleton,Minn. 1965 Jun Edited by Michael Locke Society for developmental biology.Symposia, 24 Academic press,London,1965 The Society for developmental biology formerly known as the Society for the study of development
Bioch 33.0900

SOCIETY FOR DEVELOPMENTAL BIOLOGY Reproduction molecular,subcellular and cellular :symposium Crelton,Minn. 1965 Jun Edited by Michael Locke Society for developmental biology.Symposia, 24 Academic press,New York;London,1965
Gen 34.0485

SOCIETY FOR DEVELOPMENTAL BIOLOGY The Emergence of order in developing systems Ithaca,N.Y. 1968 Jun Edited by Michael Locke Society for developmental biology. Symposia, 27 350p Academic press,New York, 1968
Bot 42.1097

SOCIETY FOR DEVELOPMENTAL BIOLOGY.SYMPOSIA, 24 Reproduction;molecular,subcellular and cellular :a symposium Carleton,Minn. 1965 Jun Society for developmental biology Edited by Michael Locke Academic press, London,1965 The Society for developmental biology formerly known as the Society for the study of development
Bioch 33.0900

SOCIETY FOR DEVELOPMENTAL BIOLOGY.SYMPOSIA, 24 Reproduction molecular,subcellular and cellular :symposium Crelton,Minn. 1965 Jun Society for developmental biology Edited by Michael Locke Academic press,New York;London, 1965
Gen 34.0485

SOCIETY FOR DEVELOPMENTAL BIOLOGY.SYMPOSIA, 25 Current status of some major problems in developmental biology :symposium Haverford, Penn. 1966 Jun Society for developmental biology Edited by Michael Locke Academic press,New York;London,1966 Formerly the Society for the study of development and growth
Gen 34.0487

SOCIETY FOR DEVELOPMENTAL BIOLOGY.SYMPOSIA, 25 Major problems in developmental biology :a symposium Haverford,Pa. 1966 Jun Society for developmental biology Edited by Michael Locke Academic press,New York;London,1966
Bioch 33.0912

SOCIETY FOR DEVELOPMENTAL BIOLOGY.SYMPOSIA, 26 Control mechanisms in developmental processes :a symposium La Jolla,Calif 1967 Jun Society for developmental biology Edited by Michael Locke Developmental biology.Supplement, 1 Academic press,New York,1967
Bioch 33.0911

SOCIETY FOR DEVELOPMENTAL BIOLOGY.SYMPOSIA, 26 Control mechanisms in developmental processes :a symposium Proceedings Society for developmental biology Edited by Michael Locke Developmental biology.Supplement, 1 Academic press,New York;London,1967
Gen 34.2153

SOCIETY FOR DEVELOPMENTAL BIOLOGY.SYMPOSIA, 26, supp.1 Control mechanisms in developmental processes :a symposium La Jolla,Calif. 1967 Jun Edited by Michael Locke Academic press,New York,1967 Formerly the Society for the study of development and growth
An 32.2535

SOCIETY FOR DEVELOPMENTAL BIOLOGY.SYMPOSIA, 27 Emergence of order in developing systems :a symposium Proceedings Society for developmental biology Edited by Michael Locke Developmental biology.Supplement, 2 Academic press,New York;London,1968
Gen 34.2154

SOCIETY FOR DEVELOPMENTAL BIOLOGY.SYMPOSIA, 27 The Emergence of order in developing systems Ithaca,N.Y. 1968 Jun Society for developmental biology Edited by Michael Locke 350p Academic press,New York,1968
Bot 42.1097

SOCIETY FOR DEVELOPMENTAL BIOLOGY.SYMPOSIA, 28 Communication in development :a symposium Boulder,Colo. 1969 Jun 16-18 Society for developmental biology Edited by Anton Lang Developmental biology.Supplement, 3 Academic press,New York;London,1969
An 32.5441

SOCIETY FOR DEVELOPMENTAL BIOLOGY.SYMPOSIA, 28 Communication in development :a symposium Proceedings Society for developmental biology Edited by A. Lang Developmental biology.Supplement, 3 Academic press,New York,1969
Gen 34.2155

SOCIETY FOR DEVELOPMENTAL BIOLOGY.SYMPOSIUM, 28 Communication in developpment :symposium Boulder,Colo. 1969 Jun 16-18 Society for developmental biology Academic press,New York,1969
Bioch 33.2255

SOCIETY FOR ENDOCRINOLOGY An International symposium on neurosecretion Proceedings Edited by H. Heller and R.B. Clark Society for endocrinology.Memoirs, 12 Academic press,London;New York,1962
Inv Med 37.0107

SOCIETY FOR ENDOCRINOLOGY Cell mechanism in hormone production and action :a symposium Proceedings London 1960 May 3-4 Edited by P.C. Williams and C.R. Austin Society for endocrinology.Memoirs, 11 Cambridge university press,Cambridge,1961
Phys 20.1363

SOCIETY FOR ENDOCRINOLOGY Cell mechanisms in hormone production and action :a symposium Proceedings London 1960 May 3-4 Edited by P.C. Williams and C.R. Austin Society for endocrinology.Memoirs,11 Cambridge university press,Cambridge,1961
Pha 16.0091

SOCIETY FOR ENDOCRINOLOGY Endocrine genetics a symposium Proceedings Cambridge 1966 Mar 29-31 Edited by S.G. Spickett and J.G.M. Shire Society for endocrinology.Memoirs, 15 Cambridge university press,Cambridge,1967
Gen 34.0555

SOCIETY FOR ENDOCRINOLOGY Endocrine genetics a symposium Proceedings Cambridge 1966 Mar 29-30 Edited by S.G. Spickett and J.G.M. Shire Society for endocrinology.Memoir, 15 Cambridge university press,Cambridge,1967
Inv Med 37.0203

SOCIETY FOR ENDOCRINOLOGY Endogenous substances affecting the myometrium :a symposium Proceedings Bristol 1965 Jul 19-20 Edited by V.R. Pickles and R.J. Fitzpatrick Society for endocrinology. Memoirs, 14 Cambridge university press, Cambridge,1966
Inv Med 37.0180

SOCIETY FOR ENDOCRINOLOGY Endogenous substances affecting the myometrium :a symposium Proceedings Bristol 1965 Jul 19-20 Edited by V.R. Pickles and R.J. Fitzpatrick Society for endocrinology. Memoirs,14 Cambridge university press, Cambridge,1966
Pha 16.0083

SOCIETY FOR ENDOCRINOLOGY Hormones and the kidney Society for endocrinology :a meeting 89th Proceedings Cambridge 1962 Sep Edited by Peter C. Williams Society for endocrinology.Memoirs, 13 Academic press, London;New York,1963
Inv Med 37.0233

SOCIETY FOR ENDOCRINOLOGY Hormones and the kidney :89th meeting Proceedings Cambridge 1962 Sep Edited by Peter C. Williams Society for endocrinology.Memoirs, 13 Academic press,London;New York,1963
Gen 34.0557

SOCIETY FOR ENDOCRINOLOGY Implantation of ova :a conference Proceedings London 1957 Nov 27 Edited by P. Eckstein Society for endocrinology.Memoirs, 6 Cambridge university press,Cambridge,1959
Phys 20.1383

SOCIETY FOR ENDOCRINOLOGY Implantation of ova :a conference Proceedings London 1957 Nov 27 Edited by P. Eckstein held at the Ciba foundation Society for endocrinology.Memoirs, 6 Cambridge university press,Cambridge,1959
Bal 39.1401

SOCIETY FOR ENDOCRINOLOGY Progress in endocrinology Proceedings Edinburgh 1959 Aug 16-20 Vol 1-2 Edited by K. Fotherby and others Society for endocrinology.Memoirs, 9-10 2 vols Cambridge university press,Cambridge,1964
Inv Med 37.0082

SOCIETY FOR ENDOCRINOLOGY Progress in endocrinology The Edinburgh meeting on endocrinology Proceedings Edinburgh 1959 Aug 16-20 Pt 1-2 Edited by K. Fotherby and others Society for endocrinology.Memoirs, 9-10 2 vols Cambridge university press,Cambridge,1960-1961
Bioch 33.0484

SOCIETY FOR ENDOCRINOLOGY Quantitative paper chromatography of steroids :a conference Proceedings London 1958 Jul 1 Edited by D. Abelson and R.V. Brooks Society for endocrinology.Memoirs, 8 Cambridge university press,Cambridge,1960
Bioch 33.2355

SOCIETY FOR ENDOCRINOLOGY Sex differentiation and development :a symposium Proceedings London 1958 Apr 10-11 Edited by C.R. Austin Society for endocrinology. Memoirs, 7 Cambridge university press, Cambridge,1960
Gen 34.0583

SOCIETY FOR ENDOCRINOLOGY Sex differentiation and development :a symposium Proceedings London 1958 Apr 10-11 Edited by C.R. Austin held at the Royal society of medicine Society for endocrinology.Memoirs, 7 Cambridge university press,Cambridge,1960
Bal 39.1402

SOCIETY FOR ENDOCRINOLOGY The Comparative endocrinology of vertebrates :a conference Proceedings Liverpool 1954 Jul 12-16 Part 1: the comparative physiology of reproduction and the effects of sex hormones in vertebrates Edited by I.Chester Jones and P. Eckstein Society for endocrinology. Memoirs, 4 Cambridge university press, Cambridge,1955
Phys 20.1378

SOCIETY FOR ENDOCRINOLOGY The Comparative endocrinology of vertebrates :a conference Proceedings Liverpool 1954 Jul 12-16 Part 2: the hormonal control of water and salt electrolyte metabolism in vertebrates Edited by I.Chester Jones and P. Eckstein Society for endocrinology.Memoirs, 5 Cambridge university press,Cambridge,1955
Phys 20.1379

SOCIETY FOR ENDOCRINOLOGY The Comparative endocrinology of vertebrates :a conference Proceedings Liverpool 1954 Jul 12-16 Pt 1: the comparative physiology of reproduction and the effects of sex hormones in vertebrates Edited by I.Chester Jones and P. Eckstein held at the University of Liverpool.Department of zoology Society for endocrinology.Memoirs, 4 Cambridge university press,Cambridge, 1955
Bal 39.1400

2013

SOCIETY FOR ENDOCRINOLOGY The Determination of adrenocortical steroids and their metabolites Proceedings London 1953 May 21 Edited by P. Eckstein and S. Zuckermann Society for endocrinology.Memoirs Dobson, London,1953
Med 36.0126

SOCIETY FOR ENDOCRINOLOGY The Determination of adrenocortical steroids and their metabolites :a conference Proceedings London 1953 May 21 Edited by P. Eckstein and S. Zuckerman Society for endocrinology. Memoirs, 2 Cambridge university press, Cambridge,1955
Radioth 35.1940

SOCIETY FOR ENDOCRINOLOGY The Gas liquid chromatography of steroids :a symposium Proceedings Glasgow 1966 Apr 4-6 Edited by J.K. Grant Society for endocrinology. Memoirs, 16 Cambridge university press, Cambridge,1967
Gen 34.0556

SOCIETY FOR ENDOCRINOLOGY The Investigation of hypothalamic-pituitary-adrenal function :a symposium Proceedings London 1967 Feb 22-23 Edited by V.H.T. James and J. Landon Society for endocrinology.Memoirs, 17 Cambridge university press,Cambridge,1968
Inv Med 37.0120

SOCIETY FOR ENDOCRINOLOGY The Investigation of hypothalamic-pituitary-adrenal function :a symposium Proceedings London 1967 Feb 22-23 Edited by V.H.T. James and J. Landon Sponsored also by the Royal society of medicine.Endocrine section Society for endocrinology.Memoirs, 17 Cambridge university press,Cambridge,1968
Phys 20.1421

SOCIETY FOR ENDOCRINOLOGY The Investigation of hypothalamic-pituitary-adrenal function :a symposium Proceedings London 1967 Feb 22-23 Edited by V.H.T. James and J. Landon Sponsored by Ciba,Horsham Society for endocrinology.Memoirs, 17 Cambridge university press,Cambridge,1968
Gen 34.0558

SOCIETY FOR ENDOCRINOLOGY The Technique and significance of oestrogen :a conference proceedings London 1954 Feb 17 Edited by P. Eckstein and S. Zuckerman Society for endocrinology.Mémoirs, 3 Cambridge university press,Cambridge,1955
Radioth 35.0045

SOCIETY FOR ENDOCRINOLOGY The Technique and significance of oestrogen determinations :a conference Proceedings London 1954 Feb 17 Edited by P. Eckstein and S. Zuckerman Society for endocrinology.Memoirs, 3 Cambridge university press,Cambridge,1955
Phys 20.1384

SOCIETY FOR ENDOCRINOLOGY The Thyroid gland : a symposium Proceedings London 1953 Feb 25 Edited by P. Eckstein and S. Zuckerman Society for endocrinology.Memoirs, 1 1953
Radioth 35.0713

2014

SOCIETY FOR ENDOCRINOLOGY 89th :a meeting Proceedings Hormones and the kidney Cambridge 1962 Sep Society for endocrinology Edited by Peter C. Williams Society for endocrinology.Memoirs, 13 Academic press,London;New York,1963
Inv Med 37.0233

SOCIETY FOR ENDOCRINOLOGY.MEMOIR, 15 Endocrine genetics a symposium Proceedings Cambridge 1966 Mar 29-30 Society for endocrinology Edited by S.G. Spickett and J. G.M. Shire Cambridge university press, Cambridge,1967
Inv Med 37.0203

SOCIETY FOR ENDOCRINOLOGY.MEMOIRS, 12 An International symposium on neurosecretion Proceedings Society for endocrinology Edited by H. Heller and R.B. Clark Academic press,London;New York,1962
Inv Med 37.0107

SOCIETY FOR ENDOCRINOLOGY.MEMOIRS, 13 Hormones and the kidney Society for endocrinology :a meeting 89th Proceedings Cambridge 1962 Sep Society for endocrinology Edited by Peter C. Williams Academic press,London;New York,1963
Inv Med 37.0233

SOCIETY FOR EXPERIMENTAL BIOLOGY Active transport and secretion Bangor 1953 Jul Society for experimental biology.Symposia, 8 Cambridge university press,Cambridge,1954
Phys 20.0958

SOCIETY FOR EXPERIMENTAL BIOLOGY Active transport and secretion :a symposium Bangor 1953 Jul Society for experimental biology. Symposia, 8 Cambridge university press, Cambridge,1954
Bioch 33.1353

SOCIETY FOR EXPERIMENTAL BIOLOGY Active transport and secretion :a symposium Papers Bangor 1953 Jul Edited by R. Brown and J.F. Danielli Society for experimental biology. Symposia, 8 Cambridge university press, Cambridge,1954
Inv Med 37.0036

SOCIETY FOR EXPERIMENTAL BIOLOGY Aspects of cell motility :a symposium Oxford 1967 Sep Society for experimental biology.Symposia, 22 Cambridge university press,Cambridge,1968
Bioch 33.1361

SOCIETY FOR EXPERIMENTAL BIOLOGY Aspects of the biology of ageing :a symposium Sheffield 1966 Sep Edited by H.W. Woolhouse Society for experimental biology.Symposia, 21 Cambridge university press,Cambridge,1967
An 32.5482

SOCIETY FOR EXPERIMENTAL BIOLOGY Aspects of the biology of ageing :a symposium Sheffield 1966 Sep 5-9 Society for experimental biology.Symposia, 21 Cambridge university press,Cambridge,1967
Bioch 33.1360

SOCIETY FOR EXPERIMENTAL BIOLOGY Biological receptor mechanisms Birmingham 1962 Sep 10-16 Society for experimental biology.Symposia, 16 Cambridge university press,Cambridge, 1962
An 32.4562

SOCIETY FOR EXPERIMENTAL BIOLOGY Biological receptor mechanisms Contributions Birmingham 1962 Sep 10-16 Edited by J.W.L. Beament Society for experimental biology. Symposia, 16 Cambridge university press, Cambridge, 1962
Psy 31.0283

SOCIETY FOR EXPERIMENTAL BIOLOGY Carbon dioxide fixation and photosynthesis Sheffield 1950 Jul Society for experimental biology.Symposia, 5 Cambridge university press,Cambridge,1951
Phys 20.0953

SOCIETY FOR EXPERIMENTAL BIOLOGY Carbon dioxide fixation and photosynthesis :a symposium Sheffield 1950 Jul Society for experimental biology.Symposia, 5 Cambridge university press,Cambridge,1951
Bioch 33.1351

SOCIETY FOR EXPERIMENTAL BIOLOGY Cell differentiation Edinburgh 1962 Sep 3-8 Society for experimental biology.Symposia, 17 Cambridge university press,Cambridge,1963
Bioch 33.1359

SOCIETY FOR EXPERIMENTAL BIOLOGY Control of organelle development :a symposium London 1969 Sep Society for experimental biology. Symposia, 24 Cambridge university press, Cambridge,1970
Bioch 33.2340

SOCIETY FOR EXPERIMENTAL BIOLOGY Control of organelle development :a symposium Proceedings London 1969 Sep 8-12 Edited by P.L. Miller Society for experimental biology.Symposia, 24 Cambridge university press,Cambridge,1970
Gen 34.2205

SOCIETY FOR EXPERIMENTAL BIOLOGY Dormancy and survival :a symposium Norwich 1968 Sep 2-6 Society for experimental biology. Symposia, 23 Cambridge university press, Cambridge,1969
Bioch 33.1362

SOCIETY FOR EXPERIMENTAL BIOLOGY Dormancy and survival :a symposium Papers Norwich 1968 Sep 2-6 Society for experimental biology.Symposia, 23 Cambridge university press,Cambridge,1969
Gen 34.2224

SOCIETY FOR EXPERIMENTAL BIOLOGY Dormancy and survival :symposium Papers Norwich 1968 Sep 2-6 Edited by H.W. Woolhouse Society for experimental biology.Symposia, 23 illus vi,598p 23cm Cambridge university press,Cambridge,1969
Sco 14.8232

SOCIETY FOR EXPERIMENTAL BIOLOGY Evolution Oxford 1952 Jul Society for experimental biology.Symposia, 7 Cambridge university press,Cambridge,1953
Phys 20.0957

SOCIETY FOR EXPERIMENTAL BIOLOGY Evolution : a symposium Papers Oxford 1952 Jul Society for experimental biology.Symposia, 7 Cambridge university press,Cambridge,1953
Psy 31.0570

SOCIETY FOR EXPERIMENTAL BIOLOGY Evolution symposium 7th Oxford 1952 Jul. Edited by R. Brown and J.F. Danielli Supported by the British council Society for experimental biology.Symposia, 7 xix,448p 25cm Cambridge university press,Cambridge, 1953
Math 3.0773

SOCIETY FOR EXPERIMENTAL BIOLOGY Evolution symposium Oxford 1952 Jul Society for experimental biology.Symposia, 7 Cambridge university press,Cambridge,1953
Gen 34.1258

SOCIETY FOR EXPERIMENTAL BIOLOGY Fibrous proteins and their biological significance :a symposium Leeds 1954 Sep Society for experimental biology.Symposia, 9 Cambridge university press,Cambridge,1955
Bioch 33.1354

SOCIETY FOR EXPERIMENTAL BIOLOGY Growth :in relation to differentiation and morphogenesis Cambridge 1947 Jul Society for experimental biology.Symposia, 2 Cambridge university press,Cambridge,1958
Phys 20.0943

SOCIETY FOR EXPERIMENTAL BIOLOGY Growth in relation to differentiation and morphogenesis. A symposium Cambridge 1947 Jul Society for experimental biology.Symposia, 2 Cambridge university press,Cambridge,1948
Bioch 33.1349

SOCIETY FOR EXPERIMENTAL BIOLOGY Growth,in relation to differentiation and morphogenesis Cambridge 1947 Jul Society for experimental biology.Symposia, 2 Cambridge university press,Cambridge,1948
Chem 18.2652

SOCIETY FOR EXPERIMENTAL BIOLOGY Homeostasis and feedback mechanisms Cambridge 1963 Sep Society for experimental biology.Symposia, 18 Cambridge university press,Cambridge,1964
Phys 20.0962

SOCIETY FOR EXPERIMENTAL BIOLOGY Mechanisms in biological competition symposium Southampton 1960 Sep Society for experimental biology.Symposia, 15 Cambridge university press,Cambridge,1961
Gen 34.1348

SOCIETY FOR EXPERIMENTAL BIOLOGY Mitochondria and other cytoplasmic inclusions : a symposium Oxford 1955 Sep Society for experimental biology.Symposia, 10 Cambridge university press,Cambridge,1957
Bioch 33.1355

SOCIETY FOR EXPERIMENTAL BIOLOGY Models and analogues in biology Bristol 1960 Sep 6-12 Edited by J.W.L. Beament Society for experimental biology.Symposia, 14 Cambridge university press,Cambridge,1960
Phys 20.0939

SOCIETY FOR EXPERIMENTAL BIOLOGY Nervous and hormonal mechanisms of integration St. Andrews 1965 Sep Society for experimental biology.Symposia, 20 Cambridge university press,Cambridge,1966
Psy 31.0137

SOCIETY FOR EXPERIMENTAL BIOLOGY Nucleic acid Cambridge 1946 Jul Society for experimental biology.Symposia, 1 Cambridge university press,Cambridge,1947
Phys 20.0942

SOCIETY FOR EXPERIMENTAL BIOLOGY Nucleic acid :a symposium Cambridge 1946 Jul Society for experimental biology.Symposia, 1 Cambridge university press,Cambridge,1947
Bioch 33.1348

SOCIETY FOR EXPERIMENTAL BIOLOGY Nucleic acid :a symposium Cambridge 1946 Jul Society for experimental biology.Symposia, 1 Cambridge university press,Cambridge,1947
Bot 42.1722

SOCIETY FOR EXPERIMENTAL BIOLOGY Nucleic acid :a symposium 1 Papers Cambridge 1946 Jul Society for experimental biology. Symposia,1 Cambridge university press, Cambridge,1947
Chem 18.1554

SOCIETY FOR EXPERIMENTAL BIOLOGY Nucleic acid :symposium Cambridge 1946 Jul Society for experimental biology.Symposia, 1 Cambridge university press,Cambridge,1947
Gen 34.0683

SOCIETY FOR EXPERIMENTAL BIOLOGY Nucleic acid symposium 1st Papers Cambridge 1946 Jul Society for experimental biology. Symposia, 1 plates 290p Cambridge university press,Cambridge,1947
Cav 7.0459

SOCIETY FOR EXPERIMENTAL BIOLOGY Physiological mechanisms in animal behaviour Cambridge 1949 Jul Edited by J.F. Danielli and R. Brown Society for experimental biology.Symposia, 4 Cambridge university press,Cambridge,1950
An 32.4327

SOCIETY FOR EXPERIMENTAL BIOLOGY Physiological mechanisms in animal behaviour Cambridge 1949 Jul Society for experimental biology.Symposia, 4 Cambridge university press,Cambridge,1950
Phys 20.0945

SOCIETY FOR EXPERIMENTAL BIOLOGY Physiological mechanisms in animal behaviour Papers Cambridge 1949 Jul Society for experimental biology.Symposia, 4 Cambridge university press,Cambridge,1950
Psy 31.0562

SOCIETY FOR EXPERIMENTAL BIOLOGY Selective toxicity and antibiotics :a symposium Edinburgh 1948 Jul Society for experimental biology.Symposia, 3 Cambridge university press,Cambridge,1949
Bioch 33.1350

SOCIETY FOR EXPERIMENTAL BIOLOGY Structural aspects of cell physiology Bristol 1951 Jul Society for experimental biology. Symposia, 6 Cambridge university press, Cambridge,1952
Phys 20.0954

SOCIETY FOR EXPERIMENTAL BIOLOGY Structural aspects of cell physiology :a symposium Bristol 1951 Jul Society for experimental biology.Symposia, 6 Cambridge university press,Cambridge,1952
Bioch 33.1352

SOCIETY FOR EXPERIMENTAL BIOLOGY Structural aspects of cell physiology :a symposium papers Bristol 1951 Jul Society for experimental biology.Symposia,6 Cambridge university press,Cambridge,1952
VA 19.0361

SOCIETY FOR EXPERIMENTAL BIOLOGY The Biological action of growth substances Cambridge 1956 Sep Society for experimental biology.Symposia, 11 Cambridge university press,Cambridge,1956
Bioch 33.1356

SOCIETY FOR EXPERIMENTAL BIOLOGY The Biological action of growth substances :a symposium Proceedings Aberystwyth 1956 Sep Society for experimental biology. Symposia, 11 Cambridge university press, Cambridge,1957
Radioth 35.0217

SOCIETY FOR EXPERIMENTAL BIOLOGY The Biological action of growth substances : symposium Aberystwyth 1956 Sep Society for experimental biology.Symposia, 11 Cambridge university press,Cambridge,1957
An 32.0089

SOCIETY FOR EXPERIMENTAL BIOLOGY The Biological replication of macromolecules :a symposium Society for experimental biology. Symposia, 12 Cambridge university press, Cambridge,1958
Bioch 33.1357

SOCIETY FOR EXPERIMENTAL BIOLOGY The Biological replication of macromolecules : symposium Society for experimental biology. Symposia, 12 Cambridge university press, Cambridge,1958
Gen 34.0736

SOCIETY FOR EXPERIMENTAL BIOLOGY The State and movement of water in living organisms Swansea 1964 Sep 8-12 Society for experimental biology.Symposia, 19 Cambridge university press,Cambridge,1965
Phys 20.0967

SOCIETY FOR EXPERIMENTAL BIOLOGY The State and movement of water in living organisms :a symposium Swansea 1964 Society for experimental biology.Symposia,19 Cambridge university press,Cambridge,1965
Pha 16.0272

SOCIETY FOR EXPERIMENTAL BIOLOGY Utilization of nitrogen and its compounds by plants :a symposium Reading 1958 Sep 15-19 Society for experimental biology.Symposia, 13 Cambridge university press,Cambridge,1959
Bioch 33.1358

SOCIETY FOR EXPERIMENTAL BIOLOGY.SYMPOSIA, 1 Nucleic acid :a symposium Cambridge 1946 Jul Society for experimental biology Cambridge university press,Cambridge,1947
Bot 42.1722

SOCIETY FOR EXPERIMENTAL BIOLOGY.SYMPOSIA, 1 Nucleic acid :a symposium Cambridge 1946 Jul Society for experimental biology Cambridge university press,Cambridge,1947
Bioch 33.1348

SOCIETY FOR EXPERIMENTAL BIOLOGY.SYMPOSIA, 1
Nucleic acid :a symposium Cambridge 1946 Jul Society for experimental biology Cambridge university press,Cambridge,1947
Radioth 35.0009

SOCIETY FOR EXPERIMENTAL BIOLOGY.SYMPOSIA, 1
Nucleic acid :symposium Cambridge 1946 Jul Society for experimental biology Cambridge university press,Cambridge,1947
Gen 34.0683

SOCIETY FOR EXPERIMENTAL BIOLOGY.SYMPOSIA, 2
Growth in relation to differentiation and morphogenesis.A symposium Cambridge 1947 Jul Society for experimental biology Cambridge university press,Cambridge,1948
Bioch 33.1349

SOCIETY FOR EXPERIMENTAL BIOLOGY.SYMPOSIA, 2
Growth,in relation to differentiation and morphogenesis Cambridge 1947 Jul Society for experimental biology Cambridge university press,Cambridge,1948
Chem 18.2652

SOCIETY FOR EXPERIMENTAL BIOLOGY.SYMPOSIA, 3
Selective toxicity and antibiotics Edinburgh 1948 Jul Society for experimental biology Cambridge university press,Cambridge,1949
Phys 20.2224

SOCIETY FOR EXPERIMENTAL BIOLOGY.SYMPOSIA, 3
Selective toxicity and antibiotics Edinburgh 1948 Jul Society for experimental biology Edited by J.F. Danielli and R. Brown Cambridge university press,Cambridge,1949
Radioth 35.1931

SOCIETY FOR EXPERIMENTAL BIOLOGY.SYMPOSIA, 3
Selective toxicity and antibiotics :a symposium Edinburgh 1948 Jul Society for experimental biology Cambridge university press,Cambridge,1949
Bioch 33.1350

SOCIETY FOR EXPERIMENTAL BIOLOGY.SYMPOSIA, 4
Physiological mechanisms in animal behaviour Cambridge 1949 Jul Society for experimental biology Edited by J.F. Danielli and R. Brown Cambridge university press, Cambridge,1950
An 32.4327

SOCIETY FOR EXPERIMENTAL BIOLOGY.SYMPOSIA, 4
Physiological mechanisms in animal behaviour Papers Cambridge 1949 Jul Society for experimental biology Cambridge university press,Cambridge,1950
Psy 31.0562

SOCIETY FOR EXPERIMENTAL BIOLOGY.SYMPOSIA, 5
Carbon dioxide fixation and photosynthesis :a symposium Sheffield 1950 Jul Society for experimental biology Cambridge university press,Cambridge,1951
Bioch 33.1351

SOCIETY FOR EXPERIMENTAL BIOLOGY.SYMPOSIA, 6
Structural aspects of cell physiology :a symposium Bristol 1951 Jul Society for experimental biology Cambridge university press,Cambridge,1952
Bioch 33.1352

SOCIETY FOR EXPERIMENTAL BIOLOGY.SYMPOSIA, 6
Structural aspects of cell physiology : symposium papers Bristol 1951 Jul Society for experimental biology Cambridge university press,Cambridge,1952
Radioth 35.0456

SOCIETY FOR EXPERIMENTAL BIOLOGY.SYMPOSIA, 7
Evolution :a symposium Papers Oxford 1952 Jul Society for experimental biology and Genetical society Cambridge university press,Cambridge,1953
Psy 31.0570

SOCIETY FOR EXPERIMENTAL BIOLOGY.SYMPOSIA, 7
Evolution symposium Oxford 1952 Jul Society for experimental biology Genetical society Cambridge university press,Cambridge, 1953
Gen 34.1258

SOCIETY FOR EXPERIMENTAL BIOLOGY.SYMPOSIA, 8
Active transport and secretion :a symposium Bangor 1953 Jul Society for experimental biology Cambridge university press,Cambridge, 1954
Bioch 33.1353

SOCIETY FOR EXPERIMENTAL BIOLOGY.SYMPOSIA, 8
Active transport and secretion :a symposium Papers Bangor 1953 Jul Society for experimental biology Edited by R. Brown and J.F. Danielli Cambridge university press, Cambridge,1954
Inv Med 37.0036

SOCIETY FOR EXPERIMENTAL BIOLOGY.SYMPOSIA, 9
Fibrous proteins and their biological significance :a symposium Leeds 1954 Sep Society for experimental biology Cambridge university press,Cambridge,1955
Bioch 33.1354

SOCIETY FOR EXPERIMENTAL BIOLOGY.SYMPOSIA, 9
Fibrous proteins and their biological significance :a symposium Papers Leeds 1954 Sep Society for experimental biology Edited by R. Brown and J.F. Danielli Cambridge university press,Cambridge,1955
An 32.5465

SOCIETY FOR EXPERIMENTAL BIOLOGY.SYMPOSIA, 10
Mitochondria and other cytoplasmic inclusions : a symposium Oxford 1955 Sep Society for experimental biology Cambridge university press,Cambridge,1957
Bioch 33.1355

SOCIETY FOR EXPERIMENTAL BIOLOGY.SYMPOSIA, 10
Mitochondria and other cytoplasmic inclusions : a symposium Papers Oxford 1955 Sep Cambridge university press,Cambridge,1957
An 32.3392

SOCIETY FOR EXPERIMENTAL BIOLOGY.SYMPOSIA, 10
Mitochondria and other cytoplasmic inclusions : a symposium Papers Oxford 1955 Sep Society for experimental biology Cambridge university press,Cambridge,1957
Radioth 35.0472

SOCIETY FOR EXPERIMENTAL BIOLOGY.SYMPOSIA, 11
The Biological action of growth substances Cambridge 1956 Sep Society for experimental biology Cambridge university press,Cambridge,1956
Bioch 33.1356

SOCIETY FOR EXPERIMENTAL BIOLOGY.SYMPOSIA, 11
The Biological action of growth substances :a symposium Proceedings Aberystwyth 1956 Sep Society for experimental biology Cambridge university press,Cambridge,1957
Radioth 35.0217

SOCIETY FOR EXPERIMENTAL BIOLOGY.SYMPOSIA, 11
The Biological action of growth substances : symposium Aberystwyth 1956 Sep Society for experimental biology Cambridge university press,Cambridge,1957
An 32.0089

SOCIETY FOR EXPERIMENTAL BIOLOGY.SYMPOSIA, 12
The Biological replication of macromolecules : a symposium Papers Aberystwyth 1956 Society for experimental biology Edited by F. K. Sanders Cambridge university press, Cambridge,1958
Radioth 35.0237

SOCIETY FOR EXPERIMENTAL BIOLOGY.SYMPOSIA, 12
The Biological replication of macromolecules : a symposium Society for experimental biology Cambridge university press,Cambridge,1958
Bioch 33.1357

SOCIETY FOR EXPERIMENTAL BIOLOGY.SYMPOSIA, 12
The Biological replication of macromolecules : symposium Society for experimental biology Cambridge university press,Cambridge,1958
Gen 34.0736

SOCIETY FOR EXPERIMENTAL BIOLOGY.SYMPOSIA, 13
Utilization of nitrogen and its compounds by plants :a symposium Reading 1958 Sep 15-19 Society for experimental biology Cambridge university press,Cambridge,1959
Bioch 33.1358

SOCIETY FOR EXPERIMENTAL BIOLOGY.SYMPOSIA, 14
Models and analogues in biology :a symposium Bristol 1960 Sep 6-12 Cambridge university press,Cambridge,1960
An 32.0058

SOCIETY FOR EXPERIMENTAL BIOLOGY.SYMPOSIA, 15
Mechanisms in biological competition symposium Southampton 1960 Sep Society for experimental biology Cambridge university press,Cambridge,1961
Gen 34.1348

SOCIETY FOR EXPERIMENTAL BIOLOGY.SYMPOSIA, 16
Biological receptor mechanisms Birmingham 1962 Sep 10-16 Society for experimental biology Cambridge university press,Cambridge, 1962
An 32.4562

SOCIETY FOR EXPERIMENTAL BIOLOGY.SYMPOSIA, 17
Cell differentiation Edinburgh 1962 Sep 3-8 Society for experimental biology Cambridge university press,Cambridge,1963
Bioch 33.1359

SOCIETY FOR EXPERIMENTAL BIOLOGY.SYMPOSIA, 17
Cell differentiation :a symposium Edinburgh 1962 Sep 3-8 Edited by G.E. Fogg Cambridge university press,Cambridge,1963
An 32.2500

SOCIETY FOR EXPERIMENTAL BIOLOGY.SYMPOSIA, 19
The State and movement of water in living organisms :a symposium Swansea 1964 Sep 8-12 Edited by G.E. Fogg Cambridge university press,Cambridge,1965
An 32.2522

SOCIETY FOR EXPERIMENTAL BIOLOGY.SYMPOSIA, 20
Nerves and hormonal mechanisms of integration : a symposium London 1965 Society for experimental biology Cambridge university press,Cambridge,1966
An 32.3785

SOCIETY FOR EXPERIMENTAL BIOLOGY.SYMPOSIA, 21
Aspects of the biology of ageing :a symposium Sheffield 1966 Sep Society for experimental biology Edited by H.W. Woolhouse Cambridge university press, Cambridge,1967
An 32.5482

SOCIETY FOR EXPERIMENTAL BIOLOGY.SYMPOSIA, 21
Aspects of the biology of ageing :a symposium Sheffield 1966 Sep 5-9 Society for experimental biology Cambridge university press,Cambridge,1967
Bioch 33.1360

SOCIETY FOR EXPERIMENTAL BIOLOGY.SYMPOSIA, 22
Aspects of cell motility :a symposium Oxford 1967 Sep Society for experimental biology Cambridge university press,Cambridge,1968
Bioch 33.1361

SOCIETY FOR EXPERIMENTAL BIOLOGY.SYMPOSIA, 22
Aspects of cell motility :a symposium Oxford 1967 Sep Society for experimental biology Cambridge university press,Cambridge,1968
An 32.5206

SOCIETY FOR EXPERIMENTAL BIOLOGY.SYMPOSIA, 22
Aspects of cell motility :a symposium Papers Oxford 1967 Sep Society for experimental biology Cambridge university press,Cambridge, 1968
Radioth 35.0473

SOCIETY FOR EXPERIMENTAL BIOLOGY.SYMPOSIA, 23
Dormancy and survival :a symposium Norwich 1968 Sep 2-6 Society for experimental biology Cambridge university press,Cambridge, 1969
Bioch 33.1362

SOCIETY FOR EXPERIMENTAL BIOLOGY.SYMPOSIA, 23
Dormancy and survival :a symposium Papers Norwich 1968 Sep 2-6 Society for experimental biology Cambridge university press,Cambridge,1969
Gen 34.2224

SOCIETY FOR EXPERIMENTAL BIOLOGY.SYMPOSIA, 23
Dormancy and survival :symposium Papers Norwich 1968 Sep 2-6 Society for experimental biology Edited by H.W. Woolhouse illus vi,598p 23cm Cambridge university press,Cambridge,1969
Sco 14.8232

SOCIETY FOR EXPERIMENTAL BIOLOGY.SYMPOSIA, 24
Control of organelle development :a symposium London 1969 Sep Society for experimental biology Cambridge university press,Cambridge, 1970
Bioch 33.2340

SOCIETY FOR EXPERIMENTAL BIOLOGY.SYMPOSIA, 24
Control of organelle development :a symposium Proceedings London 1969 Sep 8-12 Society for experimental biology Edited by P.L. Miller Cambridge university press,Cambridge, 1970
Gen 34.2205

SOCIETY FOR EXPERIMENTAL BIOLOGY.SYMPOSIA,6 Structural aspects of cell physiology :a symposium papers Bristol 1951 Jul Society for experimental biology Cambridge university press,Cambridge,1952
VA 19.0361

SOCIETY FOR EXPERIMENTAL BIOLOGY, 2 Gorwth in relation to differentiation and morphogenesis Cambridge 1947 Jul Society for experimental biology Edited by J.F. Danielli and R. Brown Cambridge university press,Cambridge,1948
An 32.5434

SOCIETY FOR EXPERIMENTAL BIOLOGY, 6 Structural aspects of cell physiology :a symposium Bristol 1951 Jul Society for experimental biology Edited by J.F. Danielli and R. Brown Cambridge university press, Cambridge,1952
An 32.5445

SOCIETY FOR EXPERIMENTAL BIOLOGY, 12 The Biological replication of macromolecules :a symposium Aberystwyth 1956 Sep Society for experimental biology Edited by K.F. Sanders Cambridge university press,Cambridge, 1957
An 32.5446

SOCIETY FOR EXPERIMENTAL BIOLOGY SYMPOSIA, 14 Models and analogues in biology Bristol 1960 Sep 6-12 Society for experimental biology vii,255p Cambridge university press,Cambridge,1960
WSM 43.2826

SOCIETY FOR EXPERIMENTAL STRESS ANALYSIS International congress on experimental mechanics 1st Proceedings New York 1961 Nov 1-3 Edited by B.E. Rossi Pergamon press,London,1963
Eng 41.2920

SOCIETY FOR EXPERIMENTAL STRESS ANALYSIS International congress on experimental mechanics 2nd Proceedings Washington,D. C. 1965 Sep 28-Oct 1 Edited by B.E. Rossi Society for experimental stress analysis, Westport,Conn.,1966
Eng 41.2921

SOCIETY FOR EXPERIMENTAL STRESS ANALYSIS Lectures and laboratory exercises on strain gage techniques Los Angeles,Calif. 1958 Aug 18-29 By William M. Murray and Peter K. Stein University of California Los Angeles, Los Angeles,Calif.,1958
Eng 41.2906

SOCIETY FOR GENERAL MICROBIOLOGY Adaptation in micro-organisms :symposium London 1953 Apr Society for general microbiology. Symposia, 3 Cambridge university press, Cambridge,1953
Gen 34.0694

SOCIETY FOR GENERAL MICROBIOLOGY Adaption in micro-organisms :a symposium Papers London 1953 Apr Edited by R. Davies and E.F. Gale Held at the Royal institution Society for general microbiology.Symposia, 3 Cambridge university press,Cambridge,1953
Bioch 33.1166

SOCIETY FOR GENERAL MICROBIOLOGY Airborne microbes :a symposium Papers London 1967 Apr Edited by P.H. Gregory and J.L. Monteith Held at the Imperial college of science and technology Society for general microbiology. Symposia, 17 Cambridge university press, Cambridge,1967
Bioch 33.1167

SOCIETY FOR GENERAL MICROBIOLOGY Autotrophic micro-organism :a symposium Papers London 1954 Apr Edited by B.A. Fry and J.L. Peel Held at the Institution of electrical engineers Society for general microbiology. Symposia, 4 Cambridge university press, Cambridge,1954
Bioch 33.1168

SOCIETY FOR GENERAL MICROBIOLOGY Bacterial anatomy :a symposium Papers London 1956 Apr Edited by E.T.C. Spooner and B.A.D. Stocker Held at the Royal institution Society for general microbiology.Symposia, 6 Cambridge university press,Cambridge,1956
Bioch 33.1169

SOCIETY FOR GENERAL MICROBIOLOGY Biochemical studies of antimicrobial drugs :a symposium Papers London 1966 Apr Edited by B.A. Newton and P.E. Reynolds Held at the Royal institution Society for general microbiology. Symposia, 16 Cambridge university press, Cambridge,1966
Bioch 33.1170

SOCIETY FOR GENERAL MICROBIOLOGY Biochemical studies of antimicrobial drugs :symposium London 1966 Apr Society for general microbiology.Symposia, 16 Cambridge university press,Cambridge,1966
Gen 34.0705

SOCIETY FOR GENERAL MICROBIOLOGY Biochemistry studies of antimicrobial drugs : symposium 16th London 1966 Apr Edited by B.A. Newton and P.E. Reynolds Society for general microbiology.Symposium,15 Cambridge university press,Cambridge,1966
Pha 16.0336

SOCIETY FOR GENERAL MICROBIOLOGY Function and structure in micro-organisms :a symposium Papers London 1965 Apr Held at the Middlesex hospital Society for general microbiology.Symposia, 15 Cambridge university press,Cambridge,1965
Bioch 33.1171

SOCIETY FOR GENERAL MICROBIOLOGY Function and structure in micro-organisms :symposium London 1965 Apr Edited by M.R. Pollock and M.H. Richmond Society for general microbiology.Symposium,15 Cambridge university press,Cambridge,1965
Pha 16.0162

SOCIETY FOR GENERAL MICROBIOLOGY Function and structure in micro-organisms :symposium London 1965 Apr Society for general microbiology.Symposia, 15 Cambridge university press,Cambridge,1965
Gen 34.0702

SOCIETY FOR GENERAL MICROBIOLOGY Mechanisms of microbial pathogenicity :a symposium Papers London 1955 Apr Edited by J.W. Howie and A.J. O'Hea Held at the Royal institution Society for general microbiology. Symposia, 5 Cambridge university press, Cambridge,1955
Bioch 33.1172

SOCIETY FOR GENERAL MICROBIOLOGY Mechanisms of microbial pathogenicity :a symposium Papers London 1955 Apr Edited by J.W. Howie and A.J. O'Hea Society for general microbiology.Symposia, 5 Cambridge university press,Cambridge,1955
Path 30.2608

SOCIETY FOR GENERAL MICROBIOLOGY Microbes and biological productivity :a symposium Papers London 1971 Apr Edited by David E. Hughes and Anthony H. Rose Society for general microbiology.Symposia, 21 Cambridge university press,Cambridge,1971
Bioch 33.2302

SOCIETY FOR GENERAL MICROBIOLOGY Microbial behaviour 'in vivo' and 'in vitro' :a symposium Papers London 1964 Apr Edited by H. Smith and Joan Taylor Held at the Royal institution Society for general microbiology.Symposia, 14 Cambridge university press,Cambridge,1964
Bioch 33.1173

SOCIETY FOR GENERAL MICROBIOLOGY Microbial behaviour 'in vivo' and 'in vitro' :a symposium Papers London 1964 Apr Edited by H. Smith and Joan Taylor Society for general microbiology.Symposia, 14 figs Cambridge university press,Cambridge,1964
Path 30.2610

SOCIETY FOR GENERAL MICROBIOLOGY Microbial behaviour in vivo and in vitro London 1964 Apr Society for general microbiology. Symposia, 14 Cambridge university press, Cambridge,1964
Gen 34.0700

SOCIETY FOR GENERAL MICROBIOLOGY Microbial classification :a symposium Papers London 1962 Apr Edited by G.C. Ainsworth and P.H.A. Sneath Held at the Royal institution Society for general microbiology.Symposia, 12 Cambridge university press,Cambridge,1962
Bioch 33.1174

SOCIETY FOR GENERAL MICROBIOLOGY Microbial classification :symposium London 1962 Apr Society for general microbiology.Symposia, 12 Cambridge university press,Cambridge,1962
Gen 34.0714

SOCIETY FOR GENERAL MICROBIOLOGY Microbial ecology :a symposium Papers London 1957 Apr Edited by R.E.O. Williams and C.C. Spicer Held at the Royal institution Society for general microbiology.Symposia, 7 Cambridge university press,Cambridge,1957
Bioch 33.1175

SOCIETY FOR GENERAL MICROBIOLOGY Microbial ecology :a symposium Papers London 1957 Apr Edited by R.E.O. Williams and C.C. Spicer Society for general microbiology. Symposia, 7 Cambridge,1957
Path 30.2585

SOCIETY FOR GENERAL MICROBIOLOGY Microbial genetics London 1960 Apr Society for general microbiology.Symposia, 10 Cambridge university press,Cambridge,1960 Two copies
Gen 34.0748

SOCIETY FOR GENERAL MICROBIOLOGY Microbial genetics :a symposium Papers London 1960 Apr Edited by W. Hayes and R.C. Clowes Held at the Royal institution Society for general microbiology.Symposia, 10 Cambridge university press,Cambridge,1960
Bioch 33.1176

SOCIETY FOR GENERAL MICROBIOLOGY Microbial genetics :a symposium Papers London 1960 Apr Edited by W. Hayes and R.C. Clowes Society for general microbiology.Symposia, 10 Cambridge university press,Cambridge,1960
Path 30.2605

SOCIETY FOR GENERAL MICROBIOLOGY Microbial growth Papers London 1969 Apr Society for general microbiology.Symposia, 19 Cambridge university press,Cambridge,1969
Path 30.2597

SOCIETY FOR GENERAL MICROBIOLOGY Microbial growth :symposium London 1969 Apr Society for general microbiology.Symposia, 19 Cambridge university press,Cambridge,1969
Gen 34.0710

SOCIETY FOR GENERAL MICROBIOLOGY Microbial reaction to environment :a symposium Papers London 1961 Apr Edited by G.C. Meynell and H. Gooder Held at the Royal institution Society for general microbiology.Symposia, 11 Cambridge university press,Cambridge,1961
Bioch 33.1177

SOCIETY FOR GENERAL MICROBIOLOGY Microbial reaction to environment :a symposium Papers London 1961 Apr Edited by G.G. Meynell and H. Gooder Society for general microbiology. Symposia, 11 pls Cambridge university press,Cambridge,1961
Path 30.2607

SOCIETY FOR GENERAL MICROBIOLOGY Molecular biology of viruses :a symposium Papers London 1968 Apr Edited by R.V. Crawford and M.G.P. Stoker Held at the Imperial college of science and technology Society for general microbiology.Symposia, 18 Cambridge university press,Cambridge,1968
Bioch 33.1178

SOCIETY FOR GENERAL MICROBIOLOGY Molecular biology of viruses :symposium London 1968 Apr Society for general microbiology. Symposia, 18 Cambridge university press, Cambridge,1968
Gen 34.0709

SOCIETY FOR GENERAL MICROBIOLOGY Organization and control in prokaryotic and eukaryotic cells :a symposium Papers Edited by H.P. Charles and B.C.J.G. Knight Held at Imperial college of science and technology Society for general microbiology. Symposia, 20 Cambridge university press, Cambridge,1970
Bioch 33.1919

SOCIETY FOR GENERAL MICROBIOLOGY Strategy of chemotherapy :a symposium Papers London 1958 Apr Edited by S.T. Cowan and Elizabeth Rowatt Held at the Royal institution Society for general microbiology.Symposia, 8 Cambridge university press,Cambridge,1958
Bioch 33.1180

SOCIETY FOR GENERAL MICROBIOLOGY Symbiotic associations :a symposium Papers London 1963 Apr Edited by P.S. Nutman and Barbara Mosse Society for general microbiology. Symposia, 13 Cambridge university press, Cambridge,1963
Bioch 33.1181

SOCIETY FOR GENERAL MICROBIOLOGY Symbiotic associations :a symposium Proceedings London Apr 1963 Society for general microbiology.Symposia, 13 Cambridge university press,Cambridge,1963
Gen 34.1355

SOCIETY FOR GENERAL MICROBIOLOGY The Molecular biology of viruses :a symposium Papers London 1968 Apr Edited by L.V. Crawford and M.G.P. Stoker Society for general microbiology.Symposia, 18 Cambridge university press for the Society of general microbiology,Cambridge,1968
Path 30.2742

SOCIETY FOR GENERAL MICROBIOLOGY The Molecular biology of viruses :a symposium Proceedings London 1968 Apr Edited by L. V. Crawford and M.G.P. Stoker Society for general microbiology.Symposia, 18 Cambridge university press,Cambridge,1968
Radioth 35.0268

SOCIETY FOR GENERAL MICROBIOLOGY The Nature of the bacterial surface :a symposium London 1949 Apr 20 Edited by A.A. Miles and N.W. Pirie illus. Blackwell,Oxford,1949
Gen 34.0697

SOCIETY FOR GENERAL MICROBIOLOGY The Nature of virus multiplication :a symposium Papers Oxford 1952 Apr Edited by Paul Fildes and W.E. Van Heyningen Held at the University of Oxford Society for general microbiology. Symposia, 2 Cambridge university press, Cambridge,1953
Bioch 33.1179

SOCIETY FOR GENERAL MICROBIOLOGY The Nature of virus multiplication :a symposium Proceedings Oxford 1952 Apr 16-17 Edited by Paul G. Fildes and W.E. Van Heyningen Society for general microbiology.Symposia, 2 Cambridge university press,Cambridge,1953
Radioth 35.0183

SOCIETY FOR GENERAL MICROBIOLOGY The Strategy of chemotherapy :a symposium Proceedings London 1958 Apr By S.T. Cowan and Elizabeth Rowatt Society for general microbiology.Symposia, 8 Cambridge university press,Cambridge,1958
Radioth 35.0225

SOCIETY FOR GENERAL MICROBIOLOGY The Strategy of chemotherapy :symposium London 1958 Apr Edited by S.T. Cowan and Elisabeth Rowatt Society for general microbiology. Symposium,8 Cambridge university press, Cambridge,1958
Pha 16.0262

SOCIETY FOR GENERAL MICROBIOLOGY Virus growth and variation :a symposium London 1958 Apr Edited by A. Isaacs and B.W. Lacey Society for general microbiology.Symposia, 9 Cambridge university press for the Society of general microbiology,Cambridge,1959
Path 30.2743

SOCIETY FOR GENERAL MICROBIOLOGY Virus growth and variation :a symposium Papers London 1959 Apr Edited by A. Isaacs and B. W. Lacey Held at the University of London Society for general microbiology.Symposia, 9 Cambridge university press,Cambridge,1959
Bioch 33.1182

SOCIETY FOR GENERAL MICROBIOLOGY 1st : symposium Proceedings Nature of the bacterial surface London 1949 Apr 20 Edited by A.A. Miles and N.W. Pirie Blackwell,Oxford,1949
Col S 12.0036

SOCIETY FOR GENERAL MICROBIOLOGY 6th : symposium Papers Bacterial anatomy London 1956 Apr Edited by E.T.C. Spooner and B.A.D. Stocker Cambridge university press,Cambridge,1956
Col S 12.0037

SOCIETY FOR GENERAL MICROBIOLOGY.SYMPOSIA, 2 The Nature of virus multiplication :a symposium Papers Oxford 1952 Apr Society for general microbiology Edited by Paul Fildes and W.E. Van Heyningen Held at the University of Oxford Cambridge university press,Cambridge,1953
Bioch 33.1179

SOCIETY FOR GENERAL MICROBIOLOGY.SYMPOSIA, 2 The Nature of virus multiplication :a symposium Proceedings Oxford 1952 Apr 16-17 Society for general microbiology Edited by Paul G. Fildes and W.E. Van Heyningen Cambridge university press,Cambridge,1953
Radioth 35.0183

SOCIETY FOR GENERAL MICROBIOLOGY.SYMPOSIA, 3 Adaptation in micro-organisms :symposium London 1953 Apr Society for general microbiology Cambridge university press, Cambridge,1953
Gen 34.0694

SOCIETY FOR GENERAL MICROBIOLOGY.SYMPOSIA, 3 Adaption in micro-organisms :a symposium Papers London 1953 Apr Society for general microbiology Edited by R. Davies and E.F. Gale Held at the Royal institution Cambridge university press,Cambridge,1953
Bioch 33.1166

SOCIETY FOR GENERAL MICROBIOLOGY.SYMPOSIA, 4 Autotrophic micro-organism :a symposium Papers London 1954 Apr Society for general microbiology Edited by B.A. Fry and J.L. Peel Held at the Institution of electrical engineers Cambridge university press,Cambridge,1954
Bioch 33.1168

SOCIETY FOR GENERAL MICROBIOLOGY.SYMPOSIA, 5 Mechanisms of microbial pathogenicity :a symposium Papers London 1955 Apr Society for general microbiology Edited by J. W. Howie and A.J. O'Hea Cambridge university press,Cambridge,1955
Path 30.2608

SOCIETY FOR GENERAL MICROBIOLOGY.SYMPOSIA, 5
Mechanisms of microbial pathogenicity :a symposium Papers London 1955 Apr Society for general microbiology Edited by J. W. Howie and A.J. O'Hea Held at the Royal institution Cambridge university press, Cambridge,1955
Bioch 33.1172

SOCIETY FOR GENERAL MICROBIOLOGY.SYMPOSIA, 6
Bacterial anatomy :a symposium Papers London 1956 Apr Society for general microbiology Edited by E.T.C. Spooner and B. A.D. Stocker Held at the Royal institution Cambridge university press,Cambridge,1956
Bioch 33.1169

SOCIETY FOR GENERAL MICROBIOLOGY.SYMPOSIA, 7
Microbial ecology :a symposium Papers London 1957 Apr Society for general microbiology Edited by R.E.O. Williams and C. C. Spicer Cambridge,1957
Path 30.2585

SOCIETY FOR GENERAL MICROBIOLOGY.SYMPOSIA, 7
Microbial ecology :a symposium Papers London 1957 Apr Society for general microbiology Edited by R.E.O. Williams and C. C. Spicer Held at the Royal institution Cambridge university press,Cambridge,1957
Bioch 33.1175

SOCIETY FOR GENERAL MICROBIOLOGY.SYMPOSIA, 8
Strategy of chemotherapy :a symposium Papers London 1958 Apr Society for general microbiology Edited by S.T. Cowan and Elizabeth Rowatt Held at the Royal institution Cambridge university press, Cambridge,1958
Bioch 33.1180

SOCIETY FOR GENERAL MICROBIOLOGY.SYMPOSIA, 8
The Strategy of chemotherapy :a symposium Proceedings London 1958 Apr By S.T. Cowan and Elizabeth Rowatt Society for general microbiology Cambridge university press,Cambridge,1958
Radioth 35.0225

SOCIETY FOR GENERAL MICROBIOLOGY.SYMPOSIA, 9
Virus growth and variation :a symposium London 1958 Apr Society for general microbiology Edited by A. Isaacs and B.W. Lacey Cambridge university press for the Society of general microbiology,Cambridge,1959
Path 30.2743

SOCIETY FOR GENERAL MICROBIOLOGY.SYMPOSIA, 9
Virus growth and variation :a symposium Papers London 1959 Apr Society for general microbiology Edited by A. Isaacs and B.W. Lacey Held at the University of London Cambridge university press,Cambridge,1959
Bioch 33.1182

SOCIETY FOR GENERAL MICROBIOLOGY.SYMPOSIA, 10
Microbial genetics London 1960 Apr Society for general microbiology Cambridge university press,Cambridge,1960 Two copies
Gen 34.0748

SOCIETY FOR GENERAL MICROBIOLOGY.SYMPOSIA, 10
Microbial genetics :a symposium Papers London 1960 Apr Society for general microbiology Edited by W. Hayes and R.C. Clowes Cambridge university press,Cambridge, 1960
Path 30.2605

SOCIETY FOR GENERAL MICROBIOLOGY.SYMPOSIA, 10
Microbial genetics :a symposium Papers London 1960 Apr Society for general microbiology Edited by W. Hayes and R.C. Clowes Held at the Royal institution Cambridge university press,Cambridge,1960
Bioch 33.1176

SOCIETY FOR GENERAL MICROBIOLOGY.SYMPOSIA, 11
Microbial reaction to environment :a symposium Papers London 1961 Apr Society for general microbiology Edited by G.C. Meynell and H. Gooder Held at the Royal institution Cambridge university press,Cambridge,1961
Bioch 33.1177

SOCIETY FOR GENERAL MICROBIOLOGY.SYMPOSIA, 11
Microbial reaction to environment :a symposium Papers London 1961 Apr Society for general microbiology Edited by G.G. Meynell and H. Gooder pls Cambridge university press,Cambridge,1961
Path 30.2607

SOCIETY FOR GENERAL MICROBIOLOGY.SYMPOSIA, 12
Microbial classification :a symposium Papers London 1962 Apr Society for general microbiology Edited by G.C. Ainsworth and P. H.A. Sneath Held at the Royal institution Cambridge university press,Cambridge,1962
Bioch 33.1174

SOCIETY FOR GENERAL MICROBIOLOGY.SYMPOSIA, 12
Microbial classification :symposium London 1962 Apr Society for general microbiology Cambridge university press,Cambridge,1962
Gen 34.0714

SOCIETY FOR GENERAL MICROBIOLOGY.SYMPOSIA, 13
Symbiotic associations :a symposium Papers London 1963 Apr Society for general microbiology Edited by P.S. Nutman and Barbara Mosse Cambridge university press, Cambridge,1963
Bioch 33.1181

SOCIETY FOR GENERAL MICROBIOLOGY.SYMPOSIA, 13
Symbiotic associations :a symposium Proceedings London Apr 1963 Society for general microbiology Cambridge university press,Cambridge,1963
Gen 34.1355

SOCIETY FOR GENERAL MICROBIOLOGY.SYMPOSIA, 14
Microbial behaviour 'in vivo' and 'in vitro' : a symposium Papers London 1964 Apr Society for general microbiology Edited by H. Smith and Joan Taylor Held at the Royal institution Cambridge university press, Cambridge,1964
Bioch 33.1173

SOCIETY FOR GENERAL MICROBIOLOGY.SYMPOSIA, 14
Microbial behaviour 'in vivo' and 'in vitro' : a symposium Papers London 1964 Apr Society for general microbiology Edited by H. Smith and Joan Taylor figs Cambridge university press,Cambridge,1964
Path 30.2610

SOCIETY FOR GENERAL MICROBIOLOGY.SYMPOSIA, 14
Microbial behaviour in vivo and in vitro London 1964 Apr Society for general microbiology Cambridge university press, Cambridge,1964
Gen 34.0700

SOCIETY FOR GENERAL MICROBIOLOGY.SYMPOSIA, 15
Function and structure in micro-organisms :a symposium Papers London 1965 Apr Society for general microbiology Held at the Middlesex hospital Cambridge university press,Cambridge,1965
Bioch 33.1171

SOCIETY FOR GENERAL MICROBIOLOGY.SYMPOSIA, 15
Function and structure in micro-organisms : symposium London 1965 Apr Society for general microbiology Cambridge university press,Cambridge,1965
Gen 34.0702

SOCIETY FOR GENERAL MICROBIOLOGY.SYMPOSIA, 16
Biochemical studies of antimicrobial drugs :a symposium Papers London 1966 Apr Society for general microbiology Edited by B. A. Newton and P.E. Reynolds Held at the Royal institution Cambridge university press, Cambridge,1966
Bioch 33.1170

SOCIETY FOR GENERAL MICROBIOLOGY.SYMPOSIA, 16
Biochemical studies of antimicrobial drugs : symposium London 1966 Apr Society for general microbiology Cambridge university press,Cambridge,1966
Gen 34.0705

SOCIETY FOR GENERAL MICROBIOLOGY.SYMPOSIA, 17
Airborne microbes :a symposium Papers London 1967 Apr Society for general microbiology Edited by P.H. Gregory and J.L. Monteith Held at the Imperial college of science and technology Cambridge university press,Cambridge,1967
Bioch 33.1167

SOCIETY FOR GENERAL MICROBIOLOGY.SYMPOSIA, 18
Molecular biology of viruses :a symposium Papers London 1968 Apr Society for general microbiology Edited by R.V. Crawford and M.G.P. Stoker Held at the Imperial college of science and technology Cambridge university press,Cambridge,1968
Bioch 33.1178

SOCIETY FOR GENERAL MICROBIOLOGY.SYMPOSIA, 18
Molecular biology of viruses :symposium London 1968 Apr Society for general microbiology Cambridge university press, Cambridge,1968
Gen 34.0709

SOCIETY FOR GENERAL MICROBIOLOGY.SYMPOSIA, 18
The Molecular biology of viruses :a symposium London 1968 Apr Society for general microbiology Cambridge university press, Cambridge,1968
PGMS 29.0416

SOCIETY FOR GENERAL MICROBIOLOGY.SYMPOSIA, 18
The Molecular biology of viruses :a symposium London 1968 Apr Society for general microbiology Edited by L.V. Crawford and M.G. P. Stoker Cambridge university press, Cambridge,1968
An 32.5380

SOCIETY FOR GENERAL MICROBIOLOGY.SYMPOSIA, 18
The Molecular biology of viruses :a symposium Papers London 1968 Apr Society for general microbiology Edited by L.V. Crawford and M.G.P. Stoker Cambridge university press for the Society of general microbiology, Cambridge,1968
Path 30.2742

SOCIETY FOR GENERAL MICROBIOLOGY.SYMPOSIA, 18
The Molecular biology of viruses :a symposium Proceedings London 1968 Apr Society for general microbiology Edited by L.V. Crawford and M.G.P. Stoker Cambridge university press, Cambridge,1968
Radioth 35.0268

SOCIETY FOR GENERAL MICROBIOLOGY.SYMPOSIA, 19
Microbial growth Papers London 1969 Apr Society for general microbiology Cambridge university press,Cambridge,1969
Path 30.2597

SOCIETY FOR GENERAL MICROBIOLOGY.SYMPOSIA, 19
Microbial growth :a symposium London 1969 Apr Society for general microbiology Edited by Pauline M. Meadow and S.J. Pirt Cambridge university press,Cambridge,1969
An 32.5379

SOCIETY FOR GENERAL MICROBIOLOGY.SYMPOSIA, 19
Microbial growth :symposium London 1969 Apr Society for general microbiology Cambridge university press,Cambridge,1969
Gen 34.0710

SOCIETY FOR GENERAL MICROBIOLOGY.SYMPOSIA, 20
Organization and control in prokaryotic and eukaryotic cells :a symposium London 1970 Apr Society for general microbiology Edited by H.P. Charles and B.C.J.G. Knight Cambridge university press,Cambridge,1970
An 32.5447

SOCIETY FOR GENERAL MICROBIOLOGY.SYMPOSIA, 20
Organization and control in prokaryotic and eukaryotic cells :a symposium Papers Society for general microbiology Edited by H. P. Charles and B.C.J.G. Knight Held at Imperial college of science and technology Cambridge university press,Cambridge,1970
Bioch 33.1919

SOCIETY FOR GENERAL MICROBIOLOGY.SYMPOSIA, 21
Microbes and biological productivity :a symposium Papers London 1971 Apr Society for general microbiology Edited by David E. Hughes and Anthony H. Rose Cambridge university press,Cambridge,1971
Bioch 33.2302

SOCIETY FOR GENERAL MICROBIOLOGY.SYMPOSIUM,8
The Strategy of chemotherapy :symposium London 1958 Apr Society for general microbiology Edited by S.T. Cowan and Elisabeth Rowatt Cambridge university press, Cambridge,1958
Pha 16.0262

SOCIETY FOR INDUSTRIAL AND APPLIED MATHEMATICS
Computers in algebra and number theory Symposium in applied mathematics Proceedings New York 1970 Mar 25-26 By Garrett Birkhoff and Marshall Hall SIAM-AMS proceedings, 4 vii,200p 26cm AMS, Providence,R.I.,1971
P Math 2.4476

SOCIETY FOR INDUSTRIAL AND APPLIED MATHEMATICS
Computing methods in optimization problems International conference on computing methods in optimization problems 2nd Papers San Remo 1968 Sep 9-13 Lecture notes in operations research and mathematical economics, 14 bibliog.,illus. Springer,Berlin,1968
Math S 3.1516

SOCIETY FOR INDUSTRIAL AND APPLIED MATHEMATICS Computing methods in optimization problems-2 International conference on computing methods in optimization problems 2nd Papers San Remo 1968 Sep 9-13 Edited by Lofti A. Zadeh and others Academic press,New York, 1969
Eng 41.5865

SOCIETY FOR INDUSTRIAL AND APPLIED MATHEMATICS Stochastic processes in mathematical physics and engineering symposium 16 proceedings New York 1963 Apr.30-May 2 American mathematical society Edited by Richard Bellman American mathematical society. Proceedings of symposia in applied mathematics, 16 viii,318p 25cm American mathematical society,Providence,R.I.,1964
Math 3.0567

SOCIETY FOR MORPHOLOGICAL RESEARCH New methods of thought and procedure Symposium on methodologies Contributions Pasadena, Calif. 1967 May 22-24 Edited by F. Zwicky and A.G. Wilson viii,338p Springer,Berlin, 1967
WSM 43.2639

SOCIETY FOR PHYSICAL CHEMISTRY Gas chromatography :a symposium 3rd Proceedings Edinburgh 1960 Jun 8-10 Institute of petroleum.Gas chromatography discussion group Edited by R.P.W. Scott xvii,466p Butterworths,London,1960
Chem E 24.0850

SOCIETY FOR RESEARCH INTO HIGHER EDUCATION Research into library services in higher education :papers presented at a conference London 1967 Nov 3 Society for research into higher education,London,1968
Eng 41.7669

SOCIETY FOR THE HISTORY OF TECHNOLOGY The Sorby centennial symposium on the history of metallurgy :a symposium Cleveland 1963 Oct 22-23 American society for metals and American institute of mining,metallurgical and petroleum engineers.Metallurgical society Edited by Cyril Stanley Smith Metallurgical society conferences, 27 Gordon and Breach, New York,1965
Met 25.2141

SOCIETY FOR THE STUDY OF DEVELOPMENT AND GROWTH Cell,organism and milieu :a symposium South Hadley,Mass. 1958 Jun 9-11 Edited by Dorothea Rudnick held at Mount Holyoke college Society for the study of development and growth.Symposia, 17 Ronald press,New York,1959
Bioch 33.1008

SOCIETY FOR THE STUDY OF DEVELOPMENT AND GROWTH Cellular mechanisms in differentiation and growth :A SYMPOSIUM Proceedings Amherst, Mass 1955 Jun 15-18 By M. Delbruck and others Edited by Dorothea Rudnick Society for the study of development and growth. Symposia, 14 illus. vii,236p Princeton university press,Princeton,N.J.,1956
Bot 42.1474

SOCIETY FOR THE STUDY OF DEVELOPMENT AND GROWTH Cellular membranes in development :a symposium Storrs,Conn. 1963 Jun 17-19 Edited by Michael Locke Society for the study of development and growth.Symposia, 22 Academic press,New York,1964
Bioch 33.0910

SOCIETY FOR THE STUDY OF DEVELOPMENT AND GROWTH Cellular membranes in development : symposium Storrs,Conn. 1963 Jun Edited by Michael Locke Society for the study of development and growth.Symposia, 22 Academic press,New York,1964
Gen 34.0480

SOCIETY FOR THE STUDY OF DEVELOPMENT AND GROWTH Cellular membranes in development : symposium Storrs,Conn. 1963 Jun 17-19 Edited by Michael Locke Society for the study of development and growth.Symposia, 22 illus. xvi,382p Academic press,New York; London,1964
Bot 42.4733

SOCIETY FOR THE STUDY OF DEVELOPMENT AND GROWTH Cytodifferentiation and macromolecular synthesis Asilomar,Calif. 1962 Jun Edited by Michael Locke Society for the study of development and growth.Symposia, 21 Academic press,New York;London,1963
Gen 34.0854

SOCIETY FOR THE STUDY OF DEVELOPMENT AND GROWTH Cytodifferentiation and macromolecular synthesis :a symposium Asilomar,Calif. 1962 Jun Edited by Michael Locke Society for the study of development and growth. Symposia, 21 Academic press,New York;London, 1963
Bioch 33.1009

SOCIETY FOR THE STUDY OF DEVELOPMENT AND GROWTH Cytodifferentiation and macromolecular synthesis :a symposium Papers Asilomar, Calif. 1962 Jun Edited by Michael Locke Society for the study of development and growth.Symposia, 21 Academic press,New York; London,1963
Bot 42.1496

SOCIETY FOR THE STUDY OF DEVELOPMENT AND GROWTH Developing cell systems and their control 1959 Edited by Dorothea Rudnick Society for the study of development and growth. Symposia, 18 Ronald press,New York,1960
Gen 34.0850

SOCIETY FOR THE STUDY OF DEVELOPMENT AND GROWTH Developing cell systems and their control : a symposium Madison,Wis. 1959 Edited by Dorothea Rudnick at the University of Wisconsin Society for the study of development and growth.Symposia, 18 Ronald press,New York,1960
Bioch 33.1010

SOCIETY FOR THE STUDY OF DEVELOPMENT AND GROWTH Developmental cytology Kingston,R.I. 1957 Jun 19-24 Edited by Dorothea Rudnick Society for the study of development and growth.Symposia, 16 Ronald press,New York, 1957
Gen 34.0849

SOCIETY FOR THE STUDY OF DEVELOPMENT AND GROWTH Dynamics of growth processes :a symposium Papers Massachusetts 1952 Jun Edited by Edgar J. Boell Society for the study of development and growth.Symposia, 11 Princeton university press,Princeton,N.J.,1954
Radioth 35.0197

SOCIETY FOR THE STUDY OF DEVELOPMENT AND GROWTH
Synthesis of molecular and cellular structure :a symposium Papers Waltham,Mass. 1960 Edited by Dorothea Rudnick Held at Brandeis university Society for the study of development and growth.Symposia, 19 Ronald press,New York,1961
Bioch 33.1012

SOCIETY FOR THE STUDY OF DEVELOPMENT AND GROWTH
The Role of chromosomes in development Amherst,Mass. 1964 Jun Edited by Michael Locke Society for the study of development and growth.Symposia, 23 illus. Academic press,New York;London,1964
Gen 34.0859

SOCIETY FOR THE STUDY OF DEVELOPMENT AND GROWTH
The Role of chromosomes in development :a symposium Amherst,Mass. 1964 Jun Edited by Michael Locke Society for the study of development and growth.Symposia, 23 Academic press,New York;London,1964 The Society for the study of development and growth known later as the Society for developmental biology
Bioch 33.1011

SOCIETY FOR THE STUDY OF DEVELOPMENT AND GROWTH. SYMPOSIA, 11 Dynamics of growth processes :a symposium Papers Massachusetts 1952 Jun Society for the study of development and growth Edited by Edgar J. Boell Princeton university press, Princeton,N.J.,1954
Radioth 35.0197

SOCIETY FOR THE STUDY OF DEVELOPMENT AND GROWTH. SYMPOSIA, 11 Dynamics of growth processes :a symposium Williamstown,Mass. 1952 Jun Edited by Edgar J. Boell Princeton university press,Princeton,N.J.,1954
An 32.2467

SOCIETY FOR THE STUDY OF DEVELOPMENT AND GROWTH. SYMPOSIA, 13 Aspects of synthesis and order in growth :a symposium Hanover,N.H. 1954 Jun 23-26 Edited by Dorothea Rudnick Princeton university press,Princeton,N.J.,1954
An 32.2482

SOCIETY FOR THE STUDY OF DEVELOPMENT AND GROWTH. SYMPOSIA, 14 Cellular mechanisms in differentiation and growth :A SYMPOSIUM Proceedings Amherst,Mass 1955 Jun 15-18 By M. Delbruck and others Society for the study of development and growth Edited by Dorothea Rudnick illus. vii,236p Princeton university press,Princeton,N.J.,1956
Bot 42.1474

SOCIETY FOR THE STUDY OF DEVELOPMENT AND GROWTH. SYMPOSIA, 14 Cellular mechanisms in differentiation and growth :a symposium Amherst,Mass. 1955 Jun 15-18 Edited by Dorothea Rudnick Princeton university press, Princeton,N.J.,1956
An 32.2478

SOCIETY FOR THE STUDY OF DEVELOPMENT AND GROWTH. SYMPOSIA, 16 Developmental cytology Kingston,R.I. 1957 Jun 19-24 Society for the study of development and growth Edited by Dorothea Rudnick Ronald press,New York, 1957
Gen 34.0849

SOCIETY FOR THE STUDY OF DEVELOPMENT AND GROWTH. SYMPOSIA, 17 Cell,organism and milieu :a symposium South Hadley,Mass. 1958 Jun 9-11 Edited by Dorothea Rudnick Ronald press,New York,1959
An 32.3226

SOCIETY FOR THE STUDY OF DEVELOPMENT AND GROWTH. SYMPOSIA, 17 Cell,organism and milieu :a symposium South Hadley,Mass. 1958 Jun 9-11 Society for the study of development and growth Edited by Dorothea Rudnick held at Mount Holyoke college Ronald press,New York, 1959
Bioch 33.1008

SOCIETY FOR THE STUDY OF DEVELOPMENT AND GROWTH. SYMPOSIA, 18 Developing cell systems and their control 1959 Society for the study of development and growth Edited by Dorothea Rudnick Ronald press,New York,1960
Gen 34.0850

SOCIETY FOR THE STUDY OF DEVELOPMENT AND GROWTH. SYMPOSIA, 18 Developing cell systems and their control :a symposium Madison,Wis. 1959 Society for the study of development and growth Edited by Dorothea Rudnick at the University of Wisconsin Ronald press,New York,1960
Bioch 33.1010

SOCIETY FOR THE STUDY OF DEVELOPMENT AND GROWTH. SYMPOSIA, 19 Synthesis of molecular and cellular structure :a symposium Papers Waltham,Mass. 1960 Society for the study of development and growth Edited by Dorothea Rudnick Held at Brandeis university Ronald press,New York,1961
Bioch 33.1012

SOCIETY FOR THE STUDY OF DEVELOPMENT AND GROWTH. SYMPOSIA, 21 Cytodifferentiation and macromolecular synthesis Asilomar,Calif. 1962 Jun Society for the study of development and growth Edited by Michael Locke Academic press,New York;London,1963
Gen 34.0854

SOCIETY FOR THE STUDY OF DEVELOPMENT AND GROWTH. SYMPOSIA, 21 Cytodifferentiation and macromolecular synthesis :a symposium Asilomar,Calif. 1962 Jun Edited by Michael Locke Academic press,London,1963
An 32.2501

SOCIETY FOR THE STUDY OF DEVELOPMENT AND GROWTH. SYMPOSIA, 21 Cytodifferentiation and macromolecular synthesis :a symposium Asilomar,Calif. 1962 Jun Society for the study of development and growth Edited by Michael Locke Academic press,New York;London, 1963
Bioch 33.1009

SOCIETY FOR THE STUDY OF DEVELOPMENT AND GROWTH. SYMPOSIA, 21 Cytodifferentiation and macromolecular synthesis :a symposium Papers Asilomar,Calif. 1962 Jun Society for the study of development and growth Edited by Michael Locke Academic press,New York;London, 1963
Bot 42.1496

SOCIETY FOR THE STUDY OF DEVELOPMENT AND GROWTH. SYMPOSIA, 22 Cellular membranes in development :a symposium Storrs,Conn. 1963 Jun 17-19 Society for the study of development and growth Edited by Michael Locke Academic press,New York,1964
Bioch 33.0910

SOCIETY FOR THE STUDY OF DEVELOPMENT AND GROWTH. SYMPOSIA, 22 Cellular membranes in development :symposium Storrs,Conn. 1963 Jun Society for the study of development and growth Edited by Michael Locke Academic press,New York,1964
Gen 34.0480

SOCIETY FOR THE STUDY OF DEVELOPMENT AND GROWTH. SYMPOSIA, 22 Cellular membranes in development :symposium Storrs,Conn. 1963 Jun 17-19 Society for the study of development and growth Edited by Michael Locke illus. xvi,382p Academic press,New York;London,1964
Bot 42.4733

SOCIETY FOR THE STUDY OF DEVELOPMENT AND GROWTH. SYMPOSIA, 22 Cellular membranes in development a symposium Proceedings Storrs, Conn. 1963 Jun Edited by Michael Locke Academic press,New York;London,1964
An 32.3295

SOCIETY FOR THE STUDY OF DEVELOPMENT AND GROWTH. SYMPOSIA, 23 The Role of chromosomes in development Amherst,Mass. 1964 Jun Society for the study of development and growth Edited by Michael Locke illus. Academic press,New York;London,1964
Gen 34.0859

SOCIETY FOR THE STUDY OF DEVELOPMENT AND GROWTH. SYMPOSIA, 23 The Role of chromosomes in development :a symposium Amherst,Mass. 1964 Jun Edited by Michael Locke Academic press,New York;London,1964
An 32.2506

SOCIETY FOR THE STUDY OF DEVELOPMENT AND GROWTH. SYMPOSIA, 23 The Role of chromosomes in development :a symposium Amherst,Mass. 1964 Jun Society for the study of development and growth Edited by Michael Locke Academic press,New York;London,1964 The Society for the study of development and growth known later as the Society for developmental biology
Bioch 33.1011

SOCIETY FOR THE STUDY OF DEVELOPMENT AND GROWTH. SYMPOSIA, 24 Reproduction:molecular, subcellular and cellular :a symposium Edited by Michael Locke Academic press,New York; London,1965
An 32.2526

SOCIETY FOR THE STUDY OF DEVELOPMENT AND GROWTH. SYMPOSIA, 25 Major problems in development biology :a symposium Edited by Michael Locke Academic press,London;New York, 1966
An 32.2529

SOCIETY FOR THE STUDY OF EVOLUTION Behaviour and evolution :conference 2nd Proceedings Princeton,N.J. 1956 Apr 30-May 5 American psychological association Edited by Anne Roe and George Gaylord Simpson 24cm New Haven, 1958
Philos 1.1827

SOCIETY FOR THE STUDY OF EVOLUTION The Problem of land connections across the south Atlantic ,with special reference to the Mesozoic :a symposium Proceedings New York 1949 Dec 28-29 Edited by Ernst Mayr American museum of natural history.Bulletin, 99,art.3 New York,1952
An 32.2158

SOCIETY FOR THE STUDY OF EVOLUTION.ANNUAL MEETING, 4TH Problem of land connections across the South Atlantic with special reference to the Mesozoic Symposium on the role of the South Atlantic basin in biogeography and evolution Proceedings New York 1949 Dec 28-29 By Maurice Ewing and others Edited by Ernst Mayr American museum of natural history.Bulletin, 99,3 American museum of natural history,New York,1952
Bot 42.2341

SOCIETY FOR THE STUDY OF FERTILITY Biology of reproduction in mammals :a symposium Proceedings Nairobi,Kenya 1968 Apr Edited by J.S. Perry and I.W. Rowlands Journal of reproduction and fertility. Supplement, 6 Blackwell,Oxford;Edinburgh, 1969
Bal 39.1404

SOCIETY FOR THE STUDY OF FERTILITY Comparative biology of reproduction in mammals symposium Proceedings London 1964 Nov 24-26 Edited by I.W. Rowlands Zoological society of London.Symposia, 15 Academic press,London,1966
An 32.3887

SOCIETY FOR THE STUDY OF FERTILITY Proceedings of the Society for the study of fertility Vol 1-10 Blackwell,Oxford, 1950 Vol 1-5:proceedings of the 1st-5th conferences of the society.Vol 6-10 called "Studies on fertility"
An 32.3791

SOCIETY FOR THE STUDY OF FERTILITY and ZOOLOGICAL SOCIETY OF LONDON Comparative biology of reproduction in mammals : international symposium Proceedings London 1964 Nov 24-26 Edited by I.W. Rowlands Zoological society of London.Symposia,15 Academic press,London,1966
VA 19.0088

SOCIETY FOR THE STUDY OF HUMAN BIOLOGY The Biology of survival :a symposium 13th Proceedings London 1964 May 7-8 Edited by O.G. Edholm Organized by the Physiological society,the society for the study of human biology,and the Zoological society of London Zoological society of London.Symposia, 13 Academic press,London, 1964
Phys 20.2132

SOCIETY FOR THE STUDY OF HUMAN BIOLOGY The Scope of physical anthropology and its place in academic studies :a symposium held at the Ciba foundation Papers London 1957 Nov 6 Edited by D.F. Roberts and J.S. Weiner Wenner-Gren foundation for anthropological research,London,1958
HE 27.0143

SOCIETY FOR THE STUDY OF HUMAN BIOLOGY.SYMPOSIA, 1 The Scope of physical anthropology and its place in academic studies :a symposium London 1957 Nov Edited by D.F. Roberts and J.S. Neiner Wenner-Gren foundation for anthropological research,London,1958
An 32.2880

SOCIETY FOR THE STUDY OF HUMAN BIOLOGY.SYMPOSIA, 2 Natural selection in human populations London 1958 Nov 8 Society for the study of human biology Edited by D.F. Roberts and G.A. Harrison Pergamon press,London,1959
Gen 34.1863

SOCIETY FOR THE STUDY OF HUMAN BIOLOGY.SYMPOSIA, 2 Natural selection in human populations a symposium London 1958 Nov 8 Edited by D.F. Roberts and G.A. Harrison Pergamon press,Oxford,1959
An 32.2881

SOCIETY FOR THE STUDY OF HUMAN BIOLOGY.SYMPOSIA, 3 Human growth :a symposium Papers London 1960 Edited by J.M. Tanner Pergamon press,Oxford,1960
An 32.2879

SOCIETY FOR THE STUDY OF HUMAN BIOLOGY.SYMPOSIA, 3 Human growth A symposium 1960 Society for the study of human biology Edited by G.M. Tanner Pergamon press,London, 1960
Gen 34.1917

SOCIETY FOR THE STUDY OF HUMAN BIOLOGY.SYMPOSIA, 4 Genetical variation in human population symposium 1961 Society for the study of human biology Edited by G.A. Harrison Pergamon press,London,1961
Gen 34.1896

SOCIETY FOR THE STUDY OF HUMAN BIOLOGY.SYMPOSIA, 4 Genetical variations in human populations :a symposium London 1960 Edited by G.A. Harrison Pergamon press, Oxford,1961
An 32.2882

SOCIETY FOR THE STUDY OF HUMAN BIOLOGY.SYMPOSIA, 5 Dental anthropology :a symposium London 1962 Edited by D.R. Brothwell Pergamon press,Oxford,1963
An 32.2883

SOCIETY FOR THE STUDY OF HUMAN BIOLOGY.SYMPOSIA, 6 Teaching and research in human biology a symposium London 1964 Apr 25 Edited by G.Ainsworth Harrison Pergamon press, Oxford,1964
An 32.2884

SOCIETY FOR THE STUDY OF HUMAN BIOLOGY.SYMPOSIA, 8 The Skeletal biology of earlier human populations :a symposium London 1965 Nov Edited by D.R. Brothwell Pergamon press, Oxford,1968
An 32.2885

SOCIETY FOR THE STUDY OF INBORN ERRORS OF METABOLISM Basic concepts of inborn errors and defects of steroid biosynthesis :a symposium 3rd Proceedings Birmingham 1965 Edited by K.S. Holt and D.N. Raine Livingstone,Edinburgh,1966
Inv Med 37.0271

SOCIETY FOR THE STUDY OF INBORN ERRORS OF METABOLISM.SYMPOSIA, 3 Basic concepts of inborn errors and defects of steroid biosynthesis Proceedings Birmingham 1965 Oct 1-2 Society for the study of inborn errors of metabolism Edited by K.S. Holt and D.N. Raine Livingstone,Edinburgh;London,1968
Gen 34.2000

SOCIETY OF AEROSPACE MATERIAL AND PROCESS ENGINEERS Filament winding conference Pasadena 1961 Mar 28-30 SAMPE,Azuza,Calif., n.d. Mimeographed
Met 25.0097

SOCIETY OF AUTOMATIVE ENGINEERS Vistas in astronautics Astronautics symposium 3rd Proceedings Los Angeles,Calif. 1960 Oct.12-14 United States.Air force.Office of scientific research Vistas in astronautics, 3 illus 266p Society of automotive engineers,New York,1960
Nap 11.0024

SOCIETY OF BIOLOGICAL PSYCHIATRY Biological psychiatry :the proceedings of the scientific sessions of the Society... San Francisco, Calif. 1958 May Vol 1 Edited by J. Masserman Grune and Stratton,New York;London, 1959
Psy 31.1624

SOCIETY OF CHEMICAL ENGINEERS,TOKYO The Preprints of the 25th anniversary congress 1961 Nov 6-14 306p Society of chemical engineers,Tokyo,1961
Chem E 24.1272

SOCIETY OF CHEMICAL INDUSTRY Industrial carbon and graphite papers read at the conference... London 1957 Sep 24-26 Society of chemical industry,London,1958
Met 25.0155

SOCIETY OF CHEMICAL INDUSTRY Modern chemistry in industry An International union of pure and applied chemistry symposium Proceedings Eastbourne 1968 Mar 11-14 Edited by J.G. Gregory London,1968
Bioch 33.1874

SOCIETY OF CHEMICAL INDUSTRY Sorption and transport processes in soils... :a symposium 1970 Society of chemical industry.Monograph, 37 Society of chemical industry,London,1970
Eng 41.8281

SOCIETY OF CHEMICAL INDUSTRY. ROAD AND BUILDING MATERIALS GROUP Symposium on particle size analysis London 1947 Feb 4 145p Institution of chemical engineers,London,1947 Supplement to Transactions,Institution of chemical engineers,vol.25,1947
Chem E 24.1431

SOCIETY OF CHEMICAL INDUSTRY.BRISTOL SECTION,and SOCIETY OF CHEMICAL INDUSTRY.COLLOID AND SURFACE CHEMISTRY GROUP Wetting :joint symposium Papers and discussion Bristol 1966 Sep 12-14 Society of chemical industry. Monographs,25 Society of chemical industry, London,1967
Col S 12.0147

SOCIETY OF CHEMICAL INDUSTRY.CHEMICAL ENGINEERING GROUP Fluidisation :a symposium Papers London 1963 Mar 11-12 Sponsored also by the Institute of petroleum 110p Society of chemical industry,London,1964
Chem E 24.1452

SOCIETY OF CHEMICAL INDUSTRY.CHEMICAL ENGINEERING GROUP The Scaling-up of chemical plant and processes :joint symposium Papers London 1957 May 28-29 By R.Edgeworth Johnstone and others Institution of chemical engineers,London,1957 Held jointly with the K.Instituut van ingenieurs and the K. Nederlandse chemische vereniging
Chem E 24.1234

SOCIETY OF CHEMICAL INDUSTRY.COLLOID AND SURFACE CHEMISTRY GROUP and SOCIETY OF CHEMICAL INDUSTRY.BRISTOL SECTION, Wetting :joint symposium Papers and discussion Bristol 1966 Sep 12-14 Society of chemical industry. Monographs,25 Society of chemical industry, London,1967
Col S 12.0147

SOCIETY OF CHEMICAL INDUSTRY.CORROSION GROUP International congress on metallic corrosion 1st London 1961 Apr 10-15 Butterworths,London,1962
Met 25.1893

SOCIETY OF CHEMICAL INDUSTRY.LONDON SECTION Detergents,wetting and emulsifying agents :a symposium Papers London 1948 Apr 5-6 48p Society of chemical industry,London,1950
Chem E 24.0923

SOCIETY OF CHEMICAL INDUSTRY.PESTICIDES GROUP Plant growth regulators :symposium Papers London 1968 Jan 8-9 Society of chemical industry.Monograph, 31 Society of chemical industry,London,1968
Bioch 33.2230

SOCIETY OF CHEMICAL INDUSTRY.PLASTICS AND POLYMERS GROUP The Physical properties of polymers :symposium Papers London 1958 Apr 15-17 Society of chemical industry. Monograph, 5 293p Society of chemical industry,London,1959
Chem E 24.1019

SOCIETY OF CHEMICAL INDUSTRY.SURFACE ACTIVITY GROUP Powders in industry :a symposium Papers London 1960 Sep 29-30 Society of chemical industry.Monograph v,447p Society of chemical industry,London,1961
Chem E 24.1445

SOCIETY OF CHEMICAL PHYSICISTS Deoxyribonucleic acid,structure,synthesis and function Annual reunion of the society of chemical physicists 11th Proceedings Col de Voza 1961 Jun 26 - Jul 1 Pergamon press, Oxford,1961
Radioth 35.0111

SOCIETY OF DIE CASTING ENGINEERS National die casting exposition and congress 1st Technical papers Detroit 1960 Nov 8-11 Society of die casting engineers,Detroit,1960
Met 25.0630

SOCIETY OF DYERS AND COLOURISTS Fibrous proteins Symposium on fibrous proteins Proceedings Leeds 1946 May 23-25 Edited by C.L. Bird Leeds,1946
Bioch 33.0503

SOCIETY OF DYERS AND COLOURISTS Fibrous proteins :a symposium Proceedings Leeds 1946 May 23-25 Edited by C.L. Bird figs, pls 27cm n.p.,c1946
Bal 44.4561

SOCIETY OF ECONOMIC PALAEONTOLOGISTS AND MINERALOGISTS Primary sedimentary structures and their hydrodynamic interpretation :a symposium papers Toronto 1964 May 18 Edited by Gerard V. Middleton Society of economic paleontologists and mineralogists.Special publication,12 bibliog. illus. map Society of economic paleontologists and mineralogists,Tulsa,Okla., 1965
Geol 8.1515

SOCIETY OF ECONOMIC PALEONTOLOGISTS AND MINERALOGISTS Finding ancient shorelines : a symposium papers Houston,Tex. 1953 Mar. 24 Edited by Jack L. Hough and Henry W. Menard Society of economic paleontologists and mineralogists.Special publication,3 illus. map Society of economic paleontologists and mineralogists,Tulsa,Oka., 1956
Geol 8.1487

SOCIETY OF ECONOMIC PALEONTOLOGISTS AND MINERALOGISTS and AMERICAN ASSOCIATION OF PETROLEUM GEOLOGISTS Classification of carbonate rocks :a symposium Papers Denver, Colo. 1961 Apr 27 Edited by William E. Ham American association of petroleum geologists. Memoir,1 Tulsa,Okla.,1962 Held under the joint auspices of the American association of petroleum geologists and the Society of economic paleontologists and mineralogists
Geog 13.0301

SOCIETY OF ENVIRONMENTAL ENGINEERS Machines for materials and environmental testing Symposium on techniques and equipment for environmental testing; symposium on developments in materials testing machine design :joint conference Manchester 1965 Sep 6-10 Under the aegis of the British national committee on materials Institution of mechanical engineers.Proceedings, 180,pt 3a I.M.E.,London,1966
Met 25.0891

SOCIETY OF EXPERIMENTAL BIOLOGY Nervous and hormonal mechanisms of integration :symposium Proceedings Edited by G.M. Hughes Society of experimental biology.Symposia,20 Cambridge university press,Cambridge,1966
Pha 16.0048

SOCIETY OF EXPERIMENTAL BIOLOGY.SYMPOSIA,20 Nervous and hormonal mechanisms of integration symposium Proceedings Society of experimental biology Edited by G.M. Hughes Cambridge university press,Cambridge,1966
Pha 16.0048

SOCIETY OF GENERAL PHYSIOLOGISTS Biogenic amines as physiological regulators :a symposium Woods Hole 1969 Aug 29-Sep 1 Edited by J.J. Blum Prentice-Hall,Englewood Cliffs,N.J.,1970
An 32.5479

SOCIETY OF GENERAL PHYSIOLOGISTS Control mechanism in respiration and fermentation :a symposium held during the meeting of the Society of general physiologists at the Marine biological laboratory Woods Hole,Mass. 1961 Sep By Harlyn O. Halvorson Edited by Barbara Wright Published for the American physiological society Ronald press,New York, 1963
Bot 42.1730

SOCIETY OF GENERAL PHYSIOLOGISTS Control mechanisms in cellular processes :a symposium Woods Hole,Mass 1960 By Sigmund R. Suskind Edited by David M. Bonner Ronald press,New York,1961 Held at the 7th annual meeting of the society
Bot 42.1492

SOCIETY OF GENERAL PHYSIOLOGISTS Molecular architecture in cell physiology :a symposium Woods Hole,Mass. 1964 Sep 8-11 Edited by Tern Hayashi and Andrew G. Szent-Gyorgyi Prentice-Hall,Englewood Cliffs,N.J.,1966 Held at the 11th annual meeting of the society
Gen 34.0668

SOCIETY OF GENERAL PHYSIOLOGISTS Subcellular particles :a symposium Woods Hole 1958 Jun 9-11 By T. Hayashi Ronald press,New York, 1959
Phys 20.1536

SOCIETY OF GENERAL PHYSIOLOGISTS Subcellular particles :a symposium Woods Hole,Mass. 1958 Jun 9-11 American physiological society Edited by Teru Hayashi Ronald press,New York, 1959
Gen 34.0828

SOCIETY OF GENERAL PHYSIOLOGISTS Subcellular particles :a symposium held during the meeting of the Society of general physiologists at the Marine biological laboratory Woods Hole,Mass. 1958 Jun 9-11 Edited by Teru Hayashi Illus The Ronald press,New York,1959
Radioth 35.0500

SOCIETY OF GENERAL PHYSIOLOGISTS Subcellular particles :a symposium held during the meeting of the Society of general physiologists at the Marine biological laboratory Woods Hole,Mass. 1958 Jun 9-11 Edited by Teru Hayashi Published for the American physiological society Ronald press,New York,1959
Bioch 33.1014

SOCIETY OF GENERAL PHYSIOLOGISTS The Cellular functions of membrane transport :a symposium Papers Woods Hole,Mass. 1963 Sep 4-7 Edited by J.F. Hoffman illus. viii,291p Prentice-Hall,New Jersey,1964 Held at the annual meeting of the society
Bot 42.1500

SOCIETY OF GENERAL PHYSIOLOGISTS The Cellular functions of membrane transport :a symposium Woods Hole,Mass. 1963 Sep 4-7 Edited by J.F. Hoffmann Prentice-Hall, Englewood Cliffs,N.J.,1964
Pha 16.0280

SOCIETY OF GENERAL PHYSIOLOGISTS The Control of nuclear activity :a symposium Papers Woods Hole,Mass. 1966 Aug 31-Sep 3 Edited by Lester Goldstein Prentice-Hall,Englewood Cliffs,N.J.,1967 Original title of the symposium 'Cytoplasmic and environmental influences on nuclear behavior'
Bot 42.4736

SOCIETY OF GENERAL PHYSIOLOGISTS The Control of nuclear activity :a symposium held under the auspices of the Society of general physiologists at its annual meeting at the Marine biological laboratory Woods Hole,Mass. 1966 Aug 31-Sep 3 Edited by Lester Goldstein Prentice-Hall,Englewood Cliffs,N.J. 1967
Bioch 33.1013

SOCIETY OF GENERAL PHYSIOLOGISTS The Specificity of cell surfaces :a symposium Woods Hole 1965 Sep 1-4 Edited by B.D. Davis and L. Warren Prentice-Hall,Englewood Cliffs,N.J.,1967
Phys 20.1506

SOCIETY OF GENERAL PHYSIOLOGISTS The Specificity of cell surfaces :a symposium Woods Hole,Mass. 1965 Sep 1-4 Edited by Bernard D. Davis and Leonard Warren Prentice-Hall,Englewood Cliffs,N.J.,1967
An 32.3436

SOCIETY OF GENERAL PHYSIOLOGISTS The Specificity of cell surfaces :a symposium Woods Hole,Mass. 1965 Sep 1-4 Edited by Bernard D. Davis and Leonard Warren Prentice-Hall,Englewood Cliffs,N.J.,1967 Held at the 12th annual meeting of the society
Gen 34.0669

SOCIETY OF GENERAL PHYSIOLOGISTS.ANNUAL MEETING 11TH Molecular architecture in cell physiology :a symposium Woods Hole,Mass. 1964 Sep 8-11 Edited by Tern Hayashi and Andrew G. Szent-Gyorgyi Under the auspices of the Society of general physiologists Prentice-Hall,Englewood Cliffs,N.J.,1966 Held at the 11th annual meeting of the society
Gen 34.0668

SOCIETY OF GENERAL PHYSIOLOGISTS.ANNUAL MEETING 12TH The Specificity of cell surfaces :a symposium Woods Hole,Mass. 1965 Sep 1-4 Edited by Bernard D. Davis and Leonard Warren Under the auspices of the Society of general physiologists Prentice-Hall,Englewood Cliffs, N.J.,1967 Held at the 12th annual meeting of the society
Gen 34.0669

SOCIETY OF GENERAL PHYSIOLOGISTS.ANNUAL MEETING 1963 The Cellular functions of membrane transport :a symposium Papers Woods Hole, Mass. 1963 Sep 4-7 Edited by J.F. Hoffman Held under the auspices of the Society of general physiologists illus. viii,291p Prentice-Hall,New Jersey,1964 Held at the annual meeting of the society
Bot 42.1500

SOCIETY OF GENERAL PHYSIOLOGISTS.ANNUAL MEETING 7TH Control mechanisms in cellular processes :a symposium Woods Hole,Mass 1960 By Sigmund R. Suskind Edited by David M. Bonner Under the auspices of the Society of general physiologists Ronald press,New York,1961 Held at the 7th annual meeting of the society
Bot 42.1492

SOCIETY OF GENERAL PHYSIOLOGISTS.ANNUAL SYMPOSIA, 6 Macromolecular complexes :a symposium Urbana,Ill. 1959 Sep Edited by M.V. Edds illus. Ronald press,New York,1961
Radioth 35.1408

SOCIETY OF INSTRUMENT TECHNICIANS Plant and process dynamic characteristics :a conference Proceedings Cambridge 1956 Apr 4-6 Academic press;Butterworths,New York;London, 1957
Eng 41.5908

SOCIETY OF INSTRUMENT TECHNOLOGY Automatic measurement of quality in process plants :a conference Proceedings Swansea 1957 Sep 23-26 xi,320p Butterworths,London,1958
Chem E 24.1160

SOCIETY OF INSTRUMENT TECHNOLOGY Joint symposium on instrumentation and computation in process development and plant design Papers London 1959 May 11-12 British conference on automation and computation Sponsored by the Institution of chemical engineers Institution of chemical engineers, London,1959 Sponsored also by the Society of instrument technology,and the British computer society
Math L 5.3290

SOCIETY OF INSTRUMENT TECHNOLOGY Joint symposium on instrumentation and computation in process development and plant design Proceedings London 1959 May 11-13 Institution of chemical engineers,London,1959
Eng 41.6011

SOCIETY OF INSTRUMENT TECHNOLOGY Plant and process dynamic characteristics :a conference Proceedings Cambridge 1956 Apr 4-6 xii, 246p Butterworths,London,1957
Chem E 24.1157

SOCIETY OF INSTRUMENT TECHNOLOGY The Use of redundancy in system design :a symposium Northampton 1964 Feb 14 Society of instrument technology,1964
Eng 41.5963

SOCIETY OF MATERIALS SCIENCE International conference on mechanical behavior of materials Proceedings Kyoto 1971 Aug 15-20 6 vols Society of materials science,Kyoto,1971
Eng 41.8524

SOCIETY OF MATERIALS SCIENCE,JAPAN Mechanical behaviour of materials International conference on 'mechanical behaviour of materials' Proceedings Kyoto 1971 Aug 15-20 Society of materials science, Kyoto,1972
Met 25.2737

SOCIETY OF MINING ENGINEERS.MINERALS BENEFICATION DIVISION Unit processes in hydrometallurgy based on an international symposium Dallas 1963 Feb 24-28 Edited by Milton E. Wadsworth and Franklin T. Davis Metallurgical society conferences, 24 Gordon and Breach,New York;London,1964
Met 25.2289

SOCIETY OF NON-DESTRUCTIVE EXAMINATION Non-destructive examination in the steel industry Conference on non-destructive examination applied to process control in the steel industry Papers and discussions Swansea 1967 Jan 3-5 Iron and steel institute. Publication, 103 illus 193p London,1967
Met 25.2601

SOCIETY OF OCCUPATIONAL MEDICINE Occupational health for the undergraduate medical student London 1970 May 14 Edited by John R. Glover Society of occupational medicine,London,c1971
PGMS 29.0667

SOCONY MOBILE OIL COMPANY.FIELD RESEARCH LABORATORY Dynamic programming Colloquium lectures in pure and applied science 8 Dallas,Texas 1963 Dec 9-13 By Rutherford Aris 310p 1965
Chem E 24.0172

SODERGARN 1965 Muscular afferents and motor control Nobel symposium 1st Proceedings Nobel foundation Almqvist and Wiksell;Wiley,Stockholm;New York,1966
Phys 20.2242

SODERGARN 1967 Fast reactions and primary processes in chemical kinetics Nobel symposium 5th Proceedings Nobel foundation Edited by Stig Claesson Wiley, New York;London,1967
Bioch 33.0328

SODERGARN 1968 Symmetry and function of biological systems at the macromolecular level Nobel symposium 11th Procedings Nobel foundation Edited by Arne Engstrom and Bror Strandberg Wiley,New York,1969
Bioch 33.0327

SODERGARN,SWEDEN 1965 Muscular afferants and motor control Nobel symposium 1st Proceedings Nobel institute for neurophysiology Edited by Ragnar Granit Wiley;Almqvist and Wiksell,New York;Stockholm, 1966
Psy 31.0250

SOFIA 1964 Symposium on interspecific hybridization in plants Edited by R. Georgieva and D. Tsikov Izdatelstvo Bolgarskoi akademii nauk,Sofia,1965 In Russian,English summaries
Gen 34.1025

SOFIA 1965 International pigment cell conference 6th International union against cancer Edited by G. Della Porta and O. Muhlbock Springer-verlag,Berlin,1966
Bioch 33.0967

SOFIA 1965 Structure and control of the melanocyte International pigment cell conference 6th Edited by G. Della Porta and O. Muhlbock Sponsored by the International union against cancer Springer, Berlin,1966
An 32.3425

SOFT MAGNETIC MATERIALS FOR TELECOMMUNICATIONS a symposium London 1952 Apr Post Office.Engineering research station Edited by C.E. Richards and A.C. Lynch Pergamon press,London,1953
Eng 41.5085

SOFT X-RAY BAND SPECTRA AND THE ELECTRONIC STRUCTURE OF METALS AND MATERIALS :a conference Proceedings Strathclyde 1967 Sep 18-21 University of Strathclyde and Royal society Edited by Derek J. Fabian Academic press,London;New York,1968
Met 25.1470

SOFTWARE ENGINEERING Symposium on computer and information sciences 3rd Proceedings Miami,Fla. 1969 Dec Vol 1-2 Edited by Julius T. Tou COINS, 3 illus 2 vols Academic press,New York,1970
Math L 5.3830

SOFTWARE ENGINEERING :a conference Report Garmisch 1968 Oct 7-11 Edited by Peter Naur and Brian Randell Sponsored by the North Atlantic treaty organization.Science committee 231p Nato,Brussels,1969
Math L 5.3490

SOFTWARE ENGINEERING TECHNIQUE :a conference Report Rome 1969 Oct 27-31 Edited by J.N. Buxton and B. Randall Sponsored by the North Atlantic treaty organization.Science committee illus. 1964p NATO science committee,Brussels,1970
Math 5.3610

SOFTWARE ENGINEERING TECHNIQUES :a conference Report Rome 1969 27-31 Oct Edited by J.N. Buxton and B. Randell Sponsored by the North Atlantic treaty organization.Science committee Illus 164p NATO science committee,Brussels,1970
Math L 5.3545

SOFTWARE SYSTEMS Joint IBM-University of Newcastle-upon-Tyne seminar Proceedings Newcastle upon Tyne 1969 Sep 8-12 University of Newcastle upon Tyne computer laboratory International business machines corporation Edited by N.S.M. Cox University of Newcastle upon Tyne,Newcastle upon Tyne,1970
Math L 5.3354

SOFTWARE WORLD Software 70 :conference Proceedings Sheffield 1970 Apr Edited by David J. Evans illus 197p Transcripta books,London,1970
Math L 5.3871

SOFTWARE 70 :conference Proceedings Sheffield 1970 Apr Software world Edited by David J. Evans illus 197p Transcripta books,London,1970
Math L 5.3871

SOIL AND ROCK MECHANICS Instruments and apparatus for soil and rock mechanics :a symposium Lafayette,Ind. 1965 Jun 13-18 American society for testing materials American society for testing materials.Special technical publication, 392 American society for testing materials,Philadelphia,Pa.,1965
Eng 41.3158

SOIL BACTERIA The Ecology of soil bacteria : an international symposium Edited by T.R.G. Gray and D. Parkinson illus. xv,681p Liverpool university press,Liverpool,1968
Bot 42.3764

SOIL DENSITY AND MOISTURE Symposium on nuclear methods for measuring soil density and moisture Atlantic City,N.J. 1960 Jun 27 American society for testing materials American society for testing materials.Special technical publication, 293 American society for testing materials,Philadelphia,Pa.,1961
Eng 41.3152

SOIL DYNAMICS Symposium on soil dynamics Atlantic City,N.J. 1961 Jun 26 American society for testing materials American society for testing materials.Special technical publication, 305 American society for testing materials,Philadelphia,Pa.,1962
Eng 41.3155

SOIL FUNGI The Ecology of soil fungi :an international symposium Edited by D. Parkinson and J.S. Waid illus. xi,324p Liverpool university press,Liverpool,1960
Bot 42.3717

SOIL MECHANICS Asian regional conference on soil mechanics and foundation engineering 1st Proceedings New Delhi 1960 Feb International society of soil mechanics and foundation engineering Indian national society of soil mechanics and foundation engineering New Delhi,1960
Eng 41.3172

SOIL MECHANICS Asian regional conference on soil mechanics and foundation engineering 2nd Proceedings Tokyo 1963 May 1-4 Vol 1-2 International society of soil mechanics and foundation engineering 2 vols 1963
Eng 41.3173

SOIL MECHANICS Australia-New Zealand conference on soil mechanics and foundation engineering 1st Proceedings Melbourne 1952 Jul University of Melbourne Institution of engineers,Australia 1952
Eng 41.3166

SOIL MECHANICS Australia-New Zealand conference on soil mechanics and foundation engineering 4th Proceedings Adelaide 1963 Aug 19-23 University of Adelaide 1963
Eng 41.3169

SOIL MECHANICS Budapest conference on soil mechanics and foundation engineering 3rd Proceedings Budapest 1968 Oct 15-18 Magyar tudomanyos akademia Akademiai kiado, Budapest,1968
Eng 41.3137

SOIL MECHANICS Congreso panamericano de mecanica de suelos y cimentaciones 1st Proceedings Mexico,D.F. 1959 Sep 7-12 Vol 1-3 Sociedad mexicana de mecanica de suelos 3 vols Mexico,D.F.,1960
Eng 41.3178

SOIL MECHANICS International conference on soil mechanics and foundation engineering 2nd Proceedings Rotterdam 1948 Jun 21-30 Vol 1-7 7 vols Rotterdam,1948
Eng 41.3139

SOIL MECHANICS International conference on soil mechanics and foundation engineering 3rd Proceedings Zurich 1953 Aug 16-27 Vol 1-3 3 vols Zurich,1953
Eng 41.3140

SOIL MECHANICS International conference on soil mechanics and foundation engineering 4th Proceedings London 1957 Aug 12-14 3 vols Butterworths,London,1957
Eng 41.3141

SOIL MECHANICS International conference on soil mechanics and foundation engineering 5th Proceedings Paris 1961 Jul 17-22 3 vols Dunod,Paris,1961
Eng 41.3142

SOIL MECHANICS International conference on soil mechanics and foundation engineering 6th Proceedings Montreal 1965 Sep 8-15 3 vols University of Toronto press,Toronto, 1965
Eng 41.3143

SOIL MECHANICS International conference on soil mechanics and foundation engineering 7th Proceedings Mexico,D.F. 1969 Vol 1-3 3 vols 1969
Eng 41.3144

SOIL MECHANICS International conference on soils mechanics and foundation engineering Proceedings Cambridge,Mass. 1936 Jun 22-26 Vol 1-3 3 vols Cambridge,Mass.,1936
Eng 41.3138

SOIL MECHANICS Problems of settlements and compressibility of soils European conference on soil mechanics and foundation engineering Proceedings Wiesbaden 1963 Vol 1-2 2 vols 1963
Eng 41.3175

SOIL MECHANICS Purdue conference on soil mechanics and its application Proceedings Lafayette,Ind. 1940 Sep 2-6 Purdue university Purdue university,Lafayette,Ind., 1940
Eng 41.3163

SOIL MECHANICS Rheology and soil mechanics : a symposium Grenoble 1964 Apr 1-8 International union of theoretical and applied mechanics Edited by J. Kravtchenko and P.M. Sirieys Springer,Berlin,1966
Eng 41.3097

SOIL MECHANICS Symposium on the use of radioisotopes in soil mechanics Cleveland, Ohio 1952 Mar 5 American society for testing materials American society for testing materials.Special technical publication, 134 American society for testing materials,Philadelphia,Pa.,1953
Eng 41.3147

SOIL MECHANICS Texas conference on soil mechanics and foundation engineering 8th Proceedings Austin,Texas 1956 Sep 14-15 University of Texas.Bureau of engineering research University of Texas,Austin,Texas, 1956
Eng 41.3164

SOIL MECHANICS AND FOUNDATION ENGINEERING Regional conference for Africa 4th Proceedings South African institution of civil engineers 3 vols Balkema,Cape Town, 1967
Eng 41.3177

SOIL MICROMORPHOLOGY International working meeting on soil micromorphology 2nd Proceedings Arnhem 1964 Sep 22-25 Edited by A. Jongerius Elsevier,Amsterdam,1964
Eng 41.3115

SOIL MICROMORPHOLOGY International working-meeting on soil micromorphology 2nd proceedings Arnhem 1964 Sep.22-25 Edited by A. Jongerius Elsevier,Amsterdam,1964
Geol 8.1511

SOIL ORGANISMS Colloquium on soil fauna and soil microflora and their relationships Proceedings Oosterbeck 1962 Sep 10-16 Edited by J. Doeksen and Joseph van der Drift North Holland,Amsterdam,1963
Bot 42.4690

The SOIL RESOURCES OF TROPICAL AFRICA London 1965 Sep 29 African studies association of the United Kingdom Edited by R.P. Moss Cambridge university press, Cambridge,1968
Geog 13.4644

SOIL STABILIZATION Conference on soil stabilization Proceedings Cambridge,Mass. 1952 Jun 18-20 Massachusetts institute of technology M.I.T.press,Cambridge,Mass.,1952
Eng 41.3134

SOIL TESTING Procedures for testing soils nomenclatures and definitions,standard methods, suggested methods American society for testing materials American society for testing materials,Philadelphia,Pa.,1958
Eng 41.3154

SOIL ZOOLOGY Easter school in agricultural science 2nd Proceedings Nottingham 1955 University of Nottingham Edited by D. Keith McE. Kevan Butterworths,London,1955
Bal 39.2006

SOIL ZOOLOGY Easter school in agricultural science 2nd Proceedings Nottingham 1955 University of Nottingham Edited by D. Keith McE. Kevan Butterworths scientific publications illus. xiv,512p Butterworths,London,1955
Bot 42.2070

SOIL ZOOLOGY Progress in soil zoology Research methods in soil zoology colloquium 1st Papers Rothamsted 1958 Jul 10-14 International society of soil science.Soil zoology committee Edited by P.W. Murphy Bibliog.,illus. Butterworths,London,1962
Bot 42.2085

SOILS Commonwealth conference on tropical and sub-tropical soils 1st proceedings Harpenden 1948 Jun Commonwealth bureau of soil science Commonwealth bureau of soil science.Technical communications,46 Commonwealth bureau of soil science,Harpenden, 1949
Geog 13.1093

SOILS Conference on soil science problems 1st proceedings Rothamsted 1930 Sep 16-18 Imperial bureau of soil science Imperial bureau of soil science.Technical communications,17 H.M.S.O.,London,1931
Geog 13.1071

SOILS International congress of soil science 3rd Oxford 1935 guidebook for the excursion round Britain Clarendon press, Oxford,1935
Geog 13.1078

SOILS Papers on soils Symposium on time rates of loading in soil tension Symposium on Atterberg limits,Session on soils and Symposium on soils for engineering purposes Atlantic City,N.J. 1959 Jun 22-23 San Francisco,Calif. 1959 Oct 16 American society for testing materials.Special technical publication, 254 American society for testing materials,Philadelphia,Pa.,1960 Papers presented at various A.S.T.M. meetings during 1959
Eng 41.3151

SOILS Pore pressure and suction in soils :a conference London 1960 Mar 30-31 International society of soil mechanics and foundation engineering.British national committee Butterworths,London,1961
Eng 41.3170

SOILS Sorption and transport processes in soils... :a symposium 1970 Society of chemical industry British society of soil science Society of chemical industry. Monograph, 37 Society of chemical industry, London,1970
Eng 41.8281

SOILS Symposium on consolidation testing of soils Atlantic City,N.J. 1951 Jun 18 American society for testing materials American society for testing materials.Special technical publication, 126 American society for testing materials,Philadelphia,Pa.,1952
Eng 41.3149

SOILS Symposium on dynamic testing of soils Atlantic City,N.J. 1953 Jul 2 American society for testing materials American society for testing materials.Special technical publication, 156 American society for testing materials,Philadelphia,Pa.,1954
Eng 41.3148

SOILS Symposium on the identification and classification of soils Atlantic City,N.J. 1950 Jun 29 American society for testing materials American society for testing materials.Special technical publication, 113 American society for testing materials, Philadelphia,Pa.,1951
Eng 41.3146

SOILS Symposium on vane shear testing of soils Atlantic City,N.J. 1956 Jun 22 American society for testing materials American society for testing materials.Special technical publication, 193 American society for testing materials,Philadelphia,Pa.,1957
Eng 41.3150

SOILS Tropical soils and vegetation :Abidjan symposium Proceedings Abidjan 1959 Oct 20-24 Unesco Jointly organized by the Commission for technical co-operation in Africa south of the Sahara Humid tropics research Unesco,Paris,1961
Geog 13.1128

SOLAR ACTIVE REGIONS Structure and development of solar active regions Budapest 1967 Sep 4-8 International astronomical union Edited by K.O. Kiepenheuer International astronomical union.Symposium, 35 Reidel,Dordrecht,1968
TA 15.0420

SOLAR ACTIVE REGIONS Structure and development of solar active regions :a symposium Budapest 1967 Sep 4-8 International astronomical union Edited by K. O. Kiepenheuer International astronomical union.Symposia, 35 Bibliog.,Illus. xvii, 608p 25cm D.Reidel,Dordrecht,1968
A Math 4.1602

SOLAR ACTIVE REGIONS Structure and development of solar active regions :a symposium Proceedings Budapest 1967 Sep International astronomical union Edited by K. O. Kiepenheuer International astronomical union.Symposium, 35 608p Reidel,Dordrecht, 1968
Obs 6.3266

SOLAR CORONA Physics of the solar corona NATO advanced study institute on physics of the solar corona Proceedings Athens 1970 Sep 6-17 North Atlantic treaty organization Edited by C.J. Macris Astrophysics and space science library, 27 illus xii,345p 24cm Reidel,Dordrecht,1971
A Math 4.1803

SOLAR CORONA Physics of the solar corona NATO advanced study institute on the physics of the solar corona Proceedings Athens 1970 Sep 6-17 North Atlantic treaty organization Edited by C.J. Macris Astrophysics and space science library, 27 345p Reidel,Dordrecht,1971
Obs 6.3607

The SOLAR CORONA :a symposium proceedings Cloudcroft,N.M. 1961 Aug.28-30 International astronomical union Edited by John W. Evans Sponsored by the Air Force Cambridge research laboratories International astronomical union.Symposium, 16 344p Academic press,New York;London, 1963
Obs 6.0564

SOLAR ECLIPSES AND THE IONOSPHERE :a symposium London 1955 Aug.22-24 Edited by W.J.G. Beynon and G.M. Brown Under the auspices of the Mixed commission on the ionosphere Pergamon press,London,1956 Vol.6 of special supplements to Journal of atmospheric and terrestrial physics
Obs 6.0176

SOLAR ECLIPSES AND THE IONOSPHERE :a symposium Proceedings London 1955 Aug 22-24 Mixed commission on the ionosphere Edited by W.J.G. Beynon and G.H. Brown Under the auspices of the International council of scientific unions Journal of atmospheric and terrestrial physics.Special supplement, 6 ix,330p Pergamon press,London;New York,1956
Nap 11.0437

SOLAR ENERGY The Conference on the use of solar energy :the scientific basis Transactions Tucson,Ariz. 1955 Oct 31-Nov 1 Vol 1-5 University of Arizona and Stanford research institute Edited by Edwin F. Carpenter and others Co-sponsored by the Association for applied solar energy illus 6 vols University of Arizona,Tucson,Ariz., 1958
Nap 11.0111

SOLAR FLARES AND SPACE RESEARCH Proceedings of the symposium held on the occasion of the 11th plenary meeting of the Committee on space research Tokyo 1968 May 9-11 Committee on space research International astronomical union Edited by C.de Jager and Z. Svestka 419p North-Holland,Amsterdam,1969 Co-sponsored by International astronomical union, International union of geodesy and geophysics and International union of radio scientists
Obs 6.3335

SOLAR MAGNETIC FIELDS Centenario della nascita di Galileo Galilei 1564-1964 4th Atti Rome 1964 Sep 14-16 Vol 2,tomo 4: atti del convegno sui campi magnetici solari e la spettroscopia ad alta resolutione. Proceedings of the meeting on solar magnetic fields and high resolution spectroscopy Galileo Galilei.Comitato nazionale per le manifestazioni celebrative Galileo Galilei. Comitato nazionale per le manifestazioni celebrative.Publicazione figs 341p Barbera,Florence,1966
WSM 43.4471

SOLAR MAGNETIC FIELDS Paris 1970 Aug 31-Sep 4 International astronomical union Edited by R. Howard International astronomical union.Symposium, 43 782p Reidel,Dordrecht,1971
Obs 6.3589

SOLAR MAGNETIC FIELDS Paris 1970 Aug 31-Sep 4 International astronomical union Edited by R. Howard International astronomical union.Symposium, 43 782p Reidel,Dordrecht,1971
TA 15.0688

SOLAR MAGNETIC FIELDS Stellar and solar magnetic fields Symposium on stellar and solar magnetic fields Munich 1963 Sep.3-10 International astronomical union Edited by R. Lust International astronomical union. Symposium, 22 460p North Holland, Amsterdam,1965
Obs 6.1148

SOLAR MAGNETIC FIELDS Stellar and solar magnetic fields :a symposium Rottach-Egern 1963 Sep.3-10 International astronomical union Edited by R. Lust International astronomical union.Symposium, 22 460p North-Holland,Amsterdam,1965
A Math 4.1160

SOLAR PHYSICS NATO advanced study institute on solar physics Proceedings Lagonissi, Athens 1965 Sep North Atlantic treaty organization.Division of scientific affairs Interscience,London,1967
Obs 6.3280

SOLAR PHYSICS The Menzel symposium on solar physics,atomic spectra and gaseous nebulae Cambridge,Mass. 1971 Apr 29 Edited by K.B. Gebbie National bureau of standards.Special publication, 353 203p National bureau of standards,Washington,D.C.,1971
Obs 6.3623

SOLAR SPECTRUM The Symposium on the solar spectrum Proceedings Utrecht 1963 Aug 26-31 Edited by C.de Jager D.Reidel,Dordrecht, 1965 Dedicated to M.G.J.Minnaert on the occasion of his 70th birthday
TA 15.0199

The SOLAR SPECTRUM symposium proceedings Utrecht 1963 Aug.26-31 Edited by C.de Jager port 417p D.Reidel,Dordrecht,1965 Proceedings dedicated to M.Minnaert
Obs 6.0876

The SOLAR SPECTRUM symposium proceedings Utrecht 1963 Aug.26-31 Rijksuniversiteit te Utrecht Edited by C.de Jager Astrophysics and space science library D. Reidel,Dordrecht-Holland,1965
A Math 4.1163

SOLAR STUDIES A Discussion on solar studies with special reference to space observations London 1970 Apr 21-22 British national committee on space research Philosophical transactions, 270A,1202 195p royal society of London,London,1971
Obs 6.3606

SOLAR SYSTEM Origin of the solar system :a conference Proceedings Goddard institute for space studies Edited by Robert Jastrow and A.G.W. Cameron 176p Academic press,New York,1963
Cav 7.2870

SOLAR SYSTEM Origin of the solar system :a conference Proceedings New York 1962 Jan 23-24 Goddard institute for space studies Edited by Robert Jastrow and A.G.W. Cameron Academic press,New York,1963
TA 15.0200

SOLAR SYSTEM Origin of the solar system conference proceedings New York 1962 Jan. 23-24 Goddard institute for space studies Edited by Robert Jastrow and A.G.W. Cameron 176p Academic press,New York,1963
Obs 6.0883

SOLAR-TERRESTRIAL PHYSICS :a symposium Review papers Belgrade 1966 Aug.29-Sep. 2 International scientific radio union Edited by J.W. King and W.S. Newman xii,390p Academic press,London;New York,1967
A Math 4.1171

SOLAR-TERRESTRIAL PHYSICS:SOLAR ASPECTS Proceedings of joint I.Q.S.Y.-Cospar symposium London 1967 Jul 17-22 Pt 1 International quiet sun year Committee on space research Edited by A.C. Stickland International quiet sun year.Annals, 4 414p MIT press,Cambridge,Mass.,1969
Obs 6.3341

SOLAR TERRESTRIAL RELATIONS Introduction to solar terrestrial relations Summer school in space physics proceedings Alpbach 1963 Jul.15-Aug.10 European space research organisation.Preparatory commission Edited by J. Ortner and H. Maseland Astrophysics and space science library ix,506p D.Reidel, Dordrecht-Holland,1965
A Math 4.1162

SOLAR TERRESTRIAL RELATIONS Introduction to solar terrestrial relations The Summer school in space physics Proceedings Alpbach 1963 Jul.15-Aug.10 Edited by J. Ortner and H. Maseland Organized by the European preparatory commission for space research Astrophysics and space science library 506p D.Reidel,Dordrecht,1965
Obs 6.1321

SOLAR TERRESTRIAL RELATIONS Introduction to solar terrestrial relations The Summer school in space physics Proceedings Alpbach 1963 Jul 15-Aug 10 European preparatory commission for space research Edited by J. Ortner and H. Maseland Astrophysics and space science library Reidel,Dordrecht,1965
TA 15.0425

SOLAR VARIATIONS,CLIMATIC CHANGE,AND RELATED GEOPHYSICAL PROBLEMS New York 1961 Jan New York academy of sciences New York academy of sciences.Annals, 95(1) New York, 1961
Bot 42.3302

SOLAR VARIATIONS,CLIMATIC CHANGE AND RELATED GEOPHYSICAL PROBLEMS :conference Papers New York 1961 Jan 24-28 New York academy of sciences,and,American meteorological society Edited by Rhodes W. Fairbridge New York academy of sciences. Annals,95,art.1 New York academy of sciences, New York,1961
Geog 13.1055

SOLAR WIND The Conference on the solar wind Proceedings Pasadena,Calif. 1964 Apr 1-4 California institute of technology.Jet propulsion laboratory Edited by Robert J. Mackin and Marcia Neugebauer JPL TR 32-630 400p pergamon press,Oxford,1966
TA 15.0201

The SOLAR WIND :a conference proceedings Pasadena,Calif. 1964 Apr.1-4 Edited by Robert J. Mackin and Marcia Neugebauer Sponsored by the California institute of technology.Jet propulsion laboratory Pergamon press,Oxford,1966
A Math 4.1165

SOLDER Symposium on solder :presented at the fifty-ninth annual meeting A.S.T.M. Atlantic city,N.J. 1956 Jun 19-20 American society for testing materials American society for testing materials.Special technical publication, 189 American society for testing materials,Philadelphia,Pa.,1957
Eng 41.4025

SOLEIL A LA RENAISSANCE:SCIENCES ET MYTHES : colloque international Brussels 1963 Apr Federation internationale des instituts et societes pour l'etude de la Renaissance and Belgique.Ministere de l'education et de la culture Universite libre de Bruxelles. Institut pour l'etude de la renaissance et de l'humanisme.Publication, 2 584p Presses universitaires de Bruxelles,Brussels;Paris, 1965 Contains Hoskin,M.and Jones,C. 'Problems in late renaissance astronomy'
WSM 43.0046

SOLID CIRCUITS AND MICROMINIATURIZATION Conference on solid circuits and microminiaturization :a conference Proceedings West Ham 1963 Jun Edited by G.W.A. Dummer Pergamon press,London,1964
Eng 41.5567

SOLID MECHANICS Midwestern conference on solid mechanics 4th proceedings Austin, Tex. 1959 Sept 9-11 University of Texas, Austin,Tex,1959
A Math 4.0368

SOLID STATE Interaction of radiation with solids Cairo solid state conference Cairo 1966 Sep 3-8 American university in Cairo Edited by Adli Bishay Plenum press,New York, 1967
TA 15.0429

SOLID STATE AND SUPERFLUID THEORY Mathematical methods in solid state and superfluid theory Scottish universities' summer school in physics 8th Papers St. Andrews 1967 Jul 31-Aug 19 Edited by R.C. Clark and G.H. Derrick Scottish universities' summer school in physics,1967 xvi,400p 25cm Oliver and Boyd,Edinburgh, 1969
Cav 7.2759

SOLID-STATE CIRCUIT CONFERENCE digest of technical papers Philadelphia,Pa. 1959 Feb.12-13 Institute of radio engineers Sponsored by American institute of electrical engineers 102p Lewis Winner,1959
Math L 5.2082

SOLID STATE DEVICES :a conference Proceedings Tokyo 1970 Sep 21-22 Japan society of applied physics Japan society of applied physics.Journal, 40, Supplement diagrs,photos 250p 26cm Japan society of applied physics,Tokyo,1970
Cav 7.2990

SOLID STATE DIFFUSION Diffusion a l'etat solide Colloque sur la diffusion a l'etat solide Saclay 1958 Jul 3-5 Centre d'etudes nuleaires,Saclay and Commissariat a l'energie atomique Centre d'etudes nucleaires de Saclay;North Holland,Gif-sur-Yvette;Amsterdam,1959
Met 25.1277

SOLID STATE OPTICS Tokyo summer lectures in theoretical physics 2nd Papers Tokyo 1966 Aug 29-Sep 3 Pt 1: dynamical processes in solid state optics University of Tokyo.Summer institute of theoretical physics Edited by Gyo Takeda and Akihito Fudjii W.A.Benjamin,New York;Tokyo,1967
Cav 7.2316

SOLID STATE OPTICS Tokyo summer lectures in theoretical physics 2nd Papers Tokyo 1966 Aug 29-Sep 3 Pt 1: dynamical processes in solid state optics University of Tokyo.Summer institute of theoretical physics Edited by Ryogo Kubo and Hiroshi Kamimura Benjamin,New York,1967
TA 15.0434

SOLID STATE PHYSICS Solid state physics in electronics and telecommunications : international conference Proceedings Brussels 1958 Jun 2-7 Vol 3,pt 1: magnetic and optical properties International union of pure and applied physics Edited by M. Desirant and J.L. Michiels Academic press,London,1960
Cav 7.0387

SOLID STATE PHYSICS Solid state physics in electronics and telecommunications : international conference Proceedings Brussels 1958 Jun 2-7 Vol 1-2,pt 1-2: semiconductors International union of pure and applied physics Edited by M. Desirant and J.L. Michiels 2 vols Academic press, London,1960
Cav 7.0388

SOLID SURFACES AND THE GAS-SOLID INTERFACE : papers presented at the Kendall award symposium honoring Stephen Brunauer Papers Kendall award symposium St.Louis,Mo. 1961 Mar 27-29 American chemical society. Division of colloid and surface chemistry Edited by Lewellyn E. Copeland and others American chemical society.Advances in chemistry series,33 American chemical society,Washington,D.C.,1961
Chem 18.0616

SOLIDIFICATION The Symposium on solidification presented at the 64th AFS castings congress and exposition Philadelphia,Pa. 1960 May 9-13 American foundrymen's society American foundrymen's society,Des Plaines,Ill.,1961
Met 25.0644

The SOLIDIFICATION OF METALS :the conference Proceedings Brighton 1967 Dec 4-7 Iron and steel institute and Institute of metals Jointly organized with the Institute of physics and the physical society and the Institution of metallurgy Iron and steel institute.Publication, 110 Iron and steel institute,London,1968
Met 25.1529

SOLIDS Atomic collision phenomena in solids : an international conference Proceedings Brighton 1969 Sep 7-12 Edited by D.W. Palmer and others Bibliog 678p 17cm North-Holland,Amsterdam,1970
Cav 7.2680

SOLIDS International symposium on the reactivity of solids 2nd Proceedings Gothenburg 1952 Jun 9-13 Under the auspices of Ingeniorsvetenskapsakademien and Chalmers tekniska hogskola 2 vols Gothenburg,1954
Min 10.0614

SOLIDS Localized excitations in solids International conference on localized excitations in solids 1st Proceedings Irvine 1967 Sep 18-22 International union of pure and applied physics National science foundation Edited by R.F. Wallis xvi,782p 26cm Plenum press,New York,1968
Cav 7.2857

SOLIDS Plasma effects in solids International conference on the physics of semiconductors 7th Paris 1964 Vol 2: plasma effects in solids International union of pure and applied physics photos, diagrms,graphs 221p 24cm Dunod,Paris, 1965 Co-sponsored by the French physical society,the French ministry for scientific research and Unesco
Cav 7.3047

SOLIDS Propieta ottiche dei solidi Scuola internazionale di fisica 'Enrico Fermi' 34 corso Rendiconti Varenna 1965 Jun 28-Jul 10 Societa italiana di fisica Edited by J. Tauc Academic press,New York;London,1966
TA 15.0443

SOLIDS Rheology of solids International symposium on high speed testing 7th Proceedings Boston,Mass. 1969 Mar 17-18 Edited by R.D. Andrews and F.R. Eirich Applied polymer symposia, 12,High speed testing, 7 Interscience,New York,1969
Met 25.2581

SOLIDS Structure and properties of solid surfaces :conference Papers Lake Geneva, Wis. 1952 Sep Edited by Robert Gomer and Cyril Stanley Smith Organized by the National research council.Committee on solids University of Chicago press,Chicago,Ill.,1953
Col S 12.0275

SOLIDS Structure et proprietes des surfaces des solides :colloque international Paris 1969 Jul 7-11 Centre national de la recherche scientifique Centre national de la recherche scientifique.Colloques internationaux, 187 Paris,1970
Met 25.2626

SOLIDS Structure of organic solids Microsymposium 'Structure of organic solids' 2nd Lectures Prague 1968 Sep 16-19 International union of pure and applied chemistry.Macromolecular division Czechoslovak chemical society In conjunction with Ceskoslovenska akademie ved illus 550p 25cm Butterworths,London,1969 Simultaneously published in 'Pure and applied chemistry',vol 18,no 4,1969
Cav 7.2659

SOLIDS The Measurement of stress and strain in solids :a conference Proceedings Manchester 1946 Jul 11-13 Institute of physics Institute of physics,London,1948
Eng 41.2898

SOLIDS The Mechanism of phase transformations in crystalline solids :an international symposium Proceedings Manchester 1968 Jul 3-5 Institute of metals Institute of metals.Monograph and report series, 33 illus,diagrs 324p 25cm London,1969
Cav 7.2666

SOLIDS Tunneling phenomena in solids NATO advanced study institute on tunneling phenomena in solids Proceedings Riso, Denmark 1967 North Atlantic treaty organization Edited by Elias Burstein and Stig Lundqvist Bibliog 580p 18cm Plenum press,New York,1969
Cav 7.2667

SOLIDS Tunneling phenomena in solids : lectures presented at the 1967 NATO advanced study institute at Riso,Denmark Riso 1967 North Atlantic treaty organization Edited by Elias Burstein and Stig Lundqvist Plenum press,New York,1969
Eng 41.8302

SOLUTION PROPERTIES OF NATURAL POLYMERS :an international symposium Proceedings Edinburgh 1967 Jul 25-28 Chemical society Chemical society.Special publication, 23 Chemical society,London,1968
Bioch 33.0632

SOLVATED ELECTRON :a symposium at the 150th meeting of the American chemical society Atlantic City 1965 Sep 15-16 American chemical society Edited by Robert F. Gould American chemical society.Advances in chemistry series, 50 American chemical society,Washington,D.C.,1965
Radioth 35.0430

SOLVAY CONFERENCE,1921,3 Atomes et electrons Conseil de physique 3eme Rapports Brussels 1921 Apr 1-6 Institut international de physique Solvay Paris,1923
Cav 7.2080

SOLVAY CONFERENCE,1927,5 Electrons et photons Conseil de physique 5eme Rapports Brussels 1927 Oct 24-27 Institut international de physique Solvay Paris,1928
Cav 7.2081

SOLVAY CONFERENCE,1933,7 Structure et proprietes des noyaux atomiques Conseil de physique 7eme Rapports Brussels 1933 Oct 22-29 Institut international de physique Solvay and Universite libre de Bruxelles Paris,1934
Cav 7.2083

SOLVAY CONFERENCE,1958,11 Structure et evolution de l'univers Conseil de physique 11eme Rapports Brussels 1958 Jun 9-13 Institut international de physique Solvay and Universite libre de Bruxelles Edited by R. Stoops Brussels,1958
Cav 7.2084

SOLVAY CONFERENCE,1961,12 Theorie quantique des champs Conseil de physique 12eme Rapports Brussels 1961 Oct 9-14 Institut international de physique Solvay and Universite libre de Bruxelles Edited by R. Stoops Interscience,New York,c 1961
13th Solvay conference issued as Conference on physics,q.v.
Cav 7.2077

SOLVAY CONFERENCE,1964,13 Structure and evolution of galaxies The Conference on physics 13th Proceedings Brussels 1964 Sep Institut international de physique Solvay and Universite libre de Bruxelles Interscience,London,1965 Other Solvay conferences issued as Conseil de physique,q.v.
Cav 7.2078

SOLVAY CONFERENCE ON CHEMISTRY 13th Proceedings Reactivity of photo-excited organic molecula Brussels 1965 Oct 25-30 International institute of chemistry,and, Universite libre de Bruxelles Interscience, London,1967
Col S 12.0568

SOLVAY CONFERENCE ON CHEMISTRY,1956 Quelques problemes de chimie minerale :conseil de chimie 10e Rapports et discussions Brussels 1956 May 22-26 Institut international de chimie Solvay.Comite scientifique R.Stoops,Brussels,1956
Chem 18.1004

SOLVAY CONFERENCE ON CHEMISTRY,1962 Transfert d'energie dans les gaz 12th Brussels 1962 Nov 5-10 Institut international de chimie Solvay Interscience; R.Stoops,New York;Brussels,c1962
Chem 18.0514

SOLVAY CONFERENCE 1958 Structure et evolution de l'univers Conseil de physique 11e Rapports Brussels 1958 Jun 9-13 Universite libre de Bruxelles Institut de physique Solvay Edited by R. Stoops 309p Brussels,1958
Obs 6.3255

SOLVAY CONFERENCE 1964 Structure and evolution of galaxies Conference on physics 13th Proceedings Brussels 1964 Sep Universite libre de Bruxelles and Institut international de physique Solvay 174p Interscience,London,1965
Obs 6.2134

SOLVENT EXTRACTION CHEMISTRY International conference on solvent extraction chemistry Proceedings Gothenburg 1966 Aug 27-Sep 1 By D. Dyrssen and others Chalmers university of technology and Goteborgs universitet North-Holland,Amsterdam,1967
Met 25.2292

SOMATIC CELL GENETICS Macy conference on genetics 4th Princeton,N.J. 1962 Oct 15-17 Josiah Macy jr.foundation Edited by Robert S. Krooth illus. University of Michigan press,Ann Arbor,Mich.,1964
Gen 34.0891

SOMATIC STABILITY IN THE NEWLY BORN :Ciba foundation symposium Ciba foundation Edited by G.E.W. Wolstenholme and M. O'Connor Churchill,London,1961
Phys 20.2127

SOME ASPECTS OF FLUID FLOW :a conference Papers,reports Leamington Spa 1950 Oct 25-28 Edited by H.R. Lang Organized by the Institute of physics viii,292p Edward Arnold,London,1951
Chem E 24.0224

SOME ASPECTS OF FLUID FLOW... :a conference Papers Leamington Spa 1950 Oct 25-28 Institute of physics Arnold,London,1951
Eng 41.6954

SOME ASPECTS OF INTERNAL IRRADIATION :a symposium Proceedings Heber,Utah 1961 May 8-11 Edited by Thomas F. Dougherty and others Pergamon press,London,1962
Radioth 35.1169

SOME ASPECTS OF THE AETIOLOGY AND BIOCHEMISTRY OF PROSTATIC CANCER Tenovus workshop 3rd Proceedings Cardiff 1970 Tenovus institute for cancer research Edited by K. Griffiths and C.G. Pierrepoint Alpha omega alpha,Cardiff,1971
Radioth 35.1928

SOME BIOCHEMICAL AND IMMUNOLOGICAL ASPECTS OF HOST-PARASITE RELATIONSHIPS :a conference New York 1963 Apr 23-25 By Clark P. Read and others New York academy of sciences Edited by T.C. Cheng New York academy of sciences.Annals, 113 p 1-510 New York,1963
Bal 39.1733

SOME CONJUGATED PROTEINS;A SYMPOSIUM The Annual conference on protein metabolism 9th six lectures New Brunswick,N.J. 1953 Jan 30-31 Rutgers university.Bureau of biological research Edited by William H. Cole Rutgers university press,New Brunswick, N.J.,1953
Bioch 33.0543

SOME FACTORS AFFECTING DRUG TOXICITY
European society for the study of drug toxicity :meeting 4th Proceedings Cambridge 1964 Jul 2-3 Edited by D.G. Davey European society for the study of drug toxicity.Proceedings,4 International congress series,81 Excerpta medica,Amsterdam, 1964
Pha 16.0326

SOME IONOSPHERIC RESULTS OBTAINED DURING THE INTERNATIONAL GEOPHYSICAL YEAR :a symposium Proceedings Brussels 1959 Sep. International scientific radio union.Special commission for the International Geophysical Year Edited by W.J.G. Beynon xi,399p Elsevier,Amsterdam,1960
Nap 11.0053

SOME MATHEMATICAL PROBLEMS IN BIOLOGY
Symposium on mathematical biology Proceedings Washington 1966 Dec 28 Edited by Murray Gerstenhaber Prepared by the American mathematical society American mathematical society.Lectures on mathematics in the life sciences, 1 Bibliog.,Illus. vi and 117p 22cm American mathematical society,Rhode Island,1968
P Math 2.3482

SORBTION-DESORBTION PHENOMENA International symposium on residual gases in electron tubes and related vacuum systems 3rd Papers Rome 1967 Mar 14-17 Section 2: sorption-desorption phenomena in high vacuum Italian society of physics Nuovo cimento.Supplement, 5,2 Editrice composori,Bologna,1967
Met 25.2366

SORBY CENTENARY CONFERENCE Papers and discussions Sheffield 1963 May 9 Iron and steel institute Sponsored by the Company of Cutlers in Hallamshire Iron and steel institute.Special report, 80 Iron and steel institute,London,1964
Met 25.2360

The SORBY CENTENNIAL SYMPOSIUM ON THE HISTORY OF METALLURGY :a symposium Cleveland 1963 Oct 22-23 American society for metals and American institute of mining, metallurgical and petroleum engineers. Metallurgical society Edited by Cyril Stanley Smith Sponsored by the Society for the history of technology Metallurgical society conferences, 27 Gordon and Breach, New York,1965
Met 25.2141

SORPTION AND TRANSPORT PROCESSES IN SOILS...
a symposium 1970 Society of chemical industry British society of soil science Society of chemical industry.Monograph, 37 Society of chemical industry,London,1970
Eng 41.8281

SOUTH AFRICAN ASSOCIATION FOR THE ADVANCEMENT OF SCIENCE :annual congress papers Karoo symposium Johannesburg 1962 South African journal of science,59,no.5 Johannesburg,1963
Geol 8.2941

SOUTH AFRICAN INSTITUTION OF CIVIL ENGINEERS
Soil mechanics and foundation engineering Regional conference for Africa 4th Proceedings 3 vols Balkema,Cape Town,1967
Eng 41.3177

SOUTH AFRICAN INSTITUTION OF CIVIL ENGINEERS DIAMOND JUBILEE CONVENTION Proceedings Johannesburg 1963 Johannesburg,1963
Eng 41.3389

SOUTH ATLANTIC BASIN Problem of land connections across the South Atlantic with special reference to the Mesozoic Symposium on the role of the South Atlantic basin in biogeography and evolution Proceedings New York 1949 Dec 28-29 By Maurice Ewing and others Edited by Ernst Mayr American museum of natural history.Bulletin, 99,3 American museum of natural history,New York, 1952
Bot 42.2341

SOUTH BIRMINGHAM TECHNICAL COLLEGE Newer engineering materials :a symposium on recent developments in new materials for use in engineering Edited by R.F. Winters Macmillan,London,1969 Based on a short course held at South Birmingham technical college
Eng 41.3447

SOUTH HADLEY,MASS. 1958 Cell,organism and milieu :a symposium Edited by Dorothea Rudnick Society for the study of development and growth.Symposia, 17 Ronald press,New York,1959
An 32.3226

SOUTH HADLEY,MASS. 1958 Cell,organism and milieu :a symposium Society for the study of development and growth Edited by Dorothea Rudnick held at Mount Holyoke college Society for the study of development and growth.Symposia, 17 Ronald press,New York, 1959
Bioch 33.1008

SOUTH KENSINGTON MUSEUM Conferences held in connection with special loan collection of scientific apparatus :physics and mechanics London 1876 420p Chapman and Hall,London, 1876
Obs 6.1966

SOUTH-CENTRAL LIBYA AND NORTHERN CHAD
Petroleum exploration society of Libya :annual field conference 8th Guidebook Libya 1966 Edited by James J. Williams Amsterdam, 1966
Geog 13.0598

SOUTH-CENTRAL LIBYA AND NORTHERN CHAD:A GUIDEBOOK
Petroleum exploration society of Libya : annual field conference 8th guidebook Libya 1966 Edited by James J. Williams Amsterdam,1966
Geol 8.4490

SOUTHAMPTON 1960 Mechanisms in biological competition symposium Society for experimental biology Society for experimental biology.Symposia, 15 Cambridge university press,Cambridge,1961
Gen 34.1348

SOUTHAMPTON 1961 On teaching mathematics Southampton mathematical conference :a report on some present-day problems in the teaching of mathematics Commonwealth library of science,technology and engineering.Mathematics division, 1 Pergamon press,Oxford,1961
Eng 41.1813

SOUTHAMPTON 1965 Applications of microelectronics :a symposium Institution of electronic and radio engineers University of Southampton.Department of electronics Sponsored also by the Institution of electrical engineers Institution of electrical engineers.Conference publication, 14 I.E.E.,London,1965 Reproduced from typescript
Eng 41.5576

SOUTHAMPTON 1966 Tall buildings with particular reference to sheer wall structures : a symposium Proceedings Edited by A. Coull and B.Stafford Smith Pergamon press,London, 1967
Eng 41.2733

SOUTHAMPTON 1968 Biodeterioration of materials:microbiological and allied aspects International biodeterioration symposium 1st Proceedings Edited by A.Harry Walters and John J. Elphick Elsevier,Amsterdam,1968
Met 25.2483

SOUTHAMPTON 1968 Biodeterioration of materials:microbiological and allied aspects International biodeterioration symposium 1st Proceedings Edited by Arthur H. Walters and John J. Elphick Bibliog.,illus. x,740p Elsevier,Amsterdam,1968
Bot 42.3773

SOUTHAMPTON MATHEMATICAL CONFERENCE :a report on some present-day problems in the teaching of mathematics On teaching mathematics Southampton 1961 Apr Commonwealth library of science,technology and engineering.Mathematics division, 1 Pergamon press,Oxford,1961
Eng 41.1813

SOUTHEASTERN CONFERENCE ON THEORETICAL AND APPLIED TECHNICS 4th Proceedings Developments in theoretical and applied mechanics New Orleans,La. 1968 Feb 29-Mar 1 Edited by D. Frederick Sponsored by Tulane university illus 637p 26cm Pergamon,Oxford,1970
A Math 4.1903

SOUTHERN ILLINOIS UNIVERSITY Teaching of probability and statistics Comprehensive school mathematics program international conference 1st Proceedings Carbondale, Ill. 1969 Mar 18-27 Edited by Lennart Rade Wiley;Almqvist and Wiksell,New York;Stockholm, 1970
An 32.5410

SOUTHERN METALS-MATERIALS CONFERENCE ON ADVANCES IN AEROSPACE MATERIALS Proceedings Orlando,Fla. 1964 Apr 16-17 American society of metals.Orlando chapter Edited by Henry M. Otte and Saul R. Locke Plenum press, New York,1965
Met 25.2282

SOUTHERN PINES,N.C. 1967 Distinguishing the health care of the aging :...seminar Proceedings American geriatrics society Edited by M.J. Lorenze American geriatrics society.Journal, 16, no 2 American geriatrics society,New York,1967
PGMS 29.0105

SOUTHWEST CENTER FOR ADVANCED STUDIES Quasi stellar sources and gravitational collapse Texas symposium on relativistic astrophysics 1st Proceedings Dallas,Texas 1963 Dec 16-18 By Ivor Robinson and others Sponsored by University of Texas 475p 25cm University of Chicago press,Chicago,Ill.,1965
Cav 7.1985

SOUTHWEST CENTER FOR ADVANCED STUDIES Quasi-stellar sources and gravitational collapse Texas symposium on relativistic astrophysics 1st proceedings Dallas,Texas 1963 Dec. 16-18 Edited by Ivor Robinson and others sponsored by University of Texas xvii,475p 25cm University of Chicago press,Chicago, 1965
Obs 6.1457

SOUTHWEST CENTER FOR ADVANCED STUDIES Quasi-stellar sources and gravitational collapse Texas symposium on relativistic astrophysics 1st proceedings Dallas,Texas 1963 Dec. 16-18 Edited by Ivor Robinson and others Sponsored by University of Texas xvii,475p 25cm University of Chicago press,Chicago, 1965
A Math 4.1161

SOUTHWESTERN METAL CONGRESS Papers New fabrication techniques American society for metals American society for metals,Cleveland, Ohio,n.d.
Met 25.0684

SOUTHWESTERN METAL CONGRESS 1ST High strength steels for aircraft Dallas,Texas 1958 May 12-16 American society for metals American society for metals,Cleveland,Ohio, 1958
Met 25.1002

SOUTHWESTERN METAL CONGRESS 1ST Sheet materials for high temperature service Papers Dallas 1957 May 12-16 American society for metals American society for metals,Cleveland,1958
Met 25.1001

SOVESHCHANII PO SPELEOLOGIV I KARSTOVEDENI Speleologiya i karstovedenie Moscow 1958 Dec. 17-18 Moskovskoe obshchestvo ispytatelei prirody Moscow,1959
Philos 1.1382

SOVESHCHANIYA PO VOPROSAM KOSMOGONII, PROISKHOZHDENIE KOSMICHESKIKH LUCHEI 3 trudy Moscow 1953 May 14-15 Edited by Viktor Amazaspovich Ambartsumyan 319p Akademia Nauk SSSR,Moscow,1954
Obs 6.0060

SOVIET AND EAST EUROPEAN AGRICULTURE Conference on Soviet agricultural affairs 2nd Papers Santa Barbara,Calif. 1965 Aug 26-28 University of California,Berkeley. Center of Slavic and east European studies Edited by Jerzy F. Karcz Russian and east European studies University of California press,Berkeley,Calif.;Los Angeles,Calif.,1967
Geog 13.2584

SOVIET ELECTROCHEMISTRY Conference on electrochemistry 4th Proceedings Moscow 1956 Oct 1-6 Vol 1-3 Translated by Consultants bureau from the Russian New York, 1961
Met 25.1821

SOVIET RESEARCH ON REMOTE CONSEQUENCES OF INJURIES CAUSED BY ACTION OF IONIZING RADIATIONS :summaries of papers presented at the conference on the remote consequences of injuries caused by the action of ionizing radiation Edited by F.G. Kortik and others Translated by Consultants bureau from the Russian Consultants bureau,New York,1957
Radioth 35.1020

SPA,BELGIUM 1958 The Prevention of accidents in childhood :report on a seminar World health organization.Regional office for Europe World health organization,Copenhagen, 1960
PGMS 29.0280

SPACE ASTRONOMY New techniques in space astronomy Munich 1970 Aug 10-14 International astronomical union Edited by F. Labuhn and R. Lust International astronomical union.Symposium, 41 418p Reidel,Dordrecht,1971
Obs 6.3590

SPACE EXPLORATION NASA-University conference on the science and technology of space exploration proceedings Chicago 1962 Nov. 1-3 Vol 1-2 United States.National aeronautics and space administration.NASA SP, 11 NASA SP 11 2 vols office of scientific and technical information, Washington,D.C.,1962
A Math 4.1157

SPACE FLIGHT AGARD-NATO specialists' meeting The Fluid dynamic aspects of space flight Marseille 1964 Apr 20-24 Vol 1-2 Agard AGARDograph, 87 2 vols Gordon and Breach,New York,1966
Eng 41.6887

SPACE OBSERVATIONS A Discussion on solar studies with special reference to space observations London 1970 Apr 21-22 British national committee on space research Philosophical transactions, 270A,1202 195p royal society of London,London,1971
Obs 6.3606

SPACE PHYSICS Earth-moon system The Dynamics of the earth-moon system conference Proceedings New York 1964 Jan.20-21 Goddard institute for space studies Edited by B.G. Marsden and A.G.W. Cameron 288p Plenum press,New York,1966
Obs 6.1169

SPACE PHYSICS Introduction to solar terrestrial relations Summer school in space physics proceedings Alpbach 1963 Jul.15-Aug.10 European space research organisation. Preparatory commission Edited by J. Ortner and H. Maseland Astrophysics and space science library ix,506p D.Reidel,Dordrecht-Holland,1965
A Math 4.1162

SPACE POWER Energy conversion for space power :a symposium Santa Monica,Calif. 1960 Sep 27-30 American rocket society Edited by Nathan W. Snyder Progress in astronautics and rocketry, 3 Academic press,New York,1961
Eng 41.4971

SPACE POWER SYSTEMS CONFERENCE Energy conversion for space power... based on a symposium Selected technical papers Santa Monica,Calif. 1960 Sep 27-30 American rocket society.Space power systems committee Edited by Nathan W. Snyder Progress in astronautics and rocketry, 3 Academic press,New York;London,1961
Met 25.2165

SPACE RESEARCH A Review of space research : symposium Report Iowa City 1962 Jun 17-Aug 10 National aeronautics and space administration Under the auspices of the Space science board National research council.Publication,1079 Washington,D.C., 1962
Obs 6.3259

SPACE RESEARCH The Significance of space research for fundamental physics :a conference Interlaken 1969 Sep 4 European space research organisation Edited by A.F. Moore and U. Hardy ESRO SO-52 175p european space research organisation,Neuilly-sur-Seine, 1971
Obs 6.3603

SPACE RESEARCH The Space science symposium 1st Proceedings Nice 1960 Jan 11-16 Edited by Hilde Kallmann-Bijl Sponsored by Committee on space research xvi,1195p North-Holland,Amsterdam,1960
Nap 11.0317

SPACE RESEARCH :symposia Vol 1- Committee on space research Edited by Hilde Kalmann-Bijl Amsterdam,1960-
Obs 6.3256

SPACE SCIENCE Moon and planets International space science :symposium 7th Papers Vienna 1966 May 10-18 International astronomical union Edited by A. Dollfus Co-sponsored by the Committee on space research 313p North-Holland, Amsterdam,1967 Papers of one session of the symposium
Obs 6.3206

SPACE SCIENCE BOARD A Review of space research :symposium Report Iowa City 1962 Jun 17-Aug 10 National aeronautics and space administration National research council.Publication,1079 Washington,D.C., 1962
Obs 6.3259

The SPACE SCIENCE SYMPOSIUM 1st Proceedings Space research Nice 1960 Jan 11-16 Edited by Hilde Kallmann-Bijl Sponsored by Committee on space research xvi,1195p North-Holland,Amsterdam,1960
Nap 11.0317

The SPACE SCIENCE SYMPOSIUM 6th Papers of a session Problems of atmospheric circulation Mar del Plata,Argentina 1965 May 11-19 Committee on space research and International union of biochemistry Edited by R.V. Garcia and T.F. Malone Co-sponsored by the International union of physiological sciences 186p Spartan books,Washington,D.C. 1966
Nap 11.0212

SPACE STRUCTURES International conference on space structures Guildford 1960 Sep University of Surrey Edited by R.M. Davies Blackwell,Oxford,1967
Eng 41.2739

SPECIAL CERAMICS :a symposium Proceedings Stoke-on-Trent 1959 Jul 13-15 By P. Popper British ceramic research association Heywood, London,1960
Met 25.0267

SPECIAL CERAMICS 4th :a symposium Proceedings Stoke-on-Trent 1967 Jul 11-13 British ceramic research association Edited by P. Popper British ceramic research association,Stoke-on-Trent,1968
Met 25.2554

SPECIAL CERAMICS 1962 2nd :a symposium Proceedings Stoke-on-Trent 1962 Jul 11-13 British ceramic research association Edited by P. Popper Academic press,London; New York,1963
Met 25.0280

SPECIAL CERAMICS 1964 British ceramic research association :symposium 3rd Proceedings Stoke-on-Trent 1964 Jul 8-10 Edited by P. Popper Academic press,London; New York,1965
Min 10.0656

SPECIAL CERAMICS 1964 3rd :a symposium Proceedings Stoke-on-Trent 1964 Jul 8-10 By P. Popper British ceramic research association Edited by P. Popper Academic press,London;New York,1965
Met 25.0281

SPECIAL CLASSIFICATION AND MULTICOLOUR PHOTOMETRY Saltsjobaden 1964 Aug 17-21 International astronomical union Edited by K. Loden International astronomical union. Symposium, 24 Academic press,London;New York,1966
TA 15.0319

SPECIAL COMMITTEE FOR THE INTERNATIONAL QUIET SUN YEAR Noctilucent clouds :international symposium Tallinn 1966 Edited by I.A. Khvostikov and G. Witt Co-sponsored by the World meteorological organization illus 235p 27cm Academy of sciences of the USSR. Soviet geophysical committee,Moscow,1967 Symposium papers in English with Russian abstracts
Sco 14.8230

SPECIALIZED TISSUES OF THE HEART Symposium on the specialized tissues of the heart Proceedings Rio de Janeiro 1960 Aug Edited by Antonio Paes de Carvalho and others Elsevier,Amsterdam,1961
An 32.3512

SPECIATION IN TROPICAL ENVIRONMENTS symposium Papers London 1968 Oct 31-Nov 1 Linnean society of London British ecological society.Tropical group Linnean society of London.Biological journal, 1,no.1-2 Academic press,London,1969
Gen 34.1298

SPECIES CONCEPT IN ITS RELATION TO THE BRITISH FLORA :a conference Report Species studies in the British flora London 1954 Botanical society of the British Isles Edited by J.E. Lousley B.S.B.I.conference reports,1954 Botanical society of the British Isles,London,1955
BG 38.1259

The SPECIES CONCEPT IN PALAEONTOLOGY :a symposium Systematics association Edited by Peter C.S. Bradley Systematics association.Publication, 2 Systematics association,London,1956
Bot 42.3421

SPECIES CONCEPT IN PALEONTOLOGY Systematics association symposium 2nd papers London 1954 May 19 Systematics association Edited by Peter C. Sylvester-Bradley Systematics association.Publication,2 Systematics association,London,1956
Geol 8.0940

The SPECIES PROBLEM :a symposium Atlanta,Ga 1955 Dec 28-29 American association for the advancement of science Edited by Ernst Mayr American association for the advancement of science.Publication, 50 American association for the advancement of science,Washington,D.C.,1957
Bot 42.3422

The SPECIES PROBLEM :a symposium presented at the Atlanta meeting of the American association for the advancement of science Atlanta,Ga. 1955 Dec 28-29 Edited by Ernst Mayr American association for the advancement of science.Publication, 50 Washington,D.C.,1957
Bal 39.0904

SPECIES STUDIES IN THE BRITISH FLORA Species concept in its relation to the British flora : a conference Report London 1954 Botanical society of the British Isles Edited by J.E. Lousley B.S.B.I.conference reports,1954 Botanical society of the British Isles,London,1955
BG 38.1259

SPECIES STUDY IN THE BRITISH FLORA The Conference on the species concept in its relation to the British flora Report London 1964 Apr 9-11 Botanical society of the British Isles Edited by J.E. Lousley Botanical society of the British Isles.B.S.B.I. conference reports, 4 London,1955
Bot 42.4709

SPECIFIC TUMOUR ANTIGENS :a symposium Sukhumi 1965 May Edited by R.J.C. Harris Organised by the International union against cancer International union against cancer. Monograph series, 2 Munksgaard,Copenhagen, 1967
Radioth 35.0901

SPECIFICATIONS ANS TESTS OF METAL CUTTING MACHINE TOOLS :conference Vol 1-2 Victoria university of Manchester.Institute of science and technology Victoria university of Manchester,Manchester,1970
Eng 41.8169

SPECIFICITE PARASITAIRE DES PARASITES DE VERTEBRES Symposium sur la specificite parasitaire des parasites de vertebres 1 er Neuchatel 1957 Universite de Neuchatel. Institut de zoologie Edited by J.G. Baer Paul Attinger,Neuchatel,1957 Title also in English.Text in English and French
Bal 39.1727

The SPECIFICITY OF CELL SURFACES :a symposium Woods Hole 1965 Sep 1-4 Edited by B.D. Davis and L. Warren Held under the auspices of the Society of general physiologists Prentice-Hall,Englewood Cliffs, N.J.,1967
Phys 20.1506

The SPECIFICITY OF CELL SURFACES :a symposium Woods Hole,Mass. 1965 Sep 1-4 Edited by Bernard D. Davis and Leonard Warren Held under the auspices of the Society of general physiologists Prentice-Hall, Englewood Cliffs,N.J.,1967
An 32.3436

The SPECIFICITY OF CELL SURFACES :a symposium Woods Hole,Mass. 1965 Sep 1-4 Edited by Bernard D. Davis and Leonard Warren Under the auspices of the Society of general physiologists Prentice-Hall,Englewood Cliffs, N.J.,1967 Held at the 12th annual meeting of the society
Gen 34.0669

SPECOLA VATICANA Stellar populations :a conference Proceedings Vatican city 1957 May 20-28 Edited by D.J.K. O'Connell Pontificiae academiae scientiarum.Scripta varia, 16 Specola vaticana.Ricerche astronomiche, 5 544p North Holland, Amsterdam,1958
TA 15.0036

SPECOLA VATICANA Stellar populations :a conference Proceedings Vatican observatory 1957 May 20-28 Edited by D.J.K. O'Connell Pontificia academia scientiarum.Scripta varia, 16 Specola vaticana.Ricerche astronomiche, 5 illus 544p North-Holland,Amsterdam, 1958
Cav 7.2390

SPECOLA VATICANA Stellar populations conferences proceedings Vatican 1957 May 20-28 Edited by D.J.K. O'Connell Pontificia academia scientiarum, 16 Specola vaticana.Ricerche astronomiche, 5 North-Holland,Amsterdam,1958
A Math 4.1150

SPECOLA VATICANA The Green flash and other low sun phenomena By D.J.K. O'Connell illus. 192p North Holland,Amsterdam,1958 With photographs (80 in col.) by Trensch,C.
TA 15.0194

SPECTRAL ANALYSIS OF TIME SERIES :advanced seminar Proceedings Madison,Wis. 1966 Oct 3-5 United States.Army.Mathematics research center and University of Wisconsin.Statistics department Edited by Bernard Harris United States.Army. Mathematics research center.Publications, 18 illus. x,319p 24cm Wiley,New York,1967
Math S 3.1456

SPECTRAL CLASSIFICATION AND MULTICOLOUR PHOTOMETRY Symposium on the spectral classification and multicolour photometry Proceedings Saltsjobaden 1964 Aug.17-21 International astronomical union Edited by K. Loden and others International astronomical union.Symposium, 24 383p Academic press, London;New York,1966
Obs 6.1092

SPECTRAL CLASSIFICATION AND MULTICOLOUR PHOTOMETRY :a symposium Saltsjobaden, Sweden 1964 Aug.17-21 International astronomical union Edited by K. Loden and others International astronomical union. Symposium, 24 383p Academic press,London; New York,1966
A Math 4.1166

SPECTROGRAPHIC ANALYSIS Congress G.A.M.S. 4th Paris 1951 June 20-22 Groupement pour l'advancement des methodes d'analyse spectrographique des produits metallurgiques 24cm Paris,1951
Philos 1.0678

SPECTROSCOPY Centenario della nascita di Galileo Galilei 1564-1964 4th Atti Rome 1964 Sep 14-16 Vol 2,tomo 4: atti del convegno sui campi magnetici solari e la spettroscopia ad alta resolutione.Proceedings of the meeting on solar magnetic fields and high resolution spectroscopy Galileo Galilei. Comitato nazionale per le manifestazioni celebrative Galileo Galilei.Comitato nazionale per le manifestazioni celebrative. Publicazione figs 341p Barbera,Florence, 1966
WSM 43.4471

SPECTROSCOPY Colloquium spectroscopicum internationale 13th Proceedings Ottawa 1967 Jun 19-23 Canadian association for applied spectroscopy Carleton university Co-sponsored by the Society for applied spectroscopy xix,460p 23cm Hilger,London, 1968
Chem 18.2845

SPECTROSCOPY Colloquium spectroscopicum internationale 14th Proceedings Debrecen 1967 Aug 7-12 Vol 1-3 Scientific society of mechanical engineers 1579p 24cm 3 vols Hilger,London,1968
Chem 18.2846

SPECTROSCOPY Colloquium spectroscopicum internationale 15th Madrid 1969 May 26-30 Vol 1-2: plenary conferences;summary of papers 22cm 2 vols Hilger,London,1969-70
Chem 18.2847

SPECTROSCOPY International conference on spectroscopy Bombay 1967 Jan 8-18 India. Department of atomic energy.Spectroscopy division 290p International council of scientific unions,Bombay,1968
TA 15.0636

SPECTROSCOPY International conference on spectroscopy at radio-frequencies :a conference Proceedings Amsterdam 1950 Sep Netherlands physical society Physica, 17,3-4 169-484p 1951
Cav 7.0055

SPECTROSCOPY Methodes nouvelles de spectroscopie instrumentale Exposes Orsay 1966 Apr 25-29 Centre national de la recherche scientifique 2nd edition Centre national de la recherche scientifique. Colloques internationaux, 161 344p CNRS, Paris,1969
Obs 6.3376

SPECTROSCOPY :a conference Report London 1962 Mar 28-31 Institute of petroleum.Hydrocarbon research group Edited by M.J. Wells Institute of petroleum,London, 1962
Chem 18.0193

SPECTROSCOPY AND MOLECULAR STRUCTURE,AND OPTICAL METHODS OF INVESTIGATING CELL STRUCTURE : a general discussion Cambridge 1950 Sep 25-28 Faraday society Faraday society. Discussions, 9 Aberdeen,1951
Bal 39.0230

SPECTROSCOPY AND MOLECULAR STRUCTURE AND OPTICAL METHODS OF INVESTIGATING CELL STRUCTURE : general discussion Cambridge 1950 Sep 25-28 Faraday society Faraday society. Discussions, 9 Faraday society,Aberdeen, 1951
Radioth 35.0453

SPECTROSCOPY IN THE METALLURGICAL INDUSTRY 2nd the symposium Papers Buxton 1962 Jul Institute of physics and the physical society.Spectroscopy group Edited by L. Bovey Hilger and Watts,London,1963
Met 25.1752

SPECTRUM FORMATION IN STARS WITH STEADY-STATE EXTENDED ATMOSPHERES Munich 1969 Apr 16-19 International astronomical union. Commission 36 Edited by H.G. Grotl and P. Wellmann International astronomical union. Commission 36.Colloquium, 2 National bureau of standards.Special publication, 332 332p National bureau of standards,Washington, D.C.,1970
Obs 6.3633

SPEECH Brain function :a conference 3rd Proceedings Los Angeles,Calif. 1963 Nov Vol 3: speech,language and communication Brain research institute Edited by Edward C. Carterette With the support of the United States.Air force.Office of scientific research UCLA forum in medical sciences, 4 University of California press,Berkeley;Los Angeles,1966
Psy 31.0458

SPELEOLOGIYA I KARSTOVEDENIE Soveshchanii po speleologiv i karstovedeni Moscow 1958 Dec. 17-18 Moskovskoe obshchestvo ispytatelei prirody Moscow,1959
Philos 1.1382

SPERMATOZOAN MOTILITY :a symposium New York 1960 Dec 29-30 American association for the advancement of science Edited by D.W. Bishop Sponsored by the American society of zoologists and the Society of general physiologists American association for the advancement of science.Publication, 72 Washington,D.C.,1962
Bal 39.0391

The SPINAL CORD :a Ciba foundation symposium Ciba foundation Churchill,London,1953
Phys 20.2005

The SPINAL CORD :a Ciba foundation symposium London 1952 Feb 26-28 Ciba foundation Edited by G.E.W. Wolstenholme Churchill, London,1953
An 32.4129

The SPIRAL STRUCTURE OF OUR GALAXY Basle 1969 Aug 29-Sep 4 International astronomical union Edited by W. Becker and G. Contopoulos International astronomical union.Symposium, 38 478p Reidel,Dordrecht,1970
Obs 6.3320

The SPIRAL STRUCTURE OF OUR GALAXY :symposium Proceedings Basle 1969 Aug 29-Sep 4 International astronomical union Edited by W. Becker and G. Contopoulos International astronomical union.Symposium, 38 478p Reidel,Dordrecht,1970
TA 15.0489

SPLINE FUNCTIONS Approximations with special emphasis on spline functions :a symposium Proceedings Madison,Wis. 1970 May 5-7 United States.Army.Mathematics research center Edited by I.J. Schoenberg Held at the University of Wisconsin United States.Army. Mathematics research center.Publications, 23 xi,488p 24cm Academic press,London;New York,1971
Math S 3.1675

SPLINE FUNCTIONS Theory and applications of spline functions :an advanced seminar Proceedings Madison,Wis. 1968 Oct 7-9 United States.Army.Mathematics research center Edited by T.N.E. Greville Held at the University of Wisconsin United States.Army. Mathematics research center.Publications, 22 xi,212p 24cm Academic press,New York; London,1969
Math S 3.1674

SPOLETO 1966 Thermoluminescence of geological materials N.A.T.O.advanced research institute on applications of thermoluminescence to geological problems Proceedings North Atlantic treaty organization Edited by D.J. McDougall Academic press,London,1968
Min 10.0628

SPORES 4 International spore conference 4th Urbana,Ill. 1968 Oct 4-6 American society for microbiology Edited by L.Leon Campbell Under the auspices of the University of Illinois American society for microbiology,1967
Gen 34.0711

SQUAW VALLEY,CALIF. 1964 Molecular biophysics International summer school of molecular biophysics Proceedings Edited by B. Pullman and M. Weissbluth Sponsored by the North atlantic treaty organization Academic press,New York,1965
Radioth 35.1418

SQUAW VALLEY,CALIF. 1964 Molecular biophysics International summer school of molecular biophysics Proceedings Edited by Bernard Pullman and Mitchel Weissbluth Sponsored jointly by the North Atlantic treaty organization and United States.Office of naval research Academic press,New York;London,1965
Bioch 33.2360

SQUAW VALLEY,CALIF. 1964 Molecular biophysics :international summer school Proceedings North Atlantic treaty organization,and,United States.Office of naval research Edited by Bernard Pullmann and Mitchel Weissbluth Academic press,New York; London,1965
Chem 18.2249

STABILITY AND DISTORTION OF VISUAL SPACE International congress on human factors in electronics :visual distortion from the engineering point of view Papers Long Beach,Calif. 1962 May 3-4 By Richard L. Gregory University of Cambridge,Cambridge, 1962
Psy 31.0382

STABILITY OF STOCHASTIC DYNAMICAL SYSTEMS : international symposium Proceedings Coventry 1972 Jul 10-14 University of Warwick.Control theory centre Edited by Ruth F. Curtain Lecture notes in mathematics, 254 ix,332p 26cm Springer,Berlin,1972
P Math 2.4569

STABILITY OF STOCHASTIC DYNAMICAL SYSTEMS Proceedings Coventry 1972 July Edited by Ruth E. Curtain illus ix,334p 25cm Springer,Berlin,1972
A Math 4.1870

STAINLESS STEELS Conference on stainless steels for the fabricator and user Proceedings Birmingham 1968 Sep 10-12 Iron and steel institute and Birmingham metallurgical association Iron and steel institute,London,1969
Met 25.2242

STANDARDIZATION IN HUMAN CYTOGENETICS Chicago conference Chicago,Ill. 1966 Sep 3-4,10 Sponsored by the National foundation-March of dimes Birth defects:original article series, 2,no.2 National foundation March of dimes,New York,1966
Gen 34.0895

STANDARDIZATION OF RADIOACTIVE WASTE CATEGORIES report of a panel Vienna 1967 Nov 6-10 International atomic energy agency International atomic energy agency.Technical reports series, 101 IAEA,Vienna,1970
Chem 18.2735

STANDING COMMITTEE ON COMBUSTION SYMPOSIA Symposium on combustion,flame and explosion phenomena 3rd Papers Madison,Wisc. 1948 Sep 7-11 Williams and Wilkins,Baltimore, Md.,1949 Chairman of committee:B.Lewis. Later volumes entitled 'Symposium (international) on combustion,flame and explosion phenomena'
Chem 18.0433

STANDING COMMITTEE ON COMBUSTION SYMPOSIA Symposium on combustion,flame and explosion phenomena abstracts of papers 3rd Madison, Wisc. 1948 Sep 7-11 University of Wisconsin,Madison,Wisc.,1948
Chem 18.0432

STANDING COMMITTEE ON COMBUSTION SYMPOSIA Symposium on combustion and flame and explosion phenomena 3rd Madison,Wis. 1948 Sep 7-11 xiii,748p Williams and Wilkins,Baltimore,Md.,1949
Chem E 24.1404

STANDING COMMITTEE ON COMBUSTION SYMPOSIA Symposium (International) on combustion (combustion and detonation waves) 4th Cambridge,Mass. 1952 Sep 1-5 xx,926p Williams and Wilkins company,Baltimore,Md., 1953
Chem E 24.1405

STANDING COMMITTEE ON COMBUSTION SYMPOSIA Symposium (international) on combustion 5th Pittsburgh,Pa 1954 Aug 30-Sep 3 combustion in engines and combustion kinetics Combustion institute v,802p Reinhold; Chapman and Hall,New York;London,1955
Chem E 24.1914

STANDING COMMITTEE ON COMBUSTION SYMPOSIA Symposium (international) on combustion,flame and explosion phenomena :combustion in engines and combustion kinetics 5th Papers Pittsburgh,Pa. 1954 Aug 30-Sep 3 Published for the Combustion institute Reinhold; Chapman and Hall,New York;London,1955 Later symposia organized by the Combustion institute
Chem 18.0435

STANDING COMMITTEE ON COMBUSTION SYMPOSIA Symposium (international) on combustion,flame and explosion phenomena (combustion and detonation waves) 4th Papers Cambridge, Mass. 1952 Sep 1-5 Held at the Massachusetts institue of technology Williams and Wilkins,Baltimore,Md.,1953 Committee chairman:B.Lewis.Earlier volumes entitled 'Symposium on combustion,flame and explosion phenomena'
Chem 18.0434

STANFORD,CALIF 1968 Computation of turbulent boundary layers :1968 AFOSR-IFP-Stanford conference Proceedings Vol 1-2 United States.Air force.Office of scientific research Edited by S.J. Kline and others Bibliog.,Illus,port 26cm 2 vols Stanford university,Stanford,1969
A Math 4.1481

STANFORD,CALIF. 1958 Symposium on naval structural mechanics 1st Proceedings United States.Office of naval research and Stanford university Edited by J.Norman Goodier and Nicholas J. Hoff International series on aeronautical sciences and space flight Pergamon,Oxford,1960
Met 25.1300

STANFORD,CALIF. 1959 Mathematical methods in the social sciences,1959 Stanford symposium 1st proceedings Stanford university Edited by Kenneth J. Arrow and others Financed by the United States.Office of naval research Stanford university.Stanford mathematical studies in the social sciences, 4 vii,365p 24cm Stanford university press,Stanford,Calif.,1960
P. Math 2.2259

STANFORD,CALIF. 1960 Creep in structures : a colloquium International union of theoretical and applied mechanics Edited by Nicholas J. Hoff Springer,Berlin,1962
Eng 41.2529

STANFORD,CALIF. 1960 Creep in structures colloquium International union of theoretical and applied mechanics Edited by Nicholas J. Hoff Springer-verlag,Berlin,1962
Met 25.0959

STANFORD,CALIF. 1960 Free radicals in biological systems Symposium on free radicals in biological systems Proceedings Stanford university Edited by M.S. Blois Academic press,New York;London,1961
Bioch 33.0309

STANFORD,CALIF. 1960 Free radicals in biological systems :symposium Proceedings Edited by M.S. Blois and others illus.,port. Academic press,New York;London,1961
Radioth 35.1407

STANFORD,CALIF. 1960 International congress for logic,methodology and philosophy of science Proceedings International union of history and philosophy of science.Division of logic,methodology and philosophy of science National academy of sciences Edited by Ernest Nagel and others ix,661p 25cm Stanford university press,Stanford,Calif.,1962
Math S 3.1853

STANFORD,CALIF. 1960 Logic,methodology and philosophy of science International congress for logic,methodology and philosophy of science Proceedings International union of history and philosophy of science.Division of logic,methodology and philosophy of science Edited by Ernest Nagel and others ix,661p Stanford university press,Stanford,Calif.,1962
WSM 43.0828

STANFORD,CALIF. 1960 The Heat transfer and fluid mechanics institute 1960 proceedings Edited by D.M. Mason and others Stanford university press,Stanford,Calif.,1960
A Math 4.0518

STANFORD,CALIF. 1962 International symposium on X-ray optics and X-ray microanalysis 3rd Proceedings Edited by H.H. Pattee,jr. and others Sponsored by Stanford university.Biophysics laboratory illus xvii,622p Academic press,New York, 1963
Cav 7.2189

STANFORD,CALIF. 1962 X-ray optics and X-ray microanalysis International symposium on x-ray microtechniques 3rd Stanford university.Biophysics laboratory Edited by H. H. Pattee and others With the financial support of the United States.Office of naval research and the National science foundation Academic press,New York;London,1963
Met 25.1379

STANFORD,CALIF. 1962 X-ray optics and x-ray microanalysis International symposium on x-ray optics and x-ray microanalysis 3rd Edited by H.H. Pattee Sponsored by Stanford university.Biophysics laboratory Academic press,New York,1963
Eng 41.4462

STANFORD,CALIF. 1963 Quantum aspects of polypeptides and polynucleotides :a symposium Papers Edited by M. Weissbluth Biopolymers. Symposia,1 Interscience,New York,1964
Pha 16.0117

STANFORD,CALIF. 1964 Joint automatic control conference 5th Preprints of conference papers Under the sponsorship of the Institute of electrical and electronics engineers.Professional technical group in automatic control Stanford,Calif.,1964
Eng 41.6014

STANFORD,CALIF. 1968 Applied mechanics International congress of applied mechanics 12th Proceedings International union of theoretical and applied mechanics Edited by H. Hetenyi and W.G. Vincenti Bibliog,illus xxiv,420p 291cm Springer,Berlin,1969
A Math 4.1722

STANFORD,CALIF. 1969 International conference on photoconductivity 3rd Proceedings International union of pure and applied physics American physical society United States.Office of naval research Edited by E.M. Pell Bibliog,diagrms,graphs xi,410p 26cm Pergamon,Oxford,1971
Cav 7.3073

STANFORD,CALIF. 1971 Symposium on operating systems principles 3rd Association for computing machinery illus 163p Association for computing machinery,New York,1971
Math L 5.3824

STANFORD RESEARCH INSTITUTE Chemical reactions in the lower and upper atmosphere :a symposium Proceedings San Francisco,Calif. 1961 Apr 18-20 390p Wiley,New York,1961
Nap 11.0929

STANFORD RESEARCH INSTITUTE Chemical reactions in the lower and upper atmosphere : an international symposium Proceedings San Francisco,Calif. 1961 Apr 18-20 Interscience,New York;London,1961
Chem 18.0476

STANFORD RESEARCH INSTITUTE High temperature technology An International symposium on high temperature technology 3rd Proceedings Pacific Grove,Calif. 1963 Sep 8-11 International union of pure and applied chemistry.Commission on high temperatures and refractories Butterworths,London,1964
Met 25.0992

STANFORD RESEARCH INSTITUTE International symposium on high temperature technology 2nd Proceedings Asilomar,Calif 1959 Oct 6-9 McGraw-Hill,New York,1960
Met 25.2535

STANFORD RESEARCH INSTITUTE The Conference on the use of solar energy :the scientific basis Transactions Tucson,Ariz. 1955 Oct 31-Nov 1 Vol 1-5 Edited by Edwin F. Carpenter and others Co-sponsored by the Association for applied solar energy illus 6 vols University of Arizona,Tucson,Ariz., 1958
Nap 11.0111

STANFORD UNIVERSITY Free radicals in biological systems Symposium on free radicals in biological systems Proceedings Stanford,Calif. 1960 Mar 21-23 Edited by M. S. Blois Academic press,New York;London,1961
Bioch 33.0309

STANFORD UNIVERSITY Mathematical methods in the social sciences,1959 Stanford symposium 1st proceedings Stanford,Calif. 1959 Jun 15-24 Edited by Kenneth J. Arrow and others Financed by the United States.Office of naval research Stanford university.Stanford mathematical studies in the social sciences, 4 vii,365p 24cm Stanford university press,Stanford,Calif.,1960
P. Math 2.2259

STANFORD UNIVERSITY Symposium on naval structural mechanics 1st Proceedings Stanford,Calif. 1958 Aug 11-14 Edited by J. Norman Goodier and Nicholas J. Hoff International series on aeronautical sciences and space flight Pergamon,Oxford,1960
Met 25.1300

STANFORD UNIVERSITY The Heat transfer and fluid mechanics institute Proceedings Seattle,Wash. 1962 Jun 13-15 University of Washington Edited by Edward Ehlers and others 293p Stanford university press, Stanford,Calif.,1962
Cav 7.2422

STANFORD UNIVERSITY.BIOPHYSICS LABORATORY International symposium on X-ray optics and X-ray microanalysis 3rd Proceedings Stanford,Calif. 1962 Aug 22-24 Edited by H. H. Pattee,jr. and others illus xvii,622p Academic press,New York,1963
Cav 7.2189

STANFORD UNIVERSITY.BIOPHYSICS LABORATORY X-ray optics and X-ray microanalysis International symposium on x-ray microtechniques 3rd Stanford,Calif. 1962 Aug 22-24 Edited by H.H. Pattee and others With the financial support of the United States.Office of naval research and the National science foundation Academic press, New York;London,1963
Met 25.1379

STANFORD UNIVERSITY.BIOPHYSICS LABORATORY X-ray optics and x-ray microanalysis International symposium on x-ray optics and x-ray microanalysis 3rd Stanford,Calif. 1962 Aug 22-24 Edited by H.H. Pattee Academic press,New York,1963
Eng 41.4462

STAR EVOLUTION Evoluzione delle stelle Scuola internazionale di fisica "Enrico Fermi" 28 corso rendiconti Varenna 1962 Aug 20-Sep.1 Societa Italiana di fisica Edited by L. Gratton 488p Academic press,New York; London,1963
Obs 6.0685

STAR EVOLUTION Evoluzione delle stelle Scuola internazionale di fisica'Enrico Fermi' 28 corso Rendiconti Varenna 1962 Aug 20-Sep 1 Societa Italiana di fisica Edited by L. Gratton Academic press,New York;London, 1963
A Math 4.1156

STARCH AND CELLULOSE Biological transformations of starch and cellulose :a symposium London 1953 Feb 21 Biochemical society Held at the London school of hygiene and tropical medicine Biochemical society. Symposia, 11 Cambridge university press, Cambridge,1953
Bioch 33.1373

STARS Low-luminosity stars Symposium on low-luminosity stars Proceedings Charlottesville,Va. 1968 University of Virginia Edited by Shiv S. Kumar 542p Gordon and Breach,New York,1969
Obs 6.3627

STARS Spectrum formation in stars with steady-state extended atmospheres Munich 1969 Apr 16-19 International astronomical union.Commission 36 Edited by H.G. Grotl and P. Wellmann International astronomical union. Commission 36.Colloquium, 2 National bureau of standards.Special publication, 332 332p National bureau of standards,Washington, D.C.,1970
Obs 6.3633

The STATE AND MOVEMENT OF WATER IN LIVING ORGANISMS Swansea 1964 Sep 8-12 Society for experimental biology Society for experimental biology.Symposia, 19 Cambridge university press,Cambridge,1965
Phys 20.0967

The STATE AND MOVEMENT OF WATER IN LIVING ORGANISMS Swansea 1964 Society for experimental biology Society for experimental biology.Symposia,19 Cambridge university press,Cambridge,1965
Pha 16.0272

The STATE AND MOVEMENT OF WATER IN LIVING ORGANISMS :a symposium Swansea 1964 Sep 8-12 Edited by G.E. Fogg Society for experimental biology.Symposia, 19 Cambridge university press,Cambridge,1965
An 32.2522

STATE COLLEGE,PA. 1950 Computing methods and the phase problem in X-ray crystal analysis :conference report Edited by Ray Pepinsky sponsored by the Rockefeller foundation Pennsylvania state college,State college,Pa.,1952
Min 10.0756

STATE OF STRESS IN THE EARTH'S CRUST :an international conference Proceedings Santa Monica,Calif. 1963 Jun 13-14 Geological society of America.Committee on rock mechanics and Rand corporation Edited by William R. Judd New York,1964
Geod 9.0143

STATE UNIVERSITY OF NEW YORK Information processing in the nervous system :a symposium Proceedings Buffalo,N.Y. 1968 Oct 21-24 Edited by K.N. Leibovic Springer,Berlin,1969
Bal 39.1019

STATE UNIVERSITY OF NEW YORK Intuitionism and proof theory :summer conference Proceedings Buffalo,N.Y. 1968 Edited by A. Kino and others Studies in logic and the foundations of mathematics viii,516p North-Holland,Amsterdam;London,1970
WSM 43.1239

STATE UNIVERSITY OF NEW YORK National conference on weather modification 1st Proceedings Albany,N.Y. 1968 Apr 28-May 1 532p American meteorological society,Boston, Mass.,1968
Nap 11.0975

STATE UNIVERSITY OF NEW YORK Set-valued mappings,selections and topological properties Proceedings Buffalo,N.Y. 1969 May 8-10 Edited by W.M. Fleischman Lecture notes in mathematics, 171 Springer,Berlin,1970
Math S 3.1657

STATE UNIVERSITY OF NEW YORK Set-valued mappings,selections and topological properties conference Proceedings Buffalo,N.Y. 1969 May 8-10 Lecture notes in mathematics, 171 x,109p 25cm Springer,Berlin,1970
P Math 2.4114

STATE UNIVERSITY OF NEW YORK University education in computing science Conference on graduate academic and related research programs in computing science Proceedings Stony Brook,N.Y. 1967 Jun Edited by Aaron Finerman Association for computing machinery. Monograph series Academic press,New York, 1968
Eng 41.2223

STATE UNIVERSITY OF NEW YORK.COLLEGE AT PLATTSBURGH Origin of anorthosite and related topics Annual George H.Hudson symposium 2nd Papers Plattsburgh,N.Y. 1966 Edited by Yngvar W. Isachsen New York state museum and science service.Memoir, 18 New York state museum and science service, Albany,N.Y.,1968
Min 10.1428

STATE UNIVERSITY OF NEW YORK.SCHOOL OF PHARMACY Immunity,cancer,and chemotherapy :basic relationships on the cellular level 1st Report Buffalo,N.Y. Edited by Enrico Mihich Academic press,New York;London,1967
Path 30.2744

STATE UNIVERSITY OF NEW YORK,BINGHAMPTON Conference on monotone mappings and open mappings 1st Proceedings Binghamton,N.Y. 1970 Oct 8-11 Edited by Louis F. McAuley xxii,422p 25cm State university of New York,Binghamton,N.Y.,1971 Proceedings dedicated to the memory of G.T.Whyburn
P Math 2.4118

STATE UNIVERSITY OF NEW YORK,BUFFALO Conference on carbon 3rd Proceedings Buffalo,N.Y. 1957 Jun 17-21 Edited by S. Mrozowski and others Co-sponsored by the United States.Office of naval research Conference on carbon.Proceedings, 2 718p Pergamon press,Oxford,1959
TA 15.0214

STATE UNIVERSITY OF NEW YORK,BUFFALO The Conference on carbon 4th Proceedings Buffalo,N.Y. 1959 Jun 15-19 Edited by S. Mrozowski and others Conference on carbon. Proceedings, 3 778p Pergamon press, Oxford,1960
TA 15.0215

STATIC ELECTRIFICATION The Institute of physics and the Physical society conference on static electrification 2nd Proceedings London 1967 May Institute of physics and the Physical society.Static electrification group Institute of physics and the Physical society.Conference series, 4 25cm London, 1967
Cav 7.2822

STATIC RELAYS SYMPOSIUM Static relays for electronic circuits Edited by Richard F. Blake illus. vii,198p 23cm Engineering publishers,Elizabeth,N.J.,1961 Based on the Static relays symposium
Math L 5.0885

STATISTICAL ASPECTS OF FATIGUE Symposium on fatigue with emphasis on statistical approach New York 1952 Jun 24 Pt 2 American society for testing materials A.S.T.M. Special technical publication American society for testing materials,Philadelphia 3, Pa.,1953
Met 25.0913

STATISTICAL ASPECTS OF FATIGUE Symposium on statistical aspects of fatigue Atlantic City, N.J. 1951 Jun 19 Pt 1 American society for testing materials A.S.T.M. Special technical publication, 121 American society for testing materials,Philadelphia,Pa., 1951
Met 25.0912

STATISTICAL COMPUTATION :a conference Proceedings Madison,Wis. 1969 Apr 28-30 Edited by Roy C. Milton and John A. Nelder Organized by the University of Wisconsin bibliog. Academic press,New York;london,1969
Math S 3.1503

STATISTICAL INFERENCE Foundations of statistical inference :a symposium Proceedings Waterloo,Ont. 1970 Mar 31-Apr 9 University of Waterloo.Department of statistics Edited by V.P. Godambe and D.A. Sprott viii,519p 25cm Holt,Toronto,1971
Math S 3.1848

STATISTICAL MECHANICS Conference on statistical mechanics and thermodynamics. Foundations and applications International union of pure and applied physics :meeting Proceedings Copenhagen 1966 Edited by Thor A. Bak table 582p 25cm Benjamin, New York,1967
Cav 7.3022

STATISTICAL MECHANICS Statistical mechanics of equilibrum and non-equilibrium The Symposium on statistical mechanics and thermodynamics Proceedings Aachen 1964 Jun 15-20 International union of pure and applied physics Edited by J. Meixner 274p North-Holland,Amsterdam,1965
Cav 7.0001

STATISTICAL METEOROLOGY National conference on statistical meteorology 1st Proceedings Hartford,Conn. 1968 May 27-29 Sponsored by the American meteorological society American meteorlogical society, Boston,Mass.,1968 Co-Sponsor:A.M.S. committee on statistical meteorology
Math S 3.1583

STATISTICAL METHOD IN INDUSTRIAL PRODUCTION conference proceedings Sheffield 1950 Sep 29-Oct 1 Royal statistical society. Industrial applications section 89p 25cm Royal statistical society,London,1951
Math 3.0821

STATISTICAL MODELS AND TURBULENCE :a symposium Proceedings San Diego,Calif. 1971 Jul 15-21 Edited by M. Rosenblatt and C. Van Atta Lecture notes in physics, 12 viii, 492p 25cm Springer,Berlin,1972
Math S 3.1832

2081

STATISTICAL THEORY OF RELIABILITY advanced seminar proceedings Madison,Wis. 1962 May 8-10 United States.Army.Mathematics research center Edited by Marvin Zelen United States army.Mathematics research center. Publications, 9 xvii,166p 23cm University of Wisconsin press,Madison,Wis., 1963
Math 3.0700

STATISTICS Bulletin de l'Institut international de statistique:compte rendus des sessions de l'Institut 19e 1930- Vol 25- Institut international de statistique v.p. 1931- Wants Vol 31:25th session
Math 3.1329

STATISTICS Foundations of statistical inference Joint statistics seminar :a discussion London 1959 Jul 27-28 By Leonard J. Savage and others Birkbeck college Imperial college of science and technology 112p Methuen;Wiley,London;New York,1964
WSM 43.1769

STATISTICS Mathematical statistics and probability Berkeley symposium on mathematical statistics and probability 5th Proceedings Berkeley,Calif. 1965 Jun-Jul Berkeley,Calif. 1966 Dec-Jan Edited by Lucien M.le Cam and Jerzy Neyman bibliog., illus. 26cm 6 vols University of California press,Berkeley,Los Angeles,Calif., 1967
Math S 3.1455

STATISTICS Mathematical statistics and probability Berkeley symposium on mathematical statistics and probability 3r proceedings Berkeley,Calif. 1954 Dec.26-31 Berkeley,Calif. 1955 Jul-Aug Vols. 1-5 University of California.Statistical laboratory Edited by Jerzy Neyman 26cm 5 vols University of California press,Berkeley; Los Angeles,Calif.,1956
Math 3.0687

STATISTICS Mathematical statistics and probability Berkeley symposium on mathematical statistics and probability 1st proceedings Berkeley,Calif. 1945 Aug.13-18 Berkeley,Calif. 1946 Jan.27-29 University of California.Statistical laboratory Edited by Jerzy Neyman viii, 501p 26cm University of California press, Berkeley;Los Angeles,Calif.,1949
Math 3.0685

STATISTICS Symposium on mathematical statistics and probability Proceedings Berkeley,Calif. 1945 Aug.13-18 and 1946 Jan.27-29 University of California. Department of mathematics Edited by Jerzy Neyman 501p University of California press, Berkeley;Los Angeles,1949
Nap 11.0048

STATISTICS AND MATHEMATICS IN BIOLOGY conference Ames,Iowa 1952 Jun-Jul Iowa state college Edited by Oscar Kempthorne and others Co-sponsored by Biometrics society xi,632p 23cm Iowa state college press,Ames,Iowa,1954
Math 3.0764

2082

STATISTICS IN THE PHYSICAL SCIENCES :a conference Proceedings Belgrade 1965 Sep 17 Under the auspices of the International association for statistics in physical sciences International association for statistics in physical sciences. Proceedings, 2 Bibliog 24cm International statistitcal institute,Belgrade, c1966 Reprinted from the Bulletin of the International statistical institute. Proceedings of the 35th session,1965
Math 3.1157

STATISTICS PROBABILITY Mathematical statistics and probability Berkeley symposium on mathematical statistics and probability 2nd proceedings Berkeley, Calif. 1950 Jul.31-Aug.12 University of California.Statistical laboratory Edited by Jerzy Neyman Also supported by the Office of naval research xi,666p 26cm University of California press,Berkeley;Los Angeles,Calif. 1951
Math 3.0686

STATISTICS PROBABILITY Mathematical statistics and probability Berkeley symposium on mathematical statistics and probability 4th proceedings Berkeley, Calif. 1960 Jun.20-Jul.30 Vols. 1-4 University of California.Statistical laboratory Edited by Jerzy Neyman 26cm 4 vols University of California press,Berkeley; Los Angeles,Calif.,1961
Math 3.0688

STATISTIS Research seminar on archaeology and related subjects day meeting on statistics and archaeology London 1964 May 30 Institute of archaeology London,1964 Unbound typescript papers
Math 3.1115

STAVANGER 1965 Peaceful uses of automation in outer space IFAC symposium on automatic control in the peaceful uses of space International federation of automatic control Edited by John A. Aseltine Plenum,New York, 1966
Eng 41.6002

STAZIONE ZOOLOGICA DI NAPOLI.PUBLICAZIONI, 21 Embriologia e genetica congress Naples 1948 Jun 1-318 Consiglio nazionale per la ricerca Per iniziativa del Centre di studio di biologia,di citologia genetica e di enzimologia Naples,1949
Gen 34.0525

STE.MARGUERITE,QUE. 1953 Brain mechanisms and consciousness :a symposium Council for international organizations of medical sciences Blackwell,Oxford,1954
Phys 20.2009

STE.MARGUERITE,QUE. 1953 Brain mechanisms and consciousness :a symposium Council for international organizations of medical sciences Edited by H.H. Jasper and others Blackwell,Oxford,1954
An 32.4625

STEAM PLANT Modern steam plant practice :a convention Proceedings The Hague 1971 Apr 28-30 Institution of mechanical engineers.Steam plant group Institut van ingenieurs.Steam plant group IMF,London,1971
Eng 41.8643

STEAM-PLANT ENGINEERING:PRESENT STATUS AND FUTURE TRENDS ;a convention Proceedings Harrogate 1963 May 2-4 Institution of mechanical engineers Institution of mechanical engineers,London,1964
Eng 41.7276

STEAM TURBINE RESEARCH Conference on steam turbine research and development Proceedings London 1953 Mar 6 Institution of mechanical engineers Institution of mechanical engineers,London,1953
Eng 41.7296

STEEL Conference on steel in architecture Proceedings London 1969 Nov 24-26 Organized by the British constructional steelwork association British constructional steelwork association,London,1970
Eng 41.0269

STEEL Continuous casting of steel Iron and steel institute :autumn general meeting Proceedings London 1964 Nov 25-26 Iron and steel institute Iron and steel institute. Special report, 89 London,1965
Met 25.2569

STEEL Joint conference on low-alloy steels organized by BISRA and the Iron and steel institute Proceedings Scarborough 1965 Apr 2-4 British iron and steel research association Iron and steel institute Iron and steel institute.Publication, 114 illus 269p London,1969
Met 25.2566

STEEL Mechanical working and steel processing committee 2nd Operating metallurgy conference 9th Proceedings Philadelphia,Pa. 1966 Dec 5-9 American institute of mining,metallurgical and petroleum engineers Metallurgical society of the AIME.Conference series, 50 Gordon and Breach,New York,1968
Met 25.2572

STEEL Non-destructive examination in the steel industry Conference on non-destructive examination applied to process control in the steel industry Papers and discussions Swansea 1967 Jan 3-5 Swansea and district metallurgical society Iron and steel institute Society of non-destructive examination Iron and steel institute. Publication, 103 illus 193p London,1967
Met 25.2601

STEEL Optimization of steel product yield Conference on optimization of steel product yield Proceedings London 1967 May 3-4 Iron and steel institute Iron and steel institute.Publication, 107 London,1967 Conference held at the Institute's annual general meeting
Met 25.2561

STEEL MAKING Symposium on steel making (acid and basic open-hearth practice) Iron and steel institute Iron and steel institute. Special report, 22 Iron and steel institute, London,1938
Met 25.0324

STEELMAKING Control of composition in steelmaking :annual general meeting Papers and discussions London 1966 May 4-5 Iron and steel institute Iron and steel institute. Special publication, 99 London,1967
Met 25.2557

STEELMAKING The Physical chemistry of steelmaking :a conference Proceedings Dedham,Mass. 1956 May 28-Jun 3 Edited by John F. Elliott Sponsored by Massachusetts institute of technology.Department of metallurgy Wiley;Chapman and Hall,New York; London,1958
Met 25.1789

STEELMAKING CONFERENCE 36TH The All-basic open-hearth furnace being also the report of the 36th steelmaking conference Leamington-Spa 1951 May 2-3 British ceramic research association and British iron and steel research association Iron and steel institute.Special report, 46 Iron and steel institute,London,1952
Met 25.0330

STEELMAKING IN THE BASIC ARC FURNACE Iron and steel institute conference :technical sessions Papers and discussions Sheffield 1964 Jul 6-10 Iron and steel institute Iron and steel institute.Special report, 87 London,1964
Met 25.2558

STEELS Conference on welding creep-resistant steels Proceedings Newcastle-upon-Tyne 1970 Feb 17-18 Welding institute Welding institute,Abington,1970 Conference director:B.Phelps
Met 25.2725

STEELS FOR REACTOR PRESSURE CIRCUITS :a symposium London 1960 Nov 30-Dec 2 Iron and steel institute Iron and steel institute.Special report, 69 Iron and steel institute,London,1961
Met 25.0414

The STEENROD ALGEBRA AND ITS APPLICATIONS :a conference to celebrate N.E.Steenrod's sixtieth birthday Proceedings Columbus, Ohio 1970 Mar 30-Apr 4 Battelle memorial institute Edited by F.P. Peterson Lecture notes in mathematics, 168 viii,317p 25cm Springer,Berlin,1970
P Math 2.4120

STELLAR AND SOLAR MAGNETIC FIELDS Symposium on stellar and solar magnetic fields Munich 1963 Sep.3-10 International astronomical union Edited by R. Lust International astronomical union.Symposium, 22 460p North Holland,Amsterdam,1965
Obs 6.1148

STELLAR AND SOLAR MAGNETIC FIELDS :a symposium Rottach-Egern 1963 Sep.3-10 International astronomical union Edited by R. Lust International astronomical union. Symposium, 22 460p North-Holland, Amsterdam,1965
A Math 4.1160

STELLAR ATMOSPHERES Aerodynamic phenomena in stellar atmospheres Symposium on cosmical gas dynamics held at the International school of physics "Enrico Fermi" 4th proceedings Varenna 1960 Aug.18-30 International astronomical union and International union of theoretical and applied mechanics Edited by R.N. Thomas and others Under the auspices of Societa Italiana di fisica International astronomical union.Symposium, 12 515p Bologna,c1961 Reprinted from Supplemento Nuovo Cimento,22,no.1,1961
A Math 4.1152

STELLAR ATMOSPHERES Aerodynamic phenomena in stellar atmospheres Symposium on cosmical gas dynamics held at the International school of physics "Enrico Fermi" 4th proceedings Varenna 1960 Aug 18-30 International astronomical union and International union of theoretical and applied mechanics Edited by R.N. Thomas and others under the auspices of the Societa italiana di fisica International astronomical union symposium, 12 515p Zanichelli,Bologna,c1961 Reprinted from the 'Supplemento del nuovo cimento',vol.22,no.1,1961
Obs 6.1736

STELLAR ATMOSPHERES Aerodynamical phenomena in stellar atmosphwre Nice 1965 Sep International astronomical union Edited by Richard N. Thomas International astronomical union.Symposium, 28 Academic press,London, 1967
TA 15.0411

STELLAR ATMOSPHERES Formation of spectrum lines Harvard-Smithsonian conference on stellar atmospheres 2nd Proceedings Cambridge,Mass. 1965 Jan 20-22 Smithsonian institution.Astrophysical laboratory and Harvard university.Astronomical observatory Smithsonian institution.Astrophysical observatory,Research in space. Special report, 174 illus 461p Cambridge,Mass.,1965
Obs 6.2121

STELLAR ATMOSPHERES Harvard-Smithsonian conference on stellar atmospheres 1st Proceedings Cambridge,Mass. 1964 Dec 20-21 Smithsonian institution.Astrophysical laboratory and Harvard university. Astronomical observatory Smithsonian institution.Astrophysical observatory.Research in space. Special report, 167 335p Cambridge,Mass.,1964
Obs 6.2122

STELLAR ATMOSPHERES Theory and observation of normal stellar atmospheres Harvard-Smithsonian conference on stellar atmospheres 3rd Proceedings Cambridge,Mass. 1968 Harvard university.Astronomical observatory Smithsonian institution.Astrophysical observatory Edited by O. Gingerich 472p Massachusetts institute of technology press, Cambridge,Mass.,1969
Obs 6.3354

STELLAR ATMOSPHERES Theorie des atmospheres stellaires :conference Saas-Fee 1971 Mar 28-Apr 3 By D. Mihalas and others 312p Observatoire de Geneve,Sauverny,1971
Obs 6.3631

STELLAR ATMOSPHERES Transfer of radiation in stellar atmospheres Colloquium on the theory of stellar atmospheres 3rd Herstmonceux 1962 Aug.15-16 International astronomical union Edited by C.de Jager and A.B. Underhill 201p Pergamon press,London, 1963 Vol 3,no.2 of Journal of quantitative spectroscopy and radiative transfer. AprJun.1963
Obs 6.0875

STELLAR ATMOSPHERES :a conference proceedings Bloomington,Ind. 1954 Oct 1-2 Edited by Marshal H. Wrubel financed by the National science foundation 183p Indiana university,Bloomington,Ind.,1955
Obs 6.1900

STELLAR CLASSIFICATION Principes fondamentaux de classification stellaire Proceedings Paris 1953 Jun 29 - Jul 4 Centre national de la recherche scientifique With the support of the Rockefeller foundation Centre national de la recherche scientifique. Colloques internationaux, 55 190p Centre national de la recherche scientifique,Paris, 1955
Obs 6.2008

STELLAR EVOLUTION Evolution stellaire avant la sequence principale Liege 1969 Jun 30-Jul 2 Societe des sciences de Liege.Memoires. 5s, 19,no 1 377p Liege,1970 Papers in English and French
TA 15.0685

STELLAR EVOLUTION Symposium on stellar evolution papers La Plata 1960 Nov. 7-11 Universidad nacional de La Plata.Observatorio astronomico Edited by J. Sahade 309p Astronomical observatory,La Plata,1962 Scientific meeting organized to celebrate the sesquicentennial of the May revolution
Obs 6.1505

STELLAR EVOLUTION :a symposium proceedings New York 1963 Nov. Edited by R.F. Stein and A.G.W. Cameron Organized by the National aeronautics and space administration.Goddard institute for space studies 464p Plenum press,New York,1966
A Math 4.1164

STELLAR EVOLUTION :a symposium proceedings New York 1963 Nov.13-15 Edited by R.F. Stein and A.G.W. Cameron Organized by Goddard institute for space studies 464p Plenum press,New York,1966
Obs 6.1677

STELLAR EVOLUTION :international conference Proceedings New York 1963 Nov 13-15 Goddard institute for space studies Edited by R.F. Stein and A.G.W. Cameron Plenum press,New York,1966
TA 15.0181

STELLAR POPULATIONS Stellar populations :a conference Proceedings Vatican observatory 1957 May 20-28 Pontificia academia scientiarum and Specola vaticana Edited by D.J.K. O'Connell Pontificia academia scientiarum.Scripta varia, 16 Specola vaticana.Ricerche astronomiche, 5 illus 544p North-Holland,Amsterdam,1958
Cav 7.2390

STELLAR POPULATIONS :a conference Proceedings Vatican city 1957 May 20-28 Specola vaticana Edited by D.J.K. O'Connell Pontificiae academiae scientiarum.Scripta varia, 16 Specola vaticana.Ricerche astronomiche, 5 544p North Holland, Amsterdam,1958
TA 15.0036

STELLAR POPULATIONS conferences proceedings Vatican 1957 May 20-28 Specola vaticana and Pontificia academia scientiarum Edited by D.J.K. O'Connell Pontificia academia scientiarum, 16 Specola vaticana.Ricerche astronomiche, 5 North-Holland,Amsterdam, 1958
A Math 4.1150

STELLAR ROTATION International astronomical union colloquium Proceedings Columbus, Ohio 1969 Sep 8-11 International astronomical union Ohio state university Edited by A. Sletterbak Co-sponsored by National science foundation port Reidel, Dordrecht,1970 Dedicated to the memory of A.J.Deutsch
Obs 6.3351

STELLAR SPECTRA Abundance determinations in stellar spectra :a symposium Utrecht 1964 Aug.10-14 International astronomical union Edited by Hans Hubenet International astronomical union.Symposium, 26 374p London;New York,1966
Obs 6.0843

STELLAR SPECTRA Abundance determinations in stellar spectra :a symposium Proceedings Utrecht 1964 Aug 10-14 International astronomical union International astronomical union.Symposium, 26 Academic press,London,1966
TA 15.0034

STELLAR SPECTRA Abundance determinations in stellar spectra :a symposium Utrecht 1964 Aug.10-14 International astronomical union Edited by Hans Hubenet International astronomical union.Symposium, 26 374p Academic press,London;New York,1966
A Math 4.1168

STELLAR SPECTRA Ultraviolet stellar spectra and related ground-based observations Lunteren 1969 Jun 24-27 International astronomical union Edited by L. Houziaux and H.E. Butler International astronomical union. Symposium, 36 361p Reidel,Dordrecht,1970
Obs 6.3321

STEREOENCEPHALOTOMY International symposium on stereoencephalotomy (stereotaxic surgery) 1st Proceedings Philadephia,Pa. 1961 Oct 11-12 Edited by E.A. Spiegel and H.T. Wycis Karger,Basle,1962
An 32.4200

STEREOLOGY International congress for stereology 2nd Proceedings Chicago,Ill. 1967 Apr 5-13 International society for stereology Edited by Hans Elias Sponsored by the National science foundation Springer, Berlin,1967
An 32.0107

STERIC ASPECTS OF THE CHEMISTRY AND BIOCHEMISTRY OF NATURAL PRODUCTS London 1959 Jun 30 Biochemical society Edited by J.K. Grant and W. Klyne Biochemical society.Symposia, 19 Cambridge university press,Cambridge,1960
Chem 18.2613

STERIC ASPECTS OF THE CHEMISTRY AND BIOCHEMISTRY OF NATURAL PRODUCTS :a symposium London 1959 Jun 30 Biochemical society Edited by J.K. Grant and W. Klyne Held at the University of London.Senate House Biochemical society.Symposia, 19 Cambridge university press,Cambridge,1960
Bioch 33.1381

STERIC COURSE OF MICROBIOLOGICAL REACTIONS London 1959 Mar 17 Ciba foundation Edited by G.E.W. Wolstenholme and C.M. O'Connor Ciba foundation study group, 2 Churchill,London,1959
PGMS 29.0540

STERIC COURSE OF MICROBIOLOGICAL REACTIONS : in honour of Dr.V.Prelog London 1959 Mar 17 Ciba foundation Edited by G.E.W. Wolstenholme and Cecilia M. O'Connor Ciba foundation study group, 2 illus Churchill,London,1959
Bioch 33.1120

STERIC EFFECTS IN CONJUGATED SYSTEMS :a symposium Hull 1958 Jul 15-17 Chemical society Edited by G.W. Gray Butterworths,London,1958
Chem 18.1687

STERILISATION AND PRESERVATION OF BIOLOGICAL TISSUES BY IONIZING RADIATIONS Panel on radiation sterilisation of biological tissues for transplantation Budapest 1969 Jun 16-20 International atomic energy agency International atomic energy agency.Panel proceedings series International atomic energy agency,Vienna,1970
Radioth 35.1915

STEROID ASSAY BY PROTEIN BINDING 2nd :a symposium Transactions Geneva 1970 Mar 23-25 Karolinska institutet World health organization Edited by E. Dicfalusy and A. Dicfalusy Karolinska symposia on research methods in reproductive endocrinology, 2 Stockholm,1970
PGMS 29.0589

STEROID BIOSYNTHESIS Basic concepts of inborn errors and defects of steroid biosynthesis :a symposium 3rd Proceedings Birmingham 1965 Society for the study of inborn errors of metabolism Edited by K.S. Holt and D.N. Raine Livingstone,Edinburgh, 1966
Inv Med 37.0271

STEROID HORMONE ADMINISTRATION Hormones, psychology and behaviour,and steroid hormone administration :two colloquia Proceedings London 1951 Apr 9-12 and London 1950 Feb 23-24 Ciba foundation Edited by G.E.W. Wolstenholme and Margaret P. Cameron Ciba foundation colloquia on endocrinology, 3 illus Churchill,London,1952
Bioch 33.0433

STEROID HORMONES A Symposium on steroid hormones Papers Madison,Wis. 1950 University of Wisconsin Edited by Edgar S. Gordon University of Wisconsin press,1950 Published in celebration of the hundredth anniversary of the founding of the University of Wisconsin
Bioch 33.0455

STEROID HORMONES Mechanism of action of steroid hormones :a conference Proceedings Dedham,Mass. 1960 Edited by Claude A. Villee and Lewis L. Engel International series of monographs on pure and applied biology Pergamon,Oxford,1961
Inv Med 37.0219

STEROID HORMONES 2nd :a symposium Proceedings Androgens in normal and pathological conditions Ghent 1965 Jun 17-19 Rijkauniversiteit te Gent Societe belge d'endrocinologie Edited by A. Vermeulen and D. Exley Excerpta medica.International congress series, 101 Excerpta medica, Amsterdam,1966
Inv Med 37.0276

STEROID HORMONES AND ENZYMES Steroid hormones and tumour growth and steroid hormones and enzymes :two colloquia Proceedings London 1950 Jul 10-12 and London 1950 Mar 8-10 Ciba foundation Edited by G.E.W. Wolstenholme Ciba foundation colloquia on endocrinology, 1 illus Churchill,London,1952
Bioch 33.0431

STEROID HORMONES AND TUMOUR GROWTH,AND STEROID HORMONES AND ENZYMES :two colloquia Proceedings London 1950 Ciba foundation Edited by G.E.W. Wolstenholme and Margaret P. Cameron Ciba foundation colloquia on endocrinology, 1 Churchill, London,1952
Phys 20.1366

STEROID HORMONES AND TUMOUR GROWTH AND STEROID HORMONES AND ENZYMES :two colloquia Proceedings London 1950 Jul 10-12 and London 1950 Mar 8-10 Ciba foundation Edited by G.E.W. Wolstenholme Ciba foundation colloquia on endocrinology, 1 illus Churchill,London,1952
Bioch 33.0431

STEROID METABOLISM AND ESTIMATION :two colloquia London 1950 Jul 31-Aug 2 London 1950 Aug 9-10 Ciba foundation Edited by G.E.W. Wolstenholme and Margaret Cameron Ciba foundation colloquia on endocrinology, 2 Churchill,London,1952
Phys 20.1367

STEROID METABOLISM AND ESTIMATION :a colloquium Proceedings London 1950 Jul 31-Aug 2 Ciba foundation Edited by G.E. W. Wolstenholme Ciba foundation colloquia on endocrinology, 2 illus Churchill,London, 1952
Bioch 33.0432

STEROIDS Biological activities of steroids in relation to cancer :a conference Proceedings Vergennes,Vt. 1959 Sep 27-Oct 2 Edited by Gregory Pincus and Erwin P. Vollmer Sponsored by the Cancer chemotherapy national service center Academic press,New York;London,1960
Radioth 35.0873

STEROIDS International congress biochemistry 2nd Paris 1952 Jul 21-27 Vol 7: symposium sur la biochemie des steroides Council for international organizations of medical sciences Societe d'edition d'enseignement superieur,Paris,1952 Text in English and French
Bioch 33.1310

STEROIDS International congress of biochemistry 4th Proceedings Vienna 1958 Sep 1-6 Vol 4: symposium 4 biochemistry of steroids International union of biochemistry Edited by E. Mosettig I.U. B.symposium series, 6 Pergamon,London,1959 Added title page in French and German.Text in English,French and German.
Bioch 33.1318

STEROIDS International congress of biochemistry 6th Abstracts New York 1964 Jul 26-Aug 1 Scheduled under the auspices of the International union of biochemistry Washington,1964
Bioch 33.1343

STEROIDS Quantitative paper chromatography of steroids :a conference Proceedings London 1958 Jul 1 Society for endocrinology Edited by D. Abelson and R.V. Brooks Society for endocrinology.Memoirs, 8 Cambridge university press,Cambridge,1960
Bioch 33.2355

STEROIDS Symposium on steroids in experimental and clinical practice Cuernavaca,Mexico 1951 Jan 15-18 Edited by Abraham White Churchill,London,1951
Radioth 35.0697

STEROIDS The Biosynthesis and secretion of adrenocortical steroids :a symposium London 1959 Feb 14 Biochemical society Edited by F. Clark and J.K. Grant Held at the University of London.Senate House Biochemical society.Symposia, 18 Cambridge university press,Cambridge,1960
Bioch 33.1380

STEROIDS IN MODERN MEDICINE :symposium Oxford 1971 Sep 3-4 Pt 1-2 Institute of hormone biology Syntex pharmaceuticals Edited by George A. Christie and Miriam Moore-Robinson Excerpta medica, Amsterdam,1972
Bioch 33.2199

STEVENS INSTITUTE OF TECHNOLOGY The Many-body problem :a symposium Proceedings Hoboken,N.J. 1957 Jan 28-29 Edited by Jerome K. Percus Interscience,New York; London,1963
Chem 18.2362

STOCHASTIC DYNAMICAL SYSTEMS Stability of stochastic dynamical systems Proceedings Coventry 1972 July Edited by Ruth E. Curtain illus ix,334p 25cm Springer, Berlin,1972
A Math 4.1870

STOCHASTIC MODELS IN MEDICINE AND BIOLOGY :a symposium Proceedings Madison,Wis. 1963 Jul 12-14 Edited by John Gurland Conducted by United States.Army.Mathematics research center United States.Army. Mathematics research center.Publications, 10 University of Wisconsin press,Madison,Wis., 1964
Gen 34.0419

STOCHASTIC OPTIMIZATION AND CONTROL :advanced seminar Proceedings Madison,Wis. 1967 Oct 2-4 United States.Army.Mathematics research center Edited by Herman Karreman United States.Army.Mathematics research center. Publications, 20 bibliog.,illus. xii,217p 23cm Wiley,New York,1968
Math S 3.1403

STOCHASTIC POINT PROCESSES :statistical analysis,theory and applications Papers Yorktown Heights,N.Y. 1971 Aug 2-6 International business machines corporation. Research center Edited by Peter A.W. Lewis xxii,894p 24cm Interscience,New York,1972
Math S 3.1803

STOCHASTIC PROCESSES IN MATHEMATICAL PHYSICS AND ENGINEERING 16 symposium proceedings New York 1963 Apr.30-May 2 American mathematical society Edited by Richard Bellman Cosponsored by the Society for industrial and applied mathematics American mathematical society.Proceedings of symposia in applied mathematics, 16 viii,318p 25cm American mathematical society,Providence,R.I., 1964
Math 3.0567

STOCHASTIC SYSTEMS Stability of stochastic dynamical systems :international symposium Proceedings Coventry 1972 Jul 10-14 University of Warwick.Control theory centre Edited by Ruth F. Curtain Lecture notes in mathematics, 254 ix,332p 26cm Springer, Berlin,1972
P Math 2.4569

STOCKHOLM Jun 27-30 Shock pathogenesis and therapy :an international symposium Edited by K.D. Bock Sponsored by Ciba Springer-verlag,Berlin,1962
PGMS 29.0080

STOCKHOLM 1909 International tuberculosis conference 8th Report International anti-tuberculosis association Edited by Gotthold Pannwitz Berlin-Charlottenburg,1910 Text in English,French,German;title also in French and German
Path 30.1024

STOCKHOLM 1938 International symposium on the chemistry of cement 2nd Proceedings Under the auspices of the ingeniorsvetenskapsakademien and Svenska cementforeningen Ingeniorsvetenskapsakademien,Stockholm,1939
Min 10.0598

STOCKHOLM 1948 Biological control Scientific basis of an international biological control organization Les Bases scientifiques d'une organisation internationale pour la lutte biologique International union of biological sciences International union of biological sciences. Serie B.Colloques, 5 Secretairiat general de I.U.I.S.B.,Paris,1949 Title also in English: The scientific bases of an international biological control organization. Text in English and French
Bal 39.3920

STOCKHOLM 1948 International congress of genetics 8th Proceedings Edited by Gert Bonnier and Robert Larsson Berlingska boktrycheriet,Lund,1949
Gen 34.1004

STOCKHOLM 1950 A Short history of botany in Sweden By R.E. Fries and others Almqvist and Wiksell,Uppsala,1950 Published on the occasion of the 7th International botanical congress
BG 38.3233

STOCKHOLM 1950 International botanical congress 7th Abstracts section for palaeobotany c1950 Duplicated.Papers in English,French and German.In folder
Bot 42.6225

STOCKHOLM 1950 International botanical congress 7th Proceedings Edited by Hugo Osvald and Ewert Aberg Almqvist and Wiksell, Uppsala,1953
Bot 42.4696

STOCKHOLM 1950 International botanical congress 7th Proposals,Abstracts International union of biological sciences International union of biological sciences, Utrecht,1950
BG 38.3128

STOCKHOLM 1950 International union of biological sciences :general assembly 10th Under the auspices of Unesco International union of biological sciences. Series A, 10 International union of biological sciences,Paris,1951
BG 38.3108

STOCKHOLM 1951 International congress of psychology 13th Proceedings and papers International union of scientific psychology Stockholm,1952 Under the presidency of professor Katz Conference languages:English, French and German
Psy 31.2536

STOCKHOLM 1952 Conference on radiobiology and radiation protection Acta radiologica, 41,fasc.1 Stockholm,1954
Radioth 35.1228

STOCKHOLM 1952 Instruments and measurements conference Transactions Svenska teknologtoreningen,Stockholm,1953
Eng 41.4142

STOCKHOLM 1955 Colloquium on fatigue Proceedings International union of theoretical and applied mechanics Edited by Waloddi Weibull and Folke K.G. Odqvist Springer,Berlin,1956
Eng 41.3754

STOCKHOLM 1955 Colloquium on fatigue Proceedings International union of theoretical and applied mechanics Edited by Waloddi Weibull and Folke K.G. Odqvist Springer-verlag,Gottingen;Heidelberg,1956 Lectures in English,French and German
Met 25.0917

STOCKHOLM 1956 Advances in radiobiology International conference on radiobiology 5th Proceedings Edited by George Carl de Hevesy and others port. Oliver and Boyd, Edinburgh;London,1957
Radioth 35.1708

STOCKHOLM 1956 Electromagnetic phemomena in cosmical physics a symposium Proceedings International astronomical union Edited by B. Lehnert In co-operation with the International union of theoretical and applied mechanics International astronomical union. Symposium, 6 544p Cambridge university press,Cambridge,1958
Obs 6.1054

STOCKHOLM 1956 Electromagnetic phenomena in cosmical physics Edited by B. Lehnert International astronomical union.Symposium, 6 Cambridge university press,Cambridge,1958
Eng 41.4475

STOCKHOLM 1956 Electromagnetic phenomena in cosmical physics International astronomical union Edited by B. Lehnert In co-operation with the International union of theoretical and applied mechanics International astronomical union.Symposium, 6 13,544p Cambridge university press, Cambridge,1958
Nap 11.0916

STOCKHOLM 1956 Electromagnetic phenomena in cosmical physics :a symposium proceedings International astronomical union Edited by B. Lehnert International astronomical union. Symposia, 6 Cambridge university press, Cambridge,1958
A Math 4.0615

STOCKHOLM 1956 Electron microscopy :a conference Proceedings Edited by F.S. Sjostrand and J. Rhodin Almqvist and Wiksell, Stockholm,1957 Some papers in German
An 32.3155

STOCKHOLM 1957 Progress in mineral dressing International mineral dressing congress Transactions Svenska Gruvforeningen and Jernkontoret Almqvist and Wiksell,Stockholm,1958
Met 25.2286

STOCKHOLM 1960 Biological structure and function I.U.B.-I.U.B.S. international symposium 1st Proceedings Vol 1 International union of biological sciences and International union of biochemistry Edited by T.W. Goodwin and O. Lindberg Academic press,London,1961
An 32.3276

STOCKHOLM 1960 Biological structure and function I.U.B.-I.U.B.S. international symposium 1st Proceedings Vol 1-2 International union of biological sciences International union of biochemistry Edited by T.W. Goodwin and O. Lindberg illus 2 vols Academic press,London;New York,1961
Bal 44.6450

STOCKHOLM 1960 Biological structure and function IUB-IUBS international symposium 1st Proceedings Vol 1-2 International union of biochemistry International union of biological sciences Edited by T.W. Goodwin and O. Lindberg 2 vols Academic press,London,1961
Gen 34.0826

STOCKHOLM 1960 Biological structure and function The I.U.B.-I.U.B.S. international symposium 1st Proceedings Vol 1-2 International union of biochemistry and International union of biological sciences Edited by T.W. Goodwin and O. Lindberg illus 2 vols Academic press,London;New York,1961
Bioch 33.0584

STOCKHOLM 1960 Biological structure and function :international symposium 1st Proceedings Vol 1-2 International union of biochemistry and International union of biological sciences Edited by T.W. Goodwin and D. Lindberg Academic press, London;New York,1961
Phys 20.1539

STOCKHOLM 1960 Biological struggle and function The I.U.B.-I.U.B.S.international symposium 1st Proceedings Vol 1-2 International union of biochemistry International union of biological sciences Edited by T.W. Goodwin and O. Lindberg illus. 2 vols Academic press,London;New York, 1961
Bot 42.1344

STOCKHOLM 1960 International geographical congress 19th Final report,bibliography Edited by Staffan Helmfrid illus 138p Stockholm,1963 Held in Norden,ie.Denmark, Finland,Iceland,Norway and Sweden
Sco 14.0133

STOCKHOLM 1960 International symposium on X-ray microscopy and X-ray microanalysis 2nd Proceedings Edited by Arne Engstrom and others illus x,542p Elsevier, Amsterdam,1960
Cav 7.2234

STOCKHOLM 1960 Recent advances in stress corrosion :a symposium Royal Swedish academy of engineering sciences.Corrosion committee Edited by Ake Bresle Almqvist and Wiksell, Stockholm,1961
Met 25.1960

STOCKHOLM 1960 X-ray microscopy and x-ray microanalysis International symposium 2nd Proceedings Edited by A. Engstrom and others Elsevier,Amsterdam,1960
Met 25.1375

STOCKHOLM 1961 International biophysics congress abstracts of contributed papers Tryckeri aktiebolaget thule,Stockholm,1961
Radioth 35.1647

STOCKHOLM 1961 International biophysics congress :abstracts of contributed papers Tryckeri Aktiebolaget,Stockholm,1961
Radioth 35.1426

STOCKHOLM 1962 International congress of mathematicians 14 proceedings International mathematical union Edited by V. Stenstrom Bibliog. 1,595p 25cm Institut Mittag-Leffler,Djursholm,Sweden,1963
P. Math 2.2601

STOCKHOLM 1962 International congress of mathematicians short communications abstracts International mathematical union 221p 24cm Stockholm,1962
P. Math 2.2602

STOCKHOLM 1962 International council of the aeronautical sciences 3rd congress Proceedings American institute of aeronautics and astronautics Spartan; Macmillan,Washington,D.C.;London,1964
Eng 41.6965

STOCKHOLM 1962 Olfaction and taste International symposium on olfaction and taste 1st Proceedings Wenner-Gren center Edited by Y. Zotterman Wenner-Gren center. International symposium.Series, 1 Pergamon press,Oxford,1963
An 32.4380

STOCKHOLM 1962 Outlines of one-hour and half-hour addresses and translations of Russian abstracts of short communications International congress of mathematicians varp 29cm Stockholm,1962
P Math 2.3005

STOCKHOLM 1965 Experimental studies and clinical experience - the assessment of risk European society for the study of drug toxicity :meeting 6th Proceedings Edited by D.G. Davey European society for the study of drug toxicity.Proceedings,6 International congress series,97 Excerpta medica,Amsterdam, 1965
Pha 16.0331

STOCKHOLM 1965 Mechanism of release of biogenic amines :international symposium Proceedings Edited by U.S. Von Euler and others Wenner-Gren center.International symposium series,5 Pergamon,Oxford,1965
Pha 16.0071

STOCKHOLM 1966 Fertility and sterility World congress on fertility and sterility 5th Proceedings Edited by Bjorn Westin and Nils Wiquist Under the supervision of the International fertility association International congress series, 133 Excerpta medica foundation,Amsterdam,1967
An 32.3899

STOCKHOLM 1966 Fertility and sterility World congress on fertility and sterility 5th Proceedings International fertility association Swedish society of obstetrics and gynaecology Edited by Bjorn Westin and Nils Wiqvist Excerpta medica.International congress series, 133 Excerpta medica, Amsterdam,1967
Inv Med 37.0230

STOCKHOLM 1966 Power systems computation conference 2nd Proceedings Pt 1-5 5 vols Stockholm,1966
Eng 41.4985

STOCKHOLM 1966 Prostaglandins Nobel symposium 2nd Proceedings Nobel foundation Edited by Sune Bergstrom and Bengt Samuelsson Wiley;Almqvist and Wiksell, New York;Stockholm,1967
Inv Med 37.0019

STOCKHOLM 1966 Structure and function of inhibitory neural mechanisms International meeting of neurobiologists 4th Proceedings Edited by C.von Euler and others Illus. ix, 563p Pergamon,Oxford,1968
Pha 16.0312

STOCKHOLM 1966 Structure and function of inhibitory neuronal mechanisms International meeting of neurobiologists 4th Proceedings Edited by C.von Euler and others Wenner-Gren center.International symposium series, 10 Pergamon press,Oxford,1968
An 32.5336

STOCKHOLM 1966 The Conference on structure and reactions of DFP sensitive enzymes Proceedings Edited by Edith Heilbronn organized by the Forsvarets forskningsanstalt, Stockholm Research institute of national defence,Stockholm,1967
Bioch 33.1065

STOCKHOLM 1967 Current problems of lower vertebrate phylogeny Nobel symposium 4th Proceedings Edited by Tor Orvig Wiley,New York,1968
Bal 39.2054

STOCKHOLM 1967 Diabetes International diabetes federation congress 6th Proceedings International diabetes federation Edited by J. Ostman and R.D.G. Milner Excerpta medica,Amsterdam,1969
Bioch 33.0462

STOCKHOLM 1969 Colloquium on spectra of meteorological variables Proceedings Vol 3: radio science Inter-union commission on radio astronomy International union of radio science.Swedish national committee illus 28cm American geophysical union,Washington,D. C.,1969
A Math 4.1745

STOCKHOLM 1969 Immunoassay of gonadotrophins ...symposium Transactions Karolinska institutet and World health organization Edited by E. Dicfalusy and A. Dicfalusy Karolinska symposia on research methods in reproductive endocrinology, 1 Stockholm,1969
PGMS 29.0182

Die STOFFPRODUKTION DER PFLANZENDECKE : vortrage und diskussionsergebnisse des internationalen okologischen symposium Stuttgart-Hohenheim 1960 May 4-7 Edited by Helmut Lieth port Fischer,Stuttgart,1962
Bot 42.1875

STOFFTRANSPORT UND STOFFVERTEILUNG IN ZELLEN HOHERER PFLANZEN :internationales symposium Schloss-Reinhardsbrunn 1968 Oct 14-19 Edited by Kurt Mothes and others Deutsche akademie der wissenschaften zu Berlin. Klasse fur medizin.Abhandlungen, 4a illus. iv,215p Akademie-verlag,Berlin,1968 Title also in English 'Transport and distribution of matter in cells of higher plants'
Bot 42.1900

STOKE-ON-TRENT 1967 Special ceramics :a symposium 4th Proceedings British ceramic research association Edited by P. Popper British ceramic research association, Stoke-on-Trent,1968
Met 25.2554

STOKE-ON-TRENT 1959 Special ceramics :a symposium Proceedings By P. Popper British ceramic research association Heywood, London,1960
Met 25.0267

STOKE-ON-TRENT 1962 Special ceramics 1962 : a symposium 2nd Proceedings British ceramic research association Edited by P. Popper Academic press,London;New York,1963
Met 25.0280

STOKE-ON-TRENT 1964 Special ceramics 1964 British ceramic research association : symposium 3rd Proceedings Edited by P. Popper Academic press,London;New York,1965
Min 10.0656

STOKE-ON-TRENT 1964 Special ceramics 1964 : a symposium 3rd Proceedings By P. Popper British ceramic research association Edited by P. Popper Academic press,London; New York,1965
Met 25.0281

STOLZENAU 1959 Internationale symposium fur vegetationskartierung Bericht Edited by Reinhold Tuxen illus. Cramer,Weinheim, 1963
Bot 42.2345

STOLZENAU 1960 Biosoziologie Internationale vereinigung fur vegetationskunde Bericht Edited by Reinhold Tuxen illus. Junk,The Hague,1965
Bot 42.4762

STOLZENAU 1961 Anthropogene vegetation Internationale vereinigung fur vegetationskunde Bericht Edited by Reinhold Tuxen illus. Junk,The Hague,1966
Bot 42.4761

STOLZENAU 1963 Pflanzensoziologie und landschaftsokologie Internationale vereinigung fur vegetationskunde Bericht Edited by Reinhold Tuxen illus. Junk,The Hague,1968
Bot 42.4765

STOLZENAU 1964 Pflanzensoziologische systematik Internationale vereinigung fur vegetationskunde Bericht Edited by Reinhold Tuxen illus. Junk,The Hague,1968
Bot 42.4764

STOMACH,INCLUDING RELATED AREAS IN THE ESOPHAGUS AND DUODINUM Hahnemann symposium 13th Edited by Charles Thompson and others Grune and Stratton,New York;London,1967
An 32.3796

STONY BROOK, N.Y. 1969 1969 number theory institute Summer institute on number theory American mathematical society Edited by Donald J. Lewis American mathematical society.Proceedings of symposia in pure mathematics, 20 xiii,451p 26cm AMS, Providence,R.I.,1971
P Math 2.4383

STONY BROOK,N.Y. 1967 University education in computing science Conference on graduate academic and related research programs in computing science Proceedings Edited by Aaron Finerman Association for computing machinery.Monograph series Bibliog.,Illus. Port. xvi,237p 24cm Academic press,New York;London,1968
P Math 2.3505

STONY BROOK,N.Y. 1967 University education in computing science Conference on graduate academic and related research programs in computing science Proceedings Edited by Aaron Finerman Held at the State university of New York Association for computing machinery.Monograph series Academic press, New York,1968
Eng 41.2223

STONY BROOK,N.Y. 1968 Galactic astronomy Edited by Hong-Yee Chiu and A. Muriel 2 vols Gordon and Breach,New York,1970
TA 15.0656

STORRS,CONN. 1963 Cellular membranes in development :a symposium Society for the study of development and growth Edited by Michael Locke Society for the study of development and growth.Symposia, 22 Academic press,New York,1964
Bioch 33.0910

STORRS,CONN. 1963 Cellular membranes in development :symposium Society for the study of development and growth Edited by Michael Locke Society for the study of development and growth.Symposia, 22 Academic press,New York,1964
Gen 34.0480

STORRS,CONN. 1963 Cellular membranes in development :symposium Society for the study of development and growth Edited by Michael Locke Society for the study of development and growth.Symposia, 22 illus. xvi,382p Academic press,New York;London,1964
Bot 42.4733

STORRS,CONN. 1963 Cellular membranes in development a symposium Proceedings Edited by Michael Locke Society for the study of development and growth.Symposia, 22 Academic press,New York;London,1964
An 32.3295

STORY OF 2ND WORLD METALLURGICAL CONGRESS held in the United States of America World metallurgical congress 2nd 1957 Oct-Nov American society for metals Edited by Kingsley W. Given 225p A.S.M.,Cleveland, Ohio,1958
Met 25.0045

STRAHLENWIRKUNG AUF MENSCHLICHE ERBANLAGEN Papers Geneva 1957 World health organization Schriften reihe strahlenschutz, 3 Gersbach,Brunswick,1958
Radioth 35.1083

STRAIN GAGE TECHNIQUES Lectures and laboratory exercises on strain gage techniques Los Angeles,Calif. 1958 Aug 18-29 By William M. Murray and Peter K. Stein University of California Los Angeles Society for experimental stress analysis University of California Los Angeles,Los Angeles,Calif., 1958
Eng 41.2906

STRANGEWAYS RESEARCH LABORATORY Tissue proteinases Wates symposium Proceedings Cambridge 1970 Apr 6-7 Royal society of London Edited by A.J. Barrett and J.T. Dingle North-Holland,Amsterdam,1971
Bioch 33.2284

STRASBOURG 1939 Le Magnetisme Travaux Universite de Strasbourg.Institut de physique and Institut international de cooperation intellectuelle Supported by Centre national de la recherche scientifique Collection scientifique,Paris,1940
Cav 7.2241

STRASBOURG 1949 Morphogenese :colloque international Centre national de la recherche scientifique Centre national de la recherche scientifique.Colloques internationaux, 28 C.N.R.S.,Paris,1951
An 32.2463

STRASBOURG 1950 Mecanisme physiologique de la secretion lactee :colloque international Centre national de la recherche scientifique Centre national de la recherche scientifique. Colloques internationaux, 32 C.N.R.S.,Paris, 1950
Phys 20.2212

STRASBOURG 1958 Chemical pathology of the nervous system International neurochemical symposium 3rd Proceedings Edited by Jordi Folch-Pi Pergamon press,London,1961
An 32.4186

STRASBOURG 1964 Hypoplasie et malformations de l'appareil genital interne de la femme Papers Societe francaise de gynecologie Edited by P. Muller Masson, Paris,1964
An 32.3832

STRASBOURG 1966 International symposium on neurosecretion 4th Edited by F. Stutinsky Springer,Berlin,1967
An 32.3779

STRASBOURG 1966-67 Seminaire de probabilites 1e Universite de Strasbourg.Institut de mathematique Lecture notes in mathematics, 39 189p 28cm Springer,Berlin,1967
Math 3.1188

STRASBOURG 1967 Regionalisation et developpement Centre national de la recherche scientifique colloques internationaux sciences humaines Centre national de la recherche scientifique Centre national de la recherche scientifique,Paris, 1968 Papers in English and French
Geog 13.7207

STRASBOURG 1967 Seminaire de probabilites 2me Universite de Strasbourg.Institut de mathematique Lecture notes in mathematics, 51 Bibliog 199p 28cm Springer,Berlin, 1968
P Math 2.2996

STRASBOURG 1967 Seminaire de probabilites 2me Universite de Strasbourg.Institut de mathematique Lecture notes in mathematics, 51 199p 28cm Springer,Berlin,1968
Math S 3.1438

STRASBOURG 1967-68 Seminaire de probabilites 3me Universite de Strasbourg.Institut de mathematique Lecture notes in mathematics, 88 Bibliog. 229p 28cm Springer,Berlin,1969
P Math 2.3517

STRASBOURG 1967-68 Seminaire de probabilites 3me Universite de Strasbourg.Institut de mathematique Lecture notes in mathematics, 88 bibliog. 229p 28cm Springer,Berlin,1969
Math S 3.1441

STRASBOURG 1968-69 Seminaire de probabilities 4me Universite de Strasbourg.Institut de mathematique Lecture notes in mathematics, 124 Bibliog. 282p 25cm Springer-Verlag,Berlin,1970
P Math 2.3518

STRASBURG 1970 Les Cultures de tissus de plantes Proceedings Centre national de la recherche scientifique Centre national de la recherche scientifique.Colloques internationaux, 193 CNRS,Paris,1971
Bioch 33.2223

STRATEGY OF CHEMOTHERAPY :a symposium Papers London 1958 Apr Society for general microbiology Edited by S.T. Cowan and Elizabeth Rowatt Held at the Royal institution Society for general microbiology. Symposia, 8 Cambridge university press, Cambridge,1958
Bioch 33.1180

The STRATEGY OF CHEMOTHERAPY London 1958 Apr Society for general microbiology Edited by S.T. Cowan and Elisabeth Rowatt Society for general microbiology.Symposium,8 Cambridge university press,Cambridge,1958
Pha 16.0262

The STRATEGY OF CHEMOTHERAPY :a symposium Proceedings London 1958 Apr By S.T. Cowan and Elizabeth Rowatt Society for general microbiology Society for general microbiology.Symposia, 8 Cambridge university press,Cambridge,1958
Radioth 35.0225

STRATFORD-UPON-AVON 1962 Techniques in endocrine research :a NATO advanced study institute Proceedings North Atlantic treaty organization Edited by Peter Eckstein and Francis Knowles Under the auspices of the University of Birmingham Academic press, London,1963
Gen 34.1964

STRATFORD-UPON-AVON 1962 Techniques in endocrine research :a NATO advanced study institute Proceedings North Atlantic treaty organization Edited by Peter Eckstein and Francis Knowles Under the auspices of the University of Birmingham Academic press, London,1963
An 32.3767

STRATFORD-ON-AVON 1947 Problems of economic reconstruction in the Far East Institute of Pacific relations conference 10th Report Institute of Pacific relations, New York,1949
Geog 13.2189

STRATFORD-UPON-AVON 1962 Techniques in endocrine research :a Nato advanced study institute Proceedings North Atlantic treaty organization Edited by P. Eckstein and F. Knowles Under the auspices of the University of Birmingham Academic press,New York,1963
Phys 20.1399

STRATHCLYDE 1967 Soft X-ray band spectra and the electronic structure of metals and materials :a conference Proceedings University of Strathclyde and Royal society Edited by Derek J. Fabian Academic press,London;New York,1968
Met 25.1470

STRATIGRAPHY Congres international de stratigraphie et de geologie du carbonifere 5e compte rendu Paris 1963 Sep 9-12 Tom 1-3 3 vols Paris,1964 Text in English,French and German
Geol 8.2988

STRATIGRAPHY Congres pour l'avancement des etudes de stratigraphie carbonifere 1er compte rendu Heerlen 1927 Jun 7-11 Edited by W.J. Jongmans Published by the geologisch-mijnbouwkundig genootschap voor Nederland en kolonien Liege,1928 Text in English,French and German
Geol 8.2984

STRATIGRAPHY Congres pour l'avancement des etudes de stratigraphie carbonifere 2e compte rendu Heerlen 1935 Sep Tom 1-3 Edited by W.J. Jongmans 3 vols Maestricht, 1937-38 Text in English,French and German
Geol 8.2985

STRATIGRAPHY Congres pour l'avancement des etudes de stratigraphie et de geologie du carbonifere 3e compte rendu Heerlen 1951 Jun 25-31 Tom 1-2 Edited by Geologisch bureau Heerlen Maestricht,1952 Text in English,French and German
Geol 8.2986

STRATIGRAPHY Congres pour l'avancement des etudes de stratigraphie et de geologie du carbonifere 4e compte rendu Heerlen 1958 Sep 15-20 Tom 1-3 3 vols Maestricht, 1960-62 Text in English,French and German
Geol 8.2987

STRATIGRAPHY Karoo symposium South African association for the advancement of science : annual congress papers Johannesburg 1962 South African journal of science,59,no.5 Johannesburg,1963
Geol 8.2941

STRATOSPHERIC AND MESOSPHERIC CIRCULATION The International symposium on stratospheric and mesospheric circulation Proceedings Berlin 1962 Aug 20-31 Institut fur meteorologie und geophysik Edited by Richard Scherhag and Gunter Warnecke Institut fur meteorologie und geophysik.Meteorologische abhandlungen,36 illus vi,644p 31cm Dietrich Reimer,Berlin,1963
Sco 14.0527

STRENGTH AND DUCTILITY OF METALS Symposium on strength and ductility of metals at elevated temperatures with particular reference to effects of notches and metallurgical changes New York 1952 Jun 23 American society for testing materials A.S.T.M.Special technical publication, 128 A.S.T.M. Philadelphia,Pa.,1953
Met 25.0979

The STRENGTH OF CONCRETE STRUCTURES :a symposium Proceedings London 1956 May Cement and concrete association Cement and concrete association,London,1958
Eng 41.2930

STRENGTH OF METALS AND ALLOYS International conference on the strength of metals and alloys Proceedings Tokyo 1967 Sep 4-8 Japan institute of metals xlvi,1049p 31cm Japan institute of metals,Sendai,1968 In commemoration of the 30th anniversary of the institute
Cav 7.3081

The STRENGTH OF METALS AND ALLOYS the international conference in commemoration of the 30th anniversary of the Japan institute of metals Proceedings Tokyo 1967 Sep 4-8 Japan institute of metals Japan institute of metals.Transactions, 9 suppl. Japan institute of metals,Tokyo,1968
Met 25.0829

STRENGTH OF METALS UNDER COMBINED STRESSES : a series of educational lectures...presented... during the twenty-second National metal congress and exhibition Cleveland,Ohio 1940 Oct 21-25 By Maxwell Gensamer American society for metals American society for metals,Cleveland,Ohio,1941
Eng 41.3742

STRENGTH OF SOLIDS A Conference on strength of solids :held at the H.H.Wills physical laboratory Report Bristol 1947 Jul 7-9 University of Bristol Edited by N.F. Mott 162p Physical sciety,London,1948
Cav 7.0346

STRENGTH OF SOLIDS :a conference Report Bristol 1947 Jul 7-9 Wills (H.H.) physical laboratory Physical society,London,1948
Met 25.1197

STRENGTHENING MECHANISMS,METALS AND CERAMICS Sagamore army materials research conference 12th Proceedings Raquette Lake,N.Y. 1965 Aug 24-27 Army materials research agency Syracuse university Syracuse university press,Syracuse,N.Y.,1966
Met 25.2746

STRESA 1960 International congress of applied mechanics 10th Proceedings Edited by F. Rolla and W.T. Koiter Elsevier, Amsterdam;New York,1962
A Math 4.1280

STRESA 1967 Continental drift,secular motion of the Pole and rotation of the earth I.A.U. symposium 32nd Papers International astronomical union and International union of geodesy and geophysics Edited by William Markowitz and B. Guinot illus 107p 25cm D.Riedel,Dordrecht,1968
Sco 14.8118

STRESA 1967 Continental drift,secular motion of the pole,and rotation of the earth International astronomical union and International union of geodesy and geophysics Edited by Wm. Markowitz and B. Guinot International astronomical union.Symposium, 32 Reidel,Dordrecht,1968
TA 15.0424

STRESA 1967 Continental drift,secular motion of the pole,and rotation of the earth : a symposium Proceedings International astronomical union International union of geodesy and geophysics Edited by W. Markowitz and B. Guinot International astronomical union.Symposium, 32 108p Reidel,Dordrecht,1968
Obs 6.3268

STRESA 1968 Studies in modern yarn production :a conference Papers Textile institute Textile institute.Annual conference, 53 Textile institute,Manchester, 1968
Eng 41.8288

STRESA,ITALY 1967 Stresa symposium on continental drift,secular motion of the pole and rotation of the earth International astronomical union and International union of geodesy and geophysics Edited by W. Markowitz and B. Guinot International astronomical union.Symposium, 32 Reidel, Dordrecht,1968
A Math 4.1308

STRESA SYMPOSIUM ON CONTINENTAL DRIFT,SECULAR MOTION OF THE POLE AND ROTATION OF THE EARTH Stresa,Italy 1967 Mar 21-25 International astronomical union and International union of geodesy and geophysics Edited by W. Markowitz and B. Guinot International astronomical union.Symposium, 32 Reidel,Dordrecht,1968
A Math 4.1308

STRESS AND STRAIN IN SOLIDS The Measurement of stress and strain in solids :a conference Proceedings Manchester 1946 Jul 11-13 Institute of physics Institute of physics, London,1948
Eng 41.2898

STRESS CORROSION CRACKING IN AIRCRAFT STRUCTURAL MATERIALS :an Agard conference Proceedings Turin 1967 Apr 18-19 North Atlantic treaty organization.Advisory group for aeronautical research and development.Structures and materials panel AGARD conference proceedings series, 18 Nato,1967
Met 25.2484

STRESS-CORROSION CRACKING OF METALS Symposium on stress-corrosion cracking of metals Philadelphia,Pa. 1944 Nov 29-Dec 1 American institute of mining and metallurgical engineers American society for testing materials American society for testing materials;American institute of mining and metallurgical engineers,Philadelphia,Pa.;New York,1945
Eng 41.3717

STRESS CORROSION TESTING :a symposium Atlantic City,N.J. 1966 Jun 26-Jul 1 American society for testing materials A.S.T. M.Special technical publication, 425 American society for testing and materials, Philadelphia,Pa.,1967
Met 25.1964

STRESS IN ROCK Determination of stress in rock :a symposium Atlantic City,N.J. 1966 Jun 26-Jul 1 American society for testing and materials American society for testing and materials.Special technical publication, 429 American society for testing materials, Philadelphia,Pa.,1967
Eng 41.3160

STRESS WAVE PROPAGATION IN MATERIALS International symposium on stress wave propagation in materials University Park,Pa. 1959 Jun 30-Jul 12 Pennsylvania state university.Department of engineering mechanics Edited by Norman Davids Sponsored by the United States.Army.Office of ordnance research Interscience,New York,1960
Eng 41.3770

STRESS WAVES International symposium on stress wave propagation in materials proceedings University Park,Pa 1959 Jun 30-Jul 2 University of Pennsylvania.Department of engineering mechanics Edited by Norman Davids Sponsored by the United States.Army. Office of ordnance research Interscience publishers,New York,1960
A Math 4.0370

STRESS WAVES IN ANELASTIC SOLIDS :symposium Providence,R.I. 1963 Apr 3-5 International union of theoretical and applied mechanics Edited by Herbert Kolsky and William Prager Springer,Berlin,1964
Eng 41.3805

STRESS WAVES IN ANELASTIC SOLIDS IUTAM symposium Proceedings Providence R.I. 1963 April 3-5 International union of theoretical and applied mechanics Edited by Herbert Kolsky and William Prager Supported by the National science foundation Springer-verlag,Berlin,1964
A Math 4.0369

STRESS-CORROSION CRACKING OF METALS Symposium on stress-corrosion cracking of metals Philadelphia,Pa. 1944 Nov 29-Dec 1 American society for testing materials and American institute of mining and metallurgical engineers.Institute of metals division A.S.T. M.,New York,1945
Met 25.1952

STRESS-CORROSION CRACKING OF TITANIUM :a symposium Seattle 1965 Oct 31-Nov 5 American society for testing materials A.S.T. M.Special technical publication, 397 American society for testing and materials, Philadelphia,Pa.,1966
Met 25.1962

STRESSES Symposium on internal stresses in metals and alloys Advance copies of papers London 1947 Oct 15-16 Institute of metals Institute of metals,London,1947
Eng 41.3768

STRESSES AND DISPLACEMENTS IN STRUCTURES Correlation between calculated and observed stresses and displacements in structures :a conference London 1955 Sep 21-22 Institution of civil engineers 2 vols Institution of civl engineers,London,1955
Eng 41.3020

STRONG,ELECTROMAGNETIC,AND WEAK INTERACTIONS International school of physics 'Ettore Majorana' 1st course proceedings Erice 1963 May-June Edited by A. Zichichi Sponsored by the European organization for nuclear research W.A.Benjamin,New York; Amsterdam,1964
A Math 4.0846

STRONG AND WEAK INTERACTIONS,PRESENT PROBLEMS International school of physics 'Ettore Majorana' 4th course proceedings Erice 1966 Jun.19-Jul.4 Edited by A. Zichichi Sponsored by the European organization for nuclear research 859p Academic press,New York,1966
A Math 4.0849

STRONG TOUGH STRUCTURAL STEELS joint conference... Proceedings Scarborough 1967 Apr 4-6 British iron and steel research association and Iron and steel institute Iron and steel institute.Publication, 104 Eyre and Spottiswoode,Margate,1967
Met 25.0428

STRONTIUM The Transfer of calcium and strontium across biological membranes :a conference Proceedings Ithaca,N.Y. 1962 May 13-16 Edited by R.H. Wasserman held at Cornell university Academic press,New York; London,1963
Bioch 33.1021

STRONTIUM METABOLISM International symposium on some aspects of strontium metabolism Proceedings Chapelcross 1966 May 5-7 Edited by J.M.A. Lenihan and others Academic press,London;New York,1967
Radioth 35.1719

STRUCTURAL ADHESIVES:THE THEORY AND PRACTICE OF GLUING WITH SYNTHETIC RESINS lectures Cambridge 1951 Aero research Lange, Maxwell and Springer,London,1951 Lectures given at a summer school on the technology of synthetic resin adhesives, Cambridge,1951
Eng 41.4022

STRUCTURAL ASPECTS OF CELL PHYSIOLOGY Bristol 1951 Jul Society for experimental biology Society for experimental biology. Symposia, 6 Cambridge university press, Cambridge,1952
Phys 20.0954

STRUCTURAL ASPECTS OF CELL PHYSIOLOGY :a symposium Bristol 1951 Jul Society for experimental biology Society for experimental biology.Symposia, 6 Cambridge university press,Cambridge,1952
Bioch 33.1352

STRUCTURAL ASPECTS OF CELL PHYSIOLOGY :a symposium papers Bristol 1951 Jul Society for experimental biology Society for experimental biology.Symposia,6 Cambridge university press,Cambridge,1952
VA 19.0361

STRUCTURAL ASPECTS OF CELL PHYSIOLOGY : symposium papers Bristol 1951 Jul Society for experimental biology Society for experimental biology.Symposia, 6 Cambridge university press,Cambridge,1952
Radioth 35.0456

STRUCTURAL DAMPING :a colloquium Papers Atlantic City,N.J. 1959 Dec Edited by Jerome E. Ruzicka Sponsored by American society of mechanical engineers.Applied mechanics division American society of mechanical engineers,New York,1959
Eng 41.6335

STRUCTURAL ENGINEERING Aluminium in structural engineering :a symposium Proceedings London 1963 Jun 11-12 Institution of structural engineers Aluminium federation Aluminium federation, London,1964
Eng 41.2694

STRUCTURAL ENGINEERS Industrialized building and the structural engineers :a symposium London 1966 May 17-19 Institution of structural engineers Institution of structural engineers,London,1966
Eng 41.3381

STRUCTURAL FATIGUE IN AIRCRAFT :a symposium American society for testing materials A. S.T.M.Special technical publication, 404 American society for testing materials, Philadelphia,Pa.,1966
Met 25.0943

STRUCTURAL INTERDEPENDENCE AND ECONOMIC DEVELOPMENT International conference on input-output techniques 3rd Proceedings Geneva 1961 Sep United Nations.Secretariat Edited by Tibor Barna and others Sponsored jointly by the Harvard economic research project Macmillan,London,1963
Geog 13.2417

STRUCTURAL LIGHTWEIGHT CONCRETE One-day symposium on structural lightweight concrete Brighton 1962 Jun 26 Vol 1-2 Reinforced concrete association 2 vols RCA, London,1963
Eng 41.2968

STRUCTURAL MECHANICS Symposium on naval structural mechanics 1st Proceedings Stanford,Calif. 1958 Aug 11-14 United States.Office of naval research and Stanford university Edited by J.Norman Goodier and Nicholas J. Hoff International series on aeronautical sciences and space flight Pergamon,Oxford,1960
Met 25.1300

STRUCTURAL METALLURGY Symposium on newer structural materials for aerospace vehicles Chicago 1964 Jun 21 American society for testing materials A.S.T.M.Special technical publication, 379 Philadelphia,Pa.,1965
Met 25.2516

STRUCTURAL PROCESSES IN CREEP :a symposium Report London 1961 May 3-4 Iron and steel institute and Institute of metals Iron and steel institute.Special report, 70 Iron and steel institute,London,1961
Met 25.2334

STRUCTURAL STEELWORK :a conference Proceedings London 1966 Sep 26-28 British constructional steelwork association British constructional steelwork association, London,1966
Eng 41.3024

STRUCTURE,FUNCTION AND EVOLUTION IN PROTEINS report of symposium Upton.N?Y. 1968 Jun 3-5 Vol 1-2 Brookhaven national laboratory.Biology department Brookhaven symposia in biology, 21 BNL 50116 2 vols brookhaven national laboratory,Upton,N.Y.,1969
Gen 34.0671

STRUCTURE,FUNCTION AND EVOLUTION IN PROTEINS : symposium Report Upton,N.Y. 1968 Jun 3-5 Brookhaven national laboratory.Biology department Brookhaven symposia in biology, 21 BNL50116(C-53) Brookhaven national laboratory.Biology department,Upton,N.Y.,1969
Bioch 33.1299

STRUCTURE AND ACTIVITY OF ENZYMES Federation of European biochemical societies symposium 1st London 1914 Mar 24 Edited by T.W. Goodwin and others Academic press,London;New York,1964
Radioth 35.0119

STRUCTURE AND ACTIVITY OF ENZYMES Edited by T.W. Goodwin and others Federation of European biochemical societies.Symposium,1 Academic press,London,1964
Pha 16.0080

STRUCTURE AND ACTIVITY OF ENZYMES :a symposium Proceedings London 1964 Mar 24 Federation of European biochemical societies Edited by T.W. Goodwin and others Federation of European biochemical societies.Symposium, 1 Academic press,London;New York,1964
Bioch 33.1404

STRUCTURE AND ACTIVITY OF ENZYMES 1 :a symposium London 1964 Mar 24 Federation of European biochemical societies Edited by T.W. Goodwin and others Academic press,London;New York,1964
Chem 18.1248

The STRUCTURE AND BIOSYNTHESIS OF MACROMOLECULES a symposium London 1961 Mar 27-28 Biochemical society Edited by D.J. Bell and J.K. Grant Held at the University of London. Senate House Biochemical society.Symposia, 21 Cambridge university press,Cambridge,1962 In commemoration of the 50th anniversary of the Biochemical society
Bioch 33.1383

The STRUCTURE AND BIOSYNTHESIS OF MACROMOLECULES a symposium London 1961 Mar 27-28 Biochemical society Edited by J.K. Grant and D.J. Bell Biochemical society symposia,21 Cambridge university press,Cambridge,1962
Chem 18.1277

STRUCTURE AND CHEMISTRY OF SOLID SURFACES International materials symposium :the structure and chemistry of solid surfaces 4th Proceedings Berkeley,Calif 1968 Jun 17-21 Lawrence radiation laboratory. Inorganic materials research division and University of California.College of chemistry Edited by Gabor A. Somorjai Inorganic materials research division series, 4 Wiley,New York,1969
Met 25.2476

STRUCTURE AND CONTROL OF THE MELANOCYTE International pigment cell conference 6th Sofia 1965 May 25-29 Edited by G. Della Porta and O. Muhlbock Sponsored by the International union against cancer Springer, Berlin,1966
An 32.3425

STRUCTURE AND DEVELOPMENT OF SOLAR ACTIVE REGIONS Budapest 1967 Sep 4-8 International astronomical union Edited by K.O. Kiepenheuer International astronomical union. Symposium, 35 Reidel,Dordrecht,1968
TA 15.0420

STRUCTURE AND DEVELOPMENT OF SOLAR ACTIVE REGIONS a symposium Budapest 1967 Sep 4-8 International astronomical union Edited by K. O. Kiepenheuer International astronomical union.Symposia, 35 Bibliog.,Illus. xvii, 608p 25cm D.Reidel,Dordrecht,1968
A Math 4.1602

STRUCTURE AND DEVELOPMENT OF SOLAR ACTIVE REGIONS a symposium Proceedings Budapest 1967 Sep International astronomical union Edited by K.O. Kiepenheuer International astronomical union.Symposium, 35 608p Reidel,Dordrecht,1968
Obs 6.3266

STRUCTURE AND EVOLUTION OF THE GALAXY North Atlantic treaty organization.Advanced study institute Proceedings Athens 1969 Sep 8-19 Edited by L.N. Mavridis Astrophysics and space science library, 22 312p S. Reidel,Dordrecht,1971
Obs 6.3456

STRUCTURE AND FUNCTION OF CONNECTIVE AND SKELETAL TISSUE :advanced study institute Proceedings St.Andrews 1964 Jun 15-25 Organized under the auspices of the North Atlantic treaty organization Butterworths, London,1965
An 32.3423

STRUCTURE AND FUNCTION OF CYTOCHROMES The Symposium on structural and chemical aspects of cytochromes Proceedings Osaka 1967 Aug 16-18 Edited by Kazuo Okunuki and others University of Tokyo press,Tokyo,1968
Bioch 33.1100

STRUCTURE AND FUNCTION OF GENETIC ELEMENTS report of a symposium Upton,N.Y. 1959 Jun 1-3 Brookhaven national laboratory. Biological department Brookhaven symposia in biology, 12 BNL 558 Brookhaven national laboratory,Upton,N.Y.,1959
Gen 34.0674

STRUCTURE AND FUNCTION OF GENETIC ELEMENTS : symposium Report Upton,N.Y. 1959 Jun 1-3 Brookhaven national laboratory.Biology department Brookhaven symposia in biology, 12 BNL558(C-29) Brookhaven national laboratory.Biology department,Upton,N.Y.,1959
Bioch 33.1291

STRUCTURE AND FUNCTION OF INHIBITORY NEURAL MECHANISMS International meeting of neurobiologists 4th Proceedings Stockholm 1966 Sep Edited by C.von Euler and others Illus. ix,563p Pergamon, Oxford,1968
Pha 16.0312

STRUCTURE AND FUNCTION OF INHIBITORY NEURONAL MECHANISMS International meeting of neurobiologists 4th Proceedings Stockholm 1966 Sep Edited by C.von Euler and others Wenner-Gren center.International symposium series, 10 Pergamon press,Oxford, 1968
An 32.5336

STRUCTURE AND FUNCTION OF POLYPEPTIDE HORMONES; INSULIN Symposium on insulin Proceedings Upton,N.Y. 1965 Nov 8-10 Edited by P.G. Katsoyannis and I.L. Schwartz Sponsored by the Brookhaven national laboratory.Medical department American journal of medicine, 40, no. 5 Yorke medical group.Publication New York,1966
Bioch 33.0557

STRUCTURE AND FUNCTION OF PROTEINS AT THE THREE-DIMENSIONAL LEVEL :a symposium Papers Cold Spring Harbor 1972 Cold Spring Harbor laboratory of quantitative biology Cold Spring Harbor symposia on quantitative biology, 36 Cold Spring Harbor,1972
Bioch 33.2334

STRUCTURE AND FUNCTION OF SUBCELLULAR COMPONENTS a symposium London 1957 Feb 23 Biochemical society Edited by E.M. Crook Biochemical society.Symposia, 16 Pl. Cambridge university press,Cambridge,1959
Radioth 35.0069

STRUCTURE AND FUNCTION OF SUBCELLULAR COMPONENTS a symposium London 1957 Feb 23 Biochemical society Edited by E.M. Crook Held at the University of London.Senate House Biochemical society.Symposia, 16 Cambridge university press,Cambridge,1959
Bot 42.1493

The STRUCTURE AND FUNCTION OF SUBCELLULAR COMPONENTS :a symposium London 1957 Feb 23 Biochemical society Edited by E.M. Crook Held at the University of London. Senate House Biochemical society.Symposia, 16 Cambridge university press,Cambridge,1959
Bioch 33.1378

STRUCTURE AND FUNCTION OF THE CEREBRAL CORTEX International meeting of neurobiologists 2nd Proceedings Amsterdam 1959 Sep 22-25 Edited by D.B. Tower and J.P. Schade Elsevier,Amsterdam,1960
An 32.4646

STRUCTURE AND FUNCTION OF THE CEREBRAL CORTEX The International meeting of neurobiologists 2nd Proceedings Amsterdam 1959 Sep 22-25 Edited by D.B. Tower and J.P. Schade Elsevier,Amsterdam,1960
Bal 39.1494

STRUCTURE AND FUNCTION OF THE ENDOPLASMIC RETICULUM IN ANIMAL CELLS The Federation of European biochemical societies meeting 4th Proceedings Oslo 1967 Jul 7 Federation of European biochemical societies Edited by F.C. Gran Organized by the Norwegian biochemical society Universitetsforlaget;Academic press,Oslo; London,1968 Symposium organizer P.N. Campbell
Bioch 33.1401

STRUCTURE AND FUNCTION OF THE EPIPHYSIS CEREBRI an international round-table conference Proceedings Amsterdam 1963 Jul 10-13 Edited by J.Ariens Kappers and J.P. Schade Progress in brain research, 10 Elsevier, Amsterdam,1964
An 32.4230

The STRUCTURE AND FUNCTION OF THE MEMBRANES AND SURFACES OF CELLS :a symposium London 1962 Mar 9 Biochemical society Edited by D.J. Bell and J.K. Grant Biochemical society.Symposia, 22 Cambridge university press,Cambridge,1963
Bioch 33.1384

The STRUCTURE AND FUNCTION OF THE MEMBRANES AND SURFACES OF CELLS :a symposium London 1962 Mar 9 Biochemical society Edited by D.J. Bell and J.K. Grant Held at the Middlesex hospital medical school Biochemical society.Symposia, 22 Cambridge university press,Cambridge,1963 Organized by J.K.Grant
Bal 39.0294

STRUCTURE AND FUNCTION OF TRANSFER RNA AND 5 S-RNA The Federation of European biochemical societies meeting 4th Proceedings Oslo 1967 Jul 5 Federation of European biochemical societies Edited by L.O. Froholm and S.G. Laland Organized by the Norwegian biochemical society Universitetsforlaget;Academic press,Oslo; London,1968 Symposium organizer H.G. Zachau
Bioch 33.1398

STRUCTURE AND FUNCTION OF TRANSFER RNA AND 5S-RNA Federation of European biochemical societies meeting 4th Proceedings Oslo 1967 Jul 3-7 Vol 3: structure and function of transfer RNA and 5S-RNA Federation of European biochemical societies Edited by L.O. Froholm and S.E. Laland Academic press, London,1968
Chem 18.2491

STRUCTURE AND FUNCTIONS OF CONNECTIVE AND SKELETAL TISSUE an advanced study institute Proceedings St.Andrews 1964 Jun 15-25 North Atlantic treaty organization.Scientific affairs division Butterworths,London,1964
Path 30.2421

STRUCTURE AND METABOLISM OF CORTICOSTEROIDS :a symposium Proceedings Paris 1963 Jul 5-6 Universite de Paris.Faculte de medicine. Laboratoire de chimie biologique Edited by Jorge R. Pasqualini and Max F. Jayle Academic press,London;New York,1964
Inv Med 37.0175

The STRUCTURE AND METABOLISM OF THE PANCREATIC ISLETS :an international Wenner-Gren symposium Proceedings Uppsala 1963 Aug 29-Sep 2 and Stockholm Wenner-Gren center Edited by S.E. Brolin and others Wenner-Gren center.International symposium series, 3 Pergamon press,Oxford,1964
Bioch 33.0485

STRUCTURE AND PROPERTIES OF ENGINEERING MATERIALS Research conference on structure and properties of engineering materials Proceedings Raleigh,N.C. 1962 Mar 12-13 North Carolina state college and United States.Army research office,Durham Materials science research, 1 Plenum press,New York, 1963
Met 25.2281

STRUCTURE AND PROPERTIES OF POROUS MATERIALS Colston research society :symposium 10th Proceedings Bristol 1958 Mar 24-27 Edited by D.H. Everett and F.S. Stone Butterworths,London,1958
Min 40 77 10.0655

The STRUCTURE AND PROPERTIES OF POROUS MATERIALS a symposium Proceedings Bristol 1958 Mar 24-27 Colston research society Edited by D.H. Everett and F.S. Stone Colston papers, 10 Butterworths,London,1958
Eng 41.3084

The STRUCTURE AND PROPERTIES OF POROUS MATERIALS a symposium Proceedings Bristol 1958 Mar 24-27 Colston research society Edited by D.H. Everett and F.S. Stone Colston papers, 10 xiv,389p Butterworths,London, 1958
Chem E 24.0854

The STRUCTURE AND PROPERTIES OF POROUS MATERIALS 10th :a symposium Proceedings Bristol 1958 Mar 24-27 Colston research society Edited by D.H. Everett and F.S. Stone Butterworths,London,1958
Chem 18.0558

STRUCTURE AND PROPERTIES OF SOLID SURFACES : a conference Lake Geneva,Wisc. 1952 Sep Edited by Robert Gomer and Cyril Stanley Smith Arranged by the National research council illus 491p University of Chicago press, Chicago,Ill.,1953
Cav 7.1344

STRUCTURE AND PROPERTIES OF SOLID SURFACES : a conference Lake Geneva,Wisc. 1952 Sep National research council Edited by Robert Gomer and Cyril Stanley Smith University of Chicago press,Chicago,1953
Met 25.1594

STRUCTURE AND PROPERTIES OF SOLID SURFACES : conference Papers Lake Geneva,Wis. 1952 Sep Edited by Robert Gomer and Cyril Stanley Smith Organized by the National research council.Committee on solids University of Chicago press,Chicago,Ill.,1953
Col S 12.0275

STRUCTURE AND PROPERTIES OF THIN FILMS An International conference on structure and properties of thin films Proceedings Bolton Landing,N.Y. 1959 Sep 9-11 General electric company Edited by C.A. Neugebauer and others John Wiley,New York,1959
Cav 7.0265

STRUCTURE AND PROPERTIES OF THIN FILMS :an international conference Proceedings New York 1959 Sep 9-11 United States.Air research and development command and General electric research laboratory Edited by C.A. Neugebauer and others Sponsored by United States.Air force.Office of scientific research Wiley,New York;London,n.d.
Met 25.1204

STRUCTURE AND PROPERTIES OF ULTRAHIGH-STRENGTH STEELS :a symposium Cleveland,Ohio 1963 Oct 22 American society for testing materials and American institute of mining, metallurgical and petroleum engineers A.S.T. M.Special technical publication, 370 American society for testing and materials, Philadelphia,Pa.,1965
Met 25.2251

STRUCTURE AND REACTIONS OF DFP SENSITIVE ENZYMES The Conference on structure and reactions of DFP sensitive enzymes Proceedings Stockholm 1966 Sep 5-7 Edited by Edith Heilbronn organized by the Forsvarets forskningsanstalt,Stockholm Research institute of national defence,Stockholm,1967
Bioch 33.1065

The STRUCTURE AND REACTIVITY OF ELECTRONICALLY-EXCITED SPECIES :a symposium Preprint of papers Ottawa 1957 Sep 5-6 Chemical institute of Canada.Physical chemistry division Part-financed by the National research council of Canada Chemical institute of Canada.Physical chemistry division,1957 Chairman:Professor K.J. Laidler
Chem 18.0227

La STRUCTURE DES SOLUTIONS SOLIDES METALLIQUES Orsay 1962 Jul 9-11 Centre national de la recherche scientifique Centre national de la recherche scientifique.Colloques internationaux, 118 Centre national de la recherche scientifique,Paris,1962 Papers in English and French
Met 25.1210

STRUCTURE ET PHYSIOLOGIE DES SOCIETES ANIMALES colloque international Paris 1950 Mar Centre national de la recherche scientifique Centre national de la recherche scientifique. Colloques internationaux, 34 C.N.R.S.,Paris, 1952
Phys 20.0540

STRUCTURE ET PROPRIETES DES SURFACES DES SOLIDES colloque international Paris 1969 Jul 7-11 Centre national de la recherche scientifique Centre national de la recherche scientifique.Colloques internationaux, 187 Paris,1970
Met 25.2626

The STRUCTURE OF CONCRETE AND ITS BEHAVIOUR UNDER LOAD :an international conference Proceedings London 1965 Sep Cement and concrete association Cement and concrete association,London,1968
Eng 41.3007

The STRUCTURE OF NUCLEIC ACIDS AND THEIR ROLE IN PROTEIN SYNTHESIS :a symposium London 1956 Feb 18 Biochemical society Edited by E.M. Crook Held at the London school of hygiene and tropical medicine Biochemical society.Symposia, 14 Cambridge university press,Cambridge,1957
Bioch 33.1376

STRUCTURE OF ORGANIC SOLIDS Microsymposium 'Structure of organic solids' 2nd Lectures Prague 1968 Sep 16-19 International union of pure and applied chemistry.Macromolecular division Czechoslovak chemical society In conjunction with Ceskoslovenska akademie ved illus 550p 25cm Butterworths,London,1969 Simultaneously published in 'Pure and applied chemistry',vol 18,no 4,1969
Cav 7.2659

The STRUCTURE OF THE EYE :the symposium held. during the 7th International congress of anatomists New York 1960 Apr 11-13 Edited by George K. Smelser Academic press, New York,1961
An 32.4773

STRUCTURES Correlation between calculated and observed stresses and displacements in structures :a conference London 1955 Sep 21-22 Institution of civil engineers 2 vols Institution of civl engineers,London, 1955
Eng 41.3020

STRUCTURES ALGEBRIQUES ET STRUCTURES TOPOLOGIQUES conferences a l'Institut Henri Poincare de la Sorbonne Paris 1956 Feb.9 and 1957 Jun.6 By H. Cartan and others Societe mathematique de France en accord avec L' Association des Professeurs de mathematique de l'enseignement public Enseignement mathematique.Monographies, 7 198p Geneva, Paris,1958
P. Math 2.0241

The STRUCTURES AND FUNCTIONS OF PROTEOLYTIC ENZYMES a discussion London 1968 Dec 5-6 Royal society of London Edited by D.C. Phillips and others Royal society of London.Philosophical transactions.Ser.B, 257,p. 63-266 Royal society,London,1970
Bioch 33.1917

STRUKTUR UND STOFFWECHSEL DES HERZMUSKELS : symposium Munster 1958 Sep 26-27 Edited by W.H. Hauss and H. Losse Thieme, Stuttgart,1959
An 32.3467

STUDIES ABOUT HUMUS Humus and plant :a symposium Prague 1961 Sep 28-Oct 6 and Brno Sep 28-Oct 6 Edited by S. Prat and V. Rypacek illus 364p Czechoslovak academy of sciences,Prague,1962
Bot 42.2108

STUDIES IN MODERN FABRICS :a conference Papers London 1970 May 26-29 Textile institute Textile institute.Annual conference, 55 Textile institute,Manchester, 1970
Eng 41.8287

STUDIES IN MODERN YARN PRODUCTION :a conference Papers Stresa 1968 May 25-28 Textile institute Textile institute. Annual conference, 53 Textile institute, Manchester,1968
Eng 41.8288

STUDIES IN SCHIZOPHRENIA :a multidisciplinary approach to mind-brain relationships New Orleans,La. 1952 Jun 11-13 Tulane university of Louisiana. Department of psychiatry and neurology Edited by Robert G. Heath,chairman Published for the Commonwealth fund Harvard university press,Cambridge,Mass.,1954 Transcript of a series of invitational meetings
Psy 28.0081

STUDIES ON ABELIAN GROUPS Symposium on the theory of Abelian groups Montpellier 1967 Jun 5-10 Edited by B. Charles Bibliog. x, 356p 25cm Dunod,Paris,1968
P Math 2.3507

STUDY GROUP ON MENTAL HEALTH ASPECTS OF THE PEACEFUL USES OF ATOMIC ENERGY Report Mental health aspects of the peaceful uses of atomic energy Geneva 1957 Oct 21-26 World health organization World health organization.Technical report series, 151 World health organization,Geneva,1958
Radioth 35.1057

STUDY GROUP ON THE EFFECTS OF RADIATION ON MEIOTIC SYSTEMS Effects of radiation on meiotic systems Vienna 1967 May 8-11 Organized by the International atomic energy agency International atomic energy agency. Panel proceedings series International atomic energy agency,Vienna,1968
Radioth 35.1201

The STUDY OF THE DISTRIBUTION OF BRITISH PLANTS a conference Report Oxford 1950 Mar 31-Apr 2 Botanical society of the British Isles Botanical society of the British Isles. B.S.B.I.conference reports, 2 London,1951
Bot 42.4707

STUDY OF TROPICAL VEGETATION Kandy symposium Proceedings Kandy 1956 Mar 19-21 Unesco. International advisory committee for humid tropics research illus. Unesco,Paris,1958
BG 38.2155

STUDY OF TROPICAL VEGETATION :a symposium Proceedings Kandy,Ceylon 1956 Mar 19-21 Unesco Jointly organized by the Government of Ceylon Humid tropics research Unesco, Paris,1958
Geog 13.1261

The STUDY OF URBAN HISTORY :international round-table conference Proceedings Leicester 1966 Sep Urban history group Edited by H.J. Dyos Arnold,London,1968
Geog 13.6762

STUDY WEEK ON NUCLEI OF GALAXIES Vatican City 1970 Apr 13-18 Edited by D.J.K. O'Connell Accademia pontifica dei nuovi lincei.Scripta varia, 35 795p North-Holland,Amsterdam,1971
Obs 6.3642

STUDY WEEK ON NUCLEI OF GALAXIES Vatican City 1971 Apr 13-18 Accademia pontifica dei nuovi lincei Edited by D.J.K. O'Connell Accademia pontifica dei nuovi lincei.Scripta varia, 35 795p North-Holland,Amsterdam, 1971
TA 15.0661

STUTTGART 1965 International vacuum congress 3rd Transactions 1: invited papers International union for vacuum science.German national committee Edited by H. Adam Pergamon,Oxford,1966 Papers mainly in English,with summaries in French and German.Title pages in English,French and German
Met 25.1691

STUTTGART 1965 International vacuum congress 3rd Transactions Vol 1: invited papers International union for vacuum science,technique and applications Edited by H. Adam 153p Pergamon press, Oxford,1966 Cover title'Advances in vacuum science and technology'
Cav 7.0015

STUTTGART 1966 1966 Intermag International magnetics conference 4th Sponsored by the Institute of electrical and electronics engineers.Magnetics group I.E.E. E.transactions on magnetics,Mag 2,no 3 I. E.E.E.,New York,1966
Eng 41.5347

STUTTGART 1969 Biochemisch nachweisbare strahlenwirkungen und deren beziehungen zur strahlentherapie :symposium der Arbeitsgemeinschaft fur strahlenbiologie Arbeitsgemeinschaft fur strahlenbiologie Edited by G.B. Gerber Thieme,Stuttgart,1970
Radioth 35.1222

STUTTGART 1969 70 jahre radiologie am Katharinen hospital der stadt :kongress der deutschen rontgengesellschaft Deutsche rontgengesellschaft Edited by Friedrich Heuck Stuttgart,1970
Radioth 35.1913

STUTTGART-HOHENHEIM 1960 Die Stoffproduktion der pflanzendecke :vortrage und diskussionsergebnisse des internationalen okologischen symposium Edited by Helmut Lieth port Fischer,Stuttgart,1962
Bot 42.1875

SUBCELLULAR COMPONENTS Structure and function of subcellular components :a symposium London 1957 Feb 23 Biochemical society Edited by E.M. Crook Held at the University of London.Senate House Biochemical society.Symposia, 16 Cambridge university press,Cambridge,1959
Bot 42.1493

SUBCELLULAR COMPONENTS The Structure and function of subcellular components :a symposium London 1957 Feb 23 Biochemical society Edited by E.M. Crook Held at the University of London.Senate House Biochemical society.Symposia, 16 Cambridge university press,Cambridge,1959
Bioch 33.1378

SUBCELLULAR COMPONENTS;PREPARATION AND FRACTIONATION :a symposium London 1967 Nov Imperial cancer research fund Edited by G.D. Birnie and Sylvia M. Fox Butterworths,London,1969
Bioch 33.1016

SUBCELLULAR PARTICLES :a symposium Woods Hole 1958 Jun 9-11 By T. Hayashi Held during the meeting of the Society of general physiologists Ronald press,New York, 1959
Phys 20.1536

SUBCELLULAR PARTICLES :a symposium Woods Hole,Mass. 1958 Jun 9-11 American physiological society Edited by Teru Hayashi Sponsored by the Society of general physiologists Ronald press,New York,1959
Gen 34.0828

SUBCELLULAR PARTICLES :a symposium held during the meeting of the Society of general physiologists at the Marine biological laboratory Woods Hole,Mass. 1958 Jun 9-11 Edited by Teru Hayashi Sponsored by the society of general physiologists Illus The Ronald press,New York,1959
Radioth 35.0500

SUBCELLULAR PARTICLES :a symposium held during the meeting of the Society of general physiologists at the Marine biological laboratory Woods Hole,Mass. 1958 Jun 9-11 Society of general physiologists Edited by Teru Hayashi Published for the American physiological society Ronald press,New York, 1959
Bioch 33.1014

SUBCELLULAR STRUCTURAL COMPONENTS Methods of separation of subcellular structural components :a symposium Louvain 1962 May 11 Biochemical society and Societe belge de biochimie Edited by J.K. Grant Held in the University of Louvain.Physiological institute Biochemical society.Symposia, 23 Cambridge university press,Cambridge,1963 Organized by C.de Duve
Bioch 33.1385

SUBMARINE GEOLOGY AND GEOPHYSICS Colston research society :symposium 17th proceedings Bristol 1965 Apr 5-9 Edited by W.F. Whittard and R. Bradshaw Butterworths,London,1965
Min 10.0573

SUBMARINE GEOLOGY AND GEOPHYSICS Colston research society symposium 17th Proceedings Bristol 1965 Apr 5-9 Edited by W.F. Whittard and R. Bradshaw Colston papers,17 Butterworths,London,1965
Geog 13.0740

SUBMARINE GEOLOGY AND GEOPHYSICS 17th :a symposium Proceedings Bristol 1965 Apr 5-9 Colston research society Edited by W.F. Whittard and R. Bradshaw Colston research society.Colston papers, 17 London, 1965
Geod 9.0580

SUBMARINE GEOLOGY AND GEOPHYSICS 17th : symposium proceedings Bristol 1965 Apr.5-9 Colston research society Edited by W.F. Whittard and R. Bradshaw Colston papers, 17 illus. map Butterworths,London,1965
Geol 8.1358

SUBUNIT STRUCTURE OF PROTEINS :biochemical and genetic aspects.Symposium Report Upton, N.Y. 1964 Jun 1-3 Brookhaven national laboratory.Biology department Brookhaven symposia in biology, 17 BNL869(C-40) Brookhaven national laboratory.Biology department,Upton,N.Y.,1964
Bioch 33.1295

SUBUNIT STRUCTURE OF PROTEINS,BIOCHEMICAL AND GENETIC ASPECTS report of symposium Upton,N.Y. 1964 Jun 1-3 Brookhaven national laboratory Brookhaven symposia in biology, 17 Brookhaven national laboratory, Upton,N.Y.,1964
Gen 34.0769

SUBVIRAL CARCINOGENESIS International symposium on tumor viruses 1st 1966 Edited by Yohei Ito Sponsored by the Aichi cancer center.Research institute and the Japanese cancer association Aichi cancer center,Nagoya,1967
Bioch 33.1191

SUCTION IN SOILS Pore pressure and suction in soils :a conference London 1960 Mar 30-31 International society of soil mechanics and foundation engineering.British national committee Butterworths,London,1961
Eng 41.3170

SUDAN AGRICULTURAL SOCIETY Agricultural development in the Sudan Philosophical society of the Sudan :annual conference 13th Proceedings and papers Khartoum 1965 Dec 3-6 Vol 1-2 Edited by D.J. Shaw maps 2 vols Philosophical society of the Sudan, Khartoum,1966 Typescript
Geog 13.4700

SUDDEN DEATH IN INFANTS Conference on causes of sudden death in infants Proceedings Seattle 1963 Sep 9-10 United States. Department of health,education and welfare Edited by Ralph J. Wedgwood and others United States.Public health service publication,1412 U.S.govt.printing office, Washington,D.C.,1965
HE 27.0158

SUKHUMI 1965 Specific tumour antigens :a symposium Edited by R.J.C. Harris Organised by the International union against cancer International union against cancer. Monograph series, 2 Munksgaard,Copenhagen, 1967
Radioth 35.0901

SULFUR IN PROTEINS :a symposium Proceedings Falmouth,Mass. 1958 May Edited by Reinhold Benesch and others Academic press, New York;London,1959
Radioth 35.0075

SULPHUR IN PROTEINS :a symposium Proceedings Falmouth,Mass. 1958 May Edited by Reinhold Benesch and others Academic press, New York;London,1959
Bioch 33.0555

SUMMER GATHERING ON FUNCTION ALGEBRAS Papers Aarhus 1969 Jul Aarhus universitet. Matematisk institut Aarhus universitet. Matematisk institut.Various publications series, 9 Bibliog. 78p 29cm University of AArhus,AArhus,August,1969
P Math 2.3451

SUMMER INSTITUTE IN NUCLEAR AND PARTICLE PHYSICS Proceedings Toronto 1967 Canadian association of physicists.Theoretical physics division Edited by B. Margolis and C.S. Lam 547p 24cm Benjamin,New York,1968
Cav 7.3100

SUMMER INSTITUTE ON NUMBER THEORY 1969 number theory institute Stony Brook, N.Y. 1969 Jul 7-Aug 1 American mathematical society Edited by Donald J. Lewis American mathematical society.Proceedings of symposia in pure mathematics, 20 xiii,451p 26cm AMS,Providence,R.I.,1971
P Math 2.4383

SUMMER INSTITUTE ON SYMBOLIC MATHEMATICAL COMPUTATION Proceedings Boston,Mass. 1968 Jun 24-Aug 16 International business machines corporation Edited by Robert G. Tobey illus 324p IBM Boston programming center,Cambridge,Mass.,1969
Math L 5.3809

SUMMER MATHEMATICAL INSTITUTE 12th Algebraic groups and discontinuous subgroups Boulder,Colo. 1965 Jul 5-Aug 6 American mathematical society Edited by Armand Borel and George D. Mostow Financed by the National science foundation American mathematical society.Proceedings of symposia in pure mathematics, 9 vii,426p 26cm American mathematical society,Providence,R.I., 1966
P Math 2.2719

SUMMER RESEARCH INSTITUTE OF THE AUSTRALIAN MATHEMATICAL SOCIETY research reports Canberra 1961 Jan. Pt. 1: probability and statistics Australian mathematical society Australian mathematical society,Canberra,1961 Typescript
Math 3.0676

SUMMER SCHOOL AND COLLOQUIUM IN MATHEMATICAL LOGIC Proceedings Logic colloquium 1969 Manchester 1969 Aug 3-23 Victoria university of Manchester Edited by R.O. Gandy and C.M.E. Yates Studies in logic and the foundations of mathematics, 61 xiv,451p 23cm North-Holland,Amsterdam,1971
P Math 2.4057

SUMMER SCHOOL IN LOGIC :N.A.T.O. advanced study institute:meeting of the association for symbolic logic Proceedings Leeds 1967 Aug 7-23 North Atlantic treaty organization Association for symbolic logic Edited by M.H. Lob Lecture notes in mathematics, 70 Bibliog. 331p 28cm Springer,Berlin,1968
P Math 2.3120

SUMMER SCHOOL IN MATHEMATICS,GEOMETRY AND TOPOLOGY proceedings Dundee 1961 Jul 10-22 Queen's college,Dundee 33cm Queen's College,Dundee,1961
P. Math 2.0292

SUMMER SCHOOL IN RADIO ASTRONOMY Papers Radio astronomy today Jodrell Bank 1962 Victoria university of Manchester Edited by H.P. Palmer and others 242p Manchester university press,Manchester,1963
TA 15.0226

SUMMER SCHOOL IN SPACE PHYSICS proceedings Introduction to solar terrestrial relations Alpbach 1963 Jul.15-Aug.10 European space research organisation.Preparatory commission Edited by J. Ortner and H. Maseland Astrophysics and space science library ix, 506p D.Reidel,Dordrecht-Holland,1965
A Math 4.1162

The SUMMER SCHOOL IN SPACE PHYSICS Proceedings Introduction to solar terrestrial relations Alpbach 1963 Jul.15-Aug.10 Edited by J. Ortner and H. Maseland Organized by the European preparatory commission for space research Astrophysics and space science library 506p D.Reidel,Dordrecht,1965
Obs 6.1321

The SUMMER SCHOOL IN SPACE PHYSICS Proceedings Introduction to solar terrestrial relations Alpbach 1963 Jul 15-Aug 10 European preparatory commission for space research Edited by J. Ortner and H. Maseland Astrophysics and space science library Reidel,Dordrecht,1965
TA 15.0425

SUMMER SCHOOL OF BOTANY Proceedings Darjeeling 1960 Jun 2-15 Ministry of scientific research and cultural affairs,New Dehli,1962
BG 38.3112

SUMMER SCHOOL ON TOPOLOGICAL ALGEBRA THEORY Bruges 1966 Sep 6-16 Bibliog 285p 27cm University of Brussels,Brussels,1966
P Math 2.2745

SUMMER SEMINAR IN APPLIED MATHEMATICS 1st Proceedings Boulder,Col. 1957 Jun 23-Jul 21 Vol 3: partial differential equations By Lipman Bers and others Edited by Alton S. Householder and others Sponsored by the American Mathematical Society Lectures in applied mathematics, 3 xiii, 343p Interscience,New York,1964
Dedicated to Arthur Norman Milgram
A Math 4.1266

SUN-EARTH ENVIRONMENT The Symposium on physical processes in the sun-earth environment Proceedings Ottawa 1959 Jul 20-21 Canada.Defence research board Edited by C. Collins DRTE no 1025 xii,392p Ottawa,1960
Nap 11.0140

SUNSPOTS Centenario della nascita di Galileo Galilei 1564-1964 4th Atti Florence 1964 Sep 9-12 Vol 2,tomo 2: atti del convegno sulle machie solari.Proceedings of the meeting on sunspots Galileo Galilei. Comitato nazionale per le manifestazioni celebrative Galileo Galilei.Comitato nazionale per le manifestazioni celebrative. Publicazioni port,figs 266p Barbera, Florence,1966
WSM 43.4470

SUPER DUTY STEELS Quality requirements of super-duty steels a technical conference Proceedings Pittsburgh 1958 May 5-6 American institute of mining,metallurgical and petroleum engineers.Committee on physical chemistry of steelmaking Edited by R.W. Lindsay Metallurgical society conferences, 3 Interscience,New York;London,1959
Met 25.0482

SUPERCONDUCTIVE TECHNIQUES Symposium on superconductive techniques for computing systems proceedings Washington,D.C. 1960 May 17-19 Sponsored by United States.Office of naval research illus. vii,413p 26cm Office of naval research,Washington,D.C.,1960
Math L 5.0859

SUPERCONDUCTIVITY International conference on low temperature physics 10th Proceedings Vol 2 a-b: superconductivity International union of pure and applied physics Akademiya nauk S.S.S.R. Edited by M.P. Malkov 22cm 2 vols
Cav 7.3031

SUPERCONDUCTORS technical sessions New York 1962 Feb 18 American institute of mining,metallurgical and petroleum engineers Edited by M. Tanenbaum and W.V. Wright Interscience,New York;London,1962
Met 25.1181

SUPERNOVA REMNANTS Pulsed-radiosources and high energy activity in supernova remnants Proceedings Rome 1969 Dec 18-20 illus 326p 27cm Accademia nazionale dei lincei, Rome,1972
A Math 4.1898

SUPERNOVAE AND THEIR REMNANTS Conference on supernovae Proceedings Greenbelt,Md. 1967 Goddard institute for space studies Edited by P.J. Brancazio and A.G.W. Cameron 240p Gordon and Breach,New York,1969
Obs 6.3361

SUPERVOLTAGE RADIATION THERAPY Roentdens, rads and riddles :a symposium on supervoltage radiation therapy Oak Ridge,Tenn. 1956 Jul 15-18 United States atomic energy commission Edited by Milton Friedman and others Held at the Oak Ridge institute of nuclear studies U. S.atomic energy commission,Washington,D.C., 1959
Radioth 35.1090

The SUPPORT OF MEDICAL RESEARCH :a symposium London 1954 Oct 4-8 Council for international organisations of medical sciences Edited by H. Himsworth and J.F. Delafresnaye Blackwell,Oxford,1956
PGMS 29.0496

The SUPRARENAL CORTEX :a symposium Proceedings Bristol 1952 Apr 1-4 Colston research society Edited by J.M. Yoffey Colston research society.Symposia, 5 Colston papers, 5 Butterworth,London, 1953
Bioch 33.0445

SURFACE ACTIVITY International congress of surface activity 2nd Proceedings London 1957 Apr 8-13 Vol 1: gas-liquid and liquid-liquid interface Edited by J.H. Schulman ix,529p Butterworths scientific publications, London,1957
Chem E 24.0935

SURFACE ACTIVITY International congress of surface activity 2nd Proceedings London 1957 Apr 8-13 Edited by J.H. Schulman 4 vols Butterworths,London,1957
Col S 12.0278

SURFACE ACTIVITY International congress of surface activity 3rd Proceedings Cologne 1960 Sep 12-17 Vol 1-4 Universitaetsdruckerei,Mainz,1961
Col S 12.0279

SURFACE AND SUBSURFACE RECONNAISSANCE Symposium on surface and subsurface reconnaissance Atlantic City,N.J. 1951 Jun 19 American society for testing materials American society for testing materials.Special technical publication, 122 American society for testing materials,Philadelphia,Pa.,1952
Eng 41.3153

SURFACE CHEMISTRY :a discussion at a joint meeting of the Societe de chimie physique and the Faraday society Papers Bordeaux 1947 Oct 5-9 Societe de chimie physique Faraday society Illus Butterworths scientific publications,London,1949
Papers in English and FRENCH
Radioth 35.0342

SURFACE OF THE MOON IAU-NASA symposium The Conference on the nature of the surface of the moon Proceedings Greenbelt,Md. 1965 Apr 15-16 Goddard space flight center and International astronomical union Edited by Wilmot N. Hess and others Sponsored bt the National aeronautics and space administration 320p Johns Hoplins press,Baltimore,1966
TA 15.0203

SURFACE PHENOMENA Cold Spring Harbor symposia on quantitative biology Papers Cold Spring Harbor 1933 Vol 1 Cold Spring Harbor biological laboratory Long Island biological association,Cold Spring Harbor,1933 Later referred to in vol.9 as "Surface phenomena"
Bioch 33.1257

SURFACE PHENOMENA Fundamental phenomena in the materials sciences :a symposium 2nd,3rd Boston,Mass. 1964,1965 Vol 2-3: surface phenomena Ilikon corporation,Natick Edited by L.J. Bonis and others 2 vols Plenum press,New York,1966
Met 25.1616

SURFACE PROTECTION AGAINST WEAR AND CORROSION two series of educational lectures... presented...during the thirty-fifth National metal congress and exposition Cleveland, Ohio 1953 Oct 19-23 By H.S. Avery and others American society for metals American society for metals,Cleveland,Ohio, 1954
Eng 41.3715

SURFACE PROTECTION AGAINST WEAR AND CORROSION... educational lectures... presented during the thirty-fifth national metal congress and exposition Cleveland 1953 Oct 19-23 By H.S. Avery and others American society for metals American society for metals, Cleveland,Ohio,1954
Met 25.1977

SURFACE STRESSING OF METALS :a series of five educational lectures...presented...during the twenty-seventh National metal congress and exposition Cleveland,Ohio 1946 Feb 4-8 By H.F. Moore and others American society for metals American society for metals, Cleveland,Ohio,1947
Eng 41.3731

SURFACE WATERS Proceedings :symposium Berkeley,Calif. 1963 Aug 19-31 World meteorological association,and,International association of scientific hydrology International association of scientific hydrology.Publication,63 illus 615p Gentbrugge,1964 Papers in English,French, German
Sco 14.0138

SURFACES AND INTERFACES 1-2 Sagamore army materials research conference 13th-14th Proceedings Raquette Lake,New York 1966-67 United States.Army materials research agency and Syracuse university Edited by John J. Burke and others Syracuse university press, New York,1967-68
Met 25.2365

SURGICAL RESEARCH SOCIETY Tools of biological research symposium 1st Proceedings London 1958 Oct 10-11 Edited by Hedley J.B. Atkins Blackwell,Oxford,1959
Gen 34.0465

SURGICAL RESEARCH SOCIETY Tools of biological research symposium 2nd Proceedings 1959 Second series Edited by Hedley J.B. Atkins illus. Blackwell, Oxford,1960
Gen 34.0466

SURGICAL RESEARCH SOCIETY Tools of biological research symposium 3rd Proceedings London 1960 Sep 30-Oct 1 Third series Edited by Hedley J.B. Atkins Blackwell,Oxford,1961
Gen 34.0467

SURTSEY BIOLOGY CONFERENCE Proceedings Surtsey research society Financial support from the United States.Office of naval research photos 48p Reykjavik,1965 Typescript
Sco 14.8155

SURTSEY RESEARCH CONFERENCE Proceedings Reykjavik 1967 Jun 25-28 Surtsey research society Sponsored by the American institute of biological sciences Surtsey research society,Reykjavik,c1967
Sco 14.7944

SURTSEY RESEARCH SOCIETY Surtsey biology conference Proceedings Financial support from the United States.Office of naval research photos 48p Reykjavik,1965 Typescript
Sco 14.8155

SURTSEY RESEARCH SOCIETY Surtsey research conference Proceedings Reykjavik 1967 Jun 25-28 Sponsored by the American institute of biological sciences Surtsey research society,Reykjavik,c1967
Sco 14.7944

SUTTON BONINGTON 1964 Physiology of the domestic fowl :a symposium Proceedings British Egg marketing board Edited by C. Horton-Smith and E.C. Amoroso British egg marketing board symposium,1 Oliver and Boyd, Edinburgh;London,1966
VA 19.0308

SUTTON BONINGTON 1965 Protein utilization by poultry :a symposium Proceedings British egg marketing board Edited by R.A. Morton and E.C. Amoroso British egg marketing board symposium,2 Oliver and Boyd, Edinburgh;London,1967
VA 19.0309

SVENSKA GRUVFORENINGEN Progress in mineral dressing International mineral dressing congress Transactions Stockholm 1957 Sep 18-21 Almqvist and Wiksell,Stockholm,1958
Met 25.2286

SWAMPSCOTT,MASS 1959 Fracture An International conference on the atomic mechanisms of fracture Proceedings National research council Edited by Benjamin Lewis Averbach and others 646p Technology press;John Wiley,Cambridge,Mass;New York,1959
Cav 7.0102

SWAMPSCOTT,MASS. 1955 Teaching of anatomy and anthropology in medical education Association of American medical colleges teaching institute 3rd Report Edited by Lura Street Jackson Association of American medical colleges,Chicago,1955
An 32.0994

SWAMPSCOTT, MASS. 1959 Atomic mechanisms of fracture :an international conference Proceedings National research council. Committee for the conference on fracture Edited by B.L. Averback and others Technology press books in science and engineering Technology press, John Wiley; Chapman and Hall, New York; London, 1959
Met 25.2323

SWAMPSCOTT, MASS. 1959 Fracture An International conference on the atomic mechanisms of fracture Proceedings Edited by B.L. Averbach and others Sponsored by the National science foundation M.I.T. press; Wiley, Cambridge, Mass.; New York, 1959
Eng 41.3767

SWANSEA 1957 Automatic measurement of quality in process plants :a conference Proceedings Society of instrument technology xi, 320p Butterworths, London, 1958
Chem E 24.1160

SWANSEA 1964 Ecology and the industrial society British ecological society symposium 5th Edited by Gordon T. Goodman and others Blackwell, Oxford, 1965
Geog 13.1216

SWANSEA 1964 Ecology and the industrial society :a symposium British ecological society Edited by Gordon T. Goodman and others British ecological society. Symposia, 5 illus. viii, 395p Blackwell scientific, Oxford, 1965
Bot 42.1974

SWANSEA 1964 The State and movement of water in living organisms Society for experimental biology Society for experimental biology. Symposia, 19 Cambridge university press, Cambridge, 1965
Phys 20.0967

SWANSEA 1964 The State and movement of water in living organisms :a symposium Edited by G.E. Fogg Society for experimental biology. Symposia, 19 Cambridge university press, Cambridge, 1965
An 32.2522

SWANSEA 1964 The State and movement of water in living organisms :a symposium Society for experimental biology Society for experimental biology. Symposia, 19 Cambridge university press, Cambridge, 1965
Pha 16.0272

SWANSEA 1965 The Fossil record a symposium with documentation Edited by W.B. Harland and others sponsored by the Geological society of London bibliog. Geological society of London, London, 1967 The papers in parts 1 and 3 were presented at a joint meeting of the Geological society of London and the Palaeontological association, Swansea, Dec 1965
Geol 8.4440

SWANSEA 1967 Non-destructive examination in the steel industry Conference on non-destructive examination applied to process control in the steel industry Papers and discussions Swansea and district metallurgical society Iron and steel institute Society of non-destructive examination Iron and steel institute. Publication, 103 illus 193p London, 1967
Met 25.2601

SWANSEA 1967 Thin walled steel structures : a symposium University of Wales Institution of structural engineers Edited by K.C. Rockey and H.V. Hill Crosby Lockwood, London, 1969
Eng 41.2747

SWANSEA 1972 NATO advanced study institute on facial structure of compact sets and applications North Atlantic treaty organization Held at University college of Swansea 147p 21cm University college of Swansea, Swansea, 1972
P Math 2.4281

SWANSEA AND DISTRICT METALLURGICAL SOCIETY Non-destructive examination in the steel industry Conference on non-destructive examination applied to process control in the steel industry Papers and discussions Swansea 1967 Jan 3-5 Iron and steel institute. Publication, 103 illus 193p London, 1967
Met 25.2601

SWEDENBORG SOCIETY International Swedenborg congress Transactions London 1910 Jul 4-8 2nd edition Swedenborg society, London, 1911
Phys 20.0723

SWEDISH CANCER SOCIETY Symposium on virus and cancer Papers Saltsjobaden 1964 Jun 8-9 Edited by H. Bergstrand and K.E. Hellstrom illus Balder, Stockholm, 1965 Symposium arranged by the Swedish cancer society in conjunction with the annual general meeting of the Unio nordica contra cancrum
Path 30.2363

SWEDISH CHEMICAL SOCIETY Thermodynamics and thermochemistry Symposium on thermodynamics and thermochemistry Plenary lectures Lund 1963 Jul 18-23 Butterworths, London, 1964
Chem E 24.0988

SWEDISH NATIONAL COMMITTEE FOR MECHANICS International symposium on fracture mechanics Kiruna 1967 Aug 7-12 Wolters-Noordhoff, Groningen, 1968
Met 25.0874

SWEDISH NATIONAL COMMITTEE FOR PHYSICS Rydberg centennial conference on atomic spectroscopy Proceedings Lund 1954 Jul 1-5 Edited by Bengt Edlen Physiographisk sallskaps i Lund. Forhandlingar. N.F., 65, 21 108p c.k.w.gleerup, Lund, 1954
Obs 6.3285

SWEDISH SOCIETY OF OBSTETRICS AND GYNAECOLOGY Fertility and sterility World congress on fertility and sterility 5th Proceedings Stockholm 1966 Jun 16-22 Edited by Bjorn Westin and Nils Wiqvist Excerpta medica. International congress series, 133 Excerpta medica, Amsterdam, 1967
Inv Med 37.0230

The SWELLING OF PROTEINS, AND ALLIED PHENOMENA London 1932 Dec 2 International society of leather trades' chemists London, 1933
Bioch 33.1486

SWISS ACADEMY OF MEDICAL SCIENCES
International symposium on chemotherapy of cancer Proceedings Lugano 1964 Apr 28-May 1 Edited by Placidus A. Plattner Sponsored by Hoffmann-La Roche and co ltd. Elsevier,Amsterdam,1964
Pha 16.0335

SWISS ACADEMY OF NATURAL SCIENCES The Development of geodesy and geophysics in Switzerland :commemorative book presented to the participants in the XIVth general assembly of the International union of geodesy and geophysics Edited by J.C. Thams illus,maps 98p Zurich,1967 In English and French, including also French t.-p.
Sco 14.0145

SWISS FEDERAL INSTITUTE OF TECHNOLOGY
Kreisel probleme :internationale symposium Vortrage Celerina,Switzerland 1962 Aug 20-23 Edited by Hans Ziegler Berlin,1963 Papers in English,French and German
Geod 9.0236

SWITCHING THEORY The International symposium on the theory of switching proceedings Cambridge,Mass. 1957 Apr 2-5 Part 1-2 Harvard university.Computation laboratory Harvard university.Computation laboratory. Annals, 29 27cm 2 vols Harvard university press,Cambridge,Mass.,1959
Math L 5.0686

SWITZERLAND 1970 International congress of biochemistry :held at Interlaken,Lucerne and Montreux 8th Abstracts International union of biochemistry Edited by J.G. Gregory Staples press,Rochester,1970
Bioch 33.2338

SYDNEY 1892 Intercolonial medical congress of Australasia 3rd session Transactions Edited by L.Ralston Huxtable Charles Potter, Sydney,1893
An 32.1141

SYDNEY 1911 Australasian medical congress 9th session Transactions Vol 1-2 2 vols Gullick,Sydney,1913
An 32.1145

SYDNEY 1951 Automatic computing machines : conference proceedings Commonwealth scientific and industrial research organization Jointly sponsored by University of Sydney.Department of electrical engineering illus. 220p 25cm Commonwealth scientific and industrial research organization,Melbourne, 1952
Math L 5.0703

SYDNEY 1955 The International wool textile research conference Proceedings Vol A-F 7 vols commonwealth scientific and industrial research organization,Melbourne, 1956 The conference took the form of a series of sessions...at Sydney,Geelong and Melbourne
Bioch 33.1481

SYDNEY 1960 Radiobiology Australasian conference on radiobiology 3rd Proceedings Australian radiation society Edited by P.L.T. Ilbery Butterworths,London, 1961
Radioth 35.1727

SYDNEY 1965 Research in chemical and extraction metallurgy Symposium on 'recent progress in research in chemical and extraction metallurgy' Proceedings Australasian institute of mining and metallurgy Edited by J.T. Woodcock and others Australasian institute of mining and metallurgy.Monograph series, 2 Melbourne, 1965 reprinted 1967
Met 25.2780

SYDNEY 1966 International congress of haematology 11th Proceedings International society of hematology Blackwell,Oxford,1966
Med 36.0203

SYDNEY 1969 Orchids under the Southern cross World orchid conference 6th Commemorative brochure 1969
BG 38.1754

SYDNEY 1969 World orchid conference 6th Proceedings Edited by Murray J.G. Corrigan and others 1971
BG 38.1753

SYLVANIA ELECTRIC PRODUCTS,INC.METALLURGICAL LABORATORIES The Physics of powder metallurgy :a symposium New York 1949 Aug 24-26 Edited by Walter E. Kingston McGraw-Hill,New York,1951
Met 25.0752

SYMBIOTIC ASSOCIATIONS :a symposium Papers London 1963 Apr Society for general microbiology Edited by P.S. Nutman and Barbara Mosse Society for general microbiology.Symposia, 13 Cambridge university press,Cambridge,1963
Bioch 33.1181

SYMBIOTIC ASSOCIATIONS :a symposium Proceedings London Apr 1963 Society for general microbiology Society for general microbiology.Symposia, 13 Cambridge university press,Cambridge,1963
Gen 34.1355

SYMBOL MANIPULATION LANGUAGES AND TECHNIQUES IFIP working conference on symbol manipulation languages Proceedings Pisa 1966 Sep 5-9 International federation for information processes Edited by Daniel G. Bobrow illus 487p North-Holland,Amsterdam,1968
Math L 5.3822

SYMBOLIC MATHEMATICAL COMPUTATION Summer institute on symbolic mathematical computation Proceedings Boston,Mass. 1968 Jun 24-Aug 16 International business machines corporation Edited by Robert G. Tobey illus 324p IBM Boston programming center, Cambridge,Mass.,1969
Math L 5.3809

SYMBOLS,UNITS AND NOMENCLATURE International conference on physics :reports on symbols, units and nomenclature approved by the general assembly of the Union London 1934 Oct.5 International union of pure and applied physics 40p Physical society,London,1935
Obs 6.1963

SYMMETRIES IN ELEMENTARY PARTICLE PHYSICS
International school of physics 'Ettore Majorana' 2nd course proceedings Erice 1964 Aug.-Sep. Edited by A. Zichichi Sponsored by the European organization for nuclear research 429p Academic press,New York;London,1965
A Math 4.0847

SYMMETRY AND FUNCTION OF BIOLOGICAL SYSTEMS AT THE MACROMOLECULAR LEVEL Nobel symposium 11th Procedings Sodergarn 1968 Aug 26-29 Nobel foundation Edited by Arne Engstrom and Bror Strandberg Wiley,New York,1969
Bioch 33.0327

SYMMETRY PRINCIPLES Coral Gables conferences on symmetry principles at high energy 4 Proceedings Coral Gables,Fla. 1967 Jan 25-27 University of Miami.Center for theoretical studies Edited by Arnold Perlmutter and Behram Kursunoglu Sponsored jointly by the United States atomic energy commission 253p Freeman,San Francisco; London,1967
A Math 4.1347

SYMMETRY PRINCIPLES AT HIGH ENERGY
Conference on symmetry principles at high energy Coral Gables,Fla. 1964 Jan 30-31 University of Miami Edited by Behram Kursunoglu and Arnold Perlmutter Freeman,San Francisco,1964
Cav 7.0543

SYMPOSIA BIOLOGICA HUNGARICA, 3
Regeneration and wound healing :symposium Budapest 1960 Nov 8-9 Edited by G. Szanto Akademiai Kiado,Budapest,1964
An 32.0093

SYMPOSIA BIOLOGICA HUNGARICA, 4
Physiologie (bewegung) der spermien :symposium Budapest 1960 Oct Edited by I. Toro Akademia Kiado,Budapest,1964
An 32.3880

SYMPOSIA BIOLOGICA HUNGARICA, 7 Callus formation :symposium on the biology of fracture healing Debrecen 1965 Jul 5-8 Edited by St. Krompecher and E. Kerner Translated by E. Kerner Akademiai Kaido, Budapest,1967
An 32.3921

SYMPOSIA CSAV The Physiology of gene and mutation expression :symposium on the mutational process Proceedings Brno 1965 Aug 4-7 Prague 1965 Aug 9-11 Edited by Margita Kohoutova and J. Hubacek Organized by the Ceskoslovenska akademie ved.Institute of microbiology Academia,Prague,1967
Gen 34.0770

SYMPOSIA IN NEUROANATOMICAL SCIENCES, 6
Neural mechanisms of the auditory and vertibular systems :a conference Proceedings Bethesda,Md. 1959 Jun 11-13 Edited by G.L. Rasmussen and W.F. Windle Sponsored by the National institute of neurological diseases and blindness illus. xiv,422p Thomas, Springfield,Ill.,1960
Phys 20.1975

SYMPOSIA IN NEUROANATOMICAL SCIENCES, 6
Neural mechanisms of the auditory and vestibular systems :a conference Bethesda,Md. 1959 Jun 11-13 Edited by Grant L. Rasmussen and William F. Windle Thomas, Springfield,Ill.,1960
An 32.4821

SYMPOSIA MATHEMATICA Meccanica non lineare e stabilita Rome 1970 Feb 23-26 Istituto nazionale di alta matematica Istituto nazionale di alta matematica.Pubblicazione, 6 395p 25cm Academic press,London;New York,1971
P Math 2.3989

SYMPOSIA MATHEMATICA Rome 1967-68 1,3
Istituto nazionale de alta matematica bibliog. 25cm 2 vols Academic press,New York;London,1969
P Math 2.3778

The SYMPOSIA OF THE GALAPAGOS INTERNATIONAL SCIENTIFIC PROJECT Galapagos Berkeley, Calif. 1964 Jan 7-18 Edited by Robert I. Bowman illus,tables xvii,318p California university press,Berkeley;Los Angeles,1966 Conference held aboard ship en route to the Islands from Jan 11-18
Bot 42.2246

SYMPOSIA ON ARCTIC BIOLOGY AND MEDICINE
Proceedings Fort Wainwright,Alaska 1962 Aug 28-30 influence of cold on host-parasite interactions By Robert I. McClaughry and others Edited by Eleanor G. Viereck Held under the auspices of the University of Alaska. Geophysical institute illus 455p 23cm Fort Wainwright,Alaska,1963
Sco 14.0882

SYMPOSIA ON ARCTIC BIOLOGY AND MEDICINE
Proceedings Fort Wainwright,Alaska 1963 Feb 17-18 4: frostbite Edited by Eleanor G. Viereck Held under the auspices of the University of Alaska.Geophysical institute illus vi,457p 23cm Arctic aeromedical laboratory,Fort Wainwright,Alaska,1964
Sco 14.0883

SYMPOSION UBER DAS NEUROVEGETATIVE SYSTEM DER GESUNDEN UND KRANKEN HAUT DES MENSCHEN Vienna 1957 May 30-Jun 1 Edited by E. Anderson and others Acta neurovegetativa, 18 Springer,Vienna,1958 In English, French and German
An 32.4046

SYMPOSION UBER KREBSPROBLEME :arbeitstagung des beratungsaussschusses fur krebsforschung beim Kultusministerium des landes Nordrhein-Westfalen Dusseldorf 1960 Jun 27-28 Edited by K.G. Ober and others Illus Springer,Berlin,1961
Radioth 35.0515

SYMPOSIUM:ATMOSPHERIC CHEMISTRY,CIRCULATION AND AEROSOLS Proceedings Visby,Sweden 1965 Aug World meteorological organization, and,Commission on atmospheric chemistry and radioactivity Tellus,18,pt 2-3 Svenska geofysiska foreningen,Stockholm,1966
Chem 18.0571

SYMPOSIUM ARISTOTELICUM 4th
Naturphilosophie bei Aristotels und Theophrast Goteborg 1966 Aug Edited by Ingemar During 292p Stiehm,Heidelberg,1969
WSM 43.0192

SYMPOSIUM DE EXPLORACION GEOQUIMICA,TOM.1-3 International geological congress 20th papers Mexico City 1956 Edited by A. Garcia Rojas and others 2cm Mexico City, 1958-60 Text in English,French,Russian and Spanish
Geol 8.3048

SYMPOSIUM DE L'ACADEMIE INTERNATIONALE DE PHILOSOPHIE DES SCIENCES Proceedings Information and prediction in science Brussels 1962 Sep 3-8 Academie internationale de philosophie des sciences Edited by S. Dockx and P. Bernays xi,272p Academic press,London;New York,1965
WSM 43.1136

SYMPOSIUM DE L'ASSOCIATION DE PSYCHOLOGIE SCIENTIFIQUE DE LANGUE FRANCAISE 2eme : la perception Textes et communications Louvain 1953 Sep 26-28 By Henri Pieron and others Association de psychologie scientifique de langue francaise Presses universitaires de France,Paris,1955
Psy 31.0887

SYMPOSIUM DE METALLURGIE SPECIALE Saclay 1957 Jun 27-28 Centre d'etudes nucleaires, Saclay Presses universitaires de France, Paris,1958
Met 25.2517

SYMPOSIUM IN APPLIED MATHEMATICS Proceedings Computers in algebra and number theory New York 1970 Mar 25-26 By Garrett Birkhoff and Marshall Hall American mathematical society Society for industrial and applied mathematics SIAM-AMS proceedings, 4 vii,200p 26cm AMS,Providence,R.I., 1971
P Math 2.4476

A SYMPOSIUM IN GENERAL BIOLOGY Papers presented Biology of phosphorus East Lansing,Mich. 1951 Apr 25-27 By G.Evelyn Hutchinson and others Edited by Lester F. Wolterink Held at Michigan state college Michigan state college press,1952
Bioch 33.0590

SYMPOSIUM IN PURE MATHEMATICS 14th Proceedings Axiomatic set theory Los Angeles,Calif. 1967 Jul 10-Aug 5 Edited by Dana S. Scott Held at the University of California Los Angeles American mathematical society.Proceedings of symposia in pure mathematics, 13,pt 1 v,474p 26cm American mathematical society,Providence,R.I., 1971
P Math 2.4061

SYMPOSIUM INTERNACIONAL DE TOPOLOGIA ALGEBRAICA Mexico City 1956 Aug Universidad nacional de Mexico.Instituto de matematicas Instituto nacional de la investigacion cientifica,Mexico Sociedad matematica mexicana xii,334p 26cm UNESCO,Mexico City,1958 In memory of Witold Hurewicz
P. Math 2.0290

SYMPOSIUM INTERNATIONAL DES SCIENCES PHYSIQUES ET MATHEMATIQUES DANS LA PREMIERE MOITIE DU XVII SIECLE Actes Pisa 1958 Jun 16-18 Vinci (Florence) Union internationale d'histoire et philosophie des sciences Academie internationale d'histoire des sciences.Collection de travaux, 11 x,278p Gruppe italiano di storia delle scienze; Hermann,Vinci (Florence);Paris,c1959
WSM 43.0066

SYMPOSIUM INTERNATIONAL D'HISTOIRE DES SCIENCES Actes Turin 1961 Jul 28-30 Comite national des celebrations du 1er centenaire de l'unite d'Italie With the participation of the Union internationale d'histoire et philosophie des sciences Academie internationale d'histoire des sciences. Collections de travaux, 14 189p Gruppe italiano di storia delle science,Vinci (Florence),1964
WSM 43.0059

SYMPOSIUM (INTERNATIONAL) ON COMBUSTION 5th Pittsburgh,Pa 1954 Aug 30-Sep 3 combustion in engines and combustion kinetics Combustion institute Under the auspices of the Standing committee on combustion symposia v,802p Reinhold;Chapman and Hall,New York; London,1955
Chem E 24.1914

SYMPOSIUM (INTERNATIONAL) ON COMBUSTION 6th New Haven,Conn. 1956 Aug 19-24 Combustion institute xxv,943p Reinhold; Chapman and Hall,New York;London,1957
Chem E 24.1407

SYMPOSIUM (INTERNATIONAL) ON COMBUSTION 10th Cambridge 1964 Aug 17-21 Combustion institute Combustion institute,Pittsburgh,Pa. 1965
Chem 18.0440

SYMPOSIUM (INTERNATIONAL) ON COMBUSTION 11th Papers Berkeley,Calif. 1966 Aug 14-20 Combustion institute Combustion institute, Pittsburgh,Pa.,1967
Chem 18.0441

SYMPOSIUM (INTERNATIONAL) ON COMBUSTION 12 Abstracts of papers Poitiers 1968 Jul 14-20 Combustion institute xxx,228p Combustion institute,Pittsburgh,Pa.,1968
Chem E 24.1408

SYMPOSIUM (INTERNATIONAL) ON COMBUSTION 13th Proceedings Salt Lake City,Utah 1970 Aug 23-29 Combustion institute xix,1190p 26cm Combustion institute,Pittsburgh,Pa., 1971
Chem 18.2863

SYMPOSIUM (INTERNATIONAL) ON COMBUSTION 7th Papers London 1958 Aug 28-Sep 3 Oxford Combustion institute Butterworths scientific publications,London, 1959 Earlier symposia entitled 'Symposium (international) on combustion,flame and explosion phenomena'
Chem 18.0437

SYMPOSIUM (INTERNATIONAL) ON COMBUSTION 8th Papers Pasadena,Calif. Aug 28-Sep 3 Combustion institute Williams and Wilkins, Baltimore,Md.,1962
Chem 18.0438

SYMPOSIUM (INTERNATIONAL) ON COMBUSTION 9th Papers Ithaca,N.Y. 1962 Aug 27-Sep 1 Combustion institute Academic press,New York; London,1963
Chem 18.0439

SYMPOSIUM (INTERNATIONAL) ON COMBUSTION,FLAME AND EXPLOSION PHENOMENA 6th Papers New Haven,Conn. 1956 Aug 19-24 Combustion institute Reinhold;Chapman and Hall,New York; London,1957 Earlier symposia organized by the Standing committee on combustion.Later symposia entitled 'Symposium (international) on combustion'
Chem 18.2602

SYMPOSIUM (INTERNATIONAL) ON COMBUSTION,FLAME AND EXPLOSION PHENOMENA 4th (combustion and detonation waves) Papers Cambridge,Mass. 1952 Sep 1-5 Standing committee on combustion symposia Held at the Massachusetts institue of technology Williams and Wilkins,Baltimore,Md.,1953 Committee chairman:B.Lewis.Earlier volumes entitled 'Symposium on combustion,flame and explosion phenomena'
Chem 18.0434

SYMPOSIUM (INTERNATIONAL) ON COMBUSTION,FLAME AND EXPLOSION PHENOMENA 5th :combustion in engines and combustion kinetics Papers Pittsburgh,Pa. 1954 Aug 30-Sep 3 Standing committee on combustion symposia Published for the Combustion institute Reinhold; Chapman and Hall,New York;London,1955 Later symposia organized by the Combustion institute
Chem 18.0435

SYMPOSIUM INTERNATIONAL SUR LA CHIMIE DES SUBSTANCES NATURELLES Chemistry of natural products The International symposium on the chemistry of natural products 5th Plenary lectures London 1968 Jul 8-13 International union of pure and applied chemistry.Division of organic chemistry and Chemical society Butterworths,London,1968 "The contents of this book appear in 'Pure and applied chemistry,vol.17, nos.3-4 (196819". Added title page in French
Bioch 33.1392

SYMPOSIUM INTERNATIONAL SUR LES ASPECTS SCIENTIFIQUES DES AVALANCHES DE NEIGE see INTERNATIONAL SYMPOSIUM ON SCIENTIFIC ASPECTS OF SNOW AND ICE AVALANCHES

SYMPOSIUM INTERNATIONAL SUR LES MAREES TERRESTRES 4th :symposium on earth tides Proceedings Brussels 1961 Jun 5-10 Edited by P. Melchior Observatoire royal de Belgique.Communication., 188,Serie geophysique, 58 R.Louis,Ixelles,i961
Geod 9.0025

SYMPOSIUM INTERNAZIONALE GENETICAE MEDICAE 1st Atti Rome 1953 Sep 6-7 1 Istituto Gregorio Mendel Edited by Luigi Gedda Analecta genetica, 1 Edizioni dell' istituto Gregorio Mendel,Rome,1954
Med 36.0148

SYMPOSIUM INTERNAZIONALE GENETICAE MEDICAE Rome 1953 Sep 6-7 3: chondrodysplasie Istituto Gregorio Mendel Edited by Luigi Gedda Analecta genetica, 1 Edizioni dell' istituto Gregorio Mendel,Rome,1954
Med 36.0150

SYMPOSIUM INTERNAZIONALE GENETICAE MEDICAE 1 Atti Rome 1953 Sep 6-7 4: il parto indolore Istituto Gregorio Mendel Edited by Luigi Gedda Analecta genetica, 1 Edizioni dell' istituto Gregorio Mendel,Rome, 1956
Med 36.0151

SYMPOSIUM INTERNAZIONALE GENETICAE MEDICAE 1st Atti Rome 1953 Sep 6-7 2: krampfbereitschaft Istituto Gregorio Mendel Edited by Luigi Gedda Analecta genetica, 1 Edizioni dell' istituto Gregorio Mendel,Rome, 1954
Med 36.0149

SYMPOSIUM OF PLASMA DYNAMICS Woods Hole,Mass. 1958 Jun 9-13 Edited by Francis H. Clauser Sponsored by United States.Air force. Office of scientific research Pergamon press; Addison-Wesley,London;Reading,Mass.,1960
Eng 41.4479

SYMPOSIUM OF PLASMA DYNAMICS proceedings Massachusetts 1968 Jun.9-13 National academy of sciences Edited by Frances H. Clauser Sponsored by the United States.Air force.Office of scientific research Pergamon press,London,1959
A Math 4.0617

SYMPOSIUM OF REAL-TIME CONTROL OF ELECTRIC POWER SYSTEMS Real-time control of electric power systems Baden 1971 Sep 27-28 Brown, Boveri and company Edited by Edmund Handschin Elsevier,Amsterdam,1972 The second Brown,Boveri symposium
Eng 41.8558

SYMPOSIUM OF RECENT ADVANCES IN THE CHEMISTRY OF NATURALLY OCCURRING PYRONES AND RELATED COMPOUNDS Proceedings Dublin 1955 Jul 12-14 University college,Dublin,and, Institute of chemistry of Ireland Royal Dublin society.Scientific proceedings,27,no 6 Royal Dublin society,Dublin,1956
Chem 18.1494

SYMPOSIUM OF THE MOLECULAR BIOLOGY OF VIRUSES Proceedings Molecular biology of viruses Edmonton,Alberta 1966 Jun 27-30 University of Alberta.Faculty of medicine Edited by John S. Colter and William Paranchych Academic press,New York;London,1967
Path 30.2745

SYMPOSIUM ON ACOUSTICAL FATIGUE Atlantic City,N.J. 1960 Jun 28 American society for testing materials A.S.T.M.Special technical publication, 284 A.S.T.M.,Philadelphia,Pa., 1961
Met 25.0929

SYMPOSIUM ON ACUTE RENAL FAILURE Papers Acute renal failure London 1963 Sep Edited by Stanley Shaldon and G.C. Cook Blackwell,Oxford,1964
Inv Med 37.0195

SYMPOSIUM ON ADVANCED MEDICINE :a conference Proceedings London 1964 Nov 16-20 Royal college of physicians of London Edited by Nigel Compston Pitman,London,1965
PGMS 29.0057

A SYMPOSIUM ON ADVANCED MEDICINE 2nd Proceedings Advanced medicine London 1965 Nov 29-Dec 3 Edited by J.R. Trounce Held at the Royal college of physicians Pitman medical,London,1966
Inv Med 37.0213

SYMPOSIUM ON ADVANCED PROBLEMS AND METHODS IN FLUID DYNAMICS 8th Proceedings Fluid dynamics transations,4 Tarda,Poland 1967 Sep 18-25 Polska akademia nauk Edited by W. Fiszdon and others bibliog.,illus. x, 812p 24cm PWN,Warsaw,1969
A Math 4.1680

SYMPOSIUM ON ADVANCES IN ELECTRON METALLOGRAPHY Papers American society for testing materials A.S.T.M.Special technical publication, 245 American society for testing materials,Philadelphia,Pa.,1958
Met 25.1362

SYMPOSIUM ON ADVANCES IN ELECTRON METALLOGRAPHY AND ELECTRON PROBE MICROANALYSIS Electron metallography and probe microanalysis 1962 Atlantic City,N.J. 1960;1961 American society for testing materials.Subcommittee on electron microstructure A.S.T.M.Special technical publication, 317 A.S.T.M., Philadelphia,Pa.,1962
Met 25.1377

SYMPOSIUM ON ADVANCES IN TECHNIQUES IN ELECTRON METALLOGRAPHY Techniques in electron metallography 1963 New York 1962 Jun 26 American society for testing materials. Subcommittee 11 on electron microstructure of metals A.S.T.M.Special technical publication, 339 American society for testing and materials,Philadelphia,Pa.,1963
Met 25.1378

SYMPOSIUM ON ADVANCES IN TRACER METHODOLOGY 6th-8th Collection of papers presented,plus other papers Edited by Seymour Rothschild Advances in tracer methodology, 2 New England nuclear corporation.Publication Plenum press,New York,1965
Bioch 33.0301

The SYMPOSIUM ON ADVANCING FRONTIERS OF LIFE SCIENCES Proceedings New Delhi 1960-61 Dec 30-Jan 1 National institute of sciences of India National institute of sciences of India.Bulletin, 19 New Delhi, 1962 Symposium held on the occasion of the Silver Jubilee of the National institute of sciences of India
Bal 39.0009

SYMPOSIUM ON AFRICAN HYDROBIOLOGY AND INLAND FISHERIES 2nd Brazzaville 1956 Jul 3-11 Scientific council for Africa south of the Sahara Published under the auspices of the Commission for technical co-operation in Africa south of the Sahara Scientific council for Africa south of the Sahara. Publication, 25 Brazzaville,1956 Title also in French:Symposium sur l'hydriobiologie et la peche en eaux douces en Afrique;text in English and French
Bal 39.1894

SYMPOSIUM ON AFRICAN HYDROBIOLOGY AND INLAND FISHERIES 2nd Brazzaville 1956 Jul 3-11 Scientific council for Africa south of the Sahara Scientific council for Africa south of the Sahara.Publications,25 Commission for technical co-operation in Africa,London,1957
Geog 13.2726

A SYMPOSIUM ON AGENTS AFFECTING FERTILITY London 1964 Biological council.Co-ordinating committee for symposia on drug action Edited by C.R. Austin and J.S. Perry Churchill,London,1965
An 32.3896

A SYMPOSIUM ON AMINO ACID METABOLISM Papers and discussions presented Baltimore,Md. 1954 Jun 14-17 Edited by William D. McElroy and H.Bentley Glass Sponsored by the McCollum-Pratt institute McCollum-Pratt institute.Contribution, 105 Johns Hopkins press,Baltimore,1955
Bioch 33.0599

SYMPOSIUM ON ANODIZING ALUMINIUM Proceedings Birmingham 1967 Apr 12-13 Aluminium federation and University of Aston in Birmingham Aluminium federation,London,1967
Met 25.2048

SYMPOSIUM ON ANTARCTIC GLACIOLOGY International association of scientific hydrology general assembly Papers Helsinki 1960 Jul 25-Aug 6 International association of scientific hydrology.Publication,55 illus 162p 24cm Gentbrugge,1961 Papers in English and French
Sco 14.0130

SYMPOSIUM ON ANTARCTIC OCEANOGRAPHY main review papers Santiago,Chile 1966 Sep 13-16 Scientific committee on antarctic research Bibliog.,Illus. xi,268p 23cm Scott polar research institute,Cambridge,1966
A Math 4.1501

SYMPOSIUM ON APPLICATIONS OF MICRO-ELECTRONICS Birmingham 1968 Mar 27 Organized by the Institution of electrical engineers Institution of electrical engineers.Conference publication, 49 Institution of electrical engineers,London,1968
Eng 41.5266

SYMPOSIUM ON APPLICATIONS OF MODERN METALLOGRAPHIC TECHNIQUES Materials engineering exposition and congress Philadelphia,Pa. 1969 Oct 13-16 American society for testing and materials American society for metals American society for testing and materials.Special technical publication, 480 ASTM,Philadelphia,Pa.,1970 Symposium presented at the exposition and congress
Met 25.2607

SYMPOSIUM ON APPROXIMATION OF FUNCTIONS 8th Proceedings Approximation of functions Warren,Mich. 1964 General motors research laboratories Edited by Henry L. Garabedian 220p Elsevier,Amsterdam,1965
TA 15.0097

SYMPOSIUM ON APPROXIMATION OF FUNCTIONS 8th proceedings Approximation of functions Warren,Mich. 1964 Aug 31-Sep 2 General motors corporation.Research laboratories Edited by Henry L. Garabedian viii,220p 25cm Elsevier publishing co.,Amsterdam,1965
P. Math 2.0978

SYMPOSIUM ON ASPECTS OF THE GENE symposium Proceedings St.Andrews 1964 Jan 9-11 University of St.Andrews.Biological society Edited by Michael D.B. Burt and Peter W. Barlow Biological journal, 4,supp. University of St.Andrews.Biological society,St. Andrews,1964
Gen 34.0762

A SYMPOSIUM ON ASTRONOMICAL OPTICS AND RELATED SUBJECTS proceedings Astronomical optics and related subjects Manchester 1955 Apr 19-22 Edited by Zdenek Kopal Under the auspices of Victoria university of Manchester.Astronomy department 428p North-Holland,Amsterdam,1956 Dedicated to the memory of Peter Yorke Millns
Obs 6.0975

SYMPOSIUM ON ATHEROSCLEROSIS Washington,D.C. 1954 Mar 22-23 National research council. Division of medical sciences Sponsored by Human factors division.Air force directorate of research and development · National research council.Publication, 338 National research council,Washington,D.C.,1955
Med 36.0266

SYMPOSIUM ON ATHEROSCLEROSIS Papers 1954 Mar 22-23 National research council.Division of medical sciences Held at the request of the United States.Air force.Directorate of research and development.Human factors division National research council. Publication, 338 National research council, Washington,D.C.,1955
Path 30.2349

SYMPOSIUM ON ATMOSPHERIC CORROSION OF NON-FERROUS METALS Atlantic City,N.J. 1955 Jun 29 American society for testing materials A.S.T. M.Special technical publication, 175 A.S.T.M. Philadelphia,Pa.,1955
Met 25.1918

A SYMPOSIUM ON ATMOSPHERIC DIFFUSION AND AIR POLLUTION proceedings Oxford 1958 Aug 24-29 International union of theoretical and applied mechanics and International union of geodesy and geophysics Edited by F. N. Frenkel and P.A. Sheppard Advances in geophysics, 6 Academic press,New York; London,1959
A Math 4.0576

SYMPOSIUM ON ATOMIC COLLISION PROCESSES Boulder summer institute for theoretical physics 11th Proceedings Boulder,Colo. 1968 Jun 17-Aug 23 University of Colorado. Department of physics and astrophysics Edited by Kalyana T. Mahanthappa and others Lectures in theoretical physics, 11c diagrms xiii,337p 23cm Gordon and Breach, London,1969 Dedicated to George Gamow
Cav 7.3010

SYMPOSIUM ON ATOMIC INTERACTIONS AND SPACE PHYSICS Autoionization,astrophysical, theoretical and laboratory experimental aspects Greenbelt,Ma. 1965 Aug.10-11 Goddard space flight center Edited by Aaron Temkin 155p Mono book corp.,Baltimore,Ma., 1966
A Math 4.0911

SYMPOSIUM ON ATOMIC INTERACTIONS AND SPACE PHYSICS Autoionization: astrophysical, theoretical and laboratory experimental aspects Papers Greenbelt,Md. 1965 Aug 10-11 Edited by Aaron Temkin Mono books, Baltimore,Md.,1966
TA 15.0479

SYMPOSIUM ON ATTERBERG LIMITS Papers on soils Symposium on time rates of loading in soil tension Symposium on Atterberg limits, Session on soils and Symposium on soils for engineering purposes Atlantic City,N.J. 1959 Jun 22-23 San Francisco,Calif. 1959 Oct 16 American society for testing materials.Special technical publication, 254 American society for testing materials, Philadelphia,Pa.,1960 Papers presented at various A.S.T.M. meetings during 1959
Eng 41.3151

SYMPOSIUM ON AUTOMATIC DEMONSTRATION Versailles 1968 Dec Edited by M. Laudet and others Organised by the Institut de recherche d'informatique et d'automatique Lecture notes in mathematics, 125 Bibliog 310p 25cm Springer-Verlag,Berlin,1970
P Math 2.3614

SYMPOSIUM ON AUTOMOBILE ENGINEERING Proceedings Advances in automobile engineering Cranfield 1963 Jul Pt 2 Edited by N.A. Carter College of aeronautics, Cranfield.Advanced school of automobile engineering Cranfield international symposium series, 5 Pergamon press,London, 1964
Eng 41.6190

SYMPOSIUM ON BASIC EFFECTS OF ENVIRONMENT ON THE STRENGTH,SCALING,AND EMBRITTLEMENT OF METALS AT HIGH TEMPERATURES Cincinnati 1955 Feb 2 American society for testing materials A.S.T.M.Special technical publication, 171 A. S.T.M.,Philadelphia,Pa.,1955
Met 25.1908

SYMPOSIUM ON BASIC MECHANISMS OF FATIGUE presented before the 61st annual meeting Boston 1958 Jun 23 American society for testing materials A.S.T.M.Special technical publication, 273 A.S.T.M.,Philadelphia,1959
Met 25.2328

SYMPOSIUM ON BASIC MECHANISMS OF FATIGUE Papers Boston,Mass. 1958 Jun 23 American society for testing materials A.S.T. M.Special technical publication, 237 A.S.T.M. Philadelphia,1959
Met 25.0925

SYMPOSIUM ON BASIC RESEARCH New York 1959 American association for the advancement of science Edited by Dael Wolfle American association for the advancement of science. Publication, 56 Washington,D.C.,1959
Eng 41.1676

A SYMPOSIUM ON BEHAVIOR OF MATERIALS IN REACTOR ENVIRONMENT Nuclear metallurgy 1956 Feb 20 By R.C. Dalzell and others American institute of mining and metallurgical engineers.Institute of metals division I.M.D. Special report series, 2 American institute of mining and metallurgical engineers,New York,1956 Lithoprinted
Met 25.1928

SYMPOSIUM ON BIOLOGICAL EFFECTS OF NEUTRON AND PROTON IRRADIATIONS Proceedings Biological effects of neutron and protein irradiations Upton,N.Y. 1963 Oct 7-11 1-2 International atomic energy agency Held at Brookhaven national laboratory International atomic energy agency.Proceedings series 2 vols International atomic energy agency,Vienna,1964
Radioth 35.1168

SYMPOSIUM ON BIOMECHANICS Papers London 1959 Apr 17 Institution of mechanical engineers London,1959
Bal 39.0061

SYMPOSIUM ON BRITTLE FAILURE OF ROTOR FORGINGS presented at special meeting for delegates to world metals congress Philadelphia,Pa. 1957 Nov 11 American society for testing materials A.S.T.M.Special technical publication, 231 American society for testing materials,Philadelphia,Pa,1957
Met 25.1325

SYMPOSIUM ON CALCIUM AND CELLULAR FUNCTION Essays London 1969 Mar 24-25 Biological council.Co-ordinating committee for symposia on drug action Edited by A.W. Cuthbert Macmillan,London,1969
Bioch 33.2264

A SYMPOSIUM ON CARBENOXOLONE SODIUM London 1967 Nov 20 Edited by J.M. Robson and F.M. Sullivan Sponsored by Berk pharmaceuticals limited Butterworths,London,1968
Surg 23.0013

SYMPOSIUM ON CATALYSIS IN PRACTICE Proceedings Harrogate 1963 Jun 20-21 Edited by J.M. Pirie The Yorkshire meeting of the Institution of chemical engineers 83p Institution of chemical engineers,London,1963
Chem E 24.0803

SYMPOSIUM ON CATECHOLAMINES 2nd Milan 1965 Jul 4-9 American society for pharmacology and experimental therapeutics Edited by George H. Acheson Held at the Istituto di ricerche farmacologiche "Mario Negri" Williams and Wilkins company, Baltimore,1966 Reprinted from the Pharmacological reviews,18,no.1
Bioch 33.0430

SYMPOSIUM ON CATECHOLAMINES 2nd Milan 1965 Jul 4-9 Edited by George H. Acheson Pharmacological reviews, 18,no.1 Williams and Wilkins,Baltimore,Md.,1966 Reprint
An 32.3777

SYMPOSIUM ON CAVITATION IN FLUID MACHINERY Chicago,Ill. 1965 Nov 7-11 American society of mechanical engineers Edited by Glenn M. Wood and others American society of mechanical engineers,New York,1965
Eng 41.6538

SYMPOSIUM ON CAVITATION RESEARCH.FACILITIES AND TECHNIQUES American society of mechanical engineers New York,1964
Eng 41.6521

SYMPOSIUM ON CHEMICAL ANALYSIS OF INORGANIC SOLIDS BY MEANS OF THE MASS SPECTROMETER Atlantic City,N.J. 1951 Jun 20 American society for testing materials.Committee E-2 on emission spectroscopy A.S.T.M.Special technical publication, 149 A.S.T.M., Philadelphia,Pa.,1951
Met 25.1723

SYMPOSIUM ON CHEMICAL-BIOLOGICAL CORRELATION 1st Washington,D.C. 1950 May 26-27 Sponsored by the Chemical-biological co-ordination centre National academy of sciences.National research council,Washington, D.C.,1951
Radioth 35.0329

SYMPOSIUM ON CHEMICAL ENGINEERING EDUCATION Discussion on papers presented Birmingham 1957 Apr 9-11 Institution of chemical engineers 1957
Chem E 24.1758

SYMPOSIUM ON CHEMICAL ENGINEERING EDUCATION Papers Birmingham 1957 Apr 9-11 Institution of chemical engineers 61p 1957
Chem E 24.1757

SYMPOSIUM ON CHEMICAL ENGINEERING IN THE METALLURGICAL INDUSTRIES Proceedings Edinburgh 1963 Sep 25-29 Institution of chemical engineers 147p Institution of chemical engineers,London,1963
Chem E 24.1131

SYMPOSIUM ON CHEMICAL PROCESS HAZARDS WITH SPECIAL REFERENCE TO PLANT DESIGN Proceedings Manchester 1960 Mar 29-31 Edited by D.M. Pirie 117p Institution of chemical engineers,London,1960
Chem E 24.1395

SYMPOSIUM ON CHEMOTHERAPY OF CANCER Oslo 1956 May 21-25 International union against cancer International union against cancer. Acta, 13,no 3 U.I.C.C.,Louvain,1957
Radioth 35.0814

SYMPOSIUM ON CHROMOSOME BREAKAGE Contributions London 1952 Jun 9-11 By C.D. Darlington and others John Innes horticultural institution Heredity, 5,supplement Oliver and Boyd,London;Edinburgh,1953
Path 30.0219

SYMPOSIUM ON CHROMOSOME MECHANICS AT THE MOLECULAR LEVEL :given at research conference for biology and medicine of the atomic energy commission Gatlinburg,Tenn. 1967 Apr 10-13 Sponsored by the Oak Ridge national laboratory.Biology division Oak Ridge national laboratory.Symposia, 20 Journal of cellular physiology, 70,suppl Wistar institute of anatomy and biology, Philadelphia,Pa.,1967
Gen 34.0846

SYMPOSIUM ON CIRCUIT ANALYSIS Proceedings Monticello,Ill. 1955 University of Illinois 1955
Eng 41.5160

A SYMPOSIUM ON CLEAR AIR TURBULENCE AND ITS DETECTION Proceedings Clear air turbulence and its detection Seattle 1968 Aug 14-16 Boeing scientific research laboratories.Flight sciences laboratory Edited by Yih-Ho Pao and Arnold Goldburg 542p 22cm Plenum press,New York,1969
Cav 7.2921

SYMPOSIUM ON CLINICAL TRIALS London 1958 Apr Royal society of medicine Edited by Edward Charles Dodds Pfizer,Tonbridge,1958
Med 36.0116

SYMPOSIUM ON COAL PREPARATION 2nd Papers Leeds 1957 Oct 21-25 University of Leeds. Department of mining xiii,513p University of Leeds,Leeds,1957
Chem E 24.1392

SYMPOSIUM ON COBALT-60 BEAM THERAPY IN MALIGNANT DISEASE Edited by S. Krishnamurthi and V. Shanta Associated printers,Madras,c1962
Radioth 35.1134

SYMPOSIUM ON CODASYL DATA BASE TASK GROUP REPORT Collected papers and discussion notes British computer society.CODASYL data base group London 1970 Oct illus 59p British computer society,London,1971
Math L 5.3861

SYMPOSIUM ON COGNITION 1st Problem solving:research,method and theory Pittsburgh,Pa. 1966 Edited by Benjamin Kleinmuntz Sponsored by the Carnegie institute of technology Wiley,New York; London,1966
Eng 41.1291

SYMPOSIUM ON COGNITION 2nd Papers Concepts and the structure of memory Pittsburgh,Pa. 1966 Apr 7-8 Edited by Benjamin Kleinmuntz Sponsored by the Carnegie institute of technology Wiley,New York;London,1967
Eng 41.0652

SYMPOSIUM ON COGNITION 3rd Papers Formal representation of human judgement Pittsburgh,Pa. 1967 Apr 13-14 Edited by Benjamin Kleinmuntz Sponsored by the Carnegie-Mellon university Wiley,New York; London,1968
Eng 41.0653

SYMPOSIUM ON COLLOID CHEMISTRY The National symposium on colloid chemistry 1st Papers and discussions Madison,Wis. 1923 Jun 12-15 University of Wisconsin Edited by J. Howard Mathews Colloid symposium monograph, 1 Department of chemistry,university of Wisconsin,Madison,Wis.,1923
Bioch 33.1469

SYMPOSIUM ON COLLOID CHEMISTRY The National symposium on colloid chemistry 2nd Papers presented Evanston,Ill. 1924 Jun 18-21 Northwestern university Edited by Harry N. Holmes Colloid symposium monograph, 2 Chemical catalog company,New York,1925
Bioch 33.1470

SYMPOSIUM ON COLLOID CHEMISTRY The National symposium on colloid chemistry 3rd Papers presented Minneapolis 1925 Jun 17-19 University of Minnesota Edited by Harry N. Holmes and Harry B. Weiser Colloid symposium monograph, 3 Chemical catalog company,New York,1925
Bioch 33.1471

SYMPOSIUM ON COLLOID CHEMISTRY The National symposium on colloid chemistry 4th Papers presented Cambridge,Mass. 1926 Jun 23-25 Massachusetts institute of technology Edited by Harry Boyer Weiser Colloid symposium monograph, 4 Chemical catalog company,New York,1926
Bioch 33.1472

SYMPOSIUM ON COLLOID CHEMISTRY The National symposium on colloid chemistry 5th Papers presented Ann Arbor,Mich. 1927 Jun 22-24 University of Michigan Edited by Harry Boyer Weiser Colloid symposium monograph, 5 Chemical catalog company,New York,1928
Bioch 33.1473

The SYMPOSIUM ON COLLOID CHEMISTRY 6th Papers presented Toronto 1928 Jun 14-16 University of Toronto Edited by Harry Boyer Weiser Colloid symposium monograph, 6 Chemical catalog company,New York,1928 Colloid symposium monograph known also as Colloid symposium annual
Bioch 33.1474

The SYMPOSIUM ON COLLOID CHEMISTRY 7th Papers presented Baltimore 1929 Jun 20-22 Johns Hopkins university Edited by Harry Boyer Weiser Colloid symposium annual, 7 Wiley;Chapman and Hall,New York;London, 1939 Colloid symposium annual known also as Colloid symposium monograph
Bioch 33.1475

The SYMPOSIUM ON COLLOID CHEMISTRY 9th Papers presented Columbus,Ohio 1931 Jun Ohio State university Edited by Harry Boyer Weiser Colloid symposium monograph, 9 Journal of physical chemistry,New York,1931
Bioch 33.1476

The SYMPOSIUM ON COLLOID CHEMISTRY 10th Papers presented Ottawa 1932 Jun Edited by Harry Boyer Weiser Colloid symposium monograph, 10 Journal of physical chemistry,New York,1932
Bioch 33.1477

The SYMPOSIUM ON COLLOID CHEMISTRY 11th Papers presented Madison,Wis. 1934 Jun Edited by Harry Boyer Weiser Colloid symposium monograph, 11 Williams and Wilkins,Baltimore,1935
Bioch 33.1478

The SYMPOSIUM ON COLLOID CHEMISTRY 12th Papers presented Ithaca,N.Y. 1935 Jun Edited by Harry Boyer Weiser Colloid symposium monograph, 12 Williams and Wilkins,Baltimore,1936
Bioch 33.1479

The SYMPOSIUM ON COLLOID CHEMISTRY 13th Papers presented St.Louis,Mo. 1936 Jun Edited by Harry Boyer Weiser Colloid symposium monograph, 13 William and Wilkins, Baltimore,1937
Bioch 33.1480

SYMPOSIUM ON COMBUSTION 9th proceedings Ithaca,New York 1962 Aug.27-Sep.1 Organized by the Combustion Institute Academic press,New York,1963
A Math 4.0970

SYMPOSIUM ON COMBUSTION,FLAME AND EXPLOSION PHENOMENA 3rd abstracts of papers Madison,Wisc. 1948 Sep 7-11 Standing committee on combustion symposia University of Wisconsin,Madison,Wisc.,1948
Chem 18.0432

SYMPOSIUM ON COMBUSTION,FLAME AND EXPLOSION PHENOMENA 3rd Papers Madison,Wisc. 1948 Sep 7-11 Standing committee on combustion symposia Williams and Wilkins, Baltimore,Md.,1949 Chairman of committee: B.Lewis.Later volumes entitled 'Symposium (international) on combustion,flame and explosion phenomena'
Chem 18.0433

SYMPOSIUM ON COMBUSTION AND FLAME AND EXPLOSION PHENOMENA 3rd Madison,Wis. 1948 Sep 7-11 Standing committee on combustion symposia xiii,748p Williams and Wilkins, Baltimore,Md.,1949
Chem E 24.1404

SYMPOSIUM ON COMPARATIVE BIOLOGY 1st Papers Comparative biochemistry of photoreactive systems Richmond,Calif. Kaiser foundation research institute Edited by Mary Belle Allen Kaiser foundation research institute. Symposia on comparative biology, 1 Academic press,New York;London,1960
Bioch 33.0334

SYMPOSIUM ON COMPOSITE MATERIALS:TESTING AND DESIGN New Orleans,La. 1969 Feb 11-13 American society for testing materials. Committee D-30 high modulus fibers and their composites A.S.T.M.Special technical publication, 460 Philadelphia,Pa.,1969 Presented at a meeting of committee D-30
Met 25.2859

SYMPOSIUM ON COMPUTER AND INFORMATION SCIENCES 3rd Proceedings Software engineering Miami,Fla. 1969 Dec Vol 1-2 Edited by Julius T. Tou COINS, 3 illus 2 vols Academic press,New York,1970
Math L 5.3830

SYMPOSIUM ON COMPUTER AUGMENTATION OF HUMAN REASONING Proceedings Washington,D.C. 1964 Jun 16-17 United States.Office of naval research Edited by Margo A. Sass and William D. Wilkinson In joint sponsorship with the Bunker-Ramo corporation Spartan books, Washington,D.C.,1965
Eng 41.2202

SYMPOSIUM ON COMPUTER AUGMENTATION OF HUMAN REASONING proceedings Washington,D.C. 1964 Jun 16-17 United States.Office of naval research Edited by Margo A. Sass and William D. Wilkinson Jointly sponsored by Bunker-Ramo corporation bibliog,illus. vi,235p 24cm Spartan books,Washington,D.C.,1965
Math L 5.1005

SYMPOSIUM ON COMPUTER AUGMENTATION OF HUMAN REASONING proceeding Washington,D.C. 1964 Jun.16-17 United States.Office of naval research Edited by Margo A. Sass and William D. Wilkinson In joint sponsorship with the Bunker-Ramo corporation vi,235p 24cm Spartan books;Macmillan,Washington,D.C.;London, 1965
Math 3.0244

SYMPOSIUM ON COMPUTER CONTROL OF NATURAL RESOURCES AND PUBLIC UTILITIES Preprints Haifa 1967 Sep 11-14 Sponsored by International federation of automatic control. Committee on applications Haifa,1967
Eng 41.5866

SYMPOSIUM ON COMPUTER PROCESSING IN COMMUNICATIONS Proceedings Computer processing in communications New York 1969 Microwave research institute Edited by Jerome Fox Microwave research institute. Symposia series, 19 illus. 850p Polytechnic institute of Brooklyn,Brooklyn,N.Y. 1970
Math L 5.3558

SYMPOSIUM ON COMPUTERS AND AUTOMATA Proceedings New York 1971 Apr 13-15 Microwave research institute Edited by Jerome Fox Microwave research institute. Symposia series, 21 illus 653p Polytechnic press,Brooklyn,N.Y.,1971
Math L 5.3635

SYMPOSIUM ON CONGESTION THEORY Proceedings Chapel Hill,N.C. 1964 Aug 24-26 Edited by Walter L. Smith and William E. Wilkinson Held at the University of North Carolina University of North Carolina.Monograph series in probability and statistics, 2 University of North Carolina press,Chapel Hill, N.C.,1965
Eng 41.2150

SYMPOSIUM ON CONGESTION THEORY proceedings Chapel Hill,N.C. 1964 Aug.24-26 University of North Carolina.Department of statistics Edited by Walter L. Smith and William E. Wilkinson Financially supported by the United States.Office of naval research North Carolina.University.Monograph series in probability and statistics, 2 xv,457p 24cm University of North Carolina press, Chapel Hill,N.C.,1965
Math 3.0568

SYMPOSIUM ON CONSOLIDATION TESTING OF SOILS Atlantic City,N.J. 1951 Jun 18 American society for testing materials American society for testing materials.Special technical publication, 126 American society for testing materials,Philadelphia,Pa.,1952
Eng 41.3149

A SYMPOSIUM ON CONTINENTAL DRIFT London 1964 Mar Royal society Edited by P.M.S. Blackett and others Royal society. Philosophical transactions,1088 Royal society,London,1965
Geog 13.5647

A SYMPOSIUM ON CONTINENTAL DRIFT Papers London 1964 Mar 19-20 By P.M.S. Blackett and others Organised for the Royal society of London Royal society of London. Philosophical transactions.Series A,258,no. 1088 vi,323p Royal society of London, London,1965
Sco 14.0260

A SYMPOSIUM ON CONTINENTAL DRIFT Papers London 1964 Mar 19-20 Royal society of London Royal society of London.Philosophical transactions.Series A, 258 London,1965 Symposium organized for the Royal society by P. M.S.Blackett,Sir Edward Bullard, and S.K. Runcorn
Geod 9.0129

A SYMPOSIUM ON CONTINENTAL DRIFT papers London 1964 Mar 19-20 By P.M.S. Blackett and others Royal society of London Royal society of London.Philosophical transactions. Series A,no.258,no.1088 map Royal society of London,London,1965
Geol 8.1262

SYMPOSIUM ON CONTROL OF ENERGY METABOLISM Control of energy metabolism A Colloquium on metabolic control and a Symposium on control of energy metabolism Proceedings Philadelphia 1965 May 20-21 Johnson research foundation Edited by Britton Chance and others Johnson research foundation. Colloquia Academic press,New York;London, 1965 "In celebration of the bicentennial of the University of Pennsylvania school of medicine"
Bioch 33.1077

SYMPOSIUM ON COPPER IN BIOLOGICAL SYSTEMS Proceedings Biochemistry of copper New York 1965 Sep 8-10 Edited by Jack Peisach and others Academic press,New York;London, 1966
Bioch 33.1068

SYMPOSIUM ON CORROSION FUNDAMENTALS :a series of lectures... Knoxsville 1955 Mar 1-3 University of Tennessee.Chemical engineering department and National association of corrosion engineers.Southeast region Edited by Anton De S. Brasunas and E. E. Stansbury University of Tennessee press, Knoxville,Tenn.,1956
Met 25.1884

SYMPOSIUM ON COSMIC RAYS,ELEMENTARY PARTICLE PHYSICS AND ASTROPHYSICS 10th Proceedings Aligarh 1967 Dec 12-16 India.Department of atomic energy.Cosmic ray committee 701p Bombay,1968
TA 15.0553

The SYMPOSIUM ON COSMICAL GAS DYNAMICS Proceedings Gas dynamics of cosmic clouds Cambridge 1953 Jul 6-11 International astronomical union and International union of theoretical and applied mechanics Edited by J.M. Burgess and H.C.van de Hulst International astronomical union.Symposium, 2 North Holland,Amsterdam,1955
TA 15.0217

SYMPOSIUM ON COSMICAL GAS DYNAMICS 2nd Gas dynamics of cosmic clouds Cambridge 1953 Jul 6-11 International union of theoretical and applied mechanics International astronomical union International astronomical union.Symposium series, 2 North-Holland,Amsterdam,1955
Eng 41.4468

SYMPOSIUM ON COSMICAL GAS DYNAMICS 3rd Proceedings Cambridge,Mass. 1957 Jun 24-29 International union of theoretical and applied mechanics International astronomical union Edited by J.M. Burgers and R.N. Thomas Reviews of modern physics, 30,no 3 American physical society,New York,1958
Eng 41.4467

SYMPOSIUM ON COSMICAL GAS DYNAMICS 3rd proceedings Cambridge,Mass. 1957 Jun.24-29 International union of theoretical and applied mechanics and International astronomical union Edited by J.M. Burgers and others Financial grant from the Smithsonian institution International astronomical union.Symposium, 8 Smithsonian Institution.Smithsonian symposium publications, 1 American physical society, New York,1958 Reviews of modern physics. Vol.30,no.3,July 1958
Obs 6.0306

SYMPOSIUM ON COSMICAL GAS DYNAMICS 4th held at the International school of physics "Enrico Fermi" proceedings Aerodynamic phenomena in stellar atmospheres Varenna 1960 Aug.18-30 International astronomical union and International union of theoretical and applied mechanics Edited by R.N. Thomas and others Under the auspices of Societa Italiana di fisica International astronomical union.Symposium, 12 515p Bologna,c1961 Reprinted from Supplemento Nuovo Cimento,22,no.1,1961
A Math 4.1152

SYMPOSIUM ON COSMICAL GAS DYNAMICS 4th held at the International school of physics "Enrico Fermi" proceedings Aerodynamic phenomena in stellar atmospheres Varenna 1960 Aug 18-30 International astronomical union and International union of theoretical and applied mechanics Edited by R.N. Thomas and others under the auspices of the Societa italiana di fisica International astronomical union symposium, 12 515p Zanichelli,Bologna,c1961 Reprinted from the 'Supplemento del nuovo cimento',vol.22,no. 1,1961
Obs 6.1736

SYMPOSIUM ON COSMICAL GAS DYNAMICS 6th Proceedings Interstellar gas dynamics Yalta 1969 Sep 8-18 International astronomical union International union of theoretical and applied mechanics International astronomical union.Symposium, 39 388p Reidel,Dordrecht,1970
Obs 6.3323

SYMPOSIUM ON COSMICAL GAS DYNAMICS 6th Proceedings Interstellar gas dynamics Yalta 1969 Sep 8-18 International union of theoretical and applied mechanics and International astronomical union Edited by H. J. Habing International astronomical union. Symposium, 39 diagrs,photos,graphs xii, 388p 24cm Reidel,Dordrecht,1970
Cav 7.2710

SYMPOSIUM ON CREEP OF CONCRETE Houston,Tex. 1964 Mar American concrete institute American concrete institute.Publication,SP-9 American concrete institute,Detroit,Mich., 1964
Eng 41.2932

SYMPOSIUM ON CYTOCHEMICAL PROGRESS IN ELECTRON MICROSCOPY Oxford 1962 Jul 2-4 Royal microscopical society Journal of the Royal microscopical society.Ser.3, 81,pt 3-4,p.107-117 London,1963
Bal 39.0235

SYMPOSIUM ON CYTOCHEMICAL PROGRESS IN ELECTRON MICROSCOPY Oxford 1962 Jul 2-4 Royal microscopical society Journal of the Royal microscopical society,1962,pt.3,4 Royal microscopical society,London,1963
An 32.3182

SYMPOSIUM ON DESIGN OF INDUSTRIAL EXPERIMENTS proceedings Experimental designs in industry Raleigh,N.C. 1956 Nov.5-9 North Carolina state college.Institute of statistics Edited by Victor Chew Supported by the United States.Air force.Office of scientific research Wiley publications in statistics Bibliog. xi,268p 28cm John Wiley and sons,New York,1958 Being parts A and B of Symposium on design of industrial experiments, proceedings,edited by Victor Chew,1957
Math 3.0583

SYMPOSIUM ON DESIGN OF INDUSTRIAL EXPERIMENTS proceedings Raleigh,N.C. 1956 Nov.5-9 North Carolina state college.Institute of statistics Edited by Victor Chew Supported by the United States.Air force.Office of scientific research Bibliog. viii,374p 27cm Institute of statistics,University of North Carolina,Raleigh,N.C.,1957
Math 3.0584

SYMPOSIUM ON DETERMINATION OF GASES IN METALS Atlantic City,N.J. 1957 Jun 18 American society for testing materials A.S.T.M. Special technical publication, 222 A.S.T.M., Philadelphia,Pa,1958
Met 25.1708

SYMPOSIUM ON DIFFERENTIATION AND GROWTH OF HAEMOGLOBIN-AND IMMUNOGLOBIN-SYNTHESISING CELLS :given at research conference for biology and medicine of the Atomic energy commission Gatlinburg,Tenn. 1966 Apr 4-7 Sponsored by the Oak Ridge national laboratory.Biology division Oak Ridge national laboratory.Biology division.Symposia, 19 Journal of cellular physiology, 67,supp. 1 Wistar institute of anatomy and biology, Philadelphia,Pa.,1966
Gen 34.0504

SYMPOSIUM ON DIGITAL COMPUTING IN THE AIRCRAFT INDUSTRY proceedings New York 1957 Jan 31-Feb 1 New York university Jointly sponsored by International business machines corporation illus. 399p 28cm International business machines corporation, New York,1957
Math L 5.0777

SYMPOSIUM ON DIRECT SHEAR TESTING OF SOILS New York 1952 Jun 26 American society for testing materials American society for testing materials.Special technical publication, 131 American society for testing materials,Philadelphia,Pa.,1953
Eng 41.3156

SYMPOSIUM ON DRUG ACTION AND MODERN THERAPY London 1961 Sep 28 Anglo-German medical society Anglo-German medical review, 1,no. 4 Schattauer,Stuttgart,1962 Text in English and German
Gen 34.1973

The SYMPOSIUM ON DRUGS AFFECTING LIPID METABOLISM Proceedings Drugs affecting lipid matabolism Milan 1960 Edited by S. Garattini and R. Paoletti illus Elsevier, Amsterdam,1961
Bioch 33.1237

SYMPOSIUM ON DYNAMIC BEHAVIOR OF MATERIALS Albuquerque,N.M. 1962 Sep 27-28 Sponsored by the American society for testing and materials American society for testing and materials.Special technical publication, 336 American society for testing and materials. Materials science series, 5 American society for testing and materials,Philadelphia, Pa.,1963
Eng 41.3788

SYMPOSIUM ON DYNAMIC TESTING OF SOILS Atlantic City,N.J. 1953 Jul 2 American society for testing materials American society for testing materials.Special technical publication, 156 American society for testing materials,Philadelphia,Pa.,1954
Eng 41.3148

SYMPOSIUM ON EFFECT OF CYCLIC HEATING AND STRESSING ON METALS AT ELEVATED TEMPERATURES Chicago 1954 Jun 17 American society for testing materials A.S.T.M.Special technical publication, 165 A.S.T.M., Philadelphia,1954
Met 25.2534

SYMPOSIUM ON EFFECT OF CYCLIC HEATING AND STRESSING ON METALS AT ELEVATED TEMPERATURES Effect of cyclic heating and shessing metals Chicago 1954 Jun 17 American society for testing materials A.S.T.M. Special technical publication, 165 A.S.T.M., Philadelphia,Pa.,1954
Met 25.0980

SYMPOSIUM ON EFFECT OF TEMPERATURE ON THE BRITTLE BEHAVIOR OF METALS WITH PARTICULAR REFERENCE TO LOW TEMPERATURES :presented at the 56th annual meeting A.S.T.M. Atlantic city,N.J. 1953 Jun 28-30 American society for testing materials American society for testing materials.Special technical publication, 158 American society for testing materials,Philadelphia,Pa.,1954
Eng 41.3750

SYMPOSIUM ON EFFECT OF TEMPERATURE ON THE BRITTLE BEHAVIOUR OF METALS WITH PARTICULAR REFERENCE TO LOW TEMPERATURES Metallic materials at low temperatures Atlantic City 1953 Jun 28-30 American society for testing materials A.S.T.M.Special technical publication, 158 American society for testing materials, Philadelphia,Pa.,1954
Met 25.0850

SYMPOSIUM ON EFFECTS OF LOW TEMPERATURES ON THE PROPERTIES OF MATERIALS :presented at a meeting of the Philadelphia district,A.S.T.M. Philadelphia,Pa. 1946 Mar 19 American society for testing materials American society for testing materials.Special technical publication, 78 American society for testing materials,Philadelphia,Pa.,1950
Eng 41.3778

SYMPOSIUM ON EFFECTS OF RADIATION AND OTHER DELETERIOUS AGENTS ON EMBRYONIC DEVELOPMENT Oak Ridge,Tenn. 1953 Apr 20-21 Sponsored by the Oak Ridge national laboratory.Biology division Wistar institute of anatomy and biology,Philadelphia,Pa.,1954 Reprinted from the Journal of cellular and comparative physiology Vol 43,Suppt 1,May 1954
An 32.2027

SYMPOSIUM ON ELECTRICAL CONDUCTIVITY IN ORGANIC SOLIDS Proceedings Durham 1960 Apr 20-22 United States.Army.Office of ordnance research United States.Office of naval research Edited by H. Kallman and M. Silver Sponsored also by United States.Air force. Office of scientific research Interscience, New York,1961
Radioth 35.1646

SYMPOSIUM ON ELECTRODE PROCESSES Transactions Philadelphia 1959 May 4-6 Electrochemical society.Theoretical electrochemistry division Edited by Ernest Yeager Wiley,New York;London,1961
Met 25.1820

SYMPOSIUM ON ELECTRODE PROCESSES Transactions Philadelphia,Pa. 1959 May 4-6 Electrochemical society.Theoretical electrochemistry division and United States.Air force.Office of scientific research Wiley,New York,1961
Chem 18.2624

SYMPOSIUM ON ELECTROFORMING - APPLICATIONS,USES, AND PROPERTIES OF ELECTROFORMED METALS presented at the 1962 committee week Dallas 1962 Feb 6-7 American society for testing materials A.S.T.M.Special technical publication, 318 A.S.T.M.,Philadelphia,Pa., 1962
Met 25.0733

SYMPOSIUM ON ELECTRON,ION AND LASER BEAM TECHNOLOGY 11th Record Electron,ion and laser beam technology 1971 Edited by R. F.M. Thornley San Francisco press,San Francisco,Calif.,1971
Eng 41.8582

SYMPOSIUM ON ELECTRON BEAM MELTING 1st Proceedings Boston,Mass. 1959 Mar 20 Edited by James S. Hetherington Sponsored by Alloyd research corporation Alloyd,Watertown, Mass.,1959
Eng 41.5613

SYMPOSIUM ON ELECTRON BEAM PROCESSES 2nd Proceedings Boston,Mass. 1960 Mar 24-25 Edited by Robert Bakish Sponsored by the Alloyd research corporation Alloyd,Cambridge, Mass.,1960
Eng 41.5614

SYMPOSIUM ON ELECTRON BEAM PROCESSES 3rd Proceedings Boston,Mass. 1961 Mar 23-24 Edited by Robert Bakish Sponsored by Alloyd electronics corporation Alloyd,Cambridge, Mass.,1961
Eng 41.5615

The SYMPOSIUM ON ELECTRON BEAM PROCESSES 3rd Proceedings Boston,Mass. 1961 Mar 23-24 Alloyd electronics corporation,Cambridge,Mass Edited by R. Bakish 379p Alloyd electronics,Cambridge,Mass.,1961
Cav 7.0113

SYMPOSIUM ON ELECTRON BEAM TECHNOLOGY 4th Proceedings Boston,Mass. 1962 Mar 29-30 Edited by R. Bakish Sponsored by the Alloyd electronics corporation Alloyd,Cambridge, Mass.,1962
Eng 41.5616

SYMPOSIUM ON ELECTRON METALLOGRAPHY Atlantic City,N.J. 1959 Jun 23 American society for testing materials A.S.T.M.Special technical publication, 262 American society for testing materials,Philadelphia,Pa.,1960
Met 25.1366

SYMPOSIUM ON ELECTRON METALLOGRAPHY Techniques of electron microscopy,diffraction and microprobe analysis Papers Atlantic City 1963 Jun 26 American society for testing materials A.S.T.M.Special technical publication, 372 Philadelphia,1964
Met 25.2495

SYMPOSIUM ON ELECTRONIC EQUIPMENT RELIABILITY 2nd Symposium digest London 1962 Oct 24-26 Institution of electrical engineers illus. 47p 28cm Institution of electrical engineers,London,1962
Math L 5.0926

SYMPOSIUM ON ELECTRONIC PHENOMENA IN CHEMISORPTION AND CATALYSIS ON SEMICONDUCTORS Moscow 1968 Jul 2-4 Edited by K. Hauffe and T. Wolkenstein De Gruyter,Berlin,1969
Chem 18.2698

SYMPOSIUM ON ELECTRONIC PHENOMENA IN CHEMISORPTION AND CATALYSIS ON SEMICONDUCTORS Moscow 1968 Jul 2-4 Edited by K. Hauffe and Th. Wolkenstein English edition De Gruyter,Berlin,1969 Held during the fourth international congress on catalysis
Met 25.2475

SYMPOSIUM ON ELEVATED TEMPERATURE STRAIN GAGES Philadelphia,Pa. 1957 Dec 4-5 American society for testing materials Sponsored by United States.Naval air material center A.S. T.M.Special technical publication, 230 American society for testing materials, Philadelphia,Pa,1958
Met 25.0985

SYMPOSIUM ON ENERGY DISPERSION X-RAY ANALYSIS X-ray and electron probe analysis Toronto,Ont. 1970 Jun 21-26 American society for testing and materials.Committee E-4 on metallography ASTM,Philadelphia,Pa., 1971 Symposium presented at the annual meeting.Co-ordinator:J.C.Russ
Met 25.2617

SYMPOSIUM ON ENGINEERING APPLICATIONS OF RANDOM FUNCTION THEORY AND PROBABILITY 1st Proceedings National science foundation Purdue university Edited by J.L. Bogdanoff and F. Kozin Wiley,New York,1963
Eng 41.5949

SYMPOSIUM ON ENGINEERING APPLICATIONS OF RANDOM FUNCTION THEORY AND PROBABILITY 1st proceedings Purdue university Edited by John L. Bogdanoff and Frank Kozin Supported by National science foundation x,421p 24cm John Wiley and sons,New York,1963
Math 3.0879

SYMPOSIUM ON ENZYME REACTION MECHANISMS Research conference for biology and medicine 12th Gatlinburg,Tenn. 1959 Apr 1-4 Sponsored by the Oak Ridge national laboratory. Biology division Journal of cellular and comparative physiology, 54,suppl.1 Wistar institute of anatomy and biology,Philadelphia, Pa.,1959
Bioch 33.1046

SYMPOSIUM ON EVALUATION OF METALLIC MATERIALS IN DESIGN FOR LOW-TEMPERATURE SERVICE presented at the 64th annual meeting Atlantic City,N.J. 1961 Jun 27-28 American society for testing materials A.S.T.M. Special technical publication, 302 A.S.T.M., Philadelphia,1962
Met 25.2335

SYMPOSIUM ON EVALUATION TESTS FOR STAINLESS STEELS Atlantic City 1949 Jun 30 American society for testing materials A.S.T.M.Special technical publication, 93 American society for testing materials, Philadelphia,Pa.,1950
Met 25.1967

A SYMPOSIUM ON EXTRACTION METALLURGY OF SOME OF THE LESS COMMON METALS Extraction and refining of the rarer metals London 1956 Mar 22-23 Institution of mining and metallurgy Institution of mining and metallurgy,London,1957
Met 25.0500

SYMPOSIUM ON FATIGUE Contributions Cranfield 1952 Mar Ergonomics research society Edited by W.F. Floyd and A.T. Welford Lewis,London,1953
Psy 31.1027

SYMPOSIUM ON FATIGUE Papers and discussions Prague 1960 Sep 8-10 Czechoslovak scientific-technical society Ceskoslovenska vedecko-technicka spolecnost,1961 Papers in English,French,German and Russian
Met 25.0936

SYMPOSIUM ON FATIGUE TESTS OF AIRCRAFT STRUCTURE: LOW-CYCLE,FULL-SCALE, AND HELICOPTERS Los Angeles 1962 Oct 1-3 American society for testing materials A.S.T.M.Special technical publication, 338 A.S.T.M.,Philadelphia,Pa., 1963
Met 25.0935

SYMPOSIUM ON FATIGUE WITH EMPHASIS ON STATISTICAL APPROACH New York 1952 Jun 24 Pt 2 American society for testing materials A. S.T.M.Special technical publication American society for testing materials,Philadelphia 3, Pa.,1953
Met 25.0913

SYMPOSIUM ON FIBROUS PROTEINS Papers Canberra 1967 Aug 14-18 Edited by W.G. Crewther Sponsored by the Australian academy of science Butterworths,Sydney,1968
Bioch 33.0512

SYMPOSIUM ON FIBROUS PROTEINS Proceedings Fibrous proteins Leeds 1946 May 23-25 Society of dyers and colourists Edited by C. L. Bird Leeds,1946
Bioch 33.0503

A SYMPOSIUM ON FLORISTICS AND STRATIGRAPHY OF GONDWANALAND Proceedings Birbal Sahni institute of palaeobotany 1964 Dec Palaeobotanical society illus. 154p 24cm Birbal Sahni institute of palaeobotany,Lucknow, 1966
Bot 42.3271

SYMPOSIUM ON FLOW VISUALIZATION :A.S.M.E. annual meeting Presentation summaries New York City 1960 Nov 30 By S.J. Kline and others American society of mechanical engineers illus var.p A.S.M.E.,New York, 1960
Chem E 24.0329

SYMPOSIUM ON FLUID DYNAMICS 5th Fluid dynamics transactions Jablonna 1961 26 Aug. 2 Sep. Vol. 1 Polska akademia nauk Edited by W. Fiszdon Pergamon press,London, 1964
A Math 4.0527

A SYMPOSIUM ON FORMATION AND EARLY DEVELOPMENT OF THE EMBRYO Beginnings of embryonic development Atlanta,Ga. 1955 Dec 27 American association for the advancement of science.Section on zoological sciences Edited by Albert Tyler and others Co-sponsored by the American society of zoologists and the Association of southeastern biologists American association for the advancement of science.Publication, 48 Washington,D.C.,1957
Bal 39.0401

SYMPOSIUM ON FOUNDATIONS OF MATHEMATICS proceedings Infinitistic methods Warsaw 1959 Sep 2-9 Polska akademia nauk.Instytut matematyczny Under the auspices of the International mathematical union 362p 26cm Pergamon press,Oxford,1961
P. Math 2.0211

SYMPOSIUM ON FRACTURE TOUGHNESS CONCEPTS FOR WELDABLE STRUCTURAL STEEL Proceedings Practical fracture mechanics for structural steel Risley 1969 Apr Edited by M.O. Dobson Published for the United Kingdom atomic energy authority Chapman and Hall, London,1970
Eng 41.3822

SYMPOSIUM ON FRACTURE TOUGHNESS CONCEPTS FOR WELDABLE STRUCTURAL STEEL Proceedings Practical fracture mechanics for structural steel Risley 1969 Apr United Kingdom atomic energy authority Edited by M.O. Dobson U.K.A.E.A. in association with Chapman and Hall,Risley,1969
Met 25.2465

SYMPOSIUM ON FREE AMINO ACIDS Amino acid pools :distribution and function of free amino acids.Symposium Proceedings Duarte,Calif. 1961 May 19-22 City of hope medical center Edited by Joseph T. Holden Under the auspices of the Institute for advanced learning in the medical sciences illus Elsevier,Amsterdam,1962 Preface:"The committee undertook the organization of the symposium under the title Conference on free amino acids
Bioch 33.0519

SYMPOSIUM ON FREE RADICALS IN BIOLOGICAL SYSTEMS Proceedings Free radicals in biological systems Stanford,Calif. 1960 Mar 21-23 Stanford university Edited by M.S. Blois Academic press,New York;London,1961
Bioch 33.0309

SYMPOSIUM ON FRETTING CORROSION New York 1952 Jun 25 American society for testing materials A.S.T.M.Special technical publication, 144 American society for testing materials,Philadelphia,Pa.,1953 Fiftieth anniversary publication
Met 25.1953

SYMPOSIUM ON FRICTION AND WEAR Proceedings Detroit,Mich. 1957 Jun 10-11 Edited by Robert Davies port Elsevier,Amsterdam,1959
Eng 41.6297

SYMPOSIUM ON FULL-SCALE TESTS ON HOUSE STRUCTURES Los Angeles,Calif. 1956 Sep 18 American society for testing materials American society for testing materials.Special technical publication, 210 American society for testing materials,Philadelphia,Pa.,1956
Eng 41.3018

SYMPOSIUM ON FULLY SEPARATED FLOWS :a conference Papers Philadelphia,Pa. 1964 May 18-20 American society of mechanical engineers Edited by Arthur G. Hansen American society of mechanical engineers,New York,1964
Eng 41.6514

SYMPOSIUM ON FUNDAMENTAL CANCER RESEARCH 12th Papers Radiation biology and cancer Houston,Tex. 1958 Anderson hospital and tumor institute University of Texas press, Austin,Tex.,1958
Phys 20.0988

SYMPOSIUM ON FUNDAMENTAL CANCER RESEARCH 16th Papers Conceptual advances in immunology and oncology Houston,Tex. 1962 Harper, New York,1962
Pha 16.0185

SYMPOSIUM ON FUNDAMENTAL PHENOMENA IN THE MATERIAL SCIENCES 1st Proceedings Boston,Mass. 1963 Feb 1 Vol 1: sintering and plastic deformation Ilikon corporation, Natick Edited by L.J. Bonis and H.H. Hausner Plenum press,New York,1964
Met 25.2280

SYMPOSIUM ON FUNDAMENTAL TOPICS IN FLUID MECHANICS AND MAGNETOHYDRODYNAMICS Proceedings Michigan State university 1962 Oct 10-11 Michigan State university Edited by Robert Wassermann and Charles P. Wells Academic press,New York,1963
Cav 7.1569

SYMPOSIUM ON GENETIC APPROACHES TO SOMATIC CELL VARIATION Gatlinburg,Tenn. 1958 Apr 2-5 Sponsored by Oak Ridge national laboratory. Biology division Oak Ridge national laboratory.Symposia, 11 Journal of cellular and comparative physiology, 52,supp.1 Wistar institute of anatomy and biology, Philadelphia,Pa.,1958
Gen 34.1094

SYMPOSIUM ON GENETIC EFFECT OF RADIATION Proceedings Misima,Japan 1960 Nov 7-9 Under the auspices of the National institute of genetics,Shizvoka Japanese journal of genetics, 36,supp. Genetics society of Japan,1961
Gen 34.1117

SYMPOSIUM ON GENETIC RECOMBINATION given at research conference for biology and medicine of the atomic energy commission Oak Ridge, Tenn. 1954 Apr 19-21 Sponsored by the Oak Ridge national laboratory.Biology division Oak Ridge national laboratory.Symposia, 7 Journal of cellular and comparative physiology, 45,supp.2 Wistar institute of anatomy and biology,Philadelphia,Pa.,1955
Gen 34.1042

SYMPOSIUM ON GENETIC RECOMBINATION given at the Research conference for biology and medicine of the Atomic Energy commission Oak Ridge,Tenn. 1954 Apr 19-21 Sponsored by Oak Ridge national laboratory.Biology division Wistar institute of anatomy and biology,Philadelphia,Pa.,1955 Reprinted from the Journal of cellular and comparative physiology,Vol.45, suppt.2,May 1955
An 32.0346

SYMPOSIUM ON GENETICS RELATED TO DENTAL HEALTH, 1 Genetics and dental health :an international symposium Proceedings Bethesda,Md. 1961 Apr 4-6 American dental association.Council on dental research Edited by Carl J. Witkop McGraw-Hill,New York,1962
Gen 34.2024

SYMPOSIUM ON GEOPHYSICAL FLUID DYNAMICS Woods Hole,Mass. 1969 Sep Woods Hole oceanographic institution Edited by Willem V. R. Malkus Bibliog,illus,port 28cm 2 vols Woods Hole oceanographic institution,Woods Hole,Mass.,1969
A Math 4.1733

SYMPOSIUM ON GIBBERELLINS Gibberellins ... The symposium...presented before the Division of agricultural and food chemistry of the 138th national meeting of the American chemical society Papers New York 1960 Sep American chemical society.Division of agricultural and food chemistry Advances in chemistry series, 28 American chemical society,Washington,1961
Bioch 33.0738

SYMPOSIUM ON GROUP THEORY abstracts of lectures Cambridge,Mass. 1963 Apr 1-3 Harvard university 59p 23cm Harvard university,Cambridge,Mass.,1963
P Math 2.2792

SYMPOSIUM ON HAEMOPOIESIS cell production and its regulation London 1960 Feb 2-4 Ciba foundation Edited by G.E.W. Wolstenholme and Maeve O'Connor Ciba foundation.Symposia Churchill,London,1960
PGMS 29.0273

SYMPOSIUM ON HIGH-ENERGY ELECTRONS Proceedings Montreux 1964 Sep 7-11 Edited by A. Zuppinger and G. Poretti Springer,Berlin, 1965
Radioth 35.1662

SYMPOSIUM ON HIGH-ENERGY NUCLEAR REACTIONS IN ASTROPHYSICS Invted papers High-energy nuclear reactions in astrophysics Philadelphia,Pa. 1967 Jan Edited by D.S.P. Shen Sponsored by American astronomical union 281p Benjamin,New York,1967 Four of the ten papers were invited after the symposium to complete the coverage
TA 15.0578

SYMPOSIUM ON HIGH-ENERGY ELECTRONS Proceedings Montreux 1964 Sep 7-11 European association of radiologists Edited by A. Zuppinger and G. Poretti Springer,Berlin, 1965
Chem 18.0123

SYMPOSIUM ON HIGH-PURITY WATER CORROSION Atlantic City,N.J. 1955 Jun 28 American society for testing materials A.S.T.M. Special technical publication, 179 American society for testing materials,Philadelphia,Pa., 1955
Met 25.1927

SYMPOSIUM ON HORMONAL CONTROL OF PROTEIN BIOSYNTHESIS Gatlinburg,Tenn. 1965 Apr 5-8 Oak Ridge national laboratory.Biology division Oak Ridge national laboratory. Annual research conference, 18 Wistar institute of anatomy and biology,Philadelphia, Pa.,1965
Gen 34.0550

SYMPOSIUM ON HORMONAL CONTROL OF PROTEIN BIOSYNTHESIS Research conference for biology and medicine 18th Gatlinburg, Tenn. 1965 Apr 5-8 Sponsored by the Oak Ridge national laboratory.Biology division Journal of cellular and comparative physiology, 66,suppl.1 Wistar institute of anatomy and biology,Philadelphia,Pa.,1965
Bioch 33.0520

SYMPOSIUM ON HUMAN PALAEOPATHOLOGY Proceedings Human palaeopathology Washington,D.C. 1965 Jan 14 National academy of sciences. Subcommittee on geographic pathology Edited by Saul Jarcho Yale university press,New Haven,Conn.,1966
An 32.3129

A SYMPOSIUM ON HYDROBIOLOGY Madison,Wis. 1940 Sep 4-6 By James G. Needham and others Held at the University of Wisconsin University of Wisconsin,Madison,Wis.,1941 Held under the auspices of the University of Wisconsin and of the Wisconsin alumni research foundation
Bal 44.6459

SYMPOSIUM ON IMAGE TUBES AND RELATED DEVICES 1st held at the Imperial college Proceedings Photo-electronic image devices London 1958 Sep 3-5 Edited by J.D. McGee and W. L. Wilcock Advances in electronics and electron physics, 12 Academic press,New York,1960
Cav 7.1456

A SYMPOSIUM ON IMMUNIZATION IN CHILDHOOD held in the Wellcome building Proceedings London 1959 May 4-6 Livingstone,London, 1960
PGMS 29.0283

SYMPOSIUM ON IMPACT TESTING Atlantic City 1955 Jun 27 American society for testing materials American society for testing materials,Philadelphia,Pa.,1955
Met 25.0902

SYMPOSIUM ON IMPACT TESTING :presented at the forty-first annual meeting A.S.T.M. Atlantic city,N.J. 1938 Jun 28 American society for testing materials American society for testing materials,Philadelphia,Pa., 1938
Eng 41.3751

SYMPOSIUM ON IMPACT TESTING :presented at the 58th A.S.T.M. annual meeting Atlantic city,N.J. 1955 Jun 27 American society for testing materials American society for testing materials.Special technical publication, 176 American society for testing materials,Philadelphia,Pa.,1956
Eng 41.3757

SYMPOSIUM ON INELASTIC SCATTERING OF NEUTRONS Inelastic scattering of neutrons Bombay 1964 Dec 15-19 Vol 1-2 International atomic energy agency Edited by M.M. Brown International atomic energy agency,Vienna,1965
Cav 7.0267

SYMPOSIUM ON INEQUALITIES 2nd Proceedings Inequalities II 1967 Aug 14-22 Edited by Oved Shisha Held at the United States air force academy xvi,439p 24cm Academic press,New York;London,1970
P Math 2.3847

SYMPOSIUM ON INEQUALITIES 3rd Proceedings Inequalities 3 Los Angeles 1969 Sep 1-9 University of California,Los Angeles Edited by Oved Shisha xxii,368p 24cm Academic press,New York;London,1972
P Math 2.4214

SYMPOSIUM ON INFINITE DIMENSIONAL TOPOLOGY Proceedings Baton Rouge,La. 1967 Mar 27-Apr 1 Edited by R.D. Anderson Annals of mathematics studies, 69 viii,300p 24cm Princeton university press;University of Tokyo press,Princeton,N.J.;Tokyo,1972
P Math 2.4112

A SYMPOSIUM ON INFORMATION THEORY Papers read Information theory London 1955 Sep 12-16 Edited by Colin Cherry Held at the Royal institution Butterworths,London,1956
Bal 39.1005

SYMPOSIUM ON INFORMATION THEORY Report of proceedings London 1950 Sep 26-29 Imperial college of science and technology. Electrical engineering department Institute of radio engineers IRE professional group on information theory.Publications committee,n.p., c1950 Foreword by W.Jackson. Duplicated
Psy 31.3221

SYMPOSIUM ON INFORMATION THEORY proceedings London 1950 Sep 26-29 Institute of radio engineers Institute of radio engineers. Professional group on information theory, 1 208p Institute of radio engineers,New York, 1953
Math L 5.2079

SYMPOSIUM ON INFORMATION THEORY symposium papers Information theory London 1955 Sep 12-16 Edited by Colin Cherry xii,401p 25cm Butterworths,London,1956
Math L 5.0760

SYMPOSIUM ON INFORMATION THEORY 1st report of proceedings, mimeograph London 1950 Sep 26-29 Edited by Willis Jackson Supported by Great Britain.Ministry of supply 206p 33cm Ministry of supply,London,1950
Math 3.0774

SYMPOSIUM ON INFORMATION THEORY 3rd papers Information theory London 1955 Sep 12-16 Edited by Colin Cherry xii,401p 25cm Butterworths,London,1956
Math 3.0761

SYMPOSIUM ON INFORMATION THEORY 4th Proceedings London 1960 Aug 29-Sep 2 Imperial college of science and technology. Electrical engineering department Edited by Colin Cherry Butterworths,London,1961
Psy 31.3386

SYMPOSIUM ON INFORMATION THEORY 4th papers Information theory London 1960 Aug.29-Sep.2 Edited by Colin Cherry xi,476p 25cm Butterworths,London,1961
Math 3.0762

SYMPOSIUM ON INFORMATION THEORY IN BIOLOGY Gatlinburg,Tenn. 1956 Oct 29-31 Edited by H.P. Yockey and others Sponsored by the Oak Ridge national laboratory London,1958 Consists of articles representing authors' results and opinions and additional papers not given at the symposium "The conference was entitled A symposium on information theory in health physics and radiobiology"
Bal 39.1006

SYMPOSIUM ON INFORMATION THEORY IN HEALTH PHYSICS AND RADIOBIOLOGY Symposium on information theory in biology Gatlinburg,Tenn. 1956 Oct 29-31 Edited by H.P. Yockey and others Sponsored by the Oak Ridge national laboratory London,1958 Consists of articles representing authors' results and opinions and additional papers not given at the symposium "The conference was entitled A symposium on information theory in health physics and radiobiology"
Bal 39.1006

A SYMPOSIUM ON INORGANIC NITROGEN METABOLISM : function of metallo-flavoproteins Papers and informal discussions presented Baltimore, Md. 1955 Jun 21-23 Edited by William D. McElroy and H.Bentley Glass Sponsored by the McCollum-Pratt institute McCollum-Pratt institute.Contribution, 125 Johns Hopkins press,Baltimore,Md.,1956
Bioch 33.0598

A SYMPOSIUM ON INORGANIC NITROGEN METABOLISM: FUNCTION OF METALLOFLAVOPROTEINS Baltimore,Md. 1955 Jun 21-23 Edited by William D. McElroy and Bentley Glass Sponsored by the McCollum-Pratt institute McCollum-Pratt institute.Contributions, 125 Johns Hopkins press,Baltimore,Md.,1956
Gen 34.0659

SYMPOSIUM ON INSULIN Proceedings Structure and function of polypeptide hormones: insulin Upton,N.Y. 1965 Nov 8-10 Edited by P.G. Katsoyannis and I.L. Schwartz Sponsored by the Brookhaven national laboratory.Medical department American journal of medicine, 40, no. 5 Yorke medical group.Publication New York,1966
Bioch 33.0557

SYMPOSIUM ON INSULIN ACTION Proceedings Insulin action Toronto 1971 Oct 25-27 Edited by Irving B. Fritz Academic press,New York,1972 Held in honour of the 50th anniversary of the discovery of insulin
Bioch 33.2192

SYMPOSIUM ON INTERACTIVE SYSTEMS FOR EXPERIMENTAL APPLIED MATHEMATICS,1967 The APL 360 terminal system By A.D. Falkoff and K.E. Iverson International business machines corporation.Watson research center RC 1922 20p 21cm new York,1968 This paper was presented at the Symposium on Interactive systems for experimental applied mathematics in Washington,D.C. Aug 26-28,1967
Math L 5.3150

SYMPOSIUM ON INTERACTIVE SYSTEMS FOR EXPERIMENTAL APPLIED MATHEMATICS 1967 An Example of the manipulation of directed graphs in the AMBIT-G programming language By Carlos Christensen CA-6711-1511 17p 28cm massachusetts computer associates,inc., Wakefield,Mass.,1967 Paper presented at the symposium on interactive systems for experimental applied mathematics.Symposium held in Washington,D.C. Aug 26-28,1967
Math L 5.3145

SYMPOSIUM ON INTERNAL STRESSES IN METALS AND ALLOYS Advance copies of papers London 1947 Oct 15-16 Institute of metals Institute of metals,London,1947
Eng 41.3768

SYMPOSIUM ON INTERNAL STRESSES IN METALS AND ALLOYS Proceedings London 1947 Oct 15-16 Institute of metals Monographs and report series, 5 vii,485p Institute of metals,London,1948
Cav 7.1454

SYMPOSIUM ON INTERSPECIFIC HYBRIDIZATION IN PLANTS Sofia 1964 Nov 10-12 Edited by R. Georgieva and D. Tsikov Izdatelstvo Bolgarskoi akademii nauk,Sofia,1965 In Russian,English summaries
Gen 34.1025

SYMPOSIUM ON ION EXCHANGE AND CHROMATOGRAPHY IN ANALYTICAL CHEMISTRY presented at the fifty-ninth annual meeting Atlantic City, N.J. 1956 Jun 18 American society for testing materials A.S.T.M.Special technical publication, 195 A.S.T.M.,Philadelphia,Pa., 1958
Met 25.1750

SYMPOSIUM ON IONIZING RADIATION ON THE NERVOUS SYSTEM Proceedings Effects of ionizing radiation on the nervous system Vienna 1961 Jun 5-9 Sponsored by the International atomic energy agency International atomic energy agency.Proceedings series International atomic energy agency,Vienna,1962
Radioth 35.1130

SYMPOSIUM ON ISOTOPE MASS EFFECTS IN CHEMISTRY AND BIOLOGY Proceedings Isotope mass effects in chemistry and biology Vienna 1963 Dec 9-13 Edited by N. Grell Held by the International union of pure and applied chemistry Pure and applied chemistry, 8,no. 3-4 Butterworths,London,1964
Radioth 35.0421

SYMPOSIUM ON LARGE-SCALE DIGITAL CALCULATING MACHINERY proceedings Cambridge,Mass. 1947 Jan 7-10 Harvard university.Computation laboratory Jointly sponsored by United States.Bureau of ordnance Harvard university. Computation laboratory.Annals, 16 illus. xxix,302p 27cm Harvard university press, Cambridge,Mass.,1948
Math L 5.0681

SYMPOSIUM ON LARGE-SCALE DIGITAL CALCULATING MACHINERY 2nd proceedings Cambridge, Mass. 1949 Sep 13-16 Harvard university. Computation laboratory Jointly sponsored by United States. Bureau of ordnance Harvard university. Computation laboratory Annals, 26 illus. xxxviii,393p 27cm Harvard university press,Cambridge,Mass.,1951
Math L 5.0684

SYMPOSIUM ON LARGE SCALE DIGITAL COMPUTING MACHINERY proceedings Lemont,Ill. 1953 Aug 3-5 Edited by J.C. Chu Sponsored by Argonne national laboratory illus. 295p 25cm Argonne national laboratory,Lemont,Ill., 1953 Photocopy
Math L 5.0737

A SYMPOSIUM ON LIGHT AND LIFE McCollum-Pratt institute Edited by William D. McElroy and Bentley Glass With support from the National science foundation McCollum-Pratt institute. Contribution,302 Johns Hopkins press, Baltimore,Md.,1961
Chem 18.0168

A SYMPOSIUM ON LIGHT AND LIFE Papers and discussions presented Baltmore,Md. 1960 Mar 28-31 Edited by William D. McElroy and H. Bentley Glass Sponsored by the McCollum-Pratt institute McCollum-Pratt institute. Contribution, 302 John Hopkins press, Baltimore,Md.,1961
Bioch 33.0600

A SYMPOSIUM ON LIGHT AND LIFE Papers and informal discussions Baltimore,Md. 1960 Mar 28-31 Edited by William D. McElroy and Bentley Glass Sponsored by the McCollum-Pratt institute McCollum-Pratt institute. Contribution, 302 Johns Hopkins university press,Baltimore,Md.,1961
Bal 44.6447

SYMPOSIUM ON LIGHT MICROSCOPY American society for testing materials 55th annual meeting Papers New York 1952 Jun 25 American society for testing materials American society for testing materials. Special publication, 143,Fiftieth anniversary publication Philadelphia,Pa.,1953
Min 10.1571

SYMPOSIUM ON LIGHT MICROSCOPY New York 1952 Jun 25 American society for testing materials A.S.T.M.Special technical publication, 143 American society for testing materials,Philadelphia,Pa.,1953
Met 25.1333

SYMPOSIUM ON LIQUIDS:STRUCTURE,PROPERTIES,SOLID INTERACTIONS proceedings Liquids: structure,properties,solid interactions Warren,Mich. 1963 Sep.5-6 Edited by Thomas J. Hughel General motors research laboratory. Symposia, 7 Elsevier,Amsterdam,1965
A Math 4.0951

SYMPOSIUM ON LISTERIC INFECTION 2nd Papers and discussions Bozeman,Montana 1962 Aug 29-31 Montana state college Edited by M.L. Gray With the support of the National institute of allergy and infectious diseases Montana state college,Bozeman,Montana,1963
Path 30.2590

SYMPOSIUM ON LOAD TESTS OF BEARING CAPACITY OF SOILS Atlantic City,N.J. 1947 Jun 16-20 American society for testing materials American society for testing materials. Special technical publication, 79 American society for testing materials,Philadelphia,Pa.,1948
Eng 41.3145

SYMPOSIUM ON LOW ENERGY ELECTRON DIFFRACTION Tucson,Ariz. 1968 Feb 4-7 American crystallographic association Edited by David H. Templeton and Gabor A. Somorjai American crystallographic association. Transactions, 4 American crystallographic association, Pittsburgh,1968
Met 25.1476

SYMPOSIUM ON LOW-LUMINOSITY STARS Proceedings Low-luminosity stars Charlottesville,Va. 1968 University of Virginia Edited by Shiv S. Kumar 542p Gordon and Breach,New York, 1969
Obs 6.3627

SYMPOSIUM ON LOW TEMPERATURE PROPERTIES OF HIGH-STRENGTH AND MISSILE MATERIALS Papers 1960 Jun 30 American society for testing materials A.S.T.M.Special technical publication, 287 A.S.T.M.,Philadelphia,1961
Met 25.0974

SYMPOSIUM ON LUBRICANTS Chicago,Ill. 1937 Mar 3 American society for testing materials American society for testing materials, Philadelphia,Pa.,1937
Eng 41.6265

SYMPOSIUM ON MACHINE INTERPRETATIONS OF PATTERSON FUNCTIONS AND ALTERNATIVE DIRECT APPROACHES and the Austin symposium on gas phase molecular structure Proceedings Austin symposium on gas phase molecular structure Austin,Texas 1966 Feb 28-Mar 2 Edited by W. P. Bradley and Harold P. Hansen American crystallographic association. Transactions, 2, 1966 Polycrystal book services,Pittsburgh,Pa. 1966
Chem 18.2633

SYMPOSIUM ON MACROMOLECULAR ASPECTS OF THE CELL CYCLE :given at research conference for biology and medicine of the atomic energy commission Gatlinburg,Tenn 1963 Apr 8-11 Sponsored by Oak Ridge national laboratory. Biology division Oak Ridge national laboratory. Symposia, 16 Journal of cellular and comparative physiology, 62,supp. 1 Wistar institute of anatomy and biology, Philadelphia,Pa.,1963
PGMS 29.0414

SYMPOSIUM ON MACROMOLECULAR ASPECTS OF THE CELL CYCLE :given at research conference for biology and medicine of the atomic energy commission Gatlinburg,Tenn. 1963 Apr 8-11 Sponsored by Oak Ridge national laboratory. Biology division Oak Ridge national laboratory. Symposia, 16 Journal of cellular and comparative physiology, 62,no.2 Wistar institute of anatomy and biology, Philadelphia,Pa.,1963
Gen 34.0842

SYMPOSIUM ON MAGNETOHYDRODYNAMICS 3rd
Magnetodynamics of conducting fluids Palo Alto,Calif. 1958 Nov 21-22 Lockheed aircraft corporation.Lockheed missiles and space division Edited by Daniel Bershader viii,145p 24cm Stanford university press, Stanford,Calif.,1959
Cav 7.2767

SYMPOSIUM ON MAGNETOHYDRODYNAMICS 3rd
Magnetohydrodynamics of conducting fluids Palo Alto,Calif. 1958 Nov 21-22 Edited by Daniel Bershader Sponsored by Lockheed aircraft corporation.Missile systems division Stanford university press,Stanford,Calif.,1959 1st symposium called 'Magnetohydrodynamics'; 2nd symposium called 'Plasma in a magnetic field',q.v.
Eng 41.4474

SYMPOSIUM ON MAGNETOHYDRODYNAMICS 4th
Plasma acceleration Palo Alto,Calif. 1959 Dec 2 Edited by Sidney W. Kash Sponsored by the Lockheed aircraft corporation.Missiles and space division Stanford university press, Stanford,Calif.,1960
Eng 41.4489

SYMPOSIUM ON MAGNETOHYDRODYNAMICS 5th
Radiation and waves Palo Alto,Calif. 1960 Dec 15-16 Edited by Morton Mitchener Sponsored by Lockheed aircraft corporation. Missile systems division Stanford university press,Stanford,Calif.,1961
Eng 41.4496

SYMPOSIUM ON MAGNETOHYDRODYNAMICS 5th
Radiation and waves in plasmas Palo Alto, Calif. 1960 Lockheed aircraft corporation. Lockheed missiles and space division Edited by Morton Mitchner 156p Stanford university press,Stanford,Calif.,1961
Cav 7.2881

SYMPOSIUM ON MAGNETOHYDRODYNAMICS 6th
Proceedings Plasma hydromagnetics Palo Alto,Calif. 1961 Dec 15-16 Edited by Daniel Bershader Sponsored by Lockheed aircraft corporation.Missile systems division Stanford university press,Stanford,Calif.,1962
Eng 41.4510

SYMPOSIUM ON MAGNETOHYDRODYNAMICS 7th
Proceedings Propagation and instabilities in plasmas Palo Alto,Calif. 1962 Dec 14-15 Edited by Walter I. Futterman Sponsored by Lockheed aircraft corporation.Missiles and space division Stanford university press, Stanford,Calif.,1963
Eng 41.4523

SYMPOSIUM ON MAJOR EFFECTS OF MINOR CONSTITUENTS ON THE PROPERTIES OF MATERIALS Atlantic City,N.J. 1961 Jun 26 American society for testing materials American society for testing and materials.Materials science series, 2 American society for testing and materials.Special technical publications, 304 American society for testing materials, Philadelphia,Pa.,1962
Eng 41.3429

SYMPOSIUM ON MAMMALIAN GENETICS AND REPRODUCTION
Gatlinburg,Tenn. 1960 Apr 4-7 Sponsored by the Oak Ridge national laboratory.Biology division Oak Ridge national laboratory. Symposia, 13 Journal of cellular and comparative physiology, 56,supp.1 Wistar institute of anatomy and biology,Philadelphia, Pa.,1960
Gen 34.1782

A SYMPOSIUM ON MAMMARY TUMORS IN MICE
National cancer institute Edited by Forest Ray Moulton American association for the advancement of science.Publication, 22 American association for the advancement of science,Washington,D.C.,1945
Gen 34.1813

SYMPOSIUM ON MATERIALS FOR GAS TURBINES :
presented at the 49th annual meeting A.S.T.M. Buffalo,N.Y. 1946 Jun 24-28 American society for testing materials American society for testing materials,Philadelphia,Pa., 1946
Eng 41.3752

SYMPOSIUM ON MATHEMATICAL BIOLOGY Proceedings
Some mathematical problems in biology Washington 1966 Dec 28 Edited by Murray Gerstenhaber Prepared by the American mathematical society American mathematical society.Lectures on mathematics in the life sciences, 1 Bibliog.,Illus. vi and 117p 22cm American mathematical society,Rhode Island,1968
P Math 2.3482

SYMPOSIUM ON MATHEMATICAL STATISTICS AND PROBABILITY Proceedings Berkeley,Calif. 1945 Aug.13-18 and 1946 Jan.27-29 University of California.Department of mathematics Edited by Jerzy Neyman 501p University of California press,Berkeley;Los Angeles,1949
Nap 11.0048

SYMPOSIUM ON MATHEMATICAL THEORY OF AUTOMATA
proceedings New York 1962 Apr 24-26 Microwave research institute Partly sponsored by American institute of electrical engineers Polytechnic institute of Brooklyn. Microwave research institute.Symposia series, 12 xix,640p 22cm Polytechnic press, Brooklyn,1962
Math L 5.0996

SYMPOSIUM ON MEASUREMENT IN UNSTEADY FLOW :
presented at the ASME hydraulic division conference Worcester,Mass. 1962 May 2-23 American society of mechanical engineers. Hydraulic division American society of mechanical engineers,New York,1962
Eng 41.6713

SYMPOSIUM ON MECHANISM OF IGNEOUS INTRUSION
Proceedings Liverpool 1969 Jan 9-11 Liverpool geological society Edited by Geoffrey Newall and Nicholas Rast Geological journal.Special issue, 2 Liverpool geological society,Liverpool,1970
Min 10.1427

SYMPOSIUM ON MECHANISMS OF IMMUNOLOGICAL TOLERANCE Proceedings Liblice 1961 Nov 8-10 Ceskoslovenska akademie ved Edited by M. Hasek and others illus. Academic press,New York,1963
BG 23.0056

SYMPOSIUM ON MEDICAL RADIOISOTOPE SCANNING Proceedings Medical radioisotope scanning Athens 1964 Apr 20-24 Held by the International atomic energy agency International atomic energy agency.Proceedings series International atomic energy agency, Vienna,1964
Radioth 35.1376

SYMPOSIUM ON MEDICINAL PLANTS :8th Pacific science congress Quezon city 1953 Nov 16-28 Pacific science association Under the auspices of the National research council of the Philippines Pacific science congress. Proceedings, 8th,4a National research council of the Philippines,Quezon city,1954
BG 38.3111

SYMPOSIUM ON METALLIC MATERIALS FOR SERVICE AT TEMPERATURES ABOVE 1600 F. Atlantic City, N.J. 1955 Jun 30 American society for testing materials A.S.T.M.Special technical publication, 174 American society for testing materials,Philadelphia,Pa.,1956
Met 25.0999

SYMPOSIUM ON METALLOGRAPHY IN COLOR (1948) presented to the 51st annual meeting Detroit 1948 Jun 21 American society for testing materials A.S.T.M.Special technical publication, 86 American society for testing materials,Philadelphia,Pa.,1949
Met 25.1342

SYMPOSIUM ON METEOR PHYSICS Proceedings Meteors Manchester 1954 Jul. Jodrell Bank experimental station Edited by T.R. Kaiser Journal of atmospheric and terrestrial physics.Special supplement, 2 204p Pergamon press,London,1955
Obs 6.0923

SYMPOSIUM ON METEORITE RESEARCH Proceedings Meteorite research Vienna 1968 Aug 7-13 International atomic energy agency Unesco Edited by Peter M. Millmann Astrophysics and space science library Reidel,Dordrecht,1969
Min 10.1442

SYMPOSIUM ON METEOROLOGY IN RELATION TO HIGH LEVEL AVIATION OVER INDIA AND SURROUNDING AREAS Papers New Delhi 1957 Dec 7 Meteorological department,India Edited by U. K. Bose Under the auspices of Indian meteorological society 159p India press, Delhi,1962
Nap 11.0420

SYMPOSIUM ON METHODOLOGIES Contributions New methods of thought and procedure Pasadena,Calif. 1967 May 22-24 California institute of technology.Office for industrial association Society for morphological research Edited by F. Zwicky and A.G. Wilson viii,338p Springer,Berlin,1967
WSM 43.2639

SYMPOSIUM ON METHODOLOGIES Contributions New methods of thought and procedure Pasadena,Calif. 1967 May 22-24 Edited by F. Zwicky and A.G. Wilson illus. viii,338p 24cm Springer,Berlin,1967
Math S 3.1389

SYMPOSIUM ON METHODS OF METALLOGRAPHIC SPECIMEN PREPARATION Atlantic City 1960 Jun 28 American society for testing materials A.S.T.M.Special technical publication, 285 American society for testing materials, Philadelphia,Pa.,1960
Met 25.1381

SYMPOSIUM ON MICROMINIATURIZATION OF ELECTRONIC ASSEMBLIES Proceedings Washington,D.C. 1958 Sep 30-Oct 1 Diamond ordnance fuze laboratories Edited by Eleonor F. Horsey and Laurence D. Shergalis Hayden,New York,1960
Eng 41.5541

SYMPOSIUM ON MICROSCOPY American society for testing materials :annual meeting 62nd papers Atlantic City,N.J. 1959 Jun 25-26 American society for testing materials.Special technical publications,257 American society for testing materials,Philadelphia,Pa.,1959
Min 10.1046

SYMPOSIUM ON MICROSEISMS Proceedings Harriman,N.Y. 1952 Sep 4-6 United States office of naval research and United States air force. Geophysical research directorate Edited by James T. Wilson and Frank Press National research council.Publication, 306 Washington,D.C.,1953
Geod 9.0376

SYMPOSIUM ON MIDDLE AND UPPER DEVONIAN STRATIGRAPHY OF PENNSYLVANIA AND ADJACENT STATES Edited by Vincent C. Shepps Pennsylvania geological survey.Bulletin.4th series,G 39 Pennsylvania geological survey, Harrisburg,Pa.,1963
Geol 8.2804

The SYMPOSIUM ON MODERN OPTICS Proceedings Modern optics New York 1967 Mar 22-24 Microwave research institute.Symposia series, 17 xxxv,804p 23cm Polytechnic press, Brooklyn,N.Y.,1967
Cav 7.2815

SYMPOSIUM ON MODERNIZATION OF MATHEMATICS TEACHING IN EUROPEAN COUNTRIES Proceedings Bucharest 1968 Sep 23-Oct 2 Unesco Bibliog 573p 24cm Editions didactiques et pedagogiques,Bucharest,1969
A Math 4.1688

SYMPOSIUM ON MOLDING MACHINES American foundrymen's society.Plant and plant equipment committee American foundrymen's society,n.p., n.d.
Met 25.0648

SYMPOSIUM ON MOLECULAR ACTION OF MUTAGENIC AND CARCINOGENIC AGENTS Gatlinburg,Tenn. 1964 Apr 6-9 Sponsored by the Oak Ridge national laboratory.Biology division Oak Ridge national laboratory.Symposia Journal of cellular and comparative physiology, 64, suppl.1 Wistar institute of anatomy and biology,Philadelphia,Pa.,1964
Gen 34.1098

SYMPOSIUM ON MOLECULAR ASPECTS OF DIFFERENTIATION given at research conference for biology and medicine of the Atomic energy commission Gatlinburg,Tenn. 1968 Apr 8-11 Sponsored by the Oak Ridge national laboratory. Biology division Oak Ridge national laboratory.Biology division.Symposia, 21 Journal of cellular physiology, 72,supp.1 Wistar institute of anatomy and biology, Philadelphia,Pa.,1968
Gen 34.0505

A SYMPOSIUM ON MOLECULAR BIOLOGY Chicago,Ill. 1956-57 Nov-Mar Edited by Raymond E. Zirkle Held under the auspices of the Chicago-Frankfurt inter-university program illus. University of Chicago press,Chicago, Ill.,1959
Gen 34.0647

A SYMPOSIUM ON MOLECULAR BIOLOGY :a seminar series and a symposium Lectures Chicago 1956-57 Nov-Mar University of Chicago Edited by Raymond E. Zirkle Under the auspices of the Chicago-Frankfurt inter-university program University of Chicago press,Chicago,1959
Bioch 33.0839

A SYMPOSIUM ON MOLECULAR BIOLOGY :a seminar series and a symposium lectures Chicago 1956-57 Nov-Mar University of Chicago Edited by Raymond E. Zirkle Under the auspices of the Chicago-Frankfurt inter-university program University of Chicago press,Chicago,1959
Bal 39.0617

SYMPOSIUM ON MONSOONS OF THE WORLD Proceedings New Delhi 1958 Feb 19-21 Meteorological office,India Under the auspices of the World meteorological organization x,270p Manager of publications,Delhi,1960
Nap 11.0284

SYMPOSIUM ON MONTE CARLO METHODS Gainesville, Fla. 1954 Mar.16-17 University of Florida. Statistical laboratory Edited by Herbert A. Meyer Sponsored by the Wright air development center Wiley publications in statistics Bibliog. xvi,382p 28cm John Wiley and sons,New York,1956
Math 3.0208

SYMPOSIUM ON MOSSBAUER EFFECT METHODOLOGY 1st Proceedings Mossbauer effect methodology New York 1965 Feb 2-5 New England nuclear corporation Edited by I.J. Gruverman Plenum press,New York,1965
Chem 18.2826

SYMPOSIUM ON MOSSBAUER EFFECT METHODOLOGY 6th Proceedings Mossbauer effect methodology New York 1970 Jan 25 New England nuclear corporation Edited by Irwin J. Gruverman viii,237p 23cm Plenum press,New York,1971
Chem 18.2805

SYMPOSIUM ON MOSSBAUER EFFECT METHODOLOGY 7th Proceedings Mossbauer effect methodology New York 1971 Jan 31 New England nuclear corporation Edited by Irwin J. Gruverman xii,308p 23cm Plenum press,London,1971
Chem 18.2961

SYMPOSIUM ON MOTOR LUBRICANTS New York 1933 Mar 8 American society for testing materials Society for automotive engineers American society for testing materials, Philadelphia,Pa.,1933
Eng 41.6264

SYMPOSIUM ON MULTIPLE-SOURCE URBAN DIFFUSION MODELS Proceedings Environmental protection 1970 United States. Environmental protection agency National air pollution control administration North Carolina consortium on air pollution Air pollution control office.Publication,AP- 86 US government printing office,Washington,D.C., 1970
Eng 41.8616

SYMPOSIUM ON MUSCLE Budapest 1966 Sep 12-16 Edited by E. Ernst and F.B. Straub Symposia biologica hungarica, 8 Akademia Kaido,Budapest,1968
An 32.4011

SYMPOSIUM ON MUSCLE RECEPTORS :a meeting held.. as part of the Golden jubilee congress of the University of Hong Kong Proceedings Hong Kong 1961 Sep University of Hong Kong Edited by David Barker Hong Kong university press,Hong Kong,1962
Phys 20.1559

SYMPOSIUM ON MUSCLE RECEPTORS :a meeting held.. as part of the golden jubilee congress of the University of Hong Kong Proceedings Hong Kong 1961 Sep Edited by David Barker Financial support provided by the University of Hong Kong Hong Kong university press,Hong Kong,1962
Bal 39.1519

SYMPOSIUM ON MUSCLE RECEPTORS Proceedings Hong Kong 1961 Sep Edited by David Barker Held as part of the Golden jubilee congress of the University of Hong Kong Hong Kong university press,Hong Kong,1962
An 32.4002

A SYMPOSIUM ON NATRIURETIC HORMONE Regulation of body fluid volumes by the kidney Smolenice castle 1969 Jun Edited by J.H. Cort and B. Lichardus Under the auspices of Ceskoslovenska akademie ved Karger,Basle, 1970
Inv Med 37.0268

SYMPOSIUM ON NATURE AND ORIGIN OF STRENGTH OF MATERIALS Properties of crystalline solids ...Symposium on recent progress in materials sciences;Symposium on nature and origin of strength of materials.Presented at the 63rd annual meeting A.S.T.M. Philadelphia,Pa. 1960 Jun 27 American society for testing materials American society for testing materials.Special technical publication, 283 American society for testing materials,Philadelphia,Pa.,1961
Eng 41.3790

SYMPOSIUM ON NAVAL HYDRODYNAMICS proceedings Washington D.C. 1956 Sep.24-28 United states.Office of naval research Edited by F. S. Sherman Sponsored by the National research council National research council. Publication, 515 National research council, Washington D.C.,1957
A Math 4.0530

SYMPOSIUM ON NAVAL HYDRODYNAMICS 5th :ship motions and drag reduction Bergen 1964 Sep 10-12 Edited by J.K. Lunde and S.W. Doroff Sponsored by the United States.Office of naval research,and Office of naval research-Department of the navy,Washington,D.C. 1965
A Math 4.1303

SYMPOSIUM ON NAVAL STRUCTURAL MECHANICS Proceedings High temperature structures and materials New York 1963 Jan 23-25 United States.Office of naval research Edited by A.M. Freudenthal and others Pergamon,Oxford,1964
Met 25.1009

SYMPOSIUM ON NAVAL STRUCTURAL MECHANICS 1st Proceedings Stanford,Calif. 1958 Aug 11-14 United States.Office of naval research and Stanford university Edited by J.Norman Goodier and Nicholas J. Hoff International series on aeronautical sciences and space flight Pergamon,Oxford,1960
Met 25.1300

SYMPOSIUM ON NAVAL STRUCTURAL MECHANICS 3rd Proceedings High temperature structures and materials New York 1963 Jan 23-25 Edited by A.M. Freudenthal and others Sponsored by the United States.Office of naval research United States.Office of naval research.Structural mechanics series Pergamon press,London,1964
Eng 41.3806

SYMPOSIUM ON NEUROENDOCRINOLOGY Advances in neuroendocrinology Miami,Fla. 1961 Dec 6-8 Edited by Andrew V. Nalbandov Supported by a grant from the National institute of neurological diseases and blindness University of Illinois press,Urbana,Ill.,1963
An 32.3766

SYMPOSIUM ON NEUROENDOCRINOLOGY Advances in neuroendocrinology Miami,Flo. 1961 Dec 6-8 Edited by A.V. Nalbandov Sponsored by the National institute of neurological diseases and blindness University of Illinois press, Urbana,Ill.,1963
Phys 20.1404

SYMPOSIUM ON NEUTRON MONITORING FOR RADIOLOGICAL PROTECTION Proceedings Neutron monitoring Vienna 1966 Aug 29-Sep 2 International atomic energy agency International atomic energy agency.Proceedings series International atomic energy agency, Vienna,1967
Radioth 35.1672

SYMPOSIUM ON NEWER METALS presented at the third Pacific area national meeting... Newer metals San Francisco,Calif. 1959 Oct 15 American society for testing materials A.S.T.M.Special technical publication, 272 A.S.T.M.,Philadelphia,Pa.,1960
Met 25.0514

SYMPOSIUM ON NEWER STRUCTURAL MATERIALS FOR AEROSPACE VEHICLES Chicago 1964 Jun 21 American society for testing materials A.S.T.M.Special technical publication, 379 Philadelphia,Pa.,1965
Met 25.2516

SYMPOSIUM ON NEWER TRENDS IN TAXONOMY Proceedings New Delhi 1966 Jan 28-30 National institute of sciences of India National institute of sciences of India. Bulletin, 34 New Delhi,1967
Bal 39.2079

The SYMPOSIUM ON NON-ELASTIC PROCESSES IN THE MANTLE Proceedings Newcastle-upon-Tyne 1966 Feb 21-25 International upper mantle committee Edited by D.C. Tozer With financial assistance of the International union of geodesy and geophysics Geophysical journal, 14;1-4 Blackwell,Oxford,1967
Geod 9.0695

A SYMPOSIUM ON NON-EQUILIBRIUM THERMODYNAMICS, VARIATIONAL TECHNIQUES AND STABILITY Chicago 1965 May 17-19 University of Chicago General motors corporation.Research laboratories Universite libre de Bruxelles Edited by Russell J. Donnelly and I. Prigogine University of Chicago,Chicago,1966
A Math 4.0712

SYMPOSIUM ON NONDESTRUCTIVE TESTING presented to the 2nd Pacific area national meeting Los Angeles 1956 Sep 17-18 American society for testing materials A.S.T.M.Special technical publication, 213 A.S.T.M. Philadelphia,1957
Met 25.1323

SYMPOSIUM ON NONDESTRUCTIVE TESTING presented to the 55th annual meeting... New York 1952 Jun 26 American society for testing materials A.S.T.M.Special technical publication, 145 A.S.T.M.,Philadelphia,1953
Met 25.2359

SYMPOSIUM ON NONLINEAR CIRCUIT ANALYSIS Proceedings New York 1953 Apr 23-24 Polytechnic institute of Brooklyn Edited by Jerome Fox Sponsored by the Microwave research institute Microwave research institute.Network symposia series, 2 Polytechnic institute of Brooklyn,New York, 1953
Eng 41.5920

SYMPOSIUM ON NONLINEAR CIRCUIT ANALYSIS Proceedings New York 1956 Apr 25-27 Polytechnic institute of Brooklyn Edited by Jerome Fox Sponsored by the Microwave research institute Microwave research institute.Network symposia series, 6 Polytechnic institute of Brooklyn,New York, 1957
Eng 41.5922

SYMPOSIUM ON NONLINEAR ESTIMATION THEORY and its applications Proceedings San Diego, Calif. 1970 Sep 21-23 United States.Air force.Office of scientific research Institute of electrical and electronics engineers.Group on automatic control vii, 297p 28cm IEEE,New York,1970
Math S 3.1698

SYMPOSIUM ON NOXIOUS EFFECTS OF LOW LEVEL RADIATION :physical elements and biological aspects Lausanne 1958 Mar 27-29 Schweizerische akademie der medizinischen wissenschaften Schwabe,Basle;Stuttgart,1958
Radioth 35.1078

SYMPOSIUM ON NUCLEAR METHODS FOR MEASURING SOIL DENSITY AND MOISTURE Atlantic City,N.J. 1960 Jun 27 American society for testing materials American society for testing materials.Special technical publication, 293 American society for testing materials, Philadelphia,Pa.,1961
Eng 41.3152

SYMPOSIUM ON NUCLEAR SEX London 1957 Sep King's college hospital medical school Edited by D.Robertson Smith and William M. Davidson Heinemann,London,1958
Gen 34.0879

SYMPOSIUM ON NUCLEAR SEX 1st Proceedings London 1957 Edited by D.Robertson Smith and William M. Davidson Heinemann,London, 1958
Inv Med 37.0197

SYMPOSIUM ON NUMERICAL ANALYSIS Numerical analysis:an introduction Birmingham 1965 Institute of mathematics and its applications Academic press,London,1966
Radioth 35.0631

SYMPOSIUM ON NUMERICAL ANALYSIS Numerical analysis:an introduction Birmingham 1965 Institute of mathematics and its applications Edited by J. Walsh bibliog.,diags. xiv, 212p 24cm Academic press,London,1966
Math L 5.0357

The SYMPOSIUM ON NUMERICAL WEATHER PREDICTION Proceedings Tokyo 1960 Nov 7-13 Meteorological agency,Japan Science council of Japan Co-sponsored by the International union of geodesy and geophysics 655p Meteorological society of Japan,Tokyo,1962
Nap 11.0923

SYMPOSIUM ON OPERATING SYSTEM PRINCIPLES Gatlinburg 1967 Oct 1-4 Association for computing machinery ACM,New York,1967
Math L 5.3488

SYMPOSIUM ON OPERATING SYSTEMS PRINCIPLES 3rd Stanford,Calif. 1971 Oct 18-20 Association for computing machinery illus 163p Association for computing machinery,New York,1971
Math L 5.3824

SYMPOSIUM ON OPTIMIZATION Nice 1969 Jun 29-Jul 5 Edited by A.V. Balakrishnan and others Lecture notes in mathematics, 132 Bibliog., Illus. 350p 26cm Springer-Verlag,Berlin, 1970
P Math 2.3719

SYMPOSIUM ON OPTIMIZATION Proceedings Nice 1969 Jun 29-Jul 5 Edited by A.V. Balakrishnan and others Lecture notes in mathematics, 132 350p 25cm Springer, Berlin,1970
Math S 3.1685

SYMPOSIUM ON ORGAN CULTURE :studies of development,function,and disease Washington,D.C. 1962 May 29-31 National cancer institute Edited by Clyde J. Dawe Sponsored by Tissue culture association and National cancer institute National cancer institute.Monograph, 11 U.S.Department of health,education,and welfare,Washington,D.C., 1963
Bioch 33.2367

SYMPOSIUM ON PARTICLE SIZE MEASUREMENT Papers American society for testing materials American society for testing materials.Special technical publication, 234 v,310p American society for testing materials,Philadelphia,Pa., 1959 Presented at the 61st annaual meeting of the American society for testing materials
Chem E 24.1440

SYMPOSIUM ON PHOTO-ELECTRIC IMAGE DEVICES 5th Proceedings Advances in electronics and electron physics Academic press,New York, 1972
Eng 41.8579

A SYMPOSIUM ON PHOTO-ELECTRONIC IMAGE DEVICES 2nd held at Imperial college Proceedings Photo-electronic image devices London 1961 Sep 5-8 Edited by J. D. McGee and others Advances in electronics and electron physics, 16 Academic press,New York,1962
Cav 7.1458

The SYMPOSIUM ON PHYSICAL PROCESSES IN THE SUN-EARTH ENVIRONMENT Proceedings Ottawa 1959 Jul 20-21 Canada.Defence research board Edited by C. Collins DRTE no 1025 xii,392p Ottawa,1960
Nap 11.0140

SYMPOSIUM ON PINK SALMON :a symposium held at the university of British Columbia Vancouver 1960 Oct 13-15 Edited by Norman J. Wilimovsky H.R.Macmillan lectures in fisheries illus,map 226p 27cm University of British Columbia institute of fisheries,Vancouver,1962
Sco 14.3615

SYMPOSIUM ON PINK SALMON Lectures Vancouver,B.C. 1960 Oct 13-15 University of British Columbia.Institute of fisheries Edited by N.J. Wilimovsky University of British Columbia.Institute of fisheries.H.R. Macmillan lectures in fisheries, 3 27cm Vancouver,B.C.,1962
Bal 44.4988

The SYMPOSIUM ON PLANETARY ATMOSPHERES Yerkes observatory 1947 Sep 8-10 University of Chicago Edited by Gerard P. Kuiper vii,366p Chicago,1949 In connection with the 15th anniversary of the Yerkes observatory
Nap 11.0340

A SYMPOSIUM ON PLANETARY ATMOSPHERES Papers Atmosphere of the earth and planets Williams Bay,Wisc. 1947 Sep 8-10 University of Chicago Edited by Gerard P. Kuiper 366p Chicago,Ill.,1949 Symposium held in connection with the 50th anniversary of the Yerkes observatory
Geod 9.0523

The SYMPOSIUM ON PLANT AND ANIMAL VIRUSES Proceedings Cuttack 1961-62 Dec 31-Jan 1 National institute of sciences of India National institute of sciences of India. Bulletin, 24,1963 National institute of sciences of India,New Delhi,1963 Held under the joint auspices of the National institute and the Council of scientific and industrial research
Bal 39.1071

SYMPOSIUM ON PLASTIC DEFORMATION OF POLYMERS Selected papers Plastic deformation of polymers New York 1969 Sep 10-11 American chemical society Dekker,New York, 1971
Met 25.2576

SYMPOSIUM ON PLASTICS Philadelphia,Pa. 1944 Feb 22-23 American society for testing and materials American society for testing materials,Philadelphia,Pa.,1944
Eng 41.3642

SYMPOSIUM ON PORCELAIN ENAMELS AND CERAMIC COATINGS AS ENGINEERING MATERIALS presented at the fifty-sixth annual meeting... American society for testing materials A. S.T.M.Special technical publication, 153 ASTM,Philadelphia,Pa.,1954
Met 25.0263

SYMPOSIUM ON POWDER METALLURGY Iron and steel institute Iron and steel institute. Special report, 38 Iron and steel institute, London,1947
Met 25.0753

SYMPOSIUM ON POWDER METALLURGY London 1954 Dec 1-2 Iron and steel institute Iron and steel institute.Special report, 58 Iron and steel institute,London,1956
Met 25.0761

SYMPOSIUM ON POWDER METALLURGY London 1954 Dec 1-2 Iron and steel institute Iron and steel institute.Special report, 58 illus. Iron and steel institute,London,1956
Eng 41.3477

SYMPOSIUM ON POWDER METALLURGY :spring meeting Buffalo,N.Y. 1943 American society for testing and materials American society for testing and materials,Philadelphia, Pa.,1943
Eng 41.3474

SYMPOSIUM ON PREVENTIVE AND SOCIAL PSYCHIATRY Washington,D.C. 1957 Apr 15-17 Walter Reed army medical center.Institute of research and National research council.Division of medical sciences US government printing office,Washington,D.C.,c1957
Psy 28.0058

SYMPOSIUM ON PRINCIPLES OF GATING presented at the 55th annual meeting Buffalo 1951 Apr 24 American foundrymen's society American foundrymen's society,Chicago,Ill., 1951
Met 25.0659

SYMPOSIUM ON PRINCIPLES OF SENSORY COMMUNICATION Contributions Sensory communication Boston,Mass. 1961 Jul 19-Aug 7 Edited by W. A. Rosenblith Held at the Massachusetts institute of technology M.I.T.press;Wiley, Boston,Mass.;New York,1961
Phys 20.2108

SYMPOSIUM ON PRINCIPLES OF SENSORY COMMUNICATION Contributions Sensory communication Cambridge,Mass. 1959 Jul 19-Aug 1 Edited by Walter Rosenblith M.I.T.press;Wiley,New York,1961
An 32.4395

The SYMPOSIUM ON PRINCIPLES OF SENSORY COMMUNICATION Contributions Sensory communication Dedham,Mass. 1959 Jul 19-Aug 1 Massachusetts institute of technology Edited by W.A. Rosenblith M.I.T.press;Wiley, New York;London,1961 Held at M.I.T.'s Endicott House,1961
Bal 39.1518

The SYMPOSIUM ON PRINCIPLES OF SENSORY COMMUNICATIONS Contributions Sensory communications Dedham,Mass. 1959 Jul 19-Aug 1 United States.Air force.Office of scientific research Edited by Walter A. Rosenblith MIT press,Cambridge,Mass.,1961
Psy 31.2842

SYMPOSIUM ON PRINTED CIRCUITS proceedings Philadelphia,Pa. 1955 Jan 20-21 Radio-Electronics-Television manufacturers association Jointly sponsored by Institute of radio engineers illus. 122p 28cm Engineering publishers,New York,1955
Math L 5.0736

SYMPOSIUM ON PROBLEMS IN ANALYSIS Papers Selected mathematical papers Princeton,N.J. 1969 Apr 1-3 Princeton university Princeton mathematical series, 31 Bibliog, port x,351p 24cm Princeton university press,Princeton,N.J.,1970 Symposium held in honour of S.Bochner
P Math 2.3992

SYMPOSIUM ON PROBLEMS OF EXTRA-GALACTIC RESEARCH Proceedings Santa Barbara,Calif. 1961 Aug 10-12 International astronomical union Edited by G.C. McVittie International astronomical union.Symposium, 15 McMillan, New York,1962
TA 15.0035

SYMPOSIUM ON PROCESS OPTIMISATION Proceedings London 1962 Jun 26 Institution of chemical engineers Edited by J.M. Pirie 70p Institution of chemical engineers,London, 1962 3rd congress of the European federation of chemical engineering
Chem E 24.1630

SYMPOSIUM ON PROGRESS IN ASTROPHYSICS Papers 1939 Feb 17 By Donald H. Menzel and others American philosophical society Franklin institute 1939 From the Proceedings of the American philosophical society,vol 81,no 2.Lacks title page
Obs 6.3230

SYMPOSIUM ON PROPERTIES OF SURFACES : presented at the fourth Pacific area national meeting Los Angeles,Calif. 1962 Oct 4 American society for testing materials American society for testing materials.Special technical publication, 340 Philadelphia,1963
Met 25.2829

SYMPOSIUM ON PROTEIN STRUCTURE Paris 1957 Jul 25-29 Edited by Albert Neuberger Sponsored by the International union of pure and applied chemistry.Protein commission Methuen;Wiley,London New York,1958
Gen 34.0667

SYMPOSIUM ON PROTEIN STRUCTURE Papers Paris 1957 Jul 25-29 Edited by Albert Neuberger Sponsored by the International union of pure and applied chemistry.Protein commission Methuen;Wiley,London;New York, 1958
Bioch 33.0535

The SYMPOSIUM ON PROTEIN STRUCTURE Proceedings Aspects of protein structure Madras 1963 Jan 14-18 University of Madras. Department of physics Edited by G.N. Ramachandran Academic press,London;New York, 1963 The Symposium on protein structure formed part of an International symposium on protein structure and crystallography organized by the University of Madras
Bioch 33.0542

SYMPOSIUM ON PSYCHOLOGICAL FACTOR ANALYSIS Uppsala 1953 Mar 17-19 Universitet i Uppsala.Institute of statistics Edited by P. Whittle Nordisk psykologi.Monograph series, 3 91p 23cm Almqvist and Wiksell, Stockholm,1953
Math 3.0777

SYMPOSIUM ON PYELONEPHRITIS Edinburgh 1966 Norwich pharmacal company and Eaton Laboratories division Livingstone,London, 1967
PGMS 29.0236

SYMPOSIUM ON QUANTITATIVE METHODS IN MORPHOLOGY held...during the eighth International congress of anatomists Proceedings Quantitative methods in morphology Wiesbaden 1965 Aug 10 Edited by Ewald R. Weibel and Hans Elias Organised by the International society for stereology Illus. Springer, Berlin,1967
An 32.0105

SYMPOSIUM ON QUESTIONS OF NUMERICAL ANALYSIS proceedings Rome 1958 Jun.30-Jul.1 Provisional international computation centre 79p Rome,1958
Math L 5.2374

SYMPOSIUM ON RADIANT ENERGY IN THE SEA Papers Helsinki 1960 Aug 4-5 International union of geodesy and geophysics Edited by N. G. Jerlov International union of geodesy and geophysics.Monograph,10 illus 115p Paris, 1961 Symposium held during the I.U.G.A. assembly at Helsinki
Sco 14.0135

SYMPOSIUM ON RADIATION EFFECTS ON METALS AND NEUTRON DOSIMETRY Los Angeles,Calif. 1962 Oct 2-3 American society for testing materials A.S.T.M.Special technical publication, 341 American society for testing and materials,Philadelphia,Pa.,1963
Met 25.1585

SYMPOSIUM ON RADIO ASTRONOMY AND THE GALACTIC SYSTEM Radio astronomy and the galactic system Noordwijk 1966 Aug 25-Sep 1 International astronomical union Edited by Hugo van Woerden Organised in co-operation with International scientific radio union International astronomical union.Symposium, 31 xviii,501p Academic press,London;New York,1967
A Math 4.1574

SYMPOSIUM ON RADIOBIOLOGY :the basic aspects of radiation effects on living systems Oberlin,Ohio 1950 Jun 14-18 Edited by James J. Nickson Sponsored by the National research council Wiley;Chapman and Hall,New York;London,1952
Radioth 35.1717

SYMPOSIUM ON RADIOGRAPHY 1st and 2nd Atlantic City 1936,1942 American society for testing materials A.S.T.M.,Philadelphia, 1943
Met 25.1317

SYMPOSIUM ON RADIOISOTOPE SAMPLE MEASUREMENT TECHNIQUES IN MEDICINE AND BIOLOGY Proceedings Radioisotope sample measurement techniques in medicine and biology Vienna 1965 May 24-25 Held by the International atomic energy agency International atomic energy agency.Proceedings series International atomic energy agency, Vienna,1965
Radioth 35.1377

SYMPOSIUM ON RADIOISOTOPES presented at the second Pacific area national meeting Los Angeles 1956 Sep 21 American society for testing materials A.S.T.M.Special technical publication, 215 A.S.T.M.,Philadelphia,1958
Met 25.1559

SYMPOSIUM ON RADIOLOGICAL HEALTH AND SAFETY IN MINING AND MILLING OF NUCLEAR MATERIALS :a symposium Proceedings Radiological health and safety in mining and milling of nuclear materials Vienna 1963 Aug 26-31 1-2 International atomic energy agency International labour organization In co-operation with the World health organization International atomic energy agency.Proceedings series International atomic energy agency, Vienna,1964
Radioth 35.1164

SYMPOSIUM ON RADIOSENSITIZERS AND RADIOPROTECTIVE DRUGS 1st Milan 1964 Edited by R. Paoletti and R. Vertus Organized by the European society for biochemical pharmacology Progress in biochemical pharmacology, 1 Karger,Basel,1965
Radioth 35.1943

SYMPOSIUM ON RADIOSOTOPES IN METALS ANALYSIS AND TESTING presented to the sixty-second annual meeting Atlantic City,N.J. 1959 Hun 22 American society for testing materials A.S.T.M.Special technical publication, 261 A.S.T.M.,Philadelphia,Pa., 1960
Met 25.1741

SYMPOSIUM ON RECENT ADVANCES IN OPTICS Proceedings Oxford 1959 Mar.20 University of Oxford 138-180p London,1959
Obs 6.1989

SYMPOSIUM ON RECENT MECHANICAL ENGINEERING DEVELOPMENTS IN AUTOMATIC CONTROL Proceedings London 1960 Jan 5-7 Institution of mechanical engineers Institution of mechanical engineers,London, 1960
Eng 41.6012

SYMPOSIUM ON RECENT PROGRESS IN MATERIALS SCIENCES Properties of crystalline solids Symposium on recent progress in materials sciences;Symposium on nature and origin of strength of materials.Presented at the 63rd annual meeting A.S.T.M. Philadelphia,Pa. 1960 Jun 27 American society for testing materials American society for testing materials.Special technical publication, 283 American society for testing materials, Philadelphia,Pa.,1961
Eng 41.3790

SYMPOSIUM ON 'RECENT PROGRESS IN RESEARCH IN CHEMICAL AND EXTRACTION METALLURGY' Proceedings Research in chemical and extraction metallurgy Sydney 1965 Feb 24-26 Australasian institute of mining and metallurgy Edited by J.T. Woodcock and others Australasian institute of mining and metallurgy.Monograph series, 2 Melbourne, 1965 reprinted 1967
Met 25.2780

SYMPOSIUM ON RECOVERY OF CELLS FROM INJURY : given at research conference for biology and medicine of the atomic energy commission Gatlinburg,Tenn. 1961 Apt 3-6 Sponsored by the Oak Ridge national laboratory.Biology division Journal of cellular and comparative physiology, 58,supp.1 Oak Ridge national laboratory.Symposia, 14 Wistar institute of anatomy and biology,Philadelphia,Pa.,1961
Gen 34.1114

SYMPOSIUM ON REGULATION OF ENZYME ACTIVITY AND SYNTHESIS IN NORMAL AND NEOPLASTIC LIVER 1st Proceedings Indianapolis,Ind. 1962 Oct 1-2 Edited by George Weber Held at Indiana university school of medicine Advances in enzyme regulation, 1 Pergamon, Oxford,1963 Technical editor Catherine E. Forrest Weber
Bot 42.1738

SYMPOSIUM ON REGULATION OF ENZYME ACTIVITY AND SYNTHESIS IN NORMAL AND NEOPLASTIC LIVER 2nd Proceedings Indianapolis 1963 Sep 30 Oct 1 Edited by George Weber Advances in enzyme regulation, 2 Pergamon,Oxford,1964
Radioth 35.0124

SYMPOSIUM ON REGULATION OF ENZYME ACTIVITY AND SYNTHESIS IN NORMAL AND NEOPLASTIC LIVER 2nd Proceedings Indianapolis,Ind. 1963 Sep 30-Oct 1 Edited by George Weber Held at Indiana university school of medicine Advances in enzyme regulation, 2 Pergamon, Oxford,1964 Technical editor Catherine E. Forrest,Weber
Bot 42.1739

SYMPOSIUM ON RELAXATION METHODS IN RELATION TO MOLECULAR STRUCTURE Lectures and papers Molecular relaxation processes Aberystwyth 1965 Jul 7-9 Chemical society Academic press,London,1966
Pha 16.0369

SYMPOSIUM ON RELAXATION METHODS IN RELATION TO MOLECULAR STRUCTURE Papers Molecular relaxation processes Aberystwyth 1965 Jul 7-9 Chemical society Chemical society special publication, 20 Academic press, London,1966
Cav 7.0005

A SYMPOSIUM ON RESPIRATORY ENZYMES Addresses Madison,Wis, 1941 Sep 11-17 By Otto Meyerhof and others held at the University of Wisconsin University of Wisconsin press, Madison,Wis.,1942
Bioch 33.1661

A SYMPOSIUM ON RESPIRATORY ENZYMES Lectures and discussions Madison,Wis. By Otto Meyerhof and others Given at the University of Wisconsin University of Wisconsin,Madison, Wis.,1942 reprinted 1948 Sponsored jointly by the universities of Wisconsin and Chicago
Bal 39.1344

SYMPOSIUM ON RHEOLOGY :presented at the ASME applied mechanics and fluids engineering conference Washington,D.C. 1965 Jun 7-9 American society of mechanical engineers Edited by A.W. Marris and J.T.S. Wang American society of mechanical engineers,New York,1965
Eng 41.4418

SYMPOSIUM ON RIBONUCLEIC ACID STRUCTURE AND FUNCTION Federation of European biochemical sciences meeting 2nd Proceedings Vienna 1965 Apr 21-24 Federation of European biochemical societies Edited by H. Tuppy Pergamon press,Oxofrd, 1966
Gen 34.0688

SYMPOSIUM ON RICE GENETICS AND CYTOGENETICS Proceedings Los Banos 1963 Feb 4-8 Sponsored by the International rice research institute Elsevier,Amsterdam,1964
Gen 34.1559

SYMPOSIUM ON ROAD VEHICLE AERODYNAMICS 1st Proceedings Road vehicle aerodynamics London 1969 Nov 6-7 Edited by A.J. Scibor-Rylski Organised by the City university. Department of aeronautics City university, London,1970 Organised in cooperation with the National physical laboratory,the Motor industry research association and the Institution of mechanical engineers
Eng 41.8336

SYMPOSIUM ON SALT AND WATER METABOLISM Papers New York 1959 Dec 11-12 Edited by Alfred P. Fishman Sponsored by the New York heart association Circulation, 21,no 5,pt 2 American heart association,New York,1959
Phys 20.1327

SYMPOSIUM ON SELECTED TOPICS IN RADIATION DOSIMETRY Proceedings Selected topics in radiation dosimetry Vienna 1960 Jun 7-11 Sponsored by the International atomic energy agency International atomic energy agency.Proceedings series International atomic energy agency,Vienna,1961
Radioth 35.1694

SYMPOSIUM ON SEMIGROUPS Proceedings Semigroups Detroit,Mich. 1968 Jun 27th-29th Wayne state university Edited by Karl W. Polley Sponsored by the National science foundation bibliog. xi,277p 24cm Academic press,New York;London,1969
P Math 2.3927

SYMPOSIUM ON SEVERAL COMPLEX VARIABLES Proceedings Park City,Utah 1970 Mar 30-Apr 3 Edited by R.M. Brooks Lecture notes in mathematics, 184 234p 25cm Springer, Berlin,1971
P Math 2.4354

SYMPOSIUM ON SHEAR AND TORSION TESTING 1960 Jun 28 American society for testing materials A.S.T.M.Special technical publication, 289 American society for testing materials,Philadelphia,Pa.,1961
Met 25.0887

SYMPOSIUM ON SHEAR TESTING OF SOILS Atlantic City,N.J. 1939 Jun 28 American society for testing materials American society for testing materials,Philadelphia,Pa.,1939
Eng 41.2988

SYMPOSIUM ON SINGULARITIES OF SMOOTH MANIFOLDS AND MAPS Liverpool singularities symposium I Proceedings Liverpool 1970 Aug -Sep University of Liverpool.Department of pure mathematics Edited by C.T.C. Wall Lecture notes in mathematics, 192 318p 25cm Springer,Berlin,1971
P Math 2.4122

SYMPOSIUM ON SINGULARITIES OF SMOOTH MANIFOLDS AND MAPS Proceedings Liverpool singularities symposium II Liverpool 1970 Aug-Sep University of Liverpool.Department of pure mathematics Edited by C.T.C. Wall Lecture notes in mathematics, 209 280p 25cm Springer,Berlin,1971
P Math 2.4123

SYMPOSIUM ON SINTER London 1953 Nov Iron and steel institute Iron and steel institute. Special report, 53 Iron and steel institute, London,1955
Met 25.0321

SYMPOSIUM ON SOIL DYNAMICS Atlantic City,N.J. 1961 Jun 26 American society for testing materials American society for testing materials.Special technical publication, 305 American society for testing materials, Philadelphia,Pa.,1962
Eng 41.3155

SYMPOSIUM ON SOILS FOR ENGINEERING PURPOSES Papers on soils Symposium on time rates of loading in soil tension Symposium on Atterberg limits,Session on soils and Symposium on soils for engineering purposes Atlantic City,N.J. 1959 Jun 22-23 San Francisco,Calif. 1959 Oct 16 American society for testing materials.Special technical publication, 254 American society for testing materials, Philadelphia,Pa.,1960 Papers presented at various A.S.T.M. meetings during 1959
Eng 41.3151

SYMPOSIUM ON SOLAR SEEING 2nd Proceedings Rome 1961 Feb 20-25 International electronic and nuclear exhibition,Rome Under the auspices of Consiglio nazionale delle ricerche 158p Rome,1962
Obs 6.2051

SYMPOSIUM ON SOLDER :presented at the fifty-ninth annual meeting A.S.T.M. Atlantic city,N.J. 1956 Jun 19-20 American society for testing materials American society for testing materials.Special technical publication, 189 American society for testing materials,Philadelphia,Pa.,1957
Eng 41.4025

A SYMPOSIUM ON SOLID AND LIQUID STATE SELENIUM PHYSICS... held at the Chemical society Proceedings Recent advances in selenium physics London 1964 June European selenium-tellurium committee Edited by Sven Walden and others Pergamon press,Oxford,1965
Cav 7.1462

The SYMPOSIUM ON SOLIDIFICATION presented at the 64th AFS castings congress and exposition Philadelphia,Pa. 1960 May 9-13 American foundrymen's society American foundrymen's society,Des Plaines,Ill.,1961
Met 25.0644

SYMPOSIUM ON SOME PROBLEMS OF NORMAL AND ABNORMAL DIFFERENTIATION AND DEVELOPMENT Proceedings Bar Harbor,Maine 1959 Jun 15-17 National cancer institute Edited by Nathan Kaliss National cancer institute.Monograph, 2 U.S.Department of health,education and welfare,Washington,D.C.,1960
Radioth 35.0808

SYMPOSIUM ON STATISTICAL ASPECTS OF FATIGUE Atlantic City,N.J. 1951 Jun 19 Pt 1 American society for testing materials A.S.T.M.Special technical publication, 121 American society for testing materials, Philadelphia,Pa.,1951
Met 25.0912

The SYMPOSIUM ON STATISTICAL MECHANICS AND THERMODYNAMICS Proceedings Statistical mechanics of equilibrum and non-equilibrium Aachen 1964 Jun 15-20 International union of pure and applied physics Edited by J. Meixner 274p North-Holland,Amsterdam,1965
Cav 7.0001

SYMPOSIUM ON STEEL MAKING (acid and basic open-hearth practice) Iron and steel institute Iron and steel institute.Special report, 22 Iron and steel institute,London, 1938
Met 25.0324

SYMPOSIUM ON STELLAR AND SOLAR MAGNETIC FIELDS Stellar and solar magnetic fields Munich 1963 Sep.3-10 International astronomical union Edited by R. Lust International astronomical union.Symposium, 22 460p North Holland,Amsterdam,1965
Obs 6.1148

SYMPOSIUM ON STELLAR EVOLUTION papers La Plata 1960 Nov. 7-11 Universidad nacional de La Plata.Observatorio astronomico Edited by J. Sahade 309p Astronomical observatory, La Plata,1962 Scientific meeting organized to celebrate the sesquicentennial of the May revolution
Obs 6.1505

A SYMPOSIUM ON STEROID HORMONES Papers Madison,Wis. 1950 University of Wisconsin Edited by Edgar S. Gordon University of Wisconsin press,1950 Published in celebration of the hundredth anniversary of the founding of the University of Wisconsin
Bioch 33.0455

SYMPOSIUM ON STEROIDS IN EXPERIMENTAL AND CLINICAL PRACTICE Cuernavaca,Mexico 1951 Jan 15-18 Edited by Abraham White Churchill,London,1951
Radioth 35.0697

SYMPOSIUM ON STRENGTH AND DUCTILITY OF METALS AT ELEVATED TEMPERATURES with particular reference to effects of notches and metallurgical changes New York 1952 Jun 23 American society for testing materials A.S.T.M.Special technical publication, 128 A.S.T.M.,Philadelphia,Pa.,1953
Met 25.0979

SYMPOSIUM ON STRESS National research council Army medical services,Washington,D.C. 1953
Med 36.0265

SYMPOSIUM ON STRESS-CORROSION CRACKING OF METALS Philadelphia,Pa. 1944 Nov 29-Dec 1 American institute of mining and metallurgical engineers American society for testing materials American society for testing materials;American institute of mining and metallurgical engineers,Philadelphia,Pa.:New York,1945
Eng 41.3717

SYMPOSIUM ON STRESS-STRAIN-TIME-TEMPERATURE RELATIONSHIPS IN MATERIALS :presented at the 65th annual meeting A.S.T.M. New York 1962 Jun 27 American society for testing and materials American society for testing and materials.Special technical publication, 325 American society for testing and materials. Materials science series, 3
Eng 41.3789

SYMPOSIUM ON STRESS-CORROSION CRACKING OF METALS Philadelphia,Pa. 1944 Nov 29-Dec 1 American society for testing materials and American institute of mining and metallurgical engineers.Institute of metals division A.S.T.M.,New York,1945
Met 25.1952

SYMPOSIUM ON STRESS-STRAIN-TIME-TEMPERATURE RELATIONSHIPS IN MATERIALS New York 1962 Jun 27 American society for testing materials A.S.T.M.Special technical publication, 325,Materials science series, 3 American society for testing and materials, Philadelphia,Pa.,1962
Met 25.1034

The SYMPOSIUM ON STRUCTURAL AND CHEMICAL ASPECTS OF CYTOCHROMES Proceedings Structure and function of cytochromes Osaka 1967 Aug 16-18 Edited by Kazuo Okunuki and others University of Tokyo press,Tokyo,1968
Bioch 33.1100

SYMPOSIUM ON SUPERCONDUCTIVE TECHNIQUES FOR COMPUTING SYSTEMS proceedings Washington,D.C. 1960 May 17-19 Sponsored by United States.Office of naval research illus. vii,413p 26cm Office of naval research,Washington,D.C.,1960
Math L 5.0859

SYMPOSIUM ON SURFACE AND SUBSURFACE RECONNAISSANCE Atlantic City,N.J. 1951 Jun 19 American society for testing materials American society for testing materials.Special technical publication, 122 American society for testing materials, Philadelphia,Pa.,1952
Eng 41.3153

SYMPOSIUM ON TECHNIQUES AND EQUIPMENT FOR ENVIRONMENTAL TESTING; SYMPOSIUM ON DEVELOPMENTS IN MATERIALS TESTING MACHINE DESIGN :joint conference Machines for materials and environmental testing Manchester 1965 Sep 6-10 Institution of mechanical engineers and Society of environmental engineers Under the aegis of the British national committee on materials Institution of mechanical engineers. Proceedings, 180,pt 3a I.M.E.,London,1966
Met 25.0891

SYMPOSIUM ON TECHNIQUES FOR ELECTRON METALOGRAPHY Atlantic City,N.J. 1953 Jun 30 American society for testing materials A.S.T.M. Special technical publication, 153 American society for testing materials,Philadelphia,Pa., 1954
Met 25.1361

SYMPOSIUM ON TECHNIQUES USED IN THE STUDY OF MICROCIRCULATION Papers Oxford 1968 Sep British microcirculation society Edited by D.R. Chambers and P.A.G. Monro British microcirculation society,Cambridge, 1969
An 32.5223

SYMPOSIUM ON TEMPERATURE 3rd Papers Washington,D.C. 1954 Oct 28-30 Edited by Hugh C. Wolfe Sponsored by the American institute of physics Reinhold,New York,1955
Eng 41.4126

SYMPOSIUM ON TESTING METAL POWDERS AND METAL POWDER PRODUCTS presented at the Cleveland spring meeting Cleveland 1952 Mar 5 American society for testing materials A.S.T.M.Special technical publication, 140 American society for testing materials, Philadelphia,Pa.,1953
Met 25.0762

SYMPOSIUM ON TESTING METAL POWDERS AND METAL POWDER PRODUCTS Papers Cleveland 1952 Mar 5 American society for testing materials American society for testing materials.Special technical publication, 140 87p American society for testing materials, Philadelphia,Pa.,1953
Chem E 24.1121

SYMPOSIUM ON TETANUS IN GREAT BRITAIN Leeds 1967 Apr 7-8 Leeds general infirmary.Tetanus unit Edited by M. Ellis c1967 Typescript
PGMS 29.0042

SYMPOSIUM ON THE ADVERSE EFFECTS OF DRUGS 1st Proceedings London 1970 Feb 4-5 Edited by Gillian C. Hanson Sponsored by Beecham research laboratories Beecham research laboratories,London,1971
PGMS 29.0673

SYMPOSIUM ON THE APPLICATION OF PRESTRESSED CONCRETE TO MACHINERY STRUCTURES Slough 1964 Jan 14-15 Prestressed concrete development group Prestressed concrete development group,London,1965
Eng 41.2948

SYMPOSIUM ON THE APPLICATION OF SWITCHING THEORY IN SPACE TECHNOLOGY SUNNYVALE,CALIF.1962 Switching and programming By A.van Wijngaarden Mathematisch centrum,Amsterdam. Rekenafdeling.Report MR, 50 16p amsterdam, 1962 Paper to be presented at the Symposium on the application of switching theory in space technology,Sunnyvale,Calif. 1962
Math L 5.2176

SYMPOSIUM ON THE APPROXIMATION OF FUNCTIONS 8th proceedings Warren,Mich. 1964 Aug.31-Sep.2 General motors corporation.Research laboratories Edited by Henry L. Garabedian bibliog. viii,220p 24cm Elsevier, Amsterdam,1965
Math L 5.0348

SYMPOSIUM ON THE ASSESSMENT OF RADIOACTIVE BODY BURDENS IN MAN Proceedings Assessment of radioactivity in man Heidelberg 1964 May 11-16 1-2 International atomic energy agency International labour organization World health organization International atomic energy agency.Proceedings series International atomic energy agency, Vienna,1964
Radioth 35.1373

SYMPOSIUM ON THE BIOLOGICAL EFFECTS OF IONIZING RADIATION AT THE MOLECULAR LEVEL Proceedings Biological effects of ionizing radiation at the molecular level Brno 1962 Jul 2-6 Held by the International atomic energy agency International atomic energy agency.Proceedings series I.A.E.A.,Vienna,1962
Radioth 35.1976

SYMPOSIUM ON THE BIOLOGY OF FRACTURE HEALING Callus formation :symposium on the biology of fracture healing Debrecen 1965 Jul 5-8 Edited by St. Krompecher and E. Kerner Translated by E. Kerner Symposia biologica hungarica, 7 Akademiai Kaido,Budapest,1967
An 32.3921

SYMPOSIUM ON THE BIOLOGY OF SKIN 14th Proceedings Aging Portland,Ore. 1964 May 9-10 Edited by William Montagna Held at the University of Oregon medical school Advances in biology of skin, 6 Oregon regional primate research center, 55 Pergamon press,Oxford,1965
An 32.4041

SYMPOSIUM ON THE BIOLOGY OF SKIN 15th Proceedings Carcinogenesis Portland,Ore. 1965 Edited by William Montagna and Richard L. Dobson Held at the University of Oregon medical school Advances in biology of skin, 7 Oregon regional primate research center, 87 Pergamon press,Oxford,1966
An 32.4042

SYMPOSIUM ON THE BIOLOGY OF SKIN 16th Proceedings Pigmentary system Portland, Ore. 1966 Edited by William Montagna and Funan Hu Held at the University of Oregon medical school Advances in biology of skin, 8 Oregon regional primate research center, 183 Pergamon press,Oxford,1967
An 32.4043

SYMPOSIUM ON THE BUSHVELD IGNEOUS COMPLEX AND OTHER LAYERED INTRUSIONS Proceedings Pretoria 1969 Jul 7-14 Geological society of South Africa Edited by D.J.L. Visser and G.von Gruenewaldt Geological society of South Africa.Special paper, 1 Geological society of South Africa,Johannesburg,1970
Min 10.1476

A SYMPOSIUM ON THE CHEMICAL BASIS OF DEVELOPMENT Baltimore,Md 1958 Mar 24-27 Edited by William D. McElroy and Bentley Glass Sponsored by the McCollum-Pratt institute Johns Hopkins press,Baltimore,Md.,1958
Bal 39.0454

A SYMPOSIUM ON THE CHEMICAL BASIS OF DEVELOPMENT Baltimore,Md. 1958 Mar 24-27 Edited by William D. McElroy and Bentley Glass Sponsored by the McCollum-Pratt institute McCollum-Pratt institute.Symposia illus. Johns Hopkins press,Baltimore,Md.,1958
Gen 34.0852

A SYMPOSIUM ON THE CHEMICAL BASIS OF DEVELOPMENT Papers and informal discussions Chemical basis of development Baltimore 1958 Mar 24-27 Edited by William D. McElroy and Bentley Glass Sponsored by the McCollum-Pratt institute McCollum-Pratt institute. Contribution, 234 Johns Hopkins press, Baltimore,1958
Bioch 33.0829

A SYMPOSIUM ON THE CHEMICAL BASIS OF HEREDITY Baltimore,Md. 1956 Jun 19-22 Edited by William D. McElroy and Bentley Glass Sponsored by the McCollum-Pratt institute McCollum-Pratt institute.Contribution, 153 Johns Hopkins press,Baltimore,Md.,1957
Gen 34.0853

A SYMPOSIUM ON THE CHEMICAL BASIS OF HEREDITY Baltimore,Md. 1956 Jun 19-22 McCollum-Pratt institute Edited by William D. McElroy and Bentley Glass Johns Hopkins press, Baltimore,Md.,1957
Bot 42.4714

A SYMPOSIUM ON THE CHEMICAL BASIS OF HEREDITY Papers and informal discussions Baltimore, Md. 1956 Jun 19-22 Edited by William D. McElroy and Bentley Glass Sponsored by the McCollum-Pratt institute McCollum-Pratt institute.Contribution, 153 Johns Hopkins press,Baltimore,Md.,1957
Bal 44.6451

A SYMPOSIUM ON THE CHEMICAL BASIS OF HEREDITY Papers and informal discussions Chemical basis of heredity Baltimore 1956 Jun 19-22 Edited by William D. McElroy and Bentley Glass Sponsored by the McCollum-Pratt institute McCollum-Pratt institute.Contribution, 153 Johns Hopkins press,Baltimore,1957
Bioch 33.0830

SYMPOSIUM ON THE CLASSIFICATION OF BRACKISH WATERS Venice 1958 Apr 8-14 By G.Sven Segerstrale International association of theoretical and applied limnology Archivio di oceanografia e limnologia, 11,suppl. Centro nazionale di studi talassografafici del consiglio nazionale delle ricerche,Venice,1959 Title also in Italian:"Simposio sulla classificazione della acque salmastre" Text in English and Italian.
Bal 39.1962

SYMPOSIUM ON THE DESIGN OF HIGH BUILDINGS Proceedings of a meeting...held as part of the Golden jubilee congress of the University of Hong Kong Hong Kong 1961 Sep University of Hong Kong Edited by Sean Mackey Hong Kong university press,Hong Kong, 1962
Eng 41.2685

SYMPOSIUM ON THE DETECTION AND USE OF TRITIUM IN THE PHYSICAL AND BIOLOGICAL SCIENCES Proceedings Tritium in the physical and biological sciences Vienna 1961 May 3-10 Vols 1-2 Sponsored by the International atomic energy agency 2 vols International atomic energy agency,Vienna,1962 Co-sponsored by the Joint commission on applied radioactivity
Radioth 35.0104

SYMPOSIUM ON THE ECOLOGY OF PELAGIC FISH SPECIES IN ARCTIC WATERS AND ADJACENT SEAS Papers Charlottenlund 1966 Sep 30-Oct 1 International council for the exploration of the sea.Distant northern seas committee and International council for the exploration of the sea.Gadoid fish committee Edited by R.W. Blacker Conseil permanent international pour l'exploration de la mer.Rapports et proces-verbaux des reunions, 158 illus,maps Host, Copenhagen,1968
Sco 14.8291

SYMPOSIUM ON THE EFFECTS OF IONIZING RADIATION ON THE NERVOUS SYSTEM Effects of ionizing radiation on the nervous system Vienna 1961 Jun 5-9 Sponsored by the International atomic energy agency International atomic energy agency,Vienna,1962
An 32.4190

SYMPOSIUM ON THE EFFECTS OF RADIATION ON CELLULAR PROLIFERATION AND DIFFERENTIATION Proceedings Effects of radiation on cellular proliferation and differentiation Monaco 1968 Apr 1-5 International atomic energy agency Organized in co-operation with the Joint commission on applied radioactivity International atomic energy agency,Vienna,1968
Radioth 35.1760

A SYMPOSIUM ON THE EFFECTS OF RADIATION ON METALS Cleveland,Ohio 1956 Oct 8 American institute of mining,metallurgical and petroleum engineers.Nuclear metallurgy committee I.M.D.Special report series, 3 Nuclear metallurgy, 3 American institute of mining,metallurgical,and petroleum engineers,New York,1956
Met 25.1576

SYMPOSIUM ON THE ENGINEERING ASPECTS OF MAGNETOHYDRODYNAMICS 2nd Proceedings Philadelphia,Pa. 1961 Mar 9-10 Edited by Clifford Mannal and Norman W. Mather Sponsored by the American institute of electrical engineers Columbia university press,New York;London,1962
Eng 41.4497

SYMPOSIUM ON THE ENGINEERING ASPECTS OF MAGNETOHYDRODYNAMICS 3rd Proceedings Engineering aspects of magnetohydrodynamics Rochester,N.Y. 1962 Mar 28-29 Edited by Norman W. Mather and George W. Sutton Gordon and Breach,New York,1964
Eng 41.4546

A SYMPOSIUM ON THE EVALUATION OF DRUG TOXICITY Macclesfield 1957 Oct 1 Imperial chemical industries.Pharmaceuticals division Edited by A.L. Walpole and A. Spinks Churchill, London,1958
Pha 16.0327

SYMPOSIUM ON THE FABRICATION OF FUEL ELEMENTS ceramic base elements ;metal base fuels and jacket components Cleveland 1958 Oct 29 American institute of mining,metallurgical and petroleum engineers.Institute of metals division Nuclear metallurgy, 5 I.M.D. Special report series, 7 Metallurgical society of American institute of mining, metallurgical and petroleum engineers,New York, 1958
Met 25.1543

SYMPOSIUM ON THE FLUID MECHANICS OF INTERNAL FLOW Warren,Mich. 1965 General motors research laboratories Edited by Gino Sovran Elsevier,Amsterdam,1967
Eng 41.6553

SYMPOSIUM ON THE GENERAL CIRCULATION OF THE OCEANS AND THE ATMOSPHERE International geodetic and geophysical union :general assembly 9th Symposium proceedings Brussels 1951 Aug International association of meteorology and atmospheric physics Unesco,Brussels,1951
Geog 13.0960

SYMPOSIUM ON THE GERM CELLS AND EARLIEST STAGES OF DEVELOPMENT Pallanza 1960 Sep 14-20 International institute of embryology A. Baselli,Milan,1961 At head of title: Istituto Lombardo accademia di scienze e lettere
Bal 39.0459

SYMPOSIUM ON THE HANDLING OF SOLIDS Proceedings London 1962 Jun 25 Institution of chemical engineers Edited by P.A. Rottenburg Institution of chemical engineers,London,1962 3rd congress of the European federation of chemical engineering
Chem E 24.1446

SYMPOSIUM ON THE HISTORY OF SCIENCE Papers Scientific change Oxford 1961 Jul 9-15 Edited by A.C. Crombie Under the auspices of the International union of the history and philosophy of science.Division of history of science Heinemann,London,1963
Geog 13.0053

SYMPOSIUM ON THE IDENTIFICATION AND CLASSIFICATION OF SOILS Atlantic City,N.J. 1950 Jun 29 American society for testing materials American society for testing materials.Special technical publication, 113 American society for testing materials, Philadelphia,Pa.,1951
Eng 41.3146

SYMPOSIUM ON THE INTERACTION BETWEEN FLUIDS AND PARTICLES Proceedings London 1962 Jun 20-22 Institution of chemical engineers illus 351p Institution of chemical engineers,London,n.d. Third congress on the European federation of chemical engineering
Chem E 24.1447

A SYMPOSIUM ON THE INTERACTION OF DRUGS AND SUBCELLULAR COMPONENTS IN ANIMAL CELLS London 1967 Biological council.Co-ordinating committee for symposia on drug action Edited by P.N. Campbell Pl. viii, 355p 21cm Churchill,London,1968
Pha 16.0252

A SYMPOSIUM ON THE INTERACTION OF DRUGS AND SUBCELLULAR COMPONENTS IN ANIMAL CELLS London 1967 Apr 10-11 Co-ordinating committee for symposia on drug action Edited by P.N. Campbell Illus Churchill,London, 1968
Radioth 35.0273

SYMPOSIUM ON THE JULY 1959 EVENTS AND ASSOCIATED PHENOMENA Helsinki 1960 Jul 28 International union of geodesy and geophysics International union of geodesy and geophysics. Monograph,7 tables,diagrs 157p 24cm I. U.G.G.,Paris,1960
Sco 14.0246

The SYMPOSIUM ON THE LESS COMMON MEANS OF SEPARATION Proceedings Birmingham 1963 Apr 24-26 Institution of chemical engineers.Midlands branch Edited by J.M. Pirie 127p I.Chem.E.,London,1964
Chem E 24.1915

The SYMPOSIUM ON THE MAGNETIC AND OTHER PECULIAR AND METALLIC-LINE A STARS Magnetic and related stars Greenbelt,Md. 1965 Nov 8-10 American astronomical society and National aeronautics and space administration Edited by Robert C. Cameron Mono book corporation, Baltimore,Md.,1967
TA 15.0310

The SYMPOSIUM ON THE MAGNETIC AND OTHER PECULIAR AND METALLIC-LINE A STARS 1st :magnetic and related stars Proceedings AAS-NASA magnetic star symposium Greenbelt,Md. 1965 Nov 8-10 and Berkeley,Calif. 1965 Dec 29 American astronomical society National aeronautics and space administration Edited by Robert C. Cameron 596p Mono book corporation,Baltimore,1967 Includes the Helen B.Warner prize lecture of the American astronomical society,by George W.Preston
Obs 6.3432

SYMPOSIUM ON THE MEASUREMENT OF VISUAL FUNCTION proceedings of Spring meeting,1965 Proceedings 1965 Armed forces-National research council committee on vision Edited by Milton A. Whitcomb and William Benson National research council,Washington,D.C.,1968
Psy 31.0427

A SYMPOSIUM ON THE MECHANISM OF CYTOPLASMIC STREAMING,CELL MOVEMENT,AND THE SALTATORY MOTION OF SUBCELLULAR PARTICLES Primitive motile systems in cell biology Princeton,N.J. 1963 Apr 2-5 Edited by Robert D. Allen and Noburo Kamiya Held at Princeton university Academic press,New York;London, 1964
Bal 39.0297

A SYMPOSIUM ON THE MECHANISM OF ENZYME ACTION Papers and discussions Baltimore 1953 Jun 16-19 Edited by William D. McElroy and Bentley Glass Sponsored by the McCollum-Pratt institute McCollum-Pratt institute. Contribution, 70 Johns Hopkins press, Baltimore,Md.,1954
Bioch 33.1059

A SYMPOSIUM ON THE MECHANISM OF ENZYME ACTIVITY McCollum-Pratt institute Edited by William D. McElroy McCollum-Pratt institute. Contribution,70 Johns Hopkins press, Baltimore,Md.,1954
Chem 18.1270

SYMPOSIUM ON THE MOON proceedings Moon Pulkovo 1960 Dec.5-11 International astronomical union Edited by Zdenek Kopal and Zdenka Kadla Mikhailov International astronomical union.Symposia, 14 map 517p Academic press,London,1962
Obs 6.0972

SYMPOSIUM ON THE MOTION OF GASEOUS MASSES OF COSMICAL DIMENSIONS Proceedings Problems of cosmical aerodynamics Paris 1949 Aug 16-19 International union of theoretical and applied mechanics International astronomical union Edited by J. M. Burgess and H.C.van de Hulst 237p Central air documents office,Dayton,Ohio,1951
Obs 6.3550

SYMPOSIUM ON THE MUTATIONAL PROCESS Symposia Csav The Physiology of gene and mutation expression :symposium on the mutational process Proceedings Brno 1965 Aug 4-7 Prague 1965 Aug 9-11 Edited by Margita Kohoutova and J. Hubacek Organized by the Ceskoslovenska akademie ved.Institute of microbiology Academia,Prague,1967
Gen 34.0770

SYMPOSIUM ON THE NATURE,OCCURRENCE,AND EFFECTS OF SIGMA PHASE presented at the fifty-third annual meeting... Atlantic City,N.J. 1950 Jun 26-27 American society for testing materials A.S.T.M.Special technical publication, 110 A.S.T.M.,Philadelphia,Pa., 1951
Met 25.0477

SYMPOSIUM ON THE NATURE OF MOLTEN SLAGS AND SALTS Physical chemistry of melts London 1952 Feb 20 Nuffield research group in extraction metallurgy Institution of mining and metallurgy,London,1953
Met 25.1787

SYMPOSIUM ON THE NUMERICAL TREATMENT OF ORDINARY DIFFERENTIAL EQUATIONS, INTEGRAL AND INTEGRO-DIFFERENTIAL EQUATIONS 4th proceedings Rome 1960 Sep.20-24 Provisional international computation centre bibliog. 679p 26cm Birkhauser,Basel,1960
Math L 5.0299

SYMPOSIUM ON THE NUMERICAL TREATMENT OF PARTIAL DIFFERENTIAL EQUATIONS WITH REAL CHARACTERISTICS proceedings Rome 1959 Jan.28-30 Provisional international computation centre 158p Birkhauser,Basel, 1959
Math L 5.2378

SYMPOSIUM ON THE ORIGIN AND DISTRIBUTION OF THE ELEMENTS Origin and distribution of the elements Paris 1967 International union of geological sciences Edited by L.H. Ahrens International series of monographs on earth sciences, 30 xvii,1178p 24cm Pergamon press,Oxford,1968
TA 15.0565

SYMPOSIUM ON THE PHYSICS AND MEDICINE OF THE UPPER ATMOSPHERE Proceedings Physics and medicine of the upper atmosphere San Antonio,Tex. 1951 Nov 6-9 Lovelace foundation for medical education and research Edited by Clayton S. White and Otis O. Benson Sponsored by School of aviation medicine. Randolph Field illus xxix,611p University of New Mexico press,Albuquerque,N.M. 1952
Nap 11.0635

SYMPOSIUM ON THE PHYSIOLOGICAL PHARMACOLOGICAL AND CLINICAL ASPECTS OF SLEEP Proceedings Sleep mechanisms Zurich 1964 Sep 18-19 Edited by K. Akert and others Progress in brain research, 18 Elsevier,Amsterdam,1965
An 32.4238

SYMPOSIUM ON THE PHYSIOLOGY AND PATHOLOGY OF HUMAN REPRODUCTION 2nd annual Proceedings Ovary Harper hospital Wayne state university Edited by Harold C. Mack Thomas,Springfield,Ill.,1968
An 32.5224

SYMPOSIUM ON THE PHYSIOLOGY OF THE INSECT CENTRAL NERVOUS SYSTEM International congress of entomology 12th Papers London 1964 Edited by J.E. Treherne and J.W.L. Beaument London,1965
Bal 39.2458

SYMPOSIUM ON THE PLACENTA :its form and functions with particular reference to the prevention of birth defects and fetal deaths New York 1964 Mar 6 Medical society of the county of New York.Special committee on infant mortality Edited by Daniel Bergsma and others Sponsored by the National Foundation-March of dimes.Great New York Chapter Birth defects original article series, 1,no.1 National foundation,New York,1965
PGMS 29.0212

SYMPOSIUM ON THE PLASMA SHEATH 2nd :its effect upon reentry communication and detection Proceedings Electromagnetic aspects of hypersonic flight Boston,Mass. 1962 Apr 10-12 Edited by Walter Rotman and others Sponsored by the Air force Cambridge research laboratories.Electromagnetic radiation laboratory Spartan books;Cleaver-Hume,Baltimore,Md.;London,1964
Eng 41.4544

SYMPOSIUM ON THE RESPIRATORY ENZYMES AND THE BIOLOGICAL ACTION OF VITAMINS The Biological action of the vitamins :a symposium Chicago 1941 Sep 15-19 University of Chicago Edited by E.A. Evans University of Chicago press,Chicago,1942 Part of the joint Symposium on the respiratory enzymes and the biological action of vitamins arranged with the University of Wisconsin to celebrate the fiftieth anniversary of the University of Chicago.The papers on Respiratory enzymes were presented at Madison,1941,Sep 11-13
Bioch 33.0343

A SYMPOSIUM ON THE ROLE OF PHOSPHOROUS IN THE METABOLISM OF PLANTS AND ANIMALS Papers and discussions Phosphorus metabolism Baltimore,Md. 1951 Jun 18-21 and Baltimore,Md. 1952 Jun 16-19 Vol 1-2 Edited by William D. McElroy and H.Bentley Glass Sponsored by the McCollum-Pratt institute McCollum-Pratt institute. Contribution, 23,36 2 vols Johns Hopkins press,Baltimore,Md.,1951-52
Bot 42.1670

A SYMPOSIUM ON THE ROLE OF PHOSPHORUS IN THE METABOLISM OF PLANTS AND ANIMALS Phosphorus metabolism Baltimore,Md. 1951 Jun 18-21 Vol 1 Edited by W.D. McElroy and B. Glass Sponsored by McCollum-Pratt institute McCollum-Pratt institute. Contributions, 23 John Hopkins press, Baltimore,Md.,1951
Radioth 35.0028

A SYMPOSIUM ON THE ROLE OF PHOSPHORUS IN THE METABOLISM OF PLANTS AND ANIMALS Phosphorus metabolism Baltimore,Md. 1951 Jun 18-21 Vol 2 Edited by W.D. McElroy and B. Glass Sponsored by the McCollum-Pratt institute McCollum-Pratt institute.Contributions, 36 John Hopkins press,Baltimore,Md.,1952
Radioth 35.0029

A SYMPOSIUM ON THE ROLE OF PHOSPHORUS IN THE METABOLISM OF PLANTS AND ANIMALS Papers and discussions presented Phosphorus metabolism Baltimore,Md. 1951 Jun 18-21 and Baltimore,Md. 1952 Jun 16-19 Vol 1-2 Edited by William D. McElroy and H. Bentley Glass Sponsored by the McCollum-Pratt institute McCollum-Pratt institute. Contribution, 23,36 2 vols Johns Hopkins press,Baltimore,Md.,1951-52
Bioch 33.0597

SYMPOSIUM ON THE ROLE OF THE SOUTH ATLANTIC BASIN IN BIOGEOGRAPHY AND EVOLUTION Proceedings Problem of land connections across the South Atlantic with special reference to the Mesozoic New York 1949 Dec 28-29 By Maurice Ewing and others Edited by Ernst Mayr American museum of natural history. Bulletin, 99,3 American museum of natural history,New York,1952
Bot 42.2341

The SYMPOSIUM ON THE SOLAR SPECTRUM Proceedings Solar spectrum Utrecht 1963 Aug 26-31 Edited by C.de Jager D. Reidel,Dordrecht,1965 Dedicated to M.G.J. Minnaert on the occasion of his 70th birthday
TA 15.0199

SYMPOSIUM ON THE SPECIALIZED TISSUES OF THE HEART Proceedings Specialized tissues of the heart Rio de Janeiro 1960 Aug Edited by Antonio Paes de Carvalho and others Elsevier, Amsterdam,1961
An 32.3512

SYMPOSIUM ON THE SPECTRAL CLASSIFICATION AND MULTICOLOUR PHOTOMETRY Proceedings Spectral classification and multicolour photometry Saltsjobaden 1964 Aug.17-21 International astronomical union Edited by K. Loden and others International astronomical union.Symposium, 24 383p Academic press, London;New York,1966
Obs 6.1092

SYMPOSIUM ON THE STRUCTURE OF LOW-MEDIUM MASS NUCLEI 3rd Lawrence,Kansas 1968 Apr 18-20 Aerospace research laboratories Edited by J.P. Davidson vi,294p 24cm University press of Kansas,Lawrence,Kansas, 1968
Cav 7.3099

SYMPOSIUM ON THE SURGERY OF CANCER The Surgical clinics of North America New York 1953 April Saunders,Philadelphia,Pa.;London, 1953
Radioth 35.0754

SYMPOSIUM ON THE THEORY OF ABELIAN GROUPS Studies on Abelian groups Montpellier 1967 Jun 5-10 Edited by B. Charles Bibliog. x, 356p 25cm Dunod,Paris,1968
P Math 2.3507

SYMPOSIUM ON THE THEORY OF FINITE GROUPS Lectures Urbana,Ill. 1967 Nov 24 American mathematical society Bibliog. 94p 25cm American mathematical society, Providence,R.I.,1969
P Math 2.3581

SYMPOSIUM ON THE THEORY OF NUMERICAL NUMBERS Proceedings Dundee 1970 Sep 15-23 Edited by John L. Morris Lecture notes in mathematics, 193 vi,152p 25cm Springer, Berlin,1971
P Math 2.4462

SYMPOSIUM ON THE THEORY OF THIN ELASTIC SHELLS proceedings Delft 1959 Aug 24-28 International union of theoretical and applied mechanics Edited by W.T. Koiter North-Holland,Amsterdam,1960
A Math 4.0366

A SYMPOSIUM ON THE ULTRASTRUCTURE OF CELLS Proceedings Electron microscopy in anatomy London 1959 Apr 16-17 Anatomical society of Great Britain and Ireland Edited by J.D. Boyd and others Edward Arnold,London, 1961
Bioch 33.2297

A SYMPOSIUM ON THE USE OF ISOTOPES IN BIOLOGY AND MEDICINE Madison,Wis. 1947 Sep 10-13 By Hans T. Clarke and others University of Wisconsin University of Wisconsin press, Madison,1948
Bioch 33.0283

SYMPOSIUM ON THE USE OF MODELS IN GEOPHYSICAL FLUID DYNAMICS 1st Proceedings Fluid models in geophysics Baltimore,Md. 1953 Sep 1-4 Edited by Robert R. Long Sponsored by United States.Office of naval research illus,diagrms v,162p U.S. Government printing office,Washington,D.C., 1956
Chem E 24.0202

SYMPOSIUM ON THE USE OF RADIOISOTOPES IN SOIL MECHANICS Cleveland,Ohio 1952 Mar 5 American society for testing materials American society for testing materials.Special technical publication, 134 American society for testing materials,Philadelphia,Pa.,1953
Eng 41.3147

The SYMPOSIUM ON THE USE OF RADIOISOTOPES IN SOIL-PLANT NUTRITION STUDIES Proceedings Radioisotopes in soil plant nutrition studies Bombay 1962 Feb 26-Mar 2 International atomic energy agency United Nations.Food and agriculture organization International atomic energy agency,Vienna,1962
Bot 42.1368

SYMPOSIUM ON THE USE OF RADIOISOTOPES IN THE STUDY OF ENDEMIC AND TROPICAL DISEASES Proceedings Radioisotopes in tropical medicine Bangkok 1960 Dec 12-16 International atomic energy agency World health organization International atomic energy agency.Proceedings series International atomic energy agency,Vienna,1962
Radioth 35.1365

SYMPOSIUM ON THE UTILIZATION OF COMPUTERS IN MATHEMATICAL RESEARCH 1966 Computers in mathematical research Edited by R.F. Churchhouse and J.-C. Herz xi,185p 23cm North-Holland,Amsterdam,1968 Mainly on a symposium on Utilization of computers in mathematical research held at the IBM world trade european education centre,Blaricum, Netherlands,August 29-31,1966
Math L 5.3116

SYMPOSIUM ON THE WELDING OF IRON AND STEEL London 1935 May 2-3 Vol 1 Iron and steel institute 2 vols Iron and steel institute,London,1935
Eng 41.3947

SYMPOSIUM ON THE WELDING OF IRON AND STEEL London 1935 May 2-3 Vol 1-2 Iron and steel institute 2 vols Iron and steel institute,London,1935
Met 25.0783

SYMPOSIUM ON THERMAL PROPERTIES Papers 3rd Thermodynamic and transport properties of gases liquids and solids Lafayette,Ind 1959 Feb 23-26 Sponsored by the American society of mechanical engineers:Heat transfer division x,472p American society of mechanical engineers;McGraw-Hill,New York,1959
Chem E 24.0984

SYMPOSIUM ON THERMODYNAMICS AND THERMOCHEMISTRY Plenary lectures Thermodynamics and thermochemistry Lund 1963 Jul 18-23 International union of pure and applied chemistry and Swedish chemical society Butterworths,London,1964
Chem E 24.0988

SYMPOSIUM ON THERMOPHYSICAL PROPERTIES 3rd Papers Advances in thermophysical properties at extreme temperatures and pressures Lafayette,Ind. 1965 Mar 22-25 Edited by Serge Gratch Sponsored by the American society of mechanical engineers American society of mechanical engineers,New York,1965
Eng 41.7121

SYMPOSIUM ON THERMOPHYSICAL PROPERTIES 2nd progress in international research on thermodynamic and transport properties Papers Princeton,N.J. 1962 Jan 24-26 American society of mechanical engineers. Standing committee on thermophysical properties American society of mechanical engineers;Academic press,New York,1962
Chem 18.0326

SYMPOSIUM ON THYROCALCITONIN AND THE C CELLS Proceedings Calcitonin London 1967 Jul 17-20 Edited by Selwyn Taylor Heinemann,London,1968
An 32.3778

SYMPOSIUM ON TIME RATES OF LOADING IN SOIL TENSION Symposium on Atterberg limits, Session on soils and Symposium on soils for engineering purposes Papers on soils Atlantic City,N.J. 1959 Jun 22-23 San Francisco,Calif. 1959 Oct 16 American society for testing materials.Special technical publication, 254 American society for testing materials,Philadelphia,Pa.,1960 Papers presented at various A.S.T.M. meetings during 1959
Eng 41.3151

SYMPOSIUM ON TIME SERIES ANALYSIS proceedings Providence,R.I. 1962 Jun.11-14 Brown university Edited by Murray Rosenblatt Also sponsored by United States.Office of naval research xiv,497p 25cm John Wiley and sons,New York,1963
Math 3.0695

The SYMPOSIUM ON TIME SERIES ANALYSIS Proceedings Providence,R.I. 1962 Jun 11-14 Brown university United States.Office of naval research Edited by Murray Rosenblatt xiv,497p 26cm Wiley,New York; London,1963
Cav 7.2744

SYMPOSIUM ON TIN New York 1952 Jun 23 American society for testing materials A.S.T.M.Special technical publication, 141 American society for testing materials, Philadelphia,Pa.,1953 Fiftieth anniversary publication. A symposium held at the 55th annual meeting of the American society for testing materials
Met 25.0567

SYMPOSIUM ON TITANIUM :presented at the second Pacific area national meeting Los Angeles,Calif. 1956 Sep 17-18 American society for testing materials American society for testing materials.Special technical publication, 204 American society for testing materials,Philadelphia,Pa.,1957
Eng 41.3598

SYMPOSIUM ON TITANIUM held at the second Pacific area national meeting Los Angeles 1956 Sep 17-18 American society for testing materials A.S.T.M.Special technical publication, 204 American society for testing materials,Philadelphia,Pa.,1957
Met 25.0571

SYMPOSIUM ON TOPOLOGICAL DYNAMICS AND ERGODIC THEORY Proceedings Lexington,Ky. 1971 Jun 3-5 University of Kentucky. Department of mathematics 80p 28cm University of Kentucky,Lexington,Ky.,1971 Duplicated notes
P Math 2.4189

SYMPOSIUM ON TRACE ANALYSIS papers New York 1955 Nov 2-4 Rockefeller foundation and Sloan-Kettering institute Edited by John H. Yoe and Henry J. Koch Wiley;Chapman and Hall,New York;London,1957
Met 25.1720

SYMPOSIUM ON TRANSMETHYLATION AND METHIONINE BIOSYNTHESIS Argonne 1964 Edited by Stanley K. Shapiro and Fritz Schlenk Sponsored by the Argonne national laboratory. Division of biological and medical research Chicago university press,Chicago,1965
Bioch 33.0607

SYMPOSIUM ON TROPICAL METEOROLOGY Proceedings Rotorua,N.Z. 1963 Nov 5-13 World meteorological organization Edited by J.W. Hutchings With the cooperation of the International union of geodesy and geophysics illus 751p New Zealand meteorological service,Wellington,N.Z.,1964
Nap 11.0650

SYMPOSIUM ON TROPICAL METEOROLOGY Proceedings Rotorua,N.Z. 1963 Nov 5-13 World meteorological organization,and,International union of geodesy and geophysics Edited by J. W. Hutchings New Zealand meteorological service,Wellington,N.Z.,1964
Geog 13.0990

SYMPOSIUM ON TURBULENCE OF FLUIDS AND PLASMAS Proceedings New York 1968 Apr 16-18 Polytechnic institute of Brooklyn Edited by Jerome Fox Co-sponsored by the Microwave research institute Microwave research institute.Symposia series, 18 512p 17cm Polytechnic press,Brooklyn,N.Y.,1969
Cav 7.2920

SYMPOSIUM ON TURBULENCE OF FLUIDS AND PLASMAS Proceedings Turbulence of fluids and plasmas New York 1968 Apr 16-18 Edited by Jerome Fox Sponsored by the Microwave research institute Microwave research institute.Symposia series, 18 bibliog., illus. 512p 22cm Polytechnic press, Brooklyn,N.Y.,1969
A Math 4.1650

SYMPOSIUM ON TURBULENCE OF FLUIDS AND PLASMAS Proceedings Turbulence of fluids and plasmas New York 1968 Apr 16-18 Polytechnic institute of Brooklyn Edited by Jerome Fox Co-sponsored by the Microwave research institute Microwave research institute.Symposia series, 18 Bibliog., Illus. xxix,511p 23cm Polytechnic press, Brooklyn N.Y.,1969
A Math 4.1508

SYMPOSIUM ON TWO-PHASE FLUID FLOW Proceedings London 1962 Feb 7 Arranged by the Institution of mechanical engineers 116p Institution of mechanical engineers,London, 1962
Chem E 24.0372

SYMPOSIUM ON TYPHOONS Proceedings Tokyo 1954 Nov 9-12 Japanese national commission for Unesco Science council of Japan Central meteorological observatory,Tokyo illus 257p Japanese national commission for Unesco,Tokyo,1955
Nap 11.0612

A SYMPOSIUM ON URANIUM AND URANIUM DIOXIDE : melting,casting,forging, rolling,coldworking, welding of uranium:preparation and fabrication of uranium dioxide for fuel element use Chicago 1957 Nov 6 American institute of mining,metallurgical and petroleum engineers. Institute of metals division Nuclear metallurgy, 4 I.M.D.Special report series, 4 The metallurgical society of American institute of mining,metallurgical,and petroleum engineers,New York,1957
Met 25.1542

SYMPOSIUM ON VANE SHEAR TESTING OF SOILS Atlantic City,N.J. 1956 Jun 22 American society for testing materials American society for testing materials.Special technical publication, 193 American society for testing materials,Philadelphia,Pa.,1957
Eng 41.3150

SYMPOSIUM ON VARIATION IN MAMMALIAN POPULATIONS Proceedings London 1969 Nov 14-15 Zoological society of London Edited by Robert J. Berry Zoological society of London. Symposia, 26 Academic press,London,1970
Gen 34.2196

SYMPOSIUM ON VIRUS AND CANCER Papers Saltsjobaden 1964 Jun 8-9 Swedish cancer society and Unio nordica contra cancrum Edited by H. Bergstrand and K.E. Hellstrom illus Balder,Stockholm,1965 Symposium arranged by the Swedish cancer society in conjunction with the annual general meeting of the Unio nordica contra cancrum
Path 30.2363

A SYMPOSIUM ON VISCOUS DRAG REDUCTION
Proceedings Viscous drag reduction Dallas,Texas 1968 Sep 24-25 Edited by C. Sinclair Wells Sponsored by United States. Office of naval research 500p 25cm Plenum press,New York,1969
Cav 7.2923

SYMPOSIUM ON VISCOUS DRAG REDUCTION
Proceedings Dallas,Texas 1968 Sep 24-25 United States.Office of naval research.Naval ship research and development center National aeronautics and space administration Edited by C.S. Wells illus xi,500p 25cm Plenum press,New York,1969
A Math 4.1736

SYMPOSIUM ON WHOLE-BODY COUNTING Proceedings
Whole-body counting Vienna 1961 Jun 12-16 International atomic energy agency. Proceedings series International atomic energy agency,Vienna,1962
Radioth 35.1361

SYMPOSIUM ON X-RAY MICROSCOPY AND MICRORADIOGRAPHY Proceedings X-ray microscopy and microradiography Cambridge 1956 Edited by V.E. Coslett and others Sponsored by the International union of pure and applied physics Academic press,New York, 1957
Radioth 35.1600

The SYMPOSIUM ON X-RAY MICROSCOPY AND MICRORADIOGRAPHY Proceedings X-ray microscopy and microradiography Cambridge 1956 Aug 16-25 Cavendish laboratory Edited by V.E. Cosslett and others Sponsored by the International union of pure and applied physics Academic press,New York,1957
Cav 7.0461

SYMPOSIUM SOBRE ALGUNOS PROBLEMAS MATEMATICOS QUE SE ESTAN ESTUDIANDO EN LATINA AMERICA
Punta del Este 1951 Dec 19-21 Universidad de la republica,Montevideo.Instituto de matematica y estadistica Bibliog 183p 24cm UNESCO para America Latina,Montevideo, 1952
P Math 2.3007

Un SYMPOSIUM SOBRE EL CRETACICO EN EL HEMISFERIO OCCIDENTAL Y SU CORRELACIO MUNDIAL papers Sistema cretacico Mexico City 1956 Tom 1-2 Edited by Lewis B. Kellum 2 vols Mexico City,1959 Prepared under the auspices of the International commission on stratigraphy and the International palaeontological union for the 20th International geological congress
Geol 8.3049

SYMPOSIUM SOBRE YACIMIENTOS DE MANGANESO
International geological congress 20th papers Mexico city 1956 Tom 1-5 Edited by Jenaro Gonzalez Reyna 2 vols Mexico city,1956 Text in English,French,German and Spanish
Geol 8.3027

SYMPOSIUM SOBRE YACIMIENTOS DE PETROLEO Y GAS
International geological congress 20th papers Mexico city 1956 Tom 1-5 Edited by Eduardo J. Guzman 2 vols Mexico city, 1956 Text in English,French and Spanish
Geol 8.3026

SYMPOSIUM SUR LA BIOCHEMIE DES STEROIDES
International congress biochemistry 2nd Paris 1952 Jul 21-27 Council for international organizations of medical sciences Societe d'edition d'enseignement superieur,Paris,1952 Text in English and French
Bioch 33.1310

SYMPOSIUM SUR LA BIOCHIMIE DE L'HEMATOPOIESE
International congress of biochemistry 2nd Paris 1952 Jul 21-27 Council for international organizations of medical sciences Societe d'edition d'enseignement superieur,Paris,1952 Text in English and French
Bioch 33.1304

SYMPOSIUM SUR LA BIOGENESE DES PROTEINES
International congress of biochemistry 2nd Paris 1952 Jul 21-27 Council for international organizations of medical sciences Societe d'edition d'enseignement superieur,Paris,1952 Text in English and French
Bioch 33.1305

SYMPOSIUM SUR LA SPECIFICITE PARASITAIRE DES PARASITES DE VERTEBRES 1 er Neuchatel 1957 Universite de Neuchatel.Institut de zoologie Edited by J.G. Baer Paul Attinger, Neuchatel,1957 Title also in English.Text in English and French
Bal 39.1727

SYMPOSIUM SUR LA SPECIFICITE PARASITAIRE DES PARASITES DE VERTEBRES 1st Universite de Neuchatel.Institut de zoologie 24cm Neuchatel,1957
Philos 1.1588

SYMPOSIUM SUR LE CYCLE TRICARBOXYLIQUE
International congress of biochemistry 2nd Paris 1952 Jul 21-27 Council for international organizations of medical sciences Societe d'edition d'enseignement superieur,Paris,1952 Text in English, French and German
Bioch 33.1306

SYMPOSIUM SUR LE METABOLISME MICROBIEN
International congress of biochemistry 2nd Paris 1952 Jul 21-27 Council for international organizations of medical sciences Societe d'edition d'enseignement superieur,Paris,1952 Text in English and French
Bioch 33.1308

SYMPOSIUM SUR LE MODE D'ACTION DES ANTIBIOTIQUES
International congress of biochemistry 2nd Paris 1952 Jul 21-27 Council for international organizations of medical sciences Societe d'edition d'enseignement superieur,Paris,1952 Text in English and French
Bioch 33.1309

SYMPOSIUM SUR LES GISEMENTS DE FER DU MONDE
International geological congress 19th Algiers 1952 Tomes 1-2 Edited by F. Blondel and L. Marvier Algiers,1952
Geol 8.3011

SYMPOSIUM SUR LES HORMONES PROTEIQUES ET DERIVEES DES PROTEINES International congress of biochemistry 2nd Paris 1952 Jul 21-27 Council for international organizations of medical sciences Societe d'edition d'enseignement superieur,Paris,1952 Text in English and French
Bioch 33.1307

SYMPOSIUM SUR LES SERIES DE GONDWANA International geological congress 19th papers Algiers 1952 Edited by Curt Teichert Algiers,1952 Text in English, French and German
Geol 8.3012

SYMPOSIUM UBER DIE MARKIERANG DER PROTEINE MITTELS RADIOAKTIVER ISOTOPEN UND DEREN ANWENDUNG IN BIOLOGIE UND MEDIZIN Bern 1964 Nov 14 Schweizerische akademie der medizinischen wissenschaften.Bulletin, 21, fasc.3-4 Schwabe,Basle,Stuttgart,1965 Parallel text in German,French and Italian
Radioth 35.1187

SYMPOSIUM "VESALE" Brussels 1964 Oct 19-24 cellule,forme et fonction Archives de biologie,Liege,1965 In English and French
An 32.3303

SYNCHRONY IN CELL DIVISION AND GROWTH Edited by Erik Zeuthen Interscience,John Wiley,New York,1964
Bal 39.0338

SYNTEX PHARMACEUTICALS Steroids in modern medicine :symposium Oxford 1971 Sep 3-4 Pt 1-2 Edited by George A. Christie and Miriam Moore-Robinson Excerpta medica, Amsterdam,1972
Bioch 33.2199

SYNTEX PHARMACEUTICALS The Treatment of carcinoma of the breast :a symposium Proceedings Cambridge 1967 Sep 9 Edited by Anthony S. Jarrett Under the auspices of the Institute of hormone biology Lewis, Cambridge,1968
Inv Med 37.0122

SYNTEX PHARMACEUTICALS The Treatment of carcinoma of the breast :a symposium Proceedings Cambridge 1967 Sep 9 Edited by Antony S. Jarrett Excerpta medica foundation,1968
PGMS 29.0460

SYNTEX PHARMACEUTICALS The Treatment of carcinoma of the breat :a symposium Proceedings Cambridge 1967 Sep 7 Edited by Antony S. Jarrett Under the auspices of the Institute of hormone biology Excerpta medica,Amsterdam,1967
Radioth 35.0926

SYNTHESIS AND METABOLISM OF ADRENOCORTICAL STEROIDS :a colloquium Proceedings London 1950 Ciba foundation Edited by W. Klyne Ciba foundation colloquia on endocrinology, 7 Churchill,London,1953
Inv Med 37.0137

SYNTHESIS AND METABOLISM OF ADRENOCORTICAL STEROIDS :a colloquium Proceedings London 1952 Jul 7-10 Ciba foundation Edited by W. Klyne and G.E.W. Wolstenholme Ciba foundation colloquia on endocrinology, 7 illus Churchill,London,1953
Bioch 33.0437

SYNTHESIS AND STRUCTURE OF MACROMOLECULES :a symposium Papers Cold Spring Harbor 1963 Cold Spring Harbor laboratory of quantitative biology Cold Spring Harbor symposia on quantitative biology, 28 Cold Spring Harbor,1963
Bioch 33.1284

SYNTHESIS AND STRUCTURE OF MACROMOLECULES :a symposium Papers Cold Spring Harbor 1963 Jun 7-13 Cold Spring Harbor laboratory of quantitative biology Cold Spring Harbor symposia on quantitative biology, 28 Cold Spring Harbor laboratory of quantitative biology,Cold Spring Harbor,1963
Radioth 35.0245

SYNTHESIS OF MOLECULAR AND CELLULAR STRUCTURE : a symposium Papers Waltham,Mass. 1960 Society for the study of development and growth Edited by Dorothea Rudnick Held at Brandeis university Society for the study of development and growth.Symposia, 19 Ronald press,New York,1961
Bioch 33.1012

SYRACUSE 1961 Celebrazioni Archimedee del secolo xx Vol. 1: conferenze generali e simposio di geometria differenziale Universita di Messina 68,190p 25cm Oderisi,Gubbio,1962
P. Math 2.2608

SYRACUSE 1961 Celebrazioni Archimedee del secolo xx Vol. 2: simposio di analisi Universita di Messina 108p 25cm Oderisi, Gubbio,1962
P. Math 2.2609

SYRACUSE 1965 Small-angle x-ray scattering the conference Proceedings American crystallographic association and Syracuse university Edited by H. Brumberger Co-sponsored by the United States Army.Research office and the National science foundation Gordon and Breach,New York,1967
Met 25.1466

SYRACUSE 1967 Population biology and evolution :an international symposium Proceedings Edited by Richard C. Lewontin Sponsored by Syracuse university,New York state science and technology foundation Syracuse university press,New York,1968
Bal 39.0940

SYRACUSE,N.Y. 1958 Recent progress in the endocrinology of reproduction conference Proceedings Edited by Charles W. Lloyd Academic press,New York;London,1959
VA 19.0282

SYRACUSE,N.Y. 1967 Population biology and evolution symposium Proceedings New York state science and technology foundation Edited by R.C. Lewontin Sponsored by Syracuse university Syracuse university press,Syracuse,N.Y.,1968
Gen 34.1070

SYRACUSE UNIVERSITY Fatigue-an interdisciplinary approach Sagamore army materials research conference 10th Raquette Lake,New York 1963 Aug 13-16 Edited by John J. Burke and others Syracuse university press,Syracuse,1964
Met 25.2330

2195

SYRACUSE UNIVERSITY Fundamentals of deformation processing Sagamore army materials conference 9th Raquette Lake,New York 1962 Aug 28-31 By Walter A. Backofen and others Co-organized and directed by the National research council Syracuse university press,Syracuse,New York,1964
Met 25.0741

SYRACUSE UNIVERSITY Population biology and evolution symposium Proceedings Syracuse,N. Y. 1967 Jun 7-9 New York state science and technology foundation Edited by R.C. Lewontin Syracuse university press,Syracuse, N.Y.,1968
Gen 34.1070

SYRACUSE UNIVERSITY Small-angle x-ray scattering :the conference Proceedings Syracuse 1965 Jun 24-26 Edited by H. Brumberger Co-sponsored by the United States Army.Research office and the National science foundation Gordon and Breach,New York,1967
Met 25.1466

SYRACUSE UNIVERSITY Strengthening mechanisms, metals and ceramics Sagamore army materials research conference 12th Proceedings Raquette Lake,N.Y. 1965 Aug 24-27 Syracuse university press,Syracuse,N.Y.,1966
Met 25.2746

SYRACUSE UNIVERSITY Surfaces and interfaces 1-2 Sagamore army materials research conference 13th-14th Proceedings Raquette Lake,New York 1966-67 Edited by John J. Burke and others Syracuse university press,New York,1967-68
Met 25.2365

SYRACUSE UNIVERSITY Ultrafine-grain metals Sagamore army materials research conference 16th Proceedings Raquette Lake,N.Y. 1969 Aug 19-22 Syracuse university press,Syracuse, N.Y.,1970
Met 25.2747

SYRACUSE UNIVERSITY.PINEBROOK CONFERENCE CENTER 1964 Cellular ultrastructure of woody plants :advanced science seminar Proceedings Edited by Wilfred A. Cote xii,603p Syracuse university press,Syracuse,N.Y.,1965
Bot 42.4732

SYRACUSE UNIVERSITY,New York state science and technology foundation Population biology and evolution :an international symposium Proceedings Syracuse 1967 Jun 7-9 Edited by Richard C. Lewontin Syracuse university press,New York,1968
Bal 39.0940

SYRACUSE UNIVERSITY RESEARCH INSTITUTE Composite materials and composite structures Sagamore ordnance materials research conference 6th Proceedings New York 1959 Aug 8-21 United States.Army.Ordnance materials research office and United States.Army.Office of ordnance research New York,1959 Mimeograph
Met 25.0093

2196

The SYSTEM OF ASTRONOMICAL CONSTANTS :a symposium proceedings Paris 1963 May 27-31 International astronomical union Edited by J. Kovalevsky International astronomical union.Symposium, 21 330p Gauthier-Villars,Paris,1965 From the Bulletin astronomique de l'observatoire de Paris,vol.25,fasc.1-3
Obs 6.0986

SYSTEM PERFORMANCE EVALUATION A.C.M.-SIGOPS workshop on system performance evaluation Cambridge,Mass. 1971 Apr 5-7 Association for computing machinery.Special interest group in operating systems illus. 378p Association for computing machinery,New York, 1971
Math L 5.3564

SYSTEM PROGRAMMING Introduction to system programming symposium proceedings London 1962 July Edited by Peter Wegner A.P.I.C. studies in data processing, 4 x,316p 23cm Academic press,London,1964
Math L 5.0947

SYSTEM SENSITIVITY AND ADAPTIVITY I.F.A.C. symposium on system sensitivity and adaptivity 2nd Preprints Dubrovnik 1968 Aug 26-31 International federation of automatic control Organised by the Yugoslav committee for electronics and automation I.F.A.C., Dubrovnik,1968
Eng 41.5889

SYSTEM UND KLASSIFIKKATION IN WISSENSCHAFT UND DOKUMENTATION :colloquium Vortrage und diskussionen Dusseldorf 1967 Apr 27-29 Edited by A. Diemer Studien zur wissenschaftstheorie, 2 183p Hain, Meisenheim am Glan,1968
WSM 43.0984

SYSTEMATICS British flowering plants and modern systematic methods The Conference on the study of critical British groups Report London 1948 Apr 9-10 Botanical society of the British Isles Edited by A.J. Wilmott Botanical society of the British Isles.B.S.B.I. conference reports, 1 London,1949
Bot 42.4706

SYSTEMATICS ASSOCIATION Aspects of Tethyan biogeography :a symposium Leicester 1966 Sep 21-23 Edited by C.G. Adams and D.V. Ager Systematics association.Publication, 7 illus. vi,336p British museum,London,1967
Bot 42.3273

SYSTEMATICS ASSOCIATION Function and taxonomic importance Systematics association symposium 3rd papers Oxford 1957 Apr 7-9 Edited by A.J. Cain Systematics association.Publication,3 Systematics association,London,1959
Geol 8.0941

SYSTEMATICS ASSOCIATION Function and taxonomic importance :a symposium Oxford 1957 Apr 7-9 By A.J. Cain Systematics association.Publication, 3 Systematics association,London,1959
Gen 34.1623

SYSTEMATICS ASSOCIATION Species concept in paleontology Systematics association symposium 2nd papers London 1954 May 19 Edited by Peter C. Sylvester-Bradley Systematics association. Publication, 2 Systematics association, London, 1956
Geol 8.0940

SYSTEMATICS ASSOCIATION Taxonomy and geography Systematics association symposium 4th Papers Oxford 1959 Mar 31-Apr 2 Edited by David Nichols Systematics association. Publication, 4 Systematics association, London, 1962
Geog 13.1162

SYSTEMATICS ASSOCIATION The Species concept in palaeontology :a symposium Edited by Peter C.S. Bradley Systematics association. Publication, 2 Systematics association, London, 1956
Bot 42.3421

SYSTEMATICS ASSOCIATION, and PALAEONTOLOGICAL ASSOCIATION Aspects of Tethyan biogeography Systematics association symposium 7th papers Leicester 1966 Sep 21-23 Edited by C.G. Adams and D.V. Ager Systematics association. Publication, 7 Systematics association, London, 1967
Geol 8.4466

SYSTEMATICS ASSOCIATION SYMPOSIUM 2nd papers Species concept in paleontology London 1954 May 19 Systematics association Edited by Peter C. Sylvester-Bradley Systematics association. Publication, 2 Systematics association, London, 1956
Geol 8.0940

SYSTEMATICS ASSOCIATION SYMPOSIUM 3rd papers Function and taxonomic importance Oxford 1957 Apr 7-9 Systematics association Edited by A.J. Cain Systematics association. Publication, 3 Systematics association, London, 1959
Geol 8.0941

SYSTEMATICS ASSOCIATION SYMPOSIUM 4th Papers Taxonomy and geography Oxford 1959 Mar 31-Apr 2 Systematics association Edited by David Nichols Systematics association. Publication, 4 Systematics association, London, 1962
Geog 13.1162

SYSTEMATICS ASSOCIATION SYMPOSIUM 4th papers Taxonomy and geography Oxford 1959 Mar 31-Apr 2 Edited by David Nichols Systematics association. Publication, 4 Systematics association, London, 1962
Geol 8.0942

SYSTEMATICS ASSOCIATION SYMPOSIUM 7th papers Aspects of Tethyan biogeography Leicester 1966 Sep 21-23 Systematics association, and, Palaeontological association Edited by C.G. Adams and D.V. Ager Systematics association. Publication, 7 Systematics association, London, 1967
Geol 8.4466

SYSTEMATICS OF TO-DAY :a symposium...in commemoration of the 250th anniversary of the birth of Carolus Linnaeus Uppsala 1957 May Edited by Olov Hedberg Uppsala universitets arsskrift, 1958:6 Uppsala, 1958
BG 38.3195

SYSTEMATICS OF TODAY :a symposium Proceedings Uppsala 1957 May 28-29 Universitet i Uppsala Edited by Olov Hedberg Uppsala universitets. Arsskrift, 6 Lundequistska bokhandeln; Otto Harrassowitz, Uppsala; Wiesbaden, 1959 In commemoration of the 250th anniversary of the birth of Carolus Linnaeus
Bot 42.3424

SYSTEMIC MYCOSES :a Ciba foundation symposium Ibadan 1967 Mar 29-31 Ciba foundation Edited by G.E.W. Wolstenholme and Ruth Porter Churchill, London, 1968
Phys 20.0998

SYSTEMIC MYCOSES :a Ciba foundation symposium in commemoration of William Balfour Baikie Ibadan 1967 Mar 29-31 Ciba foundation Edited by G.E.W. Wolstenholme and Ruth Porter illus, port Churchill, London, 1968
Path 30.2632

SYSTEMS Seminar in systems Proceedings Edited by G.D.S. MacLellan and Samuel Mercer Held at Michigan state university. College of engineering Michigan state university, East Lansing, Mich., 1958
Eng 41.6009

SYSTEMS AND COMPUTER SCIENCE Conference on systems and computer science :held at the University of Western Ontario Proceedings London, Ont. 1965 Sep 10-11 Edited by John F. Hart and Satoru Takasu Bibliog., Illus. xiv, 249p 25cm Toronto university press, Toronto, 1967
P Math 2.3494

SYSTEMS SYMPOSIUM 2nd Proceedings Views on general systems theory Cleveland, Ohio 1963 Apr Edited by Mihajlo D. Mesarovic Held at the Case institute of technology Case institute of technology. Systems research center. Publications Wiley, New York, 1964
Eng 41.5967

SYSTEMS SYMPOSIUM 3rd Proceedings Systems theory and biology Cleveland, Ohio 1966 Oct Edited by M.D. Mesarovic Sponsored by the Case institute of technology. Systems research center Springer, Berlin, 1968
Gen 34.0226

SYSTEMS SYMPOSIUM 4th Proceedings Theoretical approaches to non-numerical problem solving Cleveland, Ohio 1968 Nov 19-20 Edited by R.B. Banerji and M.D. Mesarovic Sponsored by Case western reserve university. Systems research center Lecture notes in operations research and mathematical systems, 28 bibliog., illus. vi, 466p 25cm Springer, Berlin, 1970
Math S 3.1617

SYSTEMS THEORY AND BIOLOGY Systems symposium 3rd Proceedings Cleveland, Ohio 1966 Oct Edited by M.D. Mesarovic Sponsored by the Case institute of technology. Systems research center Springer, Berlin, 1968
Gen 34.0226

SYVERTON MEMORIAL SYMPOSIUM Proceedings Analytic cell culture Detroit 1961 Jun 6-7 National cancer institute Edited by Robert E. Stevenson Sponsored by Tissue culture association and Cell culture collection committee of the viruses and cancer board, National cancer institute National cancer institute.Monograph, 7 U.S.Department of health,education and welfare,Bethesda,Md.,1962
Bioch 33.0979

SZINTIGRAPHIE UND RADIOKARDIOGRAPHIE : symposion Basel 1962 Sep 28-30 Schweizerische akademie der medizinischen wissenschaften Schwabe,Basel;Stuttgart,1963
Radioth 35.1371

TABLAS PARA EL CALCULA DE CONJUNCIONES GEOCENTRICAS DE LOS SATELITES DE JUPITER By S.G. Francos and P.Ch. Justo 27p San Fernando,1908
Obs 6.0617

TABLEAU SYNOPTIQUE DES DISPOSITIONS ADOPTEES ET RESERVEES Christiana 1914 Jun 16-Jul 30 Conference internationale du Spitsbergen 35p 33cm Grondahl,Christiana,1914
Sco 14.2859

TAGUNGSBERICHT UND WISSENSCHAFTLICHE ABHANDLUNGEN Papers Cologne 1961 May 22-26 Deutscher geographentag,Cologne Edited by Wolfgang Hartke and Friedrich Wilhelm Deutscher geographentag.Verhandlungen,33 illus,maps,15 plates 407p 26cm Franz Steiner,Wiesbaden,1962
Sco 14.2860

TAIPEI 1967 Recent developments in biochemistry colloquium Proceedings Sino-American cooperation committee Edited by Hsien-Wen Li and others Under the auspices of the Academia Sinica Academia Sinica,n.p., 1968
Gen 34.0619

TAIPEI 1967 Rice breeding with induced mutations :an FAO-IAEA research co-ordination meeting on the use of induced mutations in rice breeding Report Food and agriculture organization International atomic energy agency International atomic energy agency. Technical report series, 86 International atomic energy agency,Vienna,1968
Radioth 35.0271

TALL BUILDINGS WITH PARTICULAR REFERENCE TO SHEER WALL STRUCTURES :a symposium Proceedings Southampton 1966 Apr Edited by A. Coull and B.Stafford Smith Pergamon press,London, 1967
Eng 41.2733

TALLINN 1966 Noctilucent clouds : international symposium Special committee for the International quiet sun year and International association of meteorology and atmospheric physics Edited by I.A. Khvostikov and G. Witt Co-sponsored by the World meteorological organization illus 235p 27cm Academy of sciences of the USSR. Soviet geophysical committee,Moscow,1967 Symposium papers in English with Russian abstracts
Sco 14.8230

TAMABRA LIMESTONE OF THE POZA RICA OILFIELD International geological congress 20th papers Mexico city 1956 By A. Barnetche and L.V. Illing Mexico city,1956 Bound with two other publications of the congress
Geol 8.3028

TANANARIVE 1957 C.C.T.A.East-central and southern regional committee for geology : meeting 2nd proceedings Commission for technical co-operation in Africa south of the Sahara London,1958 Bound with proceedings of three other regional meetings
Geol 8.3117

TARDA,POLAND 1967 Fluid dynamics transations,4 Symposium on advanced problems and methods in fluid dynamics 8th Proceedings Polska akademia nauk Edited by W. Fiszdon and others bibliog.,illus. x, 812p 24cm PWN,Warsaw,1969
A Math 4.1680

TARRAGONA 1968 Present status of psychotropic drugs,pharmacological and clinical aspects International congress of the Collegium internationale neuro-psychopharmacologicum 6th Proceedings Collegium internationale neuro-psychopharmacologicum Edited by A. Cerletti and F.J. Bove Excerpta medica foundation. International congress series, 180 Excerpta medica foundation,Amsterdam,1969
Psy 31.2902

TATA INSTITUTE OF FUNDAMENTAL RESEARCH Contributions to function theory international colloquium papers presented Bombay 1960 Jan 12-19 In joint sponsorship with the Sir Dorabji Tata trust xii,231p 24cm Tata institute of fundamental research,Bombay,1960
P. Math 2.1584

TATA INSTITUTE OF FUNDAMENTAL RESEARCH Differential analysis International colloquium on differential analysis :the Bombay colloquium papers presented Bombay 1964 Jan 7-14 By M.F. Atiyah and others Jointly sponsored by the International mathematical union Tata institute of fundamental research.Studies in analysis, 2 viii,253p 25cm Oxford university press, Oxford,1964
P. Math 2.0696

TATRANSKA LOMNICA 1967 Physics and dynamics of meteors International astronomical union Edited by Lubor Kresak and Peter M. Millman International astronomical union.Symposium, 33 Reidel, Dordrecht,1968
TA 15.0422

TATRANSKA LOMNICA 1967 Physics and dynamics of meteors :a symposium Proceedings International astronomical union Edited by Lubor Kresak and Peter M. Millman International astronomical union.Symposium, 33 525p Reidel,Dordrecht,1968
Obs 6.3264

TATRANSKA LOMNICA 1967 Planetary nebulae : a symposium International astronomical union Edited by D.E. Osterbrock and C.R. O'Dell International astronomical union.Symposium, 34 Reidel,Dordrecht,1968
TA 15.0320

TATRANSKA LOMNICA 1967 Planetary nebulae : a symposium Proceedings International astronomical union Edited by D.E. Osterbrock and C.R. O'Dell International astronomical union.Symposium, 34 469p Reidel,Dordrecht, 1968
Obs 6.3265

TAVISTOCK CLINIC.CHILD DEVELOPMENT RESEARCH UNIT Determinants of infant behaviour : proceedings of a Tavistock study group on mother-infant interaction London 1959 Sep Edited by B.M. Foss Methuen;Wiley,London;New York,1961
Psy 31.1203

TAVISTOCK CLINIC.CHILD DEVELOPMENT RESEARCH UNIT Determinants of infant behaviour 2 Tavistock study group on mother-infant interaction 2nd Proceedings London 1961 Sep Edited by B.M. Foss Methuen;Wiley, London;New York,1963
Psy 31.1208

TAVISTOCK CLINIC.CHILD DEVELOPMENT RESEARCH UNIT Determinants of infant behaviour 3 Tavistock study group on mother-infant interaction 3rd Proceedings London 1963 Sep Edited by B.M. Foss Methuen;Wiley, London;New York,1965
Psy 31.1205

TAVISTOCK INSTITUTE OF HUMAN RELATIONS Exploration in group relations :a conference Report Leicester 1957 Sep By E.L. Trist and C. Sofer Leicester university press, Leicester,1959
Eng 41.0706

TAVISTOCK STUDY GROUP ON MOTHER-INFANT INTERACTION 2nd Proceedings Determinants of infant behaviour 2 London 1961 Sep Tavistock clinic.Child development research unit Edited by B.M. Foss Methuen; Wiley,London;New York,1963
Psy 31.1208

TAVISTOCK STUDY GROUP ON MOTHER-INFANT INTERACTION 3rd Proceedings Determinants of infant behaviour 3 London 1963 Sep Tavistock clinic.Child development research unit Edited by B.M. Foss Methuen; Wiley,London;New York,1965
Psy 31.1205

TAXONOMIC BIOCHEMISTRY International conference on taxonomic biochemistry, physiology and serology Lawrence,Kan. 1962 Sep Edited by Charles A. Leone xi,728p Ronald press,New York,1964
Bot 42.3437

TAXONOMY Function and taxonomic importance Systematics association symposium 3rd papers Oxford 1957 Apr 7-9 Systematics association Edited by A.J. Cain Systematics association.Publication,3 Systematics association,London,1959
Geol 8.0941

TAXONOMY Numerical taxonomy :colloquium Proceedings St.Andrews 1968 Sep University of St.Andrews Edited by Alfred J. Cole Academic press,London,1969
Bot 42.3452

TAXONOMY Symposium on newer trends in taxonomy Proceedings New Delhi 1966 Jan 28-30 National institute of sciences of India National institute of sciences of India.Bulletin, 34 New Delhi,1967
Bal 39.2079

TAXONOMY AND GEOGRAPHY Systematics association symposium 4th Papers Oxford 1959 Mar 31-Apr 2 Systematics association Edited by David Nichols Systematics association.Publication,4 Systematics association,London,1962
Geog 13.1162

TAXONOMY AND GEOGRAPHY Systematics association symposium 4th papers Oxford 1959 Mar 31-Apr 2 Edited by David Nichols Systematics association.Publication,4 Systematics association,London,1962
Geol 8.0942

TEACHING AND RESEARCH IN HUMAN BIOLOGY :a symposium London 1964 Apr 25 Edited by G.Ainsworth Harrison Society for the study of human biology.Symposia, 6 Pergamon press,Oxford,1964
An 32.2884

TEACHING COMPUTING Conference on teaching computing Proceedings Bristol 1972 Mar 28 University grants committee.Mathematical sciences sub-committee illus 165p University grants committee,London,1972
Math L 5.3878

TEACHING OF ANATOMY AND ANTHROPOLOGY IN MEDICAL EDUCATION Association of American medical colleges teaching institute 3rd Report Swampscott,Mass. 1955 Oct 18-22 Edited by Lura Street Jackson Association of American medical colleges,Chicago,1955
An 32.0994

The TEACHING OF AUTOMATIC CONTROL :a discussion Proceedings London 1962 Jan 29 Institution of mechanical engineers Institution of mechanical engineers,London, 1962
Eng 41.5814

The TEACHING OF ECOLOGY :a symposium London 1966 Apr 13-16 British ecological society Edited by J.M. Lambert British ecological society.Symposia, 7 illus. xi, 294p Blackwell scientific,Oxford,1967
Bot 42.1988

TEACHING OF ENGINEERING DESIGN Conference on the teaching of engineering design Papers and summary Scarborough 1964 Apr 1-4 Organised by the Enfield college of technology Illus Institution of engineering designers, London,1964 Co-organized by the Institution of engineering designers and the Hornsey College of art
Eng 41.0309

TEACHING OF MATHEMATICS Discussion on the teaching of mathematics... to which is added a report of the British association committee drawn up by the chairman,professor Forsyth Glasgow 1901 Sep 14 British association for the advancement of science Edited by John Perry 2nd edition Macmillan,London; New York,1902
Psy 31.3384

TEACHING OF PATHOLOGY,MICROBIOLOGY,IMMUNOLOGY, GENETICS Association of American medical colleges teaching institute 2nd Report French Lick,Ind. 1954 Oct 10-15 Edited by Lura Street Jackson Association of American medical colleges,Chicago,1955
An 32.0993

TEACHING OF PROBABILITY AND STATISTICS Comprehensive school mathematics program international conference 1st Proceedings Carbondale,Ill. 1969 Mar 18-27 Southern Illinois university Central Midwestern regional educational laboratory Edited by Lennart Rade Wiley;Almqvist and Wiksell,New York;Stockholm,1970
An 32.5410

TEACHING OF PROGRAMMING AT UNIVERSITY LEVEL Joint IBM-University of Newcastle upon Tyne seminar 3rd Proceedings Cambridge 1970 Sep 8-11 University of Newcastle upon Tyne.Computing laboratory International business machines corporation Edited by B. Shaw illus. 153p University of Newcastle, Newcastle,1971
Math L 5.3590

TECHNICAL ASPECTS OF EMULSIONS Emulsion technology,theoretical and applied :including the Symposium on technical aspects of emulsions 2nd edition expanded xiii,360p Chemical publishing company,Brooklyn,N.Y.,1946
Chem E 24.0910

TECHNICAL CONFERENCE Response of metals to high velocity deformation Estes Park,Col. 1960 Jul 11-12 American institute of mining, metallurgical and petroleum engineers Metallurgical society conferences, 9 Interscience,New York,1961
Met 25.0905

TECHNICAL CONFERENCE 1st Proceedings Flat rolled products:rolling and treatment Chicago 1959 Jan 21 American institute of mining,metallurgical and petroleum engineers. Mechanical working committee Edited by T.E. Dancy and R.L. Robinson Metallurgical society conferences, 1 Interscience,New York;London,1959
Met 25.0731

TECHNICAL CONFERENCE 3rd Proceedings Bar and allied products Pittsburgh 1961 Jan 18 American institute of mining, metallurgical and petroleum engineers. Mechanical working committee Metallurgical society conferences, 13 Interscience,New York;London,1961
Met 25.0729

TECHNICAL CONFERENCE 5th Proceedings Mechanical working of steel 1 Pittsburgh 1963 Jan 15-16 American institute of mining, metallurgical and petroleum engineers. Mechanical working committee Metallurgical society conferences, 21 Gordon and Breach, New York,1964
Met 25.0728

The TECHNICAL CONFERENCE 6th Proceedings Mechanical working of steel 2 Chicago,Ill. 1964 Jan 30-31 American institute of mining, metallurgical and petroleum engineers. Mechanical working and steel processing committee Edited by T.G. Bradbury Sponsored by the Chicago section Metallurgical society conferences, 26 Gordon and Breach,New York,1965
Met 25.2244

TECHNICAL CONFERENCE ON REFRACTORY METALS AND ALLOYS 3rd Proceedings Los Angeles 1963 Dec 9-10 American institute of mining, metallurgical and petroleum engineers. Refractory metals committee Metallurgical society conferences, 30 Gordon and Breach, New York,1966
Met 25.1017

The TECHNIQUE AND SIGNIFICANCE OF OESTROGEN :a conference proceedings London 1954 Feb 17 Society for endocrinology Royal society of medicine.Endocrinological section Edited by P. Eckstein and S. Zuckerman Society for endocrinology.Memoirs, 3 Cambridge university press,Cambridge,1955
Radioth 35.0045

TECHNIQUE EMPLOYED IN PASTURE STUDIES,BOTANICAL ANALYSES,ETC :symposium Technique of grassland experimentation in Scandinavia and Finland By G. Giobel Imperial bureau of pastures and forage crops,Herbage publication series.Bulletin, 28 Aberystwyth,1940
Bot 42.1325

TECHNIQUE OF GRASSLAND EXPERIMENTATION IN SCANDINAVIA AND FINLAND Technique employed in pasture studies,botanical analyses, etc :symposium By G. Giobel Imperial bureau of pastures and forage crops,Herbage publication series.Bulletin, 28 Aberystwyth, 1940
Bot 42.1325

TECHNIQUES FOR ELECTRON METALLOGRAPHY Symposium on techniques for electron metalography Atlantic City,N.J. 1953 Jun 30 American society for testing materials A.S.T.M.Special technical publication, 153 American society for testing materials, Philadelphia,Pa.,1954
Met 25.1361

TECHNIQUES FOR LIQUID SCINTILLATION COUNTING The Beckman summer school 2nd Proceedings 1967 Beckman instruments Glenrothes,c1967
Bioch 33.2301

TECHNIQUES FOR LIQUID SCINTILLATION COUNTING : 1967 summer school Proceedings Beckman instruments limited Beckman instruments limited,Glenrothes,n.d.
Radioth 35.1779

TECHNIQUES IN ELECTRON METALLOGRAPHY 1963 Symposium on advances in techniques in electron metallography New York 1962 Jun 26 American society for testing materials. Subcommittee 11 on electron microstructure of metals A.S.T.M.Special technical publication, 339 American society for testing and materials,Philadelphia,Pa.,1963
Met 25.1378

TECHNIQUES IN ENDOCRINE RESEARCH :a NATO advanced study institute Proceedings Stratford-upon-Avon 1962 Sep North Atlantic treaty organization Edited by Peter Eckstein and Francis Knowles Under the auspices of the University of Birmingham Academic press,London,1963
Gen 34.1964

TECHNIQUES IN ENDOCRINE RESEARCH :a NATO advanced study institute Proceedings Stratford-upon-Avon 1962 Sep North Atlantic treaty organization Edited by Peter Eckstein and Francis Knowles Under the auspices of the University of Birmingham Academic press,London,1963
An 32.3767

TECHNIQUES IN ENDOCRINE RESEARCH :a Nato advanced study institute Proceedings Stratford-upon-Avon 1962 Sep North Atlantic treaty organization Edited by P. Eckstein and F. Knowles Under the auspices of the University of Birmingham Academic press,New York,1963
Phys 20.1399

TECHNIQUES OF ELECTRON MICROSCOPY,DIFFRACTION AND MICROPROBE ANALYSIS Papers Atlantic City 1963 Jun 26 American society for testing materials A.S.T.M.Special technical publication, 372 Philadelphia,1964
Met 25.2495

Les TECHNIQUES RECENTES EN MICROSCOPIE ELECTRONIQUE ET CORPUSCULAIRE :colloque international Toulouse 1955 Apr 4-8 Centre national de la recherche scientifique Edited by Ch. Fert Centre national de la recherche scientifique.Colloques internationaux, 58 Paris,1956
Cav 7.0525

TECHNISCHE HOCHSCHULE ILMENAU.MATHEMATISCHE INSTITUT Beitrage zur graphentheorie : internationale kolloquium Vortragen Manebach 1967 May 9-12 Edited by Horst Sachs and others Bibliog.,Illus. 394p 23cm Teubner,Leipzig,1968
P Math 2.3352

TECHNOLOGICAL PLANNING ON THE CORPORATE LEVEL a conference Proceedings Boston,Mass. 1961 Sep 8-9 Harvard business school Edited by James R. Bright Harvard university, Boston,1962
Met 25.2169

TECHNOLOGY,ENGINEERING AND SAFETY International conference on the peaceful uses of atomic energy 2nd Edited proceedings Geneva 1958 Vol 2 United nations Edited by R. Hurst and others Progress in nuclear energy, 4 Pergamon press,Oxford, 1960
Met 25.1555

TECHNOLOGY AND ENGINEERING Geneva 1955 Vol 1: reactor coolants,moderators,heat transfer,reactor chemistry and corrosion of reactor materials Edited by R. Hurst and S. McLain Progress in nuclear energy, 4 Pergamon,London,1956 Includes papers presented at the United Nations conference on the peaceful uses of atomic energy
Met 25.1565

TECHNOLOGY OF COLUMBIUM (NIOBIUM) :a symposium Washington,D.C. 1958 May 15-16 Electrochemical society.Electrothermics and metallurgy division Wiley;Chapman and Hall,London,1958
Met 25.0587

The TECHNOLOGY OF ENGINEERING MANUFACTURE a conference Proceedings London 1958 Mar 25-27 Institution of mechanical engineers Institution of mechanical engineers,London,1958
Met 25.0676

The TECHNOLOGY OF MOLYBDENUM AND ITS ALLOYS :a symposium Proceedings Metal molybdenum Washington,D.C. 1956 Sep 18-19 United States.Office of naval research Edited by Julius J. Harwood American society for metals,Cleveland,Ohio,1958
Met 25.0546

TECHNOLOGY OF THE GAS-COOLED POWER REACTOR AND RELATED SUBJECTS United Nations conference on the peaceful uses of atomic energy 2nd Papers United Kingdom atomic energy authority United Kingdom atomic energy authority,Harwell,1958
Eng 41.7494

TECTONICA DE LA SIERRA MADRE ORIENTAL DE MEXICO, ENTRE TORREON Y MONTERREY Mexico City 1956 By Zoltan de Cserna Instituto nacional para la investigacion de recursos minerales de Mexico Mexico City,1956 Bound with two other publications of the congress
Geol 8.3040

TECTONICS International geological congress 20th papers Mexico City 1956 Seccion 5: relaciones entre la tectonica y la sedimentacion,tom.1-2 Edited by A. Garcia Rojas Mexico City,1957 Text in English, French,German,Russian and Spanish
Geol 8.3046

TECTONICS Stroenie i razvitie zemnoii kory conference proceedings Moscow 1963 Edited by P.N. Kropokkim Akademiya nauk SSSR, Mowcow,1964 Includes English title-page, Structure and evolution of the crust. (Proceedings of a conference on the problems of tectonics in Moscow)
Geol 8.1367

TECTONICS Voprosy sravnitelnoi tektoniki drevnikh platform conference proceedings Moscow 1963 Edited by A.A. Bogdanov and others Akademiya nauk SSSR,Moscow,1964 Includes English title-page:Problems of comparative tectonics of old platforms (Proceedings of the conference on the problems of tectonics in Moscow)
Geol 8.1366

TEDDINGTON 1953 Engineering dimensional metrology :a symposium Proceedings Vol 1-2 National physical laboratory 2 vols H.M.S.O.,London,1955
Eng 41.3927

TEDDINGTON 1953 Engineering dimensional metrology:a symposium proceedings Vols 1-2 National physical laboratory 26cm London,1955
Philos 1.0402

TEDDINGTON 1953 International symposium on automatic digital computation 3rd proceedings National physical laboratory illus vi,296p 26cm National physical laboratory,Teddington,1954
ATH L 5.0724

TEDDINGTON 1954 Creep and fracture of metals at high temperature :a symposium Proceedings National physical laboratory National physical laboratory,Teddington,Middx, 1955
Met 25.0949

TEDDINGTON 1955 Boundary layer effects in aerodynamics :a symposium 5th Proceedings National physical laboratory illus,diagrms H.M.S.O.,London,1955 Mimeograph
Chem E 24.0229

TEDDINGTON 1955 Boundary layer effects in aerodynamics: a symposium proceedings National physical laboratory. 28cm London, 1955
Philos 1.0437

TEDDINGTON 1957 Visual problems of colour : a symposium 6 Proceedings Vol 1-2 National physical laboratory 2 vols H.M.S. O.,London,1958
Psy 31.0353

TEDDINGTON 1957 Visual problems of colour : a symposium Vol 1-2 National physical laboratory National physical laboratory. Symposia, 8 2 vols H.M.S.O.,London,1958
Phys 20.1033

TEDDINGTON 1958 Mechanisation of thought processes :a symposium Proceedings Vol 1-2 National physical laboratory National physical laboratory.Symposia, 10 2 vols H. M.S.O.,London,1959
Eng 41.6010

TEDDINGTON 1958 Mechanisation of thought processes symposium proceedings National physical laboratory National physical laboratory.Proceedings of symposia, 10 23cm 2 vols H.M.S.O.,London,1959
Math L 5.0820

TEDDINGTON 1958 Mechanization of thought processes National physical laboratory symposium 10th Proceedings Vol 1-2 National physical laboratory 2 vols H.M.S. O.,London,1959
Psy 31.2019

TEDDINGTON 1961 International conference on machine translation of languages and applied language analysis proceedings Vol. 1-2 National physical laboratory National physical laboratory.Proceedings of symposia, 13 23cm 2 vols H.M.S.O., London,1962
Math L 5.0960

TEDDINGTON 1961 Sea-going qualities of ships :a seminar National physical laboratory H.M.S.O.,London,1963
Eng 41.6932

TEDDINGTON 1961 The Control of noise :a conference Proceedings National physical laboratory National physical laboratory. Symposia, 12 H.M.S.O.,London,1962
Eng 41.4218

TEDDINGTON 1962 The Accuracy of industrial measurement of length and diameter :conference Proceedings National physical laboratory National physical laboratory.Conference, 14 H.M.S.O.,London,1963
Eng 41.3933

TEDDINGTON 1963 Relation between the structure and mechanical properties of metals National physical laboratory symposium 15th Proceedings Vol 1-2 National physical laboratory 2 vols H.M.S.O.,London,1963
Met 25.1214

TEDDINGTON 1963 The Relation between the structure and mechanical properties of metals : a conference Proceedings Vol 1-2 National physical laboratory.Symposium, 15 2 vols H.M.S.O.,London,1963
Eng 41.3530

TEDDINGTON 1963 Wind effects on buildings and structures :a conference Proceedings Vol 1-2 Great Britain.Department of scientific and industrial research Great Britain.Department of scientific and industrial research.Symposium, 16 2 vols H.M.S.O.,London,1965
Eng 41.2707

TEDDINGTON 1963 Wind effects on buildings and structures :a conference Proceedings Vol 1-2 Great Britain.Department of scientific and industrial research National physical laboratory National physical laboratory.Symposium, 16 H.M.S.O.,London, 1963
Eng 41.6904

TEDDINGTON 1965 I.F.A.C.symposium on the theory of self-adaptive control systems 2nd Proceedings International federation of automatic control Edited by P.H. Hammond Plenum press,New York,1966
Eng 41.5888

TEDDINGTON 1967 Attention in neurophysiology :an international conference Proceedings National physical laboratory Edited by C.R. Evans and T.B. Mulholland port Butterworths,London,1970
Psy 31.2849

TEDDINGTON 1968 Conference on pattern recognition Organised by Institution of electrical engineers.Control and automation division Institution of electrical engineers. Conference publication, 42 Institution of electrical engineers,London,1968
Eng 41.5311

TEDDINGTON 1971 Numerical methods for unconstrained optimization Joint institute of mathematics and its application-National physical laboratory conference Proceedings National physical laboratory Institute of mathematics and its applications Edited by W. Murray xi,144p 24cm Academic press, London;New York,1972
Math S 3.1729

TEDDINGTON 1972 Conference on machine perception of patterns and pictures Proceedings Institute of physics National physical laboratory Institute of electrical engineers Institute of physics.Conference series, 13 illus ix,362p Institute of physics,London,1972
Math L 5.3882

TEICHMULLER SPACES Romanian-Finnish seminar on Teichmuller spaces and quasiformal mappings Proceedings Brasov 1969 Edited by Cabiria Andreian Cazacu 307p 21cm Publishing house of the Academy of the socialist republic of Romania, Bucharest, 1971
P Math 2.4355

TEKNILLISTEN TIETEIDEN AKATEMIA Current corrosion research in Scandinavia Scandinavian corrosion congress (NKM) : lectures 4th Helsinki 1964 Nov 24-27 Edited by Jori Larinkari and others Organized by the Scandinavian council for applied research Kemian keskusliton julkaisuja, 24 Sanoma Osakeyhtio, Helsinki, 1965
Met 25.1901

TEKNISK UKEBLAD, NORWAY Concrete shell roof construction :symposium 2nd Proceedings Oslo 1957 Jul 1-3 Teknisk ukeblad, Oslo, 1958
Eng 41.3009

TEL-AVIV 1970 International horticultural congress 18th Proceedings International society for horticultural sciences 1970
BG 38.3135

TELECOMMUNICATIONS Soft magnetic materials for telecommunications :a symposium London 1952 Apr Post Office. Engineering research station Edited by C.E. Richards and A.C. Lynch Pergamon press, London, 1953
Eng 41.5085

TELECOMMUNICATIONS Solid state physics in electronics and telecommunications : international conference Proceedings Brussels 1958 Jun 2-7 Vol 3, pt 1: magnetic and optical properties International union of pure and applied physics Edited by M. Desirant and J.L. Michiels Academic press, London, 1960
Cav 7.0387

TELECOMMUNICATIONS Solid state physics in electronics and telecommunications : international conference Proceedings Brussels 1958 Jun 2-7 Vol 1-2, pt 1-2: semiconductors International union of pure and applied physics Edited by M. Desirant and J.L. Michiels 2 vols Academic press, London, 1960
Cav 7.0388

TELEMETER MAGNETICS A High speed magnetic core storage buffer By Ben T. Goda and Waclaw Pryciak illus. 12p Telemeter magnetics, Los Angeles, Calif., 1959 Given by invitation at the Cleveland electronics conference, 1959
Math L 5.1172

TELEOLOGICAL MECHANISMS :a conference New York 1948 Oct 13 By Lawrence K. Frank and others New York academy of sciences New York academy of sciences. Annals, 50, p. 187-278 New York, 1948
Bal 39.1003

TELESCOPES Auxiliary instrumentation for large telescopes :a conference Geneva 1972 May 2-5 European southern observatory CERN Edited by S. Lanstsen and A. Reiz 525p European southern observatory, Geneva, 1972
Obs 6.3599

TELESCOPES Large telescope design :a conference Proceedings Geneva 1971 Mar 1-5 European southern observatory CERN Edited by R.M. West 499p European southern observatory, Geneva, 1971
Obs 6.3601

TELESCOPES The Construction of large telescopes :a symposium International astronomical union Edited by D.L. Crawford International astronomical union. Symposium, 27 Academic press, New York, 1966
TA 15.0311

TELESCOPES The Construction of large telescopes :a symposium Tucson, Ariz. 1965 Apr 5-12 Pasadena, Calif. International astronomical union Edited by D.L. Crawford International astronomical union. Symposium, 27 234p Academic press, New York; London, 1966
Obs 6.3203

TEMPE, ARIZ. 1961 Researches on meteorites : a symposian Proceedings University of Arizona 227p John Wiley, New York; London, 1962
Obs 6.1243

TEMPER EMBRITTLEMENT IN STEEL a symposium Philadelphia 1967 Oct 3-4 American society for testing materials. Committee A-I on steel A.S.T.M. Special technical publication, 407 A.S.T.M., Philadelphia, 1968
Met 25.0455

TEMPERATURE Temperature: its measurement and control in science and industry :a symposium 4th Proceedings Columbus, Ohio 1961 Mar 27-31 Vol 3, pt 1: basic concepts, standards and methods American institute of physics and Instrument society of America Edited by C.M. Herzfeld and Ferdinand Graft Brickwedde Co-sponsored by National bureau of standards Reinhold; Chapman and Hall, New York; London, 1962
Cav 7.0048

TEMPERATURE Temperature: its measurement and control in science and industry :a symposium 4th Proceedings Columbus, Ohio 1961 Mar 27-31 Vol 3, pt2: applied methods and instruments American institute of physics and Instrument society of America Edited by A.I. Dahl and C.M. Herzfeld Co-sponsored by National bureau of standards 1094p Reinhold publishing, New York, 1962
Cav 7.0047

TEMPERATURE Temperature: its measurement and control in science and industry :a symposium New York 1939 Nov 2-4 American institute of physics and National bureau of standards Edited by C.O. Fairchild 1362p Reinhold publishing, New York, 1941
Cav 7.0046

TEMPERATURE Temperature, its measurement and control in science and industry :a symposium Records New York City 1939 Nov 2-4 National bureau of standards Edited by C.O. Fairchild Under the auspices of the American institute of physics xiii, 1362p 23cm Reinhold publishing, New York, 1941
Sco 14.0234

TEMPERATURE Temperature,its measurement and control in science and industry :a symposium 4th Proceedings Columbus,Ohio 1961 Mar 27-31 Vol 3,1: basic concepts,standards and methods American institute of physics,and, Instrument society of America Edited by Ferdinand Graft Brickwedde With the cooperation of the National bureau of standards xvi,848p Reinhold,New York,1962 General editor:Herzfeld,Charles
Sco 14.0231

TEMPERATURE Temperature,its measurement and control in science and industry :a symposium 4th Proceedings Columbus,Ohio 1961 Mar 27-31 Vol 3,2: applied methods and instruments American institute of physics, and,Instrument society of America Edited by A.I. Dahl Co-sponsored by the National bureau of standards xiv,1094p Reinhold,New York,1962 General editor:Herzfeld,Charles
Sco 14.0233

TEMPERATURE Temperature,its measurement and control in science and industry :a symposium 4th Proceedings Columbus,Ohio 1961 Mar 27-31 Vol 3,3: Biology and medicine American institute of physics,and,Instrument society of America Edited by James D. Hardy With the cooperation of the National bureau of standards xii,683p Reinhold,New York,1962 General editor:Herzfeld,Charles
Sco 14.0232

TEMPERATURE :a symposium :its measurement and control in science and industry New York 1939 Nov 2-4 Under the auspices of the American institute of physics Reinhold, New York,1941
Eng 41.4125

TEMPERATURE :symposium :its measurement and control in science and industry New York 1939 Nov Under the auspices of American institute of physics Reinhold,New York,1941
Min 10.0599

TEMPERATURE:ITS MEASUREMENT AND CONTROL IN SCIENCE AND INDUSTRY :a symposium Papers New York 1939 Nov 2-4 American institute of physics With the cooperation of the National bureau of standards Reinhold, New York,1941
Chem 18.0636

TEMPERATURE:ITS MEASUREMENT AND CONTROL IN SCIENCE AND INDUSTRY,VOL.2 Symposium on temperature 3rd Papers Washington,D.C. 1954 Oct 28-30 Edited by Hugh C. Wolfe Sponsored by the American institute of physics Reinhold,New York,1955
Eng 41.4126

TEMPERATURE - ITS MEASUREMENT AND CONTROL IN SCIENCE AND INDUSTRY :a symposium Papers New York 1939 Nov 2-4 Under the auspices of the American institute of physics Reinhold,New York,1941
Radioth 35.1431

TEMPERATURE - ITS MEASUREMENT AND CONTROL IN SCIENCE AND INDUSTRY :a symposium Papers New York 1939 Nov 2-4 Under the auspices of the American institute of physics xiii,1362p Reinhold,New York,1941
Chem E 24.0478

TEMPERATURE - ITS MEASUREMENT AND CONTROL IN SCIENCE AND INDUSTRY 2nd :a symposium Papers New York 1939 Nov 2-4 American institute of physics and National bureau of standards Reinhold,New York,1941
Met 25.0298

TEMPERATURE - ITS MEASUREMENT AND CONTROL IN SCIENCE AND INDUSTRY 3rd : a symposium Papers Washington,D.C. 1954 Oct 28-30 Vol 2 American institute of physics and United States.Army.Office of ordnance research Edited by Hugh C. Wolfe Reinhold; Chapman and Hall,New York;London,1955
Met 25.0299

TEMPERATURE - ITS MEASUREMENT AND CONTROL IN SCIENCE AND INDUSTRY 4th Papers Columbus 1961 Mar 27-31 Vol 3,pt.2: applied methods and instruments American institute of physics Edited by A.I. Dahl Reinhold;Chapman and Hall,New York;London,1962
Met 25.0303

TEMPLE UNIVERSITY.SCHOOL OF MEDICINE The International conference on the biology of cutaneous cancer 1st Philadelphia,Pa. 1962 Apr 6-11 Edited by Frederick Urbach With financial support from Merck,Sharp and Dohme.Postgraduate program National cancer institute.Monograph, 10 US government printing office,Washington,D.C.,1963
Path 30.2423

Les TENDANCES GEOMETRIQUES EN ALGEBRE ET THEORIE DES NOMBRES colloque international Clermont-Ferrand 1964 Apr 2-9 Centre national de la recherche scientifique Edited by Marc Krasner Centre nationale de la recherche scientifique.Colloques internationaux, 143 Bibliog. 256p 25cm Centre national de la recherche scientifique, Paris,1966
P. Math 2.1775

TENOVUS INSTITUTE FOR CANCER RESEARCH Clinical management of advanced breast cancer Tenovus workshop 2nd Proceedings Cardiff 1970 Edited by C.A.F. Joslin and E. N. Gleave Alpha omega alpha,Cardiff,1971
Radioth 35.1927

TENOVUS INSTITUTE FOR CANCER RESEARCH Human adrenal gland and its relation to breast cancer Tenovus workshop 1st Proceedings Cardiff 1969 Jun 26-27 Edited by K. Griffiths and E.H.D. Cameron Alpha omega alpa,Cardiff,1969
Inv Med 37.0270

TENOVUS INSTITUTE FOR CANCER RESEARCH Human adrenal gland and its relation to breast cancer Tenovus workshop 1st Proceedings Cardiff 1969 Jun 26-27 Edited by K. Griffiths and E.H.D. Cameron Alpha omega alpha,Cardiff,1969
Bioch 33.2198

TENOVUS INSTITUTE FOR CANCER RESEARCH Prognostic factors in breast cancer :symposium Proceedings Cardiff 1967 Apr 12-14 Tenovus symposia, 1 Livingstone,Edinburgh; London,1968
Radioth 35.0935

TENOVUS INSTITUTE FOR CANCER RESEARCH Some aspects of the aetiology and biochemistry of prostatic cancer Tenovus workshop 3rd Proceedings Cardiff 1970 Edited by K. Griffiths and C.G. Pierrepoint Alpha omega alpha,Cardiff,1971
Radioth 35.1928

TENOVUS INSTITUTE FOR CANCER RESEARCH The Human adrenal gland and its relation to breast cancer Cardiff 1969 Jun 26-27 Edited by K. Griffiths and E.H.D. Cameron Tenovus workshops, 1 Alpha omega alpha publishing, Cardiff,1969
Radioth 35.0940

TENOVUS SYMPOSIA, 1 Prognostic factors in breast cancer :symposium Proceedings Cardiff 1967 Apr 12-14 Tenovus institute for cancer research Livingstone,Edinburgh; London,1968
Radioth 35.0935

TENOVUS WORKSHOP 1st Proceedings Human adrenal gland and its relation to breast cancer Cardiff 1969 Jun 26-27 Tenovus institute for cancer research Edited by K. Griffiths and E.H.D. Cameron Alpha omega alpa,Cardiff,1969
Inv Med 37.0270

TENOVUS WORKSHOP 1st Proceedings Human adrenal gland and its relation to breast cancer Cardiff 1969 Jun 26-27 Tenovus institute for cancer research Edited by K. Griffiths and E.H.D. Cameron Alpha omega alpha,Cardiff,1969
Bioch 33.2198

TENOVUS WORKSHOP 2nd Proceedings Clinical management of advanced breast cancer Cardiff 1970 Tenovus institute for cancer research Edited by C.A.F. Joslin and E.N. Gleave Alpha omega alpha,Cardiff,1971
Radioth 35.1927

TENOVUS WORKSHOP 3rd Proceedings Some aspects of the aetiology and biochemistry of prostatic cancer Cardiff 1970 Tenovus institute for cancer research Edited by K. Griffiths and C.G. Pierrepoint Alpha omega alpha,Cardiff,1971
Radioth 35.1928

TENOVUS WORKSHOPS, 1 The Human adrenal gland and its relation to breast cancer Cardiff 1969 Jun 26-27 Tenovus institute for cancer research Edited by K. Griffiths and E.H.D. Cameron Alpha omega alpha publishing,Cardiff,1969
Radioth 35.0940

TEORIA DEI GRUPPI FINITI E APPLICAZIONI : convegno internazionale Florence 1960 Apr 11-13 Unione matematica Italiana Universita di Firenze Supported by Consiglio nazionale della ricerche vii,156p 24cm Edizioni Cremonese,Rome,1960
P Math 2.2858

TEORIA DEI PLASMA Scuola internazionale di fisica 'Enrico Fermi' 25 corso Rendiconti Varenna 1962 Jul 9-21 Societa italiana di fisica Edited by N.M. Rosenbluth Academic press,New York;London,1964
Eng 41.4527

TEORIA DEL MAGNETISMO NEI METALLI DI TRANSIZIONE Scuola internazionale di fisica 'Enrico Fermi' 37 corso Rendiconti Varenna 1966 Jun 6-25 Societa italiana di fisica Edited by W. Marshall diagrs 455p 24cm Academic press,London,1967
Cav 7.2612

TEORIE ERGODICHE Scuola internationale di fisica 'Enrico Fermi' 14 corso Rendiconti Varenna 1960 May 23-31 Societa Italiana di Fisica Edited by P. Caldirola Academic press,New York;London,1961
A Math 4.0950

TEORIYA VEROYATNOSTEI.MATHMATISCHESKAYA STATISTIKA,TEORETICHESKAYA KIBERNATIKA 1969 Itogo nauki.Seriya matematika bibliog.,Illus. 104,21p Moscow,1970
P Math 2.3940

TERATOLOGY Congenital malformations :a conference on teratology papers Bethesda, Md. 1957 Apr 15-16 Published under the auspices of the Association for the aid of crippled children Pediatrics, 23,no 1,pt 2, suppt. Thomas,Springfield,Ill.,1959
An 32.2046

TERMITES Etudes sur les termites africains : colloque international Comptes rendus Leopoldville 1964 May 11-16 Universite de Louvain Edited by A. Bouillon Under the auspices of Unesco figs,pls,tables 27cm Masson,Paris,1964
Bal 44.4805

TERMITES IN THE HUMID TROPICS :symposium Proceedings New Delhi 1960 Oct 4-12 Unesco Zoological survey of India Paris, 1962
Bal 39.4249

TERPENES AND STEROLS Ciba foundation symposium on the biosynthesis of terpenes and sterols Papers London 1958 May 20-22 Ciba foundation Edited by G.E.W. Wolstenholme and Maeve O'Connor Ciba foundation symposia Churchill,London,1959
Chem 18.1511

TERPENOIDS IN PLANTS :a symposium Proceedings Aberystwyth 1966 Apr 12-14 Phytochemical group Edited by J.B. Pridham Academic press,London;New York,1967
Chem 18.1510

TESTING MATERIALS Japan congress on testing materials 6th Kyoto 1962 Oct 10111 Science council of Japan Japan society for testing materials,Kyoto,1963
Met 25.2860

TESTING MATERIALS Japanese congress on testing materials 1st-8th Proceedings 1957-65 8 vols Japan society for testing materials,Kyoto,1958-65
Eng 41.3844

TESTING OF MATERIALS AND STRUCTURES International symposium on nondestructive testing of materials and atructures Vol 1-2 International union of testing and research laboratories for materials and structures 2 vols Paris,1956
Eng 41.2833

TESTING OF SOILS Symposium on consolidation testing of soils Atlantic City,N.J. 1951 Jun 18 American society for testing materials American society for testing materials.Special technical publication, 126 American society for testing materials, Philadelphia,Pa.,1952
Eng 41.3149

TESTING OF SOILS Symposium on dynamic testing of soils Atlantic City,N.J. 1953 Jul 2 American society for testing materials American society for testing materials.Special technical publication, 156 American society for testing materials,Philadelphia,Pa.,1954
Eng 41.3148

TESTING OF SOILS Symposium on vane shear testing of soils Atlantic City,N.J. 1956 Jun 22 American society for testing materials American society for testing materials.Special technical publication, 193 American society for testing materials, Philadelphia,Pa.,1957
Eng 41.3150

TESTING TECHNIQUES FOR ROCK MECHANICS :a symposium Seattle,Wash. 1965 Oct 31-Nov 5 American society for testing and materials American society for testing and materials. Special technical publication, 402 American society for testing materials,Philadelphia,Pa., 1966
Eng 41.3161

TETANUS Symposium on tetanus in Great Britain Leeds 1967 Apr 7-8 Leeds general infirmary.Tetanus unit Edited by M. Ellis c1967 Typescript
PGMS 29.0042

TETHYS Aspects of Tethyan biogeography :a symposium Leicester 1966 Sep 21-23 Systematics association Edited by C.G. Adams and D.V. Ager Systematics association. Publication, 7 illus. vi,336p British museum,London,1967
Bot 42.3273

TEWKESBURY SYMPOSIUM ON FRACTURE 2nd Proceedings Fracture Melbourne 1969 Edited by C.J. Osborn and others Held by the University of Melbourne.Fsculty of engineering Butterworths,Sydney,1969 Financial support from the Pearson Tewkesbury bequest
Eng 41.8202

The TEWKSBURY SYMPOSIUM 1st Proceedings Fracture Melbourne 1963 Aug 26-30 University of Melbourne.Faculty of engineering Edited by C.J. Osborn University of Melbourne,Melbourne,1965
Met 25.0870

TEWKSBURY SYMPOSIUM ON FRACTURE 1st Proceedings Fracture Melbourne 1963 Aug 26-30 Edited by C.J. Osborn University of Melbourne,Melbourne,1965
Eng 41.3802

The TEWSKBURY SYMPOSIUM 2nd Proceedings Fracture Melbourne 1969 University of Melbourne.Faculty of engineering Edited by C. J. Osborn and others Butterworth,Sydney,1969
Met 25.2579

TEXAS CONFERENCE ON RELATIVISTIC ASTROPHYSICS 2ND Quasars and high energy astronomy ; including proceedings of 2nd Texas conference on relativistic astrophysics Edited by K.N. Douglas and others 485p Gordon and Breach, New York,1969 Texas conference held in 1964
TA 15.0490

TEXAS CONFERENCE ON SOIL MECHANICS AND FOUNDATION ENGINEERING 8th Proceedings Austin, Texas 1956 Sep 14-15 University of Texas. Bureau of engineering research University of Texas,Austin,Texas,1956
Eng 41.3164

TEXAS SYMPOSIUM ON RELATIVISTIC ASTROPHYSICS 1st Proceedings Quasi stellar sources and gravitational collapse Dallas,Texas 1963 Dec 16-18 By Ivor Robinson and others Southwest center for advanced studies Sponsored by University of Texas 475p 25cm University of Chicago press,Chicago,Ill.,1965
Cav 7.1985

TEXAS SYMPOSIUM ON RELATIVISTIC ASTROPHYSICS 1st proceedings Quasi-stellar sources and gravitational collapse Dallas,Texas 1963 Dec. 16-18 Southwest center for advanced studies Edited by Ivor Robinson and others sponsored by University of Texas xvii,475p 25cm University of Chicago press,Chicago, 1965
Obs 6.1457

TEXAS SYMPOSIUM ON RELATIVISTIC ASTROPHYSICS 1st proceedings Quasi-stellar sources and gravitational collapse Dallas,Texas 1963 Dec.16-18 Southwest center for advanced studies Edited by Ivor Robinson and others Sponsored by University of Texas xvii,475p 25cm University of Chicago press,Chicago, 1965
A Math 4.1161

TEXTILE INSTITUTE Studies in modern fabrics : a conference Papers London 1970 May 26-29 Textile institute.Annual conference, 55 Textile institute,Manchester,1970
Eng 41.8287

TEXTILE INSTITUTE Studies in modern yarn production :a conference Papers Stresa 1968 May 25-28 Textile institute.Annual conference, 53 Textile institute,Manchester, 1968
Eng 41.8288

TEXTILE INSTITUTE.ANNUAL CONFERENCE, 53 Studies in modern yarn production :a conference Papers Stresa 1968 May 25-28 Textile institute Textile institute, Manchester,1968
Eng 41.8288

TEXTILE INSTITUTE.ANNUAL CONFERENCE, 55 Studies in modern fabrics :a conference Papers London 1970 May 26-29 Textile institute Textile institute,Manchester,1970
Eng 41.8287

TEXTUREN IN FORSCHUNG UND PRAXIS international symposium Proceedings Clausthal-Zellerfeld 1968 Oct 2-5 Institut fur metallkunde und metallphysik der technischen universitat Clausthal Edited by J. Grewen and G. Wassermann Springer-Verlag,Berlin; Heidelberg,1969
Met 25.2363

TEXTURES IN RESEARCH AND PRACTICE Texturen in forschung und praxis international symposium Proceedings Clausthal-Zellerfeld 1968 Oct 2-5 Institut fur metallkunde und metallphysik der technischen universitat Clausthal Edited by J. Grewen and G. Wassermann Springer-Verlag,Berlin;Heidelberg, 1969
Met 25.2363

The THALAMUS 1st :international symposium Proceedings New York 1964 Edited by Dominick P. Purpura and Melvin D. Yahr Sponsored by the Parkinsons disease information and research center Columbia university press,New York;London,1966
Pha 16.0303

THE HAGUE 1963 Genetics today International congress of genetics 11th Proceedings Vol 1-3 Edited by S.J. Geerts 2 vols Pergamon press,Oxford,1963-65
Bot 42.0780

THE HAGUE 1963 Genetics today International congress of genetics 11th Proceedings Vol 1-3 Edited by S.J. Geerts 3 vols Pergamon press,Oxford,1963-65 Two copies each of vol 2 and 3
Gen 34.1007

THE HAGUE 1969 Communicable diseases: methods of surveillance :report on a seminar Convened by the World Health Organisation. Regional office for Europe Regional office for Europe,W.H.O.,Copenhagen,1969
PGMS 29.0117

THE HAGUE 1971 Modern steam plant practice a convention Proceedings Institution of mechanical engineers.Steam plant group Institut van ingenieurs.Steam plant group IME,London,1971
Eng 41.8643

THEORETICAL AND APPLIED MECHANICS International congress on theoretical and applied mechanics 8th proceedings Istanbul 1952 Aug.20-28 Vols 1-2 Istanbul universitesi.Faculty of science Edited by Kerim Erim Under the auspices of the International committee for the congresses of applied mechanics 2 vols University faculty of science,Istanbul,1953
A Math 4.0326

THEORETICAL APPROACHES TO NON-NUMERICAL PROBLEM SOLVING Systems symposium 4th Proceedings Cleveland,Ohio 1968 Nov 19-20 Edited by R.B. Banerji and M.D. Mesarovic Sponsored by Case western reserve university. Systems research center Lecture notes in operations research and mathematical systems, 28 bibliog.,illus. vi,466p 25cm Springer,Berlin,1970
Math S 3.1617

THEORETICAL BIOLOGY Towards a theoretical biology :a symposium Papers Lake Como 1967 Aug 3-12 Vol 2: sketches International union of biological sciences Edited by C.H. Waddington I.U.B.S.symposium, 2 Edinburgh university press,Edinburgh, 1969
Bal 39.0053

THEORETICAL BIOLOGY Towards a theoretical biology :an IUBS symposium Lake Como 1967 Aug 3-12 2: sketches International union of biological sciences Edited by C.H. Waddington Edinburgh university press, Edinburgh,1968
Bal 39.0017

THEORETICAL BIOLOGY Towards a theoretical biology 1.Prolegomena:an I.U.B.S. symposium Essays Bellagio 1966 Aug 28-Sep 3 International union of biological sciences Edited by C.H. Waddington 234p Edinburgh university press,Edinburgh,1968 Essays written after and in the light of the symposium
WSM 43.2897

THEORETICAL INTERPRETATION OF UPPER ATMOSPHERE EMISSIONS :a symposium Paris 1962 Jun.25-29 International astronomical union and International union of geodesy and geophysics Edited by D.P. Bates International astronomical union.Symposium, 18 264p Pergamon press,New York,1963 Reprinted from'Planetary and space science', Vol.10 1963
Obs 6.0142

THEORETICAL ORGANIC CHEMISTRY :Kekule symposium Papers London 1958 Sep 15-17 International union of pure and applied chemistry.Section of organic chemistry Organized by the Chemical society Butterworths,London,1959
Chem 18.2165

THEORETICAL ORGANIC CHEMISTRY :Kekule symposium Papers London 1958 Sep 15-17 International union of pure and applied chemistry.Section of organic chemistry,and, Chemical society Butterworths,London,1959
Chem 18.1604

THEORETICAL PHYSICS Astrophysics and the many-body problem Waltham,Mass. 1962 By E.N. Parker Brandeis university.Summer institute in theoretical physics Lectures in theoretical physics,1962,2 W.A.Benjamin,New York;Amsterdam,1963
A Math 4.1113

THEORETICAL PHYSICS High energy physics Tokyo summer lectures in theoretical physics 1st Papers Tokyo 1965 Sep 6-17 Pt 2: high energy physics University of Tokyo. Summer institute of theoretical physics Edited by Gyo Takeda Syokabo;W.A.Benjamin, Tokyo;New York,1966
Cav 7.2318

THEORETICAL PHYSICS International conference on theoretical physics Proceedings Kyoto 1953 Sep 14-24 Tokyo Science council of Japan and Kyoto university Edited by I. Imai Under the auspices of the International union of pure and applied physics 942p Tokyo,1954
Cav 7.1342

THEORETICAL PHYSICS Lectures in theoretical physics 6th Boulder 1963 Vol. 6 University of Colorado.Summer institute for theoretical physics Edited by W.E. Brittin and W.R. Chappell Sponsored by the University of Colorado.Department of physics and astrophysics Lectures in theoretical physics,Boulder.,1963,6 Colorado university press,Boulder,Colo.,1964
A Math 4.0834

THEORETICAL PHYSICS Lectures in theoretical physics 7th Boulder,Col. 1964 7a: Lorentz group University of Colorado.Summer institute for theoretical physics Edited by Wesley E. Britten and Asim O. Barut Sponsored by the University of Colorado. Department of physics and astrophysics Lectures in theoretical physics,Boulder,1964, 7a University of Colorado press,Boulder,Col., 1965 The Lorentz group symposium volume is dedicated to Professor E.Wigner
A Math 4.1335

THEORETICAL PHYSICS Lectures in theoretical physics 7th Boulder,Col. 1964 Vol. 7c: statistical physics,weak interactions, field theory University of Colorado.Summer institute for theoretical physics Edited by Wesley E. Brittin Sponsored by the University of Colorado.Department of physics and astrophysics Lectures in theoretical physics,Boulder,1964,7c University of Colorado press,Boulder,Col.,1965
A Math 4.0835

THEORETICAL PHYSICS Lectures in theoretical physics Boulder,Col. 1958-62 Vol 1-5 University of Colorado.Summer institute for theoretical physics Edited by Wesley E. Brittin and others Lectures in theoretical physics, 1-5 Interscience,New York,1959-63
Cav 7.0292

THEORETICAL PHYSICS Lectures in theoretical physics Waltham,Mass 1961 Vol. 2 By M.E. Rose and E.C.G. Sudarshan Brandeis university.Summer institute in theoretical physics Lectures in theoretical physics,1961, 2 Benjamin,New York,1962
A Math 4.0830

THEORETICAL PHYSICS Lectures in theoretical physics Waltham,Mass. 1961 Pt 1 By R.J. Eden Brandeis university.Summer institute in theoretical physics Lectures in theoretical physics,1961,1 Benjamin,New York, 1962
Cav 7.0315

THEORETICAL PHYSICS Lectures in theoretical physics Waltham,Mass. 1961 Pt 2 By M.E. Rose and E.C.G. Sudarshan Brandeis university.Summer institute in theoretical physics Lectures in theoretical physics,1961, 2 Benjamin,New York,1962
Cav 7.0316

THEORETICAL PHYSICS Lectures in theoretical physics Waltham,Mass. 1962 Vol 2: astrophysics and the many-body problem By E. N. Parker and J.S. Goldstein Brandeis university.Summer institute in theoretical physics Lectures in theoretical physics,1962, 2 Benjamin,New York,1963
Cav 7.0318

THEORETICAL PHYSICS Lectures in theoretical physics :notes from the French summer school Cargese,Corsica 1962 Jul Edited by Maurice Levy Cargese lectures in theoretical physics, 1962 W.A.Benjamin,New York,1963
Cav 7.0437

THEORETICAL PHYSICS Lectures on astrophysics and weak interactions Waltham,Mass. 1963 By S. Hayakawa and others Brandeis university.Summer institute in theoretical physics Lectures in theoretical physics,1963, 2 Brandeis university,Waltham,Mass.,1963
A Math 4.1087

THEORETICAL PHYSICS Lectures on astrophysics and weak interactions Waltham,Mass. 1963 Vol. 2 By S. Hayakawa and others Brandeis university.Summer institute in theoretical physics Lectures in theoretical physics,1963,2 Brandeis university,Waltham, Mass.,1964
A Math 4.0832

THEORETICAL PHYSICS Lectures on general relativity Waltham,Mass. 1964 By A. Trautman and others Brandeis university. Summer institute in theoretical physics Lectures in theoretical physics,1964,1 Prentice-Hall,Englewood Cliffs,N.J.,1965
Cav 7.0320

THEORETICAL PHYSICS Lectures on particle and field theory Waltham,Mass 1962 Vol 1: elementary particle physics and field theory By I. Fulton and others Brandeis university. Summer institute in theoretical physics Lectures in theoretical physics,1962,1 Benjamin,New York,1963
Cav 7.0317

THEORETICAL PHYSIÇS Lectures on particle symmetries and axiomatic field theory Papers Waltham,Mass. 1965 1: axiomatic field theory Brandeis university.Summer institute in theoretical physics Edited by M. Chretien and S. Deser Lectures in theoretical physics, 1965,1 Gordon and Breach,New York,1966
A Math 4.1334

THEORETICAL PHYSICS Lectures on particles and field theory Waltham,Mass 1964 Vol. 2 By K. Johnson and others Brandeis university.Summer institute in theoretical physics Lectures on theoretical physics,1964, 2 Prentice-Hall,New York,1965
A Math 4.0833

THEORETICAL PHYSICS Lectures on particles and field theory Waltham,Mass. 1962 Vol. 1: elementary particle physics and field theory By I. Fulton and others Brandeis university.Summer institute in theoretical physics Lectures in theoretical physics,1962, 1 Benjamin,New York,1963
A Math 4.0831

THEORETICAL PHYSICS Lectures on particles and field theory Waltham,Mass. 1964 By K. Johnson and others Brandeis university. Summer institute in theoretical physics Lectures in theoretical physics,1964,2 Prentice-Hall,Englewood Cliffs,N.J.,1965
Cav 7.0321

THEORETICAL PHYSICS Lectures on strong and electromagnetic interactions Waltham,Mass. 1963 By P.T. Matthews and others Brandeis university.Summer institute in theoretical physics Lectures in theoretical physics,1963, 1 Brandeis University,Waltham,Mass.,1964
A Math 4.0036

THEORETICAL PHYSICS Relativity,groups and topology lectures Les Houches 1963 Universite de Grenoble.Ecole d'ete de physique theorique,Les Houches Edited by C. DeWitt and B. DeWitt Universite de Grenoble.Ecole d'ete de physique theorique,Les Houches. Lectures,1963 Gordon and Breach,New York; London,1963
A Math 4.1030

THEORETICAL PHYSICS Seminar on theoretical physics Lectures Trieste 1962 Jul 16-25 Organized by the International atomic energy agency International atomic energy agency, Vienna,1963
Radioth 35.1655

THEORETICAL PHYSICS Seminar on theoretical physics lectures Trieste 1962 Jul.16-Aug. 25 Organized by the International atomic energy agency Vienna,1963 Companion volume to International summer school on nuclear theory,1962
A Math 4.0886

THEORETICAL PHYSICS Statistical physics Waltham,Mass. 1962 By G.E. Uhlenbeck and others Brandeis university.Summer institute in theoretical physics Lectures in theoretical physics,1962,3 252p W.A. Benjamin,New York,1963
Cav 7.2414

THEORETICAL PHYSICS The Seminar on theoretical physics held in Miramare Papers Trieste 1962 Jul 16-Aug 25 International atomic energy agency Edited by A. Salam STI-PUB-61 637p international atomic energy agency,Vienna,1963
Cav 7.2346

THEORETICAL PHYSICS Theoretical physics : lecture notes Typescript Waltham,Mass. 1960 By C. Moller and others Brandeis university.Summer institute in theoretical physics Edited by J. Stachel and others Lectures in theoretical physics,1960 Brandeis university,Waltham,Mass.,1960
Cav 7.0314

THEORETICAL PHYSICS Theoretical physics : lecture notes Waltham,Mass. 1960 By C. Moller and others Brandeis university.Summer institute in theoretical physics Edited by J. Stachel and others Lectures in theoretical physics,1960 Brandeis university,Waltham, Mass,c1961
A Math 4.0829

THEORETICAL PHYSICS Theoretical physics: lecture notes Waltham,Mass. 1959 By F.E. Low Brandeis university.Summer institute in theoretical physics Edited by Wayne A. Mills Lectures in theoretical physics,1959 Brandeis university,Waltham,Mass.,1959
A Math 4.0828

THEORETICAL PHYSICS Tokyo summer institute in theoretical physics 1st Tokyo 1965 Sep 6-17 Pt 2: high energy physics Edited by Gyo Takeda Tokyo summer lectures in theoretical physics,1965 iv,121p Syokabo;Benjamin,Tokyo;New York,1966
A Math 4.1346

THEORIA INCORPORATED Quantum theory and beyond :essays and discussions arising from a colloquium Cambridge 1968 Jul illus viii,345p 24cm Cambridge university press, Cambridge,1971 Based on papers read at an informal colloquium
A Math 4.1759

THEORIE DES ATMOSPHERES STELLAIRES : conference Saas-Fee 1971 Mar 28-Apr 3 By D. Mihalas and others 312p Observatoire de Geneve,Sauverny,1971
Obs 6.3631

THEORIE DES MATROIDES Rencontre Franco-britannique Proceedings Brest 1970 May 14-15 Edited by C.P. Bruter Lecture notes in mathematics, 211 108p 25cm Springer, Berlin,1970
P Math 2.4508

La THEORIE DU POTENTIEL :colloque international Orsay 1964 Jun.22-26 Centre national de la recherche scientifique Edited by M. Brelot and others Centre national de la recherche scientifique. Colloques internationaux, 146 x,312p 25cm Centre national de la recherche scientifique, Paris,1965
Math 3.0050

THEORIE DU POTENTIEL colloque international 1964 Centre national de la recherche scientifique Centre national de la recherche scientifique.Colloques internationaux 15 unbound typescript papers
Math 3.1110

THEORY AND APPLICATIONS OF SPLINE FUNCTIONS :a symposium Proceedings Madison,Wis. 1968 Oct. United States.Army.Mathematics research center United States.Army. Mathematics research center Bibliog.,Illus. xi,212p 24cm Academic press,New York; London,1969
P Math 2.3487

THEORY AND APPLICATIONS OF SPLINE FUNCTIONS : an advanced seminar Proceedings Madison, Wis. 1968 Oct 7-9 United States.Army. Mathematics research center Edited by T.N.E. Greville United States.Army.Mathematics research center.Publications, 22 Academic press,New York;London,1969
Math L 5.3443

THEORY AND FUNDAMENTAL RESEARCH IN HEAT TRANSFER American society of mechanical engineers : annual meeting Proceedings New York 1960 Nov Edited by J.A. Clark ix,220p Pergamon press,Oxford,1963
Chem E 24.0582

THEORY AND OBSERVATION OF NORMAL STELLAR ATMOSPHERES Harvard-Smithsonian conference on stellar atmospheres 3rd Proceedings Cambridge,Mass. 1968 Harvard university.Astronomical observatory Smithsonian institution.Astrophysical observatory Edited by O. Gingerich 472p Massachusetts institute of technology press, Cambridge,Mass.,1969
Obs 6.3354

THEORY AND PHENOMONOLOGY IN PARTICLE PHYSICA "ETTORE MAJORANA" International school of physics 6th course Erice 1968 Jul 13-28 Part A-B European organization for nuclear research Edited by A. Zichichi Co-sponsored by the Weizmann institute of science Bibliog.,Illus. 23cm 2 vols Academic press,New York,1969
A Math 4.1541

THEORY OF ALLOY PHASES a seminar Philadelphia 1955 Oct 15-21 American society for metals American society for metals,Cleveland,Ohio,1956 A seminar held during the 37th National metal congress and exposition
Met 25.1051

The THEORY OF ARITHMETIC FUNCTIONS :conference Proceedings Kalamazoo,Mich. 1971 Apr 29-May 1 Edited by Anthony A. Gioia and Donald L. Goldsmith Held at Western Michigan university Lecture notes in mathematics, 251 287p 25cm Springer,Berlin,1972
P Math 2.4382

THEORY OF DISTRIBUTIONS :an international summer institute Proceedings Lisbon 1964 Sep Instituto Gulbenkian de ciencia. Centro de calculo cientifico Sponsored by the North Atlantic treaty organization.Science committee 390p Instituto Gulbenkian de ciencia,Lisbon,1964
A Math 4.1265

THEORY OF DISTRIBUTIONS :international summer institute Proceedings Lisbon 1964 Sep Centro de calculo cientifico Sponsored by the North Atlantic treaty organization.Science committee Illus. xviii,390p 25cm Centro de calculo cientifico,Lisbon,1964 In English and French
P Math 2.3180

The THEORY OF ELECTRO-MAGNETIC WAVES :a symposium Proceedings New York 1950 Jun 6-8 Washington square college of arts and science New York university.Institute for mathematics and mechanics Under the auspices of Air force Cambridge research laboratories.Geophysics research directorate Interscience,New York;London,1951
Radioth 35.1507

THEORY OF GAMES:TECHNIQUES AND APPLICATIONS conference proceedings Toulon 1964 Jun.29-Jul.3 North Atlantic treaty organization.Scientific affairs committee Edited by A. Mensch xi,490p 23cm English universities press,London,1966
Math 3.0241

THEORY OF GRAPHS :international sympoaium Rome 1966 Jul International computation centre Bibliog.,Illus. xviii,416p 25cm Gordon and Breach;Dunod,New York;Paris,1967
P Math 2.3128

THEORY OF GRAPHS :colloquium Proceedings Tihany 1966 Sep Edited by P. Erdos and G. Katona Bibliog.,Illus. 270p 24cm Academic press,New York;London,1968
P Math 2.3127

THEORY OF GRAPHS :colloquium Proceedings Tihany 1968 Sep Edited by P. Erdos and G. Katona bibliog.,illus. 270p 24cm Academic press,New York,1968
Math S 3.1384

The THEORY OF GROUPS international conference proceedings Canberra 1965 Aug 10-20 Australian national university Edited by L.G. Kovacs and B.H. Neumann Bibliog xviii, 397p 24cm Gordon and Breach,New York,1967
P Math 2.2793

THEORY OF MODELS International symposium on the theory of models proceedings Berkeley, Calif. 1963 Jun 25-Jul 11 Edited by J.W. Addison and others xv,494p 23cm North-Holland,Amsterdam,1965
P. Math 2.0205

THEORY OF NUMBERS symposium Pasadena, Calif. 1963 Nov 21-22 American mathematical society Edited by Albert Leon Whiteman Under a grant from the National science foundation American mathematical society.Proceedings of symposia in pure mathematics, 8 vii,214p 26cm American mathematical society,Providence,R.I.,1965
P. Math 2.1774

The THEORY OF ORBITS IN THE SOLAR SYSTEM AND IN STELLAR SYSTEMS Thessalonika 1964 Aug 17-22 International astronomical union Edited by George Contopoulos International astronomical union.Symposium, 25 Academic press,London;New York,1966
TA 15.0332

The THEORY OF ORBITS IN THE SOLAR SYSTEM AND IN STELLAR SYSTEMS a symposium Thessaloniki 1964 Aug.17-22 International astronomical union Edited by George Contopoulos Organized in co-operation with Committee on space research International Astronomical Union.Symposium, 25 380p Academic press,London;New York,1966
Obs 6.0409

The THEORY OF ORBITS IN THE SOLAR SYSTEM AND IN STELLAR SYSTEMS :A SYMPOSIUM Thessaloniki 1964 Aug.17-22 International astronomical union Edited by George Contopoulos Organized in co-operation with Committee on space research International astronomical union.Symposium, 25 380p Academic press, London;New York,1966
A Math 4.1167

The THEORY OF ROAD TRAFFIC FLOW 2nd international symposium proceedings London 1963 Jun.25-27 Road research laboratory Edited by Joyce Almond With financial support from General motors corporation.Research laboratories ix,406p 24cm Organisation for economic co-operation and development,Paris,1965
Math 3.0922

The THEORY OF THIN ELASTIC SHELLS :a symposium Proceedings Delft 1959 Aug 24-28 International union of theoretical and applied mechanics North-Holland,Amsterdam,1960
Eng 41.2854

THEORY OF THIN SHELLS 2nd :a symposium Proceedings Copenhagen 1967 Sep 5-9 International union of theoretical and applied mechanics Edited by F.I. Niordson Springer, Berlin,1969
Eng 41.2892

THEORY OF TRAFFIC FLOW 3rd symposium proceedings Warren,Mich. 1959 Dec.7-8 Edited by Robert Herman Sponsored by General motors corporation.Research laboratories Elsevier,Amsterdam,1961
A Math 4.0949

THERMAL AND HIGH-STRAIN FATIGUE :an international conference Proceedings London 1967 Jun 6-7 Metals and metallurgy trust Organized by the Institute of metals Metals and metallurgy trust.Monograph and report series Metals and metallurgy trust, London,1967
Eng 41.3810

THERMAL AND HIGH-STRAIN FATIGUE an international conference Proceedings London 1967 Jun 6-7 Institute of metals and Iron and steel institute Metals and metallurgy trust.Monograph and report series, 32 Metals and metallurgy trust,London,1967
Met 25.0942

THERMAL CONDUCTIVITY Conference on thermal conductivity 8th Proceedings West Lafayette,Ind. 1968 Oct 7-10 Edited by C.Y. Ho and R.E. Taylor Sponsored by Purdue university.Thermophysical properties research center xx,1169p 29cm Plenum press,New York,1969
Cav 7.3075

THERMAL NEUTRON DIFFRACTION International summer school on the accurate determination of neutron intensities and structure factors Proceedings Harwell 1968 Jul 1-5 Edited by B.T.M. Willis Harwell postgraduate series Bibliog,illus xiv,229p 24cm Oxford university press,London,1970
Min 10.1524

THERMAL PROPERTIES Thermodynamic and transport properties of gases liquids and solids Symposium on thermal properties 3rd Papers Lafayette,Ind 1959 Feb 23-26 Sponsored by the American society of mechanical engineers:Heat transfer division x,472p American society of mechanical engineers;McGraw-Hill,New York,1959
Chem E 24.0984

THERMOANALYSIS OF FIBERS AND FIBER-FORMING POLYMERS :symposium Papers Atlantic City,N.J. 1965 Sep 17 American chemical society Edited by Robert F. Schwenker Applied polymer symposia, 2 Interscience, New York,1966
Met 25.2705

THERMODYNAMIC AND TRANSPORT PROPERTIES OF FLUIDS joint conference Proceedings London 1957 Jul 10-12 International union of pure and applied chemistry and Institution of mechanical engineers viii,217p Institution of mechanical engineers,London,1958
Chem E 24.0981

THERMODYNAMIC AND TRANSPORT PROPERTIES OF GASES LIQUIDS AND SOLIDS Symposium on thermal properties 3rd Papers Lafayette,Ind 1959 Feb 23-26 Sponsored by the American society of mechanical engineers:Heat transfer division x,472p American society of mechanical engineers;McGraw-Hill,New York,1959
Chem E 24.0984

THERMODYNAMICS Non-equilibrium thermodynamics variational techniques and stability :a symposium Proceedings Chicago, Ill. 1965 May 17-19 Edited by Russell J. Donnelly and others University of Chicago press,Chicago,Ill.;London,1966
Eng 41.7117

THERMODYNAMICS Statistical mechanics of equilibrum and non-equilibrium The Symposium on statistical mechanics and thermodynamics Proceedings Aachen 1964 Jun 15-20 International union of pure and applied physics Edited by J. Meixner 274p North-Holland,Amsterdam,1965
Cav 7.0001

THERMODYNAMICS AND THERMOCHEMISTRY Symposium on thermodynamics and thermochemistry Plenary lectures Lund 1963 Jul 18-23 International union of pure and applied chemistry and Swedish chemical society Butterworths,London,1964
Chem E 24.0988

THERMODYNAMICS CONFERENCE 1ST Applications of fundamental thermodynamics to metallurgical processes Conference on the thermodynamic properties of materials 1st Proceedings Pittsburgh 1964 Nov 29-Dec 1 University of Pittsburgh.Center for the study of thermodynamic properties of materials Edited by G.R. Fitterer Gordon and Breach,New York, 1967
Met 25.2443

THERMODYNAMICS IN PHYSICAL METALLURGY :a seminar held during the 31st national metal congress and exposition Cleveland,Ohio 1949 Oct 15-21 American society for metals American society for metals,Cleveland,Ohio, 1952
Met 25.2372

THERMOLUMINESCENCE OF GEOLOGICAL MATERIALS N. A.T.O.advanced research institute on applications of thermoluminescence to geological problems Proceedings Spoleto 1966 Sep 5-16 North Atlantic treaty organization Edited by D.J. McDougall Academic press,London,1968
Min 10.0628

THERMOPHYSICAL PROPERTIES Advances in thermophysical properties at extreme temperatures and pressures Symposium on thermophysical properties 3rd Papers Lafayette,Ind. 1965 Mar 22-25 Edited by Serge Gratch Sponsored by the American society of mechanical engineers American society of mechanical engineers,New York,1965
Eng 41.7121

THERMOPHYSICAL PROPERTIES Symposium on thermophysical properties progress in international research on thermodynamic and transport properties 2nd Papers Princeton,N.J. 1962 Jan 24-26 American society of mechanical engineers.Standing committee on thermophysical properties American society of mechanical engineers; Academic press,New York,1962
Chem 18.0326

THESSALONIKA 1964 The Theory of orbits in the solar system and in stellar systems International astronomical union Edited by George Contopoulos International astronomical union.Symposium, 25 Academic press,London;New York,1966
TA 15.0332

THESSALONIKI 1964 The Theory of orbits in the solar system and in stellar systems a symposium International astronomical union Edited by George Contopoulos Organized in co-operation with Committee on space research International Astronomical Union.Symposium, 25 380p Academic press,London;New York, 1966
Obs 6.0409

THESSALONIKI 1964 The Theory of orbits in the solar system and in stellar systems :a symposium International astronomical union Edited by George Contopoulos Organized in co-operation with Committee on space research International astronomical union.Symposium, 25 380p Academic press,London;New York, 1966
A Math 4.1167

THESSALONIKI 1967 Decision making and age : N.A.T.O. advanced study institute Papers North Atlantic treaty organization Edited by A.T. Welford and J.E. Birren Interdisciplinary topics in gerontology, 4 Karger,Basle;New York,c1968
Psy 31.3208

THIAMINE DEFICIENCY :biochemical lesions and their clinical sihnificance London 1966 Nov 10-11 Ciba foundation Edited by G.E.W. Wolstenholme and Maeve O'Connor Ciba foundation study group, 28 Churchill,London, 1967 In honour of Sir Rudolph Peters
Bioch 33.0574

THIN ELASTIC SHELLS Symposium on the theory of thin elastic shells proceedings Delft 1959 Aug 24-28 International union of theoretical and applied mechanics Edited by W.T. Koiter North-Holland,Amsterdam,1960
A Math 4.0366

THIN FILM PHYSICS Basic problems in thin film physics :international symposium Proceedings Clausthal-Gottingen 1965 Sep 6-11 Edited by R. Niedermayer and H. Mayer illus 756p Vandenheck and Ruprecht, Gottingen,1966
Cav 7.0512

THIN FILMS Structure and properties of thin films An International conference on structure and properties of thin films Proceedings Bolton Landing,N.Y. 1959 Sep 9-11 General electric company Edited by C.A. Neugebauer and others John Wiley,New York, 1959
Cav 7.0265

THIN FILMS Structure and properties of thin films :an international conference Proceedings New York 1959 Sep 9-11 United States.Air research and development command and General electric research laboratory Edited by C.A. Neugebauer and others Sponsored by United States.Air force. Office of scientific research Wiley,New York; London,n.d.
Met 25.1204

THIN FILMS The Use of thin films in physical investigations :a NATO advanced study institute held at the Imperial college of science and technology London 1965 Jul 19-24 Edited by J.C. Anderson Sponsored by the North Atlantic treaty organization Academic press,London,1966
Eng 41.4379

THIN FILMS papers presented at a seminar of the American society for metals American society for metals A.M.C.A.,Ohio,1964
Met 25.0175

THIN WALLED STEEL STRUCTURES :a symposium Swansea 1967 Sep 11-14 University of Wales Institution of structural engineers Edited by K.C. Rockey and H.V. Hill Crosby Lockwood, London,1969
Eng 41.2747

THOMAS GRAHAM MEMORIAL SYMPOSIUM Proceedings Diffusion processes Glasgow 1969 Sep 16 University of Strathclyde Edited by John N. Sherwood and others 2 vols Gordon and Breach,London,1971
Met 25.2597

THORIUM Metal thorium The Conference on thorium Proceedings Cleveland,Ohio 1956 Oct 11 American society for metals and United States atomic energy commission Edited by Harley A. Wilhelm American society for metals,Cleveland,Ohio,1958
Met 25.1561

THORNDIKE MEMORIAL LABORATORY Training of the physician :celebration of the Centennial of the Boston City hospital and the 40th anniversary of the Thorndike memorial laboratory Papers Boston,Mass.,1964
PGMS 29.0489

THOUGHT AND PROCEDURE New methods of thought and procedure Symposium on methodologies Contributions Pasadena,Calif. 1967 May 22-24 California institute of technology.Office for industrial association Society for morphological research Edited by F. Zwicky and A.G. Wilson viii,338p Springer,Berlin, 1967
WSM 43.2639

The THRESHOLD OF SPACE :conference on chemical aeronomy Proceedings Cambridge,Mass. 1956 Jun 25-28 Air force Cambridge research center.Geophysics research directorate Edited by M. Zelikoff Coordinated by Wentworth institute Pergamon press,London, 1957
Chem 18.0478

THROMBOSIS AND EMBOLISM International conference on thrombosis and embolism 1 Proceedings Basle 1954 Edited by Theo Koller and Willy R. Merz Benno Schwabe,Basle, 1955
Med 36.0220

THYMUS Conference on the thymus;the thymus in immunology :structure,function and role in disease Proceedings Minneapolis,Minn. 1962 Nov Edited by Robert A. Good and Ann E. Gabrielson Harper and Row,New York,1964
Med 36.0382

The THYMUS :a symposium Philadelphia 1964 Apr 29 Edited by Vittorio Defendi and Donald Metcalf Held at the Wistar institute of anatomy and biology Wistar institute symposium monograph, 2 Wistar institute press,Philadelphia,Pa.,1964
An 32.3702

The THYMUS :experimental and clinical studies:a Ciba foundation symposium Melbourne 1965 Aug 25-27 Ciba foundation Edited by G.E.W. Wolstenholme and Ruth Porter Churchill,London,1966
Phys 20.0905

The THYMUS:EXPERIMENTAL AND CLINICAL STUDIES a Ciba foundation symposium Melbourne 1965 Aug 25-27 Ciba foundation Edited by G. E.W. Wolstenholme and Ruth Porter Churchill, London,1966
An 32.3703

The THYMUS:EXPERIMENTAL AND CLINICAL STUDIES Ciba foundation symposium Melbourne 1965 Aug 25-27 Ciba foundation and Walter and Eliza Hall institute of medical research Edited by G.E.W. Wolstenholme and Ruth Porter illus Churchill,London,1966 In honour of sir Macfarlane Burnet,chairman of the conference
Path 30.2461

THYROID The Thyroid and its diseases :a conference Proceedings London 1963 Mar 15-16 Royal college of physicians of London Edited by A.Stuart Mason Pitman,London,1963
PGMS 29.0159

THYROID Thyroide et exploration thyroidienne medullo-surrenale :rapports de la vie reunion des endocrinologistes de langue francaise Paris 1959 Jun 29-Jul 1 Endocrinologistes de langue francaise Annales d'endocrinologie, 19,no.4 Masson;Doin,Paris,1959 At head of title:Colloque d'endocrinologie 1959
An 32.3755

The THYROID AND ITS DISEASES :a conference Proceedings London 1963 Mar 15-16 Royal college of physicians of London Edited by A.Stuart Mason Pitman,London,1963
PGMS 29.0159

The THYROID GLAND :a symposium Proceedings London 1953 Feb 25 Society for endocrinology and Royal society of medicine Edited by P. Eckstein and S. Zuckerman Society for endocrinology.Memoirs, 1 1953
Radioth 35.0713

THYROID HORMONES Regulation and mode of action of thyroid hormones :a colloquium Proceedings London 1956 Jun 20-22 Ciba foundation Edited by G.E.W. Wolstenholme and Elaine C.P. Millar Ciba foundation colloquia on endocrinology, 10 Churchill,London,1957
Inv Med 37.0240

THYROID HORMONES Regulation and mode of action of thyroid hormones :a colloquium Proceedings London 1956 Jun 20-22 Ciba foundation Edited by G.E.W. Wolstenholme and Elaine C.P. Millar Ciba foundation colloquia on endocrinology, 10 illus Churchill, London,1957
Bioch 33.0439

THYROID NEOPLASIA Imperial cancer research fund symposium 2nd Proceedings London 1967 Apr Edited by Stretton Young and D.R. Inman Academic press,New York,1968
An 32.3780

THYROID NEOPLASIA Imperial cancer research fund symposium 2nd Proceedings London 1967 Apr Imperial cancer research fund Edited by Stretton Young and D.R. Inman Academic press,London;New York,1968
PGMS 29.0180

THYROIDE ET EXPLORATION THYROIDIENNE MEDULLO-SURRENALE :rapports de la vie reunion des endocrinologistes de langue francaise Paris 1959 Jun 29-Jul 1 Endocrinologistes de langue francaise Annales d'endocrinologie, 19,no.4 Masson;Doin,Paris,1959 At head of title:Colloque d'endocrinologie 1959
An 32.3755

THYROTROPIN :a conference Proceedings 1962 Edited by Sidney C. Werner Sponsored by the National institutes of health.Endocrine study section Thomas,Springfield,Ill.,1963 Introduction:"This conference is the second on tropic hormones of the pituitary gland initiated by the Endocrine study section...
An 32.3762

TIHANY 1962 The Foundations of mathematics, mathematical machines and their applications colloquium Magyar tudomanyos akademia. Section of mathematics and physics Edited by Laszlo Kalmar In cooperation with the Bolyai Janos matematikai tarsulat 317p 25cm Akademiai kiado,Budapest,1965
P. Math 2.2359

TIHANY 1966 Theory of graphs :colloquium Proceedings Edited by P. Erdos and G. Katona Bibliog.,Illus. 270p 24cm Academic press, New York;London,1968
P Math 2.3127

TIHANY 1968 Theory of graphs :colloquium Proceedings Edited by P. Erdos and G. Katona bibliog.,illus. 270p 24cm Academic press, New York,1968
Math S 3.1384

TIHANY,HUNGARY 1963 Abelian groups colloquium proceedings Magyar tudomanyos akademia Edited by L. Fuchs and E.T. Schmidt 162p 24cm Akademiai kiado,Budapest,1964
P. Math 2.2028

TIME Entretiens sur le temps Cerisy-la-Salle 1964 Jul 14-23 Centre culturel international de Cerisy-la-Salle Edited by Jeanne Hersch and Rene Poirier Centre culturel international de Cerisy-la-Salle. Decade nouvelle serie, 5 351p Mouton, Paris;The Hague,1967
WSM 43.1000

TIME The Nature of time :a meeting convened by T.Gold and H.Bondi Ithaca,N.Y. 1963 May 30-Jun 1 Cornell university Edited by T. Gold xiv,248p Cornell university press, Ithaca,N.Y.,1967
WSM 43.1104

TIME The Nature of time :report of a meeting Ithaca,N.Y. 1963 May 30-Jun 1 Cornell university Edited by T. Gold and D.L. Schumacher ix,248p 23cm Cornell university,Ithaca,N.Y.,1967
Cav 7.2764

TIME-OF-FLIGHT MASS SPECTROMETRY Dynamic mass spectrometry European symposium on the time-of-flight mass spectrometry 2nd Proceedings Salford 1969 Jul 1 University of Salford Edited by D. Price and J.E. Williams Heyden,London,1970
Chem 18.2669

TIME SERIES ANALYSIS Symposium on time series analysis proceedings Providence,R.I. 1962 Jun.11-14 Brown university Edited by Murray Rosenblatt Also sponsored by United States.Office of naval research xiv, 497p 25cm John Wiley and sons,New York, 1963
Math 3.0695

TIME SERIES ANALYSIS The Symposium on time series analysis Proceedings Providence,R.I. 1962 Jun 11-14 Brown university United States.Office of naval research Edited by Murray Rosenblatt xiv,497p 26cm Wiley, New York;London,1963
Cav 7.2744

TIME-OF-FLIGHT MASS SPECTROMETRY European symposium on time-of-flight mass spectrometry 1st Proceedings Salford 1967 Jul University of Salford Edited by D. Price and J.R. Williams Pergamon,Oxford,1969
Met 25.1706

TIME-OF-FLIGHT MASS SPECTROSCOPY :based on the first European symposium Proceedings Salford 1967 Jul 3-5 University of Salford Edited by D. Price and J.E. Williams Pergamon press,Oxford,1969
Chem 18.2590

TIN Symposium on tin New York 1952 Jun 23 American society for testing materials A.S.T.M.Special technical publication, 141 American society for testing materials, Philadelphia,Pa.,1953 Fiftieth anniversary publication. A symposium held at the 55th annual meeting of the American society for testing materials
Met 25.0567

TISSUE CULTURE ASSOCIATION Cell tissue and organ culture Decennial review conference on cell tissue and organ culture 2nd Proceedings Bedford,Pa. 1966 Sep 11-15 Edited by Benton B. Westfall National cancer institute.Monograph, 26 National cancer institute,Bethesda,Md.,1967
Bioch 33.0991

TISSUE CULTURE ASSOCIATION The Chromosome: structural and functional aspects :a symposium Miami,Fla. 1965 Jun 1-4 Edited by Clyde Johnson-Dawe Williams and Wilkins,Baltimore, Md.,1965 Organized by George Yerganian
Gen 34.0783

TISSUE CULTURE ASSOCIATION The Chromosome; structural and functional aspects :a symposium Miami 1965 Jun 1-4 Vol 1 Edited by Clyde Johnson Dawe Williams and Wilkins, Baltimore,Md.,1966 Organized by George Yerganian
Bal 39.0339

TISSUE CULTURE ASSOCIATION SYMPOSIA 5th Chromosome:structural and functional aspects Miami,Fla. 1965 Jun 1-4 Edited by George Yerganian illus. xiii,107p 26cm Tissue cultural association,n.p.,1965
Bot 42.4734

TISSUE ELASTICITY :a conference Papers Hanover,N.H. 1955 Sep 1-3 Edited by John W. Remington held at Dartmouth college American physiological society,Washington,D.C., 1957
Bal 39.1204

TISSUE PROTEINASES Wates symposium Proceedings Cambridge 1970 Apr 6-7 Royal society of London Edited by A.J. Barrett and J.T. Dingle Held at the Strangeways research laboratory North-Holland,Amsterdam,1971
Bioch 33.2284

TISSUE TRANSPLANTATION International symposium on tissue transplantation Proceedings Santiago 1961 Aug 30-Sep 2 Edited by Alberto P. Cristoffanini and Gustavo Hoecker Sponsored by the Universidad de Chile Universidad de Chile,Santiago,Chile, 1962
An 32.3286

TISSUE TRANSPLANTATION The International symposium on tissue transplantation Proceedings Santiago 1961 Aug 30-Sep 2 Edited by Alberto P. Cristoffanini and Gustavo Hoecker Sponsored by Universidad de Chile Universidad de Chile,Santiago,1962
PGMS 29.0244

TITANIUM Science,technology and application of titanium International conference on titanium Proceedings London 1968 May 21-24 Institute of metals American institute of mining,metallurgical and petroleum engineers American society for metals Japan institute of metals Akademiya nauk S.S.S.R. Edited by R.I. Jaffee and N.E. Promisel Pergamon,Oxford,1970
Met 25.2852

TITANIUM Symposium on titanium :presented at the second Pacific area national meeting Los Angeles,Calif. 1956 Sep 17-18 American society for testing materials American society for testing materials.Special technical publication, 204 American society for testing materials,Philadelphia,Pa.,1957
Eng 41.3598

TITANIUM Symposium on titanium held at the second Pacific area national meeting Los Angeles 1956 Sep 17-18 American society for testing materials A.S.T.M.Special technical publication, 204 American society for testing materials,Philadelphia,Pa.,1957
Met 25.0571

TITANIUM :a symposium Report 1948 Dec 16 United States.Office of naval research Office of naval research,Washington,D.C.,1949
Met 25.0548

TITANIUM ALLOYS Applications related phenomena in titanium alloys :a symposium Los Angeles 1967 Apr 19 American society for testing materials A.S.T.M.Special technical publication, 432 A.S.T.M., Philadelphia,Pa.,1968
Met 25.0609

TJIAWI 1953 The Vegetation of the humid tropics :symposium 2nd Proceedings Unesco.Science cooperation office for south-east Asia Sponsored by the Council for sciences of Indonesia Unesco,c1953
Bot 42.4770

TOKYO 1954 Symposium on typhoons Proceedings Japanese national commission for Unesco Science council of Japan Central meteorological observatory,Tokyo illus 257p Japanese national commission for Unesco, Tokyo,1955
Nap 11.0612

TOKYO 1954 Unesco symposium on typhoons Proceedings Japanese national commission for Unesco Japanese national commission for Unesco,Tokyo,1955
Geog 13.0962

TOKYO 1955 Unesco symposium on physical oceanography Proceedings Unesco,and,Japan society for the promotion of science Unesco and Japan society for the promotion of science, Tokyo,1957
Geog 13.0875

TOKYO 1955 Nikko International symposium on algebraic number theory proceedings Science council of Japan Edited by Shokichi Iyanaga and Yukiyosi Kawada Jointly sponsored by the International mathematical union xx,265p 26cm Science council of Japan,Tokyo,1956
P. Math 2.1771

TOKYO 1957 Conference on the chemistry of muscular contraction Paper and discussions presented Committee of muscle chemistry of Japan Igaku Shoin ltd.,Tokyo;Osaka,1957
Bioch 33.2229

TOKYO 1957 Failure and defects of bridges and structures :a symposium Proceedings Japan society of civil engineers Architectural institute of Japan Japan society for the promotion of science,Tokyo, 1959
Eng 41.3015

TOKYO 1957 International geographical union regional conference Proceedings Science council of Japan,Tokyo,1959
Geog 13.5598

TOKYO 1957 Nara IGU regional conference in Japan Proceedings Science council of Japan,and,International geographical union Edited by Ryuziro Isida and others illus, maps vii,609p 26cm Tokyo,1959
Sco 14.6870

TOKYO 1957 and KYOTO 1957 The International symposium on enzyme chemistry Proceedings International union of biochemistry Organized by Science council of Japan I.U.B.Symposium series, 2 Pergamon, London,1958
Bioch 33.1053

TOKYO 1959 Prestressed structures :a symposium Proceedings Japan society of civil engineers Architectural institute of Japan Japan society for the promotion of science,Tokyo,1960
Eng 41.3016

TOKYO 1960 International congress of blood transfusion 8th Programme and abstracts International society of blood transfusion Edited by Ludwig Hollander 1960
Med 36.0201

TOKYO 1960 The Symposium on numerical weather prediction Proceedings Meteorological agency,Japan Science council of Japan Co-sponsored by the International union of geodesy and geophysics 655p Meteorological society of Japan,Tokyo,1962
Nap 11.0923

TOKYO 1960 and KYOTO 1960 World conference on earthquake engineering 2nd Proceedings Vol 1-3 Japan society of civil engineers Organised by the Science council of Japan 3 vols Science council of Japan,Tokyo,1960
Eng 41.6395

TOKYO 1962 International symposium on molecular structure and spectroscopy International union of pure and applied chemistry Science council of Japan,Tokyo, 1962
Chem 18.2639

TOKYO 1962 Kyoto 1962 Sep 7-12 The International conference on crystal lattice defects Proceedings Pt 1-3 Physical society of Japan Under the auspices of the International union of pure and applied physics Physical society of Japan.Journal, 18; Suppl.1-3 3 vols Physical society of Japan,Tokyo,1963
Cav 7.1343

TOKYO 1963 Progress in radio science 1960-63 URSI general assembly 14th Proceedings Vol 3: the ionosphere International scientific radio union. Commission on the ionosphere Edited by G.M. Brown 196p Elsevier,Amsterdam;London,1965
Nap 11.0086

TOKYO 1964 and KYOTO 1964 The Conference on photographic and spectrographic optics Proceedings International commission for optics Under the auspices of the Science council of Japan Japanese journal of applied optics, 4,suppl.1 669p Tokyo,1965
Obs 6.2050

TOKYO 1965 High energy physics Tokyo summer lectures in theoretical physics 1st Papers Pt 2: high energy physics University of Tokyo.Summer institute of theoretical physics Edited by Gyo Takeda Syokabo;W.A.Benjamin,Tokyo;New York,1966
Cav 7.2318

TOKYO 1965 International congress of physiological sciences 23rd lectures and symposia Under the auspices of the International union of physiological sciences International union of physiological sciences. Proceedings, 4 International congress series, 87 Excerpta medica,Amsterdam,1965
Phys 20.2126

TOKYO 1965 Olfaction and taste 2 Wenner-Gren center international symposium 2nd Proceedings Wenner-Gren center Edited by Takashi Hayashi Wenner-Gren center international symposium series, 8 Pergamon press,Oxford,1967
Psy 31.0296

TOKYO 1965 Tokyo summer institute in theoretical physics 1st Pt 2: high energy physics Edited by Gyo Takeda Tokyo summer lectures in theoretical physics,1965 iv,121p Syokabo;Benjamin,Tokyo;New York,1966
A Math 4.1346

TOKYO 1966 Conference on coastal engineering 10th Proceedings American society of civil engineers 2 vols American society of civil engineers,New York,1967
Eng 41.3348

TOKYO 1966 Memorial lecture meeting on the 10th anniversary of the foundation of National research institute for metals Proceedings National research institute for metals,Tokyo National research institute for metals,Tokyo, 1966
Met 25.0113

TOKYO 1966 Pacific science congress,11th Proceedings of the symposium on Pacific-Antarctic sciences :papers presented at the eleventh Pacific congress 11th Proceedings Edited by Takesi Nagata Japanese Antarctic research expedition.Scientific reports.Special issue,1 n.p.,1967
Sco 14.6039

TOKYO 1966 Tokyo summer lectures in theoretical physics 2nd Papers Pt 1: dynamical processes in solid state optics University of Tokyo.Summer institute of theoretical physics Edited by Gyo Takeda and Akihito Fudjii W.A.Benjamin,New York;Tokyo, 1967
Cav 7.2316

TOKYO 1966 Tokyo summer lectures in theoretical physics 2nd Papers Pt 1: dynamical processes in solid state optics University of Tokyo.Summer institute of theoretical physics Edited by Ryogo Kubo and Hiroshi Kamimura Benjamin,New York,1967
TA 15.0434

TOKYO 1966 Tokyo summer lectures in theoretical physics 2nd Papers Pt 2: elementary particle physics University of Tokyo.Summer institute of theoretical physics Edited by Gyo Takeda and Akihiko Fujii Tokyo; New York,1967 Course under chairmanship of Kubo,Ryogo
Cav 7.2317

TOKYO 1967 International conference on nuclear structure Physical society of Japan illus vi,755p 26cm Physical society of Japan,Tokyo,1968
Cav 7.3196

TOKYO 1967 International conference on the strength of metals and alloys Proceedings Japan institute of metals xlvi,1049p 31cm Japan institute of metals,Sendai,1968 In commemoration of the 30th anniversary of the institute
Cav 7.3081

TOKYO 1967 The Strength of metals and alloys the international conference in commemoration of the 30th anniversary of the Japan institute of metals Proceedings Japan institute of metals Japan institute of metals.Transactions, 9 suppl. Japan institute of metals,Tokyo,1968
Met 25.0829

TOKYO 1968 International congress of genetics 12th Proceedings Vol 1-3 Science council of Japan Under the auspices of the International union of biological sciences 3 vols Science council of Japan, Tokyo,1969
Gen 34.1008

TOKYO 1968 Japan national congress for applied mechanics 18th Proceedings Science council of Japan Japan national committee for theoretical and applied mechanics Bibliog,illus 188p 26cm Central scientific publishers,Tokyo,1970
A Math 4.1725

TOKYO 1968 Planetary electrodynamics International conference on the universal aspects of atmospheric electricity 4th Proceedings Vol 1-2 Edited by Samuel C. Coroniti and James Hughes 2 vols Gordon and Breach,New York;London,1969
Nap 11.0974

TOKYO 1968 Solar flares and space research Proceedings of the symposium held on the occasion of the 11th plenary meeting of the Committee on space research Committee on space research International astronomical union Edited by C.de Jager and Z. Svestka 419p North-Holland,Amsterdam,1969 Co-sponsored by International astronomical union, International union of geodesy and geophysics and International union of radio scientists
Obs 6.3335

TOKYO 1969 International conference on functional analysis and related topics Proceedings International mathematical union Science council of Japan Mathematical society of Japan xxxiii,425p 26cm University of Tokyo press,Tokyo,1970
P Math 2.4279

TOKYO 1970 Antarctic ice and water masses Proceedings Scientific committee on antarctic research Edited by George Deacon illus x,114p 23cm Scientific committee on antarctic research,1971
A Math 4.1746

TOKYO 1970 Solid state devices :a conference Proceedings Japan society of applied physics Japan society of applied physics.Journal, 40,Supplement diagrs, photos 250p 26cm Japan society of applied physics,Tokyo,1970
Cav 7.2990

TOKYO 1971 International conference on wind effects on buildings and structures 3rd Proceedings Science council of Japan Saikon,Tokyo,1971
Eng 41.8602

TOKYO 1971 Seminar on modern methods in number theory Held at the Institute of statistical mathematics,Tokyo 29cm Institute of statistical mathematics,Tokyo, 1971
P Math 2.4380

TOKYO SUMMER INSTITUTE IN THEORETICAL PHYSICS 1st Tokyo 1965 Sep 6-17 Pt 2: high energy physics Edited by Gyo Takeda Tokyo summer lectures in theoretical physics, 1965 iv,121p Syokabo;Benjamin,Tokyo;New York,1966
A Math 4.1346

TOKYO SUMMER LECTURES IN THEORETICAL PHYSICS 1st Papers High energy physics Tokyo 1965 Sep 6-17 Pt 2: high energy physics University of Tokyo.Summer institute of theoretical physics Edited by Gyo Takeda Syokabo;W.A.Benjamin,Tokyo;New York,1966
Cav 7.2318

TOKYO SUMMER LECTURES IN THEORETICAL PHYSICS 2nd Papers Tokyo 1966 Aug 29-Sep 3 Pt 1: dynamical processes in solid state optics University of Tokyo.Summer institute of theoretical physics Edited by Gyo Takeda and Akihito Fudjii W.A.Benjamin,New York;Tokyo, 1967
Cav 7.2316

TOKYO SUMMER LECTURES IN THEORETICAL PHYSICS 2nd Papers Tokyo 1966 Aug 29-Sep 3 Pt 1: dynamical processes in solid state optics University of Tokyo.Summer institute of theoretical physics Edited by Ryogo Kubo and Hiroshi Kamimura Benjamin,New York,1967
TA 15.0434

TOKYO SUMMER LECTURES IN THEORETICAL PHYSICS 2nd Papers Tokyo 1966 Aug 29-Sep 3 Pt 2: elementary particle physics University of Tokyo.Summer institute of theoretical physics Edited by Gyo Takeda and Akihiko Fujii Tokyo;New York,1967 Course under chairmanship of Kubo,Ryogo
Cav 7.2317

TOLBUTAMIDE...AFTER TEN YEARS The Brook Lodge symposium Proceedings Augusta,Mich. 1967 Mar 6-7 Edited by W.J.H. Butterfield and W.van Westering International congress series, 149 Excerpta,medica,Amsterdam,1967
Bioch 33.0425

TOLBUTAMIDE...AFTER TEN YEARS :the Brook Lodge symposium Proceedings Augusta,Mich. 1967 Mar 6-7 Edited by W.J.H. Butterfield and W.van Westering Excerpta medica international congress series,149 Bibliogs 346p Excerpta medica foundation,Amsterdam, 1967
Pha 16.0264

TOOLS OF BIOLOGICAL RESEARCH SYMPOSIUM 1st Proceedings London 1958 Oct 10-11 Surgical research society Edited by Hedley J. B. Atkins Blackwell,Oxford,1959
Gen 34.0465

TOOLS OF BIOLOGICAL RESEARCH SYMPOSIUM 2nd Proceedings 1959 Second series Surgical research society Edited by Hedley J. B. Atkins illus. Blackwell,Oxford,1960
Gen 34.0466

TOOLS OF BIOLOGICAL RESEARCH SYMPOSIUM 3rd Proceedings London 1960 Sep 30-Oct 1 Third series Surgical research society Edited by Hedley J.B. Atkins Blackwell, Oxford,1961
Gen 34.0467

TOOTH ENAMEL;ITS COMPOSITION,PROPERTIES AND FUNDAMENTAL STRUCTURE An International symposium on the composition,properties and fundamental structure of tooth enamel held at the London hospital medical college Report of the proceedings London 1964 Apr 6-7 Edited by Maurice V. Stack and Ronald W. Fearnhead Wright,Bristol,1965 Sponsored by the Wellcome trust,London;the Procter and Gamble Co., Cincinnati;Colgate-Palmolive Ltd., Great Britain;Unilever Ltd.,London
Bal 39.1298

TOPICS IN BASIC NEUROLOGY Edited by W. Bargmann and J.P. Schade Progress in brain research, 6 Elsevier,Amsterdam,1964 Lectures delivered during a symposium on "Topics in basic neurology"held as part of the Third International meeting of neurobiologists, Kiel,1962
An 32.4226

TOPOGRAPHIE ET LA GEOLOGIE DES PROFONDEURS OCEANIQUES Nice 1958 May 5-12 Centre national de la recherche scientifique. Colloques internationaux,83 C.N.R.S.,Paris, 1959
Geog 13.0734

TOPOLOGIA DIFFERENZIALE 1 ciclo Urbino 1962 Jul 2-12 By J. Cerf and others Centro internazionale matematico estivo 27cm Institute matematica dell'universita,Rome,1962
P. Math 2.0243

TOPOLOGICAL ALGEBRA Summer school on topological algebra theory Bruges 1966 Sep 6-16 Bibliog 285p 27cm University of Brussels,Brussels,1966
P Math 2.2745

TOPOLOGICAL DYNAMICS :an international symposium Fort Collins,Colo. 1967 Aug Edited by Joseph Auslander and Walter H. Gottschalk Held at the University of Colorado Bibliog.,Illus. xv,522p 23cm Benjamin,New York;Amsterdam,1968 Dedicated to Frank J.Hahn,1929-1968
P Math 2.3143

TOPOLOGICAL DYNAMICS AND ERGODIC THEORY Symposium on topological dynamics and ergodic theory Proceedings Lexington,Ky. 1971 Jun 3-5 University of Kentucky.Department of mathematics 80p 28cm University of Kentucky,Lexington,Ky.,1971 Duplicated notes
P Math 2.4189

TOPOLOGICAL PROPERTIES Set-valued mappings, selections and topological properties Proceedings Buffalo,N.Y. 1969 May 8-10 Edited by W.M. Fleischman Held at the State university of New York Lecture notes in mathematics, 171 Springer,Berlin,1970
Math S 3.1657

TOPOLOGY Algebraic geometry and topology :a symposium in honor of S.Lefschetz Princeton, N.J. 1954 Apr 8-10 Edited by R.H. Fox and others Princeton mathematical series, 12 viii,399p 24cm Princeton university press, Princeton,N.J.,1957
P. Math 2.0574

TOPOLOGY Colloque de topologie Brussels 1964 Sep 7-10 Centre belge de recherches mathematiques 235p 25cm Louvain, Paris, 1966
P. Math 2.0294

TOPOLOGY Colloque de topologie (espaces fibres) Brussels 1950 Jun 5-8 Centre Belge de recherches mathematiques 136p 24cm Centre Belge de recherches mathematiques,Brussels,1950
P Math 2.3126

TOPOLOGY Conference on general topology Proceedings Pullman,Wash. 1970 Mar 25-27 Washington state university.Department of mathematics iv,136p 27cm Washington state university.Department of mathematics, Pullman,Wash.,1970
P Math 2.4108

TOPOLOGY General topology and its relations to modern analysis and algebra Prague topological symposium 2nd Proceedings Prague 1966 Aug 30-Sep 4 Ceskoslovenska akademie ved 365p 24cm Academic press; Academia,New York;Prague,1967
P Math 2.3346

TOPOLOGY General topology and its relations to modern analysis and algebra :conference Proceedings Kanpur 1968 Oct 3-12 illus 332p 24cm Academia,Prague,1971
P Math 2.4109

TOPOLOGY General topology and its relations to modern analysis and algebra :symposium proceedings Prague 1961 Sep 1-8 Ceskoslovenska akademie ved Edited by J. Novak jointly with International mathematical union 363p 25cm Academic press,New York,1962
P. Math 2.0291

TOPOLOGY International symposium on topology and its applications Proceedings Herceg-Novi 1968 Aug 28-31 Edited by D.R. Kurepa 356p 25cm Savez matematicara,Belgrade,1969
P Math 2.4117

TOPOLOGY Lectures in topology :conference Ann Arbor 1940 Jun.24-Jul.6 University of Michigan Edited by Raymond L. Wilder and William L. Ayres vii,316p 23cm University of Michigan press,Ann Arbor, Mich., 1941
P. Math 2.0288

TOPOLOGY Summer school in mathematics, geometry and topology proceedings Dundee 1961 Jul 10-22 Queen's college,Dundee 33cm Queen's College,Dundee,1961
P. Math 2.0292

TOPOLOGY Symposium on infinite dimensional topology Proceedings Baton Rouge,La. 1967 Mar 27-Apr 1 Edited by R.D. Anderson Annals of mathematics studies, 69 viii,300p 24cm Princeton university press;University of Tokyo press,Princeton,N.J.;Tokyo,1972
P Math 2.4112

TOPOLOGY OF MANIFOLDS University of Georgia topology of manifolds institute Proceedings Athens 1969 Aug 11-22 Edited by J.C. Cantrell and C.H. Edwards xiv,514p 24cm Markham,Chicago,Ill.,1970
P Math 2.3819

TOPOLOGY OF MANIFOLDS :a conference Proceedings East Lansing,Mich. 1967 Mar 15-17 Edited by John G. Hocking Held at Michigan state university Bibliog. viii, 161p 21cm Prindle,Weber and Schmidt,Boston, Mass.,1968
P Math 2.3351

TOPOLOGY OF 3-MANIFOLDS AND RELATED TOPICS :a symposium proceedings Athens, Ga. 1961 Aug 14 - Sep 8 University of Georgia. Topology institute Edited by M.K. Fort supported by United States.Office of naval research and National science foundation viii,256p 24cm Englewood Cliffs, N.J.,1962
P. Math 2.0296

TOPOSES,ALGEBRAIC GEOMETRY AND LOGIC Conference on connections between category theory and algebraic geometry and intuitionistic logic Partial report Halifax,Nova Scotia 1971 Jan 16-19 Edited by F.W. Lawvere Lecture notes in mathematics, 274 189p 25cm Springer,Berlin,1972
P Math 2.4171

TORONTO 1928 The Symposium on colloid chemistry 6th Papers presented University of Toronto Edited by Harry Boyer Weiser Colloid symposium monograph, 6 Chemical catalog company,New York,1928 Colloid symposium monograph known also as Colloid symposium annual
Bioch 33.1474

TORONTO 1952 Association for computing machinery :meeting Proceedings illus. 160p 28cm Association for computing machinery,Washington,D.C.,1952
Math L 5.3238

TORONTO 1957 International association of scientific hydrology :general assembly 11th Proceedings Vol 4: commission on snow and ice Edited by L.J. Tison International association of scientific hydrology. Publication,46 563p 24cm Gentbrugge,1958 Papers in English,French,German,Russian
Sco 14.0125

TORONTO 1957 The Toronto meeting Transactions International union of geodesy and geophysics,and,International association of geomagnetism and aeronomy Edited by T.R. Kaiser and V. Laursen I.A.G.A.bulletin,16 x,412p 25cm North Holland,Copenhagen,1960
Sco 14.0244

TORONTO 1957 Union geodsique et geophysique internationale :assemblee generale 11th Comptes rendus 284p Paris,1958 Papers in English and French
Sco 14.7407

TORONTO 1958 Canadian conference for computing and data processing proceedings illus. xii,383p University of Toronto press,Toronto,1958
Math L 5.0845

TORONTO 1960 The Computing and data processing society of Canada conference 2nd proceedings vi,365p 25cm University of Toronto press,Toronto,1960
Math L 5.0896

TORONTO 1960 The International conference on low temperature physics 7th Proceedings International union of pure and applied physics Edited by J.M. Graham and A. C. Hollis-Hallett With the co-operation of the Canadian association of physicists University of Toronto press;North-Holland, Toronto;Amsterdam,1961
Cav 7.2330

TORONTO 1961 Nucleation and crystallization in glasses and melts :a symposium American ceramic society Edited by Margie K. Reser and others American ceramic society,Columbus,Ohio,1962
Met 25.0127

TORONTO 1964 Chemical physics of ionic solutions :a selction of invited papers and discussions presented at the international symposium of the Electrochemical society... under the chairmanship of B.E.Conway Electrochemical society Edited by B.E. Conway and R.G. Barradas J.Wiley,New York, 1966
Chem 18.0342

TORONTO 1964 International conference on electron and ion beam science and technology 1st American institute of mining, metallurgical and petroleum engineers. and Electrochemical society.Electrothermics and metallurgy divisions Edited by Robert Bakish Electrochemical society series Wiley,New York,1965
Met 25.1678

TORONTO 1964 International conference on electron and ion beam science and technology 1st Edited by Robert Bakish Sponsored by the Electrochemical society.Electrothermics and metallurgy divisions Electrochemical society series Wiley,New York,1965
Eng 41.5564

TORONTO 1964 International symposium on rarefied gas dynamics 4th Proceedings Vol 1-2 United States.Air force.Office of scientific research United States.Office of naval research National aeronautics and space administration.Institute for aerospace studies Edited by J.H.de Leeuw Advances in applied mechanics.Supplement, 3 2 vols Academic press,New York;London,1965
Eng 41.6788

TORONTO 1964 International symposium on rarefied gas dynamics 4th proceedings Vol 1-2 University of Toronto.Institute for aerospace studies Edited by J.H.de Leeuw Advances in applied mechanics.Supplement, 3 2 vols Academic press,New York,London,1965-66
A Math 4.0532

TORONTO 1964 Primary sedimentary structures and their hydrodynamic interpretation :a symposium papers Edited by Gerard V. Middleton Sponsored by the Society of economic palaeontologists and mineralogists Society of economic paleontologists and mineralogists.Special publication,12 bibliog.,illus. map Society of economic paleontologists and mineralogists, Tulsa,Okla.,1965
Geol 8.1515

TORONTO 1964 Rarefied gas dynamics : international symposium 4th Proceedings Vol 2 Edited by J.H.de Leeuw Advances in applied mechanics.Supplement, 3,vol 2 xv,608p Academic press,New York;London,1966
Chem E 24.0410

TORONTO 1965 Dolomitization and limestone diagenesis :a symposium papers Edited by Lloyd C. Pray and Raymond C. Murray Society of economic paleontologists and mineralogists. Special publication,13 illus. Society of economic paleontologists and mineralogists, Tulsa,Okla.,1965
Geol 8.1524

TORONTO 1965 Mammalian hibernation 3 International symposium on natural mammalian hibernation 3rd Proceedings Edited by Kenneth C. Fisher and others illus xiv, 534p 25cm Oliver and Boyd,Edinburgh;London, 1967
Sco 14.8231

TORONTO 1966 Determination of radial velocities and their applications International astronomical union and International council of scientific unions Edited by A.H. Batten and J.F. Heard International astronomical union.Symposium, 30 Academic press,London;New York,1967
TA 15.0307

TORONTO 1966 Determination of radial velocities and their applications :a symposium International astronomical union Edited by A. H. Batten and J.F. Heard International astronomical union.Symposium, 30 262p academic press,London,1967 publication dedicated to the memory of R.Methven Petrie
Obs 6.0147

TORONTO 1967 Summer institute in nuclear and particle physics Proceedings Canadian association of physicists.Theoretical physics division Edited by B. Margolis and C.S. Lam 547p 24cm Benjamin,New York,1968
Cav 7.3100

TORONTO 1968 International conference on cloud physics Proceedings American meteorological society 873p American meteorological society,Boston,Mass.,1969
Nap 11.0973

TORONTO 1971 Insulin action Symposium on insulin action Proceedings Edited by Irving B. Fritz Academic press,New York,1972 Held in honour of the 50th anniversary of the discovery of insulin
Bioch 33.2192

TORONTO,ONT. 1970 Symposium on energy dispersion X-ray analysis :X-ray and electron probe analysis American society for testing and materials.Committee E-4 on metallography ASTM,Philadelphia,Pa.,1971 Symposium presented at the annual meeting.Co-ordinator:J. C.Russ
Met 25.2617

TORQUAY 1863 Climate :an inquiry into the causes of its differences,and its influence on vegetable life By C. Daubeny Natural history society Bohn,Oxford,1863 Being the substance of four lectures delivered before the Natural history society
Bot 42.2042

TOUCH,HEAT AND PAIN :Ciba foundation symposium Ciba foundation Edited by A.V. S. De Reuck and J. Knight Churchill,London, 1966
Phys 20.2036

TOUCH,HEAT AND PAIN :Ciba foundation symposium Proceedings London 1965 Sep 21-23 Ciba foundation Edited by A.V.S. De Reuck and Julie Knight Churchill,London,1966
An 32.5335

TOULON 1964 Theory of games:techniques and applications conference proceedings North Atlantic treaty organization.Scientific affairs committee Edited by A. Mensch xi, 490p 23cm English universities press, London,1966
Math 3.0241

TOULOUSE 1955 Les techniques recentes en microscopie electronique et corpusculaire : colloque international Centre national de la recherche scientifique Edited by Ch. Fert Centre national de la recherche scientifique. Colloques internationaux, 58 Paris,1956
Cav 7.0525

TOURS 1959 Les Agenesies du corps calleux : rapport de neurologie presente au Congres de psychiatrie et de neurologie de langue franciase By Francis Rohmer Masson,Paris, 1959
An 32.4178

TOWARD A LESS HARMFUL CIGARETTE The World conference on smoking and health :a workshop Proceedings 1967 Sep 11-13 National cancer institute Edited by Ernest L. Wynder and Dietrich Hoffmann National cancer institute. Monograph, 28 National cancer institute, Bethesda,Md.,1968
Bioch 33.0992

TOWARDS A THEORETICAL BIOLOGY 4: essays:an IUBS symposium International union of biological sciences Edited by C.H. Waddington 299p 26cm Edinburgh university press,Edinburgh,1972
Math S 3.1866

TOWARDS A THEORETICAL BIOLOGY :a collection of essays written as a result of a symposium... Bellagio 1966 Aug 28-Sep 3 International union of biological sciences Edited by C.H. Waddington 253p 26cm Edinburgh university press,Edinburgh,1970
Math S 3.1819

TOWARDS A THEORETICAL BIOLOGY :an I.U.B.S. symposium Lake Como 1966 Aug 28-Sep 3 and Lake Como 1967 Aug 3-12 Vol 1-3 International union of biological sciences Edited by C.H. Waddington 3 vols Edinburgh university press,Edinburgh,1970
Bot 42.0320

TOWARDS A THEORETICAL BIOLOGY :an IUBS symposium Lake Como 1967 Aug 3-12 2: sketches International union of biological sciences Edited by C.H. Waddington Edinburgh university press,Edinburgh,1968
Bal 39.0017

TOWARDS A THEORETICAL BIOLOGY :a symposium Papers Lake Como 1967 Aug 3-12 Vol 2: sketches International union of biological sciences Edited by C.H. Waddington I.U.B.S.symposium, 2 Edinburgh university press,Edinburgh,1969
Bal 39.0053

TOWARDS A THEORETICAL BIOLOGY 1.Prolegomena:an I.U.B.S. symposium Essays Bellagio 1966 Aug 28-Sep 3 International union of biological sciences Edited by C.H. Waddington 234p Edinburgh university press, Edinburgh,1968 Essays written after and in the light of the symposium
WSM 43.2897

TOXICITY Selective toxicity and antibiotics Edinburgh 1948 Jul Society for experimental biology Society for experimental biology.Symposia, 3 Cambridge university press,Cambridge,1949
Phys 20.2224

TOXICITY Selective toxicity and antibiotics : a symposium Edinburgh 1948 Jul Society for experimental biology Society for experimental biology.Symposia, 3 Cambridge university press,Cambridge,1949
Bioch 33.1350

TRACE ANALYSIS Symposium on trace analysis papers New York 1955 Nov 2-4 Rockefeller foundation and Sloan-Kettering institute Edited by John H. Yoe and Henry J. Koch Wiley;Chapman and Hall,New York;London,1957
Met 25.1720

TRACE ELEMENTS Dosage des elements a l'etat de traces dans les roches et autres substances colloque Comptes-rendus Nancy 1968 Dec 4-6 Centre national de la recherche scientifique Centre national de la recherche scientifique.Colloques nationaux, 923 CNRS, Paris,1970
Min 10.1475

TRACE ELEMENTS :the conference held at Ohio agricultural experiment station Proceedings Wooster,Ohio 1957 Oct 14-16 Ohio agricultural experiment station Edited by C. A. Lamb and others Academic press,New York; London,1958
Bioch 33.0348

TRACER ELEMENTS Biological applications of tracer elements :a symposium Papers Cold Spring Harbor 1948 Cold Spring Harbor biological laboratory Cold Spring Harbor symposia on quantitative biology, 13 Long Island biological association,Cold Spring Harbor,1948
Bioch 33.1269

TRACER METHODOLOGY Annual symposium on tracer methodology 5th Proceedings Washington,D.C. 1961 Oct 20 New England nuclear corporation Packard instrument company Edited by Seymour Rothschild Advances in tracer methodology, 1 Plenum press,New York,1963 Includes selected papers from the first four annual symposia
Gen 34.0633

TRACER METHODOLOGY Symposium on advances in tracer methodology 6th-8th Collection of papers presented,plus other papers Edited by Seymour Rothschild Advances in tracer methodology, 2 New England nuclear corporation.Publication Plenum press,New York,1965
Bioch 33.0301

TRADES UNION CONGRESS Automation '65 British automation conference 1965 Eastbourne 1965 Nov 7-10 Organised by the Institution of production engineers Institution of production engineers,London, 1965
Eng 41.6018

TRAFFIC CONTROL :theory and instrumentation Papers Instrumentation requirements for traffic control systems New York 1963 Dec 16-17 Instrument society of America Edited by Thomas R. Horton Plenum press,New York, 1965
Eng 41.3310

TRAFFIC FLOW International symposium on the theory of road traffic flow London 1963 Jun 25-27 London,1963 Unbound mimeograph papers
Math 3.1113

TRAFFIC FLOW International symposium on the theory of road traffic flow London 1963 Jun 25-27 Road research laboratory Edited by Joyce Almond Organization for economic cooperation and development,Paris,1965
Eng 41.3311

TRAFFIC FLOW The Theory of road traffic flow international symposium 2nd proceedings London 1963 Jun.25-27 Road research laboratory Edited by Joyce Almond With financial support from General motors corporation.Research laboratories ix,406p 24cm Organisation for economic co-operation and development,Paris,1965
Math 3.0922

TRAFFIC FLOW Vehicular traffic science International symposium on the theory of traffic flow 3rd Proceedings New York 1965 Jun Edited by Leslie C. Edie and others Under the auspices of the Operations research society of America.Transportation science section illus x,373p 24cm American Elsevier,New York,1967
Math 3.1137

TRAINING OF THE PHYSICIAN :celebration of the Centennial of the Boston City hospital and the 40th anniversary of the Thorndike memorial laboratory Papers Thorndike memorial laboratory Boston city hospital.Harvard medical unit Boston,Mass.,1964
PGMS 29.0489

TRAITE SUR L'ANTARCTIQUE see ANTARCTIC TREATY

TRANSCRIPTION OF GENETIC MATERIAL :a symposium Papers Cold Spring Harbor 1970 Jun 4-11 Cold Spring Harbor laboratory of quantitative biology Cold Spring Harbor symposia on quantitative biology, 35 Cold Spring Harbor, 1971
Bioch 33.2335

The TRANSFER OF CALCIUM AND STRONTIUM ACROSS BIOLOGICAL MEMBRANES :a conference Proceedings Ithaca,N.Y. May 13-16 Edited by R.H. Wasserman Academic press,New York;London,1963
Inv Med 37.0227

The TRANSFER OF CALCIUM AND STRONTIUM ACROSS BIOLOGICAL MEMBRANES :a conference Proceedings Ithaca,N.Y. 1962 May 13-16 Edited by R.H. Wasserman held at Cornell university Academic press,New York;London, 1963
Bioch 33.1021

The TRANSFER OF CALCIUM AND STRONTIUM ACROSS BIOLOGICAL MEMBRANES :a conference Proceedings Ithaca,N.Y. 1962 May 13-16 Edited by R.H. Wassermann Academic press,New York,1963
Pha 16.0200

TRANSFER OF RADIATION IN STELLAR ATMOSPHERES Colloquium on the theory of stellar atmospheres 3rd Herstmonceux 1962 Aug.15-16 International astronomical union Edited by C.de Jager and A.B. Underhill 201p Pergamon press,London,1963 Vol 3,no.2 of Journal of quantitative spectroscopy and radiative transfer. AprJun.1963
Obs 6.0875

TRANSFORMATION GROUPS Conference on compact transformation groups 2nd Proceedings Amherst,Mass. 1972 Pt 1-2 Held at the University of Massachusetts Lecture notes in mathematics, 298-299 25cm 2 vols Springer,Berlin,1972
P Math 2.4110

TRANSFORMATION GROUPS Conference on transformation groups Proceedings New Orleans 1967 May 8-Jun 2 Edited by Paul S. Mostert Sponsored by the National science foundation.Advanced science seminar projects Bibliog.,Illus. xiii,456p 24cm Springer-Verlag,Berlin,1968
P Math 2.3395

TRANSFUSION SANGUINE ET ACTUALITES HEMATOLOGIQUES 1st :congres national des transfusions sanguines de France et des pays de langue francaise Algiers 1953 Apr 1-4 Edited by E. Benhamou and A. Albou Masson et compagnie,Paris,1954
Med 36.0023

TRANSITION METAL COMPOUNDS Buhl international conference on materials : transition metal compounds ;transport and magnetic properties 1st Informal proceedings Pittsburgh,Penn. 1963 Oct 31-Nov 1 Buhl foundation and Carnegie institute of technology,Pittsburgh Edited by E.R. Schatz Gordon and Breach,New York; London,1964
Met 25.2351

TRANSITIONS AND RELAXATIONS IN POLYMERS : symposium Papers Atlantic City,N.J. 1965 Sep 13-14 American chemical society Edited by Raymond F. Boyer Polymer symposia, 14 Interscience,New York,1966
Met 25.2704

TRANSITIONS INTERDITES DANS LES SPECTRES DES ASTRES Colloque international d'astrophysique 15ieme Communications Liege 1968 Jun 24-26 Societe royale des sciences de Liege.Memoires,5s, 17 410p Institut d'astrophysique,Liege,1969
TA 15.0576

TRANSMETHYLATION Symposium on transmethylation and methionine biosynthesis Argonne 1964 Edited by Stanley K. Shapiro and Fritz Schlenk Sponsored by the Argonne national laboratory.Division of biological and medical research Chicago university press, Chicago,1965
Bioch 33.0607

TRANSMISSION OF SCHIZOPHRENIA :research conference Proceedings Dorado,Puerto Rico 1967 Jun 26-Jul 1 Foundations' fund for research in psychiatry Edited by Dvid Rosenthal and Seymour S. Kety Foundations' fund for research in psychiatry.Research conference, 2 Pergamon press,Oxford,1968 reprinted 1969
PGMS 29.0566

The TRANSMISSION OF SCHIZOPHRENIA:RESEARCH CONFERENCE 2nd Proceedings Dorado, Puerto Rico 1967 Jun 26-Jul 1 Foundations fund for research in psychiatry Edited by David Rosenthal and,Seymour S. Kety Pergamon press,Oxford,1968 The second in a series of three psychiatric research conferences
Psy 28.0084

The TRANSPARANCY OF THE CORNEA :a symposium Knokke 1959 Aug 31-Sep 5 Council for international organizations of medical sciences Edited by William Stewart Duke-Elder and E.S. Perkins Blackwell,Oxford,1960
An 32.4772

TRANSPLANTATION Ciba foundation symposium on transplantation London 1961 Edited by G. E.W. Wolstenholme and Margaret P. Cameron Churchill,London,1962
An 32.3281

TRANSPLANTATION ANTIGENS AND TISSUE TYPING symposium Proceedings Charles Salt research centre 1968 Nov 11-13 Robert Jones and Agnes Hunt orthopaedic hospital management committee Edited by M.W. Elves and N.W. Nisbet Oswestry,1969
Surg 23.0031

TRANSPORT Active transport and secretion Bangor 1953 Jul Society for experimental biology Society for experimental biology. Symposia, 8 Cambridge university press, Cambridge,1954
Phys 20.0958

TRATADO ANTARTICO see ANTARCTIC TREATY

TRAUMA OF THE CENTRAL NERVOUS SYSTEM Proceedings of the Association New York 1943 Dec 17-18 Association for research in nervous and mental disease Association for research in nervous and mental disease. Research publications, 24 Williams and Wilkins,Baltimore,Md.,1945
An 32.5227

Les TRAVAUX COLLOQUE SUR LES TECHNIQUES DE CALCUL ET LES CALCULATIONS Travaux Bucharest 1967 Sep 22-26 Societate de Stiinte matematice din R.S.Romania.Bulletin mathematique, 12,1 Societate de Stiinte matematice din R.S.Romania,Bucharest,1968
Math L 5.3513

TRAVAUX TOPOGRAPHIQUES ET CARTOGRAPHIQUES International geographical congress Report Amsterdam 1938 Jul International geographical union Amsterdam,1938
Geog 13.5708

The TREATMENT OF CANCER with special reference to radiotherapy and chemotherapy Edited by J.S. Mitchell University of Cambridge. School of clinical research and postgraduate medical teaching.Course,1963 Cambridge university press,Cambridge,1965
PGMS 29.0456

The TREATMENT OF CARCINOMA OF THE BREAST :a symposium Proceedings Cambridge 1967 Sep 9 Syntex pharmaceuticals Edited by Anthony S. Jarrett Under the auspices of the Institute of hormone biology Lewis,Cambridge, 1968
Inv Med 37.0122

The TREATMENT OF CARCINOMA OF THE BREAT :a symposium Proceedings Cambridge 1967 Sep 7 Syntex pharmaceuticals Edited by Antony S. Jarrett Under the auspices of the Institute of hormone biology Excerpta medica, Amsterdam,1967
Radioth 35.0926

TREES Counting labelled trees Canadian mathematical congress 12th biennial seminar Lectures Vancouver,B.C. 1969 Canadian mathematical congress Canadian mathematical monographs, 1 x,113p 23cm Canadian mathematical congress,Montreal,Que.,1970
P Math 2.4447

TREMEZZO 1968 International course on materials science 1st Essays Accademia nazionale dei lincei.Donegani foundation Edited by Alan W. Searcy and others illus xxv,715p 24cm Interscience,New York; Chichester,1970 Being the 11th course organized by the Donegani foundation
Met 25.2593

TRENDS IN LATTICE THEORY By Garrett Birkhoff and others Edited by J.C. Abbott Van Nostrand mathematical studies, 31 Bibliog. ix,215p 20cm Van Nostrand Reinhold,New York,1970 Written versions of four talks given to a symposium on "Trends in Lattice theory" held at the U.S.Naval academy,May 1966
P Math 2.3710

TRIBOLOGY CONVENTION Douglas,I.o.M. 1971 May 12-15 Institution of mechanical engineers.Tribology group Institution of mechanical engineers.Proceedings, 19 IME, London,1971
Eng 41.8642

TRIENNIAL CONGRESS...ON THE SUBJECT OF FUNDAMENTAL RELATIONSHIPS BETWEEN ALL SECTIONS OF THE INDUSTRIAL COMMUNITY 1st Report Cambridge 1928 Jun 27-Jul 3 International association for the study and improvement of human relations and conditions in industry Secretariat I.R.I.,The Hague, 1928
Psy 31.2516

TRIESTE 1962 The Seminar on theoretical physics held in Miramare Papers International atomic energy agency Edited by A. Salam STI-PUB-61 637p international atomic energy agency,Vienna,1963
Cav 7.2346

TRIESTE 1962 Theoretical physics Seminar on theoretical physics Lectures Organized by the International atomic energy agency International atomic energy agency,Vienna,1963
Radioth 35.1655

TRIESTE 1962 Theoretical physics Seminar on theoretical physics lectures Organized by the International atomic energy agency Vienna,1963 Companion volume to International summer school on nuclear theory, 1962
A Math 4.0886

TRIESTE 1964 Plasma physics The International seminar on plasma physics proceedings International atomic energy agency.International centre for theoretical physics International atomic energy agency, Vienna,1965
A Math 4.0624

TRIESTE 1965 High-energy physics and elementary particles Seminar on elementary particles lectures International atomic energy agency International atomic energy agency.Proceedings series STIPUB117 Vienna, 1965
A Math 4.0888

TRIESTE 1966 Colloquium on late-type stars Proceedings Osservatorio astronomico di Trieste Edited by Margherita Hack 465p Trieste,1966
TA 15.0315

TRIESTE 1966 Colloquium on late-type stars proceedings Osservatorio astronomico di Trieste Edited by Margherita Hack 465p Trieste,1967
Obs 6.0712

TRIESTE 1968 Mass loss from stars Trieste colloquium on astrophysics 2nd Proceedings Edited by Margherita Hack Sponsored by the International astronomical union Astrophysics and space science library, 13 345p Reidel,Dordrecht,1969
Obs 6.3628

TRIESTE 1968 Mass loss from stars Trieste colloquium on astrophysics 2nd Proceedings Edited by Margherita Hack Sponsored by the International astronomical union Astrophysics and space science library, 13 345p Reidel,Dordrecht,1969
TA 15.0620

TRIESTE COLLOQUIUM ON ASTROPHYSICS 2nd Proceedings Mass loss from stars Trieste 1968 Sep 12-17 Edited by Margherita Hack Sponsored by the International astronomical union Astrophysics and space science library, 13 345p Reidel,Dordrecht,1969
TA 15.0620

TRIESTE COLLOQUIUM ON ASTROPHYSICS 2nd Proceedings Mass loss from stars Trieste 1968 Sep 12-17 Edited by Margherita Hack Sponsored by the International astronomical union Astrophysics and space science library, 13 345p Reidel,Dordrecht,1969
Obs 6.3628

TRINIDAD 1962 The Neotropical botany conference Proceedings Association of tropical biology.Bulletin, 1 Trinidad,1962
BG 38.3113

The TRIPLET STATE an international symposium Proceedings Beirut 1967 Feb 14-19 American university of Beirut Edited by A.B. Zahlan and others Cambridge university press, Cambridge,1967
Chem 18.2642

TRITIUM IN THE PHYSICAL AND BIOLOGICAL SCIENCES Symposium on the detection and use of tritium in the physical and biological sciences Proceedings Vienna 1961 May 3-10 Vols 1-2 Sponsored by the International atomic energy agency 2 vols International atomic energy agency,Vienna,1962 Co-sponsored by the Joint commission on applied radioactivity
Radioth 35.0104

TRITIUM IN THE PHYSICAL AND BIOLOGICAL SCIENCES The Detection and use of tritium in the physical and biological sciences symposium Proceedings Vienna 1961 May 3-10 Vol 1-2 Joint commission on applied radioactivity Sponsored by the International atomic energy agency International atomic energy agency.Proceedings series 2 vols International atomic energy agency,Vienna,1962 Papers in English and Russian.Summaries in English,French,Russian and Spanish
Gen 34.0470

TRONDHEIM 1955 International seaweed symposium 2nd Proceedings Edited by Trygve Braarud and Nils Andreas Sorensen illus. xiv,220p Pergamon,London,1956
Bot 42.4700

TROPICAL MEDICINE AND MALARIA International congress on tropical medicine and malaria 4th Proceedings Washington,D.C. 1948 May 10-18 Vol 1-2 United States. Department of State United States.Department of State.International organization and conference series, 1,5 Govt. printing office,Washington,D.C.,1948
Mol 45.0657

TROPICAL METEOROLOGY Symposium on tropical meteorology Proceedings Rotorua,N.Z. 1963 Nov 5-13 World meteorological organization Edited by J.W. Hutchings With the cooperation of the International union of geodesy and geophysics illus 751p New Zealand meteorological service,Wellington,N.Z., 1964
Nap 11.0650

TROPICAL METEOROLOGY Symposium on tropical meteorology Proceedings Rotorua,N.Z. 1963 Nov 5-13 World meteorological organization,and,International union of geodesy and geophysics Edited by J.W. Hutchings New Zealand meteorological service, Wellington,N.Z.,1964
Geog 13.0990

TROPICAL SOILS AND VEGETATION :Abidjan symposium Proceedings Abidjan 1959 Oct 20-24 Unesco Jointly organized by the Commission for technical co-operation in Africa south of the Sahara Humid tropics research Unesco,Paris,1961
Geog 13.1128

TROPICAL VEGETATION Study of tropical vegetation Kandy symposium Proceedings Kandy 1956 Mar 19-21 Unesco.International advisory committee for humid tropics research illus. Unesco,Paris,1958
BG 38.2155

TROPICS Nitrogen in the tropics A Review of nitrogen in the tropics,with particular reference to pastures :a symposium Commonwealth scientific and industrial research organisation.Division of tropical pastures Commonwealth bureau of pastures and field crops.Bulletin, 46 Commonwealth agricultural bureaux,Farnham Royal,1962
Bot 42.1784

TROPICS Recent advances in tropical ecology : a symposium Proceedings Banaras 1967 Jan 16-21 International society for tropical ecology Edited by R. Misra and B. Gopal 2 vols Vararasi,1968
Bot 42.2001

TROPOSPHERE Monograph on radio-wave propagation in the troposphere U.R.S.I. general assembly 13th Papers London 1960 Sep commission 2 on radio and troposphere International scientific radio union Edited by J.A. Saxton U.R.S.I. monographs series illus 199p Elsevier, Amsterdam,1962
Nap 11.0533

TROY,N.Y. 1961 Neutron physics :symposium Proceedings Rensselaer polytechnic institute Edited by M.L. Yeater Co-sponsored by the American nuclear society Nuclear science and technology:a series of monographs and text books, 2 xiii,303p Academic press,New York,1962
Cav 7.2543

TROY,N.Y. 1965 Interstellar grains :an international colloquium Proceedings Rensselaer polytechnic institute Edited by J. Mayo Greenberg and T.P. Roark NASA SP-140 National aeronautics and space administration, Washington,D.C.,1967
TA 15.0445

TROY,N.Y. 1970 Mathematical problems in the geophysical sciences Vol 1-2: geophysical fluid dynamics;inverse problems, dynamo theory,and tides American mathematical society Edited by W.H. Reid Lectures in applied mathematics, 13-14 illus,maps 24cm 2 vols American mathematical society,Providence,R.I.,1971
A Math 4.1742

TRUDI VSESOYUZNOGO SVEZTSDA PO LENETIKE SELEKTSII SEMENOVODSTVY I PLEMENNOMY ZHIVOTHOVODSTVY U.S.S.R.congress of genetics plant and animal breeding.Proceedings Leningrad 1929 Jan 10-16 Edited by N.I. Wulff and others 5 vols Izdanie Redaktsionnoi kollegii ceda,Leningrad, 1930
Bot 42.0654

TRUDY CHETVERTOGO VSESOYUZNOGO MATEMATICHESKOGO SEZDA Leningrad 1961 Jul 3-12 Tom 1-2 Bibliog. 27cm 2 vols Leningrad, 1963-64
P. Math 2.2606

TUBE TECHNIQUES Advances in electron tube techniques National conference on tube techniques 5th Proceedings New York 1960 Sep Edited by David Slater Sponsored by the Advisory group on electron tubes. Working group on tube techniques Pergamon press,Oxford,1961
Eng 41.5621

TUBE TECHNIQUES National conference on tube techniques 3rd Proceedings New York 1956 Sep 12-14 Sponsored by the Advisory group on electron tubes.Working group on semiconductor devices New York university press,New York,1958
Eng 41.5619

TUBE TECHNIQUES National conference on tube techniques 4th Proceedings New York 1958 Sep 10-12 Sponsored by the Advisory group on electron tubes.Working group on tube techniques New York university press,New York,1959
Eng 41.5620

TUBERCULOSIS International conference of the International union against tuberculosis 2nd Transactions London 1921 Jul 26-28 International union against tuberculosis Under the auspices of the British national association for the prevention of tuberculosis Adlard and Newman,London,1921
An 32.2990

TUBERCULOSIS International conference on tuberculosis 1st Report Berlin 1902 Oct 22-26 Central international bureau for the prevention of consumption Edited by Gotthold Pannwitz Berlin,1903 Text in English,French,German;title also in French and German
Path 30.1025

TUBERCULOSIS International tuberculosis conference 7th Report Philadelphia,Pa. 1908 Sep 24-26 International anti-tuberculosis association Edited by Gotthold Pannwitz port Internationale vereinigung gegen die tuberkulose,Berlin-Charlottenburg, 1909 Text in English,French,German;title also in French and German
Path 30.1023

TUBERCULOSIS International tuberculosis conference 8th Report Stockholm 1909 Jul 8-10 International anti-tuberculosis association Edited by Gotthold Pannwitz Berlin-Charlottenburg,1910 Text in English,French,German;title also in French and German
Path 30.1024

TUBERCULOSIS International tuberculosis conference 9th Report Brussels 1910 Oct 6-8 International anti-tuberculosis association Edited by Gotthold Pannwitz port Berlin-Charlottenburg,1911 Text in English,French,German;title also in French and German
Path 30.1026

TUBERCULOSIS Kongress zur bekampfung der tuberkulose als volkskrankheit Bericht Berlin 1899 May 24-27 Deutsches central-komite zur errichtung von heilstatten fur lungenkranke Edited by Gotthold Pannwitz Berlin,1899
Path 30.0770

TUBERCULOSIS Versammlung der tuberkulose-arzte 2te Bericht Berlin 1904 Nov 24-26 Deutsches central-komite zur errichtung von heilstatten fur lungenkranke Edited by Johannes Nietner Deutsches central-komite zur errichtung von heilstatten fur lungenkranke,Berlin,1905
Path 30.0761

TUBERCULOSIS IN THE COMMONWEALTH 1947 Commonwealth and empire health and tuberculosis conference Transactions London 1947 National association for the prevention of tuberculosis N.A.P.T.,London, 1947
HE 27.0132

TUCSON, ARIZ. 1960 Differential geometry American mathematical society Edited by Carl B. Allendoerfer With the support of the National science foundation American mathematical society.Proceedings of symposia in pure mathematics, 3 vii,200p 26cm American mathematical society,Providence,R.I., 1961
P. Math 2.0698

TUCSON,ARIZ. 1955 The Conference on the use of solar energy :the scientific basis Transactions Vol 1-5 University of Arizona and Stanford research institute Edited by Edwin F. Carpenter and others Co-sponsored by the Association for applied solar energy illus 6 vols University of Arizona,Tucson,Ariz.,1958
Nap 11.0111

TUCSON,ARIZ. 1958 Report of the round-table discussion on fast circuits of I.R.E. committee 4.10 in Tucson,Ariz. mimeograph By Wolfgang J. Poppelbaum University of Illinois.Digital computer laboratory 24p 1958
Math L 5.1678

TUCSON,ARIZ. 1965 The International conference on the physics of solids at high pressures 1st Proceedings University of Arizona Edited by C.T. Tumizuka and R.M. Emrick Jointly sponsored by the United States.Air force.Office of scientific research Academic press,New York,1965
Cav 7.2354

TUCSON,ARIZ. 1965 Pasadena,Calif. The Construction of large telescopes :a symposium International astronomical union Edited by D. L. Crawford International astronomical union. Symposium, 27 234p Academic press,New York;London,1966
Obs 6.3203

TUCSON,ARIZ. 1967 The Atmospheres of Venus and Mars :a symposium Goddard institute for space studies and Kitt Peak national observatory.Space division Edited by John C. Brandt and Michael B. McElroy Gordon and Breach,New York,1968
TA 15.0415

TUCSON,ARIZ. 1968 Symposium on low energy electron diffraction American crystallographic association Edited by David H. Templeton and Gabor A. Somorjai American crystallographic association.Transactions, 4 American crystallographic association, Pittsburgh,1968
Met 25.1476

TUCSON,ARIZ. 1970 Comets :scientific data and missions :a conference University of Arizona.Lunar and planetary laboratory Revised edition 222p University of Arizona. Lunar and planetary laboratory,Tucson,Ariz., 1972
Obs 6.3619

TUCSON,ARIZ. 1970 Dark nebulae,globules and protostars :a conference Edited by Beverly T. Lynds 150p University of Arizona press,Tucson,Ariz.
Obs 6.3638

TULANE UNIVERSITY Developments in theoretical and applied mechanics Southeastern conference on theoretical and applied technics 4th Proceedings New Orleans,La. 1968 Feb 29-Mar 1 Edited by D. Frederick illus 637p 26cm Pergamon, Oxford,1970
A Math 4.1903

TULANE UNIVERSITY Ergodic theory international symposium papers New Orleans, La. 1961 Oct.23-27 Edited by Fred B. Wright Sponsored by the National science foundation xii,316p Academic press,New York;London,1963
Math 3.0044

TULANE UNIVERSITY Serological fractions in schizophrenia :a research symposium New Orleans 1961 Jun 13-14 Edited by Robert G. Heath Harper and Row for the Commonwealth fund,New York,1963
PGMS 29.0324

TULANE UNIVERSITY OF LOUISIANA.DEPARTMENT OF PSYCHIATRY AND NEUROLOGY Studies in schizophrenia :a multidisciplinary approach to mind-brain relationships New Orleans,La. 1952 Jun 11-13 Edited by Robert G. Heath, chairman Published for the Commonwealth fund Harvard university press,Cambridge,Mass.,1954 Transcript of a series of invitational meetings
Psy 28.0081

TUMOR VIRUSES Subviral carcinogenesis International symposium on tumor viruses 1st 1966 Edited by Yohei Ito Sponsored by the Aichi cancer center.Research institute and the Japanese cancer association Aichi cancer center,Nagoya,1967
Bioch 33.1191

TUMOREN IM KINDESALTER :jahresversammlung der Schweizenschen gesellschaft fUR Padiatrie Geneva 1963 Schweizerische gesellschaft fur padiatrie Padiatrische fortbildungskurse fur die praxis, 13 Illus Schwabe,Basel, 1964
Radioth 35.0836

TUMORS OF THE ALIMENTARY TRACT IN AFRICANS : symposium Proceedings Geneva 1965 Nov 29-Dec 5 National cancer institute Edited by J.F. Murray Sponsored by the International union against cancer and the Calouste Gulbenkian foundation National cancer institute.Monograph, 25 National cancer institute,Bethesda,Md.,1967
Bioch 33.0990

TUMOUR BIOLOGY European conference on tumour biology Warsaw 1961 May 22-27 International union against cancer International union against cancer.Acta, 18, no 1-2 U.I.C;C.,Louvain,1962
Radioth 35.0817

TUNISIA 1967 Guidebook to the geology and history of Tunisia Petroleum exploration society of Libya :annual field conference 9th guidebook Edited by Lewis Martin and others Amsterdam,1967
Geol 8.4489

TUNNELING PHENOMENA IN SOLIDS   NATO advanced study institute on tunneling phenomena in solids Proceedings Riso,Denmark 1967 North Atlantic treaty organization Edited by Elias Burstein and Stig Lundqvist Bibliog 580p 18cm Plenum press,New York,1969
Cav   7.2667

TUNNELING PHENOMENA IN SOLIDS   :lectures presented at the 1967 NATO advanced study institute at Riso,Denmark Riso 1967 North Atlantic treaty organization Edited by Elias Burstein and Stig Lundqvist Plenum press,New York,1969
Eng   41.8302

TURBOMACHINERY   Internal aerodynamics (turbomachinery) :a conference Cambridge 1967 Jul 19-21 Royal society of London Institution of mechanical engineers,London, 1970
Eng   41.6829

TURBULENCE   Aircraft wake turbulence and its detection :a symposium Proceedings Seattle, Wash. 1970 Sep 1-3 Boeing company. Scientific research laboratories Edited by J. H. Olsen and others Plenum press,New York, 1971
Eng   41.8333

TURBULENCE   Mecanique de la turbulence colloque international Marseille 1961 Aug 28-Sep 2 Mecanique de la turbulence Centre national de la recherche scientifique Centre national de la recherche scientifique. Colloques internationaux, 108 Centre national de la recherche scientifique,Paris, 1962
A Math   4.0525

TURBULENCE   Statistical models and turbulence a symposium Proceedings San Diego,Calif. 1971 Jul 15-21 Edited by M. Rosenblatt and C. Van Atta Lecture notes in physics, 12 viii,492p 25cm Springer,Berlin,1972
Math S   3.1832

TURBULENCE IN GEOPHYSICS   Fundamental problems in turbulence and their relation to geophysics symposium proceedings Marseilles 1961 Sep.4-9 Vol 67 International union of geodesy and geophysics and International union of theoretical and applied mechanics Edited by Francois N. Frenkel Journal of geophysical research,Vol 67,no.8 American geophysical union, Washington D.C.,1962
A Math   4.0577

TURBULENCE OF FLUIDS AND PLASMAS   Symposium on turbulence of fluids and plasmas Proceedings New York 1968 Apr 16-18 Edited by Jerome Fox Sponsored by the Microwave research institute Microwave research institute.Symposia series, 18 bibliog.,illus. 512p 22cm Polytechnic press,Brooklyn,N.Y.,1969
A Math   4.1650

TURBULENCE OF FLUIDS AND PLASMAS   Symposium on turbulence of fluids and plasmas Proceedings New York 1968 Apr 16-18 Polytechnic institute of Brooklyn Edited by Jerome Fox Co-sponsored by the Microwave research institute Microwave research institute.Symposia series, 18 Bibliog., Illus. xxix,511p 23cm Polytechnic press, Brooklyn N.Y.,1969
A Math   4.1508

TURBULENCE OF FLUIDS AND PLASMAS   Symposium on turbulence of fluids and plasmas Proceedings New York 1968 Apr 16-18 Polytechnic institute of Brooklyn Edited by Jerome Fox Co-sponsored by the Microwave research institute Microwave research institute.Symposia series, 18 512p 17cm Polytechnic press,Brooklyn,N.Y.,1969
Cav   7.2920

TURIN 1961   Symposium international d'histoire des sciences Actes Comite national des celebrations du 1er centenaire de l'unite d'Italie With the participation of the Union internationale d'histoire et philosophie des sciences Academie internationale d'histoire des sciences. Collections de travaux, 14 189p Gruppe italiano di storia delle science,Vinci (Florence),1964
WSM   43.0059

TURIN 1967   Stress corrosion cracking in aircraft structural materials :an Agard conference Proceedings North Atlantic treaty organization.Advisory group for aeronautical research and development. Structures and materials panel AGARD conference proceedings series, 18 Nato,1967
Met   25.2484

TURIN 1967 St.Vincent   Histocompatibility testing 1967 Conference and workshop on histocompatibility testing 3r Report Edited by E.S. Curtoni and others Sponsored by the Consiglio nazionale delle ricerche Munksgaard,Copenhagen,1967
Surg   23.0032

TURIN 1968   Cranfield fluidics conference 3rd Proceedings Vol 1-3 Organised by the British hydromechanics research association 3 vols British hydromechanics research association,Cranfield,1968
Eng   41.5907

TWO-PHASE FLUID FLOW   :a symposium Proceedings 1962 Feb Institution of mechanical engineers Institution of mechanical engineers,London,1962
Eng   41.7171

TWO-PHASE FLUID FLOW   Symposium on two-phase fluid flow Proceedings London 1962 Feb 7 Arranged by the Institution of mechanical engineers 116p Institution of mechanical engineers,London,1962
Chem E   24.0372

TYPHOONS   Symposium on typhoons Proceedings Tokyo 1954 Nov 9-12 Japanese national commission for Unesco Science council of Japan Central meteorological observatory, Tokyo illus 257p Japanese national commission for Unesco,Tokyo,1955
Nap   11.0612

TYPHOONS   Unesco symposium on typhoons Proceedings Tokyo 1954 Nov 9-12 Japanese national commission for Unesco Japanese national commission for Unesco,Tokyo,1955
Geog   13.0962

U.C.L.A.CONFERENCE ON RADIOBIOLOGY 1st Proceedings Radiobiology at the intra-cellular level Catalina island 1957 Sep 9-12 University of California Edited by T.G. Hennessy and others ports. Pergamon press, London,1959
Radioth   35.1722

U.K.A.C. CONTROL CONVENTION 4th Multivariable control system design and application Manchester 1971 Sep 1-3 United Kingdom automation council Organized by the Institution of electrical engineers Institution of electrical engineers.Conference publication, 78 IEE,London,1971
Eng 41.8590

U.R.S.I. GENERAL ASSEMBLY 13th Papers Monograph on radio-wave propagation in the troposphere London 1960 Sep commission 2 on radio and troposphere International scientific radio union Edited by J.A. Saxton U.R.S.I. monographs series illus 199p Elsevier,Amsterdam,1962
Nap 11.0533

The U.S.-JAPAN JOINT CONFERENCE ON BIOLOGICAL AND BIOCHEMICAL EVALUATION OF MALIGNANCY IN EXPERIMENTAL HEPATOMAS Proceedings Biological and biochemical evaluation of malignancy in experimental hepatomas Kyoto 1965 Nov 4-5 Held under the auspices of the United States-Japan scientific cooperation program G.A.N.N.Monograph, 1 Japanese cancer association;Japanese foundation for cancer research,Tokyo,1966
Bioch 33.0952

U.S.S.R.CONGRESS OF GENETICS PLANT AND ANIMAL BREEDING.PROCEEDINGS Trudi vsesoyuznogo sveztsda po lenetike selektsii semenovodstvy i plemennomy zhivothovodstvy Leningrad 1929 Jan 10-16 Edited by N.I. Wulff and others 5 vols Izdanie Redaktsionnoi kollegii ceda, Leningrad,1930
Bot 42.0654

UCLA FORUM IN MEDICAL SCIENCES, 3 Brain and behavior Brain function conference 3rd Proceedings Los Angeles 1963 Vol 3: brain and gonadal function Brain research institute American institute of biological sciences Edited by Roger A. Gorski and Richard E. Whalen With the support of the Human ecology fund University of California press,Berkeley;Los Anjeles,1966
Psy 31.0621

UCLA FORUM IN MEDICAL SCIENCES, 7 Aggression and defence:neural mechanisms and social patterns (brain function,vol 5) Conference on brain function 5th Proceedings Pacific palisades 1965 Nov 14-17 Edited by Carmine D. Clemente and Donald B. Lindsley University of California press, Berkeley;Los Angeles,Calif.,1967
An 32.5475

UCLA FORUM IN MEDICAL SCIENCES, 11 The Interneuron :a conference Proceedings Los Angeles,Calif. 1967 Sep Edited by Mary A.B. Brazier Sponsored by the Brain research institute University of California press, Berkeley;Los Angeles,Calif.,1969
An 32.5473

ULTRA HIGH PURITY METALS :a seminar Papers Metals Park,Ohio 1961 Oct 21-22 American society for metals Edited by R.L. Smith 264p Metals Park,Ohio,1962
Cav 7.0051

ULTRA SOUND AS A DIAGNOSTIC AND SURGICAL TOOL based on the international symposium held at the Royal college of surgeons London 1962 Dec 5-6 Edited by Douglas Gordon port. Livingstone,Edinburgh;London,1964
Radioth 35.1412

ULTRA-HIGH-PURITY METALS :a seminar Papers 1961 Oct 21-22 American society for metals A.S.M.,Metals Park,Ohio,1962
Met 25.1495

The ULTRACENTRIFUGE :a conference New York 1941 Nov 14-15 By D.A. MacInnes and others New York academy of sciences.Section of physics and chemistry New York academy of sciences.Annals, 43,p 173-252 New York,1942
Bioch 33.1491

ULTRAFINE-GRAIN METALS Sagamore army materials research conference 16th Proceedings Raquette Lake,N.Y. 1969 Aug 19-22 Army materials research agency Syracuse university Syracuse university press, Syracuse,N.Y.,1970
Met 25.2747

ULTRAPURIFICATION OF SEMICONDUCTOR MATERIALS : a conference Proceedings Boston,Mass. 1961 Apr 11-13 Air force Cambridge research laboratories.Electronics research directorate Edited by M.S. Brooks and J.K. Kennedy Macmillan,New York,1962
Met 25.1149

ULTRASOUND IN BIOLOGY AND MEDICINE :a symposium Monticello,Ill. 1955 Jun 20-22 Edited by Elizabeth Kelly Sponsored by University of Illinois.Bioacoustics laboratory American institute of biological sciences. Publications, 3 American institute of biological sciences,Washington,D.C.,1957
An 32.0393

ULTRASTRUCTURE The Interpretation of ultrastructure Edited by R.J.C. Harris International society for cell biology. Symposia,1 Academic press,New York,1962 Based on a symposium held in Bern,Sept.1961
Pha 16.0008

ULTRASTRUCTURE The Interpretation of ultrastructure :symposium Bern 1961 International society for cell biology Edited by R.J.C. Harris International society for cell biology.Symposia, 1 illus Academic press,New York;London,1962
Bioch 33.0964

ULTRASTRUCTURE The Interpretation of ultrastructure :symposium Bern 1961 International society for cell biology Edited by R.J.C. Harris International society for cell biology.Symposia, 1 illus. x,438p Academic press,New York;London, 1962
Bot 42.1080

ULTRASTRUCTURE AND METABOLISM OF THE NERVOUS SYSTEM Proceedings of the Association New York 1960 Dec 9-10 Association for research in nervous and mental disease Edited by Saul R. Korey and others Association for research in nervous and mental disease.Research publications, 40 Williams and Wilkins,Baltimore,Md.,1962
An 32.4199

ULTRASTRUCTURE OF CELLS Electron microscopy in anatomy A Symposium on the ultrastructure of cells Proceedings London 1959 Apr 16-17 Anatomical society of Great Britain and Ireland Edited by J.D. Boyd and others Edward Arnold,London,1961
Bioch 33.2297

ULTRAVIOLET STELLAR SPECTRA AND RELATED GROUND-BASED OBSERVATIONS Lunteren 1969 Jun 24-27 International astronomical union Edited by L. Houziaux and H.E. Butler International astronomical union.Symposium, 36 361p Reidel,Dordrecht,1970
Obs 6.3321

UNDERWATER ACOUSTICS :an institute conducted at the Imperial college Proceedings London 1961 Jul 31-Aug 11 Imperial college of science and technology,London and Pennsylvania state university Edited by Vernon M. Albers New York,1963
Geod 9.0238

UNESCO Biological approaches to cancer chemotherapy :a symposium Louvain 1960 Jun Edited by R.J.C. Hams Academic press,London; New York,1961
Pha 16.0337

UNESCO Biological approaches to cancer chemotherapy :a symposium Louvain 1960 Jun Edited by R.J.C. Harris Academic press, London;New York,1960
Radioth 35.0238

UNESCO Biological approaches to cancer chemotherapy :a symposium Louvain 1960 Jun Edited by R.J.C. Harris Under the auspices of Unesco and the World health organization Academic press,London;New York,1961
Bioch 33.0960

UNESCO Biological organisation:cellular and sub-cellular :a symposium Proceedings Edinburgh 1957 Sep 6-10 Illus Pergamon press,London,1959
Radioth 35.0497

UNESCO Biological organisation,cellular and sub-cellular :a symposium organised on behalf of Unesco by C.H.Waddington Proceedings Edinburgh 1957 Sep 6-10 held at the University of Edinburgh Pergamon press, London,1959
Bioch 33.1019

UNESCO Biological organisation cellular and subcellular :a symposium Proceedings Edinburgh 1957 Sep 6-10 Edited by C.H. Waddington held at the University of Edinburgh Pergamon,London,1959
Bal 39.0455

UNESCO Biological organization:cellular and sub-cellular :a symposium Proceedings Edinburgh 1957 Sep 6-10 Pergamon press, London,1959
Gen 34.0829

UNESCO Biological organization:cellular and subcellular :a symposium Proceedings Edinburgh 1957 Sep 6-10 Edited by C.H. Waddington Pergamon press,London,1959
An 32.2485

UNESCO Biological organization at the cellular and supercellular level Varenna 1962 Sep 24-27 Edited by R.J.C. Harris Academic press,London;New York,1963
Bal 39.0460

UNESCO Biological organization at the cellular and supercellular level :a symposium Papers Varenna 1962 Sep 24-27 Academic press,London;New York,1963
Gen 34.0681

UNESCO Biological organization at the cellular and supercellular level :a symposium Proceedings Varenna 1962 Sep 24-27 Edited by R.J.C. Harris Academic press, London;New York,1963
An 32.3291

UNESCO Biological organization at the cellular and supercellular level :a symposium Varenna 1962 Sep 24-27 Edited by R.J.C. Harris Academic press,London;New York,1963
Bioch 33.1018

UNESCO Biological organization at the cellular and supercellular level :a symposium Varenna 1962 Sep 24-27 Edited by R.J.C. Harris Academic press,London,1963
Pha 16.0010

UNESCO Brain mechanism and learning :a symposium Montevideo 1959 Aug 2-8 Council for international organizations of medical sciences Edited by J.F. Delafresnaye Blackwell,Oxford,1960 Under the joint auspices of Unesco and WHO
Psy 31.0608

UNESCO Cellular basis and aetiology of late somatic effects of ionizing radiation :a symposium London 1962 Mar 27-30 Edited by R.J.C. Harris Academic press,London;New York,1963
Radioth 35.1750

UNESCO Curare and curare-like agents International symposium on curare and curare-like agents Proceedings Rio de Janeiro 1957 Aug 5-12 Edited by D. Bovet and others Elsevier,Amsterdam,1959
Pha 16.0243

UNESCO Etudes sur les termites africains : colloque international Comptes rendus Leopoldville 1964 May 11-16 Universite de Louvain Edited by A. Bouillon figs,pls, tables 27cm Masson,Paris,1964
Bal 44.4805

UNESCO Fine structure of cells :a symposium held at the VIIIth congress of cell biology Papers Leiden 1954 International union of biological sciences International union of biological sciences.Publications.Series B, 21 Noordhoff,Groningen,1955
Path 30.2410

UNESCO Fine structure of cells :a symposium held at the 8th congress of cell biology Leiden 1954 International union of biological sciences International union of biological sciences.Series B., 21 Noordhoff, Groningen,1955
Radioth 35.0467

UNESCO Fine structure of cells :symposium held at the VIIIth congress of cell biology Leiden 1954 International union of biological sciences.Series B, 21 Noordhoff, Groningen,1955
Phys 20.1483

UNESCO Fine structure of cells :symposium held at the 8th congress of cell biology Leiden 1954 International union of biological sciences.Publications.Series B, 21 Noordhoff,Groningen,1955
An 32.3219

UNESCO Fine structure of cells;symposium The Congress of cell biology 8th Papers Leiden 1954 International union of biological sciences International union of biological sciences.Ser.B, 21 Noordhoff, Groningen,1955
Bal 39.0283

UNESCO Immediate and low level effects of ionizing radiations :symposium Proceedings Venice 1959 Jun 22-26 International atomic energy agency Comitato nazionale per le ricerche nucleari International journal of radiation biology.Supplement Taylor and Francis,London,1960
Gen 34.2221

UNESCO Industrialization and society Conference on social implications of industrialization and technical change Papers Chicago,Ill. 1960 Sep 15-22 Edited by Bert F. Hoselitz and Wilbert E. Moore Unesco,Mouton,1963
Eng 41.1052

UNESCO Industrialization and society Conference on the social implications of industrialization and technological change Papers Chicago,Ill. 1960 Sep 15-22 University of Chicago Edited by Bert F. Hoselitz and Wilbert E. Moore Mouton,The Hague,1963
Geog 13.1608

UNESCO International colloquium on the modernization of mathematics teaching in European countries Proceedings Bucharest 1968 Sep 23-Oct 2 572p 25cm Editions didactiques et pedagogiques,Bucharest,1968
P Math 2.4020

UNESCO International conference on information processing Proceedings Paris 1959 Jun15-20 Unesco,Paris,1960
Eng 41.5999

UNESCO International conference on information processing proceedings Paris 1959 June 15-20 520p 29cm UNESCO,Paris, 1960
Math L 5.0809

UNESCO International conference on radioisotopes in scientific research 1st Proceedings Paris 1957 Sep Vol 1: research with radioisotopes in physics and industry Edited by R.C. Extermann Pergamon, London,1958
Met 25.1634

UNESCO International symposium on curare and curare-like agents Proceedings Rio de Janeiro 1957 Aug 5-12 Academia brasileira de ciencias.Conselho nacional de pesquisas and Universidade do Brasil Edited by D. Bovet and others illus Elsevier,Amsterdam, 1959 Also sponsored by the 'Istituto di sanita',of Rome.Papers in English and French
Chem 26.0103

UNESCO Marx and contemporary scientific thought The Role of Karl Marx in the development of contemporary scientific thought a symposium Papers Paris 1968 May 8-10 International council for philosophy and humanistic studies International social science council International social science council.Publications, 13 xi,612p Mouton, The Hague;Paris,1969
WSM 43.3826

UNESCO Meteorite research Symposium on meteorite research Proceedings Vienna 1968 Aug 7-13 Edited by Peter M. Millmann Astrophysics and space science library Reidel,Dordrecht,1969
Min 10.1442

UNESCO Nuclear science teaching Panel on nuclear science teaching Report Bangkok 1968 Jul 15-23 International atomic energy agency.Technical reports series, 94 International atomic energy agency,Vienna,1968
Radioth 35.1777

UNESCO Protein biosynthesis The International symposium on protein biosynthesis Proceedings Wassenaar 1960 Aug 29-Sep 2 Edited by R.J.C. Harris Under the auspices of Unesco and the Council for international organizations of medical sciences Academic press,London;New York,1961
Bioch 33.0516

UNESCO Protein biosynthesis :a symposium Wassenaar 1960 Aug 29-Sep 2 Edited by R.J. C. Harris Academic press,London;New York, 1961 Co-sponsored by the Council for international organizations of medical sciences
Radioth 35.0102

UNESCO Protein biosynthesis :a symposium Wassenaar 1960 Aug 29-Sep 2 Edited by R.J. C. Harris Sponsored also by the Council for international organizations of medical sciences Academic press,London;New York,1961
An 32.1509

UNESCO Protein biosynthesis :a symposium Wassennaar 1960 Aug 29-Sep 2 Edited by R.J. C. Harris Academic press,London;New York, 1961
Gen 34.0652

UNESCO Radio astronomy :a symposium proceedings Manchester 1955 Aug.25-27 International astronomical union Edited by H. C.van de Hulst International astronomical union.Symposium, 4 409p Cambridge university press,Cambridge,1957
Obs 6.0847

UNESCO Radioisotpoes in scientific research International conference on radioisotopes in scientific research 1st Proceedings Paris 1957 Sep 4-20 1-4 Edited by R. C. Extermann illus. 4 vols Pergamon press,London,1958
Radioth 35.1070

UNESCO Scientific problems of the humid tropical zone deltas and their implication :a symposium Proceedings Dacca 1964 Feb 24-Mar 2 Jointly sponsored by the government of Pakistan Humid tropics research Unesco, Paris,1966 Partly in French
Geog 13.5396

UNESCO Study of tropical vegetation :a symposium Proceedings Kandy,Ceylon 1956 Mar 19-21 Jointly organized by the Government of Ceylon Humid tropics research Unesco,Paris,1958
Geog 13.1261

UNESCO Symposium on modernization of mathematics teaching in European countries Proceedings Bucharest 1968 Sep 23-Oct 2 Bibliog 573p 24cm Editions didactiques et pedagogiques, Bucharest, 1969
A Math 4.1688

UNESCO Termites in the humid tropics : symposium Proceedings New Delhi 1960 Oct 4-12 Paris, 1962
Bal 39.4249

UNESCO The Initial effects of ionizing radiations on cells :a symposium Moscow 1960 Oct Edited by R.J.C. Harris Sponsored by the Akademiya nauk S.S.S.R. Academic press, London; New York, 1961
Gen 34.2222

UNESCO The Initial effects of ionizing radiations on cells :a symposium Moscow 1960 Oct Edited by R.J.C. Harris Sponsored by the Akademiya nauk S.S.S.R. Academic press, London; New York, 1961
Radioth 35.1127

UNESCO Tropical soils and vegetation : Abidjan symposium Proceedings Abidjan 1959 Oct 20-24 Jointly organized by the Commission for technical co-operation in Africa south of the Sahara Humid tropics research Unesco, Paris, 1961
Geog 13.1128

UNESCO and COMMONWEALTH SCIENTIFIC AND INDUSTRIAL RESEARCH ORGANIZATION Climatology and microclimatology :symposium Proceedings Canberra 1956 Oct Arid zone research, 11 Unesco, Paris, 1958
Geog 13.1018

UNESCO and JAPAN SOCIETY FOR THE PROMOTION OF SCIENCE Unesco symposium on physical oceanography Proceedings Tokyo 1955 Oct 19-22 Unesco and Japan society for the promotion of science, Tokyo, 1957
Geog 13.0875

UNESCO and UNITED NATIONS.ECONOMIC COMMISSION FOR LATIN AMERICA Social research and rural life in central America, Mexico and the Caribbean region :seminar Proceedings Mexico City 1962 Oct 17-27 Edited by Egbert de Vries Unesco, Paris, 1966
Geog 13.5161

UNESCO.CENTRO DE COOPERACION CIENTIFICA PARA AMERICA LATINA New research techniques in physics :a symposium Proceedings Rio de Janeiro 1952 Jul 15-29 Sao Paulo Sponsored by International union of pure and applied physics illus 447p Unesco, Rio de Janeiro, 1954 Under the auspices of Conselho nacional de pesquisas de Brasil and co-sponsored by the university of Sao Paulo
Cav 7.1977

UNESCO.INTERNATIONAL ADVISORY COMMITTEE FOR HUMID TROPICS RESEARCH Study of tropical vegetation Kandy symposium Proceedings Kandy 1956 Mar 19-21 illus. Unesco, Paris, 1958
BG 38.2155

UNESCO-MIDDLE EAST SCIENCE COOPERATION OFFICE Colloque sur la protection et la conservation de la nature dans le Proche-orient :symposium on the protection and conservation of nature in the Near East Beirut 1954 Jun 3-8 Beirut, 1954 Text in English and French
Bal 39.1787

UNESCO SYMPOSIUM ON PHYSICAL OCEANOGRAPHY Proceedings Tokyo 1955 Oct 19-22 Unesco, and, Japan society for the promotion of science Unesco and Japan society for the promotion of science, Tokyo, 1957
Geog 13.0875

UNESCO SYMPOSIUM ON TYPHOONS Proceedings Tokyo 1954 Nov 9-12 Japanese national commission for Unesco Japanese national commission for Unesco, Tokyo, 1955
Geog 13.0962

UNIO NORDICA CONTRA CANCRUM Symposium on virus and cancer Papers Saltsjobaden 1964 Jun 8-9 Edited by H. Bergstrand and K.E. Hellstrom illus Balder, Stockholm, 1965 Symposium arranged by the Swedish cancer society in conjunction with the annual general meeting of the Unio nordica contra cancrum
Path 30.2363

UNION GEOGRAPHIQUE INTERNATIONALE.COMMISSION DE GEOGRAPHIE APPLIQUEE see COMMISSION DE GEOGRAPHIE APPLIQUEE

UNION GEOGRAPHIQUE INTERNATIONALE see INTERNATIONAL GEOGRAPHICAL UNION

UNION INTERNATIONALE DE CHIMIE Conference de l'Union internationale de chimie :rapports sur les hydrates de carbone (Glucides) 10th Rapports Liege 1930 Sep 14-20 Union internationale de chimie, Paris, 1930
Chem 18.0066

UNION INTERNATIONALE DE CHIMIE PURE ET APPLIQUEE International colloquium on macromolecules Proceedings Amsterdam 1949 Sep 2-5 D.B. Centen's uitgevers-maatschappij N.V., Amsterdam, 1950
Chem 18.0503

UNION INTERNATIONALE DE MECANIQUE THEORIQUE ET APPLIQUEE Vibrations non lineaires Colloque international des vibrations non lineaires Actes Ile de Porquerolles 1951 Sep 18-21 Union internationale de radio-science France.Ministere de l'air. Publications scientifiques et techniques, 281 Information technique de l'aeronautique, Paris, 1953
Eng 41.6372

UNION INTERNATIONALE DE RADIO-SCIENCE Vibrations non lineaires Colloque international des vibrations non lineaires Actes Ile de Porquerolles 1951 Sep 18-21 Organised by the Union internationale de mecanique theorique et appliquee France. Ministere de l'air. Publications scientifiques et techniques, 281 Information technique de l'aeronautique, Paris, 1953
Eng 41.6372

UNION INTERNATIONALE DES SCIENCES BIOLOGIQUES Etudes et recherches sur les phytohormones Paris 1937 Oct 1-2 By Peter Boysen-Jensen and others League of Nations.Committee on intellectual co-operation League of Nations, Paris, 1937
Bot 42.1815

UNION INTERNATIONALE D'HISTOIRE DES SCIENCES Congres international d'histoire des sciences 9e Actes Barcelona 1959 Sep 1-7 and Madrid Vol 1 Academie internationale d'histoire des sciences.Collection de travaux, 12 733p Asociacion para la historia de ciencia espanola;Hermann,Barcelona;Paris,1960
WSM 43.0061

UNION INTERNATIONALE D'HISTOIRE DES SCIENCES Congres international d'histoire des sciences 5e Actes Lausanne 1947 Sep 30-Oct 6 Academie internationale d'histoire des sciences.Collections de travaux, 2 288p Academie internationale d'histoire des sciences;Hermann,Paris,1948
WSM 43.0060

UNION INTERNATIONALE D'HISTOIRE DES SCIENCES Congres international d'histoire des sciences et congres de la Societe internationale d'histoire de la medecine Actes Amsterdam 1950 Aug 14-21 Academie internationale d'histoire des sciences.Collection de travaux, 6 2 vols Academie d'histoire des sciences;Hermann,Paris,1951-53 The 12th congress of the Societe internationale d'histoire de la medecine is section 4 of the 6th Congres international d'histoire des sciences
WSM 43.0067

UNION INTERNATIONALE D'HISTOIRE ET PHILOSOPHIE DES SCIENCES Symposium international d'histoire des sciences Actes Turin 1961 Jul 28-30 Comite national des celebrations du 1er centenaire de l'unite d'Italie Academie internationale d'histoire des sciences.Collections de travaux, 14 189p Gruppe italiano di storia delle science,Vinci (Florence),1964
WSM 43.0059

UNION INTERNATIONALE D'HISTOIRE ET PHILOSOPHIE DES SCIENCES Symposium international des sciences physiques et mathematiques dans la premiere moitie du XVII siecle Actes Pisa 1958 Jun 16-18 Vinci (Florence) Academie internationale d'histoire des sciences.Collection de travaux, 11 x,278p Gruppe italiano di storia delle scienze; Hermann,Vinci (Florence);Paris,c1959
WSM 43.0066

UNION INTERNATIONALE POUR L'ETUDE SCIENTIFIQUE DES PROBLEMES DE LA POPULATION Congres international de la population International congress for studies on population 4th Paris 1937 July 29-Aug 1 8 vols Hermann, Paris,1938
Geog 13.1494

UNION OF SOVIET SOCIALIST REPUBLICS.SERVICE HYDRO-METEOROLOGIQUE Conference hydrologique des etats baltiques 4th Compte-rendu Leningrad 1933 Sep 6-22 Leningrad,1934
Geog 13.0767

UNION OF THE CHEMICAL SOCIETIES and INTERNATIONAL UNION OF PURE AND APPLIED CHEMISTRY Hydrogen bonding :papers held at the symposium on hydrogen bonding Papers Ljubljana 1957 Jul 29-Aug 3 Edited by D. Hadzi and H.W. Thompson Pergamon press, London,1959
Chem 18.0955

UNIONE DEGLI SCIENZATI ITALIANI Congresso scientifico italiano 5th Atti Lucca 1843 Sep 15-Sep 29 30cm Lucca,1844 Secretary general of conference:prof.Pacini
Bal 44.6331

UNIONE MATEMATICA ITALIANA Congresso dell'Unione matematica italiana 6to Atti Naples 1959 Sep 11-16 Bibliog,illus 497p 24cm Edizioni Cremonese,Rome,1960
P Math 2.3008

UNIONE MATEMATICA ITALIANA Teoria dei gruppi finiti e applicazioni :convegno internazionale Florence 1960 Apr 11-13 Supported by Consiglio nazionale della ricerche vii,156p 24cm Edizioni Cremonese,Rome,1960
P Math 2.2858

UNIONE ZOOLOGIE ITALIANA Problemi di sviluppo :un simposio Milan 1952 Sep Casa editrice ambrosiana,Milan,1954
An 32.2477

UNIONTOWN,PA. Radioactivity for pharmaceutical and allied research laboratories :a symposium Edited by Abraham Edelmann Sponsored by Nuclear science and engineering corporation Academic press,New York;London,1960
Radioth 35.1725

UNIT PROCESSES IN HYDROMETALLURGY based on an international symposium Dallas 1963 Feb 24-28 American institute of mining, metallurgical and petroleum engineers. Extractive metallurgy division and Society of mining engineers.Minerals benefication division Edited by Milton E. Wadsworth and Franklin T. Davis Metallurgical society conferences, 24 Gordon and Breach,New York; London,1964
Met 25.2289

UNITED KINGDOM 1964 Internationl geographical congress 20th Abstracts of papers Edited by F.E.Ian Hamilton Nelson, London,1964
Geog 13.5599

UNITED KINGDOM ATOMIC ENERGY AUTHORITY Human radiation cytogenetics :an international symposium Proceedings Edinburgh 1961 Oct 12-15 Edited by H.J. Evans and others North-Holland,Amsterdam,1967
Gen 34.1127

UNITED KINGDOM ATOMIC ENERGY AUTHORITY Practical fracture mechanics for structural steel Symposium on fracture toughness concepts for weldable structural steel Proceedings Risley 1969 Apr Edited by M. O. Dobson Chapman and Hall,London,1970
Eng 41.3822

UNITED KINGDOM ATOMIC ENERGY AUTHORITY Practical fracture mechanics for structural steel Symposium on fracture toughness concepts for weldable structural steel Proceedings Risley 1969 Apr Edited by M. O. Dobson U.K.A.E.A. in association with Chapman and Hall,Risley,1969
Met 25.2465

UNITED KINGDOM ATOMIC ENERGY AUTHORITY
Technology of the gas-cooled power reactor and related subjects United Nations conference on the peaceful uses of atomic energy 2nd Papers United Kingdom atomic energy authority,Harwell,1958
Eng 41.7494

UNITED KINGDOM ATOMIC ENERGY AUTHORITY.INDUSTRIAL GROUP Brittleness in metals :a conference Culcheth,Lancs 1957 Nov 1 I.G.Report 145 (RDC) Industrial group headquarters,Risley; Warrington,1959
Met 25.0859

UNITED KINGDOM ATOMIC ENERGY AUTHORITY.RESEARCH GROUP Nature of small defect clusters Consultants symposium 2nd Report Harwell 1966 Jul 4-6 Vol 1-2 Edited by M.J. Makin 2 vols Atomic energy research establishment,Harwell,1966
Met 25.1243

UNITED KINGDOM ATOMIC ENERGY AUTHORITY.RESEARCH GROUP.REPORT The Interactions between dislocations and point defects :a symposium Proceedings Harwell 1968 Jul 4-12 1-3 Atomic energy research establishment. Metallurgy division AERE-R 5944 3 vols harwell,1968
Met 25.2415

UNITED KINGDOM AUTOMATION COUNCIL Automation '65 British automation conference 1965 Eastbourne 1965 Nov 7-10 Organised by the Institution of production engineers Institution of production engineers,London, 1965
Eng 41.6018

UNITED KINGDOM AUTOMATION COUNCIL Models for decision :a conference London 1960 Oct 13-144 British computer society Operational research society English universities press, London,1965
Eng 41.1270

UNITED KINGDOM AUTOMATION COUNCIL
Multivariable control system design and application U.K.A.C. control convention 4th Manchester 1971 Sep 1-3 Organized by the Institution of electrical engineers Institution of electrical engineers.Conference publication, 78 IEE,London,1971
Eng 41.8590

UNITED KINGDOM AUTOMATION COUNCIL The Impact of users' needs on the design of data processing systems : conference Edinburgh 1964 Mar 31-Apr 3 British computer society 1964 Organized jointly by the British computer society,the British institution of radio engineers and the Institution of electrical engineers
Math L 5.3265

UNITED KINGDOM AUTOMATION COUNCIL.CONTROL CONVENTION 2nd Advances in computer control Bristol 1967 Apr 11-14 Organized by the Institution of electrical engineers. Control and automation division Institution of electrical engineers.Conference publication, 29 I.E.E.,London,1967
Eng 41.5862

UNITED NATIONS International conference on the peaceful uses of atomic energy 2nd Geneva 1958 Sep 1-13 Vol 29: chemical effects of radiation United Nations,Geneva, 1958
Met 25.1567

UNITED NATIONS International conference on the peaceful uses of atomic energy 2nd Proceedings Geneva 1958 Sep 1-13 18,21-26,29 8 vols United Nations,Geneva,1958
Radioth 35.1079

UNITED NATIONS International conference on the peaceful uses of atomic energy 2nd Proceedings Geneva 1958 Sep 1-13 Vol 31-32 2 vols United Nations,Geneva,1958
Eng 41.4466

UNITED NATIONS International conference on the peaceful uses of atomic energy Proceedings Geneva 1955 Aug 8-20 10-14 5 vols United Nations,New York,1955-56
Radioth 35.1009

UNITED NATIONS International conference on the peaceful uses of atomic energy Proceedings Geneva 1955 Aug 8-20 Vol 8: production technology of the materials used for nuclear energy 627p United nations,New York,1956
Chem E 24.1650

UNITED NATIONS International conference on the peaceful uses of atomic energy Proceedings Geneva 1955 Aug 8-20 Vol 9: reactor technology and chemical processing 771p United nations,New York,1956
Chem E 24.1651

UNITED NATIONS International conference on the peaceful uses of atomic energy Proceedings Geneva 1955 Aug 8-20 Vol 10: radioactive isotopes and nuclear radiations in medicine United nations,New York,1956
Phys 20.2093

UNITED NATIONS International conference on the peaceful uses of atomic energy Proceedings Geneva 1955 Aug 8-20 Vol 12: radioactive isotopes and ionizing radiations in agriculture,physiology and biochemistry United Nations,New York,1956
Phys 20.2244

UNITED NATIONS International conference on the peaceful uses of atomic energy 1st Proceedings Geneva 1955 Aug 8-20 Vol 6: geology of uranium and thorium United Nations,New York,1956
Min 10.0715

UNITED NATIONS International conference on the peaceful uses of atomic energy 1st Proceedings Geneva 1955 Aug 8-20 Vol 8: production technology of the materials used for nuclear nergy United Nations,New York, 1956
Min 10.0718

UNITED NATIONS Technology,engineering and safety International conference on the peaceful uses of atomic energy 2nd Edited proceedings Geneva 1958 Vol 2 Edited by R. Hurst and others Progress in nuclear energy, 4 Pergamon press,Oxford, 1960
Met 25.1555

UNITED NATIONS The International conference on the peaceful uses of atomic energy Geneva 1955 Aug 8-20 Vol 8: production technology of the materials used for nuclear energy United Nations,New York,1956
Met 25.1564

UNITED NATIONS The Use of vital and health statistics for genetic and radiation studies seminar Proceedings Geneva 1960 Sep 5-9 United Nations publications, 61,XVII.8 United Nations,New York,1962
Gen 34.1120

UNITED NATIONS United Nations international conference on the peaceful uses of atomic energy 2nd Proceedings Geneva 1958 Sep 1-13 Vol 5-7 3 vols United Nations,Geneva,1958
Met 25.1544

UNITED NATIONS.DEPARTMENT OF ECONOMIC AND SOCIAL AFFAIRS United nations regional cartographic conference for Asia and the Far East 4th Manila 1964 Nov 21-Dec 5 2: proceedings of the conference and technical papers United nations,New York,1966
Geog 13.7226

UNITED NATIONS.DEPARTMENT OF ECONOMIC AND SOCIAL AFFAIRS World population conference 1965 Belgrade 1965 Aug 30-Sep 10 4: selected papers and summaries,migration,urbanization, economic development United nations. Department of economic and social affairs,New York,1967
Geog 13.7198

UNITED NATIONS.ECONOMIC COMMISSION FOR LATIN AMERICA and UNESCO Social research and rural life in central America,Mexico and the Caribbean region :seminar Proceedings Mexico City 1962 Oct 17-27 Edited by Egbert de Vries Unesco,Paris,1966
Geog 13.5161

UNITED NATIONS.FOOD AND AGRICULTURE ORGANIZATION Radioisotopes in soil plant nutrition studies The Symposium on the use of radioisotopes in soil-plant nutrition studies Proceedings Bombay 1962 Feb 26-Mar 2 International atomic energy agency,Vienna,1962
Bot 42.1368

UNITED NATIONS.SECRETARIAT Structural interdependence and economic development International conference on input-output techniques 3rd Proceedings Geneva 1961 Sep Edited by Tibor Barna and others Sponsored jointly by the Harvard economic research project Macmillan,London,1963
Geog 13.2417

UNITED NATIONS CONFERENCE ON THE APPLICATION OF SCIENCE AND TECHNOLOGY FOR THE BENEFIT OF THE LESS DEVELOPED AREAS Report Science and technology for development Geneva 1963 Feb Vol 1-4 4 vols United Nations,New York,1963
Geog 13.2389

UNITED NATIONS CONFERENCE ON THE HUMAN ENVIRONMENT,STOCKHOLM 1972 How do you want to live?A report on the human habitat... 1972 :a study of public opinion undertaken... in connection with the United Nations conference on the human environment,Stockholm 1972 of public opinion at the request of the Secretary for the environment Great Britain. Department of the environment HMSO,London, 1972 Chairman:countess of Dartmouth
Eng 41.8609

UNITED NATIONS CONFERENCE ON THE HUMAN ENVIRONMENT,STOCKHOLM 1972 Pollution : nuisance or nemesis? Great Britain. Department of the environment HMSO,London, 1972 A report on the control of pollution, under the chairmanship of sir Eric Ashby, prepared on the occasion of the United Nations conference on the human environment,Stockholm 1972 Ashby
Eng 41.8614

UNITED NATIONS CONFERENCE ON THE HUMAN ENVIRONMENT,STOCKHOLM 1972 Sinews for survival :a report on the management of natural resources Great Britain.Department of the environment HMSO,London,1972 Undertaken in connection with the United Nations conference on the human environment, Stockholm 1972
Eng 41.8610

UNITED NATIONS CONFERENCE ON THE HUMAN ENVIRONMENT,STOCKHOLM 1972 The Human environment :the British view Great Britain. Department of the environment HMSO,London, 1972 Prepared on the occasion of the United Nations conference on the human environment,Stockholm 1972
Eng 41.8612

UNITED NATIONS CONFERENCE ON THE PEACEFUL USES OF ATOMIC ENERGY Proceedings Geneva 1959 Aug 8-20 Vol 1-15 8 vols United Nations,New York,1956 Lacks vols 6-7.10-14
Eng 41.7540

UNITED NATIONS CONFERENCE ON THE PEACEFUL USES OF ATOMIC ENERGY 2nd Papers Technology of the gas-cooled power reactor and related subjects United Kingdom atomic energy authority United Kingdom atomic energy authority,Harwell,1958
Eng 41.7494

UNITED NATIONS EDUCATIONAL,SCIENTIFIC AND CULTURAL ORGANISATION. CENTRE FOR SCIENTIFIC CO-OPERATION FOR LATIN AMERICA Algunos problemas matematicos que se estan estudiando en Latino America symposium 2 Villavicencio-Mendoza 1954 Jul.21-25 Universidad nacional,Mendoza 328p 24cm UNESCO,centro de cooperacion cientifica, Montevideo,1954
P. Math 2.2610

UNITED NATIONS EUROPEAN STUDY GROUP Report People in the countryside Leicester 1963 Edited by John Higgs NCSS 694 National council of social service,London,1966
Geog 13.1901

UNITED NATIONS INTERNATIONAL CONFERENCE ON THE PEACEFUL USES OF ATOMIC ENERGY 2nd Proceedings Geneva 1958 Sep 1-13 Vol 5-7 United Nations 3 vols United Nations,Geneva,1958
Met 25.1544

UNITED NATIONS REGIONAL CARTOGRAPHIC CONFERENCE FOR ASIA AND THE FAR EAST 4th Manila 1964 Nov 21-Dec 5 2: proceedings of the conference and technical papers United nations.Department of economic and social affairs United nations,New York,1966
Geog 13.7226

UNITED STATES 1904 International geographical congress 8th Report U.S. government printing office,Washington,D.C., 1905
Geog 13.5592

UNITED STATES.AIR FORCE Gas dynamics symposium on aerothermochemistry Proceedings Evanston,Ill 1955 Aug 22-24 Northwestern university and American rocket society Edited by Donald K. Fleming illus 284p Northwestern university,Evanston,Ill.,1956
Chem E 24.1390

UNITED STATES.AIR FORCE.AERONAUTICAL RESEARCH LABORATORY Luminescence of organic and inorganic materials :international conference New York 1962 New York university Edited by Hartmut P. Kallman and Grace Marmor Spruch xxiv,664p Wiley,New York,1962
Cav 7.1253

UNITED STATES.AIR FORCE.AEROSPACE RESEARCH LABORATORIES International symposium on multivariate analysis Proceedings Dayton, Ohio 1965 Jun 14-19 Edited by P.R. Krishnaiah Academic press,New York,1966
Eng 41.2206

UNITED STATES.AIR FORCE.DIRECTORATE OF MANAGEMENT ANALYSIS Linear programming symposium 2nd Proceedings Washington,D.C. 1955 Jan 27-29 Vol 1-2 Sponsored by the United States.Air force.Office of scientific research 24cm 2 vols National bureau of standards, Washington,D.C.,1955
Math L 5.3248

UNITED STATES.AIR FORCE.DIRECTORATE OF MATERIALS AND PROCESSES Condensation and evaporation of solids International symposium on condensation and evaporation of solids Proceedings Dayton,Ohio 1962 Sep 12-14 Edited by Emile Rutner and others Gordon and Breach,New York,1964
Met 25.1518

UNITED STATES.AIR FORCE.DIRECTORATE OF RESEARCH AND DEVELOPMENT.HUMAN FACTORS DIVISION Symposium on atherosclerosis Papers 1954 Mar 22-23 National research council.Division of medical sciences National research council.Publication, 338 National research council,Washington,D.C.,1955
Path 30.2349

UNITED STATES.AIR FORCE.EUROPEAN OFFICE OF AEROSPACE RESEARCH Optical properties and electronic structure of metals and alloys : International colloquium Proceedings Paris 1965 Sep 13-16 Institut d'optique theorique et appliquee Edited by F. Abeles North Holland,Amsterdam,1966
Cav 7.0004

UNITED STATES.AIR FORCE.OFFICE OF SCIENTIFIC RESEARCH Analysis in function space Conference on the theory and applications of analysis in function space proceedings Dedham,Mass. 1963 Jun 9-13 Massachusetts institute of technology Edited by William Ted Martin and Irving Segal vi,218p 24cm M.I.T.press,Cambridge,Mass.,1964
P. Math 2.1181

UNITED STATES.AIR FORCE.OFFICE OF SCIENTIFIC RESEARCH Analytic functions conference papers Princeton,N.J. 1957 Sep 2-14 By Rolf Nevanlinna and others Institute for advanced study Princeton mathematical series, 24 vii,197p 24cm Princeton university press,Princeton,N.J.,1960
P. Math 2.1492

UNITED STATES.AIR FORCE.OFFICE OF SCIENTIFIC RESEARCH Biophysics conference The National biophysics conference 1st Proceedings Columbus,Ohio 1957 Mar 4-6 Edited by Henry Quastler and Harold J. Morowitz Yale university press,New Haven, 1959
Bioch 33.0308

UNITED STATES.AIR FORCE.OFFICE OF SCIENTIFIC RESEARCH Brain function :a conference 3rd Proceedings Los Angeles,Calif. 1963 Nov Vol 3: speech,language and communication Brain research institute Edited by Edward C. Carterette UCLA forum in medical sciences, 4 University of California press,Berkeley;Los Angeles,1966
Psy 31.0458

UNITED STATES.AIR FORCE.OFFICE OF SCIENTIFIC RESEARCH Brain function conference 2nd Proceedings Los Angeles 1962 Vol 2: RNA and brain function memory and learning Brain research institute and American institute of biological sciences Edited by Mary A.B. Brazier U.C.L.A. forum in medical sciences, 2 University of California press, Berkeley;Los Angeles,1964
Psy 31.0246

UNITED STATES.AIR FORCE.OFFICE OF SCIENTIFIC RESEARCH Categorical algebra conference proceedings La Jolla,Calif. 1965 Jun 7-12 Edited by S. Eilenberg and others 562p 24cm Springer-Verlag,Berlin,1966
P. Math 2.2027

UNITED STATES.AIR FORCE.OFFICE OF SCIENTIFIC RESEARCH Combinatorial mathematics and its applications 2nd conference Proceedings Chapel Hill,N.C. 1970 May Edited by R.C. Bose and others 548p 25cm University of North Carolina,Chapel Hill,N.C., 1970
Math S 3.1655

UNITED STATES.AIR FORCE.OFFICE OF SCIENTIFIC RESEARCH Combinatorial mathematics and its applications Chapel Hill conference 2nd Proceedings Chapel Hill,N.C. 1970 May 8-13 Edited by R.C. Bose and others 548p 25cm University of North Carolina, Chapel Hill,N.C.,1970
P Math 2.4507

UNITED STATES.AIR FORCE.OFFICE OF SCIENTIFIC RESEARCH Computation of turbulent boundary layers :1968 AFOSR-IFP-Stanford conference Proceedings Stanford,Calif 1968 Vol 1-2 Edited by S.J. Kline and others Bibliog.,Illus,port 26cm 2 vols Stanford university,Stanford,1969
A Math 4.1481

UNITED STATES.AIR FORCE.OFFICE OF SCIENTIFIC RESEARCH Conference on complex analysis proceedings Minneapolis,Minn. 1964 Mar 16-21 Edited by A. Aeppli and others Bibliog. vii,308p 24cm Springer-Verlag,Berlin, Heidelberg,1965
P. Math 2.1582

UNITED STATES.AIR FORCE.OFFICE OF SCIENTIFIC RESEARCH Dislocations and mechanical properties of crystals :international conference Proceedings Lake Placid,N.Y. 1956 Sep 6-8 Edited by J.C. Fisher and others John Wiley,New York,1957
Cav 7.0804

UNITED STATES.AIR FORCE.OFFICE OF SCIENTIFIC RESEARCH Dislocations and mechanical properties of crystals an international conference Lake Placid 1956 Sep 6-8 Edited by J.C. Fisher and others Sponsored by United States.Air research and development command Wiley;Chapman and Hall,New York; London,1957
Met 25.1232

UNITED STATES.AIR FORCE.OFFICE OF SCIENTIFIC RESEARCH Experimental designs in industry Symposium on design of industrial experiments proceedings Raleigh,N.C. 1956 Nov.5-9 North Carolina state college.Institute of statistics Edited by Victor Chew Wiley publications in statistics Bibliog. xi, 268p 28cm John Wiley and sons,New York, 1958 Being parts A and B of Symposium on design of industrial experiments, proceedings, edited by Victor Chew,1957
Math 3.0583

UNITED STATES.AIR FORCE.OFFICE OF SCIENTIFIC RESEARCH Functional analysis and related fields :a conference held in honour of Professor Marshall Harvey Stone Proceedings Chicago,Ill. 1968 May 20-24 Edited by Felix E. Browder port. 241p 24cm Springer,Berlin,1970
P Math 2.3863

UNITED STATES.AIR FORCE.OFFICE OF SCIENTIFIC RESEARCH Fundamentals of gas-surface interactions :a symposium Proceedings San Diego,Calif. 1966 Dec 14-16 Edited by Howard M. Saltsburg and others Academic press,New York,1967
Chem 18.2696

UNITED STATES.AIR FORCE.OFFICE OF SCIENTIFIC RESEARCH Growth and perfection of crystals International conference on crystal growth Proceedings Cooperstown,N.Y. 1958 Aug 27-29 General electric company Edited by R.H. Doremus and others Wiley;Chapman and Hall,New York;London,1958
Eng 41.4438

UNITED STATES.AIR FORCE.OFFICE OF SCIENTIFIC RESEARCH Hypersonic flow research :a symposium technical papers Cambridge,Mass. 1961 Aug 16-18 American rocket society Edited by Frederick R. Riddell Progress in astronautics and rocketry, 7 Academic press,New York,London,1962
A Math 4.0533

UNITED STATES.AIR FORCE.OFFICE OF SCIENTIFIC RESEARCH Inhibition in the nervous system and gamma-aminobutyric acid :an international symposium Proceedings Duarte,Calif. 1959 May 22-24 Edited by Eugene Roberts Pergamon press,London,1960
An 32.4362

UNITED STATES.AIR FORCE.OFFICE OF SCIENTIFIC RESEARCH Inhibition in the nervous system and gamma-aminobutyric acid :an international symposium held at the City of Hope medical center Proceedings Duarte,Calif. 1959 May 22-24 Edited by Eugene Roberts and others Pergamon press,Oxford,1960
Bal 39.1471

UNITED STATES.AIR FORCE.OFFICE OF SCIENTIFIC RESEARCH International symposium on free radicals 5th Preprint of papers Uppsala 1961 Jul 6-7 1961
Chem 18.2625

UNITED STATES.AIR FORCE.OFFICE OF SCIENTIFIC RESEARCH International symposium on nonlinear differential equations and nonlinear mechanics Colorado Springs,Colo. 1961 Jul 31-Aug 4 Edited by Joseph P. LaSalle and Solomon Lefschetz Academic press,New York; London,1963
Eng 41.6385

UNITED STATES.AIR FORCE.OFFICE OF SCIENTIFIC RESEARCH International symposium on nonlinear differential equations and nonlinear mechanics proceedings Colorado Springs,Col 1961 Jul 31-Aug 4 Edited by Joseph P. La Salle and Solomon Lefschetz Academic press, New York,1963
A Math 4.0245

UNITED STATES.AIR FORCE.OFFICE OF SCIENTIFIC RESEARCH International symposium on rarefied gas dynamics 4th Proceedings Toronto 1964 Jul Vol 1-2 Edited by J.H.de Leeuw Advances in applied mechanics. Supplement, 3 2 vols Academic press,New York;London,1965
Eng 41.6788

UNITED STATES.AIR FORCE.OFFICE OF SCIENTIFIC RESEARCH International symposium on rarefied gas dynamics 6th Proceedings Cambridge,Mass. 1968 Jul Vol 1-2 Edited by Leon Trilling and Harold Y. Wachman Advances in applied mechanics.Supplement, 5 2 vols Academic press,New York,1969
Eng 41.6797

UNITED STATES.AIR FORCE.OFFICE OF SCIENTIFIC RESEARCH International symposium on the reactivity of solids 6th Proceedings Schenectady,N.Y. 1968 Aug 25-30 Edited by J.W. Mitchell and others Wiley,New York; London,1969
Met 25.2446

UNITED STATES.AIR FORCE.OFFICE OF SCIENTIFIC RESEARCH Linear programming symposium 2nd Proceedings Washington,D.C. 1955 Jan 27-29 Vol 1-2 National bureau of standards United States.Air force. Directorate of management analysis 24cm 2 vols National bureau of standards,Washington, D.C.,1955
Math L 5.3248

UNITED STATES.AIR FORCE.OFFICE OF SCIENTIFIC RESEARCH Methodologies of pattern recognition International conference on methodologies of pattern recognition Proceedings Honolulu 1968 Jan 24-26 University of Hawaii bibliog.,illus. Academic press,New York;London,1969
Math S 3.1521

UNITED STATES.AIR FORCE.OFFICE OF SCIENTIFIC RESEARCH Seminars on analytic functions conference typescript Princeton,N.J. 1957 Sep 2-14 Vols. 1-2 Institute for advanced study Edited by Marston Morse and others Bibliog. 25cm 2 vols Princeton, N.J.,1957
P. Math 2.1583

UNITED STATES.AIR FORCE.OFFICE OF SCIENTIFIC RESEARCH Sensory communications The Symposium on principles of sensory communications Contributions Dedham,Mass. 1959 Jul 19-Aug 1 Edited by Walter A. Rosenblith MIT press,Cambridge,Mass.,1961
Psy 31.2842

UNITED STATES.AIR FORCE.OFFICE OF SCIENTIFIC RESEARCH Structure and properties of thin films :an international conference Proceedings New York 1959 Sep 9-11 United States.Air research and development command and General electric research laboratory Edited by C.A. Neugebauer and others Wiley,New York;London,n.d.
Met 25.1204

UNITED STATES.AIR FORCE.OFFICE OF SCIENTIFIC RESEARCH Symposium of plasma dynamics Woods Hole,Mass. 1958 Jun 9-13 Edited by Francis H. Clauser Pergamon press;Addison-Wesley,London;Reading,Mass.,1960
Eng 41.4479

UNITED STATES.AIR FORCE.OFFICE OF SCIENTIFIC RESEARCH Symposium of plasma dynamics proceedings Massachusetts 1968 Jun.9-13 National academy of sciences Edited by Frances H. Clauser Pergamon press,London, 1959
A Math 4.0617

UNITED STATES.AIR FORCE.OFFICE OF SCIENTIFIC RESEARCH Symposium on design of industrial experiments proceedings Raleigh, N.C. 1956 Nov.5-9 North Carolina state college.Institute of statistics Edited by Victor Chew Bibliog. viii,374p 27cm Institute of statistics,University of North Carolina,Raleigh,N.C.,1957
Math 3.0584

UNITED STATES.AIR FORCE.OFFICE OF SCIENTIFIC RESEARCH Symposium on electrical conductivity in organic solids Proceedings Durham 1960 Apr 20-22 United States.Army. Office of ordnance research United States. Office of naval research Edited by H. Kallman and M. Silver Interscience,New York, 1961
Radioth 35.1646

UNITED STATES.AIR FORCE.OFFICE OF SCIENTIFIC RESEARCH Symposium on electrode processes Transactions Philadelphia,Pa. 1959 May 4-6 Wiley,New York,1961
Chem 18.2624

UNITED STATES.AIR FORCE.OFFICE OF SCIENTIFIC RESEARCH Symposium on nonlinear estimation theory and its applications Proceedings San Diego,Calif. 1970 Sep 21-23 vii,297p 28cm IEEE,New York,1970
Math S 3.1698

UNITED STATES.AIR FORCE.OFFICE OF SCIENTIFIC RESEARCH The Central nervous system and fish behavior :a conference Papers Chicago, Ill. 1967 Apr Edited by David Ingle Jointly sponsored by Perspectives in biology and medicine illus University of Chicago press,Chicago,Ill.;London,1968
Psy 31.2816

UNITED STATES.AIR FORCE.OFFICE OF SCIENTIFIC RESEARCH The Fermi surface :an international conference Proceedings Cooperstown,N.Y. 1960 Aug 22-24 General electric research laboratory Edited by W.A. Harrison and M.B. Webb John Wiley,New York, 1960
Cav 7.0541

UNITED STATES.AIR FORCE.OFFICE OF SCIENTIFIC RESEARCH The International conference on the physics of solids at high pressures 1st Proceedings Tucson,Ariz. 1965 Apr 20-23 University of Arizona Edited by C.T. Tumizuka and R.M. Emrick Academic press,New York,1965
Cav 7.2354

UNITED STATES.AIR FORCE.OFFICE OF SCIENTIFIC RESEARCH Vistas in astronautics Annual astronautics symposium 1st Edited by Morton Alperin and others Co-sponsored with General dynamic corporation.Convair division Pergamon press,London,1958
Eng 41.6880

UNITED STATES.AIR FORCE.OFFICE OF SCIENTIFIC RESEARCH Vistas in astronautics Astronautics symposium 1st proceedings San Diego 1957 Feb Edited by Morton Alperin and Marvin Stern Co-sponsored by General dynamics corporation.Convair division Vistas in astronautics, 1 International series of monographs on aeronautical sciences and space flight.Astronautics division, 1 329p Pergamon press,London,1958
Nap 11.0022

UNITED STATES.AIR FORCE.OFFICE OF SCIENTIFIC RESEARCH Vistas in astronautics Astronautics symposium 2nd Proceedings Denver,Col. 1958 Apr.28 Edited by Morton Alperin and Hollingsworth F. Gregory Co-sponsored by Institute of aeronautical sciences Vistas in astronautics, 2 International series of monographs on aeronautical sciences and space flight. Astronautics division, 2 Pergamon press, London,1959
Nap 11.0023

UNITED STATES.AIR FORCE.OFFICE OF SCIENTIFIC RESEARCH Vistas in astronautics Astronautics symposium 3rd Proceedings Los Angeles,Calif. 1960 Oct.12-14 jointly sponsored Society of automative engineers Vistas in astronautics, 3 illus 266p Society of automotive engineers,New York,1960
Nap 11.0024

UNITED STATES.AIR FORCE.OFFICE OF SCIENTIFIC RESEARCH and GENERAL ELECTRIC RESEARCH LABORATORY The Fermi surface :an international conference Proceedings Cooperstown,N.Y. 1960 Aug 22-24 J.Wiley, New York;London,1960
Chem 18.2382

UNITED STATES.AIR FORCE.OFFICE OF SCIENTIFIC RESEARCH.MATHEMATICS DIVISION Differential and combinatorial topology :a symposium in honor of Marston Morse Princeton, N.J. 1963 Edited by Stewart S. Cairns vi,265p 24cm Princeton university press,Princeton, N.J.,1965
P. Math 2.0295

UNITED STATES.AIR FORCE.OFFICE OF SCIENTIFIC RESEARCH.SOLID STATE SCIENCES DIVISION High magnetic fields International conference on high magnetic fields Proceedings Cambridge,Mass. 1961 Nov 1-4 Edited by Henry Kolm M.I.T.press;Wiley, Cambridge,Mass.;New York,1962
Eng 41.5331

UNITED STATES.AIR FORCE.SOLID STATE SCIENCES DIVISION Fatigue in aircraft structures international conference Proceedings New York 1956 Jan 30 -Feb 1 Edited by Alfred M. Freudenthal Academic press,New York,1956
Met 25.0921

UNITED STATES.AIR FORCE SYSTEMS COMMAND Matrix methods in structural mechanics :a conference Proceedings Dayton,Ohio 1965 Oct 26-28 1966
Eng 41.2732

UNITED STATES.AIR RESEARCH AND DEVELOPMENT COMMAND Conference on non-crystalline solids Alfred,N.Y. 1958 Sep 3-5 Edited by V.D. Frechette John Wiley,New York;London, 1960
Met 25.0126

UNITED STATES.AIR RESEARCH AND DEVELOPMENT COMMAND Dislocations and mechanical properties of crystals an international conference Lake Placid 1956 Sep 6-8 United States.Air force.Office of scientific research and General electric research laboratory Edited by J.C. Fisher and others Wiley;Chapman and Hall,New York;London,1957
Met 25.1232

UNITED STATES.AIR RESEARCH AND DEVELOPMENT COMMAND Fatigue in aircraft structures : an international conference New York 1956 Jan 30-Feb 1 Edited by Alfred M. Freudenthal Academic press,New York,1956
Eng 41.2772

UNITED STATES.AIR RESEARCH AND DEVELOPMENT COMMAND Growth and perfection of crystals International conference on crystal growth Proceedings Cooperstown,New York 1958 Aug 27-29 Wiley;Chapman and Hall,New York;London, 1958
Met 25.1507

UNITED STATES.AIR RESEARCH AND DEVELOPMENT COMMAND Structure and properties of thin films :an international conference Proceedings New York 1959 Sep 9-11 Edited by C.A. Neugebauer and others Sponsored by United States.Air force.Office of scientific research Wiley,New York;London,n. d.
Met 25.1204

UNITED STATES.AIR RESEARCH AND DEVELOPMENT COMMAND.MATERIALS CENTRAL Progress in very high pressure research :international conference Proceedings Bolton Landing,New York 1960 Jun 13-14 Edited by F.P. Bundy and others Wiley,New York;London,1961
Met 25.1638

UNITED STATES.AIR UNIVERSITY Matrix methods in structural mechanics :a conference Proceedings Dayton,Ohio 1965 Oct 26-28 1966
Eng 41.2732

UNITED STATES.ARMED SERVICES Lectures in applied mathematics :summer seminar proceedings Boulder,Colo. 1957 Jun.23-Jul. 19 Vol. 1: probability and related topics in physical sciences By Mark Kac American mathematical society xiii,266p 24cm Interscience publishers,London;New York,1959 With special lectures by G.E.Uhlenbeck,A.R. Hibbs,and Balth.van der Pol
Math 3.0844

UNITED STATES.ARMY.EUROPEAN RESEARCH OFFICE International symposium on free radicals 5th Preprint of papers Uppsala 1961 Jul 6-7 1961
Chem 18.2625

UNITED STATES.ARMY.HEADQUARTERS QUARTERMASTER RESEARCH AND DEVELOPMENT COMMAND Protection and functioning of the hands in cold climates :a conference Proceedings Natick,Mass. 1956 Apr 23-24 Edited by Frank R. Fisher illus vi,176p 23cm National academy of sciences,Washington,1957
Sco 14.0858

UNITED STATES.ARMY.MATHEMATICS RESEARCH CENTER Approximations with special emphasis on spline functions :a symposium Proceedings Madison, Wis. 1969 May 5-7 Edited by I.J. Schoenberg bibliog. Academic press,New York;London,1969
Math S 3.1504

UNITED STATES.ARMY.MATHEMATICS RESEARCH CENTER Approximations with special emphasis on spline functions :a symposium Proceedings Madison, Wis. 1970 May 5-7 Edited by I.J. Schoenberg Held at the University of Wisconsin United States.Army.Mathematics research center.Publications, 23 xi,488p 24cm Academic press,London;New York,1971
Math S 3.1675

UNITED STATES.ARMY.MATHEMATICS RESEARCH CENTER Asymptotic solutions of differential equations and their applications symposium proceedings Madison,Wis. 1964 May 4-6 Edited by Calvin H. Wilcox United States.Army.Mathematics research center.Publications, 13 x,249p 22cm John Wiley and sons,New York,1964 Dedicated to Professor Rudolph E.Langer
Math 3.0291

UNITED STATES.ARMY.MATHEMATICS RESEARCH CENTER
Boundary problems in differential equations symposium proceedings Madison,Wis. 1959 Apr.20-22 Edited by Rudolph E. Langer United States army.Mathematics research center. Publications, 2 x,320p 23cm University of Wisconsin press,Madison,Wis.,1960
Math L 5.0281

UNITED STATES.ARMY.MATHEMATICS RESEARCH CENTER
Electromagnetic waves :a symposium proceedings Madison,Wis. 1961 Apr.10-12 Edited by Rudolph E. Langer 396p University of Wisconsin press,Madison,Wis., 1962
Obs 6.1032

UNITED STATES.ARMY.MATHEMATICS RESEARCH CENTER
Electromagnetic waves symposium proceedings Madison,Wis. 1961 Apr. 10-12 Edited by Rudolph E. Langer University of Wisconsin press,Madison,Wis.,1962 photolith
A Math 4.0684

UNITED STATES.ARMY.MATHEMATICS RESEARCH CENTER
Error correcting codes Madison,Wis. 1968 May 6-8 Edited by Henry B. Mann United States.Army.Mathematics research center, 21 Bibliog.,Illus. xii,231p 23cm Wiley,New York,1968
P Math 2.3700

UNITED STATES.ARMY.MATHEMATICS RESEARCH CENTER
Error correcting codes :a symposium Proceedings Madison 1968 May 6-8 Edited by Henry B. Mann United States.Army. Mathematics research center.Publications, 22 Wiley,New York,1968
Math L 5.3514

UNITED STATES.ARMY.MATHEMATICS RESEARCH CENTER
Error in digital computation advanced seminar proceedings Madison,Wis. 1964 Oct.5-7 Vol. 1 Edited by Louis B. Rall United States army.Mathematics research center. Publications, 14 Bibliog. ix,324p 24cm John Wiley and sons,New York,1965
Math 3.0242

UNITED STATES.ARMY.MATHEMATICS RESEARCH CENTER
Error in digital computation symposium proceedings Madison,Wis. 1965 Apr.26-28 Vol. 2 Edited by Louis B. Rall United States army.Mathematics research center. Publications, 15 x,288p 24cm John Wiley and sons,New York,1965
Math 3.0243

UNITED STATES.ARMY.MATHEMATICS RESEARCH CENTER
Frontiers of numerical mathematics :symposium Madison,Wis. 1959 Oct 30-31 United States army.Mathematics research center.Publications, 4 xi,132p 24cm University of Wisconsin, Madison,Wis.,1960
Math L 5.3232

UNITED STATES.ARMY.MATHEMATICS RESEARCH CENTER
Graph theory and its applications :an advanced seminar Madison,Wis. 1969 Oct 13-15 bibliog.,illus. viii,262p 23cm Academic press,New York;London,1970
P Math 2.3821

UNITED STATES.ARMY.MATHEMATICS RESEARCH CENTER
Markov processes and potential theory :a symposium Proceedings Madison,Wis. 1967 May 1-3 Edited by Joshua Chover United States army.Mathematics research center. Publications, 19 Bibliog x,235p 24cm John Wiley,New York,1967
Math 3.1189

UNITED STATES.ARMY.MATHEMATICS RESEARCH CENTER
Nonlinear problems :a symposium Proceedings Madison,Wis. 1962 Apr 30-May 2 Edited by R. E. Langer bibliog.,illus. xiii,321p 24cm University of Wisconsin press,Madison,Wis., 1963
A Math 4.1615

UNITED STATES.ARMY.MATHEMATICS RESEARCH CENTER
Numerical solutions of nonlinear differential equations :advanced symposium proceedings Madison,Wis. 1966 May 9-11 Edited by Donald Greenspan United States.Army. Mathematics research center.Publications, 17 John Wiley,New York,1966
A Math 4.1220

UNITED STATES.ARMY.MATHEMATICS RESEARCH CENTER
Numerical solutions of nonlinear differential equations advanced seminar proceedings Madison,Wis. 1966 May 9-11 Edited by Donald Greenspan United States army. Mathematics research center.Publication, 17 x,343p 23cm John Wiley and sons,New York, 1966
Math L 5.0358

UNITED STATES.ARMY.MATHEMATICS RESEARCH CENTER
On numerical approximation symposium proceedings Madison,Wis. 1958 Apr. 21-23 Edited by Rudolph E. Langer United States army.Mathematics research center.Publications, 1 x,462p 24cm University of Wisconsin press,Madison,Wis.,1959
Math L 5.0267

UNITED STATES.ARMY.MATHEMATICS RESEARCH CENTER
Partial differential equations and continuum mechanics :an international conference Proceedings Madison,Wis. 1960 Jun 7-15 Edited by R.E. Langer bibliog.,illus. xv, 397p 24cm University of Wisconsin press, Madison,Wis.,1961
A Math 4.1614

UNITED STATES.ARMY.MATHEMATICS RESEARCH CENTER
Recent advances in matrix theory advanced seminar 2nd proceedings Madison,Wis. 1963 Oct.14-16 Edited by Hans Schneider United States army.Mathematics research center. Publications, 12 xi,142p 24cm University of Wisconsin press,Madison; Milwaukee,Wis.,1964
Math 3.0164

UNITED STATES.ARMY.MATHEMATICS RESEARCH CENTER
Recent advances in matrix theory seminar 2nd proceedings Madison,Wis. 1963 Oct 14-16 Edited by Hans Schneider United States army.Mathematics research center.Publications, 12 xi,142p 24cm University of Wisconsin press,Madison,Wis.,1964
P. Math 2.2087

UNITED STATES.ARMY.MATHEMATICS RESEARCH CENTER Spectral analysis of time series :advanced seminar Proceedings Madison,Wis. 1966 Oct 3-5 Edited by Bernard Harris United States.Army.Mathematics research center. Publications, 18 illus. x,319p 24cm Wiley,New York,1967
Math S 3.1456

UNITED STATES.ARMY.MATHEMATICS RESEARCH CENTER Statistical theory of reliability advanced seminar proceedings Madison,Wis. 1962 May 8-10 Edited by Marvin Zelen United States army.Mathematics research center. Publications, 9 xvii,166p 23cm University of Wisconsin press,Madison,Wis., 1963
Math 3.0700

UNITED STATES.ARMY.MATHEMATICS RESEARCH CENTER Stochastic models in medicine and biology :a symposium Proceedings Madison,Wis. 1963 Jul 12-14 Edited by John Gurland United States.Army.Mathematics research center. Publications, 10 University of Wisconsin press,Madison,Wis.,1964
Gen 34.0419

UNITED STATES.ARMY.MATHEMATICS RESEARCH CENTER Stochastic optimization and control :advanced seminar Proceedings Madison,Wis. 1967 Oct 2-4 Edited by Herman Karreman United States.Army.Mathematics research center. Publications, 20 bibliog.,illus. xii,217p 23cm Wiley,New York,1968
Math S 3.1403

UNITED STATES.ARMY.MATHEMATICS RESEARCH CENTER Theory and applications of spline functions :a symposium Proceedings Madison,Wis. 1968 Oct. United States.Army.Mathematics research center Bibliog.,Illus. xi,212p 24cm Academic press,New York;London,1969
P Math 2.3487

UNITED STATES.ARMY.MATHEMATICS RESEARCH CENTER Theory and applications of spline functions :a symposium Proceedings Madison,Wis. 1968 Oct. United States.Army.Mathematics research center Bibliog.,Illus. xi,212p 24cm Academic press,New York;London,1969
P Math 2.3487

UNITED STATES.ARMY.MATHEMATICS RESEARCH CENTER Theory and applications of spline functions : an advanced seminar Proceedings Madison, Wis. 1968 Oct 7-9 Edited by T.N.E. Greville Held at the University of Wisconsin United States.Army.Mathematics research center. Publications, 22 xi,212p 24cm Academic press,New York;London,1969
Math S 3.1674

UNITED STATES.ARMY.MATHEMATICS RESEARCH CENTER Theory and applications of spline functions : an advanced seminar Proceedings Madison, Wis. 1968 Oct 7-9 Edited by T.N.E. Greville United States.Army.Mathematics research center.Publications, 22 Academic press,New York;London,1969
Math L 5.3443

UNITED STATES.ARMY.MATHEMATICS RESEARCH CENTER and UNIVERSITY OF WISCONSIN.THEORETICAL CHEMISTRY INSTITUTE Perturbations theory and its applications in quantum mechanics an advanced seminar Papers Madison,Wisc. 1965 Oct 4-6 Edited by Calvin H. Wilcox J. Wiley,New York,1966
Chem 18.0692

UNITED STATES.ARMY.MEDICAL RESEARCH AND DEVELOPMENT COMMAND Acquisition of skills a conference Papers New Orleans,La. 1965 Mar 8-12 Edited by Edward A. Bilodeau Academic press,New York;London,1966
Psy 31.1005

UNITED STATES.ARMY.OFFICE OF ORDNANCE RESEARCH Composite materials and composite structures Sagamore ordnance materials research conference 6th Proceedings New York 1959 Aug 8-21 Arrangements by Syracuse university research institute New York,1959 Mimeograph
Met 25.0093

UNITED STATES.ARMY.OFFICE OF ORDNANCE RESEARCH Ductile chromium and its alloys :a conference.. held at the 1955 metal congress and exposition of the American society for metals American society for metals,Cleveland,Ohio, 1957
Met 25.2307

UNITED STATES.ARMY.OFFICE OF ORDNANCE RESEARCH Mechanical properties of engineering ceramics : a conference Proceedings Raleigh,N.C. 1960 Edited by W.Wurth Kriegel and Hayne Palmour Interscience,New York;London,1961
Met 25.0853

UNITED STATES.ARMY.OFFICE OF ORDNANCE RESEARCH Temperature - its measurement and control in science and industry : a symposium 3rd Papers Washington,D.C. 1954 Oct 28-30 Vol 2 Edited by Hugh C. Wolfe Reinhold;Chapman and Hall,New York;London,1955
Met 25.0299

UNITED STATES.ARMY.ORDNANCE MATERIALS RESEARCH OFFICE Composite materials and composite structures Sagamore ordnance materials research conference 6th Proceedings New York 1959 Aug 8-21 Arrangements by Syracuse university research institute New York,1959 Mimeograph
Met 25.0093

UNITED STATES.ARMY.RESEARCH OFFICE Boron-nitrogen chemistry :an international symposium Durham,N.C. 1963 Apr 23-25 American chemical society.Advances in chemistry series, 42 American chemical society,Washington,D.C., 1964 Symposium chairman:Kurt Niedenzu
Chem 18.0986

UNITED STATES.ARMY.RESEARCH OFFICE Photoelasticity :an international symposium Proceedings Chicago,Ill. 1961 Oct Edited by M.M. Frocht Pergamon press,London,1963
Eng 41.2913

UNITED STATES.ARMY MATERIALS RESEARCH AGENCY Fatigue - an interdisciplinary approach Sagamore army materials research conference 10th Proceedings Raquette Lake,N.Y. 1963 Aug 13-16 Edited by J.J. Burke and others Syracuse university press,New York,1964
Eng 41.3803

UNITED STATES.ARMY MATERIALS RESEARCH AGENCY Fatigue-an interdisciplinary approach Sagamore army materials research conference 10th Raquette Lake,New York 1963 Aug 13-16 Edited by John J. Burke and others Syracuse university press,Syracuse,1964
Met 25.2330

UNITED STATES.ARMY MATERIALS RESEARCH AGENCY
Fundamentals of deformation processing Sagamore army materials conference 9th Raquette Lake,New York 1962 Aug 28-31 By Walter A. Backofen and others Co-organized and directed by the National research council Syracuse university press,Syracuse,New York, 1964
Met 25.0741

UNITED STATES.ARMY MATERIALS RESEARCH AGENCY
Surfaces and interfaces 1-2 Sagamore army materials research conference 13th-14th Proceedings Raquette Lake,New York 1966-67 Edited by John J. Burke and others Syracuse university press,New York,1967-68
Met 25.2365

UNITED STATES.ARMY OFFICE OF ORDINANCE RESEARCH
International symposium on stress wave propagation in materials Pennsylvania state college 1959 Jun 30-Jul 2 Edited by Norman Davids Conducted by the Pennsylvania State university.Department of engineering mechanics New York,1960
Geod 9.0267

UNITED STATES.ARMY RESEARCH OFFICE,DURHAM
Boron-nitrogen chemistry :an international symposium Durham,N.C. 1963 Apr 23-25 American chemical society.Advances in chemistry series, 42 American chemical society,Washington,D.C.,1964 Symposium chairman:K.Niedenzu
Chem 18.2658

UNITED STATES.ARMY RESEARCH OFFICE,DURHAM
Environment-sensitive mechanical behaviour :a conference Proceedings Baltimore 1965 Jun 7-8 Pt 1-2 Edited by A.R.C. Westwood and N.S. Stoloff Sponsored by the Martin company research institute for advanced studies Metallurgical society conferences, 35 2 vols Gordon and Breach,New York,1966
Met 25.1963

UNITED STATES.ARMY RESEARCH OFFICE,DURHAM
Research conference on structure and properties of engineering materials Proceedings Raleigh,N.C. 1962 Mar 12-13 Materials science research, 1 Plenum press, New York,1963
Met 25.2281

UNITED STATES.BUREAU OF MINES Air pollution
Conference on air pollution Proceedings Washington,D.C. 1950 May 3-5 Edited by Louis C. McCabe Sponsored by United States. Interdepartmental committee on air pollution xiv,847p McGraw-Hill,New York,1952
Nap 11.0388

UNITED STATES.BUREAU OF ORDNANCE Symposium on large-scale digital calculating machinery 2nd proceedings Cambridge,Mass. 1949 Sep 13-16 Harvard university.Computation laboratory Harvard university.Computation laboratory Annals, 26 illus. xxxviii,393p 27cm Harvard university press,Cambridge,Mass. 1951
Math L 5.0684

UNITED STATES.BUREAU OF ORDNANCE Symposium on large-scale digital calculating machinery proceedings Cambridge,Mass. 1947 Jan 7-10 Harvard university.Computation laboratory Harvard university.Computation laboratory. Annals, 16 illus. xxix,302p 27cm Harvard university press,Cambridge,Mass.,1948
Math L 5.0681

UNITED STATES.DEPARTMENT OF HEALTH,EDUCATION AND WELFARE Sudden death in infants
Conference on causes of sudden death in infants Proceedings Seattle 1963 Sep 9-10 Edited by Ralph J. Wedgwood and others United States.Public health service publication,1412 U.S.govt.printing office, Washington,D.C.,1965
HE 27.0158

UNITED STATES.DEPARTMENT OF STATE Adoption d'un premier meridien unique et d'une heure universelle :conference Proces-verbaux Washington,D.C. 1884 Oct 1-Nov 1 216p Washington,D.C.,1884
Obs 6.3168

UNITED STATES.DEPARTMENT OF STATE
International meridian conference International conference for the purpose of fixing a prime meridian and a universal day 1st Proceedings Washington,D.C. 1884 Oct 1-Nov 1 212p Gibson,Washington,D.C.,1884 With five sheets of notes by J.C.Adams in a separate envelope
Obs 6.3169

UNITED STATES.DEPARTMENT OF THE NAVY.BUREAU OF SHIPS Zinc as a galvanic anode Navy-industry zinc symposium on cathodic protection development,application and specification 1955 Apr 21-22 U.S.Navy dept.,Washington,D.C. 1955
Met 25.2500

UNITED STATES.ENVIRONMENTAL PROTECTION AGENCY
Environmental protection Symposium on multiple-source urban diffusion models Proceedings 1970 Air pollution control office.Publication,AP- 86 US government printing office,Washington,D.C.,1970
Eng 41.8616

UNITED STATES.FEDERAL SECURITY AGENCY.OFFICE OF VOCATIONAL REHABILITATION Psycholgical diagnosis and counseling of the adult blind : selected papers from the proceedings of the University of Michigan conference for the blind Ann Arbor,Mich. 1947 Edited by Wilma Donahue and Donald Dabelstein Co-sponsored by the University of Michigan. Institute for human adjustment American foundation for the blind,New York,1950
Psy 31.2118

UNITED STATES.INTERDEPARTMENTAL COMMITTEE ON AIR POLLUTION Air pollution Conference on air pollution Proceedings Washington,D.C. 1950 May 3-5 United States.Bureau of mines Edited by Louis C. McCabe xiv,847p McGraw-Hill,New York,1952
Nap 11.0388

UNITED STATES.INTERDEPARTMENTAL COMMITTEE ON AIR POLLUTION Air pollution United States technical conference on air pollution Proceedings Washington 1950 May 3-5 xiv, 847p McGraw-Hill,New York,1952 Chairman of the committee:Louis C.McCabe
Chem E 24.1542

UNITED STATES.NATIONAL ACADEMY OF SCIENCES
Venomous and poisonous animals and noxious plants of the Pacific region :a collection of papers based on a symposium in the Public health and medical science division at the tenth Pacific science congress Papers Honolulu 1961 Aug 21-Sep 6 Pacific science association and University of Hawaii Edited by Hugh L. Keegan and W.V. MacFarlane Pergamon press.Symposium publications division, Oxford,1963
Chem 26.0448

UNITED STATES.NATIONAL AERONAUTICS AND SPACE ADMINISTRATION.NASA SP, 11 NASA-University conference on the science and technology of space exploration proceedings Chicago 1962 Nov.1-3 Vol 1-2 NASA SP 11 2 vols office of scientific and technical information,Washington,D.C.,1962
A Math 4.1157

UNITED STATES.NAVY MATHEMATICAL COMPUTING ADVISORY PANEL Advanced programming methods for digital computers symposium proceedings Washington,D.C. 1956 Jun.28-29 United States.Office of naval research United States.Office of naval research.ONR symposium reports ACR-15 83p office of naval research,Washington,D.C.,1956
Math L 5.1809

UNITED STATES.NAVY MATHEMATICAL COMPUTING ADVISORY PANEL Automatic programming for digital computers :symposium Washington,D.C. 1954 May 13-14 152p Office of naval research,Washington,D.C.,1954
Math L 5.1795

UNITED STATES.NAVY MATHEMATICAL COMPUTING ADVISORY PANEL Commercially available general-purpose electronic digital computers of moderate price symposium Washington,D.C. 1952 May 14 41p Washington,D.C.,1952
Math L 5.1790

UNITED STATES.NAVY MATHEMATICAL COMPUTING ADVISORY PANEL Managerial aspects of digital computer installations symposium papers Washington,D.C. 1953 Mar.30 36p Office of naval research,Washington,D.C.,1953
Math L 5.1792

UNITED STATES.OFFICE OF NAVAL RESEARCH
Advanced programming methods for digital computers symposium proceedings Washington, D.C. 1956 Jun.28-29 Cosponsored by United States.Navy mathematical computing advisory panel United States.Office of naval research. ONR symposium reports ACR-15 83p office of naval research,Washington,D.C.,1956
Math L 5.1809

UNITED STATES.OFFICE OF NAVAL RESEARCH
Categories of human learning :symposium on the psychology of human learning Papers Ann Arbor,Mich. 1962 Jan 31-Feb 1 Edited by Arthur W. Melton Academic press,New York; London,1964
Psy 31.1003

UNITED STATES.OFFICE OF NAVAL RESEARCH
Computing methods and the phase problem in X-ray crystal analysis :a conference Report Pennsylvania state college 1950 Apr 6-8 Pennsylvania state college,State College,Pa., 1952
Cav 7.1789

UNITED STATES.OFFICE OF NAVAL RESEARCH
Conference on carbon 3rd Proceedings Buffalo,N.Y. 1957 Jun 17-21 By S. Morzowski and others Co-sponsored by the National science foundation Pergamon press, London,1959
Met 25.0156

UNITED STATES.OFFICE OF NAVAL RESEARCH
Conference on carbon 3rd Proceedings Buffalo,N.Y. 1957 Jun 17-21 State university of New York,Buffalo Edited by S. Mrozowski and others Conference on carbon. Proceedings, 2 718p Pergamon press, Oxford,1959
TA 15.0214

UNITED STATES.OFFICE OF NAVAL RESEARCH Data processing seminar on status of digital computer and data processing developments in the Soviet Union Washington,D.C. 1958 Nov 12 179p 26cm United States Office of Naval research,Washington,D.C.,1958
Math L 5.0821

UNITED STATES.OFFICE OF NAVAL RESEARCH Fluid models in geophysics Symposium on the use of models in geophysical fluid dynamics 1st Proceedings Baltimore,Md. 1953 Sep 1-4 Edited by Robert R. Long illus,diagrms v, 162p U.S.Government printing office, Washington,D.C.,1956
Chem E 24.0202

UNITED STATES.OFFICE OF NAVAL RESEARCH High temperature structures and materials Symposium on naval structural mechanics 3rd Proceedings New York 1963 Jan 23-25 Edited by A.M. Freudenthal and others United States.Office of naval research.Structural mechanics series Pergamon press,London,1964
Eng 41.3806

UNITED STATES.OFFICE OF NAVAL RESEARCH High temperature structures and materials Symposium on naval structural mechanics Proceedings New York 1963 Jan 23-25 Edited by A.M. Freudenthal and others Pergamon,Oxford,1964
Met 25.1009

UNITED STATES.OFFICE OF NAVAL RESEARCH
International conference on high-energy physics 10th Proceedings Rochester,N.Y. 1960 Aug 25-Sep 1 University of Rochester International union of pure and applied physics Edited by E.C.G. Sudarshan and others illus xxv,890p University of Rochester,Rochester,N.Y.,1960
Cav 7.2891

UNITED STATES.OFFICE OF NAVAL RESEARCH
International conference on photoconductivity 3rd Proceedings Stanford,Calif. 1969 Aug 12-15 Edited by E.M. Pell Bibliog, diagrms,graphs xi,410p 26cm Pergamon, Oxford,1971
Cav 7.3073

UNITED STATES.OFFICE OF NAVAL RESEARCH
International symposium on rarefied gas dynamics 4th Proceedings Toronto 1964 Jul Vol 1-2 Edited by J.H.de Leeuw Advances in applied mechanics.Supplement, 3 2 vols Academic press,New York;London,1965
Eng 41.6788

UNITED STATES.OFFICE OF NAVAL RESEARCH
International symposium on rarefied gas dynamics 6th Proceedings Cambridge,Mass. 1968 Jul Vol 1-2 Edited by Leon Trilling and Harold Y. Wachman Advances in applied mechanics.Supplement, 5 2 vols Academic press,New York,1969
Eng 41.6797

UNITED STATES.OFFICE OF NAVAL RESEARCH Metal molybdenum The Technology of molybdenum and its alloys :a symposium Proceedings Washington,D.C. 1956 Sep 18-19 Edited by Julius J. Harwood American society for metals,Cleveland,Ohio,1958
Met 25.0546

UNITED STATES.OFFICE OF NAVAL RESEARCH
Microcirculation,perfusion,and transplantation of organs Conference on microcirculation, perfusion and transplantation of organs Proceedings Miami,Fla. 1969 Oct 30 Edited by Theodore I. Malinin and others Academic press,London;New York,1970
An 32.5458

UNITED STATES.OFFICE OF NAVAL RESEARCH
Millimeter waves :a symposium Proceedings New York 1959 Mr 31-Apr 2 Microwave research institute and Institute of radio engineers Polytechnic institute of Brooklyn. Microwave research institute symposia series, 9 Polytechnic press,Brooklyn,N.Y.,1960
Cav 7.1715

UNITED STATES.OFFICE OF NAVAL RESEARCH
Molecular structure and biological specificity a symposium Papers Washington,D.C. 1955 Oct 28-29 Edited by Linus Pauling and Harvey A. Itano Arranged by the American institute of biological sciences American institute of biological sciences.Publication,2 American institute of biological sciences,Washington,D. C.,1957
Chem 18.2137

UNITED STATES.OFFICE OF NAVAL RESEARCH
Molecular structure and biological specificity a symposium Washington,D.C. 1955 Oct 28-29 American institute of biological sciences Edited by Linus Pauling and Harvey A. Hana American institute of biological sciences. Publications, 2 American institute of biological sciences,Washington,D.C.,1957
An 32.3156

UNITED STATES.OFFICE OF NAVAL RESEARCH
Natural electromagnetic phenomena below 30 kilocycles NATO advanced study institute Proceedings Bad Homburg 1963 Jul 22-Aug 2 Edited by D.F. Bleil vii,470p Plenum press, New York,1964
Nap 11.0067

UNITED STATES.OFFICE OF NAVAL RESEARCH
Organization of recall International interdisciplinary conference on learning, remembering and forgetting 2nd Proceedings Princeton,N.J. 1964 Sep 27-30 New York academy of sciences and National institute of child health and human development Edited by Daniel P. Kimble New York academy of sciences.Interdisciplinary communication program,New York,1967
Psy 31.1012

UNITED STATES.OFFICE OF NAVAL RESEARCH
Plasticity Naval structural mechanics :a symposium 2nd Proceedings Rhode Island 1960 Apr 5-7 Edited by E.H. Lee and P.S. Symonds Pergamon press,New York,1960
Met 25.0845

UNITED STATES.OFFICE OF NAVAL RESEARCH
Plasticity Naval structural mechanics symposium 2nd Proceedings Providence,R. I. 1960 Apr 5-7 Edited by E.H. Lee and P.S. Symonds United States.Office of naval research.Structural mechanics series Pergamon press,London,1960
Eng 41.2479

UNITED STATES.OFFICE OF NAVAL RESEARCH
Quantum mechanical methods in valence theory : a conference Papers Shelter Island,L.I.,N. Y. 1951 Sep 8-10 National academy of sciences United States.Office of naval research.Physics branch.Publication,4 U.S. government printing office,Washington,D.C., 1952
Chem 18.2237

UNITED STATES.OFFICE OF NAVAL RESEARCH Quasi-optics :a symposium Proceedings New York 1964 Jun 8-10 Microwave research institute and Institute of radio engineers Polytechnic institute of Brooklyn.Microwave research institute symposia series, 14 Polytechnic press,Brooklyn,N.Y.,1964
Cav 7.1721

UNITED STATES.OFFICE OF NAVAL RESEARCH Self-organizing systems 1962 conference proceedings Chicago,Ill. 1962 May 22-23 Edited by Marshall C. Yovits and others Jointly sponsored by Armour research foundation ix,563p 24cm Spartan books, Washington,D.C.,1962
Math L 5.0992

UNITED STATES.OFFICE OF NAVAL RESEARCH
Semiconductor surface physics Conference on the physics of semiconductor surfaces Proceedings Philadelphia,Pa. 1956 Jun 4-6 Edited by R.H. Kingston Sponsored by the United States.Office of naval research,the University of Pennsylvania and the Lincoln laboratory University of Pennsylvania, Philadelphia,Pa.,1957
Eng 41.5399

UNITED STATES.OFFICE OF NAVAL RESEARCH
Simultaneous linear equations and the determination of Eigenvalues symposium Proceedings Los Angeles 1951 National bureau of standards Edited by L.J. Paige and Olga Taussky United States.National bureau of standards. Applied mathematics series, 29 bibliog. 125p 26cm National bureau of standards,Washington,D.C.,1953
Math L 5.0131

UNITED STATES.OFFICE OF NAVAL RESEARCH
Single-crystal films :a conference Proceedings Blue Bell,Penn. 1963 May Edited by Maurice H. Francombe and Hiroshi Sato Sponsored jointly by Philco corporation Pergamon press,Oxford,1964
Met 25.0177

UNITED STATES.OFFICE OF NAVAL RESEARCH
Surtsey biology conference Proceedings Surtsey research society photos 48p Reykjavik,1965 Typescript
Sco 14.8155

UNITED STATES.OFFICE OF NAVAL RESEARCH
Symposium on computer augmentation of human reasoning proceedings Washington,D.C. 1964 Jun.16-17 Edited by Margo A. Sass and William D. Wilkinson In joint sponsorship with the Bunker-Ramo corporation vi,235p 24cm Spartan books;Macmillan,Washington,D.C.; London,1965
Math 3.0244

UNITED STATES.OFFICE OF NAVAL RESEARCH
Symposium on computer augmentation of human reasoning Proceedings Washington,D.C. 1964 Jun 16-17 Edited by Margo A. Sass and William D. Wilkinson In joint sponsorship with the Bunker-Ramo corporation Spartan books,Washington,D.C.,1965
Eng 41.2202

UNITED STATES.OFFICE OF NAVAL RESEARCH
Symposium on computer augmentation of human reasoning proceedings Washington,D.C. 1964 Jun 16-17 Edited by Margo A. Sass and William D. Wilkinson Jointly sponsored by Bunker-Ramo corporation bibliog,illus. vi, 235p 24cm Spartan books,Washington,D.C., 1965
Math L 5.1005

UNITED STATES.OFFICE OF NAVAL RESEARCH
Symposium on congestion theory proceedings Chapel Hill,N.C. 1964 Aug.24-26 University of North Carolina.Department of statistics Edited by Walter L. Smith and William E. Wilkinson North Carolina.University. Monograph series in probability and statistics, 2 xv,457p 24cm University of North Carolina press,Chapel Hill,N.C.,1965
Math 3.0568

UNITED STATES.OFFICE OF NAVAL RESEARCH
Symposium on naval hydrodynamics proceedings Washington D.C. 1956 Sep.24-28 Edited by F. S. Sherman Sponsored by the National research council National research council. Publication, 515 National research council, Washington D.C.,1957
A Math 4.0530

UNITED STATES.OFFICE OF NAVAL RESEARCH
Symposium on naval structural mechanics 1st Proceedings Stanford,Calif. 1958 Aug 11-14 Edited by J.Norman Goodier and Nicholas J. Hoff International series on aeronautical sciences and space flight Pergamon,Oxford, 1960
Met 25.1300

UNITED STATES.OFFICE OF NAVAL RESEARCH
Symposium on superconductive techniques for computing systems proceedings Washington,D. C. 1960 May 17-19 illus. vii,413p 26cm Office of naval research,Washington,D.C.,1960
Math L 5.0859

UNITED STATES.OFFICE OF NAVAL RESEARCH
Symposium on time series analysis proceedings Providence,R.I. 1962 Jun.11-14 Brown university Edited by Murray Rosenblatt xiv,497p 25cm John Wiley and sons,New York, 1963
Math 3.0695

UNITED STATES.OFFICE OF NAVAL RESEARCH The Arctic sea ice :conference Proceedings Easton,Md. 1958 Feb 24-27 National research council.Division of earth science National research council.Publication,598 271p 28cm National academy of sciences-National research council,Washington,1958
Sco 14.0359

UNITED STATES.OFFICE OF NAVAL RESEARCH The Physics and chemistry of ceramics :a symposium Proceedings Philadelphia,Pa. 1962 May 28-30 Edited by Cyrus Klingsberg Gordon and Breach,New York;London,1963
Met 25.0277

UNITED STATES.OFFICE OF NAVAL RESEARCH The Symposium on time series analysis Proceedings Providence,R.I. 1962 Jun 11-14 Edited by Murray Rosenblatt xiv,497p 26cm Wiley,New York;London,1963
Cav 7.2744

UNITED STATES.OFFICE OF NAVAL RESEARCH
Titanium :a symposium Report 1948 Dec 16 Office of naval research,Washington,D.C.,1949
Met 25.0548

UNITED STATES.OFFICE OF NAVAL RESEARCH
Verbal behavior and learning Conference on verbal learning 2nd Proceedings Edited by C.N. Cofer and Barbara S. Musgrave McGraw-Hill,New York,1963
Psy 31.2737

UNITED STATES.OFFICE OF NAVAL RESEARCH
Vigilance :a symposium Proceedings Santa Barbara,Calif. 1961 Jan Edited by Donald N. Buckner and James J. McGrath Conducted by Human factors research McGraw-Hill series in psychology McGraw-Hill,New York,1963
Psy 31.1008

UNITED STATES.OFFICE OF NAVAL RESEARCH
Viscous drag reduction A Symposium on viscous drag reduction Proceedings Dallas, Texas 1968 Sep 24-25 Edited by C.Sinclair Wells 500p 25cm Plenum press,New York, 1969
Cav 7.2923

UNITED STATES.OFFICE OF NAVAL RESEARCH and ARCTIC INSTITUTE OF NORTH AMERICA Arctic Basin symposium Proceedings Hershey,Pa. 1962 Oct 8-11 illus,maps 313p 23cm Washington,D.C.,1963
Sco 14.5904

UNITED STATES.OFFICE OF NAVAL RESEARCH and NORTH ATLANTIC TREATY ORGANIZATION
Molecular biophysics :international summer school Proceedings Squaw Valley,Calif. 1964 Aug 17-28 Edited by Bernard Pullmann and Mitchel Weissbluth Academic press,New York;London,1965
Chem 18.2249

UNITED STATES.OFFICE OF NAVAL RESEARCH.GEOPHYSICS BRANCH Artificial stimulation of rain Conference on the physics of cloud and precipitation particles 1st Proceedings Woods Hole,Mass. 1955 Sep 7-10 Air force Cambridge research center.Geophysics research directorate Edited by Helmut Weickmann and Waldo Smith illus 442p Pergamon,London, 1957
Nap 11.0931

UNITED STATES.OFFICE OF NAVAL RESEARCH. INFORMATION SYSTEMS BRANCH Self-organizing systems :an interdisciplinary conference Proceedings Chicago,Ill. 1959 May 5-6 Edited by Marshall C. Yovits and Scott Cameron International tracts in computer science and technology and their application, 2 Pergamon press,New York, 1960
Eng 41.1162

UNITED STATES.OFFICE OF NAVAL RESEARCH.NAVAL SHIP RESEARCH AND DEVELOPMENT CENTER Symposium on viscous drag reduction Proceedings Dallas,Texas 1968 Sep 24-25 Edited by C.S. Wells illus xi,500p 25cm Plenum press, New York,1969
A Math 4.1736

UNITED STATES.OFFICE OF NAVAL RESEARCH. PHYSIOLOGICAL PSYCHOLOGY BRANCH Sensory deprivation :a symposium held at Harvard medical school Cambridge,Mass. 1958 Jun 20-21 Harvard medical school and Boston city hospital Edited by Philip Solomon and others Harvard university press,Cambridge, Mass.,1965
Psy 31.0128

UNITED STATES.OFFICE OF NAVAL RESEARCH. PSYCHOLOGICAL SCIENCES DIVISION Verbal learning and verbal behavior :a conference Proceedings New York 1959 Edited by Charles N. Cofer and Barbara S. Musgrave McGraw-Hill series in psychology McGraw-Hill, New York,1961
Psy 31.1905

UNITED STATES.OFFICE OF NAVAL RESEARCH, New perspectives in organization research : consisting of papers from a conference on research in organizations...and a seminar on the social science of organizations Pittsburgh,Pa. 1962 Jun.22-24 Pittsburgh, Pa. 1962 Jun.10-23 Edited by W.W. Cooper and others Bibliog. xxii,606p 24cm John Wiley and sons,New York,1964
Math 3.0910

UNITED STATES.OFFICE OF NAVAL RESEARCH,and Symposium on naval hydrodynamics :ship motions and drag reduction 5th Bergen 1964 Sep 10-12 Edited by J.K. Lunde and S.W. Doroff Office of naval research-Department of the navy,Washington,D.C.,1965
A Math 4.1303

UNITED STATES.PUBLIC HEALTH SERVICE Mental health in later maturity :a conference Papers Washington,D.C. 1941 May 23-24 United States.Public health service.Public health reports.Supplement, 168 US government printing office,Washington,D.C.,1942
Psy 31.3388

UNITED STATES.PUBLIC HEALTH SERVICE Renal disease and hypertension :tentn annual conference... Proceedings Denver 1971 Mar Atlantic City 1971 May Edited by Richard H. Thurm Proceedings of the annual conferences of the U.S.Public health service co-operative study (antihypertensive agents), 8 U.S.Dept. of health,education and welfare, Bethesda,1971
PGMS 29.0659

UNITED STATES.UPPER ATMOSPHERE ROCKET RESEARCH PANEL Rocket exploration of the upper atmosphere :a conference papers Oxford 1953 Aug 24-26 Edited by R.L.F. Boyd and M.J. Seaton Journal of atmospheric and terrestrial physics.Special supplement, 1 376p Pergamon press,London,1954
Nap 11.0073

UNITED STATES AIR FORCE. GEOPHYSICAL RESEARCH DIRECTORATE Symposium on microseisms Proceedings Harriman,N.Y. 1952 Sep 4-6 Edited by James T. Wilson and Frank Press National research council.Publication, 306 Washington,D.C.,1953
Geod 9.0376

UNITED STATES AIR FORCE ACADEMY Inequalities II Symposium on inequalities 2nd Proceedings 1967 Aug 14-22 Edited by Oved Shisha xvi,439p 24cm Academic press,New York;London,1970
P Math 2.3847

UNITED STATES ARMY.MATHEMATICS RESEARCH CENTER Partial differential equations and continuum mechanics :an international conference Proceedings Madison,Wisc. 1960 Jun 7-15 Edited by Rudolph E. Langer United States army.Mathematics research center.Publication, 5 Madison,Wisc.,1961
Geod 9.0202

UNITED STATES ARMY.MATHEMATICS RESEARCH CENTER Perturbation theory and its application in quantum mechanics seminar proceedings Madison,Wis. 1965 Oct 4-6 Edited by Calvin H. Wilcox Organized jointly with University of Wisconsin.Theoretical chemistry institute 428p John Wiley and sons,New York;London, 1966
A Math 4.0809

UNITED STATES ARMY.MATHEMATICS RESEARCH CENTER. Recent advances in matrix theory :advanced seminar 2nd proceedings Madison,Wis. 1963 Oct.14-16 Edited by Hans Schneider United States army.Mathematics research center. Publication, 12 xi,142p 25cm University of Wisconsin press,Madison,Wis.,1964
Math L 5.0328

UNITED STATES ARMY.MATHEMATICS RESEARCH CENTER. PUBLICATION, 14 Error in digital computation seminar proceedings Wisconsin 1964 Oct.5-7 Vols. 1-2 Edited by Louis B. Rall 2 vols John Wiley,New York,1965
A Math 4.1209

UNITED STATES ARMY.MATHEMATICS RESEARCH CENTER. PUBLICATIONS, 12 Recent advances in matrix theory seminar 2nd proceedings Madison,Wis. 1963 Oct 14-16 United States. Army.Mathematics research center Edited by Hans Schneider xi,142p 24cm University of Wisconsin press,Madison,Wis.,1964
P. Math 2.2087

UNITED STATES ARMY.MATHEMATICS RESEARCH CENTER. PUBLICATONS, 13 Asymptotic solutions of differential equations and their applications a symposium proceedings Madison,Wis 1964 Edited by C.H. Wilcox John Wiley and sons, New York,1964
A Math 4.0246

UNITED STATES ATOMIC ENERBY COMMISSION.RESEARCH CONFERENCE FOR BIOLOGY AND MEDICINE 18TH Symposium on hormonal control of protein biosynthesis Gatlinburg,Tenn. 1965 Apr 5-8 Oak Ridge national laboratory.Biology division Oak Ridge national laboratory.Annual research conference, 18 Wistar institute of anatomy and biology,Philadelphia,Pa.,1965
Gen 34.0550

UNITED STATES ATOMIC ENERGY COMMISSION A.E.C. Euratom conference on aqueous corrosion of reactor materials Brussels 1959 Oct 14-17 Office of technical services,Washington,D.C., 1959
Met 25.1930

UNITED STATES ATOMIC ENERGY COMMISSION Bioenergetics :considerations of processes of absorption stabilization, transfer and utilization.A symposium Proceedings Upton, N.Y. 1959 Oct 12-16 Brookhaven national laboratory Edited by Leroy George Augenstine Radiation research.Supplement, 2 Academic press,New York;London,1960
Bioch 33.0716

UNITED STATES ATOMIC ENERGY COMMISSION Conference on radioiodine Proceedings Chicago,Ill. 1956 Nov 5-6 Edited by D.E. Clark ACRH-100 University of Chicago, Chicago,Ill.,1956
Radioth 35.1076

UNITED STATES ATOMIC ENERGY COMMISSION Coral Gables conferences on symmetry principles at high energy 4 Proceedings Coral Gables, Fla. 1967 Jan 25-27 University of Miami. Center for theoretical studies Edited by Arnold Perlmutter and Behram Kursunoglu 253p Freeman,San Francisco;London,1967
A Math 4.1347

UNITED STATES ATOMIC ENERGY COMMISSION Lattice defects in quenched metals :an international conference Chicago 1964 Jun 15-17 Edited by R.M.J. Cotterill and others Academic press,New York;London,1965
Met 25.1240

UNITED STATES ATOMIC ENERGY COMMISSION Low level irradiation :a symposium Indianapolis, Ind. 1957 Dec 30 Edited by Austin M. Brues Co-sponsored by the Argonne national laboratory.Division of biological and medical research American association for the advancement of science.Publication, 59 American association for the advancement of science,Washington,D.C.,1959
Radioth 35.1109

UNITED STATES ATOMIC ENERGY COMMISSION Metal thorium The Conference on thorium Proceedings Cleveland,Ohio 1956 Oct 11 Edited by Harley A. Wilhelm American society for metals,Cleveland,Ohio,1958
Met 25.1561

UNITED STATES ATOMIC ENERGY COMMISSION Metallurgy information meeting 5th Proceedings Oak Ridge 1955 Apr 11-13 Pt 1-2 2 vols Washington,1960
Met 25.1552

UNITED STATES ATOMIC ENERGY COMMISSION Powder metallurgy in nuclear engineering Conference on powder metallurgy in atomic energy with additional papers Philadelphia 1955 Oct 20 By Henry H. Hausner American society for metals,Cleveland,Ohio,1958
Met 25.0767

UNITED STATES ATOMIC ENERGY COMMISSION Roentdens,rads and riddles :a symposium on supervoltage radiation therapy Oak Ridge, Tenn. 1956 Jul 15-18 Edited by Milton Friedman and others Held at the Oak Ridge institute of nuclear studies U.S.atomic energy commission,Washington,D.C.,1959
Radioth 35.1090

UNITED STATES ATOMIC ENERGY COMMISSION The International conference on high energy physics 13th Proceedings Berkeley,Calif. 1966 Aug 31-Sep 7 Edited by Margaret Alston-Garnjost Sponsored by the International union of pure and applied physics Berkeley, Calif.,1967
Cav 7.0189

UNITED STATES ATOMIC ENERGY COMMISSION.DIVISION OF TECHNICAL INFORMATION Dynamic clinical studies with radioisotopes :a symposium held at the Oak Ridge institute of nuclear studies Proceedings Oak Ridge,Tenn. 1963 Oct 21-25 Edited by R.M. Kniseley and W.N. Tauke U.S.A. E.C.,Washington,D.C.,1964
Radioth 35.1403

UNITED STATES ATOMIC ENERGY COMMISSION.RESEARCH CONFERENCE FOR BIOLOGY AND MEDICINE 11TH Symposium on genetic approaches to somatic cell variation Gatlinburg,Tenn. 1958 Apr 2-5 Sponsored by Oak Ridge national laboratory. Biology division Oak Ridge national laboratory.Symposia, 11 Journal of cellular and comparative physiology, 52,supp.1 Wistar institute of anatomy and biology, Philadelphia,Pa.,1958
Gen 34.1094

UNITED STATES ATOMIC ENERGY COMMISSION.RESEARCH CONFERENCE FOR BIOLOGY AND MEDICINE 13TH Symposium on mammalian genetics and reproduction Gatlinburg,Tenn. 1960 Apr 4-7 Sponsored by the Oak Ridge national laboratory. Biology division Oak Ridge national laboratory.Symposia, 13 Journal of cellular and comparative physiology, 56,supp.1 Wistar institute of anatomy and biology, Philadelphia,Pa.,1960
Gen 34.1782

UNITED STATES ATOMIC ENERGY COMMISSION.RESEARCH CONFERENCE FOR BIOLOGY AND MEDICINE 14TH Symposium on recovery of cells from injury : given at research conference for biology and medicine of the atomic energy commission Gatlinburg,Tenn. 1961 Apt 3-6 Sponsored by the Oak Ridge national laboratory.Biology division Journal of cellular and comparative physiology, 58,supp.1 Oak Ridge national laboratory.Symposia, 14 Wistar institute of anatomy and biology,Philadelphia,Pa.,1961
Gen 34.1114

UNITED STATES ATOMIC ENERGY COMMISSION.RESEARCH CONFERENCE FOR BIOLOGY AND MEDICINE 16TH Symposium on macromolecular aspects of the cell cycle :given at research conference for biology and medicine of the atomic energy commission Gatlinburg,Tenn 1963 Apr 8-11 Sponsored by Oak Ridge national laboratory. Biology division Oak Ridge national laboratory.Symposia, 16 Journal of cellular and comparative physiology, 62,supp.1 Wistar institute of anatomy and biology, Philadelphia,Pa.,1963
PGMS 29.0414

UNITED STATES ATOMIC ENERGY COMMISSION.RESEARCH CONFERENCE FOR BIOLOGY AND MEDICINE 16TH Symposium on macromolecular aspects of the cell cycle :given at research conference for biology and medicine of the atomic energy commission Gatlinburg,Tenn. 1963 Apr 8-11 Sponsored by Oak Ridge national laboratory. Biology division Oak Ridge national laboratory.Symposia, 16 Journal of cellular and comparative physiology, 62,no.2 Wistar institute of anatomy and biology,Philadelphia, Pa.,1963
Gen 34.0842

UNITED STATES ATOMIC ENERGY COMMISSION.RESEARCH CONFERENCE FOR BIOLOGY AND MEDICINE 17TH Symposium on molecular action of mutagenic and carcinogenic agents Gatlinburg,Tenn. 1964 Apr 6-9 Sponsored by the Oak Ridge national laboratory.Biology division Oak Ridge national laboratory.Symposia Journal of cellular and comparative physiology, 64,suppl. 1 Wistar institute of anatomy and biology, Philadelphia,Pa.,1964
Gen 34.1098

UNITED STATES ATOMIC ENERGY COMMISSION.RESEARCH CONFERENCE FOR BIOLOGY AND MEDICINE 19TH Symposium on differentiation and growth of haemoglobin-and immunoglobin-synthesising cells :given at research conference for biology and medicine of the Atomic energy commission Gatlinburg,Tenn. 1966 Apr 4-7 Sponsored by the Oak Ridge national laboratory. Biology division Oak Ridge national laboratory.Biology division.Symposia, 19 Journal of cellular physiology, 67,supp.1 Wistar institute of anatomy and biology, Philadelphia,Pa.,1966
Gen 34.0504

UNITED STATES ATOMIC ENERGY COMMISSION.RESEARCH CONFERENCE FOR BIOLOGY AND MEDICINE 20TH Symposium on chromosome mechanics at the molecular level :given at research conference for biology and medicine of the atomic energy commission Gatlinburg,Tenn. 1967 Apr 10-13 Sponsored by the Oak Ridge national laboratory. Biology division Oak Ridge national laboratory.Symposia, 20 Journal of cellular physiology, 70,suppl Wistar institute of anatomy and biology,Philadelphia,Pa.,1967
Gen 34.0846

UNITED STATES ATOMIC ENERGY COMMISSION.RESEARCH CONFERENCE FOR BIOLOGY AND MEDICINE 21ST Symposium on molecular aspects of differentiation :given at research conference for biology and medicine of the Atomic energy commission Gatlinburg,Tenn. 1968 Apr 8-11 Sponsored by the Oak Ridge national laboratory. Biology division Oak Ridge national laboratory.Biology division.Symposia, 21 Journal of cellular physiology, 72,supp.1 Wistar institute of anatomy and biology, Philadelphia,Pa.,1968
Gen 34.0505

UNITED STATES ATOMIC ENERGY COMMISSION.RESEARCH CONFERENCE FOR BIOLOGY AND MEDICINE 7TH Symposium on genetic recombination given at research conference for biology and medicine of the atomic energy commission Oak Ridge, Tenn. 1954 Apr 19-21 Sponsored by the Oak Ridge national laboratory.Biology division Oak Ridge national laboratory.Symposia, 7 Journal of cellular and comparative physiology, 45,supp.2 Wistar institute of anatomy and biology,Philadelphia,Pa.,1955
Gen 34.1042

UNITED STATES ATOMIC ENERGY COMMISSION.SYMPOSIUM SERIES, 11 Compartments pools and spaces in medical physiology :a symposium Oak Ridge, Tenn. 1966 Oct 24-27 Edited by Per.Erik E. Bergner and others Held at the Oak Ridge institute of nuclear studies U.S.atomic energy commission,1967
Radioth 35.0927

UNITED STATES CONFERENCE ON PRESTRESSED CONCRETE 1st Proceedings Cambridge,Mass. 1951 Aug 14-16 Massachusetts institute of technology Cambridge,Mass.,1951
Eng 41.3012

UNITED STATES CONTRIBUTIONS TO QUATERNARY RESEARCH International association for quaternary research congress 8th Proceedings Paris 1969 Edited by S.A. Schumm and W.C. Bradley Geological society of America.Special papers, 123 Geological society of America,Boulder,Colo.,1969
Bot 42.3249

UNITED STATES GEOLOGICAL SURVEY Geologia y depositos de carbon de la region de Sabinas, estado de Chihuahua International geological congress 20th papers Mexico City 1956 By Raymond C. Robeck and others Mexico City, 1956 Bound with two other publications of the congress
Geol 8.3041

UNITED STATES-JAPAN COMMITTEE ON SCIENTIFIC CO-OPERATION Biological and chemical aspects of oxygenases The United States-Japan symposium on oxygenases Proceedings Kyoto 1966 May 16-19 Edited by Konrad Bloch and Osamu Hayaishi Maruzen company,Tokyo,1966
Bioch 33.1102

UNITED STATES-JAPAN SCIENTIFIC COOPERATION PROGRAM Biological and biochemical evaluation of malignancy in experimental hepatomas The U.S.-Japan joint conference on biological and biochemical evaluation of malignancy in experimental hepatomas Proceedings Kyoto 1965 Nov 4-5 G.A.N.N. Monograph, 1 Japanese cancer association; Japanese foundation for cancer research,Tokyo, 1966
Bioch 33.0952

UNITED STATES-JAPAN SEMINAR ON DIFFERENTIAL AND FUNCTIONAL EQUATIONS Proceedings Minneapolis,Ma. 1967 Jun 26-30 Edited by William A. Harris and Yasutaka Sibuya Illus. xvi,585p 24cm Benjamin,New York;Amsterdam, 1967
P Math 2.3164

The UNITED STATES-JAPAN SYMPOSIUM ON OXYGENASES Proceedings Biological and chemical aspects of oxygenases Kyoto 1966 May 16-19 Edited by Konrad Bloch and Osamu Hayaishi held under the auspices of the United States-Japan committee on scientific co-operation Maruzen company,Tokyo,1966
Bioch 33.1102

UNITED STATES NATIONAL COMMITTEE FOR THE INTERNATIONAL BIOLOGICAL PROGRAM and CANADIAN NATIONAL COMMITTEE FOR THE INTERNATIONAL BIOLOGICAL PROGRAM Working party conference for the international biological program human adaptability study of circumpolar populations Report Point Barrow 1967 Nov 17-22 variousp 29cm Naval arctic research laboratory,Point Barrow, 1967
Sco 14.7529

UNITED STATES NATIONAL COMMITTEE FOR THE INTERNATIONAL GEOPHYSICAL YEAR Antarctica in the International geophysical year based on a symposium on the Antarctic Co-sponsored by the National science foundation Geophysical monograph, 1 American geophysical union. Publication, 462 illus,maps v,133p 26cm American geophysical union,Washington,D.C., 1956
Sco 14.8201

UNITED STATES NATIONAL COMMITTEE ON THEORETICAL AND APPLIED MECHANICS National congress of applied mechanics 1st proceedings Chicago 1951 June 11-16 Edited by Eli Sternberg American society of mechanical engineers,New York,1952
A Math 4.0327

UNITED STATES NATIONAL CONGRESS OF APPLIED MECHANICS 1st-5th Proceedings Edited by Eli Sternberg American society of mechanical engineers,New York,1952-66
Eng 41.1759

UNITED STATES NATIONAL CONGRESS OF APPLIED MECHANICS 2nd Proceedings Ann Arbor, Mich. Jun 14-18 Edited by Paul M. Naghdi xx,821p American society of mechanical engineers,New York,1955
Chem E 24.0231

UNITED STATES NAVAL CIVIL ENGINEERING LABORATORY Ice and snow:properties,processes,and application :a conference held at the Massachusetts Institute of technology Proceedings Cambridge,Mass. 1962 Feb 12-16 Air force Cambridge research laboratories. Terrestrial sciences laboratory Edited by W. D. Kingery illus xv,684p 24cm M.I.T. press,Cambridge,Mass.,1963
Sco 14.0269

UNITED STATES OFFICE OF NAVAL RESEARCH Symposium on microseisms Proceedings Harriman,N.Y. 1952 Sep 4-6 Edited by James T. Wilson and Frank Press National research council.Publication, 306 Washington,D.C., 1953
Geod 9.0376

UNITED STATES TECHNICAL CONFERENCE ON AIR POLLUTION Proceedings Air pollution Washington 1950 May 3-5 Sponsored by the United States.Interdepartmental committee on air pollution xiv,847p McGraw-Hill,New York,1952 Chairman of the committee:Louis C.McCabe
Chem E 24.1542

UNITES BIOLOGIQUES DOUEES DE CONTINUITE GENETIQUE colloque international Paris 1948 Jul 25-Jul 3 Centre national de la recherche scientifique Centre national de la recherche scientifique.Colloques internationaux, 8 Centre national de la recherche scientifique, Paris,1949
Gen 34.0988

UNIVERSAL ALGEBRA Conference on Universal algebra Proceedings Kingston,Ont 1969 Oct Held at Queen's university,Kingston,Ont. Queen's papers in pure and applied mathematics, 25 Bibliog. 273p 27cm Queen's university,Kingston,Ont.,1970
P Math 2.3712

UNIVERSAL DAY Adoption d'un premier meridien unique et d'une heure universelle :conference Proces-verbaux Washington,D.C. 1884 Oct 1-Nov 1 United States.Department of state 216p Washington,D.C.,1884
Obs 6.3168

UNIVERSAL DAY International meridian conference International conference for the purpose of fixing a prime meridian and a universal day 1st Proceedings Washington,D.C. 1884 Oct 1-Nov 1 United States.Department of state 212p Gibson, Washington,D.C.,1884 With five sheets of notes by J.C.Adams in a separate envelope
Obs 6.3169

UNIVERSIDAD DE CHILE International symposium on tissue transplantation Proceedings Santiago 1961 Aug 30-Sep 2 Edited by Alberto P. Cristoffanini and Gustavo Hoecker Universidad de Chile,Santiago,Chile,1962
An 32.3286

UNIVERSIDAD DE CHILE The International symposium on tissue transplantation Proceedings Santiago 1961 Aug 30-Sep 2 Edited by Alberto P. Cristoffanini and Gustavo Hoecker Universidad de Chile,Santiago,1962
PGMS 29.0244

UNIVERSIDAD DE LA REPUBLICA,MONTEVIDEO.INSTITUTO DE MATEMATICA Y ESTADISTICA Symposium sobre algunos problemas matematicos que se estan estudiando en Latina America Punta del Este 1951 Dec 19-21 Bibliog 183p 24cm UNESCO para America Latina,Montevideo,1952
P Math 2.3007

UNIVERSIDAD DE URAGUAY.FACULTAD DE MEDICINA. SERVICIO DE FISIOLOGIA OBSTETRICA Oxytocin :an international symposium Proceedings Montevideo 1959 Edited by R. Caldeyro-Barcia and H. Heller Pergamon press, London,1961
VA 19.0237

UNIVERSIDAD DE URUGUAY.FACULTAD DE MEDICINA. SERVICIO DE FISIOLOGIA OBSTETRICA Oxytocin :an international symposium Proceedings Montevideo 1959 Edited by R. Caldeyro-Barcia and H. Heller Pergamon, Oxford,1961
Pha 16.0082

UNIVERSIDAD DE URUGUAY.FACULTAD DE MEDICINA. SERVICIO DE FISIOLOGIA OBSTETRICA Oxytocin :an international symposium Proceedings Montevideo 1959 Edited by R. Caldeyro-Barcia and H. Heller Pergamon press, Oxford,1961
Phys 20.1387

UNIVERSIDAD DE URUGUAY.FACULTAD DE MEDICINA. SERVICIO DE FISIOLOGICA OBSTETRICA Oxytocin :an international symposium Proceedings Montevideo 1959 Edited by R. Caldeyro-Barcia and H. Heller Pergamon, Oxford,1961
An 32.5322

UNIVERSIDAD NACIONAL,MENDOZA Algunos problemas matematicos que se estan estudiando en Latino America symposium 2 Villavicencio-Mendoza 1954 Jul.21-25 Supported by United nations educational, scientific and cultural organisation. Centre for scientific co-operation for Latin America 328p 24cm UNESCO,centro de cooperacion cientifica,Montevideo,1954
P. Math 2.2610

UNIVERSIDAD NACIONAL DE LA PLATA.OBSERVATORIO ASTRONOMICO Symposium on stellar evolution papers La Plata 1960 Nov. 7-11 Edited by J. Sahade 309p Astronomical observatory,La Plata,1962 Scientific meeting organized to celebrate the sesquicentennial of the May revolution
Obs 6.1505

UNIVERSIDAD NACIONAL DE MEXICO.INSTITUTO DE MATEMATICAS Symposium internacional de topologia algebraica Mexico City 1956 Aug Instituto nacional de la investigacion cientifica,Mexico Sociedad matematica mexicana xii,334p 26cm UNESCO,Mexico City,1958 In memory of Witold Hurewicz
P. Math 2.0290

UNIVERSIDADE DO BRASIL International symposium on curare and curare-like agents Proceedings Rio de Janeiro 1957 Aug 5-12 Edited by D. Bovet and others Under the auspices of Unesco illus Elsevier, Amsterdam,1959 Also sponsored by the 'Istituto di sanita',of Rome.Papers in English and French
Chem 26.0103

UNIVERSIDADE DO PARANA Reciniao Brasiliera de genetica humana Parana: 1958: Nov 10-15 Edited by N. Freire-Maice and A. Quelee-Solgrado port. Universidade do Parana,1959
Gen 34.1895

UNIVERSITA DI FIRENZE Teoria dei gruppi finiti e applicazioni :convegno internazionale Florence 1960 Apr 11-13 Supported by Consiglio nazionale della ricerche vii,156p 24cm Edizioni Cremonese,Rome,1960
P Math 2.2858

UNIVERSITA DI MESSINA Celebrazioni Archimedee del secolo xx Syracuse 1961 Apr 11-16 Vol. 1: conferenze generali e simposio di geometria differenziale 68,190p 25cm Oderisi,Gubbio,1962
P. Math 2.2608

UNIVERSITA DI MESSINA Celebrazioni Archimedee del secolo xx Syracuse 1961 Apr 11-16 Vol. 2: simposio di analisi 108p 25cm Oderisi,Gubbio,1962
P. Math 2.2609

UNIVERSITA DI MILANO.ISTITUTO DI FARMACOLOGIA Immunity,cancer,and chemotherapy :basic relationships on the cellular level 1st Report Buffalo,N.Y. Edited by Enrico Mihich Academic press,New York;London,1967
Path 30.2744

UNIVERSITA DI NAPOLI.INSTITUTO DI FISICA TEORICA Lectures on field theory and the many-body problem :international spring school 1st Naples 1961 Edited by E.R. Caianiello With support of North Atlantic treaty organization Academic press,New York,1961
Cav 7.0423

UNIVERSITA DI NAPOLI.INSTITUTO DI FISICA TEORICA Lectures on the many-body problem : international spring school 2nd Naples 1962 Ravello 1963 Vol 1-2 Edited by E.R. Caianiello With support of North Atlantic treaty organization 2 vols Academic press,New York,1962-64
Cav 7.0424

UNIVERSITA DI PADUA Convegno sulla cosmologia;meeting on cosmology Atti; Proceedings Padua 1964 Sep 14-16 and Venice Quatro centenario della nascita di Galileo Galilei 1564-1964.Comitato nazionale per le manifestazioni celebrative Edited by L. Rosino Quatro centenario della nascita di Galileo Galilei.Comitato nazionale per le manifestazioni celebrative.Pubblicazioni, 2, Atti dei convegni, 3 206p Florence,1966
Obs 6.3451

UNIVERSITA DI PISA.NUCLEAR MEDICINE CENTER Labelled proteins in tracer studies Conference on problems connected with the preparation and use of labelled proteins in tracer studies Proceedings Pisa 1966 Jan 17-19 Edited by L. Donato and others EUR 2950 d,f,e European atomic energy community, Brussels,1966
Radioth 35.1752

UNIVERSITA KARLOVA Electronics and vacuum physics Czechoslovak conference 3rd Transactions Prague 1965 Sep 23-28 Ceskoslovenska akademie ved Edited by Libor Paty Academia,Prague,1967
Eng 41.5453

UNIVERSITA KARLOVA International conference on sarcoidosis 5th Proceedings Prague 1969 Jun 16-21 Edited by Ladislav Levinsky and Frantisek Macholda Univerzita Karlova, Prague,1971
PGMS 29.0666

UNIVERSITA KARLOVA Ontogenesis of the brain: the biochemical,functional and structural development of the nervous system International symposium neurontogeneticum 5th Proceedings Prague 1967 Edited by Lubor Jilek and Stanislav Trojan Universita Karlova,Prague,1968
PGMS 29.0202

UNIVERSITA KARLOVA Symposium of pediatric group working physiology 2nd Proceedings Liblice 1969 May Edited by M. Macek Universita Karlova,Prague,1970
PGMS 29.0599

UNIVERSITAT BASEL The International symposium on polarization phenomena of nucleons Basel 1960 Jul 4-8 Edited by P. Huber and K.P. Meyer Under the sponsorship of the International union of pure and applied physics Helvetica physica acta.Supplementum, 6 Birkhauser,Basel,1961
Cav 7.0135

UNIVERSITAT GOTTINGEN.ARABIDOPSIS INFORMATION SERVICE Arabidopsis research :an international symposium Report Gottingen 1965 Apr 21-24 Edited by Gerhard Robbelen Gottingen,1965 Supplement to 'Arabidopsis information service' 1965
Gen 34.1498

UNIVERSITAT KONSTANZ Pyridine-nucleotic-dependent dehydrogenases :advanced study institute Proceedings Konstanz 1969 Sep 15-20 Edited by Horst Sund Springer,Berlin, 1970
Bioch 33.2292

UNIVERSITE DE GENEVE.CLINIQUE THERAPEUTIQUE The International symposium on aldosterone Papers and discussions Geneva 1957 Jun 7-8 Edited by Alex F. Muller and Cecilia M. O'Connor J.and A.Churchill,London,1958
Bioch 33.0416

UNIVERSITE DE GRENOBLE.ECOLE D'ETE DE PHYSIQUE THEORIQUE,LES HOUCHES Geophysique exterieure Papers Les Houches 1962 Edited by Cecile De Witt and others Universite de Grenoble.Ecole d'ete de physique theorique.Les Houches. Lectures,1962 illus xiv,624p Gordon and Breach,New York,1963 Papers in English or French
Nap 11.0163

UNIVERSITE DE GRENOBLE.ECOLE D'ETE DE PHYSIQUE THEORIQUE,LES HOUCHES High energy astrophysics :lectures Les Houches 1966 Vol 3 Edited by C. Dewitt and others Supported by NATO Universite de Grenoble. Ecole d'ete de physique theorique,Les Houches. Lectures,1966 xii,449p Gordon and Breach, New York,1967
A Math 4.1562

UNIVERSITE DE GRENOBLE.ECOLE D'ETE DE PHYSIQUE THEORIQUE,LES HOUCHES Dispersion relations and elementary particles Les Houches 1960 By M.L. Goldberger Edited by C.de Witt and R Omnes Universite de Grenoble.Ecole d'ete de physique theorique,Les Houches,1960 670p Hermann,Paris,1960
A Math 4.0839

UNIVERSITE DE GRENOBLE.ECOLE D'ETE DE PHYSIQUE THEORIQUE,LES HOUCHES Geophysics:the earths environment lectures Les Houches 1962 Edited by Cecile De witt Universite de Grenoble.Ecole d'ete de physique theorique, Les Houches.Lectures,1963 Gordon and Breach, New York;London,1963
A Math 4.0578

UNIVERSITE DE GRENOBLE.ECOLE D'ETE DE PHYSIQUE THEORIQUE,LES HOUCHES High energy physics Les Houches 1965 Edited by C. DeWitt and Maurice Jacob Universite de Grenoble.Ecole d'ete de physique theorique,Les Houches,1965 509p Gordon and Breach,New York,1966
A Math 4.0864

UNIVERSITE DE GRENOBLE.ECOLE D'ETE DE PHYSIQUE THEORIQUE,LES HOUCHES Relativity,groups and topology Papers Les Houches 1963 Edited by C. DeWitt and B. DeWitt Universite de Grenoble.Ecole d'eete de physique theorique, Les Houches. Lectures,1963 929p Gordon, Breach,London,1964
TA 15.0343

UNIVERSITE DE GRENOBLE.ECOLE D'ETE DE PHYSIQUE THEORIQUE,LES HOUCHES Relativity,groups and topology lectures Les Houches 1963 Edited by C. DeWitt and B. DeWitt Universite de Grenoble.Ecole d'ete de physique theorique, Les Houches.Lectures,1963 Gordon and Breach, New York;London,1963
A Math 4.1030

UNIVERSITE DE GRENOBLE.ECOLE D'ETE DE PHYSIQUE THEORIQUE,LES HOUCHES,1965 High energy physics Les Houches 1965 Universite de Grenoble.Ecole d'ete de physique theorique,Les Houches Edited by C. DeWitt and Maurice Jacob 509p Gordon and Breach,New York,1966
A Math 4.0864

UNIVERSITE DE GRENOBLE.ECOLE D'ETUDE PHYSIQUE THEORIQUE,LES HOUCHES High energy astrophysics Lectures Les Houches 1966 Vol 1-3 Edited by C. DeWitt and others Universite de Grenoble.Ecole d'ete de physique theorique,Les Houches.Lectures,1966 Gordon and Breach,New York,1967 Lectures delivered at Les Houches,1966
TA 15.0312

UNIVERSITE DE LIEGE.INSTITUT D'ASTROPHYSIQUE Etoiles a raies d'emissions Colloque international d'astrophysique 8e Liege 14p Liege,c1957 Summary of contributions
Obs 6.3170

UNIVERSITE DE LIEGE.INSTITUT D'ASTROPHYSIQUE Instabilite gravitationnelle et formation des etoiles, des galaxies et de leurs structures caracteristiques Colloque international d'astrophysique 4e Communications Liege 1966 Jun 20-22 Institut d'astrophysique, Cointe-Sclessin,1967 Papers in English
A Math 4.1170

UNIVERSITE DE LOUVAIN Etudes sur les termites africains :colloque international Comptes rendus Leopoldville 1964 May 11-16 Edited by A. Bouillon Under the auspices of Unesco figs,pls,tables 27cm Masson,Paris, 1964
Bal 44.4805

UNIVERSITE DE MONTREAL.SEMINAIRE DE MATHEMATIQUES SUPERIEURES, 21 Logical systems containing only a finite number of symbols Montreal 1966 By Leon Henkin North Atlantic treaty organization Societe mathematique du Canada illus. 48p 28cm Presses de l'Universite de Montreal,Montreal, 1967
P Math 2.3740

UNIVERSITE DE NEUCHATEL.INSTITUT DE ZOOLOGIE Symposium sur la specificite parasitaire des parasites de vertebres 1st 24cm Neuchatel,1957
Philos 1.1588

UNIVERSITE DE NEUCHATEL.INSTITUT DE ZOOLOGIE
Symposium sur la specificite parasitaire des parasites de vertebres 1 er Neuchatel 1957 Edited by J.G. Baer Paul Attinger, Neuchatel,1957 Title also in English.Text in English and French
Bal 39.1727

UNIVERSITE DE PARIS.FACULTE DE MEDICINE. LABORATOIRE DE CHIMIE BIOLOGIQUE
Structure and metabolism of corticosteroids :a symposium Proceedings Paris 1963 Jul 5-6 Edited by Jorge R. Pasqualini and Max F. Jayle Academic press,London;New York,1964
Inv Med 37.0175

UNIVERSITE DE PARIS A LA SORBONNE Metallic solid solutions:their electronic and atomic structure :a symposium Proceedings Orsay 1962 Jul W.A.Benjamin,New York,1963
Cav 7.1737

UNIVERSITE DE STRASBOURG.INSTITUT DE MATHEMATIQUE
Seminaire de probabilites 2me Strasbourg 1967 Mar-Oct Lecture notes in mathematics, 51 Bibliog 199p 28cm Springer,Berlin,1968
P Math 2.2996

UNIVERSITE DE STRASBOURG.INSTITUT DE MATHEMATIQUE
Seminaire de probabilities 4me Strasbourg 1968-69 Lecture notes in mathematics, 124 Bibliog. 282p 25cm Springer-Verlag,Berlin,1970
P Math 2.3518

UNIVERSITE DE STRASBOURG.INSTITUT DE MATHEMATIQUE
Seminaire de probabilites 1e Strasbourg 1966-67 Nov-Feb Lecture notes in mathematics, 39 189p 28cm Springer, Berlin,1967
Math 3.1188

UNIVERSITE DE STRASBOURG.INSTITUT DE MATHEMATIQUE
Seminaire de probabilites 2me Strasbourg 1967 Mar-Oct Lecture notes in mathematics, 51 199p 28cm Springer, Berlin,1968
Math S 3.1438

UNIVERSITE DE STRASBOURG.INSTITUT DE MATHEMATIQUE
Seminaire de probabilites 3me Strasbourg 1967-68 Lecture notes in mathematics, 88 Bibliog. 229p 28cm Springer,Berlin,1969
P Math 2.3517

UNIVERSITE DE STRASBOURG.INSTITUT DE MATHEMATIQUE
Seminaire de probabilites 3me Strasbourg 1967-68 Lecture notes in mathematics, 88 bibliog. 229p 28cm Springer,Berlin,1969
Math S 3.1441

UNIVERSITE DE STRASBOURG.INSTITUT DE PHYSIQUE
Le Magnetisme Travaux Strasbourg 1939 May 21-25 Supported by Centre national de la recherche scientifique Collection scientifique,Paris,1940
Cav 7.2241

UNIVERSITE LIBRE DE BRUXELLES A Symposium on non-equilibrium thermodynamics, variational techniques and stability Chicago 1965 May 17-19 Edited by Russell J. Donnelly and I. Prigogine University of Chicago,Chicago,1966
A Math 4.0712

UNIVERSITE LIBRE DE BRUXELLES International symposium on the biology of acetabularia 1st Proceedings Brussels 1969 Jun 18-20 and Mol 1969 Jun 18-20 Edited by Jean Brachet and Silvano Bonotto Academic press, New York,1970
Bioch 33.2261

UNIVERSITE LIBRE DE BRUXELLES Proteines Conseil de chimie 9e Rapports et discussions Brussels 1953 Apr 6-14 Edited by R. Stoops Brussels,1953
Bioch 33.0522

UNIVERSITE LIBRE DE BRUXELLES Structure and evolution of galaxies Conference on physics 13th Proceedings Brussels 1964 Sep 174p Interscience,London,1965
Obs 6.2134

UNIVERSITE LIBRE DE BRUXELLES Structure and evolution of galaxies The Conference on physics 13th Proceedings Brussels 1964 Sep Solvay conference,1964,13 Interscience, London,1965 Other Solvay conferences issued as Conseil de physique,q.v.
Cav 7.2078

UNIVERSITE LIBRE DE BRUXELLES Structure et evolution de l'univers Conseil de physique 11e Rapports Brussels 1958 Jun 9-13 Edited by R. Stoops 309p Brussels,1958
Obs 6.3255

UNIVERSITE LIBRE DE BRUXELLES Structure et evolution de l'univers Conseil de physique 11eme Rapports Brussels 1958 Jun 9-13 Edited by R. Stoops Solvay conference,1958, 11 Brussels,1958
Cav 7.2084

UNIVERSITE LIBRE DE BRUXELLES Structure et proprietes des noyaux atomiques Conseil de physique 7eme Rapports Brussels 1933 Oct 22-29 Solvay conference,1933,7 Paris, 1934
Cav 7.2083

UNIVERSITE LIBRE DE BRUXELLES Theorie quantique des champs Conseil de physique 12eme Rapports Brussels 1961 Oct 9-14 Edited by R. Stoops Solvay conference,1961, 12 Interscience,New York,c 1961 13th Solvay conference issued as Conference on physics,q.v.
Cav 7.2077

UNIVERSITE LIBRE DE BRUXELLES and INTERNATIONAL INSTITUTE OF CHEMISTRY, Reactivity of photo-excited organic molecula Solvay conference on chemistry 13th Proceedings Brussels 1965 Oct 25-30 Interscience,London,1967
Col S 12.0568

UNIVERSITE LIBRE DE BRUXELLES.CENTRE D'ETUDE DE L'ENERGIE NUCLEAIRE International symposium on acetabularia 1st Proceedings Brussels 1969 Jun 18-20 Edited by Jean Brachet and Silvano Bonotto illus. xv,300p Academic press,New York;London,1970
Bot 42.3941

UNIVERSITET I OSLO Scandinavian logic symposium 2nd Proceedings Oslo 1970 Jun 18-20 Edited by J.E. Fenstad Studies in logic and the foundations of mathematics, 63 vi,405p 23cm North-Holland,Amsterdam, 1971
P Math 2.4056

UNIVERSITET I UPPSALA.INSTITUTE OF STATISTICS
Symposium on psychological factor analysis Uppsala 1953 Mar 17-19 Edited by P. Whittle Nordisk psykologi.Monograph series, 3 91p 23cm Almqvist and Wiksell, Stockholm,1953
Math 3.0777

UNIVERSITIES FEDERATION FOR ANIMAL WELFARE
Small animal anaesthesia :a symposium Proceedings London 1963 Jul Edited by Oliver Graham-Jones Pergamon press,London, 1964
Radioth 35.0270

UNIVERSITIES FEDERATION FOR ANIMAL WELFARE BRITISH SMALL ANIMALS VETERINARY ASSOCIATION
Small animal anaesthesia :a symposium Proceedings London 1963 Jul Edited by Oliver Graham-Jones Illus. Pergamon press, London,1964
VA 19.0264

UNIVERSITY COLLEGE,CORK and CHEMICAL SOCIETY
Organic reaction mechanisms :an international symposium Lectures Cork, Ireland 1964 Jul 20-25 Chemical society. Special publication,19 Chemical society, London,1965
Chem 18.1223

UNIVERSITY COLLEGE,DUBLIN and INSTITUTE OF CHEMISTRY OF IRELAND
Symposium of recent advances in the chemistry of naturally occurring pyrones and related compounds Proceedings Dublin 1955 Jul 12-14 Royal Dublin society.Scientific proceedings,27,no 6 Royal Dublin society,Dublin,1956
Chem 18.1494

UNIVERSITY COLLEGE,LONDON
Atomic collision processes The International conference on the physics of electronic and atomic collisions 3rd Proceedings London 1963 Jul 22-23 Edited by M.R.C. McDowell North-Holland,Amsterdam,1964
Cav 7.1461

UNIVERSITY COLLEGE,LONDON
The Physics of ionized gases :a conference proceedings, typescript London 1953 Apr Edited by J.B. Hasted Under the auspices of the Royal society of London.Warren research fund London,1953
A Math 4.0614

UNIVERSITY COLLEGE HOSPITAL MEDICAL SCHOOL
The Biochemical reactions of chemical warfare agents :a symposium London 1947 Dec 13 Biochemical society Edited by R.T. Williams Biochemical society.Symposia, 2 Cambridge university press,Cambridge,1948
Bioch 33.1364

UNIVERSITY COLLEGE HOSPITAL MEDICAL SCHOOL
The Relation of optical form to biological activity in the amino-acid series :a symposium London 1947 Feb 15 Biochemical society Edited by R.T. Williams Biochemical society. Symposia, 1 Cambridge university press, Cambridge,1948
Bioch 33.1363

UNIVERSITY EDUCATION IN COMPUTING SCIENCE
Conference on graduate academic and related research programs in computing science Proceedings Stony Brook,N.Y. 1967 Jun Edited by Aaron Finerman Association for computing machinery.Monograph series Bibliog. Illus.Port. xvi,237p 24cm Academic press,New York;London,1968
P Math 2.3505

UNIVERSITY EDUCATION IN COMPUTING SCIENCE
Conference on graduate academic and related research programs in computing science Proceedings Stony Brook,N.Y. 1967 Jun Edited by Aaron Finerman Held at the State university of New York Association for computing machinery.Monograph series Academic press,New York,1968
Eng 41.2223

UNIVERSITY EDUCATION IN COMPUTING SCIENCE :a conference on graduate academic and related research programs in computing science Proceedings New York 1967 Jun 5-8 Edited by Aaron Finerman Sponsored by New York state science and technology foundation ACM monograph series xvi,237p 23cm Academic press,New York,1968
Math L 5.3107

UNIVERSITY GRANTS COMMISSION
International seminar on chromosome :its structure and function Proceedings Calcutta 1968 Aug 11-13 Edited by Arunkumar Sharma and Archana Sharma Nucleus,1968,Supplementary volume Nucleus,Calcutta,1968
Gen 34.2173

UNIVERSITY GRANTS COMMITTEE.MATHEMATICAL SCIENCES SUB-COMMITTEE
Conference on teaching computing Proceedings Bristol 1972 Mar 28 illus 165p University grants committee,London,1972
Math L 5.3878

UNIVERSITY OF ABERDEEN.DEPARTMENT OF GENETICS
Human radiation cytogenetics :an international symposium Proceedings Edinburgh 1961 Oct 12-15 Edited by H.J. Evans and others North-Holland,Amsterdam,1967
Gen 34.1127

UNIVERSITY OF ADELAIDE
Australia-New Zealand conference on soil mechanics and foundation engineering 4th Proceedings Adelaide 1963 Aug 19-23 1963
Eng 41.3169

UNIVERSITY OF ALABAMA MEDICAL CENTER
Secretory mechanisms of salivary glands International conference on mechanisms of salivary secretion and their regulation Proceedings Birmingham,Ala. 1966 Aug 9-11 Edited by Leon H. Schneyer and Charlotte A. Schneyer Academic press,New York;London,1967
Bal 39.1265

UNIVERSITY OF ALASKA
Review of research on military problems in cold regions The Alaskan science conference 15th Papers Fairbanks 1964 Aug 31-Sep 4 By Charles R. Kolb and Fritz M.G. Holmstrom Under the auspices of the American association for the advancement of science 1964 Privately printed
Sco 14.0155

UNIVERSITY OF ALASKA.GEOPHYSICAL INSTITUTE Symposia on Arctic biology and medicine 1st Proceedings Fort Wainwright,Alaska 1960 Oct 9-11 1: Neural aspects of temperature regulation Edited by John P. Hannon and Eleanor Viereck illus vii,399p 23cm Arctic aeromedical laboratory,Fort Wainwright, Alaska,1961
Sco 14.0850

UNIVERSITY OF ALASKA.GEOPHYSICAL INSTITUTE Symposia on arctic biology and medicine Proceedings Fort Wainwright,Alaska 1962 Aug 28-30 influence of cold on host-parasite interactions By Robert I. McClaughry and others Edited by Eleanor G. Viereck illus 455p 23cm Fort Wainwright,Alaska,1963
Sco 14.0882

UNIVERSITY OF ALASKA.GEOPHYSICAL INSTITUTE Symposia on arctic biology and medicine Proceedings Fort Wainwright,Alaska 1963 Feb 17-18 4: frostbite Edited by Eleanor G. Viereck illus vi,457p 23cm Arctic aeromedical laboratory,Fort Wainwright,Alaska, 1964
Sco 14.0883

UNIVERSITY OF ALBERTA.FACULTY OF MEDICINE Molecular biology of viruses Symposium of the molecular biology of viruses Proceedings Edmonton,Alberta 1966 Jun 27-30 Edited by John S. Colter and William Paranchych Academic press,New York;London,1967
Path 30.2745

UNIVERSITY OF ALBERTA.FACULTY OF MEDICINE Muscle :a symposium Proceedings Edmonton, Alberta 1964 Jun 1-4 Edited by W.M. Paul and others Pergamon press,Oxford,1965 First symposium of the University of Alberta medical school held as part of the celebrations of the 50th year of this medical faculty
Bioch 33.2244

UNIVERSITY OF ALBERTA.FACULTY OF MEDICINE Muscle :symposium Proceedings Edmonton, Alta. 1964 Jun 1-4 Edited by W.M. Paul and others Pergamon press,Oxford,1965 50th anniversary of the University of Alberta medical school
An 32.4007

UNIVERSITY OF ARIZONA Researches on meteorites :a symposian Proceedings Tempe, Ariz. 1961 Mar.10 227p John Wiley,New York;London,1962
Obs 6.1243

UNIVERSITY OF ARIZONA The Conference on the use of solar energy :the scientific basis Transactions Tucson,Ariz. 1955 Oct 31-Nov 1 Vol 1-5 Edited by Edwin F. Carpenter and others Co-sponsored by the Association for applied solar energy illus 6 vols University of Arizona,Tucson,Ariz., 1958
Nap 11.0111

UNIVERSITY OF ARIZONA The International conference on the physics of solids at high pressures 1st Proceedings Tucson,Ariz. 1965 Apr 20-23 Edited by C.T. Tumizuka and R. M. Emrick Jointly sponsored by the United States.Air force.Office of scientific research Academic press,New York,1965
Cav 7.2354

UNIVERSITY OF ARIZONA.LUNAR AND PLANETARY LABORATORY Comets :scientific data and missions :a conference Tucson,Ariz. 1970 Apr 8-9 Revised edition 222p University of Arizona.Lunar and planetary laboratory,Tucson,Ariz.,1972
Obs 6.3619

UNIVERSITY OF ASTON IN BIRMINGHAM Symposium on anodizing aluminium Proceedings Birmingham 1967 Apr 12-13 Aluminium federation,London,1967
Met 25.2048

UNIVERSITY OF AUCKLAND Australasian conference on hydraulics and fluid mechanics 2nd proceedings Auckland N.Z. 1965 Dec. 6-11 Edited by C.M. Segedin Sponsored jointly with the New Zealand institution of engineers Conference organising committee, Auckland,N.Z.,1966
A Math 4.0531

UNIVERSITY OF BIRMINGHAM Techniques in endocrine research :a NATO advanced study institute Proceedings Stratford-upon-Avon 1962 Sep North Atlantic treaty organization Edited by Peter Eckstein and Francis Knowles Academic press,London,1963
Gen 34.1964

UNIVERSITY OF BIRMINGHAM Techniques in endocrine research :a NATO advanced study institute Proceedings Stratford-upon-Avon 1962 Sep North Atlantic treaty organization Edited by Peter Eckstein and Francis Knowles Academic press,London,1963
An 32.3767

UNIVERSITY OF BIRMINGHAM Techniques in endocrine research :a Nato advanced study institute Proceedings Stratford-upon-Avon 1962 Sep North Atlantic treaty organization Edited by P. Eckstein and F. Knowles Academic press,New York,1963
Phys 20.1399

UNIVERSITY OF BRISTOL A Conference on strength of solids :held at the H.H.Wills physical laboratory Report Bristol 1947 Jul 7-9 Edited by N.F. Mott 162p Physical sciety,London,1948
Cav 7.0346

UNIVERSITY OF BRISTOL Conference on the defects in crystalline solids :held at the H.H. Wills physical laboratory Report Bristol 1954 Jul By N. Bloembergen and others With co-operation of the International union of pure and applied physics.Commission on the solid state 429p Physical society,London, 1955
Cav 7.0345

UNIVERSITY OF BRISTOL Cosmic radiation based on a symposium promoted by the Colston research society 1st Papers Bristol 1948 Sep Edited by F.C. Frank and D.R. Rexworthy Colston research society.Colston papers, 1 illus viii,189p Butterworths, London,1949
Cav 7.0348

UNIVERSITY OF BRISTOL Engineering structures a symposium Papers Bristol 1949 Sep Colston research society.Research special supplement, 49 Butterworths,London,1949
Eng 41.3022

UNIVERSITY OF BRISTOL Fundamental mechanisms of photographic sensitivity A Conference on fundamental mechanisms of photographic sensitivity :held in the H.H.Wills physical laboratory Proceedings Bristol 1950 March Edited by J.W. Mitchell Butterworths, London,1951
Cav 7.0347

UNIVERSITY OF BRISTOL.DEPARTMENT OF PHYSICS A Discussion of the Faraday society 5th Papers Bristol 1949 Apr 12-14 Faraday society.Discussions, 5,1949 Butterworth, London,1959
Cav 7.0760

UNIVERSITY OF BRITISH COLUMBIA.DEPARTMENT OF COMPUTER SCIENCE Informal conference on ALGOL 68 implementations Proceedings Vancouver,B.C. 1969 Aug 29-30 Edited by J. E.L. Peck illus 119p University of British Columbia.Department of computer science,Vancouver,B.C.,1969
Math L 5.3845

UNIVERSITY OF BRITISH COLUMBIA.INSTITUTE OF FISHERIES Symposium on pink salmon Lectures Vancouver,B.C. 1960 Oct 13-15 Edited by N.J. Wilimovsky University of British Columbia.Institute of fisheries.H.R. Macmillan lectures in fisheries, 3 27cm Vancouver,B.C.,1962
Bal 44.4988

UNIVERSITY OF BUFFALO Conference on carbon 1st and 2nd Proceedings Buffalo,N.Y. 1953,1955 By S. Mrozowski and others University of Buffalo,New York,1956
Met 25.0154

UNIVERSITY OF BUFFALO Conference on carbon 3rd Proceedings Buffalo,N.Y. 1957 Jun 17-21 By S. Morzowski and others Co-sponsored by the National science foundation Pergamon press,London,1959
Met 25.0156

UNIVERSITY OF BUFFALO Conference on carbon 4th Proceedings Buffalo,N.Y. 1959 Jun 15-19 By S. Mrozowski and others Pergamon press,Oxford,1960
Met 25.0157

UNIVERSITY OF CALCUTTA International seminar on chromosome :its structure and function Proceedings Calcutta 1968 Aug 11-13 Edited by Arunkumar Sharma and Archana Sharma Nucleus,1968,Supplementary volume Nucleus, Calcutta,1968
Gen 34.2173

UNIVERSITY OF CALIFORNIA Computer graphics: utility,production,art Graphics symposium Los Angeles n.d. Edited by Fred Gruenberger Jointly sponsored by Informatics incorporated 225p 22cm Academic press, London,1967
Math L 5.3273

UNIVERSITY OF CALIFORNIA Computer graphics: utility,production,art Graphics symposium Los Angeles,Calif. 1967 Edited by Fred Gruenberger Jointly sponsored by Informatics incorporated Academic press,London,1967
Eng 41.2210

UNIVERSITY OF CALIFORNIA Computers and communications - toward a computer utility Computer communications symposium Los Angeles 1967 Mar 20-22 Edited by Fred Gruenberger 219p 23cm Prentice-Hall, Englewood Cliffs,N.J.,1968
Math L 5.3117

UNIVERSITY OF CALIFORNIA Electron microscopy and strength of crystals Berkeley international materials conference :the impact of transmission electron microscopy on theories of the strength of crystals 1st Proceedings Berkeley 1961 Jul 5-8 Edited by Gareth Thomas and Jack Washburn Interscience,New York;London,1963
Met 25.1211

UNIVERSITY OF CALIFORNIA Heat transfer and fluid mechanics institute 1958 preprints Berkeley,Calif. 1958 Jun 19-21 Stanford university press,Stanford,Calif.,1958
A Math 4.0517

UNIVERSITY OF CALIFORNIA International symposium on rarefied gas dynamics 2nd Proceedings Berkeley,Calif. 1966 Edited by L. Talbot Advances in applied mechanics.Supplement, 1 Academic press,New York;London,1961
Eng 41.6764

UNIVERSITY OF CALIFORNIA Mathematical optimization techniques symposium Santa Monica,Calif. 1960 Oct 18-20 Edited by Richard Bellman xii,346p 24cm University of California press,Berkeley,Calif.,1963
P. Math 2.2288

UNIVERSITY OF CALIFORNIA Mathematical optimization techniques symposium papers Santa Monica,Calif. 1960 Oct.18-20 Edited by Richard Bellman With assistance from the National science foundation xii,346p 24cm University of California press,Berkeley,Calif., 1963
Math 3.0238

UNIVERSITY OF CALIFORNIA On-line computing systems symposium proceedings Los Angeles, Calif. 1965 Feb 2-4 Edited by Eric Burgess Jointly sponsored by Informatics incorporated Data processing library series 152p 28cm American data processing,Detroit,Mich.,1965
Math L 5.1031

UNIVERSITY OF CALIFORNIA Radiobiology at the intra-cellular level U.C.L.A.conference on radiobiology 1st Proceedings Catalina island 1957 Sep 9-12 Edited by T.G. Hennessy and others ports. Pergamon press, London,1959
Radioth 35.1722

UNIVERSITY OF CALIFORNIA Rarefied gas dynamics The International symposium on rarefied gas dynamics 2nd Proceedings Berkeley,Calif. 1960 Aug 3-6 Edited by L. Talbot Under the sponsorship of International union of theoretical and applied mechanics Advances in applied mechanics. Supplement, 1 748p Academic press,New York,1961 Co-sponsored by National aeronautics and space administration
Cav 7.2254

UNIVERSITY OF CALIFORNIA Rarefied gas dynamics :international symposium 2nd Proceedings Berkeley,Calif. 1960 Edited by L. Talbot Advances in applied mechanics. Supplement, 1 x,748p Academic press,New York;London,1961
Chem E 24.0408

UNIVERSITY OF CALIFORNIA The Galapagos : proceedings of the symposia of the Galapagos international scientific project Proceedings Berkeley,Calif. 1964 Jan 7-8 and 1964 Jan 11-18 Edited by Robert I. Bowman illus, pls 27cm University of California press, Berkeley;Los Angeles,1966
Bal 44.4211

UNIVERSITY OF CALIFORNIA The International conference on high energy physics 13th Proceedings Berkeley,Calif. 1966 Aug 31-Sep 7 Edited by Margaret Alston-Garnjost Sponsored by the International union of pure and applied physics Berkeley,Calif.,1967
Cav 7.0189

UNIVERSITY OF CALIFORNIA The International symposium on rarefied gas dynamics 2nd proceedings Berkeley,Calif. 1960 Aug.3-6 Edited by L. Talbot Advances in applied mechanics.Supplement, 1 Academic press,New York,London,1961
A Math 4.0529

UNIVERSITY OF CALIFORNIA World conference on prestressed concrete :conference Proceedings San Francisco,Calif. 1957 Jul-Aug San Francisco,Calif.,1957
Eng 41.3011

UNIVERSITY OF CALIFORNIA World conference on shell structures Proceedings San Francisco, Calif. 1962 Oct 1-4 National research council.Publication,1187 National academy of sciences,Washington,D.C.,1064
Eng 41.3025

UNIVERSITY OF CALIFORNIA and RAND CORPORATION Proceedings of the symposium on the arctic heat budget and atmospheric circulation Lake Arrowhead 1966 Jan-Feb Edited by J.O. Fletcher Prepared for the National science foundation Memorandum RM-5233-NSF illus, diagrs 566p The Rand corporation,Santa Monica,1966
Sco 14.2407

UNIVERSITY OF CALIFORNIA.COLLEGE OF CHEMISTRY International materials symposium :the structure and chemistry of solid surfaces 4th Proceedings Berkeley,Calif 1968 Jun 17-21 Edited by Gabor A. Somorjai Inorganic materials research division series, 4 Wiley,New York,1969
Met 25.2476

UNIVERSITY OF CALIFORNIA.DEPARTMENT OF MATHEMATICS Symposium on mathematical statistics and probability Proceedings Berkeley,Calif. 1945 Aug.13-18 and 1946 Jan.27-29 Edited by Jerzy Neyman 501p University of California press,Berkeley;Los Angeles,1949
Nap 11.0048

UNIVERSITY OF CALIFORNIA.SCHOOL OF PHARMACY Challenge to pharmacy in the seventies Invitational conference on pharmacy manpower Proceedings San Francisco 1970 Sep 10-12 Edited by Joe B. Graber and Donald C. Brodie U.S.Department of health,education and welfare, Washington,D.C.,1972
PGMS 29.0674

UNIVERSITY OF CALIFORNIA.STATISTICAL LABORATORY Bernoulli 1713:Bayes 1763:Laplace 1818: anniversary volume International research seminar proceedings Berkeley,Calif. 1963 Edited by Jerzy Neyman and Lucien M. LeCam Supported by the National science foundation ix,262p 23cm Springer-verlag,Berlwn,1965
Math 3.0708

UNIVERSITY OF CALIFORNIA.STATISTICAL LABORATORY Mathematical statistics and probability Berkeley symposium on mathematical statistics and probability 2nd Proceedings Berkeley,Calif. 1950 Jul 31-Aug 12 Edited by Jerzy Neyman University of California press,Berkeley,Calif.;Los Angeles,Calif.,1951
Radioth 35.0599

UNIVERSITY OF CALIFORNIA.STATISTICAL LABORATORY Mathematical statistics and probability Berkeley symposium on mathematical statistics and probability 2nd proceedings Berkeley,Calif. 1950 Jul.31-Aug.12 Edited by Jerzy Neyman Also supported by the Office of naval research xi,666p 26cm University of California press,Berkeley;Los Angeles,Calif.,1951
Math 3.0686

UNIVERSITY OF CALIFORNIA.STATISTICAL LABORATORY Mathematical statistics and probability Berkeley symposium on mathematical statistics and probability 3rd Proceedings Berkeley,Calif. 1954 Dec 26-31 Berkeley, Calif. 1955 Jul,Aug Vol 4: contributions to biology and problems of health Edited by Jerzy Neyman University of California press,Berkeley,Calif.;Los Angeles,Calif.,1956
Radioth 35.0606

UNIVERSITY OF CALIFORNIA.STATISTICAL LABORATORY Mathematical statistics and probability Berkeley symposium on mathematical statistics and probability 4th proceedings Berkeley,Calif. 1960 Jun.20-Jul.30 Vols. 1-4 Edited by Jerzy Neyman 26cm 4 vols University of California press,Berkeley; Los Angeles,Calif.,1961
Math 3.0688

UNIVERSITY OF CALIFORNIA.STATISTICAL LABORATORY Mathematical statistics and probability Berkeley symposium on mathematical statistics and probability 3r proceedings Berkeley,Calif. 1954 Dec.26-31 Berkeley, Calif. 1955 Jul-Aug Vols. 1-5 Edited by Jerzy Neyman 26cm 5 vols University of California press,Berkeley;Los Angeles,Calif.,1956
Math 3.0687

UNIVERSITY OF CALIFORNIA.STATISTICAL LABORATORY Mathematical statistics and probability Berkeley symposium on mathematical statistics and probability 1st proceedings Berkeley,Calif. 1945 Aug.13-18 Berkeley, Calif. 1946 Jan.27-29 Edited by Jerzy Neyman viii,501p 26cm University of California press,Berkeley;Los Angeles,Calif., 1949
Math 3.0685

UNIVERSITY OF CALIFORNIA,BERKELEY.CENTER OF SLAVIC AND EAST EUROPEAN STUDIES Soviet and east European agriculture Conference on Soviet agricultural affairs 2nd Papers Santa Barbara,Calif. 1965 Aug 26-28 Edited by Jerzy F. Karcz Russian and east European studies University of California press, Berkeley,Calif.;Los Angeles,Calif.,1967
Geog 13.2584

UNIVERSITY OF CALIFORNIA,LOS ANGELES Computer graphics:utility,production,art The Graphic symposium 3rd Papers Los Angeles,Calif 1966 Edited by Fred Gruenberger 225p Thompson book co.; Academic press,Washingtom,D.C.;London,1967
TA 15.0098

UNIVERSITY OF CALIFORNIA,LOS ANGELES Inequalities 3 Symposium on inequalities 3rd Proceedings Los Angeles 1969 Sep 1-9 Edited by Oved Shisha xxii,368p 24cm Academic press,New York;London,1972
P Math 2.4214

UNIVERSITY OF CALIFORNIA,LOS ANGELES.INSTITUTE OF GEOPHYSICS 1956 :annual technical conference Symposium Rock deformation Los Angeles,Calif. 1956 Nov Edited by David Griggs and John Handin Geological society of America.Memoirs,79 New York,1960
Min 10.0348

UNIVERSITY OF CALIFORNIA LOS ANGELES Axiomatic set theory Symposium in pure mathematics 14th Proceedings Los Angeles, Calif. 1967 Jul 10-Aug 5 Edited by Dana S. Scott American mathematical society. Proceedings of symposia in pure mathematics, 13,pt 1 v,474p 26cm American mathematical society,Providence,R.I.,1971
P Math 2.4061

UNIVERSITY OF CALIFORNIA LOS ANGELES Lectures and laboratory exercises on strain gage techniques Los Angeles,Calif. 1958 Aug 18-29 By William M. Murray and Peter K. Stein University of California Los Angeles, Los Angeles,Calif.,1958
Eng 41.2906

UNIVERSITY OF CAMBRIDGE International symposium on free radicals 6th Papers Cambridge 1963 Jul 2-5 Sponsored in part by Great Britain.Air force.Office of scientific research,OAR Cambridge,1963 Mimeographed
Chem 18.0414

UNIVERSITY OF CAMBRIDGE Order of the proceedings at the Darwin celebration... :with a sketch of Darwin's life Cambridge 1909 Jun 22-24 23p Cambridge university press, Cambridge,1909
WSM 43.2842

UNIVERSITY OF CAMBRIDGE.DEPARTMENT OF RADIO-THERAPEUTICS The Cell nucleus :an informal meeting Proceedings Cambridge 1959 Aug 31-Sep 1 Faraday society Butterworths,London,1960
Bioch 33.0946

UNIVERSITY OF CAMBRIDGE.MATHEMATICAL LABORATORY Conference on high speed automatic calculating-machines report Cambridge 1949 June 22-25 Bibliog 31cm Cambridge, 1950
Philos 1.0336

UNIVERSITY OF CAMBRIDGE.MATHEMATICAL LABORATORY Conference on high-speed automatic calculating machines Report Cambridge 1949 Jun 22-25 University mathematical laboratory,Cambridge,1950
Eng 41.2175

UNIVERSITY OF CAMBRIDGE.MATHEMATICAL LABORATORY Conference on high-speed automatic calculating machines Report Cambridge 1949 Jun 22-25 University mathematical laboratory,Cambridge,1950
Math L 5.3236

UNIVERSITY OF CAMBRIDGE.SCHOOL OF CLINICAL RESEARCH AND POSTGRADUATE MEDICAL TEACHING Depression :the symposium Proceedings Cambridge 1959 Sep 22-26 Edited by E.B. Davies University of Cambridge.School of clinical research and postgraduate medical teaching.Course,1959 Cambridge university press,Cambridge,1964
PGMS 29.0330

UNIVERSITY OF CAMBRIDGE.SCHOOL OF CLINICAL RESEARCH AND POSTGRADUATE MEDICAL TRAINING Depression :a symposium Proceedings Cambridge 1959 Sep 22-26 Edited by E. Beresford Davies Cambridge university press, Cambridge,1964
Psy 31.1679

UNIVERSITY OF CHICAGO A Symposium on molecular biology :a seminar series and a symposium Lectures Chicago 1956-57 Nov-Mar Edited by Raymond E. Zirkle Under the auspices of the Chicago-Frankfurt inter-university program University of Chicago press,Chicago,1959
Bioch 33.0839

UNIVERSITY OF CHICAGO A Symposium on molecular biology :a seminar series and a symposium lectures Chicago 1956-57 Nov-Mar Edited by Raymond E. Zirkle Under the auspices of the Chicago-Frankfurt inter-university program University of Chicago press,Chicago,1959
Bal 39.0617

UNIVERSITY OF CHICAGO A Symposium on non-equilibrium thermodynamics, variational techniques and stability Chicago 1965 May 17-19 Edited by Russell J. Donnelly and I. Prigogine University of Chicago,Chicago,1966
A Math 4.0712

UNIVERSITY OF CHICAGO Atmosphere of the earth and planets A Symposium on planetary atmospheres Papers Williams Bay,Wisc. 1947 Sep 8-10 Edited by Gerard P. Kuiper 366p Chicago,Ill.,1949 Symposium held in connection with the 50th anniversary of the Yerkes observatory
Geod 9.0523

UNIVERSITY OF CHICAGO Conference on nuclear processes in geologic settings Proceedings Williamsbay 1953 Sep,21-23 Co-sponsored by National research council Williams Bay, Wisconsin,1953
Philos 1.0514

UNIVERSITY OF CHICAGO Industrialization and society Conference on the social implications of industrialization and technological change Papers Chicago,Ill. 1960 Sep 15-22 Edited by Bert F. Hoselitz and Wilbert E. Moore Sponsored jointly by Unesco Mouton,The Hague,1963
Geog 13.1608

UNIVERSITY OF CHICAGO The Biological action of the vitamins :a symposium Chicago 1941 Sep 15-19 Edited by E.A. Evans University of Chicago press,Chicago,1942 Part of the joint Symposium on the respiratory enzymes and the biological action of vitamins arranged with the University of Wisconsin to celebrate the fiftieth anniversary of the University of Chicago.The papers on Respiratory enzymes were presented at Madison,1941,Sep 11-13
Bioch 33.0343

UNIVERSITY OF CHICAGO The Central nervous system and fish behavior :symposium Papers presented Chicago 1967 Apr Edited by D. Ingle University of Chicago press,Chicago; London,1968 Sponsored jointly by the United States Air force office of scientific research, comparative physiology division of the American zoological society,and the academic journal Perspectives in biology and medicine
Bal 39.3266

UNIVERSITY OF CHICAGO The Symposium on planetary atmospheres Yerkes observatory 1947 Sep 8-10 Edited by Gerard P. Kuiper vii,366p Chicago,1949 In connection with the 15th anniversary of the Yerkes observatory
Nap 11.0340

UNIVERSITY OF CHICAGO and UNIVERSITY OF WISCONSIN A Symposium on respiratory enzymes Addresses Madison,Wisc. 1941 Sep 11-17 University of Wisconsin press,Madison, Wisc.,1942
Chem 18.1249

UNIVERSITY OF CHICAGO.CANCER TRAINING GRANT Normal and malignant cell growth :symposium Proceedings Chicago,Ill. 1968 Feb 24-25 Edited by R.J.M. Fry and others Recent results in cancer research, 17 Springer,New York,1969
Radioth 35.0865

UNIVERSITY OF CHICAGO.INSTITUTE FOR COMPUTER RESEARCH A Digitized spark chamber for automatic data retrieval By J. Bounin and others pl. 25p Chicago university, Chicago,1963?
Math L 5.1112

UNIVERSITY OF CINCINNATI Relativity Relativity conference of the Midwest Proceedings Cincinnati,Ohio 1969 Jun 2-6 Edited by M. Carmeli and others 381p Plenum press,New York,1970
TA 15.0506

UNIVERSITY OF CINCINNATI The Astrometric conference 2nd Cincinnati 1959 May 17-21 Edited by S.Aa. Strand and O.G. Franz Sponsored by the National science foundation 58p Astronomical journal,1960 Reprinted from the Astronomical journal,vol 65,no 4,May 1960
Obs 6.3243

UNIVERSITY OF COLORADO Advances in cryogenic engineering Cryogenic engineering conference 6th Proceedings Boulder,Colo. 1960 Aug 23-26 Edited by K.D. Timmerhaus Advances in cryogenic engineering, 6 Plenum press, New York,1961
Eng 41.7398

UNIVERSITY OF COLORADO Contemporary approaches to creative thinking :a symposium Boulder,Colo. 1958 Edited by Howard E. Gruber and others Atherton press;Prentice-Hall International,New York;London,1964
Eng 41.1268

UNIVERSITY OF COLORADO Lectures on fluid mechanics By Sydney Goldstein Lectures in applied mathematics:proceedings of the summer seminar,Boulder, Colo. 1957, 2 xvi,307p Interscience,New York,1960 Seminar held under the joint sponsorship of the University of Colorado, and the American mathematical society.
Chem E 24.0355

UNIVERSITY OF COLORADO Number theory conference proceedings Boulder,Colo. 1963 Aug 5-24 With the support of the National science foundation 121p 27cm University of Colorado,Boulder,Colo.,1963
P. Math 2.1772

UNIVERSITY OF COLORADO Research conference on shear strength of cohesive soils New York 1960 Boulder,Colo.,1960
Eng 41.3171

UNIVERSITY OF COLORADO Topological dynamics : an international symposium Fort Collins,Colo. 1967 Aug Edited by Joseph Auslander and Walter H. Gottschalk Bibliog.,Illus. xv, 522p 23cm Benjamin,New York;Amsterdam,1968 Dedicated to Frank J.Hahn,1929-1968
P Math 2.3143

UNIVERSITY OF COLORADO.DEPARTMENT OF PHYSICS AND ASTROPHYSICS Elementary particle physics Boulder conference on particle physics held during the first part of the 11th Boulder summer institute for theoretical physics 5th Proceedings Boulder,Colo. 1968 Jun 17-Aug 23 Vol 1-2 Edited by Kalyana T. Mahanthappa and others Lectures in theoretical physics, 11a,pt 1-2 Bibliog, diagrms xxi,629p 23cm 2 vols Gordon and Breach,New York,1969 Dedicated to George Gamow
Cav 7.3008

UNIVERSITY OF COLORADO.DEPARTMENT OF PHYSICS AND ASTROPHYSICS Lectures in theoretical physics 6th Boulder 1963 Vol. 6 University of Colorado.Summer institute for theoretical physics Edited by W.E. Brittin and W.R. Chappell Lectures in theoretical physics,Boulder.,1963,6 Colorado university press,Boulder,Colo.,1964
A Math 4.0834

UNIVERSITY OF COLORADO.DEPARTMENT OF PHYSICS AND ASTROPHYSICS Lectures in theoretical physics 7th Boulder,Col. 1964 7a: Lorentz group University of Colorado.Summer institute for theoretical physics Edited by Wesley E. Britten and Asim O. Barut Lectures in theoretical physics,Boulder,1964,7a University of Colorado press,Boulder,Col.,1965 The Lorentz group symposium volume is dedicated to Professor E.Wigner
A Math 4.1335

UNIVERSITY OF COLORADO.DEPARTMENT OF PHYSICS AND ASTROPHYSICS Lectures in theoretical physics 7th Boulder,Col. 1964 Vol. 7c: statistical physics,weak interactions, field theory University of Colorado.Summer institute for theoretical physics Edited by Wesley E. Brittin Lectures in theoretical physics,Boulder,1964,7c University of Colorado press,Boulder,Col.,1965
A Math 4.0835

UNIVERSITY OF COLORADO.DEPARTMENT OF PHYSICS AND ASTROPHYSICS Mathematical methods in theoretical physics Boulder summer institute for theoretical physics 11th Proceedings Boulder,Colo. 1968 Jun 17-Aug 23 Edited by Kalyana T. Mahanthappa and Wesley E. Brittin Lectures in theoretical physics, 11d diagrms xiv,648p 23cm Gordon and Breach, New York,1969
Cav 7.3011

UNIVERSITY OF COLORADO.DEPARTMENT OF PHYSICS AND ASTROPHYSICS Quantum fluids and nuclear matter Boulder summer institute for theoretical physics 11th Proceedings Boulder,Colo. 1968 Jun 17-Aug 23 Edited by Kalyana T. Mahanthappa and Wesley E. Brittin Lectures in theoretical physics, 11b diagrms xiv,428p 23cm Gordon and Breach, New York,1969 Dedicated to George Gamow
Cav 7.3009

UNIVERSITY OF COLORADO.DEPARTMENT OF PHYSICS AND ASTROPHYSICS Symposium on atomic collision processes Boulder summer institute for theoretical physics 11th Proceedings Boulder,Colo. 1968 Jun 17-Aug 23 Edited by Kalyana T. Mahanthappa and others Lectures in theoretical physics, 11c diagrms xiii, 337p 23cm Gordon and Breach,London,1969 Dedicated to George Gamow
Cav 7.3010

UNIVERSITY OF COLORADO.SUMMER INSTITUTE FOR THEORETICAL PHYSICS Lectures in theoretical physics 6th Boulder 1963 Vol. 6 Edited by W.E. Brittin and W.R. Chappell Sponsored by the University of Colorado.Department of physics and astrophysics Lectures in theoretical physics, Boulder.,1963,6 Colorado university press, Boulder,Colo.,1964
A Math 4.0834

UNIVERSITY OF COLORADO.SUMMER INSTITUTE FOR THEORETICAL PHYSICS Lectures in theoretical physics 7th Boulder,Col. 1964 7a: Lorentz group Edited by Wesley E. Britten and Asim O. Barut Sponsored by the University of Colorado.Department of physics and astrophysics Lectures in theoretical physics,Boulder,1964,7a University of Colorado press,Boulder,Col.,1965 The Lorentz group symposium volume is dedicated to Professor E.Wigner
A Math 4.1335

UNIVERSITY OF COLORADO.SUMMER INSTITUTE FOR THEORETICAL PHYSICS Lectures in theoretical physics 7th Boulder,Col. 1964 Vol. 7c: statistical physics,weak interactions,field theory Edited by Wesley E. Brittin Sponsored by the University of Colorado.Department of physics and astrophysics Lectures in theoretical physics, Boulder,1964,7c University of Colorado press, Boulder,Col.,1965
A Math 4.0835

UNIVERSITY OF COLORADO.SUMMER INSTITUTE FOR THEORETICAL PHYSICS Lectures in theoretical physics Boulder,Col. 1958-62 Vol 1-5 Edited by Wesley E. Brittin and others Lectures in theoretical physics, 1-5 Interscience,New York,1959-63
Cav 7.0292

UNIVERSITY OF COLORADO.SUMMER INSTITUTE ON APPLIED MATHEMATICS The General theory of quantized fields proceedings Boulder,Col. 1960 Jul.24-Aug.19 Vol. 4 By Res Jost Conducted by the American Mathematical Society Lectures in applied mathematics, 4 American mathematical society,Providence,R.I., 1965
A Math 4.0760

UNIVERSITY OF DELAWARE Delaware seminar in the philosophy of science Philosophy of science :the Delaware seminar Vol 1-: 1961- Edited by Bernard Baumrin Interscience, New York;London,1963-
WSM 43.0757

UNIVERSITY OF DENVER Energetics in metallurgical phenomena :seminar 1st-4th Proceedings Denver 1962-65 Edited by William M. Mueller Energetics in metallurgical phenomena, 1-4 Gordon and Breach,New York,1965-68
Met 25.1287

UNIVERSITY OF EDINBURGH Biological organisation,cellular and sub-cellular :a symposium organised on behalf of Unesco by C.H. Waddington Proceedings Edinburgh 1957 Sep 6-10 Unesco Pergamon press,London,1959
Bioch 33.1019

UNIVERSITY OF EDINBURGH Machine intelligence workshop 3rd;4th Proceedings 3-4 Edited by Donald Michie and Bernard Meltzer Edinburgh university press,Edinburgh,1968-69
Psy 31.3066

UNIVERSITY OF EDINBURGH.CENTRE OF AFRICAN STUDIES African primary products and international trade :international seminar Edinburgh 1964 Sep 20-23 Edited by I.G. Stewart and H. W. Ord Jointly organized by the University of Edinburgh.Department of political economy Edinburgh university press,Edinburgh,1965
Geog 13.2449

UNIVERSITY OF EDINBURGH.DEPARTMENT OF POLITICAL ECONOMY African primary products and international trade :international seminar Edinburgh 1964 Sep 20-23 University of Edinburgh.Centre of African studies Edited by I.G. Stewart and H.W. Ord Edinburgh university press,Edinburgh,1965
Geog 13.2449

UNIVERSITY OF EDINBURGH.EXPERIMENTAL PROGRAMMING UNIT Machine intelligence workshop 1st- Essays Edinburgh 1965- Edited by Donald Michie and others Oliver and Boyd; Edinburgh university press,Edinburgh,1967- The 5th contains a previously unpublished report by Alan Turing
Math S 3.1818

UNIVERSITY OF EDINBURGH.INSTITUTE OF ANIMAL GENETICS Quantitative inheritance colloquium Papers Edinburgh 1950 Apr 4-6 Edited by E.C.R. Reeve and C.H. Waddington Under the auspices of the Agricultural research council H.M.S.O.,London,1952
Gen 34.1083

UNIVERSITY OF EDINBURGH.PFIZER MEDICAL MONOGRAPHS Diabetes melitus :the opening symposium of the Pfizer foundation of the post-graduate medical school,University of Edinburgh Edinburgh 1965 Pfizer foundation Edited by L.J.P. Duncan Edinburgh university press, Edinburgh,1966
PGMS 29.0179

UNIVERSITY OF FLORIDA Methods of materials selection :a symposium Metallurgical society conferences, 40 Gordon and Breach,New York; London,1968
Met 25.2502

UNIVERSITY OF FLORIDA.CENTER FOR LATIN AMERICAN STUDIES Caribbean:Mexico today Conference on the Caribbean 14th Papers Gainesville,Fla. 1963 Dec 4-7 Edited by A. Curtis Wilgus University of Florida.Center for Latin American studies.Publications.Series 1,14 University of Florida press,Gainesville, Fla.,1964
Geog 13.5202

UNIVERSITY OF FLORIDA.COLLEGE OF ENGINEERING Deformation twinning :a conference Proceedings Gainesville 1963 Mar 21-22 In co-operation with Florida institute for continuing university studies Metallurgical society conferences, 25 Gordon and Breach, New York;London,1964
Met 25.1257

UNIVERSITY OF FLORIDA.STATISTICAL LABORATORY Monte Carlo methods symposium mimeograph Gainesville,Fla. 1954 Mar 16-17 Edited by Herbert A. Meyer Sponsored by the Wright air development center Wiley publications in statistics 382p 29cm John Wiley and sons, New York,1956
P. Math 2.2239

UNIVERSITY OF FLORIDA.STATISTICAL LABORATORY Symposium on Monte Carlo methods Gainesville, Fla. 1954 Mar.16-17 Edited by Herbert A. Meyer Sponsored by the Wright air development center Wiley publications in statistics Bibliog. xvi,382p 28cm John Wiley and sons,New York,1956
Math 3.0208

UNIVERSITY OF FLORIDA.STATISTICAL LABORATORY Symposium on Monte Carlo methods Gainesville, Fla. 1954 Mar 16-17 Edited by Herbert A. Meyer Sponsored by Wright air development center Wiley publications in statistics bibliog xvi,381p 28cm John Wiley and sons,New York,1956
Math L 5.0746

UNIVERSITY OF GEORGIA.MARINE INSTITUTE Estuaries Conference on estuaries Papers Jekyll Island,Ga. 1964 Edited by George H. Lauff American association for the advancement of science.Publications,83 New York,1967
Geog 13.0743

UNIVERSITY OF GEORGIA.MARINE INSTITUTE Estuaries :a symposium Papers Jekyll island,Ga 1964 Mar-Apr Edited by George H. Lauff American association for the advance of science.Publication, 83 Washington,D.C., 1967
Bot 42.3345

UNIVERSITY OF GEORGIA.MARINE INSTITUTE Salt marsh conference Proceedings Sapelo island, Ga. 1958 Mar 25-28 University of Georgia. Marine institute,Athens,Ga.,1959
Geog 13.0713

UNIVERSITY OF GEORGIA.TOPOLOGY INSTITUTE Topology of 3-manifolds and related topics :a symposium proceedings Athens, Ga. 1961 Aug 14 - Sep 8 Edited by M.K. Fort supported by United States.Office of naval research and National science foundation viii,256p 24cm Englewood Cliffs, N.J.,1962
P. Math 2.0296

UNIVERSITY OF GEORGIA TOPOLOGY OF MANIFOLDS INSTITUTE Proceedings Topology of manifolds Athens 1969 Aug 11-22 Edited by J.C. Cantrell and C.H. Edwards xiv,514p 24cm Markham,Chicago,Ill.,1970
P Math 2.3819

UNIVERSITY OF GLASGOW Endocrine aspects of breast cancer :a conference Proceedings Glasgow 1957 Jul 8-10 Edited by Alastair R. Currie Livingstone,Edinburgh;London,1958
Path 30.2082

UNIVERSITY OF GLASGOW Endocrine aspects of breast cancer :a conference Proceedings Glasgow 1957 Jul 8-10 Edited by Alastair R. Currie Livingstone,Edinburgh;London,1958
PGMS 29.0455

UNIVERSITY OF HAWAII Methodologies of pattern recognition International conference on methodologies of pattern recognition Proceedings Honolulu 1968 Jan 24-26 Sponsored by the United States.Air force. Office of scientific research bibliog.,illus. Academic press,New York;London,1969
Math S 3.1521

UNIVERSITY OF HAWAII Venomous and poisonous animals and noxious plants of the Pacific region :a collection of papers based on a symposium in the Public health and medical science division at the tenth Pacific science congress Papers Honolulu 1961 Aug 21-Sep 6 Edited by Hugh L. Keegan and W.V. MacFarlane Co-sponsored by the United States. National academy of sciences Pergamon press. Symposium publications division,Oxford,1963
Chem 26.0448

UNIVERSITY OF HAWAII World orchid conference 2nd Proceedings Honolulu 1957 Sep 19-23 American orchid society,Cambridge,Mass.,1958
BG 38.3117

UNIVERSITY OF HAWAII and BISHOP (BERNICE P) MUSEUM Plants and the migrations of Pacific peoples :a symposium Honolulu 1961 Aug 21-Sep 6 Edited by J. Barrau Sponsored by the National science founcation Bishop museum press,Honolulu,1963 Symposium convened at the 10th Pacific science congress
Geog 13.7202

UNIVERSITY OF HONG KONG Symposium on muscle receptors Proceedings Hong Kong 1961 Sep Edited by David Barker Hong Kong university press,Hong Kong,1962
An 32.4002

UNIVERSITY OF HONG KONG Symposium on muscle receptors :a meeting held...as part of the Golden jubilee congress of the University of Hong Kong Proceedings Hong Kong 1961 Sep Edited by David Barker Hong Kong university press,Hong Kong,1962
Phys 20.1559

UNIVERSITY OF HONG KONG Symposium on muscle receptors :a meeting held...as part of the golden jubilee congress of the University of Hong Kong Proceedings Hong Kong 1961 Sep Edited by David Barker Hong Kong university press,Hong Kong,1962
Bal 39.1519

UNIVERSITY OF HONG KONG Symposium on the design of high buildings Proceedings of a meeting...held as part of the Golden jubilee congress of the University of Hong Kong Hong Kong 1961 Sep Edited by Sean Mackey Hong Kong university press,Hong Kong,1962
Eng 41.2685

UNIVERSITY OF HONG KONG.GOLDEN JUBILEE CONGRESS Symposium on the design of high buildings Proceedings of a meeting...held as part of the Golden jubilee congress of the University of Hong Kong Hong Kong 1961 Sep University of Hong Kong Edited by Sean Mackey Hong Kong university press,Hong Kong,1962
Eng 41.2685

UNIVERSITY OF ICELAND North Atlantic biota and their history :a symposium Reykjavik 1962 Jul Sponsored by the Nato advanced institutes program pl. 430p 24cm Pergamon press,Oxford,1963
Sco 14.8334

UNIVERSITY OF ICELAND and MUSEUM OF NATURAL HISTORY,REYKJAVIK North atlantic biota and their history :a symposium Reykjavik 1962 Jul 12-25 Edited by Askell Love and Doris Love Sponsored by NATO advanced study institutes program illus. Pergamon press, Oxford,1963
Geog 13.1211

UNIVERSITY OF ILLINOIS Allerton conference on circuit and system theory 1st-6th Proceedings Monticello,Ill. 1963-68 6 vols University of Illinois,Evanston,Ill., 1963-68
Eng 41.5240

UNIVERSITY OF ILLINOIS Spores 4 International spore conference 4th Urbana,Ill. 1968 Oct 4-6 American society for microbiology Edited by L.Leon Campbell American society for microbiology,1967
Gen 34.0711

UNIVERSITY OF ILLINOIS Symposium on circuit analysis Proceedings Monticello,Ill. 1955 1955
Eng 41.5160

UNIVERSITY OF ILLINOIS.BIOACOUSTICS LABORATORY Ultrasound in biology and medicine :a symposium Monticello,Ill. 1955 Jun 20-22 Edited by Elizabeth Kelly American institute of biological sciences.Publications, 3 American institute of biological sciences, Washington,D.C.,1957
An 32.0393

UNIVERSITY OF KANSAS 1962 Cognition,theory, promise :papers read at the Martin Scheerer memorial meetings on cognitive psychology Papers Edited by Constance Scheerer Harper and Row,New York,1964
Psy 31.1919

UNIVERSITY OF KANSAS.LINDA HALL LIBRARY OF SCIENCE AND TECHNOLOGY Bibliography and natural history :a conference Essays Lawrence 1964 Jun 25-27 Edited by Thomas R. Buckman University of Kansas.Publications. Library series, 27 ix,148p University of Kansas libraries,Lawrence,Kansas,1966
WSM 43.3212

UNIVERSITY OF KENTUCKY.COLLEGE OF ARTS AND SCIENCES.DEPARTMENT OF PSYCHOLOGY Kentucky symposium Learning theory, personality theory,and clinical research :the Kentucky symposium Papers Lexington,Ky. 1953 Mar 13-14 Wiley;Chapman and Hall,New York;London,1954
Psy 31.1839

UNIVERSITY OF KENTUCKY.DEPARTMENT OF MATHEMATICS Symposium on topological dynamics and ergodic theory Proceedings Lexington,Ky. 1971 Jun 3-5 80p 28cm University of Kentucky,Lexington,Ky.,1971 Duplicated notes
P Math 2.4189

UNIVERSITY OF KENTUCKY.DEPARTMENT OF PSYCHOLOGY Learning theory,personality theory,and clinical research :the Kentucky symposium Lectures Lexington,Ky. 1953 Mar 13-14 Wiley;Chapman and Hall,New York;London,1954
Psy 28.0195

UNIVERSITY OF LEEDS Metals and enzyme activity :a symposium Leeds 1956 Jul 13 Biochemical society Edited by E.M. Crook Biochemical society.Symposia, 15 Cambridge university press,Cambridge,1958
Bioch 33.1377

UNIVERSITY OF LEEDS The Renal stone research symposium Proceedings Leeds 1968 Apr 17-20 Edited by A. Hodgkinson and B.E.C. Nordin Churchill,London,1969
PGMS 29.0240

UNIVERSITY OF LEEDS.DEPARTMENT OF MINING Symposium on coal preparation 2nd Papers Leeds 1957 Oct 21-25 xiii,513p University of Leeds,Leeds,1957
Chem E 24.1392

UNIVERSITY OF LEICESTER Exploration in group relations :a conference Report Leicester 1957 Sep By E.L. Trist and C. Sofer Leicester university press,Leicester,1959
Eng 41.0706

UNIVERSITY OF LIVERPOOL International symposium on electron and photon interactions at high energies 4th Proceedings Liverpool 1969 Sep 14-20 illus 30cm Danesbury nuclear physics laboratory,Danesbury, 1969
A Math 4.1760

UNIVERSITY OF LIVERPOOL.DEPARTMENT OF PURE MATHEMATICS Liverpool singularities symposium II Symposium on singularities of smooth manifolds and maps Proceedings Liverpool 1970 Aug-Sep Edited by C.T.C. Wall Lecture notes in mathematics, 209 280p 25cm Springer,Berlin,1971
P Math 2.4123

UNIVERSITY OF LIVERPOOL.DEPARTMENT OF PURE MATHEMATICS Symposium on singularities of smooth manifolds and maps Liverpool singularities symposium I Proceedings Liverpool 1970 Aug -Sep Edited by C.T.C. Wall Lecture notes in mathematics, 192 318p 25cm Springer,Berlin,1971
P Math 2.4122

UNIVERSITY OF LIVERPOOL.DERBY HALL The Biochemistry of fish ;a symposium Liverpool 1950 Sep 22 Biochemical society Edited by R.T. Williams Biochemical society.Symposia, 6 Cambridge university press,Cambridge,1961
Bioch 33.1368

UNIVERSITY OF LONDON Problems in eugenics International eugenics congress 1st Papers London 1912 Jul 24-30 Eugenics education society Eugenics education society, London,1912
Bot 42.6705

UNIVERSITY OF LONDON Virus growth and variation :a symposium Papers London 1959 Apr Society for general microbiology Edited by A. Isaacs and B.W. Lacey Society for general microbiology.Symposia, 9 Cambridge university press,Cambridge,1959
Bioch 33.1182

UNIVERSITY OF LONDON.SENATE HOUSE Glutathione :a symposium London 1958 Feb 15 Biochemical society Edited by E.M. Crook Biochemical society.Symposia, 17 Cambridge university press,Cambridge,1959
Bioch 33.1379

UNIVERSITY OF LONDON.SENATE HOUSE Glutathione :a symposium London 1958 Feb 15 Biochemical society Edited by E.M. Crook Biochemical society.Symposia, 17 Cambridge university press,Cambridge,1959
Radioth 35.0078

UNIVERSITY OF LONDON.SENATE HOUSE Steric aspects of the chemistry and biochemistry of natural products :a symposium London 1959 Jun 30 Biochemical society Edited by J.K. Grant and W. Klyne Biochemical society. Symposia, 19 Cambridge university press, Cambridge,1960
Bioch 33.1381

UNIVERSITY OF LONDON.SENATE HOUSE Structure and function of subcellular components :a symposium London 1957 Feb 23 Biochemical society Edited by E.M. Crook Biochemical society.Symposia, 16 Cambridge university press,Cambridge,1959
Bot 42.1493

UNIVERSITY OF LONDON.SENATE HOUSE The Biosynthesis and secretion of adrenocortical steroids :a symposium London 1959 Feb 14 Biochemical society Edited by F. Clark and J. K. Grant Biochemical society.Symposia, 18 Cambridge university press,Cambridge,1960
Bioch 33.1380

UNIVERSITY OF LONDON.SENATE HOUSE The Structure and biosynthesis of macro-molecules London 1961 Mar 27-28 Biochemical society Edited by D.J. Bell and J.K. Grant Biochemical society.Symposia, 21 Cambridge, 1962 In commemoration of the 50th anniversary of the Biochemical society
Bal 39.0515

UNIVERSITY OF LONDON.SENATE HOUSE The Structure and biosynthesis of macromolecules : a symposium London 1961 Mar 27-28 Biochemical society Edited by D.J. Bell and J.K. Grant Biochemical society.Symposia, 21 Cambridge university press,Cambridge,1962 In commemoration of the 50th anniversary of the Biochemical society
Bioch 33.1383

UNIVERSITY OF LONDON.SENATE HOUSE The Structure and function of subcellular components :a symposium London 1957 Feb 23 Biochemical society Edited by E.M. Crook Biochemical society.Symposia, 16 Cambridge university press,Cambridge,1959
Bioch 33.1378

UNIVERSITY OF MADRAS Aspects of protein structure :a symposium Proceedings Madras 1963 Jan 14-18 Edited by G.N. Ramachandran Academic press,London;New York,1963
Gen 34.0658

UNIVERSITY OF MADRAS.CENTRE OF ADVANCED STUDY IN BIOPHYSICS Conformation of biopolymers International symposium on conformation of biopolymers Papers read Madras 1967 Jan 18-21 Vol 1-2 Edited by G.N. Ramachandran Sponsored by the International union of pure and applied physics 2 vols Academic press,London;New York,1967
Bioch 33.0331

UNIVERSITY OF MADRAS.DEPARTMENT OF PHYSICS Aspects of protein structure The International symposium on protein structure and crystallography Proceedings Madras 1963 Jan 14-18 Edited by G.N. Ramachandran Academic press,London,1963
Cav 7.1495

UNIVERSITY OF MADRAS.DEPARTMENT OF PHYSICS Aspects of protein structure The Symposium on protein structure Proceedings Madras 1963 Jan 14-18 Edited by G.N. Ramachandran Academic press,London;New York,1963 The Symposium on protein structure formed part of an International symposium on protein structure and crystallography organized by the University of Madras
Bioch 33.0542

UNIVERSITY OF MADRAS.DEPARTMENT OF PHYSICS Crystallography and crystal perfection The International symposium on protein structure and crystallography Proceedings Madras 1963 Jan 14-18 session on crystallography and crystal perfection Edited by G.N. Ramachandran Academic press,London,1963
Cav 7.1494

UNIVERSITY OF MARYLAND International conference on harmonic analysis Proceedings College Park,Md. 1971 Nov 8-12 Edited by Denny Gulick and Ronald L. Lipsman Lecture notes in mathematics, 266 vi,323p 25cm Springer,Berlin,1972
P Math 2.4283

UNIVERSITY OF MARYLAND Seminar on differential equations and dynamical systems 2n Lectures College Park, Md 1970 Jul-Aug Edited by J.A. Yorke Lecture notes in mathematics, 144 viii,268p 17cm Springer-Verlag,Berlin,1970
P Math 2.3534

UNIVERSITY OF MARYLAND Several complex variables International mathematical conference Proceedings College Park,Md. 1970 Apr 6-17 1 Edited by John Horvath Lecture notes in mathematics, 155 bibliog. 214p 25cm Springer,Berlin,1970
P Math 2.3890

UNIVERSITY OF MARYLAND Several complex variables International mathematical conference Proceedings College Park,Md. 1970 Apr 6-17 2 Edited by John Horvath Lecture notes in mathematics, 185 214p 25cm Springer,Berlin,1971
P Math 2.3891

UNIVERSITY OF MELBOURNE Australia-New Zealand conference on soil mechanics and foundation engineering 1st Proceedings Melbourne 1952 Jul 1952
Eng 41.3166

UNIVERSITY OF MELBOURNE.CHEMISTRY DEPARTMENT Conference on applications of isotopes in scientific research Proceedings Melbourne 1950 Aug 14-17 University of Melbourne, Melbourne,1951
Radioth 35.1495

UNIVERSITY OF MELBOURNE.FACULTY OF ENGINEERING Fracture The Tewksbury symposium 1st Proceedings Melbourne 1963 Aug 26-30 Edited by C.J. Osborn University of Melbourne,Melbourne,1965
Met 25.0870

UNIVERSITY OF MELBOURNE.FACULTY OF ENGINEERING Fracture The Tewskbury symposium 2nd Proceedings Melbourne 1969 Edited by C.J. Osborn and others Butterworth,Sydney,1969
Met 25.2579

UNIVERSITY OF MELBOURNE.FACULTY OF ENGINEERING The Failure of metals by fatigue :a symposium Proceedings Melbourne 1946 Dec 2-6 Melbourne university press,Melbourne,1947
Met 25.0909

UNIVERSITY OF MELBOURNE.FSCULTY OF ENGINEERING Fracture Tewkesbury symposium on fracture 2nd Proceedings Melbourne 1969 Edited by C.J. Osborn and others Butterworths, Sydney,1969 Financial support from the Pearson Tewkesbury bequest
Eng 41.8202

UNIVERSITY OF MIAMI Conference on symmetry principles at high energy Coral Gables,Fla. 1964 Jan 30-31 Edited by Behram Kursunoglu and Arnold Perlmutter Freeman,San Francisco, 1964
Cav 7.0543

UNIVERSITY OF MIAMI.CENTER FOR THEORETICAL STUDIES Coral Gables conferences on symmetry principles at high energy 4 Proceedings Coral Gables,Fla. 1967 Jan 25-27 Edited by Arnold Perlmutter and Behram Kursunoglu Sponsored jointly by the United States atomic energy commission 253p Freeman,San Francisco;London,1967
A Math 4.1347

UNIVERSITY OF MIAMI.DEPARTMENT OF BIOCHEMISTRY Nucleic acid-protein interactions and nucleic acid synthesis in viral infection :symposium Miami,Fla. 1971 Jan 18-22 Miami winter symposia, 2 North-Holland,Amsterdam,1971
Bioch 33.2241

UNIVERSITY OF MIAMI SCHOOL OF MEDICINE Microcirculation,perfusion,and transplantation of organs Conference on microcirculation, perfusion and transplantation of organs Proceedings Miami,Fla. 1969 Oct 30 Edited by Theodore I. Malinin and others Academic press,London;New York,1970
An 32.5458

UNIVERSITY OF MICHIGAN Approaches to the genetic analysis of mammalian cells :Michigan conference on genetics Lectures Edited by Donald J. Merchant and James V. Neel University of Michigan press,Ann Arbor,Mich., 1962
Gen 34.0911

UNIVERSITY OF MICHIGAN Categories of human learning :symposium on the psychology of human learning Papers Ann Arbor,Mich. 1962 Jan 31-Feb 1 Edited by Arthur W. Melton Academic press,New York;London,1964
Psy 31.1003

UNIVERSITY OF MICHIGAN Decision processes seminar proceedings Santa Monica,Calif. 1952 summer Edited by R.M. Thrall and others Supported by Ford foundation viii,332p 22cm John Wiley and sons,New York,1954
Math 3.0771

UNIVERSITY OF MICHIGAN Electron and laser beam symposium 8th Proceedings Ann Arbor,Mich 1966 Apr 6-8 Edited by G.I. Haddad University of Michigan,Ann Arbor,Mich. 1966 8th in the series of annual symposia on Electron beam technology initiated in 1959 by Alloyd and held under their sponsorship for seven years.The titles of the symposia vary,therefore those catalogued have been grouped together under the keyword"Electron beams"
Eng 41.5618

UNIVERSITY OF MICHIGAN Heat transfer :a symposium Ann Arbor 1952 Edited by B.A. Uhlendorf University of Michigan.Engineering research institute special publications: symposia vii,286p Engineering research institute,Michigan,1953
Chem E 24.0544

UNIVERSITY OF MICHIGAN International conference on leukemia lymphoma Proceedings Ann Arbor 1967 Edited by Chris J.D. Zarafonetis Lea and Febiger;Kimpton, Philadelphia;London,1968
Med 36.0376

UNIVERSITY OF MICHIGAN Joint automatic control conference 8th-9th Preprints, Technical papers 1967-68
Eng 41.8208

UNIVERSITY OF MICHIGAN Lectures in topology : conference Ann Arbor 1940 Jun.24-Jul.6 Edited by Raymond L. Wilder and William L. Ayres vii,316p 23cm University of Michigan press,Ann Arbor, Mich.,1941
P. Math 2.0288

UNIVERSITY OF MICHIGAN Lectures on functions of a complex variable :conference proceedings Ann Arbor, Mich. 1953 Jun 17-Jul1 Edited by Wilfred Kaplan and others Bibliog 435p 24cm University of Michigan, Ann Arbor,Mich.,1955
P. Math 2.1533

UNIVERSITY OF MICHIGAN Midwestern conference on fluid mechanics 5th Proceedings Ann Arbor,Mich. 1957 Apr 1-2 University of Michigan.Engineering research institute publications viii,388p University of Michigan press,Ann Arbor,Mich.,1957
Chem E 24.0405

UNIVERSITY OF MICHIGAN Midwestern conference on fluid mechanics 5th Proceedings Ann Arbor,Mich. 1957 Apr 1-2 University of Michigan press,Ann Arbor,Mich.,1957
Eng 41.6958

UNIVERSITY OF MICHIGAN Proof techniques in graph theory Ann Arbor graph theory conference 2nd Proceedings Ann Arbor, Mich. 1968 Feb Edited by Frank Harary xv, 330p 24cm Academic press,New York;London, 1969
Math S 3.1658

UNIVERSITY OF MICHIGAN The National symposium on colloid chemistry 5th Papers presented Ann Arbor,Mich. 1927 Jun 22-24 Edited by Harry Boyer Weiser Colloid symposium monograph, 5 Chemical catalog company,New York,1928
Bioch 33.1473

UNIVERSITY OF MICHIGAN.ENGINEERING RESEARCH INSTITUTE The Midwestern conference on fluid mechanics 5th proceedings Ann Arbor 1957 Apr 1-2 Edited by R.C.F. Bartels University of Michigan.Engineering research institute.Publication University of Michigan press,Ann Arbor,1957
A Math 4.0516

UNIVERSITY OF MICHIGAN.INSTITUTE FOR HUMAN ADJUSTMENT Psycholgical diagnosis and counseling of the adult blind :selected papers from the proceedings of the University of Michigan conference for the blind Ann Arbor, Mich. 1947 United States.Federal security agency.Office of vocational rehabilitation and Michigan.Department of social welfare. Division of services for the blind Edited by Wilma Donahue and Donald Dabelstein American foundation for the blind,New York,1950
Psy 31.2118

UNIVERSITY OF MICHIGAN.OBSERVATORY Michigan symposium on astrophysics Proceedings Ann Arbor,Mich. 1953 Jun.29-Jul.24 Ann Arbor, 1953 Typescript
Obs 6.1978

UNIVERSITY OF MINNESOTA Midwestern conference on fluid mechanics 3rd Proceedings Minneapolis 1953 Mar 23-25 783p Minneapolis,Minn.,1953
Chem E 24.0403

UNIVERSITY OF MINNESOTA The National symposium on colloid chemistry 3rd Papers presented Minneapolis 1925 Jun 17-19 Edited by Harry N. Holmes and Harry B. Weiser Colloid symposium monograph, 3 Chemical catalog company,New York,1925
Bioch 33.1471

UNIVERSITY OF MINNESOTA.INSTITUTE OF TECHNOLOGY Midwestern conference on fluid mechanics 3rd Proceedings Minneapolis,Minn. 1953 Mar 23-25 University of Minnesota, Minneapolis,Minn.,1953
Eng 41.6953

UNIVERSITY OF MISSOURI Communication theory and research:international symposium 1st Proceedings Kansas City 1965 Mar 24-27 Edited by Lee Thayer Thomas,Springfield,Ill., 1967
Psy 31.1068

UNIVERSITY OF MONTREAL.SEMINAIRE DE MATHEMATIQUES SUPERIEURES, 8 Fondements de la geometrie algebrique moderne Montreal 1964 By Jean Dieudonne North Atlantic treaty organization Societe mathematique du Canada 151p 28cm Universite de Montreal,Montreal, 1964
P. Math 2.0507

UNIVERSITY OF MONTREAL.SEMINAIRE DE MATHEMATIQUES SUPERIEURES, 11 Homotopie et cohomologie manuscript Montreal 1964 By Beno Eckmann North Atlantic treaty organization Societe mathematique du Canada 129p 28cm Universite de Montreal,Montreal,1965
P. Math 2.0297

UNIVERSITY OF NEBRASKA.CURRENT THEORY AND RESEARCH IN MOTIVATION SYMPOSIUM, 10- Nebraska symposium on motivation Lincoln, Nebraska 1962- 1962- University of Nebraska.Department of psychology University of Nebraska press,Lincoln,Nebraska,1962-
Psy 31.1858

UNIVERSITY OF NEBRASKA.DEPARTMENT OF PSYCHOLOGY Current theory and research in motivation : a symposium Papers Lincoln,Nebraska 1953 Jan 15-16 Mar 26-27 By Judson S. Brown and others University of Nebraska press,Lincoln, Nebraska,1953
Psy 31.1857

UNIVERSITY OF NEBRASKA.DEPARTMENT OF PSYCHOLOGY Nebraska symposium on motivation Lincoln, Nebraska 1962- 1962- University of Nebraska.Current theory and research in motivation symposium, 10- University of Nebraska press,Lincoln,Nebraska,1962-
Psy 31.1858

UNIVERSITY OF NEW MEXICO Metals for supersonic aircraft and missiles Heat tolerant metals for aerodynamic applications : a conference Proceedings Alburquerque,New Mexico 1957 Jan 28-29 Edited by D.W. Grobecker American society for metals, Cleveland,Ohio,1958
Met 25.1003

UNIVERSITY OF NEW MEXICO 1955 Future of arid lands International arid land meetings Papers and recommendations Edited by Gilbert F. White American association for the advancement of science.Publications,43 American association for the advancement of science,Washington,D.C.,1956
Geog 13.1467

UNIVERSITY OF NEWCASTLE UPON TYNE Magnetism and the cosmos NATO Advanced study institute on planetary and stellar magnetism Newcastle-upon-Tyne 1965 Apr Edinburgh,1967
Geod 9.0567

UNIVERSITY OF NEWCASTLE UPON TYNE Methods in palaeomagnetism NATO Advanced study institute on procedures in palaeomagnetism Proceedings Newcastle-upon-Tyne 1964 Apr 1-10 Edited by D.W. Collinson and others Developments in solid earth geophysics, 3 Amsterdam,1967
Geod 9.0568

UNIVERSITY OF NEWCASTLE UPON TYNE.COMPUTING LABORATORY Teaching of programming at university level Joint IBM-University of Newcastle upon Tyne seminar 3rd Proceedings Cambridge 1970 Sep 8-11 Edited by B. Shaw illus. 153p University of Newcastle,Newcastle,1971
Math L 5.3590

UNIVERSITY OF NEWCASTLE-UPON-TYNE.COMPUTING LABORATORY Teaching of computer design Joint IBM-University of Newcastle-upon-Tyne seminar Proceedings Newcastle upon Tyne 1971 Sep 7-10 illus 121p University of Newcastle upon Tyne.Computing laboratory, Newcastle upon Tyne,1972
Math L 5.3723

UNIVERSITY OF NEWCASTLE UPON TYNE.SCHOOL OF PHYSICS Mantles of the earth and terrestrial planets N.A.T.O. advanced study institute :a conference Proceedings Newcastle upon Tyne 1966 Mar 30-Apr 7 Edited by S.K. Runcorn Sponsored by North Atlantic treaty organization.Science office Interscience,London,1967
TA 15.0202

UNIVERSITY OF NEWCASTLE-UPON-TYNE.SCHOOL OF PHYSICS Mantles of the earth and terrestrial planets :Nato advanced study institute Proceedings Edited by S.K. Runcorn Interscience,London,1967
Geod 9.0016

UNIVERSITY OF NEWCASTLE-UPON-TYNE Hydrogen-bonded systems :symposium on equilibria and reaction kinetics... Proceedings Newcastle-upon-Tyne 1968 Jan 10-12 Edited by A.K. Covington and P. Jones Taylor and Francis, London,1968
Chem 18.0298

UNIVERSITY OF NORTH CAROLINA Combinatorial mathematics and its applications 2nd conference Proceedings Chapel Hill,N.C. 1970 May Edited by R.C. Bose and others 548p 25cm University of North Carolina, Chapel Hill,N.C.,1970
Math S 3.1655

UNIVERSITY OF NORTH CAROLINA Combinatorial mathematics and its applications Chapel Hill conference 2nd Proceedings Chapel Hill, N.C. 1970 May 8-13 Edited by R.C. Bose and others 548p 25cm University of North Carolina,Chapel Hill,N.C.,1970
P Math 2.4507

UNIVERSITY OF NORTH CAROLINA Symposium on congestion theory Proceedings Chapel Hill, N.C. 1964 Aug 24-26 Edited by Walter L. Smith and William E. Wilkinson University of North Carolina.Monograph series in probability and statistics, 2 University of North Carolina press,Chapel Hill,N.C.,1965
Eng 41.2150

UNIVERSITY OF NORTH CAROLINA.DEPARTMENT OF MATHEMATICS Carolina conference Proceedings Chapel Hill,N.C. 1970 Mar 30-Apr 3 Edited by Donald A. Eisenman and Laird E. Taylor 107p 28cm University of North Carolina,Chapel Hill,N.C.,1970
P Math 2.4356

UNIVERSITY OF NORTH CAROLINA.DEPARTMENT OF STATISTICS Symposium on congestion theory proceedings Chapel Hill,N.C. 1964 Aug.24-26 Edited by Walter L. Smith and William E. Wilkinson Financially supported by the United States.Office of naval research North Carolina.University.Monograph series in probability and statistics, 2 xv,457p 24cm University of North Carolina press, Chapel Hill,N.C.,1965
Math 3.0568

UNIVERSITY OF NOTRE DAME International conference on sintering and related phenomena Cleveland 1965 Jun 21-23 Edited by G.C. Kuczynski and others Gordon and Breach,New York,1967
Met 25.1285

UNIVERSITY OF NOTTINGHAM Conference on anodising aluminium Proceedings Nottingham 1961 Sep 12-14 Aluminium development association,London,1961
Met 25.2047

UNIVERSITY OF NOTTINGHAM Conference on anodising aluminium Proceedings Nottingham 1961 Sep 12-14 Convened by the Aluminium development association Aluminium development association,London,1962
Eng 41.3602

UNIVERSITY OF NOTTINGHAM Control of the plant environment Easter school in agricultural science 4th Proceedings Nottingham 1957 Edited by J.P. Hudson Butterworths,London,1957
BG 38.0605

UNIVERSITY OF NOTTINGHAM Control of the plant environment Easter school in agricultural science 4th Proceedings Nottingham 1957 Edited by J.P. Hudson illus. xvi,240p Butterworths,London,1957
Bot 42.1943

UNIVERSITY OF NOTTINGHAM Digestive physiology and nutrition of the ruminant Easter school in agricultural sciences 7th Proceedings Nottingham 1960 Edited by D. Lewis Butterworth,London,1961
VA 19.0296

UNIVERSITY OF NOTTINGHAM Growth of cereals and grasses Easter school in agricultural science 12th Proceedings Nottingham Edited by F.L. Milthorpe and J.D. Ivins Bibliog.,illus,diagrs,tables xii,359p Butterworths,London,1966
Bot 42.2035

UNIVERSITY OF NOTTINGHAM Growth of leaves Easter school in agricultural science 3rd Proceedings Nottingham 1956 Edited by F. L. Milthorpe illus. x,223p Butterworths, London,1956
Bot 42.1825

UNIVERSITY OF NOTTINGHAM Measurement of grassland productivity Easter school in agricultural science 6th Proceedings Nottingham 1959 Edited by J.D. Ivins Butterworths,London,1959
Bot 42.1951

UNIVERSITY OF NOTTINGHAM Nutrition of the legumes Easter school in agricultural science 5th Proceedings Nottingham 1958 Edited by E.G. Hallsworth Butterworths,London,1958
Bioch 33.0776

UNIVERSITY OF NOTTINGHAM Reproduction in the female mammals Easter school in agricultural science 13th Proceedings Nottingham 1966 Edited by G.E. Lamming and E.C. Amoroso Butterworths,London,1967
An 32.5323

UNIVERSITY OF NOTTINGHAM Root growth Easter school in agricultural science 15th Proceedings Nottingham 1968 Edited by William J. Whittington Bibliog.,illus. xi, 450p Butterworths,London,1969
Bot 42.1437

UNIVERSITY OF NOTTINGHAM Soil zoology Easter school in agricultural science 2nd Proceedings Nottingham 1955 Edited by D. Keith McE. Kevan Butterworths,London,1955
Bal 39.2006

UNIVERSITY OF NOTTINGHAM Soil zoology Easter school in agricultural science 2nd Proceedings Nottingham 1955 Edited by D. Keith McE. Kevan Butterworths scientific publications illus. xiv,512p Butterworths,London,1955
Bot 42.2070

UNIVERSITY OF NOTTINGHAM and CHEMICAL SOCIETY Inorganic polymers :an international conference Lectures Nottingham 1961 Jul 18-21 Chemical society.Special publication, 15 Chemical society,London,1961
Chem 18.1221

UNIVERSITY OF NOTTINGHAM and CHEMICAL SOCIETY The Alkali metals :an international symposium Nottingham 1966 Jul 19-22 Chemical society.Special publication,22 Chemical society,London,1966
Chem 18.1226

UNIVERSITY OF NOTTINGHAM.FACULTY OF APPLIED SCIENCE Engineering design :papers given at the University of Nottingham, September, 1964 Edited by T.F. Roylance Pergamon press,Oxford,1966
Eng 41.0325

UNIVERSITY OF NOTTINGHAM.SCHOOL OF AGRICULTURE Growth and development of mammals Easter school in agricultural science 14th Proceedings Nottingham 1967 Edited by G. A. Lodge and G.E. Lamming Butterworths, London,1968
Gen 34.1592

UNIVERSITY OF OKLAHOMA Social psychology at the crossroads :the university of Oklahoma lectures in social psychology Papers Norman,Okla. 1950 Apr 6-11 Edited by John H. Rohrer and Mazafer Sherif Harper,New York, 1951
Psy 31.1363

UNIVERSITY OF OREGON Recent advances in invertebrate physiology :a symposium Eugene 1955 Sep Edited by B.T. Scheer and others University of Oregon publications,Eugene, Oregon,1957 Sponsored by the National science foundation,Tektronix foundation,the University of Oregon
Bal 39.2107

UNIVERSITY OF OREGON.DEPARTMENT OF GEOLOGY AND MINERAL INDUSTRIES The Andesite conference Proceedings Edited by A.R. McBirney Oregon.Department of geology and mineral industries.Bulletin, 65 Portland, Ore.,1969 Photocopy
Min 10.1477

UNIVERSITY OF OREGON MEDICAL SCHOOL Aging Symposium on the biology of skin 14th Proceedings Portland,Ore. 1964 May 9-10 Edited by William Montagna Advances in biology of skin, 6 Oregon regional primate research center, 55 Pergamon press,Oxford, 1965
An 32.4041

UNIVERSITY OF OREGON MEDICAL SCHOOL Carcinogenesis Symposium on the biology of skin 15th Proceedings Portland,Ore. 1965 Edited by William Montagna and Richard L. Dobson Advances in biology of skin, 7 Oregon regional primate research center, 87 Pergamon press,Oxford,1966
An 32.4042

UNIVERSITY OF OREGON MEDICAL SCHOOL Pigmentary system Symposium on the biology of skin 16th Proceedings Portland,Ore. 1966 Edited by William Montagna and Funan Hu Advances in biology of skin, 8 Oregon regional primate research center, 183 Pergamon press,Oxford,1967
An 32.4043

UNIVERSITY OF OTAGO MEDICAL SCHOOL.DEPARTMENT OF PSYCHOLOGICAL MEDICINE Drugs and the mind a symposium Proceedings Dunedin 1967 Jul 14-15 Edited by Basil James University of Otago medical school,Dunedin,1967
PGMS 29.0394

UNIVERSITY OF OXFORD Symposium on recent advances in optics Proceedings Oxford 1959 Mar.20 138-180p London,1959
Obs 6.1989

UNIVERSITY OF OXFORD The Nature of virus multiplication :a symposium Papers Oxford 1952 Apr Society for general microbiology Edited by Paul Fildes and W.E. Van Heyningen Society for general microbiology.Symposia, 2 Cambridge university press,Cambridge,1953
Bioch 33.1179

UNIVERSITY OF OXFORD.COMPUTING LABORATORY Advances in programming and non-numerical computation summer school proceedings Oxford 1963 Edited by L. Fox bibliog viii,218p 23cm Pergamon press,Oxford,1966
Math L 5.1026

UNIVERSITY OF OXFORD.COMPUTING LABORATORY Advances in programming and non-numerical computation summer school proceedings Oxford 1963 Edited by L. Fox viii,218p 24cm Pergamon press,Oxford,1966
Math 3.0239

UNIVERSITY OF OXFORD.COMPUTING LABORATORY Numerical solution of ordinary and partial differential equations based on a Summer school Oxford 1961 Aug-Sep Edited by L. Fox Adiwes international series in the engineering sciences Addison-Wesley series in computer science and information processing Pergamon press;Addison-Wesley,London;Reading, Mass.,1962
TA 15.0378

UNIVERSITY OF OXFORD.COMPUTING LABORATORY Numerical solution of ordinary and partial differential equations summer school Oxford 1961 Aug.-Sep. By L. Fox bibliog. ix, 509p 23cm Pergamon press,Oxford,1962
Math L 5.0308

UNIVERSITY OF OXFORD.COMPUTING LABORATORY Numerical solutions of ordinary and partial differential equations :summer school Lectures Oxford 1961 Aug-Sep Edited by L. Fox In collaboration with the University of Oxford.Delegacy for extra-mural studies Pergamon press,Oxford,1962
Cav 7.0830

UNIVERSITY OF OXFORD.DELEGACY FOR EXTRA-MURAL STUDIES Numerical solutions of ordinary and partial differential equations :summer school Lectures Oxford 1961 Aug-Sep University of Oxford.Computing laboratory Edited by L. Fox Pergamon press,Oxford,1962
Cav 7.0830

UNIVERSITY OF OXFORD.MATHEMATICS INSTITUTE Combinatorial mathematics and its applications conference Proceedings Oxford 1969 Jul 7-10 Edited by D.J.A. Welsh x,364p 24cm Academic press,London;New York,1971
P Math 2.4514

UNIVERSITY OF PENNSYLVANIA Biology of parasites;emphasis on veterinary parasites The International conference of the World association for the advancement of veterinary parasitology 2nd Proceedings Philadelphia 1965 Sep 7-9 World association for the advancement of veterinary parasitology Edited by E.J.L. Soulsby Academic press,New York;London,1966 Held.. in conjunction with Bicentennial celebrations of medical education in the United States
Bal 39.1737

UNIVERSITY OF PENNSYLVANIA Joint automatic control conference 8th-9th Preprints, Technical papers 1967-68
Eng 41.8208

UNIVERSITY OF PENNSYLVANIA Lattice dynamics The International conference on lattice dynamics Proceedings Copenhagen 1963 Aug 5-9 Edited by R.F. Wallis With the support of the International atomic energy agency 730p Pergamon press,Oxford,1965
Cav 7.0542

UNIVERSITY OF PENNSYLVANIA Photoconductivity conference Atlantic City,N.J. 1954 Nov 4-6 Edited by R.G. Breckenridge Wiley;Chapman and Hall,New York;London,1956
Eng 41.5395

UNIVERSITY OF PENNSYLVANIA Semiconductor surface physics Conference on the physics of semiconductor surfaces Proceedings Philadelphia,Pa. 1956 Jun 4-6 Edited by R. H. Kingston Sponsored by the United States. Office of naval research,the University of Pennsylvania and the Lincoln laboratory University of Pennsylvania,Philadelphia,Pa., 1957
Eng 41.5399

UNIVERSITY OF PENNSYLVANIA Single-crystal films :a conference Proceedings Blue Bell, Penn. 1963 May Edited by Maurice H. Francombe and Hiroshi Sato Sponsored jointly by Philco corporation Pergamon press,Oxford, 1964
Met 25.0177

UNIVERSITY OF PENNSYLVANIA.BICENTENNIAL CONFERENCE Cytology,genetics and evolution :university of Pennsylvania bicentennial conference By M. Demerec and others University of Pennsylvania press, Philadelphia,Pa.,1941
Bot 42.6465

UNIVERSITY OF PENNSYLVANIA.DEPARTMENT OF ENGINEERING MECHANICS International symposium on stress wave propagation in materials proceedings University Park,Pa 1959 Jun 30-Jul 2 Edited by Norman Davids Sponsored by the United States.Army.Office of ordnance research Interscience publishers, New York,1960
A Math 4.0370

UNIVERSITY OF PENNSYLVANIA.JOHNSON RESEARCH FOUNDATION Hemes and hemoprotein : colloquium 3rd Proceedings Philadelphia 1966 Apr 16-17 Edited by Britton Chance and others Academic press,New York,1966
Phys 20.1551

UNIVERSITY OF PENNSYLVANIA BICENTENNIAL CONFERENCE Cytology,genetics and evolution By M. Demerec and others University of Pennsylvania University of Pennsylvania press,Philadelphia,Pa.,1941
Gen 34.0910

UNIVERSITY OF PENNSYLVANIA BICENTENNIAL CONFERENCE Cytology,genetics and evolution By M. Demerec and others University of Pennsylvania press,Philadelphia, Pa.,1941
An 32.3347

UNIVERSITY OF PENNSYLVANIA BICENTENNIAL CONFERENCE Modern psychology By Charles S. Myers and others University of Pennsylvania press,Philadelphia,Pa.,1941
Eng 41.1542

UNIVERSITY OF PITTSBURGH.CENTER FOR THE STUDY OF THERMODYNAMIC PROPERTIES OF MATERIALS Applications of fundamental thermodynamics to metallurgical processes Conference on the thermodynamic properties of materials 1st Proceedings Pittsburgh 1964 Nov 29-Dec 1 Edited by G.R. Fitterer Gordon and Breach, New York,1967
Met 25.2443

UNIVERSITY OF PITTSBURGH.DEPARTMENT OF PSYCHOLOGY Current trends in information theory conference 7th lectures Pittsburgh,Pa. 1953 Feb.20-21 By Brockway McMillan and others Current trends in psychology series xii,188p 22cm University of Pittsburgh press,Pittsburgh,Pa.,1953 7th annual conference on Current trends in psychology
Math 3.0767

UNIVERSITY OF PITTSBURGH.DEPARTMENT OF PSYCHOLOGY Current trends in psychology and the behavioral sciences... :six lectures Lectures Pittsburgh,Pa. 1954 Mar 11-12 By John T. Wilson and others University of Pittsburgh press,Pittsburgh,1954
Psy 31.2597

UNIVERSITY OF READING Semi-conducting materials :a conference Proceedings Reading 1950 Jul 10-15 With the cooperation of the Royal society of London London,1951
Geod 9.0286

UNIVERSITY OF ROCHESTER Galileo quadricentennial Homage to Galileo Papers Rochester,N.Y. 1964 Oct 8-9 Edited by Morton F. Kaplon xii,139p MIT press, Cambridge,Mass.;London,1965
WSM 43.1985

UNIVERSITY OF ROCHESTER High energy nuclear physics Annual Rochester conference on high energy nuclear physics 3rd Proceedings Rochester,N.Y. 1952 Dec 18-20 Edited by H. Pierre Noyes and others 110p Interscience, New York,c.1953 Co-sponsored by the National science foundation
Cav 7.1986

UNIVERSITY OF ROCHESTER High energy nuclear physics Annual Rochester conference on high energy nuclear physics 4th Proceedings Rochester,N.Y. 1954 Jan 25-27 Edited by H. Pierre Noyes and others 159p Rochester,N.Y. 1954 Co-sponsored by the National science foundation.Typescript
Cav 7.1987

UNIVERSITY OF ROCHESTER High energy nuclear physics Annual Rochester conference on high energy nuclear physics 5th Proceedings Rochester,N.Y. 1955 Jan 31-Feb 2 Edited by H.Pierre Noyes and others Sponsored by the International union of pure and applied physics 196p Interscience,New York,1955 Co-sponsored by the National science foundation in cooperation with the Atomic energy commission and the United States office of naval research
Cav 7.1988

UNIVERSITY OF ROCHESTER International conference on high-energy physics 10th Proceedings Rochester,N.Y. 1960 Aug 25-Sep 1 Edited by E.C.G. Sudarshan and others Co-sponsored by the United States.Office of naval research illus xxv,890p University of Rochester,Rochester,N.Y.,1960
Cav 7.2891

UNIVERSITY OF ROCHESTER International conference on particles and fields 1967 Proceedings Rochester,N.Y. 1967 Aug 28-Sep 1 Edited by Carl Richard Hagen and others Interscience,New York,1967
A Math 4.1349

UNIVERSITY OF ROCHESTER.DEPARTMENT OF RADIOLOGY Cinefluorography Annual symposium on cinefluorography 1st Proceedings Rochester,N.Y. 1958 Nov 14-15 Edited by George H.S. Ramsey and others Thomas, Springfield,Ill.,1960
VA 19.0381

UNIVERSITY OF ROCHESTER SCHOOL OF MEDICINE AND DENTISTRY. DIVISION OF NEUROLOGY The Remote effects of cancer on the nervous system a symposium Proceedings Rochester N.Y. 1964 Sep 30-Oct 1 Edited by Russell Brain and Forbes H. Norris Contemporary neurology symposia, 1 Grune and Stratton,New York; London,1965
PGMS 29.0451

UNIVERSITY OF SALFORD Dynamic mass spectrometry European symposium on the time-of-flight mass spectrometry 2nd Proceedings Salford 1969 Jul 1 Edited by D. Price and J.E. Williams Heyden, London,1970
Chem 18.2669

UNIVERSITY OF SALFORD European symposium on time-of-flight mass spectrometry 1st Proceedings Salford 1967 Jul Edited by D. Price and J.E. Williams Pergamon,Oxford, 1969
Met 25.1706

UNIVERSITY OF SALFORD Time-of-flight mass spectroscopy :based on the first European symposium Proceedings Salford 1967 Jul 3-5 Edited by D. Price and J.E. Williams Pergamon press,Oxford,1969
Chem 18.2590

UNIVERSITY OF SHEFFIELD and CHEMICAL SOCIETY Aromaticity :an international symposium Sheffield 1966 Jul 6-8 Chemical society. Special publication,21 Chemical society, London,1967
Chem 18.1225

UNIVERSITY OF SHEFFIELD and CHEMICAL SOCIETY The Transition state :a symposium Sheffield 1962 Apr 3-4 Chemical society. Special publication,16 Chemical society, London,1962
Chem 18.1222

UNIVERSITY OF SHEFFIELD.DEPARTMENT OF FUEL TECHNOLOGY AND CHEMICAL ENGINEERING Advances in magnetohydrodynamics :a colloquium Proceedings Sheffield 1961 Oct Edited by I.A. McGrath and others Pergamon press, London,1963
Eng 41.4504

UNIVERSITY OF SHEFFIELD.DEPARTMENT OF FUEL TECHNOLOGY AND CHEMICAL ENGINEERING Advances in magnetohydrodynamics :a colloquium proceedings Sheffield 1961 Oct. Edited by I.A. McGrath and others Pergamon press, London,1963
A Math 4.0622

UNIVERSITY OF SHEFFIELD.DEPARTMENT OF METALLURGY Deformation under hot working conditions : the conference Proceedings Sheffield 1966 Jul 5-6 Edited by C.M. Sellars and W.J. McG. Tegart Iron and Steel institute,London, 1968
Met 25.2402

UNIVERSITY OF SOUTHAMPTON.DEPARTMENT OF ELECTRONICS Applications of microelectronics :a symposium Southampton 1965 Sep 21-23 Sponsored also by the Institution of electrical engineers Institution of electrical engineers.Conference publication, 14 I.E.E.,London,1965 Reproduced from typescript
Eng 41.5576

UNIVERSITY OF SOUTHERN CALIFORNIA Invariant imbedding :summer workshop Proceedings Los Angeles,Calif. 1970 Jun-Aug Edited by R.E. Bellman and E.D. Denman Lecture notes in operations research and mathematical systems, 52 148p 25cm Springer,Berlin,1971
P Math 2.4296

UNIVERSITY OF SOUTHERN CALIFORNIA.SCHOOL OF MEDICINE Conference on newer respiratory disease viruses Bethesda,Md 1962 Oct 3-5 National institutes of health National institute of allergy and infectious diseases Edited by Clayton G. Loosli American review of respiratory diseases, 88,3 pt 2 American thoracic society,1963
PGMS 29.0321

UNIVERSITY OF ST.ANDREWS Numerical taxonomy : colloquium Proceedings St.Andrews 1968 Sep Edited by Alfred J. Cole Academic press,London,1969
Bot 42.3452

UNIVERSITY OF ST.ANDREWS Pulmonary circulation and respiratory function :a symposium held at Queen's college Dundee 1955 Sep 15-16 University of St.Andrews,St. Andrews,1956
An 32.3562

UNIVERSITY OF ST.ANDREWS.BIOLOGICAL SOCIETY Symposium on aspects of the gene symposium Proceedings St.Andrews 1964 Jan 9-11 Edited by Michael D.B. Burt and Peter W. Barlow Biological journal, 4,supp. University of St.Andrews.Biological society,St. Andrews,1964
Gen 34.0762

UNIVERSITY OF STRATHCLYDE Diffusion processes Thomas Graham memorial symposium Proceedings Glasgow 1969 Sep 16 Edited by John N. Sherwood and others 2 vols Gordon and Breach,London,1971
Met 25.2597

UNIVERSITY OF STRATHCLYDE Soft X-ray band spectra and the electronic structure of metals and materials :a conference Proceedings Strathclyde 1967 Sep 18-21 Edited by Derek J. Fabian Academic press,London;New York, 1968
Met 25.1470

UNIVERSITY OF STRATHCLYDE The Engineering uses of holography :a symposium Proceedings Glasgow 1968 Sep 17-20 Edited by E.R. Robertson and J.M. Harvey Organized in association with the National physical laboratory Cambridge university press, Cambridge,1970
Eng 41.4290

UNIVERSITY OF STRATHCLYDE The Engineering uses of holography :a symposium Proceedings Glasgow 1968 Sep 17-20 Edited by Elliot R. Robertson and James M. Harvey Cambridge university press,Cambridge,1970
An 32.5364

UNIVERSITY OF SURREY International conference on space structures Guildford 1960 Sep Edited by R.M. Davies. Blackwell, Oxford,1967
Eng 41.2739

UNIVERSITY OF SUSSEX Quantum fluids The International symposium on quantum fluids Brighton 1965 Aug 16-20 Edited by D.F. Brewer viii,360p 23cm North-Holland, Amsterdam,1966
Cav 7.2935

UNIVERSITY OF SYDNEY.DEPARTMENT OF ELECTRICAL ENGINEERING Automatic computing machines : conference proceedings Sydney 1951 Aug 7-9 Commonwealth scientific and industrial research organization illus. 220p 25cm Commonwealth scientific and industrial research organization,Melbourne,1952
Math L 5.0703

UNIVERSITY OF TASMANIA Continental drift :a symposium Papers Hobart,Tasmania 1956 Mar Edited by S.Warren Carey University of Tasmania.Geology department.Symposia, 2 Hobart,1958
Geod 9.0083

UNIVERSITY OF TASMANIA.GEOLOGY DEPARTMENT 2nd symposium Papers Continental drift Hobart 1956 Mar Edited by S.Warren Carey University of Tasmania,Hobart,1958
Min 10.0561

UNIVERSITY OF TASMANIA.GEOLOGY DEPARTMENT 4th symposium Dolerite Hobart 1957 July 17-22 Hobart,1958 Typescript
Min 10.0334

UNIVERSITY OF TASMANIA.GEOLOGY DEPARTMENT. SYMPOSIA,2 Continental drift Papers Hobart 1956 Mar Edited by S.Warren Carey University of Tasmania,Hobart,1958
Geog 13.0367

UNIVERSITY OF TENNESSEE.CHEMICAL ENGINEERING DEPARTMENT Symposium on corrosion fundamentals :a series of lectures... Knoxsville 1955 Mar 1-3 Edited by Anton De S. Brasunas and E.E. Stansbury University of Tennessee press,Knoxville,Tenn.,1956
Met 25.1884

UNIVERSITY OF TEXAS Methodology in basic genetics National institutes of health. Genetics study section Edited by Walter J. Burdette Holden-Day,San Francisco,Calif., 1963 Three copies
Gen 34.1012

UNIVERSITY OF TEXAS Quasi stellar sources and gravitational collapse Texas symposium on relativistic astrophysics 1st Proceedings Dallas,Texas 1963 Dec 16-18 By Ivor Robinson and others Southwest center for advanced studies 475p 25cm University of Chicago press,Chicago,Ill.,1965
Cav 7.1985

UNIVERSITY OF TEXAS Quasi-stellar sources and gravitational collapse Texas symposium on relativistic astrophysics 1st proceedings Dallas,Texas 1963 Dec. 16-18 Southwest center for advanced studies Edited by Ivor Robinson and others xvii,475p 25cm University of Chicago press,Chicago,1965
Obs 6.1457

UNIVERSITY OF TEXAS Quasi-stellar sources and gravitational collapse Texas symposium on relativistic astrophysics 1st proceedings Dallas,Texas 1963 Dec.16-18 Southwest center for advanced studies Edited by Ivor Robinson and others xvii,475p 25cm University of Chicago press,Chicago,1965
A Math 4.1161

UNIVERSITY OF TEXAS.BUREAU OF ENGINEERING RESEARCH Texas conference on soil mechanics and foundation engineering 8th Proceedings Austin,Texas 1956 Sep 14-15 University of Texas,Austin,Texas,1956
Eng 41.3164

UNIVERSITY OF TEXAS.DEPARTMENT OF GEOLOGY,and UNIVERSITY OF TEXAS INSTITUTE OF LATIN-AMERICAN STUDIES Latin-American geology conference proceedings Austin,Tex. 1954 Mar.29-30 Edited by Fred M. Bullard map University of Texas,Austin,Tex.,1955
Geol 8.2706

UNIVERSITY OF TEXAS.DEPARTMENT OF GERMANIC LANGUAGES Introduction to modernity :a symposium on eighteenth century thought Edited by Robert Mollenauer 175p University of Texas press,Austin,Texas,1965 Essays presented in 1962 as the 4th in a series of yearly symposia
WSM 43.0649

UNIVERSITY OF TEXAS INSTITUTE OF LATIN-AMERICAN STUDIES and UNIVERSITY OF TEXAS. DEPARTMENT OF GEOLOGY, Latin-American geology conference proceedings Austin,Tex. 1954 Mar.29-30 Edited by Fred M. Bullard map University of Texas,Austin,Tex.,1955
Geol 8.2706

UNIVERSITY OF TOKYO.SUMMER INSTITUTE OF THEORETICAL PHYSICS High energy physics Tokyo summer lectures in theoretical physics 1st Papers Tokyo 1965 Sep 6-17 Pt 2: high energy physics Edited by Gyo Takeda Syokabo;W.A.Benjamin,Tokyo;New York,1966
Cav 7.2318

UNIVERSITY OF TOKYO.SUMMER INSTITUTE OF THEORETICAL PHYSICS Tokyo summer lectures in theoretical physics 2nd Papers Tokyo 1966 Aug 29-Sep 3 Pt 1: dynamical processes in solid state optics Edited by Gyo Takeda and Akihito Fudjii W.A.Benjamin, New York;Tokyo,1967
Cav 7.2316

UNIVERSITY OF TOKYO.SUMMER INSTITUTE OF THEORETICAL PHYSICS Tokyo summer lectures in theoretical physics 2nd Papers Tokyo 1966 Aug 29-Sep 3 Pt 1: dynamical processes in solid state optics Edited by Ryogo Kubo and Hiroshi Kamimura Benjamin,New York,1967
TA 15.0434

UNIVERSITY OF TOKYO.SUMMER INSTITUTE OF THEORETICAL PHYSICS Tokyo summer lectures in theoretical physics 2nd Papers Tokyo 1966 Aug 29-Sep 3 Pt 2: elementary particle physics Edited by Gyo Takeda and Akihiko Fujii Tokyo;New York,1967 Course under chairmanship of Kubo,Ryogo
Cav 7.2317

UNIVERSITY OF TORONTO The Symposium on colloid chemistry 6th Papers presented Toronto 1928 Jun 14-16 Edited by Harry Boyer Weiser Colloid symposium monograph, 6 Chemical catalog company,New York,1928 Colloid symposium monograph known also as Colloid symposium annual
Bioch 33.1474

UNIVERSITY OF TORONTO.INSTITUTE FOR AEROSPACE STUDIES International symposium on rarefied gas dynamics 4th proceedings Toronto 1964 Jul.14-17 Vol 1-2 Edited by J.H.de Leeuw Advances in applied mechanics.Supplement, 3 2 vols Academic press,New York,London,1965-66
A Math 4.0532

UNIVERSITY OF VIRGINIA Low-luminosity stars Symposium on low-luminosity stars Proceedings Charlottesville,Va. 1968 Edited by Shiv S. Kumar 542p Gordon and Breach,New York,1969
Obs 6.3627

UNIVERSITY OF WALES Thin walled steel structures :a symposium Swansea 1967 Sep 11-14 Edited by K.C. Rockey and H.V. Hill Crosby Lockwood,London,1969
Eng 41.2747

UNIVERSITY OF WARWICK.CONTROL THEORY CENTRE Stability of stochastic dynamical systems : international symposium Proceedings Coventry 1972 Jul 10-14 Edited by Ruth F. Curtain Lecture notes in mathematics, 254 ix,332p 26cm Springer,Berlin,1972
P Math 2.4569

UNIVERSITY OF WASHINGTON The Heat transfer and fluid mechanics institute Proceedings Seattle,Wash. 1962 Jun 13-15 Edited by Edward Ehlers and others Co-sponsored by Stanford university 293p Stanford university press,Stanford,Calif.,1962
Cav 7.2422

UNIVERSITY OF WASHINGTON.COLLEGE OF FISHERIES Biological and economic aspects of fisheries management :a conference Proceedings Seattle,Wash. 1959 Feb 17-19 Edited by James A. Crutchfield figures vi,160p University of Washington,Seattle,Wash.,1959
Sco 14.4563

UNIVERSITY OF WATERLOO International symposium on research in co-current gas-liquid flow Preprints Waterloo,Ontario 1968 Sep 18-19 Vol 1-2 National research council of Canada and Canadian society for chemical engineering 2 vols University of Waterloo,Waterloo,Ontario,1968 Mimeographed
Chem E 24.1520

UNIVERSITY OF WATERLOO Recent progress in combinatorics Waterloo conference on combinatorics 3rd Proceedings Waterloo, Ont. 1969 May 21-30 xiv,341p 24cm Academic press,New York;London,1969
Math S 3.1660

UNIVERSITY OF WATERLOO.DEPARTMENT OF STATISTICS Foundations of statistical inference :a symposium Proceedings Waterloo,Ont. 1970 Mar 31-Apr 9 Edited by V.P. Godambe and D.A. Sprott viii,519p 25cm Holt,Toronto,1971
Math S 3.1848

UNIVERSITY OF WESTERN AUSTRALIA.FACULTY OF ENGINEERING Hydraulics and fluid mechanics :Australasian conference 1st Proceedings Nedlands 1962 Dec 6-13 Edited by Richard Silvester illus viii, 503p Pergamon press,Oxford,1964
Chem E 24.0239

UNIVERSITY OF WISCONSIN A Symposium on hydrobiology Madison,Wis. 1940 Sep 4-6 By James G. Needham and others University of Wisconsin,Madison,Wis.,1941 Held under the auspices of the University of Wisconsin and of the Wisconsin alumni research foundation
Bal 44.6459

UNIVERSITY OF WISCONSIN A Symposium on respiratory enzymes Addresses Madison,Wis, 1941 Sep 11-17 By Otto Meyerhof and others University of Wisconsin press,Madison,Wis., 1942
Bioch 33.1661

UNIVERSITY OF WISCONSIN A Symposium on respiratory enzymes Lectures and discussions Madison,Wis. By Otto Meyerhof and others University of Wisconsin,Madison,Wis.,1942 reprinted 1948 Sponsored jointly by the universities of Wisconsin and Chicago
Bal 39.1344

UNIVERSITY OF WISCONSIN A Symposium on steroid hormones Papers Madison,Wis. 1950 Edited by Edgar S. Gordon University of Wisconsin press,1950 Published in celebration of the hundredth anniversary of the founding of the University of Wisconsin
Bioch 33.0455

UNIVERSITY OF WISCONSIN A Symposium on the use of isotopes in biology and medicine Madison,Wis. 1947 Sep 10-13 By Hans T. Clarke and others University of Wisconsin press,Madison,1948
Bioch 33.0283

UNIVERSITY OF WISCONSIN Developing cell systems and their control :a symposium Madison,Wis. 1959 Society for the study of development and growth Edited by Dorothea Rudnick Society for the study of development and growth.Symposia, 18 Ronald press,New York,1960
Bioch 33.1010

UNIVERSITY OF WISCONSIN Newcastle disease virus:an evolving pathogen :international symposium Proceedings 1963 Edited by Robert Paul Hanson University of Wisconsin press,Madison,Wis.,1964
Gen 34.0728

UNIVERSITY OF WISCONSIN Statistical computation :a conference Proceedings Madison,Wis. 1969 Apr 28-30 Edited by Roy C. Milton and John A. Nelder bibliog. Academic press,New York;london,1969
Math S 3.1503

UNIVERSITY OF WISCONSIN The National symposium on colloid chemistry 1st Papers and discussions Madison,Wis. 1923 Jun 12-15 Edited by J.Howard Mathews Colloid symposium monograph, 1 Department of chemistry,university of Wisconsin,Madison,Wis., 1923
Bioch 33.1469

UNIVERSITY OF WISCONSIN and UNIVERSITY OF CHICAGO A Symposium on respiratory enzymes Addresses Madison,Wisc. 1941 Sep 11-17 University of Wisconsin press,Madison, Wisc.,1942
Chem 18.1249

UNIVERSITY OF WISCONSIN.GRADUATE SCHOOL Non-compact groups in particle physics conference proceedings Wisconsin,Me. 1966 May 5-6 Edited by Yutze Chow 216p Benjamin,New York,1966
A Math 4.0890

UNIVERSITY OF WISCONSIN.MATHEMATICS RESEARCH CENTER Mathematics research center symposium on contributions to nonlinear functional analysis Proceedings Madison, Wisc. 1971 Apr 12-14 xii,672p 24cm Academic press,New York;London,1971
P Math 2.4288

UNIVERSITY OF WISCONSIN.MATHEMATICS RESEARCH CENTER Nonlinear functional analysis and applications Mathematics research center advanced seminar Proceedings Madison,Wisc. 1970 Oct 12-14 vii,586p 24cm Academic press,New York;London,1971
P Math 2.4286

UNIVERSITY OF WISCONSIN.PHI DELTA KAPPA CHAPTER Improving experimental design and statistical analysis Phi Delta Kappa annual symposium on educational research 7th Discussions Madison,Wisc. 1966 Rand-McNally education series Rand-McNally, Chicago,Ill.,1967
Psy 31.1123

UNIVERSITY OF WISCONSIN.STATISTICS DEPARTMENT Spectral analysis of time series :advanced seminar Proceedings Madison,Wis. 1966 Oct 3-5 Edited by Bernard Harris United States.Army.Mathematics research center. Publications, 18 illus. x,319p 24cm Wiley,New York,1967
Math S 3.1456

UNIVERSITY OF WISCONSIN.THEORETICAL CHEMISTRY INSTITUTE Perturbation theory and its application in quantum mechanics seminar proceedings Madison,Wis. 1965 Oct 4-6 United States army.Mathematics research center Edited by Calvin H. Wilcox 428p John Wiley and sons,New York;London,1966
A Math 4.0809

UNIVERSITY OF WISCONSIN.THEORETICAL CHEMISTRY INSTITUTE and UNITED STATES.ARMY. MATHEMATICS RESEARCH CENTER Perturbations theory and its applications in quantum mechanics an advanced seminar Papers Madison,Wisc. 1965 Oct 4-6 Edited by Calvin H. Wilcox J.Wiley,New York,1966
Chem 18.0692

UNIVERSITY PARK,PA 1959 International symposium on stress wave propagation in materials proceedings University of Pennsylvania.Department of engineering mechanics Edited by Norman Davids Sponsored by the United States.Army.Office of ordnance research Interscience publishers, New York,1960
A Math 4.0370

UNIVERSITY PARK,PA. 1959 International symposium on stress wave propagation in materials Pennsylvania state university. Department of engineering mechanics Edited by Norman Davids Sponsored by the United States.Army.Office of ordnance research Interscience,New York,1960
Eng 41.3770

UNIVERSITY PARK,PA. 1962 The Frontal granular cortex and behavior Pennsylvania state university Edited by K. Akert and J.M. Warren McGraw-Hill,New York;London,1964
Psy 31.0654

UNIVERSITY PARK,PA. 1962 The Frontal granular cortex and behaviour :a symposium Edited by J.M. Warren and K. Akert Supported by grants from the National science foundation McGraw-Hill,New York,1963
An 32.4656

UNIVERSITY PARK,PA. 1963 Classical conditioning :a symposium Papers Edited by William F. Prokasy Century psychology series Appleton-Century-Crofts,New York,1965 Organised by the editor
Psy 31.3015

UNIVERSITY PARK,PA. 1965 Electron and laser beam symposium 7th Proceedings Pennsylvania state university Sponsored also by the Alloyd general corporation Pennsylvania state university,University Park, Pa.,1965
Eng 41.5617

UNIVERSITY PARK,PA. 1965 International symposium on forest hydrology Proceedings National science foundation Edited by William E. Sapper and Howard W. Lull Pergamon press,Oxford,1967 A National Science Foundation advanced science seminar
Geog 13.0795

UNIVERSITY PARK,PENN. 1961 Conference on carbon 5th Proceedings Clarendon press, Oxford,1965
Met 25.0158

UNIVERSITY PARK,PENN. 1961 The Conference on carbon 5th Proceedings Vol 1 Pennsylvania state university and American carbon committee Edited by S. Mrozowski Conference on carbon.Proceedings, 4 Pergamon press,Oxford,1962 Proceedings appeared in two volumes
TA 15.0216

The UNIVERSITY TEACHING OF INTERNATIONAL RELATIONS Windsor 1950 Mar 16-20 Edited by Geoffrey L. Goodwin Held under the auspices of the International studies conference Blackwell,Oxford,1951
Geog 13.0171

UNSTEADY FLOW Symposium on measurement in unsteady flow :presented at the ASME hydraulic division conference Worcester,Mass. 1962 May 2-23 American society of mechanical engineers.Hydraulic division American society of mechanical engineers,New York,1962
Eng 41.6713

UPJOHN COMPANY Mechanism of cell and tissue damage produced by immune reactions International symposium on immunopathology 2nd Papers Brook Lodge,Mich. 1961 Edited by Pierre Grabar and Peter Miescher Schwabe,Basle;Stuttgart,1962
Path 30.2336

UPJOHN COMPANY Microbial protoplasts, spheroplasts and L-forms :a conference Proceedings Kalamazoo,Mich. 1966 Nov 10-11 Edited by Lucien B. Guze Williams and Wilkins,Baltimore,1968
Bioch 33.1142

UPPER ATMOSPHERE Physics and medicine of the upper atmosphere Symposium on the physics and medicine of the upper atmosphere Proceedings San Antonio,Tex. 1951 Nov 6-9 Lovelace foundation for medical education and research Edited by Clayton S. White and Otis O. Benson Sponsored by School of aviation medicine.Randolph Field illus xxix,611p University of New Mexico press,Albuquerque,N.M. 1952
Nap 11.0635

UPPER ATMOSPHERE Transactions of the Rome meeting International association of terrestrial magnetism and electricity a meeting Rome 1954 Sep 14-25 Edited by V. Laursen IAGA Bulletin, 15 406p Horsholm, Copenhagen,1957 The meeting took place during the 10th general assembly of the international union of geodesy and geophysics
Nap 11.0287

UPPER ATMOSPHERE EMISSIONS Theoretical interpretation of upper atmosphere emissions : a symposium Paris 1962 Jun.25-29 International astronomical union and International union of geodesy and geophysics Edited by D.R. Bates International astronomical union.Symposium, 18 264p Pergamon press,New York,1963 Reprinted from'Planetary and space science',Vol.10 1963
Obs 6.0142

UPPER MANTLE Phase transformations and the earth's interior :a conference Canberra 1969 Jan 6-10 International upper mantle committee Australiann academy of science Edited by A.E. Ringwood and D.H. Green North-Holland,Amsterdam,1970
Min 10.1459

UPPER MANTLE The Upper mantle symposium papers New Delhi 1964 Dec. International union of geological sciences,International upper mantle committee Edited by Charles H. Smith and Theodor Sorgenfrei Det Berlingske Bogtrykkeri,Copenhagen,1965
Geol 8.2191

UPPER MANTLE The World rift system symposium report Ottawa 1965 Sep 4-5 International upper mantle committee sponsored by Canada. Department of mines and technical surveys Geological survey of Canada.Paper,66-14 Ottawa,1967
Geol 8.4451

UPPER MANTLE PROJECT International symposium on geophysical theory and computers 3rd Proceedings Cambridge 1966 Jun 27-Jul 5 International upper mantle committee Sponsored by the International union of geodesy and geophysics Geophysical journal, 13;1-3 375p Blackwell,Oxford,1967
Geod 9.0694

UPPER MANTLE PROJECT The East African rift system :a seminar Papers Nairobi 1965 Apr 12-17 International upper mantle committee Sponsored by the International union of geodesy and geophysics Nairobi,1965
Geod 9.0451

UPPER MANTLE PROJECT The Symposium on non-elastic processes in the mantle Proceedings Newcastle-upon-Tyne 1966 Feb 21-25 International upper mantle committee Edited by D.C. Tozer With financial assistance of the International union of geodesy and geophysics Geophysical journal, 14;1-4 Blackwell,Oxford,1967
Geod 9.0695

UPPER MANTLE PROJECT The World rift systems : a symposium Report Ottawa 1965 Sep 4-5 National research council,Canada and International upper mantle committee Edited by T.N. Irvine Sponsored by the International union of geodesy and geophysics Geological survey of Canada.Paper, 66-14 Queen's printer,Ottawa,1967
Geod 9.0138

The UPPER MANTLE SYMPOSIUM papers New Delhi 1964 Dec. International union of geological sciences,International upper mantle committee Edited by Charles H. Smith and Theodor Sorgenfrei Det Berlingske Bogtrykkeri,Copenhagen,1965
Geol 8.2191

UPPSALA 1953 Symposium on psychological factor analysis Universitet i Uppsala. Institute of statistics Edited by P. Whittle Nordisk psykologi.Monograph series, 3 91p 23cm Almqvist and Wiksell,Stockholm,1953
Math 3.0777

UPPSALA 1957 Systematics of to-day :a symposium...in commemoration of the 250th anniversary of the birth of Carolus Linnaeus Edited by Olov Hedberg Uppsala universitets arsskrift,1958:6 Uppsala,1958
BG 38.3195

UPPSALA 1957 Systematics of today :a symposium Proceedings Universitet i Uppsala Edited by Olov Hedberg Uppsala universitets.Arsskrift, 6 Lundequistska bokhandeln;Otto Harrassowitz,Uppsala;Wiesbaden, 1959 In commemoration of the 250th anniversary of the birth of Carolus Linnaeus
Bot 42.3424

UPPSALA 1961 International symposium on free radicals 5th Preprint of papers United States.Army.European research office and United States.Air force.Office of scientific research 1961
Chem 18.2625

UPPSALA 1963 and STOCKHOLM The Structure and metabolism of the pancreatic islets :an international Wenner-Gren symposium Proceedings Wenner-Gren center Edited by S. E. Brolin and others Wenner-Gren center. International symposium series, 3 Pergamon press,Oxford,1964
Bioch 33.0485

UPSALA 1954 and STOCKHOLM 1954 Scandinavian society for clinical chemistry and clinical physiology 6th meeting Scandinavian journal of clinical and laboratory investigation, 7,supp.20 Upsala, 1955
Radioth 35.1971

UPSALA 1970 External galaxies and quasi-stellar objects International astronomical union Edited by D.S. Evans International astronomical union.Symposium, 44 549p Reidel,Dordrecht,1972
TA 15.0658

UPTAKE AND DISTRIBUTION OF ANAESTHETIC AGENTS New York 1962 Apr 23-24 Edited by E.M. Papper and Richard J. Kitz Sponsored by the National research council.Committee on anaesthesia McGraw-Hill,New York,1963
Pha 16.0043

UPTON.N?Y. 1968 Structure,function and evolution in proteins report of symposium Vol 1-2 Brookhaven national laboratory. Biology department Brookhaven symposia in biology, 21 BNL 50116 2 vols brookhaven national laboratory,Upton,N.Y.,1969
Gen 34.0671

UPTON,N.Y. 1956 Dynamics of proliferating tissues :a symposium Edited by Dorothy Price Developmental biology conference series,1956 University of Chicago press,Chicago,Ill.,1958
An 32.3285

UPTON,N.Y. 1956 Genetics in plant breeding report of symposium Brookhaven national laboratory.Biology department Brookhaven symposia in biology, 9 Brookhaven national laboratory,Upton,N.Y.,1956
Gen 34.1152

UPTON,N.Y. 1958 The Photochemical apparatus;its structure and function :a symposium Report Brookhaven national laboratory.Biology department Brookhaven symposia in biology, 11 BWL512(C-28) Brookhaven national laboratory.Biology department,Upton,N.Y.,1959
Bioch 33.1290

UPTON,N.Y. 1959 Bioenergetics : considerations of processes of absorption stabilization, transfer and utilization.A symposium Proceedings Brookhaven national laboratory Edited by Leroy George Augenstine Sponsored by the United States atomic energy commission Radiation research.Supplement, 2 Academic press,New York;London,1960
Bioch 33.0716

UPTON,N.Y. 1959 Structure and function of genetic elements :symposium Report Brookhaven national laboratory.Biology department Brookhaven symposia in biology, 12 BNL558(C-29) Brookhaven national laboratory.Biology department,Upton,N.Y.,1959
Bioch 33.1291

UPTON,N.Y. 1959 Structure and function of genetic elements report of a symposium Brookhaven national laboratory.Biological department Brookhaven symposia in biology, 12 BNL 558 Brookhaven national laboratory, Upton,N.Y.,1959
Gen 34.0674

UPTON,N.Y. 1960 Protein structure and function :a symposium Report Brookhaven national laboratory.Biology department Brookhaven symposia in biology, 13 BNL608(C-30) Brookhaven national laboratory.Biology department,Upton,N.Y.,1960
Bioch 33.1292

UPTON,N.Y. 1960 Protein structure and function :symposium Brookhaven national laboratory.Biology department Brookhaven symposia in biology,13 BNL-608 Brookhaven national laboratory,biology dep.,Upton,N.Y., 1960
Pha 16.0120

UPTON,N.Y. 1961 Fundamental aspects of radiosensitivity report of a symposium Brookhaven national laboratory.Biology department Brookhaven symposia in biology, 14 BNL 675 Brookhaven national laboratory, Upton,N.Y.,1961
Gen 34.1115

UPTON,N.Y. 1962 Enzyme models and enzyme structure :symposium Report Brookhaven national laboratory.Biology department Brookhaven symposia in biology, 15 BNL738(C-34) Brookhaven national laboratory.Biology department,Upton,N.Y.,1962
Bioch 33.1293

UPTON,N.Y. 1963 Biological effects of neutron and protein irradiations Symposium on biological effects of neutron and proton irradiations Proceedings 1-2 International atomic energy agency Held at Brookhaven national laboratory International atomic energy agency.Proceedings series 2 vols International atomic energy agency, Vienna,1964
Radioth 35.1168

UPTON,N.Y. 1963 Meristems and differentiation :symposium Report Brookhaven national laboratory.Biology department Brookhaven symposia in biology, 16 BNL805(C-38) Brookhaven national laboratory.Biology department,Upton,N.Y.,1964
Bioch 33.1294

UPTON,N.Y. 1963 Meristems and differentiation report of symposium Brookhaven national laboratory Brookhaven symposia in biology, 16 BNL 805 Brookhaven national laboratory,Upton,N.Y.,1964
Gen 34.1414

UPTON,N.Y. 1964 Subunit structure of proteins :biochemical and genetic aspects. Symposium Report Brookhaven national laboratory.Biology department Brookhaven symposia in biology, 17 BNL869(C-40) Brookhaven national laboratory.Biology department,Upton,N.Y.,1964
Bioch 33.1295

UPTON,N.Y. 1964 Subunit structure of proteins,biochemical and genetic aspects report of symposium Brookhaven national laboratory Brookhaven symposia in biology, 17 Brookhaven national laboratory,Upton,N.Y., 1964
Gen 34.0769

UPTON,N.Y. 1965 Genetic control of differentiation :report of a symposium Brookhaven national laboratory.Biology department Brookhaven symposia in biology, 18 Brookhaven national laboratory,Upton,N.Y., 1965
Gen 34.0499

UPTON,N.Y. 1965 Genetic control of differentiation :symposium Report Brookhaven national laboratory.Biology department Brookhaven symposia in biology, 18 BNL931(C-44) Brookhaven national laboratory.Biology department,Upton,N.Y.,1965
Bioch 33.1296

UPTON,N.Y. 1965 Structure and function of polypeptide hormones;insulin Symposium on insulin Proceedings Edited by P.G. Katsoyannis and I.L. Schwartz Sponsored by the Brookhaven national laboratory.Medical department American journal of medicine, 40, no. 5 Yorke medical group.Publication New York,1966
Bioch 33.0557

UPTON,N.Y. 1966 Energy conversion by the photosynthetic apparatus :symposium Report Brookhaven national laboratory.Biology department Brookhaven symposia in biology, 19 BNL989(C-48) Brookhaven national laboratory.Biology department,Upton,N.Y.,1967
Bioch 33.1297

UPTON,N.Y. 1967 Recovery and repair mechanics in radiography :symposium Report Brookhaven national laboratory.Biology department Brookhaven symposia in biology, 20 BNL50058)C-51) Brookhaven national laboratory.Biology department,Upton,N.Y.,1968
Bioch 33.1298

UPTON,N.Y. 1967 Recovery and repair mechanisms in radiobiology report of symposium Brookhaven national laboratory.Biology department Brookhaven symposia in biology, 20 BNL 50058 Brookhaven national laboratory,Upton,N.Y.,1968
Gen 34.1128

UPTON,N.Y. 1968 Structure,function and evolution in proteins :symposium Report Brookhaven national laboratory.Biology department Brookhaven symposia in biology, 21 BNL50116(C-53) Brookhaven national laboratory.Biology department,Upton,N.Y.,1969
Bioch 33.1299

URANIUM AND GRAPHITE :a symposium Proceedings London 1962 Mar 20-21 Institute of metals Institute of metals. Monograph, 27 Institute of metals,London, 1962
Met 25.1554

URBAN DIFFUSION MODELS Environmental protection Symposium on multiple-source urban diffusion models Proceedings 1970 United States.Environmental protection agency National air pollution control administration North Carolina consortium on air pollution Air pollution control office.Publication,AP-86 US government printing office,Washington, D.C.,1970
Eng 41.8616

URBAN GEOGRAPHY I.G.U.symposium in urban geography Proceedings Lund 1960 Aug 15-19 Edited by Knut Norberg Lund studies in geography.Ser.B.Human geography,24 Lunds universitet,Lund,1962 Held at the 19th International geographical congress
Geog 13.1944

URBAN MOTORWAYS :a conference British road federation British road federation, London,1957
Eng 41.3298

URBANA,ILL. 1955 Shock and vibration instrumentation;a review of the latest developments American society of mechanical engineers.Applied mechanics division New York,1956
Eng 41.6324

URBANA,ILL. 1959 Macromolecular complexes : a symposium Edited by M.V. Edds Society of general physiologists.Annual symposia, 6 illus. Ronald press,New York,1961
Radioth 35.1408

URBANA,ILL. 1961 Chemical physics of nonmetallic crystals :Northwestern university 1961 international conference Papers American institute of physics.Divisions of chemical physics and solid state physics,and, American chemical society.Division of physical chemistry Benjamin,New York,1962 First published as a supplement to the 'Journal of applied physics',January 1962
Chem 18.2385

URBANA,ILL. 1967 Symposium on the theory of finite groups Lectures American mathematical society Bibliog. 94p 25cm American mathematical society,Providence,R.I., 1969
P Math 2.3581

URBANA,ILL. 1968 Spores 4 International spore conference 4th American society for microbiology Edited by L.Leon Campbell Under the auspices of the University of Illinois American society for microbiology, 1967
Gen 34.0711

URBANA,ILL. 1972 Annual workshop on microprogramming 5th Preprints Association for computing machinery Institute of electrical and electronics engineers.Computer society Bibliog,diagrs 98p n.p.,1972 Includes 'An annotated bibliography on microprogramming,late 1969-early 1972'
Math L 5.3636

URBINO 1962 Topologia differenziale 1 ciclo By J. Cerf and others Centro internazionale matematico estivo 27cm Institute matematica dell'universita,Rome,1962
P. Math 2.0243

URINARY TRACT INFECTION National symposium on urinary tract infection 1st Proceedings London 1968 Apr Edited by Francis O'Grady and William Brumfitt Oxford medical publications Oxford university press, London,1968
Surg 23.0010

URINARY TRACT INFECTION National symposium on urinary tract infection 1st Proceedings London 1968 April Edited by Francis O'Grady and William Brumfitt Sponsored by the Beecham research laboratories Oxford university press,London,1968
PGMS 29.0238

URSI GENERAL ASSEMBLY 7th-11th Proceedings Vol 6-10 International scientific radio union 1946-54
Nap 11.0288

URSI GENERAL ASSEMBLY 13th Papers Monograph on radio noise of terrestrial origin London 1960 Sep 6 Commission 4 on radio noise of terrestrial origin International scientific radio union Edited by F. Horner URSI monographs series 202p Elsevier, Amsterdam,1962
Nap 11.0268

URSI GENERAL ASSEMBLY 13th Papers of a session Monograph on ionospheric radio London 1960 Sep. International scientific radio union Edited by W.J.G. Beynon URSI monographs series 264p Elsevier,Amsterdam; New York,1962
Nap 11.0052

URSI GENERAL ASSEMBLY 14th Proceedings Progress in radio science 1960-63 Tokyo 1963 Sep 9-20 Vol 3: the ionosphere International scientific radio union. Commission on the ionosphere Edited by G.M. Brown 196p Elsevier,Amsterdam;London,1965
Nap 11.0086

The USE OF COMPUTERS IN PRODUCTION CONTROL conference proceedings London 1959 Apr 2 Central London productivity association 69p 32cm Central London productivity association,London,1959
Math L 5.0843

The USE OF RADIOAUTOGRAPHY IN INVESTIGATING PROTEIN SYNTHESIS :a symposium Montreal 1965 Edited by C.P. Leblond and Katherine Brehme Warren Sponsored by the International society for cell biology International society for cell biology. Symposia, 4 Academic press,New York;London, 1965
Bioch 33.0286

USE OF RADIOISOTOPES AND SUPERVOLTAGE RADIATION IN RADIOTELETHERAPY;PRESENT STATUS AND RECOMMENDATION Report and background information for a study group convened by the IAEA and WHO Vienna 1959 Aug 3-5 International atomic energy agency World health organization International atomic energy agency,Vienna,1960
Radioth 35.1146

The USE OF REDUNDANCY IN SYSTEM DESIGN :a symposium Northampton 1964 Feb 14 Society of instrument technology Society of instrument technology,1964
Eng 41.5963

USE OF SECONDARY SURFACES FOR HEAT TRANSFER WITH CLEAN GASES :a symposium Proceedings 1960 Institute of mechanical engineers London,1961
Eng 41.7167

The USE OF THIN FILMS IN PHYSICAL INVESTIGATIONS a NATO advanced study institute held at the Imperial college of science and technology London 1965 Jul 19-24 Edited by J.C. Anderson Sponsored by the North Atlantic treaty organization Academic press, London, 1966
Eng 41.4379

The USE OF VITAL AND HEALTH STATISTICS FOR GENETIC AND RADIATION STUDIES seminar Proceedings Geneva 1960 Sep 5-9 United nations World health organization United Nations publications, 61, XVII.8 United Nations, New York, 1962
Gen 34.1120

The USES OF VACUUM IN METALLURGY Conference proceedings Baikov institute of metallurgy. Commission on the physico-chemical principles of steelmaking Translated by E. Bishop from the Russian Oliver and Boyd, Edinburgh; London, 1964
Met 25.1674

USPEKHI BIOLOGICHESKIKH NAUK V S.S.S.R. ZA 25 LET, 1917-42 :symposium Akademiya nauk S. S.S.R. Otdelenie biologicheskikh nauk Edited by L.A. Orbeli 356p 26cm Moscow, 1945
Philos 1.1394

USTREDNI USTAV GEOLOGICKY, PRAGUE Prager arbeitstagung uber die stratigraphie des Silurs und des Devons (1958) :a symposium papers Prague 1958 Aug.30-Sep.6 Edited by Josef Svoboda Nakladetalstvi ceskoslovenske akademie ved, Prague, 1960
Geol 8.2416

UTILIZATION OF HEAT RESISTANT ALLOYS a symposium in honor of Albert Easton White Ann Arbor 1954 Mar 11-12 American society for metals American society for metals, Cleveland, Ohio, 1921
Met 25.0998

UTILIZATION OF NITROGEN AND ITS COMPOUNDS BY PLANTS :a symposium Reading 1958 Sep 15-19 Society for experimental biology Society for experimental biology. Symposia, 13 Cambridge university press, Cambridge, 1959
Bioch 33.1358

UTRECHT 1937 Hydrophobic colloids : symposium on the dynamics of hydrophobic suspensions and emulsions Under the auspices of Nederlandsche chemische vereeniging. Colloid chemistry section Amsterdam, 1938 Reprinted from "Chemisch Weekblad" 1938
Col S 12.0100

UTRECHT 1954 Quaternary changes in level, especially in the Netherlands :a symposium Organized by the Nederlanske geologisch-mijnbouw kundig genootschap. Geologische sectie Geologie en mijnbouw, ns, 16, no.6 The Hague, 1954
Geog 13.0371

UTRECHT 1960 Concept and the role of the model in mathematics and natural and social sciences International union of history and philosophy of science colloquium Proceedings International union of history and philosophy of science. Division of philosophy of science Synthese library 194p Reidel, Dordrecht, 1961 Colloquium organized by H. Freudenthal
WSM 43.0785

UTRECHT 1960 The Concept and the role of the model in mathematics and natural and social sciences :a colloquium Proceedings International union of history and philosophy of science. Division of philosophy of science Edited by B.H. Kazemier and D. Vuysje 164p Reidel, Dordrecht, 1961
WSM 43.1797

UTRECHT 1963 Solar spectrum The Symposium on the solar spectrum Proceedings Edited by C.de Jager D.Reidel, Dordrecht, 1965 Dedicated to M.G.J.Minnaert on the occasion of his 70th birthday
TA 15.0199

UTRECHT 1963 The Solar spectrum symposium proceedings Edited by C.de Jager port 417p D.Reidel, Dordrecht, 1965 Proceedings dedicated to M.Minnaert
Obs 6.0876

UTRECHT 1963 The Solar spectrum symposium proceedings Rijksuniversiteit te Utrecht Edited by C.de Jager Astrophysics and space science library D.Reidel, Dordrecht-Holland, 1965
A Math 4.1163

UTRECHT 1964 Abundance determinations in stellar spectra :a symposium International astronomical union Edited by Hans Hubenet International astronomical union. Symposium, 26 374p London; New York, 1966
Obs 6.0843

UTRECHT 1964 Abundance determinations in stellar spectra :a symposium Proceedings International astronomical union International astronomical union. Symposium, 26 Academic press, London, 1966
TA 15.0034

UTRECHT 1964 Abundance determinations in stellar spectra :a symposium International astronomical union Edited by Hans Hubenet International astronomical union. Symposium, 26 374p Academic press, London; New York, 1966
A Math 4.1168

VACANCIES AND INTERSTITIALS IN METALS International conference on vacancies and interstitials in metals Julich 1968 Sep 23-28 Kernforschungslange Julich Arbeitsgemeinschaft metallphysik Edited by A. Seeger and others North-Holland, Amsterdam, 1970
Met 25.2414

VACANCIES AND INTERSTITIALS IN METALS : conference Proceedings Julich 1968 Sep 23-28 Edited by A. Seeger and others Bibliog 1074p 22cm North-Holland, Amsterdam, 1970
Cav 7.2686

VACANCIES AND OTHER POINT DEFECTS IN METALS AND ALLOYS a symposium Harwell, Berks 1917 Dec 10 Institute of metals Institute of metals. Monograph and report series, 23 Institute of metals, London, 1958
Met 25.1233

VACUUM DEGASSING OF STEEL Iron and steel institute annual general meeting 1965 Papers and discussions London 1965 May 5-6 Iron and steel institute Iron and steel institute.Special report, 92 Iron and steel institute,London,1965
Met 25.0372

VACUUM METALLURGY New York 1957 Jun 10-14 New York university.Department of metallurgical engineering Edited by Rointan F. Bunshah Reinhold.Chapman and Hall,New York;London,1958 Lectures presented during the course on vacuum metallurgy
Met 25.1630

VACUUM METALLURGY CONFERENCE Transactions New York 1959 Jun 1-3 New York university. Department of metallurgical engineering Edited by Rointan F. Bunshah New York university press,New York,1960
Met 25.2259

VACUUM METALLURGY SYMPOSIUM Papers Boston, Mass. 1954 Oct 6-7 Electrochemical society. Electrothermics and metallurgy division Edited by John M. Blocher Electrochemical society inc.,Boston,Mass.,1955
Met 25.2258

VACUUM PHYSICS :a symposium Birmingham 1950 Jun 27-28 Institute of physics.Midland branch Journal of scientific instruments. Supplement, 1 80,viiip Institute of physics,London,1951
Chem E 24.1353

VACUUM SCIENCE International vacuum congress 3rd Transactions Stuttgart 1965 Jun 28-Jul 2 1: invited papers International union for vacuum science.German national committee Edited by H. Adam Pergamon, Oxford,1966 Papers mainly in English,with summaries in French and German.Title pages in English,French and German
Met 25.1691

VACUUM SCIENCE AND TECHNOLOGY International vacuum congress 4th Manchester 1968 Apr 17-20 Pt 1-2 Joint British committee for vacuum science and technology International union for vacuum science, technique and applications Institute of physics and the Physical society.Conference series, 5-6 2 vols Institute of physics, London,c1968
Met 25.2834

VACUUM TECHNIQUE Meeting of the German society for vacuum technique Proceedings Heidelberg 1962 Sep 18-21 Deutsche gesellschaft fur vakuumtechnik Macmillan; Pergamon press,London;Oxford,1965
Cav 7.1062

VACUUM TECHNOLOGY International vacuum congress 3rd Transactions Stuttgart 1965 Jun 28-Jul 2 Vol 1: invited papers International union for vacuum science, technique and applications Edited by H. Adam 153p Pergamon press,Oxford,1966 Cover title'Advances in vacuum science and technology'
Cav 7.0015

VACUUM TECHNOLOGY National symposium on vacuum technology 7th Transactions Cleveland,Ohio 1960 Oct 12-14 American vacuum society Edited by C.Robert Meissner Pergamon press,Oxford,1961 Earlier symposia published by the Committee on vacuum techniques
Met 25.1669

VACUUM TECHNOLOGY National symposium on vacuum technology Transactions Chicago,Ill 1956 Oct 10-11 Committee on vacuum techniques Edited by Edmond S. Perry and John H. Durant Pergamon press,London,1957 Later symposia published by the American vacuum society
Met 25.1629

VACUUM TECHNOLOGY National vacuum symposium 8th Transactions Washington,D.C. 1961 Oct 16-19 Vol 1-2 American vacuum society and International organization for vacuum science and technology Edited by Luther E. Preuss 2 vols Pergamon press, Oxford,1962 Combined with the International congress on vacuum science and technology,2nd
Cav 7.2376

VACUUM TECHNOLOGY The National symposium on vacuum technology 7th Transactions Cleveland,Ohio 1960 Oct 12-14 American vacuum society Edited by C.Robert Meissner 427p Pergamon press,Oxford,1961 8th symposium called 'National vacuum symposium' q. v.
Cav 7.2375

VACUUMKONGRESS 1962 Vacuum technique Meeting of the German society for vacuum technique Proceedings Heidelberg 1962 Sep 18-21 Deutsche gesellschaft fur vakuumtechnik Macmillan;Pergamon press, London;Oxford,1965
Cav 7.1062

VALDUC 1967 L' Hydrogene dans les metaux : colloque Centre d'etudes de Bruyeres-le-Chatel Paris,1969
Met 25.2624

VALENCE THEORY Quantum mechanical methods in valence theory :a conference Papers Shelter Island,L.I.,N.Y. 1951 Sep 8-10 National academy of sciences With the support of the United States.Office of naval research United States.Office of naval research.Physics branch.Publication,4 U.S. government printing office,Washington,D.C., 1952
Chem 18.2237

VALUATIONS Theorie des valuations mimeograph Montreal 1964 By Paulo Ribenboim North Atlantic treaty organization Societe mathematique du Canada 313p 28cm Presses de l'universite de Montreal,Montreal, 1964
P. Math 2.1974

VANCOUVER 1960 Symposium on pink salmon :a symposium held at the university of British Columbia Edited by Norman J. Wilimovsky H. R.Macmillan lectures in fisheries illus,map 226p 27cm University of British Columbia institute of fisheries,Vancouver,1962
Sco 14.3615

VANCOUVER,B.C. 1960 Symposium on pink salmon Lectures University of British Columbia.Institute of fisheries Edited by N. J. Wilimovsky University of British Columbia. Institute of fisheries.H.R.Macmillan lectures in fisheries, 3 27cm Vancouver,B.C.,1962
Bal 44.4988

VANCOUVER,B.C. 1969 Counting labelled trees Canadian mathematical congress 12th biennial seminar Lectures Canadian mathematical congress Canadian mathematical monographs, 1 x,113p 23cm Canadian mathematical congress,Montreal,Que.,1970
P Math 2.4447

VANCOUVER,B.C. 1969 Informal conference on ALGOL 68 implementations Proceedings University of British Columbia.Department of computer science Edited by J.E.L. Peck illus 119p University of British Columbia. Department of computer science,Vancouver,B.C., 1969
Math L 5.3845

VAPOUR PHASE CHROMATOGRAPHY :a symposium Proceedings London 1956 May 30-Jun 1 Edited by D.H. Desty and C.L.A. Harbourn Sponsored by the Institute of petroleum. Hydrocarbon research group xv,435p Butterworths,London,1957
Chem E 24.0845

VAPOUR PHASE CHROMATOGRAPHY :a symposium Proceedings London 1956 May 30-Jun 1 Institute of petroleum.Hydrocarbon research group Edited by D.H. Desty and C.L.A. Harbourn Butterworths,London,1957
Chem 18.1450

VARENNA Scuola internazionale di fisica 'Enrico Fermi' Rendiconti Corso 10-14;16-: 1959- Societa italiana di fisica Nicola Zanichelli,Bologna,1960-
Cav 7.2388

VARENNA 1957 Argomenti scelti di fisica delle particelle Scuola internazionale di fisica "Enrico Fermi" 41 corso Rendiconti Societa Italiana di fisica Edited by J. Steinberger Bibliog.,Illus. vii,194p 24cm Academic press,New York,1967
A Math 4.1545

VARENNA 1960 Aerodynamic phenomena in stellar atmospheres Symposium on cosmical gas dynamics held at the International school of physics "Enrico Fermi" 4th proceedings International astronomical union and International union of theoretical and applied mechanics Edited by R.N. Thomas and others Under the auspices of Societa Italiana di fisica International astronomical union. Symposium, 12 515p Bologna,c1961 Reprinted from Supplemento Nuovo Cimento,22,no. 1,1961
A Math 4.1152

VARENNA 1960 Aerodynamic phenomena in stellar atmospheres Symposium on cosmical gas dynamics held at the International school of physics "Enrico Fermi" 4th proceedings International astronomical union and International union of theoretical and applied mechanics Edited by R.N. Thomas and others under the auspices of the Societa italiana di fisica International astronomical union symposium, 12 515p Zanichelli,Bologna, c1961 Reprinted from the 'Supplemento del nuovo cimento',vol.22,no.1,1961
Obs 6.1736

VARENNA 1960 Teorie ergodiche Scuola internationale di fisica 'Enrico Fermi' 14 corso Rendiconti Societa Italiana di Fisica Edited by P. Caldirola Academic press,New York;London,1961
A Math 4.0950

VARENNA 1961 Verifiche delle teorie gravitazionali Scuola internazionale di fisica 'Enrico Fermi' 20 corso Rendiconti Societa italiana di fisica Edited by C. Moller Academic press,New York,1962
TA 15.0084

VARENNA 1962 Biological organization at the cellular and supercellular level Edited by R.J.C. Harris Under the auspices of Unesco Academic press,London;New York,1963
Bal 39.0460

VARENNA 1962 Biological organization at the cellular and supercellular level :a symposium Edited by R.J.C. Harris Held under the auspices of Unesco Academic press, London,1963
Pha 16.0010

VARENNA 1962 Biological organization at the cellular and supercellular level :a symposium Edited by R.J.C. Harris held under the auspices of Unesco Academic press, London;New York,1963
Bioch 33.1018

VARENNA 1962 Biological organization at the cellular and supercellular level :a symposium Papers Under the auspices of Unesco Academic press,London;New York,1963
Gen 34.0681

VARENNA 1962 Biological organization at the cellular and supercellular level :a symposium Proceedings Edited by R.J.C. Harris Under the auspices of Unesco Academic press,London;New York,1963
An 32.3291

VARENNA 1962 Dispersione ed assorbimento del suono Scuola internazionale de fisica 'Enrico Fermi' 27 corso Rendiconti Societa italiana di fisica Edited by D. Sette Academic press,New York;London,1963 Papers in English and French
Chem 18.0517

VARENNA 1962 Evoluzione delle stelle Scuola internazionale di fisica "Enrico Fermi" 28 corso rendiconti Societa Italiana di fisica Edited by L. Gratton 488p Academic press,New York;London,1963
Obs 6.0685

VARENNA 1962 Evoluzione delle stelle Scuola internazionale di fisica'Enrico Fermi' 28 corso Rendiconti Societa Italiana di fisica Edited by L. Gratton Academic press, New York;London,1963
A Math 4.1156

VARENNA 1962 Teoria dei plasma Scuola internazionale di fisica 'Enrico Fermi' 25 corso Rendiconti Societa italiana di fisica Edited by N.M. Rosenbluth Academic press,New York;London,1964
Eng 41.4527

VARENNA 1964 weak interactions and high energy neutrino physics Scuola internationale di fisica 'Enrico Fermi'. 32 corso rendiconti Societa Italiana di Fisica Edited by T.D. Lee 334p Academic press,New York,1966
A Math 4.0845

VARENNA 1965 Astrofisica delle alte energie Scuola internazionale di fisica 'Enrico Fermi' 35 corso Rendiconti Societa italiana di fisica Edited by L. Gratton Academic press,New York;London,1966
TA 15.0314

VARENNA 1965 Astrofisica delle alte energie Scuola internazionale di fisica "Enrico Fermi" 35 corso rendiconti Societa Italiana di Fisica Edited by L. Gratton 463p Academic press,New York,1966
A Math 4.1169

VARENNA 1965 Propieta ottiche dei solidi Scuola internazionale di fisica 'Enrico Fermi' 34 corso Rendiconti Societa italiana di fisica Edited by J. Tauc Academic press, New York;London,1966
TA 15.0443

VARENNA 1966 Astrofisica del plasma Scuola internazionale di fisica 'Enrico Fermi' 39 corso Rendiconti Societa italiana di fisica Edited by P.A. Sturrock Academic press,New York;London,1967
TA 15.0328

VARENNA 1966 Astrofisica del plasma Scuola internazionale di fisica "Enrico Fermi" 39 corso Rendiconti Societa italiana di Fisica Edited by P.A. Sturrock xvi,364p Academic press,New York;London,1967
A Math 4.1577

VARENNA 1966 Teoria del magnetismo nei metalli di transizione Scuola internazionale di fisica 'Enrico Fermi' 37 corso Rendiconti Societa italiana di fisica Edited by W. Marshall diagrs 455p 24cm Academic press,London,1967
Cav 7.2612

VARENNA 1967 Ottica quantistica Scuola internazionale di fisica "Enrico Fermi" 42 corso Rendiconti Societa Italiana di fisica Edited by R.J. Glauber Bibliog., Illus.port. xvi,159p 25cm Academic press, New York,1968
A Math 4.1599

VARENNA 1968 Elaborazione di dati offici de parte di organism e di macchine Scuola internazionale di fisica "Enrico Fermi" 43 corso Rendaconti Societa italiana di fisica Edited by W. Reichardt Academic press,New York;London,1969
An 32.5470

VARENNA 1968 Molecular beams and reaction kinetics Scuola internazionale di fisica 'Enrico Fermi' 44 corso Rendiconti Societa italiana di fisica Edited by C. Schlier Academic press,New York,1970
Chem 18.2814

VARENNA 1969 Physics with intersecting storage rings Scuola internazionale di fisica 'Enrico Fermi' 46 corso Rendiconti Societa italiana di fisica Bibliog,photos 24cm Academic press,New York,1971
Cav 7.3012

VARENNA 1969 Questions on algebraic varieties Centro matematico estivo 3 ciclo Centro matematico estivo 343p 28cm Edizioni cremonese,Rome,1970 Conference coordinator:E.Marchionna
P Math 2.4172

VARENNA 1969 Scuola internazionale di fisica 'Enrico Fermi' 47 corso Rendiconti Societa italiana di fisica Edited by Rainer K. Sachs Academic press New Yorkp 1971cm
A Math 4.1788

VARENNA 1970 Differential games and related topics International summer school on mathematical models of action and reaction Proceedings Consiglio nazionale delle richerche Ente per gli studi monetari, bancari et finanziari 'Luigi Einaudi' Edited by H.W. Kuhn and G.P. Szego x,489p 23cm North-Holland,Amsterdam,1971
P Math 2.4574

VARENNA 1970 Differential games and related topics International summer school on mathematical models of action and reaction Proceedings Edited by H.W. Kuhn and G.P. Szego x,489p 23cm North-Holland, Amsterdam;London,1971
Math S 3.1721

VARIABLE STARS Colloquium on variable stars Non-periodic phenomena in variable stars Budapest 1968 Sep 5-9 International astronomical union.Commissions 27 and 42 International astronomical union.Colloquium, 4 490p Academic press,Budapest,1969
Obs 6.3360

VARIABLE STARS Position of variable stars in the Hertzsprung-Russell diagram Colloquium on variable stars 3rd Bamberg 1965 Aug 11-14 International astronomical union. Commissions 27 and 42 Remeis-sternwarte, Bamberg.Kleine veroffentlichungen, 4,no 40 port 304p Astronomisches institut der universitat Erlangen-Nurnberg,Bamberg,1965 I.A.U. combined colloquium of commissions 27 and 42 with the theme 'The position of stars'
TA 15.0540

VARIAN'S ANNUAL WORKSHOP ON NUCLEAR MAGNETIC RESONANCE AND ELECTRON PARAMAGNETIC RESONANCE NMR nd EPR spectroscopy Varian associates. Instrument division Pergamon press,Oxford, 1960 Papers presented at Palo Alto,Calif., October 19-23,1959
Cav 7.2389

VARIATIONS OF THE REGIME OF EXISTING GLACIERS International association of scientific hydrology :a symposium Proceedings Obergurgl 1962 Sep 10-18 Edited by W. Ward International association of scientific hydrology.Publication,58 312p Gentbrugge, 1962
Sco 14.0137

VARNA 1969 National congress of theoretical and applied mechanics 1st Proceedings Bulgarsko knizhovno druzhestvo. National committee on theoretical and applied mechanics Bibliog,illus 632p 24cm Bulgarsko knizhovno druzhestvo,Sofia,1971 Papers in English,French and Russian
A Math 4.1723

VARNA 1971 Federation of European biochemical societies meeting 7th Proceedings functional units in protein biosynthesis Federation of European biochemical societies Edited by R.A. Cox and A.A. Hadjiolov Federation of European biochemical societies.Publications, 23 Academic press,London,1972
Bioch 33.2374

VARNA 1971 Federation of European biochemical societies meeting 7th Proceedings virus-cell interactions and viral antimetabolites Federation of European biochemical societies Edited by D. Shugar Federation of European biochemical societies. Publications, 22 Academic press,London,1972
Bioch 33.2323

VASCULAR PATTERNS AS RELATED TO FUNCTIONS Conference on microcirculatory physiology and pathology 2nd Philadelphia,Pa. 1955 Apr 5 Williams and Wilkins,Baltimore,Md., 1955
An 32.3461

VATICAN 1957 Stellar populations conferences proceedings Specola vaticana and Pontificia academia scientiarum Edited by D.J.K. O'Connell Pontificia academia scientiarum, 16 Specola vaticana.Ricerche astronomiche, 5 North-Holland,Amsterdam, 1958
A Math 4.1150

VATICAN 1964 Brain and conscious experiemce :study week Pontificia academia scientiarum Edited by John Carew Eccles Springer,Berlin,1966
Phys 20.2041

VATICAN 1964 Brain and conscious experience :study week Pontificia academia scientiarum Edited by John C. Eccles Springer,Berlin,1966
Pha 16.0309

VATICAN CITY 1957 Stellar populations :a conference Proceedings Specola vaticana Edited by D.J.K. O'Connell Pontificiae academiae scientiarum.Scripta varia, 16 Specola vaticana.Ricerche astronomiche, 5 544p North Holland,Amsterdam,1958
TA 15.0036

VATICAN CITY 1970 Study week on nuclei of galaxies Edited by D.J.K. O'Connell Accademia pontifica dei nuovi lincei.Scripta varia, 35 795p North-Holland,Amsterdam, 1971
Obs 6.3642

VATICAN CITY 1971 Study week on nuclei of galaxies Accademia pontifica dei nuovi lincei Edited by D.J.K. O'Connell Accademia pontifica dei nuovi lincei.Scripta varia, 35 795p North-Holland,Amsterdam, 1971
TA 15.0661

VATICAN OBSERVATORY 1957 Stellar populations :a conference Proceedings Pontificia academia scientiarum and Specola vaticana Edited by D.J.K. O'Connell Pontificia academia scientiarum.Scripta varia, 16 Specola vaticana.Ricerche astronomiche, 5 illus 544p North-Holland,Amsterdam, 1958
Cav 7.2390

VEBLEN Oswald Analysis situs 2nd edition revised American mmathematical society. Colloquium publications, 5,pt 2 x,194p 23cm American mathematical society,New York, 1931 First edition entitled 'The Cambridge colloquium,1916.Pt 2:analysis situs'
P Math 2.2836

VEGETATION Study of tropical vegetation :a symposium Proceedings Kandy,Ceylon 1956 Mar 19-21 Unesco Jointly organized by the Government of Ceylon Humid tropics research Unesco,Paris,1958
Geog 13.1261

The VEGETATION OF THE HUMID TROPICS 2nd : symposium Proceedings Tjiawi 1953 Dec Unesco.Science cooperation office for south-east Asia Sponsored by the Council for sciences of Indonesia Unesco,c1953
Bot 42.4770

The VEGETATIVE NERVOUS SYSTEM :an investigation of the most recent advances Proceedings of the Association New York 1928 Dec 27-28 Association for research in nervous and mental disease Edited by Walter Timme and others Association for research in nervous and mental disease.Research publications, 9 Williams and Wilkins, Baltimore,Md.,1930
An 32.4884

VENICE 1958 Symposium on the classification of brackish waters By G.Sven Segerstrale International association of theoretical and applied limnology Archivio di oceanografia e limnologia, 11,suppl. Centro nazionale di studi talassografafici del consiglio nazionale delle ricerche,Venice,1959 Title also in Italian:"Simposio sulla classificazione della acque salmastre" Text in English and Italian.
Bal 39.1962

VENICE 1959 Immediate and low level effects of ionizing radiations :symposium Proceedings International atomic energy agency Comitato nazionale per le riccerche nucleari Under the auspices of Unesco International journal of radiation biology. Supplement Taylor and Francis,London,1960
Gen 34.2221

VENICE 1962 Radiation damage in solids and reactor materials :a symposium Proceedings Vol 1-3 International atomic energy agency 3 vols International atomic energy agency,Vienna,1962
Cav 7.2393

VENICE 1972 International computing symposium Proceedings Consiglio nazionale delle ricerche.Cybernetics group illus 634p Venice,1972
Math L 5.3862

VENOMS International conference on venoms 1st Papers Berkeley,Calif. 1954 Dec 27-30 American association for the advancement of science Edited by E.E. Buckley and N. Porges American association for the advancement of science.Publication, 44 des plantes et animaux indesirables
Bal 44.6436

VENUS The Atmospheres of Venus and Mars :a symposium Tucson,Ariz. 1967 Feb 28-Mar 2 Goddard institute for space studies and Kitt Peak national observatory.Space division Edited by John C. Brandt and Michael B. McElroy Gordon and Breach,New York,1968
TA 15.0415

VERANDERUNGEN DES KLIMAS SEIT DEM MAXIMUM DER LETZTEN EISZEIT Internationaler geologen kongress :eine sammlung von berichten unter mitwirking von fachenosson in verschliedenen landern 11 International geological congress.Executive committee Verlag von generalstabens litografiska anstalt,Stockholm, 1910
Bot 42.3774

VERBAL BEHAVIOR AND LEARNING Conference on verbal learning 2nd Proceedings United States.Office of naval research New York university Edited by C.N. Cofer and Barbara S. Musgrave McGraw-Hill,New York,1963
Psy 31.2737

VERBAL LEARNING AND VERBAL BEHAVIOR :a conference Proceedings New York 1959 United States.Office of naval research. Psychological sciences division and New York university.Department of psychology Edited by Charles N. Cofer and Barbara S. Musgrave McGraw-Hill series in psychology McGraw-Hill,New York,1961
Psy 31.1905

VERBAND DEUTSCHER PHYSIKALISCHER GESELLSCHAFTEN International conference on ionization phenomena in gases 5th proceedings Munich 1961 Aug 28-Sep 1 Vol. 1 Edited by H. Maecker Series in physics North Holland,Amsterdam,1962
A Math 4.0620

VEREIN DEUTSCHER INGENIEURE I.Chem.E.,VTG-VDI joint meeting on the engineering of gas-solid reactions Preprints Brighton 1968 Apr 24-26 characteristics of solid particles which are relevant to gas-solid reactions 43p Institution of chemical engineers,London, 1968 In English and German
Chem E 24.1516

VEREIN DEUTSCHER INGENIEURE I.Chem.E.,VTG-VDI joint meeting on the engineering of gas-solid reactions Preprints Brighton 1968 Apr 24-26 experiences with the operation of industrial process equipment 51p Institution of chemical engineers,London,1968 In English and German
Chem E 24.1519

VEREIN DEUTSCHER INGENIEURE I.Chem.E.,VTG-VDI joint meeting on the engineering of gas-solid reactions Preprints Brighton 1968 Apr 24-26 physical behaviour of gas-solid systems 69p Institution of chemical engineers,London,1968 In English and German
Chem E 24.1517

VEREIN DEUTSCHER INGENIEURE I.Chem.E.,VTG-VDI joint meeting on the engineering of gas-solid reactions Preprints Brighton 1968 Apr 24-26 the technical design and performance of industrial process equipment 52p Institution of chemical engineers,London, 1968 In English and German
Chem E 24.1518

VEREIN DEUTSCHER INGENIEURE The Engineering of gas-solid reactions :I.Chem.E.-VTG-VDI joint meeting Proceedings Brighton 1968 Apr 24-26 Edited by J.M. Pirie Institution of chemical engineers.Symposium series, 27 illus 255p Institution of chemical engineers,London,1968
Chem E 24.1873

VEREIN DEUTSCHER INGENIEURE.FACHGRUPPE REGELUNGSTECHNIK Regelungstechnik:moderne theorien und ihre verwendbarkeit :ein tagung Bericht Heidelberg 1956 Sep 25-29 Oldenbourg,Munich,1957
Eng 41.5842

VEREIN FUR ANGEWANDTE PSYCHOPATHOLOGIE UND PSYCHOLOGIE I.Internationale tagung fur angewandte psychopathologie und psychologie Vienna 1930 Jun 5-7 Edited by Heinz Hartmann and others Karger,Berlin,1931 From 'Abhandlungen aus der neurologie, psychiatrie,psychologie und ihre grenzbegiete', heft 61
Psy 28.0409

VEREIN FUR NATURWISSENSCHAFT ZU BRAUNSCHWEIG Versammlung Deutscher Naturforscher und Aerzte 69 Festgruss Brunswick 1897 Schulbuchhandlung,Brunswick,1897
Philos 1.0012

VERFORMUNG UND FLIESSEN DES FESTKORPERS Deformation and flow of solids :colloquium Madrid 1955 Sep 26-30 International union of theoretical and applied mechanics Edited by R. Grammel Springer,Berlin,1956 Papers in English,French and German
Met 25.1203

VERGENNES,VT. 1959 Biological activities of steroids in relation to cancer :a conference Proceedings Edited by Gregory Pincus and Erwin P. Vollmer Sponsored by the Cancer chemotherapy national service center Academic press,New York;London,1960
Radioth 35.0873

VERHANDLUNGEN DES INTERNATIONALEN BOTANISCHEN Vienna 1905 International botanical congress Edited by R.von Wettstein and others Fischer,Jena,1906 Title also in French 'Actes du congres international de botanique'
Bot 42.0647

VERIFICHE DELLE TEORIE GRAVITAZIONALI Scuola internazionale di fisica "Enrico Fermi" 20 corso Rendiconti Varenna 1961 Jun 19-Jul 1 Societa Italiana di fisica Edited by G. Moller Bibliog.,Illus. 264p 24cm Academic press,New York,1962
A Math 4.1555

VERNALIZATION AND PHOTOPERIODISM :a symposium By A.E. Murneek and others Lotsya;a biological miscellany, 1 Foreword by Kenneth V.Thimann
Bot 42.1822

VERSAILLES 1968 Symposium on automatic demonstration Edited by M. Laudet and others Organised by the Institut de recherche d'informatique et d'automatique Lecture notes in mathematics, 125 Bibliog 310p 25cm Springer-Verlag,Berlin,1970
P Math 2.3614

VERSAMMLUNG DER GESELLSCHAFT DEUTSCHER NATURFORSCHER UND AERZTE... 20th Amtlicher bericht Mainz 1842 Sep 19-26 Gesellschaft deutscher naturforscher und aerzte Edited by Johann Groser and Friedrich Karl Bruch 2 pls 28cm Kupferberg,Mainz, 1843
Bal 44.6348

VERSAMMLUNG DER TUBERKULOSE-ARZTE 2te Bericht Berlin 1904 Nov 24-26 Deutsches central-komite zur errichtung von heilstatten fur lungenkranke Edited by Johannes Nietner Deutsches central-komite zur errichtung von heilstatten fur lungenkranke,Berlin,1905
Path 30.0761

VERSAMMLUNG DEUTSCHER NATURFORSCHER UND AERTZE 69 Festschrift Brunswick 1897 Gesellschaft deutscher naturforscher und arzte Harald Bruhn,Brunswick,1897
Philos 1.0008

VERSAMMLUNG DEUTSCHER NATURFORSCHER UND AERZTE 69 Festgruss Brunswick 1897 Verein fur naturwissenschaft zu Braunschweig Schulbuchhandlung,Brunswick,1897
Philos 1.0012

VERSAMMLUNG DEUTSCHER NATURFORSCHER UND AERZTE 2ND Ueber die nationale entwicklung und bedeutung der naturwissenschaften... :rede gehalten in der zweiten allgemeinen sitzung der Versammlung deutscher naturforscher und aerzte zu Hanover am 20 September 1865 By Rudolf Virchow 17cm Hirschwald,Berlin,1865
Bal 44.0627

VERSAMMLUNG DEUTSCHER NATURFORSCHER UND AERZTE 86 Festschrift der achtundsechzigsten versammlung deutscher naturforscher und aerzte gewidmet von dem vorstand des stadtischen krankenhause Frankfurt-am-Main.Stadtisches krankenhaus Frankfurt-am-Main.Stadtisches krankenhaus.Arbeiten Frankfurt,1896
Path 30.0113

VERSAMMLUNG NORDISCHER NATURFORSCHER UND ARTZE Helsinki 1902 Jul 7-12 Verhandlungen der sektion fur anatomie,physiologie und medizinische chemie 1902
Phys 20.0160

VERTEBRATE LOCOMOTION :a symposium Proceedings London 1960 Nov 9 Zoological society of London Edited by J.E. Harris Zoological society of London.Symposia, 5 Zoological society of London,London, 1961
An 32.0459

VERTEBRATE LOCOMOTION :a symposium Proceedings London 1960 Nov 9 Zoological society of London Edited by J.E. Harris Zoological society of London.Symposia, 5 Illus. Zoological society,London,1961
VA 19.0086

VERTEBRATE LOCOMOTION 5 :a symposium Proceedings London 1960 Nov 9 Zoological society of London Edited by J.E. Harris illus,1 plate vi,132p 26cm London,1961
Sco 14.0730

VERTEBRATES Problemes actuels de paleontologie (evolution des vertebres) Paris 1966 Jun 6-11 Centre national de la recherche scientifique Centre national de la recherche scientifique.Colloques internationaux.Actes, 163 pls 27cm C.N.R.S.,Paris,1967 Organized by J.P.Lehman. Papers in English,French and German
Bal 44.4640

The VETLESEN SYMPOSIUM Galaxies and the universe New York 1966 Oct 19 Edited by Lodewijk Woltjer 112p Columbia university press,New York,1968
Obs 6.3452

VIBRATION AND NOISE IN MOTOR VEHICLES :a symposium Proceedings London 1971 Jul 6-7 Institution of emchanical engineers. Automobile division Advanced school of automotive studies IME,London,1972
Eng 41.8646

VIBRATION EFFECTS OF EARTHQUAKES ON SOILS AND FOUNDATIONS :a symposium San Francsico,Calif. 1968 Jun 23-28 American society for testing and materials American society for testing and materials.Special technical publication, 450 American society for testing materials,Philadelphia,Pa.,1969
Eng 41.3162

VIBRATION IN CIVIL ENGINEERING :a symposium Proceedings London 1965 Apr Edited by B.O. Skipp Organised by the International association for earthquake engineering Butterworths,London,1966
Eng 41.6363

VIBRATION THEORY Wave motion and vibration theory symposium Washington 1952 June 16-17 Carnegie institute of technology and American mathematical society Edited by Albert E. Heins American mathematical society.Proceedings of symposia in applied mathematics, 5 McGraw-Hill,New York,1954
A Math 4.0513

VIBRATIONS NON LINEAIRES Colloque international des vibrations non lineaires Actes Ile de Porquerolles 1951 Sep 18-21 Union internationale de radio-science Organised by the Union internationale de mecanique theorique et appliquee France. Ministere de l'air.Publications scientifiques et techniques, 281 Information technique de l'aeronautique,Paris,1953
Eng 41.6372

VICHY 1937 International congress of hepatic insufficiency and liver diseases 2: comptes rendus,discussions et communications diverses Edited by J. Aimard Wallon,Paris, 1937 Conference president:professor M. Loeper
Path 30.0948

VICK INTERNATIONAL Menthol and menthol-containing external remedies use,mode of effect and tolerance in children:international symposium Proceedings Paris 1966 Apr 27-28 Edited by F.H. Dost and B. Leiber Georg Thieme verlag,Stuttgart,1967
PGMS 29.0391

VICTORIA UNIVERSITY OF MANCHESTER Logic colloquium 1969 Summer school and colloquium in mathematical logic Proceedings Manchester 1969 Aug 3-23 Edited by R.O. Gandy and C.M.E. Yates Studies in logic and the foundations of mathematics, 61 xiv,451p 23cm North-Holland,Amsterdam,1971
P Math 2.4057

VICTORIA UNIVERSITY OF MANCHESTER Manchester university computer :inaugural conference Manchester 1951 Jul.9-12 in co-operation with Ferranti ltd pl. 40p 27cm Manchester,1957
Math L 5.1320

VICTORIA UNIVERSITY OF MANCHESTER Radio astronomy today Summer school in radio astronomy Papers Jodrell Bank 1962 Edited by H.P. Palmer and others 242p Manchester university press,Manchester,1963
TA 15.0226

VICTORIA UNIVERSITY OF MANCHESTER Radio astronomy today :international summer school Papers Jodrell Bank 1962 Edited by H.P. Palmer and others Manchester university press,Manchester,1963 Foreword by Sir Bernard Lovell
Cav 7.1751

VICTORIA UNIVERSITY OF MANCHESTER Rutherford jubilee international conference Proceedings Manchester 1961 Sep 4-8 Edited by J.B. Birks Under the sponsorship of International union of pure and applied physics 856p Haywood,London,1961 Includes commemorative session 'Rutherford at Manchester'
Cav 7.1952

VICTORIA UNIVERSITY OF MANCHESTER Rutherford jubilee international conference proceedings Manchester 1961 Sep.4-8 Edited by J.B. Birks Co-sponsored by the International union of pure and applied physics Heywood, London,1961 To commemorate the discoveries of Rutherford at Manchester
A Math 4.0826

VICTORIA UNIVERSITY OF MANCHESTER.ASTRONOMY DEPARTMENT Astronomical optics and related subjects A Symposium on astronomical optics and related subjects proceedings Manchester 1955 Apr 19-22 Edited by Zdenek Kopal 428p North-Holland,Amsterdam,1956 Dedicated to the memory of Peter Yorke Millns
Obs 6.0975

VICTORIA UNIVERSITY OF MANCHESTER.INSTITUTE OF SCIENCE AND TECHNOLOGY Specifications ans tests of metal cutting machine tools : conference Vol 1-2 Victoria university of Manchester,Manchester,1970
Eng 41.8169

VICTORIIA,B.C. 1955 British Columbia natural resources conference 8th Transactions Edited by D,B, Turner and J.D. Chapman maps 338p 25cm British Columbia natural resources conference,Victoria, B.C.,1955
Sco 14.4292

VIENNA 1905 Resultas scientifiques du congres international de botanique : wissenschaftliche ergebnisse des internationalen botanischen kongresses Wien 1905 By J.P. Lotsy Edited by R.von Wettstein and others Publications scientifiques de l'association internationale des botanishes, 1 Fischer,Jena,1906
Bot 42.0645

VIENNA 1905 Verhandlungen des internationalen botanischen International botanical congress Edited by R.von Wettstein and others Fischer,Jena,1906 Title also in French 'Actes du congres international de botanique'
Bot 42.0647

VIENNA 1930 I.Internationale tagung fur angewandte psychopathologie und psychologie Verein fur angewandte psychopathologie und psychologie Edited by Heinz Hartmann and others Karger,Berlin,1931 From 'Abhandlungen aus der neurologie,psychiatrie, psychologie und ihre grenzbegiete',heft 61
Psy 28.0409

VIENNA 1957 Symposion uber das neurovegetative system der gesunden und kranken haut des menschen Edited by E. Anderson and others Acta neurovegetativa, 18 Springer,Vienna,1958 In English, French and German
An 32.4046

VIENNA 1958 Biochemistry of lipids International conference on the biochemical problems of lipids 5th Proceedings Edited by G. Popjak Pergamon press,Oxford, 1960 Section 18 of the 4th international congress of biochemistry
Col S 12.0102

VIENNA 1958 International congress of biochemistry 4th Vol 15: biochemistry: abstracts of sectional papers and index to symposia and colloquia International union of biochemistry I.U.B.symposium series, 17 Pergamon,Oxford,1960 Added title page in French and German.Text in English,French and German
Bioch 33.1329

VIENNA 1958 International congress of biochemistry 4th Proceedings Vol 2: symposium 2 - biochemistry of wood International union of biochemistry Edited by K. Kratzi and G. Billek I.U.B.symposium series, 4 Pergamon,London,1959 Added title page in French and German.Text in English,French and German
Bioch 33.1316

VIENNA 1958 International congress of biochemistry 4th Proceedings Vol 3: symposium 3:biochemistry of the central nervous system International union of biochemistry Edited by F. Brucke I.U.B. symposium series, 5 Pergamon,London,1959 Added t.p.in French and German
Bot 42.1715

VIENNA 1958 International congress of biochemistry 4th Proceedings Vol 3: symposium 3 - biochemistry of the central nervous system International union of biochemistry Edited by Brucke I.U.B. symposium series, 5 Pergamon,London,1959 Added title page in French and German.Text in English,French and German
Bioch 33.1317

VIENNA 1958 International congress of biochemistry 4th Proceedings Vol 4: symposium 4 biochemistry of steroids International union of biochemistry Edited by E. Mosettig I.U.B.symposium series, 6 Pergamon,London,1959 Added title page in French and German.Text in English,French and German.
Bioch 33.1318

VIENNA 1958 International congress of biochemistry 4th Proceedings Vol 5: symposium 5 - biochemistry of antibiotics International union of biochemistry Edited by K.H. Spitzy and R. Brunner I.U.B. symposium series, 7 Pergamon,London,1959 Added title page in French and German.Text in English,French and German.
Bioch 33.1319

VIENNA 1958 International congress of biochemistry 4th Proceedings Vol 6: symposium 6 - biochemistry of morphogenesis International union of biochemistry Edited by W.J. Nickerson I.U.B.S.symposium series, 8 Pergamon press,London,1959
Gen 34.0482

VIENNA 1958 International congress of biochemistry 4th Proceedings Vol 6: symposium 6 - biochemistry of morphogenesis International union of biochemistry Edited by W.J. Nickerson I.U.B.symposium series, 8 Pergamon,London,1959 Added title page in French and German
Bioch 33.1320

VIENNA 1958 International congress of biochemistry 4th Proceedings Vol 7: symposium 7 - biochemistry of viruses International union of biochemistry Edited by F. Broda and W. Frisch-Niggemeyer I.U.B. symposium series, 9 Pergamon,London,1959 Added title page in French and German.Text in German and English
Bioch 33.1321

VIENNA 1958 International congress of biochemistry 4th Proceedings Vol 7: symposium 7 - biochemistry of viruses International union of biochemistry Edited by E. Broda and W. Frisch-Niggemeyer Pergamon press,London,1960
Radioth 35.0071

VIENNA 1958 International congress of biochemistry 4th Proceedings Vol 8: symposium 8 - proteins International union of biochemistry Edited by H. Neurath and H. Tuppy I.U.B.symposium series, 10 Pergamon, London,1960 Added title page in French and German.Text in English and French
Bioch 33.1322

VIENNA 1958 International congress of biochemistry 4th Proceedings Vol 8: symposium 8 - proteins International union of biochemistry Edited by H. Neurath and H. Tuppy Pergamon,London,1960
Radioth 35.0072

VIENNA 1958 International congress of biochemistry 4th Proceedings Vol 9: symposium 9 - physical chemistry of high polymers of biological interest International union of biochemistry Edited by O. Kratky I.U.B.symposium series, 11 Pergamon,London,1959 Added title page in French and German.Text in English,French and German.
Bioch 33.1323

VIENNA 1958 International congress of biochemistry 4th Proceedings Vol 9: symposium 9 - physical chemistry of high polymers of biological interest International union of biochemistry Edited by O. Kratky Pergamon press,London,1960
Radioth 35.1941

VIENNA 1958 International congress of biochemistry 4th Proceedings Vol 10: blood clotting factors By E. Deutsch Pergamon press,Oxford,1959
Med 36.0115

VIENNA 1958 International congress of biochemistry 4th Proceedings Vol 10: symposium 10 - blood clotting factors Edited by E. Deutsch I.U.B.symposium series, 12 Pergamon,London,1959 Added title page in French and German.Text in English and German
Bioch 33.1324

VIENNA 1958 International congress of biochemistry 4th Proceedings Vol 11: symposium 11 - vitamin metabolism International union of biochemistry Edited by W. Umbreit and H. Molitor I.U.B.symposium series, 13 Pergamon,London,1960 Added title page in French and German
Bioch 33.1325

VIENNA 1958 International congress of biochemistry 4th Proceedings Vol 12: colloquia International union of biochemistry Edited by H. Chantrenne and others I.U.B.symposium series, 15 Pergamon,London,1959 Added title page in French and German.Text in English,French and German
Bioch 33.1327

VIENNA 1958 International congress of biochemistry 4th Proceedings Vol 12: symposium 12 - biochemistry of insects International union of biochemistry Edited by L. Levenbook I.U.B.symposium series, 14 Pergamon,London,1959 Added title page in French and German.Text in English and German
Bioch 33.1326

VIENNA 1958 International congress of biochemistry 4th Proceedings Vol 14: transactions of the plenary sessions International union of biochemistry Edited by W. Auerswald and O. Hoffmann-Ostenhof I.U. B.symposium series, 14 Pergamon,London,1959 Added title page in French and German.Text in English,French and German.
Bioch 33.1328

VIENNA 1958 The International congress of biochemistry 4th Proceedings 13: colloquia International union of biochemistry Edited by O. Hoffmann-Osterhoff I.U.B.Symposium series, 15 Pergamon press, London,1959 Added title page in French and German
Bot 42.1696

VIENNA 1959 Controlled clinical trials :a conference Papers Convened by the Council for international organizations of medical sciences Blackwell,Oxford,1960
Radioth 35.0799

VIENNA 1959 Controlled clinical trials :a conference Papers Council for international organizations of medical sciences Blackwell,Oxford,1960 Organized under the direction of A. Bradford Hill
HE 27.0076

VIENNA 1959 Medical radioisotope scanning : a seminar Proceedings International atomic energy agency World health organization International atomic energy agency,Vienna,1959
Radioth 35.1375

VIENNA 1959 Use of radioisotopes and supervoltage radiation in radioteletherapy; present status and recommendation Report and background information for a study group convened by the IAEA and WHO International atomic energy agency World health organization International atomic energy agency,Vienna,1960
Radioth 35.1146

VIENNA 1960 Fuel element fabrication with special emphasis on cladding materials:a symposium Proceedings Vol 1-2 International atomic energy agency 2 vols Academic press,London;New York,1961
Met 25.1547

VIENNA 1960 Inelastic scattering of neutrons in solids and liquids :a symposium Proceedings International atomic energy agency Vienna,1961
Cav 7.2392

VIENNA 1960 Selected topics in radiation dosimetry Symposium on selected topics in radiation dosimetry Proceedings Sponsored by the International atomic energy agency International atomic energy agency.Proceedings series International atomic energy agency, Vienna,1961
Radioth 35.1694

VIENNA 1961 Application of isotope techniques in hydrology :a comprehensive report International atomic energy agency International atomic energy agency.Technical reports series, 11 Vienna,1962
Bot 42.3317

VIENNA 1961 Effects of ionizing radiation on the nervous system Symposium on ionizing radiation on the nervous system Proceedings Sponsored by the International atomic energy agency International atomic energy agency. Proceedings series International atomic energy agency,Vienna,1962
Radioth 35.1130

VIENNA 1961 Effects of ionizing radiation on the nervous system Symposium on the effects of ionizing radiation on the nervous system Sponsored by the International atomic energy agency International atomic energy agency,Vienna,1962
An 32.4190

VIENNA 1961 European society of haematology congress 8th Proceedings 1-2 European society of haematology Edited by Adolf Scharf 2 vols Karger,Basle; New York,1962
Med 36.0134

VIENNA 1961 International congress on mental retardation 2nd Proceedings Pt 1: organic bases and biochemical aspects of imbecility Edited by Otto Stur Karger, Basle,1963
An 32.4198

VIENNA 1961 International foundry congress 28th Congress papers International committee of foundry technical associations Verein Oesterreichischer Giesserfachleute, Vienna,1961 Papers in English,French and German
Met 25.0681

VIENNA 1961 The Detection and use of tritium in the physical and biological sciences symposium Proceedings Vol 1-2 Joint commission on applied radioactivity Sponsored by the International atomic energy agency International atomic energy agency. Proceedings series 2 vols International atomic energy agency,Vienna,1962 Papers in English and Russian.Summaries in English, French,Russian and Spanish
Gen 34.0470

VIENNA 1961 Tritium in the physical and biological sciences Symposium on the detection and use of tritium in the physical and biological sciences Proceedings Vols 1-2 Sponsored by the International atomic energy agency 2 vols International atomic energy agency,Vienna,1962 Co-sponsored by the Joint commission on applied radioactivity
Radioth 35.0104

VIENNA 1961 Whole-body counting Symposium on whole-body counting Proceedings International atomic energy agency.Proceedings series International atomic energy agency, Vienna,1962
Radioth 35.1361

VIENNA 1962 Diagnosis and treatment of radioactive poisoning Scientific meeting on the diagnosis and treatment of radioactive poisoning Proceedings World health organization International atomic energy agency International atomic energy agency. Proceedings series International atomic energy agency,Vienna,1963
Radioth 35.1368

VIENNA 1962 European congress of anaesthesiology :post-graduate courses 1st Abstracts of papers World federation of societies of anaesthesiologists English edition Wiener medizinischen akademie,Vienna, c1962
PGMS 29.0030

VIENNA 1962 Isotope techniques for hydrology International atomic energy agency International atomic energy agency.Technical reports series, 23 Vienna,1964
Bot 42.3318

VIENNA 1962 Production and use of short-lived radioisotopes from reactors Seminar on the practical applications of short-lived radioisotopes produced in small research reactors Proceedings 1-2 Held by the International atomic energy agency International atomic energy agency.Proceedings series 2 vols International atomic energy agency,Vienna,1963
Radioth 35.1369

VIENNA 1963 Isotope mass effects in chemistry and biology Symposium on isotope mass effects in chemistry and biology Proceedings Edited by N. Grell Held by the International union of pure and applied chemistry Pure and applied chemistry, 8,no. 3-4 Butterworths,London,1964
Radioth 35.0421

VIENNA 1963 Radiological health and safety in mining and milling of nuclear materials Symposium on radiological health and safety in mining and milling of nuclear materials :a symposium Proceedings 1-2 International atomic energy agency International labour organization In co-operation with the World health organization International atomic energy agency.Proceedings series International atomic energy agency, Vienna,1964
Radioth 35.1164

VIENNA 1964 Chemical effects of nuclear transformations :symposium on chemical effects associated with nuclear reactions and radioactive transformations Proceedings Vol 1-2 International atomic energy agency International atomic energy agency.Proceedings series STI-PUB 91 International atomic energy agency,Vienna,1965
Chem 18.1151

VIENNA 1964 Formal language description languages for computer programming IFIP working conference on formal language description languages proceedings International federation for information processing Edited by T.B. Steel bibliog 330p 22cm North-Holland publishing, Amsterdam,1966
Math L 5.1033

VIENNA 1964 International conference on coordination chemistry 8th Proceedings Edited by V. Gutmann c1964 Damaged copy, title page missing
Chem 18.2001

VIENNA 1964 Radioisotope techniques in the study of protein metabolism Radioisotope techniques in the study of protein metabolism : a panel International atomic energy agency International atomic energy agency.Technical reports series, 45 International atomic energy agency,Vienna,1965
Radioth 35.1150

VIENNA 1965 Antibodies to biologically active molecules Federation of European biochemical societies : meeting 2nd Proceedings Vol 1 Edited by Bernhard Cinader Pergamon,Oxford,1967
Pha 16.0190

VIENNA 1965 Applications of the Mossbauer effect in chemistry and solid-state physics : report of a panel International atomic energy agency International atomic energy agency.Technical reports series,50 International atomic energy agency,Vienna,1966
Min 10.1131

VIENNA 1965 Biophysical aspects of radiation quality Panel on biophysical aspects of radiation quality Report International atomic energy agency International atomic energy agency.Technical report series, 58 International atomic energy agency,Vienna,1966
Radioth 35.1421

VIENNA 1965 Clinical uses of whole-body counting Panel on clinical uses of whole-body counting Proceedings International atomic energy agency International atomic energy agency.Panel proceedings series International atomic energy agency,Vienna,1966
Radioth 35.1383

VIENNA 1965 Federation of European biochemical sciences meeting 2nd Proceedings symposium on ribonucleic acid structure and function Federation of European biochemical societies Edited by H. Tuppy Pergamon press,Oxofrd,1966
Gen 34.0688

VIENNA 1965 Federation of European biochemical societies meeting 2nd Proceedings Vol 1: antibodies to biologically active molecules Edited by B. Cinader Pergamon press,Oxford,1967
Phys 20.1552

VIENNA 1965 Radioisotope sample measurement techniques in medicine and biology Symposium on radioisotope sample measurement techniques in medicine and biology Proceedings Held by the International atomic energy agency International atomic energy agency.Proceedings series International atomic energy agency,Vienna,1965
Radioth 35.1377

VIENNA 1965 Recent advances in clinical neurophysiology International congress of eletroencephalography and clinical neurophysiology 6th Proceedings functions of the spinal cord; neurophysiological investigations of brain diseases;EEG in stress Edited by L. Widen Electroencephalography and clinical neurophysiology.Supplement,25 Elsevier, Amsterdam,1967
Pha 16.0304

VIENNA 1965 The Federation of European biochemical societies meeting 2nd Proceedings Vol 2: cyclitols and phosphoinositides Federation of European biochemical societies Edited by H. Kindl Pergamon press,Oxford,1966
Bioch 33.1393

VIENNA 1966 Effects of ionizing radiations on the haematopoietic tissue Panel on the effects of ionizing radiations from different sources on haematopoietic tissue Proceedings International atomic energy agency International atomic energy agency.Panel proceedings series International atomic energy agency,Vienna,1967
Radioth 35.1200

VIENNA 1966 Genetical aspects of radiosensitivity :mechanisms of repair International atomic energy agency International atomic energy agency.Proceedings series illus 175p 24cm International atomic energy agency,Vienna,1966
Bot 42.0797

VIENNA 1966 Genetical aspects of radiosensitivity:mechanisms of repair :a panel Proceedings International atomic energy agency International atomic energy agency. Panel proceedings series International atomic energy agency,Vienna,1966
Radioth 35.1194

VIENNA 1966 Moon and planets International space science :symposium 7th Papers International astronomical union Edited by A. Dollfus Co-sponsored by the Committee on space research 313p North-Holland,Amsterdam,1967 Papers of one session of the symposium
Obs 6.3206

VIENNA 1966 Mutations in plant breeding Panel on co-ordination of research on the production and use of induced mutations in plant breeding Proceedings International atomic energy agency Food and agriculture organization International atomic energy agency.Panel proceedings series 2 vols International atomic energy agency, Vienna,1966-68
Radioth 35.0261

VIENNA 1966 Neutron monitoring Symposium on neutron monitoring for radiological protection Proceedings International atomic energy agency International atomic energy agency.Proceedings series International atomic energy agency,Vienna,1967
Radioth 35.1672

VIENNA 1966 Preservation of fruit and vegetables by radiation Panel on preservation of fruit and vegetables by radiation :especially in the tropics Proceedings Organised by the Joint FAO-IAEA division of atomic energy in food and agriculture International atomic energy agency.Panel proceedings series International atomic energy agency,Vienna,1968
Radioth 35.1758

VIENNA 1966 Solid state and chemical radiation dosimetry in medicine and biology :a symposium Proceedings Edited by M.E.J. Young Organised by the International atomic energy agency International atomic energy agency.Proceedings series International atomic energy agency,Vienna,1967
Radioth 35.1673

VIENNA 1967 Biophysical aspects of radiation quality Panel on biophysical aspects of radiation quality 2nd Report International atomic energy agency International atomic energy agency.Panel proceedings series International atomic energy agency,Vienna,1968
Radioth 35.1422

VIENNA 1967 Control of livestock insect pests by the sterile-male technique Panel on the control of livestock insect pests by the sterile-male technique Proceedings Organized bythe Joint FAO-IAEA division of atomic energy in food and agriculture International atomic energy agency.Panel proceedings series International atomic energy agency,Vienna,1968
Radioth 35.0263

VIENNA 1967 Effects of radiation on meiotic systems Study group on the effects of radiation on meiotic systems Organized by the International atomic energy agency International atomic energy agency.Panel proceedings series International atomic energy agency,Vienna,1968
Radioth 35.1201

VIENNA 1967 Mutations in plant breeding II a research co-ordination meeting on the use of induced mutations in plant breeding Proceedings International atomic energy agency Food and agriculture organization International atomic energy agency.Panel proceedings series I.A.E.A.,Vienna,1968
Radioth 35.0262

VIENNA 1967 Role of computers in radiotherapy Panel on the role of computers in radiotherapy Report Organized by the International atomic energy agency International atomic energy agency.Panel proceedings series bibliog. International atomic energy agency,Vienna,1968
Radioth 35.1199

VIENNA 1967 Standardization of radioactive waste categories :report of a panel International atomic energy agency International atomic energy agency.Technical reports series, 101 IAEA,Vienna,1970
Chem 18.2735

VIENNA 1968 Enzymological aspects of food irradiation Panel on enzymological aspects of the application of ionizing radiation to food preservation Proceedings Organized by the Joint FAO-IAEA division of atomic energy in food and agriculture International atomic energy agency.Panel proceedings series International atomic energy agency,Vienna,1969
Radioth 35.1762

VIENNA 1968 Metabolism and membrane permeability of erythrocytes and thrombocytes International symposium on metabolism and membrane permeability of erythrocytes and thrombocytes 1st Edited by Erwin Deutsch and others Thieme,Stuttgart,1968 In English and German
Radioth 35.0932

VIENNA 1968 Meteorite research Symposium on meteorite research Proceedings International atomic energy agency Unesco Edited by Peter M. Millmann Astrophysics and space science library Reidel,Dordrecht,1969
Min 10.1442

VIENNA 1968 Radiation damage and sulphydryl compounds Panel on radiation damage to the biological molecular information system :with special regard to the role of SH-groups Proceedings Organized by the International atomic energy agency International atomic energy agency.Panel proceedings series International atomic energy agency,Vienna,1969
Radioth 35.1764

VIENNA 1969 Analytical control of radiopharmaceutics :a panel Proceedings Organized by the International atomic energy agency International atomic energy agency. Panel proceedings series International atomic energy agency,Vienna,1970
Radioth 35.1385

VIENNA 1969 In vitro procedures with radioisotopes in medicine :symposium Proceedings International atomic energy agency World health organization IAEA, Vienna,1970
Bioch 33.2184

VIENNA 1969 Nuclear accident dosimetry systems Panel on nuclear accident dosimetry systems Proceedings Organized by the International atomic energy agency International atomic energy agency.Panel proceedings series International atomic energy agency,Vienna,1970
Radioth 35.1384

VIERHOUTEN 1969 Pituitary,adrenal and the brain :an international conference on the pituitary-adrenal axis and the nervous system Proceedings Edited by D. De Wied and J.A.W.M. Weijnen Organized by the Rudolf Magnus institute for pharmacology Progress in brain research, 32 Elsevier,Amsterdam,1970
An 32.5235

VIEWS ON GENERAL SYSTEMS THEORY Systems symposium 2nd Proceedings Cleveland, Ohio 1963 Apr Edited by Mihajlo D. Mesarovic Held at the Case institute of technology Case institute of technology. Systems research center.Publications Wiley, New York,1964
Eng 41.5967

VIEWS ON GENERAL SYSTEMS THEORY The Second systems symposium proceedings Cleveland, Ohio 1963 Apr Case institute of technology Edited by Mihajlo D. Mesarovic xvii,178p 24cm John Wiley and sons,New York;London, 1964
Math 3.0880

VIGILANCE :a symposium Proceedings Santa Barbara,Calif. 1961 Jan United States.Office of naval research Edited by Donald N. Buckner and James J. McGrath Conducted by Human factors research McGraw-Hill series in psychology McGraw-Hill,New York,1963
Psy 31.1008

VIKING FUND Culture and personality :an interdisciplinary conference Proceedings New York 1947 Nov 7-8 Edited by S. Stansfeld Sargent and Marian W. Smith Viking fund,New York,1949
Psy 31.1931

VILLARD DE LANS 1966 Programming languages N.A.T.O. advanced study institute.Summer school North Atlantic treaty organization. Science committee Edited by F. Genuys 395p Academic press,London,1968
TA 15.0589

VILLARD-DE-LANS 1966 Programming languages N.A.T.O. advanced summer school Edited by F. Genuys Sponsored by the North Atlantic treaty organization.Science committee illus. 395p Academic press,London;New York,1968
Math L 5.3577

VILLAVICENCIO-MENDOZA 1954 Algunos problemas matematicos que se estan estudiando en Latino America symposium 2 Universidad nacional,Mendoza Supported by United nations educational,scientific and cultural organisation. Centre for scientific co-operation for Latin America 328p 24cm UNESCO,centro de cooperacion cientifica, Montevideo,1954
P. Math 2.2610

VILLEFRANCHE SUR MER 1961 Oceanographie geologique et geophysique de la Mediterrane occidentale :colloque international Travaux Centre national de la recherche scientifique Paris,1962
Geod 9.0581

VILLES ET CAMPAGNES Semaine sociologique 2nd Compte rendu Paris 1951 Mar Centre d'etudes sociologiques Edited by Georges Friedmann Ecole pratique des hautes etudes, Paris.Bibliotheque generale.Section 6 Colin, Paris,1953
Geog 13.4272

VINCI (FLORENCE) 1960 Symposium international d'histoire des sciences Actes International union of of the history and philosophy of science Academie internationale d'histoire des sciences. Collection de travaux, 13 235p Gruppe italiano di storia delle scienze,Vinci (Florence),1962
WSM 43.0063

VIRAL DISEASES The Pathogenesis and pathology of viral diseases :a symposium New York 1948 Dec 14-15 New York academy of medicine.Section on microbiology Edited by John G. Kidd New York academy of medicine. Section on microbiology.Symposia, 3 Columbia university press,New York,1950
Path 30.2554

VIRAL INFECTION Nucleic acid-protein interactions and nucleic acid synthesis in viral infection :symposium Miami,Fla. 1971 Jan 18-22 University of Miami.Department of biochemistry Papanicolaou cancer research institute Miami winter symposia, 2 North-Holland,Amsterdam,1971
Bioch 33.2241

VIRGINIA POLYTECHNIC INSTITUTE.ENGINEERING EXPERIMENT STATION. Physics of the solar system Conference on physics of the solar system and re-entry dynamics proceedings Blacksburg,Va. 1961 Jul.31-Aug.11 Part 1: physics of the solar system Virginia polytechnic institute.Engineering experiment station series, 149 Virginia polytechnic institute.Bulletin, 55,9 Virginia polytechnic institute,Blacksburg,Va.,1962
A Math 4.1155

VIRUS AND CANCER Symposium on virus and cancer Papers Saltsjobaden 1964 Jun 8-9 Swedish cancer society and Unio nordica contra cancrum Edited by H. Bergstrand and K. E. Hellstrom illus Balder,Stockholm,1965 Symposium arranged by the Swedish cancer society in conjunction with the annual general meeting of the Unio nordica contra cancrum
Path 30.2363

VIRUS AND RICKETTSIAL INFECTIONS The Dynamics of virus and rickettsial infections : an international symposium Proceedings Detroit,Mich. 1953 Oct 21-23 Henry Ford hospital Edited by Frank W. Hartmann and others illus,tables,diagrs Blakiston,New York;Toronto,1954
Path 30.2548

VIRUS-CELL INTERACTIONS AND VIRAL ANTIMETABOLITES Federation of European biochemical societies meeting 7th Proceedings Varna 1971 Sep Federation of European biochemical societies Edited by D. Shugar Federation of European biochemical societies.Publications, 22 Academic press,London,1972
Bioch 33.2323

VIRUS GROWTH AND VARIATION :a symposium London 1958 Apr Society for general microbiology Edited by A. Isaacs and B.W. Lacey Society for general microbiology. Symposia, 9 Cambridge university press for the Society of general microbiology,Cambridge, 1959
Path 30.2743

VIRUS GROWTH AND VARIATION :a symposium Papers London 1959 Apr Society for general microbiology Edited by A. Isaacs and B.W. Lacey Held at the University of London Society for general microbiology.Symposia, 9 Cambridge university press,Cambridge,1959
Bioch 33.1182

VIRUS MULTIPLICATION The Nature of virus multiplication :a symposium Papers Oxford 1952 Apr Society for general microbiology Edited by Paul Fildes and W.E. Van Heyningen Held at the University of Oxford Society for general microbiology.Symposia, 2 Cambridge university press,Cambridge,1953
Bioch 33.1179

VIRUS REPLICATION Federation of European biochemical societies meeting 4th Proceedings Oslo 1967 Jul 4 Vol 2: biochemistry of virus replication Federation of European biochemical societies Edited by S.G. Laland and L.O. Froholm Organized by the Norwegian biochemical society Universitetsforlaget;Academic press,Oslo; London,1968 Symposium organizer A.P. Nygaard
Bioch 33.1397

VIRUS VIRULENCE AND PATHOGENICITY London 1959 Jun 15 Ciba foundation Edited by G.E. W. Wolstenholme and C.M. O'Connor Ciba foundation study group, 4 Churchill,London, 1960
PGMS 29.0541

VIRUSES International congress of biochemistry 4th Proceedings Vienna 1958 Sep 1-6 Vol 7: symposium 7 - biochemistry of viruses International union of biochemistry Edited by E. Broda and W. Frisch-Niggemeyer I.U.B.symposium series, 9 Pergamon,London,1959 Added title page in French and German.Text in German and English
Bioch 33.1321

VIRUSES International congress of microbiology 6th Rome 1953 Vol 6: symposium - interaction of viruses and cells Edited by F.C. Bawden and G. Penso illus Fondazione Emanuele Paterno,Rome,1953 Title also in Italian;text in English and French
Bioch 33.1804

VIRUSES Lepetit colloquium on the biology of oncogenic viruses 2nd Proceedings Paris 1970 Nov Edited by Luigi G. Silvestri North-Holland,Amsterdam,1971
Bioch 33.2299

VIRUSES Molecular biology of viruses :a symposium Papers London 1968 Apr Society for general microbiology Edited by P. V. Crawford and M.G.P. Stoker Held at the Imperial college of science and technology Society for general microbiology.Symposia, 18 Cambridge university press,Cambridge,1968
Bioch 33.1178

VIRUSES The Ciba foundation symposium on the nature of viruses London 1956 Ciba foundation Edited by G.E.W. Wolstenholme and E.C.P. Millar Ciba foundation.Symposia Churchill,London,1957
Radioth 35.0752

VIRUSES The Molecular biology of viruses :a symposium London 1968 Apr Society for general microbiology Edited by L.V. Crawford and M.G.P. Stoker Society for general microbiology.Symposia, 18 Cambridge university press,Cambridge,1968
An 32.5380

VIRUSES :a symposium Papers Cold Spring Harbor 1953 Cold Spring Harbor biological laboratory Cold Spring Harbor symposia on quantitative biology, 18 Long Island biological association,Cold Spring Harbor,1953
Bioch 33.1274

VIRUSES,NUCLEIC ACIDS AND CANCER Annual symposium on fundamental cancer research,1963 17th Papers Houston,Tex. 1963 Anderson hospital and tumor institute Williams and Wilkins,Baltimore,Md.,1963
Radioth 35.0822

VISBY,SWEDEN 1965 Symposium:atmospheric chemistry,circulation and aerosols Proceedings World meteorological organization,and,Commission on atmospheric chemistry and radioactivity Tellus,18,pt 2-3 Svenska geofysiska foreningen,Stockholm,1966
Chem 18.0571

VISCOELASTICITY - PHENOMENOLOGICAL ASPECTS :a symposium Papers Lancaster,Pa. 1958 Apr 28-29 Edited by J.T. Bergen Sponsored by the Armstrong Cork company x,150p Academic press,New York;London,1960
Chem E 24.0413

VISCOUS DRAG REDUCTION A Symposium on viscous drag reduction Proceedings Dallas, Texas 1968 Sep 24-25 Edited by C.Sinclair Wells Sponsored by United States.Office of naval research 500p 25cm Plenum press, New York,1969
Cav 7.2923

VISCOUS DRAG REDUCTION Symposium on viscous drag reduction Proceedings Dallas,Texas 1968 Sep 24-25 United States.Office of naval research.Naval ship research and development center National aeronautics and space administration Edited by C.S. Wells illus xi,500p 25cm Plenum press,New York,1969
A Math 4.1736

VISION A Joint discussion on vision held at the Imperial college of science London 1932 Jun 3 Physical society Optical society Physical society,London,1932 Bound with 'Report of a discussion on audition' by the Physical society,q.v.
Psy 31.2610

VISTAS IN ASTRONAUTICS Annual astronautics symposium 1st United States.Air force. Office of scientific research Edited by Morton Alperin and others Co-sponsored with General dynamic corporation.Convair division Pergamon press,London,1958
Eng 41.6880

VISTAS IN ASTRONAUTICS Astronautics symposium 1st proceedings San Diego 1957 Feb United States.Air force.Office of scientific research Edited by Morton Alperin and Marvin Stern Co-sponsored by General dynamics corporation.Convair division Vistas in astronautics, 1 International series of monographs on aeronautical sciences and space flight.Astronautics division, 1 329p Pergamon press,London,1958
Nap 11.0022

VISTAS IN ASTRONAUTICS Astronautics symposium 2nd Proceedings Denver,Col. 1958 Apr.28 United States.Air force.Office of scientific research Edited by Morton Alperin and Hollingsworth F. Gregory Co-sponsored by Institute of aeronautical sciences Vistas in astronautics, 2 International series of monographs on aeronautical sciences and space flight. Astronautics division, 2 Pergamon press, London,1959
Nap 11.0023

VISTAS IN ASTRONAUTICS Astronautics symposium 3rd Proceedings Los Angeles, Calif. 1960 Oct.12-14 United States.Air force.Office of scientific research jointly sponsored Society of automative engineers Vistas in astronautics, 3 illus 266p Society of automotive engineers,New York,1960
Nap 11.0024

VISUAL INFORMATION Conferenza internazionale di informazione visiva :modalita dell'informazione visiva:ricerca scientifica e azione politica 1mo 1-2 Istituto per lo studio sperimentale di problemi sociali con tecniche filmologiche 2 vols Istituto per lo studio sperimentale di problemi sociali con tecniche filmologiche, Milan,1961 Papers in English,French and Italian
Psy 31.2545

VISUAL MECHANISMS :a symposium Chicago, Ill. 1941 Sep 24 Edited by Heinrich Kluver Biological symposia, 7 Cattell press, Lancaster,Pa.,1942
An 32.4752

VISUAL PROBLEMS OF COLOUR :a symposium Teddington 1957 Sep 23-25 Vol 1-2 National physical laboratory National physical laboratory.Symposia, 8 2 vols H. M.S.O.,London,1958
Phys 20.1033

VISUAL PROBLEMS OF COLOUR 6 :a symposium Proceedings Teddington 1957 Sep 23-25 Vol 1-2 National physical laboratory 2 vols. H.M.S.O.,London,1958
Psy 31.0353

VISUAL SEARCH TECHNIQUES :a symposium Proceedings Washington,D.C. 1959 Apr 7-8 Armed forces-National research council committee on vision Edited by Ailene Morris and E.Porter Horne National research council. Publication, 712 National research council, Washington,D.C.,1960
Psy 31.0943

The VISUAL SYSTEM:NEUROPHYSIOLOGY AND PSYCHOPHYSICS :symposium Freiburg 1960 Aug 28-Sep 3 Edited by Richard Jung and Hans Kornhuber Springer,Berlin,1961 Papers in English and German
An 32.4774

VISUAL SYSTEMS Neurophysiologie und psychophysik des visuellen systems (the visual systems:neurophysiology and psychophysics)a symposium Freiburg 1960 Aug 28-Sep 3 Edited by Richard Jung and Hans Kornhuber Springer,Berlin,1961 Papers in English, French and German
Psy 31.0355

VITAMIN B 12 COENZYMES :a conference Papers New York 1963 Apr 10-11 New York academy of sciences Edited by Harold E. Whipple New York academy of sciences.Annals, 112,p.547-921 New York,1964
Bioch 33.0565

VITAMIN B12 The Biochemistry of vitamin B12 : a symposium London 1955 Feb 19 Biochemical society Edited by R.T. Williams Held at the London school of hygiene and tropical medicine Biochemical society. Symposia, 13 Cambridge university press, Cambridge,1955
Bioch 33.1375

VITAMIN METABOLISM International congress of biochemistry 4th Proceedings Vienna 1958 Sep 1-6 Vol 11: symposium 11 - vitamin metabolism International union of biochemistry Edited by W. Umbreit and H. Molitor I.U.B.symposium series, 13 Pergamon,London,1960 Added title page in French and German
Bioch 33.1325

VLAAMSE CHEMISCHE VERENIGING Biochemical problems of lipids :an international conference 2nd Proceedings Ghent 1955 Jul 27-30 Edited by G. Popjak and E. Le Breton Butterworths,London,1956 Conference president:Professor R.Ruyssen
Chem 18.1500

VLADIVOSTOK 1966 Morskie mlekopitayushchie Sbornik uklyuchaet materialy tretego vsesoyuznogo soveshchaniya po morskim mlekopitayushchim Akademiya nauk S.S.S.R. Ministerstvo po rybnogo khozyaistva S.S.S.R. Ikhtiologicheskaya komissiya Edited by A.V.A. Arsenev and others 342p 27cm Izdatelstvo 'Nauka',Moscow,1969
Sco 14.8244

VOLCANISM International geological congress 20th papers Mexico City 1956 Seccion 1: vulcanologia del cenozoico,tom 1-2 Edited by A. Garcia Rojas and others Mexico City, 1957 Text in English,French,Russian and Spanish
Geol 8.3036

VOPROSY KHIMICHESKOI KINETIKI KATALIZA I REAKTSIONNOI SPOSOBNOSTI :doklady k vsesoyuznomu soveshchaniyu po khimicheskoi kinetike i reaktsionnoi sposobnosti Doklady Akademiya nauk S.S.S.R.Otdelenie khimicheskikh nauk Edited by V.N. Kondratev and N.M. Emanuel Izdatelstvo Akademii nauk SSSR,Moscow,1955
Chem 18.0863

VORTRAGE DES FRIDTJOF-NANSEN-GEDACHTNIS-SYMPOSIONS UBER SPITZBERGEN IN NANSENS 100. GEBURTSJAHR (GEB 10.10.1861) Stauferland-expedition,1959-60 Wurzburg 1961 Apr 3-11 Edited by Julius Budel and Alfred Wirthmann Ergebnisse der Stauferland-expedition,3 illus,7 plates 86p 25cm Franz Steiner, Wiesbaden,1965
Sco 14.2974

VORTRAGE UBER RECHENANLAGEN Gottingen 1953 Mar 19-21 Deutsche forschungsgemeinschaft Edited by L. Biermann 145p 21cm Max-Planck institut fur physik,Gottingen,1953
Math L 5.1087

VOYAGES AND EXPEDITIONS,1963-67 Ces montagnes qui flottent sur la mer By Charles Pierre Peguy Centre national de la recherche scientifique illus,maps,17 plates 314p 20cm Arthaud,1969
Sco 14.8024

VSESOYUZNAYA AKADEMIYA SELSKOKOZYAISTENNYK NAUK IMENI V.I.LENINA The Situation in biological science verbatim report Proceedings Moscow 1948 Jul 31-Aug 7 Foreign languages publishing house,Moscow,1949 Translated from the Russian
Gen 34.1210

VSESOYUZNAYA ORNITOLOGICHESKYAYA KONFERENTSIYA 11-17 SENTYABRYA,1962 3rd Materialy Lvov 1962 September 11-17 20cm Lvov,1962
Philos 1.1626

W.M.O.-S.C.A.R.-I.C.P.M.SYMPOSIUM ON POLAR METEOROLOGY Proceedings Polar meteorology Geneva 1966 Sep 5-9 World meteorological organization,and,Scientific committee on Antarctic research Also organized by the International commission on polar meteorology illus vi,540p 28cm World meteorological organization,Geneva,1967
Sco 14.7839

WAGENINGEN 1923 International conference of phytopathology and economic entomology Report Edited by T.A.C. Schoevers Committee of management,Wageningen,1923
Bot 42.0650

WAGENINGEN 1956 Chromosomes Conference on chromosomes Lectures Willink,Zwolle, 1956 Papers in English and German
An 32.3379

WAGENINGEN 1963 Barley genetics 1 International barley genetics symposium 1st Proceedings Centre for agricultural publications and documentation,Wageningen,1964
Gen 34.1560

WAGENINGEN 1965 Cell differentation and morphogenesis :international lecture course By W. Beermann and others Organized by the Landbouwhoogeschool,Wageningen North-Holland, Amsterdam,1966 "The fourth international symposium organized by the Landbouwhoogeschool"
Bal 39.0309

WAGENINGEN 1965 Cell differentiation and morphogenesis :international lecture course By W. Beermann and others Organized by the Agricultural university of Wageningen North Holland,Amsterdam,1966 The fourth international symposium organized by the Agricultural university of Wageningen
Bot 42.1098

WAGENINGEN 1965 Cell differentiation and morphogenesis :international lecture course Edited by W. Beermann and others Organized by the Agricultural university of Wageningen North-Holland,Amsterdam,1966 "The fourth international symposium organized by the Agricultural university of Wageningen"
Bioch 33.1020

WAGENINGEN 1965 Cell differentiation and morphogenesis :international lecture course Landbouwhogeschool,Wageningen Edited by W. Beermann and others North-Holland,Amsterdam, 1969
An 32.5439

WAGENINGEN 1965 Cell differentiation and morphogenesis... :international lecture course By W. Beerman Landbouwhogeschool,Wageningen North-Holland,Amsterdam,1966
Gen 34.2211

WAGENINGEN 1968 Air pollution European congress on the influence of air pollution on plants and animals 1st Proceedings Centre for agricultural publishing and documentation illus. 415p Centre for agricultural publishing and documentation, Wageningen,1969
Bot 42.2106

WAKULLA SPRINGS 1963 The Origins of prebiological systems and of their molecular matrices :a conference Proceedings Edited by Sidney W. Fox Academic press,New York; London,1965 Conducted under the auspices of the Institute for space biosciences,the Florida state university and the National aeronautics and space administration
Bal 39.0992

WAKULLA SPRINGS,FLA 1963 Origins of prebiological systems and of their molecular matrices A Conference on the origins of prebiological systems and of their molecular matrices Proceedings Institute for space biosciences Florida state university National aeronautics and space administration Edited by Sidney W. Fox Bibliog.,illus. xx, 482p Academic press,New York;London,1965
Bot 42.1364

WAKULLA SPRINGS,FLA. 1963 The Origins of prebiological systems and of their molecular matrices :a conference Proceedings Florida state university National aeronautics and space administration Edited by Sidney W. Fox Under the auspices of the Institute for space biosciences Academic press,New York;London, 1965
Gen 34.0456

WALL STRUCTURES Tall buildings with particular reference to sheer wall structures : a symposium Proceedings Southampton 1966 Apr Edited by A. Coull and B.Stafford Smith Pergamon press,London,1967
Eng 41.2733

WALTER AND ELIZA HALL INSTITUTE OF MEDICAL RESEARCH The Thymus:experimental and clinical studies Ciba foundation symposium Melbourne 1965 Aug 25-27 Edited by G.E.W. Wolstenholme and Ruth Porter illus Churchill,London,1966 In honour of sir Macfarlane Burnet,chairman of the conference
Path 30.2461

WALTER REED ARMY INSTITUTE OF RESEARCH Complexes of biologically active substances with nucleic acid and their mode of action :a research symposium Proceedings Washington, D.C. 1970 Mar 16-19 Edited by F.E. Hahn Progress in molecular and subcellular biology, 2 Springer,Berlin,1971
Bioch 33.2242

WALTER REED ARMY MEDICAL CENTER.INSTITUTE OF RESEARCH Symposium on preventive and social psychiatry Washington,D.C. 1957 Apr 15-17 US government printing office, Washington,D.C.,c1957
Psy 28.0058

WALTER REID ARMY INSTITUTE OF RESEARCH The Delayed effects of whole body radiation :a symposium Bethesda,Md. 1959 Oct 29 Johns Hopkins university.Operation research office Edited by Bernard B. Watson Johns Hopkins press,Baltimore,Md.,1960
Radioth 35.1968

WALTHAM,MASS 1961 Lectures in theoretical physics Vol. 2 By M.E. Rose and E.C.G. Sudarshan Brandeis university.Summer institute in theoretical physics Lectures in theoretical physics,1961,2 Benjamin,New York, 1962
A Math 4.0830

WALTHAM,MASS 1962 Lectures on particle and field theory Vol 1: elementary particle physics and field theory By I. Fulton and others Brandeis university.Summer institute in theoretical physics Lectures in theoretical physics,1962,1 Benjamin,New York, 1963
Cav 7.0317

WALTHAM,MASS 1964 Lectures on particles and field theory Vol. 2 By K. Johnson and others Brandeis university.Summer institute in theoretical physics Lectures on theoretical physics,1964,2 Prentice-Hall,New York,1965
A Math 4.0833

WALTHAM,MASS. 1959 Theoretical physics: lecture notes By F.E. Low Brandeis university.Summer institute in theoretical physics Edited by Wayne A. Mills Lectures in theoretical physics,1959 Brandeis university,Waltham,Mass.,1959
A Math 4.0828

WALTHAM,MASS. 1960 Synthesis of molecular and cellular structure :a symposium Papers Society for the study of development and growth Edited by Dorothea Rudnick Held at Brandeis university Society for the study of development and growth.Symposia, 19 Ronald press,New York,1961
Bioch 33.1012

WALTHAM,MASS. 1960 Theoretical physics : lecture notes By C. Moller and others Brandeis university.Summer institute in theoretical physics Edited by J. Stachel and others Lectures in theoretical physics,1960 Brandeis university,Waltham,Mass,c1961
A Math 4.0829

WALTHAM,MASS. 1960 Theoretical physics : lecture notes Typescript By C. Moller and others Brandeis university.Summer institute in theoretical physics Edited by J. Stachel and others Lectures in theoretical physics, 1960 Brandeis university,Waltham,Mass.,1960
Cav 7.0314

WALTHAM,MASS. 1961 Lectures in theoretical physics Pt 1 By R.J. Eden Brandeis university.Summer institute in theoretical physics Lectures in theoretical physics,1961, 1 Benjamin,New York,1962
Cav 7.0315

WALTHAM,MASS. 1961 Lectures in theoretical physics Pt 2 By M.E. Rose and E.C.G. Sudarshan Brandeis university.Summer institute in theoretical physics Lectures in theoretical physics,1961,2 Benjamin,New York, 1962
Cav 7.0316

WALTHAM,MASS. 1962 Astrophysics and the many-body problem By E.N. Parker Brandeis university.Summer institute in theoretical physics Lectures in theoretical physics,1962, 2 W.A.Benjamin,New York;Amsterdam,1963
A Math 4.1113

WALTHAM,MASS. 1962 Lectures in theoretical physics Vol 2: astrophysics and the many-body problem By E.N. Parker and J.S. Goldstein Brandeis university.Summer institute in theoretical physics Lectures in theoretical physics,1962,2 Benjamin,New York, 1963
Cav 7.0318

WALTHAM,MASS. 1962 Lectures on particles and field theory Vol. 1: elementary particle physics and field theory By I. Fulton and others Brandeis university.Summer institute in theoretical physics Lectures in theoretical physics,1962,1 Benjamin,New York, 1963
A Math 4.0831

WALTHAM,MASS. 1962 Statistical physics By G.E. Uhlenbeck and others Brandeis university.Summer institute in theoretical physics Lectures in theoretical physics,1962, 3 252p W.A.Benjamin,New York,1963
Cav 7.2414

WALTHAM,MASS. 1963 Lectures on astrophysics and weak interactions By S. Hayakawa and others Brandeis university. Summer institute in theoretical physics Lectures in theoretical physics,1963,2 Brandeis university,Waltham,Mass.,1963
A Math 4.1087

WALTHAM,MASS. 1963 Lectures on astrophysics and weak interactions Vol. 2 By S. Hayakawa and others Brandeis university.Summer institute in theoretical physics Lectures in theoretical physics,1963, 2 Brandeis university,Waltham,Mass.,1964
A Math 4.0832

WALTHAM,MASS. 1963 Lectures on strong and electromagnetic interactions By P.T. Matthews and others Brandeis university. Summer institute in theoretical physics Lectures in theoretical physics,1963,1 Brandeis University,Waltham,Mass.,1964
A Math 4.0036

WALTHAM,MASS. 1964 Lectures on general relativity By A. Trautman Brandeis university.Summer institute in theoretical physics Lectures in theoretical physics,1964, 1 Prentice-Hall,Englewood Cliffs,N.J.,1965
A Math 4.1025

WALTHAM,MASS. 1964 Lectures on general relativity By A. Trautman and others Brandeis university.Summer institute in theoretical physics Lectures in theoretical physics,1964,1 Prentice-Hall,Englewood Cliffs,N.J.,1965
Cav 7.0320

WALTHAM,MASS. 1964 Lectures on particles and field theory By K. Johnson and others Brandeis university.Summer institute in theoretical physics Lectures in theoretical physics,1964,2 Prentice-Hall,Englewood Cliffs,N.J.,1965
Cav 7.0321

WALTHAM,MASS. 1965 Lectures on particle symmetries and axiomatic field theory Papers 1: axiomatic field theory Brandeis university.Summer institute in theoretical physics Edited by M. Chretien and S. Deser Lectures in theoretical physics,1965,1 Gordon and Breach,New York,1966
A Math 4.1334

WALTHAM,MASS. 1968 Astrophysics and general relativity Brandeis university. Summer institute in theoretical physics 11th Proceedings Edited by Max Chretien and others 2 vols Gordon and Breach,New York, 1971
TA 15.0657

WALTHAM,MASS. 1968 Astrophysics and general relativity Brandeis university. Summer institute of theoretical physics 11th Vol 1 Edited by M. Chretien and others illus 300p 23cm 2 vols Gordon and Breach,New York,1969
A Math 4.1791

WALTHAM,MASS. 1969 Atomic physics and astrophysics Brandeis university.Summer institute in theoretical physics 12th Proceedings Edited by Max Chretien and E. Lipworth 216p Gordon and Breach,New York, 1971
TA 15.0683

WARREN,MICH. 1959 Theory of traffic flow symposium 3rd proceedings Edited by Robert Herman Sponsored by General motors corporation.Research laboratories Elsevier, Amsterdam,1961
A Math 4.0949

WARREN,MICH. 1962 Cavitation in real liquids :a symposium Proceedings General motors research laboratories Edited by Robert Davies Elsevier,Amsterdam,1964
Eng 41.6512

WARREN,MICH. 1962 Cavitation in real liquids... :a symposium 6th Proceedings Edited by Robert Davies Sponsored by General motors research laboratories port vii,189p Elsevier,Amsterdam,1964
Chem E 24.0240

WARREN,MICH. 1963 Liquids:structure, properties,solid interactions Symposium on liquids:structure,properties,solid interactions proceedings Edited by Thomas J. Hughel General motors research laboratory. Symposia, 7 Elsevier,Amsterdam,1965
A Math 4.0951

WARREN,MICH. 1963 Liquids:structure, properties,solid interactions :the symposium Proceedings General motors research laboratories Edited by Thomas J. Hughel Elsevier,Amsterdam,1965
Met 25.1514

WARREN,MICH. 1964 Approximation of functions Symposium on approximation of functions 8th Proceedings General motors research laboratories Edited by Henry L. Garabedian 220p Elsevier,Amsterdam,1965
TA 15.0097

WARREN,MICH. 1964 Approximation of functions Symposium on approximation of functions 8th proceedings General motors corporation.Research laboratories Edited by Henry L. Garabedian viii,220p 25cm Elsevier publishing co.,Amsterdam,1965
P. Math 2.0978

WARREN,MICH. 1964 Symposium on the approximation of functions 8th proceedings General motors corporation. Research laboratories Edited by Henry L. Garabedian bibliog. viii,220p 24cm Elsevier,Amsterdam,1965
Math L 5.0348

WARREN,MICH. 1965 Symposium on the fluid mechanics of internal flow General motors research laboratories Edited by Gino Sovran Elsevier,Amsterdam,1967
Eng 41.6553

WARRENTON,VA. 1963 Photosynthetic mechanisms of green plants :a symposium Papers Sponsored by the National research council.Committee on photobiology National research council.Publication,1145 Washington, 1963
Bioch 33.0774

WARRENTON,VA. 1963 Small blood vessel involvement in diabetes mellitus :conference Proceedings American institute of biological sciences Edited by Marvin D. Siperstein and others American institute of biological sciences,Washington,D.C.,1964
Bioch 33.0482

WARRENTON,VA. 1965 Future environments of North America :being the recrd of a conference convened by the Conservation Foundation Edited by F.Fraser Darling and John P. Milton Natural history press,Garden City,N.Y.,1966
Geog 13.4873

WARSAW 1953 Sesja kopernikowska... Polska akademia nauk 482p . Panstowe wydawnictwo naukowe,Warsaw,1955
WSM 43.2653

WARSAW 1958 Konferencja poswienconej omowieniu wstepnych wynikow Polskiej Wyprawy na Spitsbergen Przeland geofizyczny Rocznik, 3,pt.2 Warsaw,1960
Sco 14.2969

WARSAW 1958 Non-homogeneity in elasticity and plasticity :a symposium International union of theoretical and applied mechanics Edited by W. Olszak Pergamon press,London, 1959
Eng 41.2488

WARSAW 1959 Infinitistic methods Symposium on foundations of mathematics proceedings Polska akademia nauk.Instytut matematyczny Under the auspices of the International mathematical union 362p 26cm Pergamon press,Oxford,1961
P. Math 2.0211

WARSAW 1960 Ecological effects of biological and chemical control of undesirable plants and animal symposium International union for the conservation of nature and natural resources. Technical meeting 8th Edited by D.J. Kuenen Leiden,1961 Title also in French:Effets ecologiques du controle biologique et chimique des plantes et animaux undesirables
Bal 44.6435

WARSAW 1960 Fatigue resistance of materials and metal structural parts international conference Proceedings Institute of precision mechanics,Warsaw Edited by Alfred Buch Pergamon press;Polish scientific publishers,Oxford;Warsaw,1964
Met 25.0933

WARSAW 1961 European conference on tumour biology International union against cancer International union against cancer.Acta, 18, no 1-2 U.I.C;C.,Louvain,1962
Radioth 35.0817

WARSAW 1963 Non-classical shell problems : a symposium Proceedings International association for shell structures Edited by W. Olszak and A. Sawczuk North-Holland;PWN, Amsterdam;Warsaw,1964
Eng 41.2876

WARSAW 1965 Cracow Congres international d'histoire des sciences 11e Actes 1-6 Polska akademia nauk.Comite et institut d'histoire de la science et de la technique 6 vols Ossolineum.Maison d'edition de l'Academie polonaise des sciences,Wroclaw,1968
WSM 43.0079

WARSAW 1966 Biochemistry of blood platelets :colloquium held on the occasion of the Third meeting of the Federation of European biochemical societies Edited by E. Kowalski and S. Niewiarowski Organized by the Polish biochemical society Academic press,New York,1967
Radioth 35.0148

WARSAW 1966 The Federation of European biochemical societies meeting 3rd biochemistry of mitochondria;colloquium Federation of European biochemical societies Edited by E.C. Slater and others Organized by the Polish biochemical society Academic press,London,1967
Bioch 33.1394

WARSAW 1966 The Federation of European biochemical societies meeting 3rd genetic elements:properties and function; symposium Federation of European biochemical societies Edited by D. Shugar Organised by the Polish biochemical society Academic press,London,1967
Bioch 33.1395

WASHBURN OBSERVATORY Resonance lines in astrophysics :manuscripts presented at a conference... Boulder,Colo. 1968 Sep 9-11 480p National center for atmospheric research,Boulder,Colo.,1968
Obs 6.3349

WASHINGTON 1922 and 1928 Conference on cycles 1st-2nd Report Carnegie Institution of Washington Washington,1929
Philos 1.0096

WASHINGTON 1950 Air pollution United States technical conference on air pollution Proceedings Sponsored by the United States. Interdepartmental committee on air pollution xiv,847p McGraw-Hill,New York,1952 Chairman of the committee:Louis C.McCabe
Chem E 24.1542

WASHINGTON 1951 Frost action in soils :a symposium 30th Annual meeting Highway research board National research council. Publication,213 Highway research board. Special report 2 illus,tables,diagrs x, 385p 25cm Highway research board, Washington,D.C.,1952
Sco 14.0395

WASHINGTON 1952 Wave motion and vibration theory symposium Carnegie institute of technology and American mathematical society Edited by Albert E. Heins American mathematical society.Proceedings of symposia in applied mathematics, 5 McGraw-Hill,New York,1954
A Math 4.0513

WASHINGTON 1957 Preventive and social psychiatry :a symposium Proceedings Army institute of research 1957
Med 36.0011

WASHINGTON 1964 The Electron microprobe :a symposium Proceedings Electrochemical society.Electrothermics and metallurgy division Edited by T.D. McKinley and others Wiley,New York,1966
Met 25.2362

WASHINGTON 1966 Some mathematical problems in biology Symposium on mathematical biology Proceedings Edited by Murray Gerstenhaber Prepared by the American mathematical society American mathematical society.Lectures on mathematics in the life sciences, 1 Bibliog.,Illus. vi and 117p 22cm American mathematical society,Rhode Island, 1968
P Math 2.3482

WASHINGTON,D.C. 1884 Adoption d'un premier meridien unique et d'une heure universelle : conference Proces-verbaux United States. Department of state 216p Washington,D.C., 1884
Obs 6.3168

WASHINGTON,D.C. 1884 International meridian conference International conference for the purpose of fixing a prime meridian and a universal day 1st Proceedings United States.Department of state 212p Gibson, Washington,D.C.,1884 With five sheets of notes by J.C.Adams in a separate envelope
Obs 6.3169

WASHINGTON,D.C. 1897 Congress of American physicans and surgeons 4th Transactions New Haven,Conn.,1897
An 32.1146

WASHINGTON,D.C. 1903 Congress of American physicians and surgeons 6th triennial session Transactions New Haven,Conn.,1903
An 32.1147

WASHINGTON,D.C. 1915-16 Astronomy, meteorology and seismology The Pan American scientific congress 2nd Proceedings Pan American union Edited by Glen Levin Swiggett illus xiv,847p Government printing office, Washington,D.C.,1917
Nap 11.0838

WASHINGTON,D.C. 1935 Copper resources of the world International geological congress 16th papers Vol 1-2 2 vols Washington, D.C.,1935
Min 10.0202

WASHINGTON,D.C. 1939 Association internationale d'hydrologie scientifique : reunion Comptes rendus et rapports Tome 1-2 2 vols Washington,D.C. Vol 1 imperfect:contains only papers read at meetings of the commissions of potamology and of limnology.In English,French,German and Italian
Sco 14.0117

WASHINGTON,D.C. 1941 Mental health in later maturity :a conference Papers United States.Public health service United States. Public health service.Public health reports. Supplement, 168 US government printing office,Washington,D.C.,1942
Psy 31.3388

WASHINGTON,D.C. 1948 Tropical medicine and malaria International congress on tropical medicine and malaria 4th Proceedings Vol 1-2 United States.Department of State United States.Department of State. International organization and conference series, 1,5 Govt. printing office, Washington,D.C.,1948
Mol 45.0657

WASHINGTON,D.C. 1950 Air pollution Conference on air pollution Proceedings United States.Bureau of mines Edited by Louis C. McCabe Sponsored by United States. Interdepartmental committee on air pollution xiv,847p McGraw-Hill,New York,1952
Nap 11.0388

WASHINGTON,D.C. 1950 Symposium on chemical-biological correlation 1st Sponsored by the Chemical-biological co-ordination centre National academy of sciences.National research council,Washington,D.C.,1951
Radioth 35.0329

WASHINGTON,D.C. 1951 Characteristics and applications of resistance strain gages :a symposium Proceedings National bureau of standards National bureau of standards. Circular, 528 Washington,D.C.,1954
Eng 41.2902

WASHINGTON,D.C. 1951 Electrochemical constants NBS semicentennial symposium 7th Proceedings National bureau of standards National bureau of standards. Circular, 524 U.S.Government printing office, Washington,D.C.,1953 One of 12 semicentennial symposia
Met 25.2401

WASHINGTON,D.C. 1951 Gravity waves NBS semicentennial symposium on gravity waves 3rd Proceedings National bureau of standards National bureau of standards. Circular, 521 illus iv,287p U.S. Government printing office,Washington,D.C., 1952
Chem 8 24.0215

WASHINGTON,D.C. 1951 Gravity waves NBS semicentennial symposium on gravity waves Proceedings National bureau of standards National bureau of standards.Circular, 521 U. S.government printing office,Washington,D.C., 1952
Eng 41.6664

WASHINGTON,D.C. 1951 Mechanical properties of metals at low temperatures :NBS semi-centennial symposium By N.P. Allen and others United States.National bureau of standards United States.National bureau of standards.Circular,520 illus,tables,diagrs iv,206p 24cm Washington,D.C.,1952
Sco 14.1244

WASHINGTON,D.C. 1951 Optical image evaluation Semicentenial symposium on optical image evaluation Proceedings National bureau of standards National bureau of standards.Circular, 526 289p Washington, D.C.,1954
Obs 6.2129

WASHINGTON,D.C. 1952 Commercially available general-purpose electronic digital computers of moderate price symposium Sponsored by United States.Navy mathematical computing advisory panel 41p Washington,D.C.,1952
Math L 5.1790

WASHINGTON,D.C. 1952 International geographical union.General assembly,8th International geographical congress 17th Proceedings National research council, Washington,D.C.,1952
Geog 13.5597

WASHINGTON,D.C. 1953 Managerial aspects of digital computer installations symposium papers Organized by United States.Navy mathematical computing advisory panel 36p Office of naval research,Washington,D.C.,1953
Math L 5.1792

WASHINGTON,D.C. 1954 Automatic programming for digital computers :symposium Organized by United States.Navy mathematical computing advisory panel 152p Office of naval research,Washington,D.C.,1954
Math L 5.1795

WASHINGTON,D.C. 1954 Symposium on atherosclerosis National research council. Division of medical sciences Sponsored by Human factors division.Air force directorate of research and development National research council.Publication, 338 National research council,Washington,D.C.,1955
Med 36.0266

WASHINGTON,D.C. 1954 Symposium on temperature 3rd Papers Edited by Hugh C. Wolfe Sponsored by the American institute of physics Reinhold,New York,1955
Eng 41.4126

WASHINGTON,D.C. 1954 Temperature - its measurement and control in science and industry : a symposium 3rd Papers Vol 2 American institute of physics and United States.Army.Office of ordnance research Edited by Hugh C. Wolfe Reinhold;Chapman and Hall,New York;London,1955
Met 25.0299

WASHINGTON,D.C. 1955 Linear programming symposium 2nd Proceedings Vol 1-2 National bureau of standards United States. Air force.Directorate of management analysis Sponsored by the United States.Air force. Office of scientific research 24cm 2 vols National bureau of standards,Washington,D.C., 1955
Math L 5.3248

WASHINGTON,D.C. 1955 Molecular structure and biological specificity :a symposium American institute of biological sciences Edited by Linus Pauling and Harvey A. Hana Sponsored by the United States.Office of naval research American institute of biological sciences.Publications, 2 American institute of biological sciences,Washington,D.C.,1957
An 32.3156

WASHINGTON,D.C. 1955 Molecular structure and biological specificity :a symposium Papers United States.Office of naval research Edited by Linus Pauling and Harvey A. Itano Arranged by the American institute of biological sciences American institute of biological sciences.Publication,2 American institute of biological sciences,Washington,D.C.,1957
Chem 18.2137

WASHINGTON,D.C. 1955 Molecular structure and functional activity of nerve cells :a symposium American institute of biological sciences Edited by Robert G. Grenell and L.J. Mullins Sponsored by the United States. Office of naval research American institute of biological sciences.Publications, 1 American institute of biological sciences, Washington,D.C.,1956
An 32.4348

WASHINGTON,D.C. 1955 The Physiology of induced hypothermia :a symposium Proceedings National research council.Division of medical sciences Edited by Robert D. Dripps National research council.Publications, 451 Washington,D.C.,1956
Phys 20.0987

WASHINGTON,D.C. 1955-57 North Pacific fur seal conference Proceedings 5 vols Washington D.C.,1957 Mimeographed
Sco 14.1174

WASHINGTON,D.C. 1956 Advanced programming methods for digital computers symposium proceedings United States.Office of naval research Cosponsored by United States.Navy mathematical computing advisory panel United States.Office of naval research.ONR symposium reports ACR-15 83p office of naval research,Washington,D.C.,1956
Math L 5.1809

WASHINGTON,D.C. 1956 Metal molybdenum The Technology of molybdenum and its alloys :a symposium Proceedings United States.Office of naval research Edited by Julius J. Harwood American society for metals, Cleveland,Ohio,1958
Met 25.0546

WASHINGTON,D.C. 1957 Symposium on preventive and social psychiatry Walter Reed army medical center.Institute of research and National research council.Division of medical sciences US government printing office,Washington,D.C.,c1957
Psy 28.0058

WASHINGTON,D.C. 1958 Calcification in biological systems :a symposium presented at the Washington meeting of the American association for the advancement of science American association for the advancement of science Edited by Reidar F. Soggnaes American association for the advancement of science.Publication, 64 American association for the advancement of science, Washington,D.C.,1960
An 32.3237

WASHINGTON,D.C. 1958 Congenital heart disease :a symposium American association for the advancement of science Edited by Allan D. Bass and Gordon K. Moe American association for the advancement of science. Publications, 63 American association for the advancement of science,Washington,D.C., 1960
An 32.2063

WASHINGTON,D.C. 1958 Data processing seminar on status of digital computer and data processing developments in the Soviet Union United States.Office of Naval research 179p 26cm United States Office of Naval research, Washington,D.C.,1958
Math L 5.0821

WASHINGTON,D.C. 1958 Research in psychotherapy :a conference Proceedings American psychological association.Division of clinical psychology Edited by Eli Rubinstein and Morris B. Parloff National publishing co. Washington,D.C.,1959
Psy 31.1627

WASHINGTON,D.C. 1958 Symposium on microminiaturization of electronic assemblies Proceedings Diamond ordnance fuze laboratories Edited by Eleonor F. Horsey and Laurence D. Shergalis Hayden,New York,1960
Eng 41.5541

WASHINGTON,D.C. 1958 Technology of columbium (Niobium) :a symposium Electrochemical society.Electrothermics and metallurgy division Wiley;Chapman and Hall, London,1958
Met 25.0587

WASHINGTON,D.C. 1959 Amino acids,proteins and cancer biochemistry The Jesse P. Greenstein memorial symposium Papers American chemical society.Division of biological chemistry Edited by John T. Edsall Academic press,New York;London,1960 With a biographical article on Dr. Greenstein and a bibliography of his writings
Bioch 33.2345

WASHINGTON,D.C. 1959 Conference on experimental clinical cancer chemotherapy National cancer institute Edited by B.H. Morrison National cancer institute.Monograph, 3 U.S.Department of health,education and welfare,Washington,D.C.,1960
Radioth 35.0809

WASHINGTON,D.C. 1959 International oceanographic congress 1st Preprints of abstracts of papers American association for the advancement of science Edited by Mary Sears illus,maps 1022p 23cm Washington, D.C.,1959
Sco 14.0129

WASHINGTON,D.C. 1959 The Conference on Antarctica :conference documents.The Antarctic treaty and related papers Conference documents United States.Department of state. Publication,7060,International organization and conference series,13 78p Washington,D. C.,1960
Sco 14.6121

WASHINGTON,D.C. 1959 Visual search techniques :a symposium Proceedings Armed forces-National research council committee on vision Edited by Ailene Morris and E.Porter Horne National research council.Publication, 712 National research council,Washington,D.C. 1960
Psy 31.0943

WASHINGTON,D.C. 1960 Chemistry of cement International symposium in the chemistry of cement 4th Proceedings Vol 1-2 National bureau of standards National bureau of standards.Monograph, 43,vol 1-2 2 vols U.S.government printing office,Washington,D.C., 1962
Eng 41.3026

WASHINGTON,D.C. 1960 Chemistry of cement International symposium on the chemistry of cement 4th Proceedings Vol 1-2 National bureau of standards National bureau of standards.Monographs,43 2 vols Washington,D.C.,1962
Min 10.0647

WASHINGTON,D.C. 1960 Electrical studies on the unanesthetized brain :a symposium Edited by Estelle R. Ramey and Desmond S. O'Doherty Hoeber,New York,1960
An 32.4648

WASHINGTON,D.C. 1960 International congress of physical medicine 3rd Proceedings American congress of physical medicine and rehabilitation and American academy of physical medicine and rehabilitation Chicago,Ill.,1960
Phys 20.0990

WASHINGTON,D.C. 1960 Symposium on superconductive techniques for computing systems proceedings Sponsored by United States.Office of naval research illus. vii, 413p 26cm Office of naval research, Washington,D.C.,1960
Math L 5.0859

WASHINGTON,D.C. 1961 Advances in materials research in the N.A.T.O. nations :a symposium Proceedings North Atlantic treaty organization.Advisory group for aeronautical research and development Edited by H. Brooks and others AGARD 62 Pergamon press,Oxford, 1963
Met 25.0096

WASHINGTON,D.C. 1961 Annual symposium on tracer methodology 5th Proceedings New England nuclear corporation Packard instrument company Edited by Seymour Rothschild Advances in tracer methodology, 1 Plenum press,New York,1963 Includes selected papers from the first four annual symposia
Gen 34.0633

WASHINGTON,D.C. 1961 Large-capacity memory techniques for computing systems :a symposium Edited by Marshall C. Yovits Sponsored by the Association for computing machinery. Information systems branch Association for computing machinery.Monograph series Macmillan,New York;London,1962
Eng 41.2196

WASHINGTON,D.C. 1961 National vacuum symposium 8th Transactions Vol 1-2 American vacuum society and International organization for vacuum science and technology Edited by Luther E. Preuss 2 vols Pergamon press,Oxford,1962 Combined with the International congress on vacuum science and technology,2nd
Cav 7.2376

WASHINGTON,D.C. 1962 Advances in enzymic hydrolysis of cellular and related materials : a symposium Proceedings Edited by Elwyn T. Reese Sponsored by the American chemical society Pergamon,Oxford,1963 Including a bibliography for the years 1950-61
Bot 42.1728

WASHINGTON,D.C. 1962 Advances in glass technology International congress on glass 4th Technical papers American ceramic society Plenum press,New York,1962
Met 25.0116

WASHINGTON,D.C. 1962 International symposium on rocket and satellite meteorology Edited by H. Wexler and J.E. Caskey,jnr Sponsored by Committee on space research ix, 440p North-Holland,Amsterdam,1963
Sco 14.0516

WASHINGTON,D.C. 1962 International symposium on the use of artificial satellites for geodesy 1st Proceedings Committee on space research and International association of geodesy Edited by George Veis Co-sponsored by the International union of geodesy and geophysics North-Holland, Amsterdam,1963
Geod 9.0055

WASHINGTON,D.C. 1962 Optical processing of information :a symposium Edited by D.K. Pollock and others Cleaver-Hume,Baltimore,Md. 1963
Eng 41.5877

WASHINGTON,D.C. 1962 Reactions of coordinated ligands and homogeneous catalysis : a symposium Papers American chemical society.Division of inorganic chemistry American chemical society.Advances in chemistry series,37 American chemical society,Washington,D.C.,1963 Symposium chairman:D.H.Busch
Chem 18.1010

WASHINGTON,D.C. 1962 Symposium on organ culture :studies of development,function,and disease National cancer institute Edited by Clyde J. Dawe Sponsored by Tissue culture association and National cancer institute National cancer institute.Monograph, 11 U.S. Department of health,education,and welfare, Washington,D.C.,1963
Bioch 33.2367

WASHINGTON,D.C. 1963 Antarctic treaty meetings on telecommunications varp 37cm Washington,D.C.,1963 Unpublished conference documents
Sco 14.6118

WASHINGTON,D.C. 1963 Cell-bound antibodies a conference Proceedings National research council.Committee on tissue transplantation Edited by Bernard Amos and Hilary Koprowski Wistar institute press, Philadelphia,1963
Bioch 33.0534

WASHINGTON,D.C. 1963 Cell-bound antibodies conference Proceedings Edited by Bernard Amos and Hilary Koprowski Sponsored by the National research council.Committee on tissue transplantation Wistar institute press, Philadelphia,Pa.,1963
An 32.3296

WASHINGTON,D.C. 1963 Communication processes NATO symposium on communication processes Proceedings North Atlantic treaty organization.Advisory group on human factors Edited by Frank A. Geldard and others NATO conference series, 4 298p Pergamon press,Oxford,1965
Cav 7.2423

WASHINGTON,D.C. 1963 Ellipsometry in the measurement of surfaces and thin films symposium Proceedings National bureau of standards Edited by E. Passaglia and others National bureau of standards.Miscellaneous publication, 256 U.S.Government printing office,Washington,D.C.,1964
Met 25.1687

WASHINGTON,D.C. 1963 Ideas in modern biology International congress of zoology 16th Proceedings Edited by J.A. Moore Doubleday,NEW York,1965
Radioth 35.0259

WASHINGTON,D.C. 1963 Pavement design in frost areas :Annual meeting 42nd eleven reports 2: design considerations By James F. Haley and others Highway research board National research council.Publication,1153, Highway research record,33 illus 270p 25cm Washington,D.C.,1963
Sco 14.0935

WASHINGTON,D.C. 1964 Cornea... World congress on the cornea 1st Papers Edited by John Harry Kind and John W. McTigue Butterworths,London,1965
An 32.4777

WASHINGTON,D.C. 1964 Quantitative electron microscopy :a symposium Proceedings Armed forces institute of pathology Edited by Gunter F. Bahr and Elmar H. Zeitler Laboratory investigation, 14,no 6 Williams and Wilkins,Baltimore,Md.,1965
An 32.0162

WASHINGTON,D.C. 1964 Quantitive electron microscopy :a symposium Proceedings Edited by Gunter F. Bahr and Elmar H. Zeitler Sponsored by the Intersociety committee for research potential in pathology,inc. Laboratory investigation, 14 no.6 illus. viii,605p Baltimore,Md.,1965
Bot 42.4731

WASHINGTON,D.C. 1964 Quantitive electron microscopy :a symposium Proceedings Edited by Gunter F. Bahr and Elmar H. Zeitler held at the Armed forces institute of pathology Laboratory investigation, 14, no 6 Washington,D.C.,1966 Sponsored jointly by the Intersociety committee for research potential in pathology and the Armed forces institute of pathology
Bal 39.0239

WASHINGTON,D.C. 1964 Symposium on computer augmentation of human reasoning proceedings United States.Office of naval research Edited by Margo A. Sass and William D. Wilkinson In joint sponsorship with the Bunker-Ramo corporation vi,235p 24cm Spartan books;Macmillan,Washington,D.C.;London, 1965
Math 3.0244

WASHINGTON,D.C. 1964 Symposium on computer augmentation of human reasoning Proceedings United States.Office of naval research Edited by Margo A. Sass and William D. Wilkinson In joint sponsorship with the Bunker-Ramo corporation Spartan books, Washington,D.C.,1965
Eng 41.2202

WASHINGTON,D.C. 1964 Symposium on computer augmentation of human reasoning proceedings United States.Office of naval research Edited by Margo A. Sass and William D. Wilkinson Jointly sponsored by Bunker-Ramo corporation bibliog,illus. vi,235p 24cm Spartan books,Washington,D.C.,1965
Math L 5.1005

WASHINGTON,D.C. 1964 The Electron microprobe :a symposium Proceedings Electrochemical society Edited by T.D. McKinley and others Electrochemical society series John Wiley,New York,New York,1966
Cav 7.2425

WASHINGTON,D.C. 1965 Critical phenomena Conference on phenomena in the neighbourhood of critical points Proceedings National bureau of standards Edited by M.S. Green and J.V. Sengers National bureau of standards. Miscellaneous publication, 273 xiv,242p 27cm National bureau of standards,Washington, D.C.,1966
Cav 7.2931

WASHINGTON,D.C. 1965 Human palaeopathology Symposium on human palaeopathology Proceedings National academy of sciences. Subcommittee on geographic pathology Edited by Saul Jarcho Yale university press,New Haven,Conn.,1966
An 32.3129

WASHINGTON,D.C. 1965 Intermag conference International conference on magnetics 3rd Proceedings Spsonsored by the Institute of electrical and electronics engineers.Magnetics group Institute of electrical and electronics engineers,New York,1965
Eng 41.5345

WASHINGTON,D.C. 1965 International congress on experimental mechanics 2nd Proceedings Society for experimental stress analysis Edited by B.E. Rossi Society for experimental stress analysis,Westport,Conn., 1966
Eng 41.2921

WASHINGTON,D.C. 1965 Symposium on rheology presented at the ASME applied mechanics and fluids engineering conference American society of mechanical engineers Edited by A. W. Marris and J.T.S. Wang American society of mechanical engineers,New York,1965
Eng 41.4418

WASHINGTON,D.C. 1966 Anticholinergic drugs and brain functions in animals and man :a symposium Proceedings Edited by P.B. Bradley and M. Fink Progress in brain research, 28 Elsevier,Amsterdam,1968 Held during the 5th meeting of the Collegium Internationale Neuro-psychopharmacologicum
An 32.5232

WASHINGTON,D.C. 1966 Joint automatic control conference 7th Preprints of conference papers Sponsored by the American institute of aeronautics and astronautics Washington,D.C.,1966
Eng 41.6016

WASHINGTON,D.C. 1967 Biochemical and human factors symposium Edited by Howard Gage 174p American society of chemical engineers, New York,1967 Presented at the Biomechanical and human factors conference
Chem E 24.1803

WASHINGTON,D.C. 1967 Interactive systems for experimental applied mathematics symposium proceedings Association for computing machinery Edited by M. Klerer and J. Reinfelds Academic press,New York,1968
Math L 5.3495

WASHINGTON,D.C. 1968 Comparative morphology of hematopoietic neoplasms :a symposium Proceedings National cancer institute Edited by Carolyn H. Lingeman and F.M. Garner held at United States.Armed forces institute of pathology National cancer institute.Monograph, 32 National cancer institute,Bethesda,Md.,1969
Bioch 33.0995

WASHINGTON,D.C. 1968 International congress of physiological sciences 24th abstracts of lectures and symposia International union of physiological sciences International union of physiological sciences. Proceedings, 6 Bethesda,Md.,1968
Phys 20.2154

WASHINGTON,D.C. 1968 International congress of physiological sciences 24th abstracts of volunteer papers and films International union of physiological sciences International union of physiological sciences. Proceedings, 7 Bethesda,Md.,1968
Phys 20.2155

WASHINGTON,D.C. 1968 Neoplasms and related disorders of invertebrate and lower vertebrate animals :a symposium Proceedings National cancer institute Edited by Clyde J. Dawe and John C. Harshbarger held at the Smithsonian institution National cancer institute. Monograph, 31 National cancer institute, Bethesda,Md.,1969
Bioch 33.0994

WASHINGTON,D.C. 1969 Basic aerodynamic noise research :a conference National aeronautics and space administration Edited by Ira R. Schwartz NASA SP-207 N.A.S.A., Washington,D.C.,1969
Eng 41.4224

WASHINGTON,D.C. 1970 Complexes of biologically active substances with nucleic acid and their mode of action :a research symposium Proceedings Walter Reed army institute of research Edited by F.E. Hahn Progress in molecular and subcellular biology, 2 Springer,Berlin,1971
Bioch 33.2242

WASHINGTON,D.C. 1970 and PHILADELPHIA 1970 Renal disease and hypertension : annual conferences 9th Proceedings 7 United States.Public health service United States.Public health service.Co-operative study, 7 U.S.Department of health,education and welfare,Washington,D.C., 1970
PGMS 29.0594

WASHINGTON,D.C. 1971 American chemical society national meeting 162nd Abstracts of papers American chemical society American chemical society,Washington,D.C.,1971
Chem 18.2812

WASHINGTON,D.C.,1967 An Example of the manipulation of directed graphs in the AMBIT-G programming language By Carlos Christensen CA-6711-1511 17p 28cm massachusetts computer associates,inc.,Wakefield,Mass.,1967 Paper presented at the symposium on interactive systems for experimental applied mathematics.Symposium held in Washington,D.C. Aug 26-28,1967
Math L 5.3145

WASHINGTON,D.C.,1967 The APL 360 terminal system By A.D. Falkoff and K.E. Iverson International business machines corporation. Watson research center RC 1922 20p 21cm new York,1968 This paper was presented at the Symposium on Interactive systems for experimental applied mathematics in Washington, D.C. Aug 26-28,1967
Math L 5.3150

WASHINGTON D.C. 1956 Symposium on naval hydrodynamics proceedings United states. Office of naval research Edited by F.S. Sherman Sponsored by the National research council National research council. Publication, 515 National research council, Washington D.C.,1957
A Math 4.0530

WASHINGTON D.C. 1963 The Plasma space science symposium proceedings Catholic university of America Edited by C.C. Chang and S.S. Huang In co-operation with the National aeronautics and space administration Astrophysics and space science library, 3 D.Reidel,Dordrecht,1965
A Math 4.0625

WASHINGTON SQUARE COLLEGE OF ARTS AND SCIENCE The Theory of electro-magnetic waves :a symposium Proceedings New York 1950 Jun 6-8 Under the auspices of Air force Cambridge research laboratories.Geophysics research directorate Interscience,New York; London,1951
Radioth 35.1507

WASHINGTON STATE UNIVERSITY.DEPARTMENT OF MATHEMATICS Conference on general topology Proceedings Pullman,Wash. 1970 Mar 25-27 iv,136p 27cm Washington state university.Department of mathematics,Pullman, Wash.,1970
P Math 2.4108

WASSENAAR 1960 Protein biosynthesis The International symposium on protein biosynthesis Proceedings Edited by R.J.C. Harris Under the auspices of Unesco and the Council for international organizations of medical sciences Academic press,London;New York,1961
Bioch 33.0516

WASSENAAR 1960 Protein biosynthesis :a symposium Edited by R.J.C. Harris Under the auspice of Unesco Academic press,London; New York,1961 Co-sponsored by the Council for international organizations of medical sciences
Radioth 35.0102

WASSENAAR 1960 Protein biosynthesis :a symposium Unesco Edited by R.J.C. Harris Sponsored also by the Council for international organizations of medical sciences Academic press,London;New York,1961
An 32.1509

WASSENNAAR 1960 Protein biosynthesis :a symposium Unesco Council for international organizations of medical sciences Edited by R.J.C. Harris Academic press,London;New York, 1961
Gen 34.0652

WATER POLLUTION Air and water pollution in the iron and steel industry proceedings of the air pollution meeting 25th and 26th Sept. and the water pollution meeting 11th and 12th Dec Proceedings London 1957 Sep 25-26 London 1957 Dec 11-12 Iron and steel institute Iron and steel institute.Special report, 61 Iron and steel institute,London,1958
Met 25.0359

The WATER RELATIONS OF PLANTS :a symposium London 1961 Apr 5-8 British ecological society Edited by A.J. Rutter and F.H. Whitehead British ecological society, 3 illus. x,349p Blackwell,London,1963
Bot 42.1879

WATER REQUIREMENTS FOR KIDNEY DIALYSIS London 1969 May 28 Organised by the Elga group of companies Elga Publications,Lane End,Bucks,1969
PGMS 29.0239

WATER RESOURCES Interterritorial conference on hydrology and water resources 1st Nairobi 1950 Nov Held under the auspices of the East Africa high commission Nairobi, 1950
Geog 13.5670

WATER RESOURCES USE AND MANAGEMENT :a symposium Proceedings Canberra 1963 Sep 9-13 Australian academy of science Melbuorne university press,London;New York, 1964
Geog 13.0787

WATER STRESS IN PLANTS :a symposium Proceedings Prague 1963 Sep 30-Oct 4 Ceskoslovenska akademie ved Edited by Bohdan Slavik 318p Junk,The Hague,1965
Bot 42.1881

WATERLOO,ONT. 1968 Cocurrent gas-liquid flow International symposium on research in cocurrent gas-liquid flow Proceedings Edited by Edward Rhodes and Donald S. Scott Canadian society for chemical engineering. Symposium series, 1 illus ix,698p Plenum press,New York,1969
Chem E 24.1842

WATERLOO,ONT. 1969 Recent progress in combinatorics Waterloo conference on combinatorics 3rd Proceedings University of Waterloo xiv,341p 24cm Academic press,New York;London,1969
Math S 3.1660

WATERLOO,ONT. 1970 Foundations of statistical inference :a symposium Proceedings University of Waterloo. Department of statistics Edited by V.P. Godambe and D.A. Sprott viii,519p 25cm Holt,Toronto,1971
Math S 3.1848

WATERLOO,ONTARIO 1968 International symposium on research in co-current gas-liquid flow Preprints Vol 1-2 National research council of Canada and Canadian society for chemical engineering Held at the University of Waterloo 2 vols University of Waterloo,Waterloo,Ontario,1968 Mimeographed
Chem E 24.1520

WATERLOO CONFERENCE ON COMBINATORICS 3rd Proceedings Recent progress in combinatorics Waterloo,Ont. 1969 May 20-31 Edited by W.T. Tutte Supported by N.A.T.O. and National research council of Canada xiv, 347p 24cm Academic press,New York;London, 1969
P Math 2.3354

WATERLOO CONFERENCE ON COMBINATORICS 3rd Proceedings Recent progress in combinatorics Waterloo,Ont. 1969 May 21-30 University of Waterloo xiv,341p 24cm Academic press,New York;London,1969
Math S 3.1660

WATERWAYS AND MARITIME WORKS International engineering congress on waterways and maritime works Proceedings Glasgow 1901 Sep 3-5 Clowes,London,1902
Eng 41.3317

The WATES FOUNDATION SYMPOSIUM ON ARTERIAL CHEMORECEPTORS Proceedings Oxford 1966 Jul 18-21 Edited by R.W. Torrance Blackwell,Oxford,1968
An 32.3810

The WATES FOUNDATION SYMPOSIUM ON ARTERIAL CHEMORECEPTORS Proceedings Oxford 1966 Jul 18-21 Edited by R.W. Torrance pl. xiv,402p Blackwell scientific publications, Oxford,Edinburgh,1968
Phys 20.1968

WATES SYMPOSIUM Proceedings Tissue proteinases Cambridge 1970 Apr 6-7 Royal society of London Edited by A.J. Barrett and J.T. Dingle Held at the Strangeways research laboratory North-Holland,Amsterdam,1971
Bioch 33.2284

WAVE MOTION AND VIBRATION THEORY symposium Washington 1952 June 16-17 Carnegie institute of technology and American mathematical society Edited by Albert E. Heins American mathematical society. Proceedings of symposia in applied mathematics, 5 McGraw-Hill,New York,1954
A Math 4.0513

WAVE PROPAGATION Phase and frequency instabilities in electromagnetic wave propagation Agard Agard conference proceedings, 33 Technivision,Slough,1970 Contains the proceedings of the Electromagnetic wave propagation committee's thirteenth symposium
Eng 41.8319

WAVES ON BEACHES AND RESULTING SEDIMENT TRANSPORT Proceedings Madison,Wis. 1971 Oct Edited by R.E. Meyer illus vii,462p 23cm Academic press,New York,1972
A Math 4.1885

WAYNE STATE FUND RESEARCH RECOGNITION AWARD. ANNUAL SYMPOSIA, 1 The Cell in mitosis : a symposium Proceedings Detroit 1961 Nov 6-8 Wayne state university Edited by Laurence Levine Academic press,New York,1963
An 32.3412

WAYNE STATE FUND RESEARCH RECOGNITION AWARD. ANNUAL SYMPOSIA, 1 The Cell in mitosis annual symposium Proceedings Detroit,Mich. 1961 Nov 6-8 Wayne state university Edited by Laurence Levine illus. Academic press, New York;London,1963
Gen 34.0862

WAYNE STATE FUND RESEARCH RECOGNITION AWARD. ANNUAL SYMPOSIA, 1 The Cell in mitosis symposium Proceedings Detroit 1961 Nov 6-8 Edited by Laurence Levine Held at Wayne state university Academic press,New York; London,1963
Bioch 33.0973

WAYNE STATE UNIVERSITY Conference on training personnel for the computing machine field 1st proceedings Detroit,Mich. 1954 June 22-23 Edited by Arvid W. Jacobson Sponsored also by the Association for computing machinery,the Industrial mathematics society and the Institute of radio engineers 104p 23cm Wayne university press,Detroit, Mich.,1955
Math L 5.0730

WAYNE STATE UNIVERSITY Ovary Symposium on the physiology and pathology of human reproduction 2nd annual Proceedings Edited by Harold C. Mack Thomas,Springfield, Ill.,1968
An 32.5224

WAYNE STATE UNIVERSITY Positron annihilation conference :conference Proceedings Detroit, Mich. 1965 Jul 27-29 Edited by A.T. Stewart and L.O. Roellig xvi,438p 24cm Academic press,New York,1967
Cav 7.2855

WAYNE STATE UNIVERSITY Semigroups Symposium on semigroups Proceedings Detroit,Mich. 1968 Jun 27th-29th Edited by Karl W. Folley Sponsored by the National science foundation bibliog. xi,277p 24cm Academic press,New York;London,1969
P Math 2.3927

WAYNE STATE UNIVERSITY The Cell in mitosis annual symposium Proceedings Detroit,Mich. 1961 Nov 6-8 Edited by Laurence Levine Wayne state fund research recognition award. Annual symposia, 1 illus. Academic press, New York;London,1963
Gen 34.0862

WAYNE STATE UNIVERSITY The Cell in mitosis symposium Proceedings Detroit 1961 Nov 6-8 Edited by Laurence Levine Wayne State fund research recognition award.Annual symposia, 1 Academic press,New York;London, 1963
Bioch 33.0973

WAYNE STATE UNIVERSITY and AMERICAN CHEMICAL SOCIETY International conference on coordination chemistry Advances in the chemistry of coordination compounds 6th Proceedings Detroit,Mich. 1961 Aug 27-Sep 1 Edited by Stanley Kirschner Macmillan, New York,1961 Co-sponsored by the United States air force,and the International union of pure and applied physics
Chem 18.0991

WEAK INTERACTIONS AND HIGH ENERGY NEUTRINO PHYSICS Scuola internationale di fisica 'Enrico Fermi'. 32 corso rendiconti Varenna 1964 Jun.15-27 Societa Italiana di Fisica Edited by T.D. Lee 334p Academic press,New York,1966
A Math 4.0845

WEAR AND CORROSION Surface protection against wear and corrosion :two series of educational lectures...presented...during the thirty-fifth National metal congress and exposition Cleveland,Ohio 1953 Oct 19-23 By H.S. Avery and others American society for metals American society for metals, Cleveland,Ohio,1954
Eng 41.3715

WEATHER MODIFICATION National conference on weather modification 1st Proceedings Albany,N.Y. 1968 Apr 28-May 1 American meteorological society State university of New York 532p American meteorological society,Boston,Mass.,1968
Nap 11.0975

WEATHER MODIFICATION National conference on weather modification 2nd Proceedings Santa Barbara,Calif. 1970 Apr 6-9 National science foundation.Atmospheric sciences section American meteorological society 436p American meteorological society,Boston, Mass.,1970
Nap 11.0977

WEATHER PREDICTION The Symposium on numerical weather prediction Proceedings Tokyo 1960 Nov 7-13 Meteorological agency, Japan Science council of Japan Co-sponsored by the International union of geodesy and geophysics 655p Meteorological society of Japan,Tokyo,1962
Nap 11.0923

WEEDS The Biology of weeds :a symposium Oxford 1959 Apr 2-4 British ecological society Edited by John L. Harper British ecological society.Symposia, 1 Bibliog, maps,2 plates,tables,diagrs xv,256p Blackwell scientific,Oxford,1960
Bot 42.2032

WEIGHTS AND MEASURES La Conference generale des poids et mesures 9eme comptes rendus Paris 1948 Oct. 12-21 Comite international des poids et mesures 127p Gauthier-Villars, Paris,1949
Obs 6.2388

WEIZMANN INSTITUTE OF SCIENCE New perspectives in biology :a symposium Proceedings Rehovoth 1963 Jun 10-17 Edited by Michael Sela B.B.A.Library, 4 xviii,285p Elsevier,Amsterdam,1964 A symposium held on the occasion of the inaugration of the Ullmann institute of life sciences
Bot 42.1502

WEIZMANN INSTITUTE OF SCIENCE New perspectives in biology :a symposium Proceedings Rehovoth 1963 Jun 10-17 Edited by Michael Sela B.B.A.library, 4 Elsevier,Amsterdam,1964 A symposium held, on the occasion of the inauguration of the Ullmann institute of life sciences
Bioch 33.0606

WEIZMANN INSTITUTE OF SCIENCE Theory and phenomonology in particle physica "Ettore Majorana" International school of physics 6th course Erice 1968 Jul 13-28 Part A-B European organization for nuclear research Edited by A. Zichichi Bibliog., Illus. 23cm 2 vols Academic press,New York,1969
A Math 4.1541

WELD IMPERFECTIONS a symposium Proceedings Palo Alto 1966 Sep 19-21 Lockheed missiles and space company Edited by A.R. Pfluger and R.E. Lewis Co-sponsored by the United States.Air force.Materials laboratory and the United States office of naval research Addison-Wesley,Reading,Mass.,1968
Met 25.2319

WELDED STRUCTURES Conference on fatigue of welded structures Proceedings 1970 Vol 1-2 Welding institute Welding institute, Abington,1971 Conference director:T.R. Gurney
Eng 41.8534

WELDED STRUCTURES :a conference Proceedings London 1953 Nov 23-26 Great Britain. Ministry of works H.M.S.O.,London,1954
Eng 41.3021

WELDING Symposium on the welding of iron and steel London 1935 May 2-3 Vol 1 Iron and steel institute 2 vols Iron and steel institute,London,1935
Eng 41.3947

WELDING AND ALLIED PROCESSES IN MAINTENANCE AND REPAIR WORK Opatija 1959 International institute of welding Elsevier, Amsterdam;London,1961 Papers in English and French
Met 25.0797

WELDING AND RIVETING LARGER ALUMINIUM STRUCTURES a symposium Proceedings London 1951 Nov Aluminium development association Aluminium development association,London,1952
Eng 41.3941

WELDING IN SHIPBUILDING Joint symposium on welding in shipbuilding London 1961 Oct 30-Nov 3 Institute of welding Institute of welding,London,1962 Jointly organised by Institute of welding,Royal institution of naval architects,Institute of marine engineers, Institution of engineers and shipbuilders in Scotland and North East Coast institution of engineering and shipbuilders
Met 25.0799

WELDING INSTITUTE Conference on fatigue of welded structures Proceedings 1970 Vol 1-2 Welding institute,Abington,1971 Conference director:T.R.Gurney
Eng 41.8534

WELDING INSTITUTE Conference on weldable Al-Zn-Mg alloys Proceedings 1969 Sep 23-25 Welding institute,Abington,1970
Eng 41.8286

WELDING INSTITUTE Conference on welding creep-resistant steels Proceedings Newcastle-upon-Tyne 1970 Feb 17-18 Welding institute,Abington,1970 Conference director:B.Phelps
Met 25.2725

WELDING OF IRON AND STEEL Report on the symposium on the welding of iron and steel,May 1935 Iron and steel institute.Welding committee Iron and steel institute,London, 1936
Met 25.0782

WELDING OF IRON AND STEEL Report on the symposium on the welding of iron and steel,May 1935 Iron and steel institute.Welding committee Iron and steel institute,London, 1936
Met 25.0782

WELDING OF IRON AND STEEL Symposium on the welding of iron and steel London 1935 May 2-3 Vol 1-2 Iron and steel institute 2 vols Iron and steel institute,London,1935
Met 25.0783

WELLCOME BUILDING LONDON A Symposium on immunization in childhood held in the Wellcome building Proceedings London 1959 May 4-6 Livingstone,London,1960
PGMS 29.0283

WELLCOME HISTORICAL MEDICAL LIBRARY The History and philosophy of knowledge of the brain and its functions :an Anglo-American symposium London 1957 Jul 15-17 Edited by F.N.L. Poynter Blackwell,Oxford,1958
An 32.1649

WELLCOME INSTITUTE OF THE HISTORY OF MEDICINE Medicine and culture :a historical symposium Proceedings London 1966 Sep 27-20 Wellcome institute of the history of medicine. Publications.New series, 15 vi,321p Wellcome institute of the history of medicine, London,1969
WSM 43.3330

WELLCOME TRUST Homeostatic regulators :a Ciba foundation symposium Proceedings London 1969 Jan 28-30 Edited by G.E.W. Wolstenholme and Julie Knight Churchill, London,1969
Bioch 33.0934

WELLCOME TRUST Homeostatic regulators :a Ciba foundation symposium held jointly with the Wellcome trust London 1969 Jan 28-30 Edited by G.E.W. Wolstenholme and Julie Knight Churchill,London,1969
Phys 20.2257

WELLCOME TRUST Tooth enamel;its composition, properties and fundamental structure An International symposium on the composition, properties and fundamental structure of tooth enamel held at the London hospital medical college Report of the proceedings London 1964 Apr 6-7 Edited by Maurice V. Stack and Ronald W. Fearnhead Wright,Bristol,1965 Sponsored by the Wellcome trust,London;the Procter and Gamble Co., Cincinnati;Colgate-Palmolive Ltd.,Great Britain;Unilever Ltd., London
Bal 39.1298

WELLINGTON,N.Z. 1958 Symposium on Antarctic research New Zealand.National committee for the International geophysical year figures,tables varp Department of scientific and industrial research,Wellington, N.Z.,1958
Sco 14.6067

WEMBLEY 1924-25 Phases of modern science : collection of essays and the 'Guide to the exhibits in the science galleries' By Oliver Lodge and others Royal society of London 2nd edition 231p A.and F.Denny,London,1925 1st edition entitled 'Handbook to the exhibition of pure science arranged by the Royal society'
Nap 11.0962

WENNER-GREN CENTER Olfaction and taste International symposium on olfaction and taste 1st Proceedings Stockholm 1962 Sep Edited by Y. Zotterman Wenner-Gren center. International symposium.Series, 1 Pergamon press,Oxford,1963
An 32.4380

WENNER-GREN CENTER Olfaction and taste II International symposium on olfaction and taste 2nd Proceedings Tokyo 1965 Sep Edited by T. Hayashi Wenner-Gren center. International symposium series, 8 Pergamon press,Oxford,1967
An 32.4411

WENNER-GREN CENTER Olfaction and taste 2 Wenner-Gren center international symposium 2nd Proceedings Tokyo 1965 Sep Edited by Takashi Hayashi Wenner-Gren center international symposium series, 8 Pergamon press,Oxford,1967
Psy 31.0296

WENNER-GREN CENTER The Structure and metabolism of the pancreatic islets :an international Wenner-Gren symposium Proceedings Uppsala 1963 Aug 29-Sep 2 and Stockholm Edited by S.E. Brolin and others Wenner-Gren center.International symposium series, 3 Pergamon press,Oxford, 1964
Bioch 33.0485

WENNER-GREN CENTER.INTERNATIONAL SYMPOSIUM.SERIES, 1 Olfaction and taste International symposium on olfaction and taste 1st Proceedings Stockholm 1962 Sep Wenner-Gren center Edited by Y. Zotterman Pergamon press,Oxford,1963
An 32.4380

WENNER-GREN CENTER.INTERNATIONAL SYMPOSIUM SERIES, 3 The Structure and metabolism of the pancreatic islets :an international Wenner-Gren symposium Proceedings Uppsala 1963 Aug 29-Sep 2 and Stockholm Wenner-Gren center Edited by S.E. Brolin and others Pergamon press,Oxford,1964
Bioch 33.0485

WENNER-GREN CENTER.INTERNATIONAL SYMPOSIUM SERIES, 8 Olfaction and taste II International symposium on olfaction and taste 2nd Proceedings Tokyo 1965 Sep Wenner-Gren center Edited by T. Hayashi Pergamon press,Oxford,1967
An 32.4411

WENNER-GREN CENTER.INTERNATIONAL SYMPOSIUM SERIES, 10 Structure and function of inhibitory neuronal mechanisms International meeting of neurobiologists 4th Proceedings Stockholm 1966 Sep Edited by C.von Euler and others Pergamon press,Oxford,1968
An 32.5336

WENNER-GREN CENTER INTERNATIONAL SYMPOSIUM 2nd Proceedings Olfaction and taste 2 Tokyo 1965 Sep Wenner-Gren center Edited by Takashi Hayashi Wenner-Gren center international symposium series, 8 Pergamon press,Oxford,1967
Psy 31.0296

WENNER-GREN FOUNDATION FOR ANTHROPOLOGICAL RESEARCH African ecology and human evolution :a symposium Papers and discussions Burg-Wartenstein 1961 Jul 10-22 Edited by F.Clark Howell and Francois Bourliere illus,tables 25cm Methuen, London,1964
Bal 44.4056

WENNER-GREN FOUNDATION FOR ANTHROPOLOGICAL RESEARCH Genetic diversity and human behaviour The Burg Wartenstein symposium 27th Papers Burg 1964 Sep 16-28 behavioral consequences of genetic differences in man Edited by G.N. Spuhler Viking fund publications in anthropology, 45 Aldine, Chicago,Ill.,1967
Gen 34.2032

WENNER-GREN FOUNDATION FOR ANTHROPOLOGICAL RESEARCH Medicine and culture :a historical symposium Proceedings London 1966 Sep 27-20 Wellcome institute of the history of medicine.Publications.New series, 15 vi,321p Wellcome institute of the history of medicine,London,1969
WSM 43.3330

WENNER-GREN FOUNDATION FOR ANTHROPOLOGICAL RESEARCH Primates :stüdies in adaptation and variability Edited by Phyllis C. Jay Bibliog Holt,Rinehart and Winston,New York, 1968 Dedicated to K.R.L.Hall,and including his bibliography.Work based on the results of a symposium on primate social behaviour sponsored by the Wenner-Gren foundation for anthropological research and held at Burg Wartenstein,Austria
Psy 31.2925

WENNER-GREN FOUNDATION FOR ANTHROPOLOGICAL RESEARCH Man's role in changing the face of the earth :international symposium Princeton,N.J. 1955 Jun 16-22 Edited by William L. Thomas and others Sponsored jointly by the National science foundation University of Chicago press,Chicago,Ill.; London,1956
Geog 13.1427

WENNER-GREN FOUNDATION FOR ANTHROPOLOGY AND RESEARCH Background to evolution in Africa Systematic investigation of the African later tertiary and quaternary : symposium Burg Wartenstein 1965 Jul 14-Aug 9 Edited by Walter W. Bishop and J. Desmond Clark University of Chicago press, Chicago,Ill.,1967
Geog 13.4660

WERNER CENTENNIAL :a symposium Papers New York 1966 Sep 12-16 American chemical society American chemical society.Advances in chemistry series,62 American chemical society,Washington,D.C.,1967 Symposium chairman:G.B.Kauffman
Chem 18.0990

WEST AFRICAN MICROPALEONTOLOGICAL COLLOQUIUM 2nd Proceedings Ibadan 1966 Jun 18-Jul 1 Edited by J.E.van Hinte J.E.Brill,Leiden, 1966
Geol 8.0344

WEST END,GRAND BAHAMA 1963 Bahama international conference on burns Report Sponsored by Colonial research institute Dorrance,Philadelphia,1964
Med 36.0079

WEST HAM 1963 Solid circuits and microminiaturization Conference on solid circuits and microminiaturization :a conference Proceedings Edited by G.W.A. Dummer Pergamon press,London,1964
Eng 41.5567

WEST LAFAYETTE,IND. 1968 Conference on thermal conductivity 8th Proceedings Edited by C.Y. Ho and R.E. Taylor Sponsored by Purdue university.Thermophysical properties research center xx,1169p 29cm Plenum press,New York,1969
Cav 7.3075

The WESTERN-EUROPE CONFERENCE ON PHOTOSYNTHESIS 2nd Proceedings Currents in photosynthesis Woudschoten,Zeist 1965 Sep Edited by J.B. Thomas and J.C. Goedheer 486, 23p Donker,Rotterdam,1966
Bot 42.1890

WESTERN-EUROPE CONFERENCE ON PHOTOSYNTHESIS 2nd Proceedings Currents in photosyntheses Woudschoten,Zeist 1965 Sep 6-12 Edited by J.B. Thomas and J.C. Goedheer A.D.Donker, Rotterdam,1966
Bioch 33.0805

WESTERN METAL CONGRESS AND EXPOSITION 4TH Alloy constructional steels Los Angeles 1941 May 19-23 By Herbert J. French American society for metals,Cleveland,Ohio, 1942 With a section on corrosion in collaboration with Francis L.Laque.A series of lectures presented at the 4th Western metal congress and exposition
Met 25.0480

WESTERN METAL CONGRESS AND EXPOSITION 8TH
Zirconium and zirconium alloys Los Angeles 1953 Mar 23-27 American society for metals American society for metals,Cleveland,Ohio, 1953 A symposium on zirconium and zirconium alloys presented to members of the A. S.M. during the eighth Western metal congress and exposition,Los Angeles, march 23 to 27, 1953
Met 25.0556

WESTERN MICHIGAN UNIVERSITY Analytic theory of differentiel equations :a conference Proceedings Kalamazoo,Mich. 1970 Apr 30-May 2 Edited by P.F. Hsieh and A.W.J. Stoddart Lecture notes in mathematics, 183 bibliog. vi,225p 25cm Springer,Berlin, 1971
P Math 2.3876

WESTERN MICHIGAN UNIVERSITY Graph theory and applications :conference Proceedings Kalamazoo,Mich. 1972 May 10-13 Edited by Y. Alavi and D.R. Lick Lecture notes in mathematics, 303 viii,329p 25cm Springer, Berlin,1972
P Math 2.4111

WESTERN MICHIGAN UNIVERSITY The Theory of arithmetic functions :conference Proceedings Kalamazoo,Mich. 1971 Apr 29-May 1 Edited by Anthony A. Gioia and Donald L. Goldsmith Lecture notes in mathematics, 251 287p 25cm Springer,Berlin,1972
P Math 2.4382

WESTERN SOCIETY OF NATURALISTS Comparative physiology of carbohydrate metabolism in heterothermic animals :symposium 1959 Edited by Arthur W. Martin University of Washington press,Seattle,1961
Bal 39.1258

WESTERN SOCIETY OF NATURALISTS Contributions to marine biology :lectures and symposia given at the Hopkins marine station...at The midwinter meeting of the Western society of naturalists Pacific Grove,Calif. 1929 Dec 20-21 Stanford university press,Stanford, Calif.,1930
Bal 39.1933

WESTMINSTER 1906 National conference on infantile mortality Report of the proceedings King,Westminster,1906 President of the conference:Rt.hon.J.Burns,M.P. Chairman:E.Spicer,J.P.
Path 30.0540

WESTMINSTER HOSPITAL MEDICAL SCHOOL
Biochemical aspects of genetics :a symposium London 1949 Feb 12 Biochemical society Edited by R.T. Williams Biochemical society. Symposia, 4 Cambridge university press, Cambridge,1950
Bioch 33.1366

WETTING :joint symposium Papers and discussion Bristol 1966 Sep 12-14 Society of chemical industry.Bristol section, and,Society of chemical industry.Colloid and surface chemistry group Society of chemical industry.Monographs,25 Society of chemical industry,London,1967
Col S 12.0147

WHITE DWARFS Novae and white dwarfs Colloque international d'astrophysique Reports Paris 1939 Jul.17-23 1: observations of novae By Knut Lundmark and others Actualites scientifiques et industrielles, 901 College de France. Conferences, 13 140p Hermann,Paris,1941
Obs 6.1146

WHITE DWARFS Novae and white dwarfs Colloque international d'astrophysique reports Paris 1939 Jul.17-23 3: white dwarfs By G.P. Kuiper and others Actualites scientifiques et industrielles, 903 College de France.Conferences, 15 267p Hermann,Paris,1941
Obs 6.0996

WHITE DWARFS Novae and white dwarfs Colloque international d'astrophysique reports Paris 1939 Jul 17-23 2: nova theory supernovae By P. Swings and others Actualites scientifiques et industrielles, 902 College de France.Conferences, 14 267p Hermann,Paris,1941
Obs 6.1720

WHITE DWARFS St.Andrews 1970 Aug 11-13 Edited by W.J. Luyten International astronomical union.Symposium, 42 164p Reidel,Dordrecht,1971
TA 15.0573

WHOLE-BODY COUNTING Symposium on whole-body counting Proceedings Vienna 1961 Jun 12-16 International atomic energy agency. Proceedings series International atomic energy agency,Vienna,1962
Radioth 35.1361

WIESBADEN 1963 Problems of settlements and compressibility of soils European conference on soil mechanics and foundation engineering Proceedings Vol 1-2 2 vols 1963
Eng 41.3175

WIESBADEN 1965 International congress of anatomists 8th abstracts of papers, scientific demonstrations and films 1965 In English,French and German
An 32.1194

WIESBADEN 1965 Quantitative methods in morphology Symposium on quantitative methods in morphology held...during the eighth International congress of anatomists Proceedings Edited by Ewald R. Weibel and Hans Elias Organised by the International society for stereology Illus. Springer, Berlin,1967
An 32.0105

WILDER Raymond L. and AYRES William L. ed. Lectures in topology :conference Ann Arbor 1940 Jun.24-Jul.6 University of Michigan vii,316p 23cm University of Michigan press, Ann Arbor, Mich.,1941
P. Math 2.0288

WILDLIFE North American wildlife conference 15th Transactions San Francisco 1950 Mar 6-9 Wildlife management institute Edited by Ethel M. Quee 681p 23cm Washington,D.C.,1950
Sco 14.8304

WILDLIFE MANAGEMENT INSTITUTE North American wildlife conference 15th Transactions San Francisco 1950 Mar 6-9 Edited by Ethel M. Quee 681p 23cm Washington,D.C.,1950
Sco 14.8304

WILLIAMS BAY,WISC. 1947 Atmosphere of the earth and planets A Symposium on planetary atmospheres Papers University of Chicago Edited by Gerard P. Kuiper 366p Chicago, Ill.,1949 Symposium held in connection with the 50th anniversary of the Yerkes observatory
Geod 9.0523

WILLIAMSBAY 1953 Conference on nuclear processes in geologic settings Proceedings University of Chicago Co-sponsored by National research council Williams Bay, Wisconsin,1953
Philos 1.0514

WILLIAMSBURG,VA. 1960 and WASHINGTON D.C. Magneto-fluid dynamics International symposium on magneto-fluid dynamics proceedings Edited by F.N. Frenkel and W.R. Sears Sponsored by the International union of theoretical and applied mechanics National research council.Publication, 829 National research council,Washington,1960
A Math 4.0619

WILLIAMSBURGH,VA, 1960 and WASHINGTON,D.C. 1960 Magneto-fluid dynamics :a symposium Proceedings Edited by F.N. Frankiel and W.R. Sears Sponsored by the International union of theoretical and applied mechanics National academy of sciences.Publication, 829 American physical society,New York,1960
Eng 41.4488

WILLIAMSTOWN,MASS. 1952 Dynamics of growth processes :a symposium Edited by Edgar J. Boell Society for the study of development and growth.Symposia, 11 Princeton university press,Princeton,N.J.,1954
An 32.2467

WILLS (H.H.)PHYSICAL LABORATORY Conference on defects in crystalline solids Report Bristol 1954 July The Physcial society, London,1955
Met 25.2358

WIND EFFECTS ON BUILDINGS AND STRUCTURES International conference on wind effects on buildings and structures 3rd Proceedings Tokyo 1971 Science council of Japan Saikon,Tokyo,1971
Eng 41.8602

WIND EFFECTS ON BUILDINGS AND STRUCTURES :a conference Proceedings Teddington 1963 Jun 26-28 Vol 1-2 Great Britain. Department of scientific and industrial research Great Britain.Department of scientific and industrial research.Symposium, 16 2 vols H.M.S.O.,London,1965
Eng 41.2707

WIND EFFECTS ON BUILDINGS AND STRUCTURES :a conference Proceedings Teddington 1963 Jun 26-28 Vol 1-2 Great Britain. Department of scientific and industrial research National physical laboratory National physical laboratory.Symposium, 16 H.M.S.O.,London,1963
Eng 41.6904

WIND EFFECTS ON BUILDINGS AND STRUCTURES :a symposium Proceedings Loughborough 1968 Apr 2-4 Vol 1-2 Loughborough university of technology Edited by D.J. Johns 2 vols Loughborough,1968
Eng 41.2762

WIND EFFECTS ON BUILDINGS AND STRUCTURES : international research seminar Proceedings Ottawa 1967 Sep 11-15 Vol 1-2 National research council of Canada 2 vols University of Toronto press,Toronto,1968
Eng 41.2746

WIND LOADS Technical meeting concerning wind loads on buildings and structures Proceedings Gaithersburg,Md. 1967 Jan 27-28 National bureau of standards Edited by R.D. Marshall and H.C.S. Thom National bureau of standards.Building science series, 30 US government printing office,Washington, D.C.,1970
Eng 41.8274

WIND-SENSITIVE STRUCTURES The Modern design of wind-sensitive structures :a seminar Proceedings London 1970 Jun 18 Cement industry research and information association CIRIA,London,1971
Eng 41.8276

WINDS AND TURBULENCE IN STRATOSPHERE,MESOSPHERE AND IONOSPHERE North Atlantic treaty organization.Advanced study institute Proceedings Lindau 1966 Sep 18-Oct 1 North Atlantic treaty organization.Science committee Edited by K. Rawer 421p North Holland,Amsterdam,1968
Nap 11.0955

WINDSOR 1950 The University teaching of international relations :international meeting Edited by Geoffrey L. Goodwin Held under the auspices of the International studies conference Blackwell,Oxford,1951
Geog 13.0171

WISCONSIN 1964 Error in digital computation seminar proceedings Vols. 1-2 Edited by Louis B. Rall United States Army.Mathematics research center.publication, 14 2 vols John Wiley,New York,1965
A Math 4.1209

WISCONSIN,ME. 1966 Non-compact groups in particle physics conference proceedings University of Wisconsin.Graduate school Edited by Yutze Chow 216p Benjamin,New York,1966
A Math 4.0890

WISCONSIN ALUMNI RESEARCH FOUNDATION A Symposium on the use of isotopes in biology and medicine Addresses Madison,Wisc. 1947 Sep 10-15 University of Wisconsin press, Madison,Wisc.,1948
Chem 18.1259

WISCONSIN ALUMNI RESEARCH FOUNDATION and NATIONAL SCIENCE FOUNDATION International symposium on organometallic chemistry 2nd Abstracts of proceedings Madison,Wisc. 1965 Aug 30-Sep 3 Co-sponsored by the United States.Army research office,Durham Madison, Wisc.,1965 Mimeographed
Chem 18.1589

WOODS HOLE,MASS. 1958 Symposium of plasma dynamics Edited by Francis H. Clauser Sponsored by United States.Air force.Office of scientific research Pergamon press;Addison-Wesley,London;Reading,Mass.,1960
Eng 41.4479

WOODS HOLE,MASS. 1959 Physics of precipitation Cloud physics conference 2nd Proceedings American geophysical union.Cloud physics committee,and,National science foundation Edited by Helmut Weickmann Geophysical monograph,5 National research council.Publication,746 illus,map xii,435p 25cm American geophysical union,Washington, 1960
Sco 14.0536

WOODS HOLE,MASS. 1961 Control mechanism in respiration and fermentation :a symposium held during the meeting of the Society of general physiologists at the Marine biological laboratory By Harlyn O. Halvorson Society of general physiologists Edited by Barbara Wright Published for the American physiological society Ronald press,New York, 1963
Bot 42.1730

WOODS HOLE,MASS. 1961 North Atlantic fish marking symposium Contributions International commission for the Northwest Atlantic fisheries Edited by H.W. Graham and others International commission for the Northwest Atlantic fisheries.Special publication,4 370p ICNAF,Dartmouth,Nova Scotia,1963
Sco 14.5957

WOODS HOLE,MASS. 1961 North Atlantic fish marking symposium International commission for the northwest Atlantic fisheries International commission for the northwest Atlantic fisheries.Special publication,4 Dartmouth,N.S.,1961
Sco 14.7833

WOODS HOLE,MASS. 1963 The Cellular functions of membrane transport :a symposium Edited by J.F. Hoffmann Held under the auspices of the Society of general physiologists Prentice-Hall,Englewood Cliffs, N.J.,1964
Pha 16.0280

WOODS HOLE,MASS. 1963 The Cellular functions of membrane transport :a symposium Papers Edited by J.F. Hoffman Held under the auspices of the Society of general physiologists illus. viii,291p Prentice-Hall,New Jersey,1964 Held at the annual meeting of the society
Bot 42.1500

WOODS HOLE,MASS. 1964 Lecture notes prepared in connection with the summer institute on algebraic geometry typescript By Shreeram S. Abhyankar and others American mathematical society 29cm American mathematical society,1964
P. Math 2.0575

WOODS HOLE,MASS. 1964 Molecular architecture in cell physiology :a symposium Edited by Tern Hayashi and Andrew G. Szent-Gyorgyi Under the auspices of the Society of general physiologists Prentice-Hall, Englewood Cliffs,N.J.,1966 Held at the 11th annual meeting of the society
Gen 34.0668

WOODS HOLE,MASS. 1965 The Specificity of cell surfaces :a symposium Edited by Bernard D. Davis and Leonard Warren Held under the auspices of the Society of general physiologists Prentice-Hall,Englewood Cliffs, N.J.,1967
An 32.3436

WOODS HOLE,MASS. 1965 The Specificity of cell surfaces :a symposium Edited by Bernard D. Davis and Leonard Warren Under the auspices of the Society of general physiologists Prentice-Hall,Englewood Cliffs, N.J.,1967 Held at the 12th annual meeting of the society
Gen 34.0669

WOODS HOLE,MASS. 1966 The Control of nuclear activity :a symposium Papers Edited by Lester Goldstein Held under the auspices of the Society of general physiologists Prentice-Hall,Englewood Cliffs, N.J.,1967 Original title of the symposium 'Cytoplasmic and environmental influences on nuclear behavior'
Bot 42.4736

WOODS HOLE,MASS. 1966 The Control of nuclear activity :a symposium held under the auspices of the Society of general physiologists at its annual meeting at the Marine biological laboratory Society of general physiologists Edited by Lester Goldstein Prentice-Hall,Englewood Cliffs,N.J. 1967
Bioch 33.1013

WOODS HOLE,MASS. 1969 Symposium on geophysical fluid dynamics Woods Hole oceanographic institution Edited by Willem V. R. Malkus Bibliog,illus,port 28cm 2 vols Woods Hole oceanographic institution,Woods Hole,Mass.,1969
A Math 4.1733

WOODS HOLE,MASS. 1970 Project MAC conference on concurrent systems and parallel computation Record Massachusetts institute of technology.Project MAC bibliog.,illus. 199p Association for computing machinery,New York,1970
Math L 5.3600

WOODS HOLE MARINE BIOLOGICAL LABORATORY Control mechanism in respiration and fermentation :a symposium held during the meeting of the Society of general physiologists at the Marine biological laboratory Woods Hole,Mass. 1961 Sep By Harlyn O. Halvorson Society of general physiologists Edited by Barbara Wright Published for the American physiological society Ronald press,New York,1963
Bot 42.1730

WOODS HOLE OCEANOGRAPHIC INSTITUTION Oceanography from space :proceedings of conference on the feasibility of conducting oceanographic exploratiions from aircraft, manned orbital and lunar laboratories Proceedings Woods Hole 1964 Aug 24-28 Sponsored by National aeronautics and space administration Woods Hole oceanographic institution,Woods Hole,Mass.,1965
Sco 14.0355

WOODS HOLE OCEANOGRAPHIC INSTITUTION
Symposium on geophysical fluid dynamics Woods Hole,Mass. 1969 Sep Edited by Willem V.R. Malkus Bibliog,illus,port 28cm 2 vols Woods Hole oceanographic institution, Woods Hole,Mass.,1969
A Math 4.1733

WOOLWICH POLYTECHNIC Semiconductor devices in power engineering :a symposium Edited by J. Seymour Pitman,London,1968
Eng 41.4867

WOOSTER,OHIO 1957 Trace elements :the conference held at Ohio agricultural experiment station Proceedings Ohio agricultural experiment station Edited by C. A. Lamb and others Academic press,New York; London,1958
Bioch 33.0348

WORCESTER,MASS. 1962 Symposium on measurement in unsteady flow :presented at the ASME hydraulic division conference American society of mechanical engineers.Hydraulic division American society of mechanical engineers,New York,1962
Eng 41.6713

WORCESTER FOUNDATION FOR EXPERIMENTAL BIOLOGY
Prostaglandin :a symposium Shrewsbury 1967 Oct 16-17 Edited by Peter W. Ramwell and Jane E. Shaw Wiley,New York,1968
Inv Med 37.0185

WORK HARDENING :based on a symposium Chicago,Ill. 1966 Nov American institute of mining,metallurgical and petroleum engineers.Institute of metals division. Chemistry and physics committee Edited by J. P. Hirth and J. Weertman Metallurgical society conferences, 46 Gordon and Breach, New York,1968
Met 25.2866

WORK STUDY Conference on work study Papers Harrogate 1954 May 27-29 British institute of management British institute of management,London,1954
Eng 41.1544

WORKING CONFERENCE ON MECHANICAL LANGUAGE STRUCTURES New York 1963 Aug 14-16 Association for computing machinery Collection of 8 mimeograph abstracts of papers
Math L 5.0928

WORKING CONFERENCE ON SYMBOL MANIPULATION LANGUAGE Papers Pisa 1966 Sep 5-9 International federation for information processing 1966 Typescript papers in loose-leaf file
Math L 5.3571

WORKING OF METALS The Plastic working metals a symposium Papers and discussions Cleveland 1936 Oct 19-23 American society for metals American society for metals, Cleveland,Ohio,1936 The symposium presented before the eighteenth annual convention of the American society for metals.
Met 25.0679

WORKING PARTY CONFERENCE FOR THE INTERNATIONAL BIOLOGICAL PROGRAM human adaptability study of circumpolar populations Report Point Barrow 1967 Nov 17-22 Canadian national committee for the international biological program,and,United States national committee for the international biological program variousp 29cm Naval arctic research laboratory,Point Barrow,1967
Sco 14.7529

WORKSHOP CONFERENCE :a Serono foundation symposium Proceedings Human testis Positano 1970 Apr 23-25 Edited by Eugenia Rosemberg and C.Alvin Paulsen Advances in experimental medicine and biology, 10 Plenum,New York;London,1970
Inv Med 37.0190

WORKSHOP ON CALCULUS OF VARIATIONS AND CONTROL THEORY Control theory and the calculus of variations based on the lectures presented at the workshop on calculus of variations and control theory Edited by A.V. Balakrishnan Academic press,New York;London,1969
Math S 3.1515

WORKSHOP ON MICROPROGRAMMING 3rd Preprints Buffalo,N.Y. 1970 OCT Association for computing machinery bibliog.,illus. Association for computing machinery,New York, 1970
Math L 5.3574

WORLD ASSOCIATION FOR THE ADVANCEMENT OF VETERINARY PARASITOLOGY Biology of parasites;emphasis on veterinary parasites The International conference of the World association for the advancement of veterinary parasitology 2nd Proceedings Philadelphia 1965 Sep 7-9 Edited by E.J.L. Soulsby Held at the University of Pennsylvania Academic press,New York;London, 1966 Held...in conjunction with Bicentennial celebrations of medical education in the United States
Bal 39.1737

WORLD CLIMATE International symposium on world climate 8000 to 0 b.c. proceedings London 1966 Apr 18-19 Royal meteorological society Edited by J.S. Sawyer Royal meteorological society,London,1966
A Math 4.0579

WORLD CLIMATE FROM 8000 TO NOUGHT B.C.
International symposium on world climate 8000 to nought B.C. Proceedings London 1966 Apr 18-19 Royal meteorological society London,1966
Sco 14.0581

WORLD CONFERENCE ON EARTHQUAKE ENGINEERING 1st Proceedings Berkeley,Calif. 1956 Jun Earthquake engineering research institute Earthquake engineering research institute,San Francisco,Calif.,1967
Eng 41.6394

WORLD CONFERENCE ON EARTHQUAKE ENGINEERING 2nd Proceedings Tokyo 1960 Jul 11-18 and Kyoto 1960 Jul 11-18 Vol 1-3 Japan society of civil engineers Organised by the Science council of Japan 3 vols Science council of Japan,Tokyo,1960
Eng 41.6395

The WORLD CONFERENCE ON HISTONE BIOLOGY AND CHEMISTRY 1st Nucleohistones 1963 Apr 29 - May 3 Edited by James Bonney and Paul Ts'o Holden Day,San Francisco,Calif., 1964
Radioth 35.0121

WORLD CONFERENCE ON HISTONE BIOLOGY AND CHEMISTRY 1st Nucleohistones 1963 Apr 29-May 3 Edited by James Bonner and Paul Ts'o Holden-Day,San Francisco,Calif.,1964
Gen 34.0684

The WORLD CONFERENCE ON HISTONE BIOLOGY AND CHEMISTRY 1st Papers Nucleohistones 1963 Apr 29-May 2 Edited by James Bonner and Paul Ts'o Holden-Day,San Francisco,1964
Bioch 33.0808

WORLD CONFERENCE ON MEDICAL EDUCATION 1st Proceedings London 1953 World medical association Oxford university press,London, 1953
Med 36.0370

WORLD CONFERENCE ON MEDICAL EDUCATION 1st proceedings London 1953 Edited by Hugh Clegg Held under the auspices of the World medical association Oxford university press,London,1954
An 32.0991

WORLD CONFERENCE ON MEDICAL EDUCATION 2 Proceedings Medicine,a life-long study Chicago 1959 World medical association British medical journal,London,1961
Med 36.0371

WORLD CONFERENCE ON PRESTRESSED CONCRETE : conference Proceedings San Francisco, Calif. 1957 Jul-Aug University of California San Francisco,Calif.,1957
Eng 41.3011

WORLD CONFERENCE ON SHELL STRUCTURES Proceedings San Francisco,Calif. 1962 Oct 1-4 University of California International association for shell structures Building research advisory board National research council.Publication,1187 National academy of sciences,Washington,D.C.,1064
Eng 41.3025

The WORLD CONFERENCE ON SMOKING AND HEALTH :a workshop Proceedings Toward a less harmful cigarette 1967 Sep 11-13 National cancer institute Edited by Ernest L. Wynder and Dietrich Hoffmann National cancer institute.Monograph, 28 National cancer institute,Bethesda,Md.,1968
Bioch 33.0992

WORLD CONGRESS OF ANAESTHESIOLOGISTS 4th Proceedings Progress in anaesthesiology London 1968 Sep 9-13 By T.B. Boulton and others Excerpta Medica,International congress series, 200 Excerpta Medica Foundation,Amsterdam,1910
PGMS 29.0559

WORLD CONGRESS OF PSYCHIATRY Proceedings Montreal 1961 Jun 4-10 Vol 1-2 Canadian psychiatric association McGill university Edited by R.A. Cleghorn and others 2 vols University of Toronto press; McGill university press,Toronto;Montreal,1961 Papers in English,French,German and Spanish
PGMS 29.0322

The WORLD CONGRESS OF PSYCHIATRY 3rd Proceedings Montreal 1961 Jun 4-10 Canadian psychiatric association and McGill university 2 vols University of Toronto press;McGill university press,Toronto; Montreal,1961 Conference languages: English,French,German and Spanish
Psy 31.2540

WORLD CONGRESS ON FERTILITY AND STERILITY 5th Proceedings Fertility and sterility Stockholm 1966 Jun 16-22 Edited by Bjorn Westin and Nils Wiquist Under the supervision of the International fertility association International congress series, 133 Excerpta medica foundation,Amsterdam, 1967
An 32.3899

WORLD CONGRESS ON FERTILITY AND STERILITY 5th Proceedings Fertility and sterility Stockholm 1966 Jun 16-22 International fertility association Swedish society of obstetrics and gynaecology Edited by Bjorn Westin and Nils Wiqvist Excerpta medica. International congress series, 133 Excerpta medica,Amsterdam,1967
Inv Med 37.0230

WORLD CONGRESS ON PHYSICAL EDUCATION Report Melbourne 1956 Nov 16-21 Australian physical education association Sponsored by the Commonwealth national fitness council Australian physical education association, Melbourne,1956
HE 27.0219

WORLD CONGRESS ON THE CORNEA 1st Papers Cornea... Washington,D.C. 1964 Oct Edited by John Harry Kind and John W. McTigue Butterworths,London,1965
An 32.4777

WORLD FEDERATION OF NEUROLOGY Cerebral localization and organization :a symposium Proceedings Lisbon 1960 Oct 21-23 Edited by Georges Schattenbrand and Clinton N. Woolsey University of Wisconsin press, Madison,Wis.;Milwaukee,Wis.,1964
Psy 31.0254

WORLD FEDERATION OF NEUROLOGY Evolution of the forebrain:phylogenesis and ontology of the forebrain :a symposium Lectures Frankfurt 1965 Aug 15-19 and Sprendlingen Edited by R. Hassler and H. Stephan G.Thieme, Stuttgart,1966
Psy 31.3370

WORLD FEDERATION OF SOCIETIES OF ANAESTHESIOLOGISTS European congress of anaesthesiology :post-graduate courses 1st Abstracts of papers Vienna 1962 Sep 3-9 English edition Wiener medizinischen akademie,Vienna,c1962
PGMS 29.0030

WORLD FEDERATION OF SOCIETIES OF ANAESTHESIOLOGISTS The European congress of anaesthesiology 2nd Proceedings Copenhagen 1966 Aug 8-13 1-3 Acta anaesthesiologica Scandinavica.Supplementum, 23-25 3 vols Universitetsforlaget,Aarhus, 1966
PGMS 29.0031

WORLD HEALTH ORGANISATION.REGIONAL OFFICE FOR EUROPE   Communicable diseases:methods of surveillance :report on a seminar   The Hague 1969 May 21-30   Regional office for Europe,W. H.O.,Copenhagen,1969
PGMS   29.0117

WORLD HEALTH ORGANIZATION   Assessment of radioactivity in man   Symposium on the assessment of radioactive body burdens in man   Proceedings   Heidelberg   1964 May 11-16   1-2   International atomic energy agency. Proceedings series   International atomic energy agency,Vienna,1964
Radioth   35.1373

WORLD HEALTH ORGANIZATION   Brain mechanism and learning :a symposium   Montevideo   1959 Aug 2-8   Council for international organizations of medical sciences   Blackwell, Oxford,1961
Bal   39.1604

WORLD HEALTH ORGANIZATION   Diagnosis and treatment of radioactive poisoning   Scientific meeting on the diagnosis and treatment of radioactive poisoning   Proceedings   Vienna   1962 Oct 15-18   International atomic energy agency.Proceedings series   International atomic energy agency, Vienna,1963
Radioth   35.1368

WORLD HEALTH ORGANIZATION   Effect of radiation on human heredity :a study group   Report   Copenhagen   1956 Aug 7-11   World health organization,Geneva,1957
Radioth   35.1055

WORLD HEALTH ORGANIZATION   Genetics of the immune response :report of a WHO scientific group   Geneva   1967 Oct 2-7   World health organization.Technical report series, 402   World health organization,Geneva,1968
Gen   34.2012

WORLD HEALTH ORGANIZATION   Immunoassay of gonadotrophins ...symposium   Transactions   Stockholm   1969 Sep 23-25   Edited by E. Dicfalusy and A. Dicfalusy   Karolinska symposia on research methods in reproductive endocrinology,   1   Stockholm,1969
PGMS   29.0182

WORLD HEALTH ORGANIZATION   In vitro methods in reproductive cell biology... symposium   3rd   Transactions   Geneva   1971 Jan 25-27   Edited by E. Dicfalusy and A. Dicfalusy   Karolinska symposia on research methods in reproductive endocrinology,   3   Acta endocrinologica.Supplement, 153   Stockholm, 1971
PGMS   29.0650

WORLD HEALTH ORGANIZATION   In vitro procedures with radioisotopes in medicine : symposium   Proceedings   Vienna   1969 Sep 8-12   IAEA,Vienna,1970
Bioch   33.2184

WORLD HEALTH ORGANIZATION   Medical radiation physics   Joint IAEA-WHO expert committee on medical radiation physics   Report   Geneva 1967 Dec 12-18   World health organization. Technical report series, 390   World health organization,Geneva,1968
Radioth   35.1381

WORLD HEALTH ORGANIZATION   Medical radioisotope scanning :a seminar   Proceedings   Vienna   1959 Feb 25-27   International atomic energy agency,Vienna,1959
Radioth   35.1375

WORLD HEALTH ORGANIZATION   Mental health aspects of the peaceful uses of atomic energy   Study group on mental health aspects of the peaceful uses of atomic energy   Report   Geneva   1957 Oct 21-26   World health organization.Technical report series, 151   World health organization,Geneva,1958
Radioth   35.1057

WORLD HEALTH ORGANIZATION   Planning of radiotherapy facilities   Joint IAEA-WHO meeting on planning of radiotherapy facilities   Report   Geneva   1964 Dec 15-19   World health organization.Technical report series, 328   World health organization,Geneva,1966
Radioth   35.1060

WORLD HEALTH ORGANIZATION   Radiation-induced cancer :a symposium   Proceedings   Athens 1969 Apr 28-May 2   International atomic energy agency,Vienna,1969
Radioth   35.1917

WORLD HEALTH ORGANIZATION   Radioisotopes in tropical medicine   Symposium on the use of radioisotopes in the study of endemic and tropical diseases   Proceedings   Bangkok 1960 Dec 12-16   International atomic energy agency.Proceedings series   International atomic energy agency,Vienna,1962
Radioth   35.1365

WORLD HEALTH ORGANIZATION   Radiological health and safety in mining and milling of nuclear materials   Symposium on radiological health and safety in mining and milling of nuclear materials :a symposium   Proceedings   Vienna   1963 Aug 26-31   1-2   International atomic energy agency   International labour organization   International atomic energy agency.Proceedings series   International atomic energy agency, Vienna,1964
Radioth   35.1164

WORLD HEALTH ORGANIZATION   Steroid assay by protein binding :a symposium   2nd   Transactions   Geneva   1970 Mar 23-25   Edited by E. Dicfalusy and A. Dicfalusy   Karolinska symposia on research methods in reproductive endocrinology,   2   Stockholm, 1970
PGMS   29.0589

WORLD HEALTH ORGANIZATION   Strahlenwirkung auf menschliche erbanlagen   Papers   Geneva 1957   Schriften reihe strahlenschutz,   3   Gersbach,Brunswick,1958
Radioth   35.1083

WORLD HEALTH ORGANIZATION   The Use of vital and health statistics for genetic and radiation studies seminar   Proceedings   Geneva   1960 Sep 5-9   United Nations publications, 61,XVII.8   United Nations,New York,1962
Gen   34.1120

WORLD HEALTH ORGANIZATION Use of radioisotopes and supervoltage radiation in radioteletherapy; present status and recommendation Report and background information for a study group convened by the IAEA and WHO Vienna 1959 Aug 3-5 International atomic energy agency,Vienna,1960
Radioth 35.1146

WORLD HEALTH ORGANIZATION.EXPERT COMMITTEE ON HUMAN GENETICS.REPORT, 3 Genetic counselling :a report Geneva 1968 Sep 24-30 World health organization World health organization.Technical report series, 416 Geneva,1969
Gen 34.1970

WORLD HEALTH ORGANIZATION.REGIONAL OFFICE FOR EUROPE The Prevention of accidents in childhood :report on a seminar Spa,Belgium 1958 Jul 16-25 World health organization, Copenhagen,1960
PGMS 29.0280

WORLD HEALTH ORGANIZATION.STUDY GROUP Effect of radiation on human heredity report of a study group Copenhagen 1956 Aug 7-11 World health organization,Geneva,1957
Gen 34.1106

WORLD HEALTH ORGANIZATION.STUDY GROUP ON THE PSYCHOBIOLOGICAL DEVELOPMENT OF THE CHILD Discussions on child development : consideration of the biological,psychological and cultural approaches to the understanding of human development and behaviour :meeting Geneva 1956 Vol 4 By Jean Piaget and others Edited by J.M. Tanner and Barbel Inhelder Tavistock publications,London,1960
Psy 28.0224

WORLD HEALTH ORGANIZATION.STUDY GROUP ON THE PSYCHOBIOLOGICAL DEVELOPMENT OF THE CHILD Discussions on child development ;a consideration of the biological,psychological, and cultural approaches to the understanding of human development and behaviour Geneva, Switzerland 1953 Vol 1: proceedings Edited by J.M. Tanner and Barbel Inhelder Tavistock publications,London,1956
Psy 31.1178

WORLD HEALTH ORGANIZATION.STUDY GROUP ON THE PSYCHOBIOLOGICAL DEVELOPMENT OF THE CHILD. MEETING 1ST Discussions on child development ;a consideration of the biological, psychological,and cultural approaches to the understanding of human development and behaviour Geneva,Switzerland 1953 Vol 1: proceedings World health organization. Study group on the psychobiological development of the child Edited by J.M. Tanner and Barbel Inhelder Tavistock publications,London,1956
Psy 31.1178

WORLD HEALTH ORGANIZATION.STUDY GROUP ON THE PSYCHOBIOLOGICAL DEVELOPMENT OF THE CHILD. MEETING 4TH Discussions on child development :consideration of the biological, psychological and cultural approaches to the understanding of human development and behaviour :meeting Geneva 1956 Vol 4 By Jean Piaget and others World health organization.Study group on the psychobiological development of the child Edited by J.M. Tanner and Barbel Inhelder Tavistock publications,London,1960
Psy 28.0224

WORLD LEAGUE FOR SEXUAL REFORM Sexual reform congress 3rd Proceedings London 1929 Sep 8-14 Edited by Norman Haire World league for sexual reform.Congress, 3rd Kegan Paul,Trench,Trubner,London,1930
Psy 28.0356

WORLD LEAGUE FOR SEXUAL REFORM.CONGRESS, 3rd Sexual reform congress 3rd Proceedings London 1929 Sep 8-14 World league for sexual reform Edited by Norman Haire Kegan Paul,Trench,Trubner,London,1930
Psy 28.0356

WORLD MEDICAL ASSOCIATION Medicine,a life-long study World conference on medical education 2 Proceedings Chicago 1959 British medical journal,London,1961
Med 36.0371

WORLD MEDICAL ASSOCIATION World conference on medical education 1st Proceedings London 1953 Oxford university press,London, 1953
Med 36.0370

WORLD MEDICAL ASSOCIATION World conference on medical education 1st proceedings London 1953 Edited by Hugh Clegg Oxford university press,London,1954
An 32.0991

WORLD METALLURGICAL CONGRESS 2ND Story of 2nd world metallurgical congress held in the United States of America 1957 Oct-Nov American society for metals Edited by Kingsley W. Given 225p A.S.M.,Cleveland, Ohio,1958
Met 25.0045

WORLD METALS CONGRESS 1957 Symposium on brittle failure of rotor forgings presented at special meeting for delegates to world metals congress Philadelphia,Pa. 1957 Nov 11 American society for testing materials A.S.T.M.Special technical publication, 231 American society for testing materials, Philadelphia,Pa,1957
Met 25.1325

WORLD METEOROLOGICAL ASSOCIATION and INTERNATIONAL ASSOCIATION OF SCIENTIFIC HYDROLOGY Surface waters :symposium Proceedings Berkeley,Calif. 1963 Aug 19-31 International association of scientific hydrology.Publication,63 illus 615p Gentbrugge,1964 Papers in English,French, German
Sco 14.0138

WORLD METEOROLOGICAL ORGANIZATION Noctilucent clouds :international symposium Tallinn 1966 Special committee for the International quiet sun year and International association of meteorology and atmospheric physics Edited by I.A. Khvostikov and G. Witt illus 235p 27cm Academy of sciences of the USSR.Soviet geophysical committee,Moscow,1967 Symposium papers in English with Russian abstracts
Sco 14.8230

WORLD METEOROLOGICAL ORGANIZATION Symposium on monsoons of the world Proceedings New Delhi 1958 Feb 19-21 Meteorological office, India x,270p Manager of publications,Delhi, 1960
Nap 11.0284

WORLD METEOROLOGICAL ORGANIZATION Symposium on tropical meteorology Proceedings Rotorua,N.Z. 1963 Nov 5-13 Edited by J.W. Hutchings With the cooperation of the International union of geodesy and geophysics illus 751p New Zealand meteorological service,Wellington,N.Z.,1964
Nap 11.0650

WORLD METEOROLOGICAL ORGANIZATION and COMMISSION ON ATMOSPHERIC CHEMISTRY AND RADIOACTIVITY Symposium:atmospheric chemistry,circulation and aerosols Proceedings Visby,Sweden 1965 Aug Tellus, 18,pt 2-3 Svenska geofysiska foreningen, Stockholm,1966
Chem 18.0571

WORLD METEOROLOGICAL ORGANIZATION and INTERNATIONAL UNION OF GEODESY AND GEOPHYSICS Symposium on tropical meteorology Proceedings Rotorua,N.Z. 1963 Nov 5-13 Edited by J.W. Hutchings New Zealand meteorological service,Wellington,N.Z.,1964
Geog 13.0990

WORLD METEOROLOGICAL ORGANIZATION and SCIENTIFIC COMMITTEE ON ANTARCTIC RESEARCH Polar meteorology W.M.O.-S.C.A.R.-I.C.P.M. Symposium on polar meteorology Proceedings Geneva 1966 Sep 5-9 Also organized by the International commission on polar meteorology illus vi,540p 28cm World meteorological organization,Geneva,1967
Sco 14.7839

WORLD ORCHID CONFERENCE 2nd Proceedings Honolulu 1957 Sep 19-23 Sponsored by the University of Hawaii American orchid society, Cambridge,Mass.,1958
BG 38.3117

WORLD ORCHID CONFERENCE 3rd Proceedings London 1960 May 30-Jun 22 Sponsored by the American orchid society Royal horticultural society,London,1960
BG 38.3118

WORLD ORCHID CONFERENCE 4th Proceedings Singapore 1963 Oct illus. Straits Times press,Singapore,1964
BG 38.3119

WORLD ORCHID CONFERENCE 5th Proceedings Long beach,Calif. 1966 Apr 11-22 Edited by Lloyd R.de Garmo Fifth world orchid conference,Long beach,Calif.,1966
BG 38.3120

WORLD ORCHID CONFERENCE 6th Commemorative brochure Orchids under the Southern cross Sydney 1969 1969
BG 38.1754

WORLD ORCHID CONFERENCE 6th Proceedings Sydney 1969 Sep Edited by Murray J.G. Corrigan and others 1971
BG 38.1753

WORLD PETROLEUM CONGRESS Proceedings London 1933 Jul 19-25 Vol 2: refining, chemical and testing section Edited by A.E. Dunstan Organised by the Institution of petroleum technologists xxvi,956p World petroleum congress,London,1934
Chem E 24.1708

WORLD PETROLEUM CONGRESS 7th proceedings Mexico City 1967 Vol 2: origin of oil, geology and geophysics bibliogs,illus Elsevier publishing co ltd,Barking,1967
Geol 8.4427

The WORLD POPULATION CONFERENCE Proceedings Geneva 1927 Aug 29-Sep 3 Edited by Margaret Sanger Arnold,London,1927
Bal 39.1748

WORLD POPULATION CONFERENCE Proceedings Geneva 1927 Aug 29-Sep 3 Edited by Margaret Sanger port. Arnold,London,1927
Geog 13.1526

WORLD POPULATION CONFERENCE 1965 Belgrade 1965 Aug 30-Sep 10 4: selected papers and summaries,migration,urbanization,economic development United nations.Department of economic and social affairs United nations. Department of economic and social affairs,New York,1967
Geog 13.7198

WORLD POULTRY CONGRESS 2nd Barcelona 1924 Part 2: informations and communications Arts graphiques,Barcelona, 1924
Gen 34.1767

WORLD POWER CONFERENCE 2nd Transactions Berlin 1930 Bd 5,8 2 vols VDI verlag,Berlin,1930
Eng 41.7438

WORLD POWER CONFERENCE 4th Transactions London 1950 Jul 11-14 Vol 1-5 Edited by W.O. Skeat 5 vols Lund,Humphries, London,1952
Eng 41.7439

The WORLD RIFT SYSTEM symposium report Ottawa 1965 Sep 4-5 International upper mantle committee sponsored by Canada. Department of mines and technical surveys Geological survey of Canada.Paper,66-14 Ottawa,1967
Geol 8.4451

The WORLD RIFT SYSTEMS :a symposium Report Ottawa 1965 Sep 4-5 National research council,Canada and International upper mantle committee Edited by T.N. Irvine Sponsored by the International union of geodesy and geophysics Geological survey of Canada.Paper, 66-14 Queen's printer,Ottawa, 1967
Geod 9.0138

WORMLEY,SURREY 1954 The Commonwealth oceanographic conference Proceedings Arranged by the National oceanographic council vi,93p 25cm Cambridge university press, Cambridge,1955
Sco 14.0123

WOUDSCHOTEN,ZEIST 1965 Currents in photosyntheses Western-Europe conference on photosynthesis 2nd Proceedings Edited by J.B. Thomas and J.C. Goedheer A.D.Donker, Rotterdam,1966
Bioch 33.0805

WOUDSCHOTEN,ZEIST 1965 Currents in photosynthesis The Western-Europe conference on photosynthesis 2nd Proceedings Edited by J.B. Thomas and J.C. Goedheer 486, 23p Donker,Rotterdam,1966
Bot 42.1890

WOUND HEALING Brown university symposium on the biology of skin,1963 Proceedings Providence,R.I. 1963 Jan 26-27 Edited by William Montagna and Rupert E. Billingham Advances in biology of skin, 5 Pergamon press,Oxford,1964
An 32.4044

WOUND HEALING AND TISSUE REPAIR :a symposium New York 1956 Oct 2-4 Edited by W. Bradford Patterson Developmental biology conferences series,1956 University of Chicago press,Chicago,Ill.,1959
An 32.3161

WRIGHT AIR DEVELOPMENT CENTER Monte Carlo methods symposium mimeograph Gainesville, Fla. 1954 Mar 16-17 University of Florida. Statistical laboratory Edited by Herbert A. Meyer Wiley publications in statistics 382p 29cm John Wiley and sons,New York, 1956
P. Math 2.2239

WRIGHT AIR DEVELOPMENT CENTER Symposium on Monte Carlo methods Gainesville,Fla. 1954 Mar.16-17 University of Florida.Statistical laboratory Edited by Herbert A. Meyer Wiley publications in statistics Bibliog. xvi,382p 28cm John Wiley and sons,New York, 1956
Math 3.0208

WRIGHT AIR DEVELOPMENT CENTER Symposium on Monte Carlo methods Gainesville,Fla. 1954 Mar 16-17 University of Florida.Statistical laboratory Edited by Herbert A. Meyer Wiley publications in statistics bibliog xvi,381p 28cm John Wiley and sons,New York, 1956
Math L 5.0746

WRIGHT AIR DEVELOPMENT DIVISION.MATERIALS CENTRAL Progress in very high pressure research :an international conference Proceedings Bolton Landing,N.Y. 1960 Jun 13-14 Edited by F.P. Bundy and others 314p New York, 1961
Geod 9.0254

WRIGHT STATE UNIVERSITY International symposium on multivariate analysis Proceedings Dayton,Ohio 1968 Jun 17-22 Edited by Parachuri R. Krishnaiah xx,696p 24cm Academic press,New York;London,1969
Math S 3.1802

WROUGHT NON-FERROUS METALS The Control of quality in the production of wrought non-ferrous metals and alloys :a symposium London 1954 Apr 28 2: the control of quality in working operations Institute of metals Institute of metals.Monograph and report series, 16 Institute of metals, London,1954
Met 25.0711

WROUGHT NON-FERROUS METALS The Final forming and shaping of wrought non-ferrous metals :a symposium London 1956 Apr 12 Institute of metals Institute of metals.Monograph and report series, 20 Institute of metals, London,1956
Met 25.0709

WURZBURG 1961 Stauferland-expedition,1959-60 Fridtjof-Nansen-Gedachtnis-symposions uber Spitzbergen :in Nansens 100.Geburtsjahr Vortrage Ergebnisse der Stauferland-expedition 1959-60 Franz Steiner,Wiesbaden, 1965
Sco 14.2973

WURZBURG 1961 Stauferland-expedition,1959-60 Vortrage des Fridtjof-Nansen-gedachtnis-symposions uber Spitzbergen in Nansens 100. geburtsjahr (geb 10.10.1861) Edited by Julius Budel and Alfred Wirthmann Ergebnisse der Stauferland-expedition,3 illus,7 plates 86p 25cm Franz Steiner,Wiesbaden,1965
Sco 14.2974

X- AND GAMMA-RAY ASTRONOMY Non-solar X- and gamma-ray astronomy Rome 1969 May 8-10 International astronomical union Edited by L. Gratton International astronomical union. Symposium, 37 425p Reidel,Dordrecht,190
Obs 6.3322

X-RAY ANALYSIS Annual conference on application of X-ray analysis 18th Proceedings Denver,Colo. 1969 Aug 6-8 Denver research institute.Metallurgy division Edited by Burton L. Henke Advances in X-ray analysis, 13 Plenum press,New York,1970
Met 25.2609

X-RAY ANALYSIS Annual conference on applications of X-ray analysis 7th- Proceedings Denver,Colo. 1958- Denver research institute Edited by William M. Mueller and others Advances in X-ray analysis, 2- Plenum press,New York,1960-
Met 25.2762

X-RAY ANALYSIS Annual conference on applications of X-ray analysis 17th Proceedings Estes Park,Colo. 1968 Aug 21-23 Denver research institute.Metallurgy division Edited by Charles S. Barrett and others Advances in X-ray analysis, 12 Plenum press,New York,1969
Met 25.2824

X-RAY ANALYSIS Conference on applications of X-ray analysis 15- Proceedings Denver, Colo. 1966- Denver research institute Advances in X-ray analysis, 10- Plenum press,New York,1967-
Min 10.1549

X-RAY ANALYSIS Symposium on energy dispersion X-ray analysis :X-ray and electron probe analysis Toronto,Ont. 1970 Jun 21-26 American society for testing and materials. Committee E-4 on metallography ASTM, Philadelphia,Pa.,1971 Symposium presented at the annual meeting.Co-ordinator:J.C.Russ
Met 25.2617

X-RAY CRYSTAL ANALYSIS Computing methods and the phase problem in X-ray crystal analysis :a conference Report Pennsylvania state college 1950 Apr 6-8 Rockefeller foundation and United States.Office of naval research Pennsylvania state college, State College,Pa.,1952
Cav 7.1789

X-RAY CRYSTAL ANALYSIS Computing methods and the phase problem in x-ray crystal analysis conference report Glasgow 1960 Aug.10-12 Edited by Ray Pepinsky and others International tracts in computer science and technology and their application, 4 viii, 326p 23cm Pergamon press,Oxford,1961
Math L 5.0303

X-RAY MICROANALYSIS International symposium on X-ray microscopy and X-ray microanalysis 2nd Proceedings Stockholm 1960 Feb Edited by Arne Engstrom and others illus x, 542p Elsevier,Amsterdam,1960
Cav 7.2234

X-RAY MICROANALYSIS International symposium on X-ray optics and X-ray microanalysis 3rd Proceedings Stanford,Calif. 1962 Aug 22-24 Edited by H.H. Pattee,jr. and others Sponsored by Stanford university.Biophysics laboratory illus xvii,622p Academic press,New York,1963
Cav 7.2189

X-RAY MICROSCOPY International symposium on X-ray microscopy and X-ray microanalysis 2nd Proceedings Stockholm 1960 Feb Edited by Arne Engstrom and others illus x, 542p Elsevier,Amsterdam,1960
Cav 7.2234

X-RAY MICROSCOPY AND MICRORADIOGRAPHY Symposium on x-ray microscopy and microradiography Proceedings Cambridge 1956 Edited by V.E. Coslett and others Sponsored by the International union of pure and applied physics Academic press,New York, 1957
Radioth 35.1600

X-RAY MICROSCOPY AND MICRORADIOGRAPHY The Symposium on X-ray microscopy and microradiography Proceedings Cambridge 1956 Aug 16-25 Cavendish laboratory Edited by V.E. Cosslett and others Sponsored by the International union of pure and applied physics Academic press,New York,1957
Cav 7.0461

X-RAY OPTICS International symposium on X-ray optics and X-ray microanalysis 3rd Proceedings Stanford,Calif. 1962 Aug 22-24 Edited by H.H. Pattee,jr. and others Sponsored by Stanford university.Biophysics laboratory illus xvii,622p Academic press,New York,1963
Cav 7.2189

X-RAY OPTICS AND X-RAY MICROANALYSIS International symposium on x-ray optics and x-ray microanalysis 3rd Stanford,Calif. 1962 Aug 22-24 Edited by H.H. Pattee Sponsored by Stanford university.Biophysics laboratory Academic press,New York,1963
Eng 41.4462

X-RAY CRYSTAL ANALYSIS Computing methods and the phase problem in x-ray crystal analysis : conference 2nd report Glasgow 1960 Aug 10-12 Edited by Ray Pepinsky and others International tracts in computer science and technology and their application,4 port. Pergamon press,Oxford,1961 The first conference was held in 1950
Min 10.1207

X-RAY CRYSTALLOGRAPHY Computing methods and the phase problem in X-ray crystal analysis : conference report State college,Pa. 1950 Apr 6-8 Edited by Ray Pepinsky sponsored by the Rockefeller foundation Pennsylvania state college,State college,Pa.,1952
Min 10.0756

X-RAY MICROSCOPY AND X-RAY MICROANALYSIS International symposium 2nd Proceedings Stockholm 1960 Edited by A. Engstrom and others Elsevier,Amsterdam,1960
Met 25.1375

X-RAY OPTICS AND MICROANALYSIS Optique des rayons X et microanalyse Congres international sur l'optique des rayons X et la microanalyse 4e Orsay 1965 Sep 7-10 Edited by Raymond Castaing Hermann,Paris, 1966
Met 25.1460

X-RAY OPTICS AND X-RAY MICROANALYSIS International symposium on x-ray microtechniques 3rd Stanford,Calif. 1962 Aug 22-24 Stanford university. Biophysics laboratory Edited by H.H. Pattee and others With the financial support of the United States.Office of naval research and the National science foundation Academic press, New York;London,1963
Met 25.1379

YAKUTSK 1966 Materialy VIII vsesoyuznogo mezhduvedomstvennogo soveshchaniya po geokriologii merzlotovedeniyu Vyp 6: geomorfologicheskaya sektsiya Edited by S.P. Kachurin and G.F. Gravis 151p 22cm Knizhnoe izdatelstvo,Yakutsk,1966
Sco 14.7714

YALTA 1969 Interstellar gas dynamics Symposium on cosmical gas dynamics 6th Proceedings International astronomical union International union of theoretical and applied mechanics International astronomical union. Symposium, 39 388p Reidel,Dordrecht,1970
Obs 6.3323

YALTA 1969 Interstellar gas dynamics Symposium on cosmical gas dynamics 6th Proceedings International astronomical union International union of theoretical and applied mechanics International astronomical union. Symposium, 39 388p Reidel,Dordrecht,1970
TA 15.0541

YALTA 1969 Interstellar gas dynamics Symposium on cosmical gas dynamics 6th Proceedings International union of theoretical and applied mechanics and International astronomical union Edited by H. J. Habing International astronomical union. Symposium, 39 diagrs,photos,graphs xii, 388p 24cm Reidel,Dordrecht,1970
Cav 7.2710

YARN PRODUCTION Studies in modern yarn production :a conference Papers Stresa 1968 May 25-28 Textile institute Textile institute.Annual conference, 53 Textile institute,Manchester,1968
Eng 41.8288

YEAST METABOLISM Aspects of yeast metabolism A Guinness symposium held at the Research laboratory St.James's Gate, Dublin Discussions Dublin 1967 Edited by A.K. Mills and Hans Krebs Guinness symposia Blackwell,Oxford;Edinburgh,1968
Bioch 33.1140

YEAST METABOLISM Aspects of yeast metabolism A Guinness symposium held at the research laboratory St.James's Gate Dublin Discussions Dublin 1967 Edited by A.K. Mills and Hans Krebs Guinness symposia Blackwell,Oxford,1968
Bot 42.1751

YELLOW RUST CONFERENCE,3RD Cereal rust conferences Cambridge 1964 Jun-Jul Plant breeding institute Cambridge,1966
Bot 42.4125

YELLOW SPRINGS,OHIO 1963 Bacterial photosynthesis :a symposium Edited by Howard Gest and others Sponsored by the Charles F. Kettering research laboratory Charles F. Kettering research laboratory.Contribution, 112 illus. xvi,523p Antioch press,Yellow Springs,Ohio,1963
Bot 42.3800

YELLOW SPRINGS,OHIO 1963 Bacterial photosynthesis :a symposium Papers presented Edited by Howard Gest and others Sponsored by the Charles F.Kettering research laboratory Charles F.Kettering research laboratory. Contribution, 112 Antioch press,Yellow Springs,Ohio,1963
Bioch 33.1116

YELLOW SPRINGS,OHIO 1965 Non-heme iron proteins;role in energy conversion :a symposium Proceedings Edited by Anthony San Pietro Sponsored by the Charles F. Kettering research laboratory Charles F. Kettering research laboratory.Contribution, 201 Antioch press,Yellow Springs,Ohio,1965
Bioch 33.0546

YERKES OBSERVATORY Atmosphere of the earth and planets A Symposium on planetary atmospheres Papers Williams Bay,Wisc. 1947 Sep 8-10 University of Chicago Edited by Gerard P. Kuiper 366p Chicago,Ill.,1949 Symposium held in connection with the 50th anniversary of the Yerkes observatory
Geod 9.0523

YERKES OBSERVATORY 1947 The Symposium on planetary atmospheres University of Chicago Edited by Gerard P. Kuiper vii,366p Chicago,1949 In connection with the 15th anniversary of the Yerkes observatory
Nap 11.0340

YESHIVA UNIVERSITY Quantum mechanics : lectures By Paul A.M. Dirac Belfer graduate school of science.Monograph series, 2 Belfer graduate school of science,New York, 1964
A Math 4.0730

YONKERS 1959 Plant growth regulation International conference on plant growth regulation 4th Sponsored by the Boyce Thompson institute for plant research Iowa state university press,Ames,Iowa,1961
Bot 42.1830

YORKTOWN HEIGHTS,N.Y. 1971 Stochastic point processes :statistical analysis,theory and applications Papers International business machines corporation.Research center Edited by Peter A.W. Lewis xxii,894p 24cm Interscience,New York,1972
Math S 3.1803

YUGOSLAV COMMITTEE FOR ELECTRONICS AND AUTOMATION I.F.A.C. symposium on system sensitivity and adaptivity 2nd Preprints Dubrovnik 1968 Aug 26-31 International federation of automatic control I.F.A.C.,Dubrovnik,1968
Eng 41.5889

YUGOSLAV COMMITTEE FOR ELECTRONICS AND AUTOMATION Sensitivity methods in control theory International symposium on sensitivity analysis Dubrovnik 1964 Aug 31-Sep 5 Edited by L. Radanovic Sponsored by the International federation of automatic control. Theory committee Pergamon press,Oxford,1965
Eng 41.5835

ZEEMAN CONGRESS Papers Amsterdam 1946 Physica, 12,9-10 73p M.Nijhof,The Hague, 1946
Obs 6.2880

The ZEEMAN CONGRESS,1946 Proceedings Zeeman effect Amsterdam 1946 Sep 17-23 Edited by C.J. Bakker 221p The Hague,1946 Reprinted from 'Physica',vol 12,no 9-10
Obs 6.3192

ZEEMAN EFFECT The Zeeman congress,1946 Proceedings Amsterdam 1946 Sep 17-23 Edited by C.J. Bakker 221p The Hague,1946 Reprinted from 'Physica',vol 12,no 9-10
Obs 6.3192

ZEIST 1967 Elimination of harmful organisms from food and feed by irradiation Panel on elimination of harmful organisms from food and feed by irradiation Report Organized by the Joint FAO-IAEA division of atomic energy in food and agriculture International atomic energy agency.Panel proceedings series International atomic energy agency,Vienna,1968
Radioth 35.1761

ZESDE INTERNATIONAL BOTANISCH CONGRES Proceedings Amsterdam 1935 Sep 2-7 Vol 1-2 Edited by M.J. Sirks 2 vols Brill,Leiden,1935-36
Bot 42.0657

ZINC AS A GALVANIC ANODE Navy-industry zinc symposium on cathodic protection development, application and specification 1955 Apr 21-22 United States.Department of the navy.Bureau of ships and American zinc institute U.S. Navy dept.,Washington,D.C.,1955
Met 25.2500

ZIRCONIUM AND ZIRCONIUM ALLOYS Los Angeles 1953 Mar 23-27 American society for metals American society for metals,Cleveland,Ohio, 1953 A symposium on zirconium and zirconium alloys presented to members of the A. S.M. during the eighth Western metal congress and exposition,Los Angeles, march 23 to 27, 1953
Met 25.0556

The ZODIACAL LIGHT AND THE INTERPLANETARY MEDIUM a conference Papers Honolulu 1967 Jan 30-Feb 3 National aeronautics and space administration.Office of technology utilization NASA SP-150 430p n.a.s.a., washington,D.C.,1967
Obs 6.3425

ZONIFICACION MICROFAUNISTICA DE LAS CALIZAS CRETACICAS DEL ESTE DE MEXICO International geological congress 20th papers By Federico Bonet Associacion mexicana de geologos petroleros Mexico City, 1956
Geol 8.3050

ZOOLOGICAL NOMENCLATURE Copenhagen decisions on zoological nomenclature additions to...les regles internationales de la nomenclature zoologique Copenhagen 1953 Aug International trust for zoological nomenclature Edited by Francis Hemming International trust for zoological nomenclature,London,1953
BG 38.3107

ZOOLOGICAL SOCIETY OF JAPAN Progress in comparative endocrinology International symposium on comparative endocrinology 3rd Proceedings Oiso,Japan 1961 Jun 5-11 Edited by Kiyoshi Takewaki General and comparative endocrinology.Suppl., 1,1962 Academic press,New York,1962
Bal 39.1381

ZOOLOGICAL SOCIETY OF LONDON Biological acoustics Proceedings London 1961 May 3 Edited by P.T. Haskell and F.C. Fraser Zoological society of London.Symposia, 7 London,1962
An 32.4826

ZOOLOGICAL SOCIETY OF LONDON Comparative biology of reproduction in mammals :symposium Proceedings London 1964 Nov 24-26 Edited by I.W. Rowlands Zoological society of London.Symposia, 15 Academic press,London, 1966
An 32.3887

ZOOLOGICAL SOCIETY OF LONDON Evolutionary aspects of animal communication;imprinting and early learning :a symposium Proceedings London 1961 Nov 8-9 Zoological society of London.Symposia, 8 Zoological society of London,London,1962
An 32.4384

ZOOLOGICAL SOCIETY OF LONDON Sensory specialization in responses to environmental demands :a symposium Proceedings London 1960 Mar 4 Edited by O.E. Lowenstein Zoological society of London.Symposia, 3 Zoological society of London,London,1960
An 32.4541

ZOOLOGICAL SOCIETY OF LONDON Symposium on variation in mammalian populations Proceedings London 1969 Nov 14-15 Edited by Robert J. Berry Zoological society of London.Symposia, 26 Academic press,London, 1970
Gen 34.2196

ZOOLOGICAL SOCIETY OF LONDON The Biology of survival :a symposium 13 Proceedings Edited by O.G. Edholm illus viii,138p Academic press,London;New York,1964
Sco 14.0837

ZOOLOGICAL SOCIETY OF LONDON The Biology of survival :a symposium 13th Proceedings London 1964 May 7-8 Edited by O.G. Edholm Zoological society of London.Symposia, 13 Academic press,London,1964
Phys 20.2132

ZOOLOGICAL SOCIETY OF LONDON Vertebrate locomotion :a symposium Proceedings London 1960 Nov 9 Edited by J.E. Harris Zoological society of London.Symposia, 5 Zoological society of London,London,1961
An 32.0459

ZOOLOGICAL SOCIETY OF LONDON Vertebrate locomotion :a symposium Proceedings London 1960 Nov 9 Edited by J.E. Harris Zoological society of London.Symposia,5 Illus. Zoological society,London,1961
VA 19.0086

ZOOLOGICAL SOCIETY OF LONDON Vertebrate locomotion :a symposium 5 Proceedings London 1960 Nov 9 Edited by J.E. Harris illus,1 plate vi,132p 26cm London,1961
Sco 14.0730

ZOOLOGICAL SOCIETY OF LONDON and SOCIETY FOR THE STUDY OF FERTILITY Comparative biology of reproduction in mammals : international symposium Proceedings London 1964 Nov 24-26 Edited by I.W. Rowlands Zoological society of London.Symposia,15 Academic press,London,1966
VA 19.0088

ZOOLOGICAL SOCIETY OF LONDON.SYMPOSIA, 1 Hormones in fish :a symposium Proceedings London 1959 Oct 13 Zoological society of London Edited by I.Chester Jones Zoological society of London,London,1960
An 32.3763

ZOOLOGICAL SOCIETY OF LONDON.SYMPOSIA, 2 Cyclical activity in endocrine systems :a symposium Proceedings London 1959 Dec 8 Zoological society of London Edited by E.J.W. Barrington Zoological society of London, London,1960
An 32.3764

ZOOLOGICAL SOCIETY OF LONDON.SYMPOSIA, 3 Sensory specialization in responses to environmental demands :a symposium Proceedings London 1960 Mar 4 Zoological society of London Edited by O.E. Lowenstein Zoological society of London,London,1960
An 32.4541

ZOOLOGICAL SOCIETY OF LONDON.SYMPOSIA, 5 Vertebrate locomotion :a symposium Proceedings London 1960 Nov 9 Zoological society of London Edited by J.E. Harris Zoological society of London,London,1961
An 32.0459

ZOOLOGICAL SOCIETY OF LONDON.SYMPOSIA, 7 Biological acoustics Proceedings London 1961 May 3 Zoological society of London Edited by P.T. Haskell and F.C. Fraser London,1962
An 32.4826

ZOOLOGICAL SOCIETY OF LONDON.SYMPOSIA, 8
Evolutionary aspects of animal communication; imprinting and early learning :a symposium Proceedings London 1961 Nov 8-9 Association for the study of animal behaviour Zoological society of London Zoological society of London,London,1962
An 32.4384

ZOOLOGICAL SOCIETY OF LONDON.SYMPOSIA, 9
Comparative aspects of neurohypophysical morphology and function :a symposium Proceedings London 1962 Feb 8 Zoological society of London Edited by H. Heller Zoological society of London,1963
An 32.3765

ZOOLOGICAL SOCIETY OF LONDON.SYMPOSIA, 10
The Primates :a symposium Proceedings London 1962 Apr 12-14 Edited by John Napier and N.A. Barnicot Zoological society, London,1963
An 32.0580

ZOOLOGICAL SOCIETY OF LONDON.SYMPOSIA, 15
Comparative biology of reproduction in mammals symposium Proceedings London 1964 Nov 24-26 Society for the study of fertility Zoological society of London Edited by I.W. Rowlands Academic press,London,1966
An 32.3887

ZOOLOGICAL SOCIETY OF LONDON.SYMPOSIA, 26
Symposium on variation in mammalian populations Proceedings London 1969 Nov 14-15 Zoological society of London Edited by Robert J. Berry Academic press,London, 1970
Gen 34.2196

ZOOLOGICAL SOCIETY OF LONDON.SYMPOSIA,15
Comparative biology of reproduction in mammals international symposium Proceedings London 1964 Nov 24-26 Society for the study of fertility,and,Zoological society of London Edited by I.W. Rowlands Academic press,London,1966
VA 19.0088

ZOOLOGICAL SOCIETY OF LONDON.SYMPOSIA,5
Vertebrate locomotion :a symposium Proceedings London 1960 Nov 9 Zoological society of London Edited by J.E. Harris Illus. Zoological society,London,1961
VA 19.0086

ZOOLOGICAL SURVEY OF INDIA Termites in the humid tropics :symposium Proceedings New Delhi 1960 Oct 4-12 Paris,1962
Bal 39.4249

ZOOLOGY Ideas in modern biology International congress of zoology 16th Proceedings Washington,D.C. 1963 Aug 20-27 Edited by J.A. Moore Doubleday,NEW York,1965
Radioth 35.0259

ZOOLOGY International congress of zoology 2nd-6th,8th-9th Proceedings 1892-1913 8 vols
An 32.0447

ZUR PSYCHOANALYSE DER KRIEGSNEUROSEN :
diskussion gehalten auf dem V internationalen psychoanalytischen kongress Beitrage Budapest 1918 Sep 28-29 By Sigmund Freud and others Internationale psychoanalytische bbibliothek, 1 Internationaler psychoanelytischer verlag,Leipzig;Vienna,1919
Psy 28.0171

ZURICH 1894 International geological congress 6th guide-book livret-guide geologique dans le Jura et les Alpes de la Suisse Payot,Lausanne,1894
Geol 8.3076

ZURICH 1953 International conference on soil mechanics and foundation engineering 3rd Proceedings Vol 1-3 3 vols Zurich,1953
Eng 41.3140

ZURICH 1953 International institute of refrigeration Commissions internationales 1 et 2:meetings Proceedings problems relating to liquefaction and separation of gases... International institute of refrigeration. Bulletin.Annexe,1954-2 illus. 140p Paris, 1954 In English and French
Sco 14.0121

ZURICH 1955 Congress handbook International union of pure and applied chemistry I.U.P.A.C. organizing committee, Zurich,c1955 Abstracts in English,French, German and Italian
Chem 18.1736

ZURICH 1955 International congress of pure and applied chemistry main congress lectures and lectures in the sections 14th Lectures International union of pure and applied chemistry Birkhauser,Basle;Stuttgart,1955 Papers in English,French and German
Chem 18.1737

ZURICH 1957 International congress for psychiatry 2nd Report Vol 1-4 International organization of world congresses for psyciatry.Swiss organizing committee Edited by W.A. Stoll 4cm Fussli arts graphiques,Zurich,1959 Congress languages English,French,German,Italian and Spanish
Psy 28.0407

ZURICH 1960 Advances in aeronautical sciences International congress for the aeronautical sciences 2nd Proceedings Vol 3-4 Edited by Theodore von Karman International series in aeronautics and astronautics.Division 9:symposia, 7-8 2 vols Pergamon press,London,1962
Eng 41.6964

ZURICH 1960 Differentialgeometrie und topologie Internationales kolloquium Schweizerische mathematische gesellschaft Sponsored by the International mathematical union Enseignement mathematique.Monographies 11 159p 24cm L'enseignement mathematique universite,Geneva,1962
P. Math 2.0697

ZURICH 1964 Sleep mechanisms Symposium on the physiological pharmacological and clinical aspects of sleep Proceedings Edited by K. Akert and others Progress in brain research, 18 Elsevier,Amsterdam,1965
An 32.4238

ZURICH 1965 Festband zum 70.geburtstag von Rolf Nevanlinna kolloquium 2nd vortrage Edited by H.P. Kunzi and A. Pfluger 149p 24cm Springer-Verlag,Berlin,1966
P. Math 2.2555

ZURICH 1966 Progress in lymphology International symposium on lymphology Proceedings Thieme,Stuttgart,1967
Radioth 35.0922

ZURICH 1967 Biomechanics:technique of drawings of movement and movement analysis International seminar on biomechanics 1st Proceedings Edited by J. Wartenweiler and others Medicine and sport,2 Karger,Basle; NewYork,1968
VA 19.0128

ZURICH 1967 International union of geodesy and geophysics :general assembly 14th Abstracts of papers Vol 6: International association of scientific hydrology 1967
Sco 14.7409

ZURICH 1970 Crystal growth and epitaxy from the vapour phase International conference on crystal growth and epitaxy from the vapour phase 1st Proceedings Edited by M. Kaldis and M. Schieber North-Holland, Amsterdam,1971
Met 25.2765

ZYTO- UND HISTOCHEMIE IN DER HAMATOLOGIE Freiburger symposium :zugleich symposium der gesellschaft deutscher hamatologen 9th Freiburg 1962 Oct 25-27 Edited by Hans Merker Springer;Blackwell,Berlin;Oxford,1963
Med 36.0394

ZYTO-UND HISTO-CHEMIE IN DER HAMATOLOGIE Freiburger symposium :zugleich symposium der gesellschaft deutscher hamatologen 9th Freiburg 1962 Oct 25-27 Edited by Hans Merker Springer,Berlin,1963
An 32.3186

1939 Symposium on progress in astrophysics Papers By Donald H. Menzel and others American philosophical society Franklin institute 1939 From the Proceedings of the American philosophical society,vol 81,no 2. Lacks title page
Obs 6.3230

1948 Titanium :a symposium Report United States.Office of naval research Office of naval research,Washington,D.C.,1949
Met 25.0548

1949 The Australian environment handbook prepared for the conference on plant and animal nutrition... Commonwealth scientific and industrial research organization 2nd edition British commonwealth official scientific specialist conferences,Special agricultural conference Commonwealth scientific and industrial research organization,Melbourne,1950
Gen 34.1372

1951 NBS semicentennial symposium on energy transfer in hot gases Proceedings National bureau of standards National bureau of standards.Circular, 523 U.S.government printing office,Washington,D.C.,1954
Eng 41.7220

1953 Doklady 6 rasshirennoi sessii uchenogo soveta instituta Vyp 2-3 Nauchno-issledovatelskii institut polyarnogo zemledeliya,zhivotnovodstva i promyslovogo khozyaistva Leningrad,1953-54
Sco 14.4902

1953 Papers on non-destructive testing presented at a meeting of committee E-7... American society for testing materials. Committee E-7 on non-destructive testing American society for testing materials, Philadelphia,Pa.,1954
Met 25.1318

1954 L' Instinct dans le comportement des animaux et des hommes :colloque international Actes By M. Autuori and others Fondation Singer-Polignac Masson,Paris,1956
Psy 31.0644

1955 Boundary layer effects in aerodynamics : a symposium National physical laboratory London,1955
Eng 41.6960

1955 Materialy soveshchaniya po psikhologii Akademiya pedagogicheskikh nauk R.S.F.S.R. Institut psikhologii Edited by G.N. Voskresenskii and others Izdatekstvo Akademii pedagogicheskikh nauk RSFSR,Moscow, 1957
Psy 31.2576

1956 Resheniya mezhvedomstvennogo soveshchaniya po razrabotke unifitsirovannykh stratigraficheskikh skhem dlya Dalnego Vostoka rassmotreno i utvershdeno mezhvedomstvennym stratigraicheskim komitetom Ministerstva geologii i okhrane nedr S.S.S.R. Edited by V. N. Vereshchagin and others tables 51p 29cm Gosudarstvennoe nauchno-tekhnicheskoe izdatelstvo literatury po geologii i okhrane nedr,Moscow,1958
Sco 14.4755

1956 Rezheniya mezhvedomstevennogo soveshchaniya po razrabotke unifitsirovannykh stratigraficheskikh skhem Sibiri :rassmotreno i utverzhdeno mezhvedomstvennym stratigraficheskim komitetom Ministerstva geologii i okhrane nedr S.S.S.R. Edited by V. P. Skvortsov 91p 29cm Gosudarstvennoe nauchno-tekhnicheskoe izdatelstvo literatury po geologii i okhrane nedr,Moscow,1959
Sco 14.4753

1956 Sapropeli ikh ispolzovanie :po matrialam konferentsii Akademiya nauk Belorusskoi S.S.R.Institut torfa Izdatelstvo akademii nauk B.S.S.R.,Minsk,1958
Bot 42.6733

1957 Conference on lubrication and wear Scientific lubrication Scientific publications,Broseley,1957
Eng 41.6294

1957 Nuclear magnetic resonance :a conference Papers New York academy of sciences New York academy of sciences.Annals, 70,art 4 New York academy of sciences,New York,1958
Chem 18.2305

1957 Resheniya mezhvedomstvennogo soveshchaniya po razrabotke unifitsirovannykh stratigraficheskikh skhem dlya severo-vostoka S.S.S.R. :rassmotreno i utverzhdeno mezhvedomstvennym stratigraficheskim komitetom Ministerstva geologii i okhrane nedr S.S.S.R. Edited by N.P. Anikeev tables,maps 80p Gosudarstvennoe nauchno-tekhnicheskoe izdatelstvo literatury po geologii i okhrane nedr,Moscow,1959
Sco 14.4754

1957 World metallurgical congress 2nd Story of 2nd world metallurgical congress held in the United States of America American society for metals Edited by Kingsley W. Given 225p A.S.M.,Cleveland,Ohio,1958
Met 25.0045

1957-65 Japanese congress on testing materials 1st-8th Proceedings 8 vols Japan society for testing materials,Kyoto,1958-65
Eng 41.3844

1958 Central nervous system and behaviour Conference on the central nervous system and behaviour 1st Transactions Edited by Mary A.B. Brazier Sponsored by the Josiah Macy jr.foundation New York,1959
Phys 20.2239

1958-60 Central nervous system and behaviour The Conference on the central nervous system and behaviour 1st-3rd Transactions Edited by Mary A. Brazier . Josiah Macy jr. foundation 3 vols Josiah Macy jr. foundation,New York,1959-61
An 32.5329

1959 A Discussion on the biology of the southern cold temperate zone Royal society of London Edited by C.F.A. Pantin Royal society.Proceedings.Series B,152 plates 429-682p Royal society,London,1960
Sco 14.6173

1959 Developing cell systems and their control Society for the study of development and growth Edited by Dorothea Rudnick Society for the study of development and growth.Symposia, 18 Ronald press,New York, 1960
Gen 34.0850

1959 Tools of biological research symposium 2nd Proceedings Second series Surgical research society Edited by Hedley J. B. Atkins illus. Blackwell,Oxford,1960
Gen 34.0466

1960 Problemy geografii Sibiri i Dalnego Vostoka :itogi pervogo nauchnogo sovetshchaniya geografov Sibiri i Dalnego Vostoka 1st Institut geografii Sibiri i Dalnego Vostoka Edited by I.P. Gerosimov and others 136p Irkutskoe knizhnoe izdatelstvo, Irkutsk,1960
Sco 14.4733

1960 Recent developments in the chemistry of natural phenolic compounds :a symposium Proceedings Plant phenolics group Edited by W.D. Ollis Pergamon press,Oxford,1961
Chem 18.1495

1960 Strengthening mechanisms in solids :a seminar American society for metals American society for metals,Metals Park,Ohio, 1962
Met 25.1206

1960 Symposium on low temperature properties of high-strength and missile materials Papers American society for testing materials A.S.T.M.Special technical publication, 287 A.S.T.M.,Philadelphia,1961
Met 25.0974

1960 Symposium on shear and torsion testing American society for testing materials A.S.T.M.Special technical publication, 289 American society for testing materials, Philadelphia,Pa.,1961
Met 25.0887

1960,1963,1966 Automatic and remote control Congress of the International federation of automatic control 1st-3rd Proceedings International federation of automatic control Butterworths,London,1961-66
Eng 41.8207

1961 Conference on oceanwave spectra National academy of sciences Englewood Cliffs,N.J.,1963
Eng 41.6696

1961 Ultra-high-purity metals :a seminar Papers American society for metals A.S.M., Metals Park,Ohio,1962
Met 25.1495

1963 Annual conference on applications of x-ray analysis 12th Edited by M. Mueller and others Sponsored by the Denver research institute Advances in x-ray analysis, 7 Plenum press,New York,1964
Met 25.1445

1963 Lubrication and wear convention Proceedings Institution of mechanical engineers.Lubrication and wear group Institution of mechanical engineers,London, 1964
Eng 41.6305

1963 Newcastle disease virus:an evolving pathogen :international symposium Proceedings University of Wisconsin American association of avian pathologists Edited by Robert Paul Hanson University of Wisconsin press,Madison,Wis.,1964
Gen 34.0728

1964 Insulation of high voltages in vacuum : international symposium Massachusetts institute of technology Massachusetts institute of technology,Cambridge,1964
Eng 41.4984

1965 Recrystallization,grain growth and textures a seminar Papers American society for metals A.S.M.,Metals Park,Ohio,1966
Met 25.1269

1966 Life,land and water Environmental studies of the Glacial Lake Agassiz region :a conference Proceedings Edited by William J. Mayer-Oakes University of Manitoba. Department of anthropology.Occasional papers, 1 maps xvi,414p University of Manitoba press,Winnipeg,1967
Bot 42.2101

1967 Inequalities II Symposium on inequalities 2nd Proceedings Edited by Oved Shisha Held at the United States air force academy xvi,439p 24cm Academic press,New York;London,1970
P Math 2.3847

1967 Interdisciplinary approach to friction and wear :a symposium Proceedings National aeronautics and space administration Washington,D.C.,1968
Eng 41.6272

1967 Priroda i khozyaistvo severa :materialy pervoi nauchnoi konferentsii Kolskogo otdela geograficheskoe obshchestva S.S.S.R. Vyp 1 Geograficheskoe obshchestvo S.S.S.R. Severnyi filial Edited by I.L. Freydin and others maps 314p 29cm Akademiya nauk SSSR,Apatity,1969
Sco 14.8284

1968 Physics of hot plasmas Scottish universities' summer school 9th Proceedings Edited by B.J. Rye and J.C. Taylor illus xv,455p 25cm Oliver and Boyd,Edinburgh,1970
A Math 4.1886

1969 NUMBER THEORY INSTITUTE Summer institute on number theory Stony Brook, N.Y. 1969 Jul 7-Aug 1 American mathematical society Edited by Donald J. Lewis American mathematical society.Proceedings of symposia in pure mathematics, 20 xiii,451p 26cm AMS,Providence,R.I.,1971
P Math 2.4383

5-HYDROXYTRYPTAMINE :a symposium Proceedings London 1957 Apr 1-2 Edited by G.P. Lewis Pergamon press,London,1958
An 32.1489

5-HYDROXYTRYPTAMINE :a symposium Proceedings London 1957 Apr 1-2 Edited by G.P. Lewis Pergamon press,London,1958
Radioth 35.0788

Libraries and collections

A.Biol.	Department of Applied Biology.
A.Math.	Department of Applied Mathematics and Theoretical Physics.
An.	Department of Anatomy.
B.G.	Cory Library, Botanic Garden.
Bal.	Balfour and Newton Libraries, Department of Zoology.
Bioch.	Colman Library, Department of Biochemistry.
Bot.	Department of Botany.
Cav.	Rayleigh Library, Cavendish Laboratory, Department of Physics.
Chem.	Joint Chemical Library.
Chem.E.	Department of Chemical Engineering.
Col.S.	Oliver Gatty Library, Colloid Science Laboratory.
Eng.	Department of Engineering.
Gen.	Department of Genetics.
Geod.	Department of Geodesy and Geophysics.
Geog.	Department of Geography.
Geol.	Sedgwick Geology Library, Sedgwick Museum.
H.E.	Human Ecology Library (now incorporated in the Department of Medicine).
Hunt.	Hunterian collection, University Library (formerly in the Department of Pathology).
Inv.Med.	Department of Investigative Medicine.
Math.	Wishart Library, Statistical Laboratory.
Math.L.	University Computer Laboratory (formerly the Mathematical Laboratory).